[third edition]

Principles of Biology

Robert J. Brooker
University of Minnesota – Minneapolis

Eric P. Widmaier
Boston University

Linda E. Graham
University of Wisconsin – Madison

Peter D. Stiling
University of South Florida

PRINCIPLES OF BIOLOGY, THIRD EDITION

Published by McGraw-Hill Education, 2 Penn Plaza, New York, NY 10121. Copyright ©2021 by McGraw-Hill Education. All rights reserved. Printed in the United States of America. Previous editions ©2018 and 2015. No part of this publication may be reproduced or distributed in any form or by any means, or stored in a database or retrieval system, without the prior written consent of McGraw-Hill Education, including, but not limited to, in any network or other electronic storage or transmission, or broadcast for distance learning.

Some ancillaries, including electronic and print components, may not be available to customers outside the United States.

This book is printed on acid-free paper.

2 3 4 5 6 7 8 9 LKV 24 23 22 21 20

ISBN 978-1-260-24086-3 (bound edition)
MHID 1-260-24086-X (bound edition)
ISBN 978-1-260-70832-5 (loose-leaf edition)
MHID 1-260-70832-2 (loose-leaf edition)

Portfolio Manager: *Andrew Urban*
Product Developer: *Elizabeth M. Sievers*
Marketing Manager: *Kelly Brown*
Content Project Managers: *Jessica Portz, Tammy Juran, Sandra Schnee*
Buyer: *Sandy Ludovissy*
Designer: *David W. Hash*
Content Licensing Specialist: *Lori Hancock*
Cover Image: *©TeeJe/Getty Images*
Compositor: *MPS Limited*

All credits appearing on page are considered to be an extension of the copyright page.

Library of Congress Cataloging-in-Publication Data

Brooker, Robert J., author. | Widmaier, Eric P., author. | Graham,
  Linda E., 1946- author. | Stiling, Peter D., author.
  Principles of biology / Robert J. Brooker, University of Minnesota -
  Twin Cities, Eric P. Widmaier, Boston University, Linda E. Graham,
  University of Wisconsin - Madison, Peter D. Stiling, University of South Florida.
  Third edition. | New York, NY : McGraw-Hill Education, [2021] |
  Includes index.
  LCCN 2019017416| ISBN 9781260240863 (student edition) |
  ISBN 9781260708325 (loose-leaf edition)
  LCSH: Biology—Textbooks.
  LCC QH308.2 .B75 2101 | DDC 570—dc23
  LC record available at https://lccn.loc.gov/2019017416

The Internet addresses listed in the text were accurate at the time of publication. The inclusion of a website does not indicate an endorsement by the authors or McGraw-Hill Education, and McGraw-Hill Education does not guarantee the accuracy of the information presented at these sites.

mheducation.com/highered

Brief Contents

1. An Introduction to Biology 1

UNIT I Chemistry 20
2. The Chemical Basis of Life I: Atoms, Molecules, and Water 21
3. The Chemical Basis of Life II: Organic Molecules 37

UNIT II Cells 58
4. Evolutionary Origin of Cells and Their General Features 59
5. Membranes: The Interface Between Cells and Their Environment 98
6. How Cells Utilize Energy 121
7. How Cells Capture Light Energy via Photosynthesis 150
8. How Cells Communicate with Each Other and with the Environment 169

UNIT III Genetics 186
9. The Information of Life: DNA and RNA Structure, DNA Replication, and Chromosome Structure 187
10. The Expression of Genetic Information via Genes I: Transcription and Translation 209
11. The Expression of Genetic Information via Genes II: Non-coding RNAs 229
12. The Control of Genetic Information via Gene Regulation 248
13. Altering the Genetic Material: Mutation, DNA Repair, and Cancer 267
14. How Eukaryotic Cells Sort and Transmit Chromosomes: Mitosis and Meiosis 285
15. Transmission of Genetic Information from Parents to Offspring I: Patterns That Follow Mendel's Laws 309
16. Transmission of Genetic Information from Parents to Offspring II: Epigenetics, Linkage, and Extranuclear Inheritance 327
17. The Simpler Genetic Systems of Viruses and Bacteria 345
18. Genetic Technologies: How Biologists Study Genes and Genomes 363

UNIT IV Evolution 386
19. Evolution of Life I: How Populations Change from Generation to Generation 387
20. Evolution of Life II: The Emergence of New Species 416
21. How Biologists Classify Species and Study Their Evolutionary Relationships 433
22. The History of Life on Earth and Human Evolution 450

UNIT V Diversity 476
23. Diversity of Microbial Life: Archaea, Bacteria, Protists, and Fungi 477
24. Microbiomes: Microbial Systems on and Around Us 510
25. Plant Evolution: How Plant Diversification Changed Planet Earth 528
26. Invertebrates: The Vast Array of Animal Life Without a Backbone 552
27. Vertebrates: Fishes, Amphibians, Reptiles, and Mammals 586

UNIT VI Flowering Plants 608
28. An Introduction to Flowering Plant Form and Function 609
29. How Flowering Plants Sense and Interact with Their Environments 630
30. How Flowering Plants Obtain and Transport Water, Mineral Nutrients, and Organic Compounds 648
31. How Flowering Plants Reproduce and Develop 672

UNIT VII Animals 692
32. General Features of Animal Bodies, and Homeostasis as a Key Principle of Animal Biology 693
33. Neuroscience I: The Structure, Function, and Evolution of Nervous Systems 714
34. Neuroscience II: How Sensory Systems Allow Animals to Interact with the Environment 744
35. How Muscles and Skeletons Are Adaptations for Movement, Support, and Protection 765
36. Circulatory and Respiratory Systems: Transporting Solutes and Exchanging Gases 780
37. Digestive and Excretory Systems: Maintaining Nutrient, Water, and Energy Balance and Removing Waste 813
38. How Endocrine Systems Influence the Activities of All Other Organ Systems 848
39. The Production of Offspring: Reproduction and Development 870
40. Immune Systems: How Animals Defend Against Pathogens and Other Dangers 895
41. Integrated Responses of Animal Organ Systems to a Challenge to Homeostasis 916

UNIT VIII Ecology 930
42. Behavioral Ecology: The Struggle to Find Food and Mates and to Pass on Genes 931
43. Population Growth and Species Interactions 948
44. Communities and Ecosystems: Ecological Organization at Large Scales 965
45. Biomes: How Climate Affects the Distribution of Species on Earth 988
46. The Age of Humans 1002
47. Biodiversity and Conservation Biology 1027

About the Authors

Robert J. Brooker

Rob Brooker received his Ph.D. in genetics from Yale University in 1983. At Harvard, he studied lactose permease, the product of the *lacY* gene of the *lac* operon. He continued working on transporters at the University of Minnesota, where he is a professor in the Department of Genetics, Cell Biology, and Development. At the University of Minnesota, Dr. Brooker teaches undergraduate courses in biology and genetics. In addition to many other publications, he has written two undergraduate genetics texts: *Genetics: Analysis & Principles*, seventh edition, copyright 2021, and *Concepts of Genetics*, third edition, copyright 2019; and he is the lead author of *Biology*, fifth edition, copyright 2020, all published by McGraw-Hill Education.

Eric P. Widmaier

Eric Widmaier received his Ph.D. in 1984 in endocrinology from the University of California at San Francisco. His research focuses on the control of body mass and metabolism in mammals, the hormonal correlates of obesity, and the effects of high-fat diets on intestinal cell function. Dr. Widmaier is currently professor emeritus of biology at Boston University, where he has taught undergraduate and graduate courses in human physiology, comparative physiology, and endocrinology and recently received the university's highest honor for excellence in teaching. Among other publications, he is lead author of *Vander's Human Physiology: The Mechanisms of Body Function*, fifteenth edition, copyright 2019, and coauthor of *Biology*, fifth edition, copyright 2020, both published by McGraw-Hill Education.

Linda E. Graham

Linda Graham received her Ph.D. in botany from the University of Michigan, Ann Arbor, where she also did postdoctoral work. Her research explores the evolutionary origin of algae and land-adapted plants, focusing on cell and molecular biology as well as ecological interactions with microbes. Dr. Graham is currently a professor of botany at the University of Wisconsin–Madison. She teaches courses in microbiology and plant biology. She is the coauthor of, among other publications, *Algae*, third edition, copyright 2015, a textbook on algal biology; and *Plant Biology*, third edition, copyright 2015,

Left to right: Eric Widmaier, Linda Graham, Peter Stiling, and Rob Brooker

both published by LJLM Press. She is also a coauthor of *Biology*, fifth edition, copyright 2020, published by McGraw-Hill Education.

Peter D. Stiling

Peter Stiling obtained his Ph.D. in zoology from University College, Cardiff, United Kingdom. Subsequently, he became a postdoctoral fellow at Florida State University and later spent two years as a lecturer at the University of the West Indies, Trinidad. Dr. Stiling was formerly chair of the Department of Integrative Biology at the University of South Florida (USF) at Tampa, where he is currently an assistant vice provost for strategic initiatives and professor of biology. His research interests include plant-animal relationships and invasive species. He currently teaches biology to USF students in a London summer program, which he established in 2015. Dr. Stiling was elected an AAAS Fellow in 2012. He is also the author of *Ecology: Global Insights and Investigations*, second edition, copyright 2015, and coauthor of *Biology*, fifth edition, copyright 2020, both published by McGraw-Hill Education.

Principles of Biology: Helping Students Learn Critical-Thinking Skills

The third edition of *Principles of Biology* has been crafted with a very important goal in mind: Give students an opportunity to develop critical-thinking skills. We expect that their journey through this textbook will help them to "think like scientists" and to develop skills that are needed in many different careers in biology.

EMPHASIZING SKILLS DEVELOPMENT AND PROBLEM SOLVING

Skills! Skills! Skills! That's what instructors want their students to develop and that's what students often find the most challenging. The pedagogical features of *Principles of Biology*, 3e, are aimed at achieving this goal. Much like the way that an enzyme lowers the activation energy for a chemical reaction and thereby increases its rate, the author team has developed pedagogical features that should lower the barriers that prevent skill development and thereby allow students to develop their critical-thinking skills more quickly.

New! Critical-Thinking Questions at the End of Each Chapter

In the third edition of *Principles of Biology*, we have completely reorganized our end-of-chapter questions and have a new category of questions called **Critical-Thinking Skills.** These are largely new questions that are primarily at Bloom's levels 3 (applying) and 4 (analyzing).

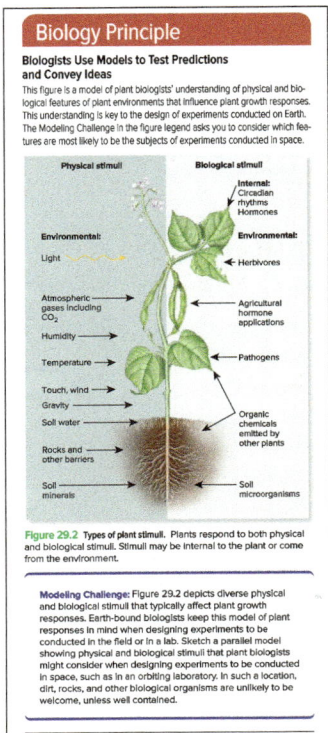

New! Modeling Challenges

A growing trend is the use of models in biology education. Students are asked to interpret models and to create models based on data or a scenario. Furthermore, using models and simulations is one of the core skills that is emphasized by "Vision and Change." The author team has added a new feature called a **Modeling Challenge,** which asks students to create a model or to interpret a model they are given. Possible answers to the Modeling Challenges are provided in Connect.

SCISKILLS

The beginning of each section of every chapter contains a set of Learning Outcomes that inform students of concepts they should understand. Many sections contain skills-based Learning Outcomes, labeled as **SCISKILLS**. These Learning Outcomes are specific to the skills that students will acquire when mastering the material and provide a specific understanding of how such skills may be assessed. SCISKILLS are mental actions such as analyzing data, forming hypotheses, making predictions, or performing calculations. These are skills that scientists generally rely on and students should practice.

5.2 Fluidity of Membranes

Learning Outcomes

1. Describe the fluidity of membranes.
2. **SCISKILLS** ▶ Predict how fluidity will be affected by changes in lipid composition.
3. **SCISKILLS** ▶ Analyze the results of experiments indicating that certain membrane proteins can diffuse laterally within the membrane.

Let's now turn our attention to the dynamic properties of membranes. Although a membrane provides a critical interface between a cell or an organelle and its environment, it is not a solid, rigid structure.

Feature Investigations

The emphasis on skills development continues in the **Feature Investigations**. Feature Investigations provide a complete description of experiments, including data analysis, so that students can understand how experimentation leads to an understanding of biological concepts.

FEATURE INVESTIGATION

Fire and Mello Showed That Double-Stranded RNA Is More Potent Than Antisense RNA in Silencing mRNA

Specific mRNAs can be targeted for translational inhibition or degradation by a mechanism involving double-stranded RNA. This mechanism was discovered during research involving plants and the nematode worm *Caenorhabditis elegans*. The study described here involved an examination of gene expression in *C. elegans*.

American biologists Andrew Fire, Craig Mello, and colleagues used *C. elegans* as their experimental organism, because it is relatively easy to inject with RNA and the expression of many of its genes had been established. In 1998, Fire and Mello investigated the effects of injected RNA on the expression of specific mRNAs. In the investigation described in Figure 11.7, we will focus on one of their experiments involving an mRNA encoded by a gene called *mex-3*. This mRNA had already been shown to be made in high amounts in early embryos of *C. elegans*.

Prior to this work, the *mex-3* gene had been identified and inserted into a plasmid. The process of inserting genes into plasmids is described in Chapter 18 (look ahead to Figure 18.2). Let's first look at the upper plasmid shown in the "Conceptual level" column in step 1. When RNA polymerase, nucleotides, and this plasmid were mixed together in a test tube, *mex-3* mRNA was made, which is called the sense strand. In living cells, the sense strand is used to make the mex-3 protein. Fire and Mello also switched the location of the promoter so that it was at the other end of the gene and signaled transcription of the opposite strand, which is called the antisense strand (see the second plasmid in the "Conceptual level" column). The sense and antisense strands are complementary to each other.

Next, they injected these RNAs into the gonads of *C. elegans*. Into some worms, they injected antisense RNA alone. Alternatively, they mixed sense and antisense RNA, which formed double-stranded RNA, and injected this double-stranded RNA into the gonads of other worms. They also used uninjected worms as controls. After injection, the RNA was taken up by eggs, which later developed into embryos. To determine the amount of *mex-3* mRNA present, Fire and Mello incubated the embryos with a probe that was complementary to

Figure 11.7 Injection of antisense and double-stranded RNA into *C. elegans* to compare their effects on mRNA silencing.

GOAL The goal was to further understand how the experimental injection of RNA was responsible for the silencing of particular mRNAs.

KEY MATERIALS The researchers used *C. elegans* as their model organism. They also had the cloned *mex-3* gene, which had been previously shown to be highly expressed in *C. elegans* embryos.

Experimental level / **Conceptual level**

1. Make sense and antisense *mex-3* RNA in vitro using cloned genes for *mex-3* with promoters on either side of the gene. RNA polymerase and nucleotides are added to synthesize the RNAs.

2. Inject either *mex-3* antisense RNA or a mixture of *mex-3* sense and antisense RNA into the gonads of *C. elegans*. This RNA is taken up by the eggs and early embryos. As a control, do not inject any RNA.

green color. Remarkably, in embryos that had received double-stranded RNA, no *mex-3* mRNA was detected! These results indicated that double-stranded RNA is more potent at silencing mRNA than is antisense RNA. In this case, the double-stranded RNA caused the *mex-3* mRNA to be degraded. Fire and Mello used the term **RNA interference (RNAi)** to describe the phenomenon in which double-stranded RNA causes the silencing of mRNA. This surprising observation led researchers to investigate the underlying molecular mechanism that accounts for this phenomenon, as described next.

6. **CONCLUSION** Double-stranded RNA is more potent at inhibiting *mex-3* mRNA than antisense RNA alone.

7. **SOURCE** Fire, A., Xu, S., Montgomery, M. K., et al. 1998. Potent and Specific Genetic Interference by Double-Stranded RNA in *Caenorhabditis elegans*. *Nature* 391: 806–811.

Experimental Questions

1. In this experiment, does the *mex-3* mRNA correspond to the sense strand or antisense strand?
2. **SCISKILLS** ▶ Explain how the sense and antisense *mex-3* RNAs were made.
3. **SCISKILLS** ▶ According to the data, which material was the most effective at causing the degradation of *mex-3* mRNA?

Quantitative Analysis

The **Quantitative Analysis** feature helps students develop their analytical skills. This feature walks them through biological concepts that have a quantitative component. The concluding Crunching the Numbers provides a sample problem to test understanding.

> **Biology Principle** | **Biology is a quantitative science**
>
> **Quantitative Analysis:** Alternative Splicing Is More Prevalent in Complex Eukaryotic Species
>
> Table 12.1 compares six species: a bacterium (*Escherichia coli*), a eukaryotic single-celled organism (yeast—*Saccharomyces cerevisiae*), a small nematode worm with about 1,000 cells (*Caenorhabditis elegans*), a fruit fly (*Drosophila melanogaster*), a flowering plant (*Arabidopsis thaliana*), and a human (*Homo sapiens*). One general trend is that less complex organisms tend to have fewer genes. For example, unicellular organisms have only a few thousand genes, whereas multicellular species have tens of thousands. However, the trend is by no means a linear one. If we compare *C. elegans* and *D. melanogaster*, the fruit fly actually has fewer genes, even though it is morphologically more complex.
>
> A second trend you can see in Table 12.1 concerns alternative splicing. This phenomenon does not occur in bacteria and is rare in *S. cerevisiae*. The frequency of alternative splicing increases from worms to flies to humans. For example, the level of alternative splicing is 10-fold higher in humans than in *Drosophila*. This trend can partially explain the increase in complexity among these species.
>
> **Crunching the Numbers:** As a rough estimate, let's suppose that the pre-mRNAs from 70% of all human genes are alternatively spliced and that an average pre-mRNA that is alternatively spliced can be spliced in five different ways. By comparison, only about 7% of the pre-mRNAs from *Drosophila* are alternatively spliced; an average pre-mRNA that is alternatively spliced can be spliced in three different ways. With a human genome size of about 22,000 protein-encoding genes and a *Drosophila* genome size of 15,600 protein-encoding genes, calculate the numbers of mature mRNAs that can be made by each species. Discuss how these numbers affect the morphological complexity of humans and *Drosophila*.

BioTIPS

In Connect, the digital partner to this textbook, we have a feature called **BioTIPS**, which is intended to help students refine problem-solving skills. Most of the **BioTIPS** are called out with icons in the textbook, but additional **BioTIPS** are included in SmartBook. The **BioTIPS** themselves are accessed through links in SmartBook. **BioTIPS** focus on 11 strategies that will help students solve problems:

1. *Make a drawing*.
2. *Compare and contrast*.
3. *Relate structure and function*.
4. *Sort out the steps in a complicated process*.
5. *Propose a hypothesis*.
6. *Design an experiment*.
7. *Predict the outcome*.
8. *Interpret data*.
9. *Use statistics*.
10. *Make a calculation*.
11. *Search the literature*.

BioTIPS will provide students with practice at applying these 11 problem-solving strategies.

A biological question related to chapter content is posed. The **BioTIPS** then walk the student through the process of answering the question. First, they help the student identify the topic of the question—what is really being asked in the question? Then they help the student collect information that was presented in the chapter that is related to the question. Finally, they help the student settle on one or more strategies that can be followed to answer the question. The answers are provided to complete the problem-solving process.

Skills! Skills! Skills!

Overall, *Principles of Biology,* third edition, has six pedagogical features that foster skill development.

1. Critical-Thinking Questions
2. Modeling Challenges
3. SCIKILLS Learning Outcomes
4. Feature Investigations
5. Quantitative Analyses
6. BioTIPS

We expect that these features will help to lower the "activation energy barrier," so that students can succeed in their skill development.

Principles of Biology: Helping Students Learn Cutting-Edge Content

We can all agree that biology textbooks cover a lot of material. To help students see the "forest and not just the trees" and to learn concepts that are essential to their future careers, the author team has developed pedagogical features to help students navigate the content in *Principles of Biology,* 3e.

WAYS TO HELP STUDENTS NAVIGATE AND LEARN THE MATERIAL IN THIS TEXTBOOK

Focus on Core Principles

Although it is intended for majors in the biological sciences, *Principles of Biology* is a shorter textbook that emphasizes core principles. Twelve principles of biology are enunciated in Chapter 1, and those principles are emphasized throughout the textbook. This emphasis is achieved in two ways. First, the principles are highlighted in selected figures in which a specific principle is illustrated.

In addition, a question at the end of each chapter is directly aimed at exploring a particular biology principle related to the content of the chapter.

Short Answer

1. Compare and contrast the different types of bonds commonly found in biological molecules.
2. **PRINCIPLES** A principle of biology is that new properties emerge from complex interactions. What examples can you cite from this chapter of emergent properties of molecules, in which atoms with one type of property combine to form molecules with completely different properties?

Learning Outcomes

Each section of every chapter begins with a set of learning outcomes. These outcomes help students understand what they should be able to do if they have mastered the material in that section.

Formative Assessment

Many instructors are devoting more class time to active learning and spending less time on directly discussing the textbook material. When students are expected to learn more textbook material on their own, it is imperative that they be given regular formative assessments, so that they can gauge whether or not they are mastering the material. Formative assessment is a major feature of this textbook and is bolstered by McGraw-Hill Connect®—a state-of-the-art digital assignment and assessment platform. In *Principles of Biology,* formative assessment is provided in multiple ways.

1. Each section of every chapter ends with a set of multiple-choice questions.
2. Many figures have Concept Check questions, so that students can determine if they understand the key points in the figure.
3. End-of-chapter questions continue to provide students with feedback regarding their mastery of the material.
4. Further assessment tools are available in Connect. Question banks, test banks, quantitative questions, and graphing interactive banks can be assigned by the professor. McGraw-Hill SmartBook® provides students with real-time assessment and feedback of their learning as they are moving through each chapter, page-by-page.

Biology Principle

The Genetic Material Provides a Blueprint That Allows Organisms to Grow, Develop, and Reproduce
The covalent linkage of a sequence of bases allows DNA to store information.

Figure 9.5 The structure of a DNA strand. Nucleotides are covalently bonded to each other in a linear manner. Notice the directionality of the strand and that it carries a particular sequence of bases. An RNA strand has a very similar structure, except the sugar is ribose rather than deoxyribose and uracil is

BioConnections and Evolutionary Connections

To help students broaden their understanding of biology, two recurring features are **BioConnections** and **Evolutionary Connections**. BioConnections are placed in key figure legends in most chapters and help students relate a topic they are currently learning to another topic elsewhere in the textbook, often in a different unit. Evolutionary Connections provide a framework for understanding how a topic in a given chapter relates to evolution, the core unifying theme in biology.

Figure 12.8 Positive control of the *lac* operon by the catabolite activator protein (CAP). When cAMP is bound to CAP, CAP binds to the DNA and causes it to bend. This bend facilitates the binding of RNA polymerase.

Image from the RCSB PDB (www.rcsb.org; H.M. Berman, et al. (2000), "The Protein Data Bank," *Nucleic Acids Research*, 28: 235–242) of PDB ID 1CGP (Schultz, S.C., Shields, G.C., Steitz, T.A.) (1991) *Science*, 253: 1001–1007.

BioConnections: Refer back to Figure 8.13. What is the function of cAMP in eukaryotic cells?

Unit Openers

Unit openers serve two purposes. They allow the student to see the big picture of the unit. In addition, the unit openers draw attention to the principles of biology that will be emphasized in that unit.

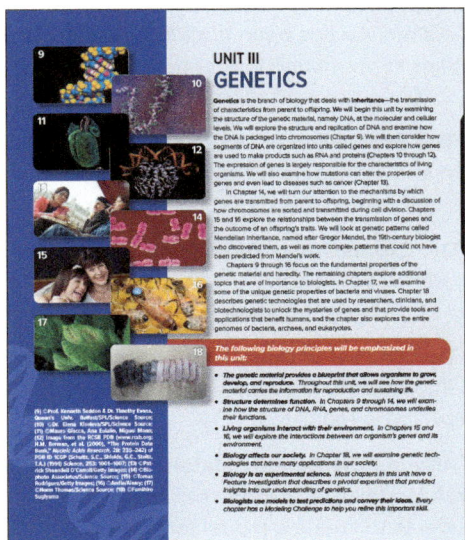

Overall, the pedagogy of *Principles of Biology* has been designed to foster student learning. Instead of being a collection of "facts and figures," *Principles of Biology* is intended to be an engaging and motivating textbook in which formative assessment and many other pedagogical features allow students to move ahead and learn the material in a productive way.

Content Changes to the Third Edition

The following are the major content changes that have occurred in the third edition.

Chapter 1 An Introduction to Biology. This chapter has a new section on the adaptations that have occurred during the evolution of a family of species known as birds-of-paradise.

Chapter 2 The Chemical Basis of Life I: Atoms, Molecules, and Water. A Modeling Challenge has been added to Figure 2.2 to help students interpret models that depict the structure of atoms.

Chapter 4 Evolutionary Origin of Cells and Their General Features. The topic of droplet organelles, which are organelles formed by liquid-liquid phase separation, has been added. An example is the nucleolus. A Modeling Challenge has been added that helps students understand how materials move through the Golgi apparatus.

Chapter 5 Membranes: The Interface Between Cells and Their Environment. The topic of osmosis has been clarified to explain why water moves in a particular direction.

Chapter 6 How Cells Utilize Energy. A new section (Section 6.9) has been added on the topic of anaerobic respiration and fermentation, and it includes two new figures.

Chapter 8 How Cells Communicate with Each Other and with the Environment. A Modeling Challenge has been added to Figure 8.9 that asks students to create a model that explains how the effects of estrogen are reversed.

Chapter 9 The information of Life: DNA and RNA Structure, DNA Replication, and Chromosome Structure. The information on chromosome structure has been updated by describing the role of SMC proteins in promoting the formation of radial loop domains (see Figure 9.21).

Chapter 10 The Expression of Genetic Material via Genes I. Transcription and Translation. New information regarding the spliceosome has been added.

Chapter 11 The Expression of Genetic Information via Genes II: Non-coding RNAs. Section 11.4 has been streamlined by focusing on microRNAs (miRNAs) and small-interfering RNAs (siRNAs).

Chapter 12 The Control of Genetic Information via Gene Regulation. Information has been added that describes how alternative splicing is regulated.

Chapter 13 Altering the Genetic Material: Mutation, DNA Repair, and Cancer. A Modeling Challenge has been added to Table 13.4 that asks students to create a model that shows how 5-bromouracil can cause a mutation.

Chapter 14 How Eukaryotic Cells Sort and Transmit Chromosomes: Mitosis and Meiosis. The topic of checkpoints has been expanded and reorganized.

Chapter 16 Transmission of Genetic Information from Parents to Offspring II: Epigenetics, Linkage, and Extranuclear Inheritance. New information has been added regarding differences in epigenetic changes between finches in rural and urban environments.

Chapter 17 The Simpler Genetic Systems of Viruses and Bacteria. Section 17.1 from the second edition is now divided into two sections in the third edition.

Chapter 18 Genetic Technologies: How Biologists Study Genes and Genomes. A few of the newer methods of DNA sequencing are briefly discussed.

Chapter 19 Evolution of Life I: How Populations Change from Generation to Generation. A Modeling Challenge has been added to Figure 19.14 that asks students to draw a series of graphs that depict the effects of diversifying selection.

Chapter 21 How Biologists Classify Species and Study Their Evolutionary Relationships. Information on the eukaryotic supergroups has been updated.

Chapter 22 The History of Life on Earth and Human Evolution. The topic of human evolution has been updated.

Chapter 23 Diversity of Microbial Life: Archaea, Bacteria, Protists, and Fungi. This chapter on the diversity of prokaryotic and eukaryotic microbial life was heavily revised in the second edition to integrate material previously covered in separate chapters. A Modeling Challenge was added to Figure 23.22 that asks the student to add endosymbiotic events to the eukaryotic phylogenetic tree.

Chapter 24 Microbiomes: Microbial Systems on and Around Us. This chapter, added to the second edition, integrates current information about microbes (archaea, bacteria, protists, and fungi) within complex organism-gene systems known as microbiomes, a major frontier of biological sciences. A Modeling Challenge was added to Figure 24.10 that asks the student to evaluate the different concepts of lichen microbiomes.

Chapter 25 Plant Evolution: How Plant Diversification Changed Planet Earth. A Modeling Challenge was added to Figure 25.10 that asks the student to sketch a graph that depicts the changes in carbon dioxide levels in Earth's atmosphere over time.

Chapter 26 Invertebrates: The Vast Array of Animal Life Without a Backbone. A new chapter-opening vignette describes how seemingly complex patterns in butterfly wings are linked to just a single gene. The implication of this work is that much of the Earth's biodiversity may be caused by relatively simple genetic changes. Several new BioConnections have been added to existing figures.

Chapter 27 Vertebrates: Fishes, Amphibians, Reptiles, and Mammals. New opening paragraphs discuss the longevity of naked mole rats. Researchers have speculated that a more detailed knowledge of naked mole rat biology may unlock the secrets of aging for humans and other animals. Speaking of aging, we discuss how Greenland sharks are probably the oldest living animals on Earth. A new section discusses the Cyclostomata, the jawless fishes. Since mammals are amniotes, a new Modeling Challenge added to Figure 27.11 asks students to draw the amniotic eggs of reptiles and eutherian mammals.

Chapter 28 An Introduction to Flowering Plant Form and Function. A Modeling Challenge was added to Figure 28.10 that asks the student to draw a model of the cell wall of genetically modified plants.

Chapter 29 How Flowering Plants Sense and Interact with Their Environments. Some new images were commissioned specifically for this textbook, to aid student comprehension of key concepts. A Modeling Challenge was added to Figure 29.2 that asks the student to plan an experiment to be performed in space examining plant stimuli.

Chapter 30 How Flowering Plants Obtain and Transport Nutrients. Some new images have been incorporated to foster student understanding. A Modeling Challenge was added to Figure 30.9 that asks the student to draw models that show how soil nutrients that are cations are affected by acid rain.

Chapter 31 How Flowering Plants Reproduce and Develop. A Modeling Challenge was added to Figure 31.10 that asks the student to draw a model that shows how synergids first interact with the egg cell and then foster fertilization when the pollen tube reaches the micropyle.

Chapter 32 General Features of Animal Bodies, and Homeostasis as a Key Principle of Animal Biology. The concept of how water distributes in an animal's body fluid compartments has been elaborated on in a new Modeling Challenge with Figure 32.13.

Chapter 33 Neuroscience I: The Structure, Function, and Evolution of Nervous Systems. The description of animal nervous systems has been reorganized under the banner of "Evolutionary Connections" to emphasize how nervous systems evolved from simple nerve nets to complex systems such as those of vertebrates. A Modeling Challenge has been added to Figure 33.8 on the topic of action potentials.

Chapter 34 Neuroscience II: How Sensory Systems Allow Animals to Interact with the Environment. New information has been added to the section on color vision in animals, including recent discoveries related to restoration of trichromatic vision in monkeys with dichromatic vision. A Modeling Challenge has been added to Figure 34.21 that asks students to predict the consequence of two different but simultaneously occurring visual disorders in humans.

Chapter 35 How Muscles and Skeletons Are Adaptations for Movement, Support, and Protection. Recent information that is revising scientists' understanding of the evolution of animal skeletons has been added. A Modeling Challenge has been added to Figure 35.8 on the cross-bridge cycle, specifically on how the cycle is arrested during rigor mortis.

Chapter 36 Circulatory and Respiratory Systems: Transporting Solutes and Exchanging Gases. The Quantitative Analysis feature that describes how blood pressure and resistance determine blood pressure has been supplemented with a Modeling Challenge that requires the student to further explore how these three important variables are related.

Chapter 37 Digestive and Excretory Systems: Maintaining Nutrient, Water, and Energy Balance and Removing Waste. The importance of an animal's microbiome to digestion is now emphasized. A more detailed description of the anatomy of alimentary canals and sphincters is included. A Modeling Challenge that relates to antidiuretic hormone has been added to Figure 37.23.

Chapter 38 How Endocrine Systems Influence the Activities of All Other Organ Systems. The complex feedback relationship between Ca^{2+} and parathyroid hormone has been edited for clarity and expanded on with a Modeling Challenge added to Figure 38.11.

Chapter 39 The Production of Offspring: Reproduction and Development. The Impact on Public Health section has been updated to reflect current statistical data on human reproductive or developmental diseases and disorders.

Chapter 40 Immune Systems: How Animals Defend Against Pathogens and Other Dangers. Recent new information regarding the evolution of toll-like receptors has been added. The structure and function of immunoglobulins have been elaborated on in a new Modeling Challenge with Figure 40.7. The two major mechanisms of cytotoxic T-cell action are now distinguished.

Chapter 41 Integrated Responses of Animal Organ Systems to a Challenge to Homeostasis. The discussion of the structure, function, and (new to this edition) evolution of baroreceptors and the baroreceptor reflex has been expanded, including a new Modeling Challenge with Figure 41.3 and new Evolutionary Connections feature.

Chapter 42 Behavioral Ecology: The Struggle to Find Food and Mates and to Pass On Genes. A new chapter-opening vignette describes how selective breeding of red foxes in Russia has produced different genetic lines of relatively tame foxes and more aggressive foxes. Genetic studies have shown how these behavioral differences are related to changes in the foxes' genomes.

Chapter 43 Population Growth and Species Interactions. New technological breakthroughs in ecology, including the use of drones and phone applications, are described. New technology has allowed ecologists to better document wildlife abundance in remote and inaccessible locations and document the spread of tree species across the United States.

Chapter 44 Communities and Ecosystems: Ecological Organization of Large Scales. A new subsection in Section 44.6, together with Figure 44.20 and Figure 44.21, describes how the biomass distribution on Earth varies across taxa and location. Trees constitute by far the highest percentage of the Earth's biomass, about 80 percent.

Content Changes to the Third Edition

Chapter 45 Biomes: How Climate Affects the Distribution of Species on Earth. This chapter now provides greatly expanded coverage of the Earth's terrestrial biomes, with new text and many new figures. A new Modeling Challenge with Figure 45.11 asks students to use temperature and precipitation data to position the chaparral biome on a Whittaker plot.

Chapter 46 The Age of Humans. This chapter highlights new research that shows how weathering of the Earth's bedrock provides about a quarter of the nitrogen circulating through the nitrogen cycle. The discussion of eutrophication has been modernized using Lake Erie and Lake Washington as examples. Also, Figure 46.25 on the collapse of the Canadian cod industry has been updated, and Section 46.7 on invasive species has been simplified.

Chapter 47 Biodiversity and Conservation Biology. Section 47.2 provides a fuller discussion of the ethical reasons for conserving biodiversity, while a new section, 47.3, discusses biodiversity and ecosystem function. Section 47.4 has simplified the debate around conservation strategies. The chapter, and the book, concludes with a new subsection on conservation and sustainability.

Strengthen Critical-Thinking Skills with Connect®

Critical-Thinking Skills

Detailed Feedback in Connect®

Learning is a process of iterative development, of making mistakes, reflecting, and adjusting over time. The question and test banks in Connect® for *Principles of Biology,* third edition, are more than direct assessments; they are self-contained learning experiences that systematically build student learning over time.

For many students, choosing the right answer is not necessarily based on applying content correctly; it is more a matter of increasing their statistical odds of guessing. A major fault with this approach is that students don't learn how to process the questions correctly, mostly because they are repeating and reinforcing their mistakes rather than reflecting and learning from them. To help students develop problem-solving skills, all higher-level Bloom's questions in Connect® are supported with hints, to help students focus on important information for answering the question. After submitting an answer, the student is given detailed feedback that walks through the problem-solving process, using Socratic questions in a decision tree–style framework to scaffold learning. Each step models and reinforces the learning process.

The feedback for each higher-level Bloom's question (Apply, Analyze, Evaluate) follows a similar process: Clarify Question, Gather Content, Choose Answer, Reflect on Process.

Unpacking the Concept

We've taken problem solving a step further. In each chapter, two higher-level Bloom's questions from the question and test banks are broken down according to the steps in the detailed feedback.

Rather than leaving it up to the student to work through the detailed feedback, we present a second version of the question in a stepwise format. Following the problem-solving steps, students need to answer questions about the problem-solving process such as "What is the key concept addressed by the question?" before answering the original question. A professor can choose which version of the question to include in the assignment based on the problem-solving skills of the students.

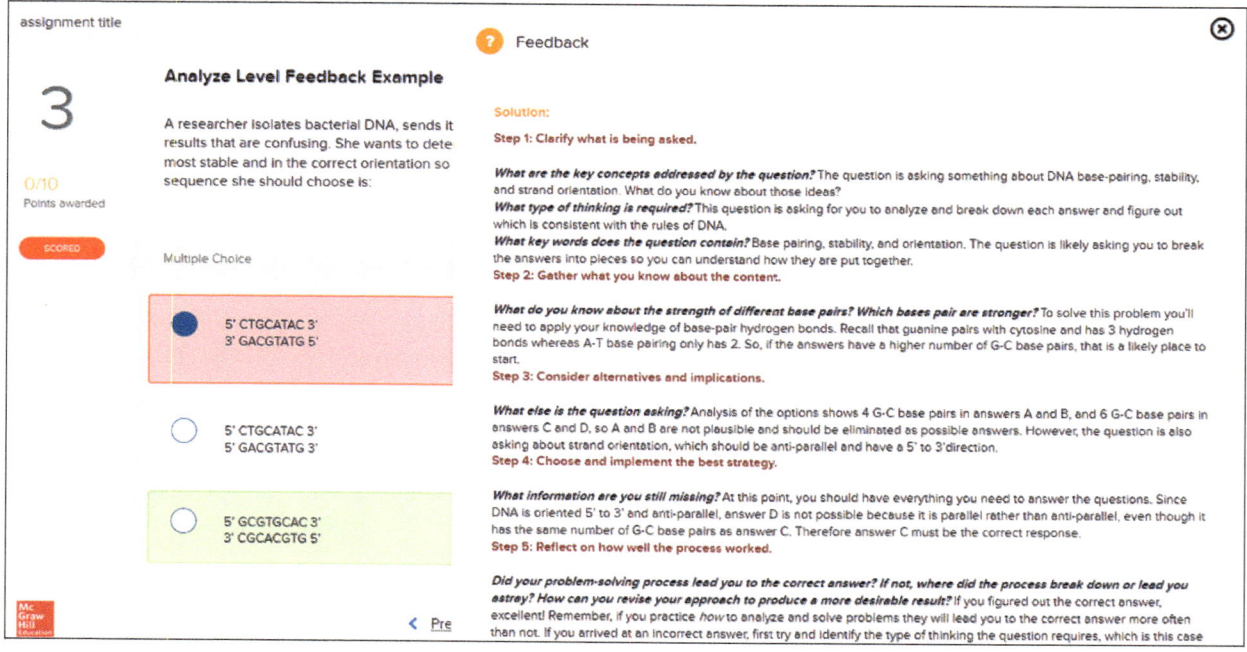

Quantitative and Data Analysis

Data and Graphing Interactives Bank

To help students develop analytical skills, Connect® for *Principles of Biology*, third edition, is enhanced with interactive data and graphing questions. Students are presented with a scientific problem and the opportunity to manipulate variables and compare different data. A series of questions follows the activity to assess if the student understands and is able to interpret the data and results.

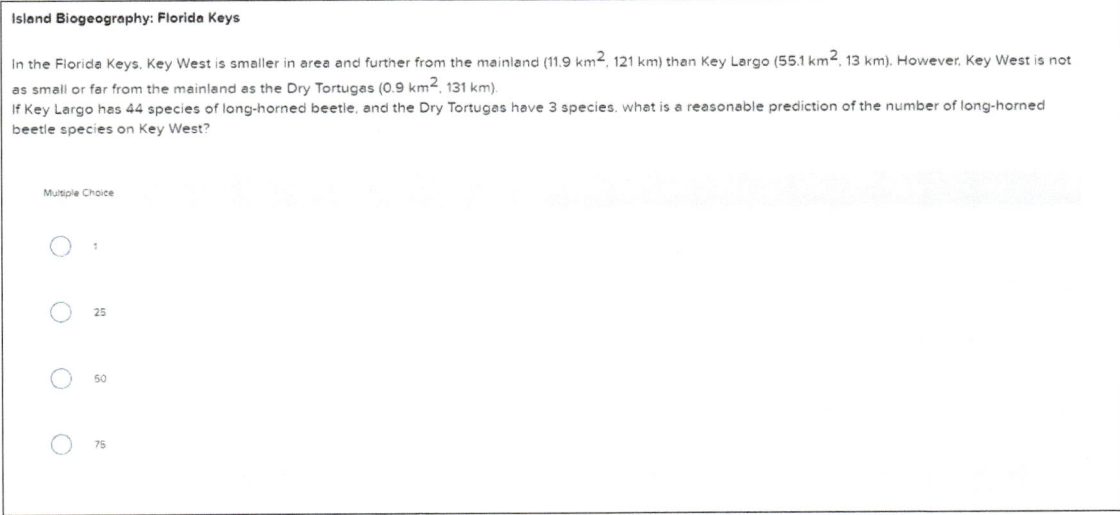

Quantitative Question Bank

The quantitative question bank contains more challenging algorithmic questions, intended to help students practice their quantitative reasoning skills. Hints and guided solution options step students through a problem before allowing them to tackle a similar problem.

FOR INSTRUCTORS

You're in the driver's seat.

Want to build your own course? No problem. Prefer to use our turnkey, prebuilt course? Easy. Want to make changes throughout the semester? Sure. And you'll save time with Connect's auto-grading too.

65%
Less Time Grading

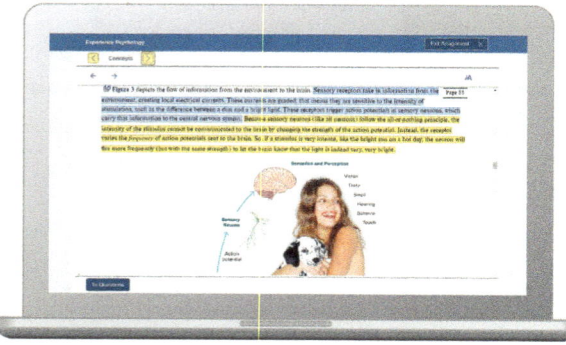

Laptop: McGraw-Hill; Woman/dog: George Doyle/Getty Images

They'll thank you for it.

Adaptive study resources like SmartBook® 2.0 help your students be better prepared in less time. You can transform your class time from dull definitions to dynamic debates. Find out more about the powerful personalized learning experience available in SmartBook 2.0 at **www.mheducation.com/highered/connect/smartbook**

Make it simple, make it affordable.

Connect makes it easy with seamless integration using any of the major Learning Management Systems—Blackboard®, Canvas, and D2L, among others—to let you organize your course in one convenient location. Give your students access to digital materials at a discount with our inclusive access program. Ask your McGraw-Hill representative for more information.

Padlock: Jobalou/Getty Images

Solutions for your challenges.

A product isn't a solution. Real solutions are affordable, reliable, and come with training and ongoing support when you need it and how you want it. Our Customer Experience Group can also help you troubleshoot tech problems—although Connect's 99% uptime means you might not need to call them. See for yourself at **status.mheducation.com**

Checkmark: Jobalou/Getty Images

FOR STUDENTS

Effective, efficient studying.

Connect helps you be more productive with your study time and get better grades using tools like SmartBook 2.0, which highlights key concepts and creates a personalized study plan. Connect sets you up for success, so you walk into class with confidence and walk out with better grades.

Study anytime, anywhere.

Download the free ReadAnywhere app and access your online eBook or SmartBook 2.0 assignments when it's convenient, even if you're offline. And since the app automatically syncs with your eBook and SmartBook 2.0 assignments in Connect, all of your work is available every time you open it. Find out more at **www.mheducation.com/readanywhere**

> *"I really liked this app—it made it easy to study when you don't have your textbook in front of you."*
>
> —Jordan Cunningham, Eastern Washington University

No surprises.

The Connect Calendar and Reports tools keep you on track with the work you need to get done and your assignment scores. Life gets busy; Connect tools help you keep learning through it all.

Calendar: owattaphotos/Getty Images

Learning for everyone.

McGraw-Hill works directly with Accessibility Services Departments and faculty to meet the learning needs of all students. Please contact your Accessibility Services office and ask them to email accessibility@mheducation.com, or visit **www.mheducation.com/about/accessibility** for more information.

Top: Jenner Images/Getty Images, Left: Hero Images/Getty Images, Right: Hero Images/Getty Images

Acknowledgments

The lives of most science-textbook authors do not revolve around an analysis of writing techniques. Instead, we are people who understand science and are inspired by it, and we want to communicate that information to our students. Simply put, we need a lot of help to get it right.

Editors are a key component that help the authors modify the content of their book so it is logical, easy to read, and inspiring. The editorial team for *Principles of Biology*, third edition, has been a catalyst that kept this project rolling. The members played various roles in the editorial process. Andrew Urban, Brand Manager for Majors Biology, did an outstanding job of overseeing the development of this new text. Elizabeth Sievers, Senior Product Developer, has been the master organizer. Liz's success at keeping us on schedule is greatly appreciated.

We would also like to acknowledge our copy editor, Deb DeBord, for keeping our grammar on track, as well as Angie Sigwarth and Jane Hoover for their excellent efforts at proofreading.

Another important aspect of the editorial process is the actual design, presentation, and layout of materials. It's confusing if the text and art aren't near each other or if a figure is too large or too small. We are indebted to the tireless efforts of Jessica Portz, Content Project Manager, and David Hash, Senior Designer, at McGraw-Hill Education. Likewise, our production company, MPS Limited, did an excellent job with the paging and art revisions.

We would like to acknowledge the ongoing efforts of the superb marketing staff at McGraw-Hill Education. Special thanks to Kelly Brown, Executive Marketing Manager, Life Sciences, for his ideas and enthusiasm for this book.

Other staff members at McGraw-Hill Education have ensured that the authors and editors were provided with adequate resources to achieve the goal of producing a superior textbook. These include G. Scott Virkler, Senior Vice President, Products & Markets; Michael Ryan, Vice President, General Manager, Products & Markets; and Betsy Whalen, Vice President, Production and Technology Services.

We would like to thank the digital authors and subject matter experts who helped in the development of the digital assets in Connect® that support *Principles of Biology*, third edition.

The authors are grateful for the help, support, and patience of their families, friends, and students: Deb, Dan, Nate, and Sarah Brooker; Maria, Caroline, and Richard Widmaier; Jim, Michael, and Melissa Graham; and Jacqui, Zoe, Leah, and Jenna Stiling.

Reviewers for *Principles of Biology*, third edition

Fernando Agudelo-Silva *College of Marin*

Sara E. Blumer-Schuette *Oakland University*

Becky Brown *College of Marin*

Silke Buschmann *Northern Arizona University*

Elizabeth C. Co *Boston University*

Janet R. Donaldson *The University of Southern Mississippi*

Jai Dwivedi *St. John's University*

Mary Ebersole *Northern Virginia Community College Manassas*

Markus Friedrich *Wayne State University*

Rohini Ganjoo *George Washington University*

Yan-Lin Guo *The University of Southern Mississippi*

Melissa A. Gutierrez *The University of Southern Mississippi*

Sarah Hosch *Oakland University*

Lisa Johansen *University of Colorado Denver*

Gerard R. Jozwiak *Oakland University*

David G. Knochel *University of Colorado Denver*

Cynthia W. Littlejohn *The University of Southern Mississippi*

Mario L. Muscedere *Boston University*

Matthew D. Overturf *University of Louisiana Monroe*

Thomas Sasek *University of Louisiana at Monroe*

Christopher Steiner *Wayne State University*

Shannon Stevenson *University of Minnesota Duluth*

Mark Sturtevant *Oakland University*

Dong-Hua Yang *St. John's University*

Xin Zhou *Northern Virginia Community College*

Rachel Zufferey *St. John's University*

CONTENTS

CHAPTER 1

An Introduction to Biology   1

1.1 Principles of Biology and the Levels of Biological Organization   2
1.2 Unity and Diversity of Life   7
Evolutionary Connections: The Study of Evolution Allows Us to Appreciate the Unity and Diversity Among Different Species   11
1.3 Biology as a Scientific Discipline   12

UNIT I   Chemistry

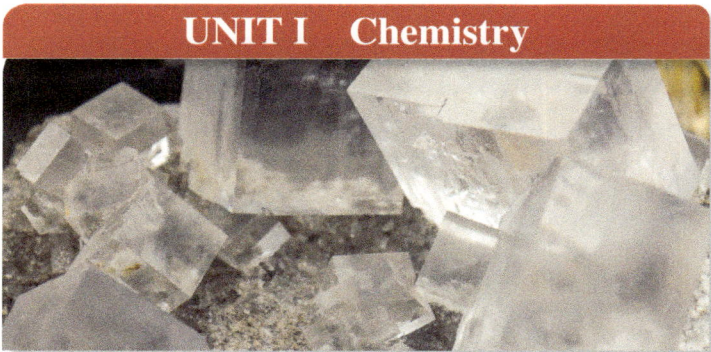

©Vladislav Gajic/Shutterstock

CHAPTER 2

The Chemical Basis of Life I: Atoms, Molecules, and Water   21

2.1 Atoms   22
2.2 Chemical Bonds and Molecules   25
2.3 Chemical Reactions   29
2.4 Properties of Water   30
Quantitative Analysis: Concentrations of Molecules in Solution Can Be Defined by Mass and Moles   31
2.5 pH and Buffers   34

CHAPTER 3

The Chemical Basis of Life II: Organic Molecules   37

3.1 The Carbon Atom and Carbon-Containing Molecules   38
3.2 Synthesis and Breakdown of Organic Molecules   40
3.3 Overview of the Four Major Classes of Organic Molecules Found in Living Cells   40
3.4 Carbohydrates   41
3.5 Lipids   44
3.6 Proteins   47

Feature Investigation: Anfinsen Showed That the Primary Structure of Ribonuclease Determines Its Three-Dimensional Structure   51
Evolutionary Connections: Proteins Contain Functional Domains   53
3.7 Nucleic Acids   54

UNIT II   Cells

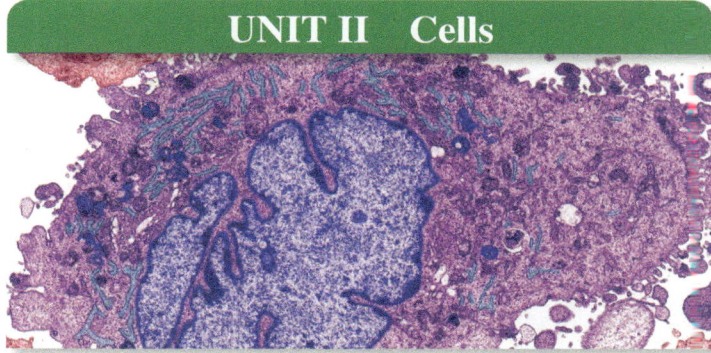

©Steve Gschmeissner/Science Source

CHAPTER 4

Evolutionary Origin of Cells and Their General Features   59

4.1 Origin of Living Cells on Earth   60
4.2 Microscopy   66
4.3 Overview of Cell Structure and Function   69
4.4 The Cytosol   74
4.5 The Nucleus and Endomembrane System   79
4.6 Semiautonomous Organelles   85
Evolutionary Connections: Mitochondria and Chloroplasts Are Derived from Ancient Symbiotic Relationships   88
4.7 Protein Sorting to Organelles   89
4.8 Extracellular Matrix and Plant Cell Walls   89
4.9 Systems Biology of Cells: A Summary   94

CHAPTER 5

Membranes: The Interface Between Cells and Their Environment   98

5.1 Membrane Structure   99
5.2 Fluidity of Membranes   100
5.3 Overview of Membrane Transport   103
5.4 Proteins That Carry Out Membrane Transport   107
Feature Investigation: Agre Discovered That Osmosis Occurs More Quickly in Cells with a Channel That Allows the Facilitated Diffusion of Water   107
5.5 Intercellular Channels   113
5.6 Exocytosis and Endocytosis   114
5.7 Cell Junctions   116

xvii

CHAPTER 6

How Cells Utilize Energy 121

- 6.1 Energy and Chemical Reactions 122
- 6.2 Enzymes 125
  Quantitative Analysis: Enzyme Function Is Influenced by Substrate Concentration and by Inhibitors 127
- 6.3 Overview of Metabolism and Cellular Respiration 129
- 6.4 Glycolysis 134
- 6.5 Breakdown of Pyruvate 137
- 6.6 Citric Acid Cycle 138
- 6.7 Oxidative Phosphorylation 138
  Feature Investigation: Yoshida and Kinosita Demonstrated That the γ Subunit of ATP Synthase Spins 143
- 6.8 Connections Among Carbohydrate, Protein, and Fat Metabolism 145
- 6.9 Anaerobic Respiration and Fermentation 146

CHAPTER 7

How Cells Capture Light Energy via Photosynthesis 150

- 7.1 Overview of Photosynthesis 151
- 7.2 Reactions That Harness Light Energy 153
- 7.3 Molecular Features of Photosystems 158
- 7.4 Synthesizing Carbohydrates via the Calvin Cycle 160
  Feature Investigation: The Calvin Cycle Was Determined by Isotope-Labeling Methods 162
- 7.5 Variations in Photosynthesis 165
  Evolutionary Connections: C_4 and CAM Plants Have Evolved a Mechanism to Minimize Photorespiration 165

CHAPTER 8

How Cells Communicate with Each Other and with the Environment 169

- 8.1 General Features of Cell Communication 170
- 8.2 Receptor Activation 173
  Quantitative Analysis: Receptors Have a Measurable Affinity for Their Ligands 174
- 8.3 Cell Surface Receptors 175
- 8.4 Intracellular Receptors 177
- 8.5 Signal Transduction and Cellular Response via an Enzyme-Linked Receptor 178
  Evolutionary Connections: Receptor Tyrosine Kinases Are Found in Choanoflagellates and Animals 178
- 8.6 Signal Transduction and Cellular Response via a G-Protein-Coupled Receptor 180

UNIT III Genetics

©Prof. Kenneth Seddon & Dr. Timothy Evans, Queen's Univ. Belfast/SPL/Science Source

CHAPTER 9

The Information of Life: DNA and RNA Structure, DNA Replication, and Chromosome Structure 187

- 9.1 Properties and Identification of the Genetic Material 188
  Feature Investigation: Avery, MacLeod, and McCarty Used Purification Methods to Reveal That DNA Is the Genetic Material 189
- 9.2 Nucleic Acid Structure 191
- 9.3 Discovery of the Double-Helix Structure of DNA 195
- 9.4 Overview of DNA Replication 197
- 9.5 Molecular Mechanism of DNA Replication 200
- 9.6 Molecular Structure of Eukaryotic Chromosomes 204

CHAPTER 10

The Expression of Genetic Information via Genes I: Transcription and Translation 209

- 10.1 Overview of Gene Expression 210
- 10.2 Transcription 211
- 10.3 RNA Modifications in Eukaryotes 213
- 10.4 Translation and the Genetic Code 216
  Feature Investigation: Nirenberg and Leder Found That RNA Triplets Can Promote the Binding of tRNA to Ribosomes 218
- 10.5 The Machinery of Translation 220
  Evolutionary Connections: Comparisons of Small Subunit rRNAs Among Different Species Provide a Basis for Establishing Evolutionary Relationships 224
- 10.6 The Stages of Translation 225

CHAPTER 11

The Expression of Genetic Information via Genes II: Non-coding RNAs 229

- 11.1 Overview of Non-coding RNAs 230

11.2 Role of Non-coding RNAs in Eukaryotic DNA Replication 234
11.3 Effects of Non-coding RNAs on Chromatin Structure and Transcription 235
11.4 Effects of Non-coding RNAs on Translation and mRNA Degradation 236
   Feature Investigation: Fire and Mello Showed That Double-Stranded RNA Is More Potent Than Antisense RNA in Silencing mRNA 237
11.5 Non-coding RNAs and Protein Sorting 239
11.6 Non-coding RNAs and Genome Defense 242
11.7 Roles of Non-coding RNAs in Human Disease and Plant Health 244

CHAPTER 12

The Control of Genetic Information via Gene Regulation 248

12.1 Overview of Gene Regulation 249
12.2 Regulation of Transcription in Bacteria 252
12.3 Regulation of Transcription in Eukaryotes: Roles of Transcription Factors 256
12.4 Regulation of Transcription in Eukaryotes: Changes in Chromatin Structure and DNA Methylation 259
12.5 Regulation of RNA Splicing and Translation in Eukaryotes 262
   Quantitative Analysis: Alternative Splicing Is More Prevalent in Complex Eukaryotic Species 264

CHAPTER 13

Altering the Genetic Material: Mutation, DNA Repair, and Cancer 267

13.1 Consequences of Mutations 268
13.2 Causes of Mutations 270
   Feature Investigation: The Lederbergs Used Replica Plating to Show That Mutations Are Random Events 271
   Quantitative Analysis: Testing Methods Determine If an Agent Is a Mutagen 274
13.3 DNA Repair 275
13.4 Cancer 277

CHAPTER 14

How Eukaryotic Cells Sort and Transmit Chromosomes: Mitosis and Meiosis 285

14.1 The Eukaryotic Cell Cycle 286
14.2 Mitotic Cell Division 291
   Evolutionary Connections: Cell Division in Bacteria Involves FtsZ, a Protein Related to Eukaryotic Tubulin 293
14.3 Meiosis and Sexual Reproduction 297
   Quantitative Analysis: Meiosis Enhances Genetic Diversity 301
14.4 Variation in Chromosome Structure and Number 303

CHAPTER 15

Transmission of Genetic Information from Parents to Offspring I: Patterns That Follow Mendel's Laws 309

15.1 Mendel's Laws of Inheritance 310
   Quantitative Analysis: A Punnett Square Is Used to Predict the Outcome of Crosses 314
15.2 Chromosome Theory of Inheritance 316
15.3 Pedigree Analysis of Human Traits 319
15.4 Variations in Inheritance Patterns and Their Molecular Basis 320
15.5 Sex Chromosomes and X-Linked Inheritance Patterns 324

CHAPTER 16

Transmission of Genetic Information from Parents to Offspring II: Epigenetics, Linkage, and Extranuclear Inheritance 327

16.1 Overview of Epigenetics 328
16.2 Epigenetics: Genomic Imprinting 329
16.3 Epigenetics: X-Chromosome Inactivation 331
16.4 Epigenetics: Effects of Environmental Agents 334
16.5 Extranuclear Inheritance: Organelle Genomes 337
   Evolutionary Connections: Chloroplast and Mitochondrial Genomes Are Relatively Small but Contain Genes That Encode Important Proteins 337
16.6 Linkage of Genes on the Same Chromosome 340
   Feature Investigation: Bateson and Punnett's Crosses of Sweet Peas Showed That Genes Do Not Always Assort Independently 340

CHAPTER 17

The Simpler Genetic Systems of Viruses and Bacteria 345

17.1 General Properties of Viruses 346
17.2 Viral Reproductive Cycles 349
17.3 Genetic Properties of Bacteria 354
17.4 Gene Transfer Between Bacteria 357
   Evolutionary Connections: Horizontal Gene Transfer Is the Transfer of Genes Between the Same or Different Species 360

CHAPTER 18

Genetic Technologies: How Biologists Study Genes and Genomes 363

18.1 Gene Cloning 364
   Quantitative Analysis: A DNA Library Is a Collection of Many Different DNA Fragments Cloned into Vectors 367
18.2 Genomics: Techniques for Studying and Altering Genomes 370

18.3 **Bacterial and Archaeal Genomes** 375
18.4 **Eukaryotic Genomes** 376
   Evolutionary Connections: Gene Duplications Provide Additional Material for Genome Evolution, Sometimes Leading to the Formation of Gene Families 378
18.5 **Repetitive Sequences and Transposable Elements** 380

UNIT IV Evolution

©Mark Dadswell/Getty Images

CHAPTER 19

Evolution of Life I: How Populations Change from Generation to Generation 387

19.1 **Overview of Evolution** 388
19.2 **Evidence of Evolutionary Change** 391
19.3 **Genes in Populations** 398
   Evolutionary Connections: Genes Are Usually Polymorphic 399
   Quantitative Analysis: The Hardy-Weinberg Equation Relates Allele and Genotype Frequencies in a Population 400
19.4 **Natural Selection** 402
   Feature Investigation: The Grants Observed Natural Selection in Galápagos Finches 406
19.5 **Genetic Drift** 409
19.6 **Migration and Nonrandom Mating** 413

CHAPTER 20

Evolution of Life II: The Emergence of New Species 416

20.1 **Identification of Species** 417
20.2 **Reproductive Isolation** 419
20.3 **Mechanisms of Speciation** 422
   Feature Investigation: Podos Found That an Adaptation for Feeding May Have Promoted Reproductive Isolation in Finches 424
20.4 **Evo-Devo: Evolutionary Developmental Biology** 427
   Evolutionary Connections: The *Hox* Genes Have Been Important in the Evolution of a Variety of Body Patterns 428

CHAPTER 21

How Biologists Classify Species and Study Their Evolutionary Relationships 433

21.1 **Taxonomy** 434
   Evolutionary Connections: Every Species Is Placed into a Taxonomic Hierarchy 435
21.2 **Phylogenetic Trees** 437
21.3 **Cladistics** 440
   Quantitative Analysis: The Principle of Parsimony Is Used to Choose from Among Possible Cladograms 443
21.4 **Molecular Clocks** 444
21.5 **Horizontal Gene Transfer** 446

CHAPTER 22

The History of Life on Earth and Human Evolution 450

22.1 **The Fossil Record** 451
   Quantitative Analysis: Radioisotopes Provide a Way to Date Fossils 452
22.2 **History of Life on Earth** 453
   Evolutionary Connections: The Origin of Eukaryotic Cells Is Hypothesized to Involve a Union Between Bacterial and Archaeal Cells 457
22.3 **Human Evolution** 463
   Evolutionary Connections: Comparing the Genomes of Humans and Chimpanzees 464

UNIT V Diversity

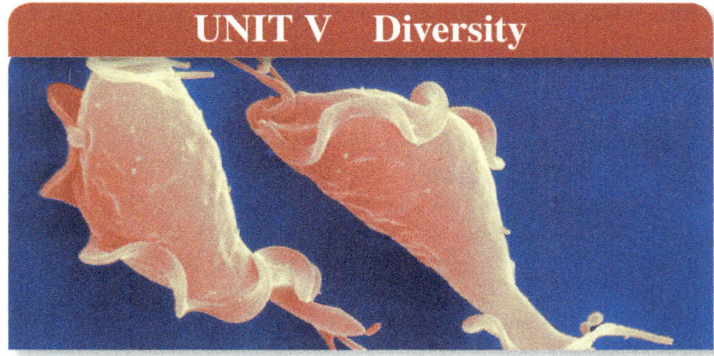

©David M. Phillips/Science Source

CHAPTER 23

Diversity of Microbial Life: Archaea, Bacteria, Protists, and Fungi 477

23.1 **Introduction to Microorganisms** 478
23.2 **Diversity and Ecological Importance of Archaea** 481
23.3 **Diversity and Ecological Importance of Bacteria** 482
23.4 **Diversity in Bacterial Cell Structure and Metabolism** 485

Contents xxi

23.5 Diversity and Ecological Importance of Protists 489
   Evolutionary Connections: Primary Plastids and Primary Endosymbiosis 494
23.6 Diversity and Ecological Importance of Fungi 499
23.7 Technological Applications of Microorganisms 507

CHAPTER 24

Microbiomes: Microbial Systems on and Around Us 510

24.1 Microbiomes: Diversity of Microbes and Functions 511
24.2 Microbiomes of Physical Systems 514
24.3 Host-Associated Microbiomes 516
   Feature Investigation: Blanton, Gordon, and Associates Found That Gut Microbiota Affect the Growth of Malnourished Children 522
24.4 Engineering Animal and Plant Microbiomes 524

CHAPTER 25

Plant Evolution: How Plant Diversification Changed Planet Earth 528

25.1 Ancestry and Diversity of Land Plants 529
25.2 An Evolutionary History of Land Plants 536
25.3 Diversity of Modern Gymnosperms 539
25.4 Diversity of Modern Angiosperms 542
   Evolutionary Connections: Flower Organs Evolved from Leaflike Structures 544
   Feature Investigation: Hillig and Mahlberg Analyzed Secondary Metabolites to Explore Species Diversification in the Genus *Cannabis* 547
25.5 Human Influences on Angiosperm Diversification 550

CHAPTER 26

Invertebrates: The Vast Array of Animal Life Without a Backbone 552

26.1 Characteristics of Animals 553
26.2 Animal Classification 554
   Evolutionary Connections: The Protostomes Consist of Two Major Clades—the Ecdysozoa and the Lophotrochozoa 557
26.3 Ctenophores: The Earliest Animals 560
26.4 Porifera: The Sponges 561
26.5 Cnidaria: Jellyfish and Other Radially Symmetric Animals 562
26.6 Lophotrochozoa: The Flatworms, Rotifers, Bryozoans, Brachiopods, Mollusks, and Annelids 564
   Quantitative Analysis: How Many Flukes? 566
26.7 Ecdysozoa: The Nematodes and Arthropods 572
26.8 Deuterostomia: The Echinoderms and Chordates 580

CHAPTER 27

Vertebrates: Fishes, Amphibians, Reptiles, and Mammals 586

27.1 Vertebrates: Chordates with a Backbone 587
27.2 Cyclostomes: Jawless Fishes 587
27.3 Gnathostomes: Jawed Vertebrates 590
27.4 Tetrapods: Gnathostomes with Four Limbs 594
   Feature Investigation: Davis and Colleagues Provided a Genetic-Developmental Explanation for Limb Length in Tetrapods 595
27.5 Amniotes: Tetrapods with a Desiccation-Resistant Egg 598
27.6 Mammals: Milk-Producing Amniotes 603

UNIT VI Flowering Plants

©George Grall/Getty Images

CHAPTER 28

An Introduction to Flowering Plant Form and Function 609

28.1 From Seed to Seed: The Life of a Flowering Plant 610
28.2 Plant Growth and Development 614
28.3 The Shoot System: Stem and Leaf Adaptations 618
   Feature Investigation: Lawren Sack and Colleagues Showed That Palmate Venation Confers Tolerance of Leaf Vein Breakage 620
28.4 Root System Adaptations 626

CHAPTER 29

How Flowering Plants Sense and Interact with Their Environments 630

29.1 Overview of Plant Behavioral Responses 631
29.2 Plant Hormones 634
   Evolutionary Connections: Plant Gibberellin Responses Evolved in a Stepwise Manner 636
29.3 Plant Responses to Light 638

29.4 Plant Responses to Gravity and Touch 641
29.5 Plant Responses to Attack 643

CHAPTER 30

How Flowering Plants Obtain and Transport Water, Mineral Nutrients, and Organic Compounds 648

30.1 Plant Nutritional Requirements 649
30.2 The Roles of Soil in Plant Nutrition 652
30.3 Transport at the Cellular Level 656
  Quantitative Analysis: The Water Potential Equation Can Be Used to Predict Cellular Water Status 660
  Evolutionary Connections: Plants Have Evolved Cellular Adaptations to Drought Stress 660
30.4 Plant Transport at the Tissue Level 661
30.5 Long-Distance Transport in Plants 663

CHAPTER 31

How Flowering Plants Reproduce and Develop 672

31.1 An Overview of Flowering Plant Reproduction 673
31.2 Flower Production, Structure, and Development 677
  Feature Investigation: Liang and Mahadevan Used Time-Lapse Video and Mathematical Modeling to Explain How Flowers Bloom 679
31.3 Male and Female Gametophytes and Double Fertilization 681
31.4 Embryo, Seed, Fruit, and Seedling Development 684
31.5 Asexual Reproduction in Flowering Plants 688
  Evolutionary Connections: Gene Expression Changes Explain the Evolution of Plantlets in *Kalanchoë* 689

UNIT VII Animals

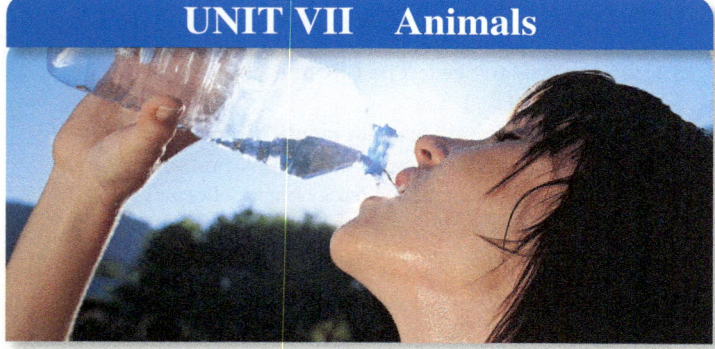

©John Rowley/Getty Images

CHAPTER 32

General Features of Animal Bodies, and Homeostasis as a Key Principle of Animal Biology 693

32.1 Organization of Animal Bodies 694
  Evolutionary Connections: Organ Development and Function Are Controlled by *Hox* Genes 699

32.2 Relationship Between Form and Function 699
32.3 General Principles of Homeostasis 700
32.4 Homeostatic Regulation of Body Temperature 703
32.5 Homeostasis of Internal Fluids 707
  Feature Investigation: Cade and Colleagues Discovered Why Athletes' Performances Wane on Hot Days 710

CHAPTER 33

Neuroscience I: The Structure, Function, and Evolution of Nervous Systems 714

33.1 Cellular Components of Nervous Systems 715
33.2 Electrical Properties of Neurons and the Resting Membrane Potential 718
  Quantitative Analysis: An Ion's Equilibrium Potential Depends on Its Concentration Gradient 720
33.3 Generation and Transmission of Electrical Signals Along Neurons 721
33.4 Communication at Synapses 726
33.5 Evolution and Development of Nervous Systems 729
  Evolutionary Connections: Animals Evolved Increasingly Complex Nervous Systems 729
33.6 Structure and Function of the Nervous Systems of Humans and Other Vertebrates 732
  Feature Investigation: Gaser and Schlaug Discovered That the Sizes of Certain Brain Structures Differ Between Musicians and Nonmusicians 737
33.7 Impact on Public Health 739

CHAPTER 34

Neuroscience II: How Sensory Systems Allow Animals to Interact with the Environment 744

34.1 Introduction to Sensation 745
34.2 Mechanoreception 746
34.3 Thermoreception and Nociception 751
34.4 Photoreception 752
  Evolutionary Connections: Color Vision Is an Ancient Adaptation in Animals 755
34.5 Chemoreception 758
  Feature Investigation: Buck and Axel Discovered a Family of Olfactory Receptor Proteins That Bind Specific Odor Molecules 759
34.6 Impact on Public Health 762

CHAPTER 35

How Muscles and Skeletons Are Adaptations for Movement, Support, and Protection 765

35.1 Types of Animal Skeletons 766
35.2 Skeletal Muscle Structure and the Mechanism of Force Generation 768

Contents xxiii

    Evolutionary Connections: Myosins Are an Ancient and Diverse Family of Proteins 771
35.3 Types of Skeletal Muscle Fibers and Their Functions 775
35.4 Impact on Public Health 777

CHAPTER 36

Circulatory and Respiratory Systems: Transporting Solutes and Exchanging Gases 780

36.1 Types of Circulatory Systems 781
    Evolutionary Connections: A Four-Chambered Heart Evolved from Simple Contractile Tubes 783
36.2 The Composition of Blood 784
36.3 The Vertebrate Heart and Its Function 786
36.4 Blood Vessels 790
36.5 Relationship Among Blood Pressure, Blood Flow, and Resistance 793
    Quantitative Analysis: Cardiac Output and Resistance Determine Blood Pressure 794
36.6 Physical Properties of Gases 796
36.7 Types of Respiratory Systems 797
36.8 Structure and Function of the Mammalian Respiratory System 800
36.9 Mechanisms of Gas Transport in Blood 803
    Quantitative Analysis: The Ability of Hemoglobin to Bind Oxygen Is Affected by Factors Such as Temperature, CO_2, and pH 804
36.10 Control of Ventilation 806
36.11 Impact on Public Health 807

CHAPTER 37

Digestive and Excretory Systems: Maintaining Nutrient, Water, and Energy Balance and Removing Waste 813

37.1 Overview of Animal Nutrition 814
37.2 General Principles of Digestion and Absorption of Nutrients 817
37.3 Vertebrate Digestive Systems 818
    Evolutionary Connections: Evolution and Genetics Explain Lactose Intolerance 824
37.4 Nutrient Use and Storage 828
37.5 Regulation of the Absorptive and Postabsorptive States in Vertebrates 830
37.6 Excretory Systems in Different Animal Groups 833
37.7 Structure and Function of the Mammalian Kidneys 836
37.8 Impact on Public Health 842
    Feature Investigation: Marshall, Warren, and Coworkers Demonstrated a Link Between Bacterial Infection and Ulcers 843

CHAPTER 38

How Endocrine Systems Influence the Activities of All Other Organ Systems 848

38.1 Types of Hormones and Their Mechanisms of Action 849
38.2 Links Between the Endocrine and Nervous Systems 852
38.3 Hormonal Control of Metabolism and Energy Balance 854
    Feature Investigation: Banting, Best, Collip, and MacLeod Were the First to Isolate Active Insulin 858
38.4 Hormonal Control of Mineral Balance 861
    Evolutionary Connections: Hormones and Receptors Evolved as Tightly Integrated Molecular Systems 863
38.5 Hormonal Control of Growth and Development 865
38.6 Hormonal Control of Reproduction 867
38.7 Impact on Public Health 867

CHAPTER 39

The Production of Offspring: Reproduction and Development 870

39.1 Overview of Sexual and Asexual Reproduction 871
    Feature Investigation: Paland and Lynch Provided Evidence That Sexual Reproduction May Promote the Elimination of Harmful Mutations in Populations 872
39.2 Gametogenesis and Fertilization 874
39.3 Human Reproductive Structure and Function 877
39.4 Pregnancy and Birth in Mammals 883
39.5 General Events of Embryonic Development 885
39.6 Impact on Public Health 891

CHAPTER 40

Immune Systems: How Animals Defend Against Pathogens and Other Dangers 895

40.1 Types of Pathogens 896
40.2 Innate Immunity 897
    Evolutionary Connections: Innate Immune Responses Require Proteins That Recognize Features of Many Pathogens 899
    Feature Investigation: Lemaitre and Colleagues Identified an Immune Function for Toll Protein in *Drosophila* 899
40.3 Adaptive Immunity in Vertebrates 902
40.4 Impact on Public Health 912

CHAPTER 41

Integrated Responses of Animal Organ Systems to a Challenge to Homeostasis 916

41.1 Effects of Hemorrhage on Blood Pressure and Organ Function 917 xxiv Contents

41.2 The Rapid Phase of the Homeostatic Response to Hemorrhage 919
  Evolutionary Connections: Baroreceptors May Have Evolved to Minimize Increases in Blood Pressure in Vertebrates 921
41.3 The Secondary Phase of the Homeostatic Response to Hemorrhage 923
41.4 Impact on Public Health 927

UNIT VIII Ecology

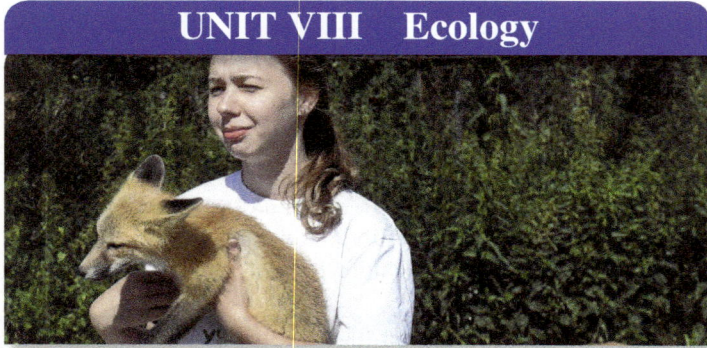

©Kirill Kukhmar/TASS/Alamy

CHAPTER 42

Behavioral Ecology: The Struggle to Find Food and Mates and to Pass on Genes 931

42.1 The Influence of Genetics and Learning on Behavior 932
  Feature Investigation: Tinbergen's Experiments Showed That Digger Wasps Learn the Positions of Landmarks to Find Their Nests 933
42.2 Communication 936
42.3 Living in Groups and Optimality Theory 938
  Quantitative Analysis: Game Theory Establishes Whether Individuals Fight or Flee 940
42.4 Altruism 941
42.5 Mating Systems 944

CHAPTER 43

Population Growth and Species Interactions 948

43.1 Measuring Population Size and Density 949
  Quantitative Analysis: Mark-Recapture Can Be Used to Estimate Population Size 950
43.2 Demography 951
43.3 How Populations Grow 954
43.4 Species Interactions 956
  Evolutionary Connections: Organisms Have Evolved Many Defenses Against Natural Enemies 959

CHAPTER 44

Communities and Ecosystems: Ecological Organization at Large Scales 965

44.1 Patterns of Species Richness and Species Diversity 966
  Quantitative Analysis: Calculating Species Diversity 968
44.2 Species Richness and Community Stability 970
44.3 Succession: Community Change 971
44.4 Island Biogeography 974
  Feature Investigation: Simberloff and Wilson's Experiments Tested the Predictions of the Equilibrium Model of Island Biogeography 977
44.5 Food Webs and Energy Flow 979
44.6 Biomass Production in Ecosystems 983

CHAPTER 45

Biomes: How Climate Affects the Distribution of Species on Earth 988

45.1 Climate and Its Relationship to Biological Communities 989
45.2 Major Biomes 995
  Evolutionary Connections: Continental Drift and Biogeography Help to Explain Species Distribution 999

CHAPTER 46

The Age of Humans 1002

46.1 Human Population Growth 1003
46.2 Global Warming and Climate Change 1005
46.3 Pollution and Human Influences on Biogeochemical Cycles 1008
  Feature Investigation: Stiling and Drake's Experiments with Elevated CO_2 Showed an Increase in Plant Growth but a Decrease in Herbivory 1009
46.4 Pollution and Biomagnification 1014
46.5 Habitat Destruction 1016
46.6 Overexploitation 1018
  Quantitative Analysis: Ecologists Make Calculations to Determine if Overexploitation is Occurring 1021
46.7 Invasive Species 1022

CHAPTER 47

Biodiversity and Conservation Biology 1027

47.1 Genetic, Species, and Ecosystem Diversity 1028
47.2 Value of Biodiversity to Human Welfare 1028
47.3 Biodiversity and Ecosystem Function 1031
47.4 Conservation Strategies 1033

Appendix A Periodic Table of the Elements A-1
Appendix B Answers to In-Chapter and End-of-Chapter Questions B-1
Index I-1

An Introduction to Biology

©TeeJe/Getty Images

The Raggiana bird-of-paradise *(Paradisaea raggiana)*, **which is the national bird of Papua New Guinea.** As discussed in Section 1.2, the striking characteristics of male birds are the result of a type of natural selection called sexual selection.

Chapter Outline

1.1 Principles of Biology and the Levels of Biological Organization

1.2 Unity and Diversity of Life

1.3 Biology as a Scientific Discipline

Assessing Your Knowledge and Skills

Biology is the study of life. The diverse forms of life found on Earth provide biologists with an amazing array of organisms to study. In many cases, the investigation of living things leads to discoveries with far-reaching benefits. Certain ancient civilizations, such as the Greeks, Romans, and Egyptians, discovered that the bark of the white willow tree (*Salix alba*) could be used to fight fever. Chemists determined that willow bark contains a substance called salicylic acid, which led to the development of the related compound acetylsalicylic acid, more commonly known as aspirin (**Figure 1.1**). Today, aspirin is taken for fever and pain relief.

As a more recent example, researchers determined that the venom from certain poisonous snakes contains a chemical that lowers blood pressure in humans. By analyzing that chemical, scientists have developed drugs called ACE inhibitors that treat high blood pressure (**Figure 1.2**).

These are just a couple of the many discoveries that make biology an intriguing discipline. The study of life not only reveals the fascinating characteristics of living species but also leads to the development of medicines that benefit the lives of people.

To make new discoveries, biologists view life from many different perspectives. What is the composition of living things? How is life organized? How do organisms reproduce? Sometimes the questions posed by biologists are fundamental and even philosophical in nature. How did living organisms originate? Can we live forever? What is the physical basis for memory? Can we save endangered species?

Future biologists will continue to make important advances. Biologists are scientific explorers looking for answers to some of life's most enduring mysteries. Unraveling these mysteries presents exciting challenges to the best and brightest minds. The rewards of a career in biology include the excitement of forging into uncharted territory, the thrill of making discoveries that can improve the health and lives of people, and the goal of trying to preserve the environment and protect endangered species. For these and many other compelling reasons, students seeking challenging and rewarding careers may wish to choose biology as a lifelong pursuit.

In this chapter, we will begin our survey of biology by examining the basic principles that underlie the characteristics of all living organisms. We will then take a deeper look at the process of evolution and how it explains the unity and diversity that we observe among living and extinct species. Finally, we will explore the general approaches that scientists follow when making new discoveries.

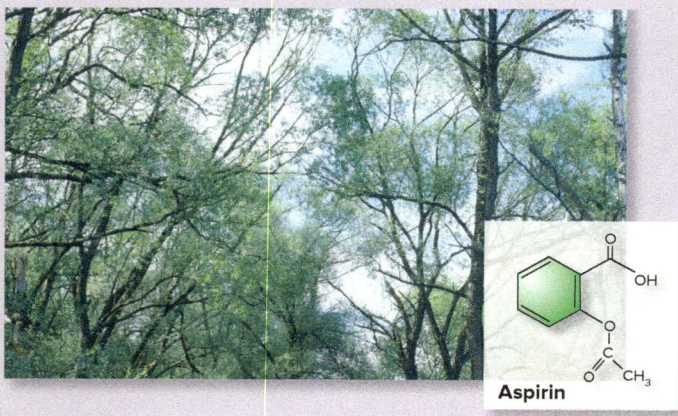

Figure 1.1 The white willow (*Salix alba*) and aspirin. Modern aspirin, acetylsalicylic acid, was developed after analysis of a chemical found in the bark of the white willow tree.
©blickwinkel/Alamy

Figure 1.2 The Brazilian arrowhead viper and an inhibitor of high blood pressure. Originally found in the venom of the Brazilian arrowhead viper (*Bothrops jararaca/jararacussa*), angiotensin-converting enzyme (ACE) inhibitors, including benazepril (Lotensin, chemical structure shown here), are commonly used to treat high blood pressure.
©Francois Gohier/Science Source

1.1 Principles of Biology and the Levels of Biological Organization

Learning Outcomes
1. Describe the principles of biology.
2. Explain how life can be viewed at different levels of biological complexity.

Because biology is the study of life, a good way to begin a biology textbook is to distinguish living organisms from nonliving objects. At first, the distinction might seem obvious. A person is alive, but a rock is not. However, the distinction between living and nonliving may seem less obvious when we consider microscopic entities. Is a bacterium alive? What about a virus or a chromosome? In this section, we will examine the principles that underlie the characteristics of all forms of life and explore other broad principles in biology. We will then consider the levels of organization that biologists study, ranging from atoms and small molecules to very large geographic areas.

The Study of Life Has Revealed a Set of Unifying Principles

In the course of studying a vast number of species, biologists have learned that a set of principles applies to all fields of biology. Twelve broad principles are described in **Figure 1.3**. The first eight principles are often used as criteria for defining the basic features of life. You will see these 12 principles at many points as you progress through this textbook. In particular, we will draw your attention to them at the beginning of each unit, and we will refer to them within particular figures in Chapters 2 through 47. It should be noted that the principles of biology are also governed by the laws of chemistry and physics, which are discussed in Chapters 2, 3, and 6.

Principle 1: Cells are the simplest units of life. The term **organism** can be applied to all living things. Organisms maintain an internal order that is separated from the environment (Figure 1.3a). The simplest unit of such organization is the cell, which we will examine in Unit II. One of the foundations of biology is the **cell theory,** which states that (1) all organisms are composed of one or more cells, (2) cells are the smallest units of life, and (3) new cells come from pre-existing cells by cell division. Unicellular organisms are composed of one cell, whereas multicellular organisms such as plants and animals contain many cells. In plants and animals, each cell has an internal order, and the cells within the organism have specific arrangements and functions.

Principle 2: Living organisms use energy. The maintenance of organization requires energy. Therefore, all living organisms acquire energy from the environment and use that energy to maintain their internal order. Cells carry out a variety of chemical reactions that are responsible for the breakdown of energy-yielding nutrients. Such reactions often release energy in a process called **cellular respiration.** The energy may be used to synthesize the components that make up

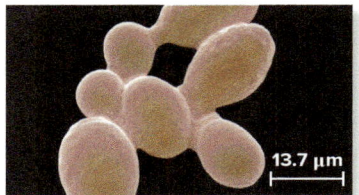

(a) **Cells are the simplest units of life:** Organisms maintain an internal order. The simplest unit of organization is the cell. Yeast cells are shown here.

(g) **Structure determines function:** In the example seen here, webbed feet (on ducks) function as paddles for swimming. Nonwebbed feet (on chickens) function better for walking on the ground.

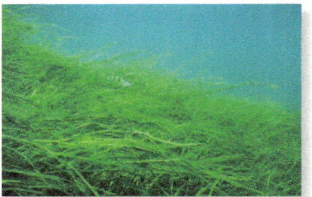

(b) **Living organisms use energy:** Organisms need energy to maintain internal order. These algae harness light energy via photosynthesis. Energy is used in chemical reactions collectively known as metabolism.

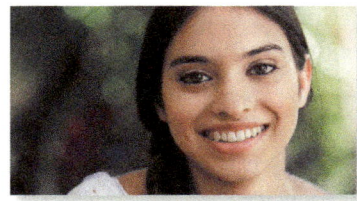

(h) **New properties of life emerge from complex interactions:** Our ability to see is an emergent property due to interactions among many types of cells in the eye and neurons that send signals to the brain.

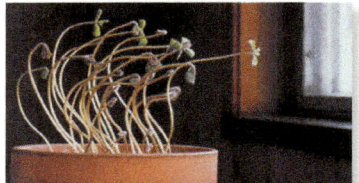

(c) **Living organisms interact with their environment:** Organisms respond to environmental changes. These plants are growing toward the light.

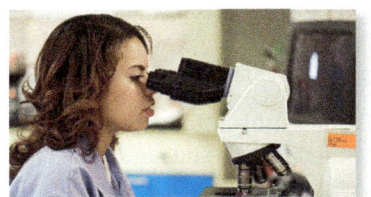

(i) **Biology is an experimental science:** The discoveries of biology are made via experimentation, which leads to theories and biological principles.

(d) **Living organisms maintain homeostasis:** Organisms regulate their cells and bodies, maintaining relatively stable internal conditions, a process called homeostasis. For example, this bird maintains its internal body temperature on a cold day.

(j) **Biology is a quantitative science:** Biologists analyze data in a quantitative way and use mathematical approaches to make predictions. For example, a Punnett square is used to predict the outcome of genetic crosses.

(e) **The genetic material provides a blueprint that allows organisms to grow, develop, and reproduce:** To sustain life over many generations, organisms must reproduce. Due to the transmission of genetic material, offspring tend to have traits like their parents.

(k) **Biologists use models and simulations to test experimental predictions and convey their ideas:** For example, molecular biologists propose models that describe the structures of proteins.

(f) **Populations of organisms evolve from one generation to the next and are related by an evolutionary history:** Populations of organisms change over the course of many generations. Evolution results in traits that promote survival and reproductive success. Evolution results in new species that are related to each other.

(l) **Biology affects our society:** Many discoveries in biology have had major effects on our society. For example, biologists developed Bt-corn, which is resistant to insect pests and is widely planted by farmers.

Figure 1.3 **Twelve principles of biology.** The first eight principles are often used as criteria for defining the basic features of life. Note: The 12 principles described here were modeled after the core concepts and core competencies described in *Vision and Change in Undergraduate Biology,* a report that was published in 2009 and organized by the American Association for the Advancement of Science. *Vision and Change* proposed several core concepts and core competencies (see http://visionandchange.org). We have slightly modified and expanded them into 12 principles to make them more accessible to beginning biology students. The five *Vision and Change* core concepts are related to our principles in the following manner: (1) evolution (principle f); (2) structure and function (principle g); (3) information flow, exchange, and storage (principle e); (4) pathways and transformations of energy and matter (principles b, c, and d); and (5) systems (principles a and h). The last four principles are modeled after four core competencies (that is, core skills) described in *Vision and Change*: ability to apply the process of science (principle i); ability to use quantitative reasoning (principle j); ability to use models and simulations (principle k); and ability to understand the relationship between science and society (principle l).

(a) ©Steve Gschmeissner/Science Photo Library/Getty Images; *(b)* ©Robert La Salle/Aqua-Photo/Alamy; *(c)* ©Cathlyn Melloan/Getty Images; *(d)* ©Cliff Keeler/Alamy; *(e)* ©Frank Krahmer/Getty Images; *(f)* ©Luiz Kagiyama/Shutterstock; *(g)* ©G.K. & Vikki Hart/Getty Images; *(h)* ©Maria Teijeiro/Getty Images; *(i)* ©Corbis/SuperStock; *(k)* ©Daniel Gage, University of Connecticut; *(l)* ©Bill Barksdale/age fotostock

BioConnections: *Look ahead to Figure 4.15. Which of these principles is this figure emphasizing?*

individual cells and living organisms. Chemical reactions involved with the breakdown and synthesis of cellular molecules are collectively known as **metabolism.** Plants, algae, and certain bacteria directly harness light energy to produce their own nutrients in the process of **photosynthesis** (Figure 1.3b). They are the primary producers of food on Earth. In contrast, some organisms, such as animals and fungi, are consumers—they must use other organisms or the products of other organisms as food to obtain energy.

Principle 3: Living organisms interact with their environment. To survive, living organisms must interact with their environment, which includes other organisms they may encounter. All organisms must respond to environmental changes. In the winter, many species of mammals develop a thicker coat of fur that protects them from the cold temperatures. Plants respond to changes in the angle of the Sun. If you place a plant in a window, it will grow toward the light (Figure 1.3c).

Principle 4: Living organisms maintain homeostasis. Although life is a dynamic process, living cells and organisms regulate their cells and bodies to maintain relatively stable internal conditions, a process called **homeostasis** (from the Greek, meaning to stay the same). The degree to which homeostasis is achieved varies among different organisms. For example, most mammals and birds maintain a relatively stable body temperature in spite of changing environmental temperatures (Figure 1.3d), whereas reptiles and amphibians tolerate a wider fluctuation in body temperature. By comparison, all organisms continually regulate their cellular metabolism so that nutrient molecules are used at an appropriate rate and new cellular components are synthesized when they are needed.

Principle 5: The genetic material provides a blueprint that allows organisms to grow, develop, and reproduce. All living organisms grow and develop specific traits. To sustain life from one generation to the next, organisms must also **reproduce,** or generate offspring (Figure 1.3e). A key feature of reproduction is that offspring tend to have characteristics that greatly resemble those of their parent(s). How is this possible? All living organisms contain genetic material composed of **deoxyribonucleic acid (DNA),** which provides a blueprint for the organization, development, and function of living things. During reproduction, a copy of this blueprint is transmitted from parent to offspring. DNA is **heritable,** which means that offspring inherit DNA from their parents.

As discussed in Unit III, **genes,** which are segments of DNA, govern the characteristics, or traits, of organisms. Most genes are transcribed into a type of **RNA (ribonucleic acid)** molecule called messenger RNA (mRNA), which is then translated into a **polypeptide** with a specific amino acid sequence. A **protein** is composed of one or more polypeptides. The structures and functions of proteins are largely responsible for the traits of living organisms.

Principle 6: Populations of organisms evolve from one generation to the next and are related by an evolutionary history. The first five characteristics of life, which we have just considered, may apply to individual organisms over the short run. Over the long run, another universal characteristic of life is **biological evolution,** or simply **evolution,** which is a heritable change in a population of organisms from generation to generation.

As a result of evolution, populations become better adapted to the environment in which they live. For example, the long snout of an anteater is an adaptation that enhances its ability to obtain food, namely ants, from hard-to-reach places (Figure 1.3f). Over the course of many generations, the fossil record indicates that the long snout occurred via biological evolution in which modern anteaters evolved from populations of organisms with shorter snouts.

Evidence from a variety of sources, including the fossil record and DNA sequences, also indicates that all organisms on Earth share a common ancestry. For example, the different species of mammals that currently exist shared a common ancestor in the past, which was also a mammal. We will discuss evolutionary relationships further in Section 1.2. In many chapters of this textbook, you will find a subsection called Evolutionary Connections, which focuses on the evolutionary aspects of the chapter's material.

Principle 7: Structure determines function. In addition to the preceding six characteristics of life, biologists have identified other principles that are important in all fields of biology. The principle that structure determines function pertains to very tiny biological molecules as well as very large biological structures. For example, at the microscopic level, a cellular protein called actin naturally assembles into structures that are long filaments. The function of these filaments is to provide support and shape to cells. At the macroscopic level, let's consider the feet of different birds (Figure 1.3g). Aquatic birds have webbed feet that function as paddles for swimming. By comparison, the feet of nonaquatic birds are not webbed and are better adapted for grasping food, perching on branches, and running along the ground. In this case, the structure of a bird's feet, webbed versus nonwebbed, is a critical feature that affects their function.

Principle 8: New properties of life emerge from complex interactions. In biology, when individual components in an organism interact with each other or with the external environment to create novel structures and functions, the resulting characteristics are called **emergent properties.** For example, the human eye is composed of many different types of cells that are organized to sense incoming light and transmit signals to the brain (Figure 1.3h). Our ability to see is an emergent property of this complex arrangement of different cell types.

Principle 9: Biology is an experimental science. We now turn to principles that involve skills that are used in the study of life. Biology is an inquiry process. Biologists are curious about the characteristics of living organisms and ask questions about those characteristics. For example, a cell biologist may wonder why a cell produces a specific protein when it is confronted with high temperature. An ecologist may ask herself why a particular bird eats insects in the summer and seeds in the winter. To answer such questions, biologists typically gather additional information and ultimately form a hypothesis, which is a proposed explanation for a natural phenomenon. The next stage is to design one or more experiments to test the validity of the hypothesis (Figure 1.3i). Like evolution, experimentation is such a key aspect of biology that most chapters of this textbook include a Feature Investigation—an actual research study that showcases the experimental approach.

Principle 10: Biology is as a quantitative science. Biologists use quantitative reasoning in a variety of ways, as you will see throughout this textbook. For example, the results of experiments are often analyzed using statistics. Also, biologists may use mathematical models to make predictions. In Chapter 15, you will see how a Punnett square is used to predict the outcome of crosses (Figure 1.3j). In Chapter 43, equations are used to predict different patterns of population growth over the course of several generations. Many chapters have a subsection that emphasizes quantitative reasoning, an important skill for careers in science and medicine. In these subsections, the quantitative nature of a given topic is described, and then students are asked to solve a problem related to that topic.

Principle 11: Biologists use models and simulations to test experimental predictions and convey their ideas. A **scientific model**, or simply a model, is a conceptual, mathematical, or physical depiction of a real-world phenomenon. Models may convey different aspects of life. For example, the model shown in Figure 1.3k is a model for the molecular structure of a protein. Why are models useful? Models are often used as a working hypothesis that helps researchers visualize or explain biological phenomena. Such models form the basis for conducting further experiments. Models are evaluated for their consistency with experimental data, which enables researchers to accept, reject, or refine them. Another useful aspect of models is that they allow biologists to convey their ideas in a relatively simple way. Most of the figures in this textbook are models that have been derived by the interpretation of experimental data.

Principle 12: Biology affects our society. The influence of biology is not confined to textbooks and classrooms. The work of biologists has far-reaching effects in our society. For example, biologists have discovered drugs that are used to treat many different human diseases. Likewise, biologists have created technologies that have many uses. Examples include the use of microorganisms to make medical products, such as human insulin, and the genetic engineering of crops to make them resistant to particular types of insect pests (Figure 1.3l).

Living Organisms Are Studied at Different Levels of Organization

The organization of living organisms can be analyzed at different levels of biological complexity, starting with the smallest level of organization and progressing to levels that are physically much larger and more complex. **Figure 1.4** depicts a scientist's view of the levels of biological organization.

1. *Atoms.* An **atom** is the smallest unit of an element that has the chemical properties of the element. All matter is composed of atoms.
2. *Molecules and macromolecules.* As discussed in Unit I, atoms bond with each other to form **molecules.** Many smaller molecules can bond together to form a large polymer called a **macromolecule.** Carbohydrates, proteins, and nucleic acids (DNA and RNA) are important macromolecules found in living organisms.
3. *Cells.* Molecules and macromolecules associate with each other to form larger structures such as cells. A **cell** is surrounded by a membrane and contains a variety of molecules and macromolecules. As noted earlier, a cell is the simplest unit of life.
4. *Tissues.* In multicellular organisms such as plants and animals, many cells of the same type associate with each other to form **tissues.** An example is muscle tissue.
5. *Organs.* In complex multicellular organisms, an **organ** is composed of two or more types of tissue. For example, the heart is composed of several types of tissue, including muscle, nervous, and connective tissue.
6. *Organism.* All living things can be called **organisms.** Biologists classify organisms as belonging to a particular **species,** which is a related group of organisms that share a distinctive form and set of attributes in nature. The members of the same species are closely related genetically. In Units VI and VII, we will examine plants and animals at the levels of cells, tissues, organs, and complete organisms.
7. *Population.* A group of organisms of the same species that occupy the same environment is called a **population.**
8. *Community.* A biological **community** is an assemblage of populations of different species that live in the same environment. The types of species found in a community are determined by the environment and by the interactions of species with each other.
9. *Ecosystem.* Researchers may extend their work beyond living organisms and also study the physical environment. Ecologists analyze **ecosystems,** which are formed by interactions between a community of organisms and its physical environment. Unit VIII considers biological organization from populations to ecosystems.
10. *Biosphere.* The **biosphere** includes all of the places on the Earth where living organisms exist. Life is found in the air, in bodies of water, on the land, and in the soil.

1.1 Reviewing the Concepts

- Biology is the study of life. Discoveries in biology help us understand how life exists, and they have many practical applications, such as the development of drugs to treat human diseases (Figures 1.1, 1.2).
- Eight principles underlie the characteristics that are common to all forms of life (Figure 1.3). All living things (1) are composed of cells as their simplest unit; (2) use energy; (3) interact with their environment; (4) maintain homeostasis; and (5) have genetic material for the development of traits and for reproduction. Also, (6) populations of organisms evolve from one generation to the next and are connected by an evolutionary history (Figure 1.3). Additional important principles of biology that apply to living organisms are that (7) structure determines function and (8) new properties emerge from complex interactions.
- Biology also involves certain skills, including (9) biology is an experimental science; (10) biology is a quantitative science; (11) models and simulations are used to test experimental predictions and convey ideas; and (12) biology influences our society.
- Living organisms can be viewed at different levels of biological organization: atoms, molecules and macromolecules, cells, tissues, organs, organisms, populations, communities, ecosystems, and the biosphere (Figure 1.4).

©Vladislav Gajic/Shutterstock
©Daniel Gage, University of Connecticut

UNIT I
CHEMISTRY

Living organisms are composed of chemicals, which are altered via chemical reactions. These reactions occur between atoms and molecules and may require or release energy. Chemical reactions and interactions between molecules play a role in virtually all aspects of a cell's activities. In order to understand how living organisms function, grow, develop, behave, and interact with their environments, we first need to understand some basic principles of atomic and molecular structure. In addition, it is necessary to understand the forces that allow atoms and molecules to interact with each other. We will begin this unit with an overview of inorganic chemistry—that is, the nature of atoms and molecules, with the exception of those that contain rings or chains of carbon. Such carbon-containing molecules form the basis of organic chemistry and are covered in Chapter 3.

The following biology principles are emphasized in this unit:

- **Living organisms use energy.** You will see how the chemical energy stored in certain bonds in molecules such as sugars and fats can be released and used by living organisms to perform numerous functions that support life, including growth, digestion, and locomotion.

- **Structure determines function.** As described in Chapters 2 and 3, the three-dimensional structure of molecules is critical to their ability to interact with and influence the function of other specific molecules.

- **The genetic material provides a blueprint that allows organisms to grow, develop, and reproduce.** Nucleic acids, the basis of inherited genetic material, are first introduced in Chapter 3.

- **New properties emerge from complex interactions.** You will learn in this unit how atoms and simple molecules are connected together to create more complex molecules with new chemical and biological properties.

- **Biologists use models to make predictions and convey their ideas.** Every chapter has a Modeling Challenge to help you refine this important skill.

The Chemical Basis of Life I: Atoms, Molecules, and Water

2

©Vladislav Gajic/Shutterstock

Crystals of rock salt. The crystals are mostly NaCl—a molecule composed of two elements—plus some impurities.

Chapter Outline

2.1 Atoms
2.2 Chemical Bonds and Molecules
2.3 Chemical Reactions
2.4 Properties of Water
2.5 pH and Buffers

Assessing Your Knowledge and Skills

Upon leaving home and beginning college, an 18-year-old student forgoes the usual well-balanced diet he was accustomed to at home in favor of large amounts of chips, crackers, pizza, burgers, and fries. By the time he graduates, he discovers at a routine physical exam that his blood pressure is a little high. One common denominator of his college diet has been the large amount of salt in his favorite foods. Table salt is a molecule made up of two elements, or atoms: sodium and chlorine. In ordinary amounts, this molecule, symbolized NaCl, is vital for all living organisms. In excess, however, it can cause numerous health problems, including, in some susceptible people, high blood pressure. An understanding of what atoms and molecules are, and how they are important for life and health, is a critical component of understanding biology.

Biology—the study of life—is founded on the principles of chemistry. All living organisms are collections of atoms and molecules bound together and interacting with each other through the forces of nature. Throughout this textbook, we will see how chemistry can be applied to living organisms as we discuss the components of cells, the functions of proteins, the flow of nutrients in plants and animals, and the evolution of new genes. This chapter lays the groundwork for understanding these and other concepts.

We begin with an overview of the structure of atoms. We next explore the various ways that atoms combine with other atoms to create molecules, looking at the different types of chemical bonds between atoms, how these bonds form, and how they determine the structures of molecules. We then examine chemical reactions between molecules. The water molecule is then highlighted, including the properties that make it a crucial component of living organisms. We conclude with a discussion about the importance of pH in the watery fluids of living organisms.

2.1 Atoms

Learning Outcomes

1. Describe the general structure of atoms.
2. Define orbital and electron shell.
3. Relate atomic structure to the periodic table of the elements.
4. **SCISKILLS** ▶ Quantify atomic mass using units of daltons and moles.
5. Explain how a single element may exist in two or more forms, called isotopes.
6. List the elements that make up most of the mass of all living organisms.

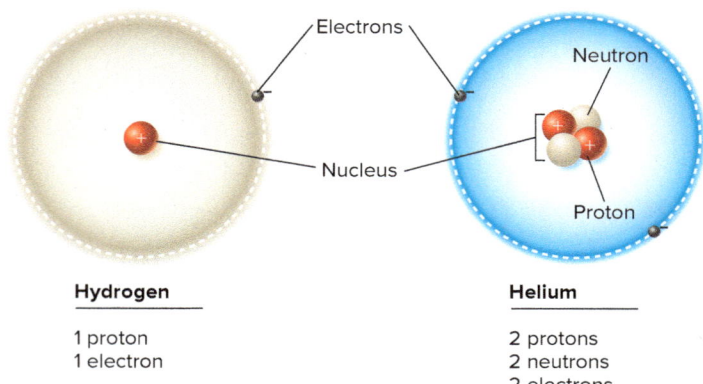

Figure 2.1 **Diagrams of two simple atoms.** These are models of the two simplest atoms, hydrogen and helium. The nucleus consists of protons and neutrons, whereas electrons are found outside the nucleus. Note: In all figures of atoms, the sizes and distances are not to scale. For example, the radius of a typical atom is roughly 10,000 times the radius of its nucleus. If an atomic nucleus were the size of a soccer ball, then the entire atom would have a radius of about 2 miles!

All organisms are composed of **matter,** which is anything that contains mass and occupies space (mass, then, is the amount of matter in any object). Matter may exist in any of three states: solid, liquid, or gas. **Atoms** are the smallest functional units of matter that form all chemical substances and ultimately all organisms; they cannot be further broken down into other substances by ordinary chemical or physical means. Atoms, in turn, are composed of different types of smaller, subatomic particles. When two or more atoms are bonded together, a **molecule** is formed. In this section, we will explore the physical properties of atoms so that we can understand how atoms combine to form molecules of biological importance.

Atoms Are Composed of Subatomic Particles

The chemicals within living organisms are composed of many different types of atoms. Each specific type of atom—nitrogen, hydrogen, oxygen, and so on—is called an **element** (or chemical element), which is defined as a pure substance made up of only one kind of atom. If you hold in your hand an ounce of pure gold, you are holding an element composed of many, many gold atoms.

Three subatomic particles—**protons (p^+), neutrons (n^0),** and **electrons (e^-)**—are found within atoms (**Figure 2.1**). The protons and neutrons are confined to a very small volume at the center of an atom, the **atomic nucleus,** whereas the electrons are found in regions at various distances from the nucleus. In most atoms, the numbers of protons and electrons are identical, but the number of neutrons may vary. Each of the subatomic particles has a different electric charge:

- Protons have one unit of positive charge.
- Electrons have one unit of negative charge.
- Neutrons are electrically neutral.

Like charges repel each other, and opposite charges attract each other. It is the opposite charges of the protons and electrons that create an atom—the positive charges in the nucleus attract the negatively charged electrons.

Because the protons are located in the atomic nucleus, the nucleus has a net positive charge equal to the number of protons it contains. The entire atom has no net electric charge, however, because the number of negatively charged electrons around the nucleus is equal to the number of positively charged protons in the nucleus.

Electrons Occupy Orbitals Around an Atom's Nucleus

At one time, scientists visualized an atom as a mini-solar system, with the nucleus being the Sun and the electrons traveling in clearly defined orbits around it. Diagrams of the two simplest atoms—hydrogen and helium—which have the smallest numbers of protons, were shown in Figure 2.1. This model of the atom is now known to be an oversimplification, because electrons do not actually orbit the nucleus in defined paths like planets around the Sun. However, this depiction of an atom remains a convenient way to diagram atoms in two dimensions.

For complex reasons associated with the physics of subatomic particles, it is impossible to predict the precise location of a given electron. We can only describe the region of space surrounding the atomic nucleus in which there is a high probability of finding that electron. These regions are called **orbitals.** A better model of an atom, therefore, is a central nucleus surrounded by cloudlike orbitals. An orbital can contain a maximum of two electrons. Consequently, any atom with more than two electrons must contain more than one orbital.

An **electron shell** is composed of one or more orbitals. The orbitals of a given electron shell have a characteristic amount of energy. **Energy** is the capacity to do work or effect a change. In biology, we often refer to various types of energy, such as light energy, mechanical energy, and chemical energy. Electrons have kinetic energy, or the energy of moving matter. Atoms with progressively more electrons have additional electron shells that are at greater and greater distances from the nucleus. These shells are numbered, with shell number 1 being closest to the nucleus. Electron shells may contain one or more orbitals, with each orbital holding up to two electrons. The first, innermost electron shell of all atoms has room for only two electrons in one orbital. The second electron shell has four orbitals holding up to four pairs of electrons, or eight electrons altogether.

Electrons are found in orbitals of varying shapes, such as spherical or propeller-type shapes. However, scientists often use simplified diagrams when depicting an atom, without showing individual orbitals within an electron shell. Figure 2.2 illustrates an example of such a depiction of a nitrogen atom. An atom of this element has seven protons and seven electrons. Two electrons fill the first shell, leaving five electrons for the second, outer shell. Two of these fill one orbital and are shown as a pair of electrons in the second shell. The other three electrons in the second shell are found singly in each of the three other available orbitals. The diagram in Figure 2.2 makes it easy to see whether the outer shell of an atom is full.

Most atoms, like nitrogen, have outer shells that are not completely filled with electrons. Atoms that have unfilled electron shells tend to share, lose, or gain electrons to fill their outer shell. Those electrons in the outermost shell are called the **valence electrons.**

Biology Principle

Biologists Use Models to Test Predictions and Convey Ideas

This figure is a model for the structure of an atom that emphasizes the locations of electrons. The Modeling Challenge in the figure legend asks you to propose a model for a different element.

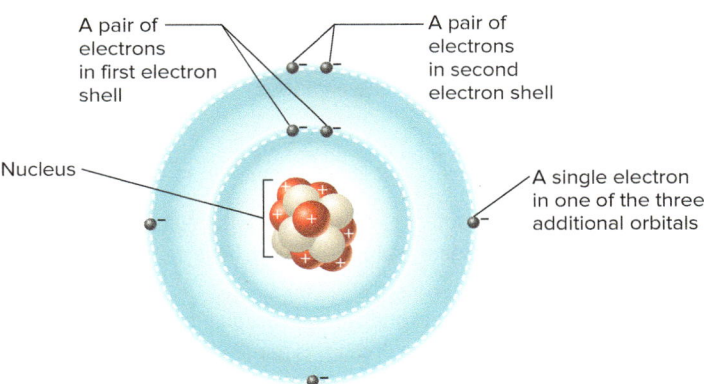

Simplified depiction of a nitrogen atom (7 electrons; 2 electrons in first electron shell, 5 in second electron shell)

Figure 2.2 Diagram showing the two electron shells of a nitrogen atom. An atom's shells fill up one by one with electrons. The first, innermost shell fills up first with two electrons. Subsequent shells gain one electron at a time until they become filled. Heavier elements contain additional shells.

Modeling Challenge: The element chlorine (Cl) has 17 electrons and 3 electron shells. As with nitrogen, the first shell is full when it contains 2 electrons, and the second shell is full when it contains 8 electrons. Draw a model for the structure of chlorine in the same format as the model in Figure 2.2.

As you will learn shortly, such electrons allow atoms to form chemical bonds with each other, in which two or more atoms become joined together to create a new substance.

Each Element Has a Unique Number of Protons

Each chemical element has a specific and unique number of protons in its nucleus that distinguishes it from other elements. The number of protons in an atom is its **atomic number.** For example, hydrogen, the simplest atom, has an atomic number of 1, corresponding to its single proton. Magnesium has an atomic number of 12, corresponding to its 12 protons. With the exception of ions, which are described later, the number of protons is identical to the number of electrons in a given atom. Therefore, the atomic number is also equal to the number of electrons in the atom, resulting in a net electric charge of zero.

Appendix A shows the periodic table of the elements, which arranges the known elements according to their atomic numbers and electron shells. A one- or two-letter symbol is used as an abbreviation for each element. The rows (known as periods) indicate the number of electron shells. For example, hydrogen (H) has one shell, lithium (Li) has two shells, and sodium (Na) has three shells. The columns (called groups) indicate the numbers of electrons in the outer shell. The outer shell of lithium (Li) has one electron, beryllium (Be) has two, boron (B) has three, and so forth. This organization of the periodic table tends to arrange elements based on similar chemical properties. For example, magnesium (Mg) and calcium (Ca) each have two electrons in their outer shell, so these two elements tend to combine with many of the same other elements. The similarities of elements within a group occur because they have the same number of electrons in their outer shells, and this gives them similar chemical bonding properties.

Atoms Have a Small but Measurable Mass

Atoms are extremely small and thus have very little mass. Protons and neutrons are nearly equal in mass, and each has more than 1,800 times the mass of an electron. Because of their tiny size relative to protons and neutrons, the mass of the electrons in an atom is ignored in calculations.

The **atomic mass** indicates an atom's mass relative to the masses of other atoms. By convention, the most common type of carbon atom, which has six protons and six neutrons, is assigned an atomic mass of exactly 12. On this scale, a hydrogen atom has an atomic mass of 1, indicating that it has 1/12 the mass of a carbon atom. A magnesium atom, with an atomic mass of 24, has twice the mass of a carbon atom.

Atomic mass is measured in units called daltons. One **dalton (Da)** equals 1/12 the mass of a carbon atom. Therefore, the most common type of carbon atom has an atomic mass of 12 Da. Because atoms such as hydrogen have a small mass, but atoms such as carbon have a larger mass, 1 gram (g) of hydrogen contains more atoms than 1 g of carbon. As first described by Italian physicist Amedeo Avogadro, 1 **mole** (mol) of any substance contains the same number of atoms as there are atoms in exactly 12 g of carbon. Twelve grams of carbon equals 1 mol of carbon, and 1 g of hydrogen equals 1 mol of hydrogen. One mol of any element contains the same number of atoms—6.022×10^{23}, which is called Avogadro's number. For example, 12 g of carbon contains 6.022×10^{23} atoms, and 1 g of hydrogen, whose atoms have 1/12 the mass of a carbon atom, contains 6.022×10^{23} atoms.

Isotopes Vary in Their Number of Neutrons

Many elements can exist in multiple forms, called **isotopes**, that differ in the number of neutrons they contain. For example, the most abundant form of the carbon atom, ^{12}C, contains six protons and six neutrons, and thus has an atomic number of 6 and an atomic mass of 12 Da. The superscript placed to the left of ^{12}C is the sum of the protons and neutrons. The rare carbon isotope ^{14}C, however, contains six protons and eight neutrons. Although ^{14}C has an atomic number of 6, it has an atomic mass of 14 Da. Nearly 99% of the carbon in living organisms is ^{12}C. Consequently, the average atomic mass of carbon is slightly greater than 12 Da because of the existence of a small amount of heavier isotopes. This explains why the atomic masses given in the periodic table do not add up exactly to whole numbers.

Isotopes of an atom have similar chemical properties but may have very different physical properties. For example, many isotopes found in nature are inherently unstable. Such unstable isotopes are called **radioisotopes**, because they lose energy by emitting subatomic particles and/or radiation, which converts them to a stable form. At the very low amounts found in nature, radioisotopes usually pose no serious threat to life, but exposure of living organisms to high amounts of radioactivity can result in the disruption of cellular function, cancer, and even death.

All Living Organisms Are Largely Composed of Four Elements

The elements oxygen, carbon, hydrogen, and nitrogen account for the majority of atoms in living organisms (Table 2.1), typically making up about 95% of an organism's mass. Much of the oxygen and hydrogen occur in the form of water, which accounts for approximately 60% of the mass of most animals and up to 95% or more in some plants. Carbon is a major building block of all living matter, and nitrogen is a vital element in all proteins. Note in Table 2.1 that although hydrogen accounts for about 63% of the atoms in the body, it makes up only a small percentage (9%) of the mass of the human body. That is because the atomic mass of hydrogen is so much lighter than that of heavier elements such as oxygen.

Other important elements in living organisms include **minerals**, which are inorganic ions or inorganic molecules that make up less than 1% of total mass. Calcium and phosphorus, for example, are important constituents of the skeletons and shells of animals. In addition, all living organisms require **trace elements**. These elements are present in extremely small quantities but are essential for normal growth and function. For example, iron plays an important role in the transport of oxygen in the blood of many animals.

2.1 Reviewing the Concepts

- Atoms are the smallest functional units of matter. Atoms are composed of protons (p^+, positive charge), electrons (e^-, negative charge), and (except for hydrogen) neutrons (n^0, electrically neutral). Electrons are found in cloudlike electron shells around the atomic nucleus (Figures 2.1, 2.2).

Table 2.1 Chemical Elements Essential for Life in Many Organisms

Element	Symbol	% Human body mass	% All atoms in human body
Most abundant in living organisms (approximately 95% of total mass)			
Oxygen	O	65	25.5
Carbon	C	18	9.5
Hydrogen	H	9	63.0
Nitrogen	N	3	1.4

Element	Symbol	Element	Symbol	Element	Symbol
Mineral elements (less than 1% of total mass)		**Trace elements (less than 0.01% of total mass)**			
Calcium	Ca	Boron	B	Manganese	Mn
Chlorine	Cl	Chromium	Cr	Molybdenum	Mo
Magnesium	Mg	Cobalt	Co	Selenium	Se
Phosphorus	P	Copper	Cu	Silicon	Si
Potassium	K	Fluorine	F	Tin	Sn
Sodium	Na	Iodine	I	Vanadium	V
Sulfur	S	Iron	Fe	Zinc	Zn

- Each element contains a unique number of protons—its atomic number.
- The atomic mass scale indicates an atom's mass relative to the mass of other atoms.
- Many atoms exist as isotopes, which differ in the number of neutrons they contain. Some isotopes are unstable radioisotopes and emit radiation.
- Oxygen, carbon, hydrogen, and nitrogen account for the majority of atoms in organisms. Living organisms require mineral and trace elements that are essential for growth and function (Table 2.1).

2.1 Testing Your Knowledge

1. Refer to the periodic table of the elements in Appendix A. Which element has 15 electrons and three electron shells?
   a. O
   b. P
   c. Na
   d. Cl
   e. Ar

2. How many electrons would be required to fill the outer electron shell of fluorine (F)?
   a. 0
   b. 1
   c. 8
   d. 7
   e. 2

2.2 Chemical Bonds and Molecules

Learning Outcomes

1. Compare and contrast the types of chemical bonds and atomic interactions that lead to the formation of molecules.
2. Explain the concept of electronegativity, and describe how it contributes to the formation of polar and nonpolar covalent bonds.
3. Describe how a molecule's shape is important to its ability to interact with other molecules.

The linkage of atoms with other atoms serves as the basis for life and gives life its great diversity. Two or more atoms bonded together make up a molecule. For example, two oxygen atoms can combine to form an oxygen molecule, represented as O_2. This representation is called a **molecular formula.** It consists of the chemical symbols for all of the atoms that are present (here, O for oxygen) and a subscript that indicates how many of those atoms are present in the molecule (in this case, two). The term **compound** refers to a molecule composed of two or more different elements. Examples include water (H_2O), with two hydrogen atoms and one oxygen atom, and the sugar glucose ($C_6H_{12}O_6$), which has 6 carbon atoms, 12 hydrogen atoms, and 6 oxygen atoms.

One of the most important features of compounds is their emergent properties. This means that the characteristics of a compound differ greatly from those of its constituent elements. Let's consider sodium as an example. Pure sodium (Na) is a soft, silvery metal that reacts violently with water to produce NaOH and hydrogen gas. By comparison, the compound that results from combining sodium with the yellow-green gas chlorine (Cl_2) is table salt (NaCl). NaCl is a white, relatively hard crystal (as seen in the chapter-opening photo) that dissolves in water. Thus, the properties of sodium in a compound can be dramatically different from its properties as a pure element.

The atoms in molecules are held together by chemical bonds. In this section, we will examine the different types of chemical bonds, how these bonds form, and how they determine the structures of molecules.

Covalent Bonds Are Formed When Atoms Share Electrons to Fill Their Outer Shells

Covalent bonds, in which atoms share a pair of electrons, can occur between atoms whose outer shells are not full. A fundamental principle of chemistry is that atoms tend to be most stable when their outer shells are filled with electrons. Figure 2.3 shows this principle as it applies to the formation of hydrogen fluoride (HF), a molecule with many important industrial and medical applications. The outer shell of a hydrogen atom is full when it contains two electrons, though a hydrogen atom has only one electron. The outer shell of a fluorine atom is full when it contains eight electrons, though a fluorine atom has only seven electrons in its outer shell. In the HF molecule, the two atoms share a pair of electrons, which spend time in the outer electron shells of both atoms. This allows both of the outer shells of those atoms to be full. Covalent bonds are strong chemical bonds, because the shared electrons behave as if they belong to each atom.

Figure 2.3 The formation of covalent bonds. In covalent bonds, electrons from the outer shells of two atoms are shared with each other in order to complete both outer shells. This simplified illustration shows hydrogen forming a covalent bond with fluorine. The number of protons in the nuclei is indicated.

Chemists sometimes depict molecules with a **structural formula** in which each covalent bond between atoms is represented by a line indicating a pair of shared electrons. For example, HF is depicted as

$$H—F$$

A molecule of water (H_2O) can be represented as

$$H—O—H$$

The structural formula of water indicates that the oxygen atom is covalently bonded to two hydrogen atoms.

Each atom forms a characteristic number of covalent bonds, which depends on the number of electrons required to fill the outer shell. The atoms of some elements important for life, notably carbon, form more than one covalent bond and become linked simultaneously to two or more other atoms. **Figure 2.4** shows the number of covalent bonds formed by the four most abundant atoms found in the molecules of living organisms—hydrogen, oxygen, nitrogen, and carbon.

For many types of atoms, their outermost shell is full when it contains eight electrons, an octet. The **octet rule** states that many atoms are most stable when they have eight electrons in their outermost electron shell. This rule applies to most atoms found in living organisms, including oxygen, nitrogen, carbon, phosphorus, and sulfur. These atoms form a characteristic number of covalent bonds to make an octet in their outermost shell (see Figure 2.4). However, the octet rule does not always apply. For example, hydrogen has an outermost shell that can contain only two electrons, not eight.

In some molecules, a **double bond** occurs when atoms share two pairs of electrons (four electrons) rather than one pair. This is the case for an oxygen molecule (O_2), which can be represented as

$$O═O$$

Another common example occurs when two carbon atoms form bonds in compounds. They may share one pair of electrons (single bond) or two pairs (double bond), depending on how many other

Figure 2.4 The number of covalent bonds formed by common elements found in living organisms. These elements form different numbers of covalent bonds due to the number of electrons in their outer shells.

covalent bonds each carbon forms with other atoms. In rare cases, carbon can even form triple bonds, in which three pairs of electrons are shared between two atoms.

Covalent Bonds May Be Polar or Nonpolar

Some atoms attract shared electrons more strongly than do other atoms. The **electronegativity** of an atom is a measure of its ability to attract electrons in a bond with another atom. Generally, electronegativity is greater in atoms that have more protons (and therefore more positive charges) but similar numbers of electron shells; thus, fluorine is more electronegative than oxygen, which is more electronegative than nitrogen (refer to Appendix A). As atoms get much larger, however, electronegativity decreases, because the distance between the nucleus (where the positive charges reside) and the outer electron shell increases. Thus, fluorine is more electronegative than chlorine.

When two atoms with different electronegativities form a covalent bond, the shared electrons are more likely at any time to be closer to the nucleus of the atom of higher electronegativity than to the nucleus of the atom of lower electronegativity. Such bonds are called **polar covalent bonds,** because the distribution of the shared electrons around the nuclei creates a polarity, or difference in electric charge, across the molecule. H_2O is a classic example of a molecule containing polar covalent bonds. Because oxygen is much more electronegative than hydrogen, the shared electrons tend to be pulled closer to the oxygen nucleus than to either of the hydrogens, resulting in a higher probability of an electron being nearer the oxygen atom at any given instant. This unequal sharing of electrons gives the molecule a region of partial negative charge (indicated by the Greek letter δ and a minus sign, $δ^-$) and two regions of partial positive charge ($δ^+$) (**Figure 2.5**).

Atoms with high electronegativity, such as oxygen and nitrogen, have a strong attraction for electrons. These atoms form polar covalent bonds with hydrogen atoms, which have low electronegativity. Examples of polar covalent bonds include O—H and N—H. In contrast, bonds between atoms with similar or identical electronegativities—for example, between two carbon atoms (C—C) or between carbon and hydrogen atoms (C—H)—are called **nonpolar covalent bonds.** Molecules containing significant numbers of polar bonds are known as **polar molecules,** whereas molecules composed predominantly of nonpolar bonds are called **nonpolar molecules.** A single molecule may have different regions with nonpolar bonds and polar bonds. As we will explore later, the physical characteristics of polar and nonpolar molecules, especially their solubility in water, are quite different.

Hydrogen Bonds and van der Waals Dispersion Forces Allow Interactions Between and Within Molecules

An important result of certain polar covalent bonds is the ability of one molecule to loosely associate with another molecule through a weak interaction called a **hydrogen bond.** A hydrogen bond forms when a hydrogen atom in one polar molecule becomes electrically attracted to an electronegative atom, such as an oxygen or a nitrogen atom, in another polar molecule. Hydrogen bonds, like those between H_2O molecules, are represented by dashed or dotted lines in diagrams to distinguish them from covalent bonds (**Figure 2.6**). A single hydrogen bond is very weak. The strength of a hydrogen bond is only a small percentage of the strength of the polar covalent bonds linking the hydrogen and oxygen within a molecule of H_2O.

Hydrogen bonds can also occur within a single, large molecule. Large molecules may have many hydrogen bonds within their structure. Collectively, many hydrogen bonds provide a strong force that helps maintain the three-dimensional structure of a molecule.

In contrast to the cumulative strength of many hydrogen bonds, the weakness of the individual bonds is also important. When an interaction between two molecules involves relatively few hydrogen bonds, such an interaction tends to be weak and readily disrupted. The reversible nature of hydrogen bonds allows molecules to interact and then to become separated again. For example, small molecules may bind to proteins called enzymes via hydrogen bonds. **Enzymes** are molecules that catalyze many biologically important chemical reactions in cells. After binding to an enzyme, a small molecule may be converted to a different molecule, which is then released.

Temporary attractive forces that are even weaker than hydrogen bonds also form between molecules. These **van der Waals dispersion forces** arise because electrons occupy orbitals in a random, probabilistic way, as mentioned previously. At any moment, the electrons in the outer shells of the atoms in a nonpolar molecule may be evenly distributed or unevenly distributed. In the latter case, a fleeting electrical attraction to other nearby molecules may arise. As with hydrogen bonds, the collective strength of these transient attractive forces between molecules can be quite strong.

Hydrogen bonds between H_2O molecules

Figure 2.6 Example of hydrogen bonds. Hydrogen bonds are important because they allow for interactions between different molecules or between atoms within a molecule. This example depicts hydrogen bonds (shown as dashed lines) between H_2O molecules. For simplicity, only a few partial charges are indicated. In this diagram, the atoms are depicted as solid shapes, which represent the outer electron shell (this is called a space-filling model for an atom).

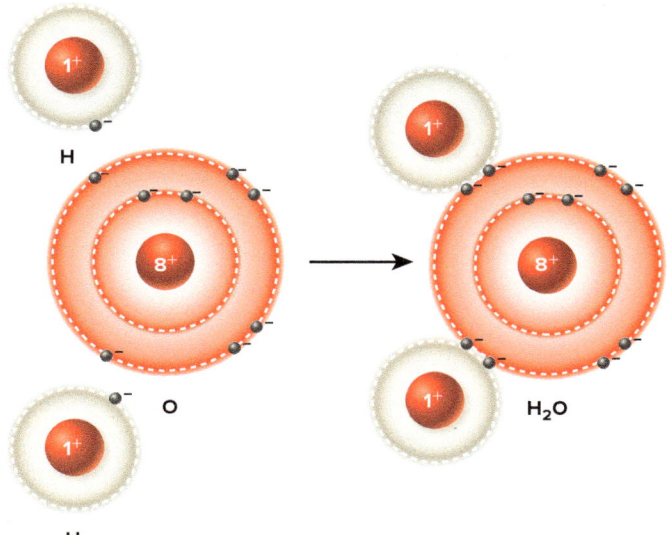

a) In H_2O, electrons are shared between an oxygen atom and two hydrogen atoms.

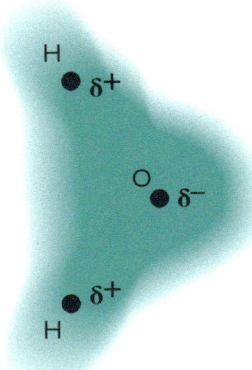

b) The shared electrons in H_2O are more likely to be near the oxygen atom at any moment, as shown in this model of a water molecule. The shading represents a rough approximation of the predicted electron densities between the three atoms. This gives oxygen a partial negative charge (δ^-) and each hydrogen a partial positive charge (δ^+).

Figure 2.5 Polar covalent bonds in the water molecule. (a) In a molecule of H_2O, two hydrogen atoms share electrons with an oxygen atom. **(b)** Because oxygen has a greater electronegativity than does hydrogen, the shared electrons are strongly attracted to the oxygen nucleus.

Ionic Bonds Involve an Attraction Between Positive and Negative Ions

Atoms are electrically neutral because they contain equal numbers of negative electrons and positive protons. If an atom or a molecule gains or loses one or more electrons, it acquires a net electric charge and becomes an **ion** (**Figure 2.7a**). For example, when a sodium atom (Na), which has 11 electrons, loses 1 electron, it becomes a sodium ion (Na^+) with a net positive charge (the superscript indicates the net charge of an ion). Ions that have a net positive charge are called **cations**. A sodium ion still has 11 protons, but only 10 electrons. A chlorine atom (Cl), which has 17 electrons, can gain an electron and become a chloride ion (Cl^-) with a net negative charge—it still has 17 protons but now has 18 electrons. Ions with a net negative charge are called **anions**.

Ions such as Na^+ and Cl^- are relatively stable because the outer electron shells are full. A sodium atom has one electron in its third (outermost) shell. If it loses this electron to become Na^+, it no longer has a third shell, and the second shell, which is full, becomes its outermost shell (see Figure 2.7). Alternatively, a Cl atom has seven electrons in its outermost shell. If it gains an electron to become a chloride ion (Cl^-), its outer shell becomes full with eight electrons. Some atoms can gain or lose more than one electron. For instance, a calcium atom, which has 20 electrons, loses 2 electrons to become a calcium ion, depicted as Ca^{2+}.

An **ionic bond** occurs when a cation binds to an anion. The right side of Figure 2.7a shows an ionic bond between Na^+ and Cl^- to form NaCl, or common table salt. NaCl often exists as crystals in which the cations and anions form a regular array. **Figure 2.7b** shows a NaCl crystal, in which the sodium and chloride ions are held together by ionic bonds.

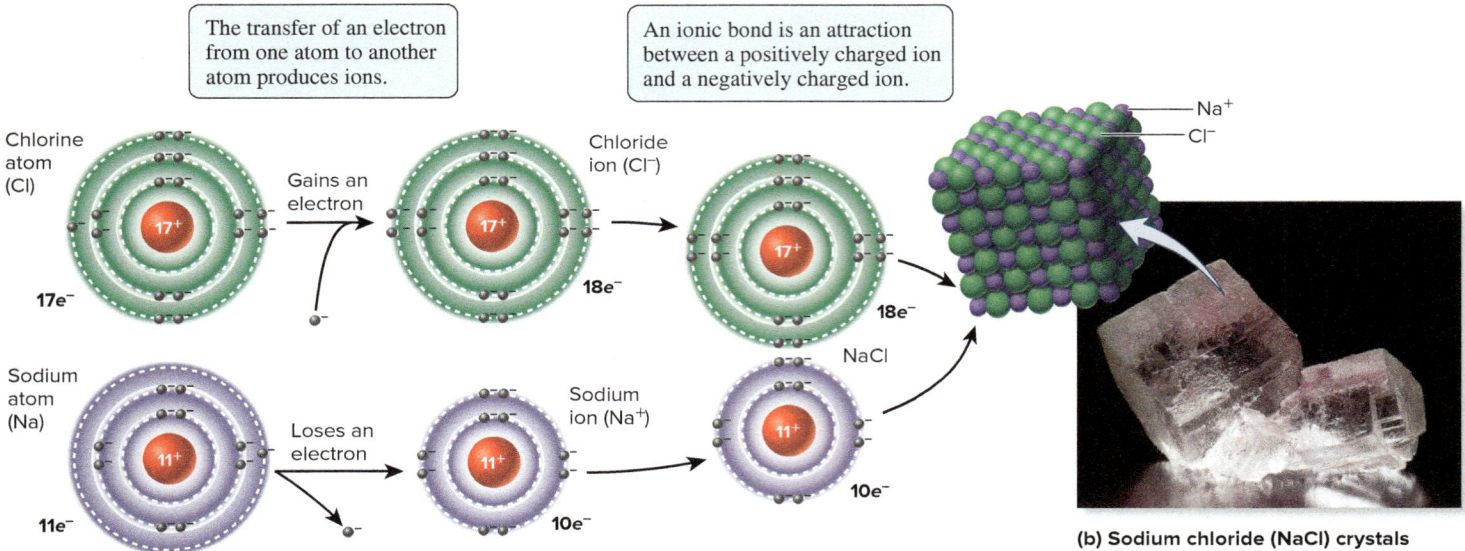

(a) Formation of ions and an ionic bond

(b) Sodium chloride (NaCl) crystals

Figure 2.7 Ionic bonding in table salt (NaCl). (a) When a sodium atom loses an electron, a sodium ion is formed, and when a chloride ion gains an electron, a chloride ion is formed. Note: Electrons do not exist free in solution; when an electron is removed from one atom, it must be transferred to some other atom. (b) In a salt crystal, a lattice is formed in which the positively charged sodium ions (Na⁺) are attracted to negatively charged chloride ions (Cl⁻).
(b) ©Charles D. Winters/Science Source

Molecules May Change Their Shapes

When atoms combine, they form molecules with various three-dimensional shapes, depending on the arrangements and numbers of bonds between their atoms. As an example, let's consider the arrangements of covalent bonds in a few simple molecules, including H_2O (**Figure 2.8**). These covalent bonds occur at defined angles between the atoms, giving the groups of atoms very specific shapes, as shown in the three examples of Figure 2.8.

Within certain limits, the shape of a molecule can change without breaking its covalent bonds. As illustrated in **Figure 2.9**, a molecule of 6 carbon atoms and 14 hydrogen atoms can assume different shapes as a result of rotations around various covalent bonds. The three-dimensional, flexible shape of molecules contributes to their biological properties. Imagine the possible variety of shape changes that could exist in very large molecules of dozens, hundreds, or even thousands of atoms, as typically occur in living organisms. In such cases, the binding of one molecule to another may affect the shape of one or both of the molecules; it is often this shape change that triggers the biological activity of a molecule in a living organism.

2.2 Reviewing the Concepts

- Two or more atoms bonded together make up a molecule. A compound is a molecule composed of two or more different elements.
- Atoms tend to form bonds that fill their outer shell with electrons. Covalent bonds, in which atoms share electrons, are very strong chemical bonds (Figures 2.3, 2.4).

Figure 2.8 Shapes of molecules. Molecules may assume different shapes, depending on the types of bonds between their atoms. The angles between bonds in groups of atoms are well defined. For example, in water at room temperature, the angle formed by the covalent bonds between the two hydrogen atoms and the oxygen atom is approximately 104.5°.

THE CHEMICAL BASIS OF LIFE I: ATOMS, MOLECULES, AND WATER

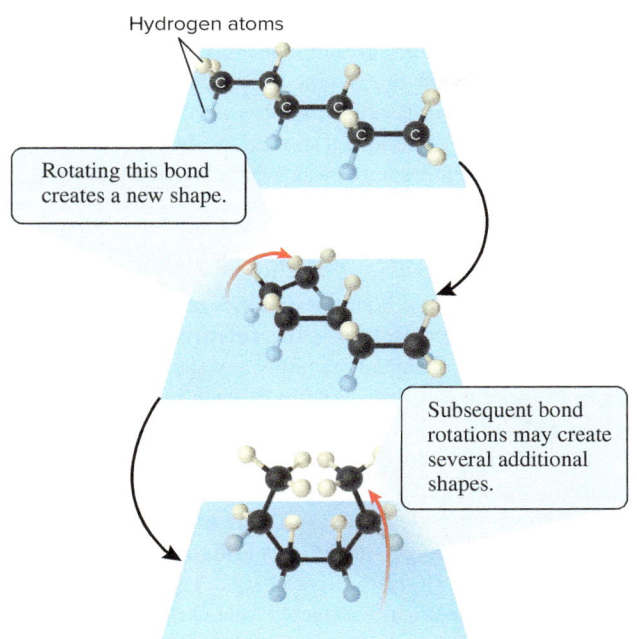

Figure 2.9 Shape changes in molecules. A single molecule may assume different three-dimensional shapes without breaking any of the covalent bonds between its atoms, as shown for a six-carbon molecule. Hydrogen atoms above the blue plane are shown in white; those below the blue plane are blue.

- The electronegativity of an atom is a measure of its ability to attract electrons in a bond with another atom. When two atoms with different electronegativities combine, they form a polar covalent bond. Polar molecules, such as H_2O, are largely composed of polar bonds, and nonpolar molecules tend to have mostly nonpolar bonds (Figure 2.5).
- One molecule may loosely associate with another molecule through weak interactions such as hydrogen bonds and van der Waals dispersion forces (Figure 2.6).
- If an atom or a molecule gains or loses one or more electrons, it acquires a net electric charge and becomes an ion. The strong attraction between two oppositely charged ions forms an ionic bond (Figure 2.7).
- The three-dimensional, flexible shapes of molecules allow them to interact and contribute to their biological properties (Figures 2.8, 2.9).

2.2 Testing Your Knowledge

1. Which of the following is a compound?
   a. Na
   b. Na^+
   c. O_2
   d. HF
   e. C

2. How many additional covalent bonds are available for each carbon in a C=C double bond?
   a. 0
   b. 1
   c. 2
   d. 3
   e. 4

2.3 Chemical Reactions

Learning Outcome

1. Relate the concept of a chemical reaction to the concept of chemical equilibrium.

Chemical Reactions Create New Compounds from Elements or Other Compounds

A **chemical reaction** occurs when one or more substances are changed into other substances by the making or breaking of chemical bonds. This can happen when two or more elements or compounds combine to form a new compound, or when one compound breaks down into two or more molecules.

Chemical reactions share many similar properties:

- They require a source of energy (heat) so that atoms and molecules can move and encounter each other.
- Many times, they must be catalyzed. A **catalyst** is an agent that speeds up the rate of a chemical reaction. An enzyme is an example of a catalyst.
- They tend to proceed in a particular direction but eventually reach a state of equilibrium.
- They usually occur in watery environments.

To understand what we mean by direction and equilibrium, let's consider a chemical reaction between methane (a component of natural gas) and oxygen. When a single molecule of methane reacts with two molecules of oxygen, one molecule of carbon dioxide and two molecules of water are produced:

$$CH_4 + 2\,O_2 \rightleftharpoons CO_2 + 2\,H_2O$$
(methane) (oxygen)   (carbon dioxide) (water)

As the reaction is written here, CH_4 and O_2 are the starting materials, or **reactants,** and CO_2 and H_2O are the **products.** The bidirectional arrows indicate that this reaction can proceed in both directions. Whether a chemical reaction is likely to proceed in a forward (left-to-right) or reverse (right-to-left) direction depends on changes in free energy, which you will learn about in Chapter 6. If we began with only CH_4 and O_2, the forward reaction would be very favorable. The reaction would produce a large amount of CO_2 and H_2O.

If the products are not converted to other molecules, chemical reactions eventually reach **chemical equilibrium,** in which the rate of the formation of products equals the rate of the formation of reactants; in other words, the concentrations of products and reactants stay the same. In the case of the reaction involving CH_4 and O_2, this equilibrium occurs when nearly all of the reactants have been converted to products.

In biological systems, many reactions do not have a chance to reach chemical equilibrium. For example, the products of a reaction may immediately be converted within a cell to different substances through a second reaction. In this case, the reactants will continue to form new products until all the reactants are used up.

One common feature of chemical reactions in living organisms is that many reactions occur in watery environments. Such chemical reactions involve reactants and products that are dissolved in water. Next, we will examine the properties of this amazing liquid and its importance to life.

2.3 Reviewing the Concepts

- A chemical reaction occurs when one or more substances are changed into different substances. All chemical reactions eventually reach chemical equilibrium, unless the products of the reaction are continually removed.

2.3 Testing Your Knowledge

1. In the reaction $CO_2 + H_2O \rightleftharpoons H_2CO_3$, chemical equilibrium will occur when
   a. the rate of CO_2 and H_2O formation equals the rate of H_2CO_3 formation.
   b. the reactants are completely converted to products.
   c. the products are completely converted to reactants.
   d. the products are converted to other substances by a second reaction.
   e. either c or d occurs.

2.4 Properties of Water

Learning Outcomes

1. Define solute and solvent.
2. Compare and contrast hydrophilic and hydrophobic substances.
3. **SCISKILLS** ▶ Calculate the molarity of a solution, and explain its meaning.
4. Describe the three states of H_2O.
5. Make a list of the roles of water that are critical for the survival of living organisms.

The bodies of all organisms are composed largely of water; most of the cells in an organism's body are not only filled with water but also surrounded by it. Up to 95% of the weight of certain plants comes from water. In humans, typically 60–70% of body weight is due to water. The brain is roughly 70% water, blood is about 80% water, and the lungs are nearly 90% water. Even our bones are about 20% water! In addition, water is an important liquid in the surrounding environments of living organisms. For example, vast numbers of species live in watery environments.

Thus far in this chapter, we have considered the features of atoms and molecules and the nature of bonds and chemical reactions between atoms and molecules. In this section, we will turn our attention to issues related to the liquid properties of living organisms and the environment in which they live. Most of the chemical reactions that occur in nature involve molecules that are dissolved in water, including those reactions that happen inside the cells of living organisms and in the spaces that surround the cells.

In this section, we will examine the properties of chemicals that influence whether they dissolve in water and consider how biologists quantify the amounts of dissolved substances. In addition, we explore some of the special properties of water that make it a vital component of living organisms and their environments.

Ions and Polar Molecules Readily Dissolve in Water

Substances dissolved in a liquid are known as **solutes,** and the liquid in which they are dissolved is the **solvent.** In all living organisms, the solvent for chemical reactions is water. Solutes dissolve in a solvent to form a **solution.** Solutions made with water are called **aqueous solutions.** To understand why a substance dissolves in water, we need to consider the chemical bonds in the solute molecule and those in water. As discussed earlier, the covalent bonds linking the two hydrogen atoms to the oxygen atom in a molecule of H_2O are polar. Therefore, the oxygen in H_2O has a slight negative charge, and each hydrogen has a slight positive charge. To dissolve in water, a substance must be electrically attracted to H_2O molecules. For example, table salt (NaCl) is a solid, crystalline substance because of the strong ionic bonds between positive sodium ions (Na^+) and negative chloride ions (Cl^-). When a crystal of sodium chloride is placed in water, the partially negatively charged oxygens of water molecules are attracted to Na^+, and the partially positively charged hydrogens are attracted to Cl^- (**Figure 2.10**). Clusters of H_2O molecules surround the ions, allowing Na^+ and Cl^- to separate from each other and enter the water—that is, to dissolve.

Generally, molecules that contain ionic and/or polar covalent bonds dissolve in water. Such molecules are said to be **hydrophilic,** which literally means water-loving. In contrast, molecules composed predominantly of carbon and hydrogen are relatively insoluble in

Figure 2.10 Table salt (NaCl crystals) dissolving in water. The ability of water to dissolve sodium chloride crystals depends on the electrical attraction between the polar water molecules and the charged sodium (Na^+) and chloride ions (Cl^-). Water molecules surround each ion as it becomes dissolved. For simplicity, the partial charges are indicated for only two water molecules.

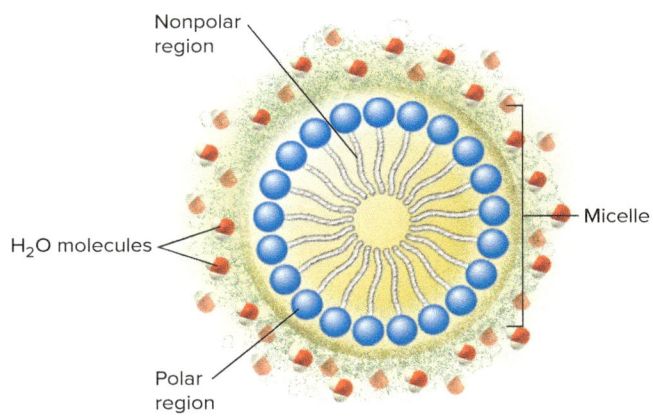

Figure 2.11 The formation of micelles by amphipathic molecules. In water, amphipathic molecules tend to arrange themselves so that their nonpolar regions are directed away from water molecules, whereas their polar regions are directed toward water molecules and can form hydrogen bonds with them.

Concept Check: *When oil dissolves in detergent micelles, where is the oil found?*

water, because carbon-carbon and carbon-hydrogen bonds are nonpolar covalent bonds. These molecules do not have partial positive and negative charges and, therefore, are not attracted to H_2O molecules. Such molecules are **hydrophobic,** meaning water-fearing. Oils are a familiar example of hydrophobic molecules. Try mixing vegetable oil with water and observe the result. The two liquids separate into an oil layer and a water layer, with very little oil dissolving in the water.

Some Molecules Have Both Hydrophilic and Hydrophobic Regions

Molecules that have both hydrophilic regions at one or more sites and hydrophobic regions at other sites are called **amphipathic** (or amphiphilic, from the Greek for both loves). When mixed with water, long, amphipathic molecules may aggregate into spheres called **micelles,** with their polar (hydrophilic) regions at the surface of each micelle, where they are attracted to the surrounding water molecules. The nonpolar (hydrophobic) ends are oriented toward the interior of the micelle (**Figure 2.11**). Such an arrangement minimizes the interaction between H_2O molecules and the nonpolar ends of the amphipathic molecules, which face inward. Nonpolar molecules can dissolve in the central nonpolar regions of these clusters and thus can exist in an aqueous environment in far higher amounts than would otherwise be possible due to their low solubility in water. Familiar examples of amphipathic molecules are those in detergents, which can form micelles that help to dissolve the oils and nonpolar molecules found in dirt. The detergent molecules in soap have polar and nonpolar ends. Oils on your skin dissolve in the nonpolar regions of the detergent, and the polar ends help the detergent rinse off in water, taking the oil with it.

In addition to micelles, amphipathic molecules may form structures consisting of double layers of molecules called bilayers. Such bilayers have two hydrophilic surfaces facing outward on both sides, in contact with water, and a hydrophobic interior facing away from water. As you will learn in Chapter 5, bilayers play a key role in cell membrane structure (look ahead to Figure 5.1).

The Amount of a Dissolved Solute per Unit Volume of Liquid Is Its Concentration

Solute **concentration** is defined as the amount of a solute dissolved in a unit volume of solution. For example, if 1 g of NaCl were dissolved in enough water to make 1 liter (L) of solution, we would say that the solute concentration is 1 g/L.

A comparison of the concentrations of two different substances on the basis of the number of grams per liter of solution does not directly indicate how many molecules of each substance are present. For example, let's compare 10 g each of glucose ($C_6H_{12}O_6$) and sodium chloride (NaCl). Because the individual molecules of glucose have more mass than those of NaCl, 10 g of glucose contains fewer molecules than 10 g of NaCl. Therefore, another way to describe solute concentration is according to the moles of dissolved solute per volume of solution. To make this calculation, we must know three things: the amount of dissolved solute, the molecular mass of the dissolved solute, and the volume of the solution.

Quantitative Analysis: Concentrations of Molecules in Solution Can Be Defined by Mass and Moles

The **molecular mass** (also referred to as molecular weight) of a molecule is equal to the sum of the atomic masses of all the atoms in the molecule. For example, using Appendix A as a guide and rounding off to the nearest whole number, glucose ($C_6H_{12}O_6$) has a molecular mass of 180 ($[6 \times 12] + [12 \times 1] + [6 \times 16] = 180$). As mentioned earlier, a mole of a substance is the amount of the substance in grams equal to its atomic or molecular mass; thus, 1 mol of glucose is 180 g of glucose. The **molarity** of a solution is defined as the number of moles of a solute dissolved in 1 L of solution. A solution containing 180 g of glucose dissolved in enough water to make 1 L is a 1 **molar** solution of glucose (1 mol/L). By convention, a 1 mol/L solution is usually written as 1 M, where the capital M stands for molar and is defined as mol/L. If 90 g of glucose (half its molecular mass) were dissolved in enough water to make 1 L, the glucose concentration would be 0.5 mol/L, or 0.5 M.

The concentrations of solutes dissolved in the fluids of living organisms are usually much less than 1 M. Many have concentrations in the range of millimoles per liter (1 mM = 0.001 M = 10^{-3} M), and others are present in even smaller concentrations—micromoles per liter (1 μM = 0.000001 M = 10^{-6} M), nanomoles per liter (1 nM = 0.000000001 M = 10^{-9} M), or less.

Crunching the Numbers: A typical concentration of glucose in the blood of many animals is around 5.5 mM. At that concentration, how many grams of glucose are present in 1 L of blood?

H_2O Exists in Three States

H_2O is an abundant compound on Earth that exists in all three states of matter—solid (ice), liquid (water), and gas (water vapor). At the temperatures found over most regions of the planet, H_2O is found primarily as a liquid in which the weak hydrogen bonds between molecules are continuously being formed, broken, and formed again as H_2O molecules move due to heat energy. If more heat is added, the speed at which H_2O molecules move increases, and hydrogen bonds are broken more readily; under such conditions, some molecules of H_2O are able to escape into the gaseous state, becoming water vapor. If the temperature falls, H_2O molecules move more slowly and consequently hydrogen bonds are broken less frequently. Larger and larger clusters of H_2O molecules are formed, until at 0°C water freezes into a crystalline matrix—ice. The H_2O molecules in ice exist in a much more orderly and highly structured arrangement than in the liquid state (**Figure 2.12**).

Changes between the solid, liquid, and gaseous states of H_2O involve an input or a release of energy. For example, when energy is supplied to raise the temperature of water to 100°C, it boils; that is, it changes from the liquid to the gaseous state—a process called vaporization. The heat required to vaporize 1 mol of any substance at its boiling point is called the substance's **heat of vaporization.** For water, this value is very high because of the hydrogen bonds between the molecules. It takes more than five times as much heat to vaporize water than it does to raise the temperature of water from 0°C to 100°C. In contrast, energy is released when water freezes to form ice. A substance's **heat of fusion** is the amount of heat energy that must be withdrawn or released from a substance to cause it to change from the liquid to the solid state. For water, this value is also high.

Another important feature for living organisms is that water has a very high **specific heat,** defined as the amount of heat energy required to raise the temperature of 1 g of a substance by 1°C (or conversely, the amount of heat energy that must be lost to lower the temperature by 1°C). The high specific heat means that it takes considerable heat to raise the temperature of water. A related concept is **heat capacity,** which is the amount of heat energy required to raise the temperature of an entire object or amount of a substance. A large beaker of pure water has a greater heat capacity than a small beaker of pure water, but the water in both beakers has the same specific heat. These properties of water contribute to the relatively stable temperatures of large bodies of water compared with inland temperatures. Most living organisms function best within a range of temperatures consistent with the liquid phase of H_2O.

The temperatures at which a solution freezes or vaporizes are influenced by the amounts of dissolved solutes. Such variations in freezing and boiling points are examples of a solution's **colligative properties,** defined as those properties that depend strictly on the total number of dissolved solute particles, not on the specific type of solute. Pure water freezes at 0°C and vaporizes at 100°C. Addition of solutes to water lowers its freezing point below 0°C and raises its boiling point to above 100°C. The presence of large amounts of solutes partly explains why the oceans do not freeze when the temperature falls below 0°C, whereas freshwater lakes and ponds do.

Figure 2.12 Structure of liquid water and ice. In water's liquid form, the hydrogen bonds between water molecules continually form, break, and re-form, resulting in a changing arrangement of molecules from instant to instant. At temperatures at or below its freezing point, H_2O forms a crystalline matrix called ice. In this solid form, hydrogen bonds are more stable.

Water Performs Many Important Tasks in Living Organisms

As noted previously, water is the primary solvent in the fluids of all living organisms. Water permits atoms and molecules to interact in ways that would be impossible in their undissolved states.

In addition to acting as a solvent, water serves many other remarkable functions that are critical for the survival of living organisms. For example, H_2O molecules participate in many chemical reactions of this general type:

$$R_1-R_2 + H-O-H \rightarrow R_1-OH + H-R_2$$

R is a general symbol used in this case to represent a group of atoms. In this reaction, R_1 and R_2 are distinct groups of atoms. On the left side of the reaction, R_1-R_2 is a compound in which these groups of atoms are connected by a covalent bond. To be converted to products, a covalent bond is broken in each reactant, R_1-R_2 and $H-O-H$; next, OH and H (from H_2O) form covalent bonds with R_1 and R_2, respectively. Reactions of this type are known as **hydrolysis reactions** (from the Greek *hydro,* meaning water, and *lysis,* meaning to break apart), because water is used to break apart another molecule (**Figure 2.13a**). As discussed in Chapter 3 and later

Biology Principle

Living Organisms Maintain Homeostasis

Healthy organisms have a normal supply of water that maintains such homeostatic functions as body temperature, ion balance, and waste levels.

(a) Water participates in chemical reactions.

(b) Water provides support. The plant on the right is wilting due to lack of water.

(c) Water is used to eliminate soluble wastes.

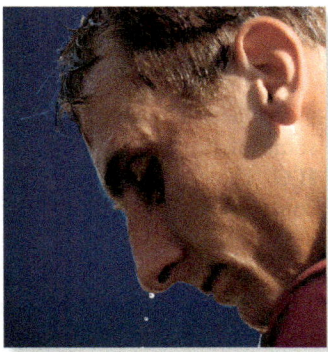

(d) Evaporation helps some animals dissipate body heat.

(e) The cohesive force of water molecules aids in the movement of water against gravity in plants.

(g) The surface tension of water explains why this water strider doesn't sink.

(f) Water in saliva serves as a lubricant during—or as shown here, in anticipation of—feeding.

Figure 2.13 Some of the amazing functions of water. In addition to acting as a solvent, water serves many crucial functions in nature.
(b) ©Aaron Haupt/Science Source; *(d)* ©Chris McGrath/Getty Images; *(e)* ©Creatas/PunchStock; *(f)* ©Gallo Images/Anthony Bannister/Getty Images; *(g)* ©Matti Suopajarvi/mattisj/Getty Images

chapters, many large molecules are broken down into smaller, biologically important units by hydrolysis reactions.

Another property of water is that it is incompressible—its volume does not significantly decrease when subjected to high pressure. This has biological importance for many organisms that use water to provide force or support. For example, water supports the bodies of some invertebrates, and it provides turgidity (stiffness) and support for plants (Figure 2.13b).

Water is also the means by which waste compounds are eliminated from an animal's body (Figure 2.13c). In mammals, for example, the kidneys filter out soluble waste products derived from the breakdown of proteins and other compounds. The filtered products remain in solution in a watery fluid, which eventually becomes urine and is excreted.

Recall from our earlier discussion of the three states of H_2O that it takes considerable energy in the form of heat to convert water into water vapor. This feature has great biological significance. Although we are familiar with the phenomenon of boiling water being converted to water vapor, water can also vaporize into the gaseous state even at ordinary temperatures. This process is known as **evaporation.** To understand this, imagine that in any volume of water at any temperature, some vibrating water molecules have higher energy than others. Those with the highest energy break their hydrogen bonds and escape into the gaseous state. During this process, energy in the form of heat is released into the environment. Evaporation is an important mechanism by which many animals cool themselves on hot days (Figure 2.13d).

The hydrogen-bonding properties of water affect its ability to form droplets and to adhere to surfaces. The phenomenon of water molecules attracting each other is called **cohesion.** Water exhibits strong cohesion due to hydrogen bonding. Cohesion aids in the movement of water through the vessels of plants (Figure 2.13e). A property similar to cohesion is **adhesion,** which is the ability of water to be attracted to, and thus adhere to, a surface that is not electrically neutral. Water tends to cling to surfaces to which it can hydrogen bond, such as a paper towel; paper is comprised of compounds such as cellulose that have an abundance of partial negative and positive charges, giving paper the ability to form hydrogen bonds. In organisms, the adhesive properties of water allow it, for example, to coat the surfaces of the digestive tract of animals and act as a lubricant for the passage of food (Figure 2.13f).

Surface tension is a measure of the attraction between molecules at the surface of a liquid. In the case of water, the attractive force between hydrogen-bonded water molecules at the interface between water and air is what causes water to form droplets. The surface water molecules attract each other into a configuration (roughly that of a sphere) that reduces the number of water molecules in contact with air. You can see this by slightly overfilling a glass with water; the water forms a rounded bulge above the rim. Likewise, surface tension allows certain insects, such as water striders, to walk on the surface of a pond without sinking (Figure 2.13g).

2.4 Reviewing the Concepts

- Water is the solvent for most chemical reactions in all living organisms. A solute dissolves in water to form a solution. Solute concentration is the amount of a solute dissolved in a unit volume of water (Figure 2.10).

Assessing Your Knowledge and Skills

Foundational Knowledge

Multiple Choice

1. _____ make up the nucleus of an atom.
   a. Protons and electrons
   b. Protons and neutrons
   c. Protons, neutrons, and electrons
   d. Neutrons and electrons
   e. Electrons

2. Living organisms are composed mainly of which atoms?
   a. calcium, hydrogen, nitrogen, and oxygen
   b. carbon, hydrogen, nitrogen, and oxygen
   c. hydrogen, nitrogen, oxygen, and helium
   d. carbon, helium, nitrogen, and oxygen
   e. carbon, calcium, hydrogen, and oxygen

3. Hydrogen bonds differ from covalent bonds in that
   a. covalent bonds can form between any type of atom, but hydrogen bonds form only between H and O.
   b. covalent bonds involve sharing of electrons, and hydrogen bonds involve complete transfer of electrons.
   c. covalent bonds result from equal sharing of electrons, but hydrogen bonds involve unequal sharing of electrons.
   d. covalent bonds involve sharing of electrons between atoms, but hydrogen bonds are the result of weak attractions between a hydrogen atom of a polar molecule and an electronegative atom of another polar molecule.
   e. covalent bonds are weak bonds that break easily, but hydrogen bonds are strong links between atoms that are not easily broken.

4. Chemical reactions in living organisms
   a. may continue to completion if the products are removed.
   b. often require a catalyst to speed them up.
   c. are in many cases reversible.
   d. occur in liquid environments, such as water.
   e. may exhibit all of the above.

5. The sum of the atomic masses of all the atoms of a molecule is its
   a. atomic weight.
   b. molarity.
   c. molecular mass.
   d. concentration.
   e. molecular formula.

Short Answer

1. Compare and contrast the different types of bonds commonly found in biological molecules.
2. **PRINCIPLES** A principle of biology is that new properties emerge from complex interactions. What examples can you cite from this chapter of emergent properties of molecules, in which atoms with one type of property combine to form molecules with completely different properties?

Critical-Thinking Skills

Multiple Choice

1. On Earth, the natural abundance of carbon isotopes is as follows: 98.95% ^{12}C; 0.63% ^{13}C; and 0.42% ^{14}C. Calculate the average atomic mass of carbon.
   a. exactly 12   c. 12.04   e. 36.04
   b. 12.01   d. 12.22

2. $6\,CO_2 + 6\,H_2O + Light \rightarrow C_6H_{12}O_6 + O_2$ is an example of a chemical reaction. We will consider this reaction in Chapter 7, where we examine photosynthesis. What is occurring during this reaction?
   a. Atoms are split and re-form as new molecules.
   b. Protons and neutrons are reorganized to form new structures.
   c. Substances are changed into other substances by the making or breaking of chemical bonds.
   d. Shells of atoms divide in order to transport protons to other atoms.
   e. Both a and b are occurring.

3. Carbon has four electrons and hydrogen has one electron in their outermost electron shells. Which of the following molecules would you predict to be stable?
   a. CH   c. CH_4 e. both c and d
   a. CH_2 d. C_2H_6

4. The molecular mass of glucose is about 180 g/mol. If 45 g of glucose is dissolved in water to make a final volume of 0.5 L, what is the molarity of the solution?
   a. 0.125 M   c. 0.5 M   e. 2.0 M
   a. 0.25 M   d. 1.0 M

5. Let's suppose a solution of pure water at 25°C contains OH^- ions at a concentration of 10^{-9} M. Calculate the pH of the solution.
   a. 3   c. 7   e. 11
   a. 5   d. 9

Short Answer

1. You will need to refer to Appendix A to answer this question. Rank the following compounds from lowest molecular mass to highest: NaCl; H_2O; $MgCl_2$; $C_6H_{12}O_6$; and $CaCl_2$.

2. In humans, the rate of breathing is controlled not by the amount of oxygen circulating in the blood but by blood pH. When the blood becomes too acidic, receptors in the brain respond to changes in blood pH and increase the breathing rate. Let's assume a person is exercising and her muscles are releasing carbon dioxide. Remind yourself how the bicarbonate pathway works, and then order the following events to show how the body maintains stable blood pH when blood carbon dioxide levels change in response to vigorous exercise: (1) Carbon dioxide combines with water to form carbonic acid; (2) pH of the blood increases; (3) pH of the blood decreases; (4) receptors in the brain detect a decrease in blood pH; (5) carbon dioxide in the blood increases due to exercise; (6) carbonic acid dissociates into water and carbon dioxide; (7) carbonic acid dissociates, forming bicarbonate and hydrogen ions; (8) an increase in breathing causes the exhalation of more carbon dioxide; (9) H^+ and bicarbonate combine to form carbonic acid.

The Chemical Basis of Life II: Organic Molecules

3

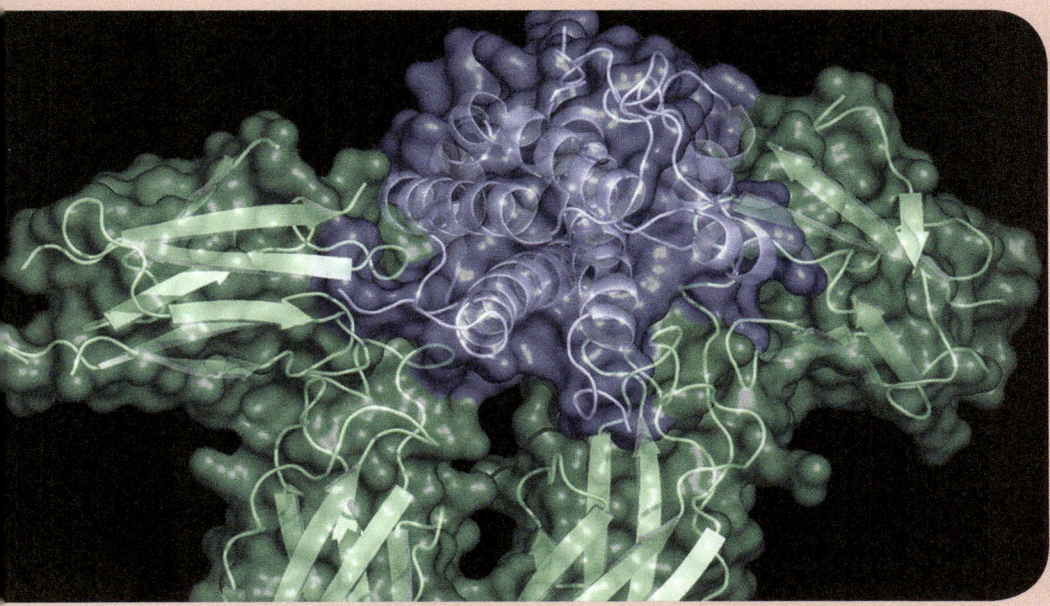
©Daniel Gage, University of Connecticut

A model showing the structure of two proteins that have bonded together. A protein is a type of organic macromolecule.

Chapter Outline

3.1 The Carbon Atom and Carbon-Containing Molecules

3.2 Synthesis and Breakdown of Organic Molecules

3.3 Overview of the Four Major Classes of Organic Molecules Found in Living Cells

3.4 Carbohydrates

3.5 Lipids

3.6 Proteins

3.7 Nucleic Acids

Assessing Your Knowledge and Skills

Of the countless possible molecules that can be produced from the known elements in nature, certain types contain carbon and are found in all forms of life. These carbon-containing molecules are collectively referred to as **organic molecules,** so named because they were first discovered in living organisms. The science of carbon-containing molecules is known as organic chemistry.

Prior to 1828, some scientists accepted a concept called "vitalism," which held that organic molecules were created by, and therefore imparted with, a vital life force that was contained within living organisms. Consequently, there was no rationale for attempting to synthesize such molecules, because they were believed to arise only through the intervention of mysterious qualities associated with life. However, in 1828, German researcher Friedrich Wöhler discovered that he had synthesized the organic molecule urea, a common waste product in animals. Wöhler's great achievement paved the way for a revolution in our understanding of the link between chemistry and biology, a discipline known as **biochemistry.**

Organic molecules include lipids and large, complex compounds called **macromolecules,** such as carbohydrates, proteins, and nucleic acids. Macromolecules differ from lipids in being long strings of repeating units of smaller molecules bonded together. In this chapter, we will survey the structures of these molecules and examine their main functions. We will begin with the element whose chemical properties are fundamental to the formation of biologically important molecules: carbon. This element provides the atomic scaffold on which life is built.

3.1 The Carbon Atom and Carbon-Containing Molecules

Learning Outcomes

1. Explain the properties of carbon that make it the chemical basis of all life.
2. Describe the variety and chemical characteristics of common functional groups of organic compounds.

In this section, we will examine the bonding properties of carbon that create biologically important organic molecules with distinct functions and shapes.

Carbon Forms Four Covalent Bonds with Other Atoms

A key property of the carbon atom is its ability to form multiple covalent bonds with other atoms, including other carbon atoms. Carbon has four electrons in its outer (second) shell, and it requires eight electrons, or four additional electrons, to fill that shell (**Figure 3.1**). In living organisms, carbon atoms most commonly form covalent bonds with other carbon atoms and with hydrogen, oxygen, nitrogen, and sulfur atoms. Bonds between two carbon atoms, between carbon and oxygen, or between carbon and nitrogen can be single or double or, in the case of certain C≡C and C≡N bonds, triple. In addition, carbon bonds may occur in configurations that are linear, ringlike, or highly branched, creating even more molecular diversity. As a result of these properties of carbon bonding, a vast number of organic compounds may be formed from only a small number of chemical elements.

As you learned in Chapter 2, bonds between atoms with similar electronegativities—for example, between two carbon atoms (C—C) or between carbon and hydrogen atoms (C—H)—are called nonpolar covalent bonds. **Hydrocarbons** are molecules with predominantly or entirely C—H and C—C bonds, and consequently are hydrophobic and poorly soluble in water. In contrast, when carbon forms polar covalent bonds with more electronegative atoms, such as oxygen or nitrogen, the resulting molecule has regions of partial negative and partial positive charges (refer back to Figure 2.5). Such a molecule is more hydrophilic and much more soluble in water, because of its electrical attraction to polar water molecules. The ability of carbon to form both polar and nonpolar bonds (**Figure 3.2**) contributes to its ability to serve as the backbone for an astonishing variety of biologically important molecules.

Another feature of carbon that is important to living organisms is that carbon bonds are stable within the large range of temperatures associated with life. This property arises in part because the carbon atom is small relative to most other atoms. Therefore, the distance between carbon atoms forming a C—C bond is short. Shorter bonds tend to be stronger and more stable than longer bonds, which form between two large atoms. For this reason, carbon bonds are compatible with what we observe about life-forms today; namely, living organisms can inhabit environments with a range of temperatures, from the Earth's frigid, icy poles to the superheated water of deep-sea vents.

Figure 3.2 Nonpolar and polar bonds in an organic molecule. Carbon can form both nonpolar and polar bonds, and both single and double bonds, as shown here in the molecule propionic acid, a common food preservative.

Carbon Atoms May Be Arranged into Functional Groups That Have Specific Properties

Aside from the simplest hydrocarbons, most organic molecules and macromolecules contain **functional groups**—groups of atoms with characteristic chemical structures and properties. Each type of functional group exhibits similar chemical properties in all molecules in which it occurs. For example, the amino group (NH_2) acts as a base. In the pH range found in living organisms, an amino group readily binds H^+ to become NH_3^+, thereby removing H^+ from an aqueous solution and increasing the pH. As discussed later in this chapter, amino groups are widely found in proteins and certain other types of organic molecules. **Table 3.1** describes examples of functional groups found in many different types of organic molecules.

3.1 Reviewing the Concepts

- Carbon can form up to four covalent bonds (polar or nonpolar) with other atoms (Figures 3.1, 3.2). A vast number of organic compounds can be formed from carbon and a relatively few other elements.
- Carbon bonds are stable at the different temperatures associated with life.
- Organic compounds may contain functional groups of atoms with characteristic properties (Table 3.1).

Figure 3.1 Model of the carbon atom. Carbon atoms have only four electrons in their outer (second) electron shell, which allows carbon to form four covalent bonds.

Table 3.1 Some Biologically Important Functional Groups That Bond to Carbon

Functional group* (with shorthand notation)	Formula	Examples of where the group is found	Properties
Amino (–NH₂)	R–NH₂ (R—N with two H)	Amino acids	Weakly basic (can accept H⁺); polar; forms part of peptide bonds
Carbonyl† (–CO)		Steroids, waxes, proteins	Polar; highly chemically reactive; forms hydrogen bonds
Ketone	R–C(=O)–R'		
Aldehyde	R–C(=O)–H		
Carboxyl (–COOH)	R–C(=O)–OH	Amino acids, fatty acids	Acidic (gives up H⁺ in water); forms part of peptide bonds
Hydroxyl (–OH)	R–OH	Steroids, alcohol, carbohydrates, some amino acids	Polar; forms hydrogen bonds with water
Methyl (–CH₃)	R–CH₃	May be attached to DNA, proteins, and carbohydrates	Nonpolar
Phosphate (–PO₄²⁻)	R–O–P(=O)(O⁻)–O⁻	Nucleic acids, ATP, phospholipids	Polar; weakly acidic and thus negatively charged at typical pH of living organisms
Sulfate (–SO₄⁻)	R–O–S(=O)(=O)–O⁻	May be attached to carbohydrates, proteins, and lipids	Polar; negatively charged at typical pH
Sulfhydryl (–SH)	R–SH	Proteins that contain the amino acid cysteine	Polar; forms disulfide bridges in many proteins

*This list contains some of the functional groups that are most important in biology. However, many more functional groups have been identified by biochemists. R and R' represent the remainder of the molecule.
†A carbonyl group is CO. In a ketone, the carbon forms covalent bonds with two other carbon atoms. In an aldehyde, the carbon is linked to a hydrogen atom.

3.1 Testing Your Knowledge

1. Carbon has an atomic number of _____ and has _____ electrons in its outer electron shell. (Refer back to Chapter 2 for help with atomic numbers.)
   a. 4 and 6
   b. 6 and 4
   c. 4 and 2
   d. 12 and 4
   e. 12 and 6

2. Hydrocarbons
   a. are molecules containing predominantly carbon, hydrogen, nitrogen, and oxygen.
   b. typically are not very soluble in water.
   c. are created when two atoms of very different electronegativities form bonds.
   d. are molecules primarily containing bonds between atoms of similar electronegativities.
   e. both b and d.

3.2 Synthesis and Breakdown of Organic Molecules

Learning Outcome
1. Explain how small organic molecules are assembled into larger ones by dehydration reactions and how hydrolysis reactions can reverse this process.

Many organic molecules are relatively small, but some are extremely large macromolecules composed of many thousands of atoms. Some important macromolecules found in living cells are formed by linking together many smaller molecules called **monomers** (meaning one part) and are thus also known as **polymers** (meaning many parts). The linkage of monomers occurs via a type of chemical reaction called a **dehydration reaction,** so named because a water molecule is released when the monomers are linked together (**Figure 3.3a**). Notice that the length of a polymer may be extended again and again by additional dehydration reactions. Some polymers can reach great lengths by this mechanism.

Polymers can be broken down into their constituent monomers by **hydrolysis reactions.** The term hydrolysis refers to the observation that a water molecule is used to break, or lyse, the linkage that holds monomers together (**Figure 3.3b**). With the exception of DNA, which is a stable polymer, the formation of other polymers in organisms, such as large carbohydrates and proteins, is generally reversible; once formed, a polymer can later be broken down.

3.2 Reviewing the Concepts
- Polymers are large macromolecules built up by dehydration reactions, in which individual monomers combine with each other. Polymers are broken down into monomers by hydrolysis reactions (Figure 3.3).

3.2 Testing Your Knowledge
1. Formation of polymers from monomers typically involves
   a. removal of a molecule of water.
   b. addition of a molecule of water.
   c. a hydrolysis reaction.
   d. a dehydration reaction.
   e. both a and d.

3.3 Overview of the Four Major Classes of Organic Molecules Found in Living Cells

Learning Outcome
1. Compare and contrast the structures and functions of carbohydrates, lipids, proteins, and nucleic acids.

(a) Polymer formation by dehydration reactions

Polymers are broken down one monomer at a time by hydrolysis reactions.

(b) Breakdown of a polymer by hydrolysis reactions

Figure 3.3 Formation and breakdown of polymers. (a) Monomers combine to form polymers in living organisms by dehydration reactions, in which a molecule of H_2O is removed each time a new monomer is added to the growing polymer. (b) Polymers can be broken down into their constituent monomers by hydrolysis reactions, in which a molecule of H_2O is added each time a monomer is released.

Table 3.2 A Comparison of the Four Types of Organic Molecules Found in Living Organisms

Type	Structure	Key functions	Examples
Carbohydrates	The general formula is $C_n(H_2O)_n$, where n is a whole number.	Simple carbohydrates are broken down to make ATP, which is used as a source of energy. Larger carbohydrates store energy or may play a structural role, as in plant cell walls. Some carbohydrates function as molecular tags, allowing recognition of specific cells and molecules.	Simple sugars, such as glucose; larger polymers, such as starch and cellulose
Lipids	Lipids are nonpolar molecules that are primarily composed of carbon and hydrogen, with some oxygen.	Lipids are a key part of cell membranes and function as hormones and in energy storage; in animals, they act as insulators and shock absorbers.	Phospholipids, estrogen, testosterone, triglycerides
Proteins	A polypeptide is a structural unit composed of a linear sequence of amino acids. A protein is a functional unit composed of one or more polypeptides.	Proteins play a key role in cell structure and carry out a diverse array of cellular functions; for example, there are proteins involved with gene expression and regulation, motor proteins, defense proteins, cell-signaling proteins, metabolic enzymes, structural proteins, and transporters.	Look ahead to Table 3.3.
Nucleic acids	A nucleic acid is a linear sequence of nucleotides; DNA is double-stranded.	DNA stores genetic information in units called genes. RNA is made from DNA and provides access to that information.	DNA and RNA

By analyzing the cells of many different species, researchers have determined that all forms of life have organic molecules and macromolecules that fall into four broad categories, based on their chemical and biological properties: carbohydrates, lipids, proteins, and nucleic acids. **Table 3.2** outlines the general structures and functions of these molecules and provides some examples. In the next sections, we will examine them in greater detail.

3.3 Reviewing the Concepts

- The four major classes of organic molecules found in living organisms are carbohydrates, lipids, proteins, and nucleic acids (Table 3.2).

3.3 Testing Your Knowledge

1. Which of the following classes of organic molecules are important for storing energy?
    a. nucleic acids (DNA and RNA)
    b. proteins
    c. carbohydrates
    d. lipids
    e. both c and d.

3.4 Carbohydrates

Learning Outcomes

1. Distinguish among different types of carbohydrate molecules, including monosaccharides, disaccharides, and polysaccharides.
2. Relate the functions of plant and animal polysaccharides to their structure.

Carbohydrates are organic molecules composed of carbon, hydrogen, and oxygen atoms. Most of the carbon atoms in a carbohydrate are linked to another carbon atom, a hydrogen atom, and an —OH functional group. However, other functional groups, such as —NH$_2$ and —COOH groups, are also found in certain carbohydrates. As discussed next, sugars are simple carbohydrates, whereas polysaccharides are large macromolecules.

Monosaccharides and Disaccharides Are Simple Carbohydrates

The simplest carbohydrates are the monomers known as **monosaccharides** (from the Greek, meaning single sugars). The most common types of monosaccharides contain five carbons,

Galactose   Glucose   Fructose

Figure 3.4 Monosaccharide structures. Glucose and galactose differ in the position of the —OH group attached to carbon atom number 4. Fructose has two carbon atoms outside the main ring structure.

called pentoses, or six carbons, called hexoses. Important pentoses are ribose ($C_5H_{10}O_5$) and the closely related deoxyribose ($C_5H_{10}O_4$), which are part of RNA and DNA molecules, respectively (described later in this chapter). The most common hexoses are galactose, glucose, and fructose (**Figure 3.4**). Note in Figure 3.4 that all three hexoses have the same chemical formula ($C_6H_{12}O_6$), but the atoms are arranged differently within the molecules. Molecules with identical formulas but different structures are called **isomers**.

The hexoses may exist in a linear structure or, more typically, in a ring. The carbon atoms in the ring are numbered by convention, as shown in Figure 3.4. The ring is formed with an oxygen atom; in glucose, the oxygen forms a linkage between carbons 1 and 5. The hydrogen atoms and the —OH groups may lie above or below the plane of the ring structure; this is one feature that distinguishes one isomer from another.

Of these three hexoses, glucose plays an especially important role in the cellular activities of living organisms. For example, glucose is very water-soluble and thus circulates in the blood or body fluids of animals, where it can be transported across cell membranes. Once inside a cell, glucose can be broken down by enzymes into smaller molecules, releasing energy that was stored in glucose's chemical bonds. This energy is then stored in the bonds of another molecule, called adenosine triphosphate, or ATP (see Chapter 6), which, in turn, provides an energy source for a variety of cellular processes.

Monosaccharides can be linked together by dehydration reactions to form **disaccharides** (meaning two sugars), which are carbohydrates composed of two monosaccharides. A familiar disaccharide is sucrose, or table sugar, which is composed of the monomers glucose and fructose (**Figure 3.5**). Sucrose is the major transport form of sugar in plants. The linking together of most monosaccharides involves the removal of an —OH group from one monosaccharide and a hydrogen atom from the other, giving rise to a molecule of H_2O and covalently bonding the two sugars together through an oxygen atom. The bond formed between two sugar molecules by such a dehydration reaction is called a **glycosidic bond**.

Polysaccharides Are Carbohydrate Polymers

When many monosaccharides are linked together to form long polymers, the products are **polysaccharides** (meaning many sugars). In living organisms, polysaccharides include

- **starch,** found in plant cells,
- **glycogen,** present in certain types of animal cells,

Biology Principle

Living Organisms Use Energy

The chemical energy stored in the bonds of glucose molecules can be harnessed by living organisms. This energy can be used to perform numerous functions that support life, including the synthesis of new molecules, growth, digestion, locomotion, and many others.

Figure 3.5 Formation of a disaccharide. Two monosaccharides can bond to each other to form a disaccharide, such as sucrose, by a dehydration reaction.

Concept Check: What type of reaction is the reverse of the one shown here, in which a disaccharide is broken down into two monosaccharides?

- **cellulose,** found in the cell walls of plant cells,
- **peptidoglycans,** found in the cell walls of certain bacteria,
- **chitin,** found in cell walls of fungi and the exoskeletons of arthropods, and
- **glycosaminoglycans,** polysaccharides found in connective tissue and surrounding cells in animals.

Starch, glycogen, and cellulose are composed of thousands of glucose molecules linked together in long chains, differing in the extent of branching and the orientation of monomers along the chain (**Figure 3.6**). The sugar monomers within other polysaccharides may have nitrogen-containing groups, such as amino groups, attached to them.

The bonds that form in polysaccharides are between specific carbon atoms of each molecule. In starch and glycogen, the bonds form between carbons 1 and 4, and between 1 and 6. The high degree

THE CHEMICAL BASIS OF LIFE II: ORGANIC MOLECULES   43

Figure 3.6 Polysaccharides that are polymers of glucose. These polysaccharides differ in their extent of branching. Note: In cellulose, the bonding arrangements cause every other glucose to be inverted with respect to its neighbors.

BioConnections: *Look ahead to Figures 4.34 and 4.35 for the role of cellulose in plant cell structure and to Figures 28.10 and 28.11 for its role in plant growth. Considering the amount of plant life on Earth, what might you conclude about the abundance of cellulose on the planet?*

of branching in glycogen contributes to its solubility in animal tissues, such as muscle. This is because the extensive branching creates a more open structure, in which many hydrophilic —OH functional groups have access to water and can hydrogen-bond with it. Starch is less branched, making it less soluble.

Starch and glycogen are used to store energy in cells. Like disaccharides, these polysaccharides can be hydrolyzed to yield monosaccharides, which are broken down to provide the energy to make ATP. Starch and glycogen are an efficient means of storing energy for those times when a plant or an animal cannot obtain sufficient energy from its environment or diet for its metabolic requirements.

Cellulose, peptidoglycans, chitin, and glycosaminoglycans, by contrast, play a structural rather than energy-storing role. Cellulose has a linear arrangement of carbon-carbon bonds and no branching (see Figure 3.6). The linear arrangement allows vast numbers of hydrogen bonds to form between cellulose molecules, which stack together in sheets and provide great strength to the structure of plant cell walls. The hydrogen bonds form between —OH groups on carbons 3 and 6.

Peptidoglycans are found in the cell walls of certain bacteria; they consist of sugars and amino acids bonded together in a strong, lattice-like arrangement that gives strength and rigidity to the cell wall. Chitin is a tough, structural polysaccharide that forms the external skeleton of insects and crustaceans (for example, shrimp and lobsters) as well as the cell walls of fungi. Glycosaminoglycans are found abundantly in cartilage—the tough, fibrous material in bone and certain other animal structures. Glycosaminoglycans are also abundant in the extracellular matrix that provides a structural framework surrounding many of the cells in an animal's body (look ahead to Figure 4.31). The nitrogen in these polysaccharides allows for additional hydrogen bonding and thus contributes to the strength of structures containing them.

3.4 Reviewing the Concepts

- Carbohydrates are composed of carbon, hydrogen, and oxygen atoms. Cells can break down glucose, an important carbohydrate, releasing energy that is then stored in the bonds of ATP.
- Carbohydrates include monosaccharides, disaccharides, and polysaccharides. Some polysaccharides store energy in cells. Other polysaccharides may serve a support or structural function (Figures 3.4, 3.5, 3.6).

3.4 Testing Your Knowledge

1. Glucose, galactose, and fructose are examples of
   a. polysaccharides.
   b. hexoses.
   c. isomers.
   d. disaccharides.
   e. both b and c.

2. _____ is a storage polysaccharide commonly found in the cells of animals.
   a. Glucose
   b. Sucrose
   c. Glycogen
   d. Starch
   e. Cellulose

3.5 Lipids

Learning Outcomes

1. List the different classes of lipid molecules important in living organisms.
2. Diagram the structure of a triglyceride, and explain how it is formed and how its structure is affected by the presence of saturated and unsaturated fatty acids.
3. Explain why some fats are solid at room temperature and others are liquid.
4. Discuss how triglycerides function as energy-storage molecules.
5. Explain why phospholipids form a bilayer when dissolved in water.
6. Describe the chemical structure of steroids, and give an example of their biological importance.

Lipids are hydrophobic molecules composed mainly of hydrogen and carbon atoms, and some oxygen. The defining feature of lipids is that they are nonpolar and therefore poorly soluble in water. Lipids include triglycerides, phospholipids, steroids, and waxes. They are not considered to be macromolecules because they are not composed of many monomers that are covalently interconnected. However, as you will see, lipids may associate with each other noncovalently to form complex structures such as those found in membranes.

Triglycerides Are Made from Glycerol and Fatty Acids

Triglycerides (commonly referred to as fats) consist of a glycerol molecule linked to three fatty acids (**Figure 3.7**). Glycerol is a three-carbon molecule with one —OH group bonded to each carbon. A fatty acid is a chain of carbon and hydrogen atoms with a carboxyl group (—COOH) at one end. Each of the —OH groups in glycerol is linked to the —COOH group of a fatty acid by the removal of a molecule of H_2O by a dehydration reaction.

The fatty acids found in triglycerides and other lipids differ with regard to their lengths and the presence or absence of double bonds (**Figure 3.8**).

- **Saturated fatty acids** are those in which all of the carbons in the hydrocarbon chain form single bonds (C—C). The term saturated indicates that each carbon has a maximal number of attached hydrogens. In other words, the carbons are saturated with respect to the number of attached hydrogens.

- Monounsaturated fatty acids contain one C=C double bond. A C=C double bond introduces a kink into the linear shape of a fatty acid.

- Polyunsaturated fatty acids contain two or more C=C double bonds.

The hydrocarbon tail of a fatty acid does not form hydrogen bonds with water and is very hydrophobic. Consequently, fatty acids tend to aggregate in solutions such that their hydrocarbon tails avoid contact with water. Saturated fatty acids can pack together more tightly than can unsaturated fatty acids because they do not have kinks in their hydrocarbon tails (see Figure 3.8), which allows increased intermolecular forces, such as van der Waals dispersion forces (see Chapter 2), between packed saturated fatty acids. As a result, it takes a greater amount of energy (heat) to melt saturated fatty acids than unsaturated ones. Animal fats generally contain a high proportion of saturated fatty acids. For example, beef fat contains high amounts of stearic acid, a saturated fatty acid with a melting point of 70°C (see Figure 3.8). When you cook a hamburger on the stove, the stearic acid and other saturated fats melt, and liquid grease appears in the frying pan. When allowed to cool to room temperature, however, the liquid grease returns to its solid form.

In contrast, triglycerides containing unsaturated fatty acids usually have low melting points and are liquids at room temperature. Such triglycerides are called oils. Triglycerides derived from plants generally contain unsaturated fatty acids. For example, olive oil contains oleic acid (see Figure 3.8), a monounsaturated fatty acid with a melting point of 16°C. Fatty acids with additional double bonds have even lower melting points; for example, linoleic acid has two double bonds and melts at −5°C. Safflower and sunflower oils contain large amounts of linoleic acid.

Like starch and glycogen, triglycerides are important for storing energy. The hydrolysis of triglycerides releases their fatty acids, which can be metabolized to provide energy to make ATP. Certain organisms, most notably mammals, have the ability to store large amounts of energy by accumulating triglycerides.

Figure 3.7 The formation of a triglyceride. The formation of a triglyceride occurs via three dehydration reactions in which fatty acids are bonded to glycerol. Note in this figure that a common shorthand notation is used for depicting fatty acids, in which a portion of the CH_2 groups forming the chain is represented as $(CH_2)_n$, where *n* in this example is 15. Thus, each fatty acid in this triglyceride has 18 carbons.

Figure 3.8 Examples of fatty acids. Fatty acids are hydrocarbon chains that have a carboxyl functional group at one end and contain either no double bonds between carbon atoms (saturated) or one or more double bonds (unsaturated). Stearic acid, for example, is an abundant saturated fatty acid in animals, whereas oleic acid is an unsaturated fatty acid found in plants. Note that the presence of a C=C double bond introduces a kink into the chain structure of oleic acid. As a consequence, unsaturated fatty acids are not able to pack together as tightly as saturated fatty acids.

Phospholipids Are Amphipathic Lipids

Phospholipids are similar in structure to triglycerides, but with one important difference. In a phospholipid, the third —OH group of glycerol is linked to a phosphate group instead of a fatty acid. In most phospholipids, a small polar or charged nitrogen-containing molecule is attached to this phosphate (**Figure 3.9a**). The glycerol backbone, phosphate group, and charged molecule constitute a polar (hydrophilic) head at one end of the phospholipid, whereas the two fatty acids form nonpolar (hydrophobic) tails at the opposite end. Recall from Chapter 2 that molecules with polar and nonpolar regions are called amphipathic molecules.

In water, phospholipids become organized into a double layer of molecules called a bilayer, with their polar heads interacting with the H_2O molecules and their nonpolar tails facing the interior, where they avoid contact with water. As you will learn in Chapter 5, this bilayer arrangement of phospholipids is critical for determining the structure of cell membranes, as shown in **Figure 3.9b**.

Steroids Contain Ring Structures

Steroids have a distinctly different chemical structure from that of the other types of lipid molecules discussed thus far. Four fused rings of carbon atoms form the general structure of all steroids. One class of steroids contains an —OH group at a particular carbon; these are called sterols and they are found in fungi, plants, and animals (**Figure 3.10**). Cholesterol, for example, is found in the cell membranes of animals, where it contributes to membrane structure and function. Cholesterol is also found in numerous animal cells, where it serves as a precursor for the synthesis of steroid hormones. These steroid hormones include estrogens and androgens, which are necessary for animal reproduction.

Waxes Are Complex Lipids That Prevent Water Loss from Organisms

Many plants and animals produce lipids called **waxes** that are secreted onto their surface, such as the leaves of plants and the cuticles of insects. Although any wax may contain hundreds of different compounds, all waxes contain one or more hydrocarbons and long structures that resemble a fatty acid attached by its carboxyl group to another long hydrocarbon chain. Most waxes are very nonpolar and therefore exclude water, providing a barrier to water loss. Some waxes may also be used as structural elements—like beeswax, which forms the honeycomb in a beehive.

3.5 Reviewing the Concepts

- Lipids are nonpolar and very insoluble in water. Major classes of lipids include triglycerides, phospholipids, steroids, and waxes.
- Triglycerides are formed by bonding glycerol with three fatty acids. In a saturated fatty acid, all the carbons are linked by single covalent bonds. Unsaturated fatty acids contain one or more C=C double bonds (Figures 3.7, 3.8).
- Phospholipids contain a polar head and two nonpolar tails, making them amphipathic (Figure 3.9).
- Steroids are constructed of four fused rings of carbon atoms. Sterols are a type of steroid and include cholesterol (Figure 3.10).
- Waxes are nonpolar and exclude water. Waxes often are found as protective coatings on leaves and the surfaces of some animals' bodies.

Biology Principle

Biologists Use Models to Test Predictions and Convey Ideas

The inset to part (b) of this figure shows a model for a portion of a phospholipid bilayer with 14 phospholipids; all of these phospholipids contain only saturated fatty acids. The Modeling Challenge in the figure legend asks you to consider the effects of unsaturated fatty acids.

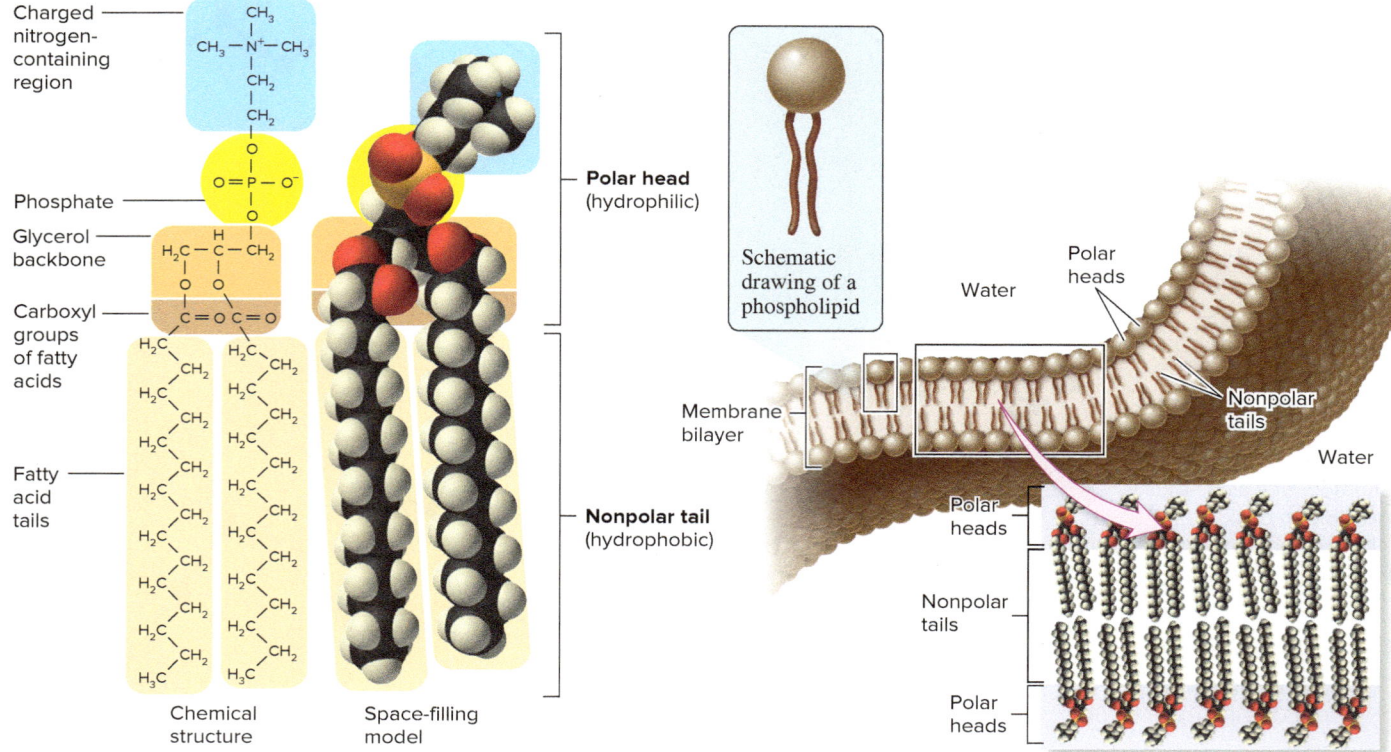

(a) Structure and model of a phospholipid

(b) Arrangement of phospholipids in a bilayer

Figure 3.9 Structure of phospholipids. (a) Chemical structure and space-filling model of phosphatidylcholine, a common phospholipid found in living organisms. Phospholipids contain both polar and nonpolar regions, making them amphipathic. The fatty acid tails are nonpolar. The rest of the molecule is polar. (b) Arrangement of phospholipids in a biological membrane, such as the plasma membrane that encloses cells. The hydrophilic region of the phospholipid faces the watery environment, whereas the hydrophobic regions associate with each other in the interior of the membrane, forming a bilayer.

Modeling Challenge: Take a look at the inset at the bottom right in Figure 3.9b and draw a modified version of this model in which 3 out of the 14 phospholipids have one monounsaturated fatty acid and 2 of them have a polyunsaturated fatty acid with two double bonds. As discussed in Chapter 5, phospholipid bilayers are more fluid when the phospholipids do not pack together tightly. Which phospholipid bilayer would you predict to be more fluid: the one in the model shown in Figure 3.9b or the modified model that you have drawn? Explain your answer.

Concept Check: When water and oil are added to a test tube, the two liquids form two separate layers (think of oil and vinegar in a bottle of salad dressing). If a solution of phospholipids was added to a mixture of water and oil, where would the phospholipids dissolve?

3.5 Testing Your Knowledge

1. Hydrolysis of a triglyceride yields
    a. 3 glycerols and 3 fatty acids.
    b. 1 glycerol and 3 fatty acids.
    c. 3 glycerols and 1 fatty acid.
    d. 3 glycerols and 6 fatty acids.
    e. a shorter triglyceride.

2. Which of the following are highly amphipathic?
    a. phospholipids
    b. triglycerides
    c. steroids
    d. waxes
    e. all of the above.

THE CHEMICAL BASIS OF LIFE II: ORGANIC MOLECULES

Figure 3.10 Structure of a common steroid. All steroids share a common ring structure, shown here. One class of steroids is the sterols, such as cholesterol.

Amino Acids Are the Building Blocks of Proteins

Proteins are polymers of **amino acids.** All amino acids contain a carbon atom, called the α-carbon, that is linked to an amino group (—NH_2) and a carboxyl group (—COOH). When an amino acid is dissolved in water at neutral pH, the —NH_2 group acts as a base and accepts a hydrogen ion, becoming positively charged. The —COOH group acts as an acid; it loses a hydrogen ion and is negatively charged at neutral pH. The α-carbon also is linked to a hydrogen atom and a side chain, which is given a general designation R.

The 20 amino acids found in proteins are distinguished by their side chains (**Figure 3.11**). The amino acids are categorized by the properties of their side chains, whether nonpolar, polar and uncharged, or polar and charged. The varying structures of the side chains are critical features of protein structure. The arrangement and chemical features of the side chains cause proteins to fold and adopt their three-dimensional shapes. In addition, certain amino side chains may be critical in protein function. For example, amino acid side chains within specific regions of enzymes are essential for enzymes to catalyze chemical reactions.

Amino acids are joined together by a dehydration reaction that links the carboxyl group of one amino acid to the amino group of another (**Figure 3.12a**). The covalent bond formed between a carboxyl group and an amino group is called a **peptide bond.** When multiple amino acids are joined by peptide bonds, the resulting molecule is called a **polypeptide** (**Figure 3.12b**). The backbone of the polypeptide in Figure 3.12 is highlighted in yellow. The amino acid side chains project from the backbone. When two or more amino acids are linked

3.6 Proteins

Learning Outcomes

1. Give examples of the general functions that are carried out by different proteins.
2. Describe how amino acids are joined to form a polypeptide.
3. Explain the four levels of protein structure.
4. Outline the factors that determine protein shape and function.
5. Define domain as it relates to protein structure.

Proteins are macromolecules composed of carbon, hydrogen, oxygen, nitrogen, and small amounts of other elements, such as sulfur. Proteins play critical roles in nearly all life processes (**Table 3.3**).

Table 3.3 Major Categories and Functions of Proteins

Category	Functions	Examples
Proteins involved in gene expression and regulation	Make mRNA from a DNA template; synthesize polypeptides from mRNA; regulate genes	RNA polymerase assists in synthesizing RNA from DNA.
Motor proteins	Initiate movement	Myosin provides the contractile force of muscles.
Defense proteins	Protect organisms against disease	Antibodies help destroy bacteria or viruses.
Metabolic enzymes	Increase rates of chemical reactions important in energy balance	Hexokinase is an enzyme involved in glucose metabolism.
Cell-signaling proteins	Enable cells to communicate with each other and to sense the environment	Notch proteins coordinate growth of cells in developing animals.
Structural proteins	Support and strengthen structures	Actin provides shape to the cytoplasm of plant and animal cells. Collagen gives strength to tendons.
Transporters	Mediate movement of solutes across membranes	Glucose transporters move glucose from outside cells to inside cells, where it can be used for energy.

Figure 3.11 The 20 amino acids found in living organisms. Amino acids have different chemical properties (for example, nonpolar versus polar) due to their different side chains. These properties contribute to the differences in the three-dimensional shapes and chemical properties of proteins, which, in turn, influence proteins' biological functions. Note: Tyrosine has both polar and nonpolar characteristics and is listed in just one category for simplicity. The common three-letter and one-letter abbreviations for each amino acid are shown in parentheses.

together, one end of the resulting molecule has a free amino group. This end is the amino end, also called the **N-terminus,** because it has a nitrogen atom. The other end of the polypeptide is called the carboxyl end, or **C-terminus,** because it has a carbon atom; the C-terminus has a free carboxyl group. As shown in **Figure 3.12c,** amino acids within a polypeptide are numbered sequentially from the N-terminus to the C-terminus.

The term polypeptide refers to a structural unit composed of a linear sequence of amino acids. A protein is a functional unit composed of one or more polypeptides that have folded and twisted into a precise three-dimensional shape. Many proteins also have carbohydrates (glycoproteins) or lipids (lipoproteins) attached to certain amino acids; these modifications impart unique functions to such proteins.

Figure 3.12 The mechanism of polypeptide formation. Polypeptides are polymers of amino acids. They are formed by linking amino acids via dehydration reactions to make peptide bonds. Every polypeptide has an amino end, or N-terminus, and a carboxyl end, or C-terminus.

Concept Check: How many H_2O molecules would be produced during the formation of a polypeptide that is 72 amino acids long?

Proteins Have a Hierarchy of Structure

Scientists describe protein structure at four progressive levels: primary, secondary, tertiary, and quaternary, shown schematically in **Figure 3.13**. Each higher level of structure depends on the preceding levels. For example, changing the primary structure may affect the secondary, tertiary, and quaternary structures.

Primary Structure The **primary structure** (see Figure 3.13) of a protein is the amino acid sequence of its polypeptide(s). The primary structures of proteins are determined by genes. As we will explore in Chapter 10, genes carry the information for the production of polypeptides with specific amino acid sequences.

Secondary Structure The amino acid sequence of a polypeptide, together with the principles of chemistry and physics, cause a protein to fold into a more compact structure. Amino acids can rotate around bonds within a polypeptide. Consequently, proteins are flexible and can fold into a number of shapes, just as a string of beads can be twisted into many configurations. Folding can be irregular, or certain regions can have a repeating folding pattern called a protein's **secondary structure**. The two basic types of a protein's secondary structure are the α helix and the β pleated sheet.

In an **α helix**, the polypeptide backbone forms a repeating helical structure that is stabilized by hydrogen bonds along the length of the backbone. As shown in Figure 3.13, the hydrogen linked to a nitrogen atom forms a hydrogen bond with an oxygen atom, which in turn is double-bonded to a carbon atom. These hydrogen bonds occur at regular intervals along the polypeptide backbone and cause the backbone to twist into a helix.

In a **β pleated sheet**, regions of the polypeptide backbone lie parallel to each other. Hydrogen bonds between a hydrogen linked to a nitrogen atom and a double-bonded oxygen form between these adjacent, parallel regions. When this occurs, the polypeptide backbone adopts a repeating zigzag, or pleated, shape.

Secondary structure contributes to the great strength of certain proteins, including those found in plant cell walls; the proteins that make up the silk webs of spiders; and collagen, the chief component of cartilage in vertebrate animals.

Some regions along a polypeptide do not assume an α helix or β pleated sheet conformation and do not have a secondary structure. These regions are sometimes called random coiled regions.

Figure 3.13 The hierarchy of protein structure. The R groups are omitted for simplicity.

Tertiary Structure As the secondary structure of a polypeptide becomes established due to the particular primary structure, additional side chains of amino acids interact. The polypeptide folds and refolds upon itself to assume a complex three-dimensional shape—its **tertiary structure** (see Figure 3.13). Tertiary structure includes all secondary structures plus any other interactions involving amino acid side chains. For some proteins, the tertiary structure is the final structure. However, as described next, other proteins are composed of two or more polypeptides.

Quaternary Structure Most proteins are composed of two or more polypeptides, which each adopt a tertiary structure and then assemble with each other (see Figure 3.13). The individual polypeptides are called **protein subunits**. Subunits may be identical polypeptides or they may be different. When proteins consist of more than one polypeptide, they are said to have **quaternary structure**. A common example is the oxygen-binding protein called hemoglobin, found in the red blood cells of vertebrate animals. As you will learn later (look ahead to Figure 36.21), four protein subunits combine to form one hemoglobin protein. Each subunit can bind a single molecule of oxygen; therefore, each hemoglobin protein can carry four molecules of oxygen in the blood.

Protein Structure Is Determined by Several Factors

Several factors determine the way proteins adopt their secondary, tertiary, and quaternary structures. The following factors are critical for protein folding and stability:

- *Hydrogen bonds.* The large number of weak hydrogen bonds within a polypeptide and between polypeptides collectively produces a strong force that promotes protein folding and stability.
- *Ionic bonds and other polar interactions.* Some amino acid side chains are positively or negatively charged. Positively charged side chains may bind to negatively charged side chains via ionic bonds.
- *Hydrophobic effect.* Some amino acid side chains are nonpolar. As a protein folds, the hydrophobic amino acids are likely to be found in the center of the protein, minimizing their contact with

water. Some proteins have stretches of nonpolar amino acids that anchor the proteins in the hydrophobic portion of membranes.
- *van der Waals dispersion forces.* Atoms within molecules have temporary attractions for each other if they are an optimal distance apart. These weak attractions are termed van der Waals dispersion forces, and they contribute to tertiary and quaternary structure (see Chapter 2).
- *Disulfide bridges.* The side chain of the amino acid cysteine (see Figure 3.11) contains a sulfhydryl group (—SH), which can react with an —SH group in another cysteine side chain. The resulting bond is a **disulfide bridge,** which links the two amino acid side chains together (—S—S—). Disulfide bridges are covalent bonds that can occur within a polypeptide or between different polypeptides.

The first four factors are also important in the ability of different proteins to interact with each other. Many cellular processes involve steps in which two or more different proteins interact with each other. For such an interaction to occur, proteins must bind to each other. Such binding is usually very specific. The shape of one protein may permit it to fit precisely into a region of another protein (**Figure 3.14**). Such **protein-protein interactions** are critically important in allowing cellular processes to occur in a series of defined steps. In addition, protein-protein interactions are important in building complicated cellular structures and in processes that provide organization and function to cells.

Biology Principle

New Properties Emerge from Complex Interactions

This principle of biology is apparent at the protein level. The primary sequence of proteins determines their final three-dimensional structures. Compare this with the chapter-opening depiction of two real proteins and the intermediate levels of protein structure shown in Figure 3.13. The three-dimensional shapes of different proteins determine their ability to interact with other molecules, including other proteins.

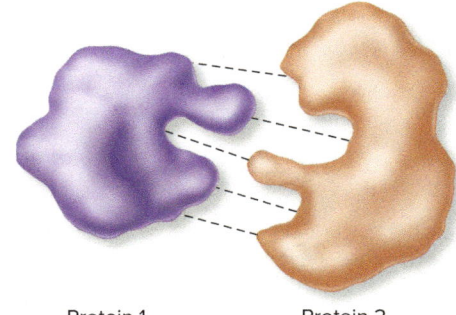

Figure 3.14 Protein-protein interaction. Two different proteins may interact with each other due to hydrogen bonding, ionic bonding, the hydrophobic effect, and van der Waals dispersion forces. Interaction is also facilitated by their respective three-dimensional shapes.

FEATURE INVESTIGATION

Anfinsen Showed That the Primary Structure of Ribonuclease Determines Its Three-Dimensional Structure

Prior to the 1960s, the mechanisms by which proteins assume their three-dimensional structures were not understood. Scientists hypothesized that correct folding required unknown cellular components or that ribosomes, the sites where polypeptides are synthesized, somehow shaped proteins during their synthesis. American biochemist Christian Anfinsen, however, postulated that the amino acid sequence of proteins is the primary factor that determines how proteins fold into their proper conformations. He hypothesized that a protein spontaneously assumes its most stable conformation based on the principles of chemistry and physics (**Figure 3.15**).

To test this hypothesis, Anfinsen studied ribonuclease, an enzyme that degrades RNA molecules. Anfinsen began with purified ribonuclease. The important point is that other cellular components were not present. Biochemists had already determined that ribonuclease has four disulfide bridges (S—S) between eight cysteines. Anfinsen exposed ribonuclease to a chemical called β-mercaptoethanol, which breaks S—S bonds, and to urea, which disrupts hydrogen bonds (H bonds) and ionic bonds. This treatment caused ribonuclease to be denatured, that is, to become unfolded. Following this treatment, he measured the ability of the enzyme to degrade RNA molecules. The enzyme had lost essentially all of its ability to degrade RNA.

Therefore, Anfinsen concluded that when ribonuclease is unfolded, or denatured, it is no longer functional.

The key step in this experiment came when Anfinsen removed the urea and β-mercaptoethanol from the solution. Because these molecules are much smaller than ribonuclease, removing them from the solution could be accomplished with a technique called size-exclusion chromatography. In size-exclusion chromatography, solutions are poured on top of a glass column of beads and allowed to flow down the column to an open collection port at the bottom. The beads in the column have microscopic pores that trap small molecules like urea and β-mercaptoethanol but permit large molecules such as ribonuclease to pass down the length of the column and out the collection port.

Using size-exclusion chromatography, Anfinsen was able to separate ribonuclease from β-mercaptoethanol and urea. He then allowed the purified enzyme to sit in water for up to 20 hours, after which he retested the ribonuclease for its ability to degrade RNA. The result revolutionized our understanding of proteins. The activity of the ribonuclease was almost completely restored! Therefore, even in the complete absence of any cellular factors or ribosomes, an unfolded protein can refold into its correct, functional structure.

Figure 3.15 Anfinsen's experiments with ribonuclease, demonstrating that the primary structure of a polypeptide plays a key role in protein folding.

HYPOTHESIS Within their amino acid sequence, proteins contain all the information needed to fold into their correct, three-dimensional shapes.

KEY MATERIALS Purified ribonuclease, RNA, denaturing chemicals, size-exclusion chromatography columns.

	Experimental level	Conceptual level	
1 Incubate purified ribonuclease in test tube with RNA, and measure its ability to degrade RNA.			Numerous H bonds and ionic bonds (not shown) and 4 S—S bonds. Protein is properly folded. (For simplicity, the three-dimensional shape is not shown; see Panel 3 for a computer model of the true structure.)
2 Denature ribonuclease by adding β-mercaptoethanol (breaks S—S bonds) and urea (breaks H bonds and ionic bonds). Measure its ability to degrade RNA.			No more H bonds, ionic bonds, or S—S bonds. Protein is unfolded.
3 Pour mixture from step 2 into a size-exclusion chromatography column. Beads in the column trap β-mercaptoethanol and urea, whereas ribonuclease flows to the bottom. Collect ribonuclease in a test tube. Allow ribonuclease to sit for up to 20 hours and then measure its ability to degrade RNA.		Beads have microscopic pores that trap β-mercapto-ethanol and urea, but not ribonuclease. Computer model of properly folded structure of ribonuclease. Renatured ribonuclease.	

4 THE DATA

5 CONCLUSION Certain proteins, like ribonuclease, can spontaneously fold into their final, functional shapes without assistance from other cellular structures or factors. (However, as described in the text, this is not true of many other proteins.)

6 SOURCE Haber, E., and Anfinsen, C. B. 1961. Regeneration of Enzyme Activity by Air Oxidation. *Journal of Biological Chemistry* 236: 422–424.

(Computer model of properly folded structure of ribonuclease) Image from the RCSB PDB (www.rcsb.org; H.M. Berman, et al. (2000), The Protein Data Bank, *Nucleic Acids Research,* 28: 235-242) of PDB ID 1A2W (Liu, Y., Hart, P.J., Schlunegger, M.P., Eisenberg, D. (1998) *PNAS,* 95: 3437-3442).

Researchers have since learned that ribonuclease's ability to refold into its functional structure does not occur for all proteins. Some proteins require assistance from enzymes and other proteins to achieve their proper folding. Even so, Anfinsen's experiments provided compelling evidence that the primary structure of a polypeptide is a key determinant of a protein's tertiary structure, an observation that earned him a share in the Nobel Prize in Chemistry in 1972.

Experimental Questions

1. What hypothesis was Anfinsen testing?
2. Why did Anfinsen use urea and β-mercaptoethanol in his experiments?
3. **SCISKILLS ▶** Explain the result that was crucial to the discovery that the tertiary structure of ribonuclease may depend entirely on the primary structure.

EVOLUTIONARY CONNECTIONS

Proteins Contain Functional Domains

Modern research has revealed that many proteins have a modular design. This means that portions within these proteins, called **domains,** have distinct structures and functions. These domains have been duplicated during evolution, so the same kind of domain is found in many different proteins. A domain that is found in different proteins has the same three-dimensional tertiary structure in all of them and performs a characteristic function.

As an example, **Figure 3.16** shows two members of a family of related proteins, known as nuclear receptors, which travel into the nucleus of animal cells and play critical roles in regulating how certain genes are turned on. These types of proteins are involved in animal development, reproduction, metabolism, and homeostasis. Nuclear receptors contain four or more domains within their structure:

- One domain found in nuclear receptors is a ligand-binding domain. These receptors are activated by a ligand, which is a molecule that binds to the receptor. For the two examples in Figure 3.16, the ligands are the steroid hormones estrogen and testosterone (a type of androgen), and the nuclear receptors are called the estrogen receptor and the androgen receptor, respectively. Though the ligand-binding domains in these two nuclear receptors have similar structures, slight structural differences allow one to bind estrogen and the other to bind testosterone (see the insets in Figure 3.16).

- A second domain, called the DNA-binding domain, binds to DNA once the receptor is activated by its ligand. The DNA-binding domains of these two receptors are sufficiently similar to enable both of them to bind to DNA but dissimilar enough that they bind to different base sequences that are located next to different genes. Therefore, estrogen and testosterone have different effects, because they regulate different sets of genes.

Figure 3.16 The domains in two related proteins called nuclear receptors. These schematic drawings do not depict the actual three-dimensional structures of the estrogen and androgen receptors but rather emphasize that both receptors are composed of a series of different domains. The insets show the actual structures of the ligand-binding domains for both receptors. Note: The estrogen and androgen receptors are each composed of two identical polypeptides. For simplicity, only one polypeptide is shown here.

- Another domain, called a nuclear localization domain, facilitates the movement of the protein into the cell nucleus, where the DNA is located.
- Once the protein binds to DNA, a fourth domain, called the activation domain, activates the transcription of the target gene.

Overall, a nuclear receptor is a single protein with multiple domains, each with a unique function.

3.6 Reviewing the Concepts

- Proteins are macromolecules that play critical roles in almost all life processes. The proteins of all living organisms are composed of the same set of 20 amino acids, corresponding to 20 different side chains (Figure 3.11, Table 3.3).
- Amino acids are joined by peptide bonds to form polypeptides. A protein is a functional unit composed of one or more polypeptides that have folded into three-dimensional shapes (Figure 3.12).
- The four levels of protein structure are primary (its amino acid sequence), secondary (α helices or β pleated sheets), tertiary (folding to assume a three-dimensional shape), and in most cases quaternary. Domains are units of structure found in many different proteins, where they serve similar functions, such as binding to DNA (Figures 3.13, 3.14, 3.15, 3.16).

3.6 Testing Your Knowledge

1. Which of the following is the primary structure of a protein?
    a. β pleated sheet
    b. its amino acid sequence
    c. its three-dimensional shape
    d. its protein subunits
    e. its random coils
2. A peptide bond is formed between
    a. amino groups of two amino acids.
    b. the carboxyl groups of two amino acids.
    c. the side groups of two amino acids.
    d. the amino group of one amino acid and carboxyl group of another amino acid.
    e. two polypeptides, which then form a protein with quaternary structure.

3.7 Nucleic Acids

Learning Outcomes

1. Describe the three components of a nucleotide.
2. Distinguish between the structures of DNA and RNA.
3. Explain how certain bases pair with others in DNA.

Nucleic acids are macromolecules that are responsible for the storage, expression, and transmission of genetic information. The expression of genetic information in the form of specific proteins determines whether an organism is a human, a fungus, an oak tree, or a bacterium. In this section, we will survey the general features of DNA and RNA, which are discussed in greater detail in Chapter 9.

Nucleotides Are the Building Blocks of DNA and RNA

The two classes of nucleic acids are **deoxyribonucleic acid (DNA)** and **ribonucleic acid (RNA).** DNA molecules store genetic information coded in the sequence of their building blocks. RNA molecules are involved in decoding this information into instructions for linking a specific sequence of amino acids to form a polypeptide.

Like other macromolecules, DNA and RNA are polymers consisting of linear sequences of repeating monomers. Each monomer, known as a **nucleotide,** has three components:

- A phosphate group
- A pentose sugar (deoxyribose or ribose)
- A single or a double ring of carbon and nitrogen atoms known as a **base** (Figure 3.17)

The nucleotides in DNA contain the five-carbon sugar **deoxyribose.** Four different nucleotides are present in DNA, corresponding to the four different bases that can be linked to deoxyribose. The **purine** bases, **adenine (A)** and **guanine (G),** have fused double rings of carbon and nitrogen atoms, and the **pyrimidine** bases, **cytosine (C)** and **thymine (T),** have a single-ring structure (Figure 3.17).

Nucleotides are covalently linked to each other to form a DNA strand or an RNA strand. The carbon atoms of the sugar are numbered 1′–5′. The phosphate groups link the 3′ carbon of one nucleotide to the 5′ carbon of the next (Figure 3.17). The linkages between sugars and phosphates form the backbone of a DNA or RNA strand with the bases projecting from the backbone.

DNA Is Composed of Two Strands That Form a Double Helix

A DNA molecule consists of two strands of nucleotides coiled around each other to form a **double helix** (Figure 3.18). The two strands are held together by hydrogen bonds between a purine base in one strand and a pyrimidine base in the opposite strand. The ring structure of each base lies in a flat plane perpendicular to the sugar-phosphate backbone, somewhat like steps on a spiral staircase.

Only certain bases can pair with others, due to the location of the hydrogen-bonding groups in the four bases (Figure 3.18). In DNA, an A in one strand is always paired with a T in another strand, and a G is paired with a C. This specificity provides the mechanism for duplicating and transferring genetic information (look ahead to Chapter 9). Two hydrogen bonds are formed between adenine and thymine (A-T pairing), and three are formed between guanine and cytosine (G-C pairing).

THE CHEMICAL BASIS OF LIFE II: ORGANIC MOLECULES

Figure 3.17 Structure of a DNA strand. Nucleotides are linked to each other to form a strand of DNA. The four bases found in DNA are shown. A strand of RNA is similar except the sugar is ribose, and uracil is substituted for thymine.

Figure 3.18 The double-stranded structure of DNA. DNA consists of two strands coiled together into a double helix. The bases form hydrogen bonds (dashed lines) in which A pairs with T, and G pairs with C.

Concept Check: *If the sequence of bases in one strand of a DNA double helix is known, can the base sequence of the opposite strand be predicted?*

RNA Is Usually Single Stranded

RNA structure differs in only a few respects from DNA. Except in some viruses, RNA consists of a single strand of nucleotides rather than a double-stranded structure. In RNA, the sugar in each nucleotide is **ribose** instead of deoxyribose. Also, the pyrimidine base thymine in DNA is replaced in RNA with the pyrimidine base **uracil (U)** (**Figure 3.19**). The other three bases—adenine, guanine, and cytosine—are found in both DNA and RNA. Certain forms of RNA called messenger RNA (mRNA), ribosomal RNA (rRNA), and transfer RNA (tRNA) are responsible for using the information contained in DNA, that is, within genes, in order to make polypeptides. This topic will be discussed in Unit III.

Figure 3.19 Uracil, a pyrimidine base found only in RNA. Uracil forms hydrogen bonds with adenine.

3.7 Reviewing the Concepts

- Nucleic acids (DNA and RNA) are responsible for the storage, expression, and transmission of genetic information. DNA and RNA consist of repeating monomers known as nucleotides, composed of a phosphate group, a five-carbon sugar, and a base (Figure 3.17).
- DNA consists of two strands of nucleotides coiled to form a double helix, held together by hydrogen bonds between a purine base in one strand and a pyrimidine base in the opposite strand. DNA molecules store genetic information coded in the sequence of their monomers (Figure 3.18).
- RNA consists of a single strand of nucleotides and includes the base uracil. RNA molecules are involved in decoding information in DNA into instructions for making polypeptides (Figure 3.19).

3.7 Testing Your Knowledge

1. _____ include adenine and guanine and have _____ rings of carbon and nitrogen atoms.
   a. Purines, double
   b. Purines, single
   c. Pyrimidines, double
   d. Pyrimidines, single

2. Which of the following is *not* found in DNA?
   a. cytosine
   b. adenine
   c. a phosphate group
   d. uracil
   e. a pentose sugar

Assessing Your Knowledge and Skills

Foundational Knowledge

Multiple Choice

1. Which of the following is a common functional group found in organic molecules?
   a. a hydrocarbon
   b. a ketone
   c. an isomer
   d. a carbon atom
   e. an oxygen atom

2. The versatility of carbon to serve as the backbone for a variety of molecules is due in part to
   a. the ability of carbon atoms to form four covalent bonds.
   b. the fact that carbon usually forms ionic bonds with many different atoms.
   c. the abundance of carbon in the environment.
   d. the ability of carbon to form single, double, and even triple bonds with many different types of atoms.
   e. both a and d

3. _____ is a polysaccharide characterized by an unbranched arrangement and a high degree of hydrogen bonding.
   a. Glucose
   b. Starch
   c. Wax
   d. Cellulose
   e. Glycogen

4. The monomers of proteins are _____, and these are linked by polar covalent bonds commonly referred to as _____ bonds.
   a. nucleotides, peptide
   b. polypeptides, peptide
   c. dipeptides, amino
   d. amino acids, peptide
   e. monosaccharides, glycosidic

5. A(n) _____ is a modular portion of a protein with a particular structure and function.
   a. peptide bond
   b. domain
   c. phospholipid
   d. amino acid
   e. monosaccharide

Short Answer

1. Explain the difference between saturated and unsaturated fatty acids as well as how the structural differences between them contribute to differences in their properties.

2. **PRINCIPLES** A principle of biology is that structure determines function. Discuss the meaning of this principle for organic molecules such as carbohydrates, lipids, and proteins.

Critical-Thinking Skills

Multiple Choice

1. Which of the following is the most hydrophobic?
   a. C_3H_7COOH
   b. CH_3OH
   c. C_2H_6
   d. $C_5H_{10}O_5$
   e. both b and d

2. A polysaccharide is composed of 74 glucose molecules. How many chemical reactions would be needed to completely break down this polysaccharide into 74 individual glucose molecules?
   a. 74 dehydration reactions
   b. 73 dehydration reactions
   c. 74 hydrolysis reactions
   d. 73 hydrolysis reactions
   e. both a and c

3. Both safflower oil and butter contain fats. However, safflower oil is liquid at room temperature and butter is solid. Why?
   a. Safflower oil has more saturated fats than butter.
   b. Safflower oil has more unsaturated fats than butter.
   c. The fatty acids of safflower oil have more double bonds than butter.
   d. The fatty acids of safflower oil have fewer double bonds than butter.
   e. Both b and c are correct.

4. The structures of three molecules are shown below. Which of them is amphipathic?

   (a)

   (b)

(c)

H-C(H)(H)-C(H)(H)-C(H)(H)-C(H)(H)-C(H)(H)-C(H)(H)-C(H)(H)-C(H)(H)-C(H)(H)-C(H)(H)-C(H)(H)-C(H)(H)-C(H)(H)-C(H)(H)-C(H)(H)-C(H)(H)-C(=O)OH

a. (a) only
b. (b) only
c. (c) only
d. (b) and (c)
e. (a) and (c)

5. In a DNA molecule, 30% of the bases are guanine (G). What percentage of the bases are adenine (A)?
   a. 30%   d. 10%
   b. 20%   e. none of the above
   c. 15%

Short Answer

1. Let's suppose that a protein's primary structure is altered such that a single amino acid was changed to a different one. How might this change affect the secondary, tertiary, or quaternary structure of the protein?

2. Let's suppose you are given two bottles that contain different types of oils. At room temperature, both of them are liquid. Describe a simple experiment to determine which one has the highest amount of unsaturated fatty acids. Explain the expected results.

Unit II
CELLS

Cell biology is the study of life at the cellular level. Although cells are the simplest units of life, they are also wonderfully complex and interesting, providing information about all living things. In this unit, Chapter 4 begins with a description of the evolutionary origin of cells and then provides an overview of their structures and functions. Chapter 5 focuses on the structure of cell membranes and the transport of substances into and out of the cell. The chapter also examines how cells are held together by cell junctions. Chapters 6 and 7 are largely devoted to metabolism—the sum of the chemical reactions in a cell or an organism. Chapter 6 explores the topic of energy and considers how enzymes facilitate chemical reactions in a cell. The chapter also examines the pathways for carbohydrate breakdown and their production of an energy intermediate called adenosine triphosphate (ATP). Chapter 7 describes how carbohydrates are made via photosynthesis by capturing the energy of sunlight. Finally, Chapter 8 considers how cells interact with their environment and with each other. The chapter examines how cells respond to signals, either those that come directly from their environment or those that are made by other cells.

The following biology principles are emphasized in this unit:

- *Cells are the simplest units of life.* Throughout this unit, we will examine how life exists at the cellular level. Chapter 4 provides an overview of cell structure and function.

- *Living organisms interact with their environment.* Chapter 5 largely focuses on the plasma membrane of the cell and how it provides an interface between the cell and its environment.

- *Living organisms use energy.* Chapters 6 and 7 address how cells utilize and store energy contained within organic molecules such as glucose.

- *Structure determines function.* Throughout this unit, we will see many examples in which the structure of proteins determines their cellular functions.

- *Biology is an experimental science.* Most chapters in this unit have a Feature Investigation describing a pivotal experiment that provided insights into the workings of cells.

- *Biologists use models to test predictions and convey their ideas.* Every chapter has a Modeling Challenge to help you refine this important skill.

(4) ©Steve Gschmeissner/Science Source; (5) ©Ramón Andrade, 3Dciencia/Science Source; (6) ©wavebreakmedia/Shutterstock; (7) ©Steven Trainoff Ph.D./Getty Images; (8) ©Dennis Kunkel Microscopy/Science Source

Evolutionary Origin of Cells and Their General Features

4

©Steve Gschmeissner/Science Source

A cell from the human liver. The cell in this micrograph was viewed by a technique called transmission electron microscopy (TEM), which is described in this chapter. The micrograph was artificially colored using a computer to enhance the visualization of certain cell structures.

Chapter Outline

- 4.1 Origin of Living Cells on Earth
- 4.2 Microscopy
- 4.3 Overview of Cell Structure and Function
- 4.4 The Cytosol
- 4.5 The Nucleus and Endomembrane System
- 4.6 Semiautonomous Organelles
- 4.7 Protein Sorting to Organelles
- 4.8 Extracellular Matrix and Plant Cell Walls
- 4.9 Systems Biology of Cells: A Summary

Assessing Your Knowledge and Skills

Cancer is a disease characterized by uncontrolled cell growth. Cancer cells form tumors that invade nearby parts of the body and may spread to distant parts via the bloodstream. About one in four Americans will eventually die from cancer. In the U.S., approximately 1.6 million people were diagnosed with cancer during 2012, and about 600,000 died from the disease. Although such numbers are sobering, there is some reason for optimism. Over the past 10 years, the death rate from cancer has consistently decreased by over 1% per year. This decrease can be attributed to a variety of factors, including lifestyle changes (for example, fewer people smoking cigarettes) and improved treatment methods.

Topics such as cancer are within the field of **cell biology**—the study of individual cells and their interactions with each other. Researchers in this field want to understand the basic features of cells and apply that knowledge in the treatment of diseases such as cystic fibrosis, sickle cell disease, and lung cancer.

The idea that organisms are composed of cells originated in the mid-1800s. German botanist Matthias Schleiden studied plant material under the microscope and was struck by the presence of many similar-looking compartments, each of which contained a dark area. Today we call those compartments cells and the dark area the nucleus. Schleiden speculated in 1838 that cells are living entities and that plants are aggregates of cells arranged according to definite laws. In 1839, German physiologist Theodor Schwann extended Schleiden's hypothesis to animals. About two decades later, German biologist Rudolf Virchow proposed that *omnis cellula e cellula*, or "every cell originates from another cell." This idea arose from his research, which showed that diseased cells divide to produce more diseased cells.

The **cell theory,** which is credited to both Schleiden and Schwann with contributions from Virchow, has three parts.

1. All living organisms are composed of one or more cells.
2. Cells are the smallest units of life.
3. New cells come only from pre-existing cells by cell division.

In this chapter, we will begin our exploration of cells with a description of hypotheses of how living cells arose on Earth. We will then consider the general features of cell structure and function. Later chapters in this unit will examine certain aspects of cell biology in greater detail.

4.1 Origin of Living Cells on Earth

Learning Outcomes

1. Outline the four overlapping stages that are hypothesized to have led to the origin of living cells.
2. List various hypotheses about how complex organic molecules formed.
3. Explain the concept of an RNA world, and describe how it could have evolved into a DNA/RNA/protein world.

As described in Chapter 1, cells are the simplest units of life. Even so, living cells are complex collections of molecules and macromolecules. DNA stores genetic information, RNA acts as an intermediary in the process of protein synthesis and plays other important roles, and proteins form the foundation for the structure and activities of living cells. Life as we know it requires this interplay among DNA, RNA, and proteins for its existence and perpetuation. On modern Earth, every living cell is made from a pre-existing cell.

But how did life get started? Because DNA, RNA, and proteins are the central players in the enterprise of life, scientists who are interested in the origin of cells have focused much of their attention on the formation of these macromolecules and their building blocks, namely, nucleotides and amino acids. To understand the origin of life, we can view the process as occurring in four overlapping stages:

- **Stage 1:** Nucleotides and amino acids were produced prior to the existence of cells.
- **Stage 2:** Nucleotides became polymerized to form RNA and/or DNA, and amino acids became polymerized to form proteins.
- **Stage 3:** Polymers became enclosed in membranes.
- **Stage 4:** Polymers enclosed in membranes acquired properties that are associated with living cells.

Researchers have followed a variety of experimental approaches to determine how life may have begun, including the synthesis of organic molecules in the laboratory without the presence of living cells or cellular material. This work has led researchers to propose a variety of hypotheses regarding the origin of cells. In this section, we will examine the origin of life at each of these stages and consider a few scientific viewpoints that wrestle with the question "How did life on Earth begin?"

Stage 1: Organic Molecules Formed Prior to the Existence of Cells

Let's begin our inquiry into the first stage of the origin of life by considering how nucleotides and amino acids may have been made prior to the existence of living cells. In the 1920s, a Russian biochemist, Alexander Oparin, and a Scottish biologist, J. B. S. Haldane, independently proposed that organic molecules such as nucleotides and amino acids arose spontaneously under the conditions that occurred on early Earth. According to this hypothesis, the spontaneous appearance of organic molecules produced what they called a "primordial soup," which eventually gave rise to living cells.

The conditions on early Earth, which were much different from today, may have been more conducive to the spontaneous formation of organic molecules. Current hypotheses suggest that organic molecules, and eventually macromolecules, formed spontaneously. The formation of such molecules is termed prebiotic (before life) or abiotic (without life) synthesis. These organic molecules accumulated slowly because there was little free oxygen gas, so they were not spontaneously oxidized, and there were as yet no living organisms, so they were also not metabolized. The slow accumulation of these molecules in the early oceans over a long period of time formed what is now called the **prebiotic soup.** The formation of this medium was a key event that preceded the origin of life.

Though most scientists agree that life originated from the assemblage of nonliving matter on early Earth, the mechanism of how and where these molecules originated is widely debated. Many intriguing hypotheses have been proposed, which are not mutually exclusive. A few of the more widely debated ideas are the reducing atmosphere hypothesis, the extraterrestrial hypothesis, and the deep-sea vent hypothesis.

Reducing Atmosphere Hypothesis Based largely on geological data, many scientists in the 1950s proposed that the atmosphere on early Earth was rich in water vapor (H_2O), hydrogen gas (H_2), methane (CH_4), and ammonia (NH_3). These components, along with a lack of atmospheric oxygen (O_2), produced a reducing atmosphere because methane and ammonia readily give up electrons to other molecules, thereby reducing them. Such oxidation-reduction reactions, or redox reactions, are required for the formation of complex organic molecules from simple inorganic molecules.

In 1953, American chemist Stanley Miller, a student in the laboratory of physical chemist Harold Urey, was the first scientist to use

Biology Principle

Biology Is an Experimental Science
By conducting experiments, researchers were able to demonstrate the feasibility of the synthesis of organic molecules prior to the emergence of living cells.

Figure 4.1 Testing the reducing atmosphere hypothesis for the origin of life—the Miller and Urey experiment.

Concept Check: *With regard to the origin of life, why are biologists interested in the abiotic synthesis of organic molecules?*

---

experimentation to test whether the prebiotic synthesis of organic molecules is possible. His experimental apparatus was intended to simulate the conditions on early Earth that were postulated in the 1950s (**Figure 4.1**). Water vapor from a flask of boiling water rose into another chamber containing hydrogen gas (H_2), methane (CH_4), and ammonia (NH_3). Miller inserted two electrodes that sent electrical discharges into the chamber to simulate lightning bolts. A condenser jacket cooled some of the gases from the chamber, causing droplets to form that fell into a trap. He then took samples from this trap for chemical analysis. In his first experiments, he observed the formation of hydrogen cyanide (HCN) and formaldehyde (CH_2O). Such molecules are precursors of more complex organic molecules. These precursors also combined to make larger molecules such as the amino acid glycine. At the end of 1 week of operation, 10–15% of the carbon had been incorporated into organic compounds. Later experiments by Miller and others demonstrated the formation of sugars, a few types of amino acids, lipids, and bases found in nucleic acids (for example, adenine).

In a study published in 2011, researchers analyzed samples that Miller had preserved from a 1958 experiment in which he used a mixture of CH_4, NH_3, hydrogen sulfide (H_2S), and carbon dioxide (CO_2). For unknown reasons, Miller had not analyzed what products were made in this experiment. When these preserved samples were analyzed using modern technology, they were found to contain 23 different amino acids and 4 amines (another type of organic molecule), more organic compounds than seen in Miller's classic experiments.

Why were these studies important? The work of Miller and Urey was the first attempt to apply scientific experimentation to the quest to understand the origin of life. Their pioneering strategy showed that the prebiotic synthesis of organic molecules is possible, although it could not prove that such synthesis did, in fact, occur. In spite of the importance of these studies, critics of the so-called reducing atmosphere hypothesis have argued that Miller and Urey were wrong about the composition of early Earth's environment. More recently, many scientists have suggested that the atmosphere on early Earth was not reducing but instead was a neutral gaseous mixture consisting mostly of carbon monoxide (CO), carbon dioxide (CO_2), nitrogen gas (N_2), and H_2O. These newer ideas are derived from studies of volcanic gas, which has much more CO_2 and N_2 than CH_4 and NH_3, and from the observation that ultraviolet (UV) radiation destroys CH_4 and NH_3, so these molecules would have been short-lived on early Earth, which had high levels of UV radiation. Nevertheless, since the experiments of Miller and Urey, many newer investigations have shown that organic molecules can be made under a variety of conditions. For example, organic molecules can be made prebiotically from a neutral gaseous mixture composed primarily of CO, CO_2, N_2, and H_2O.

Extraterrestrial Hypothesis Many scientists have argued that sufficient organic molecules may have been present in meteorites, material from asteroids and comets that reached the surface of early Earth. A significant proportion of meteorites belong to a class known as carbonaceous chondrites. Such meteorites may contain a substantial amount of organic carbon, including amino acids and nucleic acid bases. Based on this observation, some scientists have postulated that such meteorites could have transported a significant amount of organic molecules from outer space to early Earth.

Opponents of this hypothesis argue that most of this material would have been destroyed by the intense heating that accompanies the passage of large bodies through the atmosphere and their subsequent collision with the surface of the Earth. Though some organic molecules are known to reach the Earth via such meteorites, the degree to which heat would have destroyed those organic molecules remains a matter of controversy.

Deep-Sea Vent Hypothesis In 1988, German lawyer and organic chemist Günter Wächtershäuser proposed that key organic molecules may have originated in deep-sea vents, which are cracks in the Earth's surface where superheated water rich in metal ions

(a) Deep-sea vent hypothesis

(b) A deep-sea vent community

Figure 4.2 The deep-sea vent hypothesis for the origin of life. (a) Deep-sea vents are cracks in the Earth's surface that release hot gases such as hydrogen sulfide (H_2S). This heats the water near the vent and results in a gradient between the very hot water adjacent to the vent and the cold water farther from the vent. The synthesis of organic molecules can occur in this gradient. (b) Photograph of a biological community near a deep-sea vent, which includes giant (red) tube worms, crabs, and eels.
(b) ©CSSF/Ocean Networks Canada

✓ **Concept Check:** *What properties of deep-sea vents made them suitable for the prebiotic synthesis of molecules?*

and hydrogen sulfide (H_2S) mixes abruptly with cold seawater. These vents release hot, gaseous substances from the interior of the Earth at temperatures in excess of 300°C (572°F). Supporters of this hypothesis propose that biologically important molecules may have been formed in the temperature gradient between the extremely hot vent water and the cold water that surrounds the vent (**Figure 4.2a**).

Experimentally, the temperatures within this gradient are known to be suitable for the synthesis of molecules that form components of biological molecules. For example, the reaction between iron and H_2S yields pyrites and H_2 and has been shown to provide the energy necessary for the reduction of N_2 to NH_3. Nitrogen is an essential component of both nucleic acids and amino acids—the molecular building blocks of life. But N_2, which is found abundantly on Earth, is chemically inert, so it is unlikely to have given rise to life. The conversion of N_2 to NH_3 at deep-sea vents may have led to the production of amino acids and nucleic acids.

Interestingly, complex biological communities are found in the vicinity of modern deep-sea vents. Various types of fish, worms, clams, crabs, shrimp, and bacteria are found in significant abundance in those areas (**Figure 4.2b**). Unlike most other forms of life on our planet, these organisms receive their energy from chemicals in the vent and not from the Sun. In 2007, American scientist Timothy Kusky and colleagues discovered 1.43 billion-year-old fossils of deep-sea microbes near ancient deep-sea vents. This study provided more evidence that life may have originated on the bottom of the ocean. However, debate continues as to the primary way organic molecules were made prior to the existence of life on Earth.

Stage 2: Organic Polymers May Have Formed on the Surface of Clay or in Water

The preceding three hypotheses provide reasonable mechanisms by which small organic molecules could have accumulated on early Earth. Scientists hypothesize that the second stage in the origin of life was a period in which simple organic molecules polymerized to form more complex organic polymers such as DNA, RNA, or proteins. Most ideas regarding the origin of life assume that polymers with at least 30–60 monomers are needed to store enough information to make a viable genetic system. Because hydrolysis competes with polymerization, many scientists have speculated that the synthesis of polymers did not occur in a watery, prebiotic soup but instead took place on a solid surface or in evaporating tidal pools.

In 1951, Irish X-ray crystallographer John Bernal first suggested that the prebiotic synthesis of polymers took place on clay. In his book *The Physical Basis of Life*, he wrote that "clays, muds, and inorganic crystals are powerful means to concentrate and polymerize organic molecules." Many clay minerals are known to bind organic molecules such as nucleotides and amino acids. Experimentally, many research groups have demonstrated the formation of nucleic acid polymers and polypeptides on the surface of clay, given the presence of monomer building blocks. During the prebiotic synthesis of RNA, the purine bases of the nucleotides could have interacted with the silicate surfaces of the clay. Cations, such as Mg^{2+}, bound the nucleotides to the negatively charged surfaces of the clay, thereby positioning the nucleotides in a way that promoted bond formation between the phosphate of one nucleotide and the ribose sugar of an adjacent nucleotide. In this way, polymers such as RNA may have formed.

Though the formation of polymers on clay remains a reasonable hypothesis, studies by American chemist Luke Leman and his colleagues, English chemist Leslie Orgel and Iranian-American chemist M. Reza Ghadiri, indicate that polymers can also form in aqueous solutions, contrary to the prevalent view. Their work in 2004 showed that carbonyl sulfide, a simple gas present in volcanic gases and deep-sea vent emissions, can bring about the formation of peptides from amino acids under mild conditions in water. These results indicate that the synthesis of polymers could have taken place in the prebiotic soup.

Stage 3: Cell-like Structures May Have Originated When Polymers Were Enclosed by a Boundary

The third stage in the origin of living cells is hypothesized to be the formation of a boundary that separated the polymers, such as RNA, from the environment. The term **protobiont** is used to describe an aggregate of prebiotically produced molecules and macromolecules that acquired a boundary, such as a lipid bilayer, that allowed it to maintain an internal chemical environment distinct from that of its surroundings. What characteristics make protobionts possible precursors of living cells? Scientists envision the existence of four key features:

1. A boundary, such as a membrane, separated the internal contents of the protobiont from the external environment.
2. Polymers inside the protobiont contained information.
3. Polymers inside the protobiont had catalytic functions.
4. The protobionts eventually developed the capability of self-replication.

Protobionts were not capable of precise self-reproduction like living cells but could divide to increase in number. Such protobionts are thought to have exhibited simple metabolic pathways in which the structures of organic molecules were changed. In particular, the polymers inside protobionts must have gained the catalytic ability to link organic building blocks to produce new polymers. This would have been a critical step in the process that eventually provided protobionts with the ability to self-replicate. According to this scenario, metabolic pathways became more complex, and the ability of protobionts to self-replicate became more refined over time. Eventually, these structures exhibited the characteristics that we attribute to living cells. As described next, researchers have hypothesized that protobionts may have existed as coacervates or liposomes.

Coacervates In 1924, Alexander Oparin hypothesized that living cells evolved from **coacervates**, droplets that form spontaneously from the association of charged polymers such as proteins, carbohydrates, or nucleic acids surrounded by water. Their name derives from the Latin *coacervare*, meaning to assemble together or cluster. Coacervates measure 1–100 micrometers (μm) across, are surrounded by a tight skin of water molecules, and possess osmotic properties (Figure 4.3a). This skin of water allows the selective absorption of simple molecules from the surrounding medium.

Enzymes trapped within coacervates can perform primitive metabolic functions. For example, researchers have made coacervates containing the enzyme glycogen phosphorylase. When glucose-1-phosphate was made available to these coacervates, it was taken up into them, and starch was produced. The starch merged with the wall of the coacervates, and they increased in size and eventually divided into two. When the enzyme amylase was included, the starch was broken down to maltose, which was released from the coacervates.

Liposomes As a second possibility, protobionts may have resembled **liposomes**—vesicles surrounded by a lipid bilayer (Figure 4.3b). When certain types of lipids are dissolved in water, they spontaneously form liposomes. As discussed in Chapter 5, lipid bilayers are selectively permeable (look ahead to Figure 5.7),

(a) Coacervates   57 μm

(b) Liposomes   200 nm

Figure 4.3 Possible structures of protobionts. Primitive, cell-like structures such as coacervates or liposomes could have given rise to living cells. **(a)** A micrograph and an illustration of coacervates, which are droplets of charged polymers surrounded by a skin of water molecules. **(b)** An electron micrograph and an illustration of liposomes. Each liposome is made of a phospholipid bilayer surrounding an aqueous compartment.
(a) Source: A.I. Oparin. From: *The Origin of Life*, New York: Dover, 1952; *(b)* ©Mary Kraft

Concept Check: *Which protobiont seems most similar to today's cells? Explain.*

BioConnections: *Refer back to Figure 3.9. What is the physical/chemical reason phospholipids tend to form a bilayer?*

and some liposomes can even store energy in the form of an electrical gradient. Such liposomes can discharge this energy in a neuron-like fashion, showing rudimentary signs of excitability, which is characteristic of living cells.

In 2003, Danish chemist Martin Hanczyc, American chemist Shelly Fujikawa, and Canadian-American biologist Jack Szostak showed that clay can catalyze the formation of liposomes that grow and divide, a primitive form of self-replication. Furthermore, if RNA was on the surface of the clay, the researchers discovered that liposomes that enclosed RNA were formed. These experiments are compelling because they showed that the formation of membrane vesicles containing RNA molecules is a plausible explanation of the emergence of cell-like structures based on simple physical and chemical properties.

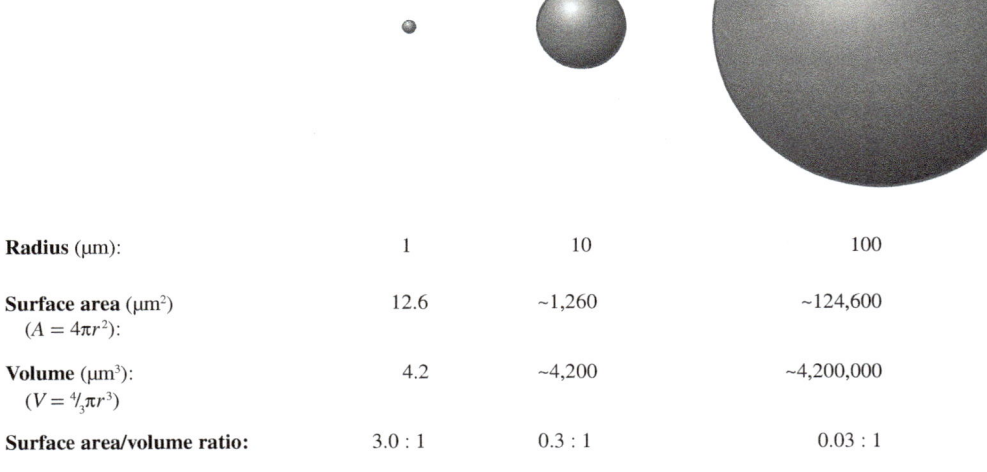

Radius (μm):	1	10	100
Surface area (μm²) $(A = 4\pi r^2)$:	12.6	~1,260	~124,600
Volume (μm³) $(V = \frac{4}{3}\pi r^3)$:	4.2	~4,200	~4,200,000
Surface area/volume ratio:	3.0 : 1	0.3 : 1	0.03 : 1

Figure 4.12 Relationship between cell size and the surface area/volume (SA/V) ratio. As cells get larger, the SA/V ratio gets smaller. Note: The three spheres shown here are not drawn precisely to scale.

BioConnections: Look ahead to Figure 36.15. How does the SA/V ratio relate to the shapes of structures involved with gas exchange?

- We can classify all forms of life into two categories based on cell structure: prokaryotic and eukaryotic cells. Bacteria and archaea are composed of prokaryotic cells with a relatively simple structure that lacks a membrane-enclosed nucleus. Structures in prokaryotic cells include the plasma membrane, cytoplasm, nucleoid, ribosomes, and cell wall. Some prokaryotic cells also have a glycocalyx (Figure 4.8).
- Differences in the proteomes of cells result in cells with dramatically different organization and structure. The proteomes of cells are largely responsible for producing the traits of organisms.
- Eukaryotic cells are compartmentalized into organelles and contain a nucleus that houses most of the DNA. Droplet organelles, such as the nucleolus, are formed by liquid-liquid phase separation. The surface area/volume (SA/V) ratio is thought to limit cell size (Figures 4.9, 4.10, 4.11, 4.12).

4.3 Testing Your Knowledge

1. Which of the following structures would *not* be found outside the plasma membrane of a prokaryotic cell?
   a. cell wall
   b. glycocalyx
   c. nucleoid
   d. pili
   e. flagella
2. A key difference between prokaryotic and eukaryotic cells is that only eukaryotic cells have
   a. a plasma membrane.
   b. genetic material (DNA).
   c. flagella.
   d. many types of membrane-bound organelles.
   e. all of the above.
3. Within a multicellular organism, differences in cell types usually arise from
   a. differences in their genomes.
   b. differences in their proteomes.
   c. differences in both their genomes and their proteomes.
   d. none of the above.

4.4 The Cytosol

Learning Outcomes

1. Identify the location of the cytosol in a eukaryotic cell, and describe its general functions.
2. Compare and contrast the three types of protein filaments that make up the cytoskeleton.
3. Explain how motor proteins interact with microtubules or actin filaments to promote cellular movements.

In Section 4.3, we focused on the general features of prokaryotic and eukaryotic cells. In the rest of this chapter, we will survey the various compartments of eukaryotic cells with a greater emphasis on structure and function. **Figure 4.13** compares four different regions in both an animal and a plant cell. We will start with the **cytosol** (shown in tan), the region of a eukaryotic cell that is outside the membrane-bound organelles but inside the plasma membrane. The other regions of the cell, which we will examine later in this chapter, include the interior of the nucleus (blue), the endomembrane system (purple and pink), and the semiautonomous organelles (orange and green). As with prokaryotic cells, the term cytoplasm refers to the region enclosed by the plasma membrane. This includes the cytosol and the organelles.

(a) Animal cell

(b) Plant cell

Figure 4.13 Compartments within (a) animal and (b) plant cells. The cytosol, which is outside the organelles but inside the plasma membrane, is shown in tan. The membranes of the endomembrane system are shown in purple (nuclear envelope and plasma membrane), pink (endoplasmic reticulum and vacuole), blue (Golgi apparatus), and dark purple (peroxisome). The interior of the nucleus is blue. Semiautonomous organelles are shown in orange (mitochondria) and green (chloroplasts).

Synthesis and Breakdown of Molecules Occur in the Cytosol

Metabolism is defined as the sum of the chemical reactions by which cells produce the materials and utilize the energy necessary to sustain life. Although many steps of metabolism also occur in cell organelles, the cytosol is a central coordinating region for many metabolic activities of eukaryotic cells. Metabolism often involves a series of steps called a metabolic pathway. Each step in a metabolic pathway is catalyzed by a specific **enzyme**—a protein that accelerates the rate of a chemical reaction. In Chapter 6, we will examine the properties of enzymes and consider a few important metabolic pathways that occur in the cytosol and cell organelles.

Some metabolic pathways involve the breakdown of a molecule into smaller components, a process termed **catabolism.** Such pathways are needed by the cell to utilize energy and to generate molecules that provide the building blocks for constructing macromolecules. Conversely, other pathways are involved in **anabolism**—the synthesis of molecules and macromolecules. For example, polysaccharides are made by linking sugar molecules. To make proteins, amino acids are covalently connected to form a polypeptide, using the information within an mRNA, a process called translation (see Chapter 10). Translation of that information occurs on ribosomes, which are found

in various locations in the cell. Some ribosomes float freely in the cytosol, others are attached to the outer membrane of the nuclear envelope or the endoplasmic reticulum membrane, and still others are found within the mitochondria or chloroplasts.

The Cytoskeleton Provides Cell Shape, Organization, and Movement

The **cytoskeleton** is a network of three different types of protein filaments: **microtubules, intermediate filaments,** and **actin filaments** (**Table 4.1**). Each type is constructed from many protein monomers. The cytoskeleton is a striking example of protein-protein interactions. The cytoskeleton is found primarily in the cytosol as well as in the nucleus along the inner nuclear membrane. Let's first consider the structure of cytoskeletal filaments and their roles in the construction and organization of cells. Later, we will examine how they are involved in cell movement.

1. **Microtubules:** Microtubules are long, hollow, cylindrical structures about 25 nm in diameter and composed of protein subunits called α- and β-tubulin. The assembly of tubulin to form a microtubule results in a polar structure with a plus end and a minus end (see Table 4.1). Microtubules grow only at the plus end but can shorten at either the plus or minus end. A single microtubule can oscillate between growing and shortening phases, a phenomenon termed **dynamic instability.** This phenomenon is important in many cellular activities, including the sorting of chromosomes during cell division.

The sites where microtubules form within a cell vary among different types of organisms. Animal cells that are not in the act of dividing contain a single structure near their nucleus called the **centrosome,** also called a microtubule-organizing center (see Table 4.1). Within the centrosome are the **centrioles,** a conspicuous pair of structures arranged perpendicularly to each other. In animal cells, microtubule growth typically starts at the centrosome in such a way that the minus end is anchored there. In contrast, most plant cells and many protists lack centrosomes and centrioles. Microtubules are created at many sites that are scattered throughout a plant cell. In plants, the nuclear membrane appears to function as a microtubule-organizing center.

Microtubules are important for cell shape and organization. Organelles such as the Golgi apparatus are attached to microtubules. In addition, microtubules are involved in the organization and movement of chromosomes during mitosis and in the orientation of cells during cell division, events we will examine in Chapter 14.

2. **Intermediate filaments:** Intermediate filaments are another class of cytoskeletal filaments found in the cells of many but not all animal species. Their name is derived from the observation that they are intermediate in diameter between actin filaments and microtubules. Intermediate filament proteins bind to each other in a staggered array to form a twisted, ropelike structure with a diameter of approximately 10 nm (see Table 4.1). Intermediate filaments function as tension-bearing fibers that help maintain cell shape and rigidity. Their lengths tend to be relatively permanent. By comparison, microtubules and actin filaments readily lengthen and shorten in cells.

Several types of proteins form intermediate filaments. Keratins are intermediate filaments in skin, intestinal, and kidney cells, where they are important for determining cell shape and providing mechanical strength. They are also a major constituent of hair and nails. As discussed later, some intermediate filaments are found inside the cell nucleus.

3. **Actin filaments:** Actin filaments are also known as **microfilaments,** because they are the thinnest cytoskeletal filaments. They are long, thin fibers approximately 7 nm in diameter. Like microtubules, actin filaments have plus and minus ends, and they are dynamic structures in which each strand grows at the plus end by the addition of actin monomers (see Table 4.1). This assembly process produces a fiber composed of two strands of actin monomers that spiral around each other.

Despite their thinness, actin filaments play a key role in cell shape and strength. They tend to be highly concentrated near the plasma membrane. In many types of cells, actin filaments support the plasma membrane and provide shape and strength to the cell. The sides of actin filaments are often anchored to other proteins near the plasma membrane, which explains why they are typically found there. The plus ends grow toward the plasma membrane and play a key role in cell shape and movement.

Motor Proteins Interact with Cytoskeletal Filaments to Promote Movements

Motor proteins are a category of proteins that use ATP as a source of energy to promote various types of movements. As shown in **Figure 4.14a**, a motor protein consists of three domains:

- **Head:** the site where ATP binds and is hydrolyzed to adenosine diphosphate (ADP) and inorganic phosphate (P_i)
- **Hinge:** the site that bends in response to ATP binding and hydrolysis. This bending is what causes movement to occur.
- **Tail:** an elongated region that is attached to other proteins or to other kinds of cellular molecules. Together, the hinge and tail make up a structure called the lever arm.

To cause movement, the head region of a motor protein interacts with a cytoskeletal filament—a microtubule or actin filament (**Figure 4.14b**). When ATP binds and is hydrolyzed, the motor protein interacts with the filament in a series of steps. The head of the motor protein is initially attached to a filament. To move forward, the head detaches from the filament, cocks forward, binds to another point farther along on the filament, and cocks backward. To picture how this works, consider the act of walking and imagine that the ground is a cytoskeletal filament, your leg is the head of the motor protein, and your hip is the hinge. To walk, you lift your leg up, you move it forward, you place it on the ground, and then you cock it backward (which propels you forward). This series of events is analogous to how a motor protein moves along a cytoskeletal filament.

Cells utilize the actions of motor proteins to promote three kinds of movements:

1. **Movement of cargo:** the filament is fixed in place and the motor protein moves a cargo from one location to another. In the example shown in **Figure 4.15a**, the tail region of a motor protein called kinesin is attached to a cargo, such as a membrane vesicle containing proteins.

(a) Three-domain structure of myosin, a motor protein

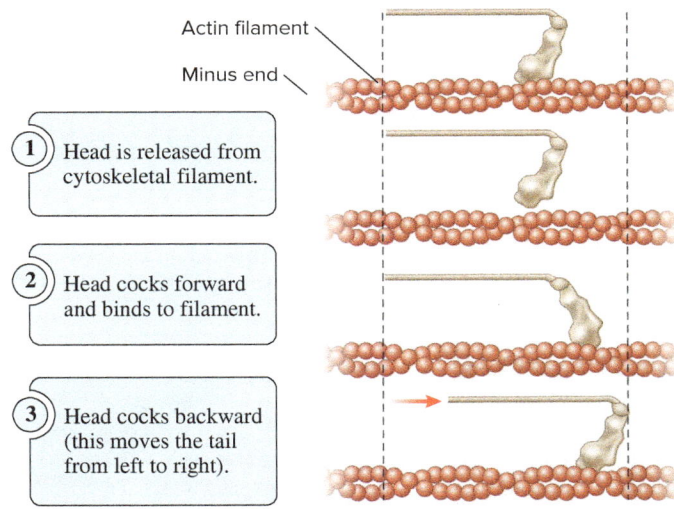

(b) Movement of a motor protein along a cytoskeletal filament

Figure 4.14 Motor proteins and their interactions with cytoskeletal filaments. The example illustrated here is the motor protein myosin (discussed in Chapter 35), which interacts with actin filaments. **(a)** Three-domain structure of myosin. **(b)** Conformational changes in a motor protein that allow it to walk along a cytoskeletal filament.

(a) Motor protein moves

(b) Filament moves

(c) Filaments bend

Figure 4.15 Three ways that motor proteins cause movements.

BioConnections: *Look ahead to Figure 35.8. Which of these three types of movements occurs during muscle contraction?*

2. **Movement of a filament:** the motor protein is fixed in place. The action of the motor protein moves the filament (Figure 4.15b). In this example, the motor protein myosin is moving an actin filament. This type of movement occurs during muscle contraction, which is described in Chapter 35.

3. **Bending of a filament:** the motor protein and filament are both fixed in place due to linking proteins, so the action of the motor protein causes the filament to bend. In this case, when motor proteins called dynein attempt to walk toward the minus end, they exert a force that causes the microtubules to bend (Figure 4.15c). This type of movement is used by cilia and flagella, which are described next.

Cilia and Flagella Allow Cells to Move

In certain kinds of eukaryotic cells, microtubules and the motor protein dynein work together to produce the bending movements of cell appendages called **flagella** and **cilia** (singular, flagellum and cilium). Flagella are usually longer than cilia, and they are typically found singly or in pairs.

Both flagella and cilia produce movement by generating bends that move along their length and push backward against the surrounding fluid. The flagellum of a sperm cell generates bends alternatively in each direction, which begin at the head and move (propagate) toward the tip of the flagellum (Figure 4.16a). By comparison, cilia are typically shorter than flagella and tend to cover all or part of the surface of a cell. Protists such as paramecia may have hundreds of cilia, which are spaced closely together and bend in a coordinated fashion to propel the organism through the water (Figure 4.16b).

Despite their differences in length, flagella and cilia have the same internal structure called the **axoneme.** The axoneme contains microtubules, the motor protein dynein, and linking proteins (Figure 4.17). In the cilia and flagella of most eukaryotic organisms, the microtubules form an arrangement called a 9 + 2 array. The outer nine are doublet microtubules, which are composed of a partial microtubule attached to a complete microtubule. Each of the two central microtubules is a single microtubule. Radial spokes project from the outer doublet microtubules toward the central pair. The microtubules in flagella and cilia emanate from **basal bodies,** which are anchored to the cytoplasmic side of the plasma membrane. At the basal body, the microtubules form a triplet structure. Much like the centrosome of animal cells, the basal bodies provide a site for microtubules to grow.

The movement of both flagella and cilia involves the propagation of a bend, which begins at the base of the structure and proceeds toward the tip (see Figure 4.16a). The bending occurs because dynein is activated to walk toward the minus end of the microtubules. ATP hydrolysis is required for this process. However, the microtubules and dynein are not free to move relative to each other because of linking proteins. Therefore, instead of freely walking along the microtubules, the dyneins exert a force that bends the microtubules (see Figure 4.15c). The dyneins at the base of the flagellum or cilium are activated first, followed by dyneins that are progressively closer to the tip. The movement of the bend from the base to the tip propels the cell.

4.4 Reviewing the Concepts

- The cytosol is a central coordinating region for many metabolic activities of eukaryotic cells, including polypeptide synthesis (Figure 4.13).
- The cytoskeleton is a network of three different types of protein filaments: microtubules, intermediate filaments, and actin filaments. Microtubules are important for cell shape, organization, and movement. Intermediate filaments help maintain cell shape and rigidity. Actin filaments support the plasma membrane and play a key role in cell strength, shape, and movement (Table 4.1).
- Motor proteins interact with cytoskeletal filaments and promote three types of movements: walking along the filament, moving the filament, and bending the filament (Figures 4.14, 4.15).
- The structure and function of cilia and flagella provide certain types of cells with motility (Figures 4.16, 4.17).

4.4 Testing Your Knowledge

1. The cytosol is the region of a eukaryotic cell that is
   a. inside the plasma membrane.
   b. inside the plasma membrane but outside the membrane-bound organelles.
   c. inside the cell nucleus.
   d. inside the plasma membrane but outside the nucleus.
   e. inside the organelles.

2. Which of the following would *not* be caused by a motor protein?
   a. movement of a cargo along a filament
   b. breakage of a filament

Figure 4.16 Bending of flagella and cilia causes cellular movements. (a) Spermatozoa (singular, spermatozoon) are sperm cells that are motile. They swim by producing repeated bends of a single, long flagellum, which move along its length. **(b)** Ciliated protozoa such as this paramecium swim by coordinated beating of many shorter cilia.
(b) ©SPL/Science Source

✓ **Concept Check:** *Describe the type of movement that occurs between the motor proteins and microtubules when flagella or cilia bend.*

(a) The bending of a sperm flagellum from the base to the tip

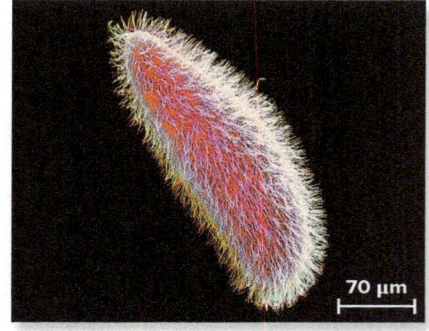

(b) Paramecium with many cilia

Figure 4.17 Structure of a eukaryotic cilium or flagellum. The structure of a cilium of a protist, *Tetrahymena thermophila* (see inset), consists of a 9 + 2 arrangement of nine outer doublet microtubules and two central microtubules. This structure is anchored to the basal body, which has nine triplet microtubules, in which three microtubules are fused together. Note: The structure of the basal body is very similar to that of centrioles in animal cells.
(top left) ©Aaron J. Bell/Science Source; *(top & bottom middle)* ©Dr. William Dentler/University of Kansas

BioConnections: Look ahead to Figure 23.15. What are some different uses of cilia and flagella among protists?

c. movement of a filament
d. bending of a filament
e. All of the above would be caused by a motor protein.

3. Which of the following is a key difference between cilia and flagella?
   a. On a single cell, cilia are more numerous than flagella.
   b. Flagella are usually longer than cilia.
   c. The internal axoneme structure is very different between cilia and flagella.
   d. Cilia contain microtubules but flagella contain actin.
   e. Both a and b are correct.

4.5 The Nucleus and Endomembrane System

Learning Outcomes

1. Describe the structure and organization of the cell nucleus.
2. Outline the structures and general functions of the components of the endomembrane system.
3. Distinguish between the rough endoplasmic reticulum and the smooth endoplasmic reticulum.

In Chapter 2, you learned that the nucleus of an atom contains protons and neutrons. In cell biology, the term **nucleus** has a different meaning. It is an organelle found in eukaryotic cells that contains most of the cell's genetic material. A small amount of genetic material is also found outside the nucleus, in mitochondria and chloroplasts.

The membranes that enclose the nucleus are part of a larger network of membranes called the **endomembrane system.** This system includes not only the nuclear envelope, which encloses the nucleus, but also the endoplasmic reticulum, Golgi apparatus, lysosomes, vacuoles, and peroxisomes. The prefix *endo* (from the Greek, meaning inside) originally referred only to these organelles and internal membranes. However, we now know that the plasma membrane is also part of this integrated membrane system (**Figure 4.18**). We will consider the major functions of the plasma membrane in Chapters 5 and 8.

Some of the membranes of the endomembrane system, such as the outer membrane of the nuclear envelope and the membrane of the endoplasmic reticulum, have direct connections to one another. Other organelles of the endomembrane system pass materials to each other via **vesicles**—small, membrane-enclosed spheres. In this section, we will examine the nucleus and survey the structures and functions of the organelles and membranes of the endomembrane system.

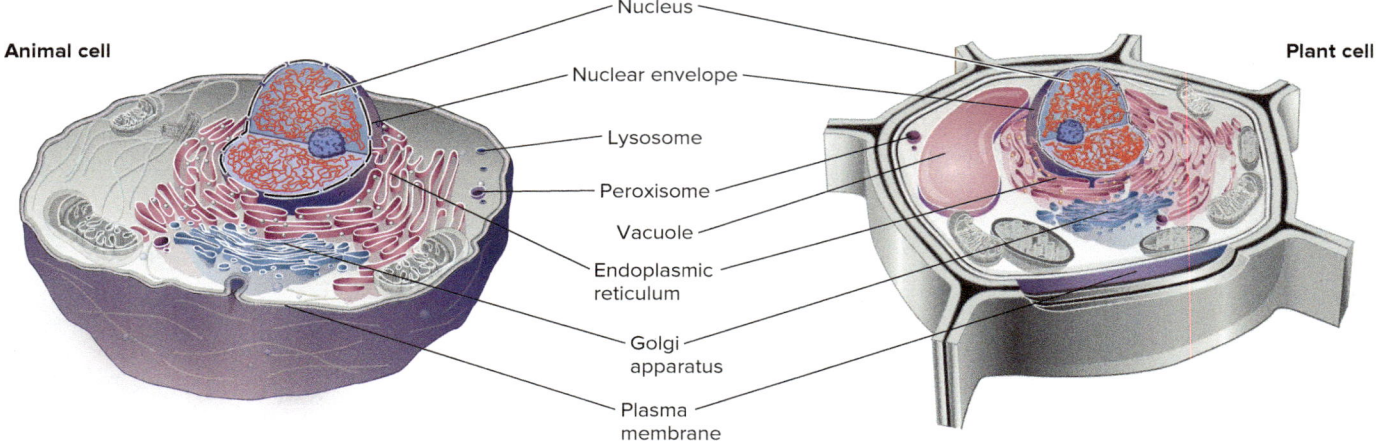

Figure 4.18 The nucleus and endomembrane system. This figure highlights the internal compartment of the nucleus (blue) and the membranes of the endomembrane system (purple, blue, and pink). The nuclear envelope is part of the endomembrane system, but the interior of the nucleus is not.

The Eukaryotic Nucleus Contains Chromosomes

The nucleus is the compartment that is enclosed by a double-membrane structure termed the **nuclear envelope** (Figure 4.19). **Nuclear pores** are formed where the inner and outer nuclear membranes make contact with each other. Nuclear pore complexes provide passageways for the movement of molecules and macromolecules into and out of the nucleus. Although cell biologists view the nuclear envelope as part of the endomembrane system, the materials within the nucleus are not.

Inside the nucleus are the chromosomes and a filamentous network of proteins called the nuclear matrix. Each **chromosome** is

Figure 4.19 The nucleus and nuclear envelope. The nuclear envelope is composed of an inner membrane and an outer membrane that meet at the nuclear pores. The inner nuclear membrane is lined with lamin proteins to form the nuclear lamina. The interior of the nucleus contains chromatin, which is attached to the internal nuclear matrix, and a nucleolus, where ribosomal subunits are assembled. *(micrographs)* ©Don W. Fawcett/Science Source

Concept Check: *What are the functions of the nuclear lamina and the internal nuclear matrix?*

composed of genetic material, namely DNA, and many types of proteins that help to compact the chromosome to fit inside the nucleus. Examples include histone proteins, described in Chapter 9. The complex formed between DNA and such proteins is termed **chromatin.** The **nuclear matrix** consists of two parts: the nuclear lamina, which is composed of intermediate filaments that line the inner nuclear membrane, and an internal nuclear matrix, which is connected to the lamina and fills the interior of the nucleus. The nuclear matrix organizes the chromosomes within the nucleus. Each chromosome is located in a distinct, nonoverlapping **chromosome territory,** which is visible when cells are exposed to dyes that label specific types of chromosomes (Figure 4.20).

The primary functions of the nucleus are the protection, organization, replication, and expression of the genetic material. These topics are discussed in Unit III. Another important function is the assembly of ribosomal subunits—cellular structures involved in producing polypeptides during the process of translation. The assembly of ribosomal subunits occurs in the **nucleolus** (plural, nucleoli), which is a droplet organelle in the nucleus. A ribosome is composed of two subunits: one small and one large (look ahead to Chapter 10, Table 10.3). Each subunit contains one or more RNA molecules and several types of proteins. Most of the RNA molecules that are components of ribosomes are made in the vicinity of the nucleolus. By comparison, the ribosomal proteins are produced in the cytosol and then imported into the nucleus through the nuclear pores. The ribosomal proteins and RNA molecules then assemble in the nucleolus to form

Figure 4.20 Chromosome territories in the cell nucleus. Chromosomes from a chicken were labeled with chromosome-specific probes. Seven types of chicken chromosomes are each stained with a different dye. Each chromosome occupies its own distinct, nonoverlapping territory within the cell nucleus.
Courtesy of Felix A. Habermann

BioConnections: Look ahead to Figure 14.7. What happens to chromosome territories during cell division?

the ribosomal subunits. Finally, the subunits exit through the nuclear pores into the cytosol, where they are needed for protein synthesis.

The Endoplasmic Reticulum Initiates the Sorting of Some Proteins and Carries Out Metabolic Functions

The **endoplasmic reticulum (ER)** is a network of membranes that form flattened, fluid-filled tubules, or **cisternae** (Figure 4.21). The terms endoplasmic (Greek, for in the cytoplasm) and reticulum (Latin, for little net) refer to the location and shape of this organelle when

Figure 4.21 Structure of the endoplasmic reticulum. (Left side) The endoplasmic reticulum (ER) is composed of a network of flattened tubules called cisternae that enclose a continuous ER lumen. The rough ER is studded with ribosomes, whereas the smooth ER lacks ribosomes. The rough ER is continuous with the outer nuclear membrane. (Right side) In this colorized TEM of the ER, the lumen is colored green and the ribosomes are purple. The cytosol is tan.
(right) ©Dennis Kunkel Microscopy/Science Source

viewed under a microscope. The term **lumen** refers to the internal space of an organelle. The ER membrane, which is continuous with the outer nuclear membrane, encloses a single compartment called the **ER lumen.** There are two distinct, but continuous, types of ER: rough ER and smooth ER.

Rough ER   The outer surface of the **rough endoplasmic reticulum (rough ER)** is studded with ribosomes, giving it a bumpy appearance when viewed under the microscope. Rough ER performs a few key functions:

- **Protein sorting:** Proteins that are destined for the ER, Golgi apparatus, lysosomes, vacuoles, plasma membrane, or extracellular environment are first directed to the rough ER. This topic is discussed later in Section 4.7.
- **Insertion of membrane proteins:** Membrane proteins with segments that span the membrane are initially inserted into the rough ER membrane as the protein is being synthesized.
- **Glycosylation:** An important function of the rough ER is the attachment of carbohydrates to proteins and lipids. This process is called **glycosylation.**

Smooth ER   The **smooth endoplasmic reticulum (smooth ER),** which lacks ribosomes, functions in diverse processes:

- **Metabolism:** The extensive network of smooth ER membranes provides a large surface area for enzymes that play important metabolic roles. In liver cells, enzymes in the smooth ER detoxify many potentially harmful organic molecules, including barbiturate drugs and ethanol (alcohol). These enzymes convert hydrophobic toxic molecules into more hydrophilic molecules, which are easily excreted from the body. Chronic alcohol consumption, as occurs in alcoholics, leads to a greater amount of smooth ER in liver cells, which increases the rate of alcohol breakdown. This explains why people who consume alcohol regularly must ingest more alcohol to experience its effects.
- **Storage of Ca^{2+}:** The smooth ER contains calcium pumps that transport Ca^{2+} into the ER lumen. The regulated release of Ca^{2+} into the cytosol is involved in many vital cellular processes, including muscle contraction in animals.
- **Lipid synthesis and modification:** The smooth ER is the primary site for the synthesis of phospholipids, which are the main lipid component of cell membranes. In addition, enzymes in the smooth ER are necessary for certain modifications of the lipid cholesterol to produce steroid hormones such as estrogen and testosterone.

The Golgi Apparatus Directs the Processing, Sorting, and Secretion of Cellular Molecules

The **Golgi apparatus** (also called the Golgi body, Golgi complex, or simply Golgi) was discovered by the Italian microscopist Camillo Golgi in 1898. It consists of a stack of flattened membranes, each of which encloses a single compartment called a **cisterna** (plural cisternae) (**Figure 4.22**). The Golgi compartments are named according to their orientation in the cell. The *cis* Golgi is near the ER membrane, the *trans* Golgi is closest to the plasma membrane, and the medial Golgi is found in the middle. Two models have been proposed to explain how materials move through the Golgi apparatus:

- *Vesicular transport model.* Materials are transported between the Golgi cisternae via membrane vesicles that bud from one compartment in the Golgi (for example, the *cis* Golgi) and fuse with another compartment (for example, the medial Golgi).
- *Cisternal maturation model.* Vesicles from the ER fuse to form a cisterna at the *cis* face; the cisterna that was previously at the *cis* face becomes a medial cisterna. Likewise, this moves medial cisternae toward the *trans* face. The cisterna at the *trans* face is lost due to the export of vesicles from its surface.

Further research will be needed to determine if either or both of these models are correct.

The Golgi apparatus performs three overlapping functions:

1. **Protein sorting:** As discussed in Section 4.7, the Golgi plays a role in the sorting of several categories of proteins.
2. **Processing:** Enzymes in the Golgi apparatus process, or modify, certain proteins and lipids. As mentioned earlier, carbohydrates can be attached to proteins and lipids in the endoplasmic reticulum. Glycosylation continues in the Golgi. For this to occur, a protein or lipid is transported via vesicles from the ER to the *cis* Golgi. Most glycosylation occurs in the medial Golgi. A second type of processing event is **proteolysis,** whereby enzymes called **proteases** cut proteins into smaller polypeptides. For example, the hormone insulin is first made as a large precursor protein termed proinsulin. In the Golgi apparatus, proinsulin is packaged with proteases into vesicles. The proteases cut out a portion of the proinsulin to create a smaller insulin molecule that is a functional hormone.
3. **Secretion:** The Golgi apparatus packages different types of materials into **secretory vesicles** that fuse with the plasma membrane, thereby releasing their contents outside the cell. Proteins destined for secretion are synthesized into the ER, travel to the Golgi, and then are transported by vesicles to the plasma membrane (see Figure 4.22). The vesicles then fuse with the plasma membrane and are secreted to the outside of the cell. The entire route is called the **secretory pathway.**

Lysosomes Are Involved in the Intracellular Digestion of Macromolecules

We now turn to another component of the endomembrane system, **lysosomes,** which are small organelles that are found in animal cells and break down molecules and macromolecules. Lysosomes contain many **acid hydrolases,** which are hydrolytic enzymes that use a molecule of water to break a covalent bond. As described in Chapter 3 (refer back to Figure 3.3b), this type of chemical reaction is called hydrolysis:

$$\text{Acid hydrolase}$$
$$R_1\text{—}R_2 + H_2O \longrightarrow R_1\text{—}OH + H\text{—}R_2$$

The acid hydrolases in a lysosome function optimally at an acidic pH. The fluid-filled interior of a lysosome has a pH of approximately 4.8.

Biology Principle

Biologists Use Models to Test Predictions and Convey Ideas

This figure is a model for the secretory pathway. As described in the text, researchers have proposed two different models for the movement of materials through the Golgi. The Modeling Challenge in the figure legend asks you to predict the movement of a fluorescently labeled particle according to each model.

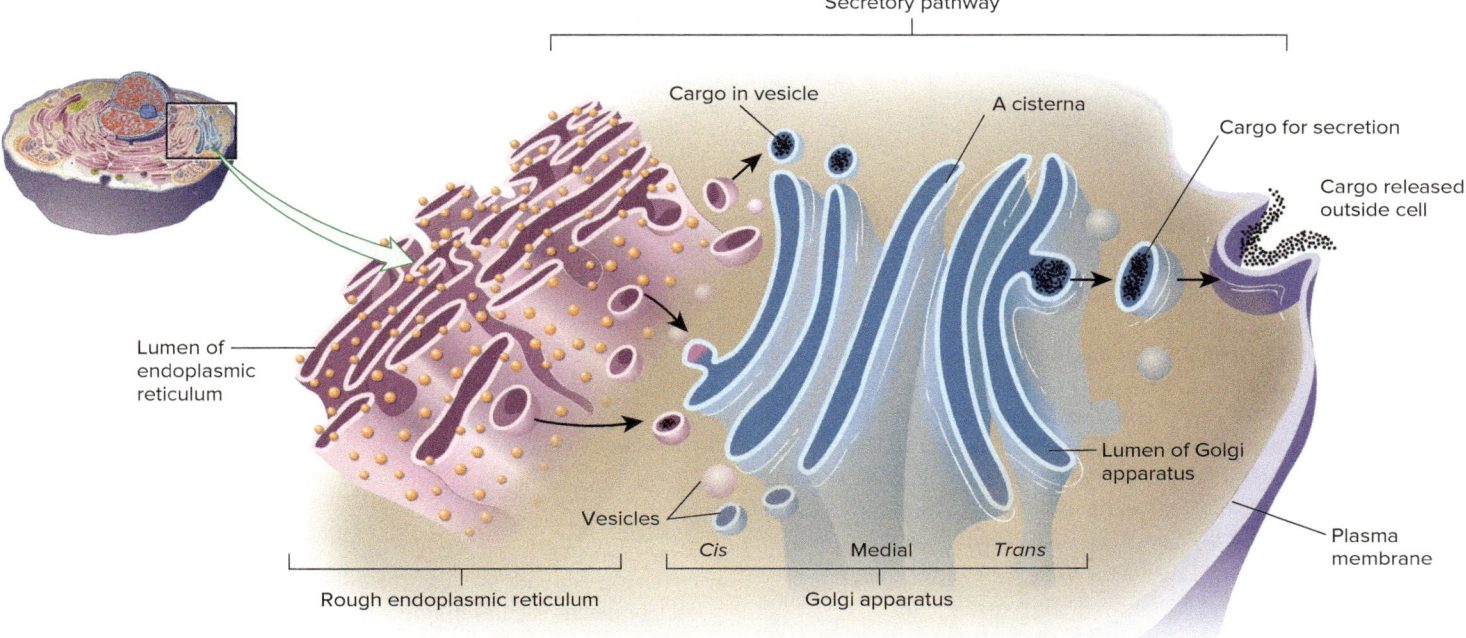

Figure 4.22 **The Golgi apparatus and secretory pathway.** The Golgi is composed of stacks of membranes that enclose separate compartments. Transport to and from the Golgi occurs via membrane vesicles. Vesicles bud from the ER and go to the Golgi, and vesicles from the Golgi fuse with the plasma membrane to release cargo to the outside. The pathway from the ER to the Golgi to the plasma membrane is termed the secretory pathway.

Modeling Challenge: Let's suppose you could inject large, fluorescently labeled particles into the *cis* Golgi. The particles are too large to be taken into membrane vesicles. They emit red fluorescence. Draw a model that shows the movement or lack of movement of red color according to the vesicular transport model and the cisternal transport model.

Concept Check: *If we consider the Golgi apparatus as three compartments (cis, medial, and trans), in what order does a protein travel through them before being secreted?*

If a lysosomal membrane breaks, releasing acid hydrolases into the cytosol, the enzymes are not very active because the cytosolic pH is neutral (approximately pH 7.0) and buffered. This prevents significant damage to the cell from lysosome breakage.

Lysosomes contain many different types of acid hydrolases that allow them to break down various complex materials, including carbohydrates, proteins, lipids, and nucleic acids. One function of lysosomes involves the digestion of substances that are taken up from outside the cell via a process called endocytosis (look ahead to Chapter 5, Figure 5.21).

Vacuoles Function in Storage, Regulation of Cell Volume, and Degradation

Vacuoles are prominent organelles in plant cells, fungal cells, and certain protists. The term vacuole (Latin, for empty space) came from early microscopic observations of these compartments. We now know that vacuoles are not empty but instead contain fluid and sometimes even solid substances. Most vacuoles are made from the fusion of many smaller membrane vesicles. In animal cells, vacuoles tend to be smaller and are more commonly used to temporarily store materials

(a) Central vacuole in a plant cell

(b) Contractile vacuoles in an algal cell

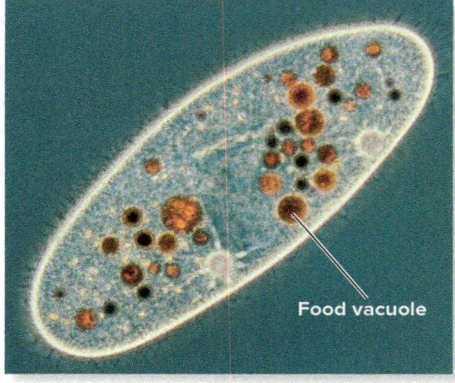
(c) Food vacuoles in a paramecium

Figure 4.23 Examples of vacuoles. Parts (a) and (b) are TEMs. Part (c) is a colorized light micrograph.
(a) ©Biophoto Associates/Science Source; *(b)* Courtesy Dr. Peter Luykx, Biology, University of Miami and Dr. David Robinson, Universität Heidelberg; *(c)* ©Dr. David Patterson/Science Source

or transport substances. In animals, such vacuoles are sometimes called storage vesicles.

The functions of vacuoles are extremely varied, and they differ among cell types and even with environmental conditions. The best way to appreciate vacuole function is to consider a few examples. Mature plant cells usually have a large **central vacuole** that occupies 80% or more of the cell volume (Figure 4.23a). The central vacuole serves two important purposes. First, it stores a large amount of water, enzymes, and inorganic ions such as calcium, as well as other materials, including proteins and pigments. Second, it performs a space-filling function. The central vacuole exerts a pressure on the cell wall, called turgor pressure. If a plant becomes dehydrated and this pressure is lost, a plant will wilt. Turgor pressure is important in maintaining the structure of plant cells and the plant itself, and it helps to drive the expansion of the cell wall, which is necessary for growth.

Certain species of protists use vacuoles to maintain cell volume. Freshwater organisms such as the alga *Chlamydomonas reinhardtii* have small, water-filled **contractile vacuoles** that expand as water enters the cell (Figure 4.23b). Once they reach a certain size, the vacuoles fuse with the plasma membrane, expelling their contents to the exterior of the cell (look ahead to Figure 5.11). This mechanism is necessary to remove excess water that continually enters the cell by diffusion across the plasma membrane.

Another function of vacuoles is degradation. Some protists engulf their food into large food vacuoles in the process of phagocytosis (Figure 4.23c). Food vacuoles contain hydrolytic enzymes that break down macromolecules within food. Macrophages, a type of cell found in animals' immune systems, engulf bacterial cells into phagocytic vacuoles, which then fuse with lysosomes, where the bacteria are destroyed.

Peroxisomes Catalyze Detoxifying Reactions

Peroxisomes, discovered by Christian de Duve in 1965, are small organelles found in all eukaryotic cells. Peroxisomes consist of a single membrane that encloses a fluid-filled lumen. A typical eukaryotic cell contains several hundred of them.

Peroxisomes catalyze a variety of chemical reactions, including some reactions that break down organic molecules and others that are biosynthetic. In mammals, large numbers of peroxisomes are found in liver cells, where toxic molecules accumulate and are broken down. A common by-product of the breakdown of toxins is hydrogen peroxide, H_2O_2:

$$RH_2 + O_2 \longrightarrow R + H_2O_2$$
(toxin)

Hydrogen peroxide has the potential to be highly toxic. In the presence of metals such as iron (Fe^{2+}), which are found naturally in living cells, H_2O_2 can be broken down to form a hydroxide ion (OH^-) and a molecule called a hydroxide free radical ($\cdot OH$):

$$Fe^{2+} + H_2O_2 \longrightarrow Fe^{3+} + OH^- + \cdot OH \text{ (hydroxide free radical)}$$

The $\cdot OH$ is highly reactive and can damage proteins, lipids, and DNA. Therefore, it is beneficial for cells to break down H_2O_2 in an alternative manner that does not form $\cdot OH$. Peroxisomes contain an enzyme called **catalase** that breaks down hydrogen peroxide to make water and oxygen gas (hence the name peroxisome):

$$2 H_2O_2 \xrightarrow{\text{Catalase}} 2 H_2O + O_2$$

A general model for peroxisome formation is shown in Figure 4.24, although the details may differ among animal, plant, and fungal cells. To initiate peroxisome formation, vesicles bud from the ER membrane and form a premature peroxisome. Following the import of additional proteins and lipids, the premature peroxisome becomes a mature peroxisome. Once the mature peroxisome has formed, it may then divide to further increase the number of peroxisomes in the cell.

4.5 Reviewing the Concepts

- The primary function of the nucleus is the organization and expression of the genetic material. A second important function is the assembly of ribosomal subunits in the nucleolus (Figures 4.18, 4.19, 4.20).

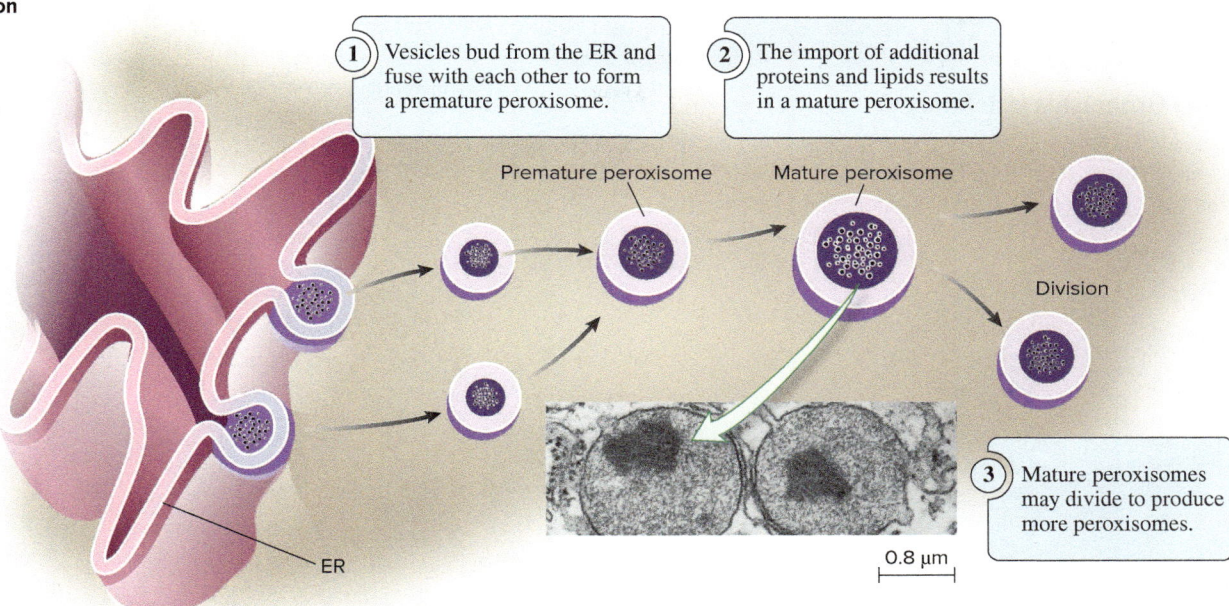

Figure 4.24 Formation of peroxisomes. The inset is a TEM of mature peroxisomes.
(inset) ©Don W. Fawcett/Science Source

- The endomembrane system includes the nuclear envelope, endoplasmic reticulum (ER), Golgi apparatus, lysosomes, vacuoles, peroxisomes, and plasma membrane. The rough endoplasmic reticulum (rough ER) plays a key role in the initial sorting of proteins. The smooth endoplasmic reticulum (smooth ER) functions in metabolic processes such as detoxification, carbohydrate metabolism, accumulation of calcium ions, and synthesis and modification of lipids. The Golgi apparatus performs three overlapping functions: processing, protein sorting, and secretion. Lysosomes degrade macromolecules and help digest substances taken up from outside the cell (endocytosis) (Figures 4.21, 4.22).
- Types of vacuoles include central vacuoles, contractile vacuoles, and food vacuoles (Figure 4.23).
- Peroxisomes catalyze a variety of chemical reactions, including those involved with the breakdown of toxic molecules such as hydrogen peroxide, and they typically contain enzymes involved in the metabolism of fats and amino acids. Peroxisomes are formed by budding from the ER, followed by maturation and division (Figure 4.24).

4.5 Testing Your Knowledge

1. Which of the following organelles is surrounded by a double membrane?
    a. nucleus
    b. endoplasmic reticulum
    c. Golgi apparatus
    d. lysosome
    e. peroxisome

2. The secretory pathway occurs in which of the following orders?
    a. Golgi to ER to plasma membrane to outside of cell
    b. ER to Golgi to plasma membrane to outside of cell
    c. plasma membrane to ER to Golgi to outside of cell
    d. ER to plasma membrane to Golgi to outside of cell
    e. none of the above

3. A characteristic of lysosomes is
    a. they have a low internal pH.
    b. they contain different acid hydrolases.
    c. they hydrolyze organic molecules into small components.
    d. they are surrounded by a single membrane.
    e. All of the above are characteristics of lysosomes.

4.6 Semiautonomous Organelles

Learning Outcomes

1. Describe the structures and general functions of mitochondria and chloroplasts.
2. **SCISKILLS** ▶ Discuss the evidence for the endosymbiosis theory.

We now turn to those organelles in eukaryotic cells that are considered semiautonomous—mitochondria and chloroplasts. These organelles grow and divide, but they are not completely autonomous because they depend on other parts of the cell for their internal components (**Figure 4.25**). For example, most of the proteins found in mitochondria are imported from the cytosol. In this section, we will survey the structures and functions of the semiautonomous organelles in

Biology Principle

Living Organisms Use Energy

Chloroplasts capture light energy and synthesize organic molecules. Mitochondria break down organic molecules and make ATP, which is used as an energy source to drive many different cellular processes.

Figure 4.25 **Semiautonomous organelles.** Mitochondria (top and bottom cell) and chloroplasts (bottom cell) are the semiautonomous organelles.

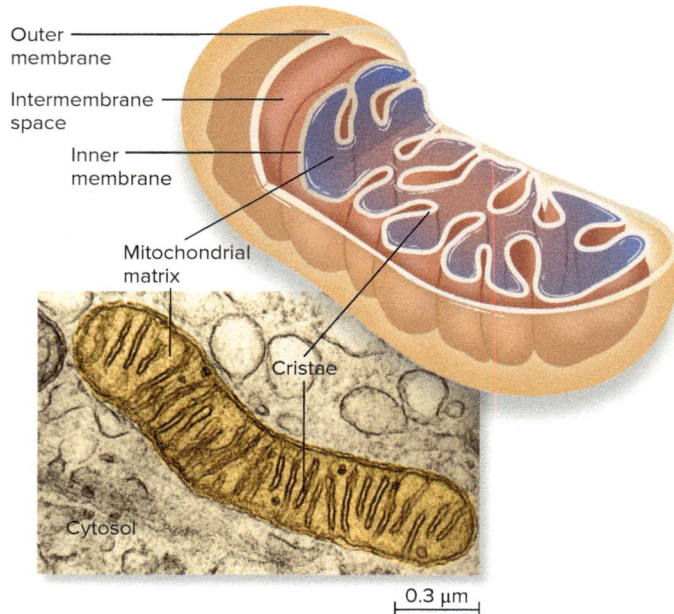

Figure 4.26 **Structure of a mitochondrion.** This figure emphasizes the membrane organization of a mitochondrion, which has an outer and an inner membrane. The invaginations of the inner membrane are called cristae. The mitochondrial matrix lies inside the inner membrane. The micrograph is a colorized TEM.
©Don W. Fawcett/Science Source

Concept Check: *What is the advantage of the mitochondrion's highly invaginated inner membrane?*

---

eukaryotic cells and consider their evolutionary origins. In Chapters 6 and 7, we will explore the functions of mitochondria and chloroplasts in greater depth.

Mitochondria Supply Cells with Most of Their ATP

Mitochondrion (plural, mitochondria) literally means thread granule, which is what mitochondria look like under a light microscope—either threadlike or granular-shaped. They are similar in size to bacteria (see Figure 4.8). A typical cell may contain a few hundred to a few thousand mitochondria. Cells with particularly heavy energy demands, such as muscle cells, have more mitochondria than others, such as skin cells. Research has shown that regular exercise increases the number and size of mitochondria in human muscle cells to meet the expanded demand for energy.

A mitochondrion has an outer membrane and an inner membrane separated by a region called the intermembrane space (**Figure 4.26**). The inner membrane is highly invaginated (folded) to form projections called **cristae.** The cristae greatly increase the surface area of the inner membrane, which is the site where ATP is made. The compartment enclosed by the inner membrane is the **mitochondrial matrix.**

The primary role of mitochondria is to make ATP. Even though mitochondria produce most of a cell's ATP, they do not create energy. Rather, their primary function is to convert chemical energy that is stored within the covalent bonds of organic molecules into a form that can be readily used by cells. Covalent bonds in sugars, fats, and amino acids store a large amount of energy. The breakdown of these molecules into simpler molecules releases energy that is used to make ATP. Many proteins in living cells use ATP as a source of energy to carry out their functions, such as muscle contraction, the uptake of nutrients, cell division, and many other cellular processes.

Mitochondria perform other functions as well. They are involved in the synthesis, modification, and breakdown of several types of cellular molecules. For example, the synthesis of certain hormones requires enzymes that are found in mitochondria. Another interesting role of mitochondria is the generation of heat in specialized fat cells known as brown fat cells. Groups of brown fat cells serve as heating pads that help to revive hibernating animals and protect sensitive areas of young animals from the cold.

Chloroplasts Carry Out Photosynthesis

Chloroplasts are semiautonomous organelles that capture light energy and use some of that energy to synthesize organic molecules

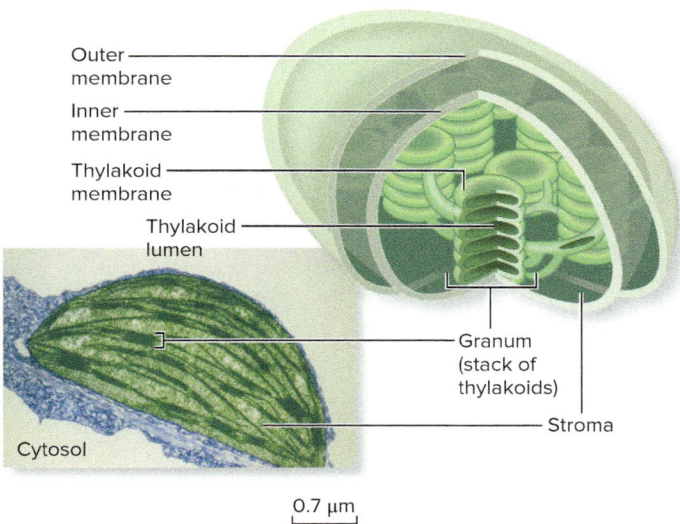

Figure 4.27 Structure of a chloroplast. Like a mitochondrion, a chloroplast is enclosed in a double membrane. In addition, it has an internal thylakoid membrane that forms flattened tubules. These tubules stack on each other to form grana. The stroma is located inside the inner membrane but outside the thylakoid membrane. The micrograph is a colorized TEM.
©Dr. Jeremy Burgess/SPL/Science Source

Mitochondria and Chloroplasts Contain Their Own Genetic Material and Divide by Binary Fission

The chromosomes found in mitochondria and chloroplasts are referred to as the **mitochondrial genome** and **chloroplast genome**, respectively, whereas the chromosomes found in the nucleus of a eukaryotic cell constitute the **nuclear genome**. Like bacterial genomes, the genomes of most mitochondria and chloroplasts are composed of a single, circular chromosome. Compared with the nuclear genome, they are very small. For example, the human genome has approximately 22,000 different protein-encoding genes, whereas the human mitochondrial genome has only about a dozen protein-encoding genes.

Mitochondria and chloroplasts increase in number via **binary fission**, or splitting in two. **Figure 4.28** illustrates the process for a mitochondrion. The mitochondrial chromosome, which is found in a region called the nucleoid, is duplicated, and the organelle divides into two separate organelles. Mitochondrial and chloroplast division are needed to maintain a full complement of these organelles when cell growth occurs following cell division. In addition, environmental

such as glucose. This process, called **photosynthesis,** is described in Chapter 7. Chloroplasts are found in the cells of plants and algae. **Figure 4.27** shows the structure of a typical chloroplast. Like a mitochondrion, a chloroplast contains an outer and inner membrane. An intermembrane space lies between these two membranes. A third membrane, the **thylakoid membrane,** forms many flattened, fluid-filled tubules that enclose a single, convoluted compartment called the thylakoid lumen. These tubules stack on top of each other to form a structure called a **granum** (plural, grana). The **stroma** is the compartment of the chloroplast that is enclosed by the inner membrane but outside the thylakoid membrane.

Chloroplasts are a specialized version of plant organelles that are more generally known as **plastids.** All plastids are derived from unspecialized **proplastids.** The various types of plastids are distinguished by their synthetic abilities and the types of pigments they contain:

- Chloroplasts, which carry out photosynthesis, contain the green pigment chlorophyll. The abundant number of chloroplasts in the leaves of plants gives them their green color.
- Chromoplasts, a second type of plastid, function in synthesizing and storing the yellow, orange, and red pigments known as carotenoids. Chromoplasts give many fruits, vegetables, and flowers their colors. In autumn, the chromoplasts also give many leaves their yellow, orange, and red colors.
- Leucoplasts typically lack pigment molecules. An amyloplast is a leucoplast that synthesizes and stores starch. Amyloplasts are common in underground plant structures such as roots and tubers.

(a) Binary fission of mitochondria

(b) Transmission electron micrographs of the process

Figure 4.28 Division of mitochondria by binary fission.
(b) ©Don W. Fawcett/Science Source

BioConnections: Look ahead to Figure 17.8. How is the process of binary fission similar to bacterial cell division, and how is it different?

conditions may influence the sizes and numbers of these organelles. For example, when plants are exposed to more sunlight, the number of chloroplasts in leaf cells increases.

EVOLUTIONARY CONNECTIONS

Mitochondria and Chloroplasts Are Derived from Ancient Symbiotic Relationships

The observation that mitochondria and chloroplasts contain their own genetic material may seem puzzling. Perhaps you might think that it would be simpler for a eukaryotic cell to have all of its genetic material in one place—the nucleus. The distinct genomes of mitochondria and chloroplasts can be traced to their evolutionary origin, which involved an ancient symbiotic association.

A symbiotic relationship occurs when two different species live in direct contact with each other. **Endosymbiosis** is a symbiotic relationship in which the smaller species—the symbiont—actually lives inside the larger species. In 1883, Andreas Schimper proposed that chloroplasts evolved from an endosymbiotic relationship between cyanobacteria (bacteria capable of photosynthesis) and eukaryotic cells. In 1922, Ivan Wallin also hypothesized an endosymbiotic origin for mitochondria. Researchers now hypothesize that mitochondria evolved from an ancient species related to a group of modern bacteria called proteobacteria, which are described in Chapter 23.

In spite of these interesting proposals, the question of whether endosymbiosis gave rise to mitochondria and chloroplasts was largely ignored until the discovery that these organelles contain their own genetic material. In 1970, the idea of endosymbiosis as the origin of mitochondria and chloroplasts was revived by Lynn Margulis in her book *Origin of Eukaryotic Cells*. During the 1970s and 1980s, the advent of molecular genetic techniques allowed researchers to analyze genes from mitochondria, chloroplasts, bacteria, and eukaryotic nuclear genomes. Researchers discovered that genes in mitochondria and chloroplasts are very similar to bacterial genes. Likewise, mitochondria and chloroplasts are strikingly similar in size and shape to certain bacterial species. These observations provided strong support for the **endosymbiosis theory,** which proposes that mitochondria and chloroplasts originated from bacteria that took up residence within primordial eukaryotic cells (**Figure 4.29**). Over the next 2 billion years, the characteristics of these intracellular bacterial cells gradually changed to those of mitochondria or chloroplasts. The origin of eukaryotic cells is discussed in more detail in Chapter 22.

A symbiotic relationship is beneficial to one or both species. According to the endosymbiosis theory, such a relationship provided eukaryotic cells with useful cellular characteristics. Chloroplasts, which were derived from cyanobacteria, have the ability to carry out photosynthesis. This benefits plant cells by giving them the ability to use the energy from sunlight. By comparison, mitochondria are thought to have been derived from a different type of bacteria known as proteobacteria. In this case, the endosymbiotic relationship enabled eukaryotic cells to synthesize greater amounts of ATP. How

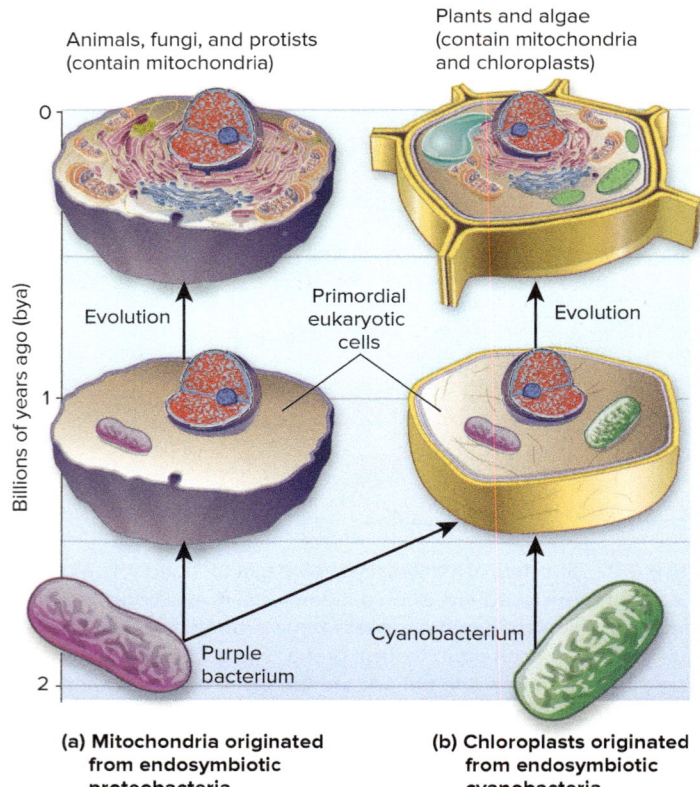

Figure 4.29 A simplified view of the endosymbiosis theory. **(a)** According to the endosymbiosis theory, modern mitochondria were derived from a type of bacteria called proteobacteria. Over the course of evolution, their characteristics changed into those found in mitochondria today. **(b)** A similar phenomenon occurred for chloroplasts, which were derived from cyanobacteria (blue-green bacteria), bacteria that are capable of photosynthesis.

Concept Check: *Discuss the similarities and differences between modern bacteria and mitochondria.*

the relationship would have been beneficial to a cyanobacterium or purple bacterium is less clear, though the cytosol of a eukaryotic cell may have provided a stable environment with an adequate supply of nutrients.

During the evolution of eukaryotic species, many genes that were originally found in the genomes of the primordial purple bacteria and cyanobacteria have been transferred from the organelles to the nucleus. This has occurred many times throughout evolution, so modern mitochondria and chloroplasts have lost most of the genes that still exist in present-day purple bacteria and cyanobacteria. Some researchers speculate that the movement of genes into the nucleus makes it easier for the cell to control the structure, function, and division of mitochondria and chloroplasts. In modern cells, hundreds of different proteins that make up these organelles are encoded by genes that have been transferred to the nucleus. These proteins are made in the cytosol and then taken up into mitochondria or chloroplasts. We will discuss this topic next.

4.6 Reviewing the Concepts

- Mitochondria and chloroplasts are considered semiautonomous organelles because they grow and divide but still depend on other parts of the cell for their internal components (Figure 4.25).
- Mitochondria produce most of a cell's ATP, which is utilized by many proteins to carry out their functions. Other mitochondrial functions include the synthesis, modification, and breakdown of cellular molecules and the generation of heat in specialized fat cells (Figure 4.26).
- Chloroplasts, which are found in the cells of plants and algae, carry out photosynthesis (Figure 4.27).
- Mitochondria and chloroplasts contain their own genetic material and divide by binary fission (Figure 4.28).
- According to the endosymbiosis theory, mitochondria and chloroplasts originated from bacteria that took up residence in early eukaryotic cells (Figure 4.29).

4.6 Testing Your Knowledge

1. Most of the ATP in a eukaryotic cell is made in the
   a. cytosol.
   b. endoplasmic reticulum.
   c. mitochondrion.
   d. chloroplast.
   e. peroxisome.

2. Chloroplasts are found in
   a. plants.
   b. algae.
   c. fungi.
   d. both a and b.
   e. all of the above.

4.7 Protein Sorting to Organelles

Learning Outcome

1. List the categories of proteins that are sorted cotranslationally and post-translationally.

As we have seen, eukaryotic cells contain a variety of membrane-bound organelles. Each protein that a cell makes usually functions within one cellular compartment or is secreted from the cell. How does each protein reach its appropriate destination? For example, how does a mitochondrial protein get sent to the mitochondrion rather than to a different organelle such as a lysosome? In eukaryotes, most proteins contain short stretches of amino acid sequences that direct them to their correct cellular location. These sequences are called **sorting signals.** Each sorting signal is recognized by specific cellular components that facilitate the proper movement of the protein carrying that signal to its correct location.

The synthesis of most eukaryotic proteins begins on ribosomes in the cytosol using mRNA, which contains the information for polypeptide synthesis (Figure 4.30). The cytosol provides amino acids, which are used as building blocks to make the proteins during translation. Cytosolic proteins lack any sorting signal, so they remain within the cytosol. By comparison, the synthesis of proteins destined for the ER, Golgi, lysosomes, vacuoles, or secretory vesicles begins in the cytosol and then halts temporarily until the ribosome has become bound to the ER membrane. After this occurs, translation resumes and the polypeptide is synthesized into the ER. Some proteins remain in the ER, and the rest are sent to the Golgi apparatus. This process is called **cotranslational sorting** because the first step in the sorting process begins while translation is occurring. These proteins eventually move to new locations via the endomembrane system. In contrast, the uptake of most proteins into the nucleus, mitochondria, chloroplasts, and peroxisomes occurs after the protein is completely made (that is, completely translated) in the cytosol. This is called **post-translational sorting** because it does not happen until translation is finished.

4.7 Reviewing the Concepts

- Proteins synthesized in eukaryotic cells are sorted to their correct cellular destination via sorting signals. The sorting process can be either cotranslational or post-translational (Figure 4.30).

4.7 Testing Your Knowledge

1. Which of the following types of protein is sorted post-translationally?
   a. a protein destined for the cytosol
   b. a protein destined for the endoplasmic reticulum
   c. a protein destined for the lysosome
   d. a protein destined for the mitochondrion
   e. a protein that is secreted from the cell

4.8 Extracellular Matrix and Plant Cell Walls

Learning Outcomes

1. Explain the functional roles of the extracellular matrix (ECM) in animals.
2. Outline the major structural components of the ECM of animals.
3. Describe the structure and function of plant cell walls.

A large portion of an animal or a plant consists of a network of material that is secreted from cells and forms a complex meshwork outside of the plasma membrane. In animals, this is called the **extracellular matrix (ECM),** whereas plant cells are surrounded by a **cell wall.** The ECM and cell walls are a major component of certain parts of animals and plants, respectively. For example, bones and cartilage in animals are composed largely of ECM, and the woody portions of plants are composed mostly of cell walls. Although the cells within wood

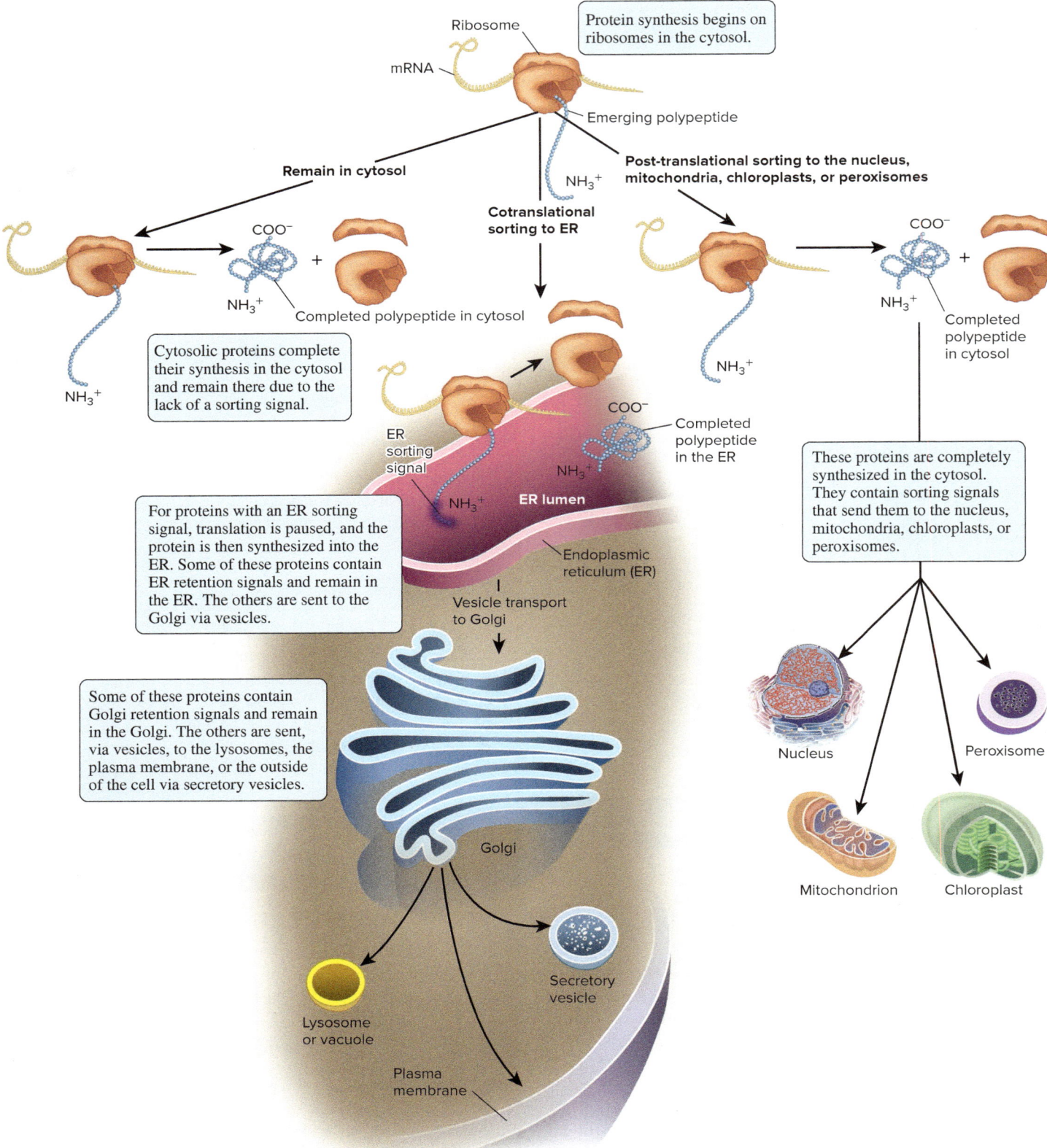

Figure 4.30 Three pathways for protein sorting in a eukaryotic cell. Proteins remain in the cytosol, are sorted into the ER (cotranslational sorting), or are sorted after they are completely synthesized (post-translational sorting).

eventually die, the cell walls they have produced remain and provide a rigid structure that supports the plant for years or even centuries.

In this section, we will begin by examining the structure and role of the ECM in animals, focusing on the functions of the major ECM components. We will then explore the structure and functions of the cell wall of plant cells.

The Extracellular Matrix in Animals Supports and Organizes Cells and Plays a Role in Cell Signaling

Unlike the cells of bacteria, archaea, fungi, and plants, the cells of animals are not surrounded by a rigid cell wall that provides structure and support. Instead, animal cells secrete materials that form an ECM, which provides support and helps to organize cells. Certain animal cells are completely embedded within an extensive ECM, whereas other cells may adhere to the ECM on only one side. **Figure 4.31** illustrates the general features of the ECM and its relationship to cells. The major macromolecules of the ECM are proteins and polysaccharides. The most abundant proteins are those that form large fibers. The polysaccharides give the ECM a gel-like character.

The ECM found in animals performs many important functions, including strength, structural support, organization, and cell signaling.

- **Strength.** The strength of the ECM in the skin of mammals prevents tearing. The ECM found in cartilage resists compression and provides protection to the joints. Similarly, the ECM protects the soft parts of the body, such as the internal organs.
- **Structural support.** The bones of many animals are composed primarily of ECM. Skeletons not only provide structural support but also facilitate movement via the functioning of attached muscles.
- **Organization.** The attachment of cells to the ECM plays a key role in the proper arrangement of cells throughout the body. In addition, the ECM binds many body parts together, such as tendons to bones.
- **Cell signaling.** A less obvious role of the ECM is cell signaling. One way that cells in multicellular organisms sense their environment is via changes in the ECM.

Adhesive and Structural Proteins Are Major Components of the ECM of Animals

In the 1850s, German biologist Rudolf Virchow suggested that all extracellular materials are made and secreted by cells. Around the same time, biologists realized that gelatin and glue, which are produced by the boiling of animal tissues, contain a common fibrous substance. This substance was named **collagen** (from the Greek, meaning glue-producing). Since that time, experimental techniques in biochemistry, microscopy, and biophysics have enabled scientists to probe the structure of the ECM. We now understand that the ECM contains a mixture of several different components, including proteins such as collagen.

The proteins found in the ECM are grouped into adhesive proteins, such as fibronectin and laminin, and structural proteins, such as collagen and elastin (**Table 4.2**).

Adhesive Proteins Fibronectin and laminin have multiple binding sites that bind to other components in the ECM, such as protein fibers and polysaccharides. The same proteins also have binding sites for receptors on the surfaces of cells. Therefore, adhesive proteins are so named because they make ECM components adhere to one another and to the cell surface. They provide organization to the ECM and facilitate the attachment of cells to the ECM.

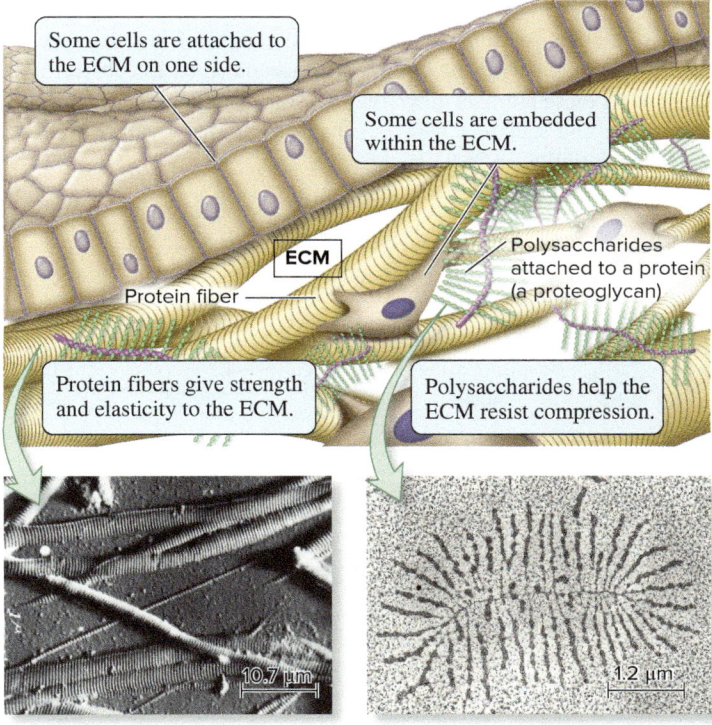

Figure 4.31 The extracellular matrix (ECM) of animal cells. The micrograph (SEM) at the bottom left shows collagen fibers, a type of protein fiber found in the ECM. The micrograph (TEM) at the bottom right shows a proteoglycan, which consists of polysaccharides attached to a protein.
(left) ©Clinical Photography, Central Manchester University Hospitals NHS Foundation Trust, UK/Science Source; *(right)* Courtesy of Dr. Joseph Buckwalter/University of Iowa

 Concept Check: *What are the four functions of the ECM in animals?*

Table 4.2	Proteins in the ECM of Animals	
General type	**Example**	**Function**
Adhesive	Fibronectin	Connects cells to the ECM and helps to organize components in the ECM
	Laminin	Connects cells to the ECM
Structural	Collagen	Forms large fibers and interconnected fibrous networks in the ECM; provides tensile strength
	Elastin	Forms elastic fibers in the ECM that can stretch and recoil

Structural Proteins and Tensile Strength Structural proteins, such as collagen, form large fibers in the ECM (see Figure 4.31). A key function of collagen is to impart tensile strength, which is a measure of how much stretching force a material can bear without tearing apart. Collagen provides high tensile strength to many parts of an animal's body. It is the main protein found in bones, cartilage, tendons, skin, and the lining of blood vessels and internal organs. In mammals, more than 25% of the total protein mass consists of collagen, much more than any other protein. Approximately 75% of the protein in mammalian skin is composed of collagen. Leather is largely a pickled and tanned form of collagen.

Structural Proteins and Elasticity Elasticity is needed in regions of the body, such as the lungs and blood vessels, that regularly expand and return to their original shape. In these places, the ECM contains elastic fibers composed primarily of the protein **elastin** (Figure 4.32). Elastin proteins form many covalent crosslinks to make a fiber with remarkable elastic properties. In the absence of a stretching force, each protein tends to adopt a compact conformation. When subjected to a stretching force, however, the compact proteins elongate, with the covalent crosslinks holding the fiber together. When the stretching force stops, the proteins naturally return to their compact conformation. In this way, an elastic fiber behaves much as a rubber band does, stretching under tension and snapping back when the tension is released.

Biology Principle

Structure Determines Function

Two structural features—the tendency of the elastin protein to adopt a compact conformation and the covalent crosslinking of multiple elastin proteins—result in fibers with the characteristic of elasticity.

Figure 4.32 Structure and function of elastic fibers. Elastic fibers are made of elastin, one type of structural protein found in the ECM surrounding animal cells.

Concept Check: *Suppose you started with an unstretched elastic fiber and treated it with a chemical that breaks the crosslinks between adjacent elastin proteins. What would happen when the fiber was stretched?*

Animal Cells Also Secrete Polysaccharides into the ECM

Polysaccharides are the second major component of the ECM of animals. As discussed in Chapter 3, polysaccharides are polymers of many simple sugars. Among vertebrates, the most abundant types of polysaccharides in the ECM are **glycosaminoglycans (GAGs).** These macromolecules are long, unbranched polysaccharides containing a repeating disaccharide unit (Figure 4.33a). GAGs are highly negatively charged molecules that tend to attract positively charged ions and water. The majority of GAGs in the ECM are linked to core proteins, forming **proteoglycans** (Figure 4.33b).

The primary function of GAGs and proteoglycans is to resist compression. Once secreted from cells, these macromolecules form a gel-like component in the ECM. How is this gel-like property important? Due to its high water content, the ECM is difficult to compress and thereby protects cells. GAGs and proteoglycans are found abundantly in regions of the body that are subjected to harsh mechanical forces, such as the joints of the human body. Two examples of GAGs are chondroitin sulfate, which is a major component of cartilage, and hyaluronic acid, which is found in the skin, eyes, and joint fluid.

Among many invertebrates, an important ECM component is **chitin,** a nitrogen-containing polysaccharide. Chitin forms the hard, protective outer covering (called an exoskeleton) of insects, such as crickets and grasshoppers, and of crustaceans, such as lobsters and shrimp. As these animals grow, they must periodically shed this rigid outer layer and secrete a new, larger one—a process called molting (look ahead to Figure 26.8).

Figure 4.33 Structures of glycosaminoglycans and proteoglycans. These macromolecules are found in the ECM, which is located outside of animal cells. **(a)** Glycosaminoglycans (GAGs) are composed of repeating disaccharide units. They range in length from several dozen to 25,000 disaccharide units. The GAG shown here is chondroitin sulfate, which is a component of cartilage. **(b)** Proteoglycans are composed of a long, linear core protein with many GAGs attached. Note that each GAG is typically 80 disaccharide units long, but only a short chain of sugars is shown in this illustration.

Concept Check: *What structural feature of GAGs gives the ECM a gel-like character?*

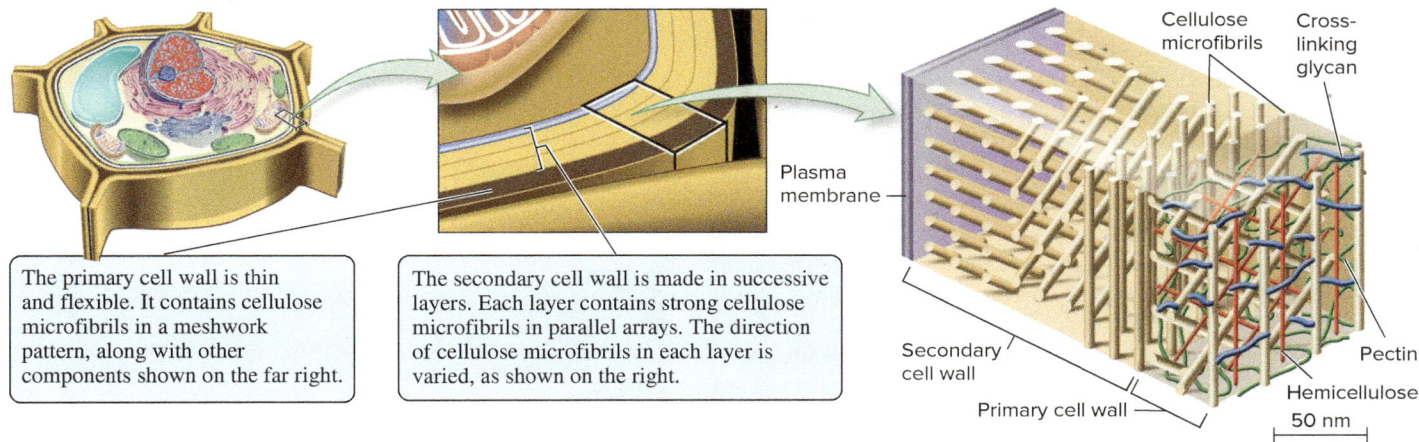

Figure 4.34 Structure of the cell wall of plant cells. The primary cell wall is relatively thin and flexible. It contains cellulose (tan), hemicellulose (red), crosslinking glycans (blue), and pectin (green). The secondary cell wall, which is produced only by certain plant cells, is made after the primary cell wall and is synthesized in successive layers.

Concept Check: With regard to cell growth, what would happen if the secondary cell wall was made too soon?

The Cell Wall of Plants Provides Strength and Resistance to Compression

Let's now turn our attention to the cell walls of plants. Plant cells are surrounded by a cell wall, a protective layer that forms outside of the plasma membrane. Like animal cells, the cells of plants are surrounded by material that provides tensile strength and resistance to compression. The cell walls of plants, however, are usually thicker, stronger, and more rigid than the ECM found in animals. Plant cell walls provide rigidity for mechanical support and play a role in the maintenance of cell shape and the direction of cell growth. As discussed in Chapter 5, the cell wall also inhibits expansion when water enters the cell, thereby preventing the cell from bursting.

The cell walls of plants are composed of a primary cell wall and a secondary cell wall (**Figure 4.34**). These walls are named according to the timing of their synthesis—the primary cell wall is made before the secondary cell wall.

Primary Cell Wall During cell division, the **primary cell wall** develops between two newly made daughter cells. It is usually very flexible and allows new cells to increase in size. The following are the main components of the primary cell wall.

- **Cellulose:** The main macromolecule of the plant cell wall is **cellulose**, a polysaccharide made of repeating molecules of glucose attached end to end. These glucose polymers associate with each other via hydrogen bonding to form microfibrils that provide great tensile strength (**Figure 4.35**). The mechanism of cellulose synthesis is described in Chapter 28 (look ahead to Figure 28.11).
- **Hemicellulose:** Hemicellulose (see Figure 4.34) is another linear polysaccharide, with a structure similar to that of cellulose, but it contains sugars other than glucose in its structure and usually forms thinner microfibrils.
- **Glycans:** Polysaccharides with branching structures are also important in cell-wall structure. The crosslinking glycans bind to cellulose and provide organization to the cellulose microfibrils.
- **Pectins:** Highly negatively charged polysaccharides, called pectins, attract water and have a gel-like character. They provide the cell wall with the ability to resist compression.

Figure 4.35 Structure of cellulose, the main macromolecule of the plant cell wall. Cellulose is made of repeating glucose units linked end to end that hydrogen-bond to each other to form microfibrils (SEM).
©SciMAT/Science Source

Secondary Cell Wall The **secondary cell wall** is synthesized and deposited between the plasma membrane and the primary cell wall (see Figure 4.34) after a plant cell matures and has stopped increasing in size. It is made in layers by the successive deposition of cellulose microfibrils and other components. Whereas the primary wall structure is relatively similar in nearly all cell types and plant species, the structure of the secondary cell wall is more variable. Some plant cells have no secondary cell wall; leaf cells that are involved in photosynthesis lack a secondary wall, allowing light to enter the cells more readily. In contrast, other cells have a thick secondary cell wall that contains components that are not found in the primary cell wall. For example, phenolic compounds called lignins, which are found in the woody parts of plants, are very hard and impart considerable strength to the secondary wall structure.

4.8 Reviewing the Concepts

- The extracellular matrix (ECM) is a network of material that forms a complex meshwork outside of animal cells; the cell wall is a similar component of plant cells.
- Proteins and polysaccharides are the major constituents of the ECM in animals. These materials are involved in strength, structural support, organization, and cell signaling (Figure 4.31).
- Adhesive proteins, such as fibronectin and laminin, help cells adhere to the ECM. Structural proteins include collagen, which forms fibers and fibrous networks that provide tensile strength, and elastin, which forms elastic fibers that stretch and recoil (Table 4.2, Figure 4.32).
- Glycosaminoglycans (GAGs) are polysaccharides of repeating disaccharide units that give a gel-like character to the ECM of animals. Proteoglycans consist of a long, linear core protein with many GAGs attached (Figure 4.33).
- Plant cells are surrounded by a cell wall composed of primary and secondary cell walls. The primary cell wall is composed largely of cellulose. The secondary cell wall is synthesized and deposited between the plasma membrane and the primary cell wall and is often thick and rigid (Figures 4.34, 4.35).

4.8 Testing Your Knowledge

1. The function of the extracellular matrix (ECM) in animals is
   a. to provide strength.
   b. to provide structural support.
   c. to organize cells and other body parts.
   d. cell signaling.
   e. all of the above.
2. The protein found in the ECM of animals that provides tensile strength and resistance to tearing when stretched is
   a. elastin.
   b. cellulose.
   c. collagen.
   d. laminin.
   e. fibronectin.
3. A major component of the plant cell wall is
   a. elastin.
   b. cellulose.
   c. collagen.
   d. laminin.
   e. fibronectin.

4.9 Systems Biology of Cells: A Summary

Learning Outcomes

1. Outline the differences in complexity among bacteria, animal cells, and plant cells.
2. Describe how a eukaryotic cell can be viewed as four interacting systems: nucleus, cytosol, endomembrane system, and semiautonomous organelles.

We conclude this chapter by reviewing cell structure and function from a perspective called **systems biology,** the study of how new properties of life arise through complex interactions of its components. The system being studied can be anything from a metabolic pathway to a cell, an organ, or even an entire organism. In this section, we will view the cell as a system. First, we will compare prokaryotic and eukaryotic cells as systems, then examine the four interconnected parts that make up the system that is the eukaryotic cell.

Bacterial Cells Are Relatively Simple Systems, Whereas Eukaryotic Cells Are a System of Four Interacting Parts

Bacterial cells are relatively small and lack the extensive internal compartmentalization characteristic of eukaryotic cells (Table 4.3). On the outside, bacterial cells are surrounded by a cell wall, and many species have flagella. Animal cells lack a cell wall, and only certain cell types have flagella or cilia. Like bacteria, plant cells also have cell walls, but the chemical composition of these walls is different from that of bacterial cells. Plant cells rarely have flagella.

As mentioned earlier in this chapter, the cytoplasm is the region of the cell enclosed by the plasma membrane. Ribosomes are found in the cytoplasm of all cell types. In bacteria, the cytoplasm is a single compartment. The bacterial genetic material, usually a single chromosome, is found in the nucleoid, which is not surrounded by a membrane. By comparison, the cytoplasm of eukaryotic cells is highly compartmentalized. The cytosol is the area that surrounds many different types of membrane-bound organelles. For example, eukaryotic chromosomes are found in the nucleus, which is surrounded by a double membrane. In addition, all eukaryotic cells have an endomembrane system and mitochondria, and plant cells also have chloroplasts.

We can view a eukaryotic cell as a system with four interacting parts: the interior of the nucleus, the cytosol, the endomembrane system, and the semiautonomous organelles (Figure 4.36). These four regions have their own structure and organization, while also playing a role in the structure and organization of the entire cell.

Table 4.3   A Comparison of Cell Complexity Among Bacterial, Animal, and Plant Cells

Structures	Bacteria	Animal cells	Plant cells
Structures on the cell surface			
Cell wall*	Present	Absent	Present
Flagella/cilia	Flagella sometimes present	Cilia or flagella present on certain cell types	Rarely present†
Plasma membrane	Present	Present	Present
Interior structures			
Cytoplasm	Usually a single compartment inside the plasma membrane	Composed of membrane-bound organelles that are surrounded by the cytosol	Composed of membrane-bound organelles that are surrounded by the cytosol
Ribosomes	Present	Present	Present
Chromosomes	Typically one circular chromosome per nucleoid; a nucleoid is not a separate compartment	Multiple linear chromosomes in the nucleus, which is surrounded by a double membrane; mitochondria also have chromosomes	Multiple linear chromosomes in the nucleus, which is surrounded by a double membrane; mitochondria and chloroplasts also have chromosomes
Endomembrane system	Absent	Present	Present
Mitochondria	Absent	Present	Present
Chloroplasts	Absent	Absent	Present

*Note that the biochemical composition of bacterial cell walls is very different from that of plant cell walls.
†Some plant species produce sperm cells with flagella, but flowering plants produce sperm within pollen grains that lack flagella.

Nucleus
- Location of most of the genome
- Gene regulation
- Organization and protection of chromosomes via the nuclear matrix

Endomembrane system
1. Nuclear envelope
   - Boundary that surrounds the nucleus
2. Endoplasmic reticulum
   - Protein secretion and sorting
   - Glycosylation
   - Lipid synthesis
   - Metabolic functions and accumulation of Ca^{2+}
3. Golgi apparatus
   - Protein secretion and sorting
   - Glycosylation
4. Lysosome/vacuoles
   - Degradation of organic molecules
   - Storage of organic molecules
   - Accumulation of water (plant vacuoles)
5. Peroxisomes
   - Breakdown of toxic molecules such as H_2O_2
   - Breakdown and synthesis of organic molecules
6. Plasma membrane
   - Uptake and excretion of ions and molecules
   - Cell signaling
   - Cell adhesion

Cytosol
- Coordination of responses to the environment
- Coordination of metabolism
- Synthesis of the proteome
- Organization and movement via cytoskeleton and motor proteins

Semiautonomous organelles
1. Mitochondria
   - Synthesis of ATP
   - Synthesis and modification of other organic molecules
   - Production of heat
2. Chloroplasts (plants and algae)
   - Photosynthesis

Figure 4.36 **The four interacting parts of eukaryotic cells: nucleus, cytosol, endomembrane system, and semiautonomous organelles.** An animal cell is illustrated here.

5

Membranes: The Interface Between Cells and Their Environment

©Ramón Andrade, 3Dciencia/Science Source

A model for the structure of aquaporin. This protein, found in the plasma membrane of many cell types, such as red blood cells and plant cells, functions as a channel that allows the rapid movement of water molecules across the membrane. A central pore in aquaporin allows water molecules to cross the membrane.

Chapter Outline

- 5.1 Membrane Structure
- 5.2 Fluidity of Membranes
- 5.3 Overview of Membrane Transport
- 5.4 Proteins That Carry Out Membrane Transport
- 5.5 Intercellular Channels
- 5.6 Exocytosis and Endocytosis
- 5.7 Cell Junctions

Assessing Your Knowledge and Skills

A membrane called the **plasma membrane** separates the internal contents of a cell from its external environment. With such a role, you might imagine that the plasma membrane would be thick and rigid. Remarkably, the opposite is true. All cell membranes, including the plasma membrane, are thin (typically 5–10 nm) and even somewhat fluid. It would take 5,000–10,000 of these membranes stacked on top of each other to equal the thickness of a piece of paper! Despite their thinness, membranes are impressively dynamic structures that effectively maintain the separation between a cell and its surroundings.

All cells have a plasma membrane that encloses the cytoplasm, and eukaryotic cells have internal membranes that form organelles (see Chapter 4). Both types are also called **biological membranes,** or simply, **membranes.** The plasma membrane and internal membranes provide interfaces where many vital cellular activities occur (Table 5.1). In this chapter, we will begin by considering the components that provide the structure and fluid properties of membranes and then explore how they are synthesized. Next, we will examine one of a membrane's primary functions—membrane transport. Biological membranes regulate the traffic of substances into and out of the cell and its organelles. This traffic occurs via simple diffusion, proteins that span the membrane, intercellular channels, exocytosis, and endocytosis. Finally, we will consider how adjacent cells are held together by cell junctions.

Table 5.1	Important Functions of Biological Membranes
Selective uptake and export of ions and molecules	
Cell compartmentalization	
Protein sorting	
Anchoring of the cytoskeleton	
Production of energy intermediates such as ATP	
Cell signaling	
Cell and nuclear division	
Adhesion of cells to each other and to the ECM	

5.1 Membrane Structure

Learning Outcomes

1. Describe the fluid-mosaic model of membrane structure.
2. Identify the three types of membrane proteins.

The two primary components of membranes are phospholipids, which form the basic matrix of a membrane, and proteins, which are embedded in the membrane or loosely attached to its surface. A third component is carbohydrates, which may be attached to membrane lipids and proteins. In this section, we will be mainly concerned with the organization of these components to form a biological membrane and their importance in the overall function of membranes.

Biological Membranes Are a Mosaic of Lipids, Proteins, and Carbohydrates

Figure 5.1 shows the biochemical organization of a membrane, which is similar in composition among all living organisms. The framework of the membrane is the **phospholipid bilayer,** which consists of two layers of phospholipids. Recall from Chapter 3 that phospholipids are **amphipathic** molecules (refer back to Figure 3.9). They have two hydrophobic (water-fearing) nonpolar tails and a hydrophilic (water-loving) polar head. The nonpolar tails of lipids are found in the interior of the membrane, and the polar heads are on the surface. Biological membranes also contain proteins, and most membranes have carbohydrates attached to lipids and proteins. Overall, the membrane is considered a mosaic of lipid, protein, and carbohydrate molecules.

The membrane structure illustrated in Figure 5.1 is referred to as the **fluid-mosaic model,** originally proposed by S. Jonathan Singer and Garth Nicolson in 1972. As discussed later, the membrane exhibits properties that resemble a fluid because lipids and proteins can move relative to each other within the membrane.

Half of a phospholipid bilayer is termed a **leaflet.** Each leaflet faces a different region. For example, the plasma membrane contains a leaflet that faces the cytosol (cytosolic leaflet) and a leaflet that faces the extracellular environment (extracellular leaflet) (see Figure 5.1). With regard to lipid composition, the two leaflets are highly asymmetric, with certain types of lipids more abundant in one leaflet than the other. A striking asymmetry occurs with glycolipids—lipids with a carbohydrate attached. These are found primarily in the extracellular leaflet. The carbohydrate portion of a glycolipid is attached to the polar head and protrudes into the extracellular environment. The types of carbohydrates on the surfaces of cells often allow cells to recognize each other, a phenomenon called cell surface recognition.

Proteins Associate with Membranes in Three Ways

Although the phospholipid bilayer forms the basic foundation of biological membranes, the protein component carries out many key functions. Some of these functions were considered in Chapter 4. Later in this chapter, we will explore how membrane proteins are involved in transporting ions and molecules across membranes and in the adhesion of cells to each other. In later chapters, we will examine how membrane proteins are responsible for other functions, including ATP synthesis (Chapter 6), photosynthesis (Chapter 7), and cell signaling (Chapter 8).

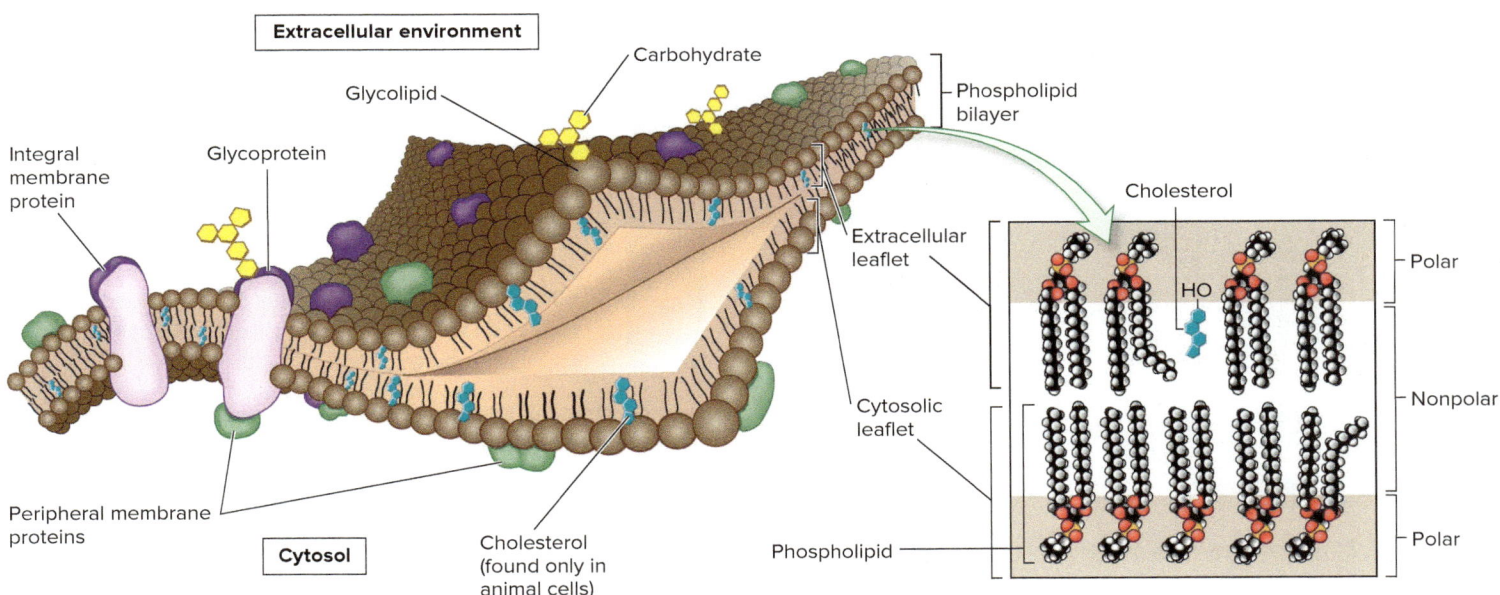

Figure 5.1 Fluid-mosaic model of membrane structure. The basic framework of a plasma membrane and other biological membranes is a phospholipid bilayer. Proteins may span the membrane and may be bound on the surface to other proteins or to lipids. Proteins and lipids that have covalently bound carbohydrates are called glycoproteins and glycolipids, respectively. The inset shows nine phospholipids and one cholesterol molecule (blue) in a bilayer, and it emphasizes the two leaflets and the polar and nonpolar regions of the bilayer. Note: A portion of the bilayer in the drawing on the left is artificially pealed apart so that you can more easily see the two leaflets.

The three types of membrane proteins are transmembrane proteins, lipid-anchored proteins, and peripheral membrane proteins. Transmembrane proteins and lipid-anchored proteins are called **integral membrane proteins,** because a portion of them is integrated into the hydrophobic region of the membrane. Each of the three types has a different way of associating with a membrane (**Figure 5.2**).

1. **Transmembrane proteins** span or traverse the membrane from one leaflet to the other. Transmembrane segments within these proteins have stretches of nonpolar amino acids that are inserted into the hydrophobic interior of the phospholipid bilayer, making it possible for a portion of the protein to traverse the membrane. In most transmembrane proteins, each transmembrane segment is folded into an α helix. Such a segment is stable in a membrane because the nonpolar amino acids interact favorably with the nonpolar tails of the lipid molecules.

2. **Lipid-anchored proteins** are proteins that associate with a membrane because they have a lipid molecule that is covalently attached to an amino acid side chain within the protein. The lipid tail is inserted into the hydrophobic portion of the membrane, thereby keeping the protein firmly attached to the membrane.

3. **Peripheral membrane proteins** are proteins that are noncovalently bound to regions of transmembrane membrane proteins that project out from the membrane and/or to the polar head groups of phospholipids (see Figure 5.1). Peripheral membrane proteins are typically bound to the membrane by hydrogen bonds or ionic bonds (or both).

5.1 Reviewing the Concepts

- A plasma membrane separates a cell from its surroundings. Biological membranes provide interfaces for carrying out vital cellular functions (Table 5.1).
- The accepted model of membranes is the fluid-mosaic model, and the basic framework of a membrane is the phospholipid bilayer. Biological membranes also contain proteins, and some membranes have attached carbohydrates (Figure 5.1).
- The three main types of membrane proteins are transmembrane proteins, lipid-anchored proteins, and peripheral membrane proteins. Transmembrane proteins and lipid-anchored proteins are classified as integral membrane proteins (Figure 5.2).

5.1 Testing Your Knowledge

1. Which of the following is *not* a characteristic of a biological membrane?
   a. contains a bilayer of phospholipids
   b. contains proteins that are inserted into the membrane
   c. contains a high percentage of water molecules
   d. has carbohydrates attached to lipids and proteins
   e. has asymmetric leaflets

2. A membrane protein that is noncovalently attached to a protein or a lipid is
   a. a transmembrane protein.
   b. a lipid-anchored protein.
   c. a peripheral membrane protein.
   d. a glycoprotein.
   e. all of the above.

Figure 5.2 Types of membrane proteins. Integral membrane proteins are of two types: transmembrane proteins and lipid-anchored proteins. Peripheral membrane proteins are noncovalently bound to the hydrophilic regions of integral membrane proteins or to the polar head groups of lipids. Inset: The protein bacteriorhodopsin contains seven transmembrane segments, depicted as cylinders, each having an α-helical structure. Bacteriorhodopsin is found in halophilic (salt-loving) archaea.

5.2 Fluidity of Membranes

Learning Outcomes

1. Describe the fluidity of membranes.
2. **SCISKILLS** ▶ Predict how fluidity will be affected by changes in lipid composition.
3. **SCISKILLS** ▶ Analyze the results of experiments indicating that certain membrane proteins can diffuse laterally within the membrane.

Let's now turn our attention to the dynamic properties of membranes. Although a membrane provides a critical interface between a cell or an organelle and its environment, it is not a solid, rigid structure. Rather, biological membranes exhibit properties of **fluidity,** which means that individual molecules remain in close association yet have the ability to readily move within the membrane. In this section, we will examine the fluid properties of biological membranes.

Membranes Are Semifluid

Though membranes are often described as fluid, it is more appropriate to say they are **semifluid,** because movements of lipids and membrane proteins occur in only two dimensions. In a fluid substance,

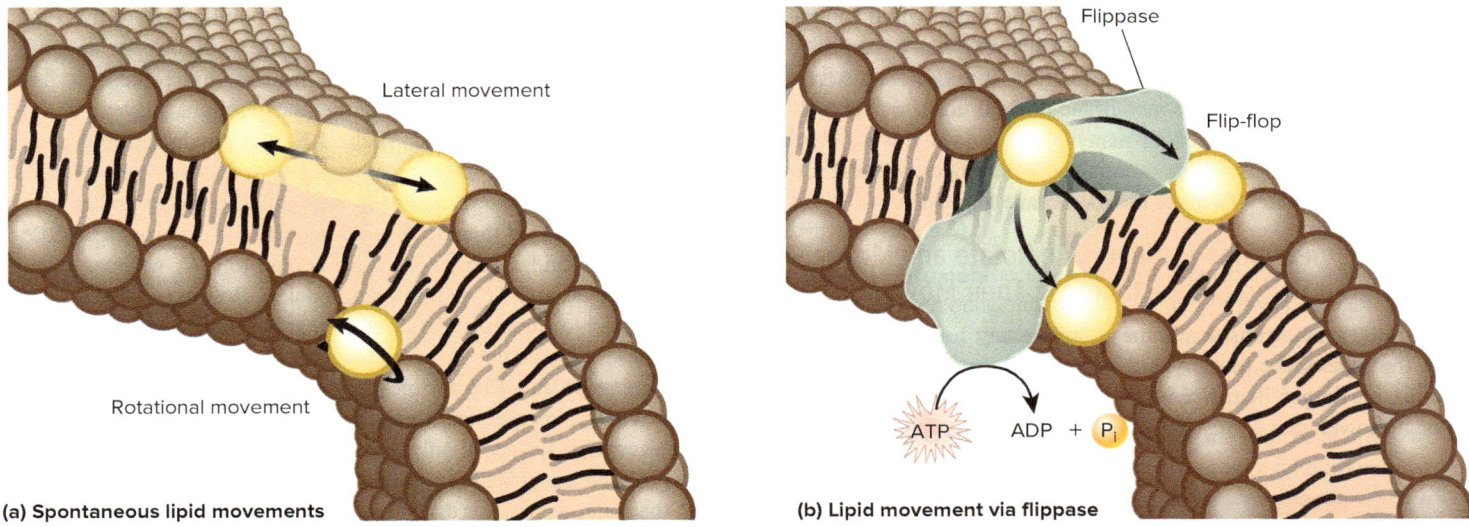

Figure 5.3 Semifluidity of the lipid bilayer. (a) Spontaneous movements in the bilayer. Lipids can rotate (that is, turn through 360°) and move laterally (for example, from left to right in the plane of the bilayer). (b) Flip-flop does not happen spontaneously, because the polar head group would have to pass through the hydrophobic region of the bilayer. Instead, the enzyme flippase uses ATP to flip phospholipids from one leaflet to the other.

Concept Check: *In an animal cell, how can changes in lipid composition affect membrane fluidity?*

molecules can move in three dimensions. By comparison, most lipids can rotate freely around their long axes and move laterally within the membrane leaflet (**Figure 5.3a**). This type of motion is considered two-dimensional, which means it occurs within the plane of the membrane. Because rotational and lateral movements keep the nonpolar tails within the hydrophobic interior, such movements are energetically favorable. At 37°C, a typical lipid molecule exchanges places with its neighbors about 10^7 times per second, and it can move several micrometers per second. At this rate, a lipid can traverse the length of a bacterial cell (approximately 1 μm) in only 1 second and the length of a typical animal cell in 10 to 20 seconds.

In contrast to rotational and lateral movements, the "flip-flop" of lipids from one leaflet to the opposite leaflet does not occur spontaneously. Flip-flop is energetically unfavorable because the polar head of a phospholipid has to travel through the hydrophobic interior of the membrane. How are lipids moved from one leaflet to the other? The transport of lipids between leaflets is due to the action of the enzyme flippase, which requires energy input in the form of ATP (**Figure 5.3b**).

Although most lipids diffuse rotationally and laterally within the plane of the lipid bilayer, researchers have discovered that certain types of lipids in animal cells tend to associate strongly with each other to form structures called lipid rafts. As the word raft suggests, a **lipid raft** is a group of lipids that float together as a unit within a larger sea of lipids. Lipid rafts have a lipid composition that differs from the surrounding membrane. For example, they usually have a high amount of cholesterol. In addition, lipid rafts may contain unique sets of lipid-anchored proteins and transmembrane proteins. The functional importance of lipid rafts is the subject of a large amount of current research. Lipid rafts may play an important role in endocytosis (discussed later in this chapter) and cell signaling.

Lipid Composition Affects Membrane Fluidity

The biochemical properties of phospholipids affect the fluidity of the phospholipid bilayer. These include the following:

- **Length of the nonpolar tails:** Shorter tails are less likely to interact with each other, which makes the membrane more fluid. The tails typically range from 14 to 24 carbon atoms, with 18 to 20 carbons being the most common.

- **Presence of double bonds:** Double bonds make a membrane more fluid. When a double bond is present, the lipid is said to be **unsaturated** with respect to the number of hydrogens that are bound to the carbon atoms (refer back to Figure 3.8). A double bond creates a kink in the lipid tail (shown in the inset to Figure 5.1), making it more difficult for neighboring tails to interact and making the bilayer more fluid.

- **Presence of cholesterol:** Cholesterol, which is found in animal cells, tends to stabilize membranes because it is a short, rigid molecule (see inset to Figure 5.1). Its effects depend on temperature. At higher temperatures, such as those observed in mammals that maintain a constant body temperature, cholesterol makes the membrane less fluid. At lower temperatures, such as icy water, cholesterol has the opposite effect. It makes the membrane more fluid and prevents it from freezing. Plant cell membranes contain phytosterols that resemble cholesterol in their chemical structure.

An optimal level of bilayer fluidity is essential for normal cell function, growth, and division. If a membrane is too fluid, which may occur at higher temperatures, it can become leaky. However, if a membrane becomes too solid, which may occur at lower temperatures, the functioning of membrane proteins will be inhibited. How can organisms cope with changes in temperature? The cells of many

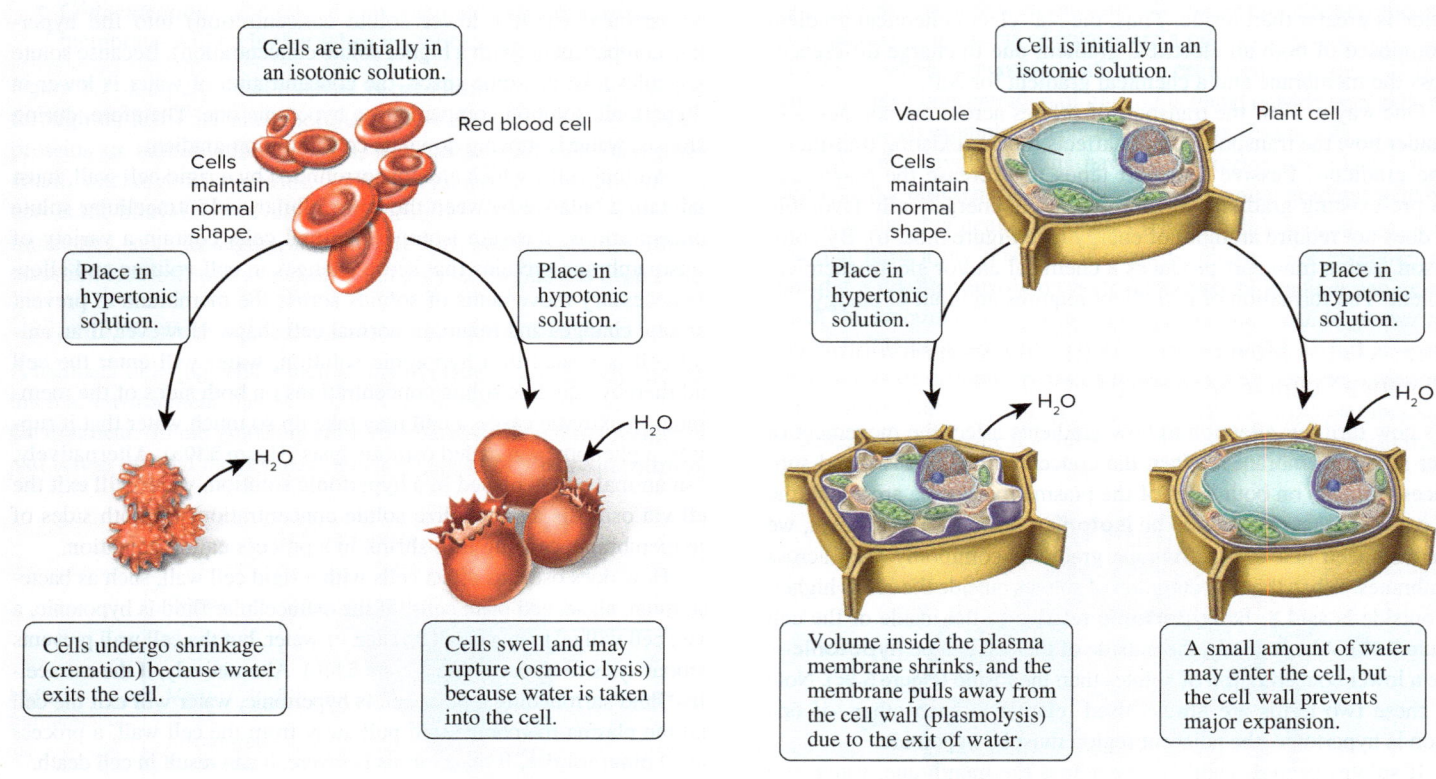

(a) Osmosis in animal cells

(b) Osmosis in plant cells

Figure 5.10 **The effects of osmosis.** **(a)** In cells that lack a cell wall, such as animal cells, osmosis may promote cell swelling and possible rupture (osmotic lysis) or shrinkage (crenation). **(b)** In cells that have a rigid cell wall, such as plant cells, a hypotonic medium causes only a minor amount of expansion, whereas a hypertonic medium causes the plasma membrane to pull away from the cell wall.

Concept Check: Let's suppose the inside of a cell has a solute concentration of 0.3 M and the outside has a concentration of 0.2 M. If the membrane is impermeable to solutes, in which direction will water move?

Biology Principle

Living Organisms Maintain Homeostasis

In this example, the paramecium maintains a relatively constant internal volume by using contractile vacuoles to remove excess water.

Figure 5.11 **The contractile vacuole in *Paramecium caudatum*.** In the upper photo, a contractile vacuole is filled with water from radiating canals that collect fluid from the cytosol. The lower photo shows the cell after the contractile vacuole has fused with the plasma membrane (which would be above the plane of this page) and has released the water from the cell.
©Michael Abbey/Science Source

5.3 Testing Your Knowledge

1. Carbon dioxide can move across a membrane without the aid of a transmembrane protein. This movement is an example of
   a. simple diffusion.
   b. facilitated diffusion.
   c. passive transport.
   d. active transport.
   e. both a and c.

2. Let's suppose the inside of a cell is hypertonic relative to the outside. If osmosis occurs, _____ will move _____ the cell to equalize solute concentrations.
   a. solute molecules, into
   b. solute molecules, out of
   c. water, into
   d. water, out of
   e. both a and d

5.4 Proteins That Carry Out Membrane Transport

Learning Outcomes

1. Outline the functional differences between channels and transporters.
2. **SCISKILLS** ▶ Analyze the results of Agre, and explain how they indicated that aquaporin is a water channel.
3. Compare and contrast uniporters, symporters, and antiporters.
4. Explain the concepts of primary active transport and secondary active transport.
5. Describe the structure and function of pumps.

Because the phospholipid bilayer is a physical barrier to the diffusion of most hydrophilic molecules and ions, cells can separate their internal contents from the external environment. However, this barrier also poses a potential problem because cells must take up nutrients from the environment and export waste products. How do cells resolve this dilemma? Over the course of millions of years, species have evolved a multitude of proteins that carry out membrane transport—transmembrane proteins that provide passageways for the movement of ions and hydrophilic molecules across the phospholipid bilayer. Such proteins play a central role in the selective permeability of biological membranes. In this section,

Figure 5.12 Mechanism of solute transport via a channel. The solute molecules are represented by purple spheres.

 Concept Check: *What is the purpose of gating?*

we will examine the two categories—channels and transporters—and see how they move solutes across the membrane.

Channels Provide Open Passageways for Solute Movement

The term **channel** refers to an open passageway for the facilitated diffusion of ions or molecules across a membrane. The type of channel shown in **Figure 5.12** is a transmembrane protein. Solutes move directly through this channel to get to the other side. When a channel is open, the transmembrane movement of solutes can be extremely rapid, up to 100 million ions or molecules per second!

Most channels are **gated,** which means they open to allow the diffusion of solutes and close to prohibit diffusion. The phenomenon of gating allows cells to regulate the movement of solutes. For example, gating may involve the direct binding of a molecule to the channel protein itself. These gated channels are controlled by the noncovalent binding of small molecules—called **ligands**—such as hormones or neurotransmitters. The ligands are often important in the transmission of signals between neurons and muscle cells or between two neurons.

FEATURE INVESTIGATION

Agre Discovered That Osmosis Occurs More Quickly in Cells with a Channel That Allows the Facilitated Diffusion of Water

As discussed earlier in this chapter, osmosis is the movement of water to balance solute concentrations. Water can cross biological membranes slowly by simple diffusion through the phospholipid bilayer. However, in the 1980s, researchers discovered that certain cell types allow water to move across the plasma membrane at a much faster rate than would be predicted by simple diffusion alone. For example, water moves very quickly across the membrane of red blood cells, which causes them to shrink and swell in response to changes in extracellular solute concentrations (see Figure 5.10a). Likewise, bladder and kidney cells, which play a key role in regulating water balance in the bodies of vertebrates, allow the rapid movement of water across their membranes. Based on these observations, researchers speculated that certain cell types might have channels in their plasma membranes that enable the rapid movement of water.

One approach to characterizing a new protein is to first identify a protein based on its relative abundance in a particular cell type and then attempt to determine the protein's function. This rationale was applied to the discovery of proteins that allow the rapid movement of water across membranes. Peter Agre and his colleagues first identified a protein that was abundant in red blood cells and kidney cells but not found in high amounts in many other cell types. Though they initially did not know the function of the protein, its physical structure was similar to other proteins that were already known to function as channels. They named this protein CHIP28, which stands for channel-forming integral membrane protein with a molecular mass of 28,000 daltons. During the course of their studies, they also identified and isolated the gene that encodes CHIP28.

In 1992, Agre and his colleagues conducted experiments to determine if CHIP28 functions in the transport of water across membranes (Figure 5.13). Because they had already isolated the gene that encodes

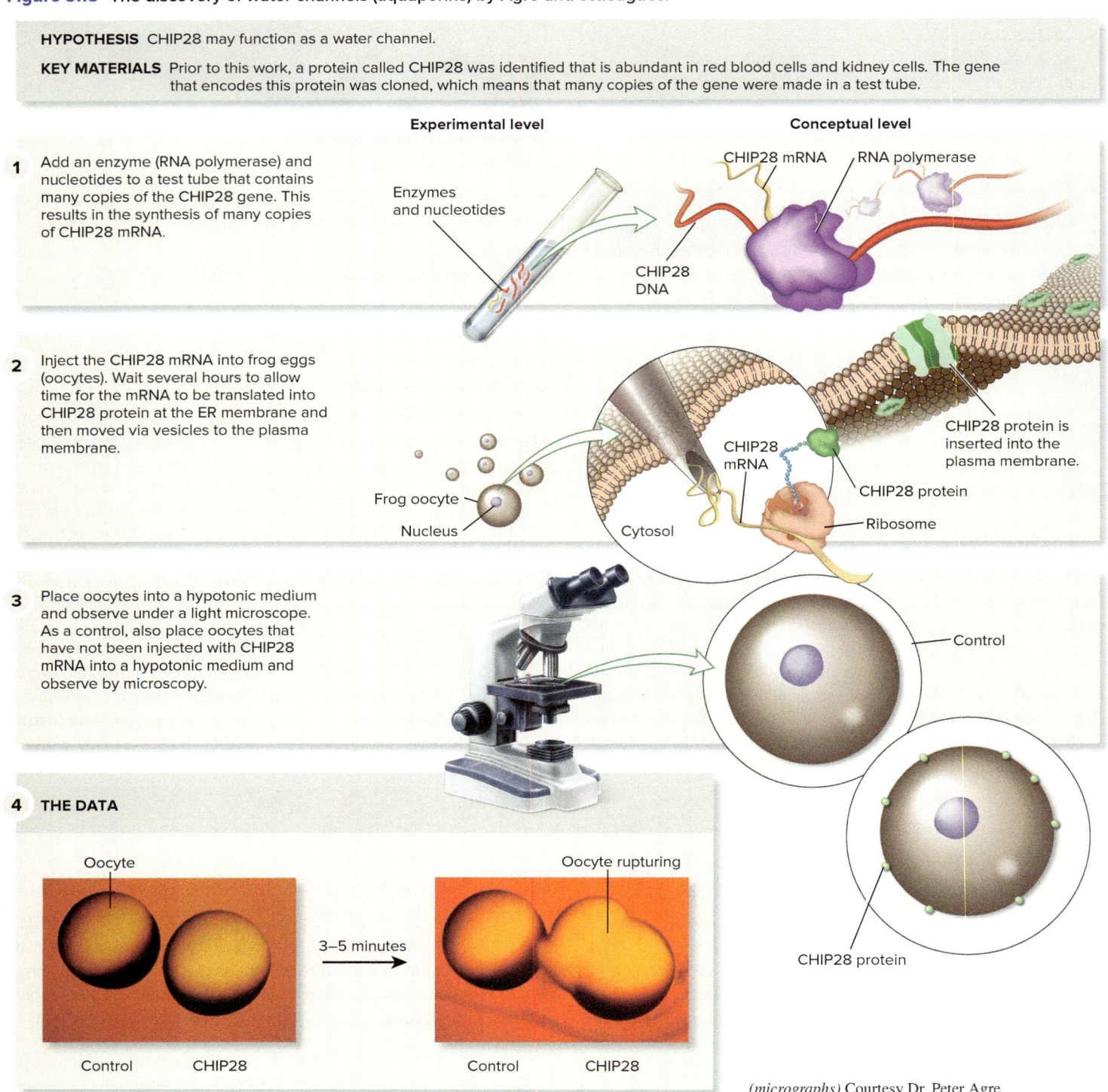

Figure 5.13 The discovery of water channels (aquaporins) by Agre and colleagues.

(micrographs) Courtesy Dr. Peter Agre

5. **CONCLUSION** The CHIP28 protein, now called aquaporin, allows the rapid movement of water across the membrane.

6. **SOURCE** Preston, G. M., Carroll, T. P., Guggino, W. B., and Agre, P. 1992. Appearance of Water Channels in *Xenopus* Oocytes Expressing Red Cell CHIP28 Protein. *Science* 256: 385–387.

CHIP28, they could make many copies of this gene in a test tube (in vitro) using gene cloning techniques (see Chapter 18). Starting with many copies of the gene in vitro, they added an enzyme to transcribe the gene into mRNA that encodes the CHIP28 protein. This mRNA was then injected into frog oocytes, chosen because these oocytes are large, are easy to inject, and lack pre-existing proteins in their plasma membranes that allow the rapid movement of water. Following injection, the mRNA was translated into CHIP28 proteins, which were inserted into the plasma membrane of the oocytes. After sufficient time had been allowed for this to occur, the oocytes were placed in a hypotonic medium. As a control, oocytes that had not been injected with CHIP28 mRNA were also exposed to a hypotonic medium.

As you can see in the data, a striking difference was observed between oocytes that expressed CHIP28 versus the control oocytes. Within minutes, oocytes that contained the CHIP28 protein were seen to swell due to the rapid uptake of water. Three to five minutes after being placed in a hypotonic medium, they actually lysed! By comparison, the control oocytes did not swell as rapidly, and they did not rupture, even after 1 hour. Taken together, these results are consistent with the hypothesis that CHIP28 functions as a channel that allows the facilitated diffusion of water across the membrane. Many subsequent studies confirmed this observation. Later, CHIP28 was renamed **aquaporin** to indicate its newly identified function of allowing water to diffuse through a channel in the membrane. More recently, the three-dimensional structure of aquaporin was determined (see chapter-opening figure). In 2003, Agre was awarded the Nobel Prize in Chemistry for this work.

Experimental Questions

1. What observations about particular cell types in the human body led to the experimental strategy in Figure 5.13?

2. What characteristics of CHIP28 made Agre and associates speculate that it may transport water? In your own words, briefly explain how they tested the hypothesis that CHIP28 has this function.

3. **SCISKILLS** ▶ Explain how the results of the experiment in Figure 5.13 support the proposed hypothesis.

Transporters Bind Their Solutes and Undergo Conformational Changes

Let's now turn our attention to a second category of proteins, known as **transporters**,* which carry out membrane transport. These transmembrane proteins bind one or more solutes in a hydrophilic pocket and undergo a conformational change that switches the exposure of the pocket from one side of the membrane to the other side (**Figure 5.14**). For example, in 1995, Robert Brooker and colleagues proposed that a transporter called lactose permease, which is found in the bacterium *Escherichia coli*, has a hydrophilic pocket that binds lactose. They further proposed that the two halves of the transporter protein come together at an interface that moves in such a way that the lactose-binding site alternates between an outwardly accessible pocket and an inwardly accessible pocket, as shown in Figure 5.14. This idea was later confirmed by studies that determined the structure of the lactose permease and related transporters.

Transporters provide the principal pathway for the cellular uptake of organic molecules, such as sugars, amino acids, and nucleotides. In animals, transporters also allow cells to take up certain hormones and neurotransmitters. In addition, many transporters play a key role in export. Waste products of cellular metabolism must be released from

---

*Transporters are also called carriers. However, this term is misleading, because transporters do not physically carry their solutes across the membrane.

Biology Principle

Structure Determines Function

Two structural features—a hydrophilic pocket and the ability to switch back and forth between two conformations—allow transporters to move ions and molecules across the membrane.

For transport to occur, a solute binds in a hydrophilic pocket exposed on one side of the membrane. The transporter then undergoes a conformational change that switches the exposure of the pocket to the other side of the membrane, where the solute is then released.

Figure 5.14 Mechanism of transport by a transporter.

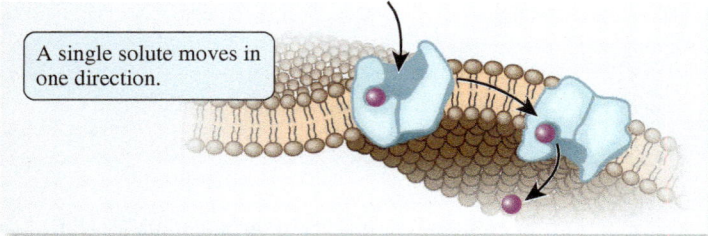
(a) Uniporter — A single solute moves in one direction.

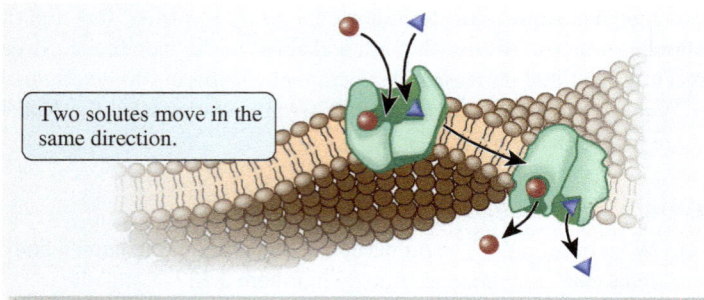
(b) Symporter — Two solutes move in the same direction.

(c) Antiporter — Two solutes move in opposite directions.

Figure 5.15 Types of transporters based on the direction of transport.

cells before they reach toxic levels; for example, a transporter removes lactic acid, a by-product of muscle cells during exercise. Other transporters, which are involved with ion transport, play an important role in regulating internal pH and controlling cell volume. Transporters tend to be much slower than channels. Their rate of transport is typically 100 to 1,000 ions or molecules per second.

Transporters are named according to the number of solutes they bind and the direction in which they transport those solutes (Figure 5.15).

- **Uniporters** bind a single ion or molecule and transport it across the membrane.
- **Symporters**, also called cotransporters, bind two or more ions or molecules and transport them in the same direction.
- **Antiporters** bind two or more ions or molecules and transport them in opposite directions.

Active Transport Is the Movement of Solutes Against a Gradient

As described earlier, active transport is the movement of a solute across a membrane against its concentration gradient—that is, from a region of lower concentration to one of higher concentration. Active transport is energetically unfavorable and requires an input of energy. Two general types are found:

- **Primary active transport** involves the functioning of a **pump**—a type of transporter that directly uses energy to transport a solute against a concentration gradient. Figure 5.16a shows a pump that uses energy in the form of ATP to transport H^+ against a gradient. Such a pump establishes an H^+ electrochemical gradient across a membrane.
- **Secondary active transport** involves the use of a pre-existing gradient to drive the active transport of another

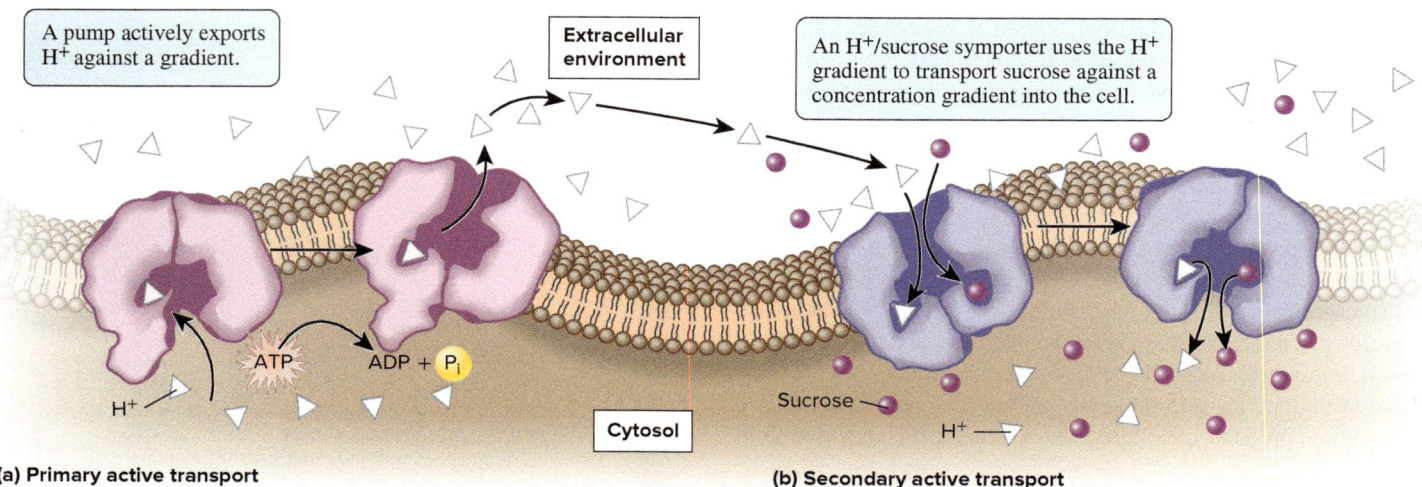

Figure 5.16 Types of active transport. (a) During primary active transport, a pump directly uses energy, in this case from ATP, to transport a solute against a concentration gradient. The pump shown here uses ATP to establish an H^+ electrochemical gradient. (b) Secondary active transport via a symporter involves the use of this gradient to drive the active transport of a solute, such as sucrose.

solute. For example, an H+/sucrose symporter uses an H+ electrochemical gradient, established by a pump, to move sucrose against its concentration gradient (**Figure 5.16b**). In this case, only sucrose is actively transported. Hydrogen ions move down their electrochemical gradient via the symporter.

Symporters enable cells to actively import nutrients against a gradient. These proteins use the energy stored in the electrochemical gradient of H+ or Na+ to power the uphill movement of organic solutes such as sugars, amino acids, and other necessary molecules. Therefore, with symporters in their plasma membrane, cells can scavenge nutrients from the extracellular environment and accumulate them to high levels within the cytoplasm. H+/solute symporters are more common in bacteria, fungi, algae, and plant cells, because H+ pumps are found in their plasma membranes. In animal cells, a pump that exports Na+ maintains a Na+ gradient across the plasma membrane. Na+/solute symporters are prevalent in animal cells.

ATP-Driven Ion Pumps Generate Ion Electrochemical Gradients

The phenomenon of active transport was discovered in the 1940s based on the study of the transport of sodium ions (Na+) and potassium ions (K+). In animal cells, the concentration of Na+ is lower inside the cell than outside, whereas the concentration of K+ is higher inside the cell than outside. After analyzing the movement of these ions across the plasma membranes of muscle cells, neurons, and red blood cells, researchers determined that the export of Na+ is coupled to the import of K+. In the late 1950s, Danish biochemist Jens Skou proposed that a single transporter is responsible for this phenomenon. He was the first to describe an ATP-driven ion pump, which was later named Na+/K+-ATPase. This pump can actively transport Na+ and K+ against their gradients by using the energy from ATP hydrolysis. The plasma membrane of a typical animal cell contains thousands of Na+/K+-ATPase pumps that maintain large concentration gradients in which the concentration of Na+ is higher outside the cell and the concentration of K+ is higher inside the cell.

Let's take a closer look at the Na+/K+-ATPase that Skou discovered. Every time one ATP is hydrolyzed, the Na+/K+-ATPase functions as an antiporter that pumps three Na+ into the extracellular environment and two K+ into the cytosol (**Figure 5.17a**). Because one cycle of pumping results in the net export of one positive charge, the Na+/K+-ATPase also produces an electrical gradient across the membrane. Because it generates an electrical gradient, it is called an **electrogenic pump.**

By studying the interactions of Na+, K+, and ATP with Na+/K+-ATPase, researchers have pieced together a molecular road map of the steps that direct the pumping of ions across the membrane (**Figure 5.17b**). Na+/K+-ATPase can alternate between two conformations, designated E1 and E2. In E1, the ion-binding sites are accessible from the cytosol—Na+ binds tightly to this conformation, whereas K+ has a low affinity. In E2, the ion-binding sites are accessible from the extracellular environment—Na+ has a low affinity, and K+ binds tightly.

The pumping mechanism of Na+/K+-ATPase occurs as follows (see Figure 5.17b):

1. Three Na+ bind to the E1 conformation from the cytosol. ATP is hydrolyzed to ADP and phosphate. Temporarily, the phosphate is covalently bound to the pump, an event called phosphorylation. The pump switches to the E2 conformation.
2. The three Na+ are released into the extracellular environment, because they have a lower affinity for the E2 conformation.
3. Two K+ bind from the outside.
4. The binding of two K+ causes the release of phosphate. The pump switches back to the E1 conformation. Because the E1 conformation has a low affinity for K+, the two K+ are released into the cytosol. Na+/K+-ATPase is now ready for another round of pumping.

Na+/K+-ATPase is a critical ion pump in animal cells because it maintains Na+ and K+ gradients across the plasma membrane. Many other types of ion pumps are also found in the plasma membrane and in organellar membranes. Ion pumps play the primary role in the formation and maintenance of ion electrochemical gradients that drive many important cellular processes (**Table 5.2**). ATP is commonly the source of energy that drives ion pumps, and cells typically use a substantial portion of their ATP to keep these pumps working. For example, neurons use up to 70% of their ATP to operate ion pumps! As discussed in Chapter 33, neurons need to maintain Na+ and K+ gradients so that they can conduct signals known as action potentials.

Table 5.2	Important Functions of Ion Electrochemical Gradients
Function	**Description**
Transport of ions and molecules	Symporters and antiporters use H+ and Na+ gradients to take up nutrients and export waste products (see Figure 5.16).
Production of energy intermediates	In the mitochondrion and chloroplast, H+ gradients are used to synthesize ATP.
Osmotic regulation	Animal cells control their internal volume by regulating ion gradients between the cytosol and extracellular fluid.
Neuronal signaling	Na+ and K+ gradients are involved in conducting action potentials, the signals transmitted by neurons.
Muscle contraction	Ca^{2+} gradients regulate the ability of muscle fibers to contract.
Bacterial swimming	H+ gradients drive the rotation of bacterial flagella.

5.4 Reviewing the Concepts

- Two classes of proteins that carry out membrane transport are channels and transporters.
- Channels form an open passageway for the facilitated diffusion of solutes across the plasma membrane. Most channels are gated, which allows cells to regulate the movement of solutes (Figure 5.12).
- An example of a channel is aquaporin, which allows the movement of water (Figure 5.13).

Biology Principle

Biologists Use Models to Test Predictions and Convey Ideas

This figure is a model for the transport of Na+ and K+ ions by an ion pump. The Modeling Challenge in the figure legend asks you to propose a model for a different ion pump that is found in the cells that line your stomach.

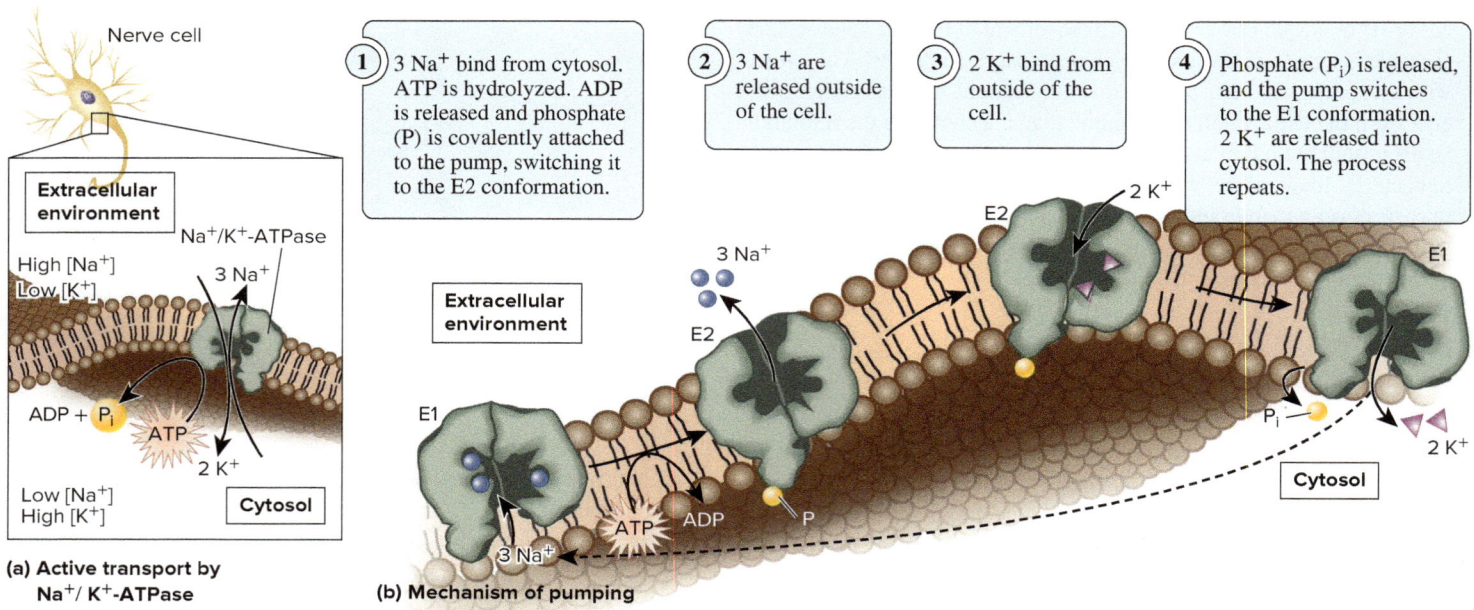

Figure 5.17 Structure and function of Na+/K+-ATPase. (a) Active transport by Na+/K+-ATPase. Each time this protein hydrolyzes one ATP molecule, it pumps out three Na+ and pumps in two K+. (b) Pumping mechanism. This figure illustrates the protein conformational changes between E1 and E2. As this occurs, ATP is hydrolyzed to ADP and phosphate. During the process, phosphate is covalently attached to the protein but is released after two K+ bind.

Modeling Challenge: The epithelial cells that line your stomach contain a gastric ATPase in their plasma membrane that is evolutionarily related to the Na+/K+-ATPase. The gastric ATPase pumps hydronium ions (H_3O^+) from the cytosol of the epithelial cells into the lumen of your stomach. The hydronium ions then dissociate into H_2O and H^+. The formation of H^+ acidifies the stomach lumen. Like the Na+/K+-ATPase, the gastric ATPase is an antiporter. It antiports a hydronium ion with a K+ ion. Each time this protein hydrolyzes one ATP molecule, it pumps one hydronium ion into the stomach lumen and one K+ into the cytosol of an epithelial cell. Draw a model for the mechanism of pumping by the gastric ATPase, using the same format as the mechanism shown in Figure 5.17b. Is the gastric ATPase an electrogenic pump?

- Transporters, which tend to function at a slower rate than channels, bind their solutes in a hydrophilic pocket and undergo a conformational change that switches the exposure of the pocket to the other side of the membrane. They can be uniporters, symporters, or antiporters (Figures 5.14, 5.15).
- Primary active transport involves pumps that directly use energy to generate a solute gradient. Secondary active transport uses a pre-existing gradient (Figure 5.16).
- Na+/K+-ATPase is an electrogenic pump that uses energy from ATP to transport ions across the membrane. Ion electrochemical gradients perform several important cellular functions (Figure 5.17, Table 5.2).

5.4 Testing Your Knowledge

1. Which of the following is *not* a characteristic of channels?
   a. They may be gated.
   b. They form an open passageway for the movement of solutes.
   c. They are usually faster than transporters.
   d. They bind their solutes in a hydrophilic pocket, much like an enzyme.
   e. They are transmembrane proteins.

2. Let's suppose Na+/K+-ATPase was in an environment in which Na+ was inside the cell, ATP was inside the cell, but no K+ was

outside the cell. Under these conditions, Na+/K+-ATPase would be stuck in which conformation?
a. E1 with phosphate covalently bound
b. E2 with phosphate covalently bound
c. E1 without phosphate covalently bound
d. E2 without phosphate covalently bound
e. either a or c

5.5 Intercellular Channels

Learning Outcome

1. Outline the structure and function of gap junctions and plasmodesmata.

Thus far, we have seen that cells produce two general types of transmembrane proteins, channels and transporters, that allow the transport of solutes across the plasma membrane. Some transport solutes into the cell and others facilitate the export of solutes. The cells of multicellular organisms also may have **intercellular channels** that allow the direct movement of substances between adjacent cells. In this section, we will examine gap junctions that often occur between adjacent animal cells and plasmodesmata that occur between plant cells.

Gap Junctions Between Animal Cells Provide Passageways for Intercellular Transport

A type of junction found between animal cells is called a **gap junction**, because a small gap occurs between the plasma membranes of cells connected by these junctions (Figure 5.18). Gap junctions are abundant in tissues and organs where cells need to communicate with each other. For example, cardiac muscle cells, which cause your heart to beat, are interconnected by many gap junctions. Because gap junctions allow the passage of ions, electrical changes in one cardiac muscle cell are easily transmitted to an adjacent cell that is connected via gap junctions. These connections are needed for the coordinated contraction of cardiac muscle cells.

In vertebrates, gap junctions are composed of many integral membrane proteins called connexins. Invertebrates have a structurally similar protein called innexin. Six connexin proteins in one vertebrate cell form a channel called a **connexon**. A connexon in one cell aligns with a connexon in an adjacent cell to form an intercellular channel (see the middle panel in Figure 5.18). The term gap junction refers to a cluster of many connexons that are close to each other in the plasma membrane and form many intercellular channels.

Gap junction channels allow the passage of ions and small molecules, including amino acids, sugars, and signaling molecules such as Ca^{2+} and cAMP, between cells. In this way, gap junctions allow adjacent cells to share metabolites and directly signal each other. However, gap junction channels are too small to allow the passage of RNA, proteins, or polysaccharides. Therefore, cells that communicate via gap junctions still maintain their own distinctive sets of macromolecules.

Plasmodesmata Are Channels Connecting the Cytoplasm of Adjacent Plant Cells

In 1879, Eduard Tangl, a Russian botanist, observed intercellular connections in the seeds of the strychnine tree and hypothesized that the cytoplasm of adjacent cells is connected by ducts in the cell walls. He was the first to propose that direct cell-to-cell communication integrates the functioning of plant cells. The ducts or intercellular channels that Tangl observed in plant cells are now known as **plasmodesmata** (singular, plasmodesma).

Plasmodesmata are functionally similar to gap junctions in animal cells, because they are open pores that allow the passage of ions and molecules between the cytosols of adjacent plant cells. However, the structure of the plasmodesmata is quite different from that of gap junctions. As shown in Figure 5.19, the plasma membrane of one cell is continuous with the plasma membrane of the adjacent cell, which forms a pore that permits the diffusion of molecules from the cytosol of one cell to the cytosol of the other. In addition to a cytosolic connection, plasmodesmata also have a central tubule, called a desmotubule, connecting the smooth ER membranes of adjacent cells.

Figure 5.18 Gap junctions between adjacent cells that line the intestine. Gap junctions form intercellular channels that allow the passage of small solutes with masses less than 1,000 Da. One connexon consists of six proteins called connexins. Two connexons align to form an intercellular channel. The micrograph shows a gap junction, which is composed of many connexons, between intestinal cells.
(photo) Courtesy Dr. Dan Goodenough/Harvard Medical School

Figure 5.19 Structure of a plasmodesma. Plasmodesmata are cell junctions connecting the cytosols of adjacent plant cells, allowing water, ions, and molecules to pass from cell to cell. At these pores, the plasma membrane of one cell is continuous with the plasma membrane of an adjacent cell. In addition, the smooth ER from one cell is connected to that of the adjacent cell via a desmotubule.
©Biophoto Associates/Science Source

BioConnections: *Look ahead to Figure 30.17. How do plasmodesmata play a role in the movement of nutrients through a plant root?*

Plant cells can alter the diameter of the channel formed by plasmodesmata. The channel can occur in closed, open, and dilated states. In the open state, plasmodesmata allow the passage of ions and small molecules, such as sugars and ATP. In this state, plasmodesmata play a similar role to gap junctions between animal cells. Plasmodesmata tend to close when a plant is wounded. The closure of plasmodesmata between adjacent cells helps to prevent the loss of water and nutrients from the wound site.

Unlike gap junctions between animal cells, plasmodesmata can dilate to allow the passage of macromolecules and even viruses between adjacent plant cells. Though the mechanism of dilation is not well understood, the wider opening of plasmodesmata is important for the passage of proteins and mRNA during plant development.

5.5 Reviewing the Concepts

- Gap junctions are intercellular channels composed of connexons, which permit the direct passage of ions and small molecules between adjacent animal cells (Figure 5.18).
- The plasma membranes and ER of adjacent plant cells are connected via plasmodesmata, which allow the passage of ions and molecules between the cytosols of adjacent plant cells (Figure 5.19).

5.5 Testing Your Knowledge

1. Which of the following is *not* a characteristic of gap junctions?
   a. They allow the passage of small molecules between adjacent cells.
   b. They are composed of membrane proteins called connexins, which form connexons.
   c. They are found between adjacent plant cells.
   d. They are found between adjacent animal cells.
   e. All of the above are correct.

5.6 Exocytosis and Endocytosis

Learning Outcomes

1. Describe the steps in exocytosis and endocytosis.
2. Compare and contrast three types of endocytosis.

We have seen that most small substances are transported via transmembrane proteins such as channels and transporters, which provide passageways for the movement of ions and molecules directly across the membrane. Eukaryotic cells have two other mechanisms, exocytosis and endocytosis, for transporting larger molecules, such as proteins and polysaccharides, and even very large particles. Both mechanisms involve the packaging of the transported substance, sometimes called the cargo, into a membrane vesicle or vacuole. Table 5.3 describes some examples.

Exocytosis During **exocytosis,** material inside the cell is packaged into vesicles and then excreted into the extracellular environment (Figure 5.20). These vesicles are usually derived from the Golgi apparatus. As a vesicles forms, a specific cargo is loaded into the interior, and a protein coat forms around the emerging vesicle. The assembly of the proteins to make a coat on the surface of the Golgi membrane causes the bud to form. Eventually, the bud separates from the membrane to form a vesicle. After the vesicle is released, the coat is shed. Finally, the vesicle fuses with the plasma membrane and releases the cargo into the extracellular environment.

Endocytosis During **endocytosis,** the plasma membrane invaginates, or folds inward, to form a vesicle that brings substances into the cell. Three types of endocytosis are receptor-mediated endocytosis, pinocytosis, and phagocytosis.

Table 5.3	Examples of Exocytosis and Endocytosis
Exocytosis	**Description**
Hormones	Certain hormones, such as insulin, are composed of polypeptides. To exert its effect, insulin is secreted into the bloodstream via exocytosis from cells of the pancreas.
Digestive enzymes	Digestive enzymes that function in the lumen of the small intestine are secreted via exocytosis from cells of the pancreas.
Endocytosis	**Description**
Uptake of vital nutrients	Many important nutrients are highly insoluble in the blood. Therefore, they are bound to proteins in the blood and then taken into cells via endocytosis. Examples include the uptake of lipids (bound to low-density lipoprotein) and iron (bound to transferrin protein).
Root nodules	Nitrogen-fixing root nodules found in certain species of plants, such as legumes, are formed by the endocytosis of bacteria. After being taken up, the bacterial cells are contained within a membrane-enclosed compartment in the root nodules.
Immune system	Cells of the immune system, known as macrophages, engulf and destroy bacteria via phagocytosis.

Figure 5.20 Exocytosis.

 What is the function of the protein coat?

- In **receptor-mediated endocytosis,** a receptor in the plasma membrane is specific for a given cargo (**Figure 5.21**). Receptor-mediated endocytosis allows cells to take up cargo that may not be very concentrated in the extracellular environment, such as cholesterol. Cargo molecules binding to their specific receptors stimulate many receptors to aggregate, and then the coat proteins bind to the membrane. The protein coat causes the membrane to invaginate and form a vesicle. Once it is released into the cell, the vesicle sheds its coat. In most cases, the vesicle fuses with an internal organelle, such as a lysosome, and the receptor releases its cargo. Then, the cargo may be directly released into the cytosol, or it may be digested into simpler building blocks before release.

- **Pinocytosis** (from the Greek, meaning cell-drinking) involves the formation of membrane vesicles from the plasma membrane

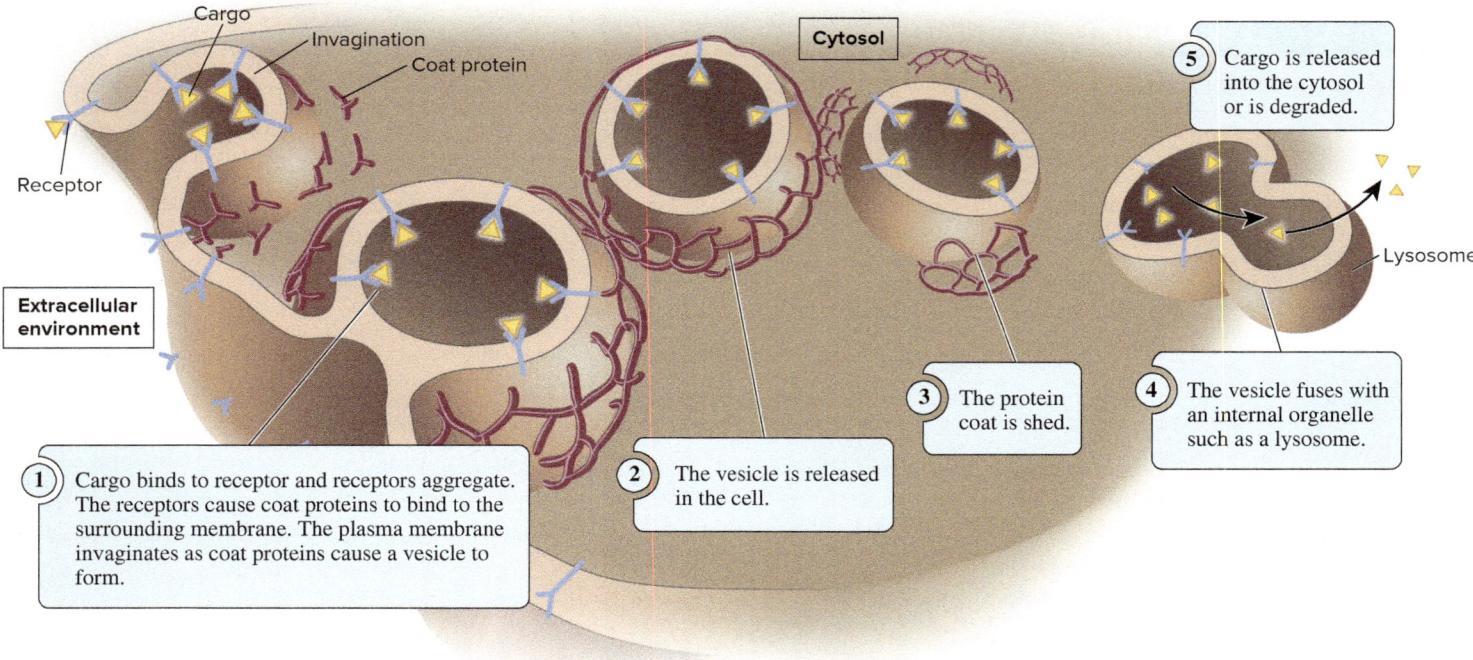

Figure 5.21 Receptor-mediated endocytosis.

as a way for cells to internalize the extracellular fluid. This allows cells to sample the extracellular solutes. Pinocytosis is particularly important in cells that are actively involved in nutrient absorption, such as cells that line the intestine in animals.

- **Phagocytosis** (from the Greek, meaning cell-eating) involves the formation of an enormous membrane vesicle called a phagosome, which engulfs a large particle such as a bacterium. Only certain kinds of cells, termed phagocytes, can carry out phagocytosis. For example, macrophages, which are cells of the immune system in mammals, kill bacteria via phagocytosis. Macrophages engulf bacterial cells into phagosomes. Once inside the cell, the phagosome fuses with a lysosome, and the digestive enzymes within the lysosome destroy the bacterium.

5.6 Reviewing the Concepts

- In eukaryotes, exocytosis and endocytosis are used to transport large molecules and particles. Exocytosis is a process in which material inside the cell is packaged into vesicles and excreted into the extracellular environment. During endocytosis, the plasma membrane folds inward to form a vesicle that brings substances into the cell. Receptor-mediated endocytosis, pinocytosis, and phagocytosis are types of endocytosis (Figures 5.20, 5.21; Table 5.3).

5.6 Testing Your Knowledge

1. The form of endocytosis that involves the formation of an enormous intracellular vesicle is called
   a. receptor-mediated endocytosis.
   b. pinocytosis.
   c. phagocytosis.
   d. all of the above.
   e. both a and b.

5.7 Cell Junctions

Learning Outcome

1. Outline the structure and function of anchoring junctions and tight junctions.

To become a multicellular organism, cells within the body must be linked to each other. In Section 5.5, we considered junctions that form intercellular channels that allow the movement of solutes between adjacent cells. In this section, we will explore junctions that function to physically adhere animal cells to each other or to the extracellular matrix (ECM).

Anchoring Junctions Link Animal Cells to Each Other and to the Extracellular Matrix (ECM)

Electron microscopy allows researchers to explore the types of junctions that occur between adjacent cells and between cells and the extracellular matrix (ECM), collectively called **anchoring junctions.** These junctions are particularly common in parts of the body where the cells are tightly connected and form linings. An example is the layer of cells that lines the small intestine. Anchoring junctions keep intestinal cells tightly adhered to one another, thereby forming a strong barrier between the lumen of the intestine and the blood. Key components of anchoring junctions that form the actual connections are integral membrane proteins called **cell adhesion molecules (CAMs).** Two types of CAMs are cadherins and integrins.

Anchoring junctions are grouped into four main categories, according to their functional roles and their connections to cellular components. Figure 5.22 shows these junctions between cells of the mammalian small intestine.

- **Adherens junctions** connect cells to each other via cadherins. In many cases, these junctions are organized into bands around cells. On the cytosolic side of the plasma membrane, adherens junctions bind to cytoskeletal filaments called actin filaments, described in Chapter 4 (refer back to Table 4.1).

- **Desmosomes** also connect cells to each other via cadherins. They are spotlike points of intercellular contact that rivet cells together. Desmosomes are connected to cytoskeletal filaments called intermediate filaments, described in Chapter 4 (refer back to Table 4.1).

- **Hemidesmosomes** connect cells to the ECM via integrins. Like desmosomes, they interact with intermediate filaments.

- **Focal adhesions** also connect cells to the ECM via integrins. On the cytosolic side of the plasma membrane, focal adhesions bind to actin filaments.

Let's now consider the molecular components of anchoring junctions. As mentioned, **cadherins** are CAMs that create cell-to-cell junctions. Two cadherin proteins, each in adjacent cells, bind to each other to promote cell-to-cell adhesion (see Figure 5.22). This binding requires the presence of calcium ions (Ca^{2+}), which change the conformation of the cadherin protein so that binding occurs. (This calcium dependence is where cadherin gets its name—$\underline{Ca}^{2+}$-dependent a$\underline{dhering}$ molecule.) The two identical cadherin proteins form what is called a dimer. On the interior of the cell, linker proteins connect cadherins to actin or intermediate filaments of the cytoskeleton. This promotes a more stable interaction between two cells, because their strong cytoskeletons are connected to each other.

Integrins, a group of cell-surface receptor proteins, are a second type of CAM, one that creates connections between cells and the ECM. Integrins do not require Ca^{2+} to function. Like cadherins, integrins also bind to actin or intermediate filaments on the cytosolic side of the plasma membrane, via linker proteins, to promote a strong association between the cytoskeleton and the ECM.

When CAMs were first discovered, researchers imagined that cadherins and integrins played only a mechanical role. In other words, their functions were described as holding cells together or to the ECM. More recently, however, experiments have shown that cadherins and integrins are also important in cell communication. The

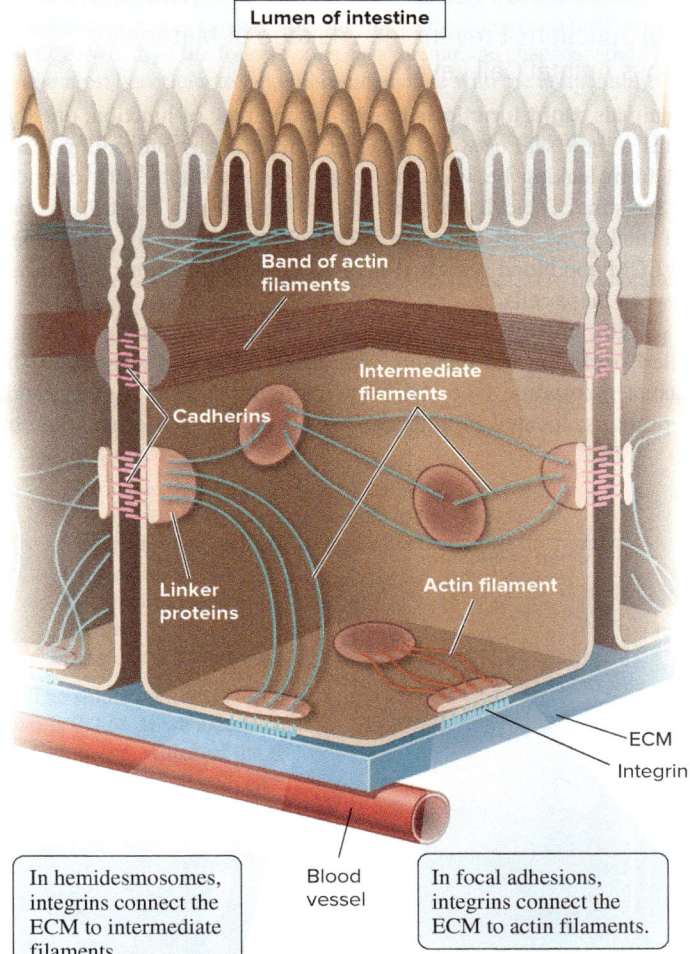

Figure 5.22 Types of anchoring junctions. This figure shows anchoring junctions in three adjacent intestinal cells. The tops of these cells face the lumen of the intestine, whereas the bottoms are adjacent to the ECM and face blood vessels.

Concept Check: *Which junctions are cell-to-cell junctions, and which are cell-to-ECM junctions?*

formation or breaking of cell-to-cell and cell-to-ECM anchoring junctions affects signal transduction pathways within the cell. Similarly, intracellular signal transduction pathways affect cadherins and integrins in ways that alter intercellular junctions and the binding of cells to ECM components.

Abnormalities in CAMs such as integrins are associated with the ability of cancer cells to metastasize, that is, to move to other parts of the body. CAMs are critical for keeping cells in their correct locations. When these adhesion molecules become defective

5. Researchers have identified a drug that prevents coat proteins in eukaryotic cells from binding to each other. If you added the drug to a sample of cells, which of the following steps in receptor-mediated endocytosis or exocytosis would be inhibited?
   a. the formation of vesicles at the plasma membrane during receptor-mediated endocytosis
   b. the uncoating of vesicles during receptor-mediated endocytosis
   c. fusion between a vesicle and a lysosome during receptor-mediated endocytosis
   d. the formation of vesicles at the Golgi membrane during exocytosis
   e. Both a and d would be inhibited.

Short Answer

1. Proteins in the plasma membrane are often the target of medicines. Discuss why you think this is the case. How would you determine experimentally that a specific membrane protein was the target of a drug?

2. Rank the following substances according to their rate of simple diffusion through an artificial phospholipid bilayer (from fastest to slowest): lactose (a disaccharide); glucose; starch; arginine; methane (CH_4); ethane (C_2H_6); and Na^+.

How Cells Utilize Energy

6

©wavebreakmedia/Shutterstock

Exercise and mitochondria. Regular exercise increases the number of mitochondria in muscle cells to meet the higher energy demands.

Chapter Outline

- 6.1 Energy and Chemical Reactions
- 6.2 Enzymes
- 6.3 Overview of Metabolism and Cellular Respiration
- 6.4 Glycolysis
- 6.5 Breakdown of Pyruvate
- 6.6 Citric Acid Cycle
- 6.7 Oxidative Phosphorylation
- 6.8 Connections Among Carbohydrate, Protein, and Fat Metabolism
- 6.9 Anaerobic Respiration and Fermentation

Assessing Your Knowledge and Skills

The human body adapts to environmental and behavioral changes. Many research studies have shown that regular exercise increases the number of mitochondria in muscle cells. Why is this change beneficial? During exercise, muscle cells need more energy. Mitochondria play a key role in the production of ATP, which is used to power a variety of processes, including muscle contraction. By increasing the number of mitochondria, muscle cells have a greater capacity to generate ATP. Exercise is a powerful inducer of mitochondrial biogenesis. Recent research has even shown that exercise may increase the number of mitochondria in brain cells. Therefore, exercise may not only enhance muscle function but also improve brain function.

We will begin this chapter with a discussion of energy and how it is used by living organisms to power a variety of processes. Much of our discussion will center on **chemical reactions,** which are processes in which one or more substances are changed into other substances (see Chapter 2). Every living cell continuously performs thousands of chemical reactions to sustain life. We will examine what factors control the direction of a chemical reaction and what determines its rate, paying particular attention to the role of enzymes.

Metabolism is the sum total of all chemical reactions that occur within a cell or an organism. Metabolism also refers to a specific set of chemical reactions occurring at the cellular level. For example, biologists may speak of sugar metabolism or fat metabolism. In this chapter, we will consider metabolism at the cellular level, examining how chemical reactions are often coordinated with each other in metabolic pathways. We will largely focus on a group of chemical reactions that carries out the breakdown of the main carbohydrate that most cells use as an energy source, the sugar glucose. As you will learn, mitochondria play a key role in this process.

6.1 Energy and Chemical Reactions

Learning Outcomes

1. Define energy, and distinguish between potential and kinetic energy.
2. State the first and second laws of thermodynamics, and discuss how they relate to living things.
3. Explain how the change in free energy determines the direction of a chemical reaction and how chemical reactions eventually reach a state of equilibrium.
4. Distinguish between exergonic and endergonic reactions.
5. Describe how cells use the energy released by the hydrolysis of ATP to drive endergonic reactions.

(a) Kinetic energy

(b) Potential energy

Figure 6.1 Examples of energy. **(a)** Kinetic energy, such as that possessed by a swinging bat, is energy associated with motion. **(b)** Potential energy is stored energy, as in a bow that is ready to shoot an arrow.
(a) ©sirtravelalot/Shutterstock; *(b)* ©B2M Productions/Getty Images

Two general factors govern the fate of a given chemical reaction in a living cell: its direction and rate. In this section, we will begin by examining the interplay of energy and the concentration of reactants as they govern the direction of a chemical reaction. You will learn that cells use energy-intermediate molecules, such as ATP, to drive chemical reactions in a desired direction.

Energy Exists in Different Forms

To understand why a chemical reaction occurs, we first need to consider **energy,** which is the ability to promote change or do work. Physicists often consider energy in two general forms: kinetic energy and potential energy (**Figure 6.1**). **Kinetic energy** is energy associated with movement, such as the movement of a baseball bat from one location to another. By comparison, **potential energy** is the energy that a substance or an object possesses due to its structure or location, such as the location of an arrow when a bow is drawn. The energy that is stored in the bonds between atoms is a type of potential energy called **chemical potential energy.** The breaking of bonds is one way that energy can be released, which is then used by living cells to perform various functions. **Table 6.1** summarizes various forms of energy that are common in biological systems.

An important phenomenon in biology is the ability of energy to be converted from one form to another. The study of energy transfers and transformations is called **thermodynamics.** Physicists have determined that four laws govern thermodynamics, the following two of which we'll study here:

1. **First law of thermodynamics.** Energy cannot be created or destroyed. However, energy can be transferred from one place to another and can be transformed from one type to another (as when, for example, chemical energy is transformed into heat).

Table 6.1	Types of Energy That Are Important in Biology	
Energy type	**Description**	**Biological example**
Light	Light is a form of electromagnetic radiation that is visible to the eye. The energy of light is packaged in photons.	During photosynthesis, light energy is captured by pigments in chloroplasts (described in Chapter 7). Ultimately, this energy is used to produce organic molecules.
Heat	Heat is the transfer of kinetic energy from one object to another or from an energy source to an object. In biology, heat is often viewed as kinetic energy that can be transferred due to a difference in temperature between two objects or locations.	Many organisms, including humans, maintain their bodies at a constant temperature. This is achieved, in part, by chemical reactions that generate heat.
Mechanical	Mechanical energy is the energy possessed by an object due to its motion or its position relative to other objects.	In animals, mechanical energy is associated with movement due to muscle contraction, such as walking.
Chemical potential	Chemical potential energy is potential energy stored in the electrons of molecules. When bonds are broken and rearranged, energy may be released.	The covalent bonds in organic molecules, such as glucose and ATP, store large amounts of energy. When bonds are broken in larger molecules to form smaller molecules, the chemical energy that is released can be used to drive cellular processes.
Electrical/ion gradient	The movement of charge or the separation of charges can provide energy. Also, a difference in ion concentration across a membrane constitutes an electrochemical gradient, which is a source of potential energy.	During a stage of cellular respiration called oxidative phosphorylation (described in this chapter), an H^+ gradient provides the energy to drive ATP synthesis.

Figure 6.2 Entropy, a measure of the disorder of a system. An increase in entropy means an increase in disorder.

Concept Check: Which do you think has more entropy, a NaCl crystal at the bottom of a beaker of water or the solution that would be formed after the Na^+ and Cl^- forming the crystal have dissolved in the water?

2. **Second law of thermodynamics.** Any energy transfer or transformation from one form to another increases the degree of disorder of a system, called **entropy** (Figure 6.2). Entropy is a measure of the randomness of molecules in a system. When a physical system becomes more disordered, the entropy increases. As the energy becomes more evenly distributed, that energy is less able to promote change or do work.

The Change in Free Energy Determines the Direction of a Chemical Reaction

Energy is required for many cellular processes, including chemical reactions, cellular movements such as those occurring in muscle contraction, and the maintenance of cell organization. To understand how organisms use energy, we need to distinguish between the energy that can be used to promote change or do work (usable energy) and the energy that cannot (unusable energy):

Total energy = Usable energy + Unusable energy

Why is some energy unusable? The main culprit is entropy. As stated by the second law of thermodynamics, energy transfers or transformations involve an increase in entropy, a degree of disorder that cannot be harnessed in a useful way. The total energy of a system is termed **enthalpy (H)**, and the usable energy—the amount of energy that is available and can be used to promote change or do work—is called the **free energy (G)**. The use of the letter G is in recognition of American physicist J. Willard Gibbs, who proposed the concept of free energy in 1878. The unusable energy is the system's entropy (S). Gibbs proposed that these three components of a system's energy are related to each other in the following way:

$$H = G + TS$$

where T is the absolute temperature in kelvins (K). Because our focus is on free energy, we can rearrange this equation as

$$G = H - TS$$

A critical issue in biology is whether a process does or does not occur spontaneously—for example, will glucose be broken down into carbon dioxide and water? Another way of framing this question is to ask "Is the breakdown of glucose a spontaneous, or favorable, reaction?" The key way to evaluate if a chemical reaction is spontaneous is to determine the free-energy change that occurs as a result of the reaction:

$$\Delta G = \Delta H - T\Delta S$$

where the symbol Δ (the Greek letter delta) indicates a change, such as before and after a chemical reaction. If a chemical reaction has a negative free-energy change ($\Delta G < 0$), the products have less free energy than the reactants, and therefore free energy is released during product formation. Such a reaction is said to be **exergonic**. Exergonic reactions are spontaneous. Alternatively, if a reaction has a positive free-energy change ($\Delta G > 0$), requiring the addition of free energy, it is termed **endergonic**. An endergonic reaction is not a spontaneous reaction.

If ΔG for a chemical reaction is negative, the reaction favors the conversion of reactants to products, whereas a reaction with a positive ΔG favors the formation of reactants. Chemists have determined free-energy changes for a variety of chemical reactions, which allows them to predict the direction of a reaction. As an example, let's consider **adenosine triphosphate (ATP)**, which is a molecule commonly used to drive cellular processes. ATP is broken down to adenosine diphosphate (ADP) and inorganic phosphate (HPO_4^{2-}, abbreviated P_i). Because water is used to remove a phosphate group, chemists refer to this reaction as the hydrolysis of ATP (Figure 6.3). For the conversion of 1 mole of ATP to 1 mole of ADP and P_i, ΔG equals –7.3 kcal/mol. Because this is a negative value, the formation of products is strongly favored.

Chemical Reactions Eventually Reach a State of Equilibrium

Even when a chemical reaction is associated with a negative free-energy change, not all of the reactants are converted to products. The reaction reaches a state of **chemical equilibrium** in which the rate of formation of products equals the rate of formation of reactants. Let's consider the generalized reaction

$$aA + bB \rightleftharpoons cC + dD$$

where A and B are the reactants, C and D are the products, and a, b, c, and d are the numbers of moles of reactants and products. The reaction reaches equilibrium, such that

$$K_{eq} = \frac{[C]^c[D]^d}{[A]^a[B]^b}$$

where K_{eq} is the equilibrium constant. Each type of chemical reaction has a specific value for K_{eq}. When K_{eq} is greater than 1, the reaction favors the formation of products; when it is less than 1, the reaction favors the formation of reactants.

Biologists make two simplifying assumptions when determining values for equilibrium constants. First, the concentration of water does not change during the reaction, and the pH remains constant at pH 7. The equilibrium constant under these conditions is designated

Biology Principle

Living Organisms Use Energy

The hydrolysis of ATP is an exergonic reaction that is used to drive many cellular reactions.

$\Delta G = -7.3$ kcal/mol

Figure 6.3 The hydrolysis of ATP to ADP and P_i. As shown in this figure, ATP has a net charge of −4, and ADP and P_i are shown with net charges of −2 each. When these compounds are shown in chemical reactions with other molecules, the net charges are indicated. Otherwise, these compounds are simply designated ATP, ADP, and P_i. At neutral pH, ADP^{2-} dissociates to ADP^{3-} and H^+.

Cells Use ATP to Drive Endergonic Reactions

In living organisms, many processes require the addition of free energy; that is, they are endergonic and do not occur spontaneously. How is this problem overcome? One strategy is to couple exergonic reactions with endergonic reactions. If an exergonic reaction is coupled with an endergonic reaction, the endergonic reaction will proceed spontaneously if the net free-energy change for both processes combined is negative.

As discussed later in this chapter, the breakdown of sugars, such as glucose, is a vital process for living organisms to utilize energy. For example, glucose is a key energy source for your brain. Certain steps in glucose metabolism are endergonic and are coupled to ATP hydrolysis. An example is described next:

$$\text{Glucose} + \text{Phosphate}^{2-} \rightarrow \text{Glucose-6-phosphate}^{2-} + H_2O$$
$$\Delta G = +3.3 \text{ kcal/mol}$$

$$ATP^{4-} + H_2O \rightarrow ADP^{2-} + P_i^{2-} \quad \Delta G = -7.3 \text{ kcal/mol}$$

Coupled reaction:

$$\text{Glucose} + ATP^{4-} \rightarrow \text{Glucose-6-phosphate}^{2-} + ADP^{2-}$$
$$\Delta G = -4.0 \text{ kcal/mol}$$

The first reaction, in which phosphate is covalently attached to glucose, is endergonic and by itself is not spontaneous. The second reaction, the hydrolysis of ATP, is exergonic. If the two reactions are coupled, the combined net free-energy change for both is negative ($\Delta G = -4.0$ kcal/mol); the coupled reaction is exergonic. In the coupled reaction, a phosphate is directly transferred from ATP to glucose, in a process called **phosphorylation.** This coupled reaction proceeds spontaneously because the net free-energy change is negative. Exergonic reactions, such as the breakdown of ATP, are commonly coupled to chemical reactions and other cellular processes that would otherwise be endergonic.

6.1 Reviewing the Concepts

- The fate of a chemical reaction is determined by its direction and rate.
- Energy, the ability to promote change or do work, exists in many forms. According to the first law of thermodynamics, energy cannot be created or destroyed, but it can be converted from one form to another. The second law of thermodynamics states that energy transfers or transformations involve an increase in entropy (Figures 6.1, 6.2; Table 6.1).
- Free energy is the amount of available energy that can be used to promote change or do work. Spontaneous or exergonic reactions, which release free energy, have a negative free-energy change, whereas endergonic reactions have a positive free-energy change (Figure 6.3).
- Chemical reactions proceed until they reach a state of chemical equilibrium, where the rate of formation of products equals the rate of formation of reactants.
- Cells couple exergonic reactions, such as the hydrolysis of ATP, to endergonic reactions, with the result that endergonic reactions proceed spontaneously.

---

with a prime symbol, K_{eq}'. If water is one of the reactants, as in a hydrolysis reaction, it is not included in the chemical equilibrium equation. As an example, let's consider the chemical equilibrium for the hydrolysis of ATP:

$$ATP^{4-} + H_2O \rightleftharpoons ADP^{2-} + P_i^{2-}$$

$$K_{eq}' = \frac{[ADP][P_i]}{[ATP]}$$

Experimentally, the value for K_{eq}' for this reaction has been determined to be approximately 1,650,000 M. Such a large value indicates that the equilibrium greatly favors the formation of products—ADP and P_i.

6.1 Testing Your Knowledge

1. When a reaction or process is exergonic,
   a. the free-energy change is negative.
   b. it is spontaneous.
   c. it does not need added energy to proceed.
   d. the entropy usually increases.
   e. all of the above are correct.
2. Cells often use ATP hydrolysis to
   a. decrease their water content.
   b. drive endergonic reactions.
   c. prevent exergonic reactions.
   d. achieve chemical equilibrium.
   e. do all of the above.

6.2 Enzymes

Learning Outcomes

1. Explain how enzymes increase the rates of chemical reactions by lowering the activation energy.
2. Describe how enzymes bind their substrates with high specificity and undergo induced fit.
3. **SCISKILLS** ▶ Analyze the velocity of chemical reactions, and evaluate the effects of competitive and noncompetitive inhibitors.
4. Explain how factors, such as nonprotein molecules or ions, temperature, and pH, influence enzyme activity.

For most chemical reactions in cells to proceed at a rapid pace, a catalyst is needed. A **catalyst** is an agent that speeds up the rate of a chemical reaction without being permanently changed or consumed by it. In living cells, the most common catalysts are **enzymes,** which are proteins. The term was coined in 1876 by German physiologist Wilhelm Kühne, who discovered trypsin, an enzyme in pancreatic juice that is needed for the digestion of food proteins. In this section, we will explore how enzymes work.

Enzymes Increase the Rates of Chemical Reactions

If a chemical reaction has a negative free-energy change, the reaction will be spontaneous; it will tend to proceed in the direction of reactants to products. Although thermodynamics governs the direction of an energy transformation, it does not determine the rate of a chemical reaction. For example, the breakdown of the molecules in gasoline to smaller molecules is an exergonic reaction. Even so, we could place gasoline and oxygen in a container and nothing much would happen (provided it wasn't near a flame). If we came back several days later, we would expect to see the gasoline still sitting there. Perhaps if we came back in a few million years, the gasoline would have been broken down. On a timescale of months or a few years, however, the chemical reaction would proceed very slowly.

In living cells, the rates of enzyme-catalyzed reactions typically occur millions of times faster than the corresponding uncatalyzed reactions. A dramatic example involves the enzyme catalase, which catalyzes the breakdown of hydrogen peroxide (H_2O_2) into water and oxygen. Catalase speeds up this reaction 10^{15}-fold faster than the uncatalyzed reaction!

Why are catalysts necessary to speed up a chemical reaction? Chemical reactions between molecules involve bond breaking and bond forming. The process of either breaking or forming a covalent bond may begin with the straining or stretching of one or more bonds in the starting molecule(s), and/or it may involve the positioning of two molecules so that they interact with each other properly. Enzymes help to facilitate these types of events. As an example, let's consider the reaction in which ATP is used to phosphorylate glucose:

$$\text{Glucose} + \text{ATP}^{4-} \rightarrow \text{Glucose-6-phosphate}^{2-} + \text{ADP}^{2-}$$

For this reaction to occur between glucose and ATP, the molecules must collide in the correct orientation and possess enough energy so that the chemical bonds can be changed. As glucose and ATP get close together, the electrons in the outer shells of their atoms repel each other. To overcome this repulsion, an initial input of energy, called the **activation energy,** is required (**Figure 6.4**).

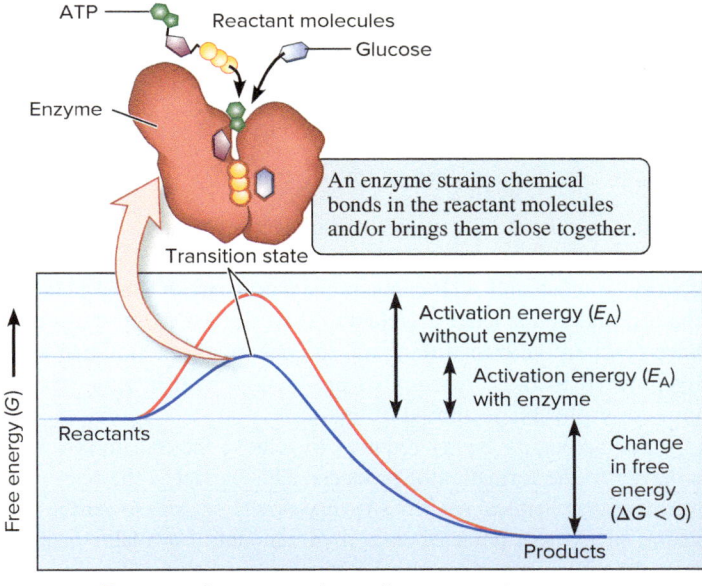

Figure 6.4 Activation energy of a chemical reaction. This figure depicts an exergonic reaction. The activation energy (E_A) is needed for molecules to achieve a transition state. One way that enzymes lower the E_A is by straining chemical bonds in the reactants so that less energy is required to attain the transition state. A second way is by binding two reactants so that they are close to each other and in a favorable orientation.

Concept Check: *How does lowering the activation energy affect the rate of a chemical reaction? How does it affect the direction?*

Biology Principle

Structure Determines Function

A key function of enzymes is their ability to bind their substrates with high specificity. This specificity is due to the structure of the enzyme's active site.

① Substrates (ATP and glucose) bind to the enzyme (hexokinase).

② Enzyme undergoes conformational change that binds the substrates more tightly. This induced fit strains chemical bonds within the substrates and/or brings them closer together.

③ Substrates are converted to products.

④ Products (ADP and glucose-6-phosphate) are released. Enzyme is ready to be reused.

Figure 6.5 The steps of an enzyme-catalyzed reaction. The example shown here involves the enzyme hexokinase, which binds glucose and ATP. The products are glucose-6-phosphate and ADP, which are released from the enzyme.

Concept Check: *During which step is the activation energy lowered?*

Activation energy (E_A) allows the molecules to get close enough to cause a rearrangement of bonds. With the input of activation energy, glucose and ATP can achieve a **transition state** in which the original bonds have stretched to their limit. Once the reactants have reached the transition state, the chemical reaction can readily proceed to the formation of products, which in this case are glucose-6-phosphate and ADP.

The activation energy required to achieve the transition state is a barrier to the formation of products. This barrier is the reason the rate of many chemical reactions is very slow. Enzymes lower the activation energy to a point where a small amount of available heat can push the reactants to a transition state.

How do enzymes lower the activation energy barrier of chemical reactions? The following are two common mechanisms:

- ***Straining reactants:*** Enzymes are large proteins that bind relatively small reactants. When bound to an enzyme, the bonds in the reactants can be strained, thereby making it easier for them to achieve the transition state (see Figure 6.4).
- ***Positioning reactants close together:*** When a chemical reaction involves two or more reactants, the enzyme provides sites in which the reactants are positioned very close to each other in an orientation that facilitates the formation of new covalent bonds.

Enzymes Recognize Their Substrates with High Specificity and Undergo Conformational Changes

Thus far, we have considered how enzymes lower the activation energy of a chemical reaction and thereby increase its rate. Let's consider some other features of enzymes that enable them to serve as effective catalysts in chemical reactions. The **active site** is the location in an enzyme where the chemical reaction takes place. The **substrates** for an enzyme are the reactant molecules that bind to an enzyme at the active site and participate in the chemical reaction. For example, hexokinase is an enzyme whose substrates are glucose and ATP (Figure 6.5). The binding between enzyme and substrate produces an **enzyme-substrate complex.**

A key feature of nearly all enzymes is their ability to bind their substrates with a high degree of specificity. For example, hexokinase recognizes glucose but does not recognize other, similar sugars such as fructose and galactose very well. In 1894, German chemist Emil Fischer proposed that the recognition of a substrate by an enzyme resembles the interaction between a lock and key: Only the correctly shaped key (the substrate) will fit into the keyhole (active site) of the lock (the enzyme). Further research revealed that the interaction between an enzyme and its substrates also involves movements or conformational changes in the enzyme itself. As shown in Figure 6.5, step 2, these conformational changes cause the substrates to bind more tightly to the enzyme—a phenomenon called **induced fit,** which was proposed by American biochemist Daniel Koshland in 1958. Only after induced fit takes place does the enzyme catalyze the conversion of reactants to products. Induced fit is a key phenomenon that lowers the activation energy.

Biology Principle	Biology is a quantitative science

Quantitative Analysis: Enzyme Function Is Influenced by Substrate Concentration and by Inhibitors

The degree of attraction between an enzyme and its substrate(s) is called the affinity of the enzyme for its substrate(s). Some enzymes have a very high **affinity,** which means they bind their substrates even when the substrate concentration is relatively low. Other enzymes have lower affinity for their substrates; the enzyme-substrate complex is likely to form only when the substrate concentration is higher.

Let's consider how biologists analyze the relationship between substrate concentration and enzyme function. In the experiment in **Figure 6.6a**, tubes labeled A, B, C, and D each contained 1 μg of enzyme, but they varied in the amount of substrate that was added. This enzyme recognizes a single substrate and converts it to a product. The samples were incubated for 60 seconds, and then the amount of product in each tube was measured. The velocity, or rate, of the chemical reaction is expressed as the amount of product produced per second. As we see in **Figure 6.6a**, the velocity increases as the substrate concentration increases but eventually reaches a plateau. Why does the plateau occur? At high substrate concentrations, nearly all of the active sites of the enzyme are occupied with substrate, so further increasing the substrate concentration has a negligible effect. At this point, the enzyme is saturated with substrate, and the velocity of the chemical reaction is near its maximal rate, called its V_{max}.

Figure 6.6a also helps us understand the relationship between substrate concentration and velocity. The K_M is the substrate concentration at which the velocity is half its maximal value. The K_M is also called the Michaelis constant, in honor of German biochemist Leonor Michaelis, who carried out pioneering work with Canadian biochemist Maud Menten regarding the study of enzymes. The K_M is a measure of the substrate concentration required for a chemical reaction to occur. An enzyme with a high K_M requires a higher substrate concentration to achieve a particular reaction velocity than does an enzyme with a lower K_M.

For an enzyme-catalyzed reaction, we can view the formation of product as occurring in two steps: (1) binding or release of substrate and (2) formation of product:

$$E + S \rightleftharpoons ES \rightarrow E + P$$

where E is the enzyme, S is the substrate, ES is the enzyme-substrate complex, and P is the product.

(a) Reaction velocity in the absence of inhibitors

(b) Competitive inhibition

(c) Noncompetitive inhibition

Figure 6.6 The relationship between velocity and substrate concentration in an enzyme-catalyzed reaction, and the effects of inhibitors. (a) In the absence of an inhibitor, the maximal velocity (V_{max}) of an enzyme-catalyzed reaction is achieved when the substrate concentration is high enough to be saturating. The K_M value for an enzyme is the substrate concentration at which the velocity of the reaction is half the maximal velocity. (b) A competitive inhibitor binds to the active site of an enzyme and raises the K_M. (c) A noncompetitive inhibitor binds to an allosteric site outside the active site and lowers the V_{max}.

Concept Check: *Enzyme A has a K_M of 0.1 mM, whereas enzyme B has a K_M of 1.0 mM. The reactions the two enzymes catalyze have the same V_{max}. If the substrate concentration was 0.5 mM, which reaction—the one catalyzed by enzyme A or the one catalyzed by enzyme B—would have the higher velocity?*

If the second step—the rate of product formation—is the slowest step, the K_M is inversely related to the affinity between the enzyme and the substrate. In such cases, enzymes with a high K_M have a low affinity for their substrates—they bind them more weakly. By comparison, enzymes with a low K_M have a high affinity for their substrates—they bind them more strongly.

Now that we understand the relationship between substrate concentration and the velocity of an enzyme-catalyzed reaction, we can explore how inhibitors may affect enzyme function. These can be categorized as reversible inhibitors that bind noncovalently to an enzyme or irreversible inhibitors that usually bind covalently to an enzyme and permanently inactivate its function.

Reversible Inhibitors Cells often use reversible inhibitors to modulate enzyme function. We will consider examples later in our discussion of glucose metabolism. **Competitive inhibitors** are molecules that bind noncovalently to the active site of an enzyme and inhibit the ability of the substrate to bind. Such inhibitors compete with the substrate in binding to the enzyme. Competitive inhibitors usually have a structure or a portion of their structure that mimics the structure of the enzyme's substrate. As seen in **Figure 6.6b**, when competitive inhibitors are present, the apparent K_M for the substrate increases—a higher concentration of substrate is needed to achieve the same rate of the chemical reaction. In this case, the effects of the competitive inhibitor can be overcome by increasing the concentration of the substrate.

By comparison, **Figure 6.6c** illustrates the effects of a **noncompetitive inhibitor.** This type of inhibitor lowers the V_{max} for the reaction without affecting the K_M. A noncompetitive inhibitor binds noncovalently to an enzyme at a location outside the active site, called an **allosteric site,** and inhibits the enzyme's function.

Irreversible Inhibitors Irreversible inhibitors usually bind covalently to an enzyme to inhibit its function. For example, some irreversible inhibitors bind covalently to an amino acid at the active site of an enzyme, thereby preventing the enzyme from catalyzing a chemical reaction. An example of an irreversible inhibitor is diisopropyl fluorophosphate (DFP). DFP is a type of nerve gas that was developed in the 1930s as a chemical weapon. This molecule binds covalently to the enzyme acetylcholinesterase, thereby preventing the proper communication between neurons and muscle cells.

Irreversible inhibition is not a common way for cells to control enzyme function. Why do cells usually control enzymes via reversible inhibitors? The answer is that a reversible inhibitor allows an enzyme to be used again, when the inhibitor concentration is lower. Being able to reuse an enzyme is energy-efficient. In contrast, irreversible inhibitors permanently inactivate an enzyme, thereby preventing its further use.

Crunching the Numbers: As we have seen, the presence of a competitive inhibitor raises the K_M for an enzyme-catalyzed reaction. The K_M value in the presence of the inhibitor is called the apparent K_M (abbreviated K_M^{app}). The relationship between the K_M and K_M^{app} is given by the following equation:

$$K_M^{app} = K_M \left(1 + \frac{[I]}{K_i}\right)$$

where K_i is a measure of the affinity between the enzyme and the inhibitor and [I] is the inhibitor concentration. As an example, let's consider hexokinase, which was featured in Figure 6.5. The K_M for glucose for a particular hexokinase found in humans is 0.15 mM. A structurally related sugar called 2-deoxyglucose acts as a competitive inhibitor with a K_i of 0.3 mM. What would be the K_M^{app} for glucose if the 2-deoxyglucose concentration was 1.0 mM?

Additional Factors Influence Enzyme Function

Enzymes, which are proteins, sometimes require the following nonprotein molecules or ions to carry out their functions:

- **Prosthetic groups** are small molecules that are permanently attached to the surface of an enzyme and aid in enzyme function.
- **Cofactors** are usually inorganic ions, such as Fe^{3+} or Zn^{2+}, that temporarily bind to the surface of an enzyme and promote a chemical reaction.
- **Coenzymes** are organic molecules that temporarily bind to an enzyme and participate in the chemical reaction that the enzyme catalyzes but are left unchanged when the reaction is completed. Some of these coenzymes are synthesized by cells, but many of them are taken in as dietary vitamins by animal cells.

The ability of enzymes to increase the rate of a chemical reaction is also affected by their environment. In particular, the temperature, pH, and ionic conditions play an important role in the proper functioning of enzymes. Most enzymes function maximally in a narrow range of temperature and pH. For example, many human enzymes work best at 37°C (98.6°F), which is normal body temperature. If the temperature is several degrees above or below this optimal temperature due to infection or environmental causes, the function of many enzymes is greatly inhibited (**Figure 6.7**). Very high temperatures may denature a protein, causing it to unfold and lose its three-dimensional shape, thereby inhibiting its function.

Enzyme function is also sensitive to pH. Certain enzymes in the stomach function best at the acidic pH found in this organ. For example, pepsin is a protease—an enzyme that digests proteins into peptides—that is released into the stomach. The optimal pH for pepsin function is around pH 2.0, which is extremely acidic. By comparison, many cytosolic enzymes function optimally at a more neutral pH, such as pH 7.2, which is the pH normally found in the cytosol of human cells. If the pH was significantly above or below this value, function would be decreased for cytosolic enzymes.

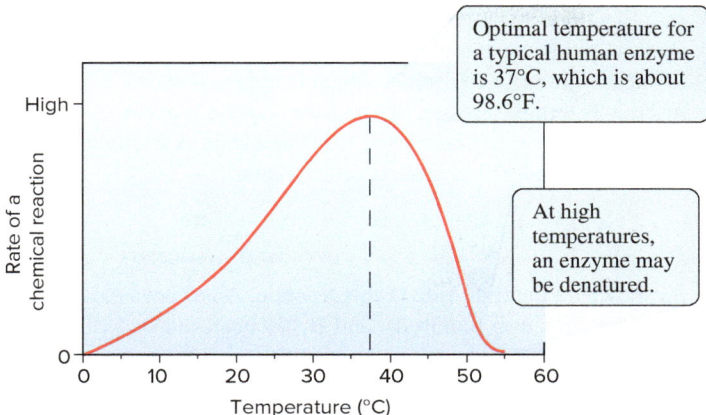

Figure 6.7 Effects of temperature on a typical human enzyme. Most enzymes function optimally within a narrow temperature range. Many human enzymes function best at 37°C—normal body temperature.

6.2 Reviewing the Concepts

- Enzymes are proteins that speed up the rate of a chemical reaction by lowering the activation energy (E_A) needed to achieve a transition state (Figure 6.4).
- Enzymes recognize reactant molecules, also called substrates, with high specificity. Conformational changes in an enzyme cause its substrate(s) to bind more tightly to it, a phenomenon called induced fit (Figure 6.5).
- Each enzyme-catalyzed reaction has a maximal velocity (V_{max}). The K_M value for an enzyme is the substrate concentration at which the velocity of the reaction is half of the maximal value. Competitive inhibitors raise the K_M for the substrate, whereas noncompetitive inhibitors lower the V_{max} (Figure 6.6).
- Enzymes sometimes need nonprotein molecules (prosthetic groups, cofactors, and coenzymes) in order to function, and other factors, including temperature and pH, can affect the proper functioning of enzymes (Figure 6.7).

6.2 Testing Your Knowledge

1. The main effect of enzymes on a chemical reaction is to
   a. lower the activation energy and thereby increase the rate of the reaction.
   b. raise the activation energy and thereby increase the rate of the reaction.
   c. lower the activation energy and thereby make the free energy more negative.
   d. raise the activation energy and thereby make the free energy more positive.
   e. do both a and c.

2. An inhibitor raises the K_M for an enzyme but has no effect on the V_{max}. This inhibitor probably binds to
   c. the active site.
   d. an allosteric site.
   e. the substrate.
   d. all of the above.
   e. both a and b.

6.3 Overview of Metabolism and Cellular Respiration

Learning Outcomes

1. Describe the concept of a metabolic pathway, and distinguish between catabolic and anabolic reactions.
2. Explain how catabolic reactions are used to generate building blocks to make larger molecules and to produce energy intermediates.
3. Define redox reaction.
4. Compare and contrast three ways that metabolic pathways are regulated.
5. Outline the four metabolic pathways needed to break down glucose to CO_2.

In the previous sections, we examined the underlying factors that govern chemical reactions and explored the properties of enzymes. In living cells, chemical reactions are coordinated with each other and often occur in a series of steps called a **metabolic pathway**, with each step catalyzed by a different enzyme (**Figure 6.8**). These pathways are categorized according to whether the reactions lead to the breakdown or the synthesis of substances. **Catabolic reactions** result in the breakdown of larger molecules into smaller ones. Such reactions are often exergonic. By comparison, **anabolic reactions** involve the synthesis of larger molecules from smaller precursor molecules. These reactions are usually endergonic and, in living cells, must be coupled to an exergonic reaction. In this section, we will survey the general features of catabolic and anabolic reactions and explore the ways in which metabolic pathways are controlled. Also, we will begin to examine cellular respiration, a process that involves four metabolic pathways.

Catabolic Reactions Recycle Organic Building Blocks and Produce Energy Intermediates Such as ATP

Catabolic reactions result in the breakdown of larger molecules into smaller ones. Such catabolic reactions have two advantages.

Recycling of Organic Building Blocks One reason to break down macromolecules is to recycle their organic molecules, which are used as building blocks to construct new molecules and macromolecules. For example, polypeptides, which make up proteins,

Figure 6.8 A metabolic pathway. In this metabolic pathway, three different enzymes catalyze the attachment of phosphate groups to various positions on a sugar molecule, beginning with a starting substrate and ending with a final product.

are composed of a linear sequence of amino acids. When a protein is improperly folded or is no longer needed by a cell, the peptide bonds between the amino acids in the protein are broken by enzymes called proteases. This process generates amino acids that can be used in the construction of new proteins.

$$\text{Protein} \xrightarrow{\text{Proteases}} \text{Many individual amino acids}$$

Breakdown of Organic Molecules to Obtain Energy A second reason for the breakdown of macromolecules and smaller organic molecules is to obtain energy that is used to drive endergonic processes in the cell. Covalent bonds store a large amount of energy. However, when cells break covalent bonds in organic molecules such as glucose, they do not directly use the energy released in this process. Instead, the released energy is stored in **energy intermediates,** molecules such as ATP, which are directly used to drive endergonic reactions in cells.

As an example, let's consider the breakdown of glucose into two molecules of pyruvate. As discussed later, the breakdown of glucose to pyruvate involves a catabolic pathway called glycolysis. Some of the energy released during the breakage of covalent bonds in glucose is harnessed to synthesize ATP. Glycolysis involves a series of steps in which covalent bonds are broken and rearranged. This process produces molecules that readily donate a phosphate group to ADP, thereby producing ATP. For example, phosphoenolpyruvate has a phosphate group attached to pyruvate. Due to the arrangement of bonds in phosphoenolpyruvate, this phosphate bond is unstable and easily broken. Therefore, the phosphate can be transferred from phosphoenolpyruvate to ADP, which is used to synthesize ATP:

This is an exergonic reaction and therefore favors the formation of products. In this step of glycolysis, the breakdown of an organic molecule, namely phosphoenolpyruvate, results in the formation of pyruvate and the synthesis of an energy intermediate, a molecule of ATP, which can then be used by a cell to drive an endergonic reaction.

Redox Reactions Involve the Transfer of Electrons

During the breakdown of small organic molecules, the removal of one or more electrons from atoms or molecules may occur. This process is called **oxidation,** because oxygen is frequently involved in chemical reactions that remove electrons from an atom or a molecule. By comparison, **reduction** is the addition of one or more electrons to an atom or a molecule. Reduction is so named because the addition of a negatively charged electron reduces the net charge of an atom or a molecule.

Electrons do not exist freely in solution. When an atom or a molecule is oxidized, the electron that is removed must be transferred to another atom or molecule, which becomes reduced. This type of reaction is termed a reduction-oxidation reaction, or **redox reaction.** An electron may be transferred from molecule A to molecule B, as shown in the following generalized equation:

$$Ae^- + B \rightarrow A + Be^-$$
$$\text{(oxidized) (reduced)}$$

As shown on the right side of this reaction, A has been oxidized (that is, had an electron removed), and B has been reduced (that is, had an electron added). In general, a substance that has been oxidized has less energy, whereas a substance that has been reduced has more energy.

During the oxidation of organic molecules such as glucose, the electrons that are removed may be used to produce energy intermediates such as NADH (**Figure 6.9**). In this process, an organic molecule is oxidized, and **nicotinamide adenine dinucleotide (NAD$^+$)** is reduced to NADH. Cells use NADH in two common ways. First, as discussed later, the oxidation of NADH is a highly exergonic reaction that can be used to make ATP. Second, NADH can donate electrons to other organic molecules and thereby energize them. Such energized molecules can more readily form covalent bonds. Therefore, NADH is often needed in anabolic reactions that involve the synthesis of larger molecules through the formation of covalent bonds between smaller molecules.

Metabolic Pathways Are Regulated in Three General Ways

The regulation of metabolic pathways is important for a variety of reasons. Catabolic pathways are regulated so that organic molecules are broken down only when they are no longer needed or when the cell requires energy. During anabolic reactions, regulation ensures that a cell synthesizes molecules only when they are needed. The regulation of catabolic and anabolic pathways occurs at the genetic, cellular, and biochemical levels.

Gene Regulation Enzymes are proteins that are encoded by genes. One way that cells control metabolic pathways is through gene regulation. For example, if a bacterial cell is not exposed to a particular sugar in its environment, it will turn off the genes that encode the enzymes that are needed to break down that sugar. Then, if the sugar becomes available, the genes are switched back on. Chapter 12 examines the steps of gene regulation in detail.

Regulation via Cell-Signaling Pathways Metabolic reactions are also coordinated at the cellular level. Cells integrate signals from their environment and adjust their metabolic pathways to adapt to those signals. As discussed in Chapter 8, cell-signaling pathways often lead to the activation of protein kinases—enzymes that covalently attach a phosphate group to a target protein. For example, when people are frightened, they secrete a hormone called epinephrine into their bloodstream. This hormone binds to the surface of muscle cells and stimulates an intracellular pathway that leads to the phosphorylation of specific enzymes involved in carbohydrate metabolism. These activated enzymes promote the breakdown of

Figure 6.9 **The reduction of NAD⁺ to produce NADH.** NAD⁺ is composed of two nucleotides: one with an adenine base and one with a nicotinamide base. The oxidation of organic molecules releases electrons. These electrons and an H⁺ bind to NAD⁺, which results in the formation of NADH. The two electrons and H⁺ are incorporated into the nicotinamide ring. Note: The actual net charges of NAD⁺ and NADH are −1 and −2, respectively. They are designated NAD⁺ and NADH to emphasize the net charge of the nicotinamide ring, which is involved in redox reactions.

Concept Check: *Which is the oxidized form, NAD⁺ or NADH?*

carbohydrates, an event that supplies the frightened individual with more energy. Epinephrine is sometimes called the fight-or-flight hormone because the added energy prepares an individual to either stay and fight or run away quickly.

Biochemical Regulation The third and very prominent way that metabolic pathways are controlled is at the biochemical level. In this case, the noncovalent binding of a molecule to an enzyme directly regulates the enzyme's function. As discussed earlier, one form of biochemical regulation involves the binding of molecules called competitive or noncompetitive inhibitors (see Figure 6.6). An example of noncompetitive inhibition is a type of regulation called **feedback inhibition,** in which the product of a metabolic pathway inhibits an enzyme that acts early in the pathway, thus preventing the overaccumulation of the product (**Figure 6.10**). This form of regulation is reversible; inhibition will occur only when the concentration of the product is high.

Many metabolic pathways use feedback inhibition as a form of biochemical regulation. In such cases, the enzyme being inhibited has two binding sites. One site is the active site, where the reactants are converted to products. In addition, enzymes controlled by feedback inhibition also have an allosteric site, where a molecule can bind noncovalently and affect the enzyme's function. The binding of a molecule to an allosteric site causes a conformational change in the enzyme that inhibits its catalytic function. For example, the conformational change may alter the structure of the active site so that reactants cannot bind. Allosteric sites are often found in the enzymes that catalyze the early steps in a metabolic pathway. Such allosteric sites typically bind molecules that are the products of the metabolic pathway. When the products bind to these sites, they inhibit

the function of these enzymes, thereby preventing the formation of too much product.

Rate-Limiting Step Cell-signaling and biochemical regulation are important, rapid ways to control chemical reactions in a cell. For a metabolic pathway composed of several enzyme-catalyzed reactions, which enzyme should be controlled? In many cases, a metabolic pathway has a **rate-limiting step,** which is the slowest step in the pathway. If the rate-limiting step is inhibited or enhanced, such changes will have the greatest influence on the formation of the product of the metabolic pathway. Rather than affecting all of the enzymes in a metabolic pathway, cell-signaling and biochemical regulation are often directed at the enzyme that catalyzes the rate-limiting step. This is an efficient, rapid way to control the amount of product of a pathway.

The Breakdown of Glucose Occurs in Four Stages

Cellular respiration is a process by which living cells obtain energy from organic molecules and release waste products. A primary aim of cellular respiration is to make adenosine triphosphate (ATP). When oxygen (O_2) is used, this process is termed **aerobic respiration.** During aerobic respiration, O_2 is used, and carbon dioxide (CO_2) is released via the oxidation of organic molecules. When we breathe, we inhale the oxygen needed for aerobic respiration and exhale CO_2, a by-product of the process. For this reason, respiration also means the act of breathing.

Different types of organic molecules, such as carbohydrates, proteins, and fats, are used as energy sources to drive aerobic respiration. In the rest of this chapter, we will largely focus on the use of glucose as an energy source for cellular respiration, as described in the following reaction:

Biology Principle

Biologists Use Models to Test Predictions and Convey Ideas

This figure is a model for the mechanism of feedback inhibition. The Modeling Challenge in the figure legend asks you to propose a model for a mutant enzyme that is not controlled by feedback inhibition.

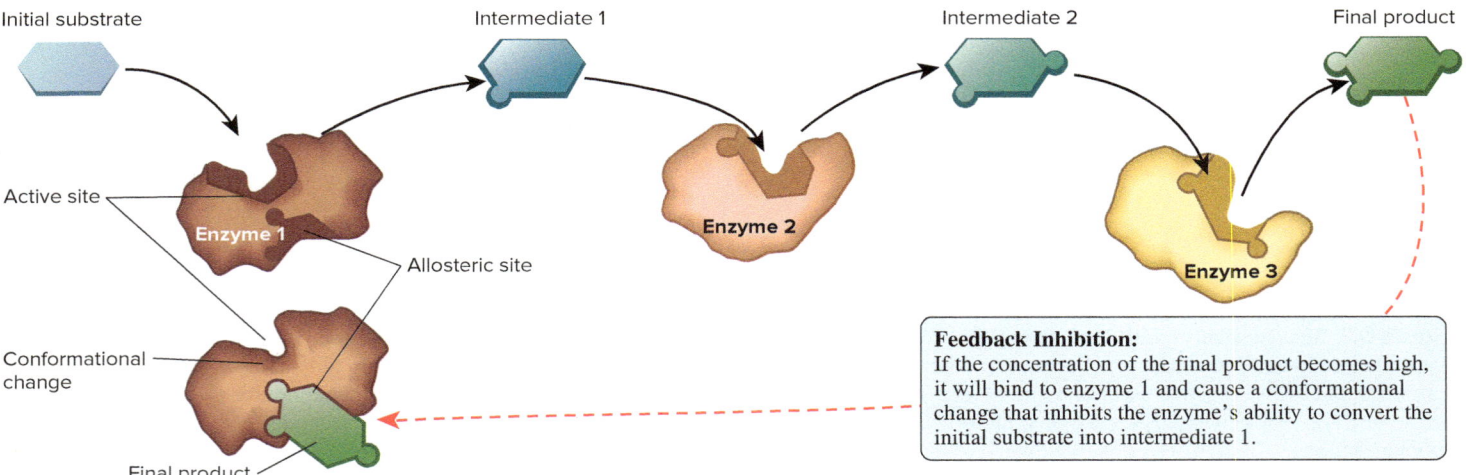

Figure 6.10 Feedback inhibition. In this process, the final product of a metabolic pathway inhibits an enzyme that functions early in the pathway, thereby preventing the overaccumulation of the product.

Modeling Challenge: Let's suppose that a mutation in the gene that encodes enzyme 1 in Figure 6.10 alters its structure in such a way that feedback inhibition no longer occurs. Draw side-by-side models that depict the structure of enzyme 1 with and without this mutation. Explain how the function of enzyme 1 differs in the normal and mutant enzymes. How will the concentration of the final product differ in a normal cell and in a cell that only makes the mutant enzyme?

$$C_6H_{12}O_6 + 6\,O_2 \rightarrow 6\,CO_2 + 6\,H_2O \qquad \Delta G = -686 \text{ kcal/mol}$$
Glucose

Certain covalent bonds within glucose store a large amount of energy. When glucose is broken down via oxidation, ultimately to CO_2 and water, a tremendous amount of free energy is released (–685 kcal/mol). Some of the energy is lost as heat, but much of it is used to make three energy intermediates: ATP, NADH, and $FADH_2$. This process involves four metabolic pathways: (1) glycolysis, (2) the breakdown of pyruvate, (3) the citric acid cycle, and (4) oxidative phosphorylation (**Figure 6.11**):

1. **Glycolysis.** In glycolysis, glucose (a compound with six carbon atoms) is broken down to two pyruvate molecules (with three carbons each), producing a net energy yield of two ATP molecules and two NADH molecules. The two ATP molecules are synthesized via **substrate-level phosphorylation,** which occurs when an enzyme directly transfers a phosphate from an organic molecule to ADP. In eukaryotes, glycolysis occurs in the cytosol.

2. **Breakdown of pyruvate to an acetyl group.** In eukaryotes, the two pyruvate molecules enter the mitochondrial matrix, where each one is broken down to an acetyl group (with two carbons each) and one CO_2 molecule. For each pyruvate broken down via oxidation, one NADH molecule is made by the reduction of NAD^+.

3. **Citric acid cycle.** Each acetyl group is incorporated into an organic molecule, which is later oxidized to liberate two CO_2 molecules. One ATP, three NADH, and one $FADH_2$ are made in this process. Because there are two acetyl groups (one from each pyruvate), the total yield is four CO_2, two ATP via substrate-level phosphorylation, six NADH, and two $FADH_2$. This process occurs in the mitochondrial matrix.

4. **Oxidative phosphorylation.** The NADH and $FADH_2$ made in the three previous stages contain high-energy electrons that can be readily transferred in a redox reaction to other molecules. Once removed from NADH or $FADH_2$, these high-energy electrons release some energy, and through an electron transport chain, that energy is harnessed to produce an H^+ electrochemical gradient. In **chemiosmosis,** energy stored in the H^+ electrochemical gradient is used to synthesize ATP from ADP and P_i. The overall process of electron transport and ATP synthesis is called oxidative phosphorylation because NADH or

Biology Principle

Living Organisms Use Energy

Molecules such as glucose store a large amount of energy. The breakdown of glucose is used to make energy intermediates, such as ATP molecules, which drive many types of cellular processes.

Figure 6.11 An overview of cellular respiration. The 30–34 ATP molecules produced via chemiosmosis represent the maximum number possible. As described later in this chapter, mitochondria may use NADH, $FADH_2$, and the H^+ electrochemical gradient for purposes other than ATP synthesis.

$FADH_2$ has been oxidized and ADP has become phosphorylated to make ATP. Approximately 30 to 34 ATP molecules can be made via oxidative phosphorylation.

In eukaryotes, oxidative phosphorylation occurs along the **cristae,** which are projections formed by the invagination of the inner mitochondrial membrane. The cristae greatly increase the surface area of the inner membrane, thereby increasing the amount of ATP that can be made. In bacteria and archaea, oxidative phosphorylation occurs along the plasma membrane because the proteins that carry out this process are located there.

6.3 Reviewing the Concepts

- Metabolism is the sum of the chemical reactions in a living organism. Metabolic pathways consist of coordinated chemical reactions that occur in steps and are catalyzed by specific enzymes (Figure 6.8).
- Catabolic reactions involve the breakdown of larger molecules into smaller ones. These reactions recycle organic molecules that are used as building blocks to make new molecules. The organic molecules are also broken down to make energy intermediates

Figure 6.15 Breakdown of pyruvate and the attachment of an acetyl group to CoA.

6.5 Testing Your Knowledge

1. Which of the following is *not* a product of the breakdown of pyruvate?
   a. ATP
   b. NADH
   c. acetyl group (attached to CoA)
   d. CO_2
   e. both a and b

6.6 Citric Acid Cycle

Learning Outcomes

1. Explain the concept of a metabolic cycle.
2. Describe how an acetyl group enters the citric acid cycle, and list the net products of the cycle.

The third stage of glucose metabolism introduces a new concept, that of a **metabolic cycle.** During a metabolic cycle, some molecules enter the cycle while others leave. The process is cyclical because it involves a series of organic molecules that are regenerated with each turn of the cycle. The idea of a metabolic cycle was first proposed in the early 1930s by German biochemist Hans Krebs. While studying carbohydrate metabolism in England, he analyzed cell extracts from pigeon muscle and determined that citric acid and other organic molecules participate in a cycle that results in the breakdown of carbohydrates to CO_2. This cycle is called the **citric acid cycle,** or the Krebs cycle, in honor of Krebs, who was awarded the Nobel Prize in Physiology or Medicine in 1953.

An overview of the citric acid cycle is shown in Figure 6.16. In the first step of the cycle, the acetyl group (with two carbons) is removed from acetyl CoA and attached to oxaloacetate (with four carbons) to form citrate (with six carbons), also called citric acid. During the cycle, two CO_2 molecules are released, and three molecules of NADH, one molecule of $FADH_2$, and one molecule of guanosine triphosphate (GTP) are made. The GTP, which is made via substrate-level phosphorylation, is used to make ATP. After a total of eight steps, oxaloacetate is regenerated so that the cycle can begin again, provided that acetyl CoA is available. Figure 6.17 shows a more detailed view of the citric acid cycle. For each acetyl group attached to CoA, the net reaction of the citric acid cycle is as follows:

$$\text{Acetyl CoA} + 2\ H_2O + 3\ NAD^+ + FAD + GDP^{2-} + P_i^{2-} \rightarrow$$

$$\text{CoA—SH} + 2\ CO_2 + 3\ NADH + FADH_2 + GTP^{4-} + 3\ H^+$$

Regulation of the Citric Acid Cycle How is the citric acid cycle controlled? The rate of the cycle is largely regulated by the availability of substrates, such as acetyl CoA and NAD^+, and by feedback inhibition. The three steps in the cycle that are highly exergonic are those catalyzed by citrate synthase, isocitrate dehydrogenase, and α-ketoglutarate dehydrogenase (see Figure 6.17). Each of these steps is rate-limiting under certain circumstances, and the way that each enzyme is regulated varies among different species. Let's consider an example. In mammals, NADH and ATP act as feedback inhibitors of isocitrate dehydrogenase, whereas NAD^+ and ADP act as activators. In this way, the citric acid cycle is inhibited when NADH and ATP levels are high, but it is stimulated when NAD^+ and ADP levels are high.

6.6 Reviewing the Concepts

- During the citric acid cycle, an acetyl group is removed from acetyl CoA and attached to oxaloacetate to make citrate. In a series of steps, two CO_2 molecules, three NADH, one $FADH_2$, and one ATP are made, after which the cycle begins again (Figures 6.16, 6.17).

6.6 Testing Your Knowledge

1. During the citric acid cycle, what happens to carbon?
   a. Organic carbon is released as inorganic carbon dioxide.
   b. Carbon is oxidized to make NADH and $FADH_2$.
   c. Carbon dioxide is used to make ATP.
   d. All of the above are correct.
   e. Only a and b are correct.

6.7 Oxidative Phosphorylation

Learning Outcomes

1. Outline how the electron transport chain produces an H^+ electrochemical gradient.
2. Explain how ATP synthase utilizes the H^+ electrochemical gradient to synthesize ATP.
3. **SCISKILLS** ▶ Analyze and interpret the results of the experiment that showed that ATP synthase is a rotary machine.

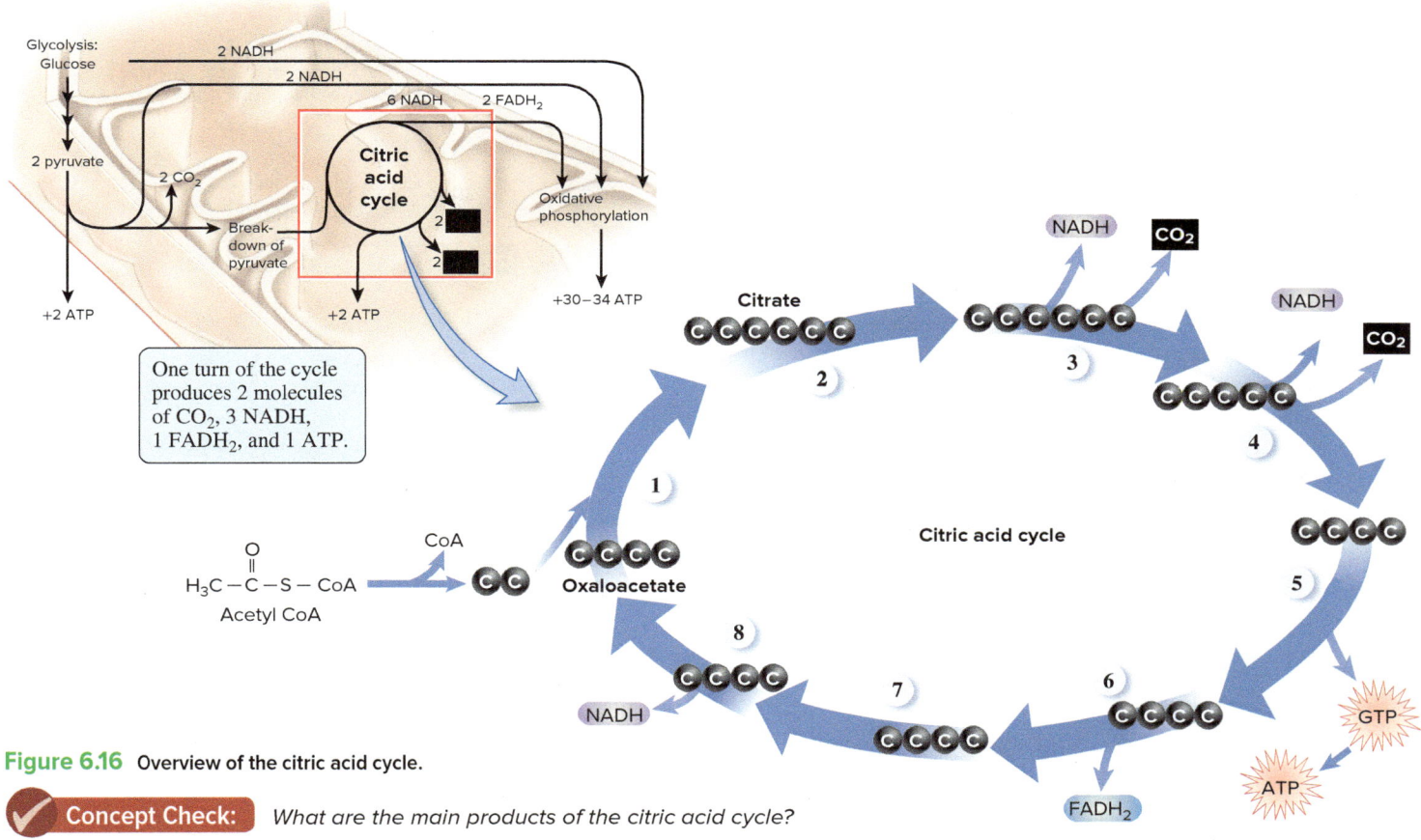

Figure 6.16 Overview of the citric acid cycle.

Concept Check: *What are the main products of the citric acid cycle?*

During the first three stages of glucose metabolism, the oxidation of glucose yields 6 molecules of CO_2, 4 molecules of ATP, 10 molecules of NADH, and 2 molecules of $FADH_2$. Let's now consider how high-energy electrons are removed from NADH and $FADH_2$ to produce more ATP. This process is called **oxidative phosphorylation.** As mentioned earlier, the term refers to the observation that NADH and $FADH_2$ have had electrons removed and have thus become oxidized, and ATP is made by the phosphorylation of ADP. In this section, we will examine how the oxidative process involves the electron transport chain, whereas the phosphorylation of ADP occurs via ATP synthase.

The Electron Transport Chain Establishes an Electrochemical Gradient

The **electron transport chain (ETC)** consists of a group of protein complexes and small organic molecules embedded in the inner mitochondrial membrane. These components are referred to as an electron transport chain because electrons are passed from one component to the next in a series of redox reactions (**Figure 6.18**). Most members of the ETC are protein complexes (designated I–IV in the figure) that have prosthetic groups. For example, cytochrome oxidase contains two prosthetic groups, each with an iron atom. The iron in each prosthetic group can readily accept and release an electron. One member of the ETC, ubiquinone (Q), is not a protein but is instead a small organic molecule that can accept and release an electron.

The red line in Figure 6.18 shows the path of electron flow. The electrons, which are originally found in NADH or $FADH_2$, are transferred to components of the ETC. The electron path is a series of redox reactions in which electrons are transferred to components with increasingly higher electronegativity. As discussed in Chapter 2, electronegativity is a measure of an atom's ability to attract electrons. At the end of the chain is oxygen, which is the most electronegative component and the final electron acceptor. The ETC is also called the **respiratory chain,** because the oxygen we breathe is used in this process. However, as discussed in Section 6.9, some microorganisms can use molecules other than O_2 as their final electron acceptor.

NADH and $FADH_2$ donate their electrons at different points in the ETC. Two high-energy electrons from NADH are first transferred, one at a time, to NADH dehydrogenase (complex I). They are then transferred to ubiquinone (Q), cytochrome b-c_1 (complex III), cytochrome c, and cytochrome oxidase (complex IV). The final electron acceptor is O_2. By comparison, $FADH_2$ transfers electrons to succinate reductase (complex II), then to ubiquinone and the rest of the chain.

As shown in Figure 6.18, some of the energy that is released during the movement of electrons is used to pump H^+ across the inner mitochondrial membrane into the intermembrane space. This active transport establishes a large H^+ **electrochemical gradient,** in which the concentration of H^+ is higher outside the mitochondrial matrix than inside and an excess of positive charge exists outside the matrix. Because hydrogen ions consist of protons, the H^+ electrochemical

140 CHAPTER 6

① The cycle begins when the acetyl group from acetyl CoA is attached to oxaloacetate to form citrate.

② In a 2-step reaction, citrate is rearranged to an isomer called isocitrate.

③ Isocitrate is oxidized to α-ketoglutarate. CO_2 is released and NADH is formed.

④ α-Ketoglutarate is oxidized as it combines with CoA to form succinyl-CoA. Once again, CO_2 is released and NADH is formed.

⑤ Succinyl-CoA is broken down to CoA and succinate. This exergonic reaction drives the synthesis of GTP from GDP and P_i. GTP can transfer its phosphate to ADP, thereby forming ATP.

⑥ Succinate is oxidized to fumarate. $FADH_2$ is made.

⑦ Fumarate combines with water to make malate.

⑧ Malate is oxidized to oxaloacetate. NADH is made. The cycle can begin again.

Figure 6.17 A detailed look at the steps of the citric acid cycle. The blue boxes indicate the location of the acetyl group, which is oxidized at step 6. (It is oxidized again in step 8.) The green boxes indicate the locations where CO_2 molecules are removed.

gradient is also called the **proton-motive force.** NADH dehydrogenase, cytochrome b-c_1, and cytochrome oxidase are H^+ pumps. The movement of electrons along the ETC is an exergonic process—it releases energy. The binding of electrons to certain proteins of the ETC enables them to actively transport H^+ out of the matrix into the intermembrane space against the H^+ electrochemical gradient.

Chemicals that inhibit the flow of electrons along the ETC can have lethal effects. For example, one component of the ETC, cytochrome oxidase (complex IV), is inhibited by cyanide. The deadly effects of cyanide ingestion occur because the ETC is shut down, preventing cells from making enough ATP for survival.

ATP Synthase Makes ATP via Chemiosmosis

The second event of oxidative phosphorylation is the synthesis of ATP by an enzyme called **ATP synthase.** The H^+ electrochemical gradient across the inner mitochondrial membrane is a source of potential energy. How is this energy used? The passive flow of H^+ back into the matrix is an exergonic process. The lipid bilayer is relatively

HOW CELLS UTILIZE ENERGY   141

1a NADH is oxidized to NAD⁺. High-energy electrons are transferred to NADH dehydrogenase. Some of the energy is harnessed to pump H⁺ into the intermembrane space. Electrons are then transferred to ubiquinone.

1b $FADH_2$ is oxidized to FAD. High-energy electrons are transferred to succinate reductase and then to ubiquinone.

2 From ubiquinone, electrons travel to cytochrome b-c_1. Some of the energy is harnessed to pump H⁺ into the intermembrane space. Electrons are transferred to cytochrome c.

3 From cytochrome c, electrons are transferred to cytochrome oxidase. Some of the energy is harnessed to pump H⁺ into the intermembrane space. Electrons are transferred to oxygen, and water is produced.

4 Steps 1–3 produce an H⁺ electrochemical gradient. As H⁺ flow down their electrochemical gradient into the matrix through ATP synthase, the energy within this gradient causes the synthesis of ATP from ADP and P_i.

Figure 6.18 Oxidative phosphorylation. This process consists of two distinct events: the electron transport chain (ETC) and ATP synthesis. The ETC oxidizes, or removes electrons from, NADH or $FADH_2$ and pumps H⁺ across the inner mitochondrial membrane. In chemiosmosis, ATP synthase uses the energy in this H⁺ electrochemical gradient to phosphorylate ADP, thereby synthesizing ATP. In this figure, an oxygen atom is represented as $\frac{1}{2} O_2$ to emphasize that the ETC reduces oxygen when it is in its molecular (O_2) form.

 Concept Check: *Explain the meaning of the name cytochrome oxidase. What might be another appropriate name?*

impermeable to H⁺. However, H⁺ can pass through the membrane-embedded portion of ATP synthase. The flow of H⁺ causes conformational changes in ATP synthase that result in the synthesis of ATP from ADP and P_i (see bottom of Figure 6.18). This is an example of an energy conversion: Energy in the form of an H⁺ electrochemical gradient is converted to chemical potential energy in ATP. The synthesis of ATP that occurs as a result of pushing H⁺ across a membrane is called chemiosmosis (from the Greek *osmos*, meaning to push). The theory behind it was proposed by Peter Mitchell, a British biochemist who was awarded the Nobel Prize in Chemistry in 1978.

Regulation of Oxidative Phosphorylation How is oxidative phosphorylation controlled? This process is regulated by a variety of factors, including the availability of ETC substrates, such as NADH and O_2, and by the ATP/ADP ratio. When ATP levels are high, ATP binds to a subunit of cytochrome oxidase (complex IV), thereby inhibiting the ETC and oxidative phosphorylation. By comparison, when ADP levels are high, oxidative phosphorylation is stimulated for two reasons: (1) ADP stimulates cytochrome oxidase, and (2) ADP is a substrate that is used (with P_i) to make ATP.

NADH Oxidation Makes a Large Proportion of a Cell's ATP

For each molecule of NADH that is oxidized and each molecule of ATP that is made, the two chemical reactions of oxidative phosphorylation can be represented as follows:

$$NADH + H^+ + \tfrac{1}{2} O_2 \rightarrow NAD^+ + H_2O$$

$$ADP^{2-} + P_i^{2-} \rightarrow ATP^{4-} + H_2O$$

The oxidation of NADH to NAD⁺ results in an H⁺ electrochemical gradient in which more hydrogen ions are in the intermembrane space than are in the matrix. The synthesis of one ATP molecule is thought to require the movement of three to four ions down their H⁺ electrochemical gradient into the matrix.

When we add up the maximal amount of ATP that can be made by oxidative phosphorylation, most researchers agree it is in the range of 30 to 34 ATP molecules for each glucose molecule that is broken down to CO_2 and H_2O. However, that maximal amount of ATP is rarely achieved, for two reasons.

- First, although 10 NADH and 2 $FADH_2$ are available to make the H⁺ electrochemical gradient across the inner mitochondrial membrane, a cell uses some of these molecules for anabolic pathways. For example, NADH is used in the synthesis of organic molecules such as glycerol (a component of phospholipids).

- Second, the mitochondrion may use some of the energy in the H⁺ electrochemical gradient for other purposes. For example, the gradient is used for the uptake of pyruvate into the matrix via an H⁺/pyruvate symporter (see Figure 6.15).

Therefore, the actual amount of ATP synthesized is usually a little less than the maximum of 30 to 34 molecules. Even so, when we compare the amount of ATP that is made by glycolysis (2), the citric acid cycle (2), and oxidative phosphorylation (30–34), we see that oxidative phosphorylation provides a cell with a much greater capacity to make ATP.

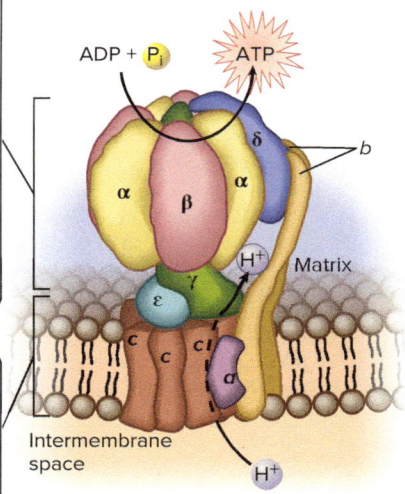

The nonmembrane-embedded portion consists of 1 ε, 1 γ, 1 δ, 3 α, and 3 β subunits. Movement of H⁺ between a *c* subunit and the *a* subunit causes the γ subunit to rotate. The rotation, in 120° increments, causes the β subunits to progress through a series of 3 conformational changes that lead to the synthesis of ATP from ADP and P_i.

The membrane-embedded portion consists of a ring of 9–12 *c* subunits, 1 *a* subunit, and 2 *b* subunits. H⁺ moves between a *c* subunit and the *a* subunit.

Figure 6.19 The subunit structure and function of ATP synthase.

Concept Check: Look at Figure 6.20. If the β subunit in the front center of this figure is in conformation 2, what are the conformations of the β subunit on the left and the β subunit in the back right?

ATP Synthase Is a Rotary Machine That Makes ATP as It Spins

The structure and function of ATP synthase are particularly intriguing and have received much attention over the past few decades. ATP synthase is a rotary machine (**Figure 6.19**). The term machine refers to the observations that it has moving parts and is powered by an energy source. The region embedded in the membrane is composed of three types of subunits called *a*, *b*, and *c*. Approximately 10 to 14 *c* subunits form a ring in the membrane. One *a* subunit is bound to this ring, and two *b* subunits are attached to the *a* subunit and protrude from the membrane. The nonmembrane-embedded subunits are designated with Greek letters. One ε and one γ subunit bind to the ring of *c* subunits. The γ subunit forms a long stalk that pokes into the center of another ring of three α and three β subunits. Each β subunit contains a catalytic site where ATP is made. Finally, the δ subunit forms a connection between the ring of α and β subunits and the two *b* subunits.

When hydrogen ions pass through a narrow channel at the contact site between a *c* subunit and the *a* subunit, a conformational change causes the γ subunit to turn clockwise (when viewed from the intermembrane space). Each time the γ subunit turns 120°, it changes its contacts with the three β subunits, which, in turn, causes the β subunits to change their conformations. How do these conformational changes promote ATP synthesis? The answer is that the conformational changes occur in a way that favors ATP synthesis and release. As shown in **Figure 6.20**, the conformational changes in the β subunits happen in the following order:

- Conformation 1: ADP and P_i bind with good affinity.
- Conformation 2: ADP and P_i bind very tightly, which strains chemical bonds so that ATP is made.

Figure 6.20 Conformational changes that result in ATP synthesis. For simplicity, the α subunits are not shown. This drawing emphasizes the conformational changes in the β subunit shown at the top. The other two β subunits also make ATP. All three β subunits alternate between three conformational states due to their interactions with the γ subunit.

- Conformation 3: ATP (and ADP and P_i) binds very weakly, and ATP is released.

Each time the γ subunit turns 120°, it causes a β subunit to change to the next conformation. After conformation 3, a 120° turn by the γ subunit returns a β subunit back to conformation 1, and the cycle of ATP synthesis can begin again. Because ATP synthase has three β subunits, each subunit is in a different conformation at any given time.

American biochemist Paul Boyer proposed the concept of a rotary machine in the late 1970s. In his model, the three β subunits alternate between three conformations, as described previously. Boyer's original idea was met with great skepticism, because the concept that part of an enzyme could spin was very novel, to say the least. In 1994, British biochemist John Walker and his colleagues determined the three-dimensional structure of the nonmembrane-embedded portion of the ATP synthase. The structure revealed that each of the three β subunits has a different conformation—one with ADP bound, one with ATP bound, and one without any nucleotide bound. This result supported Boyer's model. In 1997, Boyer and Walker shared the Nobel Prize in Chemistry for their work on ATP synthase. As described in the Feature Investigation, other researchers subsequently visualized the rotation of the γ subunit.

FEATURE INVESTIGATION

Yoshida and Kinosita Demonstrated That the γ Subunit of ATP Synthase Spins

In 1997, Masasuke Yoshida, Kazuhiko Kinosita, and colleagues set out to experimentally visualize the rotary nature of ATP synthase (**Figure 6.21**). The membrane-embedded region of ATP synthase can be separated from the rest of the protein by treating mitochondrial membranes with a high concentration of salt, releasing the portion of the protein containing one γ, three α, and three β subunits. The researchers adhered the resulting $γα_3β_3$ complex to a glass slide so that the γ subunit was protruding upward. Because the γ subunit is too small to be seen with a light microscope, the rotation of this subunit cannot be visualized directly. To circumvent this problem, the researchers attached a long, fluorescently labeled actin filament to the γ subunit via linker proteins. The fluorescently labeled actin filament is very long compared to the γ subunit and can be readily seen with a fluorescence microscope.

Because the membrane-embedded portion of the protein was missing, you may be wondering how the researchers could get the γ subunit to rotate. The answer is that they added ATP. Although the normal function of ATP synthase is to make ATP, it can also hydrolyze ATP. In other words, ATP synthase can run backwards. As shown in the data in Figure 6.21, when the researchers added ATP, they observed that the fluorescently labeled actin filament rotated in a counterclockwise direction, which is opposite the direction that the γ subunit rotates when ATP is synthesized. Actin filaments were observed to rotate for more than 100 revolutions in the presence of ATP. These results convinced the scientific community that ATP synthase is a rotary machine.

Experimental Questions

1. **SCISKILLS** ▶ The components of ATP synthase are too small to be visualized by light microscopy. For the experiment of Figure 6.21, how did the researchers observe the movement of ATP synthase?

2. **SCISKILLS** ▶ In the experiment of Figure 6.21, what observation indicated to the researchers that ATP synthase is a rotary machine? What was the control of this experiment? What did it indicate?

3. **SCISKILLS** ▶ Were the rotations seen by the researchers in the data of Figure 6.21 in the same direction as they are expected to occur in mitochondria during ATP synthesis? Why or why not?

Figure 6.21 Yoshida and Kinosita provide evidence that ATP synthase is a rotary machine.

(Images) From: H. Noji, M. Yoshida (2001), "The Rotary Machine in the Cell, ATP Synthase," *Journal of Biological Chemistry*, 276(3): 1665–1668, Fig. 2b ©2001 The American Society for Biochemistry and Molecular Biology.

6 **CONCLUSION** The γ subunit rotates counterclockwise when ATP is hydrolyzed. It would be expected to rotate clockwise when ATP is synthesized.

7 **SOURCE** Noji, H., Yasuda, R., Yoshida, M., and Kinosita, K. 1997. Direct observation of the rotation of F_1-ATPase. *Nature* 386: 299–303.

6.7 Reviewing the Concepts

- Oxidative phosphorylation involves two events: (1) The electron transport chain (ETC) oxidizes NADH or $FADH_2$ and generates an H^+ electrochemical gradient, and (2) this gradient is used by ATP synthase to make ATP via chemiosmosis (Figure 6.18).
- ATP synthase is a rotary machine. The rotation is triggered by the passage of H^+ through a channel between a c subunit and the α subunit, which causes the γ subunit to spin, resulting in three conformational changes in the β subunits that promote ATP synthesis (Figures 6.19, 6.20).
- Yoshida and Kinosita demonstrated rotation of the γ subunit by attaching a fluorescently labeled actin filament to the γ subunit via linker proteins and observing its movement during the hydrolysis of ATP (Figure 6.21).

6.7 Testing Your Knowledge

1. The electrons that travel down the electron transport chain come from
   a. NADH.
   b. $FADH_2$.
   c. O_2.
   d. ATP.
   e. either a or b.

2. The source of energy that directly drives the synthesis of ATP during oxidative phosphorylation is
   a. the oxidation of NADH.
   b. the oxidation of glucose.
   c. the oxidation of pyruvate.
   d. the H^+ gradient.
   e. the reduction of O_2.

3. The spinning of the γ subunit of ATP synthase is caused by
   a. the oxidation of NADH.
   b. the synthesis of ATP.
   c. the hydrolysis of ATP.
   d. the movement of H^+ through ATP synthase.
   e. either c or d.

6.8 Connections Among Carbohydrate, Protein, and Fat Metabolism

Learning Outcome
1. Explain how carbohydrate, protein, and fat metabolism are interconnected.

When you eat a meal, it usually contains not only carbohydrates (including glucose) but also proteins and fats. These molecules are broken down by some of the same enzymes involved with glucose metabolism. The use of the same pathways for the breakdown of sugars, amino acids, and fats makes cellular metabolism more efficient because the same enzymes can be used for the breakdown of different starting molecules.

As shown in **Figure 6.22**, proteins and fats can enter glycolysis or the citric acid cycle at different points. Proteins are first acted on

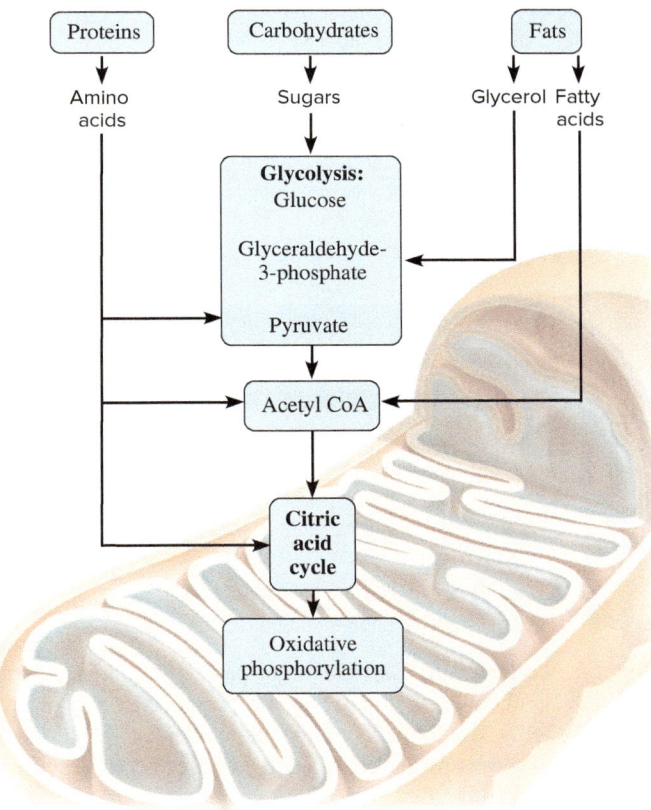

Figure 6.22 Integration of carbohydrate, protein, and fat metabolism. Breakdown products of proteins and fats can be used as fuel for cellular respiration, entering the same pathway used to break down carbohydrates.
(photo) ©Ernie Friedlander/Cole Group/Getty Images

Concept Check: What advantage does integrating protein, carbohydrate, and fat metabolism have for cells?

by enzymes, either in digestive juices or within cells, that cleave the bonds connecting individual amino acids. Breakdown products of some amino acids can enter at later steps of glycolysis, or an acetyl group can be removed from certain amino acids and become attached to CoA and then enter the citric acid cycle. Other amino acids can be modified and enter the citric acid cycle.

Fats are typically broken down to glycerol and fatty acids. Glycerol can be modified to glyceraldehyde-3-phosphate and enter glycolysis. Fatty acid tails have two carbon acetyl units removed, which bind to CoA and enter the citric acid cycle.

6.8 Reviewing the Concepts

- Proteins and fats can enter glycolysis or the citric acid cycle at different points (Figure 6.22).

6.8 Testing Your Knowledge

1. The advantage of connecting metabolic pathways for carbohydrate, protein, and fat metabolism is that
   a. more ATP is made.
   b. more NADH is made.
   c. less O_2 is consumed.
   d. the efficiency is higher because the same enzymes are used to break down different starting materials.
   e. All of the above are correct.

6.9 Anaerobic Respiration and Fermentation

Learning Outcomes

1. Describe how certain microorganisms make ATP using a final electron acceptor in the electron transport chain that is not oxygen.
2. Explain how muscle and yeast cells use fermentation to synthesize ATP under anaerobic conditions.

Thus far, we have surveyed catabolic pathways that result in the complete breakdown of glucose in the presence of oxygen. Cells also commonly metabolize organic molecules in the absence of oxygen. The term **anaerobic** is used to describe an environment that lacks oxygen. Many bacteria and archaea and some fungi exist in anaerobic environments but must oxidize organic molecules to obtain sufficient amounts of energy. Examples include microbes living in your intestinal tract and those living deep in the soil. Similarly, when a person exercises strenuously, the rate of oxygen consumption by muscle cells may greatly exceed the rate of oxygen delivery—particularly at the start of the strenuous exercise. Under these conditions, muscle cells become anaerobic and must obtain sufficient energy in the absence of oxygen to maintain their level of activity.

Cells can use two different strategies to metabolize organic molecules in the absence of oxygen. One mechanism is to use a substance other than O_2 as the final electron acceptor of the electron transport chain, a process called **anaerobic respiration.** The second approach is to produce ATP via substrate-level phosphorylation only, without any net oxidation of organic molecules, a process called **fermentation**. In this section, we will consider examples of both strategies.

Some Microorganisms Carry Out Anaerobic Respiration

At the end of the ETC, as shown earlier in Figure 6.18, cytochrome oxidase recognizes O_2 and catalyzes its reduction to H_2O. The final electron acceptor of the chain is O_2. Many species of bacteria that live under anaerobic conditions have evolved enzymes that function similarly to cytochrome oxidase but recognize molecules other than O_2 and use them as the final electron acceptor.

For example, under anaerobic conditions *Escherichia coli*, a bacterial species found in your intestinal tract, produces an enzyme called nitrate reductase. This enzyme uses nitrate (NO_3^-) as the final electron acceptor of the electron transport chain. **Figure 6.23** shows a simplified ETC in *E. coli* in which nitrate is the final electron acceptor. In *E. coli* and other bacterial species, the ETC is in the plasma membrane that surrounds the cytoplasm. Electrons travel from NADH to NADH dehydrogenase to ubiquinone to cytochrome *b* and then to nitrate reductase. At the end of the chain, NO_3^- is converted to nitrite (NO_2^-). This process generates an H^+ electrochemical gradient in three ways. First, NADH dehydrogenase pumps H^+ out of the cytoplasm. Second, ubiquinone picks up H^+ in the cytoplasm and carries it to the other side of the membrane. Third, the reduction of nitrate to nitrite consumes H^+ in the cytoplasm. The generation of an H^+ electrochemical gradient by these three processes allows *E. coli* cells to make ATP via chemiosmosis under anaerobic conditions.

Fermentation Is the Breakdown of Organic Molecules Without Net Oxidation

Many organisms, including animals and yeast, use only O_2 as the final electron acceptor of their ETCs. When confronted with anaerobic conditions, these organisms must have a different way of producing sufficient ATP. One strategy is to make ATP via glycolysis, which can occur under either anaerobic or aerobic conditions. Under anaerobic conditions, cells do not use the citric acid cycle or the ETC but rather make ATP only via glycolysis.

A key issue is that glycolysis requires NAD^+ and generates NADH. Under aerobic conditions, NADH is oxidized to NAD^+ to make more ATP. However, this cannot occur under anaerobic conditions in yeast and animals; as a result, NADH builds up and NAD^+ decreases. This is a potential problem for two reasons:

- First, at high concentrations, NADH haphazardly donates its electrons to other molecules and promotes the formation of free radicals, highly reactive chemicals that damage DNA and cellular proteins. For this reason, yeast and animal cells exposed

Figure 6.23 An example of anaerobic respiration in *E. coli*. When oxygen is absent, *E. coli* can use nitrate instead of oxygen as the final electron acceptor of the electron transport chain. This generates an H⁺ electrochemical gradient, which is used to make ATP via chemiosmosis. Note: As shown in this figure, ubiquinone picks up H⁺ on one side of the membrane and deposits it on the other side. A similar event happens during aerobic respiration in mitochondria, except that ubiquinone transfers H⁺ to cytochrome b-c$_1$, which pumps it into the intermembrane space.

to anaerobic conditions must have a way to remove the excess NADH generated from the breakdown of glucose.

- The second problem is the decrease in NAD⁺. Cells need to regenerate NAD⁺ to keep glycolysis running and make ATP via substrate-level phosphorylation.

Fermentation in Muscle Cells How do muscle cells cope with the buildup of NADH and accompanying decrease in NAD⁺? When a muscle cell is working strenuously and its environment becomes anaerobic, as in high-intensity exercise, the pyruvate from glycolysis is reduced to make lactate. (The uncharged, or protonated, form is called lactic acid.) The electrons to reduce pyruvate are derived from NADH, which is oxidized to NAD⁺ (**Figure 6.24a**). Therefore, this process decreases NADH and reduces its potentially harmful effects. It also increases the level of NAD⁺, thereby allowing glycolysis to continue. The lactate is secreted from muscle cells.

Once sufficient oxygen is restored, the lactate produced during strenuous exercise can be taken up by cells, converted back to pyruvate, and used for energy, or this lactate may be used by the liver and other tissues to make glucose.

Fermentation in Yeast Cells Yeast cells cope with anaerobic conditions differently. During wine making, a yeast cell metabolizes sugar under anaerobic conditions. The pyruvate is broken down to CO_2 and a two-carbon molecule called acetaldehyde. The acetaldehyde is then reduced by NADH to make ethanol, and NADH is oxidized to NAD⁺ (**Figure 6.24b**). Similar to lactate production in muscle cells, this process decreases NADH and increases NAD⁺, thereby preventing the harmful effects of NADH and allowing glycolysis to continue.

Fermentation is the breakdown of organic molecules to harness energy without any net oxidation (that is, without any removal of electrons). The pathways for breaking down glucose to lactate or ethanol are examples of fermentation. Although electrons are removed from an organic molecule, such as glucose, to make pyruvate and NADH, the electrons are donated back to an organic molecule in the production of lactate or ethanol. Therefore, there is no net removal of electrons from an organic molecule. Compared with oxidative phosphorylation, fermentation produces far less ATP, for two reasons. First, glucose is not oxidized completely to CO_2 and H_2O. Second, the NADH made during glycolysis cannot be used to make more ATP. Overall, the complete breakdown of glucose in the presence of oxygen yields 34–38 ATP molecules. By comparison, the anaerobic breakdown of glucose to lactate or ethanol yields only 2 ATP molecules.

6.9 Reviewing the Concepts

- Anaerobic respiration occurs in the absence of oxygen. Certain microorganisms carry out anaerobic respiration by using as the final electron acceptor of the ETC a substance other than oxygen, such as nitrate (Figure 6.23).
- During fermentation, organic molecules are broken down without any net oxidation (that is, without any net removal of electrons). Examples include lactic acid production in muscle cells and ethanol production in yeast cells (Figure 6.24).

6.9 Testing Your Knowledge

1. Compared with oxidative phosphorylation in mitochondria, anaerobic respiration in bacteria differs in that
   a. more ATP is made.
   b. ATP is made only via substrate-level phosphorylation.
   c. O_2 is converted to H_2O_2 rather than H_2O.
   d. something other than O_2 acts as a final electron acceptor of the ETC.
   e. both b and d occur.

Biology Principle

Biology affects our society

Fermentation by microorganisms is used in wine making, beer brewing, and bread making.

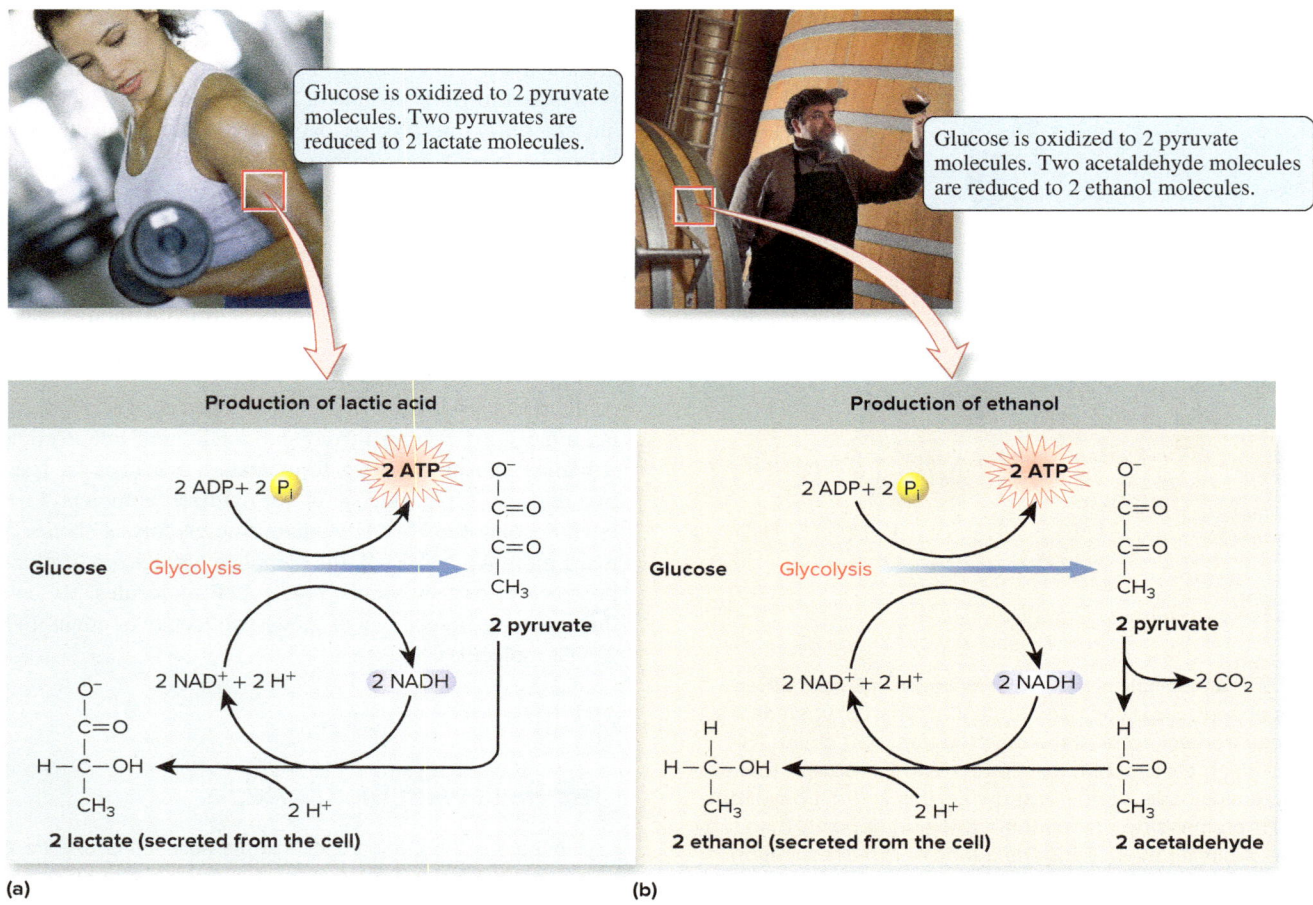

Figure 6.24 Examples of fermentation. In these examples, NADH is produced by the oxidation of an organic molecule, and then the NADH is converted back to NAD^+ when it donates electrons to a different organic molecule such as pyruvate **(a)** or acetaldehyde **(b)**.
(a) ©Purestock/SuperStock; (b) ©FreeProd/Alamy

Assessing Your Knowledge and Skills

Foundational Knowledge

Multiple Choice

1. Reactions that release free energy are
   a. exergonic.
   b. spontaneous.
   c. endergonic.
   d. endothermic.
   e. both a and b.

2. Enzymes speed up reactions by
   a. providing chemical energy to fuel a reaction.
   b. lowering the activation energy necessary to initiate the reaction.
   c. causing an endergonic reaction to become exergonic.
   d. substituting for one of the reactants necessary for the reaction.
   e. doing none of the above.

3. When NADH is converted to $NAD^+ + H^+$, NADH has been
   a. reduced.
   b. phosphorylated.
   c. oxidized.

d. decarboxylated.
e. methylated.

4. The net products of glycolysis are
   a. 6 CO_2, 4 ATP, and 2 NADH.
   b. 2 pyruvate, 2 ATP, and 2 NADH.
   c. 2 pyruvate, 4 ATP, and 2 NADH.
   d. 2 pyruvate, 2 GTP, and 2 CO_2.
   e. 2 CO_2, 2 ATP, and glucose.

5. The ability to diagnose tumors using [^{18}F]-fluorodeoxyglucose (FDG) is based on the phenomenon that most types of cancer cells exhibit higher levels of
   a. glycolysis.
   b. pyruvate breakdown.
   c. citric acid metabolism.
   d. oxidative phosphorylation.
   e. all of the above.

Short Answer

1. What causes the rotation of the γ subunit of ATP synthase? How does this rotation promote ATP synthesis?
2. **PRINCIPLES** A principle of biology is that living organisms maintain homeostasis. How is glucose breakdown regulated to maintain homeostasis? What would be some potentially harmful consequences if glucose metabolism was not regulated properly?

Critical-Thinking Skills

Multiple Choice

1. For the idealized reaction $aA + bB \rightleftharpoons cC + dD$, let's suppose that the equilibrium constant, K_{eq}, is 0.01. If the starting concentrations for A, B, C, and D are 1 M each, what would you predict based on the value of K_{eq}?
   a. The forward reaction is favored.
   b. The reverse reaction is favored.
   c. The forward reaction is fast.
   d. The reverse reaction is fast.
   e. Both b and d are correct.

2. Reaction 1 is $A + B \rightleftharpoons C + H_2O$. It is endergonic with $\Delta G = 4.6$ kcal/mol.

   Reaction 2 is $ATP + H_2O \rightleftharpoons ADP + P_i$. It is exergonic with $\Delta G = -7.3$ kcal/mol.

   What is the overall free energy for the following coupled reaction: $A + B + ATP \rightleftharpoons ADP + P_i + C$?
   a. 0 kcal/mole
   b. −2.7 kcal/mol
   c. +2.7 kcal/mol
   d. −11.9 kcal/mol
   e. +11.9 kcal/mol

3. Succinate dehydrogenase is an enzyme in the citric acid cycle that catalyzes a reaction that converts succinate to fumarate (see Figure 6.17). A molecule called malonate acts as an inhibitor of succinate dehydrogenase. The presence of malonate raises the K_M for succinate but does not affect the V_{max} of the reaction. Malonate
   a. is a competitive inhibitor.
   b. is a noncompetitive inhibitor.
   c. binds to the active site of the enzyme.
   d. binds to an allosteric site of the enzyme.
   e. Both a and c are correct.

4. An insecticide called rotenone inhibits NADH dehydrogenase, which is involved in the first step of the electron transport chain (ETC). Rotenone blocks the ability of NADH dehydrogenase to accept electrons from NADH and inhibits ATP synthesis. What is the mechanism by which rotenone inhibits ATP synthesis?
   a. Rotenone prevents the synthesis of O_2.
   b. Rotenone inhibits the formation of an H$^+$ electrochemical gradient.
   c. Rotenone directly inhibits ATP synthase.
   d. Rotenone prevents $FADH_2$ from donating electrons to the ETC.
   e. Rotenone prevents the reduction of NAD$^+$ to NADH.

5. Certain drugs act as ionophores that cause the mitochondrial membrane to be highly permeable to H$^+$. How would such drugs affect oxidative phosphorylation?
   a. Movement of electrons down the ETC would be inhibited.
   b. ATP synthesis would be inhibited.
   c. ATP synthesis would be unaffected.
   d. ATP synthesis would be stimulated.
   e. Both a and b are correct.

Short Answer

1. Explain how life can maintain its order in spite of the second law of thermodynamics. Are we defying this law?
2. Cyanide is a powerful poison that inhibits cytochrome oxidase. Explain how the effects of cyanide can be lethal.

7

How Cells Capture Light Energy via Photosynthesis

©Steven Trainoff Ph.D./Getty Images

Algae in the open ocean. Algae, such as these giant kelp, are photosynthetic organisms.

Chapter Outline

- 7.1 Overview of Photosynthesis
- 7.2 Reactions That Harness Light Energy
- 7.3 Molecular Features of Photosystems
- 7.4 Synthesizing Carbohydrates via the Calvin Cycle
- 7.5 Variations in Photosynthesis
- Assessing Your Knowledge and Skills

Nearly all of the oxygen in every breath you take is made by Earth's abundant plants, algae (see the chapter-opening photo), and cyanobacteria. We often think of photosynthesis in terms of terrestrial plants such as grasses and trees. Even so, you may be surprised to learn that nearly half of the oxygen produced globally is made by phytoplankton—a diverse mixture of aquatic and photosynthetic microorganisms that include cyanobacteria and protists such as algae. Global warming may decrease phytoplankton populations in oceans and lakes, thereby decreasing oxygen levels on a global scale.

Photosynthesis is the process in which the energy from light is captured and used to synthesize glucose and other organic molecules. Nearly all living organisms rely on photosynthesis for their nourishment, either directly or indirectly. Photosynthesis is also responsible for producing the oxygen that makes up a large portion of the Earth's atmosphere. Therefore, all aerobic organisms rely on photosynthesis for cellular respiration.

We will begin this chapter with an overview of photosynthesis as it occurs in green plants and algae. We will then explore the two stages of photosynthesis in more detail. In the first stage, called the **light reactions,** light energy is absorbed by pigments and converted to chemical energy in the form of two energy intermediates: ATP and NADPH. During the second stage, known as the **Calvin cycle,** ATP and NADPH are used to drive the synthesis of carbohydrates. We will conclude with a consideration of the variations in photosynthesis that occur in plants existing in hot and dry conditions.

7.1 Overview of Photosynthesis

Learning Outcomes

1. Write the general equations that represent the process of photosynthesis.
2. Diagram how photosynthesis powers the biosphere.
3. Describe the general structure of chloroplasts.
4. Explain how photosynthesis occurs in two stages: the light reactions and the Calvin cycle.

In the mid-1600s, a Flemish physician, Jan Baptista Van Helmont, conducted an experiment in which he transplanted the shoot of a young willow tree into a bucket of soil and allowed it to grow for 5 years. After this time, the willow tree had added 164 pounds to its original weight, but the soil had lost only 2 ounces. Van Helmont correctly concluded that the willow tree did not get most of its nutrients from the soil. From this work, he also hypothesized that the mass of the tree came from the water he had added over the 5 years. This hypothesis was partially correct, but we now know that CO_2 from the air also is a major contributor to the growth and mass of plants.

Photosynthesis is carried out by certain species of bacteria, algae, and plants. Most photosynthetic organisms use H_2O for photosynthesis, although some bacteria use hydrogen sulfide (H_2S) instead of water. When water is used, a general equation for photosynthesis is

$$CO_2 + 2\,H_2O + \text{Light energy} \rightarrow CH_2O + O_2 + H_2O$$

When the carbohydrate produced is glucose ($C_6H_{12}O_6$), we multiply each side of the equation by 6 to obtain

$$6\,CO_2 + 12\,H_2O + \text{Light energy} \rightarrow C_6H_{12}O_6 + 6\,O_2 + 6\,H_2O$$

$$\Delta G = +685 \text{ kcal/mol}$$

In this redox reaction, CO_2 is reduced during the formation of glucose, and H_2O is oxidized during the formation of O_2. Notice that the free-energy change required for the production of 1 mole of glucose from carbon dioxide and water is a whopping +685 kcal/mol! As we learned in Chapter 6, an endergonic reaction is driven forward by being coupled with an exergonic process that releases free energy. In this case, the energy from sunlight ultimately drives the synthesis of glucose.

In this section, we will survey the general features of photosynthesis as it occurs in green plants and algae. The sections that follow will examine the various steps in this process.

Photosynthesis Powers the Biosphere

The term **biosphere** refers to the places on the surface of the Earth and in the atmosphere where living organisms exist. Organisms can be categorized as autotrophs and heterotrophs.

- **Autotrophs** sustain themselves by producing organic molecules from inorganic sources. For example, plants can make glucose (an organic molecule) from inorganic CO_2 and H_2O. Autotrophs that use light as a source of energy to make organic molecules are called **photoautotrophs.** These include plants, algae, and some bacterial species such as cyanobacteria.

- **Heterotrophs** must consume food—organic molecules from their environment—to sustain life. Most species of bacteria and protists, as well as all species of fungi and animals, are heterotrophs.

Life in the biosphere is largely driven by the photosynthetic power of plants, algae, and cyanobacteria. The existence of most species relies on a key energy cycle that involves the interplay between organic molecules (such as glucose) and inorganic molecules, namely O_2, CO_2, and H_2O (**Figure 7.1**). Photoautotrophs make a large proportion of the Earth's organic molecules via photosynthesis, using light energy, CO_2, and H_2O. During this process, they also produce O_2. To supply their energy needs, both photoautotrophs and heterotrophs metabolize organic molecules via cellular respiration. As described in Chapter 6, cellular respiration generates CO_2 and H_2O and is used to make ATP. The CO_2 is released into the atmosphere and can be reused by photoautotrophs to make more organic molecules such as glucose. In this way, an energy cycle between photosynthesis and cellular respiration sustains life on our planet.

Biology Principle

Living Organisms Use Energy

Photosynthetic species capture light energy and store it in organic molecules, which are used by photosynthetic and nonphotosynthetic organisms as energy sources.

Figure 7.1 An important energy cycle between photosynthesis and cellular respiration. Photosynthesis uses light, CO_2, and H_2O to produce O_2 and organic molecules. The organic molecules are broken down to CO_2 and H_2O via cellular respiration to supply energy in the form of ATP; O_2 is reduced to H_2O.

Concept Check: Which types of organisms carry out cellular respiration? Is it heterotrophs, autotrophs, or both?

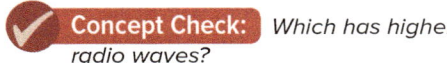

Figure 7.4 **The electromagnetic spectrum.** The bottom portion of this figure emphasizes visible light—the wavelengths of electromagnetic radiation visible to the human eye. Light in the visible portion of the electromagnetic spectrum drives photosynthesis.

Concept Check: *Which has higher energy, gamma rays or radio waves?*

Light also has properties that are characteristic of particles. Albert Einstein proposed that light is composed of discrete particles called **photons**—massless particles traveling in a wavelike pattern and moving at the speed of light (about 300 million m/sec). Each photon contains a specific amount of energy. An important difference between the various types of electromagnetic radiation, shown in Figure 7.4, is the amount of energy of the photons. Shorter-wavelength radiation carries more energy per unit of time than longer-wavelength radiation. For example, the photons of gamma rays carry more energy than those of radio waves.

The Sun radiates the entire spectrum of electromagnetic radiation, but the atmosphere prevents much of this radiation from reaching the Earth's surface. For example, the ozone layer forms a thin shield in the upper atmosphere, protecting life on Earth from much of the Sun's ultraviolet (UV) radiation. Even so, a substantial amount of electromagnetic radiation reaches the Earth's surface. The effect of light on living organisms is critically dependent on the energy of the photons that reach them. The photons in gamma rays, X-rays, and UV rays have very high energy. When molecules in cells absorb such energy, the effects can be devastating. Such radiation can cause mutations in DNA and even lead to cancer. By comparison, the energy of photons in visible light is much lower. Molecules can absorb this energy in a way that does not cause damage. Next, we will consider how molecules in living cells absorb the energy within visible light.

Pigments Absorb Light Energy

When light strikes an object, one of three things happens. First, light may simply pass through the object. Second, the object may change the path of light toward a different direction. A third possibility is that the object may absorb the light. The term **pigment** is used to describe a molecule that can absorb light energy. When light strikes a pigment, some of the wavelengths of light energy are absorbed, whereas others are reflected. For example, leaves look green to us because they reflect light energy with wavelengths in the green region of the visible spectrum. Various pigments in the leaves absorb the energy of other wavelengths. At the extremes of color reflection are white and black. A white object reflects nearly all of the visible light energy falling on it, whereas a black object absorbs nearly all of the light energy. This is why it is coolest to wear white clothes on a sunny, hot day.

What do we mean when we say that light energy is absorbed? Light energy in the visible spectrum can be absorbed by an atom when an electron is boosted to a higher energy level (**Figure 7.5**). The location in which an electron is found is called its orbital. Electrons in different orbitals possess different amounts of energy. For an electron to absorb light energy and be boosted to an orbital with a higher energy, it must overcome the difference in energy between the orbital it is in and the orbital to which it is going. For this to happen, an electron must absorb a photon that contains precisely that amount of energy.

After an electron absorbs energy, it is said to be in an excited state. Usually, this is an unstable condition. The electron can release the energy in different ways.

- An excited electron can release heat. For example, on a sunny day, the sidewalk heats up because it absorbs light energy that is released as heat.

- An electron can release energy in the form of light. Certain organisms, such as jellyfish, possess molecules that make them glow. This glowing is due to the release of light when electrons drop down to lower energy levels, a phenomenon called fluorescence.

- An excited electron can transfer its extra energy to an electron in a nearby molecule, a process called **resonance energy transfer.**

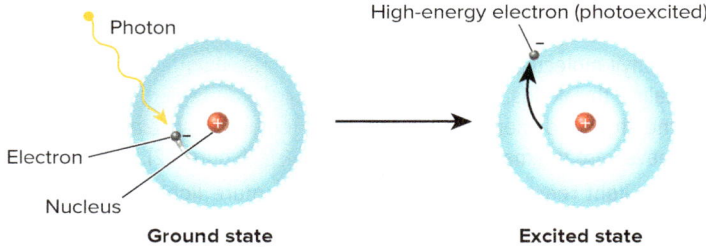

Figure 7.5 **Absorption of light energy by an electron.** When a photon of light having the correct amount of energy strikes an electron, the electron is boosted from the ground (unexcited) state to a higher energy level (an excited state). When this occurs, the electron occupies an orbital that is farther away from the nucleus of the atom. At this farther distance, the electron is held less firmly and is considered unstable.

Concept Check: *Describe four events that enable a photoexcited electron to become more stable.*

- In the case of photosynthetic pigments, another event can happen that is critical for the process of photosynthesis. Rather than releasing energy or transferring it to another molecule, an excited electron in a photosynthetic pigment is removed from that molecule and transferred to another molecule where the electron is stable. When this occurs, the energy in the electron is said to be "captured," because the electron does not readily drop down to a lower energy level and release heat or light.

Plants Contain Different Types of Photosynthetic Pigments

Plant cells that carry out photosynthesis synthesize different pigment molecules that absorb the light energy used to drive photosynthesis.

Chlorophylls Two types of chlorophyll pigments, termed **chlorophyll a** and **chlorophyll b**, are found in green plants and green algae (Figure 7.6a). In the chloroplast, both chlorophylls a and b are bound to integral membrane proteins in the thylakoid membrane, which traverse the membrane.

The chlorophylls contain a porphyrin ring and a hydrocarbon tail, also called a phytol tail. A magnesium ion (Mg^{2+}) is bound to the porphyrin ring. An electron in the porphyrin ring is able to hop from one atom in the ring to another. Because this electron isn't restricted to a single atom, it is called a delocalized electron. The delocalized electron can absorb light energy. The phytol tail in chlorophyll is a long hydrocarbon that is hydrophobic. Its function is to anchor the pigment to the surface of hydrophobic proteins within the thylakoid membrane of chloroplasts.

Carotenoids These pigments impart a color that ranges from yellow to orange to red (Figure 7.6b). Carotenoids are often the major pigments in flowers and fruits. In leaves, the more abundant chlorophylls usually mask the colors of carotenoids. In temperate climates where the leaves change colors, the quantity of chlorophyll in the leaf declines during autumn. The carotenoids become readily visible and produce the yellows, oranges, and reds of autumn foliage.

Absorption Spectra An **absorption spectrum** is a graph that plots a pigment's light absorption as a function of the light's wavelength. Each of the photosynthetic pigments shown in Figure 7.7a absorbs light in different regions of the visible spectrum. The absorption spectra of chlorophylls a and b are slightly different, though both chlorophylls absorb light most strongly in the red and violet-blue parts of the visible spectrum and absorb green light poorly. Carotenoids absorb light in the blue and blue-green regions of the visible spectrum, reflecting yellow and red.

Why do plants have different pigments? Having different pigments allows plants to absorb light at many different wavelengths. In this way, plants are more efficient at capturing the energy in sunlight. This phenomenon is highlighted in an **action spectrum**, which plots the rate of photosynthesis as a function of wavelength (Figure 7.7b). The highest rates of photosynthesis in green plants correlate with the wavelengths that are strongly absorbed by the chlorophylls and carotenoids. Photosynthesis is poor in the green region of the spectrum, because these pigments do not readily absorb this wavelength of light.

Photosystems II and I Work Together to Produce ATP and NADPH via Linear Electron Flow

As noted previously, photosynthetic organisms have the unique ability not only to absorb light energy but also to capture that energy in a stable way. Let's now consider how chloroplasts capture light energy. The thylakoid membranes of the chloroplast contain two distinct complexes of proteins and pigment molecules called **photosystem I (PSI)** and **photosystem II (PSII)** (Figure 7.8). Photosystem I was discovered before photosystem II, but photosystem II is the initial step in photosynthesis.

Events Within Photosystem II As described in steps 1a and 1b of Figure 7.8, light excites electrons in pigment molecules, such as chlorophylls, which are located in a region of PSII called the light-harvesting complex. Rather than releasing their energy in the form of heat, the excited electrons begin to follow a path shown by the red arrow.

Initially, the excited electrons move from a pigment molecule called P680 in PSII to other electron carriers called pheophytin (Pp), Q_A, and Q_B. PSII also oxidizes water, which generates O_2 and adds H^+ into the thylakoid lumen. The electrons released from the oxidized water molecules are used to replenish the electrons that exit PSII via Q_B.

Electron Transport Chain After a pair of electrons reaches Q_B, each one enters an **electron transport chain (ETC)**—a series of electron carriers—located in the thylakoid membrane. From Q_B, an electron goes to a cytochrome complex; then to plastocyanin (Pc), a small protein; and then to photosystem I. Along its journey from

(a) Chlorophylls a and b

(b) β-Carotene (a carotenoid)

Figure 7.6 Structures of pigment molecules. (a) The structure of chlorophylls a and b. As indicated, chlorophylls a and b differ only at a single site, at which chlorophyll a has a —CH_3 group and chlorophyll b has a —CHO group. **(b)** The structure of β-carotene, an example of a carotenoid. The dark green and light green areas of the structures in parts **(a)** and **(b)** are the regions where a delocalized electron can hop from one atom to another.

Biology Principle

Biologists Use Models to Test Predictions and Convey Ideas

This figure shows mathematical models in the form of graphs that describe the absorption of light and the rate of photosynthesis for photosynthetic organisms that are green. The Modeling Challenge in the figure legend asks you to propose a graph for the absorption spectrum for red algae.

(a) Absorption spectra

(b) Action spectrum

Figure 7.7 Properties of pigment function: absorption and action spectra. **(a)** These absorption spectra show the absorption of light by chlorophyll *a*, chlorophyll *b*, and β-carotene. **(b)** An action spectrum of photosynthesis depicting the relative rate of photosynthesis in green plants at different wavelengths of light.

Modeling Challenge: The absorption spectrum shown in part (a) is typical of green plants and green algae. Rhodophyta, also called red algae, contain a pigment called phycoerythrin that does not absorb light in the red region of the visible spectrum, which is why these algae appear red. Instead, phycoerythrin absorbs strongly in the blue and violet regions, and a little bit into the green region, of the spectrum. Draw an absorption spectrum for a red alga. Look back at Figure 7.4. What advantage would red algae have compared to green algae?

photosystem II to photosystem I, the electron releases some of its energy at particular steps and is transferred to the next component that has a higher electronegativity. The energy released is harnessed to pump H^+ into the thylakoid lumen.

Photosystem I and NADPH Synthesis A key role of photosystem I is to make NADPH (see Figure 7.8, steps 2 and 3). When light strikes the light-harvesting complex of photosystem I, this energy is also transferred to a reaction center in which a high-energy electron is removed from a pigment molecule, designated P700, and transferred to a primary electron acceptor. A protein called ferredoxin (Fd) can accept two high-energy electrons, one at a time, from the primary electron acceptor. Fd then transfers the two electrons to the enzyme $NADP^+$ reductase. This enzyme transfers the two electrons to $NADP^+$, which also accepts an H^+ to become NADPH. The formation of NADPH results in fewer H^+ in the stroma. The combined action of photosystem II and photosystem I is termed **linear electron flow** because the electrons move linearly from PSII to PSI and ultimately reduce $NADP^+$ to NADPH.

ATP Synthesis **Photophosphorylation** is the process of ATP synthesis in chloroplasts. The term refers to the idea that light is used to drive the phosphorylation of ADP to make ATP. As shown in Figure 7.8 (step 4), ATP is synthesized as H^+ flow through ATP synthase from the thylakoid lumen into the stroma. An H^+ electrochemical gradient is generated in three ways:

- The splitting of water, which places H^+ in the thylakoid lumen
- The movement of high-energy electrons along the ETC from photosystem II to photosystem I, which pumps H^+ into the thylakoid lumen
- The formation of NADPH, which consumes H^+ in the stroma

Products of Photosynthesis In summary, the steps of the light reactions of photosynthesis produce three chemical products: O_2, NADPH, and ATP:

1. O_2 is produced in the thylakoid lumen by the oxidation of water by photosystem II. Two electrons are removed from water, producing $2\ H^+$ and $½\ O_2$. The two electrons are transferred to $P680^+$ molecules as described later in Figure 7.10. The O_2 moves out of the chloroplast via passive diffusion.
2. NADPH is produced in the stroma using high-energy electrons that originate in the light-harvesting complex of photosystem II and are boosted a second time in photosystem I. Two high-energy electrons and one H^+ are transferred to $NADP^+$ to produce NADPH.
3. ATP is produced in the stroma via ATP synthase that uses an H^+ electrochemical gradient.

PSI Produces Only ATP via Cyclic Electron Flow

The mechanism of harvesting light energy described in Figure 7.8 is called linear electron flow because it is a linear process. This electron flow produces ATP and NADPH in roughly equal amounts. However, as we will see later, the Calvin cycle uses more ATP than NADPH. How can plant cells avoid making too much NADPH

Figure 7.8 The synthesis of ATP, NADPH, and O_2 by the concerted actions of photosystems II and I. The movement of electrons from photosystem II to photosystem I to NADPH is called linear electron flow.

Concept Check: *Are ATP, NADPH, and O_2 produced in the stroma or in the thylakoid lumen?*

and not enough ATP? In 1959, Polish-born American plant physiologist Daniel Arnon discovered a pattern of electron flow that is cyclic and generates only ATP (Figure 7.9). This process is called **cyclic photophosphorylation** because (1) the path of electrons is cyclic, (2) light energizes the electrons, and (3) ATP is made via the phosphorylation of ADP. Due to the path of electrons, the mechanism is also called **cyclic electron flow.**

When light strikes photosystem I, high-energy electrons are sent to the primary electron acceptor and then to ferredoxin (Fd). The key difference in cyclic electron flow is that the high-energy electrons are transferred from Fd to Q_B. From Q_B, the electrons then go to the cytochrome complex, then to plastocyanin (Pc) and back to photosystem I. As the electrons travel along this cyclic route, they release energy, and some of this energy is used to transport H^+ into the thylakoid lumen. The resulting H^+ gradient drives the synthesis of ATP via ATP synthase.

Cyclic electron flow is favored when the level of $NADP^+$ is low and NADPH is high. Under these conditions, there is sufficient NADPH to run the Calvin cycle. Alternatively, when $NADP^+$ is high and NADPH is low, linear electron flow is favored, so more NADPH can be made. Cyclic electron flow is also favored when ATP levels are low.

7.2 Reviewing the Concepts

- Light is a form of electromagnetic radiation that travels in waves and is composed of photons with discrete amounts of energy (Figure 7.4).
- Electrons can absorb light energy and be boosted to a higher energy level—an excited state (Figure 7.5).
- Photosynthetic pigments include chlorophylls *a* and *b* and carotenoids. These pigments absorb light energy in the visible spectrum to drive photosynthesis (Figures 7.6, 7.7).
- During linear electron flow, electrons from photosystem II follow a pathway along an electron transport chain (ETC) in the thylakoid membrane. This pathway generates an H^+ gradient that is used to make ATP. In addition, light energy striking photosystem I (PSI) boosts electrons to a very high energy level that allows the synthesis of NADPH (Figure 7.8).

Figure 7.9 **The synthesis of only ATP by cyclic photophosphorylation.** In this process, electrons follow a cyclic path that is powered by photosystem I (PSI). This produces an H⁺ electrochemical gradient, which is then used by ATP synthase to make ATP.

Concept Check: Why does cyclic photophosphorylation provide an advantage to a plant over using only linear electron flow?

- During cyclic photophosphorylation, electrons are excited in PSI and flow from ferredoxin (Fd) through the ETC and back to PSI. This cyclic electron route produces an H⁺ gradient that is used to make ATP (Figure 7.9).

7.2 Testing Your Knowledge

1. When a photoexcited electron returns to the ground state, which of the following will *not* occur?
   a. It could give off light.
   b. It could absorb a photon of light.
   c. It could give off heat.
   d. All of the above are possible.

2. During linear electron flow, the primary function of photosystem I (PSI) is to make
   a. ATP.
   b. NADPH.
   c. O_2.
   d. glucose.
   e. all of the above.

3. What is the purpose of cyclic photophosphorylation, which involves photosystem I but *not* photosystem II?
   a. production of ATP
   b. production of NADPH
   c. production of both ATP and NADPH
   d. production of O_2
   e. production of ATP and O_2

7.3 Molecular Features of Photosystems

Learning Outcomes

1. Describe the molecular mechanism by which photosystem II absorbs light energy and produces O_2.
2. Diagram the variation in the energy of an electron as it moves from PSII to NADP⁺.

The previous section provided an overview of how chloroplasts absorb light energy and produce ATP, NADPH, and O_2. As you have learned, two photosystems—PSI and PSII—play critical roles in two aspects of photosynthesis. First, both PSI and PSII absorb light energy and capture that energy in the form of excited electrons. Second, PSII oxidizes water, thereby producing O_2. In this section, we will take a closer look at how these events occur at the molecular level.

Photosystem II Captures Light Energy and Produces O_2

PSI and PSII have two main components: a light-harvesting complex and a reaction center. **Figure 7.10** shows how these components function in PSII. In 1932, American biologist Robert Emerson and an undergraduate student, William Arnold, discovered the **light-harvesting complex** in the thylakoid membrane. It is composed by a process called **resonance energy transfer.** The energy may be transferred among multiple pigment molecules until it is eventually transferred to a special pigment molecule designated P680, which is located within a site in PSII called the reaction center, where redox reactions take place. The P680 pigment is so named because it can directly absorb light at a wavelength of 680 nm. However, P680 is more commonly excited by resonance energy transfer from a chlorophyll pigment in the light-harvesting complex. In either case, when an electron in P680 is excited, it is designated P680*.

Unlike the pigments in the light-harvesting complex that undergo resonance energy transfer, P680* can actually release its high-energy electron and become P680$^+$.

$$P680^* \rightarrow P680^+ + e^-$$

One role of the reaction center is to carry out a redox reaction in which the high-energy electron from P680* is transferred to another molecule, where the electron is stable. This molecule is called the **primary electron acceptor** (see Figure 7.10). The transfer of the electron from P680* to the primary electron acceptor is remarkably fast. It occurs in less than a few picoseconds! (One picosecond equals one-trillionth of a second, or 10^{-12} sec.) Because this occurs so quickly, the excited electron does not have much time to release its energy in the form of heat or light.

When the primary electron acceptor has received this high-energy electron, the light energy has been captured and can be used to perform cellular work. As discussed earlier, the work it performs is to synthesize the energy intermediates ATP and NADPH (see Figure 7.8).

Let's now consider what happens to P680$^+$, which has given up its high-energy electron. After P680$^+$ is formed, the electron that has been removed must be replaced so that P680 can function again. Therefore, another role of the reaction center is to replace the electron that is removed when P680* becomes P680$^+$. This missing electron of P680$^+$ is replaced with a low-energy electron from water (see Figure 7.10).

$$H_2O \rightarrow 2\ H^+ + \tfrac{1}{2}\ O_2 + 2\ e^-$$
$$2\ P680^+ + 2\ e^- \rightarrow 2\ P680$$
(from water)

The oxidation of water results in the formation of oxygen gas (O_2), which many organisms use for cellular respiration. Photosystem II is the only known protein complex that can oxidize water, resulting in the release of O_2 into the atmosphere.

Electrons Vary in Their Energy Level as They Travel from Photosystem II to Photosystem I to NADP$^+$

In 1960, Robin Hill and Fay Bendall proposed that photosynthesis involves two events in which light is absorbed. According to their model, known as the **Z scheme,** an electron proceeds through a series of energy changes during photosynthesis (**Figure 7.11**). The Z refers to the zigzag shape of this energy curve. Based on our modern understanding of photosynthesis, we now know that these events involve increases and decreases in the energy of an electron

Figure 7.10 A closer look at how photosystem II harvests light energy and oxidizes water. Note: Two electrons are released during the oxidation of water, but they are transferred one at a time to P680$^+$. The primary electron acceptor is a pigment molecule named pheophytin; see Figure 7.8.

of several dozen pigment molecules that are anchored to transmembrane proteins. The role of the complex is to directly absorb photons of light. When a pigment molecule absorbs a photon, an electron is boosted to a higher energy level. As shown in Figure 7.10, the energy (not the electron itself) is transferred to adjacent pigment molecules

Figure 7.11 The Z scheme, showing the energy of an electron moving from photosystem II to NADP⁺. The oxidation of water releases two electrons that travel one at a time from photosystem II to NADP⁺. As seen here, the input of light boosts the energy of the electron twice. At the end of the pathway, two electrons are used to make NADPH.

Concept Check: *During its journey from photosystem II to NADP⁺, at what point does an electron have the highest amount of energy?*

as it moves linearly from photosystem II through photosystem I to NADP⁺.

- An electron on a nonexcited pigment molecule in the light-harvesting complex of photosystem II has the lowest energy.
- In photosystem II, light boosts such an electron to a much higher energy level.
- As the electron travels from photosystem II to photosystem I, some of the energy is released.
- The input of light in photosystem I boosts the electron to an even higher energy than it attained in photosystem II.
- The electron releases a little energy before it is eventually transferred to NADP⁺.

7.3 Reviewing the Concepts

- In the light-harvesting complex of photosystem II (PSII), pigment molecules absorb light energy that is transferred to P680 in the reaction center via resonance energy transfer. A high-energy electron from P680* is transferred to a primary electron acceptor. An electron from water is then used to replenish the electron lost from P680*. The oxidation of water produces O_2 (Figure 7.10).
- The Z scheme proposes that an electron absorbs light energy twice, at both PSII and PSI, losing some of that energy as it flows along the ETC before it is transferred to NADP⁺ (Figure 7.11).

7.3 Testing Your Knowledge

1. After a chlorophyll pigment in the light-harvesting complex of PSII absorbs a photon of light, the
   a. excited electron is transferred to the reaction center.
   b. excited electron is transferred to the electron transport chain.
   c. energy is transferred to P680 via resonance energy transfer.
   d. energy is transferred to the primary electron acceptor via resonance energy transfer.
   e. energy is released as heat, which is absorbed by the primary electron acceptor.

2. After P680* becomes P680⁺, it is converted back to P680
   a. by resonance energy transfer.
   b. by receiving an electron from the light-harvesting complex.
   c. by receiving an electron from the primary electron acceptor.
   d. by receiving an electron from water.
   e. by receiving an electron from O_2.

7.4 Synthesizing Carbohydrates via the Calvin Cycle

Learning Outcomes

1. Outline the three phases of the Calvin cycle.
2. **SCISKILLS** ▶ Analyze the results of Calvin and Benson, and explain how they identified the components of the Calvin cycle.

In the previous sections, we learned how the light reactions of photosynthesis produce O_2 and two energy intermediates, ATP and NADPH. We will now turn our attention to the second phase of photosynthesis—the Calvin cycle—in which ATP and NADPH are used to make carbohydrates. The Calvin cycle consists of a series of steps that occur in a metabolic cycle. In plants and algae, the Calvin cycle takes place in the stroma of chloroplasts. In cyanobacteria, it occurs in the cytoplasm, where the enzymes that catalyze each of the steps are located.

The Calvin cycle takes CO_2 from the atmosphere and incorporates the carbon into organic molecules, namely carbohydrates. As mentioned earlier, carbohydrates are critical for two reasons. First, they provide the precursors to make the organic molecules and macromolecules of nearly all living cells. The second key reason is the storage of energy. The Calvin cycle produces carbohydrates, which store energy. These carbohydrates are accumulated inside plant cells. The stored carbohydrates are used as a source of energy. Similarly, when an animal consumes a plant, it uses the carbohydrates as an energy source.

In this section, we will examine the three phases of the Calvin cycle. We will also explore the experimental approach of Melvin Calvin and his colleagues that enabled them to elucidate the steps of this cycle.

The Calvin Cycle Incorporates CO_2 into a Carbohydrate

The Calvin cycle, also called the Calvin-Benson cycle, was determined by chemists Melvin Calvin and Andrew Adam Benson and their colleagues in the 1940s and 1950s. This cycle requires a massive input of energy. For every 6 CO_2 molecules that are incorporated into a carbohydrate such as glucose ($C_6H_{12}O_6$), 18 ATP molecules are hydrolyzed and 12 NADPH molecules are oxidized:

$$6\ CO_2 + 12\ H_2O \rightarrow C_6H_{12}O_6 + 6\ O_2 + 6\ H_2O$$

$$18\ ATP + 18\ H_2O \rightarrow 18\ ADP + 18\ P_i$$

$$12\ NADPH \rightarrow 12\ NADP^+ + 12\ H^+ + 24\ e^-$$

Although biologists commonly describe glucose as a product of photosynthesis, glucose is not directly made by the Calvin cycle. Instead, glyceraldehyde-3-phosphate (G3P) is the product of the Calvin cycle. G3P molecules are used as starting materials for the synthesis of glucose and other molecules, including sucrose. After glucose molecules are made, they may be linked together to form a polymer of glucose called starch, which is stored in the chloroplast for later use. Alternatively, the disaccharide sucrose may be made and transported out of the leaf to other parts of the plant.

The Calvin cycle is divided into three phases: carbon fixation, reduction and carbohydrate production, and regeneration of ribulose bisphosphate (RuBP) (**Figure 7.12**).

1. **Carbon fixation:** During **carbon fixation**, CO_2 is incorporated into RuBP, a five-carbon sugar. Fixation means that the carbon has been removed from the atmosphere and incorporated into an organic molecule that is not a gas. More specifically, the product of this reaction in phase 1 is a six-carbon intermediate that immediately splits in half to form two molecules of 3-phosphoglycerate (3PG). The enzyme that catalyzes this step is named RuBP carboxylase/oxygenase, or **rubisco**. It is the most abundant protein in chloroplasts and perhaps the most abundant protein on Earth! This observation underscores the massive amount of carbon fixation that happens in the biosphere.

2. **Reduction and carbohydrate production:** In the second phase of the Calvin cycle, ATP is used to convert 3PG to 1,3-bisphosphoglycerate (1,3-BPG). Next, electrons from NADPH reduce 1,3-BPG to glyceraldehyde-3-phosphate (G3P). G3P is a carbohydrate with three carbon atoms. The key difference between 3PG and G3P is that a carbon in 3PG forms a C—O bond, whereas the analogous carbon in G3P has a C—H bond (see Figure 7.12). The C—H bond results when the G3P molecule is reduced by the addition of two electrons from NADPH. Compared with 3PG, the bonds in G3P store more energy and enable G3P to readily form larger organic molecules such as glucose.

   As shown in Figure 7.12, only some of the G3P molecules are used to make glucose or other carbohydrates. Phase 1 begins with 6 RuBP molecules and 6 CO_2 molecules. By the end of the second phase, 12 G3P molecules have been made but only 2 of these G3P molecules are used in carbohydrate production. As described next, the other 10 G3P molecules are needed to keep the Calvin cycle turning by regenerating RuBP.

3. **Regeneration of RuBP:** In the third phase of the Calvin cycle, a series of enzymatic steps converts the 10 G3P molecules into 6 RuBP molecules, using 6 molecules of ATP. After the RuBP molecules are regenerated, they serve as acceptors for CO_2, thereby allowing the cycle to continue.

As we have just seen, the Calvin cycle begins by using carbon from an inorganic source (that is, CO_2) and ends with organic molecules that will be used by the plant to make other molecules. You may be wondering why CO_2 molecules cannot be directly linked to form these larger molecules. The answer lies in the number of electrons that are located around carbon atoms. In CO_2, the carbon atom is considered electron poor. Oxygen is a very electronegative atom that monopolizes the electrons it shares with other atoms. In a covalent bond between carbon and oxygen, the shared electrons are closer to the oxygen atom.

By comparison, in an organic molecule, the carbon atom is electron rich. During the Calvin cycle, ATP provides energy and NADPH donates high-energy electrons, so the carbon originally in CO_2 has been reduced. The Calvin cycle combines less electronegative atoms with carbon atoms so that C—H and C—C bonds are formed. This allows the eventual synthesis of larger organic molecules, including glucose, amino acids, and so on. In addition, the covalent bonds within these molecules store large amounts of energy.

autoradiography.

X-ray film

Figure 7.12 The Calvin cycle. This cycle has three phases: (1) carbon fixation, (2) reduction and carbohydrate production, and (3) regeneration of RuBP.

Concept Check: *Why is NADPH needed during this cycle?*

8.1 General Features of Cell Communication

Learning Outcomes

1. Explain the two general reasons for cell signaling: responding to environmental changes and cell-to-cell communication.
2. Compare and contrast the five ways that cells communicate with each other based on the distance between them.
3. Outline the three-stage process of cell signaling.

All living organisms, including bacteria, archaea, protists, fungi, plants, and animals, must engage in cell communication to survive. Cell communication, also known as **cell signaling,** involves both incoming and outgoing signals. For example, on a sunny day, cells can sense their exposure to ultraviolet (UV) light—a physical signal—and respond accordingly. In humans, UV light acts as an incoming signal, and one way that we respond to it is to synthesize melanin, a protective pigment that helps to prevent the harmful effects of UV radiation. In addition, cells produce outgoing signals that influence the behavior of neighboring cells. Plant cells, for example, make hormones that influence the pattern of cell elongation, so the plant grows toward light. Cells of all living organisms both respond to incoming signals and produce outgoing signals. Cell communication is a two-way street.

In this section, we will begin by considering why cells need to respond to signals. We will then examine various forms of signaling that are based on the distance between the cells that communicate with each other. Finally, we will examine the main steps that occur when a cell is exposed to a signal and produces a response to it.

Cells Detect and Respond to Signals from Their Environment and from Other Cells

Before getting into the details of cell communication, let's take a general look at why cell communication is necessary.

1. Responding to Changes in the Environment Changes in the environment are a persistent feature of life, and living cells are continually faced with alterations in temperature and availability of nutrients, water, and light. A cell may even be exposed to a toxic chemical in its environment. Being able to respond to change at the cellular level is called a **cellular response.** As an example, let's consider the response of a yeast cell to glucose in its environment (**Figure 8.1**). Some of the glucose acts as a signaling molecule that binds to a receptor and causes a cellular response. In this case, the cell responds by increasing the number of glucose transporters needed to take glucose into the cell and by increasing the number of metabolic enzymes required to utilize glucose once it is inside. The cellular response allows the cell to use glucose efficiently.

2. Cell-to-Cell Communication Cells need to communicate with each other—a type of cell communication called **cell-to-cell communication.** In one of the earliest experiments demonstrating cell-to-cell communication, Charles Darwin and his son Francis studied phototropism, the phenomenon in which plants grow toward light (**Figure 8.2**). The Darwins observed that the

Biology Principle

Living Organisms Interact with Their Environment
In this example, cell signaling allows yeast cells to respond to the presence of glucose in their environment.

Due to glucose in the environment, the yeast cell has undergone a cellular response by synthesizing more glucose transporters and enzymes that are needed to metabolize glucose.

Figure 8.1 Response of a yeast cell to glucose. When glucose is absent from the extracellular environment, the cell is not well prepared to take up and metabolize this sugar. However, when glucose is present, some of that glucose binds to receptors in the cell membrane, which leads to changes in the amounts and properties of glucose transporters and metabolic enzymes so that the cell can readily use glucose.

Concept Check: *What is the signaling molecule in this example?*

actual bending occurs in a zone below the growing shoot tip. They concluded that a signal must be transmitted from the growing tip to lower parts of the shoot. Later research revealed that the signal is a molecule called auxin, which is transmitted from cell to cell. A higher amount of auxin accumulates on the nonilluminated side of the shoot. It binds to a receptor that promotes cell elongation on that side of the shoot only, thereby causing the shoot to bend toward the light source.

Cell-to-Cell Communication Occurs Between Adjacent Cells and Between Cells That Are Long Distances Apart

Organisms have a variety of mechanisms to achieve cell-to-cell communication. The mode of communication depends, in part, on the distance between the cells. One way to categorize cell signaling is by the manner in which the signal is transmitted from one cell to another. Signals are relayed between cells in five common ways, all of which involve a cell that produces the signal and a target cell that receives it (**Figure 8.3**).

1. Direct Intercellular Signaling In a multicellular organism, cells adjacent to each other may have contacts, called intercellular channels, that enable them to pass ions, signaling molecules, and

Figure 8.2 Phototropism in plants. This process involves cell-to-cell communication that leads to a shoot bending toward light just beneath its growing tip.
©Cordelia Molloy/Science Source

other materials between the cytosol of one cell and the cytosol of the other (Figure 8.3a). For example, cardiac muscle cells, which cause your heart to beat, have intercellular channels called gap junctions that allow the passage of ions needed for the coordinated contraction of these cells.

2. Contact-Dependent Signaling Some molecules are bound to the surface of a cell and provide a signal to other cells that make contact with the surface of that cell (Figure 8.3b). In the case of contact-dependent signaling, one cell has a membrane-bound signaling molecule that is recognized by a receptor on the surface of another cell. This type of cell-to-cell communication occurs, for example, when neurons (cells of the nervous system) grow and make contact with other neurons. This is important for the formation of the proper connections between neurons.

3. Autocrine Signaling In autocrine signaling, a cell secretes signaling molecules that bind to receptors on its own cell surface or on the surfaces of neighboring cells of the same cell type, stimulating a response (Figure 8.3c). One purpose of autocrine signaling is for groups of cells to sense cell density. When cell density is high, the concentration of autocrine signals is also high. In some cases, such signals will inhibit further cell growth and thereby limit cell density.

4. Paracrine Signaling In paracrine signaling, a cell secretes a signaling molecule that does not affect that cell but instead influences the behavior of target cells in close proximity (Figure 8.3d). Paracrine signaling is typically of short duration. Usually, the signal is broken down too quickly to be carried to other parts of the body and affect distant cells. A specialized form of paracrine signaling occurs in the nervous systems of animals. Neurotransmitters—molecules made in neurons—are released at the end of a neuron and traverse a narrow space called the synaptic cleft (look ahead to Chapter 33, Figure 33.12). The neurotransmitter then binds to a receptor in a target cell.

5. Endocrine Signaling In contrast to the previous mechanisms of cell-to-cell communication, endocrine signaling occurs over relatively long distances (Figure 8.3e). In both animals and plants, molecules involved in long-distance signaling are called **hormones.** They usually last longer than signaling molecules that are used in autocrine and paracrine signaling. In vertebrate animals, endocrine signaling involves the secretion of hormones into the bloodstream, which may affect virtually all cells of the body, including those that are far from the cells that secrete the signaling molecules. In vascular plants, hormones move through the vascular system as well as through adjacent cells. Some hormones are even gases that diffuse into the air. Ethylene, a gas given off by plants, plays a variety of roles, one of which is to accelerate the ripening of fruit.

Cells Usually Respond to Signals via a Three-Stage Process

Up to this point, we have learned that signals influence the behavior of cells in close proximity or at long distances, interacting with receptors to elicit a cellular response. What events occur when a cell receives a signal? In most cases, the binding of a signaling molecule to a receptor causes the receptor to activate a signal transduction pathway, which then leads to a cellular response. **Figure 8.4** diagrams the three common stages of cell signaling: receptor activation, signal transduction, and a cellular response.

Stage 1: Receptor Activation In the initial stage, a signaling molecule binds to a receptor of the target cell, causing a conformational change in the receptor that activates its function. In most cases, the activated receptor initiates a response by causing changes in a series of proteins that collectively form a signal transduction pathway, as described next.

Stage 2: Signal Transduction During signal transduction, the initial signal is converted—or transduced—to a different signal inside the cell. This process is carried out by a group of proteins that form a **signal transduction pathway.** These proteins undergo a series of changes that may result in the production of intracellular signals.

Stage 3: Cellular Response Signals cause cells to respond in several different ways. Figure 8.4 shows three common categories of proteins that are controlled by cell signaling: enzymes, structural proteins, and transcription factors.

- Many signaling molecules exert their effects by altering the activity of one or more enzymes. For example, certain hormones provide a signal that the body needs energy. These hormones activate enzymes that are required for the breakdown of molecules such as carbohydrates.

(a) Direct intercellular signaling: Signals pass through an intercellular channel from the cytosol of one cell to adjacent cells.

(b) Contact-dependent signaling: Membrane-bound signals bind to receptors on adjacent cells.

(c) Autocrine signaling: Cells release signals that affect themselves and nearby target cells.

(d) Paracrine signaling: Cells release signals that affect nearby target cells.

(e) Endocrine signaling: Cells release signals that travel long distances to affect target cells.

Figure 8.3 Types of cell-to-cell communication based on mechanism and the distance between cells.

Concept Check: Which type of signal, paracrine or endocrine, is likely to exist for a longer period of time? Explain why this longer existence is necessary.

Figure 8.4 The three stages of cell signaling: receptor activation, signal transduction, and a cellular response. Note: The examples described in this chapter involve signaling molecules that activate receptors. Some signaling molecules inhibit receptor function. Transcription factors function within the nucleus, whereas enzymes and structural proteins may function in various regions throughout the cell.

Concept Check: Explain why a signal transduction pathway is necessary for most signaling molecules to have an effect.

- Cells also respond to signals by altering the functions of intracellular structural proteins. For example, when animal cells move during embryonic development or when an amoeba moves toward food, signals play a role in the rearrangement of actin filaments, which are components of the cytoskeleton. The coordination of signaling and changes in the cytoskeleton enables a cell to move in the correct direction.
- Signaling molecules may also affect the function of **transcription factors**—proteins that regulate the transcription of genes. Some transcription factors activate gene expression. For example, when cells are exposed to sex hormones, transcription factors activate genes that change the properties of cells, which can lead to changes in the sexual characteristics of organisms.

8.1 Reviewing the Concepts

- A signal is an agent that can influence the properties of cells. A signal binds to a receptor to elicit a cellular response. Cell signaling enables cells to respond to environmental changes and to communicate with each other (Figures 8.1, 8.2).
- Cell-to-cell communication varies in the mechanism and the distance that a signal travels. Signals are relayed between cells in five common ways: direct intercellular, contact-dependent, autocrine, paracrine, and endocrine signaling (Figure 8.3).
- Cell communication is usually a three-stage process involving receptor activation, signal transduction, and a cellular response. A signal transduction pathway is a group of proteins that convert an initial signal to a different signal inside the cell (Figure 8.4).

8.1 Testing Your Knowledge

1. A general reason for cell signaling is
   a. to respond to environmental changes.
   b. cell-to-cell communication.
   c. protein degradation.
   d. all of the above.
   e. both a and b.
2. A type of cell signaling that involves signaling molecules that travel long distances is
   a. direct intercellular.
   b. contact-dependent.
   c. autocrine.
   d. paracrine.
   e. endocrine.
3. Which of the following is the correct order for the three stages of cell signaling?
   a. receptor activation → cellular response → signal transduction
   b. receptor activation → signal transduction → cellular response
   c. signal transduction → receptor activation → cellular response
   d. signal transduction → cellular response → receptor activation
   e. cellular response → signal transduction → receptor activation

8.2 Receptor Activation

Learning Outcomes

1. Explain how a signaling molecule, or ligand, binds to its receptor and causes a conformational change that activates the receptor.
2. **SCISKILLS** ▶ Calculate the affinity, measured as a dissociation constant, that a receptor has for its signaling molecule, or ligand.

In this section, we will take a closer look at receptors and their interactions with signaling molecules. Our focus will be on receptors that respond to chemical signaling molecules. Other receptors, discussed in Units VI and VII, respond to mechanical motion (mechanoreceptors), pressure (baroreceptors), temperature change (thermoreceptors), and light (photoreceptors).

Receptors Bind to Specific Signals and Undergo Conformational Changes

The ability of cells to respond to a signal usually requires precise recognition between the signal and its receptor. In many cases, the signal is a molecule, such as a steroid or a protein, which binds to the receptor. A signaling molecule binds to a receptor in much the same way that a substrate binds to the active site of an enzyme, as described in Chapter 6 (refer back to Figure 6.5). Most receptors bind to their signaling molecule with a high degree of specificity, which means a given receptor only recognizes a signaling molecule with a very particular type of structure. The signaling molecule, which is called a **ligand,** binds noncovalently to the receptor molecule. The binding occurs when the ligand and receptor happen to collide in the correct orientation and with enough energy to form a **ligand·receptor complex.**

$$[\text{Ligand}] + [\text{Receptor}] \underset{k_{\text{off}}}{\overset{k_{\text{on}}}{\rightleftharpoons}} [\text{Ligand·Receptor complex}]$$

Square brackets, [ ], indicate concentration. The value k_{on} is the rate at which binding occurs. After a complex forms between the ligand and its receptor, the noncovalent interaction between a ligand and its receptor remains stable for a finite period of time. The term k_{off} is the rate at which the ligand·receptor complex falls apart, or dissociates.

Unlike enzymes, which convert their substrates into products, receptors do not usually alter the structure of their ligands. Instead, the ligands alter the structure of their receptors, causing a conformational change (**Figure 8.5**). In this case, the binding of the ligand to its receptor changes the receptor in a way that activates its ability to initiate a cellular response. Because the binding of a ligand to its receptor is a reversible process, the ligand and receptor will also dissociate. Once the ligand is released, the receptor is no longer activated.

Figure 8.5 Receptor activation.

BioConnections: Look back at Figure 6.5. How is the binding of a ligand to its receptor similar to the binding of a substrate to an enzyme? How are they different?

Biology Principle | Biology is a quantitative science

Quantitative Analysis: Receptors Have a Measurable Affinity for Their Ligands

In general, the binding and dissociation of a ligand and its receptor occur relatively rapidly, and therefore an equilibrium is reached when the rate of formation of new ligand·receptor complexes equals the rate at which existing ligand·receptor complexes dissociate:

$$k_{on}[\text{Ligand}][\text{Receptor}] = k_{off}[\text{Ligand·Receptor complex}]$$

Rearranging the equation gives

$$\frac{[\text{Ligand}][\text{Receptor}]}{[\text{Ligand·Receptor complex}]} = \frac{k_{off}}{k_{on}} = K_d$$

K_d is called the **dissociation constant** between a ligand and its receptor. The K_d value is inversely related to the affinity between the ligand and receptor. A low K_d value indicates that a receptor has a high affinity for its ligand.

Let's look carefully at the left side of this equation and consider what it means. At a ligand concentration where half of the receptors are bound to a ligand, the concentration of the ligand·receptor complex equals the concentration of receptor that doesn't have ligand bound. At this ligand concentration, [Receptor] and [Ligand·Receptor complex] cancel out of the equation because they are equal. Therefore, at a ligand concentration where half of the receptors have bound ligand

$$K_d = [\text{Ligand}]$$

When the ligand concentration is above the K_d value, most of the receptors are likely to have ligand bound to them. In contrast, if the ligand concentration is substantially below the K_d value, most receptors will not be bound by their ligand. The K_d values for many different ligands and their receptors have been experimentally determined. How is this information useful? It allows researchers to predict when a signaling molecule is likely to cause a cellular response. If the concentration of a signaling molecule is far below the K_d value, a cellular response is not likely because relatively few receptors will form a complex with a signaling molecule.

Crunching the Numbers: Type 1 diabetes is a human disease in which the pancreas is unable to make sufficient amounts of the hormone insulin. The disorder often begins in childhood, but its onset may occur in adulthood. Type 1 diabetes affects over 1 million Americans.

After a nondiabetic person eats a meal, a proper amount of insulin is secreted into the bloodstream. Insulin binds to the insulin receptor found on the surfaces of certain cell types, such as muscle cells, and promotes the uptake of glucose from the blood into the cells. The insulin receptor on skeletal muscle cells has a K_d of approximately 0.2 nM. Keisha, who does not have type 1 diabetes, eats a meal and her blood insulin level rises to 0.45 nM. Ryan has type 1 diabetes. If he is not given insulin by injection, his blood insulin level following a meal rises to only 0.06 nM. For both Keisha and Ryan, is the percentage of insulin receptors that have insulin bound under these conditions above or below 50%? Why does Ryan have to take regular insulin injections?

8.2 Reviewing the Concepts

- The binding of a signaling molecule, also called a ligand, to a receptor is usually very specific and alters the conformation of the receptor (Figure 8.5).
- A ligand binds to a receptor with an affinity that is measured as a dissociation constant, or K_d value. The K_d value is inversely related to the affinity between the ligand and its receptor.

8.2 Testing Your Knowledge

1. When a ligand (signaling molecule) binds to a receptor,
   a. the ligand is converted to a product.
   b. the ligand forms a covalent bond with the receptor.
   c. the receptor undergoes a conformational change.
   d. the receptor is rapidly degraded.
   e. the receptor phosphorylates the ligand.

2. A receptor has a K_d for its ligand of 50 nM. At a ligand concentration of 100 nM,
   a. a small percentage of receptors will have ligand bound to them.
   b. 50% of the receptors will have ligand bound to them.
   c. most receptors will have ligand bound to them.
   d. all receptors will have ligand bound to them.
   e. none of the above will be true.

8.3 Cell Surface Receptors

Learning Outcome

1. Compare and contrast the three general types of cell surface receptors.

Most signaling molecules are either small, hydrophilic molecules or large molecules that do not readily pass through the plasma membrane of cells. Such extracellular signaling molecules bind to **cell surface receptors**—receptors embedded in the plasma membrane. A typical cell usually has dozens or even hundreds of different cell surface receptors that enable the cell to respond to different kinds of extracellular signaling molecules. By analyzing the functions of cell surface receptors from many different organisms, researchers have determined that most fall into one of three categories: enzyme-linked receptors, G-protein-coupled receptors, and ligand-gated ion channels.

1. **Enzyme-Linked Receptors** Receptors known as **enzyme-linked receptors** are found in the cells of all living species. Many human hormones bind to this type of receptor. For example, when insulin binds to an enzyme-linked receptor in muscle cells, it enhances the ability of those cells to use glucose. Enzyme-linked receptors typically have two important domains: an extracellular domain, which binds a signaling molecule, and an intracellular domain, which has a catalytic function (Figure 8.6a). When a signaling molecule binds to the extracellular domain, a conformational change is transmitted through the membrane-embedded portion of the protein that affects the conformation of the intracellular catalytic domain. In most cases, this conformational change causes the intracellular catalytic domain to become functionally active.

Most types of enzyme-linked receptors function as **protein kinases,** enzymes that transfer a phosphate group from ATP to a specific amino acid in a protein (Figure 8.6b). For example, tyrosine kinases attach phosphate to the amino acid tyrosine, whereas serine/threonine kinases attach phosphate to the amino acids serine and threonine. In the example shown in Figure 8.6b, the catalytic domain of the receptor is inactive when no signaling molecule is present. However, when a signal binds to the extracellular domain, the catalytic domain becomes activated. Under these conditions, the receptor may phosphorylate itself, or (as shown in Figure 8.6b) it may phosphorylate intracellular proteins. The attachment of a negatively charged phosphate changes the structure of a protein and can thereby alter its function. Later in this chapter, we will explore how this event leads to a cellular response, such as the activation of enzymes that affect cell function.

2. **G-Protein-Coupled Receptors** Receptors called **G-protein-coupled receptors (GPCRs)** are found in the cells of all eukaryotic species and are particularly common in animals. GPCRs typically contain seven transmembrane segments that wind back and forth through the plasma membrane. The receptors interact with intracellular proteins called **G proteins,** which are so named because of

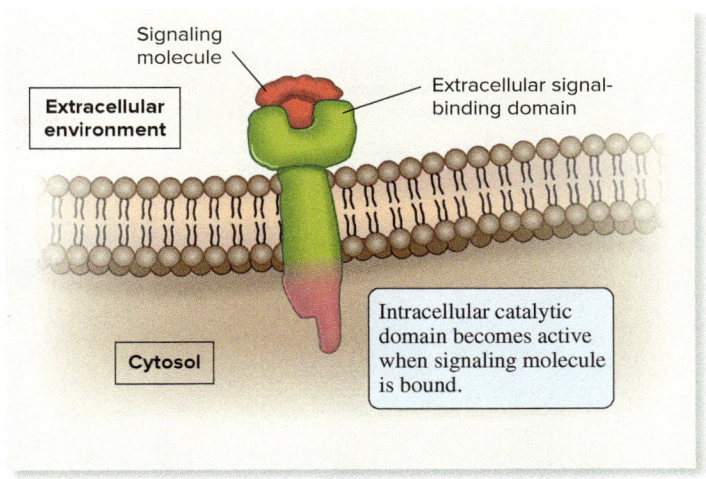

(a) **Structure of an enzyme-linked receptor**

(b) **A receptor that functions as a protein kinase**

Figure 8.6 Enzyme-linked receptors.

Concept Check: *Based on your understanding of ATP as an energy intermediate, is the phosphorylation of a protein via a protein kinase an exergonic or endergonic reaction? What is the purpose of phosphorylation?*

Figure 8.7 The activation of G-protein-coupled receptors (GPCRs) and G proteins.

Concept Check: *What has to happen before the α and β/γ subunits of the G protein can reassociate with each other?*

their ability to bind guanosine triphosphate (GTP) and guanosine diphosphate (GDP). GTP is similar in structure to ATP except it has guanine as a base instead of adenine.

Figure 8.7 shows how a GPCR and a G protein interact. At the cell surface, a signaling molecule binds to a GPCR, resulting in a conformational change that activates the receptor, causing it to bind to a G protein. The G protein, which is a lipid-anchored protein, releases GDP and binds GTP instead. The binding of GTP changes the conformation of the G protein, causing it to dissociate into an α subunit and a β/γ dimer. Later in this chapter, we will examine how the α subunit interacts with other proteins in a signal transduction pathway to elicit a cellular response.

3. Ligand-Gated Ion Channels As described in Chapter 5 (refer back to Figure 5.12), ion channels are proteins that allow the diffusion of ions across biological membranes. **Ligand-gated ion channels** are a third type of cell surface receptor found in the plasma membranes of animal, plant, and fungal cells. When signaling molecules (ligands) bind to this type of receptor, the ion channel opens and allows the flow of ions through the membrane, changing the concentration of the ions in the cell (**Figure 8.8**). In animals, ligand-gated ion channels are important in the transmission of signals between neurons and muscle cells and between two neurons.

8.3 Reviewing the Concepts

- Enzyme-linked receptors have some type of catalytic function. Many of them are protein kinases that phosphorylate proteins (Figure 8.6).
- G-protein-coupled receptors (GPCRs) interact with G proteins to initiate a cellular response (Figure 8.7).
- Ligand-gated ion channels are receptors that allow the flow of ions across cell membranes (Figure 8.8).

8.3 Testing Your Knowledge

1. When a signaling molecule binds to an enzyme-linked receptor, it causes
   a. the receptor to undergo a conformational change.
   b. the receptor to bind to a G protein.
   c. the receptor to become catalytically active.

The binding of two extracellular signaling molecules (ligands) opens the ion channel, permitting ions to pass through the membrane.

Figure 8.8 The function of a ligand-gated ion channel.

d. an ion channel to open.
   e. both a and c.
2. After a G-protein-coupled receptor is activated, what happens to the G protein that makes it active?
   a. The receptor binds to the G protein.
   b. The G protein exchanges GDP for GTP.
   c. The G protein dissociates into α and β/γ subunits.
   d. All of the above happen.
   e. None of the above happens.

8.4 Intracellular Receptors

Learning Outcome
1. Describe intracellular receptors, using the estrogen receptor as an example.

Although most receptors for signaling molecules are located in the plasma membrane, some are found inside the cell. In these cases, an extracellular signaling molecule must diffuse through the plasma membrane to gain access to its receptor. In vertebrates, receptors for steroid hormones are intracellular. As discussed in Chapter 39, steroid hormones, such as estrogens and androgens, are secreted into the bloodstream from cells of endocrine glands. The domain structure of the estrogen and androgen receptor is described in Chapter 3 (look back at Figure 3.16).

The behavior of estrogen is typical of many steroid hormones (**Figure 8.9**). Because estrogen is hydrophobic, it can diffuse through the plasma membrane of a target cell. Estrogen enters the nucleus and binds to receptor subunits there. (Other steroids bind to receptor subunits in the cytosol, then travel into the nucleus.) After this binding occurs, the estrogen receptor subunit undergoes a conformational change that enables it to form a dimer with another subunit that also has estrogen bound. The dimer, which is the active estrogen receptor, then binds to the DNA and activates the transcription of specific genes. The estrogen receptor is an example of a transcription factor—a protein that regulates the transcription of genes. The expression of specific genes changes cell structure and function in a way that results in a cellular response.

Biology Principle

Biologists Use Models to Test Predictions and Convey Ideas
This figure is a model for the mechanism of action of estrogen and the estrogen receptor. The Modeling Challenge in the figure legend asks you to propose a model for how this pathway is reversed when estrogen levels fall.

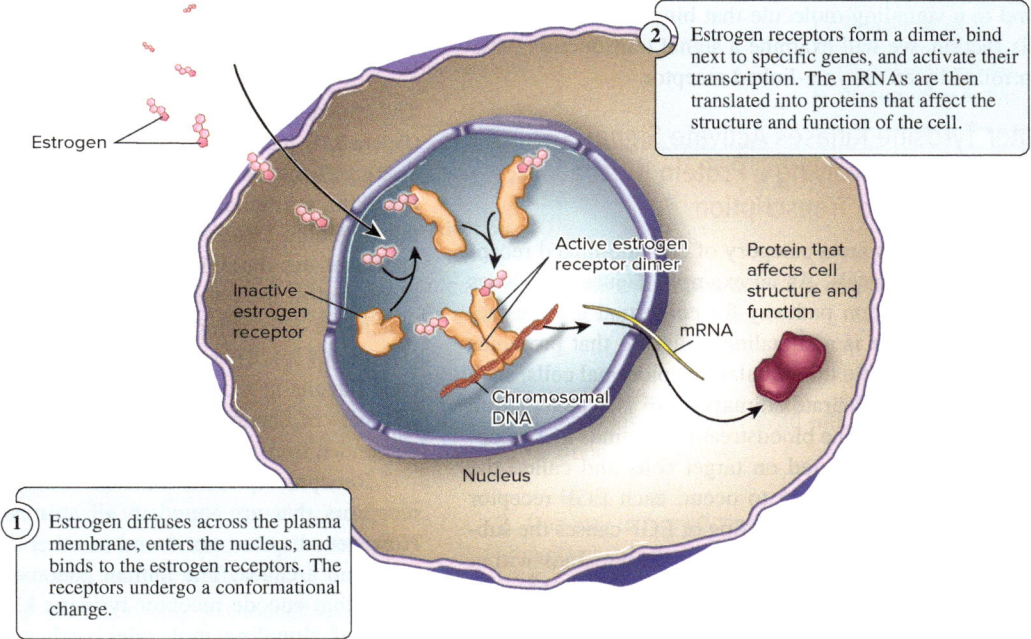

Figure 8.9 Estrogen receptor in mammalian cells. The estrogen receptor is an example of an intracellular receptor.

Modeling Challenge: The effects of hormones such as estrogen can be reversed by the actions of nucleases and proteases. These enzymes break down the mRNA and proteins, respectively, that were synthesized in response to estrogen. Create a model that describes the steps that would occur to reverse the effects of estrogen. Note: This reversal is possible because estrogen taken into a cell will eventually be broken down or converted to other steroid molecules that do not bind to the estrogen receptor.

Biology Principle

Populations of Organisms Evolve from Generation to Generation

Animals evolved from an ancestral choanoflagellate.

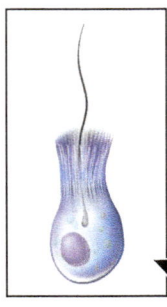

Figure 8.11 Choanoflagellates and sponges. (a) A unicellular choanoflagellate (*Monosiga brevicollis*). (b) A colonial choanoflagellate (*Proterospongia* sp.). (c) A sponge (*Ulosa stuposa*), which is a very simple animal. The inset depicts a choanocyte found within the body of a sponge.

(a, b) Mark J. Dayel (2012) Choanoflagellates and animal multicellularity, http://www.dayel.com/choanoflagellates (visited on 03/26/2013); (c) ©Wilfried Bay Nouailhat

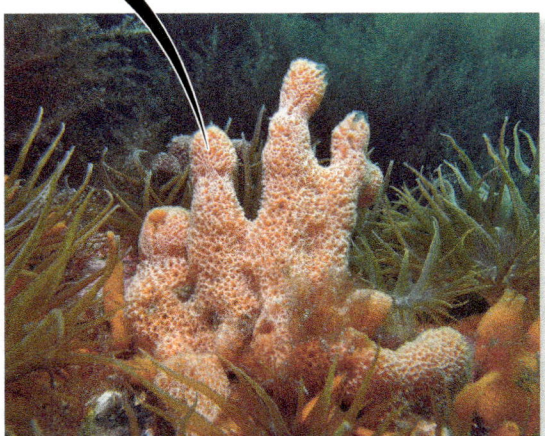

(a) A unicellular choanoflagellate
(b) A multicellular choanoflagellate
(c) A sponge

8.6 Signal Transduction and Cellular Response via a G-Protein-Coupled Receptor

Learning Outcomes

1. Describe the signal transduction pathway for signaling molecules that bind to a G-protein-coupled receptor (GPCR).
2. Explain the advantages of the production of second messengers such as cAMP.

In this section, we will examine a pathway that is controlled by a G-protein-coupled receptor (GPCR). As you will learn, this pathway involves the production of intracellular signals called second messengers.

Second Messengers Such as Cyclic AMP Are Key Components of Many Signal Transduction Pathways

Extracellular signaling molecules that bind to cell surface receptors are sometimes referred to as first messengers. After first messengers bind to receptors such as GPCRs, many signal transduction pathways lead to the production of **second messengers**—small molecules or ions that relay signals inside the cell. The signals that result in second messenger production often act quickly, in a matter of seconds or minutes, but their duration is usually short. Therefore, such signaling typically occurs when a cell needs a quick and short cellular response. Cells use several different types of second messengers, and more than one type may be used at the same time. We will focus our discussion of second messengers on cyclic AMP, which is described next.

Production of cAMP Mammalian and plant cells make several different types of G protein α subunits. One type of α subunit binds to **adenylyl cyclase**, an enzyme in the plasma membrane. This interaction stimulates adenylyl cyclase to synthesize **cyclic adenosine monophosphate (cyclic AMP, cAMP)** from ATP (**Figure 8.12**). cAMP is an example of a second messenger.

Signal Transduction Pathway and Cellular Response Let's explore a signal transduction pathway in which the G-protein-coupled receptor recognizes the hormone epinephrine (also called adrenaline). Epinephrine is produced when an individual is confronted with a stressful situation. It promotes cellular changes that help the individual deal with a perceived threat or danger. These changes improve the individual's ability to stay and fight or to

Figure 8.12 **The synthesis and breakdown of cyclic AMP.** Cyclic AMP (cAMP) is a second messenger formed from ATP by adenylyl cyclase, an enzyme in the plasma membrane. cAMP is inactivated by the action of an enzyme called phosphodiesterase, which converts cAMP to AMP.

flee. For this reason, epinephrine is sometimes called the fight-or-flight hormone. The following steps occur at the cellular level.

1. Epinephrine binds to a GPCR, which promotes the dissociation of the α subunit from the G protein (Figure 8.13).
2. The α subunit then activates adenylyl cyclase, which catalyzes the production of cAMP from ATP.
3. One effect of cAMP is to activate protein kinase A (PKA), which is composed of four subunits: two catalytic subunits that phosphorylate specific cellular proteins, and two regulatory subunits that inhibit the catalytic subunits when they are bound to those subunits. cAMP binds to the regulatory subunits of PKA. The binding of cAMP separates the regulatory and catalytic subunits, which allows each catalytic subunit to be active.
4. The catalytic subunit of PKA phosphorylates specific cellular proteins such as enzymes, structural proteins, and transcription factors, thus affecting cell structure and function.

As a specific example of a cellular response, Figure 8.14 shows how a skeletal muscle cell responds to elevated levels of epinephrine. When PKA becomes active, it phosphorylates two enzymes: phosphorylase kinase and glycogen synthase. Both of these enzymes are involved with the metabolism of glycogen, which is a polymer of glucose used to store energy. When phosphorylase kinase is phosphorylated, it becomes active. The function of phosphorylase

Figure 8.13 **A signal transduction pathway involving cAMP.** The pathway leading to the formation of cAMP and subsequent activation of PKA, which is mediated by a G-protein-coupled receptor (GPCR).

 Concept Check: *In this figure, where does the signal transduction pathway begin and end, and what is the cellular response?*

Figure 8.14 The cellular response of a skeletal muscle cell to epinephrine.

Concept Check: *Does phosphorylation activate or inhibit enzyme function?*

kinase is to phosphorylate another enzyme in the cell called glycogen phosphorylase, which then becomes active. This enzyme causes glycogen breakdown by phosphorylating glucose units at the ends of a glycogen polymer, which releases individual glucose-phosphate molecules from glycogen:

$$\text{Glycogen}_n + P_i \xrightarrow{\text{Glycogen phosphorylase}} \text{Glycogen}_{n-1} + \text{Glucose-phosphate}$$

where *n* is the number of glucose units in the glycogen polymer.

When PKA phosphorylates glycogen synthase, the function of this enzyme is inhibited rather than activated (see Figure 8.14). The function of glycogen synthase is to make glycogen. Therefore, the effect of cAMP is to prevent glycogen synthesis.

Taken together, the effects of epinephrine in skeletal muscle cells are to stimulate glycogen breakdown and inhibit glycogen synthesis. This provides these cells with more glucose molecules, which they can use to make ATP that is needed for muscle contraction. In this way, the individual is better prepared to fight or flee.

Reversal of the Cellular Response When a signaling molecule is no longer produced and its concentration falls, a larger percentage of the receptors are not bound by their ligands. When a ligand dissociates from a GPCR, the GPCR becomes deactivated. Intracellularly, the α subunit hydrolyzes its GTP to GDP and P_i, and the α subunit and β/γ dimer reassociate to form an inactive G protein (see Figure 8.7, step 3). The amount of cAMP decreases due to the action of an enzyme called **phosphodiesterase**, which converts cAMP to AMP (see Figure 8.12).

As the cAMP level falls, the regulatory subunits of PKA release cAMP, and the regulatory and catalytic subunits reassociate, thereby inhibiting PKA. Finally, enzymes called **protein phosphatases** are responsible for removing phosphate groups from proteins, which reverses the effects of PKA:

Figure 8.15 Signal amplification. An advantage of second messengers is the amplification of a signal. In this case, a single signaling molecule leads to the phosphorylation of many target proteins, perhaps hundreds or thousands.

 Concept Check: *In the case of signaling molecules such as hormones, why is signal amplification an advantage?*

The Main Advantages of Second Messengers Are Signal Amplification and Speed

The production of second messengers such as cAMP has been found to have two important advantages: signal amplification and speed. Amplification of the signal involves the synthesis of many cAMP molecules, which in turn activate many PKA proteins (**Figure 8.15**). Likewise, each PKA protein phosphorylates many target proteins in the cell to promote a cellular response.

A second advantage of second messengers such as cAMP is speed. Because second messengers are relatively small and water-soluble, they can be made rapidly and diffuse through the cytosol. For example, Brian Bacskai and colleagues studied the response of neurons to a signaling molecule called serotonin, which is a neurotransmitter that binds to a GPCR. In humans, a low serotonin level is believed to play a role in depression, anxiety, and other behavioral disorders. To monitor cAMP levels, neurons grown in a laboratory were injected with a fluorescent protein that changes its fluorescence when cAMP is made. As shown schematically in the drawing on the right in **Figure 8.16**, such cells made a substantial amount of cAMP within 20 sec after the addition of serotonin, which quickly diffused through the cell.

Figure 8.16 The rapid speed of cAMP production. The schematic drawing on the left shows a neuron prior to its exposure to serotonin, a signaling molecule; the drawing on the right shows the same cell 20 sec after exposure. Blue indicates a low level of cAMP, yellow is an intermediate level, and red/purple is a high level.

fight-or-flight hormone. Epinephrine activates enzymes that increase glycogen breakdown and inhibits enzymes that cause glycogen synthesis (Figure 8.14).
- Second messenger pathways amplify the signal and occur with great speed (Figures 8.15, 8.16).

8.6 Reviewing the Concepts

- Second messengers, such as cAMP, play a key role in signal transduction pathways, such as those that occur via G-protein-coupled receptors (Figures 8.12, 8.13).
- An example of a pathway that uses cAMP is found in skeletal muscle cells responding to elevated levels of epinephrine, the

8.6 Testing Your Knowledge

1. The enzyme that is responsible for the synthesis of cAMP is
   a. the α subunit of a G protein.
   b. adenylyl cyclase.
   c. protein kinase A.
   d. phosphodiesterase.
   e. phosphatase.

UNIT III
GENETICS

Genetics is the branch of biology that deals with **inheritance**—the transmission of characteristics from parent to offspring. We will begin this unit by examining the structure of the genetic material, namely DNA, at the molecular and cellular levels. We will explore the structure and replication of DNA and examine how the DNA is packaged into chromosomes (Chapter 9). We will then consider how segments of DNA are organized into units called genes and explore how genes are used to make products such as RNA and proteins (Chapters 10 through 12). The expression of genes is largely responsible for the characteristics of living organisms. We will also examine how mutations can alter the properties of genes and even lead to diseases such as cancer (Chapter 13).

In Chapter 14, we will turn our attention to the mechanisms by which genes are transmitted from parent to offspring, beginning with a discussion of how chromosomes are sorted and transmitted during cell division. Chapters 15 and 16 explore the relationships between the transmission of genes and the outcome of an offspring's traits. We will look at genetic patterns called Mendelian inheritance, named after Gregor Mendel, the 19th-century biologist who discovered them, as well as more complex patterns that could not have been predicted from Mendel's work.

Chapters 9 through 16 focus on the fundamental properties of the genetic material and heredity. The remaining chapters explore additional topics that are of importance to biologists. In Chapter 17, we will examine some of the unique genetic properties of bacteria and viruses. Chapter 18 describes genetic technologies that are used by researchers, clinicians, and biotechnologists to unlock the mysteries of genes and that provide tools and applications that benefit humans, and the chapter also explores the entire genomes of bacteria, archaea, and eukaryotes.

The following biology principles will be emphasized in this unit:

- **The genetic material provides a blueprint that allows organisms to grow, develop, and reproduce.** Throughout this unit, we will see how the genetic material carries the information for reproduction and sustaining life.
- **Structure determines function.** In Chapters 9 through 14, we will examine how the structure of DNA, RNA, genes, and chromosomes underlies their functions.
- **Living organisms interact with their environment.** In Chapters 15 and 16, we will explore the interactions between an organism's genes and its environment.
- **Biology affects our society.** In Chapter 18, we will examine genetic technologies that have many applications in our society.
- **Biology is an experimental science.** Most chapters in this unit have a Feature Investigation that describes a pivotal experiment that provided insights into our understanding of genetics.
- **Biologists use models to test predictions and convey their ideas.** Every chapter has a Modeling Challenge to help you refine this important skill.

(9) ©Prof. Kenneth Seddon & Dr. Timothy Evans, Queen's Univ. Belfast/SPL/Science Source; (10) ©Dr. Elena Kiseleva/SPL/Science Source; (11) ©Mauro Giacca, Ana Eulalio, Miguel Mano; (12) Image from the RCSB PDB (www.rcsb.org) H.M. Berman, et al. (2000), "The Protein Data Bank," *Nucleic Acids Research*, 28: 235–242) of PDB ID 1CGP (Schultz, S.C., Shields, G.C., Steitz, T.A.) (1991) *Science*, 253: 1001–1007; (13) ©Patrick Sheandell O'Carroll/Getty Images; (14) ©Biophoto Associates/Science Source; (15) ©Tomas Rodriguez/Getty Images; (16) ©Andia/Alamy; (17) ©Norm Thomas/Science Source; (18) ©Fumihiro Sugiyama

The Information of Life: DNA and RNA Structure, DNA Replication, and Chromosome Structure

9

©Prof. Kenneth Seddon & Dr. Timothy Evans, Queen's Univ. Belfast/SPL/Science Source

A molecular model for the structure of a DNA double helix.

Chapter Outline

9.1 Properties and Identification of the Genetic Material
9.2 Nucleic Acid Structure
9.3 Discovery of the Double-Helix Structure of DNA
9.4 Overview of DNA Replication
9.5 Molecular Mechanism of DNA Replication
9.6 Molecular Structure of Eukaryotic Chromosomes

Assessing Your Knowledge and Skills

On October 17, 2001, Mario K. was set free after serving 16 years in prison. He had been convicted of a sexual assault and murder that occurred in 1985. The charges were dropped because investigators discovered that another person, Edwin M., had actually committed the crime. How was Edwin M. identified as the real murderer? In 2001, he committed another crime, and his DNA was entered into a computer database. Edwin's DNA matched the DNA that had been collected from the victim in 1985, and other evidence was then gathered indicating that Edwin M. was the true murderer. Like Mario K., over 200 other inmates have been exonerated when DNA tests have shown that a different person was responsible for the crime.

Deoxyribonucleic acid, or **DNA,** is the genetic material that provides the blueprint to produce an organism's traits. Researchers have determined that all living organisms, including bacteria, archaea, protists, fungi, plants, and animals, use DNA as their genetic material. Viruses are not considered to be living organisms, because they do not fulfill all of the criteria described in Chapter 1 (refer back to Figure 1.3a–h). Some viruses use DNA as their genetic material, whereas others use RNA.

We will begin our survey of genetics by examining DNA at the molecular level. Once we understand how DNA works at this level, it becomes easier to see how the function of DNA controls the properties of cells and ultimately the characteristics of unicellular and multicellular organisms.

To a large extent, our understanding of genetics comes from our knowledge of the molecular structure of DNA. In this chapter, we will begin by considering some classic experiments that provided evidence that DNA is the genetic material. We will then survey the molecular features of DNA, which allows us to appreciate how DNA can store information and be accurately copied. This will allow us to explore the details of how DNA strands are used as templates for the synthesis of new DNA strands. Next, we will consider the components of ribonucleic acid (RNA), which show striking similarities to those of DNA. Finally, we will examine the molecular composition of chromosomes, where DNA is found.

9.1 Properties and Identification of the Genetic Material

Learning Outcomes

1. List the four key criteria that the genetic material must fulfill.
2. **SCISKILLS** ▶ Analyze the results of the experiments that identified DNA as the genetic material.

DNA carries the genetic instructions for the traits of all living organisms. In the case of multicellular organisms such as plants and animals, the information stored in the genetic material enables a fertilized egg to develop into an embryo and eventually into an adult organism. To fulfill its role, the genetic material must meet the following key criteria:

1. *Information.* The genetic material must contain the information necessary to construct an entire organism.
2. *Replication.* The genetic material must be accurately copied, a process known as **DNA replication.**
3. *Transmission.* After it is replicated, the genetic material can be passed from parent to offspring. It also must be passed from cell to cell during the process of cell division.
4. *Variation.* Differences in the genetic material must account for the known variation within each species and among different species.

How was the genetic material discovered? The quest to identify the genetic material began in the late 1800s, when a few scientists postulated that living organisms possess a blueprint that has a biochemical basis. In 1883, August Weismann and Karl Nägeli championed the idea that a chemical substance exists within living cells that is responsible for the transmission of traits from parents to offspring. During the next 30 years, experimentation along these lines centered on the behavior of **chromosomes,** the cellular structures that we now know contain the genetic material.

The term chromosome comes from the Greek words *chromo* and *soma,* meaning colored body, which refers to the observation of early microscopists that chromosomes are easily stained by colored dyes. By studying the transmission patterns of chromosomes from cell to cell and from parent to offspring, researchers were convinced that chromosomes carry the determinants that control the outcome of traits. The two main components of chromosomes are proteins and DNA. As we will examine in this section, experimentation revealed that the DNA component is the genetic material of living organisms.

Griffith's Bacterial Transformation Experiments Indicated the Existence of a Genetic Material

Studies in microbiology were important in developing an experimental strategy to identify the genetic material. In the late 1920s, an English microbiologist, Frederick Griffith, studied a type of bacterium known then as pneumococci and now classified as *Streptococcus pneumoniae.* Some strains of *S. pneumoniae* secrete a polysaccharide capsule, and other strains do not. When streaked on petri plates containing solid growth media, capsule-secreting strains have a smooth colony morphology. Those strains unable to secrete a capsule have a colony morphology that looks rough. In mammals, smooth strains of *S. pneumoniae* may cause pneumonia and other symptoms. In mice, such infections are usually fatal because the capsule prevents the mouse's immune system from killing the bacteria. By comparison, rough strains are not pathogenic because the immune system can destroy them.

As shown in **Figure 9.1**, Griffith injected live and/or heat-killed bacteria into mice and then observed whether or not the bacteria caused them to die. He investigated the effects of two strains of *S. pneumoniae:* type S (for smooth) and type R (for rough).

- **Step 1:** When injected into a live mouse, the type S strain killed the mouse. The capsule made by a type S strain prevents the mouse's immune system from killing the bacterial cells. Following the death of the mouse, many type S bacteria were found in the mouse's blood.

Figure 9.1 Griffith's experiments showing that genetic material can be transferred from one bacterium to another. Note: To determine if a mouse's blood contained live bacteria, a sample of blood was also applied to solid growth media. (This part of the procedure is not shown.) For steps 1 and 4, smooth bacterial colonies were observed. For steps 2 and 3, no bacterial colonies were observed because the type R cells were killed by the immune system of the mouse.

BioConnections: *Look ahead to Figure 17.10. How does bacterial transformation play a role in the transfer of genes, such as antibiotic-resistance genes, from one bacterial species to another?*

- **Step 2:** When type R bacteria were injected into a mouse, the mouse survived, and after several days, living bacteria were not found in the live mouse's blood because the mouse's immune system was able to destroy bacteria that had no capsule.
- **Step 3:** In a follow-up to these results, Griffith also heat-killed the smooth bacteria and then injected them into a mouse. As expected, the mouse survived.
- **Step 4:** A surprising result occurred when Griffith mixed live type R bacteria with heat-killed type S bacteria and then injected them into a mouse—the mouse died. The blood from the dead mouse contained living type S bacteria! How did Griffith explain these results? He postulated that a substance from dead type S bacteria transformed the type R bacteria into type S bacteria. Griffith called this process **transformation,** and he termed the unidentified material responsible for this phenomenon the transformation principle.

Let's consider what these observations mean with regard to the four criteria for the genetic material: information, replication, transmission, and variation. According to Griffith's results, the transformed bacteria had acquired the information (criterion 1) to make a capsule from the heat-killed cells. For the transformed bacteria to proliferate and thereby kill the mouse, the substance conferring the ability to make a capsule must be replicated (criterion 2) and then transmitted (criterion 3) from mother to daughter cells during cell division. Finally, Griffith already knew that variation (criterion 4) existed in the ability of his strains to produce a capsule (S strain) or not produce a capsule (R strain). In the experiment of Figure 9.1, step 4 indicated that some genetic material from the heat-killed type S bacteria had been transferred to the living type R bacteria and provided those bacteria with a new trait. At the time of his studies, however, Griffith could not determine the biochemical composition of the transforming substance.

FEATURE INVESTIGATION

Avery, MacLeod, and McCarty Used Purification Methods to Reveal That DNA Is the Genetic Material

In the 1940s, American physician Oswald Avery and American biologists Colin MacLeod and Maclyn McCarty were also interested in the process of bacterial transformation. During the course of their studies, they realized that Griffith's observations could be used as part of an experimental strategy to biochemically identify the genetic material. They asked, "What substance is being transferred from the dead type S bacteria to the live type R bacteria?"

To answer this question, Avery, MacLeod, and McCarty needed to purify the general categories of substances found in living cells. They used established biochemical procedures to purify classes of macromolecules, such as proteins, DNA, and RNA, from the type S streptococcal strain. Initially, they discovered that only the purified DNA could convert type R bacteria into type S. To further verify that DNA is the genetic material, they performed the investigation outlined in **Figure 9.2**.

Figure 9.2 The Avery, MacLeod, and McCarty experiments that identified DNA as Griffith's transformation principle—the genetic material.

HYPOTHESIS A purified macromolecule from type S bacteria, which functions as the genetic material, will be able to convert type R bacteria into type S.

KEY MATERIALS Type R and type S strains of *Streptococcus pneumoniae*.

1. Purify DNA from the type S strain. This involves breaking open cells and separating the DNA away from other components by centrifugation.

2. Mix the DNA extract with type R bacteria. Also, carry out the same steps but add the enzyme DNase, RNase, or protease to the DNA extract, which digests DNA, RNA, or proteins, respectively. As a control, don't add any DNA extract to some type R cells.

3. Allow time for the DNA to be taken up by the type R cells, converting some of them to type S.

4. Add an antibody, a protein made by the immune system of mammals, that specifically recognizes type R cells that haven't been transformed. The binding of the antibody causes the type R cells to aggregate.

5. Subject the tubes to centrifugation. The aggregated type R cells form a pellet at the bottom of the tubes, while the type S cells remain in the supernatant. Pour the supernatant onto solid growth media within petri plates. Allow time for cells to divide to form bacterial colonies.

6. **THE DATA**

A: Control
B: DNA extract
C: DNA extract + DNase
D: DNA extract + RNase — Smooth bacterial colony composed of type S cells
E: DNA extract + protease

7. **CONCLUSION** DNA is responsible for transforming type R cells into type S cells.

8. **SOURCE** Avery, O. T., MacLeod, C. M., and McCarty, M. 1944. Studies on the Chemical Nature of the Substance Inducing Transformation of Pneumococcal Types. *Journal of Experimental Medicine* 79: 137–158.

They purified DNA from the type S bacteria and mixed it with type R bacteria (step 1). After allowing time for DNA uptake into the type R bacteria, they added an antibody that aggregated any nontransformed type R bacteria, which were then removed by centrifugation. The remaining bacteria were incubated overnight on petri plates.

As a control, no DNA extract was added, and no type S bacterial colonies were observed on the petri plates (see plate A in step 6). When the researchers mixed their S strain DNA extract with type R bacteria, some of the bacteria were converted to type S bacteria (see plate B in step 6). This result was consistent with the idea that DNA is the genetic material.

Even so, a careful biochemist could argue that the DNA extract might not have been 100% pure. For this reason, the researchers realized that a small amount of contaminating material in the DNA extract, rather than the DNA, could actually be the genetic material. The most likely contaminating substances in this case would be RNA or protein.

To address this possibility, Avery, MacLeod, and McCarty treated the DNA extract with enzymes that digest either DNA (called **DNase**), RNA (**RNase**), or protein (**protease**) (see step 2). When the DNA extracts were treated with RNase or protease, the type R bacteria were still converted into type S bacteria, indicating that contaminating RNA or protein in the extract was not acting as the genetic material (see step 6, plates D and E). However, when the DNA extract was treated with DNase, it lost the ability to convert type R bacteria into type S bacteria (see plate C). Taken together, these results were consistent with the idea that DNA is the genetic material.

Experimental Questions

1. **SCISKILLS** ▶ Avery, MacLeod, and McCarty worked with two strains of *Streptococcus pneumoniae* to determine the biochemical identity of the genetic material. Explain the characteristics of the *S. pneumoniae* strains that made them particularly well suited for the researchers' experiment.

2. What is a DNA extract?

3. **SCISKILLS** ▶ In the experiment of Avery, MacLeod, and McCarty, what was the purpose of using DNase, RNase, and protease if only the DNA extract caused transformation?

9.1 Reviewing the Concepts

- All living organisms use DNA as the genetic material, but some viruses use RNA.
- The genetic material carries information to produce the traits of organisms. It is accurately replicated and transmitted from cell to cell and parent to offspring. The genetic material carries differences that explain the variation among different organisms.
- Griffith's work with type S and type R bacteria indicated the existence of a genetic material, which he called the transformation principle (Figure 9.1).
- Avery, MacLeod, and McCarty used biochemical methods to show that DNA is the genetic material (Figure 9.2).

9.1 Testing Your Knowledge

1. In the work of Griffith, the mixture of heat-killed S strain bacteria and living R strain bacteria resulted in the death of the mouse. The correct interpretation of these results is that
   a. the S strain bacteria came back to life.
   b. genetic material was transferred from the heat-killed S strain to the R strain.
   c. heat-killed S strain bacteria can kill mice.
   d. R strain bacteria can kill mice.

9.2 Nucleic Acid Structure

Learning Outcomes

1. Outline the structural features of DNA at five levels of complexity.
2. Describe the structures of nucleotides, a DNA strand, and the DNA double helix.

DNA and its molecular cousin, RNA, are known as **nucleic acids,** polymers consisting of nucleotides, which are responsible for the storage, expression, and transmission of genetic information. This term is derived from the discovery of DNA by Friedrich Miescher in 1869. He identified a novel phosphorus-containing substance from the nuclei of white blood cells found in waste surgical bandages. He named this substance nuclein. As the structure of DNA and RNA became better understood, they were found to be acidic molecules, which means they release hydrogen ions (H^+) in solution and have a net negative charge at neutral pH. Thus, the name nucleic acid was coined.

DNA is a very large macromolecule composed of smaller building blocks. As shown in **Figure 9.3**, we can consider the structural features of DNA at different levels of complexity:

1. **Nucleotides** are the building blocks of DNA.
2. A **strand** of DNA is formed by the covalent linkage of nucleotides in a linear manner.

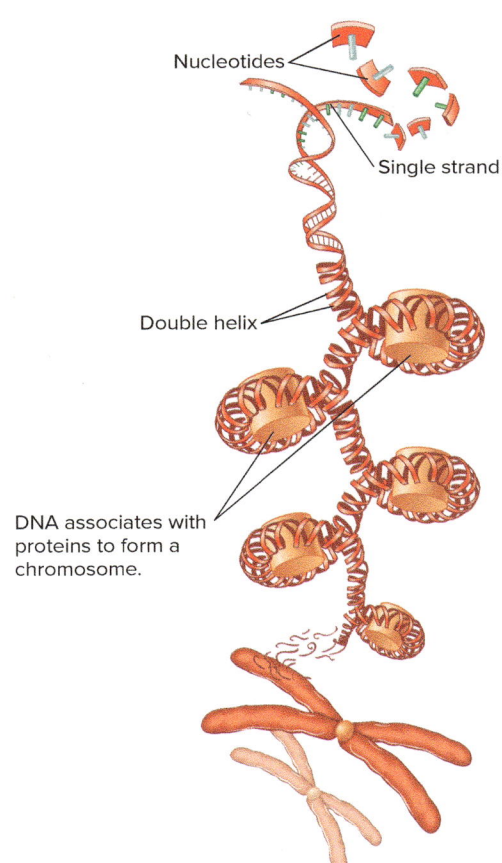

Figure 9.3 Levels of DNA structure within a chromosome.

3. Two strands of DNA hydrogen-bond with each other to form a **double helix**—two DNA strands that are twisted together to form a structure that resembles a spiral staircase, or helix.
4. In living cells, DNA is associated with an array of different proteins to form chromosomes. The association of proteins with DNA organizes the long strands into a compact structure.
5. A **genome** is the complete complement of an organism's genetic material. For example, the genome of most bacteria is a single circular chromosome, whereas eukaryotic cells have DNA in their nucleus, mitochondria, and chloroplasts.

The first two levels of complexity will be the focus of this section. Level 3 will be explored in Section 9.3, level 4 will be discussed in Section 9.6, and level 5 will be examined in Chapter 18.

Nucleotides Contain a Phosphate, a Sugar, and a Base

A nucleotide within a DNA or RNA strand has three components: a phosphate group, a pentose (five-carbon) sugar, and a nitrogen-containing base. The base and phosphate group are attached to the sugar molecule (**Figure 9.4**). The nucleotides in DNA and RNA contain different sugars:

- **Deoxyribose** is found in DNA.
- **Ribose** is found in RNA.

Five different bases are found in nucleotides, although any given nucleotide contains only one base. The five bases are subdivided into two categories due to differences in their structures (see Figure 9.4):

- **Purine** bases: **Adenine (A)** and **guanine (G)** have a double-ring structure and are found in both DNA and RNA.
- **Pyrimidine** bases: **Thymine (T), uracil (U),** and **cytosine (C)** have a single-ring structure. Cytosine is found in both DNA and RNA. Thymine is found only in DNA, whereas uracil is found only in RNA.

A conventional numbering system describes the locations of carbon and nitrogen atoms in the sugars and bases:

The prime symbol (') is used to distinguish the numbering of carbons in the sugar. The atoms in the ring structures of the bases are not given the prime designation. The sugar carbons are designated 1' (read as "one prime"), 2', 3', 4', and 5', with the carbon atoms

Figure 9.4 Nucleotides and their components. Note: This is a description of nucleotides within a DNA or RNA strand. Single nucleotides that are not within a strand may contain one, two, or three phosphate groups.

Concept Check: *Which pyrimidine(s) is (are) found in both DNA and RNA?*

numbered in a clockwise direction starting with the carbon atom to the right of the ring oxygen atom. The fifth carbon is outside the ring. A base is attached to the 1' carbon atom, and a phosphate group is attached at the 5' position. Compared with ribose, deoxyribose lacks a single oxygen atom at the 2' position; the prefix deoxy- (meaning without oxygen) refers to this missing atom.

A Strand Is a Linear Linkage of Nucleotides with Directionality

The next level of nucleotide structure is the formation of a strand of DNA or RNA in which nucleotides are covalently attached to each other in a linear fashion. **Figure 9.5** depicts a short strand of DNA with four nucleotides. The key structural features of a DNA strand are as follows:

- Nucleotides are linked together by covalent bonds—called phosphoester bonds—between phosphorus and oxygen. A sugar molecule in one nucleotide is attached to a phosphate group in the next nucleotide. Another way of viewing this

Biology Principle

The Genetic Material Provides a Blueprint That Allows Organisms to Grow, Develop, and Reproduce

The covalent linkage of a sequence of bases allows DNA to store information.

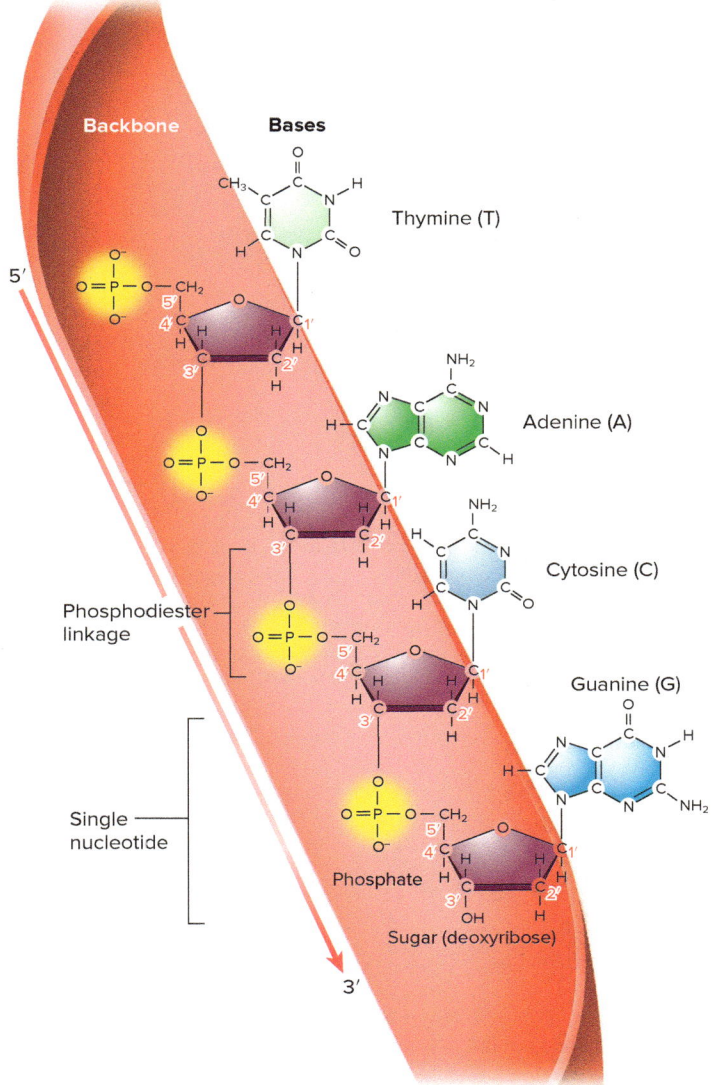

Figure 9.5 **The structure of a DNA strand.** Nucleotides are covalently bonded to each other in a linear manner. Notice the directionality of the strand and that it carries a particular sequence of bases. An RNA strand has a very similar structure, except the sugar is ribose rather than deoxyribose and uracil is substituted for thymine.

✓ **Concept Check:** *What is the difference between a phosphoester bond and a phosphodiester linkage?*

linkage is to notice that a phosphate group connects two sugar molecules. From this perspective, the linkage in DNA strands is called a **phosphodiester linkage,** which has two phosphoester bonds.

- The phosphates and sugar molecules form the **backbone** of a DNA or RNA strand, and the bases (thymine, adenine, cytosine, and guanine) project from the backbone.

- A strand has a directionality based on the orientation of the sugar molecules within that strand. In Figure 9.5, the direction of the strand is said to be 5′ to 3′ when going from top to bottom. The 5′ end of a DNA strand has a phosphate group, and the 3′ end has an —OH group. This strand is abbreviated 5′–TACG–3′ to indicate its directionality.

DNA Has a Repeating, Antiparallel, Helical Structure Formed by the Complementary Base Pairing of Nucleotides

The DNA double helix has several distinguishing features (**Figure 9.6a**):

- DNA is a double-stranded structure with the sugar-phosphate backbone on the outside and the bases on the inside. It is a helical structure, which means it is cylindrically spiral.

- The double helix is stabilized by hydrogen bonding between the bases in opposite strands to form **base pairs.**

- Base pairing is specific. An adenine (A) in one strand forms two hydrogen bonds with a thymine (T) in the opposite strand, or a guanine (G) forms three hydrogen bonds with a cytosine (C) (**Figure 9.6b**). This specificity is called the **AT/GC rule.**

- Due to the AT/GC rule, the base sequences of two DNA strands are complementary to each other. For example, if one strand has the sequence 5′–GCGGATTT–3′, the opposite strand must be 3′–CGCCTAAA–5′.

- One complete turn of the double helix is 3.4 nm in length and comprises about 10 base pairs.

- With regard to their 5′ and 3′ directionality, the two strands of a DNA double helix are **antiparallel.** If you look at Figure 9.6, one strand runs in the 5′ to 3′ direction from top to bottom, whereas the other strand is oriented 3′ to 5′ from top to bottom.

- As shown in **Figure 9.7**, two grooves, called the major groove and the minor groove, spiral around the double helix. This figure shows a space-filling model in which the atoms are depicted as spheres. It emphasizes the surface of DNA. The sugar-phosphate backbone is on the outermost surface of the double helix. The indentations where the atoms of the bases make contact with water in the surroundings are termed grooves. The **major groove** occurs where the DNA backbones of the two strands are farther apart, whereas the **minor groove** is where they are closer together.

Figure 9.6 Structure of the DNA double helix. As seen in part **(a)**, DNA is a helix composed of two antiparallel strands. Part **(b)** shows the AT/GC base pairing that holds the strands together via hydrogen bonds.

Concept Check: *If one DNA strand is 5'–GATTCGTTC–3', what is the complementary strand?*

9.2 Reviewing the Concepts

- DNA is composed of nucleotides, which are covalently linked to form DNA strands. Two DNA strands are held together by hydrogen bonds between the bases to form a double helix. DNA associates with various proteins to form a chromosome (Figure 9.3).
- Nucleotides within a DNA or RNA strand are composed of a phosphate, a sugar, and a base. The sugar can be deoxyribose (DNA) or ribose (RNA). The purine bases are adenine and guanine, and the pyrimidine bases are thymine (DNA only), uracil (RNA only), and cytosine. The atoms in a nucleotide are numbered in a conventional way (Figure 9.4).
- A strand of DNA (or RNA) has a directionality due to the orientation of the sugar molecules (Figure 9.5).
- DNA is a double helix in which the DNA strands are antiparallel and obey the AT/GC rule (Figures 9.6, 9.7).

9.2 Testing Your Knowledge

1. A nucleotide composed of deoxyribose, phosphate, and adenine could be found in
   a. DNA only.
   b. RNA only.
   c. both DNA and RNA.
   d. neither DNA nor RNA.

2. The terms 5' and 3' refer to the directionality of a DNA strand based on
   a. the orientation of phosphate molecules in the backbone.
   b. the orientation of sugar molecules in the backbone.
   c. the direction in which the bases project from the backbone.
   d. the orientation of base pairing.

3. The strands of a DNA double helix are held together by base pairing between
   a. A to G and C to T.
   b. A to C and G to T.
   c. A to T and G to C.
   d. A to A, T to T, G to G, and C to C.

Biology Principle

Biologists Use Models to Test Predictions and Convey Ideas

This figure is a space-filling model of the structure of DNA that emphasizes the major and minor grooves of DNA. The Modeling Challenge in the figure legend asks you to propose a model for the binding of a protein into a major groove.

Figure 9.7 A space-filling model of the DNA double helix. In the sugar-phosphate backbone, sugar molecules are shown in blue, and phosphate groups are yellow. The backbone is on the outermost surface of the double helix. The atoms of the bases, shown in green, are more internally located within the double-stranded structure. Notice the major and minor grooves that are formed by this arrangement.

9.3 Discovery of the Double-Helix Structure of DNA

Learning Outcome

1. Describe and interpret the work of Franklin, Chargaff, and Watson and Crick.

Now that we have an understanding of DNA structure, let's take a look back at the methods that were used to deduce its structure. X-ray diffraction was a key experimental tool that led to the discovery of the DNA double helix. When a substance is exposed to X-rays, the atoms in the substance cause the X-rays to be scattered (Figure 9.8). If the substance has a repeating structure, the pattern of scattering, known as the diffraction pattern, is related

Figure 9.8 Rosalind Franklin's X-ray diffraction of DNA fibers. The exposure of wet DNA fibers to X-rays causes the X-rays to be scattered, and the pattern of scattering is related to the position of the atoms in the DNA fibers.

Modeling Challenge: Chapter 3 describes the concept of protein domains (refer back to Figure 3.16). A domain that is commonly found in proteins that bind into the major groove of DNA is the helix-turn-helix domain, as shown to the right. An α helix within a protein is just the right width to fit into the major groove. One of the α helices within this domain is called the recognition helix because it binds to bases in the DNA. In this way, the protein is able to bind to a specific sequence of bases and thereby affect gene expression. Draw a model that shows the binding of a helix-turn-helix domain to the DNA.

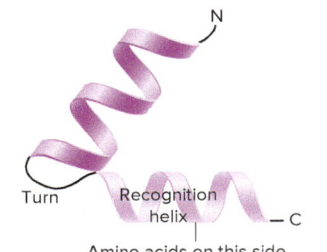

to the structural arrangement of the atoms causing the scattering. The diffraction pattern is analyzed using mathematical theory to provide information regarding the three-dimensional structure of the molecule. Rosalind Franklin, a British biophysicist working in the 1950s in the same laboratory as New Zealand-born British biophysicist Maurice Wilkins, was a gifted researcher who made marked advances in X-ray diffraction techniques involving DNA. The diffraction pattern of DNA fibers produced by Franklin suggested a helical structure with a diameter that is relatively uniform and too wide to be a single-stranded helix. In addition, the pattern provided information regarding the number of nucleotides per turn and was consistent with a 2-nanometer (nm) spacing between the strands, which corresponds to a purine (A or G) bonding with a pyrimidine (T or C).

Another piece of information that proved to be critical for the determination of the double-helix structure came from the studies of Austrian-born American biochemist Erwin Chargaff. In 1950, Chargaff analyzed the base composition of DNA that was isolated from many different species. His experiments consistently showed that the amount of adenine in each sample was similar to the amount of thymine, and the amount of cytosine was similar to the amount of guanine (Table 9.1).

In the early 1950s, more information was known about the structure of proteins than that of nucleic acids. American chemist Linus Pauling correctly proposed that some regions of proteins fold into a structure known as an α helix. To determine the structure of the α helix, Pauling built large models by linking together simple ball-and-stick units. In this way, he could see if atoms fit together properly in a complicated three-dimensional structure. This approach is still widely used today, except researchers now construct three-dimensional models on computers.

Using the ball-and-stick approach, American biologist James Watson and English biologist Francis Crick, in collaboration with Wilkins, set out to determine the structure of DNA. They assumed that nucleotides are linked together in a linear fashion and that the chemical linkage between two nucleotides is always the same. Figure 9.9 shows a ball-and-stick model that helped them deduce the structure. In a 1953 article in the journal *Nature*, Watson and Crick proposed that the structure of DNA is a double helix in which the DNA strands are antiparallel and obey the AT/GC rule. In 1962, Watson, Crick, and Wilkins were awarded the Nobel Prize in Physiology or Medicine. Unfortunately, Rosalind Franklin had died before this time, and the Nobel Prize is awarded only to living recipients.

Figure 9.9 Original ball-and-stick model of the DNA double helix proposed by Watson and Crick.
©Hulton Archive/Getty Images

Table 9.1	Base Composition of DNA from a Variety of Organisms as Determined by Chargaff			
	Percentage of bases			
Organism	Adenine	Thymine	Guanine	Cytosine
Escherichia coli (bacterium)	26.0	23.9	24.9	25.2
Streptococcus pneumoniae (bacterium)	29.8	31.6	20.5	18.0
Saccharomyces cerevisiae (yeast)	31.7	32.6	18.3	17.4
Turtle	28.7	27.9	22.0	21.3
Salmon	29.7	29.1	20.8	20.4
Chicken	28.0	28.4	22.0	21.6
Human	30.3	30.3	19.5	19.9

9.3 Reviewing the Concepts

- The X-ray diffraction data of Franklin, the biochemical data of Chargaff, and the ball-and-stick model approach of Pauling helped reveal the structure of DNA (Figure 9.8, Table 9.1).
- Watson and Crick determined that DNA is a double helix in which the DNA strands are antiparallel and obey the AT/GC rule (Figure 9.9).

9.3 Testing Your Knowledge

1. The data of Chargaff suggested that DNA has similar amounts of A and T and of G and C. This conclusion was based on
   a. a determination of the base sequences of opposite DNA strands.
   b. the composition of viral DNA.
   c. a comparison of base composition of several different species.
   d. both b and c.
2. To determine the structure of DNA, Watson and Crick used
   a. the X-ray diffraction data of Franklin.
   b. the biochemical data of Chargaff.
   c. the ball-and-stick model approach of Pauling.
   d. all of the above.

9.4 Overview of DNA Replication

Learning Outcomes

1. **SCISKILLS** ▶ Describe and interpret the experiments of Meselson and Stahl.
2. Explain how the double-stranded structure of DNA and the AT/GC rule allow DNA to be replicated semiconservatively.

The structure of DNA immediately suggested to Watson and Crick a mechanism by which DNA can be copied. They proposed that during this process, known as **DNA replication,** the original DNA strands are used as templates for the synthesis of new DNA strands. In this section, we will look at an early experiment that helped to determine the mechanism of DNA replication and then examine the structural characteristics that enable a double helix to be faithfully copied.

Meselson and Stahl Considered Three Proposed Mechanisms of DNA Replication

Researchers in the late 1950s considered three different models for the mechanism of DNA replication (**Figure 9.10**). In all of these models, the two newly made strands are called the **daughter strands,** and the original strands are the **parental strands.**

1. **Semiconservative mechanism** (Figure 9.10a): In this model, the double-stranded DNA is half conserved following the replication process. The new double-stranded DNA contains one parental strand and one daughter strand. This model is consistent with the proposal of Watson and Crick.
2. **Conservative mechanism** (Figure 9.10b): Following DNA replication, the original arrangement of parental strands is completely conserved, and the two newly made daughter strands are also together following replication.
3. **Dispersive mechanism** (Figure 9.10c): After DNA replication is completed, segments of parental DNA and newly made daughter DNA are interspersed in both strands.

In 1958, Matthew Meselson and Franklin Stahl devised an experimental approach to distinguish among these three

(a) Semiconservative mechanism. DNA replication produces DNA molecules with 1 parental strand and 1 newly made daughter strand.

(b) Conservative mechanism. DNA replication produces 1 double helix with both parental strands and the other with 2 new daughter strands.

(c) Dispersive mechanism. DNA replication produces DNA strands in which segments of new DNA are interspersed with the parental DNA.

Figure 9.10 Three proposed mechanisms for DNA replication. The strands of the original (parental) double helix are shown in red. Two rounds of replication are illustrated, with the daughter strands shown in blue.

mechanisms. An important feature of their research was the use of isotope labeling. Nitrogen, which is found in DNA, occurs in a common light (^{14}N) form and a rare heavy (^{15}N) form. Meselson and Stahl studied DNA replication in the bacterium *Escherichia coli*. They grew *E. coli* cells for many generations in a medium that contained only the ^{15}N form of nitrogen (**Figure 9.11**). This produced a population of bacterial cells in which all of the DNA was labeled with the heavy isotope. Then they switched the bacteria to a medium that contained only the ^{14}N form as its nitrogen source. The cells were allowed to divide, and samples were collected after one generation (that is, one round of DNA replication), two

Figure 9.11 The Meselson and Stahl experiment showing that DNA replication is semiconservative.

(5) ©M. Meselson, F. Stahl (1958), "The replication of DNA in *Escherichia coli*," *PNAS*, 44(7): 671–682, Fig. 4a

Concept Check: *If this experiment was conducted for four rounds of DNA replication (that is, four generations), what would be the expected fractions of light DNA and half-heavy DNA according to the semiconservative model?*

generations, and so on. Because the bacteria were doubling in a medium that contained only ^{14}N, all of the newly made DNA strands were labeled with light nitrogen, whereas the original strands remained labeled with the heavy form.

Meselson and Stahl then used centrifugation to separate DNA molecules based on differences in density. Samples were placed on top of a solution that contained a salt gradient—in this case, cesium chloride (CsCl). A double helix containing all heavy nitrogen would have a higher density and therefore travel closer to the bottom of the gradient. By comparison, if both DNA strands contained ^{14}N, the DNA would have a lower density and remain closer to the top of the gradient. If one strand contained ^{14}N and the other strand contained ^{15}N, the DNA would be half-heavy and have an intermediate density, ending up near the middle of the gradient. The DNA within the gradient was observed by exposing the gradient to UV light.

After one cell doubling (that is, one round of DNA replication), all of the DNA was half-heavy, exhibiting an intermediate density (Figure 9.11, step 5). These results are consistent with both the semiconservative and dispersive mechanisms. In contrast, the conservative mechanism predicts two different DNA bands: one of high density and one of low density. Because the DNA was found in a single half-heavy band after one doubling, the conservative model was disproved.

After two cell doublings, both light DNA and half-heavy DNA bands were observed. This result was also predicted by the semiconservative mechanism of DNA replication, because a parental strand that was light would produce all light DNA, whereas a parental strand that was heavy would produce half-heavy DNA (see Figure 9.10a). However, with the dispersive mechanism, all of the DNA strands would have been 1/4 heavy after two generations. This mechanism predicts that the heavy nitrogen would be evenly dispersed among four double helices, each strand containing 1/4 heavy nitrogen and 3/4 light nitrogen (see Figure 9.10c). This prediction did not agree with the data. Taken together, the results of the Meselson and Stahl experiment are consistent only with a semiconservative mechanism for DNA replication.

DNA Replication Proceeds According to the AT/GC Rule

As originally proposed by Watson and Crick, DNA replication relies on the complementarity of DNA strands according to the AT/GC rule. During the replication process, the two complementary strands of DNA separate and serve as **template strands** for the synthesis of daughter strands of DNA (Figure 9.12a). After the double helix has separated, individual nucleotides have access to the template strands in a region called the replication fork. First, individual nucleotides hydrogen-bond to the template strands according to the AT/GC rule: An adenine (A) in one strand bonds with a thymine (T) in the opposite strand, or a guanine (G) bonds with a cytosine (C). Next, a covalent bond is formed between the phosphate of one nucleotide and the sugar of the previous nucleotide. The end result is that two double helices are made that have the same base sequence as the original DNA molecule (Figure 9.12b). This is a critical feature of DNA

Biology Principle

Structure Determines Function

A double-stranded structure whose base sequence obeys the AT/GC rule underlies the function of DNA replication.

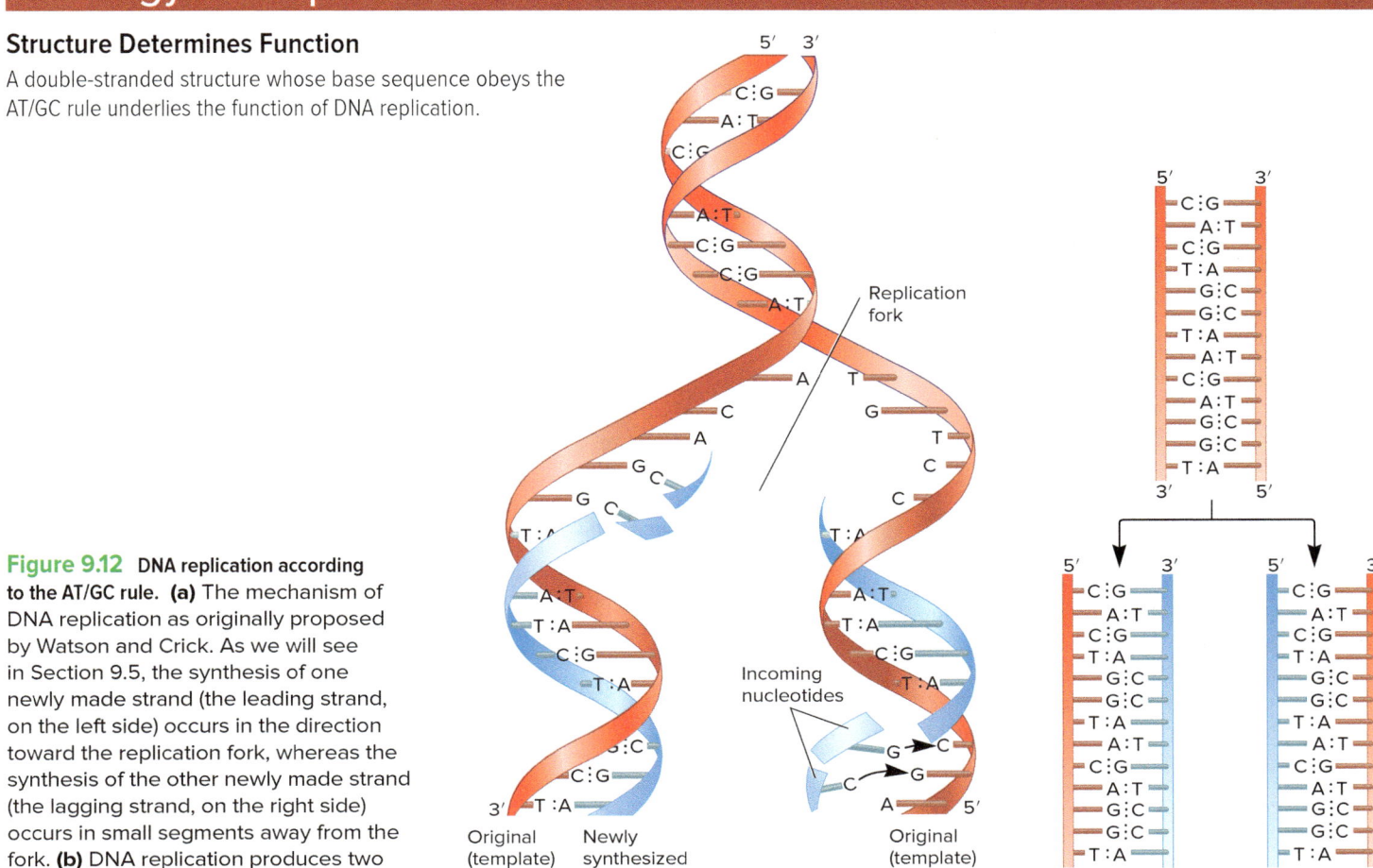

Figure 9.12 DNA replication according to the AT/GC rule. (a) The mechanism of DNA replication as originally proposed by Watson and Crick. As we will see in Section 9.5, the synthesis of one newly made strand (the leading strand, on the left side) occurs in the direction toward the replication fork, whereas the synthesis of the other newly made strand (the lagging strand, on the right side) occurs in small segments away from the fork. **(b)** DNA replication produces two copies of DNA with the same sequence of bases as the original DNA molecule.

replication, because it enables the replicated DNA molecules to retain the same information (that is, the same base sequence) as the original molecule. In this way, DNA has the remarkable ability to direct its own duplication.

9.4 Reviewing the Concepts

- Meselson and Stahl used isotope labeling with light and heavy forms of nitrogen to show that DNA is replicated by a semiconservative mechanism in which the product of DNA replication is one original strand and one new strand (Figures 9.10, 9.11).
- During DNA replication, the double-stranded DNA separates and each strand serves as a template for the synthesis of daughter strands. Nucleotides hydrogen-bond to the template strands according to the AT/GC rule: An adenine (A) in one strand bonds with a thymine (T) in the opposite strand, or a guanine (G) bonds with a cytosine. The result of DNA replication is two double helices with the same base sequence as the original DNA (Figure 9.12).

9.4 Testing Your Knowledge

1. According to Meselson and Stahl's procedure, what would be the predicted results after four generations according to the semiconservative model?
   a. ¼ half-heavy and ¾ light
   b. ⅛ half-heavy and ⅞ light
   c. $1/16$ half-heavy and $15/16$ light
   d. $1/32$ half-heavy and $31/32$ light

2. What key features of DNA structure allow it to be replicated?
   a. double-stranded and 10 base pairs per turn
   b. double-stranded with major and minor grooves
   c. double-stranded and the AT/GC rule
   d. the AT/GC rule and 10 base pairs per turn

9.5 Molecular Mechanism of DNA Replication

Learning Outcomes

1. Explain how the synthesis of new DNA strands begins at an origin of replication.
2. Describe the functions of helicase, topoisomerase, single-strand binding protein, primase, and DNA polymerase at the replication fork.
3. Outline the key differences in the synthesis of the leading and lagging strands.
4. List three reasons why DNA replication is very accurate.

Thus far, we have examined the general mechanism of DNA replication, known as semiconservative replication, and seen how DNA synthesis obeys the AT/GC rule. In this section, we will examine the details of DNA replication as it occurs inside living cells. As we will see, several different proteins are needed to initiate DNA replication and allow it to proceed quickly and accurately.

DNA Replication Begins at an Origin of Replication

An **origin of replication** is a site within a chromosome that serves as a starting point for DNA replication. At the origin, the two DNA strands unwind (Figure 9.13). DNA replication proceeds outward from two **replication forks,** a process termed **bidirectional replication.** A bacterial chromosome is relatively small and usually circular and has a single origin of replication. Bidirectional replication starts at the origin of replication and proceeds until the new strands meet on the opposite side of the chromosome. Eukaryotes have larger chromosomes that are linear. They have multiple origins of replication, so the DNA can be replicated within a reasonable length of time. The newly made strands from each origin eventually make contact with each other to complete the replication process.

DNA Replication Requires the Action of Several Different Proteins

In all living species, a set of several different proteins is involved in DNA replication. An understanding of the functions of these proteins is critical to explaining the replication process at the molecular level.

Helicase, Topoisomerase, and Single-Strand Binding Proteins: Formation and Movement of the Replication Fork To act as a template for DNA replication, the strands of a double helix must separate, and the resulting replication fork must move. As mentioned, an origin of replication serves as a site where this separation begins. Each fork then moves outward from the origin. DNA helicase, DNA topoisomerase, and single-strand binding proteins are responsible for fork formation and movement.

- **DNA helicase:** At each fork, DNA helicase binds to one DNA strand and travels in the 5′ to 3′ direction (Figure 9.14). It uses energy from ATP to break hydrogen bonds between base pairs and thereby separates the DNA strands. This keeps the fork moving forward.
- **DNA topoisomerase:** The action of DNA helicase can cause knots (called supercoils) to form just ahead of the replication fork. These knots are removed by another enzyme called DNA topoisomerase.
- **Single-strand binding proteins:** After the two DNA strands have separated, they must remain that way so they can act as templates to make complementary daughter strands. The function of single-strand binding proteins is to coat both of the single strands of template DNA and prevent them from re-forming a double helix.

Figure 9.13 The bidirectional replication of DNA. DNA replication proceeds in both directions from an origin of replication.

Figure 9.14 Proteins that facilitate the formation and movement of the replication fork.

THE INFORMATION OF LIFE: DNA AND RNA STRUCTURE, DNA REPLICATION, AND CHROMOSOME STRUCTURE   201

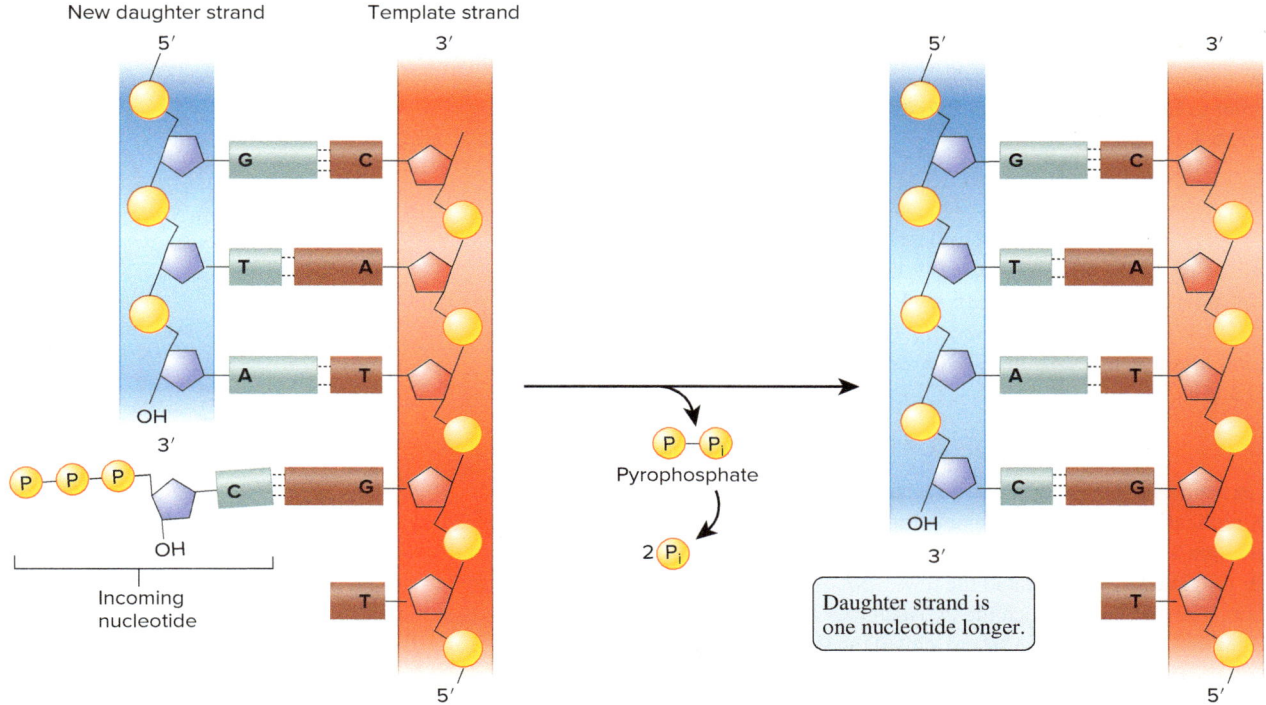

Figure 9.15 Enzymatic synthesis of DNA. **(a)** Structure of DNA polymerase. **(b)** DNA polymerase breaks the bond between the first and second phosphate in an incoming deoxynucleoside triphosphate, causing the release of pyrophosphate. This provides the energy to form a covalent bond between the resulting deoxynucleoside monophosphate and the previous nucleotide in the growing strand. The pyrophosphate is broken down to two phosphates.

DNA Polymerase and Primase: Synthesis of DNA Strands Two enzymes are needed to synthesize DNA strands during DNA replication.

- **DNA polymerase:** The enzyme called **DNA polymerase** is responsible for covalently linking nucleotides together to form DNA strands. The name indicates that this enzyme makes a polymer of DNA nucleotides.

- **DNA primase:** DNA polymerase is unable to begin DNA synthesis on a bare template strand. A different enzyme called **DNA primase** is required if the template strand is bare. DNA primase makes a complementary **primer,** which is actually a short segment of RNA, typically 10 to 12 nucleotides in length that starts, or primes, the process of DNA replication.

Let's first consider the catalytic properties of DNA polymerase, and then we will examine how DNA strands are made at the replication fork. The structure of DNA polymerase resembles a human hand with the DNA threaded through it (**Figure 9.15a**). As DNA polymerase slides along the DNA, individual nucleotides with three phosphate groups, called **deoxynucleoside triphosphates,** hydrogen-bond to the exposed bases in the template strand according to the AT/GC rule (**Figure 9.15b**). At the catalytic site, DNA polymerase breaks a bond between the first and second phosphate and then attaches the resulting nucleotide with one phosphate group (a deoxynucleoside monophosphate) to the 3′ end of a growing strand via a phosphoester bond. The breakage of the covalent bond releases pyrophosphate (PP_i); this is an exergonic reaction that provides the energy to covalently connect adjacent nucleotides. The pyrophosphate is broken down to two phosphates.

The rate of DNA synthesis is truly remarkable. In bacteria, DNA polymerase synthesizes DNA at a rate of 500 nucleotides per second, whereas eukaryotic species make DNA at a rate of about 50 nucleotides per second.

DNA polymerase has two additional enzymatic features that affect how DNA strands are made. First, DNA polymerase is unable

(a) Need for a primer

(b) 5' to 3' direction of synthesis

Figure 9.16 Enzymatic features of DNA polymerase. **(a)** DNA polymerase needs a primer to begin DNA synthesis, and **(b)** it synthesizes DNA only in the 5' to 3' direction.

to begin DNA synthesis on a bare template strand. However, if a DNA strand or an RNA primer is already attached to a template strand, DNA polymerase can elongate such a pre-existing strand by making DNA (**Figure 9.16a**). A second feature of DNA polymerase is that it can synthesize new DNA only in a 5' to 3' direction (**Figure 9.16b**).

Leading and Lagging DNA Strands Are Made Differently

Let's now examine how new DNA strands are made at replication forks. DNA replication occurs near the opening that forms each replication fork (**Figure 9.17**, step 1). The synthesis of a strand begins with an RNA primer (depicted in yellow), and the new DNA is made in the 5' to 3' direction. The manner in which the two daughter strands are synthesized is strikingly different. One strand, called the **leading strand,** is made in the same direction that the fork is moving. The leading strand is synthesized as one long, continuous molecule.

By comparison, the other daughter strand, termed the **lagging strand,** is made differently. Because DNA polymerase can synthesize DNA only in the 5' to 3' direction (Figure 9.16), the lagging strand is made as a series of small fragments that are subsequently connected to each other to form a continuous strand. These DNA fragments are known as **Okazaki fragments,** after Japanese molecular biologists Reiji and Tsuneko Okazaki, who initially discovered them in the late 1960s. The synthesis of Okazaki fragments occurs in the opposite direction from the movement of the replication fork. For example, notice that the Okazaki fragments seen in Figure 9.17, steps 2 and 3, are synthesized from left to right. As shown in Figure 9.17, step 4, the RNA primer is eventually removed, and adjacent Okazaki fragments are connected to each other to form a continuous strand of DNA.

Figure 9.17 emphasizes the way that new DNA strands are synthesized. **Figure 9.18** also includes the proteins involved in the synthesis of the leading and lagging strands in *Escherichia coli*. In this bacterium, two different DNA polymerases, called DNA polymerase I and DNA polymerase III, are primarily responsible for DNA replication. In the leading strand, DNA primase makes one RNA primer at the origin, and then DNA polymerase III attaches nucleotides in a 5' to 3' direction as it slides toward the opening of the replication fork.

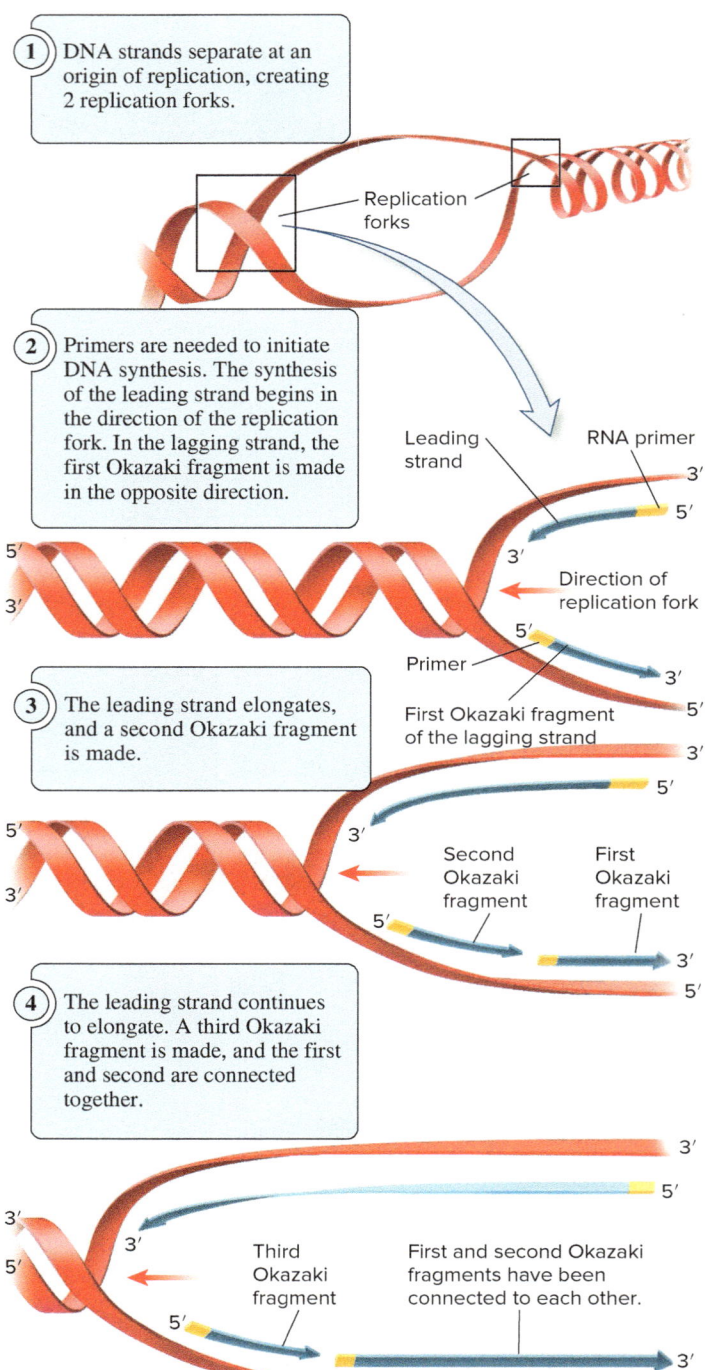

Figure 9.17 Synthesis of new DNA strands. The separation of DNA at the origin of replication produces two replication forks that move in opposite directions. New DNA strands are made near the opening of each fork. The leading strand is made continuously in the same direction the fork is moving. The lagging strand is made as small pieces (Okazaki fragments) in the opposite direction. These small pieces are then connected to each other to form a continuous lagging strand.

✓ **Concept Check:** *Which strand, the leading or lagging strand, is made discontinuously in the direction opposite the movement of the replication fork?*

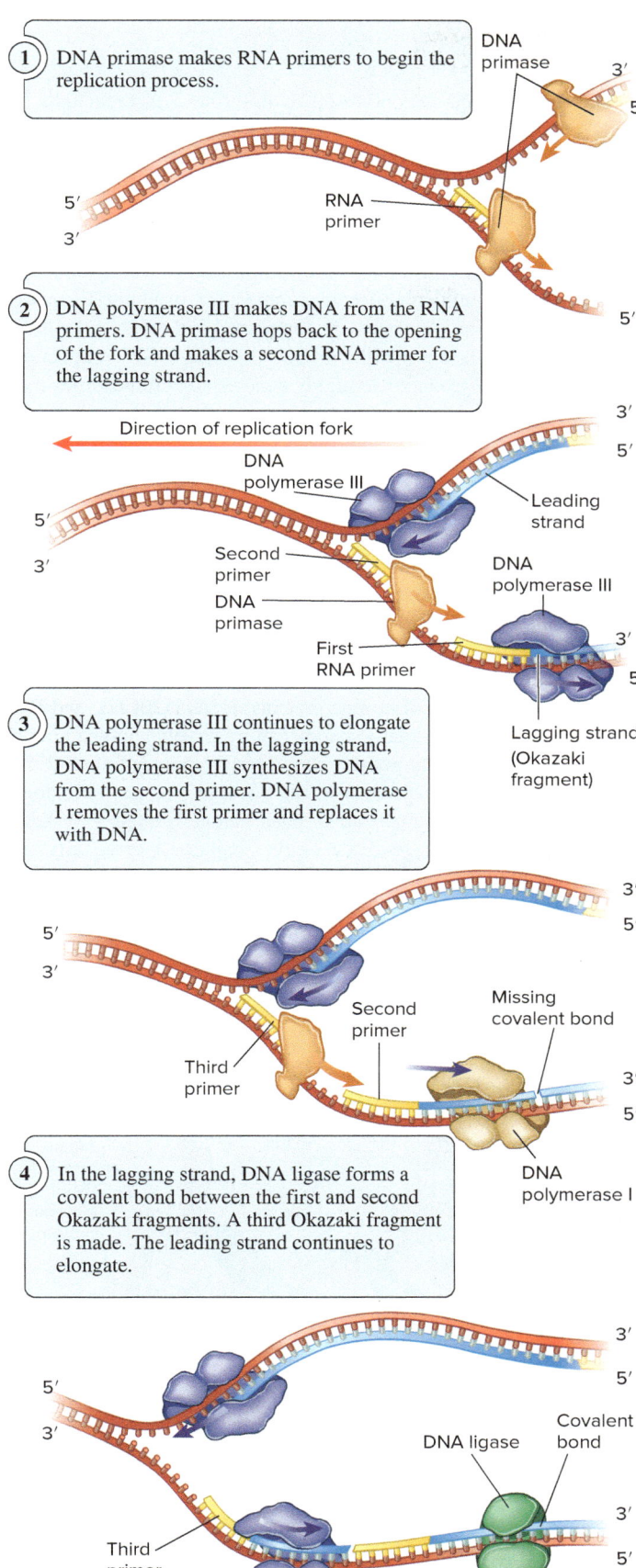

Figure 9.18 Proteins involved with the synthesis of the leading and lagging strands in *E. coli*.

In the lagging strand, DNA is also synthesized in a 5′ to 3′ direction, but in the direction away from the replication fork. Short segments of DNA are made discontinuously as a series of Okazaki fragments, each of which requires its own primer. DNA polymerase III synthesizes the remainder of the fragment (Figure 9.18, step 2). To complete the synthesis of Okazaki fragments within the lagging strand, three additional events occur: the removal of the RNA primers, the synthesis of DNA in the area where the primers have been removed, and the covalent joining of adjacent fragments of DNA (Figure 9.18, steps 3 and 4). The RNA primers are removed by DNA polymerase I, which digests the linkages between nucleotides in a 5′ to 3′ direction and fills in the vacant region with DNA. However, once the DNA has been completely filled in, a covalent bond is missing between the last nucleotide added by DNA polymerase I and the first DNA nucleotide in the adjacent Okazaki fragment. An enzyme known as **DNA ligase** catalyzes the formation of a covalent bond between these two DNA fragments to complete the replication process in the lagging strand (Figure 9.18, step 4).

DNA Replication Is Very Accurate

Although errors can happen during DNA replication, permanent mistakes are extraordinarily rare. For example, during bacterial DNA replication, only one mistake is made per 100 million nucleotides. Biologists use the term high fidelity to refer to a process that occurs with relatively few mistakes. Three factors explain such a remarkably high fidelity for DNA replication:

1. Hydrogen bonding between A and T or between G and C is more stable than hydrogen bonding between mismatched pairs of bases.
2. The active site of DNA polymerase is unlikely to catalyze bond formation between adjacent nucleotides if a mismatched base pair is formed.
3. DNA polymerase can identify a mismatched nucleotide and remove it from the daughter strand. This event, called **proofreading**, occurs when DNA polymerase detects a mismatch and then reverses its direction and digests the linkages between nucleotides at the end of a newly made strand in the 3′ to 5′ direction. Once it passes the mismatched base and removes it, DNA polymerase then continues to synthesize DNA in the 5′ to 3′ direction.

9.5 Reviewing the Concepts

- DNA synthesis occurs bidirectionally from an origin of replication. The synthesis of new DNA strands happens near each replication fork (Figure 9.13).
- DNA helicase separates DNA strands, single-strand binding proteins keep them separated, and DNA topoisomerase alleviates coiling ahead of the fork (Figure 9.14).
- Deoxynucleoside triphosphates bind to the template strands according to the AT/GC rule. DNA polymerase recognizes these deoxynucleoside triphosphates and attaches a deoxynucleoside monophosphate to the 3′ end of a growing strand (Figure 9.15).
- DNA polymerase requires a primer and synthesizes new DNA strands only in the 5′ to 3′ direction (Figure 9.16).

- The leading strand is made continuously, in the same direction the fork is moving. The lagging strand is made in the opposite direction from that of the replication fork as short Okazaki fragments that are synthesized and connected together (Figure 9.17).
- DNA primase makes one RNA primer in the leading strand and multiple RNA primers in the lagging strand. In *E. coli*, DNA polymerase III extends these primers with DNA, and DNA polymerase I removes the primers when they are no longer needed and fills in with DNA. DNA ligase connects adjacent Okazaki fragments in the lagging strand (Figure 9.18).
- DNA replication is very accurate because (1) hydrogen bonding according to the AT/CG rule is more stable than mismatched pairs, (2) DNA polymerase is unlikely to catalyze bond formation if a mismatched base pair is formed, and (3) DNA polymerase carries out proofreading.

9.5 Testing Your Knowledge

1. Which protein is needed to form a replication fork and keep it moving?
   a. DNA helicase
   b. topoisomerase
   c. single-strand binding protein
   d. All of the above are needed.

2. Which of the following is *not* a property of DNA polymerase?
   a. It cannot begin DNA synthesis on a bare template strand.
   b. It can elongate a DNA strand from a primer or pre-existing DNA strand.
   c. It can synthesize DNA in the 3′ to 5′ direction.
   d. It can synthesize DNA in the 5′ to 3′ direction.

3. Okazaki fragments are made during the synthesis of the _____ strand. They are connected together by _____.
   a. lagging, DNA ligase
   b. lagging, helicase
   c. leading, DNA ligase
   d. leading, helicase

9.6 Molecular Structure of Eukaryotic Chromosomes

Learning Outcomes

1. Describe the structure of nucleosomes and the 30-nm fiber, and how the 30-nm fiber forms radial loop domains.
2. Outline the various levels of compaction that lead to a metaphase chromosome.

We now turn our attention to the structure of eukaryotic chromosomes. A typical eukaryotic chromosome contains a single, linear, double-stranded DNA molecule that may be hundreds of millions of base pairs in length. If the DNA from a single set of human chromosomes was stretched from end to end, it would be over 1 meter in length! By comparison, most eukaryotic cells are only 10–100 micrometers (μm) in diameter, and the cell nucleus is typically about 2–4 μm in diameter. Therefore, to fit inside the nucleus, the DNA in a eukaryotic cell must be folded and compacted to a staggering degree.

The term chromosome is used to describe a discrete unit of genetic material. For example, a human somatic cell contains 46 chromosomes. By comparison, the term chromatin has a biochemical meaning. **Chromatin** is the complex of DNA and proteins that makes up eukaryotic chromosomes. As described in Chapter 11, it may also contain noncoding RNAs. Chromosomes are very dynamic structures that alternate between tight and loose compaction states. In this section, we will focus on two issues of chromosome structure. First, we will consider how chromosomes are compacted and organized within the cell nucleus. Then we will examine the additional compaction that is necessary to produce the highly condensed chromosomes that occur during cell division.

DNA Wraps Around Histone Proteins to Form Nucleosomes

Eukaryotic DNA is first compacted by wrapping itself around a group of proteins called **histones.** As shown in Figure 9.19, a repeating structural unit of chromatin is the **nucleosome,** which is 11 nm in diameter at its widest point. Each nucleosome is composed of 146 or 147 base pairs (bp) of DNA wrapped around a core of eight histone proteins called a histone octamer. Each octamer contains two molecules each of four types of histone proteins: H2A, H2B, H3, and H4. Histone proteins contain a large number of positively charged amino acids, namely lysine and arginine. The negative charges found in the phosphates of DNA are attracted to the positive charges on histone proteins. The amino terminal tail of each histone protein protrudes from the histone octamer. As discussed in Chapter 12, these tails can be covalently modified and play a key role in gene regulation.

Figure 9.19 Structure of a nucleosome. A nucleosome is composed of double-stranded DNA wrapped around an octamer of histone proteins. A linker region connects two adjacent nucleosomes. Histone H1 is bound to the linker region, as are other proteins not shown in this figure.

The nucleosomes are connected by linker regions of DNA that vary in length from 20 to 100 bp, depending on the species and cell type. A particular histone named H1 is bound to the linker region, as are other types of proteins (Figure 9.19). In electron micrographs, the overall structure of connected nucleosomes resembles beads on a string. This structure shortens the length of the DNA molecule about sevenfold.

Nucleosomes Form a 30-nm Fiber

Nucleosome units are organized into a more compact structure that is 30 nm in diameter, known as the **30-nm fiber** (Figure 9.20a). Histone H1 and other proteins are important in the formation of the 30-nm fiber, which shortens the nucleosome structure another sevenfold. In one model for the 30-nm fiber, linker regions are relatively straight and the nucleosomes zigzag back and forth (Figure 9.20b).

Chromatin Loops Are Anchored to the Nuclear Matrix

A third level of compaction involves interactions between the 30-nm fibers and proteins to form **radial loop domains** (Figure 9.21). One way that a radial loop is formed is by a category of proteins called SMC proteins. SMC stands for structural maintenance of chromosomes. An SMC protein forms a dimer that can wrap itself around two segments of DNA and thereby form a loop. These loops are often 25,000 to 200,000 bp in size.

The compaction level of chromosomes in the cell nucleus is not completely uniform. This variability can be seen with a light

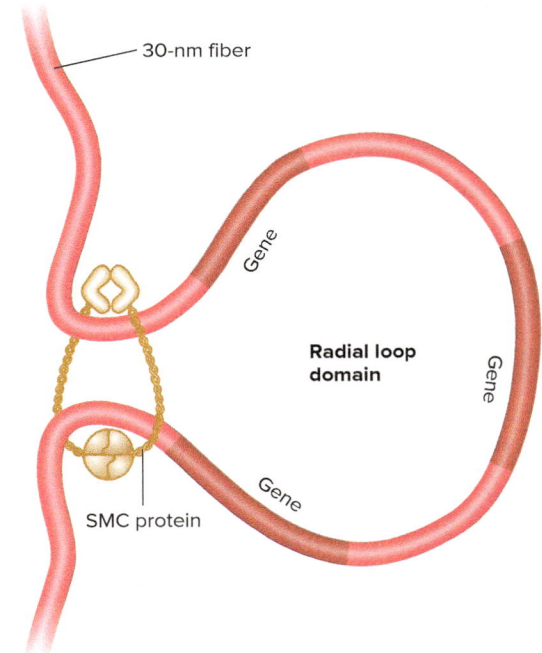

Figure 9.21 Formation of a radial loop domain by an SMC protein.

Concept Check: *Explain how an SMC protein works.*

(a) Micrograph of a 30-nm fiber

(b) Model of a 30-nm fiber

Figure 9.20 The 30-nm fiber. **(a)** A photomicrograph of the 30-nm fiber. **(b)** In this model, the linker DNA is relatively straight and the nucleosomes zigzag back and forth.
(a) ©Dr. Barbara A. Hamkalo

microscope and was first observed by German cytologist E. Heitz in 1928. He coined the term **heterochromatin** to describe the highly compacted regions of chromosomes during interphase. The less condensed regions are known as **euchromatin,** which is the form of chromatin in which the 30-nm fiber forms radial loop domains. In heterochromatin, these radial loop domains are compacted even further. In nondividing cells, most chromosomal regions are euchromatic, and these are the regions where genes are usually located. By comparison, the centromeric and telomeric regions are often heterochromatic and may not contain genes.

During Cell Division, Chromosomes Undergo Maximum Compaction

When cells prepare to divide, the chromosomes become even more compacted or condensed. Each chromosome becomes entirely condensed. This aids in their proper alignment during metaphase, which is a stage of eukaryotic cell division described in Chapter 14. Figure 9.22 illustrates the levels of compaction that contribute to the formation of a metaphase chromosome. DNA in the nucleus is always compacted by the formation of nucleosomes, 30-nm fibers, and radial loop domains (Figure 9.22a–d). In euchromatin, the 30-nm fibers are arranged in radial loop domains that are relatively loose, meaning that a fair amount of space occurs between the 30-nm fibers (Figure 9.22d). The average width of such loops is about 300 nm.

Figure 9.22 The steps in eukaryotic chromosomal compaction leading to the metaphase chromosome.
(a) ©Science Source; *(b)* ©Dr. Barbara A. Hamkalo; *(c)* Courtesy Dr. Jerome B. Rattner, Cell Biology and Anatomy, University of Calgary; *(d)* Micrograph courtesy of Dr. James R. Paulson; *(e, f)* ©Peter Engelhardt/Department of Virology, Haartman Institute

Concept Check: After they have replicated and become compacted in preparation for cell division, chromosomes are often shaped like an X, as in part (f) of this figure. Which proteins are primarily responsible for this X shape?

BioConnections: Look ahead to Figure 14.7. Why do you think it is necessary for the chromosomes to become highly compacted in preparation for cell division?

During cell division, chromosome condensation involves a much tighter packing of the loops, so little space is left between the 30-nm fibers (Figure 9.22e). All of the euchromatin becomes highly condensed, which greatly shortens the chromosomes. For a metaphase chromosome that contains two copies of the DNA (Figure 9.22f), the width averages about 1,400 nm, but a metaphase chromosome is much shorter than the same chromosome in the nucleus of a nondividing cell.

9.6 Reviewing the Concepts

- Chromosomes carry the genetic material. Chromatin is the name given to the complex of DNA, proteins, and noncoding RNAs that makes up chromosomes.
- In eukaryotic chromosomes, the DNA is wrapped around histone octamers to form nucleosomes. Nucleosomes are further compacted into 30-nm fibers (Figures 9.19, 9.20).
- A third level of compaction of eukaryotic chromosomes involves the formation of radial loop domains that are held together by SMC proteins (Figure 9.21).
- When cells prepare to divide, chromosomes become even more condensed. (Figure 9.22).

9.6 Testing Your Knowledge

1. A nucleosome is composed of
   a. 146 bp or 147 bp of DNA.
   b. an octamer of histone proteins.
   c. nuclear matrix proteins.
   d. both a and b.

2. Which of the following is the correct order for the progression of chromatin compaction seen in metaphase chromosomes?
   a. radial loop domains, further compaction of radial loop domains, nucleosomes, 30-nm fiber
   b. radial loop domains, further compaction of radial loop domains, 30-nm fiber, nucleosomes
   c. nucleosomes, 30-nm fiber, radial loop domains, further compaction of radial loop domains
   d. 30-nm fiber, nucleosomes, radial loop domains, further compaction of radial loop domains

Assessing Your Knowledge and Skills

Foundational Knowledge

Multiple Choice

1. Considering the components of a nucleotide, what component(s) is (are) always different when comparing nucleotides in a DNA strand to those in an RNA strand?
   a. phosphate group
   b. pentose sugar
   c. base
   d. both b and c
   e. a, b, and c

2. Of the following statements, which is correct when considering the process of DNA replication?
   a. New DNA molecules are composed of two completely new strands.
   b. New DNA molecules are composed of one strand from the old molecule and one new strand.
   c. New DNA molecules are composed of strands that are a mixture of sections from the old molecule and sections that are new.
   d. None of the above is correct.

3. The difference in the synthesis of the leading and lagging strands is the result of which of the following?
   a. DNA polymerase replicates only the leading strand.
   b. The two template strands are antiparallel, and DNA polymerase makes DNA only in the 5′ to 3′ direction.
   c. The lagging strand is the result of DNA breakage due to UV light.
   d. The cell does not contain enough nucleotides to make two complete strands.

4. The protein that separates DNA strands at the replication fork is
   a. DNA polymerase.
   b. DNA ligase.
   c. DNA topoisomerase.
   d. DNA helicase.
   e. DNA primase.

5. The conversion of euchromatin into heterochromatin involves
   a. the formation of more nucleosomes.
   b. the formation of less nucleosomes.
   c. greater compaction of radial loop domains.
   d. less compaction of radial loop domains.
   e. both a and c.

Short Answer

1. What are the four criteria that the genetic material must fulfill? What was Griffith's contribution to the study of DNA, and why was it important?

2. **PRINCIPLES** A principle of biology is that structure determines function. Discuss how the structure of DNA underlies different aspects of its function.

Critical-Thinking Skills

Multiple Choice

1. You have analyzed the base composition of the genetic material in a virus and found the following: 32% adenine, 17% thymine, 18% guanine, and 33% cytosine. From this analysis, which of the following is the most likely structure of the viral genome?
   a. single-stranded DNA
   b. double-stranded DNA
   c. single-stranded RNA
   d. double-stranded RNA
   e. either a or b

2. Which of the following equations would be appropriate when considering DNA base composition?
   a. %A + %T = %G + %C
   b. %A = %G
   c. %A = %G = %T = %C
   d. %A + %G = %T + %C
   e. %T = %C

3. If the sequence of a segment of a DNA strand is 5′–CGCAACTAC–3′, what is the correct sequence for the opposite strand?
   a. 5′–GCGTTGATG–3′
   b. 3′–ATACCAGCA–5′
   c. 5′–ATACCAGCA–3′
   d. 3′–GCGTTGATG–5′
   e. 5′–CGCAACTAC–3′

4. What would be the expected results for the Meselson and Stahl experiment after five generations (that is, five cell divisions)? Note: Generation zero has all heavy DNA, and all subsequent generations use light nitrogen to make new DNA strands.
   a. 1/8 heavy, 7/8 light
   b. 1/16 heavy, 15/16 light
   c. 1/32 heavy, 31/32 light
   d. 1/16 half-heavy, 15/16 light
   e. 1/32 half-heavy, 31/32 light

5. This drawing shows the synthesis of the lagging strand (on the bottom). The top strand is the template DNA. Where is DNA ligase needed?

   a. at the left arrow only
   b. at the right arrow only
   c. at both arrows
   d. at neither arrow

Short Answer

1. **SCISKILLS** ▶ Some bacterial strains exhibit resistance to killing by antibiotics. For example, certain strains of bacteria are resistant to the drug tetracycline, whereas other strains are sensitive to this antibiotic. Describe an experiment you would carry out to demonstrate that tetracycline resistance is an inherited trait carried in the DNA of the resistant strain.

2. During DNA replication, the newly made daughter strand is made according to the AT/GC rule. What critical function of DNA would be lost if the AT/GC rule was not followed during DNA replication?

The Expression of Genetic Information via Genes I: Transcription and Translation

10

©Dr. Elena Kiseleva/Science Source

An electron micrograph of many ribosomes that are translating an mRNA molecule into polypeptides. This colorized electron micrograph shows ribosomes attached to a bacterial mRNA and synthesizing polypeptides. Ribosomes are blue, mRNA is red, and emerging polypeptides are green. The complex of a single mRNA and multiple ribosomes is called a polysome.

Chapter Outline

10.1 Overview of Gene Expression
10.2 Transcription
10.3 RNA Modifications in Eukaryotes
10.4 Translation and the Genetic Code
10.5 The Machinery of Translation
10.6 The Stages of Translation
Assessing Your Knowledge and Skills

Why is knowing about our genes important? Let's consider this question with regard to heart disease. Researchers have identified dozens of genes that influence a person's predisposition to develop various types of heart disease. Some people carry variations in those genes that make it more likely for them to develop the disease. Such knowledge causes some people to take preventative measures, such as exercise and modifications to their diet. In addition, by identifying genes associated with heart disease and studying the proteins specified by them, researchers may be able to develop drugs to combat various forms of the disease.

We can broadly define a gene as a unit of heredity. Geneticists view gene function at different biological levels. In this chapter, we will explore how genes work at the molecular level. We will see how nucleotide sequences are organized to form genes and how those genes are used as a template to make RNA copies, which ultimately leads to the synthesis of functional proteins. In Chapters 15 and 16, we will examine how genes affect the traits, or characteristics, of individuals. The term **gene expression** can refer to gene function either at the molecular level or at the level of traits. In reality, the two phenomena are intricately woven together. The expression of genes at the molecular level affects the structure and function of cells, which in turn determine the traits that an organism displays.

In this chapter, we will explore the steps of gene expression as they occur at the molecular level. These steps include the use of a gene as a template to make an RNA molecule, the modifications of RNA into a functional molecule (in eukaryotes), and the use of RNA to direct the formation of a protein. This chapter focuses on the expression of genes that encode polypeptides, which are called **protein-encoding genes**. The next chapter focuses on some of the interesting functions of genes that produce **non-coding RNAs (ncRNAs)**, which are RNAs that do not encode polypeptides.

10.1 Overview of Gene Expression

Learning Outcomes

1. Outline the general steps of gene expression at the molecular level, which together constitute the central dogma.
2. Explain how proteins are largely responsible for determining an organism's characteristics.

In this section, we will begin to examine how genes work at the molecular level. As we will see, the expression of protein-encoding genes requires the process of transcription, in which DNA is used to make RNA, and the process of translation, in which RNA is used to make polypeptides. We will also consider how the synthesis of proteins is a key factor that determines the structure and function of cells and ultimately an organism's traits.

Molecular Gene Expression Involves the Processes of Transcription and Translation

At the molecular level, the general steps of gene expression occur as follows:

1. **Transcription:** The first step, known as **transcription,** produces an RNA copy of a gene, also called an RNA transcript (**Figure 10.1**). Transcription literally means the act of making a copy. Most genes, which are termed protein-encoding genes, or structural genes,* are used to make an RNA molecule that contains the information to specify a polypeptide with a particular amino acid sequence. This type of RNA is called **messenger RNA** (abbreviated **mRNA**), because its function is to carry information from the DNA to cellular components called ribosomes.

2. **Translation:** The process of synthesizing a specific polypeptide is called **translation.** The term translation is used because a sequence of bases in an mRNA is "translated" into an amino acid sequence of a polypeptide. The term **polypeptide** refers to a linear sequence of amino acids; it denotes structure. Most genes carry the information to make a particular polypeptide. By comparison, the term **protein** denotes function. Some proteins are composed of one polypeptide, whereas others are composed of two or more polypeptides.

Together, the transcription of DNA into mRNA and the translation of mRNA into a polypeptide constitute the **central dogma** of gene expression at the molecular level, which was first proposed in 1958 by Francis Crick (Figure 10.1). The central dogma applies to bacteria, archaea, and eukaryotes. However, in eukaryotes, an additional step occurs between transcription and translation. During **RNA modification,** which is described later in this chapter, the RNA transcript, termed **pre-mRNA,** is modified in ways that make it a functionally active mRNA (Figure 10.1b).

Another difference between bacteria and eukaryotes is the cellular location of transcription and translation. In bacteria, both events occur in the same compartment, namely the cytoplasm. In eukaryotes, transcription occurs in the nucleus. The mRNA then exits the nucleus through a nuclear pore, and translation occurs in the cytosol.

---

*Geneticists commonly use the term structural gene to describe all genes that encode polypeptides. Some geneticists, however, distinguish structural genes from regulatory genes—genes that encode proteins that regulate the expression of structural genes.

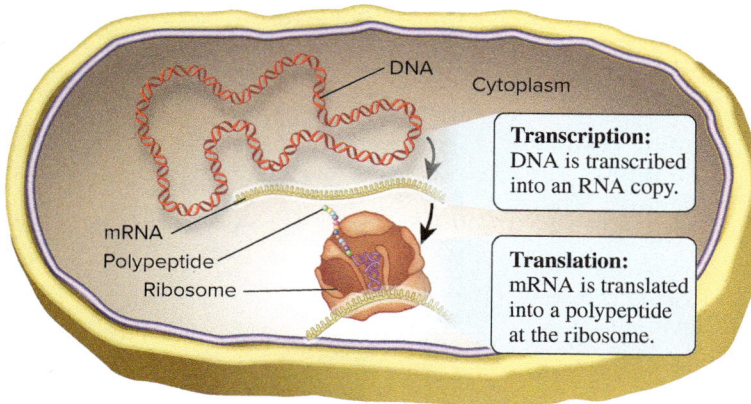

(a) Molecular gene expression in bacteria

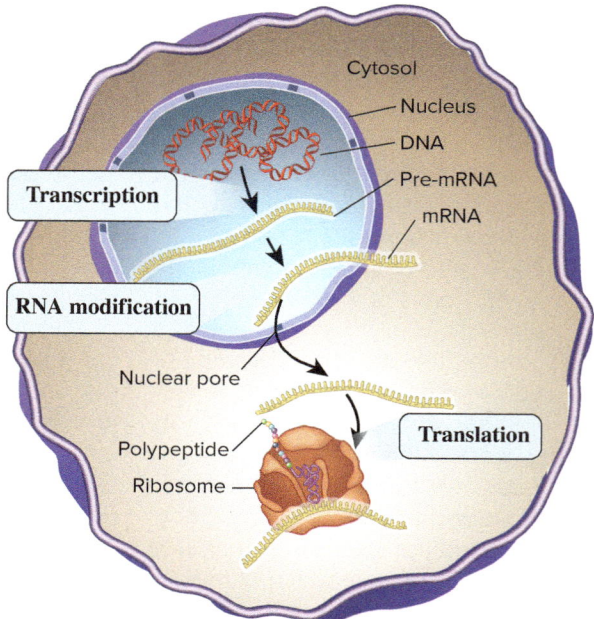

(b) Molecular gene expression in eukaryotes

Figure 10.1 The central dogma of gene expression at the molecular level. **(a)** In bacteria, transcription and translation occur in the cytoplasm. **(b)** In eukaryotes, transcription and RNA modification occur in the nucleus, whereas translation takes place in the cytosol.

 Concept Check: *What is the direction of flow of genetic information?*

Though the direction of information flow, that is, from DNA to RNA to protein, is the most common pathway, exceptions do occur. For example, as we will see in Chapter 17, certain viruses called retroviruses use RNA as a template to synthesize DNA.

The Protein Products of Genes Largely Determine an Organism's Characteristics

The genes that constitute the genetic material provide a blueprint for the characteristics of every organism. They contain the information necessary to produce an organism and allow it to interact appropriately with its environment. Each protein-encoding gene stores the information for the production of a polypeptide, which then becomes a unit within a functional protein. The activities of proteins largely determine the structure and function of cells. Furthermore, the characteristics of an organism are rooted in the activities of cellular proteins. In addition, non-coding RNAs also are important in cell structure and function. This topic is discussed in Chapter 11.

One of the main purposes of the genetic material is to encode the production of proteins in the correct cell, at the proper time, and in suitable amounts. This is an intricate task, because living cells make thousands of different kinds of proteins. Genetic analyses have shown that a typical bacterium can make a few thousand different proteins, and estimates for eukaryotes range from several thousand in simpler eukaryotes to tens of thousands in more complex eukaryotes such as humans.

10.1 Reviewing the Concepts

- At the molecular level, the central dogma states that most genes are transcribed into mRNA, and then the mRNA is translated into polypeptides. In eukaryotes, an additional step, RNA modification, occurs between transcription and translation (Figure 10.1).
- A polypeptide is a unit of structure. A protein, composed of one or more polypeptides, is a unit of function. The molecular expression of genes is an underlying factor that largely determines an organism's characteristics.

10.1 Testing Your Knowledge

1. During transcription, _____ is used as a template to make _____.
   a. DNA, a polypeptide
   b. DNA, RNA
   c. RNA, DNA
   d. RNA, a polypeptide

2. During translation, the information within _____ is used to make _____.
   a. DNA, a polypeptide
   b. DNA, mRNA
   c. mRNA, DNA
   d. mRNA, a polypeptide

3. Proteins are
   a. composed of one or more polypeptides.
   b. key components that affect the structure and function of cells.
   c. largely responsible for the outcome of an individual's traits.
   d. all of the above.

10.2 Transcription

Learning Outcomes

1. Give a molecular definition for the term gene.
2. Outline the three stages of transcription and the role of RNA polymerase in this process.
3. Compare and contrast transcription in bacteria and eukaryotes.

DNA is an information storage bank. For genes to be expressed, the information in them must be accessed at the molecular level. Rather than accessing the information directly, however, a working copy of the DNA, composed of RNA, is made. This occurs by the process of transcription, in which a DNA sequence is copied into an RNA sequence. Importantly, transcription does not permanently alter the structure of DNA. Therefore, the same DNA can continue to store information even after an RNA copy has been made. In this section, we will examine the steps necessary for genes to act as transcriptional units. We will also consider some differences in these steps between bacteria and eukaryotes.

At the Molecular Level, a Gene Is Transcribed and Produces a Functional Product

What is the definition of a gene?

> At the molecular level, a **gene** is defined as an organized unit of nucleotide sequences that enables a segment of DNA to be transcribed into RNA and ultimately results in the formation of a functional product.

When a protein-encoding gene is transcribed, an mRNA is made that specifies the amino acid sequence of a polypeptide. After it is made, the polypeptide becomes a functional product as a protein or part of a protein. The mRNA is an intermediary in polypeptide synthesis. Among all species, most genes are protein-encoding genes. However, for many genes, the functional product is the RNA itself. As mentioned, these RNAs are called non-coding RNAs (ncRNAs). Chapter 11 surveys some of the interesting functions of ncRNAs. In this chapter, we will explore the roles of two ncRNAs, because they are used during the process of translation.

- **Transfer RNA (tRNA)** is needed to make polypeptides during translation.
- **Ribosomal RNA (rRNA)** forms part of ribosomes, which provide the site where translation occurs.

A gene is composed of specific base sequences organized in a way that allows the DNA to be transcribed into RNA. **Figure 10.2** shows the general organization of sequences in a protein-encoding gene in bacteria.

- **Promoter:** a site in the DNA where transcription begins
- **Terminator:** a site in the DNA where transcription ends
- **Regulatory sequences:** sequences that function as sites for the binding of regulatory proteins that affect the rate of transcription. As discussed in Chapter 12, some regulatory proteins enhance the rate of transcription, whereas others inhibit it.

Biology Principle

The Genetic Material Provides a Blueprint for Growth, Development, and Reproduction

Transcription is the first step in accessing the information that is stored in DNA.

Figure 10.2 A bacterial protein-encoding gene as a transcriptional unit. Note: As shown later in Figure 10.4, transcription in eukaryotes first produces a pre-mRNA that is modified before translation occurs.

Concept Check: *How would removing a terminator from a gene affect transcription? Where would transcription end?*

As shown in Figure 10.2, the DNA is transcribed into mRNA from the promoter to the terminator. Within this transcribed region is the information that will specify the amino acid sequence of a polypeptide when the mRNA is translated.

During Transcription, RNA Polymerase Uses a DNA Template to Make RNA

Transcription occurs in three stages, called initiation, elongation, and termination, during which various proteins interact with DNA sequences (**Figure 10.3**).

1. The **initiation stage** is a recognition step. In bacteria, a protein called **sigma factor** binds to **RNA polymerase,** a type of enzyme that synthesizes strands of RNA. Sigma factor recognizes the base sequence of a promoter and binds there. The role of sigma factor is to enable RNA polymerase to bind to the promoter. The initiation stage is completed when the DNA strands are separated near the promoter to form an **open complex** that is approximately 10–15 bp long.

2. During the **elongation stage,** RNA polymerase synthesizes the RNA transcript. For this to occur, sigma factor is released and RNA polymerase slides along the DNA in a way that maintains the open complex as it goes. The DNA strand that is used as a template for transcription is called the **template strand.** For protein-encoding genes, the opposite DNA strand is called the **coding strand.** The coding strand has the same sequence of bases as the resulting mRNA, except that it contains thymine, whereas the mRNA contains uracil. The coding strand is so named because, like mRNA, it carries the information that codes for a polypeptide.

    During the elongation stage of transcription, nucleotides bind to the template strand and are covalently connected in the 5′ to 3′ direction (see inset of step 2, Figure 10.3). The complementarity rule used in this process is similar to the AT/GC rule of DNA replication, except that in RNA, uracil (U) substitutes for thymine (T). For example, a DNA template with the sequence 3′–TACAATGTAGCC–5′ will be transcribed into an RNA sequence reading 5′–AUGUUACAUCGG–3′. In bacteria, the rate of RNA synthesis is about 40 nucleotides per second! Behind the open complex, the DNA rewinds back into a double helix.

3. Eventually, RNA polymerase reaches a terminator, which causes it and the newly made RNA transcript to dissociate from the DNA. This event constitutes the termination stage of transcription.

Transcription in Eukaryotes Involves More Proteins

The basic features of transcription are similar among all organisms. The genes of all species have promoters, and the transcription process occurs in the stages of initiation, elongation, and termination. However, the transcription of eukaryotic genes tends to involve a greater complexity of protein components than does the transcription of bacterial genes. For example, three forms of RNA polymerase—designated I, II, and III—are found in eukaryotes. RNA polymerase II is responsible for transcribing the mRNA from eukaryotic protein-encoding genes, whereas RNA polymerases I and III transcribe non-coding genes such as the genes that encode tRNAs and rRNAs. By comparison, bacteria have a single type of RNA polymerase that transcribes all genes, though many bacterial species have more than one type of sigma factor that can recognize different promoters.

The initiation stage of transcription is also more complex in eukaryotes. Recall that in bacteria, sigma factor recognizes the promoter of a gene. By comparison, RNA polymerase II of eukaryotes always requires five transcription factors to initiate transcription. **Transcription factors** are proteins that influence the ability of RNA polymerase to transcribe genes. The roles of eukaryotic transcription factors are examined in Chapter 12.

10.2 Reviewing the Concepts

- In a gene, a site called a promoter specifies where transcription begins. A terminator specifies where transcription ends (Figure 10.2).

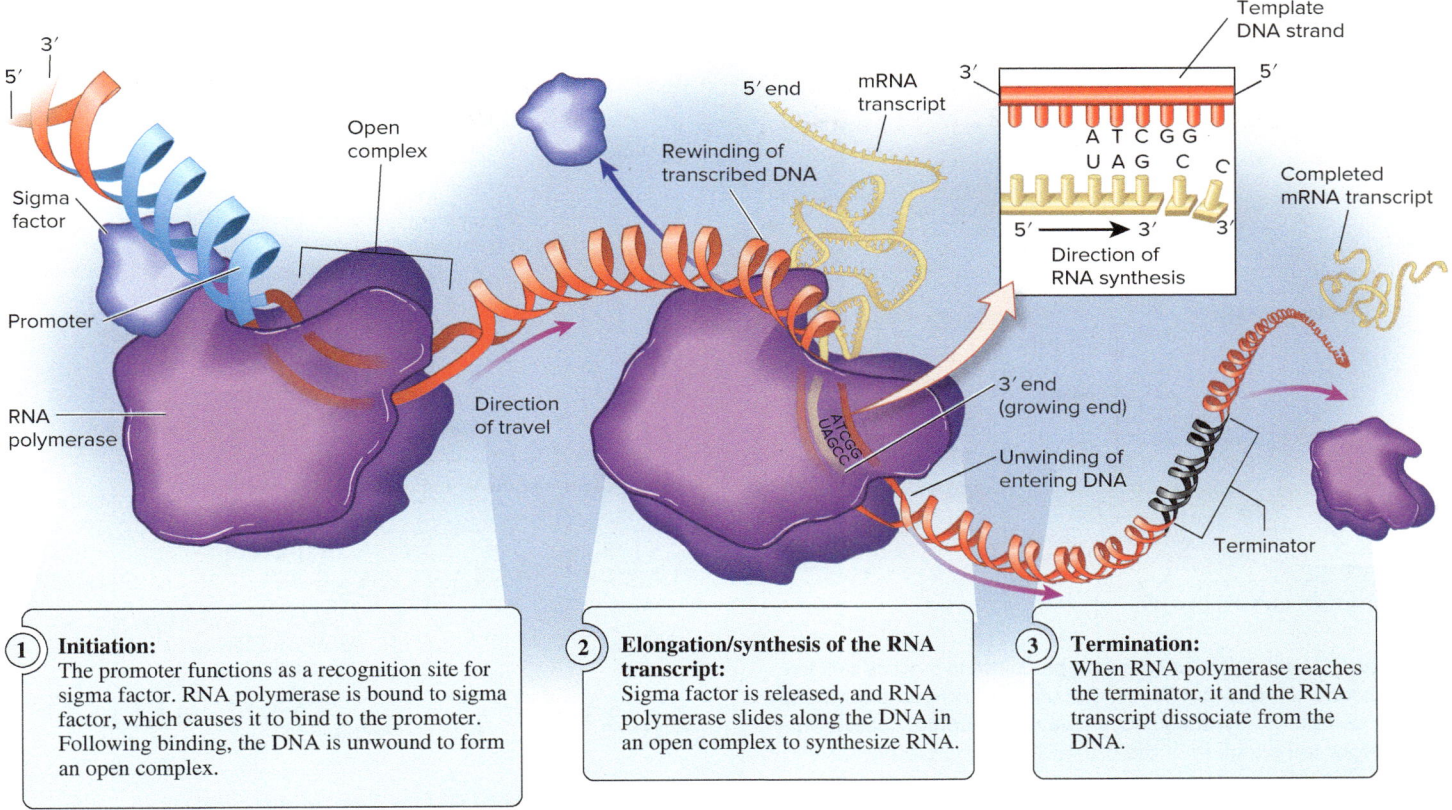

Figure 10.3 Stages of transcription. Transcription can be divided into initiation, elongation, and termination stages. The inset emphasizes the direction of RNA synthesis and the base pairing between the DNA template strand and RNA. Note: The direction of the template DNA strand is opposite to the direction of RNA synthesis.

BioConnections: *Look back at the role of DNA polymerase shown in Figure 9.15. What similarities and differences occur between the function of DNA polymerase and that of RNA polymerase?*

- In bacteria, the initiation of transcription begins when sigma factor binds to RNA polymerase and to a promoter. During elongation, synthesis of an RNA transcript occurs via base pairing of nucleotides to the template strand of DNA as RNA polymerase slides along the DNA. RNA polymerase and the RNA transcript dissociate from the DNA at the terminator (Figure 10.3).

10.2 Testing Your Knowledge

1. A region of the template strand of DNA used for transcription is 3′–AGCTTGAG–5′. What sequence of RNA would be transcribed from this region?
   a. 3′–AGCTTGAG–5′
   b. 5′–AGCTTGAG–3′
   c. 3′–UCGAACUC–5′
   d. 5′–UCGAACUC–3′

2. The stage of bacterial transcription in which sigma factor causes the binding of RNA polymerase to the promoter is
   a. initiation.
   b. elongation.
   c. termination.
   d. none of the above.

10.3 RNA Modifications in Eukaryotes

Learning Outcomes

1. Describe the addition of the 5′ cap and 3′ poly A tail to eukaryotic mRNA.
2. Outline the process of splicing that produces mature eukaryotic mRNA.

As noted previously, eukaryotic transcripts, called **pre-mRNAs**, undergo certain modifications before they exit the nucleus as functionally active **mature mRNAs**, or simply mRNAs. For example, eukaryotic pre-mRNAs are modified by the addition of caps and tails to their ends (**Figure 10.4**). In addition, many eukaryotic pre-mRNAs have coding sequences that are separated by sequences that are later removed. These intervening sequences that are not translated are called **introns**, whereas sequences contained in the mature mRNA are termed **exons**. Exons are considered to be expressed regions, because they contain the coding sequence for a polypeptide. In contrast, introns are intervening regions that are not expressed, because they are removed from the pre-mRNA. To

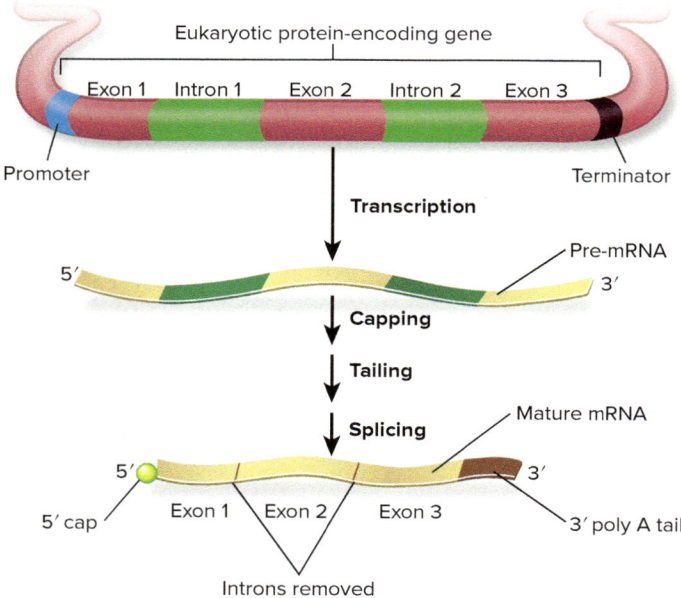

Figure 10.4 Modifications to eukaryotic pre-mRNA to produce a mature mRNA molecule. Note: Most RNA molecules are spliced after the pre-mRNA is completely synthesized. However, for some, splicing may begin before transcription of the pre-mRNA is completed.

(a) Cap structure at the 5′ end of eukaryotic mRNA

(b) Addition of a poly A tail at the 3′ end of eukaryotic mRNA

Figure 10.5 Modifications that alter the ends of mRNA in eukaryotic cells. (a) A 7-methylguanosine cap is attached to the 5′ end of a pre-mRNA. This guanine base is modified by the attachment of a methyl group. The linkage between the cap and the mRNA is a 5′ to 5′ linkage. (b) A poly A tail is added to the 3′ end of a pre-mRNA.

Concept Check: *Do the ends of protein-encoding genes have a poly T region that acts as a template for the synthesis of a poly A tail in mRNA? Explain.*

become a functional mRNA, many pre-mRNAs undergo an RNA modification known as **RNA splicing,** or simply splicing, in which introns are removed and the remaining exons are connected to each other (see Figure 10.4).

After the modifications shown in Figure 10.4 have been completed, the mRNA leaves the nucleus and enters the cytosol, where translation occurs. In this section, we will examine the molecular mechanisms that account for RNA modifications and consider why they are functionally important.

The Ends of Eukaryotic Pre-mRNAs Are Modified by the Addition of a 5′ Cap and a 3′ Poly A Tail

Mature mRNAs of eukaryotes have a modified form of guanine covalently attached at the 5′ end, an event known as **capping** (Figure 10.5a). Capping occurs while a pre-mRNA is being made by RNA polymerase, usually when the transcript is only 20 to 25 nucleotides in length. What are the functions of the cap? The 7-methylguanosine structure, called a **5′ cap,** is recognized by cap-binding proteins, which are needed for the proper exit of mRNAs from the nucleus. After an mRNA is in the cytosol, the cap structure helps to prevent its degradation and is recognized by other cap-binding proteins that enable the mRNA to bind to a ribosome for translation.

At the 3′ end, most mature eukaryotic mRNAs have a string of adenine nucleotides, typically 100 to 200 nucleotides in length, referred to as a **poly A tail** (Figure 10.5b). The poly A tail is not encoded in the gene sequence. Instead, the tail is added enzymatically after a pre-mRNA has been completely transcribed. A long poly A tail aids in the export of mRNAs from the nucleus. It also stabilizes a eukaryotic mRNA so that it can exist for a longer period of time in the cytosol.

Splicing Involves the Removal of Introns and the Linkage of Exons

Introns are found in many but not all eukaryotic genes. Splicing is less frequent among unicellular eukaryotic species, such as yeast, but is a widespread phenomenon among more complex eukaryotes. In animals and flowering plants, most protein-encoding genes have one or more introns. For example, an average human gene has about nine introns. The sizes of introns vary from a few dozen nucleotides to over 100,000! A few bacterial genes have been found to have introns, but they are rare among bacterial and archaeal species.

Introns are precisely removed from eukaryotic pre-mRNA by a complex called a **spliceosome,** which is composed of several different subunits known as snRNPs (pronounced "snurps"). Each snRNP contains small nuclear RNA and a set of proteins. Splicing occurs in a series of steps (Figure 10.6):

1. Spliceosome subunits bind to specific sequences at three locations in the intron RNA: One is termed the branch site, and

Biology Principle

Biologists Use Models to Test Predictions and Convey Ideas

This figure is a series of models for the mechanism of splicing of pre-mRNA in eukaryotes. The Modeling Challenge in the figure legend asks you to propose a model for step 2 if a mutant pre-mRNA was missing a 3′ splice site.

Figure 10.6 The splicing of a eukaryotic pre-mRNA by a spliceosome.

Modeling Challenge: Take a look at the pre-mRNA shown in Figure 10.4, which has three exons and two introns. Let's suppose this pre-mRNA had a mutation that removed the 3′ splice site from the first intron, which is called intron 1. When this happens, the spliceosome will choose the next available 3′ splice site. Draw a model in the format shown in step 2 of this figure, which shows the loop that would be formed during the splicing of this mutant pre-RNA. Label the loop with regard to which intron(s) and exon(s) it contains.

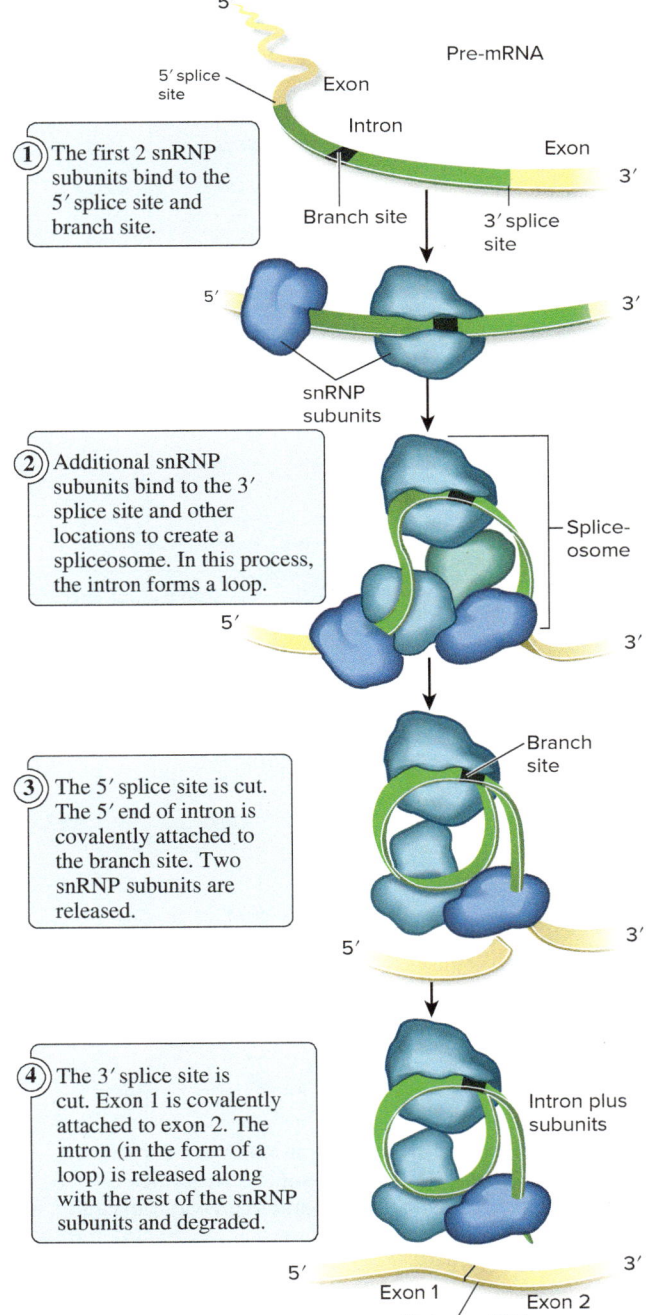

two sequences at the intron-exon boundaries are called the 5′ splice site and the 3′ splice site.

2. This binding causes the intron to loop outward, which brings the two exons close together.
3. The 5′ splice site is then cut, and the 5′ end of the intron becomes covalently attached to the branch site.
4. The 3′ splice site is cut, and the two exons are covalently attached to each other. The intron, in the form of a loop, is released and eventually degraded.

The catalytic events in steps 3 and 4 are catalyzed by an RNA component within a snRNP.

In some cases, the function of the spliceosome is regulated so that the splicing of a given mRNA can occur in two or more ways. This phenomenon, called **alternative splicing,** allows a single gene to encode two or more polypeptides with differences in their amino acid sequences. As described in Chapter 12, alternative splicing allows complex eukaryotic species to use the same gene to make different proteins at different stages of development or in different cell types.

Although primarily found in pre-mRNAs, introns occasionally occur in rRNA and tRNA molecules of certain species. These introns, however, are not removed by the action of a spliceosome. Instead, such rRNAs and tRNAs are **self-splicing,** which means the RNA itself can catalyze the removal of its own intron(s). Portions of the RNA act like an enzyme to cleave the covalent bonds at the intron-exon boundaries and connect the exons together. An RNA molecule that catalyzes a chemical reaction is termed a **ribozyme.**

10.3 Reviewing the Concepts

- In eukaryotes, transcription produces a pre-mRNA that is spliced and given a 5′ cap and a poly A tail (Figures 10.4, 10.5).
- During RNA splicing, intervening sequences called introns are removed from eukaryotic pre-mRNA by a spliceosome (Figure 10.6).

10.3 Testing Your Knowledge

1. The splicing of pre-mRNA in eukaryotes involves
   a. the removal of exons and the connection of introns.
   b. the removal of introns and the connection of exons.
   c. the removal of introns only at the 3′ end.
   d. the removal of introns only at the 5′ end.

2. The function of the 5′ cap on eukaryotic mRNA is to
   a. aid in export from the nucleus.
   b. prevent degradation.
   c. aid in translation.
   d. do all of the above.

10.4 Translation and the Genetic Code

Learning Outcomes

1. Explain how the genetic code specifies the relationship between the sequence of codons in mRNA and the amino acid sequence of a polypeptide.
2. **SCISKILLS** ▶ Analyze the experiments of Nirenberg and Leder that led to the deciphering of the genetic code.

In the two previous sections, we considered how an RNA transcript is made from DNA and how eukaryotes process that transcript. Recall that this type of RNA is called messenger RNA (mRNA), because its function is to transmit information from DNA to ribosomes, where polypeptide synthesis occurs. In this section, we will consider the next stage—translation—which is the synthesis of polypeptides using information within mRNA. We will also explore experiments that helped to crack the code.

The Genetic Code Specifies the Amino Acids Within a Polypeptide

The ability of mRNA to be translated into an amino acid sequence of a polypeptide relies on the **genetic code,** which specifies the relationship between the sequence of bases in the mRNA and the sequence of amino acids in a polypeptide. The code is read in groups of three nucleotide bases known as **codons.** The genetic code consists of 64 different codons (**Table 10.1**). The sequence of three bases in most codons specifies a particular amino acid. For example, the codon CCC specifies the amino acid proline, whereas the codon GGC encodes the amino acid glycine. From the analysis of many different species, researchers have found that the genetic code is nearly universal. Only a few rare exceptions have been discovered.

Why are there 64 codons, as shown in Table 10.1? Because there are 20 types of amino acids, at least 20 different codons are needed so that each amino acid can be specified by a codon. With four types of bases in mRNA (U, C, A, and G), a genetic code containing two bases in a codon would not be sufficient, because only 4^2, or 16, different codons would be possible. A three-base system can specify 4^3, or 64, different codons, which is more than the number of amino acids. The genetic code is said to be **degenerate,** or redundant, because more than one codon can specify the same amino acid (see Table 10.1). For example, the codons GGU, GGC, GGA, and GGG all code for the amino acid glycine. In most instances, the third base in the codon is the degenerate, or variable, base.

During Translation, mRNA Is Used to Make a Polypeptide with a Specific Amino Acid Sequence

Let's look at the organization of a bacterial mRNA to see how translation occurs (**Figure 10.7**).

- **Ribosomal-binding site:** a site that is located near the 5′ end of an mRNA and provides a location for the ribosome to bind to the mRNA.

Table 10.1 The Genetic Code*

		Second Position			
First Position	U	C	A	G	Third Position
U	UUU } Phe UUC UUA } Leu UUG	UCU } Ser UCC UCA UCG	UAU } Tyr UAC UAA Stop UAG Stop	UGU } Cys UGC UGA Stop UGG Trp	U C A G
C	CUU } Leu CUC CUA CUG	CCU } Pro CCC CCA CCG	CAU } His CAC CAA } Gln CAG	CGU } Arg CGC CGA CGG	U C A G
A	AUU } Ile AUC AUA AUG Met/start	ACU } Thr ACC ACA ACG	AAU } Asn AAC AAA } Lys AAG	AGU } Ser AGC AGA } Arg AGG	U C A G
G	GUU } Val GUC GUA GUG	GCU } Ala GCC GCA GCG	GAU } Asp GAC GAA } Glu GAG	GGU } Gly GGC GGA GGG	U C A G

*These codon sequences are in the 5′ to 3′ direction, going from left to right. Exceptions to the genetic code are rarely found among various species. For example, AUA encodes methionine in yeast and in mammalian mitochondria.

Figure 10.7 The organization of a bacterial mRNA as a translational unit. The string of colored balls represents a sequence of amino acids in a polypeptide. During and following translation, a sequence of amino acids folds into a more compact structure, as described in Chapter 3 (refer back to Figure 3.13).

Concept Check: If a mutation eliminated the start codon from a protein-encoding gene, how would the mutation affect transcription, and how would it affect translation?

- **Start codon:** the first codon that is used in translation. This codon (AUG) specifies the amino acid methionine and is only a few nucleotides from the ribosomal-binding site. The first amino acid during bacterial polypeptide synthesis is a modified form of methionine called formyl methionine, but it may be removed from the polypeptide.
- **Coding sequence:** the region that begins with the start codon and specifies the entire amino acid sequence of a polypeptide. A typical polypeptide is a few hundred amino acids in length. The coding sequence consists of a series of codons.
- **Stop codon:** the last codon, which signals the end of translation. These codons, also known as termination codons or nonsense codons, are UAA, UAG, and UGA. They do not code for an amino acid.

The start codon also defines the **reading frame** of an mRNA, which is the order in which codons are read during translation. Beginning at the start codon, each adjacent codon is read as a group of three bases, also called a **triplet,** in the 5′ to 3′ direction. For example, let's look at the following two mRNA sequences and their corresponding amino acid sequences:

	Ribosomal-binding site	Start codon

mRNA 5′–AUAAGGAGGUUACG(AUG)(CAG)(CAG)(GGC)(UUU)(ACC)–3′

Polypeptide   Met - Gln - Gln - Gly - Phe - Thr

	Ribosomal-binding site	Start codon

mRNA 5′–AUAAGGAGGUUACG(AUG)(UCA)(GCA)(GGG)(CUU)(UAC)C–3′

Polypeptide   Met - Ser - Ala - Gly - Leu - Tyr

The first sequence shows how the mRNA codons would be correctly translated into amino acids. In the second sequence, an additional U has been added to the sequence after the start codon. This shifts the reading frame, thereby changing the codons as they occur in the 5′ to 3′ direction. The polypeptide produced from this series of codons has a very different sequence of amino acids. From this comparison, we can also see that the reading frame is not overlapping, which means that each base functions within a single codon.

DNA Stores Information, Whereas mRNA and tRNA Access That Information to Make a Polypeptide

The relationships among the DNA sequence of a gene, the mRNA transcribed from the gene, and the polypeptide sequence are shown schematically in **Figure 10.8**. Recall that the template strand is used to make mRNA. The resulting mRNA strand corresponds to the coding strand of DNA, except that U in the mRNA substitutes for T in the DNA. The 5′ end of the mRNA contains an untranslated region (5′ UTR), as does the 3′ end (3′ UTR). The middle portion contains a series of codons that specify the amino acid sequence of a polypeptide.

To translate a nucleotide sequence of mRNA into an amino acid sequence, recognition occurs between mRNA and transfer RNA (tRNA) molecules. Transfer RNA, which is described in Section 10.5, functions as the "translator," or intermediary, between an mRNA codon and an amino acid. The **anticodon** is a three-base sequence in a tRNA molecule that is complementary to a codon in mRNA. Due to this complementarity, the anticodon in a tRNA and a codon in an mRNA bind to each other. Furthermore, the anticodon in a tRNA corresponds to the amino acid that it carries. For example, if the anticodon in a tRNA is 3′–AAG–5′, it is complementary to a 5′–UUC–3′ codon. According to the genetic code, a UUC codon specifies phenylalanine (Phe). Therefore, a tRNA with a 3′–AAG–5′ anticodon must carry phenylalanine.

As seen at the bottom of Figure 10.8, the direction of polypeptide synthesis parallels the 5′ to 3′ orientation of mRNA. The first amino acid is said to be at the amino end, or **N-terminus,** of the polypeptide. The term N-terminus refers to the presence of a nitrogen atom (N) at this end, whereas amino end indicates the presence of an amino group (NH_2). **Peptide bonds** connect the amino acids together. These covalent bonds form between the carboxyl group (COOH) of the previous amino acid and the amino group of the next amino acid. The last amino acid in a completed polypeptide does not have another amino acid attached to its carboxyl group. This last amino acid is said to be located at the carboxyl end, or **C-terminus.** A carboxyl group is always found at this end of the polypeptide. Note that at neutral pH, the amino group is positively charged (NH_3^+), whereas the carboxyl group is negatively charged (COO^-).

Synthetic RNA Helped Researchers Decipher the Genetic Code

Let's look at some early experiments that allowed scientists to decipher the genetic code. During the early 1960s, the genetic code was determined by the collective efforts of several researchers, including American biochemist Marshall Nirenberg, Spanish-American biochemist Severo Ochoa, and American geneticist Philip Leder. Prior to their studies, other scientists had discovered that bacterial cells can be broken open and components from the cytoplasm can synthesize polypeptides if mRNA is also present. This mixture is termed an **in vitro translation system.** Nirenberg and Ochoa made synthetic mRNA molecules using an enzyme that covalently connects nucleotides together. Using this synthetic mRNA, they then determined which amino acids were incorporated into polypeptides. For example, if a synthetic mRNA molecule had only adenine-containing nucleotides (for example, 5′–AAAAAAAAAAAAAAAAAAAA–3′), a polypeptide was produced that contained only lysine. This result indicated that the AAA codon specifies lysine.

Another method used to decipher the genetic code involved the chemical synthesis of short RNA molecules, as described in the Feature Investigation.

CHAPTER 10

Figure 10.8 Relationships among the coding sequence of a gene, the coding sequence of an mRNA, the anticodons of tRNA, and the amino acid sequence of a polypeptide. Note: The tRNAs are detached from the polypeptide, one at a time, as it being synthesized. Also, this gene does not contain any introns.

Concept Check: If an anticodon in a tRNA molecule has the sequence 3′–ACC–5′, which amino acid does it carry?

FEATURE INVESTIGATION

Nirenberg and Leder Found That RNA Triplets Can Promote the Binding of tRNA to Ribosomes

In 1964, Nirenberg and Leder discovered that RNA molecules containing three nucleotides (that is, a triplet) can cause a tRNA molecule to bind to a ribosome. In other words, an RNA triplet acts like a codon within an mRNA molecule. To establish the relationship between triplet sequences and specific amino acids, Nirenberg and Leder made triplets with specific base sequences (**Figure 10.9**). For example, in one experiment, they studied 5′–CCC–3′ triplets. This triplet was added to 20 different tubes. To each tube, they next added an in vitro translation system, which contained ribosomes and tRNAs that already had amino acids attached to them. However, each translation system had only one type of radiolabeled amino acid. One translation system had only proline that was radiolabeled, a second translation system had only radiolabeled serine, and so on.

Figure 10.9 Nirenberg and Leder's use of triplet binding to decipher the genetic code.

HYPOTHESIS	An RNA triplet can bind to a ribosome and promote the binding of the tRNA that carries the amino acid that the RNA triplet specifies.
KEY MATERIALS	The researchers made 20 in vitro translation systems, which included ribosomes, tRNAs, and 20 amino acids. The 20 translation systems differed with regard to which amino acid was radiolabeled. For example, in 1 translation system, radiolabeled glycine was added, and the other 19 amino acids were unlabeled. In another system, radiolabeled proline was added, and the other 19 amino acids were unlabeled. The in vitro translation systems also contained the enzymes that attach amino acids to tRNAs.

1. Mix together RNA triplets of a specific sequence and 20 in vitro translation systems. In the example shown here, the triplet is 5′–CCC–3′. Each translation system contained a different radiolabeled amino acid. (Note: Only 3 tubes are shown here.)

2. Allow time for the RNA triplet to bind to the ribosome and for the appropriate tRNA to bind to the RNA triplet.

3. Pour each mixture through a filter that allows the passage of unbound tRNA but does not allow the passage of ribosomes.

4. Count radioactivity on the filter.

5. **THE DATA**

Triplet	Radiolabeled amino acid trapped on the filter	Triplet	Radiolabeled amino acid trapped on the filter
5′–AAA–3′	Lysine	5′–GAC–3′	Aspartic acid
5′–ACA–3′	Threonine	5′–GCC–3′	Alanine
5′–ACC–3′	Threonine	5′–GGU–3′	Glycine
5′–AGA–3′	Arginine	5′–GGC–3′	Glycine
5′–AUA–3′	Isoleucine	5′–GUU–3′	Valine
5′–AUU–3′	Isoleucine	5′–UAU–3′	Tyrosine
5′–CCC–3′	Proline	5′–UGU–3′	Cysteine
5′–CGC–3′	Arginine	5′–UUG–3′	Leucine
5′–GAA–3′	Glutamic acid		

6. **CONCLUSION** This method enabled the researchers to identify many of the codons of the genetic code.

7. **SOURCE** Leder, P., and Nirenberg, M. W. 1964. RNA Codewords and Protein Synthesis, III. On the Nucleotide Sequence of a Cysteine and a Leucine RNA Codeword. *Proceedings of the National Academy of Sciences* 52: 1521–1529.

As shown in step 2 of Figure 10.9, the triplets became bound to the ribosomes just like mRNAs bind to a ribosome. The tRNA with an anticodon that was complementary to the added triplet would bind to the triplet, which was already bound to the ribosome. For example, when the triplet was 5′–CCC–3′, a tRNA with a 3′–GGG–5′ anticodon became bound to the triplet/ribosome complex. This tRNA carries proline.

To determine which tRNA had bound, the contents from each tube were poured through a filter that trapped the large ribosomes but did not trap tRNAs that were not bound to ribosomes (see step 3). If the tRNA carrying the radiolabeled amino acid was bound to the triplet/ribosome complex, radioactivity would be trapped on the filter. Using a scintillation counter, the researchers determined the amount of radioactivity on each filter. Because only one amino acid was radiolabeled in each in vitro translation system, they could determine which triplet corresponded to which amino acid. In the example shown here, CCC corresponds to proline. Therefore, the in vitro translation system containing radiolabeled proline showed a large amount of radioactivity on the filter. As seen in the data, by studying triplets with different sequences, Nirenberg and Leder identified many codons of the genetic code.

Experimental Questions

1. Briefly explain how a triplet mimics the role of an mRNA molecule. How was this observation useful in the study done by Nirenberg and Leder?
2. **SCISKILLS** ▶ What was the benefit of using radiolabeled amino acids in the Nirenberg and Leder experiment?
3. **SCISKILLS** ▶ Predict the results that Nirenberg and Leder would have found for the following triplets: AUG, UAA, UAG, and UGA.

10.4 Reviewing the Concepts

- Based on the genetic code, each of the 64 codons specifies a start codon (AUG, which specifies methionine), other amino acids, or a stop codon (UAA, UAG, or UGA) (Table 10.1, Figure 10.7).
- The template strand of DNA is used to make an mRNA that contains a series of codons. Recognition between an mRNA and many tRNA molecules determines the amino acid sequence of a polypeptide. A polypeptide has a directionality in which the first amino acid is at the N-terminus, or amino end, whereas the last amino acid is at the C-terminus, or carboxyl end (Figure 10.8).
- Nirenberg and Leder used the ability of RNA triplets to promote the binding of tRNAs to ribosomes as a way to identify many of the codons of the genetic code (Figure 10.9).

10.4 Testing Your Knowledge

1. The codon 5′–GAC–3′ specifies aspartic acid. What would be the anticodon of a tRNA that recognized this aspartic acid codon?
   a. 5′–GAC–3′
   b. 3′–GAC–5′
   c. 5′–CUG–3′
   d. 3′–CUG–5′
2. The reading frame of an mRNA
   a. begins with the start codon.
   b. is read as a series of adjacent codons.
   c. is not affected if a single nucleotide is added to the coding sequence.
   d. Both a and b are correct.

10.5 The Machinery of Translation

Learning Outcomes

1. Describe the structure and function of tRNA.
2. Explain how aminoacyl-tRNA synthases attach amino acids to tRNAs.
3. Outline the structural features of bacterial and eukaryotic ribosomes.
4. **SCISKILLS** ▶ Discuss how ribosomal RNA (rRNA) is used to evaluate evolutionary relationships among different species.

In the remaining sections of this chapter, we will turn our attention to the mechanism by which living cells use the genetic code to translate mRNA into polypeptides. A cell must make many different components, including mRNAs, tRNAs, ribosomes, and translation factors, in order to synthesize polypeptides (Table 10.2).

Though the estimates vary from cell to cell and from species to species, most cells use a substantial amount of their energy to translate mRNA into polypeptides. In E. coli, for example, approximately 90% of the cellular energy is used for this process. This value underscores the complexity and importance of translation in living organisms. In this section, we will focus on the components of the translation machinery.

Transfer RNAs Share Common Structural Features

To understand how tRNAs function as carriers of the correct amino acids during translation, researchers have examined their structural characteristics. The tRNAs of all species share common features.

- The two-dimensional structure of a tRNA resembles a cloverleaf. The structure has three stem-loops and a fourth stem

Table 10.2 Components of the Translation Machinery

Component	Function
mRNA	mRNA contains the information for the sequence of amino acids in a polypeptide according to the genetic code.
tRNA	A molecule with two functional sites: One—the anticodon—binds to a codon in mRNA; a second site—the 3′ single-stranded region—is where an appropriate amino acid is attached.
Ribosome	Composed of rRNA molecules and many proteins, the ribosome provides a location where mRNA and tRNA molecules can properly interact with each other. The ribosome also catalyzes the formation of covalent bonds between adjacent amino acids to make a polypeptide.
Translation factors	Proteins needed for the three stages of translation: (1) Initiation factors are required for the assembly of mRNA, the first tRNA, and ribosomal subunits. (2) Elongation factors are needed to synthesize the polypeptide. (3) Release factors are needed to recognize the stop codon and disassemble the translation machinery. Several translation factors use GTP as an energy source to carry out their functions.

Biology Principle

Structure Determines Function

The structure of tRNA has two functional sites: a 3′ single-stranded region where an amino acid is attached and an anticodon that binds to a codon in mRNA.

(a) Two-dimensional structure of tRNA

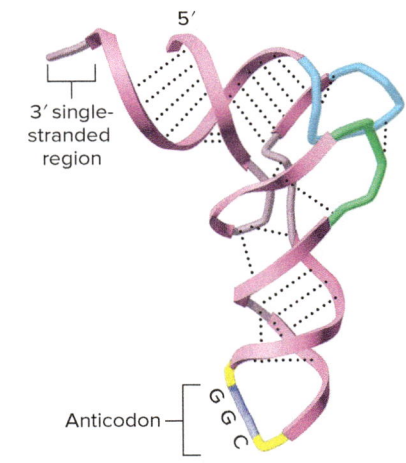

(b) Three-dimensional structure of tRNA

Figure 10.10 Structure of tRNA. (a) The two-dimensional, or secondary, structure of tRNA resembles a cloverleaf, with the anticodon in the loop of the middle stem-loop structure. The 3′ single-stranded region is where an amino acid can attach. (b) The actual three-dimensional structure of tRNA folds in on itself.

Concept Check: What is the function of the anticodon?

---

with a 3′ single-stranded region (**Figure 10.10a**). The stem in a stem-loop is a region where the RNA is double-stranded due to complementary base pairing via hydrogen bonding, whereas the loop is a region without base pairing.

- The anticodon is located in the middle loop.
- The 3′ single-stranded region is the amino acid attachment site.
- The three-dimensional structure of tRNA molecules involves additional folding of the secondary structure (**Figure 10.10b**).

The cells of every organism make many different tRNA molecules, each encoded by a different gene. A tRNA is named according to the amino acid it carries. For example, tRNA^{Ser} carries a serine.

Aminoacyl-tRNA Synthetases Charge tRNAs by Attaching an Appropriate Amino Acid

To perform its role during translation, a tRNA must have the appropriate amino acid attached to its 3′ end. The enzymes that catalyze the attachment of amino acids to tRNA molecules are known as **aminoacyl-tRNA synthetases**. Cells make 20 distinct types of aminoacyl-tRNA synthetase enzymes, with each type recognizing just 1 of the 20 different amino acids. Each aminoacyl-tRNA synthetase is named for the specific amino acid it attaches to tRNA. For example, alanyl-tRNA synthetase recognizes alanine and attaches this amino acid to all tRNAs with alanine anticodons.

Aminoacyl-tRNA synthetases catalyze chemical reactions involving an amino acid, a tRNA molecule, and ATP (**Figure 10.11**).

1. A specific amino acid and ATP bind to the enzyme.
2. The amino acid is activated by the covalent attachment of an adenosine monophosphate (AMP), and pyrophosphate (PP$_i$) is released.
3. A specific tRNA binds to the enzyme. The activated amino acid is covalently attached to the 3′ end of the tRNA molecule, and AMP is released.

Figure 10.11 Aminoacyl-tRNA synthetase charging a tRNA.

BioConnections: *Refer back to Figure 6.3, which describes the hydrolysis of ATP. Why is ATP needed to charge a tRNA?*

4. The tRNA with its attached amino acid, called a **charged tRNA,** or an **aminoacyl tRNA,** is released from the enzyme.

Aminoacyl-tRNA synthetases are amazingly accurate enzymes. The wrong amino acid is attached to a tRNA less than once in 100,000 times! The anticodon region of the tRNA is usually important for recognition by the correct aminoacyl-tRNA synthetase. In addition, the base sequences in other regions may facilitate binding to an aminoacyl-tRNA synthetase.

Ribosomes Are Composed of rRNA and Proteins

Let's now turn our attention to the **ribosome,** the site where translation takes place. Bacterial cells have one type of ribosome, which translates all mRNAs in the cytoplasm. The most abundant type of eukaryotic ribosome functions in the cytosol. In addition, mitochondria have ribosomes, and plant and algal cells have ribosomes in their chloroplasts. Unless otherwise noted, the term eukaryotic ribosome refers to ribosomes in the cytosol, not to those found in mitochondria or chloroplasts.

A ribosome is a large complex composed of structures called the large and small subunits. The term subunit is perhaps misleading, because each ribosomal subunit is itself assembled from many different proteins and one or more RNA molecules, which are called ribosomal RNAs, or simply rRNAs. In the bacterium *E. coli*, the small ribosomal subunit is called 30S, and the large subunit is 50S. The designations 30S and 50S refer to the rate at which these subunits sediment when subjected to a centrifugal force. This rate is described as a sedimentation coefficient in Svedberg units (S) in honor of Swedish chemist Theodor Svedberg, who invented the ultracentrifuge. Together, the 30S and 50S subunits form a 70S ribosome. (Svedberg units don't add up linearly, because the sedimentation coefficient is a function of both size and shape.)

The compositions of bacterial and eukaryotic ribosomes are described in **Table 10.3**. In bacteria, ribosomal proteins and rRNA molecules are synthesized in the cytoplasm, and the ribosomal subunits are assembled there as well. The synthesis of eukaryotic rRNA occurs in the nucleolus, a droplet organelle in the nucleus that is specialized for that purpose. The ribosomal proteins are made in the cytosol and imported into the nucleus. The rRNAs and ribosomal proteins are then assembled within the nucleolus to make the 40S and 60S subunits. The 40S and 60S subunits are exported into the cytosol, where they associate to form an 80S ribosome during translation.

Due to structural differences between bacterial and eukaryotic ribosomes, certain chemicals may bind to bacterial ribosomes but not to eukaryotic ribosomes, and vice versa. Some **antibiotics** (chemicals that inhibit the growth of certain microorganisms) bind only to bacterial ribosomes and inhibit translation. Examples include tetracycline, erythromycin, and chloramphenicol. Because

Table 10.3 Composition of Bacterial and Eukaryotic Ribosomes

	Small subunit	Large subunit	Assembled ribosome
Bacterial			
Sedimentation coefficient	30S	50S	70S
Number of proteins	21	34	55
rRNA	16S rRNA	5S rRNA, 23S rRNA	16S rRNA, 5S rRNA, 23S rRNA
Eukaryotic			
Sedimentation coefficient	40S	60S	80S
Number of proteins	33	49	82
rRNA	18S rRNA	5S rRNA, 5.8S rRNA, 28S rRNA	18S rRNA, 5S rRNA, 5.8S rRNA, 28S rRNA

these chemicals do not inhibit eukaryotic ribosomes, they have been effective drugs for the treatment of bacterial infections in humans and domesticated animals.

Components of Ribosomal Subunits Form Functional Sites for Translation

To understand the structure and function of the ribosome at the molecular level, researchers have determined the locations and functional roles of individual ribosomal proteins and rRNAs. In recent years, a few research groups have succeeded in purifying ribosomes and causing them to crystallize in a test tube. When researchers use X-ray diffraction to study the crystallized ribosomes, they obtain detailed information about ribosome structure. **Figure 10.12a** shows a model of a bacterial ribosome.

During bacterial translation, the mRNA lies on the surface of the 30S subunit, within a space between the 30S and 50S subunits (**Figure 10.12b**). As a polypeptide is synthesized, it exits through a hole within the 50S subunit. Ribosomes contain three discrete sites where tRNAs may be located. In 1964, James Watson proposed a two-site model for tRNA binding to the ribosome. These sites are known as the **peptidyl site (P site)** and **aminoacyl site (A site)**. In 1981, German geneticists Knud Nierhaus and Hans-Jörg Rheinberger expanded this to a three-site model (see Figure 10.12b). The third site is known as the **exit site (E site)**. In Section 10.6, we will examine the roles of these sites in the synthesis of a polypeptide.

(a) Bacterial ribosome model based on X-ray diffraction studies

(b) Schematic model for ribosome structure

Figure 10.12 Ribosome structure. (a) A model for the structure of a bacterial ribosome based on X-ray diffraction studies, showing the large and small subunits and the major binding sites. The rRNA is shown in gray (large subunit) and turquoise (small subunit), whereas the ribosomal proteins are magenta (large subunit) and dark blue (small subunit). The rRNAs form a scaffold for the binding of many ribosomal proteins. (b) A schematic model emphasizing functional sites in the ribosome and showing bound mRNA and tRNA with an attached polypeptide. The amino acids are depicted as different-colored balls to emphasize that translation produces a polypeptide with a specific amino acid sequence.
(a) ©Tom Pantages

Figure 10.13 Comparison of small subunit rRNA gene sequences from three mammalian and three bacterial species. Note the many similarities (yellow) and differences (green and red) among the sequences. The gray color indicates differences among the three bacterial species.

Concept Check: Based on the gene sequences shown here, pick two species that are closely related evolutionarily and two that are distantly related.

EVOLUTIONARY CONNECTIONS

Comparisons of Small Subunit rRNAs Among Different Species Provide a Basis for Establishing Evolutionary Relationships

Translation is a fundamental process that is vital for the existence of all living species. Research indicates that the components needed for translation arose very early in the evolution of life on our planet in ancestors that gave rise to all known living species. For this reason, all organisms have translational components that are evolutionarily related to each other. For example, the rRNA found in the small subunit of ribosomes is similar among all forms of life, though it is slightly larger in eukaryotic species (18S) than in bacterial species (16S). The gene for the small subunit rRNA (SSU rRNA) is found in all species.

How is this observation useful? One way that geneticists explore evolutionary relationships is by comparing the sequences of evolutionarily related genes. At the molecular level, gene evolution involves changes in DNA sequences. After two different species have diverged from each other during evolution, the genes of each species can accumulate changes, or mutations, that alter the sequences of those genes. After many generations, evolutionarily related species have genes that are similar but not identical to each other, because each species accumulates different mutations over time. In general, if a very long time has elapsed since two species diverged evolutionarily, their genes tend to be quite different. In contrast, if two species diverged relatively recently on an evolutionary timescale, their genes tend to be more similar.

Figure 10.13 compares a portion of the sequence of the SSU rRNA gene from three mammalian and three bacterial species. The colors highlight different types of comparisons. The bases shaded in yellow are identical in five or six species. Bases that are identical in different species are said to be **evolutionarily conserved.** Presumably, these bases were found in the primordial gene that gave rise to modern species. Perhaps because these bases may have some critical function, they have not changed over evolutionary time. Those bases shaded in green are identical in all three mammals but differ from the sequences in one or more bacterial species. Actually, if you scan the mammalian species, you may notice that all three have identical sequences in this region. The bases shaded in red are identical or very similar in the bacterial species, but they differ from the mammalian SSU rRNA genes. The sequences from *Escherichia coli* and *Serratia marcescens* are more similar to each other than the sequence from *Bacillus subtilis* is to either of them. This observation suggests that *E. coli* and *S. marcescens* are more closely related evolutionarily than either of them is to *B. subtilis*.

10.5 Reviewing the Concepts

- Translation requires mRNA, charged tRNAs, ribosomes, and many translation factors (Table 10.2).
- tRNA molecules have a two-dimensional structure resembling a cloverleaf. Two important sites are the amino acid attachment site at the 3′ single-stranded region and the anticodon, which form base pairs with a codon in mRNA (Figure 10.10).
- The enzyme aminoacyl-tRNA synthetase attaches the correct amino acid to a tRNA molecule, producing a charged tRNA (Figure 10.11).
- A ribosome is composed of a small and a large subunit, each consisting of rRNA molecules and many proteins. Bacterial and eukaryotic ribosomes differ in their molecular composition (Table 10.3).
- Ribosomes have three sites, termed the A, P, and E sites, which are locations for the binding and release of tRNA molecules (Figure 10.12).
- The gene that encodes the small subunit rRNA (SSU rRNA) has been used to determine evolutionary relationships among different species (Figure 10.13).

10.5 Testing Your Knowledge

1. The anticodon of a tRNA molecule is located
   a. at the 3′ end.
   b. in the first stem.
   c. in the middle stem.
   d. in the middle loop.

2. An aminoacyl-tRNA synthetase has binding sites for
   a. a specific tRNA.
   b. a specific amino acid.
   c. ATP.
   d. all of the above.

3. Ribosomes are composed of
   a. proteins only.
   b. rRNA only.
   c. proteins and rRNA.
   d. proteins and DNA.

10.6 The Stages of Translation

Learning Outcomes

1. Describe the three stages of translation.
2. Summarize the similarities and differences between translation in bacteria and in eukaryotes.

Like transcription, the process of translation occurs in three stages called initiation, elongation, and termination.

1. During initiation, an mRNA, the first tRNA, and the ribosomal subunits assemble into a complex.
2. In the elongation stage, the ribosome moves in the 5′ to 3′ direction from the start codon in the mRNA toward the stop codon, synthesizing a polypeptide according to the sequence of codons in the mRNA.
3. Finally, the process is terminated when the ribosome reaches a stop codon and the complex disassembles, releasing the completed polypeptide.

In this section, we will examine the three stages of translation as they occur in living cells.

Translation Is Initiated with the Assembly of mRNA, tRNA, and Ribosomal Subunits

During the **initiation stage,** a complex is formed between an mRNA molecule, the first tRNA, and the ribosomal subunits. The assembly of this complex requires the help of translation factors called **initiation factors** that facilitate the interactions between these components (see Table 10.2). The assembly also requires an input of energy. Guanosine triphosphate (GTP) is hydrolyzed by certain initiation factors to provide the necessary energy.

The initiation of translation in bacteria occurs as follows:

1. To begin assembly of the complex in bacteria, an mRNA binds to the small ribosomal subunit (**Figure 10.14**). The binding of bacterial mRNA to this subunit is facilitated by a short ribosomal-binding site near the 5′ end of the mRNA. The ribosomal-binding site is a sequence of bases that is complementary to a portion of the 16S rRNA within the small ribosomal subunit. The mRNA becomes bound to the ribosome because the ribosomal-binding site and rRNA hydrogen-bond to each other by base pairing. The start codon (AUG) is usually just a few nucleotides downstream (that is, toward the 3′ end) from the ribosomal-binding site.
2. A specific tRNA, which functions as the **initiator tRNA,** recognizes the start codon in mRNA (AUG) and binds to it. In bacteria, it carries a methionine that has been modified by the attachment of a formyl group.
3. To complete the initiation stage, the large ribosomal subunit associates with the small subunit. At the end of this stage, the initiator tRNA is located in the P site of the ribosome.

In eukaryotic species, the initiation phase of translation differs from the process in bacteria in a few ways.

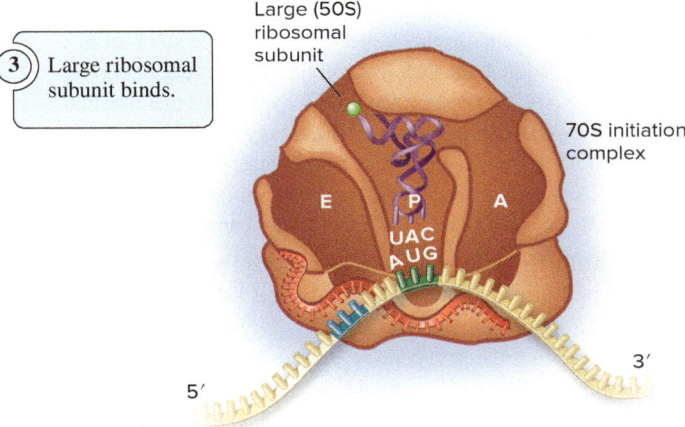

Figure 10.14 Initiation of translation in bacteria.

Concept Check: *What promotes the binding between the mRNA and the small ribosomal subunit?*

- Instead of an RNA sequence that functions as a ribosomal-binding site, eukaryotic mRNAs have a 7-methylguanosine cap (5′ cap) at their 5′ end. This 5′ cap is recognized by cap-binding proteins that promote the binding of the mRNA to the small ribosomal subunit.
- The first AUG codon is not always the start codon. In 1978, American biochemist Marilyn Kozak proposed that the small ribosomal subunit identifies a start codon by beginning at the 5′ end and then scanning along the mRNA in the 3′ direction. By analyzing the sequences of many eukaryotic mRNAs, Kozak and her colleagues discovered that the sequence around an AUG codon is important for it to be selected as the start codon. Aside

from an AUG codon itself, a guanine just past the start codon and the sequence of six bases directly upstream from the start codon are important for start codon selection.

- In eukaryotes, the initiator tRNA carries a methionine, not formyl-methionine.

Polypeptide Synthesis Occurs During the Elongation Stage

As its name suggests, the **elongation stage** involves the covalent bonding of amino acids to each other, one at a time, to produce a polypeptide. This process is similar in all species and involves several different components. Elongation occurs at a remarkable rate. Under normal cellular conditions, the translation machinery can elongate a polypeptide at a rate of 15 to 18 amino acids per second in bacteria and 6 amino acids per second in eukaryotes!

To elongate a polypeptide by one amino acid, a tRNA brings a new amino acid to the ribosome, where it is attached to the end of a growing polypeptide. In step 1 of **Figure 10.15**, translation has already proceeded to a point where a short polypeptide is attached to the tRNA located in the P site of the ribosome. This is called peptidyl tRNA. Let's consider the steps that would cause this polypeptide to become one amino acid longer.

1. **Binding of a charged tRNA:** A charged tRNA carrying a single amino acid binds to the A site. This binding occurs because the anticodon in the tRNA is complementary to the codon in the mRNA. The hydrolysis of GTP by proteins that function as **elongation factors** provides the energy for the binding of the tRNA to the A site (see Table 10.2). At this stage of translation, a peptidyl tRNA is in the P site and a charged tRNA (an aminoacyl tRNA) is in the A site. This is how the P and A sites came to be named.

2. **Peptide bond formation and peptidyl transfer:** In the second step, a peptide bond is formed between the amino acid at the A site and the growing polypeptide, thereby lengthening the polypeptide by one amino acid. As this occurs, the polypeptide is removed from the tRNA in the P site and transferred to the amino acid at the A site, an event termed a **peptidyl transfer reaction.** This reaction is catalyzed by an rRNA molecule within a region of the 50S subunit known as the peptidyltransferase center.

3. **Translocation and exit of uncharged tRNA:** The ribosome moves, or translocates, toward the 3' end of the mRNA by exactly one codon. This shifts the tRNAs in the P and A sites to the E and P sites, respectively. The uncharged tRNA exits the E site. Notice that the next codon in the mRNA (GCU in Figure 10.15) is now exposed at the unoccupied A site. At this point, a charged tRNA can enter the A site, and the same series of steps will add the next amino acid to the polypeptide.

Termination Occurs When a Stop Codon Is Reached in the mRNA

Elongation continues until a stop codon moves into the A site of a ribosome. The three stop codons, UAA, UAG, and UGA, are recognized by a protein known as a **release factor.** The three-dimensional

Figure 10.15 Elongation stage of translation in bacteria.

structure of a release factor protein mimics the structure of tRNAs, which allows it to fit into the A site.

Figure 10.16 illustrates the **termination stage** of translation.

1. A release factor binds to the stop codon at the A site. At this point, the completed polypeptide is attached to a tRNA in the P site.
2. The bond between the polypeptide and the tRNA is hydrolyzed, causing the polypeptide and tRNA to be released from the ribosome.
3. The mRNA, ribosomal subunits, and release factor dissociate.

THE EXPRESSION OF GENETIC INFORMATION VIA GENES I: TRANSCRIPTION AND TRANSLATION 227

Figure 10.16 Termination of translation in bacteria.

Table 10.4	Comparison of Bacterial and Eukaryotic Translation	
	Bacterial	**Eukaryotic**
Cellular location	Cytoplasm	Cytosol*
Ribosome composition	70S ribosomes:	80S ribosomes:
	30S subunit: 21 proteins + 1 rRNA	40S subunit: 33 proteins + 1 rRNA
	50S subunit: 34 proteins + 2 rRNAs	60S subunit: 49 proteins + 3 rRNAs
Initiator tRNA	tRNA$^{Formyl-methionine}$	tRNA^{Methionine}
Initial binding of mRNA	Requires a ribosomal-binding site	Requires a 7-methylguanosine cap
Selection of a start codon	Just downstream from the ribosomal-binding site	According to Kozak's sequences
Termination factors	Two factors: RF1 and RF2	One factor: eRF

*The components for eukaryotic translation described in this table refer to those that are used in the cytosol. Different types of ribosomes are used for translation inside mitochondria and chloroplasts.

The termination stage of translation is similar in bacteria and eukaryotes except that bacteria have two different termination factors (RF1 and RF2) that recognize stop codons, whereas eukaryotes have only one (eRF). **Table 10.4** compares some of the key differences between bacterial and eukaryotic translation.

10.6 Reviewing the Concepts

- Translation occurs in three stages called initiation, elongation, and termination.
- During the initiation stage of translation, an mRNA assembles with the ribosomal subunits and an initiator tRNA molecule, which carries formyl-methionine (in bacteria) or methionine (in eukaryotes) (Figure 10.14).
- During the elongation stage, amino acids are added, one at a time, to a growing polypeptide (Figure 10.15).
- The termination stage occurs when the binding of a release factor to a stop codon causes the release of the completed polypeptide from the tRNA and the disassembly of the mRNA, ribosomal subunits, and the release factor (Figure 10.16).
- Although translation is strikingly similar in bacteria and eukaryotes, some key differences have been observed (Table 10.4).

10.6 Testing Your Knowledge

1. A key difference in the initiation phase of translation between bacteria and eukaryotes is that in eukaryotes
   a. the 5′ cap promotes binding of mRNA to the small ribosomal subunit.
   b. the selection of the start codon depends on the sequence around it.
   c. the initiator tRNA carries methionine rather than formyl-methionine.
   d. all of the above occur.

2. A tRNA carrying a polypeptide would *never* be found in
   a. the A site.
   b. the P site.
   c. the E site.
   d. either a or c.

Assessing Your Knowledge and Skills

Foundational Knowledge

Multiple Choice

1. Which of the following best represents the central dogma of gene expression?
   a. During transcription, DNA codes for polypeptides.
   b. During transcription, DNA codes for mRNA, which codes for polypeptides during translation.
   c. During translation, DNA codes for mRNA, which codes for polypeptides during transcription.
   d. None of the above is correct.

2. The functional product of a protein-encoding gene is
   a. tRNA.
   b. mRNA.
   c. rRNA.
   d. a polypeptide.
   e. a, b, and c.

3. Which of the following is *not* a property of the genetic code?
   a. It specifies the amino acids within a polypeptide.
   b. It is composed of codons, which are specific sequences of three bases.
   c. It has a start codon, which specifies the starting point for polypeptide synthesis.
   d. It has stop codons, which specify the end of polypeptide synthesis.
   e. It determines the rate of transcription.

4. The region of a tRNA that is complementary to a codon in mRNA is
   a. the 3′ single-stranded region.
   b. the codon.
   c. the peptidyl site.
   d. the anticodon.
   e. the stem-loop.

5. The movement of the polypeptide from the tRNA in the P site to the tRNA in the A site is referred to as
   a. peptide bonding.
   b. aminoacyl binding.
   c. translation.
   d. the peptidyl transfer reaction.
   e. elongation.

Short Answer

1. What is the function of an aminoacyl-tRNA synthetase?
2. **PRINCIPLES** A principle of biology is that the genetic material provides a blueprint for growth, development, and reproduction. Explain how the information within the blueprint is accessed at the molecular level.

Critical-Thinking Skills

Multiple Choice

1. A mutation prevents a gene from being transcribed into an mRNA. The mutation most likely disrupts
   a. the promoter.
   b. the terminator.
   c. the start codon.
   d. the stop codon.
   e. both a and c.

2. Codons that are closer to the 5′ end of a mature mRNA specify amino acids that are
   a. closer to the C-terminus of the polypeptide.
   b. closer to the N-terminus of the polypeptide.
   c. spliced out and not found in the polypeptide.
   d. duplicated in the polypeptide.
   e. both a and d.

3. Let's suppose the anticodon for a tRNA that carries tyrosine is 3′–AUA–5′. A mutation occurs within the tRNA gene that encodes this tRNA and changes the anticodon to 3′–AUU–5′. The 3′–AUU–5′ anticodon is complementary to a UAA stop codon. In spite of this change, the mutant tRNA is charged with tyrosine. This mutant tRNA would do which of the following?
   a. insert tyrosines correctly during translation
   b. insert a tyrosine where a UAA stop codon is found
   c. insert isoleucines in places where tyrosines should be found
   d. terminate translation where tyrosine codons are found
   e. Both b and d would happen.

4. If a eukaryotic mRNA failed to have a cap attached to its 5′ end, what would the negative consequence(s) be?
   a. The mRNA would not properly exit the nucleus.
   b. The mRNA would not properly bind to a ribosome.
   c. The mRNA would not receive a poly A tail.
   d. The mRNA would not use the correct start codon.
   e. Both a and b are correct.

5. In which of the ribosomal sites, the A site, P site, and/or E site, could a tRNA with an attached polypeptide be found?
   a. A site only
   b. P site only
   c. E site only
   d. A and P sites
   e. P and E sites

Short Answer

1. Let's suppose a gene mutation changed the sixth codon from one that specifies lysine to a stop codon. How would this mutation affect transcription? How would it affect translation?
2. Why do you think some complexes, such as spliceosomes and ribosomes, have both protein and RNA components?

The Expression of Genetic Information via Genes II: Non-coding RNAs

11

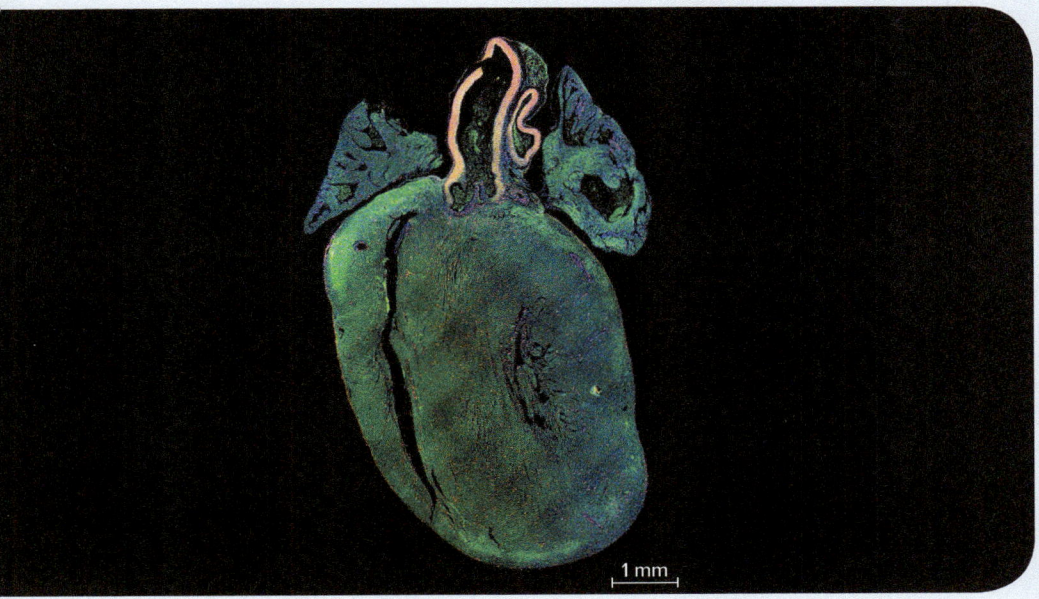

©Mauro Giacca, Ana Eulalio, Miguel Mano

Non-coding RNAs and heart repair. The muscles of the mammalian heart have poor regenerating abilities. Researchers have identified several non-coding RNAs that stimulate cardiac muscle regeneration. This heart is from a mouse that was treated with such a non-coding RNA, and it showed a significant increase in proliferating cells. This discovery may lead to new therapies to help heart attack victims regenerate cardiac muscle.

Chapter Outline

- 11.1 Overview of Non-coding RNAs
- 11.2 Role of Non-coding RNAs in Eukaryotic DNA Replication
- 11.3 Effects of Non-coding RNAs on Chromatin Structure and Transcription
- 11.4 Effects of Non-coding RNAs on Translation and mRNA Degradation
- 11.5 Non-coding RNAs and Protein Sorting
- 11.6 Non-coding RNAs and Genome Defense
- 11.7 Roles of Non-coding RNAs in Human Disease and Plant Health

Assessing Your Knowledge and Skills

People with a rare genetic disorder called cartilage-hair hypoplasia (CHH) have short stature, underdeveloped hair, and short limbs with malformations in the cartilage. In addition, they have a higher predisposition to develop lymphomas and other cancers, and they may lack a normal immune response. Though CHH was first identified by American geneticist Victor McKusick in 1965, the underlying genetic cause remained a mystery for 36 years. In 2001, the function of a single mutant gene was linked to this disease. The gene specifies an RNA molecule that does not encode a protein. Rather, the RNA made from this gene is the RNA component of RNase MRP, which is an RNA-protein complex involved in the modification of some ribosomal and mitochondrial RNAs. This was the first case in which geneticists determined that a human genetic disease was due to a mutation in a nuclear gene that is not a protein-encoding gene. By comparison, the first human genetic disease involving a protein-encoding gene was discovered nearly a century earlier! In 1909, Archibald Garrod proposed that patients with a disease called alkaptonuria, which is characterized by bluish-black discoloration of the cartilage and skin, is due to a mutation in a gene that encodes the protein homogentisic acid oxidase.

Why such a big time gap in our understanding of protein-encoding genes versus other types of genes? It's all about tools. The experimental tools to study the structure and function of proteins and to identify protein-encoding genes have been around for a long time. By contrast, the tools to study RNA structure and function and to identify genes that do not specify proteins are much more recent and are under rapid development. We are witnessing a revolution in molecular biology that is uncovering an unprecedented number of functions for RNA molecules.

In Chapter 10, we focused our attention on gene expression at the molecular level. The emphasis was on protein-encoding genes, which are transcribed into mRNA. During translation, the information within mRNAs is used to make polypeptides, which then assemble into functional proteins. The human genome has about 22,000 protein-encoding genes. In contrast, other genes are transcribed into **non-coding RNAs (ncRNAs)**, which are RNA molecules that do not encode polypeptides. As this chapter will reveal, ncRNAs perform a very diverse set of functions. In humans, the number of genes that specify ncRNAs is still difficult to measure and a matter of controversy. Estimates range from several thousand to tens of thousands.

In the past, educators have tended to emphasize proteins and DNA in the teaching of biology at the molecular level. For example, in the cell unit (Chapters 4 through 8), we saw many examples of how proteins affect cell structure and function, and the genetics unit (Chapters 9 through 18) largely focuses on the structure and function of DNA. Although DNA, RNA, and proteins are key molecular players in living cells, a historical bias has existed against RNA. With a few exceptions, the educational exploration of RNA has been limited to its role in making proteins (see Chapter 10).

The purpose of this chapter is to lessen the bias against RNA. New molecular tools have enabled researchers to discover that ncRNAs perform a spectacular array of cellular functions in bacteria, archaea, protists, fungi, plants, and animals. ncRNAs play important roles in a variety of processes, including DNA replication, chromatin modification, transcription, translation, and genome defense. In most cell types, ncRNAs are more abundant than mRNAs. For example, in a typical human cell, only about 20% of transcription involves the production of mRNAs, whereas 80% is associated with making ncRNAs! This observation underscores the importance of RNA in the enterprise of life and indicates why it deserves greater recognition and deeper study. Furthermore, abnormalities in ncRNAs are associated with a wide range of human diseases, including CHH, cancer, neurological disorders, and cardiovascular diseases. Many ncRNAs are also critical to the growth of plants, including the crop plants that are so essential to human survival.

In this chapter, we will begin with an overview of the general properties of ncRNAs, then examine specific examples of the functions they perform. We will end the chapter by considering the role of ncRNAs in different human diseases and in plant health.

11.1 Overview of Non-coding RNAs

Learning Outcomes

1. Describe the ability of ncRNAs to bind to other molecules and macromolecules.
2. Outline the general functions of ncRNAs.
3. Define ribozyme.
4. List several examples of ncRNAs, and describe their functions.

The study of ncRNAs is a rapidly expanding field, and researchers speculate that many ncRNAs have yet to be discovered. Also, due to the relative youth of this field, not all researchers agree on the names of certain ncRNAs or their primary functions. Even so, some broad themes are beginning to emerge. In this section, we will survey the general features of ncRNAs, and in later sections, we will discuss specific examples in greater detail.

ncRNAs Can Bind to Different Types of Molecules

The ability of ncRNAs to carry out an amazing array of functions is largely related to their ability to bind to different types of molecules. **Figure 11.1a** shows four common types of molecules that are recognized by ncRNAs. Some ncRNAs bind to DNA or another RNA through complementary base pairing. This allows ncRNAs to affect processes such as DNA replication, transcription, and translation. In addition, ncRNAs can bind to proteins or small molecules.

As described in Chapter 10, RNA molecules, such as tRNAs, can form stem-loop structures (refer back to Figure 10.10). Similar structures in other ncRNAs may bind to pockets on the surface of proteins, or multiple stem-loops may form a binding site for a small molecule. In some cases, a single ncRNA may contain multiple binding sites. This allows an ncRNA to facilitate the formation of a large structure composed of multiple molecules, such as an ncRNA and three different proteins, as shown in **Figure 11.1b**.

ncRNAs Can Perform a Diverse Set of Functions

In recent decades, researchers have uncovered many examples in which ncRNAs play a critical role in different biological processes. Let's first consider how ncRNAs work in a general way. The common functions of ncRNAs are the following.

Scaffold Some ncRNAs contain binding sites for multiple components, such as a group of different proteins. Much like the beams in a building, an ncRNA can act as a scaffold for the formation of a complex, as in Figure 11.1b.

Guide Some ncRNAs guide a molecule to a specific location in a cell. For example, an ncRNA may bind to a protein and guide it to a target site in the DNA that is part of a particular gene (**Figure 11.2**). This function relies on the ncRNA having multiple binding sites: one for the protein and another for the target site in the DNA.

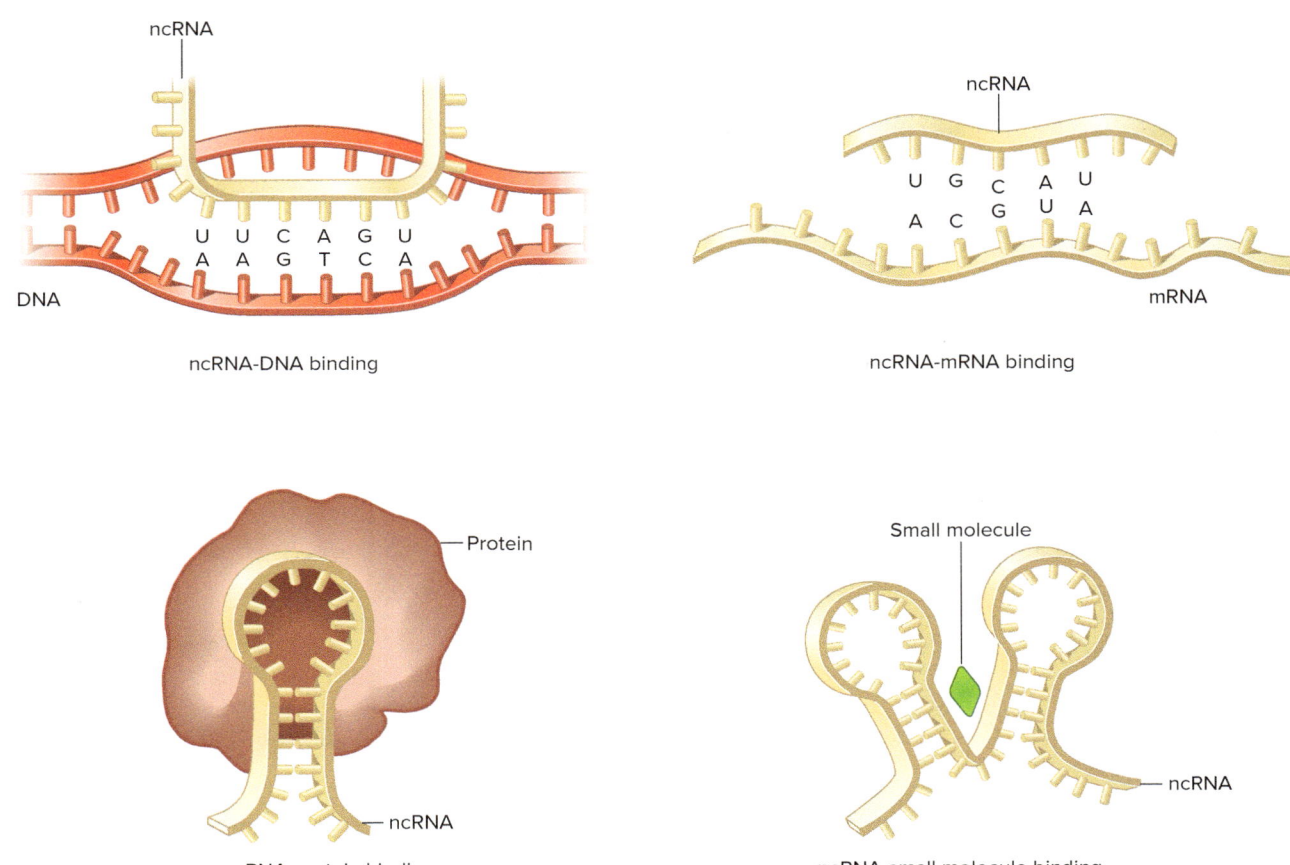

(a) Common binding interactions between ncRNAs and other molecules

(b) Multiple binding sites in a single ncRNA

Figure 11.1 Ability of ncRNAs to bind to other molecules. (a) ncRNA molecules can bind to DNA, mRNA, proteins, and small molecules. (b) Some ncRNAs have multiple binding sites for different molecules, such as proteins.

Concept Check: *Which of these binding interactions might be inhibited by the formation of a stem-loop within the ncRNA?*

Biology Principle

Biologists Use Models to Test Predictions and Convey Ideas

This figure is a model that illustrates how the structure of an ncRNA can allow it to function as a guide. The Modeling Challenge in the figure legend asks you to propose a model for a different ncRNA that functions as a guide and as a scaffold.

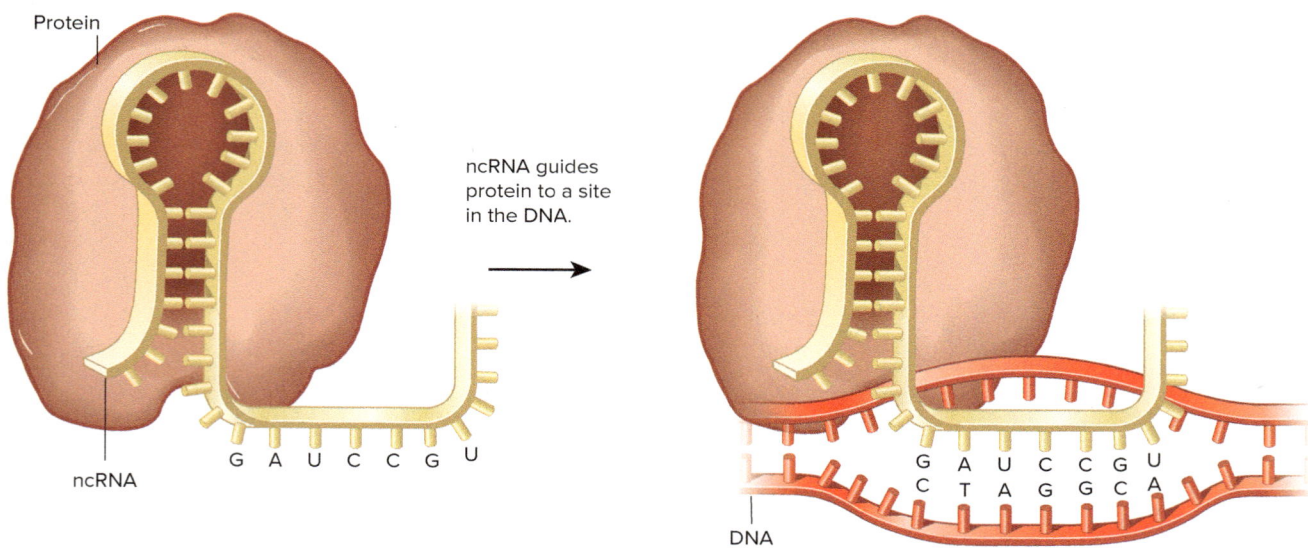

Figure 11.2 Ability of an ncRNA to function as a guide.

Modeling Challenge: Let's suppose an ncRNA functions as both a guide and a scaffold. (A scaffold is shown in Figure 11.1b.) A segment near the 5′ end of the ncRNA has the sequence CCGUUACU, and this sequence acts a guide that binds to gene X. The ncRNA also acts as a scaffold for the binding of four proteins called proteins A, B, C, and D. Draw a model that shows how this ncRNA guides the four proteins to gene X.

Alteration of Protein Function or Stability When it binds to a protein, an ncRNA can alter that protein's structure, which in turn can have a variety of effects. The binding of an ncRNA may affect

- the ability of the protein to act as a catalyst;
- the ability of the protein to bind to other molecules, such as proteins, DNA, or RNA; and
- the stability of the protein.

Ribozyme Another interesting feature of some ncRNAs is that they function as **ribozymes,** which are RNA molecules with catalytic function. For example, in Chapter 10, we saw how peptidyltransferase, which is a component of the large ribosomal subunit, catalyzes peptide bond formation during translation. An rRNA within peptidyltransferase plays the key role in this catalysis. In other words, a part of the ribosome is a ribozyme.

Blocker An ncRNA may physically prevent or block a cellular process from happening. For example, in bacteria, an antisense RNA is a type of ncRNA that is complementary to an mRNA. When an antisense RNA binds to an mRNA, it blocks the ability of a ribosome to bind to the mRNA, thereby inhibiting translation.

Decoy Some ncRNAs recognize other ncRNAs and sequester them, thereby preventing them from working. For example, a decoy ncRNA may bind to a different ncRNA called a microRNA (miRNA), which is described later in this chapter (**Figure 11.3**). The function of an miRNA is to inhibit the translation of a particular mRNA. However, if a decoy ncRNA binds to an miRNA, the miRNA is unable to carry out its function.

The Functions of Some ncRNAs Are Understood

Table 11.1 describes several examples of ncRNAs that have been well characterized. Some of these are discussed in other chapters. In the remaining sections of this chapter, we will focus on the functions of ncRNAs that are not discussed elsewhere.

Figure 11.3 Ability of an ncRNA to function as a decoy.

Table 11.1	Examples of ncRNAs		
Type of ncRNA	**Plays a role in**	**Described in**	**Description**
Telomerase RNA component (TERC)	DNA replication	This chapter	TERC facilitates the binding of telomerase to the telomere and acts as a template for DNA replication.
X inactive specific transcript (Xist RNA)	Chromatin structure, transcription	Chapter 16	Xist RNA coats one of the X chromosomes in female mammals and plays a role in its compaction and resulting inactivation.
Hox transcript antisense intergenic RNA (HOTAIR)	Chromatin structure, transcription	This chapter	HOTAIR alters chromatin structure and thereby represses transcription by guiding histone-modifying complexes to target genes.
Transfer RNA (tRNA)	Translation	Chapter 10	tRNA molecules recognize mRNA codons during translation and carry the appropriate amino acid.
Ribosomal RNA (rRNA)	Translation	Chapter 10	rRNAs are components of ribosomes, which are the site of polypeptide synthesis.
microRNA (miRNA), small-interfering RNA (siRNA)	Translation and RNA degradation	This chapter	miRNAs and siRNAs regulate the expression and degradation of mRNAs.
RNA component of signal recognition particle (SRP-RNA)	Protein sorting and secretion	This chapter	In bacteria, SRP directs the synthesis of some polypeptides to the plasma membrane. In eukaryotes, it directs polypeptide synthesis to the endoplasmic reticulum.
CRISPR RNA (crRNA)	Genome defense	This chapter	crRNA, found in bacteria and archaea, guides an endonuclease to foreign DNA, such as the DNA of a bacteriophage.

11.1 Reviewing the Concepts

- Non-coding RNAs (ncRNAs) are RNA molecules that do not encode polypeptides.
- ncRNAs bind to different types of molecules, including DNA, other RNAs, proteins, and small molecules (Figure 11.1).
- An ncRNA can provide a scaffold, act as a guide, alter protein function or stability, function as a ribozyme, function as a blocker, and/or act as a decoy (Figures 11.2, 11.3).
- ncRNAs play a role in DNA replication, chromatin structure, transcription, translation, RNA degradation, protein sorting and secretion, and genome defense (Table 11.1).

11.1 Testing Your Knowledge

1. Which of the following can bind to ncRNAs?
   a. DNA
   b. RNA
   c. proteins
   d. small molecules
   e. all of the above
2. When an ncRNA functions as a decoy, it
   a. contains binding sites for many different proteins, thereby promoting the formation of a large complex.
   b. recognizes other ncRNAs and sequesters them, thereby preventing them from working.
   c. can alter its conformation, allowing it to switch between active and inactivate conformations.
   d. may physically prevent or block a cellular process from happening.
   e. does all of the above.

11.2 Role of Non-coding RNAs in Eukaryotic DNA Replication

Learning Outcome

1. Explain how an ncRNA plays a role in the replication of telomeres in eukaryotes.

The topic of DNA replication is discussed in Chapter 9. A specialized form of DNA replication that requires the function of an ncRNA happens at the ends of eukaryotic chromosomes. These regions, called the **telomeres,** have a short nucleotide sequence that is repeated a few dozen to several hundred times (Figure 11.4). The repeat sequence shown here, 5′–GGGTTA–3′, is the sequence found in human telomeres. Flowering plants have a similar repeat sequence, but it has an extra T (5′–GGGTTTA–3′). A telomere has a region at the 3′ end that is termed a 3′ overhang, because it does not have a complementary strand.

As discussed in Chapter 9, DNA polymerase synthesizes DNA only in a 5′ to 3′ direction and requires a primer. For these reasons, DNA polymerase cannot copy the 3′ ends of double-stranded DNA.

Figure 11.4 Telomere sequences at one end of a human chromosome. The telomere sequence shown here is found in humans and other mammals. The length of the telomere and the 3′ overhang varies among different species and cell types.

BioConnections: *Refer back to Figure 9.16. Explain why DNA polymerase cannot copy the 3′ end of a linear chromosome.*

Therefore, if this replication problem were not overcome, a linear chromosome would become progressively shorter with each round of DNA replication.

In 1984, American molecular biologist Carol Greider and Australian-born American molecular biologist Elizabeth Blackburn discovered an enzyme called **telomerase** that prevents chromosome shortening by attaching many copies of a DNA repeat sequence to the 3′ ends of chromosomes (**Figure 11.5**). Telomerase contains both proteins and an ncRNA called telomerase RNA component (TERC). The lengthening of telomeres occurs in three steps.

Binding of Telomerase

TERC has a sequence that is complementary to the DNA repeat sequence. In this way, TERC acts as a guide that allows telomerase to bind to the 3′ overhang region of the telomere.

Polymerization

Following binding, TERC has a second function. This ncRNA has a sequence that functions as a template for the synthesis of a six-nucleotide sequence at the end of the DNA strand. This synthesis is called polymerization, because it is analogous to the function of DNA polymerase. Telomere lengthening is catalyzed by a protein within telomerase called telomerase reverse transcriptase (TERT). TERT's name indicates that it catalyzes the reverse of transcription; it uses an RNA template to synthesize DNA.

Translocation

Following polymerization, telomerase then moves—a process called translocation—to the new end of the DNA strand and attaches another six nucleotides to the end.

This binding-polymerization-translocation cycle occurs many times in a row, thereby greatly lengthening the 3′ end of the DNA strand in the telomere. This lengthening provides an upstream site for

THE EXPRESSION OF GENETIC INFORMATION VIA GENES II: NON-CODING RNAS   235

Figure 11.5 Mechanism of DNA replication by telomerase.

✓ **Concept Check:** *What does telomerase use as a template to make the DNA repeat sequence?*

an RNA primer to be made. DNA polymerase then synthesizes the complementary DNA strand. In this way, the progressive shortening of eukaryotic chromosomes is prevented.

11.2 Reviewing the Concepts

- The ends of linear, eukaryotic chromosomes have telomeres composed of repeat sequences. An ncRNA within telomerase called TERC guides telomerase to the telomere repeat sequence and also functions as a template for the synthesis of a six-nucleotide repeat. This happens many times in a row to lengthen one DNA strand of the telomere. This synthesis also provides a site for an RNA primer to be made, and DNA polymerase synthesizes the complementary DNA strand (Figures 11.4, 11.5).

11.2 Testing Your Knowledge

1. Which of the following is a function of TERC, the RNA component of telomerase?
    a. a guide
    b. a template
    c. a primer for DNA synthesis
    d. all of the above
    e. a and b only

11.3 Effects of Non-coding RNAs on Chromatin Structure and Transcription

Learning Outcome

1. Explain how the ncRNA known as HOTAIR plays a role in gene repression.

Hox transcript antisense intergenic RNA, referred to as HOTAIR, is a recently discovered ncRNA found in humans and other mammals that alters chromatin structure and thereby represses gene transcription. The gene that encodes HOTAIR is located within a cluster of genes called the *HoxC* genes. (*Hox* genes, which play a role in animal development, are described in Chapter 20.) HOTAIR is so named because it is transcribed from the opposite (antisense) strand to the strand used for the *HoxC* genes.

Figure 11.6 shows a simplified mechanism for the repression of gene transcription by HOTAIR. HOTAIR acts as a scaffold for the binding of two protein complexes that covalently modify histone proteins. One of these complexes binds to the 5′ end of HOTAIR, and the other binds to the 3′ end. HOTAIR then guides these complexes to a target gene by binding to a region near the gene that contains many purines, which is called a GA-rich region. For example, HOTAIR binds to a GA-rich region that is next to a *HoxD* gene. A portion of HOTAIR is complementary to this GA-rich region.

Figure 11.6 Simplified mechanism for inhibition of gene transcription by HOTAIR. This is just one proposed role of HOTAIR. This ncRNA is known to interact with other proteins as well. Note: The abbreviation HM stands for histone modification.

Concept Check: *Explain why HOTAIR binds to the target gene. Why doesn't it bind next to every gene?*

The next event involves histone modifications. As described in Chapter 12, histone modifications may affect gene transcription (look ahead to Figure 12.14). The modifications that are facilitated by HOTAIR are known to inhibit transcription. This inhibition can occur in two ways:

- The histone modifications may directly inhibit the ability of RNA polymerase to transcribe the target gene. For example, these modifications may prevent RNA polymerase from forming a preinitiation complex.
- Rather than directly affecting transcription, the histone modifications may attract other chromatin-modifying enzymes to the target gene, which would lead to further changes in chromatin structure that inhibit transcription.

Of great interest in the study of HOTAIR is its role in human disease. As discussed later in this chapter, certain types of cancer, such as breast cancer, may occur when HOTAIR is not functioning properly.

11.3 Reviewing the Concepts

- HOTAIR is an ncRNA found in humans and other mammals that regulates transcription by forming a scaffold that binds two protein complexes and guides them to particular genes. The protein complexes covalently modify histones, and these modifications inhibit transcription of the target genes (Figure 11.6).

11.3 Testing Your Knowledge

1. Which of the following is a function of the ncRNA known as HOTAIR?
   a. a decoy
   b. a scaffold
   c. a guide
   d. all of the above
   e. b and c only

11.4 Effects of Non-coding RNAs on Translation and mRNA Degradation

Learning Outcomes

1. **SCISKILLS** ▶ Analyze experimental evidence that double-stranded RNA is more potent in silencing mRNA than antisense RNA is.
2. Outline the steps of RNA interference.

In the previous section, we considered how an ncRNA can affect the process of transcription. In recent years, researchers have discovered that ncRNAs often exert their effects on RNA molecules that are already made. In this section, we will consider how ncRNAs can affect the ability of mRNAs to be translated or degraded.

FEATURE INVESTIGATION

Fire and Mello Showed That Double-Stranded RNA Is More Potent Than Antisense RNA in Silencing mRNA

Specific mRNAs can be targeted for translational inhibition or degradation by a mechanism involving double-stranded RNA. This mechanism was discovered during research involving plants and the nematode worm *Caenorhabditis elegans*. The study described here involved an examination of gene expression in *C. elegans*.

American biologists Andrew Fire, Craig Mello, and colleagues used *C. elegans* as their experimental organism, because it is relatively easy to inject with RNA and the expression of many of its genes had been established. In 1998, Fire and Mello investigated the effects of injected RNA on the expression of specific mRNAs. In the investigation described in **Figure 11.7**, we will focus on one of their experiments involving an mRNA encoded by a gene called *mex-3*. This mRNA had already been shown to be made in high amounts in early embryos of *C. elegans*.

Prior to this work, the *mex-3* gene had been identified and inserted into a plasmid. The process of inserting genes into plasmids is described in Chapter 18 (look ahead to Figure 18.2). Let's first look at the upper plasmid shown in the "Conceptual level" column in step 1. When RNA polymerase, nucleotides, and this plasmid were mixed together in a test tube, *mex-3* mRNA was made, which is called the sense strand. In living cells, the sense strand is used to make the mex-3 protein. Fire and Mello also switched the location of the promoter so that it was at the other end of the gene and signaled transcription of the opposite strand, which is called the antisense strand (see the second plasmid in the "Conceptual level" column). The sense and antisense strands are complementary to each other.

Next, they injected these RNAs into the gonads of *C. elegans*. Into some worms, they injected antisense RNA alone. Alternatively, they mixed sense and antisense RNA, which formed double-stranded RNA, and injected this double-stranded RNA into the gonads of other worms. They also used uninjected worms as controls. After injection, the RNA was taken up by eggs, which later developed into embryos. To determine the amount of *mex-3* mRNA present, Fire and Mello incubated the embryos with a probe that was complementary to

Figure 11.7 Injection of antisense and double-stranded RNA into *C. elegans* to compare their effects on mRNA silencing.

GOAL The goal was to further understand how the experimental injection of RNA was responsible for the silencing of particular mRNAs.

KEY MATERIALS The researchers used *C. elegans* as their model organism. They also had the cloned *mex-3* gene, which had been previously shown to be highly expressed in *C. elegans* embryos.

Experimental level   **Conceptual level**

1. Make sense and antisense *mex-3* RNA in vitro using cloned genes for *mex*-3 with promoters on either side of the gene. RNA polymerase and nucleotides are added to synthesize the RNAs.

2. Inject either *mex-3* antisense RNA or a mixture of *mex-3* sense and antisense RNA into the gonads of *C. elegans*. This RNA is taken up by the eggs and early embryos. As a control, do not inject any RNA.

3. Incubate and then subject early embryos to in situ hybridization. In this method, a labeled probe is added that is complementary to *mex-3* mRNA. If cells express *mex-3*, the mRNA in the cells will bind to the probe and become labeled. After incubation with a labeled probe, the cells are washed to remove unbound probe.

4. Observe embryos under the microscope.

5. **THE DATA**

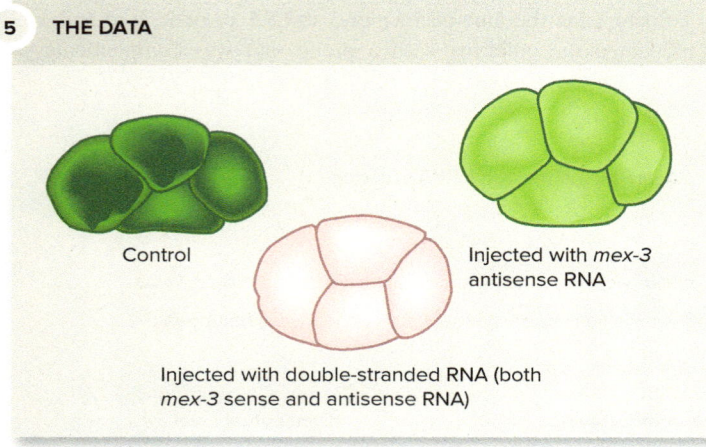

6. **CONCLUSION** Double-stranded RNA is more potent at inhibiting *mex-3* mRNA than antisense RNA alone.

7. **SOURCE** Fire, A., Xu, S., Montgomery, M. K., et al. 1998. Potent and Specific Genetic Interference by Double-Stranded RNA in *Caenorhabditis elegans*. *Nature* 391: 806–811.

mex-3 mRNA. Because the probe was labeled, any *mex-3* mRNA that became bound to the probe could be observed under the microscope. After this incubation step, any probe that was not bound to the *mex-3* mRNA was washed away.

As seen in the schematic data of Figure 11.7, the control embryos were very darkly labeled as denoted by their dark green color. These results indicated that the control embryos contained a high amount of *mex-3* mRNA, which was known from previous research. In the embryos that had received antisense RNA, *mex-3* mRNA levels were decreased, but detectable, as shown by the lighter green color. Remarkably, in embryos that had received double-stranded RNA, no *mex-3* mRNA was detected! These results indicated that double-stranded RNA is more potent at silencing mRNA than is antisense RNA. In this case, the double-stranded RNA caused the *mex-3* mRNA to be degraded. Fire and Mello used the term **RNA interference (RNAi)** to describe the phenomenon in which double-stranded RNA causes the silencing of mRNA. This surprising observation led researchers to investigate the underlying molecular mechanism that accounts for this phenomenon, as described next.

Experimental Questions

1. In this experiment, does the *mex-3* mRNA correspond to the sense strand or antisense strand?
2. **SCISKILLS** ▶ Explain how the sense and antisense *mex-3* RNAs were made.
3. **SCISKILLS** ▶ According to the data, which material was the most effective at causing the degradation of *mex-3* mRNA?

RNA Interference Is Mediated by MicroRNAs or Small-Interfering RNAs via the RNA-Induced Silencing Complex

RNA interference is found in most eukaryotic species, including animals and plants. It can arise from two sources: microRNAs and small-interfering RNAs. **MicroRNAs (miRNAs)** are ncRNAs that are transcribed from endogenous eukaryotic genes—genes that are normally found in the genome. They play key roles in regulating gene expression, particularly during embryonic development in animals and plants. Most commonly, a single type of miRNA inhibits the translation of several different mRNAs. An miRNA and an mRNA bind to each other because they have base sequences that are partially complementary. In humans, over 2,000 genes encode miRNAs. Researchers estimate that 60% of human protein-encoding genes are regulated by miRNAs.

By comparison, **small-interfering RNAs (siRNAs)** are ncRNAs that usually originate from sources that are exogenous, which means they are not normally made by cells. The siRNAs can come from viruses that infect a cell, or they might be synthesized by researchers to study gene function experimentally. In the experiment of Figure 11.7, the sense and antisense RNAs would have bonded to each other to make siRNA. In most cases, siRNAs are a perfect match to a single type of mRNA. The functioning of siRNAs is thought to play a key role in preventing certain types of viral infections. In addition, siRNAs have become important experimental tools in molecular biology.

How do miRNAs and siRNAs cause the silencing of specific mRNAs? **Figure 11.8** shows how an miRNA or an siRNA leads to RNA interference. The miRNA is first synthesized as a pri-miRNA (for primary-miRNA) in the nucleus. Due to complementary base pairing, the pri-miRNA folds into a hairpin structure (also called a stem-loop) with long, single-stranded 5′ and 3′ ends. The pri-miRNA is cleaved at both ends to form a pre-miRNA (for precursor-miRNA, not to be confused with pri-miRNA). The pre-miRNA is then exported from the nucleus.

As shown in Figure 11.8, siRNAs do not go through the processing events that occur in the nucleus. Instead, pre-siRNAs may directly enter the cytosol from outside of the cell. They may be derived from viral RNAs, or they may be made by researchers and taken up by cells. For example, in the work of Fire and Mello described in Figure 11.7, the double-stranded *mex-3* RNA is an example of a pre-siRNA. The pre-siRNA is formed from two complementary RNA molecules that base-pair with each other.

Once they are in the cytosol, both pre-miRNAs and pre-siRNAs are cut by an endonuclease called dicer (see Figure 11.8). This releases a double-stranded RNA molecule that is typically 20–25 bp long. This double-stranded RNA associates with proteins to form a complex called the **RNA-induced silencing complex (RISC)**. One of the RNA strands is degraded. The remaining single-stranded miRNA or siRNA is complementary to specific mRNAs that will be silenced. The miRNA or siRNA acts as a guide that causes RISC to recognize and bind to such mRNA molecules.

After RISC binds to an mRNA, one of two things may happen:

- RISC may inhibit translation without degrading the mRNA. This is more common for miRNAs, which often are only partially complementary to their target mRNAs.
- RISC may direct the degradation of the mRNA. One of the proteins in RISC can cleave the mRNA. This usually occurs for siRNAs that typically are a perfect match (or highly complementary) to their target mRNA.

These two effects are termed RNA interference because the miRNA or siRNA interferes with the proper expression of an mRNA. In 2006, Fire and Mello received the Nobel Prize in Physiology or Medicine for their discovery of this mechanism.

RNA interference is believed to have at least two benefits:

- This mechanism represents an important form of regulation. When genes encoding pri-miRNAs are turned on, the production of miRNAs silences the expression of specific mRNAs.
- RNA interference provides a defense against viruses. This mechanism is widely used by plants to prevent viral infections.

11.4 Reviewing the Concepts

- Fire and Mello showed that double-stranded RNA is more potent at silencing mRNA than is antisense RNA (Figure 11.7).
- RNA interference is a mechanism of mRNA silencing in which miRNA or siRNA becomes part of an RNA-induced silencing complex (RISC) that inhibits the translation of a specific mRNA or causes its degradation, respectively (Figure 11.8).

11.4 Testing Your Knowledge

1. The process of RNA interference may lead to
   a. the degradation of an mRNA.
   b. the inhibition of translation of an mRNA.
   c. the synthesis of an mRNA.
   d. all of the above.
   e. a and b only.

11.5 Non-coding RNAs and Protein Sorting

Learning Outcome

1. Describe the function of SRP, and explain the roles of SRP RNA with regard to SRP function.

To carry out their functions, proteins need to be directed to particular locations (refer back to Figure 4.30). For example, some proteins function extracellularly and need to be secreted from the cell. For such proteins to be secreted, they are first sorted to the plasma membrane in bacteria and archaea, or to the endoplasmic reticulum (ER) membrane in eukaryotic cells. This process is facilitated by an RNA-protein complex called **signal recognition particle (SRP)**. In bacteria, SRP is composed of one ncRNA and one protein. In eukaryotes, SRP is composed of one ncRNA and six different proteins.

Figure 11.9 shows how SRP works in eukaryotes. To be directed to the ER membrane, a polypeptide must contain a sorting signal

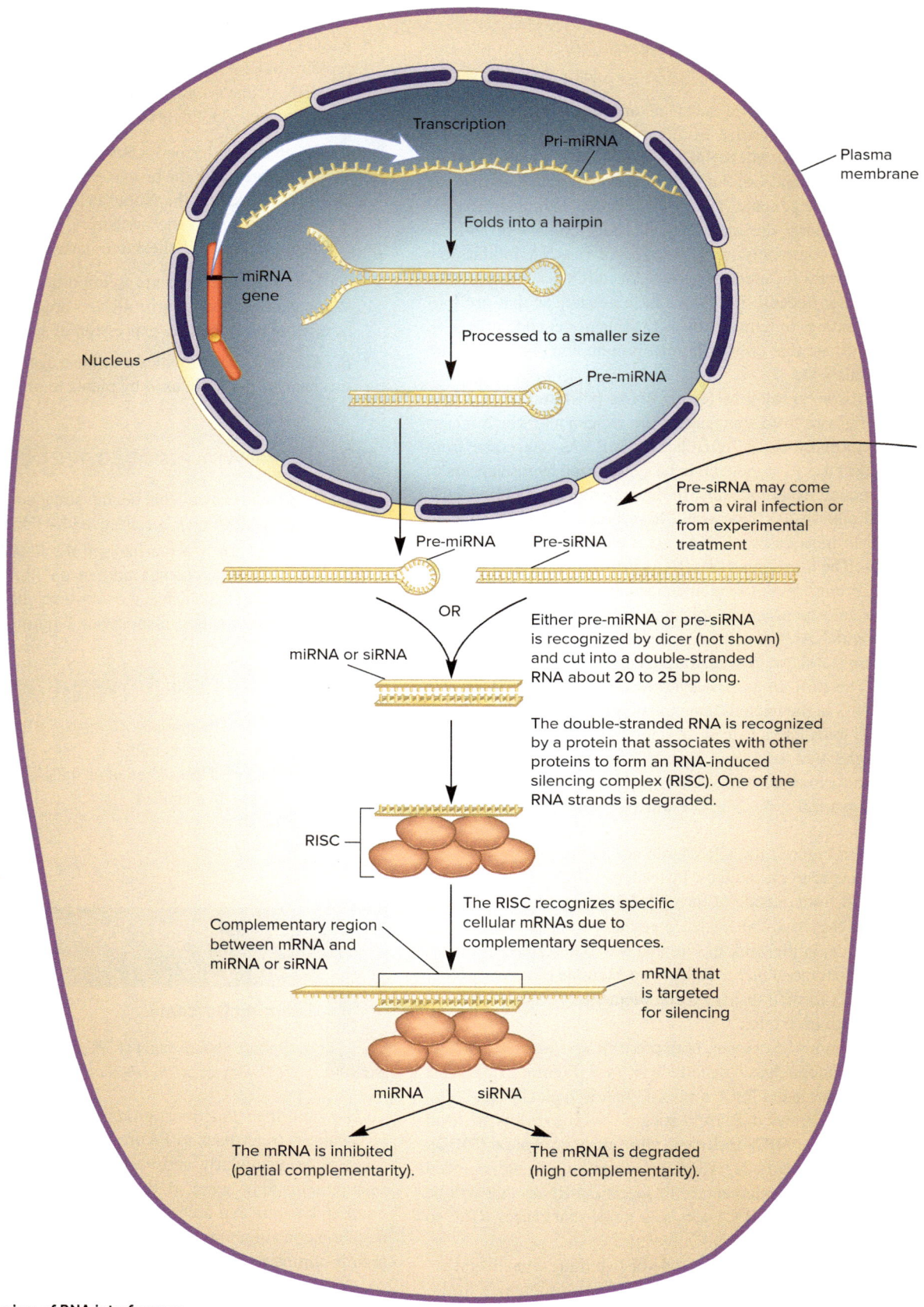

Figure 11.8 Mechanism of RNA interference.

Concept Check: *Explain why RISC binds to a specific mRNA. What type of bonding occurs?*

THE EXPRESSION OF GENETIC INFORMATION VIA GENES II: NON-CODING RNAS   241

Figure 11.9 Directing of polypeptides to the endoplasmic reticulum membrane via SRP. In eukaryotes, several categories of proteins are first directed to the ER via SRP. These include proteins that are secreted from the cell as well as proteins that are destined to stay in the ER, Golgi, lysosomes, or vacuoles. Note: SRP in eukaryotes is composed of six proteins and one ncRNA. This simplified drawing of SRP does not distinguish these seven components.

BioConnections: *Refer back to Figure 4.30. Which types of proteins need SRP to reach their proper location, and which do not?*

called an **ER signal sequence,** which is a sequence of about 6–12 amino acids that are predominantly hydrophobic and usually located near the N-terminus. As the ribosome is making the polypeptide in the cytosol, the ER signal sequence emerges from the ribosome and is recognized by a protein in SRP. The binding of SRP to the polypeptide pauses translation.

SRP then binds to an SRP receptor in the ER membrane, which docks the ribosome over a channel. For this binding to occur, proteins within SRP and the SRP receptor must also be bound by GTP. Next, these GTP-binding proteins hydrolyze their GTP, which causes the release of SRP from the SRP receptor and the polypeptide. Once SRP is released, translation resumes and the growing polypeptide is threaded through a channel to cross the ER membrane. In the case of a secreted protein, the newly made polypeptide then travels through the Golgi apparatus and then to the plasma membrane, where it is released outside of the cell.

Researchers have identified at least two key roles for SRP RNA:

1. SRP RNA provides a scaffold for the binding of SRP proteins.
2. After SRP binds to the SRP receptor in the ER membrane, the SRP RNA stimulates proteins within both SRP and the SRP receptor to hydrolyze their GTP. In other words, SRP RNA alters the structures of these proteins to enhance their GTPase activities. This stimulation is essential for the release of SRP.

11.5 Reviewing the Concepts

- Signal recognition particle (SRP), which is composed of one or more proteins and an ncRNA, plays a role in directing proteins to the plasma membrane of prokaryotic cells and to the ER membrane of eukaryotic cells (Figure 11.9).

11.5 Testing Your Knowledge

1. Which of the following is a function of SRP?
   a. pauses translation of polypeptides that have an ER signal sequence
   b. binds to an SRP receptor in the ER membrane
   c. docks the ribosome over a channel
   d. hydrolyzes GTP to promote SRP release from the SRP receptor
   e. All of the above are functions of SRP.

11.6 Non-coding RNAs and Genome Defense

Learning Outcome

1. Explain how the CRISPR-Cas system defends bacteria against bacteriophages.

Much like the immune system found in vertebrates, a system called the **CRISPR-Cas system** provides some species of bacteria and archaea with a means of defense against foreign invaders. CRISPR-Cas systems are an effective defense against bacteriophages, which are viruses that infect bacteria (discussed in Chapter 17), and transposons, which are small segments of DNA that can be inserted into the chromosomes of all species (discussed in Chapter 18). ncRNAs play a key role in CRISPR-Cas systems. About half of all bacterial species and most archaeal species have such a system. Three general types are known, designated types I, II, and III. In this section, we will focus on the type II CRISPR-Cas system and its role in providing bacteria with defense against bacteriophages.

The CRISPR-Cas System Provides Bacteria with Defense Against Bacteriophages

In 1993, Spanish microbiologist Francisco Mojica and colleagues were the first to recognize that different species of bacteria and archaea have a site in their chromosome, now called the CRISPR locus, that contains a series of repeated sequences. In 2005, by analyzing the DNA sequences of the CRISPR locus in a variety of bacterial species, Mojica, Spanish geneticist Giles Vergnaud, and Russian microbiologist Alexander Bolotin independently proposed that this locus provides protection against bacteriophage infection. The hypothesis was based on the observation that the CRISPR locus contains segments that are derived from bacteriophage DNA. Their hypothesis was confirmed in 2007 by French microbiologist Philippe Horvath and colleagues, who showed experimentally that the CRISPR-Cas locus provides defense against bacteriophage infection.

Figure 11.10a shows a common organization of the CRISPR-Cas system (also called the CRISPR locus), which has five genes: *tracr, Cas9, Cas1, Cas2,* and *Crispr*. A key feature of the *Crispr* gene is a group of clustered, regularly interspaced, short, palindromic repeats—hence the name CRISPR. The repeats within the *Crispr* gene are interspersed by short, unique sequences, which are called spacers. The CRISPR-Cas type II system also includes a gene that encodes an ncRNA called tracrRNA and a few protein-encoding CRISPR-associated genes (*Cas* genes), which are usually adjacent to the *Crispr* gene. These genes are needed to mediate the defense against bacteriophages.

The CRISPR-Cas system is considered an adaptive defense system because a bacterial cell must first be exposed to an agent, such as a bacteriophage, to elicit a response from the system. As shown in **Figure 11.10b–d**, the defense mechanism occurs in three phases.

Adaptation The process of adaptation (also called spacer acquisition) occurs after a bacterial cell has been exposed to a bacteriophage.

(a) Simplified organization of the CRISPR-Cas system in the bacterial chromosome

Figure 11.10 **The CRISPR-Cas system of genome defense in bacteria.** The system shown here is a type II system, which is found in the chromosome of certain bacterial species but not in archaea. **(a)** Organization of the CRISPR-Cas system in a bacterial chromosome. This drawing shows a typical organization, but different species have variations. **(b–d)** A simplified mechanism of the CRISPR-Cas system. The defense occurs in three phases, called adaptation, expression, and interference.

THE EXPRESSION OF GENETIC INFORMATION VIA GENES II: NON-CODING RNAS

(b) Adaptation

(c) Expression

(d) Interference

Figure 11.10 continued

Concept Check: Which component of the CRISPR-Cas system directly recognizes the bacteriophage DNA?

The proteins encoded by the *Cas1* and *Cas2* genes form a complex that recognizes the bacteriophage DNA as being foreign and cleaves it into small pieces. As shown in Figure 11.10b, a piece of bacteriophage DNA, usually between 20 and 50 bp in length, is inserted into the *Crispr* gene. The mechanism of insertion is not entirely understood. The newly inserted piece of bacteriophage DNA is called a spacer because it acts as a space between adjacent repeats. The different spacers found in the *Crispr* gene of modern bacterial species are derived from past bacteriophage infections. Each spacer provides a bacterium with defense against a particular bacteriophage. Once a bacterial cell has become adapted to a particular bacteriophage, it will pass this trait on to its daughter cells.

By cleaving the bacteriophage into pieces, the adaptation phase can protect a bacterial cell, because it cuts up the bacteriophage DNA and thereby inactivates the phage. However, a more effective way of destroying phages is provided by the expression and interference phases of this system.

Expression If a bacterial cell or one of its ancestors has already been adapted to a bacteriophage, a subsequent infection by that type of phage will result in the expression phase, in which the system gets ready for action by expressing the *Crispr, tracr,* and *Cas9* genes (Figure 11.10c). The *Crispr* gene is transcribed from a single promoter and produces a long ncRNA called pre-crRNA, which contains several repeat sequences separated by spacers. The gene encoding the tracrRNA is also transcribed, which produces many molecules of tracrRNA. As mentioned earlier, the tracrRNA is also an ncRNA. A region of the tracrRNA is complementary to the repeat sequences of the pre-crRNA. Several molecules of tracrRNA base-pair with the pre-crRNA. The pre-crRNA is then cleaved into many small molecules, now called crRNAs. Each crRNA is attached to a tracrRNA. A region of the tracrRNA is recognized by the Cas9 protein. The tracrRNA acts as a guide that causes the tracrRNA-crRNA complex to bind to a Cas9 protein.

Interference After the tracrRNA-crRNA-Cas9 complex has formed, the bacterial cell is ready to destroy the bacteriophage DNA. This phase is called interference because it resembles the process of RNA interference described earlier in this chapter (see Figure 11.8). Each spacer within a crRNA is complementary to one of the strands of a bacteriophage DNA. Therefore, the crRNA acts as a guide that causes the tracrRNA-crRNA-Cas9 complex to bind to that bacteriophage DNA (Figure 11.10d). After binding, the Cas9 protein functions as an endonuclease that breaks both strands in the bacteriophage DNA. This cleavage inactivates the phage and thereby prevents phage proliferation.

Since discovering the CRISPR-Cas system in bacteria and archaea, researchers have been able to modify certain components of this system and use them to mutate genes in living cells. We will consider this technology in Chapter 18.

11.6 Reviewing the Concepts

- The CRISPR-Cas system in bacteria and archaea provides defense against bacteriophages and transposons. The defense occurs in three phases: adaptation, expression, and interference (Figure 11.10).

11.6 Testing Your Knowledge

1. Which of the following components are needed for the adaptation phase of the CRISPR-Cas system?
   a. crRNA and Cas1
   b. crRNA and Cas2
   c. crRNA and Cas9
   d. Cas1 and Cas2
   e. tracrRNA and Cas9

2. In the CRISPR-Cas system, what does the tracrRNA bind to?
   a. crRNA and Cas1
   b. crRNA and Cas2
   c. crRNA and Cas9
   d. Cas1 and Cas2
   e. Cas1, Cas2, Cas9, and crRNA

11.7 Roles of Non-coding RNAs in Human Disease and Plant Health

Learning Outcomes

1. List examples in which ncRNAs are associated with human diseases.
2. List examples in which ncRNAs play a role in plant health.

During the past two decades, researchers have discovered that abnormalities in ncRNAs are associated with a wide range of human diseases. As mentioned at the beginning of this chapter, cartilage-hair hypoplasia (CHH) was the first human disease that was shown to be caused by an ncRNA transcribed from a nuclear gene. Since the identification of CHH-associated mutations in 2001, many ncRNAs have been shown to play a key role in human diseases. Researchers speculate that we are still seeing only the "tip of the iceberg" with regard to identifying the roles of ncRNAs in human pathology. Likewise, the impact of ncRNAs on plant health is only beginning to be appreciated but has exciting potential in the field of agriculture. In this section, we will focus on the roles of ncRNAs in the development of cancer, neurological disorders, and cardiovascular diseases, as well as their effects on plant health.

ncRNAs Play a Role in Many Forms of Cancer and Other Human Diseases

As we have seen throughout this chapter, ncRNAs play important roles in DNA replication, chromatin modification, gene transcription, mRNA translation, and protein function. When certain ncRNAs are expressed abnormally, that is, at too high or too low a level, disease conditions are known to occur. Such abnormal expression levels can be caused by mutations in specific genes or by epigenetic changes, described in Chapter 16, that alter the expression of genes that encode ncRNAs. Several examples of human diseases associated with the abnormal expression of ncRNAs are listed in **Table 11.2**.

Table 11.2 Examples of ncRNAs Associated with Human Diseases

Type of ncRNA	Disease(s)*
A group of miRNAs called the miR-200 family	Several types of cancer, including bladder cancer, melanoma, stomach cancer, and colorectal cancer
HOTAIR	Several types of cancer, including breast cancer, lung cancer, and colorectal cancer
Many miRNAs	Alzheimer's disease
Many miRNAs	Multiple sclerosis
An miRNA called miR-1	Heart arrhythmias
Several different miRNAs, including miR-10a, miR-145, and miR-143	Formation of arterial plaques

*The diseases listed in this table show an association with abnormal levels of ncRNAs. In many cases, it is not yet clear if the disease symptoms are caused, in part, by the abnormal levels of ncRNAs or if the abnormal levels are a consequence of the disease symptoms.

ncRNAs and Cancer The topic of cancer is described in Chapter 13 (look ahead to Section 13.4). The roles of ncRNAs in cancer have been most thoroughly studied with respect to miRNAs. In nearly all forms of human cancer, levels of expression of particular miRNAs differ between normal and cancer cells. In some cases, the genes that encode the miRNAs behave as tumor-suppressor genes, because a lower level of expression of particular miRNAs allows tumor growth. In other cases, the genes that encode certain miRNAs act as oncogenes; their overexpression promotes cancer.

A well-studied example of the role of miRNAs in cancer involves a group of several different miRNAs called the miR-200 family. The miR-200 family plays an essential role in tumor suppression by inhibiting metastasis—the process by which cancer cells can spread through the bloodstream to other parts of the body. Low levels of expression of miR-200 members have been associated with many types of cancer, including bladder cancer, melanoma, stomach cancer, and colorectal cancer.

Though they have been less well studied, other ncRNAs are also associated with particular types of human cancers. HOTAIR, which was discussed in Section 11.3, is an ncRNA that is highly expressed in a variety of cancers, including breast cancer, lung cancer, and colorectal cancer. When overexpressed, the gene that encodes HOTAIR behaves as an oncogene. High levels of HOTAIR expression in primary breast tumors are a significant predictor of metastasis and death. HOTAIR is known to interact with a variety of cellular components, but the mechanism by which it promotes cancer is not well understood.

ncRNAs and Neurological Disorders Many miRNAs are essential for the proper development and functioning of the nervous system. Approximately 70% of all miRNAs are expressed in the brain, and many of them are specific to neurons. miRNAs are involved in neuron growth and the overall development of the nervous system. Abnormal levels of expression of miRNAs have been associated with nearly all neurological disorders in which these ncRNAs have been investigated! Table 11.2 describes some examples in which the expression of miRNAs has been altered and associated with neurological disorders. For example, in Alzheimer's disease, abnormally expressed miRNAs are thought to be involved in down-regulating the expression of the enzyme β-secretase, which leads to the overproduction of certain β-amyloid peptides—a key feature of the disease. miRNAs are also known to control the inflammatory process that leads to the development of multiple sclerosis.

ncRNAs and Cardiovascular Diseases Abnormalities in miRNA levels have been linked to several cardiovascular diseases. A particular miRNA, called miR-1, is associated with the development of arrhythmias—irregularities in the rate or rhythm of the heartbeat. This miRNA regulates the expression of genes that encode ion channel proteins, which are important for proper signaling between cardiac muscle cells. Other miRNAs appear to play a role in vascular disease. The formation of arterial plaques is associated with abnormal expression levels of several miRNAs, including miR-10a, miR-145, and miR-143.

ncRNAs Are Essential to Plant Health

In parallel to the growing knowledge of the roles of ncRNAs in human diseases, plant biologists are discovering that abnormalities in ncRNAs play many essential roles that contribute to the health of plants. This realization is likely to have great impact in the field of agriculture as we develop methods to change the expression of ncRNAs in order to modify the characteristics of agriculturally important plants. Table 11.3 describes several ncRNAs that are known to play key roles in plant health.

11.7 Reviewing the Concepts

- Abnormalities in the expression of ncRNAs have been associated with many human diseases, including cancer, neurological disorders, and cardiovascular diseases (Table 11.2).
- The proper level of expression of ncRNAs is also important for plant health (Table 11.3).

11.7 Testing Your Knowledge

1. An miRNA is discovered to be overexpressed, and this overexpression promotes cancer. In this case, the miRNA is behaving as
   a. an oncogene.
   b. a tumor-suppressor gene.
   c. a decoy.
   d. both an oncogene and a tumor-suppressor gene.
   e. both a decoy and a tumor-suppressor gene.

Table 11.3 Importance of ncRNAs in Plant Health

Type of ncRNA*	Normal role in plant structure and function
Several miRNAs, including miR-156, miR-157, and miR-159	Control the time of year when flowering occurs
An ncRNA called COOLAIR	Promotes vernalization, the phenomenon in which certain plants will only flower after being exposed to cold winter temperatures
Two miRNAs called miR-167 and miR-397	Play a role in seed development
An miRNA called miR-402	Affects the rate of seed germination and seedling growth under stress conditions
An miRNA called miR-824	Plays a role in the development of stomata
An ncRNA called IPS1	Affects the ability of plants to cope with phosphate starvation

*Most of the examples listed are miRNAs, which are short ncRNAs. COOLAIR and IPS1 are longer ncRNAs.

Assessing Your Knowledge and Skills

Foundational Knowledge

Multiple Choice

1. Which of the following types of molecules could bind to an ncRNA due to base pairing?
   a. DNA
   b. RNA
   c. protein
   d. small molecule
   e. both a and b

2. HOTAIR causes certain genes to be repressed by facilitating
   a. the binding of a repressor protein.
   b. the release of an activator protein.
   c. the covalent modification of histones.
   d. the removal of nucleosomes.
   e. both a and c.

3. During RNA interference, what binds to an mRNA to inhibit translation?
   a. a pri-miRNA
   b. a pre-miRNA or pre-siRNA
   c. a double-stranded miRNA or double-stranded siRNA
   d. a single-stranded miRNA or single-stranded siRNA
   e. dicer

4. Cas1 and Cas2 proteins play a role during which of the following phases of genome defense?
   a. adaptation
   b. expression
   c. interference
   d. both adaptation and expression
   e. both expression and interference

5. Abnormalities in the expression of ncRNAs are associated with
   a. many forms of cancer.
   b. neurological disorders.
   c. cardiovascular diseases.
   d. all of the above.
   e. only a and b.

Short Answer

1. An ncRNA may have the following functions: scaffold, guide, alterer of protein function or stability, ribozyme, blocker, and/or decoy. Which of those functions are mediated by the following examples: HOTAIR, RNA of SRP, microRNA, crRNA? Note: A single ncRNA may have more than one function.

2. **PRINCIPLES** A principle of biology is that structure determines function. Explain how the structure of HOTAIR allows it to carry out its function.

Critical-Thinking Skills

Multiple Choice

1. TERC, which is the RNA component of telomerase, plays two key roles in the synthesis of telomere repeat sequences. During this process,
   a. TERC first acts a template, then acts as a guide.
   b. TERC first acts a guide, then acts as a template.
   c. TERC first acts a template, then acts as a decoy.
   d. TERC first acts a guide, then acts as a decoy.
   e. TERC first acts as a decoy, then acts as a template.

2. Let's suppose that a mutation in the HOTAIR gene altered the HOTAIR ncRNA in a way that prevented it from binding to GA-rich regions. This mutation would prevent the HOTAIR ncRNA from acting as a
   a. guide.
   b. scaffold.
   c. decoy.
   d. ribozyme.
   e. blocker.

3. A key difference between microRNA (miRNA) and small-interfering RNA (siRNA) is that
   a. miRNA silences the expression of mRNA, but siRNA does not.
   b. siRNA silences the expression of mRNA, but miRNA does not.
   c. miRNAs are made from endogenous genes, but siRNAs are usually taken into cells from outside sources.
   d. miRNAs usually inhibit translation, but siRNAs usually cause mRNA degradation.
   e. Both c and d are correct.

4. One of the roles of the RNA component of signal recognition particle (SRP) is to stimulate certain proteins to hydrolyze GTP. If this function of SRP RNA did not work properly, what would you expect to happen?
   a. SRP would not bind to the ER signal sequence of a polypeptide.
   b. SRP would not cause translation to pause.
   c. SRP would not bind to an SRP receptor in the ER membrane.
   d. SRP would not be released from the ER membrane.
   e. Both a and b are correct.
5. Let's suppose a bacterium is part of a lineage that has been previously exposed to bacteriophage λ. The ancestors to this bacterium were able to mount an attack against bacteriophage λ via the CRISPR-Cas9 system. If this bacterium is now exposed to bacteriophage λ, which phase(s) of the CRISPR-Cas9 system would be necessary for it to destroy that bacteriophage?
   a. adaptation only
   b. adaptation and interference
   c. expression and interference
   d. adaptation and interference
   e. adaptation, expression, and interference

Short Answer

1. The structure and function of ribosomes are described in Chapter 10. Describe three functions of ribosomal RNAs.
2. Review the concept of an RNA world described in Chapter 4, Section 4.1. Discuss which ncRNAs described in Table 11.1 may have arisen during the RNA world, and which probably arose after the modern DNA/RNA/protein world came into being.

12 The Control of Genetic Information via Gene Regulation

Image from the RCSB PDB (www.rcsb.org; H.M. Berman, et al. (2000), "The Protein Data Bank," *Nucleic Acids Research*, 28: 235–242) of PDB ID 1CGP (Schultz, S.C., Shields, G.C., Steitz, T.A.) (1991) *Science*, 253: 1001–1007.

A model for a protein that binds to DNA and regulates genes. The catabolite activator protein (CAP), shown in light purple, is binding to the DNA double helix, shown in orange. CAP, shown again in Figure 12.8, activates gene transcription when cAMP (shown in yellow) is bound to it.

Chapter Outline

- 12.1 Overview of Gene Regulation
- 12.2 Regulation of Transcription in Bacteria
- 12.3 Regulation of Transcription in Eukaryotes: Roles of Transcription Factors
- 12.4 Regulation of Transcription in Eukaryotes: Changes in Chromatin Structure and DNA Methylation
- 12.5 Regulation of RNA Splicing and Translation in Eukaryotes

Assessing Your Knowledge and Skills

Consider the muscles of a professional weight lifter and those of someone who doesn't lift weights. A weight lifter's upper arm muscles (the biceps and triceps) can attain enormous sizes, whereas the same muscles in a non–weight lifter are relatively small. How do we explain such differences? Unknowingly, when people lift weights, they are "turning on," or activating, hundreds of genes in their muscle cells. For example, genes that encode actin (a cytoskeletal filament) and myosin (a motor protein) are turned on. As discussed in Chapter 35, actin and myosin are key components that promote muscle contraction. By making more of these proteins, the weight lifter increases the mass of her or his muscles and develops muscles that can contract with greater force and duration.

At the molecular level, **gene expression** is the process by which the information within a gene is made into a functional product, such as an RNA molecule or a protein. The majority of genes in all species are regulated so that the proteins they specify are produced at appropriate times, in the appropriate cell types, and in specific amounts. The term **gene regulation** refers to the ability of cells to control the expression of their genes. By comparison, some genes have relatively constant levels of expression in all conditions over time. These are called **constitutive genes.** Frequently, constitutive genes encode proteins that are constantly required for the survival of an organism, such as certain metabolic enzymes involved with carbohydrate breakdown.

The importance of gene regulation is underscored by the number of genes devoted to this process. For example, in *Arabidopsis thaliana,* a plant that is studied by many plant geneticists, over 5% of the genome is involved with regulating gene transcription. In humans, researchers estimate that 2,600 genes encode proteins that bind to DNA and affect gene function. With a human genome size of about 22,000 protein-encoding genes, nearly 12% of them are directly involved with gene regulation.

In this chapter, we will begin with an overview that emphasizes the benefits of gene regulation and the general mechanisms that achieve such regulation in bacteria and in eukaryotes. The following sections will describe how bacteria regulate gene expression in the face of environmental change and the more complex nature of gene regulation in eukaryotes.

12.1 Overview of Gene Regulation

Learning Outcomes

1. Discuss the various ways that organisms benefit from gene regulation.
2. Identify where gene regulation can occur during the process of gene expression in bacteria and eukaryotes.

How do living organisms benefit from gene regulation? One benefit is that gene regulation conserves energy. RNAs and proteins that are encoded by genes are produced only when needed. In multicellular organisms, gene regulation also ensures that genes are expressed in the appropriate cell types and at the correct stage of development. In this section, we will examine a few examples that illustrate the important consequences of gene regulation. We will also survey the major points in the gene expression process at which genes are regulated in bacterial and eukaryotic cells.

Bacteria Regulate Genes in Response to Changes in Their Environment

The bacterium *Escherichia coli* can use many types of sugars as food sources, thereby increasing its chances of survival. With regard to gene regulation, we will focus on how it uses lactose, which is a sugar found in milk. *E. coli* carries genes that code for proteins that enable the bacterium to take up lactose from the environment and metabolize it.

Figure 12.1 illustrates the effects of lactose on the regulation of those genes. In order to utilize lactose, an *E. coli* cell requires both a transporter (lactose permease) that facilitates the uptake of lactose into the cell and an enzyme (β-galactosidase) that catalyzes the breakdown of lactose. When lactose is not present in the environment, an *E. coli* cell makes very little of these proteins. However, when lactose becomes available, the bacterium produces many more of these proteins, enabling it to readily use lactose from its environment. Eventually, all of the lactose in the environment will be used up. At this point, the genes encoding these proteins will be shut off, and most of the proteins will be degraded. Overall, gene regulation conserves

Biology Principle

Living Organisms Use Energy
Gene regulation provides a way for cells to avoid wasting energy by making proteins only when they are needed.

Figure 12.1 Gene regulation of lactose utilization in *E. coli*.

(a) Skeletal muscle cell (b) Neuron (c) Skin cell

Figure 12.2 Examples of different cell types in humans. These cells have the same genetic composition. Their unique morphologies are due to differences in the proteins they make.
(a) ©Ed Reschke/Getty Images; (b) ©McGraw-Hill Education/Al Telser; (c) Source: M. Rein/CDC

 Concept Check: *How does gene regulation underlie the different morphologies of these various types of cells?*

energy because it ensures that the proteins needed for lactose utilization are made only when lactose is present in the environment.

Eukaryotic Gene Regulation Produces Different Cell Types in a Single Organism

One of the most amazing examples of gene regulation is the phenomenon of **cell differentiation,** the process by which cells become specialized into particular types. In humans, for example, cells may differentiate into muscle cells, neurons, skin cells, or other types. **Figure 12.2** shows micrographs of three types of cells found in humans. As the micrographs show, the morphologies of these cells are strikingly different. Likewise, their functions within the body are also quite different. Muscle cells are important in body movements, neurons function in cell signaling, and skin cells form a protective outer surface to the body.

Gene regulation is responsible for producing different types of cells within a multicellular organism. The three cell types shown in Figure 12.2 contain the same **genome,** meaning they carry the same set of genes. However, their **proteomes**—the sets of proteins they make—are quite different. Certain proteins are found in particular cell types but not in others. Alternatively, whereas a protein may be present in several or even many cell types, the relative amounts of the protein may differ. The amount of a given protein depends on many factors, including how strongly the corresponding gene is turned on and how much protein is synthesized from mRNA. Gene regulation plays a major role in determining the proteome of each cell type.

Eukaryotic Gene Regulation Enables Multicellular Organisms to Proceed Through Developmental Stages

In multicellular organisms that advance through developmental stages, certain genes are expressed at particular stages of development but not at others. Let's consider an example of such gene regulation in mammals. Early stages of development occur in the uterus of female mammals. In humans, the embryonic stage lasts from fertilization to 8 weeks. During this stage, major developmental changes produce the various body parts. The fetal stage occurs from 8 weeks to birth (41 weeks). This stage is characterized by a continued refinement of body parts and a large increase in size.

The oxygen demands of a rapidly growing embryo and fetus are quite different from the needs of the mother. Gene regulation plays a vital role in ensuring that an embryo and a fetus get the proper amount of oxygen. Hemoglobin is a protein that delivers oxygen to the cells of a mammal's body. A hemoglobin protein is composed of four globin polypeptides: two encoded by one globin gene and two encoded by another globin gene (**Figure 12.3**). The genomes of mammals carry several genes (designated with Greek letters) that encode slightly different globin polypeptides. During the embryonic stage of development, the epsilon (ε)- and zeta (ζ)-globin genes are turned on. At the fetal stage, these genes are turned off, and the alpha (α)- and gamma (γ)-globin genes are turned on. Finally, at birth, the γ-globin gene is turned off, and the beta (β)-globin gene is turned on.

How do the embryo and fetus acquire oxygen from their mother's bloodstream? The hemoglobin produced during the embryonic and fetal stages has a higher binding affinity for oxygen than does the hemoglobin produced after birth. Therefore, the embryo and fetus can remove oxygen from the mother's bloodstream and use that oxygen for their own needs. This occurs across the placenta, where the mother's bloodstream is adjacent to the bloodstream of the embryo or fetus. In this way, gene regulation enables mammals to develop inside the mother's body, even though the embryo and fetus are not breathing on their own. Gene regulation ensures that the correct hemoglobin protein is produced at the right time in development.

Gene Regulation Occurs at Different Points in the Process from DNA to Protein

For protein-encoding genes, the regulation of gene expression can occur at any of the steps in the process that produces a functional protein. Bacterial gene expression can be regulated at the following stages (**Figure 12.4a**):

- **Transcription:** Gene regulation most commonly occurs at the level of transcription, which means that bacteria regulate how much mRNA is made from genes. Because transcription is the first step in gene expression, transcriptional regulation is a particularly efficient way to regulate genes because cells avoid wasting energy when the product of the gene is not needed.

Biology Principle

The genetic material provides a blueprint that allows organisms to grow, develop, and reproduce.

Gene regulation is an important process that allows organisms to proceed through developmental stages.

	Embryo	Fetus	Birth to Adult
Hemoglobin protein	2 ζ-globins 2 ε-globins	2 α-globins 2 γ-globins	2 α-globins 2 β-globins
Oxygen affinity	Highest	High	Moderate
Gene expression α-globin gene β-globin gene γ-globin gene ζ-globin gene ε-globin gene	Off Off Off On On	On Off On Off Off	On On Off Off Off

Figure 12.3 Regulation of human globin genes at different stages of development.

- **Translation:** Bacteria may also control the ability of an mRNA to be translated into protein, but this mechanism is not as prominent as transcriptional regulation.
- **Post-translation:** After a protein is synthesized via translation, its amount or function may be regulated, which is termed post-translational regulation. This form of regulation is best understood within the context of cell biology (refer back to Unit II). For example, feedback inhibition of enzymes is discussed in Chapter 6 (refer back to Figure 6.10).

In eukaryotes, gene expression can also be regulated at several stages (**Figure 12.4b**):

- **Transcription:** Transcriptional regulation is a very common form of regulation in eukaryotes.
- **RNA modification:** As discussed in Chapter 10 (refer back to Figure 10.4), eukaryotes modify their mRNA transcripts in ways that do not commonly occur in bacteria. For example, RNA splicing is a widespread phenomenon in eukaryotes. As discussed later, such events are subject to regulation.
- **Translation:** Though not as prominent as transcriptional regulation, this is a fairly common way for eukaryotes to regulate gene expression.
- **Post-translation:** This form of regulation is also prominent in eukaryotes.

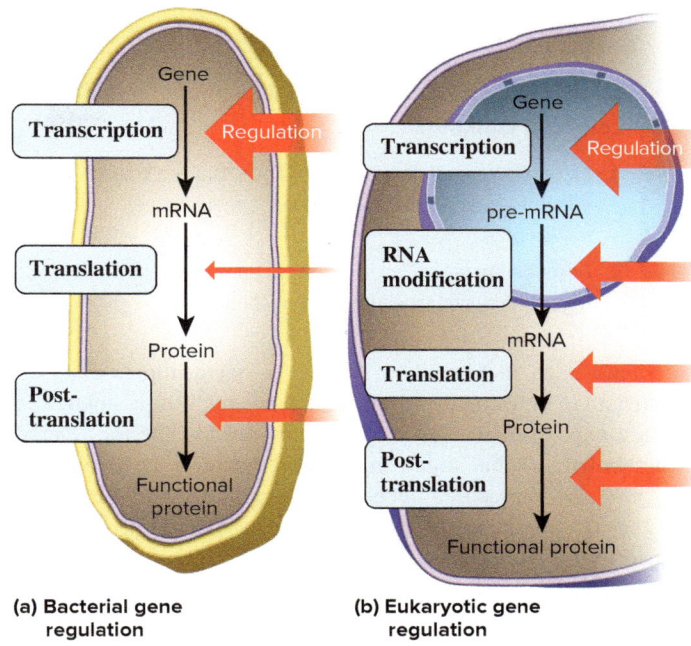

Figure 12.4 Overview of gene regulation in (a) bacteria and (b) eukaryotes. The relative widths of the red arrows indicate the prominence with which gene regulation is used to control the production of functional proteins.

12.1 Reviewing the Concepts

- Most genes are regulated so that the level of gene expression can vary under different conditions. By comparison, constitutive genes are expressed at constant levels.
- Gene regulation ensures that gene products are made only when needed. An example is the synthesis of the gene products needed for lactose utilization in bacteria (Figure 12.1).
- In eukaryotes, gene regulation leads to the production of different cell types, such as neurons, muscle cells, and skin cells. Eukaryotic gene regulation also enables gene products to be produced at different developmental stages (Figures 12.2, 12.3).
- All organisms regulate gene expression at a variety of stages, including transcription, translation, and post-translation. Eukaryotes also regulate RNA modification (Figure 12.4).

12.1 Testing Your Knowledge

1. A reason for gene regulation is
   a. to respond to environmental changes.
   b. to make proteins in specific cell types.
   c. to make proteins at specific stages of development.
   d. to do all of the above.

2. The most common point of gene regulation in bacteria is
   a. transcription.
   b. RNA modification.
   c. translation.
   d. post-translation (after a protein is made).

12.2 Regulation of Transcription in Bacteria

Learning Outcomes

1. Explain how regulatory transcription factors and small effector molecules are involved in the regulation of transcription.
2. Describe the organization of the *lac* operon and how it is regulated by negative and positive control.

As we have seen in Figure 12.1, when a bacterium is exposed to a particular nutrient in its environment, such as a sugar, the genes are expressed that encode proteins needed for the uptake and metabolism of that sugar. In addition, bacteria have enzymes that synthesize molecules such as particular amino acids. For these genes, the control of expression often occurs at the level of transcription. In this section, we will examine the underlying molecular mechanisms that bring about transcriptional regulation in bacteria.

Transcriptional Regulation Involves Regulatory Transcription Factors and Small Effector Molecules

In most cases, regulation of transcription involves the actions of **regulatory transcription factors**—proteins that bind to regulatory sequences in the DNA, which are usually in the vicinity of a promoter, and affect the rate of transcription of one or more nearby genes. These transcription factors either decrease or increase the rate of transcription of a gene (**Figure 12.5a**):

- **Repressors** are regulatory transcription factors that bind to the DNA and decrease the rate of transcription. This is a form of regulation called **negative control.**
- **Activators** are regulatory transcription factors that bind to the DNA and increase the rate of transcription, a form of regulation termed **positive control.**

In conjunction with regulatory transcription factors, molecules called **small effector molecules** often play a critical role in transcriptional regulation. A small effector molecule exerts its influence by binding to a repressor or an activator and causing a conformational change in the protein. In many cases, the effect of the conformational

(a) Actions of regulatory transcription factors

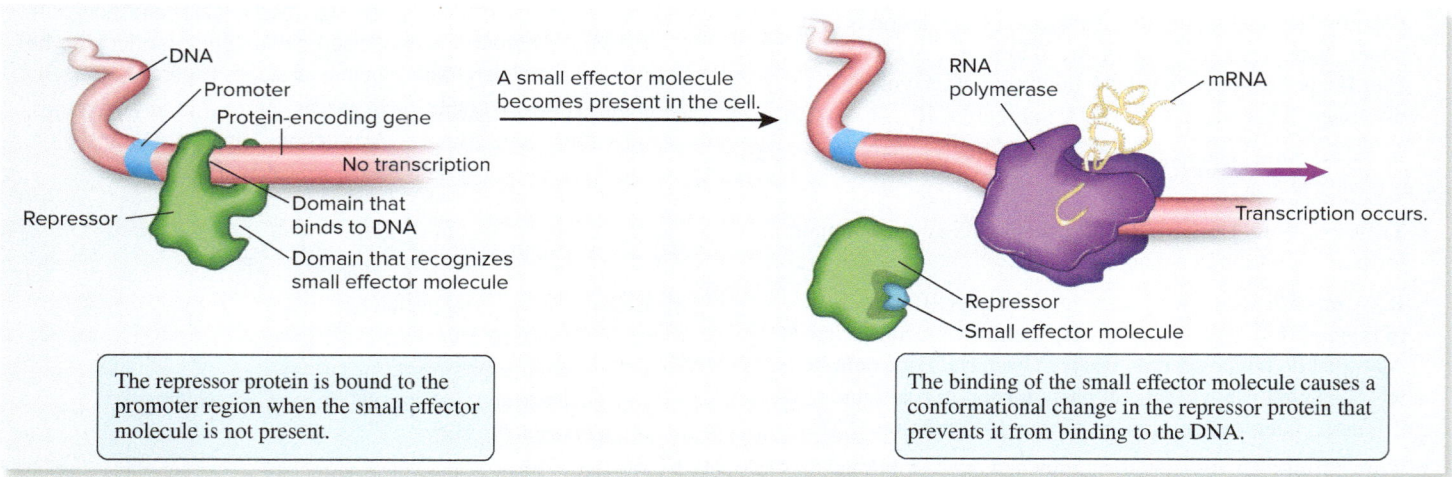

(b) Action of a small effector molecule on a repressor

Figure 12.5 Actions of regulatory transcription factors and small effector molecules. (a) Regulatory transcription factors are proteins that exert negative or positive control. (b) One way that a small effector molecule may exert its effects is by preventing a repressor protein from binding to the DNA.

change determines whether or not the protein can bind to the DNA. Figure 12.5b illustrates an example involving a repressor. When the small effector molecule is not present in the cytoplasm, the repressor binds to the DNA and inhibits transcription. However, when the small effector molecule is found in the cytoplasm, it will bind to the repressor and cause a conformational change that inhibits the ability of the protein to bind to the DNA. Transcription can occur because the repressor is not able to bind to the DNA. Repressors and activators that respond to small effector molecules have two functional regions called domains. One domain is a site where the protein binds to the DNA, whereas the other is the binding site for the small effector molecule.

The *lac* Operon Contains Genes That Encode Proteins Involved in Lactose Metabolism

In bacteria, a set of two or more protein-encoding genes may be under the transcriptional control of a single promoter. This arrangement is known as an **operon**. The genes in an operon are transcribed as a single unit, resulting in the production of a **polycistronic mRNA,** an mRNA that encodes more than one polypeptide, which become units within functional proteins. What advantage does this arrangement provide? An operon allows a bacterium to coordinate the regulation of a group of genes that encode proteins whose functions are used in a common pathway.

The genome of *E. coli* carries an operon, called the **lac operon,** that contains the genes for the proteins that allow the bacterium to metabolize lactose (see Figure 12.1). **Figure 12.6a** shows the organization of this operon as it is found in the *E. coli* chromosome, as well as the polycistronic mRNA that is transcribed from it.

- The **promoter,** *lacP*, is used to transcribe three genes: *lacZ*, *lacY*, and *lacA*.
- The *lacZ* gene encodes β-galactosidase, which is an enzyme that breaks down lactose (**Figure 12.6b**). In a side reaction, β-galactosidase also converts a small percentage of lactose into allolactose, a structurally similar sugar, or lactose analogue, which (as discussed later) functions as a small effector molecule.

(a) Organization of DNA sequences in the *lac* region of the *E. coli* chromosome

(b) Functions of lactose permease and β-galactosidase

Figure 12.6 The *lac* operon. (a) This diagram depicts a region of the *E. coli* chromosome that contains the *lacI* gene and the adjacent *lac* operon, and also shows the polycistronic mRNA transcribed from the operon. The mRNA is translated into three proteins: β-galactosidase, lactose permease, and galactoside transacetylase. (b) Lactose permease cotransports H⁺ with lactose. Bacteria maintain an H⁺ gradient across their cytoplasmic membrane that drives the active transport of lactose into the cytoplasm. β-galactosidase cleaves lactose into galactose and glucose. As a side reaction, it can also convert lactose into allolactose.

 Concept Check: *Which genes are under the control of the lac promoter?*

- The *lacY* gene encodes lactose permease, which is a membrane protein required for the transport of lactose into the cytoplasm of the bacterium.
- The *lacA* gene encodes galactoside transacetylase, which covalently modifies lactose and lactose analogues by attaching an acetyl group (—COCH$_3$). The attachment of acetyl groups to nonmetabolizable lactose analogues prevents their toxic buildup in the cytoplasm.
- The **operator,** or *lacO* site, is a regulatory sequence in the DNA. The sequence of bases at the *lacO* site provides a binding site for a repressor protein.
- The **CAP site** is a regulatory sequence in the DNA that is recognized by an activator protein.

Adjacent to the *lac* operon is the *lacI* gene, which encodes the **lac repressor.** This repressor protein is important for the regulation of the *lac* operon. The *lacI* gene, which is constitutively expressed at a fairly low level, has its own promoter called the *i* promoter. The *lacI* gene is not considered a part of the *lac* operon. Let's now take a look at how the *lac* operon is regulated by the lac repressor.

The *lac* Operon Is Under Negative Control by a Repressor Protein

In the late 1950s, the first researchers to investigate gene regulation were French biologists François Jacob and Jacques Monod at the Pasteur Institute in Paris, France. Their focus on gene regulation stemmed from an interest in the phenomenon known as enzyme adaptation, which had been identified early in the 20th century. Enzyme adaptation occurs when a particular enzyme appears within a living cell only after the cell has been exposed to the substrate for that enzyme. Jacob and Monod studied lactose metabolism in *E. coli* to investigate this phenomenon. When they exposed bacteria to lactose, the levels of lactose-utilizing enzymes in the cells increased by 1,000- to 10,000-fold. After lactose was removed, the synthesis of the enzymes abruptly stopped.

The first mechanism of regulation that Jacob and Monod discovered involved the lac repressor, which binds to the sequence of nucleotides found at the *lac* operator. Once bound, the lac repressor prevents RNA polymerase from transcribing the *lacZ, lacY,* and *lacA* genes (**Figure 12.7a**). RNA polymerase can bind to the promoter when the lac repressor is bound to the operator but is largely prevented from moving past the operator to transcribe the *lacZ, lacY,* and *lacA* genes. Therefore, if *E. coli* does not have lactose in its environment, only small amounts of lactose permease, β-galactosidase, and transacetylase are made.

Alternatively, when *E. coli* is exposed to lactose, the following events occur (**Figure 12.7b**):

1. A small amount of lactose is transported into the cytoplasm via lactose permease, and β-galactosidase converts some of it to allolactose.
2. The cytoplasmic level of allolactose gradually rises. Allolactose is an example of a small effector molecule. The lac repressor

Biology Principle

Biologists Use Models to Test Predictions and Convey Ideas

This figure is a model for the regulation of the *lac* operon by the lac repressor. The Modeling Challenge in the figure legend asks you to propose a model for a mutant lac repressor that is unable to bind to allolactose.

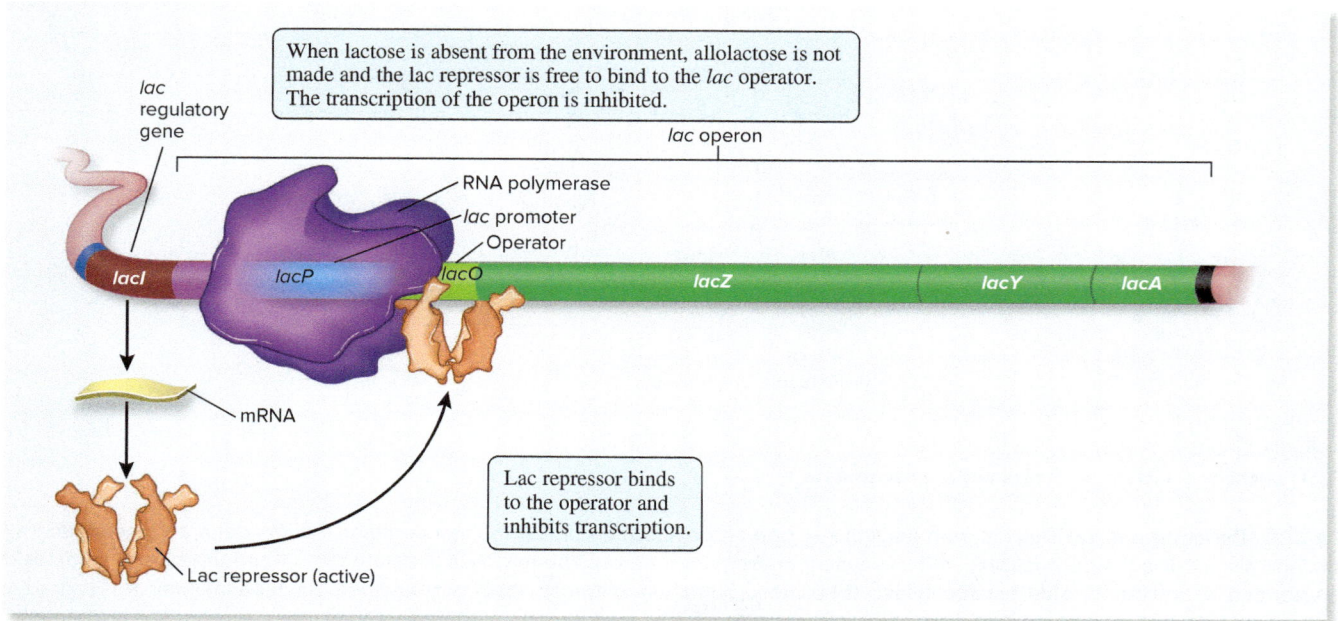

(a) Lactose absent from the environment

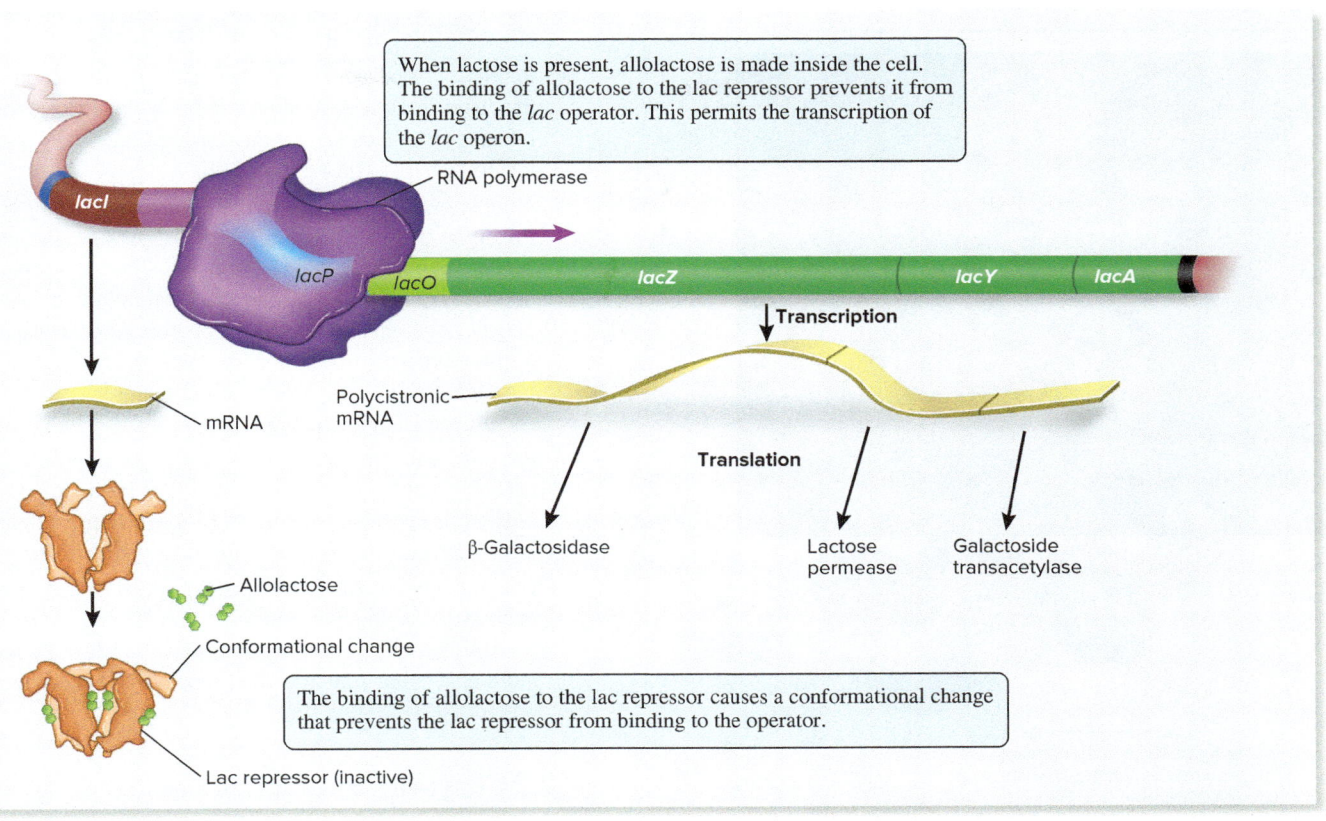

(b) Lactose present

Figure 12.7 Negative control of an inducible set of genes: function of the lac repressor in regulating the *lac* operon.

Modeling Challenge: Let's suppose that researchers identified a mutant form of the lac repressor that cannot bind to allolactose. The mutation does not affect the ability of the lac repressor to bind to the operator. Draw side-by-side models for the structures of the normal and mutant repressor in the absence of allolactose and in its presence, using the style shown in this figure. Indicate where the structure of the mutant lac repressor is altered with an arrow. In cells that make the mutant lac repressor, under what conditions would you predict that the *lac* operon would be induced?

---

protein contains four identical subunits, each of which recognizes a single allolactose molecule. When four allolactose molecules bind to the lac repressor, a conformational change occurs that prevents the repressor from binding to the operator.

3. When the lac repressor is not bound to the operator, RNA polymerase transcribes the *lacZ, lacY,* and *lacA* genes at a high rate.
4. Translation of the encoded polypeptides produces an abundant amount of the proteins needed for lactose uptake and metabolism, as described previously in Figure 12.1.

The regulation of the *lac* operon enables *E. coli* to conserve energy because lactose-utilizing proteins are made in large amounts only when lactose is present in the environment. Allolactose is an **inducer,** a small effector molecule that increases the rate of transcription, and the *lac* operon is said to be an **inducible operon.**

The *lac* Operon Is Also Under Positive Control by an Activator Protein

In addition to being under negative control by a repressor protein, the *lac* operon is also positively regulated by an activator called the **catabolite activator protein (CAP).** CAP is controlled by a small effector molecule, **cyclic AMP (cAMP),** which is produced from ATP via an enzyme known as adenylyl cyclase. When cAMP binds to CAP, the cAMP-CAP complex binds to the CAP site near the *lac* promoter (**Figure 12.8**). This causes a bend in the DNA that enhances the ability of RNA polymerase to bind to the promoter. However, the presence of glucose in the environment inhibits the production of cAMP, thereby preventing the binding of CAP to the DNA. In this way, glucose blocks the activation of the *lac* operon, thereby inhibiting transcription.

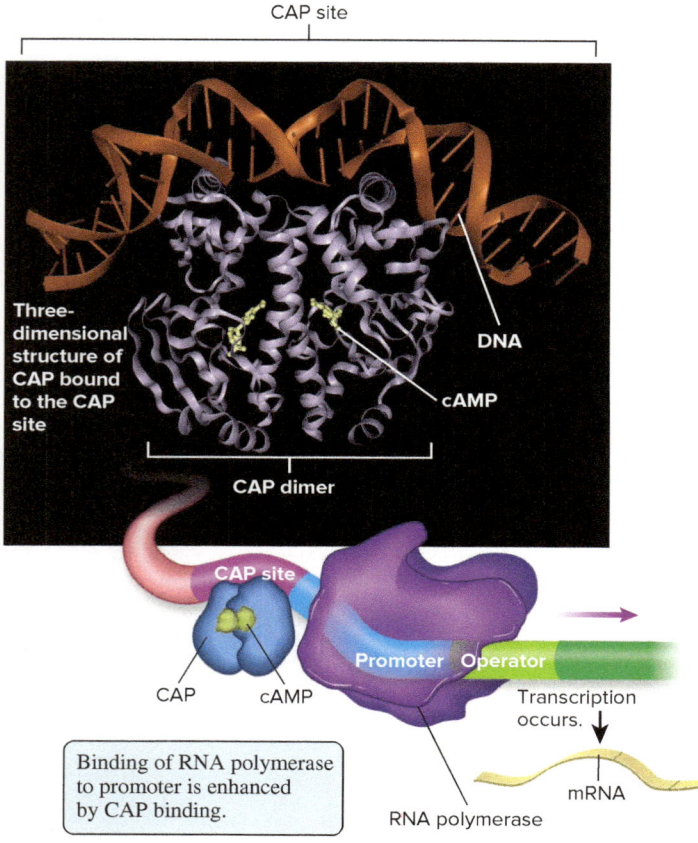

Figure 12.8 Positive control of the *lac* operon by the catabolite activator protein (CAP). When cAMP is bound to CAP, CAP binds to the DNA and causes it to bend. This bend facilitates the binding of RNA polymerase.

Image from the RCSB PDB (www.rcsb.org; H.M. Berman, et al. (2000), "The Protein Data Bank," *Nucleic Acids Research*, 28: 235–242) of PDB ID 1CGP (Schultz, S.C., Shields, G.C., Steitz, T.A.) (1991) *Science*, 253: 1001–1007.

BioConnections: *Refer back to Figure 8.13. What is the function of cAMP in eukaryotic cells?*

Figure 12.9 considers the four possible environmental conditions that an *E. coli* bacterium might experience with regard to the two sugars lactose and glucose.

- **High lactose and high glucose** (Figure 12.9a): The rate of transcription of the *lac* operon is low to moderate, because CAP does not activate transcription. Under these conditions, the bacterium primarily uses glucose rather than lactose. The bacterium conserves energy by using one type of sugar at a time.
- **High lactose and low glucose** (Figure 12.9b): The transcription rate of the *lac* operon is very high because CAP is bound to the CAP site and the lac repressor is not bound to the operator. Under these conditions, the bacterium readily metabolizes lactose.
- **Low lactose and low or high glucose** (Figure 12.9c, d): When lactose levels are low, the lac repressor prevents transcription of the *lac* operon, whether glucose levels are high or low.

12.2 Reviewing the Concepts

- Repressors and activators are regulatory transcription factors that bind to DNA and affect the transcription of genes. Small effector molecules control the ability of regulatory transcription factors to bind to DNA (Figure 12.5).
- An operon is a set of two or more protein-encoding genes controlled by a single promoter and an operator. The *lac* operon is an example of an inducible operon. The lac repressor exerts negative control by binding to the operator and preventing RNA polymerase from transcribing the operon. When allolactose binds to the repressor, a conformational change occurs that prevents the repressor from binding to the operator so that transcription can proceed (Figures 12.6, 12.7).
- Positive control of the *lac* operon occurs when the catabolite activator protein (CAP) binds to the CAP site in the presence of cAMP. This causes a bend in the DNA, which promotes the binding of RNA polymerase to the promoter (Figure 12.8).
- Glucose inhibits cAMP production, which in turn inhibits the expression of the *lac* operon, because CAP cannot bind to the CAP site. This form of regulation provides bacteria with a more efficient utilization of their resources because the bacteria preferentially use one sugar (Figure 12.9).

12.2 Testing Your Knowledge

1. Let's suppose that a mutation in the *lacI* gene prevented the lac repressor from being made. How would this mutation affect the expression of the *lac* operon?
   a. It would be expressed only in the presence of lactose in the environment.
   b. It would be expressed in the presence or absence of lactose in the environment.
   c. It would be expressed only in the absence of lactose in the environment.
   d. It would never be expressed.

12.3 Regulation of Transcription in Eukaryotes: Roles of Transcription Factors

Learning Outcomes

1. Explain the concept of combinatorial control.
2. Describe how RNA polymerase and general transcription factors initiate transcription at the core promoter.
3. Compare and contrast the roles of activators, coactivators, repressors, and TFIID in gene regulation.

Regulation of transcription in eukaryotes has some of the characteristics seen in bacteria. For example, activator and repressor proteins are involved in regulating genes by influencing the ability of RNA polymerase to initiate transcription. In addition, many eukaryotic genes

THE CONTROL OF GENETIC INFORMATION VIA GENE REGULATION

(a) Lactose high, glucose high

(b) Lactose high, glucose low

(c) Lactose low, glucose high

(d) Lactose low, glucose low

Figure 12.9 Effects of lactose and glucose on the expression of the *lac* operon.

Concept Check: What are the advantages of having both an activator and a repressor protein?

are regulated by small effector molecules. However, some important differences also occur. In eukaryotic species, genes are almost always organized individually, not in operons. In addition, eukaryotic gene regulation tends to be more intricate, because eukaryotes possess complexities not found in their bacterial counterparts. For example, eukaryotes have more complicated cell structures that contain many more proteins and a variety of cell organelles.

By studying transcriptional regulation, researchers have discovered that most eukaryotic genes, particularly those found in multicellular species, are regulated by many factors. This phenomenon is called **combinatorial control** because the combination of many factors determines the expression of any given gene. At the level of transcription, the following factors contribute to combinatorial control:

1. One or more activators may stimulate the ability of RNA polymerase to initiate transcription.
2. One or more repressors may inhibit the ability of RNA polymerase to initiate transcription.
3. The function of activators and repressors may be modulated in several ways, which include the binding of small effector molecules, protein-protein interactions, and covalent modifications.
4. Activators are necessary to alter chromatin structure in the region where a gene is located, thereby making it easier for the gene to be recognized and transcribed by RNA polymerase.
5. DNA methylation usually inhibits transcription, either by preventing the binding of an activator or by recruiting proteins that inhibit transcription.

All five of these factors may contribute to the regulation of a single gene, or possibly only three or four will play a role. In most cases, transcriptional regulation is aimed at controlling the initiation of transcription at the promoter. In this section and the following section, we will survey these basic types of gene regulation in eukaryotic species.

Eukaryotic Protein-Encoding Genes Have a Core Promoter and Regulatory Elements

To understand gene regulation in eukaryotes, we first need to consider the DNA sequences that are needed to initiate transcription. For eukaryotic protein-encoding genes, certain features are common among most promoters (**Figure 12.10**):

- **Core promoter:** The TATA box and transcriptional start site form the **core promoter.** The **transcriptional start site** is the place in the DNA where transcription actually begins. The **TATA box,** which is a 5′–TATAAA–3′ sequence, is usually about 25 bp upstream from a transcriptional start site. The TATA box determines the precise starting point for transcription. If it is missing from the core promoter, transcription may start at any of a variety of locations. The core promoter, acting alone, results in a low level of transcription that is termed **basal transcription.**

- **Regulatory elements: Regulatory elements** (or **regulatory sequences**) are DNA segments that regulate eukaryotic genes. As described later, regulatory elements are recognized by regulatory transcription factors that control the

Figure 12.12 Effect of an activator via TFIID, a general transcription factor.

- Activators and repressors may regulate RNA polymerase II by interacting with TFIID (a GTF) (**Figure 12.12**).

12.3 Testing Your Knowledge

1. The term combinatorial control refers to the observation that
   a. GTFs must combine with each other to initiate transcription.
   b. the preinitiation complex is a combination of RNA polymerase and GTFs.
   c. a combination of different factors may regulate the expression of a given gene.
   d. eukaryotic transcription relies on a combination of a core promoter and regulatory elements.

2. The core promoter for eukaryotic protein-encoding genes usually contains
   a. a transcriptional start site.
   b. a TATA box.
   c. regulatory elements.
   d. all of the above.
   e. both a and b.

a gene may be in a **closed conformation**, difficult or impossible. Transcription requires changes in chromatin structure that allow transcription factors to gain access to and bind to the DNA in the promoter region. Such chromatin, said to be in an **open conformation,** is accessible to GTFs and RNA polymerase II so that transcription can take place. In this section, we will examine how chromatin is converted from a closed to an open conformation. We will also explore how **DNA methylation**—the attachment of methyl groups to cytosine bases—affects chromatin conformation and gene expression.

Transcription Is Controlled by Changes in Chromatin Structure

In recent years, geneticists have been trying to identify the steps that promote the interconversion between the closed and open conformations of chromatin. One way to change chromatin structure is through **chromatin-remodeling complexes,** which are complexes of proteins that alter chromatin structure. Such complexes use energy from ATP hydrolysis to drive a change in the locations and/or compositions of nucleosomes, thereby making the DNA more or less amenable to transcription. Therefore, chromatin remodeling is important for both the activation and the repression of transcription.

How do chromatin-remodeling complexes change chromatin structure? Three effects are possible:

- These complexes may bind to chromatin and change the locations of nucleosomes (**Figure 12.13a**). This may involve a shift of the relative positions of a few nucleosomes or a change in the relative spacing of nucleosomes over a long stretch of DNA.

- A second effect is that chromatin-remodeling complexes may evict a histone octamer from the DNA, thereby creating gaps where nucleosomes are not found (**Figure 12.13b**).

- A third possibility is that chromatin-remodeling complexes may change the composition of nucleosomes by removing standard histone proteins from an octamer and replacing them with histone variants (**Figure 12.13c**). A **histone variant** is a histone protein that has a slightly different amino acid sequence than that of the standard histone proteins, which are described in

Figure 12.10 A common organization of sequences for the promoter of a eukaryotic protein-encoding gene. The core promoter has a TATA box and a transcriptional start site. The TATA box sequence is 5′–TATAAA–3′. However, not all protein-encoding genes in eukaryotes have a TATA box. The A highlighted in dark blue is the transcriptional start site. This A marks the site of the first adenine in the RNA transcript. The sequence that flanks the A of the transcriptional start site consists of two pyrimidines (Py_2), then a cytosine (C), then five pyrimidines (Py_5). Py stands for pyrimidine—cytosine or thymine. Regulatory elements, such as enhancers and silencers, are usually found upstream from the core promoter.

Figure 12.11 The preinitiation complex. General transcription factors (GTFs) and RNA polymerase II assemble into the preinitiation complex at the core promoter of eukaryotic protein-encoding genes.

ability of RNA polymerase to initiate transcription at the core promoter. Regulatory elements come in two general types:

1. Some regulatory elements, known as **enhancers,** play a role in the ability of RNA polymerase to begin transcription, thereby enhancing the rate of transcription. When enhancers are not functioning, most eukaryotic genes have very low levels of basal transcription.
2. Other regulatory elements, known as **silencers,** prevent transcription of a given gene when its expression is not needed. When these sequences function, the rate of

RNA polymerase II and GTFs must come together at the TATA box of the core promoter so that transcription can be initiated. A series of interactions that occurs between these proteins enables RNA polymerase II to bind to the DNA and begin the transcription process. The completed assembly of RNA polymerase II and GTFs at the TATA box is known as the **preinitiation complex** (Figure 12.11).

Activators and Repressors May Influence the Function of GTFs

In eukaryotes, regulatory transcription factors called activators and repressors bind to enhancers or silencers, respectively, and regulate the rate of transcription of genes. Activators and repressors commonly regulate the function of RNA polymerase II by affecting the function of GTFs. Figure 12.12 illustrates how an activator may work:

1. An activator binds to an enhancer. In this example, an activator also interacts with a **coactivator**—a protein that increases the rate of transcription but does not directly bind to the DNA itself.
2. The activator/coactivator complex improves the ability of a GTF called transcription factor IID (TFIID) to bind to the TATA box.
3. The function of TFIID is to promote the assembly of GTFs and RNA polymerase II into the preinitiation complex.

---

cell lineage, perhaps in the range of 10 or more, are needed for cancer to occur. Such changes involve the overexpression of oncogenes and the inactivation of tumor-suppressor genes. Many cancers begin with a benign genetic alteration that, over time and with additional genetic changes, leads to malignancy. Furthermore, a malignancy can continue to accumulate genetic changes that make it even more difficult to treat because the cells divide faster or invade surrounding tissues more readily.

As an example, let's consider lung cancer, which was briefly discussed at the beginning of this chapter. Most cancers of the lung are **carcinomas**—cancers of epithelial cells (Figure 13.14). Epithelial cells form the lining of all internal and external body surfaces. The top images in this figure show the normal epithelium found in a healthy lung. The rest of the figure shows the progression of a carcinoma that is due to mutations in basal cells, a type of epithelial cell.

1. Cancer occurs due to the accumulation of mutations in a cell lineage, beginning with an initial mutant cell, which then divides multiple times to produce a population of many daughter cells. As mutations accumulate in a lineage of basal cells, the number of these cells increases dramatically. This causes a thickening of the epithelium, a condition called hyperplasia.
2. The proliferation of the basal cells causes the loss of the ciliated, columnar epithelial cells that normally line the airways and help remove mucus and particles trapped in it from the lungs.
3. As additional mutations accumulate in this cell lineage, the basal cells develop more abnormal morphologies, a condition known as dysplasia. In the early stages of dysplasia, the abnormal basal cells are precancerous. If the source of chronic irritation (usually cigarette smoke) is eliminated, the abnormal cells are likely to disappear.
4. If smoking continues, the abnormal cells may accumulate additional genetic changes and lose the ability to stop dividing. Such cells have become cancerous—the person has basal cell carcinoma.

The basement membrane is a sheetlike layer of extracellular matrix that provides a barrier between the lung cells and the bloodstream. If the cancer cells have not yet penetrated the basement membrane, they will not have metastasized, that is, spread into the blood, bone, or other parts of the body. If the entire tumor is removed at this stage, the patient should be cured. The lower images in Figure 13.14 show a tumor that has broken through the basement membrane. The metastasis of these cells to other parts of the body will likely kill the patient, usually within a year of diagnosis.

The cellular changes that lead to lung cancer are correlated with genetic changes. These include the occurrence of mutations that create oncogenes and inhibit tumor-suppressor genes. The order of mutations is not absolute. It takes time for multiple changes to accumulate, so cancer is generally a disease of older people. Reducing your exposure to mutagens such as cigarette smoke throughout your lifetime helps minimize the risk of mutations to your genes that could promote cancer.

Figure 13.14 Progression of changes leading to lung cancer. Lung tissue is largely composed of different types of connective tissue and epithelial cells, including columnar and basal cells. A progression of cellular changes in basal cells, caused by the accumulation of mutations, leads to basal cell carcinoma, a common type of lung cancer.
©Dr. Oscar Auerbach, used by permission of the Estate of Oscar Auerbach

Normal lung epithelium

Hyperplasia

Loss of ciliated cells

Dysplasia (initially precancerous, then cancerous)

 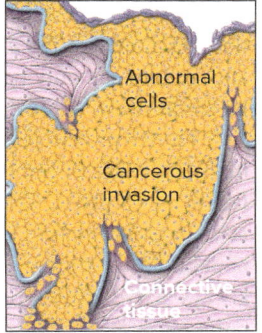
Invasive cancerous cells that can metastasize

13.4 Reviewing the Concepts

- Cancer is due to the accumulation of mutations and other genetic changes in a lineage of cells that leads to uncontrolled cell growth (Figure 13.8).
- Mutations in proto-oncogenes that result in overactivity produce cancer-causing genes called oncogenes. Oncogenes often encode proteins involved in cell-signaling pathways that promote cell division (Figures 13.9, 13.10).
- Three common types of genetic changes—namely, missense mutations, gene amplifications, and chromosomal translocations—can change proto-oncogenes into oncogenes (Figures 13.11, 13.12).
- The normal function of most tumor-suppressor genes is to negatively regulate cell division or maintain genome integrity. Loss-of-function mutations in such genes promote cancer.
- Checkpoint proteins such as p53 monitor the integrity of the genome and prevent the cell from progressing through the cell cycle if abnormalities are detected (Figure 13.13).
- Another checkpoint protein, Rb, is an inhibitor of the cell cycle, because it negatively controls E2F, a transcription factor that promotes cell division.
- Tumor-suppressor genes can be inactivated by gene mutations, chromosome loss, and changes in chromatin structure.
- Most forms of cancer, such as lung cancer, involve multiple genetic changes that lead to malignancy (Figure 13.14).

13.4 Testing Your Knowledge

1. A common way to change a proto-oncogene into an oncogene is by
   a. a missense mutation.
   b. a gene amplification.
   c. a chromosomal translocation.
   d. any of the above.

2. A common function of a tumor-suppressor gene is
   a. to maintain the integrity of the genome.
   b. to act as a negative regulator of cell division.
   c. to act as a cytoskeletal protein.
   d. Both a and b are correct.

Assessing Your Knowledge and Skills

Foundational Knowledge

Multiple Choice

1. In which of the following locations is a mutation least likely to affect gene function?
   a. promoter
   b. coding region
   c. splice junction
   d. intergenic region
   e. regulatory site

2. Mutagens can cause mutations by
   a. chemically altering DNA nucleotides.
   b. disrupting DNA replication.
   c. altering the genetic code of an organism.
   d. all of the above.
   e. a and b only.

3. The mutagenic effect of UV light is
   a. the alteration of cytosine bases to adenine bases.
   b. the formation of adenine dimers that interfere with genetic expression.
   c. the breaking of the sugar-phosphate backbone of the DNA molecule.
   d. the formation of thymine dimers that disrupt DNA replication.
   e. the deletion of thymine bases along the DNA molecule.

4. The Ames test
   a. provides a way to determine if any type of cell has experienced a mutation.
   b. provides a way to determine if an agent is a mutagen.
   c. allows researchers to experimentally disrupt gene activity by causing a mutation in a specific gene.
   d. provides a way to repair mutations in bacterial cells.
   e. does all of the above.

5. Cancer cells are said to be metastatic when they
   a. begin to divide uncontrollably.
   b. invade healthy tissue.
   c. migrate to other parts of the body.
   d. cause mutations in other healthy cells.
   e. do all of the above.

Short Answer

1. Distinguish between spontaneous and induced mutations. Which are more harmful? Which are avoidable?

2. **PRINCIPLES** A principle of biology is that the genetic material provides a blueprint that allows organisms to grow, develop, and reproduce. Explain how mutations may cause alterations to the genetic material that are detrimental to reproduction and sustaining life.

Critical-Thinking Skills

Multiple Choice

1. A mutation changes a proline codon (CCG) into a threonine codon (ACG). This is an example of
   a. a silent mutation.
   b. a missense mutation.
   c. a nonsense mutation.
   d. a frameshift mutation.
   e. both b and c.

2. Which of the following is an example of a somatic mutation?
   a. a mutation in an embryonic muscle cell
   b. a mutation in a sperm cell
   c. a mutation in an adult nerve cell
   d. a mutation in a primary oocyte that gives rise to an egg cell
   e. Both a and c are examples of somatic mutations.

3. During the Ames test, if a substance is a mutagen, there will be _____ colonies on the plates in which the cells had been exposed to the mutagen, because the mutagen converts some of the cells from _____.
   a. fewer, his$^+$ to his$^-$
   b. more, his$^+$ to his$^-$
   c. fewer, his$^-$ to his$^+$
   d. more, his$^-$ to his$^+$
   e. the same number of, his$^-$ to his$^+$

4. If a mutation eliminated the function of UvrC, which function of nucleotide excision repair would not work?
   a. sensing a damaged DNA site
   b. cutting of the damaged strand
   c. removal of the damaged strand
   d. synthesis of a new strand, using the undamaged strand as a template
   e. None of the above would occur.

5. Researchers have identified a drug that prevents the Ras protein from hydrolyzing GTP. How would this drug affect the growth factor signaling pathway? How would it affect cell growth?
   a. The signaling pathway would stay turned on and cell growth would be stimulated.
   b. The signaling pathway would stay turned on and cell growth would be inhibited.
   c. The signaling pathway would be turned off and cell growth would be stimulated.
   d. The signaling pathway would be turned off and cell growth would be inhibited.
   e. None of the above would occur.

Short Answer

1. A mutagen changes cytosine to uracil. Chemically, what is the mutagen doing to cytosine to convert it to uracil? Give an example of a mutagen that can do this.

2. Researchers have produced a *p53* knockout mouse in which both copies of the *p53* gene are defective. The knockout mouse is phenotypically normal at birth. However, it is highly sensitive to UV light. Explain the normal phenotype at birth and the sensitivity to UV light.

How Eukaryotic Cells Sort and Transmit Chromosomes: Mitosis and Meiosis

14

©Biophoto Associates/Science Source

A scanning electron micrograph of human chromosomes. These chromosomes are highly compacted and found in a dividing cell.

Chapter Outline

14.1 The Eukaryotic Cell Cycle
14.2 Mitotic Cell Division
14.3 Meiosis and Sexual Reproduction
14.4 Variation in Chromosome Structure and Number

Assessing Your Knowledge and Skills

Over 10,000,000,000,000! Researchers estimate the adult human body contains somewhere between 10 trillion and 50 trillion cells. It is almost an incomprehensible number. Even more amazing is the accuracy of the process that produces these cells. After a human sperm and egg unite, the fertilized egg goes through a long series of cell divisions to produce an adult with over 10 trillion cells. Let's suppose you randomly removed a cell from your arm and compared it with a cell from your foot. If you examined the chromosomes in both cells under the microscope, they would look identical. The DNA sequences along those chromosomes would also be the same, barring rare mutations. Similar comparisons could be made among the trillions of cells in your body. When you consider how many cell divisions are needed to produce an adult human, the precision of cell division is truly remarkable.

What accounts for this high level of accuracy? As we will examine in this chapter, **cell division**, the reproduction of cells, is a highly regulated process that distributes and monitors the integrity of the genetic material. The eukaryotic cell cycle is a series of phases needed for cell division.

The cells of eukaryotic species follow one of two different sorting processes so that new daughter cells receive the correct number and types of chromosomes. The first sorting process we will explore, called mitosis, is needed so two daughter cells receive the same amount of genetic material as the mother cell that produced them. The second sorting process we will consider, called meiosis, is needed for sexual reproduction. In meiosis, cells that have two sets of chromosomes produce daughter cells with a single set of chromosomes. Finally, we will examine variation in the structure and number of chromosomes. As you will see, a variety of mechanisms that alter chromosome structure and number can have important consequences for the affected organisms.

14.1 The Eukaryotic Cell Cycle

Learning Outcomes

1. **SCISKILLS** ▶ Describe the features of chromosomes and how sets of chromosomes are examined microscopically.
2. List the phases of the eukaryotic cell cycle.
3. Explain how cyclins and cdks work together to advance a cell through the eukaryotic cell cycle.

Life is a continuum in which new living cells are formed by the division of pre-existing cells. The Latin axiom *omnis cellula e cellula*, meaning every cell originates from another cell, was first proposed in 1858 by Rudolf Virchow, a German biologist. From an evolutionary perspective, cell division has a very ancient origin. All living organisms, from unicellular bacteria to multicellular plants and animals, have been produced by a series of repeated rounds of cell growth and division extending back to the beginnings of life nearly 4 billion years ago.

A **cell cycle** is a series of events that results in cell division. In all species, it is a highly regulated process, to ensure that cell division occurs at the appropriate time. As discussed in Chapter 17, most bacterial cells reproduce via binary fission, which is relatively simple (look ahead to Figure 17.8). The cell cycle in eukaryotes is more complex, in part, because eukaryotic cells have sets of chromosomes that need to be sorted properly. In this section, we will examine the phases of the eukaryotic cell cycle and see how the cell cycle is controlled by proteins that carefully monitor the division process to ensure its accuracy. But first we need to consider some general features of chromosomes in eukaryotic species.

Chromosomes Are Inherited in Sets and Occur in Homologous Pairs

To understand the chromosomal composition of cells and the behavior of chromosomes during cell division, scientists use microscopes to observe cells and chromosomes. **Cytogenetics** is the field of genetics that involves the microscopic examination of chromosomes. When a cell prepares to divide, the chromosomes become tightly compacted, a process that decreases their apparent length and increases their diameter. A consequence of this compaction is that distinctive shapes and numbers of chromosomes become visible under a light microscope.

Microscopic Examination of Chromosomes Figure 14.1 shows the general procedure for preparing and viewing chromosomes from a eukaryotic cell. In this example, the cells were obtained from a sample of human blood. Specifically, the chromosomes within leukocytes (white blood cells) were examined.

A photographic representation of the chromosomes, such as the photo shown in step 5 of Figure 14.1, is called a **karyotype**. A karyotype reveals the number, size, and form of chromosomes found within an actively dividing cell. The chromosomes viewed in actively dividing cells have already replicated. The two copies are still joined to each other and are referred to as a pair of **sister chromatids** (see inset to step 5 of Figure 14.1).

Sets of Chromosomes What type of information is learned from a karyotype? By studying the karyotypes of many species, scientists have discovered that eukaryotic chromosomes occur in sets. Each set is composed of several different types of chromosomes. For example, one set of human chromosomes contains 23 different types of chromosomes (see step 5 of Figure 14.1).

By convention, the chromosomes are numbered according to size, with the largest chromosomes having the smallest numbers. For example, human chromosomes 1, 2, and 3 are relatively large, whereas 21 and 22 are the two smallest. This numbering system does not apply to the **sex chromosomes,** which determine the sex of the individual. Sex chromosomes in humans are designated with the letters X and Y; females are XX and males are XY. The chromosomes that are not sex chromosomes are called **autosomes.** Humans have 22 different autosomes.

A common feature of many eukaryotic species is that most cells contain two sets of chromosomes. The karyotype shown in Figure 14.1 contains two sets of chromosomes, with 23 different chromosomes in each set. Therefore, this human cell contains a total of 46 chromosomes. Each cell has two sets because the individual inherited one set from the father and one set from the mother. When the cells of an organism carry two sets of chromosomes, that organism is said to be **diploid.** Geneticists use the letter n to represent a set of chromosomes. Diploid organisms are referred to as $2n$, because they have two sets of chromosomes. For example, humans are $2n$, where $n = 23$. Most human cells are diploid. An exception is the **gametes,** the sperm and egg cells. Gametes are **haploid,** or $1n$, which means they contain one set of chromosomes.

Homologous Pairs of Chromosomes When an organism is diploid, the members of a pair of chromosomes are called **homologs** (see inset to step 5 of Figure 14.1). The term **homology** refers to any similarity that is due to common ancestry. Pairs of homologous chromosomes are evolutionarily derived from the same chromosome. However, homologous chromosomes are not usually identical because over many generations they have accumulated some genetic changes that make them distinct.

How similar are homologous chromosomes to each other? The two chromosomes in a homologous pair are nearly identical in size and contain a very similar composition of genetic material. A particular gene found on one copy of a chromosome is also found on the homolog. However, the two homologs may vary in the way that a gene affects an organism's traits. As an example, let's consider a gene in humans called *Herc2*, which plays a major role in determining eye color. The *Herc2* gene is found on chromosome 15. One copy of chromosome 15 might carry the form of this eye color gene that confers brown eyes, whereas the gene on the homolog might confer blue eyes.

The DNA sequences on homologous chromosomes are very similar. In most cases, the sequence of bases on one homolog differs by less than 1% from the sequence on the other homolog. For example, the DNA sequence of chromosome 1 that you inherited from your mother is likely to be more than 99% identical to the DNA sequence of chromosome 1 that you inherited from your father. Nevertheless, keep in mind that the sequences are not identical. The slight differences in DNA sequence provide important variation in gene function. Again, if we use the eye color gene *Herc2* as an example, a minor

Figure 14.1 The procedure for making a karyotype. In this example, the chromosomes were treated with Giemsa stain, and the resulting dark bands are called G bands. The banding patterns make it much easier to distinguish the different chromosomes from each other.
(4) ©Véronique Burger/Science Source; (5) Courtesy of the Genomic Centre for Cancer Research and Diagnosis, CancerCare Manitoba, University of Manitoba, Winnipeg, Manitoba, Canada

Concept Check: *Researchers usually treat cells with drugs that stimulate them to divide before beginning the procedure for making a karyotype. Why is this treatment useful?*

difference in DNA sequence distinguishes two forms of the gene, brown versus blue.

The striking similarity between homologous chromosomes does not apply to the sex chromosomes (for example, X and Y in humans). These chromosomes differ in size and genetic composition. Certain genes found on the X chromosome are not found on the Y chromosome, and vice versa. The X and Y chromosomes are not considered homologous chromosomes, although they do have short regions of homology.

The Cell Cycle Is a Series of Phases That Lead to Cell Division

Eukaryotic cells that are destined to divide advance through the cell cycle, a series of changes that involves growth, replication, and division and ultimately produces new cells. **Figure 14.2** provides an overview of the cell cycle. In this diagram, the mother cell has three pairs of chromosomes, for a total of six individual chromosomes. This cell is diploid ($2n$) and contains three chromosomes per set ($n = 3$). The paternal set is shown in blue, and the homologous maternal set is shown in red.

The phases of the cell cycle are

- **G_1 phase** (first gap),
- **S phase** (synthesis of DNA, the genetic material),
- **G_2 phase** (second gap), and
- **M phase** (mitosis and cytokinesis).

The G_1 and G_2 phases were originally described as gap phases to indicate the periods between DNA synthesis and mitosis. In actively

Biology Principle

Cells Are the Simplest Units of Life

The cells of eukaryotic species are made via cell division during the eukaryotic cell cycle.

1. Prior to cell division, a mother cell in G_1 has 6 chromosomes, 2 sets of 3 chromosomes.
2. During S phase, chromosome replication produces 6 pairs of sister chromatids.
3. During G_2, the cell prepares to divide.
4. At the beginning of mitosis, the nucleus breaks apart, and replicated chromosomes condense.
5. During mitosis, sister chromatids separate into two nuclei, and then 2 cells are formed during cytokinesis.

Two daughter cells form, each containing 6 chromosomes.

Figure 14.2 **The eukaryotic cell cycle.** Dividing cells advance through a series of phases denoted G_1, S, G_2, and M. This diagram shows the progression of a cell through the cell cycle to produce two daughter cells. The original diploid cell had three pairs of chromosomes, for a total of six individual chromosomes. During S phase, these replicate to yield 12 chromatids. After mitosis is completed, the two daughter cells each contain six individual chromosomes. The width of the phases shown in this figure is not meant to reflect their actual length. G_1 is typically the longest phase of the cell cycle, whereas M phase is relatively short.

Concept Check: *Which phases make up interphase?*

---

dividing cells, the G_1, S, and G_2 phases are collectively known as **interphase**. During interphase, the cell grows and copies its chromosomes in preparation for cell division. Alternatively, cells may exit the cell cycle and remain for long periods of time in a phase called **G_0** (G zero). The G_0 phase is an alternative to proceeding through G_1. A cell in the G_0 phase has postponed division or, in the case of terminally differentiated cells (such as muscle cells in an adult animal), will never divide again. G_0 is a nondividing phase.

G_1 Phase The G_1 phase is a period when a cell may become committed to divide. Depending on the environmental conditions and the presence of signaling molecules, a cell in the G_1 phase may accumulate molecular changes that cause it to advance through the rest of the cell cycle. Cell growth typically occurs during the G_1 phase.

S Phase During the S phase, each chromosome is replicated to form a pair of sister chromatids (see Figure 14.1). When S phase is completed, a cell has twice as many chromatids as the number of chromosomes in the G_1 phase. For example, a human cell in G_1 phase has 46 distinct chromosomes, whereas the same cell in G_2 phase will have 46 pairs of sister chromatids, for a total of 92 chromatids.

G_2 Phase During the G_2 phase, a cell synthesizes the proteins necessary for chromosome sorting and cell division. Some cell growth may occur.

M Phase The first part of M phase is **mitosis,** in which one cell nucleus divides into two nuclei, distributing the duplicated chromosomes so that each daughter cell receives the same complement of chromosomes. As noted previously, a human cell in G_2 phase has 92 chromatids, which are found in 46 pairs. During mitosis, these pairs of chromatids are separated and sorted so that each daughter cell receives 46 chromosomes. In most cases, mitosis is followed by **cytokinesis,** which is the division of the cytoplasm to produce two distinct daughter cells.

Cell Cycle Length The length of the cell cycle varies considerably among different cell types, ranging from several minutes in quickly growing embryos to several months in slow-growing adult cells. For fast-dividing mammalian cells in adults, such as skin cells, the length of the cycle is often in the range of 10 to 24 hours. The various phases within the cell cycle also vary in length. G_1 is often the longest and the most variable phase, and M is usually the shortest. For a cell that divides in 24 hours, the following lengths of time for each phase are typical:

- G_1 phase: 11 hours
- S phase: 8 hours
- G_2 phase: 4 hours
- M phase: 1 hour

What factors determine whether or not a cell will divide? First, cell division is controlled by external factors, such as environmental conditions and signaling molecules. The effects of growth factors on cell division are discussed in Chapter 8 (refer back to Figure 8.10). Second, internal factors affect cell division. These include cell cycle control molecules and checkpoints, as discussed next.

Cyclins and Cyclin-Dependent Kinases Advance a Cell through the Cell Cycle

In eukaryotes, the cell cycle is a highly regulated process to ensure that cells divide at the appropriate time. Different categories of proteins are needed so that cell division occurs properly.

Cyclins and Cyclin-Dependent Kinases Proteins called **cyclins** and **cyclin-dependent kinases (cdks)** are responsible for advancing a cell through the phases of the cell cycle. Cyclins are so named because their amounts are cyclic—their amounts rise and fall during the cell cycle. To be active, the cyclin-dependent kinases controlling the cell cycle must bind to (are dependent on) a cyclin. The numbers of different types of cyclins and cdks vary from species to species.

Figure 14.3 provides a simplified description of how cyclins and cdks work together to advance a cell through the cell cycle.

1. During G_1, the amount of a cyclin termed G_1 cyclin increases in response to sufficient nutrients and growth factors.
2. G_1 cyclin binds to a cdk to form an activated G_1 cyclin/cdk complex. Once activated, cdk functions as a protein kinase that phosphorylates other proteins needed to advance the cell to the next phase in the cell cycle. For example, certain proteins involved with DNA synthesis are phosphorylated and activated, thereby allowing the cell to replicate its DNA in S phase.
3. After the cell passes into the S phase, G_1 cyclin is degraded.
4. A different cyclin, called mitotic cyclin, accumulates late in G_2. It binds to a cdk to form an activated mitotic cyclin/cdk complex. This complex phosphorylates proteins that are needed to advance the cell into M phase. After M phase is completed, mitotic cyclin is degraded.

Checkpoints Advancement through the cell cycle is a process that is highly regulated to ensure that the chromosomes are intact and that the conditions are appropriate for a cell to divide. Three critical regulatory points called **checkpoints** occur in the cell cycle of eukaryotic cells (see Figure 14.3). At these checkpoints, a variety of proteins, referred to as checkpoint proteins, control whether a cell will advance past a given checkpoint. Some checkpoint proteins function as sensors to determine if a cell is in the proper condition to divide. If these proteins sense that conditions are unfavorable, other checkpoint proteins will be activated that will prevent the cell from advancing through the cell cycle. These other checkpoint proteins often exert their effects by inhibiting cyclin-dependent kinases.

At the G_1 checkpoint, the checkpoint proteins determine if conditions are favorable for cell division. In addition, G_1-checkpoint proteins can sense if the DNA has incurred damage. What happens if DNA damage is detected? The checkpoint proteins prevent the formation of active cyclin/cdk complexes, thereby stopping the advancement of the cell cycle.

A second checkpoint exists in G_2. At this checkpoint, proteins also check for DNA damage and ensure that all of the DNA has been replicated. In addition, the G_2 checkpoint proteins monitor the levels of the proteins that are needed to advance through M phase.

A third checkpoint, called the metaphase checkpoint, is controlled by proteins that monitor the integrity of the spindle apparatus, which is involved in chromosome sorting and described later in this chapter. Metaphase is a step in mitosis during which all of the chromosomes should be attached to the spindle apparatus. If a chromosome is not correctly attached, the metaphase checkpoint proteins will stop the cell cycle. This checkpoint prevents cells from incorrectly sorting their chromosomes during division.

p53 and Rb Proteins Checkpoint proteins delay the cell cycle until problems are fixed or prevent cell division when problems cannot be fixed. Researchers have identified many different checkpoint proteins. We will consider two examples, p53 and Rb.

Figure 14.3 Checkpoints in the cell cycle. Advancement through the cell cycle requires the formation of activated cyclin/cdk complexes. Cells make different types of cyclin proteins, which are typically degraded after the cell has progressed to the next phase. The formation of activated cyclin/cdk complexes is regulated by checkpoint proteins.

 BioConnections: *Refer back to Figure 13.13. How do checkpoint proteins prevent cancer?*

p53 is so named because it is a protein with a molecular mass of 53 kDa (kilodaltons). The expression of the *p53* gene is induced when DNA is damaged. The p53 protein functions as a regulatory transcription factor that activates several different genes, leading to the synthesis of proteins that stop the cell cycle and other proteins that repair the DNA. p53 controls the ability of cells to advance past the G_1 checkpoint and the metaphase checkpoint. If the DNA is eventually repaired, a cell may later proceed through the cell cycle.

In some cases, DNA damage may be so severe that it cannot be properly repaired. When that happens, the p53 protein will activate genes that promote programmed cell death. This process, called **apoptosis,** involves cell shrinkage and DNA degradation. Enzymes known as caspases are activated during apoptosis. Caspases digest selected cellular proteins such as actin filaments, which are components of the cytoskeleton. The destruction of these microfilaments causes the cell to break into small vesicles that are eventually phagocytized by cells of the immune system. Why is apoptosis beneficial? A cell with DNA damage may function abnormally and it may even become a cancer cell. Therefore, apoptosis provides a way to eliminate such cells from the body.

Let's now turn our attention to the Rb protein, which was first identified in the study of patients with a particular type of cancer called retinoblastoma (refer back to Chapter 13). This checkpoint protein is a negative regulator of cell division. Its activity is controlled by various factors such as cell size, nutrient availability, and growth factors. When conditions are unfavorable for cell division, the Rb protein negatively controls a regulatory transcription factor called E2F that activates genes required for cell cycle advancement from G_1 to S phase. The binding of the Rb protein to E2F inhibits its activity and prevents cell division (**Figure 14.4**).

However, when conditions are favorable for cell division, cyclins bind to cyclin-dependent kinases (cdks). This binding activates the kinases, which catalyze the transfer of a phosphate to the Rb protein (see step 2 in Figure 14.4). The phosphorylated form of the Rb protein is released from E2F, thereby allowing E2F to become active. E2F primarily activates genes that are needed for DNA replication.

14.1 Reviewing the Concepts

- A micrograph that shows the alignment of chromosomes from a given cell is called a karyotype. Eukaryotic chromosomes are inherited in sets. A diploid cell has two sets of chromosomes. The members of each pair are called homologs (Figure 14.1).
- The eukaryotic cell cycle consists of four phases called the G_1 (first gap), S (synthesis of DNA), G_2 (second gap), and M phases (mitosis and cytokinesis). The G_1, S, and G_2 phases are collectively known as interphase (Figure 14.2).
- An interaction between cyclins and cyclin-dependent kinases (cdks) is necessary for cells to advance through the cell cycle. Checkpoint proteins, such as p53 and Rb, sense the integrity of the genome and the environmental conditions, respectively, and control whether or not the cell advances through the cell cycle (Figures 14.3, 14.4).

Figure 14.4 Function of the Rb protein. The Rb protein inhibits the function of E2F, which turns on genes that cause a cell to divide.

> **Concept Check:** *Would cancer occur if both copies of the Rb gene and both copies of the E2F gene were rendered inactive due to mutations?*

14.1 Testing Your Knowledge

1. A cell that is diploid has
   a. two chromosomes.
   b. pairs of homologous chromosomes.
   c. two sets of chromosomes.
   d. both b and c.

2. In eukaryotes, DNA replication produces sister chromatids. This event occurs during
   a. G_1 phase.
   b. S phase.
   c. G_2 phase.
   d. M phase.

3. A protein that drives a cell through the cell cycle is a _____. It is directly controlled by a _____.
   a. cyclin, cyclin-dependent kinase
   b. checkpoint protein, cyclin
   c. checkpoint protein, cyclin-dependent kinase
   d. cyclin-dependent kinase, cyclin

14.2 Mitotic Cell Division

Learning Outcomes
1. Explain how the replication of eukaryotic chromosomes produces sister chromatids.
2. Describe the structure and function of the mitotic spindle.
3. Outline the key events that occur during the phases of mitosis.

We now turn our attention to the process of **mitotic cell division,** during which a cell divides to produce two new cells (the daughter cells) that are genetically identical to the original cell (the mother cell). Mitotic cell division begins with mitosis—the division of one nucleus into two nuclei—and is followed by cytokinesis—in which the mother cell divides into two daughter cells.

Why is mitotic cell division important? One reason is **asexual reproduction,** a process in which genetically identical offspring are produced from a single parent. Certain unicellular eukaryotic organisms, such as baker's yeast (*Saccharomyces cerevisiae*) and the amoeba, increase their numbers in this manner. A second important reason for mitotic cell division is multicellularity. Organisms such as plants, animals, and most fungi are derived from a single cell that subsequently undergoes repeated cell divisions to become a multicellular organism. As discussed at the beginning of the chapter, adult humans are composed of trillions of cells.

In this section, we will explore how the process of mitotic cell division requires the replication, organization, and sorting of chromosomes. We will also examine how a single cell is separated into two distinct cells by cytokinesis.

In Preparation for Cell Division, Eukaryotic Chromosomes Are Replicated and Compacted to Produce Pairs Called Sister Chromatids

We will now consider how chromosomes are replicated and sorted during cell division. In Chapter 9, we examined the molecular process of DNA replication. **Figure 14.5** describes the process at the chromosomal level. Prior to DNA replication, the DNA of each eukaryotic chromosome consists of a linear DNA double helix that is found in the nucleus and is not highly compacted. When the DNA is replicated, two identical copies of the original double helix are produced. As discussed earlier, these copies, along with associated proteins, lie side-by-side and are termed sister chromatids. When a cell prepares to divide, the sister chromatids become highly compacted and readily visible under the microscope. As shown in Figure 14.5b, the two sister chromatids are tightly associated at a region called the **centromere.** In addition, the centromere serves as an attachment site for a group of proteins that form the **kinetochore,** a structure necessary for sorting the chromosomes.

The Mitotic Spindle Organizes and Sorts Chromosomes During Cell Division

What structure organizes and sorts the chromosomes during cell division? The answer is the **mitotic spindle** (**Figure 14.6**). It is composed of microtubules—long, cylindrical protein fibers that are components

Biology Principle

The Genetic Material Provides a Blueprint That Allows Organisms to Grow, Develop, and Reproduce
The process of mitosis ensures that each daughter cell receives a complete copy of the genetic material.

(a) Chromosome replication and compaction

(b) Schematic drawing of a metaphase chromosome

Figure 14.5 Replication and compaction of chromosomes into pairs of sister chromatids. (a) Chromosomal replication produces a pair of sister chromatids. While the chromosomes are elongated, they are replicated during S phase to produce two copies that are connected and lie parallel to each other. This is a pair of sister chromatids. At the start of mitosis, the sister chromatids condense into more compact structures that are easily seen with a light microscope. (b) A schematic drawing of a metaphase chromosome. This structure has two chromatids that lie side-by-side. The two chromatids are held together by cohesin proteins (not shown in this drawing). The kinetochore is a group of proteins that are attached to the centromere and play a key role during chromosome sorting.

✓ **Concept Check:** *Refer back to the karyotype in Figure 14.1. In this micrograph, is each of the 46 objects a pair of sister chromatids?*

---

of the cytoskeleton (refer back to Table 4.1). In animal cells, microtubule growth and organization start at two **centrosomes,** structures that are also referred to as microtubule-organizing centers. A single centrosome duplicates during interphase. After the two centrosomes separate from each other during mitosis, each centrosome defines a **pole** of the spindle apparatus, one within each of the future daughter cells. The centrosome in animal cells has a pair of **centrioles,** which are arranged perpendicular to each other. However, centrioles are not found in many other eukaryotic species, such as most plants, and are not required for spindle formation.

Each centrosome organizes the construction of the microtubules by rapidly polymerizing tubulin proteins. The three types of spindle microtubules are termed astral, polar, and kinetochore microtubules (). The astral microtubules, which are not attached to chromosomes, are important for positioning the spindle apparatus within the cell. The polar microtubules project into the region between the two poles. Polar microtubules that overlap with each other play a role in the separation of the two poles. Finally, the kinetochore microtubules are attached to kinetochores, which are bound to the centromere of each chromosome.

The Transmission of Chromosomes in Eukaryotes Requires a Sorting Process Known as Mitosis

During mitosis in eukaryotes, the duplicated chromosomes are distributed so that each daughter cell receives one copy of each chromosome. **Figure 14.7** depicts the process of mitosis in an animal cell, though the process is quite similar in a plant cell. Mitosis occurs as a continuum of phases known as prophase, prometaphase, metaphase, anaphase, and telophase. In the simplified diagrams along the bottom of Figure 14.7, the original cell has six chromosomes. One set of chromosomes is depicted in red, and the homologous set is blue. The different colors are intended to distinguish between maternal and paternal chromosomes, respectively.

Interphase Prior to mitosis, the cells are in interphase, which consists of the G_1, S, and G_2 phases of the cell cycle. The chromosomes, which are located in the cell nucleus, have replicated in S phase and are not condensed (Figure 14.7a). The nucleolus, which is the site where the components of ribosomes assemble into ribosomal subunits, is visible during interphase.

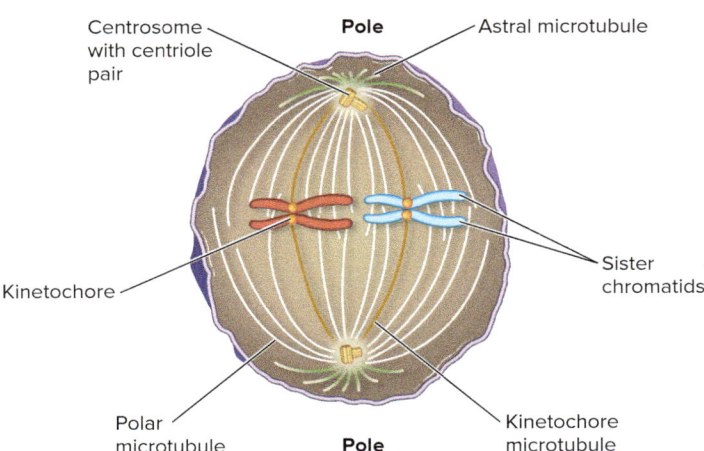

Figure 14.6 **The structure of the mitotic spindle.** The mitotic spindle in animal cells is formed by the centrosomes, which produce three types of microtubules. The astral microtubules emanate away from the centrosome and are not attached to chromosomes. The polar microtubules project into the region between the two poles. The kinetochore microtubules are attached to the kinetochores of sister chromatids. Note: For simplicity, this diagram shows only one pair of homologous chromosomes. Eukaryotic species typically have multiple chromosomes per set.

Concept Check: What are the functions of the three types of microtubules?

Prophase At the start of mitosis, in **prophase,** the chromosomes have already replicated to produce 12 chromatids, joined as 6 pairs of sister chromatids that have condensed into highly compacted structures readily visible by light microscopy (Figure 14.7b). As prophase proceeds, the nuclear envelope begins to dissociate into small vesicles. The nucleolus is no longer visible.

Prometaphase During **prometaphase,** the nuclear envelope completely fragments into small vesicles, and the mitotic spindle is fully formed (Figure 14.7c). As prometaphase advances, the centrosomes move apart and demarcate the two poles. Once the nuclear envelope has dissociated, the spindle fibers can interact with the sister chromatids. How do the sister chromatids become attached to the spindle apparatus? Initially, microtubules are rapidly formed and can be seen under a microscope growing out from the two poles. As it grows, if a microtubule happens to make contact with a kinetochore, it is said to be captured and remains firmly attached to the kinetochore. Alternatively, if a microtubule does not collide with a kinetochore, the microtubule eventually depolymerizes and retracts to the centrosome. This random process is how sister chromatids become attached to kinetochore microtubules. As the end of prometaphase nears, the two kinetochores on each pair of sister chromatids are attached to kinetochore microtubules from opposite poles. As these events are occurring, the sister chromatids are seen under the microscope to undergo jerky movements as they are tugged, back and forth, between the two poles by the kinetochore microtubules.

Metaphase Eventually, the pairs of sister chromatids are aligned in a single row along the **metaphase plate,** a plane halfway between the poles of the spindle. When this alignment is complete, the cell is in **metaphase** of mitosis (Figure 14.7d).

Anaphase During **anaphase,** the connections between the pairs of sister chromatids are broken (Figure 14.7e). Each chromatid, now an individual chromosome, is linked to only one of the two poles by one or more kinetochore microtubules. As anaphase proceeds, the kinetochore microtubules shorten, pulling the chromosomes toward the pole to which they are attached. In addition, the two poles move farther away from each other. This occurs because the overlapping polar microtubules lengthen and push against each other, thereby pushing the poles farther apart.

Telophase During **telophase,** the chromosomes have reached their respective poles and decondense. The nuclear envelope now re-forms to produce two separate nuclei. In Figure 14.7f, two nuclei that contain six chromosomes each are being produced.

Cytokinesis In most cases, mitosis is quickly followed by cytokinesis, in which the two nuclei are segregated into separate daughter cells. Whereas the phases of mitosis are similar between plant and animal cells, the process of cytokinesis is quite different. In animal cells, cytokinesis involves the formation of a **cleavage furrow,** which constricts like a drawstring to separate the cells (**Figure 14.8a**). In plants, vesicles from the Golgi apparatus move along microtubules to the center of the cell and coalesce to form a **cell plate** (**Figure 14.8b**), which then forms a cell wall between the two daughter cells.

Cytoskeletal proteins, which are discussed in Chapter 4, play key roles in the process of cytokinesis. Microtubules are important for the proper positioning of the cleavage plane in animal cells and in the formation of a cell plate in plants. In animals, actin is involved in the formation of the cleavage furrow.

What are the results of mitosis and cytokinesis? These processes ultimately produce two daughter cells with the same number of chromosomes as the mother cell. Barring rare mutations, the two daughter cells are genetically identical to each other and to the mother cell from which they were derived. The critical consequence of this sorting process is to ensure genetic consistency from one cell to the next. The development of multicellularity relies on the repeated process of mitosis and cytokinesis.

EVOLUTIONARY CONNECTIONS

Cell Division in Bacteria Involves FtsZ, a Protein Related to Eukaryotic Tubulin

As discussed in Chapter 17 (look ahead to Figure 17.8), bacteria divide by a process called binary fission. Because bacteria usually have only one type of chromosome, the process of sorting different types of chromosomes is not necessary. Even so, events during bacterial cell division may provide insights into the manner in which mitosis evolved in eukaryotes.

Biology Principle

Biologists Use Models to Test Predictions and Convey Ideas

This figure is a model for the mechanism of mitotic cell division. The Modeling Challenge in the figure legend asks you to propose a model for the outcome of mitosis if one of the pairs of chromosomes fails to disjoin properly.

Figure 14.7 **The process of mitosis in an animal cell.** The top panels are fluorescence micrographs of a newt cell advancing through mitosis. The bottom panels are schematic drawings that emphasize the sorting and separation of the chromosomes in a diploid mother cell with six chromosomes (three in each set). At the start of mitosis, these have already replicated into 12 chromatids. The final result is two daughter cells, each containing six chromosomes.

Photographs by Dr. Conly L. Rieder, Wadsworth Center, Albany, New York 12201-0509

Modeling Challenge: In Figure 14.7, the starting cell in mitosis has three homologous pairs of chromosomes that differ in length: short, medium, and long. Let's suppose that one of the long chromosomes (the blue one) fails to disjoin properly and both of the sister chromatids go into the same daughter cell. Draw a model showing the chromosomal composition of the two daughter cells at the end of mitosis. For each pair of homologs, draw one red and the other blue, as in the figure.

Prior to cell division, bacterial cells copy, or replicate, their chromosomal DNA. This produces two identical copies of the genetic material, as shown at the top of **Figure 14.9**. During binary fission, the two daughter cells become separated from each other by the formation of a septum. Recent evidence has shown that bacterial species produce a protein called FtsZ, which is important in cell division. FtsZ proteins assemble into a ring at the future site of the septum. FtsZ is thought to be the first protein to move to this division site, and it recruits other proteins that produce a new cell wall between the daughter cells.

FtsZ is evolutionarily related to the eukaryotic protein called tubulin, the main component of microtubules that are part of the mitotic spindle. In all eukaryotes, the midpoint of the mitotic spindle, which is called the metaphase plate, identifies the site for cytokinesis (see Figure 14.8). This observation indicates that tubulin is also critical for cytokinesis in eukaryotic cells.

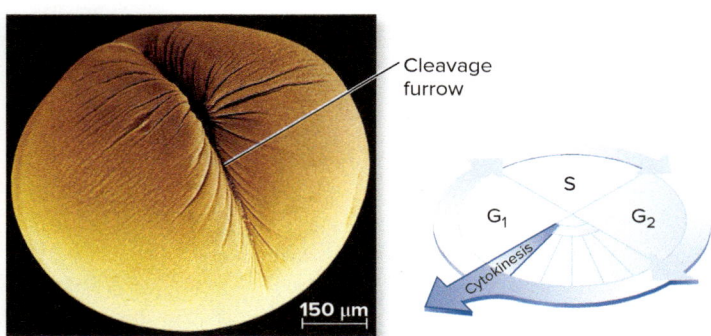

(a) Cleavage of an animal cell

(b) Formation of a cell plate in a plant cell

Figure 14.8 Micrographs showing cytokinesis in animal and plant cells.
(a) ©Don W. Fawcett/Science Source; (b) ©Carolina Biological Supply Company/Medical Images

Concept Check: What are the similarities and differences between animal and plant cells with regard to cytokinesis?

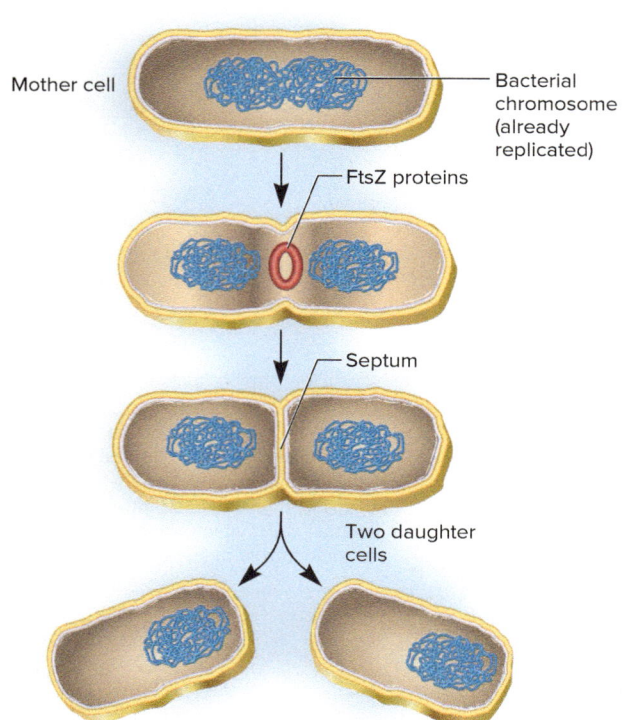

Figure 14.9 Binary fission in bacteria. Several FtsZ proteins form a ring at the future site of cell division.

14.2 Reviewing the Concepts

- In the process of mitotic cell division, a cell divides to produce two new cells (the daughter cells) that are genetically identical to the original cell.
- During S phase, eukaryotic chromosomes are replicated to produce a pair of identical sister chromatids that remain attached to each other (Figure 14.5).
- The mitotic spindle is a network of microtubules that plays a central role in chromosome sorting during cell division (Figure 14.6).
- During mitosis, the nucleus of a mother cell divides into two genetically identical nuclei. Mitosis occurs in five phases called prophase, prometaphase, metaphase, anaphase, and telophase (Figure 14.7).
- Cytokinesis occurs by a cleavage furrow in animal cells and by the formation of a cell plate in plant cells (Figure 14.8).
- FtsZ in bacteria and tubulin in eukaryotes are homologous proteins that are important in identifying the site where cell division will occur (Figure 14.9).

14.2 Testing Your Knowledge

1. Let's suppose a cell is diploid, with eight chromosomes per set. How many chromatids would be found in the cell after S phase but prior to cytokinesis?
   a. 8
   b. 16
   c. 32
   d. 64

2. Sister chromatids become attached to spindle fibers during
   a. prophase.
   b. prometaphase.
   c. metaphase.
   d. anaphase.
   e. telophase.

3. Sister chromatids separate from each other during
   a. prophase.
   b. prometaphase.
   c. metaphase.
   d. anaphase.
   e. telophase.

14.3 Meiosis and Sexual Reproduction

Learning Outcomes

1. Describe the processes of synapsis and crossing over.
2. Outline the key events that occur during the phases of meiosis.
3. Compare and contrast mitosis and meiosis, focusing on key steps that account for the different outcomes of these two processes.
4. Distinguish among the life cycles of diploid-dominant species, haploid-dominant species, and species that exhibit an alternation of generations.

We now turn our attention to **sexual reproduction**, a process in which two haploid gametes unite to form a diploid cell called a **zygote**. For multicellular species such as animals and plants, the zygote then grows and divides by mitotic cell divisions into a multicellular organism with many diploid cells.

As discussed earlier, a diploid cell contains two homologous sets of chromosomes, whereas a haploid cell contains a single set. For example, a diploid human cell contains 46 chromosomes, but a human gamete—sperm or egg cell—is a haploid cell that contains only 23 chromosomes. **Meiosis** is the process by which haploid cells are produced from a cell that was originally diploid. The term meiosis, which means to make smaller, refers to the fewer chromosomes found in cells that have undergone this process. For haploid cells to be produced, the chromosomes must be correctly sorted and distributed in a way that reduces the chromosome number to half its original diploid value. In the case of human gametes, for example, each gamete must receive 1 chromosome from each of the 23 pairs. For this to happen, two rounds of division are necessary, termed meiosis I and meiosis II (**Figure 14.10**). When a cell begins meiosis, it has chromosomes that are found in homologous pairs. When meiosis is completed, a single diploid cell with homologous pairs of chromosomes has produced four haploid cells.

In this section, we will examine the cellular events of meiosis that reduce the chromosome number from diploid to haploid. In addition, we will briefly consider how this process plays a role in the life cycles of animals, plants, fungi, and protists.

Meiosis I Separates Homologous Chromosomes

Like mitosis, meiosis begins after a cell has advanced through the G_1, S, and G_2 phases of the cell cycle. **Figure 14.11** presents simplified diagrams of a diploid cell ($2n$) with three chromosomes per set for a total of six chromosomes (as in the diagram of mitosis in Figure 14.7). Prior to meiosis, the chromosomes are replicated in S phase to produce pairs of sister chromatids. This single replication event is then followed by sequential divisions called meiosis I and II. Like mitosis, each of these processes is a continuous series of stages called prophase, prometaphase, metaphase, anaphase, and telophase. The sorting that occurs during **meiosis I** separates homologous chromosomes from each other (Figure 14.11a–e).

Prophase I During prophase I, homologous pairs of sister chromatids associate with each other, lying side by side to form a **bivalent**

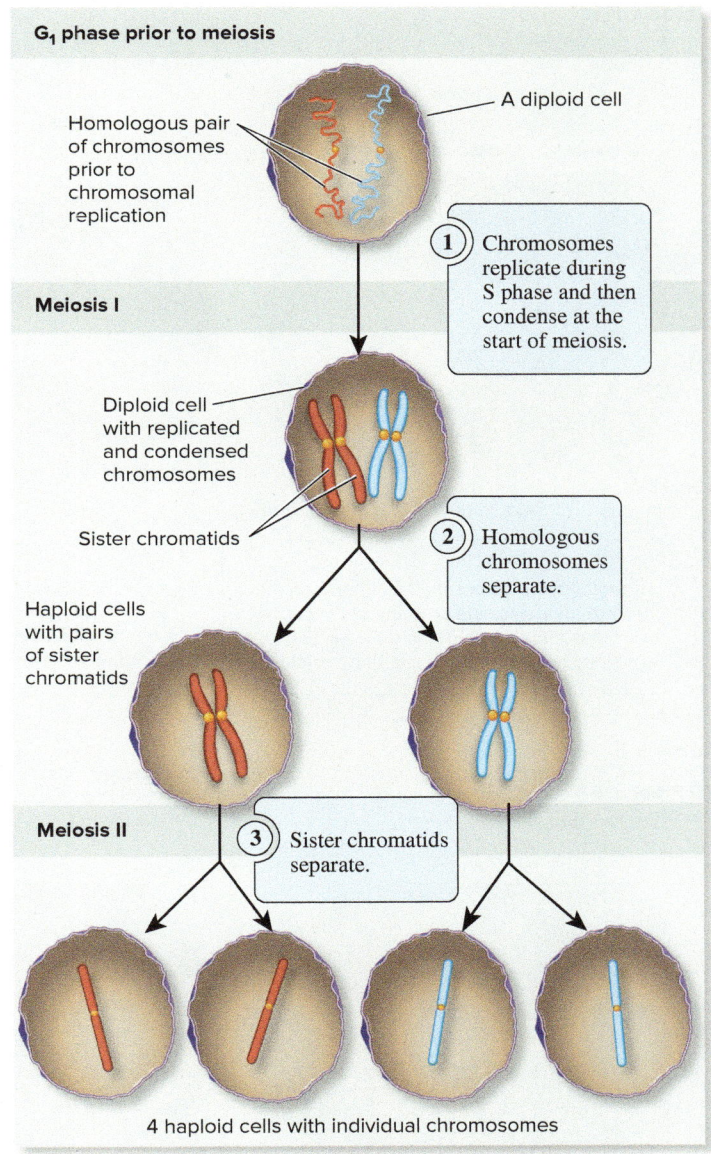

Figure 14.10 How the process of meiosis reduces chromosome number. This simplified diagram emphasizes the reduction in chromosome number as a diploid cell divides by meiosis to produce four haploid cells.

(Figure 14.11a). The process of forming a bivalent is termed **synapsis**. After synapsis, crossing over occurs, a process in which segments of chromosomes are exchanged with each other. In Figure 14.11a, see how certain red and blue segments are exchanged at a crossover site. As discussed in Chapter 16, crossing over increases the genetic variation of sexually reproducing species. During prophase I, the nuclear envelope starts to fragment into small vesicles.

Prometaphase I In prometaphase I, the nuclear envelope is completely broken down into vesicles, and the spindle apparatus is entirely formed. The sister chromatids become attached to kinetochore microtubules. However, a key difference occurs between mitosis and meiosis I. In mitosis, a pair of sister chromatids is attached to both poles via

Meiosis I

(a) Prophase I (b) Prometaphase I (c) Metaphase I

Meiosis II

(f) Prophase II (g) Prometaphase II (h) Metaphase II

1. Homologous chromosomes bind to each other to form bivalents, and crossing over occurs. Chromosomes condense, and the nuclear envelope begins to dissociate into vesicles.

2. The nuclear envelope completely dissociates into vesicles, and bivalents become attached to kinetochore microtubules.

3. Bivalents randomly align along the metaphase plate. Each pair of sister chromatids is attached to one pole.

6. Sister chromatids condense, and the spindle starts to form. The nuclear envelope begins to dissociate into vesicles.

7. The nuclear envelope completely dissociates into vesicles. Sister chromatids attach to the spindle via kinetochore microtubules.

8. Sister chromatids align along the metaphase plate. Each pair of sister chromatids is attached to both poles.

(d) Anaphase I

(e) Telophase I and cytokinesis

Cleavage furrow

④ Homologous chromosomes separate and move toward opposite poles as kinetic microtubules shorten. Polar microtubules lengthen and push the poles apart.

⑤ The chromosomes decondense, and the nuclear envelope re-forms. The 2 daughter cells are separated by a cleavage furrow.

(i) Anaphase II

(j) Telophase II and cytokinesis

Four haploid cells

⑨ Sister chromatids separate, and individual chromosomes move toward the poles as kinetochore microtubules shorten. Polar microtubules lengthen and push the poles apart.

⑩ Chromosomes decondense, and the nuclear envelope re-forms. Cleavage furrows separate the 2 cells into 4 cells.

Figure 14.11 The phases of meiosis in an animal cell.

 Concept Check: *Relative to the original cell, what is the end result of meiosis?*

kinetochore microtubules (see Figure 14.7c). In meiosis I, a pair of sister chromatids is attached to just one pole (Figure 14.11b).

Metaphase I   At metaphase I, the bivalents are organized along the metaphase plate. Notice how this pattern of alignment is strikingly different from that observed during mitosis (see Figure 14.7d). In particular, the sister chromatids are aligned in a double row rather than a single row (as in mitosis). Furthermore, the arrangement of sister chromatids within this double row is random with regard to the (red and blue) homologs. (Remember that these different colors indicate maternal and paternal chromosomes.) In Figure 14.11c, one of the red homologs is to the left of the metaphase plate and the other two are to the right, whereas two of the blue homologs are to the left of the metaphase plate and the other one is to the right. In other cells, homologs can be arranged differently along the metaphase plate (for example, three blues to the left and none to the right, or none to the left and three to the right).

Anaphase I   The segregation of homologs occurs during anaphase I (Figure 14.11d). The connections between bivalents break, but not the connections that hold sister chromatids together. Each joined pair of chromatids migrates to one pole, and the homologous pair of chromatids moves to the opposite pole, both pulled by kinetochore microtubules.

Telophase I   At telophase I (Figure 14.11e), the sister chromatids have reached their respective poles and then decondense. The nuclear envelope now re-forms to produce two separate nuclei.

If we consider the end result of meiosis I, we see that two nuclei are produced, each with half the number of sister chromatids as the starting cell in prophase; this is called a reduction division. The original diploid cell had two sets of chromosomes (three per set); its chromosomes were in homologous pairs. In contrast, the two cells produced as a result of meiosis I and cytokinesis are considered haploid—they do not have pairs of homologous chromosomes.

Meiosis II Separates Sister Chromatids

Meiosis I is followed by cytokinesis and then **meiosis II** (Figure 14.11f–j). DNA replication does not occur between meiosis I and meiosis II. The sorting events of meiosis II are similar to those of mitosis, but the starting point is different. For a diploid cell with six chromosomes, mitosis begins with 12 chromatids that are joined as six pairs of sister chromatids (see Figure 14.7). By comparison, the two cells that begin meiosis II each have six chromatids that are joined as three pairs of sister chromatids. Otherwise, the steps that occur during prophase, prometaphase, metaphase, anaphase, and telophase of meiosis II are analogous to those in a mitotic division. Sister chromatids are separated during anaphase II, unlike anaphase I, in which bivalents are separated.

Mitosis and Meiosis Differ in a Few Key Steps

How are the outcomes of mitosis and meiosis different from each other? Mitosis produces two diploid daughter cells that are genetically identical. In our example shown in Figure 14.7, the starting cell had six chromosomes (three homologous pairs of chromosomes), and both daughter cells received copies of the same six chromosomes. By comparison, meiosis reduces the number of sets of chromosomes. In the example shown in Figure 14.11, the starting cell also had six chromosomes, whereas the resulting four daughter cells have only three chromosomes. However, the daughter cells do not contain a random mix of three chromosomes. Each haploid daughter cell has one complete set of chromosomes, whereas the original diploid mother cell had two complete sets.

How do we explain the different outcomes of mitosis and meiosis? Table 14.1 emphasizes the differences between certain key steps in mitosis and meiosis that account for the different outcomes of these two processes:

- DNA replication occurs prior to mitosis and meiosis I but not between meiosis I and II.

- During prophase of meiosis I, the homologs bind to each other to form bivalents. This explains why crossing over occurs commonly during meiosis but rarely during mitosis.

- During prometaphase of mitosis and meiosis II, pairs of sister chromatids are attached to both poles. In contrast, during

Table 14.1   A Comparison of Mitosis, Meiosis I, and Meiosis II

Event	Mitosis	Meiosis I	Meiosis II
DNA replication:	Occurs prior to mitosis	Occurs prior to meiosis I	Does not occur between meiosis I and II
Synapsis during prophase:	No	Yes, bivalents are formed.	No
Crossing over during prophase:	Rarely	Commonly	Rarely
Attachment to poles at prometaphase:	A pair of sister chromatids is attached to kinetochore microtubules from both poles.	A pair of sister chromatids is attached to kinetochore microtubules from just one pole.	A pair of sister chromatids is attached to kinetochore microtubules from both poles.
Alignment along the metaphase plate:	Sister chromatids align.	Bivalents align.	Sister chromatids align.
Type of separation at anaphase:	Sister chromatids separate. A single chromatid, now called a chromosome, moves to each pole.	Homologous chromosomes separate. A pair of sister chromatids moves to each pole.	Sister chromatids separate. A single chromatid, now called a chromosome, moves to each pole.
End result when the mother cell is diploid:	Two daughter cells that are diploid	—	Four daughter cells that are haploid

meiosis I, each pair of sister chromatids (within a bivalent) is attached to a single pole.

- Bivalents align along the metaphase plate during metaphase of meiosis I, whereas sister chromatids align along the metaphase plate during metaphase of mitosis and meiosis II.
- At anaphase of meiosis I, the homologous chromosomes separate, but the sister chromatids remain together. In contrast, sister chromatid separation occurs during anaphase of mitosis and meiosis II.

Taken together, the steps of mitosis produce two diploid cells that are genetically identical, whereas the steps of meiosis involve two sequential cell divisions that produce four haploid cells that may not be genetically identical.

Biology Principle | Biology is a quantitative science

Quantitative Analysis: Meiosis Enhances Genetic Diversity

The random alignment of homologous chromosomes provides a mechanism to promote a vast amount of genetic diversity among the resulting haploid cells. Because eukaryotic species typically have many chromosomes per set, maternal and paternal homologs can be randomly aligned along the metaphase plate in a variety of ways. When meiosis is complete, it is very unlikely that any two human gametes will have the same combination of homologous chromosomes.

For any diploid species, the possible number of different, random alignments during metaphase I of meiosis equals 2^n, where n equals the number of chromosomes per set. The random alignments equal 2^n because each chromosome is found in a homologous pair and each member of the pair can align on either side of the metaphase plate. It is a matter of chance which daughter cell of meiosis I will get the maternal chromosome of a homologous pair and which will get the paternal chromosome. Because the homologs are genetically similar but not identical, the random alignment of homologous chromosomes provides a mechanism to promote a vast amount of genetic diversity among the resulting haploid cells.

Crunching the Numbers:
Humans have 23 chromosomes per set. How many possible random alignments could occur during metaphase I? How does crossing over further contribute to the genetic diversity of the resulting haploid cells?

Sexually Reproducing Species Produce Haploid and Diploid Cells at Different Times in Their Life Cycles

Let's now turn our attention to the relationship among mitosis, meiosis, and sexual reproduction in animals, plants, fungi, and protists. For any given species, the sequence of events that produces another generation of organisms is known as a **life cycle**. For sexually reproducing organisms, this usually involves an alternation between haploid cells or organisms and diploid cells or organisms (Figure 14.12).

Diploid-Dominant Species Most species of animals are diploid ($2n$), and their haploid gametes ($1n$) are considered to be a specialized type of cell. For this reason, animals are viewed as **diploid-dominant species** (Figure 14.12a). Certain diploid cells in the testes or ovaries undergo meiosis to produce haploid sperm or eggs, respectively. During fertilization, sperm and egg unite to form a diploid zygote, which then undergoes repeated mitotic cell divisions to produce a diploid multicellular organism.

Haploid-Dominant Species By comparison, most fungi and some protists are just the opposite; they are **haploid-dominant species** (Figure 14.12b). In fungi, the multicellular organism is haploid; only the zygote is diploid. During sexual reproduction, haploid cells unite to form a diploid zygote, which then immediately proceeds through meiosis to produce four haploid cells called spores. Each spore goes through mitotic cellular divisions to produce a haploid multicellular organism.

Alternation of Generations Plants and some algae have life cycles that are intermediate between diploid and haploid dominance. Such species exhibit an **alternation of generations** (Figure 14.12c). The species alternate between diploid multicellular organisms called **sporophytes** and haploid multicellular organisms called **gametophytes.** Meiosis in certain cells within the sporophyte produces haploid spores, which divide by mitosis to produce the gametophyte. Particular cells within the gametophyte differentiate into haploid gametes. Fertilization occurs between two gametes, producing a diploid zygote that then undergoes repeated mitotic cell divisions to produce a sporophyte.

Among different plant species, the relative sizes of the haploid and diploid organisms vary greatly. In mosses, the haploid gametophyte is a visible, multicellular organism, whereas the diploid sporophyte is smaller and remains attached to the haploid organism. In other plants, such as ferns (Figure 14.12c), both the diploid sporophyte and the haploid gametophyte can grow independently. The sporophyte is considerably larger and is the organism we commonly think of as a fern. In seed-bearing plants, such as roses and oak trees, the diploid sporophyte is the large multicellular plant, whereas the gametophyte is composed of only a few cells and is formed within the sporophyte.

When comparing animals, plants, and fungi, it's interesting to consider how gametes are made. Animals produce gametes by meiosis. In contrast, plants and fungi produce reproductive cells by mitosis. The gametophyte of plants is a haploid multicellular organism that is created by mitotic cellular divisions of a haploid spore. Within the multicellular gametophyte, certain cells become specialized as gametes.

14.3 Reviewing the Concepts

- The process of meiosis begins with a diploid cell and produces four haploid cells with one set of chromosomes each (Figure 14.10).

Figure 14.12 A comparison of three types of life cycles for sexually reproducing organisms.

Concept Check: *What is the main purpose of meiosis in animals? What is the main purpose of mitosis in animals?*

- Meiosis consists of two divisions—meiosis I and II—each composed of prophase, prometaphase, metaphase, anaphase, and telophase. During meiosis I, the homologs are separated into two different cells, and during meiosis II, the sister chromatids are separated into four different cells (Figure 14.11, Table 14.1).
- Animals are diploid-dominant species, whereas most fungi and some protists are haploid-dominant species. Plants alternate between diploid and haploid forms, which is called alternation of generations (Figure 14.12).

14.3 Testing Your Knowledge

1. In meiosis I, crossing over occurs during
   a. prophase.
   b. prometaphase.
   c. metaphase.
   d. anaphase.
   e. telophase.

2. The alignment of bivalents occurs during
   a. metaphase I.
   b. metaphase II.
   c. anaphase I.
   d. anaphase II.

3. A type of organism that exhibits alternation of generations is
   a. an animal.
   b. a fungus.
   c. a plant.
   d. Both fungi and plants alternate generations.

14.4 Variation in Chromosome Structure and Number

Learning Outcomes

1. Describe how chromosomes vary in size, centromere location, and number.
2. Identify the four ways that the structure of chromosomes can be changed.
3. Compare and contrast changes in the number of sets of chromosomes and changes in the number of individual chromosomes.
4. Give examples of how changes in chromosome number affect the characteristics of animals and plants.

In the previous sections of this chapter, we examined two important features of chromosomes. First we considered how chromosomes occur in sets, and second we explored two sorting processes that determine the number of sets of chromosomes following cell division. In this section, we will examine how the structures and numbers of chromosomes can vary between different species and within the same species.

Why is the study of chromosome variation important? First, geneticists have discovered that variations in chromosome structure and number can have major effects on the characteristics of an organism. We now know that several human genetic diseases are caused by such changes. In addition, changes in chromosome structure and number have been an important factor in the evolution of new species, which is a topic we will consider in Chapter 20.

Chromosome variation can be viewed in two ways. Among different species, the structure and number of chromosomes tend to vary greatly. There is also considerable variety in the size and shape of the chromosomes of a given species. On relatively rare occasions, however, the structure or number of chromosomes changes so that an individual is different from most other members of the same species. This is generally viewed as an abnormality. In this section, we will examine both normal and abnormal types of chromosome variation.

Natural Variation Exists in Chromosome Structure and Number

Before we begin to examine chromosome variation, we need to have a reference point for a normal set of chromosomes. To determine what the normal chromosomes of a species look like, a cytogeneticist microscopically examines the chromosomes from several members of the species. Chromosome composition within a given species tends to remain relatively constant. In most cases, individuals of the same species have the same number and types of chromosomes. For example, as mentioned previously, the usual chromosome composition of human cells is 46 chromosomes (two sets of 23 chromosomes). Other diploid species have different numbers of chromosomes. The dog has 78 chromosomes (two sets of 39), the fruit fly has 8 chromosomes (two sets of 4), and the tomato has 24 chromosomes (two sets of 12). When comparing distantly related species, such as humans and fruit flies, major differences in chromosome composition are observed.

The chromosomes of a given species also vary considerably in size and shape. Cytogeneticists have various ways to classify and identify chromosomes in their metaphase form. The three most commonly used features are size, location of the centromere, and banding patterns, which are revealed when the chromosomes are treated with stains. Based on centromere location, each metaphase chromosome is classified as (Figure 14.13)

- **metacentric** (centromere near the center),
- **submetacentric** (centromere off center),
- **acrocentric** (centromere near one end), or
- **telocentric** (centromere at the end).

Because the centromere is not exactly in the center of a chromosome, each chromosome has a short arm and a long arm. The short arm is designated with the letter p (for the French *petite*), and the long arm is designated with the letter q. In the case of telocentric chromosomes, the short arm may be nearly nonexistent. When a karyotype is prepared (see Figure 14.1), the chromosomes are aligned with the short arms on top and the long arms on the bottom.

Different chromosomes often have similar sizes and centromere locations. Therefore, cytogeneticists use additional methods to accurately identify each type of chromosome within a karyotype. For detailed identification, chromosomes are treated with stains to produce characteristic banding patterns. Cytogeneticists use several different staining procedures to identify specific chromosomes. An example is Giemsa stain, which produces dark G bands (see Figure 14.1). The alternating pattern of G bands is unique for each type of chromosome.

The banding pattern of eukaryotic chromosomes is useful in two ways. First, individual chromosomes can be distinguished from each other, even if they have similar sizes and centromere locations. Also,

Metacentric　　Submetacentric　　Acrocentric　　Telocentric

Figure 14.13 A comparison of centromere locations in metaphase chromosomes. Chromosomes are classified into four types—metacentric, submetacentric, acrocentric, and telocentric—on the basis of the location of the centromere.

Mutations Can Alter Chromosome Structure

Let's now consider how the structures of chromosomes can be modified by a mutation—a heritable change in the genetic material. Chromosomal mutations, which involve the breaking and rejoining of chromosomes, are categorized as deletions, duplications, inversions, and translocations (Figure 14.14). Deletions and duplications are changes in the total amount of genetic material in a single chromosome.

- When a **deletion** occurs, a segment of chromosomal material is removed (Figure 14.14a). The affected chromosome becomes deficient in a significant amount of genetic material.
- In a **duplication**, a section of a chromosome is repeated two or more times (Figure 14.14b).

What are the consequences of a deletion or duplication? The possible effects depend on their size and whether they include genes or portions of genes that are vital to the development of the organism. When deletions or duplications have an effect, they are usually detrimental. Larger changes in the amount of genetic material tend to be more harmful because more genes are missing or duplicated.

Inversions and translocations are chromosome rearrangements.

- An **inversion** is a change in the direction of the genetic material along a single chromosome. When a segment of one chromosome has been inverted, the order of G bands is opposite that of a normal chromosome (Figure 14.14c).
- A **translocation** occurs when one segment of a chromosome becomes attached to a different chromosome. In a **simple translocation**, a single piece of chromosome is attached to another chromosome (Figure 14.14d). In a **reciprocal translocation**, two nonhomologous chromosomes exchange pieces by an abnormal crossing-over event, thereby producing two chromosomes carrying translocations (Figure 14.14e).

Variation Occurs in the Number of Chromosome Sets and the Number of Individual Chromosomes

Variations in chromosome number can be categorized in two ways: variation in the number of sets of chromosomes and variation in the number of particular chromosomes within a set. The suffix -ploid or -ploidy refers to a complete set of chromosomes. Organisms that are **euploid** (the prefix eu- means true) have chromosomes that occur in one or more complete sets. For example, in a species that is diploid, a euploid organism would have two sets of chromosomes in its somatic cells. In the fruit fly (*Drosophila melanogaster*), a normal individual has eight chromosomes. The species is diploid, having two sets of four chromosomes each (Figure 14.15a).

Organisms can vary in the number of sets of chromosomes they have. For example, on rare occasions, an abnormal fruit fly can be produced with 12 chromosomes (having three sets of 4 chromosomes each) or 16 chromosomes (with four sets of 4 chromosomes each) (Figure 14.15b). Organisms with three or more sets of chromosomes are

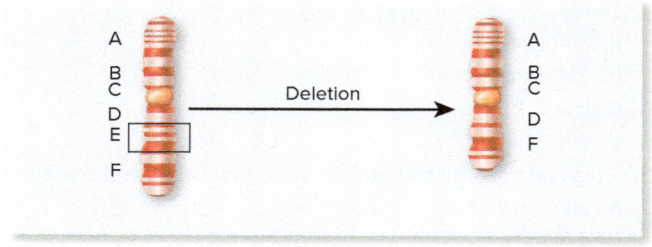

(a) Deletion: removes a segment of chromosome

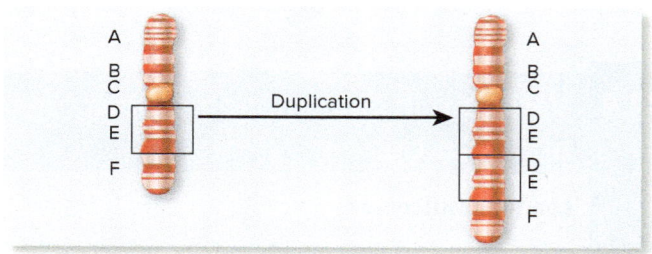

(b) Duplication: doubles a particular region

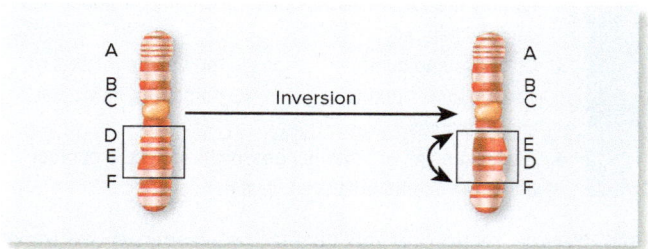

(c) Inversion: flips a region to the opposite orientation

(d) Simple translocation: moves a segment of 1 chromosome to another chromosome

(e) Reciprocal translocation: exchanges pieces between 2 different chromosomes

Figure 14.14 Types of change in chromosome structure. The letters alongside the chromosomes are placed there as frames of reference.

Concept Check: Which types of changes in chromosome structure shown here do not affect the total amount of genetic material?

(a) Normal fruit fly chromosome composition

(b) Polyploidy

(c) Aneuploidy

Figure 14.15 Types of variation in chromosome number in *Drosophila*. **(a)** The normal diploid number of chromosomes of a female fly. The X chromosome is also called chromosome 1. **(b)** Examples of polyploidy (triploid and tetraploid). **(c)** An example of aneuploidy (showing trisomy 2 and monosomy 3).

called **polyploid.** A diploid organism is referred to as $2n$, a **triploid** organism as $3n$, a **tetraploid** organism as $4n$, and so forth. All such organisms are euploid because they have complete sets of chromosomes.

A second way that chromosome number can vary is a phenomenon called **aneuploidy.** This is an alteration in the number of a particular chromosome so that the total number of chromosomes is not an exact multiple of a set (**Figure 14.15c**). For example, an abnormal fruit fly could have nine chromosomes instead of eight because it has three copies of chromosome 2 instead of the normal two copies. Instead of being perfectly diploid, the animal is $2n + 1$. Such a condition is called **trisomy,** and the animal is said to have trisomy 2, since chromosome 2 is affected. By comparison, a fruit fly could be lacking a single chromosome, such as chromosome 3, and have a total of seven chromosomes ($2n - 1$). This condition is called **monosomy,** and the animal has monosomy 3.

Variations in chromosome number usually have a significant effect on the characteristics of plants and animals. Therefore, researchers want to understand the mechanisms that cause these variations. In some cases, a change in chromosome number is the result of the abnormal sorting of chromosomes during cell division. The term **nondisjunction** refers to an event in which the chromosomes do not separate properly during cell division. Nondisjunction can occur during meiosis I or meiosis II and produces haploid cells that have too many or too few chromosomes. **Figure 14.16** illustrates the consequences of nondisjunction during meiosis I. In this case, one pair of homologs moved into the cell on the left instead of separating from each other. This results in the production of aneuploid cells, with either too many ($n + 1$) or too few ($n - 1$) chromosomes. If such a cell becomes a gamete that fuses with another gamete during fertilization, the zygote and the resulting organism will have an abnormal number of chromosomes in all of its cells.

Changes in Chromosome Number Have Important Consequences

How do changes in chromosome number affect the characteristics of animals and plants? Let's consider a few examples.

Changes in Chromosome Number in Animals In many cases, animals do not tolerate deviations from diploidy well. For example, polyploidy in mammals is generally a lethal condition. However, a few cases of naturally occurring variations from diploidy do occur in animals. Male bees, which are produced from unfertilized eggs, contain a single set of chromosomes and are therefore haploid organisms. By comparison, fertilized eggs become female bees, which are diploid. A few examples of vertebrate polyploid animals have been discovered. Interestingly, on rare occasions, animals that are morphologically very similar to each other can be found as a diploid species as well as a separate polyploid species. This situation occurs among certain amphibians and reptiles.

One important reason that geneticists are so interested in aneuploidy is its relationship to certain inherited disorders in humans. Even though most people are born with 46 chromosomes, alterations in chromosome number occur at a surprising frequency during gamete formation. About 5–10% of all fertilized human eggs result in an embryo with an abnormality in chromosome number. In most cases, these abnormal embryos do not develop properly and result in the very early miscarriage of a pregnancy, also referred to as spontaneous abortion. Approximately 50% of all miscarriages are due to alterations in chromosome number.

In some cases, an abnormality in chromosome number produces an offspring that can survive. Several human disorders are the result of abnormalities in chromosome number. The most common are trisomies of chromosome 21, 18, or 13 and abnormalities in the number of the sex chromosomes (**Table 14.2**). These syndromes are most likely due to nondisjunction. For example, Turner syndrome (X0) may occur when a gamete that is lacking a sex chromosome due to nondisjunction fuses with a normal gamete carrying an X chromosome during fertilization. By comparison, triple X syndrome (XXX) occurs when a gamete carrying two X chromosomes fuses with a gamete carrying a single X chromosome.

Most of the known trisomies involve chromosomes that are relatively small, so they carry fewer genes. Trisomies of the other human

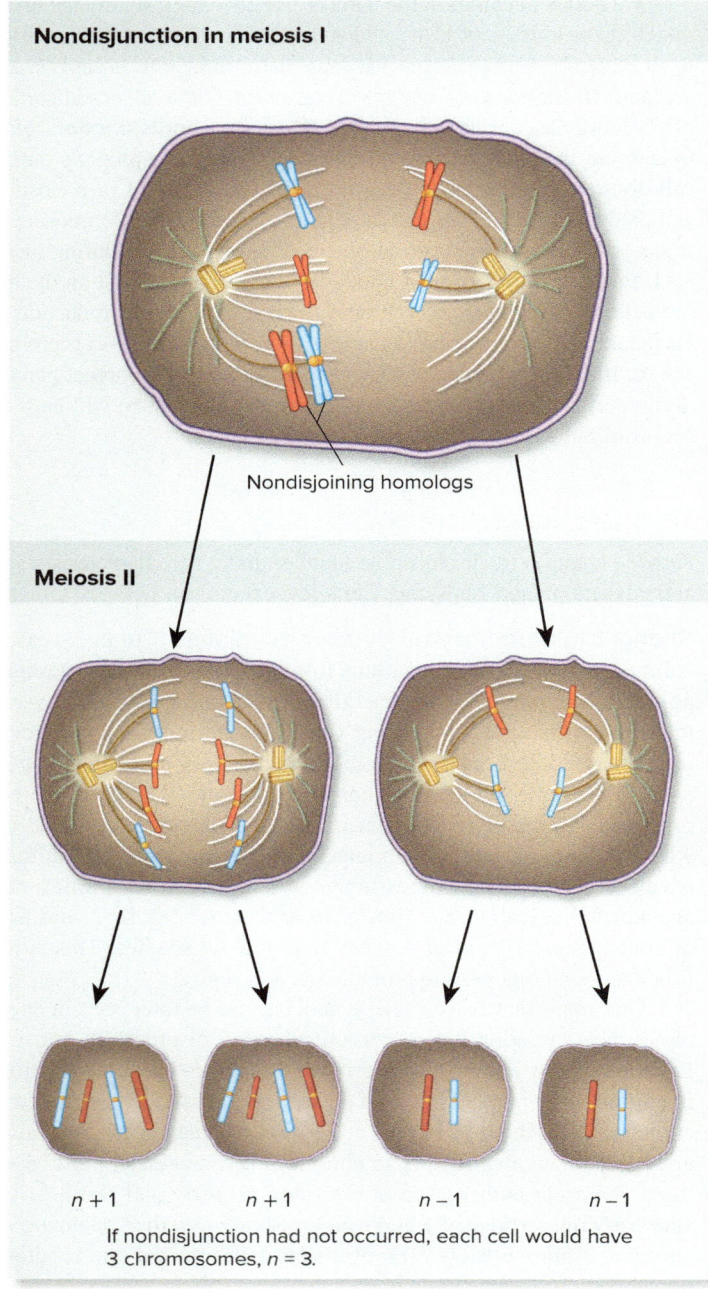

Figure 14.16 Nondisjunction during meiosis I. For simplicity, this cell shows only three pairs of homologous chromosomes. One of the three pairs did not disjoin (separate) properly, and both homologs have moved into the cell on the left. The resulting haploid cells shown at the bottom are all aneuploid, resulting in gametes with four chromosomes and two chromosomes, instead of three.
©irin-k/age fotostock

Table 14.2	Aneuploid Conditions in Humans		
Condition	Frequency (fraction of live births)	Syndrome	Characteristics
Autosomal			
Trisomy 21	1/800	Down	Mental impairment, abnormal pattern of palm creases, slanted eyes, flattened face, short stature
Trisomy 18	1/6,000	Edward	Mental and physical impairment, facial abnormalities, extreme muscle tone, early death
Trisomy 13	1/15,000	Patau	Mental and physical impairment, wide variety of defects in organs, large triangular nose, early death
Sex chromosomal			
XXY	1/1,000 (males)	Klinefelter	Sexual immaturity (no sperm), breast swelling (males)
XYY	1/1,000 (males)	Jacobs	Tall
XXX	1/1,500 (females)	Triple X	Tall and thin, menstrual irregularity
X0	1/5,000 (females)	Turner	Short stature, webbed neck, sexually undeveloped

chromosomes and most monosomies are presumed to be lethal and have been found in spontaneously aborted embryos and fetuses.

Human abnormalities in chromosome number are influenced by the age of the parents. Older parents are more likely to produce children with abnormalities in chromosome number, possibly because meiotic nondisjunction is more likely to occur in older cells. **Down syndrome,** which was first described by English physician John Langdon Down in 1866, provides an example. This disorder is caused by the inheritance of three copies of chromosome 21 (Table 14.2). The incidence of Down syndrome rises with the age of either parent. In males, however, the rise occurs relatively late in life, usually past the age when most men have children. By comparison, the likelihood of having a child with Down syndrome rises dramatically during the later reproductive years in women.

Changes in Chromosome Number in Plants In contrast to animals, plants commonly exhibit polyploidy, which is important in agriculture. In many instances, polyploid species of plants display characteristics that are helpful to humans. They are often larger in size and more robust, traits that are clearly advantageous in the production of food. For example, the species of wheat that we use to make bread, *Triticum aestivum*, is a hexaploid (with six sets of chromosomes) that arose from the union of diploid genomes from three closely related species (**Figure 14.17**). During the course of its cultivation, two diploid species must have interbred to produce a tetraploid, and then a third species interbred with the tetraploid to produce a hexaploid. Plant polyploids tend to exhibit a greater adaptability, which allows them to withstand harsher environmental conditions.

Figure 14.17 An example of a polyploid plant. Cultivated wheat, *Triticum aestivum*, is a hexaploid. It was derived from three different diploid species of grasses that originally were found in the Middle East and were cultivated by ancient farmers in that region. Modern varieties of wheat have been produced from this hexaploid species.
©irin-k/age fotostock

Although polyploidy is often beneficial in plants, aneuploidy usually has detrimental consequences for the characteristics of an organism of any eukaryotic species. Why is aneuploidy usually detrimental? To answer this question, we need to consider the relationship between gene expression and chromosome number. For many but not all genes, the level of gene expression is correlated with the number of genes per cell. Compared with a diploid cell, if a gene is carried on a chromosome that is present in three copies instead of two, approximately 150% of the normal amount of gene product is usually made. Alternatively, if only one copy of that gene is present due to a missing chromosome, only 50% of the gene product is typically made. For some genes, producing too much or too little of the gene product may not have adverse effects. However, for other genes, the over- or underexpression may interfere with the proper functioning of cells.

14.4 Reviewing the Concepts

- Chromosomes are classified as metacentric, submetacentric, acrocentric, or telocentric, based on their centromere location. Individual chromosomes can be distinguished from each other by their banding patterns after staining (Figure 14.13).
- Deletions, duplications, inversions, and translocations are ways in which mutations alter chromosome structure (Figure 14.14).
- A euploid organism has chromosomes that occur in complete sets. A polyploid organism has three or more sets of chromosomes. An organism that has one too many (trisomy) or one too few (monosomy) chromosomes is termed aneuploid. Aneuploidy can be caused by nondisjunction, an event in which the chromosomes do not separate properly during cell division. Aneuploidy in humans is responsible for several types of human genetic diseases, including Down syndrome (Figures 14.15, 14.16; Table 14.2).
- Polyploid animals are relatively rare, but polyploid plants are common and tend to be larger and more robust than their diploid counterparts (Figure 14.17).

14.4 Testing Your Knowledge

1. A chromosome in which the telomere is at one end is called
   a. metacentric.
   b. submetacentric.
   c. acrocentric.
   d. telocentric.

2. A change in chromosome structure that does not affect the total amount of genetic material is
   a. a deletion.
   b. a duplication.
   c. an inversion.
   d. a reciprocal translocation.
   e. Both c and d are correct.

3. A person who is born with Down syndrome carries three copies of chromosome 21 in his or her cells. Which of the following terms should *not* be used to describe such an individual?
   a. trisomy
   b. trisomy 21
   c. triploid
   d. aneuploid

Assessing Your Knowledge and Skills

Foundational Knowledge

Multiple Choice

1. In which phase of the cell cycle are chromosomes replicated?
   a. G_1 phase
   b. S phase
   c. M phase
   d. G_2 phase
   e. none of the above

2. Checkpoints during the cell cycle are important because they
   a. allow the organelle activity to catch up to cellular demands.
   b. ensure the integrity of the cell's DNA.
   c. allow the cell to generate sufficient ATP for cellular division.
   d. are the only times DNA replication can occur.
   e. do all of the above.

3. A replicated chromosome is composed of
   a. two homologous chromosomes held together at the centromere.
   b. four sister chromatids held together at the centromere.
   c. two sister chromatids held together at the centromere.
   d. four homologous chromosomes held together at the centromere.
   e. one chromosome with a centromere.

4. Which of the following statements accurately describes a difference between mitosis and meiosis?
   a. If the mother cell is diploid, mitosis will produce diploid daughter cells, whereas meiosis will produce haploid daughter cells.
   b. Homologous chromosomes synapse during meiosis but do not synapse during mitosis.
   c. Crossing over commonly occurs during meiosis, but it does not commonly occur during mitosis.
   d. All of the above are correct.
   e. Both a and c are correct.

5. Aneuploidy may be the result of
   a. a duplication of a region of a chromosome.
   b. an inversion of a region of a chromosome.
   c. nondisjunction during meiosis.
   d. interspecies breeding.
   e. any of the above.

Short Answer

1. Distinguish between homologous chromosomes and sister chromatids.
2. **PRINCIPLES** A principle of biology is that cells are the simplest units of life. Explain how mitosis is a key process in the formation of new cells.

Critical-Thinking Skills

Multiple Choice

1. Let's suppose you are observing cell division in cells derived from a diploid organism. You add a chemical that causes the cells to stop advancing through mitosis when they reach the end of prophase. At this point, you count the number of chromatids and determine that each cell has 48 of them. How many chromosomes were in these cells during the G_1 phase of the cell cycle?
   a. 6
   b. 12
   c. 24
   d. 48
   e. 96

2. The *Herc2* gene, which influences eye color in humans, is found on chromosome 15. Though you can't see the gene, how many copies of *Herc2* are found in the karyotype of Figure 14.1?
   a. 1
   b. 2
   c. 4
   d. 8
   e. 14

3. A student is looking at cells under the microscope. The cells are from an organism that has a diploid number of 14. One particular cell has 7 replicated chromosomes (sister chromatids) aligned in a single row at the metaphase plate. Which of the following statements accurately describes this cell?
   a. The cell is in metaphase of mitosis.
   b. The cell is in metaphase of meiosis I.
   c. The cell is in metaphase of meiosis II.
   d. All of the above are correct.
   e. Both b and c are correct.

4. During S phase of the cell cycle, a chromosome suffers a deletion that removes its centromere. When would it be most likely for this deletion to be detected, and what would be the consequences of its detection?
   a. It would be detected at the G_1 checkpoint, and cell division would be arrested.
   b. It would be detected at the G_1 checkpoint, and cell division would be accelerated.
   c. It would be detected at the G_2 checkpoint, and cell division would be arrested.
   d. It would be detected at the G_2 checkpoint, and cell division would be accelerated.
   e. It would be detected at the metaphase checkpoint, and cell division would be arrested.

5. An individual with Down syndrome has 47 total chromosomes. Such a person was produced from the union between a normal gamete and a gamete with
   a. 23 chromosomes.
   b. 24 chromosomes.
   c. 45 chromosomes.
   d. 46 chromosomes.
   e. 47 chromosomes.

Short Answer

1. Why do you think it is necessary for chromosomes to condense during mitosis and meiosis and become very compact structures? What might happen if chromosomes did not condense?
2. Normally, crossing over occurs between homologous chromosomes during prophase of meiosis I. However, sometimes mistakes happen, and crossing over occurs between two chromosomes that are *not* homologous. Describe the results of such a crossover.

Transmission of Genetic Information from Parents to Offspring I: Patterns That Follow Mendel's Laws

15

©Tomas Rodriguez/Getty Images

Parents and offspring often show a striking resemblance to each other.

Chapter Outline

- **15.1** Mendel's Laws of Inheritance
- **15.2** Chromosome Theory of Inheritance
- **15.3** Pedigree Analysis of Human Traits
- **15.4** Variations in Inheritance Patterns and Their Molecular Basis
- **15.5** Sex Chromosomes and X-Linked Inheritance Patterns

Assessing Your Knowledge and Skills

Many people are intrigued to observe that the traits of children often resemble those of their parents (see the chapter-opening photograph). Likewise, the offspring of other animals and plants typically are similar to their parents. Even so, geneticists have discovered that inheritance patterns are full of surprises. For example, if you cross a pea plant with purple flowers to a plant with white flowers, all of the offspring have purple flowers, which is the same as one of their parents. On the other hand, if you cross a four-o'clock plant with red flowers to one with white flowers, all of the offspring have pink flowers, which is not the same as either parent. A key objective of this chapter is to explain how such different outcomes can occur. The goal is to understand **inheritance**—the acquisition of traits by their transmission from parent to offspring.

Microscopic observations of chromosome transmission during mitosis and meiosis in the second half of the 19th century provided compelling evidence for **particulate inheritance**—the idea that the determinants of hereditary traits, which are now called genes, are transmitted in discrete units, or particles, from one generation to the next. Remarkably, this idea was first put forward in the 1860s by an Austrian monk who knew nothing about chromosomes or DNA (**Figure 15.1**). Gregor Mendel used quantitative analysis of carefully designed plant-breeding experiments to arrive at the concept of a **gene,** which is broadly defined as a unit of heredity.

Figure 15.1 Gregor Johann Mendel.
©Science Source

considering, the true-breeding tall plant (*TT*) is a homozygote, as is the dwarf plant (*tt*).

- A **heterozygote** carries two different alleles of a gene. Plants of the F_1 generation are heterozygotes with the genotype *Tt*. Every individual carries one copy of the tall allele (*T*) and one copy of the dwarf allele (*t*). The F_2 generation includes both homozygous individuals (homozygotes) and heterozygous individuals (heterozygotes).
- **Phenotype** refers to the characteristics of an organism that are the result of the expression of its genes. In the example in Figure 15.5a, the phenotype of one parent is tall, and the other is dwarf. Although the F_1 offspring are heterozygous (*Tt*), their phenotype is tall because each of them has a copy of the dominant tall allele. In contrast, the F_2 plants display both phenotypes in a ratio of 3:1.

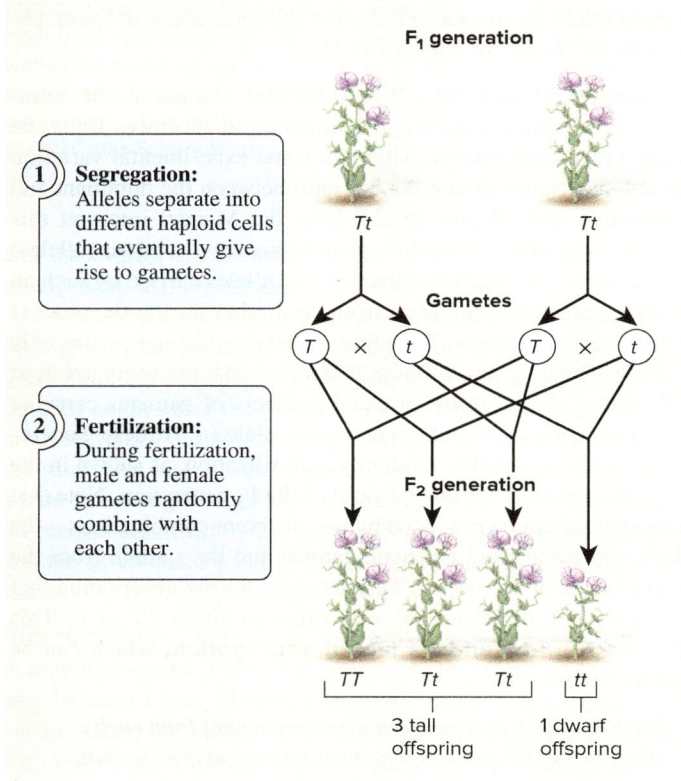

Figure 15.6 How the law of segregation explains Mendel's observed ratios. The segregation of alleles in the F_1 generation gives rise to gametes that carry just one of the two alleles. These gametes combine randomly during fertilization, producing the allele combinations *TT*, *Tt*, and *tt* in the F_2 offspring. The combination *Tt* occurs twice as often as either of the other two combinations because it can be produced in two different ways. The *TT* and *Tt* offspring are tall, whereas the *tt* offspring are dwarf.

✓ **Concept Check:** What is the ratio of the *T* allele to the *t* allele in the F_2 generation? Does this ratio differ from the 3:1 phenotype ratio? If so, explain why.

Biology Principle | Biology is a quantitative science

Quantitative Analysis: A Punnett Square Is Used to Predict the Outcome of Crosses

A common way to predict the outcome of simple genetic crosses is to make a **Punnett square,** a method originally proposed by British geneticist Reginald Punnett. To construct a Punnett square, you must know the genotypes of the parents. What follows is a step-by-step description of the Punnett-square approach, using a cross of heterozygous tall plants.

Step 1. *Write down the genotypes of both parents.* In this example, a heterozygous tall plant is crossed to another heterozygous tall plant. The plant providing the pollen is considered the male parent; the plant providing the eggs, the female parent. (In self-pollination, a single individual produces both types of gametes.)

Male parent: *Tt*
Female parent: *Tt*

Step 2. *Write down the possible gametes that each parent can make.* Remember, the law of segregation tells us that a gamete contains only one copy of each allele.

Male gametes: *T* or *t*
Female gametes: *T* or *t*

Step 3. *Create an empty Punnett square.* The number of columns equals the number of male gametes, and the number of rows equals the number of female gametes. Our example has two rows and two columns. Place the male gametes across the top of the Punnett square and the female gametes along the side.

Step 4. *Fill in the possible genotypes of the offspring by combining the alleles of the gametes in the empty boxes.*

Step 5. *Determine the relative proportions of genotypes and phenotypes of the offspring.* The genotypes are obtained directly from the Punnett square. In this example, the genotype ratio is 1 *TT* : 2 *Tt* : 1 *tt*. To determine the phenotypes, you must know which allele is dominant. For plant height, *T* (tall) is dominant to *t* (dwarf). The genotypes *TT* and *Tt* are tall, whereas the genotype *tt* is dwarf. Therefore, our Punnett square shows us that the phenotype ratio is expected to be 3 tall : 1 dwarf.

Crunching the Numbers: In pea plants, purple flowers are dominant to white. A cross is made between a heterozygous purple-flowered plant and a white plant. Set up a Punnett square, and predict the outcome of this cross.

Analyzing the Inheritance Pattern of Two Characters Demonstrated the Law of Independent Assortment

To obtain additional insights into how genes are transmitted from parents to offspring, Mendel conducted crosses in which he simultaneously followed the inheritance of two different characters. A cross of this type is called a **two-factor cross.** We will examine a two-factor cross in which Mendel simultaneously followed the inheritance of seed color and seed shape (**Figure 15.7**). He began by crossing strains of pea plants that bred true for both characters. The plants of one strain had yellow, round seeds, and the plants of the other strain had green, wrinkled seeds. He then allowed the F$_1$ offspring to self-fertilize and observed the phenotypes of the F$_2$ generation.

What are the possible patterns of inheritance for two characters?

1. The two genes may be linked in some way, so variants that occur together in the parents are always inherited as a unit. In our example, the allele for yellow seeds (*Y*) would always be inherited with the allele for round seeds (*R*), and the alleles for green seeds (*y*) would always be inherited with the allele for wrinkled seeds (*r*), as shown in Figure 15.7a.
2. The transmission of the two genes may be independent of one another, so their alleles are randomly distributed into gametes (Figure 15.7b).

By following the transmission pattern of two characters simultaneously, Mendel could determine whether the genes that control seed shape and seed color assort (are distributed) together as a unit or independently of each other.

What experimental results could Mendel predict for each of these two models? The two homozygous plants of the P generation can produce only two kinds of gametes, *YR* and *yr*, so in either case the F$_1$ offspring would be heterozygous for both genes; that is, they would have the genotypes *YyRr*. Because Mendel knew from his earlier experiments that yellow was dominant to green and round was dominant to wrinkled, he could predict that all F$_1$ plants would have yellow, round seeds. In contrast, as shown in Figure 15.7, the ratios he obtained in the F$_2$ generation would depend on whether the alleles of both genes assorted together or independently.

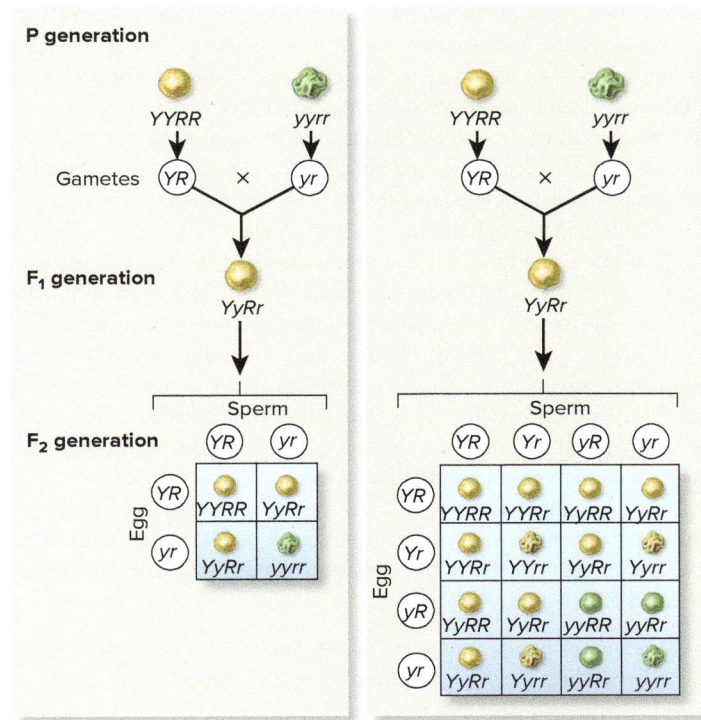

(a) Hypothesis: linked assortment

(b) Hypothesis: independent assortment

P cross	F$_1$ generation	F$_2$ generation
Yellow, round seeds × Green, wrinkled seeds	Yellow, round seeds	315 yellow, round seeds 101 yellow, wrinkled seeds 108 green, round seeds 32 green, wrinkled seeds

(c) The data observed by Mendel

Figure 15.7 Two hypotheses for the assortment of two different genes. In a two-factor cross between two true-breeding pea plants, one with yellow, round seeds and the other with green, wrinkled seeds, all F$_1$ offspring have yellow, round seeds. When the F$_1$ offspring self-fertilize, two hypotheses predict different phenotypes in the F$_2$ generation: **(a)** linked assortment, in which parental alleles stay associated with each other, or **(b)** independent assortment, in which each allele assorts independently. **(c)** Mendel's data supported the independent assortment hypothesis.

Concept Check: *What ratio of phenotypes would have occurred in the F$_2$ generation if the linked hypothesis had been correct?*

If the parental genes are linked, as in Figure 15.7a, the F$_1$ plants would produce gametes that are only *YR* or *yr*. These gametes would combine to produce offspring with the genotype *YYRR* (yellow, round), *YyRr* (yellow, round), or *yyrr* (green, wrinkled). The ratio of phenotypes would be 3 yellow, round to 1 green, wrinkled. Every F$_2$ plant would be phenotypically like one P-generation parent or the other. None would display a new combination of the parental traits. However, if the alleles assort independently, the F_2 generation would show a wider range of genotypes and phenotypes, as shown by the large Punnett square in Figure 15.7b. In this case, each F_1 parent produces four kinds of gametes—*YR, Yr, yR,* and *yr*—instead of two, so the Punnett square is constructed with four rows on each side and shows 16 possible genotypes. The F_2 generation includes plants with yellow, round seeds; yellow, wrinkled seeds; green, round seeds; and green, wrinkled seeds, in a ratio of 9:3:3:1.

The actual results of this two-factor cross are shown in the data of Figure 15.7c. Crossing the true-breeding parents produced **dihybrid** offspring—offspring that are hybrids with respect to both traits. These F_1 dihybrids all had yellow, round seeds, confirming that yellow and round are dominant traits. This result was consistent with either hypothesis. However, the data for the F_2 generation were consistent only with the independent assortment hypothesis. The F_2 offspring showed four different phenotypes in a ratio that was very close to 9:3:3:1.

In his original studies, Mendel reported that he had obtained similar results for every pair of characters he analyzed. This work gave rise to **Mendel's law of independent assortment,** which can be stated as follows:

The alleles of different genes assort independently of each other during the process that gives rise to gametes.

Independent assortment means that a specific allele for one gene may be found in a gamete regardless of which allele for a different gene is found in the same gamete. In our example, the yellow and green alleles assort independently of the round and wrinkled alleles. The union of gametes derived from F_1 plants carrying these alleles produces the F_2 genotype and phenotype ratios shown in Figure 15.7b.

As we will see in Chapter 16, not all two-factor crosses exhibit independent assortment. In some cases, the alleles of two genes that are physically located near each other on the same chromosome do not assort independently.

15.1 Reviewing the Concepts

- Mendel studied seven characters of garden peas that existed in two variants each. He allowed his pea plants to self-fertilize or conducted hybridization experiments (Figures 15.1, 15.2, 15.3, 15.4).
- Mendel determined that certain traits exist in two forms, dominant and recessive. An individual carries two genes for a given character, and genes have variant forms, which are called alleles.
- By following the inheritance pattern of a single character (using a single-factor cross) for two generations, Mendel proposed the law of segregation, which states that two alleles segregate during the process that gives rise to gametes so that every gamete receives only one allele. In a single-factor cross, a 3:1 ratio is observed between the dominant and the recessive traits in the F_2 generation (Figures 15.5, 15.6).
- The genotype is the genetic makeup of an organism. Phenotype is a description of the traits that an organism displays.
- A Punnett square is constructed to predict the outcome of a cross.
- By conducting a two-factor cross, Mendel determined the law of independent assortment, which states that the alleles of different genes assort independently of each other during gamete formation. In a two-factor cross, this yields a 9:3:3:1 ratio in the F_2 generation (Figure 15.7).

15.1 Testing Your Knowledge

1. A cross is made between a pea plant that is *Pp* and one that is *pp*, where *P* is a dominant allele for purple flowers and *p* is the recessive allele that causes white flowers. What percentage of offspring would you expect to have purple flowers?
   a. 100%   c. 50%
   b. 75%   d. 25%

2. In peas, *Y* (yellow seeds) is dominant to *y* (green seeds), and *R* (round seeds) is dominant to *r* (wrinkled seeds). A plant that is heterozygous for both genes is crossed to a plant that is homozygous for both recessive alleles. What is the expected ratio of offspring?
   a. 3 yellow/round : 1 green wrinkled
   b. 9 yellow/round : 3 yellow/wrinkled : 3 green/round : 1 green wrinkled
   c. 9 yellow/wrinkled : 3 yellow/round : 3 green/round : 1 green wrinkled
   d. 1 yellow/round : 1 yellow/wrinkled : 1 green/round : 1 green wrinkled

15.2 Chromosome Theory of Inheritance

Learning Outcomes

1. Outline the principles of the chromosome theory of inheritance.
2. Relate the behavior of chromosomes during meiosis to Mendel's two laws of inheritance.

Mendel's studies with pea plants eventually led to the concept of a gene, which is the foundation for our understanding of inheritance. However, at the time of Mendel's work, the physical nature and location of genes were a complete mystery. Several scientists, including German biologists Eduard Strasburger and Walther Flemming, observed dividing cells under the microscope and suggested that the chromosomes are the carriers of the genetic material. As we now know, the genetic material is the DNA within chromosomes. Other researchers noted that Mendel's laws of segregation and independent assortment showed a striking parallel with the segregation and sorting of chromosomes during meiosis (described in Chapter 14). Among these scientists were German biologist Theodor Boveri and American biologist Walter Sutton, who independently proposed the chromosome theory of inheritance. According to this theory, the inheritance patterns of traits can be explained by the transmission of chromosomes during meiosis and fertilization.

A modern version of the **chromosome theory of inheritance** consists of a few fundamental principles:

1. Chromosomes contain DNA, which is the genetic material. Genes are found within the chromosomes.
2. Chromosomes are replicated (copied) and passed from parent to offspring. They are also passed from cell to cell during the development of a multicellular organism.

3. The nucleus of a diploid cell contains two sets of chromosomes, which are found in homologous pairs. The maternal and paternal sets of homologous chromosomes are functionally equivalent; each set carries a full complement of genes.
4. At meiosis, one member of each homologous pair segregates into one daughter nucleus, and its homolog segregates into the other daughter nucleus. During the formation of haploid cells, the members of different homologous pairs segregate independently of each other.
5. Gametes are haploid cells that combine to form a diploid cell during fertilization, with each gamete transmitting one set of chromosomes to the offspring.

In this section, we will relate the chromosome theory of inheritance to Mendel's laws of inheritance.

Mendel's Law of Segregation Is Explained by the Segregation of Homologous Chromosomes During Meiosis

The physical location of a gene on a chromosome is called the gene's **locus** (plural, loci). As shown in **Figure 15.8**, each member of a homologous chromosome pair carries an allele of the same gene at the same locus. The individual in this example is heterozygous (Tt), so each homolog has a different allele.

How can we relate the locations of genes along chromosomes to Mendel's law of segregation? **Figure 15.9** shows a simplified example that follows a pair of homologous chromosomes through the events of meiosis. This example involves a pea plant, heterozygous for height, Tt. The top of Figure 15.9 shows the two homologous chromosomes prior to DNA replication. When a cell prepares to divide, the homologs replicate to produce pairs of sister chromatids. Each chromatid carries a copy of the allele found on the original homolog, either T or t. During meiosis I, the homologs, each consisting of two sister chromatids, pair up and then segregate into two daughter cells. One of these cells has two copies of the T allele, and the other has two copies of the t allele. The sister chromatids separate during meiosis II, which produces four haploid cells. The end result of meiosis is that each haploid cell has a copy of just one of the two original homologs. Two

Figure 15.8 A gene locus. The locus (location) of a gene is the same for each member of a homologous pair, whether the individual is homozygous or heterozygous for that gene. This individual is heterozygous (Tt) for a gene for plant height.

BioConnections: *Refer back to Section 14.3 in Chapter 14. Explain the relationship between sexual reproduction and homologous chromosomes.*

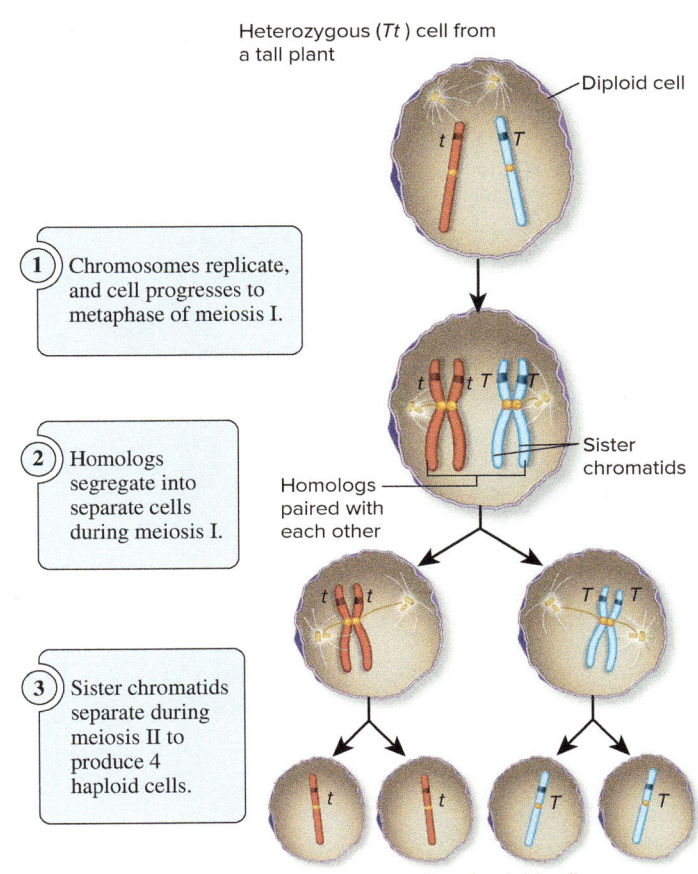

Figure 15.9 The chromosomal basis of allele segregation. This example shows a pair of homologous chromosomes in a cell of a pea plant. The blue chromosome was inherited from the male parent, and the red chromosome was inherited from the female parent. This individual is heterozygous (Tt) for a height gene. The two homologs segregate from each other during meiosis, leading to segregation of the tall allele (T) and the dwarf allele (t) into different haploid cells. Note: For simplicity, this diagram shows a single pair of homologous chromosomes, though eukaryotic cells typically have several different pairs of homologous chromosomes.

BioConnections: *When we say that alleles segregate, what does the word segregate mean? How is this related to meiosis, shown in Figure 14.11?*

of the cells have a chromosome carrying the T allele, and the other two have a chromosome carrying the t allele at the same locus. If the haploid cells shown at the bottom of Figure 15.9 combine randomly during fertilization, they produce diploid offspring with the genotype and phenotype ratios shown earlier in Figure 15.6.

Mendel's Law of Independent Assortment Is Explained by the Independent Alignment of Different Chromosomes During Meiosis

How can we relate the chromosome theory of inheritance to Mendel's law of independent assortment? **Figure 15.10** shows the alignment and segregation of two pairs of chromosomes in a pea plant. One

Biology Principle

Biologists Use Models to Test Predictions and Convey Ideas

This figure shows two models that depict how chromosomes from a cell with two pairs of chromosomes can align during metaphase of meiosis I and produce different possible haploid cells. The Modeling Challenge in the figure legend asks you to propose a series of models for a cell with three pairs of chromosomes.

1. Chromosomes replicate, and cell advances to metaphase of meiosis I. Alignment of homologs can occur in more than one way.

2. Homologs segregate into separate cells during meiosis I.

3. Sister chromatids separate during meiosis II to produce 4 haploid cells.

Figure 15.10 The chromosomal basis of independent assortment. The alleles for seed color (*Y* or *y*) and seed shape (*R* or *r*) in peas are located on different chromosomes. During metaphase of meiosis I, different arrangements of the two chromosome pairs lead to different combinations of the alleles in the resulting haploid cells. On the left, the chromosome carrying the recessive *y* allele has segregated with the chromosome carrying the dominant *R* allele. On the right, the two chromosomes carrying the dominant alleles (*Y* and *R*) have segregated together. Note: For simplicity, this diagram shows only two pairs of homologous chromosomes, though eukaryotic cells typically have several different pairs of homologous chromosomes.

Modeling Challenge: Figure 15.10 shows the possible alignments at metaphase of meiosis I for a cell that has two pairs of chromosomes. Let's suppose a cell has three pairs of chromosomes, as in Figure 14.11 in Chapter 14. Each set is represented by a different color (red or blue), and each pair is distinguished by its length. One pair is long, one is medium-sized, and one is short. Draw a series of models that show all of the possible alignments at metaphase of meiosis I for a cell that has three pairs of chromosomes. Note: Your models do not need to show the occurrence of crossovers. If the original diploid cell was heterozygous for three genes, *TtYyRr*, how many possible types of haploid cells could it produce?

pair carries the gene for seed color: The yellow allele (*Y*) is on one chromosome, and the green allele (*y*) is on its homolog. The other pair of chromosomes carries the gene for seed shape: One member of the pair has the round allele (*R*), whereas its homolog carries the wrinkled allele (*r*). Therefore, this individual is heterozygous for both genes, with the genotype *YyRr*.

When meiosis begins, the DNA in each chromosome has already replicated, producing two sister chromatids. At metaphase I of meiosis, the two pairs of chromosomes randomly align themselves along the metaphase plate. This alignment can occur in two equally probable ways, shown on the two sides of the figure. On the left, the chromosome carrying the *y* allele is aligned on the same side of the metaphase plate as the chromosome carrying the *R* allele; *Y* is aligned with *r*. On the right, the opposite has occurred: *Y* is aligned with *R*, and *y* is with *r*. In each case, the chromosomes that aligned on the same side of the metaphase plate segregate into the same daughter cell. In this way, the random alignment of chromosome pairs during meiosis I leads to the independent assortment of alleles found on different chromosomes. For two genes found on different chromosomes, each with two variant alleles, meiosis produces four allele combinations in equal numbers (*yR*, *Yr*, *YR*, and *yr*), as seen at the bottom of the figure.

If a *YyRr* plant undergoes self-fertilization, any two gametes can combine randomly during fertilization. Because four kinds of gametes are made, 4^2, or 16, allele combinations (genotypes) are possible in the offspring. These genotypes, in turn, produce four phenotypes in a 9:3:3:1 ratio, as seen in Figure 15.7. This ratio is the expected outcome when a heterozygote for two genes on different chromosomes undergoes self-fertilization.

But what if two different genes are located on the same chromosome? In this case, the transmission pattern may not conform to the law of independent assortment. We will discuss this phenomenon, known as linkage, in Chapter 16.

15.2 Reviewing the Concepts

- The chromosome theory of inheritance explains how the behavior of chromosomes during meiosis accounts for Mendel's laws of inheritance. Each gene is located at a particular site, or locus, on a chromosome (Figures 15.8, 15.9, 15.10).

15.2 Testing Your Knowledge

1. During which phase of cellular division does the event underlying Mendel's law of segregation physically occur?
   a. mitosis
   b. meiosis I
   c. meiosis II
   d. all of the above
   e. b and c only

2. Which phase of cellular division explains Mendel's law of independent assortment?
   a. mitosis
   b. metaphase of meiosis I
   c. metaphase of meiosis II
   d. anaphase of meiosis I
   e. anaphase of meiosis II

15.3 Pedigree Analysis of Human Traits

Learning Outcomes

1. **SCISKILLS** ▶ Apply pedigree analysis to deduce inheritance patterns in humans.
2. Distinguish between recessive disorders and dominant disorders.

As we have seen, Mendel conducted experiments by making selective crosses of pea plants and analyzing large numbers of offspring. Later geneticists also relied on crosses of experimental organisms, especially fruit flies (*Drosophila melanogaster*). However, geneticists studying human traits cannot use this approach, for ethical and practical reasons. Instead, human geneticists must rely on the information contained in family trees, or pedigrees. In this approach, called **pedigree analysis,** an inherited trait is analyzed over the course of a few generations in one family.

Pedigree analysis has been used to understand the inheritance of human genetic diseases that follow simple Mendelian patterns. In human populations, genes that play a role in disease may exist in different forms:

- A wild-type allele is a common allele in a population.
- A mutant allele is rare. Some mutant alleles are known to cause inherited diseases in humans.

Pedigree analysis allows us to determine whether the mutant allele is dominant or recessive and to predict the likelihood of an individual being affected.

Let's consider a recessive condition to illustrate pedigree analysis. The pedigree in **Figure 15.11** concerns a human genetic disease known as cystic fibrosis (CF), which involves a mutation in a gene that encodes the cystic fibrosis transmembrane regulator (the *CFTR* gene; refer back to Figure 1.12). Approximately 3% of Americans of European descent are heterozygous carriers of the recessive *CFTR* (disease-causing) allele. Carriers do not show disease symptoms; they are unaffected. Individuals who are homozygous for the recessive *CFTR* allele exhibit the disease symptoms, which include abnormalities of the lungs, pancreas, intestine, and sweat glands.

- The oldest generation (designated by the Roman numeral I) is at the top, with later generations (II and III) below it. Within the same generation, individuals are numbered from left to right.
- A male is represented by a square and a female by a circle.
- Couples who produce offspring are connected by a horizontal line.
- A vertical line connects parents with their offspring.
- Siblings (brothers and sisters) are placed on downward projections from a single horizontal line, from left to right in the order of their birth. For example, individuals I-1 and I-2 are the parents of individuals II-2, II-3, and II-4, who are all siblings.
- Individuals affected by the disease, such as individual II-3, are depicted by filled symbols.

(a) Human pedigree showing cystic fibrosis

○ ♀		Female
□ ♂		Male
○ □		Unaffected individual
● ■		Affected individual
◐ ◧		Presumed heterozygote (carrier)

(b) Symbols used in a human pedigree

Figure 15.11 **A family pedigree for a recessive trait.** Some members of the family in this pedigree are affected with cystic fibrosis. Unaffected individuals I-1, I-2, II-4, and II-5 are presumed to be heterozygotes (carriers) because they have produced affected offspring.

Concept Check: *Let's suppose a genetic disease is caused by a mutant allele. If two affected parents produce an unaffected offspring, can the mutant allele be recessive?*

Why does this pedigree indicate a recessive pattern of inheritance for CF? The answer is that two unaffected individuals can produce an affected offspring. For example, II-4 and II-5 have an affected offspring. Such individuals are presumed to be heterozygotes (designated by a half-filled symbol). However, the same unaffected parents can also produce unaffected offspring (depicted by an unfilled symbol), because an individual must inherit two copies of the mutant allele to exhibit the disease. A recessive mode of inheritance is also consistent with the observation that all offspring of two affected individuals are affected themselves. However, for genetic diseases that limit survival or fertility, there may rarely, if ever, be cases where two affected individuals produce offspring.

Although many of the alleles causing human genetic diseases are recessive, some are known to be dominant. Let's consider Huntington disease, a condition that causes the degeneration of brain cells involved in emotions, intellect, and movement. If you examine the pedigree shown in **Figure 15.12**, you will see that every affected individual has one affected parent. This pattern is characteristic of most dominant disorders in which a heterozygote exhibits the symptoms. However, affected parents do not always produce affected offspring. For example, II-6 is a heterozygote that has passed the normal allele to his offspring, thereby producing unaffected offspring (III-3 and III-4).

Figure 15.12 **A family pedigree for a dominant trait.** Huntington disease is caused by a dominant allele.

Concept Check: *What observation in a pedigree suggests a dominant pattern of inheritance?*

Most human genes are found on the paired chromosomes known as **autosomes,** which are the same in both sexes. Mendelian inheritance patterns involving these autosomal genes are described as autosomal inheritance patterns. Huntington disease is an example of a trait with an autosomal dominant inheritance pattern, whereas cystic fibrosis displays an autosomal recessive pattern. However, some human genes are located on sex chromosomes, which are different in males and females. These genes have their own characteristic inheritance patterns, which we will consider later in this chapter.

15.3 Reviewing the Concepts

- The inheritance patterns in humans are determined from pedigree analysis (Figures 15.11, 15.12).

15.3 Testing Your Knowledge

1. Which of the following observations in a pedigree would rule out a recessive pattern of inheritance for a human disease?
   a. An affected offspring has two unaffected parents.
   b. An unaffected offspring has two affected parents.
   c. Two affected offspring are brother and sister.
   d. An affected grandfather has an affected grandson.

15.4 Variations in Inheritance Patterns and Their Molecular Basis

Learning Outcomes

1. Relate dominant and recessive alleles and genotypes to protein function.
2. Describe pleiotropy, and explain why it occurs.
3. **SCISKILLS** ▶ Predict the outcomes of crosses that exhibit incomplete dominance.
4. Discuss how the environment plays a critical role in determining the expression of traits.

The term **Mendelian inheritance** describes the inheritance patterns of genes that segregate and assort independently. In the first section of this chapter, we considered the inheritance pattern of traits affected by a single gene that is found in two variants, one of which is dominant to the other. This pattern is called **simple Mendelian inheritance,** because the phenotype ratios in the offspring clearly conform to Mendel's laws. In this section, we will discuss the molecular basis of dominant and recessive traits and see how the molecular expression of a gene can have widespread effects on an organism's phenotype. In addition, we will examine the inheritance patterns of genes that segregate and assort independently but do not display a simple dominant/recessive relationship. This modern knowledge also sheds light on the role of the environment in producing an organism's phenotype, which we will discuss at the end of the section.

Protein Function Explains the Phenomenon of Dominance

As we discussed at the beginning of this chapter, Mendel studied seven characters that were found in two variants each (see Figure 15.2). The dominant variants are caused by the wild-type alleles for these traits in pea plants. A wild-type allele usually encodes a protein that is made in the proper amount and functions normally.

In the case of Mendel's seven characters in pea plants, the recessive alleles are due to rare mutations. By studying genes and their gene products at the molecular level, researchers have discovered that a recessive allele is often defective in its ability to express a functional protein. In other words, mutations that produce recessive alleles are likely to decrease or eliminate the synthesis or functional activity of a protein. These are called **loss-of-function alleles.**

In a simple dominant/recessive relationship, the recessive allele does not affect the phenotype of the heterozygote. A single copy of the dominant (wild-type) allele is sufficient to mask the effects of the recessive allele. How do we explain the dominant phenotype of the heterozygote? **Figure 15.13** considers the example of flower color in a pea plant. The gene encodes an enzyme (protein P) that is needed to convert a colorless molecule into a purple pigment (Figure 15.13a). The *P* allele is dominant because one *P* allele encodes enough of the functional protein—50% of the amount found in a normal homozygote—to provide a purple phenotype (Figure 15.13b). Therefore, the *PP* homozygote and the *Pp* heterozygote both make enough of the purple pigment to yield purple flowers. The *pp* homozygote cannot make any of the functional protein required for pigment synthesis, so its flowers are white.

This explanation—that 50% of the functional protein is enough—is true for many dominant alleles. In such cases, the homozygote carrying the dominant allele is making much more of the protein than necessary, so if the amount is reduced to 50%, as it is in the heterozygote, the individual still has plenty of this protein to accomplish whatever cellular function it performs. In other cases, however, an allele may be dominant because the heterozygote actually produces more than 50% of the normal amount of functional protein. This increased production is due to the phenomenon of gene regulation. The dominant allele is up-regulated in the heterozygote to compensate for the lack of function of the recessive allele.

The idea that recessive alleles usually cause a substantial decrease in the expression of a functional protein is supported by analyses of many human genetic diseases. Keep in mind that many genetic diseases are caused by rare mutant alleles. **Table 15.2** lists several examples of human genetic diseases in which a recessive allele

(a) Function of protein P

Genotype	PP	Pp	pp
Amount of functional protein P produced	100%	50%	0%
Phenotype	Purple	Purple	White

(b) Relationship between the amount of functional protein P and phenotype

Figure 15.13 How genes give rise to traits during simple Mendelian inheritance. In many cases, the amount of protein encoded by a single dominant allele is sufficient to produce the normal phenotype. **(a)** In this example, the gene encodes an enzyme that is needed to produce a purple pigment. **(b)** A plant with one or two copies of the normal allele produces enough pigment to produce purple flowers. In a *pp* homozygote, the complete lack of the normal protein (enzyme) results in white flowers.

Table 15.2	Examples of Recessive Human Genetic Diseases	
Disease	Protein encoded by the gene*	Description
Phenylketonuria (PKU)	Phenylalanine hydroxylase	Inability to metabolize phenylalanine, an amino acid. Can lead to severe mental impairment and physical degeneration. Can be prevented by following a phenylalanine-free diet beginning at birth.
Cystic fibrosis (CF)	Cystic fibrosis transmembrane conductance regulator, which is a chloride-ion transporter	Inability to regulate ion balance in epithelial cells. Leads to a variety of abnormalities, including production of thick lung mucus and chronic lung infections.
Tay-Sachs disease	Hexosaminidase A	Inability to metabolize certain lipids. Leads to paralysis, blindness, and early death.
Sickle cell disease	β globin	Anemia and blockages in blood circulation. Symptoms include swelling of the hands and feet, fatigue, jaundice, and episodes of pain.

*Individuals who exhibit the disease are homozygous for a recessive allele that results in a defect in the amount or function of the normal protein.

fails to produce a specific cellular protein in its active form. Over 9,000 human disorders are caused by mutations in single protein-encoding genes; most of these are recessive alleles. With a human genome size of about 22,000 protein-encoding genes, roughly 40% of such genes are known to cause some kind of abnormality when mutations alter the expression or functionality of their gene product.

However, it should be emphasized that not all recessive alleles are detrimental with regard to phenotype. For example, recessive alleles that result in blue eyes and light skin have spread rapidly in human populations that occupy Northern Europe (look ahead to Figure 22.23).

Recessive Alleles That Cause Diseases May Have Multiple Effects on Phenotype

Single-gene disorders, such as those described in Table 15.2, often illustrate the phenomenon of **pleiotropy,** which means that a mutation in a single gene can have multiple effects on an individual's phenotype. Pleiotropy occurs for several reasons, including the following:

1. The expression of a single gene can affect cell function in more than one way. For example, a defect in a microtubule protein may affect cell division as well as cell movement.
2. A gene may be expressed in different cell types in a multicellular organism.
3. A gene may be expressed at different stages of development.

In this genetics unit, we tend to discuss genes as they affect a single trait. This educational approach allows us to appreciate how genes function and how they are transmitted from parents to offspring. However, this focus may also obscure how amazing genes really are. In all or nearly all cases, the expression of a gene is pleiotropic with regard to the characteristics of an organism. The expression of any given gene influences the expression of many other genes in the genome, and vice versa. Pleiotropy is revealed when researchers study the effects of gene mutations.

As an example of a pleiotropic effect, let's consider cystic fibrosis (CF), which we discussed earlier as an example of a recessive human disorder (see Figure 15.11). The normal allele encodes a protein called the cystic fibrosis transmembrane regulator (CFTR), which regulates ion balance by allowing the transport of chloride ions (Cl^-) across epithelial cell membranes (refer back to Figure 1.12). The mutation that causes CF diminishes the function of this Cl^- transporter, affecting several parts of the body in different ways.

- The most severe symptom of CF is the production of thick mucus in the lungs, which occurs because of a water imbalance.
- Similarly, thick mucus can also block the tubes that carry digestive enzymes from the pancreas to the small intestine. Without these enzymes, certain nutrients are not properly absorbed into the body. As a result, persons with CF may show poor weight gain.
- In sweat glands, the normal Cl^- transporter has the function of recycling salt out of the glands and back into the skin before it can be lost to the outside world. Persons with CF have excessively salty sweat due to their inability to recycle salt back into their skin cells.

Taken together, these observations indicate that a defect in CFTR has multiple effects throughout the body.

Incomplete Dominance Results in an Intermediate Phenotype

We will now turn our attention to an example in which the alleles for a given gene do not show a simple dominant/recessive relationship. In some cases, a heterozygote that carries two different alleles exhibits a phenotype that is intermediate between those of the homozygotes that each carry two identical alleles. This phenomenon is known as **incomplete dominance.**

In 1905, Carl Correns discovered this pattern of inheritance for alleles affecting flower color in the four-o'clock plant (*Mirabilis jalapa*). **Figure 15.14** shows a cross between two four-o'clock plants: a red-flowered homozygote and a white-flowered homozygote. The

Figure 15.14 Incomplete dominance in the four-o'clock plant. When red-flowered and white-flowered homozygotes ($C^R C^R$ and $C^W C^W$) are crossed, the resulting heterozygote ($C^R C^W$) has an intermediate phenotype of pink flowers.

allele for red flower color is designated C^R, and the white allele is C^W. These alleles are designated with superscripts rather than upper- and lowercase letters because neither allele is dominant. The offspring of this cross have pink flowers—they are $C^R C^W$ heterozygotes with an intermediate phenotype. If these F_1 offspring are allowed to self-fertilize, the F_2 generation has 1/4 red-flowered plants, 1/2 pink-flowered plants, and 1/4 white-flowered plants. This is a 1:2:1 phenotype ratio, rather than the 3:1 ratio observed for simple Mendelian inheritance. What is the molecular explanation for this ratio? In this case, the red allele encodes a functional protein needed to produce a red pigment, whereas the white allele is a mutant allele that is nonfunctional. In the $C^R C^W$ heterozygote, 50% of the protein encoded by the C^R allele is not sufficient to produce the red-flower phenotype, but it does provide enough pigment to result in pink flowers.

The Environment Plays a Vital Role in the Making of a Phenotype

In this chapter, we have been mainly concerned with the effects of genes on phenotypes. In addition, phenotypes are shaped by an organism's environment. An organism cannot exist without its genes or without an environment in which to live. Both are indispensable for life. An organism's genotype provides the plan to create a phenotype, and the environment provides nutrients and energy so that the plan can be executed.

The **norm of reaction** is the phenotype range that individuals with a particular genotype exhibit under differing environmental conditions. To evaluate the norm of reaction, researchers study members of true-breeding strains that have the same genotypes and subject them to different environmental conditions. For example, **Figure 15.15** shows the norm of reaction for genetically identical plants that are homozygous for an allele that confers maximal height (like the tall allele in pea plants). Even so, the height of the plant depends on temperature. As shown in the figure, these plants attain a maximal height when raised at 75°F. At 50°F and 85°F, the plants are substantially shorter. Growth cannot occur below 40°F or above 95°F.

Biology Principle

Living Organisms Interact with Their Environment

As seen in this figure, the environment plays a key role in the outcome of traits.

Figure 15.15 The norm of reaction. The norm of reaction is the range of phenotypes that a population of organisms with a particular genotype exhibit under different environmental conditions. In this example, genetically identical plants were grown at different temperatures in a greenhouse and then measured for height.

Concept Check: *Could you study the norm of reaction in a wild population of squirrels?*

15.4 Reviewing the Concepts

- Recessive inheritance is often due to a loss-of-function allele. In many simple dominant/recessive relationships, the heterozygote has a dominant phenotype because 50% of the normal protein is sufficient to produce that phenotype (Figure 15.13).
- Mutant alleles, which are recessive, are responsible for many inherited diseases in humans (Table 15.2).
- In many cases, the effects of a mutant allele are pleiotropic, meaning the expression of the gene affects several different aspects of bodily structure and function.
- Incomplete dominance occurs when a heterozygote has a phenotype that is intermediate between that of either homozygote. This occurs because 50% of the functional protein is not enough to produce the same phenotype as a homozygote (Figure 15.14).

- Genes and the environment interact to determine an individual's phenotype. The norm of reaction is the phenotype range that individuals with the same genotype exhibit under different environmental conditions (Figure 15.15).

15.4 Testing Your Knowledge

1. In pea plants, *P* is a dominant allele that confers purple flower color and *p* is a recessive allele that confers white flowers. If the function of the flower-color gene is to make purple pigment, how do you explain the white phenotype?
   a. The *P* allele is a loss-of-function allele.
   b. The *p* allele is a loss-of-function allele.
   c. The *PP* homozygote does not make any purple pigment.
   d. Both a and b are correct.

2. The effects of a gene may be pleiotropic because
   a. the expression of a gene can affect cell function in more than one way.
   b. a gene may be expressed in different cell types in a multicellular organism.
   c. a gene may be expressed at different stages of development.
   d. All of the above are correct.

15.5 Sex Chromosomes and X-Linked Inheritance Patterns

Learning Outcomes

1. Describe different systems of sex determination in animals and plants.
2. **SCISKILLS** ▶ Predict the outcome of crosses when genes are located on sex chromosomes.
3. Explain why X-linked recessive traits are more likely to occur in males.

We will now turn our attention to genes located on **sex chromosomes.** As you learned in Chapter 14, sex chromosomes are a distinctive pair of chromosomes that are different in males and females and that determine the sex of individuals. Sex chromosomes are found in many but not all species with two sexes. The study of sex chromosomes proved pivotal in confirming the chromosome theory of inheritance. The distinctive transmission patterns of genes on sex chromosomes helped early geneticists show that particular genes are located on particular chromosomes. Later, other researchers became interested in these genes because some of them were found to cause inherited diseases in humans.

In this section, we will consider a few mechanisms by which sex chromosomes in various species determine an individual's sex. We will then explore the inheritance patterns of genes on sex chromosomes and why recessive alleles are expressed more frequently in males than in females.

In Many Species, Sex Differences Are Due to the Presence of Sex Chromosomes

Several mechanisms of sex determination have been found in different species of animals. Some examples are described in **Figure 15.16.** All of these mechanisms involve chromosome differences between the sexes, and most involve a difference in a single pair of sex chromosomes.

X-Y System In the X-Y system in mammals, the somatic cells of males have one X and one Y chromosome, whereas female somatic cells contain two X chromosomes (Figure 15.16a). For example, the 46 chromosomes carried by human cells consist of 22 pairs of autosomes and 1 pair of sex chromosomes (either XY or XX). Which chromosome, the X or Y, determines sex? In mammals, the presence of the Y chromosome causes maleness. This is known from the analysis of rare individuals that carry chromosome abnormalities. For example, mistakes that occasionally occur during meiosis may produce an individual that carries two X chromosomes and one Y chromosome. Such an individual develops into a male. A gene called the *SRY* gene located on the Y chromosome of mammals plays a key role in the developmental pathway that leads to maleness.

X-0 System The X-0 system operates in many insects (Figure 15.16b). Unlike the X-Y system in mammals, the presence of the Y chromosome in the X-0 system does not determine maleness. Females in this system have a pair of sex chromosomes and are designated XX. In some insect species that follow the X-0 system,

(a) The X-Y system in mammals

(b) The X-0 system in certain insects

(c) The Z-W system in birds

(d) The haplodiploid system in bees

Figure 15.16 Different mechanisms of sex determination in animals. The numbers shown in the circles indicate the numbers of autosomes.

✓ **Concept Check:** *If a person is born with only one X chromosome and no Y chromosome, would you expect that person to be a male or a female? Explain your answer.*

the male has only one sex chromosome, the X. In other X-0 insect species, such as *Drosophila melanogaster,* the male has both an X chromosome and a Y chromosome. In all cases, an insect's sex is determined by the ratio between its X chromosomes and its sets of autosomes. If a fly has one X chromosome and is diploid for the autosomes (2n), this ratio is 1/2, or 0.5. This fly will become a male, whether or not it receives a Y chromosome. On the other hand, if a diploid fly receives two X chromosomes, the ratio is 2/2, or 1.0, and the fly becomes a female.

Z-W System In some species, such as birds and some fish, the male carries two similar chromosomes (Figure 15.16c). This is called the Z-W system to distinguish it from the X-Y system found in mammals. The male is ZZ, and the female is ZW.

Haplodiploid System An interesting mechanism of sex determination known as the haplodiploid system is found in bees (Figure 15.16d). The male bee, or drone, is produced from an unfertilized haploid egg. Therefore, male bees are haploid individuals. Females, both worker bees and queen bees, are produced from fertilized eggs and are diploid.

Environmental Systems Although sex in many species of animals is determined by chromosomes, other mechanisms are also known. In certain reptiles and fish, sex is controlled by environmental factors such as temperature. For example, in the American alligator (*Alligator mississipiensis*), temperature controls sex development. When eggs of this alligator are incubated at 33°C, nearly all of them produce male individuals. When the eggs are incubated at a temperature significantly below 33°C, they produce nearly all females, whereas increasing percentages of females are produced above 34°C.

In Humans, X-Linked Recessive Traits Are More Likely to Occur in Males

In humans, the X chromosome is rather large and carries over 1,000 protein-encoding genes, whereas the Y chromosome is quite small and has fewer than 100. Therefore, many genes are found on the X chromosome but not on the Y; these are known as **X-linked genes.** By comparison, fewer genes are known to be Y-linked, meaning they are found on the Y chromosome but not on the X. **Sex-linked genes** are found on one sex chromosome but not on the other. Because fewer genes are found on the Y chromosome, the term usually refers to X-linked genes. In mammals, a male cannot be described as being homozygous or heterozygous for an X-linked gene, because these terms apply to genes that are present in two copies. Instead, the term **hemizygous** is used to describe an individual with only one copy of a particular gene. A male mammal is said to be hemizygous for an X-linked gene.

Many X-linked recessive alleles cause diseases in humans, and these diseases occur more frequently in males than in females. An example is the X-linked recessive disorder called classical hemophilia (hemophilia A). In individuals with hemophilia, blood does not clot properly, and a minor cut may bleed for a long time. Common accidental injuries that are minor in most people pose a threat of severe internal or external bleeding for hemophiliacs. Hemophilia A is caused by an X-linked recessive allele that encodes a defective form of a clotting protein. If a mother is a heterozygous carrier of hemophilia A, each of her children has a 50% chance of inheriting the recessive allele. The following Punnett square shows a cross between an unaffected father and a heterozygous mother. X^H designates an X chromosome carrying the functional allele, and X^{h-A} is the X chromosome that carries the recessive allele for hemophilia A.

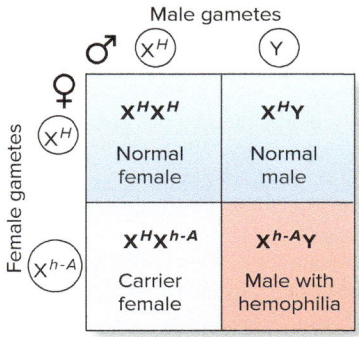

Although each child has a 50% chance of inheriting the hemophilia allele from the mother, only 1/2 of the sons will exhibit the disorder. Because a son inherits only one X chromosome, a son who inherits the recessive allele from his mother will have hemophilia. However, a daughter inherits an X chromosome from both her mother and her father. In this example, a daughter who inherits the recessive (hemophilia-causing) allele from her mother also inherits a functional allele from her father. This daughter will not have hemophilia, but if she passes the recessive allele to her son, he will have hemophilia.

15.5 Reviewing the Concepts

- Most species of animals have separate male and female sexes. In many species, sex chromosomes determine an individual's sex (Figure 15.16).
- X-linked genes are found on the X chromosome but not the Y chromosome. X-linked recessive alleles in humans cause diseases, such as hemophilia, which are more likely to occur in males.

15.5 Testing Your Knowledge

1. On rare occasions, an organism may inherit an abnormal number of sex chromosomes. If an individual is XXY, what is the phenotype in a mammal and in an insect?
   a. male mammal and male insect
   b. male mammal and female insect
   c. female mammal and male insect
   d. female mammal and female insect
   e. None of the above are correct.

2. A woman is heterozygous for an X-linked trait, hemophilia A. If she has a child with a man without hemophilia A, what is the probability that the child will be a male with hemophilia A? (Note: Assume there is a 50% chance that a child will be a male.)
   a. 100%
   b. 75%
   c. 50%
   d. 25%
   e. 0%

Assessing Your Knowledge and Skills

Foundational Knowledge

Multiple Choice

1. Experimental advantages of using pea plants included which of the following?
   a. They came in several varieties.
   b. They were capable of self-fertilization.
   c. They were easy to cross.
   d. Their anthers could be removed prior to pollination.
   e. All of the above were advantages.

2. Which of the following phenomena cannot be deduced from a single-factor cross?
   a. Mendel's law of segregation
   b. dominant/recessive relationships
   c. Mendel's law of independent assortment
   d. incomplete dominance
   e. All of the above can be deduced from a single-factor cross.

3. A pea plant has the genotype *TtRr*. The independent assortment of these two genes occurs at _____ because chromosomes with the _____ alleles line up independently of the _____ alleles.
   a. metaphase of meiosis I, *T* and *t*, *R* and *r*
   b. metaphase of meiosis I, *T* and *R*, *t* and *r*
   c. metaphase of meiosis II, *T* and *t*, *R* and *r*
   d. metaphase of meiosis II, *T* and *R*, *t* and *r*
   e. metaphase of mitosis, *T* and *t*, *R* and *r*

4. A gene that affects more than one trait is said to be
   a. dominant.
   b. wild-type.
   c. dihybrid.
   d. pleiotropic.
   e. heterozygous.

5. Genes located on one sex chromosome but not the other are said to be
   a. incompletely dominant.
   b. dominant.
   c. hemizygous.
   d. sex-linked.
   e. autosomal.

Short Answer

1. Describe the difference between genotype and phenotype. Give an example. Is it possible for two individuals to have the same phenotype but different genotypes?

2. **PRINCIPLES** A principle of biology is that living organisms interact with their environment. Discuss how the environment plays a key role in determining the outcome of an individual's traits.

Critical-Thinking Skills

Multiple Choice

1. A *Tt* plant is crossed to a *tt* plant. What is the expected outcome for the ratio of offspring from this cross?
   a. 3 tall : 1 dwarf
   b. 1 tall : 1 dwarf
   c. 1 tall : 3 dwarf
   d. 2 tall : 1 dwarf
   e. 1 tall : 2 dwarf

2. A cross is made between a pea plant that is *RrYy* and a plant that is *rrYy*. What is the predicted outcome of the seed phenotypes?
   a. 9 round, yellow : 3 round, green : 3 wrinkled, yellow : 1 wrinkled, green
   b. 9 round, green : 3 wrinkled, green : 3 round, yellow : 1 wrinkled, yellow
   c. 3 round, yellow : 3 round, green : 1 wrinkled, yellow : 1 wrinkled, green
   d. 3 round, yellow : 1 round, green : 3 wrinkled, yellow : 1 wrinkled, green
   e. 1 round, yellow : 1 round, green : 1 wrinkled, yellow : 1 wrinkled, green

3. Which of the following would *not* be observed in a pedigree if a genetic disorder was inherited in a recessive manner?
   a. Two unaffected parents have an affected offspring.
   b. Two unaffected parents have an unaffected offspring.
   c. Two affected parents have an unaffected offspring.
   d. One affected and one unaffected parent have an unaffected offspring.
   e. All of the above would be possible for a recessive disorder.

4. A woman does not show symptoms of hemophilia A. She has a son with a man with hemophilia, and that son does have hemophilia. Their next child is a daughter. What is the probability that this daughter will have hemophilia?
   a. 100%
   b. 75%
   c. 50%
   d. 25%
   e. 0%

5. A hypothetical flowering plant species produces blue, light blue, and white flowers. To determine the inheritance pattern, the following crosses were conducted, with the results indicated:

   blue × blue → all blue

   white × white → all white

   blue × white → all light blue

   What type of inheritance pattern does this represent?
   a. simple Mendelian
   b. X-linked
   c. sex-linked
   d. incomplete dominance
   e. pleiotropy

Short Answer

1. A peach without its fuzz is called a nectarine; basically, a nectarine is a smooth peach. A peach tree can produce an occasional nectarine. The trait of fuzziness is controlled by a single gene that is found in two alleles, *D* and *d*. Based on your understanding of gene function at the molecular level, do you think it is more likely that the peach is homozygous for a recessive allele or that the nectarine is homozygous for the recessive allele? Explain your choice.

2. When examining a human pedigree, what patterns do you look for to distinguish between X-linked recessive inheritance and autosomal recessive inheritance?

Transmission of Genetic Information from Parents to Offspring II: Epigenetics, Linkage, and Extranuclear Inheritance

16

©Andia/Alamy

Queen bee development. Female honeybees that are fed royal jelly throughout their entire larval stage and into adulthood develop into queen bees. The larger queen bee is shown with a blue disk labeled 68. By comparison, those larvae that do not receive this diet become smaller worker bees. These differences in development are caused by epigenetic modifications.

Chapter Outline

16.1 Overview of Epigenetics

16.2 Epigenetics: Genomic Imprinting

16.3 Epigenetics: X-Chromosome Inactivation

16.4 Epigenetics: Effects of Environmental Agents

16.5 Extranuclear Inheritance: Organelle Genomes

16.6 Linkage of Genes on the Same Chromosome

Assessing Your Knowledge and Skills

Female honeybees (*Apis mellifera*) are of two types: queen bees and worker bees (see the chapter-opening photo). Queens are larger, live for years, and produce up to 2,000 eggs each day. By comparison, the smaller worker bees are sterile, typically live only for weeks, and engage in specialized types of work, such as constructing and cleaning comb cells, nurturing larvae, guarding the hive entrance, and foraging for pollen and nectar. The striking differences between queen and worker bees are caused by differences in their diets. Certain worker bees, called nurse bees, produce a nutritive substance called royal jelly from glands in their mouths. All female larvae are initially fed royal jelly, but those that are bathed in royal jelly throughout their larval development and feed on it into adulthood become queens. Alternatively, female larvae that are weaned at an early stage of development and switched to a diet of pollen and nectar become worker bees.

The differences between queen and worker bees are not due to differences in the alleles they carry. We cannot predict the phenotype of female bees based on their genotype. Likewise, we cannot carry out crosses and predict the ratio between queen and worker bees based on Mendel's laws. This observation does not mean Mendel was wrong, however. A large number of genes do obey his laws and follow the types of inheritance patterns that we examined in Chapter 15. However, many genes, including those that control female development in honeybees, do not.

How do we distinguish genes that follow a Mendelian pattern from genes that do not? Mendelian inheritance patterns follow three general rules:

1. Except in the case of rare mutations, genes are passed unaltered from cell to cell, and from generation to generation.
2. The genes obey Mendel's law of segregation.
3. For crosses involving two or more genes, the genes obey Mendel's law of independent assortment.

For Mendelian patterns of inheritance, we must also consider other factors, such as the dominant/recessive relationship of alleles and whether or not the gene is on the X chromosome. Once these additional factors are understood, we can predict the phenotypes of offspring from their genotypes and predict the outcome of crosses based on Mendel's laws. Most genes in eukaryotic species follow a Mendelian pattern of inheritance.

In this chapter, we will examine additional types of inheritance patterns that deviate from a Mendelian pattern because one of the three rules is broken. In Sections 16.1–16.4, we will explore epigenetics, which breaks rule number 1. By epigenetic mechanisms, genes that are passed to offspring are altered, either in the gametes that produce the offspring or during embryonic development of the offspring. This alteration does not change the DNA sequence of a gene but, instead, alters the structure of chromatin and thereby affects the gene's expression. We will then examine inheritance patterns that arise because some genetic material is not located in the cell nucleus. Certain organelles, such as mitochondria and chloroplasts, contain their own genetic material. The inheritance patterns of genes within mitochondria and chloroplasts do not obey rule number 2; they do not follow the law of segregation. Finally, we will consider inheritance patterns that violate rule number 3, which is Mendel's law of independent assortment. As you will learn, some genes do not independently assort because they are located on the same chromosome.

16.1 Overview of Epigenetics

Learning Outcomes

1. Define epigenetics and epigenetic inheritance.
2. Outline the types of molecular changes that underlie epigenetic effects on gene expression.

The term epigenetics was first coined by British biologist Conrad Waddington in 1941. The prefix epi-, which means over, suggests that some types of changes in gene expression are at a level that goes beyond changes in DNA sequences. How do geneticists distinguish an epigenetic effect from other types of gene regulation, such as those described in Chapter 12 (refer back to Section 12.4)? An epigenetic effect begins with an initial event that causes a change in gene expression. For example, DNA methylation may inhibit transcription. However, for this to be an epigenetic effect, the change must be passed from cell to cell and must not involve a change in the sequence of DNA.

Thus, a key feature of an epigenetic effect is the long-term maintenance of a change in gene expression. As an example, let's consider muscle cells in humans. Some genes in the human genome should be expressed in muscle cells, such as the genes that encode the proteins called actin and myosin, which are required for muscle contraction (see Chapter 35). In embryonic muscle cells, these genes are subjected to epigenetic changes that promote their expression through adulthood. Alternatively, in other cell types, these same genes are inhibited by epigenetic changes such as DNA methylation. As the embryo grows and eventually becomes an adult, these epigenetic changes are passed from cell to cell so that adult muscle cells express actin and myosin genes at very high levels, whereas many other cell types do not.

Some epigenetic changes, such as those involving the expression or inhibition of actin and myosin genes, are relatively permanent during the life of an individual. Alternatively, other epigenetic changes may be reversible during the life of an individual or may be reversible from one generation to the next. For example, many species of flowering plants undergo **vernalization,** which is the process in which certain species of plants require an exposure to cold temperatures in order to flower. After vernalization, plants do not necessarily initiate flowering, but they acquire the ability to do so. Most commonly, plants are vernalized by exposure to cold winter temperatures and then flower the following spring or summer. Vernalization involves epigenetic changes to genes that play a role in flowering. These epigenetic changes are induced by cold winter temperatures, and they are maintained during the flowering season. However, the epigenetic changes are reversed after the flowering season is over.

Although researchers are still debating the proper definition, one way to define epigenetics is the following.

> ***Epigenetics*** *is the study of mechanisms that lead to changes in gene expression that can be passed from cell to cell and are reversible but do not involve a change in the sequence of DNA. This type of change may also be called an* ***epimutation****—a heritable change in gene expression that does not alter the sequence of DNA.*

In multicellular species that reproduce via gametes, some types of epigenetic changes are passed from parent to offspring.

- **Epigenetic inheritance** is the passage of epigenetic changes from parent to offspring via sperm or egg cells.

In Section 16.2, we will consider an example of epigenetic inheritance called genomic imprinting. However, not all epigenetic changes are inherited from parents. For example, a person may be exposed to an environmental agent in cigarette smoke that causes an epigenetic change in a lung cell that is subsequently transmitted from cell to cell and promotes lung cancer. Such a change would not be transmitted to offspring.

The molecular mechanisms of epigenetics are the subject of a large amount of recent research. The most common types of molecular changes that underlie epigenetic effects on gene expression are DNA methylation, chromatin remodeling, covalent histone modification, and the localization of histone variants (**Table 16.1**). These types of changes are also involved in transient (nonepigenetic) gene regulation that is not transmitted from cell to cell. The details of these mechanisms were examined in Chapter 12 (refer back to Section 12.4). In some cases, epigenetic changes repress the transcription of a given gene; in other cases, they activate gene transcription. In Sections 16.2–16.4, we will

Table 16.1	Chromatin Modifications That May Have an Epigenetic Effect on Gene Expression
DNA methylation	Methyl groups may be attached to cytosine bases in DNA. When this occurs near promoters, transcription is usually repressed.
Chromatin remodeling	Nucleosomes may be moved to new locations or evicted. When such changes occur in the vicinity of promoters, the level of transcription may be altered. Also, larger-scale changes in chromatin structure may occur, such as those that happen during X-chromosome inactivation in female mammals, discussed later in this chapter.
Covalent histone modification	Specific amino acid side chains in the amino terminal tails of histones can be covalently modified. For example, they can be acetylated or phosphorylated. Such modifications may repress or activate transcription.
Localization of histone variants	Histone variants may become localized to specific locations, such as near the promoters of genes, and affect transcription.

consider examples in which these types of changes have an epigenetic effect on gene expression and thereby affect an individual's phenotype.

Interestingly, a 2017 study by American biologists Sabrina McNew, Michael Skinner, and colleagues suggested a potential role of epigenetic changes in the adaptation of species to their local environments. They focused their work on two species of Darwin's finches, *Geospiza fortis* and *G. fuliginosa*. The researchers found dramatic epigenetic differences between the urban and rural populations of both species, based on DNA methylation analysis.

16.1 Reviewing the Concepts

- Epigenetics is the study of mechanisms that lead to changes in gene expression that can be passed from cell to cell and are reversible but do not involve a change in the sequence of DNA.
- The most common types of molecular changes that underlie epigenetic effects on gene expression are DNA methylation, chromatin remodeling, covalent histone modification, and the localization of histone variants (Table 16.1).

16.1 Testing Your Knowledge

1. Which of the following can have an epigenetic effect that alters gene expression?
   a. DNA methylation
   b. chromatin remodeling
   c. covalent histone modification
   d. localization of histone variants
   e. all of the above

16.2 Epigenetics: Genomic Imprinting

Learning Outcomes

1. **SCISKILLS** ▶ Predict the outcome of crosses for imprinted genes.
2. Explain the molecular basis of genomic imprinting.

The term imprinting implies a type of marking process that has a memory. For example, newly hatched birds identify marks on their parents, which allows them to distinguish their parents from other individuals. The term **genomic imprinting** refers to an analogous situation in which a segment of DNA is marked, and that mark is retained and recognized throughout the life of the organism inheriting the marked DNA. However, depending on the gene, it may be marked by females during egg formation or by males during sperm formation, but not both. This marking process, which involves epigenetic modifications, affects whether or not the gene is expressed in the offspring. Depending on how a particular gene is marked by one of the parents, the offspring expresses either the maternal or the paternal allele, but not both. Imprinted genes do not follow a Mendelian pattern of inheritance because imprinting causes the offspring to distinguish between maternal and paternal alleles. From an epigenetic perspective, the cells in the offspring are able to "remember" an event that occurred during gamete formation in their parents.

Genomic imprinting, which was discovered in the early 1980s, occurs in numerous species, including insects, plants, and mammals. The number of human genes that are imprinted is still a matter of controversy, but most estimates are in the range of a few hundred to over 1,000. In this section, we will consider an example of genomic imprinting that occurs in mammals.

For Imprinted Genes, the Gene from Only One Parent Is Expressed

One of the first imprinted genes to be identified is a gene called *Igf2*, found in mice and other mammals. This gene encodes a growth hormone called insulin-like growth factor 2 (Igf2), which is needed for proper growth. If a functional copy of this gene is not expressed, a mouse will be dwarf. The *Igf2* gene is known to be located on an autosome, not on a sex chromosome. Because mice are diploid, they have two copies of this gene, one from each parent.

Researchers have discovered mutations in the *Igf2* gene that block the function of the Igf2 hormone. Such a mutant allele is designated *Igf2*⁻. When a homozygous *Igf2 Igf2* mouse and a homozygous *Igf2*⁻ *Igf2*⁻ mouse are crossed, the results are surprising (**Figure 16.1**). If the female parent is homozygous for the mutant allele and the male parent is homozygous for the functional allele (left side), all of the offspring grow to a normal size. By contrast, if the female is homozygous for the functional allele and the male is homozygous for the mutant allele (right side), all of the offspring are dwarf. The reason this result is so surprising is that the normal-size and dwarf offspring have the same genotype (*Igf2 Igf2*⁻) but different phenotypes! In mice, the *Igf2* gene is imprinted in such a way that only the paternal allele is expressed, which means it is transcribed into mRNA. The maternal allele is not transcribed. The newborn mice shown on the left side of the photograph of Figure 16.1 are normal size, because they express a functional paternal allele. By contrast, the mice on the right side are dwarf, because they express a paternal allele that is defective and results in a nonfunctional hormone.

Figure 16.1 **An example of genomic imprinting in the mouse.** In the cross on the left, a homozygous female carrying the mutant allele *Igf2⁻* was crossed to a homozygous male with the functional *Igf2* allele. Offspring are normal size, because the paternal allele is expressed. In the cross on the right, a homozygous *Igf2 Igf2* female was crossed to a homozygous male carrying the mutant allele. In this case, offspring are dwarf because the paternal allele is defective and the maternal allele is not expressed.

Concept Check: If you crossed an *Igf2 Igf2⁻* male mouse to an *Igf2 Igf2⁻* female mouse, what would be the expected results?

DNA Methylation Affects the Transcription of an Imprinted Gene

Why is the maternal gene encoding Igf2 not transcribed into mRNA? To answer this question, we need to consider the regulation of gene transcription in eukaryotes. As discussed in Chapter 12, DNA methylation, which is the attachment of methyl (—CH₃) groups to bases of DNA, can alter gene transcription. Researchers have discovered that DNA methylation is the marking process that occurs during the imprinting of certain genes, including the *Igf2* gene. For most genes, DNA methylation silences gene expression by inhibiting the initiation of transcription or by causing the chromatin in a region to become more compact. By contrast, for a few imprinted genes, DNA methylation may enhance gene expression by attracting activator proteins to the promoter or by preventing the binding of repressor proteins.

Figure 16.2 shows the imprinting process in which a maternal gene is methylated. The left side of the figure follows the marking process during the life of a female individual; the right side follows the same process in a male. Both individuals received a methylated gene from their mother and a nonmethylated copy of the same gene from their father. Via cell division, the zygote develops into a multicellular organism. Each time a somatic cell divides, enzymes in the cell maintain the methylation of the maternal gene, but the paternal gene remains unmethylated. If methylation inhibits transcription of this gene, only the paternal copy will be expressed in the somatic cells of both the male and female offspring.

The methylation state of an imprinted gene may be altered when individuals make gametes. First, the methylation is erased (Figure 16.2, step 2). Next, the gene may be methylated again, but that depends on whether the individual is female or male. In females making eggs, both copies of the gene are methylated; in males making sperm, neither copy is methylated. When we consider the effects of methylation over the course of two or more generations, we can see how this phenomenon results in an epigenetic transmission pattern. The male in Figure 16.2 has inherited a methylated gene from his mother that is transcriptionally silenced in his somatic cells. Although he does not express this gene during his lifetime, he can pass on an active, nonmethylated copy of exactly the same gene to his offspring.

Genomic imprinting is a recently discovered phenomenon that has been shown to occur for many genes in mammals. For some genes, such as *Igf2*, the maternal allele is silenced, but for other genes, the paternal allele is silenced. Although several hypotheses have been advanced, biologists are still trying to determine whether this curious marking process confers any benefit to the individual.

16.2 Reviewing the Concepts

- As a result of genomic imprinting, offspring express either a maternal or a paternal allele, depending on how a particular gene is marked, or imprinted (Figure 16.1).
- During gamete formation, DNA methylation of an allele from one parent is a mechanism to achieve imprinting (Figure 16.2).

16.2 Testing Your Knowledge

1. A male mouse that is homozygous for the functional allele of the *Igf2* gene is mated to a female that is heterozygous, carrying one functional copy and one mutant (nonfunctional) copy of the gene. What would be the expected outcome of this cross?
   a. all normal-size offspring
   b. 1/2 normal-size and 1/2 dwarf offspring
   c. all dwarf offspring
   d. 3/4 normal-size and 1/4 dwarf offspring
   e. none of the above

Figure 16.2 Genomic imprinting via DNA methylation. The cells of the female and male offspring at the top of this figure have a methylated gene inherited from the mother and a nonmethylated version of the same gene inherited from the father. This pattern of methylation is maintained in the somatic cells of both the female and male offspring. The methylation is erased during gamete formation, but in females, the gene is methylated again at a later stage in the formation of eggs. Therefore, females always transmit a methylated, transcriptionally silent copy of this gene, whereas males transmit a nonmethylated, transcriptionally active copy.

2. Which of the following statements regarding the mechanism of genomic imprinting is *false*?
   a. Genomic imprinting begins during the process that produces gametes.
   b. Genomic imprinting is an example of epigenetic inheritance.
   c. Genomic imprinting changes the DNA sequence of a gene.
   d. Methylation of an imprinted gene occurs differently during the formation of eggs versus sperm.
   e. It is possible for an imprinted allele that is silenced in the somatic cells of one individual to be active in the somatic cells of its offspring.

16.3 Epigenetics: X-Chromosome Inactivation

Learning Outcomes

1. Explain how X-chromosome inactivation may affect the phenotype of female mammals.
2. Describe the process of X-chromosome inactivation at the cellular level.

As we have seen, genomic imprinting involves epigenetic changes that occur during gamete formation and are passed from parent to offspring. Other types of epigenetic changes begin later, such as during embryonic development, and then persist into adulthood. For example, as mentioned in Section 16.1, epigenetic changes may silence actin and myosin genes in some embryonic cells, and these changes are passed from cell to cell so that the same genes are silenced in nonmuscle cells of the adult.

In this section, we will consider an example of epigenetic silencing that begins during embryonic development in female mammals. As discussed in Chapter 15, female mammals carry two X chromosomes in their cells, whereas males carry one X and one Y. During embryonic development in female mammals, one of the X chromosomes undergoes an epigenetic change called **X-chromosome inactivation (XCI)**. This process causes that X chromosome to become highly compacted, which silences the genes that it carries. In this section, we will examine how XCI may affect a female mammal's phenotype and how the process occurs at the cellular and molecular levels.

In Female Mammals, One X Chromosome Is Inactivated in Each Somatic Cell

In 1961, British geneticist Mary Lyon proposed the epigenetic phenomenon of X-chromosome inactivation, in which one of the two copies of the X chromosome in the somatic cells of female mammals is inactivated, meaning that its genes are not expressed. XCI is based on two lines of evidence.

- In 1949, Canadian physicians Murray Barr and Ewart Bertram identified a highly condensed structure in the cells of female cats that was not found in the cells of male cats. This structure was named a **Barr body** after one of its discoverers. It is now known that the Barr body is an inactivated X chromosome (**Figure 16.3**).

Figure 16.3 X-chromosome inactivation in female mammals. (a) A Barr body is seen on the periphery of a human nucleus (during interphase) after the nucleus has been stained with a DNA-specific dye. Because it is compact, the Barr body is the most brightly stained feature. (b) The same nucleus was labeled using a yellow fluorescent probe that recognizes the X chromosome. The Barr body has undergone epigenetic changes that make it more compact than the active X chromosome, which is to the left of the Barr body.
Courtesy of I. Solovei, University of Munich (LMU)

Concept Check: *How is the Barr body different from the other X chromosome in this cell?*

- Lyon's second line of evidence was the inheritance pattern of variegated coat colors in certain female mammals. A classic case is the calico cat, which has randomly distributed patches of black and orange fur (Figure 16.4a).

How do we explain this patchwork phenotype? According to Lyon's hypothesis, the calico pattern is due to the permanent inactivation of one X chromosome in each cell that forms a patch of the cat's skin, as shown in Figure 16.4b. The gene involved is an X-linked gene that occurs as an orange allele, X^O, and a black allele, X^B. A female cat heterozygous for this gene will be calico. (The cat's white underside is due to a dominant allele of a different, autosomal gene.) At an early stage of embryonic development, one of the two X chromosomes is randomly inactivated in each of the cat's somatic cells, including those that will give rise to the hair-producing skin cells. As the embryo grows and matures, the pattern of XCI is maintained during subsequent cell divisions. Skin cells derived from a single embryonic cell in which the X^B-carrying chromosome has been inactivated produce a patch of orange fur, because they express only the X^O allele that is carried on the active chromosome. Alternatively, a group of skin cells in which the chromosome carrying X^O has been inactivated express only the X^B allele, producing a patch of black fur. If female mammals are heterozygous for X-linked genes, approximately half of their somatic cells express one allele, whereas the rest of their somatic cells express the other allele. The result is an animal with randomly distributed patches of black and orange fur. These heterozygotes are called **mosaics** because they have somatic regions that are composed of two types of cells.

Why does XCI occur? Researchers have proposed that XCI achieves **dosage compensation,** an equalization of the expression of X-linked genes in male and female mammals. The inactivation of one X chromosome in females reduces the number of expressed copies (doses) of X-linked genes from two to one, the same as expressed in males. As a result, the expression of X-linked genes in females and males is roughly equal.

The X Chromosome Has an X Inactivation Center That Controls Compaction into a Barr Body

After Lyon's hypothesis was confirmed, researchers became interested in the cellular and molecular mechanisms of XCI. The cells of humans and other mammals have the ability to count their X chromosomes and allow only one of them to remain active. Additional X chromosomes are converted to Barr bodies. In normal females, two X chromosomes are counted and one is inactivated. In normal males, one X chromosome is counted and none inactivated.

On rare occasions, people are born with abnormalities in the number of their sex chromosomes. In the disorders known as Turner syndrome, triple X syndrome, and Klinefelter syndrome, the cells inactivate the number of X chromosomes necessary to leave a single active chromosome (Table 16.2). For example, in triple X syndrome, in which an extra X chromosome is found in each cell, two X chromosomes are converted to Barr bodies. In spite of XCI, people with these three syndromes do exhibit some abnormalities. The symptoms associated with these disorders may be due to effects that occur prior to XCI or because not all of the genes on the Barr body are completely silenced.

Although the mechanism of XCI is not entirely understood at the molecular level, a short region on the X chromosome called the **X inactivation center (Xic)** is known to play a critical role. Finnish-born American geneticist Eeva Therman and German-born American geneticist Klaus Patau determined that XCI is accomplished by counting the number of Xics and inactivating all X chromosomes except for one. In cells with two X chromosomes, if one of them is missing its Xic due to a chromosome mutation, neither X chromosome will be inactivated, because only one Xic is counted. Having two active X chromosomes is a lethal condition for a human female embryo.

Table 16.2	Relationship Between XCI and the Number of X Chromosomes		
Phenotype	Chromosome composition	Number of Barr bodies	Number of active X chromosomes
Normal female	XX	1	1
Normal male	XY	0	1
Turner syndrome (female)	X0	0	1
Triple X syndrome (female)	XXX	2	1
Klinefelter syndrome (male)	XXY	1	1

Biology Principle

Biologists Use Models to Test Predictions and Convey Ideas

Part (b) in this figure is a model for the process of X-chromosome inactivation (XCI). The Modeling Challenge in the figure legend asks you to propose a model for XCI in a mammal carrying three X chromosomes.

(a) A calico cat

Figure 16.4 X-chromosome inactivation in a calico cat. (a) A calico cat. (b) The calico pattern is due to random X-chromosome inactivation in a female that is heterozygous for an X-linked gene with black (X^B) and orange (X^O) alleles. The cells at the top of this figure represent a small mass of cells making up the very early embryo. In these cells, both X chromosomes are active. At an early stage of embryonic development, one X chromosome is randomly inactivated in each cell. The initial inactivation pattern is an epigenetic change that is maintained in the descendents of each cell as the embryo matures into an adult. The pattern of orange and black fur in the adult cat reflects the pattern of XCI in the embryo.
(a) ©G.K. & Vikki Hart/Getty Images

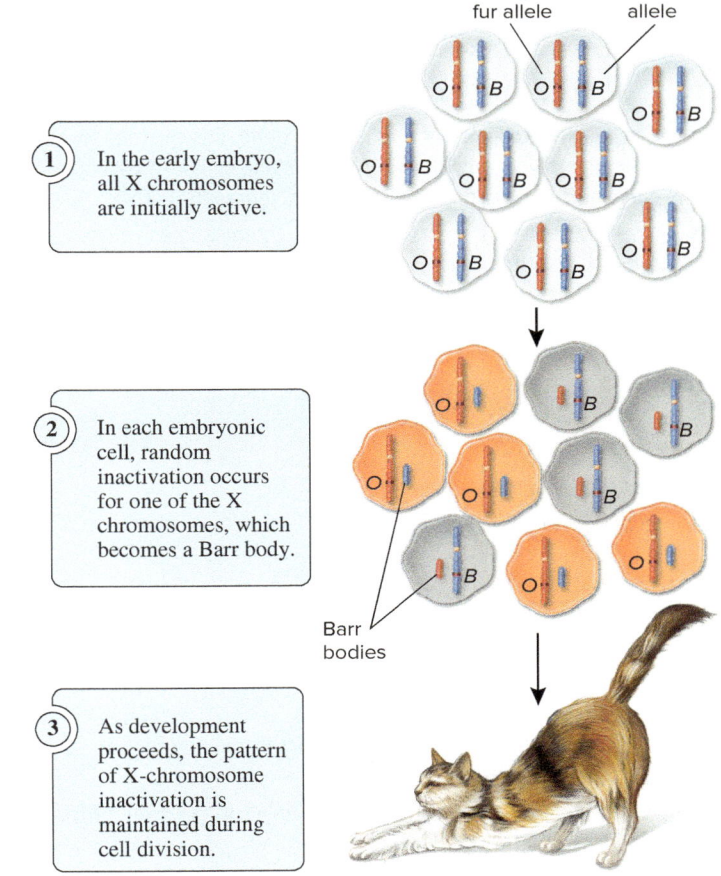

(b) Process of X-chromosome inactivation

Modeling Challenge: On rare occasions, a female mammal is born with three X chromosomes instead of the usual number of two. As described in Table 16.2, each somatic cell in such an XXX individual contains two Barr bodies following X-chromosome inactivation. Let's suppose that a gene affecting fur color occurs in three alleles: yellow (Y), orange (O), and black (B). In this mammalian species, X-chromosome inactivation occurs at the 8-cell stage. Draw an embryo with 32 cells. In each cell, color the background yellow, orange, or gray, to indicate that the yellow, orange, or gray allele is expressed, respectively. Also, as in Figure 16.4b, draw the active and inactive chromosomes.

The expression of a specific gene within the Xic is required for compaction of the X chromosome into a Barr body. This gene, discovered in 1991, is named *Xist* (for **X** **i**nactive **s**pecific **t**ranscript). The *Xist* gene product is a long RNA molecule that does not encode a protein. Instead, the role of *Xist* RNA is to coat one of the two X chromosomes during the process of XCI. After coating, proteins associate with the *Xist* RNA and cause epigenetic changes that promote the compaction of the chromosome into a Barr body. The *Xist* gene on the Barr body continues to be expressed after other genes on this chromosome have been silenced. The expression of the *Xist* gene also maintains a chromosome as a Barr body during cell division. Whenever a somatic cell divides in a female mammal, the Barr body is replicated to produce two Barr bodies.

16.3 Reviewing the Concepts

- If a female is heterozygous for an X-linked gene, this can lead to a mosaic phenotype, with some of the somatic cells expressing one allele and some expressing the other (Figures 16.3, 16.4).
- XCI in female mammals occurs when one X chromosome in every somatic cell is randomly inactivated. The counting of X inactivation centers (Xics) causes somatic cells to have only one active X chromosome (Table 16.2)

16.3 Testing Your Knowledge

1. Calico coat pattern in cats is the result of
   a. X-chromosome inactivation.
   b. X-linked inheritance.
   c. incomplete dominance.
   d. pleiotropy.
2. Which of the following statements regarding X-chromosome inactivation is *false*?
   a. Xic is the site where X-chromosome inactivation begins.
   b. Xic is where the *Xist* gene is located.
   c. The *Xist* gene on the inactivated X chromosome (the Barr body) is not expressed.
   d. The product of the *Xist* gene coats the inactivated X chromosome.

16.4 Epigenetics: Effects of Environmental Agents

Learning Outcomes

1. Explain how chemicals in the diet may affect the individual's phenotype.
2. List examples of chemicals that cause epigenetic changes that may contribute to cancer.

One of the most active fields in genetics is the study of certain environmental agents that cause epigenetic changes and thereby affect gene expression. Two areas that have received a great deal of attention are the effects of diet and the potential effects of toxic agents, such as carcinogens—cancer-causing agents. In this section, we will consider examples in which environmental agents promote epigenetic changes that affect an individual's phenotype or cause a disease such as cancer.

Chemicals in an Individual's Diet May Cause Epigenetic Changes That Affect Phenotype

At the beginning of this chapter, we considered how chemicals in royal jelly are responsible for producing queen bees (see the chapter-opening photo). Another striking example of how chemicals in the diet can promote epigenetic changes is illustrated by studies of the *Agouti* gene (designated *A*) found in mice. This gene encodes the Agouti signaling peptide that controls the deposition of yellow pigment in developing hairs. In mice that are homozygous for a functional allele, *AA*, the expression of this gene promotes the synthesis of pheomelanin, a yellow pigment. During the growth of a hair, melanocytes (pigment-producing cells) within a hair follicle initially make eumelanin, which is black. The transient expression of the *Agouti* gene causes the melanocytes to make pheomelanin. The melanocytes then revert to making black pigment. The result is a band of yellow pigment sandwiched between layers of black pigment, which gives a brown color. The yellow pigment is not synthesized near the tip of the hair, so the hair of *AA* mice is brown with black tips.

Researchers have identified several mutations that affect the expression of the *Agouti* gene. For example, mice that are homozygous for a loss-of-function allele, *aa*, have black fur because pheomelanin is not made. Alternatively, a gain-of-function mutation that causes the *Agouti* gene to be overexpressed results in a mouse with yellow fur. One such mutation is designated A^{vy}. (*A* refers to Agouti, *v* refers to viable, and *y* refers to yellow. The letter *v* is used because other mutations in the *Agouti* gene are not viable.) By characterizing the A^{vy} allele at the molecular level, researchers determined that it is due to the insertion of a new promoter next to the normal promoter of the *Agouti* gene. This new promoter is very active, which causes the overexpression of the *Agouti* gene.

An intriguing observation of mice carrying the A^{vy} allele is that they exhibit a wide phenotypic variation, ranging from yellow to mottled to pseudo-agouti (**Figure 16.5a**). Why should mice with the same genotype show such a wide range of phenotypic variation? Although the answer is not entirely understood, researchers have speculated that the new promoter in mice carrying the A^{vy} allele is very sensitive to epigenetic changes. In particular, this promoter may be more likely to be modified by DNA methylation, which would inhibit its function. Furthermore, a variety of environmental factors may cause this epigenetic change to occur. The sensitivity of this promoter to epigenetic modifications, together with variation in environmental factors, may explain the phenotypic variation seen in these mice.

A key factor that may affect epigenetic changes is diet. With regard to the A^{vy} allele, the exposure of pregnant female mice to different diets affects the phenotypes of the resulting offspring. In 2003, American geneticists Robert Waterland and Randy Jirtle conducted a study in which they investigated the effects of certain dietary supplements. Their goal was to determine if nutrients that are known to inhibit DNA methylation would alter the expression of the *Agouti* gene and thereby affect coat color. A variety of dietary factors can inhibit the enzyme DNA methyltransferase, which methylates DNA. These include folic acid, vitamin B_{12}, betaine, and choline chloride.

(a) Range in coat-color phenotypes in $A^{vy}a$ mice due to epigenetic changes

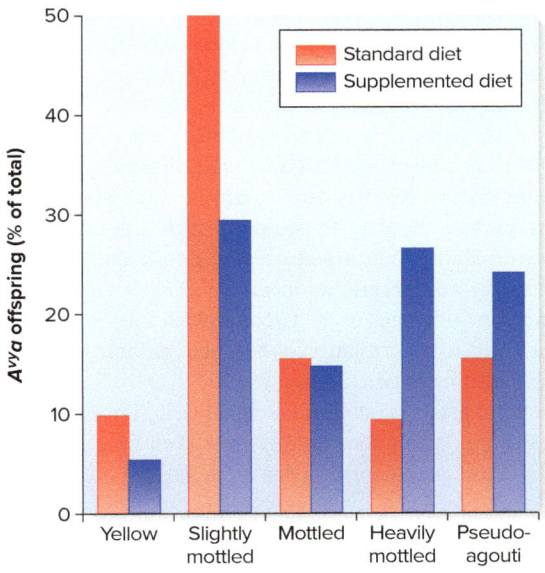

(b) Effect of diet on coat color

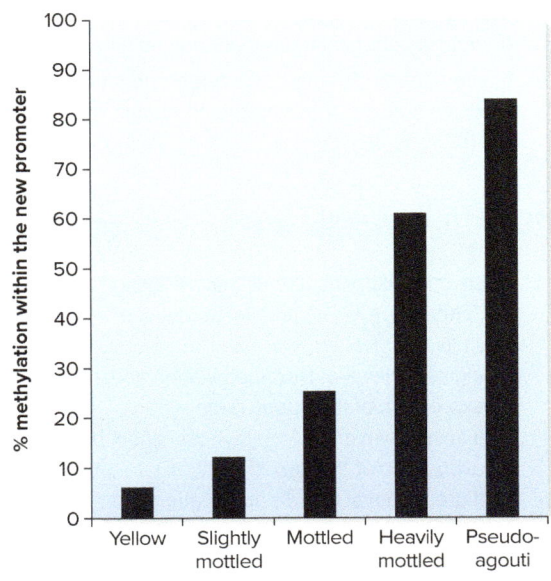

(c) Level of DNA methylation within the new promoter among mice with different coat colors

Figure 16.5 Dietary effects on coat color in mice. (a) Mice carrying the A^{vy} allele exhibit a range of phenotypes. The mice shown here are heterozygotes, $A^{vy}a$; they carry the A^{vy} allele and a loss-of-function allele (a). (b) Effects of diet supplementation on coat color. Red bars represent offspring from females given a standard diet, and blue bars represent offspring from females given a supplemented diet. (c) DNA methylation patterns among mice with different coat colors. Data in parts (b) and (c) are from Waterland, R. A., and Jirtle, R. L. 2003. Transposable Elements: Targets for Early Nutritional Effects on Epigenetic Gene Regulation. *Mol. Cell. Biol.* 23: 5293–5300. *(image & text)* Source: D.C. Dolinoy et al., "Maternal genistein alters coat color and protects A^{vy} mouse offspring from obesity by modifying the fetal epigenome," *Environ Health Perspect.* 2006 Apr, 114(4): 567–572.

Waterland and Jirtle began with female mice carrying the A^{vy} allele and divided them into a control group (which was fed a standard diet) and an experimental group (which was fed a diet supplemented with folic acid, vitamin B_{12}, betaine, and choline chloride). Both groups were fed their respective diets before and during pregnancy and up to the stage of weaning. Offspring that inherited the A^{vy} allele were then analyzed with regard to their coat color and levels of DNA methylation.

As expected, a range of coat colors was observed among the offspring (Figure 16.5b). However, the offspring of females that had been fed a supplemented diet tended to have darker coats. For example, over 25% of the offspring with heavily mottled coats had mothers that were fed a supplemented diet (blue bars), whereas less than 10% had mothers that were given a standard diet (red bars).

The coat colors of the offspring were correlated with the degree of methylation that occurred at the new promoter—offspring with darker coat color had greater levels of DNA methylation (Figure 16.5c). How do we explain these results? In the mice that are more yellow, the new promoter has undergone less DNA methylation. Therefore, the promoter remains active, leading to the transcription of the *Agouti* gene and the overproduction of yellow pigment. By contrast, the new promoter in the darker mice has undergone more methylation. Such methylation inhibits the overexpression of the *Agouti* gene and thereby prevents the overproduction of yellow pigment, resulting in darker fur.

Environmental Agents May Cause Epigenetic Changes That Are Associated with Human Diseases, Such as Cancer

One of the most active fields in genetics involves the study of epigenetic changes that contribute to human diseases. We have probably seen only "the tip of the iceberg" with respect to our understanding of this topic. Many medical studies have identified correlations between epigenetic changes and particular diseases. For example, a research study could compare one variable, such as the level of DNA methylation of a specific gene, to a second variable, such as the severity of a disease. If a high level of DNA methylation was associated with an increase in disease severity, this would be a positive correlation. Researchers analyze the data to decide if such a correlation is statistically significant. When a statistically significant correlation is obtained, how do we interpret its meaning? Such a result suggests a true **association**—changes in the two variables follow a pattern. For example, in a positive correlation, when one variable increases, the other variable also increases.

However, an association does not necessarily imply a cause-and-effect relationship. When considering the role of epigenetic changes and human disease, an association can arise in three common ways:

- The epigenetic changes directly contribute to the disease symptoms. There is a cause-and-effect relationship.
- Conversely, the disease symptoms may arise first, and then they cause subsequent epigenetic changes to happen. This is also a cause-and-effect relationship, but in the opposite direction.
- The association is indirect because a third factor is involved. For example, a toxic agent in the environment may cause a disease and also cause particular types of epigenetic changes even though those epigenetic changes do not contribute to the disease.

Correlations often identify associations between two variables. We should use caution, however, because the correlation by itself cannot prove that the association is due to cause and effect. Even so, research studies that identify associations are very useful because they provide the rationale to carry out further research to determine if a cause-and-effect relationship exists.

Researchers have identified many examples in which epigenetic changes are associated with a particular disease. These include Alzheimer's disease, cardiovascular diseases, diabetes, multiple sclerosis, and asthma. For these diseases, further research is needed to determine if these epigenetic changes are directly contributing to the disease symptoms. The role of epigenetics in disease has been most extensively studied with regard to cancer. Table 16.3 describes several examples in which an environmental factor is associated with a particular type of cancer.

In some of the examples listed in Table 16.3, scientific evidence indicates that the association is causative. For example, certain agents in tobacco smoke have been shown to cause epigenetic changes that underlie lung cancer. As described in Chapter 13, cancer-causing genes are placed into two categories: oncogenes, which are overexpressed in cancer, and tumor-suppressor genes, which are inhibited. In cases where research has shown that an environmental agent results in an epigenetic effect that contributes to cancer, it is more common for that agent to inhibit tumor-suppressor genes. However, in some cases, it may activate oncogenes. Alternatively, some of the agents listed in Table 16.3 show an association with particular cancers, but researchers are still trying to determine if the epigenetic changes caused by these agents actually promote the changes that result in cancer.

16.4 Reviewing the Concepts

- Dietary factors during early stages of development cause epigenetic changes that affect phenotype (Figure 16.5).
- Environmental agents have been shown to cause epigenetic changes that are associated with human diseases such as cancer (Table 16.3).

16.4 Testing Your Knowledge

1. When mice carrying the A^y allele exhibit a darker coat color, this phenotype is thought to be caused by dietary factors that result in
   a. a greater level of DNA methylation and a decrease in the expression of the *Agouti* gene.
   b. a lower level of DNA methylation and a decrease in the expression of the *Agouti* gene.
   c. a greater level of DNA methylation and the overexpression of the *Agouti* gene.
   d. a lower level of DNA methylation and the overexpression of the *Agouti* gene.
   e. both a and c.

Table 16.3	Environmental Agents That Are Associated with Cancer and Are Known to Cause Epigenetic Changes	
Environmental agent	**Occurrence**	**Association with particular cancers**
Polycyclic aromatic hydrocarbons	Tobacco smoke, automobile exhaust, charbroiled food	Lung, breast, stomach, and skin cancer
Benzene	Tobacco smoke, automobile exhaust	Leukemia, lymphoma, multiple myeloma
Endocrine disruptors (such as diethylstilbestrol)	Insecticides, fungicides, herbicides, some types of plastic	Breast, prostate, and thyroid cancer
Cadmium	Tobacco products, production of batteries	Lung and breast cancer
Nickel	Occupational exposure in mining, welding, and electroplating and in the manufacturing of jewelry, stainless steel, and batteries	Lung and nasal cancer
Arsenic	Lead alloy, feed additive in agriculture, insecticides	Skin, bladder, kidney, and liver cancer

2. When an association occurs between an environmental agent that causes an epigenetic change and a particular disease, this association can be due to which of the following factors?
    a. The epigenetic change directly contributes to the disease symptoms.
    b. The disease symptoms arise first, and they cause subsequent epigenetic changes to happen.
    c. The association is indirect because a third factor is involved.
    d. All of the above are possible.
    e. Only a and b are possible.

16.5 Extranuclear Inheritance: Organelle Genomes

Learning Outcomes

1. Describe the general features of mitochondrial and chloroplast genomes.
2. Explain why chloroplast and mitochondrial genes usually exhibit maternal inheritance.
3. List human diseases associated with mutations in mitochondrial genes.

In this section, we will explore inheritance patterns that violate the law of segregation. As described in Chapter 15, the segregation of genes is explained by the pairing and segregation of homologous chromosomes during meiosis (refer back to Figure 15.9). However, some genes are not found on the chromosomes in the cell nucleus, and these genes do not segregate in the same way. The transmission of genes located outside the cell nucleus is called **extranuclear inheritance.** Two important types of extranuclear inheritance patterns involve genes found in chloroplasts and mitochondria. In this section, we will examine the transmission patterns observed for genes found in the chloroplast and mitochondrial genomes and consider how mutations in these genes may affect an individual's traits.

EVOLUTIONARY CONNECTIONS

Chloroplast and Mitochondrial Genomes Are Relatively Small but Contain Genes That Encode Important Proteins

As we discussed in Chapter 4 (refer back to Figure 4.29), mitochondria and chloroplasts are found in eukaryotic cells because of an ancient endosymbiotic relationship. Because they originated from bacterial cells, these organelles contain their own genetic material, called the mitochondrial genome and chloroplast genome, respectively (**Figure 16.6**). Mitochondrial and chloroplast genomes are composed of a single, circular DNA molecule. The mitochondrial genome of many mammalian species has been analyzed and usually contains a total of 37 genes. Twenty-four genes encode tRNAs and rRNAs, which are needed for translation inside the mitochondrion, and 13 genes encode proteins that are involved in oxidative phosphorylation. As discussed in Chapter 6, the primary function of the mitochondrion is the synthesis of ATP via oxidative phosphorylation. Among different species of

(a) An animal cell

(b) A plant cell

Figure 16.6 The locations of genetic material in animal and plant cells. The chromosomes in the cell nucleus are collectively known as the nuclear genome. Mitochondria and chloroplasts have small, circular chromosomes called the mitochondrial and chloroplast genomes, respectively.

BioConnections: *Refer back to Figure 4.29. What is the evolutionary origin of mitochondria and chloroplasts in eukaryotic cells?*

flowering plants, chloroplast genomes typically contain between 100 and 200 genes. Many of these genes encode proteins that are vital to the process of photosynthesis, which is discussed in Chapter 7.

Chloroplast Genomes Are Often Maternally Inherited

One of the first experiments showing an extranuclear inheritance pattern was carried out by German botanist Carl Correns in 1909. Correns discovered that leaf pigmentation in the four-o'clock plant (*Mirabilis jalapa*) follows a pattern of inheritance that does not obey Mendel's law of segregation. Four-o'clock leaves may be green, white, or variegated. Correns observed that the pigmentation of the offspring depended solely on the pigmentation of the female parent, a phenomenon called **maternal inheritance** (Figure 16.7).

In cross 1, if the female parent providing the eggs had white leaves and the male parent providing the pollen had green leaves, all of the offspring had white leaves like the female parent. Correns also conducted a **reciprocal cross**—a cross in which the sexes and phenotypes are reversed compared to another cross. In the reciprocal cross of cross 1, the female parent had green leaves and the male parent had white leaves. Again, the offspring exhibited a phenotype like the female parent. All of the offspring of this recirpocal cross were green. In cross 2, the offspring of a variegated female parent could be green, white, or variegated. In the reciprocal cross, all of the offspring had green leaves.

What accounts for maternal inheritance? At the time, Correns did not understand that chloroplasts contain genetic material. Subsequent research identified DNA present in chloroplasts as responsible for the unusual inheritance pattern he observed. We now know that the pigmentation of four-o'clock leaves can be explained by the occurrence of genetically different types of chloroplasts in the leaf cells. As discussed in Chapter 7, chloroplasts are the site of photosynthesis, and their green color is due to the presence of the pigment called chlorophyll. Certain genes required for chlorophyll synthesis are located in the chloroplast DNA. For four-o'clock plants, the green phenotype is due to the presence of chloroplasts that have functional genes and synthesize the usual quantity of chlorophyll. The white phenotype is caused by a loss-of-function mutation in a gene within the chloroplast DNA that prevents the synthesis of most of the chlorophyll. (Enough chlorophyll is made for the plant to survive.) The variegated phenotype occurs in leaves that have a mixture of the two types of chloroplasts.

Leaf pigmentation follows a maternal inheritance pattern because the chloroplasts in four-o'clocks are transmitted only through the cytoplasm of the egg (Figure 16.8). In most species of plants, the egg cell provides most of the zygote's cytoplasm, whereas the much smaller male gamete often provides little more than a nucleus. Therefore, chloroplasts are most often inherited via the egg. Recall from Chapter 4 that chloroplasts are derived from proplastids, which can later develop into chloroplasts. In four-o'clocks, the egg cell contains several proplastids that are inherited by the offspring. The sperm cell does not contribute any proplastids. For this reason, the phenotype of a four-o'clock plant reflects the types of proplastids it inherits from the maternal parent. If the maternal parent transmits only normal proplastids, all offspring will have green leaves (Figure 16.8a). Alternatively, if the maternal parent transmits only mutant proplastids, all offspring will have white leaves (Figure 16.8b). Because an egg cell contains several proplastids, an offspring from a variegated maternal parent may inherit only normal proplastids, only mutant proplastids, or a mixture of normal and mutant proplastids. Consequently, the offspring of a variegated maternal parent can be green, white, or variegated individuals (Figure 16.8c).

How do we explain the variegated phenotype at the cellular level? This phenotype is due to events that occur after fertilization. As a zygote containing both types of proplastids grows via cellular division to produce a multicellular plant, some cells may receive mostly proplastids that develop into normal chloroplasts. Further division of these cells gives rise to a patch of green tissue. Alternatively, as a matter of chance, other cells may receive all or mostly mutant protoplastids that develop into chloroplasts that are defective in chlorophyll synthesis. The result is a patch of white tissue.

Figure 16.7 Maternal inheritance in the four-o'clock plant. In four-o'clocks, the egg contains all of the proplastids that are inherited by the offspring, so the phenotype of the offspring is determined by the maternal parent. The green phenotype is due to the presence of normal chloroplasts. The white phenotype in four-o'clocks is due to chloroplasts with a mutant allele that greatly reduces green pigment production. The variegated phenotype is due to a mixture of normal and mutant chloroplasts.

Concept Check: *In this example, where is the gene located that causes the green color of four-o'clock leaves? How is this gene transmitted from parent to offspring?*

(a) Egg cell from a maternal parent with green leaves

(b) Egg cell from a maternal parent with white leaves

(c) Possible egg cells from a maternal parent with variegated leaves

Figure 16.8 Plastid compositions of egg cells from green, white, and variegated four-o'clock plants. In this drawing of four-o'clock egg cells, normal proplastids are represented as green and mutant proplastids as white. (Note: This drawing is schematic. Proplastids do not differentiate into chloroplasts in egg cells, and they are not actually green.) (a) A green plant produces eggs carrying normal proplastids. (b) A white plant produces eggs carrying mutant proplastids. (c) A variegated plant produces eggs that may contain either or both types of proplastids.

In seed-bearing plants, maternal inheritance of chloroplasts is the most common transmission pattern. However, certain species exhibit a pattern called **biparental inheritance,** in which both the pollen and the egg contribute chloroplasts to the offspring. Others exhibit **paternal inheritance,** in which only the pollen contributes these organelles. For example, most species of pine trees show paternal inheritance of chloroplasts.

Mitochondrial Genomes Are Maternally Inherited in Humans and Most Other Species

Mitochondria are found in nearly all eukaryotic species. As with the transmission of chloroplasts in plants, maternal inheritance is the most common pattern of mitochondrial transmission in eukaryotes, although some species do exhibit biparental or paternal inheritance.

In humans, mitochondria are maternally inherited. Researchers have discovered that mutations in human mitochondrial genes cause a variety of rare diseases (Table 16.4). These are usually chronic degenerative disorders that affect organs and cells, such as the brain, eyes, heart, muscle, kidney, and endocrine glands, that require high levels of ATP. For example, Leber hereditary optic neuropathy (LHON)

Table 16.4	Examples of Human Mitochondrial Diseases
Disease	**Causes and symptomsv**
Leber hereditary optic neuropathy (LHON)	A mutation in one of several mitochondrial genes that encode electron transport proteins. The main symptom is loss of vision.
Neurogenic muscle weakness	A mutation in a mitochondrial gene that encodes a subunit of ATP synthase, which is required for ATP synthesis. Symptoms involve abnormalities in the nervous system that affect the muscles and eyes.
Maternal myopathy and cardiomyopathy	A mutation in a mitochondrial gene that encodes a tRNA for leucine. The primary symptoms involve muscle abnormalities, most notably in the heart.
Myoclonic epilepsy and ragged-red muscle fibers	A mutation in a mitochondrial gene that encodes a tRNA for lysine. Symptoms include epilepsy, dementia, blindness, deafness, and heart and kidney malfunctions.

affects the optic nerve and leads to the progressive loss of vision in one or both eyes. LHON is caused by point mutations in several different mitochondrial genes.

16.5 Reviewing the Concepts

- Mitochondria and chloroplasts are derived from an ancient endosymbiotic relationship and carry a small number of genes. The inheritance of such genes is called extranuclear inheritance (Figure 16.6).
- Chloroplasts in the four-o'clock plant are transmitted via the egg, a pattern called maternal inheritance (Figures 16.7, 16.8).
- Several human diseases are known to be caused by mutations in mitochondrial genes, which follow a maternal inheritance pattern (Table 16.4).

16.5 Testing Your Knowledge

1. In the four-o'clock plant (*Mirabilis jalapa*), leaf pigmentation is maternally inherited. A plant with green leaves is pollinated by a plant with variegated leaves. What would be the expected phenotype(s) of the offspring?
   a. all offspring with green leaves
   b. all offspring with variegated leaves
   c. some offspring with variegated, some with green, and some with white leaves
   d. mostly offspring with green leaves, but some with variegated leaves
   e. none of the above

2. Which of the following statements regarding human mitochondrial diseases is true?
   a. They are maternally inherited.
   b. They are due to mutations in the mitochondrial genome.
   c. They affect organs and cells that require high levels of ATP.
   d. All of the above are true.

16.6 Linkage of Genes on the Same Chromosome

Learning Outcomes

1. Describe how linkage violates the law of independent assortment.
2. Explain how experimental crosses can demonstrate linkage.

In the previous section, we examined extranuclear inheritance, which violates the law of segregation. In this section, we will explore inheritance patterns that violate the law of independent assortment. A typical chromosome contains hundreds or even a few thousand different genes. When two genes are close together on the same chromosome, they tend to be transmitted as a unit, a phenomenon known as **linkage.** A group of genes that usually stay together during meiosis is called a **linkage group,** and the genes in the group are said to be linked. In a two-factor cross, linked genes that are close together on the same chromosome do not follow the law of independent assortment.

In this section, we will begin by examining the first experimental cross that demonstrated linkage. This pattern was subsequently explained by Thomas Hunt Morgan, who proposed that different genes located close to each other on the same chromosome tend to be inherited together.

FEATURE INVESTIGATION

Bateson and Punnett's Crosses of Sweet Peas Showed That Genes Do Not Always Assort Independently

The first study showing linkage between two different genes involved crosses of sweet peas carried out by William Bateson and Reginald Punnett in 1905. A surprising result occurred when they conducted a two-factor cross involving flower color and pollen shape (**Figure 16.9**). In the P generation, one of the parent plants had purple flowers (*PP*) and long pollen (*LL*); the other had red flowers (*pp*) and round pollen (*ll*). As Bateson and Punnett expected, the F_1 plants all had purple flowers and long pollen (*PpLl*), because purple flowers and long pollen are dominant traits. The unexpected result came in the F_2 generation.

Figure 16.9 Bateson and Punnett's cross of sweet peas showing that independent assortment does not always occur.

4. Observe the phenotypes of the F$_2$ offspring.

Purple flowers, long pollen 15.6 : Purple flowers, round pollen 1.0 : Red flowers, long pollen 1.4 : Red flowers, round pollen 4.5

Fertilization

F$_2$ offspring having phenotypes of purple flowers with long pollen or red flowers with round pollen occurred more frequently than expected from Mendel's law of independent assortment.

5. **THE DATA**

Phenotypes of F$_2$ offspring	Observed number	Observed ratio	Expected number	Expected ratio
Purple flowers, long pollen	296	15.6	240	9
Purple flowers, round pollen	19	1.0	80	3
Red flowers, long pollen	27	1.4	80	3
Red flowers, round pollen	85	4.5	27	1

6. **CONCLUSION** The data are not consistent with the law of independent assortment.

7. **SOURCE** Bateson, William, and Punnett, Reginald C. 1911. On the Inter-relations of Genetic Factors. *Proceedings of the Royal Society of London, Series B*, 84: 3–8.

Although the F$_2$ offspring displayed the four phenotypes predicted by Mendel's laws, the observed numbers of offspring did not conform to the predicted 9:3:3:1 ratio (refer back to Figure 15.7). Rather, as seen in the data in Figure 16.9, the F$_2$ generation had a much higher proportion of the two phenotypes found in the P generation: purple flowers with long pollen, and red flowers with round pollen. How did Bateson and Punnett explain these results? They suggested that the transmission of flower color and pollen shape was somehow coupled, so these traits did not always assort independently. Although the law of independent assortment applies to many other genes, in this example, the hypothesis of independent assortment was rejected.

After rejecting the law of independent assortment for the flower color and pollen shape genes, Bateson and Punnett did not offer an alternative hypothesis to explain their results. We now know their results can be explained by linkage: The genes for flower color and pollen shape are closely linked on the same chromosome.

Experimental Questions

1. What hypothesis was Bateson and Punnett testing when conducting the crosses of sweet peas?
2. **SCISKILLS** ▶ What were the expected results of Bateson and Punnett's cross of F$_1$ plants?
3. **SCISKILLS** ▶ A statistic called the chi square (χ^2) test can be used to evaluate whether observed data are consistent with predicted data based on a hypothesis. This statistic is described in a statistics primer in Connect. Conduct a χ^2 square test using the data in this experiment. Based on your results, would you accept or reject an independent assortment hypothesis for these data?

Linkage and Crossing Over Produce Nonrecombinant and Recombinant Types

As mentioned, Bateson and Punnett realized their results did not conform to Mendel's law of independent assortment. However, they did not know why the genes were not assorting independently. A few years later, Thomas Hunt Morgan obtained similar ratios while studying the transmission pattern of genes in *Drosophila*. Like Bateson and Punnett, Morgan observed many more F$_2$ offspring with the combination of traits found in the P generation than predicted on the basis of independent assortment. To explain his data, Morgan proposed three ideas:

1. When different genes are located on the same chromosome, the traits determined by those genes are more likely to be inherited together. This violates the law of independent assortment.

2. Due to crossing over during meiosis, homologous chromosomes can exchange pieces of chromosomes and create new combinations of alleles.
3. The likelihood of a crossover occurring in the region between two genes depends on the distance between the two genes. Crossovers between homologous chromosomes are much more likely to occur between two genes farther apart along a chromosome than are crossovers of two genes that are closer together.

To illustrate the first two ideas, **Figure 16.10** considers a series of crosses involving two genes linked on the same chromosome in *Drosophila*. The two genes are located on an autosome, not on a sex

Figure 16.10 Linkage and recombination of alleles. In the P generation, an experimenter crossed $b^+b^+c^+c^+$ and *bbcc* flies to produce F_1 heterozygotes. F_1 females were then testcrossed to *bbcc* males. F_2 nonrecombinants have phenotypes that are the same as the P generation, whereas recombinants have a new combination of traits that is different from the P generation. The large number of nonrecombinant phenotypes in the F_2 generation suggests that the two genes are linked on the same chromosome. F_2 recombinant phenotypes occur because the alleles can be rearranged by crossing over. Note: The b^+ and c^+ alleles are dominant, and the *b* and *c* alleles are recessive.

Concept Check: *In which fly or flies did crossing over occur to produce the recombinant offspring of the F_2 generation?*

BioConnections: *Refer back to Figure 14.11. When does crossing over occur?*

chromosome. The P generation cross is between flies that are homozygous for alleles that affect body color and wing shape. The female is homozygous for the dominant wild-type alleles that produce gray body color (b^+b^+) and straight (normal) wings (c^+c^+); the male is homozygous for recessive mutant alleles that produce black body color (bb) and curved wings (cc). The symbols for the genes in Drosophila are different from what you learned earlier in the chapter for other organisms. In Drosophila, gene names are based on the name of the mutant allele; the dominant wild-type allele is indicated by a superscript plus sign ($^+$). The chromosomes next to the flies in Figure 16.10 show the arrangement of these alleles. If the two genes are on the same chromosome, we know the arrangement of alleles in the P generation flies because these flies are homozygous for both genes ($b^+b^+c^+c^+$ for one parent and $bbcc$ for the other parent). In the P generation female on the left, b^+ and c^+ are linked, whereas b and c are linked in the male on the right.

Let's now look at the outcome of the crosses in Figure 16.10. As expected, the F_1 offspring (b^+bc^+c) all had gray bodies and straight wings, confirming that these are the dominant traits. In the next cross, F_1 females were mated to males that were homozygous for both recessive alleles ($bbcc$). The purpose of the experiment, which is called a **testcross,** is to determine if the genes for body color and wing shape are linked. If the genes are on different chromosomes and assort independently, this cross will produce F_2 offspring with the four possible phenotypes in a 1:1:1:1 ratio. The observed numbers clearly conflict with this prediction. The two most abundant phenotypes are those with the combinations of characteristics in the P generation: gray bodies and straight wings or black bodies and curved wings. These offspring are termed **nonrecombinants,** because this combination of traits has not changed from the P generation. The smaller number of offspring that have a combination of traits not found in the P generation—gray bodies and curved wings or black bodies and straight wings—are called **recombinants.**

How do we explain the occurrence of recombinants when genes are linked on the same chromosome? As shown beside the flies of the F_2 generation in Figure 16.10, each recombinant individual has a chromosome that is the product of a crossover. The crossover occurred while the F_1 female fly was making egg cells.

As shown below, four different egg cells are possible:

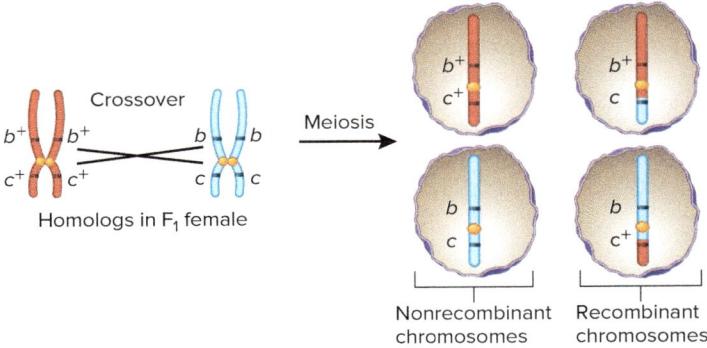

Due to crossing over, two of the four egg cells produced by meiosis have recombinant chromosomes. How are recombinant offspring produced? In the testcross, each of the male fly's sperm cells carries a chromosome with the two recessive alleles. An egg with the recombinant chromosome carrying the b^+ and c alleles produces an F_2 offspring with a gray body and curved wings. If the egg contains the recombinant chromosome carrying the b and c^+ alleles, F_2 offspring will have a black body and straight wings. Therefore, crossing over in the F_1 female can explain the occurrence of both types of F_2 recombinant offspring.

16.6 Reviewing the Concepts

- When two different genes are close together on the same chromosome and tend to transmitted as a unit, they are said to be linked. Linked genes tend to be inherited as a unit, unless crossing over separates them (Figures 16.9, 16.10).

16.6 Testing Your Knowledge

1. Two genes in fruit flies are linked on the same chromosome. An AABB individual is crossed to an aabb individual. Female offspring, which are AaBb, are then crossed to aabb males. Which offspring from this second cross do you expect to be the most common?
   a. Aabb and aaBb
   b. AaBb and aaBb
   c. AaBb and aabb
   d. Aabb and aabb

2. Genes that are linked do not conform to
   a. Mendel's law of segregation.
   b. Mendel's law of independent assortment.
   c. both of Mendel's laws.
   d. the chromosome theory of inheritance.

Assessing Your Knowledge and Skills

Foundational Knowledge

Multiple Choice

1. Which of the following is an example of an epigenetic change that alters gene expression?
   a. chromatin remodeling
   b. covalent histone modification
   c. localization of histone variants
   d. DNA methylation
   e. all of the above

2. In mice, the allele of the *Igf2* gene that is inherited from the mother is never expressed in her offspring. This happens because the *Igf2* gene from the mother
   a. always undergoes a mutation that inactivates its function.
   b. is deleted during oogenesis.
   c. is deleted during embryonic development.
   d. is not transcribed in the somatic cells of her offspring.
   e. All of the above are correct.

3. The marking process for genomic imprinting initially occurs during
   a. gametogenesis.
   b. fertilization.
   c. embryonic development.
   d. adulthood.
   e. both b and c.

4. In many organisms, organelles, such as the mitochondria, are transmitted only via the egg. This phenomenon is known as
   a. biparental inheritance.
   b. paternal inheritance.
   c. X-linked inheritance.
   d. maternal inheritance.
   e. both c and d.
5. Extranuclear inheritance occurs because
   a. certain genes are found on the X chromosome.
   b. chromosomes in the nucleus may be transferred to the cytoplasm.
   c. some cellular organelles contain DNA.
   d. the nuclear membrane breaks down during cell division.
   e. of both a and c.

Short Answer

1. Define epigenetics. Are all epigenetic changes passed from parent to offspring? Explain.
2. **PRINCIPLES** A principle of biology is that the genetic material provides a blueprint for growth, development, and reproduction. Explain how epigenetics affects that blueprint.

Critical-Thinking Skills

Multiple Choice

1. A female mouse that is $Igf2\ Igf2^-$ is crossed to a male that is also $Igf2\ Igf2^-$. The expected outcome for the phenotypes of the offspring for this cross is
   a. all normal size.
   b. all dwarf.
   c. 1 normal size : 1 dwarf.
   d. 3 normal size : 1 dwarf.
   e. 1 normal size : 3 dwarf.
2. According to Lyon's hypothesis, the patchwork pattern on a calico cat can be explained by which of the following?
   a. One of the X chromosomes is converted to a Barr body in somatic cells of female mammals.
   b. One of the X chromosomes is converted to a Barr body in all cells of female mammals.
   c. Both of the X chromosomes are converted to Barr bodies in somatic cells of female mammals.
   d. Both of the X chromosomes are converted to Barr bodies in all cells of female mammals.
   e. One of the X chromosomes is lost in the somatic cells of female mammals.
3. A female mouse that is homozygous for the A^{vy} allele is mated to a male that is homozygous for a loss-of-function (a) allele. How would you expect the diet of this female during pregnancy to affect her offspring?
   a. Dietary agents that promote a greater level of DNA methylation would produce offspring with more yellow fur.
   b. Dietary agents that promote a lower level of DNA methylation would produce offspring with more yellow fur.
   c. Dietary agents that promote a greater level of DNA methylation would produce offspring with darker fur.
   d. Dietary agents that promote a lower level of DNA methylation would produce offspring with darker fur.
   e. Both b and c are correct.
4. Environmental agents may cause epigenetic changes that alter the expression of specific genes. Which of the following epigenetic changes would be expected to promote cancer?
   a. a change that resulted in the overexpression of a tumor-suppressor gene
   b. a change that resulted in the inhibition of a tumor-suppressor gene
   c. a change that resulted in the overexpression of an oncogene
   d. a change that resulted in the inhibition of an oncogene
   e. both b and c
5. Which of the following examples is an epigenetic change that is *not* caused by environmental agents?
   a. variation in coat color in mice due to variation in expression of the A^{vy} allele
   b. X-chromosome inactivation
   c. effects of royal jelly on the formation of queen bees
   d. maternal inheritance
   e. Neither b nor d are caused by environmental agents.

Short Answer

1. Certain agents found in cigarette smoke are known to cause the silencing of a tumor-suppressor gene called *p53*, which is described in Chapter 13 (refer back to Figure 13.13). This silencing promotes lung cancer. However, by performing DNA sequencing, researchers have determined that the DNA sequence for this gene is sometimes not altered, even though it is silenced. Give two possible explanations for these results.
2. 5-Azocytidine is a chemical that inhibits DNA methyltransferase, which is the enzyme that methylates DNA. If 5-azocytidine was fed to female mice during pregnancy, explain how you think it would affect the coat color of offspring carrying the A^{vy} allele.

The Simpler Genetic Systems of Viruses and Bacteria

17

A tobacco plant infected with tobacco mosaic virus (TMV). Note the mottled lesions on the surface of the leaves.

Chapter Outline

17.1 General Properties of Viruses
17.2 Viral Reproductive Cycles
17.3 Genetic Properties of Bacteria
17.4 Gene Transfer Between Bacteria
Assessing Your Knowledge and Skills

©Norm Thomas/Science Source

We often view viruses as agents that infect people and cause illness. Many people don't realize that viruses infect all types of organisms, including bacteria and plants. The first virus to be discovered, the tobacco mosaic virus (TMV), infects tobacco and many other plant species, causing mosaic-like patterns in which normal-colored patches are interspersed with light green or yellowish patches on the leaves (see the chapter-opening photo). TMV damages leaves, flowers, and fruit but almost never kills the plant.

In 1883, German chemist Adolf Mayer determined that this disease could be spread by spraying the sap from one plant onto another. By subjecting this sap to filtration, Russian scientist Dmitri Ivanovski demonstrated that the disease-causing agent was too small to be a bacterium. At first, some researchers suggested the agent was a chemical toxin. However, Dutch botanist Martinus Beijerinck ruled out this possibility by showing that sap continued to transmit the disease after many plant generations. A toxin would have been diluted after many generations, but Beijerinck's results indicated the disease agent was multiplying in the plant.

In this chapter, we will first turn our attention to the genetic analyses of viruses. Once a cell is infected, the genetic material of a virus orchestrates a series of events that ultimately leads to the production of new virus particles. We will consider the biological complexity of viruses and explore viral reproductive cycles.

In the remaining sections of this chapter, we will examine the bacterial genome and the methods used in its investigation. Like their eukaryotic counterparts, bacteria have genetic differences that affect their cellular traits, and techniques of modern microbiology make many of these differences, such as sensitivity to antibiotics and differences in nutritional requirements, easy to study.

Although bacteria reproduce asexually by cell division, their genetic diversity is enhanced by the phenomenon known as gene transfer, in which genes are passed from one bacterial cell to another. Like sexual reproduction in eukaryotes, gene transfer enhances the genetic diversity observed among bacterial species. In this chapter, we will explore three interesting ways that bacteria transfer genetic material.

17.1 General Properties of Viruses

Learning Outcome

1. Compare and contrast types of viruses with regard to their host range, structure, and genomes.

Viruses are nonliving particles with nucleic acid genomes. Why are viruses considered nonliving? The answer is that they do not exhibit all of the properties associated with living organisms (refer back to Figure 1.3). Viruses are not composed of cells, and by themselves they do not use energy or carry out metabolism, maintain homeostasis, or even reproduce. A virus or its genetic material must be taken up by a living cell to replicate.

Because of viruses' ability to cause disease, microbiologists, geneticists, and molecular biologists have taken a great interest in the structure, genetic composition, and replication of viruses. In this section, we will discuss the structure and genetic composition of viruses.

Viruses Are Remarkably Varied, Despite Their Simple Structure

A **virus** is a small, infectious particle that consists of nucleic acid enclosed in a protein coat. Researchers have identified and studied over 4,000 different types of viruses. Although all viruses share some similarities, such as small size and the reliance on a living cell for replication, they vary greatly in their characteristics, including their host range, structure, and genome composition. Some of the major differences are described next, and the characteristics of selected viruses are shown in Table 17.1.

Differences in Host Range A cell that is infected by a virus is called a **host cell,** and a species that can be infected by a specific virus is called a host species for that virus. Viruses differ greatly in their **host range**—the number of species and cell types they can infect. Table 17.1 lists a few examples of viruses with widely different ranges of host species. For example, TMV has a broad host range. It is known to infect over 150 different species of plants.

Table 17.1 Hosts and Characteristics of Selected Viruses

Virus or group of viruses	Host	Effect on host	Nucleic acid*	Genome size (kb)†	Number of genes†
Phage λ	*E. coli*	Can exist harmlessly in the host cell or cause lysis	dsDNA	48.5	36
Phage T4	*E. coli*	Causes lysis	dsDNA	169	288
Tobacco mosaic virus (TMV)	Many plants	Causes mottling and necrosis of leaves and other plant parts	ssRNA	6.4	6
Baculoviruses	Insects	Most baculoviruses are species specific; they usually kill the insect.	dsDNA	133.9	154
Influenza virus	Mammals	Causes classical "flu," with fever, cough, sore throat, and headache	ssRNA	13.5	11
Epstein-Barr virus	Humans	Causes mononucleosis, with fever, sore throat, and fatigue	dsDNA	172	80
Adenovirus	Humans	Causes respiratory symptoms and diarrhea	dsDNA	34	35
Herpes simplex type II	Humans	Causes blistering sores around the genital region	dsDNA	158.4	77
HIV (type I)	Humans	Causes AIDS, an immunodeficiency syndrome eventually leading to death	ssRNA	9.7	9

*The abbreviations ss and ds refer to single-stranded and double-stranded, respectively.
†Several of the viruses listed in this table are found in different strains that vary with regard to genome size and number of genes. The numbers reported in this table are typical values. The abbreviation kb refers to kilobase, which equals 1,000 bases.

Table 17.2 Examples of Human Viruses

Area of body infected	Virus	Name of disease
Brain and central nervous system	Flavivirus	Yellow fever
	Rhabdovirus	Rabies
Skin	Herpes simplex I	Cold sores
	Variola virus	Smallpox
Respiratory tract	Influenza virus	Flu
	Rhinovirus	Common cold
Immune system	Human immunodeficiency virus (HIV)	Acquired immunodeficiency syndrome (AIDS)
	Epstein-Barr virus	Mononucleosis
Digestive system	Hepatitis B virus	Viral hepatitis
	Norwalk virus	Viral gastroenteritis
Reproductive system	Herpes simplex II	Genital herpes
	Papillomavirus	Genital warts, cervical cancer
Blood	Ebola virus	Hemorrhagic fever
	Hantavirus	Hemorrhagic fever with renal syndrome

By comparison, other viruses have a narrow host range, with some infecting only a single species. Furthermore, a virus may infect only a specific cell type in a host species. **Table 17.2** describes some viruses that infect human cells in certain parts of the body and thereby cause disease.

Structural Differences Viruses are so small that they cannot be resolved by even the best light microscope. Although the existence of viruses was postulated in the 1890s, viruses were not observed until the 1930s, when the electron microscope was invented. Viruses range in size from about 20 to 400 nm in diameter (1 nm [nanometer] = 10^{-9} m). For comparison, a typical bacterium is 1,000 nm in diameter, and the diameter of most eukaryotic cells is 10 to 1,000 times that of a bacterium. Adenoviruses, which cause infections of the respiratory and gastrointestinal tracts, have an average diameter of 75 nm. Over 50 million adenoviruses could fit into an average-size human cell.

What are the common structural features of all viruses? As shown in **Figure 17.1**, all viruses have a protein coat called a **capsid,** which encloses a genome consisting of one or more molecules of nucleic acid. Capsids are composed of one or several different protein subunits called capsomers. Capsids have a variety of shapes, including helical and polyhedral. Figure 17.1a shows the structure of TMV, which has a helical capsid made of identical capsomers. Figure 17.1b shows an influenza virus, which has a polyhedral capsid. Many viruses that infect animal cells, such as influenza virus, have a **viral envelope** enclosing the capsid. The envelope consists of a lipid bilayer that is derived from a cellular membrane of the host cell and is embedded with virally encoded spike glycoproteins.

In addition to encasing and protecting the genetic material, the capsid and envelope enable viruses to infect their hosts. In many viruses, spike glycoproteins (Figure 17.1b) help the virus bind to the surface of a host cell. Viruses that infect bacteria, called **bacteriophages,** or simply **phages,** may have more complex protein coats, with accessory structures used for anchoring the virus to a host cell and transferring the viral nucleic acid into the bacterial cytoplasm (Figure 17.1c). For example, the tail fibers of such bacteriophages are needed to attach the virus to the bacterial cell wall.

Genome Differences The genetic material in a virus is called a **viral genome.** The composition of viral genomes varies markedly among different types of viruses, as indicated by the examples in Table 17.1. The nucleic acid of some viruses is DNA, whereas in others it is RNA. These are referred to as DNA viruses and RNA viruses, respectively. It is striking that some viruses use RNA for their genome, whereas all living organisms use DNA. In some viruses, the nucleic acid is single-stranded, whereas in others, it is double-stranded. The genome can be linear or circular, depending on the type of virus. Some kinds of viruses have more than one copy of the genome.

Viral genomes also vary considerably in size, ranging from a few thousand to more than a hundred thousand nucleotides in length (see Table 17.1). For example, the genome of TMV is only 6,400 base pairs in length and contains only six genes. Other viruses, particularly those with a complex structure, such as phage T4, contain many more genes. These extra genes encode many different proteins that are involved in the formation of the elaborate structure shown in Figure 17.1c.

17.1 Reviewing the Concepts

- Viruses are nonliving particles that do not exhibit all of the properties associated with living organisms. A virus or its genetic material must be taken up by a living cell to replicate.
- Viruses vary with regard to their host range, structure, and genome composition (Tables 17.1, 17.2; Figure 17.1).

17.1 Testing Your Knowledge

1. Viruses differ with regard to their
   a. host range.
   b. structure.
   c. use of DNA or RNA as the genetic material.
   d. all of the above.

Biology Principle

Biologists Use Models to Test Predictions and Convey Ideas

The four drawings along the right side of this figure are simplified models for the structures of four different viruses. The Modeling Challenge in the figure legend asks you to propose a model for how the virus shown in part (c) injects its DNA into a host cell.

(a) Tobacco mosaic virus, a nonenveloped virus with a helical capsid

(b) Influenza virus, an enveloped virus with spikes

(c) T4, a bacteriophage

Figure 17.1 Variations in the structure of viruses, as shown by transmission electron microscopy (TEM). All viruses contain nucleic acid (DNA or RNA) surrounded by a protein capsid. They may or may not have an outer envelope surrounding the capsid. (a) Tobacco mosaic virus (TMV) has a capsid made of 2,130 identical protein subunits, helically arranged around a strand of RNA. (b) Influenza virus has a polyhedral capsid that encloses its RNA genome. It also has an envelope composed of a lipid bilayer and spike glycoproteins. The envelope is obtained from the host cell when the virus buds from the plasma membrane. (c) Some bacteriophages, such as T4, have protein coats with accessory structures such as tail fibers that facilitate invasion of a bacterial cell.
(a) ©Science Source; *(b)* ©Chris Bjornberg/Science Source; *(c)* ©Omikron/Science Source

Modeling Challenge: During the infection process of T4 bacteriophage, the T4 DNA is injected into the bacterial cytoplasm, but most of the T4 proteins remain on the surface of the bacterium. First, the tips of the tail fibers bind to the outer membrane of the bacterial cell. Next, the base plate makes contact with the surface of the cell, and then the shaft contracts. During this contraction, proteins within the shaft poke a hole that penetrates into the bacterial cytoplasm. After this hole is formed, the phage DNA leaves the phage head and travels through the shaft and into the cytoplasm. Draw a model that shows this process. Your model should show three or four steps. Note: You may want to look ahead to Figure 17.2a for help.

17.2 Viral Reproductive Cycles

Learning Outcomes

1. List the six steps in viral reproductive cycles, and distinguish between the lysogenic and lytic cycles.
2. Describe how emerging viruses such as HIV arise and spread through populations.
3. Explain how an understanding of virus structure and reproduction can aid in the development of drugs to combat viruses.

When a virus infects a host cell, the expression of viral genes leads to a series of steps, called a **viral reproductive cycle,** which results in the production of new viruses. The details of the steps differ among various types of viruses, and even the same virus may have the capacity to follow alternative cycles. Even so, by studying the reproductive cycles of hundreds of different viruses, researchers have determined that the viral reproductive cycle consists of five or six basic steps.

Viral Reproductive Cycles Consist of Common Steps

To illustrate the general features of viral reproductive cycles, Figure 17.2 considers these steps for two types of viruses. Figure 17.2a shows the cycle of phage λ (lambda), a bacteriophage with double-stranded DNA as its genome, and Figure 17.2b depicts the cycle of HIV, an enveloped animal virus containing single-stranded RNA. The descriptions that follow compare the reproductive cycles of these two very different viruses.

Step 1: Attachment In the first step of a viral reproductive cycle, the virus attaches to the surface of a host cell. This attachment is usually specific for one or just a few types of cells because proteins in the virus recognize and bind to specific molecules on the cell surface. In the case of phage λ, the phage tail fibers bind to proteins in the outer bacterial cell membrane of *E. coli* cells. In the case of HIV, spike glycoproteins in the viral envelope bind to protein receptors in the plasma membrane of human blood cells called helper T cells.

Step 2: Entry After attachment, the viral genome enters the host cell. Attachment of phage λ stimulates a conformational change in the phage coat proteins, so the shaft contracts, and the phage injects its DNA into the bacterial cytoplasm. In contrast, the envelope of HIV fuses with the plasma membrane of the host cell, so both the capsid and its contents are released into the cytosol. Some of the HIV capsid proteins are then removed by host cell enzymes, a process called uncoating. This releases two copies of the viral RNA and molecules of reverse transcriptase and integrase into the cytosol. As discussed shortly, these enzymes are needed for step 3.

The genome of some viruses, including both phage λ and HIV, can integrate into a chromosome of the host cell. For such viruses, the cycle may proceed from step 2 to step 3, as described next, delaying the production of new viruses. Alternatively, the cycle for phage λ may proceed directly from step 2 to step 4 and quickly lead to the production of new viruses.

Step 3: Integration Viruses capable of integration carry a gene that encodes an enzyme called **integrase.** For integration of phage λ to occur, this gene is expressed soon after entry and the integrase protein is made. Integrase cuts the host's chromosomal DNA and inserts the viral genome into the chromosome. In the case of phage λ, the double-stranded DNA that entered the cell can be directly integrated into the double-stranded DNA of the chromosome. Once integrated, the phage DNA in a bacterium is called a **prophage.** When a bacterial cell divides, the prophage DNA is copied and transmitted to daughter cells along with the bacterial chromosomal DNA. The phase of a viral reproductive cycle when a prophage is integrated into a bacterial chromosome is called the **lysogenic cycle.** New phages are not made during the lysogenic cycle, and the host cell is not destroyed. On occasion, a prophage can be excised from the bacterial chromosome to end the lysogenic cycle. An excised prophage may then proceed to the lytic cycle.

How can an RNA virus integrate its genome into the host cell's DNA? For this to occur, the viral genome must be copied into DNA. HIV accomplishes this by means of a viral enzyme called **reverse transcriptase,** which is carried within the capsid and released into the host cell along with the viral RNA. Reverse transcriptase uses the viral RNA strand to make a complementary copy of DNA. The complementary DNA is then used as a template to make double-stranded viral DNA. This process is called reverse transcription because it is the reverse of the usual transcription process, in which a DNA strand is used to make a complementary strand of RNA. The viral double-stranded DNA enters the host cell nucleus and is inserted into a host chromosome via integrase. Like reverse transcriptase, integrase is carried within the capsid and released into the host cell during uncoating. Once integrated, the viral DNA in a eukaryotic cell is called a **provirus.** Viruses that follow this mechanism are called **retroviruses.**

Step 4: Synthesis of Viral Components The production of new viruses by a host cell involves the synthesis of new copies of the viral genome and the viral proteins that make up the protein coat. A prophage must be excised as described in step 3 before the synthesis of new viral components can occur. Following excision, host cell enzymes make many copies of the phage DNA and transcribe the genes within these copies into mRNA. Host cell ribosomes translate this viral mRNA into viral proteins. The expression of phage genes also leads to the degradation of the host chromosomal DNA.

In the case of HIV, the provirus DNA is not excised from the host chromosome. Instead, it is transcribed in the nucleus to produce many copies of viral RNA. These viral RNA molecules enter the cytosol, where they are used to make viral proteins and serve as the genome for new viral particles.

Step 5: Viral Assembly After all of the necessary components have been synthesized, they must be assembled into new viruses.

Figure 17.2 Comparison of the steps of two viral reproductive cycles. (a) The reproductive cycle of phage λ, a bacteriophage with a double-stranded DNA genome. (b) The reproductive cycle of HIV, an enveloped animal virus with a single-stranded RNA genome.

BioConnections: *Refer back to Figure 5.20. How does the release of HIV resemble exocytosis?*

THE SIMPLER GENETIC SYSTEMS OF VIRUSES AND BACTERIA 351

4. Synthesis of viral components: In the lytic cycle, phage DNA directs the synthesis of viral components. During this process, the phage DNA circularizes, and the host chromosomal DNA is degraded.

5. Viral assembly: Phage components are assembled with the help of noncapsid proteins to make many new phages.

6. Release: The viral enzyme called lysozyme causes cell lysis, and new phages are released from the broken cell.

4. Synthesis of viral components: Proviral DNA directs the synthesis of viral components.

5. Viral assembly: Capsid proteins enclose 2 RNA molecules and molecules of reverse transcriptase and integrase. Capsid assembles with spike glycoproteins during budding.

6. Release: Virus buds from the plasma membrane of the host cell and is released. The new viral envelope is derived from a portion of the host cell plasma membrane.

Some viruses with simple structures self-assemble—viral components spontaneously bind to each other to form a complete virus particle. An example of a self-assembling virus is TMV, which we examined earlier (see Figure 17.1a). TMV capsid proteins assemble around a TMV RNA molecule, which becomes trapped inside the hollow capsid.

Other viruses, including the two shown in Figure 17.2, do not self-assemble. The assembly of phage λ requires the help of non-capsid proteins not found in the completed phage particle. Some of these noncapsid proteins function as enzymes that modify capsid proteins, whereas others serve as scaffolding for the assembly of the capsid.

The assembly of HIV occurs in two stages. First, capsid proteins assemble around two molecules of viral RNA and molecules of reverse transcriptase and integrase. Next, the newly formed capsid acquires its outer envelope in a budding process. This second phase of assembly occurs during step 6, as the virus is released from the cell.

Step 6: Release The last step of a viral reproductive cycle is the release of new viruses from the host cell. The release of bacteriophages is a dramatic event. Because bacteria are surrounded by a rigid cell wall, the phages must burst, or lyse, their host cell to escape. After the phages have been assembled, a phage-encoded enzyme called lysozyme digests the bacterial cell wall, causing the cell to burst. Lysis releases many new phages into the environment, where they can infect other bacteria and begin the cycle again. Collectively, steps 1, 2, 4, 5, and 6 are called the **lytic cycle,** because they lead to cell lysis.

The release of enveloped viruses from an animal cell is far less dramatic. This type of virus escapes by a mechanism, called budding, that does not lyse the cell. In the case of HIV, a newly assembled virus particle associates with a portion of the plasma membrane containing HIV spike glycoproteins. The membrane enfolds the viral capsid and eventually buds from the surface of the cell. This is how the virus acquires its envelope, which is a piece of host cell membrane studded with viral glycoproteins.

Latency in Bacteriophages During the lysogenic cycle, viruses integrate their genomes into a host chromosome (see step 3). In some cases, the resulting prophage or provirus may remain inactive, or **latent,** for a long time. Most of the viral genes are silent during latency, and the viral reproductive cycle does not advance to step 4. When a lysogenic bacterium prepares to divide, it copies the prophage DNA along with its own DNA, so each daughter cell inherits a copy of the prophage. A prophage can be replicated repeatedly in this way without killing the host cell or producing new phage particles.

For phage λ, environmental conditions influence whether or not the viral DNA is integrated into a host chromosome and how long the virus remains in the lysogenic cycle. If nutrients are readily available, phage λ usually proceeds directly to the lytic cycle after its DNA enters the cell. Alternatively, if nutrients are in short supply, the lysogenic cycle is favored because sufficient material may not be available to make new viruses. If more nutrients become available later, the prophage may become activated. At this point, the viral reproductive cycle switches to the lytic cycle, and new viruses are made and released.

Latency in Human Viruses Latency among human viruses occurs in two ways. For HIV, latency occurs because the virus has integrated into the host genome and may remain dormant for a long time. Alternatively, the genomes of other viruses can exist as an **episome**—a genetic element that replicates independently of the chromosomal DNA but also can occasionally integrate into chromosomal DNA. Examples of viral genomes that exist as episomes include different types of herpesviruses that cause cold sores (usually herpes simplex type I), genital herpes (usually herpes simplex type II), and chickenpox (varicella-zoster). A person infected with a given type of herpesvirus may have periodic outbreaks of disease symptoms when the virus switches from the latent, episomal form to the active form that produces new virus particles.

As an example, let's consider the herpesvirus called varicella-zoster. The initial infection by this virus causes chickenpox, after which the virus may remain latent for many years as an episome. If the varicella-zoster virus becomes active and starts making new virus particles, it can cause a disease called shingles. Shingles begins as a painful rash that eventually erupts into blisters. The blisters follow the path of the neurons that carry the latent varicella-zoster virus. The blisters often form a ring across the back of the patient's body. The term shingles is derived from a Latin word meaning girdle, referring to the observation that the blisters girdle a part of the body.

Emerging Viruses, Such as HIV and Zika Virus, Have Arisen Recently and May Rapidly Spread Through a Population

A key reason researchers are interested in viral reproductive cycles is the ability of many viruses to cause diseases in humans and other hosts. **Emerging viruses** are those that have arisen recently or are likely to have a greater probability of causing infection. They typically arise via mutations in pre-existing viruses. Emerging viruses may lead to significant loss of human life and often cause public alarm. New strains of influenza virus arise fairly regularly due to mutations. One example is the strain H1N1, also called swine flu. In the U.S., despite attempts through vaccination to minimize deaths, over 30,000 people die annually from influenza.

Another example of an emerging virus is Zika virus, a type of flavivirus. It is an enveloped virus with a genome composed of single-stranded RNA. Its name comes from the Zika Forest of Uganda, where the virus was first isolated in 1947. Zika virus is primarily spread by mosquitoes of the genus *Aedes*. The most common symptoms of a Zika infection are fever, rash, joint pain, and conjunctivitis. In most people, the illness is usually mild, but in rare cases, Zika infections in adults can cause a more serious illness called Guillain-Barré syndrome. In addition, Zika virus infection during pregnancy can result in serious brain abnormalities, including the condition microcephaly, in which the infant's head is smaller than normal. The Zika virus has spread globally from Africa into Asia,

South America, and North America. Though predictions for infection rates vary, some epidemiologists expect that millions of people will become infected with the virus in the coming years.

During the past few decades, the most devastating example of an emerging virus is **human immunodeficiency virus (HIV)**, the causative agent of **acquired immunodeficiency syndrome (AIDS)**. HIV is primarily spread by sexual contact between infected and uninfected individuals, but it can also be spread by the sharing of needles among intravenous drug users and passed from infected mother to unborn child. Since AIDS was first recognized in 1981, the total number of AIDS deaths has been nearly 40 million, making it one of the most deadly diseases in human history. More than 0.6 million of these deaths have occurred in the U.S. In 2018, about 37 million people were living with HIV; approximately 3 million of them were infected that year. In that year, nearly 2 million died from AIDS. Worldwide, nearly 1 in every 100 people between ages 15 and 49 is infected. In the U.S., about 50,000 new HIV infections occur each year, 70% of which are in men and 30% in women.

The devastating effects of AIDS result from viral destruction of helper T cells, a type of white blood cell that plays an essential role in the immune system of mammals, which is discussed in Chapter 40. Figure 17.3 shows HIV virus particles invading a helper T cell, which interacts with other cells of the immune system to facilitate the production of antibodies and other molecules that target and kill foreign invaders of the body. When large numbers of helper T cells are destroyed by HIV, the function of the immune system is seriously compromised, and the individual becomes susceptible to diseases, called opportunistic infections, that would not normally occur in a healthy person. For example, *Pneumocystis jiroveci*, a fungus that causes a type of pneumonia, is easily destroyed by a healthy immune system. However, in people with AIDS, *P. jiroveci* pneumonia can be fatal.

An insidious feature of HIV replication, which is summarized in Figure 17.2b, is that reverse transcriptase, the enzyme that copies the RNA genome into DNA, lacks a proofreading function. In Chapter 9, we learned that DNA polymerase can identify and remove mismatched nucleotides in newly synthesized DNA. Because reverse transcriptase lacks this function, it makes more errors and thereby tends to create mutant strains of HIV. This high rate of mutation undermines the body's ability to combat HIV because mutant strains may be resistant to the body's defenses. In addition, mutant strains of HIV may be resistant to antiviral drugs, which are described next.

Drugs Have Been Developed to Combat the Proliferation of HIV

A compelling reason to understand the reproductive cycle of HIV and other disease-causing viruses is that such knowledge may be used to develop drugs that stop viral proliferation. For example, in the U.S., the highest rate of AIDS-related deaths was approximately 17 per year per 100,000 people in 1994 and 1995. The current rate is about 4 to 5 deaths per year per 100,000 people, owing in part to the use of antiviral drugs. These drugs inhibit viral proliferation, although they cannot eliminate the virus from the body.

One approach to the design of antiviral treatments is to develop drugs that specifically bind to proteins encoded by the viral genome. For example, azidothymidine (AZT) mimics the structure of a normal nucleotide and binds to the enzyme reverse transcriptase. In this way, AZT inhibits reverse transcription, thereby inhibiting viral replication. Another way to combat HIV involves the use of antiviral drugs that inhibit proteases, enzymes that are needed during the assembly of the HIV capsid. An HIV protease cuts capsid proteins, which makes them smaller and able to assemble into a capsid structure. If the protease does not function, the capsid will not assemble, and new HIV particles will not be made. Several drugs known as protease inhibitors have been developed that bind to HIV protease and inhibit its function.

A major challenge in AIDS research is to discover drugs that inhibit viral proteins without also binding to host cell proteins and inhibiting normal cellular functions. A second challenge is to develop drugs to which mutant strains will not become resistant. As mentioned, HIV readily accumulates mutations during viral replication. A current strategy is to treat HIV patients with a "cocktail" of three or more HIV drugs, making it less likely that any mutant strain will overcome all of the inhibitory effects.

17.2 Reviewing the Concepts

- The viral reproductive cycle consists of a series of steps: attachment, entry, integration, synthesis of components, assembly, and release. Some bacteriophages follow two different reproductive cycles: the lytic cycle and the lysogenic cycle (Figure 17.2).
- Emerging viruses, such as human immunodeficiency virus (HIV), cause significant loss of human life. HIV, the causative agent of the disease AIDS, is a retrovirus whose reproductive cycle involves the integration of the viral genome into a chromosome in the host cell. HIV accomplishes this by means of a viral enzyme called reverse transcriptase. Drugs to combat HIV have been developed that specifically inhibit viral proteins (Figure 17.3).

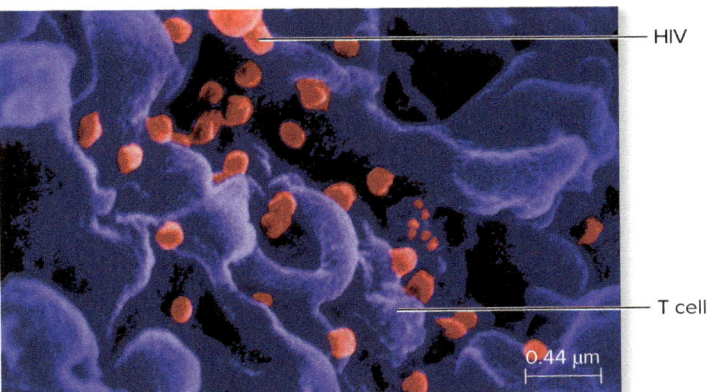

Figure 17.3 **Micrograph of HIV invading a human helper T cell.** This is a colorized scanning electron micrograph (SEM). The surface of the T cell is purple, and HIV particles are red.
Source: Cynthia Goldsmith/CDC

17.2 Testing Your Knowledge

1. The step in the viral cycle that follows attachment is
   a. entry.
   b. integration into the host genome.
   c. synthesis of new viral components.
   d. release.

17.3 Genetic Properties of Bacteria

Learning Outcomes
1. Outline the key features of a bacterial chromosome.
2. Describe the two processes that compact the bacterial chromosome.
3. Outline the structure and functions of plasmids.
4. Diagram the process of cell division in bacteria.

Bacterial cells may exist as single units or remain associated with each other after cell division, forming pairs, chains, or clumps. Bacteria are widespread on Earth, and numerous species are known to cause various types of infectious diseases. We will begin this section by exploring the structure and replication of the bacterial genome and the organization of DNA sequences along a bacterial chromosome. We will then examine how the bacterial chromosome is compacted to fit inside a bacterium and how it is transmitted during asexual reproduction.

Bacteria Typically Have Circular Chromosomes That Carry a Few Thousand Genes

The genes of bacteria are found within bacterial chromosomes.

- Although a bacterial cell usually has a single type of chromosome, it may have more than one copy of that chromosome. The number of copies depends on the species and on growth conditions, but a bacterium typically has one to four identical chromosomes.

- Each bacterial chromosome is tightly packed within a distinct **nucleoid** of the cell (**Figure 17.4**). Unlike the eukaryotic nucleus, the nucleoid is not a separate cellular compartment bounded by a membrane. The DNA in the nucleoid is in direct contact with the cytoplasm of the cell.

- Bacterial chromosomes contain molecules of double-stranded DNA along with many different proteins. They are usually circular and typically only a few million base pairs (bp) long. For example, the chromosome of *Escherichia coli* has approximately 4.6 million bp, and the *Haemophilus influenzae* chromosome has roughly 1.8 million bp.

- A typical bacterial chromosome contains a few thousand unique genes that are found throughout the chromosome. Gene sequences, primarily those that encode proteins, account for the largest part of bacterial DNA.

Figure 17.4 Nucleoids within the bacterium *Bacillus subtilis*. In the light micrograph shown here, the nucleoids are fluorescently labeled and seen as purplish-blue irregularly shaped areas within the bacterial cytoplasm. Two or more nucleoids are usually found within each cell.
©M. Wurtz/Biozentrum, University of Basel/Science Source

Concept Check: *How many nucleoids are in this bacterial cell?*

BioConnections: *Refer back to Figures 4.8 and 4.9. How is a nucleoid different from a nucleus found in a eukaryotic cell?*

- Other nucleotide sequences in the chromosome play a role in DNA replication, gene expression, and chromosome structure. One of these sequences is the origin of replication, which is a few hundred base pairs long. Bacterial chromosomes have a single origin of replication that functions as an initiation site for the assembly of several proteins that are required for DNA replication.

The Formation of Chromosomal Loops and DNA Supercoiling Make the Bacterial Chromosome Compact

Bacterial cells are much smaller than most eukaryotic cells (refer back to Figure 4.8). *E. coli* cells, for example, are approximately 1 μm wide and 2 μm long. To fit within a bacterial cell, the DNA of a typical bacterial chromosome must be compacted about 1,000-fold. How does this occur? The compaction of a bacterial chromosome, shown in **Figure 17.5**, occurs by two processes: the formation of loop domains and DNA supercoiling.

Unlike eukaryotic DNA, bacterial DNA is not wound around histone proteins to form nucleosomes. However, the binding of proteins to bacterial DNA is important in the formation of **loop domains**—chromosomal segments that are folded into loops. As seen in Figure 17.5, DNA-binding proteins anchor the bases of the loops in place. The number of loops varies according to the size of a bacterial chromosome and the species. The *E. coli* chromosome has a few hundred loop domains, each with about 10,000 bp. This looping compacts the circular chromosome about 10-fold.

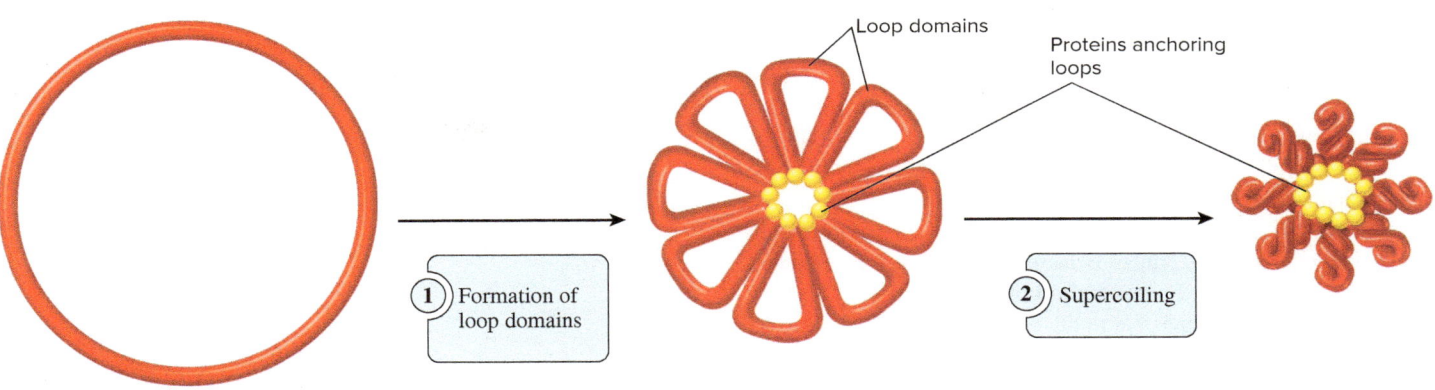

Figure 17.5 **The compaction of a bacterial chromosome.** To compact the large, circular chromosome, segments are organized into smaller loop domains by binding to proteins at the bottoms of the loops. These loops are made more compact by DNA supercoiling. Note: This is a simplified drawing; bacterial chromosomes typically have a few hundred loop domains.

Concept Check: *Describe how the loop domains are held in place.*

DNA supercoiling is the second important compaction process for the bacterial chromosome. Because DNA is a long, thin molecule, twisting can dramatically change its conformation. This compaction process is similar to what happens to a rubber band if you twist it in one direction. Because the two strands of DNA already coil around each other, the formation of additional coils due to twisting is referred to as supercoiling. Bacterial enzymes called topoisomerases twist the DNA and control the degree of DNA supercoiling.

Plasmids Are Small, Circular Pieces of Extrachromosomal DNA

In addition to chromosomal DNA, bacterial cells commonly contain **plasmids,** small, circular pieces of DNA that exist separately from the bacterial chromosome (**Figure 17.6**). Plasmids occur naturally in many strains of bacteria and some eukaryotic cells, such as yeast. The smallest plasmids consist of just a few thousand base pairs and carry only a gene or two. The largest are in the range of 100,000 to 500,000 bp and carry several dozen or even hundreds of genes. A plasmid has its own origin of replication that allows it to be replicated independently of the bacterial chromosome. The DNA sequence of the origin of replication influences how many copies of the plasmid are found within a cell. Some origins are said to be very strong because they result in many copies of the plasmid, perhaps as many as 100 per cell. Other origins of replication have sequences that are much weaker, so the number of copies is relatively low, such as one or two per cell.

Why do bacteria have plasmids? Certain genes within a plasmid usually provide some type of growth advantage to the cell or may aid in survival under certain conditions. By studying plasmids in many different species, researchers have discovered that most plasmids fall into five categories:

1. Resistance plasmids, also known as R factors, contain genes that confer resistance against antibiotics and other types of toxins.
2. Degradative plasmids carry genes that enable the bacterium to digest and utilize an unusual substance. For example, a degradative plasmid may carry genes that allow a bacterium to digest an organic solvent such as toluene.
3. Col-plasmids contain genes that encode colicins, which are proteins that kill other bacteria.
4. Virulence plasmids carry genes that turn a bacterium into a pathogenic strain.
5. Fertility plasmids, also known as F factors, allow bacteria to transfer genes to each other, as described later in this chapter.

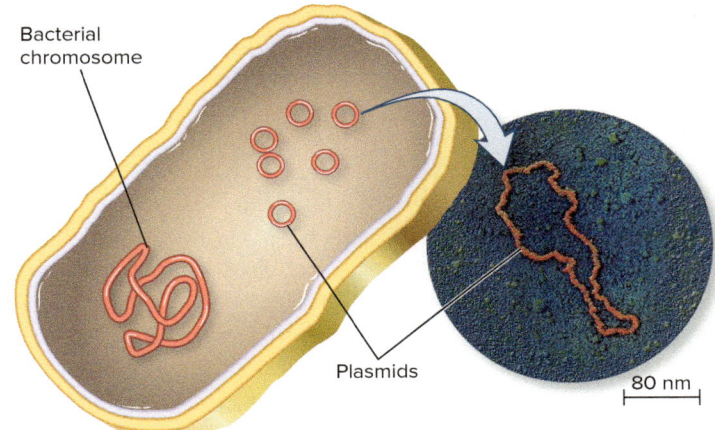

Figure 17.6 **Plasmids in a bacterial cell.** Plasmids are small, circular DNA molecules that exist independently of the bacterial chromosome.
©Stanley Cohen/Science Source

Concept Check: *Describe the similarities and differences between a bacterial chromosome and a plasmid.*

On occasion, a plasmid may integrate into the bacterial chromosome. Such plasmids, which can integrate or remain independent of the chromosome, are termed episomes.

Most Bacteria Reproduce by Binary Fission

Thus far, we have considered the genetic material of bacteria and the compaction of the bacterial chromosome to fit inside the cell. Let's now turn our attention to the process of bacterial cell division. The capacity of bacteria to divide is quite astounding. The cells of some species, such as *E. coli*, can divide every 20–30 minutes. When placed on a solid growth medium in a petri dish, an *E. coli* cell and its daughter cells undergo repeated cellular divisions and form a group of genetically identical cells called a **bacterial colony** (Figure 17.7). Starting with a single cell that is invisible to the naked eye, a visible bacterial colony containing 10 to 100 million cells forms in less than a day!

Cell division of most bacterial species occurs by a process called **binary fission,** during which a cell divides into two daughter cells. Figure 17.8 shows this process for a cell with a single chromosome. Before it divides, the cell replicates its DNA. This produces two identical copies of the chromosome. Next, the cell's plasma membrane is drawn inward and deposits new cell-wall material, separating the two daughter cells. Each daughter cell receives one of the copies of the original chromosome. Therefore, except when a mutation occurs, each daughter cell contains an identical copy of the mother cell's genetic material.

Plasmids replicate independently of the bacterial chromosome. During binary fission, the plasmids are distributed to the daughter cells, so each daughter cell usually receives one or more copies of the plasmid.

17.3 Reviewing the Concepts

- Bacteria typically have a single type of circular chromosome found in the nucleoid of the cell. The chromosome contains many genes and one origin of replication (Figure 17.4).
- The bacterial chromosome is made more compact by the formation of loop domains and by DNA supercoiling (Figure 17.5).
- Plasmids are small, circular DNA molecules that exist independently of the bacterial chromosome (Figure 17.6).
- When placed on solid growth media, a single bacterial cell will divide many times to produce a colony composed of many cells (Figure 17.7).
- Bacteria reproduce by asexual reproduction in a process called binary fission, during which a cell divides to form two daughter cells (Figure 17.8).

17.3 Testing Your Knowledge

1. Which of the following would *not* be found in a typical bacterial chromosome?
   a. a few thousand genes
   b. one origin of replication
   c. a centromere
   d. All of the above would be found in a bacterial chromosome.
2. Bacterial chromosomes are made more compact by
   a. attachment to the plasma membrane.
   b. the formation of loop domains.
   c. DNA supercoiling.
   d. both b and c.
3. Which of the following is a characteristic of plasmids?
   a. They have an origin of replication.
   b. They are usually circular.
   c. They may carry genes, such as antibiotic-resistance genes.
   d. They exist independently of the bacterial chromosome.
   e. All of the above are correct.

Figure 17.7 Growth of a bacterial colony. Through successive cell divisions, a single bacterial cell of *E. coli* forms a genetically identical group of cells called a bacterial colony.
©Dr. Jeremy Burgess/SPL/Science Source

Concept Check: *Suppose a bacterial strain divides every 30 minutes. If a single cell is placed on a plate, how many cells will be in the colony after 16 hours?*

Biology Principle

The Genetic Material Provides a Blueprint That Allows Organisms to Grow, Develop, and Reproduce

After cell division, a bacterial cell increases in size.

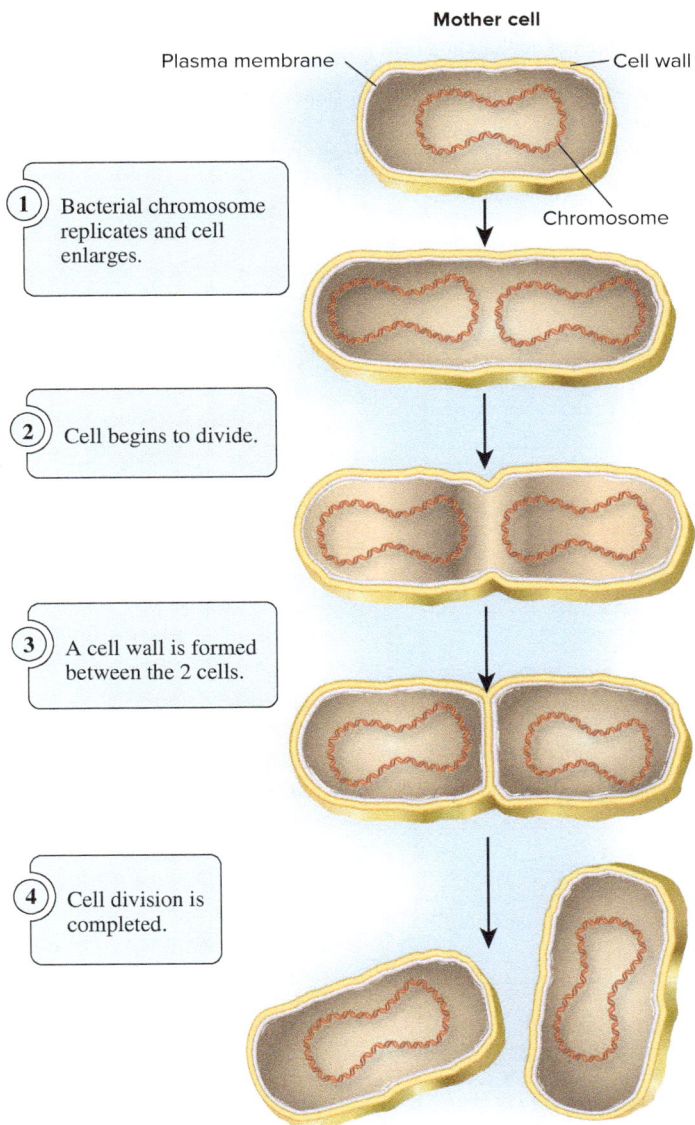

Figure 17.8 Bacterial cell division. Bacteria reproduce by binary fission. Before a bacterium divides, the bacterial chromosome is replicated to produce two identical copies. These two copies segregate from each other during cell division, with one copy going to each daughter cell.

17.4 Gene Transfer Between Bacteria

Learning Outcomes

1. Compare and contrast the three forms of gene transfer between bacteria: conjugation, transformation, and transduction.
2. Describe the process of horizontal gene transfer.

Even though bacteria reproduce asexually, they exhibit a great deal of genetic diversity. Within a given bacterial species, the term **strain** refers to a lineage that has genetic differences compared with another lineage. For example, one strain of *E. coli* may be resistant to an antibiotic, whereas another strain may be sensitive to the same antibiotic. How does genetic diversity arise in an asexual species? It comes primarily from two sources. First, mutations occur that alter the bacterial genome and affect the traits of bacterial cells. Second, diversity arises from **gene transfer,** in which genetic material is transferred from one bacterial cell to another. Through gene transfer, genetic variation that arises in one bacterium can spread to other strains and even to other species. For example, an antibiotic-resistance gene may be transferred from a resistant strain to a sensitive strain.

Gene transfer occurs in three ways: conjugation, transformation, and transduction (**Table 17.3**).

- **Conjugation** involves a direct physical interaction between two bacterial cells. During conjugation, one bacterium acts as a donor and transfers DNA to a recipient cell.
- **Transformation** is a process in which DNA is released into the environment and taken up by another bacterial cell.
- **Transduction** occurs when a bacteriophage infects a bacterial cell and then a newly made bacteriophage transfers some of that cell's DNA to another bacterium.

These three types of gene transfer have been extensively investigated in research laboratories, and their molecular pathways continue to be studied with great interest. In this section, we will examine these mechanisms in greater detail.

During Conjugation, DNA Is Transferred from a Donor Cell to a Recipient Cell

In the early 1950s, American microbiologists Joshua and Esther Lederberg, Irish physician William Hayes, and Italian geneticist Luca Cavalli-Sforza independently discovered that certain bacterial strains can donate genetic material during conjugation. For example, about 5% of *E. coli* strains found in nature act as donor strains. Further research showed that a strain that cannot act as a donor may acquire this ability after being mixed with a donor strain. Hayes correctly proposed that donor strains contain a type of plasmid called a fertility factor, or **F factor,** that can be transferred to recipient strains. Also, other donor *E. coli* strains were later identified that transfer portions of the bacterial chromosome at high frequencies. After a segment of the chromosome is transferred, it then inserts, or recombines, into the chromosome of the recipient cell. Such donor strains were named

Table 17.3	Mechanisms of Gene Transfer Between Bacterial Cells
Mechanism	**Description**
Conjugation:	Conjugation requires direct contact between a donor cell and a recipient cell. The donor cell transfers a strand of DNA to the recipient. In the example shown here, the DNA in a plasmid is transferred to the recipient cell. The end result is that both donor and recipient cells have a plasmid.
Transformation:	A fragment of DNA from a donor cell is released into the environment. This may happen when a bacterial cell dies. This DNA fragment is taken up by a recipient cell, which incorporates the DNA into its chromosome.
Transduction: 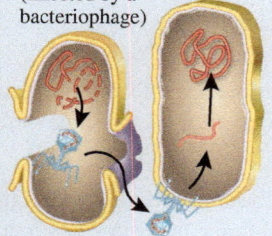	When a bacteriophage infects a donor cell, it causes the bacterial chromosome of the donor cell to break up into fragments. A fragment of bacterial chromosomal DNA is incorporated into a newly made bacteriophage. The bacteriophage then transfers this fragment of DNA to a recipient cell.

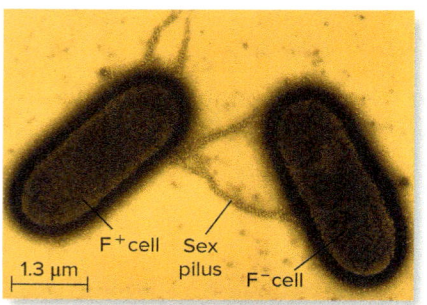

(a) Micrograph of conjugating cells

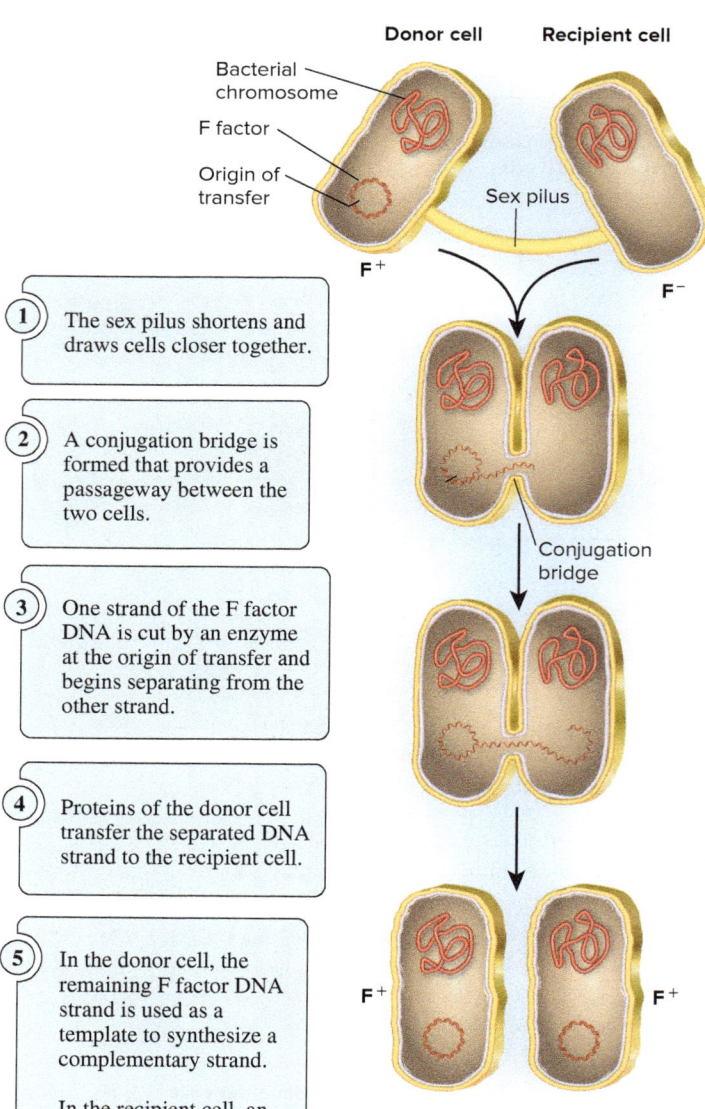

1. The sex pilus shortens and draws cells closer together.
2. A conjugation bridge is formed that provides a passageway between the two cells.
3. One strand of the F factor DNA is cut by an enzyme at the origin of transfer and begins separating from the other strand.
4. Proteins of the donor cell transfer the separated DNA strand to the recipient cell.
5. In the donor cell, the remaining F factor DNA strand is used as a template to synthesize a complementary strand.

   In the recipient cell, an enzyme joins the ends of the transferred DNA strand, and the complementary strand is made. Each cell now has a double-stranded circular F factor.

(b) Transfer of an F factor during conjugation

Hfr (for **H**igh **f**requency of **r**ecombination). In our discussion, we will focus on donor strains that carry F factors.

The micrograph in **Figure 17.9a** shows two conjugating *E. coli* cells. The cell on the left is designated F^+, meaning that it has an F factor. The F^+ cell, also called the donor cell, makes **sex pili** (singular, pilus) that bind specifically to recipient cells, which are designated

Figure 17.9 Bacterial conjugation. (a) A micrograph of two *E. coli* cells that are conjugating. The cell on the left, designated F^+, is the donor; the cell on the right, designated F^-, is the recipient. The two cells make contact via a sex pilus made by the F^+ cell. (b) The transfer of an F factor during conjugation. At the end of conjugation, both the donor cell and the recipient cell are F^+.
(a) ©Dr. L. Caro/SPL/Science Source

Concept Check: *If a donor cell has only one F factor, explain how the donor and recipient cells both contain one F factor following the transfer of an F factor during conjugation.*

F⁻ cells. The F factor found in F⁺ cells carries several genes that are required for conjugation, such as those required for sex pili formation. Also, some F factors, called F′ factors (F prime factors) carry additional bacterial genes that are not needed for conjugation.

Figure 17.9b describes the events that occur during conjugation in *E. coli*. The process is similar in other bacteria that are capable of conjugating, although the details vary somewhat from one species to another.

1. In *E. coli* and some other species, F⁺ cells make very long pili that attempt to make contact with nearby F⁻ cells. Once contact is made, the pili shorten, drawing the donor and recipient cells closer together.
2. A conjugation bridge is formed that provides a direct passageway for DNA transfer.
3. One strand of F factor DNA is cut at the origin of transfer and separates from the other strand.
4. Proteins of the donor cell transfer the separated DNA strand through the conjugation bridge into the recipient cell. The other strand remains in the donor cell.
5. In the donor cell, the complementary strand is synthesized, thereby restoring the F factor DNA to its original double-stranded condition. In the recipient cell, the two ends of the newly acquired F factor DNA strand are joined to form a circular molecule, and its complementary strand is synthesized to produce a double-stranded F factor.

If conjugation is successful, the end result is that the recipient cell has acquired an F factor, converting it from an F⁻ to an F⁺ cell. The genetic composition of the donor strain has not been changed.

In Transformation, Bacteria Take Up DNA from the Environment

In contrast to conjugation, the process of gene transfer known as bacterial transformation does not require direct contact between bacterial cells. Frederick Griffith first discovered this process in 1928 while working with strains of *Streptococcus pneumoniae* (refer back to Figure 9.1).

How does a bacterial cell become transformed? First, it imports a strand of DNA from the environment. This DNA strand, which is typically derived from a dead bacterial cell, may then insert or recombine into the bacterial chromosome. The live bacterium then carries genes from the dead bacterium—the live bacterium has been transformed.

Not all bacterial strains have the ability to take up DNA. Those that do have this ability are described as naturally **competent,** and they have genes that encode proteins called competence factors. Competence factors facilitate the binding of DNA fragments to the bacterial cell surface, the uptake of DNA into the cytoplasm, and the incorporation of the imported DNA into the bacterial chromosome. Temperature, ionic concentrations, and the availability of nutrients affect whether or not a bacterium will be competent to take up genetic material.

In recent years, biologists have unraveled some of the steps that occur when competent bacterial cells are transformed by taking up genetic material from the environment. In the example shown in Figure 17.10, the DNA released from a dead bacterium carries a gene, tet^R, that confers resistance to the antibiotic tetracycline.

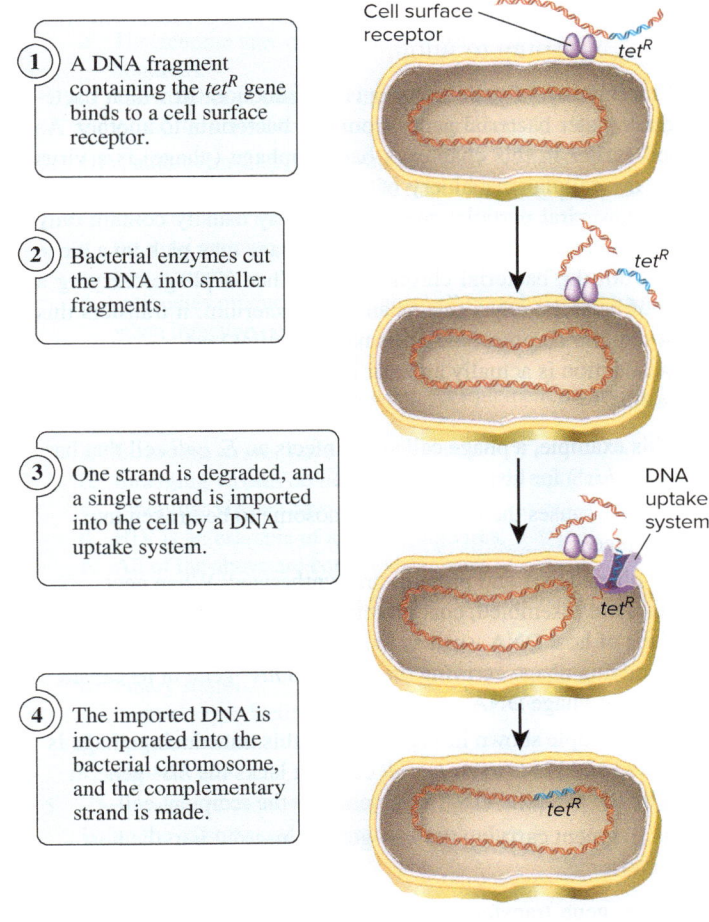

1. A DNA fragment containing the tet^R gene binds to a cell surface receptor.
2. Bacterial enzymes cut the DNA into smaller fragments.
3. One strand is degraded, and a single strand is imported into the cell by a DNA uptake system.
4. The imported DNA is incorporated into the bacterial chromosome, and the complementary strand is made.

Transformed cell that is resistant to the antibiotic tetracycline

Figure 17.10 Bacterial transformation. This process has transformed a bacterium that was sensitive to the antibiotic tetracycline into one that is resistant to this antibiotic.

BioConnections: *Refer back to Figures 9.1 and 9.2. How did the phenomenon of transformation allow researchers to demonstrate that DNA is the genetic material?*

1. A large fragment of the DNA binds to a cell surface receptor on the outside of a bacterial cell that is sensitive to tetracycline.
2. Enzymes secreted by the bacterium cut this large fragment into fragments of DNA small enough to enter the cell.
3. One of the two strands of a fragment of DNA containing the tet^R gene is degraded. The other strand enters the bacterial cytoplasm via a DNA uptake system that transports the DNA across the plasma membrane.
4. Finally, the imported single-stranded DNA strand is incorporated into the bacterial chromosome, and the complementary strand is synthesized.

Following transformation, the recipient cell has been transformed from a tetracycline-sensitive cell to a tetracycline-resistant cell.

c. the virus recognizes and binds to specific molecules on the surface of the host cell.
d. the host cell produces channel proteins that provide passageways for viruses to enter the cytoplasm.
e. the virus releases specific proteins that make holes in the membrane large enough for the virus to enter.

3. Genetic diversity is *not* increased in bacterial populations by
   a. binary fission.
   b. mutation.
   c. transformation.
   d. transduction.
   e. conjugation.

4. Let's suppose that F^+ and F^- cells grow at the same rate. If you mix together an equal number of F^+ and F^- cells, how would you expect the proportions to change over time? In other words, do you expect an increase in the relative proportion of F^+ or of F^- cells?
   a. The F^+ cells would become more prevalent over time.
   b. The F^- cells would become more prevalent over time.
   c. The relative proportions of the F^+ and F^- cells would stay the same.
   d. The F^+ cells would initially become more prevalent but later decrease.
   e. The F^- cells would initially become more prevalent but later decrease.

5. Which of the following is an example of horizontal gene transfer?
   a. the transfer of a gene from one strain of *E. coli* to a different strain via conjugation
   b. the transfer of a gene from one strain of *E. coli* to a different strain via transduction
   c. the transfer of an antibiotic-resistance gene from *E. coli* to *Salmonella typhimurium* via transformation
   d. the transfer of an antibiotic-resistance gene from one strain of *Salmonella typhimurium* to a different strain via transformation
   e. All of the above are examples of horizontal gene transfer.

Short Answer

1. Certain drugs to combat human viral diseases affect spike glycoproteins in the viral envelope. Discuss how you think such drugs may prevent viral infection.

2. Conjugation is sometimes called "bacterial mating." Discuss how conjugation is similar to sexual reproduction in eukaryotes and how it is different.

Genetic Technologies: How Biologists Study Genes and Genomes

18

©Fumihiro Sugiyama

Gene editing. Japanese researchers used an exciting new genetic approach, called CRISPR-Cas technology, to silence a gene in mice that encodes tyrosinase, an enzyme needed for pigment production. In the seven albino newborn mice on the left, both copies of this gene have been silenced. In the ones with a mottled appearance, partial silencing was achieved. The gene was not silenced in the black newborn on the right. The method of changing a gene's sequence, in this case to silence it, is called gene editing.

Chapter Outline

- **18.1** Gene Cloning
- **18.2** Genomics: Techniques for Studying and Altering Genomes
- **18.3** Bacterial and Archaeal Genomes
- **18.4** Eukaryotic Genomes
- **18.5** Repetitive Sequences and Transposable Elements

Assessing Your Knowledge and Skills

Jon was born with hemophilia A, which is a blood clotting disorder. As discussed in Chapter 15, hemophilia is inherited as an X-linked recessive trait. Jon's blood is unable to clot properly because he is missing a protein, called factor VIII, that is needed in a pathway required for normal blood clotting. Fortunately, Jon can take injections of purified factor VIII and thereby minimize the harmful effects of hemophilia. You might be surprised to learn that purified factor VIII is not obtained from human blood. Instead, it's made by cells grown in the laboratory. These cells have been genetically modified to synthesize factor VIII in large amounts. This process is just one example of how researchers have been able to apply **recombinant DNA technology**—the use of laboratory techniques to bring together fragments of DNA from multiple sources—to benefit humans.

In the early 1970s, the first successes in making recombinant DNA molecules were accomplished independently by two groups at Stanford University: David Jackson, Robert Symons, and Paul Berg; and Peter Lobban and A. Dale Kaiser. Both groups were able to isolate and purify pieces of DNA in a test tube and then covalently link two or more DNA fragments. Once inside a host cell, the recombinant molecules were replicated to produce many identical copies. The process of making multiple copies of a particular gene is known as **gene cloning.** In the first section of this chapter, we will explore recombinant DNA technology and gene cloning, techniques that have enabled geneticists to probe the relationships between gene sequences and phenotypic consequences.

In the second section, we will consider the topic of **genomics**—the molecular analysis of the entire genome of a species. In recent years, molecular techniques have progressed to the point where researchers can study the structure and function of many genes as large, integrated networks and can make changes to the genome in order to silence a particular gene (see the chapter-opening photo) or alter its DNA sequence in a specific way.

In the remaining sections of this chapter, we will consider the genome characteristics of bacteria, archaea, and eukaryotes. In many species, such as multicellular plants and animals, the genome has a high abundance of repetitive sequences—sequences that are repeated multiple times in the same genome. We will examine how such repetitive sequences increase in number.

18.1 Gene Cloning

Learning Outcomes

1. Outline the steps of gene cloning using vectors.
2. Distinguish between a genomic library and a cDNA library.
3. Explain how gel electrophoresis is used to separate DNA fragments.
4. Describe the steps of gene cloning using polymerase chain reaction (PCR).

As already mentioned, gene cloning is the process of making many copies of a particular gene. Why is gene cloning useful? **Figure 18.1** provides an overview of the steps and goals of gene cloning. The process is usually done with one of two goals in mind. One is that a researcher or clinician wants many copies of the gene, perhaps to study the DNA directly or to use the DNA as a tool. For example, geneticists may want to determine the sequence of a gene from a person with a disease to see if the gene carries a mutation. Alternatively, the goal may be to obtain a large amount of the gene product, such as a specific protein. Along these lines, biochemists use gene cloning to obtain large amounts of proteins to study their structure and function. In recent years, gene cloning has provided the foundation for critical technical advances in a variety of disciplines, including molecular biology, genetics, cell biology, biochemistry, microbiology, and medicine. In this section, we will examine the procedures that are used in gene cloning.

Step 1: Vector DNA and Chromosomal DNA Are the Starting Materials for Gene Cloning

One way to carry out gene cloning uses a type of DNA known as a **vector** (from the Latin, for one who carries) (see Figure 18.1). Vector DNA acts as a carrier of the DNA segment that is to be cloned. In cloning experiments, a vector may carry a small segment of chromosomal DNA, perhaps only a single gene. By comparison, a chromosome carries up to a few thousand genes. When a vector is introduced into a living cell, it can replicate, and so the DNA that it carries is also replicated. This produces many identical copies of the inserted gene.

The vectors commonly used in gene cloning experiments were originally derived from two natural sources: plasmids or viruses.

- **Plasmids** are small, circular pieces of DNA that are found naturally in many strains of bacteria and exist independently of the bacterial chromosome. Commercially available plasmids have been genetically engineered for effective use in cloning experiments. They contain unique sites into which geneticists can easily insert pieces of chromosomal DNA.

- **Viral vectors** are derived from viruses, which can infect living cells and propagate themselves by taking control of the host cell's metabolic machinery. When a chromosomal gene is inserted into a viral vector, the gene is replicated whenever the viral DNA is replicated. Therefore, viruses can be used as vectors to carry other pieces of DNA.

The second material necessary for cloning a gene is the gene itself, which we will call the gene of interest. The source of the gene is the chromosomal DNA that carries the gene. The preparation of chromosomal DNA involves breaking open cells and extracting and purifying the DNA using biochemical separation techniques such as chromatography and centrifugation.

Step 2: Cutting Chromosomal and Vector DNA into Pieces and Linking Them Together Produces Recombinant Vectors

The second step in a gene cloning experiment is the insertion of the gene of interest into the vector (see Figure 18.1, step 2). How is this accomplished? DNA molecules are cut and pasted into vectors to produce recombinant vectors. To cut DNA, researchers use enzymes known as **restriction enzymes.** These enzymes, which were discovered by Swiss geneticist Werner Arber and American microbiologists Hamilton Smith and Daniel Nathans in the 1960s and 1970s, are made naturally by many different species of bacteria. Restriction enzymes protect bacterial cells from invasion by viruses by degrading the viral DNA into small fragments. Several hundred different restriction enzymes from various bacterial species have been identified and are commercially available to molecular biologists.

The restriction enzymes used in cloning experiments bind to a specific base sequence and then cleave the DNA backbone at two defined locations, one in each strand. Most restriction enzymes recognize sequences that are palindromic, which means a sequence is identical when read in the opposite directions on the two strands (**Table 18.1**). For example, the sequence recognized by the restriction enzyme *Eco*RI is 5′–GAATTC–3′ in the top strand shown in Table 18.1. Read in the opposite direction in the bottom (complementary) strand, this sequence is the same: 5′–GAATTC–3′. Certain restriction enzymes are useful in cloning because they digest DNA into fragments with single-stranded ends (termed sticky ends) that hydrogen-bond to other DNA

Biology Principle

Biology Is an Experimental Science
The technique of gene cloning allows researchers to study genes and gene products in greater detail.

Gene cloning is done to achieve one of two main goals:

Producing large amounts of DNA of a specific gene	Expressing the cloned gene to produce the encoded protein
Examples	*Examples*
• Cloned genes provide enough DNA for DNA sequencing. The sequence of a gene can help us understand how a gene works and identify mutations that cause diseases.	• Large amounts of the protein can be purified to study its structure and function.
	• Cloned genes can be introduced into bacteria or livestock to make pharmaceutical products such as insulin.
• Cloned DNA can be used as a probe to identify the same gene or similar genes in other organisms.	• Cloned genes can be introduced into plants and animals to alter their traits.
	• Cloned genes can be used to treat diseases—a clinical approach called gene therapy.

Figure 18.1 Overview of gene cloning. The process of gene cloning is used to produce large amounts of a gene or its protein product.

---

fragments that are cut with the same enzyme and thus have complementary sequences.

Figure 18.2 shows the action of a restriction enzyme (*Eco*RI) and the insertion of a gene into a vector. This vector, which is a plasmid, carries the *amp*^{R} and *lacZ* genes, whose useful functions will be discussed later. *Eco*RI binds to specific sequences in both the vector and chromosomal DNA. It then cleaves the DNA backbones, producing DNA fragments with sticky ends (Figure 18.2, step 1). The sticky ends of a piece of chromosomal DNA and the vector DNA can hydrogen-bond with each other (Figure 18.2, step 2), a process called annealing. However, this interaction is not stable, because it involves only a few hydrogen bonds between complementary bases. To establish a permanent connection between two DNA fragments, the sugar-phosphate backbones of the DNA strands must be covalently linked, or ligated. This linkage is catalyzed by DNA ligase (Figure 18.2, step 3). Recall from Chapter 9 that DNA ligase is an enzyme that catalyzes the formation of a covalent bond between the sugar-phosphate backbones of two adjacent DNA fragments.

In some cases, the two ends of the vector simply ligate back together, restoring it to its original circular structure; this forms a recircularized vector. In other cases, a fragment of chromosomal DNA may become ligated to both ends of the vector. When this happens, a segment of chromosomal DNA has been inserted into the vector. The result is a vector containing a piece of chromosomal DNA, which is called a **recombinant vector.** Such a vector is ready to be cloned. A recombinant vector may contain the gene of interest, or it may contain a different piece of chromosomal DNA.

Table 18.1	Examples of Restriction Enzymes Used in Gene Cloning	
Restriction enzyme*	Bacterial source	Sequence recognized†
EcoRI	Escherichia coli (strain RY13)	5′–GAATTC–3′ 3′–CTTAAG–5′
SacI	Streptomyces achromogenes	5′–GAGCTC–3′ 3′–CTCGAG–5′

*Restriction enzymes are named according to the species in which they are found. The first three letters are italicized because they indicate the genus and species names. Because a species may produce more than one restriction enzyme, the enzymes are designated I, II, III, and so on to indicate the order in which they were discovered in a given species.
†The arrows show the locations in the upper and lower DNA strands where the restriction enzymes cleave the DNA backbone.

Step 3: The Uptake of Recombinant Vectors into Host Cells Followed by the Growth of Those Cells Achieves Gene Cloning

The third step in gene cloning is the actual cloning of the gene of interest. In this step, the goal is for the recombinant vector carrying the desired gene to be taken up by bacterial cells that have been treated with agents that render them permeable to DNA molecules (**Figure 18.3**). The process, called transformation, is described in Chapter 17 (refer back to Figure 17.10). Bacterial cells with the ability to take up DNA from the extracellular environment are called **competent.** After the bacteria are combined with a mixture of recircularized vectors and recombinant vectors, some bacterial cells take up a single vector, although most cells fail to take up any vector. The bacteria are then inoculated onto petri plates containing a bacterial growth medium and ampicillin.

In the experiment shown in Figure 18.3, the bacterial cells were originally sensitive to ampicillin. The vector, which is a plasmid, already carries an antibiotic-resistance gene, called the *amp^R* gene. What is the purpose of this gene in a cloning experiment? Such a gene is called a **selectable marker** because the presence of the antibiotic selects for the growth of cells expressing the *amp^R* gene. The *amp^R* gene encodes an enzyme known as β-lactamase that degrades the antibiotic ampicillin, which normally kills bacteria. Bacteria that have not taken up a plasmid are killed by the antibiotic. In contrast, any bacterium that has taken up a plasmid carrying the *amp^R* gene grows and divides many times to form a visible bacterial colony containing tens of millions of cells. Because each cell in a single colony is derived from the same original cell that took up a single plasmid, all cells within a colony contain the same type of plasmid DNA.

The experimenter also needs a way to distinguish bacterial colonies that contain cells with a recombinant vector from those containing cells with a recircularized vector. In a recombinant vector, a piece of chromosomal DNA has been inserted into a region of the vector that contains the *lacZ* gene, which encodes the enzyme β-galactosidase. The insertion of chromosomal DNA into the vector disrupts the *lacZ* gene, thereby preventing the synthesis of β-galactosidase. By comparison, a recircularized vector has

Figure 18.2 Step 2 of gene cloning: the actions of a restriction enzyme and DNA ligase to produce a recombinant vector.

Concept Check: *In the procedure shown in this figure, has the gene of interest been cloned?*

BioConnections: *Refer back to Figure 17.6. What are the general properties of plasmids?*

a functional *lacZ* gene. The functionality of *lacZ* can be determined by adding to the growth medium a colorless compound, X-Gal, which is cleaved by β-galactosidase into a blue dye. Bacteria grown in the presence of X-Gal form blue colonies if they produce a functional β-galactosidase enzyme, and white colonies if they do not. In this experiment, therefore, bacterial colonies containing recircularized vectors form blue colonies, whereas colonies containing recombinant vectors carrying a segment of chromosomal DNA are white.

After a bacterial cell has taken up a recombinant vector, two subsequent events lead to the production of many copies of that vector. First, when the vector has a highly active origin of replication, the bacterial host cell produces many copies of the recombinant vector per cell. Second, the bacterial cells divide approximately every 20 minutes. Following overnight growth, a population of many millions of bacteria is obtained from a single cell. For example, a bacterial colony may comprise 10 million cells, with each cell containing 50 copies of the recombinant vector. Therefore, this bacterial colony has 500 million copies of the cloned gene!

Biology Principle | Biology is a quantitative science

Quantitative Analysis: A DNA Library Is a Collection of Many Different DNA Fragments Cloned into Vectors

In a typical cloning experiment, such as the one described in Figures 18.2 and 18.3, the treatment of chromosomal DNA with restriction enzymes actually yields many different DNA fragments. Therefore, after the DNA fragments are ligated individually to vectors, a researcher has a collection of many recombinant vectors, with each vector containing a particular fragment of chromosomal DNA. A collection of recombinant vectors containing DNA fragments of a given organism is known as a **DNA library** (Figure 18.4). Researchers make DNA libraries using the methods shown in Figures 18.2 and 18.3 and then use those libraries to obtain clones that carry a gene of interest.

Two types of DNA libraries are commonly made. When the inserts are derived from chromosomal DNA, the library is called a **genomic library.** Alternatively, researchers can isolate mRNA and use the enzyme reverse transcriptase, which is described in Chapter 17, to make DNA molecules using mRNA as a starting material. Such DNA is called **complementary DNA, or cDNA.** A **cDNA library** is a collection of recombinant vectors that have inserts derived from cDNA.

In the case of eukaryotic genes, an important experimental advantage of cDNA is that it lacks introns—intervening sequences that are not translated into proteins. Because introns can be quite large, it is much simpler for researchers to insert cDNAs into vectors rather than chromosomal DNA segments if they want to focus their attention on the coding sequence of a gene. In addition, because bacteria do not splice out introns, using cDNAs provides an advantage if researchers want to express the gene of interest in bacteria.

Crunching the Numbers: The restriction enzyme *Eco*RI recognizes a 6-bp sequence (see Table 18.1). Assuming that this

Figure 18.3 Step 3 of gene cloning: introduction of a recombinant vector into a host cell. For cloning to occur, a recombinant vector is introduced into a host cell, which copies the vector and divides to produce many cells. This produces many copies of the gene of interest.

Concept Check: *In this cloning experiment, what is the purpose of having the lacZ gene in the vector?*

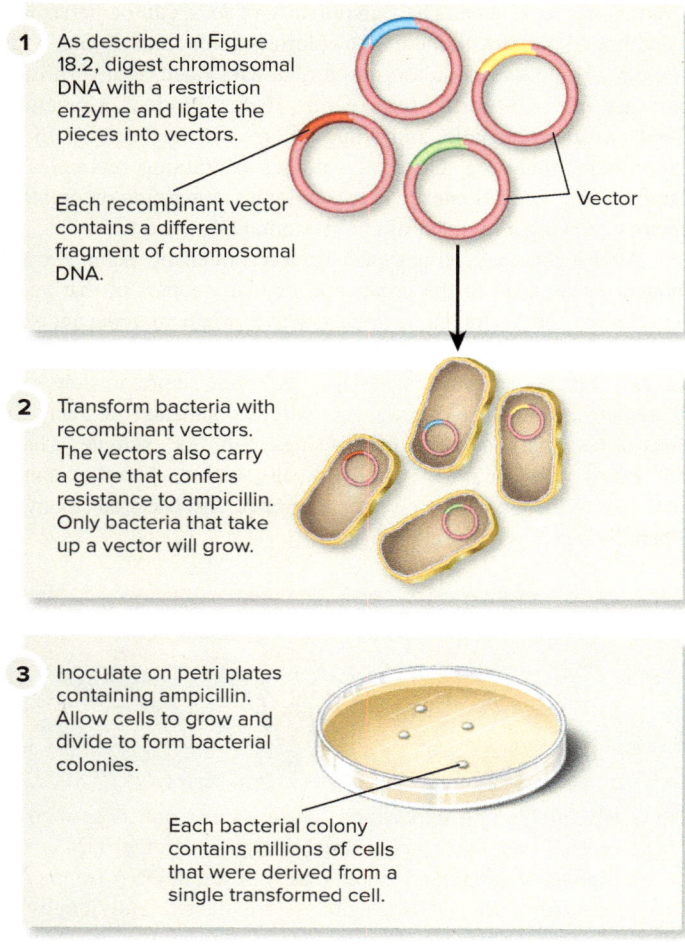

Figure 18.4 A DNA library. Each colony in a DNA library contains a vector with a different piece of chromosomal DNA.

sequence is found randomly in the human genome, how often (on average) will this restriction enzyme cut the chromosomal DNA? With a human genome size of approximately 3 billion base pairs, how many DNA fragments will be generated by cutting with *Eco*RI?

Figure 18.5 Separation of molecules by gel electrophoresis. In this example, samples containing many fragments of DNA are loaded into wells at the top of the gel and then subjected to an electric current that causes the fragments to move toward the positively charged electrode at the bottom of the gel. This separates the fragments according to their masses, with the smaller DNA fragments near the bottom of the gel.

Concept Check: *One DNA fragment contains 600 bp and another has 1,300 bp. Following electrophoresis, which will be closer to the bottom of the gel?*

Gel Electrophoresis Separates Macromolecules, Such as DNA Fragments

Gel electrophoresis is a technique for separating macromolecules, such as DNA and proteins, as they migrate through a gel. This method is often used to evaluate the results of a cloning experiment. For example, gel electrophoresis is used to determine the sizes of DNA fragments that have been inserted into recombinant vectors.

Gel electrophoresis can separate molecules based on their charge, size/length, and mass. In the example shown in **Figure 18.5**, gel electrophoresis is used to separate different fragments of chromosomal DNA based on their masses. The flat slab of semisolid gel has depressions at the top called wells, where samples are added. Electrodes are located at each end of the gel. An electric current is applied to the gel, which causes charged molecules, either proteins or nucleic acids, to migrate from the top of the gel toward the bottom—a process called electrophoresis. DNA is negatively charged and moves toward the positively charged electrode, which is at the bottom in this figure. As gel electrophoresis occurs, the DNA fragments are separated into distinct bands within the gel. Smaller DNA fragments move more quickly through the gel than larger ones in a given amount of time and therefore are located closer to the bottom of the gel than the larger ones. The fragments in each band can then be stained with a dye for identification.

Polymerase Chain Reaction (PCR) Is Also Used to Make Many Copies of DNA

As we have seen, one method of cloning involves an approach in which the gene of interest is inserted into a vector, introduced into a host cell, and then propagated. Another cloning technique, in which DNA is copied without the aid of vectors and host cells, is a process called **polymerase chain reaction (PCR),** which was developed by American biochemist Kary Mullis in 1985 (**Figure 18.6**).

The goal of PCR is to make many copies of DNA from a defined region, perhaps encompassing a gene or part of a gene. Several reagents are required for the synthesis of DNA. First, two different primers are needed that are complementary to sequences at each end of the DNA region to be amplified. These primers are usually about 20 nucleotides long. One primer is called the forward primer, and the other is the reverse primer. PCR also requires all four deoxynucleoside triphosphates (dNTPs) and a heat-stable form of DNA polymerase called *Taq* polymerase. *Taq* polymerase is isolated from the bacterium *Thermus aquaticus*, which lives in hot springs and can tolerate temperatures up to 95°C. A heat-stable form of DNA polymerase is necessary because PCR is conducted at high temperatures that would inactivate DNA polymerase from most other bacteria.

To make copies of a DNA region, the following three steps occur:

1. A sample of chromosomal DNA, called the template DNA, is heated to separate (denature) the DNA into single-stranded molecules.
2. The primers bind to the DNA as the temperature is lowered (see Figure 18.6). The binding of the primers to the specific sites in the template DNA is called primer annealing.

Figure 18.6 **Polymerase chain reaction (PCR).** During each PCR cycle, the steps of denaturation, primer annealing, and primer extension take place. The net result of PCR is the synthesis of many copies of DNA in the region that is flanked by the two primers. To conduct this type of PCR experiment, researchers must have prior knowledge about the base sequence of the template DNA so they can make primers with base sequences that are complementary to the ends of the template DNA. Note: The temperatures shown in this figure are approximate and may vary depending on the primer sequence and length.

Concept Check: *Why do the primers used in PCR bind specifically to the primer-annealing sites?*

BioConnections: *Refer back to Figure 9.16. Why are primers needed in a PCR experiment?*

With each successive cycle, the relative amount of DNA fragments that end exactly at both primer sites (marked *) increases. Therefore, after many cycles, the vast majority of DNA fragments contain only the region that is flanked by the 2 primer sites.

3. After the primers have annealed, the temperature is slightly raised and *Taq* polymerase uses dNTPs to catalyze the synthesis of complementary DNA strands, thereby doubling the amount of DNA in the region that is flanked by the primers. This step is called primer extension, because the length of the primers is extended by the synthesis of DNA.

The sequential process of denaturation followed by primer annealing and then primer extension is repeated many times in a row. This method is called a chain reaction because the products of each step are used as reactants in subsequent steps. A device that controls the temperature and automates the timing of each step, known as a thermocycler, is used to carry out PCR. The PCR technique can amplify a sample of DNA by a staggering amount. After 30 cycles of denaturation, primer annealing, and primer extension, a DNA sample will have increased by 2^{30}, approximately a billionfold, in a few hours!

18.1 Reviewing the Concepts

- Recombinant DNA technology is the use of laboratory techniques to bring together DNA fragments from two or more sources. Gene cloning, the process of making multiple copies of a gene, is used to obtain many copies of a particular gene or large amounts of the protein encoded by the gene (Figure 18.1).
- In one method of gene cloning, both a vector and chromosomal DNA are cut with restriction enzymes. The DNA fragments hydrogen-bond to each other at their sticky ends, and the pieces are covalently linked together via DNA ligase, producing recombinant vectors. When a recombinant vector is introduced into a bacterial cell, the cell replicates the vector and divides to produce many cells (Figures 18.2, 18.3).
- A collection of recombinant vectors, each with a particular piece of chromosomal DNA, is introduced into bacterial cells to produce a DNA library. If the DNA molecules are derived from mRNA, this creates a complementary DNA (cDNA) library (Figure 18.4).
- Gel electrophoresis is used to separate macromolecules by applying an electric current that causes them to move through a gel matrix. Gel electrophoresis typically separates molecules according to their charge, size/length, and mass (Figure 18.5).
- Polymerase chain reaction (PCR) is an alternative technique for gene cloning that does not involve vectors and host cells. Primers that flank the region of DNA to be amplified are required (Figure 18.6).

18.1 Testing Your Knowledge

1. In a cloning experiment using plasmid vectors, a restriction enzyme is used
   a. to cut the chromosomal DNA into many fragments.
   b. to open up the vector at a single location.
   c. to connect a chromosomal DNA fragment to a vector.
   d. for both a and b.

2. A cDNA library is a collection of recombinant vectors that
   a. carry fragments of chromosomal DNA.
   b. carry DNA segments derived from RNA.
   c. carry only a gene of interest.
   d. carry DNA segments from many different species.

3. The three steps in a PCR cycle occur in which of the following orders?
   a. primer annealing—primer extension—denaturation
   b. denaturation—primer annealing—primer extension
   c. denaturation—primer extension—primer annealing
   d. primer extension—primer annealing—denaturation

18.2 Genomics: Techniques for Studying and Altering Genomes

Learning Outcomes

1. Distinguish between genomics and functional genomics.
2. Outline the steps of DNA sequencing using the dideoxy chain-termination method.
3. Explain what a DNA microarray is and how it is used to identify the genes expressed by a sample of cells.
4. Describe how CRISPR-Cas technology can be used to edit genes.

As discussed throughout Unit III, the genome is the complete genetic composition of a cell, an organism, or a species. As genetic technology has progressed over the past few decades, researchers have gained an increasing ability to analyze the composition of genomes as a whole unit. The term genomics refers to the molecular analysis of the entire genome of a species. Segments of chromosomes are cloned and analyzed in progressively smaller pieces, the locations of which are known on the intact chromosomes. This is the mapping phase of genomics. The mapping of a genome ultimately progresses to the determination of the complete DNA sequence, which provides the most detailed description available of an organism's genome at the molecular level. By comparison, **functional genomics** studies the expression of a genome. For example, functional genomics can be used to analyze which genes are turned on or off in normal versus cancer cells. In this section, we will consider a few of the methods that are used in genomics and functional genomics.

The Dideoxy Chain-Termination Method Is Used to Determine the Base Sequence of DNA

The term **DNA sequencing** refers to a procedure that is aimed at determining the base sequence of DNA. Scientists can learn a great deal about the function of a gene if its nucleotide sequence is known. For example, the investigation of genetic sequences has been vital in our understanding of the genetic basis of human diseases.

One type of DNA sequencing, developed in 1977 by English biochemist Frederick Sanger and colleagues, is known as the **dideoxy chain-termination method,** or more simply,

dideoxy sequencing. Dideoxy sequencing is based on knowledge of DNA replication. As described in Chapter 9, DNA polymerase connects adjacent deoxynucleoside triphosphates (dNTPs) by catalyzing a covalent linkage between the phosphate group at the 5′ position on an incoming nucleotide and the —OH group at the 3′ position on the growing strand. Chemists, however, can synthesize nucleotides,

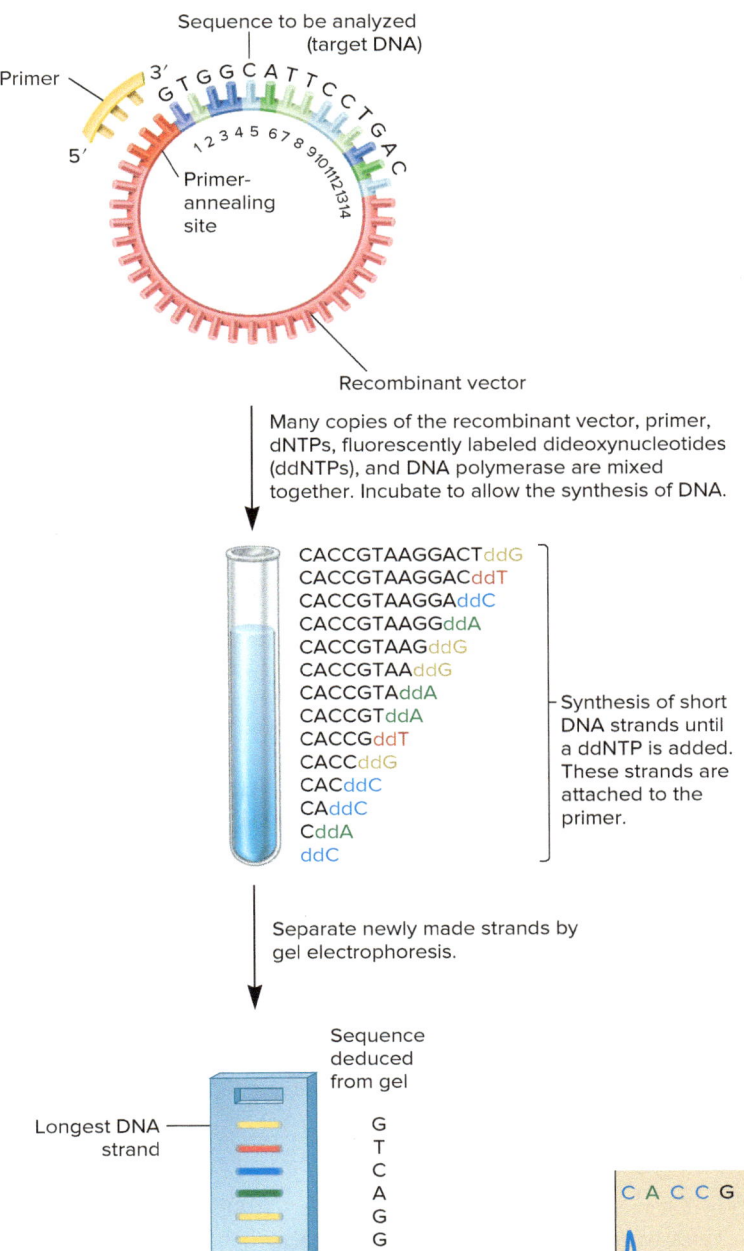

(a) Dideoxy DNA sequencing method

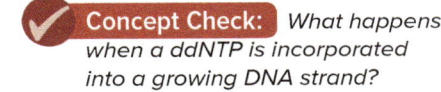

(b) Output from automated sequencing

Figure 18.7 **Structure of a dideoxynucleotide triphosphate.** This figure shows the structure of dideoxyguanosine triphosphate (ddGTP). It has a hydrogen, shown in red, instead of a hydroxyl group at the 3′ position. The prefix dideoxy- means that the sugar has two (di) missing (de) oxygens (oxy) compared with ribose, which has —OH groups at both the 2′ and 3′ positions.

called dideoxynucleoside triphosphates (ddNTPs), that are missing the —OH group at the 3′ position (Figure 18.7). What happens if a ddNTP is incorporated during DNA replication? If a ddNTP is added to a growing DNA strand, the strand can no longer grow, because the 3′ —OH group, the site of attachment for the next nucleotide, is missing. This ending of DNA synthesis is called chain termination.

Before discussing the steps of this DNA sequencing procedure, let's first consider the DNA segment that is analyzed in such an experiment. The segment of DNA to be sequenced, the target DNA, is obtained in large amounts by using the gene cloning techniques described earlier in this chapter. In Figure 18.8a, the target DNA has been inserted into a vector next to a primer-annealing site, the place where a primer will bind. The target DNA is initially double-stranded, but Figure 18.8a shows the DNA after it has been denatured into a single strand by heat treatment.

Let's now examine the steps involved in DNA sequencing.

1. Many copies of single-stranded template DNA are placed into a tube and mixed with primers that bind to the primer-annealing site. DNA polymerase and all four types of dNTPs are also added. In addition, a low concentration of each of the four possible dideoxynucleoside triphosphates—ddGTP, ddATP, ddTTP, and ddCTP—is added. Each type of ddNTP is tagged with a different-colored fluorescent molecule; typically, ddA is green, ddT is red, ddG is yellow, and ddC is blue.

2. The tube is then incubated to allow DNA polymerase to synthesize strands that are complementary to the target DNA

Figure 18.8 **Dideoxy sequencing of DNA. (a)** The procedure of dideoxy sequencing uses fluorescently labeled ddNTPs. **(b)** This method uses a fluorescence detector that measures the four kinds of ddNTPs as they emerge from the gel.

Concept Check: What happens when a ddNTP is incorporated into a growing DNA strand?

sequence. However, addition of a ddNTP causes DNA synthesis to terminate early, preventing further elongation of the strand. For example, let's consider ddTTP. Synthesis of new DNA strands stops at the sixth or thirteenth position after the annealing site if a ddTTP, instead of a dTTP, is incorporated into the growing DNA strand (Figure 18.8a). This means the target DNA has a complementary A at the sixth and thirteenth positions. Eventually, a set of fluorescently tagged strands results, with the color of the ddNTP representing the last nucleotide added to the strand.

3. After the samples have been incubated for several minutes, the newly made DNA strands are separated according to their lengths by subjecting them to gel electrophoresis. This can be done using a slab gel, as shown in Figure 18.8a, or more commonly by running them through a gel-filled capillary tube. The shorter strands move to the bottom of the gel more quickly than the longer ones. Electrophoresis is continued until each band emerges from the bottom of the gel, where a laser excites the fluorescent dye. A fluorescence detector records the amount of fluorescence emission at four wavelengths, corresponding to the four dyes.

An example of a printout from a fluorescence detector is shown in Figure 18.8b. The peaks of fluorescence correspond to the DNA sequence that is complementary to the target DNA. The heights of the fluorescent peaks are not always the same, because ddNTPs get incorporated at some sites more readily than at others. Note that although ddG is usually labeled with a yellow dye, it is converted to black ink on the printout for ease of reading. Though improvements in automated sequencing continue to be made, a typical sequencing run can provide a DNA sequence that is approximately 700 to 900 bases long, perhaps even longer.

Researchers are also developing alternative methods to the dideoxy chain-termination method in order to sequence DNA. These methods have largely superceded the dideoxy sequencing methods on sequencing projects in which the goal is to sequence an entire genome. They are sometimes called next-generation sequencing methods because they came after the dideoxy sequencing method. Though a description of next-generation sequencing methods is beyond the scope of this textbook, various innovations have made them faster and less expensive than dideoxy sequencing. These include the following:

- Some of the newer sequencing approaches rely on optical detection methods that make it unnecessary to perform gel electrophoresis.
- Some methods are able to sequence single DNA molecules, which makes it unnecessary to make many copies of the template DNA.
- In nanopore sequencing, a strand of DNA is passed through a protein nanopore in response to an electric current. The current is changed as the bases G, A, T and C pass through the pore in different combinations, allowing the DNA sequence to be read.

A Microarray Can Identify Which Genes Are Transcribed by a Cell

Let's now turn our attention to functional genomics. Researchers have developed an exciting technology, called a **DNA microarray,** that is used to monitor the expression of thousands of genes simultaneously.

A DNA microarray is a small silica, glass, or plastic slide that is dotted with many different sequences of single-stranded DNA, each corresponding to a short sequence within a known gene. Each spot contains multiple copies of a known DNA sequence. For example, one spot in a microarray may correspond to a sequence within the β-globin gene; another might correspond to a different gene, such as a gene that encodes a glucose transporter. A single slide contains tens of thousands of different spots in an area the size of a postage stamp. These microarrays are typically produced using a technology that "prints" spots of DNA sequences onto a slide, similar to the way that an inkjet printer deposits ink on paper.

What is the purpose of using a DNA microarray? In the experiment shown in **Figure 18.9**, the goal is to determine which genes are transcribed into mRNA from a particular sample of cells. In other words, which genes in the genome of these cells are expressed? To conduct this experiment, the mRNA was isolated from the cells and then used to make fluorescently labeled cDNAs. The labeled cDNAs were then incubated with a DNA microarray. The single-stranded DNA in the microarray corresponds to the coding strand—the strand that has a sequence that is similar to mRNA. Those cDNAs that are complementary to the DNAs in the microarray hybridize, thereby remaining bound to the microarray. The array is washed and then analyzed using a microscope equipped with a computer that scans all of the spots and generates an image of their relative fluorescence.

If the fluorescence intensity in a spot is high, a large amount of cDNA was in the sample that hybridized to the DNA at this location. For example, if the β-globin gene was expressed in the cells being tested, a large amount of cDNA for this gene would be made, and the fluorescence intensity for that spot would be high. Because the DNA sequence of each spot is already known, a fluorescent spot identifies a cDNA that is complementary to that DNA sequence. Furthermore, because the cDNA was generated from mRNA, this technique identifies genes that have been transcribed in a particular cell type under a given set of conditions. However, the amount of protein encoded by an mRNA may not always correlate with the amount of mRNA, due to variation in the rates of mRNA translation and protein degradation.

Thus far, the most common use of DNA microarrays is to study gene expression patterns. In addition, the technology of DNA microarrays has found several other important uses, described in **Table 18.2**.

Genes in Living Cells Can Be Altered Using CRISPR-Cas Technology

In Chapter 11, we considered how the CRISPR-Cas system provides bacteria with a defense against bacteriophages (refer back to Figure 11.10). Researchers realized that the components of this system can be used to alter genes, an approach called **gene editing.** This method is named **CRISPR-Cas technology.** Researchers have made a modification to the natural system to make it efficient for gene editing. They create a single RNA in which tracrRNA and crRNA are covalently linked to each other (**Figure 18.10a**). This is called the single guide RNA (sgRNA). The spacer region of the sgRNA is designed to be complementary to one of the DNA strands of a target gene that a researcher

GENETIC TECHNOLOGIES: HOW BIOLOGISTS STUDY GENES AND GENOMES 373

Figure 18.9 Identifying transcribed genes within a DNA microarray. In this simplified example, only three cDNAs specifically hybridize to spots on the microarray (see step 4). Those genes were expressed in the cells from which the mRNA was isolated. In an actual experiment, the array typically contains hundreds or thousands of different cDNAs and tens of thousands of different spots. *(bottom)* ©Alfred Pasieka/Science Source

Concept Check: *If a fluorescent spot appears on a microarray, what information does this provide regarding gene expression?*

Table 18.2	Applications of DNA Microarrays
Application	**Description**
Cell-specific gene expression	A comparison of microarray data using cDNAs derived from mRNAs of different cell types can identify genes that are expressed in a cell-specific manner.
Gene regulation	Because environmental conditions play an important role in gene regulation, a comparison of microarray data using cDNAs derived from mRNAs isolated from cells exposed to two different environmental conditions may reveal genes that are induced under one set of conditions and repressed under another set.
Elucidation of metabolic pathways	Genes that encode proteins that participate in a common metabolic pathway are often expressed together and can be revealed from a microarray analysis.
Tumor profiling	Different types of cancer cells exhibit striking differences in their gene expression profiles, which can be revealed by a DNA microarray analysis. This approach is gaining use as a tool to classify tumors that are sometimes morphologically indistinguishable.
Genetic variation	A mutant allele may not hybridize to a spot on a microarray as well as a wild-type allele. Therefore, microarrays are gaining use as a tool for detecting genetic variation. This application has been used to identify disease-causing alleles in humans and to identify mutations that contribute to quantitative traits in plants and other species.
Microbial strain identification	Microarrays can distinguish between closely related bacterial species and subspecies.

wants to edit. The sgRNA binds to Cas9 and guides it to the target gene. Cas9 then makes a double-strand break in this gene.

Following this break, two different DNA repair events are possible. If the break is repaired by a process called end joining, the gene may incur a small deletion (see the left side of **Figure 18.10b**). This deletion may inactivate the gene, particularly if it causes a frameshift mutation in the coding sequence. Alternatively, a researcher can add a double-stranded segment of DNA, called the donor DNA, that is homologous to the region where the break occurs (see the right side of Figure 18.10b). This homologous DNA is made synthetically and is designed to carry a particular mutation, such as a point mutation, that the researcher wants to introduce. A different DNA repair system swaps in the donor DNA by a double crossover. In this way, researchers can introduce a specific mutation into a gene.

The experiment described in Figure 18.10b can be performed on different cell types and even on whole organisms. By studying how a gene mutation affects cell structure and function or by examining the phenotypic effects of a mutation in a whole organism (see the chapter-opening photo), researchers often gain a better understanding of how genes function.

18.2 Reviewing the Concepts

- Genomics is the molecular analysis of an entire genome of a species; functional genomics studies the expression of a genome.
- Dideoxy sequencing uses ddNTPs to determine the base sequence of a segment of DNA (Figures 18.7, 18.8).

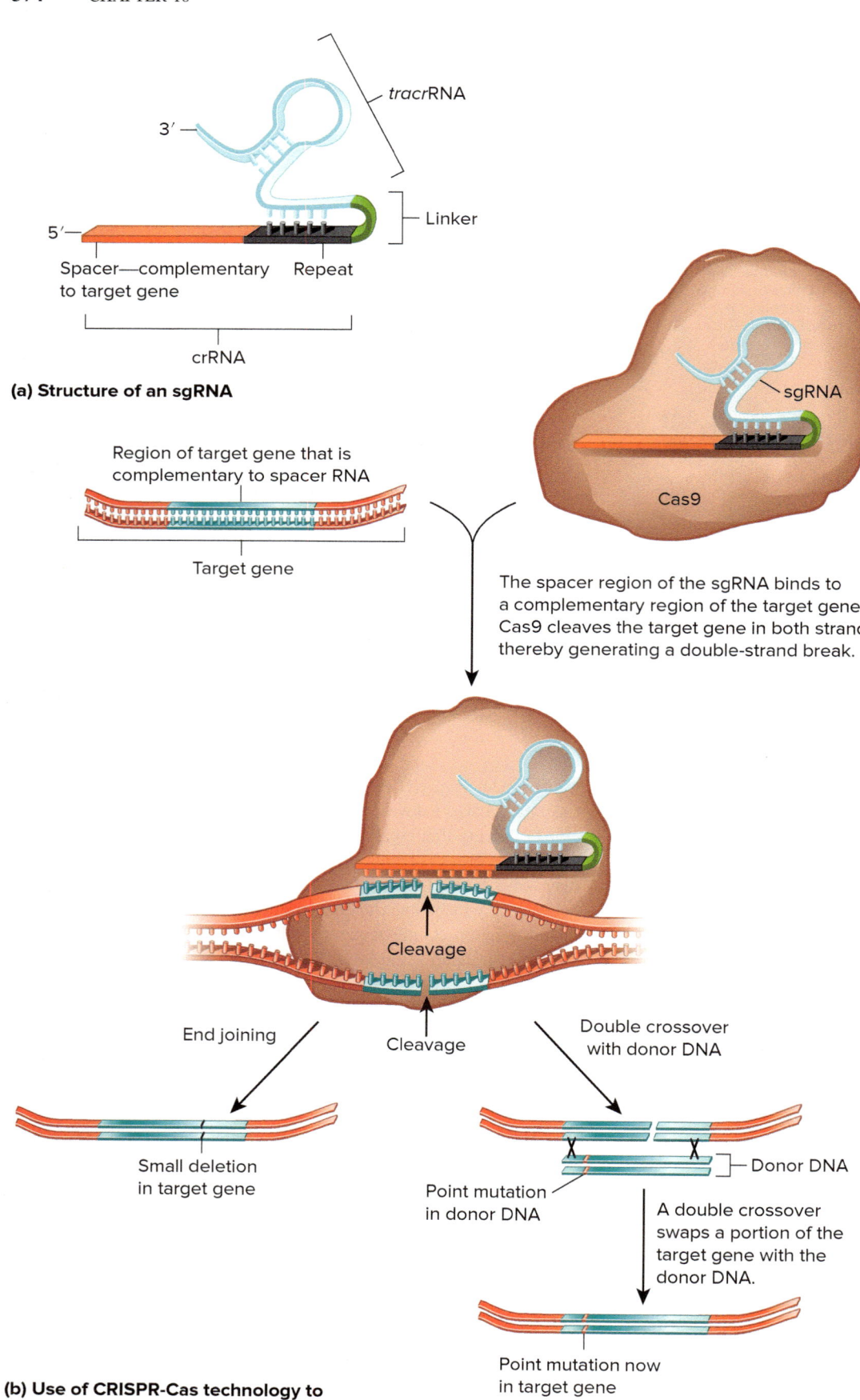

Figure 18.10 The use of CRISPR-Cas technology for gene editing. **(a)** Structure of an sgRNA, which is composed of a crRNA connected to a tracrRNA via a linker. **(b)** Use of CRISPR-Cas technology. On the left side, the gene has been repaired by end joining and suffers a small deletion. This deletion may result in gene inactivation. On the right side, a double crossover has occurred that swaps in the donor DNA. The target gene now carries a point mutation.

Concept Check: *How is the sgRNA different from certain components of the bacterial defense system described in Chapter 11 (refer back to Figure 11.10)?*

- A DNA microarray is a small silica, glass, or plastic slide dotted with different sequences of single-stranded DNA, each corresponding to a short sequence within a known gene. It is used to study gene expression patterns (Figure 18.9, Table 18.2).
- CRISPR-Cas technology can be used to edit genes (Figure 18.10).

18.2 Testing Your Knowledge

1. In dideoxy sequencing, what is a key property of a dideoxy nucleotide?
   a. It stops the further growth of a DNA strand.
   b. It is fluorescently labeled.
   c. It cannot be incorporated into a growing DNA strand.
   d. All of the above are correct.
   e. Both a and b are correct.

2. A microarray allows researchers to
   a. clone genes.
   b. make recombinant DNA molecules.
   c. study the expression of many genes simultaneously.
   d. make mutations within genes.
   e. do all of the above.

18.3 Bacterial and Archaeal Genomes

Learning Outcome

1. List the key characteristics of bacterial and archaeal genomes.

The past decade has seen remarkable advances in our overall understanding of the entire genome of many species. As genetic technology has progressed, researchers have gained an increasing ability to analyze the composition of genomes as a whole unit. The complete DNA sequence is now known for many species, providing the most detailed description available of an organism's genome at the molecular level. In this section, we will survey the sizes and composition of genomes in selected species of bacteria and archaea.

Studying the Genomes of Bacteria and Archaea Has Important Applications

Why are researchers interested in the genomes of bacteria and archaea?

1. Bacteria cause many diseases that affect humans as well as other animals and plants. Studying the genomes of bacteria reveals important clues about the process of infection, which may also help us find ways to combat bacterial infections.
2. Some bacterial species exhibit mutualistic relationships in which they rely on other bacterial species for growth. Sequencing the genomes of mutualistic bacteria may help to reveal the reasons for their interdependence.
3. The knowledge that is obtained by studying bacterial and archaeal genomes often applies to larger and more complex organisms.
4. A fourth reason is evolution. The origin of the first eukaryotic cell probably involved a union between an archaeal cell and a bacterial cell, as we will explore in Chapter 22. The study of bacterial and archaeal genomes helps us understand how all living species evolved.
5. Bacteria are often used as tools in research, as discussed in Section 18.1. A better understanding of their genomes can make them more effective tools.

The Genomes of Bacteria and Archaea Typically Consist of a Circular Chromosome with a Few Thousand Genes

Geneticists have made great progress in the study of bacterial and archaeal genomes. The genomes of thousands of bacterial and archaeal species have been sequenced and analyzed. The chromosomes of bacteria and archaea are usually a few million base pairs in length. Genomic researchers refer to 1 million base pairs as 1 megabase pair, abbreviated 1 Mb. Most bacteria and archaea contain a single type of chromosome, though multiple copies may be present in a single cell. However, some bacteria are known to have more than one type of chromosome. For example, *Vibrio cholerae*, the bacterium that causes the diarrheal disease cholera, has two different chromosomes in each cell; one has 2.9 Mb and the other 1.1 Mb.

Bacterial and archaeal chromosomes are usually circular. For example, the two chromosomes in *V. cholerae* are circular, as is the single type of chromosome found in *E. coli*. However, linear chromosomes are found in some species, such as *Borrelia burgdorferi*, the bacterium that causes Lyme disease, the most common tick-borne disease in the U.S. Certain bacterial species even contain both linear and circular chromosomes. *Agrobacterium tumefaciens*, which infects plants and causes crown gall tumors, has one linear chromosome (2.1 Mb) and one circular chromosome (3.0 Mb).

Table 18.3 compares the sequenced genomes from several bacterial and archaeal species. These genomes range in size from 1.7 to 5.2 Mb. The total number of genes is correlated with the total genome size. Roughly 1,000 genes are found for every 1 Mb of DNA. Compared with eukaryotic genomes, bacterial and archaeal genomes are less complex. Their chromosomes lack centromeres and telomeres and have a single origin of replication. Also, chromosomes of bacteria and archaea have relatively little repetitive DNA, whereas repetitive sequences, which are discussed later in this chapter, are often abundant in eukaryotic genomes.

In addition to one or more chromosomes, bacteria may have plasmids, circular pieces of DNA that exist independently of the bacterial chromosome (refer back to Figure 17.6). Plasmids are typically small, ranging in length from a few thousand to tens of thousands of base pairs, though some can be quite large, with hundreds of thousands of base pairs. The various functions of plasmids were described in Chapter 17, and their use as vectors in gene cloning was discussed in Section 18.1.

18.3 Reviewing the Concepts

- The genome is the complete genetic makeup of a cell, an organism, or a species.

Table 18.3 Examples of Bacterial and Archaeal Genomes That Have Been Sequenced*

Species	Genome size (Mb)†	Number of genes‡	Description
Methanobacterium thermoautotrophicum	1.7	1,921	An archaeal species that produces methane
Haemophilus influenzae	1.8	1,753	One of several different bacterial species that cause respiratory illness and meningitis
Sulfolobus solfataricus	3.0	3,032	An archaeal species that metabolizes sulfur-containing compounds
Lactobacillus plantarum	3.3	3,052	A type of lactic acid–producing bacterium used in the production of cheese and yogurt
Mycobacterium tuberculosis	4.4	4,294	The bacterium that causes the respiratory disease tuberculosis
Escherichia coli	4.6	4,377	A naturally occurring intestinal bacterium; certain strains cause human illness
Bacillus anthracis	5.2	5,439	The bacterium that causes the disease anthrax

*Bacterial and archaeal species often exist in different strains that may differ slightly in their genome size and number of genes. The data are from common strains of the indicated species. The species shown in this table have only one type of chromosome.
†Mb equals 1 million base pairs, or a megabase pair.
‡The number of genes is an estimate based on the analysis of genome sequences.

- Bacterial and archaeal genomes are typically a single circular chromosome with a few million base pairs of DNA. Such genomes usually have a few thousand different genes. In addition to one or more chromosomes, bacteria often have plasmids (Table 18.3).

18.3 Testing Your Knowledge

1. Which of the following is *not* a reason to study bacterial and archaeal genomes?
   a. to better understand how bacteria and archaea cause disease
   b. to better understand the biology of more complex organisms
   c. to better understand evolution
   d. to make them more effective tools in research and biotechnology
   e. All of the above are good reasons.

18.4 Eukaryotic Genomes

Learning Outcomes

1. Describe the key features of eukaryotic genomes.
2. Explain how gene duplications occur and lead to the formation of gene families.
3. List the goals and results of the Human Genome Project.

In the previous section, we examined bacterial and archaeal genomes. We now turn to eukaryotes, which include protists, fungi, animals, and plants. As we will see, their genomes are larger and more complex than those of their bacterial and archaeal counterparts. We will also examine how the duplication of genes can lead to families of related genes and will survey the goals and results of the Human Genome Project, which was aimed at mapping and sequencing the human genome.

Studying the Genomes of Eukaryotes Has Important Applications

Why are researchers interested in the genomes of eukaryotes? Motivation to sequence eukaryotic genomes comes from four main sources:

1. The availability of genome sequences makes it easier for researchers to identify and characterize the genes of model organisms. This was the impetus for genome projects involving baker's yeast (*Saccharomyces cerevisiae*), the fruit fly (*Drosophila melanogaster*), a nematode worm (*Caenorhabditis elegans*), the flowering plant called thale cress (*Arabidopsis thaliana*), and the mouse (*Mus musculus*).
2. Studying eukaryotic genomes enables researchers to gather more information for identifying and treating human diseases. Researchers hope that knowing the DNA sequence of the human genome will help to identify genes in which mutation plays a role in disease.
3. Sequencing the genomes of agriculturally important species can lead to new strains of livestock and crop species with improved traits.
4. Biologists are increasingly relying on genome sequences as a way to establish evolutionary relationships.

The Nuclear Genomes of Eukaryotes Are Sets of Linear Chromosomes That Vary Greatly in Size and Composition Among Different Species

As discussed in Chapter 14 (refer back to Figure 14.1), the genome located in the nucleus of eukaryotic species is usually found in sets of linear chromosomes. In humans, for example, one set contains 23 linear chromosomes—22 autosomes and 1 sex chromosome, X or Y. In addition, certain organelles in eukaryotic cells contain a small amount of DNA, including the mitochondrion, which plays a role in ATP synthesis, and the chloroplast (found in plants and algae), which carries out photosynthesis. The genetic material in these organelles is referred to as the mitochondrial or the chloroplast genome to distinguish it from the nuclear genome, which is located in the cell nucleus. In this chapter, we will focus on the nuclear genome of eukaryotes.

Sizes of Nuclear Genomes In the past decade or so, the DNA sequence of entire nuclear genomes has been determined for hundreds of eukaryotic species, including several dozen mammalian genomes. Examples are shown in **Table 18.4**.

Relationship Between Genome Sizes and Repetitive Sequences Eukaryotic genomes are generally larger than bacterial and archaeal genomes, in terms of both the number of genes and genome size. The genomes of simpler eukaryotes, such as yeast, carry several thousand different protein-encoding genes, whereas the genomes of more complex eukaryotes contain tens of thousands (see Table 18.4). Note that the number of genes is not the same as genome size. When we speak of genome size, we mean the total amount of DNA, often measured in megabase pairs. The relative sizes of nuclear genomes vary dramatically among different eukaryotic species (**Figure 18.11a**). In general, increases in the amount of DNA are correlated with increases in cell complexity and body complexity. For example, yeasts have smaller genomes than animals.

However, major variations in genome size are often observed among species that are similar in form and function. For example, the total amount of DNA found within different species of amphibians varies over 100-fold. As another example, let's consider two closely related species of the plant called the globe thistle, *Echinops bannaticus* and *Echinops nanus* (**Figure 18.11b, c**). These species have similar numbers of chromosomes, but *E. bannaticus* has nearly double the amount of DNA that *E. nanus* has. What is the explanation for the larger genome of *E. bannaticus*? The genome of *E. bannaticus* is not likely to contain twice as many genes. Rather, its genome composition includes many repetitive sequences, which are short DNA sequences that are present in many copies throughout the genome. Repetitive sequences are often abundant in eukaryotic species. We will examine the characteristics of such repetitive sequences in Section 18.5.

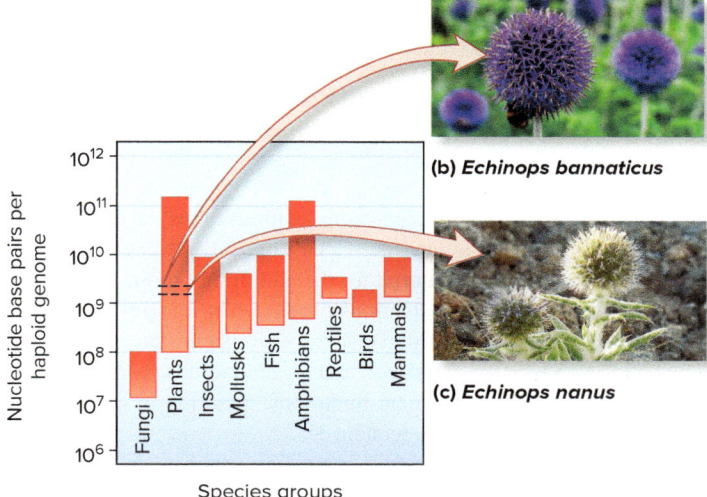

Figure 18.11 Genome sizes among selected groups of eukaryotes.
(a) Genome sizes among various groups of eukaryotes are shown on a log scale. As an example for comparison, two closely related species of globe thistle are pictured. These species have similar characteristics, but *Echinops bannaticus* **(b)** has nearly double the amount of DNA that *E. nanus* **(c)** has, as a result of the accumulation of repetitive DNA sequences.
(b) ©The Picture Store/SPL/Science Source; (c) ©Photo by Michael Beckmann, Institute of Geobotany and Botanical Garden, Halle, Germany

Concept Check: What are two reasons that the groups of species shown in part (a) vary in their total amount of DNA?

Table 18.4 Examples of Eukaryotic Nuclear Genomes That Have Been Sequenced

Species	Nuclear genome size (Mb)	Number of protein-encoding genes	Description
Saccharomyces cerevisiae (baker's yeast)	12.1	~6,600	One of the simplest eukaryotic species, which has been extensively studied by researchers trying to understand eukaryotic molecular biology
Caenorhabditis elegans (a nematode worm)	100	~20,000	A model organism used to study animal development
Drosophila melanogaster (fruit fly)	180	~15,000	A model organism used to study many genetic phenomena, including development
Arabidopsis thaliana (thale cress)	120	~26,000	A model organism studied by plant biologists
Oryza sativa (rice)	440	~40,000	A cereal grain that has a relatively small genome and is very important worldwide as a food crop
Mus musculus (mouse)	2,500	~21,000	A model mammalian organism used to study genetics, cell biology, and development
Homo sapiens (humans)	3,200	~22,000	Our own genome, the sequencing of which will help to elucidate our understanding of inherited traits and aid in the identification and treatment of diseases

Note: The genome size refers to the number of megabase pairs (Mb) in one set of chromosomes. For species with sex chromosomes, it would include both sex chromosomes.

EVOLUTIONARY CONNECTIONS

Gene Duplications Provide Additional Material for Genome Evolution, Sometimes Leading to the Formation of Gene Families

Let's now turn our attention to gene duplications, a way of increasing the number of genes in a genome. Gene duplications are important because they provide raw material for the addition of more genes into a species' genome. Such duplications produce **homologous genes,** two or more genes that are derived from the same ancestral gene (**Figure 18.12a**). Over the course of many generations, each version of the gene accumulates different mutations, resulting in genes with similar but not identical DNA sequences.

How do gene duplications occur? One mechanism that produces gene duplications is a misaligned crossover (**Figure 18.12b**). In this example, two homologous chromosomes have paired with each other during meiosis, but the homologs are misaligned. A crossover produces one chromosome with a gene duplication, one with a gene deletion, and two normal chromosomes. Each of these chromosomes is segregated into different haploid cells. If a haploid cell carrying the chromosome with the gene duplication participates in fertilization with another gamete, an offspring with a gene duplication is produced. In this way, gene duplications can form and be transmitted to future generations.

During evolution, gene duplications can occur several times. Two or more homologous genes within a single species are also called paralogous genes, or **paralogs.** Multiple gene duplications followed by the accumulation of mutations in each paralog result in a **gene family**—a group of paralogs that carry out related functions. A well-studied example is the globin gene family found in animals. The globin genes encode polypeptides that are subunits of proteins that function in oxygen binding. Hemoglobin, which is made in red blood cells, carries oxygen throughout the body. In humans, the globin gene family is composed of 14 paralogs that were originally derived from a single ancestral globin gene (**Figure 18.13**). According to an evolutionary analysis, the ancestral globin gene duplicated between 500 and 600 mya. Since that time, additional duplication events and chromosomal rearrangements have occurred to produce the current number of 14 genes on three different human chromosomes. Four of these are pseudogenes—genes that have been produced by gene duplication but have accumulated mutations that make them nonfunctional, so they are not transcribed into RNA.

The accumulation of different mutations in the various family members has produced globins that are specialized in their function.

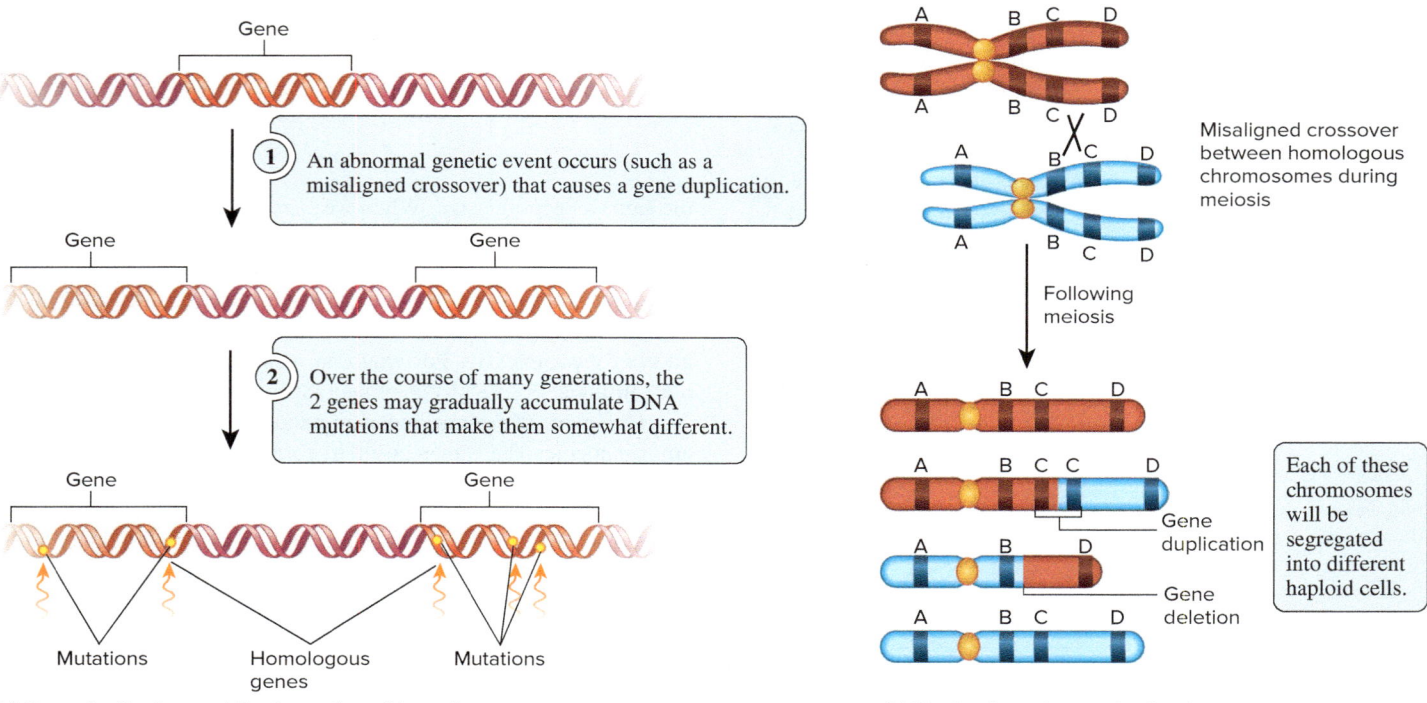

Figure 18.12 Gene duplication and the evolution of homologous genes. (a) A gene duplication produces two copies of the same gene. Over time, these copies accumulate different random mutations, which results in homologous genes with similar but not identical DNA sequences. (b) A mechanism of gene duplication. If two homologous chromosomes misalign during meiosis, a crossover may produce a chromosome with a gene duplication.

Biology Principle

Populations of Organisms Evolve from One Generation to the Next and Are Related by an Evolutionary History

In this example, evolution involves two different types of genetic changes. First, a gene duplication occurs; second, each copy of the gene accumulates different mutations.

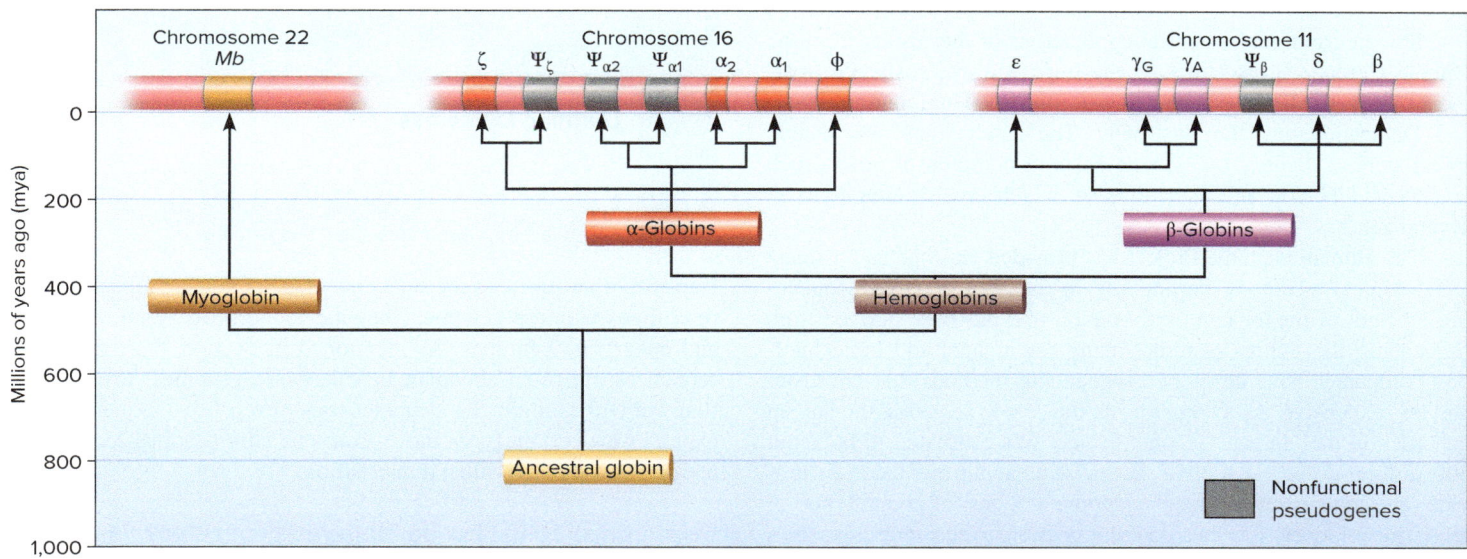

Figure 18.13 The evolution of the globin gene family in humans. The globin gene family evolved from a single ancestral globin gene.

BioConnections: *Refer back to Figure 12.3. How do the proteins encoded by different members of the globin gene family vary in their affinity for oxygen?*

For example, myoglobin binds and stores oxygen in muscle cells, whereas the hemoglobins bind and transport oxygen via red blood cells. Also, different globin genes are expressed during different stages of development. The zeta (ζ)-globin and epsilon (ε)-globin genes are expressed very early in embryonic life. During the second trimester of gestation, the alpha (α)-globin and gamma (γ)-globin genes are turned on. Following birth, the γ-globin genes are turned off, and the β-globin gene is turned on. These differences in the expression of the globin genes reflect the differences in the oxygen transport needs of humans during the embryonic, fetal, and postpartum stages of life (refer back to Figure 12.3).

The Human Genome Project Has Stimulated Genomic Research

Before ending our discussion of genomes, let's consider the **Human Genome Project,** a research effort to identify and map all human genes. Scientists had been discussing how to undertake this project beginning in the mid-1980s. In 1988, the National Institutes of Health (NIH) in Bethesda, Maryland, established an Office of Human Genome Research, with James Watson as its first director.

The Human Genome Project officially began on October 1, 1990, and was largely finished by the end of 2003. It was an international consortium that included research institutions in the U.S., the U.K., France, Germany, Japan, and China. From its outset, the Human Genome Project had the following goals:

- *To obtain the DNA sequence of the entire human genome.* The first draft of a nearly completed DNA sequence was published in February 2001, and a second draft was published in 2003. The entire human genome is approximately 3.2 billion base pairs in length.

- *To identify all human genes.* This involved mapping the locations of genes throughout the entire genome.

- *To develop technology for the generation and management of human genome information.* Some of the efforts of the Human Genome Project have involved improvements in molecular genetic technology, such as gene cloning and DNA sequencing. The Human Genome Project has also developed computer tools that allow scientists to easily access up-to-date information from the project and analytical tools to interpret genomic information.

- *To analyze the genomes of model organisms.* These include *E. coli, S. cerevisiae, D. melanogaster, C. elegans, A. thaliana,* and *M. musculus.*

- *To develop programs focused on understanding and addressing the ethical, legal, and social implications of the results obtained from the Human Genome Project.* The Human Genome Project raised many ethical issues regarding genetic information and genetic engineering. Who should have access to genetic information? Should employers, insurance companies, law enforcement agencies, and schools have access to our genetic makeup?

Some current and potential applications of the results from the Human Genome Project include the improved diagnosis and treatment of genetic diseases such as cystic fibrosis, Huntington disease, and Duchenne muscular dystrophy. The results may also enable researchers to identify the genetic basis of common disorders such as cancer, diabetes, and heart disease, which involve alterations in several genes.

The Human Genome Project has provided an impetus for many technological advances, particularly in the area of DNA sequencing. In light of the amazing information that has been derived from sequencing the human genome, many researchers and biotechnology companies have developed sequencing methods that are faster and less expensive. As a comparison, the cost to sequence the human genome via the Human Genome Project was estimated to be about 300 million dollars. By 2019, newer sequencing methods can provide the sequence of a person's genome at a cost of about $1000 or less! This relatively low cost makes it possible to sequence a person's genome as a diagnostic procedure for the purposes of identifying, treating, and/or preventing many different kinds of diseases.

18.4 Reviewing the Concepts

- The nuclear genomes of eukaryotic species are composed of sets of linear chromosomes with a total length of several million to billions of base pairs. They typically contain several thousand to tens of thousands of genes. Genome sizes vary greatly among eukaryotic species (Figure 18.11, Table 18.4).
- One way that the number of genes in a genome can increase is via gene duplication. Gene duplication may occur as a result of a misaligned crossover during meiosis. This is one mechanism that can produce a gene family, two or more homologous genes in a species that have related functions (Figures 18.12, 18.13).
- The Human Genome Project, an international effort to map and sequence the entire human genome, was completed by an international consortium in 2003.

18.4 Testing Your Knowledge

1. The sizes of eukaryotic genomes vary because
   a. more complex eukaryotes tend to have more genes.
   b. the amount of repetitive sequences can vary greatly among different species.
   c. the diameter of the cell nucleus may be different.
   d. of both a and b.

2. The members of a gene family
   a. are called paralogs.
   b. have similar but not identical DNA sequences.
   c. usually carry out similar functions.
   d. All of the above are characteristics of gene family members.

18.5 Repetitive Sequences and Transposable Elements

Learning Outcomes

1. Describe the characteristics of repetitive sequences.
2. Name the two major types of transposable elements, and explain how they move about the genome.

Repetitive sequences are segments of DNA that are repeated multiple times within a genome. They are found in bacterial, archaeal, and eukaryotic genomes. As mentioned in Section 18.4, repetitive sequences are often abundant in eukaryotic genomes. In this section, we will examine the characteristics of repetitive sequences and explore how certain types move from one chromosomal location to another by a process called transposition.

Segments of DNA May Be Moderately or Highly Repetitive

Repetitive sequences fall into two broad categories, moderately and highly repetitive. Sequences that are repeated a few hundred to several thousand times are called **moderately repetitive sequences.** In some cases, these sequences are multiple copies of the same gene. For example, the genes that encode ribosomal RNA (rRNA) are found in many copies. The cell needs a large amount of rRNA for its cellular ribosomes. This is accomplished by having and expressing multiple copies of the genes that encode rRNA. In addition, other types of functionally important sequences can be moderately repetitive. For example, multiple copies of origins of replication are found in eukaryotic chromosomes. Other moderately repetitive sequences may play a role in the regulation of gene transcription and translation.

Highly repetitive sequences are those that are repeated tens of thousands to millions of times throughout the genome. Each copy of a highly repetitive sequence is relatively short, ranging from a few to several hundred nucleotides in length.

Some highly repetitive sequences are clustered together in a tandem array, in which a very short nucleotide sequence is repeated many times in a row. In *Drosophila*, for example, 19% of the chromosomal DNA is highly repetitive DNA found in tandem arrays. An example is shown here:

AATATAATATATAATATAATATAATATAT
TTATATTATATATTATATTATATTATATA

In this particular tandem array, two related sequences, AATAT and AATATAT (in the top strand), are repeated multiple times. Highly repetitive sequences, which contain tandem arrays of short sequences, can be quite long, sometimes more than 1 million bp in length!

Other highly repetitive sequences are interspersed throughout the genome. A widely studied example is the *Alu* family of sequences found in humans and other primates. The *Alu* sequence is approximately 300 bp long. This sequence derives its name from the observation that it contains a site for cleavage by a restriction enzyme known as *Alu*I. It represents about 10% of the total human DNA and occurs (on average) approximately every 5,000–6,000 bases. Evolutionary studies suggest that the *Alu* sequence arose 65 million years ago (mya) from a section of a single ancestral gene known as the 7SL RNA gene. Remarkably, over the course of 65 million years, the *Alu* sequence has been copied and inserted into the human genome so often that it now occurs more than 1 million times. The mechanism for the proliferation of *Alu* sequences will be described later in this section.

Figure 18.14 shows the composition of the classes of DNA sequences found in the nuclear genome of humans. Surprisingly, exons, the coding regions of protein-encoding genes, and the genes that give rise to rRNA and tRNA make up only about 2% of our genome! The other 98% is composed of non-coding sequences. Though we often think of genomes as being the repository of sequences that code for proteins, most eukaryotic genomes are largely composed of other types of sequences. Intron DNA makes up about 24% of the human genome, and unique non-coding DNA constitutes 15%. Repetitive DNA makes up 59% of the DNA in the genome.

Scientists once described non-coding sequences as "junk DNA" because it was believed to have no biological function. To address this question, the Encyclopedia of DNA Elements (ENCODE) Project was started in 2003. This undertaking has involved more than 440 researchers in 32 laboratories. The goal of the ENCODE Project is to identify functional sequences in the human genome. In 2012, the researchers announced that they were able to assign function to approximately 80% of the human genome. Surprisingly, much of the previously described "junk DNA" appears to play a role in a complex network of gene regulation, a topic discussed in Chapters 11 and 12.

Transposable Elements Move from One Chromosomal Location to Another

As we have seen, repetitive sequences can be moderately or highly repetitive, and they can be found in tandem arrays or interspersed throughout a genome. Some types of repetitive sequences, which are interspersed in genomes, are **transposable elements (TEs)**—DNA segments that can move throughout the genome. The *Alu* sequence discussed earlier is an example of a TE. The process in which a TE moves to a new site in a genome is called **transposition.** TEs range from a few hundred to several thousand base pairs in length. They have sometimes been referred to as "jumping genes," because they are inherently mobile.

American cytogeneticist Barbara McClintock first identified TEs in the late 1940s from her studies with corn plants (**Figure 18.15**). She identified a segment of DNA that could move into and out of a gene that affected the color of corn kernels, producing a speckled appearance. Since that time, biologists have discovered many different types of TEs in nearly all of the species they have examined.

Though McClintock identified TEs in corn in the late 1940s, her work was met with great skepticism because many researchers had trouble believing that DNA segments could be mobile. The advent of molecular technology in the 1960s and 1970s allowed scientists

Biology Principle

The Genetic Material Provides the Blueprint for Growth, Development, and Reproduction

Only a small percentage of the human genome is involved with encoding the proteins that are largely responsible for human traits.

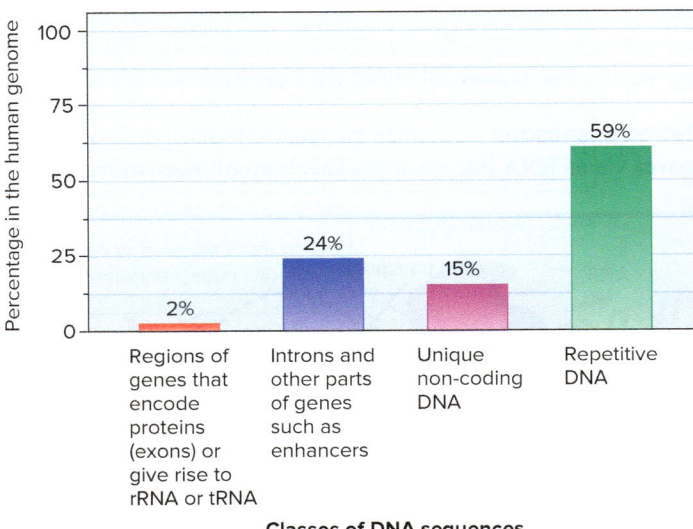

Figure 18.14 The composition of DNA sequences in the nuclear genome of humans. Only about 2% of our genome codes for proteins. Most of our genome is made up of repetitive sequences.

(a) Barbara McClintock

(b) Speckled corn kernels caused by transposable elements

Figure 18.15 Barbara McClintock, who discovered transposable elements (TEs). As shown in part (b), when a TE is found within a pigment gene in corn, its frequent movement disrupts the gene, causing the kernel color to be speckled.

(a) ©Topham/The Image Works; *(b)* ©Kenneth Keifer/Getty Images

(a) Bulldog

(b) Greyhound

(c) Dachshund

Figure 19.7 **Common breeds of dogs that have been obtained by selective breeding.** By selecting individuals carrying the alleles that influence traits desirable to humans, dog breeders have produced breeds with distinctive features. The dogs in this figure carry the same kinds of genes (for example, genes that affect their size, shape, and fur color). However, the alleles for many of these genes are different among these dogs, thereby allowing breeders to select for or against them and produce breeds with strikingly different phenotypes.
(a) ©WilleeCole/Getty Images; *(b)* ©Martin Rugner/age fotostock; *(c)* ©Juniors Bildarchiv/age fotostock

All dogs are members of the same species, *Canis lupus*, subspecies *familiaris*, so they can interbreed to produce offspring. Selective breeding can dramatically modify the traits in a species. When you compare certain breeds of dogs (for example, a greyhound and a dachshund), they hardly look like members of the same species! A study in 2007 by American geneticist Nathan Sutter and colleagues indicated that the size of dogs may be determined by alleles of the *Igf1* gene that encodes a growth hormone called insulin-like growth factor 1. A particular allele of this gene was found to be common to all small breeds of dogs but nearly absent from very large breeds, suggesting that this allele is a major contributor to body size in small breeds of dogs.

A Comparison of Homologies Shows Evolution of Related Species from a Common Ancestor

Let's now consider other widespread observations of the process of evolution among different species. In biology, the term **homology** refers to a similarity that occurs due to descent from a common ancestor. Two species may have a similar trait because the trait was originally found in a common ancestor. As described next, such homologies may involve anatomical, developmental, or molecular features.

Anatomical Homologies An examination of the limbs of modern vertebrate species reveals similarities that indicate the same set of bones has undergone evolutionary changes, becoming modified to perform different functions in different species. As seen in **Figure 19.8**, the forelimbs of vertebrates have a strikingly similar pattern of bone arrangements. These are termed **homologous structures**—structures that are similar to each other because they evolved from a structure in a common ancestor. The forelimb has developed different functions among various vertebrates, including grasping, walking, flying, swimming, and climbing. The process of evolution explains how these animals have descended from a common ancestor and how natural selection has resulted in modifications to the structure of the original set of bones in ways that ultimately allowed them to be used for several different functions.

Biology Principle

Structure Determines Function

These homologous sets of bones have evolved into somewhat different structures, which result in differences in their functions in humans, turtles, bats, and whales.

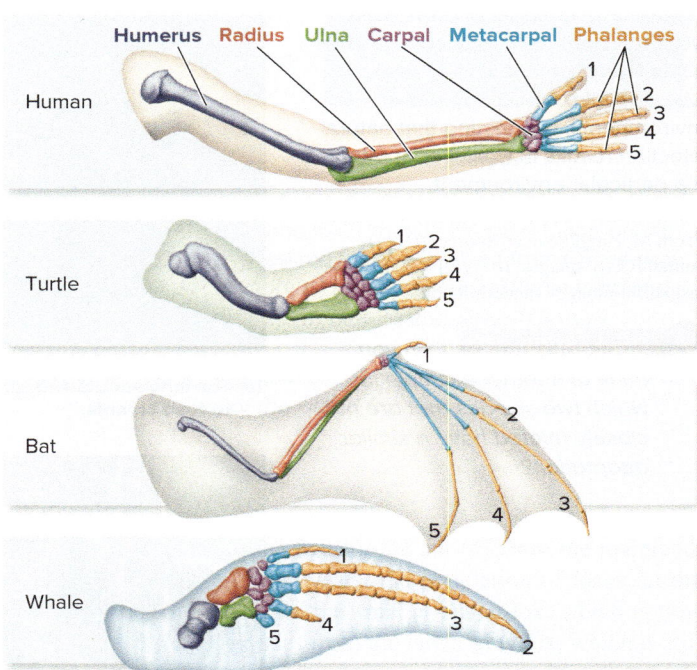

Figure 19.8 **An example of anatomical homology: homologous structures found in vertebrates.** The same set of bones is found in the human arm, turtle arm, bat wing, and whale flipper, although their relative sizes and shapes differ significantly. This homology suggests that all of these animals evolved from a common ancestor.

Table 19.3	Examples of Vestigial Structures in Animals
Organism	Vestigial structure(s)
Human	Bony tail in embryo and muscles to wiggle ears in adult
Boa constrictor	Skeletal remnants of hip and hind leg bones
Whale	Skeletal remnants of a pelvis
Manatee	Fingernails on the flippers

Another result of evolution is the phenomenon of **vestigial structures,** anatomical features that have little or no current function but resemble structures of their presumed ancestors (Table 19.3). An interesting case is found in humans. People have a complete set of muscles for moving their ears, even though most people are unable to do so. By comparison, many modern mammals can move their ears to determine the direction of sounds, and presumably this was an important trait in a distant human ancestor.

Why would organisms have structures that are no longer useful? Within the context of evolutionary theory, vestigial structures are evolutionary relics. Natural selection maintains functional structures in a population of individuals. However, if a species changes its lifestyle, so that the structure loses its purpose, the selection that would normally keep the structure in a functional condition is no longer present. When this occurs, the structure may degenerate over the course of many generations due to the accumulation of mutations that limit its size and shape. Natural selection may eventually eliminate such traits due to the inefficiency and cost of producing unused structures.

Developmental Homologies Species that differ substantially at the adult stage often bear striking similarities during early stages of embryonic development. These temporary similarities are called developmental homologies. In addition, evolutionary history is revealed during development in certain organisms, such as vertebrates. For example, if we consider human development, several features are seen in the embryo that are not present at birth. Human embryos have rudimentary gill ridges, like a fish embryo, even though human embryos receive oxygen via the umbilical cord. The presence of gill ridges indicates that humans evolved from an aquatic species that had gill slits. A second observation is that human embryos have a bony tail. It is difficult to see the advantage of such a structure in utero but easier to understand its presence assuming that an ancestor of the human lineage possessed a tail. These observations, and many others, illustrate that closely related species share similar developmental pathways.

Molecular Homologies A similarity between organisms at the molecular level due to descent from a common ancestor is called a **molecular homology.** Compelling molecular evidence indicating that modern life-forms are derived from an interrelated group of common ancestors is revealed by analyzing genetic sequences. The same type of gene is often found in diverse organisms. Furthermore, the degree of similarity between genetic sequences from different species reflects the evolutionary relatedness of those species. As an example, let's consider the *p53* gene, which encodes the p53 protein—a checkpoint protein of the cell cycle (refer back to Figure 14.3). Figure 19.9 shows a short amino acid sequence that makes up part of the p53

	Short amino acid sequence within the p53 protein	Percentages of amino acids in the whole p53 protein that are identical to human p53
Human (*Homo sapiens*)	Val Pro Ser Gln Lys Thr Tyr Gln Gly Ser Tyr Gly Phe Arg Leu Gly Phe Leu His Ser Gly Thr	100
Rhesus monkey (*Macaca mulatta*)	Val Pro Ser Gln Lys Thr Tyr His Gly Ser Tyr Gly Phe Arg Leu Gly Phe Leu His Ser Gly Thr	95
Green monkey (*Cercopithecus aethiops*)	Val Pro Ser Gln Lys Thr Tyr His Gly Ser Tyr Gly Phe Arg Leu Gly Phe Leu His Ser Gly Thr	95
Rabbit (*Oryctolagus cuniculus*)	Val Pro Ser Gln Lys Thr Tyr His Gly Asn Tyr Gly Phe Arg Leu Gly Phe Leu His Ser Gly Thr	86
Dog (*Canis lupus familiaris*)	Val Pro Ser Pro Lys Thr Tyr Pro Gly Thr Tyr Gly Phe Arg Leu Gly Phe Leu His Ser Gly Thr	80
Chicken (*Gallus gallus*)	Val Pro Ser Thr Glu Asp Tyr Gly Gly Asp Phe Asp Phe Arg Val Gly Phe Val Glu Ala Gly Thr	53
Channel catfish (*Ictalurus punctatus*)	Val Pro Val Thr Ser Asp Tyr Pro Gly Leu Leu Asn Phe Thr Leu His Phe Gln Glu Ser Ser Gly	48
European flounder (*Platichthys flesus*)	Val Pro Val Val Thr Asp Tyr Pro Gly Glu Tyr Gly Phe Gln Leu Arg Phe Gln Lys Ser Gly Thr	46
Congo puffer fish (*Tetraodon miurus*)	Val Pro Val Thr Thr Asp Tyr Pro Gly Glu Tyr Gly Phe Lys Leu Arg Phe Gln Lys Ser Gly Thr	41

Figure 19.9 An example of genetic homology: a comparison of a short amino acid sequence within the p53 protein from nine different animals. This figure compares a short region of the p53 protein, a checkpoint protein that plays a role in preventing cancer. Amino acids are represented by three-letter abbreviations. The orange-colored amino acids in the sequences are identical to those in the human sequence. The numbers in the right column give the percentage of amino acids within the whole p53 protein that are identical with the human p53 protein, which is 393 amino acids in length. For example, 95% of the amino acids, or 373 of 393, are identical between the p53 sequences found in humans and the one in rhesus monkeys.

Concept Check: In the sequence shown in this figure, how many amino acid differences occur between the following pairs: rhesus and green monkeys, Congo puffer fish and European flounder, and rhesus monkey and Congo puffer fish? What do these differences tell you about the evolutionary relationships among these four species?

protein in a variety of species, including five mammals, one bird, and three fish. The top sequence is the human p53 sequence, and the right column gives the percentages of amino acids within each animal's entire sequence that are identical to those in the entire human sequence. Amino acids in the other species that are identical to those in humans are highlighted in orange. The sequences from the two monkeys are the most similar to those in humans, followed by the other two mammalian species (rabbit and dog). The three fish sequences are the least similar to the human sequence, but the fish sequences tend to be similar to each other.

Taken together, the data shown in Figure 19.9 illustrate two critical points about evolution at the gene level. First, specific genes are found in a diverse array of species such as mammals, birds, and fishes. Second, the sequences of closely related species tend to be more similar to each other than they are to distantly related species.

19.2 Reviewing the Concepts

- Evidence of evolutionary change is found in studies of the fossil record, biogeography, convergent evolution, selective breeding, and homologies (Table 19.2).
- Fossils provide evidence of evolutionary change in a series of related organisms. The fossil record often reveals transitional forms that link past ancestors to modern species (Figures 19.3, 19.4).
- Biogeography provides information on the geographic distribution of related species. When populations become isolated on islands or continents, they often evolve into new species (Figure 19.5).
- In convergent evolution, independent adaptations result in similar characteristics, or analogous structures, because different species occupy similar environments (Figure 19.6).
- Selective breeding, the selecting and breeding of individual organisms having desired traits, is a human-driven form of selection (Figure 19.7).
- Homologies are similarities that occur due to descent from a common ancestor. Homologous structures, such as the set of bones in the forelimbs of vertebrates, and vestigial structures, structures that were functional in an ancestor but no longer have a useful function in modern species, are evidence of evolutionary change. Homologies can also be seen during embryonic development and at the molecular level (Figures 19.8, 19.9; Table 19.3).

19.2 Testing Your Knowledge

1. Which of the following provides evidence of evolutionary change?
   a. an analysis of fossils of the whale lineage
   b. the identification of transitional species in the fossil record
   c. the identification of island species that are closely related to mainland species
   d. the similarity of homologous genes
   e. all of the above

2. Homologous traits show similarities because the species exhibiting those traits
   a. occupy similar environments.
   b. evolved from a common ancestor.
   c. evolved in a similar environment.
   d. are geographically close to each other.

19.3 Genes in Populations

Learning Outcomes
1. Define gene pool.
2. Distinguish between allele frequency and genotype frequency.
3. **SCISKILLS** ▶ Use the Hardy-Weinberg equation to calculate allele and genotype frequencies in a given population.
4. List the conditions that must be met for a population to be in Hardy-Weinberg equilibrium.
5. Describe the factors that cause microevolution to occur.

Population genetics is an extension of our understanding of Darwin's theory of natural selection, Mendel's laws of inheritance, and newer studies in molecular genetics. All of the alleles for every gene in a given population make up the **gene pool.** Each member of the population receives its genes from its parents, which, in turn, are members of the gene pool. Individuals that reproduce contribute to the gene pool of the next generation. Population geneticists study the genetic variation within a gene pool and how such variation changes from one generation to the next. The emphasis is often on understanding the variation in alleles among members of a population. In this section, we will examine some of the general features of populations and gene pools.

Populations Are Dynamic Units

A population is a group of individuals of the same species that occupy the same environment at the same time and, for sexually reproducing organisms, can interbreed with one another. Certain species occupy a wide geographic range and are divided into discrete populations due to geographic isolation. For example, distinct populations of a given species may be located on different sides of a physical barrier, such as a mountain range.

Populations may change in size and geographic location from one generation to the next. As the size and location of a population change, its genetic composition generally changes as well. Some of the genetic changes involve adaptation, in which a population becomes better suited to its environment, consisting of individuals that are more likely to survive and successfully reproduce. For example, a population of mammals may move from a warmer to a colder geographic location. Natural selection may favor those individuals in the population that have thicker fur, which over time will produce a population of individuals that are better insulated against the colder temperatures.

EVOLUTIONARY CONNECTIONS

Genes Are Usually Polymorphic

The term **polymorphism** (from the Greek, meaning many forms) refers to the presence of two or more variants or traits for a given character within a population. **Figure 19.10** illustrates a striking example of polymorphism in the elder-flowered orchid (*Dactylorhiza sambucina*). Throughout the range of this species in Europe, both yellow- and red-flowered individuals are prevalent.

Polymorphism of a character is usually due to the existence of two or more alleles of a gene that influences the character. Geneticists also use the term polymorphism to describe the variation in the DNA sequence of genes. A gene that commonly exists as two or more alleles in a population is a **polymorphic gene.** To be considered polymorphic, a gene must exist in at least two different forms or alleles, and each allele must occur at a frequency that is greater than 1%. By comparison, a **monomorphic gene** exists predominantly as a single allele in a population. When 99% or more of the alleles of a given gene are identical in a population, the gene is considered to be monomorphic.

What types of molecular changes cause genes to be polymorphic? A polymorphism may involve various types of changes, such as a deletion of a significant region of the gene, a duplication of a region, or a change in a single nucleotide. This last type of variation is called a **single-nucleotide polymorphism (SNP).** SNPs (pronounced "snips") are the smallest type of genetic variation that can occur within a given gene and the most common. For example, the allele that causes the recessive disorder called sickle cell disease involves a single-nucleotide change in the human β-globin gene, which encodes a subunit of the oxygen-carrying protein called hemoglobin. The non-disease-causing allele and the sickle cell allele represent a SNP of the β-globin gene:

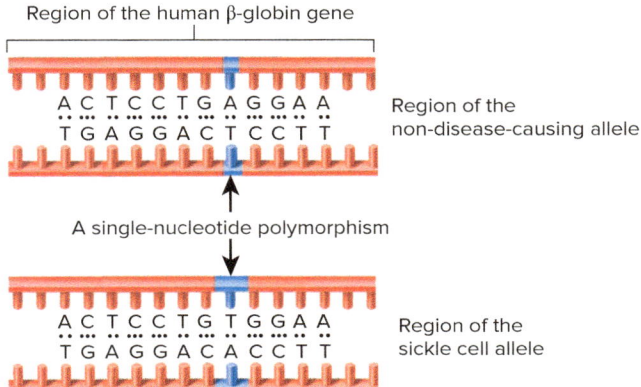

In this single-nucleotide polymorphism, the single-nucleotide A (in the top strand) of the non-disease-causing allele changes to a T (in the sickle cell allele).

SNPs represent 99% of all variation in human DNA sequences that occurs among people. In human populations, a gene that is 2,000–3,000 bp in length, on average, contains 10 different SNPs. Likewise, SNPs with a frequency of 1% or more are found very frequently among genes of nearly all species. Polymorphism is the norm for relatively large, healthy populations of nearly all species, as evidenced by the occurrence of SNPs within most genes.

Population Genetics Is Concerned with Allele and Genotype Frequencies

One approach to analyzing genetic variation in populations is to consider the frequency of specific alleles and genotypes in a quantitative way. Two fundamental calculations are central to population genetics: **allele frequency** and **genotype frequency.** Allele and genotype frequencies are defined as follows:

$$\text{Allele frequency} = \frac{\text{Number of copies of a specific allele in a population}}{\text{Total number of all alleles for that gene in the population}}$$

$$\text{Genotype frequency} = \frac{\text{Number of individuals with a particular genotype in a population}}{\text{Total number of individuals in the population}}$$

Although allele and genotype frequencies are related, make sure you clearly distinguish between them. As an example, let's consider a population of 100 four-o'clock plants (*Mirabilis jalapa*) with the following genotypes:

49 red-flowered plants with the genotype $C^R C^R$

42 pink-flowered plants with the genotype $C^R C^W$

9 white-flowered plants with the genotype $C^W C^W$

When calculating an allele frequency for a diploid species, remember that homozygous individuals have two copies of a given allele, whereas heterozygotes have only one. For example, in tallying

Figure 19.10 An example of polymorphism: the two color variations found in the orchid *Dactylorhiza sambucina*.
©Paul Harcourt Davies/SPL/Science Source

However, due to their low rate of occurrence, mutations by themselves do not have a major effect in changing allele frequencies in a population over time. They do not significantly disrupt Hardy-Weinberg equilibrium.

The second phenomenon required for evolution to occur is one or more mechanisms that alter the prevalence of a given allele or genotype in a population. These mechanisms are natural selection, genetic drift, migration, and nonrandom mating (see Table 19.4). Over the course of many generations, these mechanisms may promote widespread genetic changes in a population. In the remainder of this chapter, we will examine how natural selection, genetic drift, migration, and nonrandom mating affect genetic variation from one generation to the next.

19.3 Reviewing the Concepts

- Population genetics is the study of genes and genotypes in a population. A population is a group of individuals of the same species that occupy the same environment at the same time and can interbreed if the species reproduces sexually. All of the alleles for every gene in a population constitute the population's gene pool.
- Polymorphism, which is very common in nearly all populations, is the presence of two or more variants of a character in a population. A monomorphic gene exists as a single allele (comprising more than 99% of the alleles of the gene) in a population (Figure 19.10).
- Allele frequency is the number of copies of a specific allele divided by the total number of all alleles for that gene in a population. Genotype frequency is the number of individuals with a given genotype divided by the total number of individuals in a population.
- The Hardy-Weinberg equation ($p^2 + 2pq + q^2 = 1$) predicts that allele and genotype frequencies will remain in equilibrium if no new mutations arise, no natural selection occurs, the population size is very large, migration does not occur, and mating is random (Figure 19.11).
- Sources of new genetic variation include random gene mutations, gene duplications, and horizontal gene transfer. Natural selection, genetic drift, migration, and nonrandom mating may alter allele and genotype frequencies and cause a population to evolve (Table 19.4).

19.3 Testing Your Knowledge

1. The frequency of individuals in a population that exhibit a rare recessive disorder is 0.0025. The value of 0.0025 is an example of
   a. an allele frequency.
   b. a genotype frequency.
   c. a phenotype frequency.
   d. all of the above.
   e. both b and c.

2. Hardy-Weinberg equilibrium assumes
   a. a very large population with no random fluctuations (no genetic drift).
   b. no natural selection.
   c. random mating.
   d. no new mutations.
   e. all of the above.

19.4 Natural Selection

Learning Outcomes

1. Explain how natural selection results in a population that is better adapted to its environment and more successful at reproduction.
2. **SCISKILLS** ▶ Calculate the fitness values of given genotypes.
3. Compare and contrast four different types of natural selection.
4. Define sexual selection, and distinguish between intrasexual and intersexual selection.

As discussed earlier in this chapter, natural selection is the process by which individuals with certain heritable traits tend to survive and reproduce at higher rates than those without those traits. As a result, favorable heritable traits become more common while detrimental heritable traits become less common in a population over many generations. Keep in mind that natural selection itself is not evolution. Rather, it is a key mechanism that causes evolution to happen. Over time, natural selection results in **adaptations**—changes in populations of living organisms that increase their ability to survive and reproduce in a particular environment. In this section, we will examine various ways that natural selection produces such adaptations.

Natural Selection Favors Individuals with Greater Reproductive Success

Reproductive success is the likelihood of an individual contributing fertile offspring to the next generation. Natural selection occurs because some individuals in a population have greater reproductive success than other individuals. Those individuals having heritable traits that favor reproductive success are more likely to pass such traits to their offspring. Reproductive success is commonly attributed to two categories of traits:

1. Certain characteristics make organisms better adapted to their environment and therefore more likely to survive to reproductive age. Therefore, natural selection favors individuals with characteristics that provide a survival advantage.
2. Reproductive success may involve traits that are directly associated with reproduction, such as the abilities to find a mate and produce viable gametes and offspring. Traits that enhance the ability of individuals to reproduce, such as brightly colored plumage in male birds, are often subject to natural selection.

A modern description of the principles of natural selection can relate our knowledge of molecular genetics to the process of evolution:

1. Within a population, allelic variation arises from random mutations that cause differences in DNA sequences. A mutation that creates a new allele may alter the amino acid sequence of the encoded protein. This, in turn, may alter the function of the protein.
2. Some alleles encode proteins that enhance an individual's survival or reproductive capability over other members of the population. For example, an allele may produce a protein that is more efficient at a higher temperature, conferring on the individual a greater probability of survival in a hot climate.
3. Individuals with beneficial alleles are more likely to survive and reproduce and thereby contribute their alleles to the gene pool of the next generation.
4. Over the course of many generations, allele frequencies of many different genes may change through natural selection, thereby significantly altering the characteristics of a population. The net result of natural selection is a population that is better adapted to its environment and more successful at reproduction.

Fitness Is a Quantitative Measure of Reproductive Success

To continue our discussion of natural selection, we need to consider the concept of **fitness,** which is the relative likelihood that one genotype will contribute to the gene pool of the next generation compared with other genotypes. Although this property often correlates with physical fitness, the two ideas should not be confused. Fitness is a measure of reproductive success. An extremely fertile individual may have a higher fitness than a less fertile individual that appears more physically fit.

To examine fitness, let's consider an example of a hypothetical gene existing in alleles A and a. We assign fitness values to each of the three possible genotypes according to their relative reproductive success. For example, let's suppose the average reproductive successes of the three genotypes are

AA produces 5 offspring

Aa produces 4 offspring

aa produces 1 offspring

By convention, the genotype with the highest reproductive success is given a fitness value of 1.0. Fitness values are denoted by the variable w. The fitness values of the other genotypes are assigned values relative to this 1.0 value.

Fitness of AA: $w_{AA} = 1.0$

Fitness of Aa: $w_{Aa} = 4/5 = 0.8$

Fitness of aa: $w_{aa} = 1/5 = 0.2$

Variation in fitness occurs because certain genotypes result in individuals that have greater reproductive success than individuals with other genotypes.

Natural Selection Follows Different Patterns

By studying species in their native environments, population geneticists have discovered that natural selection can occur in several ways. In most of the examples described next, natural selection leads to adaptations that make certain members of a species more likely to survive to reproductive age.

Directional Selection
During **directional selection,** individuals at one extreme of a phenotypic range have greater reproductive success in a particular environment. Different phenomena may initiate the process of directional selection. A common reason for directional selection is that a population may be exposed to a prolonged change in its living environment. Under the new environmental conditions, the relative fitness values may change to favor one genotype, which will promote the elimination of other genotypes. As an example, let's suppose a population of finches on a mainland already has genetic variation that affects beak size (see Figure 19.2). A small number of these birds migrate to an island where the seeds are generally larger than on the mainland. In this new environment, birds with larger beaks have a higher fitness because they are better able to crack open the larger seeds and thereby survive to reproduce. Over the course of many generations, directional selection produces a population of finches carrying alleles that promote larger beak size.

Another way that directional selection may arise is that a new allele may be introduced into a population by mutation, and the new allele may confer a higher fitness in individuals that carry it (**Figure 19.12**). What are the long-term effects of directional selection? If the homozygote carrying the favored allele has the highest fitness value, directional selection may cause this favored allele to eventually predominate in the population, perhaps even leading to a monomorphic gene.

Stabilizing Selection
A type of natural selection called **stabilizing selection** favors the survival of individuals with intermediate phenotypes and selects against those with extreme phenotypes. Stabilizing selection tends to decrease genetic diversity, because it eliminates alleles that cause extreme phenotypes. An example of stabilizing selection involves clutch size (number of eggs laid) of birds, which was first studied by British biologist David Lack in 1947. Under stabilizing selection, birds that lay too many or too few eggs per nest have lower fitness values than do those that lay an intermediate number. When a bird lays too many eggs, many offspring die due to inadequate parental care and food. In addition, the strain on the parents themselves may decrease their likelihood of survival and consequently their ability to produce more offspring. Having too few offspring, however, does not contribute many individuals to the next generation. Therefore, the most successful parents are those that produce an intermediate clutch size. In the 1980s, Swedish evolutionary biologist Lars Gustafsson and his colleagues examined the phenomenon of stabilizing selection in the collared flycatcher (*Ficedula albicollis*) on the Swedish island of Gotland. They discovered that Lack's hypothesis concerning an optimal clutch size appears to be true for this species (**Figure 19.13**).

Diversifying Selection
Diversifying selection (also known as disruptive selection) favors the survival of two or more different genotypes that produce different phenotypes. In diversifying selection,

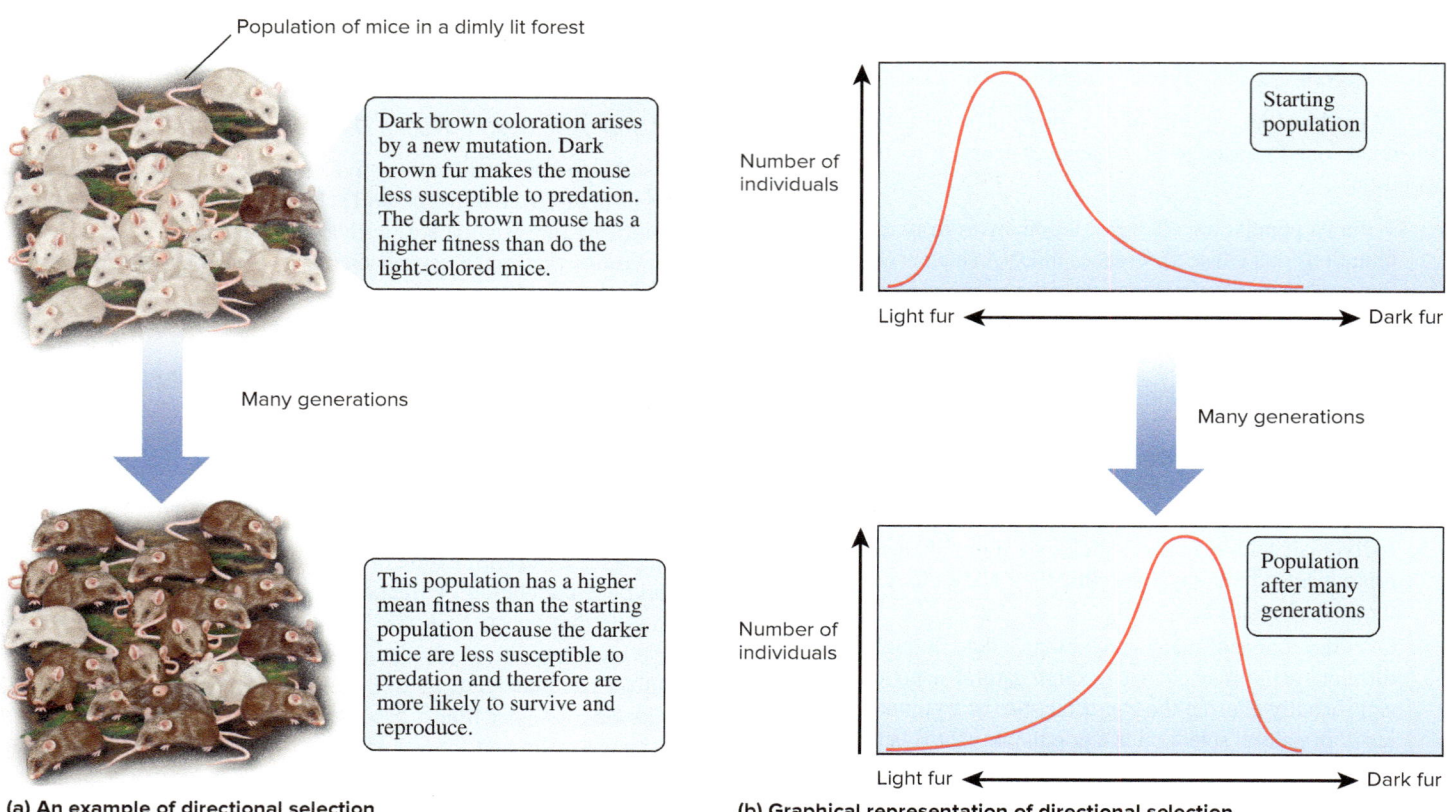

Figure 19.12 Directional selection. This pattern of natural selection selects for one extreme of a phenotype that confers the highest fitness in the population's environment. **(a)** In this example, a mutation causing darker fur arises in a population of mice. This new genotype confers higher fitness, because mice with darker fur can better evade predators and are more likely to survive and reproduce. Over many generations, directional selection favors the prevalence of individuals with darker fur. **(b)** These graphs show the change in fur color phenotypes before and after directional selection.

Concept Check: Let's suppose the climate on an island abruptly changed so that the average temperature is 10°C higher. The climate change is permanent. How would directional selection affect the genetic diversity in a population of mice on the island (1) over the short run and (2) over the long run?

the fitness values of a particular genotype are higher in one environment and lower in a different one, whereas the fitness values of the second genotype vary in an opposite manner. Diversifying selection is likely to occur in populations that occupy heterogeneous environments, so some members of the species are more likely to survive in each type of environmental condition.

An example of diversifying selection involves colonial bentgrass (*Agrostis capillaris*) (**Figure 19.14**). In certain locations where this grass is found, such as near abandoned mines in Wales, places occur where the soil is contaminated with high levels of heavy

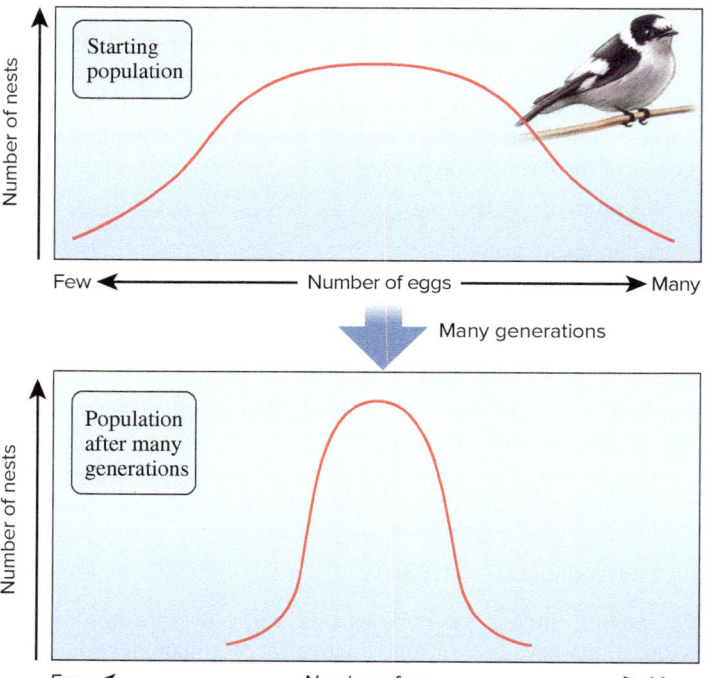

Figure 19.13 Stabilizing selection. In this pattern of natural selection, the extremes of a phenotypic distribution are selected against. Those individuals with intermediate traits have the highest fitness. These graphs show the results of stabilizing selection on clutch size in a population of collared flycatchers (*Ficedula albicollis*). This process results in a population with less diversity and more uniform traits.

Concept Check: Why does stabilizing selection decrease genetic diversity?

Biology Principle

Biologists Use Models to Test Predictions and Convey Ideas

The graphs along the right side of this figure are mathematical models that depict how the relative numbers of individuals with certain phenotypes may change over the course of many generations. The Modeling Challenge in the figure legend asks you to propose graphical models for a scenario involving three different alleles for a given gene in a population.

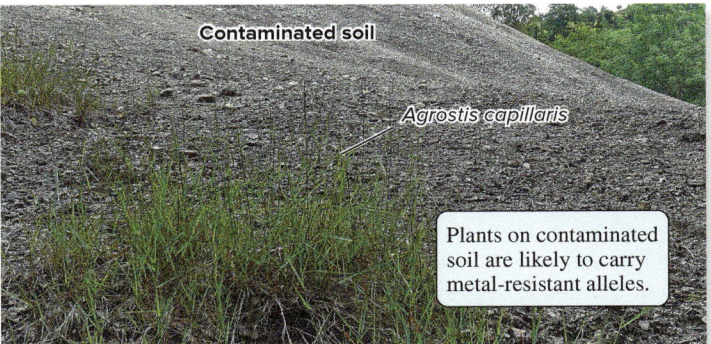

(a) Growth of *Agrostis capillaris* on contaminated soil

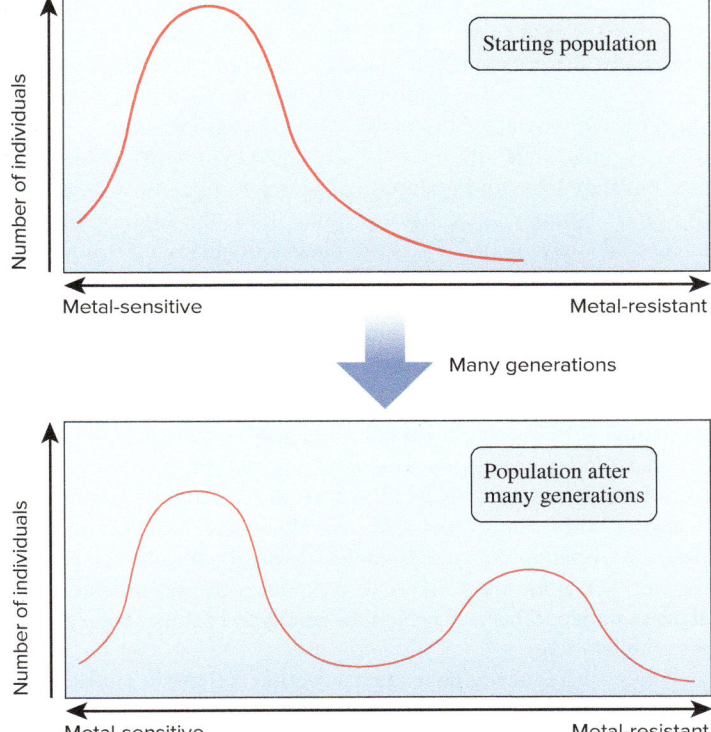

(b) Graphical representation of diversifying selection

Figure 19.14 Diversifying selection. This pattern of natural selection selects for two different phenotypes, each of which is most fit in a particular environment. **(a)** In this example, random mutations have resulted in metal-resistant alleles in colonial bentgrass (*Agrostis capillaris*) that allow it to grow on contaminated soil. In uncontaminated soils, the grass does not show metal tolerance. The existence of both metal-resistant and metal-sensitive alleles in the population is an example of diversifying selection due to heterogeneous environments. **(b)** Graphs showing the change in phenotypes in this bentgrass population before and after diversifying selection.

(a) Courtesy Mark MacNair, University of Exeter

Modeling Challenge: Let's suppose that a gene in a rodent species exists in three alleles that confer long hair, medium-length hair, or short hair. Hair length is positively correlated with survival in a cold climate. The rodent species occupies a wide geographic range in which the average temperature varies. About 70% of the region is fairly cold year-round, 20% has a medium temperature, and 10% is hot. If the starting population was composed predominantly of short-haired individuals, draw a series of two graphs in the style of Figure 19.14 that depict the effects of diversifying selection after many generations.

metals due to mining. The relatively recent metal contamination has selected for the proliferation of mutant strains of *A. capillaris* that are tolerant of the heavy metals (see Figure 19.14a). Genetic changes enable these mutant strains to grow on contaminated soil but tend to inhibit their growth on normal, noncontaminated soil. These metal-resistant plants often grow on contaminated sites that are close to plants that grow on uncontaminated land and do not show metal tolerance.

Balancing Selection Contrary to a popular misconception, natural selection does not always cause the elimination of "weaker," or less-fit, alleles. **Balancing selection** is a type of natural selection that maintains genetic diversity in a population. Over many generations, balancing selection results in a **balanced polymorphism,** in which two or more alleles are kept in balance and therefore are maintained in a population over many generations.

How does balancing selection maintain polymorphism? Population geneticists have identified two common ways that balancing selection occurs. First, for genetic variation involving a single gene, balancing selection can favor the heterozygote over either corresponding homozygote. This phenomenon is called **heterozygote advantage.** Heterozygote advantage sometimes explains the persistence of alleles that are deleterious in a homozygous condition.

A classic example of heterozygote advantage involves the H^S allele of the human β-globin gene. A homozygous $H^S H^S$ individual has sickle cell disease, which causes the red blood cells to have a deformed, sickle shape. Sickle-shaped cells deliver less oxygen to the body's tissues and can block the flow of blood through the vessels. The $H^S H^S$ homozygote has a lower fitness than a homozygote with two copies of the more common β-globin allele, $H^A H^A$. Heterozygotes, $H^A H^S$, do not typically show symptoms of the disease, but they have an increased resistance to malaria. Compared with $H^A H^A$ homozygotes, heterozygotes have the highest fitness because they have a 10–15% better chance of surviving if infected by the malarial parasite *Plasmodium falciparum*. Therefore, the H^S allele is maintained in populations where malaria is prevalent, even though the allele is detrimental in the homozygous state (**Figure 19.15**).

Negative frequency-dependent selection is a second way that natural selection produces balanced polymorphism. In this pattern of natural selection, the fitness of a genotype decreases when its frequency becomes higher. In other words, common individuals have a lower fitness, and rare individuals have a higher fitness. Therefore, common individuals are less likely to reproduce, whereas rare individuals are more likely to reproduce, thereby producing a balanced polymorphism in which no genotype becomes too rare or too common.

Negative frequency-dependent selection is thought to maintain polymorphisms among species that are preyed upon by predators. Research has shown that certain predators form a mental "search image" for their prey, which is usually based on the common type of prey in an area. A prey that exhibits a rare polymorphism that affects its appearance is less likely to be recognized by the predator. For example, a prey that is a different color from most other members of its species may not be readily recognized by the predator. Such relatively rare organisms are subject to a lower rate of predation. This type of selection maintains polymorphism among certain prey populations.

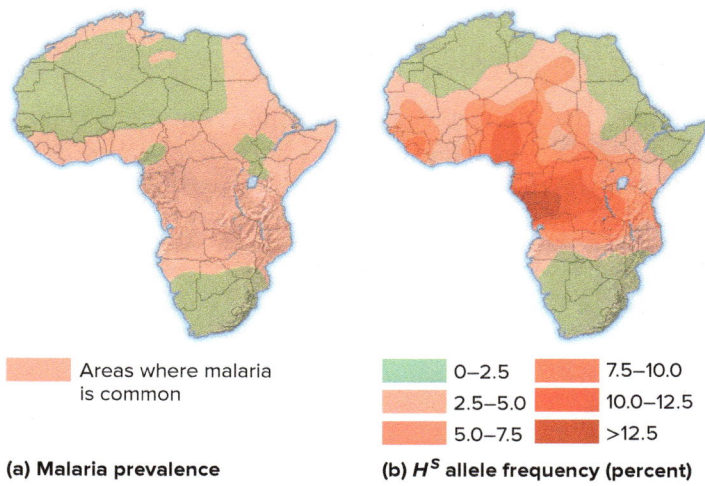

(a) Malaria prevalence
(b) H^S allele frequency (percent)

Figure 19.15 Balancing selection and heterozygote advantage. **(a)** The geographic prevalence of malaria in Africa. **(b)** The frequency of the H^S allele of the β-globin gene in the same area. In the homozygous condition, the H^S allele causes sickle cell disease. This allele is maintained in human populations in areas where malaria is prevalent, because the heterozygote ($H^A H^S$) has a higher fitness than either of the corresponding homozygotes ($H^A H^A$ or $H^S H^S$).

Concept Check: *If malaria was eradicated, what would you expect to happen to the frequencies of the H^A and H^S alleles over the long run?*

FEATURE INVESTIGATION

The Grants Observed Natural Selection in Galápagos Finches

Let's now turn to an example that shows natural selection occurring as a recent event. Since 1973, British evolutionary biologists Peter Grant, Rosemary Grant, and their colleagues have studied natural selection in finches found on the Galápagos Islands. For over 40 years, the Grants have focused much of their work on one of the Galápagos Islands known as Daphne Major (**Figure 19.16a**). This small island (0.34 km²) has a moderate degree of isolation (it is 8 km from the nearest island), an undisturbed habitat, and a resident population of *Geospiza fortis*, the medium ground finch (**Figure 19.16b**).

To study natural selection, the Grants observed various traits in finches over the course of many years. One trait they observed was beak size. The medium ground finch has a relatively small, crushing beak, allowing it to feed more easily on small, tender seeds (see Table 19.1). The Grants quantified beak size among the medium ground finches of Daphne Major by carefully measuring beak depth—a measurement of the beak from top to bottom (**Figure 19.17**). The small size of the island made it possible for them to measure a large percentage of birds and their offspring. During the course of their studies, they compared the beak depths of parents and offspring by examining many broods over

Biology Principle

Populations of Organisms Evolve from One Generation to the Next and Are Related by an Evolutionary History

On this island, the Grants analyzed how beak size may change from one generation to the next.

(a) Daphne Major

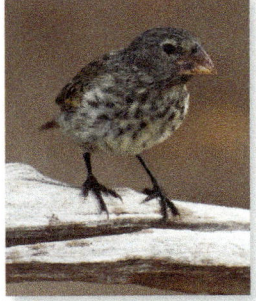
(b) Medium ground finch

Figure 19.16 Daphne Major and a resident finch. (a) Daphne Major, one of the Galápagos Islands. (b) One of the medium ground finches (*Geospiza fortis*) that populate this island.
(a) ©Worldwide Picture Library/Alamy; *(b)* ©Ralph Lee Hopkins/Getty Images

By measuring many birds every year, the Grants were able to assemble a detailed portrait of natural selection in action. In the study shown in Figure 19.17, they measured beak depth from 1976 to 1978. In the wet year of 1976, the plants of Daphne Major produced an abundance of the small, tender seeds that these finches could easily eat. However, a severe drought occurred in 1977. During this year, the plants on Daphne Major tended to produce few of the smaller seeds, which the finches rapidly consumed. Therefore, the finches resorted to eating larger, drier seeds, which are harder to crush. As a result, birds with larger beaks were more likely to survive and reproduce because they were better at breaking open the large seeds. As shown in the data, the average beak depth of birds in the population increased substantially, from 8.8 mm in predrought offspring to 9.8 mm in postdrought offspring.

How do we explain these results? Birds with larger beaks were more likely to survive and pass this trait to their offspring. Overall, these results illustrate the power of natural selection to alter the features of a trait—in this case, beak depth—in a given population over time. This is an example of directional selection.

Experimental Questions

1. What features of Daphne Major made it a suitable field site for studying the effects of natural selection?

2. **SCISKILLS** ▶ Why is beak depth in finches a good trait for a study of natural selection? What environmental conditions were important in allowing the Grants to collect information concerning natural selection?

3. **SCISKILLS** ▶ Analyze the results of the Grants' study, and explain what they mean with regard to natural selection and evolution.

several years and found that the depth of the beak was transmitted from parents to offspring, indicating that differences in beak depths are due to genetic differences in the population. In other words, they found that beak depth is a heritable trait.

Figure 19.17 The Grants' investigation of natural selection in the medium ground finch.

HYPOTHESIS Dry conditions produce larger seeds and may result in larger beaks in succeeding generations of *Geospiza fortis* due to natural selection.

KEY MATERIALS A population of *G. fortis* on the Galápagos Island called Daphne Major.

		Experimental level	Conceptual level
1	In 1976, measure beak depth in parents and offspring of the species *G. fortis*.	Capture birds and measure beak depth.	This is a way to measure a trait that may be subject to natural selection.
2	Repeat the procedure on offspring that were born in 1978 and had reached mature size. A drought had occurred in 1977 that caused plants on the island to produce mostly large, dry seeds and relatively few small seeds.	Capture birds and measure beak depth.	This is a way to measure a trait that may be subject to natural selection.

3 THE DATA

4 CONCLUSION
Because a drought produced larger seeds, birds with larger beaks were more likely to survive and reproduce. The process of natural selection produced postdrought offspring that had larger beaks compared to predrought offspring.

5 SOURCE
Grant, B. Rosemary, and Grant, Peter R. 2003. What Darwin's Finches Can Teach Us About the Evolutionary Origin and Regulation of Biodiversity. *Bioscience* 53: 965–975.

Sexual Selection Is a Type of Natural Selection Pertaining to Traits That Are Directly Involved with Reproduction

Thus far, we have largely focused on examples of natural selection that produce adaptations for survival in particular environments. Now let's turn our attention to a form of natural selection called **sexual selection**, in which individuals with certain traits are more likely to engage in successful reproduction than other individuals. Darwin originally described sexual selection as "the advantage that certain individuals have over others of the same sex and species solely with respect to reproduction."

In many species of animals, sexual selection affects the characteristics of males more intensely than those of females. Unlike females, which tend to be fairly uniform in their reproductive success, male success tends to be more variable, with some males mating with many females and others not mating at all. Sexual selection results in the prevalence of traits, called secondary sex characteristics, that favor reproductive success. The process can result in **sexual dimorphism**—a significant difference between the morphologies of the two sexes within a species.

Sexual selection operates in one of two ways.

- In **intrasexual selection**, members of one sex, usually males, directly compete with each other for the opportunity to mate with individuals of the opposite sex. Examples of traits that result from intrasexual selection in animals include horns in male sheep, antlers in male moose, and the enlarged claw of male fiddler crabs (**Figure 19.18a**). In fiddler crabs (*Uca paradussumieri*), males enter the burrows of females that are ready to mate. If another male attempts to enter the burrow, the male already inside stands in the burrow shaft and blocks the entrance with his enlarged claw. Males with the largest claws are more likely to be successful at driving off their rivals and being able to mate; therefore, they are more likely to pass on their genes to future generations.

- In **intersexual selection**, also called mate choice, members of one sex, usually females, choose their mates from individuals of the other sex on the basis of certain desirable characteristics. This type of sexual selection often results in showy characteristics in males. **Figure 19.18b** shows a classic example that involves the Indian peafowl (*Pavo cristatus*). Male peacocks have long and brightly colored tail feathers, which they fan out as a mating behavior. Female peahens select among males based on feather color and pattern, as well as the physical prowess of the display.

Many animals have secondary sex characteristics, and evolutionary biologists generally agree that sexual selection is responsible for such traits. But why should males compete, and why should females be choosy? Researchers have proposed various hypotheses to explain the underlying mechanisms. One possible reason is related to the different investments of females and males in the production of offspring. Females may expend a great deal of energy producing eggs and, in mammals, providing nutrients to unborn offspring during internal gestation. In some animal species, the female is the primary caregiver, whereas the male plays a minor role. In such species, mating behavior may influence the fitness of both males and females. Males increase their fitness by mating with multiple females. This increases their likelihood of passing their genes on to the next generation. By comparison, females may produce relatively fewer offspring, and their reproductive success may not be limited by the number of available males. In these circumstances, females will have higher fitness if they choose males that are good defenders of their territory and have alleles that confer a survival advantage to their offspring.

Figure 19.18 Examples of the results of sexual selection, a type of natural selection. (a) An example of intrasexual selection. The enlarged claw of the male fiddler crab is used in direct male-to-male competition. In this photograph, a male inside a burrow is extending its claw out of the burrow to prevent another male from entering and mating with the female. (b) An example of intersexual selection. Female peahens choose male peacocks based on the males' colorful and long tail feathers and the robustness of their display.
(a) ©Gerald Cubitt; *(b)* ©Topham/TopFoto/The Image Works

✓ **Concept Check:** *Male birds of many species have loud and elaborate courtship songs. Is this trait likely to be involved in intersexual or intrasexual selection? Explain.*

(a) Intrasexual selection

(b) Intersexual selection

19.4 Reviewing the Concepts

- Natural selection is the process by which individuals with certain heritable traits that favor survival and reproduction tend to become more prevalent in a population. Fitness, the relative likelihood that a genotype will contribute to the gene pool of the next generation, is a measure of reproductive success.
- Directional selection is the process by which one extreme of a phenotypic distribution is favored (Figure 19.12).
- Stabilizing selection is the process by which an intermediate phenotype is favored (Figure 19.13).
- Diversifying selection is the process by which two or more phenotypes are favored. An example is a population that occupies a diverse environment (Figure 19.14).
- Balancing selection maintains balanced polymorphism in a population. Examples of the ways in which balancing selection occurs include heterozygote advantage and negative frequency-dependent selection (Figure 19.15).
- A study by the Grants showed natural selection occurring in a population of finches (Figures 19.16, 19.17).
- Sexual selection is a form of natural selection in which individuals with certain traits are more likely than others to engage in successful reproduction.
- In intrasexual selection, members of one sex compete with each other for the opportunity to mate with individuals of the opposite sex. In intersexual selection, members of one sex choose their mates on the basis of certain desirable characteristics (Figure 19.18).

19.4 Testing Your Knowledge

1. In a population of fish, body coloration varies from a light shade, almost white, to a very dark shade of green. Changes in the environment that result in decreased predation of individuals with the lightest coloration is an example of _____ selection.
   a. diversifying
   b. stabilizing
   c. directional
   d. balancing

2. Considering the same population of fish described in question 1, if the stream environment included several areas of sandy, light-colored bottom areas and a lot of dark-colored vegetation, both the light- and dark-colored fish would have selective advantage and increased survival in certain places. This type of scenario could explain the occurrence of which type of natural selection?
   a. diversifying
   b. stabilizing
   c. directional
   d. balancing

3. The enlarged claw of male fiddler crabs is an example of
   a. intrasexual selection.
   b. intersexual selection.
   c. both intrasexual and intersexual selection.
   d. neither intrasexual nor intersexual selection.

19.5 Genetic Drift

Learning Outcomes

1. Define genetic drift, and explain its effects on allele frequencies over time.
2. Compare and contrast the bottleneck and founder effects.
3. Explain how neutral mutations can spread through a population.

Thus far, we have focused on natural selection as a mechanism that promotes widespread genetic changes in a population. Let's now turn our attention to a second important way the gene pool of a population can change. In the 1930s, American geneticist Sewall Wright played a large role in developing the concept of **genetic drift** (also called random genetic drift), which refers to changes in allele frequencies due to random chance. The term genetic drift is derived from the observation that allele frequencies may "drift" randomly from generation to generation as a matter of chance.

Changes in allele frequencies due to genetic drift happen regardless of the fitness of individuals that carry those alleles. For example, an individual with a high fitness value may, by chance, not encounter a member of the opposite sex. Likewise, random chance can influence which alleles happen to be carried in the gametes that fuse with each other in a successful fertilization. In this section, we will examine how genetic drift alters allele frequencies in populations.

Genetic Drift Has a Greater Effect in Small Populations

What are the effects of genetic drift? Over the long run, genetic drift favors either the elimination (frequency of 0%) or the fixation (frequency of 100%) of an allele in a population. However, the number of generations it takes for an allele to be lost or fixed greatly depends on the population size. Figure 19.19 illustrates the potential consequences of genetic drift in one large ($N = 1,000$) and two small ($N = 10$) populations. This simulation involves the frequency of hypothetical B and b alleles of a gene for fur color in a population of mice—B is the black allele, and b is the white allele.

At the beginning of this simulation, which ran for 50 generations, all three populations had identical allele frequencies: $B = 0.5$ and $b = 0.5$. In the small populations, the allele frequencies fluctuated substantially from generation to generation. Eventually, in one population, the b allele was eliminated; in another, it was fixed at 100%. These small populations then consisted of only black mice or white mice, respectively. At this point, the gene became monomorphic, and allele frequencies could no longer fluctuate due to genetic drift.

By comparison, the frequencies of B and b in the large population fluctuated much less. The relative effect of random chance, also termed random sampling error, is much smaller when the sample size is large. Nevertheless, genetic drift can eventually lead to allele elimination or fixation even in large populations, but this will take many more generations to occur than it does in small populations.

In nature, genetic drift may rapidly alter allele frequencies when the size of a population dramatically decreases. Two examples of this phenomenon are the bottleneck effect and the founder effect, which are described next.

Bottleneck Effect A population can be dramatically reduced in size by events such as earthquakes, floods, drought, and human destruction of habitat. These occurrences may eliminate most members of the population without regard to their genetic composition. The population is said to have passed through a bottleneck. The change in allele frequencies of the resulting population due to genetic drift is called the **bottleneck effect.** Some alleles may be overrepresented, whereas others may even be eliminated. Such changes may happen for two reasons. First, the surviving members may have allele frequencies that differ from those of the original population that was much larger. Second, as we saw in Figure 19.19, genetic drift acts more quickly to reduce genetic variation when the population size is small. Eventually, a population that has gone through a bottleneck may regain its original size. However, the new population is likely to have less genetic variation than the original one.

A hypothetical example of the bottleneck effect is shown with a population of frogs in Figure 19.20. In this example, a starting population of frogs is found in three phenotypes: yellow, dark green, and striped. Due to a bottleneck caused by a drought, the dark green variety is lost from the population.

As a real-life example, the northern elephant seal (*Mirounga angustirostris*) has lost much of its genetic variation. This was caused by a bottleneck in which the population decreased to approximately

Figure 19.19 Genetic drift and population size. This graph shows three hypothetical simulations of genetic drift and its effect on small and large populations of black (B allele) and white (b allele) mice. In all cases, the starting allele frequencies are $B = 0.5$ and $b = 0.5$. The red lines illustrate two populations of mice in which $N = 10$; the blue line shows a population in which $N = 1,000$.

In a large population, many more generations are required before an allele is eliminated or fixed.

Biology Principle

Populations of Organisms Evolve from One Generation to the Next and Are Related by an Evolutionary History
Genetic drift randomly changes allele frequencies and (in the long run) leads to a loss or fixation of an allele.

Figure 19.20 A hypothetical example of the bottleneck effect. This example involves a population of frogs in which a drought dramatically reduced population size, resulting in a bottleneck. The bottleneck reduced the genetic diversity in the population.

Concept Check: *How does the bottleneck effect undermine the efforts of conservation biologists who are trying to save species nearing extinction?*

① The starting population includes 3 phenotypes of frogs: yellow, dark green, and striped.

② A drought causes a bottleneck in which the population size is decreased and the dark green phenotype is lost.

③ The population size recovers, but genetic variation is decreased, and only 2 phenotypes are left.

---

20 to 30 surviving members in the 1890s due to hunting. The species has rebounded in numbers to over 100,000, but the bottleneck effect reduced its genetic variation to very low levels.

Founder Effect Another common phenomenon in which genetic drift may rapidly alter allele frequencies is the **founder effect.** This occurs when a small group of individuals separates from a larger population and establishes a colony in a new location. For example, a few individuals from a large population on a continent may move to an island and become the founders of an island population. The founder effect differs from the bottleneck effect in that it occurs in a new location, although both effects are related to a reduction in population size. The founder effect has two important consequences.

- First, the founding population, which is relatively small, is expected to have less genetic variation than the larger original population from which it was derived.

- Second, as a matter of chance, the allele frequencies in the founding population may differ markedly from those of the original population.

Population geneticists have studied many examples in which isolated populations were founded via colonization by members of another population. For example, in the 1960s, American geneticist Victor McKusick studied allele frequencies in the Amish of Lancaster County, Pennsylvania. At that time, this group included about 8,000 people, descended from just three couples that immigrated to the U.S. in 1770. Among this population of 8,000, a genetic disease known as the Ellis–van Creveld syndrome (a recessive form of dwarfism) was found at a frequency of 0.07, or 7%. By comparison, this disorder is extremely rare in other human populations, even the population from which the founding members originated. Evidence suggests that the high frequency in the Lancaster County population can be traced to one couple, one of whom carried the mutated gene that causes the syndrome.

Genetic Drift Plays an Important Role in Promoting Genetic Change

In 1968, Japanese evolutionary biologist Motoo Kimura proposed that much of the DNA sequence variation seen in genes in natural populations is the result of genetic drift rather than natural selection. Genetic drift is a random process that does not preferentially select for any particular allele—it can alter the frequencies of both beneficial and deleterious alleles. Much of the time, genetic drift promotes **neutral variation**—changes in genes and proteins that do not have an effect on reproductive success.

According to Kimura, most variation in DNA sequences is due to the accumulation of neutral mutations that have attained high frequencies in a population via genetic drift. This is called the **neutral theory of evolution.** For example, a new mutation within a gene that changes a glycine codon from GGG to GGC would not affect the amino acid sequence of the encoded protein. Both genotypes may be equal in fitness. However, such new mutations can spread throughout a population due to genetic drift (Figure 19.21). Kimura agreed with Darwin that natural selection is responsible for adaptive changes in a species during evolution. The long snout of an anteater is the result of natural selection. His main idea is that much of the variation in DNA sequences is explained by neutral variation rather than adaptive variation.

The sequencing of genomes from many species is consistent with Kimura's theory. When researchers examine changes of the coding sequence within protein-encoding genes, nucleotide substitutions are more prevalent in the third base of a codon than in the first or second base. Mutations in the third base are often neutral; that is, they do not change the amino acid sequence of the protein (refer back to Table 10.1). In contrast, random mutations at the first or second base are more likely to be harmful than beneficial and tend to be eliminated from a population via natural selection.

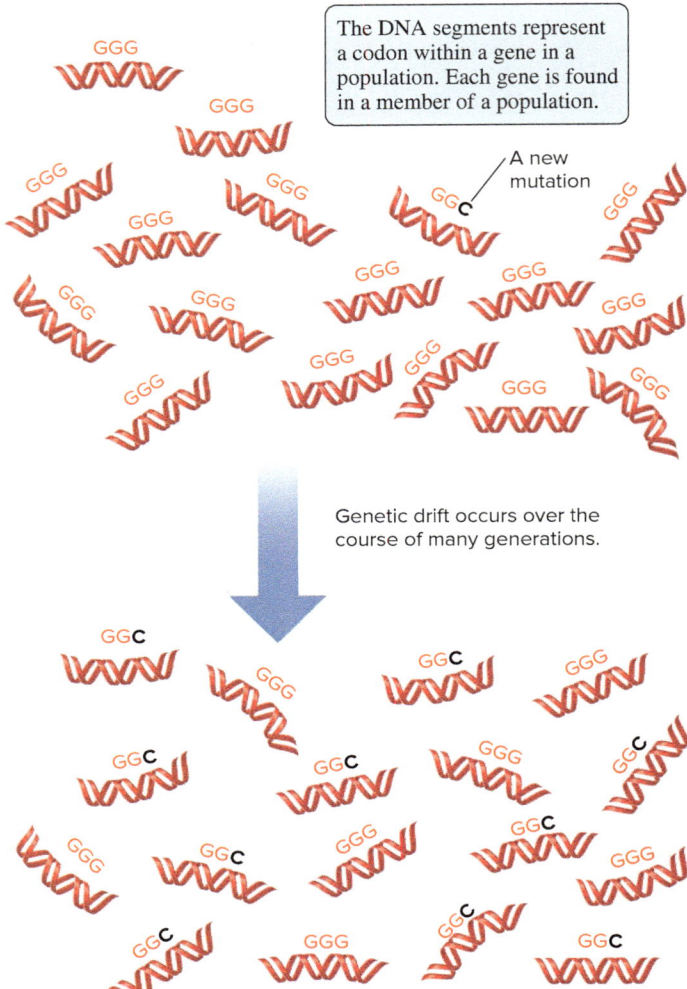

Figure 19.21 Non-Darwinian evolution in a population. In this example, a mutation within a gene changes a glycine codon from GGG to GGC, which does not affect the amino acid sequence of the encoded protein. Each gene shown represents a copy of the gene in a member of a population. Over the course of many generations, genetic drift may cause this neutral allele to become prevalent in the population, perhaps even monomorphic.

BioConnections: *Refer back to the genetic code described in Table 10.1. Describe three different genetic changes that you would expect to be neutral.*

19.5 Reviewing the Concepts

- Genetic drift involves changes in allele frequencies over time due to random chance. It occurs more rapidly in small populations and leads to either the elimination or the fixation of alleles (Figure 19.19).
- In the bottleneck effect, an environmental event dramatically reduces a population size and the allele frequencies of the resulting population change due to genetic drift (Figure 19.20).
- The founder effect occurs when a small population moves to a new geographic location and genetic drift alters the genetic composition of that population.
- Kimura proposed the neutral theory of evolution, in which genetic drift promotes the accumulation of neutral genetic changes in genes and proteins that do not affect reproductive success. Much of the variation in DNA sequences in populations appears to be the result of genetic drift rather than natural selection (Figure 19.21).

19.5 Testing Your Knowledge

1. A hurricane kills 98% of a population of cranes. The population eventually rebounds to its original size, but the amount of genetic variation in the rebounded population is significantly less. This is an example of
  a. the founder effect.
  b. the bottleneck effect.
  c. diversifying selection.
  d. both a and b.

2. Kimura's proposal regarding neutral mutations differs from Darwinian evolution in that
  a. natural selection does not exist.
  b. most of the genetic variation in a population is due to neutral mutations, which do not affect reproductive success.
  c. neutral variation alters survival and reproductive success.
  d. neutral mutations are not affected by population size.
  e. Both b and c are correct.

19.6 Migration and Nonrandom Mating

Learning Outcomes

1. Describe how gene flow affects genetic variation in neighboring populations.
2. Define inbreeding, and explain how it may have detrimental consequences.

Thus far, we have considered how natural selection and genetic drift operate as key mechanisms that cause evolution to happen. In addition, migration between neighboring populations and nonrandom mating may influence genetic variation and the relative proportions of genotypes. In this section, we will explore how these mechanisms work.

Migration Between Two Populations Tends to Increase Genetic Variation

Earlier in this chapter, we considered the founder effect, in which migration to a new location by a relatively small group can result in a population with an altered genetic composition due to genetic drift. In addition, migration between two different established populations can alter allele frequencies. As an example, let's consider two populations of a particular species of deer that are separated by a mountain range running north and south (**Figure 19.22**). On rare occasions, a few deer from the western population may travel through a narrow pass between the mountains and become members of the eastern population. If the two populations are different with regard to genetic variation, this migration will alter the frequencies of certain alleles in the eastern population. Of course, this migration could occur in the opposite direction as well and would then affect the western population. This transfer of alleles into or out of a population, called **gene flow**, occurs whenever individuals move between populations having different allele frequencies.

What are the consequences of migration?

- First, migration tends to reduce differences in allele frequencies between neighboring populations. Population geneticists can evaluate the extent of migration between two populations by analyzing the similarities and differences between their allele frequencies. Populations that frequently mix their gene pools via migration tend to have similar allele frequencies, whereas the allele frequencies of isolated populations tend to differ due to the effects of natural selection and genetic drift.
- Second, migration generally increases genetic diversity within populations. As discussed earlier in this chapter, new mutations are relatively rare events. Therefore, a new mutation may arise in only one population, and migration may then introduce this new allele into a neighboring population.

Nonrandom Mating Affects the Relative Proportion of Homozygotes and Heterozygotes in a Population

As mentioned earlier, one of the conditions required to establish Hardy-Weinberg equilibrium is random mating, which means that members of a population choose their mates irrespective of their genotypes or phenotypes. In many species, including human populations, this condition is violated. Such **nonrandom mating** takes different forms. Assortative mating occurs when individuals with similar phenotypes are more likely to mate. If the similar phenotypes are due to similar genotypes, assortative mating tends to increase the proportion of homozygotes and decrease the proportion of heterozygotes in the population. The opposite situation, where dissimilar phenotypes mate preferentially, causes heterozygosity to increase.

Another form of nonrandom mating involves the choice of mates based on their genetic history rather than their phenotypes. Individuals may choose a mate that is part of the same genetic lineage. The mating of two genetically related individuals, such as cousins, is called **inbreeding.** This sometimes occurs in human societies and is more likely to take place in nature when population size becomes very small.

In the absence of other evolutionary factors, nonrandom mating does not affect allele frequencies in a population. However, it will alter the balance of genotypes predicted by the Hardy-Weinberg equation. As an example, let's consider a human pedigree involving a mating between cousins (**Figure 19.23**). Individuals III-2 and III-3 are cousins and have produced the daughter labeled IV-1. She is said to be inbred, because her parents are genetically related. The parents of an inbred individual have one or more common ancestors. In the pedigree of Figure 19.23, I-2 is the grandfather of both III-2 and III-3.

Inbreeding increases the relative proportions of homozygotes and decreases the likelihood of heterozygotes in a population. Why

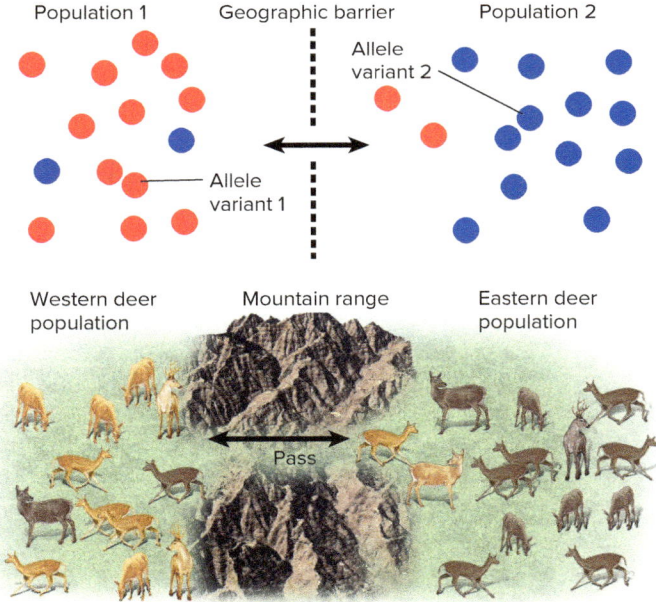

Figure 19.22 Migration and gene flow. In this example, two populations of a deer species are separated by a mountain range. On rare occasions, a few deer from one population travel through a narrow pass and become members of the other population. If the two populations differ in regard to genetic variation, this migration will alter the frequencies of alleles in the populations.

Concept Check: *How does migration affect the genetic compositions of populations?*

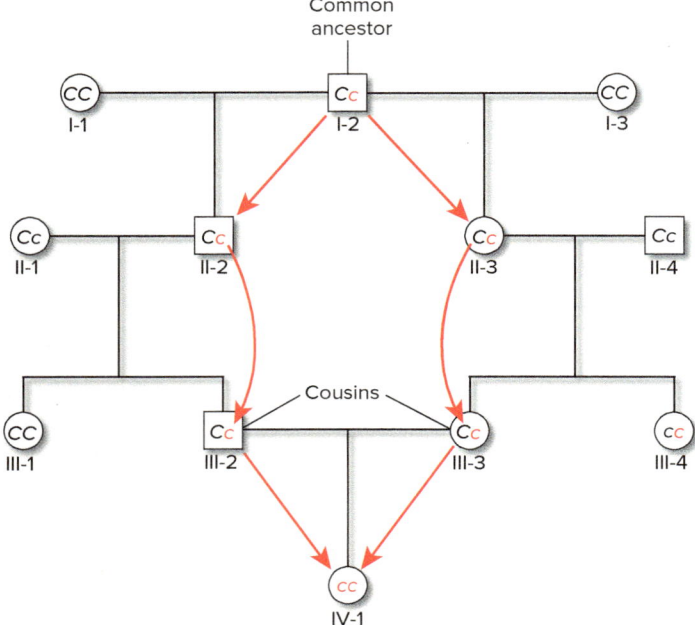

Figure 19.23 A human pedigree containing inbreeding. The parents of individual IV-1 are genetically related (cousins), and therefore individual IV-1 is inbred. Inbreeding increases the likelihood that an individual will be homozygous for any given gene. The red arrows show how IV-1 could become homozygous by inheriting the same allele (*c*) that the common ancestor (I-2) passed to both of her parents.

 BioConnections: *Many inherited human diseases show a recessive pattern of inheritance (refer back to Table 15.2). Explain whether inbreeding would increase or decrease the likelihood of such diseases.*

does this happen? An inbred individual has a higher chance of being homozygous for any given allele because the same allele for that gene could be inherited twice from a common ancestor. For example, individual I-2 is a heterozygote, *Cc*. The *c* allele could pass from I-2 to II-2 to III-2 and finally to IV-1 (see red arrows in Figure 19.23). Likewise, the *c* allele could pass from I-2 to II-3 to III-3 and then to IV-1. Therefore, IV-1 has a chance of being homozygous because she inherited both copies of the *c* allele from a common ancestor to both of her parents. Inbreeding does not favor any particular allele—it does not favor *c* over *C*—but it does increase the likelihood that an individual will be homozygous for any given gene.

Although inbreeding by itself does not affect allele frequencies, it may have negative consequences with regard to recessive alleles. Rare recessive alleles that are harmful in the homozygous condition are found in nearly all natural populations. Such alleles do not usually pose a problem because heterozygotes carrying a rare recessive allele are also rare, making it very unlikely that two such heterozygotes will mate with each other. However, related individuals share some of their genes, including recessive alleles. Therefore, if inbreeding occurs, homozygous offspring are more likely to be produced. For example, rare recessive diseases in humans are more frequent when inbreeding occurs.

19.6 Reviewing the Concepts

- Gene flow occurs when individuals migrate between populations with different allele frequencies. It reduces differences in allele frequencies between populations and enhances genetic diversity (Figure 19.22).
- Inbreeding, a form of nonrandom mating in which genetically related individuals have offspring with each other, tends to increase the proportion of homozygotes relative to heterozygotes (Figure 19.23).

19.6 Testing Your Knowledge

1. If the members of one population frequently migrated to a neighboring population and vice versa, which of the following would you expect to be correct?
   a. The two populations would tend to have similar allele frequencies.
   b. The two populations would tend to be more genetically diverse than if no migration had occurred.
   c. The two populations would be more susceptible to genetic drift.
   d. All of the above are correct.
   e. Only a and b are correct.

2. Populations that experience inbreeding may also experience
   a. a decrease in fitness due to an increased frequency of recessive genetic diseases.
   b. an increase in fitness due to increases in heterozygosity.
   c. very little genetic drift.
   d. no apparent change.
   e. increased mutation rates.

Assessing Your Knowledge and Skills

Foundational Knowledge

Multiple Choice

1. A change in one or more heritable characteristics of a population that occurs from one generation to the next is called
   a. natural selection.
   b. sexual selection.
   c. population genetics.
   d. evolution.
   e. inheritance of acquired characteristics.

2. Homology occurs because
   a. different species occupy similar environments.
   b. different species evolved from a common ancestor.
   c. new mutations modify existing structures.
   d. populations evolve from one generation to the next.
   e. of both a and b.

3. Which of the following is an example of a developmental homology seen in human embryonic development and other vertebrate species that are not mammals?
   a. gill ridges
   b. umbilical cord

c. tail
d. Both a and c are correct.
e. All of the above are correct.

4. Population geneticists are interested in the genetic variation in populations. The most common type of genetic change that causes polymorphism in a population is
   a. a deletion of a gene sequence.
   b. a duplication of a region of a gene.
   c. a rearrangement of a gene sequence.
   d. a change in a single nucleotide.
   e. an inversion of a segment of a chromosome.

5. Which of the following statements regarding mutations is true?
   a. Mutations are not important in evolution.
   b. Mutations provide the source for genetic variation, but other evolutionary factors are more important in determining allele frequencies in a population.
   c. Mutations occur at such a high rate that they promote major changes in the gene pool from one generation to the next.
   d. Mutations are of greater importance in smaller populations than in larger ones.
   e. Mutations are of greater importance in larger populations than in smaller ones.

Short Answer

1. What is convergent evolution? How do the pairs of organisms shown in Figure 19.6 support the theory of descent with modification?

2. **PRINCIPLES** A principle of biology is that populations of organisms evolve from one generation to the next and are related by an evolutionary history. Explain how the homologous forelimbs of vertebrates indicate that populations evolve from one generation to the next.

Critical-Thinking Skills

Multiple Choice

1. Which of the following is *not* an example of homology?
   a. the similarity in the β-globin gene sequence among different species of mammals
   b. the similarity in the structure of ribosomes found in all modern species
   c. the similarity in bone structure in the limbs of vertebrate species
   d. the long snouts and tongues of the giant anteater and the echidna that allow them to feed on ants
   e. All of the above are examples of homology.

2. In the Hardy-Weinberg equation, which term(s) would be used to calculate the frequency of individuals that do *not* exhibit a recessive disease but are carriers of a recessive allele?
   a. q
   b. p^2
   c. $2pq$
   d. q^2
   e. both b and d

3. A gene exists in two alleles, *B* and *b*. The fitness values are 0.7 for *BB*, 1.0 for *Bb*, and 0.4 for *bb*. This scenario is likely to result in
   a. directional selection.
   b. stabilizing selection.
   c. diversifying (disruptive) selection.
   d. balancing selection.
   e. both c and d.

4. A population of mice occupies a heterogeneous environment in which the fitness of some genotypes is higher in one area of the environment while the fitness of other individuals is higher in another area. This scenario is likely to result in
   a. directional selection.
   b. stabilizing selection.
   c. diversifying (disruptive) selection.
   d. balancing selection.
   e. both a and d.

5. A catastrophic earthquake killed about 90% of the maple trees in a river valley. There is no known genetic variation in maple trees that confers a survival advantage in the face of an earthquake. Eventually, the maple tree population recovered to its original numbers, but researchers have determined that certain traits, such as resistance to the damaging effects of some herbivores, has been lost from the rebounded population. The most likely explanation for this observation is that the earthquake caused _____ and then _____ eliminated certain alleles from the population.
   a. a bottleneck effect, directional selection
   b. a bottleneck effect, diversifying (disruptive) selection
   c. a bottleneck effect, genetic drift
   d. a heterogeneous environment, diversifying (disruptive) selection
   e. a heterogeneous environment, genetic drift

Short Answer

1. The percentage of individuals exhibiting a recessive disease in a population is 0.04, which is 4%. Assuming Hardy-Weinberg equilibrium, what percentage of individuals are expected to be heterozygous carriers?

2. The term natural selection is sometimes confused with the term evolution. Discuss the meanings of these two terms. Explain how the processes are different and how they are related to each other.

20

Evolution of Life II: The Emergence of New Species

Two different species of zebras. Grevy's zebra (*Equus grevyi*) is shown on the left, and Grant's zebra (*E. quagga boehmi*), which has fewer and thicker stripes, is shown on the right.

Chapter Outline

20.1 Identification of Species
20.2 Reproductive Isolation
20.3 Mechanisms of Speciation
20.4 Evo-Devo: Evolutionary Developmental Biology

Assessing Your Knowledge and Skills

©Frederic B. Siskind

The origin of living organisms has been described by philosophers as the great "mystery of mysteries." Perhaps that is why so many different views have been put forth to explain the existence of living species. At the time of Aristotle (4th century B.C.E.), most people believed that some living organisms came into being by spontaneous generation—the idea that nonliving materials could give rise to living organisms. For example, it was commonly believed that worms and frogs could arise from mud, and mice could come from grain. By comparison, many religious teachings contend that species were divinely made and have remained the same since their creation. In contrast to these ideas, Charles Darwin proposed the scientific theory of evolution by descent with modification. According to this view, species arise by the modification of existing species.

This chapter builds in an exciting way on the information that we considered in previous chapters. In Chapter 4, we examined how the first primitive cells may have arisen in an RNA world. In Chapter 19, we surveyed the tenets on which the theory of evolution is built and explored **microevolution**—evolution on the small scale of changes in allele frequencies in a population. In this chapter, we will consider evolution on a larger scale. **Macroevolution** refers to evolutionary changes that produce new species and groups of species.

To biologists, the concept of a **species** refers to a group of related organisms that share a distinctive set of attributes in nature. Members of the same species share an evolutionary history, which makes them more genetically similar to each other than they are to members of a different species. You may already have an intuitive sense of this concept. It is obvious that zebras and mice are different species. However, as discussed in the first section of this chapter, the distinction between different, closely related species is often blurred in natural environments. Two closely related species may look very similar, as the chapter-opening photo illustrates.

In this chapter, we will also focus on the mechanisms that promote the formation of new species, a phenomenon called **speciation.** Such macroevolution typically occurs by the accumulation of microevolutionary changes, those that occur in single genes. We will also consider how variations in the genes that control development play a role in the evolution of new species.

20.1 Identification of Species

Learning Outcomes

1. Outline the characteristics that biologists use to distinguish different species.
2. Describe different species concepts.

How many different species are on Earth? The number is astounding. A study done by American biologist E. O. Wilson and colleagues in 1990 estimated the known number of species at approximately 1.4 million. Currently, about 1.8 million species have been identified and catalogued. However, a vast number of species have yet to be classified. This is particularly true among bacteria and archaea, which are difficult to categorize into distinct species. Also, new invertebrate and even vertebrate species are still being found in the far reaches of pristine habitats. Common estimates of the total number of species range from 5 to 50 million! Many estimates suggest about 10 million.

When studying natural populations, evolutionary biologists are often confronted with situations in which some differences between two populations are apparent, but it is difficult to decide whether the two populations truly represent separate species. When two or more geographically restricted groups of the same species display one or more traits that are somewhat different but not enough to warrant their placement into different species, biologists sometimes classify such groups as **subspecies.** Similarly, many bacterial species are subdivided into **ecotypes.** Each ecotype is a genetically distinct population adapted to its local environment. In this section, we will consider the characteristics that biologists examine when deciding if two groups of organisms constitute different species.

Each Species Is Established Using Characteristics and Histories That Distinguish It from Other Species

As mentioned, a species is a group of organisms that share a distinctive set of attributes in nature. In the case of sexually reproducing species, members of one species usually cannot successfully interbreed with members of other species. Members of the same species share an evolutionary history that is distinct from those of other species. Although this may seem like a reasonable way to characterize a given species, biologists would agree that distinguishing between species is a more difficult undertaking. What criteria do we use to distinguish species? How many differences must exist between two populations to classify them as different species? Such questions are often difficult to answer.

The characteristics that a biologist uses to identify a species depend, in large part, on the species in question. For example, the traits used to distinguish insect species are quite different from those used to identify different bacterial species. The relatively high level of horizontal gene transfer among bacteria presents special challenges in the grouping of bacterial species. Among bacteria, it is sometimes very difficult and perhaps arbitrary to divide closely related organisms into separate species.

The most commonly used characteristics for identifying species are morphological traits, the ability to interbreed, molecular features, ecological factors, and evolutionary relationships. A comparison of these concepts will help you appreciate the various approaches that biologists use to identify the bewildering array of species on our planet.

Morphological Traits One way to establish that a population constitutes a unique species is based on the physical characteristics of its members. Organisms are classified as the same species if their anatomical traits appear to be very similar. Likewise, microorganisms can be classified according to morphological traits at the cellular level. By comparing many different morphological traits, biologists may be able to decide that certain populations constitute a unique species.

Although an analysis of morphological traits is a common way for biologists to establish that a particular group constitutes a species, this approach has drawbacks. First, researchers may have difficulty deciding how many traits to consider. In addition, quantitative traits, such as size and weight, that vary in a continuous way among members of the same species are not easy to analyze. Another drawback is that the degree of dissimilarity that distinguishes different species may not show a simple relationship. The members of the same species sometimes look very different, and conversely, members of different species sometimes look remarkably similar to each other. For example, **Figure 20.1a** shows two different frogs of the species *Dendrobates tinctorius,* commonly called the dyeing poison frog. This species exists in many different-colored morphs, which are individuals of the same species that have noticeably dissimilar appearances. In contrast, **Figure 20.1b** shows two different species of frogs, the northern leopard frog (*Rana pipiens*) and the southern leopard frog (*Rana utricularia*), which look fairly similar.

Reproductive Isolation Why would biologists describe two types of organisms, such as the northern leopard frog and southern leopard frog, as being different species if they are morphologically similar? One reason is that biologists have discovered that the two species of frogs are unable to breed with each other in nature. Therefore, a second way of identifying a species is by the ability of its members to interbreed. In the late 1920s, geneticist Theodosius Dobzhansky proposed that each species is reproductively isolated from other species. Such **reproductive isolation** prevents one species from successfully interbreeding with other species. In 1942,

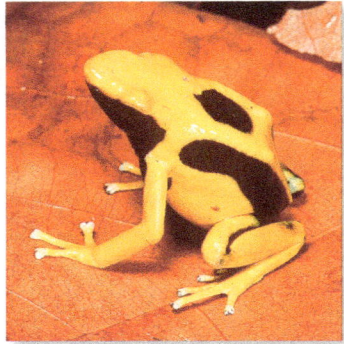

(a) Frogs of the same species

(b) Frogs of different species

Figure 20.1 Difficulties of using morphological traits to identify species. In some cases, members of the same species appear quite different. In other cases, members of different species look very similar. **(a)** Two frogs of the same species, the dyeing poison frog (*Dendrobates tinctorius*). **(b)** Two different species of frog, the northern leopard frog (*Rana pipiens,* left) and the southern leopard frog (*Rana utricularia,* right).
(a left) ©Mark Smith/Science Source; *(a right)* ©Pascal Goetgheluck/ardea.com; *(b left)* ©Ron Erwin/Getty Images; *(b right)* ©Robin Chittenden/Alamy

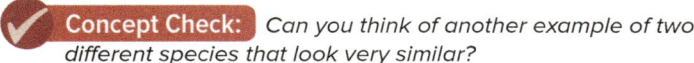

Concept Check: *Can you think of another example of two different species that look very similar?*

German evolutionary biologist Ernst Mayr expanded on the ideas of Dobzhansky to provide a definition of a species. According to Mayr, a key feature of sexually reproducing species is that, in nature, the members of one species have the potential to interbreed with one another to produce viable, fertile offspring but cannot successfully interbreed with members of other species.

Reproductive isolation has been used to distinguish many plant and animal species, especially those that look alike but do not interbreed. Even so, this criterion suffers from four main problems.

- In nature, it may be difficult to determine if two populations are reproductively isolated, particularly if the populations have nonoverlapping geographic ranges.
- Biologists have noted many cases in which two different species can interbreed in nature yet consistently maintain themselves as separate species. For example, different species of yucca plants, such as *Yucca pallida* and *Yucca constricta,* do interbreed in nature yet typically maintain populations with distinct characteristics. For this reason, they are viewed as distinct species.
- Another drawback of reproductive isolation is that it does not apply to asexual species such as bacteria and some species of plants and fungi.
- A fourth drawback is that reproductive isolation cannot be applied to extinct species.

For these reasons, reproductive isolation has been primarily used to distinguish closely related species of modern animals and plants that reproduce sexually.

Molecular Features Molecular features are now commonly used to determine if two different populations are different species. Evolutionary biologists often compare DNA sequences within genes, gene order along chromosomes, chromosome structure, and chromosome number in order to identify similarities and differences among different populations. For example, researchers may compare the DNA sequence of the *16S rRNA* gene between different bacterial populations as a way of determining if two populations represent different species. When the sequences are very similar, such populations will probably be judged to be the same species. However, it may be difficult to draw the line when separating groups into different species. How much difference must be present for species to be considered separate? Is a 2% difference in their genome sequences sufficient to warrant placement into two different species, or do we need a 5% difference?

Ecological Factors A variety of factors related to an organism's habitat may be used to distinguish one species from another. For example, certain species of warblers are distinguished by the habitat in which they forage for food. Some species search the ground for food, others forage in bushes or small trees, and some primarily forage in tall trees. Such habitat differences are used to distinguish different species that look morphologically similar.

Many bacterial species have been categorized as distinct based on ecological factors. Bacterial cells of the same species are likely to use the same types of resources (such as sugars and vitamins) and grow under the same types of conditions (such as particular temperature and pH ranges). However, a drawback of this approach is that different groups of bacteria sometimes display very similar growth characteristics, and even the same species may show great variation in the growth conditions it will tolerate.

Evolutionary Relationships In Chapter 21, we will examine the methods used to produce "evolutionary trees," diagrams that describe the relationships between ancestral species and modern species. In some cases, such relationships are based on an analysis of the fossil record. For example, we will consider how the fossil record was used to construct a tree that shows the ancestors that led to modern horse species. Alternatively, another way of establishing evolutionary relationships is by the analysis of DNA sequences. Researchers obtain samples of cells from different individuals and compare the genes within those cells to see how similar or different they are.

Species Concepts Emphasize Particular Features to Define and Distinguish Species

A **species concept** is a way of defining the concept of a species and/or of providing an approach to distinguish one species from another. However, even Darwin realized the difficulty in defining a species. In

1859, he said, "No one definition [of species] has as yet satisfied all naturalists; yet every naturalist knows vaguely what he means when he speaks of a species." Since 1942, over 20 different species concepts have been proposed by various evolutionary biologists. Let's consider a few of the more popular ones.

- Ernst Mayr proposed one of the first species concepts, called the **biological species concept,** which defines a species as a group of individuals whose members have the potential to interbreed with one another in nature to produce viable, fertile offspring but cannot successfully interbreed with members of other species.
- According to the **evolutionary lineage concept** proposed by American paleontologist George Gaylord Simpson in 1961, species should be defined based on the separate evolution of lineages. A **lineage** is a series of species that forms a line of descent, with each new species the direct result of speciation from an immediate ancestral species.
- The **ecological species concept,** described by American evolutionary biologist Leigh Van Valen in 1976, defines each species based on an ecological niche, which is the unique set of habitat resources that a species requires, as well as its influence on the environment and other species.
- In 1998, American zoologist Kevin de Queiroz suggested that there is only a single general species concept, which generally concurs with Simpson's evolutionary lineage concept. According to de Queiroz's **general lineage concept,** each species is a population of an independently evolving lineage. Each species has evolved from a specific series of ancestors and, as a consequence, forms a group of organisms with a particular set of characteristics. Multiple criteria are used to determine if a population is part of an independent evolutionary lineage, and thus a species, which is distinct from others. These include key factors from other species concepts, such as reproductive isolation and ecological niches. Because of its generality, the general lineage concept has received significant support.

20.1 Reviewing the Concepts

- A species is a group of related organisms that share a distinctive set of attributes in nature. Speciation is the process by which new species are formed. Macroevolution refers to evolutionary changes that produce new species and groups of species.
- Various characteristics, including morphological traits, reproductive isolation, molecular features, ecological factors, and evolutionary relationships, are used to identify species (Figure 20.1).

20.1 Testing Your Knowledge

1. According to Mayr's biological species concept, a species is defined by
    a. ecological factors.
    b. morphological factors.
    c. the lineage to which it belongs.
    d. reproductive isolation.
    e. all of the above.

2. According to de Queiroz's general lineage concept, each species
    a. is a population of an independently evolving lineage.
    b. occupies its own niche.
    c. is reproductively isolated.
    d. All of the above are correct.
    e. Both a and b are correct.

20.2 Reproductive Isolation

Learning Outcome

1. Compare and contrast prezygotic and postzygotic isolating mechanisms.

Thus far, we have considered various ways of differentiating species. From the discussion, you may have realized that the identification of a species is not always a simple matter. The phenomenon of reproductive isolation has played a major role in the way biologists study plant and animal species, partly because it identifies a possible mechanism for the process of forming new species. For this reason, much research has been done to try to understand **reproductive isolating mechanisms,** the mechanisms that prevent interbreeding between different species. Why do reproductive isolating mechanisms occur? Populations do not intentionally erect these reproductive barriers. Rather, reproductive isolation is a consequence of genetic changes that usually occur because a species becomes adapted to its own particular environment. The view of evolutionary biologists is that reproductive isolation typically evolves as a by-product of genetic divergence. Over time, as a species evolves its own unique characteristics, some of those traits are likely to prevent breeding with other species.

Reproductive isolating mechanisms fall into two categories: **prezygotic isolating mechanisms,** which prevent the formation of a zygote, and **postzygotic isolating mechanisms,** which block the development of a viable and fertile individual after fertilization has taken place. Figure 20.2 summarizes some of the more common ways that reproductive isolating mechanisms prevent reproduction between different species. When two species do produce offspring, such an offspring is called an **interspecies hybrid.**

Prezygotic Isolating Mechanisms We will consider five types of prezygotic isolating mechanisms.

1. **Habitat isolation:** One obvious way to prevent interbreeding is for members of different species to never come in contact with each other. This phenomenon, called habitat isolation, may involve a geographic barrier to interbreeding.
2. **Temporal isolation:** In temporal isolation, species happen to reproduce at different times of the day or year. In the northeastern U.S., for example, the two most abundant field crickets, *Gryllus veletis* and *Gryllus pennsylvanicus* (spring and fall field crickets, respectively), do not differ in song or habitat and are morphologically very similar (**Figure 20.3**). *G. veletis* matures in the spring, whereas *G. pennsylvanicus* matures in the fall, which minimizes interbreeding between the two species.

Species 1   Species 2

Prezygotic isolating mechanisms

Habitat isolation: Species occupy different habitats, so they never come in contact with each other.

Temporal isolation: Species have different mating or flowering seasons or times of day or become sexually mature at different times of the year.

Behavioral isolation: Sexual attraction between males and females of different animal species is limited due to differences in behavior or physiology.

↓ Attempted mating

Mechanical isolation: Morphological features such as size and incompatible genitalia prevent 2 members of different species from interbreeding.

Gametic isolation: Gametic transfer takes place, but the gametes fail to unite with each other. This can occur because the male and female gametes fail to attract, because they are unable to fuse, or because the male gametes are inviable in the female reproductive tract of another species. In plants, the pollen of one species usually cannot generate a pollen tube to fertilize the egg cells of another species.

↓ Fertilization

Postzygotic isolating mechanisms

Hybrid inviability: The egg of one species is fertilized by the sperm from another species, but the fertilized egg fails to develop past the early embryonic stages.

Hybrid sterility: An interspecies hybrid survives, but it is sterile. For example, the mule, which is sterile, is produced from a cross between a male donkey (*Equus asinus*) and a female horse (*Equus caballus*).

Hybrid breakdown: The F_1 interspecies hybrid is viable and fertile, but succeeding generations (F_2, and so on) become increasingly inviable. This is usually due to the formation of less-fit genotypes by genetic recombination.

↓ Interspecies hybrid

Figure 20.2 **Reproductive isolating mechanisms.** These mechanisms prevent successful breeding between different species. They can occur prior to fertilization (prezygotic) or after fertilization (postzygotic).

BioConnections: *Refer back to Figure 19.18b. Is female choice an example of a prezygotic or postzygotic isolating mechanism?*

(a) Spring field cricket (*Gryllus veletis*)   **(b)** Fall field cricket (*Gryllus pennsylvanicus*)

Figure 20.3 **Temporal isolation.** Interbreeding between these two species of crickets does not usually occur because *Gryllus veletis* matures in the spring, whereas *Gryllus pennsylvanicus* matures in the fall.
(a) ©C. Allan Morgan/Getty Images; *(b)* ©Bryan E. Reynolds

Concept Check: *Is temporal isolation an example of a prezygotic or a postzygotic isolating mechanism?*

3. **Behavioral isolation:** In the case of animals, mating behavior and anatomy often play key roles in promoting reproductive isolation. An example is found between the western meadowlark (*Sturnella neglecta*) and eastern meadowlark (*Sturnella magna*) (**Figure 20.4**). In the zone of overlap, very little interspecies mating takes place, largely due to differences in their songs, which enable meadowlarks to recognize potential mates as members of their own species.

4. **Mechanical isolation:** Morphological features such as size or incompatible genitalia prevent two species from interbreeding. For example, male dragonflies use a pair of special appendages to grasp females during copulation. When a male tries to mate with a female of a different species, his grasping appendages do not fit her body shape.

5. **Gametic isolation:** Two species may attempt to interbreed, but the gametes fail to unite in a successful fertilization event. In flowering plants, gametic isolation is commonly associated with pollination. Plant fertilization is initiated when a pollen grain lands on the stigma of a flower and sprouts a pollen tube that ultimately reaches an egg cell. When pollen is released from a plant, it could be transferred to the stigma of many different plant species. In most cases, when a pollen grain lands on the stigma of a different species, either it fails to generate a pollen tube or the tube does not grow properly and thus does not reach the egg cell.

Postzygotic Isolating Mechanisms Let's now turn to postzygotic mechanisms of reproductive isolation, of which there are three common types.

1. **Hybrid inviability:** This occurs when an egg of one species is fertilized by a sperm from another species, but the fertilized egg cannot develop past early embryonic stages.

2. **Hybrid sterility:** An interspecies hybrid may be viable but sterile. A classic example of hybrid sterility is the mule, which is produced by a mating between a male donkey (*Equus asinus*) and a female horse (*Equus ferus caballus*) (**Figure 20.5**). All male mules and most female mules are sterile. Why are mules usually

Biology Principle

Populations of Organisms Evolve from One Generation to the Next and Are Related by an Evolutionary History

One of the evolutionary changes that took place in these two species of meadowlarks is that their mating songs became different.

(a) Western meadowlark (*Sturnella neglecta*)

Figure 20.4 Behavioral isolation. (a) The western meadowlark (*Sturnella neglecta*) and (b) eastern meadowlark (*Sturnella magna*) are very similar in appearance. The red region in this map shows where the two species' ranges overlap. However, very little interspecies mating takes place due to differences in their songs.
(a) ©Rod Planck/Science Source; *(b)* ©Ron Austing/Science Source

 ×

Male donkey (*Equus asinus*)   Female horse (*Equus ferus caballus*)

Mule

Figure 20.5 Hybrid sterility. When a male donkey (*Equus asinus*) mates with a female horse (*Equus ferus caballus*), their offspring is a mule, which is usually sterile.
(top left) ©Mark Boulton/Science Source; *(top right)* ©MyLoupe/UIG/Getty Images; *(bottom)* ©Stephen L. Saks/Science Source

Concept Check: *Is hybrid sterility an example of a prezygotic or a postzygotic isolating mechanism?*

---

sterile? Two reasons explain the sterility. Because the horse has 32 chromosomes per set and a donkey has 31, a mule inherits 63 chromosomes (32 + 31). Due to the uneven number, not all of the chromosomes can pair evenly. Also, the chromosomes of the horse and donkey have structural differences, which either prevent them from pairing correctly or lead to chromosomal abnormalities if crossing over occurs during meiosis. For these reasons, mules usually produce inviable gametes.

3. **Hybrid breakdown:** Interspecies hybrids may be viable and fertile, but the subsequent generation(s) may harbor genetic abnormalities that are detrimental. This mechanism, called hybrid breakdown, can be caused by changes in chromosome structure. The chromosomes of closely related species may have structural differences from each other, such as inversions. In hybrids, a crossover may occur in the region that is inverted in one species but not the other. This will produce gametes with too little or too much genetic material. Such hybrids often have offspring with developmental abnormalities.

Postzygotic isolating mechanisms tend to be uncommon in nature compared with prezygotic mechanisms. Why are postzygotic mechanisms rare? One explanation is they are more costly in terms of energy and resources used. For example, a female mammal uses a large amount of energy to produce an offspring that is sterile. Evolutionary biologists hypothesize that natural selection has favored prezygotic isolating mechanisms because they do not waste a lot of energy.

20.2 Reviewing the Concepts

- Reproductive isolating mechanisms prevent two different species from breeding with each other (Figure 20.2).
- Prezygotic isolating mechanisms include habitat isolation, temporal isolation, behavioral isolation, mechanical isolation, and gametic isolation (Figures 20.3, 20.4).
- Postzygotic isolating mechanisms include hybrid inviability, hybrid sterility, and hybrid breakdown (Figure 20.5).

20.2 Testing Your Knowledge

1. A mechanism of reproductive isolation that does *not* prevent fertilization from happening is called
   a. a prezygotic isolating mechanism.
   b. a postzygotic isolating mechanism.
   c. Both a and b prevent fertilization from happening.
   d. Both a and b do not prevent fertilization from happening.

2. Which of the following is *not* an example of a prezygotic isolating mechanism?
   a. Two species breed at different times of the year.
   b. An egg from one species is fertilized by the sperm of another, but the embryo dies at an early stage.
   c. Two different bird species have different mating calls.
   d. Two different species are geographically isolated from each other.
   e. All of the above are prezygotic isolating mechanisms.

20.3 Mechanisms of Speciation

Learning Outcomes

1. Describe how allopatric speciation can occur and how it can lead to adaptive radiation.
2. Outline two different mechanisms of sympatric speciation.

Speciation, the formation of a new species, is caused by genetic changes in a particular group that make it different from the species from which it was derived. As discussed in Chapter 19, mutations in genes can be acted on by natural selection and other evolutionary mechanisms to alter the genetic composition of a population. New species commonly evolve in this manner. In addition, interspecies matings, changes in chromosome number, and horizontal gene transfer may also contribute to the formation of new species. In all of these cases, the underlying cause of speciation is the accumulation of genetic changes that ultimately promote enough differences that a population can be recognized as a unique species.

Even though genetic changes account for the phenotypic differences observed among living organisms, such changes do not fully explain the existence of many distinct species on our planet. Why does life often diversify into the more or less discrete populations that we recognize as species? Two main explanations have been proposed:

1. In some cases, speciation may occur due to abrupt events, such as changes in chromosome number, that cause reproductive isolation.
2. More commonly, species arise as a consequence of adaptation to different ecological niches. For sexually reproducing organisms, reproductive isolation is typically a by-product of that adaptation.

Depending on the species involved, one or both factors may play a dominant role in the formation of new species. In this section, we will consider how reproductive isolating mechanisms and adaptation to particular environments are critical aspects of the speciation process.

Geographic and Habitat Isolation Can Promote Allopatric Speciation

Cladogenesis is the splitting or diverging of a population into two or more species. In the case of sexually reproducing organisms, the process of cladogenesis requires that gene flow becomes interrupted between two or more populations, limiting or eliminating reproduction between members of those populations. **Allopatric speciation** (from the Greek *allos*, meaning other, and the Latin *patria*, meaning homeland) is the most prevalent way for cladogenesis to occur. This form of speciation occurs when a population becomes isolated from other populations and evolves into one or more new species. Typically, this isolation may involve a geographic barrier such as a large area of land or body of water.

In some cases, geographic separation may be caused by slow geological events that eventually produce large geographic barriers. For example, a mountain range may emerge and split one species that occupies the lowland regions, or a creeping glacier may divide a population. **Figure 20.6** shows an interesting example in which geological

Biology Principle

Populations of Organisms Evolve from Generation to Generation

These two species of fish look similar because they share a common ancestor that existed in the fairly recent past.

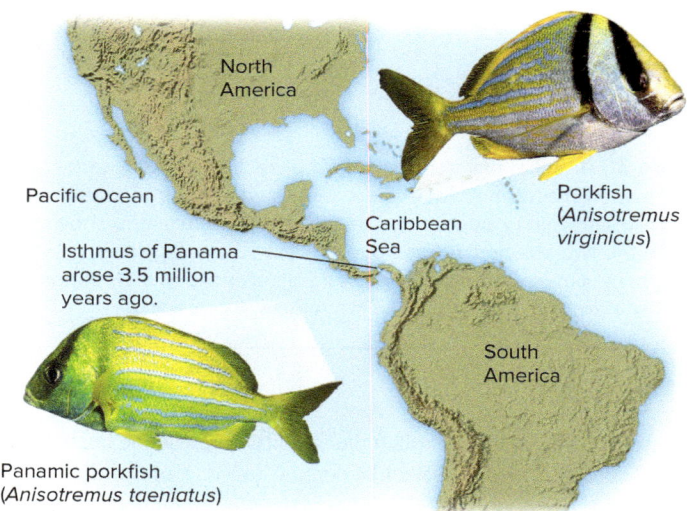

Figure 20.6 Allopatric speciation. An ancestral fish population was split into two by the formation of the Isthmus of Panama about 3.5 mya. Since that time, different genetic changes occurred in the two populations. These changes eventually led to the formation of different species: The Panamic porkfish (*Anisotremus taeniatus*) is found in the Pacific Ocean, and the porkfish (*Anisotremus virginicus*) is found in the Caribbean Sea.
(left) ©Hal Beral/V&W/imagequestmarine.com; *(right)* ©Amar and Isabelle Guillen/Guillen Photography/Alamy

separation promoted speciation. A fish called the Panamic porkfish (*Anisotremus taeniatus*) is found in the Pacific Ocean, whereas the porkfish (*Anisotremus virginicus*) is found in the Caribbean Sea. These two species were derived from an ancestral species whose population was split by the formation of the Isthmus of Panama about 3.5 mya. Before that event, the waters of the Pacific Ocean and Caribbean Sea mixed freely. Since the formation of the isthmus, the two populations have been geographically isolated and have evolved into distinct species.

Allopatric speciation can also occur when a small population moves to a new location that is geographically isolated from the main population. The Hawaiian Islands are a showcase of this type of allopatric speciation. The islands' extreme isolation coupled with their phenomenal array of ecological niches has enabled a small number of founding species to evolve into a vast assortment of different species. Biologists have investigated several examples of **adaptive radiation**, in which a single ancestral species has evolved into a wide array of descendant species that differ in their habitat, form, or behavior.

As shown in **Figure 20.7**, an example of adaptive radiation is seen with a family of birds called honeycreepers (Drepanidinae). Researchers estimate that the honeycreepers' ancestor arrived in Hawaii 3–7 mya. This ancestor was a single species of finch, possibly a Eurasian rosefinch (genus *Carpodacus*) or, less likely, the North American house finch (*Carpodacus mexicanus*). At least 54 different species of honeycreepers, many of which are now extinct, evolved from this founding population to fill available niches in the islands' habitats. Natural selection resulted in the formation of many species with different feeding strategies. Seed eaters have stouter, stronger bills capable of cracking tough husks. Insect-eating honeycreepers have thin, warbler-like bills adapted for picking insects from foliage or strong, hooked bills to root out wood-boring insects. The curved bills of nectar-feeding honeycreepers enable them to extract nectar from the flowers of Hawaii's endemic plants.

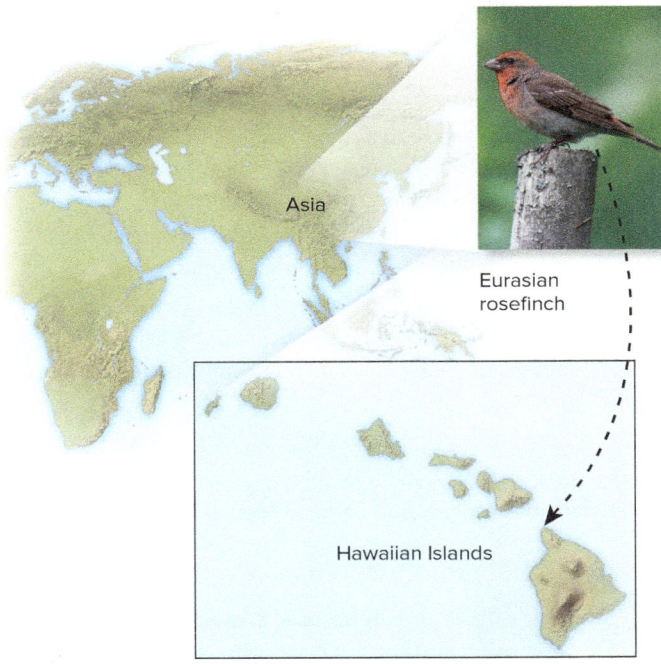

(a) Migration of ancestor to the Hawaiian Islands

(b) Examples of Hawaiian honeycreepers

Figure 20.7 Adaptive radiation. (a) The honeycreepers' ancestor is believed to be related to a Eurasian rosefinch that arrived on the Hawaiian Islands approximately 3–7 mya. Since that time, at least 54 different species of honeycreepers (Drepanidinae) have evolved on the islands. (b) Adaptations to feeding have produced honeycreeper species with notable differences in beak morphology.
(a) (Eurasian rosefinch) ©FLPA/Alamy; *(b) (Palila), (Nihoa finch), (Maui Alauahio),* and *(Akohekohe)* ©Jack Jeffrey Photography; *(Akikiki)* and *(I'iwi)* ©Jim Denny

BioConnections: *Refer back to Figure 19.14. Discuss how diversifying selection played a role in the diversity of honeycreepers on the Hawaiian Islands.*

FEATURE INVESTIGATION

Podos Found That an Adaptation for Feeding May Have Promoted Reproductive Isolation in Finches

In 2001, American evolutionary biologist Jeffrey Podos analyzed the songs of Darwin's finches on the Galápagos Islands to determine how environmental adaptation may contribute to reproductive isolation. As in honeycreepers, the differences in beak sizes and shapes among the various species of finches are adaptations to different feeding strategies. Podos hypothesized that changes in beak morphology might also affect the songs that the birds produce, thereby having the potential to affect mate choice. The components of the vocal tract of birds, including the trachea, larynx, and beak, work collectively to produce a bird's song. Birds actively modify the shape of their vocal tracts during singing, and beak movements are normally very rapid and precise.

Podos focused on two aspects of a bird's song. The first feature is the frequency range, which is a measure of the minimum and maximum frequencies in a bird's song, measured in kilohertz (kHz). The second feature is the trill rate. A trill is a series of notes or group of notes repeated in succession. **Figure 20.8** shows a graphical depiction of the songs of Darwin's finches. As you can see, the song patterns of these finches are quite different from each other.

To quantitatively study the relationship between beak size and song, Podos first captured male finches on Santa Cruz, one of the Galápagos Islands, and measured their beak sizes (**Figure 20.9**). The birds were banded and then released into the wild. The banding provided a way of identifying the birds whose beaks had already been measured. After release, the songs of the banded birds were recorded on a tape recorder, and their range of frequencies and trill rate were analyzed. Podos then compared the data for the Galápagos finches to a large body of data that had been collected on many other bird species. This comparison was used to evaluate whether beak size—in this case, beak depth, the measurement of the beak from top to bottom, at its base—constrained the frequency range and/or the trill rate of the finches.

The results of this comparison are shown in the data of Figure 20.9. As seen here, the relative constraint on vocal performance became higher as the beak depth became larger. This means that birds with larger beaks had a narrower frequency range and/or a slower trill rate. Podos proposed that as jaws and beaks became adapted for strength to crack open larger, harder seeds, they became less able to perform the rapid movements associated with certain types of songs. In contrast, the finches with smaller beaks adapted to probe for insects or eat smaller seeds had less constraint on their vocal performance. From the perspective of speciation, the changes observed in song patterns for the Galápagos finches could have played an important role in promoting reproductive isolation, because song pattern is an important factor in mate selection in birds. Therefore, a by-product of beak adaptation for feeding is that it also appears to have affected song pattern, possibly promoting reproductive isolation and eventually the formation of distinct species.

Experimental Questions

1. What did Podos hypothesize regarding the effects of beak size on a bird's song? How could changes in beak size and shape lead to reproductive isolation among the finches?

2. **SCISKILLS** ▶ How did Podos test the hypothesis that beak morphology caused changes in the birds' songs?

3. **SCISKILLS** ▶ Analyze the results of Podos' study, and explain whether they support his original hypothesis. What is meant by a by-product of adaptation, and how does it apply to this particular study?

Figure 20.8 **Differences in the songs of Galápagos finches.** These spectrograms depict the frequency of each bird's song over time, measured in kilohertz (kHz). The songs are produced in a series of trills that have a particular pattern and occur at regular intervals. Notice the differences in frequency and trill rate between different species of birds.

Figure 20.9 Study by Podos investigating the effects of beak depth on song among different species of Galápagos finches.

HYPOTHESIS Changes in beak morphology that are an adaptation for feeding may also affect the songs of Galápagos finches and thereby lead to reproductive isolation between species.

KEY MATERIALS This study was conducted on finch populations of the Galápagos Island of Santa Cruz.

Experimental level — **Conceptual level**

1. Capture male finches and measure their beak depth. Beak depth is measured at the base of beak, from top to bottom.

This is a measurement of phenotypic variation in beak size.

2. Band the birds and release them back into the wild.

Banding allows identification of birds with known beak depths.

3. Record the bird's songs on a tape recorder.

This is a measurement of phenotypic variation in song.

4. Analyze the songs with regard to frequency range and trill rate.

The frequency range is the value between high and low frequencies. The trill rate is the number of repeats per unit time.

5. **THE DATA**

The data for the Galápagos finches were compared to a large body of data that had been collected on many other bird species. The relative constraint on vocal performance is higher if a bird has a narrower frequency range and/or a slower trill rate. These constraints were analyzed with regard to each bird's beak depth.

6. **CONCLUSION** Larger beak size, which is an adaptation to cracking open large, hard seeds, constrains vocal performance. This may affect mating song patterns and thereby promote reproductive isolation and, in turn, speciation.

7. **SOURCE** Podos, Jeffrey. 2001. Correlated Evolution of Morphology and Vocal Signal Structure in Darwin's Finches. *Nature* 409: 185–188.

Sympatric Speciation Occurs When Populations Are in Direct Contact

Sympatric speciation (from the Greek *sym*, meaning together) occurs when members of a species that are within the same range diverge into two or more different species even though there are no physical barriers to interbreeding. Although sympatric speciation is believed to be less common than allopatric speciation, particularly in animals, evolutionary biologists have discovered several ways in which it can occur. These include polyploidy and adaptation to local environments.

Polyploidy A type of genetic change that can cause immediate reproductive isolation is **polyploidy,** in which an organism has more than two sets of chromosomes. Plants tend to be more tolerant of changes in chromosome number than animals. For example, many crops and decorative species of plants are polyploid. How does polyploidy occur? One mechanism is complete nondisjunction of chromosomes, which increases the number of chromosome sets in a given species (autopolyploidy). Such changes can result in an abrupt sympatric speciation. For example, nondisjunction could produce a tetraploid plant with four sets of chromosomes from a species that was diploid with two sets. A cross between a tetraploid and a diploid produces a triploid offspring with three sets of chromosomes. Triploid offspring are usually sterile because an odd number of chromosomes cannot be evenly segregated during meiosis. This hybrid sterility causes reproductive isolation between the tetraploid and diploid species.

Another mechanism that leads to polyploidy is interspecies breeding. For example, interbreeding between two different species may produce an **allodiploid,** an organism that has one set of chromosomes from two different species. This term refers to the occurrence of chromosome sets (ploidy) from the genomes of different (allo-) species. Diploid species that are close evolutionary relatives can sometimes interbreed and produce allodiploid offspring. An organism containing two sets of chromosomes from two different species, for a total of four sets, is called an **allotetraploid.** An allotetraploid can occur as a result of nondisjunction in an allodiploid organism. This can also abruptly lead to reproductive isolation, because the allotetraploid would be reproductively isolated from related diploid species.

Polyploidy is so frequent in plants that it is a major mechanism of their speciation. In ferns and flowering plants, about 40–70% of the species are polyploid. By comparison, polyploidy can occur in animals, but it is much less common. For example, less than 1% of reptiles and amphibians are polyploids derived from diploid ancestors. The reason polyploidy is not usually tolerated in animals is not understood.

Adaptation to Local Environments In some cases, populations occupying different local environments that are continuous with each other may diverge into different species. An early example of this type of sympatric speciation was described by American biologists Jeffrey Feder, Guy Bush, and colleagues. They studied the North American apple maggot fly (*Rhagoletis pomenella*). This fly originally fed on native hawthorn trees. However, the introduction of apple trees approximately 200 years ago provided a new local environment for this species. The apple-feeding populations of this species develop more rapidly because apples mature more quickly than hawthorn fruit. The result is partial temporal isolation, which is an example of a prezygotic isolating mechanism. Although the two populations—those that feed on apple trees and those that feed on hawthorn trees—are considered subspecies, evolutionary biologists speculate they may eventually become distinct species due to reproductive isolation and the accumulation of independent mutations.

American entomologist Sara Via and colleagues have studied the beginnings of sympatric speciation in pea aphids (*Acyrthosiphon pisum*), a small, plant-eating insect. Pea aphids in the same geographic area can be found on both alfalfa (*Medicago sativa*) and red clover (*Trifolium pratenae*) (Figure 20.10). Although pea aphids on these two host plants look identical, they show significant genetic differences and are highly ecologically specialized. Pea aphids that are found on alfalfa exhibit a lower fitness when transferred to red clover, whereas pea aphids found on red clover exhibit a lower fitness when transferred to alfalfa. The same traits involved in this host specialization cause these

Biology Principle

Populations of Organisms Evolve from One Generation to the Next and Are Related by an Evolutionary History

Populations of pea aphids are evolving based on preference for different food sources—alfalfa or red clover. The populations may eventually evolve into separate species.

Figure 20.10 Pea aphids, a possible example of sympatric speciation in progress. Some pea aphids prefer alfalfa, whereas others prefer red clover. These two populations may be in the process of sympatric speciation.
©Dr. Sara Via, Department of Biology and Department of Entomology, University of Maryland

Concept Check: *How may host preference eventually lead to speciation?*

two groups of pea aphids to be substantially reproductively isolated. Taken together, the observations of the North American apple maggot fly, pea aphids, and other insect species suggest that diversifying selection (described in Chapter 19) occurs because some members within the same range evolve to feed on a different host. This may be an important mechanism of sympatric speciation among insects.

20.3 Reviewing the Concepts

- Allopatric speciation occurs when a population becomes isolated from other populations and evolves into one or more new species. When speciation from a single ancestral species occurs multiple times, the process is called adaptive radiation (Figures 20.6, 20.7).
- Podos hypothesized that changes in beak depth, associated with adaptation for feeding, promoted reproductive isolation by altering the song patterns of finches (Figures 20.8, 20.9).
- Sympatric speciation involves the formation of different species in populations that are not geographically isolated from one another. Polyploidy and adaptation to local environments are mechanisms that promote sympatric speciation (Figure 20.10).

20.3 Testing Your Knowledge

1. Which of the following is an example of allopatric speciation?
   a. A few birds fly to a distant island, and after many generations a new bird species is formed.
   b. A diploid species becomes a new tetraploid species due to nondisjunction.
   c. A species diverges into two different species due to adaptation to local environments.
   d. All of the above are examples of allopatric speciation.

2. Which of the following would be the most likely to promote an abrupt speciation event?
   a. formation of a new species on an island
   b. adaptation of a species to local environments
   c. formation of a polyploid species from a single diploid species
   d. formation of an allotetraploid from two diploid species
   e. both c and d

20.4 Evo-Devo: Evolutionary Developmental Biology

Learning Outcomes

1. Describe how the spatial expression of genes, such as *BMP4* and *Gremlin*, affects pattern formation.
2. Explain the relationship between the number of *Hox* genes and the body pattern of an animal species.
3. Outline how differences in the growth rates of body parts can change the characteristics of species.

In biology, **development** refers to a series of changes in the state of a cell, a tissue, an organ, or an organism. Development is the process that gives rise to structures and functions of living organisms. In recent years, many evolutionary biologists have begun to investigate how genetic variation produces species and groups of species with novel shapes and forms. The underlying reasons for such changes are often rooted in the developmental pathways that control an organism's morphology.

Evolutionary developmental biology (referred to as **evo-devo**) is an exciting and relatively new field of biology that compares the development of different organisms in an attempt to understand ancestral relationships between organisms and the mechanisms that bring about evolutionary change. During the past few decades, developmental geneticists have gained a better understanding of biological development at the molecular level. Much of this work has involved the discovery of genes that control development in model organisms. As the genomes of more organisms have been analyzed, researchers have become interested in the similarities and differences that occur between closely related and distantly related species. The field of evolutionary developmental biology has arisen out of this interest. In this section, we will see that proteins that control developmental changes, such as cell-signaling proteins and transcription factors, often play a key role in promoting the morphological changes that occur during evolution.

The Spatial Expression of Genes That Affect Development Has a Dramatic Effect on Phenotype

Genes that play a role in development influence cell division, cell migration, cell differentiation, and cell death. The interplay among these four processes produces an organism with a specific body pattern, a process called **pattern formation.** As you might imagine, developmental genes are very important to the phenotypes of individuals. They affect traits such as the shape of a bird's beak, the length of a giraffe's neck, and the size of a plant's flower. In recent years, the study of development has revealed that developmental genes are key players in the evolution of many types of traits. Changes in such genes affect traits that can be acted on by natural selection. Furthermore, variation in the expression of these genes may be commonly involved in the acquisition of new traits that promote speciation.

As an example, let's compare pattern formation of a chicken's foot with that of a duck's foot. Two different patterns occur: a nonwebbed pattern, in which the digits are not interconnected, and a webbed pattern, in which the digits are connected by sheets of skin. Developmental biologists have discovered that the morphological differences between a nonwebbed and a webbed pattern are due to the differential expression of two different cell-signaling proteins called bone morphogenetic protein 4 (BMP4) and gremlin. The *BMP4* gene is expressed throughout the developing limb of both the chicken and the duck; this is shown in Figure 20.11a, in which the BMP4 protein is stained purple. The BMP4 protein causes cells to undergo programmed cell death. The gremlin protein, which is stained brown in Figure 20.11b, inhibits the function of BMP4, thereby allowing cells to survive. In the developing chicken limb, the *Gremlin* gene is expressed throughout the limb, except in the regions between each digit. Therefore, in these regions, the cells die, and a chicken develops a nonwebbed foot (Figure 20.11c). By comparison, in the duck, *Gremlin* is expressed throughout the entire limb, including the interdigit regions, and the duck develops a webbed foot. Interestingly, researchers have been able to introduce

(a) BMP4 protein levels—similar expression in chicken and duck

Future interdigit regions

(b) Gremlin protein levels—not expressed in interdigit region in chicken

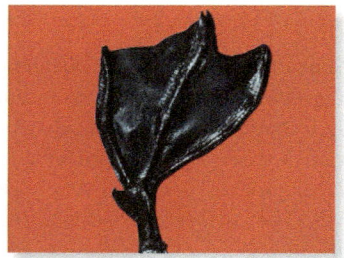

(c) Comparison of a chicken foot and a duck foot

Figure 20.11 **The role of cell-signaling proteins in the pattern formation of birds' feet.** This figure shows how changes in developmental gene expression can affect webbing between the toes. (a) Expression of the *BMP4* gene in the developing limbs. BMP4 protein is stained purple here and is expressed throughout the limb. (b) Expression of the *Gremlin* gene in the developing limbs. Gremlin protein is stained brown here. Note that *Gremlin* is not expressed in the interdigit regions of the chicken but is expressed in these regions of the duck. Gremlin inhibits BMP4, which causes programmed cell death. (c) Because BMP4 is not inhibited in the interdigit regions in the chicken, the cells in this region die, and the foot is not webbed. By comparison, inhibition of BMP4 in the interdigit regions in the duck results in a webbed foot. *(a) (left) (right)* Courtesy Ed Laufer; *(b–c) (left) (right)* Courtesy of Dr. J.M. Hurle. Originally published in *Development*. 1999 Dec. 126(23): 5515–5522

 Concept Check: *What would you expect to happen to the pattern of the feet of ducks if the Gremlin gene was underexpressed?*

gremlin protein into the interdigit regions of developing chicken limbs. This produces a chicken with webbed feet!

How are these observations related to evolution? During the evolution of birds, genetic variation arose such that some individuals expressed the *Gremlin* gene in the regions between each digit, but others did not. This variation determined whether or not a bird's feet were webbed. In terrestrial settings, having nonwebbed feet is an advantage because it enables the individual to hold on to perches, run along the ground, and snatch prey. Therefore, natural selection would favor nonwebbed feet in terrestrial environments. This process explains the occurrence of nonwebbed feet in chickens, hawks, crows, and many other terrestrial birds. In aquatic environments, however, webbed feet are an advantage because they act as paddles for swimming, so genetic variation that produced webbed feet in aquatic birds would have been acted on by natural selection. Over time, this gave rise to the webbed feet now found in a wide variety of aquatic birds, including ducks, geese, and penguins.

EVOLUTIONARY CONNECTIONS

The *Hox* Genes Have Been Important in the Evolution of a Variety of Body Patterns

The study of developmental genes has revealed interesting trends among large groups of species. *Hox* genes are found in nearly all species of animals, indicating that they originated very early in animal evolution. They specify the fate of a particular segment or region of the body. Developmental biologists have hypothesized that variation in the *Hox* genes has spawned the formation of many new body patterns. As shown in **Figure 20.12**, the number and arrangement of *Hox* genes vary considerably among different types of animals. Sponges, the simplest of animals, have at least one gene that is homologous to *Hox* genes. Insects typically have nine or more *Hox* genes. In most cases, multiple *Hox* genes occur in a cluster, lying close to each other along a chromosome. In mammals, *Hox* gene clusters have been duplicated twice during the course of evolution to form four clusters, all slightly different, containing a total of 39 genes.

Researchers propose that increases in the number of *Hox* genes have been instrumental in the evolution of many animal species whose body structures show greater complexity. To understand how, let's first consider *Hox* gene function. All *Hox* genes encode transcription factors that act as master control proteins for directing the formation of particular regions of the body. Each *Hox* gene controls a hierarchy of many regulatory genes that regulate the expression of genes encoding proteins that ultimately affect the morphology of the organism. The evolution of complex body patterns is associated with an increase not only in the number of regulatory genes—as evidenced by the increase in *Hox* gene complexity during evolution—but also in genes that encode proteins that directly affect an organism's form and function.

How would an increase in *Hox* genes enable more complex body patterns to evolve? Part of the answer lies in the spatial expression of the *Hox* genes. Different *Hox* genes are expressed in different regions of the body along the body axis called the anteroposterior axis, which runs from the head toward the abdomen (**Figure 20.13**). Therefore, an increase in the number of *Hox* genes allows each of these master control genes to become more specialized in the region that it controls. In fruit flies, one segment in the middle of the body can be controlled by a particular *Hox* gene and forms wings and legs, whereas a segment in the head region can be controlled by a different *Hox* gene and develops antennae. Therefore, research suggests that one way for new,

Figure 20.12 *Hox* gene number and body complexity in different types of animals. Researchers speculate that the duplication of *Hox* genes and *Hox* gene clusters played a key role in the evolution of more complex body patterns in animals. A correlation is observed between increasing numbers of *Hox* genes and increasing complexity of body structure. The different colors of *Hox* genes correspond to genes that are most closely related in different animal groups.
*Note: Sponges, which are the simplest animals, have no true tissues. They do not have true *Hox* genes, though they have an evolutionarily related gene called an NK-like gene. Some species of sponges have more than one copy of this gene. Ctenophores, the earliest diverging animals, do not possess *Hox* genes.

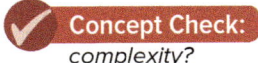 **Concept Check:** *What is the relationship between the total number of Hox genes in an animal species and its morphological complexity?*

more complex body patterns to evolve is by increasing the number of *Hox* genes, thereby making it possible to form many specialized parts of the body that are organized along a body axis.

Three lines of evidence support the idea that increases in *Hox* gene number have been instrumental in the evolution and speciation of animals with more complex body patterns.

- *Hox* genes are known to control the fate of regions along the anteroposterior axis.
- As described in Figure 20.12, a general trend is observed in which animals with a more complex body structure tend to have more *Hox* genes and *Hox* clusters in their genomes than do simpler animals.
- A comparison of *Hox* gene evolution and animal evolution reveals striking parallels. Researchers have analyzed *Hox* gene sequences among modern species and made estimates regarding the timing of past events. Though the date is difficult to precisely pinpoint, the first *Hox* gene arose well over 600 mya. In addition, gene duplications of this primordial gene produced clusters of *Hox* genes in other species. Clusters such as those found in modern insects were likely to be present approximately 600 mya. A duplication of a *Hox* cluster is estimated to have occurred around 520 mya.

Estimates of *Hox* gene origins correlate with major diversification events in the history of animals. As described in Chapter 22, the Cambrian period, stretching from 543 to 490 mya, saw a great diversification of animal species. This diversification occurred after the *Hox* cluster was formed and was possibly undergoing its first duplication to produce two *Hox* clusters. Also, approximately 420 mya, a second duplication produced species with four *Hox* clusters. This event preceded the proliferation of tetrapods—vertebrates with four limbs—that occurred during the Devonian period, approximately 417–354 mya. Modern tetrapods have four *Hox* clusters. This

(a) *Drosophila*

(b) Mouse

Figure 20.13 Expression of *Hox* genes along the anteroposterior axis in animals. The colors shown in this figure correspond to the gene colors in Figure 20.12, from left to right.

second duplication may have been a critical event that led to the evolution of complex terrestrial vertebrates with four limbs, such as amphibians, reptiles, and mammals.

Variation in Growth Rates Can Have a Dramatic Effect on Morphology

As shown in **Figure 20.14**, another way that developmental genes influence morphology is by controlling the relative growth rates of different

Biology Principle

Biologists Use Models to Test Predictions and Convey Ideas

This figure is a model for the differences in head development between a human and a chimpanzee due to heterochrony. The Modeling Challenge in the figure legend asks you to propose a series of models that compare development in different breeds of dogs.

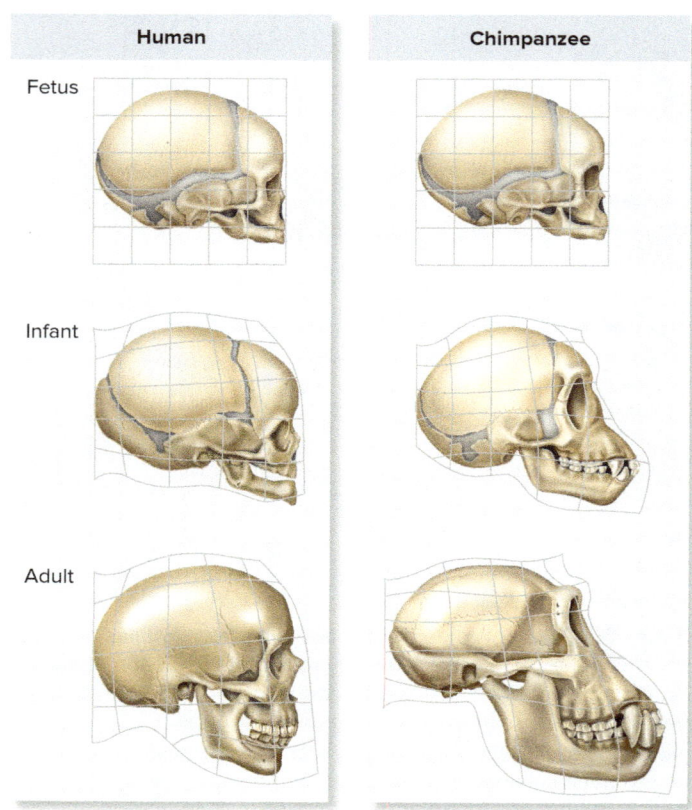

Figure 20.14 Heterochrony. Due to heterochrony, one region of the body may grow faster than another during development in different species. The skulls of adult chimpanzees and humans have different shapes, even though their fetal skull shapes are similar.

Modeling Challenge: As discussed in Chapter 19, selective breeding has created different breeds of dogs that look strikingly different. For example, Figure 19.7 shows a greyhound (b) and a dachshund (c). Some of their differences are due to heterochrony. The drawing in the upper right of this Modeling Challenge is that of a dog at an early embryonic stage. At this stage, a greyhound and a dachshund look fairly similar. Make a series of drawings for a greyhound and a dachshund that begins with a fetal dog and shows a newborn pup and an adult animal. Next to your drawings, explain how heterochrony plays a role.

parts of the body during development. The term **heterochrony** refers to differences among species in the rate or timing of developmental events. The speeding up or slowing down of growth appears to be a common occurrence in evolution and leads to different species with striking morphological differences. With regard to the pace of evolution, such changes may rapidly lead to the formation of new species.

As an example, Figure 20.14 compares the progressive growth of human and chimpanzee skulls. At the fetal stage, the sizes and shapes of the skulls look fairly similar. However, after this stage, the relative growth rates of certain regions become markedly different, thereby affecting the shape and size of the adult skull. In the chimpanzee, the jaw region grows faster, giving the adult chimpanzee a much larger and longer jaw. In the human, the jaw grows more slowly, and the region of the skull that surrounds the brain—the cranium—grows faster. The result is that adult humans have a smaller jaw but a larger cranium.

20.4 Reviewing the Concepts

- Evolutionary developmental biology (evo-devo) compares the development of different species in order to understand ancestral relationships and the mechanisms that bring about evolutionary change. These changes often involve variation in the expression of cell-signaling proteins and transcription factors.
- The spatial expression of genes that affect development can affect phenotypes dramatically, as shown by the expression of the *BMP4* and *Gremlin* genes in birds, resulting in nonwebbed or webbed feet (Figure 20.11).
- An increase in the number of *Hox* genes played an important role in the evolution of more complex body patterns in animals (Figures 20.12, 20.13).
- A difference in the relative growth rates of body parts among different species, called heterochrony, can have a major effect on morphology (Figure 20.14).

20.4 Testing Your Knowledge

1. Which of the following species would you expect to have the largest number of *Hox* genes?
   a. nematode worm
   b. fruit fly
   c. bird
   d. mammal
   e. All of the above have the same number of *Hox* genes.

Assessing Your Knowledge and Skills

Foundational Knowledge

Multiple Choice

1. Macroevolution comprises evolutionary changes that
   a. occur in multicellular organisms.
   b. produce new species and groups of species.
   c. occur over long periods of time.
   d. cause changes in allele frequencies.
   e. occur in large mammals.

2. The ecological species concept classifies a species based on
   a. morphological characteristics.
   b. reproductive isolation.
   c. the niche the organism occupies in the environment.
   d. genetic relationships between an organism and its ancestors.
   e. both a and b.

3. A large number of honeycreeper species on the Hawaiian Islands is an example of
   a. adaptive radiation.
   b. genetic drift.
   c. stabilizing selection.
   d. horizontal gene transfer.
   e. microevolution.

4. The key difference between allopatric and sympatric speciation is
   a. how fast they occur.
   b. whether speciation involves geographic separation.
   c. the role of sexual selection in the speciation process.
   d. the role of reproductive isolation in the speciation process.
   e. All of the above are differences between the types of speciation.

5. Researchers suggest that an increase in the number of *Hox* genes
   a. caused reproductive isolation in all cases.
   b. could explain the evolution of color vision.
   c. facilitated the evolution of more complex body patterns in animals.
   d. decreased the number of body segments in insects.
   e. All of the above are correct.

Short Answer

1. What is the key difference between prezygotic and postzygotic isolating mechanisms? Give an example of each type. Which type is more costly from the perspective of energy?

2. **PRINCIPLES** A principle of biology is that populations of organisms evolve from one generation to the next and are related by an evolutionary history. Describe one example in which genes that control development played an important role in the evolution of different species.

Critical-Thinking Skills

Multiple Choice

1. Which of the following could be used to classify species according to the biological species concept?
   a. the song patterns among birds
   b. the ability of bacteria to use particular nutrients in the environment
   c. the ability of frogs to interbreed with each other
   d. differences in lineages among primates
   e. both a and b

2. Which of the following is an example of a postzygotic isolating mechanism?
   a. incompatible genitalia
   b. different mating seasons
   c. incompatible gametes
   d. a mountain range separating two populations
   e. the failure of fertilized eggs to develop normally

3. Two different species of desert plants produce flowers only after a rainstorm, which is a very infrequent event. After a rainstorm, one of the species flowers a week later, and the other species flowers a month later. Pollination occurs within 3 days of flowering. Even though the two plant species are closely related evolutionarily, they are reproductively isolated from each other. Based on their flowering behavior, these two species exhibit a _____ isolating mechanism that is due to _____.
   a. prezygotic, habitat isolation
   b. prezygotic, temporal isolation
   c. prezygotic, mechanical isolation
   d. prezygotic, gametic isolation
   e. postzygotic, hybrid inviability

4. Let's suppose one species of fish found in freshwater lakes in the midwestern U.S. evolved into two different species because some members tend to eat insect larvae that are near the surface of the water, whereas other members eat algae growing along the shore. This would be an example of _____ speciation due to _____.
   a. allopatric, adaptation to local environments
   b. allopatric, a geographic separation
   c. allopatric, slow geological events
   d. sympatric, adaptation to local environments
   e. sympatric, a geographic separation

5. Evolutionary changes in the rate or timing of developmental events (heterochrony) could explain
   a. why mammals are more complex than fruit flies.
   b. why human skulls and jaws are different sizes than those of chimpanzees.
   c. why some birds have webbed feet and others do not.
   d. how new species arise due to adaptation to local environments.
   e. how polyploidy can promote speciation.

Short Answer

1. Compare and contrast different mechanisms of allopatric and sympatric speciation. How are genetic changes related to these two general forms of speciation?

2. Discuss the type of speciation (allopatric or sympatric) that is most likely to occur under each of the following conditions:
   a. A pregnant female rat is transported by an ocean liner to a new continent.
   b. A meadow containing several species of grasses is exposed to a pesticide that promotes nondisjunction.
   c. In a very large lake containing several species of fishes, the water level gradually falls over the course of several years. Eventually, the large lake becomes subdivided into smaller lakes, some of which are connected by narrow streams.

How Biologists Classify Species and Study Their Evolutionary Relationships

21

Darwin's bark spider, *Caerostris darwini* (left side), which was discovered in 2010. The right side shows an example of the enormous web that a female spider of this species can weave.

Chapter Outline

- 21.1 Taxonomy
- 21.2 Phylogenetic Trees
- 21.3 Cladistics
- 21.4 Molecular Clocks
- 21.5 Horizontal Gene Transfer

Assessing Your Knowledge and Skills

©Photos M. Kuntner

New species are discovered on a regular basis. In 2010, researchers reported the discovery of a new species of spider in Madagascar, which they named Darwin's bark spider, *Caerostris darwini*. As seen in the chapter-opening photo, this species builds the largest orb-style webs that are known. Such webs have been found spanning rivers, streams, and lakes, reaching up to 25 m in length. The silk spun by these spiders is thought to be the toughest biological material ever studied. It is over 10 times stronger than Kevlar—the synthetic fiber used in bullet-proof vests, bicycle tires, and racing sails!

The rules for the classification of newly described species, such as Darwin's bark spider, are governed by the discipline of taxonomy (from the Greek *taxis*, meaning order, and *nomos*, meaning law). **Taxonomy** is the science of describing, naming, and classifying **extant** species, those that still exist today, as well as **extinct** species, those that have died out.

Taxonomy results in the ordered division of species into groups based on similarities and dissimilarities in their characteristics. This task has been ongoing for over 300 years. In the mid- to late 1600s, naturalist John Ray made the first attempt to broadly classify all known forms of life. Ray's ideas were later extended by naturalist Carolus Linnaeus in the mid-1700s, an achievement considered by some to be the official birth of taxonomy.

Systematics is the study of biological diversity and the evolutionary relationships among organisms, both extant and extinct. In the 1950s, German entomologist Willi Hennig proposed that evolutionary relationships should be inferred from features shared by descendants of a common ancestor. Researchers now try to place new species into taxonomic groups based on evolutionary relationships with other species. In addition, previously established taxonomic groups are revised as new data shed light on evolutionary relationships.

In this chapter, we will begin with a discussion of taxonomy and the concept of taxonomic groups. We will then examine how biologists use systematics to determine evolutionary relationships among organisms, looking in particular at how those relationships are portrayed in diagrams called phylogenetic trees. We will then explore how analyses of morphological data and molecular genetic data are used to understand the evolutionary history of life on Earth.

21.1 Taxonomy

Learning Outcomes

1. Identify the three domains of life.
2. Outline the hierarchy of groupings in taxonomy.
3. Explain how species are named using binomial nomenclature.

A hierarchy is a system of organization that classifies entities into successive levels. In biological taxonomy, every species is placed into several nested groups within a hierarchy. For example, a leopard and a fruit fly are both classified into a large group known as animals, though they differ in many traits. By comparison, leopards and lions are placed together into a group with a smaller number of species called felines (more formally named Felidae), which are predatory cats. The felines are a subset of the animal group, which has species that share many similar traits. The species that are placed together into small taxonomic groups are likely to share many of the same characteristics. In this section, we will consider how biologists use a hierarchy to group similar species.

Living Species Are Subdivided into Three Domains of Life

Modern taxonomy places species into progressively smaller hierarchical groups. Each group at any level is called a **taxon** (plural, taxa). In the late 1970s, based on information in the sequences of genes, American biologist Carl Woese proposed the idea of creating a category called a **domain.** Under this system, all forms of life are grouped within three domains: **Bacteria, Archaea,** and **Eukarya** (**Figure 21.1**). The terms Bacteria and Archaea are capitalized when referring to the domains, but they are not capitalized when referring to individual species. A single bacterial cell is called a bacterium, and a single archaeal cell is an archaeon.

The domain Eukarya formerly consisted of four kingdoms called Protista, Fungi, Plantae, and Animalia. Researchers later discovered that protists do not constitute a separate kingdom but, instead, are a very broad collection of species. Taxonomists now subdivide the eukaryotic domain into seven groups called supergroups. In the taxonomy of eukaryotes, a **supergroup** lies between a domain and a kingdom (Figure 21.1). As discussed in Chapter 23, all seven supergroups contain a distinctive group of protists. In addition, the

Figure 21.1 A classification system for living and extinct organisms. All organisms are grouped into three domains: Bacteria, Archaea, and Eukarya. Eukaryotes are divided into seven supergroups. These seven supergroups encompass most eukaryotes. However, several small eukaryotic lineages are classified outside of the seven supergroups.

BioConnections: *Refer back to Figure 4.8. Which of the three domains contains organisms with prokaryotic cells?*

Table 21.1 Distinguishing Cellular and Molecular Features of Domains Bacteria, Archaea, and Eukarya*

Characteristic	Bacteria	Archaea	Eukarya
Chromosomes	Usually circular	Circular	Usually linear
Nucleosome structure	No	No	Yes
Chromosome segregation	Binary fission	Binary fission	Mitosis/meiosis
Introns in genes	Rarely	Rarely	Commonly
Ribosomes	70S	70S	80S
Initiator tRNA	Formyl-methionine	Methionine	Methionine
Operons	Yes	Yes	No
Capping of mRNA	No	No	Yes
RNA polymerases	One	Several	Three
Promoters of structural genes	−35 and −10 sequences	TATA box	TATA box
Cell compartmentalization	No	No	Yes
Membrane lipids	Ester-linked	Ether-linked	Ester-linked

*The descriptions in this table represent the general features of most species in each domain. Some exceptions are observed. For example, certain bacterial species have linear chromosomes, and operons occasionally are found in eukaryotes, such as the nematode worm *Caenorhabditis elegans*.

kingdoms named Fungi and Animalia are within the supergroup Opisthokonta, because they are closely related to other protists in this supergroup. Kingdom Plantae is within the supergroup called Land Plants and Relatives. Green plants are closely related to green algae, which are protists in this supergroup. Table 21.1 compares a variety of molecular and cellular characteristics among the domains Bacteria, Archaea, and Eukarya.

EVOLUTIONARY CONNECTIONS

Every Species Is Placed into a Taxonomic Hierarchy

Why is it useful to categorize species into groups? The three domains of life contain millions of different species. Subdividing them into progressively smaller taxonomic groups makes it easier for biologists to appreciate the relationships among such a large number of species. The order of hierarchy is

domain > supergroup > kingdom > phylum (plural, phyla) > **class > order > family > genus** (plural, genera) > **species**

Each of these taxa contains progressively fewer species that are more similar to each other than they are to the members of the taxa above them in the hierarchy. For example, the taxon Animalia, which is at the kingdom level, has a larger number of fairly diverse species than does the class Mammalia, which contains fewer species that are relatively similar to each other.

To further understand taxonomy, let's consider the classification of a species such as the gray wolf (*Canis lupus*) (**Figure 21.2**). The classification of this species is as follows:

- Domain Eukarya (several million species): The gray wolf is a eukaryote. See Table 21.1 for a description of eukaryotes.
- Supergroup Opisthokonta (over 1 million species): As described in Chapter 23, this supergroup includes the animal and fungal kingdoms and related protists.
- Kingdom Animalia (over 1 million species): Animals are multicellular and heterotrophic, which means they must consume other organisms or parts of other organisms to survive. Their cells lack a cell wall. Except for sponges, animal bodies contain tissues that have specialized cell types. Examples are muscle and nervous tissue.
- Phylum Chordata (~50,000 species): Animals in this phylum all have four common features at some stage of their development. These are a notochord (a cartilaginous rod that runs along the back), a tubular nerve or spinal cord located above the notochord, gill slits or arches, and a postanal tail.
- Class Mammalia (5,513 species): Two distinguishing features are hair, which helps the body maintain a constant body temperature, and mammary glands, which produce milk to nourish the young.
- Order Carnivora (282 species): Carnivores typically have sharp claws, prominent canine teeth, and binocular vision. They usually follow a diet that primarily consists of meat.
- Family Canidae (34 species): All species in the family Canidae are doglike animals. This family is composed of different species of wolves, jackals, foxes, wild dogs, and the coyote.
- Genus *Canis* (7 species): This genus is composed of closely related, doglike species, which include four species of jackals, the coyote, and two types of wolves. The species *Canis lupus* encompasses several subspecies, including the domestic dog (*Canis lupus familiaris*).

Binomial Nomenclature Is Used to Name Species

As originally advocated by Linnaeus, **binomial nomenclature** is the standard format for naming species. The scientific name of every species has two names, its genus name and its unique, specific epithet. An example is the gray wolf, *Canis lupus*. The genus name is always capitalized, but the specific epithet is not. Both names are italicized. After the first mention, the genus name is often abbreviated to a single letter. For example, we write that *Canis lupus* is the gray wolf, and in subsequent sentences, the species is referred to as *C. lupus*.

When naming a new species, genus names are always nouns or treated as nouns, whereas species epithets may be either nouns or adjectives. The names often have a Latin or Greek origin and refer to characteristics of the species or to features of its habitat. For example, the genus name of the gray wolf, *Canis*, is from the Latin *canis*, meaning dog, and the species name, *lupus*, is from the Latin *lupus*,

Biology Principle

Populations of Organisms Evolve from One Generation to the Next and Are Related by an Evolutionary History

A goal of taxonomy is to relate the diversity of species according to their evolutionary relationships.

Taxonomic group	Gray wolf found in	Number of current species
Domain	Eukarya	~4–10 million
Supergroup	Opisthokonta	>1 million
Kingdom	Animalia	>1 million
Phylum	Chordata	~50,000
Class	Mammalia	5,513
Order	Carnivora	282
Family	Canidae	34
Genus	*Canis*	7
Species	*lupus*	1

Figure 21.2 A taxonomic classification of the gray wolf (*Canis lupus*).

 Concept Check: Which group is broader, a phylum or a family?

---

meaning wolf. However, the choice of name for a species can be lighthearted. For example, there is a beetle named *Agra vation*.

21.1 Reviewing the Concepts

- Taxonomy is the field of biology concerned with describing, naming, and classifying extant and extinct species. Systematics is the study of biological diversity and the classification of evolutionary relationships among species, both extant and extinct.
- Taxonomy places all living organisms into progressively smaller hierarchical groups called taxa (singular, taxon). The broadest groups are the three domains, called Bacteria, Archaea, and Eukarya, followed by supergroups, kingdoms, phyla, classes, orders, families, genera, and species (Figures 21.1, 21.2; Table 21.1).
- Binomial nomenclature is the standard format for naming species that provides each species with a genus name and a species epithet.

21.1 Testing Your Knowledge

1. The three domains of life are
   a. Animals, Bacteria, and Archaea.
   b. Eukarya, Prokaryotes, and Viruses.
   c. Eukarya, Bacteria, and Archaea.
   d. Protists, Animals, and Fungi.

2. Which of the following is a correct order of taxa?
   a. supergroup > domain > kingdom > phylum > class > order > family > genus > species
   b. domain > supergroup > kingdom > phylum > class > order > family > genus > species
   c. domain > supergroup > kingdom > class > phylum > order > family > genus > species
   d. supergoup > domain > kingdom > phylum > class > family > order > genus > species

21.2 Phylogenetic Trees

Learning Outcomes

1. Explain the concept of phylogeny and its basis for the construction of phylogenetic trees.
2. Explain the process of cladogenesis, the primary way that new species arise.
3. Describe how morphological and genetic homologies are used to construct phylogenetic trees.

As mentioned, systematics is the study of biological diversity and evolutionary relationships. By studying the similarities and differences among species, biologists can construct a **phylogeny**—the evolutionary history of a species or group of species. To propose a phylogeny, biologists use the tools of systematics. For example, the classification of the gray wolf in Figure 21.2 is based on systematics. Therefore, one use of systematics is to place species into taxa and to understand the evolutionary relationships among different taxa.

In this section, we will consider the features of diagrams, or "trees," that describe the evolutionary relationships among various species, both extant and extinct. As you will learn, such trees are usually based on morphological or genetic data.

A Phylogenetic Tree Depicts Evolutionary Relationships Among Species

A **phylogenetic tree** is a diagram that describes the evolutionary relationships among various species, based on the information available to and gathered by systematists. A phylogenetic tree should be viewed as a hypothesis that is proposed, tested, and later refined as additional data become available. Let's look at what information a phylogenetic tree contains and the form in which it is presented. **Figure 21.3** shows a hypothetical phylogenetic tree of the relationships among various flowering plant species, in which the species are labeled A through K. The vertical axis represents time, and the oldest species is at the bottom.

New species can be formed by **anagenesis,** in which a single species evolves into a different species. However, the primary way that new species arise is by **cladogenesis,** in which a species diverges into two or more species. Most evolutionary biologists focus only on cladogenesis in the construction of their trees. The branch points in a phylogenetic tree, also called **nodes,** indicate times when cladogenesis has occurred. For example, approximately 12 mya,

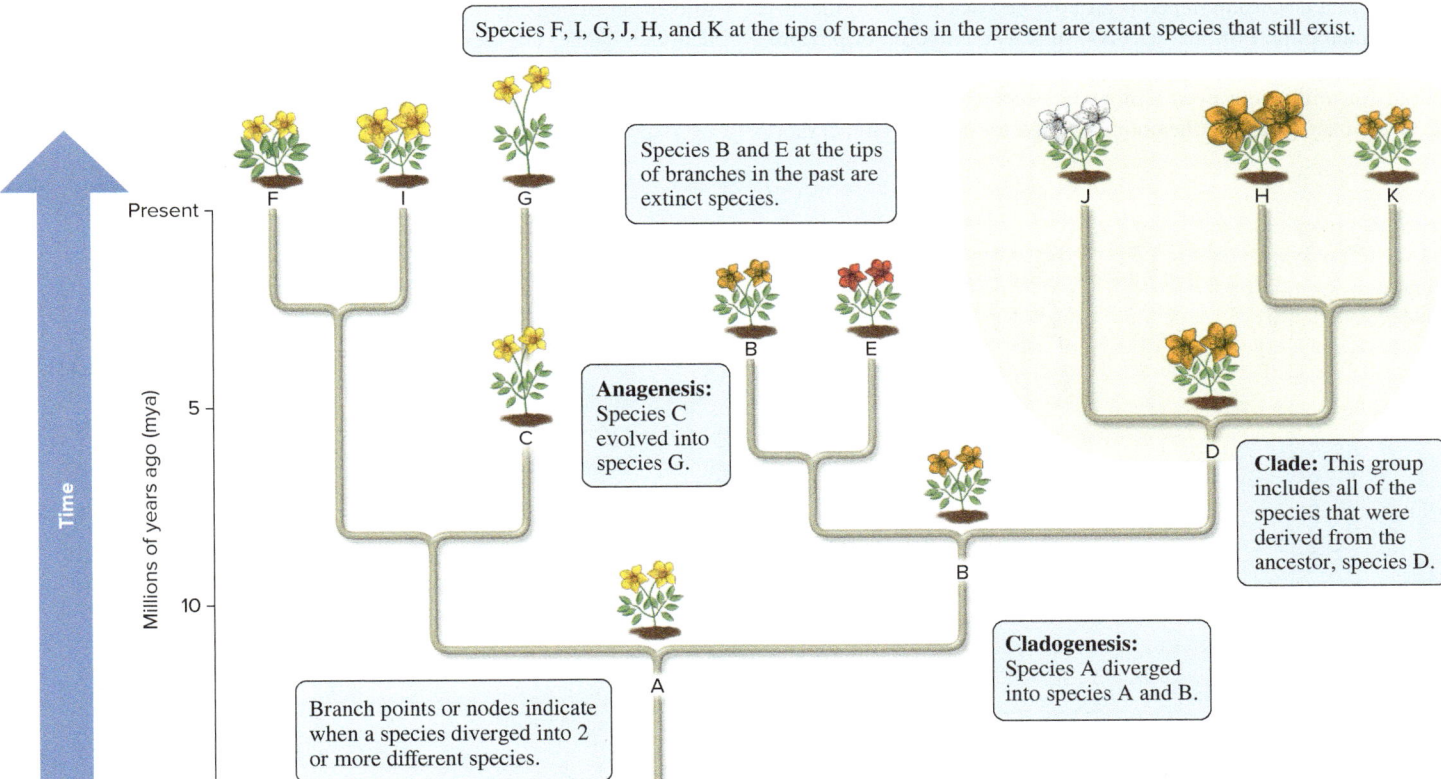

Figure 21.3 How to read a phylogenetic tree. This hypothetical tree shows the proposed relationships among various plant species. Species are placed into clades, groups of organisms containing an ancestral organism and all of its descendants. Note: Anagenesis is a possible way for a new species to arise, but cladogenesis is the primary mechanism.

 Concept Check: *Can two different species have more than one common ancestor?*

species A diverged into species A and species B. Figure 21.3 also shows anagenesis in which species C evolved into species G. The tips of branches may represent species that became extinct in the past, such as species B and E, or living species, such as F, I, G, J, H, and K, which are at the top of the tree. Species A and D are also extinct but gave rise to species that are still in existence.

A **clade** consists of a common ancestral species and all of its descendant species. For example, the group highlighted in light green in the upper right of Figure 21.3 is a clade derived from the common ancestral species labeled D. Likewise, the entire tree forms a clade, with species A as a common ancestor. Therefore, smaller and more recent clades are nested within larger clades that have older common ancestors.

Systematics Constructs Taxonomic Groups Based on Evolutionary Relationships

A key goal of modern systematics is to create taxonomic groups that reflect evolutionary relationships. Systematics attempts to organize species into clades so that each group includes an ancestral species and all of its descendants. A **monophyletic group** is a taxon that is a clade. Ideally, every taxon, whether it is a domain, supergroup, kingdom, phylum, class, order, family, or genus, should be a monophyletic group.

What is the relationship between a phylogenetic tree and taxonomy? The relationship depends on how far back we go to identify a common ancestor. For broader taxa, such as a kingdom or phylum, the common ancestor existed a very long time ago, on the order of hundreds of millions or even billions of years ago. For smaller taxa, such as a family or genus, the common ancestor occurred much more recently, on the order of millions or tens of millions of years ago. This concept is shown in a schematic way in Figure 21.4. This small, hypothetical kingdom is a clade that contains 64 living species. (Actual kingdoms are obviously larger and exceedingly more complex.) The diagram emphasizes the taxa that contain the species designated number 43. The common ancestor that gave rise to this kingdom existed approximately 1 bya. Over time, more recent species arose that subsequently became the common ancestors to the phylum, class, order, family, and genus that contain species number 43.

The Study of Systematics Is Usually Based on Morphological or Genetic Homology

As discussed in Chapter 19, the term **homology** refers to a similarity that occurs due to descent from a common ancestor. Such features are said to be homologous. For example, the arm of a human, the wing of a bat, and the flipper of a whale are homologous structures (refer back to Figure 19.8). Similarly, genes found in different species are homologous if they have been derived from the same ancestral gene (refer back to Figure 19.9).

In systematics, researchers identify homologous features that are shared by some species but not by others, which allows them to group species based on shared similarities. Researchers usually study homology by examining morphological features—those related to the structure of an organism—or they analyze genetic data. In addition, the data they gather are viewed in light of geographic data. Many organisms do not migrate extremely long distances. Species that are closely related evolutionarily are relatively likely to inhabit neighboring or overlapping geographic regions, though many exceptions occur.

Figure 21.4 Schematic relationship between a phylogenetic tree and taxonomy, when taxonomy is correctly based on evolutionary relationships. The shaded areas highlight the kingdom, phylum, class, order, family, and genus for species number 43. All of the taxa are clades. Broader taxa, such as phyla and classes, are derived from more ancient common ancestors. Smaller taxa, such as families and genera, are derived from more recent common ancestors. These smaller taxa are subsets of the broader taxa.

Concept Check: *Which taxon would have a more recent common ancestor, a phylum or an order?*

Analysis of Morphological Features The first studies in systematics focused on morphological features (also called anatomical features) of extinct and living species. Morphological traits continue to be widely used in systematic studies, particularly in those studies pertaining to extinct species and those involving groups that have not been extensively studied at the molecular level. To establish evolutionary relationships based on morphological features, many traits have to be analyzed to identify similarities and differences.

By studying morphological features of extinct species in the fossil record, paleontologists can propose phylogenetic trees that chart the evolutionary lineages of species, including those that still exist. In this approach, the trees are based on morphological features that change over the course of many generations. As an example, **Figure 21.5** depicts a current hypothesis of the evolutionary changes that led to the development of the modern horse. This figure shows representative species from various genera. Many morphological features were used to create this tree. Because hard parts of the body are more commonly preserved in the fossil record, this tree is largely based on the analysis of skeletal changes in foot structure, lengths and shapes of various leg bones, skull shape and size, and jaw and teeth. Over an evolutionary timescale, the accumulation of many genetic changes has had a dramatic effect on the species' characteristics. In

Biology Principle

Structure Determines Function

The changes in structural features during horse evolution are related to changes in their functional needs. During the course of their evolution, horse populations shifted from feeding on leaves in forested regions to feeding on abrasive grasses in more wide-open spaces.

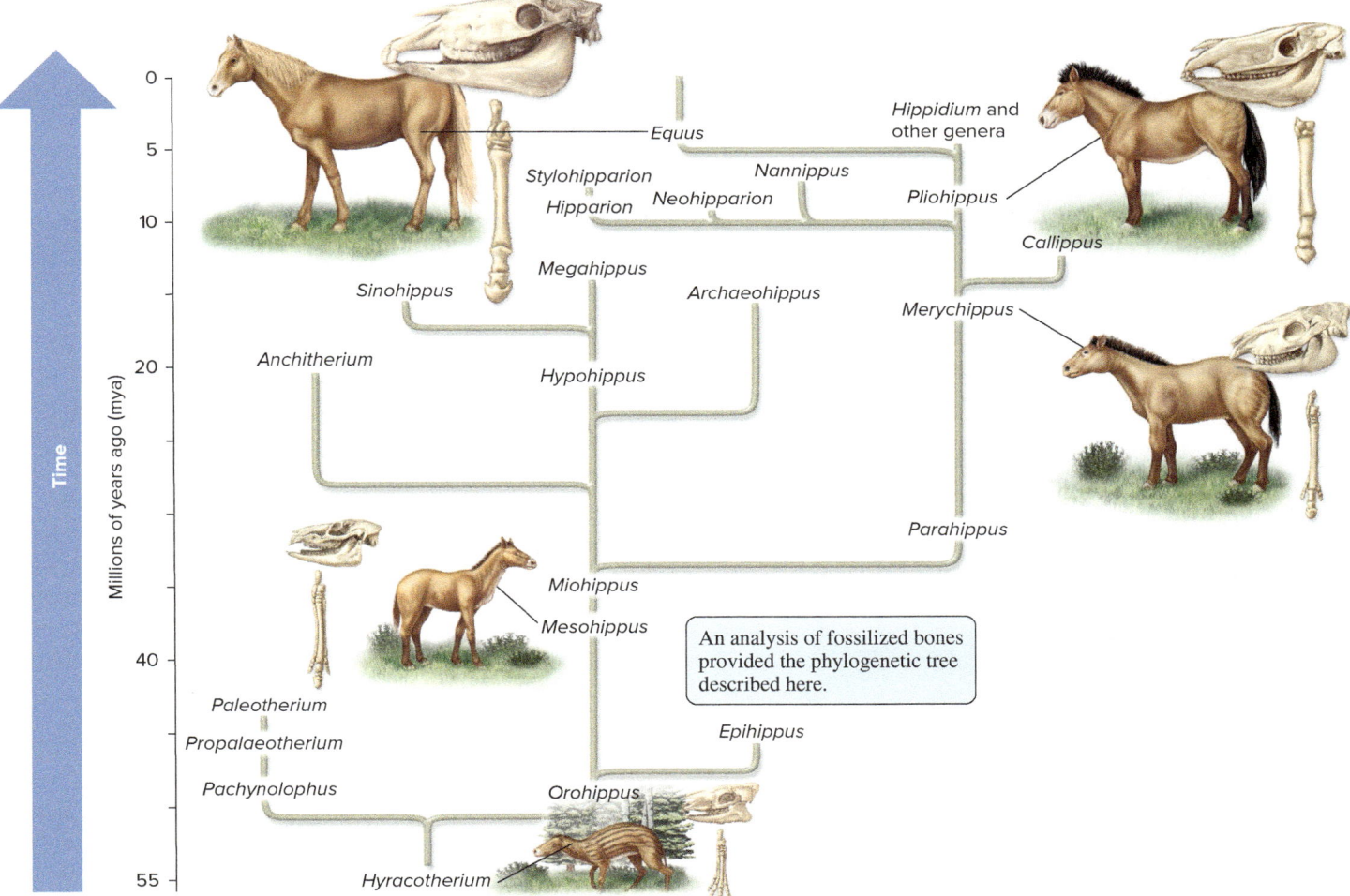

Figure 21.5 Evolution of the horses. An analysis of morphological traits was used to produce this phylogenetic tree, which shows the evolutionary history of the modern horse. As shown next to some of the genera, three important morphological changes that occurred were a trend toward larger size, fewer toes, and teeth suited for grazing.

the genera depicted in this figure, a variety of morphological changes occurred, such as an increase in size, a reduction in the number of toes, and modifications in the jaw and teeth consistent with a dietary shift from tender leaves to fibrous grasses.

How do evolutionary biologists explain these changes in horses' traits? The changes can be attributed to natural selection, which acted on existing variation and resulted in adaptations to changes in climate. Over North America, where much of horse evolution occurred, changes in climate caused large areas of dense forests to be replaced with grasslands. The increase in size and changes in foot structure enabled horses to escape predators more easily and travel greater distances in search of food. The changes seen in horses' teeth are consistent with a shift from eating the tender leaves of bushes and trees to eating grasses and other, more abrasive types of vegetation that require more chewing.

A potential drawback of using morphological features for constructing phylogenetic trees is that similar morphological features among different species may be due to convergent evolution. As described in Chapter 19, convergent evolution results in analogous structures—characteristics that arose independently in different lineages because the species evolved in similar environments. For example, the giant anteater and the echidna have long snouts and tongues that enable these animals to feed on ants (refer back to Figure 19.6). These traits are not derived from a common ancestor. Rather, they arose independently during evolution due to adaptation to similar environments. The use of analogous structures in systematics can cause errors if a researcher assumes that a particular trait arose only once and that all species having the trait are derived from a common ancestor.

Molecular Systematics The field of **molecular systematics** involves the analysis of genetic data, such as DNA sequences or amino acid sequences, to identify and study genetic homologies and propose phylogenetic trees. In 1963, Austrian biologist Emile Zuckerkandl and American chemist Linus Pauling were the first to suggest that molecular data could be used to establish evolutionary relationships. How can a comparison of genetic sequences help to establish evolutionary relationships? As discussed later in this chapter, DNA sequences change over the course of many generations due to the accumulation of mutations. Therefore, when homologous sequences in different species are compared, DNA sequences from closely related species are more similar to each other than they are to sequences from distantly related species.

21.2 Reviewing the Concepts

- The evolutionary history of a species or group of species is its phylogeny. A phylogenetic tree is a diagram that describes the phylogeny of particular species and should be viewed as a hypothesis that is proposed, tested, and refined as more data become available (Figure 21.3).
- A key goal of systematics is to construct taxa and phylogenetic trees based on evolutionary relationships. Smaller taxa, such as families and genera, are derived from more recent common ancestors than are broader taxa, such as kingdoms and phyla (Figure 21.4).
- Both morphological and genetic data are used to propose phylogenetic trees (Figure 21.5).

21.2 Testing Your Knowledge

1. A phylogenetic tree is
    a. a hypothesis about the evolutionary relationships among different species.
    b. a diagram that depicts the evolutionary relationships among different species.
    c. a diagram that should be organized according to the grouping of species into clades.
    d. all of the above.

2. Earlier in this chapter, we considered the taxonomy of the gray wolf. Of the following taxa, which one would have a common ancestor that existed the longest time ago?
    a. kingdom Animalia
    b. phylum Chordata
    c. order Carnivora
    d. genus *Canis*

3. To construct a phylogenetic tree, a researcher compares the DNA sequence of a gene that encodes the cytoskeletal protein called tubulin in many different species. This is an example of using
    a. morphological homology.
    b. genetic homology.
    c. the fossil record.
    d. both a and b.

21.3 Cladistics

Learning Outcomes

1. Distinguish between shared ancestral characters and shared derived characters.
2. Explain the use of cladistics to construct a phylogenetic tree.
3. **SCISKILLS** ▶ Describe how the principle of parsimony is used to choose among phylogenetic trees.

Cladistics is the classification of species based on evolutionary relationships. A cladistic approach constructs phylogenetic trees by considering the possible pathways of evolutionary changes that involve characteristics that are shared or not shared among various species. In this section, we will consider how the cladistic approach is used to produce phylogenetic trees known as **cladograms.**

Species Differ with Regard to Ancestral and Derived Characters

A cladistic approach compares homologous features, also called **characters,** which may exist in two or more **character states.** For example, among different species, a front limb, which is a character, may exist in different character states such as a wing, an arm, or a flipper. The various character states are either shared or not shared by different species.

To understand the cladistic approach, let's take a look at a simplified phylogeny (**Figure 21.6**). We can place the species that currently

HOW BIOLOGISTS CLASSIFY SPECIES AND STUDY THEIR EVOLUTIONARY RELATIONSHIPS   441

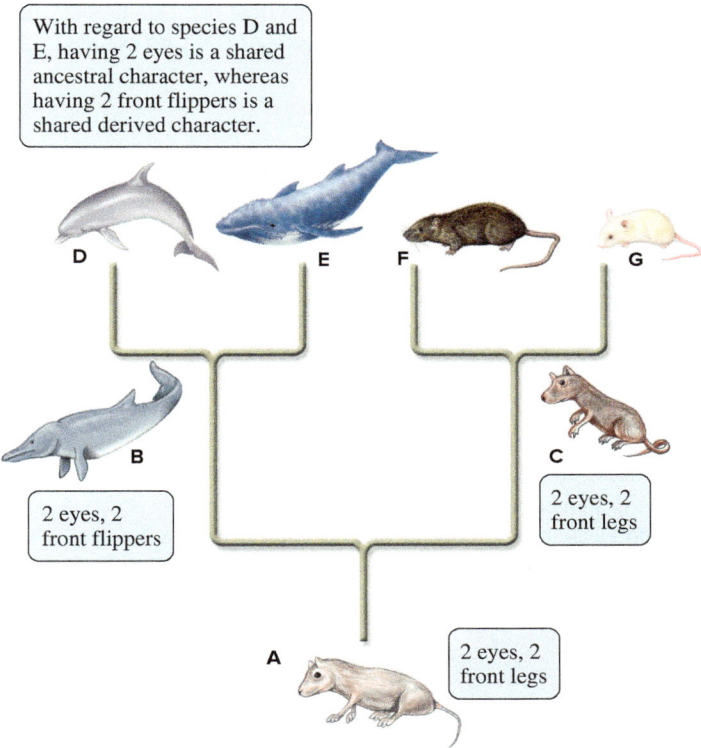

Figure 21.6 A comparison of shared ancestral characters and shared derived characters.

complex than the front limbs of ancestral species A (see Figure 21.6), which were limbs with individual toes. Derived characters can be similar in complexity, less complex, or more complex than ancestral characters.

A Cladistic Approach Produces a Cladogram Based on Shared Derived Characters

To illustrate how shared derived characters are used to propose a cladogram, **Figure 21.7a** compares several characters among five species of animals. The cladogram shown in **Figure 21.7b** is consistent with the distribution of shared derived characters among these species. A branch point is where two species differ in a character. The oldest common ancestor, which is now extinct, had a notochord and was an ancestor to all five species. Vertebrae are a shared derived character of the

exist into two groups: D and E, and F and G. The most recent common ancestor to D and E is B, whereas species C is the most recent common ancestor to F and G. With these ideas in mind, let's focus on the front limbs (flippers versus legs) and eyes.

A character that is shared by two or more different taxa and inherited from ancestors older than their last common ancestor is called a **shared ancestral character.** Such characters are viewed as being older—ones that occurred earlier in evolution. With regard to species D, E, F, and G, having two eyes is a shared ancestral character. It originated prior to species B and C.

By comparison, a **shared derived character** is a character that is shared by two or more species or taxa and originated in their most recent common ancestor. With regard to species D and E, having two front flippers is a shared derived character that originated in species B, their most recent common ancestor (see Figure 21.6). Shared derived characters are more recent traits on an evolutionary timescale than are shared ancestral characters. For example, among mammals, only some species have flippers, such as whales and dolphins. In this case, flippers were derived from the two front limbs of an ancestral species. The word derived indicates that evolution involves the modification of traits in pre-existing species. In other words, populations of organisms with new traits are derived from changes in pre-existing populations. The basis of the cladistic approach is to analyze many shared derived characters among groups of species to deduce the pathway that gave rise to those species.

The terms ancestral and derived do not indicate the complexity of a character. For example, the flippers of a dolphin do not appear more

	Lancelet	Lamprey	Salmon	Lizard	Rabbit
Notochord	Yes	Yes	Yes	Yes	Yes
Vertebrae	No	Yes	Yes	Yes	Yes
Hinged jaw	No	No	Yes	Yes	Yes
Tetrapod	No	No	No	Yes	Yes
Mammary glands	No	No	No	No	Yes

(a) Characteristics among species

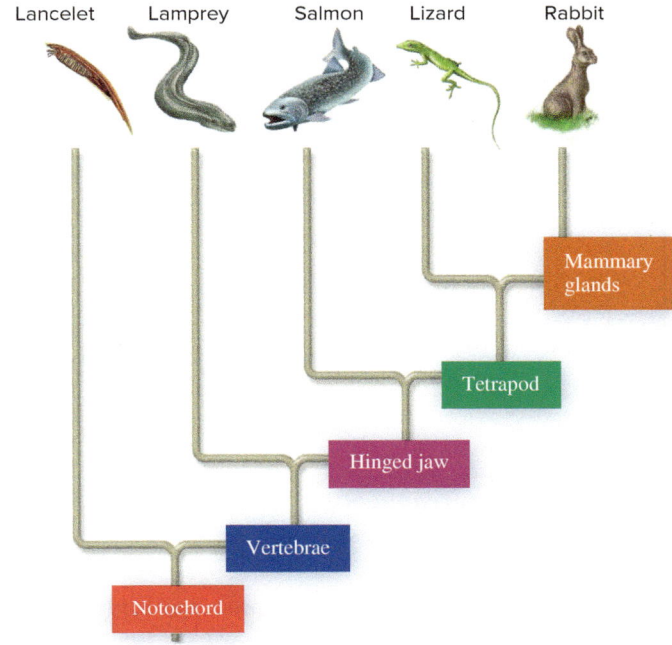

(b) Cladogram based on morphological traits

Figure 21.7 Using shared ancestral characters and shared derived characters to propose a phylogenetic tree. **(a)** A comparison of characteristics among these species. **(b)** This phylogenetic tree illustrates both shared ancestral and shared derived characters in a cladogram of five animal species.

Concept Check: *What shared derived character is common to the salmon, lizard, and rabbit, but not the lamprey?*

lamprey, salmon, lizard, and rabbit, but not the lancelet, which is an invertebrate. By comparison, a hinged jaw is a shared derived character of the salmon, lizard, and rabbit, but not of the lamprey or lancelet.

In a cladogram, an **ingroup** is the group whose evolutionary relationships we wish to understand. By comparison, an **outgroup** is a species or group of species that is assumed to have diverged before the species in the ingroup. An outgroup lacks one or more shared derived characters that are found in the ingroup. A designated outgroup can be closely related or more distantly related to the ingroup. In the tree shown in Figure 21.7, if the salmon, lizard, and rabbit are an ingroup, the lamprey is an outgroup. The lamprey has a notochord and vertebrae but lacks a character shared by the ingroup, namely a hinged jaw. Thus, for the ingroup, the notochord and vertebrae are shared ancestral characters, whereas the hinged jaw is a shared derived character not found in the outgroup.

Likewise, the concept of shared derived characters can apply to molecular data, such as a gene sequence. Let's consider an example to illustrate this idea. Our example involves molecular data obtained from seven different hypothetical plant species called A–G. In these species, a homologous region of DNA was sequenced as shown here:

```
         1 2 3 4 5 6 7 8 9 10

A:       GATAGTACCC
B:       GATAGTTCCC
C:       GATAGTTCCG
D:       GGTATTACCC
E:       GGTATAACCC
F:       GGTAGTACCA
G:       GGTAGTACCC
```

The cladogram of **Figure 21.8** is a hypothesis of how these DNA sequences arose. In this case, a mutation that changes the sequence of nucleotides is comparable to a modification of a character. For example, let's designate plant species D as an outgroup and species A, B, C, F, and G as the ingroup. In this case, a G (guanine) at the fifth

Biology Principle

Biologists Use Models to Test Predictions and Convey Ideas

A cladogram is a model that depicts evolutionary relationships. In this case, the cladogram is based on DNA sequence data. The Modeling Challenge in the figure legend asks you to propose a simpler cladogram based on five short DNA sequences.

Figure 21.8 **The use of shared derived characters applied to molecular data.** This phylogenetic tree illustrates a cladogram involving homologous gene sequences found in seven hypothetical plant species. Mutations that alter a ancestral DNA sequence are shared among certain species but not others. Note: A, T, G, and C refer to nucleotide bases, and the numbers refer to the position of the base in the nucleotide sequences. For example, A6 refers to an adenine at the sixth position.

Modeling Challenge: Below are short sequences from the same gene in five different species. The first sequence is the proposed ancestral sequence, and the other four are derived from this sequence. Construct a cladogram based on these sequences.

Ancestral sequence: ACCGTAC; species A: ATCGAAG; species B: ACCGAAC; species C: ATCGATG; and species D: ATCGAAC.

position is a shared derived character. The genetic sequence carrying this G is derived from an older ancestral sequence.

In this textbook, most phylogenetic trees are rooted, which means that a single node at the bottom of the tree represents a common ancestor of all species or groups of species in the tree. A method for rooting trees is the use of a noncontroversial outgroup. Such an outgroup typically shares enough morphological traits and/or DNA sequence similarities with the members of the ingroup to allow a comparison between the ingroup and outgroup. Even so, the outgroup must be noncontroversial in that it has enough distinctive differences with the ingroup to be considered a clear outgroup. For example, if the ingroup is a group of mammalian species, an outgroup might be a reptile species.

Biology Principle | Biology is a quantitative science

Quantitative Analysis: The Principle of Parsimony Is Used to Choose from Among Possible Cladograms

Systematists are confronted with the task of choosing the phylogenetic tree that is most likely to be correct, among many possible trees. Although different methods may be used for choosing among possible cladograms, one approach is to assume that the best hypothesis is the one that requires the fewest evolutionary changes to explain the differences among observed characters. This concept, called the **principle of parsimony,** states that the preferred hypothesis is the one that is the simplest for all the characters and their states. For example, if two species possess a tail, we initially assume that a tail arose once during evolution and that both species have descended from a common ancestor with a tail. Such a hypothesis is simpler, and more likely to be correct, than assuming that tails arose twice during evolution and that the tails in the two species are not due to descent from a common ancestor.

The principle of parsimony can also be applied to genetic data, in which case the most likely hypothesis is the one requiring the fewest base changes. Let's consider a hypothetical example involving a DNA sequence from four taxa (A–D), where A is presumed to be the outgroup.

```
            1 2 3 4 5

    A:      GTACA (outgroup)
    B:      GACAG
    C:      GTCAA
    D:      GACCG
```

Given that B, C, and D are the ingroup, three hypotheses for phylogenetic trees are shown in **Figure 21.9**, although more are possible. Tree 1 requires seven mutations, and tree 2 requires six, whereas tree 3 requires only five. Therefore, tree 3 requires the smallest number of mutations and is considered the most parsimonious. Based on the principle of parsimony, the tree with the fewest base changes is the hypothesis that is the most likely to accurately reflect the evolutionary history of the ingroup in question. In practice, when researchers have multiple sequences that are longer than the ones shown here, computer programs are used to find the most parsimonious tree.

Crunching the Numbers: According to the principle of parsimony, which of the phylogenetic trees in Figure 21.9 would you consider to be the least likely? Explain why.

Figure 21.9 Using the principle of parsimony and molecular genetic data to choose a phylogenetic tree. Shown are three possible phylogenetic trees for the evolution of a short DNA sequence (although many more are possible). Changes in nucleotide sequence are indicated for each tree. For example, T2 → A means that the second base, a T, was changed to an A. According to the principle of parsimony, tree number 3 is the most likely choice because it requires only five mutations.

21.3 Reviewing the Concepts

- In the cladistic approach to creating a phylogenetic tree, also called a cladogram, species are grouped together according to shared derived characters, characters shared by two or more species or taxa and originating in their most recent common ancestor (Figure 21.6).
- An ingroup is the group whose evolutionary relationships are of interest, whereas an outgroup is a species or group of species that lacks one or more shared derived characters found in the ingroup. A comparison of the ingroup and outgroup is used to determine which character states are ancestral or derived (Figures 21.7, 21.8).
- The cladistic approach produces many possible cladograms. The most likely phylogenetic tree may be chosen by applying the principle of parsimony (Figure 21.9).

21.3 Testing Your Knowledge

1. In a cladogram, the branch points, or nodes, are determined by differences in
   a. species complexity.
   b. shared derived characters.
   c. genetic homology but not molecular homology.
   d. molecular homology but not genetic homology.

2. With regard to choosing the most likely cladogram, explain how the principle of parsimony is used.
   a. The tree that has the minimum number of branch points is chosen.
   b. The tree that has the maximum number of branch points is chosen.
   c. The tree that has the fewest evolutionary changes is chosen.
   d. The tree that has the most evolutionary changes is chosen.

21.4 Molecular Clocks

Learning Outcomes

1. Explain how molecular clocks are used in the dating of evolutionary events.
2. **SCISKILLS** ▶ Compare and contrast the use of different genes to produce phylogenetic trees.

As we have seen, a phylogenetic tree describes the evolutionary relationships among various species. Researchers are interested not only in the most likely pathway of evolution (the branches of the trees) but also in the timing of evolutionary change (the lengths of the branches). How can researchers determine when different species diverged from each other in the past? As shown in Figure 21.5, the fossil record can sometimes help researchers apply a timescale to a phylogeny.

Another way to infer the timing of past events is by analyzing genetic sequences. The **neutral theory of evolution** proposes that most genetic variation that exists in populations is due to the accumulation of neutral mutations—changes in genes and proteins that are not acted upon by natural selection. The reasoning behind this concept is that favorable mutations are likely to be very rare, and detrimental mutations are likely to be eliminated from a population by natural selection. A large body of evidence supports the idea that much of the genetic variation observed in living species is due to the accumulation of neutral mutations. From an evolutionary point of view, if neutral mutations occur at a relatively constant rate, they can serve as a **molecular clock** to measure evolutionary time. In this section, we will consider the concept of a molecular clock and its application in phylogenetic trees.

The Timing of Evolutionary Change May Be Inferred from Molecular Clock Data

Figure 21.10 illustrates the concept of a molecular clock. The graph's vertical axis is the number of base-pair differences in a homologous gene between different pairs of species. The horizontal axis plots the amount of time that has elapsed since each pair of species shared a common ancestor.

As an example, let's suppose a researcher compares a gene sequence that is 500 bp long. Between species A and species B, this sequence might differ at 10 places and be identical at 490 places. By comparison, the 500-bp sequence might differ at 20 places between species A and species C and be the same at 480 places. Such a result is consistent with the idea that species A and species B share a more

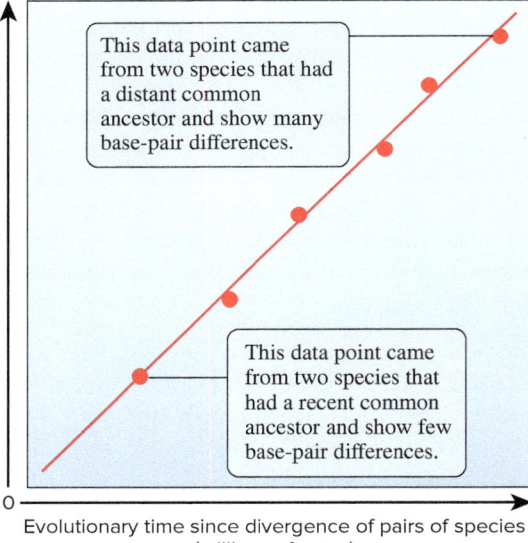

Figure 21.10 A molecular clock. According to the concept of a molecular clock, neutral mutations accumulate at a relatively constant rate over evolutionary time. In a comparison of the same homologous gene between pairs of different species, those species that diverged more recently tend to have fewer differences than do those whose common ancestor occurred in the distant past.

BioConnections: *Look back at Table 10.1, which shows the genetic code. Propose a mutation that would change the sequence of a codon and be neutral.*

recent common ancestor than do species A and species C. The explanation for this phenomenon is that the gene sequences of various species accumulate independent mutations after they have diverged from each other. A longer period of time since their divergence allows for a greater accumulation of mutations, which makes their sequences of base pairs more different.

Figure 21.10 suggests a linear relationship between the number of base-pair changes and the time of divergence. For example, a linear relationship predicts that a pair of species that has, say, 20 differences in a given gene sequence would have a common ancestor that is roughly twice as old as that of a pair showing 10 differences. Although actual data sometimes show a relatively linear relationship over a defined time period, evolutionary biologists do not think that molecular clocks are perfectly linear over very long periods of time. Several factors can contribute to nonlinearity of molecular clocks. These include differences in the generation times of the species being analyzed and variation in the mutation rates of genes between different species.

To obtain reliable data, researchers must calibrate their molecular clocks. How much time does it take to accumulate a certain percentage of base-pair changes? To perform such a calibration, researchers must have information regarding the date when two species diverged from a common ancestor. Such information could come from the fossil record, for instance. The genetic differences between those species are then divided by the amount of time since their last common ancestor to calculate a rate of evolutionary change.

As an example of clock calibration, let's consider primates. The fossil evidence suggests that humans and chimpanzees diverged from a common ancestor approximately 6 mya. The percentage of base-pair differences between the mitochondrial DNA of humans and chimpanzees is 12%. From these data, the molecular clock for changes in mitochondrial DNA sequences of primates is calibrated at roughly 2% base-pair changes per million years.

Different Genes Are Analyzed to Study Phylogeny and Evaluate the Timing of Evolutionary Change

For evolutionary comparisons, the DNA sequences of many genes have been obtained from a wide range of sources. Many different genes have been used to propose phylogenetic trees and evaluate the timing of past events. For example, the gene that encodes an RNA found in the small ribosomal subunit (SSU rRNA) has been commonly used in evolutionary studies. As noted in Chapter 10, the gene for SSU rRNA is found in the genomes of all living organisms. Therefore, its function must have been established at an early stage in the evolution of life on this planet, and its sequence has changed fairly slowly. Furthermore, SSU rRNA is a rather large molecule, so it contains a large amount of sequence information. This gene has been sequenced for thousands of different species (refer back to Figure 10.13). Slowly changing genes such as the gene that encodes SSU rRNA are useful for evaluating distant evolutionary relationships, such as comparing higher taxa. For example, SSU rRNA data can be used to place eukaryotic species into their proper phyla or orders.

Other genes have changed more rapidly during evolution because of a greater tolerance of neutral mutations. For example, the mitochondrial genome and DNA sequences within introns can more easily incur neutral mutations (compared to the coding sequences within exons), and so their sequences change more frequently during evolution. Rapidly changing DNA sequences have been used to study recent evolutionary relationships, particularly among eukaryotic species such as large animals that have longer generation times and tend to evolve more slowly. In these cases, slowly evolving genes may not be very useful for establishing evolutionary relationships because two closely related species are likely to have identical or nearly identical DNA sequences for such genes.

Figure 21.11 shows a simplified phylogeny of closely related species of primates. This tree was proposed by comparing DNA sequence changes in the gene for cytochrome oxidase subunit II, one of several subunits of cytochrome oxidase, a protein in the mitochondrial inner membrane that is involved in cellular respiration. This gene tends to change fairly rapidly on an evolutionary timescale. The vertical scale of Figure 21.11 represents time, and the branch points that are labeled with letters represent common ancestors. Let's take a look at three branch points (labeled A, D, and E) and relate them to the accumulation of neutral mutations.

- *Ancestor A:* This ancestor diverged into two species that ultimately gave rise to siamangs and the other five species. Since this divergence, there has been a long time (approximately 23 million years) for the siamang genome to accumulate a relatively high number of random neutral changes that would be different from the random changes that have occurred in the genomes of the other five species (see the yellow bar in Figure 21.11). Therefore, the gene in the siamangs is fairly different from the genes in the other five species.

- *Ancestor D:* This ancestor diverged into two species that eventually gave rise to humans and chimpanzees. This divergence occurred a moderate time ago, approximately 6 mya, as illustrated by the red bar. The differences in gene sequences between humans and chimpanzees are relatively moderate.

- *Ancestor E:* This ancestor diverged into two species of chimpanzees. Since the divergence of species E into two species, approximately 3 mya, the time for the molecular clock to "tick" (that is, accumulate random mutations) is relatively short, as depicted by the green bar in Figure 21.11. Therefore, the two existing species of chimpanzees have fewer differences in their gene sequences than are found between other primates.

21.4 Reviewing the Concepts

- The neutral theory of evolution proposes that most genetic variation in a population is due to the accumulation of neutral mutations. Assuming that neutral mutations occur at a relatively constant rate, genetic data can serve as a molecular clock to measure evolutionary time (Figure 21.10).
- Slowly changing genes are useful for analyzing distant evolutionary relationships, whereas rapidly changing genes are used to analyze more recent evolutionary relationships, particularly among eukaryotes that have long generation times and evolve more slowly (Figure 21.11).

Figure 21.11 **The use of DNA sequence changes to study primate evolution.** This phylogenetic tree, which shows relationships among closely related species of primates, is based on a comparison of mitochondrial gene sequences encoding the protein cytochrome oxidase subunit II.

Concept Check: *Which pair of species would be expected to have fewer genetic differences: orangutans and gorillas or gorillas and humans?*

21.4 Testing Your Knowledge

1. When the same homologous gene is compared between closely related species (for example, a lion and a tiger) versus distantly related species (for example, a lion and a turtle),
   a. the genes between *closely* related species have *fewer* differences because their common ancestor occurred more recently.
   b. the genes between *distantly* related species have *fewer* differences because their common ancestor occurred more recently.
   c. the genes between *closely* related species have *more* differences because their common ancestor occurred more recently.
   d. the genes between *distantly* related species have *more* differences because their common ancestor occurred more recently.

2. An analysis of the number of neutral genetic changes among different species can be used to estimate the dates of past evolutionary events, such as the divergence of one species into two species. For this type of estimate to be valid,
   a. researchers must have a way to calibrate their clock, such as using information from the fossil record.
   b. the rate of neutral mutations must be relatively constant over a given time period.
   c. the gene sequences that are analyzed must be relatively short.
   d. Both a and b must be true.

21.5 Horizontal Gene Transfer

Learning Outcome

1. Explain how horizontal gene transfer affects evolution and the relationships among different taxa.

Thus far, we have considered various ways to construct phylogenetic trees, which describe the relationships between ancestors and their descendants. The type of evolution depicted in the phylogenetic trees in previous sections, which involves changes in groups of species due to descent from a common ancestor, is sometimes called vertical evolution. Since the time of Darwin, vertical evolution has been the traditional way biologists have viewed the evolutionary process. However, over the past couple of decades, researchers have come to realize that evolution is not so simple. In addition to vertical evolution, horizontal gene transfer has also played a significant role in the phylogeny of living species.

Horizontal gene transfer (also called lateral gene transfer) is any process in which an organism incorporates genetic material from another organism without being the offspring of that organism. This phenomenon has reshaped the way biologists view the evolution of species. Horizontal gene transfer has played a major role in the evolution of many species. As discussed in Chapter 17 (refer back to Table 17.3), bacteria can transfer genes via conjugation, transformation, and transduction. Bacterial gene transfer can occur between strains of the same species or, occasionally, between cells of different bacterial species. The transferred genes may encode proteins that provide a survival advantage, such as resistance to antibiotics or the ability to metabolize an organic molecule in the environment. Horizontal gene transfer is also fairly common among certain unicellular eukaryotes. However, its relative frequency and importance in the evolution of multicellular eukaryotes remain difficult to evaluate.

Scientists have debated the role of horizontal gene transfer in the earliest stages of evolution, prior to the divergence of the bacterial and archaeal domains. The traditional viewpoint was that the three domains of life—Bacteria, Archaea, and Eukarya—arose from a single type of prokaryotic (or pre-prokaryotic) cell called the universal ancestor. However, genomic research has suggested that horizontal gene transfer may have been particularly common during the early stages of evolution on Earth, when all species were unicellular. Horizontal gene transfer may have been so prevalent that the universal ancestor may have actually been an ancestral community of cell lineages that evolved as a whole. If that was the case, the tree of life cannot be traced back to a single prokaryotic ancestor.

Figure 21.12 illustrates a schematic scenario for the evolution of life on Earth that includes the roles of both vertical evolution and

Figure 21.12 A web of life. This phylogenetic tree shows not only the vertical evolution of life on Earth but also the contribution of horizontal gene transfer. In this scenario, horizontal gene transfer was prevalent during the early stages of evolution, when all organisms were unicellular, and continues to be a prominent factor in the speciation of bacteria and archaea. Note: This tree is schematic. Also, while the introduction of chloroplasts into the eukaryotic domain is shown as a single event, such events have occurred multiple times and by different mechanisms.

 Concept Check: *How does the phenomenon of horizontal gene transfer muddle the concept of monophyletic groups?*

horizontal gene transfer. This has been described as a "web of life" rather than a "tree of life." In this scenario, instead of a universal ancestor, a community of primitive cells frequently transferred genetic material in a horizontal fashion. Horizontal gene transfer was also prevalent during the early evolution of bacteria and archaea, and when eukaryotes first emerged as unicellular species. In living bacteria and archaea, it remains a prominent way to foster evolutionary change. By comparison, the region of the diagram that contains eukaryotic species has a more treelike structure. Researchers have speculated that multicellularity and sexual reproduction have presented barriers to horizontal gene transfer in most eukaryotes. For a gene to be transmitted to eukaryotic offspring, it would have to be transferred into a eukaryotic cell that is a gamete or a cell that gives rise to gametes. Horizontal gene transfer has become less common in eukaryotes, though it does occur occasionally.

21.5 Reviewing the Concepts

- Horizontal gene transfer is the phenomenon in which an organism incorporates genetic material from another organism without being the offspring of that organism. Due to the prevalence of horizontal gene transfer, the tree of life may more accurately be described as a web of life (Figure 21.12).

21.5 Testing Your Knowledge

1. Which of the following is an example of horizontal gene transfer?
   a. a mating between two different species, such as a horse and a donkey, to produce a mule
   b. the passage of genes from a mother cell to two daughter cells during cell division
   c. the transfer of an antibiotic-resistance gene from one bacterial species to a different species
   d. Both a and c are examples of horizontal gene transfer.

Assessing Your Knowledge and Skills

Foundational Knowledge

Multiple Choice

1. Which of the following is the correct order of taxa used to classify organisms?
   a. kingdom, supergroup, domain, phylum, class, order, family, genus, species
   b. domain, supergroup, kingdom, class, phylum, order, family, genus, species
   c. supergroup, domain, kingdom, phylum, class, family, order, genus, species
   d. domain, supergroup, kingdom, phylum, class, order, family, genus, species
   e. kingdom, supergroup, domain, phylum, order, class, family, species, genus

2. Which of the following characteristics is *not* shared by bacteria, archaea, and eukaryotes?
   a. DNA is the genetic material.
   b. Messenger RNA encodes the information to produce proteins.
   c. All cells are surrounded by a plasma membrane.
   d. The cytoplasm is compartmentalized into organelles.
   e. Both a and d are not shared by bacteria, archaea, and eukaryotes.

3. Which of the following occur(s) at branch points, or nodes, in a phylogenetic tree?
   a. anagenesis
   b. cladogenesis
   c. horizontal gene transfer
   d. both a and b
   e. both b and c

4. A taxon composed of all species derived from a common ancestor is referred to as
   a. a phylum.
   b. a monophyletic group or clade.
   c. a genus.
   d. an outgroup.
   e. any of the above.

5. A goal of modern taxonomy is to
   a. classify all organisms based on morphological similarities.
   b. classify all organisms into monophyletic groups.
   c. classify all organisms based solely on genetic similarities.
   d. determine the evolutionary relationships only between similar species.
   e. None of the above is a goal of modern taxonomy.

Short Answer

1. What is a molecular clock? How is it used in depicting phylogenetic trees?

2. **PRINCIPLES** A principle of biology is that populations of organisms evolve from one generation to the next and are related by an evolutionary history. What are some advantages and potential pitfalls of using changes in morphological features to construct phylogenetic trees?

Critical-Thinking Skills

Multiple Choice

Use the phylogenetic tree shown below to answer questions 1, 2, and 3. Modern species are labeled with letters, and nodes (branch points) are labeled with numbers.

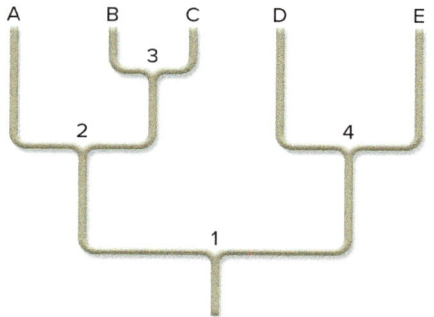

1. Which species are the most closely related?
   a. A, B, and C
   b. A and B
   c. B and C
   d. C and D
   e. D and E

2. A common ancestor to species A, B, and C is located at which node(s)?
   a. 1
   b. 2
   c. 3
   d. 4
   e. both 1 and 2

3. Assuming that this phylogenetic tree was based on an analysis of homology, such as the sequence of homologous genes among these species, which of the following groups constitute a clade?
   a. B and C
   b. D and E
   c. A, B, and C
   d. B, C, D, and E
   e. The species in a, b, and c constitute a clade.

4. According to the phylogenetic tree shown below, what is a shared derived character for primates, rodents, crocodiles, and birds?

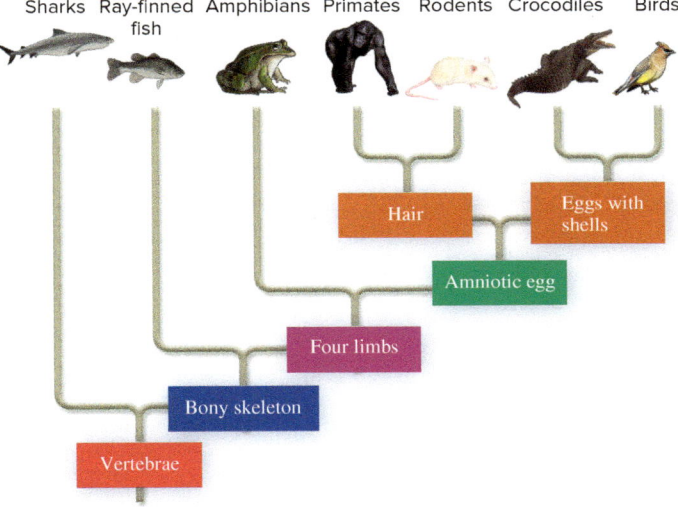

   a. four limbs
   b. amniotic egg
   c. hair
   d. eggs with shells
   e. Both a and b are correct.

5. If you were going to use a cladistics approach to construct a phylogenetic tree of doglike animals, which of the following would be the best outgroup?
   a. wolf
   b. tiger
   c. domestic dog
   d. hyena
   e. jackal

Short Answer

1. Refer back to the data found in Figure 19.9 in Chapter 19. Construct a phylogenetic tree based on these data and on your knowledge of the relatedness of these species.

2. Discuss how taxonomy is useful. Make a list of some practical applications that are derived from taxonomy.

22

The History of Life on Earth and Human Evolution

©George Bernard/SPL/Science Source

A fossil fish. This 50-million-year-old fossil of a unicorn fish (*Naso rectifrons*) is an example of the many different kinds of organisms that have existed during the history of life on Earth.

Chapter Outline

22.1 The Fossil Record
22.2 History of Life on Earth
22.3 Human Evolution

Assessing Your Knowledge and Skills

The amazing origin of the universe is difficult to comprehend. Astronomers propose that the universe began with a cosmic explosion called the Big Bang about 13.8 billion years ago (bya), after which the first clouds of the elements hydrogen and helium were formed. Over a long time period, gravitational forces collapsed these clouds to create stars that converted hydrogen and helium into heavier elements, including carbon, nitrogen, and oxygen, which are the atomic building blocks of life on Earth. These elements were returned to interstellar space by exploding stars called supernovas, which created clouds in which simple molecules such as water, carbon monoxide, and hydrocarbons formed. The clouds then collapsed to make a new generation of stars and solar systems.

Our solar system began about 4.6 bya after one or more local supernova explosions. According to one widely accepted scenario, hundreds of planetesimals (small celestial bodies, such as asteroids) occupied the region where Venus, Earth, and Mars are now found. The Earth, which is estimated to be 4.55 billion years old, grew from the aggregation of such planetesimals over a period of 100 to 200 million years. For the first half billion years or so after its formation, the Earth was too hot to allow liquid water to accumulate on its surface. By 4 bya, the Earth had cooled enough for the outer layers of the planet to solidify and for oceans to begin to form.

The period between 4.0 and 3.5 bya marked the emergence of life on our planet. The first forms of life that we know about produced microscopic fossils, the preserved remains of organisms that existed in the past. These fossils, estimated to be about 3.5 billion years old, resemble modern cyanobacteria, which are photosynthetic bacteria (**Figure 22.1**). Researchers cannot travel back through time and observe how the first life-forms came into being. However, plausible hypotheses regarding how life first arose have emerged from our understanding of modern life.

Chapter 4 discusses hypotheses for how living cells could have originated from organic molecules,

(a) Fossil prokaryote

(b) Modern cyanobacteria

Figure 22.1 Earliest fossils and living cyanobacteria. (a) A fossilized prokaryote about 3.5 billion years old that is thought to be an early cyanobacterium. (b) A modern cyanobacterium, which has a similar morphology. Cyanobacterial cells connect to each other to form chains, as shown here.
(a) ©J.W. Schopf; (b) ©Michael Abbey/Science Source

beginning with an RNA world (see Section 4.1). This chapter emphasizes when and how particular forms of life arose after the emergence of the first primitive cells. We will begin by considering the methods that researchers use to analyze and date fossils, which are the remains of past life-forms. We will then consider how fossils, such as the one shown in the chapter-opening photo, have provided biologists with evidence of the history of life on Earth from its earliest beginnings to the present day. The last section focuses on one of the most interesting stories of evolution, which is the lineage that gave rise to modern humans.

22.1 The Fossil Record

Learning Outcomes

1. Describe how fossils are formed.
2. **SCISKILLS** ▶ Explain how radiometric dating is used to estimate the age of a fossil.
3. List several factors that affect the completeness of the fossil record.

Let's begin by turning our attention to a process that has given us a window into the history of life over the past 3.5 billion years. **Fossils** are the preserved remains of past life on Earth. They can take many forms, including bones, shells, and leaves, and the impressions of cells or other evidence, such as footprints or burrows. Scientists who study fossils are called **paleontologists** (from the Greek *palaios*, meaning ancient). Because our understanding of the history of life is derived primarily from the fossil record, it is important to appreciate how fossils are formed and dated and to understand why the fossil record cannot be viewed as complete.

Fossils Are Formed Within Sedimentary Rock

How are fossils usually formed? Most fossils are found in sedimentary rocks that were formed from particles of older rocks broken apart by water or wind. These particles—in the form of gravel, sand, or mud—settle and bury living and dead organisms at the bottoms of rivers, lakes, and oceans. Over time, more particles pile up, and sediments at the bottom of the pile eventually become compressed into rock. Most fossils are formed when organisms are buried quickly, and then during the process of sedimentary rock formation, their hard parts are gradually replaced over millions of years by minerals, producing a recognizable representation of the original organism (see, for example, the chapter-opening photo).

The relative ages of fossils can sometimes be revealed by their locations in sedimentary rock formations. Because sedimentary rocks are formed by small particles settling in layers, the layers are piled one on top of the other. In a sequence of layered rock, the lower layers are usually older than the upper layers. Paleontologists often study changes in life-forms over time by studying the fossils in layers from bottom to top (**Figure 22.2**). The more ancient life-forms are found in the lower layers, and newer species are found in the upper ones. However, such an assumption can occasionally be misleading when geological processes, such as folding, have flipped the layers.

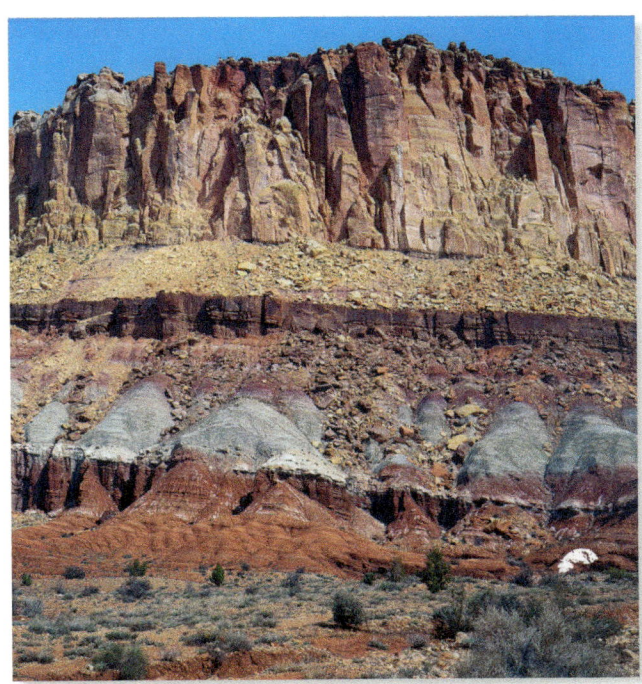

Figure 22.2 An example of layers of sedimentary rock that contain fossils.
©Simon Fraser/SPL/Science Source

✓ **Concept Check:** Which rock layer in this photo is most likely to be the oldest?

Biology Principle | Biology is a quantitative science

Quantitative Analysis: Radioisotopes Provide a Way to Date Fossils

A common way to estimate the age of a fossil is by analyzing the decay of radioisotopes within the accompanying rock, a process called **radiometric dating**. The atoms of certain elements can exist in multiple forms, called isotopes, that differ in the number of neutrons they contain. A radioisotope is an unstable isotope of an element that decays spontaneously, releasing radiation at a constant rate. The **half-life** is the length of time required for a radioisotope to decay to exactly one-half of its initial quantity. Each radioisotope has its own unique half-life (**Figure 22.3a**). Within a sample of rock, scientists can measure the amount of a given radioisotope as well as the amount of the decay product—the isotope that is produced when the original isotope decays. For dating geological materials, several types of isotope decay patterns are particularly useful, including carbon-14 to nitrogen-14 and uranium-235 to lead-207 (**Figure 22.3b**).

To determine the age of a rock using radiometric dating, paleontologists need to have a way to set the clock—extrapolate back to a starting point at which a rock did not have any amount of the decay product. Except for fossils less than 50,000 years old, in which carbon-14 (^{14}C) dating can be employed, fossil dating is not usually conducted on the fossil itself or on the sedimentary rock in which the fossil is found. Most commonly, igneous rock—rock formed through the cooling and solidification of lava—in the vicinity of the sedimentary rock is dated. Why is igneous rock chosen? One reason is that igneous rock derived from an ancient lava flow initially contains uranium-235 (^{235}U) but no lead-207 (^{207}Pb). The ultimate decay product of ^{235}U is ^{207}Pb. By comparing the relative proportions of ^{235}U and ^{207}Pb in a sample, the age of igneous rock can be accurately determined.

The relationship between a radioisotope and the process of decay is represented by the following equation:

$$N = N_0 e^{-(0.693t/T_{1/2})}$$

where

- N is the number of atoms of a radioisotope after a certain time period.
- N_0 is the number of atoms of that radioisotope that were originally present prior to any decay.
- e is the natural logarithm.
- t is the time period during which decay has occurred.
- $T_{1/2}$ is the half-life of the radioisotope.

Crunching the Numbers: A paleontologist discovered a fossil of a previously unidentified reptile. A sample of nearby igneous rock that weighed 1 kg contained 0.11 mg of uranium-235 and 0.035 mg of lead-207. Estimate the age of this newly discovered fossil. (Note: Uranium and lead have different molecular masses, so you first need to calculate the number of uranium and lead atoms in this sample.)

Biology Principle

Biologists Use Models to Test Predictions and Convey Ideas

The graph shown in part (a) is a mathematical model that depicts, in a general way, the decay of a radioisotope and the production of its decay product. The Modeling Challenge in the figure legend asks you to modify this graph to depict the decay of 2 mg of carbon-14.

(a) Decay of a radioisotope

Radioisotope	Decay product	Half-life (years)	Useful dating range (years)
Carbon-14	Nitrogen-14	5,730	100–50,000
Potassium-40	Argon-40	1.3 billion	100,000–4.5 billion
Rubidium-87	Strontium-87	47 billion	10 million–4.5 billion
Uranium-235	Lead-207	710 million	10 million–4.5 billion
Uranium-238	Lead-206	4.5 billion	10 million–4.5 billion

(b) Radioisotopes that are useful for geological dating

Figure 22.3 Radiometric dating of fossils. **(a)** A rock can be dated by measuring the relative amounts of a radioisotope and its decay product within the rock. **(b)** These five radioisotopes, each of which has a useful dating range, are commonly analyzed for the dating of fossils.

Modeling Challenge: Add new labels to the x- and y-axes of the graph in Figure 22.3a so that it depicts the decay of 2 mg of carbon-14. Also, change the general labels (Buildup of decay product, and Decay of radioisotope) next to the blue and red lines so they apply to carbon-14.

Table 22.1 Factors That Affect the Fossil Record

Factor	Description
Anatomy	Organisms with hard body parts, such as animals with a skeleton or thick shell, are more likely to be preserved than are organisms composed of only soft tissues.
Size	The fossil remains of larger organisms are more likely to be found than those of smaller organisms.
Number	Species that existed in greater numbers or over a larger area are more likely to be preserved within the fossil record than those that existed in smaller numbers or in a smaller area.
Environment	Inland species are less likely to become fossilized than are those that lived in a marine environment or near the edge of water because sedimentary rock is more likely to be formed in or near water.
Time	Species that lived relatively recently or existed for a long time are more likely to be found as fossils than species that lived very long ago or for a relatively short time.
Geological processes	Due to the chemistry of fossilization, certain organisms are more likely to be preserved than are other organisms.
Paleontology	Certain types of fossils may be more interesting to paleontologists. In addition, a significant bias exists with regard to the locations where paleontologists search for fossils. For example, they tend to search in regions where other fossils have already been found.

Several Factors Affect the Completeness of the Fossil Record

The fossil record should not be viewed as a complete and balanced representation of the species that existed in the past. Several factors affect the likelihood that extinct organisms have been preserved as fossils and will be identified by paleontologists (Table 22.1).

- Certain organisms are more likely than others to become fossilized. Organisms with hard shells or bones tend to be overrepresented in the fossil record.
- Factors such as anatomy, size, numbers of the species, and the environment and time in which they lived play important roles in determining the likelihood that organisms will be preserved in the fossil record.
- Geological processes may favor the fossilization of certain types of organisms.
- Unintentional biases arise that are related to the efforts of paleontologists. For example, scientific interests may favor searching for and analyzing certain species over others. Many paleontologists have been greatly interested in finding the remains of dinosaurs.

Although the fossil record is incomplete, it has provided a wealth of information regarding the history of the types of life that existed on Earth. The rest of this chapter will survey the emergence of life-forms from 3.5 bya to the present.

22.1 Reviewing the Concepts

- Fossils, which are preserved remnants of past life-forms, are formed in sedimentary rock. Our understanding of the history of life on Earth is derived primarily from the fossil record (Figure 22.2).
- Radiometric dating is one way of estimating the age of a fossil (Figure 22.3).
- Several factors affect the likelihood that organisms have been preserved in the fossil record (Table 22.1).

22.1 Testing Your Knowledge

1. Fossils are typically formed
   a. within igneous rock.
   b. within sedimentary rock.
   c. when dead organisms are exposed to intense sunlight.
   d. when dead organisms are left undisturbed for long periods of time.
   e. when dead organisms are quickly frozen.

2. Which of the following may account for biases in the fossil record?
   a. Certain organisms are more likely than others to become fossilized.
   b. Various factors, such as anatomy and number, affect the likelihood that organisms will be preserved in the fossil record.
   c. Geological processes may favor the fossilization of certain types of organisms.
   d. Unintentional biases arise that are related to the efforts of paleontologists.
   e. All of the above may cause biases.

22.2 History of Life on Earth

Learning Outcomes

1. List the types of environmental changes that have affected the history of life on Earth.
2. Describe the cell structure and energy utilization of the first living organisms that arose during the Archaean eon.
3. Explain how the origin of eukaryotic cells involved a union between bacterial and archaeal cells.
4. Describe the key features of multicellular organisms, which arose during the Proterozoic eon.
5. Outline the major events and changes in species diversity during the Paleozoic, Mesozoic, and Cenozoic eras.

The oldest fossils that have been discovered thus far were from single-celled organisms, which were preserved approximately 3.5 bya. In this section, we will begin with a brief description of the geological changes on Earth that have affected the emergence of new forms of life and then examine some of the major changes in life that have occurred since it began.

Many Environmental Changes Have Occurred Since the Origin of the Earth

The **geological timescale** is a timeline of the Earth's history and major events from its origin approximately 4.55 bya to the present (Figure 22.4). This timeline is subdivided into four eons—the Hadean, Archaean, Proterozoic, and Phanerozoic—and then further

Figure 22.4 The geological timescale and an overview of the history of life on Earth.

subdivided into eras. The first three eons are collectively known as the Precambrian because they preceded the Cambrian era, a geological era that saw a rapid increase in the diversity of life. We will examine these time periods later in this section.

The changes that occurred in species over the past 4 billion years are the result of two interactive processes. First, as discussed in Chapters 19–21, genetic changes in organisms can affect their characteristics. Such changes can influence the ability of organisms to survive and reproduce in their native environment. Second, the environment on Earth has undergone dramatic changes that have profoundly influenced the types of organisms that have existed during different periods of time. In some cases, an environmental change might allow new types of organisms to flourish. Alternatively, environmental changes can result in **extinction**—the complete loss of a species or group of species. The major types of environmental changes are described next.

Changes in Temperature During the first 2.5 billion years of its existence, the surface of the Earth gradually cooled. However, during the last 2 billion years, the Earth has undergone major fluctuations in temperature, producing periods of widespread glaciation, called Ice Ages, that alternate with warmer periods. Furthermore, the temperature on Earth is not uniform, which produces environments, such as tropical rain forests and arctic tundra, where the temperatures are quite different.

Changes in Atmosphere Composition The chemical composition of the gases surrounding the Earth has changed substantially over the past 4 billion years. One notable change has been the amount of oxygen. Prior to 2.4 bya, relatively little oxygen gas was in the atmosphere, but at that time, levels of oxygen in the form of O_2 began to rise significantly. The emergence of organisms capable of photosynthesis added oxygen to the atmosphere. Our current atmosphere contains about 21% O_2. Increased levels of oxygen are thought to have played a key role in various aspects of the history of life, including the following:

- The origin of many animal body plans coincided with a rise in atmospheric O_2.
- The conquest of land by arthropods (about 410 mya) and a second conquest by arthropods and vertebrates (about 350 mya) occurred during periods in which O_2 levels were high or increasing.
- Increases in animal body sizes are associated with higher O_2 levels.

Higher levels of O_2 could have contributed to these events because higher O_2 levels may enhance the ability of animals to carry out aerobic respiration. These events are discussed later in this chapter and in more detail in Unit V.

Shifting of Landmasses As the Earth cooled, landmasses formed that were surrounded by bodies of water. This produced two different environments: terrestrial and aquatic. Furthermore, over the course of billions of years, Earth's major landmasses, known as the continents, have shifted their positions, changed their shapes, and separated from each other. This phenomenon, called **continental drift,** is shown in **Figure 22.5**.

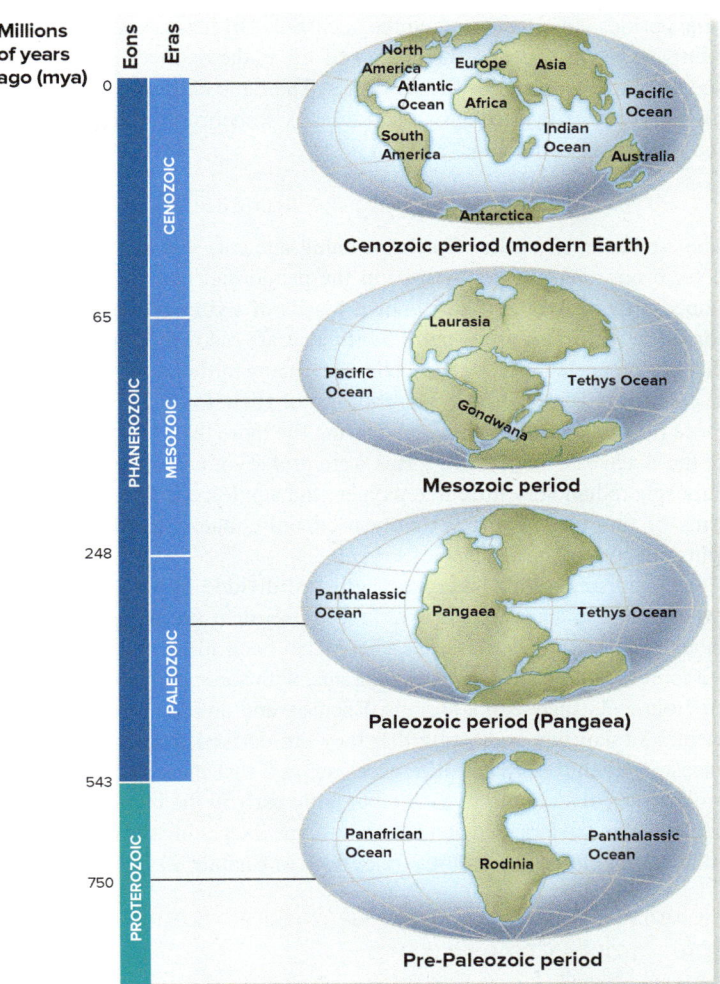

Figure 22.5 Continental drift. The relative locations of the continents on Earth have changed dramatically over time.

Floods and Glaciations Catastrophic floods have had major effects on the organisms in the flooded regions. Glaciers have periodically moved across continents and altered the composition of species on those landmasses.

Volcanic Eruptions The eruptions of volcanoes harm organisms in the vicinity of an eruption, sometimes causing extinctions. In addition, volcanic eruptions in the oceans lead to the formation of new islands. Massive eruptions may also spew so much debris into the atmosphere that they affect global temperatures and limit solar radiation, which restricts photosynthetic production.

Meteorite Impacts During its long history, the Earth has been struck by many meteorites. Large meteorites have significantly affected Earth's environment.

The effects of one or more of these events have sometimes caused large numbers of species to go extinct at the same time, which is called a **mass extinction.** Five large mass extinctions occurred near the ends of the Ordovician, Devonian, Permian, Triassic, and Cretaceous periods, respectively. The boundaries between geological time periods are often based on the occurrence of mass extinctions. A recurring pattern seen in the history of life is the extinction of some species and the emergence of new ones. The rapid extinction of many modern species due to human activities is sometimes referred to as the sixth mass extinction.

Prokaryotic Cells Arose During the Archaean Eon

The Archaean (from the Greek, meaning ancient) was an eon when diverse microbial life flourished in the primordial oceans. As mentioned previously, the oldest known fossils of living cells were preserved in rocks that are about 3.5 billion years old (see Figure 22.1), though scientists postulate that cells arose many millions of years prior to this time. Based on the morphology of their fossilized remains, these first cells were prokaryotic. During the more than 1 billion years of the Archaean eon, all life-forms were prokaryotic. Because Earth's atmosphere had very little free oxygen, the single-celled microorganisms of this eon almost certainly used only anaerobic respiration, which occurs in the absence of oxygen.

Organisms with prokaryotic cells are divided into two groups: bacteria and archaea. Bacteria are more prevalent on modern Earth, though many species of archaea have also been identified. Archaea are found in many different environments, with some found in extreme environments such as hot springs. Bacteria and archaea share fundamental similarities, indicating that they are derived from a common ancestor. Even so, certain differences suggest that these two types of prokaryotes diverged from each other quite early in the history of life. In particular, bacteria and archaea show some interesting differences in cellular and molecular features (refer back to Chapter 21, Table 21.1).

Biologists Are Undecided About Whether Heterotrophs or Autotrophs Came First

An important factor that greatly influenced the emergence of new species was the availability of energy. As we learned in Unit II, all organisms require energy to survive and reproduce. Organisms follow two strategies to obtain energy.

- **Heterotrophs** derive their energy from the chemical bonds within the organic molecules they consume. Because the most common sources of organic molecules today are other organisms, heterotrophs typically consume other organisms or materials from other organisms.

- **Autotrophs** directly harness energy from either light or inorganic molecules. Among modern species, plants are an important example of autotrophs. Plants can directly absorb light energy and use it (via photosynthesis) to synthesize organic molecules such as glucose. On modern Earth, heterotrophs ultimately rely on autotrophs for the production of food.

Were the first forms of life heterotrophs or autotrophs? The answer is not resolved. Some biologists have speculated that autotrophs, such as those living near deep-sea vents, may have arisen first. These organisms would have used chemicals that were made near the vents as an energy source to make organic molecules. Alternatively, many scientists have hypothesized that the first living cells were heterotrophs. They reason that it would have been simpler for the first primitive cells to use the organic molecules in the prebiotic soup as a source of energy.

(a) Fossil stromatolite

(b) Modern stromatolites

Figure 22.6 Fossil and modern stromatolites: evidence of early autotrophs. Each stromatolite is a rocklike structure, typically 1 meter in diameter. **(a)** Section of a fossilized stromatolite. These layers are mats of mineralized cyanobacteria, one layer on top of the other. The existence of fossil stromatolites provides evidence of early autotrophic organisms. **(b)** Modern stromatolites that have formed in western Australia.
(a) ©Dirk Wiersma/SPL/Science Source; *(b)* ©Horst Mahr/age fotostock

If heterotrophs came first, why were cyanobacteria, which are autotrophs, preserved in the earliest fossils rather than heterotrophs? One possible reason is related to their manner of growth. Certain cyanobacteria promote the formation of a layered structure called a **stromatolite** (Figure 22.6a). The aquatic environment where these cyanobacteria survive is rich in minerals such as calcium. The cyanobacteria grow in large mats that form layers. As they grow, they deplete the carbon dioxide (CO_2) in the surrounding water. This causes calcium carbonate in the water to gradually precipitate over the bacterial cells, calcifying the older cells in the lower layers into rock and trapping grains of sediment. Newer cells produce a layer on top. Over time, many layers of calcified cells and sediment are formed, thereby producing a stromatolite. This process still occurs in places such as Shark Bay in western Australia, which is renowned for the stromatolites along its beaches (Figure 22.6b).

The emergence and proliferation of ancient cyanobacteria had two critical consequences. First, the autotrophic nature of these bacteria enabled them to produce organic molecules from CO_2. This prevented the depletion of organic foodstuffs that would have been exhausted if only heterotrophs existed. Second, cyanobacteria produce O_2 as a waste product of photosynthesis. During the Archaean and Proterozoic eons, the activity of cyanobacteria led to the gradual rise in atmospheric O_2 discussed earlier. The increase in O_2 spelled doom for many anaerobic species, which became restricted to a few anoxic (without oxygen) environments, such as deep within the ground. However, O_2 enabled the emergence of new bacterial and archaeal species that used aerobic (with oxygen) respiration (refer back to Chapter 6). In addition, aerobic respiration is likely to have played a key role in the emergence and eventual explosion of eukaryotic life-forms, which typically have high energy demands. These eukaryotic life-forms are described next.

EVOLUTIONARY CONNECTIONS

The Origin of Eukaryotic Cells Is Hypothesized to Involve a Union Between Bacterial and Archaeal Cells

Eukaryotic cells arose during the Proterozoic eon, which began 2.5 bya and ended 543 mya (see Figure 22.4). The manner in which the first eukaryotic cells originated is not entirely understood. In modern eukaryotic cells, genetic material is found in three distinct organelles. All eukaryotic cells contain DNA in the nucleus and mitochondria, and plant and algal cells also have DNA in their chloroplasts. To address the issue of the origin of eukaryotic species, scientists have examined the DNA sequences found in these three organelles. They have concluded that the nuclear, mitochondrial, and chloroplast genomes appear to be derived from once-separate cells that came together.

Nuclear Genome Both bacteria and archaea contributed substantially to the nuclear genome of eukaryotic cells. Eukaryotic nuclear genes encoding proteins involved in metabolic pathways and lipid biosynthesis appear to be derived from ancient bacteria, whereas genes involved with transcription and translation appear to be derived from an archaeal ancestor.

To explain the origin of the nuclear genome, several hypotheses have been proposed. The most widely accepted hypothesis involves a symbiotic relationship between ancient bacteria and archaea. **Endosymbiosis** is a relationship in which a smaller organism (the endosymbiont) lives inside a larger organism (the host).

Researchers have suggested that an archaeal species evolved the ability to invaginate its plasma membrane, which could have two results (**Figure 22.7**). First, it could eventually lead to the formation of an extensive internal membrane system and enclose the genetic material in a nuclear envelope. Second, the ability to invaginate the plasma membrane provides a mechanism for taking up materials from the environment via endocytosis, which is described in Chapter 5. In the scenario described in Figure 22.7, an ancient archeon (an archaeal cell) engulfed a bacterium via endocytosis, maintaining the bacterium in its cytoplasm as an endosymbiont. Over time, many genes from the

Figure 22.7 Possible endosymbiotic relationships that gave rise to the first eukaryotic cells.

BioConnections: *Refer back to Figure 5.21. Explain how endocytosis played a role in endosymbiosis.*

bacterium were transferred to the archaeal host cell, and the resulting genetic material eventually became the nuclear genome.

Mitochondrial and Chloroplast Genomes Mitochondria found in eukaryotic cells are likely derived from a bacterial species that resembles modern proteobacteria, a diverse group of bacteria that carry out oxidative phosphorylation to make ATP. One possibility is that an endosymbiotic event involving an ancestor of this bacterial species produced the first eukaryotic cell and that the mitochondrion is a remnant of that event. Alternatively, endosymbiosis may have produced the first eukaryotic cell, and then a subsequent endosymbiosis resulted in mitochondria (see Figure 22.7). DNA-sequencing data indicate that chloroplasts were derived from a separate endosymbiotic relationship between a primitive eukaryotic cell and a cyanobacterium. As discussed in Chapter 23, plastids, such as chloroplasts, have arisen independently on several occasions via primary, secondary, and tertiary endosymbiosis (look ahead to Figure 23.22).

Multicellular Eukaryotes and the Earliest Animals Arose During the Proterozoic Eon

The first multicellular eukaryotes are thought to have emerged about 1.5 bya, in the middle of the Proterozoic eon. The oldest fossil evidence for multicellular eukaryotes was an organism that resembled modern red algae; this fossil was dated at approximately 1.2 billion years old.

Simple multicellular organisms are believed to have originated in one of two ways. One mechanism is that many individual cells can find each other and aggregate to form a colony. Cellular slime molds are examples of modern organisms in which groups of single-celled organisms can come together to form a small multicellular organism. According to the fossil record, such organisms have remained very simple for hundreds of millions of years.

Alternatively, another way that multicellularity can occur is when a single cell divides and the resulting cells stick together. This pattern occurs in many simple multicellular organisms, such as algae and fungi, as well as in species with more complex body plans, such as plants and animals. Biologists cannot be certain whether the first multicellular organisms arose by an aggregation process or by cell division and adhesion. However, the development of complex multicellular organisms now occurs by cell division and adhesion.

An interesting example showing changes in the level of complexity from unicellular organisms to more complex multicellular organisms is found among evolutionarily related species of volvocine green algae. These algae exist as unicellular species, as small clumps of cells of the same cell type, or as larger groups of cells with two distinct cell types. **Figure 22.8** compares four species of volvocine algae. *Chlamydomonas reinhardtii* is a unicellular alga (Figure 22.8a). It is called a biflagellate, because each cell has two flagella. *Gonium pectorale* is a multicellular organism composed of eight cells (Figure 22.8b). This simple multicellular organism is formed from a single cell by cell division and adhesion. All of the cells in this species are biflagellate. Other volvocine algae have evolved into larger

Biology Principle

New Properties Emerge from Complex Interactions
The formation of different cell types is an emergent property of multicellularity.

   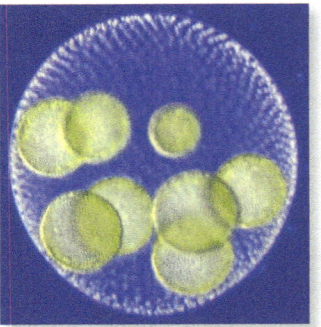

Flagella

3 μm

(a) *Chlamydomonas reinhardtii*, a unicellular alga

10 μm

(b) *Gonium pectorale*, composed of 16 identical cells

30 μm

(c) *Pleodorina californica*, composed of 64 to 128 cells, has 2 cell types, somatic and reproductive

100 μm

(d) *Volvox aureus*, composed of about 1,000 to 2,000 cells, has 2 cell types, somatic and reproductive

Figure 22.8 Variation in the level of multicellularity among volvocine algae.
(a) Courtesy of Dr. Barbara Surek, Culture Collection of Algae at the University of Cologne (CCAC); *(b)* ©Bill Bourland/Maple Ferryman; *(c)* ©Bowen Jiang; *(d)* ©Roland Birke/Getty Images

and more complex organisms. *Pleodorina californica* has 64–128 cells (Figure 22.8c), and *Volvox aureus* has about 1,000–2,000 cells (Figure 22.8d). A feature of these more complex organisms is that they have two cell types: somatic and reproductive cells. The somatic cells are biflagellate cells, but the reproductive cells are not. *V. aureus*, which is larger, has a higher percentage of somatic cells than *P. californica*.

Overall, an analysis of these four species of algae illustrates three important principles found among complex multicellular species:

1. Multicellular organisms arise from a single cell that divides to produce daughter cells that adhere to one another.
2. The daughter cells can follow different fates, thereby producing multicellular organisms with different cell types.
3. As organisms get larger, a greater percentage of their cells tend to be somatic cells. The somatic cells carry out the activities required for the survival of the multicellular organism, whereas the reproductive cells are specialized for the sole purpose of producing offspring.

Toward the end of the Proterozoic eon, the first multicellular animals emerged. The first animals were invertebrates—animals without a backbone. Most animals, except for organisms such as sponges and jellyfish, exhibit bilateral symmetry—a two-sided body plan with a right and a left side that are mirror images. Because each side of the body has appendages such as legs, one advantage of bilateral symmetry is that it facilitates locomotion. Bilateral animals also have anterior and posterior ends, with the mouth at the anterior end. In 2004, Chinese paleontologist Jun-Yuan Chen, American paleobiologist David Bottjer, and their colleagues discovered a fossil in southern China, which they described as the earliest known ancestor of animals with bilateral symmetry. This small animal, with a shape like a flattened helmet, is barely visible to the naked eye (Figure 22.9). The fossil, called *Vernanimalcula guizhouena,* is approximately 580–600 million years old. However, the interpretation that *V. guizhouena* was the earliest ancestor of animals with bilateral symmetry remains controversial and is under active investigation.

Phanerozoic Eon: The Paleozoic Era Saw the Diversification of Invertebrates and the Colonization of Land by Plants and Animals

The proliferation of multicellular eukaryotic life has been extensive during the Phanerozoic eon, which started 543 mya and extends to the present day. Phanerozoic means well-displayed life, referring to the abundance of fossils of plants and animals that have been identified from this eon. As shown in Figure 22.4, the Phanerozoic eon is subdivided into three eras: Paleozoic, Mesozoic, and Cenozoic. Because these eras are relatively recent and researchers have identified many fossils from them, each is further subdivided into periods. We will consider each era with its associated conditions and prevalent forms of life separately.

The term Paleozoic means ancient animal life. The Paleozoic era covers approximately 300 million years, from 543 to 248 mya, and is subdivided into six periods: Cambrian, Ordovician, Silurian, Devonian, Carboniferous, and Permian. Periods are usually named after regions where rocks and fossils of that age were first discovered.

Cambrian Period (543–490 mya)
The climate in the Cambrian period was generally warm and wet, with no evidence of ice at the poles. During this time, the diversity of animal species increased rapidly, an event called the **Cambrian explosion.** However, recent evidence suggests that many types of animal groups present during the Cambrian period actually arose prior to this time.

- By the middle of the Cambrian period, all of the existing major types of invertebrates were present, plus many others that no longer exist.
- Examples of animals that first appeared in the Cambrian period and still exist include echinoderms (sea urchins and starfish), arthropods (insects, spiders, and crustaceans), mollusks (clams and snails), chordates (organisms with a dorsal nerve chord), and vertebrates (animals with backbones).

The cause of the Cambrian explosion is not understood. Because it occurred shortly after marine animals evolved shells, some scientists have speculated that the changes observed in animal species may have allowed them to exploit new environments. Alternatively, others have suggested that the increase in diversity may be related to atmospheric oxygen levels. During this period, oxygen levels were increasing, and perhaps more complex body plans became possible only after the atmospheric oxygen surpassed a certain threshold. In addition, as atmospheric oxygen reached its present levels, an ozone (O_3) layer was produced that screens out harmful ultraviolet radiation, thereby allowing complex life to live in shallow water and eventually on land. Another possible contributor to the Cambrian explosion was an "evolutionary arms race" between interacting species. The ability of predators to capture prey and the ability of prey to avoid predators may have been major factors in the diversification of animals into many different species.

Figure 22.9 Fossil of an early invertebrate animal showing bilateral symmetry. This fossil of an early animal, *Vernanimalcula guizhouena*, dates from 580 to 600 mya.
(right) We thank Prof. Jun-Yuan Chen for permission to use this image

✓ **Concept Check:** *Name three other species that exhibit bilateral symmetry.*

Ordovician Period (490–443 mya) As in the Cambrian period, the climate of the early and middle parts of the Ordovician period was warm, and the atmosphere was moist.

- A diverse group of hard-shelled marine invertebrates, including trilobites and brachiopods, appeared in the fossil record (Figure 22.10). Marine communities consisted of invertebrates, algae, early jawless fishes (a type of early vertebrate), mollusks, and corals.
- Early land plants and arthropods may have first invaded the land during this period.

Toward the end of the Ordovician period, the climate changed rather dramatically. Large glaciers formed, which drained the relatively shallow oceans, causing the water levels to drop. This resulted in a mass extinction in which as much as 60% of the existing marine invertebrates became extinct.

Silurian Period (443–417 mya) In contrast to the dramatic climate changes observed during the Ordovician period, the climate during the Silurian period was relatively stable. The glaciers largely melted, which caused the ocean levels to rise.

- No new major types of invertebrate animals appeared during this period, but significant changes were observed among existing vertebrate and plant species.
- Many new types of fishes appeared in the fossil record.
- Coral reefs made their first appearance during this period.
- The Silurian marked a major colonization of land by terrestrial plants and animals. For this to occur, certain species evolved adaptations, such as a hard cuticle, that prevented them from drying out.
- Ancestral relatives of spiders and centipedes became widespread.
- The earliest vascular plants, which have tissues that are specialized for the transport of water, sugar, and salts throughout the plant body, arose during this period.

Devonian Period (417–354 mya) In the Devonian period, generally dry conditions occurred across much of the northern landmasses. However, the southern landmasses were mostly covered by cool, temperate oceans.

- The Devonian saw a major increase in the number of terrestrial plant species. At first, the vegetation consisted primarily of small plants, only a meter tall or less. Later, ferns, horsetails, and seed plants, such as gymnosperms, also emerged. By the end of the Devonian, the first trees and forests were formed.
- A major expansion of terrestrial animals also occurred. Insects first appeared in the fossil record, and other invertebrates became plentiful. In addition, the first tetrapods—vertebrates with four legs—are believed to have arisen in the Devonian. Early tetrapods included amphibians, which lived on land but required water in which to lay their eggs.
- In the oceans, many types of invertebrates flourished, including brachiopods, echinoderms, and corals. This period is sometimes called the Age of Fishes, as many new types of fishes emerged.

During a period of approximately 20 million years near the end of the Devonian period, a prolonged series of extinctions eliminated many marine species. The cause of this mass extinction is not well understood.

Carboniferous Period (354–290 mya) The term Carboniferous refers to the deposits of coal, a sedimentary rock primarily composed of carbon, that were formed during this period. The Carboniferous period had the ideal conditions for the subsequent formation of coal. It was a cooler period, and much of the land was covered by forest swamps. Coal was formed over many millions of years from compressed layers of rotting vegetation.

- Very large plants and trees became prevalent. For example, tree ferns such as *Psaronius* grew to a height of 15 meters or more (Figure 22.11).

(a) Trilobite

(b) Brachiopod

Figure 22.10 Shelled invertebrate fossils of the Ordovician period. Trilobites existed for millions of years before becoming extinct about 250 mya. Many species of brachiopods exist today.
(a) ©Francois Gohier/Science Source; *(b)* ©kavring/Shutterstock

THE HISTORY OF LIFE ON EARTH AND HUMAN EVOLUTION

Figure 22.11 A giant tree fern, *Psaronius*, from the Carboniferous period. This genus became extinct during the Permian. The illustration is a re-creation based on fossil evidence. The inset shows a fossilized section of the trunk.
(right) ©Natural History Museum, London/SPL/Science Source

- The first flying insects emerged. Giant dragonflies with a wingspan of over 2 feet inhabited the forest swamps.
- Terrestrial vertebrates also became more diverse.
- Amphibians were very widespread.
- The amniotic egg was one of the most important evolutionary innovations of the Carboniferous period. This innovation was critical for the emergence of reptiles, tetrapods that had the ability to lay eggs on land. In reptiles, the amniotic egg was covered with a leathery or hard shell, which prevented the desiccation of the embryo inside.

Permian Period (290–248 mya) At the beginning of the Permian, continental drift had brought much of the Earth's total land together into a supercontinent known as Pangaea (see Figure 22.5). The interior regions of Pangaea were dry, with large seasonal fluctuations.

- The forests of fernlike plants were replaced with gymnosperms, seed-bearing land plants. Species resembling modern conifers first appeared in the fossil record.
- Amphibians were prevalent, but reptiles became the dominant vertebrate species.

At the end of the Permian period, the largest known mass extinction in the history of life on Earth occurred; 90–95% of marine species and a large proportion of terrestrial species were eliminated. The cause of the Permian extinction is the subject of much research and controversy. One possibility is that glaciation destroyed the habitats of terrestrial species and lowered ocean levels, which would have caused greater competition among marine species. Another hypothesis is that enormous volcanic eruptions in Siberia produced large ash clouds that abruptly changed the climate on Earth.

Phanerozoic Eon: The Mesozoic Era Saw the Rise and Fall of the Dinosaurs

The Permian extinction marks the division between the Paleozoic and Mesozoic eras. Mesozoic means middle animals. It was a time that saw great changes in animal and plant species. This era is sometimes called the Age of Dinosaurs, because those animals flourished during this time. The climate during the Mesozoic era was consistently hot, and terrestrial environments were relatively dry. Little, if any, ice was found at either pole. The Mesozoic is divided into three periods: Triassic, Jurassic, and Cretaceous.

Triassic Period (248–206 mya)

- Reptiles were plentiful in this period, including new groups such as crocodiles and turtles.
- The first dinosaurs, which were a diverse group of reptiles that typically had an erect posture and sometimes attained a very large size, emerged during the middle of the Triassic. The first mammals, such as the small *Megazostrodon* also appeared (**Figure 22.12**).
- Gymnosperms were the dominant land plants.

Volcanic eruptions near the end of the Triassic are thought to have caused global warming, resulting in mass extinctions that eliminated many marine and terrestrial species.

Jurassic Period (206–144 mya)

- Gymnosperms, such as conifers, continued to be the dominant vegetation.
- Mammals were not prevalent.

Figure 22.12 *Megazostrodon*, the first known mammal of the Triassic period. The illustration is a re-creation based on fossilized skeletons. The *Megazostrodon* was 10 to 12 cm long.

BioConnections: *Look ahead to Table 27.1. What are the common characteristics of mammals?*

Figure 22.13 A fossil of an early, birdlike animal, *Archaeopteryx lithographica*, which emerged in the Jurassic period.
©Jason Edwards/Getty Images

- Dinosaurs and other reptiles continued to be the dominant land vertebrates. Modern birds are descendants of a dinosaur lineage called theropod (meaning beast-footed) dinosaurs. *Tyrannosaurus rex* is one of the best known theropod dinosaurs. An early, birdlike animal, *Archaeopteryx lithographica* (Figure 22.13), emerged in the Jurassic period. However, paleontologists are debating whether or not *Archaeopteryx* is a true ancestor of modern birds.

Cretaceous Period (144–65 mya)

- On land, dinosaurs continued to be the dominant animals.
- The earliest flowering plants, called angiosperms, which form seeds within a protective chamber, emerged and began to diversify.

The end of the Cretaceous period witnessed another mass extinction, which brought an end to many previously successful groups of organisms. Except for the lineage that gave rise to birds, dinosaurs abruptly died out, as did many other species. As with the Permian extinction, the cause or causes of this mass extinction are still debated. One plausible hypothesis suggests that a large meteorite hit the region that is now the Yucatan Peninsula of Mexico, lifting massive amounts of debris into the air and thereby blocking the sunlight from reaching the Earth's surface. Such a dense haze could have cooled the Earth's surface by 11–15°C (20–30°F). Evidence also points to huge volcanic eruptions as a contributing factor to this mass extinction.

Phanerozoic Eon: Mammals and Flowering Plants Diversified During the Cenozoic Era

The Cenozoic era spans the most recent 65 million years. It is divided into two periods: Tertiary and Quaternary. In many parts of the world, tropical conditions were replaced by a colder, drier climate. During this time, mammals became the largest terrestrial animals, which is why the Cenozoic is sometimes called the Age of Mammals. However, the Cenozoic era also saw an amazing diversification of many types of organisms, including birds, fishes, insects, and flowering plants.

Tertiary Period (65–1.8 mya)

- On land, the mammals that survived from the Cretaceous period began to diversify rapidly during the early part of the Tertiary period.
- Angiosperms became the dominant land plants, and insects became important for their pollination.
- Fishes also diversified, and sharks became abundant.
- Toward the end of the Tertiary period, about 7 mya, hominoids came into existence. **Hominoids** include humans, chimpanzees, gorillas, orangutans, and gibbons, plus all of their recent ancestors. The subset of hominoids called **hominins** includes modern humans, extinct human species (for example, of the *Homo* genus), and our immediate ancestors.

Quaternary Period (1.8 mya–Present)
Periodic Ice Ages have been prevalent during the last 1.8 million years, covering much of Europe and North America.

- This period has witnessed the widespread extinction of many species of mammals, particularly larger ones.
- Certain species of hominins became increasingly more like modern humans. *Homo sapiens*—modern humans—first appeared about 200,000 years ago. The evolution of hominins is discussed in more detail in Section 22.3.

22.2 Reviewing the Concepts

- The geological timescale, which is divided into four eons and many eras and periods, charts the major events that occurred during the history of life on Earth (Figure 22.4).
- Both the emergence of new species and mass extinctions are correlated with changes in temperature, the amount of O_2 in the atmosphere, landmass locations, floods and glaciation, volcanic eruptions, and meteorite impacts (Figure 22.5).
- During the Archaean eon, bacteria and archaea arose. The proliferation of cyanobacteria led to a gradual rise in O_2 levels (Figure 22.6).
- Eukaryotic cells arose during the Proterozoic eon. This origin involved a union between bacterial and archaeal cells that is hypothesized to have been endosymbiotic. The origins of mitochondria and chloroplasts were also the result of endosymbiosis (Figure 22.7).
- Multicellular eukaryotes arose about 1.5 bya during the Proterozoic eon. Multicellularity now occurs via cell division and the adherence of the resulting cells to each other. A multicellular organism can have multiple cell types. The first animal showing bilateral symmetry emerged toward the end of the Proterozoic eon (Figures 22.8, 22.9).
- The Phanerozoic eon is subdivided into the Paleozoic, Mesozoic, and Cenozoic eras. During the Paleozoic era, invertebrates greatly diversified, particularly during the Cambrian explosion, and the land became colonized by plants and animals. Terrestrial vertebrates, including tetrapods, became more diverse (Figures 22.10, 22.11).
- Mammals, dinosaurs, and birds emerged during the Mesozoic era. Dinosaurs were particularly prevalent during the Jurassic period (Figures 22.12, 22.13).

- During the Cenozoic era, mammals diversified, and flowering plants became the dominant plant species. The first hominoids emerged approximately 7 mya. *Homo sapiens*, our species, first appeared about 200,000 years ago.

22.2 Testing Your Knowledge

1. A heterotroph is an organism that
   a. only consumes autotrophs.
   b. only consumes other heterotrophs.
   c. derives its energy from the chemical bonds within organic molecules.
   d. directly harnesses energy from inorganic sources.
2. Eukaryotic cells were originally derived from
   a. an endosymbiotic relationship between archaeal and bacterial cells.
   b. an endosymbiotic relationship between different archaeal cells.
   c. an endosymbiotic relationship between different bacterial cells.
   d. a single archaeal species.
   e. a single bacterial species.
3. Multicellularity in most modern species occurs because
   a. individual cells find each other and aggregate.
   b. a single cell undergoes multiple divisions and the cells adhere to one another.
   c. Both a and b are correct.
   d. Neither a nor b is correct.

22.3 Human Evolution

Learning Outcomes

1. List the common characteristics of primates, and describe their evolutionary relationships.
2. Explain how human species evolved from other primate species, and describe how they spread across the Earth.
3. Provide examples of how populations of *Homo sapiens* are still evolving.
4. Compare and contrast modern human variation at the phenotype and genotype levels.

Hardly a topic in biology has evoked more interest or public debate than human evolution. The question of "where did we come from" has been considered by people for thousands of years. In this section, we will tackle this question from an evolutionary perspective.

We will begin with an overview of primate evolution and explore how humans are evolutionarily related to their closest nonhuman relatives. We will then take a closer look at the evolutionary events that gave rise to modern humans. As you will see, many extinct species of humans have existed, including the Denisovans, which were discovered as recently as 2010! Finally, we will turn our attention to the level of existing genetic variation among modern humans and consider whether humans are still evolving.

Primates Evolved from a Tree-Dwelling Species and Exhibit a Distinctive Set of Characteristics

Primates are primarily tree-dwelling species that evolved from a group of small, arboreal, insect-eating mammals about 85 mya, before dinosaurs went extinct. Primates have several defining characteristics, mostly relating to their tree-dwelling nature:

- **Binocular vision.** Primates have forward-facing eyes that are positioned close together on a flattened face. Swinging from branch to branch requires accurate judgment of distances; this movement is facilitated by binocular vision in which the fields of vision for both eyes overlap, producing a single image.
- **At least some digits with flat nails instead of claws.** This feature is believed to aid in the manipulation of objects.
- **Grasping hands.** All primates have grasping hands, a characteristic that enables them to hold on to branches. The monkeys, gibbons, orangutans, gorillas, chimpanzees, and humans also possess an opposable thumb, a thumb that can be placed opposite the fingers of the same hand, which gives them a precision grip and enables the manipulation of small objects. The same primates, except humans, also have opposable big toes.
- **Large brain.** Acute vision and other senses enhancing the ability to move quickly through the trees require the efficient processing of large amounts of information. As a result, primate brains are relatively large for their body sizes and are well developed.
- **Complex social behavior and well-developed parental care.** Compared to other mammals, primates have a tendency toward complex social behavior and relatively long parental care of their offspring.

Some of these characteristics are found in other animals. For example, binocular vision occurs in owls and some other birds, grasping hands are found in raccoons, and relatively large brains occur in marine mammals. Primates are defined by possessing the whole suite of these characteristics.

Primates are classified in multiple ways. Taxonomists divide them into two broad groups: the strepsirrhini and the haplorrhini (**Figure 22.14**). The **strepsirrhini** are smaller species, such as bush babies, lemurs, and pottos. These are generally nocturnal and smaller-brained primates with eyes positioned a little more toward the sides of their heads (**Figure 22.15a**). The strepsirrhini are named for their wet noses with no fur at the tip. The **haplorrhini** have dry noses and fully forward-facing eyes. This group consists of the tarsiers and the larger-brained and diurnal **anthropoidea:** monkeys and the **hominoidea** (gibbons, orangutans, gorillas, chimpanzees, and humans) (**Figure 22.15b, c**). A key feature of anthropoidea is an opposable thumb, which makes it easier to grasp and handle objects.

What differentiates monkeys from hominoids? Most monkeys have tails, but hominoids do not. In addition, hominoids have more mobile shoulder joints, broader rib cages, and a shorter spine. These features aid in brachiation, a swinging movement in trees. Hominoids also possess relatively long limbs and short legs and, with the exception of gibbons, are much larger than monkeys. The 20 species of hominoids are split into two groups: the lesser apes (family Hylobatidae), or the gibbons; and the greater apes (family

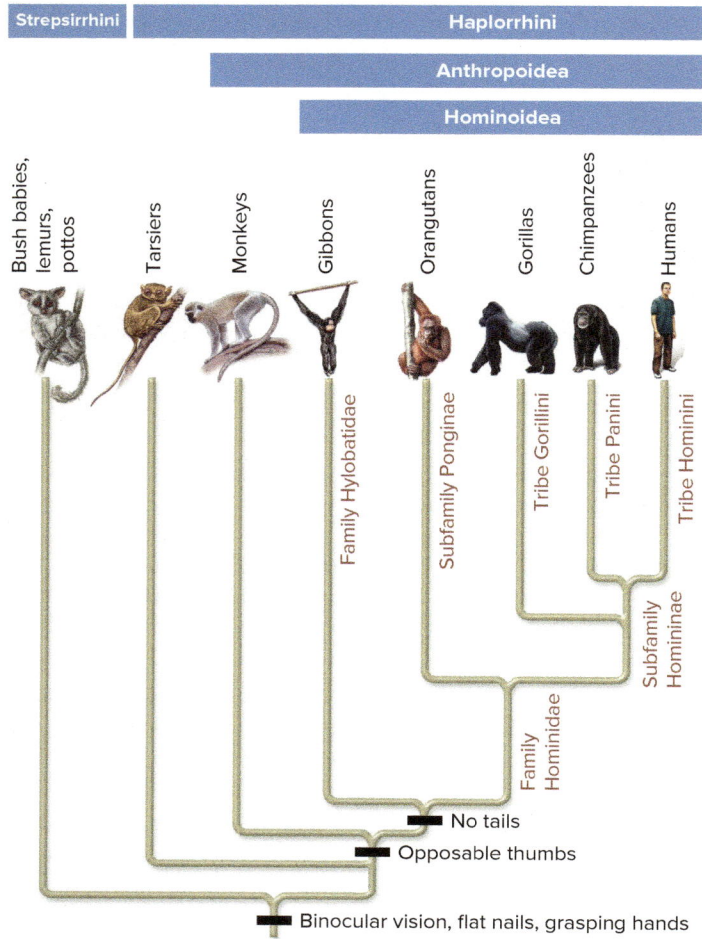

Figure 22.14 Evolutionary tree of the primates.

published an initial sequence of the chimpanzee genome. The draft sequence followed the 2003 publication of the human genome (see Chapter 18) and allowed scientists to make detailed comparisons between the two species. These comparisons revealed that the sequences of base pairs of both species' genomes differ by only 1.23%, which is 10 times less than the difference between the mouse and rat genomes. Comparisons of human and chimpanzee proteomes also showed that 29% of all proteins are identical, with most others differing by one or two amino acid substitutions.

Many of the genetic differences between chimps and humans result from chromosome inversions and duplications. Over 1,500 inversions occur between the chimp and human genomes. Although many inversions are in the non-coding regions of the genome, the DNA in these regions may regulate the expression of the genes in the coding regions. Duplications and deletions are also common. For example, humans have lost a gene called *caspase-12*, which in other primates may protect against Alzheimer's disease.

Some interesting genetic differences were apparent between chimps and humans even before their entire genomes were sequenced. In 2002, Swedish molecular geneticist Svante Pääbo discovered differences between humans and chimps in a gene called *FOXP2*, which plays a role in speech development. Proteins encoded by this gene differ in just 2 amino acids of a 715-amino-acid sequence. Researchers propose that the mutations in this gene have been crucial for the development of human speech.

In 2006, a team led by American geneticist David Reich discovered that the human X chromosome diverged from the chimpanzee X chromosome about 1.2 million years more recently than the other chromosomes. This indicated to the researchers that the human and chimp lineages split apart, then began interbreeding before diverging again. If so, the interbreeding explains why many fossils appear to exhibit traits of both humans and chimps, because they may actually have been human-chimpanzee hybrids.

Hominidae), or the orangutans, gorillas, chimpanzees, and humans (**Figure 22.16**). The lesser apes are strictly arboreal, whereas the greater apes often descend to the ground to feed.

Although humans are closely related to chimpanzees and gorillas, they did not evolve directly from them. Rather, all hominoid species share a common ancestor. Recent molecular studies show that gorillas, chimpanzees, and humans are more closely related to one another than they are to gibbons and orangutans, so scientists have split the family Hominidae into groups, including the subfamily Ponginae (orangutans) and the subfamily Homininae (gorillas, chimpanzees, and humans and their ancestors) (see Figure 22.14). In turn, the Homininae are split into three tribes: the Gorillini (gorillas), the Panini (chimpanzees), and the Hominini (humans and their ancestors).

EVOLUTIONARY CONNECTIONS

Comparing the Genomes of Humans and Chimpanzees

A male chimp called Clint, which lived at a primate research center in Atlanta, provided the DNA used to sequence the chimp genome. In 2005, the Chimpanzee Sequencing and Analysis Consortium

Bipedalism Is a Distinguishing Feature of Humans

About 7 mya in Africa, a lineage that led to modern humans diverged from other primate lineages. The evolution of humans should not be viewed as a neat, stepwise progression from one species to another. Rather, human evolution, like the evolution of all species, can be visualized more like a tree, with one or two **hominin** species—members of the Hominini tribe—likely coexisting at the same point in time, with some eventually going extinct and some giving rise to other species (**Figure 22.17**).

The key characteristic differentiating hominins from other apes is that hominins walk on two feet; that is, they are **bipedal**. At about the time when hominins diverged from other ape lineages, the Earth's climate had cooled, and the forests of Africa gave way to grassy savannas. Bipedal locomotion and an upright stance may have been advantageous in allowing hominins to peer over the tall grasses of the savanna to see predators or prey.

Bipedalism is correlated with important anatomical changes in hominins. First, the opening of the skull where the spinal cord enters shifted forward, causing the spine to be more directly underneath the

(a) Strepsirrhini (lesser bush baby) (b) Anthropoidea (capuchin monkey) (c) Hominoidea (white-handed gibbon)

Figure 22.15 Primate classification. The primates are divided into two groups: **(a)** the strepsirrhini (smaller, nocturnal species such as this bush baby) and the haplorrhini (larger, diurnal species). Haplorrhini comprise **(b)** the monkeys and tarsiers, such as this capuchin monkey (*Cebus capucinus*), and **(c)** the hominoids, species such as this white-handed gibbon (*Hylobates lar*).
(a) ©David Haring/DUPC/Getty Images; (b) ©Brand X Pictures/PunchStock; (c) ©catherinka/123RF

  Concept Check: *What are the defining features of primates?*

(a) Gorilla (*Gorilla gorilla*)   (b) Chimpanzee (*Pan troglodytes*)   (c) Human (*Homo sapiens*)

Figure 22.16 Members of the family Hominidae. (a) Gorillas, the largest of the living primates, are ground-dwelling herbivores that inhabit the forests of Africa. **(b)** Chimpanzees are smaller, omnivorous primates that also live in Africa. The chimpanzees are close living relatives of modern humans. **(c)** Humans are also members of the family Hominidae, as is the orangutan.
(a) ©iStockphoto/Getty Images; (b) ©imageBROKER/Alamy; (c) ©Tetra Images/Getty Images

head. Second, the hominin pelvis became broader to support the additional weight. And third, the lower limbs, used for walking, became relatively larger than those in other apes. These are the types of anatomical changes paleontologists look for in the fossil record to determine whether fossil remains are hominin.

The earliest known hominin, *Sahelanthropus tchadensis*, was discovered in Central Africa in 2002. Fossils of this species are dated at 7 million years old. The evolutionary relationship between *S. tchadensis* and later hominin species is unclear. Another early group of hominins included several species of the genus *Australopithecus*,

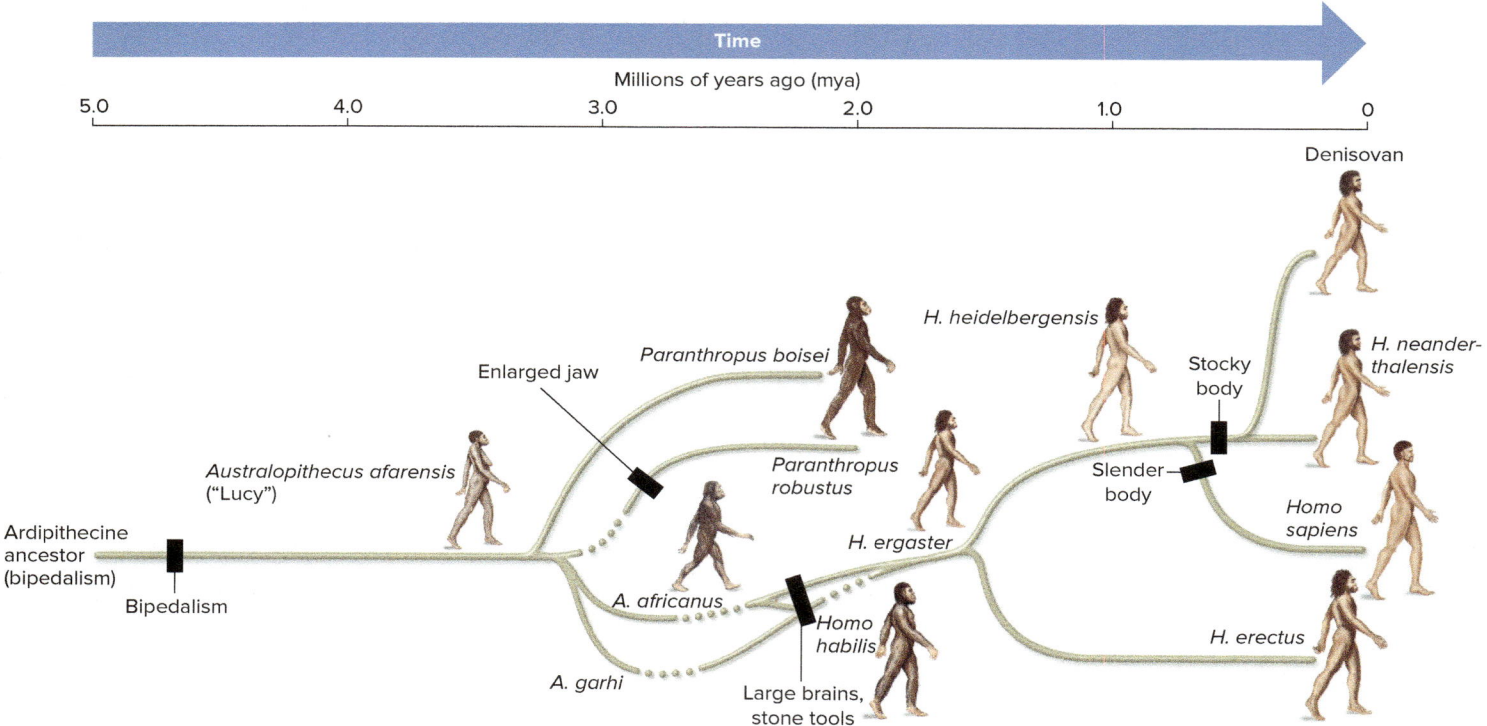

Figure 22.17 A possible scenario for human evolution. In this human family tree (based on the work of Donald Johanson and others), several hominin species lived during the same time period as others, but only one lineage gave rise to modern humans (*Homo sapiens*).

which first emerged in Africa about 4 mya. As shown in Figure 22.17, *Australopithecus afarensis* is generally regarded as the direct ancestor of most hominin species, but it could be a close relative of an unknown species that was the direct ancestor. From there, the evolution of different human species is still debated. It is generally agreed that two genera evolved from *Australopithecus*: the more robust *Paranthropus* and the genus *Homo*. The early stages of the evolution of *Homo* species, and their differentiation from at least two possible *Australopithecus* species, have not yet been determined with great certainty. However, the later divergence of various *Homo* species is better understood.

Australopithecus and *Paranthropus* Are Early Human Genera

Let's consider some of the general features of the two early human genera, *Australopithecus* and *Paranthropus*. In 1924, the first fossil was found in South Africa for a member of the genus *Australopithecus* (from the Latin *austral,* meaning southern, and the Greek *pithecus,* meaning ape). Since then, hundreds of fossils of this group have been unearthed all over southern and eastern Africa, the areas where fossil deposits are best exposed to paleontologists. This group was widespread, with at least six species.

In 1974, American paleontologist Donald Johanson discovered the skeleton of a female *A. afarensis* in the Afar region of Ethiopia and dubbed her Lucy. (The Beatles' song "Lucy in the Sky with Diamonds" was playing in the camp the night when Johanson was sorting the unearthed bones.) Over 40% of the skeleton had been preserved, enough to provide a good idea of the physical appearance of australopithecines. Compared with modern humans, they were relatively small, about 1–1.5 m in height and approximately 18 kg in weight (Figure 22.18). Females were much smaller than males, a characteristic known as sexual dimorphism. Examination of the bones revealed that *A. afarensis* walked on two legs. They had a facial structure and brain size (about 500 cubic centimeters [cm^3]) similar to those of a chimp.

In the 1930s, the remains of bigger-boned hominins were found. Two of the larger species now considered to be a separate genus, *Paranthropus,* weighed about 40 kg and lived during the same time period as australopithecines and early *Homo* species. *Paranthropus* were vegetarians with enormous jaws used for grinding tough roots and tubers. Both *Paranthropus* species died out rather suddenly about 1.5–2.0 mya.

Although *Australopithecus africanus* was thought to have evolved slightly later than *A. afarensis,* its bones were found much earlier than those of *A. afarensis*. In the 1920s, Australian anthropologist Raymond Dart described *A. africanus* from infant bones discovered in a cave in Taung, South Africa. The well-preserved skull was small but well rounded, unlike the skulls of chimpanzees and gorillas. Also, the positioning of the head on the vertebral column indicated bipedalism. These observations suggested to Dart that he had found a transitional form between apes and humans. However, it would take another 20 years and the discovery of more fossils to convince the scientific world to support Dart's view. In 1996, remains of

Figure 22.18 **A modern woman compared to an australopithecine.** Compared with modern humans, australopithecines, as illustrated by this reconstruction of the famous fossil Lucy, were much smaller and lighter.

another species, *Australopithecus garhi,* were also found in the Afar region. They were a surprise in that the teeth suggested similarities with *Paranthropus boisei.* Garhi means surprise in the local Afar language. The proposal that both *A. garhi* and *A. africanus* are ancestors of modern humans has been the subject of much debate. They have been viewed as either dead-end cousins or the ancestors of the first members of the genus *Homo.*

The Genus *Homo* Includes Modern Humans and Their Most Recent Relatives

Seven different species in the *Homo* genus are shown in Figure 22.17. Let's consider how they are evolutionarily related to each other.

Homo habilis In the 1960s, British paleontologist Louis Leakey found hominin fossils estimated to be about 2 million years old in Olduvai Gorge, Tanzania. Two particularly interesting observations about these fossils stand out. First, reconstruction of the skull showed a brain size of about 680 cm^3, larger than that of *Australopithecus.* Second, the fossils were found with a wealth of stone tools. As a result, Leakey assigned the fossils to a new species, *Homo habilis,* from the Latin, meaning handy man. The discovery of several more *Homo* fossils followed, but there have been no extensive finds, as there were with Lucy. This makes it difficult to determine which *Australopithecus* lineage gave rise to the *Homo* lineage (see Figure 22.17), and paleontologists remain divided on this point.

Homo habilis lived alongside *Paranthropus* in East Africa but had much smaller jaws and teeth, indicating that it probably ate large quantities of meat. The smaller jaw provided more space in the skull for brain development (look ahead to Figure 35.7). *Homo habilis* probably scavenged most of its meat from the kills of large predators. A meatier diet is easier to digest and is rich in nutrients and calories. The human brain uses a lot of energy, 20% of the body's total energy production. The meat-eating habit thus helped propel the evolution of increasing brain size in humans. Cut marks on animal bones of the period reveal that early humans used stone tools to smash open bones and extract the protein-rich bone marrow.

Homo ergaster Although the evidence is not entirely clear, *H. habilis* probably gave rise to one of the most important species of *Homo, Homo ergaster. Homo ergaster* was a hominin that arose in Africa about 2 million years ago; it had a human-looking face and skull, with downward-facing nostrils. *Homo ergaster* was also a tool user, and the tools, such as hand axes, were larger and more sophisticated. *Homo ergaster* evolved in a period of global cooling and drying that reduced tropical forests even more and promoted savanna conditions. Hairlessness and the regulation of body temperature through sweating may also have evolved at this time as adaptations to the sunny environment. A leaner body shape was evident. We know this from so-called Turkana boy, a fossil teenage boy found in Kenya in 1984. Though only 13 years old, scientists predict he would have been about 185 cm (6 ft 1 in.) when adult, much the same height as the Masai tribesmen that inhabit the area today. A dark skin probably protected *H. ergaster* from the Sun's rays. The pelvis had narrowed, promoting efficiencies in walking upright, and the size of the brain and hence the skull had increased, which may have produced more difficulty in childbirth. Mothers had to push increasingly large-brained infants through a narrowed pelvis. Researchers think that as a result, the human gestation period was shortened. Earlier birth led to prolonged care of human infants compared with that in other apes. Prolonged child care required well-nourished mothers, who would have benefited from the support of their male partner and other members of a social group. Some anthropologists have suggested this was the beginning of the family.

H. ergaster is thought to have given rise to many species, including *Homo erectus, Homo heidelbergensis, Homo neanderthalensis,* and *Homo sapiens.* A possible timeline and geographic locations for these species are shown in **Figure 22.19.** *Homo ergaster* probably was the first type of human to leave Africa, as similar bones have been found in the Eurasian country of Georgia. *H. ergaster* is believed to be a direct ancestor of modern humans, with *H. heidelbergensis* viewed as an intermediary step. Living at the same time as *H. heidelbergensis* was another descendent of *H. ergaster, H. erectus,* though some researchers consider *H. ergaster* and *H. erectus* as the same species. We will treat them as separate species here.

Homo erectus *H. erectus* was a large hominin, as large as a modern human but with heavier bones and a smaller brain capacity of between 750 and 1,225 cm^3 (modern human brain size is about 1,350 cm^3). Fossil evidence shows that *H. erectus* was a social species that used tools, hunted animals, and cooked over fires. The meat-eating habit may have sparked the migration of *H. erectus.* Carnivores tend to have larger ranges than similar-sized herbivores, because the food sources of carnivores (prey) are usually scarcer per unit area. *H. erectus* spread out of Africa soon after the species appeared, over a million years ago, and fossils have been found as far away as China and Indonesia. The first fossil was found by Dutch physician Eugene Dubois in 1891 on the Indonesian island of Java. Stone tools are rarely found in these Asian sites, suggesting *H. erectus* based their technology on other

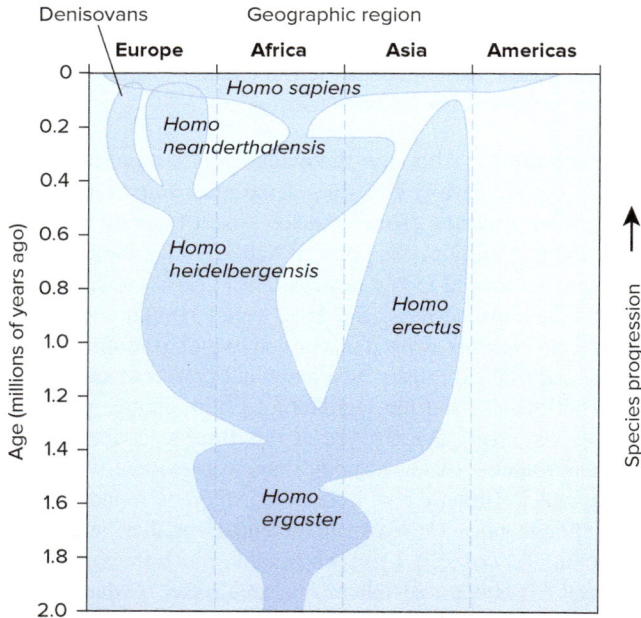

Figure 22.19 One view of the temporal and geographic evolution of hominid populations. Though not shown in this figure, the range of the Denisovans and *Homo neanderthalensis* also extended into Asia.

materials, such as bamboo, which was abundant at that time. Bamboo is strong yet lightweight and could have been used to make spears. These humans may even have used rafts to take to the seas. *H. erectus* went extinct about 100,000 years ago, for reasons that are unclear but may be related to the spread of *H. sapiens* into its range.

Homo heidelbergensis *H. heidelbergensis* was similar in body form to modern humans. Large caches of this species' bones were found in Spain, at the bottom of a 14-m (45-ft) shaft known as La Sima de Los Huesos (the pit of bones). Similar remains were also found at Boxgrove in England. Shinbones recovered from Boxgrove suggest males stood around 180 cm (6 ft) and weighed 88 kg (196 pounds). Skulls were large, with brain volumes from 1,100 to 1,400 cm^3, similar to modern humans. Animal bones from these sites showed cut marks from stone blades beneath tooth marks from carnivores. This indicated that humans were killing large prey before scavengers arrived. Horses, giant deer, and rhinoceroses were common prey and would have required much skill and cooperation to hunt. *H. heidelbergensis* gave rise to three species, *Homo neanderthalensis*, the Denisovans, and *Homo sapiens*. However, some scientists view these as three subspecies. In the discussion that follows, we will consider them to be three different species.

Homo neanderthalensis *H. neanderthalensis* was named for the Neander Valley of Germany, where the first fossils of its type were found. In the Pleistocene epoch, glaciers were locked in a cycle of advance and retreat, and the European landscape was often covered with snow. Over the course of many generations, the more slender body form of *H. heidelbergensis* evolved into a shorter, stockier build that was better equipped to conserve heat; we now call this type of human Neanderthal. Neanderthals also possessed a more massive skull and larger brain size than modern humans, about 1,450 cm^3, perhaps associated with their bulk. Males were about 168 cm (5 ft 6 in.) in height and would have been very strong by modern standards. They had a large face with a prominent bridge over the eyebrows, a large nose, and no chin (**Figure 22.20a**). They lived predominantly in Europe, with a range extending to the Middle East and Asia (**Figure 22.20b**). Their stocky and muscular physique was well suited to the rigors of cold climates and hunting prey. Paleontologists have found a high rate of head and neck injuries in Neanderthal bones, similar to that seen in present-day rodeo riders. This suggests that close encounters with large prey often resulted in blows that knocked the hunters off their feet. The hyoid bone, which holds the larynx (voice box) in place, was well developed, suggesting speech was used. The Neanderthals went extinct about 40,000 to 30,000 years ago.

Denisovans In 2010, scientists sequenced DNA from a fossilized pinky finger of a young female found in Denisova Cave in Siberia, Russia, and found it to be genetically distinct from the DNA of *H. neanderthalensis* and *H. sapiens*. Because it is such a recent discovery, its species name is yet to be agreed upon. The common name given to this type of human is a Denisovan. Using carbon radioisotope dating, the fossil was estimated to be about 400,000 years old. Scientists now hypothesize that Denisovans were a sister group to the Neanderthals, diverging from the Neanderthals about 400,000 years ago. It is difficult to reconstruct the physical traits of Denisovans, because only a finger bone and a few molar teeth have been discovered thus far. Even so, the finger bone was unusually broad and robust, suggesting the Denisovans were similar in build to the Neanderthals. Like the Neanderthals, the Denisovans are thought to have gone extinct about 40,000 to 30,000 years ago.

Homo sapiens *H. sapiens* (from the Latin, meaning wise man) is our own species. *H. sapiens* is a slender, lighter-weight species with a slightly smaller brain capacity than that of the Neanderthals. Researchers hypothesize a variety of reasons *H. sapiens* thrived while the Neanderthals disappeared, including possessing a more efficient body type with lower energy needs, increased longevity, and differences in social structure and cultural adaptations.

Evidence Suggests That *Homo sapiens* Arose in Africa and Then Migrated to Other Parts of the World

Two models have been proposed to explain where the species of modern humans, *Homo sapiens*, arose. The first model, the out-of-Africa hypothesis, suggests that *H. sapiens* evolved in Africa from *H. heidelbergensis*. Some members of *H. sapiens* later migrated to other parts of the world and gradually replaced species such as *H. erectus* and *H. neanderthalensis*. An alternative hypothesis, called the multiregional hypothesis, which is not widely accepted, suggests that human groups have evolved from *H. ergaster* populations in a number of different parts of the world. According to this hypothesis, gene flow between neighboring populations prevented the formation of several different *H. sapiens* species.

In 1987, American evolutionary biologists Rebecca Cann, Allan Wilson, and colleagues analyzed the sequences of mitochondrial

(a) An adult male Neanderthal

(b) Geographic range of Neanderthals

Figure 22.20 Neanderthals. **(a)** Artist's rendition of a Neanderthal human. Neanderthals were shorter and stockier than modern humans, with larger elbow and ankle joints, shorter forearms, and a larger, broader rib cage. **(b)** The range of Neanderthals was confined to Europe and western Asia, with a northerly limit that corresponds to about 50° north, the southern limit of glaciation. Total population size may have been only 70,000 at its peak.

DNA that was collected from many people from around the world. By comparing these sequences to each other, they were able to conclude that *H. sapiens* arose in Africa about 200,000 years ago. These results and those of subsequent studies support the out-of-Africa hypothesis. According to this hypothesis, *H. ergaster* arose in Africa and spread to Asia and Europe. Later, *H. erectus* diverged from *H. ergaster* in Africa and spread into Asia, and *H. neanderthalensis* diverged from *H. heidelbergensis* in Europe. Both of these species and the Denisovans became extinct as *H. sapiens* migrated across the world. Some researchers have suggested that the extinction of these other human species may have occurred because they were outcompeted by *H. sapiens*. However, further research is needed to confirm or refute that idea.

The analysis of DNA sequences from modern humans across the world can also be used to construct a map that describes the migration of humans out of Africa. **Figure 22.21** describes such a map. However, the time periods should be considered as approximate. As we gather more data from people in various regions of the world, this map is under frequent revision and is more complicated than the simplified map shown in Figure 22.21. Modern humans spread first into the Middle East and Asia, then later into Europe and Australia, and finally across the Bering Strait to the Americas.

Much remains to be resolved in our understanding of human evolution, and new data provide paleontologists with information to revise their hypotheses. For example, in 2004, the remains of a small human on the Indonesian island of Flores were discovered and given the name *Homo floresiensis*, nicknamed hobbits by the media. Many species—for example, deer and elephants—develop into small forms in insular situations, so hobbit humans seemed plausible. Since then, many researchers have suggested these people were modern humans who were suffering from a genetic disorder. Even modern humans on Flores are pygmies. Pathological dwarfism would have made these people even smaller. Only *H. sapiens* tools have been found at the area where the bones occur, suggesting these individuals were indeed dwarf forms of modern humans. A 2014 study showed that the small brain size of one of the *H. floresiensis* fossils was in the range predicted for an individual with Down syndrome. The authors went on to argue that *H. floresiensis* is an invalid species.

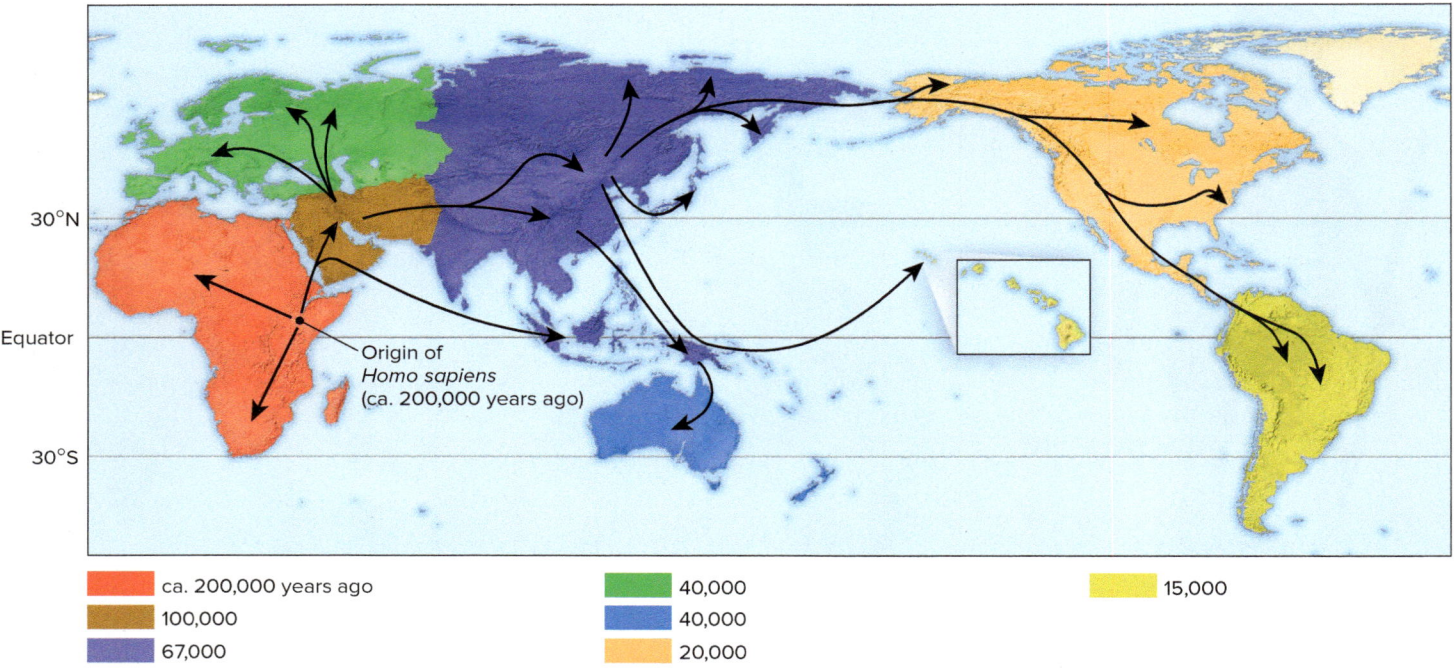

Figure 22.21 **A simplified model for the origin and spread of *Homo sapiens* throughout the world.** This map, based on differences of mtDNA throughout current members of the world's population, suggests *Homo sapiens* originated in East Africa. About 100,000 years ago, the species spread into the Middle East and from there to Europe, Asia, Australia, and the Americas.

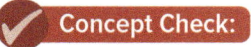 **Concept Check:** *List the locations in their proper order through which humans had to migrate to become established in South America.*

Human Evolution Has Involved Interbreeding Among Closely Related Species

Because the remains of Neanderthals and Denisovans are less than 100,000 years old, researchers have been able to extract DNA from their fossils and compare their sequences to each other and to *H. sapiens*. Detailed comparisons of Neanderthal, Denisovan, and modern human genomes have revealed interbreeding among the three species as *H. sapiens* spread into Europe and Asia. In 2018, paleogeneticists (scientists who study the DNA of ancient remains) Viviane Slon and Svante Pääbo analyzed the DNA of a female who lived about 90,000 years ago and determined that she was the offspring of a female Neanderthal and a male Denisovan. This child is the first direct evidence that interbreeding occurred among different human groups.

The genomes of modern humans of African descent contain little or no DNA that is derived from Neanderthals or Denisovans. However, 1 to 4% of the DNA from a person of European descent is derived from Neanderthals, and 4 to 6% of the DNA from a person of Southeast Asia descent is derived from Denisovans. These results indicate that *H. sapiens* ancestors interbred with Neanderthals and Denisovans as *H. sapiens* spread across Europe and Asia. Interestingly, some people of African descent carry very small amounts of Neanderthal DNA. How is this possible? By analyzing the DNA sequences of certain African people, researchers speculate that some *H. sapiens* from Europe may have migrated back to Africa about 3,000 years ago and interbred with a few isolated African populations of *H. sapiens*.

When analyzing the human genome on a population level, at least 20% of the Neanderthal genome is found in the genome of modern humans of European descent. No one individual has all 20%; rather, any given person of European descent has about 1 to 4%. Researchers have speculated that certain Neanderthal genes may have provided a survival advantage, and that may explain why they have been retained in particular human populations. For example, a group of related proteins called keratins form filaments that play a role in the formation of human skin, hair, and nails. Such filaments may differ among populations that have evolved in a warm climate versus a cold one. Some alleles of keratin genes encode keratins that provide better insulation, a trait that is advantageous in cold climates. In people of European descent, genes that encode keratins are often of Neanderthal origin. This observation is consistent with the idea that some Neanderthal alleles may have helped European *H. sapiens* adapt to colder European environments.

On the downside, scientists also speculate that Neanderthals carried an allele of a gene involved with fatty acid uptake that helped them store fat better than *H. sapiens*. Such a gene was an advantage for a Neanderthal lifestyle in which hunter-gatherers gorged on prey and then went for days without eating. However, in modern humans who overeat, fat storage can put them at risk for developing various diseases, such as type 2 diabetes. In certain human populations, particularly Native Americans and those of Latin American descent, this Neanderthal allele is fairly common. Some people are heterozygous, carrying one allele derived from the Neanderthal genome and one from the *H. sapiens* genome. Such individuals are 20% more likely to develop type 2 diabetes compared to people who are homozygous for the *H. sapiens* allele. Furthermore, people who are homozygous for the Neanderthal allele are 40% more likely to develop diabetes.

Are Human Populations Still Evolving?

As discussed in Chapter 19, natural selection is the process in which individuals with greater reproductive success are more likely to pass their genes to future generations. Natural selection results in evolution. (Other processes, such as genetic drift, can also promote evolutionary change.) One factor that often plays a role in natural selection is the environment. Various types of environmental factors such as temperature, predators, and food sources can affect reproductive success and thereby cause a population to evolve in a particular direction. For example, a prolonged decrease in temperature may favor the survival of mammals with thicker fur, which will increase in frequency over the course of many generations.

Modern humans have great control over their environment. They live in dwellings where they can control the temperature, they can largely avoid predators, and they usually don't rely entirely on locally grown food for their survival. For such reasons, the impact of the natural environment on the evolution of human populations may have lessened compared to humans that lived long ago. Even so, human evolution via natural selection is still occurring. For example, we continue to evolve genetic resistance to infectious diseases. The bubonic plague of the 14th century killed about one-third of the Asian and European populations, yet many people survived and passed on alleles that confer greater resistance to this serious disease. Similarly, alleles that confer resistance to malaria are becoming more common in certain African populations.

A classic example of recent human evolution is the ability to digest lactose, a sugar found in milk. In most human populations, the ability to readily digest lactose is lost after the age when babies are weaned from their mother's milk. After weaning, lactose becomes indigestible to most people and they suffer bloating, abdominal cramps, flatulence, diarrhea, nausea, or vomiting if they eat or drink dairy products. However, the ability to consume dairy products may have provided a survival advantage as people began to domesticate cows, sheep, and goats. In human populations where such domestication took place, the ability to digest lactose, called lactose tolerance, is expected to increase in frequency due to natural selection if it provided people with greater reproductive success.

Recent studies suggest that mutations that confer lactose tolerance arose several thousand years ago in a few different places. The frequency of this mutation increased dramatically over the past few thousand years in populations where dairy products are commonly consumed. The genetic mutation for digesting lactose after weaning is now carried by most northern Europeans. Lactose intolerance is less common in this geographic region (**Figure 22.22**). For this population, the trait is linked to a single mutation that affects the expression of the lactase gene, which encodes an enzyme that is needed to digest lactose. The mutation prevents the gene from being turned off after weaning. Lactose tolerance has also been found in other populations, such as Africa and the Middle East, but these mutations occurred independently of the one that is usually found in lactose-tolerant people of European descent.

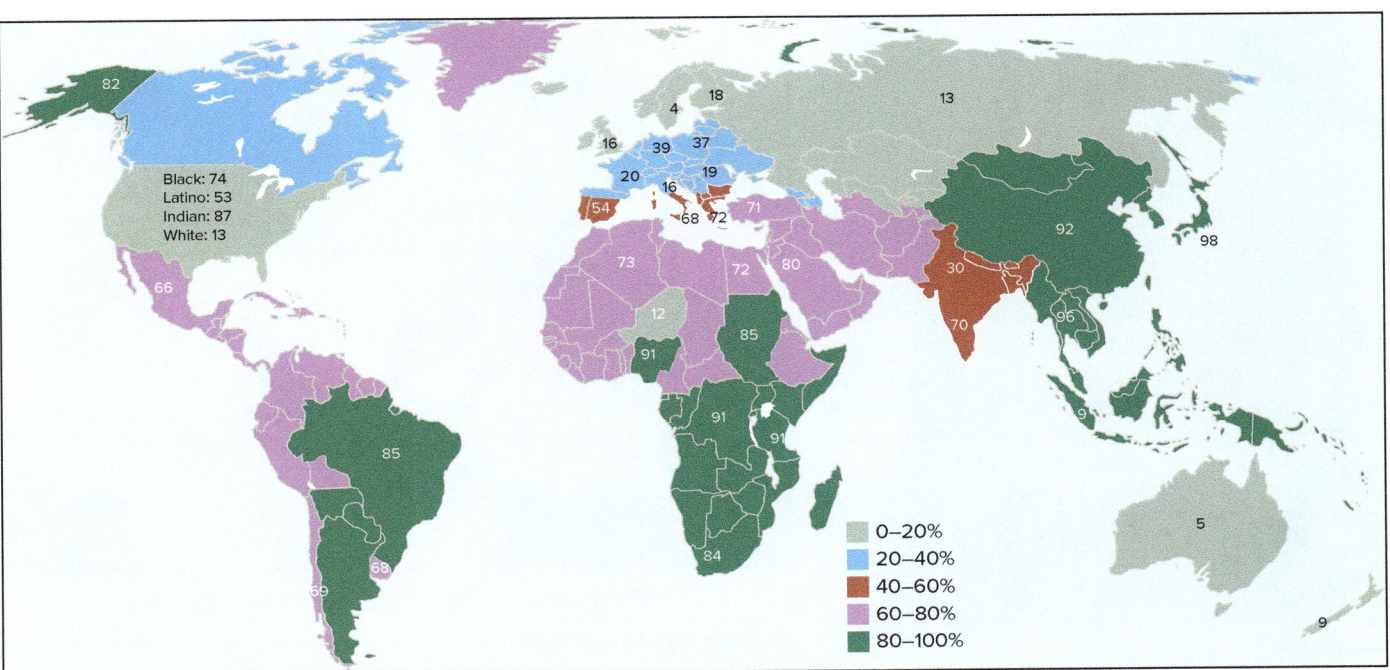

Figure 22.22 **The frequency of lactose intolerance in human populations.** This map emphasizes lactose intolerance, which is the inability to digest lactose after weaning. In contrast, most northern Europeans can digest the sugar lactose found in dairy products; they are lactose tolerant.

 Concept Check: *Explain why the frequency of lactose intolerance is high in some places and low in others.*

As another example, eye color in certain human populations has also changed in recent times. Over 10,000 years ago, all or nearly all humans had brown eyes. About 10,000 years ago, someone who lived near the Baltic Sea acquired a mutation in a gene called *Herc2* that resulted in blue eyes in the homozygous state. In 2008, researchers discovered that this mutation decreases the expression of the *Oca2* gene, which encodes a protein that is needed for the production of melanin pigment in the iris and other parts of the body. The frequency of this allele increased, especially in northern Europe (Figure 22.23), and by about 3,000 years ago, blue eyes had spread across Europe.

Why did the frequency of blue eyes increase in such a rapid fashion in human populations? The dramatic rise in blue eye color suggests that natural selection was playing a role, but researchers have yet to come up with a definitive answer regarding its selective advantage. One possible explanation has to do with vitamin D deficiency. Vitamin D is an important human vitamin that requires skin exposure to the UV rays in sunlight in order to be made. People living in northern latitudes are exposed to much less sunlight compared to those living nearer the equator, putting them at greater risk for vitamin D deficiency. A decrease in melanin synthesis not only affects eye color (brown to blue) but also results in lighter skin, which more easily absorbs UV rays. Therefore, one hypothesis for the spread of blue eyes (and lighter skin) through the human population is that it may have enabled humans to avoid the harmful health effects of vitamin D deficiency, which includes weakness and bone abnormalities. In this scenario, natural selection acted on skin color, and the eye color phenotype increased due to its association with a lighter skin color.

Modern Humans Show Relatively Small Genetic Variation Yet Exhibit a Significant Amount of Phenotypic Variation

Looks can be deceiving. If you take a plane ride around the world and make stops in Japan, Nigeria, Norway, Brazil, and Australia, you might get the impression that the human species is genetically diverse (Figure 22.24). The physical differences among humans in many parts of the globe are striking. To assess the level of genetic variation, the **1000 Genomes Project** was launched in January 2008; it is an international research effort to establish the level of human genetic variation. By 2015, this project had sequenced the genomes of over 2,500 people from around the world and compared those sequences to each other. The results indicate that, genetically speaking, humans are very, very similar to each other. Our level of genetic variation is lower than most species of mammals whose genomes have been sequenced, and it is even lower than the variation within certain fruit fly species. This may seem surprising; if you compared two fruit flies of the same species to each other, they would probably look very similar!

Most human genetic variation is in the form of single-nucleotide polymorphisms (SNPs), which are places where two humans' genomes differ at a single base pair (see Chapter 19). DNA sequencing allows researchers to identify sites where two people's genomes differ. The

(a) Blue eye color

(b) Percentage of people with blue eye color in Europe and surrounding regions

Figure 22.23 Blue eye color and its spread throughout Europe. (a) In humans, blue eye color is due to a lack of melanin in the iris. This mutation probably appeared only about 10,000 years ago. (b) Many people in western and northern Europe have blue eyes due to the spread of this single mutation that originated in someone living near the Baltic Sea. Note: The indicated percentages include blue eye color and other light colors of eyes, such as green.

(a) ©harpazo_hope/Getty Images

Figure 22.24 Examples of human phenotypic variation in different parts of the world. The women seen here are from (a) Japan, (b) Nigeria, (c) Norway, (d) Brazil, and (e) Australia. They are genetically very similar, even though they look somewhat different.
(a) ©William Perugini/123RF; *(b)* ©Filipe Frazao/Shutterstock; *(c)* ©Andrea Magugliani/Alamy; *(d)* ©Daniel Ernst/Alamy; *(e)* ©David Freund/Getty Images

data have revealed that greater than 99.9% of those differing sites are due to SNPs. If you compared your DNA with that of any unrelated person, you would probably find about 4.5 million SNP sites where your SNP was different from his or hers. This might sound like a lot, but consider that your genome is 3 billion base pairs long. If 4.5 million SNP differences occur between two people, this amounts to a genome difference of only 0.15%. The difference is much less than 1%!

Furthermore, this very low value of genetic variation is not greatly affected by the pairs of people who are analyzed. Let's suppose you compared the SNPs between two people of Japanese descent and compared the SNPs between a person of Japanese descent and a person of northern European descent. Both pairs would show a low level of SNP variation, which would be around 0.15%. The value for the first pair (Japanese-Japanese) would probably be a little lower (say, 0.14%), and the other pair (Japanese–northern European) might be a little higher (say, 0.16%), but both pairs would be remarkably similar. Genetic variation in human populations is very low and most of it occurs within all populations. Relatively little additional genetic variation is observed when comparing individuals from populations that are geographically separated.

Why is the genetic diversity of our species low? Various factors are involved. First, *Homo sapiens* has not been around that long. Even though 200,000 years might seem like a long time, humans have a long generation time, and evolution occurs over the course of generations. Second, until recently, human populations have been relatively small. About 10,000 years ago, the human population size is estimated to have been about 5 million. Small populations tend to be less genetically diverse than larger ones. Although the human population has grown enormously over the past few centuries, this expansion is very recent on an evolutionary timescale.

How do we explain the disparity between a low level of genetic variation and a seemingly higher level of phenotypic diversity? The answer is not entirely understood, but it may be related to the traits that influence our perception of diversity. Some of the traits that are most visually obvious, such as eye and skin color, may be dramatically influenced by natural selection even though they involve changes in a relatively small number of genes. For example, we have already considered how blue eye color and light skin color spread rapidly throughout Europe, and this phenotypic change involved a mutation in a single gene. This observation underscores how our visual perception of diversity may be biased by traits not rooted in major genetic differences.

22.3 Reviewing the Concepts

- Many defining characteristics of primates relate to their tree-dwelling nature; these include grasping hands, large brain, nails instead of claws, and binocular vision (Figures 22.14–22.16).
- About 7 mya in Africa, a lineage that led to humans began to separate from other primate lineages. A key characteristic of hominins (extinct and modern humans) is bipedalism. Human evolution can be visualized as a tree, with a few hominin species coexisting at the same point in time, some eventually going extinct, and some giving rise to other species (Figures 22.17–22.20).
- Data from the sequencing of human mitochondrial DNA suggest that *H. sapiens* originated in East Africa. From there, *H. sapiens* spread to Asia and then to all other parts of the globe (Figure 22.21).
- Human evolution has involved interbreeding between closely related species, such as *H. sapiens* and the Neanderthals and the Denisovans.
- Evidence that human populations are still evolving includes traits such as lactose tolerance and blue eye color (Figures 22.22, 22.23).
- Modern humans show relatively little genetic variation yet exhibit a significant amount of phenotypic variation (Figure 22.24).

22.3 Testing Your Knowledge

1. Which of the following is *not* a trait found in some or all primates?
   a. binocular vision
   b. opposable thumb
   c. flat nails
   d. large brain
   e. All of the above are found in some or all primates.

2. Some modern humans have a small amount of Neanderthal DNA. This observation suggests that
   a. *H. sapiens* and *N. neanderthalensis* interbred with each other.
   b. *H. sapiens* and *N. neanderthalensis* were not derived from a common ancestor.
   c. *H. sapiens* incurred new mutations that made them more like *N. neanderthalensis*.
   d. All of the above are correct.
   e. Only a and b are correct.

Assessing Your Knowledge and Skills

Foundational Knowledge

Multiple Choice

1. The movement of landmasses that have changed their positions, shapes, and association with other landmasses is called
   a. glaciation.
   b. Pangaea.
   c. continental drift.
   d. biogeography.
   e. geological scale.

2. Paleontologists estimate the dates of fossils by
   a. the layer of rock in which the fossils are found.
   b. the analysis of radioisotopes found in nearby igneous rock.
   c. the complexity of the body pattern of the organism.
   d. all of the above.
   e. a and b only.

3. According to the endosymbiosis theory, eukaryotic cells came into existence by
   a. endocytosis of an archaeal cell by a bacterial cell.
   b. endocytosis of a bacterial cell by an archaeal cell.
   c. fusion of an archaeal cell and a bacterial cell.
   d. fusion of a bacterial cell and an archaeal cell.
   e. a and c only.

4. The earliest fossils of vascular plants were formed during the _____ period.
   a. Ordovician
   b. Silurian
   c. Devonian
   d. Triassic
   e. Jurassic

5. Which of the following statements regarding modern humans, *H. sapiens,* is *false*?
   a. Some modern humans have a small amount of DNA that is derived from Neanderthals.
   b. Some modern humans have a small amount of DNA that is derived from Denisovans.
   c. *H. sapiens* probably arose in Africa.
   d. Modern humans are very genetically diverse compared to most other species.
   e. Modern humans are still evolving.

Short Answer

1. How are the ages of fossils determined? In your answer, you should discuss which types of rocks are analyzed and explain the concepts of radiometric dating and half-life.

2. **PRINCIPLES** Two principles of biology are (1) living organisms interact with their environment and (2) populations of organisms evolve from one generation to the next and are related by an evolutionary history. Describe two examples in which changes in the global climate affected the evolution of species.

Critical-Thinking Skills

Multiple Choice

1. You are analyzing a fossil using radiometric dating. The rock that surrounds the fossil contained about 18 mg of potassium-40 when it formed. The rock now contains 4.5 mg of potassium-40. Using the table in Figure 22.3b, what is the best estimate of the fossil's age?
   a. 1.3 billion years
   b. 2.6 billion years
   c. 3.9 billion years
   d. 5.2 billion years
   e. 10.4 billion years

2. Most biologists agree that mitochondria were acquired by eukaryotic cells before chloroplasts. What evidence is consistent with this idea?
   a. All eukaryotes have mitochondria, but only plants and some algae have chloroplasts.
   b. Mitochondria are derived from proteobacteria, whereas chloroplasts are derived from cyanobacteria.
   c. Mitochondira are autonomous, whereas chloroplasts are semiautonomous.
   d. Photosynthesis cannot occur without mitochondria.
   e. Both b and d are correct.

3. Which of the following evolutionary innovations was advantageous for survival in a terrestrial environment?
   a. amniotic eggs in certain groups of animals
   b. seeds in certain groups of plants
   c. a cell wall in certain groups of fungi
   d. water channels in certain groups of protists
   e. Only a and b are correct.

4. Which of the following events would be least likely to cause a mass extinction?
   a. glaciation
   b. volcanic eruptions
   c. the formation of polyploid plants
   d. changes in temperature
   e. meteorite impacts

5. Which of the following is *not* an example indicating that the human population is still evolving?
   a. resistance to diseases that have caused a great loss of human lives
   b. variation in eye color
   c. the presence of DNA in human mitochondria
   d. lactose tolerance in some people
   e. All of the above are examples that humans are still evolving.

Short Answer

1. How was the phenomenon of endosymbiosis important in the evolution of the first eukaryotic cells?

2. Discuss the factors that have contributed to the dramatic changes in life-forms since the origin of life on Earth about 3.5 to 4 billion years ago.

UNIT V
DIVERSITY

Biological diversity encompasses the variety of living things that exist on Earth now, as well as all the life-forms that lived in the past. Knowing about the many different kinds of modern organisms helps us to understand how life-forms are structured in ways that allow them to function differently in nature (described in Units VI and VII) and how species interact with each other and with their environments (described in Unit VIII—Ecology).

Unit V begins with microorganisms, which include bacteria, archaea, diverse lineages of protists, and fungi, Earth's oldest, simplest, and most numerous life-forms; Chapter 23 reveals the diversity of microorganisms that influence our lives. In Chapter 24, we will learn how microbes occur in complex systems known as microbiomes, which influence our environments and health. Chapter 25 explores the diversification of plants, a process that changed features of planet Earth, culminating in seed plants that are vital sources of human food, fiber, and medicine. The simplest animals, the invertebrates, are covered in Chapter 26. More complex animals, including mammals, are the focus of Chapter 27.

The following biology principles are emphasized in this unit:

- **Cells are the simplest units of life.** Microbial species commonly have bodies consisting of only one or a few cells, but some bacteria, many fungi and protists, and all plants and animals are composed of many cells of specialized structure and function.

- **Living organisms interact with their environment.** Ancient microorganisms and plants dramatically changed the composition of Earth's atmosphere, and their modern descendants continue to influence today's atmosphere and climate.

- **Populations of organisms evolve from one generation to the next and are related by an evolutionary history.** This concept, emphasized throughout the unit, helps us to understand the origin of Earth's biodiversity.

- **Structure determines function.** The chapters in this unit provide many examples of ways in which evolutionary change in molecular, cellular, or body structure has allowed organisms to achieve new and diverse functions.

(23) ©David M. Phillips/Science Source;
(24) ©RuslanDashinsky/Getty Images;
(25) ©Linda Graham; (26) ©InVision_Photography/Getty Images;
(27) ©Neil Bromhall/Shutterstock

Diversity of Microbial Life: Archaea, Bacteria, Protists, and Fungi

23

©David M. Phillips/Science Source

The protist *Trichomonas vaginalis* is one type of microorganism that causes infection of the human urogenital tract.

Chapter Outline

23.1 Introduction to Microorganisms
23.2 Diversity and Ecological Importance of Archaea
23.3 Diversity and Ecological Importance of Bacteria
23.4 Diversity in Bacterial Cell Structure and Metabolism
23.5 Diversity and Ecological Importance of Protists
23.6 Diversity and Ecological Importance of Fungi
23.7 Technological Applications of Microorganisms

Assessing Your Knowledge and Skills

Millions of people around the world suffer from infections with urogenital tract microorganisms, which include prokaryotic bacteria as well as eukaryotic protists and fungi, such as yeasts. *Trichomonas vaginalis*, for example, is a sexually transmitted protist that causes disease symptoms when it invades the human genitourinary tract. In this location, *T. vaginalis* consumes bacteria and host epithelial and red blood cells, as well as carbohydrates and proteins released from damaged host cells. An estimated 170 million cases of infection with this organism occur each year around the globe, and infections can predispose humans to other diseases. *T. vaginalis* has an undulating membrane and flagella that allow it to move over mucus-coated skin (see the chapter-opening photo). *Giardia intestinalis,* which infects the human intestinal tract, is another example of a protist whose flagella are critical to the ability to cause disease. Many types of disease-causing bacteria and fungi likewise may inhabit human bodies.

These examples illustrate how knowledge of microbial diversity and structure is essential in medicine, and such knowledge is also important in comprehending the many ecological roles that Earth's microorganisms play. By influencing global environment, many microbes that do not cause disease nevertheless influence human health and sustainability. In this chapter, we will survey microbial diversity, including the prokaryotic domains Archaea and Bacteria, as well as diverse eukaryotic protists and fungi.

23.1 Introduction to Microorganisms

Learning Outcomes

1. Define the terms microorganism, archaea, bacteria, protist, and fungi.
2. Draw a diagram that illustrates horizontal gene transfer, and explain how it benefits the recipient organism.

Microorganisms, also known as microbes, are organisms that are typically so small in size that they can be seen only with the use of a microscope. Although a few are microscopic in size, most animals and plants grow large enough to be seen without a microscope. In this chapter, we will focus on groups that are primarily microscopic:

- The prokaryotic domain Archaea (informally known as archaea)
- The prokaryotic domain Bacteria (informally, bacteria)
- Diverse phyla of protists and fungi having eukaryotic cells, which are classified in the domain Eukarya, together with animals and plants

Recall that prokaryotic cells lack many of the features that occur in eukaryotic cells. Examples of distinctive eukaryotic cell features include nuclei having an envelope with complex pores, mitochondria, and other membrane-bound organelles. Comparative study of modern microbes helps to reveal how eukaryotic cells first arose from prokaryotic ancestors.

Archaea and Bacteria Are Ancient Lineages That Continue to Be Abundant Today

The domains Archaea and Bacteria are both monophyletic groups, meaning they each have a single common ancestor. Although archaea and bacteria together are sometimes referred to as the "prokaryotes," this combination does not represent a monophyletic group; Archaea are more closely related to the domain Eukarya (**Figure 23.1**).

Archaea and bacteria include Earth's smallest known cells and are the most abundant organisms in the world. About half of Earth's total biomass consists of an estimated 10^{30} individual bacteria and archaea. Just a pinch of garden soil can contain 2 billion prokaryotic cells, and about a million occur in 1 mL of seawater. Archaea and bacteria live in nearly every conceivable habitat, including extremely

Figure 23.1 Bacteria, Archaea, and Eukarya are the three domains of life on Earth. Only 10 of about 100 known bacterial phyla are shown, for simplicity. Note the close relationship between Archaea and Eukarya. Horizontal gene transfer among organisms belonging to the same or different domains is a key evolutionary process.

Biology Principle

Cells Are the Simplest Units of Life

In the cases of unicellular bacteria and archaea, a single cell represents an entire organism, but some bacteria have multicellular bodies whose cells are interconnected and may be specialized.

(a) A unicellular bacterium, *Vibro parahaemoliticus*, with flagella

(b) A multicellular cyanobacterium, *Cylindrospermum*, with specialized cells: heterocyte and akinete

Figure 23.2 Bacterial cells. (a) *Vibrio parahaemoliticus*, a unicellular bacterium that causes seafood poisoning, artificially colored. (b) *Cylindrospermum*, a multicellular blue-green bacterium with specialized cells: pale heterocytes that foster nitrogen-fixation and larger, orange-coated akinetes that may survive stressful conditions. The blue-green color of the other cells is the natural color of photosynthetic light-harvesting pigments.
(a) ©Dennis Kunkel Microscopy/Science Source; *(b)* ©Lee W. Wilcox

Protists Are Early-Diverging Eukaryotes Displaying Diverse Structural Types

The term protist comes from the Greek word *protos,* meaning first, reflecting fossil and molecular evidence that protists were Earth's first eukaryotes. **Protists** are diverse eukaryotic organisms that cannot be classified into the kingdom Fungi, Plantae, or Animalia. Protists that lack chloroplasts are informally termed protozoa (Figure 23.3a) or fungus-like protists (Figure 23.3b), whereas protists having one or more chloroplasts are informally termed algae (Figure 23.3c). Although most protists are microscopic in size (see Figure 23.3a, b), many can be observed without using a microscope, examples being the multicellular marine photosynthetic protists commonly known as seaweeds (see Figure 23.3c).

Fungi Form a Kingdom of Diverse, Plastidless Eukaryotes

Fungi are eukaryotic organisms that lack plastids, so they are not photosynthetic and live by absorbing organic food. Some fungi are unicellular, and others occur as microscopic threads known as hyphae (Figure 23.4). Although most of the body of most fungi is microscopic in size, the hyphae of many species can merge to form structures visible with the naked eye—mushrooms are examples. Another distinctive feature of fungi is that most species produce cell walls containing the polysaccharide chitin.

Horizontal Gene Transfer Fosters Microbial Evolution

Archaea, bacteria, protists, and fungi often live in communities that occupy the same types of habitats. Living closely together in communities fosters the exchange of genetic material between microbial species, an important evolutionary process known as horizontal gene transfer. **Horizontal gene transfer (HGT)** is the process in which an organism receives genetic material from another organism without being the offspring of that organism. This process contrasts with vertical gene transfer, which occurs from parent to progeny. Horizontal gene transfer is increasingly recognized as an important evolutionary mechanism. Recipient organisms benefit from this process by acquiring new metabolic capabilities.

Horizontal gene transfer is particularly common among archaea and bacteria (see Figure 23.1). For example, at least 17% of the genes present in the common human gut bacterium *Escherichia coli* came from other bacteria. The study of nearly 200 genomes has revealed that about 80% of bacterial and archaeal genes have been involved in horizontal transfer at some point in their history. Genes also move among the bacterial, archaeal, and eukaryotic domains. For example, about a third of the genes present in the archaeal species *Methanosarcina mazei* originally came from bacteria. Bacterial genes have also been transferred to many types of eukaryotic protists, fungi, and plants. Horizontal gene transfer also occurs among different fungal species, as well as between fungi and protists, animals, and plants. Such gene transfer increases the genetic diversity of microorganisms and other living things, influencing their evolution.

Horizontal gene transfer can occur between different prokaryotic species via both transduction and transformation, as discussed in Chapter 17. Horizontal gene transfer also occurs when eukaryotes feed or when microbes invade the cells of other organisms, known as hosts.

hot or salty waters that support no other life, and they are Earth's most ancient organisms, having originated more than 3.4 bya. Their great age and varied habitats have resulted in extraordinarily high metabolic diversity. Such high metabolic diversity explains why modern archaea and bacteria play diverse ecological roles and can be used in many technological applications. Prokaryotic organisms are often single-celled (Figure 23.2a), though many achieve larger body size by aggregating cells into colonies or rows of cells known as **filaments.** Some bacteria display the defining features of true multicellularity, such as intercellular connections and specialized cells (Figure 23.2b).

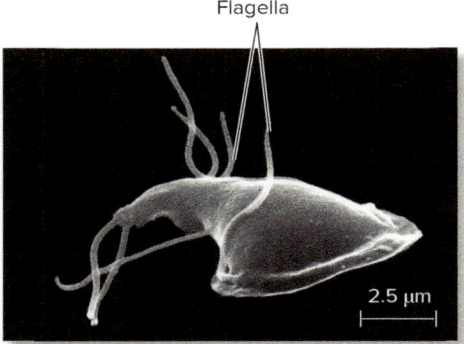
(a) A single-celled protist, the flagellate protozoan, *Giardia*

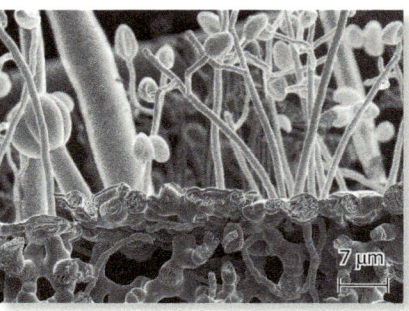
(b) *Phytophthora infestans*, a fungus-like protist whose filaments grow within and emerge from a plant leaf

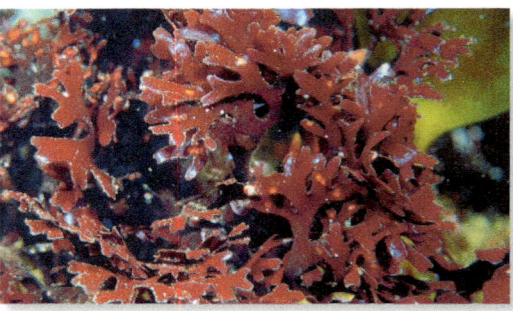
(c) *Chondrus crispus*, a multicellular algal protist

Figure 23.3 Protist bodies take diverse forms that can be unicellular or multicellular. (a) Scanning electron microscopic (SEM) view of the single-celled flagellate protozoan, *Giardia*, which infects the human gut. (b) SEM view of microscopic filaments produced by the fungus-like protist *Phytophthora infestans*, which causes disease of wild and crop plants. (c) An edible multicellular red alga that is conspicuous to the unaided eye, *Chondrus crispus*. Note: SEM images are made with a scanning electron microscope (SEM) that employs electrons rather than visible light, with the result that cellular structures do not normally appear in color.
(a) Source: Dr. Stan Erlandsen, Dr. Dennis Feely/CDC; *(b)* ©Andrew Syred/Science Source; *(c)* ©Andrew J. Martinez/Science Source

(a)

(b)

Figure 23.4 Fungal bodies can be unicellular or composed of microscopic filaments known as hyphae. (a) Scanning electron microscopic (SEM) view of the single-celled yeast fungus *Saccharomyces cerevisiae*, important in the brewing and baking industries. (b) SEM of *Trichoderma* hyphae attacking the wider hyphae of another fungus.
(a) ©Mediscan/Alamy; *(b)* ©Biophoto Associates/Science Source

23.1 Reviewing the Concepts

- Prokaryotic Bacteria (informally, bacteria) and Archaea (informally, archaea) form two of the three domains of life. Prokaryotic species are often single-celled, though cellular aggregations are common, and examples of multicellularity occur (Figures 23.1, 23.2).
- Protists are classified into the domain Eukarya and occur as single eukaryotic cells or more complex bodies; multicellularity is common, and some protists, such as photosynthetic seaweeds, are quite large (Figure 23.3).
- Fungi form a eukaryotic kingdom of non-photosynthetic unicellular or filamentous organisms that usually have chitin in their cell walls (Figure 23.4).
- Horizontal gene transfer (HGT) commonly occurs among prokaryotic species and between domains of life, fostering evolutionary diversification.

23.1 Testing Your Knowledge

1. Imagine that you have found a new type of organism that is composed of several cells but can only be observed by using a microscope. Is it possible that the cells of this organism have prokaryotic organization?
   a. Yes; although many prokaryotes are single-celled, some prokaryotes are multicellular.
   b. No; so far as is known, prokaryotes occur only as single cells.

2. What process can serve as a mechanism of horizontal gene transfer?
   a. viral transduction
   b. transformation
   c. feeding by eukaryotes
   d. invasion of host cells by microbes
   e. All of the above are correct.

23.2 Diversity and Ecological Importance of Archaea

Learning Outcomes

1. Describe the evolutionary and ecological importance of archaea.
2. List the features of archaea that enable them to grow in extreme habitats.

From an evolutionary perspective, archaea are important because they share a number of features with eukaryotes, suggesting common ancestry. For example, histone proteins are typically associated with the DNA of both archaea and eukaryotes, but such proteins are absent from most bacteria. Archaea and eukaryotes also share more than 30 ribosomal proteins that are not present in bacteria, and archaeal RNA polymerases are closely related to their eukaryotic counterparts. These similarities help evolutionary biologists learn how the information systems of eukaryotes evolved.

Archaea are also ecologically important. Though many archaea occur in soils and surface ocean waters of moderate conditions, diverse archaea occupy habitats with very high salt content, acidity, methane levels, or temperatures that would kill most bacteria and eukaryotes. Organisms that occur primarily in extreme habitats are known as **extremophiles** (meaning lovers of extreme conditions). One example is *Methanopyrus*, which grows best at deep-sea thermal vent sites where the temperature is 98°C and is thus known as a hyperthermophile (meaning lover of high temperatures). At this temperature, the proteins of most organisms would denature, but those of *Methanopyrus* are resistant to such damage. Archaea help biologists to better understand the origin of life, the origin of eukaryotes, how life on Earth has evolved in extreme environments, and what kinds of extraterrestrial life might exist.

The domain Archaea includes the phyla Lokiarchaeota, Korarchaeota, Thaumarchaeota, Crenarchaeota, and Euryarchaeota (see Figure 23.1). Lokiarchaeota are known for their particularly close evolutionary relationship to Eukarya. Thaumarchaeota species that oxidize ammonia are important in global nitrogen cycling. The early-diverging Euryarchaeota include some hyperthermophiles, diverse methane producers, and extreme halophiles—species able to grow in higher than usual salt concentrations.

Archaea possess distinctive membrane lipids, which are formed with ether bonds; in contrast, ester bonds characterize the membrane lipids of bacteria and eukaryotes (**Figure 23.5**). Ether-bonded membranes are resistant to damage by heat and other extreme conditions, which helps explain why many archaea are able to grow in extremely harsh environments. Also note that archaea use isoprene chains instead of fatty acid chains to build membranes. Although some archaea lack cell walls, most possess a wall composed of protein (which differs from the chemical composition of most bacterial cell walls described in Section 23.4).

Figure 23.5 Structure of bacteria and archaea. (a) Bacteria and (b) archaea both have prokaryotic cell structure, but (c) bacterial membrane lipids are formed with ester linkages, whereas (d) archaeal membrane lipids feature ether linkages, which are thought to be more stable under extreme environmental conditions. As shown in transmission electron microscopic (TEM) images (a) and (b), most bacteria feature a cell wall made of a material known as peptidoglycan, which is often enclosed by an outer envelope, whereas archaea lack these features. Most archaea have outer coverings made of protein.
(a) ©Linda Graham; *(b)* ©Eye of Science/Science Source

Archaea are also ecologically important as producers of methane (CH_4)—the major component of natural gas. Methane is a greenhouse gas that increases global warming over 20 times more per molecule than does CO_2. In recent years, the level of CH_4 has been increasing in Earth's atmosphere as the result of human activities. Several groups of anaerobic archaea convert CO_2, methyl groups, or acetate to CH_4 and release CH_4 from their cells into the atmosphere. Methane-producing archaea live in swampy wetlands, the bottoms of lakes, deep-sea habitats, and the digestive systems of animals such as cattle and humans.

23.2 Reviewing the Concepts

- Domain Archaea, particularly the phylum Lokiarchaeota, is more closely related to Domain Eukarya than either is to Domain Bacteria.
- Many representatives of Domain Archaea occur in extremely hot, salty, or acidic habitats. Ether-linked membrane lipids are among the features of archaea that enable their survival in extreme habitats (Figure 23.5).
- Several groups of archaea that occupy watery habitats generate the powerful greenhouse gas methane (CH_4).

23.2 Testing Your Knowledge

1. What features are characteristic of archaea but not bacteria?
   a. Archaea possess membrane lipids linked by ether bonds—not ester bonds, as in bacteria.
   b. Archaea typically have cell walls composed of protein—not peptidoglycan, as in most bacteria.
   c. The DNA of archaea is bound to histone proteins, as in eukaryotes but not bacteria.
   d. All of the above are correct.
   e. None of the above is correct.
2. How does methane production by certain archaea contribute directly to global climate change?
   a. Methane has a cooling effect on Earth's climate.
   b. Methane in Earth's atmosphere increases rainfall.
   c. Methane is a powerful greenhouse gas that helps to warm Earth's climate.
   d. Methane in Earth's atmosphere increases the chance of fire.
   e. None of the above is correct.

23.3 Diversity and Ecological Importance of Bacteria

Learning Outcomes

1. Explain the evolutionary and ecological importance of cyanobacteria.
2. List some examples of proteobacteria important in the environment.
3. Explain how populations of bacteria increase and how some bacteria survive under stressful conditions.

Table 23.1 Representative Bacterial Phyla

Phyla	Characteristics
Deinococcus-Thermus	Extremophiles. The genus *Deinococcus* is known for high resistance to ionizing radiation, and the genus *Thermus* inhabits hot springs. *Thermus aquaticus* has been used in commercial production of *Taq* polymerase enzyme used in the polymerase chain reaction (PCR), an important procedure in molecular biology laboratories.
Chloroflexi	Known as the green nonsulfur bacteria. Chloroflexi conduct photosynthesis without releasing oxygen (anoxygenic photosynthesis).
Firmicutes	Diverse Gram-positive bacteria, some of which produce endospores. The disease-causing *Clostridium difficile* is an example.
Actinobacteria	Gram-positive bacteria producing branched filaments; many form spores. *Mycobacterium tuberculosis*, the agent of tuberculosis in humans, is an example. Actinobacteria are notable antibiotic producers; over 500 different antibiotics are known from this group.
Cyanobacteria	The oxygen-producing photosynthetic bacteria. Photosynthetic pigments include chlorophyll *a* and blue-green or red accessory pigments. Cyanobacteria occur as unicells, colonies, unbranched filaments, and branched filaments. Many of the filamentous species produce specialized cells: dormant akinetes and heterocytes in which nitrogen fixation occurs. In waters having excess nutrients, cyanobacteria produce blooms and may release toxins harmful to the health of humans and wild and domesticated animals.
Spirochaetes	Motile bacteria having distinctive corkscrew shapes, with flagella held close to the body. Spirochaetes include the pathogens *Treponema pallidum*, the agent of syphilis, and *Borrelia burgdorferi*, which causes Lyme disease.
Bacteroidetes	Includes representatives of diverse metabolism types. Some are common in the human intestinal tract, and others are primarily aquatic.
Chlamydiae	Notably tiny, obligate intracellular parasites. Some cause eye disease in newborns or sexually transmitted diseases.
Planctomycetes	Reproduce by budding rather than binary fission. Cell walls lack peptidoglycan; the cytoplasm may contain nucleus-like bodies, as in *Gemmata obscuriglobus*.
Proteobacteria	A very large group of Gram-negative bacteria, collectively having high metabolic diversity. Proteobacteria includes many species important in medicine, agriculture, and industry.

Molecular studies suggest the existence of nearly 100 bacterial phyla, though many are poorly known. Though some members of Domain Bacteria live in extreme environments, most favor moderate conditions. The characteristics of 10 prominent bacterial phyla (see Figure 23.1) are briefly summarized in **Table 23.1**. Among these, the Cyanobacteria and the Proteobacteria are particularly diverse and relevant to eukaryotic cell evolution, global ecology, and human affairs.

Cyanobacteria Are Photosynthetic Bacteria That Produce Oxygen and Play Other Important Ecological Roles

The phylum Cyanobacteria contains photosynthetic bacteria that are abundant in fresh waters, oceans, and wetlands and on the surfaces of arid soils. Cyanobacteria are named for blue-green (cyan) coloration conferred by photosynthetic pigments that help chlorophyll absorb light energy (see Figure 23.2b). Cyanobacteria are the only bacteria known to generate oxygen as a product of photosynthesis. Ancient relatives of modern cyanobacteria produced Earth's first oxygen-rich atmosphere, and modern cyanobacteria continue to generate atmospheric oxygen. The chloroplasts of eukaryotic algae and plants evolved from cyanobacteria, a process that involved massive horizontal transfer of bacterial genes into eukaryotic nuclear genomes.

In phosphorus-rich bodies of water such as lakes, some species of cyanobacteria grow rapidly into large, visible populations (blooms) that color the water blue-green, or cyan. The individual cells release small amounts of toxins that help to keep small aquatic animals from eating them, but when blooms occur, toxins can rise to levels that harm humans, pets, livestock, and wildlife. Consequently, public health authorities often warn people not to swim in waters with visible blue-green blooms or allow pets and livestock to drink such water. People can help to prevent the formation of cyanobacterial blooms by reducing the input of phosphorus-rich fertilizers, manure, and sewage into bodies of water.

Despite the harmful effects of some species, cyanobacteria provide important benefits to humans and other organisms, in addition to producing atmospheric oxygen. Cyanobacteria generate diverse organic compounds, including antibiotics and sticky mucilage that helps to stabilize soils and aquatic sediments. Many cyanobacteria also have the ability to convert abundant but inert atmospheric nitrogen gas into ammonia, which algae and plants can use to synthesize amino acids and proteins. This process, known as **nitrogen fixation**, enriches nutrient-poor soils, particularly wet paddy fields where rice is grown in many regions of the world, thereby helping to provide food for billions of people. Cyanobacteria illustrate the wide range of body diversity found among bacterial phyla:

- Unicells (Figure 23.6a) are single-celled bodies.
- Colonies are groups of cells held together by a thick, gluey substance called mucilage (Figure 23.6b).
- Filaments are cells that are attached end-to-end (Figure 23.6c); some filaments are branched (Figure 23.6d). Some filamentous

cyanobacteria display the hallmarks of multicellularity that also mark multicellular plants and animals: cell-to-cell attachment, specialized cells, intercellular chemical communication, and programmed cell death.

Proteobacteria Do Not Produce Oxygen but Play Other Ecologically Important Roles

Although some proteobacteria are photosynthetic, these species do not produce oxygen as do the cyanobacteria. Certain proteobacteria play an important role by using oxygen to consume methane, thereby reducing the impact of this greenhouse gas. Methane-destroying proteobacteria often live in close association with photosynthetic plants and algae, which provide the oxygen the bacteria need to oxidize methane. The ancient bacteria that evolved into mitochondria were closely related to modern Proteobacteria. Though proteobacteria share molecular and cell-wall features, this phylum displays amazing diversity of form and metabolism. The class Alphaproteobacteria includes nitrogen-fixing species that associate with plants, thereby fostering plant growth. Some alphaproteobacterial species destroy methane, thereby reducing the potential environmental impacts of this greenhouse gas.

Bacterial Reproduction Helps to Explain Ecological Function

Bacteria lack eukaryote-type sexual reproduction involving specialized gametes, gamete fusion (syngamy), and meiosis, though they can exchange some genes by conjugation, transformation, and transduction (described in Chapter 17). Some bacteria reproduce asexually by budding, producing small progeny cells from larger parental cells. However, most bacteria reproduce asexually by division of a single cell into two equal progeny cells, a process known as **binary fission** (Figure 23.7a).

Binary fission is the basis of a widely used method for detecting and counting bacteria in food, water samples, or patient fluids. Microbiologists who study the spread of disease need to quantify bacterial cells in samples taken from the environment. Medical

| 0.1 mm | 250 μm | 700 μm | 50 μm |

(a) Unicells   **(b) Colony of cells**   **(c) Unbranched filaments**   **(d) Branched filaments**

Figure 23.6 Major body types found in the phylum Cyanobacteria. **(a)** The genus *Chroococcus* occurs as unicells. **(b)** The genus *Merismopedia* forms a flat colony of cells held together by mucilage. **(c)** The genus *Oscillatoria* exists as unbranched filaments. **(d)** The genus *Stigonema* forms a branched filament having a mucilage sheath; sunscreen compounds that protect the cells from damage by ultraviolet (UV) radiation cause the brown color of the sheath.
(a) ©Linda Graham; *(b)* ©Michael Abbey/Science Source; *(c)* ©Sinclair Stammers/Science Source; *(d)* ©Lee W. Wilcox

(a) Bacterium undergoing binary fission

(b) Colonies developed from single cells

(c) Bacteria stained with fluorescent DNA-binding dye

Figure 23.7 Bacteria can be visualized and counted by means of colonies formed by repeated binary fission or by using fluorescent dye. *(a)* ©SCIMAT/Science Source; *(b)* ©Linda Graham; *(c)* ©Lee W. Wilcox

technicians often need to count bacteria in body fluid samples to assess the level of infection. However, because bacterial cells are small and often unpigmented, they are difficult to view and count directly. One way that microbiologists count bacteria is to place a measured volume of sample into plastic dishes filled with a semisolid nutrient medium. Bacteria in the sample undergo repeated binary fission to form colonies of cells visible to the unaided eye. Because each colony is assumed to represent a single cell that was present in the original sample, the number of colonies in the dish approximates the number of living bacteria in the original sample (**Figure 23.7b**). Bacteria that cannot be cultivated in the laboratory can, however, be observed and counted with the use of a fluorescent dye that binds bacterial DNA (**Figure 23.7c**).

Some bacteria produce thick-walled cells that are able to survive unfavorable conditions in a dormant state. These specialized cells develop when bacteria have experienced stress, such as low nutrients or unfavorable temperatures, and are able to germinate into metabolically active cells when conditions improve. For example, when winter approaches, aquatic filamentous cyanobacteria often produce **akinetes**—large, thick-walled, food-filled cells (see Figure 23.2b). Akinetes are able to survive winter at the bottoms of lakes, and they produce new filaments in spring when they are carried by water currents to the brightly lit surface. Persistence of such akinetes, followed by their growth into filaments by means of binary fission, explains how harmful cyanobacterial blooms can develop year after year in overly fertile lakes.

Endospores (**Figure 23.8**) are produced inside the cells of some bacterial species. DNA and other materials become enclosed within a tough coat and then are released when the enclosing cell dies and breaks down. Bacterial endospores can remain alive, though in a dormant state, for long periods, then reactivate when conditions are suitable. The ability to produce endospores allows some Gram-positive Firmicutes bacteria to cause serious diseases. For example, *Bacillus anthracis* causes the disease anthrax, a potential agent in bioterrorism and germ warfare. Most cases of human anthrax result when endospores of *B. anthracis* enter breaks in the skin, causing skin infections that are relatively easily cured by antibiotic treatment. But sometimes the endospores are inhaled or consumed in undercooked, contaminated meat, potentially causing more serious illness or death. *Clostridium botulinum* can contaminate improperly canned food that has not been heated to temperatures high enough to destroy its tough endospores. When the endospores germinate and bacterial cells grow in the food, they produce deadly botulinum toxin, as well as NH_3 and CO_2 gas, which causes can lids to bulge. If humans consume the food, the toxin causes botulism, a severe type of food poisoning that can lead to respiratory and muscular paralysis. The toxin is so potent that only 400 g of *C. botulinum* toxin would be sufficient to kill every human on Earth. That being the case, it is interesting that people use a commercial preparation of botulinum toxin in the form of Botox, which is injected into the skin to paralyze facial muscles, thereby reducing the appearance of wrinkles.

Clostridium tetani produces a nerve toxin that causes lockjaw, also known as tetanus, when bacterial cells or endospores from soil enter wounds. The ability of the genera *Bacillus* and *Clostridium* to produce resistant endospores helps to explain their widespread presence in nature and their effect on humans.

Figure 23.8 Specialized bacterial cells capable of dormancy. An endospore with a resistant wall develops within the cytoplasm of the pathogen *Clostridium difficile*.
©Dr. Kari Lounatmaa/Science Source

Concept Check: *How do endospores influence the ability of some bacteria to cause disease?*

23.3 Reviewing the Concepts

- Domain Bacteria includes 50 or more phyla, including Cyanobacteria and Proteobacteria (Table 23.1).
- Cyanobacteria were ancestral to all modern chloroplasts, are the only prokaryotic organisms that produce oxygen as a result of photosynthesis, sometimes form harmful bloom populations in overly fertile waters, and illustrate prokaryotic body types (Figure 23.6).
- Proteobacteria do not produce oxygen and include environmentally important species.
- Bacteria reproduce by binary fission or budding, and some produce akinetes or endospores that allow them to survive under unfavorable conditions (Figures 23.7, 23.8).

23.3 Testing Your Knowledge

1. What feature of cyanobacteria can be harmful to humans and other animals?
   a. oxygen production
   b. toxin production
   c. production of blue-green photosynthetic pigments
   d. carbon fixation
   e. nitrogen fixation

2. What essential role do some alphaproteobacteria play in the lives of plants?
   a. They were precursors of plastids.
   b. They were precursors of nuclei.
   c. Some are able to convert atmospheric nitrogen gas into ammonia, which plants use to make amino acids and proteins.
   d. Some evolved into the endomembrane system.
   e. All of the above are correct.

23.4 Diversity in Bacterial Cell Structure and Metabolism

Learning Outcomes

1. Discuss structural adaptations that have increased the complexity of prokaryotic cells.
2. Describe the structural differences between Gram-positive and Gram-negative bacterial cells.
3. Explain how mucilage influences bacterial behavior.
4. List the different means by which prokaryotic cells can move.

Even though bacteria have a much simpler cellular organization than do eukaryotes, many bacteria display structural adaptations that increase their complexity. Cyanobacteria and other photosynthetic bacteria, for example, are able to use light energy to produce organic compounds because their cells contain large numbers of thylakoids—flattened, tubular membranes that grow inward from the plasma membrane (**Figure 23.9a**). The extensive membrane surface of the thylakoids bears large amounts of chlorophyll and other components required for photosynthesis. This explains why thylakoids are also

Thylakoids provide a greater surface area for chlorophyll and other molecules involved in photosynthesis.

The gas vesicles buoy this photosynthetic organism to the lighted water surface, where it often forms conspicuous scums.

(a)

(b)

(c)

Figure 23.9 Bacterial internal cell structures (a) Photosynthetic thylakoid membranes and numerous gas vesicles found in a cell of the aquatic cyanobacterial genus *Microcystis*. (b) Magnetosomes found in the spirillum *Magnetospirillum magnetotacticum*. Each iron-rich magnetite crystal in the internal row is enclosed by a membrane derived from the plasma membrane. (c) A nucleus-like structure in *Gemmata obscuriglobus*, a species of planctomycete bacteria. Although the nuclear envelope–like structure lacks pores, it encloses the bacterial cell's DNA, known as a nucleoid.
(a) ©Norma Lang; (b) ©Dr. Richard P. Blakemore, University of New Hampshire; (c) ©Richard Webb and John Fuerst, The University of Queensland, Australia/Centre for Microscopy and Microanalysis

abundant in plant chloroplasts, which descended from cyanobacterial ancestors. Thylakoids enable photosynthetic bacteria and chloroplasts to take maximum advantage of light energy in their environments.

In other bacteria, plasma membrane ingrowth has generated additional intriguing adaptations—magnetosomes and nucleus-like bodies—that are sometimes described as bacterial organelles. Magnetosomes are tiny crystals of an iron mineral known as magnetite, each surrounded by a membrane. These structures occur in the bacterium *Magnetospirillum* and related genera (**Figure 23.9b**). In each cell, about 15 to 20 magnetosomes occur in a row, together acting as a compass needle that responds to the Earth's magnetic field. Magnetosomes help the bacteria to orient themselves in space and thereby locate the submerged, low-oxygen habitats they prefer. Plasma membrane ingrowths may also surround cellular DNA, producing nucleus-like structures within bacteria of the phylum Planctomycetes (**Figure 23.9c**). Common in aquatic habitats, such bacteria are important to evolutionary biologists who are interested in the evolutionary origin of the eukaryotic nucleus.

Prokaryotic Cells Vary in Shape

Bacterial cells occur in five common shapes (**Figure 23.10**):

- Spheres, known as **cocci,** maximize surface area/volume ratio, thereby enhancing exchange of materials with the environment.
- Elongate rods, called **bacilli,** are able to store more nutrients than some other shapes.
- Comma-shaped cells are called **vibrios.**
- Spiral-shaped cells that are flexible are known as **spirochaetes.**
- Spiral-shaped cells that are rigid are termed **spirilli.**

Cytoskeletal proteins similar to those present in eukaryotic cells control these cell shapes. For example, helical strands of an actin-like protein are responsible for the rod shape of bacilli; if this protein is not produced, bacilli become spherical in shape.

Bacterial Cells Vary in Cell-Wall Structure

Most prokaryotic cells possess a rigid cell wall outside the plasma membrane. Cell walls maintain cell shape and help protect against

(a) Gram-positive bacteria (b) Gram-negative bacteria

Figure 23.11 Gram-positive and Gram-negative bacteria. **(a)** *Streptococcus pneumoniae,* a member of the phylum Firmicutes, stains positive (purple) with the Gram stain. **(b)** *Escherichia coli,* a member of the Proteobacteria, stains negative (pink) when the Gram stain procedure is applied.
(a) ©CNRI/Science Source; *(b)* ©Lee W. Wilcox

attack by viruses or predatory bacteria. Cell walls also help bacteria and other microbes avoid lysing in hypotonic conditions, when the solute concentration is higher inside the cell than outside. The structure and composition of bacterial cell walls are medically important.

A polymer known as **peptidoglycan** is an important component of most bacterial cell walls. Peptidoglycan is composed of carbohydrates that are crosslinked by peptides. Bacterial cell walls occur in two major forms, which differ in peptidoglycan thickness, the presence or absence of a membrane occurring outside the peptidoglycan layer, staining properties, and response to antibiotics. Bacteria having these chemically different walls are called Gram-positive or Gram-negative bacteria, after the staining process used to distinguish them (**Figure 23.11**). The stain is named for its inventor, Danish scientist Hans Christian Gram.

Gram-positive bacteria classified in the phyla Firmicutes and Actinobacteria have walls with a relatively thick peptidoglycan layer (**Figure 23.12a**). By contrast, the cell walls of cyanobacteria, proteobacteria, and other Gram-negative species have a thinner peptidoglycan layer and are enclosed by a thin, outer envelope whose outer

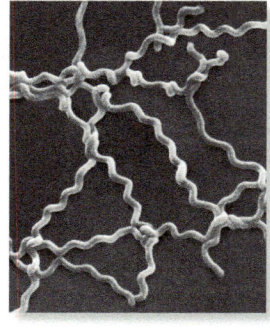

Sphere-shaped cocci
(*Lactococcus lactis*)

Rod-shaped bacilli
(*Lactobacillus plantarum*)

Comma-shaped vibrios
(*Vibrio cholerae*)

Spiral-shaped spirochaetes
(*Leptospira* sp.)

Figure 23.10 Major types of prokaryotic cell shapes. Scanning electron microscopic (SEM) views, which naturally lack color.
(1) ©SCIMAT/Science Source; *(2, 4)* ©Dennis Kunkel Microscopy/Science Source; *(3)* ©Media for Medical/UIG/Getty Images

(a) Gram-positive: thick peptidoglycan cell-wall layer, no outer envelope

(b) Gram-negative: thinner peptidoglycan cell-wall layer, with outer envelope

Figure 23.12 Cell-wall structures of Gram-positive and Gram-negative bacteria. (a) The structure of the cell wall of Gram-positive bacteria. (b) The structure of the cell wall and lipopolysaccharide envelope typical of Gram-negative bacteria.

leaflet is rich in **lipopolysaccharides** (Figure 23.12b). This outer envelope of Gram-negative bacteria is a lipid bilayer but is distinct from the plasma membrane. Peptidoglycan and lipopolysaccharides can affect disease symptoms, the composition of vaccines, and bacterial responses to antibiotics.

The lipopolysaccharide-rich outer membrane of Gram-negative bacteria helps them to resist the entry of some antibiotics and can contain proteins that help disease-causing bacteria to attach to target cells. However, this outer envelope also impedes the secretion of proteins from bacterial cells into the environment, a process that normally allows cells to communicate with each other. Gram-negative bacteria have adapted to the presence of an outer membrane by evolving several types of protein systems that function in secretion. In some disease-causing bacteria, these secretion systems have evolved into weapons used to attack plant or animal cells.

Distinguishing Gram-positive from Gram-negative bacteria is an important factor in choosing the best antibiotics for treating infectious diseases. For example, Gram-positive bacteria are typically more susceptible than Gram-negative bacteria to penicillin and related antibiotics, because these antibiotics interfere with the synthesis of peptidoglycan, which Gram-positive bacteria require in larger amounts. For this reason, penicillin or related antibiotics such as methicillin are widely used to treat infections caused by Gram-positive bacteria. However, it is of societal concern that some strains of Gram-positive bacteria have become resistant to some antibiotics, an example being <u>m</u>ethicillin-<u>r</u>esistant *Staphylococcus aureus*, or MRSA. The evolution of antibiotic resistance in bacteria can often be traced to a horizontal gene transfer event.

Slimy Mucilage Often Coats Cellular Surfaces

Many bacteria exude a coat of slimy mucilage, also called a capsule, a glycocalyx, or an extracellular polymeric substance. Mucilage, which varies in consistency and thickness, is largely composed of hydrated polysaccharides and protein. A capsule helps some pathogenic bacteria evade the defense system of their host. You may recall that Frederick Griffith discovered the transfer of genetic material while experimenting with capsule-producing pathogenic strains and capsule-less, nonpathogenic strains of the bacterium *Streptococcus pneumoniae*. The immune system cells of mice are able to destroy this bacterium only if it lacks a capsule.

Mucilage plays many additional roles: holding cells together closely enough for chemical communication and DNA exchange to occur, helping aquatic species to float in water, binding mineral nutrients, helping cells to stick to surfaces where they may form coatings known as **biofilms** (see Chapter 24), and repelling attack. Pigmented mucilage sheaths (see Figure 23.6d) coat some bacterial filaments, where they help to prevent UV damage.

Bacteria Move by a Variety of Cellular Adaptations

Many bacteria have structures at the cell surface or within cells that enable them to change position in their environment, a process known as motility. Motility allows bacteria and other microbes to respond to chemical signals emitted from other cells and to move to favorable conditions within gradients of light, gases, or nutrients.

- Internal gas vesicles (see Figure 23.9a) help cyanobacteria to float into well-illuminated waters that allow photosynthesis to occur. Cyanobacteria filled with such vesicles can collect at the water surface, forming noticeable scums.

- Threadlike cell surface structures known as pili (singular, pilus) allow some species to twitch or glide across surfaces and may aid reproduction.

- Bacterial flagella (see Figure 23.2a) allow cells to move by twitching, gliding, or swimming in liquids at rates of more than 150 μm per second and thus can foster the spread of infection through an animal or plant body.

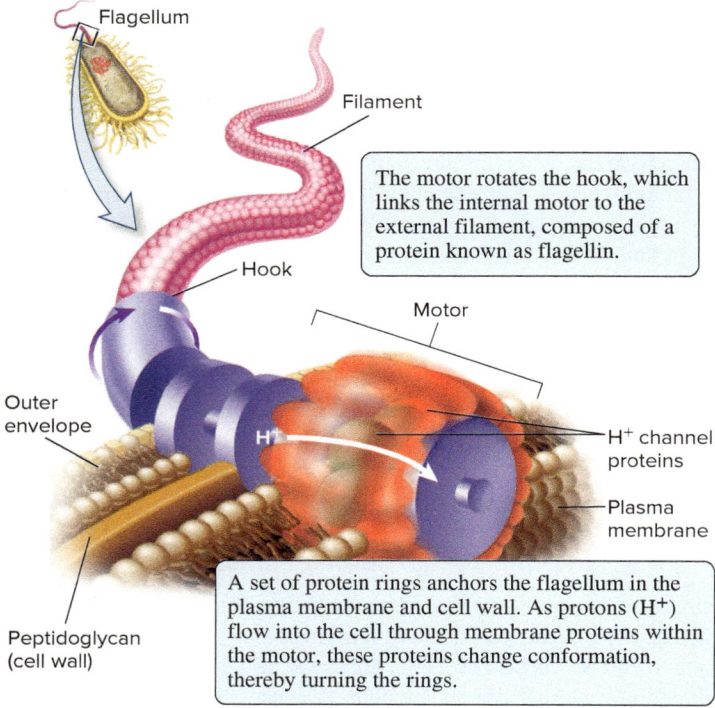

Figure 23.13 Diagram of a bacterial flagellum, showing a filament, hook, and motor.

Concept Check: Does the filament move more like the arms of a human swimmer or the shaft of a boat propeller?

Bacterial flagella (singular, flagellum) differ from eukaryotic flagella in several ways. Although bacterial flagella are largely built of about 30 types of proteins, they lack a plasma membrane covering, an internal cytoskeleton of microtubules, and motor proteins—all features that characterize eukaryotic flagella. Unlike eukaryotic flagella, prokaryotic flagella do not repeatedly bend and straighten. Instead, prokaryotic flagella spin, propelled by molecular machines composed of a filament, hook, and motor that work together, somewhat like a boat's outboard motor and propeller (Figure 23.13). Lying outside the cell, the long, stiff, curved filament acts as a propeller. The hook links the filament with the motor, which contains a set of protein rings at the cell surface. Hydrogen ions (protons), which have been pumped out of the cytoplasm, usually via the electron transport system, diffuse back into the cell through channel proteins within the motor. This proton flow powers the turning of the hook and filament at rates of hundreds of revolutions per second. (Archaea may have flagella that also rotate but are much thinner than bacterial flagella, composed of different proteins, and powered differently—by the hydrolysis of ATP.)

Bacteria Can Be Grouped by Type of Nutrition and Response to Oxygen

Microbes can be classified according to their energy source, carbon source, response to oxygen, and presence of specialized metabolic processes such as nitrogen fixation.

- **Autotrophs** (from the Greek, meaning self-feeders) are organisms that are able to produce all or most of their own organic compounds from inorganic sources. Autotrophs fall into two categories: photoautotrophs and chemolithoautotrophs.
- **Photoautotrophs** such as cyanobacteria use light as a source of energy for the synthesis of organic compounds from CO_2 and H_2O or from H_2S.
- **Chemolithoautotrophs** use energy obtained by chemical modifications of inorganic compounds to synthesize organic compounds. Such chemical modifications include nitrification (the conversion of ammonia to nitrate) and the oxidation of sulfur, iron, or hydrogen.
- **Heterotrophs** (from the Greek, meaning other feeders) are organisms that require at least one organic compound, and often more, from their environment.
- **Photoheterotrophs** are able to use light energy to generate ATP, but they must take in organic compounds from their environment as a source of carbon.
- **Chemoheterotrophs** must obtain organic molecules both for energy and as a carbon source. Fungi, archaea, and many bacteria are chemoheterotrophs. Among the many types of bacterial chemoheterotrophs is the Gram-positive species *Propionibacterium acnes,* which contributes to the skin condition acne, affecting up to 80% of adolescents in the U.S. The genome sequence of *P. acnes* has revealed numerous genes that allow it to break down skin cells and consume the products.

Bacterial species also differ in their need for and responses to oxygen. These variations are important to the ecological or medical roles of microbes.

- **Obligate aerobes** require O_2 in order to survive.
- **Obligate anaerobes,** such as the bacterial Firmicutes genus *Clostridium,* are poisoned by O_2. People suffering from gas gangrene caused by *Clostridium perfringens* or related species are often treated by placement in a chamber with a high oxygen content (called a hyperbaric chamber), which kills the organisms and deactivates the toxins.
- **Aerotolerant anaerobes** do not use O_2, but they are not poisoned by it, either. These organisms obtain their energy by fermentation or anaerobic respiration, which uses electron acceptors other than oxygen in electron transport processes. Anaerobic metabolic processes include denitrification (the conversion of nitrate into N_2 gas) and the reduction of manganese, iron, and sulfate, which are all important in the Earth's cycling of minerals.
- **Facultative anaerobes** can use O_2 via aerobic respiration, obtain energy via anaerobic fermentation, or use inorganic chemical reactions to obtain energy, shifting between modes depending on environmental conditions.

Many cyanobacteria and some other prokaryotic species are able to convert the abundant atmospheric nitrogen gas into a reduced form that algae and plants can use to produce amino acids and proteins. Oxygen poisons the principal enzyme involved in the nitrogen fixation process, so many cyanobacteria accomplish nitrogen fixation within specialized cells known as heterocytes, where oxygen concentrations are low (see Figure 23.2b). Some cyanobacteria that lack heterocytes accomplish nitrogen fixation at night, when photosynthetic oxygen production does not occur.

Cyanobacteria and other autotrophic bacteria, together with photosynthetic eukaryotes—plants and algae—are important **producers,** organisms that use photosynthesis to synthesize the organic compounds used by other organisms for food. **Decomposers,** also known as saprobes, include heterotrophic microorganisms (as well as animals). Bacteria, together with fungi, play important ecological roles by breaking down dead organisms and organic matter, releasing minerals for uptake by living things.

23.4 Reviewing the Concepts

- Intracellular structures such as thylakoids, gas vesicles, magnetosomes, and nucleus-like features are examples of prokaryotic cell structure complexity (Figure 23.9).
- Major types of bacterial cell shapes are spherical cocci, rod-shaped bacilli, comma-shaped vibrios, and coiled spirochaetes and spirilli (Figure 23.10).
- Most bacterial cell walls contain peptidoglycan, which is composed of carbohydrates crosslinked by peptides. Gram-positive bacterial cells have thick peptidoglycan walls, whereas Gram-negative cells have less peptidoglycan in their walls and are enclosed by an outer lipopolysaccharide envelope (Figures 23.11, 23.12).
- Motility enables bacteria to change positions within their environment, which aids in locating favorable conditions for growth. Gas vesicles, pili, and flagella are structures that enable motility (Figure 23.13).
- Bacteria display diverse types of metabolism, ways in which they interact with oxygen and obtain energy and carbon.

23.4 Testing Your Knowledge

1. Which structure is associated with motility in one or more types of bacteria?
   a. flagellum
   b. peptidoglycan
   c. lipopolysaccharide envelope
   d. mucilage
   e. All of the above are correct.

23.5 Diversity and Ecological Importance of Protists

Learning Outcomes

1. List three features that define protists.
2. Distinguish among algae, protozoa, and fungus-like protists.
3. Define the terms phytoplankton, periphyton, flagellate, ciliate, and amoeba.
4. Describe a distinctive structural characteristic for each of seven eukaryotic supergroups.

The term protist comes from the Greek word *protos,* meaning first, reflecting the observation that protists were Earth's first eukaryotes. Protists are eukaryotes that are not classified in the plant, animal, or fungal kingdom. Protists display two additional common characteristics: They are most abundant in moist habitats, and most of them are microscopic in size. Despite their small size, protists have a greater influence on global ecology and human affairs than most people realize. For example, the photosynthetic protists known as algae generate at least half of the oxygen in the Earth's atmosphere and produce organic compounds that feed marine and freshwater animals. The oil that fuels our cars and industry is derived from pressure-cooked algae that accumulated on the ocean floor over millions of years. Because fossil oil deposits are becoming depleted, algae are being engineered into systems for producing renewable biofuels.

Protists also include some parasites that cause serious human illnesses, such as *Trichomonas vaginalis,* shown in the chapter-opening photo. Another example is the waterborne protist *Cryptosporidium parvum,* which in 1993 sickened 400,000 people in Milwaukee, Wisconsin, costing $96 million in medical expenses and lost work time. Species of the related protist *Plasmodium,* which is carried by mosquitoes in many warm regions of the world, cause the disease malaria. Every year, about 200 million people become ill with malaria, and about 500,000 die of this disease.

Protists Can Be Informally Labeled According to Their Ecological Roles

Protists are often labeled according to their ecological roles and occur in three major types: algae, protozoa, and fungus-like protists (see Figure 23.3). All of these protists occur widely in aquatic habitats.

- **Algae** (singular, alga, from the Latin, meaning seaweeds) are protists that are generally photoautotrophic, meaning that most possess chlorophyll and other photosynthetic pigments and can produce organic compounds from inorganic sources by means of photosynthesis (**Figure 23.14**). In addition to organic compounds, photosynthetic algae produce oxygen. Diverse types of algae are the main sources of oxygen and organic food for organisms that live in aquatic habitats. Despite the common occurrence of photosynthesis, algae do not form a monophyletic group descended from a single common ancestor.

- **Protozoa** (from the Greek, meaning first life) are heterotrophic protists that cannot produce their own organic food and must obtain it by feeding. Protozoa feed by absorbing small organic molecules or by ingesting bacteria or small eukaryotes as prey. For example, the protozoa known as ciliates may consume smaller photosynthetic algae known as diatoms (see Figure 23.14). Like the algae, the protozoa do not form a monophyletic group.

- **Fungus-like protists** are a nonmonophyletic assemblage of organisms having bodies, nutrition, or reproduction mechanisms similar to those of the true fungi (see Section 23.6). Even so, fungus-like protists are not closely related to fungi; some of their similar features may arise from the horizontal transfer of genes from fungi. Water molds, some of which cause diseases of fish, and *Phytophthora infestans,* which causes diseases of many wild and crop plants, are examples of fungus-like protists (see Figure 23.3b).

Biology Principle

Living Organisms Use Energy

The organic components of these golden diatoms are digested by the ciliate and used as food, and indigestible diatom cell walls are excreted.

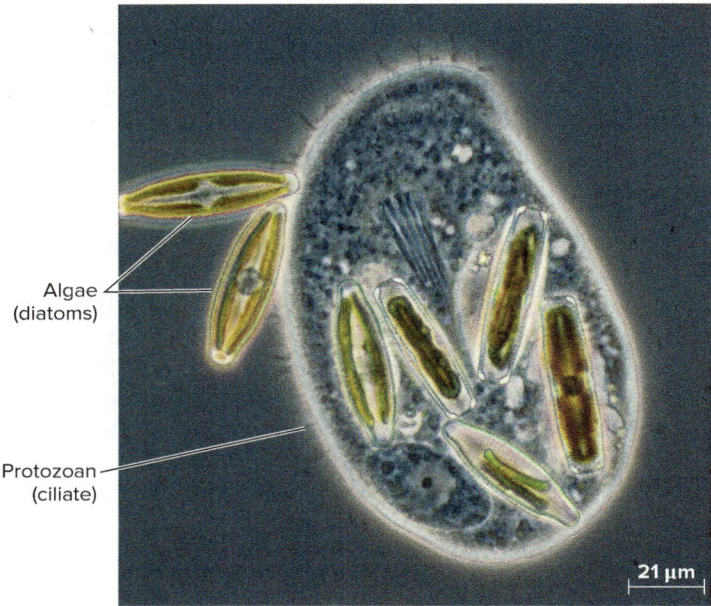

Figure 23.14 A heterotrophic protozoan feeding on photosynthetic algae. The ciliate shown here has consumed several oil-rich, golden-pigmented, silica-walled algal cells known as diatoms. Diatom cells that have avoided capture glide nearby.
©Dr. Hilda Canter-Lund/FBA

Protists Can Be Informally Labeled According to Their Type of Motility

Microscopic protists have evolved diverse ways to propel themselves in moist environments of diverse types. Swimming by means of flagella or cilia and amoeboid movement are major types of protist motility.

- **Flagellates** are the many types of photosynthetic and heterotrophic protists that are able to swim because they produce one or a few eukaryotic flagella (Figure 23.15a). Recall that eukaryotic flagella are cellular extensions whose movement is based on interactions between microtubules and the motor protein dynein. Eukaryotic flagella rapidly bend and straighten, thereby pulling or pushing cells through the water. Flagellates are typically composed of one or only a few cells and are small—usually from 2 to 20 μm long. Some flagellate protists are sedentary, living attached to underwater surfaces. Sedentary protists use flagella to collect bacteria and other small particles for food. Macroalgae and other immobile protists often produce small, flagellate reproductive cells that allow these protists to mate and disperse to new habitats.

- **Ciliates** are protists that use many tiny, hairlike surface extensions, known as **cilia**, to move. Cilia are structurally similar to eukaryotic flagella but are shorter and more abundant on cells (Figure 23.15b). Having many cilia allows ciliates to achieve relatively large cell sizes (up to several millimeters in length) while remaining buoyant in water. In some cases, cilia have become aggregated into bristles that function as limbs to accomplish movement across surfaces.

- **Amoebae** are protists that move by extending cytoplasm into lobes, known as pseudopodia (from the Greek, meaning false feet) (Figure 23.15c). Once these pseudopodia move toward a food source or other stimulus, the rest of the cytoplasm flows after them, thereby changing the shape of the entire organism as it creeps along.

Protists Can Be Informally Labeled According to Their Habitats

Although protists occupy nearly every type of moist habitat, including animal bodies, they are particularly common and diverse in oceans, lakes, wetlands, and rivers. Even extreme aquatic environments such as Arctic and Antarctic ice and acidic hot springs serve as habitats for some protists. In such places, protists may swim or float in open water or live attached to surfaces such as rocks or beach sand. As noted earlier, these different habitats influence protist structure and size. When humans pollute natural waters, harmful or nuisance

(a) A cryptomonad with eukaryotic flagella

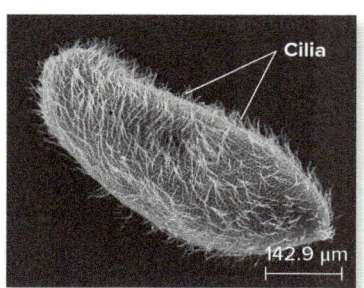

(b) A protzoan ciliate covered by cilia

(c) An amoeba extending many pseudopodia

Figure 23.15 SEMs of three protists: a flagellate, a ciliate, and an amoeba. (a) This cryptomonad, with two flagella, is an example of a flagellate. (b) The ciliate *Paramecium*, showing numerous cilia on the cell surface. (c) An amoeba of the genus *Pelomyxa*, showing pseudopodia.
(a, b) ©Dennis Kunkel Microscopy/Science Source; (c) ©Steve Gschmeissner/Science Source

growths of protists may result, but it is also the case that protists provide important ecological services.

- Planktonic protists swim or float in fresh or salt water, together with planktonic bacteria, viruses, and small animals. The photosynthetic protists in plankton are known as **phytoplankton** (plantlike plankton). Staying afloat is particularly important for phytoplankton, which need light for photosynthesis, so small size and other buoyancy adaptations are common. Coastal aquatic environments that have become polluted with too many nutrients may foster the growth of toxic phytoplankton into large populations known as blooms or red tides (Figure 23.16a).

- Periphytic protists attach themselves to underwater surfaces such as rocks, sand, and plants in microbial communities known as **periphyton.** Because sinking is not a problem for attached protists, their bodies can be larger than those of the plankton. In water that is polluted with excessive nutrients such as phosphorus, growths of periphytic protists may become conspicuous though are not toxic (Figure 23.16b). Periphytic protists provide key ecological benefits—food, oxygen, and habitat—to beneficial microbes and small animals.

- **Seaweeds,** also known as macroalgae, are marine photosynthetic protists that are large enough to see with the unaided eye. Seaweeds usually grow attached to underwater surfaces such as rocks, sand, docks, ship hulls, or offshore oil platforms. Kelp forests, communities dominated by tree-sized seaweeds, serve as refuges for valued aquatic animals such as sea otters (Figure 23.16c). Kelp forests also generate large amounts of organic carbon that enter ocean food chains, and they are harvested for essential industrial products. Recent increases in the frequency and duration of marine heat waves worldwide have caused extensive losses of kelp forests and their ecological services.

Eukaryotic Supergroups Include Protists and Related Plant, Animal, and Fungal Kingdoms

At one time, protists were classified into a single kingdom. However, modern phylogenetic analyses based on comparative analysis of DNA sequences and cellular features reveal that protists do not form a monophyletic group (Figure 23.17). That is because some protists are more closely related to plants, animals, or fungi than to other protists.

Over billions of years of Earth's history, evolution has generated diverse lineages of eukaryotes. Many more eukaryotic branches exist than are shown in the streamlined diagram of Figure 23.17. (Also, look ahead to Table 23.2.)

Concepts of protist evolution and relationships have been changing as new information becomes available. Even so, molecular and cellular data reveal that many protist phyla can be classified within seven eukaryotic **supergroups** that each display distinctive features (see Figure 23.17). All of the eukaryotic supergroups include phyla of protists; some, in fact, contain only protist phyla. The supergroup Opisthokonta includes the multicellular animal and fungal kingdoms and related protists, whereas another supergroup includes the multicellular plant kingdom and the protists most closely related to plants. The study of protists helps to reveal how multicellularity originated in

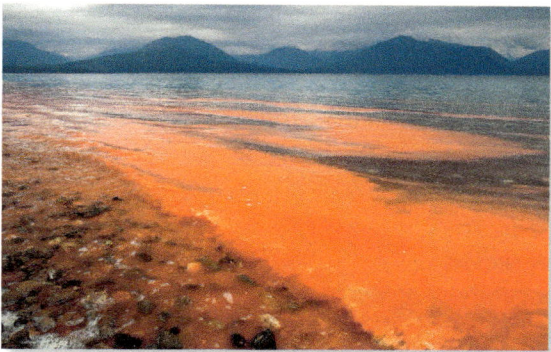

(a) Coastal red tide of phytoplankton

(b) Periphyton in shallow, nearshore water

(c) Offshore kelp forest

Figure 23.16 Protists can be classified according to habitats. **(a)** In polluted waters, phytoplankton—the swimming or floating photosynthetic plankton—can form large populations known as blooms or red tides, which produce harmful toxins. **(b)** Nutrient-rich waters foster conspicuous nearshore growths of periphyton, diverse communities of microorganisms that attach to substrata. Here, the periphyton is dominated by bushy, nontoxic green algae that harbor beneficial bacteria, protists, fungi, and small animals. **(c)** Kelp forests are extensive growths of tree-sized brown seaweeds that grow in deeper offshore waters and serve as habitat for many other organisms of importance to humans.

(a) ©Purestock/Alamy; *(b)* ©Linda Graham; *(c)* ©Jeff Rotman/Science Source

Biology Principle

Populations of Organisms Evolve from Generation to Generation

Figure 23.17 Phylogenetic diagram showing relationships among seven major eukaryotic supergroups. All eukaryotic supergroups include protists, and some supergroups include only protists. Relationships of some protists to these supergroups are still unclear.

animals, fungi, and plants. The following survey of eukaryotic supergroups reveals the ecological importance of modern protists.

A Feeding Groove Characterizes Many Protists Classified in the Excavates

The protist supergroup known informally as the excavates (named for a feeding groove that appears to be "excavated" into the cell) originated very early among eukaryotes, so this supergroup is important in understanding the early evolution of eukaryotes. Many excavates, such as the genus *Jakoba,* and other protists display a feeding groove (Figure 23.18a). The feeding groove is an important adaptation that allows these single-celled organisms to ingest small particles of food in their aquatic habitats. Once food particles are collected within the feeding groove, they are then taken into cells by a type of endocytosis known as **phagocytosis** (from the Greek, meaning cellular eating) (see Figure 23.18b). During phagocytosis, a vesicle of plasma membrane surrounds each food particle and pinches off within the cytosol. Enzymes within these food vesicles break the food particles down into small molecules that, upon their release into the cytosol, can be used for energy.

Phagocytosis is also the basis for an important evolutionary process known as **endosymbiosis,** a symbiotic association in which a smaller species known as the **endosymbiont** lives within the body of a larger host species. Phagocytosis provides a way for protist cells that function as hosts to take in diverse types of prokaryotic or eukaryotic cells. If not digested, ingested cells may become endosymbionts that confer valuable traits. Examples include endosymbiotic proteobacteria that evolved into mitochondria and endosymbiotic cyanobacteria that evolved into to chloroplasts. Invasion of cells by microbes that become endosymbionts is a major way in which many new genes can be horizontally transferred into protist cells.

The supergroup informally called excavates includes euglenoid flagellates (Figure 23.19). Many euglenoids lack plastids, but others contain green plastids acquired by endosymbiosis. Photosynthetic

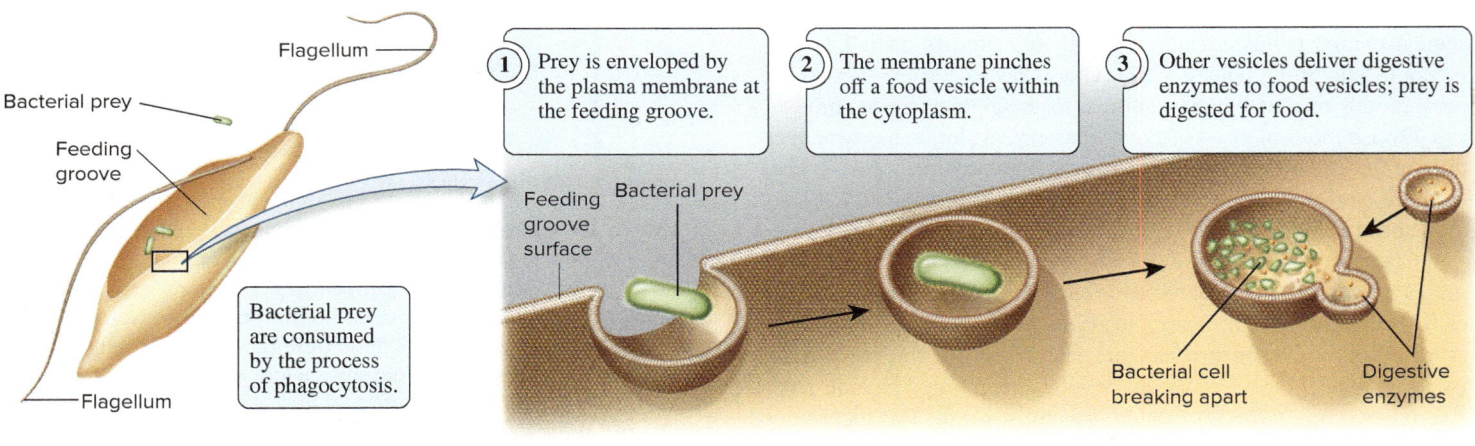

(a) *Jakoba* with feeding groove

(b) The process of phagocytosis

Figure 23.18 Excavates. (a) The flagellate *Jakoba intestinalis* is an example of the supergroup excavates. (b) Many protists display a characteristic feeding groove that functions in phagocytosis.

Figure 23.19 A euglenoid, a common excavates protist that contains plastids.
©Gerd Guenther/SPL/Science Source

euglenoids display conspicuous particles of carbohydrate and light-sensing systems that include red eyespots, which enable the cells to swim to favorable light environments.

Some plastidless members of excavates are parasites, organisms that live and reproduce within a living host organism. Examples include the human parasites *Trichomonas vaginalis* (see the chapter-opening photo) and *Giardia intestinalis* (previously known as *G. lamblia*) (see Figure 23.3a). *G. intestinalis* causes giardiasis, an intestinal infection that can result from drinking untreated water or from unsanitary conditions in day-care centers. Nearly 300 million human infections occur every year, and the disease also harms young farm animals, dogs, and cats, as well as wild animals. In the animal body, flagellate cells cause disease and produce tough, infectious stages known as cysts that are transmitted in feces and can survive several weeks outside a host. When an animal ingests as few as 10 of these cysts, within 15 minutes stomach acids induce the flagellate stage to develop and adhere to cells of the small intestine.

Land Plants and Related Algae Share Similar Genetic Features

Genetic analyses indicate that the supergroup that includes land plants also encompasses several protist phyla. Together, land plants and closely related green algae form a lineage informally known as streptophytes (see Chapter 25), and other green algae are classified in the phylum Chlorophyta. Rhodophyta, informally known as red algae, and the phylum Cryptophyta, informally cryptomonads, are related to green algae and land plants (see Figure 23.15a).

Green Algae Diverse structural types of green algae (**Figure 23.20**) occur in fresh water, in the ocean, and on land. Some green algae are planktonic, some occur as attached periphyton, and others are larger seaweeds. Most green algae are photosynthetic, and their cells contain the same types of plastids and photosynthetic pigments that are present in land plants (Kingdom Plantae) (see Chapter 25). Some green algae are useful as model lab organisms and others provide resources for aquatic animals. Studies of chemical traces in ancient rocks indicate that green algae played an important role in the early diversification of animal life.

Red Algae Most species of red algae are multicellular marine seaweeds, some of which people harvest for food or making industrial products (see Figure 23.3c). Fossils indicate that red algae have existed for more than a billion years. The red appearance of

Figure 23.20 Green algal body diversity fosters survival in aquatic habitats. **(a)** The single-celled, flagellate genus *Chlamydomonas* occurs in open water plankton. *Chlamydomonas* is also an important model lab organism. **(b)** The genus *Monactinus* is composed of several cells that are associated to form a colony whose lacy star shape helps to keep this photosynthetic protist afloat in brightly illuminated surface waters. Colonies can also avoid being eaten by predators that consume single-celled prey. **(c)** The filamentous genus *Desmidium* occurs as a twisted row of cells, another way to achieve larger body size while remaining afloat. **(d)** The large-celled, branched, filamentous, attached body of *Cladophora* is too large for most aquatic animals to consume. **(e)** The seaweed genus *Acetabularia* represents many amazing species of tropical green algae whose bodies are the size of a dandelion plant or larger yet are composed of only one extremely large cell.
(a) ©Brian P. Piasecki; *(b)* ©Roland Birke/Medical Images; *(c–e)* ©Linda Graham

(a) Mycelium growing in liquid medium

(b) Mycelium growing on flat, solid medium

Figure 23.32 Fungal shape shifting. (a) When a mycelium, such as that of this *Rhizoctonia solani*, is surrounded by food substrate in a liquid medium, it will grow into a spherical form. **(b)** When the food supply is limited to a two-dimensional supply, as shown by *Neotestudina rosatii* in a laboratory dish, the mycelium will form a disc. Likewise, distribution of the food substrate determines the mycelium shape in nature.
(a) ©Agriculture and Agri-Food Canada, Southern Crop Protection and Food Research Centre, London ON; *(b)* Source: CDC

exclusively by asexual means. DNA-sequencing studies have revealed that many types of modern fungi that reproduce only asexually have evolved from ancestors that had both sexual and asexual reproduction.

Many fungi produce asexual spores known as **conidia** (from the Greek *konis*, meaning dust) at the tips of hyphae (**Figure 23.33**). When they land on a favorable substrate, conidia germinate into a new mycelium that produces many more conidia. The green molds that form on citrus fruits are familiar examples of conidial fungi. A single fungus can produce as many as 40 million conidia per hour over a period of 2 days.

Figure 23.33 Asexual reproductive cells of fungi. SEM of the asexual spores (conidia) of *Aspergillus versicolor*, which causes skin infections in burn victims and lung infections in AIDS patients. Each of these small cells is able to detach and grow into an individual that is genetically identical to the parent fungus and so is able to grow in similar conditions.
©Dennis Kunkel Microscopy/Science Source

✓ **Concept Check:** *How might you try to protect a burn patient from infection by a conidial fungus?*

Because asexual fungi can spread so rapidly, they are responsible for costly fungal food spoilage, allergies, and diseases. Medically important fungi that reproduce primarily by asexual means include the athlete's foot fungus (*Epidermophyton floccosum*) and the infectious yeast (*Candida albicans*). **Yeasts** are unicellular fungi of various lineages. Asexual reproduction in some yeasts occurs by budding, the production of a new cell on the surface of a larger parental cell (see Figure 23.4a).

Fungi Have Distinctive Sexual Reproductive Processes

As is typical for eukaryotes, the fungal sexual reproductive cycle involves the union of gametes, the formation of zygotes, and the process of meiosis. The fungal life cycle is generally haploid-dominant, as is the life cycle for many protists.

Fungal Gametes and Mating Early-diverging fungi that live in the water produce flagellate sperm that swim to nonmotile eggs, as do animals, many protists, and some plants. By contrast, the gametes of terrestrial fungi are cells of hyphal branches rather than distinguishable male and female gametes. Fungal mycelia occur in multiple mating types that differ biochemically, controlled by particular genes. During fungal sexual reproduction, hyphal branches of different but compatible mycelia are attracted to each other by secreted peptides, and when hyphae have grown sufficiently close together, they fuse. The mating of hyphal branches represents an adaptation to terrestrial life.

Fruiting Bodies As we have seen, under appropriate environmental conditions, such as seasonal change, a mated mycelium may produce a fruiting body such as a mushroom (see Figure 23.30). Fruiting bodies typically disperse haploid spores that can grow into haploid mycelia. If a haploid mycelium encounters hyphae of an appropriate mating type, hyphal branches will fuse and start the sexual cycle over again. The structures of fruiting bodies vary in ways that reflect different adaptations that foster spore dispersal by wind, rain, or animals.

- Mature puffballs have delicate surfaces upon which just a slight pressure causes the spores to puff out into wind currents (**Figure 23.34a**). Birds' nest fungi form characteristic, egg-shaped spore clusters. Raindrops splash on these clusters and disperse the spores.

- The fruiting bodies of stinkhorn fungi smell and look like rotting meat (**Figure 23.34b**), which attracts carrion flies. The flies land on the fungi to investigate the potential meal and then fly away, in the process dispersing spores that stick to their bodies.

- The fruiting bodies of fungal truffles (**Figure 23.34c**) are unusual in being produced underground. Truffles have evolved a spore dispersal process that depends on animals that eat fungi. Mature truffles emit an odor that attracts wild pigs and dogs, which break up the fruiting bodies while digging for them, thereby dispersing the spores. Collectors use trained, leashed pigs or dogs to locate valuable truffles in forests for the market.

Many fungal fruiting bodies such as truffles and morels are edible, and several species of edible fungi are cultivated for human

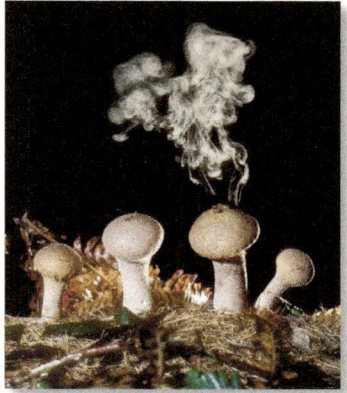
(a) Fruiting bodies adapted for dispersal of spores by wind

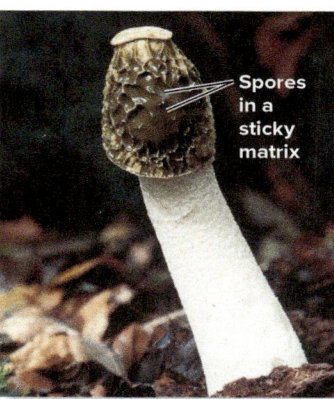
(b) Fruiting body adapted for dispersal of spores by insects

(a) *Amanita*

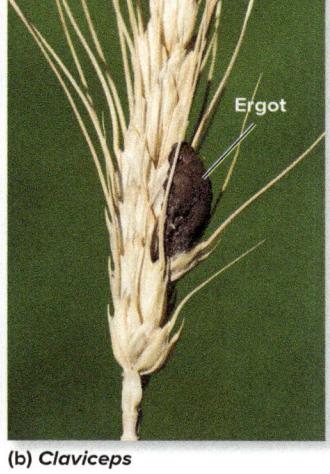
(b) *Claviceps*

Figure 23.35 Toxic or hallucinogenic fruiting bodies. **(a)** Common in conifer forests, *Amanita muscaria* is both toxic and hallucinogenic. Ancient people used this fungus to induce spiritual visions and to reduce fear during raids. This fungus produces a toxin, amanitin, which specifically inhibits RNA polymerase II of eukaryotes. **(b)** The fungus *Claviceps purpurea* infects rye and other grasses, producing hard masses of mycelia known as ergots in place of some of the grains (fruits). Ergots such as the one illustrated produce alkaloids related to LSD and thus cause psychotic delusions in humans and animals that consume products made with infected rye.
(a) ©MyLoupe/UIG CALC/Getty Images; *(b)* ©Adelheid Nothegger/imageBROKER/Superstock

 BioConnections: *Refer back to Figure 12.11, which illustrates the cellular role of RNA polymerase II in eukaryotes. What effect would the amanitin toxin have on human cells?*

(c) The black truffle *Tuber melanosporum*, an ascomycete fungus, highly valued for its taste

Figure 23.34 Fruiting body adaptations that foster spore dispersal. **(a)** When disturbed by wind gusts or animal movements, spores puff from fruiting bodies of the puffball fungus (*Lycoperdon perlatum*). **(b)** The fruiting bodies of stinkhorn fungi, such as this *Phallus impudicus,* smell and look like dung or rotting meat. This attracts flies, which come into contact with the sticky fungal spores, thereby dispersing them. **(c)** The black truffle *Tuber melanosporum*, an ascomycete fungus, produces fruiting bodies underground. The smell of mature truffle fruiting bodies attracts animals that dig them up, thereby dispersing spores.
(a) ©Chris Howes/Wild Places Photography/Alamy; *(b)* ©Bob Gibbons/ardea.com; *(c)* ©Nacivet/Getty Images

consumption. However, the bodies of many other fungi produce toxic substances that may deter animals from consuming them. For example, several fungi that attack stored grains, fruits, and spices produce aflatoxins that cause liver cancer and are a major health concern worldwide. The forest mushroom *Amanita virosa* (**Figure 23.35a**), known as the "destroying angel," contains a powerful toxin that can cause liver failure so severe that death may ensue unless a liver transplant is performed. Each year, many people in North America are poisoned when they consume similarly toxic mushrooms gathered in the wild. There is no reliable way for nonexperts to distinguish poisonous from nontoxic fungi; it is essential to receive instruction from an expert before foraging for mushrooms in the woods. Therefore, many authorities recommend that it is better to search for mushrooms in the grocery store than in the wild.

Several types of fungal fruiting bodies produce hallucinogenic or psychoactive substances. As in the case of fungal toxins, fungal hallucinogens may have evolved as herbivore deterrents, but humans have inadvertently experienced their effects. For example, *Claviceps purpurea*, which causes a disease of rye crops and other grasses known as ergot, produces a psychogenic compound related to lysergic acid diethylamide (LSD) (**Figure 23.35b**). Some experts speculate that cases of hysteria, convulsions, infertility, and a burning sensation of the skin that occurred in Europe during the Middle Ages and that were attributed to witchcraft resulted from ergot-contaminated rye used in foods. Another example of a hallucinogenic fungus is the "magic mushroom" *Psilocybe*, which is used in traditional rituals in some cultures. Like ergot, the magic mushroom produces a compound similar to LSD. Consuming hallucinogenic fungi is risky because the amount used to achieve psychoactive effects is dangerously close to a poisonous dose.

Decomposer and Predatory Fungi Play Important Ecological Roles

Decomposer fungi are essential components of the Earth's ecosystems. Together with bacteria, they decompose dead organisms and wastes, preventing the buildup of organic debris in ecosystems. For example, only certain bacteria and fungi can break down cellulose and lignin, the main components of wood. Decomposer fungi and bacteria are Earth's recycling engineers. They release CO_2 into the air and minerals into the soil and water, making these essential nutrients available to plants and algae.

More than 200 species of predatory soil fungi use special adhesive or nooselike hyphae to trap tiny soil animals, such as nematodes, and absorb nutrients from their bodies (**Figure 23.36**). Such fungi help to control populations of nematodes, some of which attack plant roots. Other fungi obtain nutrients by attacking insects, and certain fungal species have been used as biological control agents to kill black field crickets, red-legged earth mites, and other pests.

Pathogenic Fungi Cause Plant and Animal Diseases

One of the most important ways in which fungi affect humans is by causing diseases that may infect us and valued wild and domesticated plants and animals. Such disease-causing microbes are called pathogens. Five thousand fungal species are known as plant pathogens because they cause serious crop diseases that reduce yield. Plant pathogenic fungi use the organic compounds absorbed from plants to grow, attack additional plant cells, and produce reproductive spores capable of infecting more plants. Wheat rust is an example of a common crop disease caused by fungi (**Figure 23.37**). Rusts are named for the reddish spores that emerge from the surfaces of infected plants. To control the spread of crop pathogens from one country to another, agricultural customs inspectors closely monitor the entry of plants, soil, foods, and other materials that might harbor fungi.

Athlete's foot and ringworm are common human skin diseases caused by several types of fungi known as dermatophytes because they colonize the human epidermis. *Pneumocystis jiroveci* and *Cryptococcus neoformans* are fungal pathogens that infect individuals with weakened immune systems, such as those with AIDS, sometimes causing death. *Blastomyces dermatitidis*, which causes the

Figure 23.36 A predatory fungus. The fungus *Arthrobotrys anchonia* traps nematode worms in hyphal loops that suddenly swell in response to the animal's presence. Fungal hyphae then grow into the worm's body and digest it.
©Science Source

Biology Principle

Biology Affects Our Society

Recent analyses indicate that environmental changes linked to human activities correlate with increases in the incidence of new fungal pathogen infestations that threaten human health and agricultural sustainability.

Figure 23.37 Wheat rust. The plant pathogenic fungus *Puccinia graminis* grows within the tissues of wheat plants, using plant nutrients to produce rusty streaks of red spores that erupt at the stem and leaf surface where spores can be dispersed. Red spore production is but one stage of a complex life cycle involving several types of spores. Rusts infect many other crops in addition to wheat, causing immense economic damage.
©Nigel Cattlin/Science Source; *(Inset)* ©Herve Conge/ISM/Medical Images

disease blastomycosis; *Coccidioides immitis*, the cause of coccidioidomycosis; and *Histoplasma capsulatum*, the agent of histoplasmosis, are fungal pathogens that affect the lungs and may spread to other parts of the body, causing severe illness. Pathogenic chytrids that cause devastating skin infections in amphibians are examples of fungi that reduce populations of wild animals.

Fungi Form Beneficial Associations with Other Species

Although some fungi are harmful, others occur in beneficial partnerships with other organisms, including animals, plants, algae, and even bacteria. In the next chapter, we will see how fungi function in microbial systems, such as their associations with plant roots known as mycorrhizae and with bacteria and algae to form lichens.

23.6 Reviewing the Concepts

- Fungi form a monophyletic kingdom that is part of the eukaryotic supergroup Opisthokonta (Figure 23.28).
- Major fungal lineages include early-diverging, aquatic-adapted chytrids and later-diverging, land-adapted Mucoromycota, Glomeromycota, Ascomycota, and Basidiomycota (Figure 23.29).

- Most fungi produce filamentous hyphae that have absorptive nutrition and form bodies known as mycelia that can take different shapes (Figures 23.30, 23.31, 23.32).
- Fungi spread rapidly in the environment by means of asexual conidia and sexual spores produced in fruiting bodies of diverse types (Figures 23.33, 23.34, 23.35).
- Fungi play important ecological roles as decomposers, predators, and pathogens and by forming beneficial associations with other organisms (Figures 23.36, 23.37).

23.6 Testing Your Knowledge

1. How does asexual reproduction benefit fungi?
   a. Asexual reproduction allows fungi to spread rapidly in a favorable environment without expending resources required for mating and the formation of fruiting bodies.
   b. Asexual reproduction fosters genetic diversity.
   c. Asexual reproduction allows fungi to produce fruiting bodies in which meiosis occurs.
   d. Asexual reproduction does not benefit fungi.
   e. None of the above is correct.

23.7 Technological Applications of Microorganisms

Learning Outcome

1. List some examples of ways in which bacteria, protists, and fungi are used in human technology.

Bacteria, protists, and fungi are used in diverse technological applications.

- The food industry uses bacteria to produce chemical changes in food that improve consistency or flavor. Cheese makers add pure cultures of certain bacteria to milk. The bacteria consume milk sugar (lactose) and produce lactic acid, which aids in curdling the milk.
- The chemical industry produces enzymes, vinegar, amino acids, vitamins, insulin, vaccines, antibiotics, and other useful pharmaceuticals by growing particular bacteria in giant vats. Industrially grown bacteria produce the antibiotics streptomycin, tetracycline, kanamycin, gentamycin, bacitracin, polymyxin-B, and neomycin.
- The hot springs bacterial species *Thermus aquaticus* is a source of a form of DNA polymerase widely used in biology laboratories to amplify DNA in the polymerase chain reaction (PCR).
- Vast accumulations of the silica-rich walls of ancient diatoms, known as diatomite or diatomaceous earth, are mined for use in reflective paint and other industrial products.
- Diverse, photosynthetic brown seaweeds (Phaeophyceae) are sources of polysaccharides known as alginates that are used to make diverse industrial products.
- Some red seaweeds are harvested for extraction of food industry emulsifiers known as carageenans and for production of agar and agarose, which are widely used in biology laboratories.
- Bacterial and algal cells can be used as chemical factories. By genetically modifying microbial genomes, microorganisms can be made to produce particular useful compounds, including pharmaceuticals and renewable biofuels.
- The ability of some microorganisms to break down organic compounds, precipitate metals from mine waste, or absorb water pollutants such as phosphorus makes them very useful in treating wastewater, industrial discharges, and oil spills. This process, known as bioremediation, is used to reduce levels of harmful materials in the environment.
- A variety of industrial processes employ fungi to convert inexpensive organic compounds into valuable materials such as citric acid used in the food industry; glycerol; antibiotics such as penicillin; and cyclosporine, a drug widely used to prevent rejection of organ transplants.
- Enzymes extracted from fungi are used to break down plant materials for renewable bioenergy production.
- In the food industry, fungi are used to produce the distinctive flavors of blue cheese and other cheeses. Other fungi secrete enzymes that are used in the manufacture of protein-rich tempeh and other food products from soybeans.
- The brewing and winemaking industries find yeasts essential, and the baking industry depends on the yeast *Saccharomyces cerevisiae* for bread production.
- The yeast *S. cerevisiae* is also widely used as a model organism for fundamental biological studies of heterotrophic eukaryotes. Yeasts are useful in the laboratory because they have short life cycles, they are easy and safe for lab workers to maintain, and their genomes resemble those of animals.

23.7 Reviewing the Concepts

- Many bacteria, protists, and fungi are useful in industrial and other applications to make food products, medicines, or renewable biofuels; to clean up polluted environments; and to use as valuable laboratory resources.

23.7 Testing Your Knowledge

1–4. Match each of the microbes listed below with an appropriate technological application.
1. the protist stramenopiles known as diatoms
2. the fungus *Saccharomyces cerevisiae*
3. *Thermus aquaticus*
4. red algae

a. heterotrophic eukaryotic model for lab studies
b. source of commercial DNA polymerase
c. ingredient in reflective paint
d. agar is extracted for use in microbiology labs

Assessing Your Knowledge and Skills

Foundational Knowledge

Multiple Choice

1. The bacterial phylum that typically produces oxygen gas as the result of photosynthesis is
   a. the proteobacteria.
   b. the cyanobacteria.
   c. the Gram-positive bacteria.
   d. all of the above.
   e. none of the above.

2. Gram staining is a procedure that microbiologists use to
   a. determine if a bacterial strain is a pathogen.
   b. determine if a bacterial sample can break down oil.
   c. infer the structure of a bacterial cell wall and bacterial response to antibiotics.
   d. count bacteria in medical or environmental samples.
   e. do all of the above.

3. The structures that enable some Gram-positive bacteria to remain dormant for extremely long periods of time are known as
   a. akinetes.
   b. endospores.
   c. biofilms.
   d. lipopolysaccharide envelopes.
   e. pili.

4. Protists that move around by means of cilia are the
   a. cryptomonad algae.
   b. euglenoids.
   c. ciliate protozoa.
   d. red algae.
   e. green algae.

5. Fungal cells differ from animal cells in that fungal cells
   a. lack ribosomes, though these are present in animal cells.
   b. lack mitochondria, though these occur in animal cells.
   c. generally have chitin-containing cell walls, whereas animal cells lack rigid walls.
   d. lack cell walls, whereas animal cells possess walls.
   e. None of the above is correct.

Short Answer

1. Explain why many microbial populations are able to grow rapidly, thereby influencing the rate of food spoilage, infection, and the formation of harmful blooms in aquatic habitats.

2. **PRINCIPLES** A principle of biology is that biology affects our society. How are the endospores produced by certain bacteria important in human life?

Critical-Thinking Skills

Multiple Choice

1. Considering gene transfer from new endosymbionts to the host nucleus, how many different genomes are potentially represented in protists having tertiary plastids?
   a. at least 3
   b. at least 5
   c. at least 7
   d. at least 15
   e. at least 20

2. Prokaryotic species, protists, and fungi occur in diverse shapes and sizes. As you learned in Chapter 4, cells are generally microscopic in size because spherical cells that are smaller in diameter have a higher surface area/volume ratio than are spherical cells of larger diameter, and a high surface area/volume ratio aids the import of materials and export of wastes. Use this principle to arrange the following types of common prokaryotes or protists from smallest to largest total body surface area/volume ratio: giant kelp, *E. coli*, yeast.
   a. *E. coli*, yeast, giant kelp
   b. giant kelp, yeast, *E. coli*
   c. *E. coli,* yeast, giant kelp
   d. *E.* coli, giant kelp, yeast
   e. yeast, giant kelp, *E. coli*

3. Consider the hypotheses listed below. Which if any of them is a reasonable and testable hypothesis to answer the following question: In an aquatic environment, how might larger body size—conferred by larger cells in the cases of unicellular species or by multicellularity— be advantageous to photosynthetic prokaryotes or protists?
   a. Larger bodies might convert more sunlight energy into organic compounds.
   b. Larger bodies might be able to store larger amounts of essential mineral nutrients when these are most abundant in the surrounding environment.
   c. Larger bodies might be less vulnerable to being eaten by small herbivores (organisms that consume photosynthetic prey).
   d. All of the above are reasonable hypotheses.
   e. None of the above is a reasonable hypothesis.

4. How might you test the hypothesis that a fungus-like protist acquired the capacity to take up and metabolize a compound (that is also food for fungi) by means of horizontal gene transfer from a fungal species?
   a. You might compare the phylogenetic relationships of relevant genes that occur in the genomes of fungi and fungus-like protists.
   b. You might test the ability of a fungus-like protist that lacks the trait to express the trait after using genetic engineering to incorporate one or more relevant fungal genes into the protist genome.
   c. You might compare the amino acid sequences of proteins needed for expression of the trait in fungi with those of fungus-like protists.
   d. All of the strategies listed above are reasonable.
   e. None of the strategies listed above is reasonable.

5. New methods are needed to reduce methane emissions into Earth's atmosphere. What materials would you need to construct a technological system that employs biological organisms to remove methane?
   a. You would need one or more strains of methane-oxidizing bacteria capable of growing in an industrial system and would need to provide the bacteria with optimal oxygen levels and an optimal source of organic food.
   b. You would need to cultivate one or more strains of methane-producing bacteria, provide the methanogens with a source of organic food, and devise a way to keep oxygen out of the system.
   c. You would need to acquire one or more strains of nitrogen-fixing bacteria, come up with a system to provide the bacteria with both methane and fixed nitrogen, and determine the optimal source of organic food.

d. You would need to employ one or more strains of cyanobacteria and provide the bacteria with a source of light and oxygen so that they could generate their own organic carbon food source.
e. You would need to cultivate one or more strains of eukaryotic algae and provide them with optimal levels of oxygen and light.

Short Answer

1. Imagine that you have been asked to help determine the species of cellular microbe that might cause a particular human infectious disease, with the ultimate goal of developing treatment options. Describe how you might use information provided in this chapter to start the process of pathogen type determination.

2. Imagine that you lead a team that is working on identifying the microbes responsible for extensive toxic algal blooms that occur in coastal water and inland lakes. Describe how you might make these determinations.

24

Microbiomes: Microbial Systems on and Around Us

Nutrition can affect the microbiomes of infants and young children, which in turn affect proper growth.

Chapter Outline

- 24.1 Microbiomes: Diversity of Microbes and Functions
- 24.2 Microbiomes of Physical Systems
- 24.3 Host-Associated Microbiomes
- 24.4 Engineering Animal and Plant Microbiomes

Assessing Your Knowledge and Skills

©Ruslan Dashinsky/Getty Images

Ideally, all of the world's children would have good food sufficient to start healthy lives. Unfortunately, malnutrition has left nearly 180 million children around the world stunted in their growth. Public health experts have tried to use dietary supplements to restore normal growth, a strategy that does not always work. New research on microbes living in the bodies of these children has revealed why. Stunted children retain an infantile set of gut microbes, in contrast to children whose more mature collection of gut microbes stimulates growth hormones. This understanding may help people to devise new ways to improve the health of millions of children.

Chapter 23 introduced the diversity of microorganisms—archaea, bacteria, protists, and fungi—that influence our health and environments in many ways. This chapter builds on that foundation, describing how diverse microbes occur together in microbiomes, complex systems associated with particular environments, such as the human body. Medical aspects of the human microbiome are often in the news. Oceans, ice, fresh waters, soils, and the bodies of other organisms likewise have distinctive microbiomes.

Because microbes are generally small, microbiomes are often inconspicuous. New molecular approaches described in this chapter have enabled biologists to study microbiome composition and functions. Using these new methods, medical scientists are identifying new ways in which people can improve their health. Agricultural scientists are discovering new strategies for engineering crop microbiomes to promote plant health. Exploring the microbiomes of other organisms and environments has revealed previously unrecognized global ecological effects. In this chapter, we will examine the properties of microbiomes, which have become an important and fast-growing area of investigation in modern biology.

24.1 Microbiomes: Diversity of Microbes and Functions

Learning Outcomes

1. Define microbiome.
2. List some reasons microbiomes are considered to be complex systems.
3. Discuss how biologists use ribosomal RNA gene sequences and whole metagenomic sequencing to catalog diverse types of microbes in a microbiome.
4. Explain how biologists identify microbiome functions.

Microbes include millions of species of archaea, bacteria, protists, and fungi that play important ecological roles worldwide (see Chapter 23). A **microbiome** is a particular assemblage of microbes and genes that occurs in a defined environment (**Figure 24.1**).

Microbiomes Are Complex Biological Systems

Microbiomes are complex systems, in part, because they often contain thousands of different microbial species that are challenging to identify. Microbiome studies often reveal new types of microbes that have not been studied in the lab and, so, have not even been formally named. Diverse species of bacteria or archaea dominate some microbiomes (as in Figure 24.1), whereas other microbiomes include many types of eukaryotic protists and fungi, in addition to prokaryotic species (**Figure 24.2**). Some microbiomes also contain a surprising number of microscopic invertebrate animals, and viruses may also be common.

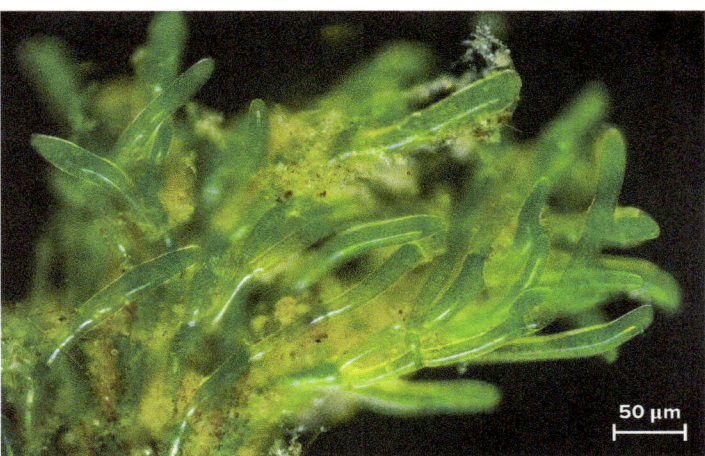

Figure 24.2 **A complex microbiome that includes diverse prokaryotic and eukaryotic species.** This light micrograph shows that the branched green alga *Cladophora* provides extensive surface area for attachment of bacteria (too small to distinguish by light microscopy) and larger eukaryotic microbes such as white ciliates and golden diatoms. Photosynthetic members of the microbiome provide organic food and oxygen to heterotrophic members, which return the favor by providing autotrophs with essential vitamins and minerals.
©Lee W. Wilcox

Figure 24.1 **A microbiome dominated by diverse bacterial types.** This microbiome, viewed by SEM, occurs on the surface of a multicellular, photosynthetic cyanobacterium. Spherical, rod-shaped, and elongate objects are bacteria of diverse types held together in a biofilm by delicate mucilage threads. Photosynthesis by the cyanobacterium provides the heterotrophic bacteria with organic molecules and oxygen.
©Linda Graham

These diverse life-forms carry out many types of metabolism that influence the environments and other members of the same microbiome. Microbes communicate with each other by means of chemical or electrical signals that biologists are just beginning to understand. In these ways, microbiomes resemble culturally diverse human groups whose social networks and responses to outside influences are complex. Identifying what species occur together, how microbes affect each other and their environments, and the effects of environmental change are major goals of microbiome research.

Determining the species composition, functions, and responses of microbiomes is challenging because microbes are so small and difficult to distinguish. For example, different bacterial species often have similar body structure such as single round or rod-shaped cells about 1 μm in diameter (see Figure 24.1). Likewise, the bodies of millions of fungal species are composed of thin hyphae that often look alike, even with the use of a microscope (see Chapter 23).

For these reasons, biologists commonly use genetic differences to distinguish and identify microbial species and genes present in a complex microbiome, as well as to infer their functions. In addition to cataloging the types of microbes and genes present in a microbiome, biologists examine proteins and metabolites.

Microbiome Complexity May Be Evaluated by Analyzing Genes That Encode Ribosomal RNA

We will first consider how biologists catalog organisms in a particular microbiome by using gene sequences that encode ribosomal RNA. All of Earth's living things, including microbes, produce proteins by means of ribosomes, which contain ribosomal RNA (rRNA). Ribosomal RNA is so important to living things that rRNA structure

1. Obtain a sample to analyze.

Sample from living organism

Sample of seawater

2. Extract the DNA from the sample. This yields a collection of many DNA fragments from many different microbial species.

DNA fragments

3.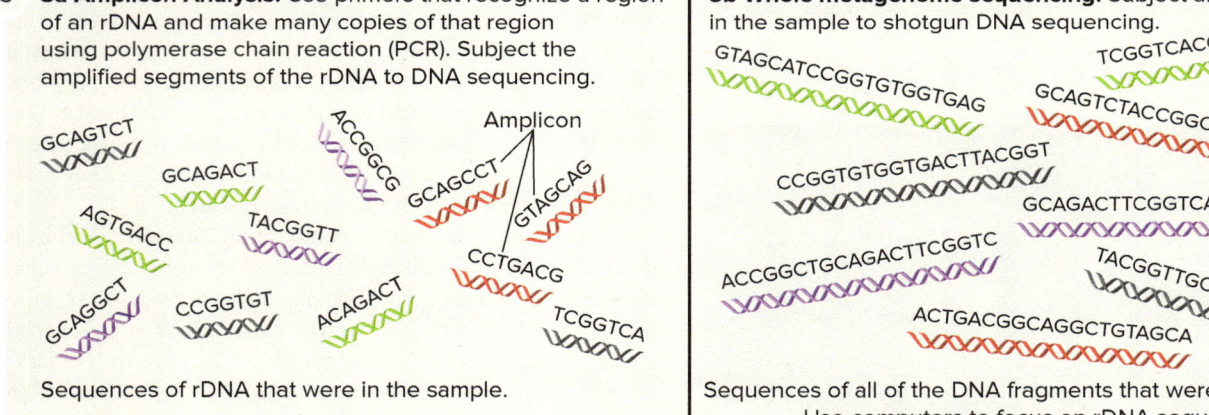

3a Amplicon Analysis: Use primers that recognize a region of an rDNA and make many copies of that region using polymerase chain reaction (PCR). Subject the amplified segments of the rDNA to DNA sequencing.

Sequences of rDNA that were in the sample.

3b Whole metagenome sequencing: Subject all of the DNA in the sample to shotgun DNA sequencing.

Sequences of all of the DNA fragments that were in the sample. Use computers to focus on rDNA sequences.

4. Compare the DNA sequences obtained in step 3 to DNA sequences in a database. The database sequences are already known to come from particular microbial species. This allows researchers to match the sequences obtained in step 3 to known sequences. For example, if a DNA sequence obtained in step 3a matches an rDNA sequence from the bacterium *E. coli*, this result indicates that *E. coli* was in the microbiome of that sample.

Figure 24.3 Characterization of microbiomes via amplicon analysis or whole metagenome sequencing.
(photo) © Goodshoot/Alamy

tends to be conserved in evolution. Likewise, the sequence of bases of rRNAs within a given species or closely related species is highly conserved. As described in Chapter 10, changes in the sequences of rRNAs can be used to evaluate evolutionary relationships (refer back to Figure 10.13). The genes that encode rRNAs are called rDNAs. Differences in rDNA sequences are used to identify and classify the microbes present in a microbiome, even if it includes thousands of microbial species.

Biologists usually begin a microbiome study by obtaining a sample and then extracting the DNA (Figure 24.3). As seen in step 3a, polymerase chain reaction (PCR) can be used to copy a particular region of an rDNA. This process yields many copies of that region from many different species that are in the sample. These copied rDNA regions are known as **amplicons.** Amplifying the DNA may be needed to generate sufficient DNA for sequencing (described in Chapter 18).

The amplicons are then subjected to DNA sequencing, which yields the base sequence of each rDNA type that was amplified in the original sample. Biologists use computers to compare the DNA sequences of each amplicon to reference sequences in databases. These reference sequences come from microbes whose names and metabolic functions are already known. This relatively inexpensive way to examine the diversity of known microbes in a microbiome is called **amplicon analysis.**

In amplicon analysis, researchers often use PCR primers that amplify a region of *16S rDNA*, a gene that encodes 16S rRNA. 16S rRNA is found in the small ribosomal subunit of bacteria and archaea. Gene sequences that encode 16S rRNA are commonly used to classify prokaryotic species because such genes are distinctive for prokaryotic organisms.

In eukaryotes, the small ribosomal subunit rRNA is longer and designated 18S rRNA. DNA sequences encoding this 18S rRNA are used to identify and classify eukaryotic microbiome components: protists, fungi, or small animals and plants. Other types of nucleic acid sequences are employed to detect and identify viruses. Used together, these identification methods reveal the diversity of known microbes that are present in a microbiome.

Microbiome Complexity May Also Be Analyzed by Whole Metagenomic Sequencing

An alternative method for characterizing microbial diversity is to obtain base sequences of all the DNA present in a sample, a process known as **whole metagenomic sequencing (WMS)** (see Figure 24.3, step 3b). A **metagenome** is defined as the genomes of all the organisms present in a sample. WMS is carried out using **shotgun sequencing,** so named because the process generates many small pieces of DNA, reminding people of the tiny metal pellets of shot sprayed by a shotgun blast. This approach does not have to focus on the sequencing of one particular gene, such as the *16S rRNA* gene.

Instead, many fragments of DNA are randomly sequenced, and then biologists use computers to identify places where the ends of DNA fragments have the same DNA sequences (Figure 24.4). These overlapping regions allow researchers to align the DNA fragments into longer sequences known as contiguous sequences, or **contigs.** This process, known as metagenome assembly, provides greater amounts of information needed for more detailed classification when sequence comparisons are made to taxonomic reference databases.

If metagenomic sequences are relatively long, or if a microbiome contains relatively few microbes, it may be possible to use computer methods to assemble contigs into whole microbial genomes (see Chapter 18). A number of microbial species were discovered in this way and even today are known only from their genomic sequence. Some experts consider that one goal of WMS should be to assemble entire microbe genome sequences, a process known as **genome-centric metagenomics.**

If sufficient DNA has been analyzed, WMS and computer methods can be used to obtain multiple ribosomal markers (16S plus 23S rRNA gene sequences from prokaryote taxa, and 18S plus 28S rRNA gene sequences from eukaryotes) from the same sample. By contrast, amplicon analyses typically focus on the amplification of a particular gene from a selected group of species. For example, researchers may focus on 16S rRNA gene amplicons, which may identify some, but not all, of the known prokaryotic taxa in a given sample. For this reason, many experts consider that the term microbiome should be limited to microbial communities characterized by WMS. The term **microbiota** is commonly used to describe collections of microbial life cataloged by limited amplicon analysis, such as when only 16S rDNA sequence data are used.

Certain Functions Within Microbiomes Are Determined by Identifying Protein-Encoding Genes

When analyzing microbiomes by WMS, another goal is to find and classify protein-encoding genes, providing a deeper view of microbial function. To gain more information about microbiome function, biologists may look within whole metagenomic sequence data for particular protein-coding genes that indicate specialized microbial functions. Amplicon analyses may also be employed to infer the presence of specific genes linked to particular functions. Three examples are described next.

Nitrogen Fixation One important microbial function is nitrogen fixation, the process in which atmospheric nitrogen gas is reduced to form ammonia. Nitrogen fixation is environmentally important and is a key process by which industry produces crop fertilizer. Only certain prokaryotic species have the natural ability to accomplish nitrogen fixation (see Chapter 23). Crop plants and other photosynthetic organisms require ammonia or another source of

Figure 24.4 A comparison of two DNA fragments that contain an overlapping region. A contig is a series of DNA fragments that contain overlapping regions.

```
                                    Overlapping region
TTACGGTACCAGTTACAAATTCCAGACCTAGTACC
AATGCCATGGTCAATGTTTAAGGTCTGGATCATGG
                         GACCTAGTACCGGACTTATTCGATCCCCAATTTTGCAT
                         CTGGATCATGGCCTGAATAAGCTAGGGGTTAAAACGTA
```

fixed nitrogen to make amino acids, chlorophyll, and other essential molecules. To obtain evidence for microbiome bacteria that are able to fix nitrogen, scientists may identify gene sequences known to encode enzymes essential for nitrogen fixation. One such gene is *nifH*, an indicator of nitrogen fixation. Such genes are known as **marker genes,** because they "mark" the occurrence of a particular function—in this case, nitrogen fixation.

Methane Oxidation Additional marker genes encode subunits of the enzyme methane monooxygenase (MMO). This enzyme uses oxygen gas to oxidize the greenhouse gas methane, important in global carbon cycles and climate warming. The presence of these genes in a microbiome indicates that a function of the microbiome is to oxidize methane.

Metabolites Some microbes produce very specific compounds that are not produced by most microbial species. These compounds are called metabolites, because they are the products of metabolic pathways. Examples include certain vitamins and toxins. In many cases, previous research has identified the enzymes that are needed to produce a particular metabolite. When WMS identifies genes that encode these enzymes, this result indicates that one function of the microbiome is to produce that metabolite. One example of a metabolite is vitamin B_{12}, which some bacteria produce and some other organisms require. The presence of genes that encode the enzymes needed to synthesize vitamin B_{12} indicates that a microbiome may provide this service.

The Analysis of mRNAs, Proteins, and Metabolites Provides Additional Information About Microbiome Function

Catalogs of microbial species and genes obtained by amplicon analysis or WMS don't reveal which genes were actually being transcribed and which transcripts were being translated at the time a sample was collected.

Metatranscriptome To learn which mRNAs were present in a microbiome at the time of sampling, biologists analyze transcriptomes. A **transcriptome** is a collection of all the mRNA sequences produced by a single organism under defined conditions. A **metatranscriptome** is a collection of all the mRNAs present in an environmental sample, that is, all of the mRNAs produced by all of the organisms sampled at a particular time.

Metaproteome Biologists sometimes use the number of mRNA sequences of a particular type to infer abundance or activity level of a translated protein. However, mRNA abundance is influenced by the extent to which microbes were actively growing when they were collected, the lifetime of a particular transcript, and how often that transcript is translated. A proteome analysis can provide more direct information about what proteins are present in a particular microbiome. A proteome analysis, accomplished by chromatographic and spectroscopy methods, reveals which proteins are present in a particular sample. The **metaproteome** is all the proteins produced by all the members of a microbiome.

Meta-metabolome Because proteins, even if present, might not be functionally active, researchers may analyze the products of metabolism. **Metabolomes** are collections of information about the types and abundances of molecules produced by metabolism, such as sugars or fatty acids, by a single organism. A **meta-metabolome** provides similar information for an entire microbiome.

24.1 Reviewing the Concepts

- Microbiomes are complex biological systems that are difficult to characterize using a microscope alone (Figures 24.1, 24.2).
- The members of microbiomes are usually identified using genetic techniques such as amplicon analysis or whole metagenomic sequencing (WMS) (Figures 24.3, 24.4).
- The analysis of protein-encoding genes can reveal important functions of microbiomes.
- Additional information regarding microbiome function can be obtained by characterizing the metatranscriptome, metaproteome, and meta-metabolome, which are all the types of RNAs, proteins, and metabolic products, respectively, that are present in a microbiome.

24.1 Testing Your Knowledge

1. A key difference between an amplicon analysis and whole metagenome sequencing (WMS) is that
   a. only an amplicon analysis involves DNA sequencing.
   b. only WMS involves DNA sequencing.
   c. an amplicon analysis focuses on specific genes, such as a gene that encodes rRNA, whereas WMS considers all genes present in a microbiome.
   d. a WMS focuses on specific genes, such as a gene that encodes rRNA, whereas an amplicon analysis considers all genes present in a microbiome.
   e. Both a and c are correct.

2. All of the proteins produced by a microbiome are called a
   a. transcriptome.
   b. metatranscriptome.
   c. metaproteome.
   d. meta-metabolome.
   e. Both b and c are correct.

24.2 Microbiomes of Physical Systems

Learning Outcomes

1. List some types of microbiomes that are found within physical systems.
2. Explain how microbiome analyses can help monitor environments for microbial activities that affect human health.

Thus far, our discussion of microbiomes as complex systems, and our survey of the methods that biologists use to study microbiome composition and function, provides the foundation for surveying the diversity of Earth's microbiomes. Some microbiomes are found within **physical systems,** which include oceans, ice, fresh waters, and soils. For example, biologists may study the microbiome in a sample of seawater or in a block of ice. Alternatively, other microbiomes are associated with living organisms known as hosts. In this section, we will consider microbiomes within physical systems, and then Section 24.3 will explore their occurrence within or on the surfaces of hosts.

Microbiomes Are Abundant in the World's Oceans and in Ice

Although you might imagine that animals such as fish and whales dominate Earth's oceans, in fact, microbes are far more numerous. Oceans occupy 71% of Earth's surface and a volume of 1.37 billion cubic kilometers, with a concentration of 10^4–10^6 microbial cells per milliliter; the number of microbes in 1 liter of seawater reaches into the billions. That's a lot of microbes!

Together, ocean microbes represent an immense amount of genetic and functional diversity that influences the entire planet. For example, floating or swimming cyanobacteria and algae (see Chapter 23) are so abundant in Earth's oceans that these photosynthetic microbes produce about half of Earth's annual organic carbon and oxygen. Other ocean microbes play essential roles in degrading organic molecules and recycling dissolved minerals, a process essential to ocean productivity. The cyanobacterial genus *Synechococcus* and its phages, together with stramenopile, alveolate, and rhizarian protists (see Chapter 23), are key to the movement of organic molecules to deep-ocean waters. Retention of carbon in the deep ocean for long periods affects global climate and is the mechanism by which extensive ocean oil and methane (fossil-fuel) deposits form.

Biologists have recently used genetic techniques (described in Section 24.1) to catalog viruses, prokaryotic species, and small eukaryotes from 68 ocean locations worldwide. By also monitoring the physical features of these places, they have discovered that water temperature is a major factor influencing the composition of ocean microbiomes, raising questions about the impact of global climate change.

Earth's icy environments—collectively known as the **cryosphere**—likewise contain a surprisingly large number of microbes, an estimated 10^{25}–10^{28} cells. The sampling of microbes from sea ice and glaciers is challenging, so adventurous biologists use ice-breaking ships, helicopters, planes, tractors, drilling rigs, and remotely operated vehicles to access polar oceans, sea ice sheets, snow, and glaciers (**Figure 24.5a**). Modern cryosphere microbiomes provide insight into periods of the Proterozoic eon when the entire planet Earth was engulfed by ice. Biologists are also intrigued by the possibility that Earth's cold microbiomes might resemble life on cold or icy planetary bodies such as Mars, Jupiter's moon Europa, and Saturn's moon Enceladus.

Microbiome studies of cold habitats have revealed surprisingly diverse types of microbiota that colorize their otherwise white environments. Beneath floating sheets of sea ice live conspicuous brown diatom growths that dangle into the cold ocean. These photosynthetic algae supply organic molecules and oxygen to heterotrophic bacteria; ciliate, flagellate, and foraminiferan protists (see Chapter 23);

(a) Researchers sampling microbiomes of the cryosphere

(b) A glacier colored by algae in ice microbiomes

(c) Blood Falls in Antarctica colored red by iron released by bacteria living in ice microbiomes

Figure 24.5 Microbiomes within and on ice. (a) Researchers obtaining samples of ice for microbiome analysis. **(b)** The color of the ice seen here is due to the presence of algae. **(c)** Blood Falls in Antarctica. The red color is from iron that is released by bacteria, which are part of a microbiome within the ice.
(a) Courtesy Cody S. Sheik; *(b)* ©Jason Edwards/National Geographic Image Collection; *(c)* Source: Peter Rejcek, NSF

and small animals. Algal cells on the surfaces of glaciers can be so abundant that they color the ice green, red, yellow, purple, or gray, in this way influencing the amount of energy that glaciers reflect into space (Figure 24.5b). Blood Falls in Antarctica gets its dramatic red color from dissolved iron released from subsurface minerals by iron-metabolizing bacteria of ice microbiomes (Figure 24.5c).

Microbiomes Are Found in Fresh Water and Within the Soil

Freshwater and soil microbiomes are important to diverse human concerns, including drinking water safety and agricultural production. *16S rDNA* and other genes are commonly analyzed to detect infectious or toxic organisms in the water used for drinking and recreation. Experts are particularly concerned about the effects of global climate warming on freshwater and soil cyanobacteria, because these organisms grow abundantly in warmer temperatures, particularly where humans have polluted environments with excess minerals.

Some abundant cyanobacteria produce persistent and potent toxins that harm people and wildlife. For example, the cyanobacterial genus *Microcystis* (Figure 24.6) produces more than 100 different chemical forms of the toxin **microcystin,** which binds to and inhibits eukaryotic phosphatases, enzymes that cells use to remove phosphate from organic compounds. The inability to remove phosphate interferes with many cellular processes, including cell signaling (see Chapter 8). By this mechanism, microcystin can cause severe liver damage and hemorrhage in mammals, and chronic exposure to microcystin is associated with high rates of human liver cancer. Humans and other animals become exposed to microcystin when large numbers of cyanobacteria occur in water used for drinking or swimming, a growing public health concern. Molecular methods are increasingly used to monitor aquatic microbiomes for cyanobacterial species likely to produce toxins. Because the genes encoding all enzymes required for the biosynthesis of microcystin and other cyanobacterial toxins are known, metagenomic data can be analyzed for the presence of these genes and thereby detect the potential for toxin production.

Soil microbiomes are key to our ability to grow crops, because various soil microbes foster or decrease plant health (see Chapter 23). A single gram of soil contains as many as 50,000 bacterial species and diverse fungi, many of which affect plants in some way. Soil microbiomes are important for understanding how terrestrial plants acquire microbiomes, described in Section 24.3.

24.2 Reviewing the Concepts

- Seawater serves as a physical system for microbiomes, which include important photosynthetic microbes such as cyanobacteria and algae.
- Biologists also explore the microbiomes found within ice. These microbiomes can affect the color of ice and its ability to absorb light (Figure 24.5).
- Microbiomes are also found in freshwater habitats and may elicit human health concerns, for example, due to toxins produced by *Microcystis* (Figure 24.6).
- Soils contain microbiomes rich in bacteria and fungi that affect the growth of crops and wild plants.

24.2 Testing Your Knowledge

1. Which of the following is *not* a microbiome of a physical system?
   a. a microbiome in a block of ice
   b. a microbiome in the human gut
   c. a microbiome in a soil sample
   d. a microbiome in a sample of seawater
   e. All of the above are examples of microbiomes of physical systems.

24.3 Host-Associated Microbiomes

Learning Outcomes

1. Define the terms holobiont and hologenome.
2. Describe a few examples of ways in which animal and plant hosts acquire microbiomes.
3. Make a list of the benefits that are derived from the associations between microbiomes and their hosts.
4. **SCISKILLS** ▶ Analyze the evidence that the human microbiome plays a role in health.

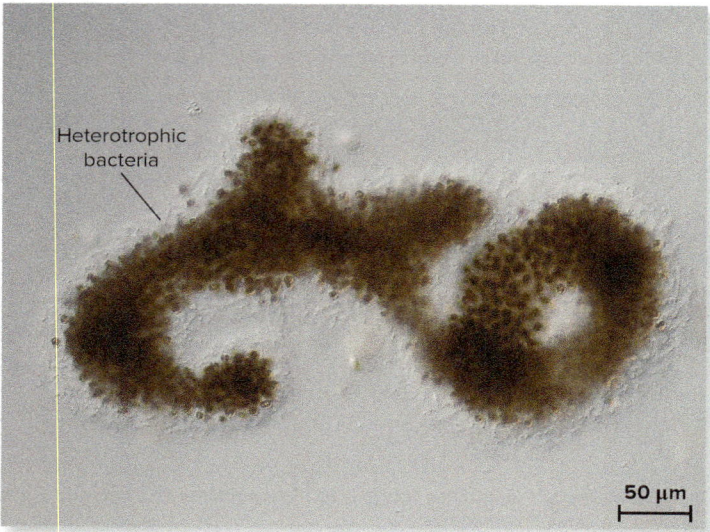

Figure 24.6 The colony-forming cyanobacterial genus *Microcystis* with a surface microbiome containing colorless heterotrophic bacteria. *Microcystis* cells look dark because they contain many light-refracting gas vesicles that aid in floatation. *Microcystis* and its microbiome produce toxins that may occur in drinking water and are harmful to human health.
©Lee W. Wilcox

Many people have heard news media stories about the human microbiome, which includes thousands of different microbial species that live in various locations within and on the surface of the human body. Consequently, humans are said to be microbiome **hosts.** The human

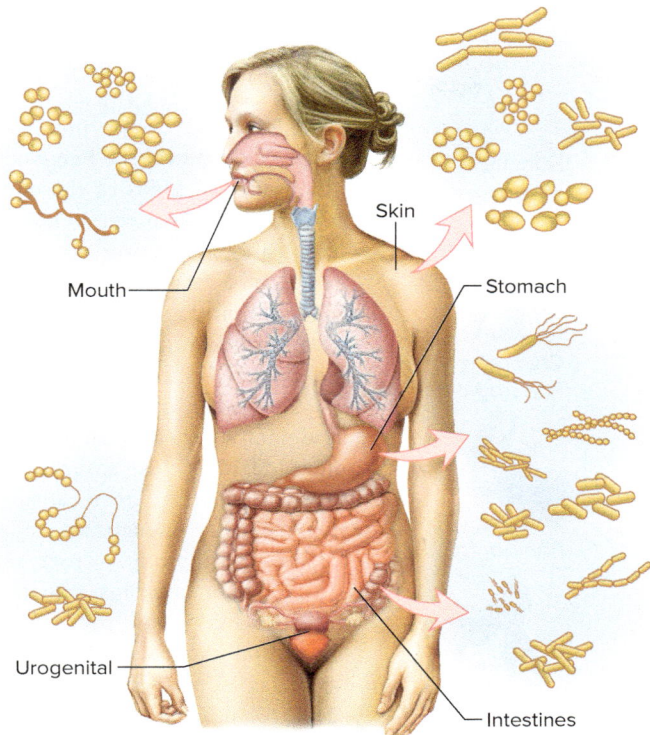

Figure 24.7 The human microbiome includes diverse bacteria, but also some archaea, protists, and fungi.

- Amoebozoan protists coat valuable bacteria with a protein that protects the bacteria within protist cells and during transmission to the next generation.
- Termites use specific behaviors to transfer among themselves microbes they need to break down plant materials into food.
- Other insects coat the casings of their eggs with bacteria; when the young hatch, they become inoculated with beneficial microbes.
- Mammals, including humans, transmit important microbes as the young transit the birth canal.
- Though seeds or spores may transmit some bacteria from plant parents to progeny, seedlings commonly acquire microbiota from the surrounding soil and air, using organic secretions determined by host genetics to attract and sustain beneficial microbes.

This list indicates that both host genetics and the environment help to determine the types of microbes present in microbiomes. Additional evidence for the importance of host genetics is that closely related tropical plants have microbiomes more similar to each other than to the microbiomes of more distantly related plant species. The importance of host genetics is also illustrated by comparative studies of the gut microbiomes of certain primates, including African apes (chimpanzees, bonobos, and gorillas) and their close relative, humans (see Chapter 22). The gut microbiomes of African apes and those of humans contain similar genera of bacteria and archaea, but in different proportions (Figure 24.8).

After humans diverged from African apes, the African-ape diet remained plant-rich, whereas the human diet became increasingly animal-rich. These dietary differences affected the environment of gut microbiomes, which were exposed to different types of nutrients. For example, compared to African apes, humans have more gut microbes from the Bacteroidetes phylum, such as the genus *Bacteroides*. This feature is associated with diets rich in animal fats and proteins. In contrast to apes, humans have lower numbers of the archaeon genus *Methanobrevibacter*, which degrades complex plant polysaccharides to methane. Humans also have lower amounts of *Fibrobacter*, a bacterial genus common in ape microbiomes, where it helps to break down the plant foods that apes consume. Microbiomes vary among modern human societies that have different dietary habits, as well as in groups whose diets regularly change with the seasons.

body serves as an environment for such microbes, some of which provide us with nutritional or protective benefits. Our microbes, in turn, receive benefits from us, such as organic molecules that serve as microbial food (Figure 24.7). Likewise, the bodies of other animals, plants, fungi, protists, and even bacteria support complex microbiomes that promote both host and microbial survival. In this section, we will examine the composition and functional properties of host-associated microbiomes.

Microbiomes Extend Host Capabilities and Are Acquired by Their Hosts in Different Ways

The combination of a host organism and its microbiome is known as a **holobiont**. The host and microbiome genomes together are called the **hologenome**. Microbiomes contribute many more genes to the hologenome than do their hosts. For example, the human genome contains about 22,000 protein-encoding genes, whereas the human microbiome is estimated to have a few million! In this way, microbiomes extend host metabolic capabilities and influence host evolution. Chemical signals produced by the host act on particular microbes that serve as information hubs, transmitting information to the broader microbial community. Microbiomes thereby function as complex biological networks.

Having a functionally useful microbiome aids the survival of the young and thereby increases fitness. The acquisition of a microbiome by its host can occur in a variety of ways.

Bacterial, Protist, and Fungal Microbiomes Perform Important Functions

Thus far, we have considered some general properties of host-associated microbiomes, primarily using examples from animals and plants. Bacteria, protists, and fungi also serve as hosts for microbiomes that carry out important functions.

Bacterial Microbiomes Certain species of bacteria, such as cyanobacteria, can produce relatively large, multicellular bodies that host microbiomes that include many other types of bacteria (see Figure 24.1). Smaller, heterotrophic bacteria live on surfaces

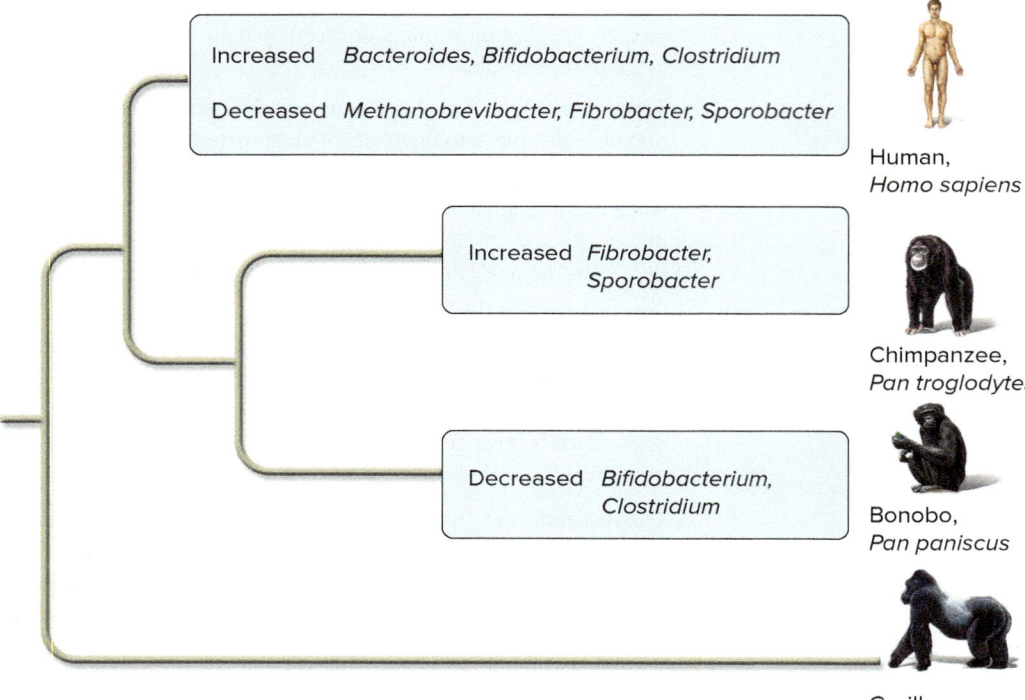

Figure 24.8 Changes in gut microbiomes during the evolution of African apes and humans. The changes seen in humans, chimpanzees, and bonobos are relative to the gut microbiome of gorillas.
Source: Moeller, Andrew H., et al., "Rapid Changes in the Gut Microbiome During Human Evolution," *Proceedings of the National Academy of Sciences*, vol. 111, 2014, 16431–16435.

of the cyanobacterium *Microcystis*, previously mentioned for its importance as a toxin producer in freshwater microbiomes (see Figure 24.6). Surface bacteria benefit from the oxygen and organic molecules produced by this photosynthetic host. WMS shows that *Microcystis*-associated bacteria synthesize vitamin B_{12}, which the cyanobacterial host requires but cannot produce. The surface bacteria also produce toxins different from those generated by the cyanobacterial host.

Protist Microbiomes Microbiomes associated with protist hosts are likewise ecologically important. The green alga *Cladophora*, which grows along freshwater and marine shorelines worldwide (see Figure 23.16b), provides an example. WMS and SEM reveal that in fresh waters this highly branched green protist (see Figure 24.2) operates something like a large hotel by hosting hundreds of bacterial, protist, and fungal microbes, as well as at least a hundred species of microscopic animals. Bacteria attach themselves to the algal cell wall by secreting mucilage to form a **biofilm,** which is a group of microbes that stick together (**Figure 24.9**). In addition to a congenial living space, the photosynthetic host provides room service in the form of oxygen and organic molecules. Bacterial guests include diverse and numerous methane-oxidizers, which reduce the amount of the greenhouse gas methane that enters Earth's atmosphere from aquatic environments. Some bacterial guests pay their hotel bill by producing vitamin B_{12}, which the algal host requires, as do the methane-oxidizers. These and many other microbiome functions are important on a global scale, because the algal host has such a large surface area and is so globally widespread and abundant.

Fungal Microbiomes Fungi occur in thousands of kinds of the microbial partnerships known as **lichens.** These partnerships were originally thought to be relatively simple, composed of a particular fungus species and a particular green algal and/or cyanobacterial species. Lichens are now known to represent complex microbiomes that may include multiple fungal species, several types of photosynthetic green algae or cyanobacteria (sometimes both), heterotrophic protists representing diverse supergroups, and many types of heterotrophic bacteria (and perhaps some archaea). The heterotrophic lichen partners depend upon organic carbon and fixed nitrogen provided by photosynthetic partners, which benefit from inorganic nutrients and water that spongy fungal hyphae absorb from the environment. Partners also benefit from fungus-produced sunscreen pigments that protect DNA and the photosynthetic apparatus of autotrophs from sunlight damage. Bacterial components fix nitrogen, help break down harmful organic compounds, and play other roles.

Figure 24.9 A biofilm composed of numerous microbes attached to a surface by mucilage. This is an SEM of a biofilm on the outer cell wall of the green protist *Cladophora*, which is also shown in a light micrograph in Figure 24.2. Biofilms occur on many other biological and nonliving surfaces.
©Linda Graham

Lichen bodies exhibit different morphological forms. Many lichens attach to an underlying surface, though others live unattached. Three common types are crustose, foliose, and fruticose lichens.

- Crustose lichens are flat and adhere tightly to an underlying surface (Figure 24.10a).
- Foliose lichens are flattened and leaflike (Figure 24.10b).
- Fruticose lichens grow upright (Figure 24.10c) or hang down from tree branches.

Light microscopy views of lichen slices reveal a distinct layer of photosynthetic green algae or cyanobacteria close to the upper lichen surface, with fungal hyphae often making up much of the lichen body (Figure 24.10d). Microscopy and WMS analysis have shown that diverse bacteria and protists may live on lichen surfaces. Lichen body structure differs dramatically from that of fungal components grown separately, indicating that other members of the microbiome may influence lichen form.

Lichens often grow on rocks, buildings, tombstones, tree bark, soil, and other surfaces that easily become dry. When water is not available, lichen metabolism is inactive until moisture returns. For this reason, lichens often grow very slowly, persisting through stressful times. Fossil remains indicate that lichens have existed for hundreds of millions of years. Modern lichens can be very old; some are estimated to be thousands of years old. Lichens occur in diverse types of habitats, including some of the most extreme terrestrial sites on Earth—deserts, mountaintops, and the Arctic and Antarctic. In locations where plant vegetation is sparse, lichens may serve as a food source for reindeer and other hardy organisms. Though unpalatable, lichens are not toxic to humans and have served as survival foods for indigenous peoples in times of shortages.

Soil building is another important lichen function. Lichen acids help to break up the surfaces of rocks, beginning the process of soil development. Lichens having nitrogen-fixing cyanobacterial partners (see Chapter 23) can also increase soil fertility. One study showed that such lichens release 20% of the nitrogen they fix into the environment, where it is available for uptake by plants.

Microbiomes Aid Plant Growth

Comparative WMS studies of the microbiomes of modern bryophytes and related green algae that model earliest plants (see Chapter 25) indicate that plants have hosted beneficial microbiomes throughout their evolutionary history. Microbiomes occurring on both aboveground and subterranean parts of modern plant bodies aid their growth in many ways.

Aboveground Plant Microbiomes Leaf surfaces commonly host as many as 10^7 bacterial cells per square centimeter, and microbes occur within leaf and stem tissues. Globally, leaf microbe numbers are estimated at 10^{26}. Leaves of tropical forest trees may host more than 400 types of bacteria, and those of temperate forest trees likewise support microbes whose diversity correlates with photosynthetic production.

Subterranean Root Microbiomes Plant roots and other subterranean parts acquire beneficial bacterial and fungal partners from a more diverse assortment of microbes in the soil close to plant root

Biology Principle

Biologists Use Models to Test Predictions and Convey Ideas

(a) Crustose lichen

(b) Foliose lichen

(c) Fruticose lichen

(d) Microscopic view of a cross section of a lichen

Figure 24.10 Lichen structure. (a) An orange-colored crustose lichen grows tightly pressed to the substrate. (b) The flattened, leaf-shaped genus *Umbilicaria* is a common foliose lichen. (c) The highly branched genus *Cladonia* is a common fruticose lichen. (d) A handmade thin slice of *Umbilicaria* viewed with a light microscope reveals that green-pigmented photosynthetic algae occur just beneath a thin upper layer of brown-pigmented fungal hyphae. A cottony mass of pale fungal hyphae occurs in lower regions of the lichen body.
(a) ©Perry Mastrovito/Getty Images; *(b, d)* ©Lee W. Wilcox; *(c)* ©Ed Reschke/Getty Images

Modeling Challenge: An older concept of lichens is a partnership between a photosynthetic microbe (green algal and/or cyanobacterial species) that exchanges oxygen and organic food for protection and inorganic materials provided by a particular fungus. Modern microbiome studies show that lichens actually have much more taxonomic and functional diversity than previously recognized. Make two diagrams that show how our understanding of lichens as microbiomes has changed over time, using arrows and labels to indicate exchanges of key materials.

surfaces, a region called the **rhizosphere**. Legumes and some other plants are known to form partnerships with soil bacteria that provide fixed nitrogen, greatly aiding plant growth (see Chapter 30). Certain fungi have long been important components of plant root microbiomes, because fungal hyphae efficiently absorb minerals from the soil and transport these essential nutrients to plant partners. Such partnerships are known as mycorrhizae.

Mycorrhizae Beneficial associations that occur between the hyphae of certain fungi and the roots of most seed plants are known as **mycorrhizae** (from the Greek, meaning fungus roots). Modern fungus-root associations are very important in nature and agriculture; more than 80% of terrestrial plants form mycorrhizae. Plants that have mycorrhizal partners receive an increased supply of water and mineral nutrients, primarily phosphate, copper, and zinc. They do so because an extensive fungal mycelium is able to absorb minerals from a much larger volume of soil than can roots alone (Figure 24.11). Added together, the branches of a fungal mycelium in 1 m^3 of soil can reach 20,000 km in total length. Experiments have shown that mycorrhizae greatly enhance plant growth compared with plants lacking fungal partners. In return, plants provide fungi with organic food molecules, sometimes contributing as much as 20% of their photosynthetic products.

The two most common types of mycorrhizae are endomycorrhizae, which occur within root tissue, and ectomycorrhizae, which coat roots. **Endomycorrhizae** (from the Greek *endo*, meaning inside) are partnerships between plants and fungi in which the fungal hyphae penetrate the spaces between root cell walls and plasma membranes and grow along the outer surface of the plasma membrane. In such spaces, endomycorrhizal fungi often form highly branched, bushy arbuscules (from the word arbor, referring to their treelike shape). As the arbuscules develop, the root plasma membrane also expands. Consequently, the arbuscules and the root plasma membranes surrounding them have a very high surface area that facilitates rapid and efficient exchange of materials: Minerals flow from fungal hyphae to root cells, whereas organic food molecules move from root cells to hyphae. These fungus-root associations are known as **arbuscular mycorrhizae**, abbreviated **AM** (Figure 24.12). AM fungi are associated with apple

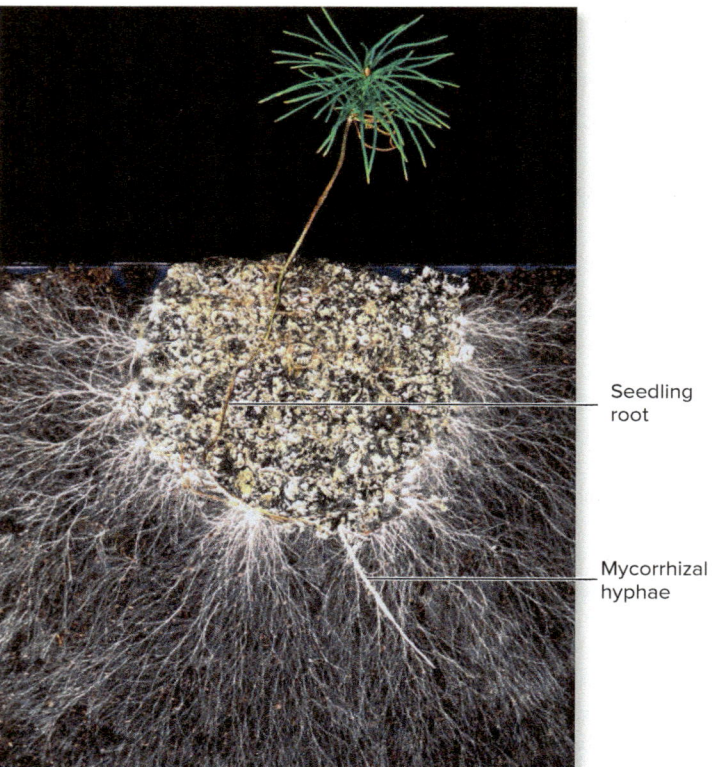

Figure 24.11 Tree seedling with mycorrhizal fungi. Hyphae of a mycorrhizal fungus extend farther into the soil than do the seedling's roots, helping the plant to obtain water and mineral nutrients.
©Dr. D.P. Donelley and Prof. J.R. Leake, University of Sheffield, Department of Animal & Plant Sciences

(a) Micrograph of arbuscular mycorrhizae

(b) Hyphae growing between cell walls and plasma membranes

Figure 24.12 Endomycorrhizae. (a) Light micrograph showing black-stained AM fungi within the roots of the forest herb *Asarum canadensis*. Endomycorrhizal fungal hyphae penetrate plant root cell walls, then branch into the space between root cell walls and plasma membranes. (b) Diagram showing the position of highly branched arbuscules. Hyphal branches, or arbuscules, are found on the surface of the plasma membrane, which becomes highly invaginated. The result is that both hyphae and plant membranes have very high surface areas.
(a) ©Mark Brundrett

(a) Ectomycorrhizal fruiting body

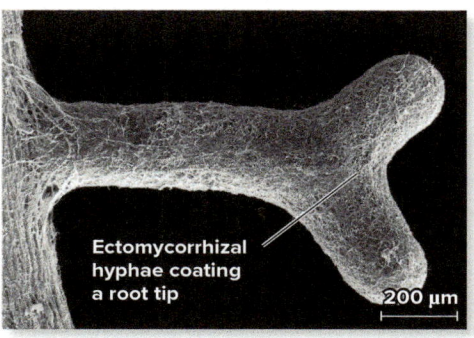
(b) SEM of ectomycorrhizal hyphae

(c) Hyphae invading intercellular spaces

Figure 24.13 Ectomycorrhizae. (a) The fruiting structure of the common forest fungus *Laccaria bicolor*. This is an ectomycorrhizal fungus that is associated with tree roots. **(b)** SEM showing ectomycorrhizal fungal hyphae of *L. bicolor* covering the surfaces of young *Pinus resinosa* root tips. **(c)** Diagram showing that the hyphae of ectomycorrhizal fungi do not penetrate root cell walls but grow within intercellular spaces. In this location, fungal hyphae are able to obtain organic food molecules produced by plant photosynthesis.
(a) ©Jacques Landry, Mycoquebec.org; *(b)* Courtesy of Dr. R. Larry Peterson and Hugues Massicotte

Concept Check: What benefits do plants obtain from the association with fungi?

and peach trees, coffee shrubs, and many herbaceous plants, including legumes, grasses, tomatoes, and strawberries.

Ectomycorrhizae (from the Greek *ecto*, meaning outside) are partnerships between temperate forest trees and soil fungi, particularly basidiomycetes. The fungi that engage in such associations are known as ectomycorrhizal fungi (**Figure 24.13a**). The hyphae of ectomycorrhizal fungi coat tree-root surfaces (**Figure 24.13b**) and grow into the spaces between root cells but do not penetrate the cell membrane (**Figure 24.13c**). Some species of oak, beech, pine, and spruce trees will not grow unless their ectomycorrhizal partners are also present. Mycorrhizae are thus essential to the success of commercial nursery tree production and reforestation projects.

Animal Microbiomes Serve Many Useful Functions

Animal microbiomes contain viruses, archaea, bacteria, fungi, protists, and microscopic animals. These species affect animal health and may play important environmental roles or have medical applications. Termites, for example, are globally important in recycling plant biomass, a function enhanced by hundreds of gut bacterial species. Tunicates, an early-diverging lineage of chordates (see Chapter 26), occur worldwide as colonies of animals that filter microbial food from seawater. The guts of tunicates contain complex microbiomes, including beneficial cyanobacteria that synthesize sterols useful to the animal and nonphotosynthetic bacteria that produce defensive molecules. Such microbiomes are potential sources of new antibiotics that control disease-causing microbes without harming the animal host.

Microbiomes are likewise important to mammals, such as the brown bear, *Ursus arctos*. Brown bears are notable for their seasonal lifestyle, which involves fat accumulation in summer, with reliance on that fat during hibernation in winter. Ribosomal DNA amplicon analysis (see Figure 24.3) reveals that the microbiome in a bear's gut differs in winter versus summer. Winter microbiomes are less diverse and include higher levels of the bacterial phylum Bacteroidetes, which is associated with the breakdown of lipids and proteins. Bacteria linked to diets rich in plant fiber are also lower in winter than in summer, when bear diets include plants.

The most intensively studied animal microbiome is that of humans, the subject of several large scientific projects involving many biologists. The Human Microbiome Project, for example, characterized the microbes of 18 body sites on 300 healthy U.S. adults, finding that humans have distinctive microbiomes in the gut, vagina, urogenital tract, mouth, nose, skin, and teeth (see Figure 24.7). The Human Microbiome Project and other studies have shown that human bodies host 100 trillion microbes that make up 1–3% of our body weight! Microbiota associated with humans affect our health in many ways:

- The microbiomes of teeth form biofilms, known as plaque, that are detrimental to dental health.
- Up to 240 bacterial genera can be associated with human skin alone, performing beneficial functions. Some species within skin microbiomes break down dead skin, and others help to prevent infections or transform skin oil into natural moisturizer.
- Microbes of the digestive system, particularly the gut, play important roles that start in early life. *Bifidobacterium longum* subspecies *infantis*, for example, is the most prevalent microbe in the gut of healthy infants. This bacterium has genes encoding proteins that bind, import, and metabolize milk polysaccharides into short fatty acids such as acetate. Some of these fatty acids serve as food for infant colon cells, aid the baby's immune system, and reduce gut pH, which deters some disease microbes. The best foods for *B. longum* are polysaccharides that are abundant in human milk but rare or absent in that of other animals. Breastfeeding thereby fosters the growth of these beneficial microbes.

FEATURE INVESTIGATION

Blanton, Gordon, and Associates Found That Gut Microbiota Affect the Growth of Malnourished Children

As noted at the beginning of this chapter, information about the human microbiome has the potential to improve child growth. In a study published in 2016, Laura Blanton, Jeffrey Gordon, and colleagues described the effects of differences in gut microbiota on the growth of malnourished children. These researchers began their work by analyzing *16S rDNA* to determine how gut microbiota change during the first 32 months of life in malnourished children from the same locale (**Figure 24.14**). Some of these malnourished children appeared healthy based on their growth, but others were stunted in their growth and were underweight. The results revealed that the healthy children had gut microbiota that changed over time so that the older children had a different, more mature gut microbiota than did the infants. By contrast, the children who were stunted in growth retained an immature microbiota (see left side of data of Figure 24.14).

Figure 24.14 Impact of the gut microbiota on growth.

HYPOTHESIS Microbiota from children who were stunted in their growth due to malnourishment will impair the growth of mice.

KEY MATERIALS Fecal samples from healthy and stunted children; germ-free mice, which are mice, that have been raised in an aseptic environment and do not have any microbes in or on their bodies.

Experimental level

Conceptual level

1. Obtain fecal samples from many healthy and stunted children of the same locale from birth to 32 months of age.

2. Perform an amplicon analysis on the fecal samples from birth to 32 months old.

See Figure 24.3.

This method reveals which prokaryote species are in the gut microbiota and how they change over the course of 32 months.

3. Using a tube, introduce fecal samples from healthy and stunted children into the gut of 5-week-old germ-free mice.

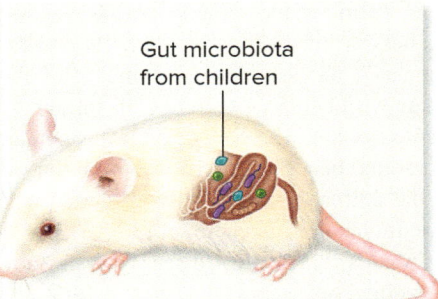

Gut microbiota from children

4. Give the mice a diet that corresponds to a diet that the children eat.

The two different types of gut microbiota may affect the ability of the mice to gain weight.

5 Monitor the weight of the mice over a 5-week period.

Received fecal material from a healthy donor

Received fecal material from a stunted donor

Weight gain is a measure of how the gut microbiota affect growth.

6 **THE DATA**

From step 2:

Healthy children: — From birth to 32 months → Microbiota changed over 32 months

Stunted children: — From birth to 32 months → Microbiota stayed similar over 32 months

From step 5:
- Mice that received fecal transplant from healthy children
- Mice that received fecal transplant from stunted children

% Initial weight vs. Days after the fecal transplant was established in the mouse

7 **CONCLUSION** The gut microbiota from stunted donors impaired the growth of mice.

8 **SOURCE** Blanton, Laura B., et al. 2016. Gut Bacteria That Prevent Growth Impairments Transmitted by Microbiota from Malnourished Children. *Science* 351: 830.

In the next phase of their work (see step 3 of Figure 24.14), these biologists transplanted the microbiota of healthy or stunted children into separate sets of microbe-free (germ-free) mice so that the effects on growth could be monitored in ways that would have been difficult to accomplish with children. After this fecal transplantation, the mice were fed germ-free food of the same type eaten by the children. This experiment revealed that the mice having microbiota transplanted from the stunted children gained less weight than the mice that received microbiota transplants from healthy children, even though the mice consumed the same amount of food (see right side of data of Figure 24.14). Though not shown in Figure 24.14, these researchers also used magnetic resonance imaging technology to determine lean and fat body mass, as well as micro-CT (computerized tomography) to evaluate the femur bone structure of the mice. The results of these imaging studies also showed that the gut microbiota played an important role in proper growth.

In further work that is not shown in Figure 24.14, the researchers took a closer look at the composition of the microbiota to identify potential microbes that were responsible for better growth. By performing an amplicon analysis, they determined that the microbiota of the mice that had received fecal transplants from healthy donors tended to be dominated by a few bacterial species—*Ruminococcus gnavus, Clostridium symbiosum*, and a few others. To determine if any of these species were responsible for greater growth, researchers grew cultures of these bacterial species in the laboratory and then introduced them into mice that had received fecal transplants from stunted children. When these bacterial species were transplanted into the smaller mice, they grew larger! The researchers then characterized the gut microbiota from these mice and determined that they were dominated by *Ruminococcus gnavus* and *Clostridium symbiosum*. Metabolomic studies revealed that these two bacterial species had the effect of decreasing amino acid oxidation and increasing amino acid incorporation into proteins that contributed to increases in body mass. In this way, it was inferred that the microbes helped malnourished children to better use amino acids for growth.

Experimental Questions

1. Why did investigators feed experimental mice germ-free food?
2. **SCISKILLS** ▶ Explain how the microbiota from healthy or stunted children affected the growth of mice.

24.3 Reviewing the Concepts

- The combination of a host organism and its microbiome (or microbiota) is known as a holobiont. The host and microbiome (or microbiota) genomes together are the hologenome (Figure 24.7).
- Hosts acquire microbiomes or microbiota in different ways, and these associations may change as species diverge from each other during evolution (Figure 24.8).
- A variety of species—including other microbes, plants, fungi, and animals—can act as microbiome or microbiota hosts. The associations result in an astonishing array of benefits for both the hosts and their microbial partners (Figures 24.9–24.13).
- Blanton, Gordon, and colleagues found that gut microbiota affect the growth of children, and this observation was confirmed by studies in mice (Figure 24.14).

24.3 Testing Your Knowledge

1. Which of the following statements regarding genome size is true?
   a. The human genome is larger than the metagenome of the human microbiome.
   b. The human genome is smaller than the metagenome of the human microbiome.
   c. The human genome is about the same size as the metagenome of the human microbiome.
2. Mycorrhizae are associations between
   a. plant roots and protists.
   b. plant roots and fungi.
   c. plant roots and bacteria.
   d. plant shoots and protists.
   e. plant shoots and fungi.

24.4 Engineering Animal and Plant Microbiomes

Learning Outcomes

1. Explain why people want to engineer animal and plant microbiomes.
2. Discuss how biologists produce synthetic microbiomes.
3. Describe artificial selection methods for identifying the most beneficial microbiomes for hosts.

Because microbiomes affect the health of animals and plants, researchers are investigating the potential of altering microbiomes to benefit humans, domesticated animals, and crop plants. Manipulating the composition of a microbiome to improve host characteristics is known as **microbiome engineering.** Much animal microbiome engineering research is done on mice or other laboratory animals, but it eventually may be applied to humans.

Can Aid Human Health Transplantion of Gut Microbiomes

As we saw in Figure 24.14, the gut microbiome can have an important impact on health. For this reason, fecal transplants are performed to introduce the entire gut microbiome of a healthy animal host into the gut of an unhealthy host. However, because thousands of microbial species and many complex interactions may be involved, the biological basis for success or failure of this treatment process can be difficult to determine. Alternatively, biologists perform experiments to determine more precisely which particular microbiome members are consistently associated with host health. As shown by the studies of Blanton, Gordon, and colleagues, described earlier, comparing the microbiota of healthy hosts with those of unhealthy hosts helps to reveal these key beneficial species.

Additional information about healthy human microbiomes is gained by studying humans in nonindustrial societies. For example, the microbiomes of isolated Amerindians having no previous contact with Westerners have the highest known diversity of gut and skin bacteria and microbiome functions. By contrast to humans living in industrialized societies, isolated Amerindians have lower diversity of bacterial species from the phylum Bacteroidetes and higher diversity of other important microbiome species. Such differences indicate that humans living in industrialized countries have undergone shifts in their microbiomes that are implicated in gastrointestinal disorders, obesity, and autoimmune disease. For example, by comparison with isolated populations, people in industrialized countries don't consume enough dietary fiber, composed of complex plant carbohydrates. Studies of mice show that the chronic lack of dietary fiber reduces gut microbiome diversity.

Researchers use such information to assemble **synthetic microbiomes** by mixing cultures of beneficial microbial species. The effects on host health are determined by implanting these synthetic microbiomes into germ-free hosts, also known as axenic or gnotobiotic hosts, such as mice, rats, or guinea pigs. The microbe combinations associated with the most positive effects on host health may be considered for use as treatments in humans.

Can Improve Human and Plant Microbiomes Probiotic Treatment

A **probiotic treatment** involves the introduction of a single microbial strain or a few strains into the microbiome of a host to promote host health. For example, the bacterial genus *Lactobacillus* that normally occurs in the human vagina has been used to treat vaginal conditions that result from the loss of the normal microbiome. Plant microbiomes likewise naturally contain microbes that aid plant health and protect against pests and disease, and therefore can be administered as probiotics. For example, in a model laboratory system containing the plant *Arabidopsis thaliana* and the bacterium *Pseudomonas simiae*, the microbe stimulates the plant to produce a transcription factor that causes roots to secrete specific molecules. These compounds help the plant to take up iron and to deter the growth of fungal disease agents. Adding probiotic microbes has been used to promote crop growth.

Can be Used to Engineer Microbiomes Artificial Selection

Yet another engineering approach relies on artificial selection, which is a human-controlled form of natural selection (see Chapter 19). In this process, biologists engineer communities by artificial selection of

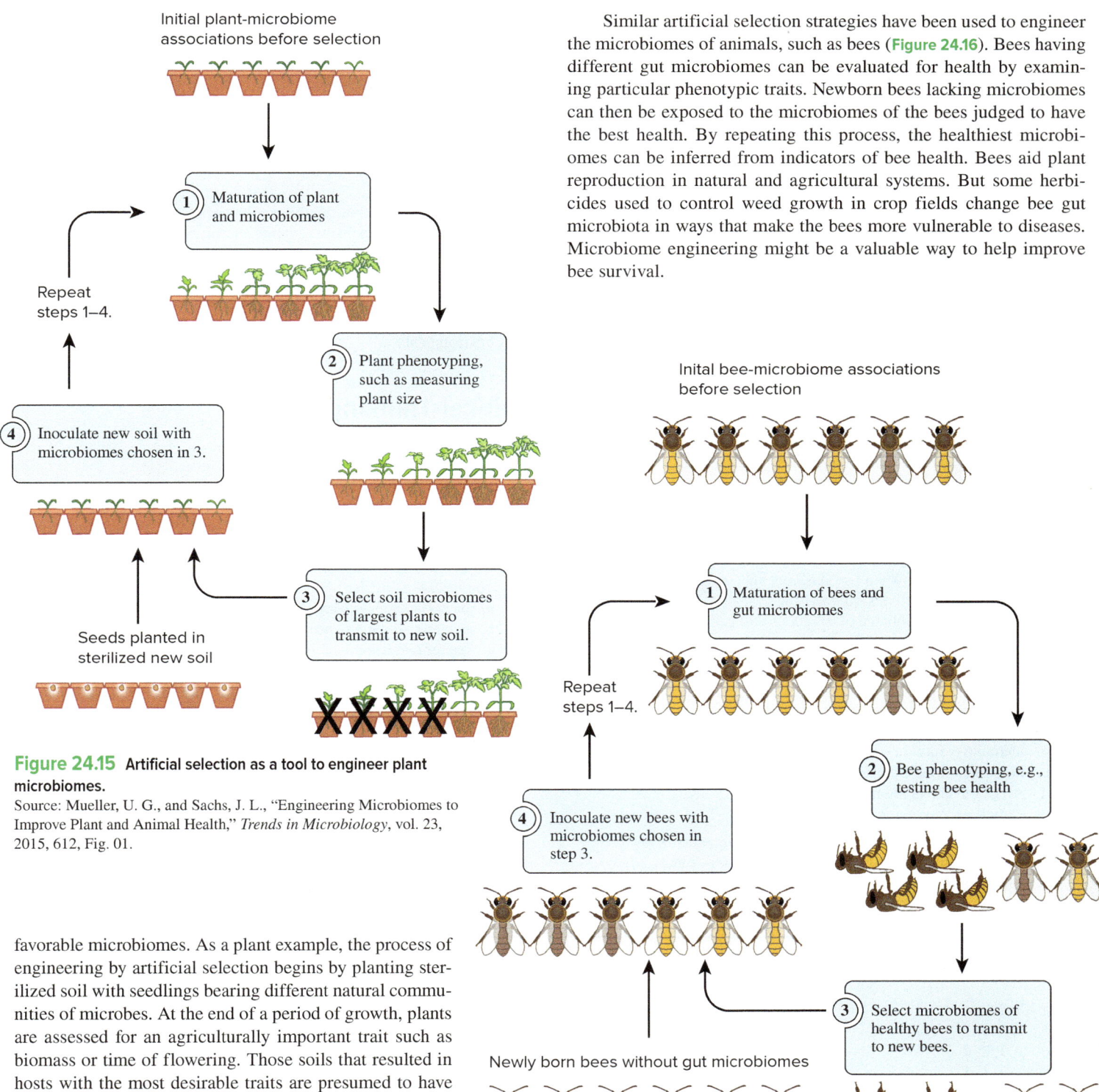

Figure 24.15 Artificial selection as a tool to engineer plant microbiomes.
Source: Mueller, U. G., and Sachs, J. L., "Engineering Microbiomes to Improve Plant and Animal Health," *Trends in Microbiology*, vol. 23, 2015, 612, Fig. 01.

Similar artificial selection strategies have been used to engineer the microbiomes of animals, such as bees (Figure 24.16). Bees having different gut microbiomes can be evaluated for health by examining particular phenotypic traits. Newborn bees lacking microbiomes can then be exposed to the microbiomes of the bees judged to have the best health. By repeating this process, the healthiest microbiomes can be inferred from indicators of bee health. Bees aid plant reproduction in natural and agricultural systems. But some herbicides used to control weed growth in crop fields change bee gut microbiota in ways that make the bees more vulnerable to diseases. Microbiome engineering might be a valuable way to help improve bee survival.

favorable microbiomes. As a plant example, the process of engineering by artificial selection begins by planting sterilized soil with seedlings bearing different natural communities of microbes. At the end of a period of growth, plants are assessed for an agriculturally important trait such as biomass or time of flowering. Those soils that resulted in hosts with the most desirable traits are presumed to have acquired microbes that produce the most beneficial microbiomes. The microbiomes from the soils of the plants with favorable phenotypes are selected and used to inoculate soils for the next generation of plants. In this way, scientists use plant characteristics to select the most beneficial microbiomes (Figure 24.15). Amplicon analysis or WMS is then used to determine the composition of the winning microbiome.

Figure 24.16 Artificial selection as a tool to engineer animal microbiomes. Engineering the bee microbiome provides an example.
Source: Mueller, U. G., and Sachs, J. L., "Engineering Microbiomes to Improve Plant and Animal Health," *Trends in Microbiology*, vol. 23, 2015, 612, Fig. 01.

24.4 Reviewing the Concepts

- Researchers are exploring a variety of approaches for engineering microbiomes, including the transplantation of gut microbiomes, the use of probiotics, and artificial selection (Figures 24.15, 24.16).

24.4 Testing Your Knowledge

1. The introduction of one or a few strains of beneficial microbes into the microbiome of a host is called
   a. gut microbiome transplantation.
   b. probiotics.
   c. artificial selection.
   d. All of the above are correct.

Assessing Your Knowledge and Skills

Foundational Knowledge

Multiple Choice

1. A microbiome is
   a. an interaction between two different species of microorganisms.
   b. an environment that is microscopic.
   c. a particular assemblage of microbes and genes that occurs in a defined environment.
   d. the entire genetic makeup of a particular microorganism.
   e. both a and d.

2. An example of a microbiome function is
   a. nitrogen fixation.
   b. methane oxidation.
   c. the production of particular metabolites, such as vitamin B_{12}.
   d. All of the above are microbiome functions.
   e. Only b and c are microbiome functions.

3. What would be the order of the four DNA fragments below if they were aligned into a contig? With the exception of black, DNA regions with the same sequence are shown in the same color.

   Fragment 1:
   Fragment 2:
   Fragment 3:
   Fragment 4:

   a. 1, 2, 3, 4
   b. 2, 3, 4, 1
   c. 1, 3, 2, 4
   d. 3, 2, 1, 4
   e. b or d

4. Which of the following is a microbiome of a physical system?
   a. a microbiome on the surface of a leaf
   b. a microbiome in the human gut
   c. a microbiome in a soil sample
   d. a microbiome in a sample of human saliva
   e. All of the above are correct.

5. The combination of a host organism and its microbiome is known as
   a. a microbiome.
   b. a holobiont.
   c. a metagenome.
   d. a metabolome.
   e. both a and c.

Short Answer

1. Describe examples of holobionts that represent each of the following host organisms: cyanobacteria, protists, fungi, plants, animals.
2. **PRINCIPLES** A principle of biology is that living organisms interact with their environment. Describe an example of how a plant microbiome might influence plant interactions with the environment.

Critical-Thinking Skills

Multiple Choice

1. Imagine that you are asked to plan a study to use rDNA amplicons to compare microbiota in members of a particular human population. For this study, you will need to determine amplicon primer sequences specific to the taxonomic groups you would expect to find. Which cluster of potential microbiota groups would you need to consider?
   a. bacteria and some archaea
   b. bacteria, some archaea, and protists
   c. bacteria and some archaea, protists, and fungi
   d. bacteria, some archaea, protists, fungi, and viruses

2. If you had a whole metagenomic sequencing (WMS) data set from which to make a list of all the known types of *nifH* genes (markers of nitrogen fixation) that are present in a particular water sample, which groups of taxa would you expect to find represented?
   a. alphaproteobacteria and cyanobacteria
   b. green algae and red algae
   c. ascomycete and basidiomycete fungi
   d. all of the above
   e. none of the above

3. How might you test the competing hypotheses that (1) the bacterial microbiota of a particular holobiont are determined primarily by the environment or (2) microbiota are primarily determined by host traits that reflect evolutionary effects, such as production of chemical secretions that attract particular bacteria?
   a. You might compare the microbiota of holobionts collected from environments that differ in some major way, because if the first hypothesis is valid, microbiota might co-vary with environments.
   b. You might compare the microbiota of holobionts with that of the surrounding environment, because substantial similarity would support the first hypothesis.
   c. You might compare the microbiota of closely related holobionts, because similarity of holobiont microbiota would help to support the alternate hypothesis.
   d. All of the above hypothesis-testing processes would be useful.
   e. None of the above processes would be helpful.

4. Let's assume that a particular microbiome is known to undergo changes in taxonomic diversity in response to a particular environmental change. You want to know how microbiome proteins change. Which of the methods listed below could you employ?
   a. You could obtain whole metagenomic sequences from DNA extracted from samples taken before and after environmental change, then look for sequences related to known proteins.
   b. You could obtain RNA from samples taken before and after environmental change, then look for differences related to known proteins.
   c. You could extract and analyze as many proteins as possible from samples taken before and after environmental change, then compare lists.
   d. You could try to detect as many metabolites as possible in the samples.
   e. You could perform all of the above.
5. Although many people now know that humans and other animals depend on healthy microbiomes, the existence of other types of microbiomes might not be so widely recognized. Which of the following statements correctly describes the resources a type of non-animal host could provide to diverse microbial species?
   a. Colonial and filamentous cyanobacteria provide attachment space, organic food, and oxygen.
   b. Filamentous green algae provide extensive attachment area, organic food, and oxygen.
   c. Fungi in the form of lichens provide protection in the form of chemical compounds that may reduce edibility or light damage.
   d. Plants of diverse types provide habitat, organic carbon, and oxygen.
   e. All of the above statements are correct.

Short Answer

1. Imagine that you are part of a team that aims to improve the health of domesticated honeybees. What microbiome engineering process might you use?
2. Algal growths on polar ice surfaces are known to reduce ice reflectivity, thereby potentially affecting climate on a global basis. Let's say that your collaborators have made collections of algae from polar ice and you lead a lab team responsible for determining what types of algae are present in the ice samples. What technological expertise would members of your team need to have?

25

Plant Evolution: How Plant Diversification Changed Planet Earth

©Linda Graham

The drought-resistant water fern *Marsilea drummondii* in the Australian outback.

Chapter Outline

25.1 Ancestry and Diversity of Land Plants
25.2 An Evolutionary History of Land Plants
25.3 Diversity of Modern Gymnosperms
25.4 Diversity of Modern Angiosperms
25.5 Human Influences on Angiosperm Diversification

Assessing Your Knowledge and Skills

During an expedition to the desert outback of Australia in which the goal was to learn more about ancient events in the evolutionary history of land plants, a team of plant biologists was amazed to find extensive growths of the so-called water fern *Marsilea drummondii* (chapter-opening photo). They were surprised because most ferns live in relatively moist environments, including the humid interiors of buildings where ferns are often used as ornamentals. As indicated by their common name, most water ferns live in very moist places indeed. But some fern species, including *M. drummondii*, are so resistant to drying that they can live in arid places. This is possible because this fern's leaves are fuzzy with surface hairs that act as a blanket to retard drying and excessive heating. The leaves are also able to fold themselves, thereby reducing surface area from which water can evaporate. The plant can become so dry that it crunches underfoot; it is not dead but is instead living in a state of reduced metabolism until sporadic rains stimulate active photosynthesis and growth. With these adaptations, *M. drummondii* became sufficiently abundant that aboriginal Australians, who call this plant "nardoo," have long gathered its small, brown, food-rich reproductive structures, using grinding stones to make flour for cakes.

Nardoo is just one of the hundreds of thousands of plants that have become adapted in diverse ways to life on land, even very arid places. As is illustrated by aboriginal Australians' use of nardoo and developed societies' reliance on agricultural crops, humans depend on this plant diversity for food and many other materials. The importance of plants in Earth's ecology is illustrated by a recent global comparison of the amount of carbon in living things. Plants contain 450 gigatons (GT) of carbon, greatly exceeding the amounts present in any other kind of life, with bacteria next at 70 GT and animals at 2 GT. In this chapter, we will survey the phyla of modern plants and explore how plant evolution has affected Earth's atmosphere, climate, and soils and the diversity of many other organisms.

25.1 Ancestry and Diversity of Land Plants

Learning Outcomes

1. Name several characteristics unique to land plants.
2. Compare and contrast the features of vascular and nonvascular plants.
3. **SCISKILLS** ▶ Explain how cuticles, epidermal stomata, and internal conduction systems work together to maintain stable water content in vascular plants.

Several hundred thousand modern species are formally classified into the kingdom Plantae, informally known as the land plants or simply plants (Figure 25.1). Although some species live in watery places, plants can be defined as multicellular, mostly photosynthetic eukaryotes that are adapted in many ways for life on land. Molecular and other evidence indicates that the plant kingdom evolved from green algal ancestors similar to particular modern algae. Together, plants and their closest green algal relatives are known as **streptophytes** (from Latin words meaning twisted and plant) (Figure 25.2). Land plants are distinguished from closely related algae by the presence of traits that foster survival in terrestrial conditions, which are drier, sunnier, hotter, colder, and less physically supportive than aquatic habitats. Likewise, the first land animals had to acquire similar structural and reproductive adaptations.

Reflect Adaptation to Terrestrial Life Distinctive Features of the Land Plants

Land plants display distinctive features that represent early adaptations to the land habitat:

- The bodies of all land plants are primarily composed of **tissues,** defined as close associations of cells. Bodies composed of tissues have lower surface area/volume ratios than simpler green algal bodies and, thus, less readily lose water in dry air by evaporation.
- Land plant tissues arise from one or more actively dividing cells that occur at growing tips, forming **apical meristems.** The tissue-producing apical meristems of land plants produce relatively thick, robust bodies able to withstand drought and mechanical stress. Related green algae have simpler growth and body structure.
- The life cycle of land plants, known as **alternation of generations,** involves two types of multicellular bodies. A haploid generation known as the gametophyte produces eggs and/or sperm; a diploid generation known as the sporophyte produces spores by meiosis. Related algae possess a simpler life cycle in which a single-celled zygote is the only diploid stage.
- Land plant embryos are young, multicellular sporophytes that are produced by mitotic divisions of zygotes. Plant embryos, like those of mammals, depend on maternal tissues for food during early development. Embryos are not present in related green algae but are such a basic feature of land plants that the latter are also known as embryophytes.
- Tough-walled plant **spores** allow land plants to disperse offspring through dry air. Plant spores are tough because their walls contain

Figure 25.1 A temperate rain forest containing diverse plant phyla in Olympic National Park in Washington state.
©Craig Tuttle/Getty Images

a protective coating known as **sporopollenin.** This coating retards drying and microbial attack. By contrast, related green algae protect zygotes with sporopollenin, but when such zygotes undergo meiosis to produce flagellate spores for dispersal in water, these spores are not sporopollenin-coated and thus cannot survive long in air.

- Specialized structures called **gametangia** generate, protect, and disperse land plant gametes; tough **sporangia** likewise produce, protect, and disperse the spores of land plants.

Modern Land Plants Can Be Classified into Nine Phyla

Plant biologists use molecular and structural information to classify plants into phyla and organize phyla into an evolutionary sequence (Figure 25.2). The modern plant phyla are listed below by their informal names, together with a brief description of their evolutionary, societal, or ecological importance.

- The structurally simple, early-diverging plant groups informally known as mosses, liverworts, and hornworts serve as useful models of the earliest land plants. Mosses play an important role in Earth's climate now and have likely been doing so for hundreds of millions of years.
- Lycophytes are the earliest-diverging group of vascular land plants—those having conduction systems that also provide skeletal strength. In the ancient past, lycophytes were more abundant, much larger in size, and more diverse than modern representatives.
- Pteridophytes include many modern species of ferns that are particularly common in tropical habitats and are valued for horticultural plantings. Like the lycophytes, pteridophytes do not produce seeds but display several adaptations that foster stable tissue-water content, a feature critical to the ability to survive through dry periods.

Biology Principle

Populations of Organisms Evolve from Generation to Generation

This figure shows evolutionary divergence times indicated by molecular clock and some fossil evidence, suggesting when plant groups may first have arisen.

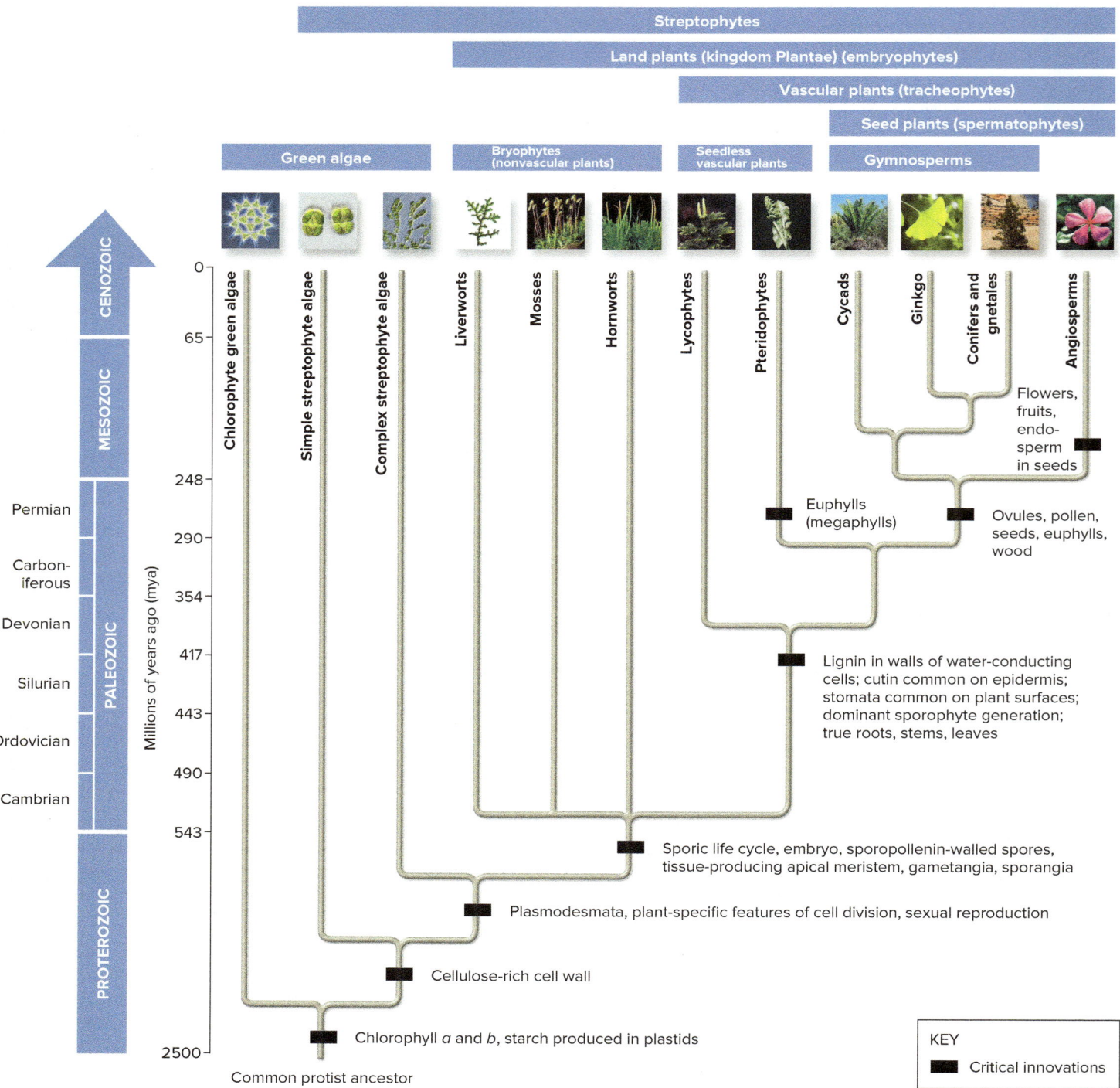

Figure 25.2 Evolutionary relationships of the modern plant phyla. Land plants gradually acquired diverse structural, biochemical, and reproductive adaptations to land that are known as critical innovations.

(1) ©Roland Birke/Medical Images; (2) ©the CAUP image database, http://botany.natur.cuni.cz/algo/database; (3–6) ©Lee W. Wilcox; (7) ©Ed Reschke/Getty Images; (8) ©Patrick Johns/Corbis/VCG/Getty Images; (9) ©Philippe Psaila/SPL/Science Source; (10) ©Toth Tamas/Shutterstock; (11) ©imageBROKER/Alamy; (12) ©Gallo Images/Getty Images

- Cycads are an early-diverging lineage of seed plants and, so, are important in understanding how seeds first evolved. Bearing spectacular, large leaves, cycads are widely planted for their horticultural beauty.
- Ginkgos are seed plants that were more diverse in the past than today. Modern ginkgos are valued as hardy and attractive streetside trees.
- Conifers, the cone-bearing seed plants, are particularly important to humans as sources of wood and other commercial materials. Conifer forests occur at higher latitudes and altitudes than forests of other types.
- Flowering plants, also known as angiosperms, are the plants upon which humans and many other animals depend for food and other resources.

Mosses, Liverworts, and Hornworts Are the Simplest Land Plants

Mosses, liverworts, and hornworts (**Figures 25.3, 25.4**) are Earth's simplest land plants. There are about 12,000 or more species of mosses, 6,500 species of modern liverworts, and 100 species of hornworts. Collectively, mosses, liverworts, and hornworts are known informally as the **bryophytes** (from the Greek *bryon*, meaning moss, and *phyton*, meaning plant). It is unclear if bryophytes all derive from a single common ancestor (see Figure 25.2), but the term bryophyte is useful for expressing common structural, reproductive, and ecological features. Bryophytes are relatively small in stature and are most common and diverse in moist habitats. Although many bryophytes possess a simple type of conducting tissue, this tissue is not tough enough to also provide structural support. By contrast, the tougher and more extensive conducting tissues of vascular plants support taller stature.

Bryophytes require moist habitats in order to accomplish sexual reproduction because they produce flagellate sperm that must swim through water films to fertilize eggs. Recently, Todd Rosenstiel, Sarah Eppley, and associates discovered that at least some bryophytes emit perfume-like chemical attractants that lure small animals such as mites and springtails, which then help to disperse bryophyte sperm. This was a surprising discovery because animal-aided sexual reproduction was previously regarded as a seed plant trait.

Bryophytes are also distinctive in having a dominant (larger and longer-lived) gametophyte generation; the bryophyte sporophyte is a relatively small and short-lived life phase that remains dependent on the maternal gametophyte and does not branch (see Figure 25.3b, Figure 25.4). By contrast, as described next, vascular plant sporophytes become independent of maternal gametophytes and these sporophytes branch, giving rise to larger bodies; the gametophytes are quite small and inconspicuous.

Lycophytes and Pteridophytes Are Vascular Plants That Do Not Produce Seeds

Vascular plants, which include all groups of modern plants except bryophytes, possess a conduction system that also provides structural

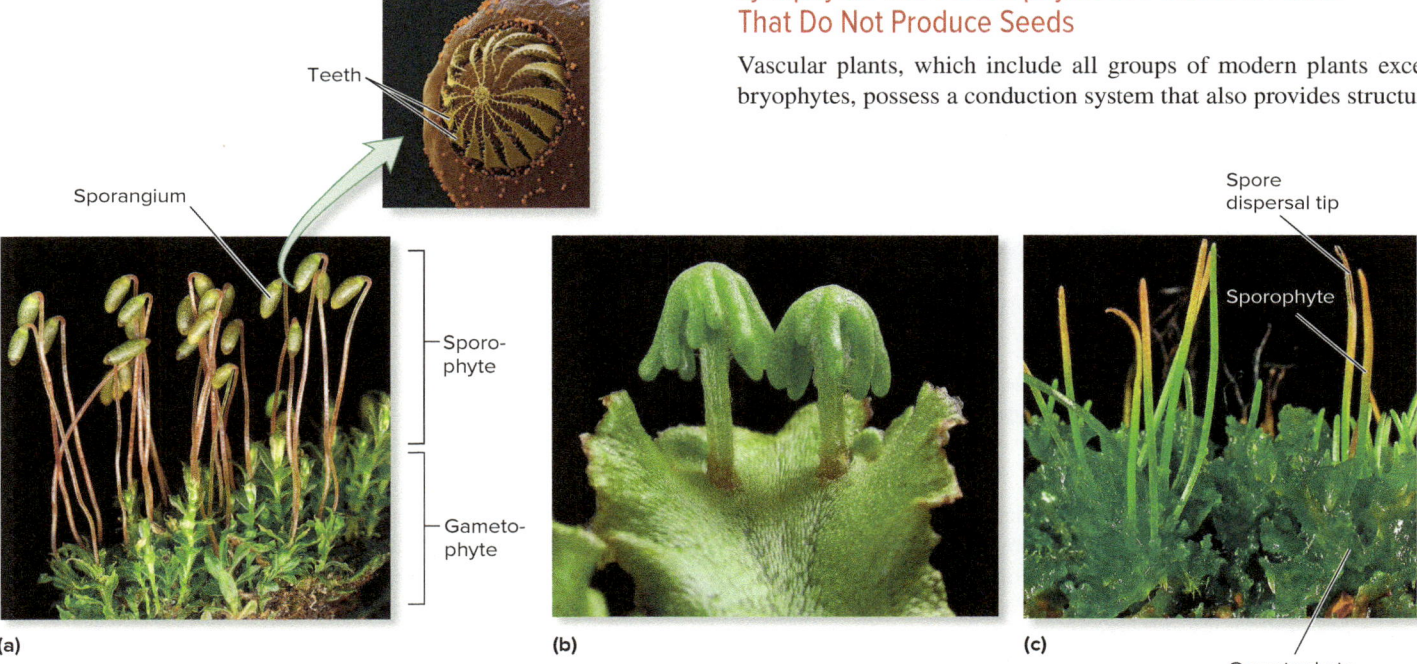

Figure 25.3 Moss, liverwort, and hornwort representatives. **(a)** The moss genus *Mnium* has a conspicuous upright, leafy, green gametophyte that generates eggs and sperm and a less noticeable attached, unbranched sporophyte that bears a spore-producing sporangium at its tip. Inset: This artificially colored SEM shows that the tips of moss sporangia often produce a number of tooth-shaped structures separated by spaces. Such tips function something like a salt shaker to sprinkle spores into the wind over a period of time, rather than releasing spores all at once. **(b)** The liverwort *Marchantia polymorpha* has a flat body that produces umbrella-shaped structures bearing egg-shaped sporophytes on the undersides. When sporophytes mature, they release tough-walled spores into the air. **(c)** Hornwort gametophytes grow close to the ground and produce upright sporophytes that release spores from tips into the air.
(a–c) ©Lee W. Wilcox; *(a inset)* ©Eye of Science/Meckes/Ottawa/Science Source

Concept Check: *Why do you think liverworts, mosses, and hornworts produce their reproductive spores on raised structures?*

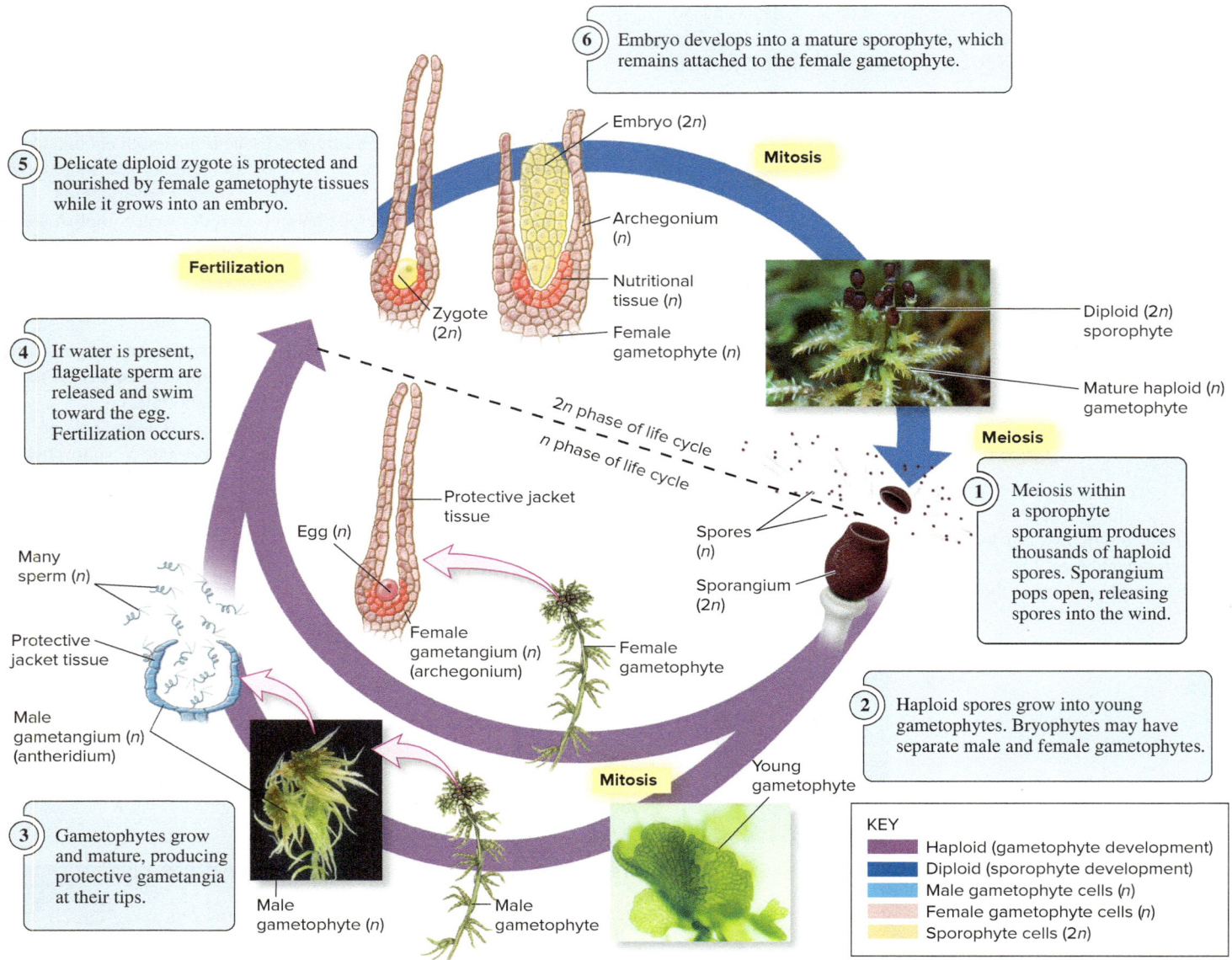

Figure 25.4 The life cycle of the early-diverging peat moss genus *Sphagnum*. The life cycle of this common peat moss illustrates reproductive adaptations commonly found in bryophytes that were likely also present in early land plants. *(top inset)* ©Larry West/Science Source; *(bottom insets)* ©Linda Graham

support. Fossils indicate that vascular plants have been important to Earth's ecology for hundreds of millions of years. Molecular data demonstrate that the lycophytes are the oldest phylum of living vascular plants and that pteridophytes are the next oldest living vascular plant phylum (see Figure 25.2). In the past, diverse lycophytes included tall trees, but now only about 1,000 relatively small species exist (**Figure 25.5a**). Pteridophytes diversified more recently, and there are about 12,000 modern species, including horsetails, whisk ferns, and other ferns (**Figure 25.5b–d**). Because the lycophytes and pteridophytes diverged prior to the origin of seeds, they are known as the seedless vascular plants.

Together, lycophytes, pteridophytes, and seed-producing vascular plants are also known as the **tracheophytes,** named for **tracheids,** a type of specialized conducting cell. Tracheids occur in specialized tissues, known as **xylem,** that conduct water and minerals throughout the plant body. The xylem also provides structural support, allowing vascular plants to grow taller than nonvascular plants. This support function arises from the presence of a compression- and decay-resistant, waterproofing material known as **lignin,** which occurs in the cell walls of tracheids and some other types of plant cells. A second type of vascular tissue known as **phloem** conducts a watery sap containing organic molecules such as sugar throughout the plant body. Xylem and phloem, together known as **vascular tissues,** occur in the plant organs: stems, roots, and leaves.

Tracheophyte **stems** are branching organs that bear reproductive sporangia and leaves. Most vascular plants also produce **roots**—branching organs specialized for uptake of water and minerals from the soil—and **leaves** that generally have a photosynthetic function.

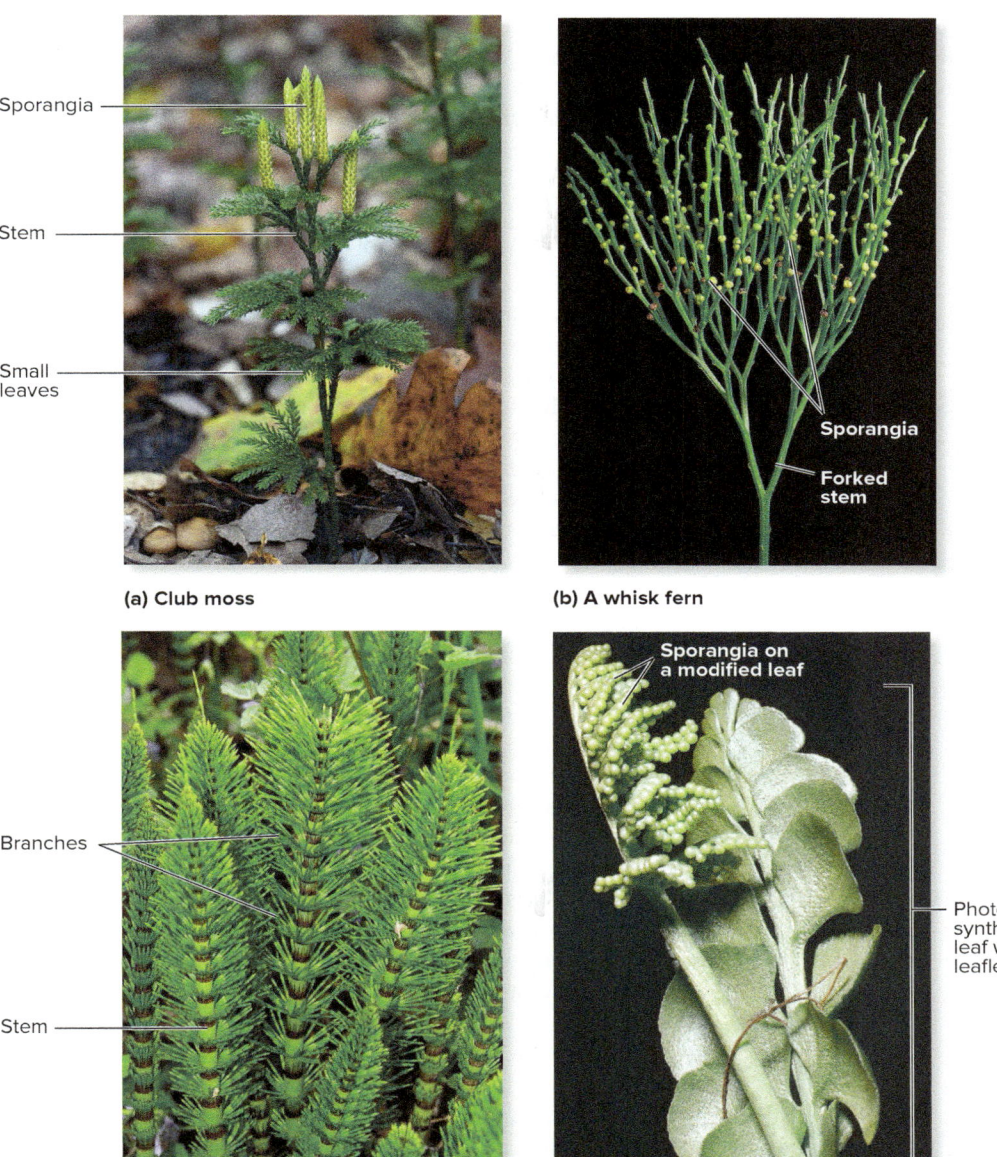

Figure 25.5 Representative seedless vascular plants. (a) The lycophyte *Lycopodium obscurum*. The sporophyte stems bear many tiny leaves, and leaves bearing sporangia generally occur in club-shaped clusters. For this reason, lycophytes are informally known as club mosses or spike mosses, though they are not true mosses. The gametophytes of lycophytes are small structures that often occur underground, where they are better protected from drying. **(b)** The leafless, rootless, green stems of the whisk fern (*Psilotum nudum*) branch by forking and bear many clusters of yellow sporangia that disperse spores via wind. The gametophyte of this plant is a tiny, pale structure that lives underground in a symbiotic partnership with fungi. **(c)** Another pteridophyte is the giant horsetail (*Equisetum telmateia*), which displays branches in whorls around the green stems. The leaves of this plant are tiny, light brown structures that encircle stems and branches at intervals. This plant produces spores in cone-shaped structures, and the wind-dispersed spores grow into small, green gametophytes. **(d)** The early-diverging fern *Botrychium lunaria*, showing a green, photosynthetic leaf with leaflets and a modified leaf that bears many round sporangia.
(a–b) ©Lee W. Wilcox; *(c)* ©Andrea Innocenti/REDA&CO/UIG/Getty Images; *(d)* ©Patrick Johns/Getty Images

Like stems, roots and leaves contain both xylem and phloem conducting tissues. Together, the branching vascular tissues of an entire vascular plant are known as the vascular system. The vascular system of a plant functions much like the vascular system of an animal; both systems carry nutrients and other materials throughout the body.

In relatively dry habitats, lycophytes, pteridophytes, and other vascular plants are able to grow to larger sizes and remain metabolically active for longer periods than can bryophytes, allowing tracheophytes to exploit a wider spectrum of land habitats. Vascular plants have this advantage because they are better able to maintain stable internal water content by means of several adaptations, including a vascular system, a waxy cuticle, and stomata.

- A vascular system allows water and minerals taken up by roots to move through stems and leaves, and it allows sugar produced in leaves (and sometimes stems) to move into locations where the sugar is metabolized to produce ATP. For more information about the structure and function of plant vascular tissues, see Chapter 30.

- A protective waxy **cuticle** present on most plant surfaces contains a polyester polymer known as cutin that deters pathogen attack and wax that helps to prevent drying (**Figure 25.6a**). A simpler cuticle that retards drying in some bryophytes is composed of materials that were ancestral to the cutin and lignin produced by vascular plants.

- The surface tissue of vascular plant stems and leaves contains many **stomata** (singular, stoma or stomate)—pores that can open and close (**Figure 25.6b**). Stomata allow plants to take in the carbon dioxide needed for photosynthesis and release oxygen to the air while conserving water. When the environment is moist, the pores open, allowing photosynthetic gas exchange to occur. When the environment is very dry, the pores close, thereby reducing water loss by evaporation into the air. For more information about stomata, see Chapter 30.

Biology Principle

Living Organisms Maintain Homeostasis

The structures illustrated in this figure explain how vascular land plants maintain homeostasis of water content.

(a) Stem showing tracheophyte adaptations

(b) Close-up of stomata

Figure 25.6 Tracheophyte adaptations for transporting and conserving water. (a) A cross section through a stem of the whisk fern *Psilotum nudum*. When viewed with fluorescence microscopy and illuminated with violet light, an internal core of xylem tracheids glows yellow, as does the surface cuticle. (b) Surface pores associated with specialized cells—the complexes known as stomata—allow for gas exchange between plant and atmosphere. This photo naturally lacks color because it was made with a scanning electron microscope (SEM), which uses electrons rather than visible light to form magnified images.
(a) ©Linda Graham; (b) ©Martha Cook

Lycophytes and pteridophytes have a life cycle dominated by the diploid sporophyte, which is large enough to produce many spores. Clusters of sporangia are often visible on the undersides of fern leaves (Figure 25.7). The fern gametophyte is small but can be seen with the unaided eye and, so, is widely used in biology classes to demonstrate the plant life cycle.

Gymnosperms and Angiosperms Are the Modern Seed Plants

Among the vascular plants, the seed plant phyla known as gymnosperms and angiosperms dominate most modern landscapes. Because both groups produce seeds, gymnosperms and angiosperms together are known as the **spermatophytes. Seeds** are complex structures having specialized tissues that protectively enclose embryos and contain stores of carbohydrate, lipid, and protein that enable embryos to grow and develop. Seed plants produce **pollen,** small air- or animal-borne spores that contain and protect microscopic male gametophytes. The ability to produce seeds and pollen helps to free seed plants from reproductive limitations experienced by the seedless plants, revealing why in many places seed plants are the dominant plants on Earth today.

The modern seed plant phyla commonly known as cycads, ginkgos, and conifers (including a group known as the gnetophytes) are collectively known as gymnosperms (Figure 25.8a–c). The term **gymnosperm** comes from the Greek, meaning naked seeds, reflecting the observation that gymnosperm seeds are not enclosed within fruits.

The **angiosperms** (Figure 25.8d) produce seeds and pollen but are distinguished from gymnosperms by the presence of flowers, fruits, and a specialized seed tissue known as endosperm. A **flower** is a short stem bearing reproductive organs that are specialized in ways that enhance seed production. **Fruits** are structures that develop from flowers, enclose seeds, and foster seed dispersal in the environment. The term angiosperm comes from the Greek, meaning enclosed seeds, reflecting the observation that the flowering plants produce seeds within fruits. **Endosperm** is a nutritive seed tissue that increases the efficiency with which food is stored in the seeds of flowering plants, and endosperm provides a large proportion of human food such as in cereal crops. Angiosperms are integral components of the diet of many animals.

25.1 Reviewing the Concepts

- Plants are multicellular eukaryotic organisms that are mostly photosynthetic and display many adaptations to life on land. Land plants evolved from green algal ancestors.
- The modern plant kingdom consists of nine phyla, informally known as the mosses, liverworts, hornworts, lycophytes, pteridophytes, cycads, ginkgos, conifers, and angiosperms (Figures 25.1, 25.2).
- The moss, liverwort, and hornwort phyla are together known informally as the bryophytes because they share some structural, reproductive, and ecological features (Figures 25.3, 25.4).
- Lycophytes, pteridophytes, and other vascular plants (tracheophytes) generally possess stems, roots, and leaves having vascular systems composed of conductive phloem and xylem tissues (Figures 25.5, 25.6, 25.7).

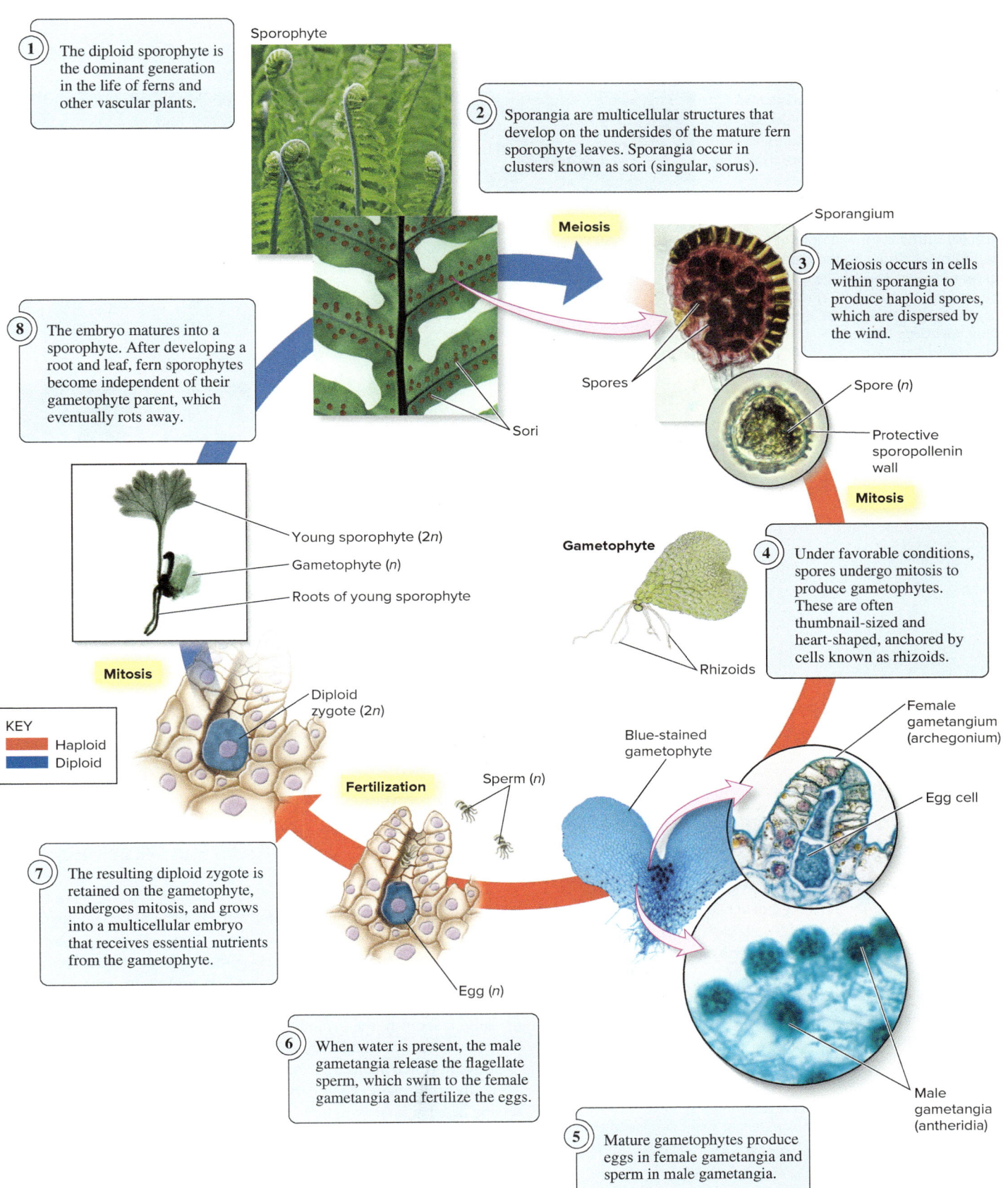

Figure 25.7 The life cycle of a typical fern. The fern life cycle is often used to illustrate plant alternation of generations because both sporophyte and gametophyte are large enough for people to see with the unaided eye.
(Sporophyte) ©Ernst Kucklich/Getty Images; *(Sori)* ©Carolina Biological Supply Company/Medical Images; *(Sporangium, spore)* ©Linda Graham; *(Gametophyte, female and male gametangium, blue-stained gametophyte)* ©Lee W. Wilcox; *(Young sporophyte)* ©Mike Peres RBP SPAS/CMSP/Newscom

Figure 25.8 Representative seed plants. **(a)** Palmlike foliage and conspicuous, seed-producing cones are features of most cycads. **(b)** *Ginkgo biloba* has leaves with forked veins and seeds with fleshy, foul-smelling seed coats. **(c)** An example of a conifer, the pine (*Pinus*). **(d)** Citrus is an example of a flowering plant, showing the distinctive features of flowers and fruits.
(a) ©Ed Reschke/Getty Images; *(b)* ©Topic Photo Agency Inc./age fotostock; *(c)* ©Lee W. Wilcox; *(d)* ©Bill Ross/Getty Images

(a) Cycad

(b) *Ginkgo*

(c) Conifer

(d) Flowering plant

- Cycads, ginkgos, and conifers (the gymnosperms) and flowering plants (the angiosperms) are the seed plants (Figure 25.8).

25.1 Testing Your Knowledge

1. Judging from Figures 25.2, 25.4, and 25.7, which of the following statements most accurately describes how plant gametophytes have changed through time?
   a. The first land plant gametophytes were, like those of modern bryophytes, larger and more independent than the sporophytes.
   b. The first land plant gametophytes, like those of modern bryophytes, had about the same size and form as the earliest plant sporophytes.
   c. The first land plant gametophytes were smaller than the earliest land plant sporophytes.
   d. The earliest type of plant life cycle, represented by modern bryophytes, was similar to that of related green algae, in which the only diploid cells are zygotes.
   e. The earliest type of plant life cycle involved the dispersal of small gametophytes within pollen, as in modern seed plants.

2. Place the plant structures in correct order, starting with earliest evolved and ending with most recently evolved.
   a. flower, seed, embryo, tracheid
   b. tracheid, embryo, seed, flower
   c. embryo, tracheid, seed, flower
   d. embryo, seed, tracheid, flower
   e. seed, embryo, tracheid, flower

25.2 An Evolutionary History of Land Plants

Learning Outcome

1. Describe some ways in which ancient plants changed Earth's environment in ways that affected the diversification of other living things.

A billion years ago, Earth's terrestrial surface was comparatively devoid of life. Green or brown crusts of cyanobacteria most likely grew in moist places, but there was probably little soil, no land plants, and no animal life. The origin of the first land plants was key to development of the first substantial soils, the rise to modern levels of atmospheric oxygen, the evolution of modern plant communities, and the colonization of land by animals. In this section, we will explore some ways in which plants transformed Earth's physical environment, especially atmospheric chemistry, and how the rise of seed plants greatly impacted animal evolution.

How can we know about such ancient events as the origin and diversification of land plants? One line of information comes from comparing molecular and other features of modern plants. For example, the genome sequence of the moss *Physcomitrella patens,* first reported in 2007, reveals the presence of genes that aid heat and drought tolerance, which are especially useful in the terrestrial habitat. Plant fossils, the preserved remains of plants that lived in earlier times, provide another line of information. Tough plant materials such

Figure 25.9 Fossil of *Pseudosalix handleyi*, an angiosperm.
©Photo by Steven R. Manchester, University of Florida. Courtesy Botanical Society of America, St. Louis, MO, www.botany.org

Concept Check: *What biochemical components of plants favor the formation of fossils?*

as sporopollenin, cutin, and lignin do not readily decay and, so, foster the formation of plant fossils (**Figure 25.9**). The study of fossils and our increasing knowledge of the molecular, structural, and functional features of modern plants have revealed an amazing story—how plants gradually acquired adaptations that enabled success on the land.

Past and Present Seedless Plants Have Transformed Earth's Ecology

One important way in which plants have transformed Earth's ecology is by storing CO_2 as decay-resistant organic compounds. This process reduces atmospheric levels of CO_2 and the buried organic materials may be transformed into coal, one type of fossil fuel. The transformation of CO_2 gas into long-lived organic compounds is known as carbon sequestration. Modern societies are interested in methods of reducing atmospheric CO_2 as a way of slowing the rate of global climate warming. Fostering plant growth is one way to accomplish this important goal.

Today, and in the ancient past, bryophytes such as the abundant peat moss genus *Sphagnum* (see Figure 25.4) have played an important role in carbon sequestration. *Sphagnum* moss produces so much decay-resistant tissue that in many places, dead moss has accumulated over thousands of years, forming deep peat deposits. People have harvested moss peat to use as fuel, to improve garden soil texture, and for other purposes. However, by converting enormous amounts of atmospheric CO_2 into peat, the peat mosses serve humanity in a much more important way—by helping to reduce the climate impacts of greenhouse gases. Experts are concerned that large regions of the world where peat mosses grow are being destroyed by land use alterations, overharvesting of peat, and other environmental changes that could reduce the ability of peat moss to moderate Earth's climate. One of the most important things people can do to slow climate change is to conserve the peat mosses and their habitats.

Fossils tell us that extensive forests dominated by tree-sized lycophytes, ferns, and early seed plants occurred in widespread, swampy regions during the warm, moist Carboniferous period (354–290 mya) (**Figure 25.10**). Plants that died and fell into oxygen-poor sediments didn't completely decay and were eventually transformed into coal. Experts estimate that such plants sequestered 10^{18} mol of atmospheric carbon into coal every million years during this time period. Much of today's coal is derived from the abundant remains of ancient plants, explaining why the Carboniferous period is commonly known as the Coal Age. During this time, plant photosynthesis and sequestration of plant-produced carbon compounds caused atmospheric CO_2 to fall dramatically and oxygen to rise to its highest known level. The removal of large amounts of the greenhouse gas CO_2 from the atmosphere by Coal Age plants had a cooling effect on the climate of the next period, which also became drier because cold air holds less moisture than warm air. This global change in climate fostered the rise of early seed plants and associated animal life, including the dinosaurs.

An Ancient Cataclysm Marked the Rise of Angiosperms

Diverse phyla of gymnosperms dominated Earth's vegetation through the Mesozoic era (248–65 mya), which is sometimes called the Age of Dinosaurs. In addition, fossils provide evidence that early mammals and flowering plants existed in the Mesozoic. Gymnosperms and early angiosperms were probably sources of food for early mammals as well as for herbivorous dinosaurs.

One fateful day about 65 mya, disaster struck from the sky, causing a dramatic change in the types of plants and animals that dominated terrestrial ecosystems. That day, at least one large meteorite crashed into the Earth near the present-day Yucatán Peninsula in Mexico. This episode is known as the **Cretaceous-Paleogene event** (also sometimes referred to as the K/T event). The impact, together with substantial volcanic activity that also occurred at this time, is thought to have produced large amounts of ash, smoke, and haze that dimmed the Sun's light long enough to kill many of the world's plants. Many types of plants became extinct, though some survived and their descendants persist to the present time. With a severely reduced food supply, most dinosaurs were also doomed, the exceptions being their descendants, the birds. The demise of the dinosaurs left room for birds and mammals to expand into many kinds of terrestrial habitats formerly inhabited by dinosaurs.

After the Cretaceous-Paleogene event, ferns dominated long enough to leave large numbers of fossil spores, and then surviving groups of flowering plants began to diversify into the space left by the extinction of previous plants. The rise of angiosperms fostered the diversification of beetles and other types of insects that associate with modern plants. Because most mammals, including humans, rely directly or indirectly upon food that angiosperms produce, the rise of angiosperms was also key to the evolutionary diversification of humans and other mammals. Diversification of angiosperms was also critical to the origin of agriculture, upon which most modern humans depend for food, shelter, and many other materials.

Biology Principle

Biologists Use Models to Test Predictions and Convey Ideas

Giant dragonfly — Giant horsetail (pteridophyte) — Giant lycophyte

Figure 25.10 Reconstruction of a Carboniferous (Coal Age) forest. This ancient swampy forest was dominated by tree-sized, giant lycophytes and pteridophytes, which later contributed to the formation of large coal deposits. Sequestration of enormous amounts of plant-produced carbon into coal reduced levels of atmospheric carbon dioxide and allowed atmospheric levels of oxygen to rise. Higher atmospheric oxygen levels allowed the survival of giant insects such as dragonflies.
©Lee W. Wilcox

Concept Check: *Why did giant dragonflies occur during the Carboniferous, but not now?*

Modeling Challenge: Modeling the extent of carbon sequestration into peat and coal in Earth's past is key to comprehending how plants (and the actions of modern humans) affect global atmospheric conditions and climate. Sketch a curve that illustrates how carbon dioxide levels in Earth's atmosphere changed from the time of early, low-growing, bryophyte-like plants to the Carboniferous.

25.2 Reviewing the Concepts

- Paleobiologists and plant evolutionary biologists infer the history of land plants by analyzing the molecular features of modern plants and by comparing the structural features of fossil and modern plants (Figure 25.9).
- Ancient and modern plants transformed Earth's ecology by altering atmospheric chemistry and climate, as well as by influencing the evolutionary diversification of animals (Figure 25.10).

25.2 Testing Your Knowledge

1. Plants lack bones, but they have left many fossils. What decay-resistant plant material is largely responsible for the occurrence of plant fossils?
   a. lignin in the xylem of the plant water-conducting system
   b. cutin produced on plant surfaces
   c. sporopollenin produced on spore surfaces
   d. All of the above are correct.
   e. None of the above is correct.

2. How have plants influenced the evolution of animals?
   a. The success of the earliest land plants started the process by which terrestrial environments became able to support early land animals.
   b. The photosynthetic activity of abundant Coal Age plants decreased the amount of CO_2 and increased the amount of O_2 in Earth's atmosphere, causing global climate changes that influenced animal traits.
   c. Diverse phyla of gymnosperms provided food and habitat key to the Mesozoic dominance of dinosaurs.
   d. The diversification of angiosperms following the Cretaceous-Paleogene event fostered the diversification of insects and other animal lineages.
   e. All of the above are correct.

25.3 Diversity of Modern Gymnosperms

Learning Outcome

1. List three gymnosperm phyla, and describe their importance to humans.

Figure 25.11 shows our current understanding of the relationships among modern seedless and seed plants. Seed plants include three phyla of gymnosperms and the flowering plants, also known as angiosperms. Gymnosperms appear to have originated from ancestral seedless plants that had the capacity to produce wood, a tough material composed of xylem tissue, allowing such plants to grow to tree height. Tall height allows trees to gain an advantage in capturing light energy for photosynthesis. Wood is a common trait among gymnosperms, and diverse angiosperms that grow to tree-size inherited the capacity to produce wood. Fossils indicate that several gymnosperm groups that once existed have since become extinct. Such extinctions may explain why it remains unclear which gymnosperm phylum is most closely related to angiosperms. We next survey the distinctive features of three modern gymnosperm phyla: cycads, *Ginkgo,* and conifers.

Cycads Are Endangered in the Wild but Are Widely Used as Ornamentals

Cycads (see Figure 25.8a) are regarded as the earliest-diverging modern gymnosperm phylum, originating more than 300 mya. Although cycads are an ancient group, molecular clock analyses indicate that modern cycad species are about 12 million years old. Nearly 300 cycad species occur today, primarily in tropical and subtropical regions. However, many species of cycads are rare, and their tropical forest homes are increasingly threatened by human activities. Many cycads are listed as endangered, and commercial trade in cycads is regulated by the Convention on International Trade in Endangered Species of Wild Fauna and Flora (CITES), a voluntary international agreement between governments to protect such species.

The structure of cycads is so interesting and attractive that many species are cultivated for use in outdoor plantings or as houseplants. Cycads display spreading, palmlike leaves (cycad comes from a Greek word, meaning palm). Mature leaves of the African cycad *Encephalartos laurentianus* can reach an astounding 8.8 m in length! Individual cycad plants produce conspicuous, conelike structures that produce either pollen or **ovules,** structures that contain female gametophyte tissues, including egg cells (see Figure 25.8a). When mature, both types of reproductive structures emit odors that attract beetles. These insects carry pollen to ovules, where the pollen produces tubes that deliver sperm to eggs. Fertilization of an egg to form a zygote triggers development of the ovule into a seed.

Ginkgo biloba Is the Last Survivor of a Once Diverse Group

The beautiful tree *Ginkgo biloba* (see Figure 25.8b) is the single remaining species of a phylum that was much more diverse during the Age of Dinosaurs. Today, *G. biloba* may be nearly extinct in the wild; widely cultivated modern *Ginkgo* trees are descended from seeds produced by a tree found in a remote Japanese temple garden and taken to Europe by 17th-century explorers.

G. biloba trees are widely planted along city streets because they are ornamental and they tolerate cold, heat, and pollution better than many other trees. In addition, these trees are long-lived—individuals can live for more than a thousand years and grow to 30 m in height. Individual trees have sex chromosomes much like those of humans, so they produce either ovules and seeds or pollen. Seed embryos are protected by a fleshy, bad-smelling outer seed coat and a hard inner seed coat. For streetside or garden plantings, people usually select the pollen-producing trees to avoid the stinky seeds.

Conifers Are the Most Diverse Modern Gymnosperm Lineage

The conifers (see Figure 25.8c) are a lineage of trees named for their seed cones, of which pinecones are familiar examples. Modern conifer families include more than 50 genera. Conifers are particularly common in mountain and high-latitude forests and are important sources of wood and paper pulp.

Conifers produce simple pollen cones and more complex, ovule-bearing cones (**Figure 25.12**). The pollen cones of conifers

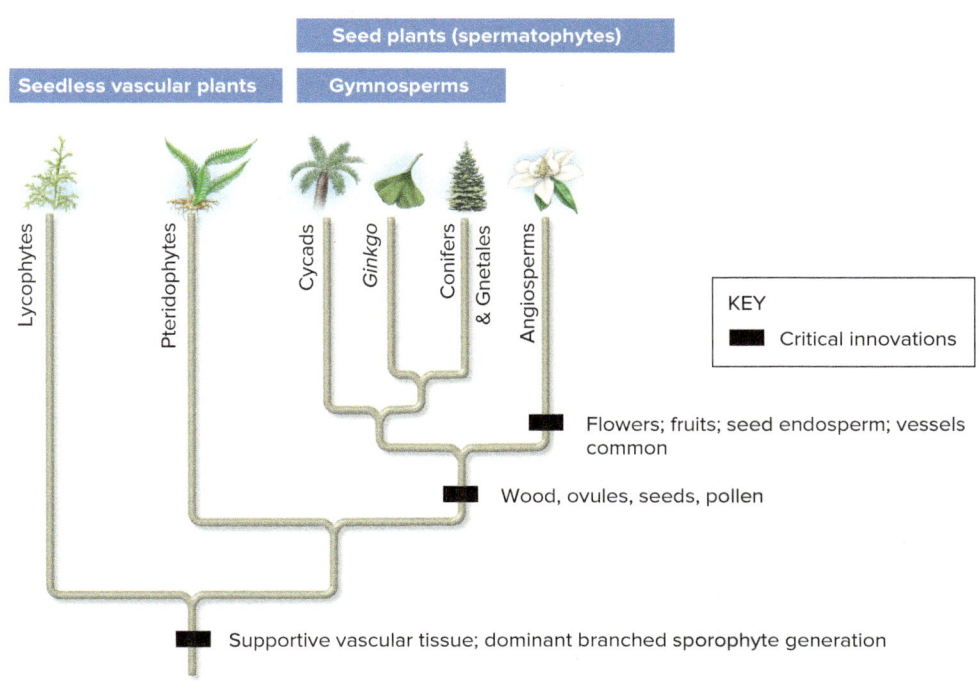

Figure 25.11 A phylogeny of modern seedless and seed plants.

Biology Principle

The Genetic Material Provides a Blueprint That Allows Organisms to Grow, Develop, and Reproduce

This diagram illustrates the entire seed-to-seed growth and development cycle of conifers.

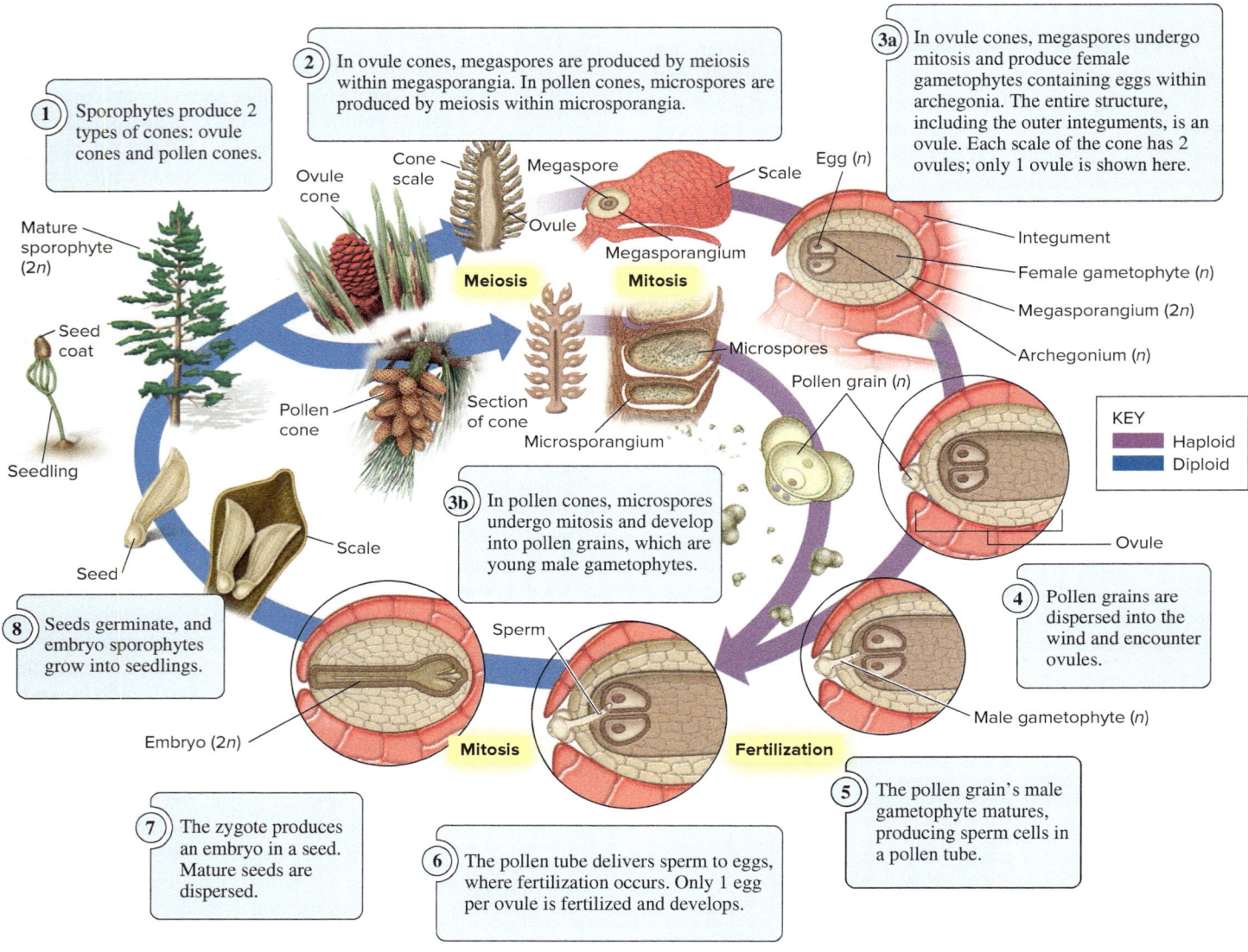

Figure 25.12 The life cycle of the genus *Pinus*.

bear many leaflike structures, each bearing a sporangium in which meiosis occurs and pollen grains develop. This pollen-producing sporangium is called a microsporangium because the pollen grains develop from small spores known as microspores. The ovule cones are composed of many short branch systems that bear ovules, within which female gametophyte tissues produce eggs.

When conifer pollen is mature, it is released into the wind, which transports pollen to ovules. When released from pollen tubes, sperm fuse with eggs, generating zygotes that grow into the embryos within seeds. Altogether, it takes nearly 2 years for pine (the genus *Pinus*) to complete the processes of male and female gamete development, fertilization, and seed development. The seeds of pine and some other conifers develop wings that aid in wind dispersal. Other conifers, such as yew and juniper, produce seeds or cones with bright-colored, fleshy coatings that are attractive to birds, which help to disperse the seeds.

Many conifers occur in cold climates and thus display numerous adaptations to such environments. Their conical shapes and flexible branches help conifer trees to shed snow, preventing heavy snow accumulations from breaking branches. People who use conifers in landscape plantings also value these traits. Conifer leaf shape and structure are adapted to resist damage from drought that occurs in both summer and winter, when liquid water is scarce. Conifer leaves are often scalelike or needle-shaped; these shapes reduce the area of leaf surface from which water can evaporate. A thick, waxy cuticle coats conifer leaf surfaces, retarding water loss and preventing attack by disease organisms. Many conifers are evergreen; that is, their leaves live for more than 1 year before being shed and are not all shed during the same season. Retaining leaves through winter helps conifers to start up photosynthesis earlier than deciduous trees, which in spring must replace leaves lost during the previous autumn. Evergreen leaves thus provide an advantage in the short growth season of alpine or high-latitude environments.

The conifer phylum also includes the Gnetales, an order of three genera, *Gnetum*, *Ephedra*, and *Welwitschia*, which feature distinctive adaptations:

- *Gnetum* is unusual among modern gymnosperms in having broad leaves similar to those of many tropical plants (Figure 25.13a). Such leaves foster light capture in the dim forest habitat. More than 30 species of the genus *Gnetum* occur as vines, shrubs, or trees.

- *Ephedra*, native to arid regions of the southwestern U.S., has tiny, brown, scalelike leaves and green, photosynthetic stems (Figure 25.13b). These adaptations help to conserve water by preventing water loss that would otherwise occur from the surfaces of larger leaves. Like all plants, *Ephedra*

Biology Principle

Biology Affects Our Society

The modern decongestant drug pseudoephedrine is based on the chemical structure of ephedrine, which was named for and originally obtained from *Ephedra*. Pseudoephedrine sales are now restricted in many places because this compound can be used as a starting point for the synthesis of illegal drugs. Ephedrine has also been used to enhance sports performance, a practice that has elicited medical concern.

(a) Genus *Gnetum*

(b) *Ephedra californica*

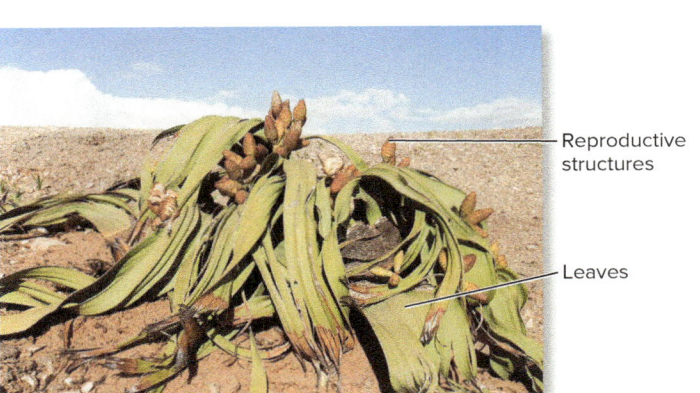

(c) *Welwitschia mirabilis*

Figure 25.13 Gnetales. (a) A tropical plant of the genus *Gnetum*, displaying broad leaves and reproductive structures. (b) *Ephedra californica* growing in deserts of North America, showing minuscule, brown leaves on green, photosynthetic stems and reproductive structures. (c) *Welwitschia mirabilis* growing in the Namib Desert of southwestern Africa, showing long, wind-shredded leaves and reproductive structures.
(a) ©Robert & Linda Mitchell; (b) ©2004 James M. Andre; (c) ©Wildlife GmbH/Alamy

produces **secondary metabolites,** compounds that are not essential for basic cellular structure and growth but that aid plant structure, protection, or reproduction. Plant secondary compounds influence humans in many ways (see Section 25.4), and *Ephedra*'s secondary compounds have long been known to affect humans. For example, early settlers of the western U.S. used *Ephedra* to treat colds and other medical conditions.

- *Welwitschia* has only one living representative species. *Welwitschia mirabilis* is a strange-looking plant that grows in the coastal Namib Desert of southwestern Africa, one of the driest places on Earth (Figure 25.13c). A long taproot anchors a stubby stem that barely emerges from the ground. Two very long leaves grow from the stem but are rapidly shredded by the wind into many strips. The plant is thought to obtain most of its water from coastal fog that accumulates on the leaves, explaining how *W. mirabilis* can grow and reproduce in such a dry place.

25.3 Reviewing the Concepts

- The seed plants (also called spermatophytes) consist of the gymnosperms, with exposed seeds, and the angiosperms, with seeds enclosed in fruits (Figures 25.11, 25.12).
- The diversity of modern gymnosperms includes three modern phyla that are all important to human life: cycads, *Ginkgo biloba*, and the conifers (including Gnetales) (Figure 25.13).

25.3 Testing Your Knowledge

1. In what way are gymnosperms different from angiosperms?
   a. Gymnosperms possess embryos.
   b. Gymnosperms possess lignified xylem.
   c. Gymnosperms possess seeds.
   d. Gymnosperms possess wood.
   e. Gymnosperms lack flowers, fruit, and seed endosperm.
2. Which of the following plant traits represents an example of plant adaptations that are useful to humans?
   a. lignin-walled xylem
   b. stems, leaves, and roots
   c. seed endosperm
   d. secondary metabolites
   e. All of the above are correct.

25.4 Diversity of Modern Angiosperms

Learning Outcomes

1. List four flower organs and their functions, and explain how each flower part may have first evolved.
2. Describe how diversification of flowers and fruits enhances seed production and dispersal.
3. Name three major types of angiosperm secondary metabolites and how these affect animals.

The flowering plants, including hundreds of thousands of species, are the most diverse group of modern land plants. The evolutionary success of angiosperms is based upon traits that are absent or rare in other plants—flowers, fruits, and endosperm. Recall that a flower is a short stem bearing reproductive organs, fruits develop from flowers, and endosperm is a nutritive seed tissue. Flowers foster seed production, fruits favor seed dispersal, and endosperm food helps embryos within seeds to grow into seedlings. Angiosperms also possess more efficient water-conducting systems than do other plants. In addition to tracheids, most angiosperms also possess water-conducting **vessels,** which are wider than tracheids and therefore increase the efficiency of water flow through plants. Though similar structures occur in some seedless plants and gymnosperms, these vessels are thought to have evolved independently. Angiosperms also possess vascular systems that branch more densely than do those of other plants, a trait that allows flowering plants to more effectively conduct materials through the body.

More than 124 mya, an extinct group of gymnosperms gave rise to the angiosperms—the flowering plants. Charles Darwin famously referred to the origin of the flowering plants as "an abominable mystery," one that has not been fully solved even today. Modern plant evolutionary biologists continue to study the evolutionary origin of distinctive angiosperm traits, such as the flower, that help to explain why flowering plants are so diverse.

Flowers Include Stamens and/or Carpels and Often Petals and Sepals

Flowers are produced at stem tips and may contain four types of organs: sepals, petals, pollen-producing stamens, and ovule-producing carpels (Figure 25.14). These flower organs are supported by tissue located at the tip of a flower stem. The functioning of several genes that control flower organ development explains why carpels are the

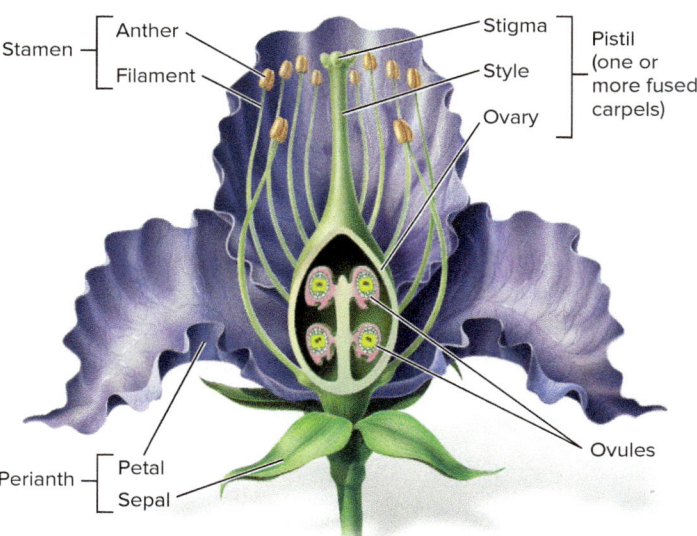

Figure 25.14 Generalized flower structure. Although flowers are diverse in size, shape, and color, they commonly have the parts illustrated here.

most central flower organs, why stamens surround carpels, and why petals and sepals are the outermost flower organs.

Many flowers produce attractive **petals** that play a role in **pollination,** the transfer of pollen among flowers of similar type. **Sepals** of many flowers are green and form the outer layer of flower buds. The sepals of some flowers, such as tulip and lily, look similar to petals, in which case both sepals and petals are known as tepals. All of a flower's sepals and petals are collectively known as the perianth. Most flowers produce one or more **stamens,** the structures that produce and disperse pollen. Most flowers also contain a single or multiple **carpels,** structures that produce ovules. Some flowers produce only a single carpel, others display several separate carpels, and many possess several carpels that are fused together into a compound structure. Both a single carpel and fused carpels are referred to as a **pistil** (from the Latin *pistillum,* meaning pestle) because it resembles the device people use to grind materials to powder in a mortar (Figure 25.14). Pistil structure can be divided into three regions having distinct functions. The topmost portion of the pistil, known as the stigma, receives and recognizes pollen of the appropriate species or genotype. The elongate middle portion of the pistil is called the style. The lowermost portion of the pistil is the ovary, which encloses and protects ovules.

During the flowering plant life cycle (Figure 25.15), the stigma allows pollen of appropriate genetic type to germinate, producing a long pollen tube that grows through the style. The pollen tube delivers two sperm cells to ovules. In the distinctive angiosperm process known as **double fertilization,** one sperm nucleus fuses with the egg nucleus to form a zygote, and the other sperm nucleus fuses with the nuclei of other haploid cells of the female gametophyte. The latter is the first step in the development of a characteristic angiosperm nutritive tissue known as endosperm. Fed by the endosperm, the zygote develops into an embryo, and the ovule develops into a seed. Ovaries (and sometimes additional flower parts) develop into fruits.

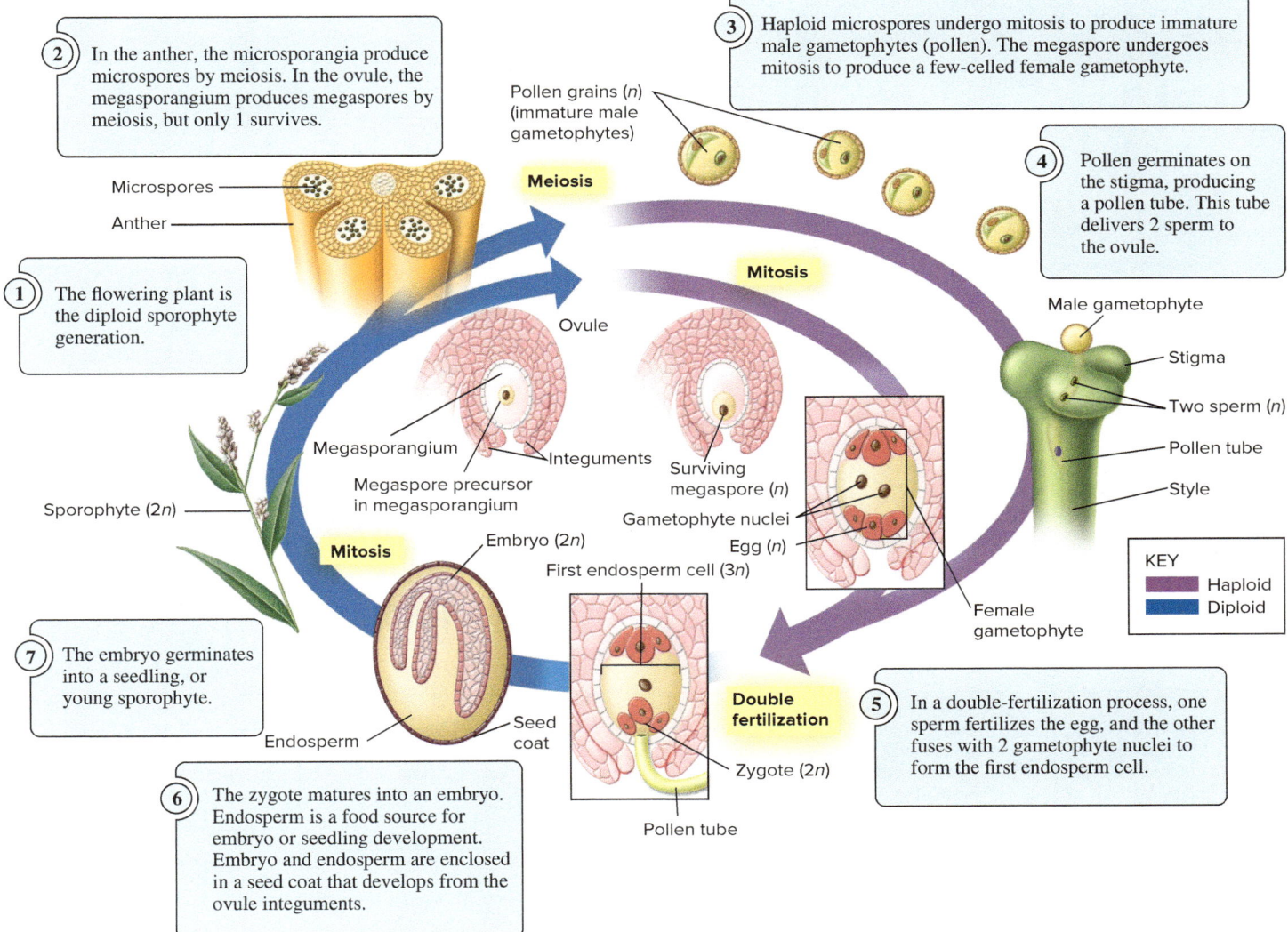

Figure 25.15 The life cycle of a flowering plant, illustrated by the genus *Polygonum*. Flowering plant life cycles differ in the length of the cycle and the number of cells and nuclei in the female gametophyte, the seven cells and eight nuclei of *Polygonum* being common.

EVOLUTIONARY CONNECTIONS

Flower Organs Evolved from Leaflike Structures

Petals and sepals are often similar to leaves in shape and structure of their vascular system, features that suggest the evolution of petals and sepals from leaves. Structural comparison and molecular data indicate that stamens and pistils likewise originated from leaflike structures. Angiosperm stamens resemble leaflike structures of gymnosperms that produce the microspores that develop into pollen. Plant evolutionary biologists hypothesize that over time, the microspore-producing tissues became restricted to the stamen tips, generating anthers on a supporting filament (**Figure 25.16a**). Plant evolutionary biologists likewise hypothesize that carpels are homologous to leaflike structures of gymnosperms that bear ovules on their surfaces. The carpels of some early-diverging modern plants are leaflike structures that fold over ovules, with the carpel edges stuck together by secretions (**Figure 25.16b**). Most modern flowers produce carpels whose edges have fused together into a tube whose lower portion (ovary) encloses and protects ovules, improving plant fitness. The first flowers arose when early stamens, carpels, and perianth parts aggregated into a single structure.

(a) Stamen evolution

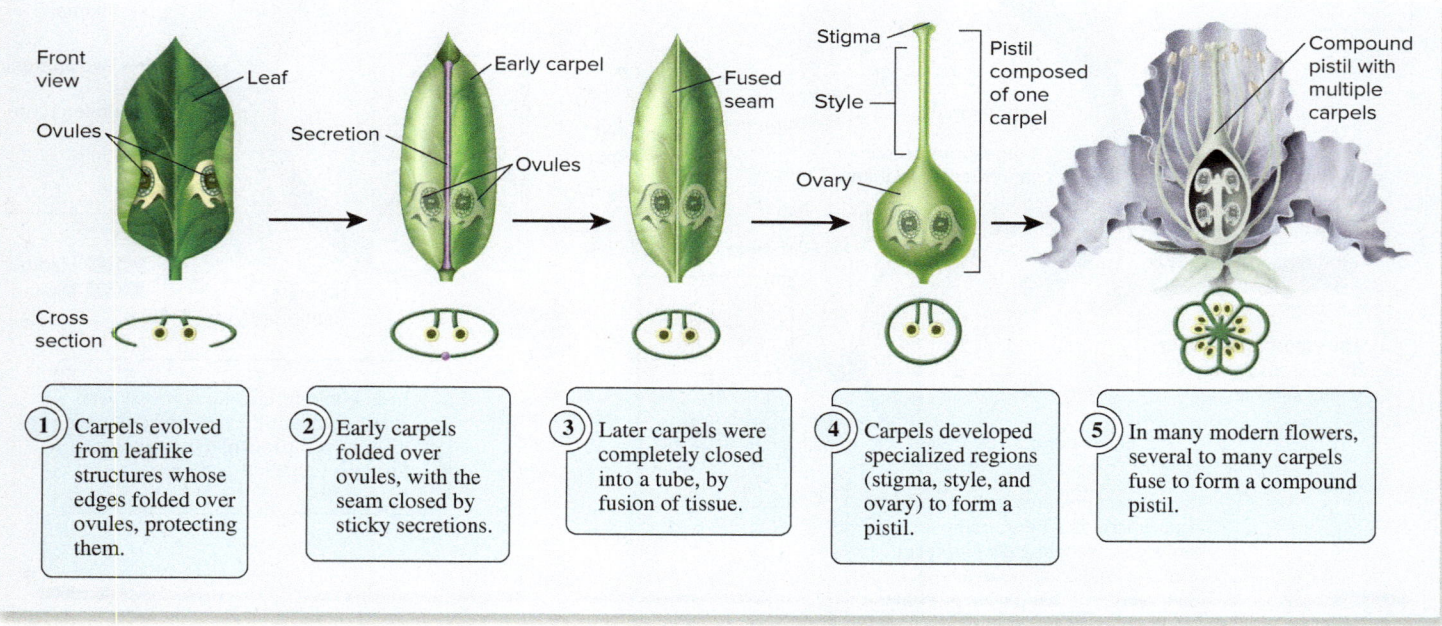

(b) Carpel and pistil evolution

Figure 25.16 Hypothetical evolutionary origin of stamens, carpels, and pistils. Plant biologists test these models by searching for new fossils or generating additional molecular data.

PLANT EVOLUTION: HOW PLANT DIVERSIFICATION CHANGED PLANET EARTH 545

Figure 25.17 A phylogeny showing the major modern angiosperm lineages.

Flowering Plants Diversified into Several Lineages, Including Monocots and Eudicots

Figure 25.17 presents our current understanding of the relationships among modern angiosperm groups. According to gene-sequencing studies, the earliest-diverging modern angiosperms are represented by a single species called *Amborella trichopoda*, a shrub that lives in forests on the South Pacific island of New Caledonia. Later-diverging groups of angiosperms include water lilies, the star anise plant, and other close relatives (Figure 25.17). Magnoliids, represented by the genus *Magnolia*, are the next-diverging group. Magnoliids are closely related to two very large and diverse angiosperm lineages: the **monocots** and the **eudicots.**

Monocots and eudicots are named for differences in the number of embryonic leaves called cotyledons. Monocot embryos possess one cotyledon, whereas eudicots possess two cotyledons. Also, monocots typically have petals, stamens, or other floral parts numbering three or some multiple of three, whereas eudicot flower parts tend to occur in fours, fives, or a multiple of four or five. Tulips, daffodils, and irises are examples of monocots, whereas roses, snapdragons, and daisies are examples of eudicots.

Flower Diversification Has Fostered Efficient Seed Production

During the diversification of flowering plants, flower evolution has involved several types of changes that foster the transfer of pollen from one plant to another. Effective pollination is essential to efficient seed production because it minimizes the amount of energy plants must expend to accomplish sexual reproduction. The fusion of flower organs, clustering of flowers into groups, and reduction of the perianth are some examples of changes leading to effective pollination.

- Many flowers have fused petals that form floral tubes. Such tubes tend to accumulate sugar-rich nectar that provides a reward for **pollinators,** animals that transfer pollen among plants. The diameters of floral tubes vary among flowers and are evolutionarily tuned to the feeding structures of diverse animals, which range from the narrow tongues of butterflies to the wider bills of nectar-feeding birds (**Figure 25.18**). Nectar-feeding bats stick their heads into even larger, tubular flowers to lap up nectar with their tongues.

(a) Zinnia flower and butterfly (b) Hibiscus flower and hummingbird (c) Saguaro cactus flower and bat

Figure 25.18 Flowers whose perianths form nectar-containing floral tubes of different widths that accommodate different pollinators. **(a)** This zinnia is composed of an outer rim of showy flowers and a central disc of narrow, tubular flowers that produce nectar. Butterflies, but not other pollinators, are able to reach the nectar by means of narrow tongues. **(b)** The hibiscus flower produces nectar in a floral tube whose diameter corresponds to the dimensions of a hummingbird bill. **(c)** The saguaro cactus (*Carnegiea gigantea*) flower forms a floral tube that is wide enough for nectar-feeding bats to get their heads inside. The cactus flower has been drawn here as if it were transparent, to illustrate bat pollination.

- Many plants produce **inflorescences,** groups of flowers tightly clustered together, which occur in several types. The zinnia features a type of inflorescence in which many small flowers are clustered into a head (Figure 25.18a). The flowers at the center of a zinnia head function in reproduction and lack showy petals, but flowers at the rim have showy petals that attract pollinators. Flower heads allow pollinators to transfer pollen among a large number of flowers at the same time.
- The grass family features flowers with few or no perianths, which explains why grass flowers are not showy. This adaptation fosters pollination by wind, since petals would get in the way of such pollen transfer. Since grasses often grow in dense populations, wind pollination can be very effective.

Diverse Types of Fruits Function in Seed Dispersal

Fruits are structures that develop from ovary walls in diverse ways that aid the dispersal of the enclosed seeds. Seed dispersal helps to prevent seedlings from competing with their larger parents for scarce resources such as water and light. Dispersal of seeds also allows plants to colonize new habitats.

- Many mature angiosperm fruits, such as cherries, are attractively colored, soft, juicy, and tasty (Figure 25.19a). Such fruits are adapted to attract animals that consume the fruits, digest the outer portion as food, and eliminate the seeds after traveling some distance, thereby dispersing them. Hard seed coats prevent such seeds from being destroyed by the animal's digestive system.
- Strawberries are aggregate fruits, many fruits that all develop from a single flower having multiple pistils (Figure 25.19b). The ovaries of these pistils develop into tiny, single-seeded, yellow fruits on a strawberry surface; the fleshy, red, sweet portion of a strawberry develops from tissue that supported the flower. Aggregate fruits allow a single animal consumer, such as a bird, to disperse many seeds at the same time.
- Pineapples (Figure 25.19c) are juicy, multiple fruits that develop when many ovaries of an inflorescence fuse together. Such multiple fruits are larger and attract relatively large animals that have the ability to disperse seeds over long distances.
- The plant family informally known as legumes is named for its distinctive fruits, dry pods that split open down both sides when seeds are mature, thereby releasing them (Figure 25.19d).
- Nuts and grains are additional examples of dry fruits. Grains are the characteristic single-seeded fruits of cereal grasses such as rice, corn (maize), barley, and wheat. Other plants produce dry fruits with surface burrs that attach to animal fur.

Angiosperms Produce Diverse Secondary Metabolites That Play Important Roles in Structure, Reproduction, and Protection

As noted in the discussion of the gymnosperm *Ephedra* (see Section 25.3), secondary metabolites are organic compounds that are not essential for basic cell structure and growth but aid survival, structure, and reproduction. Certain prokaryotes, protists, and fungi, as well as some animals and all plants, produce secondary metabolites, but

(a) **A fleshy fruit (cherry)**

(b) **An aggregate fruit (strawberry)**

(c) **A multiple fruit (pineapple)**

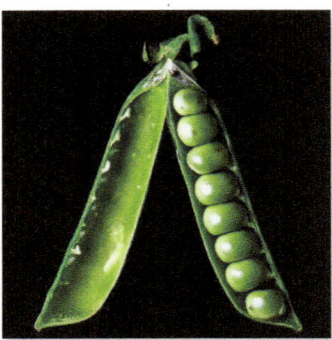

(d) **Legumes with dry pods (peas)**

Figure 25.19 Representative fruit types. (a) The cherry is an example of a fleshy fruit that is adapted to attract animals that consume the fruits and excrete the seeds. **(b)** Strawberry is an aggregate fruit, consisting of many tiny, single-seeded fruits produced by a single flower. The fruits are embedded in the surface of tissue that originally supported the flower. **(c)** Pineapple is a large, multiple fruit formed by the aggregation of smaller fruits, each produced by one of the flowers in an inflorescence. **(d)** Peas produce legumes, fruits that open on two sides to release seeds.
(a–c) ©Lee W. Wilcox; *(d)* ©Gloomerique/Getty Images

these compounds are most diverse in the angiosperms. Most of the 100,000 or so known types of secondary metabolites are produced by flowering plants. The diversification of these compounds has strongly influenced flowering plant evolution, as well as everyday human life. Three major classes of plant secondary metabolites occur: (1) terpenes and terpenoids, (2) phenolics, and (3) alkaloids (Figure 25.20).

- About 25,000 types of plant terpenes and terpenoids are derived from different arrangements of the simple hydrocarbon gas isoprene. Taxol, used in the treatment of cancer, is a terpene, as are citronella and a variety of other compounds that repel insects. Rubber, turpentine, rosin, and amber are complex terpenoids that likewise serve important roles in plant biology, as well as having useful human applications.
- Phenolic compounds are responsible for some flower and fruit colors as well as the distinctive flavors of cinnamon, nutmeg, ginger, cloves, chilies, and vanilla. Phenolics absorb ultraviolet radiation, thereby preventing damage to cellular DNA. They also help to defend plants against insects and disease microbes.

(a) Natural rubber produced by *Hevea brasiliensis* is an example of a complex terpene.

(b) Capsaicin extracted from capsicum pepper is an example of a phenolic compound.

(c) Caffeine produced by *Coffea arabica* is an example of an alkaloid.

Figure 25.20 Major types of plant secondary metabolites. Note that the chemistry of plant secondary metabolites differs from that of the primary compounds produced by all cells. The production by plants of terpenes, phenolics, and alkaloids helps to explain how plants survive and reproduce, and why plants are useful to humans in so many ways.
(a) ©samuraigirl/Shutterstock; *(b)* ©Jonathan Buckley/GAP Photo/Getty Images; *(c)* ©Science Photo Library/Alamy

Some phenolic compounds found in tea, red wine, grape juice, and blueberries are antioxidants that detoxify free radicals, highly reactive by-products of normal metabolism. Such phenolics thereby prevent cellular damage in the plant, as well as in animals that consume them.

- Alkaloids are nitrogen-containing secondary metabolites that often have potent effects on the animal nervous system.

Plants produce at least 12,000 types of alkaloids, and certain species produce many alkaloids. Caffeine, nicotine, morphine, ephedrine, cocaine, and codeine are examples of alkaloids that influence the physiology and behavior of humans, so their use and abuse are of societal concern. Like flowers and fruit structures, secondary metabolites are useful in distinguishing among Earth's hundreds of thousands of flowering plant species.

FEATURE INVESTIGATION

Hillig and Mahlberg Analyzed Secondary Metabolites to Explore Species Diversification in the Genus *Cannabis*

The genus *Cannabis* has long been a source of hemp fiber used for ropes and fabric. People have also used *Cannabis* (also known as marijuana) in traditional medicine and as a hallucinogenic drug. *Cannabis* produces tetrahydrocannabinol (THC), a type of alkaloid called a cannabinoid. THC and other cannabinoids are produced in glandular hairs that cover most of the *Cannabis* plant's surface but are particularly rich in leaves located near the flowers. THC mimics compounds known as endocannabinoids, which are naturally produced and act in the animal brain and elsewhere in the body. THC affects humans by binding to receptor proteins in plasma membranes in the same way as natural endocannabinoids. Cancer and AIDS patients sometimes choose to use cannabis to reduce nausea and stimulate their appetite, counteracting the side effects of cancer treatment and the wasting that accompanies both diseases.

Because humans have subjected cultivated *Cannabis* plants to artificial selection for so long, plant biologists have been uncertain how cultivated *Cannabis* species are related to those in the wild. In the past, plants cultivated for drug production were often identified as *Cannabis indica*, whereas those grown for hemp were typically known as *Cannabis sativa*. However, these species are difficult to distinguish on the basis of structural features, and the relevance of these names to wild cannabis was unknown. At the same time, species identification has become important for biodiversity studies, agriculture, and law enforcement. For these reasons, plant biologists Karl Hillig and Paul Mahlberg hypothesized that ratios of THC to another cannabinoid known as cannabidiol (CBD) might aid in defining *Cannabis* species and identifying plant samples at the species level, as shown in **Figure 25.21**. Like THC, CBD has important applications in modern medicine.

To test their hypothesis, the investigators began by collecting *Cannabis* fruits from nearly a hundred diverse locations around the world and then growing these plants from seed under uniform conditions in a greenhouse. The investigators next extracted cannabinoids, analyzed them by means of gas chromatography (a laboratory technique used to identify components of a mixture), and determined the ratios of THC to CBD. The results, published in 2004, suggested that the wild and cultivated *Cannabis* samples evaluated in this study could be classified into two species: *C. sativa*, displaying relatively low THC levels, and *C. indica*, having relatively high THC levels. As a result of this work, ecologists, agricultural scientists, and forensic scientists can reliably use ratios of THC to CBD to classify samples. Similar studies of plant secondary metabolites offer the benefit of uncovering potential new medicinal compounds or other applications of significance to humans.

Experimental Questions

1. **SCISKILLS** ▶ Designing an experiment requires a plan to achieve an adequate number of samples in order to allow statistical analysis. Hillig and Mahlberg obtained nearly a hundred *Cannabis* fruit samples from around the world. Why were so many samples needed?

2. Why did Hillig and Mahlberg collect samples from the leaves growing nearest the flowers?

Figure 25.21 Hillig and Mahlberg's analysis of secondary metabolites in the genus *Cannabis*.
Photos: *(1)* ©Phil Schermeister/Getty Images; *(2)* ©Matthew Kellett/Alamy

4 THE DATA

Cannabis plants isolated from diverse sources worldwide formed 2 groups—those having relatively high THC to CBD ratios and those having lower THC to CBD ratios.

Plants having low THC to CBD ratios, often used as hemp fiber sources, corresponded to the species *C. sativa*.

Plants having high THC to CBD ratios, often used as drug sources, corresponded to the species *C. indica*.

5 CONCLUSION
Differing cannabinoid ratios support a concept of 2 *Cannabis* species.

6 SOURCE
Hillig, K. W., and Mahlberg, P. G. 2004. A Chemotaxonomic Analysis of Cannabinoid Variation in *Cannabis* (Cannabaceae). *American Journal of Botany* 91: 966–975.

Table 25.1 provides a summary of the characteristics of modern plants and their closest algal relatives, together known as the streptophytes.

25.4 Reviewing the Concepts

- Flowers foster seed production and are adapted in various ways that aid pollination. Flower parts likely evolved from leaflike structures (Figures 25.14, 25.15, 25.16).
- The two largest and most diverse lineages of flowering plants are the monocots and eudicots (Figure 25.17).
- Flower diversification involved evolutionary changes such as the fusion of petals, clustering of flowers into inflorescences, and a reduced perianth, which improve pollination effectiveness and seed production (Figure 25.18).
- Fruits are structures that enclose seeds and aid in their dispersal (Figure 25.19).
- Angiosperms produce three main groups of secondary metabolites—(1) terpenes and terpenoids, (2) phenolics, and

Table 25.1 Distinguishing Features of Modern Streptophyte Algae and Land Plants

Streptophyte Algae
  Primarily aquatic habitat; zygotic life cycle; embryos and tough-walled spores absent

LAND PLANTS (EMBRYOPHYTES)
  Primarily terrestrial habitat; alternation of two multicellular generations—diploid sporophyte and haploid gametophyte; multicellular embryos are nutritionally dependent on maternal gametophyte for at least some time during development; spore-producing sporangia; gamete-producing gametangia; tough-walled spores

Nonvascular plants (Bryophytes) (**liverworts, mosses, hornworts**)
  Dominant gametophyte generation; supportive, lignin-containing vascular tissue absent; true roots, stems, leaves absent; sporophytes unbranched and unable to grow independently of gametophytes

VASCULAR PLANTS (TRACHEOPHYTES) (lycophytes, pteridophytes, spermatophytes)
  Dominant sporophyte generation; lignin-walled water-conducting tissue—xylem; specialized organic food-conducting tissue—phloem; sporophytes branched and eventually become able to grow independently of gametophytes; all have stems; most also have leaves and roots
  Lycophytes Leaves generally small with a single, unbranched vein (lycophylls); sporangia borne on sides of stems

PTERIDOPHYTES + SEED PLANTS
  Pteridophytes Leaves generally large with extensively branched vein system; sporangia borne on leaves; seeds absent

SEED PLANTS (SPERMATOPHYTES)
  Seeds present; leaves have branched vein systems

GYMNOSPERMS (**cycads, ginkgos, conifers**)
  Flowers and fruits absent; seed food stored in female gametophyte, endosperm absent
  Angiosperms (flowering plants)
  Flowers and fruits present; seed food stored after fertilization in endosperm formed by double fertilization

Key: Phyla; LARGER MONOPHYLETIC CLADES (synonyms). All other classification terms are not clades.

25.4 Testing Your Knowledge

1. A plant produces very small, petal-less flowers that are clustered into a dangling inflorescence. Among the choices given, which pollination agent is most likely?
   b. butterfly
   c. hummingbird
   d. bat
   e. wind
   f. water

2. A plant produces juicy, sweet, red fruits having relatively small seeds. Among the choices given, which seed dispersal agent is most likely in nature?
   a. butterfly
   b. bird
   c. bear
   d. wind
   e. water

25.5 Human Influences on Angiosperm Diversification

Learning Outcome
1. Explain how humans created the domesticated wheat, corn, and rice grain crops widely planted around the world today.

By means of the process known as **domestication,** which involves artificial selection for traits desirable to humans, ancient humans transformed wild plant species into new crop species. Cultivated bread wheat (*Triticum aestivum*) was probably among the earliest food crops, having originated more than 8,000 years ago in what is now southeastern Turkey and northern Syria. Bread wheat originated by a series of steps from wild ancestors (*Triticum boeoticum* and *Triticum dicoccoides*). Wheat, corn, and rice produce grains (a type of fruit) at the tip of a stem called the ear. Among the earliest changes that occurred during wheat domestication was the loss of shattering, the process by which ears of wild grain crops break apart and disperse their grains. A mutation probably caused the ears of some wheat plants to remain intact, a trait that is disadvantageous in nature but beneficial to humans. Nonshattering ears would have been easier for humans to harvest than normal ears. Early farmers probably selected seed stock from plants having nonshattering ears and other favorable traits such as larger grains. These ancient artificial selection processes, together with modern breeding efforts, explain why cultivated wheat differs from its wild relatives in shattering and other properties. The accumulation of these trait differences explains why cultivated and wild wheat plants are classified as different species.

About 9,000 years ago, people living in what is now Mexico domesticated a native grass known as teosinte (of the genus *Zea*), producing a new species, *Zea mays*, known as corn or maize. The evidence for this pivotal event includes ancient ears that were larger than wild ones and distinctive fossil pollen. Modern ears of corn are much larger than those of teosinte, with many more rows and larger and softer corn grains, and modern corn ears do not shatter, as do those of ancestral teosinte (**Figure 25.22**).

Molecular analyses indicate that domesticated rice (*Oryza sativa*) originated from ancestral wild species of grasses (*Oryza nivara* and/or *Oryza rufipogon*). As in the cases of wheat and corn, domestication of rice involved loss of ear shattering, in this case resulting from a key amino acid substitution. Ancient humans might have unknowingly selected for this mutation while gathering rice from wild populations, because the mutants would not so easily have shed grains during the harvesting process. Eventually, the nonshattering mutant became a widely planted crop throughout Asia, and today it is the food staple for millions of people.

Although humans generated these and other new plant species, in modern times humans have caused the extinction of plant species as the result of habitat destruction and other threats to species. Protecting biodiversity will continue to challenge humans as populations and demands on the Earth's resources increase.

Biology Principle

Biology Affects Our Society
The domestication of corn from a wild grass to one of the world's largest production crops is an amazing feat of artificial selection.

Figure 25.22 Ears and grains of modern corn and its ancestor, teosinte. This illustration shows that domesticated corn ears are much larger than those of the ancestral grass teosinte. In addition, corn fruits are softer and more edible than the grains of teosinte, which are enclosed in a hard casing.

Concept Check: *In what other way do corn ears differ from those of teosinte?*

25.5 Reviewing the Concepts

- Humans have produced new crop species by domesticating wild plants. The process of domestication involved artificial selection for traits such as nonshattering ears of wheat, corn, and rice (Figure 25.22).

25.5 Testing Your Knowledge

1. What feature of wild food did humans alter during crop domestication?
   a. fruit size
   b. fruit softness
   c. seed dispersal
   d. All of the above are correct.

Assessing Your Knowledge and Skills

Foundational Knowledge

Multiple Choice

1. The simplest and most ancient of modern land plants are probably
   a. the ferns.
   b. the cycads.
   c. the liverworts and mosses.
   d. the angiosperms.
   e. none of the above.

2. Plants possess a life cycle that involves alternation of two multicellular generations: the gametophyte and
   a. the lycophyte.
   b. the bryophyte.
   c. the spermatophyte.
   d. the lignophyte.
   e. the sporophyte.

3. A waxy cuticle is an adaptation that
   a. helps to prevent water loss from tracheophytes.
   b. helps to prevent water loss from algae.
   c. helps to increase water loss from tracheophytes.
   d. aids in water transport within the bodies of vascular plants.
   e. does all of the above.

4. The primary function of a fruit is to
   a. provide food for the developing seed.
   b. provide food for the developing seedling.
   c. foster pollen dispersal.
   d. foster seed dispersal.
   e. do none of the above.

5. How did human domestication affect the transformation of wild grain plants into modern crops?
   a. Domestication reduced ear shattering.
   b. Domestication increased ear size.
   c. Domestication increased grain edibility.
   d. All of the above are correct.
   e. None of the above is correct.

Short Answer

1. Explain the process of the Carboniferous period that generated massive coal deposits, which are mined today.
2. **PRINCIPLES** A principle of biology is that structure determines function. Explain how the materials cutin, lignin, and sporopollenin aid terrestrial plant survival and foster the formation of plant fossils.

Critical-Thinking Skills

Multiple Choice

1. Which sequence of plant critical adaptations reflects the order of their appearance in time?
   a. embryos, vascular tissue, seeds, flowers
   b. vascular tissue, embryos, flowers, seeds
   c. vascular tissue, seeds, embryos, flowers
   d. seeds, embryos, flowers, vascular tissue
   e. seeds, vascular tissue, embryos, flowers

2. Which of the following environmental conditions was likely experienced by early Coal Age plants to a greater degree than by plants of modern, pre-industrial times, and how did these conditions influence plant growth?
   a. high atmospheric carbon dioxide levels that fostered plant photosynthesis
   b. low temperatures that retarded plant photosynthesis
   c. herbivory by large populations of large animals that reduced plant photosynthetic surface area
   d. All of the above are correct.
   e. None of the above is correct.

3. If you could use a time machine to follow the diversification of land plants, starting in the early Paleozoic, which sequence of events would you likely witness?
   a. diversification of the modern phyla of seedless vascular plants, followed by diversification of bryophyte phyla that occur today (mosses, liverworts, and hornworts), and then seed plant diversification
   b. diversification of bryophyte phyla that occur today, followed by the origin of modern seedless vascular plant phyla, then seed plant diversification
   c. diversification of early seed plants, followed by diversification of seedless plants and finally the origin of modern bryophytes
   d. any of the above scenarios, given what we know about the fossil record of plants and molecular data
   e. none of the scenarios, given the available evidence

4. Given what we know about the evolutionary sequence represented by modern plant phyla, if you were in charge of a large, expensive project to compare plant genetic systems, which modern plant group would be a good place to start since its system is the simplest?
   a. flowering plants
   b. conifers
   c. ferns
   d. one or more bryophyte phyla
   e. No group of modern plants is likely to have a simpler genetic system than any other.

5. Imagine that you are appointed to lead an effort to use plants to reduce atmospheric levels of the greenhouse gas carbon dioxide. Which of the following steps might you reasonably recommend?
   a. Preserve modern moss-dominated peatlands, which store organic carbon as peat over long time periods.
   b. Restore peatlands from which peat has been harvested for agricultural or other uses.
   c. Preserve forests dominated by vascular land plants, because such plants take up and store carbon as tough organic compounds for long time periods.
   d. Restore forests dominated by vascular land plants.
   e. All of the above are steps that could reduce atmospheric carbon dioxide.

Short Answer

1. How have flowering plants been able to diversify into so many species?
2. Why are some fruits useful for human food?

26 Invertebrates: The Vast Array of Animal Life Without a Backbone

Butterflies and moths constitute over 10% of all known biodiversity. This tropical butterfly, *Heliconius erato*, the red postman, advertises its distastefulness to predators with bright warning colors. Recent studies on *H. erato* and other species show how spatial shifts in expression of a single gene, *WntA*, cause pattern-shape evolution.

Chapter Outline

- 26.1 Characteristics of Animals
- 26.2 Animal Classification
- 26.3 Ctenophores: The Earliest Animals
- 26.4 Porifera: The Sponges
- 26.5 Cnidaria: Jellyfish and Other Radially Symmetric Animals
- 26.6 Lophotrochozoa: The Flatworms, Rotifers, Bryozoans, Brachiopods, Mollusks, and Annelids
- 26.7 Ecdysozoa: The Nematodes and Arthropods
- 26.8 Deuterostomia: The Echinoderms and Chordates

Assessing Your Knowledge and Skills

©InVision_Photography/Getty Images

With about 180,000 named species, butterflies and moths constitute over 10% of all known biodiversity. Many of the species differ in wing size and coloration, from tiny species less than 5 mm in wingspan to the Atlas moth, *Atticus atlas*, and Queen Alexandria's bird wing butterfly, *Ornithoptera alexandre*, both with a wingspan of over 25 cm. Wing patterns, which differ widely among species, are used in nature for camouflage, to indicate toxicity and deter predators (see the chapter-opening photo), and to attract mates. An understanding of how such diversity is created in butterflies and moths helps us to explain the origins of diversity across all insects, the world's most diverse taxonomic class of animals.

French evolutionary biologist Arnaud Martin and his colleagues are using the cutting-edge gene editing tool CRISPR, which we discussed in Chapter 18, to understand how butterfly wings are shaped and colored. They found that a single gene (*WntA*) is responsible for wing color patterns across distantly related species, suggesting that biodiversity results from changes in a single developmental gene.

Animals constitute the most species-rich kingdom. Many different kinds of animals and their products are part of our diet. In addition, humans enjoy many animal species as companions and depend on other species to test life-saving drugs. However, other animals threaten our food supply and transmit deadly diseases.

Since the time of Carolus Linnaeus in the 1700s, scientists have classified animals based on their morphology, that is, on their physical structure. In the 1990s, animal classifications based on similarities in DNA and rRNA sequences became more common. Quite often, classifications based on morphology and those based on molecular data were similar, but some important differences arose. In this chapter, we will begin by defining the key characteristics of animals and then take a look at the major types of invertebrates. In Chapter 27, we will look specifically at the vertebrates, animals more familiar to us, including fish, amphibians, reptiles, and mammals. Throughout Chapters 26 and 27, we will explore how new molecular data have enabled scientists to revise and refine the animal phylogenetic tree.

26.1 Characteristics of Animals

Learning Outcome

1. List the key characteristics of animals that distinguish them from other organisms.

The Earth contains a dazzling diversity of animal species, living in environments from the deep sea to the desert and exhibiting an amazing array of characteristics. Most animals move and eat multicellular prey, and therefore they are loosely differentiated from species in other kingdoms. However, a single definition of an animal is difficult because they are so diverse; biologists can find exceptions to nearly any given characteristic. Even so, the following features can help us broadly characterize the group we call animals.

Cell Structure

- Animals are multicellular.
- Animal cells lack cell walls and are flexible. This flexibility facilitates movement.
- Animal cells gain structural support from an extensive extracellular matrix (ECM) that forms strong fibers outside the cell (refer back to Figure 4.31).
- A group of unique cell junctions—anchoring and tight junctions—plays an important additional role in holding animal cells in place and allowing communication between cells (refer back to Section 5.7).

Mode of Nutrition

- Animals are heterotrophs; that is, they ingest other organisms or their products to sustain life. Many different modes of feeding exist among animals, including suspension feeding—filtering food out of the surrounding water; bulk feeding—eating large food pieces; and fluid feeding—sucking plant sap or animal body fluids (**Figure 26.1**).

Movement

- Muscle tissue is unique to animals, and most animals are capable of some type of locomotion in order to acquire food or escape predators.
- Nervous tissue is also unique to animals, and a nervous system coordinates movement.

Genomes

- Most animals possess *Hox* genes, which function in patterning the body axis (refer back to Figures 20.12 and 20.13).
- Animals have very similar genes that encode small ribosomal subunit (SSU) rRNA (refer back to Figure 10.13).

Reproduction and Development

- Nearly all members of the animal kingdom reproduce sexually, whereby a small, mobile sperm unites with a much larger egg to form a fertilized egg, or zygote. Fertilization can occur internally, which is common in terrestrial species, or externally, which is more common in aquatic species.
- Similarly, embryos develop inside the mother or outside in the mother's environment. A particularly unusual developmental phenomenon is the occurrence of **metamorphosis**, by which an organism changes from a juvenile to an adult form—for example, when a caterpillar changes into a butterfly.

26.1 Reviewing the Concepts

- Animals constitute a very species-rich kingdom, with a number of characteristics that most of them have in common, including multicellularity, flexible cells, an extracellular matrix (ECM), and unique cell junctions, in addition to heterotrophic feeding, internal digestion, and the possession of nervous and muscle tissues.
- Many different feeding modes exist among animals, including suspension feeding, bulk feeding, and fluid feeding (Figure 26.1).
- Most animals reproduce sexually with embryos developing internally, inside the mother's body, or externally, in the environment.

26.1 Testing Your Knowledge

1. Which of the following is *not* a distinguishing characteristic of animals?
   a. the capacity to move at some point in their life cycle
   b. possession of cell walls
   c. multicellularity
   d. heterotrophy
   e. All of the above are characteristics of animals.

(a) (b) (c)

Figure 26.1 Modes of animal nutrition. (a) Suspension feeders, such as these tube worms, filter food particles from the water column. **(b)** Bears and other bulk feeders tear off large pieces of their food and chew it or swallow it whole. **(c)** Fluid feeders, such as these aphids, suck fluid from their food source.
(a) ©Wolfgang Herath/imageBROKER/age fotostock; *(b)* ©Enrique R. Aguirre Aves/Getty Images; *(c)* ©Bartomeu Borrell/age fotostock

26.2 Animal Classification

Learning Outcomes

1. Discuss why choanoflagellates are believed to be the closest living relatives of animals.
2. Describe how symmetry, germ layers, and embryonic development form the basis of animal classification.
3. Describe how differences in body cavities and segmentation may also be used to distinguish groups of animals.
4. **SCISKILLS** ▶ Draw an animal phylogeny based solely on body plans, and contrast it with the animal phylogeny based on body plans and molecular data.

Although animals constitute an extremely diverse kingdom, most biologists agree that the kingdom is monophyletic, meaning that all taxa have evolved from a single common ancestor. Today, scientists recognize about 35 animal phyla. At first glance, many of these phyla seem so distantly related to one another (for example, chordates and jellyfish) that making sense of this diversity with a classification scheme seems nearly impossible. However, over the course of centuries, scientists have come to some basic conclusions about the evolutionary relationships among animals. In this section, we will explore the major features of animal body plans that form the basis of animal phylogeny (Figure 26.2).

Animals Evolved from a Choanoflagellate-like Ancestor

Animal life on Earth has evolved over hundreds of millions of years. Some scientists suggest that changing environmental conditions, such as a buildup of dissolved oxygen and minerals in the ocean or an increase in atmospheric oxygen, eventually permitted higher metabolic rates and increased the activity of a wide range of animals. Others suggest that with the development of sophisticated locomotor skills, a wide range of predators and prey evolved, leading to an

Biology Principle

Populations of Organisms Evolve from One Generation to the Next and Are Related by an Evolutionary History

All animals are believed to be derived from a choanoflagellate-like ancestor.

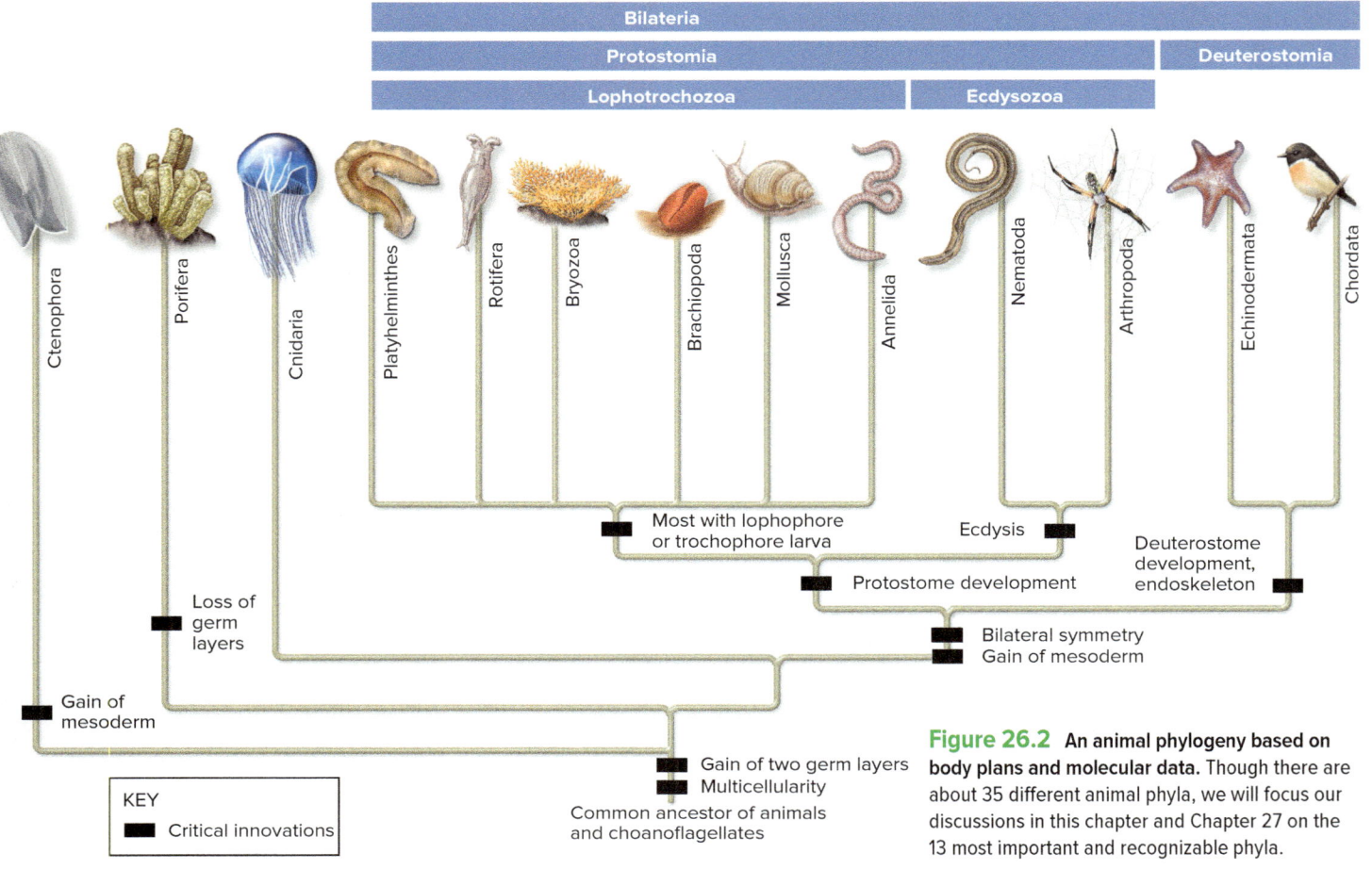

Figure 26.2 An animal phylogeny based on body plans and molecular data. Though there are about 35 different animal phyla, we will focus our discussions in this chapter and Chapter 27 on the 13 most important and recognizable phyla.

evolutionary "arms race" in which predators evolved powerful weapons and prey evolved more powerful defenses against them. Such adaptations would have led to a proliferation of different lifestyles and taxa. Finally, the evolution of *Hox* genes may have resulted in an increase in diversity.

With the monophyletic nature of the animal kingdom in mind, scientists have attempted to characterize the organism from which animals most likely evolved. According to research, the closest living relative of animals is believed to be a flagellated protist known as a choanoflagellate. Choanoflagellates are tiny, single-celled or colonial organisms, each with a single flagellum surrounded by a collar composed of cytoplasmic tentacles (refer back to Figure 23.27).

Animal Phyla Have Broad Differences Related to Body Plan, Germ Layers, and Features of Embryonic Development

Biologists traditionally classified animal diversity in terms of the following three main morphological and developmental features of animal body plans:

1. Type of body symmetry
2. Number of germ layers
3. Specific features of embryonic development

We will discuss each of these major features of animal body plans next.

Symmetry Animals may be categorized according to their type of symmetry. Symmetry is the existence of balanced proportions of the body on either side of a median plane. Some of the earliest-diverging animals, such as sponges, were asymmetric, meaning they had no plane of symmetry (**Figure 26.3a**). Radially symmetric animals can be divided equally by any longitudinal plane passing through the central axis (**Figure 26.3b**). Such animals are often circular or tubular in shape, with a mouth at one end, and include ctenophores and cnidarians.

Bilaterally symmetric animals, the **Bilateria,** can be divided along a vertical plane at the midline to create two halves (**Figure 26.3c**).

Thus, a bilateral animal has a left side and a right side, which are mirror images, as well as a **dorsal** (upper) and a **ventral** (lower) side, which are not identical, and an **anterior** (head) and a **posterior** (tail) end. Bilateral symmetry is strongly correlated with both the ability to move through the environment and **cephalization**—the localization of sensory structures at the anterior end of the body. Such abilities allow animals to encounter their environment initially with their head, which is best equipped to detect and consume prey and to detect and respond to predators and other dangers. Most animals are bilaterally symmetric.

Germ Layers Fertilization of an egg by a sperm creates a diploid zygote. The zygote then undergoes **cleavage**—a succession of rapid cell divisions with no significant growth that produces a hollow sphere of cells called a **blastula.** In all animals except the sponges, the growing zygote develops different layers of cells, called **germ layers,** during a process known as **gastrulation** (**Figure 26.4**). In gastrulation, an area in the blastula folds inward, or invaginates, creating in the process a structure called a **gastrula.** The inner layer of cells becomes the **endoderm,** which lines the primitive digestive tract. The outer layer, or **ectoderm,** covers the surface of the embryo and differentiates into the epidermis and nervous system.

A key difference between Bilateria and most other animals is that the Bilateria develop a third layer of cells, termed the **mesoderm,** between the ectoderm and endoderm. Mesoderm forms the muscles and most other organs between the digestive tract and the ectoderm (look ahead to Figure 39.17). Because the Bilateria have these three distinct germ layers, they are often referred to as **triploblastic,** whereas the cnidarians, which have only ectoderm and endoderm, are termed **diploblastic.** Interestingly, the mesoderm of the earliest-diverging animals, the ctenophores, probably originated independently of the mesoderm found in bilaterians.

Specific Features of Embryonic Development in the Bilateria A key feature of embryonic development is the development of a mouth

(a) Porifera: no symmetry

(b) Cnidaria: radial symmetry

(c) Bilateria: bilateral symmetry

Figure 26.3 Early divisions in animal phylogeny. Animals can be categorized based on symmetry: **(a)** the absence of symmetry (Porifera; the sponges). Most individual sponges consist of multiple tubes so that any line of symmetry will not divide the animal into two equal halves. **(b)** radial symmetry (the cnidarians), or **(c)** bilateral symmetry (Bilateria; all other animals).
(a) ©Wild Horizons/UIG/Getty Images; *(b)* ©Gavin Parsons/Getty Images; *(c)* ©Jens Kuhfs/Getty Images

 Concept Check: *Look back at Figure 26.2. What other animals, apart from cnidarians, are radially symmetric?*

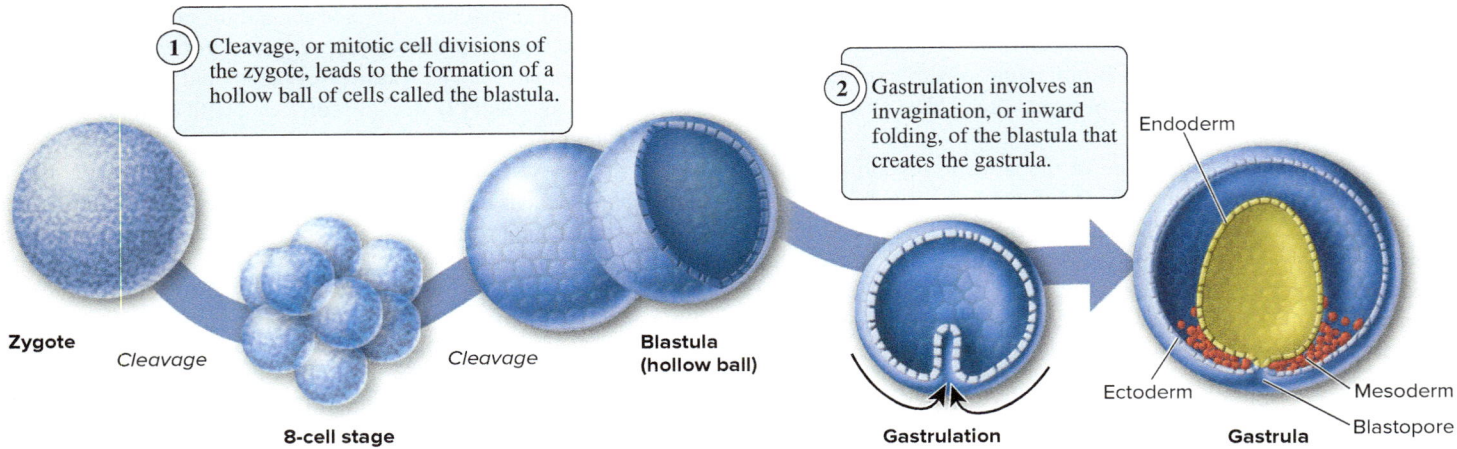

Figure 26.4 Formation of germ layers.

BioConnections: Refer back to Figure 21.7. Is the existence of three germ layers in the Bilateria a shared ancestral character or a shared derived character?

and an anus (Figure 26.5a). In gastrulation, the endoderm forms an indentation, the **blastopore,** which is the opening of the digestive tract to the outside. In **protostomes** (from the Greek *protos*, meaning first, and *stoma*, meaning mouth) (see Figure 26.2), the blastopore becomes the mouth. If an anus is formed in a protostome species, it develops from a secondary opening. In contrast, in **deuterostomes** (from the Greek *deuteros*, meaning second), the blastopore becomes the anus, and the mouth develops from a secondary opening.

Protostomes and deuterostomes also differ in the cleavage stage of embryonic development. As mentioned, the earliest stage of embryonic development involves a process known as cleavage. Protostome development is generally characterized by so-called **determinate cleavage,** in which the fate of each embryonic cell is determined very early (Figure 26.5b). If one of the cells is removed from a four-cell protostome embryo, neither the single cell nor the remaining three-cell mass can form a viable embryo, and development is halted. In contrast, deuterostome development in most species is characterized by **indeterminate cleavage,** in which each cell produced by early cleavage retains the ability to develop into a complete embryo. For example, when one cell is excised from a four-cell sea urchin embryo, both the single cell and the remaining three can go on to form viable embryos. Other embryonic cells compensate for the missing cells. In human embryos, if individual embryonic cells separate from one another early in development, identical twins can result.

Another distinguishing feature of the early bilaterian embryo is the cleavage pattern (Figure 26.5c). In **spiral cleavage,** the planes of cell cleavage are oblique to the vertical axis of the embryo, resulting in an arrangement in which newly formed upper cells lie centered between the underlying cells. Many protostomes, including mollusks and annelid worms, exhibit spiral cleavage. The coiled shells of some mollusks result from spiral cleavage. In **radial cleavage,** the cleavage planes are either parallel or perpendicular to the vertical axis of the embryo. This results in tiers of cells, one directly above the other. All deuterostomes exhibit radial cleavage, as do insects and nematodes, suggesting it may have been an ancestral condition.

Additional Morphological Criteria Distinguish the Bilateria

In older phylogenetic trees of animal life, classification was also based on additional morphological features, such as the possession of a fluid-filled body cavity called a **coelom** or the presence of body segmentation. More recent molecular data suggest that although these features are helpful in describing differences in animal structure, they are not as reliable in shedding light on the evolutionary history of animals as previously believed.

Body Cavity A body cavity is an internal space within an animal that houses internal organs. In many animals, the body cavity is completely lined with mesoderm and is called a true coelom. Animals with a true coelom, such as annelids, arthropods, and chordates, are termed **coelomates** (Figure 26.6a). If the fluid-filled cavity is not completely lined by tissue derived from mesoderm, it is known as a pseudocoelom (Figure 26.6b). Animals with a pseudocoelom, including rotifers and nematodes, are termed **pseudocoelomates.** Some animals, such as flatworms, lack a fluid-filled body cavity and are termed **acoelomates** (Figure 26.6c). Instead of fluid, this region contains mesenchyme, a tissue derived from mesoderm.

In some soft-bodied invertebrates, such as earthworms, the coelom functions as a **hydrostatic skeleton**—a fluid-filled body cavity surrounded by muscles that gives support and shape to the body. Muscle contractions at one part of the body push this fluid toward another part of the body. This type of movement can best be observed

Figure 26.5 Differences in embryonic development between protostomes and deuterostomes. See Figure 26.2 for a comparison of protostome and deuterostome groups. **(a)** In protostomes, the blastopore becomes the mouth. In deuterostomes, the blastopore becomes the anus. **(b)** Protostomes have determinate cleavage, whereas deuterostomes have indeterminate cleavage. **(c)** Many protostomes have spiral cleavage, and all deuterostomes have radial cleavage. The dashed arrows indicate the direction of cleavage.

in an earthworm. In some organisms, such as insects and mollusks, the fluid in the body cavity also acts as a simple circulatory system. Finally, the fluid is relatively incompressible and therefore cushions internal organs such as the heart and intestinal tract in all coelomate animals, helping to prevent injury from external forces.

Segmentation Another feature of the animal body plan is the presence or absence of segmentation. In segmentation, the body is divided into regions called segments. Segmentation is most obvious in the annelids, or segmented worms, but it is also evident in arthropods and chordates (**Figure 26.7**). In annelids, most segments are similar in structure and contain the same set of blood vessels, nerves, and muscles. However, some segments may differ, such as those containing the sex organs. In chordates, we can see segmentation in the backbone and muscles. The advantage of segmentation is that it allows specialization of body regions and the appendages used for movement. Many insects have wings and three pairs of legs, whereas centipedes have no wings and many legs. Crabs, lobsters, and shrimp have highly specialized thoracic appendages that aid in movement and feeding (look ahead to Figure 26.30).

EVOLUTIONARY CONNECTIONS

The Protostomes Consist of Two Major Clades—the Ecdysozoa and the Lophotrochozoa

In 1997, American molecular biologists Anna Marie Aguinaldo, James Lake, and colleagues analyzed the relationships between protostome clades by sequencing the complete gene that encodes small subunit (SSU) rRNA from a variety of representative taxa. Total genomic DNA was isolated using standard techniques and amplified by the polymerase chain reaction (PCR) (refer back to Figure 18.6). PCR fragments were then subjected to DNA sequencing, a technique also described in Chapter 18, and the evolutionary relationships among 50 species were examined. The resulting data indicated the existence of a new clade of molting animals, the **Ecdysozoa** (pronounced ek-die-so-zo-ah), consisting of the nematodes and arthropods. This study represented a scientific breakthrough because it underscored the value of molecular phylogenies. According to molecular evidence, the other major protostome clade is the **Lophotrochozoa** (pronounced low-fo-tro-ko-zo-ah), which encompasses the mollusks, the annelids, and several other phyla (see Figure 26.2). When some morphologists reviewed their data in light of this new information, they found there was also morphological support for these new groupings.

The Ecdysozoa is so named because all of its members secrete a nonliving cuticle, an external skeleton (exoskeleton); think of the hard shell of a beetle or that of a crab. As these animals grow, the exoskeleton becomes too small, and the animal molts, or breaks out of its old exoskeleton, and secretes a newer, larger one (**Figure 26.8**) This molting process is called **ecdysis,** hence the name Ecdysozoa. Although this group was named for this morphological characteristic, it was first strongly supported as a separate clade by molecular evidence such as similarities in DNA.

Although the Lophotrochozoa clade was organized primarily through the analysis of molecular data, its name stems from two morphological features seen in many organisms of this clade. The lopho part is derived from the **lophophore,** a horseshoe-shaped crown of tentacles used for feeding that is present on some phyla in this clade, such as the rotifers, bryozoans, and brachiopods (**Figure 26.9a**). The trocho part refers to the **trochophore larva,** a distinct larval stage characterized by a band of cilia around its middle that is used for swimming (**Figure 26.9b**). Trochophore larvae are found in several Lophotrochozoa phyla, such as annelid worms and mollusks, indicating their similar ancestry. Other members of the clade, such as the platyhelminthes, have neither of these morphological features and are classified as lophotrochozoans based strictly on molecular data.

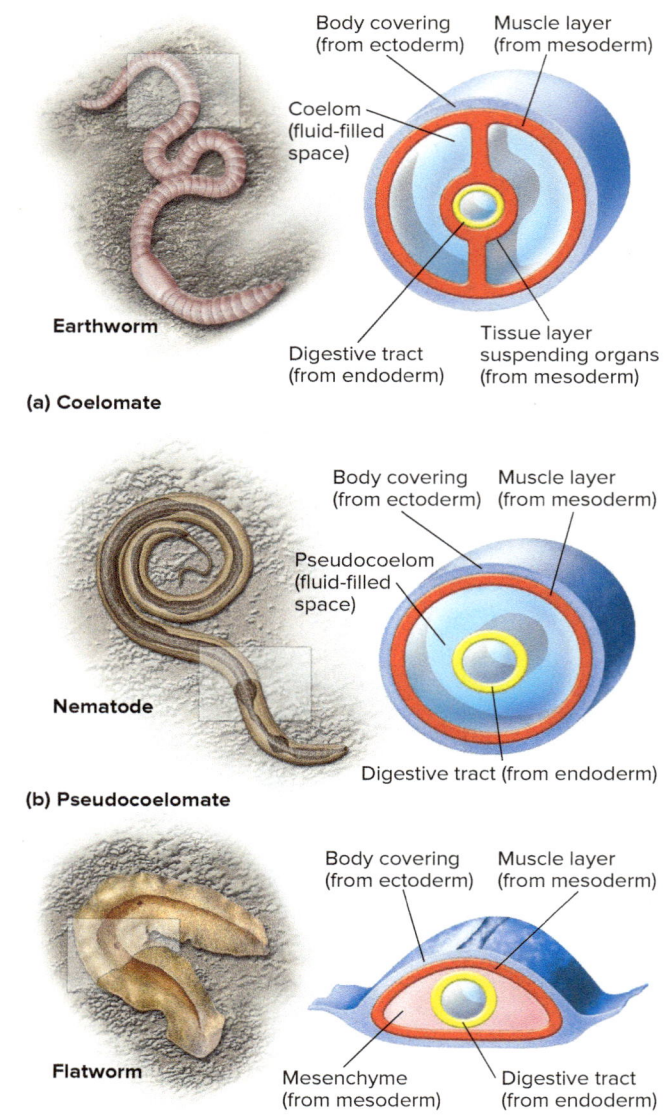

Figure 26.6 The three basic body cavities of bilaterally symmetric animals. Cross sections of each animal are shown on the right.

Concept Check: *What advantages does a coelom confer for movement?*

In earthworms, each ring is a distinct segment.

Lobsters have developed specialized appendages on many segments.

Fishes exhibit segmentation in their muscles and backbone.

Figure 26.7 Segmentation. Annelids, arthropods, and chordates all exhibit segmentation.

BioConnections: *Look ahead to Figure 26.15b. Are cestodes segmented?*

Biology Principle

The Genetic Material Provides a Blueprint That Allows Organisms to Grow, Develop, and Reproduce

For animals with exoskeletons, growth and development necessitate molting.

Figure 26.8 Ecdysis. This dragonfly, emerging from a discarded exoskeleton, is a member of the Ecdysozoa—a clade of animals exhibiting ecdysis, the periodic shedding (molting) and re-formation of the exoskeleton.
©Dwight Kuhn

BioConnections: *What two phyla have endoskeletons? Look ahead to Figure 35.1.*

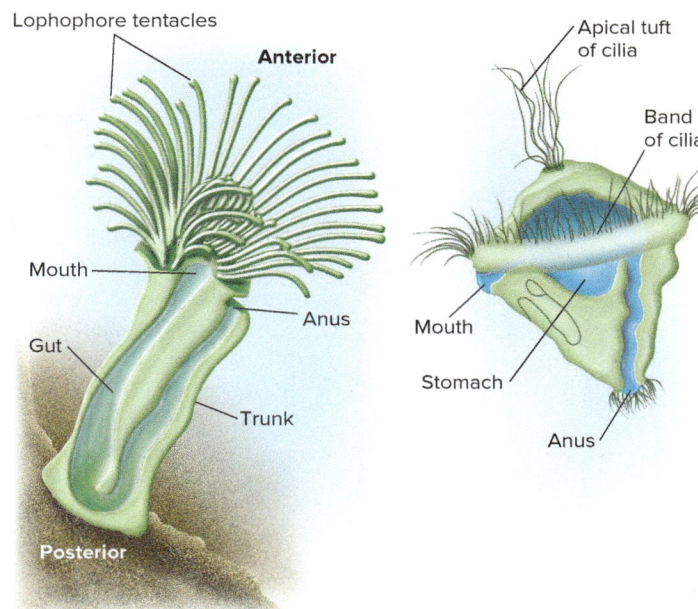

(a) Lophophore of a phoronid worm   **(b) Trochophore larva**

Figure 26.9 Characteristics of the Lophotrochozoa. (a) A lophophore, a crown of ciliated tentacles, generates a current to bring food particles into the mouth. **(b)** The trochophore larval form is found in several animal lineages.

- Recent molecular studies propose a division of the protostomes into two major clades: the Ecdysozoa and the Lophotrochozoa. Members of the Ecdysozoa secrete and periodically shed a nonliving cuticle, typically an exoskeleton, or external skeleton (Figure 26.8). Most members of the Lophotrochozoa are distinguished by two morphological features: the lophophore, a crown of tentacles used for feeding, and the trochophore larva, a distinct larval stage (Figure 26.9).

26.2 Testing Your Knowledge

1. In protostomes,
   a. the blastopore becomes the mouth.
   b. the blastopore becomes the anus.
   c. development is characterized by indeterminate cleavage.
   d. Both a and c are correct.
   e. Both b and c are correct.

2. In triploblastic animals, the inner lining of the digestive tract is derived from the
   a. ectoderm.   d. pseudocoelom.
   b. mesoderm.   e. coelom.
   c. endoderm.

3. Pseudocoelomates
   a. lack a fluid-filled cavity.
   b. have a fluid-filled cavity that is completely lined with mesoderm.
   c. have a fluid-filled cavity that is partially lined with mesoderm.
   d. have a fluid-filled cavity that is not lined with mesoderm.
   e. have an air-filled cavity that is partially lined with mesoderm.

26.2 Reviewing the Concepts

- Animals can be categorized according to their type of symmetry, whether asymmetric (the sponges), radial (the cnidarians and ctenophores), or bilateral (Bilateria, all other animals) (Figure 26.3).
- The Cnidaria have two embryonic cell layers (germ layers), the endoderm and the ectoderm, whereas the Bilateria and Ctenophora have a third germ layer termed the mesoderm, which develops between the endoderm and the ectoderm (Figure 26.4).
- Animals are also classified according to patterns of embryonic development. In protostomes, the blastopore becomes the mouth; in deuterostomes, the blastopore becomes the anus (Figure 26.5). Most protostomes have spiral cleavage, and all deuterostomes have radial cleavage.

26.3 Ctenophores: The Earliest Animals

Learning Outcome

1. Outline the unique features of ctenophores.

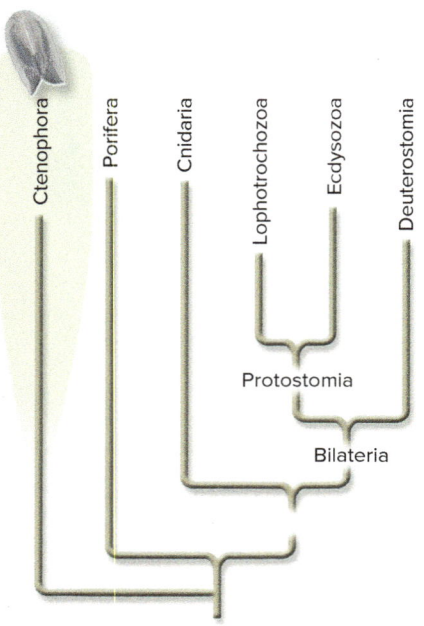

Ctenophores, also known as comb jellies, constitute the earliest-diverging animal lineage. Comb jellies are a small phylum of about 200 species, all of which are marine and look very much like jellyfish (Figure 26.10). They have eight rows of cilia on their surfaces that resemble combs. Based on morphological data, the ctenophores were once classified as close relatives to the cnidarians, which include jellyfish and corals. However, recent molecular analyses have placed them further away from the cnidarians on the evolutionary tree.

The name Ctenophora (pronounced teen-o-for′-ah) comes from the Greek *ktenos*, meaning comb, and *phora*, meaning bearing. The coordinated beating of the cilia propels a ctenophore. Averaging about 1–10 cm in length, comb jellies are probably the largest animals to use cilia for locomotion.

Comb jellies possess two long tentacles but lack stinging cells. Instead, they have cells on the tentacles that secrete a sticky substance onto which small prey adhere. The tentacles are then drawn over the mouth. Digestion occurs in a body cavity called a **gastrovascular cavity**, and waste and water are eliminated through two anal pores. Prey are generally small and may include tiny crustaceans called copepods and small fishes.

Comb jellies are often transported around the world in ships' ballast water. *Mnemiopsis leidyi*, a ctenophore species native to the Atlantic coast of North and South America, was accidentally introduced into the Black and Caspian Seas in the 1980s. With a plentiful food supply and a lack of predators, *Mnemiopsis* underwent a population explosion and ultimately devastated the local fishing industries.

All ctenophores are **hermaphrodites** (from the Greek, for the god Hermes and the goddess Aphrodite), possessing both ovaries and testes, and gametes are shed into the water to unite and eventually form a free-swimming larva that is very similar in form to the adult. Nearly all ctenophores exhibit **bioluminescence**, a phenomenon that results from chemical reactions that give off light rather than heat. Individuals are particularly evident at night, and ctenophores that wash up on shore can make the sand or mud appear luminescent.

Like jellyfish, ctenophores have both muscle and nerve cells organized as a diffuse net centralized at an elementary brain. However, the nervous system uses different neurotransmitters than those in bilaterians and jellyfish and has different types of synapses. The presence of muscle cells originating from mesoderm suggests that ctenophores share a three-germ-layer embryonic structure with bilaterians. Even so, recent analysis of the genome of the ctenophore *Mnemiopsis leidyi* suggests that ctenophores lack many of the genes involved in specifying bilaterian mesoderm. In addition, ctenophores lack *Hox* genes and possess a ctenophore-specific cleavage program. Finally, many bilaterian neuron-specific genes are absent or not expressed in ctenophores. These findings argue against a linear march of evolutionary forms from more simple animals such as sponges to complex bilaterians. Instead, evolutionary studies suggest that ctenophores were the earliest animals to diverge from a choanoflagellate-like ancestor that was multicellular and had a simple nervous system. Later, the ctenophores evolved their own unique way of forming mesoderm, which is different from other animal groups.

Figure 26.10 A ctenophore. Ctenophores are called comb jellies because the rows of cilia on their surfaces resemble combs.
©Tobias Bernhard/Getty Images

26.3 Reviewing the Concepts

- Ctenophores are the earliest-diverging animals (Figure 26.10). They possess a unique nervous system and mesoderm germ layer. They are predatory, possess a complete gut, and use cilia to propel themselves.

26.3 Testing Your Knowledge

1. Ctenophores evolved
   a. after sponges.
   b. after cnidarians.
   c. after bilaterians.
   d. before sponges.
   e. before choanoflagellates.

26.4 Porifera: The Sponges

Learning Outcomes
1. Outline the body plan and unique characteristics of sponges.
2. Describe how sponges defend themselves against predators.

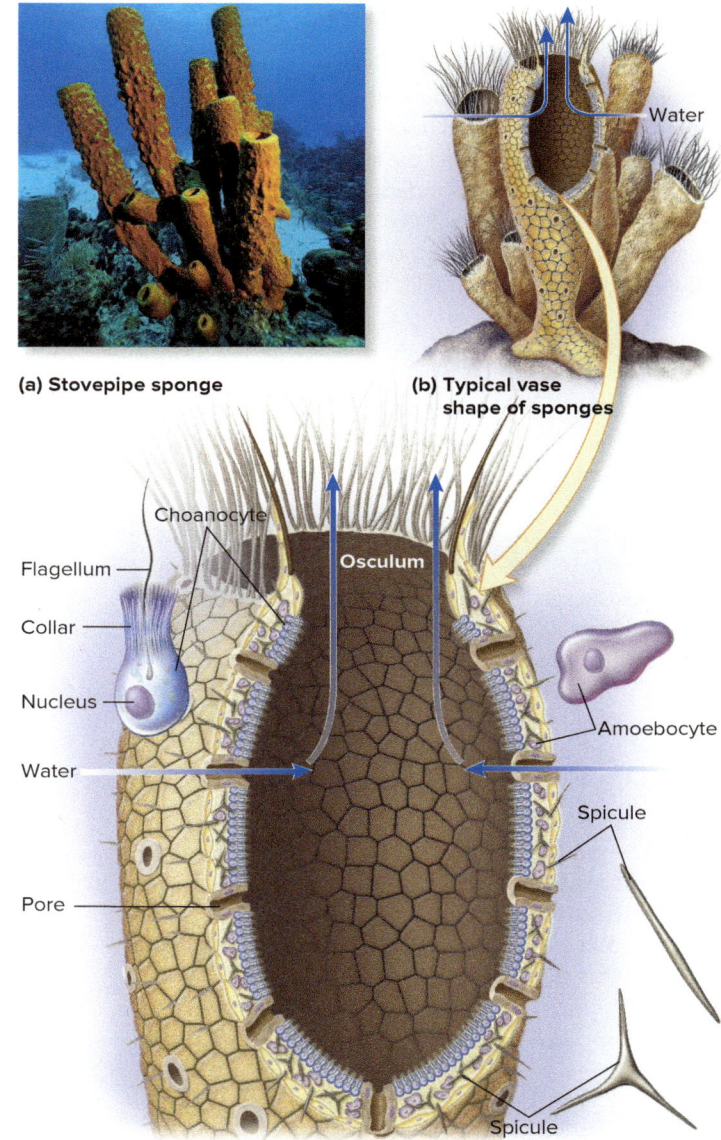

(a) Stovepipe sponge

(b) Typical vase shape of sponges

(c) Cross section of sponge morphology

Figure 26.11 (a) The stovepipe sponge (*Aplysina archeri*) is common on Caribbean reefs. (b) Many sponges have a vaselike shape. (c) A cross section reveals that sponges are truly multicellular animals, having various cell types but no distinct tissues.
(a) ©Norbert Probst/age fotostock

Members of the phylum Porifera (from the Latin, meaning pore bearers) are commonly referred to as sponges. Sponges have lost their germ layers and lack true tissues—groups of cells that have a similar structure and function. However, sponges are multicellular and possess several types of cells that perform different functions. Even though sponges carry most of the genes that are needed for a functioning nervous system, they have lost the ability to produce neurons during evolution. Biologists have identified approximately 8,000 species of sponges, the vast majority of which are marine. Sponges range in size from only a few millimeters across to more than 2 m in diameter. Smaller sponges may be radially symmetric, but most have no apparent symmetry. Some sponges have a low, encrusting growth form, whereas others grow tall and erect (Figure 26.11a). Although adult sponges are sessile—that is, anchored in place—the larvae are free-swimming.

Choanocytes Help Circulate Water

The body of most sponges looks similar to a vase pierced with small holes, or pores (Figure 26.11b). Water is drawn through these pores into a central cavity and flows out through the large opening at the top, called the osculum. The water is brought into the pores by the beating action of the flagella of the **choanocytes**, or collar cells, that line the central cavity (Figure 26.11c). In the process, the choanocytes trap and eat small particulate matter and tiny plankton. Also lining the central cavity are mobile cells called amoebocytes that absorb food from choanocytes, digest it, and carry the nutrients to other cells. Sponges are unique among the major animal phyla in using intracellular digestion, the uptake of food particles by cells, as a mode of feeding.

Sponges Have Mechanical and Chemical Defenses Against Predators

Some amoebocytes can form tough skeletal fibers that support the body. In many sponges, this skeleton consists of sharp **spicules** formed of protein, calcium carbonate, or silica. For example, some

Concept Check: *Since sponges are soft and sessile, why aren't they eaten by other organisms?*

deep-ocean species, called glass sponges, are distinguished by needle-like silica spicules that form elaborate, lattice-like skeletons. The presence of such tough spicules may help to explain why predation of sponges is rare. Other sponges have fibers of a tough protein called **spongin** that lend skeletal support. Spongin skeletons are still commercially harvested and sold as bath sponges. Many species produce toxic defensive chemicals, some of which are thought to have possible antibiotic and anti-inflammatory effects in humans.

Sponges Reproduce Sexually and Asexually

Sponges reproduce through both sexual and asexual means. Most sponges are hermaphrodites and thus can produce both sperm and eggs. Gametes are derived from amoebocytes or choanocytes. The eggs remain in the mesohyl and the sperm are released into the water and carried by water currents to fertilize the eggs of neighboring sponges. Zygotes develop into flagellated swimming larvae that eventually settle on a suitable substrate to become sessile adults. In asexual reproduction, a small fragment, or bud, may detach and form a new sponge.

26.4 Reviewing the Concepts

- The phylum Porifera, or sponges, lack true tissues but are multicellular animals possessing several types of cells (Figure 26.11).
- Porifera are asymmetric marine filter feeders.

26.4 Testing Your Knowledge

1. How do sponges protect themselves from predators?
    a. They are protected by silica spicules.
    b. They are protected by toxic defensive chemicals.
    c. They are eaten, and the leftover cells reaggregate into new, smaller sponges.
    d. Both a and b are correct.
    e. All of the above are correct.
2. Choanocytes are
    a. a group of protists believed to have given rise to animals.
    b. specialized cells of sponges that trap and eat small particles.
    c. cells that make up the gelatinous layer in sponges.
    d. cells of sponges that transfer nutrients to other cells.
    e. cells that form spicules in sponges.

26.5 Cnidaria: Jellyfish and Other Radially Symmetric Animals

Learning Outcomes

1. Compare and contrast the two body forms of cnidarians.
2. Describe how cnidarians defend themselves.

The members of the phylum Cnidaria (from the Greek *knide*, meaning nettle, and *aria*, meaning related to; pronounced nid-air′-e-ah) are mostly found in marine environments, although a few are freshwater species. Cnidaria includes hydra, jellyfish, box jellies, sea anemones, and corals. The cnidarians have only two embryonic germ layers: the ectoderm and the endoderm. A gelatinous substance called the **mesoglea** connects the two layers. In jellyfish, the mesoglea is enlarged and forms a transparent jelly, whereas in hydra and corals, the mesoglea is very thin (**Figure 26.12**). Most cnidarians have tentacles around the mouth that aid in prey detection and capture.

The phylum Cnidaria consists of four classes: Hydrozoa (including the Portuguese man-of-war), Scyphozoa (jellyfish), Anthozoa (sea anemones and corals), and Cubozoa (box jellies). The distinguishing characteristics of these classes are described in **Table 26.1**.

The Cnidarians Exist in Two Different Body Forms

Most cnidarians may exist as two different body forms and associated lifestyles: the sessile **polyp** or the motile **medusa** (Figure 26.12b). For example, corals and sea anemones exhibit only the polyp form, and jellyfish exist predominantly in the medusa form.

The polyp form has a tubular body with an opening at the oral (top) end that is surrounded by tentacles and functions as both mouth and anus (Figure 26.12a). The aboral (bottom) end is attached to the substrate. Polyps exist colonially, as in corals, or alone, as in sea anemones. Corals take dissolved calcium and carbonate ions from seawater and precipitate them as limestone underneath their bodies. In some species, this leads to a buildup of limestone deposits. As each successive generation of polyps dies, the limestone remains in place, and new polyps grow on top. Thus, huge underwater limestone deposits called coral reefs are formed (look ahead to Figure 45.14b). The largest of these is Australia's Great Barrier Reef, which stretches over 2,300 km. Many other extensive coral reefs are known, including the reef system along the Florida Keys. All coral reefs occur in warm water, generally between 20°C and 30°C.

The free-swimming medusa form has an umbrella-shaped body with an opening that serves as both mouth and anus on the concave underside, which is surrounded by tentacles (Figure 26.12b). Medusae possess simple sense organs near the bell margin, including organs of equilibrium called statocysts and photosensitive organs known as ocelli. When one side of the bell tips upward, the statocysts on that side are stimulated, and muscle contraction is initiated to reorient the medusa so that the tentacles are pointing downward. The ocelli allow medusae to position themselves in particular light levels.

Cnidarians Have Specialized Stinging Cells

One of the unique and characteristic features of the cnidarians is the existence of stinging cells called **cnidocytes**, which function in defense or the capture of prey (**Figure 26.13a**). Cnidocytes contain **nematocysts**, powerful capsules with an inverted coiled and barbed thread. Each cnidocyte has a hairlike trigger called a **cnidocil** on its surface. When the cnidocil is touched or a chemical stimulus is detected, the nematocyst is discharged; its filament penetrates the prey and injects a small amount of toxin. Small prey are immobilized and passed into the mouth by the tentacles. After discharge, the cnidocyte is absorbed, and a new one grows to replace it. The

Biology Principle

Structure Determines Function

The inverted, umbrella-shaped medusae are free-swimming, whereas the tubular, polyp forms are sedentary.

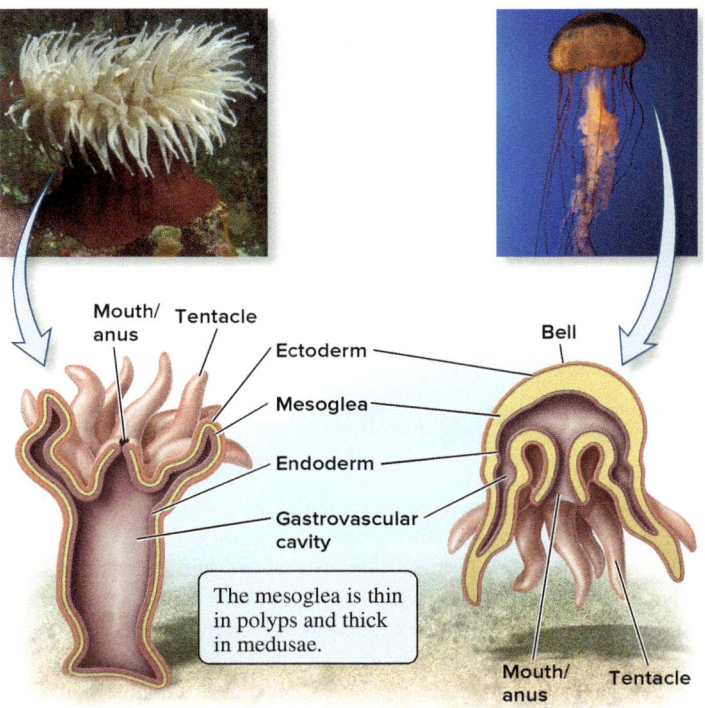

(a) Polyp (b) Medusa

Figure 26.12 Polyp and medusa forms of cnidarians. Both (a) polyp and (b) medusa forms have two layers of cells: an outer (ectoderm) and an inner (endoderm). In between is a layer of mesoglea, which is thin in polyps and thick in medusae.
(a) Source: Linda Snook, NOAA/CBNMS; (b) ©Kick Images/Getty Images

Concept Check: What are the dominant life stages of the following types of cnidarians: box jellies, corals, and the Portuguese man-of-war?

Table 26.1	Main Classes and Characteristics of the Cnidaria	
	Class and examples (est. no. of species)	**Class characteristics**
	Hydrozoa: Portuguese man-of-war, hydra, some corals (2,700)	Mostly marine; exist in polyp and medusa stages; often colonial, sometimes with individuals developing distinct morphology for prey capture, reproduction, and flotation
	Scyphozoa: jellyfish (200)	All marine; exist in medusa stage; large (up to 2 m)
	Anthozoa: sea anemones, sea fans, most corals (6,000)	All marine; exist in polyp stage; many are colonial with similar-looking individuals
	Cubozoa: box jellies, sea wasps (20)	All marine; exist in medusa stage; box-shaped

nematocysts of most cnidarians are not harmful to humans, but those on the tentacles of the larger jellyfish and the Portuguese man-of-war (**Figure 26.13b**) can be extremely painful and even fatal.

26.5 Reviewing the Concepts

- Cnidaria includes jellyfish, box jellies, sea anemones, corals, and the Portuguese man-of-war (Table 26.1). Cnidarians have only two embryonic germ layers, the ectoderm and endoderm, with a gelatinous substance (mesoglea) connecting the two layers.
- Cnidarians exist in two forms: polyp and medusa. A characteristic feature of cnidarians is their stinging cells, or cnidocytes, which function in defense and prey capture (Figures 26.12, 26.13).

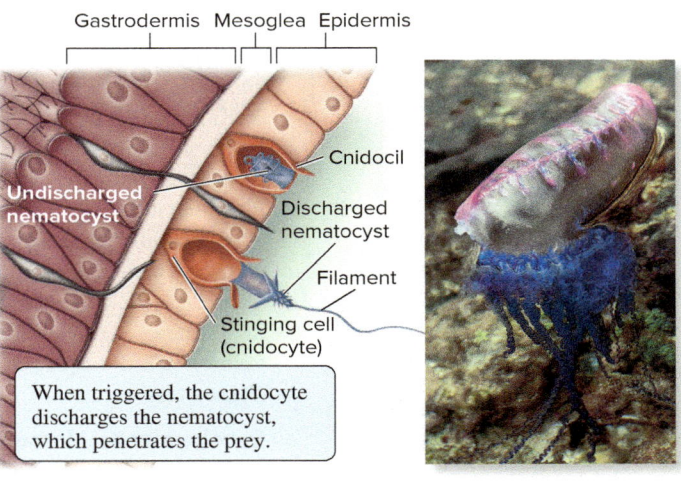

(a) Cnidocytes (b) Portuguese man-of-war

Figure 26.13 Specialized stinging cells of cnidarians, called cnidocytes.
(a) Cnidocytes, which contain stinging capsules called nematocysts, are situated in the tentacles. (b) The Portuguese man-of-war (*Physalia physalis*) employs cnidocytes that can be lethal to humans.
(b) ©Wild Horizon/Getty Images

26.5 Testing Your Knowledge

1. Cnidarians defend themselves using
   a. a nerve net.
   b. bioluminescence.
   c. nematocysts.
   d. medusae.
   e. cnidocils.

26.6 Lophotrochozoa: The Flatworms, Rotifers, Bryozoans, Brachiopods, Mollusks, and Annelids

Learning Outcomes

1. Describe the unique features of platyhelminthes, rotifers, bryozoans, and brachiopods.
2. Outline the main features and list the main classes of the mollusks.
3. List the advantages of segmentation in the annelids.

As noted earlier, morphological and molecular data are consistent with three clades of bilateral animals: the Lophotrochozoa and the Ecdysozoa (collectively known as the Protostomia, or protostomes) and the Deuterostomia (deuterostomes). In this section, we will explore the distinguishing characteristics of the Lophotrochozoa, a diverse group that includes taxa that possess either a lophophore (a crown of ciliated tentacles, seen in Bryozoa and Brachiopoda) or a distinct larval stage called a trochophore (Mollusca and Annelida). Also included in this clade are the Platyhelminthes (some of which have trochophore-like larvae) and the Rotifera (which have a lophophore-like feeding device), both of which share molecular similarities with the other members of the Lophotrochozoa.

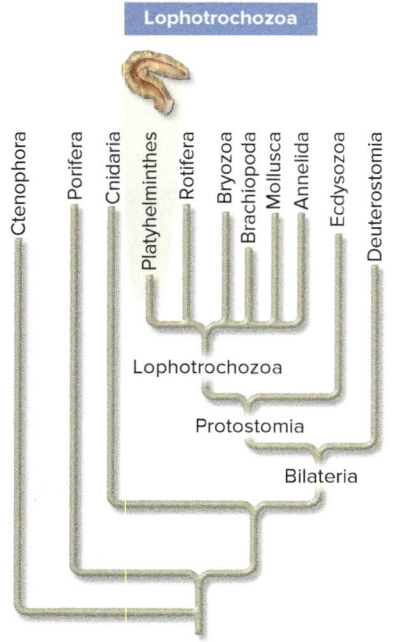

The Phylum Platyhelminthes Consists of Flatworms with No Coelom

Platyhelminthes (from the Greek *platy*, meaning flat, and *helminth*, meaning worm), or flatworms, were among the first animals to develop an active, predatory lifestyle. Platyhelminthes, and indeed most animals, possess mesoderm and are bilaterally symmetric, with a head bearing sensory appendages, a feature called cephalization (Figure 26.14).

Development of Mesoderm

The flatworms are believed to be the first bilaterian animals

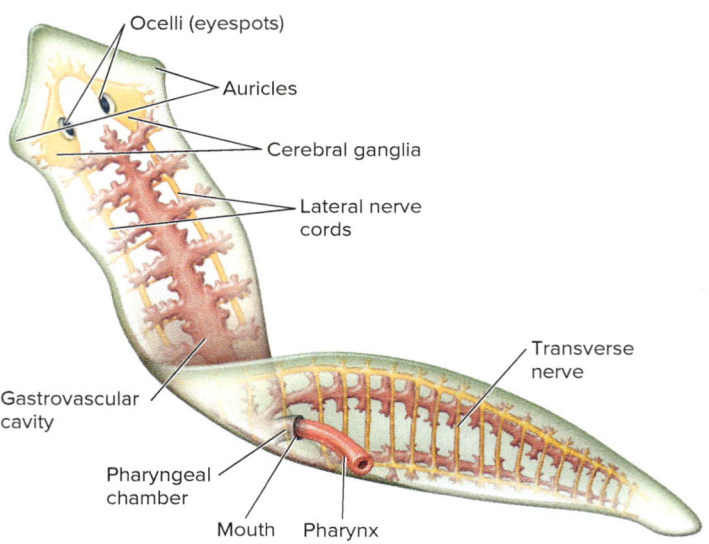

Figure 26.14 Body plan of a flatworm. Flatworm morphology is represented by a planarian, a member of the class Turbellaria.

  Concept Check: *How do flatworms breathe?*

to develop three distinctive embryonic germ layers—ectoderm, endoderm, and mesoderm—with mesoderm replacing the simpler, gelatinous mesoglea of cnidarians. As such, they are said to be triploblastic. The muscles in flatworms, which are derived from mesoderm, are well developed. The development of mesoderm was therefore a critical evolutionary innovation in bilaterian animals, leading to complex movement and the development of more sophisticated organs. Flatworms lack a fluid-filled body cavity in which the gut is suspended and, instead, mesoderm fills the body spaces around the gastrovascular cavity; hence, they are described as acoelomates (see Figure 26.6c). Flatworms also lack a specialized respiratory or circulatory system and must respire by diffusion. Thus, no cell can be too far from the surface of the body, making a flattened shape necessary. The digestive system of flatworms is incomplete, with only one opening, which serves as both mouth and anus, as in cnidarians. Most flatworms possess a muscular pharynx that may be extended through the mouth. The pharynx opens to a gastrovascular cavity, where food is digested. In large flatworms, the gastrovascular cavity is highly branched to distribute nutrients to all parts of the body. The incomplete digestive system of flatworms prevents continuous feeding. Some flatworms are predators, but many species invade other animals as parasites.

Cephalization At the anterior end of some free-living flatworms are light-sensitive eyespots, called ocelli, as well as chemoreceptive and sensory cells that are concentrated in organs called auricles. A pair of **cerebral ganglia,** which are clusters of nerve cell bodies, receives input from photoreceptors in eyespots and sensory cells. From the ganglia, a pair of lateral nerve cords running the length of the body allows rapid movement of information from anterior to posterior. In addition, transverse nerves form a nerve net on the ventral surface, similar to that of cnidarians (look ahead to Figure 33.14a, b). Thus, flatworms retain the cnidarian-style nervous system while possessing the beginnings of the more centralized type of nervous system seen throughout much of the rest of the animal kingdom.

In Platyhelminthes, reproduction is either sexual or asexual. Most species are hermaphroditic but do not fertilize their own eggs. Flatworms can also reproduce asexually by splitting into two parts, with each half regenerating the missing fragment.

Flatworm Diversity The four classes of flatworms are the Turbellaria, Monogenea, Trematoda (flukes), and Cestoda (tapeworms) (Table 26.2).

- Turbellarians are the only free-living class of flatworms and are widespread in lakes, ponds, and marine environments (Figure 26.15a).
- Monogeneans are relatively simple external parasites with just one host species (a fish).
- Trematodes and cestodes are internally parasitic in humans and therefore are of great medical and veterinary importance. They possess a variety of attachment organs, such as hooks and suckers, that enable them to remain embedded within their hosts. For example, cestodes attach to their host by means of an organ at the head end called a scolex (Figure 26.15b). They have no mouth or gastrovascular cavity and absorb nutrients across the body surface.

Reproduction in Flatworms Flatworm life cycles can be complex. Cestodes often require two separate vertebrate host species: one host, such as pigs or cattle, to begin their life cycle and another host, a primate such as a human, to complete their development. Behind the scolex in cestodes is a long ribbon of identical "segments" called proglottids (Figure 26.15b). These are not true body segments but are replicates of sex organs that produce thousands of eggs. The

Table 26.2 Main Classes and Characteristics of Platyhelminthes

Class and examples (est. no. of species)	Class characteristics
Turbellaria: planarian (3,000)	Mostly marine; free-living flatworms; predators or scavengers
Monogenea: flukes (1,000)	Marine and freshwater; usually external parasites of fish; simple life cycle (no intermediate host)
Trematoda: flukes (11,000)	Internal parasites of vertebrates; complex life cycle with several intermediate hosts
Cestoda: tapeworms (5,000)	Internal parasites of vertebrates; complex life cycle, usually with one intermediate host; no digestive system; nutrients absorbed across epidermis

Biology Principle

Biology Affects Our Society

About 1% of U.S. cattle are infected by beef tapeworms. Consuming beef that is not sufficiently well cooked can lead to infection by these parasites. There are at least 1,000 hospitalizations a year in the U.S. for tapeworm infection, most as a result of eating inadequately cooked pork.

(a)

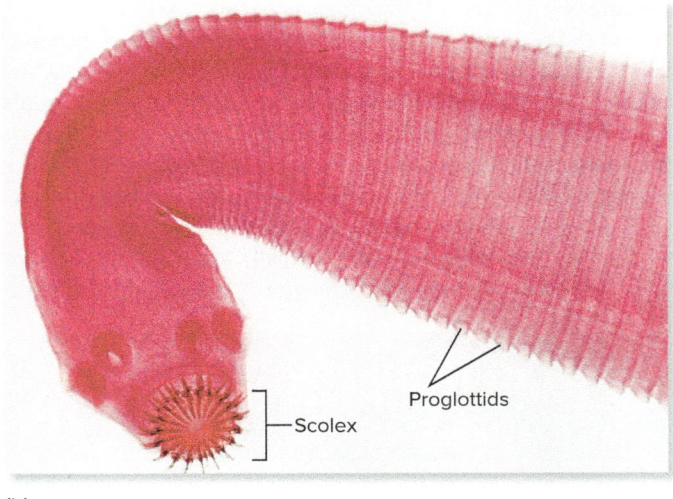
(b)

Figure 26.15 Flatworms. (a) Many free-living marine turbellarians are brightly colored such as this racing stripe flatworm, *Pseudoceros bifurcus,* from Bali, Indonesia. **(b)** The tapeworm *Taenia pisiformis* is a member of the class Cestoda. Note the tiny hooks and suckers that make up the scolex. Beyond the scolex, each segment is a proglottid filled with eggs.
(a) ©Wolfgang Poelzer/Wa./age fotostock; *(b)* ©Biophoto Associates/Science Source

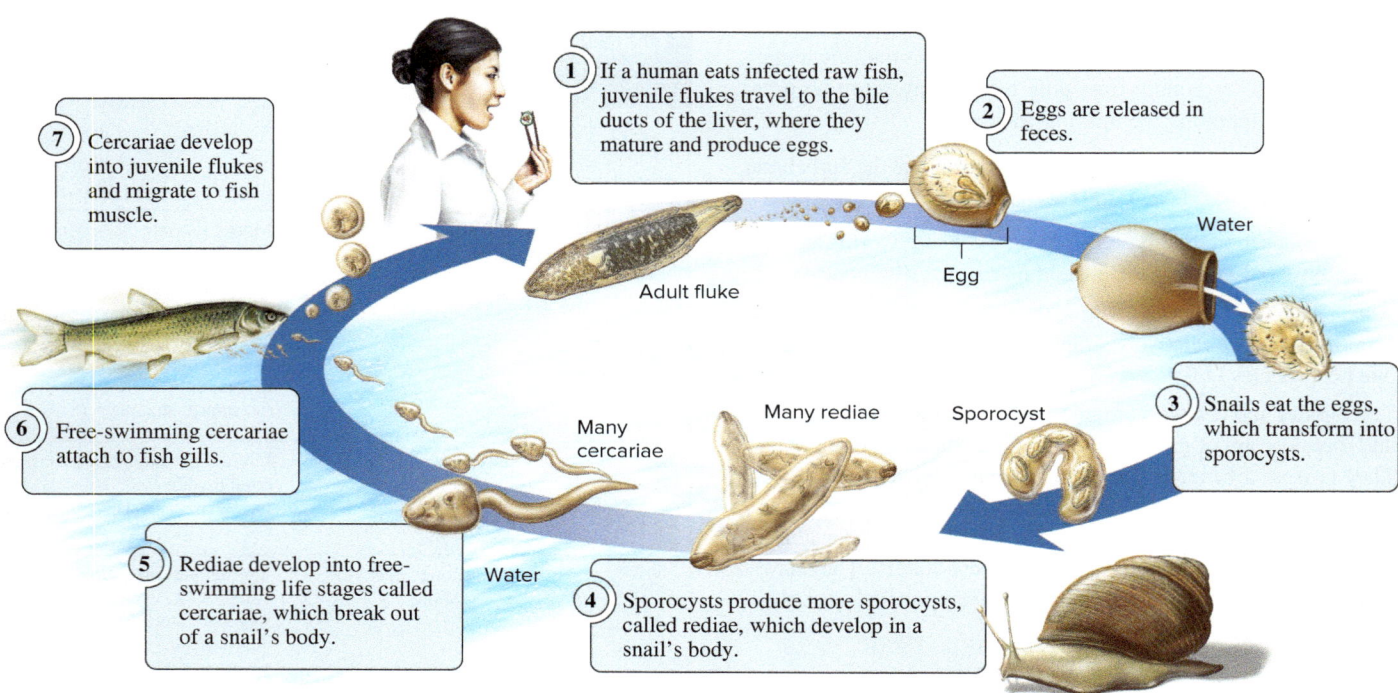

Figure 26.16 **The complete life cycle of a trematode.** This figure shows the life cycle of the Chinese liver fluke (*Clonorchis sinensis*).

proglottids are continually shed in the host's feces. Human feces passed out into the ground are eaten with grass by pigs or cattle. Many tapeworms are then ingested by humans when they consume undercooked, infected meat—hence the value of thoroughly cooking meat.

The life cycle of trematodes is even more complex than that of cestodes, involving multiple hosts. The first host, called the intermediate host, is usually a mollusk, and the final host, or definitive host, is usually a vertebrate, but often a second or even a third intermediate host is involved (**Figure 26.16**). Because the flukes multiply in each host, prodigious numbers of offspring can be produced.

Blood flukes, genus *Schistosoma*, are the most common parasitic trematodes infecting humans; they cause the disease known as schistosomiasis. Over 200 million people worldwide, primarily in tropical Asia, Africa, and South America, are infected with this disease. The inch-long adult flukes can live for years in human hosts, and the release of eggs may cause chronic inflammation and blockage in many organs. Untreated schistosomiasis can lead to severe damage to the liver, intestines, and lungs and can eventually lead to death. Sewage treatment and access to clean water can greatly reduce infection rates.

Biology Principle | Biology is a quantitative science
Quantitative Analysis: How Many Flukes?

Many parasites face the challenge of reproducing and having their offspring find a new host. How can offspring relocate to a new host when their parent is locked up inside the host's body? As noted previously, in tapeworms, each proglottid contains thousands of eggs. This ensures that vast numbers of eggs pass out of the host's body in the hope of finding a new host. Flukes, such as Chinese liver fluke (*Clonorchis sinensis*), have multiple host species. Eggs, on release from the human host, pass into water, where snails eat them (Figure 26.16, steps 2 and 3). Eggs develop into sporocysts inside the snail, and each sporocyst produces more sporocysts, called rediae. The rediae produce cercariae, which break out of the snail's body to attach to fish. The cercariae eventually lodge in fish muscles, which are then consumed by humans. The chances of infection at each of these stages are very low, so flukes produce huge numbers of offspring to compensate. We can appreciate the vast reproductive effort involved by calculating the number of offspring at each stage:

- Average number of eggs produced per fluke per day = 28,000
- Average length of life of adult fluke = 1,800 days (about 5 years)
- Number of sporocysts produced per lifetime = 28,000 × 1,800 = about 50 million
- Average number of cercariae produced per sporocyst = 200
- Total number of cercariae produced per lifetime = 200 × 50 million = 10 billion

Crunching the Numbers: Let's suppose a tapeworm sheds about 5 proglottids per day. Each proglottid contains about 50,000 eggs, and most tapeworms live in colonies of about 10 per host. How many eggs are shed by each host per day?

Members of the Phylum Rotifera Have a Pseudocoelom and a Ciliated Crown

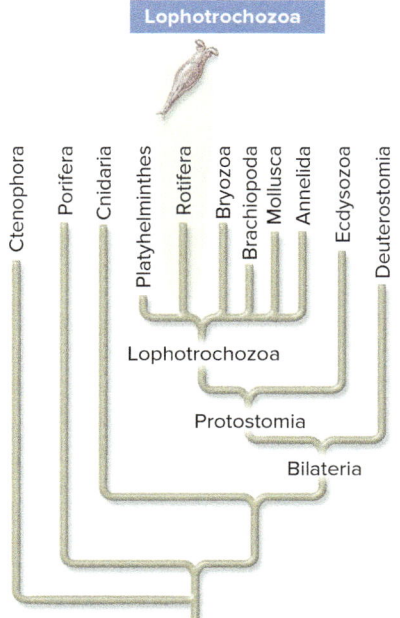

Members of the phylum Rotifera (from the Latin *rota*, meaning wheel, and *fera*, meaning to bear) get their name from their ciliated crown, or **corona,** which, when beating, looks similar to a rotating wheel (**Figure 26.17**). Most rotifers are microscopic animals, usually less than 1 mm long, and some have beautiful colors. The corona creates water currents that propel the animal through the water and that waft small, planktonic organisms or decomposing organic material toward the mouth. Rotifers have an alimentary canal, a digestive tract with a separate mouth and anus, which means they can feed continuously. There are about 2,200 species of rotifers, most of which inhabit fresh water, with a few marine and terrestrial species. Most often, they are bottom-dwelling organisms, living on the pond floor or along lakeside vegetation.

Figure 26.17 *Philodina roseola*, a common rotifer, showing the crown of cilia (corona).
©Wim van Egmond/SPL/Science Source

Bryozoa and Brachiopoda Have a Lophophore for Feeding and Gas Exchange

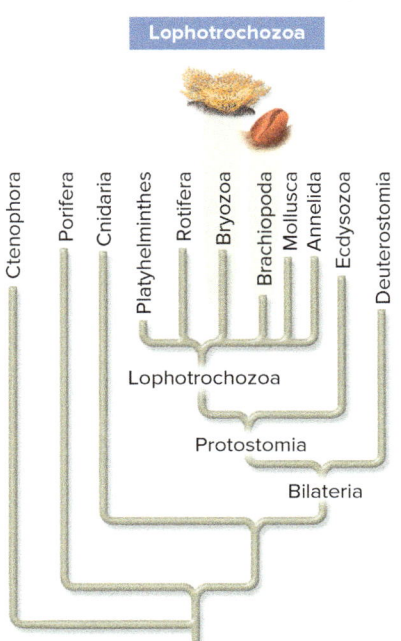

The Bryozoa and the Brachiopoda both possess a lophophore, a ciliary feeding device (see Figure 26.9a), and a true coelom (see Figure 26.6a). The lophophore is a circular fold of the body wall bearing tentacles that draw water toward the mouth. Because a thin extension of the coelom penetrates each tentacle, the tentacles also serve as a respiratory organ. Gases diffuse across the tentacles and into or out of the coelomic fluid and are carried throughout the body.

Bryozoans (from the Greek *bryon*, meaning moss, and *zoon*, meaning animal) are small, colonial animals, most of which are less than 0.5 mm long, found encrusted on rocks in shallow aquatic environments (**Figure 26.18a**). For this reason, bryozoans have been important reef-builders. Bryozoans date back to the Paleozoic era, and thousands of fossil forms have been discovered and identified.

Brachiopods (from the Greek *brachio*, meaning arm, and *podos*, meaning foot) are small, 1–8-cm, bottom-dwelling, marine organisms with two shell halves, much like clams (**Figure 26.18b**). Although they are now a relatively small group, with about 300 living species, brachiopods flourished in the Paleozoic and Mesozoic eras—about 30,000 fossil species have been identified.

Mollusca Is a Large Phylum Containing Snails, Slugs, Clams, Oysters, Octopuses, and Squids

Mollusks (from the Latin *mollis*, meaning soft) constitute a very large phylum, with over 100,000 living species, including organisms as diverse as snails, clams and oysters, octopuses and squid, and chitons. They are an ancient group, as evidenced by the classification of about 35,000 fossil species. Mollusks have a considerable economic, aesthetic, and ecological importance to humans. Many, including scallops, oysters, clams, and squids, serve as sources of food. A significant industry involves the farming of oysters to produce cultured pearls, and rare and beautiful mollusk shells are extremely valuable to collectors. Snails and slugs can damage vegetables and ornamental plants, and boring mollusks can penetrate wooden ships and wharfs. Mollusks are intermediate hosts to many parasites, and several invasive species have become serious pests. For example, populations of the zebra mussel (*Dreissena polymorpha*) have been introduced

(a) A bryozoan

(b) A brachiopod, the northern lamp shell

Figure 26.18 Bryozoans and brachiopods.
(a) Bryozoans are colonial animals that reside inside a nonliving case.
(b) Brachiopods, such as this northern lamp shell (*Terebratulina septentrionalis*), in Nova Scotia, Canada, have dorsal and ventral shells.
(a) ©Hecker/Sauer/age fotostock; *(b)* ©Andrew J. Martinez/Science Source

Concept Check: What are the two main functions of the lophophore?

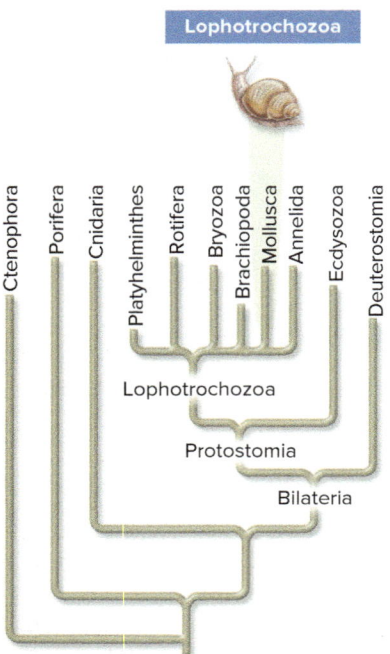

into North America from Asia via ballast water from transoceanic ships. Since their introduction, they have spread rapidly throughout the Great Lakes and an increasing number of inland waterways, adversely affecting native organisms and clogging water intake valves of municipal water-treatment plants around the lakes.

Key Features of Mollusks

One common feature of the mollusks is their soft body, which in many species is found under a protective external shell. Most mollusks are marine, although some have colonized fresh water. Many snails and slugs have moved onto land but survive only in humid areas and where the calcium necessary for shell formation is abundant in the soil. The ability to colonize freshwater and terrestrial habitats has led to a diversification of mollusk body plans. Thus, in the amazing diversity of mollusks we see how organismal diversity is related to environmental diversity.

Although great variation in morphology occurs between classes, mollusks have a basic body plan consisting of three parts (**Figure 26.19**).

- A muscular **foot** is usually used for movement, and a **visceral mass** containing the internal organs rests atop the foot.
- The **mantle**, a fold of skin draped over the visceral mass, secretes a shell in those species that form shells.
- The **mantle cavity** houses delicate **gills**, filamentous organs that are specialized for gas exchange. A continuous current of water, often induced by cilia present on the gills or by muscular pumping, flushes out the wastes from the mantle cavity and brings in new, oxygen-rich water.

Most mollusks have an **open circulatory system** with a heart that pumps body fluid, called hemolymph, through vessels and into sinuses. Sinuses are the open, fluid-filled cavities between the internal organs. The organs and tissues are therefore continually bathed in hemolymph. The sinuses coalesce to form an open cavity known as the hemocoel (blood cavity). From these sinuses, the hemolymph drains into vessels that take it to the gills and then back to the heart.

The mollusk's mouth may contain a **radula**, a unique, protrusible, tonguelike organ that has many teeth and is used to eat plants, scrape food particles off rocks, or, if the mollusk is predatory, bore into shells of other species and tear flesh (Figure 26.19, inset). Other mollusks, particularly bivalves, have lost their radula and are filter feeders that strain water brought in by ciliary currents.

Mollusk Reproduction Most mollusk species have separate sexes, although some are hermaphroditic. Gametes are usually released into the water, where they mix and fertilization occurs. In some snails, however, fertilization is internal, with the male inserting sperm directly into the female. Internal fertilization was a key evolutionary development, enabling some snails to colonize land, and can be considered a critical innovation that fostered extensive adaptive radiation. In many species, reproduction involves the

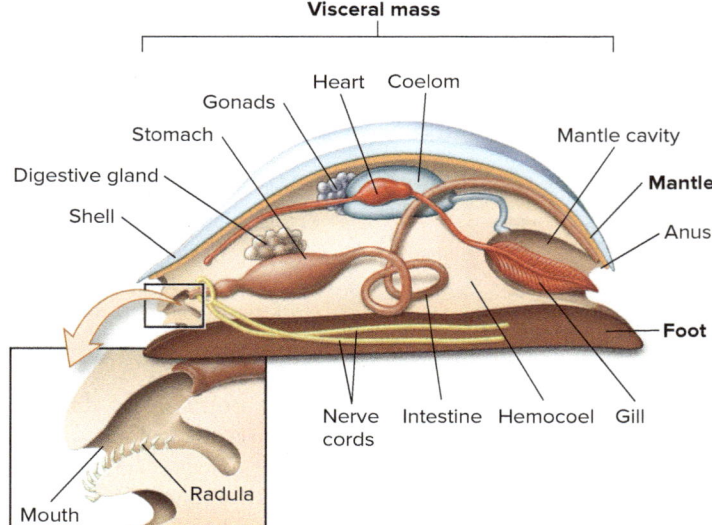

Figure 26.19 The mollusk body plan. The generalized body plan of a mollusk includes the characteristic foot, mantle, and visceral mass.

BioConnections: Does the mollusk heart pump blood? Look ahead to Figure 36.1 for more information.

production of a trochophore larva that develops into a free-swimming larva with a rudimentary foot, shell, and mantle.

The Major Molluscan Classes

Of the eight molluscan classes, the four most common are the Bivalvia (clams and mussels), Polyplacophora (chitons), Gastropoda (snails and slugs), and Cephalopoda (octopuses, squids, and nautiluses) (Table 26.3).

- Bivalves are freshwater or marine mollusks whose bodies are enclosed within a hinged shell of two valves, or halves. Prominent members of this class include oysters, clams, mussels, and scallops (Figure 26.20a).

- Polyplacophora are marine mollusks with a shell composed of eight separate plates (Figure 26.20b). Chitons are common in the intertidal zone, an area above water at low tide and under water at high tide, and they creep along when covered by the tide. Feeding occurs by scraping algae off rock surfaces. When the tide recedes, the muscular foot holds the chiton tight to the rock surface, preventing desiccation.

- The class Gastropoda (from the Greek *gaster*, meaning stomach, and *podos*, meaning foot) is the largest group of mollusks and encompasses about 75,000 living species, including snails, periwinkles, and limpets (Figure 26.20c). Most

Table 26.3 Main Classes and Characteristics of Mollusks

Class and examples (est. no. of species)	Class characteristics
Bivalvia: clams, mussels, oysters (30,000)	Marine or freshwater; shell with two halves or valves; primarily filter feeders with siphons
Polyplacophora: chitons (860)	Marine; eight-plated shell
Gastropoda: snails, slugs, nudibranchs (75,000)	Marine, freshwater, or terrestrial; most with coiled shell, but shell absent in slugs and nudibranchs; radula present
Cephalopoda: octopuses, squids, nautiluses (780)	Marine; predatory, with tentacles around mouth, often with suckers; shell often absent or reduced; closed circulatory system; jet propulsion via siphon

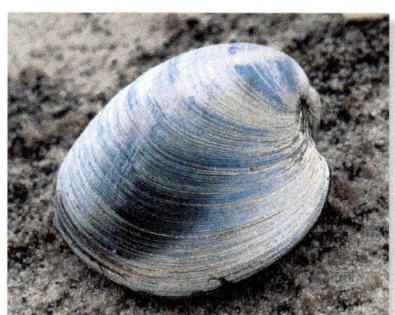
(a) A Quahog clam, class Bivalvia

(b) A chiton, class Polyplacophora

(c) A snail, class Gastropoda

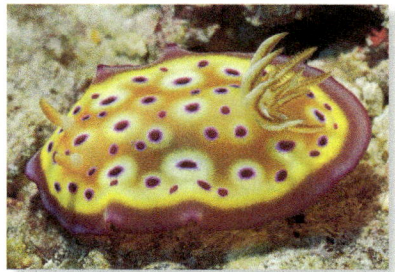
(d) A sea slug, class Gastropoda

(e) A blue-ringed octopus, class Cephalopoda

Figure 26.20 Mollusks. (a) A bivalve shell, class Bivalvia, with growth rings. Quahog clams (*Mercenaria mercenaria*) can live over 20 years. (b) A chiton (*Tonicella lineata*), a polyplacophoran with a shell made up of eight separate plates. (c) A gastropod tree snail (*Liguus fasciatus*) from the Florida Everglades, showing its characteristic coiled shell. (d) A nudibranch (*Chromodoris kuniei*). The nudibranchs are a gastropod subclass whose members have lost their shell altogether. (e) The highly poisonous blue-ringed octopus (*Hapalochlaena lunulata*) is a cephalopod.
(a) ©Andrew J. Martinez/Science Source; *(b)* ©Kjell B. Sandved/Science Source; *(c)* ©imageBROKER/Alamy; *(d)* ©Hal Beral/Getty Images; *(e)* ©Yusran Abdul Rahman/Shutterstock

Concept Check: *Why are mollusks important to humans?*

gastropods have a one-piece shell, into which the animal can withdraw to escape predators, However, the class also includes species such as slugs and nudibranchs, whose shells have been greatly reduced or completely lost during their evolution (Figure 26.20d). Although gastropods usually occupy marine or freshwater habitats, some species, including snails and slugs, have also colonized land.

The 780 species of Cephalopoda (from the Greek *kephalo*, meaning head, and *podos*, meaning foot) are the most morphologically complex of the mollusks and indeed among the most complex of all invertebrates. Most are fast-swimming marine predators that range from organisms just a few centimeters in size to the colossal squid (*Mesonychoteuthis hamiltoni*), which is known to reach over 13 m in length and 495 kg (1,091 lb) in weight. A cephalopod's mouth is surrounded by many long arms commonly armed with suckers. All cephalopods have a beaklike jaw that allows them to bite their prey, and some, such as the blue-ringed octopus (*Hapalochlaena lunulata*), deliver a deadly poison through their saliva (Figure 26.20e).

The foot of some cephalopods has become modified into a muscular siphon. Water drawn into the mantle cavity is quickly expelled through the siphon, propelling the organisms forward or backward in a kind of jet propulsion. Such vigorous movement requires powerful muscles and a very efficient circulatory system to deliver oxygen and nutrients to the muscles. Cephalopods are the only mollusks with a **closed circulatory system,** in which blood flows throughout an animal entirely within a series of vessels. One of the advantages of this type of system is that the heart can pump blood through the tissues rapidly, making oxygen more readily available. The blood of cephalopods contains the copper-rich protein hemocyanin for transporting oxygen. Less efficient than the iron-rich hemoglobin of vertebrates, hemocyanin gives the blood a blue color.

The Phylum Annelida Consists of the Segmented Worms

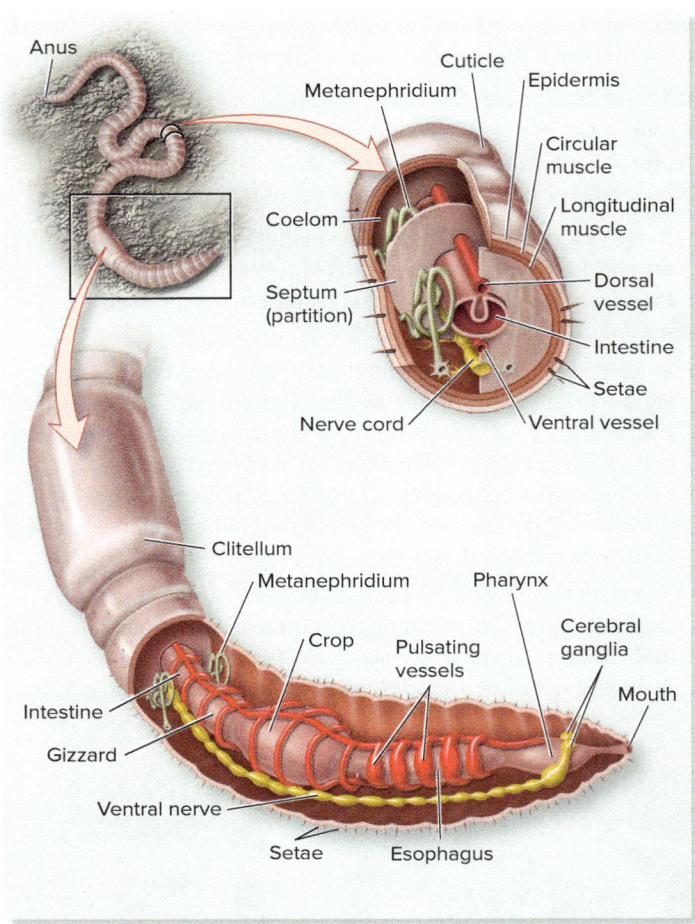

Figure 26.21 **The segmented body plan of an annelid, as illustrated by an earthworm.** The segmented nature of the worm is apparent internally as well as externally. Individual segments are separated by septa.

Concept Check: *What are some of the advantages of segmentation?*

Annelids are a large phylum with about 18,000 species of segmented worms. The members include free-ranging marine worms, tube worms, the familiar earthworm, and leeches. They range in size from less than 1 mm to enormous Australian earthworms that can reach a length of 3 m.

Key Features of Annelids

The phylum name Annelida is derived from the Latin *annulus*, meaning little ring. Each ring is a distinct segment of the annelid's body; adjacent segments are separated by septa (Figure 26.21). Segmentation in the adult confers three advantages:

- Many components of the body are repeated in each segment, including blood vessels, nerves, and excretory and reproductive organs. If the components in one segment fail, those of another segment will still function.

- Annelids possess a fluid-filled coelom that acts as a hydrostatic skeleton. In unsegmented coelomate animals, muscle contractions can distort the entire body during movement. However, such distortion is minimized in segmented animals, which allows for more effective locomotion over solid surfaces.

- Segmentation permits specialization of some segments, especially at the annelid's anterior end. Animals with more complex body plans tend to produce a greater variety of specialized segments.

All annelids except the leeches have chitinous bristles, called **setae,** on each segment. In some annelids, these are situated on fleshy, footlike **parapodia** (from the Greek, meaning almost feet) that are pushed into the substrate to provide traction during movement. In others, the setae are held closer to the body. Many annelid species burrow into soil or into muddy marine sediments and extract nutrients from ingested soil or mud. Some annelids also feed on dead or living vegetation, whereas others are predatory or parasitic.

Annelids have a nervous system with a pair of cerebral ganglia that connect to a subpharyngeal ganglion (Figure 26.21). From there, a large ventral nerve cord runs down the entire length of the body. The ventral nerve cord is unusual because it contains a few, very large nerve cells called **giant axons** that facilitate high-speed neuronal conduction and rapid responses to stimuli.

Annelids have an internal transport system in which both the circulatory system and the coelomic fluid carry nutrients, wastes, and respiratory gases. The circulatory system is closed, with dorsal and ventral vessels connected by pairs of pulsating vessels. The blood of most annelid species contains the respiratory pigment hemoglobin. Respiration occurs directly through the permeable skin surface, which restricts annelids to moist environments. The digestive system is complete and unsegmented, with many specialized regions: mouth, pharynx, esophagus, crop, gizzard, intestine, and anus.

Reproduction in Annelids Sexual reproduction in annelids involves two individuals, often of separate sexes but sometimes hermaphrodites, which exchange sperm via internal fertilization. In some species, asexual reproduction by fission occurs, in which the posterior part of the body breaks off and forms a new individual.

The Major Annelid Groups Evidence published by German evolutionary biologist Torsten Struck and colleagues in 2011 suggests that the phylum Annelida contains two major groups: the Errantia, which are active, free-ranging worms, and the Sedentaria, which are burrowers or parasites.

Members of the Errantia have many long setae bristling out of their body and are supported on footlike parapodia (**Figure 26.22a**). Most of them are free-ranging predators with well-developed eyes and powerful jaws. Many are brightly colored. In turn, most species are important prey for fishes and crustaceans.

In the Sedentaria, setae are in close proximity to the body wall, which facilitates better anchorage in tubes and burrows. Their more sedentary lifestyle is associated with reductions in head appendages. Within the Sedentaria, three types of lifestyles are apparent. Tube worms are marine sedentarians that exhibit beautiful tentacle crowns for filtering food items, such as plankton, from the water column (**Figure 26.22b**). The bulk of these worms remains hidden in a tube deep in the mud or sand.

Earthworms ingest soil and leaf tissue to extract nutrients and in the process create burrows in the earth (**Figure 26.22c**). As plant material and soil pass through the earthworm's digestive system, they are finely ground in the gizzard into smaller fragments. The biologist Charles Darwin, who penned the first detailed study of earthworm ecology, wrote, "All the fertile areas of this planet have at least once passed through the bodies of earthworms."

Leeches are usually found in freshwater environments, but there are also some marine species as well as terrestrial species that inhabit warm, moist areas such as tropical forests. Leeches have a fixed number of segments, usually 34, though in most species the septa have disappeared. Most leeches feed as blood-sucking parasites of vertebrates. They have powerful suckers at both ends of the body, and the anterior sucker is equipped with razor-sharp jaws that can bore or slice into the host's tissues. The salivary secretion of leeches (hirudin) acts as an anticoagulant to stop the prey's blood from clotting and an anesthetic to numb the pain. Leeches can suck up to several times their own weight in blood. They were once used in the medical field in the

(a) Marine worm

(b) Tube worm

(c) Earthworm

(d) Leech

Figure 26.22 Annelids. **(a)** This free-ranging marine worm from Indonesia is a member of the group Errantia. Members of the group Sedentaria include **(b)** tube worms, **(c)** earthworms, and **(d)** leeches. This species, *Hirudo medicinalis,* is sucking blood from a hematoma, a swelling of blood that can occur after surgery.
(a) ©WaterFrame/Alamy; (b) ©J.W. Alker/imageBROKER/age fotostock; (c) ©Colin Varndell/Getty Images; (d) ©St. Bartholomew's Hospital/Science Source

practice of bloodletting, the withdrawal of often considerable quantities of blood from a patient in the erroneous belief that this would prevent or cure illness and disease. Even today, leeches may be used after surgeries (Figure 26.22d). In these cases, the blood vessels are not fully reconnected and excess blood accumulates, causing a swelling called a hematoma. The accumulated blood blocks the delivery of new blood and stops the formation of new vessels. The leeches remove the accumulated blood, and new capillaries are more likely to form.

26.6 Reviewing the Concepts

- Most taxa in the Lophotrochozoa possess either a lophophore or trochophore larva. Platyhelminthes, or flatworms, are regarded as the first animals to have the organ-system level of organization (Figure 26.14; Table 26.2). Flukes and tapeworms are internally parasitic, with complex life cycles (Figures 26.15, 26.16).
- Rotifers get their name from a ciliated crown, or corona, which, when beating, looks similar to a rotating wheel (Figure 26.17).
- The bryozoa and brachiopods both possess a lophophore, a ciliary feeding structure (Figure 26.18).
- The mollusks, which constitute a large phylum with over 100,000 diverse living species, have a basic body plan with three parts—a foot, a visceral mass, and a mantle—and usually an open circulatory system (Figures 26.19, 26.20). The four most common mollusk classes are the Bivalvia (clams and mussels), Polyplacophora (chitons), Gastropoda (snails and slugs), and Cephalopoda (octopuses, squids, and nautiluses) (Table 26.3).
- Annelids are a large phylum with two main groups: Errantia, which includes free-ranging marine worms, and Sedentaria, which includes tube worms, earthworms, and leeches (Figure 26.22). Segmentation, in which the body is divided into compartments, is a critical evolutionary innovation in the annelids.

26.6 Testing Your Knowledge

1. Platyhelminthes possess
   a. a lophophore.
   b. a coelom.
   c. mesoderm.
   d. a muscular foot.
   e. giant axons.
2. All mollusks possess
   a. a radula.
   b. an open circulatory system.
   c. a soft body.
   d. an external shell.
   e. segmentation.
3. In lophotrochozoans, all species have
   a. an incomplete digestive system, which prevents continuous feeding.
   b. an open circulatory system.
   c. a segmented body.
   d. molecular similarities in their DNA.
   e. a coelom.

26.7 Ecdysozoa: The Nematodes and Arthropods

Learning Outcomes

1. List the distinguishing characteristics of nematodes.
2. Describe the arthropod body plan and its major features.
3. Give examples of the arthropod subphyla Chelicerata, Myriapoda, Hexapoda, and Crustacea.
4. List the features that help account for the diversity of insect species.

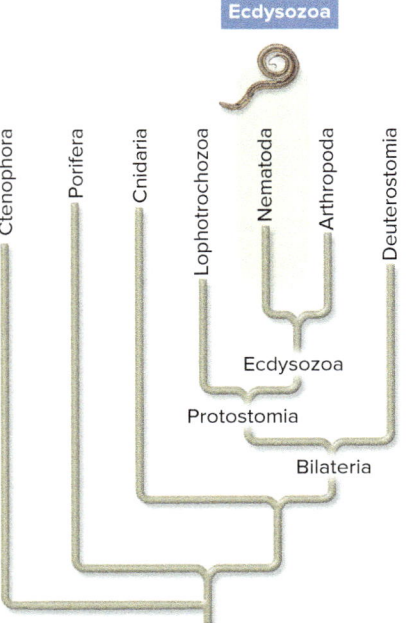

The Ecdysozoa is the sister group to the Lophotrochozoa. Although the separation is supported by molecular evidence, the Ecdysozoa is named for a process called ecdysis, or the periodic molting of the exoskeleton (see Figure 26.8). All ecdysozoans possess a **cuticle**, a nonliving covering, such as an exoskeleton, that both supports and protects the animal. Once formed, however, the cuticle typically cannot increase in size, which restricts the growth of the animal inside. The solution for growth is the formation of a new, softer cuticle under the old one. The old one then splits open and is sloughed off, allowing the new, soft cuticle to expand to a bigger size before it hardens.

Where the cuticle is thick, as in arthropods, it impedes the diffusion of oxygen across the skin. Such species acquire oxygen by lungs, gills, or a set of branching, air-filled tubes called tracheae. A variety of appendages specialized for locomotion evolved in many species, including legs for walking or swimming and wings for flying. The ability to shed the cuticle opened up developmental options for the ecdysozoans. For example, many species undergo a complete metamorphosis, changing from a wormlike larva into a winged adult. Another significant adaptation is the development of internal fertilization, which permitted species to live in dry environments.

Because of these innovations, ecdysozoans are an incredibly successful group. In this section, we will consider the two most common ecdysozoan phyla: the nematodes and the arthropods. The grouping of nematodes and arthropods is a relatively new concept supported by molecular data and implies that the process of molting arose only once in animal evolution. In support of this, certain hormones that stimulate molting have been discovered to exist only in both nematodes and arthropods.

The Phylum Nematoda Consists of Small Pseudocoelomate Worms Covered by a Tough Cuticle

The nematodes (from the Greek *nematos*, meaning thread), also called roundworms, are small, thin worms that range from less than 1 mm to about 5 cm (Figure 26.23), although some parasitic species measuring 1 m or more have been found in the placenta of sperm whales. Nematodes are ubiquitous organisms that exist in nearly all habitats, from the poles to the tropics. They are found in the soil, in both freshwater and marine environments, and inside plants and animals as parasites. A shovelful of soil may contain a million nematodes. Over 25,000 species are known, but there are probably at least four times as many undiscovered species.

Key Features of Nematodes Nematodes have several distinguishing characteristics. A tough cuticle covers the body. The cuticle is secreted by the epidermis and is made primarily of **collagen,** a structural protein also present in vertebrates (refer back to Figure 4.31). The cuticle is shed periodically as the nematode grows. The pseudocoelom functions as both a fluid-filled skeleton and a circulatory system. Diffusion of gases occurs through the cuticle. Nematodes have a complete digestive tract composed of a mouth, a pharynx, an intestine, and an anus. The mouth often contains sharp, piercing organs called **stylets,** and the muscular pharynx functions to suck in food.

Nematode Reproduction Nematode reproduction is usually sexual, with separate males and females, and fertilization takes place internally. Females are usually larger than males and can produce prodigious numbers of eggs—in some cases, over 100,000 per day. Development in some nematodes is easily observed because the organism is transparent and the generation time is short. For these reasons, the small, free-living nematode *Caenorhabditis elegans* has become a model organism for researchers to study. This nematode has 1,090 somatic cells, but 131 die, leaving exactly 959 cells. The cells die via genetically controlled programmed cell death, or apoptosis. Many diseases in humans, including acquired immunodeficiency syndrome (AIDS), cause extensive apoptosis, whereas others, such as cancer and autoimmune diseases, reduce apoptosis so that cells that should die do not. Researchers are studying the process of apoptosis in *C. elegans* in the hope of finding treatments for these and other human diseases.

Parasitic Nematodes A large number of nematodes are parasitic in humans and other vertebrates.

- The large roundworm *Ascaris lumbricoides* is a parasite of the small intestine that can reach up to 30 cm in length. Over a billion people worldwide carry this parasite, although infections are most prevalent in tropical or developing countries. Eggs pass out in feces and can remain viable in the soil for years, although they require ingestion before hatching into an infective stage.

- Hookworms (*Necator americanus*), so named because their anterior end curves dorsally like a hook, are also parasites of the human intestine. The eggs pass out in feces, and recently hatched hookworms can penetrate the skin of a host's foot to establish a new infection. In areas with modern plumbing, these diseases are uncommon.

- Pinworms (*Enterobius vermicularis*), although a nuisance, have relatively benign effects on their hosts. The rate of infection in the U.S., however, is staggering: 30% of children and 16% of adults are believed to be hosts. Adult pinworms live in the large intestine and migrate to the anal region at night to lay their eggs, which causes intense itching. The resultant scratching can spread the eggs from the hand to the mouth.

- In the tropics, some 250 million people are infected with *Wuchereria bancrofti*, a fairly large (100 mm) worm that lives in the lymphatic system, blocking the flow of lymph and, in rare cases, causing elephantiasis, an extreme swelling of the legs and other body parts (Figure 26.24). Females release tiny, live young called microfilariae, which are transmitted to new hosts via mosquitoes.

Figure 26.23 Scanning electron micrograph of a nematode within a plant leaf.
©Biophoto Associates/Science Source

Concept Check: *Both nematodes and annelids are wormlike in appearance. How are they different?*

The Phylum Arthropoda Contains the Spiders, Millipedes and Centipedes, Insects, and Crustaceans—Species with Jointed Appendages

The arthropods (from the Greek *arthron*, meaning joint, and *podos*, meaning foot) constitute perhaps the most successful phylum on Earth. About three-quarters of all described living species present on Earth are arthropods, and scientists have estimated they are also numerically common, with an estimated 10^{18} (a billion billion) individual organisms. The huge success of the arthropods, in terms of their sheer numbers and diversity, is related to features that permit these animals to live in all the major habitats on Earth, from the poles to the tropics and from marine and freshwater habitats to dry land. Such features include an exoskeleton, segmentation, and jointed appendages.

Key Features of Arthropods The body of a typical arthropod is covered by a hard cuticle, an **exoskeleton** (external skeleton) made of layers of chitin and protein. The exoskeleton provides protection and a point of attachment for muscles, all of which are internal. The exoskeleton is also relatively impermeable to water, a feature that may have

Figure 26.24 Elephantiasis in a human leg. The disease is caused by the nematode parasite *Wuchereria bancrofti*, which lives in the lymphatic system and blocks the flow of lymph.
©Noah Seelam/Stringer/Getty Images

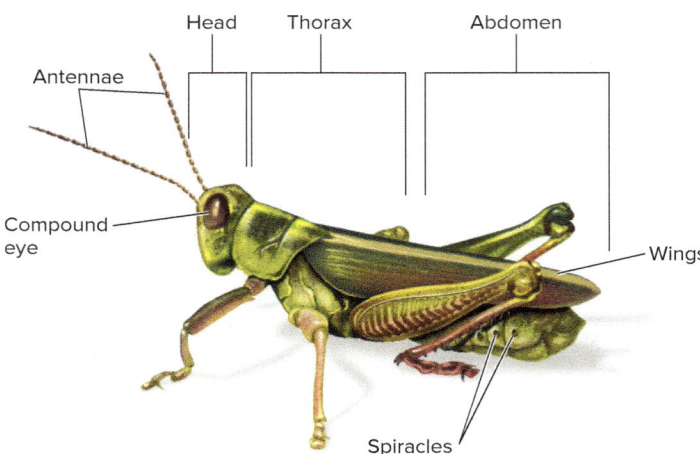

Figure 26.25 Body plan of an arthropod, as represented by a grasshopper.

hemolymph flows back into the heart via pores, called ostia, that are equipped with valves.

Because the cuticle impedes the diffusion of gases through the body surface, arthropods possess special organs that permit gas exchange. Aquatic arthropods have feathery gills, and terrestrial species have a highly developed **tracheal system** (look ahead to Figure 36.17). On the body surface, pores called **spiracles** provide openings to a series of finely branched air tubes within the body called trachea. The tracheal system delivers oxygen directly to tissues and cells.

enabled many arthropods to conserve water and colonize land. From this point of view, the development of a hard cuticle was a critical innovation. It also reminds us that the ability to adapt to diverse environmental conditions can lead to increased organismal diversity.

Arthropods are segmented, and many of the segments bear jointed appendages, which are used for such functions as walking, swimming, sensing, breathing, food handling, and reproduction. In many orders, the body segments have become fused into functional units, or **tagmata**, such as the head, thorax, and abdomen of an insect (**Figure 26.25**). Cephalization is extensive, and arthropods have well-developed sensory organs, including organs of sight, touch, smell, hearing, and balance.

Like most mollusks, arthropods have an open circulatory system (look ahead to Figure 36.1a), in which hemolymph is pumped from a tubelike heart into the aorta or short arteries and then into open sinuses from where the gases and nutrients diffuse into tissues. The

Arthropod Diversity Arthropod classification is complex and ongoing. Although many classifications have been proposed, a 1995 study of the mitochondrial DNA of arthropod species by American geneticist Jeffrey Boore and colleagues suggests a phylogeny with five main subphyla: one now-extinct subphylum, Trilobita (trilobites), and four living subphyla—Chelicerata (spiders and relatives), Myriapoda (millipedes and centipedes), Hexapoda (insects and relatives), and Crustacea (crabs and relatives) (**Table 26.4**). Molecular evidence thus suggests insects are more closely related to crustaceans than they are to spiders or to millipedes and centipedes. A group known as the Pancrustacea contains the insects and crustaceans. The extinct Trilobites were more closely related to the Chelicerates and Myriapods.

Modeling Challenge: Draw a phylogeny showing the relatedness of these six taxa: trilobites, chelicerates, myriapods, hexapods, crustaceans, and pancrustaceans.

Table 26.4 Main Subphyla and Characteristics of Arthropods

Subphyla and examples (est. no. of species)	Class characteristics
Chelicerata: spiders, scorpions, mites, ticks, horseshoe crabs, sea spiders (74,000)	Body usually with cephalothorax and abdomen only; six pairs of appendages, including four pairs of legs, one pair of fangs, and one pair of pedipalps; terrestrial; predatory or parasitic
Myriapoda: millipedes, centipedes (13,000)	Body with head and highly segmented trunk. In millipedes, each segment with two pairs of walking legs; terrestrial; herbivorous. In centipedes, each segment with one pair of walking legs; terrestrial; predatory, poison jaws.
Hexapoda: insects such as beetles, butterflies, flies, fleas, grasshoppers, ants, bees, wasps, termites, springtails (>1 million)	Body with head, thorax, and abdomen; mouthparts modified for biting, chewing, sucking, or lapping; usually with two pairs of wings and three pairs of legs; mostly terrestrial, some freshwater; herbivorous, parasitic, or predatory
Crustacea: crabs, lobsters, shrimp (45,000)	Body of two to three parts; three or more pairs of legs; chewing mouthparts; usually marine

Subphylum Chelicerata: Spiders, Scorpions, and Mites and Ticks Chelicerata consists mainly of the class Arachnida, which contains predatory spiders and scorpions as well as the ticks and mites, some of which are blood-sucking parasites that feed on vertebrates. All species have a body consisting of two tagmata: a fused head and thorax (called a cephalothorax) and an abdomen. All species also possess six pairs of appendages: the chelicerae, or fangs; a pair of pedipalps, which have various sensory, predatory, or reproductive functions; and four pairs of walking legs.

Spider fangs are supplied with venom from poison glands. Most spider bites are harmless to humans, although they are very effective in immobilizing and/or killing their insect prey. Venom from some species, including the black widow (*Latrodectus mactans*) (Figure 26.26a) and the brown recluse (*Loxosceles reclusa*), are potentially, although rarely, fatal to humans. The toxin of the black widow is a neurotoxin, which interferes with the functioning of the nervous system, whereas that of the brown recluse is hemolytic, meaning it destroys red blood cells around the bite. After the spider has subdued its prey, it pumps digestive fluid into the tissues via the fangs and sucks out the partially digested meal.

Spiders have abdominal silk glands, called spinnerets, and many spin webs to catch prey. The silk is a protein that stiffens after extrusion from the body because the mechanical shearing causes a change in the organization of the amino acids. Silk is stronger than steel of the same diameter and more elastic than Kevlar, the material used in bulletproof vests. Each spider family constructs a characteristic size and style of web and can do it perfectly on the first attempt, indicating that web spinning is an innate (instinctual) behavior (see Chapter 42).

Scorpions (order Scorpionida) are generally tropical or subtropical animals that feed primarily on insects, though they may eat spiders and other arthropods as well as smaller reptiles and mice. Their pedipalps are modified into large claws, and their abdomen tapers into a stinger, which is used to inject venom. Although the venom of most North American species is generally not fatal to humans, that of the *Centruroides* genus from deserts in the U.S. Southwest and Mexico can be deadly. Fatal species are also found in India, Africa, and other countries. Unlike spiders, which lay eggs, scorpions bear live young, which the mother then carries around on her back until they have their first molt (Figure 26.26b).

In mites and ticks (order Acari), the two main body segments (cephalothorax and abdomen) are fused and appear as one large segment. Many mite species are free-living scavengers that feed on dead plant or animal material. Other mites are serious pests on crops, and some, like chiggers (*Trombicula alfreddugesi*), are parasites of humans that can spread diseases such as typhus (Figure 26.26c). Chiggers are parasites only in their larval stage. Chiggers do not bore into the skin; their bite and salivary secretions cause skin irritation. *Demodex brevis* is a hair-follicle mite that is common in animals and humans. The mite is estimated to be present on over 90% of adult humans. Although the mite causes no irritation in most humans, *Demodex canis* causes the skin disease known as mange in domestic animals, particularly dogs.

Ticks are larger organisms than mites, and all are ectoparasitic, feeding on the body surface of vertebrates. Their life cycle includes attachment to a host, sucking blood until they are replete and dropping off the host to molt (Figure 26.26d). Ticks can carry a huge variety of viral and bacterial diseases, including Lyme disease, a global bacterial disease so named because it was first found in the town of Lyme, Connecticut, in the 1970s. Infection rates in the U.S. are highest in the Northeast, mid-Atlantic, and upper Midwest regions.

Subphylum Myriapoda: The Millipedes and Centipedes
Myriapods have one pair of antennae on the head and three pairs of appendages that are modified as mouth parts, including mandibles that act as jaws. The millipedes and centipedes, both wormlike arthropods with legs, are among the earliest terrestrial animal phyla known. Millipedes (class Diplopoda) have two pairs of legs per segment, as their class name denotes (from the Latin *diplo*, meaning two, and *podos*, meaning feet)—not 1,000 legs, as their common name suggests (Figure 26.27a). They are slow-moving, herbivorous creatures that eat decaying leaves and other plant material. When threatened, the millipede's response is to roll up into a protective coil. Many millipede species also have glands on their underside that can eject a variety of toxic, repellent secretions. Some millipedes are brightly colored, warning potential predators that they can protect themselves.

27

Vertebrates: Fishes, Amphibians, Reptiles, and Mammals

©Neil Bromhall/Shutterstock

A naked mole rat (*Heterocephalus glaber*). Do naked mole rats defy the normal laws of aging?

Chapter Outline

- 27.1 Vertebrates: Chordates with a Backbone
- 27.2 Cyclostomes: Jawless Fishes
- 27.3 Gnathostomes: Jawed Vertebrates
- 27.4 Tetrapods: Gnathostomes with Four Limbs
- 27.5 Amniotes: Tetrapods with a Desiccation-Resistant Egg
- 27.6 Mammals: Milk-Producing Amniotes

Assessing Your Knowledge and Skills

Naked mole rats (*Heterocephalus glaber*) are special animals. These burrowing rodents, native to arid areas of Africa, live in social groups but, as we will see in Chapter 42, only one female, the queen, mates and produces offspring. These rats' bodies are resistant to pain, and they can survive up to 18 minutes without oxygen.

Recently, researchers have reported that naked mole rats defy the normal laws of aging. The researchers have been studying this species for many years at Calico, a life science company that former Google board member Arthur Levin founded to investigate aging. Here, South African–born physiologist Rochelle Buffenstein has maintained colonies for many years. Based on their size, naked mole rats shouldn't live beyond about 6 years. Mice in captivity live 4 years at most. However, many naked mole rats live beyond 30 years. In 2018, Buffenstein reported that unlike other animals, the risk of dying doesn't seem to increase as naked mole rats grow older. This finding has stimulated more research as scientists strive to unlock the secrets of aging.

What are possible explanations for the longevity of naked mole rats? Their body temperature is a low 32°C, which may help them avoid accumulated molecular damage. Also, they have better DNA damage repair than most other mammals, which renders them virtually cancer-free. They appear to have no problems with their heart or bones as they age. Queens do not go through menopause and continue to reproduce into their thirties. Naked mole rats have a lot to tell us about aging.

The naked mole rat is an example of a **vertebrate** (from the Latin *vertebratus*, meaning joint of the spine), an animal with a backbone. Vertebrates range in size from tiny fishes weighing 0.1 g to huge whales with weights over 100,000 kg. They occupy nearly all of Earth's habitats, from the deepest depths of the oceans to mountaintops and the sky beyond. Throughout history, humans have depended on many vertebrate species for their welfare by domesticating species such as horses, cattle, pigs, sheep, and chickens; using skin and fur for clothes; and keeping countless species, including cats and dogs, as pets. Many vertebrate species are the subjects of conservation efforts, as we will see in Chapter 47.

In Chapter 26, we discussed two chordate subphyla: the cephalochordates (lancelets) and urochordates (tunicates).

The third subphylum of chordates, the Vertebrata, with about 66,000 species, is by far the largest and most dominant group of the phylum. In this chapter, we will explore the characteristics of vertebrates and the evolutionary development of the major vertebrate clades, including fishes, amphibians, reptiles, and mammals.

27.1 Vertebrates: Chordates with a Backbone

Learning Outcome
1. List the main distinguishing characteristics of vertebrates.

Our current understanding of the relationships between vertebrate groups is shown in **Figure 27.1**. Nested within the vertebrates are various clades based on morphological characteristics. The vertebrates retain all the chordate characteristics outlined in Chapter 26, as well as possessing several additional traits, including the following:

1. **Vertebral column.** In vertebrates, the notochord is replaced by a bony or cartilaginous column of interlocking **vertebrae** that provides support and protects the nerve cord, which lies within its tubelike structure.
2. **Cranium.** The anterior end of the nerve cord is expanded to form a more developed brain encased in a protective bony or cartilaginous housing called the **cranium**.
3. **Endoskeleton of cartilage or bone.** The cranium and vertebral column are parts of the endoskeleton, the living skeleton of vertebrates that forms within the animal's body. The endoskeleton contains living cells that secrete the skeleton, which grows with the animal, unlike the nonliving exoskeleton of arthropods.

Although these are the main distinguishing characteristics of vertebrates, there are others. For example, vertebrates have multiple clusters of *Hox* genes, compared with the single cluster of *Hox* genes in tunicates and lancelets. These additional gene clusters are believed to have permitted increasingly complex morphologies beyond those possessed by invertebrate chordates. Vertebrates also possess a great diversity of internal organs, including a liver, kidneys, endocrine glands, and a heart with at least two chambers. The liver is unique to vertebrates, and the vertebrate kidneys, endocrine system, and heart are more complex than are analogous structures in invertebrate taxa.

Although these features are exhibited in all vertebrate classes, some classes evolved critical innovations that helped them to succeed in specific environments, such as on land or in the air. For example, birds developed feathers and wings, structures that enable most species to fly. Each of the vertebrate classes is distinctly different from the others, as outlined in **Table 27.1**. One of the earliest innovations was the development of jaws. All vertebrates except some early-diverging fishes possessed jaws. Today, the only jawless vertebrates are hagfish and lampreys, which are described in Section 27.2.

27.1 Reviewing the Concepts
- Vertebrates have several characteristic features, including a vertebral column, a cranium, an endoskeleton of cartilage or bone, multiple clusters of *Hox* genes, and complex internal organs (Figure 27.1, Table 27.1).

27.1 Testing Your Knowledge
1. Which of the following is *not* a defining characteristic of vertebrates?
   a. great diversity of organs
   b. hinged jaw
   c. multiple clusters of *Hox* genes
   d. vertebral column
   e. endoskeleton

27.2 Cyclostomes: Jawless Fishes

Learning Outcome
1. Describe the two classes of existing jawless vertebrates.

Although many species of jawless fishes are known from the fossil record, most had become extinct by the end of the Devonian period. Today, the only jawless vertebrates are hagfish and lampreys, together known as the Cyclostomata. Cyclostomes, or "circle mouths," are eel-like animals that do not possess jaws. Sequencing of RNA libraries in 2010, together with genomic surveys, yielded strong support that Cyclostomata is monophyletic. In addition, hagfish and lampreys share a very distinct type of immune system. In this section, we will examine the characteristics of hagfish and lampreys.

Ancestral vertebrate

The Hagfish Are the Simplest Living Cyclostomes

The hagfish are entirely marine cyclostomes that lack eyes, jaws, fins, and even vertebrae

Biology Principle

Biologists Use Models to Test Predictions and Convey Ideas.
A phylogenetic tree, or cladogram, is a model of how life on Earth evolved among a given group of species.

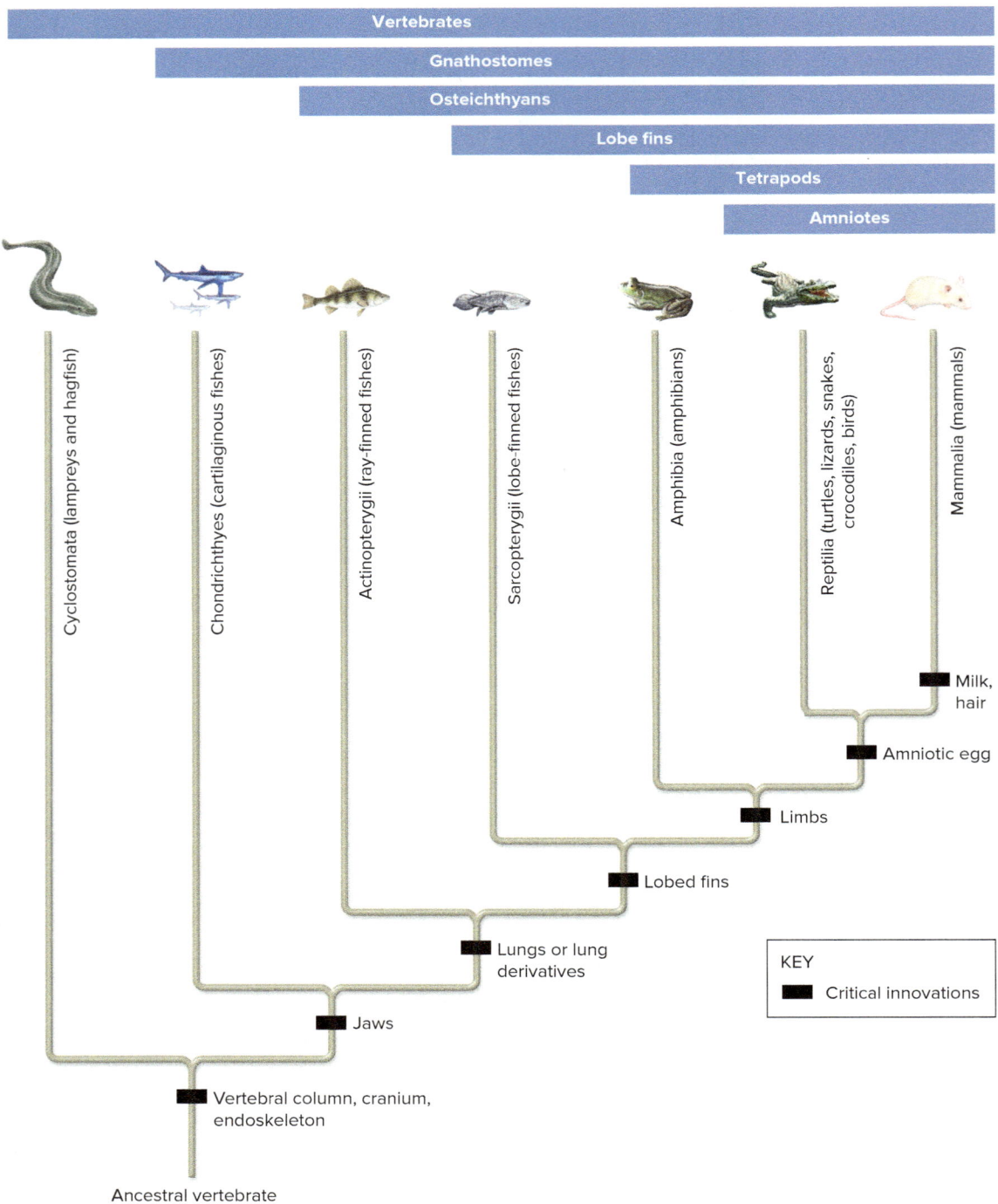

Figure 27.1 The major clades of vertebrates. Critical innovations of each group are shown. The nested labels across the top of the figure refer to groupings within the vertebrates.

Table 27.1 The Main Clades and Characteristics of Living Vertebrates

Clade	Examples (approx. no. of species)	Main characteristics
Cyclostomata	Lampreys, hagfish (100)	Jawless fishes; no appendages
Chondrichthyes	Sharks, skates, rays (970)	Fishes with cartilaginous skeleton; teeth not fused to jaw; no swim bladder; well-developed fins; internal fertilization; single blood circulation
Actinopterygii	Ray-finned fishes, most bony fish (31,800)	Fishes with ossified skeleton; gill opening covered by operculum; fins supported by rays, fin muscles within body; swim bladder often present; mucous glands in skin
Sarcopterygii	Lobe-finned fishes, of which coelacanths (2) and lungfishes (6) are the only living members	Fishes with ossified skeleton; bony extensions, together with muscles, project into pectoral and pelvic fins
Amphibia	Frogs, toads, salamanders (7,600)	Adults able to live on land; fresh water needed for reproduction; development usually involving metamorphosis from tadpoles; adults with lungs and double blood circulation; moist skin; shell-less eggs
Testudines	Turtles (346)	Body encased in hard shell; no teeth; head and neck retractable into shell; eggs laid on land
Squamata	Lizards, snakes (9,900)	Lower jaw not attached to skull; skin covered in scales
Crocodilia	Crocodiles, alligators (25)	Four-chambered heart; large, aquatic predators; parental care of young
Aves	Birds (10,425)	Feathers; hollow bones; air sacs; reduced internal organs; endothermic; four-chambered heart
Mammalia	Mammals (5,500)	Mammary glands; hair; specialized teeth; enlarged skull; external ears; endothermic; four-chambered heart; highly developed brains; diversity of body forms

(Figure 27.2). The hagfish skeleton consists largely of a notochord and a cartilaginous skull that encloses the brain. The lack of a vertebral column leads to extensive flexibility. So how can hagfish be vertebrates without a vertebral column? The strong molecular support for a cyclostome clade suggests hagfish anatomy has degenerated to a remarkable degree and only the cranium and diversity of organs provide evidence of vertebrate ancestry.

Hagfish live in the cold waters of northern oceans, close to the muddy bottom, feeding on marine worms and other invertebrates. Essentially blind, hagfish have a very keen sense of smell and are attracted to dead and dying fish, to which they attach themselves via toothed plates on the mouth. The powerful tongue then rasps off pieces of tissue. Though the hagfish cannot see approaching predators, it has special glands that produce copious amounts of slime. When provoked, the hagfish's slime production increases dramatically, enough to potentially distract predators or coat their gills and interfere with breathing. Hagfish can sneeze to clear their nostrils of their own slime.

Figure 27.2 A hagfish, *Myxine glutinosa*.
©Pat Morris/ardea.com

The Lampreys Are Eel-like Animals That Lack Jaws

Lampreys are similar to hagfish in that they lack both a hinged jaw and true appendages. However, lampreys do possess a notochord surrounded by a cartilaginous rod, representing a rudimentary vertebral column. Lampreys are found in both marine and freshwater environments. Marine lampreys are parasitic as adults. They grasp other fish with their circular mouth (Figure 27.3a) and rasp a hole in the fish's side, sucking out blood, tissue, and fluids (Figure 27.3b). Males and females of all species spawn in freshwater streams, and the resultant larval lampreys burrow into the sand or mud, much like lancelets (refer back to Figure 26.34a), emerging to feed on small invertebrates or detritus at night. This stage can last for 3 to 7 years, at the end of which time the larvae metamorphose into adults. In most freshwater species, the adults do not feed at all but quickly mate and die. Marine species migrate from fresh water back to the ocean, until they return to fresh water to spawn and then die.

27.2 Reviewing the Concepts

- Early-diverging vertebrates lacked jaws. Today, the only jawless vertebrates are the hagfish and lampreys (Figures 27.2, 27.3).

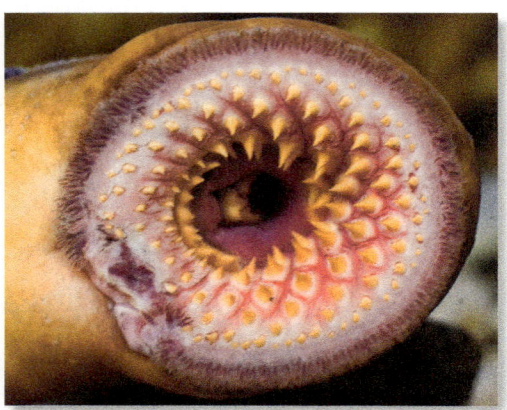
(a) Jawless mouth of a sea lamprey

(b) A sea lamprey feeding

Figure 27.3 The lamprey, a modern jawless fish. (a) The sea lamprey (*Petromyzon marinus*) has a circular, jawless mouth. (b) A sea lamprey feeding on a fish.
(a) ©R. Duran/Getty Images; (b) ©Jacana/Science Source

27.2 Testing Your Knowledge

1. Which of the following organisms are considered cyclostomes?
   a. tunicates
   b. hagfish
   c. lampreys
   d. both a and b
   e. both b and c

27.3 Gnathostomes: Jawed Vertebrates

Learning Outcomes

1. Describe how jaws evolved.
2. Discuss the distinguishing features of sharks.
3. List the three features that distinguish bony fishes from cartilaginous fishes.
4. Outline the differences between the ray-finned fish and the lobe-finned fish.

All vertebrate species that possess jaws are called **gnathostomes** (from the Greek, meaning jaw mouth) (see Figure 27.1). Gnathostomes are a diverse clade of vertebrates that includes fishes, amphibians, reptiles, and mammals. The earliest-diverging gnathostomes were fishes. Biologists have identified about 32,800 species of living fishes with jaws, more than all other species of vertebrates combined. Most are aquatic, gill-breathing species that usually possess fins and a scaly skin. The three separate clades of jawed fishes are the Chondrichthyes (cartilaginous fishes), Actinopterygii (ray-finned fishes), and Sarcopterygii (lobe-finned fishes). Each has distinguishing characteristics (Table 27.1).

A Hinged Jaw Was a Critical Innovation That Aided in Feeding

A hinged jaw was a significant evolutionary development. It enabled an animal to grip its prey more firmly, which may have increased its rate of capture, and to attack larger prey species, thus increasing its potential food supply. Accompanying the jawed mouth was the development of more sophisticated head and body structures, including appendages called fins. Gnathostomes also possess two additional *Hox* gene clusters beyond those in the cyclostomes (bringing their total to four or more), which likely led to increased morphological complexity.

The hinged jaw developed from the gill arches, cartilaginous or bony rods that help to support gills. Similarities between cells that make up jaws and gill arches support this view. Primitive jawless fishes had nine gill arches surrounding the eight gill slits (Figure 27.4a). During the late Silurian period (about 417 mya), some of these gill arches became modified. The first and second gill arches were lost, and the third and fourth pairs evolved to form the jaws (Figure 27.4b, c). This is how evolution typically works; body features do not appear de novo; instead, existing features become modified to serve other functions.

Biology Principle

Structure Determines Function

The development of a jaw increased the predatory capabilities of gnathostomes.

(a) Primitive jawless fishes

(b) Early jawed fishes (placoderms)

(c) Modern jawed fishes (cartilaginous and bony fishes)

Figure 27.4 The evolution of the vertebrate jaw. (a) Primitive fishes and modern jawless fishes, such as lampreys, have nine cartilaginous gill arches that support eight gill slits. (b) In early jawed fishes such as the placoderms, the first two pairs of gill arches were lost, and the third pair became modified to form a hinged jaw. This left six gill arches (4–9) to support the remaining five gill slits, which were still used in breathing. (c) In modern jawed fishes, the fourth gill arch also contributes to jaw support, allowing stronger, more powerful bites to be delivered.

Chondrichthyans Are Fishes with Cartilaginous Skeletons

Members of the class Chondrichthyes (the **chondrichthyans**)—sharks, skates, and rays—are called cartilaginous fishes because their skeleton is composed of flexible cartilage rather than bone. The cartilaginous skeleton is considered a derived character, meaning that the ancestors of the chondrichthyans had bony skeletons but members of

this class subsequently lost this feature. This hypothesis is reinforced by the observation that during development, the skeleton of most vertebrates is cartilaginous and then it becomes bony (ossified) as a hard calcium-phosphate matrix replaces the softer cartilage. A change in the developmental sequence of the cartilaginous fishes is believed to prevent the ossification process.

Key Features of Chondrichthyans

All chondrichthyans are denser than water, which means that they will sink if they stop swimming. Many sharks never stop swimming and maintain buoyancy via the use of their fins and a large, oil-filled liver. Perhaps the most important fin for propulsion in sharks is the large and powerful caudal fin, or tail fin, which, when swept from side to side, thrusts the fish forward at great speed (**Figure 27.5a**). The paired pelvic fins (at the back) and pectoral fins (at the front) act like flaps on airplane wings, allowing the shark to dive deeper or rise to the surface. They also aid in steering. In addition, the dorsal fin (on the shark's back) acts as a stabilizer to prevent the shark from rolling in the water as the tail fin pushes it forward. During swimming, water continually enters the shark's mouth and is forced over the gills, allowing the shark to extract oxygen and breathe.

Skates and rays are essentially flattened sharks that cruise along by using hugely expanded pectoral fins but may also rest for long periods on the ocean floor (**Figure 27.5b**). How, then, do skates and rays breathe when they rest on the ocean floor? These species, and a few sharks such as the nurse shark, use a muscular pharynx and jaw muscles to pump water over the gills.

Sharks were among the earliest fishes to develop teeth, which evolved from rough scales on the skin. Although sharks' teeth are very sharp and hard, they are not set into the jaw, as are human teeth. Therefore, they break off easily and are continually replaced, row by row (**Figure 27.5c**). In chondrichthyans and all species of fishes, the heart consists of two chambers, an atrium and a ventricle, that contract in sequence. All fishes employ what is known as a single circulation, in which blood is pumped from the heart to capillaries in the gills to collect oxygen, and then it flows through arteries to the tissues of the body before returning to the heart (look ahead to Figure 36.2a).

Chondrichthyan Senses

Sharks have a powerful sense of smell, facilitated by sense organs in the nostrils (sharks and other fishes do not use nostrils for breathing). They can see well but cannot distinguish colors. All jawed fishes have a row of microscopic organs in the skin, arranged in a line that runs laterally down each side of the body, which can sense movements in the surrounding water.

Figure 27.5 Cartilaginous fishes. (a) The silvertip shark (*Carcharhinus albimarginatus*) is one of the ocean's most powerful predators. (b) Stingrays are essentially flattened sharks with very large pectoral fins. (c) Close-up of the mouth of a sand tiger shark (*Carcharias taurus*), showing rows of teeth. (d) This mermaid's purse (egg pouch) of a dogfish shark (*Scyliorhinus canicula*) is entwined in vegetation to keep it stationary.
(a) ©Valerie & Ron Taylor/ardea.com; (b) ©Bill Curtsinger/National Geographic/Getty Images; (c) ©Jeff Rotman/naturepl.com; (d) ©Oxford Scientific/Getty Images

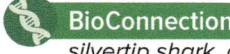**BioConnections:** Look ahead to Figure 36.16. Gills, like those seen along the side of the silvertip shark, are organs with a large surface area. Though this makes gills suitable for gas exchange between blood and water, other materials may move across gills, too. What problems do you think this would cause?

This system of sense organs, known as the **lateral line**, senses pressure waves and sends nerve signals to the brain.

Reproduction in Chondrichthyans

Fertilization is internal in chondrichthyans, with the male transferring sperm to the female via a pair of claspers, modifications of the pelvic fins. Some shark species are **oviparous**; that is, they lay eggs, often inside a protective pouch called a mermaid's purse (Figure 27.5d). In **ovoviparous** species, the eggs are retained within the female's body, but there is no placenta to nourish the young. A few species are **viviparous**; that is, the eggs develop within the uterus, receiving nourishment from the mother via a placenta. Both ovoviparous and viviparous sharks give birth to live young.

Sharks never stop growing, so the largest individuals are the oldest. Radiocarbon dating of the eye lenses of large Greenland sharks, *Somniosus microcephalus*, revealed the age at sexual maturity to be at least 156 years and showed an average life span of 272 years; the largest animal, at 502 cm long, was determined to be 392 ± 120 years old, making this species the longest-lived vertebrate known.

The Earliest-Diverging Osteichthyans Are Fishes with Bony Skeletons

Unlike the cartilaginous fishes, all other gnathostomes have a bony skeleton and belong to the clade known as Osteichthyes. This term means bony fish and was originally proposed for just that group. With the advent of modern phylogenetic systematics, however, the term **osteichthyans** has expanded to include all vertebrates with a bony skeleton, including tetrapods (see Figure 27.1). Bony fishes are the most numerous of all types of fishes, with more species (about 31,830) than any other. Most biologists now recognize two living clades, the Actinopterygii (ray-finned fishes) and the Sarcopterygii (lobe-finned fishes).

Key Features of Bony Fishes

The skin of bony fishes, unlike the rough skin of sharks, is covered by a thin epidermal layer containing glands that produce mucus, an adaptation that reduces drag during swimming. Just as in the cartilaginous fishes, water is drawn over the gills for breathing, but in bony fishes, a protective flap called an **operculum** covers the gills (look ahead to Figure 36.16). Muscle contractions around the gills and operculum draw water through the gills so that bony fishes do not need to swim continuously in order to breathe.

Some early bony fishes lived in shallow, oxygen-poor waters and developed lungs as an embryological offshoot of the pharynx. These fish could rise to the water surface and gulp air. As discussed later, modern lungfishes can breathe in this manner. In most bony fishes, the lungs evolved into a **swim bladder,** a gas-filled, balloon-like structure that helps the fish to remain buoyant in the water even when it is completely stationary. Some fishes, known as bimodal breathers, can breathe through their gills and by gulping air, absorbing oxygen through their digestive tracts. For example, Siamese fighting fish (*Betta splendens*, known as betta), a popular freshwater aquarium fish, is a bimodal breather that is relatively easy to care for, since it can survive without an air pump in its aquarium.

Bony fishes have colonized nearly all aquatic habitats. Following the cooling of the newly formed planet Earth, water condensed into rain and over a vast period of time filled what are now the oceans. Later, as water evaporated from the oceans and sodium, potassium, and calcium were added via runoff from the land, the oceans became salty. Therefore, most fishes probably evolved in freshwater habitats and secondarily became adapted to marine environments. This, of course, required the development of physiological adaptations to the different osmotic problems seawater presents compared with fresh water.

Reproduction in Bony Fishes Reproductive strategies of bony fishes vary tremendously, but most species reproduce via external fertilization, with the female shedding her eggs and the male depositing sperm on top of them. Although adult bony fishes can maintain their buoyancy, their eggs tend to sink. This helps to explain why many species spawn in shallow, more oxygen- and food-rich waters and why coastal areas are important fish nurseries.

Actinopterygii, the Ray-Finned Fish The most species-rich clade of bony fishes is the Actinopterygii, or **ray-finned fishes.** In Actinopterygii, the fins are supported by thin, bony, flexible rays and are moved by muscles on the interior of the body (**Figure 27.6**). The class has a diversity of forms, from large, predatory moray eels to delicate sea dragons. Whole fisheries are built around the harvest of such ray-finned species as cod, anchovies, and salmon.

Sarcopterygii, the Lobe-Finned Fish In the Sarcopterygii, or **lobe-finned fishes,** the fins are supported by skeletal extensions of the pectoral and pelvic areas that are moved by muscles within the fins. The Actinistia (coelacanths) and Dipnoi (lungfishes) are both considered Sarcopterygii. The name lobe fins used to refer solely to the lobe-finned fishes, but because it has become clear that terrestrial vertebrates (tetrapods) evolved from such fishes, the definition of the group has been expanded to include both lobe-finned fishes and tetrapods (see Figure 27.1).

The Actinistia, or coelacanths, were a very successful group in the Devonian period, but all fishes of the group were once believed to have died off at the end of the Mesozoic era (some 65 mya). You can, therefore, imagine the scientific excitement when in 1938 a modern coelacanth was discovered as part of the catch of a boat fishing off the coast of South Africa (**Figure 27.7a**). Another species was found more recently in Indonesian waters.

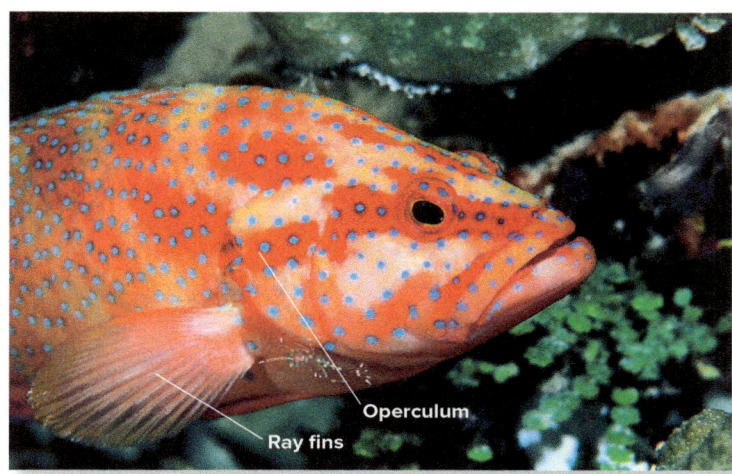

Figure 27.6 An actinopterygian fish (ray-finned fish). Black tip grouper (*Epinephelus fasciatus*), Maldives, Indian Ocean.
©Ian Cartwright/Getty Images

Concept Check: *What features distinguish ray-finned fishes from sharks?*

The Dipnoi, or **lungfishes,** like the coelacanths, are also not currently a very species-rich group, having just three genera and six species (**Figure 27.7b**). Lungfishes live in oxygen-poor freshwater swamps and ponds. They have both gills and lungs, the latter of which enable them to come to the surface and gulp air. In fact, lungfish will drown if they are unable to breathe air. Because they also have muscular lobe fins, they are often able to traverse long distances across land if their ponds dry out.

The morphological features of coelacanths, lungfishes, and primitive terrestrial vertebrates, together with the similarity of their nuclear genes, suggest to many scientists that lobe-finned ancestors gave rise to three lineages: the coelacanths, the lungfishes, and the tetrapods. In the next section, we will examine the characteristics of tetrapods in more detail.

27.3 Reviewing the Concepts

- A critical innovation in vertebrate evolution is the hinged jaw, which first developed in fishes. Gnathostomes are vertebrate species that possess a hinged jaw (Figure 27.4).
- The chondrichthyans (sharks, skates, and rays) have a skeleton composed of flexible cartilage and powerful appendages called fins. They are active predators with acute senses and were among the earliest fishes to develop teeth (Figure 27.5).
- Bony fishes consist of the Actinopterygii (ray-finned fishes, the most species-rich clade) and the lobe-finned fishes, which include the Actinistia (coelacanths) and the Dipnoi (lungfishes). In Actinopterygii, the fins are supported by thin, flexible rays and moved by muscles inside the body (Figure 27.6).

(a) A coelacanth

(b) Australian lungfish

Figure 27.7 The Sarcopterygii (lobe-finned fish). (a) An actinisian, the coelacanth (*Latimeria chalumnae*). (b) A Dipnoi, the Australian lungfish (*Neoceratodus forsteri*).
(a) ©Peter Scoones/Science Source; (b) ©Oceans-Image/Photoshot/Avalon

 Concept Check: *How are lungfishes similar to coelacanths?*

- The lobe fins comprise the lobe-finned fishes (Actinistia and Dipnoi) and the tetrapods. In the lobe-finned fishes, the fins are supported by extensions of the pectoral and pelvic areas and are moved by their own muscles (Figure 27.7).

27.3 Testing Your Knowledge

1. In modern jawed fishes, which gill arches have been modified to form the jaws?
   a. 1 and 2
   b. 2 and 3
   c. 3 and 4
   d. 4 and 5
   e. 1 and 9

2. Which of the following is *not* a morphological feature of sharks?
   a. paired pelvic fins
   b. paired pectoral fins
   c. a cartilaginous skeleton
   d. a swim bladder
   e. multiple rows of teeth

3. Which type of fish is a lobe fin?
   a. moray eel
   b. sea dragon
   c. lamprey
   d. stingray
   e. lungfish

27.4 Tetrapods: Gnathostomes with Four Limbs

Learning Outcomes

1. Describe adaptations that were required for the transition to life on land.
2. List the different amphibian orders, and describe the features that differentiate them.
3. **SCISKILLS** ▶ Explain how relatively simple experiments with mice showed that mutations in just two genes can cause large changes in limb development.

During the Devonian period (from about 417 to 354 mya) (refer back to Figure 22.4), a diversity of plants and animals colonized the land. The plants served as both a source of oxygen and a potential food source for animals that ventured out of the aquatic environment. Terrestrial arthropods and vertebrates appeared during the Devonian period.

The transition to life on land involved a large number of adaptations. Paramount among these were adaptations that prevented desiccation, or drying out, and made locomotion and reproduction on land possible. In this section, we will begin by outlining the development of early **tetrapods**, vertebrate animals having four legs or leglike appendages, and the characteristics of their immediate descendants, the amphibians.

The Origin of Tetrapods Involved the Development of Four Limbs

The fossil record of the Devonian period revealed the evolution of sturdy lobe-finned fishes to fishes with four limbs. The abundance of light and nutrients in shallow waters encouraged a profusion of plant life and the invertebrates that fed on them. The development of lungs enabled lungfishes to colonize these productive yet often oxygen-poor waters. Here, the ability to move in shallow water clogged with plants and debris was more vital than the ability to swim swiftly through open water and may have favored the progressive development of sturdy limbs. As an animal's weight began to be borne more by the limbs, the vertical column strengthened, and hip bones and shoulder bones were braced against the backbone for added strength. Such modifications are the result of changes in the expression of genes, especially *Hox* genes, as described in the Feature Investigation.

FEATURE INVESTIGATION

Davis and Colleagues Provided a Genetic-Developmental Explanation for Limb Length in Tetrapods

The development of limbs in tetrapods was a vital step that allowed animals to colonize land. The diversity of vertebrate limb types is amazing, from fins in fish and marine mammals to different wing types in bats and birds to legs and arms in primates. Early in vertebrate evolution, an ancestral gene complex was duplicated twice to give rise to four groups of genes, called *HoxA, B, C,* and *D,* which control limb development. In 1995, American molecular biologist Mario Capecchi, a co-winner of the 2007 Nobel Prize in Physiology and Medicine, and colleagues analyzed the effects of mutations in specific *Hox* genes that are responsible for determining limb formation in mice. The vertebrate forelimb is divided into three zones: humerus (upper arm); radius and ulna (forearm); and carpals, metacarpals, and phalanges (digits). The researchers had no specific hypothesis in mind; their goal was to understand the role of *Hox* genes in limb formation. As described in **Figure 27.8**, they began with strains of mice carrying loss-of-function mutations in *HoxA-11* or *HoxD-11* that, on their own, did not cause dramatic changes in limb formation. They bred the mice and obtained offspring carrying

Figure 27.8 Relatively simple changes in *Hox* genes control limb formation in tetrapods.

GOAL To determine the role of *Hox* genes in limb development in mice.

KEY MATERIALS Mice with individual mutations in *HoxA-11* and *HoxD-11* genes.

	Experimental level	Conceptual level
1	Breed mice with individual mutations in *HoxA-11* and *HoxD-11* genes. (The *A* and *D* refer to wild-type alleles; *a* and *d* are mutant alleles.) The mice bred were heterozygous for both genes (*AaDd*).	Based on previous studies, researchers expect mutant mice to produce viable offspring, perhaps with altered limb morphologies.
2	Using molecular techniques described in Chapter 18, obtain DNA from the tail and determine the genotypes of offspring.	The resulting genotypes occur in Mendelian ratios, generating mice with different combinations of wild-type and mutant alleles.
3	Stain the skeletons and compare the limb characteristics of the wild-type mice (*AADD*) to those of strains carrying mutant alleles in one or both genes.	Mutant mice may have altered bone morphologies.

4 THE DATA

Genotype	Carpal bone fusions (% of mice showing the fusion)			
	Normal (none fused)	NL fused to T	T fused to P	NL fused to T and P
AADD	100	0	0	0
AaDD	100	0	0	0
aaDD	33	17	50	0
AADd	100	0	0	0
AAdd	0	17	17	67
AaDd	17	17	33	33

5 CONCLUSION
Relatively simple mutations involving two genes can cause large changes in limb development.

6 SOURCE
Davis, A.P., et al. 1995. Absence of Radius and Ulna in Mice Lacking *HoxA-11* and *HoxD-11*. *Nature* 375: 791–795.

one, two, three, or four loss-of-function mutations. Double mutants (*aadd*) exhibited dramatically different phenotypes not seen in mice homozygous for the individual mutations. The radius and ulna were almost entirely eliminated.

As seen in the data, the mutations affected the formation of limbs. For example, the wrist contains seven bones: three proximal carpals—called navicular lunate (NL), triangular (T), and pisiform (P)—and four distal carpals (d1–d4). In mice with the genotypes *aaDD* and *AAdd*, the proximal carpal bones are usually fused together. Individuals having one recessive allele (*AADd* and *AaDD*) do not show this defect, but individuals having two recessive alleles (*AaDd*) often do. Therefore, any two mutant alleles (either from both *HoxA-11* and *HoxD-11* or one from each locus) cause carpal fusions. Deformities became even more severe with three mutant alleles (*Aadd* or *aaDd*) or four mutant alleles (*aadd*) (data not shown in the figure). Thus, scientists have shown that relatively simple mutations can control relatively large changes in limb development.

Experimental Questions

1. What was the purpose of the study conducted by Davis and colleagues?
2. How were the researchers able to study the effects of individual genes?
3. **SCISKILLS** ▶ Summarize the results of the experiment and how they relate to limb development in vertebrates.

Amphibian Limbs and Lungs Are Adaptations to a Semiterrestrial Lifestyle

Amphibians (from the Greek, *amphibios*, meaning both ways of life) live in two worlds: They have successfully invaded the land, but most must return to the water to reproduce. By the middle of the Carboniferous period (about 320 mya), species similar to modern amphibians had become common in the terrestrial environment. With a bonanza of terrestrial arthropods to feast on, the amphibians became numerous and species-rich, and often large in body size. The mid-Permian period (some 260 mya) is sometimes known as the Age of Amphibians. However, most of the large amphibians became extinct at the end of the Permian period. This was the largest known mass extinction in Earth's history, with the extinction of 90–95% of marine species and a large proportion of terrestrial species. Most surviving amphibians were smaller organisms resembling modern species.

Ancestral vertebrate

Key Features of Amphibians One of the first challenges terrestrial animals had to overcome was breathing air when on land. Amphibians breathe in several different ways. They can use the same technique as lungfishes, opening their mouths to let in air. Alternatively, they can take in air through their nostrils. They then close their nostrils and close and raise the floor of the mouth, creating a positive pressure that pumps air into the lungs. This method of breathing is called **buccal pumping.** In addition, the skin of amphibians is much thinner than that of fishes, and amphibians absorb oxygen from the air directly through their outer moist skin or through the skin lining of the inside of the mouth.

Because the skin of amphibians is so thin, the animals face the problem of desiccation on land. As a consequence, even

(a) Gelatinous mass of amphibian eggs

(b) Tadpole

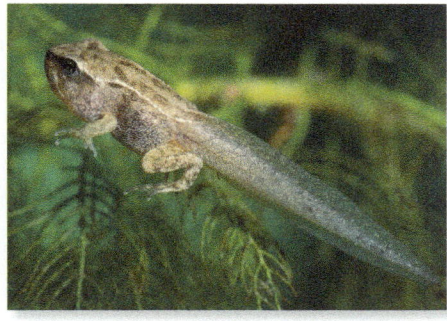
(c) Tadpole undergoing metamorphosis

Figure 27.9 Amphibian development in the European common frog (*Rana temporaria*). **(a)** Amphibian eggs are laid in gelatinous masses in water. **(b)** The eggs develop into tadpoles, aquatic herbivores with a fishlike tail that breathe through gills. **(c)** During metamorphosis, the tadpole loses its gills and tail and develops limbs and lungs.
(*a*) ©Don Vail/Alamy; (*b*) ©Nature Picture Library/Alamy; (*c*) ©NHPA/Photoshot/Avalon

 Concept Check: *Do all amphibians produce tadpoles?*

amphibian adults are more abundant in damp habitats, such as swamps or rain forests, than in dry areas. Also, most amphibians cannot venture too far from water because their larval stages are still aquatic.

Reproduction in Amphibians In frogs and toads, fertilization is generally external, with males shedding sperm over the gelatinous egg masses laid by the females in water (**Figure 27.9a**). The fertilized eggs lack a shell and would quickly dry out if exposed to the air for long. They soon hatch into tadpoles (**Figure 27.9b**)—small, fishlike animals that lack limbs and breathe through gills. As the tadpole nears the adult stage, the tail and gills are resorbed, and limbs and lungs appear (**Figure 27.9c**). Such a dramatic change from a juvenile to an adult body form is known as **metamorphosis.** A few species of amphibians do not require water to reproduce. These species are ovoviparous or viviparous—retaining the eggs in the reproductive tract and giving birth to live young.

Modern Amphibians Include a Variety of Frogs, Toads, Salamanders, and Caecilians

Approximately 7,600 living amphibian species are known, and the vast majority of these, some 6,700 species, are frogs and toads of the order Anura (from the Greek, meaning tail-less ones) (**Figure 27.10a**). The other two orders are the Apoda (from the Greek, meaning legless ones), the wormlike caecilians; and the Urodela (from the Latin, meaning tailed ones), the salamanders.

Adult anurans are carnivores, eating a variety of invertebrates by catching them on a long, sticky tongue. In contrast, their aquatic larvae (tadpoles) are primarily herbivores. In addition to secreting mucus, which keeps their skin moist, some frogs can also secrete poisonous chemicals that deter would-be predators. Some amphibians advertise the poisonous nature of their skin with warning coloration. Global warming is currently threatening many anurans with extinction.

Caecilians are a small order of about 170 species of legless, nearly blind amphibians (**Figure 27.10b**). Most are tropical and burrow in forest soils, but a few live in ponds and streams.

(a) Tree frog

(b) A caecilian

(c) Mud salamander

Figure 27.10 Amphibians. **(a)** Most amphibians are frogs and toads of the order Anura, including this red-eyed tree frog (*Agalychnis callidryas*). **(b)** The order Apoda includes wormlike caecilians such as this species from Ecuador, *Siphonops annulatus*. **(c)** The order Urodela includes species such as this mud salamander (*Pseudotriton montanus*).
(*a*) ©Gregory G. Dimijian/Science Source; (*b*) ©Danita Delimont/Alamy; (*c*) ©Gary Meszaros/Science Source

BioConnections: *Why are some frogs, like the red-eyed tree frog, brightly colored rather than camouflaged?*

The more than 550 species of salamanders possess legs and a long tail and have a more elongate body than anurans (**Figure 27.10c**). Like frogs, many salamanders have colorful skin patterns that advertise their distastefulness to predators. A few species, such as Cope's giant salamander (*Dicamptodon copei*), retain the gills and tail fins characteristic of the larval stage into adulthood, and they mature sexually in the larval stage, a phenomenon known as paedomorphosis.

27.4 Reviewing the Concepts

- Fossils record the evolution of lobe-finned fishes to fishes with four limbs. Research has shown that relatively simple mutations can control large changes in limb development (Figure 27.8).
- Amphibians live on land but return to the water to reproduce. In frogs and toads, the larval stage undergoes metamorphosis, losing gills and tail and gaining lungs and limbs (Figure 27.9).
- The majority of amphibians belong to the order Anura (frogs and toads). Other orders are the Apoda (caecilians) and Urodela (salamanders) (Figure 27.10).

27.4 Testing Your Knowledge

1. The transition of tetrapods from water to life on land required
   a. the prevention of desiccation.
   b. internal fertilization.
   c. the ability to breathe air.
   d. both a and c.
   e. all of the above.
2. Which of the following amphibian orders have legs?
   a. salamanders
   b. frogs and toads
   c. caecilians
   d. both a and b
   e. all amphibian orders

27.5 Amniotes: Tetrapods with a Desiccation-Resistant Egg

Learning Outcomes

1. Diagram the structure of the amniotic egg.
2. Identify the critical innovations of the amniotes.
3. Describe the distinguishing features of the major amniote clades.
4. List the features that enable birds to fly.

Although amphibians live successfully in a terrestrial environment, they must lay their eggs in water or in a very moist place so that the shell-less eggs do not dry out on exposure to air. Thus, a critical innovation in animal evolution was the development of a shelled egg that sheltered the embryo from desiccation on land. A shelled egg containing fluids was like a personal, enclosed pond for each developing individual. Such an egg evolved in the common ancestor of turtles, lizards, snakes, crocodiles, birds, and mammals—a group of tetrapods collectively known as the **amniotes.** The amniotic egg permitted animals to lay their eggs in a dry place so that reproduction was no longer tied to water. It was truly a critical innovation, untethering animals from water in much the same way as the development of seeds liberated plants from water (see Chapter 25).

Over time, the amniotes became very diverse in species and morphology. Mammals are considered amniotes, too, because even though most of them do not lay eggs, they retain other features of amniotic reproduction. In this section, we will begin by discussing the structure of the amniotic egg and other adaptations that permitted animal species to become fully terrestrial. We will then examine the biology of the reptiles, the first group of vertebrates to fully exploit land.

The Amniotic Egg and Other Innovations Permitted Life on Land

The **amniotic egg** (**Figure 27.11**) contains the developing embryo and the four separate extraembryonic membranes that it produces:

1. The innermost membrane is the **amnion,** which protects the developing embryo in a fluid-filled sac called the amniotic cavity.
2. The **yolk sac** encloses a stockpile of nutrients, in the form of yolk, for the developing embryo.
3. The **allantois** functions as a disposal sac for metabolic wastes.
4. The **chorion,** along with the allantois, provides gas exchange between the embryo and the surrounding air.

Surrounding the chorion is the albumin, or egg white, which also stores nutrients. The **shell** provides a tough, protective covering that is not very permeable to water and prevents the embryo from drying out. However, the shell remains permeable to oxygen and carbon dioxide so that the embryo can breathe. In birds, this shell is hard and calcareous, whereas in reptiles and early-diverging mammals such as the platypus and echidna, it is soft and leathery. In most mammals, however, the embryos embed into the wall of the uterus and receive their nutrients directly from the mother.

Along with the amniotic egg, other critical innovations that enabled the conquest of land by animals include the following:

- **Desiccation-resistant skin.** Whereas the skin of amphibians is moist and aids in respiration, the skin of amniotes is thicker and water-resistant and contains keratin, a tough protein. As a result, most gas exchange takes place through the lungs.
- **Thoracic breathing.** Amphibians use buccal pumping to breathe, contracting the mouth to force air into the lungs. In contrast, amniotes use **thoracic breathing,** in which coordinated contractions of muscles expand the rib cage, creating a negative pressure to suck air in and then forcing it out later. This results in a greater volume of air being displaced with each breath than with buccal pumping.

Biology Principle

Biologists Use Models to Test Predictions and Convey Ideas

This figure is a model for the structure of the amniotic egg. The Modeling Challenge in the figure legend asks you to draw two models of amniotic eggs, one for a reptile and one for a mammal.

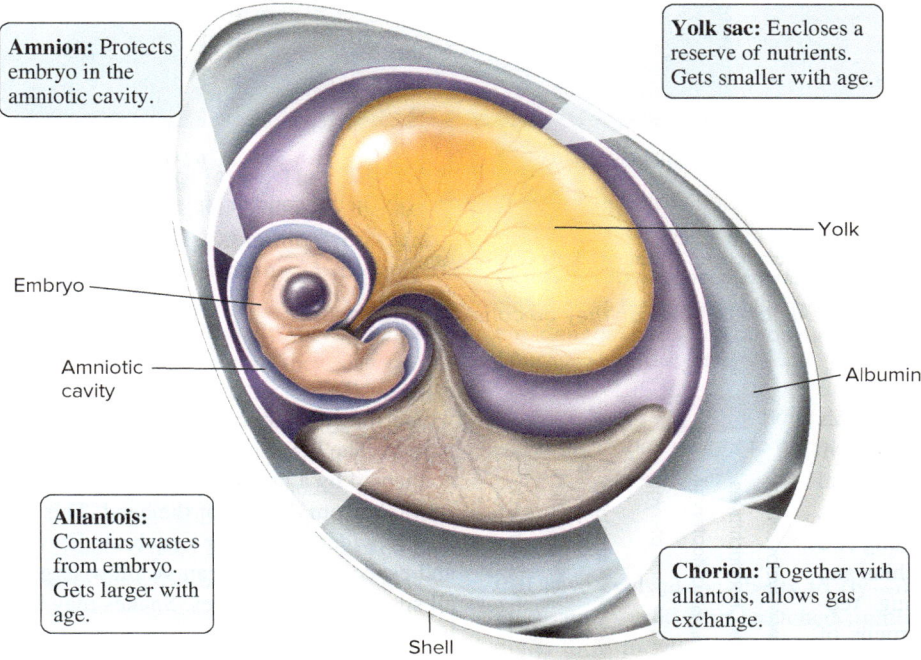

Figure 27.11 The amniotic egg.

Modeling Challenge: Even though most mammals do not lay eggs, they are considered amniotes because they retain most features of amniotic reproduction. A developing mammalian embryo in the mother's uterus is associated with the same extraembryonic membranes found in a traditional amniotic egg: the yolk sac, chorion, allantois, and amnion, together with an amniotic cavity. Draw the amniotic eggs of a reptile and a eutherian mammal. Label all of these features in each of your figures: yolk sac, chorion, allantois, amnion, embryo, and amniotic cavity. Draw the embryo about the same size in each figure, but vary the size of each of the four extraembryonic membranes as you see fit. Note: A shell does not surround the mammalian embryo. Also, the mammalian embryo is connected to the placenta via an umbilical cord.

 Concept Check: What are the other critical innovations of amniotes?

- **Water-conserving kidneys.** The ability to concentrate waste prior to elimination and thus conserve water is an important role of the amniotic kidneys.
- **Internal fertilization.** Because sperm cannot penetrate a shelled egg, fertilization occurs internally, within the female's body before the shell is secreted. In internal fertilization, the male of the species often uses a copulatory organ (penis) to transfer sperm into the female reproductive tract.

Reptiles Include Turtles, Lizards, Snakes, Crocodilians, and Birds

Early amniote ancestors gave rise to all the modern amniotes we know today, from lizards and snakes to birds and mammals. The traditional view of amniotes identified three living classes: the reptiles (turtles, lizards, snakes, and crocodilians), birds, and mammals. As we will see later in this chapter, modern systematists now consider birds to into the bones—that expand and contract when a bird inhales and exhales while the lungs remain stationary. Air is therefore being constantly moved across the lungs during inhalation and exhalation. Although making bird breathing very efficient, this process also makes birds especially susceptible to airborne toxins (hence, the utility of the canary in the coal mine; the bird's death signaled the presence of harmful carbon dioxide or methane gas that was otherwise unnoticed by miners).

3. **Reduction of organs.** Some organs are reduced in size or are lacking altogether in birds, which reduces the total mass of a bird's body. For example, birds have only one ovary and can carry relatively few eggs. As a result, they lay fewer eggs than most other reptile species. In fact, the gonads of both males and females are reduced, except during the breeding season, when they increase in size. Most birds also lack a urinary bladder. In addition, their lack of teeth has reduced weight at the head end.

4. **Lightweight bones.** Most bird bones are thin and hollow and are crisscrossed internally by thin strips of bone to give them a strong but lightweight honeycomb structure.

Other bird characteristics are related to flight. Flight requires good vision, and bird vision is the best in the vertebrate world. Because birds need a rapid metabolism and the quick production of adenosine triphosphate (ATP) to fuel flight, body temperatures are generally 40–42°C, considerably warmer than the human body's average of 37°C. Like mammals, birds also have a double circulation and a four-chambered heart that ensures rapid blood circulation.

Birds generally have high energy needs, and most birds are carnivores, eating insects or other invertebrates. However, some birds, such as parrots, eat nutrient-rich fruits and seeds. Birds feed in many different ways, which is reflected in the variety of their beak shapes (**Figure 27.16**).

Bird reproduction involves parental care. Eggs need to be kept warm for successful development, which entails sitting on the nest until the eggs have hatched, a process called brooding. Often, the males and females take turns brooding so that one parent can feed and maintain its strength. Picking successful partners is important, and birds often engage in complex courtship rituals (look ahead to Figure 42.17).

With about 10,425 species, birds are the most species-rich group of terrestrial vertebrates. Despite this diversity, birds lack the variety of body shapes that exists in the other class of vertebrates, the mammals: Some mammals can swim, some can fly, some walk on four legs, and some walk on only two legs. Most birds fly, and therefore most have the same general body shape.

27.5 Reviewing the Concepts

- The amniotic egg permitted animals to become fully terrestrial. Other critical innovations included desiccation-resistant skin, thoracic breathing, water-conserving kidneys, and internal fertilization (Figure 27.11).

Biology Principle

Structure Determines Function
Each of these beak shapes permits a different method of feeding.

(a) Cracking beak

(e) Nectar-feeding beak

(b) Scooping beak

(c) Tearing beak

(d) Probing beak

(f) Sieving beak

Figure 27.16 A variety of bird beaks. Birds have evolved a variety of beak shapes used in different types of food gathering. **(a)** Hyacinth macaw (*Anodorhynchus hyacinthinus*)—cracking. **(b)** White pelican (*Pelecanus onocrotalus*)—scooping. **(c)** Verreaux's eagle (*Aquila verreauxii*)—tearing. **(d)** American avocet (*Recurvirostra americana*)—probing. **(e)** Anna's hummingbird (*Calypte anna*)—nectar feeding. **(f)** Roseate spoonbill (*Platalea ajaja*)—sieving.
(a) ©B.G. Thomson/Science Source; *(b)* ©Jean-Claude Canton/Bruce Coleman Inc./Photoshot/Avalon; *(c)* ©Morales/age fotostock; *(d)* ©Photo 24/Getty Images; *(e)* ©Shailendra Pradhan/Getty Images; *(f)* ©Mervyn Rees/Alamy

- Living reptilian clades include the Testudines (turtles), Squamata (lizards and snakes), Crocodilia (crocodiles), and Aves (birds) (Figures 27.12–27.14).
- The fossil bird *Archaeopteryx lithographica* helps to trace a lineage from dinosaurs to birds (Figure 27.15).
- The four key characteristics of birds are feathers, air sacs, reduced organs, and a lightweight skeleton. Birds are the most species-rich clade of terrestrial vertebrates. The diversity of bird beaks reflects the varied methods they use for feeding (Figure 27.16).

27.5 Testing Your Knowledge

1. The membrane of the amniotic egg that serves as a site for waste storage is the
   a. amnion.
   b. yolk sac.
   c. allantois.
   d. chorion.

2. Which of the following is a critical innovation of amniotes?
   a. buccal pumping
   b. moist skin that aids in respiration
   c. external fertilization
   d. water-conserving kidneys
   e. bony skeleton

3. Which of the following features is associated with the ability of birds to fly?
   a. scaly legs
   b. feathers
   c. air sacs
   d. double circulation
   e. Both b and c are correct.

27.6 Mammals: Milk-Producing Amniotes

Learning Outcomes

1. Identify the four features that distinguish mammals from other vertebrate clades.
2. Describe the main orders of mammals.

Mammals evolved from amniote ancestors earlier than birds. About 225 mya, the first mammals appeared in the mid-Triassic period (refer back to Figure 22.4). The extinction of the dinosaurs in the Cretaceous period, some 65 mya, paved the way for mammals to increase in diversity of size and species. Today, biologists have identified about 5,500 species of mammals, from fishlike dolphins to birdlike bats and from small insectivores such as shrews to large herbivores such as giraffes and elephants. The range of sizes and body forms of mammals is unmatched by any other living vertebrate group, and mammals are prime illustrations of the concept that organismal diversity is related to environmental diversity. In this section, we will outline the features that distinguish mammals from other taxa. Chapter 22 also considers the evolution of humans and other primates in greater detail (refer back to Section 22.3).

Mammals Have Mammary Glands, Hair, Specialized Teeth, and an Enlarged Skull

Four characteristics distinguish mammals: the possession of mammary glands, hair, specialized teeth, and an enlarged skull.

- **Mammary glands.** Mammals, or the clade Mammalia (from the Latin *mamma*, meaning breast), are named after the female's distinctive **mammary glands,** which secrete milk. Newborn mammals suckle this fluid, which helps to promote rapid growth.

- **Hair.** All mammals have hair, although some have more than others. Whales have hair in utero, but adults are hairless or retain only a few hairs on their snout. Mammals are endothermic, and their hair is an efficient insulator. Hair can also take on functions other than insulation. Many mammals, including cats, dogs, walruses, and whales, have sensory hairs called vibrissae (**Figure 27.17a**). Hair can be of many colors, to allow mammals to blend into their background (**Figure 27.17b**). In some cases, as in porcupines and hedgehogs, the hairs become long, stiffened, and sharp (quills) and serve as a defense mechanism (**Figure 27.17c**).

- **Specialized teeth.** Mammals are the only vertebrates with highly differentiated teeth—incisors, canines, premolars, and molars—that are adapted for different types of diets (**Figure 27.18**).

- **Enlarged skull.** The mammalian skull differs from other amniote skulls in several ways. First, the brain is enlarged and is contained within a relatively large skull. Second, mammals have a single lower jawbone, unlike reptiles, whose lower jaw is composed of multiple bones. Third, mammals have three bones in the middle ear, as opposed to reptiles, which have one bone in the middle ear.

Mammals Are Morphologically Diverse and Occupy a Wide Range of Environments

Modern mammals are incredibly diverse in size and lifestyle (**Table 27.2**). They vary in size from tiny insect-eating bats, weighing in at only 2 g, to leviathans such as the blue whale, the largest animal ever known, which tips the scales at 100 tonnes (over 200,000 lb). Mammalian orders are divided into two distinct subclasses. The subclass Prototheria contains only the order Monotremata, or **monotremes,** which are found in Australia and New Guinea. There are only five species of monotremes:

(a) Sensory hairs

(b) Camouflaged coat

(c) Defensive quills

Figure 27.17 Mammalian hair. (a) The sensory hairs (vibrissae) of the walrus (*Odobenus rosmarus*). **(b)** The camouflaged coat of a bobcat (*Lynx rufus*). **(c)** The defensive quills of the crested porcupine (*Hystrix africaeaustralis*).
(a) ©National Geographic Image Collection/Alamy; *(b)* ©Charles Krebs/Getty Images; *(c)* ©Anthony Bannister/Science Source

(a) Biting teeth

(b) Grinding teeth

(c) Gnawing teeth

(d) Tusks   (e) Grasping teeth

Figure 27.18 Mammalian teeth. Mammals have different types of teeth, according to their diet. **(a)** The wolf has long canine teeth that bite and tear its prey. **(b)** The deer has a long row of molars that grind plant material. **(c)** The beaver, a rodent, has long, continually growing incisors used to gnaw wood. **(d)** The elephant's incisors are modified into tusks for defense and combat. **(e)** Dolphins and related mammals that are fish or plankton feeders have numerous small teeth used to grasp prey.
(a) ©Photo 24/Getty Images; *(b)* ©Sam Camp/Getty Images; *(c)* ©Mauritius Images GmbH/Alamy; *(d)* ©johan63/Getty Images; *(e)* ©Jim Watt/Getty Images

the duck-billed platypus (Figure 27.19a) and four species of echidna, a spiny animal resembling a hedgehog. Monotremes are early-diverging mammals that lay eggs rather than bear live young, lack a placenta, and have mammary glands with poorly developed nipples. The mothers incubate the eggs, and upon hatching, the young simply lap up the milk as it oozes onto the fur.

The subclass Theria, which contains all remaining mammals, is divided into two clades, the Metatheria and the Eutheria. The clade Metatheria, or the **marsupials,** is a group of seven orders, with about 280 species, including the rock wallaby pictured in Figure 27.19b. Once widespread, members of this order are now largely confined to Australia, although some marsupials exist in South America, and one species—the opossum—is found in North America. Fertilization is internal, and reproduction is viviparous in marsupials. Marsupials have a relatively simple, short-lived placenta that nourishes the embryo. Unlike other mammals, however, marsupials are extremely

Table 27.2 The Main Orders of Mammals, in Order of Species-Richness

Order	Examples (approx. no. of species)	Main characteristics
Rodentia	Mice, rats, squirrels, beavers, porcupines (2,337)	Plant eaters; gnawing habit, with two pairs of continually growing incisor teeth
Chiroptera	Bats (1,157)	Insect or fruit eaters; small; have ability to fly; navigate by sonar; nocturnal
Eulipotyphla	Shrews, moles, hedgehogs (462)	Insect eaters; primitive placental mammals
Primates	Monkeys, apes, humans (437)	Opposable thumb; binocular vision; large brain
Carnivora	Cats, dogs, weasels, bears, seals, sea lions (295)	Flesh-eating mammals; canine teeth
Artiodactyla	Deer, antelopes, cattle, sheep, goats, camels, pigs (240)	Herbivorous hoofed mammals, usually with two toes, hippopotamus and others with four toes; many with horns or antlers
Diprotodontia	Kangaroos, koalas, opossums, wombats (147)	Marsupials mainly found in Australia
Lagomorpha	Rabbits, hares (92)	Powerful hind legs; rodent-like teeth
Cetacea	Whales, dolphins (84)	Marine fish or plankton feeders; front limbs modified into flippers; no hind limbs; little hair except on snout
Perissodactyla	Horses, zebras, tapirs, rhinoceroses (18)	Hoofed herbivorous mammals with odd-numbered toes, one (horses) or three (rhinoceroses)
Monotremata	Duck-billed platypuses, echidna (5)	Egg-laying mammals found only in Australia and New Guinea
Proboscidea	Elephants (3)	Long trunk; large, upper incisors modified as tusks

small when they are born (often only 1–2 cm) and make their way to a ventral pouch called a marsupium for further development.

All the other mammalian orders are members of the clade Eutheria and are considered **eutherians,** or placental mammals, such as the gray langur monkeys shown in Figure 27.19c. Eutherians have a long-lived and complex placenta, compared with that of marsupials. In eutherians, fertilization is internal and reproduction is viviparous, but the developmental period, or gestation, of the fetus is prolonged.

27.6 Reviewing the Concepts

- The distinguishing characteristics of mammals are mammary glands, hair, specialized teeth, and an enlarged skull (Figures 27.17, 27.18).

- Two subclasses of mammals exist: the Prototheria (monotremes) and the Theria (the live-bearing mammals). The live-bearing mammals consist of the Metatheria (marsupials) and Eutheria (placental mammals) (Table 27.2, Figure 27.19).

27.6 Testing Your Knowledge

1. Mammals can be distinguished from other vertebrate classes by their
   a. amniotic eggs.
   b. mammary glands.
   c. warm-bloodedness.
   d. reduction of organs.
   e. thoracic breathing.

Figure 27.19 Diversity among mammals.
(a) Prototherians, such as this duck-billed platypus (*Ornithorhynchus anatinus*), lay eggs, lack a placenta, and possess mammary glands with poorly developed nipples.
(b) Metatherians, or marsupials, such as this rock wallaby (*Petrogale assimilis*), feed and carry their developing young, or "joeys," in a ventral pouch. **(c)** Gestation lasts longer in eutherians, and their young are more developed at birth, as illustrated by this gray langur (*Presbytis entellus*).
(a) ©Dave Watts/naturepl.com; *(b)* ©Nigel Pavitt/Getty Images; *(c)* ©Education Images/Getty Images

BioConnections: Look back at Figure 19.6a. What is another species of prototherian?

(a) Prototherian (duck-billed platypus)

(b) Metatherian (rock wallaby)

(c) Eutherian (gray langur)

Assessing Your Knowledge and Skills

Foundational Knowledge

Multiple Choice

1. To which of the following clades does a jawed fish belong?
    a. gnathostomes
    b. lobe fins
    c. tetrapods
    d. amniotes
    e. cyclostomes

2. Organisms that lay eggs are said to be
    a. oviparous.
    b. ovoviviparous.
    c. viviparous.
    d. placental.
    e. none of the above.

3. The earliest-diverging tetrapods were
    a. coelacanths.
    b. lungfish.
    c. amphibians.
    d. reptiles.
    e. mammals.

4. Which of the following clades is *not* considered part of the reptiles?
    a. turtles
    b. crocodiles
    c. birds
    d. mammals
    e. Both c and d are correct.

5. What evolutionary trait do frogs and other amphibians lack (but reptiles have) that continues their association with water?
    a. lungs
    b. pulmonary veins
    c. amniotic egg
    d. legs

Short Answer

1. **PRINCIPLES** A principle of biology is that structure determines function. How is vertebrate movement similar to arthropod movement, and how is it different?
2. Why aren't all reptiles endothermic, given that both birds and mammals are?

Critical-Thinking Skills

Multiple Choice

1. Your lab group is asked to identify the taxonomic group of a particular aquatic animal. You identify three characteristics—a bony skeleton, an operculum, and a swim bladder—and therefore classify the animal as a member of the
   a. class Myxini.
   b. lampreys.
   c. class Chondrichthyes.
   d. bony fishes.
   e. amphibians.
2. Let's suppose that an injury causes a frog to lose its ability to perform buccal pumping. For this frog to survive, what must happen to compensate for this loss?
   a. The pumping of blood from the atria to the ventricle must increase.
   b. The exchange of gases through the skin must increase.
   c. The pumping of blood from the ventricle to the body must increase.
   d. The diffusion of wastes through the kidneys must increase.
   e. The pumping of blood from the ventricle to the atria must increase.
3. One of the students in your study group is struggling with the idea that mammals and vertebrates are separate taxonomic designations. To illustrate the two terms, you describe an animal that has the expected mammalian characteristics. What should you include in your description of this animal that would make it clear that the animal is specifically a mammal and not just a vertebrate?
   a. clusters of *Hox* genes
   b. cranium
   c. bony skeleton
   d. hair
   e. vertebral column
4. What support can be provided for the argument that hagfish are not fish?
   a. They are not truly aquatic.
   b. They lack a cranium.
   c. They lack jaws.
   d. They lack gills.
   e. They lack an amniotic egg.
5. What was the most significant finding of Davis and colleagues' experiment on limb development?
   a. that mice could survive despite inheriting multiple loss-of-function mutations
   b. that the limbs of mice could exhibit deformities of fused bones
   c. that as few as two mutations inherited simultaneously could affect the morphology of mouse limbs to a large degree
   d. that observed deformities in mouse limbs were proportional to the number of loss-of-function mutations that were inherited
   e. that young mice inherited characteristics of their parents

Short Answer

1. The hinged jaw was a critical evolutionary innovation. Explain how the evolution of hinged jaws changed the nature of species interactions.
2. You have the pleasure of observing wood frog tadpoles in a pond on campus. You see them make the transition from gilled larvae into adults with lungs. In order to make this amazing transformation from an entirely aquatic life to one that includes time on land, your tadpoles will face a series of obstacles, not unlike those that the first vertebrate inhabitants of land faced. Discuss the major obstacles involved in a transition from living in water to spending some of the time on land.

UNIT VI
FLOWERING PLANTS

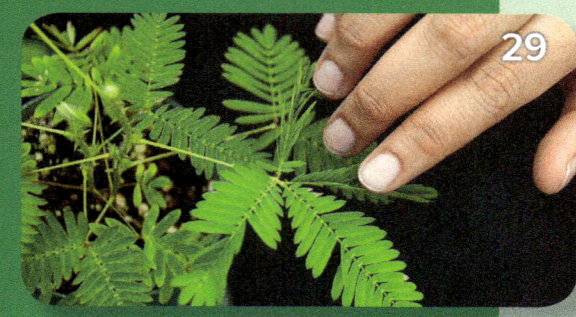

Flowering plants, also known as the angiosperms, are essential to the lives of humans and most other organisms on Earth. Flowering plants provide most of our food, either directly in the form of vegetables and fruits (which include grains) or indirectly as animal fodder. Cotton, linen, and other fibers that we use for clothing and wood that we use for construction and fuel—as well as powerful cancer drugs and many other medicines—come from the flowering plants. Unit VI reveals molecular, biochemical, structural, evolutionary, and ecological features of the hundreds of thousands of flowering plants that support Earth's life.

Unit VI begins with Chapter 28, which provides an overview of flowering plant structure and function, focusing on the seed-to-seed life cycle. By comparing plant bodies with those of animals, you will learn how plants are constructed and how they grow. Building on this background, Chapter 29 explains the genetic and physiological bases of plant responses to external stimuli, such as day length, that are mediated by internally produced hormones. In this chapter, you will learn that plants, like animals, have evolved sophisticated sensory systems that monitor environmental conditions and allow plants to respond in predictable ways. Chapter 30 explains the nutritional requirements of plants and how plants transport materials, a deep understanding of which is critical to human agriculture and our ability to feed increasing populations of humans. Chapter 31 focuses on the molecular and cellular bases of plant reproduction, the process that generates seeds and fruits. This chapter ties together key concepts presented throughout Unit VI: the seed-to-seed life cycle of plants and their distinctively structured bodies, plant development and growth in response to environmental and hormonal influences, and plant acquisition and transport of materials that support growth and reproduction.

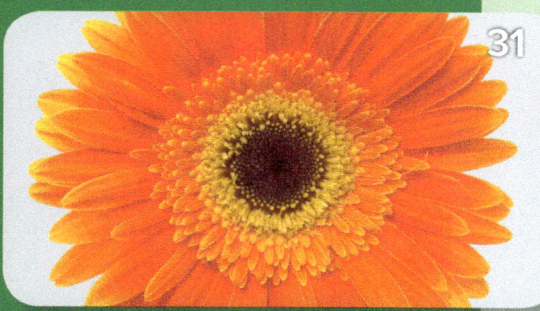

The following biology principles are emphasized in this unit:

- ***Living organisms use energy.*** *Though sunlight is the major source of energy for photosynthetic plants, hundreds of flowering plant species as well as diverse tissues occurring in all flowering plants are heterotrophic and thus require a supply of organic food as a source of energy.*

- ***Living organisms interact with their environment.*** *Some of the ways in which plants detect and respond to light and other environmental factors are similar to those operating in microbes or animals, but others are distinctive.*

- ***Living organisms maintain homeostasis.*** *As is the case for animals, it is essential for flowering plants to maintain body water content and energy balance within tolerance limits, or they will die.*

- ***The genetic material provides a blueprint that allows organisms to grow, develop, and reproduce.*** *Some of the cellular and molecular bases of plant growth and development are also features of microbes and animals, but plants display some distinctive modes of growth and development.*

- ***Structure determines function.*** *Differences in the structures of animal and plant bodies—and structural differences among plants—correspond with differences in function.*

(28) ©George Grall/Getty Images; (29-30) ©Linda Graham; (31) ©Vidok/Getty Images

An Introduction to Flowering Plant Form and Function

28

©George Grall/Getty Images

Venus flytrap. This carnivorous plant has distinctive leaves whose colors attract unsuspecting insects, then fold with a quick snap. As their bodies decompose, the trapped insects supply carnivorous plants with essential mineral nutrients.

Chapter Outline

28.1 From Seed to Seed: The Life of a Flowering Plant

28.2 Plant Growth and Development

28.3 The Shoot System: Stem and Leaf Adaptations

28.4 Root System Adaptations

Assessing Your Knowledge and Skills

Many types of fascinating carnivorous (meat-eating) flowering plants occur abundantly in wetlands around the world, even though the soils in these places are often infertile. Using modified leaves that quickly fold shut (see the chapter-opening photo) or that have other unusual shapes, carnivorous plants trap insects and other small animals. As the trapped animals die, the plant leaves absorb essential nutrients released from the animal bodies, allowing plant growth and reproduction by flowers and seeds. The example of carnivorous plants reveals that plant leaves are not just photosynthetic organs but can have additional functions. Many other plants produce leaves, stems, or roots whose distinctive structures likewise foster survival under challenging circumstances.

Chapter 25 provided an overview of modern plant evolution and diversity, which includes bryophytes, lycophytes, ferns, gymnosperms, and angiosperms—the flowering plants. The next four chapters focus on the flowering plants, the most species-rich modern plant group and the one most important in human agriculture. This chapter provides an overview of the principles of plant growth, development, and reproduction. These principles are important to plant biologists because they underlie the ability of humans to produce agricultural crops and understand how plants function in nature. We will begin by considering the life of a flowering plant, from seed germination to seedling growth into mature plants that are capable of producing the next generation.

28.1 From Seed to Seed: The Life of a Flowering Plant

Learning Outcomes

1. List ways in which flowering plant reproduction and growth differ from those of animals.
2. Describe how the two major types of flowering plants—monocots and eudicots—differ in form.
3. Distinguish among annual, biennial, and perennial plants.

Several major events punctuate the lives of flowering plants, also known as the **angiosperms**:

- When seeds germinate, dormant embryos wake to metabolic activity and begin the process of seedling development.
- Seedlings grow and develop into mature sporophyte bodies capable of reproduction. Growth is an increase in weight or size, and development is a series of changes in the state of a cell, a tissue, an organ, or a body.
- As they develop, the mature sporophytes of all flowering plants display some architectural features in common, though plant species vary widely in organ and body structure and life span.
- Flowers produce gametophytes that engage in sexual reproduction, and fruits disperse the next generation of seeds.

Seedlings Develop from Embryos in Seeds

Seeds are reproductive structures produced by flowering plants and other seed plants, usually as the result of sexual reproduction. Seeds allow plants to survive periods unsuitable for growth, remaining in a dormant condition until environmental conditions improve. Seeds contain embryos that develop into young plants—seedlings—when the seeds germinate. Embryos and seedlings are young **sporophytes**, the larger, spore-producing generation of flowering plants (Figure 28.1). By contrast, a microscopic alternating life phase known as the **gametophyte**, which occurs in flowers, has the function of sexual reproduction.

In addition to an embryo, plant seeds also contain a supply of stored food. The embryo and stored food are enclosed by a tough, protective seed coat (Figure 28.2a). The seed coat protects the delicate plant embryo during the dispersal of seeds from parent plants into the environment. Dispersed seeds may remain dormant in the soil—sometimes for long periods—but germinate when temperature, moisture, and light conditions are favorable. Such conditions activate embryo metabolism, in which stored food is respired for energy needed for growth and development. Enlarging plant embryos break the seed coat and grow into seedlings (Figure 28.2b). If sufficient resources such as water and minerals are available, and if they are able to avoid being killed by disease organisms or being eaten, seedlings develop into mature plants (Figure 28.2c). The seedling stage is often the most vulnerable phase of a plant's life.

Figure 28.1 The plant sexual cycle. The sexual cycle of all flowering plants involves alternation of diploid sporophyte and haploid gametophyte generations. In flowering plants, the sporophyte is the dominant, conspicuous generation, whereas the tiny gametophytes are mostly hidden within flowers. The bean plant shown here produces both male and female gametophytes. Embryos form when eggs produced by female gametophytes are fertilized by sperm produced by male gametophytes. Embryos, which are the young sporophytes of the next generation, are formed within seeds and dispersed from plants.

BioConnections: Refer back to Figure 25.4 to see the life cycle of a typical bryophyte, Figure 25.7 to see the life cycle of a typical fern, Figure 25.12 to see the life cycle of a typical gymnosperm, and Figure 25.15 to see an angiosperm life cycle. How does the life cycle of the flowering plant differ from those of the moss and the fern?

Biology Principle

The Genetic Material Provides a Blueprint That Allows Organisms to Grow, Develop, and Reproduce

The plant *Arabidopsis thaliana,* shown here, is widely used as a model organism for understanding the genetics of plant structure and development.

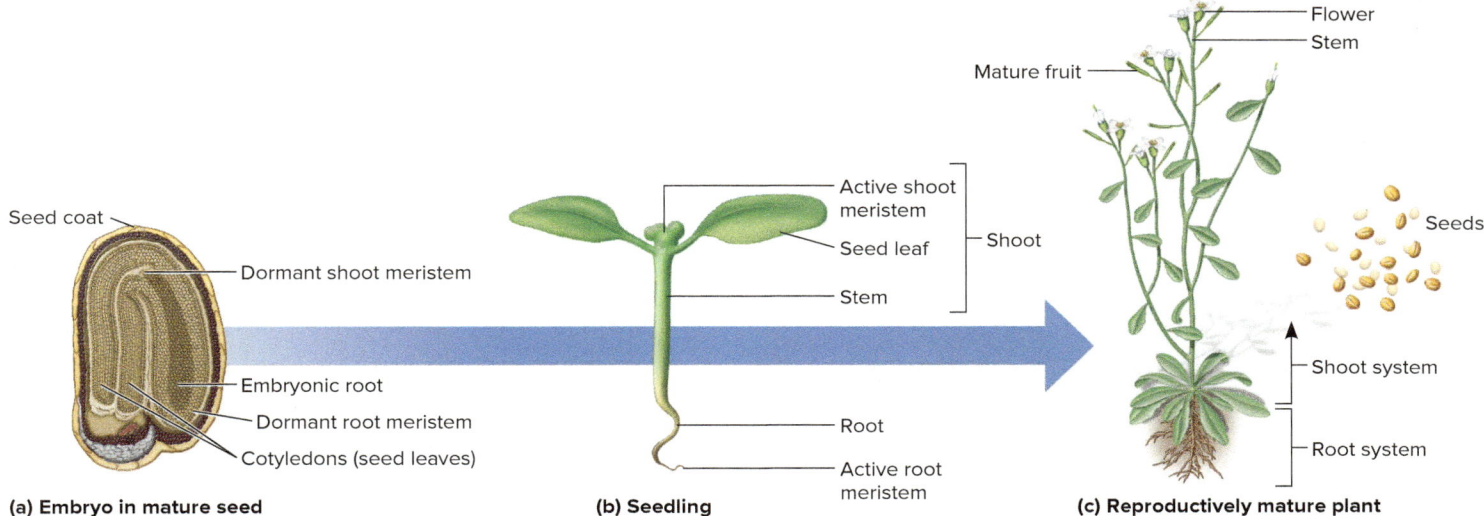

Figure 28.2 The seed-to-seed life of flowering plants. **(a)** Seed embryos possess embryonic leaves, known as cotyledons; a dormant shoot meristem; an embryonic root; and a dormant root meristem. **(b)** When seeds germinate, the shoot and root meristems become active. Meristem activity allows the embryonic root to produce the seedling root and the young shoot of the seedling to grow and produce leaves. **(c)** Reproductively mature plants have branched shoot and root systems and bear flowers and fruits that disperse seeds.

The angiosperm plant body is composed of only three types of organs: stems, leaves, and roots. **Stems** produce leaves and branches and bear the reproductive structures of mature plants. **Leaves** are flattened structures that emerge from stems and are often specialized in ways that enable photosynthesis. Stems and leaves together make up the plant **shoot** (Figure 28.2b). Mature plants often possess multiple stems bearing many leaves, which together form the shoot system. **Roots** provide anchorage in the soil and foster efficient uptake of water and minerals. The aggregate of a plant's roots make up the root system (Figure 28.2c).

Plant seedlings and mature plants produce new tissues in specific areas called meristems. A **meristem** (from the Greek *merizein*, meaning to divide) is a region of undifferentiated cells that produces new tissues by cell division. A dormant meristem occurs at the shoot and root tips of seed embryos, and these meristems become active in seedlings (Figure 28.2a, b). In mature plants, active meristems occur at each stem and root tip. Such meristems are known as shoot and root apical meristems because shoot and root tips are called apices. The term shoot apical meristem is commonly referred to with the acronym SAM, and the root apical meristem is the RAM.

Mature Sporophytes Develop from Seedlings

As seedlings develop into mature sporophytes, the aboveground shoot typically becomes green and photosynthetic and thus able to produce organic food. Photosynthesis powers the transformation of seedlings into mature plants. The development of mature plants encompasses both vegetative growth, a process that increases the size of the shoot and root systems, and reproductive development. Vegetative growth and reproductive development involve **organ systems,** structures that are composed of more than one organ. Branches, buds, flowers, seeds, and fruits are organ systems, analogous to organ systems such as the circulatory, skeletal, and reproductive systems in the animal body. The hierarchy of structure in a mature plant, ranging from specialized cells, tissues, organs, and organ systems to root and shoot systems, is shown in **Figure 28.3**.

Vegetative Growth Vegetative growth is the growth of nonreproductive parts of the plant body. During their growth, plant shoots produce **buds**—miniature shoots—each having a dormant shoot apical meristem. A bud is an example of an organ system because it contains more than one organ. Scaly, modified leaves protect the bud contents. Under favorable conditions, the bud scales fall off, and the buds open. Newly opened vegetative buds display young leaves on a short shoot. The shoot apical meristem then becomes active, producing new stem tissue and leaves. In this way, buds generate leafy branches that contribute to vegetative growth. By contrast, during reproduction, floral buds are produced; these open to produce flowers, modified shoots that do not contribute to vegetative growth.

Vegetative shoots often display **indeterminate growth,** meaning that apical meristems continuously produce new stem tissues and leaves as long as conditions remain favorable. This process explains how very

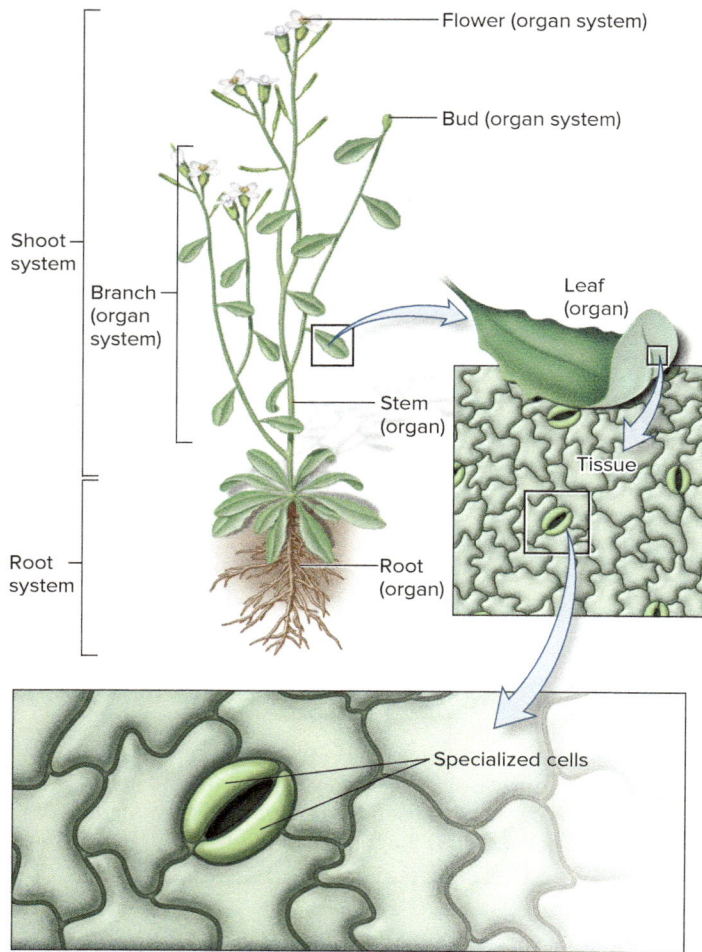

Figure 28.3 Levels of biological organization in a plant. Flowering plant sporophyte bodies consist of a root system and a shoot system. Shoot systems produce organ systems such as buds, flowers, fruits, and seeds, which are composed of organs, tissues, and specialized cells. Root systems are composed of organs, tissues, and specialized cells.

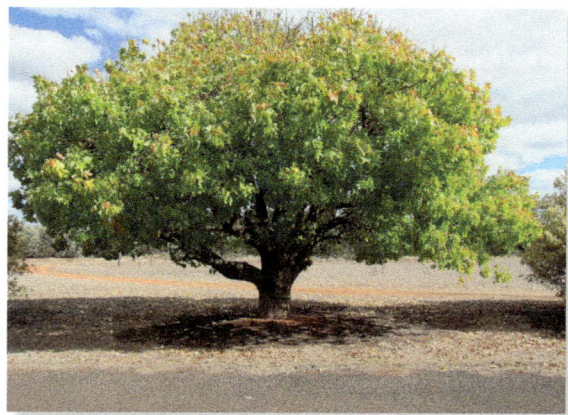

(a) *Brachychiton*, a tree native to Australia

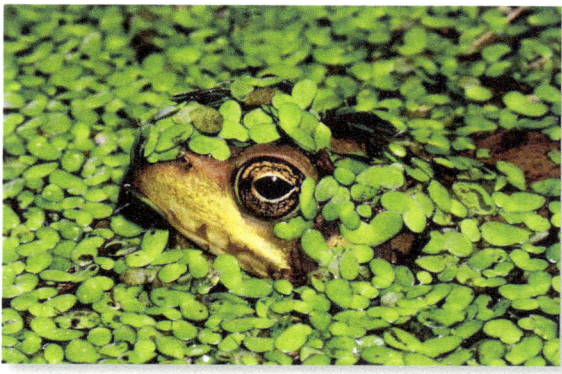

(b) *Lemna* (common duckweed)

Figure 28.4 Plants display indeterminate growth but vary in size. (a) A large, woody angiosperm has the same indeterminate growth pattern as (b) tiny, floating duckweed plants, each smaller than the eye of the frog hiding in the water beneath.
(a) ©Linda Graham; *(b)* ©David Davis/Science Source

large plants, such as trees, can develop from seedlings (Figure 28.4a). However, plant size is also under genetic control, so some plants remain small even when they are mature. The tiny, floating plants of *Lemna* species, commonly known as duckweeds, which sometimes cover the surfaces of ponds in summer, are examples of plants whose small size is genetically determined (Figure 28.4b). Indeterminate growth allows plants to adapt their vegetative body structure to environmental conditions. By contrast, animal bodies and flowers display **determinate growth,** which is growth of limited duration.

Reproductive Development Under favorable conditions, mature plants produce reproductive structures: flowers, seeds, and fruits. **Flowers** and floral buds are reproductive shoots that develop when shoot apical meristems produce flower parts instead of new stem tissues and leaves. Flower development occurs under the control of several genes whose roles are well understood. The gene *LEAFY*, for example, acts in the SAM to flip the molecular switch from vegetative to floral development, and interactions of *APETALA*, *PISTILLATA*, *AGAMOUS*, and *SEPALLATA* genes control the development of flower parts.

In contrast to shoots, which often show indeterminate growth, flowers are produced by determinate growth. A floral shoot no longer produces new stem growth or leaves. Therefore, vegetative growth and reproductive development are alternative processes. In order to flower, a plant must give up some of its potential to continue vegetative growth.

Flower tissues enclose and protect tiny male and female gametophytes during their growth and development (see Figure 28.1). Female gametophytes produce eggs within structures known as ovules, produced in the ovary of a flower pistil (refer back to Figure 25.15). Male gametophytes begin their development within pollen grains produced in the anthers of a flower stamen. Pollen is dispersed to the flower pistil, whereupon pollen grains may germinate, producing a tube that delivers sperm to eggs. Fertilization generates zygotes, which develop into embryos, and triggers the process by which ovules develop into seeds and flower parts develop into fruits. **Fruits** thus enclose seeds and function in seed dispersal. Flowers and fruits are features that distinguish flowering plants from other plant phyla (see Chapter 25).

Flowering Plants Vary in Structure and Life Span

With some exceptions, flowering plants occur in two groups, informally known as the **eudicots** and the **monocots**. These groups take their names from the number of seed leaves (cotyledons) that are present on seed embryos. For example, bean plants and relatives, which possess two (*di*) seed leaves, are examples of eudicots. Most woody trees, shrubs, and vines are also eudicots. Corn, which has only one (*mono*) seed leaf, is an example of a monocot, as is tiny duckweed (Figure 28.4b). Eudicots and monocots also vary in the structure of other organs and organ systems. For example, eudicot flowers typically have petals and other parts numbering four, five, or a multiple of those numbers, whereas monocot flower parts usually occur in threes or a multiple of three. Stems, roots, leaves, and pollen of eudicots and monocots also vary in distinctive ways, as shown in Table 28.1.

The life span of a flowering plant can vary from a few weeks to many years.

- Plants that die after producing seed during their first year of life are known as **annuals.** Corn and the common bean are examples of annual crops whose nutrient-rich seeds are harvested within a few months after planting and must be replanted at the beginning of each new growing season.
- Plants that do not reproduce during the first year of life but may reproduce within the following year are known as **biennials.** Such plants often store food in fleshy roots during the first year of growth, and this food fuels reproduction during the second or a later year of life. Humans use some of these fleshy roots for food, including carrots, parsnips, and sugar beets.
- Trees are examples of **perennials,** plants that live for more than 2 years, often producing seed each year after they reach reproductive maturity.

Table 28.1 Distinguishing Features of Eudicots and Monocots, Two Major Groups of Flowering Plants

Feature	Eudicots	Monocots
Number of seed leaves (cotyledons)	Two	One
Number of flower parts	Usually four, five, or multiples of these	Usually three or multiple of three
Stem vascular bundles	Arranged in a ring	Scattered
Root system	Branched taproot	Fibrous; adventitious
Leaf venation	Netted or branched	Often parallel
Pollen	Three pores or slits	One pore or slit

28.1 Reviewing the Concepts

- Seed embryos, seedlings, and mature plants are stages of the sporophyte generation in the flowering plant sexual cycle; tiny gametophytes develop and grow within flowers (Figure 28.1).
- Plant organs are composed of tissues that contain specialized cells. The basic plant organs are stems, leaves, and roots. Shoot systems include stems and stem branches, and stems produce leaves, buds, flowers, and fruits. Root systems include one or more main roots with branches. Buds, flowers, fruits, and seeds are organ systems, composed of more than one organ (Figures 28.2, 28.3).
- The two major groups of flowering plants—eudicots and monocots—differ in the structure of their seed embryos, flowers, stems, roots, leaves, and pollen. Plants can be categorized according to their life span as annuals, biennials, and perennials (Figure 28.4, Table 28.1).

28.1 Testing Your Knowledge

1. Which of the following statements represents plant vegetative growth rather than plant reproduction?
   a. A plant produces new root tissue from a root apical meristem.
   b. A plant produces new shoot tissue from a shoot apical meristem.
   c. A vegetative bud opens to generate a new leafy shoot.
   d. All of the above are correct.
   e. None of the above is correct.

2. How do annual plants differ from biennials and perennials?
   a. An annual plant has a seed-to-seed life span of a year or less, whereas the life spans of biennials and perennials are longer.
   b. An annual plant reproduces in the second year of its life span, whereas biennials and perennials reproduce in the first year of life.

Modeling Challenge: Figure 28.10 is a model depicting the processes involved in plant cell-wall expansion. Let's suppose that the genes encoding expansin proteins were inactivated using genetic technology, such as CRISPR-Cas (refer back to Figure 18.10). Draw a model showing the expected spatial relationship between crosslinking polysaccharides and cellulose microfibrils in the genetically modified plants, and use this diagram to predict the phenotype of such plants.

The direction in which a plant cell expands depends on the arrangement of cellulose microfibrils in its cell wall, which is in turn determined by the orientation of cytoplasmic microtubules. These microtubules are thought to influence the positions of cellulose-synthesizing protein complexes located in the plant plasma membrane. The protein complexes connect sugars to form cellulose polymers, spinning cellulose microfibrils onto the cell surface to form the cell wall. As a result, cell-wall cellulose microfibrils encircle cells in the same orientation as underlying cytoplasmic microtubules (**Figure 28.11**). Because cellulose microfibrils do not extend lengthwise, plant cell walls expand more easily in a direction perpendicular to them. Microtubules control not only the direction of cell expansion but also the plane of cell division, which is likewise critical to normal plant form.

28.2 Reviewing the Concepts

- Apical-basal polarity and radial symmetry are distinctive architectural features of plants. Plant growth occurs by the addition of new cells and by cell expansion (Figures 28.5, 28.6).
- Shoot apical meristems contain undifferentiated stem cells and produce primary meristems that generate new cells, thereby increasing plant length and producing tissues and organs: stems, roots, and leaves (Figures 28.7, 28.8; Table 28.2).
- Plant cells are able to expand by water uptake into vacuoles and under conditions that result in loosening of cell-wall components. The direction in which plant cells expand is determined by the arrangement of cell-wall cellulose microfibrils, which is in turn influenced by the orientation of microtubules in the nearby cytoplasm (Figures 28.9, 28.10, 28.11).

28.2 Testing Your Knowledge

1. Which feature distinguishes plant growth and development from those of animals?
   a. apical-basal polarity that is maintained throughout life
   b. increase in length by the activity of apical meristems
   c. maintenance of a population of undifferentiated stem cells within meristems
   d. expansion of cells by water uptake and in controlled directions
   e. All of the above are correct.

2. How does a gene that is closely related to the animal tumor-suppressor gene *Rb* function in plants?
   a. The plant gene controls the ability to detect and respond to light.
   b. The plant gene is essential to normal cell cycle function of plant stem cells.
   c. Mutations in the plant gene cause a form of plant cancer.
   d. All of the above are correct.
   e. None of the above is correct.

28.3 The Shoot System: Stem and Leaf Adaptations

Learning Outcomes

1. Discuss why plant shoots are said to have a modular structure.
2. Explain why leaves having different shapes and vein patterns exist in nature.
3. Compare the structure and function of the conducting tissues xylem and phloem.
4. Create a drawing showing how bark and wood are produced in woody plants.

As we have seen, the shoot system includes all of a plant's stems, branches, leaves, and buds. The shoot system also produces flowers and fruits when the plant has reached reproductive maturity. In these ways, the shoot system is key to plant growth, photosynthesis, and reproduction.

Shoot Systems Have a Modular Structure

More than 200 years ago, German writer, politician, and scientist Johann Wolfgang von Goethe realized that plants are modular organisms, composed of repeated units. Shoots are notably modular (**Figure 28.12**). Each shoot module consists of four parts: a stem node, an internode, a leaf, and an axillary bud. A **node** is the stem

Figure 28.11 Control of the direction of plant cell expansion by microfibrils and microtubules. Plant cells enlarge in the direction perpendicular to encircling cell-wall cellulose microfibrils, which run parallel to the orientation of underlying cytoplasmic microtubules.

AN INTRODUCTION TO FLOWERING PLANT FORM AND FUNCTION **619**

(a) Modular structure of herbaceous shoot

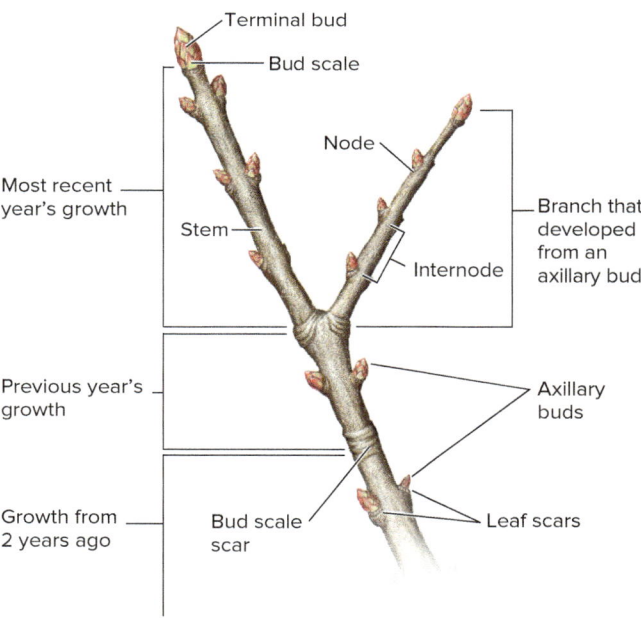

(b) Modular structure of woody shoot in winter

Figure 28.12 The modular organization of plant shoots. (a) The top end of a herbaceous stem showing the shoot modules. Each module consists of a node with its associated leaf and axillary bud and an internode. **(b)** The modular organization shown by a woody stem as it appears during winter. Axillary buds lie above the scars left by leaf fall. Regions between successive sets of bud scale scars mark each year's growth.

Concept Check: *If a twig has five sets of bud scale scars, how old is the twig likely to be?*

region from which one or more leaves emerge. An **internode** is the region of stem between adjacent nodes. Differences in numbers and lengths of internodes help to explain why plants differ in height (see Figure 28.4). Each time a young leaf is produced at the SAM, a new meristem develops in the upper angle formed where the leaf emerges from the stem. This angle is known as an axil (from the Greek *axilla*, meaning armpit), and the meristem formed there is called an axillary meristem. Such axillary meristems generate **axillary buds,** which bear a SAM at their tips and can produce leafy branches known as lateral shoots or flowers.

Hormones and MicroRNAs Influence Leaf Development

As mentioned, leaf primordia are surface bumps of tissue that develop at the sides of a SAM (see Figure 28.6). Production of such leaf primordia is under the control of a hormone known as **auxin.** In general, **hormones** are signaling molecules that are produced at one site and exert their effects on distant target cells (plant hormones are described in more detail in Chapter 29). The outermost epidermal layer of cells at shoot tips produces auxin, which moves from cell to cell by means of specific membrane transport proteins. Auxin accumulates in particular locations because cells of the shoot apex differ in their ability to import and export auxin. When auxin accumulates in a particular apical region, the hormone causes expression of the gene encoding expansin to increase. When expansin loosens cell walls (see Figure 28.10), the cells expand by taking up water, thereby forming a tissue bulge—a leaf primordium. The development of leaf primordia depletes auxin from nearby tissue, with the result that the next leaf primordium will develop in a different place on the shoot apex, where the auxin level is higher. Such changes in auxin concentration on the surface of the shoot explain why leaf (and flower) primordia develop in spiral or whorled patterns around the shoot tip. The youngest leaf primordia occur closest to the shoot tip, and successively older leaf primordia occur on the sides of the shoot tip (see Figure 28.6).

Hormones also influence the transformation of primordia into leaves. The cells of leaf primordia do not produce a protein known as KNOX, which is produced by other shoot meristematic cells. KNOX is a transcription factor, a protein that regulates gene transcription. The absence of KNOX proteins induces leaf primordia to produce a plant hormone known as **gibberellic acid.** This hormone stimulates both cell division and cell enlargement, causing young leaves to grow.

Other molecules produced by a SAM direct leaf flattening and differentiation of the upper and lower leaf tissues (see Figure 28.8c). A particular set of genes expressed in young primordia causes uppermost leaf tissue layers to develop, and a different set of genes specifies development of lower leaf tissue layers. This differential gene expression is caused by differences in the location of a type of microRNA (miRNA), a small RNA molecule that silences the expression of pre-existing mRNAs. The miRNA accumulates in the lower regions of a primordium. In this location, the miRNA interrupts the expression of genes that would otherwise stimulate development of upper leaf tissues while allowing expression of the genes specifying lower leaf tissue development. Other types of miRNA molecules influence leaf shape, which helps to explain the diversity of leaf shapes produced by different plant species.

(a) Eudicot stem with simple leaf and pinnate venation
(b) Monocot stem and leaf
(c) A compound leaf
(d) A simple leaf and palmate venation

Figure 28.13 Examples of variation in leaf form. **(a)** A simple eudicot leaf, showing blade, petiole, and axillary bud. This leaf has a pinnate venation pattern. **(b)** The leaf of a monocot, showing parallel veins. The base of the leaf encircles the stem. **(c)** A compound leaf divided into leaflets. **(d)** A simple leaf having palmate venation.
(a–d) ©Lee W. Wilcox

Leaf Shape and Surface Features Reflect Adaptation to Environmental Stress

As previously noted, leaf flatness facilitates solar energy collection, and thinness helps leaves to avoid overheating. Leaf shape and surface features also reflect adaptation to stressful environmental conditions.

Leaf Shape The flattened portion of a leaf is known as the leaf **blade**. In beans and most other eudicots, blades are attached to the stem by means of a stalk known as a **petiole**, and an axillary bud occurs at the junction of the stem and petiole (Figure 28.13a). In contrast, corn and other monocots have leaf blades that grow directly from the stem, encircling it to form a leaf sheath (Figure 28.13b).

Leaf shape can be simple or compound, each having particular advantages. Simple leaves have only one blade, though the edges may be smooth, toothed, or lobed. Simple leaves are advantageous in shady environments because they provide maximal light absorption surface, but they can overheat in sunny environments. As an evolutionary response to heating stress, the blades of some leaves have become highly dissected into leaflets. Such leaves are known as compound leaves (Figure 28.13c). Leaflets can be distinguished from leaves because leaflets lack axillary buds at their bases. Compound leaves are common in hot environments because leaflets foster heat dissipation.

Leaf Vein Patterns Leaf vein patterns are known as venation. Eudicot leaves occur in two major venation forms. They may have a single main vein from which smaller lateral veins diverge in a feather-like pattern known as **pinnate** venation (Figure 28.13a). Alternatively, several main veins may spread from a common point on the petiole, like the fingers of your hand, a pattern known as **palmate** venation (Figure 28.13d). In eudicot leaves, small veins connect in a netted pattern, but most monocot leaves have a distinctive parallel venation (Figure 28.13b; see Table 28.1).

FEATURE INVESTIGATION

Lawren Sack and Colleagues Showed That Palmate Venation Confers Tolerance of Leaf Vein Breakage

In 2008, American plant structure expert Lawren Sack and associates studied the adaptive value of leaf venation patterns by comparing water conduction after injury in pinnately and palmately veined leaves (Figure 28.14). Leaves with palmate venation have several main veins, whereas leaves with pinnate venation have a single main vein. The investigators hypothesized that the multiple main veins of palmate leaves could confer greater tolerance of vein breakage of the type that would occur during mechanical injury or insect damage; if one main vein were damaged, water flow could continue through the other main veins. To test the hypothesis, the investigators experimentally cut a main vein in the leaves of several plants belonging to seven different plant species: four having pinnately veined leaves and three having palmately veined leaves. They conducted the experiments in vivo—that is, in the plant's natural forest environment. After the wounds had healed, the investigators measured the extent of water flow within the leaves at two or three places on each leaf. They found that across all species examined, palmately veined leaves tolerated the disruption in water flow better than pinnately veined leaves. Although palmate venation provides redundancy in case a main vein becomes damaged, it is more costly in terms of materials needed to construct the additional main veins. Hence, leaves with pinnate venation are less costly to produce and work well when the potential for vein damage is low.

Experimental Questions

1. Why did Sack and associates conduct their studies of palmate venation on plants growing in a forest rather than in a greenhouse?
2. **SCISKILLS** ▶ Why did Sack and colleagues use splints on uncut control leaves as well as for experimentally cut leaves?
3. **SCISKILLS** ▶ Why did Sack and associates measure leaf water conduction at two or more places on each leaf?

Figure 28.14 Sack and colleagues investigated the function of palmate venation.

HYPOTHESIS Palmate venation provides vascular redundancy, which allows leaves to tolerate vein breakage.

KEY MATERIALS Seven species of trees or shrubs at Harvard Forest, Petersham, MA.

	Experimental level	Conceptual level
1. Identify 7 species, 4 with pinnately veined leaves and 3 with palmately veined leaves. For each species, the researchers analyzed 10 leaves on 3 different plants.	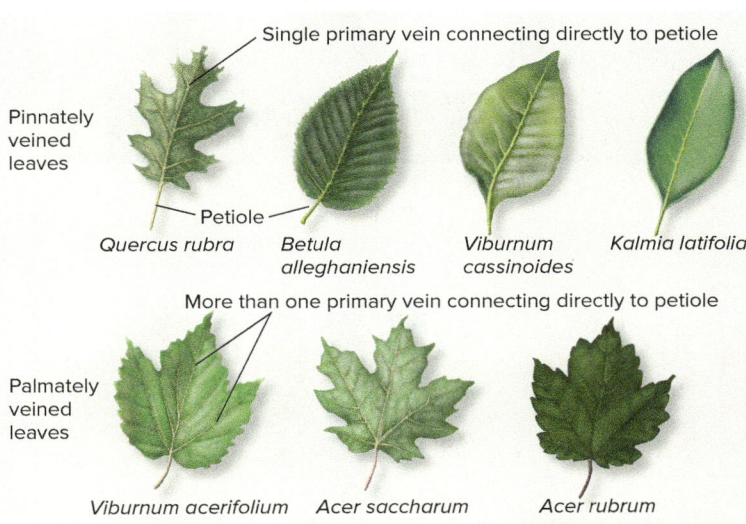	Locate pinnate and palmate leaves for comparison.
2. With the leaf still attached to the plant, use a scalpel to cut across 1 primary vein in 5 experimental leaves but not 5 control leaves.		This procedure initially cuts off water supply via 1 primary vein.
3. Cover the cuts with medical tape on both top and bottom of leaves. Tape same area of uncut control leaves.		The tape prevents infection. Controls: Tape is applied to controls for experimental consistency.
4. Fold cardboard over base of cut leaves, forming a splint. Splint the same area of uncut control leaves.		The cardboard prevents leaf collapse. Control: Cardboard is applied to controls for experimental consistency.
5. Measure water conduction 2–9 weeks after treatment, in 2 regions (A, B) of each pinnate leaf and in 3 regions (A, B, C) of each palmate leaf.		After 2–9 weeks, the cuts had healed. Measurement of water conduction will determine the effect of the vein cut.

6 THE DATA

* = Statistically significant difference from control

7 CONCLUSION Palmately veined leaves did not suffer as much conduction loss from a primary vein cut as did pinnately veined leaves.

8 SOURCE Sack, L., et al. 2008. Leaf Palmate Venation and Vascular Redundancy Confer Tolerance of Hydraulic Disruption. *Proceedings of the National Academy of Sciences of the U.S.* 105: 1567–1572.

Leaf Surface Features   Leaf surfaces also show adaptive features. Leaf epidermal cells secrete a cuticle composed of protective wax and polyester compounds. The cuticle helps plants to avoid drying in the same way that enclosure in waxed paper keeps food moist. Plants that grow in very arid climates often have thick cuticles, whereas plants native to moist habitats typically have thinner cuticles.

Some leaf epidermal cells may differentiate into spiky or hairlike projections known as **trichomes** (Figure 28.15). Blankets of trichomes offer protection from excessive light, UV radiation, extreme air temperature, excess water loss, or attack by herbivores—animals that consume plant tissues. Broken trichomes of the stinging nettle, for example, release a caustic substance that irritates animals' skin, causing them to avoid these plants. Leaf epidermal cells include pairs of specialized guard cells located on either side of stomata (Figure 28.15). These **guard cells** allow stomata to be open during moist conditions and to close when conditions are dry, thereby preventing plants from losing too much water. The opening and closing of stomata are described in more detail in Chapter 30.

Figure 28.15   Leaf surface features.   These features, viewed by SEM (with artificial colors), include cuticular wax, trichomes, and stomatal guard cells.
©Eye of Science/Science Source

AN INTRODUCTION TO FLOWERING PLANT FORM AND FUNCTION   623

(a) Tendrils

(b) Bud scales

(c) Bracts

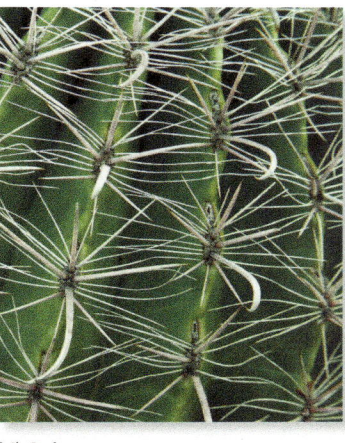
(d) Spines

Figure 28.16 Examples of modified leaves. (a) The tendrils of an American vetch plant are modified leaves that help the plant attach to a trellis. (b) Bud scales, such as those on this sycamore bud, are modified leaves that protect buds from winter damage. (c) The attractive red bracts of poinsettia are modified leaves that function like flower petals to attract pollinator insects to the small flowers. (d) Spines on surfaces of cacti such as the giant saguaro are modified leaves that function in defense.
(a) ©Ed Reschke/Getty Images; *(b)* ©McGraw-Hill Education/John Farmar; *(c)* ©Steve Terrill/Getty Images; *(d)* ©Don Paulson Photography/Purestock/SuperStock

Concept Check: *Because cactus leaves are so highly modified for defense that they cannot effectively accomplish photosynthesis, how do cacti obtain organic compounds?*

Modified Leaves Perform Diverse Functions

Though most leaves function primarily as photosynthetic organs, some plants produce leaves that are modified in ways that allow them to play other roles. For example, threadlike tendrils that help some plants attach to a supporting structure are modified leaves or leaflets (**Figure 28.16a**). The tough scales that protect buds on plants, such as the sycamore, from winter damage are modified leaves (**Figure 28.16b**). Poinsettia "petals" are actually modified leaves known as bracts, which are larger and more brightly colored than the flowers they surround and help to attract pollinators (**Figure 28.16c**). Cactus spines are modified leaves that have taken on a defensive role, leaving photosynthesis to the cactus stem (**Figure 28.16d**).

Primary and Secondary Vascular Tissues

Stems, leaves, roots, buds, flowers, and fruits all contain vascular tissues, in the form of xylem and phloem, which conduct water, minerals, and organic compounds. **Herbaceous plants** such as corn and bean produce mostly primary vascular tissues. In contrast, **woody plants** produce both primary and secondary vascular tissues. A comparison of primary and secondary vascular tissues will aid in understanding their roles.

Primary Vascular Tissues
Primary vascular tissues occur in primary xylem and phloem, described in more detail in Chapter 30. Primary xylem is a complex tissue containing several cell types (Table 28.2): unspecialized parenchyma cells, stiff fibers that provide structural support, and two types of cells that facilitate water transport. Water transport cells of angiosperms include narrower tracheids and wider vessel elements. Arranged in pipeline-like arrays, tracheids and vessel elements conduct water, along with dissolved minerals and certain organic compounds (**Figure 28.17**). Mature tracheids and vessel elements are no longer living cells, and the absence of cytoplasm facilitates water flow. During development, these cells lose their cytoplasm by the

Biology Principle

Structure Determines Function
Because lignin strengthens water-conducting cells in xylem tissues, these cells do not change shape as large volumes of water move through them.

Figure 28.17 Water-conducting cells of the xylem. In this thinly sliced portion of a stem, the red-stained, lignin-impregnated walls of narrow tracheids and wider vessel elements can be distinguished.
©Lee W. Wilcox

process of programmed cell death, a process that resembles apoptosis in animals. However, the cell walls of tracheids and vessel elements don't easily break down or collapse because they are impregnated with the tough polymer lignin (see Chapter 25). The rigid cell walls of tracheids and vessel elements not only foster water conduction but also help to support the plant body. Like the walls of plastic plumbing pipes, lignin provides the hydrophobic surface needed for water movement, as well as the strength to support trees weighing more than 2,000 metric tons.

Primary phloem, which is composed of living cells, transports organic compounds such as sugars and certain minerals in a watery solution. Phloem tissue has sieve-tube elements, thin-walled cells that are arranged end to end to form pipelines (**Figure 28.18**). Pores in the end walls of sieve-tube elements allow solutions to move from one cell to another. Phloem tissue also includes companion cells that aid sieve-tube element metabolism, supportive fibers, and parenchyma cells (Table 28.2). Phloem fibers are tough-walled sclerenchyma cells that are surprisingly long (20–50 mm) and valued for their high strength. The phloem fibers of hemp (*Cannabis sativa*), flax (*Linum usitatissimum*), jute (*Corchorus capsularis*), kenaf (*Hibiscus cannabinus*), and ramie (*Boehmeria nivea*) are commercially important in the production of rope, textiles, and paper.

Secondary Vascular Tissues Woody plants begin life as herbaceous seedlings that possess only primary vascular tissues. But as these plants mature, they produce secondary vascular tissues and bark. Secondary vascular tissues are composed of secondary xylem and secondary phloem. **Secondary xylem** is also known as **wood**, a component of plants that plays many important roles in human life. Wood is composed of about 25% lignin, 45% cellulose, and 25% other polysaccharides that are together known as hemicelluloses.

Secondary phloem is the **inner bark**. **Outer bark** is composed of protective layers of mostly dead cork cells that cover the outside of woody stems and roots. Therefore, bark includes both inner bark (secondary phloem) and outer bark (cork). Woody plants produce secondary vascular tissues by means of **secondary meristems,** also known as lateral meristems, which form rings of actively dividing cells that encircle the stem. The two types of secondary meristems are vascular cambium and cork cambium, which are derived from primary meristems.

The secondary meristem known as **vascular cambium** is a ring of dividing cells that produces secondary xylem to its interior and secondary phloem to its exterior (**Figure 28.19**). Secondary xylem conducts most of a woody plant's water and minerals. Cell divisions that occur in secondary meristems increase the girth of woody stems. During each new growing season, the vascular cambium produces new cylinders of secondary xylem and secondary phloem. In temperate trees, each year's addition of new secondary xylem forms growth rings that can be observed on the cut stem surfaces (**Figure 28.20**). The growth rings of secondary xylem surround the remains of the primary xylem and a central cylinder of parenchyma cells known as the pith. If environmental conditions favor plant growth, the growth rings formed at that time will be wider than those formed during stressful conditions. Climatologists use growth ring widths in samples of old wood to deduce past climatic conditions, and archaeologists use growth ring data to determine the age of wood constructions and artifacts left by ancient cultures.

Secondary xylem may transport water for several years, but usually only the current year's production of secondary phloem is active in food transport. This is because thin-walled sieve elements typically live for only a year. Thus, only a thin layer of phloem, the inner bark, is responsible for most of the sugar transport in a large tree. Deep abrasion of tree bark may damage this thin phloem layer, disrupting a tree's food transport.

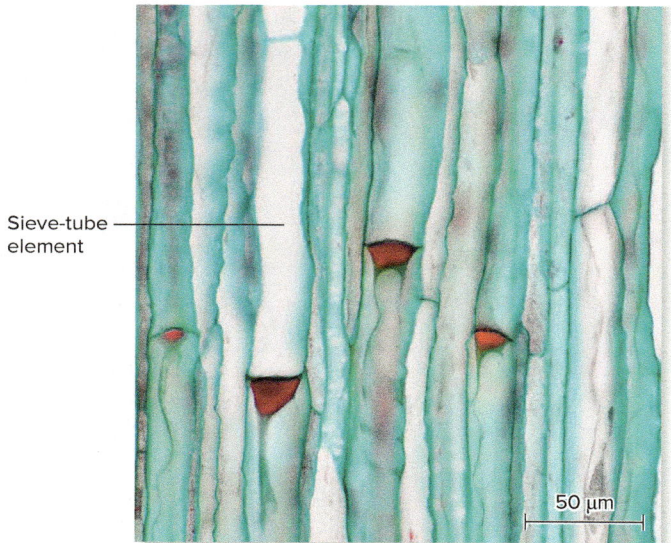

Figure 28.18 Food-conducting cells of the phloem. This thinly sliced portion of a stem shows green-stained, thin-walled sieve-tube elements that conduct watery solutions of certain minerals and organic compounds such as sugar. The red-stained material is a protein that seals wounded phloem cells, thereby preventing loss of fluids, much like blood clotting in humans.
©Lee W. Wilcox

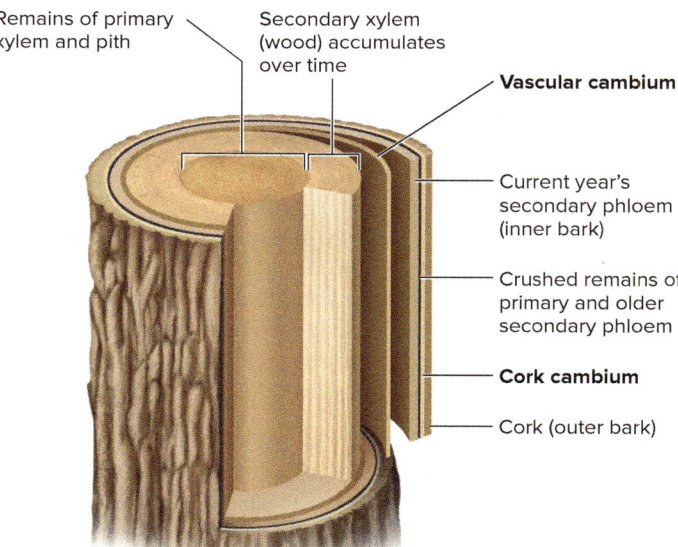

Figure 28.19 Formation of wood and bark by secondary (lateral) meristems. The vascular cambium is a thin cylinder of tissue that produces a thick cylinder of wood (secondary xylem) toward the inside of the stem and a thinner cylinder of inner bark (secondary phloem) toward the outside of the stem. The cork cambium forms an outer coating of protective cork (outer bark).

Figure 28.20 **The anatomy of a tree trunk.** Each year, a new cylinder of wood is produced; this yearly wood production appears as annual rings on the cut surface of a woody stem. The stem partial cross section at left is from a 1-year-old stem having a single secondary xylem growth ring. The stem partial cross section at right is from a stem that has more wood growth rings.
©Lee W. Wilcox

Concept Check: *Why do tree trunks have a thicker layer of wood (secondary xylem) than of inner bark (secondary phloem)?*

As a young woody stem begins to increase in diameter, its thin epidermis eventually ruptures and is replaced by outer bark, which is composed of protective cork tissues. Cork is produced by a secondary meristem called the **cork cambium,** another ring of actively dividing cells. The cork cambium surrounds the secondary phloem (see Figures 28.19 and 28.20). Together, the cork cambium, layers of cork tissue produced by the cambium, and associated parenchyma cells are known as a **periderm.** The outer bark becomes thicker as woody stems accumulate multiple periderm layers. The outer bark surface is often interrupted by passages known as **lenticels** that allow inner stem tissues to accomplish gas exchange.

Cork cells are dead when mature, and their walls are layered with suberin, a material that helps to prevent both attack by microbial pathogens and water loss from the stem surface. Cork tissues also produce tannins, compounds that protect against pathogens by inactivating their proteins. The cracked surfaces of tree trunks are dead cork tissues of the outer bark. Commercial cork is sustainably harvested from the cork oak tree (*Quercus suber*) for production of flooring material, bottle stoppers, and other items.

Modified Stems Display Diverse Forms and Functions

Stems mostly grow upright because light is required for photosynthesis. But some stems, known as rhizomes, occur underground and grow horizontally. For example, potato tubers are the swollen, food-storing tips of rhizomes. Grass stems also grow horizontally, as either rhizomes just beneath the soil surface or stolons, which grow along the soil surface. The leaves and reproductive shoots of grasses grow upward from the point where they are attached to these horizontal stems. Grass blades continue to elongate from their bases even if you cut their tips off, explaining why lawns must be mowed repeatedly during the growing season. The horizontal stems of grasses are adaptations that help to protect vulnerable shoot apical meristems against natural hazards such as fire and grazing animals.

28.3 Reviewing the Concepts

- Shoots are modular systems; each module includes a node, an internode, a leaf, and an axillary bud. Axillary buds develop from axillary meristems and may grow into new branches (Figure 28.12).
- Variations in leaf structure reflect adaptations that aid photosynthesis or protect against stress (Figures 28.13, 28.14, 28.15).
- Herbaceous plants are those whose stems produce little or no wood and are mostly composed of primary vascular tissues in the form of primary xylem and primary phloem (Figures 28.16, 28.17, 28.18).
- In addition to primary tissues, woody plants—trees, shrubs, and woody vines—possess secondary meristems that produce wood and bark. The vascular cambium produces secondary xylem (wood) and secondary phloem (inner bark). The cork cambium produces cork tissues that form outer bark (Figures 28.19, 28.20).

28.3 Testing Your Knowledge

1. Imagine that you are asked to draw a typical plant stem. Where would you correctly draw buds?
   a. anywhere along the stem
   b. anywhere on the surfaces of the internodes
   c. on the surfaces of leaves
   d. at any junction of leaf and stem
   e. only at the junction of the topmost leaf and the stem
2. What function might a leaf serve?
   a. photosynthesis
   b. attachment of the plant to a supporting structure
   c. protection
   d. defense
   e. All of the above are leaf functions.

28.4 Root System Adaptations

Learning Outcomes

1. List ways that root structure has been modified in different plants in response to different habitats.
2. Create a drawing that shows how root systems and branch roots develop.

Roots play the essential roles of absorbing water and minerals, anchoring plants in soil, and storing nutrients. The external form of roots varies among flowering plants, reflecting adaptation to particular life spans or habitats. In contrast, root internal structure is more uniform.

Modified Roots Display Diverse External Forms and Functions

In addition to the underground taproot systems of eudicots and the fibrous root systems of monocots (see Table 28.1), several types of root modification provide adaptive advantages in particular habitats. For example, corn and many other plants produce supportive prop roots from the lower portions of their stems. Many tropical trees grow in such thin soils that the trees are vulnerable to being blown down in windstorms. Such trees often produce dramatic aboveground buttress roots that help to keep trees upright (**Figure 28.21a**). Many mangrove trees that grow along tropical coasts produce pneumatophores (Greek, meaning breath bearers), roots that grow upward into the air (**Figure 28.21b**). Functioning as snorkels, pneumatophores absorb oxygen-rich air, which diffuses to submerged roots growing in oxygen-poor sediments. This is necessary because, like animal cells, root cells require a supply of oxygen in order to produce ATP. Roots use this ATP to power root growth and the uptake of mineral nutrients (see Chapter 30).

(a) Buttress roots

—Pneumatophores

(b) Pneumatophores

Figure 28.21 Modified aboveground roots. **(a)** Buttress roots help to keep tropical trees such as this *Ficus macrophylla* from toppling in windstorms. **(b)** Pneumatophores produced by mangroves are roots that extend upward into the air. These aerial roots take up air and then transmit it to underwater roots that grow in oxygen-poor sediments.
(a) ©Rob Broek/Getty Images; *(b)* ©Peter E. Smith/Natural Sciences Image Library

Root Internal Growth and Tissue Specialization Occur in Distinct Zones

Studies of gene expression in roots of the model plant *Arabidopsis thaliana* reveal remarkably complex internal structure, with at least 15 distinct cell types. For our purposes, a simpler microscopic examination of root internal structure reveals three major zones: a root apical meristem (RAM) protected by a root cap, a zone of root elongation, and a zone of maturation in which specialized cells can be observed (**Figure 28.22**). However, comparisons of gene transcription patterns show that these three zones are not sharply delineated but instead reflect a gene expression gradient. Higher expression of genes involved in development occurs at the root tip, and genes that function in differentiation are expressed at higher levels toward the zone of maturation.

- The root apical meristem (RAM) contains stem cells that surround a tiny region of cells that rarely divide, known as the quiescent center. Signals emanating from the quiescent center keep nearby stem cells in an undifferentiated state. Root stem cells farther away from the quiescent center produce new cells in multiple directions. (1) Toward the root tip, stem cells produce columella cells that sense gravity and touch, which helps roots to extend downward into the soil and around obstacles such as rocks. (2) At the sides of the quiescent center, stem cells produce the protective root cap and epidermal cells. Root tip epidermal cells secrete a sticky substance called mucigel that lubricates root growth through the soil. (3) Toward the shoot, stem cells generate cells that become internal root tissues such as xylem, phloem, endodermis, and cortex. Root tip damage reduces the ability of root systems to grow.

- Above the RAM lies the **zone of elongation,** in which cells extend by water uptake, thereby dramatically increasing root length.

- The **zone of maturation,** located above and overlapping with the root zone of elongation, is where most root cell differentiation and tissue specialization occur. Specialized root tissues include (1) mature vascular tissues at the root core, (2) an enclosing cylinder of cells known as the pericycle that produces branch roots (**Figure 28.23**), (3) another cell cylinder called the endodermis (meaning inside skin) that helps roots to avoid conducting toxic materials, (4) relatively unspecialized parenchyma cells forming a starch-storing cortex, and (5) epidermal cells at the root surface.

- **Root hairs** are specialized epidermal cells that can be as long as 1.3 cm, about the width of your little finger, but are only 10 μm in diameter (**Figure 28.24**). Their small diameter allows root hairs to obtain water and minerals from soil pores that are too narrow for even the smallest roots to enter. Root hair plasma membranes are rich in transport proteins that use ATP or an ion electrochemical gradient to selectively absorb materials from the soil. Damage to root hairs reduces a plant's ability to take up essential resources.

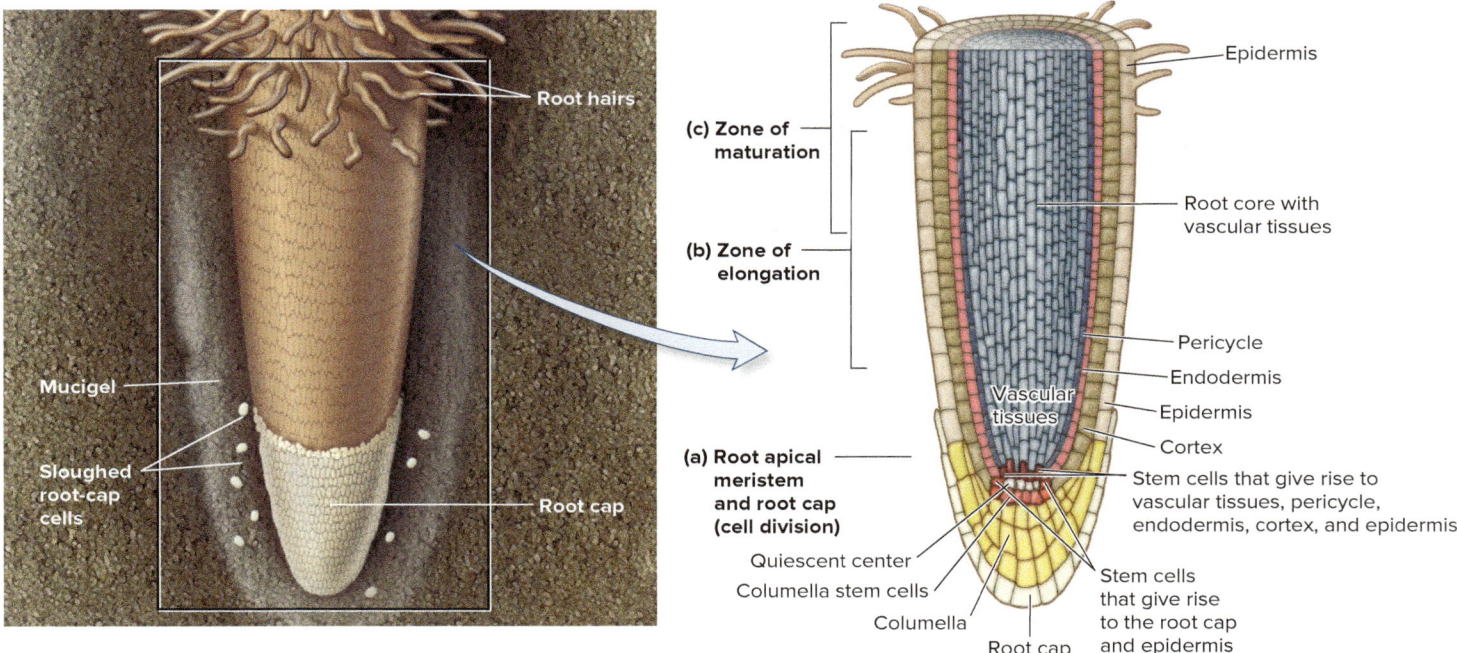

Figure 28.22 Three zones of root growth. A longitudinal view of a typical root reveals three major zones: **(a)** a root apical meristem region that includes stem cells, a quiescent zone, columella, and a root cap; **(b)** a zone of elongation; and **(c)** a zone of maturation, characterized by specialized cells and tissues, including epidermal root hairs, cylinders of endodermis and pericycle tissue, and a core of vascular tissue.

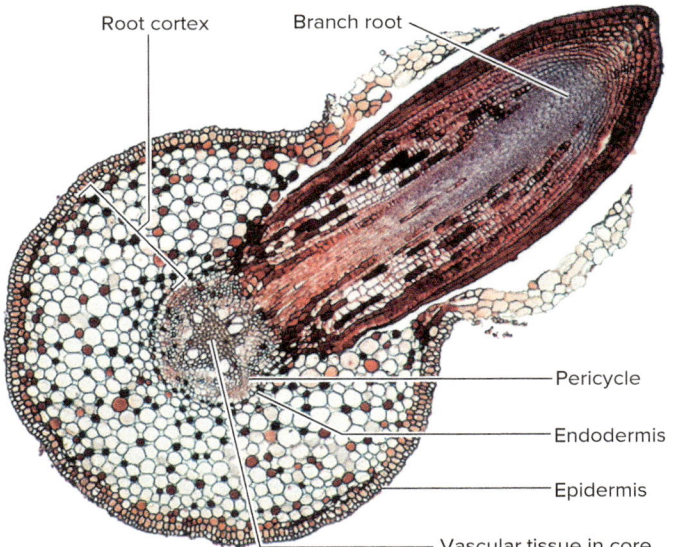

Figure 28.23 Cross section of a mature root. This stained light micrograph shows the epidermis and cortex of a root surrounding a central core of vascular tissue. An inner cortex layer is the endodermis, which surrounds a cylinder of meristematic pericycle tissue. The pericycle has produced a young branch root that has grown through the cortex and the epidermis.
©Lee W. Wilcox

 Why are lateral roots produced in this way?

Figure 28.24 Root hairs. Root hairs are so essential to root function that a young seedling, shown here, produces many of them, as do older plants.
©Nigel Cattlin/Alamy

28.4 Reviewing the Concepts

- Roots occur in multiple forms that reflect adaptation to environmental conditions (Figure 28.21).
- The internal organization of roots is comparatively uniform, and three major zones can be recognized on microscopic examination: the root apical meristem and root cap, a zone of cell and root elongation, and a zone of tissue maturation. Features of the mature root include an inner core of vascular tissue, a pericycle that produces lateral (branch) roots, an endodermis that functions in mineral selection, and epidermal root hairs that aid nutrient uptake (Figures 28.22, 28.23, 28.24).

28.4 Testing Your Knowledge

1. Which of the following structures is a type of root?
   a. endodermis
   b. epidermis
   c. cortex
   d. pneumatophore
   e. none of the above

2. Which of the following series of words most accurately describes the external structure of a root, starting at the tip and moving toward the shoot?
   a. zone bearing root hairs, zone lacking root hairs
   b. quiescent center, columella, stem cells
   c. root cap, zone of elongation, zone of maturation
   d. conducting tissues, endodermis, pericycle
   e. none of the above

Assessing Your Knowledge and Skills

Foundational Knowledge

Multiple Choice

1. Where would you look to find the gametophyte generation of a flowering plant?
   a. at the shoot apical meristem
   b. at the root apical meristem
   c. in seeds
   d. in flower parts
   e. Flowering plants lack a gametophyte generation.

2. Which type of plant is most likely to have food-rich roots that are useful as human food?
   a. an annual
   b. a biennial
   c. a perennial
   d. a centennial
   e. a plant that grows along coastal shorelines

3. Which of the following describes the distinctive architectural features of plants?
   a. radial symmetry and apical-basal polarity
   b. bilateral symmetry and apical-basal polarity
   c. radial symmetry and absence of apical-basal polarity
   d. bilateral symmetry and absence of apical-basal polarity
   e. absence of symmetry and absence of apical-basal polarity

4. Which is the most accurate description of how plants grow?
   a. by the addition of new cells at meristems that include stem cells
   b. by cell enlargement as the result of water uptake
   c. by both the addition of new cells and cell expansion
   d. by addition of fat cells
   e. All of the above are correct.

5. Where would you look for leaf primordia?
   a. at a vegetative shoot tip
   b. at the root apical meristem
   c. at the vascular cambium
   d. at the cork cambium
   e. in a floral bud

Short Answer

1. What would be the consequences if leaves were radially symmetric (shaped like spheres or cylinders)?
2. **PRINCIPLES** A principle of biology is that structure determines function. Why are most tall plants woody, rather than herbaceous?

Critical-Thinking Skills

Multiple Choice

1. An elementary school teacher wants to help her students learn about tree trunk structure by observing a stump. Which structure(s) would the children be able to see without difficulty, and which structure(s) might the students have difficulty locating without the teacher helping them by drawing on the stump with a marking pen?
   a. The children would be able to distinguish all of these structures: secondary phloem and xylem, annual rings, vascular and cork cambia, and outer bark.
   b. The children would be able to easily recognize vascular cambium, secondary phloem, and cork cambium without the teacher's markings, but not secondary xylem, annual rings, or outer bark.
   c. The children would be able to easily distinguish outer bark, secondary xylem, and secondary phloem, but not annual rings or the vascular or cork cambia unless the teacher made markings.
   d. The children would be able to distinguish secondary xylem (wood), annual rings of secondary xylem, and outer bark but not the vascular cambium, secondary phloem, or cork cambium unless the teacher made markings.
   e. The children would likely not be able to distinguish features other than outer bark.

2. A large population of beetles has fed on and thereby destroyed about half of the terminal and axillary buds produced by a tree during one summer. What effects would you expect to see the following spring?
   a. Spring leaf production would be lower than in previous years.
   b. Flowers would be more sparse than in previous years.
   c. Fewer later shoots (newest branches) would occur in spring, by comparison to previous years.
   d. All of the above are reasonable expectations.
   e. The tree is not likely to suffer any impacts of bud loss during the previous summer.

3. Water and cellular cytoplasm are more dense than air. In a well-hydrated plant, which leaf tissue(s) is (are) probably the least dense, and why?
   a. the upper epidermis, because a waxy cuticle is present
   b. the lower epidermis, because stomata are normally present
   c. the palisade parenchyma, because elongate cells are pressed closely together
   d. the spongy parenchyma, because this tissue contains air spaces
   e. the vascular tissues, because tracheids and vessels are normally filled with water and solutes

4. Imagine that some precious plants have accidentally been treated with chemicals that destroy the underground root system. How might these plants respond to this treatment and survive?
   a. New adventitious roots might be able to form on stem cuttings that have been placed in water.
   b. Some of the plants might be able to produce adventitious roots from leaves.
   c. If the plants are monocots, they might regenerate a new root system from stem tissue. Monocots normally produce mature root systems in this way, because the monocot embryonic root dies soon after seed germination.
   d. All of the above are reasonable scenarios.
   e. None of the above is likely to happen.

5. During its development, a water-conducting tracheid elongates in a direction parallel to the shoot or root axis. Based on this information, what can you say about the orientation of cellulose cell-wall microfibrils and cytoplasmic microtubules in this developing tracheid?
   a. The microfibrils will be oriented perpendicularly (at right angles) to the long axis of the developing tracheid, encircling it, but the cytoplasmic microtubules will be oriented parallel to the direction in which the tracheid is elongating.
   b. Microfibrils and microtubules will both be oriented perpendicularly (at right angles) to the elongating axis of the tracheid.
   c. Microfibrils and microtubules will both be oriented parallel to the direction of tracheid elongation.
   d. Microfibrils will be oriented parallel to the direction of tracheid elongation, but microtubules will be perpendicular (at right angles) to both the microfibrils and the elongating tracheid.
   e. None of the above is correct.

Short Answer

1. Water lilies produce large leaves that float on water surfaces. These leaves are connected to bottom-growing rhizomes (horizontal stems like those produced by many grasses) by long petioles. A friend asks you why water lilies keep growing densely around his fishing dock even when he regularly cuts the leaf stalks. How would you answer?

2. Mangroves provide valuable ecological services in coastal areas by moderating wave effects and serving as nurseries for wildlife. What would happen to these valuable plants if pneumatophores were destroyed?

29

How Flowering Plants Sense and Interact with Their Environments

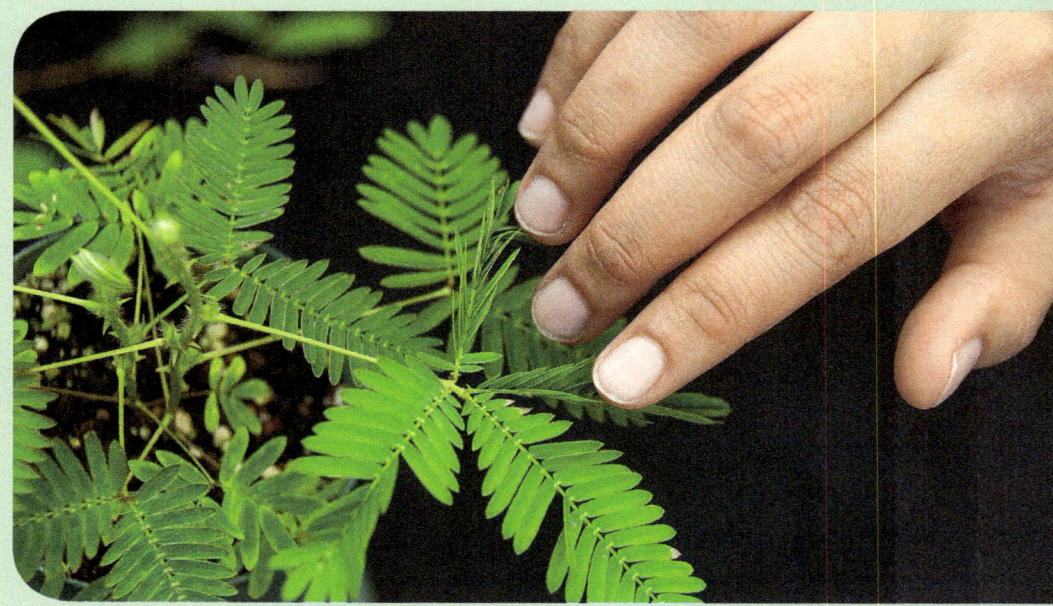
©Linda Graham

Behavior of the sensitive plant, *Mimosa pudica*. When touched, the leaves of the sensitive plant quickly fold, and only slowly unfold.

Chapter Outline

- 29.1 Overview of Plant Behavioral Responses
- 29.2 Plant Hormones
- 29.3 Plant Responses to Light
- 29.4 Plant Responses to Gravity and Touch
- 29.5 Plant Responses to Attack

Assessing Your Knowledge and Skills

If you touch a leaf of a sensitive plant (*Mimosa pudica*), in less than a second the leaf will fold, only slowly returning to its original flat position (see the chapter-opening photo). Such rapid leaf folding is an example of plant behavior, in this case a response to touch. Experts are not completely sure why sensitive plants fold their leaves when touched. One possibility is that leaf folding in response to the touch of wind reduces the amount of leaf surface exposed to evaporation, thereby helping the plant to retain water. Sensitive plant leaf folding is one of the fastest of plant behavioral responses, but there are many more. Plants also respond to changes in the Sun's position, the period of daily illumination, gravity, attack by animals and disease-causing microorganisms, and other environmental stimuli. This chapter provides examples of ways in which understanding plant behavior has been key to increasing agricultural productivity and protecting natural ecosystems for human benefit.

We will begin with a survey of the diverse types of internal and external stimuli that induce plant behavior and review how cells perceive and respond to stimuli by means of signal transduction pathways. Next, we will focus on plant hormones, which are major types of internal mobile molecules that influence plant behavior. Finally, we will consider how responses to light, gravity, touch, and attack foster plant survival and reproduction.

29.1 Overview of Plant Behavioral Responses

Learning Outcomes
1. List examples of plant responses to biological and physical stimuli.
2. Describe the three stages of cell signaling.

You might not have noticed that plants display behavior, possibly because much of it occurs at a slower pace than humans easily observe. But plants, like all living things, respond to various types of stimuli, which is a definition of behavior. Examples of plant behavior include plant movements, some types of which were described in 1880 by Charles Darwin and his son Francis in their book *The Power of Movement in Plants*. Modern time-lapse photography, which makes plant behavior more obvious to humans, reveals that most plants are constantly in motion, bending, twisting, or rotating in dancelike movements known as nutation (Figure 29.1). Many other intriguing examples of plant behavior exist:

- Plant shoots typically grow toward light and against the pull of gravity, and most roots grow toward water and in the same direction as the gravitational force.
- Seeds may germinate when they detect the presence of sufficient light and moisture for successful seedling growth.
- Plants produce flowers, fruit, and seeds only at the season most favorable for reproductive success.
- Plants take protective actions when they sense attack by disease-causing microbes or hungry animals, thereby preventing excessive damage to their own bodies or those of neighboring plants.

Figure 29.1 Example of plant movement. Sixteen superimposed photographs of a shoot of the honeysuckle vine *Lonicera japonica*, taken over a period of 2 hours, reveal the circular movement known as nutation.
©Digital Photography by Ash Kaushesh, University of Central Arkansas, Conway, Arkansas 72035 USA. Courtesy Botanical Society of America, St. Louis, MO, www.botany.org

Plant Behavior Involves Responses to Biological and Physical Stimuli

Most people are aware that both biological and physical stimuli influence animal behavior. Bird nesting behavior in spring, for example, involves hormonal changes triggered by seasonal conditions. Plants likewise respond to biological and physical stimuli (Figure 29.2).

Biological Stimuli Biological stimuli can originate from inside or outside the plant body. Plants respond to two types of internal stimuli: internal biological clocks and mobile chemical signals. Internal biological clocks, based on **circadian rhythms,** occur not only in plants but also in animals and other organisms. The word circadian comes from the Latin words meaning about and day. Circadian rhythms evolved under the influence of Earth's rotation, which causes the regular alternation of night and day. Leaf movements, flower opening, fragrance emission, and many other behaviors result from the operation of circadian rhythms.

Like animals, plants also respond to mobile chemical signals that are produced within the body and move from one location to another, acting at very low concentrations. In plants, these chemical signals may move short distances from cell to cell or long distances within the vascular system. Mobile plant chemical signals include transcription factor proteins and microRNAs as well as chemically different compounds known as **hormones.** Some plant hormones are similar to particular animal hormones. One example is the plant hormone jasmonic acid, which is chemically similar to prostaglandins produced by animals. The gaseous hormone nitric oxide (NO) functions in both animals and plants. Other hormones are distinctive to plants. As in the case of animal hormones, plant hormones often interact with each other and with external biological and physical stimuli to maintain a stable internal environment (homeostasis) and enable progress through the life stages described in Chapters 28 and 31.

Other biological stimuli arise from outside the plant body; these include

- herbivores, animals that consume plant parts;
- airborne or soil pathogens, disease-causing microbes;
- beneficial microorganisms; and
- chemicals emitted from neighboring plants.

Physical Stimuli Plants sense and respond to many types of physical stimuli (Figure 29.2). Physical stimuli in natural plant environments include light, atmospheric gases such as CO_2 and water vapor, temperature, touch, wind, gravity, soil water, rocks and other barriers to root growth, and soil minerals. Crop plants also respond to applications of agricultural chemicals. That plants have evolved such a broad array of sensory capacity is not surprising, because all of the listed environmental influences affect plant survival and reproduction.

Sensing and Responding to Environmental Stimuli Though plants lack the specialized sense organs typical of animals, receptor molecules located in plant cells sense stimuli and cause responses. When many cells of a tissue receive and respond to the same

> # Biology Principle
>
> **Biologists Use Models to Test Predictions and Convey Ideas**
>
> This figure is a model of plant biologists' understanding of physical and biological features of plant environments that influence plant growth responses. This understanding is key to the design of experiments conducted on Earth. The Modeling Challenge in the figure legend asks you to consider which features are most likely to be the subjects of experiments conducted in space.
>
>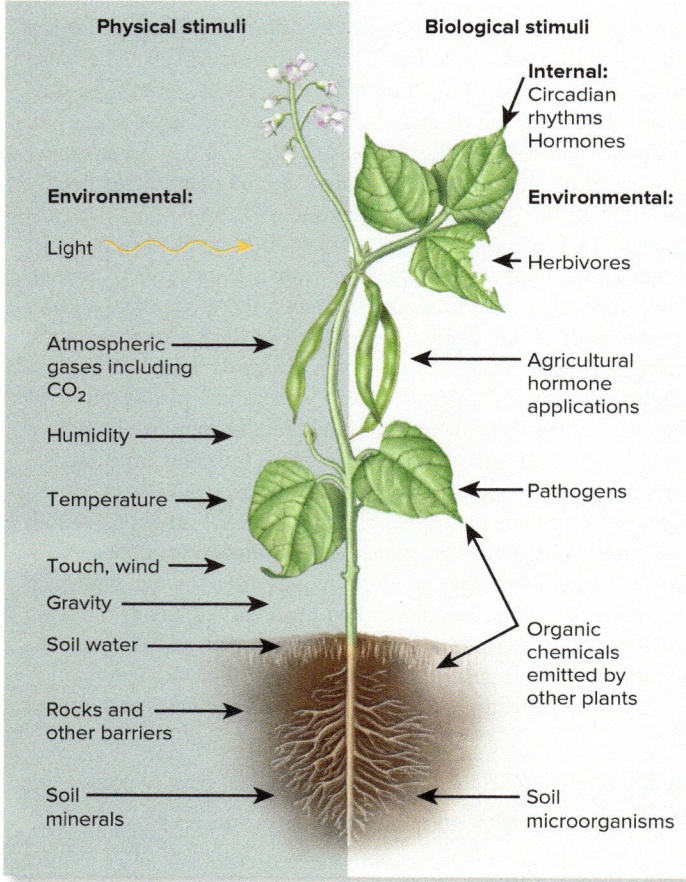
>
> **Figure 29.2** Types of plant stimuli. Plants respond to both physical and biological stimuli. Stimuli may be internal to the plant or come from the environment.

> **Modeling Challenge:** Figure 29.2 depicts diverse physical and biological stimuli that typically affect plant growth responses. Earth-bound biologists keep this model of plant responses in mind when designing experiments to be conducted in the field or in a lab. Sketch a parallel model showing physical and biological stimuli that plant biologists might consider when designing experiments to be conducted in space, such as in an orbiting laboratory. In such a location, dirt, rocks, and other biological organisms are unlikely to be welcome, unless well contained.

biological or physical stimuli, entire organs or plant bodies display behavior. For example, plants tend to grow toward a light source. This process, known as positive **phototropism,** involves both a cellular perception of light and a growth response of stem tissue to an internal chemical signal. In general, a tropism is a growth response that depends on a stimulus that occurs in a particular direction. In the case of phototropism, cells in plant shoot tissues that face a light source sense the light and cause a plant hormone known as auxin to move across the stem into cells not facing the light. In stem tissues not facing the light, the inward movement of auxin influences gene expression, causing cells to expand by taking up water (refer back to Figure 28.9). Because cells not facing the light expand more than the cells facing the light, the stem bends toward light. Bending toward light enables the plant to acquire more light energy for photosynthesis than if bending did not occur. Phototropism is only one of many examples of ways in which plants perceive their physical environments and respond to them.

We will next review how plant cells more generally receive signals and transmit them intracellularly. This process, known as cell communication (also called cell signaling), occurs in all cells.

Plant Cell Communication Involves Receptors, Messengers, and Effectors

Cell communication is the process in which a cell perceives a physical or chemical signal, thereby switching on an intracellular pathway that leads to a cellular response. Cell communication often involves three types of molecules and processes:

1. Receptors that may become activated in the process known as receptor activation
2. Second messengers that transmit signals in the process known as signal transduction
3. Effector molecules that cause a cellular response

Receptors are sensor proteins that become activated when they receive a specific type of signal (**Figure 29.3**). Receptors occur in diverse cellular locations. Some plant defense receptors are located within the plasma membrane, light receptors may occur in the cytosol, and auxin receptors are found in the nucleus. Some activated receptors directly generate a response, such as an increased flow of ions across a membrane. In contrast, many activated receptors first bind to molecules that initiate an intracellular communication pathway. For example, the binding of certain defense hormones to plasma membrane receptors results in the intracellular production of second-messenger molecules that transmit the signal.

Second messengers transmit messages from many types of activated receptors. Cyclic AMP (cAMP), inositol trisphosphate (IP_3), and calcium ions (Ca^{2+}) are major types of second messengers in animal and plant cells. Ca^{2+} is a particularly common messenger in plant cells. Touch and various other stimuli cause Ca^{2+} to flow from storage sites in the endoplasmic reticulum (ER) lumen into the cytosol. Calmodulin or other calcium-binding proteins then bind the Ca^{2+}. Calcium binding alters the structure of the proteins, causing them to interact with other cell proteins or to alter enzymatic function.

Effectors are molecules that directly influence cellular responses. In plants, calcium-dependent protein kinases (CDPKs)

Biology Principle

Cells Are the Simplest Units of Life

Internal and environmental stimuli are received and elicit responses at the cellular level.

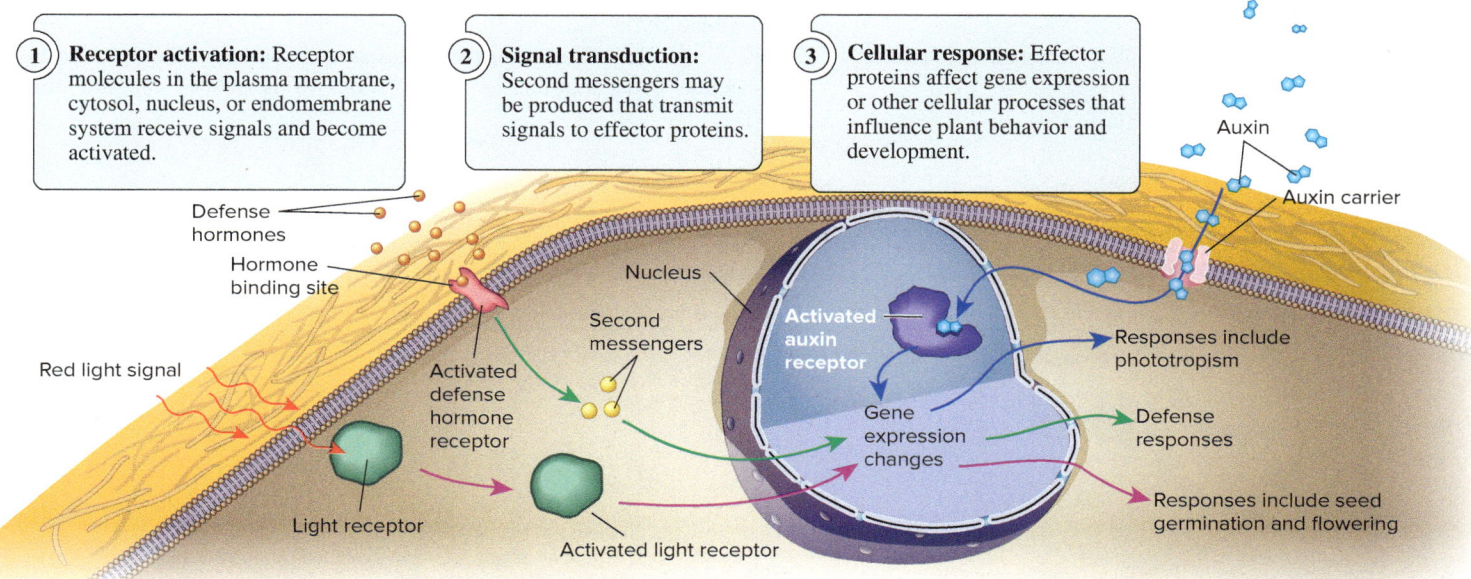

Figure 29.3 An overview of plant cell communication. Plant cells respond to hormonal signals produced within the plant body, as well as to environmental stimuli. Three different signal transduction processes are shown here: one initiated when a compound senses light, one initiated when a defense hormone binds to a plasma membrane receptor, and one initiated when the plant hormone auxin binds a receptor located in the nucleus.

BioConnections: *Refer back to Figure 8.4, which shows an example of a cell-communication process in an animal cell. Which of the plant processes illustrated above is most similar?*

---

are particularly important effector molecules. Cell communication ends when an effector causes a cellular response, such as opening or closing an ion channel or switching the transcription of particular genes on or off. A single activated receptor can dispatch many second-messenger molecules, which in turn can activate scores of effectors, leading to many molecular responses within a single cell.

29.1 Reviewing the Concepts

- Plants sense and respond to diverse biological and physical signals and thus display behavior (Figures 29.1, 29.2).
- During the process of cell communication, cellular receptors respond to environmental physical and biological stimuli as well as to internal hormonal signals. The communication process involves receptor activation, signal transduction by messengers, and a cellular response (Figure 29.3).
- Cellular responses include changes in gene expression and ion channels that influence plant growth, development, reproduction, defense, and movements such as leaf folding.

29.1 Testing Your Knowledge

1. Which of the following processes most accurately reflects cell communication during the plant behavior known as phototropism?
   a. In a stem, plant cells that face a light source sense this physical signal, causing the movement of the hormone auxin across the stem to cells not facing the light; cells not facing the light respond to auxin entry by expanding, causing the stem to bend toward the light.
   b. In a stem, plant cells not facing a light source cause auxin to move across the stem to cells facing the light; in response to auxin, cells facing the light contract, causing the stem to bend toward the light.
   c. In a stem, plant cells facing a light source cause auxin to move across the stem to cells not facing the light; the contraction of cells not facing the light causes the stem to bend away from the light.
   d. In a stem, plant cells not facing a light source cause auxin to move across the stem to cells facing the light; cells facing the light expand, causing the stem to bend away from the light.
   e. None of the above is correct.

29.2 Plant Hormones

Learning Outcomes

1. Explain what plant hormones are.
2. Discuss the general mechanism by which plant hormones affect gene expression.
3. List some ways in which plant hormones are relevant to agriculture.

Plant hormones are chemical signals transported within the plant body; when taken up by target cells, the signals elicit responses. Plant hormones are small molecules synthesized in metabolic pathways that also make amino acids, nucleotides, sterols, or secondary metabolites. In this section, we will focus on the major plant hormones: auxins, cytokinins, gibberellins, ethylene, abscisic acid, and brassinosteroids (**Table 29.1**).

Individual plant hormones often have multiple effects, and different concentrations or combinations of hormones can produce

Table 29.1 The Major Types of Plant Hormones

Type of plant hormone	Chemical structure of an example	Functions (a partial list)
Auxins	Indoleacetic acid (IAA)	Establish apical-basal polarity, induce vascular tissue development, mediate phototropism, promote formation of roots, inhibit leaf and fruit drop, and stimulate fruit development
Cytokinins	Zeatin	Promote cell division, influence cell specialization and plant aging, activate secondary meristem development, promote adventitious root growth, and promote shoot development in plant tissue culture
Gibberellins	Gibberellic acid	Stimulate cell division and cell elongation, stimulate stem elongation and flowering, and promote seed germination
Ethylene	Ethylene ($H_2C=CH_2$)	Promotes seedling growth, induces fruit ripening, plays a role in leaf and petal aging and drop, and coordinates defenses against osmotic stress and pathogen attack
Abscisic acid	Abscisic acid	Slows or stops metabolism during environmental stress, induces bud and seed dormancy, prevents seed germination in unfavorable conditions, and promotes stomatal closing
Brassinosteroids	Brassinolide	Promote cell expansion, stimulate shoot elongation, retard leaf drop, stimulate xylem development, and promote stress responses

distinct growth or developmental responses. Several plant hormones are known to act by removing gene repressors, thereby allowing gene expression to occur. This general mechanism, which is akin to releasing the brake on a car, allows hormones to cause relatively rapid responses. A closer look at the major types of plant hormones reveals their multifaceted roles.

Auxins Are the Master Plant Hormones

Plants produce several types of **auxins,** which are often considered to be the master plant hormones because they, often working with other hormones, influence plant structure, development, and behavior in many ways. Indoleacetic acid (IAA) is one plant auxin (Table 29.1), but other natural and artificial compounds have similar structures and effects. In this section, we will refer to this family of related compounds simply as auxin.

Auxin exerts so many effects because it promotes the expression of diverse genes, together known as **auxin-responsive genes.** Under low auxin conditions, proteins called Aux/IAA repressors prevent plant cells from expressing these genes. The repressor proteins prevent gene expression by binding to activator proteins at gene promoters. When the auxin concentration is high enough, auxin molecules glue repressors onto a protein complex called TIR1, which causes the breakdown of the repressors. Free of the repressors, the activator proteins enhance the expression of auxin-responsive genes.

Auxin Transport The way in which auxin is transported into and out of cells is integral to its effects in plants. Auxin is produced in apical shoot tips and young leaves, and it is directionally transported from one living parenchyma cell to another. Auxin in an uncharged form (IAAH) may enter cells from intercellular spaces by means of diffusion; however, the negatively charged form (IAA⁻) requires the aid of a plasma membrane protein known as the **auxin influx carrier (AUX1).** Several types of proteins, called PIN proteins, transport auxin out of cells. They are named for the pin-shaped shoot apices of plants having mutations in *PIN* genes. Because they transport auxin out of cells, PIN proteins are called **auxin efflux carriers.** They are necessary because in the cytoplasm, auxin occurs as a charged ion that does not readily diffuse out of cells.

In shoots, AUX1 is located at the apical ends of cells, whereas PIN proteins often occur at the basal ends (Figure 29.4a). This polar distribution of auxin carriers explains polar auxin movement (also

(a) Cellular mechanism of auxin transport

(b) Auxin transport throughout a plant

Figure 29.4 Auxin transport. (a) Polar and lateral auxin movement is controlled by the distribution of auxin efflux carriers located in the plasma membrane. When efflux carriers primarily occur at the basal ends of cells, auxin will flow downward. Auxin may flow laterally when auxin efflux carriers occur at the sides of cells. (b) In a whole plant, auxin primarily flows downward from shoot tips to root tips, where it then flows upward for a short distance.

Concept Check: *How could auxin carriers be organized to allow auxin to move upward in roots?*

called **polar transport**) in which auxin primarily flows downward in shoots and into roots (**Figure 29.4b**). However, the locations of auxin carriers can also change within cell plasma membranes, allowing lateral (Figure 29.4b) or upward transport of auxin. Differences in the presence and positions of auxin carrier proteins explain variations in auxin concentration within plants. By measuring the local auxin concentration, plant cells determine their position within the plant body and respond by dividing, expanding, or specializing.

Auxin Effects In nature, auxin influences plants throughout their lifetimes. In addition to mediating phototropism, auxin establishes the apical-basal polarity of seed embryos, induces vascular tissue to differentiate, promotes formation of roots, and stimulates fruit development. Many of auxin's effects are of practical importance to humans:

- Auxin treatment is used to produce some types of seedless fruit.
- Growers may use auxin sprays to retard premature fruit drop in orchards.
- To foster growth of whole plants from stem cuttings, people often dip cut stems into powdered auxin to stimulate root development.
- To make many clones of a valuable plant, in a process called plant tissue culture, growers first culture cells taken from the plant, grow many cultures from these cells, then induce mature plants to develop from the cell cultures by using a hormone treatment that includes auxin. This process can be used to generate many plants for market or to help preserve rare species.

Cytokinins Stimulate Cell Division

Like auxins, the plant hormones known as **cytokinins** play varied and important roles throughout the lives of plants. One example of a cytokinin is the compound zeatin, named for the corn genus *Zea* from which zeatin can be obtained (Table 29.1). The name cytokinins reflects these hormones' major effect—an increase in the rate of plant cytokinesis, or cell division. Root tips are major sites of cytokinin production, but shoots and seeds also make this plant hormone. Transported in the xylem to meristems and other plant parts, cytokinins bind to receptors thought to be located in the plasma membrane. At shoot and root tips, cytokinins influence meristem size, stem cell activity, and vascular tissue development. Cytokinins are also involved in root and shoot growth and branching, the production of flowers and seeds, and leaf aging. Plant tissue culture depends on hormone treatment with cytokinin, in addition to auxin and gibberellins.

Gibberellins Stimulate Cell Division and Elongation

The **gibberellins** (also known as gibberellic acids) (Table 29.1) are another family of plant hormones. Gibberellins are produced in apical buds, roots, young leaves, and seed embryos. In addition to promoting shoot development in plant tissue culture, gibberellins interact with light and other hormones to foster seed germination and enhance stem elongation and flowering in nature. Gibberellins also retard leaf and fruit aging. These multiple effects largely arise from gibberellin's stimulatory effects on cell division and elongation.

More than a hundred different forms of gibberellin have been found. Many kinds of dwarf plants are short because they produce less gibberellin than taller varieties of the same species. The dwarf strain of pea plants Mendel used in some of his breeding experiments is an example. When dwarf varieties of plants are experimentally sprayed with gibberellin, their stems grow to normal heights. However, dwarf wheat and rice crops are valued in agriculture because they can be more productive and less vulnerable to storm damage than taller varieties. Since the discovery of gibberellin, plant scientists have determined how this plant hormone works at the molecular level and how gibberellin regulation of plant growth evolved.

EVOLUTIONARY CONNECTIONS

Plant Gibberellin Responses Evolved in a Stepwise Manner

In flowering plants, gibberellin works by helping to liberate repressed transcription factors, allowing gene transcription to proceed. When gibberellin is absent, certain proteins (known as DELLA proteins) bind to particular transcription factor proteins needed for the expression of gibberellin-responsive genes. In this way, DELLAs function as brakes that restrain cell division and expansion. When sufficient gibberellin is present, it releases the brakes by binding to a protein known as GID1 protein. Gibberellin binding to GID1 causes DELLAs to release transcription factor proteins and fosters the destruction of DELLAs. In the absence of DELLA proteins, transcription factor proteins are able to bind the promoter regions of gibberellin-responsive genes, allowing their expression. As a result, cell division and expansion occur, leading to growth.

In 2007, plant scientists Yuki Yasumura, Nicholas Harberd, and their colleagues reported that the components of the gibberellin-DELLA mechanism in flowering plants arose from features present in more ancient lycophyte and bryophyte lineages (described in Chapter 25). Modern bryophytes and lycophytes possess GID1 as well as DELLA and the transcription factor proteins it binds, but these substances do not affect plant growth. This observation suggests that DELLA-mediated repression of plant growth evolved after the divergence of lycophytes but prior to the appearance of the first gymnosperms and angiosperms (**Figure 29.5**). Though the necessary components were present earlier, they did not assemble into a growth regulation system until relatively late in plant evolutionary history.

Ethylene's Effects Include Cell Expansion

The plant hormone **ethylene** is particularly important in coordinating plant developmental and stress responses. Ethylene is a simple

Biology Principle

Populations of Organisms Evolve from One Generation to the Next and Are Related by an Evolutionary History

The gibberellin-DELLA system that controls the growth of flowering plants evolved in a step-by-step fashion.

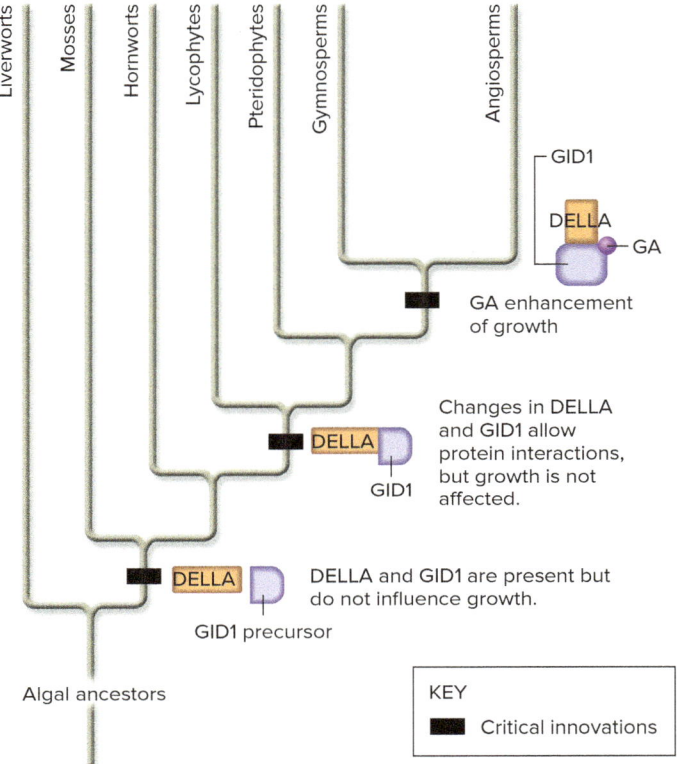

Figure 29.5 **Evolution of the gibberellin-DELLA system.** Although DELLA and GID1 proteins are present in a bryophyte, they do not interact. In lycophytes, DELLA binds to the GID1 protein, a process that also occurs in seed plants, but the lycophyte DELLA system does not influence growth. Gymnosperms and angiosperms have a fully functional system in which gibberellin binds to GID1, thereby also binding to DELLA proteins. After this, gibberellin-responsive gene expression occurs. This pattern suggests that the process by which gibberellin operates in flowering plants evolved from components present in earlier-diverging plant lineages.

Figure 29.6 Seedling growth showing the triple response to ethylene.
©Nigel Cattlin/Alamy

Concept Check: *What adaptive advantage does this seedling behavior provide?*

hydrocarbon gas produced during seedling growth, flower development, and fruit ripening (Table 29.1). In the root tip, ethylene determines how many stem cells remain inactive and how many cells undergo divisions. This hormone also plays important roles in defense against osmotic stress and pathogen attack, as well as leaf and petal aging and drop. As a gas, ethylene is able to diffuse through the plasma membrane and cytosol to bind to ethylene receptors localized in the ER. When activated by ethylene binding, these receptors inactivate a particular protein, enabling the transcription of various genes.

One effect of ethylene is to increase the disorder of microtubules within cells, thereby causing random orientation of cell-wall microfibrils. As a result, cells exposed to ethylene tend to expand in all directions rather than elongating (refer back to Figure 28.11). Ethylene's effects on cell expansion explain this hormone's important role in dicot seedling growth. The tender apical meristems of seedlings could be easily damaged during their growth through crusty soil. Ethylene helps seedlings to avoid such damage by inducing what is known as the triple response (Figure 29.6). First, ethylene prevents the seedling stem and root from elongating. Second, the hormone induces the stem and root to swell radially, thereby increasing in thickness. Together, these responses strengthen the seedling stem and root. Third, the seedling stem bends so that embryonic leaves and the delicate meristem grow horizontally rather than vertically. The bent portion of the stem, known as a hook, then pushes up through the soil. The hook forms as the result of an imbalance of auxin across the stem axis, which causes cells on one side of the stem to elongate faster than cells on the other side. Ethylene drives this auxin imbalance.

Knowledge of the effects of ethylene on fruit has been very useful commercially. Ripe fruit can be easily damaged during transit, but tomatoes and apples can be picked before they ripen for transport with minimal damage. At their destination, such fruit can be ripened by treatment with ethylene. However, fruit that becomes overripe may exude ethylene, which hastens ripening in nearby, unripe fruit. For this reason, fruit that must be stored for extended periods is kept in ethylene-free environments.

Several Hormones Help Plants to Cope with Environmental Stresses

Several plant hormones share the property of helping plants to respond to environmental stresses such as flooding, drought, high salinity, cold, heat, and attack by disease-causing microorganisms and animal herbivores. These protective hormones include the major plant hormones known as **abscisic acid (ABA)** and **brassinosteroids** (Table 29.1). Additional protective hormones are salicylic acid (SA), whose chemical structure is similar to that of aspirin, and a peptide known as systemin. The fragrant compound jasmonic acid and the gas nitric oxide (NO) are also protective plant hormones.

Abscisic Acid Abscisic acid (ABA) was named at a time when plant biologists thought that it played a role in leaf or fruit drop, also known as abscission. Later, they discovered that ethylene actually causes leaf and fruit abscission, whereas abscisic acid slows or stops plant metabolism when growing conditions are poor. Other important effects of ABA include the following:

- **Induction of bud and seed dormancy.** In preparation for winter, ABA stimulates the formation of tough, protective scales around the dormant buds of perennial plants. Seed coats of apple, cherry, and other plants also accumulate ABA, which prevents seeds from germinating unless temperature and moisture conditions are favorable for seedling growth. Dormant buds and seeds resume growth only when specific environmental signals reveal the onset of conditions suitable for survival.
- **Stomatal closure.** Water-stressed roots produce ABA, which is then transported to shoots, where (together with ABA produced by water-stressed leaf mesophyll) it helps to prevent evaporation of water from leaf surfaces by inducing leaf pores (stomata) to close.

Brassinosteroids Named after the cruciferous plant genus *Brassica* (which includes cabbage and broccoli), from which they were first identified, brassinosteroids occur in seeds, fruit, shoots, leaves, and flower buds of all types of plants. Brassinolide is an example of a brassinosteroid (Table 29.1). Brassinosteroids are chemically related to animal steroid hormones, such as the sex hormones testosterone and estrogen. In plants, brassinosteroids induce vacuole water uptake and influence enzymes that alter cell-wall carbohydrates, thereby fostering cell expansion. Mutations that affect brassinosteroid synthesis cause plants to exhibit dwarfism. Such plants have small, dark green cells because their tissues are unable to expand. Brassinosteroids also impede leaf drop, help grass leaves to unroll, and stimulate xylem development. They can be applied to crops to help protect plants from heat, cold, high salinity, and herbicide injury.

29.2 Reviewing the Concepts

- Plant hormones interact with environmental stimuli to control plant development, growth, and behavior (Table 29.1).
- Auxin plays an important role in many aspects of plant behavior, including phototropism. The position of auxin influx and efflux carriers determines the direction of auxin transport (Figure 29.4).
- The major effect of cytokinins is to increase the rate of cell division. Auxin and cytokinin are used in the process of plant tissue culture.
- Gibberellin function illustrates the general principle that plant hormones often act to release cellular brakes on gene expression. The flowering plant gibberellin-DELLA system evolved from components present in earlier-diverging plant lineages (Figure 29.5).
- The gaseous hormone ethylene plays an important role in seedling development (Figure 29.6).
- Abscisic acid (ABA) and brassinosteroids help plants to respond to environmental stress.

29.2 Testing Your Knowledge

1. If you were asked to draw a picture showing how several plant hormones affect gene expression, which of the following auto parts would be most useful as a component of your drawing?
   a. crankshaft
   b. tires
   c. axle
   d. brakes
   e. steering wheel

2–6. Match the major plant hormones with a major function.
   2. auxins
   3. cytokinins
   4. gibberellin
   5. ethylene
   6. abscisic acid

   a. promote cell division and shoot development in tissue-cultured plants
   b. mediate phototropism
   c. promote seed germination
   d. promote stomatal closing
   e. induce fruit ripening

29.3 Plant Responses to Light

Learning Outcome

1. Describe how phytochrome allows plants to respond to light, including day length.

How do buried seeds know if enough light is available to support seedling photosynthesis? How do plants know to flower at times of the year that are most beneficial for achieving pollination or seed dispersal? The answer in both cases is that plants possess cellular systems for measuring light and determining the season of the year. Because of these systems, plant seeds germinate only when there is sufficient light for seedling growth, and flowering occurs during the most advantageous season.

Plants Detect Light and Measure Day Length

A plant's ability to measure and respond to day length, a process called **photoperiodism,** is based on the presence of several types of light receptors within cells. Such light sensors are known as photoreceptors; in this section, we will focus on the particular type known as phytochrome.

Phytochrome is a red- and far-red-light receptor that influences many plant processes. This light sensor operates much like a light switch, flipping back and forth between two conformations: P_r, which absorbs red light, and P_{fr}, which absorbs only far-red light (light at the far-red end of the visible spectrum) (Figure 29.7). When red light is abundant, as in full sunlight, P_r absorbs red light and changes to P_{fr}, which activates cellular responses such as seed germination. When left in the dark for a long period, P_{fr} slowly transforms into P_r. The phytochrome light switch controls (1) seed germination, (2) seasonal flowering, and (3) plant response to shading.

1. The role of phytochrome as a plant "light switch" has been shown experimentally in studies of lettuce-seed germination. Researchers have found that water-soaked lettuce seeds will not germinate in darkness, but they will germinate if exposed to as little as 1 minute of red light (Figure 29.8). This amount of light exposure is sufficient to transform a critical amount of P_r to the active P_{fr} conformation, which stimulates germination. However, if this brief red-light treatment is followed by a few minutes of treatment with far-red light, the lettuce seeds will not germinate. This short period of far-red illumination is enough to convert seed P_{fr} back to the inactive P_r conformation.

   The most recent light exposure determines whether the phytochrome occurs in the active or inactive conformation. In nature, if seeds are close enough to the surface that their phytochrome is switched on by red light, the seeds will germinate. But if seeds are buried too deeply for red light to penetrate, they will not germinate. In this way, seeds can sense if they are close enough to the surface to begin the germination process.

2. Phytochromes also play a critical role in photoperiodism. Flowering plants can be classified as long-day, short-day, or day-neutral plants. When scientists named these groups, they thought that plants measured the amount of daylight; researchers later discovered that plants actually measure night length.

   - Lettuce, spinach, radish, beet, clover, gladiolus, and iris are examples of **long-day plants** because they flower in spring or early summer, when the night period is shorter (and thus the day length is longer) than a defined period (Figure 29.9).
   - Asters, strawberries, dahlias, poinsettias, potatoes, soybeans, and goldenrods are examples of **short-day plants** because they flower only when the night length is longer than a defined period. Such night lengths occur in late summer, fall, or winter, when days are short (Figure 29.9).

Figure 29.7 How phytochrome acts as a molecular light switch.

Concept Check: What kind of light does the active conformation of phytochrome absorb, and what kind of change does such absorption cause?

Figure 29.8 How phytochrome influences seed germination.
(1–5) ©Linda Graham

 Concept Check: *Describe the change in phytochrome that would occur if a deeply buried seed were uncovered enough to receive sunlight.*

In darkness, seeds do not germinate because phytochrome remains in the inactive P_r conformation.

Even a brief exposure to red light generates the active P_{fr} conformation of phytochrome, allowing seeds to germinate.

Exposure to far-red light after red-light exposure converts active P_{fr} to inactive P_r, so seeds do not germinate.

Exposure to red light after far-red light switches phytochrome back to the active P_{fr} conformation, so seeds germinate.

The most recent light exposure determines whether phytochrome occurs in the active P_{fr} or in the inactive P_r conformation. If in the latter, most seeds do not germinate.

- Roses, snapdragons, cotton, carnations, dandelions, sunflowers, tomatoes, and cucumbers flower regardless of the night length, as long as day length meets the minimal requirements for plant growth, and thus they are known as **day-neutral plants.**

   Ornamental plant growers manipulate night length to produce flowers for market during seasons when they are not naturally available. For example, chrysanthemums are short-day plants that usually flower in the fall, but growers use light-blocking shades to increase night length in order to produce flowering chrysanthemums at any season.

3. Phytochrome also mediates plant responses to shading. These responses include the extension of leaves from shady portions of a dense tree canopy into the light, and growth that allows plants to avoid being shaded by neighboring plants. These growth responses occur by the elongation of shoot internodes. Leaves detect shade as an increased proportion of far-red light to red light. This means that more of the phytochrome in shaded leaves is in the inactive (P_r) state than is the case for leaves in sunlight. Activated phytochrome (P_{fr}) inhibits the growth of shoot internodes, but phytochrome in the inactivated state does not, so branches bearing shaded leaves extend toward sunlight.

29.3 Reviewing the Concepts

- Light-sensitive pigments such as phytochrome allow plants to respond to light stimuli and influence seed germination, control of seasonal flowering, and response to shading (Figures 29.7, 29.8, 29.9).

29.3 Testing Your Knowledge

1–4. Imagine a garden that has been planted with irises, dahlias, and carnations and in which goldenrods also occur as weeds. Match each plant with the seasonal period during which blooming will occur.

1. iris (a long-day plant)
2. dahlia (a short-day plant)
3. carnation (a day-neutral plant)
4. goldenrod

a. spring or early summer when nights are relatively short
b. anytime that light amounts meet plant growth needs
c. late summer, fall, or winter when days are short
d. none of the above

Figure 29.9 Flowering and photoperiodism. Iris is a long-day plant that flowers in response to the short nights of late spring and early summer, whereas goldenrod is a short-day plant that flowers in response to the longer nights of autumn. The length of night is the critical factor, as shown by the effects of light flashes.
(top left) (bottom left) ©Ray Bulson/Newscom; *(top right) (bottom right)* ©Lee W. Wilcox; *(middle left)* ©Garden Picture Library/Getty Images; *(middle right)* ©Comstock Images/PictureQuest

Concept Check: *What would happen if you gave these plants a brief exposure to darkness in the middle of the daytime?*

29.4 Plant Responses to Gravity and Touch

Learning Outcomes
1. Discuss how plant shoots and roots respond to gravity.
2. Describe how plant roots and leaves respond to touch.

Have you ever wondered what causes plant stems to generally grow upward and roots downward? The upward growth of shoots and the downward growth of roots are behaviors known as **gravitropism,** growth in response to the force of gravity. Shoots are said to be negatively gravitropic because they often grow in the direction opposite to gravitational force. If a potted plant is turned over on its side, the shoot will eventually bend and begin to grow vertically again (**Figure 29.10**). Most roots are said to be positively gravitropic because they grow in the same direction as the gravitational force.

Both roots and shoots detect gravity by means of starch-heavy plastids known as **statoliths,** which are located in specialized

Figure 29.10 Negative gravitropism in a shoot. This tomato shoot system has resumed upward growth after being placed on its side. Upward growth started about 4 hours after the plant was turned sideways; this photo was taken 20 hours later. Shoots sense gravity by means of starchy statoliths present in endodermal tissue.
©Lee W. Wilcox

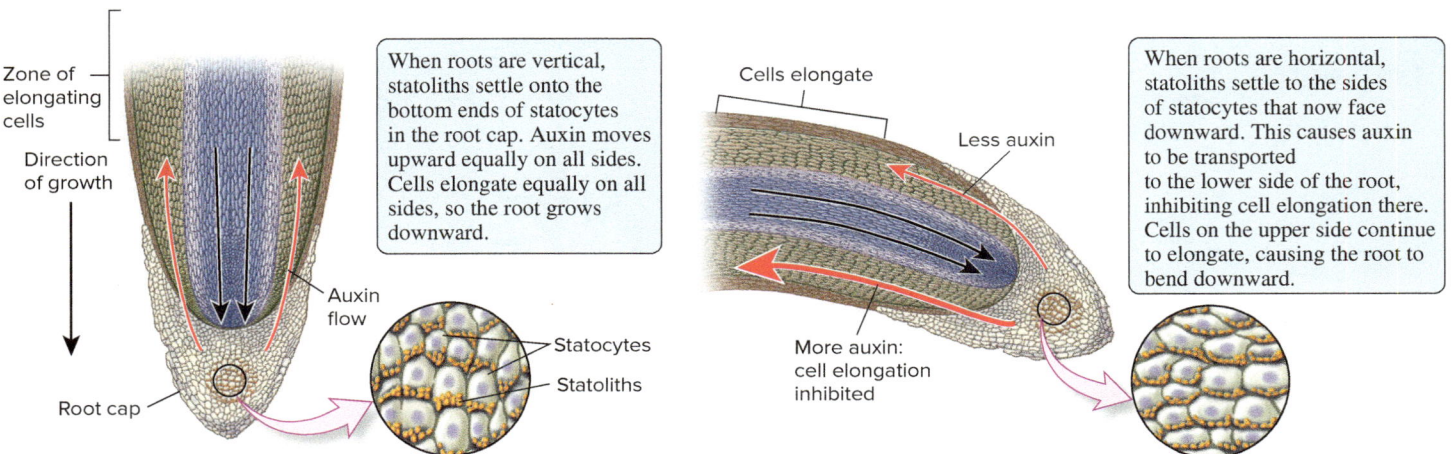

Figure 29.11 Positive gravitropism in a root. Root-tip cells sense gravity by means of starchy statoliths present in statocytes at the center of the root cap.

Concept Check: *Is there any other environmental signal that roots can use to achieve downward growth?*

gravity-sensing cells called statocytes. In shoots, statocytes are located in a tissue known as the endodermis, which forms a sheath around vascular tissues. Although roots also possess an endodermis, in roots, gravity-sensing cells primarily occur in the center of the root cap (refer back to Figure 28.22).

Gravity causes the relatively heavy statoliths to sink, which causes changes in calcium ion messengers that affect the direction of auxin transport. This process induces changes in the direction of shoot or root growth. For example, in a root that becomes oriented horizontally, the statoliths are pulled by gravity to the lower sides of statocytes (**Figure 29.11**). The change in statolith position causes auxin to move to cells on the lower sides of roots. In roots, auxin inhibits cell elongation (in contrast to its action in shoots). Therefore, root growth slows on the lower side, whereas cell elongation continues normally on the upper side. This process causes the root to bend so that it eventually grows downward again.

Recent studies suggest that gravity responses are related to touch responses. For example, when roots encounter rocks or other barriers to their downward growth in the soil, they display a touch response that temporarily supersedes their response to gravity. Such roots grow horizontally until they get around the barrier, whereupon downward growth in response to gravity resumes.

Plant shoots and leaves may also respond to touch. Vines, for example, may produce tendrils that wind around or clasp supporting structures. Wind also induces shoot touch responses. In very windy places, trees tend to be shorter than normal, giving them the advantage of being less likely to blow over than are taller trees. In the laboratory, plant scientists have simulated natural touch responses by rubbing plant stems and found that this treatment can result in shorter plants. Touch causes the release of calcium ion messengers that influence gene expression.

Rapid responses to touch include trapping of prey animals by the leaves of insectivorous plants (see Chapter 28's opening photo), and leaf folding by the sensitive plant (see this chapter's opening photo). Sensitive plant leaves undergo changes in the water content of cells within a structure known as a **pulvinus** (plural, pulvini), a swelling located at the base of attachment of each pair of leaflets in complex leaves. A pulvinus consists of a thick layer of parenchyma cells that surrounds a core of vascular tissue (**Figure 29.12**). When the leaflet of a sensitive plant is touched, an electrical signal known as an action potential opens ion channels in parenchyma cells near the lower surfaces of the pulvini. These cells expel potassium (K^+) and chloride (Cl^-) ions, causing water to flow out and the cells to become flattened. This bends the leaflets together, starting the leaflet-folding process. The action potential generated at the touch site also flows through the leaflet, causing many or all of the leaflets to also bend, with the result that the entire leaf folds. Reversal of this process allows the leaf to unfold. Action potentials are likewise important in the transmission of nerve signals by animals, described in Chapter 33.

Some plant movements that are unrelated to touch, including sleep movements and Sun tracking, also result from the action of pulvini. Sleep movements are changes in leaf position that occur in response to day-night cycles. Sun tracking is the movement of leaves or flowers in response to the Sun's position.

29.4 Reviewing the Concepts

- Plant shoots and roots respond to gravity (in the process of gravitropism) by means of starch-heavy statoliths located within statocytes (Figures 29.10, 29.11).
- Rapid touch responses, such as leaf folding in sensitive plants, depend on changes in the water content of cells in structures known as pulvini (Figure 29.12).

Figure 29.12 Leaf folding in the sensitive plant: how pulvini change the positions of leaflets. The electrical signals known as action potentials result from the rapid flow of ions through membrane ion channels. Action potentials spread from one plant cell to another through plasmodesmata. Swelling of cells on one side of the pulvinus and loss of water from cells on the other side cause bending.

29.4 Testing Your Knowledge

1. If you turn an evenly illuminated potted tomato plant onto its side, after some period of time the shoot will turn upward and the roots will turn downward. Which description of this plant growth response is most accurate?
   a. The shoot displays phototropism, and the root displays a touch response.
   b. The shoot displays a touch response, and the root displays negative phototropism.
   c. The shoot displays negative gravitropism, and the root displays positive gravitropism.
   d. The shoot displays positive gravitropism, and the root displays negative gravitropism.
   e. None of the above is accurate.

29.5 Plant Responses to Attack

Learning Outcomes

1. Discuss how plants sense and respond to attack by herbivorous animals.
2. Discuss how plants sense and respond to attack by disease-causing microbes.

Plants are vulnerable to attack by herbivores, animals that consume plants. Plants are also vulnerable to pathogens—disease-causing microorganisms. Structural barriers such as cuticles, trichomes, and outer bark help to reduce infection and herbivore attack (refer back to Figure 28.15). These barriers, together with chemical defense compounds, explain why remarkably little natural vegetation is lost to herbivore or pathogen attack. However, agricultural crops can be more vulnerable to attack than their wild counterparts. This is because some protective adaptations have been lost during crop domestication as the result of genetic changes that increase edibility. For this reason, crop scientists are particularly interested in understanding plant defense, with the goal of being able to breed or genetically engineer crop plants that are better protected from pests.

Plants Defend Themselves Against Herbivores

Herbivory defense compounds include diverse secondary metabolites—terpenes and terpenoids, phenolics, and alkaloids (refer back to Figure 25.20)—but also poisonous hydrogen cyanide and other molecules. Some of these substances act directly on herbivores, making plants taste bad so that herbivores learn to avoid them. Other chemical compounds function indirectly, on the principle that "the enemy of my enemy is my friend." For example, when attacked by insect caterpillars, cruciferous plants release terpenoids that attract the bodyguard wasp (*Cotesia rubecula*), which attacks the caterpillars.

Plants can also signal to healthy plant parts that other tissues have suffered herbivore attack, allowing the healthy parts to mount defensive responses. The model plant *Arabidopsis thaliana*, for example, employs the amino acid glutamate, glutamate receptors, and calcium ions for long-distance communication, in a process that is intriguingly similar to glutamate-mediated neurotransmission in mammals. Glutamate is the most widespread excitatory neurotransmitter found in animal nervous systems (see Chapter 33). The plant glutamate signaling process starts within seconds after attack. Glutamate released from wounded cells moves through the plant vascular system at about 1 mm/s, causing calcium to enter unharmed cells. The

Figure 29.13 **Plant responses to herbivore attack.** **(1)** Leaf damage induces local defensive responses, including production of compounds that travel through the plant and cause defensive responses elsewhere. **(2a)** In addition, attacked plants may release volatile compounds, such as jasmonic acid. These volatile compounds foster the development of defenses in undamaged parts of the same plant. **(2b)** Such volatile compounds may also induce defenses in neighboring plants, which then become less vulnerable to herbivores. Volatile compounds may also attract predators that feed on the herbivore attacker.

Concept Check: *What advantage would a predator obtain by responding to the volatile signals emitted from herbivore-damaged plants? What adaptive advantage could a plant obtain by attracting the enemies of its attackers?*

calcium induces healthy cells to accumulate the hormone jasmonic acid (JA), which induces defensive responses.

Plants may also signal herbivore damage to neighboring plants, allowing them to prepare for attack. For example, when insects wound tomato plants, damaged cells release the peptide hormone systemin. This hormone induces undamaged cells to produce JA, which travels in the phloem or through the air to undamaged parts of the plant (**Figure 29.13**). Airborne compounds released from herbivore-damaged plants may also attract the enemies of insect attackers. Nearby plants may detect these airborne signals and respond by similarly arming themselves against attack (Figure 29.13, step 2b).

Plants Defend Themselves Against Pathogen Attack

Every year, about 15% of worldwide crop production is lost to disease caused by certain bacteria, fungi, protists, or viruses. During their evolution, plants have acquired mechanisms for detecting and responding to disease-causing microbes, but microbes can rapidly evolve new disease processes. Agricultural scientists aim to understand how pathogenic microbes attack plants and how plants are able to prevent disease, in order to develop disease-resistant crop varieties.

Plant pathogens produce compounds known as **elicitors,** which promote the infection of plant cells and tissues (**Figure 29.14**). Some bacteria inject elicitors into plant cells by means of syringe-like systems (Figure 29.14). Fungal pathogens often deliver elicitors into plant cells and absorb nutrients from them by means of specialized penetration structures. In response to elicitors, plant cells have evolved multiple lines of defense:

1. An early line of defense consists of plasma membrane receptors that specifically bind microbe-associated molecules, such as bacterial lipopolysaccharides or the fungal cell-wall compound chitin (see Chapter 23). Such plant plasma membrane receptors, which are typically protein kinases, allow plant cells to sense disease-causing microbes. Researchers have discovered that animal cells use homologous receptor molecules to sense and respond to pathogens.

2. A subsequent line of defense occurs in the cytosol. Here, microRNAs help to destroy the nucleic acids of invading viruses, in processes similar to those present in animal cells. In addition, receptors that occur in the cytosol allow plant cells to recognize injected elicitors. The binding of receptors with elicitors triggers the production of chemical defense signals: hydrogen peroxide (H_2O_2), the gas nitrous oxide (NO), and mobile hormones. Defense responses include deposition of polymers that strengthen plant cell walls at the site of attack, programmed death of infected cells as a way of limiting the spread of infection, and production of compounds that kill pathogens (Figure 29.14).

3. The model plant *Arabidopsis thaliana* produces extracellular vesicles that deliver small RNA molecules into the cells of

Figure 29.14 Pathogen/plant interactions and the hypersensitive response to pathogen attack.
(inset) ©G.R. "Dick" Roberts/Natural Sciences Image Library

nearby fungal pathogens before a disease process starts. Within the fungal cells, the small RNAs silence fungal genes needed for plant attack, thereby preventing disease.

Many types of **resistance genes (R genes)** encode plant disease receptor proteins. If a plant is genetically unable to produce a receptor that can recognize a pathogen's presence or elicitor, disease may result. By contrast, plants successfully resist disease when the product of a dominant *R* gene (a receptor) recognizes a pathogen's presence. Comparative genome analysis shows that plant *R* genes evolved early, in algal relatives of the first land plants, indicating the importance of such genes. Diverse alleles of *R* genes occur in modern plant populations, providing plants with a large capacity to cope with different types of pathogens. These receptors, together with plant responses to pathogen attack—the hypersensitive response and systemic acquired resistance—represent the plant immune system.

The Hypersensitive Response to Pathogen Attack

The plant **hypersensitive response (HR)** is a local reaction to pathogen attack that limits the progression of disease (Figure 29.14, steps 3 and 4). There are several components to the hypersensitive response:

- Plant-produced hydrogen peroxide (H_2O_2) can kill pathogens and helps to strengthen the cell wall by deposition of cross-binding polymers.
- The gaseous hormone NO works with H_2O_2 to stimulate the synthesis of defensive secondary metabolites, the hormone salicylic acid (SA), and tough lignin in the cell walls of nearby tissues. Recall that lignin is a plant cell-wall polymer that helps cell walls to resist microbial attack and compression and that has waterproofing qualities (see Chapters 25 and 28).
- NO also induces programmed cell death, which deprives pathogens of food, helping to limit the spread of disease. Necrotic spots are brown patches on plant organs that reveal where disease pathogens have attacked plants and where plant tissues have battled back.
- Salicylic acid or the related compound methyl salicylate sends alarm signals to other parts of the plant, which respond by preparing defenses.

Systemic Acquired Resistance

Whereas the hypersensitive response is a local reaction, **systemic acquired resistance (SAR)** is an immune response of the whole plant induced by a pathogen attack (Figure 29.15). SAR immunizes the plant not only from the inducing pathogen but also from many others. The SAR system uses the same type of long-distance signaling process that is associated with herbivore attack (see Figure 29.13):

- At or near a wound, systemin induces the production of jasmonic acid (JA) in nearby vascular tissues.
- JA is then transported throughout the plant, inducing defenses.
- In addition, the hormone salicylic acid can be converted to volatile methyl salicylate, which diffuses into the air

Figure 29.15 Systemic acquired resistance to pathogen attack.

✓ **Concept Check:** *In what way is systemic acquired resistance to pathogens similar to plant responses to herbivore attack?*

surrounding a plant, inducing resistance in noninfected tissues. These signals involve changes in transcription of up to 10% of plant genes, including plant homologs of the *BRCA* genes associated with human breast cancer.

- Responding plant tissues may produce defensive enzymes, which can break down pathogen cell walls, or they may generate tannins, which are toxic to microorganisms. Experimental infection of *Arabidopsis* leaves with the bacterial pathogen *Pseudomonas syringae* led to induction of SAR within 6 hours.

29.5 Reviewing the Concepts

- Plants cope with biological stresses such as herbivore and pathogen attack by means of structural and chemical adaptations. Injured plant parts produce hormones that signal other parts of the same plant and nearby plants to produce defensive responses (Figure 29.13).
- Plant pathogens begin their attack by producing elicitor compounds. If an attacked plant can produce a compound that interferes with that elicitor, encoded by a resistance (*R*) gene, then

the plant can resist infection by that pathogen. The hypersensitive response (HR) is a local defensive response to pathogen attack, whereas systemic acquired resistance (SAR) is a whole-plant immune response (Figures 29.14, 29.15).

29.5 Testing Your Knowledge

1. Which of the following substances is involved in plant defensive responses to pathogen attack?
   a. systemin
   b. jasmonic acid
   c. glutamate
   d. hydrogen peroxide
   e. All of the above are correct.

Assessing Your Knowledge and Skills

Foundational Knowledge

Multiple Choice

1. The major types of plant hormones include
   a. cyclic AMP, IP_3, and calcium ions.
   b. calcium, CDPKs, and DELLA proteins.
   c. auxin, cytokinin, and gibberellin.
   d. chlorophyll, carotenoids, and phytochrome.
   e. statoliths, pulvini, and parenchyma.

2. Phototropism is
   a. the production of flowers in response to a particular day length.
   b. the production of flowers in response to a particular night length.
   c. the growth response of a plant, an organ system, or an organ to directional light.
   d. the growth response of a plant, an organ system, or an organ to gravity.
   e. the growth response of a plant, an organ system, or an organ to touch.

3. What is the most accurate order of events during signal transduction?
   a. first, receptor activation; then, messenger signaling; and last, an effector response
   b. first, an effector response; then, messenger signaling; and last, receptor activation
   c. first, messenger signaling; then, receptor activation; and last, an effector response
   d. first, an effector response; then, receptor activation; and last, messenger signaling
   e. None of the above is accurate.

4. Which of the plant hormones is known as the "master hormone," and why?
   a. cytokinin, because many plant functions require cell division
   b. gibberellins, because growth is essential to many plant responses
   c. abscisic acid, because it is necessary for leaf and fruit drop
   d. brassinosteroids, because water uptake is so fundamental to plant growth
   e. auxin, because there are many different auxin-responsive genes

5. Gaseous hormones are able to enter cells without requiring special membrane transporter systems. Which of the major plant hormones is a diffusible gas?
   a. auxin
   b. gibberellin
   c. cytokinin
   d. ethylene
   e. abscisic acid

Short Answer

1. Why can plants be said to display behavior?
2. **PRINCIPLES** A principle of biology is that living organisms interact with their environment. The plant environment includes physical and biological components such as disease-causing pathogenic microbes. Why do plants produce so many types of pathogen resistance (*R*) genes?

Critical-Thinking Skills

Multiple Choice

1. Photoreceptors, such as phytochrome, allow plant cells to detect and respond to light of particular wavelengths. Photoreceptors are so important to photosynthetic organisms that phytochrome evolved even before the earliest plants, indicated by the light responses of particular green algae known to be closely related to land plants. Which wavelength changes would you expect to affect these algal relatives most strongly?
   a. change from a violet-light to an ultraviolet-light environment and the reverse change
   b. change from a red-light to a green-light environment and the reverse change
   c. change from a red-light to a far-red-light environment and the reverse change
   d. change from a yellow-light to a blue-light environment and the reverse change
   e. change from an orange-light to a green-light environment and the reverse change

2. Which of the following plant responses is consistent with gravity effects?
   a. A tomato shoot is blown to a horizontal position by strong wind but over time assumes a normal vertical orientation.
   b. A plant cutting develops adventitious roots that grow downward into a fluid.
   c. A plant root that has been growing horizontally to get around a rock starts growing downward again.
   d. All of the above are correct.
   e. None of the above is correct.

3. A team of plant biologists is given the task of understanding how an important crop plant interacts with a particular fungal pathogen (disease agent). Which of the following possibilities should the team investigate?
   a. The plant is able to recognize and respond to the presence of chitin, an elicitor specific to fungi.
   b. The fungus uses specialized penetration structures to attack the plant.
   c. The plant secretes vesicles containing small RNA molecules that interfere with the expression of fungal genes essential to disease progression.
   d. All of the above should be investigated.
   e. None of the above should be investigated.

4. One way that biologists investigate gravity's effects on biological organisms is to take (usually small) representatives into gravity-free space, to the International Space Station, for example. Let's consider germinating lettuce seedlings growing in flat dishes that are attached vertically to a common support, with light coming from all directions. How would you expect the seedlings to respond?
   a. Most seedlings would generate embryonic roots in the direction of the support structure.
   b. Most seedlings would produce embryonic roots in the direction away from the support structure.
   c. Most seedlings would not grow in space.
   d. Seedlings would produce embryonic roots in various directions.
   e. None of the above is correct.

5. Imagine that you are able to use genetic technology to decrease or completely eliminate the expression of genes involved in the production of the gas NO in a plant species that serves as a model system for studying the effects of plant pathogens. What effects would you expect to observe?
   a. The genetically altered plants would not be able to undergo programmed cell death to halt the growth of pathogens.
   b. The engineered plants would produce higher than normal amounts of salicylic acid in cells attacked by the pathogen.
   c. The altered plants would produce higher than typical amounts of lignin.
   d. The modified plants would be able to tolerate higher levels of pathogens than would wild type plants.
   e. All of the above would be expected.

Short Answer

1. Why are most wild plants distasteful, and some even poisonous, to people?
2. How could you increase the resistance of a crop plant species to a particular type of herbivore?

30

How Flowering Plants Obtain and Transport Water, Mineral Nutrients, and Organic Compounds

Arid land vegetation in the Mojave Desert of the U.S.

Chapter Outline

30.1 Plant Nutritional Requirements
30.2 The Roles of Soil in Plant Nutrition
30.3 Transport at the Cellular Level
30.4 Plant Transport at the Tissue Level
30.5 Long-Distance Transport in Plants
Assessing Your Knowledge and Skills

©Linda Graham

Millions of people live in arid lands. Arid lands are places where natural and crop plant growth is limited by water availability, as shown in the chapter-opening photo. Past indigenous societies invented clever agricultural practices that effectively supported relatively low human populations. But more recently, increasing human populations and global climate change have challenged the ability of people to survive in dry lands. Famines can arise when droughts decrease food production. To aid food production everywhere, and particularly in dry lands, biologists work closely with climate and soil scientists to better understand how plants interact with water and other materials in the soil and atmosphere.

One amazing discovery resulting from such multidisciplinary work is that plants actually increase local rainfall by moving huge amounts of water from the soil to the atmosphere. Plants take up CO_2 from the air and absorb water and dissolved mineral nutrients from the soil for use in photosynthesis. Plant photosynthesis generates organic food that plants use for growth and reproduction, and that also supports herbivorous animals. For every CO_2 molecule that plants convert to organic food during the process of photosynthesis, plants transport 500 water molecules from the soil through their bodies, releasing the water to the air as vapor! Because rain is condensed water vapor, plants actually generate much of the precipitation they and humans need, particularly in arid regions. In addition, by exuding water vapor, plants cool themselves and the surrounding air. Consequently, when vegetation is lost as the result of livestock overgrazing and other effects of burgeoning human populations, local rainfall decreases and temperatures increase. Agriculture and human life then become even more challenging, particularly in arid regions. Dryland agricultural concerns illustrate how people benefit from understanding the processes that plants use to obtain and transport water, minerals, and organic compounds.

30.1 Plant Nutritional Requirements

Learning Outcomes
1. List the major nutritional resources that green plants need for healthy growth.
2. Distinguish between plant macronutrients and micronutrients.

Plants, like animals, require **nutrients**—substances that are metabolized by or incorporated into an organism. In addition to light, photosynthetic plants require water (H_2O), carbon dioxide (CO_2), and more than a dozen elements to produce organic food by means of photosynthesis. CO_2 is primarily absorbed from air, whereas H_2O and elements in the form of mineral ions—such as potassium (as K^+), nitrogen (as NH_4^+ or NO_3^-), and calcium (as Ca^{2+})—are primarily taken up from soil (**Figure 30.1**).

The chemical elements required by plants play many roles in plant metabolism, often functioning as enzyme cofactors (**Table 30.1**). Elements that are generally required in amounts of at least 1 g/kg of plant dry matter are known as **macronutrients.** In contrast, elements that are needed in amounts at or less than 0.1 g/kg of plant dry mass are known as **micronutrients,** or trace elements. As is also the case for animals, an insufficient amount of essential nutrients can cause deficiency symptoms. Chlorosis, a yellowing of plant tissue resulting from an insufficiency of one or more minerals needed to make chlorophyll, is an example of a plant nutrient deficiency symptom (**Figure 30.2**). Because insufficient amounts of water, light, carbon dioxide, and other mineral nutrients can limit the extent of green plant growth, these resources are known as limiting factors.

Table 30.1	Plant Essential Mineral Nutrients			
Element (chemical symbol)	Percent of plant dry mass	Major source	Form taken up by plants	Function(s)
Macronutrients				
Carbon (C)	45	Air	CO_2	Component of all organic molecules
Oxygen (O)	45	Air, soil, water	CO_2, O_2, H_2O	Component of all organic molecules
Hydrogen (H)	6	Water	H_2O	Component of all organic molecules; protons used in chemiosmosis and cotransport
Nitrogen (N)	1.5	Soil	NO_3^-, NH_4^+	Component of proteins, nucleic acids, chlorophyll, coenzymes, and alkaloids
Potassium (K)	1.0	Soil	K^+	Essential role in cell ionic balance
Calcium (Ca)	0.5	Soil	Ca^{2+}	Component of cell walls; messenger in signal transduction
Magnesium (Mg)	0.2	Soil	Mg^{2+}	Component of chlorophyll; activates some enzymes
Phosphorus (P)	0.2	Soil	HPO_4^{2-}	Component of nucleic acids, ATP, phospholipids, and some coenzymes
Sulfur (S)	0.1	Soil	SO_4^{2-}	Component of proteins, some coenzymes, and defense compounds
Micronutrients				
Chlorine (Cl)	0.01	Soil	Cl^-	Required for water splitting in photosystem; cell ion balance
Iron (Fe)	0.01	Soil	Fe^{3+}, Fe^{2+}	Enzyme cofactor; component of cytochromes
Manganese (Mn)	0.005	Soil	Mn^{2+}	Enzyme cofactor
Boron (B)	0.002	Soil	$B(OH)_3$	Enzyme cofactor; component of cell walls
Zinc (Zn)	0.002	Soil	Zn^{2+}	Enzyme cofactor
Sodium (Na)	0.001	Soil	Na^+	Required to generate PEP in C_4 and CAM plants
Copper (Cu)	0.0006	Soil	Cu^+, Cu^{2+}	Enzyme cofactor
Molybdenum (Mo)	0.00001	Soil	MoO_4^{2-}	Enzyme cofactor
Nickel (Ni)	0.000005	Soil	Ni^{2+}	Enzyme cofactor

Figure 30.1 The major types of plant nutrients and their sources.

Figure 30.2 Chlorosis as a symptom of mineral deficiency. This camellia plant is suffering from an iron deficiency, as revealed by its yellow leaves, a symptom known as chlorosis. Other mineral deficiencies can also cause chlorosis.
©Geoff Kidd/SPL/Science Source

Water Supplies Hydrogen and Plays Other Essential Roles

Though plants vary in water content, water typically makes up about 90% of the weight of living plants. Although many types of plants living in arid lands display adaptations that allow them to survive for extended periods in nearly dry conditions, all plants require an adequate supply of water for active metabolism and growth. Water is essential to plants for several reasons:

- Water acts as a nutrient by supplying most of the hydrogen atoms for incorporation into organic compounds. Hydrogen from water is incorporated into plants when CO_2 is reduced during the process of photosynthesis.
- Water is the solvent for other mineral nutrients and is the main transport medium in plants, allowing the movement of minerals and other solutes throughout the plant body via the vascular tissues.
- Evaporative loss of water from plant surfaces helps to cool plants and their environments.
- Cytoplasmic and vacuolar water help to support plants by maintaining hydrostatic pressure on the cell wall.

Light Provides Photosynthetic Plants with Energy

Photosynthetic plants require light energy for the formation of the covalent bonds of organic compounds that make up the plant's body. Green plants' use of light energy as an essential resource parallels animals' nutritional requirements for organic food as a source of chemical energy. Remarkably, several hundred species of flowering plants have lost photosynthetic capacity; for these

Biology Principle

Living Organisms Use Energy
Monotropa uniflora is one of many types of nongreen heterotrophic flowering plants that lack photosynthetic capacity and therefore must absorb organic compounds for use as an energy source. In contrast, nearby autotrophic green plants use sunlight as a source of energy.

Figure 30.3 The heterotrophic flowering plant Indian pipe (*Monotropa uniflora*).
©Robert Ziemba/age fotostock

heterotrophic plants, essential nutrients include absorbed organic compounds that replace light as a source of energy. An example is the ghostly white Indian pipe (*Monotropa uniflora*), which grows in forests (**Figure 30.3**).

For photosynthetic plants, the amount of light available may be less than or greater than optimal, and different plant species are adapted to cope with varying amounts of light and shade:

- In forests, light availability limits the growth of tree seedlings and other small plants that are shaded by the leafy tree canopy overhead. Forest seedlings often rely on organic food stored within the seed or supplied by underground networks of mycorrhizal fungi (refer back to Figure 24.11), which provide organic food tapped from the roots of mature trees. Forest saplings may not be able to grow maximally until a light gap appears in the leafy canopy.
- Conversely, plants growing in deserts or on mountains often experience light so intense that it can damage their photosynthetic components. Plants have evolved light-absorbing

carotenoid pigments and other adaptations that help them to cope with too much light.

- Plants can adjust the thickness of leaves in response to variations in light availability. Some plants produce sun-leaves in sunnier parts and shade-leaves in more shaded parts. Sun-leaves have a thicker layer of chlorophyll-rich mesophyll and are able to harvest more of the bright sunlight that penetrates deeply into the leaf. In contrast, shade-leaves have a thinner mesophyll layer, lacking enough light to supply a thicker layer of photosynthetic tissue (Figure 30.4).

Carbon Dioxide Provides Most Plant Dry Mass

Light provides the energy for plant photosynthesis, but most plant dry mass originates from carbon dioxide (CO_2). Most plants obtain CO_2 gas from the atmosphere by absorption through **stomata,** pores that occur in the plant epidermis, which plants can close to retain water or open to allow the entry of CO_2 needed for photosynthesis. Under experimental conditions, plant photosynthetic rates increase with CO_2 concentration until the Calvin cycle enzyme rubisco has become fully supplied—that is, saturated with CO_2. In nature, plants are often unable to obtain enough CO_2 for maximal photosynthesis. This is partly because the modern atmospheric concentration of CO_2 is only 390 μL/L (a small fraction of atmospheric content), whereas considerably more CO_2 would be required to maximize photosynthesis in most plants.

Climate plays an important role in the ability of plants to acquire CO_2. Plants that experience hot, dry conditions are particularly vulnerable to low CO_2 levels. When their stomata close to conserve water, the plants cannot absorb CO_2, which limits photosynthesis. About 30,000 species of C_4 and CAM plants, common in arid lands, possess various types of structural and biochemical adaptations that help them to cope with CO_2 limitation by improving their CO_2 absorption (see Chapter 7). You may recall that C_4 photosynthesis typically involves two types of leaf cells; mesophyll cells use a special enzyme (PEP carboxylase) to trap CO_2 into a four-carbon compound that moves into bundle-sheath cells that surround leaf veins (vascular bundles). Within the bundle-sheath cells, CO_2 is released and utilized by the Calvin cycle to produce organic compounds. This system allows C_4 plants to keep the Calvin cycle supplied with CO_2 even when stomata must be partially closed to conserve water. The desert plant *Bienertia cycloptera* is remarkable because C_4 photosynthesis occurs within single cells. Plastids at the cell periphery use PEP carboxylase to trap CO_2, thereby functioning like mesophyll cells of most C_4 plants, and plastids clustered near the nucleus function like the bundle-sheath cells of most C_4 plants (Figure 30.5). Plant biologists study diverse C_4 plants to gain insight into ways that crop plants might be engineered to increase food production.

When grown in greenhouses and supplied with sufficient light, moisture, and air that has been enriched with CO_2, tomatoes, cucumbers, leafy vegetables, and some other crops can double their growth rate. This observation has stimulated plant biologists to perform experiments designed to indicate how wild and crop plants respond to human-generated increases in Earth's atmospheric CO_2 content.

In addition to light, CO_2, and water, plants require additional elements that occur naturally in soil or that can be added in the form of fertilizers (Table 30.1). Section 30.2 focuses on the roles of soil in plant nutrition.

Biology Principle

Structure Determines Function

These micrographs of stained leaf slices reveal the thinner layer of photosynthetic tissues (mesophyll) and greater amount of air spaces in **(a)** shade-leaves compared with **(b)** sun-leaves.

(a) Shade-leaf

(b) Sun-leaf

Figure 30.4 Shade- and sun-leaves.
(a–b) ©Lee W. Wilcox

30.1 Reviewing the Concepts

- Terrestrial green plants require H_2O, light energy, CO_2, and mineral ions such K^+, NO_3^-, and Ca^{2+} absorbed from soil. Deficiency symptoms such as chlorosis signal the lack of sufficient nutrients (Figures 30.1, 30.2; Table 30.1).
- Like animals and fungi, diverse nongreen flowering plants obtain chemical energy by metabolizing organic compounds absorbed from their environment (Figure 30.3).
- Plants display adaptations that allow them to cope with resource levels that are not optimal. Sun- and shade-leaves are adaptations to variation in light availability; C_4 and CAM photosynthesis represent adaptation to arid environments. (Figures 30.4, 30.5).

Figure 30.5 A CO_2-acquisition adaptation. The desert plant *Bienertia cycloptera* is able to conduct C_4 photosynthesis within the confines of a single cell. In this fluorescence photograph of a *B. cycloptera* cell, chlorophyll fluoresces red, indicating plastid location. Plastids at the cell periphery function like the mesophyll cells of most C_4 plants, which take up CO_2 and use it to produce four-carbon organic acids that eventually move into the cytoplasm. In contrast, the red *B. cycloptera* plastids clustered near the green-stained nucleus function much like the bundle-sheath cells of most C_4 plants, in which CO_2 (released from organic acids absorbed from the cytoplasm) is used in the Calvin cycle to make sugar.
Confocal fluorescence micrograph courtesy Simon D.X. Chuong

30.1 Testing Your Knowledge

1. Animals require O_2 and food in the form of organic compounds that supply energy and provide essential minerals, such as calcium. What essential resource do green plants require?
   a. water as a source of H and O
   b. light as a source of energy
   c. CO_2 as a source of carbon for making organic compounds
   d. elements in the form of mineral ions such as K^+, NH_4^+ or NO^{3-}, and Ca^{2+}
   e. All of the above resources are required.

2. Which environment is likely to display more types of plants that possess C_4 photosynthesis?
   a. cool, moist environment
   b. hot, dry environment
   c. cool, dry environment
   d. hot, moist environment
   e. All of the above are correct.

30.2 The Roles of Soil in Plant Nutrition

Learning Outcomes
1. Discuss the advantages of using soils high in organic matter for plant cultivation.
2. Explain the advantages of loam soils for plant cultivation.
3. List reasons that plants require a source of fixed nitrogen.

In addition to light, CO_2, and water, plants require additional elements that occur naturally in soil or that can be added in the form of fertilizers (Table 30.1). Soil is an essential resource for most wild and cultivated plants, providing water and other essential nutrients. For this reason, extensive loss of soil by wind and water erosion is of wide concern. Soils vary greatly in fertility—their ability to support plant growth—because they vary in degree of aeration, water-holding capacity, and mineral content. These properties arise from differences in the physical structure and chemical features of soils.

Soils Have Both a Physical and Chemical Structure

Natural soils display layers, the topsoil being the uppermost layer and the one most important to plant growth (**Figure 30.6a**). The remains of plants that have recently died and other organisms form a layer of organic litter above the topsoil. Many of the inorganic minerals and organic materials that enrich high-quality topsoil arise from the activities of microorganisms that decompose the litter. In this way, the minerals contained in living plants are eventually recycled to subsequent generations. Beneath the topsoil lie layers called subsoil and soil base, which are largely composed of mineral materials. Bedrock is the bottom layer that supports the soil layers. Plant roots play an important role in conveying deep-lying minerals to the surface, thereby helping to enrich the topsoil.

Soil layers vary in composition and thickness, depending on various factors, including climate, vegetation, bedrock type, and human influences. For example, natural grasslands produce deep topsoil rich in both organic matter and minerals useful to plants, with the result that grassland soils have been transformed into croplands in many regions of the world (**Figure 30.6b**). In dramatic contrast, tropical rain forests often have only thin layers of topsoil; their low-fertility soils are composed mostly of inorganic materials of types that are not useful to plants (**Figure 30.6c**). Understanding the roles of soil organic matter and inorganic soil components is essential to maintaining and improving soil fertility.

Soil Organic Matter Soil organic matter, also known as humus, is largely derived from plant detritus, the dead and decaying remains of plants, although animal wastes and decayed animal bodies also contribute to the organic content of soils. Maintaining or improving soil organic matter provides many benefits:

- Organic-rich soils, containing 8% or so organic matter, are less likely to erode—wash or blow away with water or wind.
- Soil organic matter binds mineral nutrients, thereby fostering soil fertility.
- Soil organic matter softens the soil's consistency, thereby fostering plant root growth and farmers' ability to cultivate the soil.

(a) Soil structure
(b) Thick layer of topsoil in cropland
(c) Thin layer of topsoil in rain forest

Figure 30.6 **The structural layers of soil.** **(a)** Diagram showing general soil structure. **(b)** A vertical view of an agricultural soil, showing a relatively deep layer of dark topsoil. **(c)** A vertical view of a tropical rain forest soil, showing a thin layer of dark topsoil.
(b) Courtesy of C.A. Stiles, USDA-NRCS; (c) ©Ruddy Gold/age fotostock

- Gardeners often produce organic-rich compost by layering vegetable waste from the kitchen and yard waste with soil, as well as watering and turning the pile occasionally to introduce the water and oxygen necessary for decomposition. Mature compost is then mixed with garden soil to improve the soil's fertility.
- After harvesting a crop such as corn, farmers often leave the plant stubble to help hold soil in place and then plough the rotting remains back into the soil to increase its organic content.

Inorganic Soil Components Inorganic materials in soil are derived from the physical and chemical breakdown of rock by (1) cycles of freezing and thawing and (2) organic acids exuded from plant roots or lichens. Inorganic materials needed by plants, such as K^+, NH_4^+ or NO_3^-, and Ca^{2+} (Table 30.1), can be lost from soil by **leaching,** the dissolution and removal of minerals as water percolates through materials. Heavy rainfall can reduce the fertility of soils by leaching large amounts of nutrients from them. For example, high annual rainfall in tropical regions causes extensive leaching of minerals, explaining why tropical soils can be low in nutrients. In arid lands, irrigation of crops with ion-laden water, followed by evaporation, can increase soil saltiness to a degree that inhibits plant growth.

Soil Types Soils can be classified according to their relative content of coarse and fine inorganic materials, classified as sand, silt, or clay (**Figure 30.7**):

- Sand grains range from 20 to 2,000 micrometers (μm) in diameter.
- Particles of silt range from 2 to 20 μm in diameter.
- Clay particles are smaller than 2 μm in diameter.

Because of their size differences, sand, silt, and clay particles confer different properties on soils, allowing them to be classified as sandy, silty-clay, or loam soils.

- Soils that contain 45% or more sand and 35% or less clay are classified as sandy soils. The relatively large size and irregular shapes of sand particles allow air and water to move rapidly through sandy soils, which are said to be porous (**Figure 30.8**). Sandy soils are well aerated, which is favorable to the growth of plant roots, which require oxygen for cellular respiration. However, sandy soils hold less water than the same volume of clay, and rapid percolation of water through sandy soils both reduces the amount of water available to the roots and leaches minerals from the soil.
- Silty-clay soils contain more than 35% of silt and clay particles. Silt and clay particles fit closely together, so such soils are less porous than sandy soils. Water percolates less easily through silty-clay soils, which therefore retain more minerals than do sandy soils. In addition, clay particles have negative charges on their surfaces that electrostatically bind positively charged ions

Figure 30.7 The relative sizes of inorganic soil components.

Figure 30.8 The movement of air and water in sandy soil.

Concept Check: *In which type of soil is nutrient leaching a more common problem: sandy or clay soil?*

(cations) such as K^+, NH_4^+, and Ca^{2+}. Because these and some other soil cations are plant nutrients, their retention benefits plant growth.

- Loam soils contain a balanced mixture of sand, silt, and clay and thus combine the aeration benefits of sandy soils with the water- and cation-retention properties of silty-clay soils. Hence, loam soil is considered ideal for the cultivation of many plants.

Cation Exchange To become available to plants, cations must be detached from clay particles. Plant roots often exude organic acids that help to release cations. Recall that an acid is a molecule that releases hydrogen ions (H^+, protons) into solutions. Hydrogen ions released into soil water are able to replace mineral cations on the surfaces of clay particles in a process known as **cation exchange** (**Figure 30.9**). Cation exchange releases cations into soil water, making them available for uptake by plant roots. However, free ions are also more easily washed out of soil.

If the H^+ concentration becomes too high, large numbers of mineral ions are released and can be leached from soil by heavy rainfall. Such leached minerals may include metals such as aluminum (Al^{3+}) that would otherwise be bound in soil. These processes explain how acid rain causes loss of soil fertility and pollution of streams with toxic substances such as aluminum that can harm human health. Humans foster acid rain by burning fossil fuels and setting fire to woodlands, processes that add acidic contaminants to the atmosphere.

The Role of Fertilizers Soil additions known as fertilizers enhance plant growth by providing essential elements that are otherwise absent or present in less than optimal amounts. Fertilizers occur in organic and inorganic forms:

- Organic fertilizers are those in which most of the minerals are bound to organic molecules and are thus released relatively slowly. Manure and compost are examples of organic fertilizers that play an important role in organic farming. Organic crops are grown without the use of inorganic fertilizers, growth substances, and pesticides.

- Inorganic fertilizers consist largely of more water-soluble inorganic minerals, which are immediately useful to plants but can more easily leach from soils by heavy rainfall. Nitrogen (N), phosphorus (P), and potassium (K) are the mineral nutrients that most frequently limit crop growth and are common components of commercial inorganic fertilizers. Such fertilizers are available in different ratios of minerals that are optimal for different types of plants.

Even though fertilizers can improve the growth of crops, excessive application of fertilizers to fields and lawns is undesirable because (1) fertilizers are costly and (2) minerals not taken up by plant roots are easily washed away by rain into waterways. In aquatic ecosystems, excess nutrients can fuel large growths of algae and aquatic plants that can harm other aquatic life. For example, large areas of the Gulf of Mexico and other coastal regions are now known as "dead zones" because microbial decomposition of large algal populations depletes oxygen from the water, suffocating the animal life. These large populations of algae are fostered by fertilizers that wash from farm soils into rivers, such as the Mississippi River, and then drain into coastal oceans. By the overuse of fertilizers, humans are directly responsible for the occurrence of coastal "dead zones." More careful application of fertilizers and planting stream and river edges with vegetation that helps to absorb mineral nutrients can help to reduce or prevent "dead zones."

Plants Require Fixed Nitrogen

As we have seen, plants require a variety of mineral nutrients obtained from the soil (Table 30.1). In this section, we will focus on nitrogen as an important example. Nitrogen availability frequently limits plant growth in nature and in crop fields because plants require large amounts of N to synthesize amino acids, nucleotides, chlorophyll, and alkaloids and other cellular materials.

Plants take up nitrogen in a chemically combined form, ammonium (NH_4^+) or nitrate (NO_3^-), and such combined forms of nitrogen are known as **fixed nitrogen.** New fixed nitrogen can be added to soils by the action of lightning, fire, and air pollution, as well as biological and industrial nitrogen fixation. **Nitrogen fixation** is the process by which atmospheric N_2 is combined with hydrogen (that is, reduced) to produce ammonium (NH_4^+). Most of the fixed nitrogen in soils is produced by biological nitrogen fixation, which is performed in nature only by certain prokaryotic species. These species possess

Biology Principle

Biologists Use Models to Test Predictions and Convey Ideas

Part (b) of this figure is a model for the process of cation exchange that occurs on the surface of clay particles in soil. The Modeling Challenge in the figure legend asks you to draw two models that compare cation exchange before and after a public policy change related to acid rain.

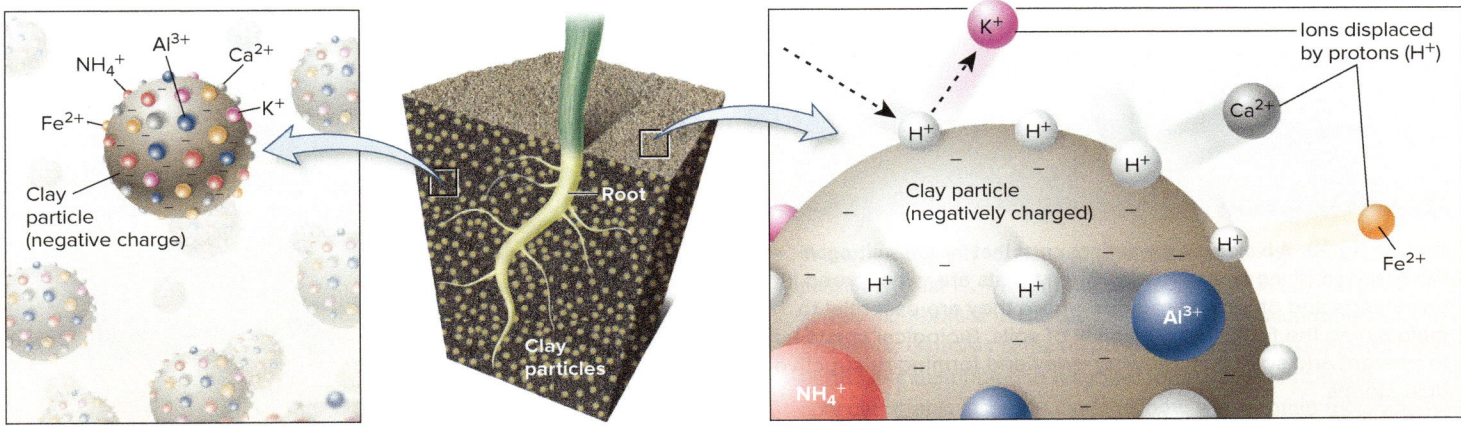

(a) Electrostatic attraction between clay particles and mineral ions

(b) Cation exchange

Figure 30.9 Cation binding and exchange. (a) Clay and organic particles in the soil display negative electrostatic surface charges that bind cations. Bound cations include not only plant mineral nutrients such as ammonium (NH_4^+) but also cations that are not plant nutrients, such as aluminum ion (Al^{3+}). (b) Cation exchange occurs when protons (H^+) displace other cations, releasing them to soil water. This process makes cations more available for uptake by plant roots, but it also increases the potential for cations to leach away during rains or floods.

Modeling Challenge: Acid precipitation, commonly known as acid rain, is linked to human health issues as well as declining forest health in some regions of the world. Plant health can be adversely affected by the accumulation of Al^{3+} in plant tissues and a loss of soil fertility in the form of essential mineral cations. The 1970 federal Clean Air Act, which mandated a reduction of acid-forming sulfur compounds from power plant emissions into the atmosphere, was influenced by scientific studies showing that acid rain is associated with decline in forest health. Consider the diagrammatic model of cation exchange shown in Figure 30.9b, and draw a model showing how soil clay particle associations with cations that are plant nutrients (for example, NH_4^+) might change after an increase in acid rain. Then draw another model that shows how these associations would change after a decrease in acid rain as a consequence of the Clean Air Act. Discuss how the differences between your two models would affect forest health.

---

an enzyme known as nitrogenase that reduces atmospheric N_2 at environmental temperatures and pressures. Nitrogen fertilizers applied to crops are produced by industrial nitrogen fixation, a process that requires very high temperatures and pressures and thus consumes large amounts of energy.

Biological Nitrogen Fixation by Bacteria Nitrogen-fixing bacteria include many types of cyanobacteria, which are photosynthetic organisms that occur in oceans, lakes, and other aquatic systems (see Chapter 23). Certain nitrogen-fixing cyanobacteria live symbiotically within the leaves of water ferns that naturally occur in wet rice paddies. Fixed nitrogen produced by these cyanobacteria fertilizes not only the water ferns but also the rice plants, thereby helping to feed millions of people who depend on rice as their staple food. Nitrogen-fixing cyanobacteria also occur in surface soil crusts that are common in arid lands; fixed nitrogen produced by soil crusts enhances soil fertility (**Figure 30.10**).

Various types of nonphotosynthetic nitrogen-fixing bacteria also enrich water and soil. Nitrogen-fixing bacterial symbionts are taken up and fostered within special nodules on the roots of legume plants such as beans, peas, and peanuts (**Figure 30.11**). This close association allows the bacteria to transfer fixed nitrogen directly to plant cells and explains why food and forage legumes are valued for their high protein content. Biologists have been trying to incorporate nitrogen-fixing symbionts or genes into cereal crops, which lack such effective nitrogen-fixation capacity.

Biology Principle | Biology is a quantitative science

Quantitative Analysis: The Water Potential Equation Can Be Used to Predict Cellular Water Status

How can researchers use water potential concepts to predict the water status of plant cells?

Crunching the Numbers: Imagine that you are able to use specialized equipment to measure the solute potential (ψ_S) of a plant cell as -0.15 MPa and measure the pressure potential (ψ_P) as $+0.15$ MPa. Use the simple form of the water potential equation, $\psi_W = \psi_S + \psi_P$, to calculate this cell's water potential. From that number, is this cell turgid, plasmolyzed, or flaccid?

EVOLUTIONARY CONNECTIONS

Plants Have Evolved Cellular Adaptations to Drought Stress

Plants native to cold, dry, or saline environments have evolved diverse adaptations that allow them to cope with low water availability in these habitats. Plant cells may increase their cytosolic concentration of the amino acid proline, sugars, or sugar alcohols such as mannitol. This solute concentration increase draws water into cells. By increasing the concentration of solutes inside cells, cold-resistant plants prevent water from moving out of their cells when ice crystal formation in intercellular spaces lowers the water potential outside the cells. The additional solutes also lower the freezing point of the cytosol, in the same way that adding antifreeze to a car's radiator in winter keeps the radiator fluid from freezing.

Arid-land plants can often survive in a nearly dry state for as much as 10 months of the year, growing and reproducing only after the rains come. The cytosol of such desiccation-tolerant plants is typically rich in sugars that bind to phospholipids to form a glasslike structure. This helps to stabilize their cellular membranes. Plant cells under water stress may also increase the number of plasma membrane aquaporins, allowing cells to absorb water quickly when it becomes available.

Forest plants native to moister habitats have likewise evolved traits that aid survival in drought. A recent study of 40 forests revealed that water potential is an important factor in predicting how forest plant species are likely to respond to climate change.

Relative Water Content Measurements Reveal Whole Plant Adaptation to Water Stress

Researchers measure a property known as **relative water content (RWC)** in studies of natural plant adaptation to cold, drought, or salt stress and in agricultural research for developing drought-tolerant crops. RWC can be used to predict a plant's ability to recover from the wilted condition. An RWC of less than 50% spells death for most plants. However, some plants, particularly those of arid habitats, can tolerate lower water content for substantial time periods, an evolutionary adaptation to water stress. A standard method for determining RWC involves three simple weight measurements:

- Sample tissue taken from a plant under a given set of conditions is first weighed to obtain the fresh weight.
- Then the sample is completely hydrated in water within an enclosed, lighted chamber until constant turgid weight is achieved.
- Finally, the sample is dried to a constant dry weight.
- RWC is calculated as $100 \times$ [(fresh weight minus dry weight) divided by (turgid weight minus dry weight)].

Comparisons of the RWC of plants sampled from or grown in environments varying in water availability, followed by assessment of the plants' ability to recover from water stress, reveal that some plants have evolved in ways that allow them to survive for longer periods in arid conditions.

30.3 Reviewing the Concepts

- Plant root hairs take up minerals via plasma membrane transporters, a process that involves ATP hydrolysis, and water flows into solute-rich root cells by osmosis. Short- and long-distance transport allow these materials to move from roots to stems and leaves. Organic compounds produced by photosynthetic tissues move to nonphotosynthetic tissues by short- and long-distance transport (Figures 30.12, 30.13, 30.14).
- Turgor pressure increases as water enters plant cells because cell walls restrict the extent to which cells can swell. A cell that is so full of water that the plasma membrane presses closely against the cell wall is turgid. A cell that contains so little water that the plasma membrane pulls away from the cell wall is plasmolyzed. A flaccid cell has a water content between these extremes (Figure 30.15).
- Solute potential (ψ_S) and pressure potential (ψ_P) arising from the presence of a cell wall are major factors affecting cellular water potential (ψ_W). Water moves from a region of higher water potential to a region of lower water potential (Figure 30.16).
- The measurement known as relative water content (RWC) can be used to examine the drought recovery capacities of whole plants.

30.3 Testing Your Knowledge

1. Assuming equally thick cell walls of the same composition, which plant cell would have the highest turgor pressure?
    a. a plant cell that displays a lower concentration of solutes than its environment
    b. a plant cell that displays the same concentration of solutes as its environment
    c. a plant cell that displays a higher concentration of solutes than its environment
    d. All of these plant cells would have the same turgor pressure.

2. A plant is able to recover from a relative water content (RWC) of 40%. To what type of environment is this plant most likely adapted?
   a. an area of tropical rain forest that receives plentiful rainfall year round
   b. an area of temperate forest that receives plentiful rainfall during the growing season
   c. an area of grassland that receives relatively low levels of rainfall that is evenly distributed throughout the growing season
   d. an area of arid desert in which water is not available during much of the year
   e. All of the above are correct.

30.4 Plant Transport at the Tissue Level

Learning Outcomes

1. Distinguish among transmembrane transport, symplastic transport, and apoplastic transport.
2. Explain how the root endodermis functions as a diffusion barrier.

Now that we have reviewed how water, dissolved minerals, and organic compounds enter or leave plant cells, we are prepared to consider short-distance transport within and among nearby plant tissues. Understanding tissue-level transport is an essential step toward comprehending how plants accomplish long-distance transport. Tissue-level transport occurs in three forms: transmembrane transport, symplastic transport, and apoplastic transport (Figure 30.17).

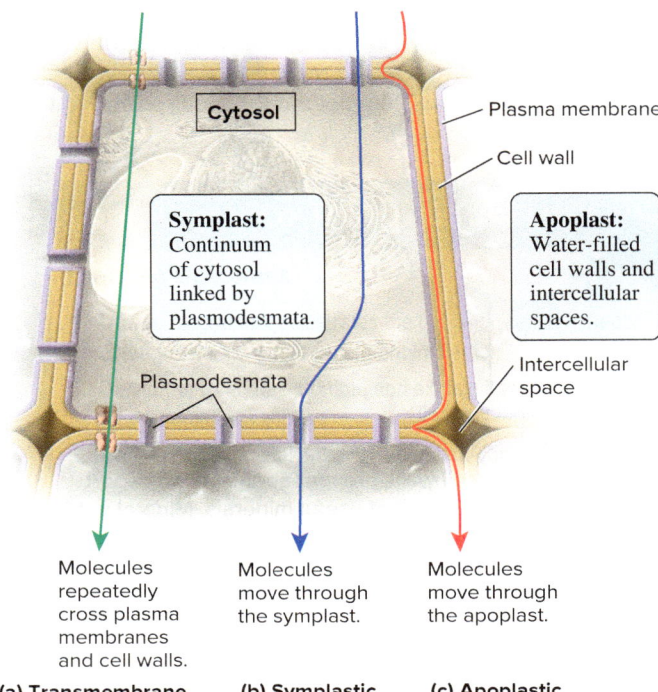

Figure 30.17 Three routes of tissue-level transport in plants: (a) transmembrane, (b) symplastic, and (c) apoplastic.

Transmembrane transport involves the export of a material from one cell via membrane proteins, followed by import of the same substance by an adjacent cell (Figure 30.17a). One prominent example of transmembrane transport is the movement of the plant hormone auxin downward in shoots. Auxin travels from one parenchyma cell to another with the aid of carrier proteins (refer back to Figure 29.4a). This process explains how auxin produced in one part of the plant body can influence more distant tissues.

- **Symplastic transport** is the movement of a substance from the cytosol of one cell to the cytosol of an adjacent cell via membrane-lined channels called plasmodesmata (Figure 30.17b). Plasmodesmata are large enough in diameter to allow the transport of proteins and nucleic acids, as well as smaller molecules. Together, all of a plant's protoplasts (the cell contents without the cell walls) and plasmodesmata form the **symplast.** Symplastic transport has the potential to move molecules widely among the cells and tissues of the plant body.

- **Apoplastic transport** is the movement of water and solutes along cell walls and the spaces between cells. Both symplastic and apoplastic transport play important roles in the transport of mineral nutrients through the outer tissues of roots (Figure 30.18).

The Root Endodermis Plays Key Roles

Water and dissolved minerals can move apoplastically into root epidermal and cortex tissues. However, apoplastic movement of water and minerals stops at the **endodermis,** a term meaning inside skin. In roots, the endodermis is a thin cylinder of tissue whose close-fitting cells and specialized cell walls form a barrier to diffusion between the cortex and the central core of vascular tissue. Root endodermal cells are able to function as diffusion barriers because their walls possess ribbon-like strips of waterproof suberin, composed of wax and phenolic polymers. Suberin is also abundant in cork, the protective outer bark of trees and shrubs (see Chapter 28). Known as **Casparian strips,** these suberin ribbons wrap all the way around each endodermal cell, forming a tight seal (Figure 30.19).

The plant endodermis is functionally analogous to animal epithelial tissues whose cellular tight junctions likewise form diffusion barriers. Because of the Casparian strips, materials in the root apoplast cannot penetrate farther into the root unless endodermal cells transport the materials into their cytosol, a process that requires specific transporter proteins. In this way, the root endodermis prevents harmful solutes such as toxic metal ions from moving through the apoplast to vascular tissues and being transported to the shoot. In addition, the endodermis prevents xylem solutes such as K^+, NO_3^-, or Ca^{2+} from flowing back into root cortex or soil.

30.4 Reviewing the Concepts

- Plant tissue-level transport occurs in three forms. Transmembrane transport involves the movement of materials from one cell to another from intercellular spaces, across plasma membranes, and into the cytosol. Symplastic transport allows materials to move from one cell to another through the symplast, the cells' cytosol and plasmodesmata, without crossing plasma and close a pore. As guard cells become fully turgid, their volume expands by 40–100%. This expansion does not occur evenly, however, because the innermost cell walls are thicker and less extensible than are other parts of the cell walls. In addition, the guard cells expand primarily in the lengthwise direction because bands of radially oriented cellulose microfibrils prevent the guard cells from expanding laterally (Figure 30.23b). Thus, when the guard cells are turgid, a stomatal pore opens between them, allowing air exchange with the leaf's spongy mesophyll. Conversely, when the guard cells lose their turgor, their volume decreases, and the stomatal pore closes.

What causes the change in guard cell turgor? In flowering plants, stomata often open early in the morning, in response to sunlight. This response makes sense, given that light, water, and carbon dioxide are all required for photosynthesis. Blue light stimulates H^+-ATPase proton pumps, leading to guard cell uptake of ions, especially potassium (K^+), and other solutes. As a result of increases in solute concentrations inside guard cells, osmotic water uptake occurs via plasma membrane aquaporins, resulting in cell expansion and stomatal opening (Figure 30.24a).

At night, the reverse process closes flowering plant stomata. Potassium (K^+) and other solutes are pumped out of guard cells, causing water to exit and deflate guard cells, resulting in pore closure. Flowering plants also close stomata during the daytime under conditions of water stress, a process mediated by the stress hormone abscisic acid (ABA). Water stress causes a 50-fold increase in ABA, which is transported in the xylem sap to guard cells. There, ABA binds to a membrane receptor, which elicits a Ca^{2+} second messenger, causing the guard cells to lose solutes and deflate (Figure 30.24b).

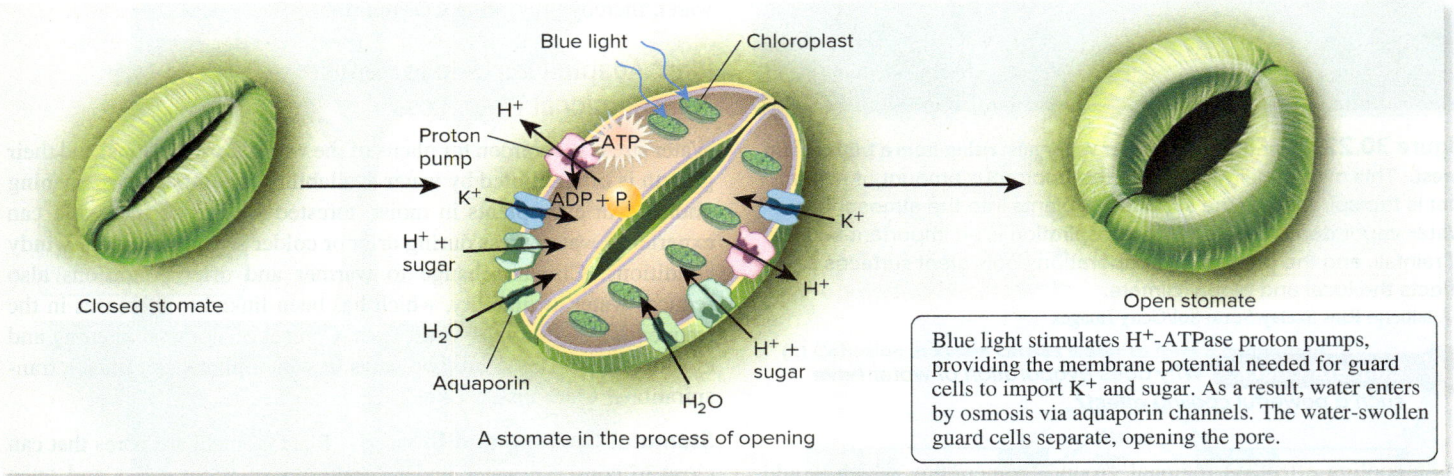

(a) The process of stomatal opening

(b) The process of stomatal closing

Figure 30.24 How stomata of flowering plants open and close. (a) Angiosperm stomata usually open in response to sunlight. (b) Angiosperm stomata usually close in response to lack of sunlight. Stomata can also close during the day under conditions of water stress, which induces plants to produce more of the hormone abscisic acid (ABA). Guard cell plasma membranes possess ABA receptors, which receive the drought signal.

Leaf Abscission Angiosperm trees and shrubs of seasonally cold habitats experience water stress every winter, when evaporation from plant surfaces occurs but soil water is frozen and therefore unavailable for uptake by roots. Under these conditions, air bubbles can form in the xylem, preventing bulk water transport, much as the introduction of air into an animal's circulatory system can block conduction. Desert plants also experience water stress conditions, but at less predictable times and for much of the year. Many types of plants that live in cold climates or deserts are adapted to cope with water stress by dropping their leaves, a process known as leaf **abscission.** Dropping leaves lets these plants avoid dangerously low water potentials. Leaf abscission also reduces the amount of root mass that plants must produce to obtain water under arid conditions. The ocotillo (*Fouquieria splendens*) of North American deserts can produce leaves after sporadic rains and then drop leaves as a direct response to drought as many as six times a year (**Figure 30.25**).

The sugar maple (*Acer saccharum*) is an example of the many types of trees or shrubs of temperate forests that drop their leaves each autumn and are known as deciduous plants. During their evolution, temperate-zone angiosperm trees and shrubs have acquired the genetic capacity to use seasonal changes to predict the onset of cold, dry winter conditions and respond with preemptive leaf abscission. Leaf abscission reduces the potential for evaporative loss of water from a perennial plant during the period when soil water is frozen and thus cannot be taken up by roots.

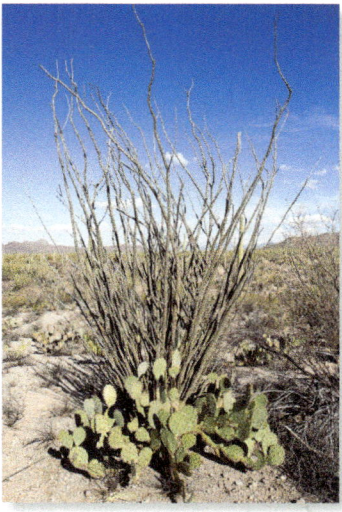

(a) Ocotillo with leaves **(b) Ocotillo without leaves**

Figure 30.25 Leaf abscission as a drought adaptation. The ocotillo (*Fouquieria splendens*), a plant native to North American deserts, is known for its ability to respond to intermittent rain and drought by producing and dropping leaves multiple times within a year.
(a) ©Dan Suzio/Science Source; *(b)* ©Matthew Heinrichs/Alamy

Concept Check: Why doesn't the ocotillo drop its leaves at a single predictable time each year, as do temperate angiosperm trees and shrubs?

Leaf abscission is a highly coordinated developmental process. The hormone ethylene stimulates an abscission zone to develop at the bases of leaf petioles (**Figure 30.26a**). The abscission zone contains two types of tissues: a separation layer of short, thin-walled cells and an underlying protective layer of suberin-impregnated cork cells (**Figure 30.26b**). As noted in the discussion of endodermal Casparian strips, suberin contains both waterproofing wax and phenolic polymers that retard microbial attack. As the abscission layer develops across the vein linking the petiole with the stem, the layer eventually cuts off the water supply to the leaf. Chlorophyll in the leaves degrades, revealing colorful orange and yellow carotenoid and xanthophyll pigments that were hidden beneath. In addition, some plants synthesize red and reddish blue pigments in response to changing environmental conditions. The presence of these pigments explains colorful autumn vegetation in temperate zones. Enzymes eventually break down the cell-wall components of the separation layer, causing the petiole to break from the stem. The underlying protective layer forms a leaf scar that seals the wound, helping to protect the plant stem from water loss and pathogen attack.

Transport in the Phloem Occurs by Pressure Flow

As we have seen, transport of water and dissolved minerals occurs within dead, empty xylem tracheids and vessels by suction, explained by the cohesion-tension theory. Movement of materials in the watery sap of living phloem tissues involves a contrasting process that is explained by the **pressure-flow hypothesis.** This hypothesis states that sugars flow from the places where sugars are made to sites where sugars are utilized, by means of differences in turgor pressure. Leaf mesophyll tissue makes sugars and is known as a sugar source. Cells that receive and utilize sugar, referred to as sinks, occur in the non-photosynthetic tissues of roots and stems, as well as flowers, fruits, and seeds.

The process by which sugars made in leaf mesophyll cells are moved into the living cells of phloem is known as sugar loading and often requires energy input in the form of ATP (**Figure 30.27**). The movement of sugar into living phloem cells causes water to also enter by osmosis, with the result that the pressure of phloem sap rises. This pressure causes phloem contents to flow from regions of higher solute concentration (leaf veins) to regions of lower solute concentrations (such as roots, flowers, fruits, and seeds). Phloem bulk flow is based on positive pressure, much as you would exert when squirting water from a syringe by pushing on the plunger (Figure 30.27). The water that enters sugar-rich leaf phloem comes from nearby xylem, explaining why vascular bundles that make up the plant conduction system contain both xylem and phloem tissues (see Chapter 28).

The living food-conducting components of phloem are known as sieve-tube elements. Joined end-to-end, sieve-tube elements form long sieve tubes whose structure fosters the flow of sap for long distances through the plant body. During their development, sieve tubes arise by the mitotic division of precursor cells that also produce narrow cells known as companion cells (refer back to Figure 28.18). Sieve-tube elements and neighboring companion cells remain connected by plasmodesmata, channels that run through cell walls. As they mature, sieve-tube elements lose their nucleus, but

Biology Principle

Living Organisms Maintain Homeostasis

Water content homeostasis is fostered by leaf drop in response to, or in anticipation of, arid conditions or cold temperatures.

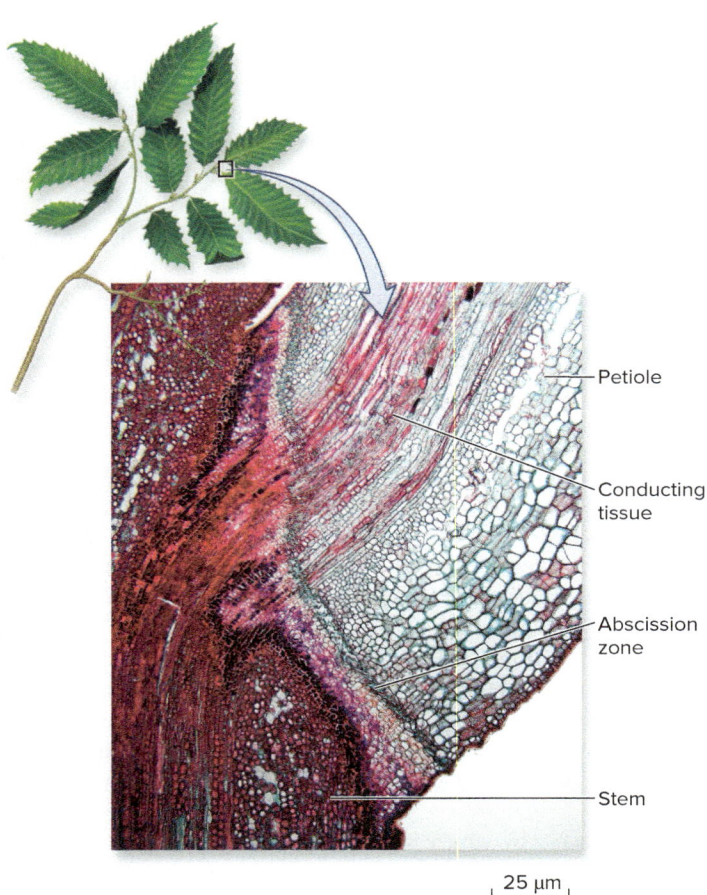

(a) LM of leaf abscission zone at the junction of a petiole and stem, stained with dyes

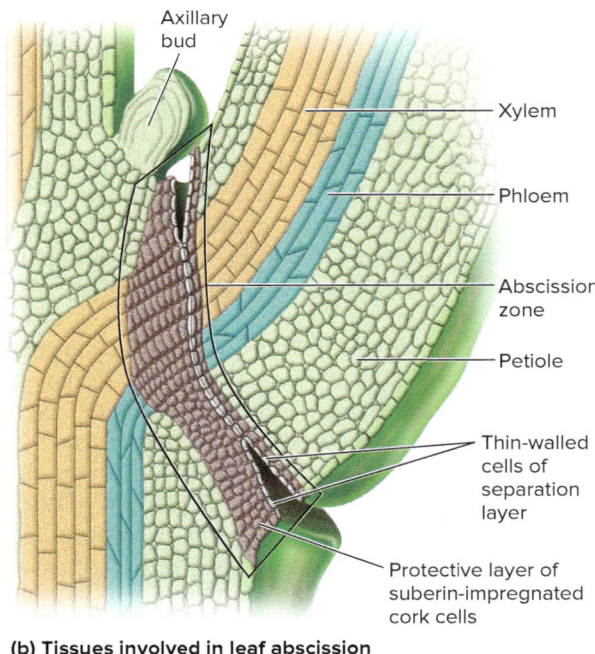

(b) Tissues involved in leaf abscission

Figure 30.26 Leaf abscission. Leaf abscission helps to maintain plant hydration.
(a) ©Lee W. Wilcox

---

companion cells retain a nucleus. Consequently, sieve-tube elements become dependent upon a neighboring companion cell for essential messenger RNA (mRNA) and proteins, which are supplied via plasmodesmata.

An example of sieve-tube function can be observed when a plant's conducting system is damaged. In this case, a wound response occurs that involves a protein known as P protein (for phloem protein). Aggregates of this protein accumulate along the end walls of damaged sieve-tube elements, preventing loss of phloem sap (refer back to Figure 28.18). Such protein aggregates function much like a clot that helps reduce blood loss from wounded animals. P protein also binds to the cell walls of pathogens, thereby helping to prevent infection at wounds. Companion cells provide either P protein mRNA or the protein itself to sieve-tube elements. Plants then deposit the carbohydrate callose as a sealant at the wound site.

Companion cells also play an essential role in loading sugar into the phloem. Although glucose and some other monosaccharides can occur in phloem, the more stable disaccharide sucrose is the main form in which most plants transport sugar over long distances. Two types of phloem loading occur: symplastic and partly apoplastic. Many woody plants transport sucrose from sugar-producing cells of the leaf mesophyll to companion cells and then to sieve-tube elements via plasmodesmata, a process known as symplastic phloem loading (**Figure 30.28a**). The advantage of symplastic loading is that it does not require ATP; by moving through plasmodesmata, sugar does not have to cross plasma membranes.

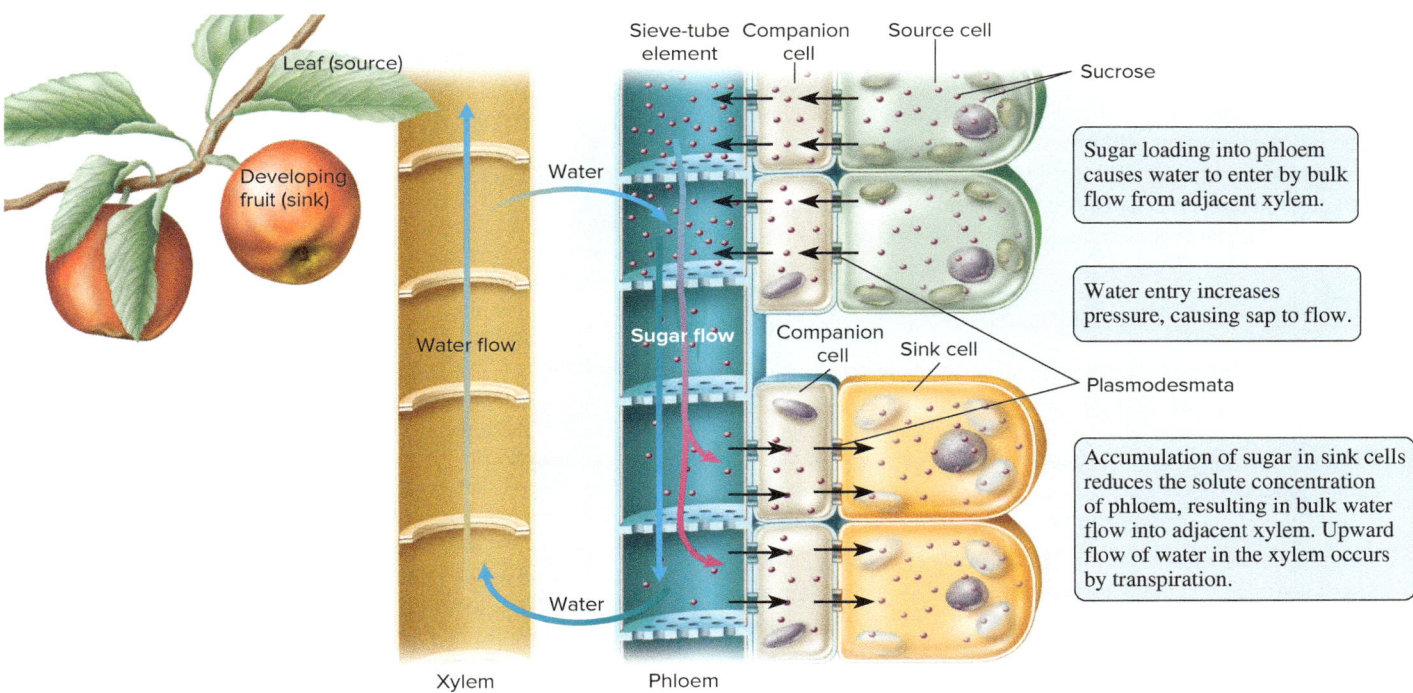

Figure 30.27 Pressure-flow hypothesis for phloem transport. Water flows by osmosis into solute-rich source phloem tissues from nearby xylem. The increased water pressure pushes solute-rich phloem sap toward sink tissues, which are lower in solute concentration. This process explains why phloem transport from sources to sinks is referred to as pressure-flow. At sink tissues, sugars are unloaded from the phloem, thereby decreasing phloem solute concentration, and these processes allow water to flow back into xylem. The role of water in the phloem pressure-flow process explains why phloem and xylem tissues occur close together in vascular bundles that form the plant conduction system.

Figure 30.28 Symplastic and partly apoplastic phloem loading. Variations in sugar loading that occur among plants illustrate important transport concepts and are important in crop production.

In contrast, most herbaceous plants, including important crop plants and the model plant *Arabidopsis*, load sugar into sieve-tube elements or companion cells from intercellular spaces, often against a concentration gradient. ATP must be used to move the sugar across a plasma membrane into a companion cell or sieve-tube element (Figure 30.28b). Therefore, this second type of phloem loading is partly apoplastic and partly a transmembrane process. Understanding how sugar-loading processes differ among plants is essential to improving crop production.

30.5 Reviewing the Concepts

- Water and solutes move for long distances by bulk flow within the xylem and phloem. Plant tracheids, vessels, and sieve-tube elements are structured in ways that foster bulk flow. Bulk flow of water upward in xylem is powered by the water pressure difference between moist soil and drier air, the latter resulting from solar heating (Figures 30.20, 30.21).
- The cohesion-tension theory proposes that as the result of water's cohesion and the tension exerted on water at the plant's surface by evaporation, a continuous stream of water can be pulled up through the plant body from the soil, into roots, through stems, and into leaves. Transpiration, the evaporative loss of water from plant surfaces, strongly influences local and global climate (Figure 30.22).
- Regulation of stomata helps plants to prevent excessive water loss by transpiration. Expansion of guard cells causes stomata to open, allowing CO_2 intake. Guard cell deflation causes pores to close, limiting water loss (Figures 30.23, 30.24).
- Plants under existing or predicted water stress often drop their leaves in a process known as abscission, an adaptive response that lets plants avoid very low water potentials (Figures 30.25, 30.26).
- The pressure-flow hypothesis helps to explain the movement of phloem sap as a process driven by differences in turgor pressure that occur between a sugar source (for example, leaves) and a sugar sink (for example, developing fruit). Phloem loading, the movement of sugars into sieve-tube elements for long-distance transport, occurs by symplastic or partly apoplastic transport (Figures 30.27, 30.28).

30.5 Testing Your Knowledge

1. What structural feature of xylem tracheids and vessel elements fosters bulk flow of water?
   a. absence of cytoplasmic contents, resulting from controlled cell death
   b. arrangement of cells end-to-end to form long arrays
   c. cell walls coated with lignin
   d. All of the above are correct.
   e. None of the above is correct.

2. Under what environmental condition are plant stomata most likely to be open?
   a. at night
   b. on a hot, dry, windy day
   c. in the daytime under humid conditions
   d. All of the above are correct.

Assessing Your Knowledge and Skills

Foundational Knowledge

Multiple Choice

1. Which of the following can limit plant growth in nature?
   a. sunlight
   b. water
   c. carbon dioxide
   d. fixed nitrogen
   e. all of the above

2. In what form do plants take up most soil minerals?
   a. as ions dissolved in water
   b. as neutral salts
   c. as mineral-clay complexes
   d. linked to particles of organic carbon
   e. None of the above is correct.

3. Soil organic matter provides the benefit of
   a. allowing water to percolate rapidly through soil.
   b. making soil softer in consistency.
   c. increasing the aluminum content of soil.
   d. causing minerals to be leached more rapidly from soil.
   e. None of the above is correct.

4. How do plants accomplish tissue-level transport?
   a. transmembrane transport of solutes from one cell to another
   b. symplastic transport of materials from one cell to another via plasmodesmata
   c. apoplastic transport of water and dissolved solutes through cell walls and intercellular spaces
   d. by all of the above
   e. by none of the above

5. A root endodermis is
   a. an innermost layer of cortex cells that each display characteristic Casparian strips.
   b. a layer of cells just inside the epidermis of a root.
   c. a layer of cells just outside the epidermis of a root.
   d. a group of specialized cells within the root epidermis.
   e. none of the above.

Short Answer

1. Why is it a bad idea to overfertilize your houseplants? If the amount recommended on the package is good, wouldn't more be better?

2. **PRINCIPLES** A principle of biology is that biology affects our society. Why should farmers (those barely able to grow enough crops to feed themselves) in arid lands prevent livestock from grazing natural vegetation to the point that it disappears?

Critical-Thinking Skills

Multiple Choice

1. Which of the following statements explains how water enters root cells from the soil?
   a. Roots accumulate sugars from shoots, thereby increasing root cell ability to absorb water by osmosis.

b. Roots actively pump water from the soil using the chemical energy of ATP.
c. Water enters the spongy spaces between cells and within cell walls.
d. Membrane-level transport of ions from soil into root-hair cells increases the osmotic flow of water into them.
e. Both c and d are correct.

2. What structural features of guard cells foster their ability to form an open pore in plant epidermal surfaces?
   a. thickened inner cell walls and radially oriented microfibrils
   b. thickened outer cell walls and radially oriented microfibrils
   c. thickened inner cell walls and longitudinal microfibrils
   d. thickened outer cell walls and longitudinal microfibrils
   e. uniform thickness of cell walls and randomly arranged microfibrils

3. Which of the following statements describes how a substance is produced that plugs wounded sieve-tube elements, thereby preventing the leakage of phloem sap?
   a. Sieve-tube elements produce C protein.
   b. Xylem tracheids and vessels produce C protein.
   c. Companion cells aid production of P protein.
   d. Guard cells control production of P protein.
   e. None of the above is correct.

4. Which scenario is most closely related to sugar loading and long-distance sugar transport in a plant?
   a. Dry air causes a suction effect that draws water from plant surfaces by evaporation.
   b. The uptake of mineral ions from moist soil causes water to osmotically flow into a plant root.
   c. A long period of dry weather causes a desert plant to respond by losing its leaves.
   d. A tree with lush foliage is loaded with developing fruit.
   e. Cold, dry conditions cause air bubbles to form in vessels.

5. In addition to the macronutrients and micronutrients that all plants require, additional minerals are thought to improve plant growth. For example, some plants accumulate silicon (Si), though this element is not considered an essential nutrient because its absence does not disrupt the plant life cycle. Silicon is thought to confer protective functions in plants that accumulate this mineral. Which soil type likely contains the highest levels of Si?
   a. loam
   b. silty-clay
   c. sandy
   d. organic-rich compost
   e. All of the above contain equivalent levels of Si.

Short Answer

1. Imagine that you are part of a team assigned to determine what temperature and moisture conditions best suit a new crop plant so that it can be recommended to farmers in appropriate climate regions. What features of the crop plant might you investigate?

2. Why should a farmer try to avoid overfertilizing farm fields that lie near clean trout streams?

31
How Flowering Plants Reproduce and Develop

The Gerbera daisy, *Gerbera hybrida*.

Chapter Outline

31.1 An Overview of Flowering Plant Reproduction

31.2 Flower Production, Structure, and Development

31.3 Male and Female Gametophytes and Double Fertilization

31.4 Embryo, Seed, Fruit, and Seedling Development

31.5 Asexual Reproduction in Flowering Plants

Assessing Your Knowledge and Skills

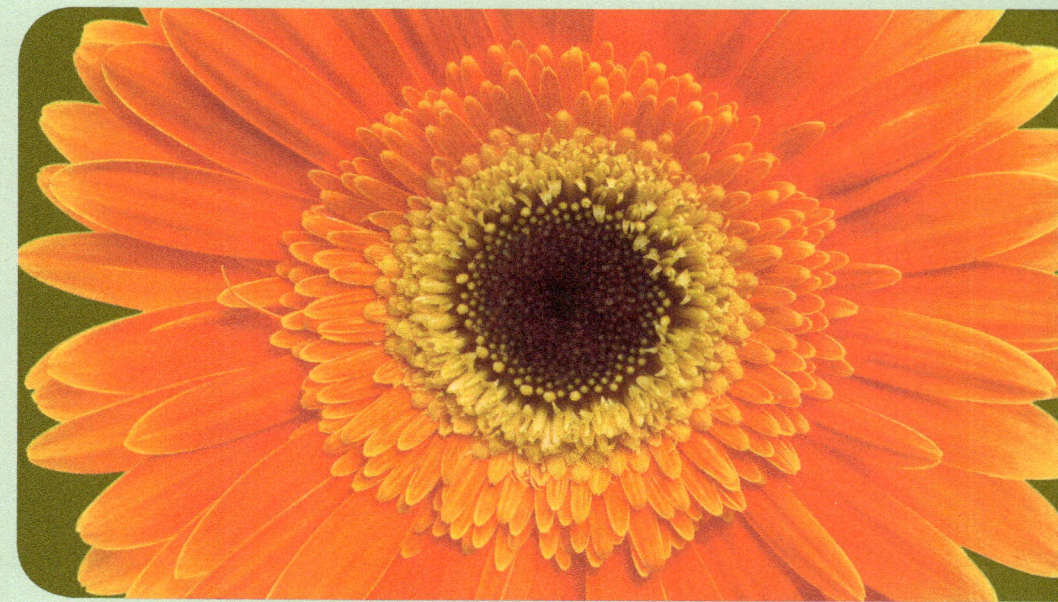

©Vidok/Getty Images

A type of daisy known as the *Gerbera* (chapter-opening photo) is one of the most popular plants you might find at the florist or garden shop. As is the case for other daisies and sunflowers, the attractive "flowers" of a *Gerbera* are actually inflorescences, hundreds of small flowers clustered together. The centermost disk flowers of a *Gerbera* produce sexual organs, pollen, and seeds, but the outermost ray flowers do not. Instead, their petals are fused together to form large, colorful flags that attract bees, butterflies, or birds. These animals accomplish cross-pollination, the exchange of pollen between *Gerbera* flowers of different genetic types. As a result of cross-pollination, the next generation of seeds will be genetically diverse, fostering species survival and evolution. Both the inner disk and outer ray flowers of a Gerbera daisy play essential roles in the plant's reproductive success.

The Gerbera daisy is also a popular model organism, which molecular biologists use to explore the genetic basis for flower traits. Molecular biologists have discovered how *Gerbera* produces both disk and ray flowers in the same inflorescence. During development, a particular transcription factor is expressed less strongly at the inflorescence center than at the margins, where the protein starts a developmental process that shapes the beautiful ray flowers and cancels their production of sexual organs. Less affected by the protein, the central disk flowers produce seeds within fruits that aid seed dispersal.

Flowers, seeds, and fruits are defining reproductive features of the diverse types of flowering plants that are found in nature and that humans rely upon for food and many other products. This chapter begins with an overview of the sexual reproductive cycle of flowering plants. This background allows a closer look at flower structure and development and some of the genes that control flower production and appearance. We will also see how dandelions and some other interesting plants reproduce without using the sexual process.

31.1 An Overview of Flowering Plant Reproduction

Learning Outcomes

1. Explain how the sexual life cycle of flowering plants differs from that of earlier-evolved mosses.
2. List the four organs found in many flowers and the functions of each organ.
3. Describe the phenomenon of double fertilization.
4. List the parts of an angiosperm seed, and explain how each part is produced and functions.

Most flowering plants display **sexual reproduction**, the process by which two gametes fuse to produce offspring that have unique combinations of genes. Flowering plants, also known as angiosperms, inherited their sexual life cycle, known as **alternation of generations** (**Figure 31.1**), from ancestors extending back to the earliest land plants (see Chapter 25). Though all plants share the same basic life cycle, flowering plants display unique reproductive features.

Alternation of Generations Is the Sexual Cycle of Flowering Plants

All groups of land plants produce two alternating, multicellular life cycle stages, in essence, two distinct types of plant body. These two life cycle stages are the diploid, spore-producing **sporophyte** and the haploid, gamete-producing **gametophyte.** In all groups of plants, haploid spores are typically produced by diploid sporophytes as the result of meiosis. These spores undergo mitotic cell divisions to produce multicellular gametophytes. Certain cells within the gametophytes differentiate into gametes. Thus, meiosis does not directly generate the gametes of plants, in contrast with animals. The processes of meiosis and fertilization form the transitions between the plant sporophyte and gametophyte life stages and link them in a cycle (Figure 31.1) (refer back to Figure 25.15). The life cycles of bryophytes (refer back to Figure 25.4), ferns (refer back to Figure 25.7), and gymnosperms (refer back to Figure 25.12) also exhibit alternation of generations.

During the evolutionary diversification of land plants, the sporophyte generation has become larger and more complex, whereas the gametophyte generation has become smaller and less complex. To illustrate this, we can compare the life cycle stages of mosses to those of angiosperms. Mosses diverged early in the history of land plants, whereas angiosperms appeared much later. During the intervening time, the relative sizes and dependence of the sporophyte and gametophyte generations changed dramatically. Moss sporophytes are small structures that always grow attached to larger, photosynthetic gametophytes, because moss sporophytes are incapable of independent life and depend on gametophytes to supply them with essential nutrients (**Figure 31.2a**).

In contrast, flowering plant sporophytes are notably larger and more complex than gametophytes. A tall oak tree, for example, is a single sporophyte. However, oak gametophytes are few-celled, microscopic structures that develop and grow within flowers (**Figure 31.2b**). In addition, photosynthetic oak seedlings and trees grow independently, but nonphotosynthetic oak gametophytes depend completely on the sporophyte for their nutrition. A closer look at flower structure will help you to gain a more complete understanding of angiosperm gametophyte structure and function.

Flowers Produce and Nurture Male and Female Gametophytes

A **flower** is defined as a reproductive shoot, a stem that produces reproductive organs instead of leaves. Flowers are organ systems because several different organs typically occur within a flower (see Chapter 25). Flower organs are produced by shoot apical meristems much like those that generate leaves and are thought to have evolved from leaflike structures by descent with modification (refer back to Figure 25.16).

Most flowers contain four types of organs: sepals, petals, stamens, and carpels (**Figure 31.3**). **Sepals** often protect the unopened flower bud. **Petals** usually attract insects or other animals for pollen transport. **Stamens** and **carpels** each produce distinctive types of spores by the process of meiosis. From these spores, tiny, multicellular gametophytes develop, and certain cells of these gametophytes become specialized gametes.

Stamens Stamens produce **male gametophytes** and foster their early development. Most stamens display an elongate stalk, known as a **filament,** which is topped by an anther. Filaments contain vascular tissue that delivers nutrients from the parental sporophyte

Figure 31.1 Alternation of generations, the plant life cycle.

BioConnections: *Refer back to Figure 25.15 to see a more detailed illustration of the flowering plant life cycle. How can you recognize and where can you find the gametophyte generation of a flowering plant?*

(a) Gametophyte-dominant bryophyte (moss)

(b) Sporophyte-dominant flowering plant (oak)

Figure 31.2 Evolutionary shift in plant life cycle stage dominance. (a) In mosses, the gametophyte is the dominant life cycle stage, and the sporophyte depends on the gametophyte for resources. (b) In flowering plants such as oak trees, the sporophyte life cycle stage is dominant. Microscopic flowering plant gametophytes develop and grow within flower tissues and depend completely on the sporophyte generation.

Concept Check: *What advantages do flowering plants obtain by having such small and dependent gametophytes?*

(a) Flower parts

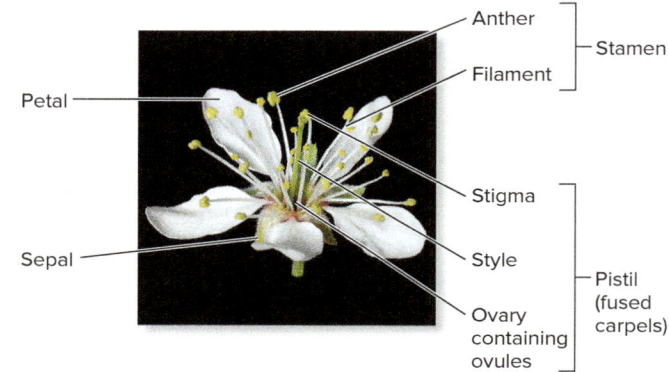

(b) *Prunus americana* (plum)

Figure 31.3 The structure of a typical flower.
(b) ©Lee W. Wilcox

Concept Check: *Do all flowers have all of the structures illustrated here?*

to the anthers. Each **anther** is a group of four **sporangia**—structures in which spores are produced. Within the anther's sporangia, many diploid cells undergo meiosis, each producing four tiny, haploid spores. Because they are so small, the spores produced within anthers are known as **microspores.** Immature male gametophytes, known as **pollen grains,** develop from microspores. The term pollen comes from a Latin word meaning fine flour, reflecting the small size of pollen grains. Pollen grains are eventually dispersed through pores or slits in the anthers. At the time of dispersal, the pollen grain is a two- or three-celled, immature male gametophyte produced by mitotic division. During a later phase of development, a mature male gametophyte produces **sperm** cells.

Carpels Carpels are vase-shaped structures that produce, enclose, and nurture **female gametophytes.** Carpels contain veins of vascular tissue that deliver nutrients from the parent sporophyte to the developing gametophytes. The term **pistil** (named for its resemblance to the pestle used to grind materials to a powder) refers to a single carpel or several fused carpels (Figure 31.3). The topmost portion of a pistil, known as a stigma (Greek, meaning mark), receives pollen grains. The style is the middle portion of the pistil, and an ovary is at the bottom of the pistil. The **ovary** produces and nourishes one or more ovules. An **ovule** consists of a spore-producing sporangium and enclosing tissues known as integument. Within an ovule, a diploid cell produces four haploid **megaspores** by meiosis; megaspores are larger than the microspores produced in anthers, and megaspores have a different fate. Within each group of four megaspores, three die, which allows the surviving one to gain more nutrients. This survivor generates a female gametophyte by mitosis. The female gametophytes of flowering plants typically consist of seven cells, one of which is the female gamete, the **egg**

cell. The female gametophytes of flowering plants also include a large, central cell that typically has two nuclei, as well as antipodal cells, whose function is not well understood. This basic information about male and female gametophytes will help you to understand how they function to produce a young sporophyte within a seed.

Fertilization Triggers the Development of Embryonic Sporophytes, Seeds, and Fruits

In flowering plants, fertilization leads to the production of a young sporophyte that lies within a seed, completing the life cycle (Figure 31.4). Prior to fertilization, pollen grains released from anthers first find their way to the stigma of a compatible flower, the process known as pollination. **Self-pollination** occurs when pollen from the anthers of a flower is transferred to the stigma of the same flower or between flowers of the same plant. **Cross-pollination** occurs when a stigma receives pollen from a different plant of the same species. Flowers are adapted for effective pollination by diverse pollination mechanisms. Many flowers, including the Gerbera daisy, are attractive to insects or other animals that transport pollen, whereas oak flowers and those of some other angiosperms are adapted for pollen transport by wind. A few plants move pollen by means of water currents.

Pollen Germination When pollen grains land on the stigma, the stigma may function as a "gatekeeper," allowing only pollen of

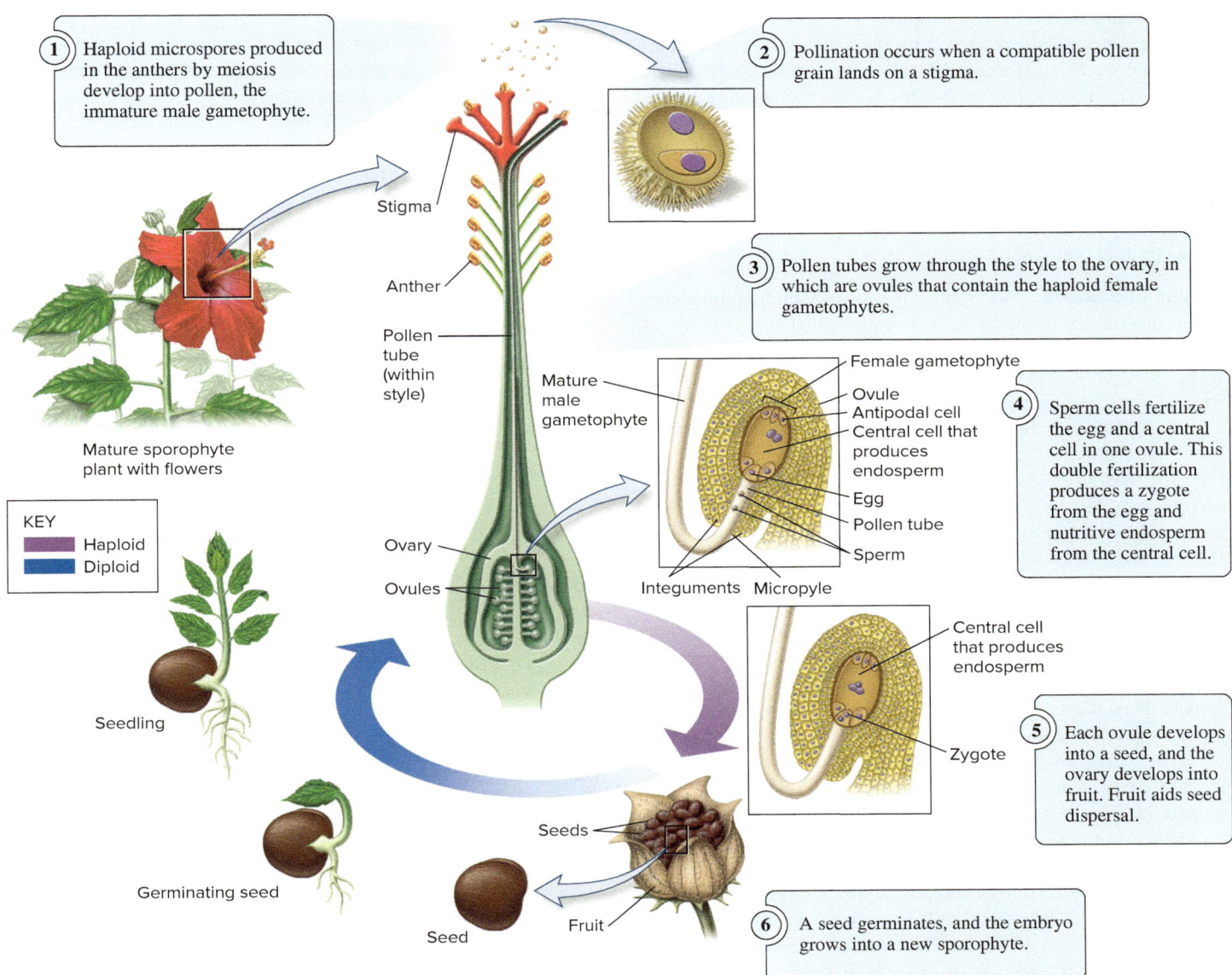

Figure 31.4 The life cycle of a flowering plant. The plant reproductive cycle is illustrated here by hibiscus.

Concept Check: *What advantage does the hibiscus flower gain by clustering its stamens around the pistil?*

appropriate genotype to germinate. During germination, a pollen grain produces a long, thin **pollen tube** that contains two sperm cells. The pollen tube grows through the style toward the ovary. Upon reaching the ovules, the pollen tube grows through an opening in the ovule and delivers sperm to the female gametophyte (Figure 31.4). These sperm unite with haploid cells of the female gametophyte in the process of **fertilization.** Note that pollination and fertilization are distinct processes in flowering plants.

Double Fertilization Angiosperms display a phenomenon known as **double fertilization.** In this process, two different fertilization events occur. One of the two sperm cells delivered by a pollen tube fertilizes the egg cell, thereby forming a diploid **zygote.** This zygote may develop by mitotic division into a young sporophyte, known as an **embryo.** Fertilization thus begins a new cycle of alternation between sporophyte and gametophyte generations. The other sperm delivered by the same pollen tube fuses with the two nuclei in the central cell of the female gametophyte (Figure 31.4). The cell formed by this second fertilization undergoes mitosis, eventually producing a nutritive tissue known as the **endosperm.** The embryo and the endosperm are essential parts of maturing seeds. Fertilization not only starts the development of zygotes into embryos but also triggers the transformation of ovules into seeds and ovaries into fruits. Embryo, seed, and fruit development occur at the same time.

Embryos and Seeds An embryo is a young, multicellular diploid sporophyte that develops from a single-celled zygote by mitosis. Because they are not yet capable of photosynthesis, embryos depend on organic food and other materials supplied by sporophytes. Therefore, embryo development occurs within developing seeds located in a flower ovary. Seeds develop from fertilized ovules. Each developing seed contains an embryo and nutritive endosperm tissue, enclosed and protected by a **seed coat** that develops from the ovule integuments. When embryos and the seed coat have fully matured, they undergo drying, and the seed enters a phase of metabolic slowdown known as dormancy. Fully mature, dormant seeds are ready to be dispersed.

Fruit and Seed Dispersal A **fruit** is a structure that encloses and helps to disperse seeds (Figure 31.4). Seed dispersal benefits plants by reducing competition for resources among seedlings and parental plants. Seed dispersal also allows plants to colonize new sites.

Fruits develop from the flower's ovary and sometimes include other flower parts. Young fruits bearing immature seeds are typically small and green. While embryos and seeds are developing, the fruit also matures. The ovary wall changes into a fruit wall known as a **pericarp** (from the Greek, meaning surrounding the fruit). Mature fruits vary greatly among plant species in size, shape, color, and water content. These variations represent adaptations for seed dispersal in different ways. Fruit variation is also extremely important to wild animals and in human agriculture. For example, most fruit crops are juicy and sweet, with relatively small seeds. In nature, these features foster dispersal by birds and other animals that feed on such fruits.

Seed Germination and Seedlings If a dispersed seed encounters favorable conditions, including sufficient sunlight and water, the seed will undergo seed **germination.** During seed germination, the embryo absorbs water, becomes metabolically active, and grows out of the seed coat, producing a seedling. In a favorable environment, the seedling may grow into a mature sporophyte capable of producing flowers. In the next section, we will focus on flower production, structure, and development.

31.1 Reviewing the Concepts

- Flowering plants display a sexual life cycle known as alternation of generations, a characteristic of all land plant groups, including early-diverging mosses. The gamete-producing male and female gametophytes of flowering plants are very small and depend entirely on nurturing sporophytic tissues. Plant gametes arise by the process of mitosis (Figures 31.1, 31.2).
- Flowers are reproductive shoots that develop from a shoot apical meristem. The role of flowers is to promote seed production. A flower shoot generally produces four types of organs: sepals, petals, stamens, and carpels. Stamens produce and disperse pollen grains, which are immature male gametophytes. Carpels produce, enclose, and nurture female gametophytes in ovules (Figure 31.3).
- A mature male gametophyte, or pollen tube, produces and delivers two sperm to ovules within the ovary. Flowering plants display double fertilization: One of the two sperm released from a pollen tube combines with an egg cell to form a zygote while the other fuses with two nuclei located in a central cell of the female gametophyte, producing the first cell of endosperm tissue. Seed endosperm is a nutritive tissue that supports development of an embryonic sporophyte (Figure 31.4).
- Seeds are reproductive structures containing a dormant embryo enclosed by a protective seed coat that develops from ovule integuments. Fruits are structures that contain seeds and foster seed dispersal. Like flowers and endosperm, fruits are unique features of flowering plants.

31.1 Testing Your Knowledge

1. How does the sexual life cycle of a flowering plant differ from the life cycle of an animal?
    a. Two multicellular bodies alternate in the sexual life cycles of flowering plants, but not in the sexual cycles of animals.
    b. Meiosis generates spores in plants, whereas meiosis generates gametes in animals.
    c. The gamete-producing plant gametophyte is haploid, whereas a diploid animal produces gametes.
    d. All of the above are correct.
    e. None of the above is correct.
2. Which of the following is a unique combination of reproductive features found in flowering plants?
    a. flowers, fruits, and endosperm
    b. stems, roots, and leaves
    c. alternation of generations and sporophytes
    d. gametophytes, eggs, sporangia, and spores
    e. All of the above are correct.

31.2 Flower Production, Structure, and Development

Learning Outcomes

1. Give an example of how gene expression affects flowering time or flower appearance.
2. List examples of complete, incomplete, perfect, imperfect, radially symmetric, and bilaterally symmetric flowers and inflorescences.
3. **SCISKILLS** ▶ Explain recent research findings on how flowers bloom.

Figure 31.5 The occurrence of flower parts in concentric whorls.

Flowers are essential sources of food for many animal pollinators. As the result of coevolutionary relationships with such animals, flowers occur in a spectacular array of colors and forms that attract particular animal pollinators. Flowers also attract humans because we possess sensory systems much like those of animal pollinators. We give bouquets to show love and appreciation; decorate homes, workplaces, and objects with flowers; display flower arrangements on ceremonial occasions; and make perfume from flowers. Consequently, many types of flowers are grown for the florist and perfume industries. Flowers are necessary for the production of grain and other fruit crops. For these and other reasons, biologists investigate how flower development is controlled by environmental signals and changes in gene expression.

Environmental Signals Interact with Genes to Control Flower Production

You've probably noticed that plants flower at particular times of the year. How do plants know when to flower? Flowering time is controlled by the integration of environmental information such as temperature and day length (photoperiod) with hormonal influences and circadian rhythms (see Chapter 29). These stimuli are perceived and integrated by companion cells in leaf phloem, which then make and transport a mobile protein known as FT (flowering time) to the shoot apex. In this location, FT interacts with other proteins, initiating the process by which a leaf-producing apical meristem transforms into a reproductive meristem.

Some plants—such as winter wheat—are planted and sprout in fall, are dormant in winter, and then flower the following spring. In these plants, a special transcription factor represses flowering genes, thereby preventing flowers from appearing too soon. Exposure to cold winter conditions causes the production of a small RNA molecule that silences the transcription factor, allowing these plants to flower when the appropriate day length occurs in spring. The process by which cold exposure allows plants to flower in spring is known as vernalization.

Developmental Genes Control Flower Structure

Organ identity genes specify the four basic flower organs: sepals, petals, stamens, and carpels. Other genes determine flower shape, color, odor, or grouping into bunches known as inflorescences, such as those of the *Gerbera*.

The Genetic Basis of Flower Organ Identity Sepals, petals, stamens, and carpels occur in four concentric rings known as whorls. Sepals (collectively known as the calyx) form the outermost whorl, and petals (together known as the corolla) form an adjacent whorl type. Stamens (together, the androecium) create a third type of whorl, and carpels (the gynoecium) form the innermost whorl (**Figure 31.5**). The **perianth** consists of the calyx plus the corolla.

Gene classes referred to as A, B, C, and E encode transcription factors that control the production and arrangement of these flower organ whorls. The expression of A genes is required for the formation of sepals and petals, B gene expression is needed for petal and stamen development, and C genes specify stamen and carpel production. Consequently, the effects of the A, B, and C genes in determining flower organs are together known as the ABC model of floral organ identity. More recently, a fourth class of genes, the E genes, have been recognized as essential for the formation of petals, stamens, and carpels. Gymnosperms possess genes similar to the class B and C genes of flowering plants, suggesting an early stage in the evolution of the ABCE system.

Variation in Number of Whorls Flowers that possess all four types of flower whorls—calyx, corolla, androecium (stamens), and gynoecium (one or more carpels)—are known as **complete flowers.** In contrast, flowers that lack one or more flower whorls are described as **incomplete flowers.** Flowers having both stamens and carpels are said to be **perfect flowers,** whereas flowers lacking stamens or carpels are described as **imperfect flowers.** An imperfect flower that produces only carpels is known as a carpellate flower (or pistillate flower). Imperfect flowers that produce only stamens are described as staminate flowers.

Corn produces both imperfect staminate and carpellate flowers on an individual plant (**Figure 31.6**). The flowers of corn start to develop as perfect flowers, but in carpellate flowers, the stamens stop developing. After pollination and fertilization, each carpellate flower produces one of the kernels on a cob of corn. In contrast, staminate flowers of corn, which are found in corn tassels, produce the pollen. Corn produces staminate and carpellate flowers on the same plant. Holly and willow are examples of the many types of plants that produce staminate and carpellate flowers, though on separate plants.

Variation in Flower Organ Number In addition to variation in whorls, flowers vary in the number of organs. You may recall that flowering plants occur in two major groups differing in flower structure and other features: the eudicots and the monocots (refer

- After pollination, interactions between proteins of pistil cells and those of pollen determine pollen germination.
- Germinated pollen delivers two sperm to female gametophytes by means of a long pollen tube. One sperm nucleus fuses with the egg to produce a zygote, the first cell of a new sporophyte generation. The other sperm nucleus fuses with the two nuclei of the central cell, generating the first cell of the nutritive endosperm tissue (Figure 31.11).

31.3 Testing Your Knowledge

1. Put the events in the life of a pollen grain in the correct order:
   I. Microspore cytoplasm divides by mitosis to produce a two-celled male gametophyte, protected by a tough pollen wall.
   II. Mature pollen is transported, most commonly by animal pollinators or wind, to the stigma of a flower pistil.
   III. Diploid cells of sporangia located within flower anthers undergo meiosis to produce haploid microspores.
   IV. A compatible pollen grain takes up water and germinates: The tube cell produces the pollen tube and the generative cell has divided to produce two sperm cells.
   V. Double fertilization occurs, leading to the formation of a zygote and the first endosperm cell.
   VI. At the ovule, the pollen tube bursts, propelling sperm toward the egg and central cell.
   a. I, II, III, IV, V, VI
   b. III, I, II, IV, VI, V
   c. I, II, III, VI, V, IV
   d. V, IV, III, II, I, VI
   e. None of the above is correct.

31.4 Embryo, Seed, Fruit, and Seedling Development

Learning Outcomes
1. Discuss the major stages in angiosperm embryo development.
2. List examples of environmental and internal factors that influence seed germination.

Seeds and fruits are major components of plant reproduction and are essential to the nutrition of animals, including humans. Seeds contain dormant plant embryos that may, under favorable conditions, develop into seedlings. As noted earlier, fruits aid seed dispersal, which allows plants to colonize new sites. Embryos, seeds, and fruits mature simultaneously, and their development is coordinated by hormonal signals, which were introduced in Chapters 28 and 29. Seedling development is also hormonally regulated.

Embryos Develop from Zygotes

Fueled by endosperm nutrients, angiosperm embryos undergo development in a series of stages known as **embryogenesis** (Figure 31.12). Sometime within a period of days to several weeks following fertilization, a zygote begins to divide. At this point, a zygote is blanketed with a layer of callose, which helps to seal it off from the environment, thereby fostering embryo-specific gene expression. The zygote's first cell division is unequal, producing a smaller cell and a larger cell (Figure 31.12, step 1). This unequal division helps to establish the apical-basal (top-bottom) polarity of the embryo, which persists through the life of the plant. The smaller cell develops into

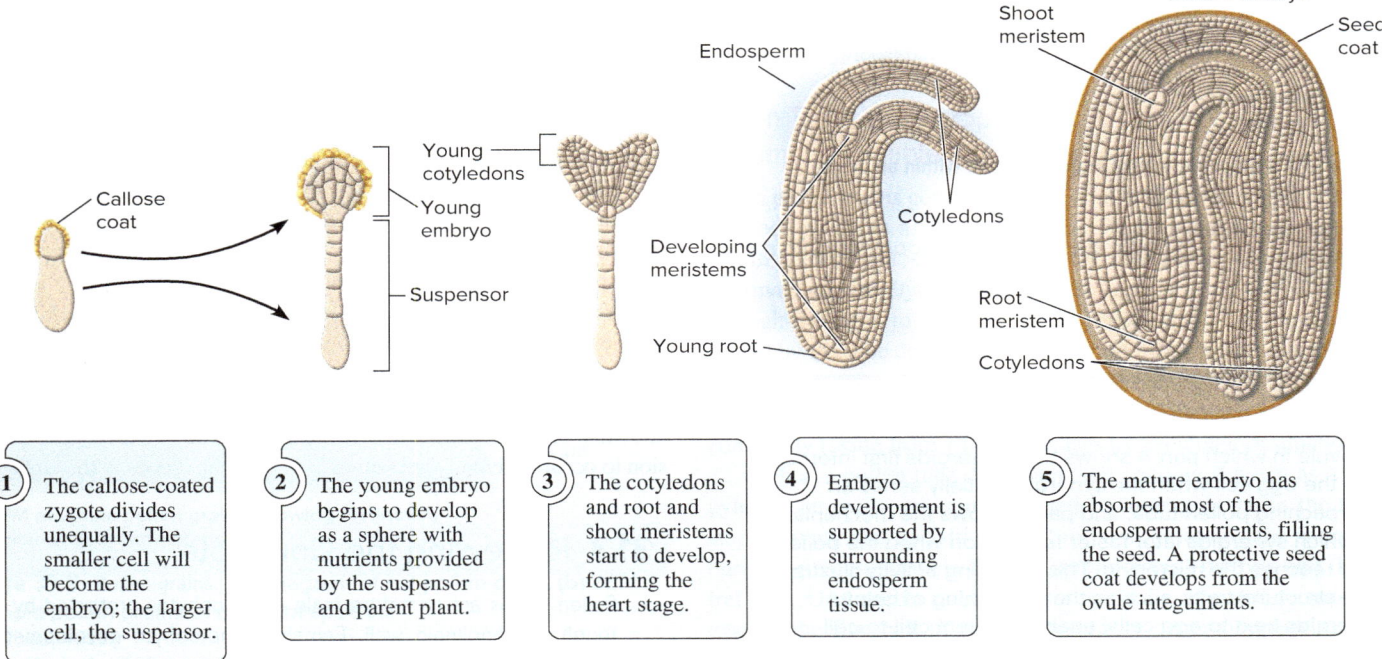

1. The callose-coated zygote divides unequally. The smaller cell will become the embryo; the larger cell, the suspensor.
2. The young embryo begins to develop as a sphere with nutrients provided by the suspensor and parent plant.
3. The cotyledons and root and shoot meristems start to develop, forming the heart stage.
4. Embryo development is supported by surrounding endosperm tissue.
5. The mature embryo has absorbed most of the endosperm nutrients, filling the seed. A protective seed coat develops from the ovule integuments.

Figure 31.12 Embryogenesis in the eudicot *Arabidopsis*.

the embryo, whose radial symmetry is established at this point and continues in adult *Arabidopsis* plants. The larger cell develops into a short chain of cells, called a suspensor, that is anchored near the micropyle at the ovule entrance (Figure 31.12, step 2). These suspensor cells channel nutrients and hormones into the young embryo, which absorbs them at its surfaces.

Young eudicot embryos are spherical, but they soon become heart-shaped as the seedling leaves, called **cotyledons,** start to develop (Figure 31.12, step 3). At this point, auxin and a mobile protein transcription factor are involved in establishing the young shoot and root at the apical and basal poles, respectively. Eudicot embryos such as *Arabidopsis* then become torpedo-shaped, and as the cotyledons grow, they often curl to fit within the developing seed (Figure 31.12, step 4). In contrast, mature monocot embryos are cylindrical, with a single cotyledon and a side notch where the apical meristem forms.

Mature Seeds Contain Dormant Embryos

As they mature, seeds undergo changes leading to dormancy, an adaptation that prevents them from germinating when environmental conditions are not suitable for seedling growth. During this process, embryos become dry and thus able to survive in the absence of water. Seed maturation includes transformation of the ovule's integuments into a tough seed coat (Figure 31.12, step 5). The seed coat restrains seedlings from growing and prevents the entry of water and oxygen, which maintains low seed metabolism. In addition, the coats of some seeds are darkly colored with pigments that may help to prevent damage by UV radiation or microbial attack. Another change leading to seed dormancy is gradual, controlled loss of water from the embryo and other seed tissues. As a result, the water content of dispersed seeds is only 5–15%. Abscisic acid (ABA) is a hormone that induces the activity of genes that help embryo tissues to survive the drying process.

The structure of mature monocot and eudicot seeds differs. Within eudicot seeds such as beans and peas, mature embryos often display an epicotyl. An **epicotyl** is the portion of an embryonic stem having a first bud with two tiny leaves; this bud is located above the point of attachment of the cotyledons (Figure 31.13a). The **hypocotyl** is the portion of an embryonic stem located below the point of attachment of the cotyledons. An embryonic root, the radicle, extends from the hypocotyl. Much of the endosperm has been absorbed into the large cotyledons. In contrast, mature monocot embryos, such as those of corn, feature an epicotyl with a first bud enclosed in a protective sheath known as the coleoptile. The young monocot root is enclosed within a protective envelope known as the coleorhiza (Figure 31.13b). When the embryo-containing seeds of flowering plants are dry and ready for dispersal, they are released from the plant while enclosed in a fruit or released when the fruit breaks open.

Fruits Develop from Ovaries and Other Flower Parts

All fruits develop from ovaries and sometimes other flower parts. They occur in diverse forms that aid seed dispersal. Some fruits are dry, whereas others are moist and juicy; some open to release seeds, and others do not. Fruits also display a wide variety of sizes, colors,

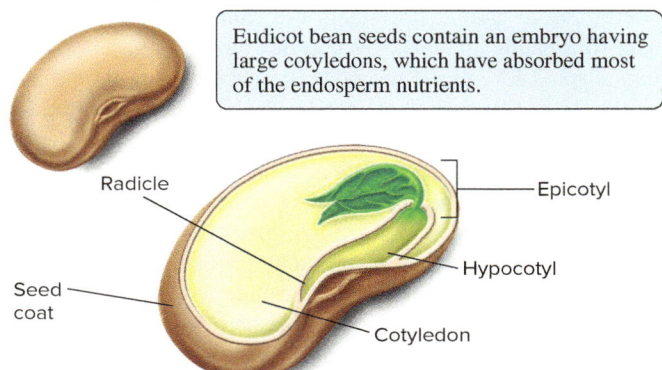

(a) **Eudicot bean seed, showing embryo with epicotyl, hypocotyl, and radicle**

(b) **Monocot corn seed, showing an embryo protected by coleoptile and coleorhiza**

Figure 31.13 Structure of mature angiosperm seeds and embryos.

Concept Check: *Why do mature seeds of eudicots lack extensive amounts of endosperm?*

and fragrances. These variations result from differences in the process of fruit development. Plant hormones, including auxin, gibberellic acid, and cytokinins, control this transformation. ABA stimulates cell expansion, and ethylene influences fruit ripening. For instance, ethylene helps to ripen nuts, a type of dry fruit, by inducing plasma membranes to rupture, causing water loss.

Under the influence of plant hormones, the pericarp (ripened ovary wall) of peaches, plums, and related fruits swells and softens, and orange or red chromoplasts replace green chloroplasts. As fruits mature, the outer protective cuticle often becomes very thick, contributing to peel toughness, which helps to prevent microbe attack. In addition, many maturing fruits increase their sugar and acid content, which produces the distinctive tastes of ripe fruit. Many fruits also produce fragrant volatile compounds.

Differences in the shape, color, fragrance, and moisture content of wild fruits reflect evolutionary adaptation for effective seed dispersal. Whereas many fruits and seeds are dispersed by wind or water or by attaching to animal fur, others use fruit color and fragrance to attract fruit-eating animals. Blackberries provide a good example of fruits adapted for animal dispersal. Blackberry flowers produce many separate pistils, each containing a single ovule (Figure 31.14a). Following pollination and fertilization, the ovary of each pistil develops into a sweet, juicy fruitlet containing a single seed. As the individual fruitlets develop, they fuse together at the sides. Consequently, the many fruitlets produced by a single blackberry flower are dispersed together (Figure 31.14b). Attracted by the color, birds consume the whole aggregate and excrete the seeds, thereby dispersing many at a time. Many other types of fruits occur, and these likewise represent adaptations that foster seed dispersal (refer back to Figure 25.19). Although a fruit is usually defined as a mature ovary containing seeds, commercial seedless fruits such as watermelon are produced by genetic modification or treatment with artificial auxin.

Environmental and Internal Factors Influence Seed Germination

Seeds vary greatly in their ability to germinate after dispersal. Small seeds such as those of dandelions and lettuces germinate quickly if light is available. Other seeds require a period of dormancy before germination occurs. Some seeds can remain dormant for amazingly long time periods. For example, a lotus (*Nelumbo nucifera*) seed collected from a lakebed in China germinated at the age of 1,300 years, as determined by radiocarbon dating. In 2005, plant scientists germinated a 2,000-year-old date seed found in Israel.

Water is generally required to rehydrate seeds so that embryos can resume their metabolic activity. Water absorption also swells seeds, helping to break the seed coat and allowing embryonic organs to emerge. In some cases, rainfall of sufficient duration to leach germination-inhibiting compounds out of seeds is required. The optimal temperature for germination of most seeds lies between 25°C and 31.25°C (77°F and 86.25°F). This explains why gardeners wait until the soil is warm before planting seeds outdoors in spring. However, some seeds need a period of cold treatment or seed coat abrasion before they will germinate. Such physical stimuli induce the activity of more than 2,000 genes associated with seed germination.

When grass seeds rehydrate, the young shoot secretes the hormone gibberellic acid from the seed cotyledon into the outermost endosperm layer, known as the aleurone. In response, the aleurone secretes digestive enzymes into the central endosperm, releasing sugars from stored starch (Figure 31.15). The seedling uses these sugars for growth. This highly coordinated process allows grass seeds to germinate quickly when it rains, an advantage in arid grassland habitats. Humans also use this basic process of germination to make beer. In the process known as malting, beer brewers apply gibberellic acid to barley seeds to induce them to germinate simultaneously. The barley seeds are then baked at a high temperature to stop germination, a process that produces malt. Brewers then treat malt with water and heat, add the dried flowers of the hop plant (the genus *Humulus*), and add yeasts to ferment the plant sugars to alcohol.

Once seeds have germinated, plants vary in the process by which the embryonic shoot emerges. When bean and onion seeds

Biology Principle

Living Organisms Interact with Their Environment

This principle is illustrated by the structure of blackberry fruits, which evolved as an effective way to achieve dispersal by birds.

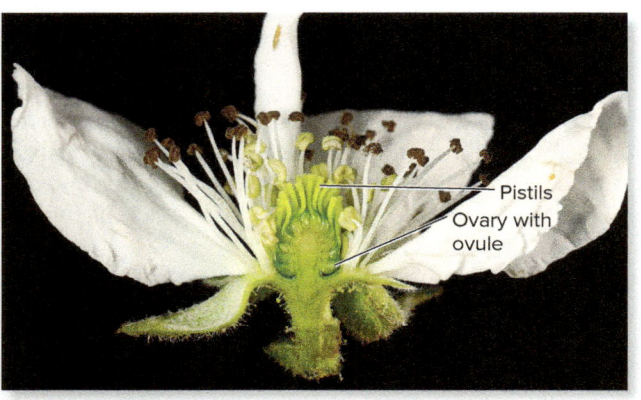

(a) *Rubus allegheniensis* (common blackberry) flower

(b) Blackberry fruit

Figure 31.14 Blackberry flower and fruit. (a) Each of the many separate pistils in a blackberry flower is able to produce a single one-seed fruit (called a fruitlet) if fertilization occurs. (b) Together, the individual fruitlets of the blackberry compose an aggregate fruit that allows many seeds to be efficiently dispersed at the same time by the same animal agent.
(a-b) ©Lee W. Wilcox

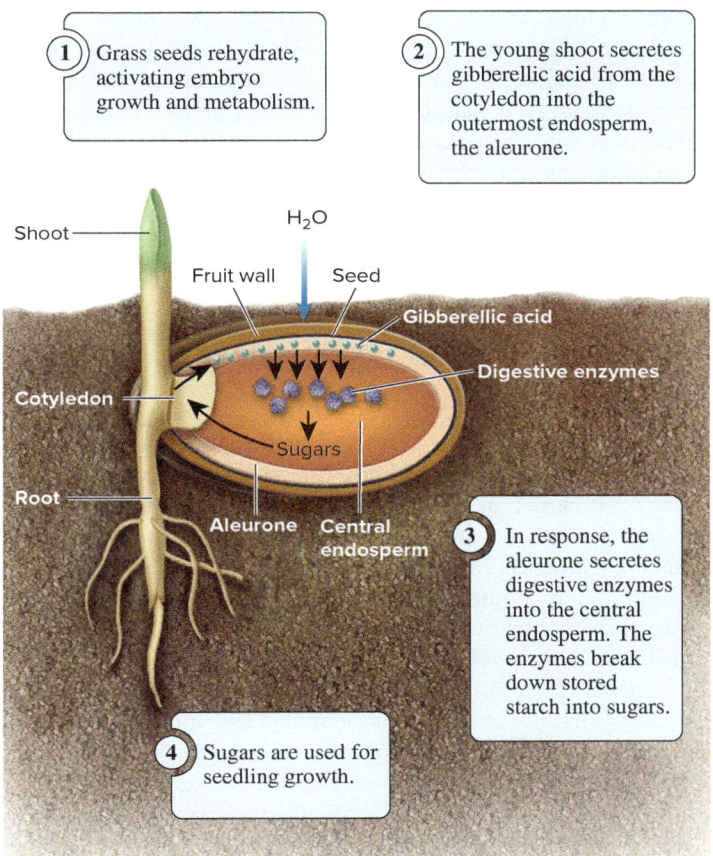

Figure 31.15 Germination of grass seeds.

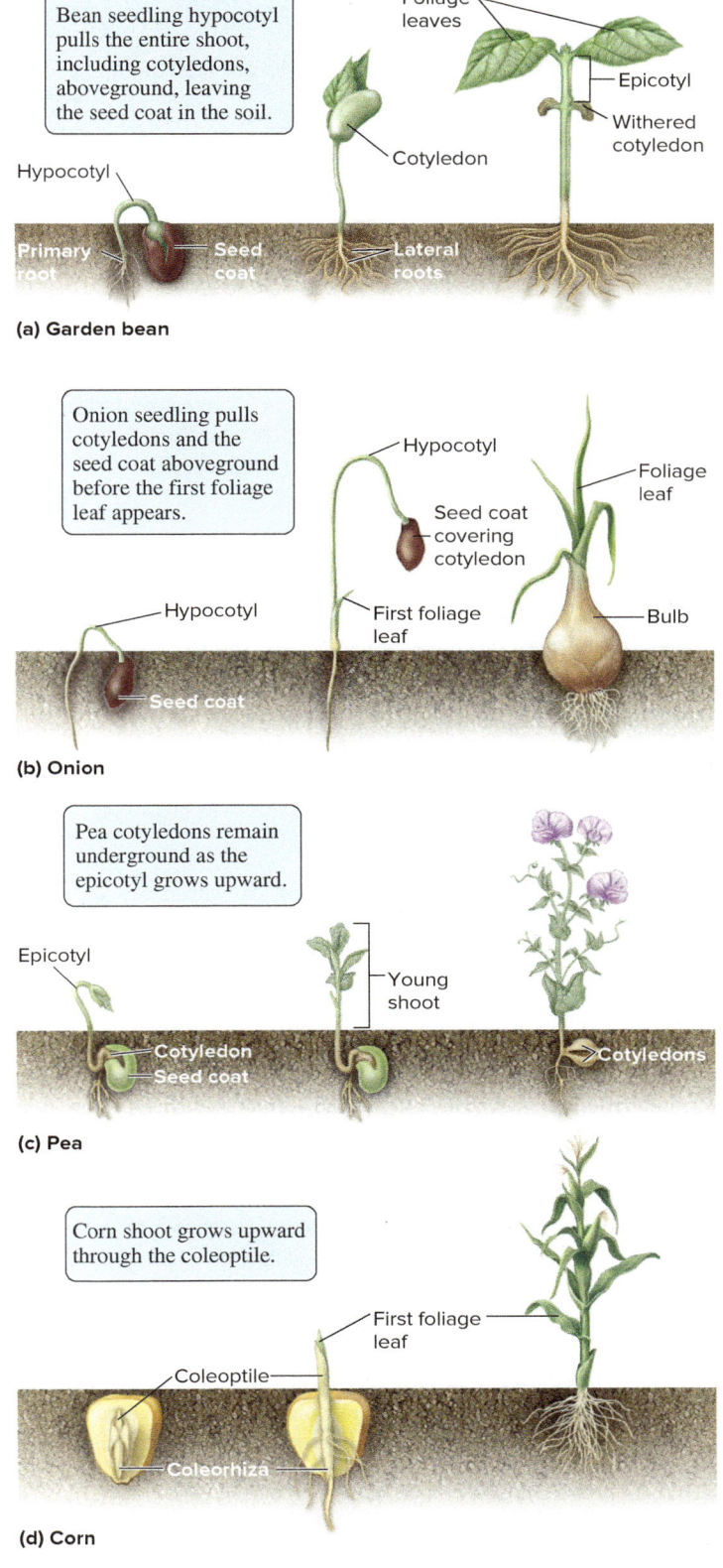

Figure 31.16 Variations in seed germination and seedling growth patterns.

germinate, the hypocotyl forms a hook that first breaches the soil surface and then straightens, thereby pulling the rest of the seedling and cotyledons aboveground (Figure 31.16a, b). In contrast, when pea seeds germinate, the epicotyl forms a hook that pulls the shoot tip out of the ground, leaving the cotyledons beneath the soil surface (Figure 31.16c). The tough hooks bear the brunt of passage through hard surface soil crusts, thereby protecting the delicate shoot tips. The shoot tips of corn seedlings and those of other grasses are protected by the **coleoptile**, a protective tube that encloses the first foliage leaves (Figure 31.16d).

31.4 Reviewing the Concepts

- Unequal division of a zygote leads to development of an embryo and nutritive suspensor. (Figures 31.12, 31.13).
- Mature seeds contain embryos that become dry and are protected by desiccation-resistance proteins and a tough seed coat. These adaptations enable seeds to withstand long periods of dormancy, germinating only when conditions are favorable for seedling survival. Mature fruits develop from ovaries and aid in seed dispersal (Figure 31.14).
- Seed germination is influenced by environmental and internal factors (Figures 31.15, 31.16).

31.4 Testing Your Knowledge

1. Put the following events in the life of an embryo flowering plant in the correct order:
   I. The zygote uses endosperm nutrients to undergo mitotic division to form a smaller cell that develops into the embryo and a larger cell that becomes the nutrient-channeling suspensor.
   II. In maturing seeds, embryos dry and become dormant.
   III. Young embryos take up nutrients and divide mitotically to form cotyledons, young shoot, and young root.
   IV. Under favorable light and moisture conditions, seeds germinate and embryos grow into seedlings.
   V. Mature fruits, often with the aid of wind or animals, disperse mature seeds containing dry embryos into the environment.
   a. I, II, III, IV, V
   b. I, III, II, V, IV
   c. I, II, III, V, IV
   d. V, IV, II, III, I
   e. None of the above is correct.

31.5 Asexual Reproduction in Flowering Plants

Learning Outcomes
1. Explain the benefits of asexual reproduction to flowering plants.
2. List several examples of ways in which plant asexual reproduction is important to agriculture.

Many plants rely on sexual reproduction. However, a wide variety of angiosperms reproduce primarily by asexual means, and other plants commonly utilize both sexual and asexual reproduction. **Asexual reproduction** is the production of new individuals from a single parent without the occurrence of fertilization. Although sexual reproduction provides beneficial genetic variation, asexual reproduction can be advantageous in other ways. For example, asexual reproduction maintains favorable gene combinations that allow faster population growth in stable environments. Asexual reproduction is also advantageous in stressful habitats where pollinators or mates can be rare, because it allows a single individual to start a new population. Finally, asexual reproduction allows some plants to persist for very long periods of time. Among the oldest known plants are creosote bushes that are asexual clones of a parent that grew from a seed about 12,000 years ago!

The artificial propagation of plants from cuttings is a form of asexual reproduction that is widely used commercially and by home gardeners. In this section, we will explore the three main mechanisms of plant asexual reproduction: production of plant clones from nonreproductive plant organs, somatic embryogenesis, and apomixis.

Asexual Reproduction Generates Plant Clones from Organs

Roots, stems, and leaves are plant organs that can function as asexual reproductive structures. For example, root sprouts, such as those produced by aspens, can generate entire groves of genetically identical trees. Sucker shoots, such as those appearing at the bases of banana plants and date palms, and the pieces of tuber bearing "eyes" (which are buds) that develop into potato plants are examples of vegetative plant organs that have agricultural importance. Attractive horticultural varieties of African violets and other plants can be propagated from leaf cuttings. Such asexual offspring grow into adult plants that have the same valued properties as their parent plant. In contrast, growing these plants from seed would produce diverse progeny, not all of which would have economically prized properties.

Somatic Embryogenesis Is the Production of Plant Embryos or Plantlets from Body (Somatic) Cells

Somatic embryos can develop from many types of plant body cells, known as somatic cells. Unlike the sexually produced embryo described earlier in Figure 31.12, somatic embryos do not have endosperm and do not develop within a seed. Even so, somatic embryos produce root and shoot systems and can develop into mature plants. Somatic embryogenesis occurs naturally in citrus, mango, onion, and tobacco plants, but agricultural scientists have made use of it as well.

In the 1950s, British plant biologist F. C. Steward and associates were the first researchers to successfully clone a complex organism, carrot plants, by means of somatic embryogenesis. The biologists extracted differentiated cells from carrot roots, grew the cells in conditions that caused some cells to lose their specialized properties and develop into embryos, and then cultivated each embryo in conditions that favored development into a mature carrot plant. Many types of plants are now cloned by somatic embryogenesis, which allows commercial growers to produce large numbers of genetically identical individuals.

The leaves of the common houseplant *Kalanchoë daigremontiana* produce many tiny plantlets at their edges via somatic embryogenesis. When these detach, they are able to take root and grow into new individuals (**Figure 31.17**). Recent molecular studies have revealed that evolutionary changes in embryogenesis genes are involved in this process.

Figure 31.17 Asexual reproduction via somatic embryogenesis. Expression of embryogenesis genes causes the leaves of this *Kalanchoë* plant to produce small plantlets around the edges. When mature, these plantlets drop off and, under the right conditions, grow into new plants.
©Lee W. Wilcox

EVOLUTIONARY CONNECTIONS

Gene Expression Changes Explain the Evolution of Plantlets in *Kalanchoë*

Kalanchoë daigremontiana is informally known as mother of thousands because it produces many plantlets at the edges of leaves. American plant biologists Helena Garces, Neelima Sinha, and their colleagues investigated the evolution of this type of asexual reproduction by studying genes that are involved in the development of organs and embryos in four species of *Kalanchoë*. In addition to *K. daigremontiana*, they investigated *K. marmorata*, which does not produce plantlets; *K. pinnata*, which produces plantlets only under certain stressful conditions; and *K. gastonis-bonnieri*, which produces plantlets both normally and when stressed. In leaves of these species, the biologists looked for expression of the gene *STM*, which encodes a key regulator of leaf production at the shoot meristem. They found that *STM* is expressed in cells at the leaf margins of all *Kalanchoë* species that produce leaf plantlets, but not the species that do not produce plantlets. Then the investigators checked for the expression in leaves of two genes (*LEC1* and *FUS3*) that are involved in embryo development. They discovered that these embryo-linked genes were expressed only in the leaf margins of species that normally form plantlets, not the species in which plantlet formation is induced by stress.

In a survey of a larger number of species, these investigators also discovered that *Kalanchoë* species that normally produce plantlets have an altered LEC1 protein. The normal form of LEC1 protein is essential to the process by which embryos become dry and thus tolerant of arid conditions. LEC1 is therefore necessary for the production of viable seeds, those able to germinate. *Kalanchoë* species that produce plantlets only under stressful conditions and those that do not produce plantlets at all were able to produce viable seeds. By contrast, the seeds of plantlet-producing species were not viable.

Together, these data allowed the investigators to infer that the evolution of plantlet formation began when certain leaf cells of some species gained the ability to function like a shoot meristem, thereby producing structures resembling small shoots. In some of the descendants of these species, normal LEC1 function in seeds was lost, as was the ability to produce viable seeds. Some species adapted by expressing the embryo-development process in leaf margin cells, a process that allowed them to produce plantlets. Such plantlets are not affected by loss of LEC1 function because, unlike seeds, they do not undergo a drying process during development.

Apomixis Is Seed Production Without Fertilization

Apomixis (from the Greek, meaning away from mixing—that is, genetic mixing) is a natural asexual reproductive process in which fruits and seeds are produced in the absence of fertilization. By contrast, somatic embryogenesis does not involve fruit and seed production. More than 300 species of flowering plants, including hawkweeds, dandelions, and some types of citrus, are able to reproduce asexually by apomixis. Dandelions and some other apomictic plants require pollination to stimulate seed development, but others do not. Agricultural scientists are interested in apomixis as a potential method for producing genetically uniform seeds, propagating hybrids, and removing the need for fertilization in crop plants.

31.5 Reviewing the Concepts

- Asexual reproduction is the production of new individuals from a single parent without the occurrence of fertilization. Plant clones can develop asexually from roots, stems, and leaves.
- Somatic embryogenesis is the production of embryos or plantlets from individual body cells (Figure 31.17).
- Apomixis is a mechanism by which some plants produce seeds and fruits from flowers without the occurrence of fertilization.

31.5 Testing Your Knowledge

1. What is a reasonable way to try to produce many clones (copies) of a particularly valuable plant?
   a. You can try planting cuttings from the valuable plant.
   b. You can try to grow cells from the valuable plant under conditions that allow somatic embryogenesis in carrot, as in F. C. Steward's cloning experiments.
   c. You can try to genetically alter the plant so that seeds and fruits are produced by apomixis, as occurs naturally in hawkweeds and dandelions.
   d. All of the above are reasonable.
   e. Plant cloning has not yet been achieved.

Assessing Your Knowledge and Skills

Foundational Knowledge

Multiple Choice

1. Where do the pollen grains of flowering plants develop?
   a. in the anthers of a flower
   b. in the carpels of a flower
   c. while being dispersed by wind, water, or animals
   d. within ovules
   e. within pistils

2. Where do mature male gametophytes of flowering plants primarily develop?
   a. in the anthers of a flower
   b. in the carpels of a flower
   c. while being dispersed by wind, water, or animals
   d. within ovules
   e. on the surfaces of leaves

3. Where would you find female gametophytes of a flowering plant?
   a. in the anthers of a flower
   b. at the stigma of a pistil
   c. in the style
   d. within ovules in a flower's ovary
   e. in structures that are dispersed by wind, water, or animals

4. From what structure does the fleshy, outer part of a juicy fruit primarily develop?
   a. the style
   b. a stamen filament
   c. the ovary wall
   d. a group of fused sepals
   e. the stigma

5. Which plant hormone is key to the process by which seed endosperm is digested to produce sugars that can be used by seedlings to start growth (and is valued by brewers for the process of making beer)?
   a. gibberellic acid
   b. auxin
   c. cytokinin
   d. abscisic acid
   e. ethylene

Short Answer

1. Why do pollen grain walls contain sporopollenin?
2. **PRINCIPLES** A principle of biology is that populations of organisms evolve from one generation to the next and are related by an evolutionary history. Modern flowering plants likely possessed a single common ancestor. Why, then, do modern flowers occur in such a diversity of shapes and colors?

Critical-Thinking Skills

Multiple Choice

1. How does double fertilization occur in flowering plants?
   a. The two sperm in a pollen tube fertilize the two egg cells present in each female gametophyte.
   b. One of the two sperm in a pollen tube fertilizes the single egg in a female gametophyte, and the other fuses with the two nuclei present in the central cell.
   c. Two sperm, one contributed by each of two different pollen tubes, fertilize the two egg cells in a single female gametophyte.
   d. Two sperm contributed by separate pollen tubes enter a single female gametophyte; one of the sperm fertilizes the egg cell; the other fertilizes the central cell.
   e. None of the above is correct.

2. A seed is
   a. an embryo produced by the fertilization of an egg, which is protected by a seed coat that develops from the ovule's integuments.
   b. a structure that germinates to form a seedling under favorable environmental conditions.
   c. an embryo produced by asexual apomixis that is enclosed by a seed coat.
   d. all of the above.
   e. none of the above.

3. What are the likely chemical composition and source of the chemical stimulus of flowering that is transported to the shoot apical meristem?
   a. The hormone auxin is produced at the shoot apex.
   b. Stimulated by environmental signals, leaf vein companion cells produce FT protein, which is then transported in the phloem to the shoot apex.
   c. Pollen development involves the carbohydrate callose.
   d. Roots take up potassium cations from the soil and transport these ions into the shoot.
   e. All of the above are chemical stimuli and sources key to the process by which shoot apical meristems make the developmental switch from vegetative growth (new stem and leaf tissues) to flower production.

4. How many whorls of organs occur in complete flowers, and are complete flowers always perfect?
   a. 2, no
   b. 4, yes
   c. 6, no
   d. 8, yes
   e. 10, no

5. If an ovary contains eight ovules, how many seeds can result if pollen tubes, each containing two sperm, reach all eight ovules?
   a. 1
   b. 4
   c. 8
   d. 16
   e. more than 20

Short Answer

1. Observe or view images of orchid flowers. Are orchid flowers bilaterally symmetric or radially symmetric? Are orchid flowers more likely to be pollinated by wind or by animals? What gene might be involved in the production of orchid flower shape?
2. Why are seed coats often tough?

UNIT VII
ANIMALS

In Chapters 26 and 27, we considered the amazing diversity of animal life. Despite the obvious differences between animals, they are nevertheless linked by fundamental similarities, which we will explore in this unit. The basic features of animal bodies, the relationship between structure and function, and the ability of animals to maintain homeostasis will be introduced in Chapter 32. Chapters 33 and 34 will describe major principles of nervous systems in animals. The ability of animals to move through their environments will be covered in Chapter 35. Circulatory, respiratory, digestive, excretory, and endocrine systems will then be covered in Chapters 36–38. Chapters 39 and 40 will cover animal reproduction, development, and immune systems. In the last chapter in this unit (41), a synthesis of a whole-body response to a major homeostatic challenge that involves all the organ systems is explored.

Biologists study the various organ systems in animals at two levels. Anatomy is the study of the structure (or form) of an animal's body, and physiology is the study of how body structures function. Collectively, anatomy and physiology form the foundation of animal biology. We will begin the unit, therefore, with an examination of the structure and organization of animal bodies.

The following biology principles are emphasized in this unit:

- ***Living organisms maintain homeostasis.*** This key principle of biology is related to all aspects of animal biology and is covered throughout the unit.

- ***Living organisms use energy.*** We will see how many of the processes that achieve and maintain homeostasis require energy; for example, it takes energy to regulate body temperature (Chapter 32), to obtain oxygen and maintain blood flow (Chapter 36), and to maintain stable concentrations of sodium and other ions in body fluids (Chapter 37).

- ***The genetic material provides a blueprint that allows organisms to grow, develop, and reproduce.*** In Chapters 38 and 39, we will explore the factors that regulate how animals reproduce, how an embryo develops into a mature animal, and how animals grow in size.

- ***Living organisms interact with their environment.*** How animals sense changes in their external environment, such as light and dark, smells, and vibrations, will be covered in Chapter 34.

- ***Structure determines function.*** The relationship between structure and function in animals will be evident at multiple levels: cells, tissues, organs, organ systems, and whole bodies. This will be particularly evident in the discussion of muscles and skeletons (Chapter 35).

- ***Populations of organisms evolve from one generation to the next and are related by an evolutionary history.*** This principle will be explored in the Evolutionary Connections features throughout the unit. As one example, the evolution of a four-chambered heart is described in Chapter 36.

- ***Biology affects our society.*** At the ends of Chapters 33 to 40 are sections entitled Impact on Public Health. These sections relate what you've learned in a chapter to important facets of human disease, including their costs to society.

- ***Biology is a quantitative science.*** This principle will be evident in the discussion of how oxygen is transported in the body fluids of animals (Chapter 36) and how animal nervous systems function (Chapter 33).

- ***Biology is an experimental science.*** Several chapters include a Feature Investigation that describes an experiment notable for its innovation and importance and for moving the study of animal biology forward.

(32) ©John Rowley/Getty Images; (33) Courtesy Ann-Shyn Chiang, Tsing Hua Chair Professor/Brain Research Center & Institute of Biotechnology/National Tsing Hua University; (34) ©Gustavo Mazzarollo/Getty Images; (35) ©Don W. Fawcett/Science Source; (36) ©SPL/Science Source; (37) ©Richard Hutchings/PhotoEdit; (38) ©Sovereign/ISM/Medical Images; (39) ©Robert F. Sisson/National Geographic/Getty Images; (40) ©SPL/Science Source; (41) ©BSIP SA/Alamy

General Features of Animal Bodies, and Homeostasis as a Key Principle of Animal Biology

32

©John Rowley/Getty Images

Perspiring and drinking water are both mechanisms that help achieve homeostasis—a stable internal body environment.

Chapter Outline

- **32.1** Organization of Animal Bodies
- **32.2** Relationship Between Form and Function
- **32.3** General Principles of Homeostasis
- **32.4** Homeostatic Regulation of Body Temperature
- **32.5** Homeostasis of Internal Fluids

Assessing Your Knowledge and Skills

In 2002, a healthy young woman died shortly after completing the Boston Marathon. The cause of death was traced to an imbalance in the concentrations of ions such as sodium (Na^+) in her body fluids. This occurred because the woman drank an extreme excess of water before and during the race, thereby diluting the concentrations of Na^+ and other ions in her body fluids. When this happens, water moves by osmosis into cells, causing them to swell. Whereas some animal cells can withstand considerable swelling, the cells of a human's brain cannot. Swelling of brain cells increases pressure in the brain, because the skull prevents the brain from expanding. The pressure in turn can squeeze and compress local blood vessels. This decreases blood flow in the brain, depriving it of the oxygen it needs to function. Brain damage, coma, and death can result. This tragic event underscores the importance of matching fluid and ion (salt) intake with the body's requirements, since too much or too little of either can have serious consequences.

The process by which the different aspects of an animal's internal environment are maintained within normal limits, even in the face of changes in the external surroundings, is known as **homeostasis** (from the Greek *homoios,* meaning similar, and *stasis,* meaning to stand still). Before you can fully appreciate the importance of homeostasis to animals and how it is achieved, however, you first need to understand some basic features of animal bodies.

We will begin this chapter with a discussion of the organization of cells into tissues, and tissues into organs. We will look at tissues and organs from the perspective of the whole animal, examining how the properties of life arise from the complex interactions of its components. Next, we will discuss a principle of biology that also helps you to understand homeostasis, namely that structure determines function. We will then link these principles together with a detailed look at what homeostasis means for different animals, how it may be challenged, and how it is restored or maintained.

32.1 Organization of Animal Bodies

Learning Outcomes

1. List the different types of animal tissue, providing general functions and specific examples of each.
2. Name the various organ systems found in many animals, list the major components of each, and describe their general functions.

All animals share similarities in the ways in which they exchange materials with their surroundings, obtain energy from organic nutrients, synthesize complex molecules, detect and respond to signals in their immediate environment, and reproduce. Animals typically begin life as a single cell—most commonly, a fertilized egg—which divides to create two cells, each of which divides in turn, resulting in four cells, and so on. If cell multiplication were the only event occurring, the end result would be a spherical mass of identical cells. As we will see in Chapter 39, however, cells become specialized during development to perform a particular function (that is, they differentiate). Examples of differentiated cells are muscle and blood cells. As Figure 32.1 shows for the urinary system of mammals, the cells of an animal's body are organized into progressively more complex structures, including tissues, organs, and organ systems. In this section, we will explore this organization in greater detail.

Specialized Cells Are Organized into Tissues

A **tissue** is an association of many cells having a similar structure and function. The tissues in a typical animal's body can be classified into four types, according to the cell types that constitute them and the functions they perform: muscle, nervous, epithelial, and connective tissues (Figure 32.1). Within each of these functional categories, subtypes of tissues perform variations of that function, as illustrated by the three types of muscle tissue.

Muscle Tissues **Muscle tissues** consist of cells specialized to shorten, or contract, generating the mechanical forces that may

- produce body movements,
- decrease the diameter of a tube, or
- exert pressure on a fluid-filled cavity.

Three types of muscle tissue are found in animals (Figure 32.2). **Skeletal muscles** are generally attached to bones in vertebrates

Biology Principle

New Properties of Life Emerge from Complex Interactions

Individually, none of the four tissues that constitute a bladder or kidney could perform the functions of those organs, but when combined in precise ways, the result is a functional organ system capable of removing soluble wastes from the fluids of an animal's body.

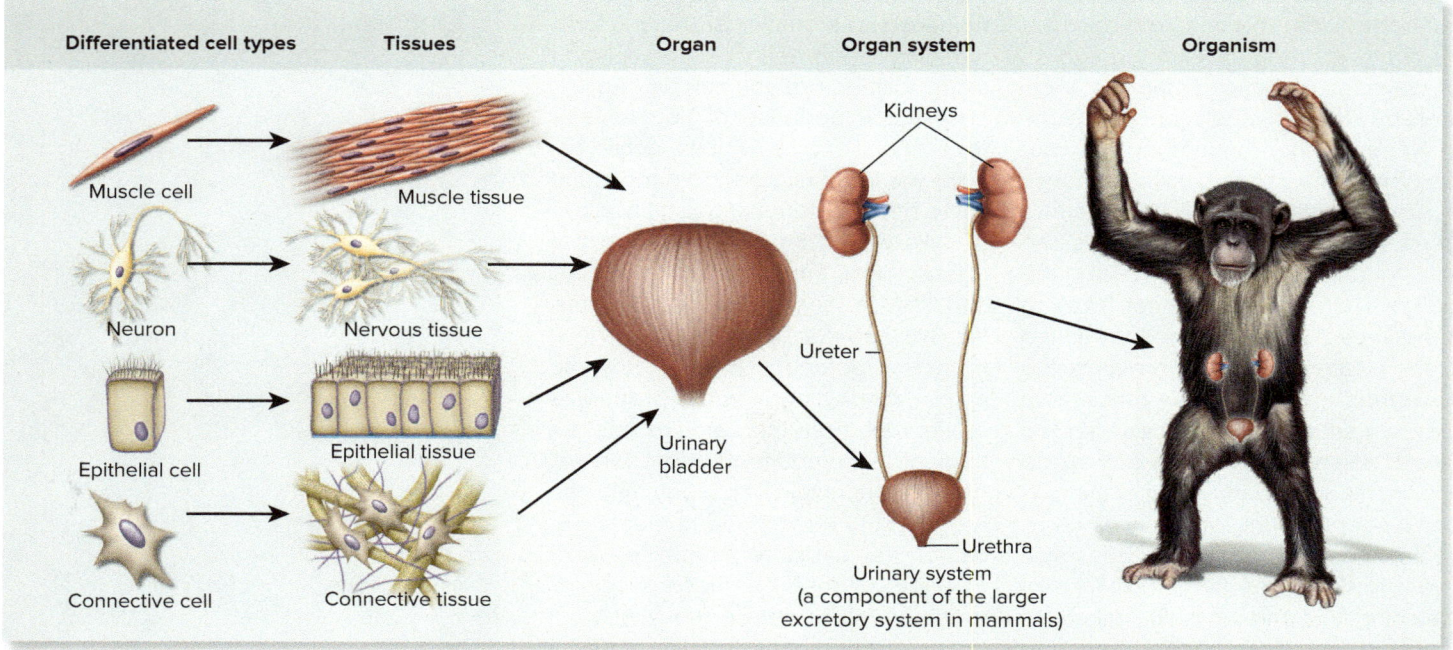

Figure 32.1 The internal organization of cells, tissues, organs, and an organ system in a mammal. Most animals share the same four tissue types.

via bundles of collagen fibers called tendons, and to the exoskeleton of invertebrates. When skeletal muscles are stimulated by signals from the nervous system, they generate force that leads to the contraction of the muscle (see Chapter 35). Contraction of these muscles may be under voluntary control and can produce the types of movements required for locomotion, such as extending limbs or flapping wings.

Smooth muscles surround hollow tubes and cavities inside the body's organs, such that their contraction or relaxation can control the movement of the contents of those organs. For example, the contraction or relaxation of smooth muscle in the lung tubes called bronchioles regulates the rate at which air can flow through those tubes. Contraction of all smooth muscle is involuntary—that is, it occurs without conscious control.

In the third type of muscle tissue, **cardiac muscle**, physical and electrical connections between individual cells enable many cells to contract almost in unison. This type of muscle tissue is found only in the heart, where it provides the force that generates pressure sufficient to propel blood through an animal's body. As with smooth muscle, contraction of cardiac muscle is involuntary.

Nervous Tissues **Nervous tissues** are complex networks of cells called **neurons** that are specialized to receive, generate, and conduct electrical signals from one part of an animal's body to another part (**Figure 32.3**). An electrical signal produced in one neuron may stimulate other neurons to initiate new electrical signals, stimulate muscle cells to contract, or stimulate glandular cells to secrete substances into the animal's body fluids. Thus, nervous tissue provides a means of controlling many diverse activities in an animal's body.

Epithelial Tissues **Epithelial tissues** are sheets of densely packed cells that cover the body or individual organs and line the interiors of various cavities inside an animal's body. Epithelial

Biology Principle

Living Organisms Interact with Their Environment
Animals interact with their environment in many ways; these interactions largely depend on the use of nervous tissue, which allows animals to sense and respond to changes in the environment.

Figure 32.3 **Nervous tissue in the brain of a vertebrate.** Nervous tissue consists of neurons with extensive cell-to-cell contacts, as shown in this confocal micrograph of a section from a human brain. The cells are labeled with fluorescent markers; different colors signify different depths in the section.
©Dr. Gopal Murti/Science Source

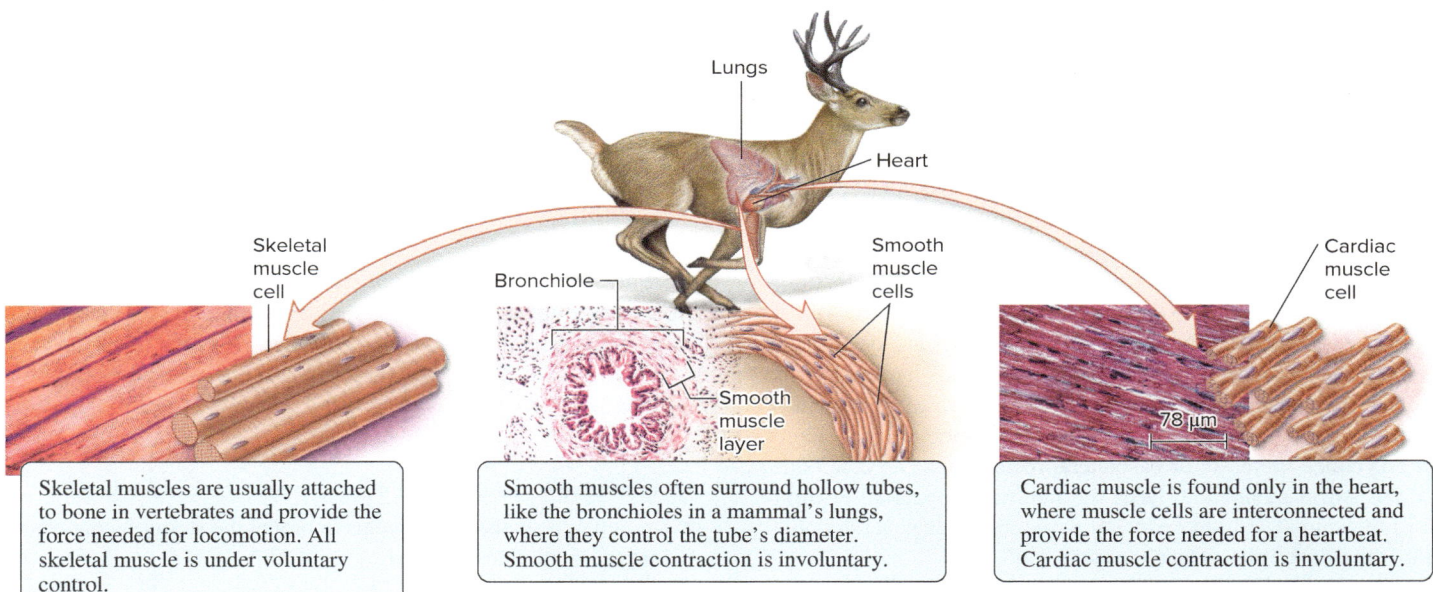

Figure 32.2 **Three types of muscle tissue: skeletal, smooth, and cardiac.** All three types produce force, but they differ in their appearance and in their locations within animals' bodies.
(left) ©Michael Abbey/Science Source; *(middle)* ©Sinclair Stammers/Science Source; *(right)* ©Biophoto Associates/Getty Images

Figure 32.4 Examples of epithelial tissue. The types of epithelial tissue are distinguished by their structure. Epithelial tissue is used to construct body coverings and the protective sheets that line and cover hollow tubes and cavities. The epithelial cells that make up epithelial tissues have apical and basal (or basolateral) membranes. The apical side typically faces the exterior of the body or the lumen of a structure such as the intestines. The shapes and sizes of these cells determine their properties. For example, ciliated epithelial cells in the nasal passage help to trap and filter out airborne particles that could damage the lungs.

cells are specialized to protect structures and to secrete and transport ions and organic molecules. For example, epithelial tissue can invaginate (fold inward) to form sweat glands that secrete water and ions onto the surface of an animal's skin.

Epithelial cells come in a variety of shapes, including cuboidal (cube-shaped), squamous (flattened), and columnar (elongated). These three types of epithelial cells are typically arranged in epithelial tissues as simple epithelia (one layer), stratified epithelia (multiple layers), pseudostratified epithelia (one layer, but with nuclei located in such a way that it appears stratified), or transitional epithelia (multiple layers with the ability to expand and contract) (**Figure 32.4**).

Regardless of their shape, organization into tissues, or location, all epithelial cells are asymmetric, or polarized. This means that one side of such a cell is anchored to or faces an extracellular matrix (ECM) called the basal lamina, or basement membrane; this side of the cell is called the basal or basolateral membrane. The other side, called the apical membrane, faces the internal or external environment of the animal. Within epithelial tissue, the cells generally align such that their membranes face the same way.

Connective Tissues As their name implies, **connective tissues** connect, surround, anchor, and support the structures of an animal's body. In many cases, connective tissues are composed of a relatively sparse number of cells within a noncellular matrix of some type. Connective tissues include adipose (fat-storing) tissue, bone, cartilage, loose connective tissue, dense connective tissue, and blood, a type of fluid connective tissue (**Figure 32.5**). Many of the types and functions of proteins that make up connective tissue were covered in Chapter 4.

Different Tissue Types Combine to Form Organs and Organ Systems

An **organ** is composed of two or more kinds of tissues arranged in various proportions and patterns, such as sheets, tubes, layers, bundles, or strips. The different tissues interact with each other through direct contact and secretions. An example is the vertebrate stomach (**Figure 32.6**), which consists of

- an outer layer and inner layer of epithelial tissue,
- connective tissue that covers and cements the organ together,
- smooth muscle tissue that helps to propel the stomach contents into the small intestine, and
- nervous tissue that regulates smooth muscle activity.

In an **organ system,** different organs work together to perform an overall function or functions. In the example just described, the stomach is part of the digestive system, along with other structures,

Figure 32.5 Examples of connective tissue in mammals. Connective tissue connects, surrounds, anchors, and supports other tissues and may exist as a suspension of cells in a fluid (blood), clumps of cells (fat), or tough, rigid material (bone and cartilage). The samples have been stained or the micrographs have been computer-colorized to reveal connective tissue.
(Blood) ©Dennis Kunkel Microscopy/Science Source; *(Adipose tissue)* ©Ed Reschke/Getty Images; *(Bone)* ©Tissuepix/Science Source; *(Cartilage)* ©Victor P. Eroschenko; *(Loose connective tissue, Dense connective tissue)* ©McGraw-Hill Education/Al Telser

such as the mouth, esophagus, small and large intestines, and anus. In another example, the urinary system (which is part of the larger excretory system) in mammals was shown in Figure 32.1. The organ systems commonly found in animals are listed in **Table 32.1**.

Organ systems should not be considered as functioning in isolation. Instead, they influence and depend on each other in many ways. For example, signals from the nervous, endocrine, and circulatory systems all strongly influence how much water the vertebrate urinary system retains in the body as it forms urine, an adaptation that can be lifesaving under certain circumstances.

The spatial arrangement of organs into organ systems is part of the overall body plan of animals. Organ systems develop at specific times and locations within the body and along the anteroposterior body axis (the formation of the "head" and "tail" of the body), as do other structures, such as limbs, tentacles, antennae, and other animal appendages. Scientists have long questioned how the layout of animal bodies is determined during the period when an embryo is developing. Remarkably, organ development in animals appears to be under the control of a highly conserved family of body-plan genes with homologs in all animals, as described next.

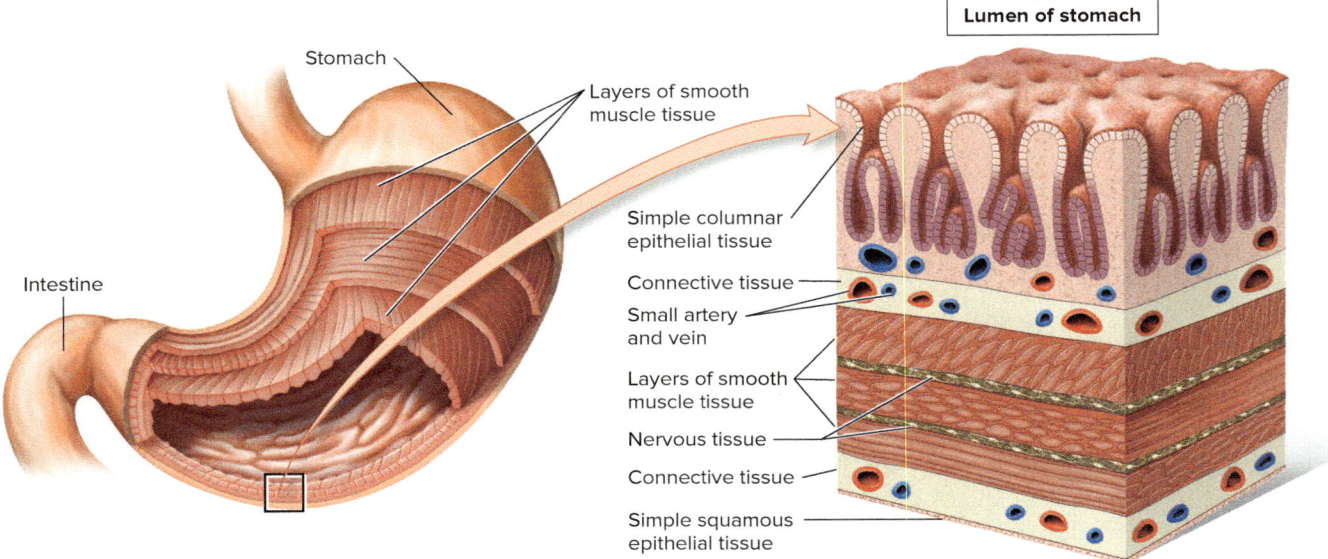

Figure 32.6 The vertebrate stomach as an example of an organ composed of all four tissue types. In this illustration, the thickness and appearance of the layers of nervous tissue have been considerably exaggerated for visual clarity. Lumen refers to the cavity of a structure, in this case the stomach.

Concept Check: Do animal organs necessarily contain equal proportions of all four tissue types?

BioConnections: Look ahead to Figure 37.6 and compare the structure of the small intestine in that figure with the structure of the stomach shown in this figure. What can you conclude about common functions of the stomach and intestine from this comparison?

Table 32.1 Organ Systems Found in Animals

Organ system	Major components*	Major functions
Circulatory	Contractile element (heart or vessel); distribution network (blood vessels); blood	Distributes solutes such as nutrients and gases to all parts of an animal's body
Digestive	Ingestion structures (mouth, mouthparts); storage structures (crop, stomach); digestive and absorptive structures (stomach, intestines); egestion structures (rectum, anus); accessory structures (pancreas, gallbladder, liver, salivary glands)	Breaks complex molecules from food into absorbable units; absorbs organic and inorganic nutrients and water; eliminates solid wastes
Endocrine	All glands, organs, tissues, or cells that secrete hormones	Regulates and coordinates growth, development, metabolism, mineral balance, water balance, blood pressure, behavior, and reproduction
Excretory	All organs, including respiratory structures (such as gills and lungs), kidneys, and related structures that are involved in removing soluble wastes from the body; the vertebrate urinary system is a part of the excretory system and includes the kidneys, ureters, bladder, and urethra	Eliminates soluble metabolic wastes; regulates body fluid volume and solute concentrations
Immune and lymphatic	Circulating leukocytes; lymphatic vessels and lymph nodes or other structures with clusters of immune cells	Defends against pathogens, cancer cells, and toxins
Integumentary	Body surfaces (skin)	Protects from dehydration and injury; provides a barrier against pathogens; in some animals, plays a role in regulation of body temperature
Musculoskeletal	Force-producing structures (muscles); support structures (bones, cartilage, exoskeleton); connective structures (tendons, ligaments)	Produces locomotion; generates force; propels materials through body organs; supports body
Nervous	Sensory and processing organs (brain); structures for signal delivery (spinal cord, nerves, ganglia, sense organs)	Regulates and coordinates movement, sensation, organ functions, behavior, and learning
Reproductive	Gonads and associated structures	Produces gametes (sperm and egg); in some animals, provides nutritive environment for embryo and fetus
Respiratory	Gas-exchange structures (gills, skin, tracheae, lungs)	Exchanges oxygen and carbon dioxide with environment; regulates blood pH

*Selected examples only; these do not necessarily pertain to all animals.

EVOLUTIONARY CONNECTIONS

Organ Development and Function Are Controlled by *Hox* Genes

Hox genes are a family of ancient, highly conserved genes that are found in all animals. The evolution of *Hox* genes was a major event that led to the precise timing and spatial patterning of the anteroposterior body axis during animal development. For example, these genes determine the number and position of legs and wings in *Drosophila* and other insects. *Hox* genes play a similar role in determining the spatial patterning of the vertebrate body and appendages (refer back to Figures 20.12 and 20.13). Recently, however, scientists have begun exploring the role of *Hox* genes in the development and spatial patterning of the organs that make up animals' organ systems.

By generating mutant mice that fail to express one or more *Hox* genes (the genes are said to have been knocked out), researchers have discovered that these genes have the important function of determining where within the vertebrate body particular organs form. For example, one member of the *Hox* gene family in mice has been found to be important for development of anterior parts of the body, including the neck. When this gene is knocked out, mouse embryos show defects in neck structure, such as abnormal blood vessels. Also, the organs within the neck—including the thymus, thyroid, and other glands—do not develop normally or even fail to form at all. Experimental deletion of *Hox* genes associated with other body segments does not affect the development and function of neck glands and organs. Likewise, investigators have uncovered vital roles of different *Hox* genes in lung development within the thorax and the proper positioning and development of the vertebrate kidneys in the abdomen.

Of particular interest is the discovery that *Hox* genes are important not only for spatial patterning of organs but also for their growth, development, and function. For instance, certain *Hox* genes help to determine the branching patterns of the airways of the lungs, the size of the lungs, and the ability of the lungs to produce secretions that are important for breathing air after birth. Other *Hox* genes control cell proliferation, shape changes, apoptosis, cell migration, and cell-to-cell adhesion within various organs. This is also true in invertebrates, such as the leech, where *Hox* genes are first expressed during organ formation, and in *Drosophila,* where the final shape and size of the heart are partly controlled by *Hox* genes. In all animals, therefore, this evolutionarily ancient gene family plays different but broadly related roles in how an animal's body and organs form.

32.1 Reviewing the Concepts

- Groups of cells with similar properties form tissues, which combine with other types of tissues to form organs, which are functionally linked to form organ systems (Figure 32.1). The four types of tissue are muscle, nervous, epithelial, and connective tissues (Figures 32.2, 32.3, 32.4, 32.5). An organ is composed of two or more kinds of tissue. In an organ system, different organs work together to perform an overall function (Figure 32.6, Table 32.1).
- The development, spatial positioning, and function of many body organs are under the control of highly conserved *Hox* genes in animals.

32.1 Testing Your Knowledge

1. Which of the following statements is *false?*
   a. All organs consist of at least two different tissue types.
   b. Epithelial tissue is specialized to initiate and conduct electrical signals.
   c. Cardiac muscle is found only in the heart.
   d. Blood is a type of fluid connective tissue.
   e. The final locations of the various organs in an animal's body are under genetic control.

2. Match the representative functions with the organ system.

supports the body and produces locomotion	digestive system
absorbs nutrients into the body	circulatory system
regulates growth, metabolism, reproduction, mineral balance	integumentary system
distributes solutes throughout the body	endocrine system
provides a barrier against pathogens entering the body	musculoskeletal system

32.2 Relationship Between Form and Function

Learning Outcomes

1. Provide an example of how the structure of an animal's tissues or organs can help to predict their function.
2. **SCISKILLS** ▶ Describe the quantitative relationship between the surface area of an object and its volume, and explain the importance of this relationship to animal form and function.

A key principle of biology throughout this unit is that structure (form) determines function. Consider, for instance, the similarities in the finger-like projections of the small intestine of a human, the skin folds of some high-altitude frogs, the cellular extensions of the neurons of a mouse, and the feathery antennae of a moth (**Figure 32.7**). What do these structures have in common? They are all adaptations that increase surface area, which often predicts a function related to transport, absorption, or detection of environmental stimuli. In the examples shown in Figure 32.7, these features maximize the intestine's ability to absorb nutrients, the skin's ability to obtain oxygen from the environment, a neuron's ability to form contacts with many other cells, and an antenna's ability to detect airborne chemicals. A large increase in surface area of a structure is an important adaptation,

Figure 32.7 Examples of structures in which extensive surface area is important for function. A large surface area allows **(a)** high rates of transport of nutrients across the intestine of a human, **(b)** increased diffusion of oxygen across the folds of skin of a frog living at high altitude in Lake Titicaca, where O_2 is less available, **(c)** extensive communication between neurons in a mouse's brain, and **(d)** detection of airborne chemicals by moth antennae.
(a) ©Biophoto Associates/Science Source; *(b)* ©Dante Fenolio/Science Source; *(c)* ©Thomas Deerinck, NCMIR/Science Source; *(d)* ©Anthony Bannister/NHPA/Science Source

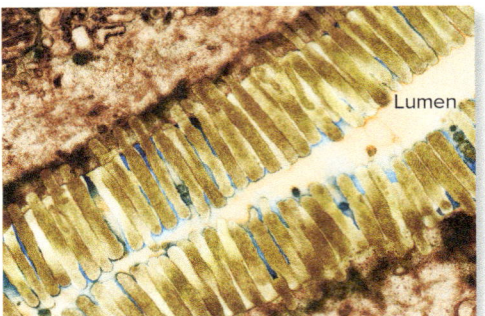
(a) Human intestine (longitudinal section)

(b) Frog skin

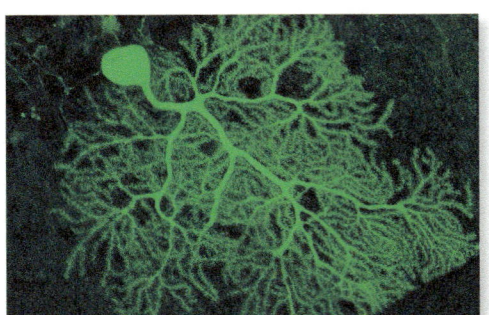
(c) Mouse brain neuron stained with a fluorescent marker

(d) Moth antennae

 Concept Check: Is having a large surface area important only for animals, or might it also provide advantages to other living organisms?

but it comes at the expense of greatly increasing a structure's volume unless the shape of the structure is changed. The reason is that as an object enlarges, its volume grows relatively more than its surface area; surface area increases by the power of 2, whereas volume increases by the power of 3 (refer back to Figure 4.12).

The challenge of packaging an extensive surface area into a confined space is overcome by changes in shape (see, for example, the way the inner surface of the intestine in Figure 32.7a folds inward to form finger-like extensions). The relationship between a structure's surface area and its volume is called the **surface area/volume (SA/V) ratio**. A high SA/V ratio is ideal for exchange of heat, solutes, and water across a surface without contributing greatly to the mass or volume of a body part. This concept will appear throughout this unit as we explore the ways in which animals obtain energy, regulate their metabolism and body temperature, obtain oxygen, and eliminate wastes.

When all of an animal's organ systems function properly and body fluid levels and solute composition are maintained within normal limits, an animal is generally considered to be in a healthy condition. In the rest of this chapter, we will examine the process of achieving and maintaining this condition, which is known as homeostasis.

32.2 Reviewing the Concepts

- The structure of an animal's tissues and organs is related to the function of those structures. A large surface area maximizes the ability of a tissue or an organ to absorb nutrients, exchange oxygen and carbon dioxide with the environment, communicate with other cells, or receive sensory information from the environment (Figure 32.7).

32.2 Testing Your Knowledge

1. Which of the following statements is *false*?
   a. The more folded a structure is, the greater its surface area will generally be.
   b. Increasing surface area of a structure is a useful adaptation for processes such as transporting solutes.
   c. When the linear dimensions of a structure increase, so do both the surface area and the volume of that structure.
   d. A structure's surface area and volume usually change in opposite directions; as one increases, the other decreases.

32.3 General Principles of Homeostasis

Learning Outcomes

1. Explain how the concept of homeostasis applies to the internal environment of animals.
2. Distinguish between conforming and regulating as mechanisms for maintaining homeostasis.
3. List several variables that are regulated within a homeostatic range in vertebrate animals.
4. Name the components of a homeostatic control system, and describe the importance of each to the regulation of an animal's internal environment.
5. **SCISKILLS** ▶ Contrast negative feedback, positive feedback, and feedforward regulation, and explain how they do or do not contribute to the maintenance of homeostasis in animals.

The environmental conditions in which organisms—including all animals—live are rarely, if ever, constant. For example, animals may be exposed to fluctuations in air and water temperatures, nutrient and water availability, the presence of external threats, and oxygen availability. Any one of these environmental changes can be harmful or even fatal if an organism is unable to respond appropriately. However, as you might expect from the incredible diversity of environments in which they exist, animals can adjust in many ways to their surroundings and thrive. As noted earlier, the process of maintaining a relatively stable internal environment despite changes in the external surroundings is known as homeostasis, which is an adaptation that is essential for an animal's health and survival. In this section, we will examine several of the most important features of homeostasis, including how it is achieved and maintained.

Some Animals Conform to External Environments; Others Regulate Their Internal Environments

Some animals conform to their environments so that some feature of their internal environment matches that of their external surroundings. A marine crab, for example, has about the same solute concentration in its body fluids as is found in seawater. Likewise, the body temperatures of many fishes and aquatic invertebrates match the temperatures of the surrounding waters. Energetically, conforming is a cheap strategy for survival. A small fish would have to expend a great deal of energy to maintain its body temperature at, say, 37°C when swimming in the waters near Antarctica. However, because conformers do not actively adjust their internal body composition, they are often restricted to living in environments that are relatively unchanging.

Other animals homeostatically regulate their body temperature and the composition of their fluids and solutes, allowing them to differ from the external environment. Regulating the internal environment requires considerable energy. For this reason, homeostasis for regulators comes with a high energy price tag. However, the price paid by regulators makes it possible for them to inhabit a wide range of environments, even if those environments fluctuate significantly, something that is difficult for conformers.

Vertebrates Maintain Most Physiological Variables Within a Narrow Range

In vertebrates, the common physiological variables are usually maintained within a certain range, despite fluctuating external environmental conditions. Examples of such variables include the concentrations of blood-borne solutes such as minerals, glucose, and oxygen; the pH of the body fluids; blood pressure; and energy stores. At first glance, homeostasis may appear to be a state of stable balance of physiological variables. However, this simple description does not capture the scope of homeostasis. For example, no physiological function is constant for very long, which is why we call them variables. Some variables may fluctuate around an average value during the course of a single day yet still be considered in balance.

Homeostasis is a dynamic process, not a static one. It is not an equilibrium but rather a steady-state. The difference is that a steady-state requires a constant input of energy to be maintained, whereas an equilibrium does not. To use one of the examples just mentioned, maintaining homeostasis of oxygen content in an animal's blood requires regular energy input in order to power the muscles of breathing.

Consider an example of a physiological variable in your own body. Normally, blood sugar (glucose) remains at fairly steady and predictable concentrations in any healthy individual. After a meal, however, the concentration of glucose in your blood can increase quickly, especially if you have just eaten something sweet. Conversely, if you skip a few meals, your blood glucose concentration may decrease slightly. Such fluctuations above and below the normal value might suggest that the blood glucose concentration is not homeostatically controlled, but this is incorrect. Once the blood glucose concentration increases or decreases, homeostatic mechanisms restore it back toward normal. In the case of glucose, the endocrine and nervous systems are primarily responsible for this adjustment, but in other examples, a wide variety of control systems may be initiated. In later chapters, we will see how every organ and tissue of an animal's body contribute to homeostasis, sometimes in multiple ways and usually in concert with each other.

Homeostasis, then, does not imply that a given physiological function or variable is rigidly constant. Instead, homeostasis means that a variable fluctuates within a certain normal range and that if disturbed from that range, compensatory mechanisms restore the variable toward normal.

Homeostatic Control Systems Maintain the Internal Environment

The activities of cells, tissues, and organs must be regulated and coordinated with each other so that any change in the extracellular fluid—a part of the body's internal environment—initiates a response to correct the change. These compensating regulatory responses are performed by homeostatic control systems. A **homeostatic control system** must have several components, including

- a **set point**, which is the normal value for a controlled variable;
- a **sensor,** which monitors the level or activity of a particular variable;
- an **integrator,** which compares signals from the sensor with the set point; and
- an **effector,** which compensates for any deviation between the actual value and the set point.

Figure 32.8 shows an example of a homeostatic control system that regulates body temperature in mammals. This system is somewhat analogous to the heating system of a home. In that case, a sensor and an integrator within the thermostat compare the actual room temperature with the set point temperature that was determined by setting the thermostat to a given temperature. If the room temperature becomes cooler than the thermostat setting, the effector (furnace) is activated and adds heat to the room. In a mammal, the sensors are temperature-sensitive neurons in the skin and a part of the brain, whereas the integrator is within another brain structure. Signals from the brain are sent along nerves to the effectors, which include skeletal muscles. If body temperature decreases, the muscles contract vigorously in response to these signals,

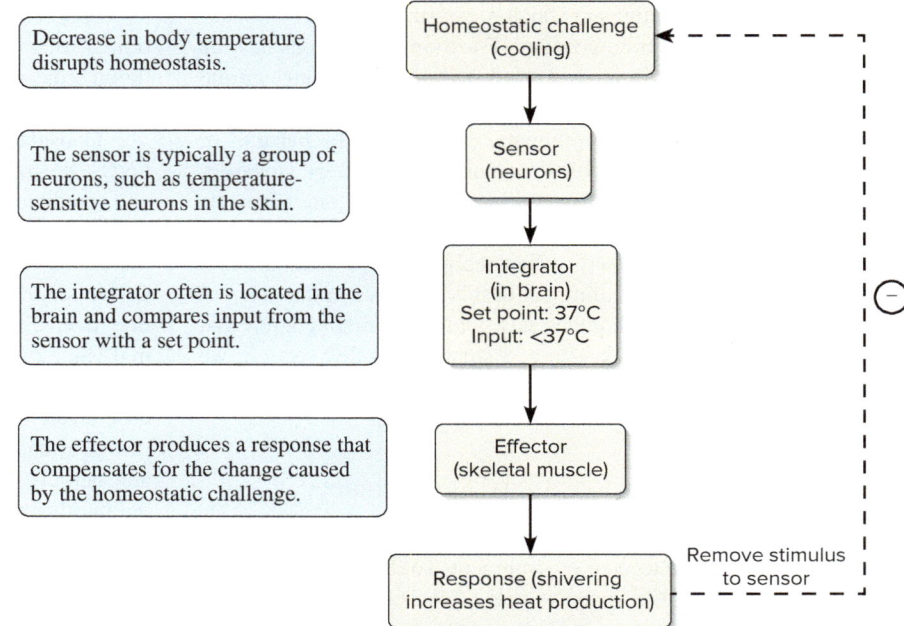

Figure 32.8 An example of a homeostatic control system. The sensor, integrator, and effector for responding to a decrease in body temperature are shown. Negative feedback (dashed line and minus symbol) prevents the response from continuing above normal. Different homeostatic control systems have different sensors and effectors.

BioConnections: *Refer back to Figure 6.10. Can negative feedback occur at a molecular level and contribute to homeostasis?*

resulting in shivering—a major way in which mammals generate heat. We will discuss other heat-conserving and heat-generating mechanisms that contribute to this important homeostatic control system in Section 32.4.

Negative Feedback Is a Key Feature of Homeostasis

A fundamental feature of homeostasis is that disturbances to a physiological variable, when they occur, are minimized and corrected. The temperature-regulation system just described is an example of a **negative feedback loop,** in which a change in the variable being regulated brings about responses that move the variable in the opposite direction. Thus, a decrease in body temperature leads to responses that increase body temperature—that is, move it back toward its original value. This removes the input to the sensor in the control system, and the effectors (in this example, skeletal muscles that cause shivering) are no longer stimulated.

Negative feedback prevents homeostatic responses from overcompensating. Whereas it is beneficial to initiate warming mechanisms when an animal's body temperature decreases, it would not be beneficial to continue to generate heat beyond the normal, healthy range.

Not all forms of feedback contribute to homeostasis. In some cases, a **positive feedback loop** may accelerate a process (think of an avalanche that begins with a small snowball rolling down a steep hill). Once begun, a positive feedback loop has no obvious means of stopping, since a change in a variable or process leads to events that amplify that change. Look again at Figure 32.8. Imagine what would happen if shivering not only generated heat but also stimulated the sensor neurons to send additional stimulatory inputs to the effectors. Heat production would continue in such a circumstance without any means of stopping, until a potentially dangerous or even life-threatening body temperature was reached. This is contrary to the principle of homeostasis, in which large fluctuations in a variable are minimized and reversed. Not surprisingly, therefore, positive feedback is far less common than negative feedback in animals. However, it does have important roles in certain processes, such as ovulation and birth in mammals, blood clotting, and the generation of action potentials in neurons, all of which will be described in later chapters.

Feedforward Regulation Prepares for an Upcoming Challenge to Homeostasis

In animals with well-developed nervous systems, homeostasis is aided by **feedforward regulation,** the process in which an animal's body prepares for a change in some variable before it even occurs. For example, consider the increase in heart rate that occurs prior to an increase in an animal's activity (think of a racehorse at the starting gate). Another classic example of feedforward regulation is that demonstrated by Russian physiologist Ivan Pavlov in the 1890s. When a hungry dog (or human!) sees or smells food, it begins to salivate and its stomach starts churning and producing acid. Salivation and activity of the stomach are important components of the digestive process, yet at this stage the animal has not actually eaten any food. Instead, its digestive system is already preparing for the arrival of food in order to maximize digestive efficiency, speed the flow of nutrients into the blood, and minimize the time required for active cells to replenish energy stores. Therefore, feedforward regulation helps to prepare an animal for an ensuing challenge, such as the requirement for increased activity or the assimilation of nutrients.

32.3 Reviewing the Concepts

- Homeostasis is the process of maintaining a relatively stable internal environment despite changes in the external environment. Some animals conform to their environment, but many others homeostatically regulate internal processes in response to changes in their environment.

- Homeostatic control systems coordinate homeostasis. Negative feedback loops minimize changes in a variable and prevent homeostatic responses from overcompensating. Feedforward regulation prepares the body for an upcoming challenge to homeostasis (Figure 32.8).

32.3 Testing Your Knowledge

1. The normal value for a controlled variable such as body temperature is called the
   a. set point.
   b. integrator.
   c. effector.
   d. feedback.
   e. sensor.

2. Which of the following statements regarding negative feedback is *false*?
   a. It helps regulate variables such as body temperature.
   b. It explains why an animal may salivate when it sees food.
   c. It is a major feature of homeostatic control systems.
   d. It prevents homeostatic responses from overcompensating.
   e. Both b and d are false.

32.4 Homeostatic Regulation of Body Temperature

Learning Outcomes

1. Provide examples of how changes in temperature affect chemical reactions, protein function, and membrane structure.
2. Define metabolic rate, and describe how it is related to heat production and body temperature.
3. List and define the four terms used to categorize organisms based on their source of heat and ability to maintain body temperature.
4. Identify the four main mechanisms animals use to exchange heat with the environment.
5. Define several mechanisms by which animals can alter their rate of heat gain or loss.

As we have seen, the control of body temperature is an example of homeostasis. Look at the image of the young woman in the chapter-opening photo. What is happening at that moment? It is a hot day, and the woman is perspiring and thirsty; it appears that she has just been exercising outdoors. Under such conditions, her body must work to prevent her body temperature from increasing, because high body temperature can destroy enzymes and damage cellular membranes. One way some animals, including humans, eliminate heat from the body is to perspire. Recall from Chapter 2 that water has a high heat of vaporization, and therefore the evaporation of water from the body surface helps to eliminate body heat. In this section, we will begin by discussing why body temperature homeostasis is important for the health and survival of animals. Next, we will examine how metabolic rate is related to body temperature. We will then consider the homeostatic mechanisms by which animals gain or lose heat and how these mechanisms may be regulated.

Temperature Affects Chemical Reactions, Protein Structure and Function, and Membrane Structure

Most animals can survive only in a relatively narrow range of body temperatures. Mammals, for example, typically have a resting body temperature of 35–38°C (95–100°F). In humans, a body temperature of 41°C (106°F) may result in organ damage (notably, the brain), and a body temperature of 42–43°C (107–109°F) is usually fatal. Temperature affects three vital aspects of animals' bodies:

- chemical reactions (and the enzymes that catalyze them),
- protein function, and
- membrane structure.

First, chemical reactions depend on temperature. Heat accelerates the motion of molecules, so as an animal's body temperature increases, the rates at which the molecules in its body move and encounter each other also increase. Consequently, the rate of most chemical reactions in animals doubles or even triples for every 10°C increase in body temperature. Low temperatures slow down chemical reactions, making it harder for an animal to remain active and carry out functions such as digestion, reproduction, and immunity. An effect on an animal's immunity is particularly important, as many vertebrates become susceptible to disease when their body temperatures are decreased for long periods.

A second effect of temperature is that very high temperature can cause proteins to become denatured (refer back to Figure 6.7); that is, they lose the normal three-dimensional structure that is crucial to their ability to function properly. Denaturation occurs because the bonds that form tertiary and quaternary protein structures result from weak interactions, such as hydrogen bonds, and can be disrupted by heat. Denaturation of enzymes, in particular, is especially serious because of the major role they have in the metabolism of nutrients for the generation of ATP.

A third effect of temperature is on the structures of plasma membranes and the membranes of intracellular organelles. At low temperatures, membranes become less fluid and more rigid, partly due to changes in membrane phospholipids. Rigid membranes are less able to perform biological functions, such as transporting ions and nutrients. Alternatively, if the temperature becomes too high, membranes can become leaky.

Heat Production Is Related to Metabolic Rate

The amount of energy an organism uses in a given period of time to power its activities is called its **metabolic rate.** The greater the metabolic rate, the more heat an animal generates. This heat arises from chemical reactions and, in turn, speeds up the rate of other chemical reactions.

The breakdown of organic molecules liberates energy in their chemical bonds that can be transferred to and stored in the bonds in ATP (see Chapter 6). This is the energy that cells use to perform various biological activities such as muscle contraction, active transport, and molecular synthesis. We refer to these activities as work. Not all of the energy is used to do work, however. Some of it appears as heat, which contributes to an animal's body temperature or is dissipated to the environment. Biologists have historically quantified the energy of metabolism in calories. A **calorie** is approximately the amount of heat required to raise the temperature of 1 gram of water by 1 degree Celsius. Most biological activities, however, require much greater amounts of energy than a calorie, and consequently the more common unit of measurement is the kilocalorie (1,000 calories, or 1 **kcal**). (In food labeling, 1 Calorie with a capital C is the same as 1 kcal.) Biologists often measure and compare the metabolic rates of different animals to learn, for example, how some animals are capable of hibernating, how an animal's body temperature influences its metabolic rate, and how hormones and other factors alter an animal's metabolism.

The most common measure used to compare the metabolic rates of different species is the **basal metabolic rate (BMR)**. BMR can be thought of as the metabolic cost of living in a resting animal under controlled conditions. In mammals, most of it can be attributed to the routine functions of the heart, liver, kidneys, and brain. To compare the BMRs of different animals, however, a distinction must be made between animals that generate their own internal heat and those that do not (or do so less well). **Endotherms** are animals that generate their own internal heat through their metabolism. By contrast, **ectotherms** are animals that rely on heat from the external environment to warm themselves (Figure 32.9).

In addition, biologists define **homeotherms** as animals capable of maintaining their body temperature within a narrow range, whereas **heterotherms** have body temperatures that may undergo considerable variation (Figure 32.9). Most animals can be categorized as either endotherms or ectotherms and as either homeotherms or heterotherms. Generally, birds and mammals are endothermic and homeothermic. Other vertebrates and most invertebrates are ectothermic and heterothermic.

Not all animals, however, can be classified into the same two categories at all times. Hibernating mammals, for example, are endotherms and homeotherms, except during the transition from fall to winter and again from winter to spring, when they are endotherms and heterotherms. Generally, though, birds and mammals can quickly adjust the body's mechanisms for retaining or releasing heat so that body temperature remains within a narrow homeostatic range. This regulation provides the advantage that the body's chemical reactions occur at optimal rates even when the environment imposes extreme challenges. The metabolic rate of a resting mammal, for example, is roughly six times greater than that of a comparably sized reptile. A suddenly awakened mammal is instantly capable of intense activity even on a winter's day, but an icy-cold reptile could be at the mercy of a predator because of the time required to warm itself in order to flee.

Endothermy does have three major disadvantages, however.

- **Requirement for larger amounts of energy.** First, to produce sufficient heat by metabolic processes, endotherms must consume larger amounts of food to provide the nutrients used by cells in the synthesis of ATP, during which heat is generated. Endotherms with high BMRs, such as shrews, must eat almost continuously and may die if deprived of food for as little as a day. By contrast, many ectotherms, such as snakes, can go for weeks without eating.

- **Risk of overheating.** Second, endotherms have a greater risk of overheating during periods of intense activity, even in cold weather.

- **Loss of body fluids.** Third, as described shortly, one mechanism for preventing overheating is the evaporation of body fluids by, for example, perspiration; this can result in an imbalance in water levels in an animal's body.

Animals Exchange Heat with the Environment in Four Ways

Body temperature homeostasis is aided through the exchange of heat between an animal's body and its environment. The surface of an animal's body can lose or gain heat from the external environment via four mechanisms: radiation, evaporation, conduction, and convection.

- **Radiation** is the emission of electromagnetic waves from the surface of an object or organism. The rate of emission is determined by the temperature of the radiating surface. Thus, if the body surface is warmer than the environment, the body loses heat at a rate that depends on the temperature difference. If the outside temperature is warmer than body temperature, the body gains heat by radiation—for instance, from sunlight. We can observe radiated heat from an animal's body with imaging devices that detect infrared light, the wavelength at which thermal energy is radiated from animal bodies (Figure 32.10).

Figure 32.9 Body temperature and environmental temperature in endotherms and ectotherms. In this idealized example, the rabbit is also a homeotherm because its body temperature doesn't change; the lizard is also a heterotherm because its body temperature changes considerably.

Concept Check: Into which thermoregulatory categories do humans fit?

Figure 32.10 Visualization of heat exchange in an ectotherm and an endotherm. Thermal-imaging cameras can detect heat radiated from an animal's body. Note the warm skin of the endotherm (the human) and the cold body surface of the ectotherm (the tarantula), even though both animals are at the same environmental temperature of about 25°C.
©Nutscode/T Service/Science Source

Concept Check: Which animal in this figure is a heterotherm?

- **Evaporation** is the conversion of water from the liquid to the gaseous state. Animals can lose body heat through evaporation of water from the skin and membranes lining the respiratory tract, including the surface of the tongue. A large amount of energy in the form of heat is required to transform water from liquid to gas. Whenever water vaporizes from the body's surface, the heat required to drive the process is conducted from the surface, thereby cooling the animal.

- **Conduction** is a process whereby the body surface loses or gains heat through direct contact with cooler or warmer objects (see, for example, how in Figure 32.10 the edges of the tarantula's legs are warmer than the rest of its body). The greater the temperature difference, the greater is the rate of heat transfer. Different materials have different abilities to absorb heat, however. Water has a higher specific heat (see Chapter 2) than air, meaning that at any temperature, water will retain greater amounts of heat than will air. Consequently, aquatic animals in water that is 10°C lose considerably more heat in a given time than terrestrial animals lose in air that is 10°C. On a hot day, terrestrial animals can efficiently lose heat by immersing themselves in water.

- **Convection** is the transfer of heat by the movement of air or fluid next to the body. For example, the air close to an endotherm's body is heated by conduction. Because warm air is less dense than cold air, the warm air near the body is forced upward and carries away heat by convection. Convection is aided by creating currents of air around an animal's body. Humans do this by sitting near fans, but other animals can create cooling air currents themselves, as when an elephant waves its ears.

The four mechanisms of heat transfer just described can be homeostatically regulated in animals, such that heat is retained within the body at some times and lost at other times, as we will see next.

Several Mechanisms Can Alter Rates of Heat Gain or Loss in Endotherms

For purposes of temperature control, think of an endotherm's body as a central core surrounded by a shell consisting of skin and subcutaneous (just below the skin) tissue. Depending on the species, the temperature of the central core of endotherms is regulated at approximately 35–42°C (95–108°F), but the temperature of the outer surface of the skin varies considerably. If the skin were a perfect insulator, the body would never lose or gain heat by conduction. The skin of animals does not insulate completely, however, so the temperature of its outer surface generally is somewhere between that of the external environment and that of the core. Only in animals that store large amounts of subcutaneous fat (blubber) does the body surface provide considerable insulation. In endotherms without blubber, the main form of insulation is a covering of hair, fur, or feathers, which traps heat from the body in a layer of warm air near the skin, reducing heat loss by conduction. Let's take a look at four mechanisms that different endotherms use to regulate how much heat is gained or lost between the environment and their bodies.

Changes in Skin Blood Flow Rather than acting as an insulator, as in some animals, in most endotherms the skin functions as a variable heat exchanger that can be adjusted to increase or decrease heat loss from the body (**Figure 32.11**). In such animals, blood vessels just beneath the surface of the skin dilate (widen) on hot days. This increases the flow of warm blood to the skin and helps to dissipate heat to the environment across the skin surface. These same vessels constrict (get narrower) on cold days to retain body heat. Signals from the nervous system regulate the dilation and constriction of these blood vessels. Diving birds and diving mammals are good examples of animals that use this mechanism. Ducks, seals, and walruses greatly reduce the amount of blood flowing to the skin when they dive in cold waters. This allows them to retain body heat that would otherwise be conducted into the water.

Countercurrent Exchange Many endotherms and ectotherms regulate heat loss to the environment through **countercurrent exchange,** in which heat is transferred between fluids flowing in opposite directions. Countercurrent exchange regulates heat loss to the environment by returning heat to the body's core and keeping the core much warmer than the extremities. In endotherms, countercurrent exchange occurs primarily in the extremities—the flippers of dolphins, for example, or the legs of birds and certain other terrestrial animals (**Figure 32.12a**). As warm blood travels from the body core through arteries down a bird's leg, heat moves by conduction from the artery to adjacent veins carrying cooler blood in the opposite direction (**Figure 32.12b**). By the time the arterial blood reaches the tip of the leg, its temperature has decreased considerably, reducing the amount of heat lost to the environment and returning the heat to the body's core.

Evaporative Heat Loss Recall that some animals can lose body heat through evaporation of water from the skin and membranes

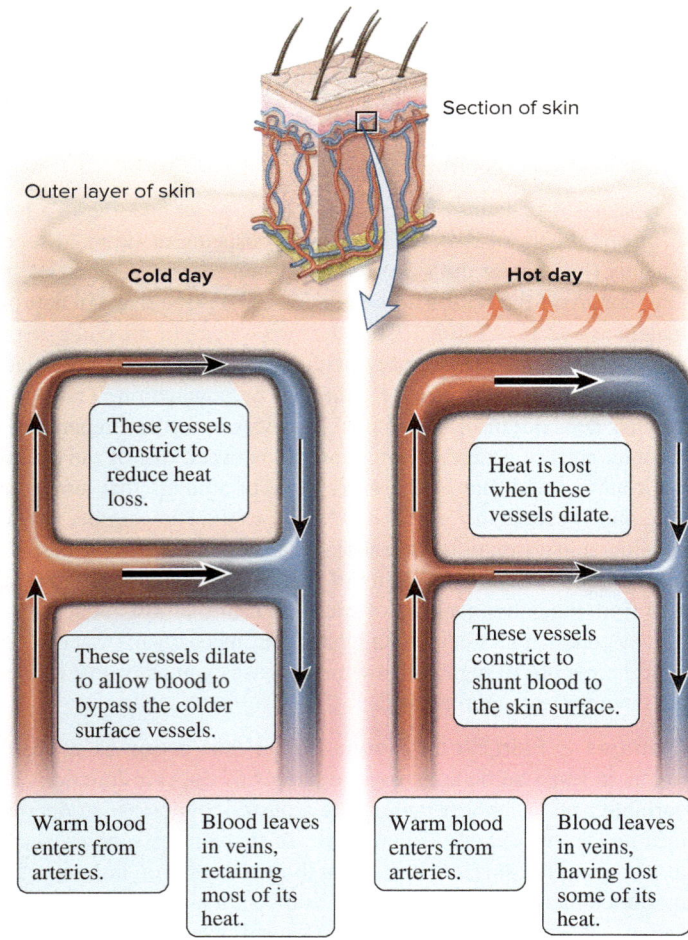

Figure 32.11 **Regulation of heat exchange in the skin.** As shown in this schematic illustration, the skin functions as a variable heat exchanger. The arrows in the blood vessels indicate direction and relative amount of blood flow.

(a) **Countercurrent exchange of heat in the leg of an endotherm**

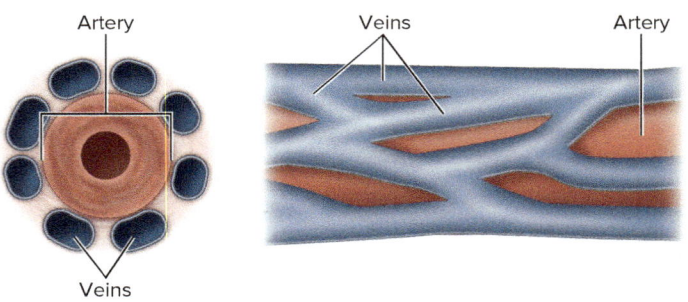

(b) **Cross section and surface view of veins covering an artery**

Figure 32.12 **Countercurrent exchange.** **(a)** Countercurrent exchange retains heat in the leg of an endotherm such as this bird. Black arrows in vessels indicate direction of blood flow. Red arrows indicate the direction of heat transfer. **(b)** A depiction of the arrangement of veins surrounding an artery in a wading bird's leg. The artery is almost completely covered by overlying veins, allowing efficient heat exchange between the vessels.

Concept Check: *In which direction does heat transfer in this example, from artery to veins or from veins to artery?*

lining the respiratory tract. Heat exchange in some mammals can be regulated by changing the rate of water evaporation through perspiration (as shown in the chapter-opening photo). Nerves to the sweat glands stimulate the production of sweat, a dilute solution containing sodium chloride.

In endotherms that lack sweat glands, such as birds, or those that have very few glands, such as dogs and cats, panting (short, rapid breaths with the mouth open) promotes evaporation of water from the tongue surface. Panting has advantages over sweating, because only water but no salt is lost, and panting provides the air current that promotes heat exchange by convection. However, the surface area of the mouth and tongue is relatively small, which limits the rate at which heat can be eliminated.

Behavioral Adaptations Behavioral mechanisms can also alter heat loss by radiation, conduction, and evaporation. For example, on hot days birds may ruffle their feathers and raise their wings, whereas many mammals will reduce their activity and spread their limbs; these postural changes increase the surface area available for heat transfer. Terrestrial endotherms seek shade, partially immerse themselves in water, or burrow into the ground when the Sun is high. Animals that neither sweat nor pant can still benefit from evaporative heat loss. The evaporation of fluids deposited on the body surface by licking the skin or splashing the skin with water also draws heat from the body.

Similarly, animals respond to cold temperatures with numerous behavioral adaptations. Huddling in groups, curling up into a ball, hunching the shoulders, burying the head and feet in feathers, and similar maneuvers reduce the surface area exposed to a cold environment and decrease heat loss by radiation and conduction.

Muscle Activity and Brown Adipose Tissue Metabolism Increase Heat Production

We have discussed how heat is gained or lost to the environment and how heat can be retained by reducing blood flow to the skin on a cold day. Body temperature, however, is a balance between these factors and heat production. Changes in muscle activity constitute a major mechanism for the control of heat production in endotherms.

When an endotherm is exposed to low environmental temperatures, core body temperature begins to decrease. The primary response to decreasing temperatures is to decrease the flow of blood to regions that permit the conduction of heat. If this does not adequately reduce heat loss, skeletal muscle contraction is increased. This may lead to shivering, which consists of rapid skeletal muscle contractions without any locomotion. Virtually all of the energy liberated by the chemical reactions that power the contracting muscles appears as internal heat, a process known as **shivering thermogenesis.** Many birds that remain in cold climates during the winter shiver almost continuously.

In many mammals, chronic cold exposure also induces **nonshivering thermogenesis,** an increase in the metabolic rate and therefore heat production that is not due to increased muscle activity. Nonshivering thermogenesis occurs primarily in **brown adipose tissue** (also called brown fat), a specialized tissue that occurs in small or young mammals in which the shivering mechanism is not present or not yet fully developed. Brown adipose tissue is responsive to hormones and signals from the nervous system, which are activated when body temperature decreases. Unlike ordinary white adipose tissue, which stores energy in the form of fat, brown adipose tissue metabolizes fat and generates heat as a by-product.

32.4 Reviewing the Concepts

- The amount of energy an organism expends to power its activities is called its metabolic rate. In the resting state, this is called the basal metabolic rate (BMR). The greater the BMR of an animal, the more heat the animal generates.
- Ectotherms depend on external heat sources to warm their bodies, whereas endotherms use their own metabolically generated heat to warm themselves. Homeotherms maintain their body temperature within a narrow range, but the body temperature of heterotherms varies with environmental conditions (Figure 32.9).
- The surface of an animal's body can lose or gain heat from the environment via four mechanisms: radiation, evaporation, convection, and conduction (Figure 32.10).
- The skin can function as a variable heat exchanger; blood vessels near the skin surface dilate to dissipate heat or constrict to retain it. Heat exchange can also be controlled by countercurrent exchange, perspiration, and behavioral mechanisms (Figures 32.11, 32.12).
- Muscle activity (shivering thermogenesis) and brown adipose tissue metabolism (nonshivering thermogenesis) increase the production of heat.

32.4 Testing Your Knowledge

1. Animals that have body temperatures that are maintained within a narrow range are
   - a. endotherms.
   - b. ectotherms.
   - c. homeotherms.
   - d. heterotherms.
   - e. both b and d.

2. The rate of heat loss in a mammal is regulated by
   - a. the degree of blood flow at the surface of the skin.
   - b. the amount of perspiration.
   - c. behavioral adaptations.
   - d. air currents near the animal's body.
   - e. all of the above.

32.5 Homeostasis of Internal Fluids

Learning Outcomes

1. Describe how the bodies of animals are organized into fluid compartments.
2. **SCISKILLS** ▶ Predict the outcomes of imbalances of water and ions on animals' function and survival.
3. Describe what osmolarity is and how it relates to the movement of water and cell shape.
4. **SCISKILLS** ▶ Predict how water will move between body fluid compartments if solute concentrations differ between the compartments.
5. Describe the factors that result in obligatory exchanges of water and ions between an animal's body and its environment.

Look again at the chapter-opening photo of a young woman who has been exercising on a hot day. As she perspires, she loses heat, but this comes with a cost because of the water and ions she loses. Animal bodies are composed in large part of water, which is contained in body fluid compartments. Within the fluids are many salts, such as NaCl and KCl. Salts dissociate in solution into ions. (Because ions from dissolved salts are electrically charged, they are sometimes referred to as electrolytes.) In this section, we will first examine how water is distributed in animal bodies and why water and ion homeostasis is so important. We will then explore some of the challenges that must be overcome to maintain that homeostasis.

Body Fluids Are Distributed into Fluid Compartments

Up to now, we have examined organ systems and their component structures without any reference to the other major feature of animal bodies, namely water. All animal bodies are composed primarily of water. This water is distributed within fluid compartments (**Figure 32.13**). Most of the water in an animal's body is contained inside its cells; this fluid compartment is called **intracellular fluid** (from the Latin *intra,* meaning inside of). The rest of the water in its body exists outside the cells and is therefore called **extracellular fluid** (from the Latin *extra,* meaning outside of). In vertebrates, extracellular fluid includes the fluid (noncellular) part of blood, called **plasma,** and the fluid that fills the spaces that surround cells, called **interstitial fluid** (from the Latin *inter,* meaning between). The interstitial fluid and the plasma are separated by the wall of blood vessels. Water can move between body fluid compartments by the process of osmosis, as described in Chapter 5. This has important consequences for a variety of processes that occur in animals, such as the control of the volume of blood in an animal's body. For example, if the solute

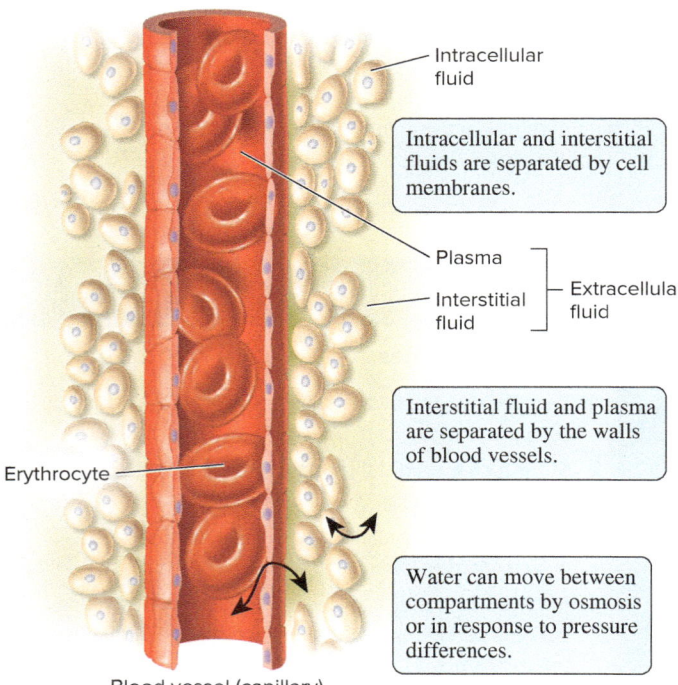

Figure 32.13 Fluid compartments in a typical vertebrate. Most of the fluid within an animal's body exists within cells (intracellular fluid). Extracellular fluid is that portion of the body's fluid that lies outside cells (interstitial fluid) or within blood vessels (plasma), such as the capillary shown here. The majority of extracellular fluid is in the interstitial compartment. Arrows indicate directions of water movement between adjacent compartments.

Concept Check: What would happen to the distribution of water in the fluid compartments of an animal's body if a blood vessel were damaged and leaked its contents internally?

concentration of an animal's blood were to increase (perhaps due to ingestion of a salty meal), water would move by osmosis from the interstitial fluid into the plasma. This would enlarge the blood volume and could cause an increase in blood pressure.

The solute composition of the extracellular fluid is very different from that of the intracellular fluid. Maintaining differences in solute composition across the plasma membrane is an important way in which animal cells regulate their own activity. For example, differences in ion concentrations on either side of the plasma membrane of a cell create the potential for moving electrical charges across the membrane; this is a key mechanism by which cells of the nervous system communicate with each other. The mechanisms by which solutes move across plasma membranes were also described in Chapter 5.

A Balance of Water and Ions Is Critical for Survival

Maintenance of normal levels of body water is of great importance for all animals for several reasons (refer back to Figure 2.13). Water is the solvent that permits dissolved solutes to participate in chemical reactions. It also directly participates in hydrolysis reactions, and it

is the transport vehicle that takes oxygen and nutrients to cells and removes wastes generated by metabolism.

When an animal's water volume is decreased below the normal range, we say the animal is dehydrated. In terrestrial animals, dehydration may occur if sufficient drinking water is not available or when water is lost by evaporation (through perspiring or panting). Dehydration can be a serious, potentially life-threatening condition. For example, because blood is roughly 50% water (plasma), blood volume tends to decrease in dehydrated animals. Decreased blood volume compromises the ability of the circulatory system to move nutrients and wastes throughout the body and, in endotherms, to assist in the regulation of body temperature on hot days.

Ion balance is also very important for animals. A change of only a few percentage points in the extracellular concentration of K^+, for example, can trigger changes in nerve, heart, and skeletal muscle function by altering membrane potentials (see Chapter 33). Other ions, such as Ca^{2+}, Mg^{2+}, PO_4^{3-}, and SO_4^{2-}, also participate in various biological activities. These functions include serving as cofactors for enzyme activation, participating in bone formation, forming part of the extracellular matrices around cells, and activating cellular events such as exocytosis and muscle contraction. An imbalance in any of these ions can seriously disrupt cellular activities. Dehydration is one way in which such an imbalance might arise. That is because when the amount of water in a body fluid compartment decreases, the concentrations of solutes in the remaining water increase.

Water moves between adjacent body compartments by osmosis (refer back to Figure 5.10). Changes in the ion concentration in one compartment will lead to changes in fluid distribution between the compartments. These changes can cause cells to shrink or swell. If, for example, the ion concentration of the extracellular fluid increases, water moves by osmosis from inside cells to the extracellular fluid, causing the cells to shrink. Shrinking or swelling of cells in the brain, heart, and other vital organs can rupture plasma membranes, leading to cell death.

The solute concentration of an aqueous solution is known as the solution's **osmolarity,** expressed as milliosmoles/liter (mOsm/L). The number of dissolved solute particles determines a solution's osmolarity. For example, a 150 mM NaCl solution has an osmolarity of 300 mOsm/L, because each NaCl molecule dissociates into two ions, one Na^+ and one Cl^- ($2 \times 150 = 300$). For reference, the value of 300 mOsm/L is well within the range of typical osmolarities of animal body fluids. Solutions greater than about 300 mOsm/L are said to be hyperosmotic. Those below about 300 mOsm/L are hypoosmotic. Solutions that are the same osmolarity as that of an animal's body fluids are said to be iso-osmotic.

Many Exchanges of Ions and Water with the Environment Are Obligatory

Many vital processes—eliminating nitrogenous wastes, obtaining oxygen and eliminating carbon dioxide (CO_2), consuming and metabolizing food, and regulating body temperature—have the potential to disturb ion and water balance (**Figure 32.14**). Therefore, these processes generally require additional energy expenditure to minimize or reverse the disturbance. Exchanges of ions and water with the environment that occur as a consequence of such vital processes are called obligatory exchanges (because the animal is "obligated" to make them).

Modeling Challenge: Figure 32.13 illustrates how fluids are distributed in the body of a vertebrate. In the simplified model depicted below, typical relative volumes are indicated by the width of each body fluid compartment, and solutes are indicated by dots. Note that each compartment has a similar solute concentration. Let's suppose that solute was added to the interstitial fluid, thereby doubling the solute concentration. Draw a new model that approximates the relative amounts of fluid in each body fluid compartment following this introduction of additional solute to the interstitial fluid. Assume that the solute cannot move between compartments, but water can. To create this model, add more dots to the interstitial fluid and then shift the vertical lines separating the compartments to the left or right. Keep the total volume (that is, the sum of the three compartments) the same as shown here.

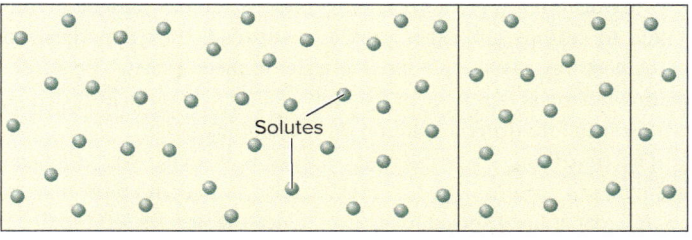

Intracellular fluid | Interstitial fluid | Plasma

Distributions of fluids in a vertebrate

Respiration: Water vapor exits during breathing.

Metabolism: Cells produce H_2O during metabolism.

Removal of wastes: Ions and H_2O are lost in feces and urine.

Food ingestion: Food introduces salt and H_2O.

Body temperature regulation: H_2O is lost by evaporation.

Figure 32.14 Types of obligatory ion and water exchanges in a terrestrial animal. Obligatory exchanges with the environment occur as the result of necessary life processes.

Concept Check: *Can animals completely avoid all the losses resulting from obligatory exchanges?*

Exchanges Due to Elimination of Nitrogenous Wastes When carbohydrates and fats are metabolized by animal cells, the major waste product is CO_2, which is exhaled or diffuses across the body surface. By contrast, proteins and nucleic acids contain nitrogen; when these molecules are broken down and metabolized, nitrogenous wastes are generated. **Nitrogenous wastes** are molecules that include nitrogen from amino groups (—NH_2). These wastes are toxic at high concentrations and must be eliminated from the body, but unlike carbon dioxide gas, they cannot be eliminated by exhaling. As we will see in Chapter 37, the excretion of nitrogenous wastes is carried out by excretory organs and may require body water.

Exchanges Due to Respiration The requirements for both respiration and water and ion balance present different challenges to air- and water-breathing animals. To ventilate its lungs, an air-breathing animal moves air into and out of its airways. Water vapor in the mouth, nasal cavity, and upper airways exits the body with each exhalation. As an animal becomes more active, breathing becomes deeper and more rapid, which in turn increases the rate of water loss from the body. Therefore, respiration in animals with lungs is associated with significant water loss, as you can observe in cold weather when you can "see your breath."

In water-breathing animals, the challenge of water and ion balance is more complex, because such animals move water, not air, over their respiratory organs (gills). Gills, like all respiratory organs, are thin structures with large amounts of surface area and an extensive network of blood vessels. Although these features make gills suitable for gas exchange by diffusion between the blood and the surrounding water, they also make them suitable for ion and water movement by diffusion and osmosis, respectively. This can create imbalances in the volume and osmolarity of fluids in an animal's body.

Exchanges Due to Feeding Because foods contain salts and water, eating also involves obligatory exchanges of these substances. Some plant products are over 95% water by weight, and other foods may contain high amounts of Na^+ or other ions. Therefore, the type of diet an animal consumes determines how much salt and water it ingests.

When food molecules are metabolized to provide energy that will be stored in the chemical bonds of ATP, oxygen captures electrons and combines with hydrogen ions, thereby making water (refer back to Figure 6.18). This water is called metabolic water to indicate its origin and, in some animals, is all the additional water they need to survive.

Exchanges Due to Evaporation of Water As we have seen, endotherms use body water to cool off. For example, sweating and panting are used to cool the body. These behaviors use the evaporation of water to draw heat out of the body. In the process, however, the animal loses water and, in sweat, some ions.

Other than perspiration and panting, very little water is gained or lost directly across the body surface of most terrestrial vertebrates (with the exception of amphibians), because their skin is impermeable to water. In invertebrates, the rate of water loss across the body surface depends on whether the animal is soft-bodied, like worms, or covered in a waxy, water-impermeable cuticle, like most insects.

The significance of obligatory exchanges and their effects on homeostasis was dramatically illustrated by a long-term investigation by a research team at the University of Florida, as described next. Their discovery would lead to a revolution in our understanding of exercise physiology and ion and water homeostasis in humans.

FEATURE INVESTIGATION

Cade and Colleagues Discovered Why Athletes' Performances Wane on Hot Days

On a typically hot summer day in the mid-1960s in Gainesville, Florida, the University of Florida football team was practicing in full equipment. The players were becoming dehydrated and, unbeknownst to them, the osmolarity of their body fluids was increasing as their bodies produced copious amounts of sweat in an effort to maintain body temperature. The athletes became aware of two things. First, they discovered that they did not need to urinate for long periods after a strenuous practice session and, second, their performance on the field suffered as they became increasingly fatigued and more susceptible to severe muscle cramps. Occasionally, players required medical treatment for their symptoms. In extreme cases, athletes exercising in these conditions have been known to develop seizures—uncontrolled activity of neurons in the brain. This situation did not escape the notice of the team physicians and, notably, university faculty member and kidney specialist Robert Cade.

Many of the symptoms experienced by the players could be readily explained. The fatigue was directly related to loss of water from the body, which put a strain on the circulatory system and reduced the amount of blood flow to muscles and other organs. It was worsened by a decrease in blood glucose concentration that was not compensated for during the long periods of strenuous activity. The muscle cramps and even occasional seizures arose from an imbalance in extracellular ions—notably, Na^+ and K^+—which are secreted outside the body by sweat glands in the process of perspiration. Sweat is a hypo-osmotic solution relative to body fluids, and consequently sweating results in an imbalance in extracellular ion concentrations. In the athletes, this caused a change in the electrical potential across muscle and neuronal cell membranes, which triggered the symptoms. Finally, the decreased urine production is one of the body's mechanisms for reducing fluid loss when body water is decreasing.

Cade and his colleagues rejected what was then the prevailing view that drinking any fluids during heavy exercise somehow contributed to cramps and other problems. Instead, they hypothesized that the best way to maintain ion and water homeostasis in a profusely sweating person is to restore to the body exactly what was lost; that is, the person should drink a solution that resembles sweat!

The first thing Cade did was analyze how much Na^+, K^+, and other ions are present in sweat. Fortunately, he had an abundance of human sweat at his disposal to analyze. Once the players left the field, their jerseys were wrung out, and the composition of the collected sweat was determined with an ion analyzer, the flame spectrophotometer shown in Figure 32.15. The concentrations were then compared with known values of ion concentrations in human blood. Today, we

Figure 32.15 Cade and colleagues discovered a way to improve athletic performance and prevent ion and water imbalance during strenuous exercise.

HYPOTHESIS Athletic performance can be enhanced by maintaining the body's ion and H_2O balance during exercise.

KEY MATERIALS Supply of human sweat for analysis, ion analyzer, salt solution.

1. Obtain human sweat from exercising athletes.

2. Analyze composition of sweat using a flame spectrophotometer, which measures ion concentrations. Prepare artificial solution that mimics composition of sweat. Compare composition of both sweat and artificial solution to known ion concentration in human blood.

3. Add flavoring and sugar to artificial solution.

Sugar improves flavor and provides energy.

4. Provide freshman team with the artificial solution and varsity B-team with water. Hold scrimmage.

5. **THE DATA**

6. **CONCLUSION** Replacement of fluid with solute concentrations similar to those found in human sweat improves athletic performance compared with water replacement alone.

7. **SOURCE** Most of the original studies described here were published in expanded form in a later report. See Cade, R., et al. 1972. Effect of Fluid, Electrolyte, and Glucose Replacement During Exercise on Performance, Body Temperature, Rate of Sweat Loss, and Compositional Changes of Extracellular Fluid. *Journal of Sports Medicine and Physical Fitness* 12: 150–156.

know that the composition of human sweat can change under certain conditions and can vary among people, but Cade's results were typical. The athletes' sweat contained mostly Na^+, K^+, and Cl^- at concentrations that indicated the solution was dilute compared with blood. Once Cade completed this analysis, he simply prepared an artificial solution of a composition similar to human sweat. The next step was to have the players drink the solution before and during the practice sessions and games. Improving its taste—adding some lemon flavoring and sugar—removed any inhibitions the players may have had about drinking it while also providing an energy boost.

For the first trial, Cade gave the solution to the freshman players during an intrasquad scrimmage against the more experienced varsity B-team, whose members received only pure water to drink as a control. At first, the freshman team appeared overmatched by the B-team, as might be expected. In the second half of the scrimmage, however, the freshman team vastly outperformed the more experienced players and did not suffer the characteristic late-game fatigue the B-team experienced. Based on this test, the varsity A-team was given a similar solution to drink the next day during a game against a heavily favored opponent, whom they beat handily on a hot 39°C (102°F) day.

In subsequent years, Cade and other researchers conducted carefully controlled experiments with humans and laboratory animals to confirm that a balanced solution of ions similar to that present in sweat effectively improves exercise performance and decreases the possibility of dehydration and its consequences. Because the solution was envisioned as "aid" for the team known as the University of Florida "Gators," the drink eventually came to be called Gatorade.

The story of Gatorade is one of a practical application based on solid scientific principles of osmolarity and ion and water homeostasis. You can now understand why drinking a dilute salt solution (in any of numerous sports drinks) during strenuous exercise is better than drinking water. Although drinking pure water prevents dehydration, drinking water in excess will decrease plasma ion concentrations below normal. In other words, it will replace one type of imbalance with another, as happened to the marathoner in the chapter introduction.

Experimental Questions

1. What symptoms are sometimes seen in athletes after prolonged, strenuous exercise, particularly in hot weather? How are these symptoms related to water loss during exercise, and what did Cade and his colleagues hypothesize about this?
2. **SCISKILLS ▶** How did the researchers test their hypothesis?
3. **SCISKILLS ▶** Propose a hypothesis: Would drinking a sports drink like Gatorade be beneficial for someone at rest or someone exercising only lightly?

32.5 Reviewing the Concepts

- Body fluids are contained in intracellular and extracellular compartments. In vertebrates, the extracellular compartment is made up of plasma and interstitial fluid (Figure 32.13).
- Exchanges of ions and water with the environment resulting from vital processes, such as respiration or the elimination of wastes, are called obligatory exchanges (Figure 32.14).
- The solute concentration of a solution of water is known as the solution's osmolarity.
- Cade and colleagues discovered that fluid replacement during exercise is particularly beneficial if the fluid contains solutes at concentrations similar to those in sweat (Figure 32.15).

32.5 Testing Your Knowledge

1. Which of the following statements is true of body fluid compartments?
   a. Most body water is in cells.
   b. Water cannot move between compartments.
   c. The osmolarities of the fluids in each compartment are usually very different from each other.
   d. The extracellular compartment consists of the plasma and interstitial fluid.
   e. Both a and d are true.

2. Which of the following statements is *false*?
   a. Sweat has a lower solute concentration than internal body fluids such as blood.
   b. Some animals can survive without drinking water.
   c. A solution of 1,000 mOsm/L is hyperosmotic relative to the body fluids of many animals.
   d. The rate of water loss in invertebrates depends in part on the type of body surface they have.
   e. A 300 mOsm/L solution of NaCl contains 300 mM Na^+ and 300 mM Cl^-.

Assessing Your Knowledge and Skills

Foundational Knowledge

Multiple Choice

1. Tissue that is specialized to conduct electrical signals from one structure in the body to another structure is _____ tissue.
   a. epithelial
   b. nervous
   c. connective
   d. muscle

2. Structures composed of two or more tissue types arranged in various proportions and patterns are
   a. cells.
   b. organs.
   c. tissue systems.
   d. tissues.
   e. organ systems.

3. Adapting to changes in the external environment and maintaining internal variables within normal ranges is called
   a. equilibrium.   d. set point.
   b. homeostasis.   e. negative control.
   c. positive feedback.

4. When examining the structure of many animal organs,
   a. it is apparent that an organ's surface area increases more than its volume as an organ enlarges.
   b. a general function can sometimes be predicted based on the organ's structural features.
   c. different tissues do not come into contact with each other within an organ.
   d. all four tissue types are equally represented in all organs.
   e. None of the above is correct.

5. Most of the water in an animal's body
   a. lacks any type of ions or other solutes.
   b. is found in the spaces between cells.
   c. is contained inside cells.
   d. is located in the extracellular fluid.
   e. is unable to move between body compartments.

Short Answer

1. Describe the components of a homeostatic control system.
2. Define negative feedback loop and positive feedback loop, and describe how they differ.

Critical-Thinking Skills

Multiple Choice

1. Which of the following events would be counter to the concept of homeostasis?
   a. blood vessels dilating in the skin of an endotherm on a cold day
   b. an animal feeling thirsty after a meal
   c. a person drinking a bottle of soda pop when he or she is not thirsty
   d. an endotherm perspiring on a hot day
   e. both a and c

2. A man eats several slices of pizza topped with pepperoni and sausage, which are very salty. What do you expect will happen?
   a. The sodium concentration of his extracellular fluids will decrease.
   b. Water will move by osmosis out of his cells into his extracellular fluid.
   c. His cells will swell.
   d. His cells will shrink.
   e. Both b and d are correct.

3. A person consumes a very sugary meal. After a short time, her blood glucose concentration increases from its normal resting value of 100 mg/dL to 210 mg/dL. (A deciliter, dL, is equal to 100 milliliters and is often used in quantifying glucose concentration in the blood.) A little while later, it drops to 50 mg/dL. Which of the following statements is true?
   a. She is experiencing normal glucose homeostasis.
   b. She is unable to produce compensating mechanisms for the sugar that enters her blood.
   c. She has some compensating mechanisms, but they are not functioning homeostatically.

4. Refer to Figure 32.7a. What would happen if the dimensions of the finger-like projections shown in that figure were to double in size?
   a. Their surface area would double.
   b. Their surface area would increase eightfold.
   c. Their volume would double.
   d. Their volume would increase eightfold.
   e. Their ratio of surface area/volume would increase.

5. Unlike most vertebrates, frogs have skin that is relatively permeable to water. When a frog is submerged in a freshwater pond, which typically has an osmolarity of less than about 50 mOsm/L, what do you predict will happen?
   a. The frog will lose body water to the pond.
   b. There will be no net gain or loss of water.
   c. The frog will gain water across its skin.
   d. The solute concentration of the frog's body fluids will increase.
   e. Both a and d will occur.

Short Answer

1. Would you expect all animals to have a similar set point for a given homeostatic variable? Explain why or why not.
2. **PRINCIPLES** Two principles of biology are that living organisms use energy and living organisms maintain homeostasis. How are energy use and the maintenance of homeostasis related?

33

Neuroscience I: The Structure, Function, and Evolution of Nervous Systems

Courtesy Ann-Shyn Chiang, Tsing Hua Chair Professor/Brain Research Center & Institute of Biotechnology/National Tsing Hua University

Three-dimensional image of the brain of a fruit fly. The brains of animals are organized into anatomical structures with specialized functions. For example, the structures colored green in this image are important for learning and memory, the purple structures are important for olfaction, and the large, orange structures on either side are involved in processing visual information.

Chapter Outline

33.1 Cellular Components of Nervous Systems

33.2 Electrical Properties of Neurons and the Resting Membrane Potential

33.3 Generation and Transmission of Electrical Signals Along Neurons

33.4 Communication at Synapses

33.5 Evolution and Development of Nervous Systems

33.6 Structure and Function of the Nervous Systems of Humans and Other Vertebrates

33.7 Impact on Public Health

Assessing Your Knowledge and Skills

After learning that she was going to lose her job, a 42-year-old woman began losing interest in her daily activities. A mother of two, she found it increasingly difficult to find the energy or motivation to care for the children, whereas in the past she had been an attentive, involved, and nurturing parent. Her relationship with her husband deteriorated as she found herself spending more and more time isolated and unwilling even to engage in ordinary conversation. Eventually, simply getting out of bed in the morning became overwhelming. As her condition worsened, she finally agreed to see a physician, who diagnosed her with major depressive disorder, commonly called depression. This disorder affects how brain cells signal each other through the release of chemical messengers called **neurotransmitters.** In depression, certain neurotransmitters that are important for maintaining a positive mood may be deficient. The woman was treated, therefore, with a drug that increases the brain concentration of one of these neurotransmitters. After several months, she began to feel "like her old self" again and eventually the medication was withdrawn. Her case illustrates just how crucial it is that a proper balance of signaling molecules be present in the human brain.

The **nervous system** is composed of coordinated groups of cells that sense internal and environmental changes and transmit signals that enable animals to respond in appropriate ways. Nervous systems provide animals with advantages that promote survival. For example, they help animals to avoid predation and other environmental dangers; form social bonds that enhance the chances of survival for the individual and the group; and even perform the complex tasks of thinking, learning, remembering, and, as illustrated by the woman with depression, maintaining a stable mood.

Nervous systems are composed of circuits of **neurons**—highly specialized cells that communicate with each other and with other types of cells by electrical or chemical signals. In many animals, neurons become organized into a central-processing area of the nervous system called a **brain** (see the chapter-opening photo). The brain sends commands to other parts of the body through **nerves**—bundles of neuronal cell extensions encased in connective tissue that project to and from various tissues and organs.

We will begin by investigating the structural and functional features of neurons that make them suited for rapid communication. We will then examine neurotransmitters, how nervous systems are organized, how they evolved, and how the human brain is affected by disease.

33.1 Cellular Components of Nervous Systems

Learning Outcomes

1. Explain the difference between the central nervous system (CNS) and the peripheral nervous system (PNS).
2. List the cellular components of the nervous system, and describe the function of each cell type.
3. **SCISKILLS** ▶ Identify the different parts of a typical neuron, and explain how the structure of each relates to its function.
4. Describe the functional relationships among sensory neurons, motor neurons, and interneurons.
5. **SCISKILLS** ▶ Explain what a reflex is, and propose reasons why reflexes are important in animals.

The organization of the nervous system permits extremely rapid responses to changes in an animal's external or internal environment. In many animals, the **central nervous system (CNS)** consists of a brain and a nerve cord, which in vertebrates extends from the brain through the vertebral column and is called the **spinal cord**. The **peripheral nervous system (PNS)** consists of all neurons and projections of their plasma membranes that are outside of but connect with the CNS, such as projections that end on muscle and gland cells. In certain invertebrates with simpler nervous systems, the distinction between the CNS and PNS is less clear or not present.

In this section, we will survey the general properties of the cells that are common to all nervous systems.

Cells of the Nervous System Are Specialized to Receive and Send Signals

The major signaling cells of nervous systems are neurons. All animals except sponges have neurons, which comprise the following parts (**Figure 33.1a, b**):

- A **cell body** contains the cell nucleus and other organelles.
- **Dendrites** (from the Greek word *dendron*, meaning tree) are projections of the cell body with numerous, branching extensions that provide a large surface area for contacts with other neurons. Electrical and chemical messages from other neurons are received by the dendrites and from there move toward the cell body (Figure 33.1c). Some messages also arrive directly at the cell body.
- An **axon** is an extension of the cell body that transmits outgoing signals along its length and eventually to other cells. A typical neuron has a single axon.
- An **axon hillock** is the part of the axon closest to the cell body and is important in the initial generation of the electrical signals that travel along an axon.
- **Axon terminals** are the ends of axons and are the sites at which electrical or chemical messages are communicated to other cells, such as other neurons or muscle cells.

Within an animal's body, axons from many neurons are arranged in parallel bundles to form nerves, which are covered by a protective layer of connective tissue (Figure 33.1d). Nerves enter and leave the CNS and transmit signals in both directions between the PNS and the CNS.

Nervous systems also contain **glia**, which are non-neuronal cells that surround, support, and protect the neurons. In vertebrates, specialized glial cells wrap around certain axons of both the CNS and the PNS at regular intervals to form an insulating layer called a **myelin sheath** (Figure 33.1). The sheath is periodically interrupted by noninsulated gaps called nodes of Ranvier. As we will see later, myelin and the nodes of Ranvier increase the speed at which electrical signals travel along an axon.

Neurons Are Specialized for Different Functions

Neurons can be categorized into three main types: sensory neurons, motor neurons, and interneurons. The structures of each type reflect their specialized functions.

Sensory Neurons As their name suggests, **sensory neurons** detect or sense information from the outside world, such as light, odors, touch, or heat. In addition, they detect internal body conditions such as blood pressure or body temperature. Sensory neurons are also called afferent (from the Latin, meaning to bring toward) neurons because they transmit information from the periphery to the CNS.

Figure 33.1 Structure and basic function of a typical vertebrate neuron and associated glial cells. **(a)** A stained neuron seen at high magnification (confocal fluorescence microscopy). **(b)** A diagrammatic representation of a peripheral neuron with associated glial cells. The glial cells wrap their membranes around the axon at regular intervals, creating a myelin sheath that is interrupted by bare patches called nodes of Ranvier. **(c)** The structures involved in signaling by a neuron. Signals flow in the direction shown. **(d)** Cross section of a nerve, as seen in a false-color SEM. Axons are light green, surrounded by myelin sheaths (red). Connective tissue fibers (purple) and a portion of a blood vessel (yellow) can also be seen.
(a) ©James Cavallini/Medical Images; *(d)* ©Steve Gschmeissner/Science Source

Many sensory neurons have a long axon that branches into two main processes, with the cell body in between (Figure 33.2a).

Motor Neurons **Motor neurons** transmit signals away from the CNS and elicit some type of response that depends on the cells receiving the signals (Figure 33.2b). They are so named because one type of response they cause is movement. In addition, motor neurons may cause other effects such as the secretion of hormones from endocrine glands. Because they transmit signals away from the CNS, motor neurons are also called efferent (from the Latin, meaning to carry from) neurons. Like sensory neurons, motor neurons tend to have long axons but do not branch into two main processes.

Interneurons A third type of neuron, called the **interneuron**, forms connections between other neurons in the CNS. The signals sent between interneurons are critical in the interpretation of information that the CNS receives, as well as the response that it may elicit. Interneurons tend to have many dendrites, and their axons are typically short and highly branched (Figure 33.2c). This arrangement allows interneurons to form complex connections with many other cells.

Reflex Arcs As a way to understand the interplay among sensory neurons, interneurons, and motor neurons, let's consider a simple example in which these types of neurons form interconnections with each other. Neurons transmit signals to each other through a

Biology Principle

Cells Are the Simplest Units of Life

Cells are the simplest units of an organ system, including the nervous system. However, the variety and complexity of cells within an organ system can be great, as shown here.

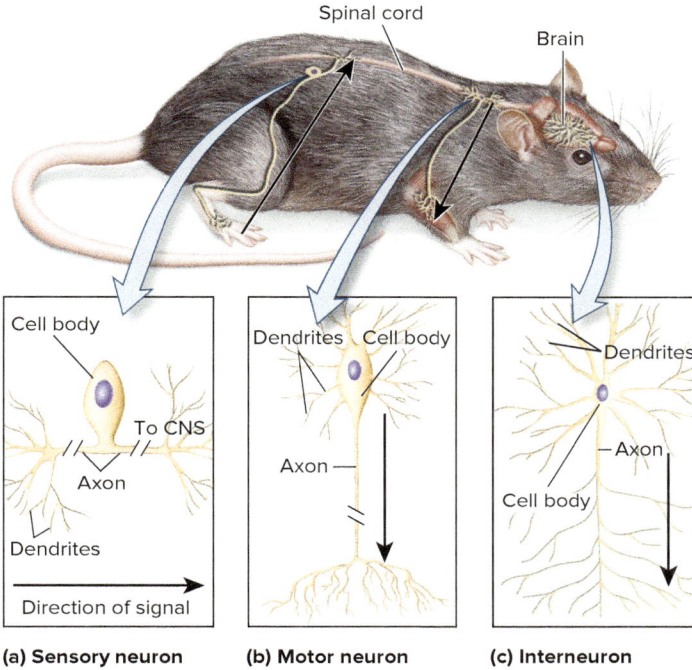

Figure 33.2 Types of neurons. (a) Vertebrate sensory neurons have an axon that bypasses the cell body and projects to the CNS. (b) Motor neurons transmit signals away from the CNS and usually have long axons that enable them to act on distant cells. In parts (a) and (b), the hatch marks indicate that the axons are not to scale but are actually much longer than shown. (c) Interneurons are usually short neurons that connect two or more other neurons within the CNS. Although short, the axons and dendrites may have extensive branches, allowing them to receive many inputs and transmit signals to many neurons. The black arrow indicates the direction that a signal travels along the axon.

---

series of connections that form a circuit. An example of a simple circuit is a **reflex arc,** which allows an organism to respond rapidly to inputs from sensory neurons and which consists of only a few neurons (Figure 33.3). For example, the stimulus from sensory neurons that sense a tap to the patellar tendon is sent to the CNS, but there is little or no interpretation of the signal. Typically, very few, if any, interneurons are involved, as in the example shown in Figure 33.3. A signal is then transmitted to motor neurons, the axons of which leave the CNS and elicit a response, in this case a knee jerk. Such a response is very quick and automatic.

Reflexes are among the evolutionarily oldest and most important features of nervous systems, because they allow animals to respond quickly to potentially dangerous or otherwise important events. For instance, many vertebrates will immediately cringe, jump, leap, or take flight in response to a loud noise, which could represent sudden danger. Other reflexes, such as the patellar tendon reflex just described, are important for postural changes and locomotion. There are a great many examples of reflexes, and their importance is evident from the observation that they exist in abundance in nearly all animals.

33.1 Reviewing the Concepts

- The central nervous system (CNS) is composed of a brain and a nerve cord. The peripheral nervous system (PNS) consists of all neurons and their projections that are outside of and connect with the CNS. Nerves transmit signals between the PNS and CNS.
- In neurons, signals flow from dendrites to the cell body and then to the axon and axon terminals. The types of neurons include sensory neurons, motor neurons, and interneurons. A neuron's structure is a reflection of its function (Figures 33.1, 33.2). Glia are non-neuronal cells that surround neurons and perform numerous supportive functions.
- The most basic neuronal circuit is a reflex arc, which allows rapid responses to inputs from sensory neurons and consists of one or a few sensory and motor neurons (Figure 33.3).

33.1 Testing Your Knowledge

1. Which of the following statements is *false*?
   a. The CNS of vertebrates comprises the brain and spinal cord.
   b. Certain glial cells produce a myelin sheath that wraps around axons.
   c. A typical reflex arc consists of only a small number of both sensory and motor cells.
   d. Motor neurons are afferent neurons that convey information into the CNS.
   e. Interneurons are found in the CNS and are specialized for communication with many other cells.
2. Signals arrive at a neuron at the
   a. axon.
   b. axon terminal.
   c. dendrites and cell body.
   d. axon hillock.
   e. myelin sheath.

Figure 33.3 A reflex arc. The knee-jerk response is an example of a reflex. A tap below the kneecap (also known as the patella) stretches the patellar tendon, which acts as a stimulus for a sensory neuron. This initiates a reflex arc that activates (+) a motor neuron that causes the extensor muscle on top of the thigh to contract, which extends the leg. At the same time, a different interneuron inhibits (−) the motor neuron of the flexor muscle, causing it to relax so that it does not bend the leg.

Concept Check: *Animals have many types of reflexes. Once initiated, most reflexes occur to completion. Do you think that in some cases they may be overridden or partially suppressed?*

33.2 Electrical Properties of Neurons and the Resting Membrane Potential

Learning Outcomes

1. Explain what a membrane potential is.
2. Outline how the resting membrane potential is established and maintained.
3. Describe what is meant by an electrochemical gradient.
4. **SCISKILLS** ▶ Write the Nernst equation, and use it to predict the direction in which different ions move across a plasma membrane.

Neurons use electrical signals called action potentials (described in Section 33.3) to communicate with other neurons, muscle cells, or gland cells. These signals involve changes in the amount of electric charge across a neuron's plasma membrane. In this section, we will first examine the electrical and chemical (concentration) gradients across the plasma membrane of neurons. In subsequent sections, we will explore how these gradients provide a way for neurons to generate and conduct action potentials.

Neurons Establish Differences in Ion Concentration and Electric Charge Across Their Membranes

Like all cell membranes, the plasma membrane of a neuron acts as a hydrophobic barrier that separates charges. The charges are positively and negatively charged ions. Ion concentrations differ between the interior and exterior of the cell. Such differences create a charge imbalance across the plasma membrane; this produces an electrical potential measurable in units called millivolts (mV). Analogous to a battery, neurons have negative and positive poles, in this case the intracellular and extracellular surfaces, respectively, of the plasma membrane. For this reason, a neuron is said to be electrically polarized. The difference between the electric charges along the intracellular and extracellular surfaces of a cell membrane is called a membrane potential. The **resting membrane potential** is the membrane potential of a cell that is not producing action potentials; the cell is considered "at rest" and is not at that moment sending signals to other cells.

Let's examine how the resting membrane potential is established and maintained. When investigators first measured the resting membrane potential of a typical neuron, they registered a voltage of about −70 millivolts (mV) inside the cell with respect to the outside. This means that the interior of the resting cell had a more negative charge than the exterior (**Figure 33.4a**). Negative ions within the cell and positive ions outside the cell attract each other and line up along the membrane. Although there are more positive charges along the outside surface of a neuron and more negative charges inside, the actual number of ions that contribute to the resting membrane potential is extremely small compared with the total number of ions inside and outside the cell. The ions that are most critical for establishing the resting potential are Na^+ and K^+.

The following are major factors responsible for the resting membrane potential (**Figure 33.4b**).

1. **Establishment of ion concentration gradients.** The Na^+/K^+-ATPase pumps within the plasma membrane continually transport Na^+ out of the cell and K^+ into the cell (refer back to Figure 5.17). The Na^+/K^+-ATPase pump transports three Na^+ out

Biology Principle

Living Organisms Use Energy

The Na+/K+-ATPase pump uses the energy stored in ATP to actively transport ions across the neuronal plasma membrane. Consequently, establishing and maintaining resting membrane potentials accounts for a significant fraction of the energy expended per day in an animal.

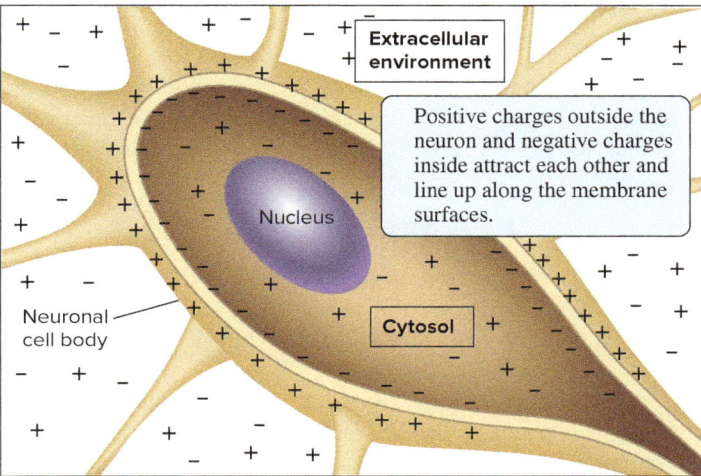

(a) Distribution of charges across the neuronal plasma membrane

(b) Two major factors that influence the resting membrane potential

Figure 33.4 The resting membrane potential. The slight excess of negative charges inside and positive charges outside the cell membrane is shown in part **(a)**. In part **(b)**, the major factors that contribute to this charge distribution are shown: the Na+/K+-ATPase pump that establishes ion concentration gradients and leak ion channels that permit diffusion of Na+ but especially K+ across the membrane.

of the cell for every two K+ it moves into the cell. Consequently, the pump contributes modestly to a charge difference across the plasma membrane. More importantly, however, it establishes concentration gradients for Na+ and K+ by actively transporting them across the membrane in opposite directions, such that more Na+ is outside the cell and more K+ is inside. The charge difference and the concentration gradients will contribute to movement of these ions across the membrane, which in turn will create the membrane potential.

2. **Membrane permeability to different ions.** The plasma membrane contains ion channels that affect the permeability of the membrane to Na+ and K+. These channels are relatively specific for Na+ or K+, and they influence the resting membrane potential by allowing the diffusion of these ions across the plasma membrane. They are referred to as ungated, or leak, channels that are open in a resting, unstimulated cell. In most neurons, there are anywhere from 10 to 100 times more leak K+ channels than leak Na+ channels, so at rest the membrane is more permeable to K+ than to Na+. As you might predict from the concentration gradients, when the cell is at rest, Na+ tends to diffuse into the cell and K+ diffuses out of the cell through these channels. The greater number of leak channels for K+ means that diffusion of K+ will make a more significant contribution to establishing a membrane potential than will diffusion of Na+. The movement of positively charged K+ out of the cell will make the interior of the cell negatively charged with respect to the outside. This is the resting membrane potential. Let's now see how this happens.

An Electrochemical Gradient Affects the Diffusion of Ions Across a Membrane

Both the membrane potential and the concentration of ions influence the direction of ion diffusion across a membrane. The direction in which an ion diffuses depends on the **electrochemical gradient** for that ion, which is the combined effect of both an electrical and a concentration (chemical) gradient (refer back to Figure 5.8). **Figure 33.5** considers the concept of an electrochemical gradient for K+. This hypothetical situation illustrates two compartments containing solutions separated by a membrane that permits only the diffusion of K+. Figure 33.5a illustrates an electrical gradient. In this case, the concentration of K+ is equal across the membrane, but the concentrations of other ions such as Na+ and Cl− are unequal on opposite sides of the membrane and thereby produce an electrical gradient. Because K+ is positively charged, it is attracted to the side of the membrane with more negative charge.

Figure 33.5b shows a concentration gradient in which K+ concentration is higher on one side than the other, but there is no electrical gradient. In this scenario, K+ diffuses from a region of high to one of low concentration.

Finally, Figure 33.5c shows an electrochemical gradient. The electrical gradient would favor the movement of K+ from left to right, but the concentration gradient would favor movement from right to left. If perfectly balanced, these opposing forces can create an electrochemical equilibrium in which there is no net diffusion of K+ in either direction. The membrane potential at which this occurs for a particular ion at a given concentration gradient is referred to as that ion's **equilibrium potential**. The equilibrium potential for a given ion can be calculated under a variety of starting conditions, as described next.

(a) Electrical gradient, no concentration gradient for K⁺

(b) Concentration gradient for K⁺, no electrical gradient

(c) Electrochemical gradient, K⁺ equilibrium

Figure 33.5 Electrical and concentration gradients. This hypothetical example depicts two chambers separated by a membrane that is permeable only to K⁺. **(a)** In this example, the compartments initially contain equal concentrations of K⁺, but an electrical gradient exists due to an unequal distribution of Na⁺ and Cl⁻. Potassium ions are attracted to the greater amount of negative charge on the right side of the membrane. **(b)** In this case, there is initially no electrical gradient across the membrane, and the left compartment contains a lower concentration of K⁺ than in the right compartment. Under these conditions, K⁺ diffuses down its concentration gradient from right to left. **(c)** This example illustrates opposing electrical and chemical (concentration) gradients. The right compartment contains a higher concentration of K⁺, and the left side has a greater net amount of positive charge. These gradients balance each other, so no net movement of K⁺ occurs and a state of electrochemical equilibrium is reached.

✓ **Concept Check:** In part (b), does the diffusion of K⁺ down its concentration gradient result in an electrical gradient? What will eventually stop the net diffusion of K⁺?

BioConnections: What are some biological functions of electrochemical gradients in living organisms? Refer back to Table 5.2 for help.

Biology Principle | Biology is a quantitative science

Quantitative Analysis: An Ion's Equilibrium Potential Depends on Its Concentration Gradient

Given the charge difference across a resting neuron's membrane, is it possible to predict the direction that an ion will diffuse across the membrane at any given concentration gradient for that ion? In other words, can we compare the relative strengths of the electrical and concentration gradients and predict their net effect? By measuring the membrane potential of isolated neurons in the presence of changing concentrations of extracellular ions, German neuroscientist Walther Nernst deduced a mathematical formula that relates electrical and concentration gradients to each other. This formula, now called the **Nernst equation,** gives the equilibrium potential for an ion at any given concentration gradient. For monovalent cations such as Na⁺ and K⁺ at 37°C, the Nernst equation can be expressed as

$$E = 60 \text{ mV} \log_{10} ([X_{extracellular}]/[X_{intracellular}])$$

where E is the equilibrium potential; $[X]$ is the extracellular or intracellular concentration of an ion; and 60 mV is a value that depends on temperature, the valence of the ion, and other factors. (For anions, the value is −60 mV.)

The Nernst equation allows researchers to predict when an ion is or is not at electrochemical equilibrium. To understand the usefulness of this equation, consider two examples. First, let's suppose the resting membrane potential of a given neuron is observed to be −88.6 mV and the K⁺ concentration is 5 mM outside and 150 mM inside the neuron. If we enter these concentrations into the Nernst equation,

$$E = 60 \text{ mV} \log_{10} (5/150) = 60 \text{ mV} (-1.48)$$
$$= -88.6 \text{ mV}$$

This value is a good approximation of a typical equilibrium potential for K⁺ in many neurons. Under these conditions, the K⁺ equilibrium potential calculated by the Nernst equation exactly equals the actual resting membrane potential in our example. Therefore, K⁺ is in electrochemical equilibrium, and no net diffusion of K⁺ occurs, even if many K⁺ channels are open.

Now let's suppose that the resting membrane potential of another neuron is −70 mV, and the Na⁺ concentration is 100 mM outside and 10 mM inside the neuron. If we enter these concentrations into the Nernst equation, we get

$$E = 60 \text{ mV} \log_{10} (100/10) = 60 \text{ mV} (1)$$
$$= 60 \text{ mV}$$

This turns out to be a typical value for the equilibrium potential for Na⁺ in many neurons. At a resting membrane potential of −70 mV, as in this example, the value of +60 mV tells us that Na⁺ is not in electrochemical equilibrium. When the equilibrium potential for a given ion—calculated by the Nernst equation—and the resting membrane potential do not match, there is a driving force for that ion to diffuse across the membrane. In this example, Na⁺ will diffuse into the cell, because both the electrical and concentration gradients favor an inward flow. However, if the actual membrane potential were +60 mV instead of −70 mV, Na⁺ would be in equilibrium at those concentrations, and therefore there would be no net diffusion of Na⁺ in either direction.

Typically, the resting membrane potential of neurons is not at the equilibrium potentials for either K⁺ or Na⁺ but is somewhere in between. Because neuronal membranes generally have so many more leak K⁺ channels than Na⁺ ones, the actual resting membrane potential is typically much closer to the equilibrium potential for K⁺ than to that for Na⁺. Therefore, these ions will move across the plasma membrane if there are open ion channels. The diffusion of a charged ion across the plasma membrane down its electrochemical gradient results in an electric current (a current is any net flow of a charged particle in a given direction). This small current provides the necessary electrical signal that neurons use to communicate with one another, as we will see in the next section.

Crunching the Numbers: A certain neuron has an intracellular concentration of K⁺ of 145 mM and an extracellular concentration of 6.5 mM. The resting membrane potential is −70 mV. What is the equilibrium potential for K⁺ under these conditions? Is K⁺ at its equilibrium potential in this neuron? If not, in which way will K⁺ diffuse if the membrane has open K⁺ channels?

33.2 Reviewing the Concepts

- The resting membrane potential of a neuron is determined by the distribution and relative permeabilities of ions across the plasma membrane (Figure 33.4).
- Diffusion of ions through ion channels occurs as a result of an electrochemical gradient across the membrane (Figure 33.5).
- The Nernst equation gives the equilibrium potential for an ion at any given concentration gradient. Differences between this value and the resting membrane potential determine whether and in which direction an ion will diffuse across a membrane.

33.2 Testing Your Knowledge

1. The concentrations of anion X inside and outside a neuron are 94 mM and 19 mM, respectively. Calculate its equilibrium potential.
   a. 112.5 mV
   b. 1.08
   c. −1.08
   d. 41.7 mV
   e. −41.7 mV

2. The resting membrane potential occurs because
   a. the membrane is leaky to Na⁺ and K⁺.
   b. the concentrations of Na⁺ and K⁺ are unequal across the membrane.
   c. the Na⁺/K⁺-ATPase pump moves two K⁺ into the cell for every three Na⁺ it moves out of the cell.
   d. Both a and b are correct.
   e. Answers a, b, and c are all correct.

33.3 Generation and Transmission of Electrical Signals Along Neurons

Learning Outcomes

1. Distinguish between ligand-gated and voltage-gated ion channels.
2. Compare and contrast graded potentials and action potentials.
3. **SCISKILLS** ▶ Describe the steps involved in the generation of an action potential, beginning with the spread of excitation to the axon hillock.
4. Explain how myelination influences the propagation of an action potential.

Communication between neurons begins when one cell receives a stimulus and in turn sends an electrical signal along its axon via currents generated by ion movements. This signal then influences the next neuron in a pathway. In this section, we will survey the amazing ability of neurons to generate the electrical changes that produce action potentials and see how those potentials are transmitted along an axon.

Signaling by a Neuron Occurs Through Changes in the Membrane Potential

Recall that a neuron is polarized because of the separation of charges across its plasma membrane. Changes in the membrane potential, therefore, are changes in the degree of polarization. **Depolarization** occurs when the cell membrane becomes less polarized—that is, when it becomes less negative inside the cell relative to the outside. When a neuron is stimulated by an electrical or chemical stimulus, one or more types of ion channels open; this may, for example, result in the diffusion of Na⁺ into the cell down its electrochemical gradient, thereby bringing positive charge into the cell. This diffusion of Na⁺ makes the new membrane potential somewhat less negative than the original resting membrane potential. In contrast, **hyperpolarization** occurs when the cell membrane becomes more polarized than at rest. For example, an increase in the rate of diffusion of K⁺ out of the cell would make the charge along the inside of the cell membrane more negative than it normally is when at the resting membrane potential.

Although all cells in an animal's body have a membrane potential, neurons are called excitable cells because they can change their resting membrane potential and generate electrical signals. This is accomplished by another class of ion channels called gated ion channels, because they open and close in a manner analogous to a gate in a fence (Figure 33.6). **Voltage-gated ion channels** open and close in response to changes in voltage (that is, charge difference) across the membrane. **Ligand-gated ion channels** open or close when ligands, such as neurotransmitters, bind to them on their extracellular surface. The opening and closing of voltage-gated and ligand-gated ion channels are responsible for two types of changes in a neuron's membrane potential: graded potentials and action potentials.

Figure 33.6 Examples of gated ion channels. (a) A voltage-gated ion channel allows ions to diffuse into the cell. These channels open or close depending on changes in charge (voltage) across the membrane. (b) A ligand-gated ion channel opens or closes in response to ligand binding. In the example here, the binding of a neurotransmitter opens the channel.

BioConnections: *Are ions the only substances that can move through channels in plasma membranes? Refer back to Figure 5.13 to be reminded of an important molecule that diffuses through membranes via channels.*

Graded Potentials Vary in Size Depending on the Strength of a Stimulus

A **graded potential** is any depolarization or hyperpolarization that varies with the strength of a stimulus. A large change in membrane potential occurs when a strong stimulus opens many ion channels, whereas a weak stimulus causes a small change because only a small number of ion channels are opened (**Figure 33.7**).

Graded potentials occur locally where an electrical or chemical stimulus opens gated ion channels on a region of the plasma membrane of dendrites or the cell body. From this initial site of stimulation, a graded potential spreads a small distance across nearby regions of the plasma membrane. When the stimulus stops, the gated ion channels close, and ion pumps restore the ion concentration gradients. This allows the resting membrane potential to return to normal as Na^+ and K^+ diffuse through their leak channels.

Graded potentials occur on all neurons and are particularly important for the function of sensory neurons, which must distinguish

Figure 33.7 Graded membrane potentials. Within a limited range, the change in the membrane potential occurs in proportion to the intensity of the stimulus. If the membrane potential becomes less negative following a stimulus, the resulting change is called a depolarization. If large enough, a depolarization can elicit an action potential. If the membrane potential becomes more negative, the change is called hyperpolarization. A cell that is hyperpolarized is less likely to elicit an action potential.

between strong and weak stimuli reaching the organism from the environment. In addition, however, graded potentials can act as triggers for the second type of electrical signal, the action potential.

Action Potentials Transmit Electrical Signals Along the Length of Axons

As stated earlier, an **action potential** is the electrical event that transmits a signal along an axon. In contrast to a graded potential, an action potential is always a large depolarization. Once an action potential has been triggered, it occurs in an all-or-none manner. In other words, under normal circumstances it cannot be graded; a cell cannot have a partial action potential. Unlike a graded potential, an action potential is conducted, or propagated, along an axon, regenerating itself as it travels. Action potentials travel rapidly down an axon to the axon terminals, where they initiate a response at a junction with another cell.

Figure 33.8 shows the electrical changes that happen in a localized region of an axon when an action potential is occurring. Voltage-gated Na^+ and K^+ channels are both present in very high numbers from the axon hillock to the axon terminals. An action potential begins when a cell receives a strong stimulus that results in a graded potential that is large enough to spread from the cell body to the axon hillock and depolarize the membrane there. This opens some voltage-gated Na^+ channels, depolarizing the membrane further to a value called the threshold potential (step 2 in Figure 33.8). The **threshold potential** is the membrane potential, typically around -55 to -50 mV, at which sufficient voltage-gated Na^+ channels are open that the all-or-none phase of the action potential begins.

The opening of voltage-gated Na^+ channels involves a change in the conformation of the membrane-spanning region of the channel (Figure 33.8). When the channel opens, Na^+ rapidly diffuses into the cell down its electrochemical gradient (recall that the equilibrium potential

Biology Principle

Biologists Use Models to Test Predictions and Convey Ideas

This figure is a model for an action potential in an axon of a neuron. The model includes a depiction of when different gated ion channels open and close during an action potential. The Modeling Challenge in the figure legend asks you to predict how membrane potential would change in a neuron that was stimulated in the presence of an inhibitor of voltage-gated Na⁺ channels.

Figure 33.8 Changes that occur during an action potential. The movement of ions across the plasma membrane of an axon first depolarizes and then repolarizes the cell. These changes in the membrane potential are caused by the opening and closing of voltage-gated Na⁺ and K⁺ channels. The values given for membrane potential are representative and not necessarily the same in all neurons.

Modeling Challenge: Figure 33.8 is a model of a typical action potential in an axon of a neuron. Note how certain stages of the action potential result from the opening and closing of specific ion channels. In some animals, such as members of the family Tetraodontidae (puffer fish and others), symbiotic bacteria in the animal's microbiome produce a compound called tetradotoxin (TTX) that is stored safely in the animal's liver, gonads, or other organs. It is highly toxic when ingested by other animals, however, and consequently TTX is a defense against predation. TTX acts by binding to and blocking only the voltage-gated Na⁺ channels in the plasma membranes of neurons, thereby preventing the influx of Na⁺ that is normally associated with action potentials. Draw a graphical model of the change, if any, in membrane potential of an axon when a neuron is stimulated to the threshold potential in the presence of TTX. Use the same axes and scales as shown in Figure 33.8.

for Na$^+$ is about +60 mV). This influx of charges further depolarizes the cell, causing even more voltage-gated Na$^+$ channels to open, resulting in the spike in membrane potential that characterizes the action potential. This positive feedback process is so rapid that the membrane potential reaches its peak positive value in less than 1 millisecond!

As the membrane potential becomes more and more depolarized, a second, delayed conformational change in the voltage-gated Na$^+$ channel blocks further flow of Na$^+$ into the cell. This conformational change involves the **inactivation gate** (step 3 in Figure 33.8), a string of amino acids that juts out from the channel protein into the cytosol. The inactivation gate swings into the opening of the channel, thereby preventing any further movement of Na$^+$ through the channel into the cell. Thus, the inactivation gate terminates the depolarization phase of an action potential; this gate does not swing out of the channel until the membrane potential returns to a value near the resting potential.

Voltage-gated K$^+$ channels are also opened by the change in voltage to the threshold potential, but they open more slowly, about the time the Na$^+$ channels become inactivated. When K$^+$ channels open, some K$^+$ ions diffuse out of the cell, moving down their electrochemical gradient, and the membrane potential becomes repolarized. The membrane has a brief period of hyperpolarization while the K$^+$ channels remain open and the membrane potential approaches the equilibrium potential for K$^+$, which is more negative than the resting potential (step 4 in Figure 33.8). The repolarization then closes the voltage-gated K$^+$ channels. The membrane returns to the resting potential as the Na$^+$/K$^+$-ATPase pump restores the original ion concentration gradients (step 5 in Figure 33.8) and Na$^+$ and K$^+$ diffuse only through their leak channels. At this stage, both the Na$^+$ and K$^+$ voltage-gated channels have the ability to reopen if the membrane potential is again depolarized to the threshold potential by a new stimulus.

While the voltage-gated Na$^+$ channels are open, and then while they are inactivated, the membrane at that region is in its **absolute refractory period** (Figure 33.8, step 3), during which time that portion of membrane is unresponsive to another stimulus. As the cell repolarizes, there is a brief period when the inactivation gate is released and the voltage-gated Na$^+$ channels revert to the closed state, but the voltage-gated K$^+$ channels are still open. During this time, the membrane is in its **relative refractory period** and becomes even more polarized than the resting membrane potential (Figure 33.8, step 4). A new action potential may be generated at this time, but only in response to an unusually large stimulus. The reason is that the continued K$^+$ efflux through the open voltage-gated K$^+$ channels would oppose some of the Na$^+$ entry that normally occurs during the beginning of a new action potential. Therefore, the refractory periods place a limit on how many action potentials can be generated in a given time. The refractory periods also allow the brain to distinguish signals arriving from weak (few action potentials/sec) and strong (many action potentials/sec) stimuli; only the latter can "break through" the relative refractory period.

Action Potentials Are Propagated Along the Axon to the Axon Terminals

Let's now consider how action potentials are propagated along an axon from the axon hillock to the axon terminals (Figure 33.9). Each action potential in a membrane results in Na$^+$ entering the cell and diffusing along the membrane. The absolute number of ions that cross the membrane in a single action potential is actually extremely small and does not alter the concentration gradient in a measurable way. Nonetheless, the charges carried with the Na$^+$ are sufficient to open nearby voltage-gated Na$^+$ channels in an adjacent strip of membrane along the axon toward the axon terminals. (Na$^+$ that diffuses in the other direction can be ignored, as that portion of the membrane is in its absolute refractory period.) A new action potential is then generated, which again depolarizes the next strip of membrane, and so on. In this way, the sequential opening of Na$^+$ channels along the axon membrane conducts a wave of depolarization from the axon hillock to the axon terminals.

An action potential can be conducted down the axon at a speed as fast as 100 m/sec or as slow as 1 cm/sec. The speed is determined by two factors: the axon diameter and the presence or absence of a myelin sheath. The axon diameter influences the rate at which incoming ions can spread along the inner surface of the plasma membrane. The flow of ions meets less resistance in a wide axon than it does in a narrow axon. Therefore, in a wider axon, the action potentials propagate faster. The exceptionally large axons of squids and lobsters, for example, conduct action potentials very rapidly, allowing the animals to move quickly when threatened.

Myelination also increases the speed at which action potentials travel along an axon. An insulating myelin sheath reduces the leakage of charge across the membrane of an axon (see Figure 33.1). The exposed nodes of Ranvier are the only areas of myelinated axons that have sufficient numbers of voltage-gated Na$^+$ channels to generate an action potential (**Figure 33.10**). When Na$^+$ ions diffuse into the cell at one node, the charge spreads through the cytosol, causing the Na$^+$ channels at the next node to open, where an action potential is regenerated. This type of conduction is called **saltatory conduction** (from the Latin *saltare*, meaning to leap) because the action potential seems to "jump" from one node to the next. In reality, action potentials do not "jump" from node to node but instead are regenerated anew at each node.

Saltatory conduction speeds up the conduction process because it takes less time for charge to travel from node to node, regenerating new action potentials only at the nodes, than it would if each tiny strip of membrane generated action potentials all along the length of the axon. Saltatory conduction is an adaptation that permits fast propagation of signals without a requirement for increasing the diameter of axons. It occurs primarily in long axons in the PNS and CNS; very short neurons such as interneurons are typically not myelinated.

33.3 Reviewing the Concepts

- Gated ion channels enable a cell to rapidly change its membrane potential and form graded potentials (Figures 33.6, 33.7).
- Graded potentials that reach the threshold potential will trigger an action potential, an event that carries an electrical signal along an axon. Axon diameter and myelination influence the speed at which an action potential propagates (Figures 33.8, 33.9, 33.10).

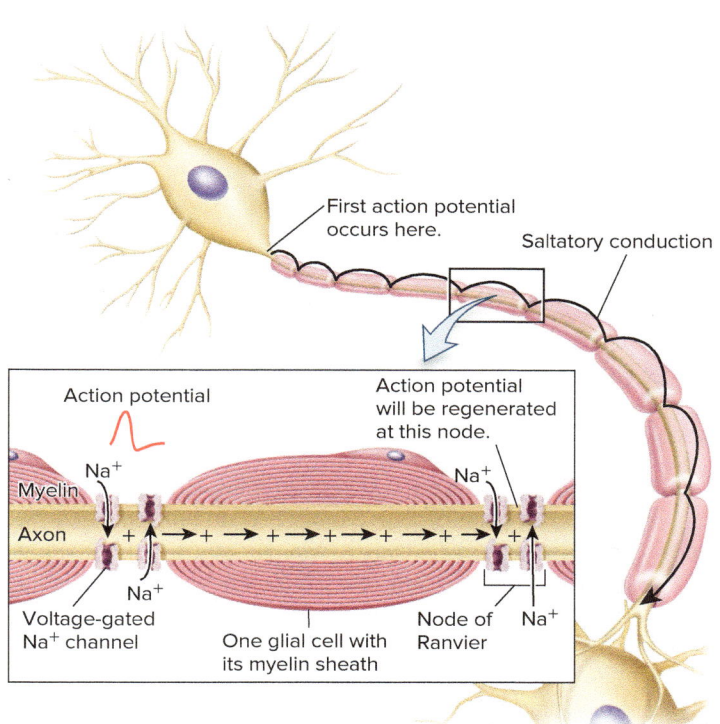

Figure 33.10 Saltatory conduction along a myelinated axon. Action potentials are generated only at the nodes of Ranvier, which lack a surrounding sheath of myelin.

Concept Check: *What is the main advantage of saltatory conduction?*

33.3 Testing Your Knowledge

1. Which of the following statements is *false*?
   a. Graded potentials may die out or trigger an action potential.
   b. Action potentials arise at the axon hillock and move toward the axon terminals.
   c. The diameter of an unmyelinated axon is directly related to the speed at which action potentials travel along its length.
   d. Saltatory conduction occurs in all axons.
   e. Action potentials are all-or-none, but graded potentials can have different sizes.

2. Action potentials
   a. require a threshold potential around −55 to −50 mV in order to occur.
   b. have many different shapes and sizes, depending on how a cell was stimulated.
   c. occur only in myelinated neurons.
   d. occur whenever a neuron is sufficiently hyperpolarized.
   e. are limited in their frequency by inactivation gates of K^+ channels.

Figure 33.9 Propagation of the action potential along an axon. All Na^+ channels shown refer to voltage-gated channels.

1. Unstimulated neuron; Na^+ channels are closed.
2. Threshold potential is reached at axon hillock. Na^+ channels open, and an action potential is generated. Na^+ diffuses along the axon.
3. Entry of Na^+ depolarizes membrane and opens channels toward the axon terminal, generating new action potentials. Previously opened Na^+ channels are inactivated.
4. Continued entry of Na^+ depolarizes the membrane and opens channels farther down the axon. Previously opened Na^+ channels are inactivated. Previously inactivated Na^+ channels switch to a closed state when K^+ channels (not shown) open and restore the resting potential.
5. Process continues and action potential moves down axon.

KEY: Closed Na^+ channels | Open Na^+ channels | Inactivated Na^+ channels

33.4 Communication at Synapses

Learning Outcomes
1. Describe the structural features of a synapse.
2. **SCISKILLS** ▶ Predict the effects of excitatory postsynaptic potentials (EPSPs) and inhibitory postsynaptic potentials (IPSPs) on the ability of neurons to generate an action potential.
3. List the classes of neurotransmitters, and briefly describe their generalized functions.
4. Describe the two general types of postsynaptic membrane receptors.

Neurons communicate with other cells at a **synapse,** which is a junction where an axon terminal meets another neuron or other cell. In this section, we will examine the structure and function of different types of synapses, including the role that neurotransmitters have in neuronal communication.

Synapses May Be Electrical or Chemical

A synapse includes an axon terminal of the neuron that is sending the signal, the nearby plasma membrane of the receiving cell, and often a synaptic cleft, or extracellular space between the two cells. By definition, the presynaptic cell sends the signal, and the postsynaptic cell receives it. A given cell may be presynaptic to one cell and postsynaptic to another (**Figure 33.11**).

Figure 33.11 Presynaptic and postsynaptic cells. The black arrows show the direction of signal transmission from one neuron to the next. Note that neuron 2 is postsynaptic with respect to neuron 1 and presynaptic with respect to neuron 3.

Two types of synapses occur in animals: electrical and chemical. In the first type, the **electrical synapse,** electric current passes directly from the presynaptic to the postsynaptic cell via gap junctions (refer back to Figure 5.18). The most well-studied examples occur in certain aquatic invertebrates. In those animals, electrical synapses occur where a group of neurons must fire extremely quickly and synchronously, such as when a crayfish must coordinate a number of muscles to dart away from a predator. An advantage of electrical synapses is their great speed of transmission.

The second type of synapse is a **chemical synapse,** in which a neurotransmitter is released from an axon terminal into the synaptic cleft. The neurotransmitter diffuses to the membrane of the postsynaptic cell, where it interacts with receptor proteins and triggers a change in the membrane potential of the postsynaptic cell. As a result of the time required for these events, chemical synapses are slower than electrical ones. However, unlike electrical synapses, chemical synapses allow for complex modulation of the response of the postsynaptic cell, as we will see next.

In Chemical Synapses, Neurotransmitters Trigger Changes in Membrane Potential of Postsynaptic Cells

Figure 33.12 shows the steps that occur when two neurons communicate via a chemical synapse. The cytosol of an axon terminal of a presynaptic cell contains vesicles—small, membrane-enclosed packets, each containing thousands of molecules of neurotransmitters. The membranes of axon terminals contain voltage-gated Ca^{2+} channels. When an action potential arrives at an axon terminal, these channels open, allowing Ca^{2+} ions to diffuse down their electrochemical gradient into the cell. Calcium ions bind to a protein associated with the vesicle membrane. This triggers exocytosis, in which the vesicle fuses with the presynaptic axon terminal membrane, thereby releasing its neurotransmitter molecules into the synaptic cleft. The neurotransmitter molecules then diffuse across the cleft and bind to receptor proteins in the postsynaptic cell membrane.

The binding of neurotransmitter molecules to their receptors opens or closes ion channels, thereby changing the membrane potential of the postsynaptic cell. Some neurotransmitters are called excitatory, because they depolarize the postsynaptic membrane; the response is an **excitatory postsynaptic potential (EPSP).** An EPSP is a graded potential that brings the membrane potential closer to the threshold potential that would trigger an action potential. Two ways in which an EPSP can be generated are by the opening of Na^+ channels in a postsynaptic membrane and by the closing of K^+ channels. In both cases, positive charges accumulate inside the cell. Conversely, an inhibitory neurotransmitter usually hyperpolarizes the postsynaptic membrane, producing an **inhibitory postsynaptic potential (IPSP).** An IPSP is a graded potential that brings the postsynaptic cell's membrane potential farther away from the threshold for producing an action potential. A common way in which an IPSP is formed is by the opening of Cl^- channels; the electrochemical gradient for Cl^- is such that it usually favors diffusion of Cl^- into neurons, thus bringing negative charge into the cell. Hyperpolarization is a major way in which neurons are inhibited; see, for example, the motor neuron in the reflex arc shown in Figure 33.3.

Figure 33.12 Structure and function of a chemical synapse. In response to an action potential, Ca^{2+} enters the presynaptic neuron axon terminal. This results in vesicle fusion with the plasma membrane, releasing neurotransmitter molecules into the synaptic cleft. The neurotransmitter molecules then bind to receptors in the plasma membrane of the postsynaptic cell. This causes ligand-gated ion channels to open or close, which in turn changes the membrane potential of the postsynaptic cell. In this example, Na^+ channels open, which will create an EPSP in the postsynaptic cell.

Concept Check: What is the benefit of having degrading enzymes in the synaptic cleft break down neurotransmitter molecules?

To end the synaptic signal, neurotransmitter molecules in the synaptic cleft are broken down by enzymes or transported back into the terminal of the presynaptic cell and repackaged into vesicles for reuse. The latter event, called reuptake, is an efficient mechanism for recapturing unused neurotransmitters that were released into the synaptic cleft. These mechanisms also prevent a postsynaptic cell from being chronically stimulated.

The effect of activating a single synapse is usually far too weak to elicit an action potential in the postsynaptic neuron. A neuron may receive inputs from many synapses on its dendrites and cell body. Some of these synapses may release neurotransmitters onto the neuron at the same or nearly the same time. When two or more EPSPs are generated at one time in a postsynaptic cell, their depolarizations sum together. The resulting larger depolarization may bring the membrane potential at the axon hillock to the threshold potential, initiating an action potential. Conversely, if EPSPs and IPSPs arrive together at a postsynaptic cell, the two types of signals may cancel each other out, and no action potential is elicited. This is a major advantage of chemical synapses; responses of postsynaptic cells can be modified to reflect the sum of many different inputs arriving at once or in close succession. It is also a way in which an animal's brain can determine the relative importance of competing stimuli. Imagine a hungry fish confronted with a worm and a predator. IPSPs sent to hunger-sensitive neurons can override the activation of those cells, whereas other neurons that respond to danger cues may receive only EPSPs, so the fish ignores the worm and seeks safety.

Several Classes of Neurotransmitters Elicit Responses in Postsynaptic Cells

Neuroscientists have identified more than 100 different neurotransmitters in animals. Generally, neurotransmitters are categorized by structure. The changing balance between excitatory and inhibitory neurotransmitters controls the state of nervous system circuits at any one time. To understand how neurotransmitters work, imagine driving a car with one foot on the gas pedal (which you can think of here as an EPSP) and one foot on the brake (an IPSP). To speed up, you can press down on the gas pedal, ease up on the brake, or both, whereas to slow down, you can do the opposite. Although you would never drive a car like this, all nervous systems operate in this way, with combined excitatory and inhibitory actions of neurotransmitters. This creates opportunities to fine-tune signaling between neurons.

The different types of neurotransmitters found in animals fall into the following classes:

- **Acetylcholine.** Acetylcholine is released at synapses where a neuron contacts skeletal or cardiac muscle. It is also released at synapses within the brain and on many gland cells. Depending on where it acts, it may be excitatory or inhibitory. For example, it stimulates contraction of skeletal muscle but inhibits cardiac muscle activity.
- **Biogenic amines.** The biogenic amines are compounds containing amine groups that are derived from amino acids. They include the catecholamines—dopamine, norepinephrine, and epinephrine—and serotonin and histamine. The catecholamines are formed from the amino acid tyrosine, serotonin from tryptophan, and histamine from histidine. They have widespread physiological effects such as control of blood pressure, metabolism, mood, attention, behavior, and learning.
- **Amino acids.** The amino acids glutamate, aspartate, glycine, and γ-aminobutyric acid (GABA) function as neurotransmitters. Glutamate is the most widespread excitatory neurotransmitter found in animal nervous systems, and GABA is the most common inhibitory neurotransmitter.
- **Neuropeptides.** Neuropeptides are short polypeptides that can be excitatory or inhibitory. Neuropeptides are often called neuromodulators, because they can modulate the response of a postsynaptic neuron to other neurotransmitters. For example, a neuropeptide released from a presynaptic cell may stimulate an increase in the number of receptors for another neurotransmitter on the postsynaptic cell, which makes the cell more responsive to that neurotransmitter.
- **Gases.** Certain gases such as nitric oxide (NO) are produced locally in many tissues, including some neurons where they function as neurotransmitters. Unlike other neurotransmitters, they are not sequestered into vesicles but are released from terminals by diffusion. They also differ by not acting on postsynaptic receptors but rather by diffusing into the postsynaptic cell and acting on intracellular enzymes. These enzymes may alter membrane potential in a postsynaptic neuron, or in non-neuronal cells they may affect such actions as smooth muscle contraction.

Postsynaptic Membrane Receptors Determine the Type of Response to Neurotransmitters

As you have just learned, in some cases the same neurotransmitter can have both excitatory and inhibitory effects depending on the cell upon which it acts. The response of the postsynaptic cell depends on the type of receptor present in the postsynaptic membrane. The two major types of postsynaptic membrane receptors are ionotropic and metabotropic, and many neurotransmitters act on both (**Figure 33.13**).

Figure 33.13 The two major categories of postsynaptic receptors. **(a)** Ionotropic receptors are ligand-gated channels with several subunits. Neurotransmitters bind to ionotropic receptors and directly open ion channels in the membrane. **(b)** Metabotropic receptors are G-protein-coupled receptors. Neurotransmitters bind to metabotropic receptors and initiate a signaling pathway that typically opens or closes nearby ion channels (refer back to Figures 8.7 and 8.13).

Ionotropic receptors are ligand-gated ion channels that open when they bind a neurotransmitter (Figure 33.13a). When neurotransmitter molecules bind to these receptors, the conformation of the receptors changes, allowing ions to flow through the channels to cause an EPSP or an IPSP. Acetylcholine and amino acids bind to ionotropic receptors. Ionotropic receptors have a quaternary structure (see Chapter 3) and are composed of multiple subunits that form the receptor's channel.

Metabotropic receptors (Figure 33.13b) are G-protein-coupled receptors (GPCRs) (refer back to Figures 8.7 and 8.13). They do not form a channel but instead are coupled to an intracellular signaling pathway that initiates changes in the postsynaptic cell. A common response is the phosphorylation of specific ion channels in the plasma membrane. In Chapter 34, we will see that metabotropic receptors are particularly important in activating sensory cells that respond to visual and other stimuli.

33.4 Reviewing the Concepts

- In an electrical synapse, an electric current is conducted from one cell to another via gap junctions. In a chemical synapse, a neurotransmitter carries the signal from the presynaptic to the postsynaptic cell. The response in the postsynaptic cell at a chemical synapse may be inhibitory (IPSP) or excitatory (EPSP). Many EPSPs generated at one time can sum together and bring the membrane potential to the threshold potential, initiating an action potential (Figures 33.11, 33.12).
- Chemical classes of neurotransmitters found in animals include acetylcholine, biogenic amines, amino acids, neuropeptides, and gaseous neurotransmitters. The two types of receptors for neurotransmitters are ionotropic and metabotropic receptors (Figure 33.13).

33.4 Testing Your Knowledge

1. A certain medical condition is associated with too little serotonin being released by neurons of the brain. Which of the following could theoretically reverse this problem by increasing how much serotonin is present in synaptic clefts?
   a. eating foods high in tyrosine content
   b. taking a drug that binds to and inhibits serotonin receptors on the postsynaptic cell
   c. taking a drug that prevents reuptake of serotonin
   d. taking a drug that prevents influx of Ca^{2+} into the presynaptic cell
   e. taking a drug that stimulates serotonin-degrading enzymes on the postsynaptic cell membrane
2. Which of the following is a true statement?
   a. An EPSP may be caused by opening K^+ channels in a postsynaptic membrane.
   b. An IPSP may be caused by closing K^+ channels in a postsynaptic membrane.
   c. It is possible for certain neurotransmitters to be either excitatory or inhibitory in different cells.
   d. Metabotropic receptors are also ligand-gated ion channels.
   e. One way for a neurotransmitter to depolarize a postsynaptic cell is to activate ionotropic receptors that are linked to Cl^- channels.

33.5 Evolution and Development of Nervous Systems

Learning Outcomes

1. List some different types of nervous systems found in animals.
2. Describe the three basic divisions of the vertebrate brain.
3. Discuss anatomical changes that accompanied the evolution of brain complexity.

Studying the evolution and development of animal nervous systems will help you understand how particular nervous systems are adapted to different functions. The characteristics of an animal's nervous system determine the behaviors that an animal displays. In this section, we will survey the nervous systems of animals and examine the brains of vertebrates in greater detail.

The Evolution of Nervous Systems Enabled Animals to Sense and Respond to Changes in the Environment

Precisely when nervous systems first arose and whether or not the nervous systems of most or all animals can be traced back to a common ancestor are questions of active investigation. For example, recent genetic studies have uncovered remarkable similarities in the genes that regulate neuronal development in bilaterally symmetric animals. Those studies suggest that the patterning of nervous system development in bilateral animals can be traced to a common ancestor that lived more than 500 mya!

EVOLUTIONARY CONNECTIONS

Animals Evolved Increasingly Complex Nervous Systems

Today, all multicellular animals except sponges have a nervous system. Recently, however, researchers have discovered that even sponges express dozens of genes that are similar to genes expressed in vertebrate neurons. The functions of the sponge genes are uncertain, but the proteins encoded by them interact with each other in ways that are similar to certain vertebrate synaptic proteins.

The simplest nervous systems found in animals are the **nerve nets** present in the radially symmetric cnidarians (**Figure 33.14a**). The neurons in this type of nervous system are arranged in a network of connections between the inner and outer body layers of the animals. Activation of neurons in any one region leads to excitation that spreads in all directions at once, activating contractile cells to contract. This causes the animal to move or swim, enabling it to capture food or evade a predator. Recent research has identified regions of specialized function in the nerve nets of some cnidarians, such as local sensory neurons in the outer body wall. These findings push the origins of specialized nervous system structures and functions further back in evolutionary time than previously thought.

Sea stars and other echinoderms have a slightly more complex nervous system, with a nerve ring around the mouth that is connected to radial nerves extending into the limbs (**Figure 33.14b**). This arrangement allows the mouth and limbs to operate independently.

During the evolution of animals, more complex body types have been associated with a concentration of sense organs at the anterior end of the body, along with an increasingly complex brain that controls sensory and motor functions of the entire body. Within the brain, neuronal pathways provide the integrative functions necessary for an animal to make more sophisticated responses to its environment. Brains are found in all vertebrates and most invertebrates, and they are composed of more than one anatomical and functional region with considerable synaptic complexity.

The first animals to have evolved a brain as defined by regionalization and synaptic complexity are the platyhelminthes such as *Planaria*. In these animals, different regions of the brain appear to integrate inputs from sense organs, such as the eyes, and control motor outputs such as movements involved in swimming (**Figure 33.14c**).

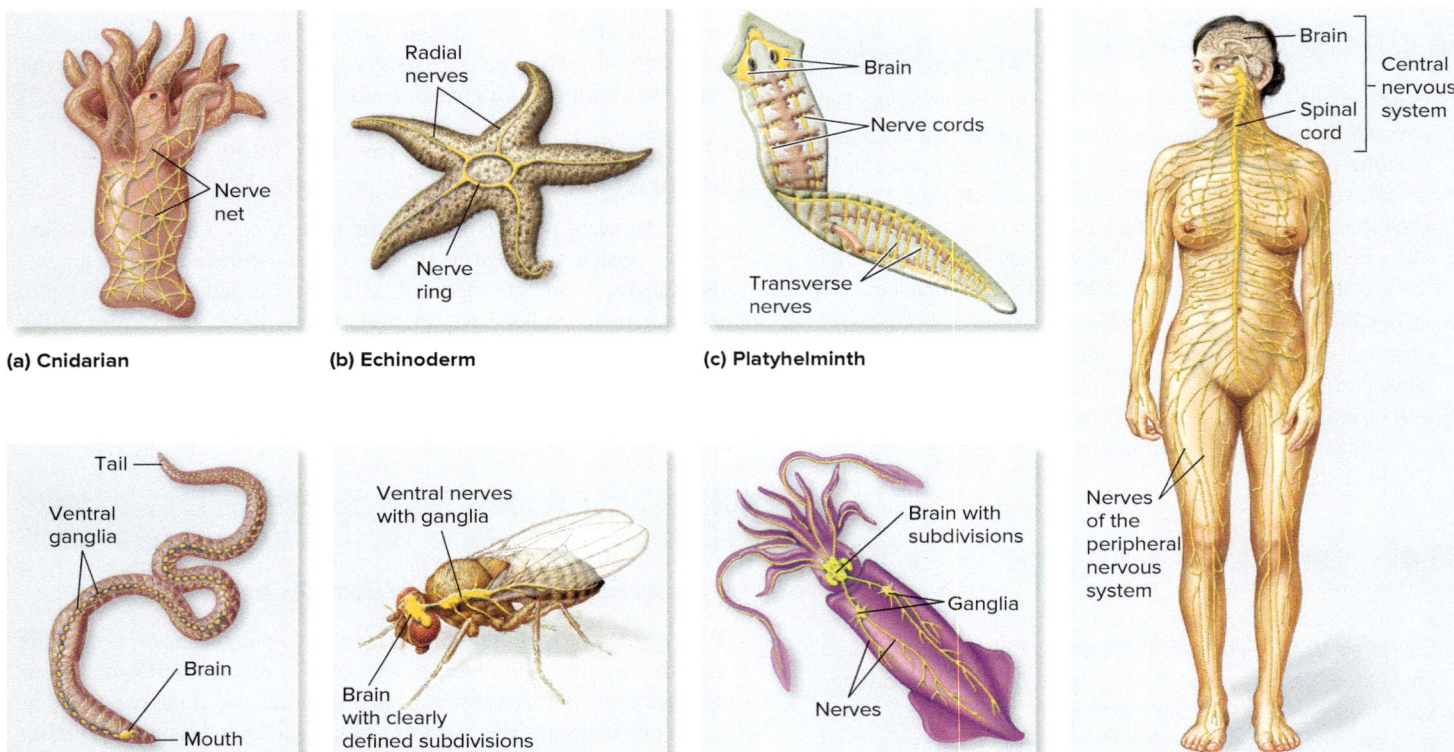

Figure 33.14 Representative nervous systems throughout the animal kingdom.

BioConnections: Nervous systems are one of the defining features of animals. In Chapter 26, you learned that several other features define animals and distinguish them from other living organisms. What are some of these features?

Two **nerve cords** extend along the ventral surface of the animal and are connected to each other by short nerves.

In annelids, ganglia and nerves are present at each body segment, where they coordinate local sensory and motor activities (**Figure 33.14d**). **Ganglia** are collections of neuron cell bodies with limited processing ability, limited synapses, and few to no subdivisions like those found in a brain. Ganglia are present in most animals and often coordinate local signaling in a body part.

In *Drosophila* (**Figure 33.14e** and see the chapter-opening photo), the brain has several subdivisions with distinct, well-defined functions, such as a region devoted to learning and memory. Additional limited processing of sensory and motor functions occurs within ganglia.

In the simpler types of mollusks, such as snails, the nervous system has features similar to those of annelids. Some mollusks, such as the squid and octopus, have brains with well-developed subdivisions that allow these animals to coordinate the complex visual and motor behaviors necessary for their predatory lifestyle (**Figure 33.14f**).

In the embryos of chordates, a dorsally located nerve cord is present that, in vertebrates, develops into the brain and spinal cord. Together, the brain and spinal cord constitute the CNS in vertebrates (**Figure 33.14g**). Nerves from the PNS route information into and out of the CNS at separate segments along the spinal cord.

Brains of Vertebrates Have Three Basic Divisions

As evident from **Figure 33.15**, the vertebrate brain develops as basically a linear structure with expansions along its length. Fossils of jawless fishes that lived 400 mya show that their brains were organized into the three basic divisions that have been retained in all modern vertebrates: the **hindbrain, midbrain,** and **forebrain.**

The vertebrate hindbrain coordinates many basic reflexes, motor functions, and other bodily functions, such as breathing, that maintain the normal homeostatic processes of an animal. The midbrain processes several types of sensory inputs, including vision, olfaction (smell), and audition (hearing), and has a role in alertness. The forebrain initiates motor functions and processes sensory inputs. Many forebrain functions are attributed to a region called the **cerebrum,** the surface layer of which is especially developed in mammals and is called the **cerebral cortex.** As you will learn later, in humans this region is critical for thought, reasoning, and learning, among other functions. Other structures of the forebrain lie beneath the cerebral cortex and will be discussed later in this chapter.

Evolution of Increased Brain Complexity Involved a Larger, Highly Folded Cerebrum

With the increased complexity of the brain, the size of the cerebrum also increased, making up a greater proportion of the brain. In mammals, many of the important functions of the cerebrum are carried

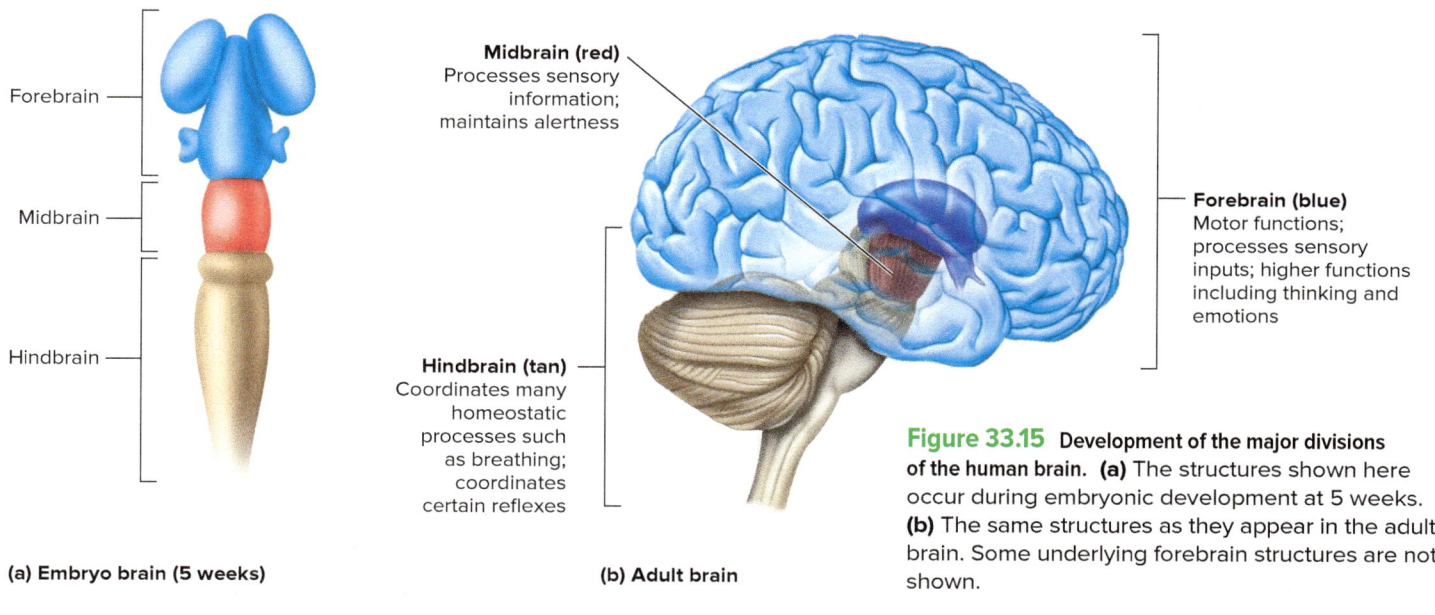

Figure 33.15 Development of the major divisions of the human brain. (a) The structures shown here occur during embryonic development at 5 weeks. (b) The same structures as they appear in the adult brain. Some underlying forebrain structures are not shown.

out by the cerebral cortex. Therefore, the increased complexity of the brain in these animals is also correlated with an increased surface area of the cerebral cortex. During the evolution of mammals, this increase in surface area occurred more rapidly than an expansion in the size of the skull. How could this occur? The answer is that the external surface of the cerebral cortex in animals with increasingly complex brains is highly convoluted, forming many folds and grooves. Compare the relatively smooth-looking surface of the cerebral cortex of a rat, for example, with the highly folded one of the human in **Figure 33.16**.

Brain mass and the amount of folding are correlated with more complex behaviors. Greater size and increased folding provide a larger number of neurons and synapses and make more surface area available for cell-to-cell signaling. This facilitates greater processing and interpretation of information. Even so, it would be wrong to assume that an animal with a small brain is profoundly limited in its behavioral repertoire. A bat with a 0.9-g brain and an elephant with a 2,500-g brain can both perform an astounding variety of complex behaviors, such as navigating across great distances and maintaining sophisticated social hierarchies with other members of their species.

33.5 Reviewing the Concepts

- Nervous systems in animals range from nerve nets in cnidarians to increasingly complex systems with nerve cords and ganglia and a brain (Figure 33.14).
- In all vertebrates, the three major divisions of the brain are the hindbrain, midbrain, and forebrain (Figure 33.15).
- Increased mass and folding of the brain's cerebral cortex allow for expansion of regions associated with conscious thought, reasoning, and learning (Figure 33.16).

Biology Principle

Structure Determines Function

The increased folding of the cerebral cortex of certain mammals increases the surface area of this part of the brain, allowing for more neuronal connections and complex behaviors, including conscious thought.

Figure 33.16 The degree of cerebral cortex folding in different mammalian species. The brains are not shown to scale.

33.5 Testing Your Knowledge

1. Match the nervous system feature with the animal.
   - nerve net
   - two ventral nerve cords
   - nerve ring and radial nerves
   - brain with several well-defined subdivisions
   - sea star (echinoderm)
   - cnidarian
   - *Planaria*
   - *Drosophila*

2. Match the nervous system structure with its function(s).
   - cerebral cortex
   - midbrain
   - ganglia
   - hindbrain
   - control of homeostatic functions such as breathing
   - thinking and learning
   - local signaling in a body part
   - sensory processing and alertness

33.6 Structure and Function of the Nervous Systems of Humans and Other Vertebrates

Learning Outcomes

1. Outline the anatomical organization of the vertebrate nervous system.
2. Describe the organization of the peripheral nervous system.
3. Distinguish between the somatic and autonomic branches of nervous systems.
4. Briefly describe the major structures and functions of the human brain.

The nervous systems of vertebrates are amazingly complex—in humans, the brain alone has over 100 billion neurons and an even greater number of glial cells. Moreover, complexity is defined by more than just numbers of cells. Within the brain, for example, are enormous numbers of connections between neurons—a single neuron in the cerebellum may have as many as 100,000 synapses with other cells! In this section, we will examine the structure and function of the vertebrate nervous system, with special attention to the human nervous system.

Information Is Conveyed Within and Between the CNS and PNS by Nerves and Tracts

In vertebrates, the CNS and PNS are connected anatomically and functionally by nerves (**Figure 33.17**). Nerves that enter and exit the brain are called **cranial nerves;** those that enter and exit through the spinal cord are called **spinal nerves.**

Information may flow in both directions within nerves between the CNS and PNS. For example, suppose you accidentally lean against a newly painted fence. Neuronal endings in your skin, which are part of the PNS, transmit tactile (touch) information directly into the spinal cord via spinal nerves. From there, the information travels to your brain, where the sensation is analyzed and identified as something sticky. Signals are sent from your brain, down your spinal cord, and out through motor neurons in spinal nerves to your muscles, causing you to move away.

Information is also conveyed from region to region within the CNS. In that case, axons occur in bundles known as **tracts.** One of the most obvious characteristics of the CNS is that some parts look white, and others appear gray, leading to the terms white matter and gray matter (**Figure 33.18**). The white matter gets its appearance from myelin; it consists of myelinated axons bundled together in large numbers to form tracts. The gray matter is darker in appearance and consists of neuronal cell bodies, dendrites, and unmyelinated axons. In the spinal cord, for example, the gray matter is located in the center and forms two dorsal horns and two ventral horns (Figure 33.18). Each dorsal horn connects to a dorsal root, which is part of a nerve. Dorsal roots receive incoming information from sensory neurons of the PNS. The ventral horn connects to the ventral root, which transmits outgoing information through motor neurons.

Unlike the PNS, the CNS is encased in protective structures, including bone (the skull and vertebrae) and sheathlike membranes called meninges. The meninges are protective structures encasing a watery solution called cerebrospinal fluid, which absorbs physical shocks to the CNS that result from sudden movements, injuries, or blows to the head.

The PNS Consists of the Somatic and Autonomic Nervous Systems

The PNS of vertebrates consists of two major functional and anatomical divisions: the somatic nervous system and the autonomic nervous system.

Somatic Nervous System The primary function of the **somatic nervous system** is to sense the external environment and activate skeletal muscles. The afferent sensory neurons of the somatic nervous system receive stimuli, such as heat, light, odors, chemicals (in food), sounds, and touch, and transmit signals to the CNS. Efferent motor neurons of the somatic nervous system activate skeletal muscles. The cell bodies of these motor neurons are located within the CNS. The axons from these cells leave the spinal cord and project directly onto skeletal muscle cells, where they release acetylcholine at the synapse. Acetylcholine stimulates the muscle cells to contract.

Autonomic Nervous System The **autonomic nervous system** regulates homeostasis and organ function. For example, it is involved in regulating heart rate, blood pressure, glucose homeostasis, and digestion. For the most part, the autonomic nervous system is not subject to voluntary control. We usually cannot consciously change our blood glucose concentration, for example.

The efferent pathways of the autonomic nervous system involve two neurons. The cell body of the first neuron is within the CNS and synapses on a second neuron in ganglia outside the spinal cord; these ganglia, therefore, are part of the PNS. This second neuron sends its axon to smooth muscles, cardiac muscle, and glands.

Biology Principle

Populations of Organisms Evolve from One Generation to the Next and Are Related by an Evolutionary History

The similarities between the nervous systems of these two very different-looking animals point to a common ancestor.

(a) The human nervous system

(b) The amphibian nervous system

Figure 33.17 Organization of the vertebrate nervous system. The CNS consists of the brain and spinal cord, both of which are encased in bone (not shown). The PNS includes cranial nerves, ganglia, and spinal nerves, which carry information to and from the CNS, and many other neurons throughout the body. Note the similarities between two widely divergent vertebrates, **(a)** humans and **(b)** frogs.

The motor nerves of the autonomic system are subdivided into the sympathetic and parasympathetic divisions (**Figure 33.19**). Both divisions typically act on the same organs and usually have opposing actions. The **sympathetic division** is responsible for rapidly activating systems that prepare the body for danger or stress. Imagine, for example, the physiological responses that would occur if you were hiking and came upon a dangerous animal. You would react instinctively to prepare yourself to confront (fight) or avoid (flight) a potentially life-threatening situation. This is the **fight-or-flight response,** and it is found to varying degrees in all vertebrates. It is characterized by increased heart rate; stronger pumping action of the heart; relaxed airways (in humans and many other air breathers) for taking fast, deep breaths; inhibition of digestive activity; increased blood flow to skeletal muscles; and increased secretion by the liver of glucose into

Figure 33.18 Gray matter and white matter in the CNS. The gray matter is composed of groups of cell bodies, dendrites, and unmyelinated axons. The white matter consists of tracts of myelinated axons.

Biology Principle

Living Organisms Maintain Homeostasis

The autonomic nervous system provides an excellent illustration of a major way in which homeostasis is maintained. Many organs in the body are controlled in opposite ways by the two divisions of this branch of the nervous system. Therefore, the function of these structures can be modulated in two directions; for example, heart rate can be increased or decreased to match an animal's immediate metabolic requirements.

Figure 33.19 Sympathetic and parasympathetic divisions of the autonomic nervous system. For simplicity, only some of the major functions of each division are shown in this figure. The sympathetic and parasympathetic divisions tend to have opposing effects, and most parts of the body receive inputs from both divisions. Neurons from the sympathetic division make connections with a chain of ganglia, many of which are alongside the spinal cord. Neurons of the parasympathetic division make connections in ganglia near or in their targets (for clarity, the ganglia are shown with only one representative synapse).

Figure 33.20 Major structures of the human brain. An overview of the human brain, showing several internal structures. The epithalamus is obscured by the limbic system in this illustration. Only a portion of the limbic system is shown. The basal nuclei are not shown for clarity; they are located beneath the cerebral cortex in the region of the limbic system and thalamus. The midbrain, pons, and medulla oblongata collectively constitute the brainstem.

Concept Check: Are the three major divisions of hindbrain, midbrain, and forebrain unique to humans? To mammals in general?

the blood. All of these actions are mediated in part by the release of norepinephrine from axon terminals of neurons of the sympathetic nervous system.

The **parasympathetic division** of the autonomic nervous system is involved in maintaining and restoring bodily functions. It is particularly active during restful periods or after a meal, which is why it is sometimes said to mediate the **rest-and-digest response.** Neurons of the parasympathetic division promote digestion and absorption of food from the intestines, slow the heart rate, and decrease the amount of fuel supplied to the blood from the liver and adipose tissue. These effects are all mediated by acetylcholine, the neurotransmitter of the parasympathetic division.

The Human Hindbrain Is Important for Homeostasis and Essential Bodily Functions

Let's turn our attention to a more detailed analysis of the structure and function of the human brain (**Figure 33.20**). We will begin with the three major structures of the hindbrain, which is the evolutionarily oldest part of the brain and contains structures that control the basic processes that sustain life.

Medulla Oblongata The **medulla oblongata** regulates numerous homeostatic processes and, so, is essential for life. It is involved in the control of heart rate, blood pressure, breathing, digestion, glucose balance, and swallowing, among many other functions.

Cerebellum The **cerebellum** sits dorsal to the medulla oblongata and receives sensory inputs from the cerebral cortex. It also receives inputs from the spinal cord and inner ears that convey information about the position of the limbs and head, thereby helping to maintain balance and coordinate hand-eye movements. In addition, the cerebellum synchronizes fine motor activities such as texting or touching the fingers to the tip of the nose with eyes closed. When the cerebellum is damaged or injured, such as in an accident, a person finds it difficult to maintain balance and coordination.

Pons The **pons** sits anterior to the medulla oblongata and ventral to the cerebellum. Large tracts pass through the pons into and out of the cerebellum, which allows the pons to serve as a relay between the cerebellum and other areas of the brain. In addition to this integrative motor function, the pons has a very important role in regulating the unconscious control of breathing.

The Midbrain Processes Sensory Inputs and Maintains Alertness

The human midbrain lies anterior to the pons. It receives sensory inputs from visual, olfactory, and auditory pathways. It supplies tracts that pass this information to other parts of the brain for further processing and interpretation. As one example, the midbrain is responsible for activating neural reflexes that change the diameter of the pupil of the eye in response to a change in the amount of ambient light.

Together, the midbrain, medulla oblongata, and pons constitute the **brainstem.** All three parts of the brainstem contain nuclei (collections of many neurons in one location that have a similar function) that form a network of neuronal connections called the **reticular formation.** Axons from the brainstem reticular formation nuclei travel in tracts that extend throughout the CNS. The reticular formation controls consciousness, alertness, sleep, arousal, and essential functions such as regulation of the heart. Because of the importance of the brainstem's functions, damage to it is catastrophic and may result in coma or death.

The Human Forebrain Is Responsible for Motor, Sensory, and Higher Functions

The human forebrain consists of the **diencephalon**—which in turn encompasses the thalamus, hypothalamus, and epithalamus—and the cerebrum. The cerebrum includes the cerebral cortex, basal nuclei, and limbic system (**Figure 33.21**; also see Figure 33.20).

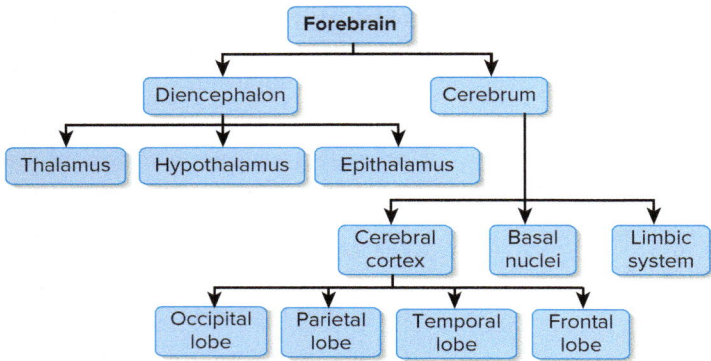

Figure 33.21 Relationships among major structures that make up the human forebrain.

Concept Check: Is the occipital lobe part of the diencephalon or cerebrum?

Figure 33.22 The four lobes of the cerebral cortex as seen on the right hemisphere, as well as some major functions they control.

Diencephalon (Thalamus, Hypothalamus, and Epithalamus)

The **thalamus** has a major role in relaying sensory information to appropriate parts of the cerebrum. It receives input from all sensory systems except the olfactory system. One type of processing performed by neurons in the thalamus is that of filtering out sensory information in a way that allows us to pay attention to important cues while temporarily ignoring less important ones. This filtering mechanism begins in the reticular formation, which relays information about sensory inputs to the thalamus. Together, these brain regions permit selective attention to certain stimuli. A good example of this is a new parent's ability to sleep through a thunderstorm but awaken immediately to the cry of a baby in the next room.

The **hypothalamus**, located beneath the thalamus at the bottom of the forebrain, controls functions of the digestive and reproductive systems, body temperature, circadian rhythms, and certain drives related to homeostasis such as appetite and thirst.

The **epithalamus** is a collection of structures that have several functions in the interpretation of olfactory and visceral inputs. It also includes the pineal gland, an endocrine gland that secretes a hormone called melatonin into the blood. In humans, melatonin appears to function primarily in the control of daily cycles such as our sleep/wake cycle.

Cerebrum (Cerebral Cortex, Basal Nuclei, and Limbic System)

The cerebrum is divided into two halves (hemispheres) connected to each other by a large tract called the corpus callosum.

The cerebral cortex is the surface layer that covers the cerebrum and comprises four lobes in each hemisphere: the frontal, parietal, occipital, and temporal lobes (**Figure 33.22**). A few general functions of each lobe are listed here:

- The **frontal lobe** is important for the conscious motor commands that initiate movement. It is also chiefly responsible for such higher functions as making decisions, controlling impulses, planning for the future, and exhibiting judgment; some aspects of short-term memory; and conscious thought and social awareness.

- The **parietal lobe** receives and interprets sensory input from somatic pathways, including touch from the surface of the body. This lobe also has an important role in our ability to use visual cues to orient ourselves in space.

- The **occipital lobe** receives visual information and controls many aspects of visual perception and color recognition.

- The **temporal lobe** is necessary for language, hearing, and some types of memory.

The **basal nuclei** are a group of forebrain nuclei that surround the thalamus and lie beneath the cerebral cortex. The basal nuclei are involved in planning, learning, and fine-tuning movements. They also function via a complex circuitry to smoothly initiate or inhibit movements. Parkinson's disease is a relatively common neurological disorder that affects the basal nuclei and results in numerous and severe motor difficulties.

The **limbic system** is a collection of evolutionarily older cerebral structures that form an inner layer near the base of the forebrain (see Figure 33.20). Historically, some neuroscientists have included parts of the diencephalon as part of the limbic system because of very extensive connections between these regions. The limbic system is primarily involved in the formation and expression of emotions, and it has a role in learning, memory, and the perception and recognition of smells. The expression of emotions occurs early in childhood before the more advanced functions of the cerebrum are evident. Thus, even very young babies can express fear, distress, and anger as well as bond emotionally with their parents.

A major part of the limbic system is the **hippocampus,** which includes several layers of cells that are connected together in a complex circuit (see Figure 33.20). Its main function appears to be establishing memories for spatial locations, facts, and sequences of events. Damage to certain parts of the hippocampus in humans results in an inability to form new memories, a devastating condition that prevents the recognition of other people or even an awareness of daily events.

(a) Brain activity of a person thinking about a task that requires finger movements

(b) Brain activity when the same person is performing this task

Figure 33.23 Exploring the functional activity of brain regions using an fMRI scan. Red color indicates higher O_2 use; both hemispheres are shown.
(a & b: brain on left) ©Science Photo Library/Alamy; *(a & b: brain on right)* ©Simone Brandt/Getty Images

Concept Check: *What can we conclude from part (a) about the metabolic cost of thinking?*

Imaging Studies Help Scientists Understand the Functions of Brain Regions

A key way in which scientists have learned about the functions of the parts of the brain is through the use of imaging techniques. These include a technique called **magnetic resonance imaging (MRI)**, which can detect the size and shape of individual brain structures, as well as abnormal structures such as brain tumors. A related tool is **functional MRI (fMRI)**, which measures the level of activity of different brain regions at any moment (**Figure 33.23**).

The use of fMRI has revealed many fascinating aspects of the activities of different brain regions, notably in people who have suffered brain damage or loss of sensory inputs. For example, individuals who are blind from birth might be predicted to have occipital lobes that are less functional or active than are those in sighted persons (recall that the occipital lobes have the major role in visual perception). However, the work of American neuroscientist Harold Burton and others has revealed with fMRI that the occipital lobes of blind persons are active but have become adapted to other sensory functions such as tactile signals from the fingers, including those arising from using Braille, a tactile reading system. Amazingly, this reassignment of occipital function occurs to some extent even in individuals who lose their vision later in life. Most likely, this does not represent a new function of the occipital lobes but rather an enhancement of an existing function that remains relatively minor in sighted persons.

fMRI is also revealing differences in brain structure and function in individuals due to the types of activities in which they regularly engage, as described next.

FEATURE INVESTIGATION

Gaser and Schlaug Discovered That the Sizes of Certain Brain Structures Differ Between Musicians and Nonmusicians

MRI and fMRI have been extremely useful in revealing which brain areas are involved in a particular function. Such research has also shown that the human brain is surprisingly adaptable. A number of studies have been carried out on musicians, because they practice extensively throughout their lives, enabling researchers to study the effects of repeated use on brain function.

In 2003, American neuroscientist Christian Gaser and German neuroscientist Gottfried Schlaug used MRI to examine the sizes of brain structures in three groups of people—professional musicians, amateur musicians, and nonmusicians (**Figure 33.24**). Individuals were assigned to each group based on their reported history of musical training: (1) professional musicians with over 2 hours of concentrated musical practice time each day, (2) amateur musicians who played a musical instrument regularly but not professionally (practicing about 1 hour/day), and (3) those who had never played a musical instrument. The researchers hypothesized that repeated exposure to musical training would increase the size of brain areas associated with visual, motor, and auditory skills, because all of these activities are used to read, hear, make, and interpret music. As seen in the data in part 4, the brain areas involved in hearing, moving the fingers, and coordinating movements with vision and hearing were larger in professionals than in amateur musicians and were larger in amateurs than in nonmusicians. The region of the brain that controls finger movements was particularly well developed in the professional musicians, an interesting finding because all of the musicians in this study played keyboard instruments such as the piano.

In subsequent studies, Schlaug and colleagues examined the brains of people of different ages who either did or did not have musical training beginning in childhood. The researchers were able to correlate the degree of training with the sizes of different brain regions. They identified one particular region within the temporal lobe that was highly correlated with the extent of musical training in the course of one's life. This region, called the planum temporale, is believed to be especially important for recognizing and interpreting sound, identifying its source, and translating auditory signals (such as music) into motor processes (such as playing an instrument or humming a melody).

In a related study, American neuroscientists Vincent Schmithorst and Scott Holland used fMRI to determine if musicians' brains were activated differently than nonmusicians' brains when they heard music. These researchers found that one area of the

Figure 33.24 Gaser and Schlaug's study of the size of visual, motor, and auditory nuclei in the brains of musicians and nonmusicians.

HYPOTHESIS Musical training is associated with structural differences in the brain.
KEY MATERIALS Volunteer subjects with different degrees of musical training.

	Experimental level	Conceptual level
1	Establish 3 groups of subjects with different musical backgrounds. Interview subjects for musical history and assign to 1 of 3 groups.	**Controls (nonmusicians)** No musical training. **Amateur musicians** Play an instrument about 1 hr/day and are not employed as musicians. **Professional musicians** Employed as musicians and practice their instrument > 2 hr/day.
2	Control for possible factors that might affect results. Make group assignments in a way that matches subjects' age and other characteristics.	(see table below)

	Controls	Amateurs	Professionals
Age (average)	23	26	27
Type of instrument	None	Keyboard	Keyboard
Mental IQ	119	123	118
Sex	Male	Male	Male

3. Perform MRI and calculate volume of gray matter in different brain regions.

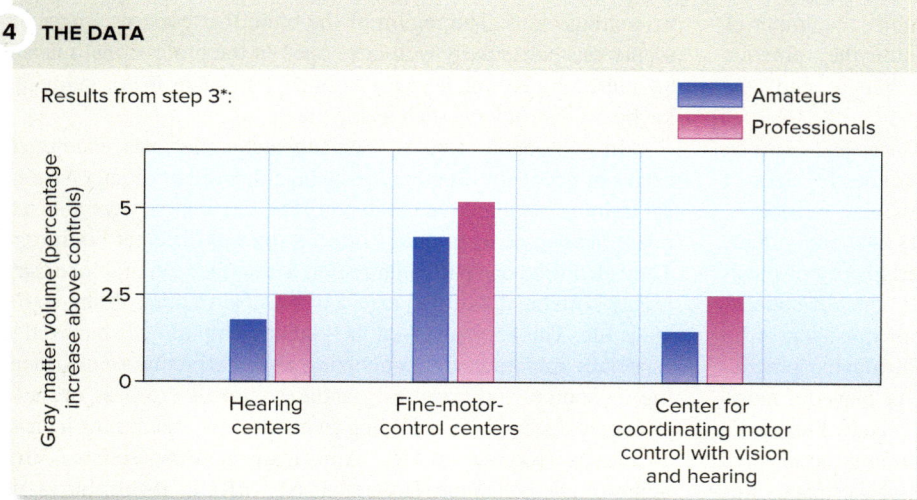

Posterior → Anterior

Regions of brain involved with a particular skill

4. **THE DATA**

Results from step 3*:
(Bar graph showing Gray matter volume, percentage increase above controls, for Amateurs (blue) and Professionals (pink) in three regions)
- Hearing centers: Amateurs ~1.7, Professionals ~2.5
- Fine-motor-control centers: Amateurs ~4.2, Professionals ~5.2
- Center for coordinating motor control with vision and hearing: Amateurs ~1.5, Professionals ~2.5

*Controls are not shown separately because the data are expressed relative to controls.

5. **CONCLUSION** Musical training is associated with increased volumes of brain regions involved in hearing, fine-motor control, and the coordination of motor and sensory information.

6. **SOURCE** Gaser, C., and Schlaug, G. 2003. Brain Structures Differ Between Musicians and Non-musicians. *Journal of Neuroscience* 23: 9240–9245.

cerebral cortex was selectively activated by melodies only in musicians. This study differed from that of Gaser and Schlaug, because Schmithorst and Holland examined the activity of brain areas as well as their sizes. Their results showed that listening to music activates certain neurons and pathways in the brains of musicians but not in nonmusicians.

The human studies of Gaser, Schlaug, Schmithorst, and Holland have not determined the underlying reason(s) for increased sizes of certain brain structures. One possibility is that people with increased sizes of these regions are more likely to become musicians. Alternatively, musical training may actually cause certain regions of the brain to grow larger and alter their neuronal connections.

In other research involving experimental animals, groups of animals have been randomly separated into those learning a task versus controls (that do not learn the task). Such experiments have shown increases in the size of brain regions that are associated with learning and memory. The increased size may result from the formation of new synapses, the growth of blood vessels to the region, and/or the production of more glial cells.

Experimental Questions

1. What was the hypothesis proposed by Gaser and Schlaug? How did Gaser and Schlaug test this hypothesis? What were the results of their experiment?
2. **SCISKILLS** ▶ How did the research of Schmithorst and Holland differ from that of Gaser and Schlaug? Were their results generally supportive of Gaser and Schlaug's hypothesis?
3. **SCISKILLS** ▶ Based on the results described here and what you have learned in this chapter, predict general differences that might be found between the brain of a professional athlete, such as a tennis player, and that of a nonathlete. What about the brain of a person who has been deaf from birth compared to that of a person with normal hearing?

33.6 Reviewing the Concepts

- In humans and other vertebrates, the brain and spinal cord make up the central nervous system (CNS). The neurons and all axons outside the CNS, including the cranial and spinal nerves, constitute the peripheral nervous system (PNS). The CNS receives sensory input from the PNS, and the PNS receives commands from the CNS (Figure 33.17).
- The CNS is composed of gray and white matter and is protected by meninges and cushioned by cerebrospinal fluid (Figure 33.18).
- The PNS consists of the somatic nervous system, which senses external environmental conditions and controls skeletal muscles, and the autonomic nervous system, which controls homeostatic functions (Figure 33.19).
- The human brain consists of the hindbrain, which controls the basic processes that sustain life; the midbrain, which processes several types of sensory inputs, including vision, olfaction, and audition; and the forebrain, which is made of the diencephalon (thalamus, hypothalamus, and epithalamus) and the cerebrum (cerebral cortex, limbic system, and basal nuclei) (Figures 33.20, 33.21, 33.22).
- Imaging techniques such as MRI and fMRI allow researchers to examine the structure and activity of the brain (Figures 33.23, 33.24).

33.6 Testing Your Knowledge

1. The division of the nervous system that controls voluntary muscle movement is
   a. the autonomic nervous system.
   b. the sensory division.
   c. the somatic nervous system.
   d. the parasympathetic division.
   e. the sympathetic division.
2. Match the human brain region with its representative function(s).

thalamus	making decisions and exercising judgment
hypothalamus	controlling breathing/relay between cerebellum and rest of brain
basal nuclei	filtering incoming sensory information
cerebral cortex	fine-tuning, starting, or inhibiting movement
pons	controlling body temperature and circadian rhythms

33.7 Impact on Public Health

Learning Outcomes

1. Describe how disorders of neurotransmission can affect human health.
2. Describe the cellular changes that occur in Alzheimer's disease and its impact on public health.
3. List several recreational and illicit drugs and their effects on the nervous system.

When neurons fail to develop properly or their function becomes impaired, the consequences can be devastating—affecting mood, behavior, and even the ability to think or move. Over 100 neurological disorders have been identified in humans. Most neurological disorders can be classified into several broad groups (**Table 33.1**). These disorders collectively affect hundreds of millions of people around

Table 33.1 Diseases or Disorders Affecting the Human Central Nervous System

Category	Examples
Demyelinating	Multiple sclerosis (damage to the myelin sheath around neurons in the central nervous system)
Headache	Migraine (severe, recurring headache)
Infectious	Meningitis (inflammation of the meninges), encephalitis (inflammation of the brain)
Injury-related	Brain and spinal cord injuries due to accidents
Mood	Major depressive disorder, bipolar disorder
Movement	Parkinson's disease (death of neurons in the midbrain and basal nuclei)
Neurodegenerative	Alzheimer's disease
Seizure	Epilepsy
Sleep	Sleep apnea (in one form, the brain fails to regulate breathing during sleep)
Tumors	Glioma (a tumor arising from glial cells)

the world. In addition, all so-called recreational drugs, including alcohol and tobacco (and in some states marijuana), as well as illicit drugs such as cocaine, heroin, LSD, and methamphetamine, exert their effects by altering neurotransmission, as summarized in **Table 33.2**. The use of these drugs, therefore, can result in symptoms similar to those of neurological disorders. Certain therapeutic drugs also exert their effects by altering neurotransmission—for example, to correct a neurotransmitter deficiency.

Disorders of Neurotransmission Can Impact Mood

Several neurological disorders result from disrupted neurotransmission between cells. Genetic processes involved in the synthesis of neurotransmitters or malfunction of synaptic events can increase or decrease activity at synapses, which in turn affects emotions and behavior. The most common mood disorder is **major depressive disorder,** or depression. This illness results in prolonged periods of sadness, despair, and most commonly a profound lack of interest in daily activities. Depression affects 5–12% of men and 10–25% of women at some time during their lives. This condition is thought to result from decreased activity of synapses that release biogenic amines such as serotonin, which changes neuronal activity within specific areas in the brain that are involved in processing mood and emotion. Drugs used to treat depression include the **selective serotonin reuptake inhibitors (SSRIs),** such as Prozac, Zoloft, and Lexapro, which reduce the reuptake of serotonin into the presynaptic terminal after it is released. As a result, serotonin accumulates in the synaptic cleft, counteracting the presumed deficit that causes the alteration in mood. Patients taking SSRIs often report a greater interest in daily activities and a gradual return to normal, as was the case for the woman in the chapter-opening story.

Alzheimer's Disease Impairs Memory, Thought, and Language

Alzheimer's disease (AD) is the leading cause of dementia worldwide. It is characterized by a loss of memory and cognitive function. AD is a progressive, incurable disease that begins with small memory lapses, leading in later stages to problems with language and abstract thinking and finally to loss of normal motor control and eventual death. The disease typically appears after age 65, though some inherited forms can strike people in their thirties and forties.

Although psychological testing can help to diagnose AD, a definitive diagnosis is possible only after death when the brain is examined microscopically. Brains of AD patients show two noticeable changes: the existence of plaques and neurofibrillary tangles. Plaques are extracellular deposits of a misfolded protein, β-amyloid, that forms large, sticky aggregates. These plaques were first noted in 1906 by German physician Alois Alzheimer, after whom the disease is named. Neurofibrillary tangles are intracellular, twisted accumulations of cytoskeletal fibers. Scientists are unsure how these changes influence intellectual function and memory. AD is also associated with the degeneration and death of neurons, particularly in the hippocampus and parietal lobes, which is why it is considered a neurodegenerative disease.

Researchers have identified variation in a few genes whose products are associated with the likelihood of developing AD later in life, but the underlying changes that result in the expression of these and other possible AD-related genes are still the subject of considerable research. Although genetics undoubtedly has a role in AD, it is not the only possible cause. For example, when one identical twin develops AD, the other appears to be at increased risk but does not always develop the disease, even if he or she survives to a very old age. Moreover, evidence suggests that severe head injuries, metabolic diseases such as diabetes, bacterial infections (notably from gum disease), and heart and blood vessel disease may increase a person's risk to develop AD in later life.

Currently, AD cannot be prevented or cured. However, three major clinical approaches are being tested to slow down its progression. These are designed to (1) induce a person's immune system to destroy β-amyloid as soon as it is formed, (2) prevent the formation of β-amyloid with drugs that block its synthesis, or (3) prevent the accumulation of β-amyloid into large aggregates using antiaggregation drugs. These approaches hold promise but remain unproven.

Until a cure for AD is found, its impact on public health remains enormous. Currently, about 4–5 million Americans have AD, a number that is expected to grow to nearly 16 million by 2050. The prevalence of the disease is about 3% for people between the ages of 65 and 74, and 25–50% for people older than 85. Estimated costs associated with providing health care and housing for AD patients (30% of whom live in nursing homes), as well as lost productivity in the workplace, total a staggering $100 billion per year in the U.S. alone. This number will rise substantially now that the oldest members of the population spike known as the baby boom generation have passed the age of 70.

Table 33.2 Representative Effects of Drugs on Neurotransmitter Action, Mood, and Behavior

Name of drug*	Actions on neurotransmission	Effects on mood or behavior	Effects of abuse or overdose
Illicit or Recreational[†]			
Alcohol (ethanol)	Enhances inhibitory GABA transmission; increases dopamine transmission; inhibits glutamate transmission	Relaxation; euphoria; sleepiness	Liver damage; brain damage
Amphetamines: "uppers," "crystal meth," "speed"	Stimulate the release of dopamine and norepinephrine	Euphoria; increased activity	High blood pressure; psychosis
Cocaine	Blocks norepinephrine and dopamine reuptake	Intense euphoria followed by depression	Convulsions; hallucinations; death from overdose
Lysergic acid diethylamide (LSD)	Binds to serotonin receptors	Hallucinations; sensory distortions	Unpredictable and irrational behavior
Marijuana (tetrahydrocannabinol)	Binds to receptors for natural cannabinoids	Increased sense of well-being; decreased short-term memory; decreased goal-directed behavior; increased appetite	Delusions; paranoia; confusion
Narcotics: heroin, morphine, oxycodone	Bind to opiate receptors	Pain relief; euphoria; sedation	Slowed breathing; death from overdose
Nicotine	Initially stimulates but then depresses activity in adrenal medulla and neurons in the peripheral nervous system; increases dopamine in brain	Increased attention; decreased irritability	Heart disease; lung disease
Phencyclidine (PCP or "angel dust")	Blocks channel for excitatory amino acid neurotransmitters; increases dopamine activity	Violent behavior; feelings of power; numbness; disorganized thoughts	Psychosis; convulsions; coma; death
Therapeutic			
Antianxiety drugs: benzodiazepines (for example, Xanax, Valium, Librium, Rohypnol ["date rape drug," "roofies"])	Bind to GABA receptors and increase inhibitory neurotransmission	Relieve anxiety	Drowsiness; memory loss in some cases
Antidepressants: tricyclic antidepressants (for example, Elavil, Anafranil)	Block the reuptake of norepinephrine from synapses	Relieve depression and obsessive-compulsive disorder	Drowsiness; confusion
Antipsychotic drugs: phenothiazines (for example, Thorazine, Mellaril, Stelazine), atypical antipsychotics (Abilify, Risperdal)	Block dopamine receptors	Ease schizophrenic symptoms	Decreased control of movement
Monoamine oxidase inhibitors (for example, Parnate, Nardil)	Block the breakdown of biogenic amine neurotransmitters	Relieve depression	Liver damage; hyperexcitability
Selective serotonin reuptake inhibitors (SSRIs) (for example, Prozac, Zoloft, Paxil, Lexapro)	Block the reuptake of serotonin from synapses	Relieve depression and obsessive-compulsive disorder	Insomnia; anxiety; headache; sexual dysfunction

*The capitalized drug names represent only some of the brand name varieties available in each category. In some cases, generic versions are also available.
[†]Some of these drugs have some therapeutic value under certain conditions.

Many Illicit Drugs Disrupt Normal Neurotransmission

Many illicit drugs work at the synapse to either enhance or interfere with the normal mechanisms of neurotransmission (Table 33.2). In the presynaptic terminal, such drugs can decrease neurotransmitter release by decreasing Ca^{2+} entry into the cell and preventing the exocytosis of vesicles containing stored neurotransmitters. In the synaptic cleft, drugs can slow the rate at which the neurotransmitter is broken down into an inactive form or taken back up into the presynaptic neuron, thereby prolonging the action of the neurotransmitter in the synaptic cleft. Some substances act on the postsynaptic membrane either by preventing the neurotransmitter from binding to its receptor or by acting as a substitute for the neurotransmitter by stimulating its receptor.

In effect, these drugs produce changes or imbalances in neurotransmission similar to those observed in some neurological disorders. These substances can increase activity, alter mood, and produce hallucinations. They can also have life-threatening effects and be highly addictive.

Some drugs, such as cocaine, block the removal of the amine neurotransmitters dopamine and norepinephrine from the synaptic cleft by preventing their reuptake into the presynaptic terminal. Opioids and marijuana mimic the actions of biological substances already in the brain, binding to receptors on the postsynaptic membrane. With these drugs, the resulting effects are much stronger than are the effects of natural neurotransmitters. It is no surprise that many of these drugs are termed mind-altering. They do, after all, change the ways in which neurons communicate with each other.

33.7 Reviewing the Concepts

- Disorders of the human central nervous system can be placed into several broad categories. Most neurological conditions can be classified as disorders of either neurotransmission or conduction. Mood disorders caused by disrupted neurotransmission include major depressive disorder (Table 33.1).
- Alzheimer's disease is a progressive disorder characterized by the formation of plaques and neurofibrillary tangles in brain tissue.
- Drugs used in the treatment of neurological disorders and many recreational and illicit drugs usually alter neurotransmission (Table 33.2).

33.7 Testing Your Knowledge

1. _____ is a progressive disease that causes a loss of memory and intellectual and emotional function.
   a. Major depressive disorder
   b. Parkinson's disease
   c. Amnesia
   d. Alzheimer's disease
   e. Stroke
2. Therapeutic drugs that reduce symptoms of depression may act
   a. by decreasing exocytosis of neurotransmitters in the brain.
   b. by increasing serotonin levels in brain synapses.
   c. by decreasing the formation of plaques and neurofibrillary tangles.
   d. by blocking calcium entry into presynaptic neurons.
   e. by decreasing acetylcholine levels in brain synapses.

Assessing Your Knowledge and Skills

Foundational Knowledge

Multiple Choice

1. In vertebrates, the brain and spinal cord constitute
   a. the peripheral nervous system.
   b. the efferent division of the nervous system.
   c. the central nervous system.
   d. the autonomic nervous system.
   e. the central and peripheral nervous systems.
2. Which part of a neuron contains voltage-gated Ca^{2+} channels and stored neurotransmitters?
   a. the myelin sheath
   b. the axon hillock
   c. the dendrites
   d. the axon terminals
   e. the cell body
3. The myelin sheath
   a. is produced by neurons in the peripheral nervous system.
   b. is only formed around neurons in the brain.
   c. is present on all neurons.
   d. is generally present around long axons in either the CNS or the PNS.
   e. significantly slows transmission along neurons.
4. Neurons that function mainly in connecting other neurons in the central nervous system are
   a. sensory neurons.
   b. efferent neurons.
   c. motor neurons.
   d. afferent neurons.
   e. interneurons.
5. The difference in charges across the plasma membrane of an unstimulated neuron is called
   a. an EPSP.
   b. the resting membrane potential.
   c. an IPSP.
   d. the graded potential.
   e. the action potential.

Short Answer

1. Describe the differences between graded and action potentials.
2. **PRINCIPLES** A principle of biology is that new properties emerge from complex interactions. How is this principle evident in the structure and function of animal nervous systems?

Critical-Thinking Skills

Multiple Choice

1. Imagine that several EPSPs arrive at a postsynaptic cell at the same location in quick succession. What might happen as a result?
   a. Their depolarizations will sum, and this might produce an action potential.
   b. Each EPSP will create a graded potential, some of which will depolarize the cell and some of which will hyperpolarize the cell.
   c. Each EPSP will cause a separate action potential.
   d. They will cancel each other out, leading to no action potentials.
   e. Their depolarizations will sum, but this cannot produce an action potential.
2. Assume that an experimental drug is available that could prevent the inactivation gates of voltage-gated Na^+ channels in a neuron from closing. How would the shape of an action potential be affected in a neuron that was electrically stimulated to threshold potential in the presence of this drug?
   a. The rising phase of the action potential would be less steep.
   b. The repolarization phase would occur more gradually, and the cell might never return to its resting potential.

c. The peak of the action potential would be lower, and it would return much more quickly to baseline.
   d. There would be a second peak occurring shortly after the first.
3. From what you've learned about factors that can modify the speed of transmission of an action potential along an axon, in which axons would the speed be predicted to be the greatest?
   a. wide, unmyelinated axons
   b. narrow, unmyelinated axons
   c. wide, myelinated axons
   d. narrow, myelinated axons
4. A frog is confronted by a snake. Which of the following events, which are mediated in part by neurons, would *not* be expected to occur in the frog?
   a. increased heart rate
   b. increased breathing rate
   c. increased blood flow to skeletal muscles
   d. a rapid decrease in the blood glucose concentration
   e. a rapid decrease in blood flow to the intestines
5. A particular neuron has a resting membrane potential of −64 mV, and its equilibrium potential for Na^+ is +59 mV. In which direction will Na^+ diffuse if Na^+ channels in the membrane are open?
   a. Na^+ will move into the cell.
   b. Na^+ will move out of the cell.
   c. Na^+ will move equally in both directions.

Short Answer

1. Scientists have determined that the average brain mass of *Homo neanderthalensis* (Neanderthals) was likely greater than that of modern humans (*Homo sapiens*). Does this mean that Neanderthals were more intelligent and capable of more complex behaviors than modern humans?
2. In certain diseases, such as kidney failure, the Na^+ concentration in an animal's extracellular fluid can increase. What effect might a high extracellular Na^+ concentration have on neurons?

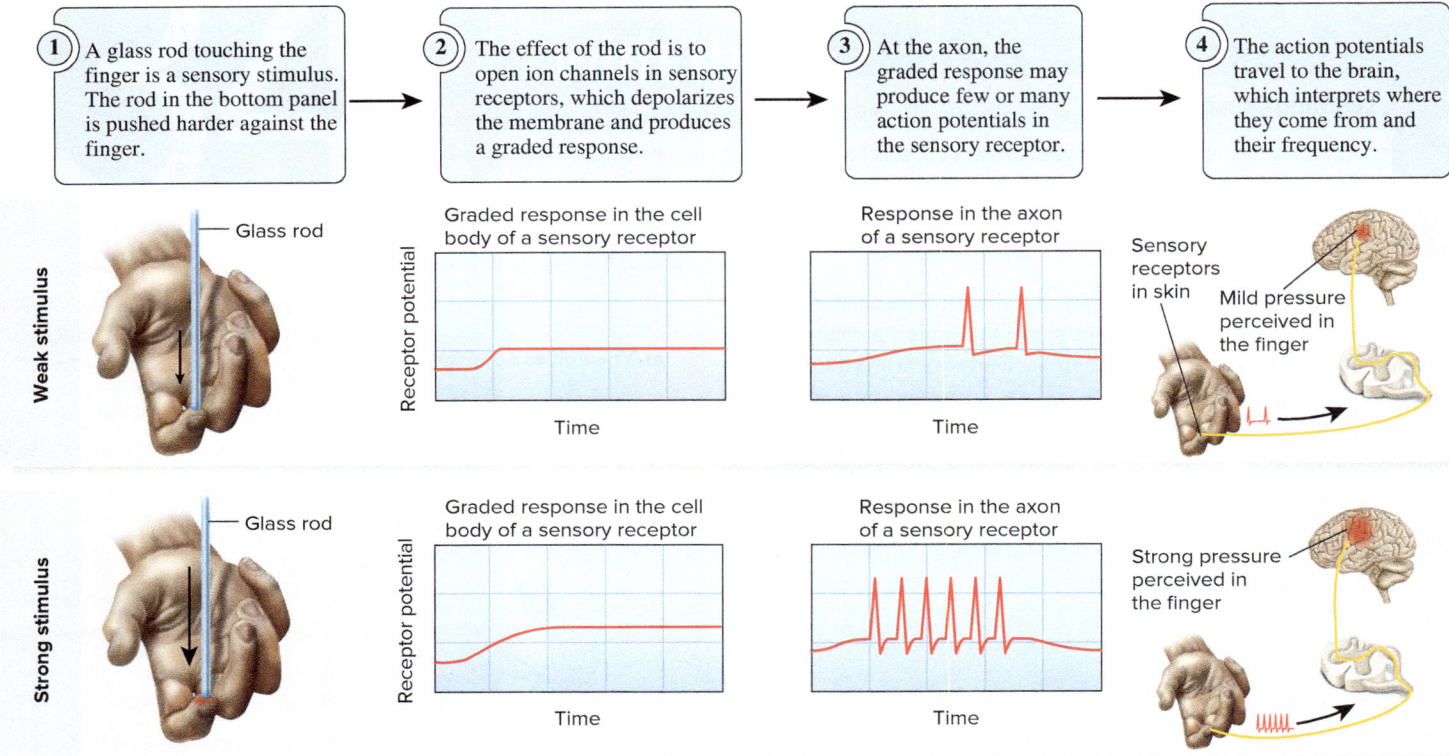

Figure 34.2 Transduction of a sensory stimulus of two different intensities. In this example, the sensory receptor is a neuron. Note the larger graded response following the stronger stimulus. The perception of stimulus strength occurs in the brain.

generated by visual sensory receptors in the eye are transmitted along a neural pathway that sends action potentials into areas of the brain that are devoted to processing vision. For this reason, the brain interprets such signals as visual stimuli. The brain can separate and identify each sense because each sense uses its own dedicated pathway.

Sensory receptors can be divided into general types based on the type of stimulus to which they respond. Each type uses a different mechanism to detect stimuli and to transmit the information to different regions of the CNS. In the following sections of this chapter, we will examine the different types of stimuli and the structures and functions of different sensory receptors.

34.1 Reviewing the Concepts

- Sensory transduction is the process by which incoming stimuli are converted to neural signals. Perception is an awareness of sensations. Sensory receptors are neurons or specialized epithelial cells that respond to stimuli and begin the process of sensory transduction (Figures 34.1, 34.2).

34.1 Testing Your Knowledge

1. The process by which incoming stimuli are converted into neural signals is called
   a. perception.
   b. sensory transduction.
   c. sensory reception.
   d. sensory perception.
   e. perceptual transduction.

34.2 Mechanoreception

Learning Outcomes

1. Describe what mechanoreceptors are and the various sensory abilities they impart to animals.
2. Outline the structure of the mammalian ear, and explain how mechanical forces move through it.
3. Describe how body position and movement are detected by sense organs.

Mechanoreceptors are cells that detect physical stimuli such as touch, pressure, stretch, sound, and movement. Physically touching or deforming a mechanoreceptor cell opens ion channels in its plasma membrane (Figure 34.2). Some mechanoreceptors are neurons that send action potentials to the CNS in response to physical stimuli. Other mechanoreceptors are specialized epithelial cells that contain hairlike structures that bend in response to mechanical forces.

Skin Receptors Detect Touch and Pressure

Several types of receptors in the skin of many animals detect touch, deep pressure, or the bending of hairs on the skin. Some of these specialized receptors consist of neuronal dendrites covered in dense connective tissue. When these structures become deformed (for example, by pressure on the skin), the neurons generate action potentials that travel to the brain. In mammals, these receptors are located at different depths below the surface of the skin, which makes them suitable for responding to different types of stimuli, such as soft or hard touch. Other skin mechanoreceptors located in the hair follicles respond to movements of hairs and whiskers.

Stretch Receptors Detect Expansion

One type of mechanoreceptors called **stretch receptors** are neuron endings commonly found in organs that can be distended, or stretched, such as a stomach, urinary bladder, and skeletal muscles. Stretch receptors have been best studied in crustaceans and mammals. In decapod crustaceans such as crabs and lobsters, for example, stretch receptors in muscles of the tail, abdomen, and thorax relay signals to the brain regarding the positions in space of different body parts. This information allows the animal to coordinate complex motor functions, such as walking backward or sideways. In another example, when the mammalian stomach stretches after a meal, the stretch receptors in the stomach are deformed, causing them to become depolarized and send action potentials to the brain. The brain interprets the signals as fullness, which inhibits appetite.

Hair Cells Are Mechanoreceptors That Detect Sound and Movement

Thus far, we have considered skin and stretch mechanoreceptors, which are neurons that detect touch, pressure, and stretch. Other mechanoreceptors are specialized epithelial cells called **hair cells,** which have deformable projections termed **stereocilia** that detect movement. The stereocilia are different from true cilia (refer back to Figure 4.17) because they do not contain motor proteins in their structure. They are displaced by movements of fluid or other physical stimuli (**Figure 34.3**). Stereocilia allow hair cells to be responsive to movements made by an animal.

Hair cells contain ion channels that open or close when the stereocilia bend, thereby changing the cell's membrane potential. When the plasma membrane depolarizes, voltage-gated Ca^{2+} channels open, triggering the release of neurotransmitter molecules from the hair cells. The neurotransmitter then binds to receptor proteins in adjacent sensory neurons and can elicit action potentials that travel to the brain. When unstimulated, the stereocilia are not bent, and the hair cells release only a small amount of neurotransmitter onto nearby sensory neurons, resulting in a resting rate of action potentials in the sensory neurons. In the example shown in Figure 34.3, bending of the stereocilia in one direction in response to fluid movement increases the release of neurotransmitter from the hair cell, exciting the sensory receptors. Bending in the other direction decreases the release of the same neurotransmitter, inhibiting the sensory receptors. The result is an increase or a decrease, respectively, in the rate of action potentials produced in the sensory neurons.

Hair cells provide a rich array of sensory capabilities in many animal species. For example, they are found in the hearing and equilibrium (balance) organs of many invertebrates and vertebrates, where they detect sound or changes in head position. They are also found along the body surface of fishes and some amphibians, where they detect external water currents that inform an animal about nearby predators or of the animal's own movements.

Hearing Involves the Reception of Sound Waves

The sense of hearing, called **audition,** is the ability to detect and interpret sound. This sense is critical for the survival and reproduction of many types of animals. For example, many animals locate their young by hearing its calls, or they use sound to attract a mate. Hearing is also important for detecting the approach of danger—a predator, a thunderstorm, an automobile—and locating its source.

Sound consists of oscillating pressure waves; when we hear a sound, we are actually detecting waves of air pressure. The distance from the peak of one wave to the peak of the next is a wavelength. The

Figure 34.3 The response of hair cells to mechanical stimulation. The stereocilia of these hair cells are hairlike projections of the plasma membrane that bend in response to fluid movement outside the cell.

 How are stereocilia different from cilia, which are described in Chapter 4?

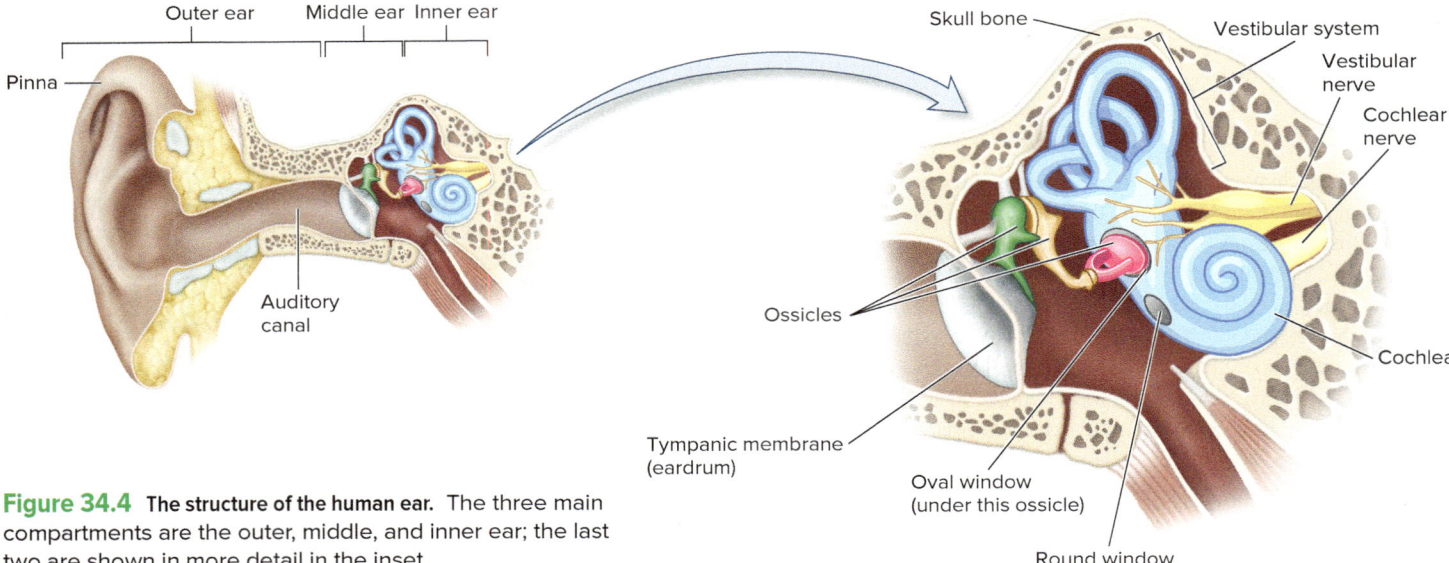

Figure 34.4 **The structure of the human ear.** The three main compartments are the outer, middle, and inner ear; the last two are shown in more detail in the inset.

number of complete wavelengths that occur in 1 sec is called the frequency of the sound (1 wave/sec = 1 Hertz, or Hz). The wavelength and frequency of sound waves impart certain characteristics to the stimulus. Short wavelengths have high frequencies that are perceived as a high pitch or tone, and long wavelengths have lower frequencies and a lower pitch. The human hearing range is 20–20,000 Hz.

The sense of hearing is present in vertebrates and to a limited extent in arthropods. Hearing, however, is especially well developed in birds and mammals. We will turn now to a detailed discussion of the mammalian ear and the mechanism by which it detects sound, including the importance of hair cells in this mechanism.

Structure of the Mammalian Ear The mammalian ear has three main compartments (**Figure 34.4**). The first is the outer ear, which consists of the external ear, or pinna (plural, pinnae), and the auditory canal. The middle ear is separated from the outer ear by the tympanic membrane (eardrum). It contains three small bones called ossicles that link movements of the eardrum with another membrane, called the oval window. The oval window separates the middle ear from the inner ear. Within the inner ear is the **cochlea**—a coiled chamber of bone containing hair cells and the membrane-like round window. The structures in the cochlea generate the signals that travel via a nerve called the cochlear nerve to auditory centers of the brain. Also present in the inner ear is the **vestibular system** which functions in balance and communicates with the brain via the vestibular nerve. We will examine this system later.

Generation of Electrical Signals in the Mammalian Ear To understand how mammals hear, let's first consider how mechanical forces are transmitted through the ear. The outer ear collects sound waves and funnels them to the tympanic membrane, causing it to vibrate back and forth (**Figure 34.5**). The ossicles transfer the vibration of the tympanic membrane to the oval window, which vibrates against the cochlea. This vibration sends pressure waves through two fluid-filled passages called the vestibular and tympanic canals. The waves eventually strike the round window, where they dissipate.

Along the way, however, the waves cause the vibration of a sheath-like structure called the **basilar membrane**, which is formed from elastic fibers tensed across a duct called the cochlear duct. The basilar membrane is stiff near the oval window and gets progressively less stiff toward its end. The fluid pressure waves interact with a portion or all of the basilar membrane, depending on the frequency of the sound that initiated them; for example, only the lowest frequency sound waves push fluid along the entire route shown in Figure 34.5.

Within the cochlea, mechanical vibrations are changed, or transduced, into electrical signals. This happens in a structure called the **organ of Corti,** which rests on top of the basilar membrane along its length. To understand how this works, look at a cross section through the cochlea (**Figure 34.6**). The organ of Corti contains supporting cells and rows of hair cells. The stereocilia of the hair cells are embedded in a gelatinous membrane called the tectorial membrane. The back-and-forth vibration of the basilar membrane bends the stereocilia in one direction and then the other. When bent in one direction, the hair cells depolarize and release neurotransmitter onto sensory neurons, which then send action potentials to the brain. When bent in the other direction, the hair cells hyperpolarize and stop releasing neurotransmitter. In this way, the frequency of action potentials generated by the sensory neurons is determined by the up-and-down vibration of the basilar membrane.

The Sense of Balance Is Mediated by Statocysts in Invertebrates and by the Vestibular System in Vertebrates

Let's now turn our attention to another form of mechanoreception, the sense of balance, also called equilibrium. This includes an animal's ability to sense the position, orientation, and movement of its body. Being able to sense body position is vital for the survival of animals. This is how a lobster, for example, rights itself when flipped over by a predator or how a bird maintains its balance while flying.

Statocysts Many aquatic invertebrates have sensory organs called **statocysts** that send information to the brain about the position of the animal in space (**Figure 34.7**). Statocysts are small,

NEUROSCIENCE II: HOW SENSORY SYSTEMS ALLOW ANIMALS TO INTERACT WITH THE ENVIRONMENT 749

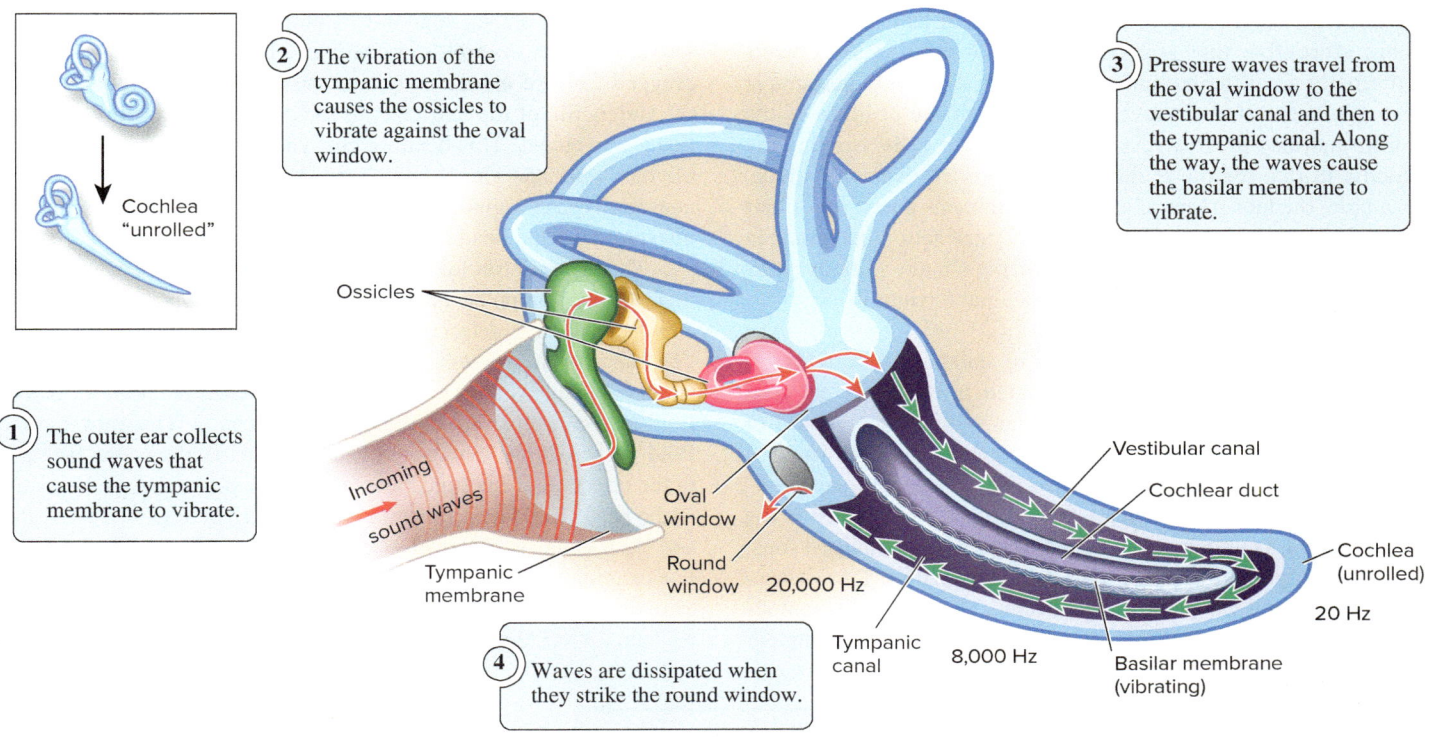

Figure 34.5 Movement of sound waves through the human ear.

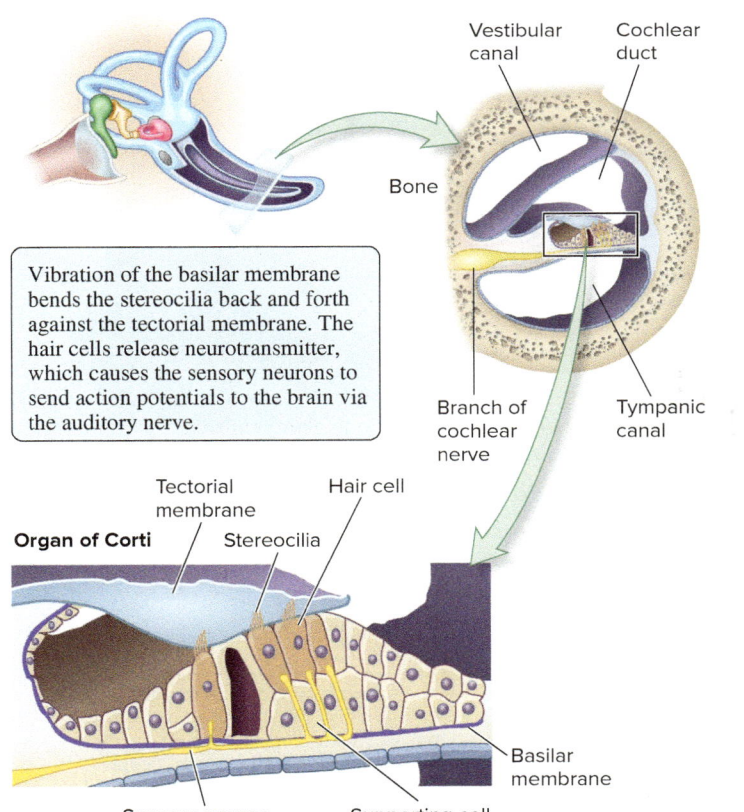

Figure 34.6 Transduction of mechanical vibrations to action potentials in the organ of Corti.

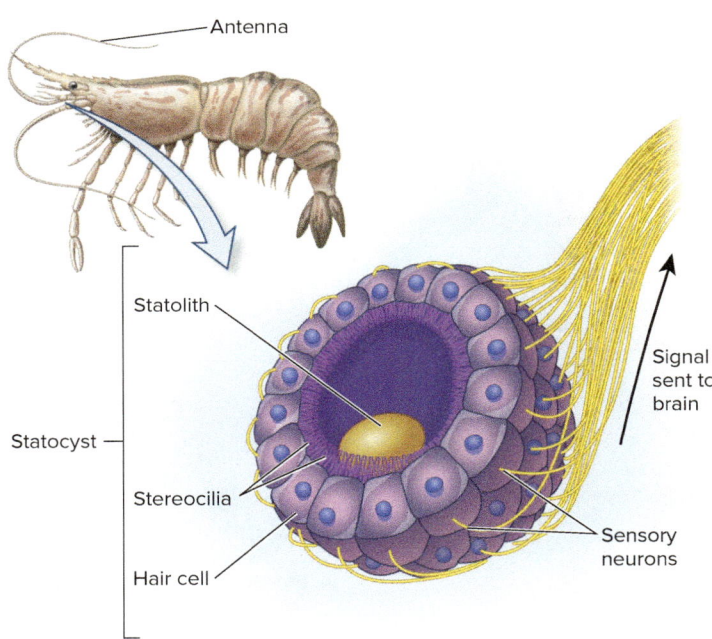

Figure 34.7 **Sensing of balance in an invertebrate.** Statocysts located near the antennae consist of a sphere of sensory hair cells surrounding one or more statoliths. When the animal moves, gravity shifts the statolith and stimulates the hair cells beneath it.

BioConnections: *Are statoliths unique to animals? Refer back to Figure 29.11 for a hint.*

round structures consisting of an outer sphere of hair cells and one or more **statoliths,** tiny granules of sand or other dense objects that normally sit at the bottom of the statocyst when the animal is at rest. When the animal moves, gravity alters the position of the statoliths and they move across some of the hair cells, bending the stereocilia in that region. If the animal is turned over on its side, for example, the movement of statoliths stimulates a new set of hair cells to release neurotransmitter. This generates action potentials in sensory neurons that inform the brain of the change in body position and trigger a reflex that allows the animal to right itself.

The Vestibular System The organ of balance in vertebrates, known as the vestibular system, is located in the inner ear next to the cochlea (Figure 34.8). The vestibular system is composed of a series of fluid-filled sacs and tubules that provide information about linear or rotational movements. The utricle and saccule—the two sacs nearest the cochlea—detect linear movements of the head (Figure 34.8a), such as those that occur when an animal runs, jumps, or changes its posture. The hair cells within these structures are embedded in a gelatinous substance that contains granules of calcium carbonate called **otoliths** (from the Latin, meaning ear stones), which are analogous to statoliths. When the head moves forward, the heavy otoliths are temporarily "left behind" as they are dragged forward more slowly, and the weight of the otoliths bends the stereocilia of the hair cells in the direction opposite that of the linear movement. This bending changes the membrane potential of the hair cells and alters the electrical responses of nearby sensory neurons. These signals are sent to the brain, which uses them to interpret how the head has moved.

Three **semicircular canals** connect to the utricle at bulbous regions called ampullae (singular, ampulla). The function of the semicircular canals is to detect rotational motions of the head (Figure 34.8b). The hair cells in the semicircular canals are embedded in the ampullae in a gelatinous cone called the cupula. When the head moves, the fluid in the canal shifts in the opposite direction. This movement of fluid pushes on the cupula and bends the stereocilia of the hair cells in the direction of the fluid flow, which is opposite that of the motion of the head. The three canals are oriented at right angles to each other, and each canal is maximally sensitive to motion in its own plane. For example, the canal that is oriented horizontally responds most to rotations such as shaking the head "no," whereas the other canals respond to "yes" motions or to tipping the ear to the shoulder. Overall,

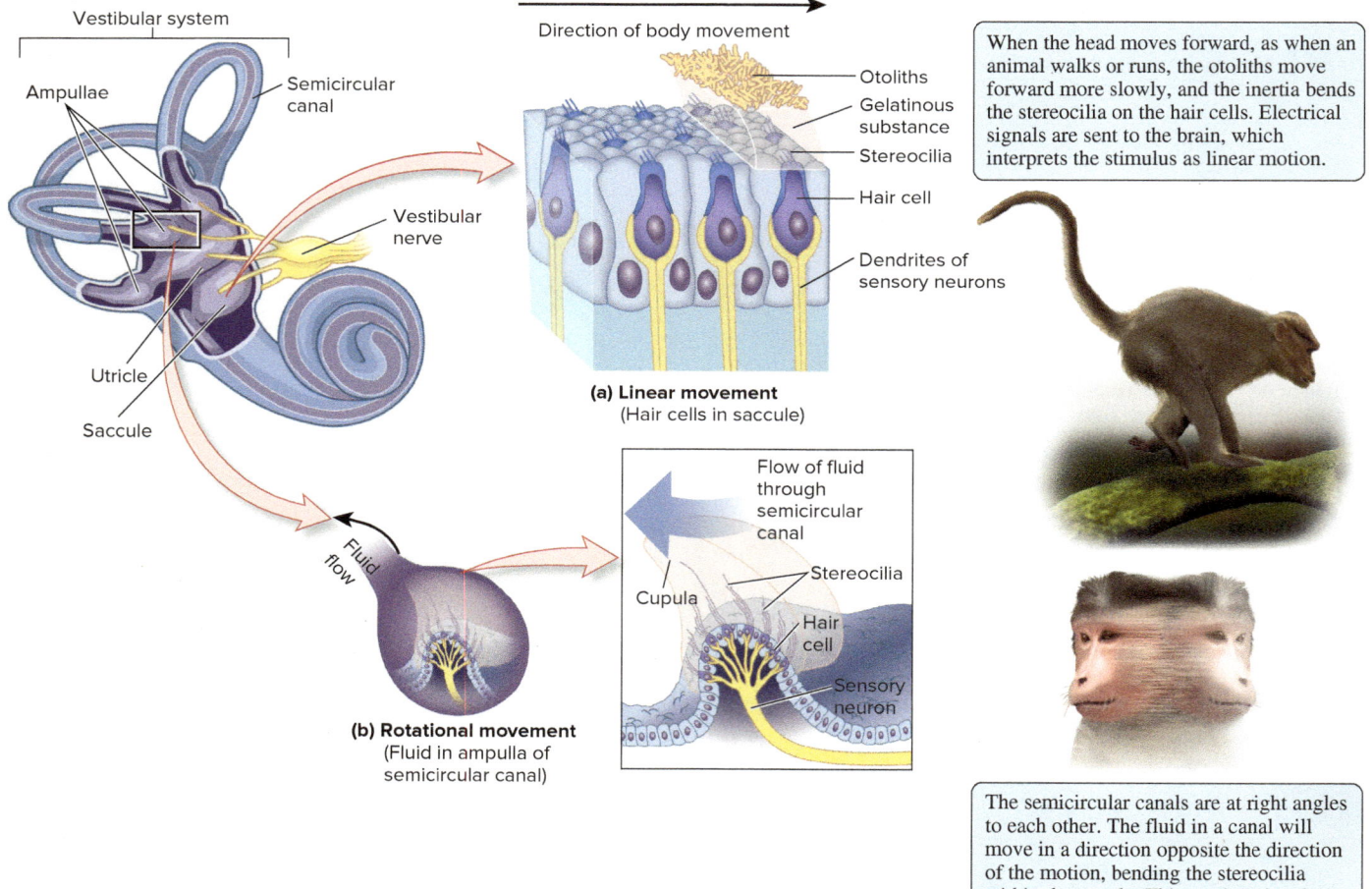

Figure 34.8 The vertebrate vestibular system.

by comparing the signals from the three canals, an animal's brain can interpret the motion of its head in three dimensions.

The vestibular system of vertebrates provides conscious information about body position and movement. It also supplies unconscious information for reflexes that maintain normal posture, control head and eye movements, and assist in locomotion. Researchers have discovered clear correlations between the types of locomotion an animal engages in and the size of its semicircular canals relative to its body mass. Among primates, for instance, agile animals with jerky forms of locomotion, such as leaping tarsiers, have much larger canals than lorises, which are quadripedal and move slowly.

34.2 Reviewing the Concepts

- Mechanoreceptors respond to physical stimuli such as touch, pressure, stretch, movement, and sound.
- Hair cells have stereocilia that respond to movements of fluid or other stimuli; these cells release neurotransmitters that can trigger action potentials in adjacent sensory neurons (Figure 34.3).
- The mammalian ear has three main compartments: the outer, middle, and inner ear. Sound (pressure) waves are transmitted through the outer and middle ear to the cochlea of the inner ear, where the hair cells are located and signals are generated that travel to the brain for interpretation (Figures 34.4, 34.5, 34.6).
- Statocysts in certain invertebrates allow an animal to sense its body position or equilibrium. The vestibular system in vertebrates allows animals to sense linear and rotational movement (Figures 34.7, 34.8).

34.2 Testing Your Knowledge

1. Which of the following statements about hair cells is true?
   a. They are sensory neurons.
   b. They have true cilia on their surface.
   c. They are important for audition and balance in many animals.
   d. They are only found in invertebrates.
   e. They are only found in vertebrates.

2. Vibrations of the _____, located within the _____, bend stereocilia in the mammalian ear and activate sensory neurons.
   a. tympanic membrane, cochlear duct
   b. round window, middle ear
   c. basilar membrane, organ of Corti
   d. ossicles, outer ear
   e. ampullae, semicircular canals

34.3 Thermoreception and Nociception

Learning Outcome

1. **SCISKILLS** ▶ Distinguish between thermoreception and nociception, and propose reasons that may explain why they are vital to the safety and survival of an animal.

The perception of temperature and pain enables animals to respond effectively to potentially dangerous changes in their environments. These sensory stimuli are related to each other in that their receptors are located in some of the same areas, share similar physical features, and under certain conditions result in similar perceptions.

Thermoreceptors Detect Temperature

An ability to sense heat is important for animals because their body temperature is affected by the external temperature. This is particularly true for ectotherms, animals that rely on heat from the environment to warm themselves. As you learned in Chapter 32, most animals can survive only if their body temperature remains within certain limits, because cell membranes and proteins function optimally only within a particular temperature range. Thermoreceptors respond to cold or hot temperatures by activating or inhibiting enzymes within their plasma membranes, which in turn alters membrane ion channels. There are two types of thermoreceptors: those that respond to hot and those that respond to cold. Both of these types of sensory receptors are neurons. Thermoreceptors are often linked with reflexive behaviors, such as the increased blinking and tearing of the eyes of some mammals in cold weather or the withdrawal of a foot or hand from a hot surface.

In addition to skin receptors that sense the outside temperature, endotherms have thermoreceptors in the brain that detect changes in core body temperature. Activation of skin or brain thermoreceptors triggers physiological and behavioral adjustments that help to maintain temperature homeostasis. These adjustments include changes in skin blood flow, shivering, and behaviors such as seeking shade or sunlight.

A fascinating example of thermoreceptors is found in a few families of snakes, including boas, pythons, and pit vipers (a group that includes copperheads and rattlesnakes). These snakes can localize prey in the dark with thermoreceptors located in pits on each side of the head (**Figure 34.9**). Within the pit, a thin, nerve-rich, heat-sensitive membrane becomes activated in response to heat emitted as infrared waves by live animals. When the snake detects the heat of the animal, it localizes the prey by moving its head back and forth until both pits detect the same intensity of heat. This indicates that the prey is centered in front of the snake.

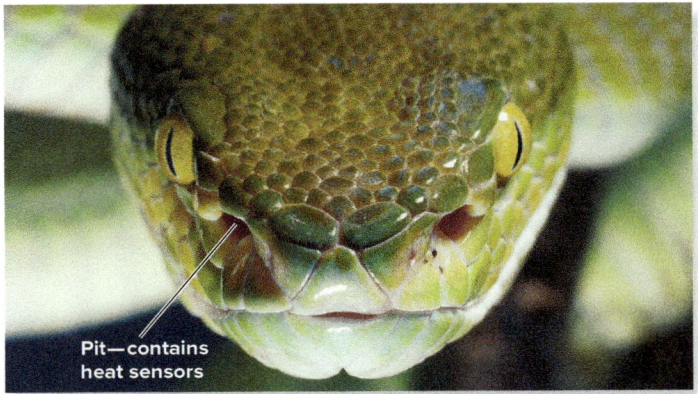

Figure 34.9 Heat sensing in a reptile. Sensory pits enable a white-lipped pit viper (*Cryptelytrops albolabris*) to detect and localize the heat given off by its prey.
©Daniel Heuclin/Science Source

Concept Check: What advantage is provided by the presence of sensory pits on both sides, rather than just one side, of a pit viper's head?

Nociceptors Warn of Pain

Nociceptors are neurons in the skin and internal organs. They respond to tissue damage or to stimuli that are about to cause tissue damage. Nociceptors are unusual because they can respond not only to external stimuli, such as extreme temperatures, but also to internal stimuli, such as molecules released into the extracellular space from injured cells. Damaged cells release a number of substances, including acids and small signaling molecules called prostaglandins, that cause inflammation and make nearby nociceptors more sensitive to painful stimuli. Anti-inflammatory drugs such as aspirin and ibuprofen reduce pain by preventing the production of prostaglandins.

Signals arising from nociceptors travel to the CNS and reach the cerebrum, where the type or cause of the pain is interpreted. The signals are also sent to the limbic system, which holds memories and emotions associated with pain, and to the reticular formation in the brainstem. The reticular formation increases alertness and arousal—an important response to a painful stimulus (see Chapter 33 for a discussion of brain regions and their functions).

The sense of pain—nociception—is one of the most important of all the senses, as was noted in the chapter-opening description of a young child with congenital insensitivity to pain. It informs an animal whether it has been injured and triggers behavioral responses that protect it from further danger. Although in many cases we cannot know whether or how animals perceive pain, nociceptors have been identified in all classes of vertebrates and in many invertebrates.

34.3 Reviewing the Concepts

- Thermoreceptors in the skin and brain allow an animal to sense external and internal temperatures, respectively. Nociceptors in the skin and internal organs sense pain, an important survival adaptation in animals (Figure 34.9).

34.3 Testing Your Knowledge

1. Which of the following statements is *false*?
   a. Nociceptors and thermoreceptors are sensory neurons.
   b. Thermoreceptors are found in the skin and the brain of some animals and are of two types: hot- and cold-sensing thermoreceptors.
   c. Signals from painful stimuli are not sent to the CNS but are interpreted and acted upon entirely by neurons within the PNS.
   d. Thermoreceptors are often linked with reflexive behaviors.
   e. Nociceptors may respond not only to pain but also to certain other things such as extreme temperature.

34.4 Photoreception

Learning Outcomes

1. Describe the structures of the various types of visual organs in animals.
2. Compare and contrast the structure and function of rods and cones, and explain how they detect and respond to light.

A type of sensory reception called electromagnetic reception is widespread in animals. It includes the ability of different animals to detect magnetic or electric fields, such as the ability of some sharks and rays to detect the tiny electric fields generated by the beating of the hearts of their prey. The most common type of electromagnetic sensing, however, is photoreception. Visual systems employ specialized neurons called **photoreceptors** that detect light. Some of the properties of light were described in Chapter 7. In this section, we will examine the variety of animal visual organs that sense light and send signals to the brain, where the image is interpreted. We will then take a detailed look at the structure and function of the vertebrate eye.

Compound Eyes Are Found in Many Invertebrates

The ability to sense light is an ancient adaptation found even in many unicellular organisms, where it may provide a selection advantage for photosynthesis or protection. In some cases, as in certain flatworms, an animal may have simple eyespots that contain a small number of light-sensitive molecules but lack the ability to discern the direction of a light source or to interpret a visual image.

In contrast, arthropods and some annelids have image-forming **compound eyes** (Figure 34.10a), which consist of hundreds to as many as 10,000 light detectors called ommatidia (singular, ommatidium) (Figure 34.10b). Each ommatidium makes up one facet of the eye. Within the ommatidium, a two-part **lens**—composed of an outer cornea and an inner crystalline cone—focuses light onto a long, central structure called a rhabdom (Figure 34.10c). The rhabdom is a column of light-sensitive microvilli that project from the cell membranes of the photoreceptor cells of the ommatidium (Figure 34.10d). The light-sensitive molecules required for vision are located in the microvilli; the extensive surface area of the microvilli provides the eye with increased sensitivity to light. Pigment cells surrounding the photoreceptor cells absorb excess light, thereby isolating each ommatidium from its neighbors.

Each ommatidium senses the intensity and color of light. Individually, each ommatidium receives light from only a very narrow field, but collectively they provide animals with a wide viewing area. In addition, animals with compound eyes are very sensitive to the movement of objects as they travel across the viewing areas of successive ommatidia. By combining the different inputs from neighboring ommatidia, the compound eye is believed to form a mosaic-type image that is interpreted by the brain. Animals such as bees and fruit flies, with large numbers of ommatidia, presumably have sharper vision than those with fewer sensory cells, such as grasshoppers. Behavioral studies have shown, however, that the resolving power of even the best compound eye is considerably less than that of the single-lens eye, which we will consider next.

Vertebrates and Some Invertebrates Have a Single-Lens Eye

Single-lens eyes are found in vertebrates and in some mollusks, such as squids and octopuses, and in some annelids. In such eyes, different patterns of light emitted from or reflected off objects in the animal's field of view are transmitted through a small opening, or pupil, through a lens, to a sheetlike layer of photoreceptors called the **retina** at the back of the eye (Figure 34.11a). The light inputs form a visual image of the environment on the retina. The activation of these photoreceptors triggers electrical changes in neurons that pass out of

Figure 34.10 The compound eye of insects. (a) Close-up of the eyes of a fruit fly (*Drosophila melanogaster*). (b) Each eye has approximately 1,000 ommatidia, which form a sheet on the surface of the eye. (c) Each ommatidium has a lens that directs light to the photosensitive rhabdom. (d) Extending from each photoreceptor cell and forming the rhabdom are many light-sensitive microvilli.
(a) ©Eye of Science/Science Source

Figure 34.11 The vertebrate single-lens eye. (a) Major structural features of the human eye. (b) Changes in lens shape help the eye to focus at different distances. When an object is near, the ciliary muscles contract and the lens becomes rounder, causing light to bend more. When the object is far away, the ciliary muscles relax and the lens flattens.

the eye through the optic nerve, carrying the signals to the brain. The brain then interprets the visual image that was transmitted.

As illustrated in Figure 34.11a, the vertebrate eye has an outer sheath called the sclera (the white of the eye). At the front of the eye, the sclera is continuous with a thin, clear layer known as the **cornea.** The iris is a circle of pigmented smooth muscle responsible for eye color, with the pupil in its center. The size of the pupil changes when the muscles of the iris reflexively relax or contract to allow more or less light to enter the eye. Between the sclera and the retina is a layer of blood vessels called the choroid. Within the eye are two chambers containing fluids—the aqueous and vitreous humors—that help to maintain eye shape.

Because light radiates in all directions from a light source, light must be bent (refracted) inward toward the photoreceptors at the back of the eye. This is accomplished by the cornea and the lens. Whenever light passes from one medium to another medium of a different density, light waves will bend (try looking at a pencil in a glass partly filled with water). The cornea initially refracts the light. The light then passes through the lens, where it is refracted again and focused onto the layer of photoreceptors, the retina, at the back of the eye. The sharpest images are those that form directly at the back, at the macula; this region includes the fovea, which contains a particularly high density of photoreceptors. The bending of the incoming light results in an upside-down and laterally inverted image on the retina, but the brain adjusts for this, and the image is perceived correctly (Figure 34.11b).

The lens is adjusted to focus light that comes from different distances. In the mammalian eye, the lens changes shape to become more or less round depending on whether the object being looked at is near or far. Contraction and relaxation of the ciliary muscles adjust the lens according to the angle at which light enters the eye. When the lens is stretched, it flattens, and light passing through it bends less than when it is round (Figure 34.11b).

Rods and Cones Are Photoreceptor Cells

We will turn now to the structure and function of the photoreceptors. Many animals, including humans, have two types of photoreceptors with names that are derived from their shapes: **rods** and **cones** (Figure 34.12). Rods and cones are cells with three functional parts. The first, the outer segment, contains folds of membranes that form stacks of discs. These discs contain the molecules that absorb light. The second, the inner segment, contains the cell nucleus and other cytoplasmic organelles. The third functional part is the synaptic terminal. Rods and cones do not have axons but have synaptic terminals with vesicles that contain neurotransmitter. These terminals synapse with neurons within the retina.

Rods are extremely sensitive to low-intensity light, but they do not discriminate different colors. Rods are used mostly at night, and they send signals to the brain that generate a black-and-white visual image. Unlike rods, cones are sensitive to wavelengths of light that allow animals to perceive color. Cones function best in daylight and are used by diurnal vertebrate species and by some insects such as the honeybee, which can detect the yellow color of pollen. Compared with rods, the human retina has fewer cones, but they are very densely clustered in and around the fovea; their density provides for sharp vision when we look directly at an object. Cones are less sensitive to light than rods, but this is less critical in daylight because the amount of light reaching the eyes at this time far exceeds what is needed to stimulate any photoreceptor cell.

Rods and Cones Contain Visual Pigments That Absorb Light

Visual pigments are the molecules that absorb light; they are embedded in the disc membranes of the outer segments of rods and cones (Figure 34.13). These pigments consist of two components bonded together. The first component is **retinal,** a derivative of vitamin A that is capable of absorbing energy from light. The second component of visual pigments is a protein called **opsin,** of which there are several types. Each type binds with retinal in different ways, which alters the wavelength of light that retinal absorbs.

Rods have only one visual pigment, called **rhodopsin** (inset, Figure 34.13). Cones may contain any one of several types of visual pigments called cone pigments. The structural differences of the opsins in different cone pigments determine the wavelengths of light that the retinal in a given cone can absorb; in humans, these correspond to red, green, or blue light.

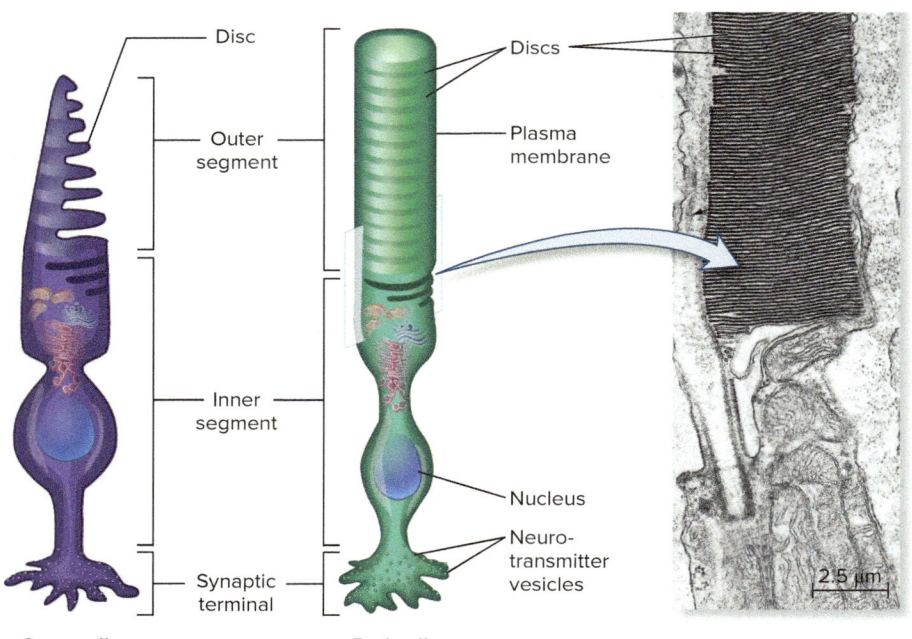

Figure 34.12 Structure of cones and rods. The illustration shows the structure of a cone and a rod, and the transmission electron micrograph shows a section of a rod. Note the stacks of multiple membranous discs in the outer segment of each cell.
(right) ©Don W. Fawcett/Science Source

Biology Principle

Structure Determines Function

The stack of discs in a photoreceptor's outer segment is an excellent example of a structural adaptation that increases surface area without significantly increasing volume, thereby allowing for greater light-capturing ability and improving the function of the cell as a sensory cell.

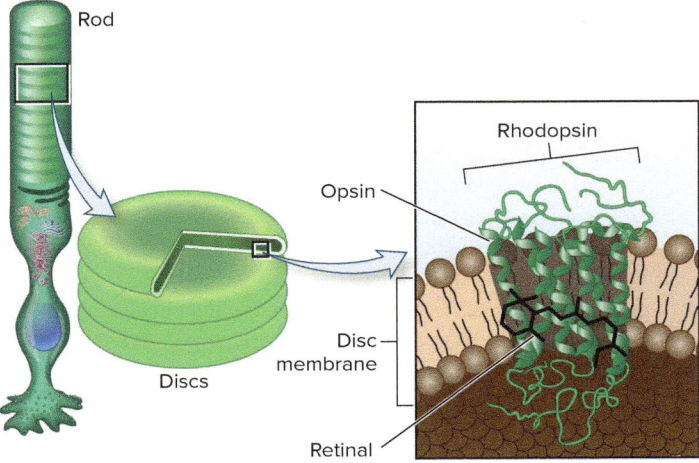

Figure 34.13 A visual pigment. The visual pigment rhodopsin spans the membrane of each disc in a rod photoreceptor. It is composed of a membrane protein, opsin, that is bound to a molecule of retinal, a derivative of vitamin A that is capable of absorbing energy from light.

EVOLUTIONARY CONNECTIONS

Color Vision Is an Ancient Adaptation in Animals

Color vision requires the presence of at least two types of opsins with optimal sensitivities to light of different wavelengths. Genome analyses and behavioral and neurophysiological testing of many animals have confirmed that color vision is widespread across invertebrate and vertebrate taxa, and it arose early in animal evolution. It is debatable why color vision provided a selective advantage to animals. Some investigators believe that it produced more acute vision due to an overall improved contrast and was not originally related to such commonly observed phenomena today as colorful plumage displays or brightly colored flowers. Coloration in plants may have evolved in response to the color sensitivity in animals, not vice versa.

Evolutionary studies indicate that the primordial color sensitivity was to short wavelengths in the ultraviolet/blue region of visible light (refer back to Figure 7.4). From there, additional opsins evolved with sensitivities to medium and longer wavelengths. Most vertebrates have four types of color-sensitive photoreceptors. Some insects, including the butterfly *Papilio xuthus,* have up to 6 types of light receptors, and certain crustaceans (for example, mantis shrimp) have 12! Such animals presumably see shades of colors we cannot. A well-studied model is the honeybee, which has four color-sensitive photoreceptors that respond to wavelengths ranging from 300 nm to 600 nm (refer back to Figure 7.4). Honeybees can perceive ultraviolet light that is not visible to humans. A rainbow would look very different to a honeybee than it does to humans, extending past the blue edge and stopping short of the red edge.

An intriguing exception to vertebrates are most mammals and, in particular, primates. Most mammals today are dichromatic; that is, they have only two types of opsins and thus see a more limited color palette than do other vertebrates. One possible explanation for dichromatic vision may be related to the evolution of mammals. The earliest mammals were probably nocturnal animals. Recall that cones are less sensitive to low levels of light and therefore color vision is limited at night. For this reason, natural selection may have favored a loss of some color-sensitive photoreceptors, thereby allowing for more rods and improving night vision. Primates, however, are an exception, because a third opsin reappeared in the course of primate evolution, one that is sensitive to middle (green) wavelengths. It has been postulated that this may have imparted an advantage that allowed fruit- and leaf-eating primates to better discern orange, red, and yellow fruits against a background of green leaves.

Most humans have normal trichromatic color vision. However, problems in color vision may result from defects in the cone pigments arising from mutations in the opsin genes. The most common is red-green color blindness, which occurs predominantly in men (1 in 12 males compared with 1 in 200 females). Individuals with red-green color blindness either lack the red or green cone pigments entirely or, more commonly, have one or both of them in an abnormal form. In one form, for example, an abnormal green pigment responds to red light as well as green, making it difficult to discriminate between the two colors.

Researchers have determined that color blindness results from a recessive mutation in one or more genes encoding the opsins. Genes encoding the red and green opsins are located on the X chromosome, but the gene encoding the blue opsin is located on a different chromosome. In males, the presence of only one X chromosome means that a single recessive allele from the mother results in red-green color blindness, even though the mother herself may not be color blind (**Figure 34.14**).

Although there is no cure or treatment for color blindness in humans, in 2009 American neuroscientist Katherine Mancuso and colleagues were able to restore trichromatic color vision in adult squirrel monkeys that had been dichromatic (red-green color blind) since birth. This was

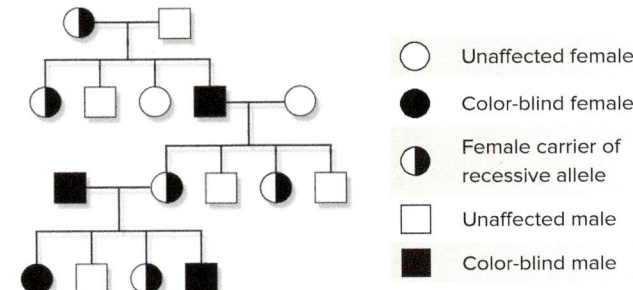

Figure 34.14 Color blindness. An example of a pedigree for red-green color blindness showing all possible offspring.

Concept Check: *Why is red-green color blindness relatively rare in females?*

accomplished using gene therapy in which the missing opsin gene was introduced into each eye, where it was then expressed in existing photoreceptors. Behavioral testing revealed that the monkeys could, for the first time, distinguish red and green colors. Incredibly, the visual centers of the adult brain were able to perceive this new input immediately, despite not having been "wired" for red-green signals during early life.

Photons Change Photoreceptor Activity by Altering the Conformation of Visual Pigments

Photoreceptors differ from other sensory receptor cells because at rest in the dark their membrane is slightly depolarized, whereas in response to a light stimulus it becomes hyperpolarized rather than further depolarized (Figure 34.15). In the dark, the cell membranes of the outer segments of resting cells are highly permeable to Na^+. These ions diffuse into the cytosol of the cell through open Na^+ channels in the outer segment membrane. The Na^+ channels are opened by intracellular cyclic guanosine monophosphate (cGMP). In the dark, cytosolic concentrations of cGMP are high, keeping Na^+ channels open and depolarizing the cell. This depolarization results in a continuous release of the neurotransmitter glutamate from the synaptic terminal of the photoreceptor. The photoreceptor synapses with a postsynaptic cell that is the next neuron in the visual pathway. This initiates a series of events within the retina that is interpreted by the brain as an absence of light. In contrast, when exposed to light, the Na^+ channels in the outer segment membranes of the photoreceptor close. The resulting decrease in Na^+ concentration leads to a hyperpolarization of the cell. In response, the release of glutamate is stopped. This results in a series of cellular activations within the retina and brain that is interpreted as a visual image.

Let's take a more detailed look at the steps in the signal transduction pathway that allow a photoreceptor to respond to light (Figure 34.16).

1. When the photoreceptor is exposed to light, the retinal within the visual pigment absorbs a photon, the smallest unit of electromagnetic energy, including light (see Chapter 7).

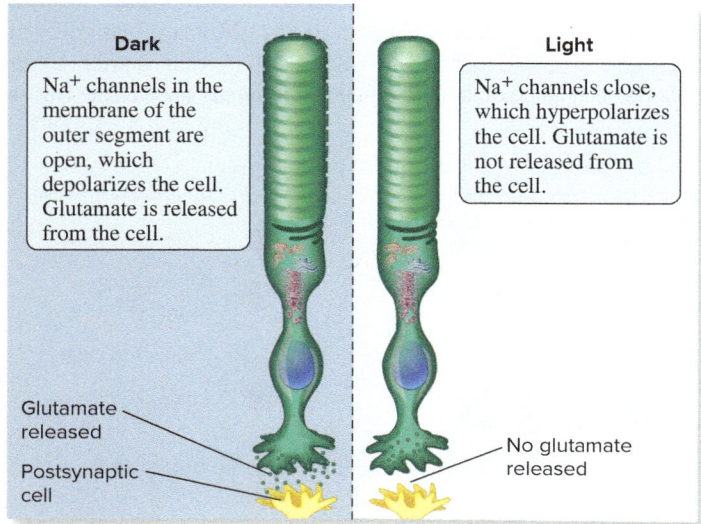

Figure 34.15 Membrane potential response of a photoreceptor to dark and light.

The energy of the photon alters the form of retinal from *cis*-retinal to *trans*-retinal, an isomer with a slightly different conformation due to a rotation at one of the molecule's double bonds (Figure 34.16).

2. This change results in retinal briefly dissociating from the opsin protein. This causes a conformational change in the opsin that, in turn, activates a G protein called transducin, located in the disc membrane. (Thus, opsins belong to the class of proteins called G-protein-coupled receptors.)

3. The activated transducin activates another disc protein, a phosphodiesterase enzyme that is specific for cGMP. When phosphodiesterase is activated, it decreases the concentration of cytosolic cGMP by converting that molecule to GMP.

4. The action of phosphodiesterase results in the closure of Na^+ channels in the outer segment membrane. Remember that in the dark these channels are kept open by intracellular cGMP. However, as the amount of cGMP in the cytosol decreases, the Na^+ channels of the outer segment membranes close, and there is no longer any net diffusion of Na^+ into the cell. The membrane potential of the cell becomes less positive than it was in the dark. Therefore, the response of the cell is a hyperpolarization that is proportional to the intensity of the light.

As a result of hyperpolarization, there is a decrease in glutamate release from the photoreceptor (see Figure 34.15), ultimately leading to a visual image. The sequential activation of enzymes following activation of a single photoreceptor results in an amplification of the original signal (refer back to Figure 8.15). Because of this property, many animals are able to detect extremely low levels of light.

The Visual Image Is Refined in the Retina

We will now turn our attention to the neural pathway through which the visual signal travels to reach the brain. The vertebrate retina has several layers of cells (Figure 34.17). The photoreceptors (rods and cones) are positioned at the back of the retina. The next layer of the retina contains bipolar cells, so named because one end (or "pole") of the cell synapses with the photoreceptors, and the other end relays responses to the front layer of cells, the ganglion cells. The ganglion cells send their axons into the optic nerves, which carry the information to visual centers of the brain.

The pathway for light reception begins at the rods and cones. When the photoreceptors absorb photons, the cells hyperpolarize and no longer release glutamate onto bipolar cells. Glutamate normally inhibits bipolar cells, and thus its removal leads to their depolarization. The depolarized bipolar cells release their neurotransmitters onto ganglion cells, stimulating action potentials in those cells.

Signals from the ganglion cells travel along pathways that reach several parts of the brain, including the thalamus, brainstem, cerebellum, and cerebral cortex. Some of these brain regions are involved in functions such as coordinating an animal's movement with its environment or eliciting emotional states that are relevant to what an animal is seeing. The cerebral cortex responds to characteristics of the visual scene such as whether something is moving, how far away it is, how one color compares with another, and the nature of the image (for example, a face). The cortex does not form a picture in the brain but forms a spatial and temporal pattern of electrical activity that is perceived as an image.

Figure 34.16 Signal transduction pathway in a photoreceptor (rod) cell in response to light.

1. In the disc membrane, the *cis*-retinal in the visual pigment absorbs a photon of light, isomerizing it to *trans*-retinal.

2. The change to *trans*-retinal causes a conformational change in opsin that activates transducin, a G protein in the disc membrane.

3. Transducin activates a phosphodiesterase, which converts cGMP into GMP.

4. The concentration of cGMP in the cytosol decreases, causing cGMP to dissociate from Na⁺ channels in the outer segment membrane. The Na⁺ channels then close, and the cell hyperpolarizes.

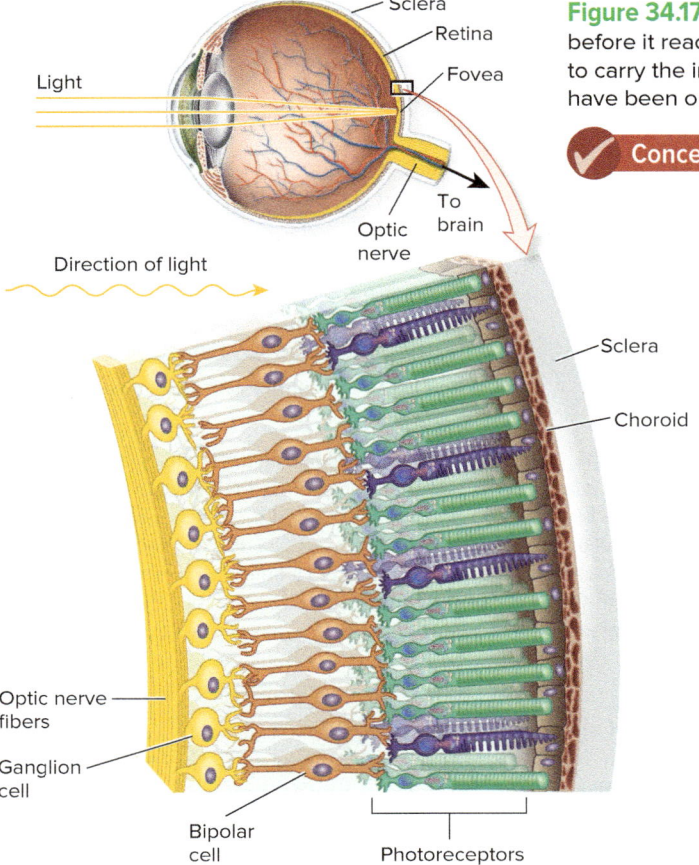

Figure 34.17 **The arrangement of cells in the retina.** Light passes through layers of cells before it reaches the photoreceptors. The ganglion cells generate action potentials to carry the information to the brain via the optic nerves. Additional structural features have been omitted for clarity.

Concept Check: *What type of cells form synapses with photoreceptors?*

34.4 Reviewing the Concepts

- The compound eye found in many invertebrates consists of many ommatidia that focus light (Figure 34.10).
- The single-lens eye is found in vertebrates and certain invertebrates. These eyes form a visual image of the environment on the retina (Figure 34.11).
- Rods and cones are photoreceptors found in the vertebrate eye. Only cones detect color. The visual pigment in rods and cones consists of retinal and a protein called opsin. Red-green color blindness is due to a defect in a type of opsin found in cone pigments (Figures 34.12, 34.13, 34.14).
- The absorption of photons by retinal triggers a signal transduction pathway that causes hyperpolarization of the photoreceptor (Figures 34.15, 34.16).
- The retina is composed of layers of cells, including photoreceptors, that receive light input and convey it to the visual centers of the brain via the optic nerve (Figure 34.17).

34.4 Testing Your Knowledge

1. Which of the following statements is true of compound eyes?
   a. They do not contain a lens or lenslike structure.
   b. They cannot sense color.
   c. They have one ommatidium.
   d. They are found in insects and many vertebrates.
   e. They probably have less resolving power than single-lens eyes.
2. Which of the following statements is *false*?
   a. Transducin is activated by an opsin.
   b. The cytosolic concentration of cGMP in a photoreceptor decreases upon light exposure.
   c. Photoreceptors of both compound and single-lens eyes have specializations that increase surface area.
   d. The sequence of electrical signaling in the vertebrate eye is from photoreceptor to ganglion cell to bipolar cell and then to the optic nerve.
   e. Color blindness in humans is an X-linked disorder.

34.5 Chemoreception

Learning Outcomes

1. **SCISKILLS** ▶ Describe olfaction and gustation in animals, and propose reasons for why these senses are advantageous.
2. Explain how olfactory receptors respond to the binding of odor molecules.
3. Outline how receptor cells within taste buds respond to the binding of food molecules.

(a) **Olfactory receptor cells in olfactory epithelium**

(b) **Cilia from dendrite with odor molecule receptors**

Figure 34.18 Olfactory structures in the human nose. Odor molecules dissolve in a layer of mucus that coats the olfactory receptor cells. The molecules bind to receptor proteins in the membranes of cilia that extend from the olfactory receptor cells. Action potentials in the olfactory receptor cells are conducted to cells in the olfactory bulb, and from there to the brain for interpretation. Basal cells periodically differentiate into new olfactory receptor cells, replacing dead or damaged cells.

Chemoreception includes the senses of smell (**olfaction**) and taste (**gustation**), both of which involve detecting chemicals in the environment or food. These chemicals bind to sensory receptors called **chemoreceptors,** which in turn initiate electrical responses that travel to the brain. In terrestrial animals, airborne molecules that bind to olfactory receptors must be small enough to be carried in the air and into the nose. Taste molecules can be larger because they are conveyed in food and liquid.

Taste and smell are important mechanisms by which many species sense and learn about their environment. The two senses are closely related. In fact, the distinction is largely meaningless for aquatic animals, because for them all chemoreception comes through the water. Even in terrestrial animals, about 80% of the perception of taste is actually due to the activation of olfactory receptors. This is why food loses its flavor when the sense of smell is impaired, such as when you have a cold. In this section, we will explore chemoreception in mammals, since they provide well-studied examples.

Mammalian Olfactory Receptors Respond to the Binding of Odor Molecules

The olfactory sensitivity of mammals varies widely depending on their supply of olfactory sensory receptor cells, which ranges from 5 or 6 million in humans to 100 million in rabbits and 220 million in dogs. Olfactory sensory receptors are neurons located in the epithelial tissue at the upper part of the nasal cavity (**Figure 34.18a**). These cells are surrounded by two additional cell types: supporting cells and basal cells. Supporting cells provide physical support for the sensory receptors. In humans, basal cells differentiate into new sensory receptors every 30–60 days, replacing those that have died.

Olfactory sensory receptors have dendrites from which long, thin cilia extend into a mucous layer that covers the epithelium. These cells do not function like the mechanoreceptor hair cells of the auditory and vestibular systems; the cilia are different from hair cell stereocilia, which bend. Instead, olfactory sensory receptor cells have receptor proteins located in the plasma membranes of the cilia (**Figure 34.18b**). Airborne molecules dissolve in the mucous layer and bind to these receptor proteins. When an odor molecule binds to its receptor protein, it initiates a signal transduction pathway that opens Na^+ channels in the plasma membrane; this permits Na^+ to diffuse into the cell due to the ion's electrochemical gradient. The subsequent depolarization results in action potentials that are transmitted to the next series of cells, which are located in the olfactory bulbs of the brain. The olfactory bulbs, which are part of the limbic system, are a collection of neurons that act as an initial processing center of olfactory information and relay it to the cerebral cortex for further processing and interpretation.

The relative size of the olfactory bulbs indicates the importance of olfaction to an animal. In humans, the olfactory bulbs make up only about 5% of the weight of the brain, whereas in nocturnal animals, like rats and mice, they can constitute as much as 20%. Even with their relatively limited olfactory sensitivity, however, humans appear to have the capacity to detect up to 10,000 different odors, and other mammals are thought to respond to many more. Recent research suggests that this number could theoretically be as great as 1 trillion! Exactly how it is possible to detect such a large number of odors remained a mystery until 1991, when two scientists uncovered the molecular basis of olfaction.

FEATURE INVESTIGATION

Buck and Axel Discovered a Family of Olfactory Receptor Proteins That Bind Specific Odor Molecules

How does the olfactory system discriminate among thousands of different odors? American neuroscientists Linda Buck and Richard Axel set out to study this question. When they began, two hypotheses were proposed to explain this phenomenon. One possibility was that many different types of odor molecules might bind to one or just a few types of receptor proteins, with the brain responding differently depending on the number or distribution of the activated receptors. The second hypothesis was that olfactory sensory receptor cells can make many different types of receptor proteins, each type binding a particular odor molecule or group of related odor molecules.

To begin their study, Buck and Axel assumed that olfactory receptor proteins would be highly expressed in the olfactory sensory receptor cells, but not in other parts of the body. Based on previous work, they also postulated that the receptor proteins would be members of the large family of G-protein-coupled receptors (GPCRs). As shown in **Figure 34.19**, they isolated olfactory sensory receptor cells from rats and then used a homogenizer to break open the cells to release their mRNA. The mRNA was purified and used to make complementary DNA (cDNA) using the enzyme reverse transcriptase. This generated a large collection of cDNAs, representing all of the genes that were expressed in the receptor cells at the time of mRNA collection.

To determine if any of the cDNAs encoded GPCRs, the researchers used primers that recognized conserved regions within previously identified genes that encoded GPCRs. A conserved region is a DNA sequence that rarely changes among different members of a gene family (see Chapter 18 for a description of gene families). The primers were used in a polymerase chain reaction (PCR) to amplify cDNAs that encoded GPCRs. This produced many PCR products, which were then subjected to DNA sequencing. Buck and Axel identified at least 100 different genes, each encoding a GPCR with a slightly different amino acid sequence as predicted from the DNA sequence. Significant variability in the predicted amino acid sequences between the olfactory GPCRs was found in the putative ligand-binding region of the molecules (believed to be transmembrane domains 3–5). Further research demonstrated that these genes were expressed in olfactory cells, but not in other parts of the body. These results were consistent with the second hypothesis—namely, that organisms produce a large number of distinct olfactory receptor proteins, each type binding a particular odor molecule or a group of related odor molecules.

Since these studies, researchers have determined that this family of olfactory genes in mammals is surprisingly large. In humans, roughly 400 genes encode olfactory receptor proteins. Each olfactory receptor cell is thought to express only one type of GPCR that recognizes its own specific odor molecule or group of closely related molecules. Most odors an animal encounters, however, are due to multiple chemicals that activate many different types of odor receptors at the same time. We perceive odors based on the combination of receptors that become activated. This ability to combine different inputs allows animals such as humans to perceive thousands of different odors despite having only a few hundred different olfactory receptor proteins.

The research of Buck and Axel explained, in part, how animals detect a myriad of odors. In 2004, the two neuroscientists received the Nobel Prize in Physiology or Medicine for this pioneering work.

Figure 34.19 Buck and Axel identified olfactory receptor proteins in olfactory receptor cells.

HYPOTHESES 1. Many different types of odor molecules bind to just a few types of receptor proteins. 2. Odor molecules are detected by many specific olfactory receptor proteins belonging to the family of G-protein-coupled receptors (GPCRs).

KEY MATERIALS Laboratory rats (*Rattus norvegicus*), PCR reagents, DNA-sequencing gels.

Experimental level | **Conceptual level**

1. Dissect and homogenize olfactory epithelium from laboratory rats.

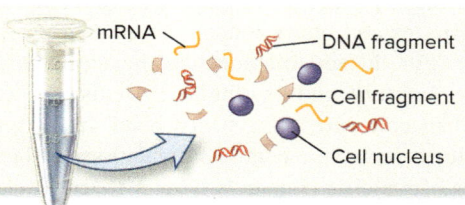

2. Purify mRNA. Make cDNA (described in Chapter 18) from the mRNA, using reverse transcriptase.

Many double-stranded cDNAs

3. Add primers that bind specifically to genes that encode GPCRs. Subject to PCR as described in Chapter 18.

Primers will hybridize only with cDNA that codes for proteins in the GPCR family and amplify those genes. Many different PCR products are obtained, each corresponding to a different gene.

4. Subject each PCR product to DNA sequencing, also described in Chapter 18.

Output from automated sequencing (example)
GTGGACTTAATGCA

Different GPCRs will have slightly different DNA sequences.

5. **THE DATA**

At least 100 different GPCRs were uniquely expressed in olfactory sensory receptor cells. Analysis of their predicted amino acid sequences revealed significant variability in the putative ligand-binding regions (transmembrane domains 3–5).

○ Amino acids that were the same in all the olfactory GPCRs.
● Amino acids that were different among the olfactory GPCRs.

Transmembrane regions shown as cylinders (labeled 1–7)

6. **CONCLUSION** Olfactory receptor cells express many different receptor proteins that account for an animal's ability to detect a wide variety of odors.

7. **SOURCE** Buck, L., and Axel, R. 1991. A Novel Multigene Family May Encode Odorant Receptors: A Molecular Basis for Odor Recognition. *Cell* 65: 175–187.

Experimental Questions

1. What were the two major hypotheses to explain how animals discriminate between different odors? How did Buck and Axel test the hypothesis of multiple olfactory receptor proteins?
2. **SCISKILLS ▶** Of the two hypotheses proposing how animals discriminate between different odors (see question 1), which one was supported by the results of this experiment? With the evidence presented by Buck and Axel, what is the current hypothesis regarding the discrimination of odors in animals?
3. **SCISKILLS ▶** Predict the survival or reproductive advantages that may be provided to animals by the ability to sense thousands or more different types of odors.

Taste Buds Detect Food Molecules

Taste buds are structures containing cells that detect particular molecules in food. Humans have about 9,000 taste buds, whereas other mammals, such as pigs and rabbits, can have double that number. The bumps that you see on your tongue are not the taste buds but are papillae, elevated structures that collect food molecules in depressions called taste pores. These pores contain many taste buds that include the sensory receptor cells along with supporting cells (**Figure 34.20**). The tips of the sensory receptor cells have microvilli that extend into the taste pore. Here, molecules in food that have dissolved in saliva bind to receptor proteins. This binding triggers intracellular signals that alter ion permeability and membrane potentials. The sensory receptor cells then release neurotransmitters onto underlying sensory neurons. Action potentials travel from these neurons to the thalamus and other regions of the forebrain, where the taste is perceived.

There are a number of different types of taste cells, and they are distributed on the tongue in different areas, but these areas overlap considerably. Each type of cell has a specific transduction mechanism that allows it to detect specific chemicals present in the foods and fluids an animal ingests. Their activation results in the perception of sweet, sour, salty, and bitter tastes. A fifth taste called umami (after the Japanese word for delicious) is associated with the presence in ingested food of glutamate and other similar amino acids and is usually described as imparting enhanced flavor to food. It may account for the widely recognized effect of monosodium glutamate (MSG) in enhancing the flavor of food.

The senses of taste and smell are enhanced when we are hungry, a phenomenon that most likely occurs in other animals as well. Once we have eaten, we are less aware of the smell and taste of food. The importance of this for survival is clear: A hungry animal needs to eat. An enhanced sense of smell aids in locating food, and a heightened sense of taste encourages an animal to eat. Afterward, these senses become temporarily dampened so as not to distract the animal from its other needs. The mechanism by which these changes occur is uncertain, but it may involve a temporary alteration in the number of smell and taste receptors or in their ability to bind ligands.

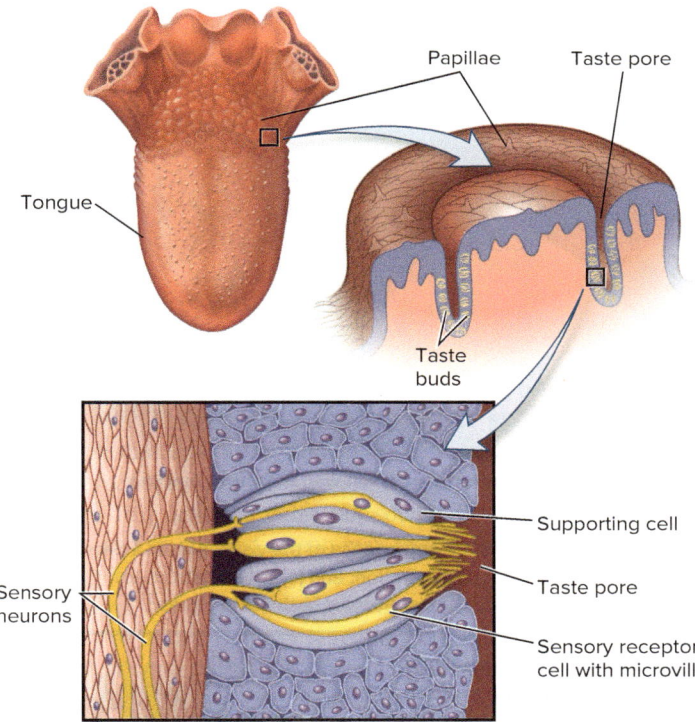

Figure 34.20 Structures involved in the sense of taste. This sense occurs in taste buds, which contain the sensory receptor cells that respond to dissolved food molecules.

Concept Check: *Where are the receptor proteins that bind to food molecules located?*

34.5 Reviewing the Concepts

- In mammals, olfactory sensory receptors are neurons in the nasal cavity. They have cilia with receptor proteins that bind specific odor molecules. The binding initiates a signal that is sent to the brain. Buck and Axel discovered that olfactory sensory receptors have many different types of receptor proteins for odor molecules (Figures 34.18, 34.19).
- Taste buds contain chemosensory cells that detect molecules in food. There are different types of taste cells, the activation of which results in the perception of five tastes: sweet, sour, salty, bitter, and umami (Figure 34.20).

34.5 Testing Your Knowledge

1. Which of the following is *not* true of chemoreception?
   a. It is not unique to mammals.
   b. It includes gustation and olfaction.
   c. The sensory cells are activated by the bending of stereocilia.
   d. Some chemoreceptors are located within taste buds.
   e. It requires receptor proteins.
2. Olfactory sensory cells in humans
   a. are much greater in number than in many other mammals, including dogs.
   b. have a relatively short lifetime and cannot be replaced.

2. _____ sense pain, _____ sense heat or cold, and _____ sense touch.
   a. Mechanoreceptors, thermoreceptors, nociceptors
   b. Nociceptors, thermoreceptors, mechanoreceptors
   c. Nociceptors, thermoreceptors, olfactory receptors
   d. Mechanoreceptors, nociceptors, stretch receptors
   e. Nociceptors, photoreceptors, mechanoreceptors

3. The sensory receptors for audition (hearing) are located in
   a. the organ of Corti.
   b. the middle ear.
   c. the utricle and saccule.
   d. the tympanic membrane.
   e. the outer ear.

4. Which of the following statements is (are) true?
   a. A loss of taste buds would reduce a person's sense of smell.
   b. Olfactory sensory receptors have a limited lifetime and can't be replaced.
   c. Chemoreception only occurs in terrestrial animals.
   d. The relative size of the olfactory bulbs is a good predictor of a mammal's ability to sense odors.
   e. There are typically only a few types of odor receptor proteins in animals.

5. Cone pigments detect different wavelengths of light due to
   a. their location in the retina.
   b. the amount of light they absorb.
   c. the type of retinal they have.
   d. the type of opsin protein they have.
   e. interactions with bipolar cells.

Short Answer

1. Distinguish between sensory transduction and perception.
2. What advantage is provided to an animal by having semicircular canals oriented in three different planes?

Critical-Thinking Skills

Multiple Choice

1. In an experiment to test the function of statocysts, researchers replaced the statoliths of a crayfish with iron filings and then held a magnet directly above the animal's head. Normally, statoliths sit at the bottom of a statocyst unless an animal moves or turns over. What do you think happened in response to the magnet?
   a. The animal turned in circles.
   b. The animal flipped itself over.
   c. The animal began moving backwards.
   d. The animal fell over onto its side.

2. Certain fishes, such as *Astyanax mexicanus*, have evolved an unusually large number of taste buds, large olfactory processing centers in the brain, and extraordinary sensitivity to touch but little or no visual capability. Which of the following environments would you predict such animals inhabit?
   a. tidal pools
   b. dark caves
   c. high-altitude mountain lakes
   d. shallow rivers
   e. coral reefs

3. Which situation would be expected to be associated with the depolarization of rods and cones?
   a. the release of neurotransmitters from bipolar cells
   b. an animal emerging from a burrow into bright sunshine
   c. a person sleeping at night
   d. an increase in signaling in ganglion cells
   e. Both a and d are correct.

4. A 5-year-old girl is tested for color vision in kindergarten and found to be red-green color blind. Her mother is color blind. Which of the following statements regarding her father and siblings is true?
   a. Her father is not color blind.
   b. All of her sisters are color blind, but her brothers are not.
   c. Her father is color blind, and so are all of her sisters and brothers.
   d. Half of the girls in her family and all of the boys are color blind.
   e. There is a 25% chance that any of her siblings is color blind.

5. Which of the following series of events that occur in the vertebrate eye is *not* in the correct order?
   a. arrival of a photon, isomerization of retinal, dissociation of retinal from opsin
   b. hyperpolarization of photoreceptor cell, activation of bipolar cell, activation of ganglion cell
   c. dissociation of retinal from opsin, activation of phosphodiesterase, activation of transducin
   d. activation of phosphodiesterase, decrease in cytosol concentration of cGMP, closure of Na^+ channels

Short Answer

1. Despite the obvious differences in their appearance, try to find structural and functional similarities between the compound and single-lens eyes found in animals.
2. **PRINCIPLES** A principle of biology is that living organisms interact with their environment. This is certainly true with respect to animals' use of their sensory systems. Of the ways in which you use your senses to interact with the environment, which do you think is the least important for human survival, and why?

How Muscles and Skeletons Are Adaptations for Movement, Support, and Protection

35

©Don W. Fawcett/Science Source

False-color scanning electron micrograph of a neuromuscular junction in a mammal (neuronal structures in yellow; muscle cells in red).

Chapter Outline

35.1 Types of Animal Skeletons

35.2 Skeletal Muscle Structure and the Mechanism of Force Generation

35.3 Types of Skeletal Muscle Fibers and Their Functions

35.4 Impact on Public Health

Assessing Your Knowledge and Skills

An active 24-year-old woman began experiencing fatigue after even simple activities and found that she needed to curtail her regular gym workouts. Around the same time, she noticed that she had difficulty swallowing, and one of her eyelids began to droop. Alarmed by these symptoms, she visited her physician. After a series of tests, it was determined that she had developed the disease called myasthenia gravis. This disease ranges in severity, but at its worst, it can make chewing, swallowing, talking, and even breathing difficult. It affects an estimated 10,000–30,000 Americans, usually not becoming apparent until adolescence or later. It cannot be cured, but a variety of medical treatments are available that can decrease its severity.

Myasthenia gravis is a disease of skeletal muscle, the structure and function of which are subjects of this chapter. In this disease, the body's immune system inactivates some of the receptor proteins on skeletal muscle cells. (This type of disease is called an autoimmune disease, which you will learn more about in Chapter 40.) These receptor proteins bind the neurotransmitter acetylcholine (ACh), which is released by motor neurons onto skeletal muscle cells, stimulating the contraction of those cells. Consequently, myasthenia gravis results in a degree of paralysis that depends on the severity of the disease. For this reason, the woman in this example was given a medication that increases the concentration of ACh in the synapses on her skeletal muscle cells. She will have to take this medication for the rest of her life.

All muscles are composed of highly specialized cells that have the ability to contract. The three types of muscle—skeletal, smooth, and cardiac—were introduced in Chapter 32. In this chapter, you will learn about the structure and function of skeletal muscle, the mechanisms by which skeletal muscle controls movements such as lifting an object, and locomotion—the movement of an animal from place to place. For skeletal muscles to produce their effects, however, they usually must exert a force on an animal's skeleton. We will begin the chapter, therefore, with an overview of animal skeletons. Then we will examine the structure and function of skeletal muscle and how the interaction of two muscle proteins—actin and myosin—produces muscle contraction. We will conclude with a consideration of important bone and muscle diseases in humans.

35.1 Types of Animal Skeletons

Learning Outcomes

1. Distinguish between exoskeletons and endoskeletons.
2. List the advantages and disadvantages of an exoskeleton.
3. List the major functions of the vertebrate skeleton.
4. Describe the composition of vertebrate bone.

A broad definition of a **skeleton** is a body structure that serves one or more functions related to support, protection, and locomotion. The first skeletons are believed to have evolved about 550 mya in a small marine invertebrate. This appears to have coincided with a change in the chemistry of the Earth's oceans, including an increased availability of dissolved calcium and other minerals that could be used to harden structures of an animal's body. In this section, we will examine the two major types of skeletons found in animals: exoskeletons (found in many invertebrates) and endoskeletons (found in all chordates and a few invertebrates).

Exoskeletons Are on the Outside of Animal Bodies

Many invertebrates have a hard outer casing or shell that covers all or part of their body surface. Bivalve mollusks, for example, are covered by a tough shell made partly of calcium carbonate. The body surface of arthropods is covered by layers of the polysaccharide chitin and minerals, including calcium. The term **exoskeleton** generally refers to any tough exterior body covering of these types in an invertebrate (**Figure 35.1a**). An exoskeleton surrounds most or all of the body surface and provides one or more of the following: support for the body, protection from the environment (for example, acting as a barrier to dessication in terrestrial species), protection from predators, and protection for internal organs.

Exoskeletons are often tough and durable and, in arthropods, are segmented to allow for flexibility and movement. They provide a support mechanism for the attachment of muscles that participate in movement or locomotion. In arthropods, the exoskeleton must be periodically shed, regrown, and strengthened again to allow the animal to grow, a process called ecdysis, or molting (Figure 35.1a). One disadvantage of such exoskeletons is that when an animal is molting, its new exoskeleton is temporarily soft, making the animal more vulnerable to predators and the environment.

Exoskeletons vary enormously in their complexity, thickness, and durability. The differences in exoskeletons are usually adaptations that enhance an animal's survival. Think, for example, of the difference between the body surfaces of a butterfly and a lobster. A butterfly's exoskeleton must be light enough for the animal to fly, whereas the thick, tough exoskeleton of the lobster provides a very effective defense against predators in the sea. Arthropods are a good example of the benefits of exoskeletons. These animals are among the most successful of all animal phyla living today, having survived for hundreds of millions of years and inhabiting nearly every possible ecological niche on the planet. To some degree, the amazing success of arthropods can be attributed to the exoskeleton.

Endoskeletons Are Internal Support Structures

An **endoskeleton** is an internal, mineralized structure derived from mesoderm (a germ layer in bilateria) (refer back to Figure 26.4) that provides support and protection, including in some cases to internal

Biology Principle

The Genetic Material Provides a Blueprint That Allows Organisms to Grow, Develop, and Reproduce

Part of the growth and development of animals involves their skeletons, as shown in this figure. For animals with exoskeletons, however, this growth may create brief periods when the animal is particularly vulnerable to predation.

(a) An arthropod next to its recently shed exoskeleton

(b) Echinoderm (sea star) and (right) SEM of its endoskeleton

Figure 35.1 Types of skeletons. (a) Exoskeleton. An arthropod's skeleton covers and protects its body, but it must be periodically shed and replaced to allow growth, as shown in this photo. **(b)** Endoskeleton. Echinoderms such as the sea star have endoskeletons of bony plates made of calcium carbonate (inset). Vertebrate endoskeletons will be described later.
(a) ©Tom McHugh/Science Source; *(b left)* ©Georgette Douwma/Science Source; *(b right)* ©The Trustees of the Natural History Museum, London/Alamy

✓ **Concept Check:** *Unlike exoskeletons, endoskeletons do not provide protection for the body surface of animals. How can the lack of an external skeleton be an advantage for animals?*

organs such as those in the thorax of vertebrates. Unlike exoskeletons, endoskeletons do not protect the outer surface of the body and do not need to be shed as an animal grows. Minerals such as calcium, magnesium, phosphate, and carbonate supply the hardening material that gives the skeleton its firm structure.

Various types of internal support structures are found in some sponges. These are spiky networks of proteins and minerals (refer back to Figure 26.11) but are not true endoskeletons. The endoskeletons of echinoderms consist of mineralized, platelike structures. Beneath the body surface of echinoderms are arrays of mineralized plates made largely of calcium carbonate. These plates extend into the spines and arms that radiate from the main body, and they serve primarily a support function (Figure 35.1b). Vertebrate endoskeletons, by contrast, are composed of cartilage, bone, or both and have important functions beyond providing support for an animal's body. Next, we will consider the structure and function of the vertebrate skeleton in greater detail.

The Vertebrate Skeleton Performs Several Important Functions

Vertebrate skeletons are composed of either cartilage—as in cartilaginous fishes (sharks, rays, and skates)—or both cartilage and bone (refer back to Figure 32.5), as in bony fishes, amphibians, reptiles, birds, and mammals. **Bone** is a living, dynamic type of dense connective tissue made up of cells and both organic and mineral components. Let's consider the vertebrate bony skeleton as a whole before examining bones in greater detail.

In the vertebrate skeleton, bones are connected in ways that allow for support, protection, and movement. The vertebrate skeleton is often considered in two parts, called the axial and appendicular skeletons (Figure 35.2). The axial skeleton is composed of the bones that form the main longitudinal axis of an animal's body, including the skull, vertebrae, sternum, and ribs. The appendicular skeleton consists of the limb bones and the bones that connect them to the axial skeleton. A joint is formed where two or more bones come together. Some joints permit free movement (for example, the joints in human shoulders). Other joints allow no movement (fused joints, like those interlocking the skull bones) or allow only limited movement (such as those of the vertebral column).

The skeleton of vertebrates serves several functions in addition to support, protection, and movement. For example, blood cells and platelets, the latter of which help blood to clot (see Chapter 36), are formed within the soft, fatty interior (called the marrow) of certain bones, including the ilia, the vertebrae, and the ends of the femurs. In addition, homeostasis of the concentrations of calcium and phosphate ions in body fluids is achieved in large part through exchanges of these ions between bone and blood.

Bone Consists of a Mixture of Organic and Mineral Components

Within bone are cells that form bone—called osteoblasts and osteocytes—and cells that break it down—called osteoclasts. The organic part of bone is secreted by osteoblasts and osteocytes and consists of the protein collagen. This protein has a unique triple helical structure that gives bone both strength and flexibility (refer back to Figure 4.31). The collagen forms a matrix that becomes woven with minerals. The mineral component is composed of a crystalline

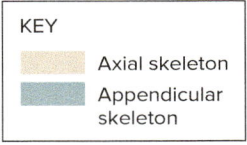

Figure 35.2 The adult human skeleton. This diagram shows the axial (beige) and appendicular (green) parts of the skeleton in an adult human, an animal with an endoskeleton. The adult human skeleton consists of 206 separate bones.

mixture of primarily Ca^{2+} and PO_4^{3-} plus a few other ions that provide rigidity. These ions must be obtained in an animal's diet, absorbed into the blood, and deposited in bone.

Bones cannot move by themselves but instead provide the scaffold upon which skeletal muscles act to cause body movement. We will turn now to a discussion of skeletal muscle and the mechanism by which it generates force.

35.1 Reviewing the Concepts

- Skeletons are structures that provide support and protection and function in locomotion.
- Two major types of skeletons are found in animals. Exoskeletons are protective external structures that typically must be shed and regrown to accommodate an animal's growth. Endoskeletons are internal structures that grow with the animal but do not protect its body surface (Figure 35.1). Vertebrate endoskeletons are made up of the axial and appendicular skeletons (Figure 35.2).

- In addition to the functions of support, protection, and locomotion, the vertebrate skeleton produces blood cells and constitutes a reservoir for ions crucial to homeostasis.

35.1 Testing Your Knowledge

1. A disadvantage of exoskeletons is that
   a. they cannot protect an animal's internal organs.
   b. they must be periodically shed, leaving the animal in a vulnerable state.
   c. they do not provide any flexibility for an animal to move easily.
   d. they are easily damaged, soft structures.
   e. they cannot protect the outside of the body surface.

2. Endoskeletons
   a. are found in all arthropods.
   b. do not protect the outer surface of an animal's body.
   c. must be periodically shed and regrown.
   d. are found in certain invertebrates but not in vertebrates.
   e. are only found in vertebrates.

35.2 Skeletal Muscle Structure and the Mechanism of Force Generation

Learning Outcomes

1. List the three types of muscle found in vertebrates and their locations in the body.
2. Identify the structural components of a muscle down to the level of the sarcomere.
3. Outline the sliding-filament mechanism of muscle contraction.
4. Explain how tropomyosin and troponin help to regulate muscle contraction.
5. Describe how electrical excitation and muscle contraction are linked by events at the neuromuscular junction.

Vertebrates have three types of muscle, which are classified according to their structure, function, and control mechanisms, as first introduced in Chapter 32:

- **Cardiac muscle** is found only in the heart and provides the force required for the heart to pump blood. Cardiac muscle will be discussed in Chapter 36.
- **Smooth muscle** surrounds and forms part of the lining of hollow organs and tubes, including those of the digestive tract, urinary bladder, uterus, blood vessels, and airways. The contraction and relaxation of smooth muscle cells are not under voluntary control. Instead, as described in Chapter 33, they are largely controlled by the autonomic nervous system.
- **Skeletal muscle** is found throughout the body and is directly involved in movements and locomotion. Let's take a detailed look at the structure and function of skeletal muscle and its cells.

A Skeletal Muscle Is a Contractile Organ That Supports and Moves Bones

A skeletal muscle is a grouping of cells—called **muscle fibers**—bound together in bundles by a succession of connective tissue layers (**Figure 35.3**). Skeletal muscles are usually linked to bones by collections of collagen fibers known as tendons. The transmission of force from contracting muscle to bone can be likened to a number of people pulling on a rope attached to a heavy object. Each person corresponds to a single muscle fiber, the rope corresponds to the tendons, and the bone is the heavy object.

When the force generated by a contracting muscle is great enough, the bone to which it is connected moves as the muscle shortens. A contracting muscle exerts only a pulling force, so as the muscle shortens, the attached bones are pulled toward or away from each other. Muscles that bend a limb at a joint (that is, reduce the angle between two bones) are called **flexors,** whereas muscles that straighten a limb (increase the angle between two bones) are called **extensors.** Two or more muscles that produce oppositely

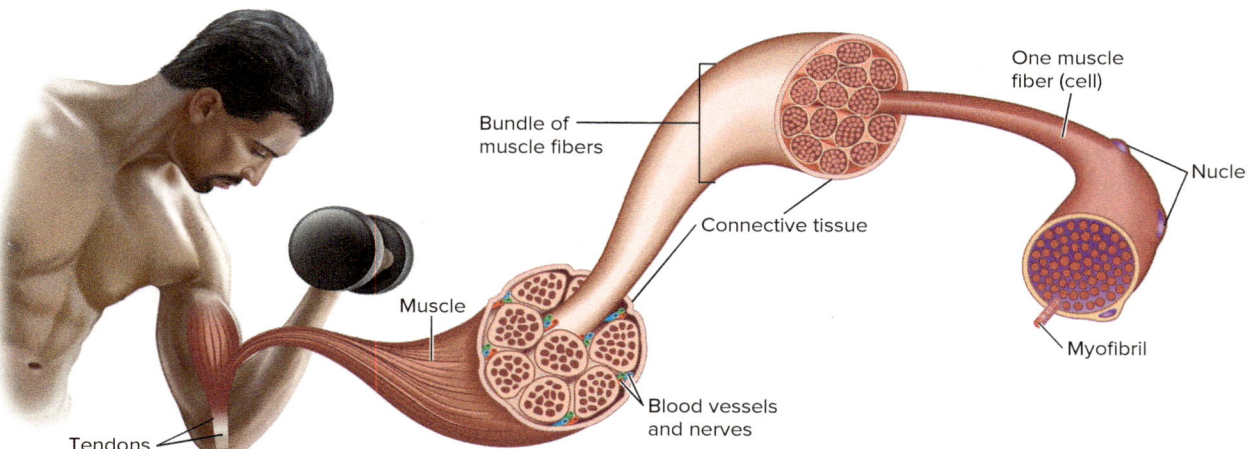

Figure 35.3 Skeletal muscle structure. Skeletal muscles attach to bone by tendons, which are bundles of collagen fibers. Each muscle consists of bundles of muscle fibers (muscle cells) bound together by connective tissue. Myofibrils are the contractile elements of muscle fibers.

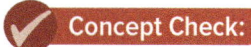 **Concept Check:** What would happen to the ability of a muscle to move a bone if its tendon were torn—for example, due to injury?

Figure 35.4 Actions of flexors and extensors. The figure shows how skeletal muscles cause flexion or extension of a limb. When the flexor muscle contracts, the extensor relaxes, and vice versa.

> ## Biology Principle
> ### Cells Are the Simplest Units of Life
> A muscle consists of many cells called muscle fibers, each containing a highly ordered arrangement of intracellular filaments. As the cells change their shape by shortening, the shape of the entire muscle also changes.

directed movements at a joint are known as **antagonists.** For example, in Figure 35.4, we can see that contraction of the hamstrings (along with relaxation of the quadriceps) flexes the leg at the knee joint, whereas contraction of the quadriceps (with relaxation of hamstrings) causes the leg to extend. Both antagonistic muscles exert only a pulling force when they contract; where their tendons connect to the shin bones determines whether the leg is flexed or extended.

Muscle Fibers Contain Myofibrils Composed of Arrays of Filaments

Each skeletal muscle fiber arises from several cells that fuse to form a single mature cell (see Figure 35.3). Each fiber contains numerous cylindrical bundles known as **myofibrils** (Figure 35.5). Myofibrils extend from one end of the fiber to the other and are linked to the tendons at the ends of the fiber. Each myofibril contains numerous functional structures called thick and thin filaments. These filaments are arranged in a repeating pattern running the length of the myofibril. One complete unit of this repeating pattern is known as a **sarcomere** (from the Greek *sarco*, meaning muscle, and *mer*, meaning part). The **thick filaments** are composed almost entirely of the motor protein **myosin,** which uses ATP as a source of energy. The **thin filaments,** which are about half the diameter of the thick filaments, contain the cytoskeletal protein **actin** and associated proteins. Because the arrangement of thick and thin filaments looks like a series of light and dark bands when viewed by a microscope, skeletal muscle is also known as striated muscle. The names and features of the bands and other components of a sarcomere are as follows (Figure 35.5):

- **A band.** This band is formed by the thick filaments located in the middle of each sarcomere, where their orderly, parallel arrangement produces a wide, dark band. A portion of the thin filaments overlaps the thick filaments in this band.
- **Z line.** This is a network of proteins to which thin filaments are attached. Two successive Z lines define the boundaries of one sarcomere.

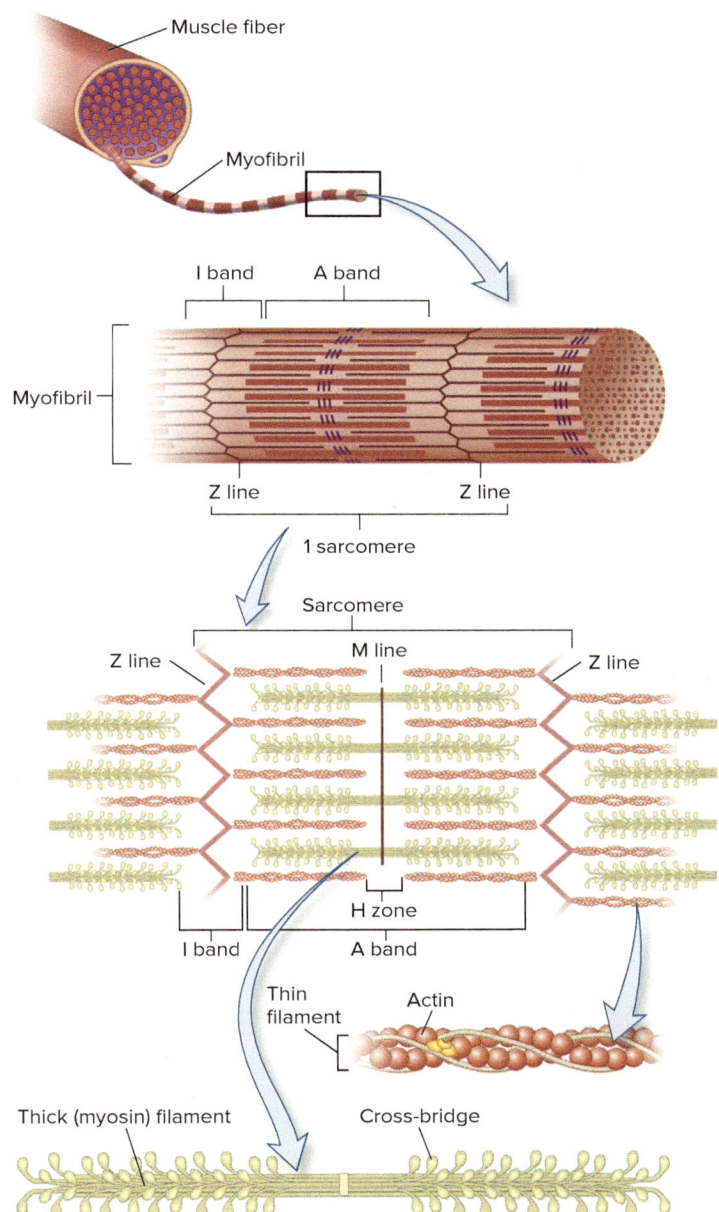

Figure 35.5 Myofibril structure. Each muscle fiber consists of numerous myofibrils containing thick and thin filaments. Each repeating unit of the arrayed filaments constitutes a sarcomere, composed largely of the proteins actin and myosin.

- **I band.** This band lies between the A bands of two adjacent sarcomeres. Each I band contains those portions of the thin filaments that do not overlap the thick filaments, and each I band is bisected by a Z line.
- **H zone.** This is a narrow region in the center of the A band. It corresponds to the space between the two sets of thin filaments in each sarcomere.
- **M line.** This band is in the center of the H zone, and it corresponds to proteins that link the central regions of adjacent thick filaments.

The spaces between overlapping thick and thin filaments are bridged by projections known as **cross-bridges**, which are regions of myosin molecules that extend from the surface of the thick filaments toward the thin filaments (Figure 35.5).

Skeletal Muscle Shortens When Thin Filaments Slide Past Thick Filaments

Movement requires shortening of muscles to pull against attached tendons and bones. However, a muscle can generate force, or contract, without producing movement. Holding a heavy weight at a constant position, for example, requires muscle contraction, but not muscle shortening. Therefore, as used in muscle physiology, the term contraction refers to activation of the cross-bridges within muscle fibers, which initiates the generation of force. When these activating mechanisms are turned off, contractions end, allowing the muscle fiber to relax.

Let's look at how the structure of the thin and thick filaments allows them to move (**Figure 35.6**). In thin filaments, actin proteins form polymers that are arranged into two intertwined, helical chains. These chains are closely associated with two proteins, called tropomyosin and troponin, that have important functions in regulating contraction that will be described in detail later.

Myosin proteins have a three-domain structure composed of two intertwined tails, two hinges, and two heads (Figure 35.6a). The hinges are flexible regions that connect the heads to the tails. Each filament is made of many myosin proteins associated in a parallel array, with the hinges and heads extending out to the sides, forming cross-bridges. Each head contains two binding sites: one for actin and one for ATP. The hydrolysis of ATP by the ATPase activity of myosin provides the energy for the cross-bridge to move via a bending motion at the hinge.

How a muscle fiber actually shortens is known as the **sliding-filament mechanism** of muscle contraction (Figure 35.6b). In this mechanism, the sarcomeres shorten, but neither the thick nor the thin filaments change in length. Instead, the thick filaments remain stationary while the thin filaments slide, pulling on the Z lines and shortening the sarcomere. The sliding-filament movement is propelled by the myosin cross-bridges. During shortening, each cross-bridge attaches to an actin molecule in a thin filament and moves in a motion much like your fist bending at the wrist. Because of the opposing orientation of the thick filaments, the movement of a cross-bridge forces the thin filaments toward the center of the sarcomere (the M line in the H zone), thereby narrowing the H zone and shortening the sarcomere. One stroke of a cross-bridge produces only a very small movement of a thin filament relative to a thick filament. As long as a muscle fiber continues to be stimulated to contract, however, each cross-bridge repeats its motion many times, resulting in continued sliding of the thin filaments. Thus, the ability of a muscle fiber to generate force and movement depends on the amount of interaction between actin and myosin.

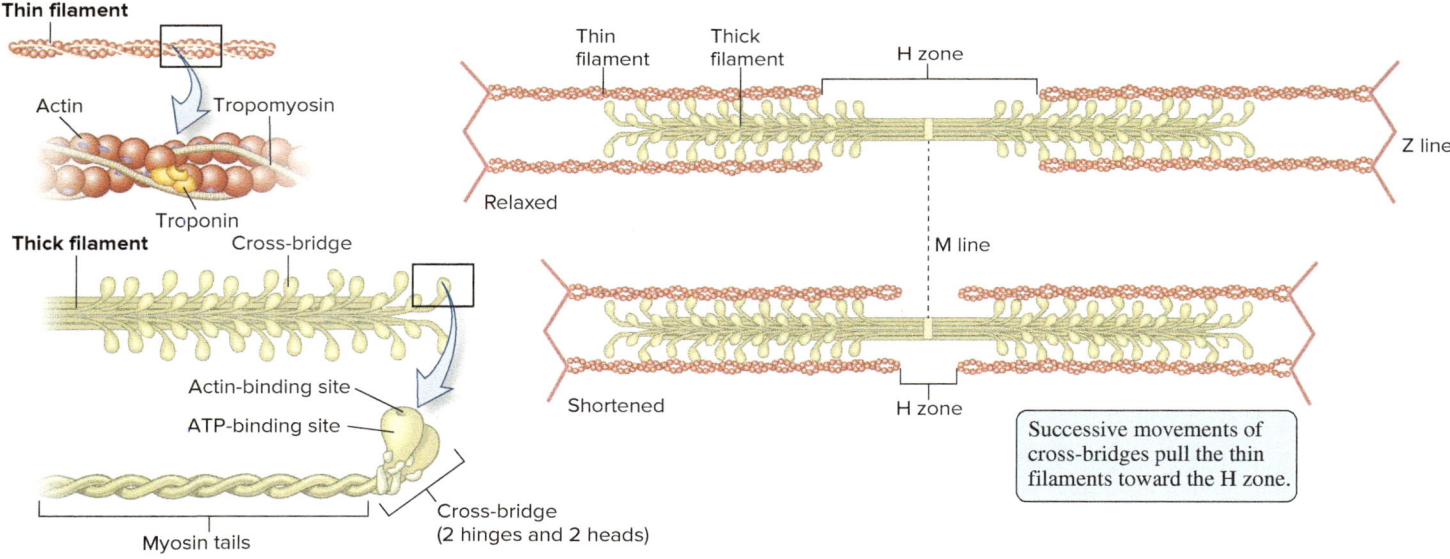

Figure 35.6 Structure and function of the filaments and the sliding-filament mechanism. **(a)** Thin filaments are composed of two intertwined actin molecules and their associated proteins, tropomyosin and troponin. Thick filaments are made of the protein myosin, which has two intertwined tails, two hinges, two heads, an actin-binding site, and an ATP-binding site. The end of a myosin molecule that contains the actin-binding sites is bent at an angle to form the cross-bridge. **(b)** When the cross-bridges on myosin molecules bind to actin, the thin filaments are pulled toward the H zone, shortening the sarcomere. The sliding of thin filaments past the overlapping thick filaments shortens the sarcomere but does not change the lengths of the filaments themselves.

EVOLUTIONARY CONNECTIONS

Myosins Are an Ancient and Diverse Family of Proteins

The protein myosin was first discovered in extracts of frog leg skeletal muscle in the 19th century by German physiologist Wilhelm Kühne. We now know that eukaryotic genomes encode a family of related myosin proteins. Myosins are among the most ancient of eukaryotic proteins and are found throughout the animal kingdom. Small differences in the sequences of genes that arose from a single primordial myosin gene have led to differences in function and tissue-specific expression of various myosin proteins. All eukaryotic myosins, however, share the common feature of possessing binding sites for both actin and ATP in their head domains.

The greatest diversity among myosin proteins evolved in the tail regions. This allowed myosins to evolve different functions in different cell types. In skeletal muscle, as we have seen, myosin is important for generating force, but in other types of cells, myosins function in organelle trafficking, cell locomotion, mitosis, signal transduction, and other activities. Many of these functions are found in nearly all eukaryotes.

Interestingly, one member of the large family of myosin-encoding genes, called *MYH16*, encodes a myosin protein that is expressed mainly in the muscles of the jaw in primates. In 2004, American physician researcher Hansell Stedman and colleagues discovered that this gene, although present in humans, did not encode a functional protein because of a mutation that deleted two base pairs from it. This mutation was present in 100% of people tested worldwide but was not found in the eight nonhuman primate species tested, which included chimpanzees.

Based on genomic comparisons and estimates of genetic divergence among species, the researchers estimated that the mutation occurred approximately 2.4 mya. Significantly, the genus *Homo*, with its smaller jaw and larger braincase, is also thought to have first appeared about 2.4 mya, leading Stedman to suggest that the loss of the MYH16 protein led to the smaller, less muscular jaws characteristic of modern humans (**Figure 35.7**). Because jaw muscles are attached to skull bones, massive jaw muscles may have placed a mechanical constraint on skull growth in early hominins (as well as in modern nonhuman primates). Smaller jaw-closing muscles would place less constraint on bone formation of the outside of the skull (such as the illustrated anchor points obscuring the sutures or growth plates). This in turn would prevent the early cessation of brain growth seen in the nonhuman primates. It may also have eliminated a major constraint on the evolution of the larger brains seen in modern humans.

Stedman's hypothesis is compelling, given that the myosin mutation affects the jaw-closing muscles and is present only in humans. Another research group analyzed gene sequences made available through the Chimpanzee Sequencing and Analysis Consortium—an effort to determine the complete DNA sequence of the chimpanzee genome—and concluded that the mutation may have arisen earlier, perhaps 4.0 to 5.4 mya, before the appearance of small-jawed species of *Homo* in the fossil record. Although scientists are certain that a human-specific mutation in *MYH16* arose at some point in the evolution of hominins, whether the mutation facilitated expansion of the cranium, which in turn may have permitted enlargement of the human brain, remains an intriguing question.

The Cross-Bridge Cycle Requires ATP and Ca^{2+}

Let's now turn our attention to how actin and myosin interact to promote muscle contraction and shortening. The sequence of events that occurs between the time when a cross-bridge binds to a thin filament and when it is set to repeat the process is known as a **cross-bridge cycle.**

Cross-bridge cycling occurs when the cytosolic Ca^{2+} concentration exceeds a critical threshold. This usually occurs when neural input stimulates processes that result in the release of Ca^{2+} from intracellular storage sites (described in detail shortly). In other words, the contraction of skeletal muscle fibers is under nervous control.

Figure 35.8 illustrates the events that occur during the four steps of the cross-bridge cycle: (1) cross-bridge binding, (2) power stroke (moving), (3) detaching, and (4) resetting. As the cycle begins, the myosin cross-bridges are in an energized state, which is produced by the hydrolysis of their bound ATP to ADP and inorganic phosphate (P_i). The ADP and P_i remain for the moment bound to the cross-bridge. The sequence of storage and release of energy by myosin is analogous to the operation of a mousetrap: Energy is first stored in the trap by cocking the spring (ATP hydrolysis) and is then released by the springing of the trap (head binds to actin and moves in power stroke). Let's look at each step of the process more closely.

Step 1: Ca^{2+} concentration increases, triggering the cross-bridge to bind to actin. When the Ca^{2+} concentration in the cytosol increases, an energized myosin cross-bridge along with its associated ADP and P_i bind to an actin molecule on a thin filament.

Step 2: Release of inorganic phosphate (P_i) fuels the power stroke: The cross-bridge and thin filaments move. The binding of an energized myosin cross-bridge to actin in step 1 triggers the release of P_i. This release causes a conformational change in the hinge domain of the myosin protein. This change in conformation causes the cross-bridge to rotate toward the M line in the H zone at the center of the sarcomere as myosin returns to a lower energy conformation. This step is known as the **power stroke,** which moves the actin filament. At this time, ADP is released from the cross-bridge.

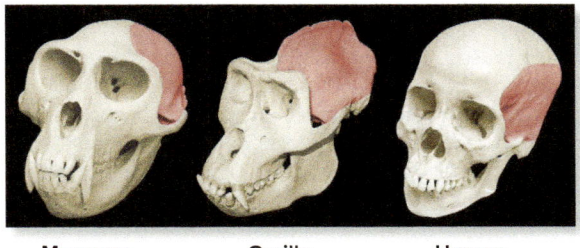

Figure 35.7 Comparison of jaw size and area of muscle attachment (shown in red) in modern primates (not to scale).
(macaque) ©Bone Clones, Inc./Science Source; *(gorilla)* ©Sabena Jane Blackbird/Alamy; *(human)* ©Elizaveta Galitckaia/Alamy

Biology Principle

Biologists Use Models to Test Predictions and Convey Ideas

This figure is a model for the mechanism of cross-bridge cycling in skeletal muscle, which includes a depiction of two ways that ATP functions during each cycle. The Modeling Challenge in the figure legend asks you to predict how the cycle might be affected by the change in cellular ATP content that occurs during rigor mortis.

1. **Binding:** When the Ca^{2+} concentration is high, energized cross-bridge can bind to actin. (ADP and P$_i$ are already bound to the cross-bridge.)

2. **Power stroke:** Release of P$_i$ causes cross-bridge to move toward the H zone of the sarcomere. This power stroke moves the actin filament toward the H zone. ADP is then released.

3. **Detaching:** ATP binds to myosin, causing the cross-bridge to detach from the actin filament.

4. **Resetting:** Hydrolysis of ATP to ADP + P$_i$ provides energy, which causes the cross-bridge to move away from the H zone. ADP and P$_i$ remain bound to the re-energized cross-bridge. Cycle can begin again.

Step 3: ATP binds to myosin, causing the cross-bridge to detach. During the power stroke, myosin is bound very firmly to actin; this linkage must be broken to allow the cross-bridge to be re-energized and the cycle to be repeated. The binding of a new molecule of ATP to the myosin cross-bridge breaks the link between actin and myosin. The dissociation of actin and myosin by ATP is an example of allosteric regulation of protein activity. When ATP binds at one site on myosin (called an allosteric site) (refer back to Figure 6.6c), myosin's affinity for actin decreases. ATP is not hydrolyzed in this step. Instead, ATP functions at this stage as an allosteric modulator of the myosin head, weakening the binding of myosin to actin and leading to their dissociation.

Step 4: ATP hydrolysis re-energizes and resets the cross-bridge. After actin and myosin dissociate, the ATP bound to myosin is hydrolyzed. This hydrolysis energizes myosin, causing it to reset to the conformation that allows actin binding. If Ca^{2+} is still available at this time, the cross-bridge can reattach to a new actin molecule in the thin filament, and the cross-bridge cycle will repeat, causing the muscle fiber to shorten further.

As we have seen, ATP performs two different roles in the cross-bridge cycle. First, the energy released from ATP hydrolysis provides the energy for cross-bridge movement. Second, ATP binding breaks the link formed between actin and myosin during the cycle, causing the cross-bridge to detach, allowing a new cycle to start. Although the precise mechanisms may differ slightly between vertebrates and invertebrates, biologists think that all skeletal muscle functions according to steps similar to those just described.

The Regulation of Muscle Contraction by Ca^{2+} Is Mediated by Tropomyosin and Troponin

How does the presence of Ca^{2+} in the cytosol of muscle cells regulate the cycling of cross-bridges? The answer requires a closer look at the two additional thin filament proteins mentioned earlier—tropomyosin and troponin.

Tropomyosin proteins are arranged end to end along the actin thin filament (**Figure 35.9a**). In the absence of Ca^{2+}, they partially cover the myosin-binding site on each actin molecule, thereby preventing cross-bridges from making contact with actin. Each tropomyosin

---

Modeling Challenge: Rigor mortis is the condition that occurs shortly after death, in which skeletal muscles stiffen. This happens because, as cells die, their ability to produce ATP decreases and eventually stops, and consequently the cytosolic concentration of ATP declines significantly. Figure 35.8 illustrates the cross-bridge cycle in skeletal muscle. Draw a new model of the cycle, beginning with step 1 and continuing as far as you think it will progress in rigor mortis.

Figure 35.8 The four steps of cross-bridge cycling in skeletal muscle. The mechanism by which Ca^{2+} permits actin to bind to energized myosin, and how the concentration of Ca^{2+} in the cytosol is regulated, will be discussed later.

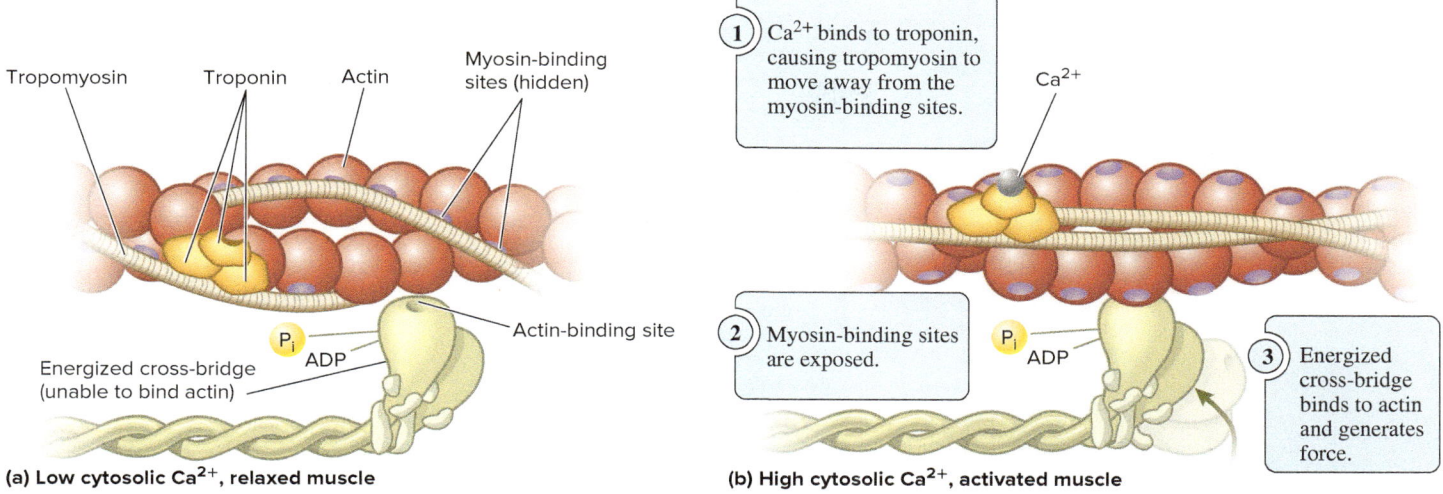

Figure 35.9 Function of Ca²⁺, tropomyosin, and troponin in cross-bridge cycling. **(a)** When the concentration of cytosolic Ca²⁺ is low, the myosin-binding sites on actin are blocked by tropomyosin. **(b)** When the concentration of cytosolic Ca²⁺ increases, Ca²⁺ binds to troponin, which in turn causes tropomyosin to move away from the myosin-binding sites on actin.

is held in this blocking position by **troponin**, a smaller, globular-shaped protein with three subunits that is bound to both tropomyosin and actin. In this way, troponin and tropomyosin block access to myosin-binding sites on actin molecules in the relaxed muscle fiber.

One of the three subunits of troponin is capable of binding Ca²⁺. When Ca²⁺ enters the cytosol, it binds to troponin. The binding of Ca²⁺ produces a change in the shape of troponin, which—through troponin's linkage to tropomyosin—causes tropomyosin to move away from the myosin-binding site on each actin molecule (Figure 35.9b). This movement allows cross-bridge cycling to occur. Conversely, release of Ca²⁺ from troponin reverses the process, blocking the myosin-binding site and stopping contractile activity.

Thus, the concentration of cytosolic Ca²⁺ determines whether or not Ca²⁺ is bound to troponin molecules, which in turn determines the number of actin sites available for cross-bridge binding. The cytosolic concentration of Ca²⁺, however, is very low in resting muscle. Let's see how the Ca²⁺ concentration is increased so that contraction can occur.

Ca²⁺ Concentration and Contraction of Skeletal Muscle Are Coupled with Electrical Excitation

Vertebrate skeletal muscle is electrically excitable tissue. Like neurons, skeletal muscle cells generate and propagate action potentials in response to an appropriate stimulus. Also like neurons, electrical excitation is the event that triggers an increase in cytosolic Ca²⁺. In a neuron, the increased concentration of Ca²⁺ in an axon terminal causes exocytosis of neurotransmitter. In skeletal muscle, however, Ca²⁺ causes contraction of a muscle fiber.

This sequence of events by which an action potential in the plasma membrane of a muscle fiber leads to cross-bridge activity is called **excitation-contraction coupling**. The increased Ca²⁺ concentration continues to activate the contractile apparatus long after electrical activity in the membrane has ceased.

The source of the increased cytosolic Ca²⁺ that occurs following a muscle action potential is the muscle fiber's **sarcoplasmic reticulum**, which acts as a Ca²⁺ reservoir. The sarcoplasmic reticulum, which is a specialized form of the endoplasmic reticulum found in other cells, is composed of interconnected, sleevelike compartments and sacs around each myofibril (Figure 35.10). Separate tubular structures, the **transverse tubules (T-tubules)**, are invaginations of the plasma membrane that lie close to the sarcoplasmic reticulum. The T-tubules course around each myofibril and conduct action potentials from the outer surface of the muscle fiber inside to the myofibrils. An action potential causes the opening of Ca²⁺ channels in the sarcoplasmic reticulum, which allows Ca²⁺ to diffuse into the cytosol and bind to troponin, initiating cross-bridge cycling.

A contraction continues until Ca²⁺ is removed from troponin and the cytosol. This is achieved by ATP-dependent Ca²⁺ transporters in the sarcoplasmic reticulum that restore the Ca²⁺ concentration in the cytosol back to its resting concentration. If this did not occur, a muscle fiber would not be able to repeatedly contract and relax.

Electrical Stimulation of Skeletal Muscle Fibers Occurs at the Neuromuscular Junction

We have seen that an action potential in the plasma membrane of a skeletal muscle fiber is the signal that triggers contraction. How are these action potentials initiated? The mechanism by which action potentials are initiated in a skeletal muscle involves stimulation by a motor neuron. These are neurons whose cell bodies are located in the central nervous system (CNS), with axons that transmit signals away from the CNS to skeletal muscle cells.

The site where a motor neuron's axon terminals synapse with a muscle fiber is known as a **neuromuscular junction** (Figure 35.11 and see the chapter-opening photo). These terminals release the neurotransmitter acetylcholine (ACh) into the synaptic cleft. The region of the muscle fiber plasma membrane that lies directly under an axon terminal is called the motor end plate; it is folded into junctional folds, where the ACh receptors are located. These folds increase the total surface area available for the membrane to respond to ACh.

Figure 35.10 Structure and function of the sarcoplasmic reticulum, transverse tubules (T-tubules), and myofibrils in a skeletal muscle fiber.

BioConnections: *What did you learn earlier about voltage-gated Ca^{2+} channels in cell-to-cell communication in animals? Refer back to Figure 33.12 for help.*

Figure 35.11 **The neuromuscular junction.** The structure of a neuromuscular junction. Action potentials in the motor neuron cause exocytosis of ACh-containing synaptic vesicles. ACh binds to receptor proteins in the plasma membrane of the junctional folds of the skeletal muscle fiber. This initiates Na^+ entry and, consequently, an action potential in the muscle fiber. Excess ACh is removed by the enzyme acetylcholinesterase.

Concept Check: *How does Na^+ enter the muscle fiber after the ACh receptor protein is activated?*

BioConnections: *What is the function of the myelin indicated in this figure? Refer back to Figure 33.10 for help.*

When an action potential in a motor neuron arrives at an axon terminal, it triggers the release of stored ACh, which diffuses across the synaptic cleft and binds to receptor proteins in the junctional folds of the muscle fiber. The ACh receptor is a ligand-gated ion channel (see Chapters 5 and 33). The binding of ACh opens the channel. Na^+ diffuses down its electrochemical gradient into the muscle fiber, which causes the muscle fiber to depolarize. The depolarization results in an action potential that spreads along the membrane of the muscle fiber and through the T-tubules.

Overstimulation of a muscle fiber is prevented by the action of the enzyme **acetylcholinesterase.** Acetylcholinesterase acts in the synapse to break down excess ACh to inactive molecules, ending neurotransmission (Figure 35.11). This inactivation of ACh allows a muscle cell to be stimulated over and over again in a regulated manner. To realize how important this is, imagine if the skeletal muscles that control your breathing were to contract as you inhaled, and then could not relax because the ACh was not removed from the synapse. Exhaling and a subsequent breath would be impossible.

35.2 Reviewing the Concepts

- A skeletal muscle is a grouping of cells, called muscle fibers, bound together by connective tissue layers. Within each fiber are myofibrils, which contain thick and thin filaments arranged in sarcomeres. Regions of the thick filaments that extend toward the thin filaments are called cross-bridges (Figures 35.3, 35.4, 35.5).
- During muscle contraction, the sarcomeres shorten by a process known as the sliding-filament mechanism. The thick filaments remain stationary while the thin filaments slide past them, propelled by the action of cross-bridges (Figure 35.6).
- Myosins are an ancient family of related proteins. Mutations in a myosin gene expressed in jaw muscles of primates may have contributed to the evolution of larger brain size in humans (Figure 35.7).
- ATP has two major functions in cross-bridge cycling; it provides the energy for cross-bridge movement, and it breaks the link between actin and myosin so that another cycle of shortening can occur (Figure 35.8).
- The binding of Ca^{2+} to troponin allows tropomyosin to move away from the myosin-binding sites, initiating cross-bridge binding (Figure 35.9).
- The source of Ca^{2+} involved in a skeletal muscle fiber's action potential is the fiber's sarcoplasmic reticulum. Transverse tubules conduct action potentials from the plasma membrane to the interior of the cell, allowing Ca^{2+} to be released from the sarcoplasmic reticulum to the cytosol (Figure 35.10).
- Electrical stimulation of skeletal muscle occurs at a neuromuscular junction, at which a motor neuron's axon terminals synapse with a muscle fiber (Figure 35.11).

35.2 Testing Your Knowledge

1. Which of the following statements is *false*?
   a. A muscle fiber is a collection of cells embedded in connective tissue.
   b. A sarcomere contains both actin and myosin arranged in parallel.
   c. The function of Ca^{2+} in contraction is to bind to tropomyosin.
   d. Myofibrils are individual muscle cells.
   e. Choices a, c, and d are false.
2. Which of the following statements is true?
   a. During the cross-bridge cycle, sarcomeres are lengthened as the filaments slide past each other.
   b. Action potentials are conducted from the interior of muscle cells to the plasma membrane via T-tubules.
   c. Troponin and tropomyosin are bound to myosin; their removal from myosin permits the filaments to slide past each other.
   d. Acetylcholine receptors are located in junctional folds at the motor end plate of muscle cells.
   e. The actions of acetylcholine are stimulated further by the enzyme acetylcholinesterase.

35.3 Types of Skeletal Muscle Fibers and Their Functions

Learning Outcomes

1. Outline the general characteristics of the three types of skeletal muscle fibers.
2. Explain how muscles adapt to exercise.

Now that you have learned about the structural characteristics of skeletal muscle and the events that initiate and produce force generation, let's explore how skeletal muscle is adapted to meet the varied functional demands of vertebrates.

Animals use skeletal muscle for a wide variety of activities—locomotion, stretching, chewing, breathing, and balance, to name a few. Therefore, it is not surprising that not all skeletal muscle fibers share identical mechanical and metabolic characteristics. In this section, we will consider how different types of fibers can be classified on the basis of their rates of shortening (as either fast or slow) and the way in which they produce the ATP required for contraction (as oxidative or glycolytic).

Skeletal Muscle Fibers Are Adapted for Different Types of Movement

Different muscle fibers contain forms of myosin that differ in the maximal rates at which they can hydrolyze ATP. This, in turn, determines the maximal rates of cross-bridge cycling and muscle shortening. Fibers containing myosin with low ATPase activity (that is, slow rates of hydrolyzing ATP) are called **slow fibers.** Those containing myosin with higher ATPase activity are classified as **fast fibers.** Although the rate of cross-bridge cycling is about four times faster in fast fibers than in slow fibers, the maximal force produced by both types of cross-bridges is approximately the same.

The second means of classifying skeletal muscle fibers is based on the type of metabolic pathways available for synthesizing ATP. Fibers that contain numerous mitochondria and have a high capacity for oxidative phosphorylation (refer back to Figure 6.18) are classified as **oxidative fibers.** Most of the ATP production by such fibers depends on blood flow delivering oxygen and nutrients to the muscle. These fibers contain large amounts of myoglobin, an oxygen-binding protein that increases the availability of oxygen in the fiber by providing an intracellular reservoir of oxygen. The benefit of oxidative muscle fibers is they can maintain sustained action over a long period of time without fatigue.

By contrast, **glycolytic fibers** have few mitochondria but possess a high concentration of the enzymes involved in glycolysis (refer back to Figure 6.12) and large stores of glycogen, a polymer that acts as a cellular reservoir of glucose. Corresponding to their limited use of oxygen, these fibers contain little myoglobin.

On the basis of these two characteristics, at least three major types of skeletal muscle fibers can be distinguished:

1. **Slow-oxidative fibers** have low rates of myosin ATPase activity but have the ability to make large amounts of ATP. These fibers are used for prolonged, regular types of movement, such as the steady, nonstop swimming of tuna or the maintenance of posture in a human. Long-distance runners have a high proportion of these fibers in their leg muscles. These types of activities require muscles that do not fatigue easily and are not needed for intense bursts of activity.
2. **Fast-oxidative fibers** have high myosin ATPase activity and can make large amounts of ATP. Like slow-oxidative fibers, these fibers do not fatigue quickly and can be used for long-term actions. They are also particularly suited for rapid actions, however, such as those of the throat muscles of songbirds to produce trilling sounds or those of the muscles that shake a rattlesnake's tail to generate the fast clicking.
3. **Fast-glycolytic fibers** have high myosin ATPase activity but cannot make as much ATP as oxidative fibers, because their source of ATP is glycolysis; thus, they fatigue much more quickly than oxidative fibers. They also tend to be larger cells with greater amounts of myosin and actin. These fibers are best suited for rapid, short-lasting but powerful actions. An example is the sudden burst of flight of chickens or turkeys when eluding a predator; these relatively large birds can quickly take flight but cannot sustain it for long.

Different muscle groups within an animal's body have different proportions of each type of fiber interspersed with one another. Many activities require the action of all three types of fibers at once. This is important when you consider the wide range of animal activities related to locomotion alone, including walking, climbing, running, swimming, flying, crawling, and jumping. Depending on the needs of an animal at any given moment, the motor nerve inputs can be adjusted to stimulate different ratios of fiber types. For example, when a crab uses its pincers to grab prey, fast-glycolytic muscles quickly snap the claws closed, and slow-oxidative fibers maintain a tight grip for as long as required.

Muscles Adapt to Exercise

The regularity with which a muscle is used, as well as the duration and intensity of the activity, affects the properties of the muscle. For example, the pectoral (flight) muscles of migratory birds increase substantially in size as migration commences. Whether or not an activity includes resistance also causes changes to skeletal muscle. In humans, increased amounts of resistance exercise—such as weightlifting—result in hypertrophy (increase in size) of muscle fibers. Normally, the number of fibers in a muscle does not change much throughout adult life. The increases in muscle size that occur with resistance exercise result primarily from increases in the size of each fiber. These fibers undergo an increase in fiber diameter due to the increased synthesis of actin and myosin filaments, which form more myofibrils.

Physical activity of relatively low intensity but long duration—popularly called aerobic exercise in humans—increases the number of mitochondria in the fibers that are required in this type of activity. In addition, the number of blood vessels around these fibers increases to supply the greater energy demands of active muscle. All of these changes increase muscle endurance.

By contrast, short-duration, high-intensity exercise, such as weightlifting, primarily affects fast-glycolytic fibers, which are used during strong contractions. In addition, this type of exercise stimulates the synthesis of glycolytic enzymes, thereby increasing the glycolytic capacity of the muscle. The results of such high-intensity exercise are the increased strength and enlarged muscles of a conditioned weight lifter. Such muscles, although very powerful, have little capacity for endurance and therefore fatigue rapidly.

35.3 Reviewing the Concepts

- Slow-oxidative muscle fibers have low rates of myosin ATPase activity and are used for prolonged, regular activities. Fast-oxidative fibers have high rates of myosin ATPase activity and are suited for rapid, long-term actions. Fast-glycolytic fibers have high myosin ATPase activity but cannot make as much ATP as oxidative fibers; they are best suited for rapid, intense, short-term actions.
- Increased amounts of exercise affect the properties of muscle and can result in hypertrophy.

35.3 Testing Your Knowledge

1. Muscles that contain large amounts of fast-glycolytic fibers
    a. are suited for prolonged, steady activity such as running a marathon.
    b. are ideal for the flight muscles of birds during long migrations.
    c. are suited for relatively brief, intense activity such as that of a predator chasing a prey for a few seconds.
    d. have low amounts of myosin ATPase activity.
    e. can make more ATP than can muscles containing oxidative fibers.

35.4 Impact on Public Health

Learning Outcomes

1. Describe the causes of the bone diseases rickets and osteoporosis and their impacts on public health.
2. Describe the general features of muscular dystrophy.

In this chapter, we have seen how skeletal muscles and, in vertebrates, bones work together to enable animals to move around in their environments. A number of diseases affect bone structure and function in humans. Bone disease may involve defects in either the mineral or organic components of bone. Poor bone formation and structure may result from inadequate nutrition, hormonal imbalances, aging, or skeletal muscle atrophy, to name just a few of the common causes.

In addition, many diseases or disorders directly affect the contraction of skeletal muscle. Some of them are temporary and not serious, such as muscle cramps, whereas others, such as the disease muscular dystrophy, are chronic and severe. Also, some muscle diseases result from defects in parts of the nervous system that control contraction of the muscle fibers rather than from defects that originate in the fibers themselves. One example is amyotrophic lateral sclerosis (also known as Lou Gehrig's disease), in which motor neurons die, leading to the atrophy of the muscles those neurons normally stimulate. Still other diseases that affect skeletal muscle function result when normal processes go awry. For example, the immune system that normally protects the body from foreign pathogens may turn on itself, as occurred in the woman with myasthenia gravis described in the chapter-opening story. This section looks at a few of these conditions in more detail.

Rickets and Osteoporosis Affect the Bones of Millions of People

Bone diseases are fairly common, particularly among individuals older than age 50. Two major abnormalities can occur in bone. The first is improper mineral deposition in bone, usually due to inadequate dietary Ca^{2+} intake or inadequate absorption of Ca^{2+} from the small intestine. Without adequate minerals, bone becomes soft and easily deformed, as occurs in the weight-bearing bones of the legs of children with **rickets** (or **osteomalacia,** as it is called in adults) (**Figure 35.12a**). These disorders are prevented or treated with vitamin D supplements, because a hormone derived from this vitamin is the most important factor in promoting the absorption of ingested Ca^{2+} from the small intestine.

The second major abnormality is a more common disease called **osteoporosis,** in which both the mineral and organic (protein) portions of bone are decreased (**Figure 35.12b**). This disease, which affects four times as many women as men, occurs when the normal balance between bone formation and bone breakdown is disrupted, resulting in decreased bone density. Osteoporosis is the most prevalent bone disease in the U.S., affecting an estimated 15–30 million individuals. It results in annual national expenditures of approximately $15–20 billion in hospital and other medical costs.

Most commonly, osteoporosis results from hormonal imbalances. Some hormones—for example, estrogen—are required to maintain bone mass. When estrogen concentrations decrease after menopause (the time when a woman's reproductive cycles cease), bone density

(a) X-ray image of leg bones of a child with rickets

(b) Histologic appearance of normal bone (top) and bone from a person with osteoporosis (bottom)

Figure 35.12 Human bone diseases.
(a) ©Dr. LR/Science Source; *(b)* ©Tim Arnett, University College London

may decline, increasing the risk of bone fractures. By contrast, some hormones—such as parathyroid hormone (see Chapter 38)—act to demineralize bone, releasing Ca^{2+} into the extracellular fluid as part of the way in which the body normally maintains Ca^{2+} homeostasis in the blood. If such hormones are present in excess, however, they can cause sufficient demineralization of bone to result in osteoporosis. This may happen in rare cases when the glands that make these hormones malfunction and overproduce the hormones.

Another cause of osteoporosis is prolonged disuse of muscles. In ways that are not completely clear, the force produced by active skeletal muscle contractions helps to maintain bone mass. When muscles are not or cannot be used—due to paralysis or long-term immobilizing illnesses or even during prolonged space flight in which the decreased gravity reduces the need for muscles to work hard—bone mass declines.

The severity of osteoporosis can be minimized with adequate Ca^{2+} and vitamin D intake and weight-bearing exercise programs. In some cases, postmenopausal women may be given estrogen to replace what their bodies are no longer producing, but this therapy is controversial due to the potential adverse effects of estrogen on cardiovascular health and possible increased risk of developing breast cancer. Drugs that inhibit bone-resorbing osteoclasts are also commonly used to minimize bone damage due to this disease.

Muscular Dystrophy Is a Rare Genetic Disease That Causes Muscle Degeneration

The group of diseases collectively called muscular dystrophy affects 1 of every 3,500 American males; it is much less common in females. **Muscular dystrophy** is associated with the progressive degeneration of skeletal and cardiac muscle fibers, weakening the muscles and leading ultimately to death from heart failure and other causes. The signs and symptoms become evident at about 2 to 6 years of age, and most affected individuals do not survive beyond the age of 30.

The most common form of muscular dystrophy, called **Duchenne muscular dystrophy,** is an X-linked recessive disorder resulting from a defective gene on the X chromosome. Because females have two X chromosomes, and males only one, a heterozygote female with one abnormal and one normal allele will not generally develop Duchenne muscular dystrophy. If she passes the abnormal allele to a son, however, he will have the disease. The affected gene normally encodes a protein known as dystrophin, which is absent or defective in patients with the disease. Dystrophin is a large protein that links cytoskeletal proteins to the plasma membrane. Scientists think it is involved in maintaining the structural integrity of the plasma membrane in muscle fibers. When dystrophin is absent or dysfunctional, the plasma membrane of muscle fibers is disrupted, causing extracellular fluid to enter the cell. Eventually, the cell ruptures and dies.

One area of interest in the treatment of muscular dystrophy involves a protein called myostatin, which is one of a large number of proteins that control muscle growth in mammals. Myostatin functions to prevent overgrowth of skeletal muscle by inhibiting stem cell maturation into new muscle fibers, and preventing excessive growth of mature fibers. Researchers have identified animals with mutations in the myostatin gene that result in the production of an inactive protein; such animals show astonishing muscle development (**Figure 35.13**). Whether or not targeting the myostatin protein will prove successful in reversing muscle loss in muscular dystrophy and other diseases remains to be determined.

Figure 35.13 An animal with a mutation in the gene that encodes the protein myostatin, which resulted in inactive myostatin protein and overdevelopment of skeletal muscle.
©Bruce Stotesbury/ZumaPress.com/Newscom

35.4 Reviewing the Concepts

- Rickets in children (osteomalacia in adults) is characterized by soft, deformed bones, usually resulting from insufficient dietary intake or absorption of Ca^{2+}. In osteoporosis, bone density is reduced when bone formation fails to keep pace with normal bone breakdown (Figure 35.12).
- Muscular dystrophy is an ultimately fatal genetic disease associated with the progressive degeneration of skeletal and cardiac muscle fibers (Figure 35.13).

35.4 Testing Your Knowledge

1. Which of the following statements is (are) *false*?
    a. Rickets is a disease of the organic component of bone in children.
    b. Osteoporosis is characterized by thickened bones in adults.
    c. In osteomalacia, bone hardens and becomes filled with excess mineral deposits.
    d. Osteoporosis is typically caused by excessive intake of Ca^{2+} and vitamin D.
    e. All of the above are false.
2. Duchenne muscular dystrophy is
    a. primarily characterized by bone loss due to Ca^{2+} depletion.
    b. an X-linked disorder, making it more common in females.
    c. due to a disruption of the plasma membranes of muscle fibers.
    d. caused by increased amounts of the protein dystrophin.
    e. characterized by the degeneration of smooth muscle.

Assessing Your Knowledge and Skills

Foundational Knowledge

Multiple Choice

1. Which two minerals are the major components of bone?
    a. Na^+ and K^+
    b. Na^+ and Ca^{2+}
    c. Ca^{2+} and K^+
    d. Ca^{2+} and PO_4^{2-}
    e. PO_4^{2-} and Mg^{2+}
2. Which if any of the following is *not* a function of the vertebrate skeleton?
    a. structural support
    b. protection of internal organs
    c. calcium reserve
    d. blood cell production
    e. All of the above are functions of the vertebrate skeleton.
3. The function of ATP during muscle contraction is to
    a. cause an allosteric change in myosin so that it detaches from actin.
    b. provide the energy necessary for the movement of the cross-bridge.
    c. expose the myosin-binding sites on the thin filaments.
    d. do all of the above.
    e. do a and b only.

4. During cross-bridge cycling, Ca^{2+} binds to _____, which results in _____.
   a. tropomyosin, detachment of the cross-bridge from actin
   b. myosin, binding of the cross-bridge to actin
   c. actin, binding of the cross-bridge to myosin
   d. troponin, exposure of the myosin-binding sites on actin
   e. troponin, exposure of the actin-binding sites on tropomyosin

5. Stimulation of a muscle fiber by a motor neuron occurs at
   a. the neuromuscular junction.
   b. the transverse tubules.
   c. the myofibril.
   d. the sarcoplasmic reticulum.
   e. the cell body of the motor neuron.

Short Answer

1. List and briefly describe the steps in the cross-bridge cycle.
2. Discuss the structure and function of the two major types of skeletons found in animals, giving examples of animals in which such skeletons are found.

Critical-Thinking Skills

Multiple Choice

1. A person takes her cat to a veterinarian because the cat has recently been dragging one of its hind legs. She fears that the cat may be developing paralysis (inability to move a skeletal muscle). The veterinarian performs a certain test and concludes that the cat's leg muscles are normal and healthy. Which of the following possibilities might account for the cat's inability to move its leg?
   a. There is an abnormally high amount of acetylcholine being released at the neuromuscular junctions in the affected leg.
   b. There are more than usual numbers of Na^+ channels in the axon terminals of the motor neurons leading to the fibers of the affected leg.
   c. There has been a change in the proportion of different types of muscle fibers in the affected leg.
   d. The motor neurons that stimulate the fibers of the affected leg have been damaged and are no longer functional.
   e. The motor end plates on muscle fibers in the affected leg contain less acetylcholinesterase than normal.

2. Which of the following statements is *false*?
   a. A monkey uses flexors when bringing a piece of fruit to its mouth.
   b. A tetrapod uses both flexors and extensors in its legs to walk up a steep hill.
   c. Actin is a single, three-domain protein composed of actin monomers, tropomysin, and troponin.
   d. Given the similar structure and amino acid sequences of many eukaryotic myosins, it is likely that many functions of myosin are similar across taxa.
   e. Experimental stimulation (with, for example, an electrode) of a motor neuron that innervates a skeletal muscle fiber will cause the fiber to contract.

3. Recent evidence suggests that rudimentary mineralized skeletons first evolved in marine organisms at a time when there was little or no predation in the Earth's oceans. Therefore, it is believed that skeletons were not selected for the protection of soft-bodied animals from predators. Which of the following possibilities is most likely to have contributed to the evolution of skeletons in early marine animals?
   a. A skeleton would have allowed animals to grow much larger, since only animals with a hard skeleton can achieve sizes larger than, say, a small worm.
   b. Mineralized structures would have provided a soft-bodied animal with stability on the ocean floor and protection from the elements (such as rocks).
   c. A skeleton would have provided a source of energy for an animal.
   d. A skeleton would have provided buoyancy for animals to help them to travel long distances in the ocean.

4. An elderly woman's bone density is determined to be lower than normal. Which of the following might theoretically be responsible for this finding?
   a. She has a disease of her parathyroid glands that causes them to produce parathyroid hormone in abnormally large amounts.
   b. Her diet is chronically deficient in foods containing Ca^{2+}.
   c. Her ovaries are producing more estrogen than normal.
   d. She has been taking large amounts of vitamin D supplements.
   e. Both a and b are correct.

5. Which of the following statements is true?
   a. It is impossible for a female to have Duchenne muscular dystrophy.
   b. Weight training will increase the proportion of slow-oxidative muscle fibers in a person's muscles.
   c. The flight muscles of a bird that migrates hundreds of miles a day and the leg muscles of a human marathon runner would be expected to have similar proportions of muscle fiber types.
   d. In a neuromuscular junction, electrical signals move from a muscle fiber to a motor neuron.
   e. Of the three types of muscle (smooth, cardiac, skeletal), only skeletal muscle is specialized to contract.

Short Answer

1. Explain the link between electrical excitation and skeletal muscle contraction. Which of these two events *must* occur first, and why?
2. **PRINCIPLES** A principle of biology is that living organisms use energy. How is that principle illustrated in this chapter, particularly with regard to the mechanisms by which skeletal muscle cells contract?

36

Circulatory and Respiratory Systems: Transporting Solutes and Exchanging Gases

©SPL/Science Source

Image from cardiac angiography of a human heart. Contrast dye injection made the coronary blood vessels (in red) easier to visualize.

Chapter Outline

- 36.1 Types of Circulatory Systems
- 36.2 The Composition of Blood
- 36.3 The Vertebrate Heart and Its Function
- 36.4 Blood Vessels
- 36.5 Relationship Among Blood Pressure, Blood Flow, and Resistance
- 36.6 Physical Properties of Gases
- 36.7 Types of Respiratory Systems
- 36.8 Structure and Function of the Mammalian Respiratory System
- 36.9 Mechanisms of Gas Transport in Blood
- 36.10 Control of Ventilation
- 36.11 Impact on Public Health

Assessing Your Knowledge and Skills

An overweight 62-year-old woman decided to get herself "back in shape" after years of a sedentary lifestyle. Rather than gradually increasing her activity level, however, she embarked on a rigorous, demanding exercise program at a local gym. While running at 7 mph on a 10% inclined treadmill, she felt an acute, crushing sensation in her chest. She became extremely anxious and stopped exercising, but the pain did not go away. She felt light-headed, short of breath, and nauseated and called for help. An ambulance arrived, and the paramedic determined from her vital signs that the woman was experiencing a heart attack. She was transported to the emergency department of a local hospital and remained there for further treatment. Eventually, she was released and placed on several medications to control her blood pressure, prevent blood clots from forming, and improve the ability of the healthy parts of her heart to function. She also received counseling from a nutritionist and a physical therapist, who provided her with a sensible plan for improving her overall health.

Cardiovascular disease—disease of the heart or blood vessels—is the leading cause of death in the U.S. and in much of the rest of the developed world. In the U.S. alone, an estimated 80 million or more people have one or more diseases that affect the heart and blood vessels. About 1–1.5 million of those individuals will have a heart attack in the next year, and between 400,000 and 600,000 of them will die as a result.

The heart is a muscular pump that requires considerable nutrients and oxygen to sustain its unceasing muscular effort. The chapter-opening photo shows the extensive network of blood vessels coursing through a typical mammalian heart. Should any of those vessels become diseased, the regions of the heart supplied by them can die. That is what happens during a heart attack. When this occurs, the damaged heart may no longer be able to generate sufficient force to pump blood to all the cells of the body.

Circulatory systems transport necessary materials to all the cells of an animal's body and transport waste products away from the cells so that they can be released into the environment. Two of the most important solutes carried in the blood are oxygen (O_2) and carbon dioxide (CO_2). Oxygen enters an animal's body via a respiratory organ such as gills or lungs. It then dissolves into body fluids and is distributed to cells. As described in Chapter 6, O_2 is used by mitochondria in the formation of ATP. One of the waste products of those reactions is CO_2, which diffuses out of cells. In vertebrates, CO_2 is returned by the circulatory system to the respiratory organ and is then released into the environment. The process of moving O_2 and CO_2 in opposite directions between the environment, body fluids, and cells is called **gas exchange,** or **respiration.** (This differs from cellular respiration, in which cells generate energy from metabolism and produce ATP, as described in Chapter 6.) A **respiratory system** includes all of an animal's structures that contribute to gas exchange.

This chapter deals with the mechanisms that control the circulation of body fluids and the means by which animals obtain and transport O_2, rid themselves of CO_2, and cope with the challenges imposed by changing metabolic demands.

36.1 Types of Circulatory Systems

Learning Outcomes

1. Compare and contrast open and closed circulatory systems.
2. Describe the differences between single and double circulations in vertebrates.

Apart from the simple type of circulation that exists in certain invertebrates that have a gastrovascular cavity (refer back to Figure 26.12), two basic types of animal circulatory systems exist: open systems and closed systems. In this section, we will compare and contrast some of the key features of the two major types of circulatory systems found in animals.

In Open Circulatory Systems, Hemolymph Enters the Body Cavity

As illustrated in **Figure 36.1a**, an **open circulatory system** is characterized by a fluid that is pumped by one or more contractile hearts into the body cavity (hemocoel) of an animal. Therefore, the fluid in the blood vessels and the interstitial fluid that surrounds cells mingle in one large, body compartment, rather than being located in separate body compartments. The mixed fluid is called **hemolymph.** Open circulatory systems are found in many invertebrates. Nutrients and wastes are exchanged by diffusion between the hemolymph and body cells, and the hemolymph is eventually returned to the heart after collecting into vessels or directly through small openings called ostia. Oxygen and CO_2 are not transported in hemolymph; in animals with open circulatory systems, these gases are exchanged with the environment by a different mechanism, which we will examine later.

(a) An open circulatory system

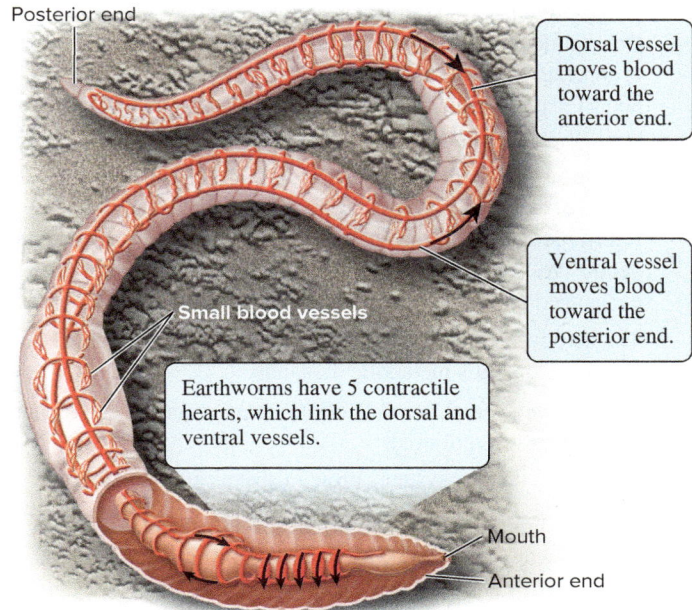

(b) A closed circulatory system

Figure 36.1 Types of circulatory systems. (a) In open circulatory systems, one or more muscular, tubular hearts pump hemolymph through open-ended vessels, where it percolates throughout the body. In arthropods such as this honeybee, hemolymph re-enters the heart through ostia. The arrows show the movement of the hemolymph. **(b)** In closed circulatory systems, such as that of this earthworm, blood remains within vessels and hearts, and it recirculates without emptying into the body cavity. Arrows indicate the direction of blood flow.

Concept Check: *It is incorrect to think of open circulatory systems as "primitive" compared with closed systems. Why?*

Open circulatory systems have certain advantages. They are metabolically inexpensive because it does not require much energy to pump hemolymph into the hemocoel; hemolymph simply drains into the hemocoel through open-ended vessels. In addition, open circulatory systems can adapt to changes in an animal's metabolic demands. As an insect takes flight, for example, its flight muscles contract more forcefully and rapidly, which acts to expand and compress the animal's thorax. This movement helps to propel hemolymph more rapidly throughout the hemocoel and into and out of the hearts. In other words, as the animal's physical activity increases, its circulation becomes more efficient, recharging the metabolically active cells with nutrients. The ability to adjust circulation to meet an animal's requirements is one of the most important features of any circulatory system.

Despite the phenomenal success of species with open circulatory systems, this type of circulation has certain limitations. For instance, because the hemolymph empties in bulk into the hemocoel, it cannot be selectively delivered to individual regions of the body that may have increased metabolic requirements relative to other regions. By contrast, during periods of increased physical activity in vertebrates, larger amounts of blood are delivered to skeletal muscles and away from less metabolically active structures, an extremely useful adaptation that results in greater endurance. This selective delivery is possible because these animals have a closed circulatory system, as we will explore next.

In Closed Circulatory Systems, Blood Remains in Blood Vessels at All Times

In a **closed circulatory system,** blood and interstitial fluid are physically separated and differ in their components and solute composition. Closed circulatory systems evolved independently several times and are found in earthworms, cephalopods (squids and octopuses), and all vertebrates (**Figure 36.1b**). Despite some differences in structure among species, closed circulatory systems share certain key features, which will be described throughout the rest of this chapter:

- Closed circulatory systems transport **blood**—a fluid connective tissue containing a mixture of cells and solutes under pressure that is generated by one or more contractile, muscular hearts. Both O_2 and CO_2 are transported in the blood of closed circulatory systems.
- Except when injury occurs, blood remains at all times within the blood vessels that distribute it throughout the body. Vessels called **arteries** carry blood away from the heart, and **veins** return blood back to the heart. **Capillaries** link the arterial and venous vessels and are the sites where fluid and solutes are exchanged between blood and interstitial fluid.
- The activity of closed circulatory systems can be adjusted to match an animal's metabolic demands.
- Within limits, closed circulatory systems generally can repair themselves when injured.
- Closed circulatory systems grow in size as an animal grows.

A closed circulatory system offers several advantages. First, animals can grow to a larger size, because blood can be directed to every cell of an animal's body, no matter how large the body. Nearly all body cells are within one or two cell widths of a blood vessel. Second, blood flow can be selectively increased or decreased to supply different parts of the body with the precise amount of blood needed at any given moment. After a meal, for example, more blood can be directed to the intestines to absorb nutrients, and if an animal's body temperature should increase above normal for some reason, additional blood can be routed to the body surface to dissipate heat to the environment. An animal with an open circulatory system cannot make these adjustments.

The closed circulatory system of vertebrates can be divided into two major arrangements: single circulation and double circulation. Single circulations are seen in fishes, and double circulations are seen in crocodiles, birds, and mammals. Amphibians and most reptiles have an intermediate type of circulation that combines features of both.

Single Circulation: Fishes In the single circulation of fishes, the heart has a single filling chamber—called an **atrium** (plural, atria)—to collect blood from the tissues, and a single exit chamber—a **ventricle**—to pump blood out of the heart (**Figure 36.2a**). The atrium and ventricle are separated by a one-way valve that prevents the backflow of blood. Arteries carry blood away from the heart to the gills, which pick up O_2 by diffusion from the water and unload CO_2 into the water. The freshly oxygenated blood then circulates through other arteries to the rest of the body. There, O_2 and nutrients are delivered to cells, and CO_2 diffuses from cells into the blood. Finally, the partially deoxygenated blood is returned to the heart via veins, where it is pumped back to the gills for another load of oxygen.

An important feature of all respiratory surfaces is that they function best when the blood flowing through them is maintained at a low pressure. A low pressure facilitates the exchange of gases. A fish heart does not generate high pressure when it pumps blood to the gills, thereby facilitating the exchange of gases and protecting the gills from mechanical damage. As blood flows through vessels, however, it tends to lose some pressure along the way as it gets distributed into more and more capillaries and veins. Thus, the pressure of blood leaving the gills will be even lower than it was upon entering the gills. This limits the rate at which oxygenated blood can be delivered to the cells of a fish's body.

Double Circulation: Crocodiles, Birds, and Mammals In a double circulation, oxygenated and deoxygenated blood are completely separated into two distinct circuits, the **systemic circulation** and the **pulmonary circulation** (**Figure 36.2b**). All animals with double circulations have four-chambered hearts: a left and right atrium and a left and right ventricle. The two atria are separated from their respective ventricles by one-way valves, similar to the situation in the heart of fishes.

The pulmonary circulation receives deoxygenated blood from the right ventricle and delivers oxygenated blood from the lungs to the left atrium, which then passes it on to the left ventricle. The left ventricle pumps blood to all the body's tissues via the systemic circulation. The systemic circulation then returns the partly deoxygenated blood from the tissues to the right atrium, which then sends it to the right ventricle, and the cycle starts again. The two ventricles fill with blood and then contract simultaneously with each other. One-way valves between each ventricle and the arteries bringing blood to the pulmonary or systemic circulation prevent back-flow of blood after it leaves a ventricle.

A major advantage of a double circulation is that the two ventricles can function as if they were, in effect, two hearts, each with its own ability to pump blood under different pressures. The blood from the right ventricle can be pumped under low pressure to the lungs, which, as mentioned earlier, is important for optimal function of respiratory structures. In a double circulation, however, once the blood picks up O_2 from the lungs, it can be returned to the left side of the heart. The left ventricle is more muscular than the right ventricle and therefore generates higher pressure when pumping blood. Thus, the oxygenated blood

CIRCULATORY AND RESPIRATORY SYSTEMS: TRANSPORTING SOLUTES AND EXCHANGING GASES

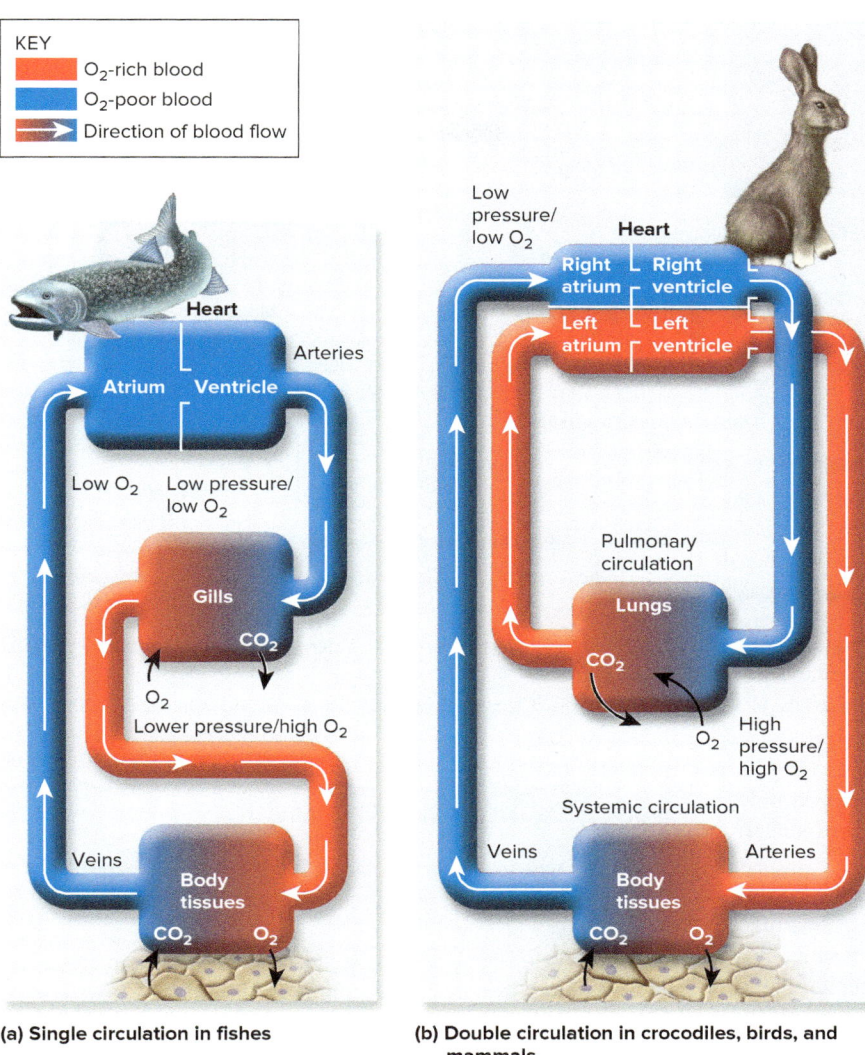

(a) Single circulation in fishes

(b) Double circulation in crocodiles, birds, and mammals

Figure 36.2 Representative vertebrate circulatory systems. (a) Fishes have a single circulation in which blood is pumped from the heart to the gills, from which it circulates to the rest of the body tissues and then finally returns to the heart. **(b)** Crocodiles, birds, and mammals have a double circulation, in which oxygenated blood is pumped under high pressure to the body's tissues and deoxygenated blood is pumped under low pressure to the lungs. Valves are represented by openings between atria and ventricles, and between ventricles and their respective arteries.

leaving the left side of the heart has sufficiently high pressure to reach all the cells of the animal's body. Consequently, O_2 and nutrients can be delivered throughout the body at a high rate, even to regions above the heart, such as the head, where the blood flow must work against gravity. This is no small feat, considering the distance blood must travel in some large animals. The left ventricle of a giraffe, for example, is particularly muscular compared with that of other animals.

EVOLUTIONARY CONNECTIONS

A Four-Chambered Heart Evolved from Simple Contractile Tubes

The heart is the first vertebrate organ to become functional during embryonic development, and the factors responsible for its development have been extensively studied. Interestingly, at least one of these developmental factors has also recently shed light on how a heart develops four chambers in crocodiles, birds, and mammals.

The first heartlike organ appears to have evolved at least 500 mya. Ancestral hearts may have resembled those observed today in cephalochordates and urochordates—a simple, linear tube with wavelike contractile properties but no valves or chambers (refer back to Figure 26.35a, b). A valvelike structure first appears in the tubular hearts of some arthropods, such as *Drosophila,* but such hearts do not have clearly defined chambers. Analyses of animal genomes have identified numerous genes that are critical for the development of the chambered vertebrate heart, including a core set of five highly conserved genes. Interestingly, however, one or more of these five genes are also expressed in linear tube hearts such as those just described and even in contractile structures in animals without hearts. For example, cnidarians do not have a heart or blood vessels, but they have muscle cells in their contractile bell. Contractions of these muscle cells not only produce locomotion but also in the process circulate water through the animal's gastrovascular cavity. The muscle cells express at least three of these heart developmental genes. Nematodes also do not have a heart but instead have a contractile pharynx that serves part of the same function. The pharyngeal muscles of these animals express four of the five key genes for heart development that are found in all vertebrates. Thus, the genes required for the formation of a contractile, multichambered heart arose well before such hearts appeared in animals.

Why are some vertebrate hearts four-chambered but others are not? In a 2009 study, American developmental biologist Benoit Bruneau and colleagues compared the expression of one of the major heart developmental genes, called *Tbx5*, in the hearts of amphibians, reptiles, birds, and mammals. (The *Tbx5* gene encodes Tbx5 protein, a transcription factor.) The researchers discovered that the spatial expression of *Tbx5* became increasingly restricted in the hearts of animals displaying greater separation of the ventricles into two chambers (**Figure 36.3**). Amphibians have two atria but only a single ventricle; in these animals, *Tbx5* was expressed throughout the developing ventricle. In lizards, in which a septum only partially divides the ventricle into two chambers, *Tbx5* expression was absent in a portion of the right half of the ventricle. In birds and mammals, with fully partitioned ventricles, Tbx5 protein was not made in the developing right ventricle but was present in large amounts in the left ventricle. This finding suggests that a gradient of Tbx5 protein is required for the ventricle to form two chambers and thereby a complete double circulation.

The investigators also showed that if *Tbx5* expression was experimentally induced in the heart of an embryonic mouse in a pattern

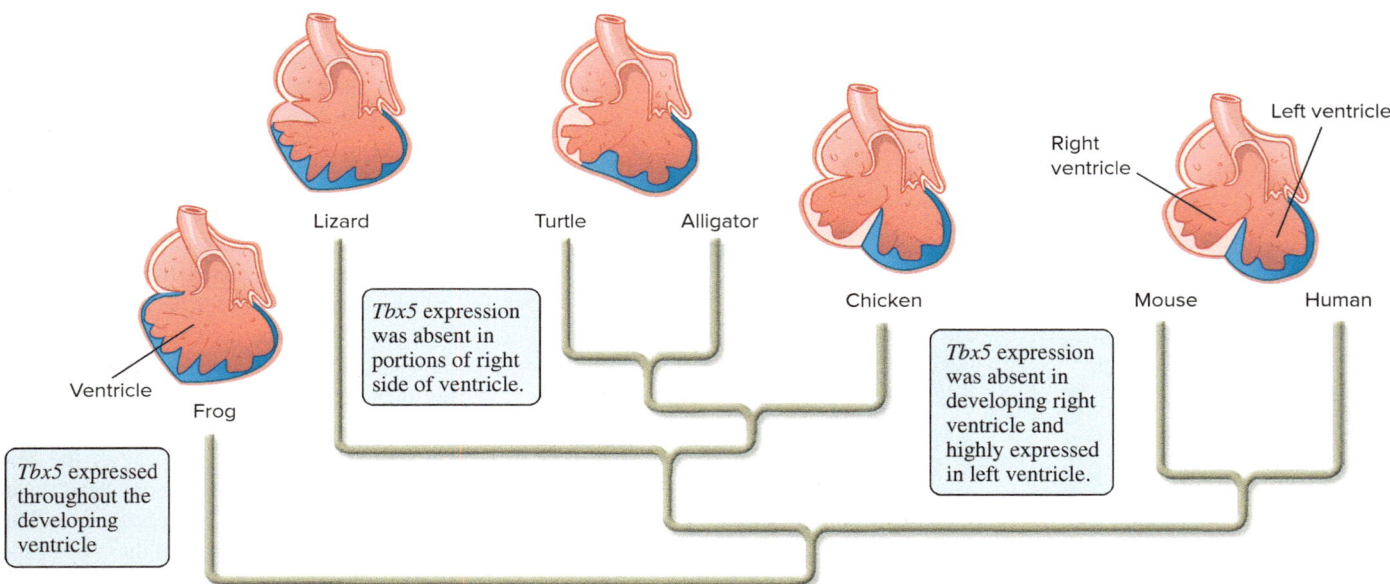

Figure 36.3 Expression of the *Tbx5* gene in embryonic vertebrate hearts. The blue color indicates where in the developing heart muscle the *Tbx5* gene was expressed. Note: Structural details have been omitted for simplicity.
Source: Koshiba-Takeuchi, Kazuko et al., "Reptilian Heart Development and the Molecular Basis of Cardiac Chamber Evolution," *Nature*, vol. 461, September 3, 2009, 95–98.

that mimicked that seen in lizards, the mouse heart developed only three clearly defined chambers—there was no distinction between left and right ventricles!

From these and other studies, it appears that the expression of a core set of genes is vital for the formation of a heart or heartlike structure and that a subset of genes, like *Tbx5*, is critical for the division of the heart into chambers. The most common congenital organ defects in humans are those associated with the heart. The relatively high incidence of heart-related birth defects makes an understanding of the genetic control of the growth and development of the heart a great concern in medicine today.

36.1 Reviewing the Concepts

- Circulatory systems transport necessary materials to all cells of an animal's body and transport waste products away from cells. The two major types of circulatory systems are open and closed circulatory systems.
- An open circulatory system does not transport O_2 and CO_2. It contains three components: hemolymph, vessels, and one or more hearts. The vessels open into the animal's body cavity. In a closed circulatory system, blood and interstitial fluid are physically separated by blood vessels (Figure 36.1).
- The closed circulatory systems of vertebrates transport O_2 and CO_2 and may be single or double circulations. In a double circulation, blood circulates at two different pressures through the systemic circulation and pulmonary circulation (Figures 36.2, 36.3).

36.1 Testing Your Knowledge

1. Which of the following statements is true?
   a. Open circulatory systems cannot adapt to changes in an animal's activity.
   b. Open circulatory systems are found in certain invertebrates and a small number of vertebrates.
   c. Closed circulatory systems may be either single or double circulations.
   d. An advantage of open circulatory systems is the ability to distribute blood to body parts in proportion to each part's metabolic requirements at a given moment.
   e. Animals with single circulations generally have higher blood pressures than those with double circulations.

2. Fishes have _____ and _____.
   a. one atrium, one ventricle
   b. one atrium, two ventricles
   c. two atria, one ventricle
   d. two atria, two ventricles

36.2 The Composition of Blood

Learning Outcomes

1. List the four components of blood, and describe the composition and functions of each.
2. Describe the process of blood clotting in mammals.

Blood is the transport medium of animals with closed circulatory systems. It moves necessary materials—including nutrients and gases

Figure 36.4 Components of blood. When a blood sample is centrifuged, it forms three visible layers: plasma, leukocytes, and erythrocytes. Leukocytes, shown in the scanning electron micrograph, are the white blood cells that make up part of the immune system. Erythrocytes are red blood cells, which carry oxygen. Note: The leukocyte layer is enlarged and not to scale, for illustrative purposes.
(right) ©Power and Syred/Science Source

 BioConnections: Leukocytes are part of the immune system of animals. Are immune defenses unique to animals? (Refer back to Figures 29.13 through 29.15 for a hint.)

such as O_2—to all cells and takes away waste products, including CO_2 and other breakdown products of metabolism. What are the components of blood that allow it to perform its functions?

Blood Is Composed of Cells Suspended in Plasma

Blood consists of cells and, in mammals, membrane-enclosed fragments of cytosol, suspended in a solution containing dissolved nutrients, proteins, gases, and other molecules and ions. If we collect a blood sample and spin it in a centrifuge, the blood separates into three visible layers (**Figure 36.4**). Let's take a look at each of these.

Plasma The top layer of the centrifuged blood sample is a yellowish solution called **plasma** (Figure 36.4). Plasma typically makes up about half of the total volume of blood in most vertebrates. It is made up of water and dissolved nutrients, oxygen, waste products of metabolism, and many other molecules released by cells, such as hormones and proteins. These proteins serve numerous functions, such as helping in the formation of blood clots, which seal off wounds to blood vessels.

Leukocytes Beneath the plasma in the centrifuged blood sample is a narrow, white layer of **leukocytes,** also known as white blood cells (Figure 36.4). Leukocytes develop from a specialized connective tissue (the marrow) of certain bones in vertebrates. Although there are several types—which we will describe further in Chapter 40—all leukocytes perform vital functions that defend the body against infection and disease and thus are key components of an animal's immune system.

Erythrocytes The bottom visible layer of the centrifuged blood sample consists of **erythrocytes,** also called red blood cells because of their color (Figure 36.4). The term **hematocrit** refers to the volume of blood (expressed as a percentage) that is composed of erythrocytes, usually between about 35% and 65% among vertebrates. Erythrocytes serve the critical function of transporting oxygen throughout the body. There are approximately a thousand times more erythrocytes than leukocytes in the circulation. Like leukocytes, erythrocytes are derived from cells in the bone marrow. In most vertebrates, mature erythrocytes retain their nuclei and other cellular organelles, but in all mammals (and a few species of fishes and amphibians), the nuclei are lost upon maturation. The lack of a nucleus and many other organelles in the mammalian erythrocyte contributes to its characteristic biconcave shape (Figure 36.4) and is believed to increase the cell's oxygen-carrying capacity. The biconcave shape of the mammalian erythrocyte increases its surface area relative to the flattened disc or oval shape present in most other vertebrates. This is believed to increase the efficiency of gas exchange between the erythrocyte and the body fluids.

Oxygen is poorly soluble in plasma. Consequently, the amount of oxygen that dissolves in plasma usually cannot support a vertebrate's basal metabolic rate, let alone more strenuous activity. Within the cytosol of erythrocytes, however, are large amounts of the protein **hemoglobin,** which can reversibly bind to oxygen. Hemoglobin greatly increases the reservoir of oxygen in the blood and enables animals to be more active. In a later section, we will consider the mechanisms by which hemoglobin binds and releases oxygen.

Platelets Vertebrate blood has a fourth component, not visible in a centrifuged sample of blood, called **platelets** in mammals and thrombocytes in other vertebrates. Platelets, which are derived from bone marrow cells, are cell fragments that lack a nucleus; they serve a crucial role in the formation of blood clots that limit blood loss after injury. The formation of a blood clot requires several steps, two of which include platelets (**Figure 36.5**). First, when a blood vessel is injured, platelets secrete substances that cause the platelets to clump together and bind to collagen fibers in the surrounding connective tissue at the wound site; this forms a plug that prevents continued blood loss. Second, other platelet secretions interact with plasma proteins to cause the precipitation from solution of a fibrous protein called fibrin. Fibrin forms a meshwork of threadlike fibers, which wrap around and between platelets and erythrocytes, enlarging and thickening the plug to form a clot. Blood clotting begins within seconds and helps to prevent injured animals from bleeding to death. Eventually, the body absorbs the clot as the injured vessel heals.

Biology Principle

Living Organisms Maintain Homeostasis

Because the blood carries with it the nutrients, hormones, proteins, and other products that must circulate throughout an animal's body, a loss of blood due to injury can be catastrophic unless the wound is healed and the blood loss is stopped. In this way, blood clotting is a key feature that contributes to the ability of animals to maintain homeostasis.

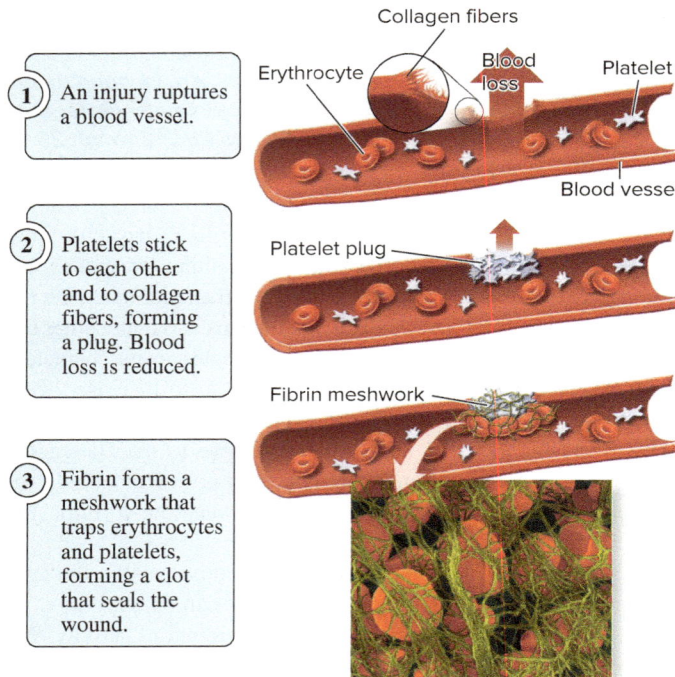

Figure 36.5 Platelets and the process of blood clot formation. A blood clot forms in two major steps: A platelet plug reduces initial blood loss, and a fibrin clot then seals the wound. An example of a fibrin clot is shown in the scanning electron micrograph.
(bottom) ©Dennis Kunkel Microscopy/Science Source

36.2 Reviewing the Concepts

- Blood is a fluid connective tissue consisting of cells suspended in a solution containing dissolved nutrients, proteins, gases, and other molecules. Blood has four components: plasma, leukocytes, erythrocytes, and (in vertebrates) platelets (or thrombocytes) (Figure 36.4).
- Platelets contribute to clot formation in two ways: by binding together to form part of the clot itself and by secreting substances that lead to precipitation of the protein fibrin (Figure 36.5).

36.2 Testing Your Knowledge

1. In an animal whose blood is 53% plasma by volume,
   a. the hematocrit is 53%.
   b. the hematocrit is 47%.
   c. the erythrocytes make up about 47% of the volume.
   d. the leukocytes make up about 47% of the volume.
   e. both b and c are correct.

2. Which of the following statements is *false*?
   a. Platelets are nucleated cells that participate in clot formation.
   b. In mammals, erythrocytes are non-nucleated cells that contain hemoglobin.
   c. Erythrocytes contain a reservoir of oxygen carried in the blood.
   d. Leukocytes are key components of the immune system of some animals.
   e. Platelets form clots both by clumping together with other platelets and by inducing the precipitation of fibrin.

36.3 The Vertebrate Heart and Its Function

Learning Outcomes

1. Describe the structure of the vertebrate heart.
2. **SCISKILLS** ▶ Outline the sequence of events of the cardiac cycle, and draw conclusions regarding the functional relationship between electrical and contractile events.
3. **SCISKILLS** ▶ Interpret the meaning of the tracings on an electrocardiogram.

As described earlier, in animals with a double circulation, such as mammals, the heart has a left atrium and left ventricle and a right atrium and right ventricle (**Figure 36.6**). The two sides of the heart are physically separated by a muscular and fibrous wall called a septum. Blood enters the atria from veins from the systemic or pulmonary circulation. The superior and inferior venae cavae return blood from the body, and the left and right pulmonary veins return blood from the lungs. The left ventricle ejects blood into the **aorta,** which branches to other arteries that distribute the blood into the systemic circulation. The right ventricle ejects blood into the pulmonary trunk, which divides into **pulmonary arteries** that lead to the right and left lungs.

Between the atria and ventricles are one-way valves called **atrioventricular (AV) valves** that control the movement of blood between the chambers. Blood normally can flow in only one direction through a valve. Blood moves down a pressure gradient through the AV valves from the atria into the ventricles. One-way valves are also found between each ventricle and the artery that a ventricle sends blood to (either the aorta or the pulmonary trunk); these are known as **semilunar valves.** The right ventricle pumps deoxygenated blood to the lungs through one semilunar valve, and the left ventricle pumps oxygenated blood to the rest of the body through another semilunar valve.

CIRCULATORY AND RESPIRATORY SYSTEMS: TRANSPORTING SOLUTES AND EXCHANGING GASES   **787**

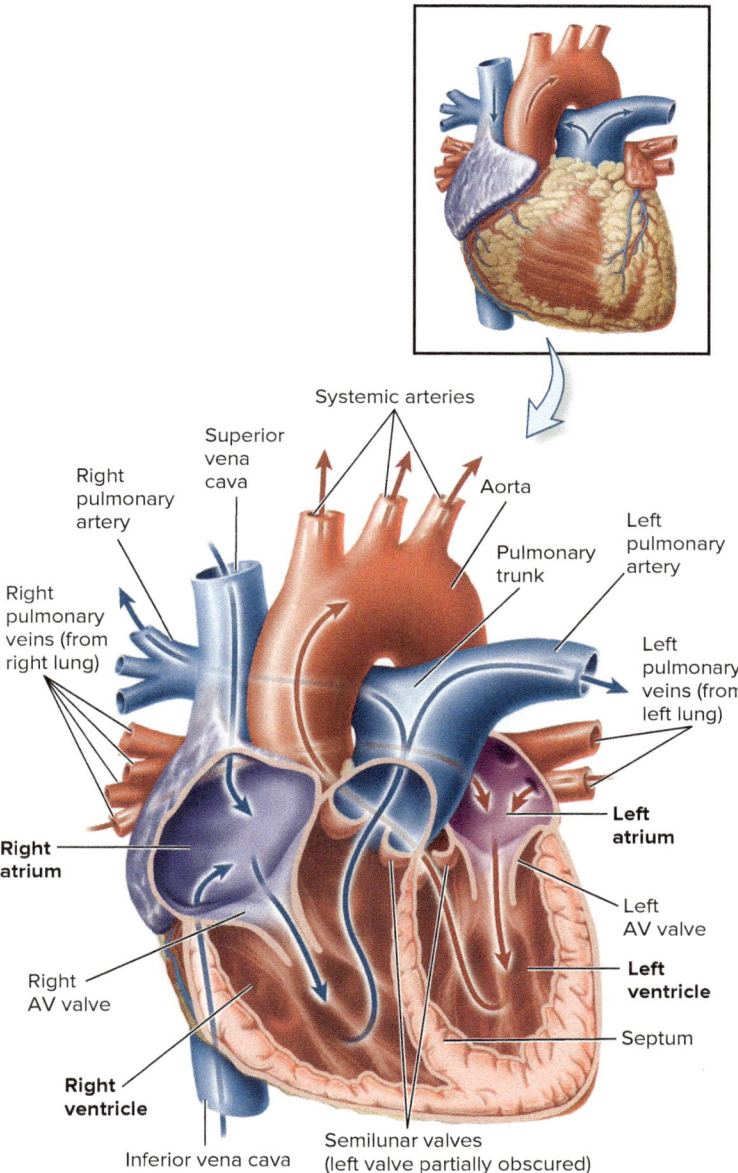

Figure 36.6 The mammalian heart and circulation. This cutaway illustration of a human heart shows the major blood vessels entering and leaving the heart, as well as the relationships of the four chambers (see the inset for the intact heart). Regions that contain oxygenated blood are shown in red, deoxygenated in blue. Note that the pulmonary veins carry oxygenated blood because they return blood from the lungs, whereas veins from the systemic circulation carry deoxygenated blood.

✓ **Concept Check:** *Which vessels in this illustration carry fully oxygenated blood?*

Vertebrates Have a Myogenic Heart

What causes the heart to beat so steadily? Animals cannot consciously initiate heart contractions. The beating of the heart is initiated either by nerves or by intrinsic activity of the heart muscle cells themselves. Many arthropods, for example, have a neurogenic heart that will not beat unless it receives regular electrical impulses from the nervous system. Many mollusks and all vertebrates, however, have a myogenic heart, in which the signaling mechanism that initiates contraction resides within the cardiac muscle itself.

In vertebrate myogenic hearts, cardiac muscle is distinguished by the interconnectedness between individual cardiac muscle cells, or cardiac myocytes. Each myocyte has membrane extensions that form interlocking networks with other myocytes. Within these networks, or intercalated discs, as they are called, are many gap junctions (refer back to Figure 5.18) that electrically couple the myocytes. The large number of gap junctions permits the rapid spread of electric current from cell to cell so that all parts of the heart are rapidly stimulated virtually as one. This electric current is the signal that increases the intracellular concentration of Ca^{2+} within the myocytes. The increase in Ca^{2+} triggers contraction in a manner that is similar in many ways to what occurs in skeletal muscle (see Chapter 35).

Myogenic hearts are electrically excitable and generate their own action potentials. The rate and forcefulness of the beating of myogenic hearts are, however, regulated by the nervous system. For example, the autonomic nervous system can increase heart rate in vertebrates during the fight-or-flight response (see Chapter 33). Nonetheless, myogenic hearts will continue to beat on their own if dissected out of an animal and placed in a nutrient bath, even without any input from nerves.

Excitation of the Heart Begins in the Atria and Spreads to the Ventricles

As with skeletal muscle (see Chapter 35), contraction of the vertebrate heart cannot occur until the muscle has been electrically excited. The electrical excitation of the vertebrate heart has two phases: atrial and ventricular. In atrial excitation, electrical signals are generated within the wall of the right atrium at the **sinoatrial (SA) node,** or **pacemaker** (Figure 36.7). The SA node is a collection of modified myocytes that have an inherently unstable resting membrane potential. Ion channels in the membranes of these cells are opened spontaneously and allow the influx of positively charged ions into the cytosol, thereby depolarizing the cell. These depolarizations produce action potentials in the SA node cells with a range of frequencies that are characteristic for a given species; the frequency determines an animal's heart rate.

After the SA node cells generate action potentials, the potentials quickly spread across one or both atria through the gap junctions described earlier. The action potentials trigger an influx of Ca^{2+} into the muscle cell cytosol, which activates contraction. Because the impulses spread very rapidly across the atria, both atria contract together almost as if they were one large muscle cell. Atrial contraction assists in providing a final push of blood through the open AV valves into the ventricles.

To begin ventricular excitation, action potentials initiated in the SA node first reach another node of specialized myocytes, the atrioventricular node. The **atrioventricular (AV) node** is located at the junction between the atria and ventricles and conducts the electrical signals from the atria to the ventricles. This is important, because the atria and ventricles are separated by a connective tissue septum

Figure 36.7 Electrical activity in the mammalian heart and the cardiac cycle. The spread of electrical activity, which is shown in yellow, begins in the sinoatrial (SA) node and quickly spreads through the atria to the atrioventricular (AV) node. Branches from the AV node transmit electrical activity throughout the ventricles via conducting fibers. The atria and ventricles do not contract until they have become electrically excited. Changes in pressure gradients between the atria and ventricles, and between the ventricles and the aorta or pulmonary trunk, are the forces that open or close the four heart valves. Note: The left semilunar valve is only partly visible in this orientation.

that does not contain gap junctions. Like the SA node, the AV node is electrically excitable, but its cells require a longer time to become excited than do the cells of the SA node. This electrical delay allows time for the atria to contract before the ventricles do.

Fibers branching from the AV node spread electrical impulses along a conducting system of cardiac muscle cells that branch throughout the muscular walls of the ventricles. This conducting system ensures that all of the ventricular muscle gets depolarized quickly and nearly simultaneously (Figure 36.7). In this way, the entire mass of both ventricles contracts in a coordinated way in response to depolarization.

The Cardiac Cycle Has Two Phases: Diastole and Systole

Each beat of the vertebrate heart requires the coordinated activities of the atria and ventricles. The contraction and relaxation events that produce a single heartbeat are known as the **cardiac cycle,** which can be divided into two phases (Figure 36.7). In the first phase, **diastole,** the ventricles are relaxed and fill with blood coming from the atria through the open AV valves. At the end of diastole, the atria are electrically stimulated, causing them to contract, which provides a boost of additional blood to the ventricles. In the second phase, **systole,** the ventricles contract and eject the blood through the open semilunar valves.

The valves open and close (**Figure 36.8**) as a result of pressure gradients established between the atria and ventricles, and between the ventricles and the arteries that leave them. During diastole, when the ventricles are in a relaxed state, pressure in the ventricles is lower than in the atria and the arteries. As a result, the AV valves are open during that time. Once the ventricles are electrically excited, they begin contracting, and the pressure within the ventricles rapidly increases. This marks the beginning of systole. When the pressure in the ventricles exceeds that in the atria, the AV valves are forced closed. Because the ventricles are thicker and stronger than the atria, ventricular contractions generate much greater pressure. The pressure exerted by the ventricles squeezes the blood upward, causing the AV valves to shut; this prevents blood from flowing backward into the atria. Closure of the AV valves makes the "lub" sound of the familiar "lub-dub" heard through a stethoscope. Ventricular pressure continues to increase as the ventricles contract. When the pressure in the ventricles exceeds the pressure in the arteries, the semilunar valves are forced open (Figure 36.8b). Blood is then ejected through the open semilunar valves into the arteries.

The pressure in the arteries increases during systole because they are expanding with the blood that was pumped into them. At this time, electrical stimulation of the ventricles ends, and therefore they begin to return to their resting, relaxed state. The pressure in the relaxing ventricles eventually falls below that in the arteries. The high pressure in the arteries causes backflow of blood toward the ventricles, which closes the semilunar valves and thereby prevents blood from flowing backwards into the heart. Closure of the semilunar valves creates the second heart sound, "dub," heard through a stethoscope. Throughout systole, meanwhile, the atria continue to fill

CIRCULATORY AND RESPIRATORY SYSTEMS: TRANSPORTING SOLUTES AND EXCHANGING GASES   789

(a) Semilunar valves in nearly closed position

(b) Semilunar valves in opened position

Figure 36.8 Appearance and function of heart valves. The heart valves consist of two or three flaps of tissue that open in one direction. Shown in this illustration are the semilunar valves of a human heart.

Concept Check: Which heart chambers pump blood through the semilunar valves?

with blood, which raises the pressure in the atria. Soon the pressure in the atria exceeds that in the relaxing ventricles, and the AV valves open again, bringing a new volume of blood into the ventricles and starting a new diastole.

Blood pressure, which is defined as the force exerted by blood on the walls of blood vessels, is highest in the arteries during systole and lowest during diastole. For this reason, blood pressure is measured with two numbers, the systolic and diastolic pressures. For historical reasons, the units are usually given in millimeters of mercury (mmHg). Of vertebrates, blood pressures are highest in mammals and birds, and lowest in fishes. For example, a typical healthy blood pressure in humans is around 120/80 mmHg (systolic/diastolic); most fishes have values 5–10 times lower.

An ECG Tracks Electrical Events During the Cardiac Cycle

An **electrocardiogram (ECG or EKG,** with the latter abbreviation from the original German spelling) is a medical and veterinary test used to investigate the function of the heart. An ECG is a record of the summed electrical potentials between various points on the body. These potentials arise from the electrical signals generated by myocytes during the cardiac cycle. Sensitive electrodes are placed on the surface of the body to monitor the wave of electricity initiated by the

Figure 36.9 An electrocardiogram (ECG). Electrodes placed on the body surface detect electrical impulses occurring in the heart, and an ECG is a recording of them. The resultant waveform is a useful indicator of cardiac health. Note that the ventricular wave (QRS complex) is taller than the atrial wave (P wave). This is because the ventricles are larger than the atria and generate more electrical activity.

Concept Check: Why is it possible to detect electrical changes in the heart using electrodes placed on the surface of the skin? (Hint: Think about the composition of body fluids and their ability to conduct electricity.)

SA node, which spreads in sequence through the atria, AV node, and ventricles. This procedure works because the body fluids that surround the heart conduct electricity, even the very weak impulses generated by a beating heart.

The tracing of an ECG reveals several waves of electrical excitation known as the P, QRS, and T waves (**Figure 36.9**):

- The P wave begins when the SA node generates action potentials and ends when the two atria are completely depolarized. In the cardiac cycle, the P wave is followed by atrial contraction, when the ventricles are in diastole and filling with blood through the open AV valves.

- The QRS wave, or complex, begins when the branches from the AV node electrically excite the ventricles and ends when both ventricles depolarize completely. In the cardiac cycle, the QRS complex is followed by ventricular contraction during systole.

- The T wave results from the repolarization of the ventricles back to their resting state. In the cardiac cycle, this causes ventricular relaxation as diastole begins again. (No wave is visible for atrial repolarization because it occurs simultaneously with the large QRS complex.)

The ECG record displays both the amplitude (strength) of the electrical signal and the direction in which it is moving within the chest. From this information, physicians and veterinarians can determine whether a person's or an animal's heart signals have a normal frequency, strength, duration, and pattern.

36.3 Reviewing the Concepts

- Blood enters the atria from veins of the systemic or pulmonary circulation, flows down a pressure gradient through the AV valves into the ventricles, and is pumped out through the semilunar valves into the arteries of the systemic and pulmonary circulations (Figure 36.6).
- Excitation of the atria begins at the sinoatrial (SA) node; excitation of the ventricles begins at the atrioventricular (AV) node. The electrical impulses spread through the atria and ventricles and stimulate the heart muscle to contract. The cardiac cycle has two phases: diastole and systole (Figures 36.7, 36.8).
- The electrical activity of the heart can be monitored using an electrocardiogram (Figure 36.9).

36.3 Testing Your Knowledge

1. The AV valves of the vertebrate heart
   a. are open during systole when the ventricles contract.
   b. conduct the electric current from the atria to the ventricles.
   c. open so that blood moves from ventricles to atria.
   d. create the second ("dub") heart sound when they close.
   e. are open during diastole.

2. The pace of the heart rate in vertebrates is set by
   a. the AV node.
   b. the SA node.
   c. the rate of electrical conduction along fibers that branch from the AV node.
   d. the QRS complex.
   e. the opening of the AV valves.

36.4 Blood Vessels

Learning Outcome

1. Explain the structural and functional distinctions among arteries, arterioles, capillaries, venules, and veins.

Now that we have discussed the vertebrate heart, let's take a more detailed look at the vessels that transport blood throughout the body. **Figure 36.10a** provides a simplified view of the route that blood flows in a closed circulatory system. Blood is pumped by the heart to large arteries and then flows to branching small arteries and eventually to the smallest arteries, which are called arterioles. Arterioles bring blood to the smallest vessels, the capillaries. Gas and nutrient exchange occurs between the blood in the capillaries and the cells surrounding the capillaries. From capillaries, the blood then flows back to the heart through the smallest veins, called venules, to small veins and finally to large veins.

Arteries Distribute Blood to Organs and Tissues

Arteries are thick-walled vessels that consist of layers of smooth muscle and connective tissue wrapped around a single-celled inner layer of specialized epithelial cells called an endothelium. The endothelium forms a smooth lining in contact with the blood (**Figure 36.10b**). Because thick layers of tissue surround the endothelium, most dissolved substances cannot diffuse across arteries. Instead, arteries act as conducting tubes that distribute blood leaving the heart to all the organs and tissues of an animal's body.

In vertebrates, the walls of the largest arteries, such as the aorta, also contain one or more layers of elastin, a protein with elastic properties (refer back to Figure 4.32). As the aorta stretches to accommodate blood arriving from the heart, the elastin layers also stretch. The thick layers of tissue in the aorta and other large arteries prevent them from stretching more than a small amount. When the heart relaxes as it readies for another beat, the elastin layers in the aorta and largest arteries recoil to their original state, something like the release of a stretched rubber band. The recoiling vessels generate a force on the blood within them; this helps to prevent blood pressure from decreasing too much while the heart is refilling during diastole.

Arterioles Distribute Blood to Capillaries

As arteries carry blood away from the heart, they branch repeatedly and become smaller in diameter. Eventually, the vessels are little more than a single-celled layer of endothelium surrounded by one or two layers of smooth muscle and connective tissue (Figure 36.10b). These are the arterioles, which deliver blood to regions of the body in proportion to metabolic demands. The adjustment of blood flow is accomplished by changing the diameter of arterioles. They widen, or dilate, in areas of high metabolic activity and narrow, or constrict, in inactive regions. Arterioles dilate when their smooth muscle cells relax, and arterioles constrict when these cells contract.

Capillaries Are the Site of Gas and Nutrient Exchange

Arterioles branch into tiny, thin-walled capillaries. Capillaries are tubes composed of a single-celled layer of endothelium resting on a layer of extracellular matrix termed a basal lamina (refer back to Figure 32.4). Capillaries are the narrowest blood vessels in the body; essentially every cell in an animal's body is near one. The diameter of a capillary is about the same as the width of erythrocytes, which move through the capillaries in single file (**Figure 36.11**).

Solutes diffuse readily between the plasma in a capillary and the surrounding interstitial fluid and intracellular fluid. Nutrients, hormones, and other solutes diffuse from the plasma and into cells. Waste products generated by cells diffuse in the opposite direction. Gases are exchanged across capillaries between the blood and the environment via a respiratory organ, such as the lungs, and between the blood and tissues (**Figure 36.12**).

Blood enters capillaries under pressure that is created by the beating of the heart. This pressure forces some of the fluid in blood out through tiny openings in capillary walls into the interstitial fluid. These openings are wide enough to permit water with its solutes, including O_2, to leave the capillary, but not cells and most proteins. The movement of fluid between capillaries and interstitial fluid is important for maintaining a normal distribution of water between the body fluid compartments.

If the fluid that leaves a capillary were to remain in the interstitial fluid, the volume of plasma in the blood would decrease, and the interstitial fluid would swell. Most of the fluid that leaves at the beginning of a capillary, however, is recaptured at the capillary's end. Any excess fluid is picked up by structures known as lymphatic vessels and later returned to veins. (The lymph system will be described in Chapter 40.)

CIRCULATORY AND RESPIRATORY SYSTEMS: TRANSPORTING SOLUTES AND EXCHANGING GASES

(a) The sequence of blood flow in a closed circulatory system

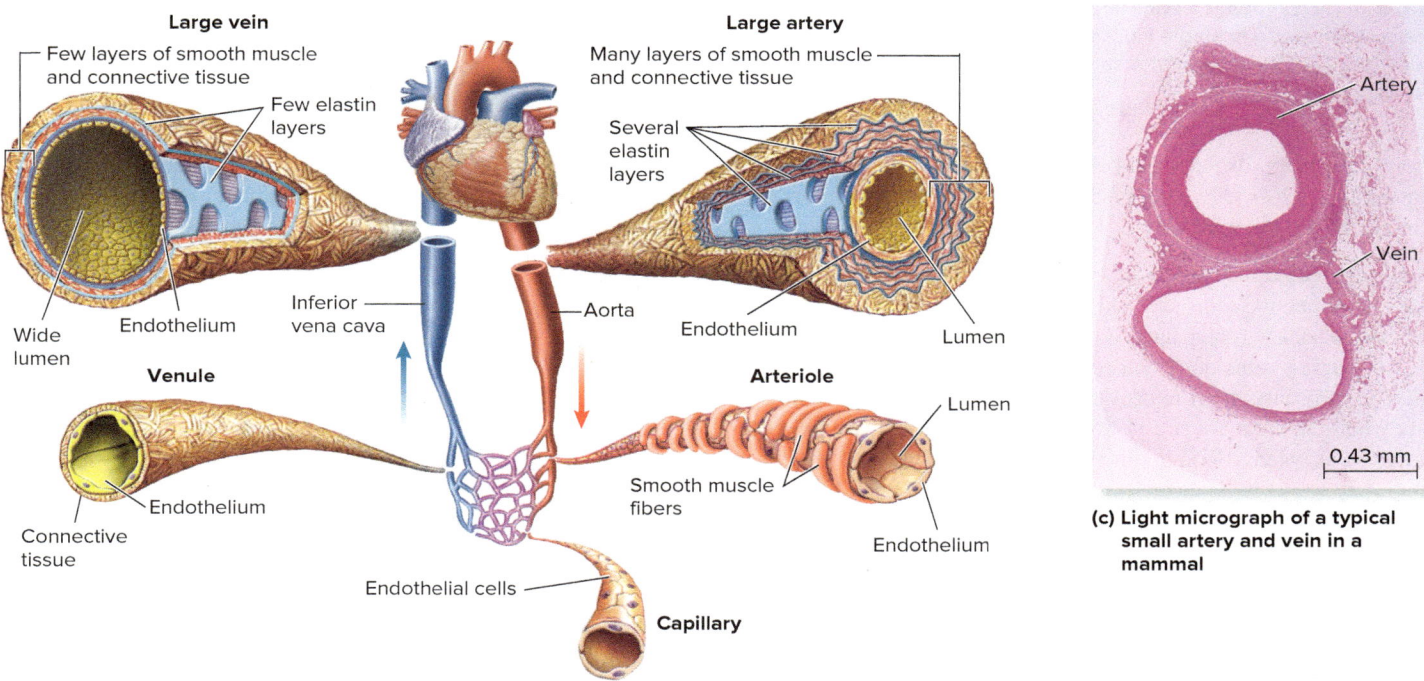

(b) The structure of blood vessels in a closed circulatory system

Figure 36.10 Comparative features of blood vessels. (a) Overview of blood flow through vessels in a closed circulatory system. Regions of gas exchange with the environment (such as the lungs) have been omitted for simplicity. (b) Blood vessel structure. Sizes are not drawn to scale. (c) Light micrograph of a small artery near a vein. Note the difference between the two vessels in wall thickness and lumen diameter.
(c) ©Carolina Biological Supply Company/Medical Images

Venules and Veins Return Blood to the Heart

Once blood travels through capillaries, picking up any substances secreted from the cells of the body, it enters the venules, which are small, thin-walled extensions of capillaries. The venules empty into veins that return blood to the heart. The walls of veins are much thinner, less muscular, and more easily distended, or stretched, than those of arteries (see Figure 36.10b, c).

By the time blood has reached the veins, the blood pressure has become very low. Veins can fill with considerable volumes of blood, particularly veins in the lower parts of an animal's body, such as the legs, where gravity tends to cause blood to pool. However, several factors assist the low-pressure blood on its way toward the heart even when it flows against gravity:

- Neurotransmitters released by neurons of the sympathetic nervous system stimulate contraction of smooth muscles of leg veins. This compresses the veins and helps to force blood back to the heart.
- Activity of skeletal muscles in the limbs assists the return of venous blood to the heart by squeezing the veins that travel through them (**Figure 36.13**).
- One-way valves inside veins ensure that blood returning from below the heart moves in only one direction, toward the heart (Figure 36.13). The valves can open toward the heart, but not in the other direction. When a leg vein, for example, is compressed by the two mechanisms just described, the blood forces open the valves and blood moves in one direction. When the vein is relaxed, the valves close again, preventing blood from flowing backward, away from the heart. (By contrast, veins located

Figure 36.11 Erythrocytes moving through a capillary. Note that the cells move along in single file, seen here through a light microscope. The diameter of the capillary is approximately 7 μm.
©Ed Reschke/Getty Images

above the heart, like those in the necks of bipeds and some quadrupeds, lack valves because gravity is sufficient to return blood toward the heart.)

Many people can easily observe the effects of gravity on venous blood flow. When the arms are held down by the sides, the veins are visible on the backs of the hands. When the arms are raised above the head, the bulging veins quickly lose blood and become less visible. When blood from the veins returns to the heart, it must travel against gravity when the arms are at a person's sides and with gravity when the arms are elevated. Blood drains from veins much more effectively when gravity works in its favor.

36.4 Reviewing the Concepts

- Arteries carry blood away from the heart. Arterioles distribute blood to capillaries, which are the site of gas and nutrient exchange. Blood from the capillaries flows into venules and then into veins, which return it to the heart (Figures 36.10, 36.11, 36.12).
- Muscle activity and one-way valves help move venous blood toward the heart against gravity (Figure 36.13).

36.4 Testing Your Knowledge

1. The vessels that mediate fluid and solute exchange are the _____, which then empty into _____.
   a. veins, the heart
   b. venules, capillaries
   c. capillaries, venules
   d. capillaries, arterioles
   e. arterioles, capillaries

2. Which of the following characteristics is true of arterioles?
   a. They have elastic properties.
   b. They consist of an endothelium plus one or two layers of smooth muscle.
   c. They affect local blood flow by dilating or constricting.
   d. They empty into venules.
   e. Both b and c are correct.

Biology Principle

Living Organisms Maintain Homeostasis
Keep this principle in mind as you read the rest of this chapter. The exchange of O_2 and CO_2 illustrated here is absolutely essential for all homeostatic processes in animals, in part because oxygen is required for the production of much of an animal's ATP. The ATP, in turn, provides the energy required to sustain processes that contribute to homeostasis.

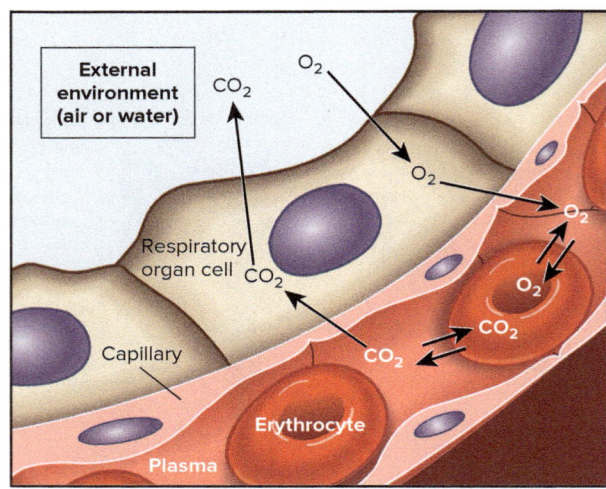

(a) Gas exchange between the environment and a respiratory organ

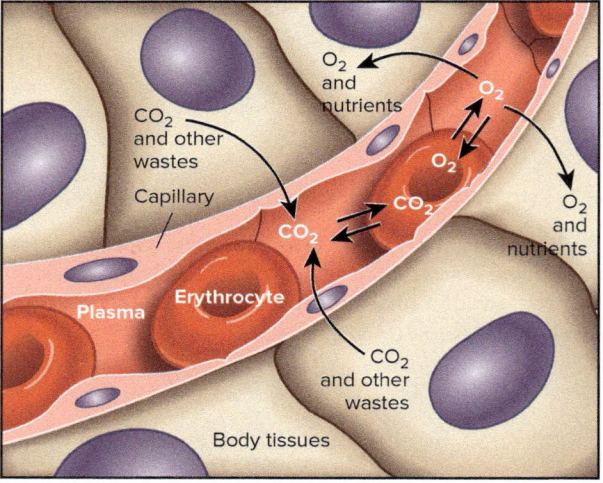

(b) Gas, nutrient, and waste exchange between the blood and tissues

Figure 36.12 Overview of gas and other solute exchange between capillaries and cells. (a) In vertebrates, oxygen (O_2) diffuses from the environment across the cells of a respiratory organ into the blood (plasma and erythrocytes); carbon dioxide (CO_2) diffuses in the opposite direction. (b) From the blood, O_2 and nutrients diffuse into tissue cells, where they are used during the synthesis of ATP and other activities. Cells generate CO_2 and other waste products, which diffuse out of cells and into blood. Note: Interstitial fluid is omitted for simplicity.

CIRCULATORY AND RESPIRATORY SYSTEMS: TRANSPORTING SOLUTES AND EXCHANGING GASES   793

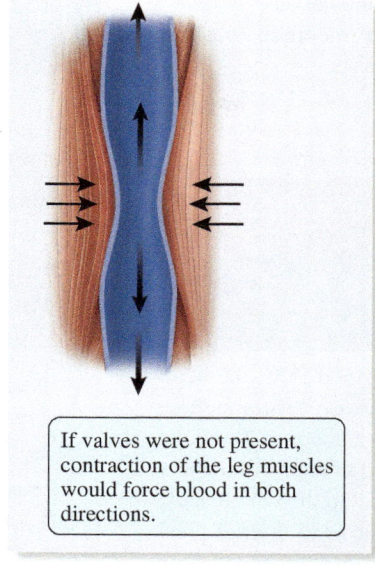

(a) Vein with one-way valves   (b) Vein without valves

Figure 36.13 One-way valves in veins. Valves are typically present in the limbs, as shown in this dog's leg. **(a)** In some veins, one-way valves assist the return of blood to the heart against the force of gravity. **(b)** Blood moving in a vein without valves would flow in both directions if the vein were compressed.

Concept Check: *Unlike most mammals, giraffes have one-way valves in the veins of their long necks. Which way do you think the valves open, toward the heart or toward the head, and why? (Hint: Search online for a picture of a giraffe drinking from a body of water.)*

36.5 Relationship Among Blood Pressure, Blood Flow, and Resistance

Learning Outcomes

1. **SCISKILLS** ▶ Be able to relate blood pressure, blood flow, and resistance in a mathematical way.
2. Distinguish between the effects of vasodilation and vasoconstriction on blood flow.
3. **SCISKILLS** ▶ Predict the effects of changes in resistance and cardiac output on blood pressure.

Blood pressure—the force exerted by blood on vessel walls—is responsible for blood flow, the movement of blood through the vessels. It is not the same in all regions of an animal's body, however, because of resistance. **Resistance (R)** refers to the tendency of blood vessels to slow the flow of blood through their lumens. The relationship among blood pressure, blood flow, and resistance can be considered on two levels—local and systemic—as we will see next.

Blood Pressure, Blood Flow, and Resistance Are Mathematically Related

The relationship among blood pressure, blood flow, and resistance is stated by Poiseuille's law, which was derived in the 1840s by Jean Marie Louis Poiseuille, a French physician and physiologist. His law is simplified here:

$$\text{Flow}(F) = \Delta\text{Pressure}(P)/\text{Resistance}(R)$$

Stated mathematically, blood flow through a blood vessel is directly proportional to the difference (Δ) in pressure of the blood between the beginning and end of the vessel, and inversely proportional to the resistance created by that vessel. The equation can be rearranged as $\Delta P = F \times R$, which demonstrates that blood pressure depends on both blood flow and resistance. Poiseuille's law applies to blood flow through a single vessel, an organ, or the entire body.

Resistance to Flow Depends on the Radius of Arterioles

Changes in arteriolar resistance are the major mechanism for increasing or decreasing blood flow to a region. The relationship between arteriolar radius and resistance is not linear. Resistance is inversely proportional to the radius of the vessel raised to the fourth power: $R \propto 1/r^4$, where $\propto$ means "proportional to" and r is the radius of the arteriolar lumen. Let's consider an arteriole with a radius that increases by a factor of 2. This would occur if the smooth muscles of the arteriole relaxed sufficiently to allow the vessel to dilate and double its original radius. Because resistance is inversely proportional to the fourth power of vessel radius, an increase in radius by a factor of 2 will result in a decrease in resistance of 2^4, or 16-fold.

Vasodilation refers to an increase in blood vessel radius, and **vasoconstriction** refers to a decrease in blood vessel radius. The signals that control arteriolar radius come from three sources:

- **Local factors.** Locally, metabolic by-products such as carbon dioxide, lactic acid, and other substances secreted by metabolically active tissues cause nearby arterioles to vasodilate, thereby decreasing their resistance to blood flow. This permits more blood flow to the active region, facilitating oxygen and nutrient delivery and waste removal.

- **Hormones.** Hormones secreted by glands throughout the body can also regulate arteriolar radius. For example, during the fight-or-flight response, some hormones cause the arterioles that deliver blood to the small intestine to vasoconstrict, routing blood away from the intestine and to more vital areas such as skeletal muscle.

- **Nervous system.** Signals from the autonomic nervous system cause contraction or relaxation of arteriolar smooth muscle; this effect is similar to that described for hormones.

Biology Principle: Biology is a quantitative science

Quantitative Analysis: Cardiac Output and Resistance Determine Blood Pressure

Because blood vessels provide resistance to blood flow, the heart must beat forcefully enough to overcome that resistance throughout the whole body. **Cardiac output (CO)** is the amount of blood the heart pumps per unit of time, usually expressed in units of liters per minute (L/min). Poiseuille's law can be adapted to the whole body. Because venous blood pressure is so low relative to arterial pressure, it can be ignored in this context and the Δ symbol can drop out. In this case, then, pressure is simply arterial blood pressure (BP) recorded using a blood pressure cuff, flow refers to CO, and resistance refers to the total resistance provided by all the arterial vessels of the systemic circulation (**total peripheral resistance**, or **TPR**). Thus, we have BP = CO × TPR. The values of CO and TPR determine the pressure the blood exerts in the arterial vessels of a closed circulatory system.

Cardiac output depends on the size of an animal's heart, how often it beats each minute, and how much blood it ejects with each beat. Each beat, or stroke, of the heart ejects an amount of blood known as the **stroke volume (SV)**, which is roughly proportional to the size of the heart (Table 36.1). Thus, if we know the stroke volume of a heart and can measure the heart rate (HR, the number of beats per minute, or bpm), we can determine the CO. Simply put, CO = SV × HR. As one example from Table 36.1, the CO of a blue whale is

100 L/beat × 10 beats/min = 1,000 L/min

To put that number in perspective, 1,000 L/min is roughly equivalent to 250 gal/min, or 25 ordinary fish aquaria!

Of course, the CO of a blue whale is far greater than that of a human, a dog, or a shrew. The heart of a typical shrew, for example, is the size of a small pea, whereas the heart of a blue whale is as large as a cow. Typically, heart size varies in proportion to body mass within a given class of vertebrates, with birds and mammals having larger hearts than similarly sized fishes, amphibians, or reptiles. Note in Table 36.1 that heart rate decreases as mammals get larger, but heart mass and stroke volume increase roughly in proportion with body mass. Smaller animals have smaller hearts and, therefore, smaller stroke volumes. However, small animals have faster heart rates than those of large animals. A shrew's resting heart rate may be more than 800 bpm, whereas a blue whale's heart may beat only 10 times per minute (although the volume of blood ejected with each of those beats is enormous!). The faster heart rates of small animals give them a greater cardiac output than predicted from the size of their hearts, which helps them to meet the extraordinary oxygen and nutrient demands of their relatively high metabolisms.

The greater the cardiac output and resistance to blood flow, the higher the blood pressure will be. Imagine that the circulatory system is like a faucet (the heart) connected to a garden hose (the arteries and arterioles) (Figure 36.14). If the faucet is fully open (analogous to maximal cardiac output) and the hose is not blocked (analogous to low resistance), the amount of water rushing into the hose will be high, and so will the water pressure, representing blood pressure. If the faucet is only partially open, the water pressure will be lower. However, now imagine that the faucet is partially open but a region of the hose is kinked, or constricted, representing a region of high resistance. In that case, the pressure of the water in the hose will increase between the faucet and the point where the hose is kinked, and it will decrease beyond the constriction, as will the flow of water.

Arterial blood pressure, therefore, is a function of how hard the heart is working and how constricted or dilated the various arterioles are. Blood pressure must be high enough for blood to reach all body tissues even at the farthest extremities, but not high enough to damage blood vessels or force excess plasma out of capillaries.

Crunching the Numbers: Using the data in Table 36.1 and a graphing program (such as Excel) on your computer, create a plot of body mass (x-axis) versus heart mass (y-axis). Now create another plot, of body mass versus heart rate. How do the graphs differ? Is one more linear than the other? Based on these relationships, can you determine which variable—heart mass or heart rate—makes the most significant contribution to the relatively greater cardiac output in smaller animals?

Table 36.1 Comparative Features of Representative Mammalian Hearts*

Animal	Body mass (kg)	Heart mass (kg)	Stroke volume (L)	Heart rate (bpm)†
Shrew‡	0.0024 (2.4 g)	0.000035 (35 mg)	0.000008 (8 μL)	835
Rat	0.20	0.001	0.0002	360
Rabbit	2	0.012	0.0013	189
Small dog	5	0.030	0.007	120
Large dog	30	0.180	0.040	88
Human	75	0.380	0.075	70
Horse	450	3.50	0.90	38
Elephant	4,000	25	4.0	25
Blue whale	100,000	600	100	10

*Values are based on average body masses and resting conditions. In some cases, stroke volumes are estimates based on heart size.
†bpm = beats per minute.
‡The shrew reported here is the Etruscan shrew, one of the smallest known mammals. Its heart is somewhat larger than would be predicted for its body mass. Note its heart rate; at 835 bpm, the heart beats 14 times per second!

Biology Principle

Biologists Use Models to Test Predictions and Convey Ideas

This figure is a model for how flow and resistance affect the pressure of water leaving a hose. The model is an analogy for blood flowing through a blood vessel. The Modeling Challenge below asks you to predict how a different combination of an open faucet and a kinked hose would affect the flow of water from the end of the hose.

(a) Maximal cardiac output with low resistance

(b) Moderate cardiac output with low resistance

(c) Moderate cardiac output with high resistance

Figure 36.14 The relationship among cardiac output, resistance, and blood pressure. A hose analogy shows the way in which cardiac output (the faucet) and resistance (constriction of the hose) affect blood pressure.

36.5 Reviewing the Concepts

- Blood pressure, which is the force exerted by blood on the walls of blood vessels, is responsible for moving blood through the vessels. Resistance is the tendency of blood vessels to slow the flow of blood through their lumens.
- Local blood flow through a vessel is directly proportional to the pressure of the blood entering the vessel and inversely proportional to the resistance created by that vessel. Arteriole radius is the major factor that regulates resistance.
- Cardiac output overcomes resistance to generate systemic blood pressure and depends on the size of an animal's heart, how often it beats each minute, and how strongly it contracts with each beat (Table 36.1).
- Cardiac output (CO) and total peripheral resistance (TPR) determine blood pressure (BP): BP = CO × TPR (Figure 36.14).

36.5 Testing Your Knowledge

1. The radius of an arteriole decreases 50%. What will happen to the flow of blood through that arteriole?
   a. It will decrease by 50%.
   b. It will decrease to 1/4 of normal.
   c. It will decrease to 1/16 of normal.
   d. It will decrease by 25%.
   e. It will decrease by a factor of 8.

2. An animal's heart rate increased but its blood pressure did not change. Theoretically, what could explain this?
   a. Its stroke volume also increased.
   b. Its total peripheral resistance increased.
   c. Its stroke volume decreased.
   d. Its peripheral resistance decreased.
   e. Either c or d could explain this.

Modeling Challenge: The models of water flow through a tube shown in Figure 36.14 demonstrate the effects of flow (how fully opened is the faucet) and resistance (how kinked is the hose) on the pressure and flow of water leaving a hose. Partially kinking a hose decreases water flow from it and increases pressure before the kink. Opening the faucet more fully increases the rate of water flow. One scenario not shown in this figure is the combination of a fully opened faucet with a hose that is partly but not completely kinked. In other words, it combines features of panels (a) and (c) of Figure 36.14. Draw a model of such a scenario. As you do this, keep in mind that the faucet is analogous to cardiac output and the kink in the hose is analogous to decreasing the diameter of blood vessels.

36.6 Physical Properties of Gases

Learning Outcomes
1. List the relative amounts of the gases that make up the majority of the air that animals breathe.
2. Define partial pressure, and explain how it is calculated.
3. List three factors that determine how much gas dissolves in a solution.

We will now turn our attention to the process of gas exchange between the environment and the fluids of an animal's body. Air is composed of about 21% O_2, 78% nitrogen (N_2), and roughly 1% CO_2 and other gases. From a respiratory standpoint, N_2 can usually be ignored because it plays no role in energy production, nor is it created as a waste product of metabolism. In many animal species, gas exchange involves an interplay between the circulatory system and the respiratory system. The exchange of O_2 and CO_2 depends on the solubility of these gases in water and their rates of diffusion. In this section, we will begin by examining some of the basic properties of the two major gases that are important in respiration.

Gases Exert Pressure, Which Depends on Altitude

The gases in the air exert pressure on the body surfaces of animals, although the pressure is not perceptible (unless it changes suddenly, as when your ears "pop" in a descending airplane). This pressure is called **atmospheric pressure.** It can be measured by noting how high a column of mercury is forced upward by the air pressure in a device called a mercury manometer. One of the traditional units of gas pressure, therefore, is millimeters of mercury (mmHg)—the same as for blood pressure.

At sea level, atmospheric pressure is 760 mmHg. It decreases at higher elevations because the gravitational pull of the Earth decreases. Consequently, gases expand and fewer gas molecules are in a given volume of air at high altitude. To visualize why gas pressure decreases at higher altitudes, think about how hydrostatic pressure decreases from the ocean floor to the ocean surface. Now, imagine that you are at the bottom of an "ocean" of air. The closer you get to the "ocean's" surface (the top of the atmosphere), the lower the pressure exerted on your body.

Atmospheric pressure is the sum of the pressures exerted by all the gases in air, in exact proportion to their amounts. The individual pressure of each gas is its **partial pressure,** symbolized by a capital P and a subscript depending on the gas. At sea level, the partial pressure of oxygen (P_{O_2}) in the air we breathe is 21% of the atmospheric pressure at sea level, or 160 mmHg (0.21 × 760 mmHg). The percentage of oxygen and other gases in air remains the same regardless of altitude, but the lower the atmospheric pressure, the lower the partial pressure of oxygen in air. In Denver, for example, where the atmospheric pressure is only about 640 mmHg, the P_{O_2} is 21% of 640 mmHg, or 134.4 mmHg.

All gases diffuse from regions of higher pressure to regions of lower pressure. The partial pressure of oxygen in the environment is what provides the driving force for its diffusion from air or water across an animal's respiratory surface and into its blood. Consequently, the rate of oxygen diffusion into the blood of a terrestrial animal decreases when the animal moves from sea level to a higher altitude, where the P_{O_2} is lower. Thus, humans who climb to high altitudes often need to breathe from a tank containing gas that is enriched in oxygen to compensate for the lower P_{O_2}.

Pressure, Temperature, and Other Solutes Influence the Solubility of Gases

Gases dissolve in water, including fresh water, seawater, and all animal body fluids. Gases such as O_2 exert their biological effects while in solution. However, most gases dissolve rather poorly in water. There is less O_2 in a given volume of water than in air, for example, which places constraints on how active many aquatic organisms can be. Among the factors that influence the solubility of a gas in water, the following are particularly important.

- **Pressure of the gas.** As the pressure of a gas that comes into contact with water increases, more of that gas will dissolve, up to a limit that is specific for each gas at a given temperature. The partial pressure of gas in water is given in the same units as atmospheric pressure (mmHg). For example, if water is in contact with air that has a P_{O_2} of 160 mmHg, the P_{O_2} of the water is also 160 mmHg.

- **Temperature of water.** More gas dissolves in a given volume of cold water than in warm water. At higher temperatures, gases in solution have more thermal energy and are therefore more likely to escape from the liquid. This means that all other things being equal (for example, density of plant life and other organisms), animals inhabiting warm waters generally have less O_2 available to them than do animals in cold waters.

- **Presence of other solutes.** Ions and other solutes decrease the amount of gas that dissolves in water. Thus, a given volume of blood—which contains many solutes—dissolves less O_2 than does pure water. Likewise, less O_2 dissolves in seawater than in fresh water at any given temperature and pressure. Animals living in cold, freshwater lakes, therefore, usually have more O_2 available to them than animals living in warm, salty seas.

36.6 Reviewing the Concepts

- The partial pressure of oxygen (P_{O_2}) in the environment provides the driving force for its diffusion. Atmospheric pressure decreases at higher elevations.
- Three factors—the pressure of the gas, temperature of the water, and presence of any other solutes—affect the solubility of a gas in water.

36.6 Testing Your Knowledge

1. In which solution would you expect the most oxygen to be dissolved?
   a. a cold sea
   b. a lake in winter
   c. a warm sea
   d. a very warm, freshwater pond

36.7 Types of Respiratory Systems

Learning Outcomes

1. Describe the different ways in which animals obtain oxygen from the environment.
2. **SCISKILLS** ▶ Diagram the countercurrent exchange that occurs in gill ventilation.
3. State Boyle's law, and explain how it relates to ventilation in air-breathing vertebrates.

Animals exchange oxygen and carbon dioxide with their surroundings with at least one of four major types of gas-exchange organs: the body surface, gills, tracheae, or lungs. **Ventilation** is the process of bringing oxygenated water or air into contact with a gas-exchange (respiratory) organ. In this section, we will examine the mechanisms of ventilation used by animals with different respiratory systems.

Some Animals Exchange Gases Across the Body Surface

In those invertebrates that are only a few cell layers thick, such as cnidarians and platyhelminthes, O_2 and CO_2 can diffuse directly across the body surface (refer back to Figures 26.12 and 26.14). In this way, O_2 reaches all the interior cells, in some cases without any specialized circulatory system.

Even in some vertebrates, the body surface may be permeable to gases. Amphibians have unusually moist, permeable skin. On land, amphibians rely primarily on their lungs, but when they are under water and thus cannot breathe air, O_2 and CO_2 can diffuse across the skin. In some species, skin folds increase surface area and thereby increase the rates of diffusion of gases (refer back to Figure 32.7b). The ability to exchange gases across the skin is an adaptation that permits amphibians to spend prolonged times under water.

Water-Breathing Animals Use Gills for Gas Exchange

Water-breathing animals use specialized respiratory structures called **gills,** which can be either external or internal. External gills are uncovered extensions from the body surface, found in many invertebrates and the larval forms of some amphibians. Internal gills, which occur in fishes, are enclosed in a protective cavity.

External Gills External gills vary widely in appearance, but all have a large surface area, often in the form of elaborate projections (**Figure 36.15**). In many cases, external gills are ventilated by being moved back and forth through the water. The ability to move external gills is particularly important for sessile invertebrates, which must otherwise rely on sporadic local water currents or the muscular efforts of their bodies to create local currents for ventilation.

Despite the success of marine invertebrates, external gills have several limitations. First, they are unprotected and therefore are susceptible to damage from the environment. Second, because water is much denser than air, considerable energy is required to continually wave the gills back and forth through the water (think of the difference between waving your hands through air and waving them through water). Finally, their appearance and motion may draw the attention of predators.

Biology Principle

Structure Determines Function

Note how the gills of this nudibranch have extensive surface modifications to increase the total area available for gas exchange. This is not unique to external gills; all gas-exchange surfaces have structural modifications that maximize their function.

Figure 36.15 **Example of an animal with external gills.** The gills of this nudibranch have extensive projections to increase surface area, which facilitates oxygen diffusion from the surrounding water.
©Hal Beral/V&W/The Image Works

Internal Gills By contrast, fishes have internal gills, which are covered by a bony plate called an operculum (**Figure 36.16a**). Fish gills are confined within the opercular cavity, the space beneath the operculum, which protects the gills and helps to streamline the body.

The main support structures of gills are the gill arches, from which project gill filaments composed of numerous platelike structures called **lamellae** (**Figure 36.16b**). Blood vessels run the length of the filaments. Oxygen-poor blood travels through a vessel called the afferent vessel along one side of the filament, and oxygen-rich blood travels through another vessel called the efferent vessel along the other side. Within the lamellae are numerous capillaries, all oriented with blood flowing from the oxygen-poor vessel to the oxygen-rich one.

Water enters a fish's mouth and flows between the lamellae in the opposite direction from blood flowing through the lamellar capillaries. This arrangement of water and blood flow is an example of **countercurrent exchange** (refer back to Figure 32.12). As oxygenated water encounters the lamellae, it comes into close proximity to blood in the gill capillaries. Recall that O_2 diffuses along a pressure gradient from a region of higher O_2 pressure to one of lower O_2 pressure. Thus, O_2 diffuses from the water into the capillaries of the lamellae. As water continues to flow across the lamellar surface, it encounters regions of capillaries that have not yet picked up O_2—in other words, even as O_2 begins to diffuse from the water into the gill capillaries, a sufficient pressure gradient remains along the lamellae to permit diffusion of more of the remaining O_2 from the water. This

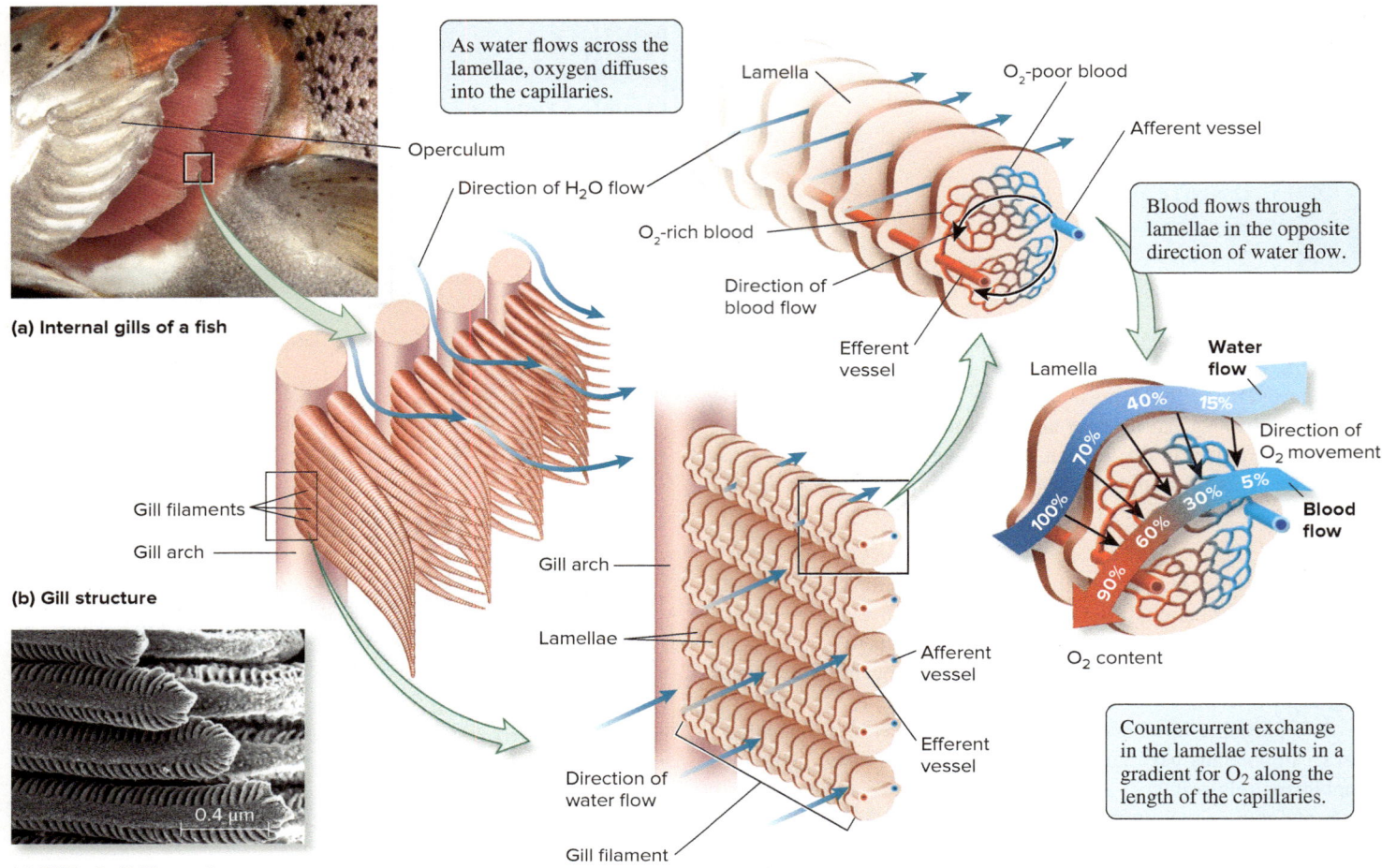

Figure 36.16 Structure of fish gills. **(a)** The operculum, which has been lifted up in this photo, protects the gills underneath. **(b)** The gills are composed of gill arches, from which numerous pairs of filaments arise. Thin, platelike lamellae are arrayed along the filaments. Blood flows through the capillaries of the gill filaments in the direction opposite from water flowing between lamellae, a mechanism called countercurrent exchange. **(c)** Several filaments with their lamellae, as revealed in a scanning electron micrograph (SEM).
(a) ©Sarah Ahrens/Alamy; *(c)* ©Electron Microscopy Unit, Royal Holloway, University of London

Concept Check: Why do fishes die when out of water? (Hint: Consider what happens if you hold two thin, wet paper towels together.)

BioConnections: You have learned about countercurrent exchange already in another context (refer back to Figure 32.12). How does it relate to heat exchange in vertebrates?

is an extremely efficient way to remove as much O_2 from the water as possible before the water passes out of the operculum.

Different fishes can ventilate their gills in three ways:

- Actively drawing water in through the mouth and out the operculum
- Swimming with the mouth open so that water continually moves across the gills
- Facing into a current of water while resting, but keeping the mouth open

Each of these ventilation mechanisms is a flow-through system—water moves only in one direction so that the gills are constantly in contact with fresh, oxygenated water. This improves gas exchange.

Insects Use Tracheal Systems to Exchange Gases with the Air

The delicate nature of gill lamellae generally makes them unsuitable for gas exchange in air. Air-breathing probably evolved as an adaptation in aquatic animals inhabiting regions that were subject to periodic drought.

One of the major mechanisms that animals evolved to breathe air is the **tracheal system** found in insects. Along the surface of an insect's body are tiny openings to the outside called **spiracles.** Arising from the spiracles are sturdy tubes called **tracheae** (singular, trachea) (**Figure 36.17**). Tracheae branch extensively into ever-smaller tubes called tracheoles, which eventually become small enough that their tips contact virtually every cell in the body. At their tips, tracheoles are filled with a small amount of fluid. Air flowing into the tracheoles

CIRCULATORY AND RESPIRATORY SYSTEMS: TRANSPORTING SOLUTES AND EXCHANGING GASES

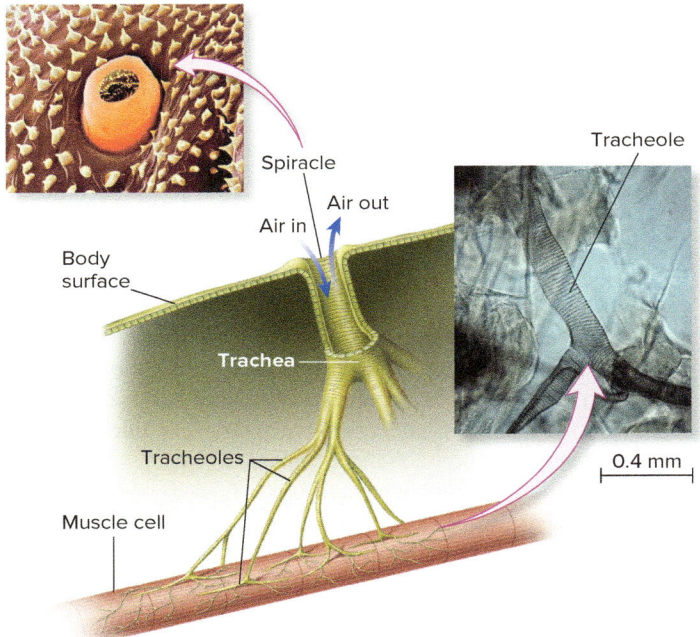

Figure 36.17 **The tracheal system of insects.** Air enters holes on the body surface called spiracles. Oxygen diffuses directly from the fluid-filled tracheole tips to cells that come into contact with the tips. The circulatory system plays no role in gas exchange in insects. The micrographs show a single spiracle and a branching tracheole.
(left) ©Microfield Scientific Ltd/Science Source; *(right)* ©Ed Reschke/Getty Images

Concept Check: *How might the structure of the respiratory system of insects be related to their relatively small size?*

comes into contact with this fluid. Oxygen from the air dissolves in the fluid, and from there it diffuses across the tracheoles and into nearby cells. Carbon dioxide diffuses in the opposite direction, from cells into the tracheoles and from there to the environment.

When an insect's oxygen demands increase due to increased activity, muscular movements of its abdomen and thorax draw air into and out of the tracheae a little like a bellows. An insect's muscles and tracheal system match ventilation with the animal's exercise intensity and O_2 requirements. This is particularly important in flying insects, which have very great metabolic demands.

As discussed earlier, the open circulatory system of insects does not participate in gas exchange. Oxygen diffuses directly from air to trachea to tracheoles and finally to body cells. This mechanism of ventilation and O_2 delivery is very effective. The relative metabolic rate of insect flight muscles is among the highest known of any tissue in any animal, and the tracheal system supplies enough O_2 to meet those enormous demands.

Air-Breathing Vertebrates Use Lungs to Exchange Gases

Except for some amphibians such as lungless salamanders, all air-breathing, terrestrial vertebrates use lungs to bring O_2 into the circulatory system and remove CO_2. **Lungs** are internal, paired structures that arise from the pharynx during embryonic life. All lungs receive partially deoxygenated blood from the heart and return fully oxygenated blood back to the heart.

Most vertebrates ventilate their lungs by a process called **negative pressure filling,** in which the pressure of air in the lungs is decreased below that of the environment. Boyle's law— named after the 17th-century Irish chemist Robert Boyle—states that the pressure and volume of a gas are inversely related (**Figure 36.18**). For example, when the size of a chamber in which a gas is contained is increased, the

Figure 36.18 **Boyle's law.** The volume and pressure of a gas are inversely related; that is, when the volume is increased, the pressure is decreased, and vice versa. This relationship creates the gas pressure gradients that ventilate the vertebrate lungs.

Concept Check: *As lungs expand during inhalation, what happens to the pressure of the air inside them?*

pressure of the gas in that container decreases. By contrast, if the chamber size is decreased, the pressure of the gas inside increases. By expanding its lungs (inhaling), an animal creates a pressure gradient for air to move from the atmosphere (higher pressure) into its lungs (lower pressure). When an animal exhales, the lungs become compressed, increasing the pressure of the air inside them; this causes air to leave the lungs.

To understand the mechanisms that achieve negative pressure filling, we will next examine the structure and function of the mammalian respiratory system as a model.

36.7 Reviewing the Concepts

- Ventilation is the process of bringing oxygenated water or air into contact with a respiratory organ.
- The body surface is permeable to gases in some invertebrates and in amphibians. Water-breathing animals use external or internal gills for gas exchange (Figures 36.15, 36.16).
- In insects, air in the tracheoles comes into contact with fluid at the tracheole tips. Oxygen from the air dissolves in this fluid and diffuses across the tracheole wall and into nearby cells (Figure 36.17).
- Air-breathing vertebrates generally use lungs to obtain O_2 and eliminate CO_2. All lungs receive deoxygenated blood from the heart and return oxygenated blood to the heart. Most vertebrates fill their lungs by negative pressure filling, which follows the principles of Boyle's law (Figure 36.18).

36.7 Testing Your Knowledge

1. Boyle's law states that
   a. the pressure and volume of a gas are directly and positively related.
   b. the pressure exerted by a gas will increase when its volume decreases.
   c. increasing the volume of a gas increases its pressure.
   d. decreasing the volume of a gas decreases its pressure.
   e. Both c and d are correct.
2. Which of the following statements is *false*?
   a. The insect respiratory system can adapt to increased metabolic demands of the animal.
   b. The tracheal system of insects is not sufficient to support the oxygen demands of very active tissues.
   c. Exchanging gases across the body surface is a feature found in certain invertebrates and some vertebrates.
   d. External gills are more susceptible to damage than are internal gills.
   e. The flow of blood and water through the gill lamellae follows a countercurrent arrangement.

36.8 Structure and Function of the Mammalian Respiratory System

Learning Outcomes

1. Describe the components of the mammalian respiratory system and the structure of the mammalian lung.
2. Outline the process of ventilation in mammalian lungs.

In this section, we will examine the structures of the mammalian respiratory system and the mechanisms by which mammals ventilate their lungs.

During Ventilation, Air Flows Through a Series of Branching Tubes

In mammals, the respiratory system includes the nose, the mouth, a series of branching tubes, lungs, and the muscles and connective tissues that encase these structures within the thoracic (chest) cavity (**Figure 36.19a**). When humans and other mammals breathe, air first enters the nose and mouth, where it is warmed and humidified. These processes protect the lungs from drying out. While in the nose, the air is partially purified as it flows over a coating of sticky mucus in the nasal cavity. The mucus and hairs in the nasal cavity trap some of the larger dust and other particles that are inhaled with air. These are then removed by the body's immune cells or swallowed.

The inhaled air from the mouth and nose converges at the back of the throat, or **pharynx,** a common passageway for air and food. From there, air passes through the **larynx,** which contains the vocal cords. Air flows from the larynx into the **trachea,** a tube that leads to the lungs.

The trachea is partially ringed by cartilage that provides rigidity and ensures that the trachea always remains open. Inhaled air flows down the trachea as it branches into two smaller tubes, called **bronchi** (singular, bronchus), which lead to each lung. The bronchi branch repeatedly into smaller and smaller tubes, eventually becoming thin-walled **bronchioles** surrounded by circular rings of smooth muscle (**Figure 36.19b**). Bronchioles can dilate or constrict in a manner analogous to that of arterioles, the small blood vessels that deliver blood to capillaries (see Figure 36.10).

The **alveoli** (singular, alveolus) are the saclike regions of the lungs where gas exchange occurs (**Figure 36.19c**). The alveoli are highly adapted for gas exchange and consist of two major types of cells. Gases diffuse across type I cells, whereas type II cells are secretory cells (described later). The alveoli are only one cell thick and resemble extremely thin sacs, appearing like bunches of grapes on a stem. Deoxygenated blood pumped from the right ventricle of the heart flows to the many capillaries surrounding the alveoli. Oxygen diffuses from the lumen of each alveolus across the type I alveolar cells, through the interstitial space outside the cells, and into the capillaries (Figure 36.19b). Carbon dioxide diffuses in the opposite direction. The oxygenated blood from the lungs then flows to the left atrium of the heart and from there enters the left ventricle, where it is pumped out through the aorta to the systemic circulation.

The Pleural Sacs Protect the Lungs

The lungs are soft, delicate tissues that could easily be damaged by the surrounding bone, muscle, and connective tissue of the thorax if not protected. Each lung is encased in a **pleural sac,** a double layer of thin, moist connective tissue. Between the two layers is a microscopically thin layer of water that acts as a lubricant and makes the two tissue layers adhere to each other.

In addition to protecting the lungs, the inner pleural sac adheres to its lung, and the outer pleural sac adheres to the chest wall. In this way, movements of the chest wall result in similar movements of the lungs. This is important because the lungs are not muscular and,

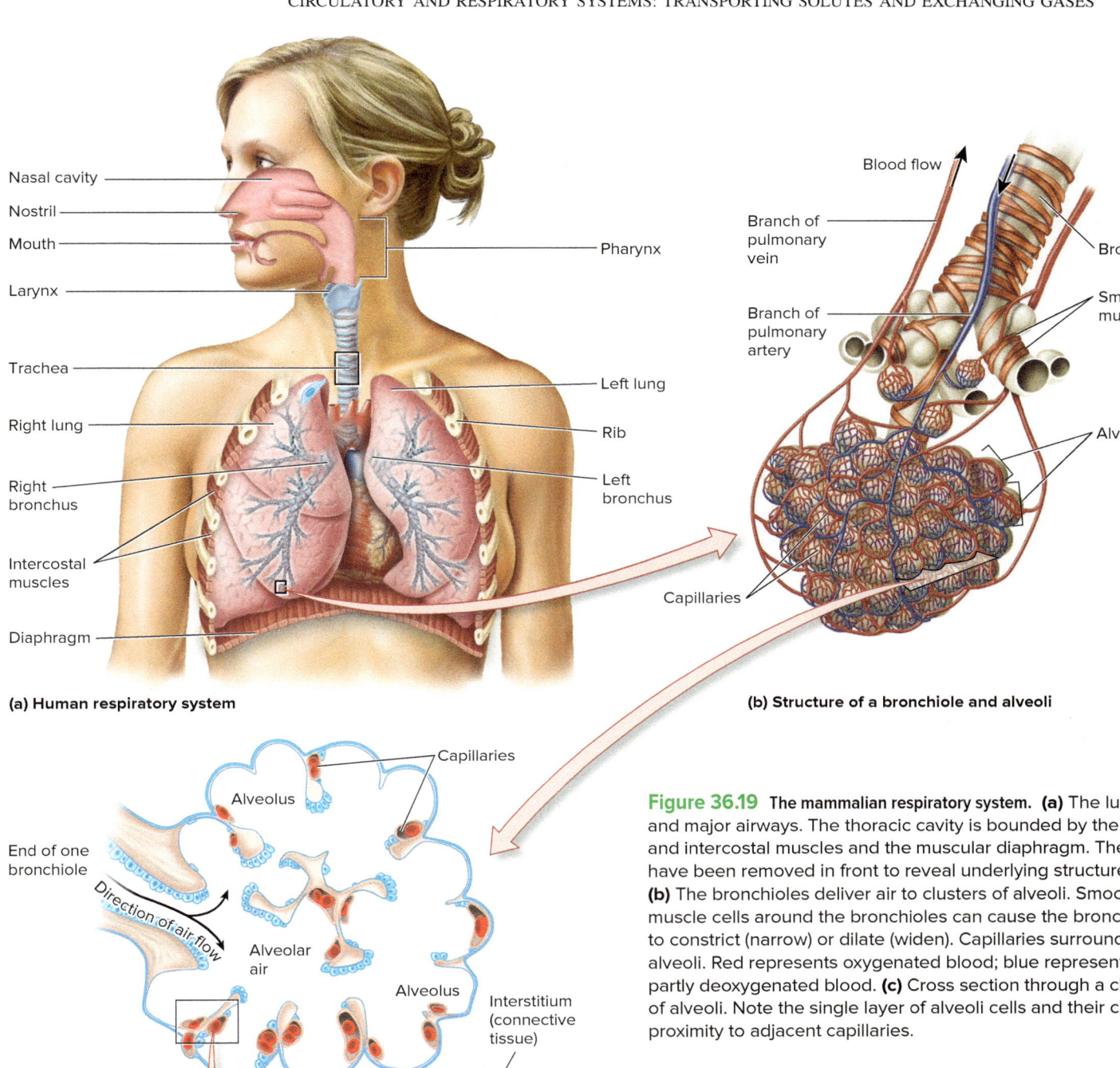

Figure 36.19 The mammalian respiratory system. (a) The lungs and major airways. The thoracic cavity is bounded by the ribs and intercostal muscles and the muscular diaphragm. The ribs have been removed in front to reveal underlying structures. (b) The bronchioles deliver air to clusters of alveoli. Smooth muscle cells around the bronchioles can cause the bronchioles to constrict (narrow) or dilate (widen). Capillaries surround the alveoli. Red represents oxygenated blood; blue represents partly deoxygenated blood. (c) Cross section through a cluster of alveoli. Note the single layer of alveoli cells and their close proximity to adjacent capillaries.

The Lungs Expand by Negative Pressure Filling

The way in which you inflate a balloon, by forcing air from your mouth into the balloon, is called positive pressure filling. By contrast, as mentioned earlier, the mechanism by which mammals and many other vertebrates ventilate their lungs is called negative pressure filling. In this process, the volume of the lungs expands, creating a decreased pressure that draws air into the lungs (see Figure 36.18). The process differs in some ways among classes of vertebrates, but in mammals, the work is provided by the intercostal muscles, which surround and connect the ribs in the chest, and a large muscle called the **diaphragm** (Figure 36.19a), which divides the thoracic cavity from the abdomen.

so, cannot inflate themselves. Instead, as we will see, the lungs are inflated by the expansion of the thoracic cavity, which results from the contraction of muscles in the thorax.

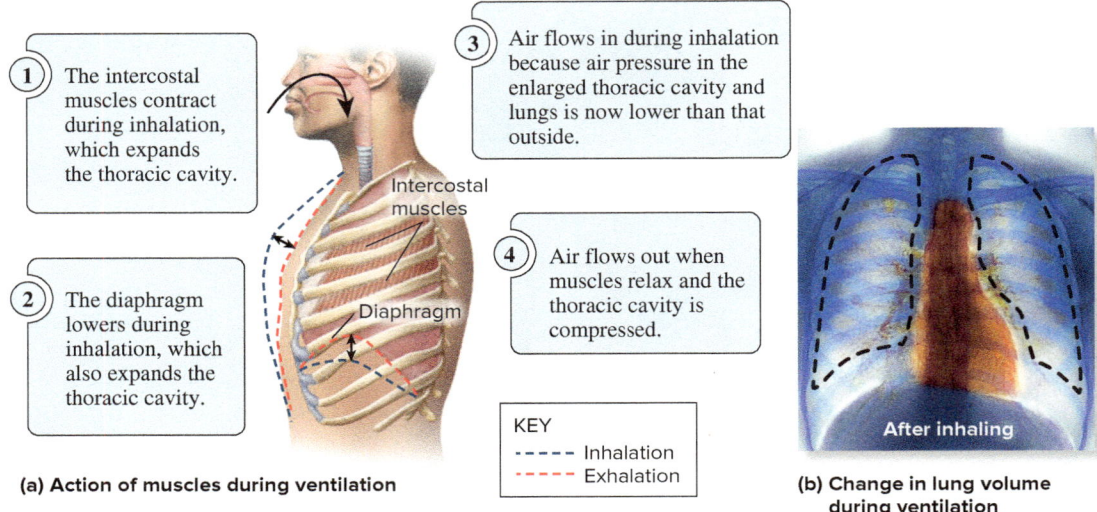

Figure 36.20 Ventilation of the mammalian lung by negative pressure filling. (a) The intercostal muscles contract, which expands the chest cavity by moving the ribs up and out. The diaphragm also contracts, causing it to pull downward, further expanding the cavity. The muscular efforts of inhalation require energy, whereas the return to the resting state by exhaling occurs primarily by recoil. (b) X-ray image of the chest of an adult man after inhaling. The volume of the lungs after exhaling is superimposed using dashed lines to illustrate the relative change in lung volume.
(b) ©Pr. M. Brauner/Science Source

Let's follow the process when a mammal, such as a human, ventilates his or her lungs (Figure 36.20). At the start of a breath, the diaphragm contracts, pulling downward. This downward pull enlarges the thoracic cavity. Simultaneously, the intercostal muscles contract, moving the chest upward and outward, which also helps to enlarge the thoracic cavity. Recall that the pleural sacs adhere the lungs to the chest wall, so as the chest expands, the lungs expand with it. According to Boyle's law, as the volume of the lungs increases, the pressures of the gases within them must decrease. In other words, the pressure in the lungs becomes negative with respect to the outside air. Air, therefore, flows down its pressure gradient from outside the mouth and nose, into the lungs.

Once the lungs are inflated with air, the chest muscles and diaphragm relax and recoil back to their original positions as a person exhales. This movement compresses the lungs, which raises the pressure inside them according to Boyle's law. As a result, air is forced out of the airways. Whereas inhaling requires the expenditure of significant amounts of energy, exhaling is mostly passive and normally does not require much energy. This is possible because the lungs and chest have large numbers of elastin fibers, which, as mentioned earlier in the case of arteries, have elastic recoil properties.

Mammals Breathe by Tidal Ventilation

When mammals exhale, air leaves via the same route that it entered during inhalation, and no new oxygen is delivered to the airways at that time. This type of breathing is called **tidal ventilation** (think of air flowing into and out of the lungs, like the ebb and flow of ocean tides). Tidal ventilation is less efficient than the unidirectional, flow-through system of fishes in which gills are always exposed to oxygenated water during all phases of the respiratory cycle, but it is sufficient for air-breathers because of the greater amount of oxygen in air compared to water.

As you likely know from experience, the lungs are neither fully inflated nor fully deflated at rest. For example, you could easily take a larger breath than normal if you wished, or exhale more than the usual amount of air. The volume of air that is normally breathed in and out at rest is the tidal volume, about 0.5 L in an average-sized human. Lung size and tidal volume are proportional to body size, both among humans and between species. A 6-foot-tall adult human, for example, has a larger tidal volume than a 4-foot-tall child because the adult has larger lungs. Similarly, horses have larger tidal volumes than humans, and humans have larger tidal volumes than dogs.

During exertion, the lungs can be inflated further than the resting tidal volume to provide additional oxygen. Likewise, the lungs can be deflated beyond their normal limits at rest, by exerting a strong effort during exhalation. The lungs never fully deflate, however, partly because they are held open by their adherence to the chest wall. Maintenance of partial inflation is important for a simple reason. Think again of our analogy of a balloon. It is much easier to fill a balloon that is already partly inflated than it is to inflate a completely empty balloon. The same is true of the lungs. The most difficult breath is the very first one that a newborn mammal takes—the only time its lungs are ever completely empty of air.

Surfactant Facilitates Lung Inflation by Decreasing Surface Tension

Like all cells, those that make up the lining of the alveoli are surrounded by extracellular fluid. This fluid layer is where gases dissolve. Unlike other internal body cells, however, alveolar cells come into contact with air, creating an air/liquid interface along the inner surface of the alveoli. This results in surface tension within the alveoli. Surface tension results from the attractive forces between water molecules at the air/liquid interface and partly explains why droplets of water form beads. This tension produces a force that tends to collapse alveoli as water molecules lining its surfaces are attracted to each other. If many or all of the alveoli were to collapse, however, the amount of surface area available for gas exchange in the lungs would be greatly reduced. What prevents them from collapsing? The type II cells of the alveoli produce **surfactant,** a mixture of proteins and amphipathic lipids (that is, lipids with both polar and nonpolar regions), and secrete it into the alveolar lumen. Surfactant forms a barrier between the air and the fluid layer inside the alveoli. This barrier reduces surface tension in the alveolar walls, allowing them to remain open.

36.8 Reviewing the Concepts

- The mammalian respiratory system includes the nose, mouth, airway tubes, lungs, muscles, and connective tissues that encase these structures within the thoracic cavity (Figure 36.19).
- Mammals ventilate their lungs by negative pressure filling. The work is provided by the intercostal muscles and diaphragm. Surfactant forms a barrier between air and the fluid layer of the alveoli, helping to keep them open (Figure 36.20).

36.8 Testing Your Knowledge

1. The _____ separate(s) the thoracic cavity from the abdomen and _____ during exhalation.
   a. pleural sacs, relax
   b. pleural sacs, contract
   c. diaphragm, relaxes
   d. diaphragm, contracts
   e. intercostal muscles, contract
2. The tidal volume
   a. decreases with age from childhood to adulthood.
   b. is about 5 L in a typical adult.
   c. is the maximum volume of air that can be inhaled in one breath.
   d. is the minimum volume of air that can be inhaled in one breath.
   e. is the volume of air normally inhaled and exhaled in one breath.

36.9 Mechanisms of Gas Transport in Blood

Learning Outcomes

1. Describe the characteristics of respiratory pigments.
2. **SCISKILLS** ▶ Analyze the quantitative relationship between oxygen and hemoglobin, and explain how certain factors can modify the shape of the dissociation curve.
3. Write the reversible reaction between hemoglobin and O_2.
4. Describe the ways in which CO_2 is carried in blood.

The amount of O_2 that can be dissolved in the body fluids is not sufficient to sustain life in most animals. In nearly all animals, therefore, the amount of O_2 in the body fluids must be increased above that which can be physically dissolved in water. This is made possible because of the widespread occurrence of oxygen-binding proteins, which increase the total reservoir of O_2 available to cells. In this section, we will examine the structure, function, and evolution of these molecules as well as how CO_2 is transported.

Oxygen Binds to Respiratory Pigments

The oxygen-binding proteins that have evolved in animals are called **respiratory pigments** because they have a color (typically blue or red). In vertebrates, the pigments are contained within erythrocytes, whereas many invertebrates have these pigments in their hemolymph. Respiratory pigments are proteins containing one or more metal atoms that bind to oxygen. In vertebrates and many marine invertebrates, the metal is iron (Fe^{2+}). As stated earlier, hemoglobin is the major iron-containing pigment and gives blood its red color. In decapod crustaceans, arachnids, and many mollusks, the metal is copper (Cu^{2+}). The copper-containing pigment **hemocyanin** gives the blood or hemolymph a bluish tint.

Hemoglobin gets its name because it is a globular protein—which refers to the shape and water solubility of a protein—and because it contains a chemical group called a heme in its core. An atom of iron is bound within the heme group. In vertebrates, each molecule of hemoglobin consists of four polypeptide subunits, each with its own iron atom to which a molecule of O_2 can bind. Thus, a single hemoglobin molecule can bind up to four molecules of oxygen (**Figure 36.21**).

Respiratory pigments share certain characteristics that make them ideal for transporting O_2. First, they all bind O_2 with high affinity. Second, the binding between the pigment and O_2 is noncovalent

Biology Principle

Structure Determines Function

The precise quaternary structure of hemoglobin permits its association with heme groups, which contain the iron atoms that bind oxygen molecules. Without a correct structure, the ability of hemoglobin to bind oxygen would be compromised.

Figure 36.21 Hemoglobin. Erythrocytes contain large amounts of the protein hemoglobin. Oxygen binds reversibly to iron atoms in the heme portion of each subunit of hemoglobin.

and reversible. Clearly, it would be of no benefit for a protein to bind O_2 if it could not later unload the O_2 to cells that require it. The reaction that describes the reversible binding of O_2 to hemoglobin (Hb) is

$$Hb + O_2 \rightleftharpoons HbO_2$$

where Hb is hemoglobin, HbO_2 is oxyhemoglobin (hemoglobin with bound O_2), and the double arrows ($\rightleftharpoons$) indicate that the reaction is reversible.

The amount of pigment present in the blood is sufficient to provide enough O_2 to meet most oxygen demands, except for those that occur under the most strenuous exertion. In humans, for example, the presence of hemoglobin gives blood about 45 times more O_2-carrying capacity compared with plasma alone.

The Amount of Oxygen Bound to Hemoglobin Depends on the P_{O_2} of Blood

The partial pressure of O_2 (P_{O_2}) in blood is a measure of its dissolved concentration. When P_{O_2} is high, more O_2 binds to hemoglobin, whereas fewer O_2 molecules are bound when P_{O_2} is low. **Figure 36.22** shows the relationship between O_2 binding and P_{O_2}, known as an **oxygen-hemoglobin dissociation curve,** for humans. At a P_{O_2} of 100 mmHg, which is typical of the oxygenated blood leaving the lungs, nearly every hemoglobin molecule is bound to four O_2 molecules (at the far right in the graph). It is essentially 100% saturated with O_2 at that P_{O_2}. The P_{O_2} of blood leaving the tissue capillaries of other parts of the body is lower because some of the oxygen has left the blood and been used by cells. At rest, the average P_{O_2} of blood capillaries in these other parts of the body is typically around 40 mmHg. At this P_{O_2}, hemoglobin releases some O_2 molecules, decreasing to about 75% saturation with O_2. During strenuous exercise, P_{O_2} in the capillaries drops even further (as low as 20 mmHg); consequently, during exercise hemoglobin releases even more O_2 and becomes less saturated. In this way, hemoglobin performs its role of O_2 delivery. In the lungs, it binds O_2, and elsewhere it releases O_2 as required.

The curve in Figure 36.22 is not linear but S-shaped (sigmoidal). This shape results because the subunits of hemoglobin cooperate with each other in binding O_2. Once a molecule of O_2 binds to one subunit's iron atom, the shape of the entire hemoglobin protein changes, making it easier for a second O_2 to bind to the next subunit, and so on. Thus, the relationship between P_{O_2} and the amount of O_2 bound to hemoglobin becomes very steep in the portion of the curve that represents the pressures that occur in the tissue capillaries throughout the body. This steepness allows O_2 release from hemoglobin to be very sensitive to even small decreases in P_{O_2} generated by the diffusion of O_2 from capillaries to cells. The curve levels off at high oxygen pressures as 100% saturation is approached.

> **Biology Principle** | **Biology is a quantitative science**
>
> **Quantitative Analysis:** The Ability of Hemoglobin to Bind Oxygen Is Affected by Factors Such as Temperature, CO_2, and pH
>
> One of the remarkable features of the oxygen-hemoglobin binding relationship is that it is influenced by metabolic waste products such as CO_2 and H^+ and by heat (temperature). **Figure 36.23a** shows three curves, one obtained under normal resting conditions and the others in the presence of low or high levels of CO_2. Carbon dioxide binds to amino acids in the hemoglobin protein (not to the iron, as O_2 does), and when it does, it decreases the affinity of hemoglobin for O_2. Note how an increase in CO_2 shifts the curve to the right, such that at any P_{O_2}, less O_2 is bound to hemoglobin. Another way of saying this is that at any P_{O_2}, more O_2 has been released from hemoglobin, thus becoming available to cells. A similar shift in the curve occurs with an increase in acidity (decreased pH), because H^+ can also bind to hemoglobin and alter its oxygen-carrying capacity; the effect of CO_2 and H^+ on the oxygen-hemoglobin dissociation curve is known as the **Bohr effect.** Elevated temperature also reduces the affinity of hemoglobin for O_2, resulting in a right-shifted curve.
>
> Cells generate each of these products—CO_2, H^+, and heat—when they are actively metabolizing nutrients such as glucose. The metabolic products enter the surrounding capillaries and diffuse into erythrocytes, where they alter the shape of hemoglobin, causing it to release more of its O_2 than would normally occur at that P_{O_2}. This phenomenon is a way in which individual body tissues obtain more O_2 from the blood to match their higher metabolic demands. Thus, when an animal increases its physical activity, the skeletal muscles generate more CO_2, H^+, and heat than do some other tissues. Therefore, the hemoglobin in muscle capillaries releases more O_2 to muscle cells compared to the hemoglobin in the capillaries of these other tissues.

Figure 36.22 **The human oxygen-hemoglobin dissociation curve.** Depending on the partial pressure of oxygen (P_{O_2}), oxygen is either loaded onto hemoglobin, as in the lungs, or unloaded from hemoglobin, as in the rest of the body tissues. When P_{O_2} is high, more Fe^{2+} atoms are bound to O_2, and therefore the hemoglobin is more saturated with O_2.

(a) Shifts in the Oxygen-Hemoglobin dissociation curve

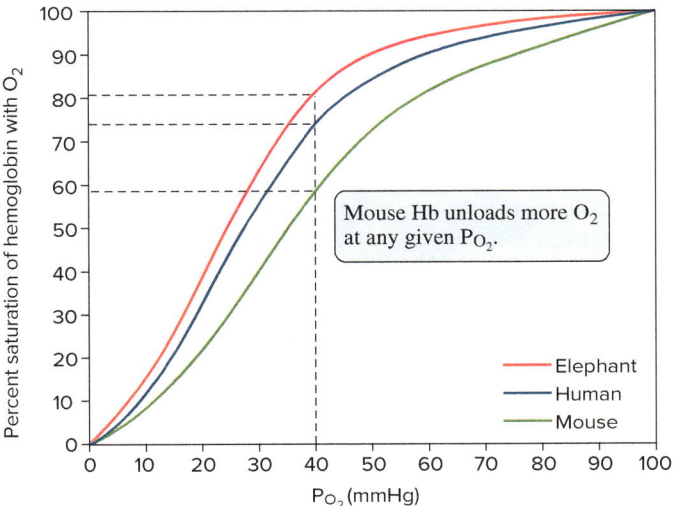

(b) Oxygen-Hemoglobin dissociation curves of different animals

Figure 36.23 Changes in oxygen-hemoglobin dissociation curves under different conditions and among different species. (a) Increasing or decreasing the amounts of CO_2 or H^+ (pH), or the temperature of the blood, shifts the oxygen-hemoglobin dissociation curve. Metabolically active tissues generate more of these products. The change in affinity of hemoglobin for oxygen (O_2) allows different tissues to obtain O_2 in proportion to their metabolic requirements. (b) Oxygen-hemoglobin dissociation curves for three mammals with low (elephant), moderate (human), or high (mouse) relative metabolic rates. For any P_{O_2}, such as the one selected in the graph (40 mmHg), which is typical of the P_{O_2} of tissue capillaries, less O_2 is bound to mouse hemoglobin than to human or elephant hemoglobin, and less O_2 is bound to human hemoglobin than to elephant hemoglobin. Therefore, O_2 is unloaded from hemoglobin more readily in smaller animals.

The shift in the oxygen-hemoglobin dissociation curve occurs in all classes of vertebrates (although not in all species), but it has different magnitudes in different species. Not surprisingly, perhaps, animals with relatively high metabolic rates, such as mice, show a greater Bohr effect than do animals with lower relative metabolic rates. Recall that these same waste products of metabolism also cause local vasodilation of arterioles. The more metabolically active a tissue is, therefore, the more blood flow it receives, which means more oxygen-bound hemoglobin. Moreover, the oxygen is unloaded from hemoglobin more readily due to the shift in the curve. This is an excellent example of how adaptive changes in circulatory and respiratory functions often are complementary.

Regardless of their sensitivity to CO_2 and other factors, the hemoglobins of metabolically active animals also have a lower affinity for oxygen (**Figure 36.23b**) even in the absence of a greater than normal amount of CO_2, H^+, and heat. In small, active animals, the curves are displaced to the right relative to the human curve (in other words, the hemoglobin P_{50} value is higher in small animals than it is in larger animals with lower relative metabolic rates). In contrast, larger animals with lower relative metabolic rates, such as the elephant, have curves shifted to the left compared with that of humans. At high oxygen pressures, such as those that occur in the lungs, these shifts have little relevance because nearly all the hemoglobin is bound to oxygen at those pressures. At lower P_{O_2} values however, such as those that would occur in the capillaries of metabolically active tissues, the difference in the curves becomes significant. For example, look at the three curves at a P_{O_2} of 40 mmHg, a typical value found in tissues that are using oxygen at a resting rate. The mouse hemoglobin has less oxygen bound (its hemoglobin is less saturated) at that pressure than does the human hemoglobin, which in turn has less O_2 bound than does the elephant hemoglobin. In other words, the mouse hemoglobin has released more of its O_2 to the active tissues than have the other animal hemoglobins. The different properties of hemoglobin among species result from changes in the sequence of hemoglobin, the result of numerous evolutionary changes in the genes encoding hemoglobin subunits in vertebrates.

Crunching the Numbers: Refer to Figure 36.23a. In which case would hemoglobin be more saturated: at a P_{O_2} of 35 mmHg and with increased CO_2, or at a P_{O_2} of 30 mmHg but with decreased CO_2? Now refer to Figure 36.23b. What percent saturation of hemoglobin would you estimate for a small mammal, such as a rabbit, at a P_{O_2} of 30 mmHg?

Carbon Dioxide Is Transported in the Blood in Three Forms

As with oxygen, only a limited amount of CO_2 physically dissolves in blood. An additional amount of CO_2 is carried bound to hemoglobin, as noted earlier. The majority of CO_2, however, is converted into highly soluble bicarbonate ions (HCO_3^-). This conversion is achieved by the following reactions, where H_2CO_3 is a short-lived compound called carbonic acid that immediately dissociates to an H^+ and an HCO_3^-:

$$CO_2 + H_2O \rightleftharpoons H_2CO_3 \rightleftharpoons H^+ + HCO_3^-$$

Note that one H^+ is formed for every CO_2 that enters this reaction. The resulting pH change is what makes CO_2 a dangerous waste product,

3. Arrange the following blood vessels in sequence from those with the highest expected pressure to the lowest, for a mammal such as a horse: (1) a large artery in the chest, (2) a vein in the neck, (3) a leg vein, (4) a large artery in the neck. Assume that the horse is standing and at rest.
   a. 1, 2, 3, 4
   b. 1, 4, 3, 2
   c. 4, 1, 2, 3
   d. 3, 2, 1, 4
   e. 4, 1, 3, 2

4. According to the following reaction, what would happen if a person at rest held his or her breath for a minute or so?

$$CO_2 + H_2O \rightleftharpoons H_2CO_3 \rightleftharpoons H^+ + HCO_3^-$$

   a. The arterial level of CO_2 would decrease.
   b. The arterial level of CO_2 would increase and that of H^+ would decrease.
   c. The blood would become more alkaline.
   d. The arterial level of CO_2 would increase and the blood would become more acidic.
   e. The arterial levels of H^+ and HCO_3^- would both decrease.

5. What would happen to an animal's blood pressure if its cardiac output increased by a factor of 2 and its total peripheral resistance decreased by 50%?
   a. It would decrease 16-fold.
   b. It would decrease 8-fold.
   c. It would increase 1.5-fold.
   d. It would increase 24-fold.
   e. It would not change.

Short Answer

1. Explain why it is an advantage for CO_2, H^+, and heat to be major factors in decreasing the affinity of hemoglobin for oxygen.
2. **PRINCIPLES** A principle of biology is that structure determines function. How is this principle related to the hemoglobin molecule?

Digestive and Excretory Systems: Maintaining Nutrient, Water, and Energy Balance and Removing Waste

37

©Richard Hutchings/PhotoEdit

A balanced meal containing many nutrients, including carbohydrates, lipids, proteins, vitamins, minerals, and water.

Chapter Outline

37.1 Overview of Animal Nutrition
37.2 General Principles of Digestion and Absorption of Nutrients
37.3 Vertebrate Digestive Systems
37.4 Nutrient Use and Storage
37.5 Regulation of the Absorptive and Postabsorptive States in Vertebrates
37.6 Excretory Systems in Different Animal Groups
37.7 Structure and Function of the Mammalian Kidneys
37.8 Impact on Public Health

Assessing Your Knowledge and Skills

A 37-year-old man visited his physician because of chronic heartburn. The man was obese, weighing 247 pounds and standing 5 feet, 8 inches tall. He was in considerable pain, particularly after eating. His physician determined that his heartburn was directly related to his obesity. The distension of his abdomen and his practice of eating large meals increased the likelihood of stomach acid being regurgitated into the esophagus—the tube that carries food from the pharynx to the stomach. The reflux of acid into the esophagus creates the symptoms of heartburn and can lead to ulcers. The man was counseled by a nutritionist regarding a sensible weight loss program and was treated with a medication that decreases acid production in the stomach. Fortunately, weight loss can reverse most of the man's symptoms. Unfortunately, however, although this person initially lost about 10% of his body weight within a short time, he regained it all within 2 years, at which time his condition worsened.

This case illustrates several topics covered in this chapter, including the importance of sufficient and sensible consumption of nutrients and the control of digestive processes such as acid production. In animals, a **nutrient** is any organic or inorganic substance that is consumed and is required for survival, growth, development, tissue repair, or reproduction; the process of consuming and using nutrients is called **nutrition.**

In this chapter, we will begin by looking at the types of nutrients that animals require. Next, we will discuss the diverse ways in which animals obtain, digest, and absorb those nutrients and how these processes are regulated. We will then consider how energy homeostasis is maintained while an animal is fasting and how various excretory organs participate in removing waste products from an animal's body. We will next consider some major features of the vertebrate and mammalian kidney and examine how the kidneys eliminate soluble wastes and regulate homeostasis. We will conclude by examining the natures of some common gastrointestinal and kidney disorders in humans.

37.1 Overview of Animal Nutrition

Learning Outcomes

1. List the major categories of nutrients consumed by animals, and describe some of their general functions.
2. Identify four groups of essential nutrients, listing several examples of each.
3. **SCISKILLS** ▶ Predict some of the consequences of a deficiency of a given nutrient.
4. Describe some examples of the diversity of animal feeding habits.

Food processing in animals occurs in four phases: ingestion, digestion, absorption, and egestion (**Figure 37.1**). **Ingestion** is the act of taking food into the body via a structure such as a mouth. From there, the food moves into a digestive cavity of some sort or, as shown for the vertebrate in Figure 37.1, a specialized tube called an **alimentary canal.** If the nutrients in the food are in a form that cannot be directly used by cells, the nutrients must be broken down into smaller molecules—a process known as **digestion.** This is followed by the process of **absorption,** in which ions, water, and small molecules diffuse or are transported from the digestive cavity into an animal's circulatory system or body fluids. **Egestion** (or defecation) is the process by which animals pass undigested material out of the body.

Animals require both organic and inorganic nutrients. Organic nutrients fall into five categories: carbohydrates, proteins, lipids, nucleic acids, and vitamins. These provide energy and the building blocks of new molecules or serve as coenzymes in many enzymatic reactions. Inorganic nutrients include water and minerals such as calcium, copper, and iron. Minerals serve many functions, including acting as cofactors in enzymatic reactions, binding oxygen, and other processes.

In this section, we will examine the various nutrients consumed by animals and some of the major functions of these molecules. We will then see that despite the enormous number of animal species and the highly varied environments in which they live, the ways in which animals obtain and ingest food can be classified into just a few categories.

Animals Require Nutrients for Energy and the Synthesis of New Molecules

Although different species consume a wide variety of foods, all animals require the same fundamental organic molecules (**Table 37.1**; see also Chapter 3 for a discussion of the chemical nature of organic molecules).

Ingested organic molecules are used for two general purposes: to provide energy and to make new molecules. As described in Chapter 6, the bonds of organic molecules may be enzymatically broken down to release energy; this energy can be used in the synthesis of ATP, a primary energy source for all cells. Indirectly, therefore, organic molecules provide the energy required for most of the chemical reactions that occur in animals' bodies. In addition, organic molecules serve as building blocks for synthesizing new cellular molecules. For example, the muscles of animals use amino acids obtained from food to make the specialized proteins that allow their muscle fibers to contract.

Essential Nutrients Must Be Obtained from the Diet

Animal cells can synthesize many organic molecules, but certain compounds cannot be synthesized from any ingested or stored precursor molecule. These **essential nutrients** must be obtained in the diet in their complete form. The essential nutrients can be classified into four groups:

- **Essential amino acids** are required in the diet of many but not all animals. In humans, the essential amino acids include isoleucine, leucine, lysine, methionine, phenylalanine, histidine, threonine, tryptophan, and valine (refer back to Figure 3.11 for the structures of the 20 amino acids in animals). Without a recurring supply of essential amino acids, protein synthesis in each cell in an animal's body would slow down or stop. Because

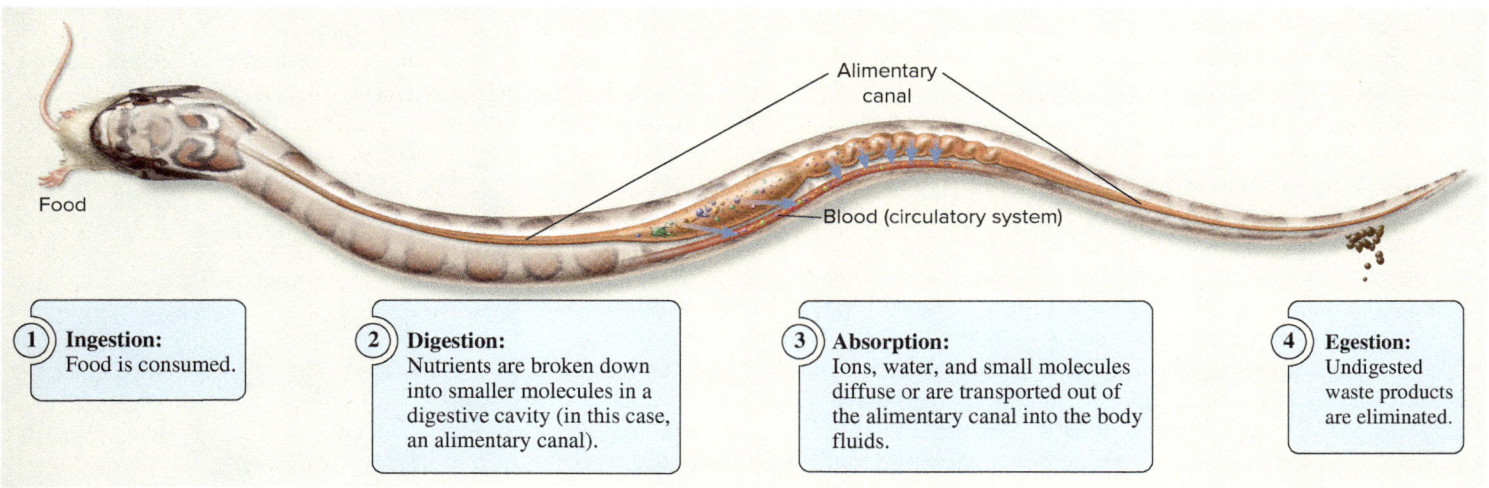

Figure 37.1 An overview of the four phases of food processing in animals, as shown in a snake.

① **Ingestion:** Food is consumed.

② **Digestion:** Nutrients are broken down into smaller molecules in a digestive cavity (in this case, an alimentary canal).

③ **Absorption:** Ions, water, and small molecules diffuse or are transported out of the alimentary canal into the body fluids.

④ **Egestion:** Undigested waste products are eliminated.

Table 37.1 Major Nutrients, Their Dietary Sources, and Some of Their Functions and Symptoms of Deficiency

Class of nutrient	Dietary sources	Functions in vertebrates	Symptoms of deficiency in humans
Carbohydrates	All food sources, especially starchy plants	Energy source; component of some proteins; source of carbon	Muscle weakness; weight loss
Proteins	All food sources, especially meat, legumes, cereals, and roots	Provide amino acids to make new proteins; build muscle; some amino acids used as energy source	Weight loss; muscle loss; weakness; weakened immune system; increased likelihood of infections
Lipids	All food sources, especially fatty meats, dairy products, and plant oils	Major component of cell membranes; energy source; thermal insulator; building blocks of some hormones	Hair loss; dry skin; weight loss; hormonal and reproductive disorders
Nucleic acids	All food sources	Provide sugars, bases, and phosphates that can be used to make DNA, RNA, and ATP	None; components of nucleic acids can be synthesized by cells from amino acids and sugars

all 20 amino acids are readily available in all meats but not as abundant in plants, some vegetation-eaters, such as cows, have evolved the capacity to synthesize the essential amino acids.

- **Essential fatty acids** are certain unsaturated fatty acids (refer back to Figure 3.8), such as linoleic acid, that cannot be synthesized by animal cells. Linoleic acid is vital to an animal's health because it is converted in cells to another fatty acid—arachidonic acid—which is the precursor required for production of several compounds important in many aspects of animal physiology. Such compounds include the prostaglandins, which function in pain, blood clotting, and smooth muscle contraction. Unsaturated fatty acids are abundant in plants, which provide a dietary source of essential fatty acids for many animals. Animals that never consume plants, however, may obtain their essential fatty acids from fishes or from the adipose (fat) tissue of birds and mammals.

- **Vitamins** are important organic nutrients that serve as coenzymes for many metabolic and biosynthetic reactions. Water-soluble vitamins, such as vitamin C, are not stored in an animal's body and must be regularly ingested. Fat-soluble vitamins, such as vitamin A, are stored in adipose tissue. Table 37.2 summarizes the vitamins required by animals, their dietary sources, some of their important functions, and health consequences associated with their deficiencies in humans.

- **Minerals** are inorganic ions required by animals for normal functioning of cells. Minerals such as iron and zinc are required as cofactors for or constituents of some enzymes and other proteins. Other minerals, such as calcium, are required for bone, muscle, and nervous system function. Still others—notably, sodium and potassium—contribute to changes in electrical differences across plasma membranes and therefore are especially critical for heart, skeletal muscle, and neuronal activity. Not all minerals are used the same way or at the same rate by all animals. For instance, the major way in which oxygen is transported in the body fluids of animals is by binding to proteins that contain one or more metal atoms in their structure. In some invertebrates, copper is the metal atom that binds oxygen, whereas in all vertebrates and most other invertebrates, iron serves this function.

Animals Have Diverse Feeding Habits

Animals share many of the same nutritional requirements, but they obtain and process those nutrients in diverse ways. Several factors determine how and when an animal obtains food, including its energetic demands, whether or not it is capable of locomotion, its local environment, and the structure and function of its digestive system. The digestive systems of **herbivores,** animals that normally eat only vegetation, contain microorganisms that assist in the digestion of cellulose, the main macromolecule of plant cell walls. By contrast, **carnivores** are primarily adapted to consume animal flesh or fluids, and **omnivores,** such as humans, eat both plant and animal products.

Although broadly useful, the three dietary categories just described are limited when considering the diversity of animal feeding habits. For example, some animals eat protists, algae, and/or fungi as a major source of food. Other animals are almost strictly carnivores at one time of year but herbivores at other times. Many nonmigratory birds, for example, feed on insects and worms during the summer but switch to eating whatever vegetation, buds, or seeds they can find during the winter. Animals like these are said to be opportunistic; they have a strong preference for one type of food but can adjust their diet if the need arises.

An animal's life stage may also influence its diet. Mammals begin life as milk-drinkers and later switch to consuming plants, animals, or both. Caterpillars eat leaves, but after metamorphosis, most species of moths and butterflies are strictly fluid-drinkers, typically consuming the nectar in flowers. Tadpoles are mostly herbivores, whereas frogs eat insects. This type of food resource preference may reduce competition between the different life stages of a species.

The three major categories also do not indicate what type of plant or animal is consumed. Grasses, cereals, and fruits are all types of plant matter, but each has a different energy and nutrient content. Some carnivores eat only flesh, whereas others drink only the blood of other animals.

Regardless of what an animal eats, the useful parts of the eaten material must be digested into molecules that its cells can absorb. Next, we will turn to a discussion of how animals digest their food and absorb the nutrients.

Table 37.2 Vitamins Required by Animals*

Class of nutrient	Dietary sources	Functions in vertebrates	Symptoms of deficiency in humans
Water-soluble vitamins			
Biotin	Liver; legumes; soybeans; eggs; nuts; mushrooms; some green vegetables	Coenzyme for gluconeogenesis and fatty acid and amino acid metabolism	Skin rash; nausea; loss of appetite; mental disorders (depression or hallucinations)
Folic acid	Green vegetables; nuts; legumes; whole grains; organ meats (especially liver, kidney, heart)	Coenzyme required for synthesis of nucleic acids	Anemia (a lower than normal number of erythrocytes in the blood); depression; birth defects
Niacin	Legumes; nuts; milk; eggs; meat	Involved in many oxidation-reduction reactions	Skin rashes; diarrhea; mental confusion; memory loss
Pantothenic acid	Nearly all foods	Part of coenzyme A, which is involved in numerous synthetic reactions, including formation of cholesterol	Burning sensation in hands and feet; gastrointestinal symptoms; depression
Vitamin B_1 (thiamine)	Meats; legumes; whole grains	Coenzyme involved in metabolism of sugars and some amino acids	Beriberi (muscular weakness, anemia, heart problems, loss of weight)
Vitamin B_2 (riboflavin)	Dairy foods; meats; organ meats; cereals; some vegetables	Respiratory coenzyme; required for metabolism of fats, carbohydrates, proteins	Seborrhea (excessive oil secretion from skin glands resulting in skin lesions)
Vitamin B_6 (pyridoxine)	Meats; liver; fish; nuts; whole grains; legumes	Coenzyme for over 100 enzymes that participate in amino acid metabolism, lipid metabolism, heme synthesis	Seborrhea; nerve disorders; depression; confusion; muscle spasms
Vitamin B_{12}	Meats; liver; eggs; some shellfish; dairy foods	Required for red blood cell formation, DNA synthesis, nervous system function	Anemia; nervous system disorders leading to sensory problems; balance and gait problems; loss of bladder and bowel control
Vitamin C (ascorbic acid)	Citrus fruits; green vegetables; tomatoes; potatoes	Antioxidant and free radical scavenger; aids in iron absorption; helps to maintain healthy connective tissue and gums	Scurvy (connective tissue disease associated with skin lesions, weakness, poor wound healing, tooth decay); bleeding gums
Fat-soluble vitamins			
Vitamin A (retinol)	Liver; green and yellow vegetables; some fruits in small amounts	Component of visual pigments; regulatory molecule affecting transcription; important for reproduction and immunity	Night blindness due to loss of visual ability; skin lesions; impaired immunity
Vitamin D	Fish oils; fish; egg yolk; liver; synthesized in skin via sunlight	Required for calcium and phosphorus absorption from intestines; bone growth	Rickets (weakened, deformed bones) in children; osteomalacia (weak bones) in adults
Vitamin E	Meats; vegetable oils; grains; nuts; seeds; small amounts in some fruits and vegetables	Antioxidant; inhibits prostaglandin synthesis	Unknown, possibly skeletal muscle atrophy; peripheral nerve disorders
Vitamin K	Legumes; green vegetables; some fruits; some vegetable oils (olive oil, soybean oil); liver; synthesized by bacteria in the large intestine	Component of blood clotting mechanism	Reduced blood clotting ability

*Not all animals require each of these vitamins. For example, most vertebrates with the exception of fishes and a few mammals (including humans) can synthesize vitamin C.

BioConnections: *All living organisms require nutrients. What are some common nutrients required by organisms other than animals? See Table 30.1 for help.*

37.1 Reviewing the Concepts

- The four phases of animal nutrition are ingestion, digestion, absorption, and egestion. Animals require organic nutrients—carbohydrates, proteins, lipids, nucleic acids, and vitamins—and inorganic nutrients (water and minerals) (Figure 37.1).
- Essential nutrients consist of vitamins, minerals, and those amino acids and fatty acids that cannot be synthesized by animals (Tables 37.1, 37.2).
- Herbivores eat only vegetation, carnivores consume animal flesh or fluids, and omnivores eat both plant and animal products.

37.1 Testing Your Knowledge

1. Which of the following is *not* an organic nutrient?
   a. lipid
   b. vitamin
   c. mineral
   d. carbohydrate
   e. protein
2. Which of the following statements is *false*?
   a. Water is an inorganic nutrient.
   b. All nutrients can be synthesized in animal cells.
   c. Copper and potassium are examples of minerals consumed by animals.
   d. In addition to other functions, carbohydrates, lipids, and proteins can all supply energy.
   e. An animal's diet depends in some cases on its life stage.

37.2 General Principles of Digestion and Absorption of Nutrients

Learning Outcomes

1. Describe the general structure of an alimentary canal.
2. **SCISKILLS** ▶ Make predictions about the functions of parts of the alimentary canal based on their structural adaptations.
3. Distinguish between passive and active absorption of food.

Once food has been ingested, some of the nutrients from the food must be broken down (digested) so that they can be absorbed by the cells of the digestive system. In this section, we will examine some of the major principles of digestion and absorption in animals, beginning with where digestion takes place.

Digestion Usually Occurs Extracellularly

Food is digested either inside cells (intracellularly) or outside cells (extracellularly). Intracellular digestion occurs only in some very simple invertebrates such as sponges and single-celled organisms and to a limited extent in cnidarians. It involves using phagocytosis to bring food particles directly into a cell, where the food is segregated from the rest of the cytoplasm in food vacuoles. Once inside these vacuoles, macromolecules in food are broken down by hydrolytic enzymes into monomers (the building blocks of polymers), which then are moved out of the vacuole to be used directly by that cell. Intracellular digestion cannot support the metabolic demands of an active animal for long, because only tiny bits of food can be engulfed and digested at one time. This form of digestion also does not provide a mechanism for storing large quantities of food so that an animal can digest it slowly while going about its other activities.

Most animals digest food extracellularly in a cavity of some sort. Extracellular digestion protects the interior of the cells from the actions of hydrolytic enzymes and allows animals to consume larger prey or vegetation. Food enters the digestive cavity, where it is stored, slowly digested, and absorbed gradually over long periods of time, ranging from hours (for example, after a human eats a pizza) to weeks (after a python eats a gazelle).

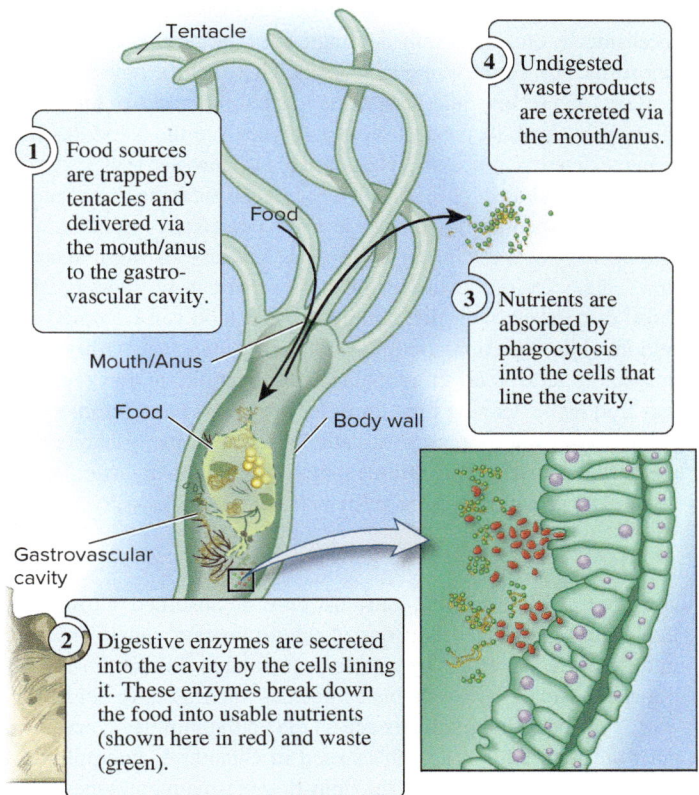

Figure 37.2 Extracellular digestion in a gastrovascular cavity. In animals with a gastrovascular cavity, such as the cnidarian illustrated here, most digestion occurs extracellularly within the cavity.

 Concept Check: *What is an advantage of extracellular digestion?*

In the simplest form of extracellular digestion—seen in invertebrates such as cnidarians and flatworms—the digestive cavity is known as an incomplete digestive cavity because it has one opening that serves as both an entry (mouth) and an exit (anus) port (**Figure 37.2**). This type of digestive cavity is also called a **gastrovascular cavity** because not only does digestion occur within it but fluid movements in the cavity also serve as a circulatory—or vascular—system that distributes digested nutrients throughout the animal's body. Food within a gastrovascular cavity is partially digested by enzymes that are secreted into the cavity by the cells lining the cavity. As the food particles become small enough, they are engulfed by the lining cells and further digested intracellularly. Undigested material that remains in the gastrovascular cavity is expelled.

Most Digestive Cavities Are Tubes with Specialized Regions and Openings at Opposite Ends

In contrast to the gastrovascular cavities of simple invertebrates, the digestive systems of all other animals consist of a single elongated tube, usually called an alimentary canal, with an opening at each end and through which food passes from one end to the other (see Figure 37.1). Along its length, the tube contains smooth muscle as part of its structure.

Contractions of this muscle churn up the ingested food so that it is mechanically broken into smaller fragments. The canal is lined on its interior surface by a layer of epithelial cells. These cells synthesize digestive enzymes and secrete other factors into the hollow cavity, or lumen, of the alimentary canal; they also secrete hormones into the blood that help to regulate digestive processes. The cells are also involved with transporting digested material out of the canal, where it can enter the body fluids.

The alimentary canal has several specialized regions along its length that vary according to species. Because of these specializations, digestive processes requiring low pH can be segregated from those requiring higher pH, and undigested food can be stored in one region while digestion continues in another area. The ability of some animals to store food in a stomach, for example, allows them to eat less frequently, leaving time for other activities. In addition, a diverse collection of microorganisms, called a microbiome, inhabits the alimentary canal. The microbiome plays a key role in digestion and even affects the growth of young animals (see Chapter 24).

Absorption of Food May Be Passive or Active

Once food is digested, the nutrients must be absorbed across the epithelium of specialized portions of the alimentary canal. This occurs in different ways, either by simple or facilitated diffusion or by active transport. Small, hydrophobic molecules such as fatty acids diffuse down concentration gradients across the epithelium. By contrast, minerals are hydrophilic; as you learned in Chapter 5, hydrophilic substances do not readily cross the lipid-rich plasma membranes of cells. Instead, minerals are usually actively transported across the membranes of epithelial cells by ATP-dependent ion transporters. In other cases, small, hydrophilic organic nutrients are transported by facilitated diffusion or secondary active transport, usually with Na^+ (refer back to Figure 5.16b for a description of secondary active transport).

After nutrients enter the epithelial cells of the alimentary canal, the cells use some of the nutrients for their own requirements. In animals with a closed circulatory system, most of the nutrients are transported across the basolateral surface of the epithelial cells (the side opposite the lumen of the canal). From there, the nutrients can diffuse into nearby blood or lymphatic vessels and circulate to the other cells of the body. Thus, nutrients enter the alimentary canal in food, are digested within the canal into smaller molecules that can be transported into epithelial cells, and from there are released into the blood, where they can reach all of the body's cells. In the special case of water, osmotic gradients established by the transport of ions and other nutrients out of the epithelial cells draw water by osmosis from the canal, across the epithelium, and from there into the blood.

The mechanisms that activate and control the digestive and absorptive functions of the alimentary canal have been extensively studied in vertebrates and have great importance for human health. In the next section, we will explore the structure and function of the digestive systems of vertebrates.

37.2 Reviewing the Concepts

- Intracellular digestion occurs in single-celled organisms and some simple invertebrates. Most animals digest food via extracellular digestion. Flatworms and cnidarians have a gastrovascular cavity with one opening that serves as both entry and exit ports (Figure 37.2).
- Nearly all other animals have an alimentary canal, open at both ends and segregated into specialized regions, through which food passes from one end to the other.
- Once food has been digested, the nutrients must be absorbed via diffusion, facilitated diffusion, or active transport.

37.2 Testing Your Knowledge

1. Which of the following statements is correct?
    a. Intracellular digestion commonly occurs in most vertebrates.
    b. Absorption of all nutrients requires active transport.
    c. Alimentary canals have two openings, whereas gastrovascular cavities have only one.
    d. Extracellular, but not intracellular, digestion requires enzymes.
    e. Most minerals are absorbed by simple diffusion.

37.3 Vertebrate Digestive Systems

Learning Outcomes

1. Describe the structure of vertebrate digestive systems.
2. Describe how food moves through the alimentary canal and how different regions contribute to the processes of digestion and absorption.
3. **SCISKILLS** ▶ Predict problems that might arise if saliva or acid cannot be secreted into the mouth or stomach, respectively.
4. Outline how microorganisms can help to digest cellulose in ruminants and other herbivores.
5. **SCISKILLS** ▶ Propose a hypothesis explaining why most human adults are lactose-intolerant to some degree.
6. Describe the mechanisms of digestion and absorption of carbohydrates, proteins, and fats in vertebrates and the importance of enzymes in these processes.
7. Explain why some nutrients do not require digestion prior to being absorbed.
8. Name three major hormones important for the regulation of digestion in vertebrates, and describe the function of each.

The vertebrate **digestive system** consists of the alimentary canal—also known as the **gastrointestinal (GI) tract**—plus several accessory organs (**Figure 37.3**). The human GI tract consists of the mouth, pharynx, esophagus, stomach, small and large intestines, and anus. The accessory structures, not all of which are found in all vertebrates, are the tongue, teeth, salivary glands, liver, gallbladder, and pancreas. The differences between the digestive systems of various vertebrates reveal much about their respective feeding behaviors and mechanisms. In this section, we will look at some of the most important differences as we discuss the form and function of each part of the vertebrate digestive system.

Figure 37.3 A vertebrate digestive system, that of a human. This figure shows the organs of the gastrointestinal tract (labeled in black) and the major accessory organs (labeled in red). Not all vertebrates share identical features of the digestive system; for example, many birds and some mammals lack a gallbladder.

The Vertebrate Alimentary Canal Is Divided into Functional Regions

The alimentary canal is one continuous tube with structural and functional specializations along its length. In some places, the specialized regions are bounded by a circular ring of smooth muscle called a sphincter. Contraction of the muscle closes the sphincter, whereas relaxation of the muscle opens it. In this way, the different functions of an alimentary canal can be isolated from each other. For example, storage of food can occur in one region while digestion of food occurs in another.

The first section of the alimentary canal, at the anterior end, functions primarily in the ingestion of food. It contains the mouth, salivary glands, pharynx (throat), and esophagus. The middle section, which functions in the storage and initial digestion of food, generally contains one or more food-storage or digestive organs, including the crop, gizzard, and stomach, depending on the species. The middle section also contains the upper part of the small intestine—where most of the digestion and absorption of food take place—and accessory structures that connect with the intestine, including the pancreas, liver, and gallbladder. The final, posterior part of the GI tract functions in some absorption and the elimination of nondigestible waste. It consists of the remainder of the small intestine and, in most vertebrates other than fishes, a large intestine. Undigested material is defecated through an opening termed an **anus,** or in many amphibians, reptiles, and birds, a cloaca (a common opening for the digestive and urogenital tracts).

From the middle of the esophagus to the anus or cloaca, the GI tract has the same general structure, with a lumen that is lined by a layer of epithelial cells. Included in the epithelial layer are secretory cells that release a protective coating of mucus into the lumen of the tract, and for this reason the epithelial layer together with some underlying tissue is also referred to as the **mucosa.** Other cells in the mucosa release hormones into the blood in response to the presence of food; these hormones control certain aspects of the digestive process. Passing through the mucosa are ducts from secretory glands that release acid, enzymes, water, and ions into the lumen. From the stomach onward, the epithelial cells are linked along the edges of their apical surfaces by tight junctions (refer back to Figure 5.23) that prevent digestive enzymes and undigested food from moving between the cells and out of the alimentary canal. In some places, such as the small intestine, the apical surface of the mucosa is highly convoluted, a feature that increases the surface area available for digestion and absorption.

The epithelial cell layer is surrounded by layers of tissue made up of smooth muscles, neurons, connective tissue, and blood vessels. The neurons are activated by signals coming from regions of the central nervous system that respond to the sight and smell of food. The neurons are also activated directly by the presence of food in the gastrointestinal tract. Contraction of the smooth muscles is controlled by these neurons and results in mechanical mixing of the contents within the stomach and intestine. This helps to speed up digestion and brings digested foods into contact with the epithelium to facilitate absorption.

Food Processing and Some Carbohydrate Digestion Begin in the Mouth

Ingestion and the start of digestion begin once food enters the mouth. In terrestrial vertebrates, the presence of food stimulates salivary glands in and around the mouth, cheeks, tongue, and throat to produce a flow of saliva—a watery fluid containing proteins, mucus, and antibacterial substances. Fishes, which lack true salivary glands, secrete mucus from specialized cells in their mouth and pharynx. Saliva has several functions, not all of which pertain to all vertebrates:

- It moistens and lubricates food to facilitate swallowing.
- It dissolves food particles to facilitate the ability of sensory cells in taste buds to taste food.
- It kills ingested bacteria with a variety of antibacterial compounds, including antibodies.
- It initiates digestion of carbohydrates through the action of a secreted enzyme called salivary amylase.

Digestion is the least important of these functions in most vertebrates, only a few of which produce salivary amylase. In humans and other primates, salivary amylase is present and accounts for a very small percentage of total carbohydrate digestion. Depending on how much salivary amylase you produce, you may be able to detect the action of amylase in your own saliva with a simple test. If you chew on a starchy soda cracker and leave it in your mouth for a while, you may notice that it begins to taste sweet. This is because some of the polysaccharides (polymers of glucose) in the cracker have been partially digested to sweet-tasting disaccharides.

The other functions of saliva, however, are very important. For example, imagine trying to chew and swallow food with a completely dry mouth. Also, the antibiotic properties of saliva help to keep the mouth free of infection. In people who have had cancerous salivary glands removed, the teeth and gums often become so diseased that tooth loss occurs.

Peristalsis Moves Swallowed Food Through the Esophagus to a Storage Organ

As food is swallowed, it moves into the next segments of the alimentary canal, the **pharynx** (throat) and **esophagus.** These structures, although not contributing to digestion or absorption, serve as a pathway to storage organs such as the stomach.

In the pharynx, swallowing begins as a voluntary action but continues in the esophagus by the process of **peristalsis**—rhythmic, spontaneous waves of smooth muscle contraction that begin near the mouth and end at the stomach. When most vertebrates eat, the mouth and stomach are roughly horizontal with respect to each other. In fact, the head of a terrestrial grazing animal is at times lower than its stomach when swallowing. The wavelike action of peristalsis ensures that swallowed food is moved toward the stomach and does not remain in

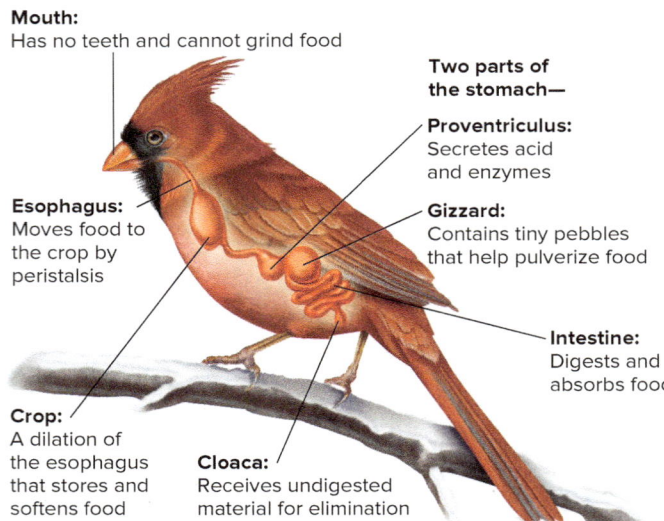

Figure 37.4 The alimentary canal of birds. The avian alimentary canal contains specialized regions for storing and softening food (the crop) and pulverizing food (the gizzard). The gizzard and proventriculus constitute the stomach. Undigested material is eliminated through the cloaca.

the esophagus or even move backward into the mouth if the head is lowered.

In some animals, food does not move directly from the esophagus to the stomach but first enters a storage organ called a **crop,** which is a dilation of the lower esophagus (Figure 37.4). Crops are found in most birds (and many invertebrates, including insects and some worms). Food is stored and softened by watery secretions in the crop, but little or no digestion occurs there. Because they process large amounts of tough food, birds that eat primarily grains and seeds have larger crops than do birds that eat primarily insects and worms. The material that birds regurgitate to their young comes from the crop.

Food Processing Continues and Protein Digestion Begins in the Stomach

Once food passes through the esophagus (and crop, if present), it reaches the stomach. A **stomach** is a muscular, saclike organ that most likely evolved as a means of storing food. In addition to its storage function, the muscular nature of the stomach allows it to mechanically break up large chunks of food into smaller, more easily digestible fragments. The stomach also partially digests some of the macromolecules in food and regulates the rate at which its contents empty into the small intestine.

Glands within the stomach wall known as gastric glands secrete hydrochloric acid (HCl) and an inactive molecule called pepsinogen into the stomach lumen. The material within the stomach does not usually flow backward into the esophagus, because of the presence of a sphincter at the place where the esophagus and stomach join.

The acid serves four major functions in the stomach lumen:

- It converts pepsinogen into an active enzyme called **pepsin,** which begins the digestion of protein. (Stomach gland cells do not make pepsin itself because, if they did, it would digest the cells' own proteins.)

- It kills many of the microorganisms that may have been ingested with food.
- It helps to dissolve the particulate matter in food.
- It alters the ionization of polar molecules, especially proteins. This disrupts the structural framework of the tissues in food and makes the proteins more accessible to the digestive actions of pepsin. (In contrast to proteins, no significant digestion of other organic molecules occurs in the stomach.)

After a period of time, the pepsin and HCl in the stomach reduce food particles to a pulpy fluid called **chyme.** Acid production is controlled by the nervous and endocrine systems. One hormone that is particularly important in this regard is **gastrin.** Molecules in chyme stimulate cells in the stomach epithelium to release gastrin into the blood. Gastrin then acts on acid-producing cells to increase the rate of HCl secreted into the stomach lumen roughly in proportion to the amount of protein ingested in a meal. The presence of gastrin also stimulates smooth muscle contraction in the stomach, which mixes the contents of the chyme with the acid and helps to move chyme into the small intestine. With the exception of water, very little or none of the products of chyme can cross the epithelium of the stomach; therefore, little or no absorption of nutrients occurs in the stomach.

In birds, the stomach is divided into two parts: the proventriculus and the gizzard (Figure 37.4). The **proventriculus** is the glandular portion of the stomach that secretes acid and pepsinogen. Partially digested and acidified food then moves to the **gizzard,** a muscular structure present in all birds and some other vertebrates and invertebrates. The gizzard has a rough inner lining that grinds food into smaller fragments. In birds, the gizzard often contains swallowed sand and tiny stones. The gritty sand and stones take the place of teeth and help to mash and grind ingested food. Grain-eating birds, particularly chickens and other fowl, many passerines (or perching birds), and pigeons and doves, generally have more muscular gizzards than do insectivorous birds, because of the difficulty in breaking down plant matter.

The Stomach of Some Herbivores Is Part of a Complex Structure That Allows the Animals to Process Cellulose

Cellulose, the main macromolecule of plant cell walls (refer back to Figures 3.6, 4.34, 28.10, and 28.11), is a very important part of the diet of herbivores and many omnivores. In human nutrition, cellulose is also called fiber or roughage and is useful in helping to eliminate solid wastes. The human digestive system is not equipped with the enzymes to digest cellulose, and it therefore passes through the system intact. **Ruminants** (sheep, goats, llamas, and cows) also lack the enzymes, but they are able to digest cellulose with the help of their microbiome. Microorganisms living within their digestive tracts break down the cellulose into monosaccharides that can be absorbed along with other by-products of microbial digestion, such as fatty acids and some vitamins. In this way, bacteria and protists predigest the food, and the animal absorbs the breakdown products.

Figure 37.5 **Digestive system of a ruminant.** Ruminants have a complex arrangement of three modified pouches arising from the esophagus: the rumen, reticulum, and omasum. The rumen and reticulum act as storage and processing sites (in large ruminants, the rumen may store up to 95 L of undigested food); the omasum absorbs some water and ions. Digestion by acid and pepsin takes place in the abomasum, which connects with the intestines.

Ruminants have a system of several chambers, beginning with three outpouchings of the lower esophagus composed of the rumen, reticulum, and omasum, followed by a true stomach called an abomasum (**Figure 37.5**). The rumen and reticulum contain the microorganisms that digest cellulose, and the omasum absorbs some of the water and ions released from the chewed and partially digested food. The tough, partially digested food (called the cud) is occasionally regurgitated, rechewed, and swallowed again.

Eventually, the partially digested food, the microorganisms, and the useful by-products of microbial digestion reach the abomasum, which secretes the acid and proteases typical of the vertebrate stomach. From the abomasum, the material passes to the intestines, where digestion and absorption are completed. The microorganisms are killed by the stomach acid, digested, or eliminated. Some microorganisms remain in the rumen and quickly multiply to replenish their populations, ensuring that a well-balanced, mutualistic relationship is maintained.

Most Digestion and Absorption Occur in the Small Intestine

Most digestion of food and absorption of nutrients occur in the **small intestine**—the portion of the alimentary canal that leads from the stomach to the large intestine (or, in animals without a large intestine, directly to the anus or cloaca). Hydrolytic enzymes break down macromolecules of organic nutrients into monomers. Some of these hydrolytic enzymes are located in the apical membrane of the intestinal epithelial cells where they face the intestinal lumen; others are secreted by the pancreas and enter the lumen through a duct.

A sphincter between the stomach and the beginning of the small intestine prevents the enzymes and their digestion products from moving back into the stomach. This sphincter opens only long enough to permit a small amount of chyme to enter the small intestine. After the chyme is digested and moved along the intestine, the sphincter relaxes again and another bolus of chyme enters from the stomach.

The products of digestion are absorbed across the epithelial cells and enter the blood. Vitamins and minerals, which do not require enzymatic digestion, are also absorbed in the small intestine. Water is absorbed by osmosis from the small intestine in response to the movement of nutrients across the intestinal epithelium.

The ability of the small intestine to carry out the bulk of digestion and absorption is aided by mucosal infoldings and specializations along its length. Finger-like projections known as **villi** (singular, villus) extend into the lumen of the vertebrate small intestine (**Figure 37.6**).

Biology Principle

Structure Determines Function

Note the arrangement of muscle layers in the small intestine. Some are oriented in circles around the lumen, whereas others are lengthwise. This structure allows the intestine to be squeezed and shortened, not necessarily at the same time, maximizing the mixing and churning of the lumen contents. Also, the structure of the villi and microvilli provides extensive surface area in a compact volume, increasing the efficiency of digestion and absorption.

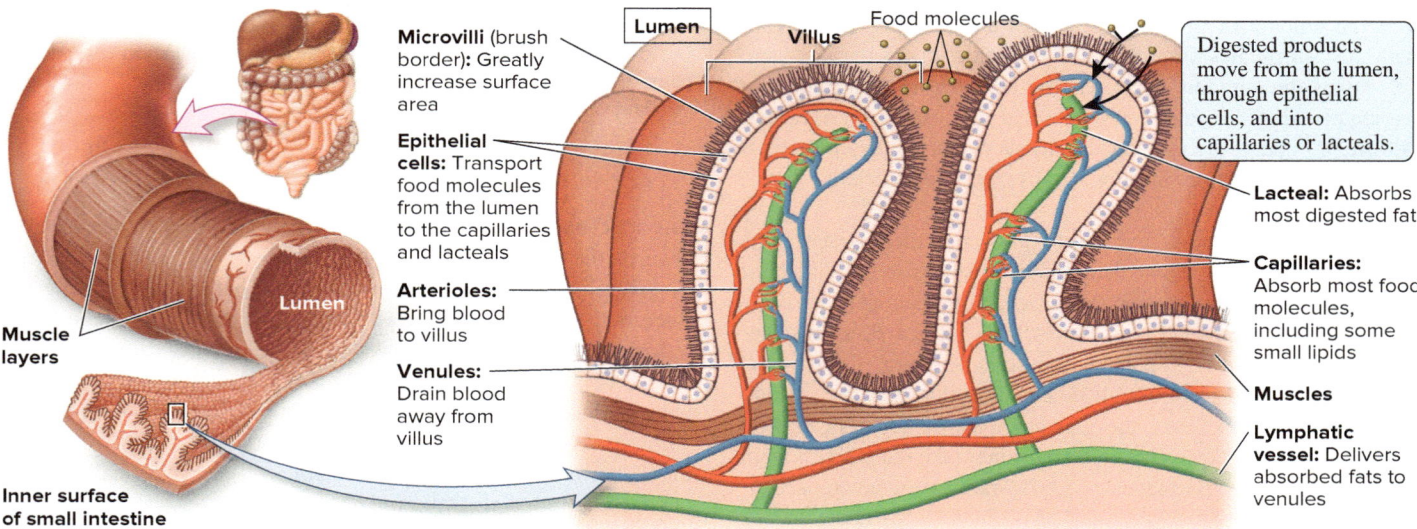

(a) Structure of the small intestine

(b) Colorized SEM of the interior of a small intestine

Figure 37.6 **The specialized arrangement of tissues in the small intestine.** (a) The small intestine is folded into numerous villi, which increase the surface area for digestion and absorption. Within each villus are capillaries and lymphatic vessels (lacteals), into which absorbed nutrients are transported. The epithelial cells of the villi have extensions from their surface called microvilli. The microvilli constitute the brush border of the intestine and greatly add to the total surface area. (b) SEM of the interior of the small intestine, showing numerous microvilli forming the brush border.
(b) ©Steve Gschmeissner/Science Source

The surface of each villus is covered with a layer of epithelial cells with plasma membranes that have many small projections called **microvilli,** known collectively as a **brush border.** The combination of folded mucosa, villi, and microvilli increases the small intestine's surface area enormously. This means that the likelihood of an ingested food particle encountering a digestive enzyme and being absorbed across the epithelium is very high, so digestion and absorption proceed more rapidly.

The center of each intestinal villus is occupied by capillaries, the smallest blood vessels in the body, and by a type of vessel called a lacteal, which is part of the lymphatic system (Figure 37.6). Most nutrients are absorbed directly into the capillaries and from there into venules, which eventually empty into veins. Most of the fat absorbed in the small intestine, however, exists as bulky, protein-coated particles that are too large to enter capillaries. Consequently, absorbed fat enters the larger, wider lacteals. Material absorbed by the lacteals eventually empties into large veins in the circulatory system.

The Pancreas Secretes into the Small Intestine Substances That Aid Digestion

As the chyme moves through the small intestine, the **pancreas** secretes substances that flow via ducts into the first portion of the intestine, the **duodenum** (Figure 37.7). The pancreas has numerous functions, but here we will focus on those that are directly involved in digestion.

The arrival of chyme into the duodenum triggers the release of two polypeptide hormones from intestinal epithelial cells. The first, cholecystokinin (CCK), enters the blood and stimulates the pancreas to secrete into the intestine digestive enzymes that break down large molecules—proteins, lipids, carbohydrates, and nucleic acids—into smaller units. It also relaxes a sphincter that permits the entry of material from the pancreatic duct into the intestine to be synchronized with the presence of chyme in the intestine. The second hormone, secretin, stimulates other cells of the pancreas to secrete bicarbonate ions (HCO_3^-) into the small intestine. The HCO_3^- neutralizes the acidity of chyme, which would otherwise inactivate the pancreatic enzymes in the small intestine and could damage the intestinal epithelium. Thus, through the actions of CCK and secretin, the secretion of digestive enzymes and HCO_3^- into the intestine is precisely coordinated with the arrival of chyme.

The digested nutrients, along with water, are absorbed across the plasma membranes of the brush border cells. Peristalsis slowly propels the remaining contents through the posterior portions of the small intestine, called the jejunum and the ileum, where further absorption occurs.

Carbohydrates Are Digested by Amylase and Brush Border Enzymes

In omnivores such as humans, most of the ingested carbohydrates are the polysaccharides starch and cellulose from plants as well as glycogen from animals (refer back to Figures 3.4–3.6 for the structures of carbohydrates). The remainder consists of simple carbohydrates, such as the monosaccharides fructose and glucose in fruit and disaccharides (combinations of two monosaccharides) such as lactose in milk. Humans also add the disaccharide sucrose (table sugar) to their food. Certain other animals consume sucrose from sources such as maple sap and sugarcane.

As mentioned earlier, depending on the species, a very small amount of carbohydrate is digested in the mouth by salivary amylase. Almost all glycogen and starch digestion, however, takes place in the small intestine due to the action of amylase secreted into the intestine by the pancreas. The products of digestion by amylase are molecules of the disaccharide maltose (Figure 37.8).

Maltose, along with any ingested sucrose and lactose, is further broken down into monosaccharides—fructose, glucose, and galactose—by enzymes located on the brush border of the small intestine epithelial cells. The monosaccharides are then absorbed into the epithelial cells either by facilitated diffusion (fructose) or by secondary active transport coupled to sodium ions (glucose and galactose). Monosaccharides then leave the epithelial cells by way of facilitated diffusion through transporters located in the basolateral membrane of the epithelial cells and enter the blood. The bloodstream distributes the monosaccharides and other absorbed nutrients to the cells of the body.

Do you feel ill after drinking milk or consuming certain dairy products? If so, you may be among the vast majority of people who cannot adequately digest lactose, the chief disaccharide in milk. A small percentage of humans, however, retain the ability to fully digest lactose throughout life. Next, we'll examine this phenomenon along with its genetic and evolutionary basis.

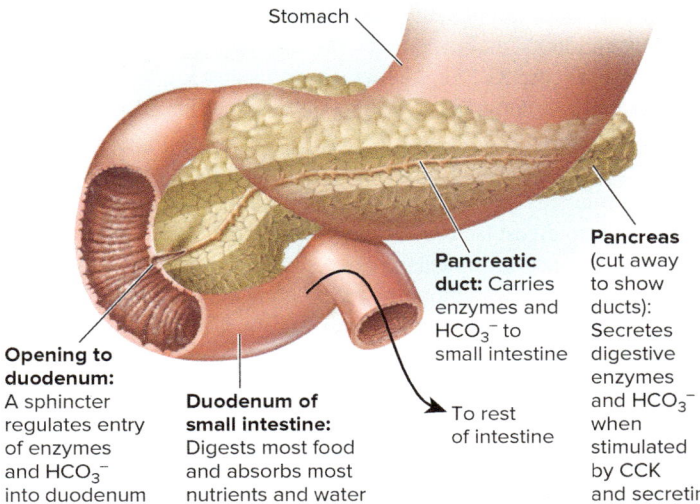

Figure 37.7 **The structures and functions of the vertebrate pancreas and small intestine.** The position of the stomach is shown for orientation.

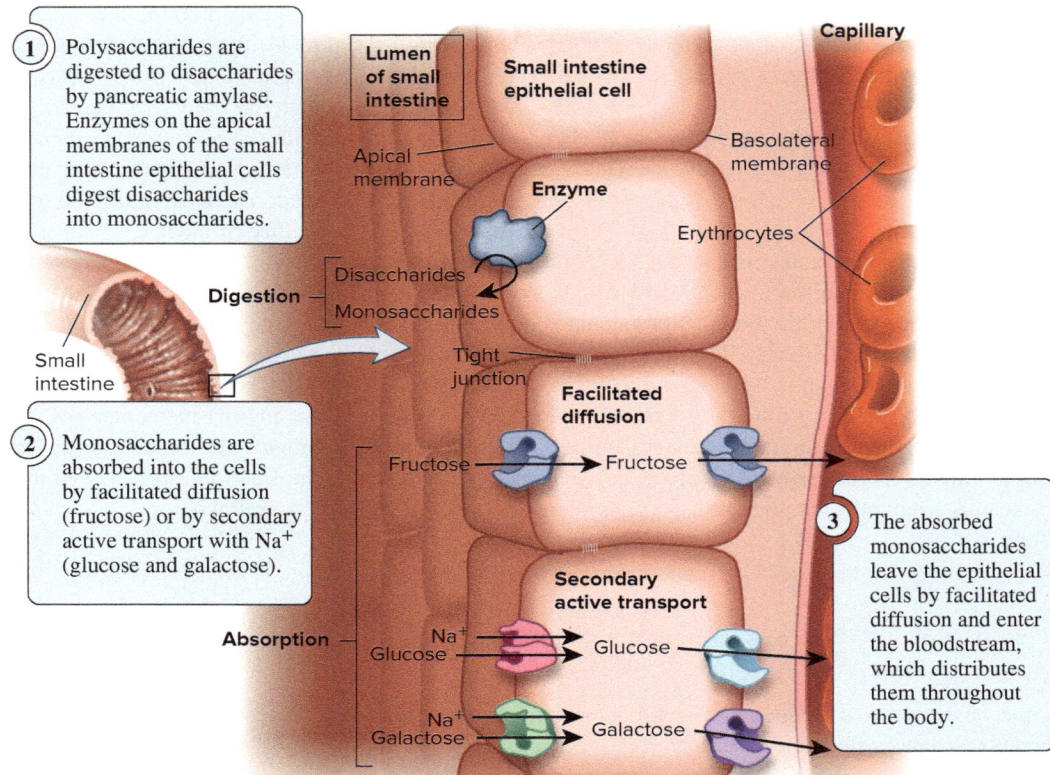

Figure 37.8 Digestion and absorption of carbohydrates in the small intestine. In this and subsequent figures, the microvilli (making up the brush border) are omitted for simplicity.

Concept Check: The absorption of many nutrients requires active transport. What can we conclude from that about the energetic cost of absorbing the nutrients from a meal?

BioConnections: Are the transport processes shown here, including facilitated diffusion and active transport, unique to animal cells? Look back at Figure 30.13 for help.

EVOLUTIONARY CONNECTIONS

Evolution and Genetics Explain Lactose Intolerance

One of the defining features of mammals is that they produce and consume milk, a nutrient-rich solution containing all the nutrients required by mammalian offspring. With rare exceptions, the milk of all mammals contains the dissacharide lactose. Lactose is digested by the brush border enzyme lactase, which cleaves lactose into glucose and galactose, which can then be absorbed into the blood. Humans with lactose intolerance cannot adequately digest lactose because their lactase is inactive, absent, or present in small amounts.

Milk is the sole food of most mammals shortly after birth and the primary food for various lengths of time thereafter until weaning, the transition from consuming mother's milk to eating a diet of solid foods. Once weaned, mammals never again drink milk, except, of course, for humans.

The only dietary source for lactose is milk, and therefore it is not surprising that older mammals lose the ability to digest this disaccharide. This occurs because the gene that encodes the enzyme lactase is shut off around the age of weaning or shortly thereafter. The developmental mechanisms that turn off lactase production and activity are not firmly understood, but they are known to involve decreased transcription of the lactase-encoding gene.

If an adult mammal were to drink milk, the undigested lactose would accumulate in the intestines. As a result, water that normally would be absorbed by osmosis when monosaccharides are formed from lactose would also remain in the intestines. Microorganisms in the large intestine can digest some of the lactose for their own use and, in the process, release by-products such as hydrogen and other gases. The combination of water retention and bacterial action results in GI symptoms such as diarrhea, gas, and cramps.

About 90% of the human population cannot fully digest lactose after early childhood and experience these symptoms to some degree (refer back to Figure 22.22). In other words, lactose intolerance is not a disorder; it is a common condition that evolved in all mammals, including humans.

Why, then, are some adults able to consume dairy products without getting ill? Researchers have discovered that the ability of human adults to digest lactose is clearly linked to genetic background. Lactose intolerance in humans is an example of a polymorphism, a genetic trait that varies among people. Other than individuals who trace their ancestry to northern Europe and a few isolated regions in Africa, nearly every population of humans shows a considerable degree of lactose intolerance, reaching nearly 100% in most of south and east Asia. One hypothesis for this phenomenon appears to be a behavioral and cultural change that occurred in Neolithic times, when certain populations domesticated cattle and added cow's milk to their diet. Adults who were able to digest lactose—due to a mutation affecting the expression of the lactase-encoding gene—enjoyed a selective advantage and tended to thrive and pass on their genes more frequently than those whose digestive systems could not handle milk.

Recently, investigators have uncovered several single-nucleotide changes located in regulatory sites that control the expression of the lactase-encoding gene in humans. Interestingly, it appears that the nucleotide changes make these regulatory sites less susceptible to epigenetic modification. It is believed that without epigenetic modification, the lactase-encoding gene remains active. These changes, therefore, are associated with lactase expression that continues long after weaning. People carrying these mutations are lactose-tolerant. By comparison, adults who lack these mutations can consume dairy products only in small amounts or not at all because they do not produce sufficient amounts of the lactase enzyme. These observations suggest that natural selection may have resulted in certain populations carrying mutations that caused the lactase-encoding gene to continue to be expressed long after weaning. Eventually, a substantial percentage of people in those regions of the world retained sufficient intestinal lactase to use milk as a staple food.

Most Protein Digestion Occurs in the Small Intestine by the Actions of Proteases

As mentioned earlier, protein digestion begins in the stomach. The partially digested protein in the chyme coming from the stomach is further digested in the small intestine by several proteases, including **trypsin** (Figure 37.9). The pancreas secretes this enzyme as an inactive precursor called trypsinogen, which prevents the active enzyme from digesting the pancreas itself. Once trypsinogen enters the small intestine, it is cleaved into the active molecule trypsin by another enzyme, called enterokinase (also called enteropeptidase), located on the apical membranes of the intestinal cells.

The polypeptide fragments produced by trypsin and other proteases are further digested into individual amino acids by the actions of proteases located on the brush border in the small intestine. These proteases cleave off one amino acid at a time from the N-terminus and C-terminus of polypeptides. Individual amino acids coupled to sodium ions then enter the epithelial cells by secondary active transport. Amino acids leave these cells and enter the blood by facilitated diffusion across the basolateral membrane. Like carbohydrates, proteins are almost completely digested and absorbed in the duodenum.

Lipid Digestion and Absorption in the Small Intestine Are Facilitated by Bile

Most ingested lipid is in the form of triglycerides, which are too large to cross the intestinal epithelium and must be broken down. This process occurs almost entirely in the small intestine. The major digestive enzyme in this process is **pancreatic lipase,** secreted with the other digestive enzymes by the pancreas into the small intestine (see Figure 37.7). The lipase reaction catalyzes the splitting of bonds in a triglyceride, producing two free fatty acids and a monoglyceride:

1 Triglyceride → 2 Free fatty acids + 1 Monoglyceride

Triglycerides, however, are poorly soluble in water and aggregate into large lipid droplets, as you can see if you shake a salad dressing made of oil and vinegar. Because lipase is a water-soluble enzyme, its digestive action in the small intestine can take place only at the surface layer of a lipid droplet. The rate of digestion is substantially increased by the process of emulsification, which disrupts the large lipid droplets into many tiny droplets, increasing their total surface area and exposure to the lipase. This occurs partly by the muscular contractions of the stomach and small intestine, which mechanically break apart the fat droplets.

To prevent the smaller droplets from coalescing, however, something else is needed. The liver secretes a mixture of substances called bile (Figure 37.10). **Bile** contains HCO_3^-, cholesterol, phospholipids, a number of organic wastes, and cholesterol-derived substances called bile salts. The HCO_3^-, like that from the pancreas, helps to neutralize acid (H^+) from the stomach, and the phospholipids and bile salts coat the small lipid droplets, thereby preventing them from coalescing into large droplets. They can do this because they are amphipathic molecules. Bile salts and phospholipids have nonpolar regions that dissolve in the lipid droplets and polar regions that face the watery environment of the

Figure 37.9 Digestion of proteins and absorption of amino acids in the small intestine.

(1) Trypsinogen secreted by the pancreas is converted to trypsin by the enzyme enterokinase. Trypsin and other proteases cleave proteins into small polypeptides.

(2) Proteases on the apical membrane of epithelial cells cleave polypeptides into amino acids.

(3) Amino acids are absorbed by secondary active transport and facilitated diffusion.

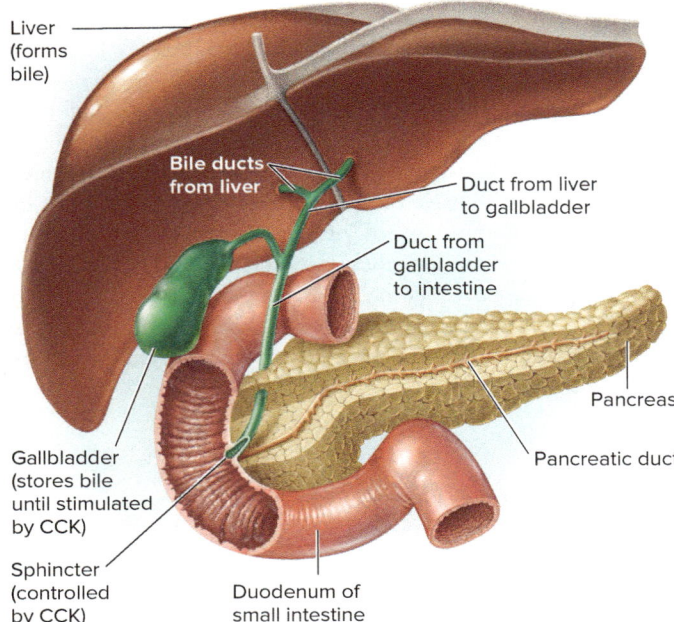

Figure 37.10 Sites of production, storage, and secretion of bile. The connection of the pancreatic duct with the intestine is via the same sphincter shown here for bile but is omitted for clarity.

intestinal lumen. The outward-facing polar regions are charged and so nearby droplets repel each other.

The liver secretes bile into small ducts that join to form one large duct that delivers the bile to the **gallbladder,** a small sac underneath the liver found in many vertebrates, including humans. During a meal, the smooth muscles in the gallbladder contract, injecting the bile solution into another portion of the large duct that empties into the small intestine through the same opening as the pancreatic duct (Figure 37.10). The gallbladder, therefore, is a storage organ that allows the release of large amounts of bile to be precisely timed to the consumption of fats. This timing is under the control of the same hormone, CCK, that you learned earlier is linked with the arrival of chyme into the small intestine. In addition to its actions on the pancreas, CCK also stimulates the gallbladder to contract and relaxes the sphincter that controls the flow of bile into the intestine.

Although bile aids the digestion of fats, absorption of the poorly soluble products of the lipase reaction is facilitated by a second action of phospholipids and bile salts, the formation of **micelles** (Figure 37.11a). Micelles consist of bile salts, phospholipids, fatty acids, and monoglycerides clustered together in very tiny droplets. Micelles continuously break apart and re-form near the epithelium of the intestine. When a micelle breaks down, the small lipid molecules are slowly released into solution and diffuse across the intestinal epithelium. As the lipids diffuse into epithelial cells, micelles release

Figure 37.11 Formation of micelles and absorption of lipids in the small intestine. Note: Droplets, triglycerides, and other molecules are not drawn to scale.

Concept Check: *What advantage is provided by resynthesizing absorbed fatty acids and monoglycerides into triglycerides in the epithelial cells?*

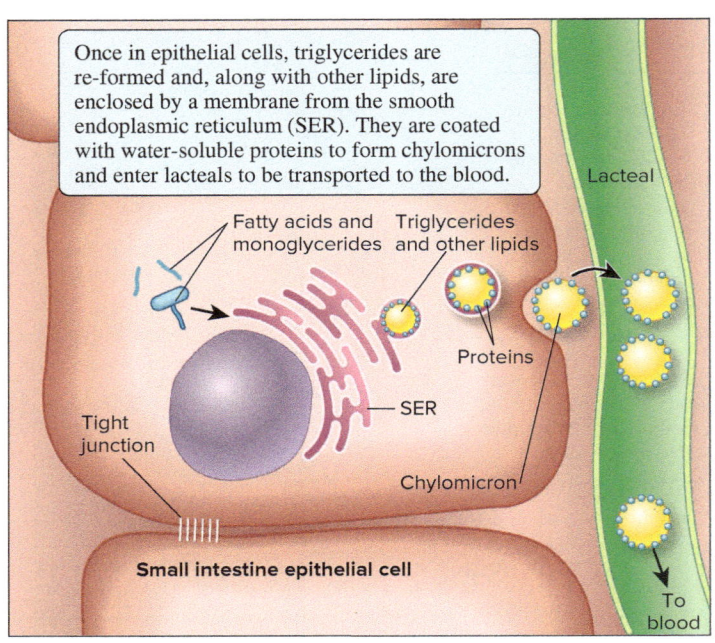

(a) **Digestion of emulsified fats into micelles, and absorption into intestinal cells**

(b) **Absorption of triglycerides via the formation and release of chylomicrons** more lipids into the aqueous phase of chyme. Thus, the micelles keep most of the insoluble fat digestion products in small, soluble aggregates while gradually releasing very tiny quantities of lipids that diffuse into the intestinal epithelium. Note that it is not the micelle that is absorbed but rather the individual lipid molecules that are released from the micelle.

During their passage through the epithelial cells, fatty acids and monoglycerides are resynthesized into triglycerides in the smooth endoplasmic reticulum (SER). This process decreases the concentration of cytosolic free fatty acids and monoglycerides in the epithelial cells and, so, maintains a diffusion gradient for these molecules from the lumen into the cells. The resynthesized triglycerides aggregate into large droplets called **chylomicrons.** The chylomicrons are coated with proteins that perform an emulsifying function similar to that of bile salts (**Figure 37.11b**), and they keep the absorbed fats in a soluble form in the blood. In addition to triglycerides, chylomicrons contain any absorbed phospholipids, cholesterol, and fat-soluble vitamins.

Chylomicrons are released by exocytosis from the epithelial cells and enter lacteals. The fluid from the lacteals eventually empties into a large vein and from there into the general blood circulation, which delivers the chylomicrons to adipose cells, where the fat can be stored.

The Large Intestine Concentrates Undigested Material, Which Is Then Egested via the Anus

Once food has been digested and the nutrients have been absorbed, the remnants of chyme pass to the large intestine. The size of the large intestine varies greatly among different vertebrates, and the organ is vestigial or even absent in many animals, notably fishes. Its first portion, the cecum, forms a small pouch from which extends the appendix, a finger-like projection having no certain essential function in humans but possibly contributing to the body's immune defense mechanisms. The next part of the large intestine in humans and other mammals is called the **colon.** Much of the colon in humans is S-shaped, forming the sigmoid colon, which empties into the **rectum,** a short segment of the large intestine that ends at the anus. The characteristic appearance of the mammalian colon is not widely found in other vertebrates, most of whom have a simple, straight large intestine.

The primary functions of the large intestine are to store and concentrate fecal material before defecation and to absorb some of the remaining ions and water that were not absorbed in the small intestine. Because most substances are absorbed in the small intestine, only a small volume of water and ions, along with undigested material (feces), is passed on to the large intestine. **Defecation** occurs when contractions of the rectum and the relaxation of associated sphincter muscles expel the feces through the final portion of the canal, the anus (see Figure 37.3).

Located within the microbiome of the large intestine are large populations of various bacteria, some of which provide benefits to animals. For example, some bacteria in the large intestine release by-products of metabolism such as certain vitamins, which can be absorbed across the intestinal epithelium. Although this source of vitamins generally provides only a small part of an animal's normal daily requirement, it may make a significant contribution when dietary vitamin intake is low. In humans, vitamin deficiencies may even occur when certain antibiotics that kill these species of bacteria are used. Another bacterial product is gas (flatus), which is a mixture of nitrogen, carbon dioxide, hydrogen, methane, and hydrogen sulfide. Some foods (for example, beans) contain large amounts of certain carbohydrates that cannot be digested by intestinal enzymes in humans but are readily metabolized by bacteria in the large intestine. Bacterial processing of these undigested polysaccharides produces these gases, except for nitrogen, which is derived from swallowed air. Much of the fecal mass is actually comprised of these bacteria, which keeps their populations in check.

BIO TIPS ONLINE

37.3 Reviewing the Concepts

- The vertebrate digestive system consists of the alimentary canal plus associated glands and organs (Figure 37.3).
- Saliva facilitates swallowing, dissolves food particles, and initiates digestion. Peristalsis moves food through the pharynx and esophagus to the stomach (Figure 37.4).
- The stomach stores food, secretes acid, partially digests proteins, and regulates the rate at which chyme empties into the small intestine. Ruminants have a stomach that is associated with several chambers that facilitate cellulose digestion (Figure 37.5).
- Nearly all digestion and absorption occur in the small intestine. Villi and microvilli increase the small intestine's surface area and maximize the efficiency of digestion and absorption. The pancreas secretes digestive enzymes and a bicarbonate-rich fluid that neutralizes the acidic chyme (Figures 37.6, 37.7).
- In the small intestine, carbohydrates are digested to monomers by amylase and brush border enzymes and are then transported into the blood. Proteins are broken down to smaller fragments in the stomach by pepsin and further broken down in the small intestine into amino acids by proteases. Lipid digestion and absorption occur entirely in the small intestine by the action of pancreatic lipase (Figures 37.8, 37.9, 37.10, 37.11).
- The large intestine concentrates undigested material, which is then egested through the anus or cloaca.

37.3 Testing Your Knowledge

1. Which is *not* part of the vertebrate alimentary canal?
   a. salivary glands
   b. pharynx
   c. esophagus
   d. stomach
   e. rectum

2. CCK does all *except* which of the following?
   a. It stimulates the pancreas to secrete digestive enzymes into the small intestine.
   b. It stimulates the production of bile by the liver.
   c. It stimulates contraction of the gallbladder.
   d. It relaxes the sphincter that connects the pancreatic and gallbladder ducts to the small intestine.

3. Choose the *incorrect* statement.
   a. Trypsin digests proteins in the small intestine.
   b. Enzymes on the brush border of the small intestine complete the process of carbohydrate digestion.
   c. Fatty acids are absorbed across the epithelium of the small intestine.
   d. Approximately 10% of the world's population is lactose-intolerant.
   e. Bile is a mixture of amphipathic compounds, wastes, cholesterol, and HCO_3^-.

37.4 Nutrient Use and Storage

Learning Outcomes

1. Describe the processes by which glucose, triglycerides, and amino acids are absorbed during the absorptive state.
2. Outline the two major ways that vertebrates can maintain the concentration of glucose in their blood during the postabsorptive state.

Once nutrients have been ingested, they are either used or stored. The metabolism of nutrients can be divided into two alternating phases: The **absorptive state** occurs when ingested nutrients enter the blood from the GI tract. The **postabsorptive state** occurs when the GI tract is empty of nutrients and an animal's own stores must supply energy. During the absorptive state, some ingested nutrients supply the immediate energy requirements of the body. The rest are added to the body's energy stores to be called upon during the next postabsorptive state.

Total body energy stores are adequate for the average human to withstand a fast of several weeks. By contrast, some animals can barely survive a single missed meal—particularly if they have low energy reserves—because their relative metabolic requirements are much greater than our own. In this section, we will focus on nutrient absorption, use, and storage in vertebrates, with a closer look at mammals.

In the Absorptive State, Nutrients Are Absorbed and Used for Energy or Stored

The categories of nutrients that are absorbed during the absorptive state include carbohydrates, lipids, proteins, nucleic acids, vitamins, minerals, and water. We will look in depth at the first three of these. Figure 37.12 gives an overview of what happens to these nutrients, as summarized here:

- **Absorbed carbohydrates.** The chief carbohydrate monomer absorbed from the GI tract of vertebrates is glucose. Much of the absorbed glucose enters all cells and is enzymatically broken down, releasing energy that can be used in the synthesis of ATP. If more glucose is absorbed than is immediately required, a portion of the excess is incorporated into glycogen, primarily in the liver and skeletal muscle. The remainder is broken down to provide sources of carbon for building triglycerides in adipose cells.

- **Absorbed triglycerides.** Triglycerides are packaged into chylomicrons, which enter lymph and, from there, the blood circulation (see Figure 37.11). As blood moves through adipose tissue, an enzyme there releases the fatty acids from the triglycerides in the chylomicrons. The fatty acids then diffuse into adipose cells and combine with glycerol to resynthesize

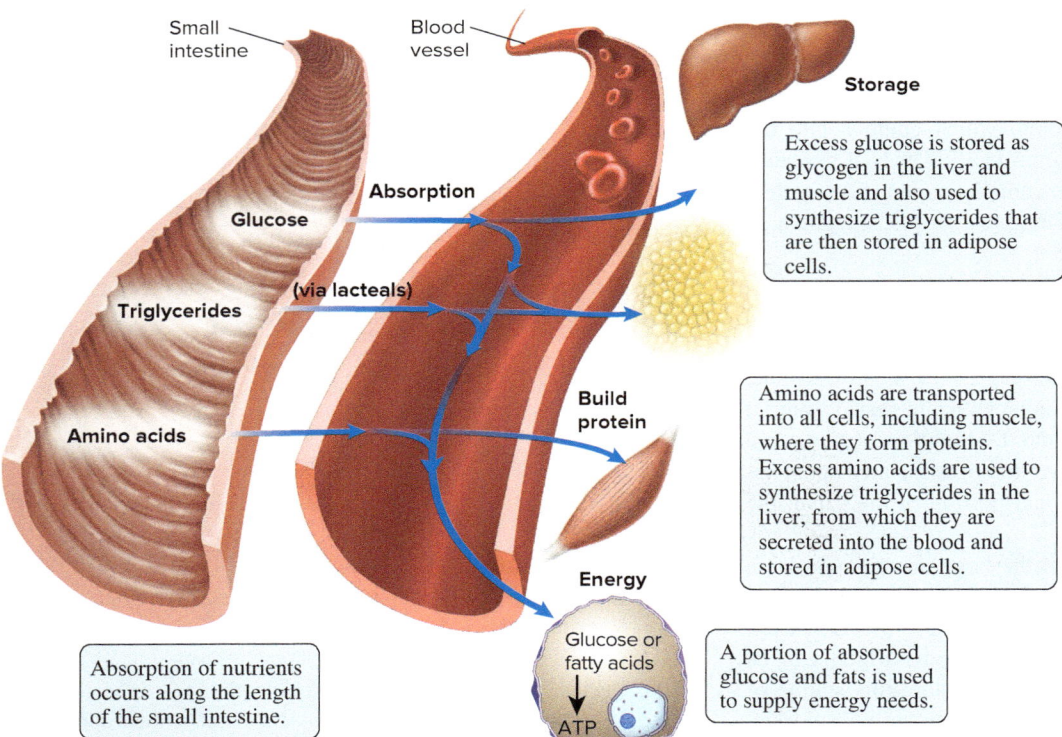

Figure 37.12 Events of the absorptive state. The products of digestion are absorbed into the blood along the length of the small intestine. These nutrients are broken down for immediate energy demands or they are incorporated into macromolecules and used for energy storage, as with glycogen, or cell function, as with proteins.

triglycerides. These triglycerides are stored in adipose cells until the body requires additional energy.

- **Absorbed amino acids.** Amino acids are taken up from the blood by all cells of an animal's body, and used there to synthesize proteins. In vertebrates, excess amino acids are enzymatically broken down in liver cells and the products are used to synthesize fatty acids, which then are incorporated into triglycerides. The triglycerides are then released into the blood and taken up and stored in adipose cells. Thus, an excess in the diet of any of these nutrients—carbohydrates, triglycerides, or proteins—will result in fat deposition in adipose cells.

In the Postabsorptive State, Stored Nutrients Are Released and Used

As the postabsorptive state begins, synthesis of glycogen and triglycerides slows and the breakdown of these substances begins. During this state, some of the macromolecules formed during the absorptive state are broken down to supply monomers that can be used for energy. No glucose is available to be absorbed from the intestines during this time, yet the blood glucose concentration must be maintained because the cells of the central nervous system (CNS) normally can use only glucose for energy. A large decrease in blood glucose concentration can disrupt CNS functions, ranging from subtle impairment of mental function to seizures, coma, or even death.

The events that maintain blood glucose concentration fall into two categories: (1) reactions that provide glucose to the blood and (2) cellular use of fatty acids for energy.

Glucose from Glycogen and Other Sources Vertebrates can produce glucose during the postabsorptive state in two major ways: (1) by breaking down glycogen into glucose and (2) by making new glucose from noncarbohydrate sources. In the first, the glycogen that was formed during the absorptive state can be broken back down into molecules of glucose by hydrolysis, a process known as **glycogenolysis** (Figure 37.13a). This process occurs primarily in the liver, after which the glucose is released into the blood, where it can be made available to all cells. Muscle glycogen is also broken down into glucose by glycogenolysis, but this glucose is used exclusively by muscle cells and not secreted into the blood.

The amount of liver glycogen available to provide glucose during the postabsorptive state varies among animals, but it is generally sufficient to maintain blood glucose concentration for only a brief time, such as an overnight fast in a human. Therefore, a second mechanism for maintaining blood glucose is required if the postabsorptive state continues. In the process of **gluconeogenesis** (literally, creation of new glucose), enzymes in the liver synthesize glucose from noncarbohydrate precursors; the glucose is then secreted into the blood (Figure 37.13b). This process occurs in all vertebrates but appears to be especially important in mammals.

A major precursor for gluconeogenesis is glycerol, which is released from triglycerides in adipose tissue by the breakdown process called **lipolysis.** In lipolysis, lipase enzymes within adipose cells hydrolyze triglycerides into fatty acids and glycerol, both of which enter the blood. The fatty acids diffuse into cells, where they are used as an alternative to glucose as an energy source (except in the CNS, which continues to require glucose). The glycerol is taken up by the

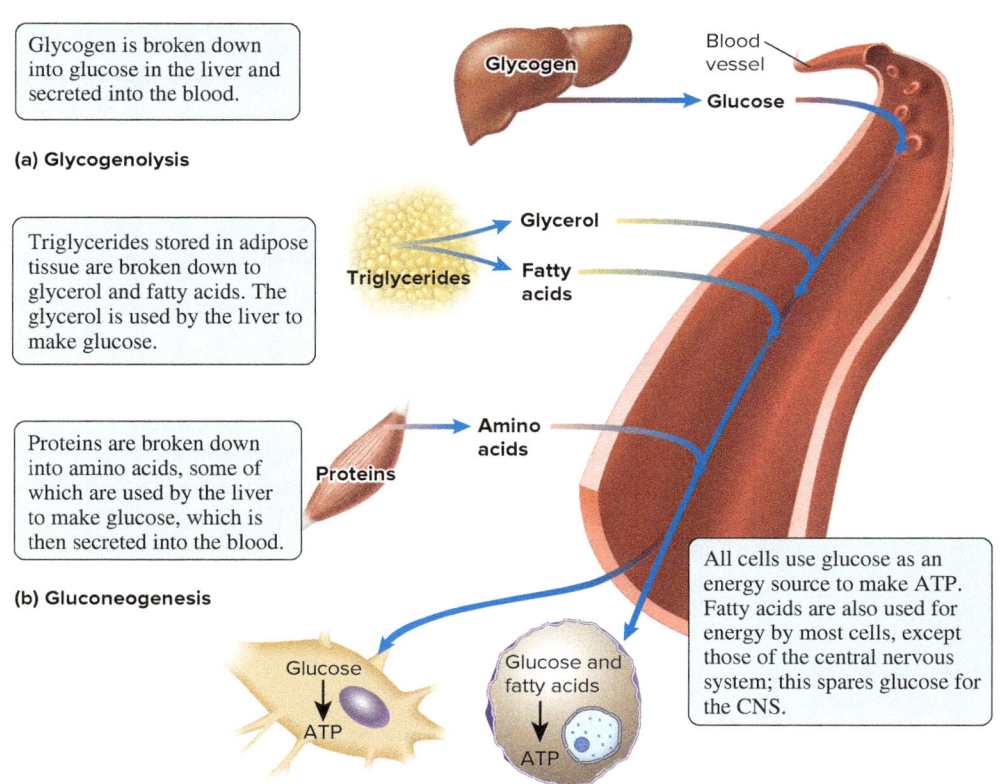

Figure 37.13 Events of the postabsorptive state. Macromolecules formed during the absorptive state are broken down to supply smaller molecules that can be used for energy. **(a)** This begins with glycogenolysis—the breakdown of glycogen into glucose. **(b)** In gluconeogenesis, the breakdown products of triglycerides and proteins (namely, glycerol, fatty acids, and amino acids) can be used for energy or can be used in the synthesis of glucose in the liver.

liver, where enzymes use it to synthesize glucose by gluconeogenesis; the glucose is then released back into the blood.

When the postabsorptive state continues for an extended period of time—as when an animal fails to find food—protein becomes an important source of gluconeogenesis. Large quantities of protein in muscle and other tissues can be broken down by hydrolysis to amino acids without serious tissue damage or loss of function. Certain of the amino acids that enter the blood are taken up by the liver. In the liver, the amino group is removed from each amino acid, and the remaining molecule is used to synthesize glucose by a stepwise series of enzyme-catalyzed reactions. This process, however, has limits. Continued protein loss can result in the death of cells throughout the body because they depend on proteins for such vital processes as plasma membrane function, enzymatic activity, and the formation of organelles.

The combined effects of glycogenolysis, gluconeogenesis, and the provision and use of fatty acids as an alternative energy source are so effective that after several days of fasting, the concentration of glucose in the blood of animals such as humans decreases by only a small percentage.

37.4 Reviewing the Concepts

- In the absorptive state, ingested nutrients are entering the blood from the alimentary canal. In the postabsorptive state, the canal is empty of nutrients and the body's own stores must supply energy.
- Much of the absorbed glucose immediately enters cells and is enzymatically broken down, providing energy required to synthesize ATP. Most absorbed triglycerides are stored in adipose cells until the body requires additional energy. Amino acids are taken up by all body cells and used to synthesize proteins (Figure 37.12).
- The events that maintain blood glucose concentration in the postabsorptive state fall into two categories: (1) glycogenolysis and gluconeogenesis, which provide glucose to the blood, and (2) cellular use of fatty acids for energy, which spares glucose for use by the nervous system (Figure 37.13).

37.4 Testing Your Knowledge

1. During the absorptive state, an animal is
   a. fasting.
   b. relying entirely on stored molecules for energy.
   c. absorbing nutrients from a recently ingested meal.
   d. metabolizing lipids stored in adipose tissue to supply ATP to its cells.
   e. doing both a and b.

2. Gluconeogenesis
   a. occurs when the liver synthesizes glucose from noncarbohydrate precursors.
   b. is the process by which glycogen is broken down to glucose.
   c. occurs primarily when an animal is in the absorptive state.
   d. occurs when triglycerides are being formed and stored in adipose cells.
   e. occurs exclusively in skeletal muscle.

37.5 Regulation of the Absorptive and Postabsorptive States in Vertebrates

Learning Outcomes

1. **SCISKILLS** ▶ Predict how the blood concentration of insulin will change as the blood concentration of glucose changes, and vice versa.
2. Describe how the use and storage of glucose are regulated.
3. Describe the effect of exercise on energy demands, and explain the mechanisms that are activated to meet those demands.

Tight control mechanisms are required to maintain homeostatic levels of energy-providing molecules in an animal's blood. These controls come in two forms: endocrine and nervous. Cells of the endocrine system produce long-distance signaling molecules called hormones. Several hormones work together with signals arising from the nervous system. In this section, we will see that one common function of the endocrine and nervous systems is to regulate the processes of glycogenolysis and gluconeogenesis so that glucose is made available to cells at all times.

Insulin Is a Key Regulator of Metabolism

In vertebrates, the blood concentration of **insulin,** a polypeptide hormone made by the pancreas, increases during the absorptive state and decreases during the postabsorptive state. Insulin controls metabolism primarily by regulating blood glucose concentration. It does this by promoting the transport of glucose from extracellular fluid into cells, where it can be used for metabolism. Glucose is a polar molecule that cannot cross plasma membranes without the aid of a transporter protein. Insulin increases glucose uptake by binding to a cell surface receptor and stimulating an intracellular signaling pathway. This pathway increases the availability of transport proteins called glucose transporters (GLUTs) in the plasma membrane (**Figure 37.14**).

Prior to insulin exposure, most GLUTs are located within preformed vesicles stored in the cytosol of cells. When a cell is stimulated by insulin, the vesicles (along with their GLUTs) fuse with the plasma membrane, making more GLUTs available to transport glucose into the cell. Consequently, insulin functions to lower blood glucose concentration because it increases the uptake of glucose into cells.

Insulin exerts its effects mainly on skeletal muscle cells and adipose cells, because these cells have insulin receptors in their plasma membranes. Cells of the brain contain GLUTs that are always present in the plasma membrane, even in the absence of insulin. This ensures that brain cells will obtain glucose at all times (recall that neurons of the CNS cannot normally use any alternative energy source).

Blood Glucose Concentration Is Maintained Within a Homeostatic Range

The blood glucose concentration of animals is controlled by a system of checks and balances. In the absorptive state, after a meal has been eaten, blood glucose concentration increases; in the postabsorptive

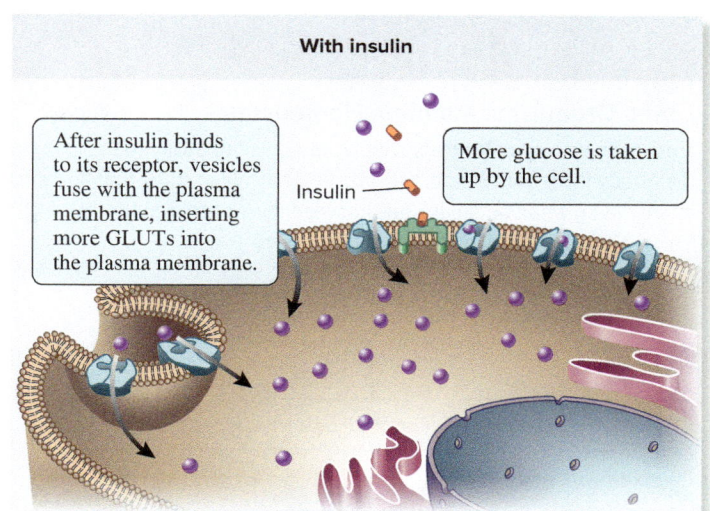

Figure 37.14 The effect of insulin on glucose transport (GLUT) proteins. In skeletal muscle and adipose cells, insulin stimulates the fusion of intracellular vesicles containing GLUT proteins with the plasma membrane, where they facilitate glucose uptake into the cell.

Concept Check: What benefit is there to having GLUT proteins premanufactured and stored in intracellular vesicles?

BioConnections: In what other contexts does the fusion of intracellular vesicles with a cell's plasma membrane occur? (Refer back to Figure 5.21 for help.)

state, depending on how long it lasts, the concentration decreases. What homeostatic mechanisms keep blood glucose concentration within a normal, homeostatic range?

Increased Glucose in the Absorptive State The primary factor controlling the secretion of insulin from the pancreas is the blood glucose concentration itself (Figure 37.15a). An increase in glucose directly stimulates the cells of the pancreas to secrete insulin in proportion to the amount of glucose in the blood. Later, after insulin promotes glucose uptake into its target cells, the resulting decrease in blood glucose concentration removes the signal for insulin secretion. This is an example of a negative feedback loop (refer back to Figure 32.8).

Decreased Glucose in the Postabsorptive State Several factors act in concert to prevent blood glucose from decreasing below the normal homeostatic range, even during a short fast. Otherwise, glucose could decrease so much that there would not be enough glucose to keep the cells of the CNS alive.

If for any reason—such as a more prolonged fast—blood glucose concentration decreases below the normal homeostatic range for an animal, neurons that respond to changes in the extracellular concentration of glucose are activated within the hypothalamus (Figure 37.15b). Signals from the hypothalamus then stimulate the production of glucose-elevating factors. These include numerous hormones, notably **glucagon,** another polypeptide hormone secreted by the pancreas. Glucagon stimulates the processes of glycogenolysis and gluconeogenesis in the liver. In addition, the neurotransmitter norepinephrine, released from neurons of the autonomic nervous system, stimulates adipose tissue to release fatty acids into the blood. The fatty acids diffuse across plasma membranes and provide another source of energy for the synthesis of ATP. The overall effect is to increase the blood concentrations of glucose and fatty acids during the postabsorptive state or during a prolonged fast.

More Energy Is Required During Exercise

We think of exercise as something humans do for fun or fitness, but in its broadest sense, **exercise** can be defined as any physical activity that increases an animal's metabolic rate (see Chapter 32). Generally, an animal becomes physically active to seek something, such as food, shelter, or a mate, or to elude something, such as a predator or a storm. The types of activities animals engage in can be quite varied. When a cheetah sprints after a small antelope, for example, the activity of both predator and prey is brief and intense, perhaps lasting only a few seconds. By contrast, a migrating bird may fly continuously for hours a day over a span of weeks.

For any type of activity or exercise, including those that result from an animal's behavioral responses to stress, nutrients must be available to provide the energy required for such things as skeletal muscle contraction, increased heart and lung activity, and increased activity of the nervous system. These energy-providing nutrients include glucose and fatty acids as well as the muscle's own glycogen. The liver supplies the blood with the glucose used during physical activity by glycogenolysis and gluconeogenesis. This occurs even in the absorptive state, and thus blood glucose concentration increases above normal when an animal is active at such times. In addition, an increase in adipose tissue lipolysis releases fatty acids into the blood, which provides an additional source of energy for exercising muscle.

These events are mediated by the same hormones and nerves responsible for the regulation of the postabsorptive state. For example, inputs from the autonomic nervous system to the pancreas

Biology Principle

Living Organisms Maintain Homeostasis

Most animals cannot feed constantly, and thus the ability to tap into stored reserves of energy allows them to maintain the concentration of glucose in their body fluids within a homeostatic range. This ensures that homeostasis of body processes can be maintained because all cells are supplied with energy even when an animal has not recently ingested nutrients.

(a) When blood glucose concentration is high   (b) When blood glucose concentration is low

Figure 37.15 Maintenance of blood glucose concentration within a normal range.

Concept Check: *What are the sources of the glycerol, fatty acids, and amino acids used to make glucose?*

---

inhibit insulin secretion during exercise or acutely stressful situations. Consequently, glucose transport into muscle and adipose cells is decreased, which tends to increase blood glucose concentration (this is part of the fight-or-flight response described in Chapter 33). Because the brain does not depend on insulin for glucose transport across neuronal membranes, as described earlier, more of the body's supply of glucose is available to the brain at such times, whereas other cells can use fatty acids for energy. Therefore, the body uses all available forms of energy in response to fasting, exercise, and stress.

a decrease in glucose concentration removes the signal for secretion. If blood glucose concentration decreases, glucose-monitoring regions in the hypothalamus stimulate the production of glucose-elevating factors such as glucagon and norepinephrine (Figure 37.15).

- Exercise is any type of physical activity that increases an animal's metabolic rate. Exercise activates pathways similar to those of the postabsorptive state to meet an animal's increased metabolic demands.

37.5 Reviewing the Concepts

- Hormones and the nervous system maintain homeostatic levels of energy sources in the blood. Insulin acts in muscle and adipose tissue to facilitate the diffusion of glucose from blood into the cell cytosol via glucose transporters (GLUTs) (Figure 37.14).
- In vertebrates, an increase in blood glucose concentration in the absorptive state stimulates the pancreas to secrete insulin;

37.5 Testing Your Knowledge

1. Insulin primarily regulates blood glucose concentration by
   a. stimulating the recruitment of GLUT proteins to the plasma membrane.
   b. stimulating gluconeogenesis.
   c. suppressing glucose uptake by muscle tissue.

d. stimulating the release of glucose from glycogen reserves in the liver.
e. inhibiting the synthesis of new GLUT proteins.

2. Which of the following increase the concentration of glucose in the blood during exercise?
   a. insulin and glucagon
   b. glucagon and norepinephrine
   c. norepinephrine and GLUTs
   d. glucagon and GLUTs
   e. insulin and norepinephrine

37.6 Excretory Systems in Different Animal Groups

Learning Outcomes

1. Describe the forms of nitrogenous wastes generated by animals.
2. Describe the general processes of filtration, reabsorption, secretion, and excretion.
3. Compare and contrast different invertebrate osmoregulatory organs.
4. Outline the general structural and functional features of kidneys that are common to all vertebrates.

We will turn now to an examination of the mechanisms by which non-useful products of metabolism are removed from an animal's body.

Broadly defined, an **excretory system** includes all of an animal's organs (such as gills, lungs, skin, and kidneys) that remove soluble wastes from the body and regulate ion and water balance. Respiratory structures and their ability to remove CO_2—the major waste product of metabolism—were covered in Chapter 36. In this section, we will focus on excretory organs that remove nitrogenous and other soluble wastes and, in the process, help to regulate ion and water balance. Most such excretory organs contain tubular structures lined with epithelial cells that have the capacity to actively transport ions and other substances across their membranes. Wastes are excreted out of the body by means of these tubes.

Some animals have considerable ability to regulate the rate at which waste is excreted and the amount of water that is lost in the process. For example, even though a thirsty mammal on a hot, sunny day must continue to rid its body of soluble waste products generated by its metabolism, it must also conserve water. In this section, we will consider the structure and function of invertebrate and vertebrate excretory organs. First, however, we will examine the major forms of nitrogenous wastes generated in animals due to the metabolism of proteins and nucleic acids.

Nitrogenous Wastes Are Found in Three Major Forms

Nitrogenous wastes are usually produced in one of three forms—ammonia (or ammonium ions), urea, or uric acid (**Figure 37.16**), depending on the species and the environment in which it lives.

- **Ammonia (NH_3)** and **ammonium ions (NH_4^+)** are the most toxic of the nitrogenous wastes because they disrupt

Biology Principle

Living Organisms Use Energy

Energy from ATP is required to metabolize nitrogenous wastes into urea and uric acid. Many animals, therefore, must expend energy each day to rid themselves of nitrogenous wastes.

Animal group	Most aquatic animals	Mammals, most amphibians, some marine fishes, some reptiles, and some terrestrial invertebrates	Birds, insects, and most reptiles
Major form of nitrogenous waste	Ammonia (NH_3) or ammonium ions (NH_4^+)	Urea	Uric acid
Energy required for production	None	Moderate	High
Amount of body water required for excretion	High	Moderate	Low
Toxicity of waste	High	Low	Low

Figure 37.16 Nitrogenous wastes produced by different animal groups. The three major forms of nitrogenous wastes, which are derived from the breakdown of proteins or nucleic acids, have different properties.

37.7 Testing Your Knowledge

1. Most useful solutes are reabsorbed in the _____, but additional fine-tuning of ion reabsorption and secretion occurs in the _____.
   a. renal corpuscle, loop of Henle
   b. collecting duct, renal corpuscle
   c. glomerulus, collecting duct
   d. proximal tubule, distal tubule and cortical collecting duct
   e. glomerulus, proximal tubule

2. A deficiency of aldosterone would be expected to result in
   a. increased plasma Na^+ and decreased plasma K^+ concentrations.
   b. decreased plasma Na^+ and decreased plasma K^+ concentrations.
   c. decreased plasma Na^+ and increased plasma K^+ concentrations.
   d. increased plasma Na^+ and increased plasma K^+ concentrations.

37.8 Impact on Public Health

Learning Outcomes

1. Describe the symptom called heartburn, and list some of its causes.
2. Identify the primary causes of ulcers.
3. Describe several common health issues related to diseases of the kidneys.
4. **SCISKILLS** ▶ Diagram the process of hemodialysis, and predict its strengths and limitations.

As we have seen, the functioning of the vertebrate digestive system is extraordinarily fine-tuned by the brain, nerves, and hormones. When people are well nourished, therefore, you might assume that digestive problems would be rare. However, each year in the U.S. alone, gastrointestinal complaints account for approximately 31 million visits to doctors and 10 million visits to hospitals and emergency rooms. In this section, we will revisit the problem experienced by the individual in the chapter-opening case and examine the causes and treatments of kidney disease in humans.

Excess Stomach Acid Production Can Lead to Heartburn

Approximately one in four people in the U.S. suffers at some time from **heartburn,** defined as a painful or burning sensation in the esophagus. As mentioned earlier, very little of the stomach contents moves backward into the esophagus, largely because a muscular sphincter at the juncture between the esophagus and the stomach prevents it. This sphincter relaxes when we swallow food, then closes again once food passes into the stomach. However, under some conditions, the sphincter either does not close entirely or is forced open by the pressure of material in the stomach. When this happens, the acid in the stomach enters the esophagus and irritates nerve endings there, causing the symptoms of heartburn. The medical term for the condition that causes these symptoms is gastroesophageal reflux, or simply acid reflux.

Many circumstances may contribute to acid reflux. Overeating, for example, can enlarge the volume of the stomach to the point of forcing its contents back through the esophageal sphincter. Lying down after a big meal removes the effect of gravity on food in the stomach and may allow some acid to leak backward. Acid reflux is also associated with consumption of acidic food such as citrus fruits; food and drink that contain chemicals that can relax the sphincter between the esophagus and stomach (for example, alcohol, coffee, and chocolate); and fatty foods (which take longer to digest than other foods and therefore remain in the stomach longer, delaying the emptying of stomach acid). One common cause of acid reflux is pregnancy. Toward the end of pregnancy, the growing fetus pushes up on the abdominal contents, which tends to force material from the stomach into the esophagus.

Common antacids contain calcium carbonate, which buffers the acid in the stomach and esophagus. In severe cases, acid reflux can sufficiently damage the walls of the esophagus to cause a chronic cough and pain or can even perforate the esophagus. Antacids may not be sufficient to treat these patients, who are instead given drugs that inhibit the stomach's ability to produce acid. These include common brand name drugs such as Zantac, Prilosec, and Nexium, some of which are among the most widely prescribed medications in the U.S. Left untreated, acid reflux disease can lead to cancer of the esophagus.

Erosion of the Walls of the Alimentary Canal Causes an Ulcer

Erosion of any portion of the wall of the alimentary canal due to any cause is called an **ulcer.** Most ulcers occur in the stomach (gastric ulcers) and the duodenum but occasionally also in the lower esophagus, because these sites have the greatest exposure to acid (**Figure 37.24**). If an ulcer is left untreated, the contents of the lumen may leak into the surrounding body cavity, where enzymes and acids from the stomach can do considerable damage. As many as 20 million Americans have an ulcer. Each year, over 30,000 patients require surgery to repair tissue damaged by ulcers, and around 6,000 die due to complications from the disease.

Acid is essential for ulcer formation, but we now know that it is not usually the primary cause. Many patients with ulcers have perfectly normal rates of acid production. Contrary to popular belief, stress or spicy food is probably not a major cause of ulcers, either. So what is the main cause of ulcers? In the 1980s, a team of Australian scientists demonstrated that many ulcers actually arise from a bacterial infection. Let's take a look at their revolutionary discovery, for which they earned the Nobel Prize in Physiology and Medicine in 2005.

Figure 37.24 Gastric and duodenal ulcers. **(a)** Common locations of ulcers in humans. **(b)** Illustration of a gastric ulcer eroding through the mucosal layer. **(c)** Photograph of an actual gastric ulcer that has penetrated through the stomach wall.
(c) ©Javier Domingo/Medical Images

(c) Human gastric ulcer

FEATURE INVESTIGATION

Marshall, Warren, and Coworkers Demonstrated a Link Between Bacterial Infection and Ulcers

For many years, the conventional wisdom regarding ulcers was that they occurred as a consequence of overproduction of stomach acid and that this, in turn, was due to factors such as stress, spicy foods, and smoking. In the early 1980s, Barry Marshall and J. Robin Warren had obtained evidence that some individuals with gastritis (inflammation of the stomach lining) or ulcers had active colonies of a bacterium later identified as *Helicobacter pylori* in their stomachs. This was a startling observation, because it had been assumed that no organism could survive the harsh pH and enzyme environment of the stomach. In preliminary studies, Marshall and Warren further observed that treatment of such individuals with compounds that kill bacteria could reduce colonies of *H. pylori* and provide relief from gastritis. Based on these observations, they hypothesized that symptoms of gastritis and ulcers in certain individuals were caused by *H. pylori* infection and that eradication of the infection would cure the disease (**Figure 37.25**).

To test their hypothesis, the investigators recruited individuals who were being treated for ulcers at a local clinic. First, the presence of *H. pylori* and an ulcer was confirmed with visual inspection, biopsies, enzymatic tests, and bacterial cultures in vitro. On the basis of these exams and tests, 100 patients were chosen for the study. They were assigned to one of four groups based on the treatments they were to be given: antacid plus placebo; antacid plus antibiotic; bismuth plus placebo; or bismuth plus antibiotic. Bismuth is known to have antibacterial properties, but its mechanism of action is unclear. The antacid used in the study is a drug that temporarily relieves the symptoms of an ulcer by neutralizing acid; it does not cure ulcers, however, and therefore served as a control.

Figure 37.25 Marshall, Warren, and coworkers demonstrate that *H. pylori* infection is a cause of ulcers in humans. (Note: In the original study, cited below, the authors initially misidentified the species of bacteria as *Campylobacter pylori*.)

HYPOTHESIS *H. pylori* infection is a cause of ulcers in humans.

KEY MATERIALS Endoscope, bacterial culture plates and medium, histologic stains, biopsy instruments.

Experimental level | **Conceptual level**

1. Confirm presence of ulcer and *H. pylori* in human subjects. — Use endoscopy to visualize the ulcer. Remove a tissue sample for analysis (biopsy), and check for presence of *H. pylori* by culture and enzyme analysis. If the cultures are *H. pylori*, then extracts of them should contain specific enzymes (measurable in a sensitive assay with a spectrophotometer).

2. Randomly assign subjects to 4 treatment groups.
   - Treatment Group 1: Antacid + Placebo
   - Treatment Group 2: Antacid + Antibiotic
   - Treatment Group 3: Bismuth + Placebo
   - Treatment Group 4: Bismuth + Antibiotic

   ↓ 10 weeks
   Re-evaluate for ulcers and *H. pylori*
   ↓ 1 year
   Follow-up exams

   Group 1 is a control, because antacids help the symptoms of ulcers, but do not cure them. A placebo is an ineffective treatment with no function. An antibiotic is a compound that kills bacteria. Bismuth is an element that also has bacteria-killing properties.

3. **THE DATA**

4. **CONCLUSION** Infection with *H. pylori* is a significant cause of ulcers in humans.

5. **SOURCE** Marshall, B. J., et al. 1988. Prospective Double-Blind Trial of Duodenal Ulcer Relapse After Eradication of *Campylobacter pylori*. *The Lancet* 332: 1437–1442.

(inset) ©Juergen Berger/Science

After 10 weeks, the patients were examined again. The results were striking: The antacid by itself or with the antibiotic had little effect on the elimination of *H. pylori* infection. Bismuth cleared the infection from 27% of patients, and bismuth plus the antibiotic cleared it from 74%. Interestingly, the use of antibiotic alone led to antibiotic-resistant strains of *H. pylori*, accounting for the failure of the antibiotic on its own to eliminate the infection. The presence of bismuth somehow prevented this resistance from occurring.

A trend was observed in which patients with cured ulcers were more likely to no longer have *H. pylori* infections. As seen in step 3, 92% of those individuals with no more evidence of infection had healed ulcers, compared to only 61% of those who still had signs of infection. Finally, a 1-year follow-up of those subjects whose ulcers had fully healed after 10 weeks (regardless of treatment) was performed. A combined bismuth/antibiotic treatment 1 year previously had reduced the rate of relapses (new ulcer formation) to only 25% compared with a high of 92% in the antacid/placebo group, a statistically highly significant result.

The investigators concluded that *H. pylori* infection is a significant cause of ulcers in humans. The mechanism seems to involve toxic by-products released by the bacterium and the body's inflammatory reaction to the microbe. These responses erode the mucous layer in the stomach and cause acid-producing cells to become overactive. Today, patients with ulcers can be easily tested for the presence of colonies of this bacterium in their stomach, and combined treatment with bismuth and one or two different antibiotics is remarkably effective at curing the disease.

Experimental Questions

1. What hypothesis was tested in this experiment, and what background information led to that hypothesis?
2. Did the results support the hypothesis? Do the results indicate that *H. pylori* infection is the only cause of ulcers in people?
3. What treatment therapy would you recommend for someone with a confirmed presence of *H. pylori* and an ulcer? From the data, can you conclude that some ulcers heal on their own?

Kidney Disease Disrupts Homeostasis

Diseases and disorders of the kidneys are a major cause of illness in the human population. Approximately 20 million people in the U.S. suffer from kidney disease, with about 15,000–20,000 individuals receiving kidney transplants each year. Many diseases affect the kidneys. Diabetes, bacterial infections, allergies, congenital defects, kidney stones (accumulation of mineral deposits in nephron tubules), tumors, and toxic chemicals are some sources of kidney damage or disease. A buildup of pressure due to obstruction of the urethra or a ureter may damage one or both kidneys and increase the likelihood of a bacterial infection and eventual renal failure. The symptoms of renal failure include fatigue, weakness, swelling of the abdomen and other regions, changes in urine amount and composition, anemia, ion imbalances, nausea, and muscle disturbances. These symptoms are similar regardless of the cause of the disease, and all stem from the condition known as **uremia,** the retention of urea and other waste products in the blood.

Assuming that a person with poorly functioning kidneys continues to ingest a normal diet containing the usual quantities of nutrients and salts, what problems might arise? Potentially toxic waste products that would normally enter the nephron tubules by filtration instead build up in the blood, because kidney damage significantly reduces the number of functioning nephrons. In addition, the usual excretion of K^+ in the urine is impaired because too few nephrons remain capable of normal tubular secretion of this ion. Increased K^+ in the blood is an extremely serious condition, because of the importance of stable extracellular concentrations of K^+ to the control of heart and neuron function.

The kidneys are still able to perform their homeostatic functions to a degree as long as at least 20% or so of the nephrons are functioning normally. The remaining nephrons undergo alterations in function—filtration, reabsorption, and secretion—to partially compensate for the missing nephrons. For example, each remaining nephron increases its rate of K^+ secretion so that the total amount of K^+ excreted in the urine can be maintained at or near normal levels. The kidneys' regulatory abilities are limited, however. If, for example, someone with severe renal disease were to eat a diet high in K^+, the remaining nephrons would not likely be able to excrete enough K^+ to prevent its concentration from increasing in the extracellular fluid.

Kidney Disease May Be Treated with Hemodialysis or Transplantation

Diseased kidneys may eventually reach a point that they can no longer excrete and reabsorb water and solutes at rates that maintain homeostasis or excrete waste products as fast as they are produced. In such cases, doctors must use various procedures that artificially perform the kidneys' filtering functions.

The most important of these procedures is **hemodialysis.** The general term dialysis means to separate substances in solution using a porous membrane. In hemodialysis, blood from an artery is purified by redirecting it through a dialysis machine, called a dialyzer (**Figure 37.26**). Within the dialyzer, blood flows through cellophane tubing that is surrounded by a special dialysis fluid. The tubing is highly permeable to most solutes but relatively impermeable to protein and completely impermeable to blood cells. These characteristics are designed to be quite similar to those of the kidneys' own glomerular capillaries. The dialysis fluid has ion concentrations similar to those in plasma but contains no urea or other substances that are to be completely removed from the blood.

As blood flows through the tubing in the dialyzer, small solutes diffuse out into the dialysis fluid until an equilibrium is reached. If, for example, the patient's plasma K^+ concentration is above normal, K^+ diffuses out of the blood across the cellophane tubing and into the dialysis fluid. Similarly, waste products and excess amounts

Figure 37.26 Simplified diagram of hemodialysis. The dialyzer is composed of many strands of very thin, sievelike tubing. In the dialyzer, blood within the dialysis tubing and the dialysis fluid bathing the tubing move in opposite directions (a countercurrent), which maximizes diffusion of substances out of the blood. The dialyzer provides a large surface area for diffusion of waste products out of the blood and into the dialysis fluid.

BioConnections: *Note that modern medical technology is using a feature found in many animals—countercurrent exchange. What other examples of countercurrent exchange have you learned about? Refer back to Figures 32.12 and 36.16 for help.*

of other substances also diffuse into the dialysis fluid and thus are eliminated from the body. Note in Figure 37.26 that blood and dialysis fluid move in opposite directions through the dialyzer. This pattern of movement establishes an artificial countercurrent exchange system that increases the efficiency with which solutes are removed from the blood. The dialyzed blood is then returned to one of the patient's veins through another tube that leaves the dialyzer. Each year, nearly 400,000 Americans undergo some type of hemodialysis.

The treatment of choice for most patients with permanent renal disease is kidney transplantation. Rejection of the transplanted kidney by the recipient's body is a potential problem, but great strides have been made in reducing the frequency of rejection. Many people who might benefit from a transplant, however, do not receive one, because the number of people needing a transplant far exceeds the number of donors. Currently, the major source of kidneys for transplanting is recently deceased persons. However, donation from a living, related donor has become more common, particularly with improved methods for preventing rejection. As noted earlier, the mammalian kidney can perform its functions with only a fraction of its nephrons intact. Due to this large safety margin, a person who donates one of his or her kidneys can function quite normally with only one kidney.

37.8 Reviewing the Concepts

- Heartburn (acid reflux) and ulcers are common disorders affecting millions of people, involving several regions of the GI tract. Many cases of ulcer are due to an infection with the *H. pylori* bacterium (Figures 37.24, 37.25).
- The symptoms of renal malfunction from any cause are similar and stem from uremia—the retention of urea and other waste products in the blood.
- Treatments for kidney disease include hemodialysis, in which a machine called a dialyzer removes wastes from the blood and returns the purified blood to the body, and kidney transplantation (Figure 37.26).

37.8 Testing Your Knowledge

1. Which of the following statements is *false*?
   a. Gastroesophageal reflux can result from eating a high-fat meal.
   b. Gastroesophageal reflux is common in late pregnancy.
   c. Gastroesophageal reflux is made worse by lying down after a large meal.
   d. The primary cause of ulcers is overproduction of acid due to stress and spicy foods.
   e. Ulcers may occur in regions of the alimentary canal other than the stomach.

2. Hemodialysis
   a. is the treatment of choice for people with advanced renal disease.
   b. directly mixes a person's blood with the dialysis fluid.
   c. can lead to a loss of blood cells.
   d. normalizes concentrations of small solutes in the blood by a diffusion mechanism.
   e. is only useful for removing nitrogenous wastes from the blood.

Assessing Your Knowledge and Skills

Foundational Knowledge

Multiple Choice

1. _____ is the process of enzymatically breaking down large molecules into smaller molecules that can be used by cells.
   a. Absorption
   b. Secretion
   c. Ingestion
   d. Digestion
   e. Egestion

2. Which of the following statements is correct?
   a. Hydrolytic enzymes are released from the gallbladder into the small intestine.
   b. Pancreatic enzyme release is stimulated by an increase in CCK in the blood.
   c. The pancreas secretes CCK through a sphincter into the small intestine.
   d. All vertebrates have a gallbladder.
   e. Secretin stimulates the release of bile from the liver.

3. Which of the following statements regarding the vertebrate stomach is *false*?
   a. Its cells secrete the protease pepsin in its active form.
   b. It is a saclike organ that may have evolved to store food.
   c. Its cells secrete hydrochloric acid.
   d. It is the initial site of protein digestion.
   e. Little or no absorption of nutrients occurs there.

4. In many vertebrates, bile is produced by the _____ and stored in the _____.
   a. liver, small intestine
   b. gallbladder, liver
   c. pancreas, small intestine
   d. small intestine, gallbladder
   e. liver, gallbladder

5. Which of the following statements is true of the large intestine?
   a. It contains microorganisms that may produce useful by-products that can be absorbed.
   b. It stores and concentrates fecal material.
   c. Its cells absorb ions and water that remain in chyme after it leaves the small intestine.
   d. It varies considerably in size and may even be absent in some vertebrates.
   e. All of the above are true of the large intestine.

Short Answer

1. List and describe the three major processes involved in urine production. Does each of these processes occur for every substance that enters an excretory organ, such as a kidney?

2. Briefly discuss the parts and functions of the nephron in the mammalian kidney. Which functions occur in multiple regions of the nephron, and which occur in only one portion?

Critical-Thinking Skills

Multiple Choice

1. Some people mistakenly believe that drinking water to excess is healthy. If a person drinks more water than is needed at a given time, what would happen as a result?
   a. The glomerular filtration rate (GFR) would not change but more water would be reabsorbed.
   b. The concentration of ADH in the blood would decrease and the GFR would increase.
   c. The concentration of ADH in the blood would increase and the GFR would increase.
   d. The GFR would decrease and the concentration of ADH would not change.
   e. The concentration of ADH in the blood would increase and the GFR would decrease.

2. Theoretically, which of the following could increase the rate of filtration in a mammalian kidney?
   a. a greater number of fenestrations in afferent arterioles
   b. a decrease in the pressure of blood entering a glomerulus
   c. dilation of the efferent arterioles
   d. a shorter loop of Henle
   e. both a and b

3. A health food store advertises "CCK Powder" as an aid to digestion, weight control, and good health. It is to be taken orally by dissolving the polypeptide hormone in water. Why will this be ineffective?
   a. CCK normally inhibits digestion.
   b. CCK will inhibit pancreatic enzyme secretion.
   c. CCK will inhibit saliva production.
   d. Pepsin and other proteases will digest CCK in the alimentary canal.
   e. CCK in the solution will break down the stomach mucosa.

4. Recall from Chapter 36 that arterioles are small arteries that can widen (dilate) or narrow (constrict), thereby greatly altering how much blood moves through them at any time. If the afferent arterioles in the renal glomeruli were to dilate, which of the following events do you think would happen?
   a. The GFR would increase.
   b. The GFR would decrease.
   c. The rate of urine production would increase.
   d. The volume of fluid in the body would decrease.
   e. Answers a, c, and d are correct.

5. Which of the following situations would be associated with the stimulation of glycogenolysis?
   a. a student suddenly realizing that he or she is late for an exam
   b. a lion that has just finished a meal
   c. a mouse running for cover from an owl
   d. a bat leaving its roost and starting its nightly search for food
   e. a, c, and d

Short Answer

1. Explain the functions of the crop and gizzard in birds. Propose a reason for why humans did not evolve a crop or gizzard. Smooth, polished stones have been found in the stomach region of fossilized skeletons of ancient sauropod dinosaurs. What does this suggest about the alimentary canal of such animals?

2. **PRINCIPLES** A principle of biology is that living organisms maintain homeostasis. The mammalian kidneys are an excellent example of this principle. Briefly, how many homeostatic processes can you describe in which the kidneys play a role?

38 How Endocrine Systems Influence the Activities of All Other Organ Systems

©Sovereign/ISM/Medical Images

A section through a human brain, highlighting the pituitary gland and its connection to the hypothalamus (white). Both of these structures secrete numerous hormones. (Image is a three-dimensional MRI.)

Chapter Outline

- 38.1 Types of Hormones and Their Mechanisms of Action
- 38.2 Links Between the Endocrine and Nervous Systems
- 38.3 Hormonal Control of Metabolism and Energy Balance
- 38.4 Hormonal Control of Mineral Balance
- 38.5 Hormonal Control of Growth and Development
- 38.6 Hormonal Control of Reproduction
- 38.7 Impact on Public Health

Assessing Your Knowledge and Skills

A fit, 23-year-old woman became alarmed when her friends repeatedly commented on her recent weight loss and voracious appetite. She had attributed these changes to her overall efforts to keep fit with regular exercise. However, despite her appetite, she continued losing weight until she reached a body weight of 102 pounds, down from her normal weight of about 122 pounds. At this point, she visited her doctor, who ordered a blood test for thyroid hormone.

In animals, a **hormone** is a signaling molecule produced by cells and secreted into the extracellular fluid. Unlike paracrine signals, which act on nearby cells, hormones enter the blood or hemolymph, where they travel to one or more distant target tissues to alter their functions. Thyroid hormone is made by cells of the thyroid gland. One of its important functions is the stimulation of metabolic rate. In this woman's case, the concentration of thyroid hormone in her blood was discovered to be greatly elevated, which increased her metabolic rate and led to her weight loss. Her appetite increased but could not fully compensate for her increased metabolism. It was later determined that she had a disease that was caused by cells of her thyroid gland overproducing thyroid hormone. She was treated with a procedure that destroys the gland, and as a result, she will have to take a daily thyroid hormone pill for the rest of her life to normalize her hormone levels. As you will learn in this chapter, there are many examples of hormone disease in humans and other animals, and many situations in which hormones are used to treat disease.

A chief function of hormones is to maintain homeostasis. Not surprisingly, therefore, hormones affect a wide range of animal bodily functions, including digestion, blood pressure, fluid and mineral balance, reproduction, and metabolism.

Although hormones are made by cells in nearly all the organs of an animal's body, hormone-producing cells are often found in specialized glands, called **endocrine glands,** the primary function of which is hormone synthesis and secretion. Examples include the pituitary gland, highlighted in the chapter-opening photo, and the thyroid gland. Collectively, all the endocrine glands and other organs containing hormone-secreting cells constitute an animal's **endocrine system.**

In this chapter, we will first examine the chemical nature of hormones and their mechanisms of action. We will then discuss how the endocrine and nervous systems interact, and the ways in which hormones influence homeostasis and other functions. We will conclude with a discussion of how hormones are used therapeutically in humans, as well as misused in certain situations.

38.1 Types of Hormones and Their Mechanisms of Action

Learning Outcomes

1. List and give examples of the three structural classes of hormones.
2. Identify the cellular locations of receptors for lipid-soluble hormones and for water-soluble hormones.
3. Describe the factors that regulate the concentration of a hormone in the blood.

Figure 38.1 is an overview of the major vertebrate endocrine structures and their hormones. Take a moment and review this figure as preparation for the rest of the chapter. In this section, we will examine some of the general characteristics of hormones, how they act, and how their concentrations in the blood are maintained.

Hormones Are Classified by Their Structures

Hormones fall into three broad classes:

- Amine hormones are synthesized from an amino acid—either tyrosine or tryptophan—and include epinephrine, norepinephrine, dopamine, thyroid hormone, and melatonin.
- Polypeptide hormones are the most abundant class of hormones. They are synthesized like any other polypeptide—by gene transcription and translation of mRNA. With the one exception of thyroid hormone (described later), amine hormones and polypeptide hormones are packaged into secretory vesicles that provide a reservoir of stored hormone that is available for immediate release by exocytosis when required.
- Steroid hormones are enzymatically synthesized from cholesterol, as depicted in **Figure 38.2**. Unlike amines and the polypeptide hormones, steroid hormones are not packaged into secretory vesicles because the hydrophobic steroid could diffuse across lipid membranes. Instead, steroid hormones are made on demand and immediately diffuse out of the cell.

Hormones Act Through Plasma Membrane or Intracellular Receptors

The amine and polypeptide hormones are generally water-soluble, and the steroid hormones are lipid-soluble. An exception is thyroid hormone. Although derived from an amino acid, it is sufficiently modified that it is mostly lipid-soluble. Water-soluble hormones cannot cross plasma membranes, and therefore they reversibly bind to a receptor protein on the surface of a target cell. Steroid hormones and thyroid hormone, however, diffuse across a target cell's plasma membrane and bind to a receptor protein located in either the cytosol or the nucleus. Hormones that act by binding to plasma membrane receptors tend to elicit fast responses, whereas those that act via intracellular receptors generally take longer.

The general mechanisms by which chemical messengers such as hormones act were covered in Chapter 8. Water-soluble hormones activate intracellular signaling pathways such as the cAMP pathway, which rapidly influences the activity of numerous proteins in a given target cell (refer back to Figure 8.13). By contrast, steroid and thyroid hormones, when bound to their receptors, act as transcription factors to regulate gene expression (refer back to Figure 8.9). Steroid and thyroid hormones can influence several genes within a single cell or in different cells. In these ways, one hormone can exert a variety of actions throughout an animal's body.

Hormone Concentrations Depend on Rates of Synthesis and Removal

When hormones are produced by endocrine cells in animals with closed circulatory systems, the hormones are released into the interstitial fluid surrounding the cells; from there, the hormones diffuse into the blood. Although most hormones are present at all times in the blood, their concentrations can increase when necessary. This can be accomplished by changing the rate of hormone secretion or by changing the rates at which hormones are inactivated or removed from the blood. The latter is important because although hormones carry out vital functions, excessive stimulation of cells by hormones often produces detrimental effects. Therefore, once a hormone has performed its functions, it is usually prevented from exerting its effects indefinitely. This is accomplished by negative feedback (see Chapter 32), by removal of the hormone from the circulation by the liver or kidneys, or by receptor-mediated endocytosis of a hormone bound to a membrane receptor.

Generally, these processes ensure that hormone concentrations in the blood remain within a normal range but have the capacity to be increased beyond that range if required. One of the ways in which changes in hormone concentrations are initiated is through sensory input to an animal's brain. As we will see next, the nervous and endocrine systems are functionally linked in many animals, including all vertebrates.

Hypothalamus:
Secretes several neurohormones that stimulate or inhibit anterior pituitary gland function. Synthesizes two neurohormones that are stored in and released from the posterior pituitary.

Anterior pituitary gland:
Produces 6 hormones with diverse actions (ACTH, FSH, LH, GH, PRL, TSH).

Posterior pituitary:
In mammals, secretes oxytocin, which stimulates uterine contractions during birth and milk secretion after birth; also secretes antidiuretic hormone, which increases water reabsorption in the kidneys.

Pineal gland:
Makes melatonin, which regulates daily rhythms and seasonal reproduction in some mammals.

Adrenal glands (medulla and cortex)

Medulla (not visible):
Makes epinephrine and norepinephrine, which mediate the fight-or-flight response.

Cortex:
Makes mineralocorticoids (such as aldosterone), which regulate Na^+ and K^+ balance in the blood; makes glucocorticoids (such as cortisol), which regulate growth, metabolism, development, immune function, and the body's response to stress; makes some androgens, which control reproduction.

Parathyroid glands (behind the thyroid):
Make parathyroid hormone, which increases blood Ca^{2+}.

Thyroid:
Makes thyroid hormone, which regulates metabolic rate, growth, and differentiation; makes calcitonin, which functions in Ca^{2+} homeostasis in some species.

Liver and kidneys:
Secrete erythropoietin, which regulates production of erythrocytes; also convert vitamin D to an active hormone.

Stomach and small intestine:
Secrete gastrin and other hormones that facilitate digestion and regulate pancreatic activity.

Pancreas:
Makes insulin, which decreases blood glucose, and glucagon, which increases blood glucose.

Ovaries (in females):
Produce estrogens and progesterone, which control female reproduction.

Testes (in males):
Produce androgens, such as testosterone, which control male reproduction.

Adipose tissue:
Produces hormones such as leptin which regulate appetite and metabolic rate.

Figure 38.1 Overview of the endocrine system in vertebrates, as seen in a human. This figure shows the major endocrine glands and selected other structures that constitute the human endocrine system, and it gives some major functions of the hormones produced by those glands. All mammals share similar endocrine structures. Other vertebrates have many of the same glands, but their structures can vary considerably.

BioConnections: *Are hormones unique to animals? Refer back to Table 29.1 for help.*

38.1 Reviewing the Concepts

- The endocrine glands and other organs with hormone-secreting cells constitute an animal's endocrine system. Hormones include amines, polypeptides, and steroids. Water-soluble hormones act on receptor molecules located in the plasma membrane, whereas lipid-soluble hormones act on intracellular receptors (Figures 38.1, 38.2).
- Concentrations of hormones can be increased or decreased by changing the rate of hormone secretion and the rates at which hormones are inactivated or removed from the blood.

38.1 Testing Your Knowledge

1. _____ hormones are hydrophilic, whereas _____ hormones diffuse through plasma membranes.
    a. Steroid, thyroid
    b. Steroid, all amine
    c. Polypeptide, steroid and thyroid
    d. Thyroid, polypeptide
    e. Polypeptide, all amine

2. The synthesis of which class(es) of hormone requires transcription and translation?
    a. amine hormones
    b. polypeptide hormones
    c. steroid hormones
    d. a and b only
    e. a, b, and c

Biology Principle

Structure Determines Function

Notice how slight changes in the arrangements of functional groups (and, in one case, an open ring) attached to the four-ring backbone create products with completely different functions. Keep in mind that the three-dimensional shape of a molecule may be changed much more by such chemical modifications than is visible in a two-dimensional depiction.

Figure 38.2 Synthesis and functions of the major steroid hormones in animals.

38.2 Links Between the Endocrine and Nervous Systems

Learning Outcomes

1. Describe the anatomical connections between the hypothalamus and pituitary gland.
2. Describe the roles of the hypothalamus and anterior pituitary gland in the regulation of endocrine function.
3. **SCISKILLS** ▶ Predict what would happen to the secretion of each hormone of the anterior pituitary gland if the portal circulatory connection between that gland and the hypothalamus were severed.
4. List some of the major functions of the hormones of the posterior pituitary.

A key feature of the endocrine system in most animals is that the concentrations of many hormones in the extracellular fluid rise and fall in response to changes in an animal's environment. For this to happen, the cells of the endocrine system must receive signals that indicate environmental changes. A sensory signal must be detected by a sensory receptor, and this information must then be converted into an endocrine response. Electrical signals are transmitted from the sensory receptors and are eventually conveyed to the hypothalamus. In this section, we will explore how the hypothalamus and pituitary gland function in linking the nervous and endocrine systems of vertebrates.

The Hypothalamus and Pituitary Gland Are Physically Connected

As described in Chapter 33, the **hypothalamus** is a collection of nuclei located at the ventral surface of the vertebrate brain (**Figure 38.3a**).

Figure 38.3 Relationship between the hypothalamus and the pituitary gland.

BioConnections: *Recall the definition of neurotransmitter from Chapter 33 and refer back to Figure 33.12. What determines whether a molecule such as dopamine (Table 38.1) is considered a neurotransmitter or a neurohormone?*

Table 38.1 Hormones of the Anterior Pituitary Gland and Hypothalamus

Anterior pituitary gland hormone	Stimulatory neurohormone from hypothalamus	Inhibitory neurohormone from hypothalamus	Major functions
Adrenocorticotropic hormone (ACTH)	Corticotropin-releasing hormone (CRH)	None known	Stimulates adrenal cortex to make glucocorticoids
Follicle-stimulating hormone (FSH)	Gonadotropin-releasing hormone (GnRH)	None known	Stimulates germ cell development and sex steroid production in gonads
Growth hormone (GH)	Growth hormone–releasing hormone (GHRH)	Somatostatin	Promotes growth; regulates glucose and fatty acid balance in blood
Luteinizing hormone (LH)	GnRH	None known	Stimulates release of eggs in females; stimulates sex steroid production from gonads
Prolactin (PRL)	TRH and other factors have been suggested as stimulators of PRL.	Dopamine	Stimulates milk formation in mammals; participates in mineral balance in other vertebrates
Thyroid-stimulating hormone (TSH)	Thyrotropin-releasing hormone (TRH)	None known	Stimulates thyroid gland to make thyroid hormones

Neurons in these nuclei are connected to an endocrine gland sitting directly below the hypothalamus, called the **pituitary gland** (see the white structure in the chapter-opening photo). The pituitary gland in humans is made up of two lobes, the anterior and posterior lobes. Some vertebrate species have an intermediate lobe that is believed to be vestigial in others, including primates. In those species that have one—for example, fishes and amphibians—hormones of the intermediate lobe function in such processes as skin color changes that help the animal to blend in better with its environment.

The hypothalamus and pituitary are connected by a thin piece of tissue called the **infundibular stalk** and by specialized blood vessels called portal veins. **Portal veins** differ from ordinary veins in that they not only collect blood from venules arising from capillaries—as all veins do—but they also drain into a second set of capillaries, instead of returning the blood directly to the heart. The second set of capillaries drain into venules and then veins that return blood to the heart. This arrangement of blood vessels bypasses the general circulation and allows the hypothalamus to communicate directly with the anterior lobe of the pituitary gland (often simply called the anterior pituitary gland).

As described in Chapter 33, the hypothalamic nuclei are vital for regulating such diverse functions as reproduction, body temperature homeostasis, circadian rhythms, appetite, thirst, metabolism, and responses to stress. The hypothalamus has such wide-ranging effects in part because it acts as a master control, signaling the pituitary gland when to produce and secrete one or more of its several hormones. However, the hypothalamus communicates differently with the anterior and posterior lobes of the pituitary gland. Let's look at the interaction between the hypothalamus and anterior pituitary gland first.

The Hypothalamus and Anterior Pituitary Gland Have Integrated Functions

Within the different nuclei of the hypothalamus are neurons that synthesize a class of hormones called neurohormones. A **neurohormone** is any hormone that is made in and secreted by neurons, typically in the hypothalamus. All neurohormones are either amines or polypeptides. The axon terminals from the hypothalamus end next to capillaries, into which they secrete the neurohormones (**Figure 38.3b**). The neurohormones are delivered directly from the hypothalamus to the cells of the anterior pituitary gland via the portal veins.

In response to these hypothalamic neurohormones, cells of the anterior pituitary gland synthesize several hormones. **Table 38.1** lists the six major anterior pituitary hormones that have well-defined functions in vertebrates and the neurohormones that either stimulate or inhibit them. The stimulatory action of several of the neurohormones led to their also being known as hypothalamic-releasing hormones, because they cause the release, or secretion, of other hormones from the anterior pituitary. The six major hormones of the anterior pituitary gland are secreted into the general blood circulation, where they act on other endocrine glands or structures.

The Posterior Pituitary Contains Axon Terminals from Hypothalamic Neurons That Store and Secrete Oxytocin and Antidiuretic Hormone

The posterior pituitary of mammals has a blood supply, but in contrast to the anterior pituitary gland, it is not connected to the hypothalamus by portal veins and does not receive neurohormones from the hypothalamus. The posterior pituitary is not actually a gland but is instead an extension of the hypothalamus that lies in close contact with the anterior pituitary gland (Figure 38.3a). Axons from neurons of the hypothalamus extend into the posterior pituitary. In mammals, the axon terminals in the posterior pituitary store one of two polypeptide hormones—oxytocin or antidiuretic hormone. When the hypothalamus receives information that these hormones are required, they are released directly from the neuron axon terminals into the blood.

Oxytocin increases in the blood of pregnant mammals just prior to birth. It stimulates contractions of the smooth muscles in the uterus, which facilitates the birth process and shortly afterward helps to expel the placenta. Oxytocin also stimulates the secretion of milk from the mammary glands of lactating females. When the mother's nipples are stimulated by the suckling of a newborn, sensory neurons transmit a signal from there to the mother's hypothalamus, which stimulates the cells that produce and secrete oxytocin. Oxytocin stimulates smooth

4 Remove pancreas from a second group of dogs.	The concentration of glucose in the blood increases due to diabetes. Glucose appears in the urine, a sign of diabetes.

5 Inject either purified extract or control solution into the diabetic dogs. Determine the levels of glucose in the blood and in the urine.	The control solution contained all of the chemicals used to prepare the extract but did not include any material from a pancreas.

6 THE DATA

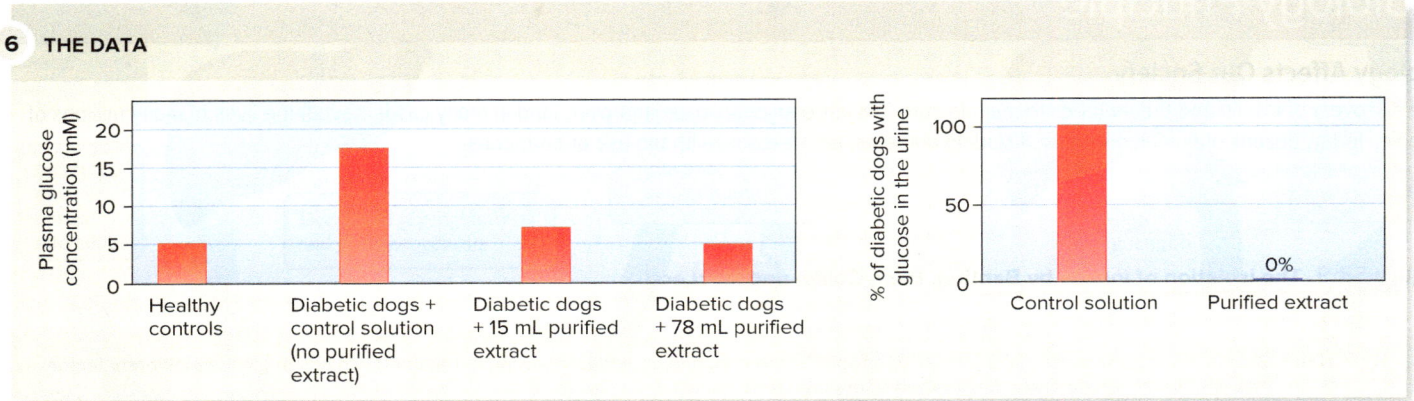

7 CONCLUSION Atrophy of the exocrine portion of the pancreas eliminated digestive enzymes that would have degraded the glucose-lowering factor (insulin) during its purification from the pancreas. Insulin was effective in reversing the major symptoms of diabetes in dogs.

8 SOURCE Banting, F. G., and Best, C. H. 1922. The Internal Secretion of the Pancreas. *Journal of Laboratory and Clinical Medicine* 7: 251–266.

---

had previously been induced by surgically removing the entire pancreas. The researchers discovered that the extract briefly kept the diabetic dogs alive. However, the researchers were unable to isolate sufficient quantities of the active factor in the extract to keep the experiments going longer than a day or so, nor were they able to obtain sufficiently pure factor to prevent side effects such as infection and fever.

This is where Collip's expertise came in. Collip developed a method for removing contaminating proteins from the extract by adding alcohol to the acid (Figure 38.9, step 3). Different proteins will precipitate (come out of solution) in response to different concentrations of alcohol. Using a concentration too low to cause the active factor to precipitate, Collip was able to remove most of the contaminating proteins, leaving behind a much more purified and safe extract.

The extract was then filtered to remove debris and further extracted to remove lipids. In this step, a hydrophobic solvent was added to the partially purified extract; the lipids in the extract dissolved in the solvent, which could then be removed. This was done because the researchers correctly predicted that the factor was a polypeptide and, by removing lipids, they would obtain a more purified preparation. Finally, the alcohol was evaporated, which concentrated the extract and increased its potency. When increasing amounts of this purified factor were injected into a second group of diabetic dogs, their blood glucose concentrations decreased and even returned to normal, and there was no longer any glucose in their urine. Diabetic dogs that received only a control solution that did not contain the factor continued to show a high glucose concentration in the blood and urine, compared with healthy animals.

Subsequent improvements in their procedures enabled the researchers to obtain larger amounts of the factor and to more accurately assess its potency without using the ligation procedure. The first successful test in humans of the extract, which they eventually named insulin, came in 1922 on a 14-year-old boy in Toronto, who was seriously ill from T1DM. The success of the team in rapidly isolating insulin and proving its effectiveness was so immediately significant that Banting and MacLeod were awarded the 1923 Nobel Prize in Medicine or Physiology.

Experimental Questions

1. How did Banting and Best propose to obtain the glucose-lowering factor produced by the pancreas?
2. What was Collip's contribution to the isolation of insulin?
3. **SCISKILLS** ▶ Propose a reason it was necessary to inject insulin as opposed to, for example, giving it in the form of an oral solution.

38.3 Reviewing the Concepts

- The thyroid gland makes thyroid hormone, which contains iodine. A major action of thyroid hormone in adult animals is to stimulate energy use by many different cell types (Figures 38.4, 38.5).
- The adrenal glands produce glucocorticoids, which increase blood glucose concentration. The endocrine pancreas produces insulin and glucagon, which have opposite effects on blood glucose concentration (Figures 38.6, 38.7, 38.8).
- Groundbreaking research by Banting, Best, Collip, and MacLeod identified and isolated insulin for therapeutic use in treating diabetes mellitus (Figure 38.9).

38.3 Testing Your Knowledge

1. Which of the following statements is true of the adrenal glands?
   a. They produce insulin.
   b. They produce hormones that control ion balance and glucose homeostasis.
   c. They produce only steroid hormones.
   d. They are inhibited by the hormone ACTH.
   e. They chiefly regulate calcium ion balance.

2. Type _____ diabetes mellitus is the most common form in humans, and it is associated with _____ .
   a. 1, immune disease
   b. 2, immune disease
   c. 1, obesity
   d. 2, obesity

38.4 Hormonal Control of Mineral Balance

Learning Outcomes

1. Describe the hormonal control of Ca^{2+} homeostasis.
2. Explain the regulation in the kidneys of water and Na^+ and K^+ balance by antidiuretic hormone and aldosterone.

All animals must maintain a balance in their cells and fluids of minerals such as Ca^{2+}, Na^+, and K^+. These ions participate in many functions common throughout much of the animal kingdom. For example, the movement of ions across plasma membranes determines the electrical properties of neurons and muscle cells. In this section, we will see how hormones maintain a homeostatic balance of these ions in an animal's body.

Vitamin D and Parathyroid Hormone Regulate the Concentration of Ca^{2+} in the Blood

Calcium ions have critical roles in neuronal transmission, heart function, muscle contraction, and numerous other events. Therefore, the concentration of Ca^{2+} in the blood is among the most tightly regulated variables in an animal's body.

Calcium is obtained from the diet and absorbed by the small intestine. Like all hydrophilic substances, Ca^{2+} cannot readily cross plasma membranes, including those of the epithelial cells of the small intestine, and thus its transport must be regulated. This process is controlled by the action of a derivative of **vitamin D.** In most mammals that receive regular exposure to sunlight, skin cells produce vitamin D from a precursor called 7-dehydrocholesterol, a reaction that requires the energy of ultraviolet light (**Figure 38.10**). In addition, many animals obtain vitamin D from food or, in humans, from milk and other foods supplemented with the vitamin. This supplementation is especially important for people who live at latitudes that receive little sunlight for much or part of the year.

Before it can act, vitamin D must first be modified by two enzymes that each add one hydroxyl group to specific carbon atoms. The first enzyme acts in the liver and the second in the kidneys. The final active product is called **1,25-dihydroxyvitamin D.** This molecule is a hormone because it is released into the blood by an organ—the kidneys—and then acts on a distant target tissue—the small intestine. The major function of 1,25-dihydroxyvitamin D in the intestine is to activate expression of a Ca^{2+} active transporter in the intestinal epithelium, thereby stimulating the absorption of Ca^{2+}. The Ca^{2+} then enters the blood, which delivers the ion to tissues for such activities as building bone and maintaining nerve, muscle, and heart functions.

When Ca^{2+} is not present in the diet, or when 1,25-dihydroxyvitamin D is not formed in normal amounts because of insufficient exposure to sunlight, the blood concentration of Ca^{2+} does not normally decrease significantly. The concentration does not drop because of the actions of another hormone called **parathyroid hormone (PTH).** This hormone is secreted from several small glands in the neck region of all tetrapods called parathyroid glands, which in humans are located at the back of the thyroid gland (**Figure 38.11a**). Cells of

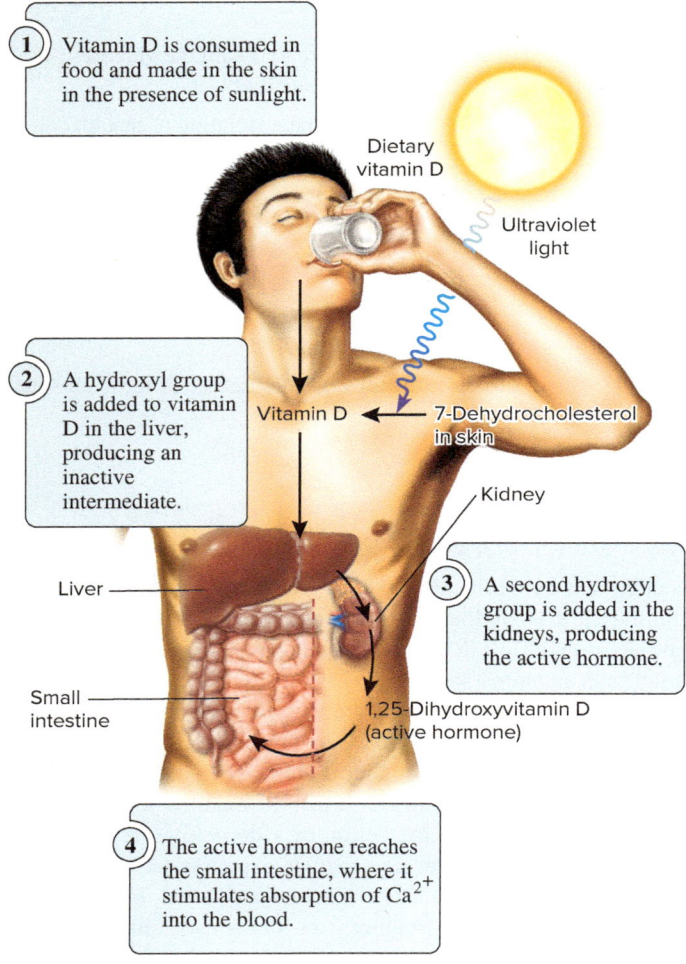

Figure 38.10 Synthesis of the active hormone formed from vitamin D that is ingested or produced from a precursor in the skin.

Concept Check: *Would you predict that all mammals synthesize the active form of vitamin D using the energy of sunlight?*

the parathyroid glands express plasma membrane receptors that bind extracellular Ca^{2+}. The binding of Ca^{2+} to these receptors inhibits the secretion of PTH. Thus, a decrease in the blood concentration of Ca^{2+} leads to less Ca^{2+} available to bind its membrane receptor; this, in turn, releases the inhibition of PTH. Consequently, the secretion of PTH into the blood increases.

PTH has three important actions that increase the concentration of Ca^{2+} in an animal's blood:

- It stimulates reabsorption of Ca^{2+} from the kidney filtrate so that less Ca^{2+} is lost in the urine.
- It activates the kidney enzyme required for the synthesis of 1,25-dihydroxyvitamin D.
- It acts on bone to stimulate the activity of cells that dissolve the mineral part of bone.

The action of PTH on bone releases Ca^{2+}, which enters the blood (**Figure 38.11b**). Typically, only a tiny fraction of the total Ca^{2+} in bone is removed in this way. Thus, besides providing a skeletal framework for the vertebrate body, bone also serves as an important reservoir of Ca^{2+}. PTH also acts to increase reabsorption of Ca^{2+} from the filtrate in the kidneys, so less Ca^{2+} is excreted in the urine. Without PTH, Ca^{2+} homeostasis is not possible. Complete absence of PTH is fatal in humans and other mammals.

In addition to 1,25-dihydroxyvitamin D and PTH, another polypeptide hormone called **calcitonin** has a role in Ca^{2+} homeostasis in some vertebrates, notably fishes and some mammals. Calcitonin is secreted by cells in the thyroid gland. In many respects, its function is opposite that of PTH. Calcitonin promotes excretion of Ca^{2+} into the urine and deposition of Ca^{2+} into bone, thereby decreasing blood Ca^{2+} concentration. This function is especially important in marine fishes, because of the high Ca^{2+} content of seawater and the entry of this ion from the water into the body fluids of the animals.

Two Hormones Regulate the Extracellular Fluid Concentrations of Na⁺ and K⁺ in Vertebrates

Like Ca^{2+}, concentrations of Na^+ and K^+ in the body fluids of most animals are tightly regulated, because these ions are the basis of membrane potential formation and action potential generation (see Chapter 33), among other functions. Like Ca^{2+}, Na^+ and K^+ are ingested in the diet and excreted in the urine in vertebrates at rates that maintain homeostatic concentrations in the body fluids. Unlike Ca^{2+}, however, no large reservoirs exist in the body for Na^+ and K^+. One of the key mechanisms for regulating the concentrations of these ions in the blood is altering the rate of Na^+, K^+, and water reabsorption from the urine as it is being formed in the kidneys. This is accomplished in large part by the actions of two hormones: ADH and aldosterone (**Figure 38.12**), both of which exert their effects on the kidneys.

The vertebrate kidneys normally reabsorb most Na^+ and K^+ from the fluid that is filtered through the kidney glomeruli (refer back to Figures 37.20 through 37.22). However, dietary and other changes can alter the concentrations of these ions in the blood. When this happens, the kidneys work to restore the ions to their normal concentrations. For example, if the blood Na^+ concentration increases above normal, the osmolarity of the blood will increase. Osmoreceptors in the brain detect this and stimulate ADH secretion from the posterior pituitary. Recall from Chapter 37 that ADH acts on the kidneys to reabsorb water from the forming urine. Increasing the amount of ADH available to act on the kidneys results in less water being excreted into the urine and more being reabsorbed into the blood. The reabsorption of water into the blood decreases the blood Na^+ concentration back toward normal.

In addition, the synthesis and secretion of the adrenal steroid hormone **aldosterone** are directly inhibited in response to an increase in Na^+ concentration. Normally, aldosterone increases Na^+ reabsorption in the kidneys. Therefore, a decrease in aldosterone in the blood results in a greater amount of Na^+ excretion into the urine than normal (refer back to Figure 37.22). Aldosterone also regulates

Biology Principle

Biologists Use Models to Test Predictions and Convey Ideas

Part (b) of this figure is a model of the relationship between the concentrations of Ca^{2+} and PTH in the blood of a vertebrate—in this example, a human. The model depicts how a change in the concentration of Ca^{2+} leads to a change in PTH secretion that restores the ion to normal. The Modeling Challenge in the figure legend asks you to predict how artificial or disease-related changes in one or more of the hormones depicted in this model can affect thyroid size and secretion.

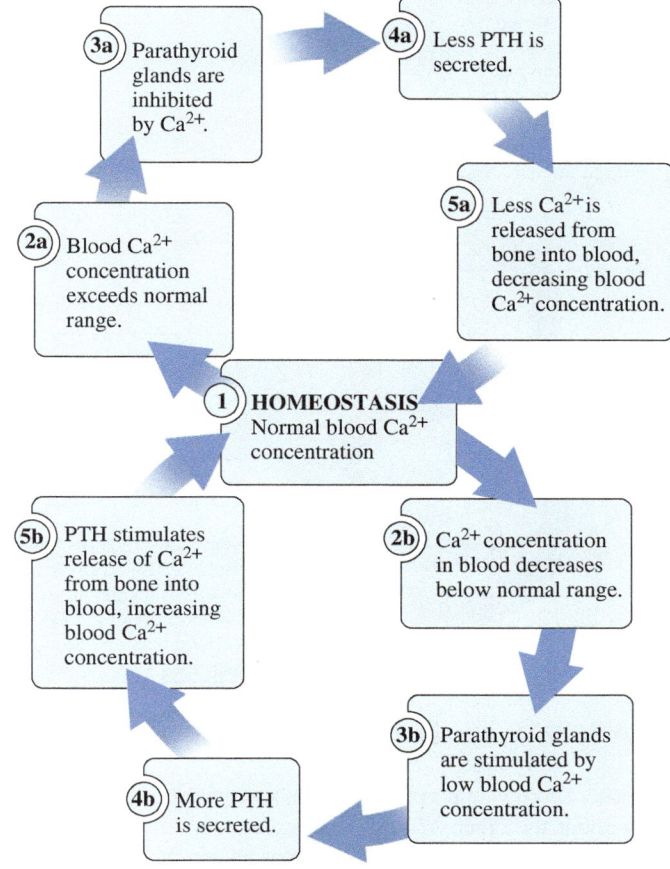

Figure 38.11 The role of parathyroid hormone (PTH) in Ca^{2+} homeostasis. (a) The four parathyroid glands are located behind the thyroid. (b) The action of PTH: Steps 2a–5a occur when Ca^{2+} is in excess in the blood; steps 2b–5b occur when the concentration of Ca^{2+} in the blood is below normal.

(a) Location of the parathyroid glands

(b) Homeostatic control of blood Ca^{2+} concentration

Modeling Challenge: A type of endocrine disorder in many vertebrates, including humans, occurs when target cells lose their ability to respond to a particular hormone. For example, target cells may lose their ability to express a functional receptor for a particular hormone. Consider the bottom part of the model in Figure 38.11b, with the steps labeled 1 through 5b. Revise that model to show how Ca^{2+} homeostasis would be affected if bone cells were unable to respond to PTH.

the extracellular fluid concentration of K^+. If the K^+ concentration increases above normal, aldosterone release is increased and this stimulates secretion of K^+ into the urine.

Aldosterone and 1,25-dihydroxyvitamin D are both steroid hormones, as is cortisol, described earlier. Their structures are similar but distinct (see Figure 38.2). These characteristics have enabled biologists to use steroid hormones as models for understanding how signaling molecules and their receptors may have coevolved, as described next.

EVOLUTIONARY CONNECTIONS

Hormones and Receptors Evolved as Tightly Integrated Molecular Systems

We have seen that hormones act by binding to receptor proteins located on the cell surface or in the cytosol or nucleus. This raises an intriguing question: Which evolved first, a given hormone or its

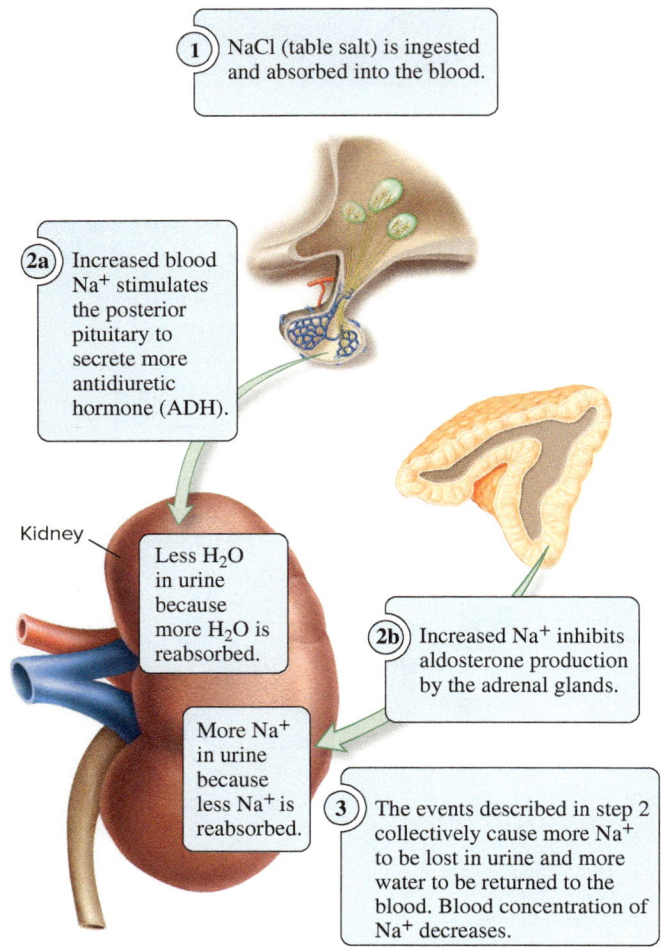

Figure 38.12 An example of Na⁺ balance achieved by hormones.

receptor? There would appear to be no selection pressure to evolve one without the other. Without a receptor, a hormone has no function, and without a hormone, its receptor has no function. Hormones and their receptors function as tightly integrated molecular systems. Each molecule depends on the other for biological activity. How such systems could have evolved, or whether they evolved together, has long been a major puzzle. However, recent research has generated intriguing hypotheses about the evolution of these signaling systems.

American biologist Joseph Thornton and colleagues addressed this puzzle by examining how two structurally related steroid hormones evolved separate activities and distinct receptors. Aldosterone, as noted, is a steroid hormone made by the adrenal cortex. Because it regulates the balance of two minerals (Na⁺ and K⁺) in the body, it is known as a mineralocorticoid. Cortisol is another steroid hormone made by the adrenal cortex and is known as a glucocorticoid, because one of its major functions is to regulate glucose balance. The actions of these hormones are mediated by intracellular receptors known as the mineralocorticoid receptor (MR) and the glucocorticoid receptor (GR), respectively. Phylogenetic analysis of the gene sequences that encode these two receptors suggests that they arose at least 450 mya, after the evolution of the first vertebrates, by duplication of an ancient gene that encoded a corticoid receptor (CR).

The researchers analyzed the known sequences of the genes for the MR and GR of many vertebrate species, including the most ancient vertebrates, to deduce a theoretical sequence of an ancestral CR. They then synthesized this CR and tested its ability to bind aldosterone and cortisol. They discovered that the CR was capable of binding both hormones, particularly aldosterone. This finding was surprising, because aldosterone is only present in tetrapods, which arose around 300 mya, long after the proposed gene duplication event that created the MR. Therefore, it appears that a receptor with high affinity for aldosterone was present long before animals acquired the capacity to synthesize aldosterone. The receptor seems to have evolved to bind other steroids. Around 450 mya, the gene for the CR duplicated, and then the two resulting genes seem to have further mutated such that one receptor gained high affinity primarily for mineralocorticoids and the other for glucocorticoids.

These studies suggest that an animal's ability to respond to aldosterone evolved long before aldosterone did, because a high-affinity receptor was already in place in animals. When aldosterone appeared in tetrapods, it used the MR derivative of the ancestral CR. In this example, the answer to "which came first" appears to be that the receptor evolved first and the hormone later.

38.4 Reviewing the Concepts

- The concentration of Ca²⁺ in the blood is tightly regulated. 1,25-Dihydroxyvitamin D and parathyroid hormone (PTH), produced by the parathyroid glands, regulate the blood concentration of Ca²⁺ (Figures 38.10, 38.11).
- The blood concentrations of Na⁺ and K⁺ are regulated in large part by the actions of ADH, which increases water retention in the body and therefore controls ion concentration indirectly, and aldosterone, which increases Na⁺ reabsorption and K⁺ secretion into the urine (Figure 38.12).
- In at least one case, it appears that a hormone's receptor evolved before the hormone itself.

38.4 Testing Your Knowledge

1. Which of the following statements is correct?
   a. Aldosterone decreases the sodium ion concentration in the blood.
   b. 1,25-Dihydroxyvitamin D acts on intestinal cells to increase Ca²⁺ transport into the blood.
   c. Parathyroid hormone acts on intestinal cells to increase Ca²⁺ transport into the blood.
   d. ADH promotes loss of water from the body via the urine.
   e. Aldosterone stimulates reabsorption of K⁺ from the urine.

2. A person takes a daily supplement containing far more Ca²⁺ than she needs. Which of the following might ensue?
   a. The concentration of PTH in her blood will decrease.
   b. The amount of 1,25-dihydroxyvitamin D in her blood will increase.
   c. Her kidneys will reabsorb more Ca²⁺ than usual.
   d. The concentration of PTH in her blood will increase, and that of 1,25-dihydroxyvitamin D will decrease.
   e. Both a and c will occur.

38.5 Hormonal Control of Growth and Development

Learning Outcome

1. Identify and describe the functions of the hormones involved in the regulation of growth and development in animals.

Hormones have a central role in controlling growth and development in animals. Growth can occur slowly over long periods or in brief spurts, with some animals exhibiting both types of growth. Although growth is determined by many factors, notably adequate nutrition, it is regulated in large part by the endocrine system.

Growth is distinguished from development, which is the process by which cells form tissues with specific functions, and tissues develop into larger and more complex structures. In this section, we will examine how both processes depend in part on the endocrine system.

Vertebrates Require a Balance of Several Hormones for Normal Growth

Several hormones stimulate growth in vertebrates. The anterior pituitary gland produces **growth hormone (GH),** which is under dual stimulatory and inhibitory control of the hypothalamus (see Table 38.1). GH binds to receptors on cells in the liver and stimulates the liver to produce another hormone, called **insulin-like growth factor-1 (IGF-1).** In mammals, IGF-1 stimulates the elongation of bones by binding to receptors on certain bone cells. This stimulation occurs most notably during puberty, when mammals become reproductively mature. Growth at that time is further accelerated by the steroid hormones of the gonads, resulting in the rapid pubertal growth spurt. Eventually, however, the gonadal hormones cause the growth regions of bone to seal, preventing any further bone elongation.

In humans, if the pituitary gland fails to make adequate amounts of GH during childhood, the concentrations of GH and IGF-1 in the blood will be lower than normal. In such cases, growth is stunted, resulting in one of the major causes of short stature. Individuals with this condition can be treated with injections of recombinant human GH and will grow to relatively normal height, as long as treatment begins before puberty is completed so that the bones can still elongate.

Hormones are also required during fetal life for development of the brain, lungs, and other organs into fully mature, functional structures. For example, cortisol from the mammalian fetal adrenal gland is vital for proper lung formation. In humans, premature babies born before the adrenal glands have matured have lungs that are not capable of inflating normally; such babies require hospitalization to survive.

Thyroid hormone influences development among all vertebrates. In amphibians, for example, thyroid hormone promotes metamorphosis—notably, in tadpoles, where it promotes the resorption of the tail and development of the legs (**Figure 38.13**). This effect can be demonstrated by experimentally decreasing or increasing the concentration of thyroid hormone in the blood of a tadpole. Decreasing the thyroid hormone concentration results in a tadpole that does not transform into a frog. Increasing the thyroid hormone above normal results in a tiny froglet in which metamorphosis occurs sooner than normal.

Many Invertebrates Grow and Develop in Spurts Under the Control of Three Hormones

Like vertebrate endocrine pathways, hormonal control systems in invertebrates often involve multiple glands and neural structures acting in concert. In insects, for example, the endocrine system is critical for the growth and development of larvae and their eventual development into pupae (**Figure 38.14**). In the case of larval growth and metamorphosis, specialized neurosecretory cells in the brain periodically secrete a hormone called **prothoracicotropic hormone (PTTH),** which then stimulates a pair of endocrine glands called the prothoracic glands. These glands, located in the thorax, synthesize and secrete into the hemolymph a steroid hormone called **20-hydroxyecdysone.**

Figure 38.13 **Effects of thyroid hormones on animal development.** Experimental manipulation of thyroid hormone concentration can slow down or accelerate the development of tadpoles.

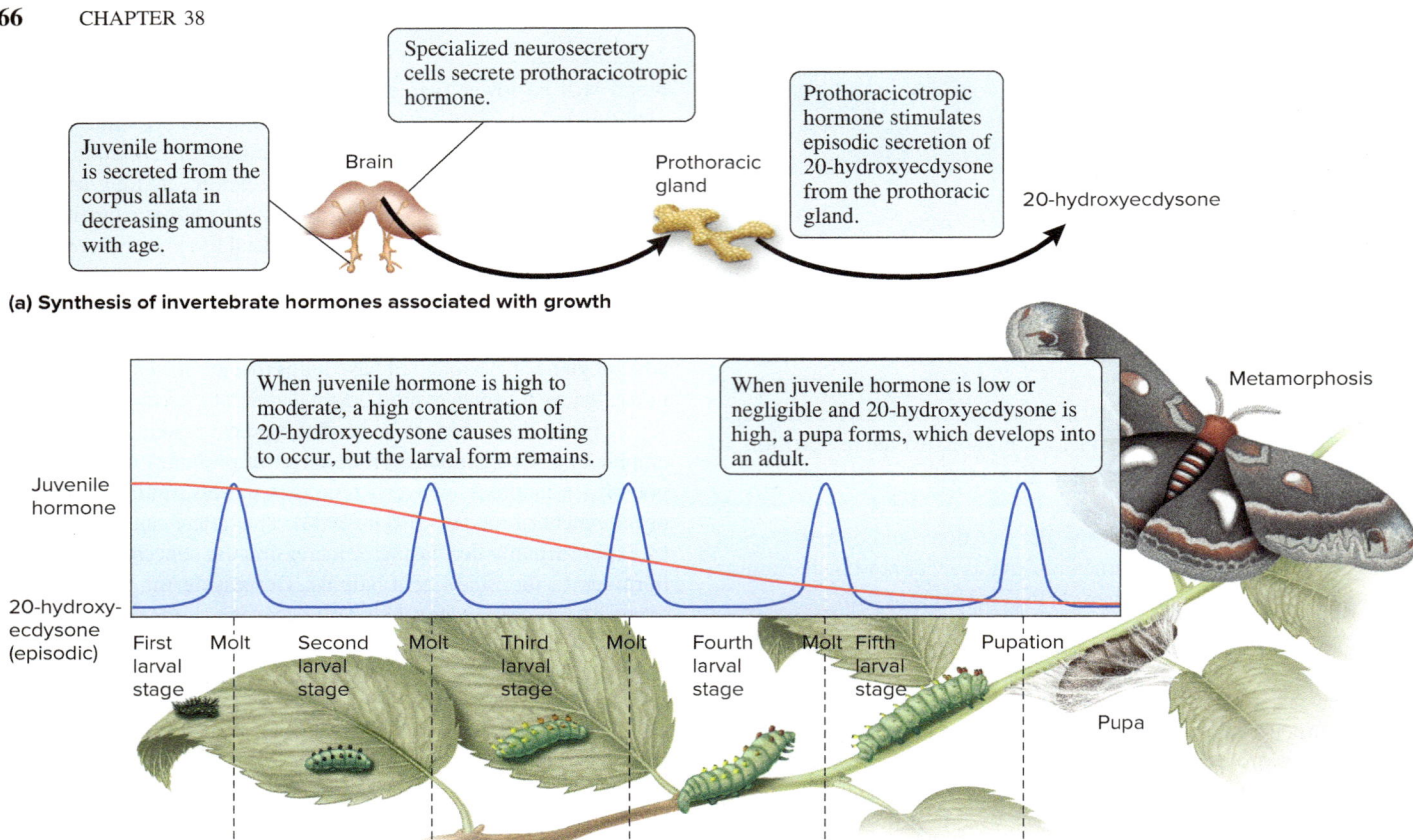

Figure 38.14 Hormonal control of insect development. The development of insects requires the coordinated actions of three hormones. Note: The relative concentrations and secretion patterns of 20-hydroxyecdysone and juvenile hormone in this figure are schematic and not representative of all insects.

Concept Check: *In what part of a cell would you predict the receptor for the steroid hormone 20-hydroxyecdysone is located?*

This hormone is secreted only periodically. In response to each burst of secretion, the larva undergoes a rapid development and molts (sheds its cuticle). It then begins a new developmental period until it molts again in response to another episode of secretion. The molting process is known as ecdysis, from which the name of the hormone is derived (refer back to Figure 26.8).

Throughout larval development, paired neurosecretory structures behind the brain, termed the corpus allata, secrete another hormone, a protein called **juvenile hormone (JH)**. JH determines the character of the molt that is induced by 20-hydroxyecdysone, by acting to prevent metamorphosis into an adult (hence the name juvenile, which reflects the fact that JH fosters the larval stage). As the larva ages, however, the amount of JH it produces gradually declines until the concentration is nearly zero. During this time, 20-hydroxyecdysone continues to be secreted periodically. The decline in JH below a certain concentration results in the transition from larva to pupa in response to a burst of 20-hydroxyecdysone; the near absence of JH is a prerequisite for the final step of metamorphosis into an adult.

38.5 Reviewing the Concepts

- Hormones play a crucial role in regulating growth and development. In vertebrates, normal growth depends on a balance among growth hormone (GH), insulin-like growth factor-1 (IGF-1), and gonadal hormones. Thyroid hormones affect development among all vertebrates (Figure 38.13).
- In insects, prothoracicotropic hormone (PTTH), 20-hydroxyecdysone, and juvenile hormone (JH) control the growth of larvae and their development into pupae (Figure 38.14).

38.5 Testing Your Knowledge

1. Some children with short stature can be effectively treated with
   a. cortisol.
   b. growth hormone.
   c. drugs that inhibit IFG-1 secretion.
   d. prothoracicotropic hormone.
   e. thyroid hormone.

2. _____ is a steroid hormone that _____ molting in insects.
   a. Prothoracicotropic hormone, stimulates
   b. 20-Hydroxyecdysone, stimulates
   c. Juvenile hormone, inhibits
   d. 1,25-Dihydroxyvitamin D, stimulates
   e. 20-Hydroxyecdysone, inhibits

38.6 Hormonal Control of Reproduction

Learning Outcomes

1. List the male and female sex steroids, their sites of production, and their general functions in reproductive development.
2. **SCISKILLS** ▶ Propose a mechanism by which hormones can link nutrition with reproduction.

The topic of reproduction is covered in detail in Chapter 39. In all vertebrates and probably most invertebrates, reproduction is closely linked with endocrine function. In this section, we will briefly examine the most common reproductive hormones and some of their major actions in male and female animals.

The Gonads Secrete Sex Steroids That Influence Most Aspects of Reproduction

Hormones produced by the gonads of animals have central roles in nearly all aspects of reproduction, from reproductive behaviors to the ability to produce offspring. In males, the gonads are called testes. They produce and house the male gametes, called sperm cells. In addition, the testes produce several related steroid hormones, collectively termed **androgens.** In many vertebrates, including humans, the primary androgen is **testosterone.** The gonads of females are called the ovaries, which contain the female gametes, or egg cells. Other cells within the ovaries secrete two major steroid hormones: **progesterone** and a family of hormones called **estrogens.** The major estrogen in many vertebrates, including humans, is **estradiol.** Collectively, the male and female gonadal hormones are referred to as the sex steroids.

The sex steroids are responsible for male- or female-specific reproductive changes associated with courtship and mating. In vertebrates, androgens from the developing testes are required for development of the male phenotype. Sex steroids are also chiefly responsible for the development of secondary sex characteristics in each sex, which in humans include growth of the appropriate external genitals, growth of the breasts in women, development of facial hair in men, and distribution and amount of fat and muscle in the body. Finally, sex steroids are required for the maturation of gametes, the transition of young animals to reproductive maturity, and the ability of females to produce young.

The ability of the testes and ovaries to produce sex steroids depends on the presence of gonadotropins. These hormones are secreted by the anterior pituitary gland, are the same in both sexes, and include **follicle-stimulating hormone (FSH)** and **luteinizing hormone (LH)** (Table 38.1). Therefore, identical anterior pituitary gland hormones control the gonads in both male and female animals.

Nutrition and Reproduction in Mammals Are Linked Through Hormones

An interesting feature of the endocrine control of reproduction is the observation that puberty is delayed in mammals, including humans, that are very undernourished. Similarly, fertility—the ability to produce offspring—is reduced in women and other adult female mammals under such conditions. This makes sense, because supporting the nutritional demands of a growing fetus would be difficult without adequate nutrition and energy stores. In such a case, it is more advantageous for an animal to delay reproduction until sufficient food is available.

How does the brain of a female mammal determine when sufficient energy is stored in her body to support pregnancy? The answer appears to be partly the result of hormonal signals, notably the polypeptide hormone **leptin.** Leptin is secreted by adipose cells in proportion to the amount of fat stored in the cells. For example, if the amount of adipose tissue in a female mammal's body increases, so does the concentration of leptin in her blood. The increase in leptin is detected by a nucleus in the brain that controls appetite and metabolism. In this case, appetite is suppressed and metabolism is increased. Conversely, a decrease in the amount of body fat leads to a decrease in the blood concentration of leptin and removal of the signal to the brain. This results in an increased appetite and decreased metabolism. The former encourages an animal to find food, and the latter conserves vital stores of energy.

Leptin also is known to stimulate the synthesis and secretion of reproductive hormones such as FSH and LH by the anterior pituitary gland. Consequently, undernourishment that causes a decrease in leptin also results in decreased production of reproductive hormones, thereby contributing to a temporary loss of fertility. When adipose stores are sufficient, there is more leptin available and the secretion of FSH and LH return to normal. Therefore, leptin acts as a link between energy stores and the reproductive system.

38.6 Reviewing the Concepts

- Hormones produced by the gonads play vital roles in nearly all aspects of reproduction. The ability of the testes and ovaries to produce the sex steroids depends on the gonadotropins, FSH and LH. The hormone leptin links nutritional status with reproduction.

38.6 Testing Your Knowledge

1. Which of the following statements is correct?
   a. Leptin inhibits reproduction by inhibiting FSH and LH.
   b. Progesterone and estradiol are examples of androgens.
   c. FSH and LH stimulate the gonads in both males and females.
   d. Undernourishment stimulates leptin production, which in turn inhibits reproduction.
   e. Both a and d are correct.

38.7 Impact on Public Health

Learning Outcomes

1. Name several hormones used therapeutically and the disease or health problem for which each is prescribed.
2. **SCISKILLS** ▶ Predict the risks associated with the misuse of hormones and with exposure to endocrine disruptors in the environment.

The list of endocrine-related diseases in the human population is lengthy, ranging from relatively common disorders such as diabetes and thyroid disease to rare ones that may affect only 1 in every

100,000 or more individuals. Hormones can be used to treat many of these diseases, but they can also be misused in ways that cause health problems. Also, a wide range of industrial chemicals with hormone-like actions are thought to cause disruption to the endocrine system. In this section, we will consider how hormones are used therapeutically as well as misused; then we will look at an example of how environmental factors that alter the function of the endocrine system can have important public health implications.

Hormones Are Used to Treat Millions of People

Since the advent of synthetic hormone production, doctors have been able to provide individuals with hormones that their own bodies may fail to produce. For example, millions of people self-administer insulin each day to treat diabetes or take thyroid pills to supplement an insufficient level of thyroid hormone. Many women are treated with gonadotropin hormones to increase fertility or take estrogen to control the symptoms of menopause. A growing number of men are administered androgens if the levels produced by their bodies decline significantly in later life. Other examples of therapeutic hormone use include recombinant human growth hormone to treat abnormally short stature in prepubertal children, epinephrine inhalers to treat asthma, and glucocorticoids to treat inflammation, lung disease, and skin disorders, to name just a few.

Hormone Misuse Can Have Disastrous Consequences

The advent of hormone therapy has been of enormous benefit to society. However, the availability of hormones has also led to their misuse. Some individuals, typically those in competitive sports, self-administer hormones such as androgens. This practice may increase muscle mass and improve athletic performance, but serious side effects may occur. For instance, androgen administration has been linked to extremely aggressive behavior and to the development of cardiovascular disease and heart attacks, skin problems, and certain cancers. Women using androgens run similar health risks as men, as well as develop masculinizing traits, including increased growth of body hair and thinning scalp hair.

Another example of misuse occurs when hormones are used to boost the number of erythrocytes in order to increase the oxygen-carrying capacity of the blood. This has become common among some athletes participating in long-distance aerobic activities, such as competitive cycling and cross-country skiing. The hormone used is **erythropoietin (EPO),** a polypeptide that acts by stimulating the maturation of erythrocytes in the bone marrow and blood, thereby increasing hematocrit.

EPO is normally made by the liver and kidneys in response to any situation where additional blood cells are required, such as following blood loss. When used abusively, however, the number of erythrocytes can become great enough that the blood becomes significantly more viscous than normal. This increased viscosity strains and damages the heart, which must work harder to pump the thickened blood. Since the 1990s, the international cycling community has been rocked by an alarming number of professional cyclists who died of heart attacks in the prime of their lives. These individuals had been using EPO to gain an unfair and, as it turns out, unwise advantage over their peers. Testing continues to detect injected EPO in the blood of some cyclists and other endurance athletes.

Synthetic Compounds May Act as Endocrine Disruptors

A recent and disturbing phenomenon is the growing occurrence of so-called **endocrine disruptors,** chemicals that interrupt the normal function of hormones in the body. These chemicals, which are found in lakes, streams, ocean water, and soil exposed to pollution runoff, are derived in many cases from industrial waste. They are also found in many common household products, including nearly all plastic goods. Many have molecular structures that resemble a hormone sufficiently to bind to its receptors. A common type of disruptor is recognized by estrogen receptors. If these compounds make their way into drinking water or food, they can exert estrogen-like actions or can inhibit the actions of the body's own estrogen. Such effects can lead to significant consequences for fertility and the development of embryos and fetuses.

The extent of the risk from endocrine disruptors is hotly debated, but the number of mature, functional germ cells produced in animals as diverse as mollusks and humans has declined dramatically in the past 50 years in the U.S. In addition, researchers throughout the world have noted feminization of freshwater fishes downstream of wastewater facilities. For example, male fishes exposed to such conditions during development show an increased production of proteins normally made by females bearing eggs. They also show changes in gonadal structures such that there is a resemblance to the female structures. Further research is being conducted to gain more information on the consequences of these contaminants for animal and human endocrine systems.

38.7 Reviewing the Concepts

- Hormones are used therapeutically to treat a variety of human disorders, including diabetes, infertility, growth disorders, asthma, and inflammation.
- Androgen misuse can cause health risks such as cardiovascular disease, skin problems, cancer, infertility in men, and masculinizing traits in women. Erythropoietin misuse can seriously damage the heart.
- Endocrine disruptors may bind to hormone receptors in animals' bodies, including humans. In some cases, they may exert estrogen-like actions or inhibit the actions of the body's own estrogen.

38.7 Testing Your Knowledge

1. Hormones are often misused. Which of the following might result if a man injects himself with androgens in an effort to increase muscle mass?
   a. low blood sugar
   b. heart disease
   c. increased fertility
   d. immune suppression
   e. testes enlargement due to increased LH and FSH

Assessing Your Knowledge and Skills

Foundational Knowledge

Multiple Choice

1. Which of the following is the defining feature of hormones?
   a. They are only produced in endocrine glands.
   b. They are secreted into the blood, where they may reach one or more distant target cells, thereby altering cell function throughout the body.
   c. They are released only by neurons but are not the same as neurotransmitters.
   d. They are never released by neurons, but only by epithelial cells.
   e. They are secreted into ducts, where they diffuse to another, nearby gland or other structure.

2. Steroid hormones are synthesized from _____ and bind _____.
   a. proteins, membrane receptors
   b. fatty acids, membrane receptors
   c. tyrosine, intracellular receptors
   d. proteins, intracellular receptors
   e. cholesterol, intracellular receptors

3. Which of the following statements about polypeptide hormones is *false*?
   a. They bind to receptors located on cell membranes.
   b. Most of them are lipid-soluble.
   c. They are the most abundant class of hormones.
   d. They normally activate second messengers.
   e. Their binding to receptors is reversible.

4. Which of the following statements about the control of hormones is *false*?
   a. Many are regulated by negative feedback.
   b. In some cases, they may be controlled by changes in blood concentrations of certain nutrients such as glucose.
   c. In some cases, they may be controlled by changes in blood concentrations of certain minerals such as Ca^{2+}.
   d. Their blood concentrations are controlled by their rates of synthesis and degradation.
   e. The rate of degradation of a hormone may go up or down, but synthesis is always constant.

5. In many invertebrates, the molting of larvae is stimulated by
   a. growth hormone.
   b. cortisol.
   c. juvenile hormone.
   d. 20-hydroxyecdysone.
   e. aldosterone.

Short Answer

1. Distinguish between type 1 and type 2 diabetes mellitus.
2. Briefly describe the general role of hormones in animal development.

Critical-Thinking Skills

Multiple Choice

1. A wolf has not been able to obtain food to eat for several days. Which hormone or hormones might be present in this animal's blood in a higher concentration than normal?
   a. insulin
   b. glucagon
   c. cortisol
   d. leptin
   e. Both b and c are correct.

2. An electrode (a fine metal wire that conducts electricity) is surgically implanted into the hypothalamus of a rat and connected to a stimulating device. Stimulation of the electrode sends depolarizing current into the neurons of the hypothalamus, thereby activating the neurons. Which of the following events might be expected to occur?
   a. The rat will experience a decreased rate of urine production.
   b. The blood pressure of the rat will gradually decrease.
   c. The rat will experience diuresis (increased loss of water in the urine).
   d. The total body fluid volume of the rat will increase.
   e. Both a and d will occur.

3. In some parts of the world, the iodine content of the soil is extremely low or absent. Humans who rely almost exclusively on crops grown in this soil for their diet may be expected to experience which of the following symptoms?
   a. increased secretion of TRH, decreased secretion of TSH, decreased T_3 concentration in the blood, and a goiter
   b. decreased secretion of TRH and TSH, decreased T_3 concentration in the blood, and a goiter
   c. increased T_3 concentration in the blood, decreased TRH and TSH secretion, and no goiter
   d. decreased T_3 concentration in the blood, increased TRH and TSH secretion, and a goiter
   e. decreased secretion of TRH, increased secretion of TSH, decreased T_3 concentration in the blood, and a goiter

4. House pets such as dogs may develop tumors in the anterior pituitary gland. One such tumor arises from cells that synthesize and secrete the hormone ACTH. These tumor cells become autonomous; that is, the cells produce ACTH without requiring any input from the hypothalamus, and they lose their ability to be inhibited by negative feedback. In such a case, predict which of the following will occur.
   a. The animal's bones will thicken.
   b. The animal will become more susceptible to disease because of suppression of its immune system.
   c. The blood glucose concentration of the animal will decrease below normal.
   d. Hypothalamic secretion of CRH will increase.
   e. The animal's muscle mass will slowly increase.

5. A 6-year-old child is very short for his age and is not growing at a normal rate. His pediatrician obtains a sample of his blood for testing. The test reveals that the child has normal levels of growth hormone (GH) in his blood. Which of the following statements could explain his failure to grow?
   a. His bone cells do not express receptors for insulin-like growth factor-1 (IGF-1).
   b. His liver cells do not express receptors for GH.
   c. His hypothalamus secretes too little somatostatin.
   d. His hypothalamus secretes more growth hormone–releasing hormone than normal.
   e. Both a and b could explain his failure to grow.

Short Answer

1. What advantage is provided by an adipose-derived signaling molecule such as leptin, in terms of an animal's body mass, metabolism, and reproduction?
2. **PRINCIPLES** A principle of biology is that living organisms maintain homeostasis. How do the opposing actions of insulin and glucagon cooperate to maintain glucose homeostasis? When are these hormones released into the blood? What would happen to a nonfasting mammal if it were injected with a high dose of glucagon?

39

The Production of Offspring: Reproduction and Development

Most animals reproduce by sexual reproduction. These oysters are releasing their sperm and eggs into the water, where they will combine and form new organisms. This is an example of sexual reproduction, which favors genetic variation in species.

Chapter Outline

- 39.1 Overview of Sexual and Asexual Reproduction
- 39.2 Gametogenesis and Fertilization
- 39.3 Human Reproductive Structure and Function
- 39.4 Pregnancy and Birth in Mammals
- 39.5 General Events of Embryonic Development
- 39.6 Impact on Public Health

Assessing Your Knowledge and Skills

©Robert F. Sisson/National Geographic/Getty Images

An 18-year-old woman joined her college's cross-country track team after competing for 2 years in high school. Gradually, her training increased from about 25 km/week to a very demanding 90 km/week. Recognizing the increased ability and competitiveness of college athletes relative to those at her high school, and at the urging of her track coach, she restricted her daily food intake to become as lean as possible. As her training and diet restriction progressed, she became increasingly preoccupied with monitoring what she was eating. Eventually, her weight dropped from 59 kg to 46 kg. This was followed by the cessation of her menstrual periods for several months. Alarmed that something was wrong, she visited her physician. After a battery of tests ruled out a variety of possible causes for missing her periods, she was diagnosed with secondary amenorrhea. This is a general term for the absence of menstrual cycles in a person who had previously been cycling normally. In her case, it was clear that the cause of her condition was undernourishment combined with intense exercise. She was encouraged to receive mental health counseling in conjunction with her parents and coach in order to address her eating disorder, and to work with a dietitian to tailor an appropriate meal plan to support her desired level of activity.

Secondary amenorrhea is very common among elite female athletes in sports and activities like cross-country track, gymnastics, and ballet, where physical activity is demanding and a lean body mass is encouraged. It is a warning sign that a person is overdoing things and needs to correct a nutritional imbalance. When a woman's body is stressed in this way, her reproductive function may cease. From a physiological perspective, this makes sense, as the likelihood of supporting a healthy pregnancy in such circumstances is greatly reduced and the body requires whatever energy is available just to survive.

This woman's case illustrates the delicate control of **reproduction**—the processes by which organisms produce offspring. The biological mechanisms that favor successful reproduction in the animal kingdom are extraordinarily diverse. Many of the observable differences in animal behavior and anatomy are the result of adaptations that increase an animal's chances of reproducing.

In this chapter, we will examine the diverse means of reproduction throughout the animal kingdom, including asexual and sexual reproduction. We will consider the control mechanisms of reproduction in mammals, including humans. We then will cover some of the general events in embryonic development and will conclude with a discussion of some key issues related to fertility in the human population today, including those that affected the young woman described in this introduction.

39.1 Overview of Sexual and Asexual Reproduction

Learning Outcomes

1. Describe several differences between sexual and asexual reproduction.
2. **SCISKILLS** ▶ Propose a hypothesis explaining why most animals reproduce by sexual reproduction.

Sexual reproduction in animals is the production of a new individual by the joining of two haploid reproductive cells called **gametes**—typically one from each parent. This produces offspring that are genetically different from both parents. **Asexual reproduction** in animals occurs when offspring are produced from a single parent without the fusion of gametes. The offspring are clones of the parent organism. In this section, we will consider general features of sexual and asexual reproduction, including the advantages and disadvantages of each.

Sexual Reproduction Occurs in Most Animals

Sexual reproduction occurs in most invertebrate and vertebrate species and involves the joining of two gametes. The gametes are spermatozoa (usually shortened to **sperm**) from the male and **egg cells,** or **ova** (singular, ovum), from the female. (In some species, sperm and egg cells are produced by a single organism called a hermaphrodite.) When a sperm unites with an egg—the process termed **fertilization**—each haploid gamete contributes its set of chromosomes to produce a diploid cell called a fertilized egg, or **zygote**. As the zygote undergoes cell divisions and begins to develop, it is called an **embryo**.

Asexual Reproduction Takes Several Forms

Asexual reproduction occurs in some invertebrates and in a small number of vertebrate species. These animals reproduce in one of three ways: budding, regeneration, or parthenogenesis. **Budding,** which is seen in cnidarians and a few other invertebrate species, occurs when a portion of the parent organism pinches off to form a new, complete individual (**Figure 39.1a**). In this process, cells from the parent undergo mitosis and differentiate into specific types of structures before the new individual breaks away from the parent. At any one time, a parent organism may have one, two, or multiple buds forming simultaneously. Budding continues throughout such an animal's lifetime.

Certain species of sponges, echinoderms, and worms reproduce by **regeneration,** or fragmentation, in which a complete organism can be formed from a fragment of the animal's body. In some sea

(a) Budding

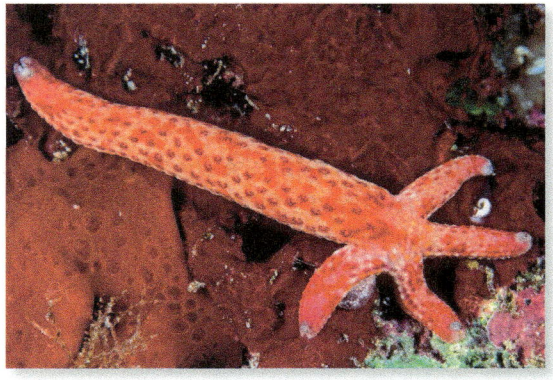

(b) Regeneration

Figure 39.1 Examples of asexual reproduction. (a) *Hydra* with a single bud. (b) A starfish regenerating a new body from a single arm.
(a) ©Clouds Hill Imaging Ltd./Getty Images; *(b)* ©WaterFrame/Alamy

BioConnections: *Is asexual reproduction unique to animals? Look back at Figure 31.17 for some help.*

stars and starfish, for example, an arm removed by injury or predation can grow into an entirely new individual (**Figure 39.1b**). Similarly, a flatworm bisected into two pieces can regenerate into two distinct individuals.

Parthenogenesis is the development of offspring from an unfertilized egg. Animals produced by parthenogenesis are usually diploid but in some cases are haploid. This form of asexual reproduction occurs in several invertebrate classes and in a few species of fishes and reptiles.

Sexual and Asexual Reproduction Have Advantages and Disadvantages

Asexual reproduction provides a relatively simple way for an organism to produce many copies of itself, whereas sexual reproduction usually requires two individuals to produce offspring. What are the advantages and disadvantages of each method?

Asexual reproduction has certain advantages over sexual reproduction. First, an animal can reproduce asexually even if it is isolated from others of its own species—for example, because the animal is nonmotile or because it rarely encounters another member of its species. Second, individuals can reproduce rapidly at any time because they need not seek out, attract, and mate with an individual of the opposite sex. Asexual reproduction, therefore, is an effective way of generating large numbers of offspring. Although many kinds of animals reproduce asexually, it is more prevalent in species that live in very stable environments, where there is little selection pressure for genetic diversity in the population.

Compared with asexual reproduction, sexual reproduction is associated with unique costs. Two types of gametes (sperm and eggs) must be made (often in large numbers), males and females require specialized body parts to mate with each other, and the two sexes must be able to find each other. However, given that the vast majority of eukaryotic species reproduce sexually, the following question has intrigued biologists since the time of Darwin: What is the advantage of sexual reproduction for the perpetuation of a species?

In the context of species survival, the major difference between asexual and sexual reproduction is that sexual reproduction allows for greater genetic variation due to genetic recombination. Only certain alleles from each parent are passed on, and when a set of genes from one parent mixes with a different set from the other parent, the offspring are never exactly like either of their parents. Thus, a key consequence of sexual reproduction is increased genetic variation between successive generations. One prevalent hypothesis about the advantage of sexual reproduction is that it allows a more rapid adaptation to environmental changes than does asexual reproduction.

In particular, sexual reproduction allows alleles within a species to be redistributed via crossing over and independent assortment across many generations. As a result, some offspring carry combinations of alleles that promote survival and reproduction, whereas other offspring may carry less favorable combinations. As described in Chapter 19, natural selection can favor those combinations that promote greater reproductive success while eliminating offspring with lower fitness. By comparison, the alleles of asexual organisms are not reassorted from generation to generation. As a result, it is more difficult to accumulate potentially beneficial alleles within individuals of these species. Also, as described next, sexual reproduction may facilitate the elimination of harmful alleles from a population.

FEATURE INVESTIGATION

Paland and Lynch Provided Evidence That Sexual Reproduction May Promote the Elimination of Harmful Mutations in Populations

Evolutionary biologists have suggested that the inability of asexual species to reassort alleles may be a key disadvantage for them compared with sexually reproducing species. To investigate this question, American geneticists Susanne Paland and Michael Lynch studied the persistence of mutations in populations of *Daphnia pulex*, a freshwater organism commonly known as the water flea. The researchers chose this organism because some natural populations reproduce asexually, and others reproduce sexually.

In their experiment, shown in **Figure 39.2**, Paland and Lynch studied the sequences of several mitochondrial genes in 14 sexually reproducing and 14 asexually reproducing populations of *D. pulex*. The researchers hypothesized that asexual populations would be less able to eliminate harmful mutations. As discussed in Chapter 15, random gene mutations that change the amino acid sequence of an encoded protein are much more likely to be harmful than beneficial. The alleles of sexually reproducing populations can be reassorted from generation to generation, thereby producing succeeding generations in which the detrimental alleles are lost from the population.

As shown in step 4 of Figure 39.2, the researchers discovered that both the sexual and asexual populations could eliminate highly deleterious mutations. Organisms harboring such mutations probably died and did not reproduce. In addition, both the sexual and asexual populations retained mildly deleterious and neutral mutations. However, moderately deleterious mutations were eliminated from the sexual populations but not from the asexual ones. One interpretation of these data is that sexual reproduction allows for the reassortment of beneficial and detrimental alleles, making it easier for sexually reproducing populations to eliminate those mutations that are moderately detrimental.

Experimental Questions

1. **SCISKILLS** ▶ How did Paland and Lynch test the hypothesis that sexual reproduction reduced the frequency of deleterious mutations in a population?

2. **SCISKILLS** ▶ Approximately how many times more often did deleterious mutations persist in the asexually reproducing populations compared to the sexually reproducing ones?

3. What is the proposed evolutionary benefit of sexual reproduction?

Figure 39.2 Paland and Lynch demonstrated the importance of sexual reproduction in reducing the frequency of harmful genetic mutations.

HYPOTHESIS Sexual reproduction allows for greater mixing of alleles of different genes and thereby may prevent the accumulation of detrimental alleles in a population.

KEY MATERIALS The researchers collected samples of *Daphnia pulex* from many natural populations. A total of 14 sexual populations and 14 asexual populations were studied.

Experimental level | Conceptual level

1. Isolate mitochondrial DNA from members of 28 populations of *D. pulex*. This involves breaking open cells and extracting the DNA (refer back to Chapter 18).

2. Amplify regions of mitochondrial genes, using PCR. Subject the regions to DNA sequencing. The techniques of PCR and DNA sequencing are described in Chapter 18 (refer back to Figures 18.6 and 18.8).

3. Using computer technology, align the sequences and determine the number of DNA changes that would cause amino acid substitutions. These amino acid changes were categorized as those that would be highly deleterious, moderately deleterious, mildly deleterious, or neutral for protein function. Compare the sexual and asexual populations for the persistence of these types of changes.

4. **THE DATA**

Results from step 3:

Types of amino acid substitutions (The amino acid substitutions were due to rare mutations that occurred in the natural populations of *D. pulex*.)	% of total amino acid substitutions	Allowed to persist in Sexual populations	Allowed to persist in Asexual populations
Highly deleterious	73.2	No	No
Moderately deleterious	13.3	No	Yes
Mildly deleterious	4.4	Yes	Yes
Neutral	9.1	Yes	Yes

5. **CONCLUSION** Moderately deleterious mutations are less likely to persist in populations of animals that reproduce sexually.

6. **SOURCE** Paland, S., and Lynch, M. 2006. Transition to Asexuality Results in Excess Amino Acid Substitutions. *Science* 311: 990–992.

39.1 Reviewing the Concepts

- Asexual reproduction occurs when offspring are produced from a single parent, without the fusion of genetic material from two parents (Figure 39.1).
- Sexual reproduction is the production of a new individual by the joining of two haploid gametes: a sperm from the male and an egg from the female.
- Paland and Lynch showed that sexual reproduction was more effective than asexual reproduction in eliminating deleterious mutations from a population of *Daphnia pulex* (Figure 39.2).

39.1 Testing Your Knowledge

1. Choose the correct statement.
   a. The majority of animals reproduce by asexual reproduction.
   b. Together, budding, regeneration, and parthenogenesis account for reproduction in nearly all invertebrates.
   c. A zygote is the haploid organism produced by the union of a sperm and an egg.
   d. Potentially beneficial alleles are more likely to accumulate in species that reproduce sexually than in asexually reproducing species.
   e. In asexual reproduction, gametes from each parent join to form a zygote.

39.2 Gametogenesis and Fertilization

Learning Outcomes

1. Outline the processes of spermatogenesis and oogenesis.
2. **SCISKILLS** ▶ Propose reasons why internal fertilization is advantageous.

This section covers how and where gametes are formed, as well as how two gametes join to form a new organism.

Sperm and Eggs Are Produced During the Process of Gametogenesis

In most invertebrates and all vertebrates, male and female gametes are formed within the **gonads**—the **testes** (singular, testis) in males and the **ovaries** (singular, ovary) in females. The formation of gametes—**gametogenesis**—begins with cells called germ cells, which multiply by mitosis, resulting in diploid cells (carrying two sets of chromosomes, denoted as $2n$) called **spermatogonia** (singular, spermatogonium) in males and **oogonia** (singular, oogonium) in females (**Figure 39.3**). Some of these cells become **primary spermatocytes** or **primary oocytes** that may begin the process of meiosis. (See Chapter 14 for a review of the processes of mitosis and meiosis.) Until this point, the development of sperm is similar to that of eggs. From this point on, gametogenesis differs between the two types of gamete.

Spermatogenesis The formation of haploid sperm from a diploid germ cell is called **spermatogenesis**. As shown in Figure 39.3a, primary spermatocytes begin this process by undergoing the first of two meiotic divisions (meiosis I). In the primary spermatocyte, meiosis I produces two haploid (n) cells called **secondary spermatocytes.** These cells also undergo meiosis (meiosis II), producing four haploid **spermatids** that eventually differentiate into mature haploid sperm cells. Gametogenesis in males, therefore, results in four gametes from each spermatogonium.

The most striking change in each spermatocyte as it differentiates into a sperm is the formation of a flagellum, also called the tail (Figure 39.3b). The movements of the tail require energy and make the sperm motile. The sperm also has a head region, which contains the nucleus that carries the chromosomes. At the tip of the head is a special structure termed the **acrosome,** which contains hydrolytic enzymes that help break down the protective outer layers surrounding an egg cell. The head and tail are separated by a midpiece that typically contains as many as several dozen mitochondria, depending on the species, that produce the ATP required for tail movements.

Oogenesis Whereas spermatogenesis produces four gametes from each primary spermatocyte, gametogenesis in the female, called **oogenesis,** results in the production of a single gamete from each primary oocyte (Figure 39.3c). In mammals, oogenesis begins in the female embryo. Oogonia divide by mitosis to form primary oocytes that enter meiosis I but stop the process, remaining in an arrested state after birth. Meiosis I does not resume until puberty, the time after which a mammal first becomes capable of reproducing. Beginning at puberty, meiosis I is completed in some primary oocytes, producing from each one a haploid **secondary oocyte** plus a smaller cell, called a polar body, that eventually degenerates. Meiosis II begins in the secondary oocyte but stops at metaphase. Once the secondary oocyte is released from the ovary in the process of **ovulation,** it can become fertilized if sperm are available. Meiosis II is not complete until the oocyte is fertilized by a sperm. If the oocyte does not encounter a sperm, it never undergoes the second meiotic division. Once a haploid egg nucleus fuses with a haploid sperm nucleus, a diploid zygote is produced.

Each oocyte undergoes growth and development within an ovarian structure called a follicle. In mammals, a layer of glycoproteins called the **zona pellucida** surrounds the surface of the secondary oocyte in the follicle. In turn, a layer of cells called the cumulus mass, which provides protection and nutritive support, is outside the zona pellucida. An outer cellular layer known as the theca produces hormones that control oocyte growth (Figure 39.3d). These layers have important roles in fertilization, as discussed later.

Fertilization Involves the Union of Sperm with Egg

Fertilization is the process by which the haploid male and female gametes unite and become a diploid zygote. Several important cellular and molecular processes must occur before the nuclei of the gametes can fuse. The mechanism by which the egg and sperm make contact has been studied extensively in sea urchins, and evidence suggests that a similar process occurs in humans and other mammals. Sea urchin eggs emit chemical attractants that bind to nearby sperm. This increases cellular respiration (that is, the breakdown of nutrients to synthesize ATP) within the sperm, which helps to increase sperm motility. The sperm then swim toward the egg by following the concentration gradient of the attractants.

Biology Principle

Structure Determines Function

The structure of a sperm cell is a good example of this principle of biology at the cellular level. Notice how each of the three major structural elements of the sperm cell is suited for a particular function. When all three elements function together, the cell is capable of fertilizing an egg.

(a) Spermatogenesis

(c) Oogenesis

(b) Mature human sperm

(d) Mature human follicle and oocyte

Figure 39.3 Gametogenesis and gametes in males and females. (a) In the process of spermatogenesis, male diploid (2n) germ cells undergo two meiotic divisions to produce mature haploid (n) sperm. (b) The characteristic head, midpiece, and flagellum (tail) of a mature human sperm, as seen in a drawing and accompanying scanning electron micrograph (SEM). (c) The process of oogenesis in females, which produces a haploid secondary oocyte that enters but does not complete meiosis II until it is fertilized. (d) Mature follicle and oocyte. The drawing depicts a secondary oocyte within its follicle; the SEM shows an isolated human oocyte covered by its zona pellucida and remnants of the cumulus mass.

(b) ©Eye of Science/Science Source; (d) ©P.M. Motta & G. Familiari/Univ. "La Sapienza"/Science Source

Biology Principle

Biologists Use Models to Test Predictions and Convey Ideas

This figure is a model of fertilization. The model depicts how the nucleus of a sperm (with its DNA) enters an egg. The Modeling Challenge in the figure legend asks you to predict how the genetic material in the nucleus of the resulting zygote will appear, using the conventions for chromosomes depicted earlier in Figure 39.3.

(a) Acrosomal reaction

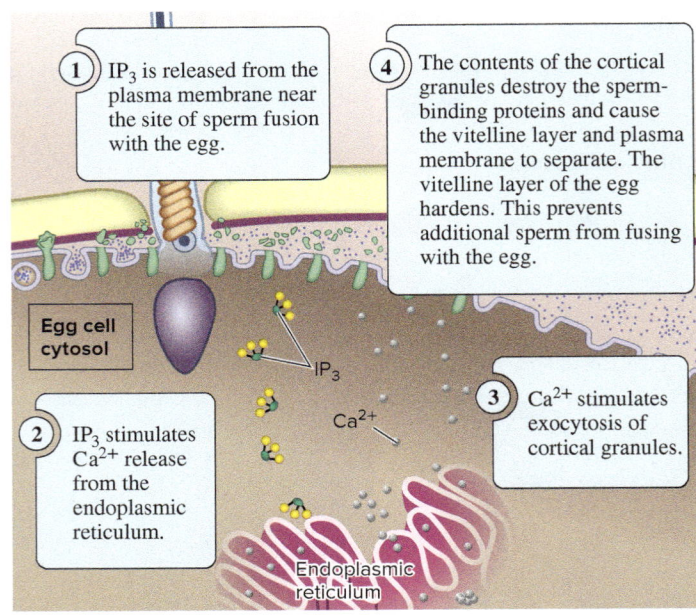

(b) Cortical reaction

Figure 39.4 The events during fertilization of an egg by a sperm. **(a)** Acrosomal reaction. The contact of a sperm with an egg initiates a series of events that permits the head of the sperm to bind to the plasma membrane of the egg. **(b)** Cortical reaction. Sperm fusion leads to an increased concentration of cytosolic Ca^{2+} that ultimately causes the vitelline layer of the egg to harden, preventing additional sperm from fusing.

Modeling Challenge: Figure 39.4 depicts how the nucleus of a sperm enters the cytosol of an egg. After the cortical reaction, the sperm and egg nuclei will eventually fuse, forming a zygote. Draw two models that depict the chromosomes before and after this nuclear fusion event. In both models, the chromosomes should be shown in red and blue, as in Figure 39.3. Assume that each gamete has only two chromosomes, and distinguish between them by drawing one longer than the other. In the first model, show an egg cell with its nucleus containing (red) chromosomes, and a sperm nucleus with (blue) chromosomes in the egg cytosol, but not yet fused with the nucleus of the egg. In the second model, depict a zygote after fusion of the two nuclei. For simplicity, omit all other intracellular and membrane features. If necessary, refer back to Figures 14.10 and 14.11 for help.

For a sperm to physically contact an egg, it must first penetrate the layers surrounding the egg's plasma membrane (**Figure 39.4**). In sea urchins, as in many animals, the sperm must penetrate a jelly-like layer consisting of glycoproteins and polysaccharides before contacting the plasma membrane of the egg. The sperm is able to do this because of the **acrosomal reaction,** in which hydrolytic enzymes are released from the acrosome onto the jelly coat of the egg (Figure 39.4a). These enzymes dissolve a localized region of the jelly coat, allowing the sperm head to bind to proteins in the egg's plasma membrane. Binding is followed by fusion of the sperm head membrane with the egg membrane, and shortly thereafter by penetration of the sperm head and release of its nucleus into the egg.

The acrosomal reaction is followed by the **cortical reaction** (Figure 39.4b). Normally, the cytosolic Ca^{2+} concentration in eggs, as in most cells, is kept low by several mechanisms, including one in which Ca^{2+} is transported by an ATP-dependent pump out of the cytosol and into the endoplasmic reticulum. When the sperm binds to the egg, a small signaling molecule called inositol trisphosphate (IP_3) is released from the region of the plasma membrane nearest to the sperm entry point. IP_3 then binds to nearby sites on the endoplasmic reticulum (ER) and opens Ca^{2+} channels. Within seconds after a sperm cell binds to an egg, Ca^{2+} is released from the lumen of the ER and into the cytosol.

The release of Ca^{2+} in the cortical reaction has several important effects. First, it causes exocytosis of membrane-bound vesicles called cortical granules in the egg's cytosol. The granules release enzymes and other substances that inactivate the sperm-binding proteins on the plasma membrane. In addition, the outer coating of the egg cell, known as the vitelline layer in sea urchins or the zona pellucida in vertebrates (see Figure 39.3d), becomes hardened and begins to separate from the plasma membrane. These events create a barrier to more sperm fusing with the egg. Additionally, the burst of cytosolic Ca^{2+} leads to the activation of molecular signaling pathways that initiate the first cell cycle and triggers an increase in protein synthesis and metabolism within the egg cell.

Shortly afterward, the nucleus of the sperm fuses with the nucleus of the egg, creating a diploid zygote. The first cell division of the zygote occurs approximately 90 minutes after fertilization in sea urchins and amphibians, but it can take up to 24 hours in mammals.

In a Given Species, Fertilization Occurs Either Outside or Inside the Female

For sperm to fertilize eggs, the two gametes must physically come into contact. This can occur either outside or inside the female's body. When this occurs outside of the female, the process is called **external fertilization.** This type of fertilization occurs in aquatic environments, when eggs and sperm are released into the water close enough for fertilization to occur. The aqueous environment protects the gametes from drying out. Animals that reproduce by external fertilization show species-specific behaviors that help to bring the eggs and sperm together. For instance, very soon after a female fish lays her eggs, a male deposits his sperm cells in the water nearby, where they spread over the clump of eggs. The fertilized eggs then develop outside the mother's body.

Although the aqueous environment protects against desiccation, eggs can be eaten by predators, washed away by currents, or subjected to potentially lethal changes in water temperature or other variables such as pH or oxygen. Such environmental challenges have provided selection pressure for some species, including many aquatic or amphibious animals, to release very large numbers of eggs and sperm at once—as many as several million in some species, such as the oysters in the chapter-opening photo.

In contrast to external fertilization, most terrestrial animals and some aquatic animals reproduce by **internal fertilization,** in which sperm are deposited within the reproductive tract of the female during the act called copulation. Internal fertilization protects the gametes and zygotes from dessication, environmental hazards, and predation and guarantees that sperm are placed and remain in very close proximity to eggs. Once fertilization occurs within the female, the zygotes then develop into offspring.

The behaviors and anatomical structures involved in achieving internal fertilization are extremely varied among species. Typically, mating involves accessory sex organs, which are reproductive structures other than the gonads. The external accessory sex organs involved in copulation include certain types of genitalia (for example, the **penis** and the **vagina**), which are used to physically join the male and female so that sperm can be deposited directly into the female's reproductive tract. A penis or an analogous structure is present in most insects, reptiles, some species of birds (ratites and waterfowl), and all mammals. However, males of other vertebrate species—including most birds—that reproduce by internal fertilization lack a structure that can be inserted into the female, instead depositing sperm in the female by cloacal contact. The cloaca is a common opening for the reproductive, digestive, and excretory systems in these animals.

39.2 Reviewing the Concepts

- Male and female gametes are formed within the gonads—the testes in males and the ovaries in females. Within the ovaries, each oocyte undergoes growth and development within a follicle before it leaves the ovary in a process called ovulation (Figure 39.3).
- Major events in fertilization include the acrosomal and cortical reactions (Figure 39.4).
- In external fertilization, sperm and eggs are released into an aquatic environment, where they unite. In internal fertilization, sperm are deposited within the reproductive tract of the female during copulation or by cloacal contact.

39.2 Testing Your Knowledge

1. In _____, gametes are released in large numbers in a watery environment.
   a. copulation
   b. internal fertilization
   c. external fertilization
   d. the cortical reaction
   e. spermatogenesis

2. Meiosis I produces haploid cells. These are _____ in males and _____ in females.
   a. sperm, oogonia
   b. primary spermatocytes, primary oocytes
   c. spermatogonia, primary oocytes
   d. secondary spermatocytes, secondary oocytes
   e. spermatids, zygotes

39.3 Human Reproductive Structure and Function

Learning Outcomes

1. Describe the structures of the human male reproductive tract.
2. Outline the process of hormonal control of the male reproductive system.
3. Describe the structures of the human female reproductive tract.
4. Diagram the events of the ovarian cycle.
5. Outline the process of hormonal control of the reproductive cycle of the human female.
6. Explain how maternal hormones prepare the uterus to accept the embryo.

We will turn now to a detailed look at the human reproductive system. For both sexes, we will begin with a description of the anatomy of the reproductive system, including the gonads and the accessory sex structures. We will then examine the hormones that control the production of the gametes and the preparation for and establishment of pregnancy.

The Human Male Reproductive Tract Is Specialized for Production and Ejaculation of Sperm

The external structures of the male reproductive tract—the genitalia—consist of the penis and the scrotum, the sac that contains the testes and holds them outside the body cavity, where the temperature is better suited for spermatogenesis (Figure 39.5).

Each testis is composed of tightly packed **seminiferous tubules** encased in connective tissue. Surrounding the tubules are Leydig cells—scattered endocrine cells that secrete the steroid hormone testosterone. Spermatogenesis begins at puberty and continues throughout life. It occurs all along the walls of the seminiferous tubules. Cells at the earliest stages of spermatogenesis—the spermatogonia—are located nearest the outer surface of the wall. Cells of more advanced stages are located progressively inward, such that the sperm are released directly into the tubule lumen. Cells within the seminiferous tubules are continuously developing from spermatogonia into spermatocytes and eventually to sperm, so at any one time, all types of cells are present along the seminiferous tubule. Support cells, called Sertoli cells, surround the developing spermatogonia and spermatocytes, providing them with nutrients and protection and playing a role in facilitating their maturation into sperm.

Sperm moving out of the seminiferous tubules empty into the epididymis, a coiled, tubular structure located on the surface of the

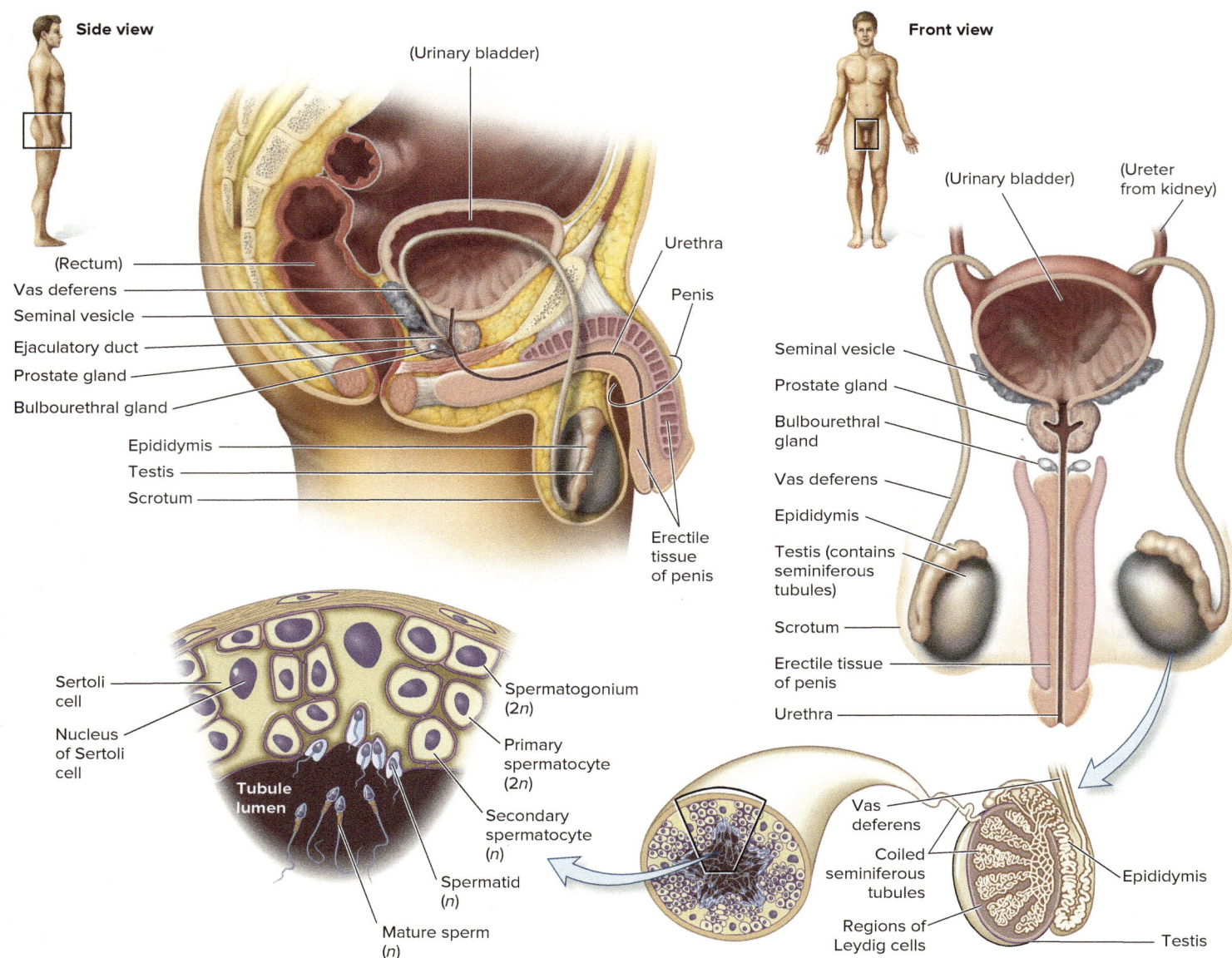

Figure 39.5 Male reproductive structures in humans. Side and front views of the male reproductive system (nonreproductive structures are identified in parentheses for orientation purposes). The enlargement (at bottom) shows the internal structure of a testis and associated structures and indicates the stages of spermatogenesis.

testis (Figure 39.5). Here the sperm complete their differentiation by becoming motile and gaining the capacity to fertilize an egg.

Sperm leave the epididymis through the vas deferens, a muscular tube leading to the ejaculatory duct, which then connects to the urethra (Figure 39.5). The urethra originates at the bladder and extends to the end of the penis. In males, the urethra not only conducts urine but also carries **semen,** a mixture containing fluid and sperm, that is released during **ejaculation**—the movement of semen through the urethra by contraction of muscles at the base of the penis.

The liquid components of semen are important for the survival and movement of sperm through the female reproductive tract. This liquid is formed by three accessory glands—the seminal vesicles, the prostate gland, and the bulbourethral glands—which secrete substances into the urethra to mix with the sperm. The seminal vesicles secrete the monosaccharide fructose, the main nutrient for sperm, as well as other factors that enhance sperm motility and survival. The prostate gland and bulbourethral glands secrete a thin, alkaline fluid that protects sperm from acidic fluids in the urethra and within the female reproductive tract.

Introduction of sperm into the female reproductive system during copulation is made possible by erection of the penis. Erection occurs when blood fills spongy erectile tissue located along the length of the penis (Figure 39.5). Sexual arousal stimulates release of the gaseous neurotransmitter nitric oxide (NO) in the penis, which causes local arteries and arterioles to vasodilate (widen). The pressure of the blood flowing into the widened arteries and arterioles constricts nearby veins, causing a reduction in venous drainage from the penis, engorging it with blood. After ejaculation, NO release is decreased, causing a reversal of the vascular changes responsible for erection.

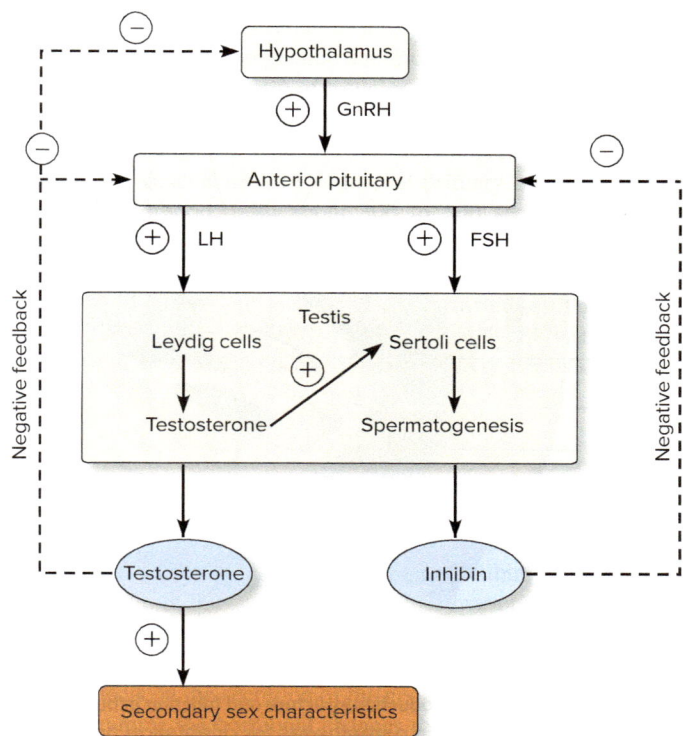

Figure 39.6 The hormonal control of male reproduction. In response to LH, Leydig cells in the testes secrete testosterone, which, along with FSH, acts on Sertoli cells to facilitate spermatogenesis. Negative signs indicate inhibitory effects via negative feedback, and plus signs indicate stimulatory effects.

Male Reproductive Function Requires the Actions of Hormones

Recall from Chapter 38 that the hypothalamus is a structure on the ventral surface of the brain that synthesizes neurohormones, including gonadotropin-releasing hormone (GnRH) (refer back to Table 38.1). GnRH stimulates the anterior pituitary gland to release two gonadotropins: luteinizing hormone (LH) and follicle-stimulating hormone (FSH).

In males, LH stimulates the Leydig cells of the testes to produce androgens, particularly testosterone (**Figure 39.6**). Testosterone stimulates the growth of the male reproductive tract and the genitalia during development and puberty, as well as the development of male secondary sex characteristics. In humans, these include facial hair growth, increased muscle mass, and deepening of the voice.

The other pituitary gonadotropin, FSH, functions along with testosterone to stimulate spermatogenesis. FSH does this by stimulating the activity of the Sertoli cells within the seminiferous tubules (see Figures 39.5 and 39.6). As stated earlier, Sertoli cells provide the nutritional and structural support necessary for development of the sperm. They also respond to testosterone produced by the Leydig cells, by stimulating mitosis and meiosis of the germ cells associated with them in the tubules.

Production of sperm and testosterone is kept in check by negative feedback mechanisms that control the amount of gonadotropins produced (Figure 39.6). Testosterone at high concentrations inhibits the secretion of GnRH from the hypothalamus, so both LH and FSH are inhibited when the blood concentration of testosterone is high. Testosterone also directly inhibits LH secretion by the anterior pituitary gland. In addition, when Sertoli cells are activated by FSH, they secrete a protein hormone called inhibin, which enters the blood and inhibits further secretion of FSH. These feedback mechanisms maintain homeostatic concentrations of FSH and LH in the blood.

Before puberty, LH is not released in sufficient amounts to stimulate significant testicular production of testosterone, and the reproductive system is quiescent—spermatogenesis does not occur. Although the mechanisms that initiate puberty in mammals are still not completely understood, research has shown that increased GnRH production at that time initiates increased LH and FSH secretion from the pituitary. The testosterone induced by LH stimulates the development of adult male characteristics. Testosterone is also responsible for an increased sex drive (libido) at this time.

The Human Female Reproductive Tract Is Specialized for Production and Fertilization of the Egg and Development of the Embryo

The female genitalia are composed of the larger, hair-covered outer folds called the labia majora and smaller, inner folds, called the labia minora, which surround the external opening of the reproductive tract (**Figure 39.7**). At the anterior part of the labia minora is the clitoris,

Neurulation Involves Formation of the Central Nervous System and Segmentation of the Body

By studying development in several different vertebrate species, researchers are beginning to understand some of the fundamental steps in the formation of the central nervous system (CNS)—the brain and spinal cord—in vertebrates. The multistep embryological process responsible for initiating CNS formation is called **neurulation** (Figure 39.19). Neurulation occurs just after gastrulation and involves formation of the **neural tube** from ectoderm located dorsal to the notochord. All neurons and their supporting cells in the CNS originate from neural precursor cells derived from the neural tube.

Neural Tube Formation Neurulation in vertebrates occurs in several major steps, as shown in Figure 39.19:

1. In the first step, ectoderm overlaying the notochord thickens by the elongation of cells in the dorsal region to form the neural plate, with adjacent regions that will eventually form the epidermis and a structure called the neural crest (discussed shortly). The neural plate then elongates, resulting in the formation of a single, dorsal, elongated epithelial cell layer that is aligned with the animal's anteroposterior axis.

2. Next, a column of cells along the midline of the neural plate—the medial hinge point—initiates folding of the neural plate, leading to the formation of the neural groove.

3. After folding, in the next step, columns of cells in the dorsal lateral hinge points undergo morphological changes that lead to convergence of the two sides of the neural groove and generation of a tubelike structure that is not yet sealed on the dorsal surface.

4. In the next step of neurulation, called fusion (Figure 39.19, step 4), the dorsal-most cells on either side of the neural tube are released from adjacent ectoderm and make contact with each other, thereby closing the neural tube. At the same time, ectoderm on either side of the neural tube moves toward the centerline, then up and over the neural tube, where it forms the dorsal epidermis of the embryo.

Neural Crest Formation Another important group of cells that arises during neurulation is the **neural crest,** which is unique to vertebrates. It consists of cells that originate from the ectoderm overlaying

Figure 39.19 Neurulation and the beginning of neural crest formation in vertebrates. The major steps of neurulation are (1) thickening and elongation; (2) folding, which creates the neural groove; (3) convergence, in which the neural tube begins to take shape; and (4) fusion, in which the neural tube is completed. In a later event, cells migrate away from the neural crest to form several other structures, including the neurons of the peripheral nervous system.

(1–4) Courtesy Kathryn Tosney/University of Miami

the dorsal surface of the newly formed neural tube and that migrate to other regions of the embryo (Figure 39.19, step 4). Once these cells reach their final destination in the embryo, they differentiate into a variety of cell types different from those that arise from the rest of the ectoderm. All neurons and supporting cells of the peripheral nervous system in vertebrates are derived from neural crest cells. In addition, the neural crest gives rise to skeletal and cartilaginous structures in the head and face, melanocytes (specialized cells that provide pigmentation to the skin of vertebrates), the medulla of the adrenal glands, and connective tissue in numerous organs, notably the heart.

Organogenesis Is the Process of Organ Formation

As described in Chapter 32, organs are specialized structures that consist of arrangements of two or more tissue types. Most organs, such as the kidney, contain all four tissue types: nervous, muscle, epithelial, and connective tissue (refer back to Figure 32.1). The developmental event in which cells and tissues form organs is called **organogenesis.** Each germ layer gives rise to particular types of cells found within different organs (see Figure 39.17).

Many organs begin to form during or just after neurulation. However, these organs become functional at different times during development. For example, the heart is the first functional organ to form in the vertebrate embryo. It begins to beat and pump blood by 2.5 days after fertilization in chicks, 9 days in mice, and about 22 days in humans. By contrast, the lungs of mammals do not acquire the ability to function until shortly before birth. As we saw in Chapter 32, the development of different organs in animals is controlled by genes in the embryo—notably, the *Hox* genes. *Hox* genes are important for establishing structures along the anteroposterior axis. Many of the genes controlling the processes of gastrulation, neurulation, and organogenesis encode secreted proteins or growth factors that induce cells in their local vicinity to differentiate along a specific developmental pathway. For example, the notochord produces many signaling proteins that help to establish tissue patterns in the embryo. Proteins produced within it induce segment-specific expression of the *Hox* genes in subsequent stages of development.

39.5 Reviewing the Concepts

- The process by which a fertilized egg is transformed into an organism with distinct physiological systems and body parts is called embryonic development. The process by which different cells within a developing organism acquire specialized forms and functions, due to the expression of cell-specific genes, is called cellular differentiation.
- Development in many animals, including vertebrates, involves five general events: fertilization, cleavage, gastrulation, neurulation, and organogenesis (Figure 39.13).
- Cleavage involves cell divisions without cell growth and results in daughter cells called blastomeres. Cleavage-stage embryos in triploblasts have animal and vegetal hemispheres. The hemispheres determine, in part, the future anteroposterior and dorsoventral axes of the embryo (Figure 39.14).
- Incomplete, or meroblastic, cleavage occurs in vertebrates whose eggs contain large amounts of yolk. Complete, or holoblastic, cleavage occurs in animals whose eggs have smaller amounts of yolk. In mammals, cleavage occurs in the oviduct and the embryo implants in the uterine wall (Figures 39.15, 39.16).
- During gastrulation, the hollow ball of cells that makes up a blastula or blastocyst develops into a gastrula, containing the three germ layers: endoderm, mesoderm, and ectoderm. Each germ layer gives rise to specific structures (Figures 39.17, 39.18).
- Neurulation is the multistep embryological process responsible for initiating formation of the nervous system. Neurulation involves formation of the neural tube and neural crest (Figure 39.19).
- Organogenesis, the developmental event during which cells and tissues form organs, begins during or just after neurulation; however, organs become functional at different times during development.

39.5 Testing Your Knowledge

1. The three germ layers of triploblasts are established during
   a. fertilization.
   b. cleavage.
   c. blastula formation.
   d. gastrulation.
   e. neurulation.

2. Choose the correct sequence of events.
   a. cleavage → gastrulation → organogenesis → neurulation
   b. neurulation → cleavage → organogenesis → gastrulation
   c. cleavage → gastrulation → neurulation → organogenesis
   d. organogenesis → neurulation → gastrulation → cleavage

39.6 Impact on Public Health

Learning Outcomes

1. Explain the most common causes of human infertility.
2. Compare the different types of birth control, and explain their mechanisms of pregnancy prevention.

Approximately 5–10% of individuals of reproductive age in the U.S. are infertile; that is, they cannot reproduce. In men and women alike, fertility can be decreased by a variety of factors. In this section, we will discuss some of the common causes of **infertility**—the inability of a man to produce sufficient numbers or quality of sperm to impregnate a woman, or the inability of a woman to become pregnant or maintain a pregnancy. We will then conclude by examining the methods in use today to prevent pregnancy.

Infertility May Result from Disease, Developmental Disorders, Inadequate Nutrition, and Stress

As many as 75% of infertility cases have some identifiable cause, ranging from disease to toxin exposure.

Disease Primary among the diseases that affect fertility are sexually transmitted diseases (STDs). For example, some STDs may cause blockage in the ducts of the testes, thus preventing normal sperm transport, or they can cause permanent damage to the oviducts, uterus, and surrounding tissues. The Centers for Disease Control and Prevention (CDC) put the number of reported cases of STDs in the U.S. at about 2.3 million in 2017. Most of these cases are treatable if caught early. In addition, however, the CDC estimates that more than 20,000 women become infertile each year due to undiagnosed STD infection. In most cases, the site of irreversible damage is localized to the oviducts.

Developmental Disorders Developmental disorders are conditions that either are present at birth or arise during childhood and adolescence. In some developmental disorders that affect fertility, inherited mutations of genes that encode enzymes involved in the biosynthesis of reproductive hormones cause abnormal expression of those genes. The result is either too much or too little of one or more of these hormones, notably estradiol or testosterone. Other developmental disorders that compromise fertility include malformations of the cervix or oviducts.

Inadequate Nutrition and Other Stressors Adequate nutrition is required for normal growth and development of all parts of the body, including the reproductive system. Because the reproductive system is not essential for an individual's survival, it often becomes inactive when nutrients are chronically scarce—such as during starvation—and temporary infertility results. In this way, precious stores of energy in the body are preserved for vital functions, such as those of the brain and heart.

Malnutrition can also delay sexual development and puberty during late childhood and early adolescence. Undernourished children may begin puberty several years later than normal. The brains of all mammals, including humans, contain a center that monitors the body's fat stores. One of the triggers that initiate puberty may be a signal—such as the hormone leptin—from adipose tissue to the brain (see Chapter 38). Very low fat stores in undernourished girls, for example, result in decreased leptin in the blood; this low level of leptin signals the brain that the body does not contain sufficient energy to support the energetic demands of pregnancy, and thus puberty is delayed.

Starvation or poor nutrition is considered a type of stress, defined in biology as any real or perceived threat to an animal's homeostasis. Physical and psychological forms of stress can and do affect fertility in humans. Many immediately nonessential functions, including the maintenance of menstrual cycles in women, can be suppressed by chronic stress. The reproductive consequences of stress are much greater in females than in males, likely because only females bear the energetic cost of pregnancy. Interestingly, from a reproduction viewpoint, the human body responds to long-term strenuous physical exercise in the same way that it responds to long-term stress. This is why many young ballerinas and gymnasts experience delayed puberty and why female long-distance runners (like the one in the chapter introduction) may have menstrual cycles that are abnormal or absent (secondary amenorrhea).

Other Causes When the causes of infertility cannot be determined, a variety of factors come under suspicion. Among these possible causes are the ingestion of toxins (for example, certain heavy metals such as cadmium and lead), tobacco smoking, marijuana use, and injuries to the gonads.

Treatments Among several currently available treatments for increasing the likelihood of pregnancy in infertile couples are hormone therapy for the woman to increase egg production and a collection of procedures known as **assisted reproductive technologies (ART)**. In the most common ART procedure, in vitro fertilization, sperm and eggs collected from a man and a woman are placed together in culture dishes. Once the sperm have fertilized the eggs and the resulting zygotes have undergone several cell divisions, one or more of the embryos are inserted into a woman's uterus with the goal that one will implant. When this procedure was first used in 1978, the children born as a result came to be known as "test-tube babies." Since then, several million children have been born using this technology.

Contraception Usually Prevents Pregnancy

The use of methods to prevent fertilization or the implantation of a fertilized egg is termed **contraception.** Methods of contraception can be either permanent or temporary.

Permanent Methods The permanent forms of contraception surgically prevent the transport of gametes through the reproductive tract (**Figure 39.20a**). **Vasectomy** is a surgical procedure in men that severs the vas deferens, thereby preventing the release of sperm at ejaculation (however, semen is still released). In women, **tubal ligation** involves the cutting and sealing of the oviducts. This procedure prevents the movement of the egg from the oviduct into the uterus. These procedures are considered permanent, because it is difficult—sometimes impossible—to reverse the surgery.

Temporary Methods Temporary methods of preventing fertilization include barrier methods, which prevent sperm from reaching an egg (**Figure 39.20b**). Barrier methods include **vaginal diaphragms,** which are placed in the upper part of the vagina just prior to intercourse and block the movement of sperm to the cervix, and **condoms,** which are sheathlike membranes worn over the penis that collect the ejaculate. In addition to their contraceptive function, condoms significantly reduce the risk of STDs such as HIV infection, syphilis, gonorrhea, chlamydia, and herpes. Other types of contraception do not reduce this risk.

Another temporary form of contraception involves synthetic hormones. Oral contraceptives (birth control pills) are synthetic forms of estradiol and progesterone, which are taken by mouth to prevent ovulation in women by inhibiting pituitary LH and FSH release (recall the feedback actions of estradiol and progesterone). The hormones in these pills also affect the composition of cervical mucus such that sperm cannot easily pass through it into the uterus. In addition to the oral route, hormones can be administered by injections, skin patches, and vaginal rings.

Another temporary method of contraception involves placement in the uterus of an **intrauterine device (IUD),** a small object that prevents fertilization in part by inhibiting sperm movement and survival in the uterus. Some IUDs may also induce local inflammation in the endometrium and make it less likely that a blastocyst could implant there. Although not as widely used as other contraceptives in the U.S., IUDs are the most commonly used means of contraception by women worldwide because of their effectiveness and simplicity of use.

In addition to the contraceptive methods used before or during intercourse, any of several emergency contraception pills can be taken by a woman within 72 hours of intercourse. These pills include a high dose of progestin (a synthetic form of progesterone), a high dose of combined estrogen and progestin, and a dose of ulipristal acetate, a

THE PRODUCTION OF OFFSPRING: REPRODUCTION AND DEVELOPMENT 893

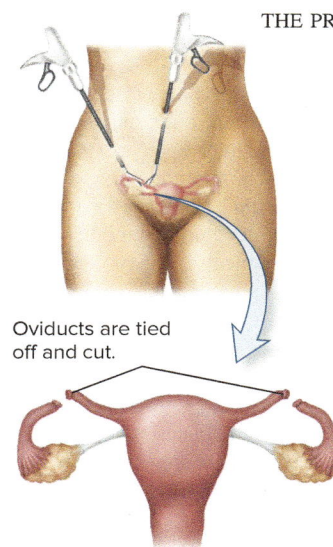

Vasectomy (<1.0%)   **Tubal ligation (<1.0%)**
(a) Permanent methods

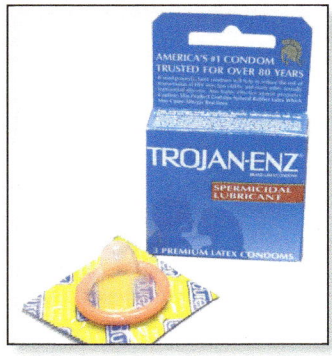

Diaphragm (5–20%)   **Condoms (male) (2–15%)**

Oral contraceptive (1–2%)   **Intrauterine device (IUD) (1–2%)**
(b) Temporary methods

Figure 39.20 Examples of contraceptive methods. These methods may be used by men or women to **(a)** permanently or **(b)** temporarily prevent pregnancy. The estimated first-year failure rates for each method are given in parentheses (collected from data published by the U.S. Food and Drug Administration and other organizations). A failure rate of 10% means that 10 of every 100 women using that method of contraception will become pregnant in the first year of use. The large failure rate for the use of condoms and diaphragms is due to improper use of these devices by many people. Female condoms are also available and have a failure rate of approximately 20%.
(*b diaphragm, condoms, oral contraceptive*) ©McGraw-Hill Education/Jill Braaten

substance that acts on the progesterone receptor. These hormones and chemicals prevent ovulation and impede the ability of sperm to reach and fertilize an egg.

Used prior to the advent of modern contraception, and still in use today by individuals who prefer not to use contraceptives, is the rhythm method, which involves abstaining from sexual intercourse near the time of ovulation. Its main drawback is the difficulty in precisely pinpointing the time of ovulation, which can occur at any time within a roughly 2-week window of the 28-day cycle. Predicting the time of ovulation is also difficult in that several of the detectable changes characteristic of the midpoint of the ovarian cycle—including a small rise in body temperature and changes in the characteristics of the cervical mucus—occur only after ovulation. These drawbacks explain why the rhythm method has a relatively high failure rate.

39.6 Reviewing the Concepts

- Infertility is the inability of a man to produce sufficient numbers or quality of sperm to impregnate a woman, or the inability of a woman to become pregnant or to maintain a pregnancy. A primary cause of infertility is STDs.
- The use of procedures to prevent fertilization or the implantation of a fertilized egg is termed contraception. Methods of contraception include vasectomy and tubal ligation, vaginal diaphragms, condoms, oral contraceptives, and IUDs (Figure 39.20).

39.6 Testing Your Knowledge

1. Which of the following statements is true about the use of contraceptives?
   a. They all greatly reduce the likelihood of spreading or contracting an STD.
   b. They include barrier methods, which prevent sperm from reaching an egg.
   c. There are surgical methods available for men but not for women.
   d. The most common form of contraception worldwide is the vaginal diaphragm.
   e. They are all approximately 100% effective.

Assessing Your Knowledge and Skills

Foundational Knowledge

Multiple Choice

1. Which of the following is an advantage of sexual reproduction?
   a. necessity to locate a mate
   b. increased energy expenditure in producing large numbers of gametes that may not be used in reproduction
   c. increased genetic variation
   d. decreased genetic variation
   e. increased adaptability to relatively nonchanging, stable environments

2. Cells of the neural crest
   a. give rise to the central nervous system.
   b. originate from ectoderm.
   c. migrate to different areas of the body and differentiate into a variety of cells, including neurons of the peripheral nervous system.
   d. do all of the above.
   e. do b and c only.
3. Compared with external fertilization, in internal fertilization,
   a. male gametes have a higher chance of coming into close proximity to female gametes.
   b. gametes are less protected against predation or other harmful environmental factors.
   c. gametes typically become dessicated.
   d. gametes come into contact only outside the mother's reproductive tract.
   e. mating must occur in watery environments.
4. During the human ovarian cycle, ovulation is stimulated by
   a. a decrease in FSH secretion.
   b. an increase in progesterone secretion.
   c. an increase in LH secretion.
   d. the presence of semen in the vagina.
   e. a decrease in estradiol concentration in the bloodstream.
5. A major function of FSH is to
   a. stimulate the development of the gonads during early development.
   b. stimulate spermatogenesis in males and oocyte maturation in females.
   c. increase the secretion of testosterone by the testes.
   d. regulate the secretion of the bulbourethral glands.
   e. inhibit the activity of Sertoli cells in the testes.

Short Answer

1. In what ways does the hypothalamus control vertebrate reproduction?
2. Describe the major events in animal development, beginning with cleavage.

Critical-Thinking Skills

Multiple Choice

1. Although gametogenesis differs in several important ways in males and females, in what way is it similar in both sexes?
   a. The primary spermatocyte and primary oocyte both eventually produce four gametes.
   b. The formation of gametes in both sexes involves mitosis and meiosis.
   c. The gametes in both sexes have a flagellum.
   d. The gametes that are produced are diploid.
   e. In both sexes, meiosis is not completed until fertilization occurs.

2. A relatively common complication of pregnancy in older horses and some other mammals is prolonged retention of the placenta and fluids in the uterus after birth has occurred. This can lead to discomfort and infection if not treated. Which procedure do you think would be suitable for a veterinarian to employ in such a situation?
   a. injecting the mother with oxytocin after birth
   b. injecting the mother with drugs that inhibit the synthesis of prostaglandins at the onset of labor
   c. treating the mother with a drug that blocks estrogen receptors for a week or so prior to the expected due date
   d. injecting the mother with inhibitors of both estrogen and prostaglandins after birth
   e. administering oxytocin receptor blockers to the mother before and after birth
3. The common forms of birth control pills in humans typically contain estrogen and progesterone. They act in women to suppress ovarian function. Theoretically, which of the following could be used as a means of birth control in *either* sex?
   a. a drug that prevents GnRH from binding to its receptors
   b. a drug that stimulates FSH and LH such that they are always at high concentrations in the blood
   c. a drug that permanently shuts off secretion of inhibin
   d. a drug that only blocks the LH surge but otherwise does not alter LH or FSH secretion
   e. a drug that mimics the actions of FSH
4. A principle of biology is that living organisms maintain homeostasis. Which of the following does *not* conform to that principle?
   a. the production of sperm in male animals
   b. the negative feedback effect of inhibin on gonadotropins
   c. the estrogen-dependent LH surge in a female mammal
   d. the gradual growth of an egg within an ovary
   e. the suppression of fertility in a female during times of undernourishment
5. A young woman has not menstruated for 3 months. She confirms with a pregnancy test that she is not pregnant. Which of the following might account for the woman's amenorrhea?
   a. She is experiencing chronic and severe emotional difficulties in a relationship.
   b. She has an eating disorder that greatly restricts her food intake.
   c. She has recently embarked on a very strenuous exercise program to train for a marathon.
   d. She lives close to a factory that makes batteries and other metal-containing products.
   e. All of the above are correct.

Short Answer

1. What disadvantages are associated with external fertilization, and what is one major way in which some animals overcome those disadvantages?
2. **PRINCIPLES** A principle of biology is that living organisms use energy. What are some of the energy costs associated with sexual reproduction? What outweighs those costs and accounts for the observation that most animals reproduce sexually?

Immune Systems: How Animals Defend Against Pathogens and Other Dangers

40

©SPL/Science Source

In an immune system response, a macrophage engulfs numerous rod-shaped bacteria. (Image is a false-color SEM.)

Chapter Outline

40.1 Types of Pathogens
40.2 Innate Immunity
40.3 Adaptive Immunity in Vertebrates
40.4 Impact on Public Health
Assessing Your Knowledge and Skills

A 50-year-old man in the northwestern U.S. was bitten while trying to remove a dead rodent from the mouth of a stray cat that his family had recently taken in and cared for. Within a few days, the man developed a fever, aches and pains, bleeding sores in his mouth and nose, and patches of dying, blackened skin. After he was admitted to a hospital, tests revealed the presence of the infectious bacterium *Yersinia pestis*, the organism that causes bubonic plague. This disease has caused upwards of 100 million deaths since the 14th century, and 1,000–3,000 cases still occur worldwide each year. The bacterium typically enters through the skin, normally an animal's first line of defense against infectious agents. In humans, it then enters various immune structures, where it multiplies rapidly and causes severe inflammation. Treatment of the disease is with a class of drugs called antibiotics, which kill bacteria. Such treatment is successful roughly 85–90% of the time, but prior to the advent of antibiotics, as many as 90% of individuals died when infected with *Y. pestis*.

All animals must defend themselves against various threats to their internal environment, including the invasion of potentially harmful microorganisms such as bacteria, the presence of foreign molecules such as the products of microorganisms, and the presence of abnormal cells such as cancer. The ability of an animal to ward off these threats—an animal's **immunity**, or immune defenses—is the subject of this chapter. The cells and organs within an animal's body that contribute to immune defenses collectively constitute an animal's **immune system**.

Immune defenses are usually considered as two types: innate and adaptive. **Innate immunity** refers to defenses that are present at birth and that act against any foreign material in much the same way regardless of the specific identity of the invader. Thus, innate immunity of animals is sometimes referred to as nonspecific immunity. An example of an innate immune response is seen in the chapter-opening photo, in which an immune cell known as a macrophage is engulfing numerous bacteria. All animals have innate immune defenses.

In contrast, **adaptive immunity** develops only after the body is exposed to foreign substances, which is why it is also sometimes called acquired immunity. This type of immunity is characterized by the ability of certain cells of the immune system to recognize a particular foreign substance and initiate a response that targets that specific substance. Another distinctive feature of adaptive immunity is that repeated exposure to a foreign substance elicits greater and greater defense responses. In contrast, in innate immunity, each exposure to the foreign material elicits the same magnitude of defense responses. Adaptive immunity has been identified in all vertebrates except for the jawless fishes and has not been unequivocally identified in invertebrates.

We will begin with a brief overview of the types of infectious agents that cause disease in animals. We will then consider the mechanisms that provide animals with innate and adaptive defenses. We will conclude with a discussion of the public health implications of some immunity-related conditions in humans.

40.1 Types of Pathogens

Learning Outcome

1. List the three main types of pathogens that elicit immune responses and how each damages its host organism.

All animals are susceptible to disease, often with catastrophic consequences. In recent years, outbreaks of diseases such as Ebola in primates, hoof-and-mouth disease in domestic farm animals, white nose syndrome in bats, West Nile virus in birds, and white plague disease in coral have devastated large populations of these animals worldwide. To prevent or minimize disease, an animal's immune defenses must protect against a variety of foreign materials, but most important among them are disease-causing **pathogens,** which include microorganisms and viruses. Both terrestrial and aquatic animals encounter each of the three major types of pathogens: certain bacteria, viruses, and eukaryotic parasites.

Bacteria Are Prokaryotes Responsible for Many Animal Diseases

Bacteria are single-celled prokaryotic organisms that lack a true nucleus (refer back to Figure 4.8). Bacteria can either damage tissues at an open wound site of an animal or release toxins that enter the blood and disrupt functions in other parts of the body. In humans, bacteria are responsible for many diseases and infections, including typhoid fever, strep throat, skin infections, middle ear infections, and food poisoning. The major ways in which bacteria gain entry into an animal's body are through direct bodily contact, open wounds, inhalation through the respiratory tract, and ingestion of food or water that has been contaminated by fecal matter. This last situation may arise because many infectious bacteria enter the intestines of animals and are excreted in feces, which may be deposited near food or water sources used by animals.

Viruses May Cause Illness or Cancer

All viruses contain nucleic acid (DNA or RNA) within a protein coat (see Figure 17.1). Unlike bacteria, viruses are not living organisms and lack the metabolic machinery to synthesize the proteins they require to replicate themselves. Instead, they infect a host cell and use its biochemical and genetic machinery, including nucleotides and energy sources, to make more viruses. The viral nucleic acid directs the host cell to synthesize the proteins required for viral replication.

After entering a cell, some viruses, such as those that cause the common cold, multiply rapidly, kill the cell, and then infect other cells. Other viruses can lie dormant within host cells before suddenly undergoing rapid replication, which causes cell damage or death. Finally, certain viruses can transform their host cells into cancerous cells. In humans, viruses are responsible for a great variety of illnesses, including some sexually transmitted diseases. Viruses typically enter an animal's body through the respiratory tract or through an open wound.

Eukaryotic Parasites Enter an Animal's Body in Several Ways and Cause Widespread Illness

Eukaryotic parasites—whether protists, fungi, or worms—damage a host by using the host's nutrients for their own growth and reproduction or by secreting toxic chemicals. In humans, parasites account for an enormous number of cases of disease annually. Parasitic infections may enter a host through the bite of an infected insect, as in malaria; by ingestion of food or water containing parasitic organisms, such as roundworms; or in some cases by penetration of the skin, as with blood flukes.

40.1 Reviewing the Concepts

- Three major types of pathogens elicit an immune response: bacteria, viruses, and eukaryotic parasites. Pathogens enter the body through direct contact, open wounds, ingestion, insect bites, and the respiratory tract.

40.1 Testing Your Knowledge

1. Which pathogens are *not* living organisms and require a host cell's genetic machinery to reproduce themselves?
   a. bacteria
   b. roundworms
   c. eukaryotic parasites
   d. viruses
   e. the pathogens that cause such illnesses as middle ear infections and strep throat

40.2 Innate Immunity

Learning Outcomes

1. Identify the general characteristics of innate defense mechanisms in animals.
2. Explain how an animal's body surface provides protection from pathogens.
3. Describe the process of phagocytosis, and explain its importance in innate immune responses.
4. List the cell types involved in innate immunity, and identify the functions of each.
5. Outline the sequence of events in the inflammatory response.
6. Describe the role of Toll proteins and Toll-like receptors in innate immunity.

Innate immune defenses protect against foreign cells or substances without having to recognize the invaders' specific identities. This type of defense mechanism is called innate because animals inherit the ability to perform these protective functions and because this type of immunity does not require prior exposure to pathogens. Instead of distinguishing among foreign materials, innate defenses recognize some general, conserved property marking the invader as foreign, such as a particular class of carbohydrate or lipid present in the cell walls of many different kinds of microbes.

In this section, we will first consider defenses at the body surfaces. We will then examine the actions of phagocytic cells, the response to injury known as inflammation, and various proteins secreted by cells of the innate immune system that facilitate the destruction of pathogens.

The Body Surface Is an Initial Line of Defense

Although not part of innate immunity, it is worth noting that an animal's initial defenses against pathogens are the cellular, anatomical, and chemical barriers provided by a surface exposed to the external environment. Very few microorganisms can penetrate the intact skin or body surface of most animals, particularly the tough, thick, or scaly skin characteristic of many vertebrates or the sturdy exoskeleton of many arthropods. In addition, glands in the body surfaces of many invertebrates and vertebrates secrete a variety of antimicrobial molecules, including mild acids and enzymes, such as lysozyme, that destroy bacterial cell walls.

The secretions from cells in the mucous membranes lining the respiratory and upper digestive tracts of many vertebrates also contain antimicrobial molecules. Equally important, mucus is sticky—pathogens that become stuck in it are prevented from penetrating the mucous membrane barrier and entering the internal environment of the body.

If, however, a pathogen is able to penetrate a barrier and gain entry into an animal's internal tissues and fluids, innate defense mechanisms are activated. These mechanisms are mediated by several types of cells that reside in the body fluids and tissues, as described next.

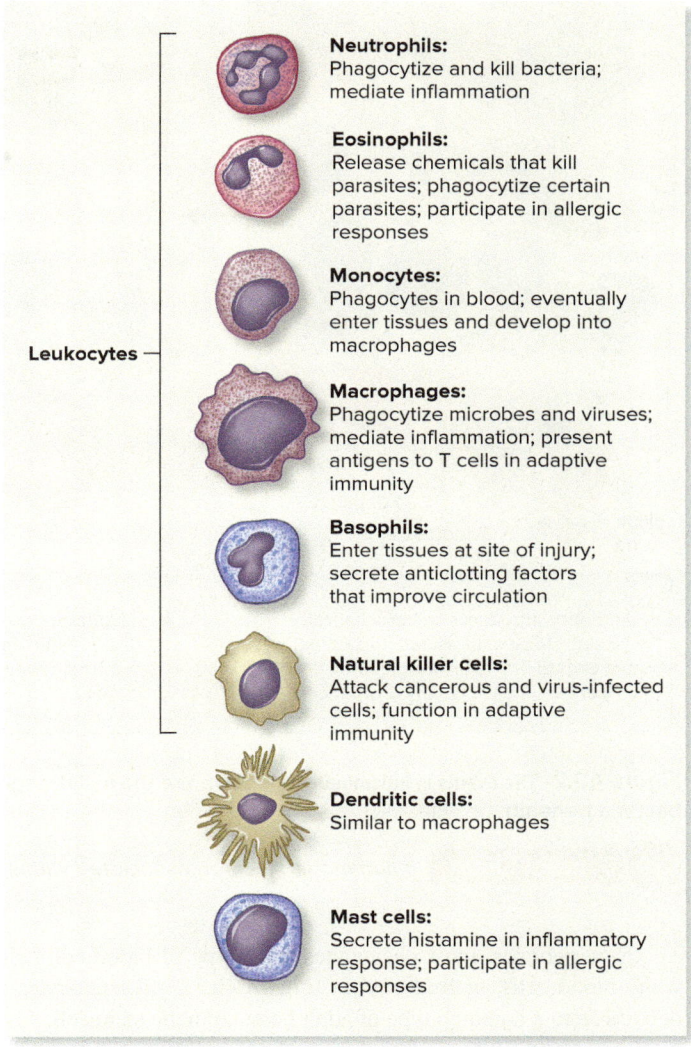

Figure 40.1 Cells involved in innate immunity in vertebrates. Six of these types of cells are leukocytes, and several are phagocytes. Some of the cells also participate in adaptive immunity.

BioConnections: *Refer back to Figure 36.4. How do the numbers of leukocytes in blood compare with the number of erythrocytes?*

In Innate Immunity, Phagocytic Cells Defend Against Pathogens That Enter the Body

Several different types of cells have key functions in innate immunity (**Figure 40.1**). Many of these cells are **phagocytes**—cells capable of phagocytosis. **Phagocytosis** is a type of endocytosis in which a cell engulfs a large particle, such as a bacterium, by forming an internal vesicle called a phagosome. Within the phagosome, enzymes and oxidizing compounds such as hydrogen peroxide begin the process of destroying the bacterium or other engulfed matter. Phagocytes are found in extracellular fluids and within various tissues and organs. They are present in all classes of animals and are among the most ancestral forms of immune defenses.

Figure 40.2 **The events in inflammation.** Shown are the initial stages of inflammation in response to a penetrating wound that introduces bacteria beneath the skin.

Concept Check: *Inflammation is often associated with swelling of the inflamed area. Might this swelling have an adaptive value?*

In vertebrates, most phagocytes are a type of blood cell called white blood cells, or **leukocytes** (Figure 40.1). All leukocytes are derived from a common type of adult hematopoietic stem cell, which in mammals and birds is found in the bone marrow. The leukocytes involved in innate immunity include neutrophils, eosinophils, monocytes, macrophages, basophils, and natural killer (NK) cells. In addition to leukocytes, dendritic cells and mast cells are also derived from bone marrow stem cells and have important functions in innate immunity. As we will see later, some of these cells also participate in certain features of adaptive immunity.

Inflammation Is an Innate Response to Infection or Injury

Inflammation is an innate, local response to infection or injury. The functions of inflammation are to destroy or inactivate pathogens and to clear the infected region of dead cells and other debris. Some of the key cells involved in this process are neutrophils, macrophages, and mast cells.

The events of inflammation are induced and regulated by chemical mediators. These include a family of proteins called **cytokines** that function in both innate and adaptive immune defenses. Cytokines provide a chemical communication network that synchronizes the components of the immune response. Most cytokines are secreted by more than one type of immune system cell and by certain nonimmune cells such as endothelial cells.

The sequence of events in a typical inflammatory response to a bacterial infection is summarized in **Figure 40.2**. A tissue injury, such as that caused by a splinter, begins the inflammation process, which results in the familiar signs and symptoms of local redness, swelling, heat, and pain. Substances secreted into the extracellular fluid from injured tissue cells, mast cells, and neutrophils contribute to the inflammatory response. For example, histamine from mast cells and nitric oxide (NO) from endothelial cells cause vasodilation of the small blood vessels in the infected and damaged area. This increases blood flow to the inflamed area, which accounts for the redness and heat, speeds the delivery of beneficial proteins and leukocytes, and increases local metabolism to facilitate healing. In addition, the vessels become leakier as they dilate, which helps the neutrophils and plasma proteins that participate in inflammation to gain entry to the interstitial fluid. The swelling in an inflamed area also results from this increased leakiness of blood vessels.

Once neutrophils and macrophages arrive at the site of an injury, they begin the process of phagocytosis and killing of the bacteria. The neutrophils and macrophages also release substances into the extracellular fluid that can destroy bacteria even before phagocytosis occurs.

Inflammation or illness may be accompanied by **fever,** an increase in an animal's body temperature that results from an infection. In mammals, cytokines released into the circulation by activated macrophages act within the hypothalamus to raise the body's set point for temperature. The result is an increase in heat generation (for example, by shivering). Studies of animal models indicate that a small or moderate increase in temperature stimulates leukocyte activity and proliferation and provides a less hospitable environment for at least some types of pathogens.

Defenses Against Pathogens Include Interferons and Complement Proteins

In addition to phagocytes and the inflammatory response, at least two other types of innate defenses against invading pathogens can also be involved. These defenses allow for extracellular destruction of pathogens without prior exposure to them. The first of these innate defenses is used against viruses. **Interferons** are proteins that generally inhibit viral replication inside host cells. In response to viral infection, many cell types produce interferons and secrete them into the extracellular fluid. When the interferons bind to plasma membrane receptors on the secreting cell and on other cells, each cell synthesizes a variety of proteins that interfere with the ability of the viruses to replicate. Interferons are not specific. Many kinds of viruses induce interferon synthesis, and the same interferons, in turn, can inhibit the multiplication of many different kinds of viruses.

A second type of innate defense is provided by a family of circulating proteins known as **complement.** Inactive complement proteins normally circulate in the blood at all times and are activated by contact with the surface of many different microbes. Activation of complement proteins results in a cascade of events. Among their many actions, complement proteins stimulate the release of histamine from mast cells, thereby increasing the permeability of local blood vessels, as described earlier. Five of the active proteins generated in the complement cascade form a multiunit protein called the **membrane attack complex (MAC)**, which, by embedding itself in a pathogen's plasma membrane or envelope, creates porelike channels. Water and ions enter the pathogen through the channels, and the pathogen is destroyed.

EVOLUTIONARY CONNECTIONS

Innate Immune Responses Require Proteins That Recognize Features of Many Pathogens

As we have seen, innate immunity often depends on an immune cell's recognition of some general molecular feature common to many types of microbes or viruses. It is easy to imagine the selective advantage for animals provided by their ability to recognize such pathogen features, or **pathogen-associated molecular patterns (PAMPs),** as they are known. Until 1985, however, it was not known how that recognition was accomplished. At that time, German biologist Christiane Nüsslein-Volhard and American biologist Eric Wieschaus were interested in how animal embryos differentiate into adults. In the course of their studies, they discovered a protein they named Toll (now called Toll-1) that was required for the proper dorsoventral orientation of the body of the fruit fly (*Drosophila melanogaster*). For their investigations of embryonic development and discovery of Toll proteins, Nüsslein-Volhard and Wieschaus were awarded the 1995 Nobel Prize in Physiology or Medicine.

In a landmark study in 1996, it was discovered that Toll-1 also confers on adult flies the ability to fight off fungal infections (see the following Feature Investigation). It later became clear that a family of Toll proteins exists in animals from nematodes to mammals, including humans, and these proteins are found in the plasma membranes and endosomal membranes of macrophages and dendritic cells, among others.

In vertebrates, one function of Toll proteins is to recognize and bind to ligands containing PAMPs, such as lipopolysaccharide and other common bacterial lipids and carbohydrates, viral and bacterial nucleic acids, and a protein found in the flagellum of many bacteria. PAMPs are molecular features that are generally considered to be vital to the survival of a particular pathogen. When one of these ligands binds a Toll protein on the plasma membrane of an immune cell, second messengers are generated in the cell, triggering the secretion of inflammatory mediators such as various cytokines. These in turn stimulate the activity of other leukocytes involved in the innate immune response. Some of these mediators also activate cells involved in the adaptive immune response. Because many of the Toll proteins are plasma membrane-bound, bind extracellular ligands, and induce second-messenger formation, they are referred to as receptors, and the family of proteins is now known as **Toll-like receptors (TLRs).** Despite this, not all TLRs generate intracellular signals when bound to a ligand. For example, some TLRs induce attachment of a microbe to a macrophage, thereby facilitating phagocytosis and subsequent destruction of the microbe.

The importance of TLRs in mammals has been demonstrated in mice with a mutated, nonfunctional form of one member of the family called Toll-4. These mice are hypersensitive to the effects of injections of lipopolysaccharide (to mimic a bacterial infection) and are less able to ward off bacterial infection. In humans, recent studies suggest that certain naturally occurring variants in a specific TLR are associated with increased risk of certain diseases.

TLRs are currently an active area of investigation among biologists because of their significance in the immune response. Certain domains of these receptors have also been identified in plants, where they may also be involved in disease resistance. In 2018, Israeli microbiologist Shany Doron and colleagues reported that the intracellular domain of the vertebrate TLRs is similar to that present in newly discovered defense proteins in prokaryotes. The genes that encode these proteins appear to confer nonspecific protection for bacteria against a wide variety of phages. Therefore, TLRs may be among the first immune defense mechanisms to have evolved in living organisms.

FEATURE INVESTIGATION

Lemaitre and Colleagues Identified an Immune Function for Toll Protein in *Drosophila*

As you've just learned, a Toll protein was initially identified as a critical molecule directing the proper dorsoventral development of *Drosophila* embryos. Interestingly, however, certain aspects of the sequence and function of Toll share similarities with at least one important immune protein found in vertebrates. For example, both Toll and the receptor for the cytokine interleukin-1 are transmembrane proteins with similar amino acid sequences in their cytosolic domains. Moreover, both proteins activate a similar intracellular signaling pathway once they have bound an extracellular ligand. These findings led to the hypothesis proposed by French microbiologist Bruno Lemaitre and coworkers that Toll protein may have a role in immunity in *Drosophila*.

To test this hypothesis, the researchers compared wild-type flies with mutant flies in which the *Toll* gene was underexpressed due to a mutation (**Figure 40.3**). They then infected the flies with a fungal pathogen by pricking them with a needle that had been dipped into a concentrated solution of fungal spores. Survival of the flies was

Figure 40.3 Lemaitre and colleagues identified a function for Toll protein in immunity in *Drosophila*.
(top right) ©Daniel L. Geiger/SNAP/Alamy

4 THE DATA

Survival test results: (Wild-type/water, Mutant/water, Wild-type/*A. fumigatus*, Mutant/*A. fumigatus*)

Northern blot results: Drosomycin mRNA

5 CONCLUSION Toll protein receptor is essential for survival of *Drosophila* exposed to a common fungal pathogen and for induction of antifungal genes.

6 SOURCE Lemaitre, B., Nicolas, E., Michaut, L., Reichhart, T., and Hoffmann, J. A. 1996. The Dorsoventral Regulatory Gene Cassette Spätzle/Toll/Cactus Controls the Potent Antifungal Response in *Drosophila* Adults. *Cell* 86: 973–983.

monitored over the next several days and compared with the survival of wild-type flies that had normal levels of Toll protein. What they found was remarkable: Within days of being infected with the fungus *Aspergillus fumigatus*, 100% of the mutant flies died, but only about 30% of the wild-type flies died (step 4 of Figure 40.3). The mutant flies were observed to be covered in germinating hyphae of the fungus, clearly demonstrating that the flies were unable to fight the infection. Moreover, analysis of mRNA from the treated and untreated flies revealed that in contrast to the wild-type flies, the infected mutant flies were unable to increase expression of a key antifungal gene (known to encode a polypeptide called drosomycin).

Since then, these and other researchers have identified much of the mechanism by which Toll promotes immunity to certain microbes in *Drosophila* and, presumably, other insects. Activation of Toll protein induces the expression of several antimicrobial genes. Unlike vertebrate TLRs, however, in *Drosophila* Toll does not directly recognize PAMPs associated with microbial membranes. Rather, microbial infection induces activation of extracellular signaling molecules in *Drosophila* that, in turn, bind to Toll located on cell membranes. This binding induces intracellular signals that activate gene expression. Despite this difference, the identification of an immune function of an ancestral Toll protein paved the way for a clearer understanding of TLRs and innate immunity in all animals; in 2011, Jules Hoffmann received a share of the Nobel Prize in Physiology or Medicine for this pioneering research.

Experimental Questions

1. What led the investigators to hypothesize that Toll protein may serve an immune function in *Drosophila*?
2. Is *Drosophila* Toll protein a receptor that recognizes PAMPs?
3. **SCISKILLS** ▶ Based on the results of their survival studies of infected flies, was the researchers' hypothesis supported?

40.2 Reviewing the Concepts

- Phagocytes, many of which are leukocytes, play an important role in innate immunity (Figure 40.1).
- Inflammation is an innate, local response to infection or injury characterized by local redness, swelling, heat, and pain. The events of inflammation are induced and regulated by chemical mediators called cytokines (Figure 40.2).
- Antimicrobial proteins include interferons and complement proteins. Activation of the complement proteins results in the formation of a membrane attack complex (MAC).
- Toll-like receptors are evolutionarily ancient proteins that recognize PAMPs (Figure 40.3).

40.2 Testing Your Knowledge

1. An example of a PAMP is
   a. lipopolysaccharide.
   b. Toll-like receptor.
   c. Toll protein.
   d. interleukin-1.
   e. complement protein.

2. During inflammation,
   a. histamine is released from neutrophils.
   b. histamine causes local vasoconstriction.
   c. NO causes local vasodilation.
   d. mast cells secrete complement proteins.
   e. an animal's set point for body temperature is lowered.

40.3 Adaptive Immunity in Vertebrates

Learning Outcomes

1. Describe the major classes of lymphocytes.
2. Outline the three stages of the adaptive immune response.
3. Compare and contrast humoral immunity with cell-mediated immunity.
4. **SCISKILLS** ▶ Explain how the structures of different immunoglobulins allow predictions to be made about their functions.
5. Explain how antibody structure and genetic rearrangement lead to the diverse set of antibodies found in mammals.
6. List the steps of a typical humoral immune response.
7. **SCISKILLS** ▶ Predict the benefit of secondary immune responses, and explain the function of memory cells in these responses.

In adaptive immunity, cells of the immune system first encounter and later recognize a specific foreign cell or protein to be destroyed, as opposed to recognizing some general conserved feature of pathogens. Any molecule that can trigger an adaptive immune response is called an **antigen.** An antigen, therefore, is any molecule that the host does not recognize as self. Most antigens are either proteins or very large polysaccharides. Antigens include the protein coats of viruses, bacterial surface proteins, specific macromolecules on pollens and other allergens, membrane proteins of cancerous cells or transplanted cells, toxins, and vaccines.

Adaptive immunity is found in all jawed vertebrates. Jawless fishes possess some but not all of the features of adaptive immunity found in other vertebrates. Adaptive immunity is considered to be absent in invertebrates. We will consider adaptive immune responses in the context in which they are best understood—the jawed vertebrates.

Adaptive Immune Defense Mechanisms Include Lymphoid Organs, Tissues, and Cells

The cells responsible for adaptive immunity are a type of leukocyte called **lymphocytes.** At any given time, some lymphocytes circulate in the blood, but most of them reside in a group of organs and tissues that constitutes the **lymphatic system** (Figure 40.4a). The system is composed primarily of a network of lymphatic vessels. These vessels drain the fluid known as lymph from the interstitial fluid. The lymph is transported through lymphatic vessels to various immune organs before eventually returning to the circulatory system.

Primary Lymphoid Organs The two major adaptive immune structures in vertebrates are grouped into primary and secondary lymphoid organs. The primary lymphoid organs are the structures in which lymphocytes differentiate into mature immune cells. The primary lymphoid organs are the thymus and, in birds and mammals, the bone marrow. In animals without extensive bone marrow, specialized regions of other organs such as the kidney and liver may serve as primary lymphoid organs.

Secondary Lymphoid Organs The primary lymphoid organs supply mature lymphocytes to secondary lymphoid organs, where the lymphocytes multiply and function. In humans, for example, secondary lymphoid organs include the lymph nodes (Figure 40.4b), the spleen, the tonsils, and scattered lymphocyte accumulations in the linings of the intestinal, respiratory, genital, and urinary tracts.

Destruction of or damage to a primary lymphoid organ typically results in a severe inability to fight off infections. The loss of any of the secondary lymphoid organs, although not as serious, nevertheless increases the risk of infections throughout an animal's life. For example, the spleen must occasionally be surgically removed in humans due to injury or disease. Such individuals must be carefully monitored for the rest of their lives because of their increased vulnerability to infection.

Recirculation of Lymphocytes After leaving the bone marrow or thymus, lymphocytes circulate between the secondary lymphoid organs, the blood, the lymph, and all the tissues of the body. Lymphocytes from all the secondary lymphoid organs continually leave those structures and enter the blood. Simultaneously, some circulating lymphocytes leave venules all over the body to enter the interstitial fluid. From there, they re-enter lymphatic vessels and are carried back to secondary lymphoid organs. This constant recirculation of lymphocytes increases the likelihood that any given lymphocyte will encounter an antigen it is specifically programmed to recognize.

Lymphocytes Include B Cells and T Cells

Different kinds of lymphocytes participate in coordinated adaptive immune responses (Figure 40.5). In addition to NK cells, described previously, two major types of lymphocytes are B cells and T cells. **B cells** were first observed to mature in an avian organ called the bursa of Fabricius—and therefore were named B cells. In mammals, B cells mature within bone marrow. If stimulated by antigen, some B cells differentiate further into **plasma cells,** which synthesize and secrete **antibodies,** proteins that bind to and help to destroy foreign molecules or pathogens, as described later. This type of response is called **humoral immunity.**

T cells get their name because they mature within the thymus. T cells may directly or indirectly kill infected, mutated, or transplanted cells. T cells include a few distinct types of lymphocytes, including two called cytotoxic T cells and helper T cells. **Cytotoxic T cells** travel to the locations of their targets, bind to those targets by

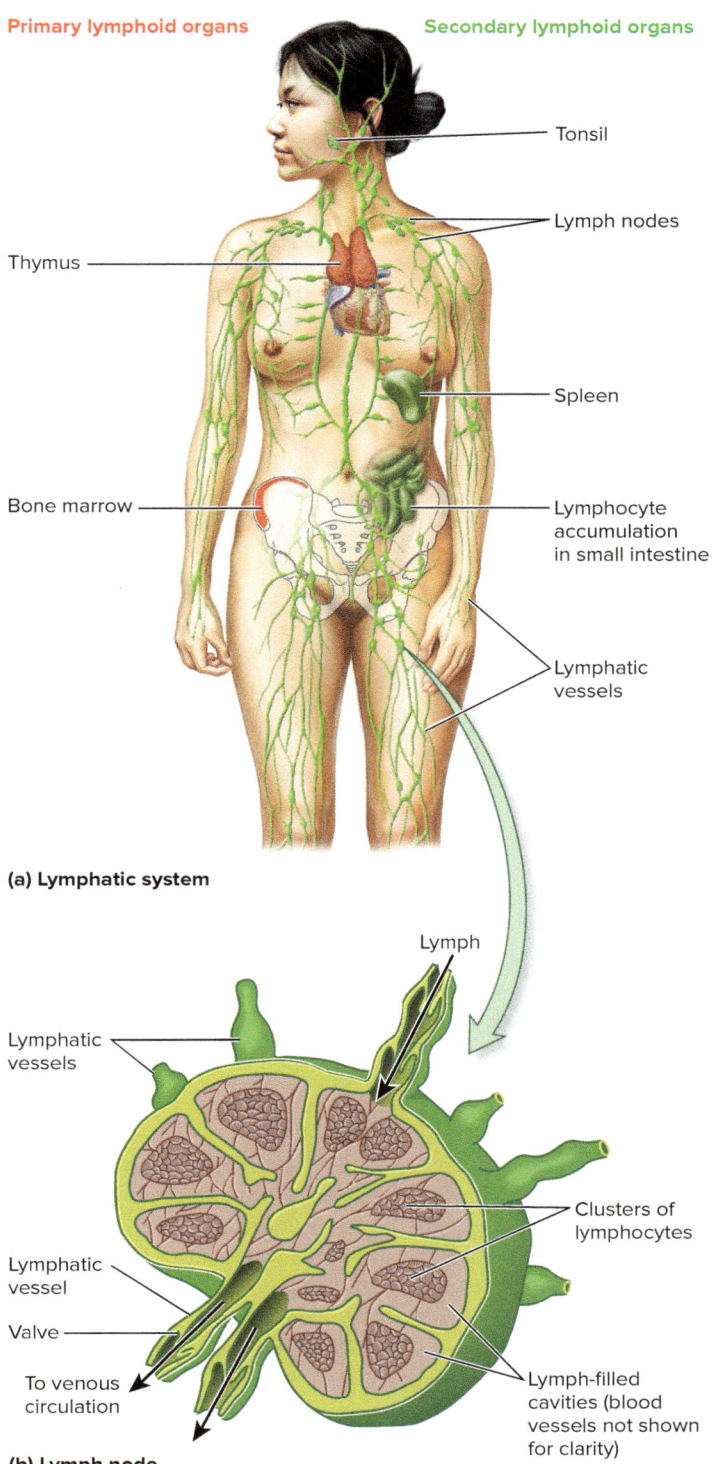

(a) Lymphatic system

(b) Lymph node

Figure 40.4 The lymphatic system in humans. (a) The major components of the human lymphatic system. Primary lymphoid organs are shown in red, and secondary lymphoid organs are shown in green. In adult humans, the primary lymphoid organs in bone are found in the sternum, the ribs, parts of the skull, small regions of the femur and humerus, and, as shown here, the pelvis. (b) The structure of a lymph node. Lymph nodes occur along the course of lymphatic vessels, which drain interstitial fluid from tissues and return it to the venous circulation. Within a lymph node, lymph percolates through open cavities containing clusters of lymphocytes.

 Concept Check: *From where does lymph arise?*

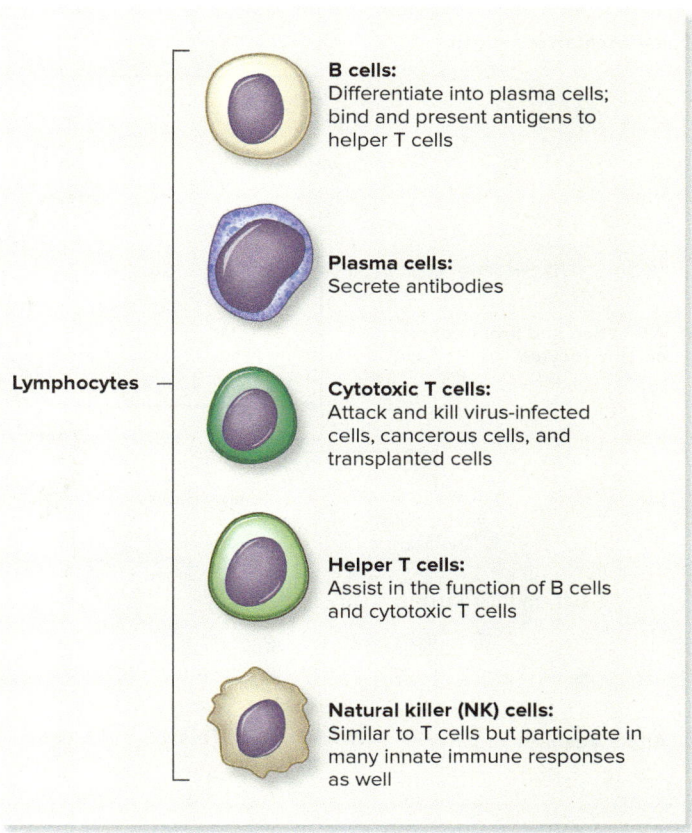

Figure 40.5 Cells involved in adaptive immunity in vertebrates. Some important functions of each type of lymphocyte are noted. The shape and color conventions shown in this figure will be used throughout this chapter.

recognizing an antigen, and destroy the targets via secreted chemicals. In addition, responses mediated by cytotoxic T cells are directed against body cells that have become cancerous or infected by pathogens. (NK cells also destroy such cells by secreting toxic chemicals.) This type of immune response is called **cell-mediated immunity.**

As their name implies, **helper T cells** do not themselves function as effector cells. Instead, they assist in the activation and function of B cells and cytotoxic T cells. With only a few exceptions, B cells and cytotoxic T cells cannot function adequately unless first stimulated by cytokines secreted from helper T cells.

A third type of T cell, which we won't consider further here, is the regulatory T cell. These cells function as a check on the activity of the other two classes of T cells. Among other things, they are believed to be important for preventing autoimmunity, a condition in which an animal's immune system fails to properly distinguish self from nonself.

Humoral and Cell-Mediated Immune Responses Occur in Three Stages

In both humoral immunity and cell-mediated immunity, an adaptive immune response can usually be divided into three stages (**Figure 40.6**): recognition of antigen, activation and proliferation of lymphocytes, and attack against a recognized antigen.

Figure 40.6 **The three stages of an adaptive immune response.** All three cell types in stage 1 recognize the same antigen. Helper T cells secrete cytokines that activate B cells and cytotoxic T cells, as indicated by the ⊕ symbols. Both B cells and T cells undergo cell division to form clones when activated, and in both cases, some of the cloned cells are set aside as memory cells to fight off a future infection of the same type.

Concept Check: Which cell type participates in both humoral and cell-mediated immunity?

Stage 1: Recognition of antigen. During its development, each lymphocyte synthesizes a type of membrane receptor that can bind to a specific antigen. If the lymphocyte subsequently encounters that antigen, the antigen becomes bound to the receptor. This specific binding is the immunological meaning of the word recognize. Antigens that bind to a lymphocyte receptor are said to be recognized by the lymphocyte as being nonself molecules. The ability of lymphocytes to distinguish one antigen from another, therefore, is determined by the nature of their plasma membrane receptors. Each lymphocyte is specific for just one type of antigen.

Stage 2: Activation and proliferation of lymphocytes. In the second stage, the binding of an antigen to a receptor on a lymphocyte activates that lymphocyte. Upon activation, the lymphocyte undergoes multiple cycles of cell division. The result is the formation of many identical cells, called clones, that express the same receptor as the receptor that first recognized the antigen. The molecular nature of the antigen determines which individual lymphocytes will be activated to form clones. This process requires the function of helper T cells, which divide when activated and then secrete the cytokines that promote cell division. Some of the cloned lymphocytes become effector cells—the plasma cells and cytotoxic T cells that carry out the attack response—whereas others are stored as **memory cells,** which remain poised to recognize the antigen if it returns in the future.

In a typical human, the size of the lymphocyte population is staggering. Over 100 million different lymphocytes, each with the ability to recognize a unique antigen, are estimated to exist in a person's immune system. This vast population explains why our bodies are able to recognize so many different antigens as foreign and eventually destroy them.

Stage 3: Attack against antigen. In the third stage, the effector cells attack all antigens of the kind that initiated the immune response. In humoral immunity, plasma cells secrete antibodies, which recruit and guide other molecules and cells that perform the actual attack. In cell-mediated immunity, cytotoxic T cells directly attack and kill the cells bearing the antigens.

Once the attack is successfully completed, the great majority of plasma cells and cytotoxic T cells that participated in it die by apoptosis. However, memory cells persist even after the immune response has been successfully completed, so they can recognize and fight off any future infection bearing the same type of antigen. Let's look at each of these three stages in more detail, first for humoral immunity and then for cell-mediated immunity.

Biology Principle

Biologists Use Models to Test Predictions and Convey Ideas

This figure is a model for the structure of an immunoglobulin. The model depicts how different domains of an immunoglobulin have characteristic structures, notably the variable regions of the protein that bind to a specific antigen. The Modeling Challenge in the figure legend asks you to predict the shape of a variable region of an immunoglobulin that binds a different antigen than the one modeled in the figure.

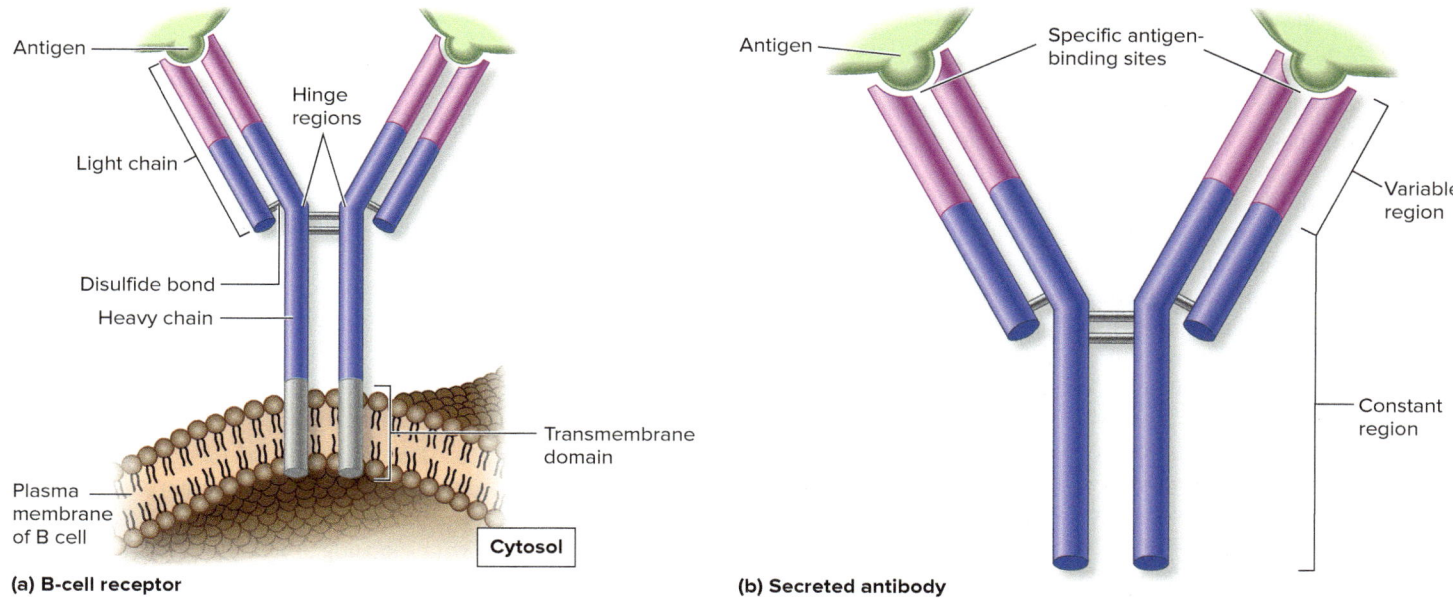

Figure 40.7 Immunoglobulins. **(a)** B-cell receptor and **(b)** secreted antibody. Immunoglobulins are composed of polypeptides known as heavy chains and light chains, held together by disulfide bonds. B-cell receptors are anchored to the plasma membrane by a transmembrane domain, whereas antibodies are secreted into the extracellular fluid. Within both types of immunoglobulins, the constant regions of the heavy and light chains (shown in blue) have identical amino acid sequences. In contrast, the antigen-binding sites formed by the light- and heavy-chain variable regions (shown in purple) have unique amino acid sequences. This gives the variable region a unique shape and is what imparts to each receptor or antibody its specificity for a particular antigen.

Modeling Challenge: The structures of proteins determine their functions. This is true of all immunoglobulins. Using the model of a secreted antibody shown in Figure 40.7b as a reference, draw a new model of an antibody that binds to the antigen depicted to the right. Binding site refers to the part of the antigen that is recognized by the antibody.

In Humoral Immunity, B Cells Produce Immunoglobulins That Serve as Receptors or Antibodies

In stage 1 of humoral immunity (Figure 40.6, left side), B cells recognize antigens with the help of B-cell receptors. When B cells are activated by cytokines in stage 2, they proliferate and differentiate into plasma cells, which secrete antibodies into the extracellular fluid. These proteins travel all over the body to reach and bind antigens identical to those that stimulated their production. In some cases, the binding of an antibody to an antigen simply prevents that antigen from infecting a host cell, in a process called neutralization. For example, many viruses are prevented from infecting host cells in this way. In other cases, antibodies bound to an antigen guide an attack that eliminates the antigens or the cells bearing them, a process we will discuss in more detail later. Antibody-mediated responses are the major defense against bacteria, viruses, and other pathogens in the extracellular fluid and against toxin molecules.

Structure and Function of Immunoglobulins B-cell receptors and the antibodies secreted by plasma cells share many structural and functional similarities (**Figure 40.7**). They are both members of a family of proteins called **immunoglobulins (Ig)**. However, they show some important differences. B-cell receptors have a transmembrane domain that anchors them in the plasma membrane of the B cell (Figure 40.7a). Antibodies are soluble proteins that are secreted from plasma cells (Figure 40.7b). Interestingly, B-cell receptors and the antibodies made by plasma cells are encoded by the same genes. In plasma cells, the pre-mRNA is alternatively spliced, a phenomenon described in Chapter 12 (refer back to Figure 12.17), so the transmembrane domain is not present in the protein. For this reason, the B-cell receptor in a particular B cell and secreted antibodies from the resulting plasma cells recognize the same antigen.

Mammals have five classes of immunoglobulins, designated IgM, IgG, IgA, IgE, and IgD, which are involved in different types of functions. For example, the most abundant immunoglobulins in mammals are IgM and IgG. Together, these two immunoglobulin classes provide the bulk of adaptive immunity against bacteria and viruses in the extracellular fluid. IgA antibodies are secreted by plasma cells in the linings of structures exposed to the environment. For example, IgA molecules secreted into saliva help to keep animals' mouths relatively free of pathogens. IgE antibodies participate in defenses against multicellular eukaryotic parasites. They also attach to mast cell membranes, where they have a role in allergic responses. Finally, the functions of IgD are unclear, but these immunoglobulins may contribute to B-cell activation.

Each immunoglobulin molecule comprises four interlinked polypeptides: two heavy chains and two light chains (Figure 40.7b). Hinge regions provide the molecule with flexibility. One portion of an immunoglobulin chain is called the **constant region.** The amino acid sequence of the constant region is identical for all immunoglobulins of a given class and distinguishes the classes from each other. The constant regions are important for the binding of antibodies to immune cells and to complement proteins, and thus they contribute to the eventual destruction of antigen that is bound to an antibody. A defining feature of immunoglobulins, however, is their **variable region,** which gets its name because its sequence varies among different B cells. The variable region is the site that specifically recognizes a particular antigen.

Recombination of Immunoglobulin Genes The amino acid sequences of the variable regions vary widely from immunoglobulin to immunoglobulin. The enormous number of variable sequences results in countless unique structures of immunoglobulins. Thus, each of the five classes of immunoglobulins contains up to millions of unique molecules, each capable of combining with only one specific antigen or, in some cases, with several antigens with very similar structures. However, vertebrate genomes generally have 20,000 or so protein-encoding genes in their genomes. This raises an intriguing question. How can millions of different immunoglobulin proteins be produced if there are presumably only a small number of immunoglobulin genes? The answer is that the genes undergo a unique process involving gene rearrangements.

Along the length of a typical human immunoglobulin gene are many gene segments that encode a piece of the final immunoglobulin protein. Each segment along the length of the gene is associated with recognition sequences that bind enzymes that are expressed only in developing lymphocytes. These enzymes randomly cut out segments, which are then recombined in different ways, resulting in a new, permanent immunoglobulin gene for a given B cell.

The process of gene recombination results in each lymphocyte within an individual's body producing a unique type of immunoglobulin. Because an animal makes hundreds of millions of different lymphocytes, the immune system can produce antibodies capable of recognizing an incredibly diverse array of antigens. Nearly any foreign antigen that is taken into the body is recognized by some lymphocytes in this large population.

Activated B Cells Produce Plasma Cells That Secrete Antibodies

In stages 2 and 3 of the humoral immune response, B cells are activated and differentiate into plasma and memory cells. The plasma cells then secrete antibodies that attack the antigen that was detected (see Figure 40.6, left side). In this section, we will take a closer look at these processes.

Clonal Selection B cells are activated by a specific antigen, with the aid of a helper T cell (a process we will discuss later). When an antigen-stimulated lymphocyte divides and replicates itself, the progeny of this lymphocyte—all of which express the same receptor—are clones. The process by which these clones are formed is called **clonal selection** (Figure 40.8). This term emphasizes that lymphocyte proliferation is selected by exposure to an antigen.

Antibodies Facilitate Destruction of Pathogens via Opsonization The antibodies secreted from the plasma cells circulate through the lymphatic system and the blood. Eventually, the antibodies combine with the antigen that initiated the immune response. These antibodies then direct the attack against the pathogen to which they are now bound. Thus, immunoglobulins play two distinct roles in humoral immune responses. First, during antigen recognition, immunoglobulins (B-cell receptors) on the surface of B cells bind to antigen brought to them. Second, antibodies secreted by the resulting plasma cells bind to pathogens bearing the same antigens, marking them as the targets to be destroyed.

Instead of directly destroying the pathogens, antibodies bound to antigen on the pathogen surface inactivate the pathogens in various ways. Antibodies may physically link the pathogens to phagocytes (neutrophils and macrophages), complement proteins, or NK cells. This linkage—called **opsonization**—triggers the attack mechanism and ensures that only the pathogens, and not nearby body cells, are destroyed.

In a second mechanism, antibodies that recognize toxins produced by bacterial pathogens in the extracellular fluid bind to the toxins, thereby preventing them from harming susceptible body cells. The antibody-antigen complexes that are formed are then destroyed by phagocytes.

In a similar way, antibodies that recognize certain viral surface proteins bind to the viruses in the extracellular fluid, preventing them from attaching to the plasma membranes of potential host cells. As with bacterial toxins, the antibody-virus complexes that are formed are subsequently phagocytized.

In Cell-Mediated Immunity, T Cells Recognize Antigens Complexed with Self Proteins

In cell-mediated immunity, T cells recognize and are activated by antigens (see Figure 40.6, right side). T-cell receptors for antigens have specific regions that differ from one T cell to another. As shown

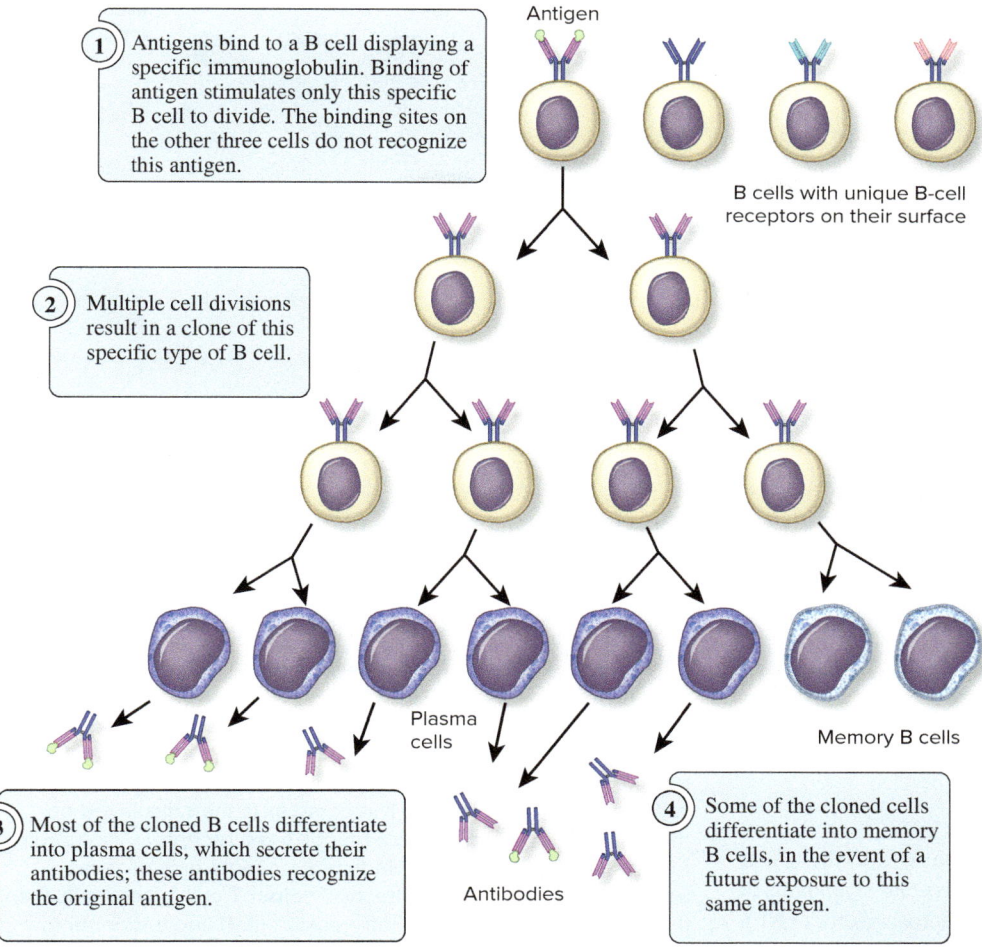

Figure 40.8 Clonal selection. In this example, a B cell with a specific immunoglobulin on its surface recognizes an antigen and is stimulated to divide into a clone of identical cells.

in **Figure 40.9**, T-cell receptors are composed of two polypeptides, each with a variable and a constant region, along with a transmembrane domain. The variable regions recognize an antigen. As in B-cell development, multiple DNA rearrangements occur during T-cell maturation, leading to millions of distinct types of T cells, each with a receptor of unique specificity. For T cells, this maturation occurs as they develop in the thymus.

A T-cell receptor cannot bind to an antigen unless the antigen is already complexed with a protein that is found on the surface of a host cell, such as a macrophage. Such proteins are called **major histocompatibility complex (MHC)** proteins. In many but not all vertebrates, two major classes of MHC proteins are known, called class I and class II MHC proteins. In humans, class I MHC proteins are found on the surface of all body cells except erythrocytes (that is, all nucleated cells). Class II MHC proteins are found only on the surface of macrophages, B cells, and dendritic cells.

To recognize an antigen, the two different types of T cells have different MHC requirements. Cytotoxic T cells require that antigen be associated with class I MHC proteins, whereas helper T cells require an association with class II MHC proteins. One reason for this difference stems from the presence of different proteins on the surfaces of T cells. Helper T cells can be identified by a unique membrane protein called CD4, and cytotoxic T cells are identified by a membrane protein known as CD8. CD4 binds to class II MHC proteins, whereas CD8 binds to class I MHC proteins.

Antigen Presentation to Helper T Cells How do foreign antigens become complexed with MHC proteins on the surface of the body's own cells? The answer involves the mechanism known as antigen presentation. As previously noted, helper T cells can bind antigen only when the antigen appears on the plasma membrane of a host cell complexed with the cell's class II MHC proteins. Cells bearing these complexes, therefore, function as **antigen-presenting cells (APCs)**. Because only macrophages, B cells, and dendritic cells express class II MHC proteins, only these cells can function as APCs for helper T cells.

Let's consider the function of macrophages as APCs for helper T cells (**Figure 40.10**). After a protein antigen has been phagocytized by a macrophage in an innate immune response, it is partially broken down into smaller polypeptide fragments by the macrophage's proteolytic enzymes within intracellular vesicles called endosomes. The resulting digested fragments then bind in the endosomes to class II MHC proteins synthesized by the macrophage. Each fragment-MHC complex is then transported to the plasma membrane, where it is displayed on the cell surface. A specific helper T-cell receptor then binds this entire complex on the cell surface of the macrophage. The CD4 protein helps to link the two cells. What is complexed to MHC proteins and presented to the helper T cells is not the intact antigen but instead a polypeptide fragment of the antigen—called an antigenic determinant, or **epitope.**

B cells process antigen and present it to helper T cells in essentially the same way as macrophages do. The ability of B cells to present antigen to helper T cells is a second function of B cells in response to antigenic stimulation, in addition to their differentiation into antibody-secreting plasma cells.

The binding between the helper T-cell receptor and the complex of antigen fragment and class II MHC protein on an APC is the essential antigen-specific event in helper T-cell activation. However, by itself this specific binding does not result in helper T-cell activation. In addition, interactions occur between nonantigenic proteins on the surfaces of the attached helper T cell and the APC. These interactions constitute a necessary costimulus for helper T-cell activation (Figure 40.10).

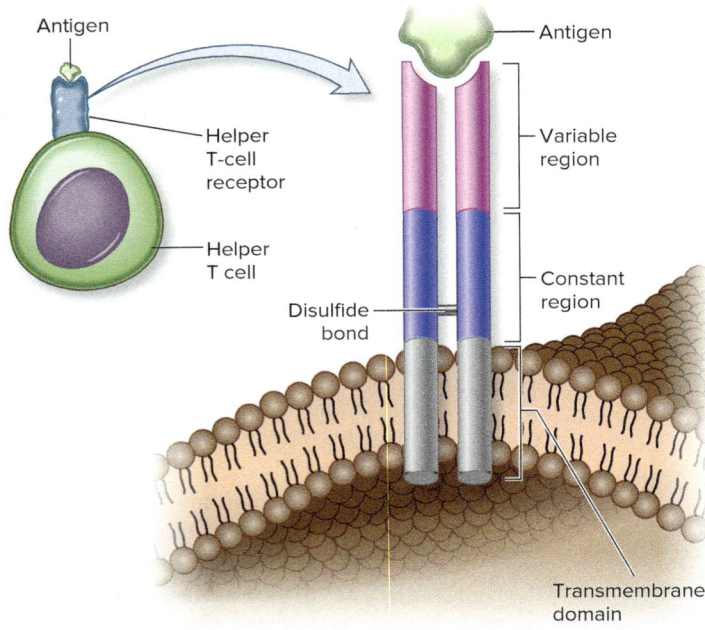

Figure 40.9 Structure of a T-cell receptor in the plasma membrane.

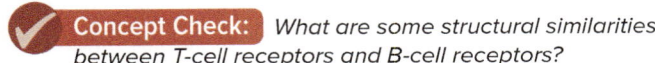 **Concept Check:** *What are some structural similarities between T-cell receptors and B-cell receptors?*

Figure 40.10 **Antigen presentation and helper T-cell activation.** In the initial events in helper T-cell activation, an antigen fragment is complexed with a class II MHC protein within an antigen-presenting cell such as a macrophage. The complex is then displayed on the cell surface and binds to a helper T-cell receptor. Also required for T-cell activation are the binding of nonantigenic proteins between the APC and the attached helper T cell (a costimulus) and the actions of the cytokines interleukin 1 (IL-1) and tumor necrosis factor (TNF).

Finally, the antigen-dependent binding of the APC to the helper T cell plus the costimulus induces the APC to secrete large amounts of two cytokines—interleukin 1 (IL-1) and tumor necrosis factor (TNF). These molecules also stimulate the attached helper T cell. Thus, the APC participates in the activation of a helper T cell in three ways:

1. Presenting antigen
2. Providing a costimulus
3. Secreting cytokines

Activated helper T cells then secrete cytokines of their own that stimulate B cells and cytotoxic T cells.

Helper T Cells and B-Cell Activation Now we can reconsider how B cells are activated by helper T cells. This process begins when a helper T cell specific for a particular antigen binds to a complex of that antigen and a class II MHC protein on an APC, thus activating the helper T cell. Along with other signals, this binding induces the activated helper T cell to divide. Some of the resulting activated helper T cells then bind to B cells that display the same antigen on their surfaces. This binding, along with additional cytokines, stimulates the B cell to go through the process of clonal selection. Thus, helper T cells are so named because their secretions help to activate B cells that have bound antigen, in addition to their participation in the activation of cytotoxic T cells and antigen presentation.

Antigen Presentation to Cytotoxic T Cells Unlike helper T cells, cytotoxic T cells require class I MHC proteins for activation. This distinction helps to explain the major function of cytotoxic T cells—destruction of any of an animal's own altered cells that have become cancerous or infected with viruses. The crucial point is that the antigens that complex with class I MHC proteins typically arise within body cells and are thus endogenous antigens.

How do antigens that activate cytotoxic T cells arise? In viral infections, once a virus has entered a host cell, the expression of viral genes results in the synthesis of viral proteins, which are foreign to the cell. Cancer cells accumulate mutations, some of which alter the amino acid sequence of proteins. Such abnormal proteins act as antigens.

In both virus-infected and cancerous cells, cytosolic enzymes hydrolyze some of the endogenously produced antigenic proteins into polypeptide fragments, which are transported into the endoplasmic reticulum. There, the fragments are complexed with the host cell's class I MHC proteins and then shuttled by exocytosis to the plasma membrane, where a cytotoxic T cell specific for the antigen/MHC protein complex can bind to it. Once binding occurs, cytotoxic T cells release chemicals that kill the infected or cancerous cell, as discussed next.

Activated Cytotoxic T Cells Kill Infected or Cancerous Cells

So far, this section has described how immune responses provide long-term defenses against bacteria, viruses, and individual foreign molecules that enter the body's extracellular fluid. We will now

IMMUNE SYSTEMS: HOW ANIMALS DEFEND AGAINST PATHOGENS AND OTHER DANGERS 909

Figure 40.11 Summary of events in the killing of virus-infected cells by cytotoxic T cells. The sequence is similar for cancerous cells attacked by a cytotoxic T cell. Not shown are the formation of memory T cells, which like memory B cells remain poised for the possibility of a future infection with the same antigen.

examine how an animal's own cells that have become infected by viruses or transformed into cancerous cells are destroyed by cell-mediated immune responses (see Figure 40.6, right side).

What is the value of destroying virus-infected host cells? First, and most importantly, such destruction prevents cells from making more viruses. Second, for cells that already are making mature viruses, it results in the release of the viruses into the extracellular fluid, where they can be neutralized by circulating antibody.

Functions of Cytotoxic T Cells A typical cytotoxic T-cell response triggered by viral infection of a vertebrate's body cells is summarized in **Figure 40.11**. The response triggered by a cancerous cell would be similar. A virus-infected cell produces foreign proteins, viral antigens that are processed and presented on the plasma membrane of the cell complexed with class I MHC proteins. Cytotoxic T cells specific for the particular antigen bind to the complex (Figure 40.11, step 1). As with B cells, binding to antigen alone does not cause activation of the cytotoxic T cell.

Macrophages phagocytize extracellular viruses (or, in the case of cancer, antigens released from the surface of cancerous cells) and then process and present antigen, in association with class II MHC proteins, to the helper T cells (step 2). The activated helper T cells then bind to cytotoxic T cells. They also release IL-2 and other cytokines, which stimulate the proliferation of helper T cells as well as cytotoxic T cells. Why is proliferation important if a cytotoxic T cell has already located and bound to its target? The answer is that there is rarely just one virus-infected or cancerous cell. By expanding the population of cytotoxic T cells capable of scanning the entire body and recognizing the particular antigen, the likelihood is greater that an appropriate cytotoxic T cell will encounter other virus-infected or cancerous cells.

The cytotoxic T cells specific for that virus then find and bind to other virus-infected cells (step 3). There are two major ways in which cytotoxic T cells kill their target cells. The first way involves the release by exocytosis of secretions from the T cell. The second mechanism involves binding of a membrane protein on the T cell to

Figure 40.12 Summary of events in a typical humoral immune response in a mammal. Most of the events depicted occur within a lymph node.

surface receptors on the target cell. Both mechanisms induce apoptosis in the target cell. Let's examine the first mechanism in detail.

Each cytotoxic T cell releases the contents of its secretory vesicles directly into the extracellular space between itself and the target cell to which it is bound (thereby ensuring that other, nearby host cells will not be killed). These vesicles contain proteases, as well as a protein called perforin, which is similar in structure to the proteins of the complement system's membrane attack complex. Perforin inserts into the target cell's membrane and forms pores ("perforations") in it (step 4). This allows proteases secreted by a cytotoxic T cell to enter the attacked cell and causes apoptosis. The cell also takes in water through the pores, which causes it to burst. The cytotoxic T cell is not harmed by this process and can then continue to kill other virus-infected cells (step 5). Target cell killing by activated cytotoxic T cells occurs by several mechanisms, but this is one of the most important.

Viruses can be eliminated from an animal's body in two ways: through the humoral actions of antibodies in body fluids and through the cell-mediated killing of virus-infected cells. Although cytotoxic T cells have an important role in the attack against such cells, they are not the only mechanisms. NK cells also destroy virus-infected and cancerous cells by secreting toxic chemicals. As mentioned earlier in this chapter, NK cells can recognize general features on the surface of such cells and participate in innate immunity. In addition, in a cell-mediated adaptive immune response, NK cells can be linked to such target cells by antibodies and then can destroy them by the release of toxic molecules.

Summary: Example of an Adaptive Immune Response

Let's bring together our discussions of the adaptive immune system by looking in detail at one example. One classic example is the humoral immune response that results in the destruction of bacteria. The sequence of events, which is quite similar to the humoral response to a virus in the extracellular fluid, is summarized in **Figure 40.12**. For this example, we will consider the response in mammals, in which lymph nodes are present. Many features of the response, however, are similar in other vertebrates.

This process starts the same way as for innate responses, with the bacteria penetrating one of the body's linings through an injury and entering the interstitial fluid (Figure 40.12, step 1). The bacteria then move with lymph into the lymphatic system and are carried to lymph nodes (step 2). Within the lymph nodes, both a macrophage and a B cell recognize one of the bacteria as foreign and bind to it.

B-cell activation usually requires activation of helper T cells. The helper T cell binds to a complex of processed antigen and class II MHC protein on the macrophage (an APC). The macrophage phagocytizes the bacterium, hydrolyzes its proteins into peptide fragments, complexes the fragments with class II MHC proteins, and displays the complexes on its surface. Once a helper T cell specific for the complex binds to it, the helper T cell becomes activated. The macrophage helps this process in two other ways: It provides a costimulus, and it secretes the cytokines IL-1 and TNF.

IL-1 and TNF stimulate the helper T cell to secrete another cytokine, IL-2. IL-2 stimulates the activated helper T cell to divide, which leads eventually to the formation of a clone of activated helper T cells (step 4). The activated helper T cells bind to B cells and secrete IL-2 and other cytokines (step 5). Certain of these cytokines provide the additional signals that are usually required to activate nearby antigen-bound B cells to proliferate (step 6). These cells differentiate into memory cells, which help to ward off possible future attacks by the same antigen, and plasma cells, which then secrete specific antibodies (step 7). The antibodies enter the blood and bind to bacterial cells, which are then destroyed (step 8).

B Cells and T Cells That Recognize Self Molecules Must Be Killed or Inhibited

As we have seen, the lymphocytes responsible for the adaptive immune response in vertebrates are very capable killers of pathogens—so capable, in fact, that it raises a question: Why don't these cells attack and kill normal self cells? In other words, how does an animal's adaptive immune system distinguish between self and nonself components and develop what is called **immune tolerance,** or tolerance of an animal's own proteins and other molecules?

Recall that the huge diversity of lymphocyte receptors is ultimately the result of multiple, random DNA recombination processes. It is virtually certain, therefore, that every animal possessing adaptive immune defenses would have lymphocytes with receptors that could bind to that individual's own proteins. The continued existence and functioning of such lymphocytes would be disastrous, because such binding would launch an immune attack against all body cells expressing these proteins.

At least two mechanisms explain why individuals normally lack active lymphocytes that respond to self components. First, during early development in vertebrates, T cells are exposed to a wide mix of self proteins in the thymus. Those T cells with receptors capable of binding self proteins are destroyed by apoptosis in a process termed **clonal deletion.** The second mechanism, termed **clonal inactivation,** occurs outside the thymus and causes potentially self-reacting T cells to become nonresponsive. B cells undergo similar processes.

The mechanisms by which these two events occur are still under investigation. Occasionally, however, these mechanisms fail, and an animal's immune cells attack the body's own cells, resulting in an autoimmune disease. **Autoimmune diseases** are conditions in which the body's normal state of immune tolerance somehow breaks down, with the result that both humoral and cell-mediated immunity attacks are directed against certain of the body's own cells. A growing number of human diseases, such as multiple sclerosis, rheumatoid arthritis, systemic lupus erythematosus, and type 1 diabetes mellitus, are being recognized as autoimmune in origin.

Immunological Memory Is an Important Feature of Adaptive Immunity

As you have learned, the adaptive immune response to a given antigen depends on whether or not an animal's body has previously been exposed to that antigen. Consider, for example, the humoral immune response. In mammals, antibody production in response to the first contact with an antigen occurs slowly, over a few weeks. This response to an initial antigen exposure is termed a **primary immune response** (Figure 40.13). Any subsequent infection by the same pathogen elicits an immediate and heightened production of additional specific antibodies against that particular antigen, a reaction termed a **secondary immune response.**

In the case of humoral immunity, this secondary response occurs more quickly, is stronger, and lasts longer because memory B cells that were produced in response to the initial antigen exposure are quickly stimulated to multiply and differentiate into thousands of plasma cells. These cells then produce large amounts of specific antibodies. The immune system's ability to produce this secondary response is called immunological memory.

Immunological memory also occurs with cell-mediated immunity and memory T cells. It explains why animals are able to fight off many illnesses to which they have been previously exposed, such as many common childhood diseases in humans. It is also the basis for the efficacy of vaccinations. In **vaccinations,** small quantities of living, dead, or altered pathogens; small quantities of toxins; or harmless antigenic molecules derived from a microorganism or its toxin are injected into the body so as to induce a primary immune response, including the production of memory cells. Subsequent natural exposure to the immunizing antigen results in

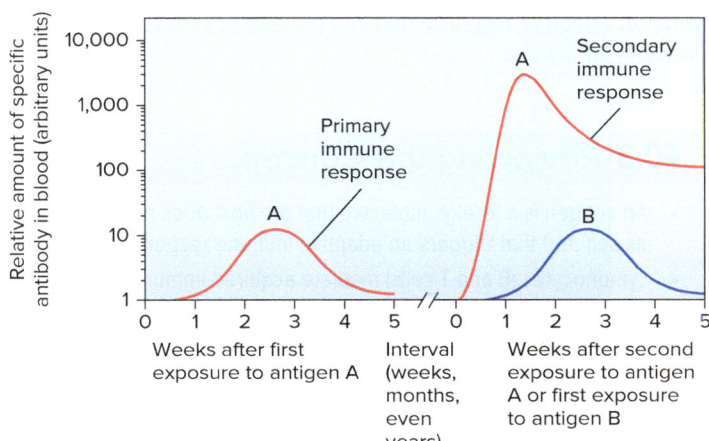

Figure 40.13 **Primary and secondary immune responses.** In a primary response, as shown on the left of this graph, an initial exposure to an antigen produces modest levels of specific antibody over a period of weeks. In a secondary response, subsequent exposure to the same antigen results in greater antibody production that occurs more rapidly and lasts longer than a primary response. (Note that the scale of the y-axis is logarithmic.) The secondary response is specific for that antigen. Exposure at that time to a different antigen for the first time produces the usual primary response.

Concept Check: *What is the advantage of a secondary immune response?*

41

Integrated Responses of Animal Organ Systems to a Challenge to Homeostasis

A person donating blood. Donations such as these improve the chances for survival of individuals who have lost significant amounts of blood due to injury.

Chapter Outline

- 41.1 Effects of Hemorrhage on Blood Pressure and Organ Function
- 41.2 The Rapid Phase of the Homeostatic Response to Hemorrhage
- 41.3 The Secondary Phase of the Homeostatic Response to Hemorrhage
- 41.4 Impact on Public Health

Assessing Your Knowledge and Skills

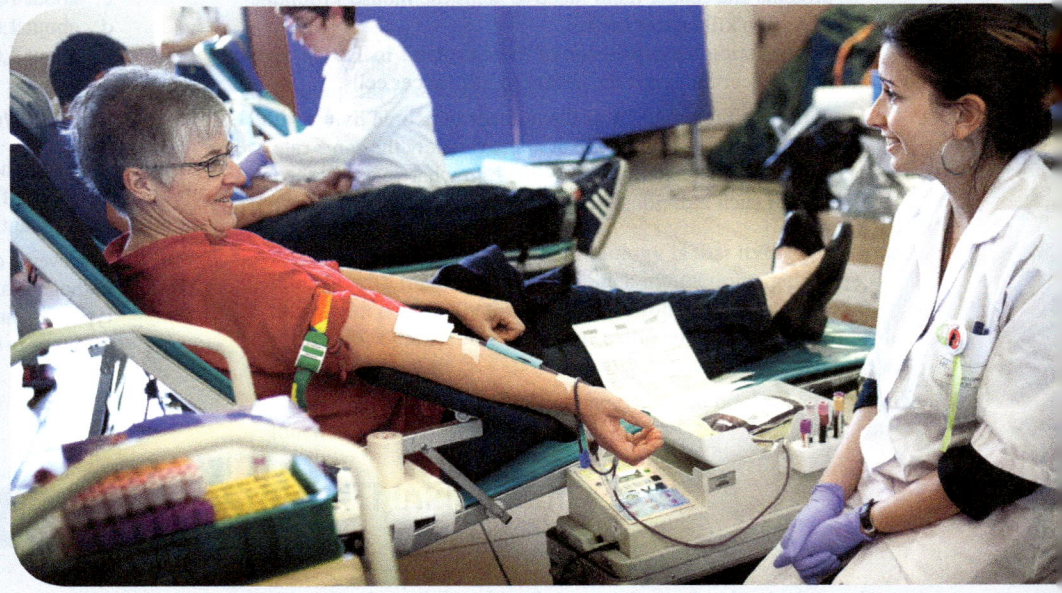
©BSIP SA/Alamy

Shortly after the start of a high school soccer match, a 16-year-old girl falls to the ground after tripping over an opponent's foot. A teammate running behind the girl cannot stop herself in time and accidentally kicks the girl with a hard, cleated shoe in the lower back, near the region of the girl's left kidney. The injury is not terribly painful, and the girl is able to get up on her own and walk off the field, where she sits down and rests. By the end of the match, however, she is complaining of feeling "a little woozy." She becomes agitated yet remains alert and responsive. Nonetheless, her concerned coach has her transported by ambulance to a local hospital. A paramedic records her blood pressure as 108/66 mmHg—which, according to the girl's mother, is a little lower than usual. Her heart rate is 101 beats per minute (bpm), which is abnormally rapid. She is breathing at a rate of 20 breaths/minute, which is above normal.

The kick she received has ruptured one or more large blood vessels, causing her to bleed internally. The loss of blood through a ruptured blood vessel is called a **hemorrhage.** Without prompt treatment, a significant hemorrhage runs the risk of leading to a serious and potentially life-threatening condition known as **shock,** in which the ability of the circulatory system to provide nutrients and oxygen to vital organs decreases to the point that those organs can no longer function properly and their cells begin to die.

One of the principles of biology that you learned in Chapter 1 is that living organisms maintain homeostasis. Throughout this unit, you have seen numerous examples of how this principle applies to animal biology. Recall from Chapter 32 that organ systems in animals do not function in isolation. That is, changes in the activity of one organ system often result in changes in other organ systems. Many times, two or more organ systems function together to control an important variable such as blood pressure. In this chapter, we will take a detailed look at how a significant challenge to homeostasis results in changes in the activities of all organ systems, which function together to meet the challenge of restoring homeostasis. Our example of a homeostatic challenge will be hemorrhage, which may be internal, as in the girl just described, or external, as when you cut yourself and bleed to the outside. Keep in mind, however, that there are many homeostatic challenges that result in the coordinated activation of more than one organ system; hemorrhage is just one well-studied example. The nature or magnitude of the responses may differ for different situations, but the general concept of an integrated, whole-body response remains the same.

Hemorrhage can have life-threatening consequences if not compensated for by homeostatic mechanisms. A significant loss of blood decreases an animal's blood pressure, which slows the delivery of nutrients and oxygen to vital organs such as the brain and heart. The result can be catastrophic and even fatal if pressure is not corrected. The compensatory mechanisms for hemorrhage have been studied in all vertebrates but most thoroughly in mammals, and thus we will use mammals as our example in this chapter. We will begin by describing these compensatory mechanisms and then, at the end of the chapter, return to the case of the young soccer player and discover how an understanding of homeostatic control mechanisms helped her recover.

41.1 Effects of Hemorrhage on Blood Pressure and Organ Function

Learning Outcomes

1. Explain why a decrease in blood pressure can be dangerous.
2. **SCISKILLS** ▶ Predict changes that may occur to an animal's blood pressure when its blood volume is increased or decreased.

In any type of hemorrhage, no matter how small or large, one of the first responses of an animal's body is sealing the wound. This is accomplished by the clotting mechanisms described in Chapter 36 (refer back to Figure 36.5). For the purpose of our discussion, we will assume that an animal suffering a hemorrhage has normal clotting mechanisms and that they are initiated immediately following the hemorrhage and continue for some time until the wound is closed. In this section, we will focus on understanding why hemorrhage is so dangerous and how blood volume and pressure are linked.

Decreased Blood Pressure May Lead to Cell Death

Blood circulates in the body under pressure, as described in Chapter 36. The source of that pressure is the beating of the heart. Blood flows through arteries when it leaves the heart and pushes against the inner surfaces of the vessels. This force is what we refer to when we measure a person's blood pressure using an arm cuff, for example. In most mammals, blood pressure is roughly similar to that of humans, but different animals face different challenges in getting blood to all parts of the body. Think of the distance blood must travel from the heart to the brain in a giraffe. Even in humans, the distance from the heart to the brain results in a decrease in blood pressure as the blood flows upward against gravity. You may have experienced a sense of light-headedness on occasion when standing suddenly; it takes a second or two to generate the pressure required to overcome the effect of gravity on the blood circulating to your head. This temporary light-headedness illustrates the importance of blood pressure. All cells require uninterrupted delivery of nutrients and oxygen to perform the metabolic and other activities required for optimal functioning and survival. Very active cells, such as those of the brain, are the first to show signs of malfunction when the rate of nutrient and oxygen delivery is decreased.

Because of its vital importance for cell function, let's focus for the moment on oxygen delivery. The rate of oxygen delivery to any tissue in an animal's body depends on at least four factors, which were described in detail in Chapter 36. These include

- the rate of blood flow to a region, which in turn is influenced by the degree to which vessels are dilated or constricted,
- the number of erythrocytes in a given volume of blood,
- the percent saturation of hemoglobin in the erythrocytes, and
- the ability of hemoglobin to release its oxygen to cells.

Following a hemorrhage, the first of these factors—blood flow—is immediately affected as blood pressure begins to decrease. Very quickly, unless compensatory events occur, this alone can be sufficient to cause widespread cell death. Adjustments to all four factors, however, can help to prevent this from happening.

Blood Volume and Blood Pressure Are Linked

Refer again to the chapter-opening photo of a person donating blood. This procedure removes about 500 mL of blood, which is roughly 10% of the total blood volume of a typical adult human. This controlled procedure has been adapted by investigators in understanding how an animal's body responds to a hemorrhage. For instance, using a thin, flexible, hollow tube (called a cannula) inserted into a vein of an experimental animal to withdraw carefully controlled amounts of blood, investigators can determine the relationship between the degree of hemorrhage and its effects on different organ systems. Then, investigators can monitor how these organ systems respond in an integrated, coordinated way—first, to prevent blood pressure from continuing to decrease and, second, to help restore blood volume and pressure to normal.

The results of such experiments indicate that mammals can usually cope well with a 10% hemorrhage, as humans can when donating

Figure 41.4 **Changes in the distribution of total blood volume after hemorrhage.** Blood is redistributed away from areas that are not immediately important so that the heart and brain can still receive sufficient blood despite hemorrhage. The idealized values shown here are based on research derived from numerous mammalian species.

of norepinephrine receptors. Binding to one type of receptor causes contraction of the muscles, resulting in vasoconstriction of the vessel. This vasoconstriction occurs in the skin, skeletal muscles, and digestive system, among others. By contrast, when norepinephrine binds to the other type of norepinephrine receptor found in vessels of the heart and brain, it helps to keep those vessels dilated to ensure sufficient blood flow to those regions (**Figure 41.4**). This redistribution of blood away from organs with less immediately vital functions to those with more important functions is an effective means of partially compensating for a decreased total blood volume.

In addition to their actions on the arterioles, the neurons of the sympathetic nervous system also have endings on the smooth muscles of many of the large veins, such as those in the legs. When stimulated by norepinephrine, these smooth muscles contract, and this helps to squeeze blood up through the veins toward the heart. Recall from Chapter 36 that venous blood pressure is very low compared to arterial pressure. Compressing the veins in this way helps to provide an important boost of blood from the venous reservoir.

There is an additional benefit from the increased venous blood returning to the heart. In the early 20th century, German physiologist Otto Frank and British physiologist Ernest Starling, working independently to investigate the mechanisms by which heart muscle cells contract, recognized an intriguing property of the heart muscle. Within limits, if the ventricles of the heart are expanded (stretched), they subsequently contract with greater force than normal. In other words, the more blood that enters the heart before a beat, the more the ventricles will be stretched and the greater will be the force of the beat. This can be thought of as somewhat analogous to stretching a rubber band before releasing it. The more you stretch the rubber band, the more forcefully it snaps back to its original length. When the amount of blood returning from the veins to the heart is increased after hemorrhage, the contractile force of the heart chambers is increased, making them more effective in delivering a greater cardiac output. This phenomenon is now known as the **Frank-Starling Law of the Heart** and is very important in the early response to blood loss. It occurs together with the baroreceptor reflex.

The Respiratory System Aids in Circulating Blood and Delivering Oxygen

In addition to the nervous system, the respiratory system also participates in the overall response to hemorrhage. When oxygen delivery to any region of an animal's body is decreased, the cells in that region depend more on fermentation to produce the ATP required for survival. A by-product of fermentation is lactic acid, which is released into the blood and decreases the blood pH. Recall from Chapter 36 that chemoreceptors are present in certain blood vessels of the circulatory system of animals. In mammals, the chemoreceptors are associated with the same vessels that contain the baroreceptors, but in this case they are sensitive not to pressure but to changes in the amount of oxygen, carbon dioxide, and acidity of the blood. A buildup of lactic acid (decreased pH) stimulates the chemoreceptors, which then activate associated neurons that send signals to the brainstem centers that control breathing (refer back to Figure 36.24). Breaths become deeper and more rapid, thereby maximizing the saturation of hemoglobin with oxygen.

In addition, the mechanical action of the chest and diaphragm creates a siphon-like effect that draws blood up from veins below the diaphragm, allowing greater venous return to the heart. For example, during inhalation, the pressure in the thorax decreases as the chest expands while the abdominal pressure increases due to the diaphragm contracting and moving downward. This creates a pressure gradient that helps to draw blood from the abdomen into the thorax and then to the heart. This thoraco-abdominal pump—or respiratory pump, as it is sometimes called—is another important way in which the Frank-Starling effect is generated. The faster the rate and the deeper each breath, the more significant the Frank-Starling effect becomes.

The acidic pH of the blood has another benefit. As you learned in Chapter 36 (refer back to Figure 36.23), a decreased pH results in a decreased affinity of hemoglobin for oxygen (recall that this is called the Bohr effect). The lactic acid generated by poorly oxygenated tissues causes hemoglobin to unload a greater amount of oxygen, which is an important benefit that helps to offset the decreased oxygen delivery to the tissues resulting from decreased blood flow.

41.2 Reviewing the Concepts

- Baroreceptors are pressure-sensitive neuron endings associated with certain large blood vessels. In the baroreceptor reflex, a change in blood pressure is rapidly compensated for by a change in cardiac output and in peripheral resistance in many blood vessels (Figure 41.3).
- The nervous system mediates a redistribution of blood after hemorrhage such that the vital organs continue to receive as much blood as possible (Figure 41.4).
- The respiratory system contributes to venous return via the chemoreceptors and the thoraco-abdominal pump.

41.2 Testing Your Knowledge

1. After a moderate hemorrhage, an animal's cardiac output is found to have decreased from its normal value of 4.6 L/min to 3.7 L/min. The animal's heart rate at that time is 99 beats per minute. What is its stroke volume at that time?
   a. 366 mL/beat
   b. 0.0366 L/beat
   c. 89.1 mL/beat
   d. 3.66 L/beat
   e. 3.66 beats/L

2. Which of the following would be expected to elicit a baroreceptor reflex?
   a. a person jumping out of bed in the morning
   b. a squirrel escaping from a hawk, but not until being cut by the bird's talons
   c. a giraffe lifting its head from drinking at a lake to scan the environment for predators
   d. an experimental animal infused intravenously with a large volume of fluid
   e. All of the above are correct.

41.3 The Secondary Phase of the Homeostatic Response to Hemorrhage

Learning Outcomes

1. Describe how and why water moves between body fluid compartments after a hemorrhage.
2. Explain how the endocrine and urinary systems help to restore blood volume.
3. Describe the mechanism by which hematocrit first decreases and then is returned to normal, as well as how this mechanism depends on the endocrine and muscularskeletal systems.
4. List some changes in the functions of the reproductive, integumentary, and digestive systems that contribute to a restoration of homeostasis after hemorrhage.

After the baroreceptor reflex and venous return contribute to stabilizing the blood pressure, the next phase of the compensatory response to the homeostatic challenge of blood loss is to restore fluid volume and the number of circulating erythrocytes. In this section, we will examine a special way in which fluid compartments in an animal's body can be redistributed to increase blood volume and how the endocrine and urinary systems contribute to the long-term restoration of pressure and hematocrit.

Water Moves from Interstitial Fluid to the Plasma After a Hemorrhage

Recall from Figure 32.13 that water is found in three compartments of an animal's body: within cells (intracellular fluid), around cells (interstitial fluid), and in the blood (plasma). Water will move between compartments if there is an osmotic gradient or if there is a difference in fluid pressure. Ernest Starling determined that water moves across a capillary between the plasma and interstitial fluid compartments due to the difference between these two forces (**Figure 41.5**). In mammals, plasma and interstitial fluid are very similar in composition, except that much more protein is found in plasma than in interstitial fluid. This difference creates an osmotic force that tends to draw water into a capillary. By contrast, the fluid (hydrostatic) pressure in a capillary is much higher than in the surrounding interstitial fluid, and this pressure difference tends to force fluid out of the capillary. The net effect of these forces, now called **Starling forces,** is that fluid leaves at the beginning of a capillary due to hydrostatic pressure and then, as the pressure decreases along the length of the vessel, re-enters at the other end of the capillary due to osmosis.

After a hemorrhage occurs, the delicate balance between the pressure difference and osmotic gradient is altered because the blood pressure decreases. The net effect of the change in Starling forces is that more water enters the capillary by osmosis than is forced out of the capillary by hydrostatic pressure. This shift in fluid from the interstitial fluid to the blood helps to restore blood volume. This effect takes some time to develop and is not as rapid as the baroreceptor reflex, but it is a very important part of the longer-term response to hemorrhage.

What effect does this redistribution of fluid have on the hematocrit—the percentage of blood volume that is occupied by erythrocytes (refer back to Figure 36.4)? For a few hours after a hemorrhage, the hematocrit does not change, because whole blood is lost from the injured vessel. There are fewer total erythrocytes but also less total fluid in proportion, so the hematocrit is initially unchanged. However, it will eventually decrease below normal as the blood becomes diluted by the movement of fluid into capillaries due to Starling forces. As you have learned throughout this unit, homeostatic responses to one challenge often produce new challenges, which must in turn be addressed at a later time. In this case, helping to restore blood volume so that pressure can be maintained is of more immediate importance than is bringing the hematocrit back to normal. As we will see next, hematocrit is eventually restored during later stages of the secondary phase.

Hematocrit Changes in a Biphasic Manner After a Hemorrhage

Recall that initially hematocrit does not change when blood is lost following a hemorrhage. As time goes on, the blood becomes more dilute (as just described) and the hematocrit decreases below normal (**Figure 41.6**). This decrease must eventually be corrected so that oxygen transport in the blood can be returned to normal.

Erythrocytes mature in the marrow of certain bones of the mammalian skeleton. Stem cells in the bones follow a path of differentiation characterized by the loss of the nucleus and other organelles. This differentiation is completed in the blood once the cells leave the bone marrow. The hormone erythropoietin (EPO) controls the rate of this differentiation and increases the survival of the developing cells in the marrow. EPO is made primarily by the kidneys. Whenever the total amount of hemoglobin in the blood is decreased below normal for any reason (such as a decreased hematocrit), the EPO-producing cells of the kidneys are activated to produce and secrete EPO into the blood. The effect of the hormone on the number of circulating

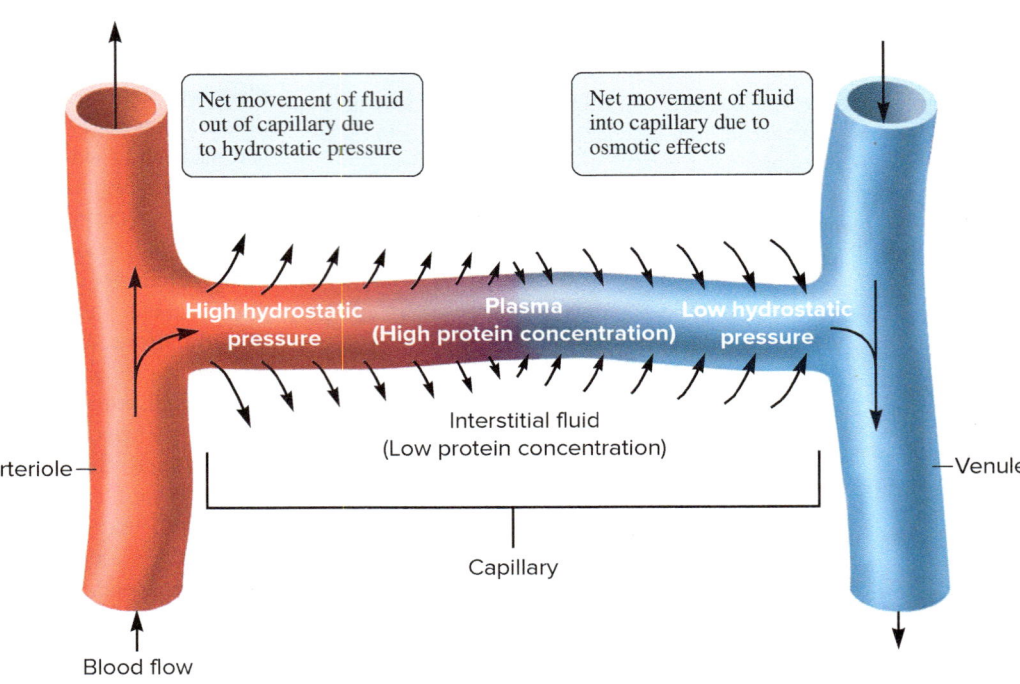

Figure 41.5 **Starling's forces in a capillary in the absence of a hemorrhage.** Capillaries are the smallest blood vessels and allow the movement of water and solutes between the plasma and interstitial fluid. Near the arteriole, the pressure of blood moving through the capillary forces a protein-free filtrate of plasma out of the capillary, but near the venule, the osmotic effect of proteins in the blood draws fluid into the capillary. When blood volume is lowered, as in a hemorrhage, a change in the balance between these two forces will change the net flow of fluid. Hemorrhage decreases blood pressure, resulting in a greater net movement of fluid into the capillary from the interstitial fluid to the plasma. This diagram does not show it, but some fluid that exits a capillary is not recaptured but, instead, enters a nearby lymphatic vessel.

Concept Check: *What would happen to fluid moving across a capillary if the protein concentration of plasma is decreased?*

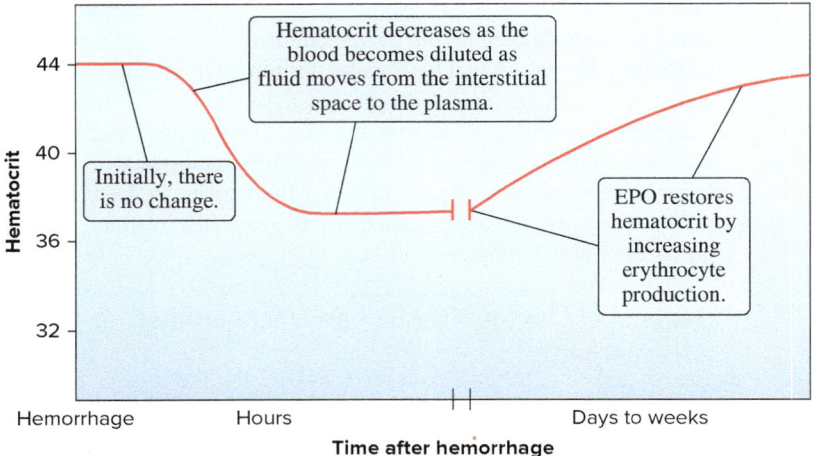

Figure 41.6 **The effect of fluid shifts on hematocrit of the blood following hemorrhage.** Note that hematocrit decreases as fluid moves from the interstitial fluid into the blood vessel. This is later compensated for by the production of more erythrocytes in the bone marrow and their release into the blood under the control of the hormone erythropoietin (EPO).

erythrocytes is not apparent for at least several hours, and it does not fully restore hematocrit for up to several days or even weeks, depending on the magnitude of the hemorrhage.

Interestingly, some mammals such as horses, seals, and pigs contain a large reservoir of mature erythrocytes in their spleen, an organ of the immune system involved in the production of leukocytes (refer back to Figure 40.4). In response to activation by the sympathetic nervous system, smooth muscle cells in the spleen contract, causing it to eject its reservoir of erythrocytes into the circulation. This effect happens much more quickly than the slow EPO-induced restoration of hematocrit and provides a clear survival advantage after hemorrhage. Under normal conditions, the spleen response gives such animals a boost in the oxygen-carrying capacity of the blood during periods of increased activity or, in the case of seals, when diving under water for long periods.

The Endocrine and Urinary Systems Function Together to Retain Fluid in the Body

An animal that has lost significant blood would seem to benefit from decreasing the amount of fluid that is lost via the urine. On the other hand, the kidneys serve the important function of removing soluble wastes from the blood, which requires water for excretion. This is another example of how the response to a homeostatic challenge may result in new challenges that must be met at a later time.

The sympathetic nervous system decreases the amount of plasma that is filtered into the kidney nephrons by mediating the vasoconstriction of the renal afferent arterioles (**Figure 41.7a** and refer back to Figure 37.20). With less blood entering the glomerular capillaries, less fluid is filtered into the nephron tubules. This conserves fluid in the blood but also limits the kidneys' ability to remove soluble wastes. The accumulation of wastes in the blood needs to be corrected eventually, but in the short term, the net effect of helping to minimize water loss in the urine may be lifesaving. In addition, as the limited volume of filtrate that does get formed moves through the

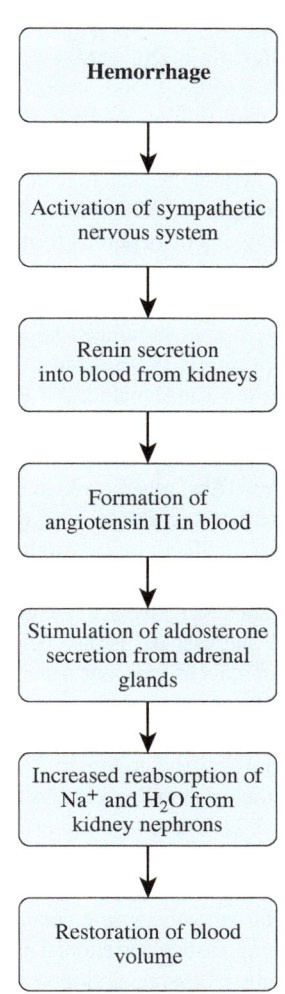

(a) Effect of vasodilation and vasoconstriction of afferent arteriole on the amount of filtrate formed in a nephron

(b) Actions of hormones on kidney nephrons leading to the increased return of Na⁺ and H₂O to the blood after hemorrhage

(c) Stimulation of aldosterone secretion and its effect on blood volume

Figure 41.7 A nephron of the mammalian kidney, with its associated afferent and efferent arterioles. (a) Schematic illustration of the effect of changing the diameter of afferent arterioles on the filtration of fluid out of a capillary. The afferent arteriole brings blood into a glomerulus associated with a kidney nephron. The relative amount of filtrate (red arrows) in the proximal tubule of the nephron is indicated by the thickness of the arrow. Vasoconstriction of an afferent arteriole will decrease the amount of fluid that gets filtered into the tubule, thereby preserving fluid in the blood. (b) Additional changes along the distal tubule of the nephron and collecting ducts promote greater reabsorption of Na⁺ and water into the blood. The result is a decreased volume of urine that is more concentrated than normal, as well as a return of additional fluid to the circulation. (c) Production of the hormones angiotensin II (AII) and aldosterone, along with some major actions. Intermediate steps in AII synthesis are omitted for simplicity.

nephron tubules, greater amounts of Na⁺ and water are reabsorbed from the filtrate and returned to the blood, making the urine more concentrated and retaining more fluid in the body (**Figure 41.7b**).

These effects on urine and blood volume are mediated by hormones produced by the endocrine system. Aldosterone, a steroid hormone produced by the adrenal glands, is secreted whenever blood volume is decreased. It acts on distal regions of nephrons to increase Na⁺ reabsorption, which creates an osmotic gradient that draws water from the filtrate in the tubules into the blood (refer back to Figure 37.22).

The synthesis of aldosterone is stimulated by another hormone, called angiotensin II (AII), which is synthesized in the blood whenever blood pressure is decreased. The signal for AII formation, in turn, is an enzyme called renin, which is released into the blood by kidney cells whenever blood pressure is low and the sympathetic nervous system is activated (**Figure 41.7c**). Thus, a hemorrhage elicits a sequence of events that leads to the synthesis of aldosterone, thereby helping to minimize the volume of urine and increase the volume of the blood.

In addition to stimulating aldosterone production, AII is also a potent vasoconstrictor of arterioles and directly helps to increase blood pressure by that mechanism. In some animals, AII is also believed to stimulate thirst and thus encourages animals to seek out water to drink. This is beneficial because the ingested water will be absorbed into the blood.

The other major hormone that helps to retain water in the body is antidiuretic hormone (ADH) (refer back to Figure 37.23). ADH stimulates cells in the tubules of the mammalian kidneys to reabsorb water (but not ions) into the blood, concentrating the urine. The secretion of ADH is triggered by signals from the baroreceptors that activate the hypothalamic neurons that synthesize ADH. Like AII, ADH also has direct actions on vascular smooth muscle cells, causing them to contract and thereby promoting vasoconstriction. Thus, AII and ADH increase blood pressure directly by vasoconstriction and indirectly (via aldosterone, in the case of AII) by helping to retain water in the body.

Other Organ Systems Are Affected by Hemorrhage and May Help in the Compensatory Response

As we have seen, the circulatory, nervous, respiratory, endocrine, and urinary systems have a variety of important functions that contribute to an animal's ability to survive a serious hemorrhage. In addition, all other organ systems are affected to some degree by hemorrhage, and they all are involved, at least indirectly, in the overall compensatory response that restores blood volume and blood pressure.

You have already learned how the respiratory system contributes to the return of blood from the veins to the heart. The muscular-skeletal system may also assist the flow of blood from the limbs to the heart. Recall that one-way valves within limb veins help to move blood against gravity toward the heart (refer back to Figure 36.13). Each time the skeletal muscles of the limbs contract, the veins coursing through those muscles are squeezed and blood is pushed through the valves. The anxiety that may be triggered by the injury and hemorrhage may encourage such muscular activity, and this has the benefit of contributing to the Frank-Starling effect.

The integumentary system also contributes indirectly as the arterioles that normally deliver blood to the skin constrict, redirecting blood to vital organs. This also happens to blood vessels throughout much of the alimentary canal of the digestive system. In addition, the smooth muscles of the canal itself relax, due to inhibitory actions of the sympathetic nervous system. This effect occurs even if an animal has just eaten a meal prior to the hemorrhage, and it results in a considerable energy savings. In such emergency situations, energy supplies tend to be spared for the heart, the brain, and a few other organs, such as the placenta in a pregnant female mammal.

In major homeostatic challenges such as hemorrhage, energy is also redirected away from the reproductive system. At such times, it is more important that the individual survives so that it can reproduce another day.

Finally, even the immune and lymphatic systems respond to hemorrhage. You have learned that a major immune system organ, the spleen, may contribute to the maintenance of a normal hematocrit in certain mammals. In addition, if the injury has resulted from an external wound, part of the response to hemorrhage is to prevent infection. The cells and secretions of the innate immune response trigger a local inflammatory reaction (refer back to Figure 40.2) at the site of the wound, and the adaptive immune response responds to any specific pathogens that have entered the circulation or local tissues.

Table 41.1 summarizes some of the major events of the homeostatic response to hemorrhage in a mammal. Not all homeostatic challenges engage such widespread responses, but typically the activities of more than one organ system are affected directly or indirectly. There are many common examples of this fundamental feature of homeostasis. For example, when you are ill, you may have a fever caused by secretions of the cells of the immune system. Fever may contribute to fighting off pathogens. However, fever also induces sweating, which in turn decreases the amount of fluid and alters the concentrations of Na⁺ and other ions in the blood. The urinary and endocrine systems help to restore fluid and ion balance. If you miss a meal or two, the endocrine and nervous systems together help to maintain glucose homeostasis in the blood. If you move to a high altitude or go mountain-climbing, the respiratory system helps to maintain the amount of oxygen in your blood by stimulating your breathing rate. That, however, decreases the amount of carbon dioxide in the blood, which in turn increases the pH of the blood. The urinary system will then help to restore pH to normal by excreting bicarbonate ions in the urine and reabsorbing more hydrogen ions into the blood. These and many other examples demonstrate once again that organ systems do not function in isolation but are integrated with each other to ensure homeostasis of all of an animal's physiological functions.

41.3 Reviewing the Concepts

- After hemorrhage, fluid flows from the interstitial space to the plasma, which helps to restore blood volume (Figure 41.5).
- Hematocrit decreases as fluid enters the plasma due to Starling forces, then is slowly restored to normal by the action of erythropoietin on immature erythrocytes in bone marrow (Figure 41.6).
- Antidiuretic hormone and aldosterone (stimulated by angiotensin II) act on kidney tubules to help retain water and ions in the blood, thereby concentrating and decreasing the amount of urine produced and increasing blood volume (Figure 41.7).
- All organ systems are affected by hemorrhage to some degree and directly or indirectly contribute to the restoration of homeostasis.

Table 41.1 Summary of Key Homeostatic Responses Following a Hemorrhage

Variable	Rapid responses	Secondary responses
Rate of blood flow to organs	Decreased to most organs, but less so to heart and brain	Restored to normal as fluid shifts into the plasma and the kidneys reabsorb water
Heart rate (HR)	Increased	Returns to normal once pressure is capable of sustaining itself
Stroke volume (SV)	Decreased due to less venous return, but the decrease is minimized by increased force of contraction of the heart	Increased toward normal due to Frank-Starling effect and sympathetic nervous system (norepinephrine)
Cardiac output (HR × SV)	Maintained near normal by baroreceptor reflex if hemorrhage is not severe, but decreased in severe hemorrhage	Slowly returned to normal or even increased above normal as venous return increases
Hematocrit	Not usually affected in a rapid response, except in animals with contractile spleens that release erythrocytes into the bloodstream	Decreased due to fluid moving into blood, but eventually restored by erythropoietin
Breathing rate and depth	Increased (improves venous return)	Return to normal once pressure is normal and blood pH is restored
Hemoglobin (Hb) saturation	Increased toward 100% by hyperventilation	Restored to the normal range
Unloading of oxygen from Hb	Increased by Bohr effect due to lactic acid	Restored to normal once blood flow to organs returns to normal and pH is normalized
Urine production	Decreased due to vasoconstriction of afferent arterioles	Further decreased due to effects of ADH and aldosterone

41.3 Testing Your Knowledge

1. The initial response to hemorrhage is mediated largely by the _____ system(s); the later responses are mediated largely by the _____ system(s).
   a. nervous, immune
   b. nervous and circulatory, endocrine and urinary
   c. urinary and endocrine, respiratory
   d. urinary and endocrine, nervous
   e. respiratory, nervous

2. _____ increases hematocrit, _____ increases the reabsorption of only water from the renal filtrate, and _____ increases the reabsorption of both Na$^+$ and water.
   a. Aldosterone, antidiuretic hormone (ADH), erythropoietin (EPO)
   b. EPO, ADH, aldosterone
   c. EPO, aldosterone, ADH
   d. ADH, EPO, aldosterone
   e. ADH, aldosterone, EPO

41.4 Impact on Public Health

Learning Outcomes

1. Explain the relation between the degree of hemorrhage and its consequences in a human.
2. Describe procedures used to help restore normal circulatory function in humans after a large hemorrhage.

Let's now return to the case of the girl described at the beginning of the chapter. Her initial symptoms were light-headedness, agitation without any sign of mental confusion, rapid breathing rate, a slightly lower blood pressure than normal for her, and a very rapid heart rate. All of these suggested significant internal bleeding due to the rupture of one or more blood vessels near one of her kidneys. Once in the ambulance, therefore, she was placed on her back with her feet slightly elevated. Gravity helped to return blood from her leg veins to the right atrium of her heart. This effect of gravity was important because the amount of blood ejected with each heartbeat is roughly proportional to the amount of blood that fills the heart (recall the Frank-Starling Law of the Heart). She was anxious but still conscious and alert, suggesting that only a moderate hemorrhage had occurred.

A decision was made to not infuse whole blood but instead to simply supply fluids. A cannula was placed into a vein in her hand and attached to a solution of ions, which was allowed to rapidly enter her circulation at an initial rate of about 1 liter/10 minutes. The fluid was approximately iso-osmotic with her blood. Recall from Chapter 32 (and refer back to Figure 5.10) that cells can be deformed and even killed when exposed to extracellular fluid with an osmolarity that is very different from normal.

After sufficient fluids were infused, the girl's blood pressure rose to 114/71 mmHg, which was within the normal range for her. Her heart rate was still somewhat elevated at 84 bpm, compared to her usual resting rate of about 72 bpm; this, however, could be attributed mainly to her continued anxiety, not to the baroreceptor reflex. Her breathing rate by this time was a steady 15 breaths per minute, normal for a girl her age, and her hematocrit was 34 (normal is about 38–46 for females). A blood sample was drawn to test for indicators of kidney damage. Later analyses of the sample, along with a CT scan of her thorax and abdomen, revealed that the kidneys and other major organs were functioning properly and undamaged, and thus it appeared that fluid replacement was sufficient in this case. Her hematocrit was low for some time, but it eventually recovered due to the actions of erythropoietin.

Had the hemorrhage been more severe—for example, resulting in a loss of greater than 20% of her total blood volume—the outcome *ethos*, meaning habit or manner), focused on specific genetic and physiological mechanisms of behavior. For example, we can hypothesize that male deer rut or fight with other males in the fall because a change in day length stimulates the eyes, brain, and pituitary gland and triggers hormonal changes in their bodies. These factors, which directly affect behavior, are called **proximate causes.** However, we can also hypothesize that male deer fight to determine which deer get to mate with the most female deer and pass on their genes. This hypothesis leads to a different answer from the one that is concerned with changes in day length. This answer focuses on the adaptive significance of fighting to the deer—that is, on the effect of a particular behavior on reproductive success. These factors are called **ultimate causes** of behavior.

In this chapter, we will explore the role of both proximate and ultimate causes of behavior. We will begin by exploring how behavior is determined, which involves the roles of both genetics and the environment. We will consider how different behaviors include movement, communication, and group living. Finally, we will examine whether an organism can truly behave in a way that benefits others at a cost to itself and how behavior shapes various mating systems.

42.1 The Influence of Genetics and Learning on Behavior

Learning Outcomes

1. Describe the differences among innate behavior, classical conditioning, and operant conditioning.
2. Describe how memory is used in local movement.

As we will see in this section, behavior is controlled by both genetics and the environment. Genetics allows the evolutionary responses of ancestors to selection to be incorporated into current behavior, while responses to the environment permit animals the opportunity to change their behavior over the short term.

Innate Behavior Patterns Are Genetically Programmed

Behaviors that seem to be genetically programmed are referred to as **innate** (also called instinctual). For example, a spider will spin a specific web without ever seeing a member of its own species build one. A classic example of innate behavior is the egg-rolling response in geese (**Figure 42.1**). If an incubating goose notices an egg out of the nest, she will extend her neck toward the egg, get up, and then roll the egg back to the nest using her beak. Such behavior improves the survival of offspring. Eggs that roll out of the nest get cold and fail to hatch. Geese that fail to exhibit the egg-rolling response pass on fewer of their genes to future generations. Interestingly, any round object, from a wooden egg to a volleyball, can elicit the egg-rolling response. Objects that elicit such a behavioral response are called **releasers**.

Learning Occurs When Animals Modify Their Behavior Based on Past Experience

Although many of the behavioral patterns exhibited by animals are innate, animals can modify their behavior based on previous experience, a process that involves learning. Perhaps the simplest form of learning is **habituation,** in which an organism learns to ignore a repeated stimulus. For example, animals in African safari parks become habituated to the presence of vehicles containing tourists. Habituation can be a problem at airports, however, where birds eventually ignore the alarm calls designed to scare them away from the runways.

Habituation is a form of nonassociative learning, which is learning without a positive or negative reinforcement. A positive or negative association between a stimulus and a response is termed **associative learning.** The two main types of associative learning are classical conditioning and operant conditioning.

In **classical conditioning,** an involuntary response comes to be associated with a stimulus that did not originally elicit the response. This type of learning was investigated by Russian psychologist Ivan Pavlov. In his original experiments in the 1920s, Pavlov restrained a hungry dog in a harness and presented small portions of food at regular intervals. The dog salivated whenever it smelled the food. Pavlov then began to sound a metronome when presenting the food. Eventually, the dog salivated at the sound of the metronome, whether or not the food

Figure 42.1 Innate behavior. A female goose retrieves an egg that has rolled outside the nest through a set sequence of movements. The goose completes the entire sequence even if a researcher takes the egg away before the goose has rolled it back to the nest.

(a) Blue jay eating a monarch   (b) Vomiting reaction

Figure 42.2 Operant conditioning, also known as trial-and-error learning. **(a)** A young blue jay will eat a monarch butterfly, not knowing that it is noxious. **(b)** After the first experience of eating a monarch and vomiting, a blue jay will avoid the insects in the future.

Concept Check: What is the difference between operant conditioning and classical conditioning?

BioConnections: Look ahead to Figure 43.12f. How might operant conditioning relate to the similar appearance of king snakes and coral snakes?

was present. Classical conditioning is widely observed in animals; for example, many insects quickly learn to associate certain flower odors with nectar rewards. In some humans, the sound of a dentist's drill is enough to produce a feeling of uneasiness, tension, and sweaty palms.

In **operant conditioning,** an animal's behavior is reinforced by a consequence, either a reward or a punishment. Operant conditioning, also called trial-and-error learning, is common in animals. Often, it is associated with negative rather than positive reinforcement. For example, toads eventually refuse to strike at insects that sting, such as wasps and bees, and birds learn to avoid bad-tasting butterflies (**Figure 42.2**).

Behavior is also affected by **spatial learning,** in which specific landmarks are learned to construct cognitive maps, mental representations of the spatial relationships of objects, to aid in local movements. For example, in the fall, some birds known as nutcrackers (genus *Nucifraga*) store thousands of seeds in many underground caches over a 35-km^2 area. The birds relocate many of these caches in the winter but leave alone the caches of other nutcrackers. As long ago as the 1930s, Dutch-born ethologist Niko Tinbergen showed how the female digger wasp uses landmarks to relocate her nests, as described next.

FEATURE INVESTIGATION

Tinbergen's Experiments Showed That Digger Wasps Learn the Positions of Landmarks to Find Their Nests

In the sandy, dry soils of Europe, the solitary female digger wasp (*Philanthus triangulum*) digs four or five nests in which to lay her eggs. Each nest stretches obliquely down into the ground for 40–80 cm. After digging the nesting holes, the wasp performs a sequence of apparently genetically programmed events. She catches and stings a honeybee, which paralyzes it; returns to the nest; drags the bee into the nest; and lays an egg on it. The egg hatches into a larva, which then feeds on the paralyzed bee. However, the larva needs to ingest five to six bees before it is fully developed. This means the wasp must catch and sting four or five more bees for each larva. She can carry only one bee at a time. After each visit, the wasp must seal the nest with soil to prevent the bee from being stolen, find another bee, relocate the nest, open it, and add the new bee. How does the wasp relocate the nest after spending considerable time away? Niko Tinbergen observed the wasps hover and fly around the nest each time they took off. He hypothesized that they were learning the nest position by creating a mental map of the landmarks in the area.

To test his hypothesis, Tinbergen experimentally adjusted the landmarks around the burrow that the wasps might be using as cues (**Figure 42.3**). First, he put a ring of pinecones around the nest entrance to train the wasp to associate the pinecones with the nest. Then, when the wasp was out hunting, he moved the circle of pinecones a distance

Figure 42.3 How Tinbergen discovered the digger wasp's nest-locating behavior.

Concept Check: How would you test what type of spatial landmarks are used by female digger wasps?

HYPOTHESIS Digger wasps (*Philanthus triangulum*) use visual landmarks to locate their nests.

STARTING LOCATION The female digger wasp excavates an underground nest, to which she returns daily, bringing food to the larvae located inside.

1 Place a ring of pinecones around the nest to train the wasp to associate pinecones with the nest.

the scout traces a figure 8, in the middle of which she waggles her abdomen and produces bursts of sound. Again, the other bees maintain contact with her. The amazing part of the waggle dance is that the angle at which the central part of the figure 8 deviates from the vertical direction of the comb represents the same angle at which the food source deviates from the point at which the Sun hits the horizon (Figure 42.6c). The direction is always up-to-date, because the bee adjusts the dance as the Sun moves across the sky.

42.2 Reviewing the Concepts

- Communication is a form of behavior. The use of different forms of communication between organisms depends on the environment in which they live.
- Chemical communication often involves marking territories; auditory and visual forms of communication are used to attract or defend mates; and tactile communication is used to establish bonds between group members (Figures 42.5, 42.6).

42.2 Testing Your Knowledge

1. Most terrestrial mammals are nocturnal. This means they are likely to communicate primarily via what means?
   a. visual
   b. auditory
   c. chemical
   d. both a and b
   e. both b and c

42.3 Living in Groups and Optimality Theory

Learning Outcomes
1. Outline the costs and benefits of living in groups.
2. Outline the costs and benefits of defending a territory.
3. **SCISKILLS** ▶ Explain how game theory is used to determine winners between contestants.

Some of the more complex animal behavior occurs when individuals live together in groups such as flocks or herds. When is it best to live alone, and when does it benefit an individual to live in a group? One way to approach this question is to assess the costs and benefits involved. Although group living increases competition for food and the spread of disease, it also has benefits that compensate for the costs involved. Many of these benefits relate to group defense against predators and locating and securing food sources. Group living can reduce predator success in at least two ways: through increased vigilance and through protection in numbers.

Living in Large Groups May Reduce the Risk of Predation Because of Increased Vigilance

For many predators, success depends on the element of surprise. If an individual prey is alerted to an attack, the predator's chance of success is lowered. A woodpigeon (*Columba palumbus*) in a flock takes to the air when it spots a goshawk (*Accipiter gentilis*). Once one pigeon takes flight, the other members of the flock are alerted and follow suit. If each individual in a group occasionally scans the environment for predators, the larger the group, the less time an individual forager needs to devote to vigilance and the more time it can spend feeding. This explanation is referred to as the **many-eyes hypothesis** (Figure 42.7). Of course, cheating is a possibility, because some birds might never look up, relying on others to keep watch while they keep feeding. However, the individual that happens to be scanning when a predator approaches is most likely to escape, whereas cheaters are more likely to get caught and not pass their genes to future generations.

Living in Groups Offers Protection by the Selfish Herd

Group living also provides protection in sheer numbers. Typically, predators take one prey animal per attack. In any given attack, an individual antelope in a herd of 100 has a 1 in 100 chance of being selected, whereas a single individual has a 1 in 1 chance. Large herds

Figure 42.7 Living in groups and the many-eyes hypothesis. The larger the number of woodpigeons, the less likely an attack will be successful.

Concept Check: *With regard to attacks by predators, what other advantage do large groups provide to individuals?*

may be attacked more frequently than a solitary individual, but a herd is unlikely to attract 100 times more attacks than an individual, often because of the territorial nature of predators. Furthermore, large numbers of prey are able to defend themselves better than single individuals, which usually choose to flee. For example, groups of nesting black-headed gulls mob a crow, thereby reducing the crow's ability to steal the gulls' eggs.

Research has shown that within a group, each individual can minimize the danger to itself by choosing the location that is as close to the center of the group as possible. This was the subject of a famous paper, "The Geometry of the Selfish Herd," by the British evolutionary biologist W. D. Hamilton. The explanation of this type of defense is that predators are likely to attack prey on the periphery because they are easier to isolate visually. Many animals in herds tend to bunch close together when they are under attack, making it physically difficult for the predator to get to the center of the herd.

Predator Success May Be Increased in Groups

Group living may allow predators such as lions and wolves the opportunity to locate and take down prey of a disproportionately large size, which a single predator would be unable to handle. Furthermore, as we saw with honeybees, when animals search for food within a group, each individual may be able to exploit the discovery of others.

Interestingly, the actual size of predator groups may be different from the optimal size. Imagine that a solitary killer whale hunting for a seal has a given payoff for its efforts (Figure 42.8). Here, payoff is defined as the amount of energy gained by eating the prey minus the amount of energy expended in capturing it. If another killer whale joins in the hunt, the seal is more easily caught, and despite having to share the prey, payoff for either individual increases. A third killer whale increases the payoff even more. Three whales is the optimal group size, for which the payoff to each individual is maximized. When the group size becomes four or five, then the average payoff to each individual decreases, because the food has to be shared among more individuals. However, living in a group of four or five is still a better option than hunting alone, so whales continue to join groups of two, three, or four other whales. When the group size is six or more, a whale would do better on its own than in the group. In this example, the most commonly observed or stable group size is five individuals. Paradoxically, the optimal group size of three, which affords the greatest payoff to each whale, is not equivalent to the stable group size.

Some Animals Defend a Territory

Some animals prefer living a solitary existence to living in a herd. Many of these animals actively defend a **territory**, a fixed area in which an individual excludes other members of its own species, and sometimes other species, by aggressive behavior or territory marking. The primary benefit of a territory is that it provides exclusive access to a particular resource, whether it be food, mates, or sheltered nesting sites. Large territories provide more of a resource but may

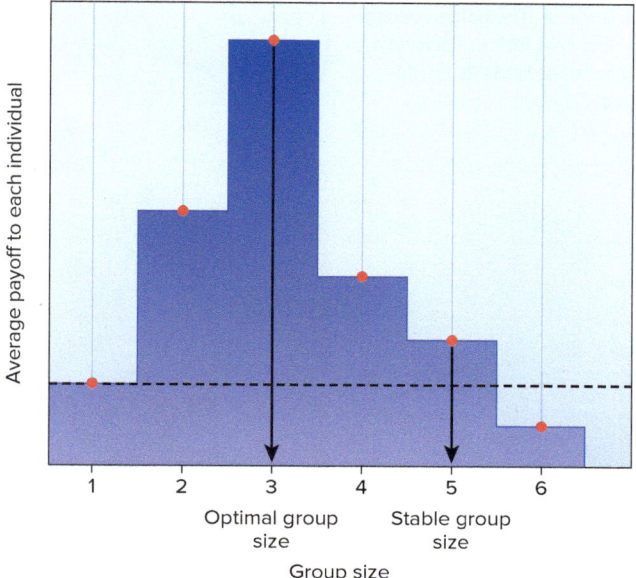

Figure 42.8 Optimal and stable group size. The payoff to a group of two or three predators is greater than for a solitary predator. The highest payoff is for a group size of three, which is the optimal size. However, other solitary individuals would benefit from joining the group, and group size increases to five individuals. In a group size of six, any one individual would do better on its own. The stable group size is then five.

be costly to defend, whereas small territories that are less costly to defend may not provide enough resources.

The best size of a territory may be predicted by **optimality theory**, which states that an animal should behave in a way that maximizes the benefits of a behavior minus its costs. In this case, the benefits are the nutritional or caloric values of the food items within the territory, and the costs are patrolling and defending the territory. When the difference between the energetic benefits of food gathering and the energetic costs of maintaining a territory is maximized, an organism is optimizing its behavior. Optimality theory can also be used in decisions about food gathering and when it may be profitable to remain at a patch to look for food versus when is better to look for a completely new patch.

In studies of the territorial behavior of the golden-winged sunbird (*Nectarinia reichenowi*) in East Africa, American ornithologists Frank Gill and Larry Wolf measured the energy content of nectar as the benefit of maintaining a territory and compared it with the energy costs of activities such as perching, flying, and fighting (Figure 42.9). Defending the territory ensured that other sunbirds did not take nectar from available flowers, thus increasing the amount of nectar in each flower. In defending a territory, the sunbird gained 780 kilocalories (kcal) a day in extra nectar content. However, the sunbird also spent 728 kcal in defense of the territory, yielding a net gain of 52 kcal a day and making territorial defense advantageous.

Defending a territory can be dangerous, and biologists have long been interested in determining the outcome of contests between individuals. American mathematician John Nash developed **game theory** to establish the outcome of contests in which an individual's choice

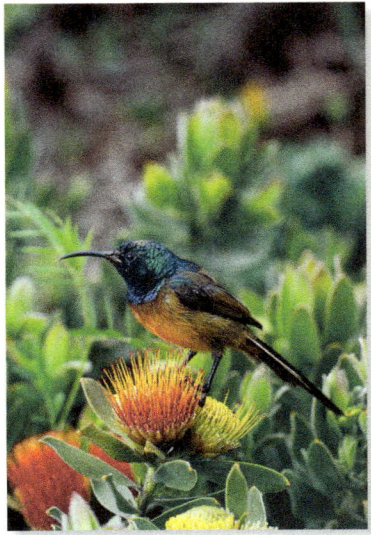

Figure 42.9 The golden-winged sunbird of East Africa (*Nectarinia reichenowi*) defends its territory.
©Tony Camacho/Science Source

Figure 42.10 Hawk and Dove strategies. In these dueling meerkats, the individual on the left is escalating the fight, behaving in a Hawklike manner, and the individual on the right is in a submissive posture, a Dovelike strategy.
©SCFotos/Stuart Crump Visuals/Alamy

of strategy depends on the actions of other individuals. Originally developed to understand human economic behavior, game theory has been used to help describe the behavior of animals in competitive situations, as described next.

Biology Principle | Biology is a quantitative science

Quantitative Analysis: Game Theory Establishes Whether Individuals Fight or Flee

Imagine a game in which individuals of the same species engage in contests over resources. The possible behaviors for any given individual are to display, presenting itself as a contestant; to fight to win, attacking until the opponent retreats; and to retreat, giving up the resource to the opponent. Some individuals, the "Hawks," always fight to win, risking injury to themselves in the process. "Doves" initially perform aggressive displays when encountering another Dove but always retreat before injury occurs. When a Hawk encounters a Dove, the Dove does not display, but instead assumes a submissive posture (**Figure 42.10**).

Let's assign the winner of a contest +50 and the loser 0. The cost of injury is −100 because the injured player may not be able to compete again for a long time or may die, and the cost of spending time in a display is −10. It is assumed that Hawks and Doves reproduce in proportion to their scores. What happens when Hawks fight other Hawks or Doves, or when Doves fight Doves? We can analyze such a contest by constructing a two-by-two matrix with the average rewards for four possible types of encounter (**Table 42.1**). In a population of all Hawks, the chance of winning is 50%. Half the population wins and gets a reward of 50, and half the population loses and gets −100. The average is ½(50) + ½(−100) = −25. Thus, any individual playing Dove would do better because, when a Dove meets a Hawk, it gets 0 in every encounter, since it retreats immediately, which is not very good but still better than −25. Therefore, Hawk is not always the best strategy. In a population of all Doves, the chance of winning is 50%. Half of the population wins and gets a reward of 40 (the reward minus the cost of displaying), and half of the population loses and gets −10 (the cost of display); the score, on average is ½(50 − 10) + ½(−10) = +15. Even though the average score is positive, Dove is not always the best strategy, either. In this population, any individual Hawk would do very well, and the Hawk strategy would soon spread because when a Hawk meets a Dove, it gets +50 in every encounter.

Although all individuals in a population will not always adopt a Hawk or Dove strategy, a mixture of Hawk and Dove strategies is more likely. In such a population, a stable equilibrium is reached in which the average reward for a Hawk is equal to the average reward for a Dove. For the rewards we have specified, the stable mixture can be calculated as follows. If h is the proportion of Hawks in the population, the proportion of Doves is $(1 − h)$. The average reward for a Hawk, H, is the reward for each type of fight multiplied by the probability of meeting each type of contestant.

Therefore, the average reward for a Hawk will be

$$H = -25h + 50(1 - h)$$

Table 42.1 Average Rewards to the Attacker in the Game Between Hawk and Dove

Attacker	Opponent	
	Hawk	Dove
Hawk	½(50) + ½(−100) = −25	+50
Dove	0	½(50 − 10) + ½(−10) = +15

Costs and rewards: Winner +50   Injury −100

Loser 0   Display −10

For Doves, the average reward, D, will be

$$D = 0h + 15(1 - h)$$

At the stable point, H is equal to D. When H = D, the proportion of Hawks $(h) = 7/12$, and the proportion of Doves $(1 - h) = 5/12$. You can verify this as follows:

$$H = -25(7/12) + 50(5/12) = -175/12 + 250/12 = 75/12$$
$$D = 0 (7/12) + 15(5/12) = 0 + 75/12 = 75/12$$

This stable point could be achieved in two ways. First, the population could consist of 7/12 of individuals who always played Hawk and 5/12 who always played Dove. Alternatively, the population could consist of individuals who all adopted a mixed strategy, playing Hawk in 7/12 contests and Dove in the other 5/12.

Despite the simplicity of these models, some important conclusions can be drawn from them:

1. Fighting strategy is frequency dependent: It depends on what other animals are doing. A Hawk strategy is good for an individual in a population of Doves but not in one of Hawks. Territory owners may be more likely to behave like Hawks than challengers.

2. The strategies employed by individuals are dependent on the values of rewards. If the rewards are changed, the proportion of individuals playing Hawk and Dove strategies will change.

3. The frequency of Hawk behavior increases as the payoff increases. For males, one of the biggest payoffs is the opportunity to mate with a female because failure to do so is equivalent to genetic death. Fights over females are therefore commonly severe and sometimes fatal, and Hawk strategies dominate.

In the real world, animals also adjust their strategies according to the vigor of their opponents. It would be pointless to adopt a Hawk strategy against a bigger opponent, even for a territory owner. Players may use a strategy that follows the rule "if larger, behave like a Hawk; if smaller, behave like a Dove; if equally matched, adopt the Hawk strategy if a territory owner."

Crunching the Numbers: What is the average reward to a Hawk in a Hawk-versus-Hawk confrontation, where the reward to the winner is +30 and all other rewards are the same as in Table 42.1?

42.3 Reviewing the Concepts

- Many benefits of group living relate to defense against predators, offering protection through sheer numbers. Groups of predators may enjoy greater success hunting in a pack than alone (Figures 42.7, 42.8).
- Some animals prefer a solitary existence over group living and may maintain exclusive territories. The size of a territory, a fixed area in which an individual or a group excludes other members of its own species, tends to be optimized according to the costs and benefits involved (Figure 42.9).
- Game theory can be used to determine the strategies used by animals in contests over territories, food, or mates (Figure 42.10, Table 42.1).

42.3 Testing Your Knowledge

1. In nature, the most commonly observed number of individuals in a group is generally
   a. <100.
   b. >10.
   c. the optimal group size.
   d. the stable group size.
   e. the selfish herd size.

2. What is the main advantage of defending a territory?
   a. exclusive access to mates
   b. exclusive use of resources
   c. reduced conflict with others
   d. reduced transmission of disease
   e. reduced predation pressure

3. In any given population, the most commonly observed fighting strategy is
   a. Hawk.
   b. Dove.
   c. a mixture of Hawk and Dove.
   d. if smaller, behave like a Hawk.
   e. if a territory owner, behave like a Dove.

42.4 Altruism

Learning Outcomes

1. **SCISKILLS** ▶ Evaluate the arguments for and against the concept of group selection.
2. Describe how the concept of kin selection can explain altruistic behavior.
3. **SCISKILLS** ▶ Predict which types of relatives are more or less likely to be the recipients of apparently altruistic acts.
4. Explain why eusociality is a form of altruism.

As discussed in Chapter 19, each species exhibits adaptations that favor reproductive success. Via natural selection, organisms that have favorable adaptations are more likely to reproduce and pass their genes to future generations. However, ecologists have identified many examples in which some individuals forgo reproduction or exhibit behaviors that put them at risk, apparently for the benefit of the group. **Altruism** is a behavior that appears to benefit others at a cost to oneself. In this section, we will consider why altruistic behavior takes place.

Kin Selection Is Explained by Hamilton's Rule

We are not surprised if parents forgo food for themselves while passing food on to their offspring, because all offspring have copies of their parents' genes. In taking care of their young, parents are actually caring for copies of their own genes. Such behavior is known as **kin selection**, selection for behavior that lowers an individual's own survival and reproductive capabilities but enhances the survival and reproductive success of a relative. As we will see later, kin selection can operate via many different relatives, not just parents and their offspring.

The probability that any two individuals will share a copy of a particular gene is a quantity called the **coefficient of relatedness** (*r*). During meiosis in a diploid species, any given copy of a gene has a 50% chance of segregating into an egg or sperm. A mother and father are, on average, related to their children by the amount $r = 0.5$, because half of a child's genes come from its mother and half from its father. By similar reasoning, brothers, sisters, or brothers and sisters are related by the amount $r = 0.5$ (they share half their mother's genes and half their father's); grandchildren and grandparents, by 0.25; and cousins, by 0.125 (Figure 42.11). In 1964, British evolutionary biologist W. D. Hamilton realized the implication of the coefficient of relatedness for the evolution of altruism. An organism can pass on its genes not only through having offspring but also through ensuring the survival of siblings, nieces, nephews, and cousins. This means an organism has a vested interest in protecting its brothers and sisters, and even the offspring of its parents' siblings.

The term **inclusive fitness** is used to designate the total number of copies of genes passed on through one's relatives, as well as one's own reproductive output. Hamilton proposed that an altruistic gene is favored by natural selection when

$$rB > C$$

where *r* is the coefficient of relatedness of the donor (the altruist) to the recipient, *B* is the benefit received by the recipient of the altruism,

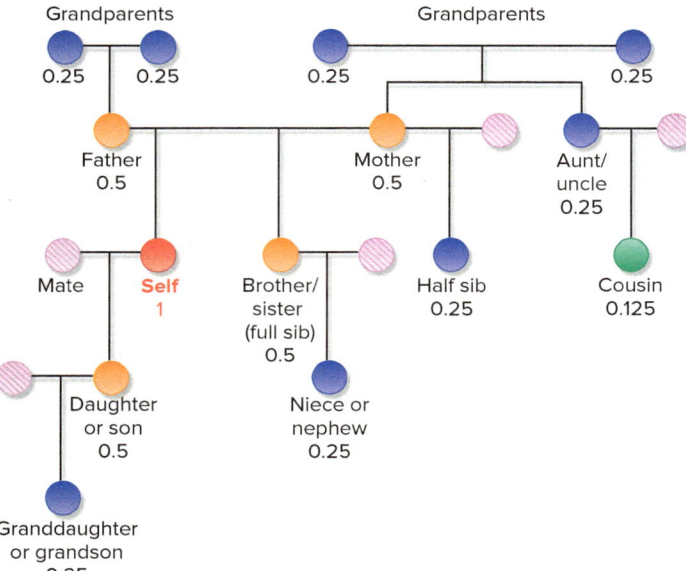

Figure 42.11 Degree of genetic relatedness to self in a diploid organism. Pink, hatched circles represent completely unrelated individuals.

Concept Check: With regard to kin selection, would it be better to sacrifice your life to save one sister or nine cousins?

Figure 42.12 Altruistic behavior. *Datana ministra* caterpillars exhibit a bright, striped warning pattern to advertise their bad taste to predators.
©Peter Stiling

Concept Check: Why do these caterpillars congregate in clusters?

and *C* is the cost incurred by the donor. This relationship is known as **Hamilton's rule**.

Let's examine a situation involving altruism within a group of animals. Many insect larvae, especially caterpillars, are soft-bodied creatures. They rely on a bad taste or toxin to deter predators and may advertise this condition with bright warning colors. For example, noxious *Datana ministra* caterpillars, which feed on oaks and other trees, have bright red and yellow stripes and adopt a specific posture with head and tail ends upturned when threatened (Figure 42.12). Unless it is born with an innate avoidance of this prey type, a predator has to kill and eat one of the caterpillars in order to learn to avoid similar individuals in the future. It is of no personal benefit to a single unlucky caterpillar to be killed. However, animals with warning colors often aggregate in kin groups because they hatch from the same egg mass. In this case, the death of one individual is likely to benefit its siblings, which are less likely to be attacked by the same predator in the future, and thus its genes will be preserved. This explains why the genes for bright color and a warning posture are successfully passed on from generation to generation. In a case where $r = 0.5$, $B = 50$, and $C = 1$, the benefit of 25 is greater than 1, so the genes for this behavior are favored by natural selection. Hamilton's rule has also been supported in humans; social network analysis has shown that in times of severe emotional or financial crises, individuals have felt they were more likely to receive important support from closer relatives than distant relatives.

A common example of altruism in social animals occurs when a sentry raises an alarm call in the presence of a predator. This behavior has been observed in Belding's ground squirrels (*Spermophilus beldingi*). The squirrels feed in groups, with certain individuals acting as sentries and watching for predators. As a predator approaches, the sentry typically gives an alarm call, and the group members retreat into their burrows. Similar behavior occurs in prairie dogs (*Cynomys* spp.) (Figure 42.13). In drawing attention to itself, the caller is at a higher risk of being attacked by the predator. However, in many groups, those closest to the sentry are most likely

Biology Principle

The Genetic Material Provides a Blueprint That Allows Organisms to Grow, Develop, and Reproduce

The similarities in DNA between kin promote behavior in which some animals act to save the lives of their close relatives.

Figure 42.13 Alarm calling, an example of kin selection. This prairie dog sentry is emitting an alarm call to warn other individuals, which are often close kin, of the presence of a predator. It is believed that by doing so, the sentry draws attention away from the others but becomes an easier target itself.
©Danita Delimont/Alamy

Figure 42.14 A naked mole rat colony (*Heterocephalus glaber*). In this mammalian species, most females do not reproduce; only the queen (shown underneath her babies) has offspring.
©Neil Bromhall/Science Source

to be offspring or brothers or sisters; thus, the altruistic act of alarm calling is reasoned to be favored by kin selection.

BIO TIPS ONLINE

Altruism in Eusocial Animals Arises Partly from Genetics and Partly from Lifestyle

Perhaps the most extreme form of altruism is the evolution of sterile castes in social animals, in which the vast majority of females, known as workers, rarely reproduce themselves but instead help one reproductive female (the queen) to raise offspring, a phenomenon called **eusociality**. In insects, the explanation of eusociality lies partly in the particular genetics of most social insect reproduction. Females develop from fertilized eggs and are diploid, the product of fertilization of an egg by a sperm. Males develop from unfertilized eggs and are haploid.

Such a system of sex determination is called the **haplodiploid system** (refer back to Figure 15.16). If they have the same parents, each daughter receives an identical set of genes from her haploid father. The other half of a female's genes comes from her diploid mother, so the coefficient of relatedness (r) of sisters is 0.50 (from father) + 0.25 (from mother) = 0.75. The result is that females are more closely related to their sisters (0.75) than they would be to their own offspring (0.50). This suggests that it is evolutionarily more advantageous for females to stay in the nest or hive and care for other female offspring of the queen, which are their full sisters.

Elegant though these types of explanations are, they do not explain the whole picture. Large eusocial colonies of termites exist, but termites are diploid, not haplodiploid. How do we account for the existence of eusociality? For termites, the coefficient of relatedness among sisters is 0.5 rather than 0.75, so it shouldn't matter to females whether they rear siblings or daughters of their own. Rather than proposing kin selection as a simple explanation, some ecologists have suggested that particular lifestyles of some animals, as well as genetics, promote eusociality. In the 1970s, a eusocial mammal, the naked mole rat (*Heterocephalus glaber*) was found living in arid areas of Africa in large, underground colonies where only one female, the queen, mates with different males and produces offspring (**Figure 42.14**). Like termites, the species is diploid. The large nests, or burrows, of the mole rat are enclosed, making it difficult for individuals to escape and form their own colony. Many females are forced to live together, making it possible for the queen to suppress reproduction in other females. She does this by producing a pheromone in her urine that is passed around the colony by grooming.

42.4 Reviewing the Concepts

- Altruism is behavior that benefits others at a cost to oneself. Most apparently altruistic acts are associated with outcomes beneficial to those most closely related to the individual, a concept termed kin selection (Figures 42.11, 42.12, 42.13).
- In eusociality, a phenomenon in which most individuals of a species do not reproduce, altruism may arise partly from the unique genetics of the animals and partly from lifestyle (Figure 42.14).

42.4 Testing Your Knowledge

1. In diploid organisms, a mother is related to her daughter by the amount $r =$
   a. 0.
   b. 0.25.
   c. 0.5.
   d. 0.75.
   e. 1.0.

2. The existence of worker ants that do not mate can be explained by
   a. eusociality.
   b. a haplodiploid mating system.
   c. the enclosed nature of their nests.
   d. both a and b.
   e. both b and c.

42.5 Mating Systems

Learning Outcomes

1. Compare and contrast promiscuous, monogamous, polygynous, and polyandrous mating systems.
2. **SCISKILLS** ▶ Predict the circumstances that favor polygyny.

In nature, males produce millions of sperm, but females produce far fewer eggs. It would seem that the majority of males are superfluous, because one male could easily fertilize all the females in a local area. If one male can mate with many females, why in most species does the sex ratio remain at approximately 1 to 1? The answer lies with natural selection. Let's consider a hypothetical population that contains 10 females to every male; each male mates, on average, with 10 females. A parent whose children were exclusively sons could expect to have 10 times the number of grandchildren than a parent with the same number of daughters. Under such conditions, natural selection would favor the spread of genes for male-producing tendencies, and males would become prevalent in the population. If the population were mainly males, females would be at a premium, and natural selection would favor the spread of genes for female-producing tendencies. Such constraints operate on the numbers of both male and female offspring, keeping the sex ratio at about 1:1. This idea was developed in 1930 by the British geneticist Ronald Fisher and has come to be known as Fisher's principle.

Even though the sex ratio is fairly even in most species, that doesn't mean that one female always mates with one male, or vice versa. Four different types of mating systems occur in nature (**Figure 42.15**). In some species, mating is promiscuous, with each female and each male mating with multiple partners within a breeding season. In monogamy, each individual mates exclusively with one partner over at least a single breeding cycle and sometimes for longer. In contrast, polygamy describes a mating pattern in which either males or females mate with more than one partner in a breeding season. There are two types of polygamy. In polygyny (Greek, for many females), one male mates with more than one female, but females mate only with one male. In polyandry (Greek, for many males), one female mates with several males, but males mate with only one female. In this section, we will examine these four types of mating systems.

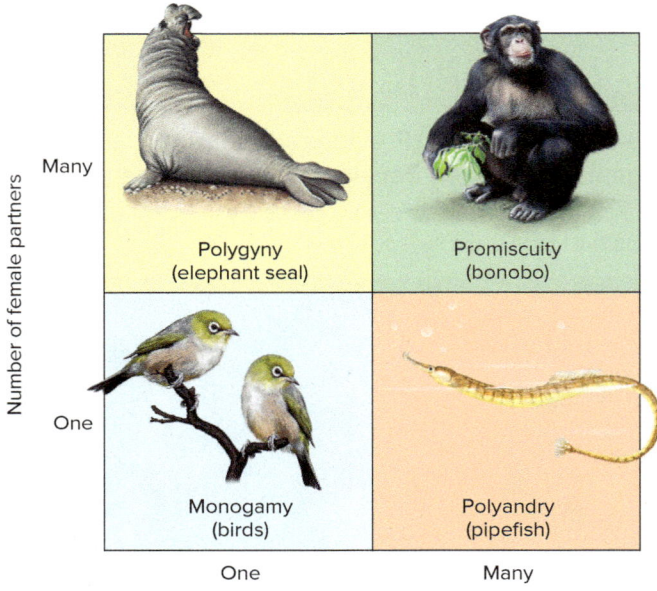

Figure 42.15 The four animal mating strategies.

In Promiscuous Mating Systems, Each Male or Female Mates with Multiple Partners

Chimpanzees and bonobos are somewhat **promiscuous;** each male mates with many females, and vice versa. Here, sex has a social as well as a reproductive function, allowing for the alleviation of conflict within the social group. However, females that maximize the genetic diversity of their offspring are also more likely to have at least some offspring that will survive in a changing world. Intertidal and terrestrial mollusks are also usually promiscuous. Individuals copulate with several partners, and eggs are fertilized with sperm from several individuals. These mollusks are slow moving and they risk desiccation when searching for a mate. The risk of not finding a mate is believed to promote promiscuous mating.

In Monogamous Mating Systems, Males and Females Are Paired for at Least One Reproductive Season

In **monogamy,** each individual mates exclusively with one partner over at least a single breeding cycle and sometimes for longer. Males and females do not exhibit much **sexual dimorphism**—a pronounced difference in the morphologies of the two sexes within a species—and are generally similar in body size and structure (**Figure 42.16a**). Several hypotheses have been developed to explain the existence of monogamy. The first is the **mate-guarding hypothesis,** which suggests that males stay with a female to protect her from being fertilized by other males. Such a strategy may be advantageous when receptive females are widely scattered and difficult to find.

The **male-assistance hypothesis** maintains that males remain with females to help them rear their offspring. Monogamy is common among birds, where about 70% of the pairings remain intact during at least one breeding season. According to the male-assistance hypothesis, monogamy is prevalent in birds because eggs and chicks take a considerable amount of parental care. Most eggs need to be incubated

(a) Monogamous species (b) Polygynous species (c) Polyandrous species

Figure 42.16 Sexual dimorphism in body size and mating system. **(a)** In monogamous species, such as these Manchurian cranes (*Grus japonensis*), males and females do not exhibit pronounced sexual dimorphism and appear very similar. **(b)** In polygynous species, such as elk (*Cervus canadensis*), males are bigger than females and have large horns, with which they engage other males in combat over females. **(c)** In polyandrous species, females are usually bigger, as with these golden silk spiders (*Nephila clavipes*).
(a) ©Masahiro Iijima/ardea.com; *(b)* ©Wildlife GmbH/Alamy; *(c)* ©Millard H. Sharp/Science Source

continuously if they are to hatch, and chicks require almost continual feeding. It is therefore in the male's best interest to help raise his young, because he would have few surviving offspring if he did not.

The **female-enforced monogamy hypothesis** suggests that females stop their male partners from being polygynous. Male and female burying beetles (*Nicrophorus defodiens*) work together to bury small, dead animals, which provide a food resource for their developing offspring. Males release pheromones to attract other females to the site. However, while an additional female might increase the male's fitness, the additional developing offspring might compete with the offspring of the first female, decreasing her fitness. As a result, on sensing these pheromones, the first female interferes with the male's attempts at signaling, preserving the monogamous relationship.

In Polygynous Mating Systems, One Male Mates with Many Females

In **polygyny,** one male mates with more than one female in a single breeding season. Physiological constraints often dictate that female organisms must care for the young. Because of these constraints, at least in many organisms with internal fertilization such as mammals and some fishes, males are able to mate with and then desert several females. Polygynous systems are therefore associated with uniparental care of young, with males contributing little. Sexual dimorphism is typical in polygynous mating systems, with males developing a larger body size to boost success in competition over mates (**Figure 42.16b**). Sexual selection is prevalent in polygynous mating systems, with females choosing the most attractive males (intersexual selection) or mating with the most competitively superior males (the "winners" of intrasexual selection) (refer back to Figure 19.18). Several types of polygyny are known, including resource-based polygyny, harem mating, and communal courting.

Resource-Based Polygyny When some critical resource is patchily distributed and in short supply, certain males may dominate the resource and breed with more than one visiting female. The major source of nestling death in the lark bunting (*Calamospiza melanocorys*), which lives in North American grasslands, is overheating from too much exposure to the Sun. Prime territories are therefore those with abundant shade, and some males with shaded territories attract two females, even though the second female can expect no help from the male in the process of rearing young. Males in some exposed territories remain bachelors for the season. From the dominant male's point of view, polygyny is advantageous; from the female's point of view, there may be costs. Although by choosing a dominant male, a female may be gaining access to good resources, she may have to share these resources with other females.

Harem Mating Sometimes males defend a group of females without commanding a resource-based territory. This pattern is more common where females naturally congregate in groups or herds, perhaps to avoid predation, as with horses, zebras, and some deer, and where space is limited, as with southern elephant seals, which mate on beaches. Usually, the largest and strongest males command most of the matings, but defending the harem is usually so exhausting that males may manage to remain the strongest male for only a year or two.

Communal Courting Polygynous mating can occur where neither resources nor groups of females are defended. In some instances, particularly in birds and mammals, males display in designated communal courting areas called **leks** (**Figure 42.17**). Females go to these areas specifically to find a mate, and they choose a prospective mate after the males have performed elaborate displays. Most females seek to mate with the best male, so a few of the flashiest males perform the vast majority of the matings. For example, at a lek of the white-bearded manakin (*Manacus manacus*) of South America, one male accounted for 75% of the 438 matings when there were as many as 10 males. A second male mated 56 times (13% of matings), but 6 others mated a total of only 10 times.

Figure 42.17 Male birds at a lek. Black grouse (*Tetrao tetrix*) congregate at a moorland lek in Scotland in April. Females visit the leks, and males display to them.
©Chris Knights/ardea.com

In Polyandrous Mating Systems, One Female Mates with Many Males

Polyandry, in which one female mates with several males, is rarer than polygyny. Nevertheless, it occurs in many species of mammals, birds, fishes, and insects. Sexual dimorphism is present, with the females being the larger of the sexes (**Figure 42.16c**). In the Arctic tundra, the summer season is short but very productive, providing a bonanza of insect food for 2 months. The productivity of the breeding grounds of the spotted sandpiper (*Actitis macularia*) is so high that the female becomes rather like an egg factory, laying up to five clutches of four eggs each in 40 days. Her reproductive success is limited not by food but by the number of males she can find to incubate the eggs, and females compete for males, defending territories where the males sit.

Polyandry is also seen in some species where egg predation is high and males are needed to guard the nests. For example, in the pipefish (*Syngnathus typhle*), males have brood pouches that provide eggs with safety and a supply of oxygen- and nutrient-rich water. Females produce enough eggs to fill the brood pouches of two males and may mate with more than one male. Insects show widespread polyandry, because they can store sperm from several males for extended periods of time.

Ultimately, as we have seen, most behaviors have evolved to maximize an individual's reproductive output. In a successful group of individuals, this leads to population growth. But we are not knee deep in black grouse or spotted sandpipers, so there must be some constraints on reproductive output. In Chapter 43, we will turn to the topic of population ecology and explore how populations grow and the factors that limit their growth.

42.5 Reviewing the Concepts

- Four types of mating systems are found among animals: promiscuity, monogamy, polygyny, and polyandry. The relative body size of males and females depends on their mating system (Figures 42.15, 42.16).
- Polygynous mating can often occur in situations when males dominate a resource, defend groups of females, or display in common courting areas called leks (Figure 42.17).

42.5 Testing Your Knowledge

1. Fisher's principle suggests that the sex ratio of males to females in a population should tend toward
   a. 100:1.
   b. 10:1.
   c. 1:1.
   d. 1:10.
   e. 1:100.

2. In polyandrous mating systems,
   a. males are usually bigger than females.
   b. females are usually bigger than males.
   c. males and females are of roughly equal size.

Assessing Your Knowledge and Skills

Foundational Knowledge

Multiple Choice

1. Certain behaviors seem to have very little environmental influence. Such behaviors are the same in all individuals regardless of the environment and are referred to as _____ behaviors.
   a. genetically programmed
   b. instinctual
   c. innate
   d. All of the above are correct.
   e. Only b and c are correct.

2. Nutcrackers are able to relocate caches of seeds by means of
   a. innate behavior.
   b. imprinting.
   c. classical conditioning.
   d. operant conditioning.
   e. cognitive maps.

3. The modification of behavior based on prior experience is called
   a. habituation.
   b. learning.
   c. association.
   d. adjustment behavior.
   e. innate.

4. When an individual behaves in a way that reduces its own fitness but increases the fitness of others, the organism is exhibiting
   a. kin selection.
   b. group selection.
   c. altruism.
   d. selfishness.
   e. ignorance.

5. In a polygynous mating system,
   a. one male mates with one female.
   b. one female mates with many different males.
   c. one male mates with many different females.
   d. many different females mate with many different males.

Table 43.2 Summary of the Types of Species Interactions

Nature of interaction	Species 1*	Species 2*
Competition	−	−
Amensalism	−	0
Mutualism	+	+
Commensalism	+	0
Predation, herbivory, parasitism, parasitoidism	+	−

*+ = positive effect; − = negative effect; 0 = no effect.

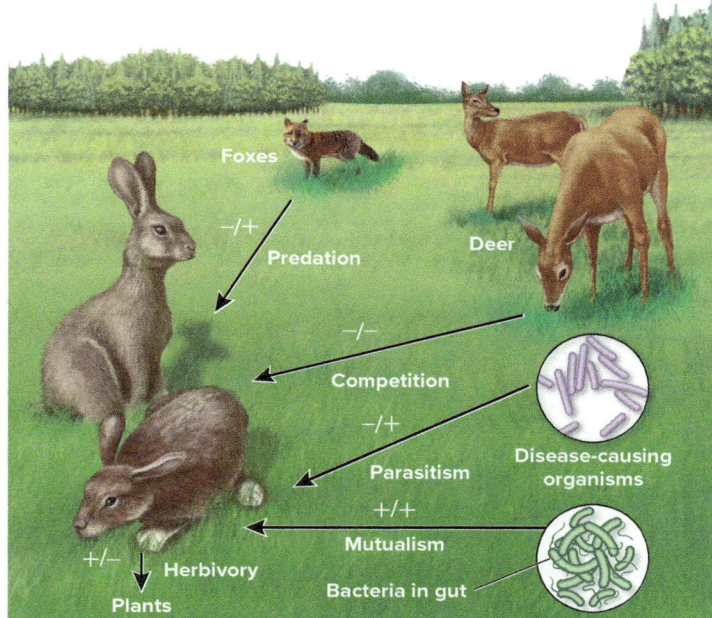

Figure 43.8 Species interactions. These rabbits can interact with a variety of species, experiencing predation by foxes, competition with deer for food, and parasitism from various disease-causing organisms. Herbivory occurs when rabbits feed on various plants. Finally, mutualism occurs between the rabbit and gut microbes that aid in the digestion of cellulose-rich plant material. The effects of each species on the other are shown by the terms assigned to the arrows, as discussed in the text.

any of the eight forms listed in **Table 43.2**. **Competition** is an interaction that affects both species negatively (−/−), as both species compete over food or other resources. Sometimes, however, an interaction is quite one-sided, being detrimental to one species and neutral to the other; such an interaction is called **amensalism** (−/0). **Mutualism** is an interaction in which both species benefit (+/+), whereas **commensalism** benefits one species and is neutral to the other (+/0). Finally, four types of interactions are positive for one species but detrimental to the other (+/−):

- **Predation,** in which a predator feeds on prey and causes its rapid death
- **Herbivory,** in which an animal feeds on a plant
- **Parasitism,** in which a parasite establishes a relatively long-term relationship with its host, which may or may not lead to the host's death
- **Parasitoidism,** in which an organism lays eggs in a host and the resulting larvae remain in the host and usually kill it

To illustrate how species interact in nature, let's consider a rabbit population in a grassland community (**Figure 43.8**). To determine what factors influence the size and density of the rabbit population, we need to understand each of its possible species interactions. For example, the rabbit population might be limited by the quality or availability of food. It is also likely that other species, such as deer, use the same resource and thus compete with the rabbits. The rabbit population might be limited by predation from foxes or by the virus that causes the disease myxomatosis, which is usually spread by fleas and mosquitoes. Finally, gut microbes help the rabbits to digest cellulose-rich plant materials and receive nutrients for themselves.

Competition Can Negatively Affect the Population Growth of Competing Species

The influence of competition on population growth was illustrated by the Russian microbiologist Georgyi Gause, who, in 1934, studied competition among three protist species, *Paramecium aurelia*, *P. bursaria,* and *P. caudatum*, all of which feed on bacteria and yeast, which in turn feed on an oatmeal medium in a culture tube in the laboratory. More bacteria occurred in the oxygen-rich upper part of the culture tube, more yeast in the oxygen-poor lower part of the tube. Because each *Paramecium* species was a slightly different size, Gause calculated population growth as a combination of numbers of individuals per milliliter of solution multiplied by their unit volume to give a population volume for each species. When they were grown separately, the population volume of all three *Paramecium* species followed a logistic growth pattern (**Figure 43.9a**). When Gause cultured *P. caudatum* and *P. aurelia* together, *P. caudatum* eventually went extinct (**Figure 43.9b**). Both species used bacteria as food, but *P. aurelia* grew at a rate six times faster than *P. caudatum*.

However, when Gause cultured *P. caudatum* and *P. bursaria* together, neither went extinct (**Figure 43.9c**). The population volume of each was much less than when they were grown alone, because some competition occurred between them. Gause discovered, however, that *P. bursaria* was better able to use the yeast in the lower part of the culture tubes. *P. bursaria* have tiny green algae inside them, which produce oxygen and allow *P. bursaria* to thrive in the lower oxygen levels at the bottom of the tubes.

From these experiments, Gause concluded that two species with exactly the same requirements cannot live together in the same place and use the same resources—that is, occupy the same **niche.** His conclusion was later termed the **competitive exclusion principle.** The

(a) Each *Paramecium* species grown alone

(b) Competition between *P. aurelia* and *P. caudatum*

(c) Competition between *P. caudatum* and *P. bursaria*

Figure 43.9 Competition among *Paramecium* species. (a) When each of three species (*Paramecium aurelia*, *P. bursaria*, and *P. caudatum*) is grown alone, its population grows according to the logistic model. **(b)** When *P. aurelia* is grown with *P. caudatum*, the density of *P. aurelia* is lower than when grown alone, and *P. caudatum* goes extinct. **(c)** When *P. caudatum* is grown with *P. bursaria*, the population densities of both are lowered, but the species coexist.

term **resource partitioning** describes the differentiation of niches, both in space and in time, that enables similar species to coexist in a community, just like two species of *Paramecium* feeding in different parts of a culture tube.

By reviewing studies that have investigated competition in nature, scientists can see how frequently it occurs and in what particular circumstances it is most important. In a 1983 review of field studies by American ecologist Joseph Connell, competition was found in 55% of 215 species surveyed, demonstrating that it is frequent in nature. Generally in studies of single pairs of species utilizing the same resource, competition is almost always reported, whereas in studies involving more species, the frequency of competition decreases. Why should this be the case? Imagine a resource, such as a series of different-sized grains, being fed upon by four types of species—ants, beetles, mice, and birds (**Figure 43.10a**). The ants feed on the smallest grain, the beetles and mice on the grain of intermediate sizes, and the birds on the largest. If only adjacent species competed with each other, competition would be expected only between the ant and beetle, the beetle and mouse, and the mouse and bird. Thus, competition would be found in only three out of the six possible species pairs (50%). Naturally, the percentage would vary according to the number of species on the resource spectrum. If only three species occurred along the spectrum, we would expect competition in two of the three pairs (67%). If just two species used the resource spectrum, however, we would expect competition in almost 100% of the cases (**Figure 43.10b**).

Some other general patterns were evident from Connell's review. Plants showed a high degree of competition, perhaps because they are rooted in the ground and cannot easily escape or perhaps because they are competing for the same set of limiting nutrients—water, light, and minerals. Marine organisms tended to compete more than terrestrial ones, most likely because many of the species studied lived in the intertidal zone and were attached to the rock face, in a manner similar to the rooting of plants. Because the area of the rock face is limited, competition for space is quite important.

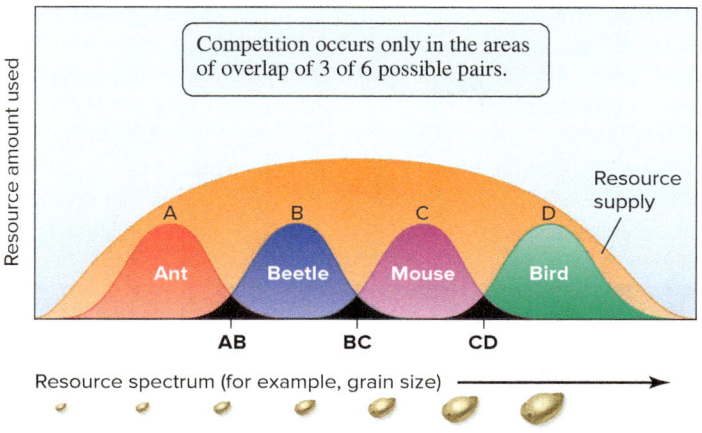
(a) Competition among 4 species for a resource

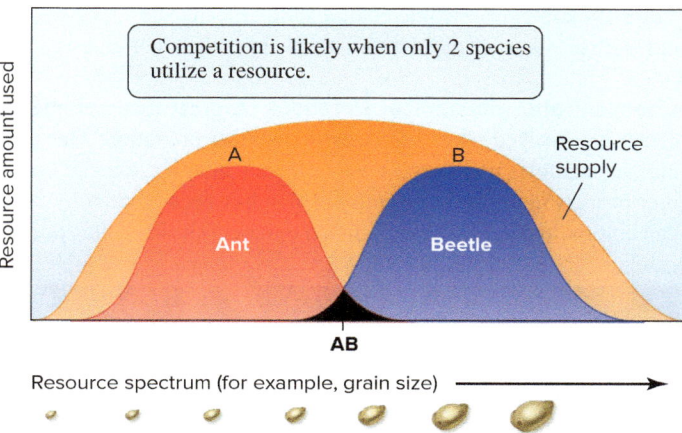
(b) Competition among 2 species for a resource

Figure 43.10 The frequency of competition according to the number of species involved. (a) Resource supply and utilization curves of four species, A, B, C, and D, along the spectrum of a hypothetical resource such as grain size. If competition occurred only between species with adjacent resource utilization curves, competition would be expected between three of the six possible pairings: A and B, B and C, and C and D. (b) When only two species utilize a resource set, competition is nearly always expected between them.

Concept Check: *If five species utilized the resource set in part (a), what percentage of the interactions would be competitive?*

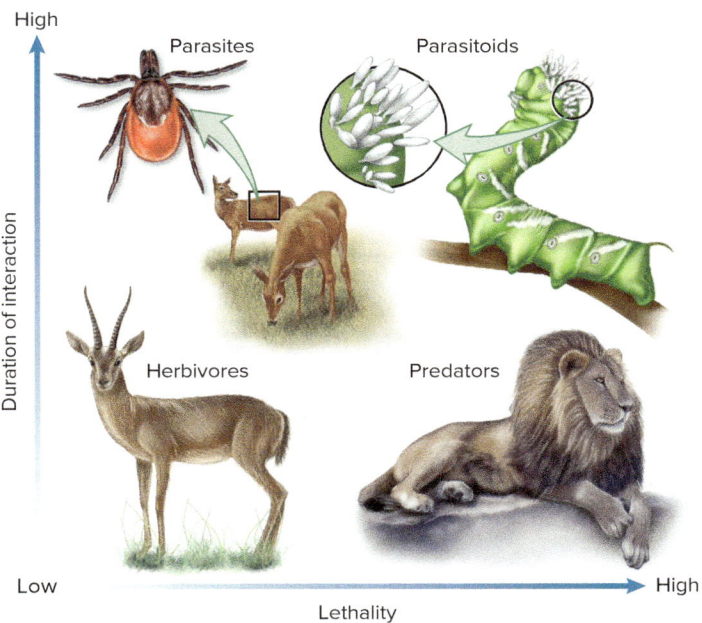

Figure 43.11 Possible interactions between populations. Lethality represents the probability that an interaction results in the death of the prey. Duration represents the length of the interaction between the consumer and the prey.

Concept Check: *Where do omnivores fit in this figure?*

BioConnections: *Look back at Figure 27.3. Where do lampreys fit in this figure?*

Predators, Herbivores, Parasites, and Parasitoids Interact with Their Prey or Host Species in Different Ways

Predation, herbivory, parasitism, and parasitoidism are interactions that have a positive effect for one species and a negative effect for the other. These categories of species interactions can be classified according to how lethal they are for the prey or host and the length of association between the consumer and prey (**Figure 43.11**). Each has particular characteristics that set it apart. Herbivory usually involves nonlethal feeding on plants, whereas predation generally results in the death of the prey. Parasitism, like herbivory, is typically nonlethal and differs from predation in that the adult parasite typically lives and reproduces for long periods in or on the living host (refer back to Figure 26.16). Parasitoids are insects that lay eggs in a host that the resulting larvae remain with and almost always kill. For example, a parasitoid female wasp inserts its eggs into a host, such as a caterpillar, and the resulting larvae consume the insides of the caterpillar and eventually kill it. The larvae emerge to create white pupae on the outside of the caterpillar. They thus have features in common with parasites (a long association) and predators (lethality). Parasitoids are common in the insect world and are often used in the biological control of pests, such as caterpillars.

EVOLUTIONARY CONNECTIONS

Organisms Have Evolved Many Defenses Against Natural Enemies

The variety of strategies that organisms have evolved to avoid being eaten suggests that predators, herbivores, parasites, and parasitoids, also known as **natural enemies,** are a strong selective force.

Common strategies include chemical and mechanical defenses, camouflage, displays of intimidation, mimicry, and armor and weaponry.

Chemical and Mechanical Defenses A great many animal species have evolved chemical defenses against predation. One of the classic examples of a species with a chemical defense is the bombardier beetle (*Stenaptinus insignis*), which has been studied by American entomologist Tom Eisner and coworkers. These beetles possess a reservoir of hydroquinone and hydrogen peroxide in their abdomen. When threatened, they eject the chemicals into an "explosion chamber," where the subsequent release of oxygen causes the whole mixture to be violently ejected as a hot spray (about 88°C, or 190°F) that can be directed at the beetle's attackers (**Figure 43.12a**). Many other arthropods, such as millipedes, also have chemical sprays, and the phenomenon is also found in vertebrates, as anyone who has had a close encounter with a skunk can testify. Often associated with a

(a) This bombardier beetle (*Stenaptinus insignis*) directs its hot, stinging spray at a forceps "attacker."

(b) Aposematic coloration advertises the poisonous nature of this blue poison arrow frog (*Dendrobates azureus*) from South America.

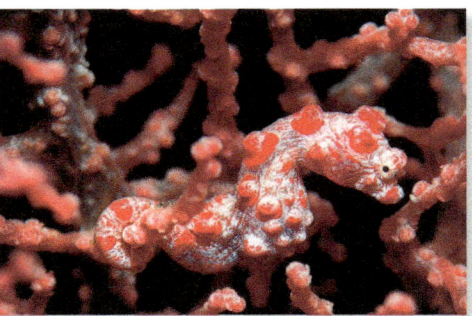

(c) Camouflage allows this pygmy sea horse (*Hippocampus bargibanti*) from Bali to blend in with its background.

(d) In a display of intimidation, this porcupine fish (*Diodon hystrix*) puffs itself up to look threatening to its predators.

(e) Müllerian mimicry. Viceroy (left) and monarch (right) butterflies are both noxious and have similar color patterns.

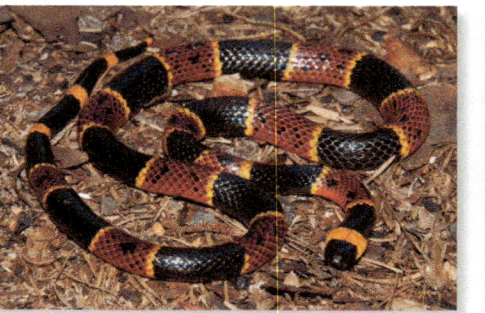

(f) In this example of Batesian mimicry, an innocuous scarlet king snake (*Lampropeltis elapsoides*) (left) mimics the venomous eastern coral snake (*Micrurus fulvius*) (right).

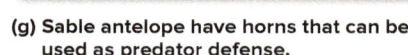

(g) Sable antelope have horns that can be used as predator defense.

Figure 43.12 Antipredator adaptations.

(a) ©CB2/ZOB/Supplied by WENN/Newscom; *(b)* ©Hans D. Dossenbach/ardea.com; *(c)* ©Thomas Aichinger/V&W/The Image Works; *(d)* ©Paul Springett 01/Alamy; *(e left)* ©Rick & Nora Bowers/Alamy; *(e right)* ©Eric Carr/Alamy; *(f left)* ©Suzanne L. & Joseph T. Collins/Science Source; *(f right)* ©Jason Ondreicka/Alamy; *(g)* ©cd123/123RF

BioConnections: *Refer back to Figure 42.2. Do predators learn to avoid bad-tasting prey by habituation, classical conditioning, or operant conditioning?*

chemical defense is an **aposematic coloration,** or warning coloration, which advertises an organism's unpalatable taste or poisonous nature. For instance, many tropical frogs have bright warning coloration that calls attention to their harmful effects, which are due to toxins in their skin (Figure 43.12b).

Many unusual and powerful chemicals are present in plants—such as nicotine in tobacco, morphine in poppies, cocaine in coca, and caffeine in coffee. These compounds are not part of the primary metabolic pathway that plants use to obtain energy and are therefore referred to as **secondary metabolites.** Most of these chemicals are bitter tasting or toxic, thereby deterring herbivores from feeding on the plants. The staggering variety of secondary metabolites in plants, an estimated 200,000 identified so far, may be a testament to the large number of organisms that feed on plants. In an interesting twist, many of these compounds have medicinal properties that have proved to be beneficial to humans. In addition to containing chemical compounds, many plants have an array of mechanical defenses, such as thorns, hairs, and spines.

Camouflage Camouflage is the blending of an organism with the background of its habitat and is a common method of avoiding detection (refer back to Figure 27.17b). For example, many grasshoppers are green and blend in with the foliage on which they feed. Stick insects mimic branches and twigs with their long, slender bodies. In most cases, these animals stay perfectly still when threatened, because movement alerts a predator. Camouflage is prevalent in the vertebrate world, too. Many sea horses adopt a body shape and color pattern similar to the environment in which they are found (Figure 43.12c).

Displays of Intimidation Some animals put on displays of intimidation in an attempt to discourage predators. For example, a cat arches its back, a frilled lizard extends it collar, and a porcupine fish inflates itself when threatened in order to appear larger (Figure 43.12d). All of these animals have evolved displays to deceive potential predators about the ease with which they can be eaten.

Mimicry There are two major types of mimicry. In **Müllerian mimicry,** two or more toxic species converge to look the same, thus reinforcing the basic distasteful design. The viceroy butterfly (*Limenitis archippus*) and the monarch butterfly (*Danaus plexippus*) are examples of Müllerian mimicry. Both species are unpalatable and look similar, but the viceroy can be distinguished from the monarch by a black line that crosses its hindwings (Figure 43.12e).

Batesian mimicry is the mimicry of an unpalatable species (the model) by a palatable one (the mimic). Some of the best examples involve flies, especially hoverflies of the family Syrphidae, which are striped black and yellow and resemble stinging bees and wasps but are themselves harmless. Among vertebrates, the nonvenomous scarlet king snake (*Lampropeltis elapsoides*) mimics the venomous eastern coral snake (*Micrurus fulvius*), thereby gaining protection from would-be predators (Figure 43.12f).

Armor and Weaponry The shells of tortoises and turtles are an effective means of defense against most predators, as are the quills of porcupines (refer back to Figure 27.17c). Though many animals developed horns and antlers for sexual selection, these projections can also be used in defense against predators (Figure 43.12g). Many invertebrate species have powerful claws, pincers, or, in the case of scorpions, venomous stingers that can be used in defense as well as offense.

Natural Enemies Can Have a Dramatic Effect on Prey Populations

Despite this impressive array of defenses, many studies have shown a strong effect of natural enemies on their prey populations. Invasive species—which are species introduced into a new area, usually by humans, that tend to spread rapidly—provide particularly striking examples of the effects of natural enemies.

The prickly pear cactus (*Opuntia stricta*) was imported from South America to Australia in the 19th century as a food source for an insect being used to make a bright red dye for the British. The cactus, however, quickly became invasive and established itself as a major pest of rangeland. Then, in the 1920s, the small cactus moth (*Cactoblastis cactorum*), which feeds on *Opuntia* cactus, was introduced to Australia from South America and soon controlled the cactus populations. This reuniting of an invasive plant with its natural enemy successfully saved hundreds of thousands of acres of valuable rangeland from being overrun by the cacti (Figure 43.13).

Biology Principle

Biology Affects Our Society

Control of prickly pear by the cactus moth cleared hundreds of thousands of hectares of cacti, allowing sheep to graze the area and farming to thrive.

(a) Before biological control   (b) After biological control

Figure 43.13 Successful biological control of prickly pear cactus. The prickly pear cactus (*Opuntia stricta*) in Chinchilla, Australia, **(a)** before and **(b)** after control by the cactus moth (*Cactoblastis cactorum*).

Source: State Library of Queensland, neg. no.'s API-101-01-0001r and API-101-01-0002r

4. Which pattern is found when a population exhibits a constant per capita rate of increase?
   a. A
   b. B
   c. C
   d. D

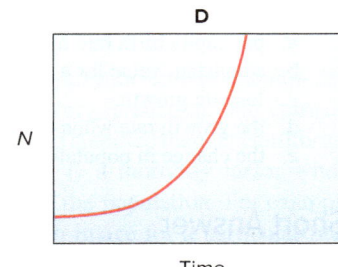

5. In a growing population, the youngest age classes have
   a. the highest number of deaths per year.
   b. the smallest number of deaths per year.
   c. the most individuals.
   d. the most individuals and the largest number of deaths per year.
   e. the fewest individuals.

Short Answer

1. Using the logistic equation, calculate population growth when $K = 1,000$, $N = 500$, and $r = 0.1$ and when $K = 1,000$, $N = 100$, and $r = 0.1$. Compare the results with those shown in Section 43.3, where $K = 1,000$, $N = 900$, and $r = 0.1$. At which time is population growth highest as a percentage of the population?

2. You survey an annually breeding butterfly population and determine that it contains 1,000 females. The next year, you count 1,200 females in the same area. What is the finite rate of increase, λ? What will the population be 5 years from now if the rate of increase is the same each year?

Communities and Ecosystems: Ecological Organization at Large Scales

44

©Gary Braasch/Corbis/Getty Images

Mount St. Helens erupting in 1980. Ecologists have been monitoring gradual change in the species composition of the area since the disturbance.

Chapter Outline

44.1 Patterns of Species Richness and Species Diversity

44.2 Species Richness and Community Stability

44.3 Succession: Community Change

44.4 Island Biogeography

44.5 Food Webs and Energy Flow

44.6 Biomass Production in Ecosystems

Assessing Your Knowledge and Skills

At 8:32 a.m. on May 18, 1980, Mount St. Helens, in the Washington Cascades, erupted. The blast felled trees over a 600-km² area, and the landslide that followed destroyed everything in its path, killing nearly 60 people. For almost 40 years, ecologists have studied the recovery of plant and animal species in that area and have noted how the appearance of some species facilitated the recovery of others.

The assemblage of many populations that live in the same place at the same time is known as a **community**. **Community ecology** explores the factors that influence the number and abundance of species in a community. In this chapter, we will begin by considering patterns of species richness and discuss how biodiversity is a measure of both species richness and the distribution of individuals among species. We will examine why, on a global scale, the number of species is usually greatest in the tropics and declines toward the poles. Next, we will discuss why the recovery of communities following a disturbance such as a fire or a volcanic eruption tends to occur in a predictable sequence—a process termed succession—which may be determined by the balance between the rates of immigration of new species and the extinction of existing ones.

The term **ecosystem** was first used in 1935 by British plant ecologist A. G. Tansley to describe the system formed by the interaction between a community of organisms and its physical environment. **Ecosystem ecology** addresses the flow of energy and the production of biomass, which is the total mass of living matter in a given area, usually measured in grams or kilograms per square meter. In Section 44.5, we will explore energy flow, the movement of energy through an ecosystem. In examining energy flow, we will consider the complex networks of feeding relationships among species, which are represented by food webs. Finally, we will explore how ecologists measure biomass within ecosystems.

44.1 Patterns of Species Richness and Species Diversity

Learning Outcomes

1. Identify the latitudinal gradient of species richness.
2. List and describe three hypotheses for observed patterns of species richness.
3. **SCISKILLS** ▶ Calculate the Shannon diversity index.
4. **SCISKILLS** ▶ Calculate the effective number of species.

Community ecology addresses which factors influence the number of different species in a community, or **species richness.** Globally, the number of species of most taxa varies along a latitudinal gradient, generally increasing from polar to temperate areas and reaching a maximum in the tropics. For example, the species richness of North American birds increases from Arctic Canada to Panama (**Figure 44.1**). A similar pattern exists for mammals, amphibians, reptiles, and plants. Although the latitudinal gradient of species richness is an important pattern, species richness is also influenced by topographical variation. More mountains mean more hilltops, valleys, and differing habitats; thus, the number of birds is greater in the mountainous U.S. West than in the East. Species richness is also reduced by the peninsular effect, in which the number of species decreases as a function of distance from the main body of land.

Figure 44.1 Species richness of birds in North America. The values indicate the numbers of different species in a given area. Contour lines show equal numbers of bird species, with colors indicating incremental changes. Note the pronounced latitudinal gradient toward the tropics and the high diversity in California and the Rocky Mountains, regions of considerable topographical variation and habitat diversity.

Many hypotheses for the latitudinal gradient in species richness have been advanced. In this section, we will consider three hypotheses for patterns of species richness. The key factors are evolutionary time, area, and productivity; each factor can influence species richness to some degree. Although they are discussed separately here, these three hypotheses are not mutually exclusive. All three factors can contribute to patterns of species richness.

The Species-Time Hypothesis Suggests That Communities Diversify with Age

Many ecologists argue that communities diversify, or gain species, with time. Tropical communities are usually older than temperate communities, because the species in temperate regions are periodically wiped out by glaciers. The **species-time hypothesis** proposes that temperate regions have less-rich communities than tropical ones because they are younger. According to this idea, Ice Ages have driven many species extinct in temperate regions and it takes time for remaining species to evolve and diversify in those regions after the glaciers have retreated. More than half of the families of flowering plants have no temperate representatives.

In support of the species-time hypothesis, British ecologist H. John Birks found a significant correlation between the numbers of species of insects on various British trees and the length of time these tree species have been present in Britain (**Figure 44.2a**). Many of the tree species in Britain are relatively recent colonists, having appeared following the departure of the glaciers that covered most of the islands after the last Ice Age. Birks used radiocarbon dating of pollen collected from deep lake sediments to estimate the length of time a tree species had been present in Britain. No tree species had been present for longer than 13,000 years. He then gathered information on numbers of insect species present on trees from lists provided by other experts who had been examining the insect fauna of trees in Britain for many years. The significant relationship between pollen age and total number of insects indicated that older tree species support more insect species.

However, ecologists recognize drawbacks to the species-time hypothesis. For example, this hypothesis may help to explain variations in the species richness of terrestrial organisms, but it has limited applicability to marine organisms. Although we might not expect terrestrial species, particularly plants, to redistribute themselves quickly following a glaciation—especially if there is a physical barrier, like the English Channel, to overcome—there seems to be no reason that marine organisms couldn't relatively easily shift their distribution patterns during glaciations, yet the latitudinal gradient of species richness still exists in marine habitats.

The Species-Area Hypothesis Suggests That Large Areas Support More Species

The **species-area hypothesis** proposes that larger areas contain more species than smaller areas because they can support larger populations and a greater range of habitats. Much evidence supports the species-area hypothesis. For example, in 1974, American ecologist Donald Strong showed that insect species richness on tree species in Britain was better correlated with the area over which a tree species could be found than with time of habitation since the last Ice Age (**Figure 44.2b**). The points cluster more tightly around the line of best fit. Even relatively newly introduced species, such as apple trees,

Biology Principle

Biology Is a Quantitative Science

Ecologists have estimated both the number of years tree species have been present in Britain and their area of coverage, in order to determine which of these abiotic variables best explains the number of insect species each tree species supports.

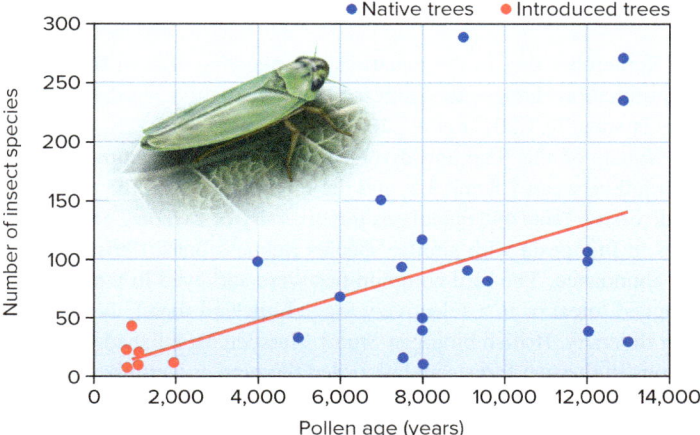

(a) Insect species richness increases on older tree species.

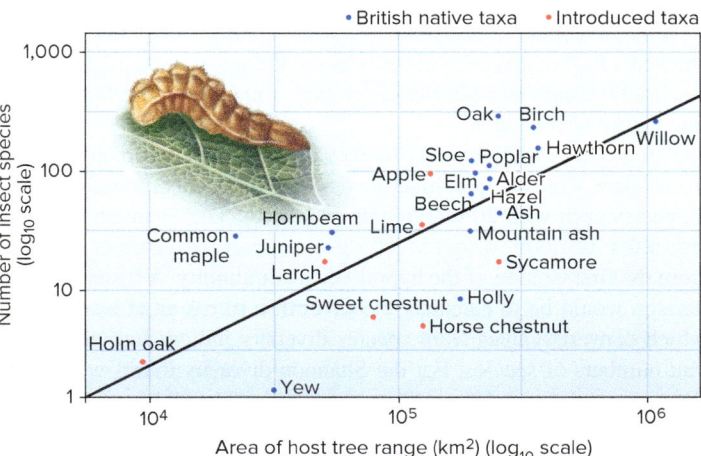

(b) Insect species richness increases on more widely occurring tree species.

Figure 44.2 Relationship between species richness on British host trees and both time and area. **(a)** Insect species richness is greater on tree species that have been present in Britain for longer time periods, which supports the species-time hypothesis. **(b)** A positive correlation is also found between insect species richness and the area of the host tree's range, in square kilometers (km^2), which supports the species-area hypothesis. Note the log scales in part (b).

supported a large number of insect herbivores if they were planted over a wide area. The observation that the number of species tends to increase with increasing area is called the **species-area effect**.

The large, climatically similar area of the tropics has been proposed as a reason that the tropics have high species richness. However, the species-area hypothesis seems unable to explain why, if increased richness is linked to increased area, more species are not found in certain regions such as the vast contiguous landmass of Asia. Furthermore, although tundra may be the world's largest biome in terms of landmass, it has low species richness. Finally, the largest marine system, the open ocean, which has the greatest volume of any habitat, has fewer species than tropical nearshore waters, which have a relatively small volume.

The Species-Productivity Hypothesis Suggests That Greater Productivity Results in More Species

The **species-productivity hypothesis** proposes that greater production by plants results in greater overall species richness. An increase in plant productivity, the total mass of plant material produced over time, leads to an increase in the number of herbivores and hence an increase in the number of predator, parasite, and scavenger species. Productivity itself is influenced by factors such as temperature and rainfall, because many plants grow better where it is warm and wet. For example, in 1987, Canadian biologist David Currie and colleagues showed that the species richness of trees in North America is best predicted by the **evapotranspiration rate,** the rate at which water moves into the atmosphere through the processes of evaporation from the soil and other surfaces and the transpiration of plants, which are influenced by the amount of solar energy (**Figure 44.3**).

Figure 44.3 Tree species richness in North America. Contour lines show equal numbers of tree species, with colors indicating incremental changes. Tree species richness and evapotranspiration rates are highest in the Southeast.

Concept Check: Why doesn't the species richness of trees increase in mountainous areas of the West, as it does for birds?

44.2 Species Richness and Community Stability

Learning Outcome

1. Describe the diversity-stability hypothesis, and explain the evidence supporting it.

A community is considered to be stable when little to no change can be detected in the number of species and their abundance over a given time period. Such a community is also said to be in equilibrium. Community stability is an important concept for ecologists. A decrease in the stability of a community over time may alert ecologists to a possible problem. For example, an invasive species such as the zebra mussel, *Dreissena polymorpha*, in the Great Lakes can cause undesirable reductions in the numbers of native invertebrates living there. In this section, we will explore the question of whether communities with higher species richness are more stable than communities with lower richness, using evidence from a field experiment.

The Diversity-Stability Hypothesis States That Species-Rich Communities Are More Stable Than Those with Fewer Species

The link between species richness and stability was first explicitly proposed by British ecologist Charles Elton in the 1950s. He suggested that a disturbance in a species-rich community would be cushioned by large numbers of interacting species and would not produce as drastic an effect as would a disturbance in a species-poor community. For example, an introduced predator or parasite would be more likely to cause extinctions in a species-poor community compared to a more species-rich community, where its effects would be buffered by interactions with more species. Elton argued that outbreaks of pests are often found on cultivated land or land disturbed by humans, both of which are species-poor communities with few naturally occurring species. His idea became known as the **diversity-stability hypothesis**.

However, some ecologists challenged Elton's association of diversity with stability. They pointed out many examples of introduced species that have assumed pest proportions in species-rich areas, including rabbits in Australia and pigs in North America. They also suggested that disturbed or cultivated land may suffer from pest outbreaks not because of its low number of species but because invasive species often have no natural enemies in the new environment. In contrast, native species and their natural enemies have had long associations. For example, in Europe, coevolved predators, such as foxes, prevent rabbit populations from increasing to pest proportions. Research was needed to determine if a link existed between species richness and stability.

Field Studies Have Linked Stability to Species Richness

In 1996, American ecologist David Tilman reported the relationship between species richness and community stability from an 11-year study of 207 grassland plots in Minnesota that varied in their species richness. At the end of every year, he measured the biomass of every

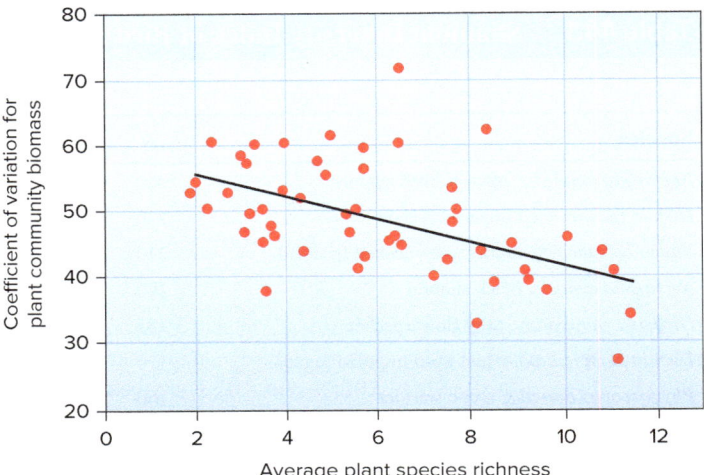

Figure 44.4 Biomass variation and species richness. Tilman's 11-year study of grassland plots in Minnesota revealed that year-to-year variability in community biomass was lower in species-rich plots. Each dot represents an individual plot. Only the plots from one field are graphed.

plant species in each plot. He then analyzed how much this biomass varied from year to year through a statistical calculation called the coefficient of variation, which is a measure of relative variability. (It is the ratio of the standard deviation to the mean.) Less variation in biomass indicates community stability. Year-to-year variation in plant community biomass was significantly lower in plots with greater plant species richness (Figure 44.4). These results showed that higher richness enhances community stability.

Tilman suggested that more species-rich communities are more likely to contain disturbance-resistant species that in the event of a disturbance, could grow in number and compensate for the loss of disturbance-sensitive species. For example, when a change in climate such as drought decreased the abundance of competitively dominant species that thrived under normal conditions, drought-resistant species increased in mass and replaced them. Thus, declines in the number of sensitive species were compensated by increases in other species, which acted to stabilize total community biomass.

Although ecologists recognize a link between species richness and community stability, they are also aware that over long periods of time, communities may experience severe disturbances. The change in the composition and structure of communities that follows occurs in a predictable way, termed succession, which is described next.

44.2 Reviewing the Concepts

- Community stability is an important concept in ecology. The diversity-stability hypothesis maintains that species-rich communities are more stable than communities with fewer species. Tilman's field studies, which showed that year-to-year variation in plant biomass decreased with increasing species richness, established a link between diversity and stability (Figure 44.4).

44.2 Testing Your Knowledge

1. Which evidence suggests that more diverse communities are more stable than less diverse communities?
   a. Agricultural land, with fewer species, undergoes less frequent pest outbreaks than natural prairies containing more species.
   b. In Australia, introduced rabbits achieved pest proportions. In Europe, coevolved predators such as foxes prevent rabbit populations from reaching pest proportions.
   c. Long-term studies of American grasslands show that fields with higher numbers of plant species vary less in biomass from year to year.
   d. Both a and b are correct.
   e. Both b and c are correct.

2. Ecologists began to question Elton's link of increased stability to increased diversity because
   a. cultivated land undergoes few outbreaks of pests.
   b. highly disturbed areas have high numbers of species.
   c. pest outbreaks are caused by lack of long associations with natural enemies, not because they occur in simple systems.
   d. All of the above are correct.

44.3 Succession: Community Change

Learning Outcomes
1. Distinguish between primary and secondary succession.
2. Compare and contrast facilitation, inhibition, and tolerance as mechanisms of succession.

The term **succession** describes the gradual and continuous change in species composition of a community following a disturbance. **Primary succession** refers to succession on a newly exposed site that has no biological legacy in terms of plants, animals, or microbes, such as bare ground caused by a volcanic eruption or sediment created by the retreat of glaciers. In primary succession on land, the plants must often build up the soil, and thus a long time—even hundreds of years—may be required for the process. Only a tiny proportion of the Earth's surface is currently undergoing primary succession, including the area around Mount St. Helens, on new lava flows around the volcanoes in Hawaii and off the coast of Iceland, and behind retreating glaciers in Alaska and Canada.

Secondary succession refers to succession on a site that previously supported life but has undergone a disturbance such as a fire, tornado, hurricane, or flood. In terrestrial areas, soil is already present. Clearing a natural forest and farming the land for several years is an example of a severe forest disturbance that does not kill all native species. Some plants and many soil bacteria, nematodes, and insects are still present. Secondary succession occurs if farming has ended. The secondary succession in abandoned farmlands (also called old fields) can lead to a pattern of vegetation quite different from one that develops after primary succession following glacial retreat. For example, the plowing and added fertilizers, herbicides, and pesticides may have caused substantial changes in the soil of an old field, allowing species that require a lot of nitrogen to colonize. These species would not be present for many years in newly created glacial soils.

American plant ecologist Frederic Clements is often viewed as the founder of successional theory. His work in the early 20th century emphasized succession as proceeding through several stages to a distinct end point, or **climax community**. Although disturbance can return a community from a later stage to an earlier stage, generally the community progresses in one direction. Clements's depiction of succession focused on a process termed facilitation, but two other mechanisms of succession—inhibition and tolerance—have since been described. In this section, we will examine the evidence for each of them.

Facilitation Assumes That Each Invading Species Creates a More Favorable Habitat for Succeeding Species

A key assumption of Clements's view of succession was that each colonizing species makes the local environment a little different, such as a little shadier or a little richer in soil nitrogen, so that it becomes more suitable for other species, which then invade and outcompete the earlier residents. This process, known as **facilitation**, continues until the most competitively dominant species have colonized, when the community is at climax. The composition of the climax community for any given region is determined by climate, soil conditions, and frequency of disturbance.

Succession following the gradual retreat of Alaskan glaciers is often used as a specific example of facilitation as a mechanism of succession. Over the past 200 years, the glaciers in Glacier Bay have undergone a dramatic retreat of nearly 100 km (**Figure 44.5**). Succession in Glacier Bay has followed a distinct pattern of vegetation. As glaciers retreat, they leave moraines—deposits of stones, pulverized rock, and debris that serve as soil.

In Alaska, the bare soil has a low nitrogen content and little organic matter. In the pioneer stage, the soil is first colonized by a black crust of cyanobacteria, mosses, lichens, horsetails (*Equisetum variegatum*), and the occasional river beauty (*Epilobium latifolium*) (**Figure 44.6a**). Because the cyanobacteria are nitrogen fixers, the soil nitrogen increases a little, but soil depth and litterfall (fallen leaves, twigs, and other plant material) are still minimal. At this stage, there are instances of seeds and seedlings of dwarf shrubs of the rose family, commonly called mountain avens (*Dryas drummondii*); alders (*Alnus sinuata*); and spruce but they are rare. After about 40 years, mountain avens dominate the landscape (**Figure 44.6b**). Soil nitrogen increases, as do soil depth and litterfall, and alder trees begin to invade.

At about 60 years, alders form dense, close thickets (**Figure 44.6c**). Alders have nitrogen-fixing bacteria that live mutualistically in their roots and convert nitrogen from the air into a biologically useful form. Soil nitrogen dramatically increases, as does litterfall. Sitka spruce trees (*Picea sitchensis*) begin to invade at about this time. After about 75 to 100 years, the spruce trees begin to overtop the alders, shading them out. The litterfall is still high, and the large volume of needles turns the soil acidic. The shade causes competitive exclusion of many of the original understory species, including alder, and only mosses carpet the ground. At this stage, seedlings of western hemlock (*Tsuga*

Figure 44.5 The degree of glacier retreat at Glacier Bay, Alaska, since 1794. (a) Primary succession begins on the bare rock and soil evident at the edges of the retreating glacier. (b) The lines reflect the position of the glacier in 1794 and its subsequent retreat northward.

(a) ©Charles D. Winters/Science Source

Concept Check: Why might ecologists think of walking the coastline of Glacier Bay as the equivalent of walking back in time?

heterophylla) and mountain hemlock (*Tsuga mertensiana*) may also occur. After 200 years, a mixed spruce-hemlock climax forest occupies the location (Figure 44.6d).

Other studies also provide evidence of facilitation. Early primary succession on Mount St. Helens shows that decomposition of fungi allows mosses and other fungi to colonize the soil, providing evidence of facilitation. Succession on sand dunes also supports the facilitation model. Pioneer plant species stabilize the sand dunes and facilitate the establishment of subsequent plant species. The foredunes, those nearest the shoreline, are the most frequently disturbed and are maintained in a state of early succession, whereas more stable communities develop farther away from the shoreline.

Succession also occurs in aquatic communities. Although soils do not develop in marine environments, facilitation may still occur when one species enhances the quality of settling and establishment sites for another species. When experimental test plates used to measure settling rates of marine organisms were placed in Delaware Bay, researchers discovered that certain cnidarians enhanced the attachment of tunicates, and both facilitated the attachment of mussels, the dominant species in the community. In this experiment, the smooth surface of the test plates prevented many species from colonizing, but once the surface became rougher, because of the presence of the cnidarians, many other species were able to colonize. In a similar fashion, early colonizing bacteria, which create biofilms on rock surfaces, can facilitate succession of other organisms.

Inhibition Implies That Early Colonists Can Prevent Later Arrivals from Replacing Them

Although data on succession in some communities fit the facilitation model, researchers have proposed alternative hypotheses concerning how succession may operate. In the mechanism known as **inhibition**, early colonists prevent colonization by other species. For example, removing the litter of *Setaria faberi*, an early successional plant species in New Jersey old fields, causes an increase in the biomass of a later species, *Erigeron annuus*. The release of toxic compounds from decomposing *Setaria* litter or physical obstruction by the litter itself blocks the establishment of *Erigeron*. Without the litter present, however, *Erigeron* dominates and reduces the biomass of *Setaria*.

Inhibition has been seen as the primary method of succession in the marine intertidal zone, where space is limited. In this habitat, early successional species are at a great advantage in maintaining possession of valuable space. In 1974, American ecologist Wayne Sousa created an environment for testing how succession works in the intertidal zone by scraping rock faces clean of all algae or putting out fresh boulders or concrete blocks. The first colonists of these areas were the green algae of the genus *Ulva*. By removing *Ulva* from the substrate, Sousa showed that the large red alga *Chondracanthus canaliculatus* was able to colonize more quickly (Figure 44.7). The results of Sousa's study indicate that early colonists can inhibit rather than facilitate the invasion of subsequent colonists. Succession may eventually occur because early colonizing species, such as *Ulva*, are more susceptible than later successional species, such as *Chondracanthus*, to the rigors of the physical environment and to attacks by herbivores, such as crabs (*Pachygrapsus crassipes*).

Tolerance Suggests That Early Colonists Neither Facilitate nor Inhibit Later Colonists

In 1977, researchers Joseph Connell and Ralph Slatyer proposed a third mechanism of succession, which they termed **tolerance.** In this process, any species can start the succession, but the eventual climax community is reached in a somewhat orderly fashion. The species that establish themselves and remain do not change the environment in ways that either facilitate or inhibit subsequent colonists. Species have differing tolerances to the intensity of competition as more species accumulate. Relatively competition-intolerant species are more successful early in succession, when the intensity of competition is low

Figure 44.6 The pattern of primary succession at Glacier Bay, Alaska. **(a)** The first species to colonize the bare ground following retreat of the glaciers are small species such as cyanobacteria, moss, and lichens. **(b)** Mountain avens (*Dryas drummondii*) is a flowering plant common in the *Dryas* stage. **(c)** Soil nitrogen and litterfall increase rapidly as alder (*Alnus sinuata*) invades. Note also the appearance of a few spruce trees higher up the valley. **(d)** Sitka spruce (*Picea sitchensis*) and hemlock (*Tsuga heterophylla*) trees make up a climax spruce-hemlock forest at Glacier Bay, with moss carpeting the ground. Two hundred years ago, a glacier occupied this spot.
(a) ©Leon Werdinger/Alamy; *(b)* ©James Hager/age fotostock; *(c)* ©Howie Garber/AccentAlaska.com; *(d)* ©Craig Lovell/Eagle Visions P/Newscom

and resources are abundant. Relatively competition-tolerant species appear later in succession and at climax. Connell and Slatyer found the best evidence for the tolerance model in American plant ecologist Frank Egler's earlier work on floral succession. In the 1950s, Egler showed that succession in plant communities is determined largely by species that already exist in the ground as buried seeds or old roots. Whichever species germinates first or regenerates from roots initiates the succession sequence. Germination or root regeneration, in turn, depends on the timing of a disturbance. For example, an early-season tree fall would promote early-germinating species to grow in the subsequent light gap, whereas a late-season tree fall would promote the growth of late-germinating species. As succession proceeds, earlier germinating or regenerating species may be outcompeted by different species.

The Three Models of Succession Differ in How One Species Affects Colonization by a Different Species

The key distinction between the three models of succession is in the manner in which one species affects the colonization of another species. In the facilitation model, species replacement is facilitated by previous colonists; in the inhibition model, it is inhibited by the action of previous colonists; and in the tolerance model, species may be affected by previous colonists, but they do not require them (**Figure 44.8**).

44.3 Reviewing the Concepts

- Succession is the gradual and continuous change in community structure over time. Primary succession is succession on a newly exposed site with no prior biological legacy; secondary succession is succession on a site that previously supported life but has undergone a disturbance (Figures 44.5, 44.6).

- Three mechanisms have been proposed for succession. In facilitation, each species facilitates or makes the environment more suitable for subsequent species. In inhibition, initial species inhibit later colonists. In tolerance, any species can start the succession, and species replacement is unaffected by previous colonists (Figures 44.7, 44.8).

Biology Principle

Biology Is an Experimental Science
The results of Sousa's study indicate that early colonists can inhibit rather than facilitate the invasion of subsequent colonists.

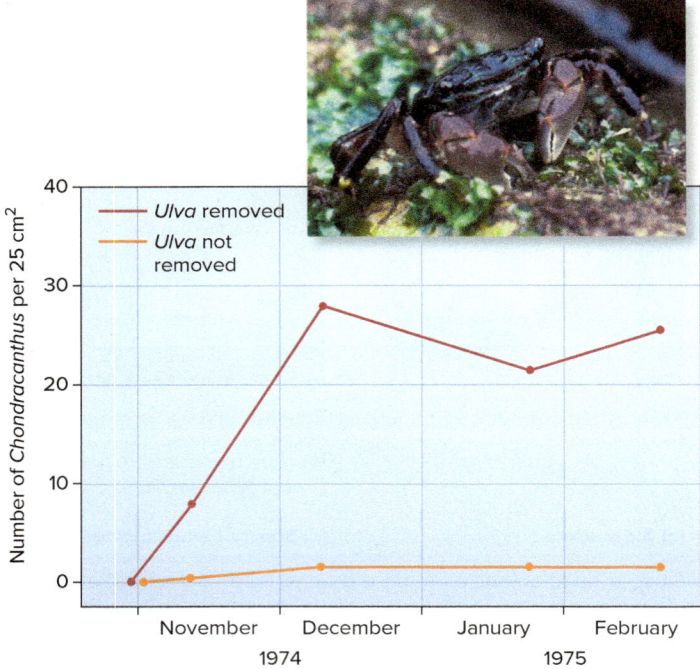

Figure 44.7 Inhibition as a method of succession in the marine intertidal zone. Removing *Ulva* from intertidal rock faces allowed colonization by *Chondracanthus*. The inset shows *Ulva* on a rock face with the striped shore crab *Pachygrapsus crassipes,* an herbivore.
©Wayne Sousa/University of California, Berkeley

44.3 Testing Your Knowledge

1. Which of the following is an example of secondary succession?
   a. plants growing in cracks on the pavement of a quiet street
   b. the recovery of vegetation following the 2004 Indonesian tsunami
   c. the colonization of new sand by beach plants
   d. the recovery of forests following a wildfire
   e. Both b and d are correct.

2. In New England salt marshes, *Spartina* grass stabilizes the soil and reduces water velocity, which enables other seedlings to emerge. This is an example of
   a. primary succession.
   b. facilitation.
   c. tolerance.
   d. inhibition.
   e. a climax community.

Figure 44.8 Three models of succession. A, B, C, and D represent four stages. D represents the climax community. An arrow indicates "is replaced by," and + = facilitation, − = inhibition, and 0 = no effect. The facilitation model is the classic model of succession. In the inhibition model, early-arriving species outcompete later-arriving species. In the tolerance model, the structure of the community depends on which species gets there first. The colored arrows show that succession may bypass some stages in the tolerance model.

Concept Check: *Inhibition implies that competition exists between species, with early-arriving species tending to outcompete later arrivals, at least for a while. Does competition or mutualism feature more prominently in facilitation?*

3. A tree falls in a forest in spring, and flowers germinate in the light gap. Following a tree fall in autumn, different species of flowers germinate in the light gap. This illustrates the principle of
   a. facilitation.
   b. tolerance.
   c. inhibition.
   d. primary succession.
   e. climax communities.

44.4 Island Biogeography

Learning Outcomes

1. **SCISKILLS** ▶ Interpret a graph that illustrates the equilibrium model of island biogeography.
2. List the three predictions of the island biogeography model, and discuss how well the evidence supports them.

In some newly formed habitats such as volcanic islands, succession may be affected not only by facilitation, inhibition, or tolerance but also by the ability of species from neighboring areas (such as a mainland) to colonize isolated areas (such as an island). In these cases, species richness is affected by the distance of isolated habitats from a source of potential colonists from neighboring areas and by the size of the areas to be colonized. In the 1960s, American ecologists Robert MacArthur and E. O. Wilson developed a comprehensive model to explain the process of succession on new islands, where a gradual buildup of species proceeds from a sterile beginning. Their model, termed the **equilibrium model of island biogeography,** holds that the number of species on an island tends toward an equilibrium number that is determined by the balance between two factors: immigration rates and extinction rates. Their model has been applied not just to newly formed oceanic islands but also to virtual islands, such as mountains surrounded by deserts, lakes surrounded by dry land, or conservation areas surrounded by agricultural land or urban landscapes. In this section, we will explore island biogeography to investigate how well the model's predictions are supported by data.

The Island Biogeography Model Suggests That During Succession, Species Gains from Immigration Are Balanced by Losses from Extinction

MacArthur and Wilson's model of island biogeography suggests that species repeatedly arrive on an island and either thrive or become extinct. The rate of immigration of new species is highest when no species are present on the island. As the number of species accumulates, the immigration rate decreases, since subsequent immigrants are more likely to represent species already present on the island. The rate of extinction is low at the time of first colonization, because few species are present and many have large populations. With the addition of new species, the population sizes of some species diminish, so the probability of extinction increases. Over time, the number of species tends toward an equilibrium, $\hat{S}$, in which the rates of immigration of new species equal the rates of extinction of existing species. Species may continue to arrive and go extinct, but the number of species on the island remains approximately the same.

MacArthur and Wilson reasoned that when plotted graphically, both the immigration and extinction lines would be curved, for several reasons (**Figure 44.9a**). First, species arrive on islands at different rates. Some organisms, including plants with seed-dispersal mechanisms and winged animals, are more mobile than others and arrive quickly. Other organisms arrive more slowly. This pattern causes the immigration curve to start off steep but get progressively shallower. On the other hand, extinctions rise at accelerating rates, because as later species arrive, competition increases and more species are likely to go extinct. The strength of the island biogeography model was that it generated several testable predictions:

1. **Species-area relationships.** The number of species should increase with increasing island size (area), the species-area effect (see Figure 44.2b). Extinction rates would be lower on larger islands because population sizes should be larger and less susceptible to extinction (**Figure 44.9b**).
2. **Species-distance relationships.** The number of species should decrease with increasing distance of the island from the mainland—that is, from the **source pool,** the pool of potential species available to colonize the island. Immigration rates should be greater on islands near the source pool because species do not have as far to travel (Figure 44.9b).

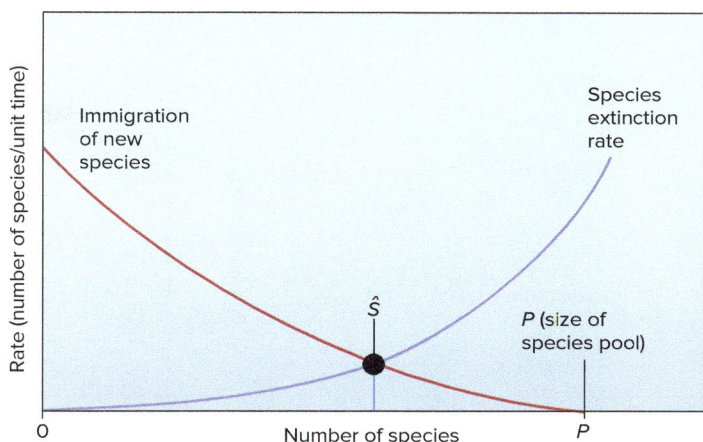

(a) Effects of immigration and extinction on species number

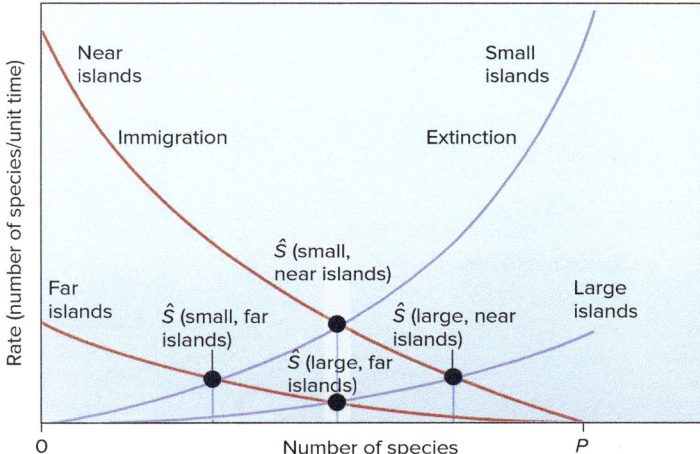

(b) Added effects of island size and proximity to the mainland on species number

Figure 44.9 MacArthur and Wilson's equilibrium model of island biogeography. (a) The interaction of immigration rate and extinction rate produces an equilibrium number of species on an island, $\hat{S}$. $\hat{S}$ can vary from 0 species to P species, the total number of species available to colonize. (b) $\hat{S}$ varies according to the island's size and distance from the mainland. An increase in distance (near to far) lowers the immigration rate. An increase in island area (small to large) lowers the extinction rate.

Concept Check: *Can you think of a scenario where there would be large numbers of species on a small island?*

BioConnections: *Look ahead to Figure 47.7. How might the model of island biogeography be useful in the design of nature reserves?*

3. **Species turnover.** The turnover of species should be considerable. Whereas the number of species on an island might remain relatively constant, the composition of the species should vary over time as new species colonize the island and others become extinct.

Let's examine the three predictions of the island biogeography model and see how well the data support each one.

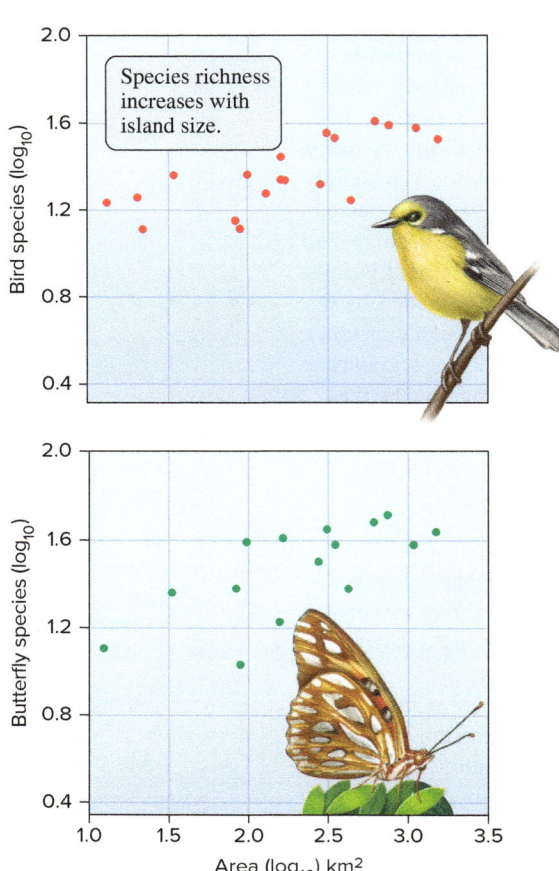

Figure 44.10 Species richness and island size. **(a)** The Lesser Antilles extend from Anguilla in the north to Grenada in the south. **(b)** On these islands, the number of bird and butterfly species increases with the area of an island. Note that these relationships are traditionally plotted on a double logarithmic scale, a so-called log-log plot, in which the horizontal axis is the logarithm to the base 10 of the area and the vertical axis is the logarithm to the base 10 of the number of species. A linear plot of the area versus the number of species would be difficult to produce because of the wide range of area and species richness involved. Logarithmic scales condense this variation to manageable limits.

Concept Check: *Calculate the approximate change in bird species richness across islands in the Lesser Antilles*

Species-Area Relationships The West Indies has traditionally been a key location for ecologists studying island biogeography. The physical geography and the plant and animal life of the islands are well known. Furthermore, all of the Lesser Antilles, from Anguilla in the north to Grenada in the south, enjoy a similar climate and are surrounded by deep water (**Figure 44.10a**). In 1999, American ecologists Robert Ricklefs and Irby Lovette summarized the available data on the species richness of four groups of animals—birds, bats, reptiles and amphibians, and butterflies—across 19 islands that varied in area over two orders of magnitude (13–1,510 km²). In each case, a positive correlation occurred between area and species richness (**Figure 44.10b**).

Species-Distance Relationships In studies of the numbers of lowland forest bird species in Polynesia, MacArthur and Wilson found that the number of species decreased with the distance from the source pool of New Guinea (**Figure 44.11**). They expressed the richness of bird species on the islands as a percentage of the number of bird species found on New Guinea. A significant decline in this percentage was observed with increasing distance. More distant islands were inhabited by lower numbers of species than nearer islands. This research substantiated the prediction of species richness declining with increasing distance from the source pool.

Species Turnover Studies involving species turnover on islands are difficult to perform because detailed and complete species lists are needed over long periods of time, usually many years and often decades. The lists that do exist are often compiled in a casual way and are not usually suitable for comparison with more modern data. In 1980, British ecologist Francis Gilbert reviewed 25 investigations carried out to analyze turnover and found a lack of this type of rigor in nearly all of them. Furthermore, most of the observed turnover in these studies, usually less than 1% per year, or less than one species per year, appeared to be due to immigrants that never became established rather than to the extinction of well-established species. More recent studies have revealed similar findings, suggesting that the rates of turnover are low rather than high, giving little conclusive support to the third prediction of the equilibrium model of island biogeography. Even the most rigorous study, by E. O. Wilson and his student Dan Simberloff, showed negligible turnover, as described next.

COMMUNITIES AND ECOSYSTEMS: ECOLOGICAL ORGANIZATION AT LARGE SCALES

(a) New Guinea and neighboring islands

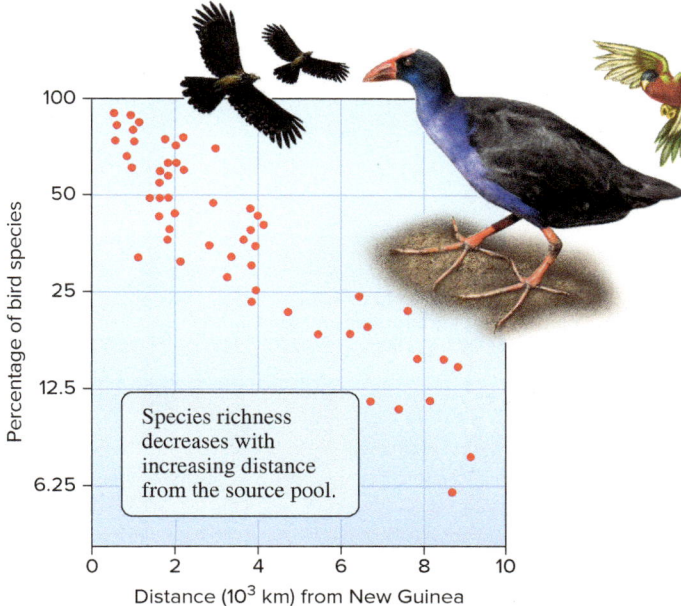

(b) Relationship between species richness and distance from source

Figure 44.11 Species richness and distance from the source pool. (a) Map of Australia, New Guinea, and these Polynesian Islands: New Caledonia, Fiji Islands, Cook Islands, Marquesas Islands, Pitcairn, and Easter Island. (b) The number of bird species on the islands decreases with increasing distance from the source pool, New Guinea. The species richness is expressed as the percentage of bird species on New Guinea.

FEATURE INVESTIGATION

Simberloff and Wilson's Experiments Tested the Predictions of the Equilibrium Model of Island Biogeography

In the 1960s, American ecologists Daniel Simberloff and E. O. Wilson conducted possibly the most rigorous test of the equilibrium model of island biogeography ever performed, using islands in the Florida Keys. First, they surveyed small red mangrove (*Rhizophora mangle*) islands, 11–25 m in diameter, taking a census of the numbers of all their terrestrial arthropods. Then they enclosed each island with a plastic tent and had the islands fumigated with methyl bromide, a short-acting insecticide, to kill all arthropods on them. The tents were removed, and periodically thereafter Wilson and Simberloff surveyed the islands to examine the recolonization rates. At each survey, they counted all the species present, noting any species not there at the previous census and the absence of others that were previously there but had presumably gone extinct (see the data of **Figure 44.12**). In this way, they estimated turnover of species on islands.

Figure 44.12 Simberloff and Wilson's experiments on the equilibrium model of biogeography.
(photos) Courtesy Dr. Daniel Simberloff, University of Tennessee-Knoxville

HYPOTHESIS The island biogeography model predicts higher species richness for islands closer to the mainland and significant turnover of species on islands.

STARTING LOCATION Mangrove islands in the Florida Keys.

	Experimental level	Conceptual level
1	Take initial census of all terrestrial arthropods on 4 mangrove islands. Erect a framework over each mangrove island.	Each mangrove island is isolated.

2. Cover the frameworks with tents and fumigate with methyl bromide to kill all arthropod species.

Methyl bromide is a short-acting insecticide that at low levels will not kill plant life.

3. Remove the tents and conduct censuses every month to monitor recolonization of arthropods and to determine extinction rates.

Mangrove islands are recolonized.

4. **THE DATA** Island E2 was closest to the mainland and supported the highest number of species both before and after fumigation. E3 and ST2 were at an intermediate distance from the mainland, and E1 was the most distant.

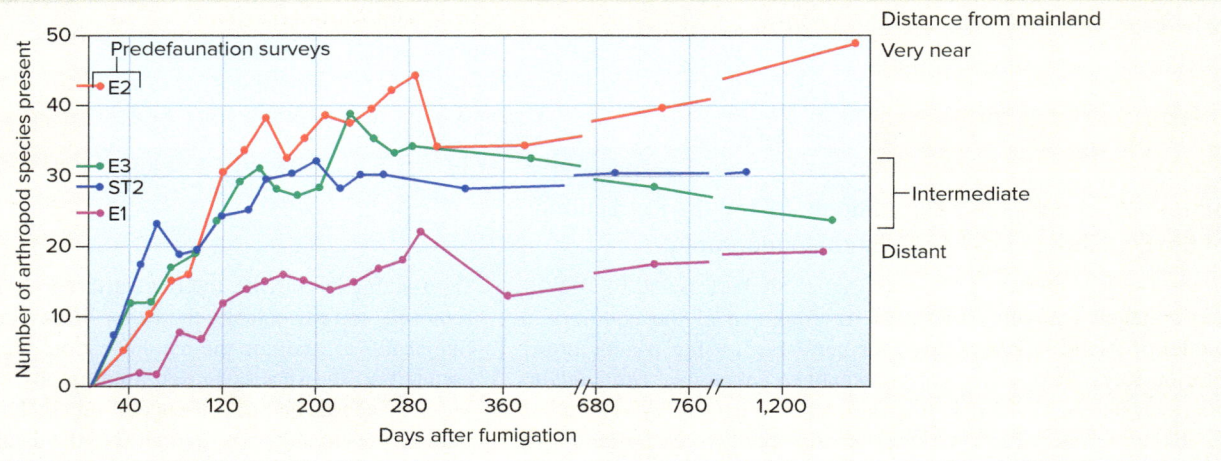

5. **CONCLUSION** Island distance from the mainland influences species richness on mangrove islands in the Florida Keys. However, species turnover is minimal, and species richness changes little following initial recolonization.

6. **SOURCE** Simberloff, D. S. 1978. Colonization of Islands by Insects: Immigration, Extinction and Diversity, pp. 139–153 in L. A. Mound and N. Waloff (eds.), *Diversity of Insect Faunas*. Blackwell Scientific Publications, Oxford, U.K.

After 250 days, all but one of the islands had a number of arthropod species similar to the number observed before fumigation, even though population densities were still low. The data indicated that recolonization rates were higher on islands nearer to the mainland than on more distant islands—as the island biogeography model predicts. However, the data, which consisted of lists of species on islands before and after extinctions, provided little support for the prediction of substantial turnover. Rates of turnover were low, only 1.5 extinctions per year, compared with the 15–40 species found on the islands within a year. Simberloff and Wilson concluded that turnover probably involves only a small subset of transient or less common species, with the more common species remaining permanent after colonization.

Experimental Questions

1. What was the purpose of Simberloff and Wilson's study?
2. Why did the researchers conduct a thorough species survey of arthropods before experimental removal of all the arthropod species?
3. **SCISKILLS** ▶ What did the researchers conclude about the relationship between island proximity to the mainland and species richness and turnover?

The equilibrium model of island biogeography has stimulated much research to confirm the strong effects of area and distance on species richness. However, species turnover appears to be low rather than considerable, which suggests that succession on most islands is a fairly orderly process. This conclusion means that colonization is not a random process and that the same species seem to colonize first and other species gradually appear in the same order.

The principles of island biogeography have been applied to wildlife preserves, which are essentially islands in a sea of developed land consisting of agricultural fields or urban sprawl. Conservationists have therefore used the model of island biogeography in the design of nature preserves, a topic we will return to in Chapter 47.

44.4 Reviewing the Concepts

- In the equilibrium model of island biogeography, the number of species on an island tends toward an equilibrium number determined by the balance between immigration and extinction rates (Figure 44.9).
- The model predicts that the number of species increases with increasing island size, that the number of species decreases with distance from the source pool, and that turnover is high. Support exists for the first two predictions of the model, but experiments on islands in the Florida Keys refuted the third prediction (Figures 44.10, 44.11, 44.12).

44.4 Testing Your Knowledge

1. Which of the following is part of the original MacArthur-Wilson theory of island biogeography?
   a. $\hat{S}$ is increased by distance from the source pool.
   b. $\hat{S}$ is decreased by island size.
   c. $\hat{S}$ is a balance between immigration and extinction.
   d. Island size influences immigration rates.
   e. Distance from source pool influences extinction rates.

2. Why are the immigration and extinction lines in the MacArthur-Wilson model both curved?
   a. Species arrive at different rates.
   b. Some organisms are more mobile than others.
   c. Competition increases as more species arrive.
   d. Later-arriving species tend to be better competitors.
   e. All of the above are true.

44.5 Food Webs and Energy Flow

Learning Outcomes

1. Compare and contrast food chains and food webs, and distinguish primary producers and primary, secondary, and tertiary consumers.
2. List and describe the different types of ecological pyramids.

We will now turn our attention to ecosystem ecology, which studies the movement of energy and materials through organisms and their communities. Key factors that affect species richness are the amount of available energy and the feeding relationships among organisms. These feeding relationships can be characterized by an unbranched **food chain**—a linear depiction of energy flow, with each organism feeding on and deriving energy from the preceding organism. In this section, we will consider the unidirectional flow of energy in a food chain and examine a food web, a more complex model of interconnected food chains. We will then explore two of the most important features of food webs—chain length and the pyramid of numbers.

The Main Trophic Levels Within Food Chains Consist of Primary Producers, Primary Consumers, and Secondary Consumers

Each level in a food chain is called a **trophic level** (from the Greek *trophos*, meaning feeder), with different species feeding at different levels. In a food-chain diagram, an arrow connects each trophic level with the one above it (Figure 44.13). Food chains typically consist of organisms that obtain energy in different ways. **Autotrophs** harvest light or chemical energy and store that energy in carbon compounds. Most autotrophs, which include plants, algae, and photosynthetic bacteria, use sunlight for this process. These organisms, called **primary producers,** form the base of the food chain. They produce the energy-rich organic molecules upon which nearly all other organisms depend.

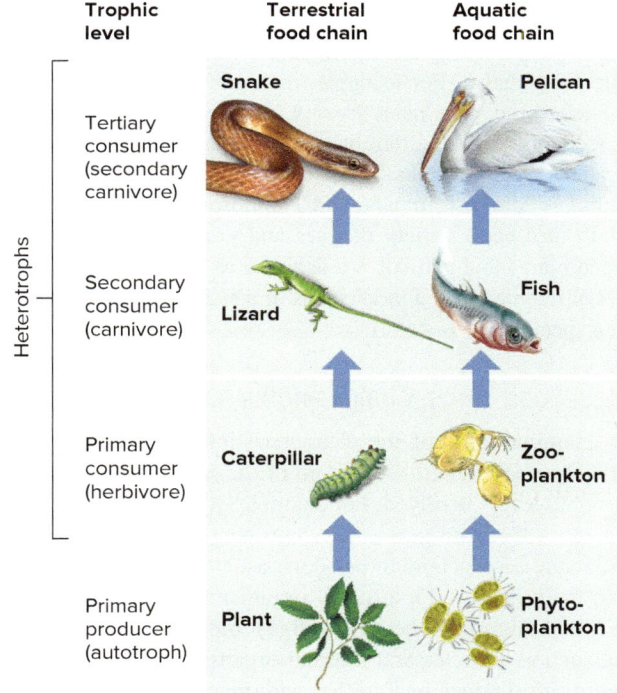

Figure 44.13 Food chains. Two examples of the flow of food energy up the trophic levels: a terrestrial food chain and an aquatic food chain.

BioConnections: *In these two food chains, plants and protists (phytoplankton) are the producers. Refer back to Section 23.4. What other organisms are producers and could also support food chains?*

Organisms that consume organic molecules from their environment to sustain life and thus receive their nutrition by eating other organisms or products of organisms are termed **heterotrophs.** Heterotrophs that obtain their food by consuming primary producers are termed **primary consumers** (also called herbivores) and include most protists, most animals, and even some plants such as mistletoe, which is parasitic on other plants. Organisms that eat primary consumers are **secondary consumers** (also called carnivores). Organisms that feed on secondary consumers are **tertiary consumers,** and so on. Thus, energy enters a food chain through primary producers, via photosynthesis, and is passed up the food chain to primary, secondary, and tertiary consumers (Figure 44.13).

At each trophic level, many organisms die before they are eaten. Most energy from the first trophic level, such as plants, goes unconsumed by herbivores. Instead, unconsumed plants die and decompose in place. This material, along with dead remains of animals and waste products, is called **detritus.** Consumers that get their energy from detritus, called **detritivores,** break down dead organisms from all trophic levels. For example, carrion beetles feed on the dead bodies of other animals. Some detritivores, which do not ingest their food but feed by absorbing it on a molecular scale, are known as decomposers. The most important are fungi and bacteria. In terrestrial systems, detritivores probably consume 80–90% of plant matter, with different groups such as earthworms and fungi working in concert to extract most of the energy. Detritivores may, in turn, support a community of predators that feed on them.

In nature, feeding relationships are usually more complex than simple food chains. For example, many different herbivore species may feed on the same plant species. Also, each species of herbivore may feed on several different plant species. For instance, on the African savanna, cheetahs, lions, and hyenas all eat a variety of prey, including wildebeest, impala, and Thompson's gazelle. These herbivores in turn eat a variety of trees and grasses. Such relationships between plants and animals are depicted as a **food web,** a complex model of interconnected food chains in which there are multiple links among species (**Figure 44.14**).

In Most Food Webs, Chain Lengths Are Short

Let's examine some of the characteristics of food webs in more detail. The chain length is the sum of the number of links between the trophic levels involved. For example, if a lion feeds on a zebra, and a zebra feeds on grass, the chain length is two. In many food webs, chain lengths tend to be short, usually five or fewer, because of two main factors. First, many organisms cannot digest all their prey. They take only the easily digestible plant leaves or animal tissue such as muscles and internal organs, leaving the hard wood or energy-rich bones behind. Second, much of the energy assimilated by animals is used in maintenance and is lost from the organism as heat. Both of these factors acting together means that on average only about 10% of the available energy is transferred from one trophic level to another. Because energy is lost at each link, after a few links, most of the available energy has been expended and relatively little energy is available for higher trophic levels (**Figure 44.15**).

Ecological Pyramids Describe the Distribution of Numbers, Biomass, or Energy Between Trophic Levels

The abundance of organisms, biomass, or available energy at each trophic level of a food web can be expressed graphically as an ecological pyramid. One of the best-known ecological pyramids, described by British ecologist Charles Elton in 1927, is the **pyramid of numbers,** in which the number of individuals decreases at each trophic level, with a large number of individuals at the base and fewer individuals at the top. For example, in a grassland, there may be millions of individual plants per hectare, hundreds of thousands of insects that feed on the plants, tens of thousands of spiders feeding on the insects, and one or two birds that feed on the spiders (**Figure 44.16a**).

Ecologists have, however, discovered some exceptions to this pyramid. One single producer such as an oak tree can support hundreds of herbivorous beetles, caterpillars, and other primary consumers, which in turn may support thousands of insect predators and parasitoids. This is called an inverted pyramid of numbers.

One way to reconcile this apparent exception is to weigh the organisms in each trophic level, creating a **pyramid of biomass. Biomass** is the total mass of living matter in a given area or volume, for example, grams per square meter. For example, an oak tree weighs more than all of its herbivores and predators combined. American ecologist Howard Odum measured the pyramid of biomass for a freshwater ecosystem, Silver Springs, in Florida (**Figure 44.16b**). Beds of eelgrass (genus *Sagittaria*) and attached algae make up most of the producers. Insects, snails, herbivorous fishes, and turtles eat the producers. Other fishes form the secondary and tertiary consumers. Odum also noted the presence of fungi and bacteria, which were involved in decomposition on all trophic levels.

Another way is to express the pyramid in terms of production rate. The **pyramid of energy** shows the rate of energy production rather than biomass (**Figure 44.16c**). The laws of thermodynamics ensure that the highest amounts of energy are found at the lowest trophic levels. The energy pyramid for Silver Springs also shows that large amounts of energy pass through decomposers, despite their relatively small biomass.

44.5 Reviewing the Concepts

- Ecosystem ecology concerns the movement of energy and materials through organisms and their communities. Organisms that obtain energy from light or chemicals are primary producers (or autotrophs). Organisms that feed on primary producers are called primary consumers (or herbivores). Organisms that feed on primary consumers are called secondary consumers (or carnivores). Consumers that get their energy from the remains and waste products of organisms are called detritivores (Figures 44.13, 44.14).

- Food webs are complex models of interconnected food chains in which multiple links occur between species. Food webs tend to have five or fewer links between top and bottom trophic levels. Energy conversions are not 100% efficient, and usable energy is lost within each trophic level and from one trophic level to the next (Figures 44.15, 44.16).

Biology Principle

Biologists Use Models to Test Predictions and Convey Ideas

This figure is a model for the structure of an African savanna food web, which consists of a composite of many different food chains. The Modeling Challenge in the figure legend asks you to provide examples of food chains that are contained within this African savanna food web.

Figure 44.14 A food web from an African savanna ecosystem. Each trophic level is occupied by different species. Generally, each species feeds on, or is fed upon by, more than one species.

Modeling Challenge: Take another look at Figure 44.14 and draw the five different food chains that are associated with hyenas. Be sure to label each trophic level as secondary carnivore, primary carnivore, herbivore, or autotroph. For the purposes of this challenge, ignore any interactions with wild dogs.

Concept Check: At which trophic level do decomposers feed?

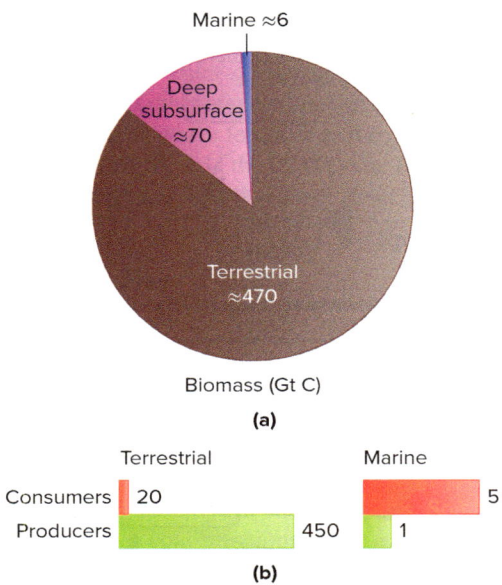

Figure 44.21 Biomass distributions across different habitats. **(a)** Across marine, terrestrial and subsurface, **(b)** across producers and consumers in different habitats.

It is estimated that humans have reduced the biomass of wild land mammals seven-fold and marine mammals five-fold over the past 50,000 to 3,000 years. Fish constitute the largest biomass of wild animals. Livestock biomass is greater than humans and much more than wild mammals. Humans have also reshaped the total quantity of carbon sequestered by plants as forests have been replaced by crops. Such changes have reduced total plant biomass two-fold relative to the start of human civilization, but crop biomass accounts for only about 2% of plant biomass.

Across different environments, land biomass at 470 Gt C is almost two orders of magnitude higher than the 6 Gt C of marine biomass (**Figure 44.21a**) despite oceans covering 71% of the Earth's surface and occupying a huge volume. Most plant biomass is concentrated on land because plants constitute only a small fraction of marine biomass as green algae and seagrasses. On the other hand, most animal biomass occurs in marine environments and most bacterial biomass occurs in subsurface environments, in deep aquifers and far below the seabed, >8 m below ground.

In looking at biomass across trophic levels, the biomass of primary producers on land dwarfs that of consumers (**Figure 44.21b**). However, in marine environments, primary production is only 1 Gt C, more than two orders of magnitude less than that on land. Marine primary production supports about 5 Gt C of consumer biomass, resulting in an inverted pyramid of biomass, described earlier. This is possible because primary producers have a rapid turnover of biomass, usually on the order of days, whereas consumer biomass turns over much more slowly, on the order of years for many fish.

44.6 Reviewing the Concepts

- Net primary production (NPP) is gross primary production minus the energy released during respiration via photosynthetic organisms. NPP in terrestrial ecosystems is limited primarily by temperature and the availability of water and nutrients. In aquatic ecosystems, NPP is limited mainly by the availability of light and nutrients (Figures 44.17, 44.18).
- Secondary production is limited by available primary production, but most primary production goes to detritivores (Figure 44.19).
- On a global scale, by far the highest percentage of biomass on Earth is attributable to plants, with bacteria a distant second. By contrast, animal biomass is minuscule. The amount of biomass on land greatly exceeds that in marine systems or that deep underground (Figures 44.20, 44.21).

44.6 Testing Your Knowledge

1. Net primary production is
   a. the energy that passes from plants to herbivores.
   b. gross primary production minus the energy used by herbivores.
   c. the energy fixed in photosynthesis.
   d. the energy fixed in photosynthesis minus the energy used in the respiration of photosynthetic organisms.
   e. gross primary production minus the energy used by consumers.

2. Primary production in terrestrial systems is often limited by
   a. temperature.
   b. water.
   c. nutrients
   d. both b and c.
   e. a, b, and c.

3. Most gross primary production is used
   a. in plant respiration.
   b. by herbivores.
   c. by bacteria.
   d. by nematodes.
   e. by both c and d.

4. Most of the Earth's biomass is found
   a. in the oceans.
   b. as bacteria.
   c. as plants.
   d. as animals.
   e. deep underground.

Assessing Your Knowledge and Skills

Foundational Knowledge

Multiple Choice

1. Autotrophic organisms are
   a. primary consumers.
   b. secondary consumers.
   c. tertiary consumers.
   d. primary producers.
   e. decomposers.

2. Which organisms are the most important consumers of energy in a Georgia salt marsh?
   a. *Spartina* grass and algae
   b. insects
   c. spiders
   d. crabs
   e. bacteria

3. The mechanism of succession in which early colonizers exclude subsequent colonists from moving into a community is referred to as
   a. facilitation.
   b. competitive exclusion.
   c. secondary succession.
   d. inhibition.
   e. natural selection.

4. On which type of island would you expect species richness to be greatest?
   a. small, near mainland
   b. small, distant from mainland
   c. large, near mainland
   d. large, distant from mainland
   e. Species richness is equal on all these types of islands.

5. Based on your knowledge of patterns of species richness, a large, tropical continent would likely have _____ species richness.
   a. low
   b. uneven
   c. high
   d. intermediate
   e. variable

Short Answer

1. At what trophic level does a carrion beetle feed?
2. **PRINCIPLES** A principle of biology is that biology is an experimental science. In the nutrient-poor heathlands of Europe, scotch heather (*Calluna vulgaris*) and cross-leaved heath (*Erica tetralix*) are gradually replaced by variegated purple moor grass (*Molinia caerulea*) and wavy hair grass (*Deschampsia flexuosa*). Adding *Calluna* litter or nitrogen fertilizer speeds up this process. Explain this phenomenon, and state which mechanism of succession is supported.

Critical-Thinking Skills

Multiple Choice

1. The process of primary succession occurs
   a. around a recently erupted volcano.
   b. on a newly plowed field.
   c. on a hillside that has suffered a mudslide.
   d. on a recently flooded riverbank.
   e. on none of the above.

2. You discover a narrow, subterranean river in the outback of Australia and estimate that it is at least 50 million years old. You gather a crew of scientists to perform species sampling of the river and apply metagenomics techniques to calculate microbial diversity. The results indicate that although short in length, the river is extremely diverse. You even get to name a new species of shrimp after yourself! You are asked to explain the high diversity of the subterranean river. What hypothesis would you cite?
   a. the species-time hypothesis
   b. the species-area hypothesis
   c. the intermediate disturbance hypothesis
   d. the species-productivity hypothesis

3. In the figure, each letter represents a species within a community and the arrows represent energy flow though the ecosystem. Which species is most likely a primary producer?

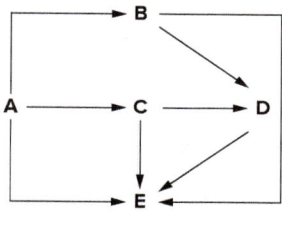

   a. A
   b. B
   c. C
   d. D
   e. E

4. How is it possible for a community to have low species diversity but high species richness?
   a. The individuals have a perfectly even distribution among the species.
   b. Most of the individuals belong to a single species.
   c. There are one or two rare species.
   d. There are low numbers of individuals in each species.

5. A hectare of wheat fixes 32,000 kg of carbon per year through photosynthesis and uses 7,000 kg of carbon per year for cellular respiration. The net primary productivity of the wheat field is therefore _____ kg of carbon/hectare/year.
   a. 39,000
   b. 37,000
   c. 27,000
   d. 25,000
   e. 4,600

Short Answer

1. Forest A has 5 tree species with 100 individuals. Forest B has 5 tree species with 10 individuals. What is the Shannon diversity index for each forest? Which forest has the highest diversity? What does this exercise tell you about the limitations of the Shannon diversity index?

2. Calculate the species diversity of the following four communities. Which community has the highest diversity? What is the maximum diversity each community could have?

Community	Relative abundance of species			H_s	Maximum possible diversity
	Species 1	Species 2	Species 3		
1	90	10	—		
2	50	50	—		
3	80	10	10		
4	33.3	33.3	33.3		

45 Biomes: How Climate Affects the Distribution of Species on Earth

African savanna, Masai Mara, Kenya.

Chapter Outline

45.1 Climate and Its Relationship to Biological Communities

45.2 Major Biomes

Assessing Your Knowledge and Skills

©AGF/UIG/Getty Images

For most areas of the world, we need only see a picture for a few seconds to know where we are: a tropical rain forest, a hot desert, or the African savanna. This is because the plants and animals in these large-scale communities, called **biomes,** are characteristic of the climate in which they live. In this chapter, we will examine how and why climate varies across the Earth and how climate affects the global distribution of biome types. First, we will discuss how temperature differences on Earth arise and how these differences affect air circulation and precipitation patterns. We will see how the tilt of the Earth causes seasonality and how elevation and proximity to water can affect local climate. In Section 45.2, we will examine how variations in temperature and rainfall patterns produce different biome types.

Humans are fundamentally changing the appearance of traditional biomes on Earth as they plant crops, cut down old-growth forests, and urbanize areas. Some recent studies have suggested that human-dominated areas now cover more of the Earth's land surface than natural ones do. To evaluate the effects of agriculture, deforestation, and urbanization at large scales, ecologists need to understand the factors that naturally limit the distribution of certain large-scale communities, such as tropical forests or temperate prairies.

45.1 Climate and Its Relationship to Biological Communities

Learning Outcomes
1. Explain how differences in global temperatures drive atmospheric circulation.
2. Discuss how the tilt and rotation of the Earth influence weather patterns.
3. Explain how both mountains and proximity to large bodies of water can change local temperature and precipitation patterns.
4. Explain how waves and tides are generated in marine systems.

Temperature and precipitation are the most important components of **climate**, the prevailing weather pattern in a given region. The distribution and abundance of organisms are primarily influenced by these factors. Therefore, to understand the patterns of abundance of life on Earth, ecologists need to study the global climate. In this section, we will examine global climate patterns, focusing on how temperature variation drives atmospheric circulation and how features such as elevation and landmass can alter these patterns. Knowing how and why climate changes around the world enables us to understand and predict the occurrence of biomes, large-scale terrestrial communities such as tropical rain forests and hot deserts.

Atmospheric Circulation Is Driven by Differences in Global Temperatures

Substantial differences in temperature occur over the Earth, mainly due to latitudinal variations in the incoming solar radiation. At higher latitudes, such as northern Canada and Russia, the Sun's rays hit the Earth obliquely and are spread out over more of the planet's surface than they are in equatorial areas (**Figure 45.1**). More heat is also lost

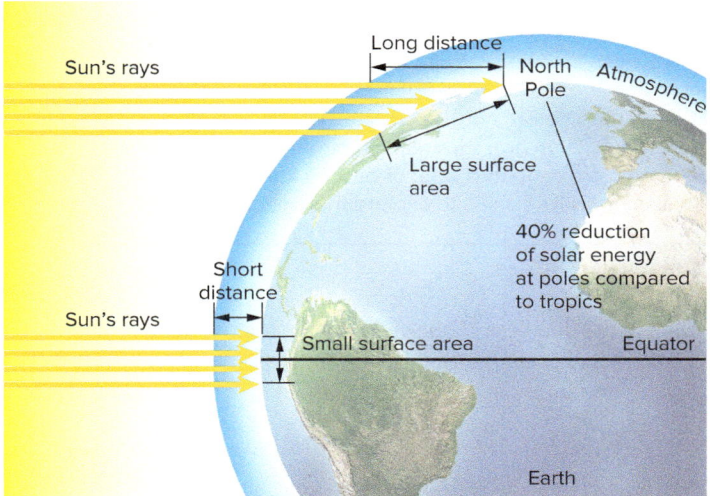

Figure 45.1 **The intensity of solar radiation at different latitudes.** In polar areas, the Sun's rays strike the Earth at an oblique angle and deliver less energy than at tropical locations. In tropical areas, the energy is concentrated over a smaller surface and travels a shorter distance through the atmosphere.

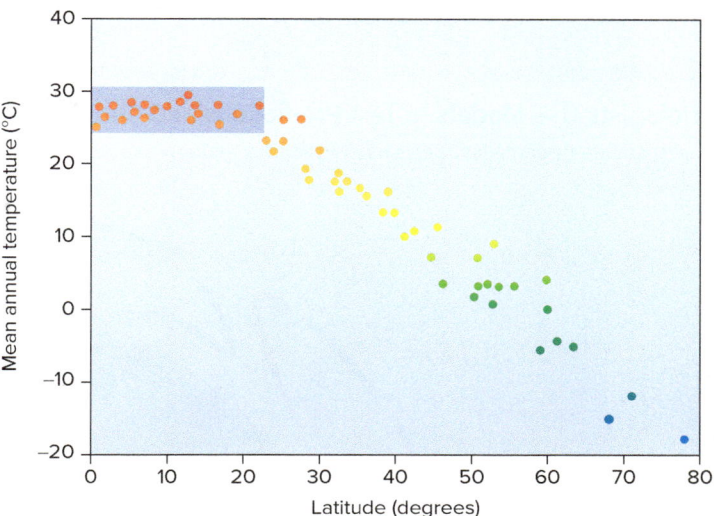

Figure 45.2 **Variation of the Earth's temperature.** The temperatures shown in this figure were measured at moderately moist continental locations of low elevation.

Concept Check: *Why is there a wide band of similar temperatures at the tropics? (See the orange dots highlighted in the purple rectangle.)*

in the atmosphere of higher latitudes, because the Sun's rays travel a greater distance through the atmosphere, allowing more heat to be dissipated by cloud cover. The result is that 40% less solar energy strikes polar latitudes than equatorial areas. Generally, temperatures increase as the amount of solar radiation increases (**Figure 45.2**). However, in the tropics, both cloudiness and rain reduce average temperature, so temperatures do not continue to increase toward the equator.

Global patterns of atmospheric circulation and precipitation are influenced by solar energy. High temperatures at the equator cause the surface equatorial air to heat up and rise vertically into the atmosphere. As the warm air rises away from its source of heat, it cools and becomes less buoyant, but the cool air does not sink back to the surface, because of the warm air behind it. The warm air rising near the equator forms towers of cumulus clouds that provide rainfall, which in turn maintains the lush vegetation of the equatorial rain forests. As the upper flow of air moves toward the poles, it begins to subside, or fall back to Earth, at about 20–30° north and south of the equator. These **subsidence zones** are areas of high pressure and are the sites of the world's hot deserts, because the subsiding air is relatively dry, having released all of its moisture over the equator (**Figure 45.3**).

From the centers of the subsidence zones, the surface flow splits into two directions, one of which flows toward the poles and the other toward the equator. The equatorial flow from both hemispheres meets near the equator in a region called the intertropical convergence zone (ITCZ). In the three-cell model, the circulation between 20–30° and 45–55° latitude is opposite that of the cell nearest the equator because the net surface flow is poleward. Additional zones of high precipitation occur in this cell, usually between 45° and 55°. In the final circulation cell, at the poles, the air has cooled and descends, but it

Biology Principle

Biologists Use Models to Test Predictions and Convey Ideas
The modified three-cell model is used to predict temperature and rainfall patterns on Earth and the positions of the biomes they create.

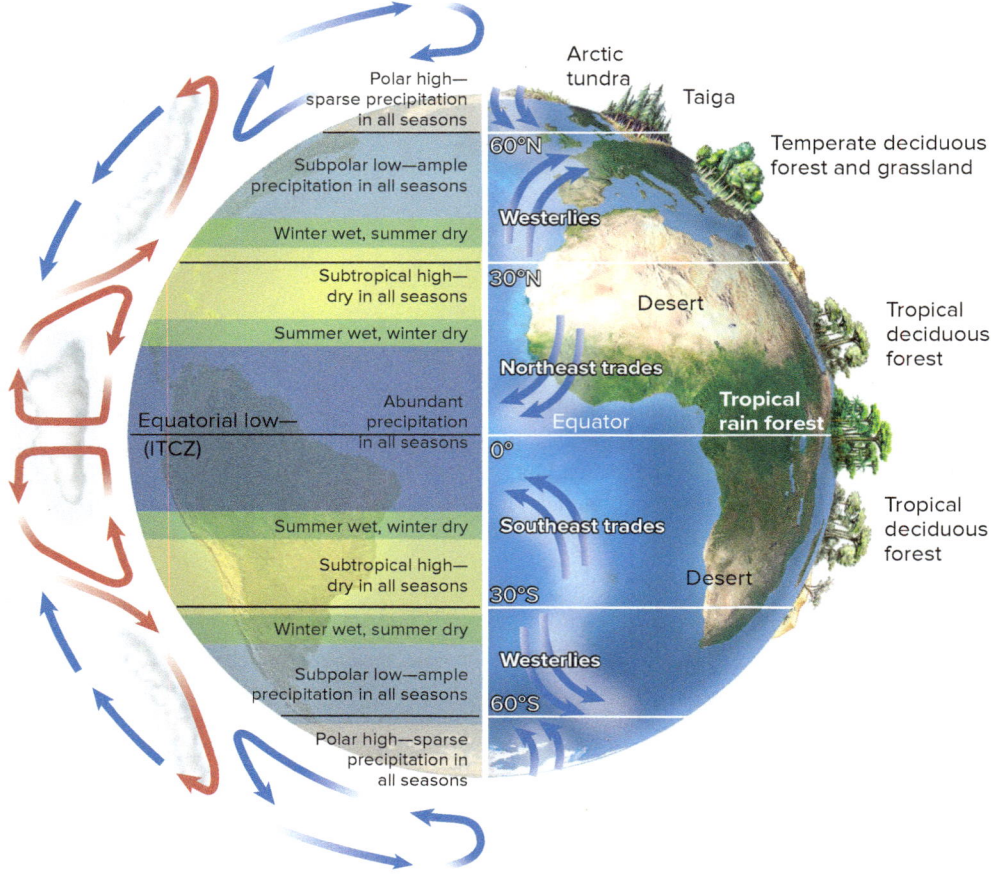

Figure 45.3 Global circulation based on a modified three-cell model. Warm, moist air rises in the equatorial zone where rain is frequent year round. Tropical forests exist mainly in a band around the equator, where it is hot and rainy. At around 20° to 30° north and south, the descending air is hot and dry, and deserts exist. A secondary zone of precipitation exists at around 45° to 55° north and south, where temperate forests are located. The polar regions are generally cold and dry. The term high refers to areas of high pressure resulting from falling air. Low refers to areas of low pressure resulting from rising air. Northerly or southerly air movements are deflected west or east by the rotation of the Earth and are shown as westerlies or so-called northeast/southeast trade winds, which allowed sailing ships to explore the world.

---

has little moisture left, explaining why many high-latitude regions are desert-like.

The distributions of the major biomes are largely determined by temperature differences and the rainfall patterns they generate. Hot, tropical forest blankets the tropics, where rainfall is high. At about 20–30° latitude, the air cools and descends, but it is without moisture, so the hot deserts occur around that latitude. The middle cell of the circulation model shows that at about 45° to 55° latitude, the air has warmed and gained moisture, so it ascends, dropping rainfall over the wet, temperate deciduous forests of North America and western Europe in the Northern Hemisphere and New Zealand and Chile in the Southern Hemisphere. In the high-latitude regions of the Northern and Southern Hemispheres (above 60° latitude), only short tundra vegetation can be supported.

The Tilt and Rotation of the Earth Also Affect Climate

The three-cell model provides a good understanding of global circulation, but it is oversimplified. Overlaid on these patterns is a seasonality caused by the tilt of the Earth. The Earth's axis of rotation is tilted at 23.5° from the vertical for the full 365 days the Earth takes to orbit the Sun (**Figure 45.4**). The solar equator, the area receiving the most solar energy, varies seasonally and reaches 23.5° north on June 21

BIOMES: HOW CLIMATE AFFECTS THE DISTRIBUTION OF SPECIES ON EARTH 991

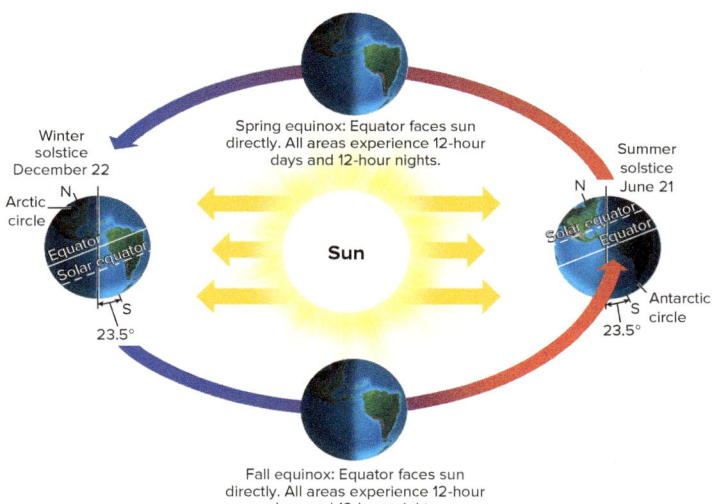

Figure 45.4 Seasonality. The constant 23.5° tilt of the Earth's axis causes changes in the solar energy striking the Earth during its 365-day procession around the Sun. The times of the solstices and equinoxes are shown for the Northern Hemisphere and are reversed for the Southern Hemisphere.

and 23.5° south on December 21. In the Northern Hemisphere, these dates are called the summer and winter solstices, from the Latin *sol*, sun, and *sistit*, stands. For several days before and after each solstice, the noontime elevation of the Sun appears in the same place. On March 21 and September 22, the so-called spring and autumn equinoxes, all locations in the Northern and Southern Hemispheres have approximately equal amounts of solar radiation. This means that for half of the year the Northern Hemisphere receives more solar energy, and for the other half of the year the Southern Hemisphere receives more solar energy. At 60° north, during the northern winter, temperatures in Siberia may average only −12°C, whereas in the summer they may average 16°C, a difference of 28°C. In contrast, tropical temperatures vary relatively little, perhaps 2–3°C, year round. Southern Hemisphere temperatures also vary seasonally, but the large expanses of open water moderate the temperature extremes.

Associated with these seasonal changes in temperature are changes in day length and precipitation. The ITCZ follows the solar equator (**Figure 45.5**). Thus, as the solar equator moves seasonally, the band of rainfall associated with it drifts north or south of the equator, depending on the season. The result is a seasonality of tropical rainfall between about 20° north and 20° south of the equator. Two rainy seasons of equal length exist at the equator, both about 3 to 4 months long, as the ITCZ drifts north or south. At the solar equator, there is just one rainy season of about 4–5 months. As a result of changes in latitude, there is much global variation in precipitation patterns.

In addition to seasonal changes in temperature, the wind direction is deflected by the rotation of the Earth, a phenomenon called the Coriolis force, after the French scientist Gaspard Gustave de Coriolis. Imagine a rocket fired from the North or South Pole toward

Figure 45.5 Seasonal variations in the intertropical convergence zone. The intertropical convergence zone is a belt of low pressure that circles the Earth near the equator, where trade winds converge and heavy rainfall is frequent.

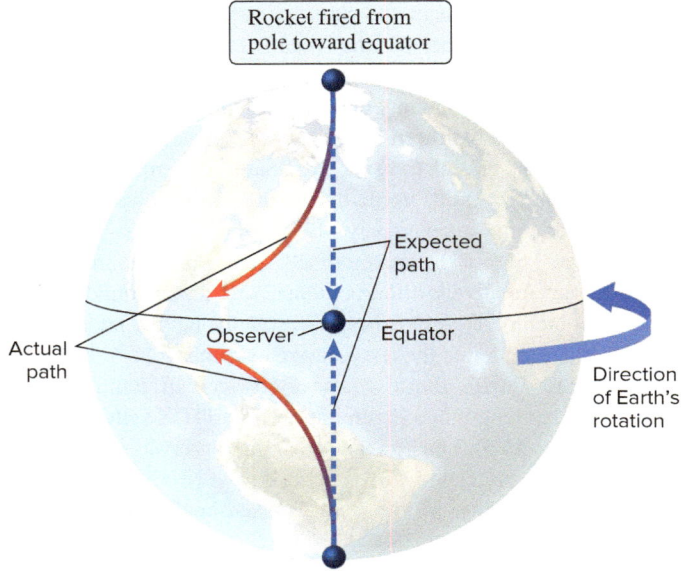

Figure 45.6 Diagrammatic representation of the Coriolis force. Even though a rocket fired from the North Pole flies due south, by the time it reaches the equator its intended target would have moved and the landing point would be at a more westward spot.

the equator. By the time the rocket reaches the equator, the target will have moved to the east. The rocket always travels due south or north, but to an observer at the equator its path bends some distance to the west before the rocket lands (Figure 45.6). A similar phenomenon occurs with winds and aircraft, though pilots compensate for Coriolis forces when flying. The opposite phenomenon happens when a rocket is fired from the equator either north or south. The Coriolis force deflects wind directions east or west, gives spin to storm systems, and is a reason hurricanes spin counterclockwise in the Northern Hemisphere but clockwise in the Southern Hemisphere.

Elevation and Proximity of a Landmass to Water Affect Local Climates

Thus far, we have considered climate on a global scale. The geographic features of a landmass can also have an important effect on the local climate in that area. For example, the elevation of a region greatly influences its temperature range. On mountains, temperatures decrease with increasing elevation. This decrease is a result of a process known as **adiabatic cooling,** in which increasing elevation leads to a decrease in air temperature caused by pressure changes. When air is blown across the Earth's surface and up over mountains, it expands because of the reduced pressure. As it expands, it cools at a rate of about 10°C for every 1,000 m in elevation, as long as no water vapor or cloud formation occurs. (Adiabatic cooling is also the process applied in the function of a refrigerator, in which refrigerant gas cools as it expands coming out of the compressor.) A vertical ascent of 600 m produces a temperature change roughly equivalent to that brought about by an increase in latitude of 1,000 km. This explains why mountaintop vegetation, even in tropical areas, can have the characteristics of a colder biome.

Mountains can also influence patterns of precipitation. Colder air holds less moisture than warmer air. Thus, when warm, moist air encounters the windward side of a mountain, it flows upward and cools, releasing precipitation in the form of rain or snow. On the side of the mountain sheltered from the wind (the leeward side), drier air descends, producing what is called a **rain shadow,** an area where precipitation is noticeably less (Figure 45.7). For example, the western side of the Cascade Range in Washington State receives more than 500 cm of annual precipitation, whereas the eastern side receives only 50 cm. The Gobi Desert in central Asia lies in the rain shadow of the

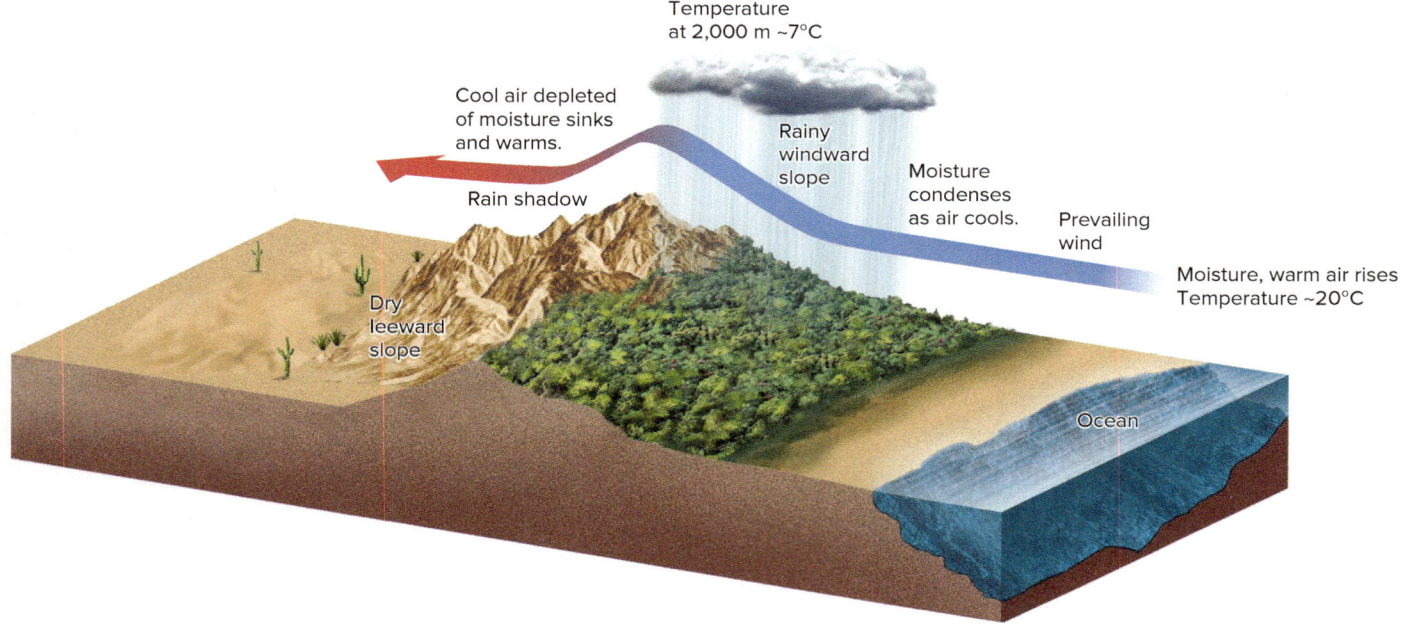

Figure 45.7 The influence of elevation on climate.

(a) Sea breeze

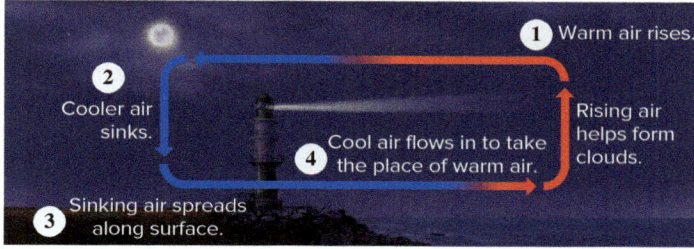

(b) Land breeze

Figure 45.8 Proximity to water affects climate. (a) When the landmass is warmer than the ocean, a sea breeze is created, bringing in cooler, moisture-laden air and sometimes fog. **(b)** When the sea is warmer than the land, a land breeze results.

Himalayas, and the Atacama Desert in Chile and Peru is in the rain shadow of the Andes.

The proximity of a landmass to a large body of water can also affect regional climate because land heats and cools more quickly than the sea. Recall from Chapter 2 that water has a very high specific heat—the amount of energy required to raise the temperature of 1 gram of a substance by 1°C. The specific heat of the land is much lower than that of the water, allowing the land to warm more quickly than water. During the day, the warmed air rises and cooler air flows in to replace it. This pattern creates onshore sea breezes in coastal areas (**Figure 45.8a**). At night, the land cools quicker than the sea, and so the pattern is reversed, creating offshore breezes (**Figure 45.8b**). The sea, therefore, has a moderating effect on the temperatures of coastal regions, especially islands. The climates of coastal regions may differ markedly from those of their climatic zones. Many never experience frost, and fog is often evident. Thus, along coastal areas, vegetation patterns may differ from those in areas farther inland. In fact, some areas of the U.S., including Florida, would be deserts, were it not for the warm water of the sea and the moisture-laden clouds that form above them.

Ocean Currents, Waves, and Tides Also Affect Climate

Together with the rotation of the Earth, winds also create ocean currents. The major ocean currents act as "pinwheels" between continents, running clockwise in the ocean basins of the Northern Hemisphere and counterclockwise in those of the Southern Hemisphere (**Figure 45.9**). The Gulf Stream, equivalent in flow to 50 times the world's major rivers combined, takes warm water from the Caribbean and the U.S. coasts across the Atlantic Ocean, where it combines with the North Atlantic Drift to moderate the climate of Europe. The Humboldt Current takes cool conditions to the western coast of South America and almost to the equator, and the California Current takes cooler climate to the Hawaiian Islands.

In addition to currents, wind also creates waves, which can range in size from small ripples to huge swells. As the wind blows, the

Figure 45.9 Major ocean currents of the world. The red arrows represent warm water; the blue arrows, cold water.

(a) Spring tide

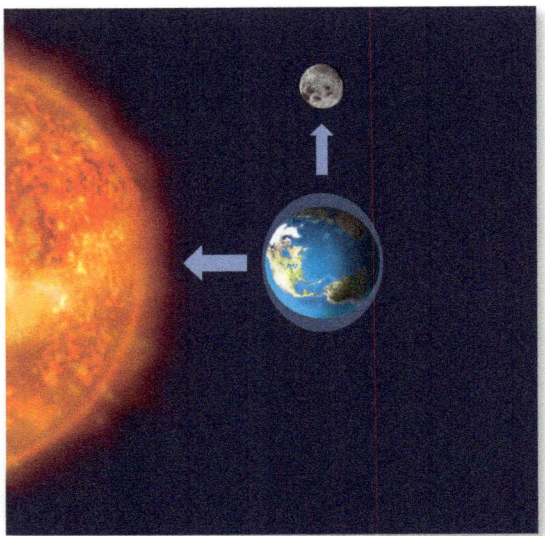
(b) Neap tide

Figure 45.10 Tide formation. (a) Spring tides form when the Sun, Earth, and Moon are in alignment. **(b)** Neap tides result when the Sun, Moon, and Earth are at a 90° angle.

Concept Check: *Which celestial body influences tides the most, the Sun or the Moon?*

friction between the air and the water creates small ripples. Once the ripples have formed, the wind has something to push against and the waves may increase in size. Four factors influence wave size: wind speed, fetch, duration of time of wind, and water depth. Fetch is the distance of open water over which the wind has blown. Shorter fetches in lakes reduce wave size, whereas long fetches of open ocean may allow the formation of 4- to 5-m waves.

Coastal areas are also influenced by tides, the rise and fall of sea levels. The gravitational pull of the Moon and the Sun causes tides. The gravitational force of the Sun is 179 times stronger than that of the Moon, but because the Sun is 389 times more distant, its effect is weaker. The tidal force of the Moon is 2.2 times greater than that of the Sun. As the Earth turns, each area of the globe is close to the Moon once a day. At the equator, oceans are pulled toward the Moon at this time, creating high tides at the equator and low tides at higher latitudes (**Figure 45.10**). Similarly, when an ocean is on the opposite side of the Earth away from the Moon, the tide is high. This is because the Earth is itself pulled more toward the Moon at this point, leaving the water behind, causing the water to rise relative to the Earth. Thus, most areas of the Earth have two tides per day.

The tidal range, the difference between the high and low tides, varies over a 2-week cycle. When the Sun, Moon, and Earth form a straight line, a condition known as syzygy, the gravitational pull of the Sun reinforces that of the Moon and the tidal range is maximal (Figure 45.10a). This is called a spring tide, because the ocean's surface appears to spring upward, and it occurs at around the time of a new or full Moon. When the Sun, Earth, and Moon are at a 90° angle, the gravitational forces of the Sun partially cancel those of the Moon, and the tidal range is minimal (Figure 45.10b). This is termed a neap tide, and it occurs when the Moon is in its first or third quarter.

45.1 Reviewing the Concepts

- Differences in global temperatures are caused by variations in incoming solar radiation and patterns of atmospheric circulation (Figures 45.1, 45.2, 45.3).
- The tilt of the Earth causes seasonality, including variations in temperature and rainfall. The rotation of the Earth affects wind direction (Figures 45.4, 45.5, 45.6).
- Elevation and the proximity of a landmass to a large body of water can affect climate (Figures 45.7, 45.8).
- Ocean currents, waves, and tides also affect climate and local environmental conditions (Figures 45.9, 45.10).

45.1 Testing Your Knowledge

1. The world's major subsidence zones occur at
   a. latitudes 0° and 45–55°.
   b. latitude 20–30° and the poles.
   c. latitude 0° and the poles.
   d. latitudes 20–30° and 45–55°.
   e. latitudes 0° and 20–30°.

2. Which of the following is *not* one of the four factors that influence wave size?
   a. duration time of wind
   b. water temperature
   c. water depth
   d. length of fetch
   e. wind speed

45.2 Major Biomes

Learning Outcomes

1. **SCISKILLS** ▶ Predict how changes in temperature and rainfall affect the distribution patterns of terrestrial biomes.
2. Describe how the distribution of species on Earth has been affected by continental drift.
3. Explain the concept of biogeographic regions.

As mentioned at the beginning of this chapter, a biome is a large-scale community characteristic of the climate in which it is found. In this section, we will survey the various types of terrestrial and aquatic biomes. We will also consider how continental drift and biogeography help to explain species distribution.

Ecologists Recognize 10 Different Terrestrial Biomes Based on Precipitation and Temperature

Many types of classification schemes are used for mapping the geographic locations of terrestrial biomes, but one of the most useful was developed by American ecologist Robert Whittaker, who classified biomes according to the physical factors of average annual precipitation and temperature (**Figure 45.11**). This classification scheme, commonly known as a Whittaker plot, recognizes 10 terrestrial biomes distributed widely across the Earth (**Figure 45.12**).

Tropical Rain Forest Tropical rain forests are hot and wet year round, with many different plant species (**Figure 45.13a**). The upper layer of forest foliage, or canopy, is dense and little light reaches the forest floor, so the ground cover is sparse. Animal life is diverse with many species of insects, reptiles, amphibians, and mammals, though large mammals are not common. Mimicry and bright, protective coloration that warns of bad taste or toxins are common. Tropical forests cover much of northern South America, Central America, western and central Africa, Southeast Asia, and various islands in the Indian and Pacific Oceans.

Tropical Deciduous Forest Tropical deciduous forests have a distinct 2- to 3-month (or longer) dry season when many trees shed their leaves (**Figure 45.13b**). An understory of herbs and grasses may grow at this time. Where the dry season is 6–7 months long, shorter, thorny plants such as acacia trees grow. The diversity of animal life is high, with species such as antelopes, wild pigs, tigers, and monkeys. Much of India consists of tropical deciduous forests, as do parts of Brazil, Mexico, Africa, and Australia.

Temperate Rain Forest Temperate rain forest is a relatively uncommon biome restricted to coastal locales such as the northwestern North American coast from northern California to Alaska, western Chile, parts of Norway, and Australia and New Zealand, where the nearby oceans greatly influence the climate. Large coniferous trees such as hemlocks, firs, and spruces thrive there and animal life is diverse (**Figure 45.13c**).

Temperate Deciduous Forest Temperatures in temperate deciduous forests often fall below freezing in winter, so snow may occur. Much of eastern Europe, eastern Asia, and the eastern United States

Biology Principle

Biologists Use Models to Test Predictions and Convey Ideas

A Whittaker plot, like that shown in Figure 45.11, is a model that gives a good visual representation of the biome types associated with different temperatures and precipitation patterns on Earth. The Modeling Challenge in the figure legend asks you to expand this Whittaker plot by including the chaparral biome.

Figure 45.11 The relationship between the world's terrestrial biome types and temperature and precipitation patterns.

Modeling Challenge: Some authorities recognize a biome type called chaparral, which occurs in regions of cool winters and warm summers averaging 10–25°C annually. Annual precipitation is generally 30–50 cm per year. This biome type is found in southern California, the Mediterranean region of Europe, South Africa, and southwestern Australia. Using the precipitation amounts and temperatures just described, draw the position of the chaparral biome on the Whittaker plot shown in Figure 45.11, and comment on its position relative to neighboring biomes. Based on its position, what vegetation types might be present in a chaparral?

BioConnections: Look back at Figure 7.16. In which biome type are CAM plants most likely to be found?

consists of temperate deciduous forests where oak, maple, and hickory trees are common (**Figure 45.13d**). Leaves are shed in the fall and reappear in the spring, when herbaceous plants flower before the light is reduced by the forest canopy. Mammals such as squirrels, wolves, bobcats, foxes, bears, and mountain lions occur, but reptiles

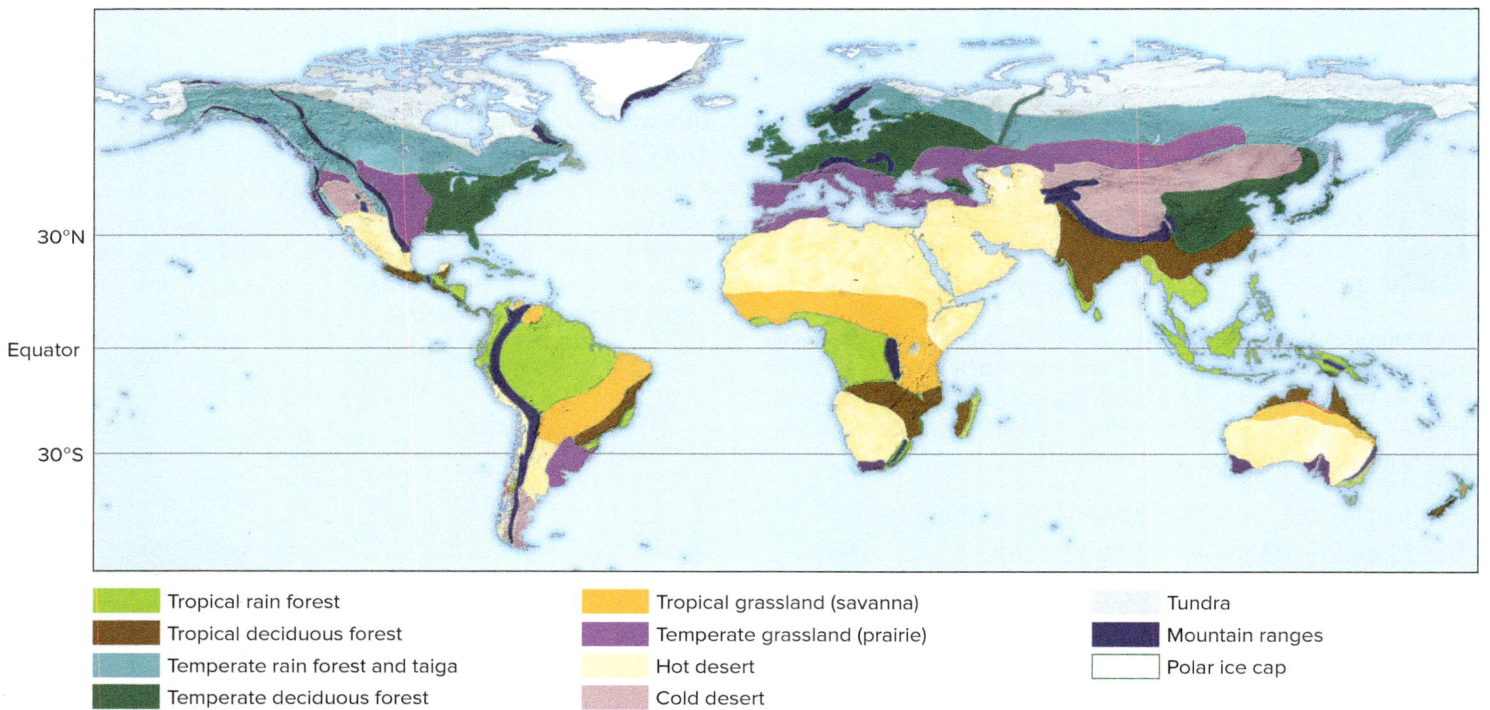

Figure 45.12 Geographic location of terrestrial biomes. The distribution patterns of taiga and temperate rain forest are combined because of their similarity in tree species and because temperate rain forest is actually limited to a very small area.

 Concept Check: *In a globally warmed world, what biome might expand into areas currently occupied by tundra?*

are relatively less common. In the cold months, many species hibernate and some birds migrate.

Temperate Coniferous Forest (Taiga) Vast tracts of the temperate coniferous biome, often known by its Russian name of taiga, occur in North America and Russia. In the Southern Hemisphere, little land area occurs at latitudes where extensive taiga could exist. Temperatures are very cold, often below freezing, for long periods of time, and snow is common. Most of the trees are evergreens or conifers with conical shapes to reduce bough breakage from heavy loads of snow (**Figure 45.13e**). The thick canopy and abundant pine needles greatly reduce the number of understory plants. Reptiles and amphibians are rare and mammals are heavily furred.

Tropical Grassland (Savanna) Tropical grassland, or savanna, exists in Africa, South America, and northern Australia where rainfall is seasonal, or limited to about 50–130 cm per year, insufficient to support forests (**Figure 45.13f**). Fire may be prevalent and most plants have extensive root systems that enable them to resprout quickly after a fire. The world's greatest assemblages of large mammals occur in savannas. Herds of antelope, zebra, and wildebeest are found together with their associated predators: lion, leopard, cheetah, and hyena.

Temperate Grassland (Prairie) Temperate grasslands, or prairies, also have limited rainfall but winter temperatures often fall below freezing. Included are the North American prairies, the steppes of Russia, the pampas of Argentina, and the South African veldt. Large grazing mammals such as bison and pronghorn feed on the grasses in North American prairies. Wild horses are common in Asia, as are kangaroos in Australia (**Figure 45.13g**).

Hot Desert Rainfall is generally less than 30 cm a year in deserts. In hot deserts, temperatures may vary from as high as 50°C in the day to below freezing at night. Annual plants grow only when there is rain. Succulents, such as saguaro cactus and other barrel cacti, store water. Other shrubs drop their leaves in the dry season (**Figure 45.13h**). All plants produce numerous small seeds to conserve water. Seed eaters such as ants, birds, and rodents are common, as are the reptiles that prey on them. Deserts are found at latitudes of 20–30° north and south of the equator and include the Sahara of North Africa, the Kalahari and Namib of southern Africa, the Atacama of Chile, the Sonoran of northern Mexico, and the Simpson of Australia.

Cold Desert Cold deserts are also dry but temperatures commonly average only around freezing in the winter. Plant life is sparse and most species are deciduous and spiny (**Figure 45.13i**). Many small herbivores live in burrows to escape the cold. Large cold deserts include the Great Basin Desert in North America, the Patagonia Desert in eastern Argentina, and the Gobi Desert in central Asia.

Tundra Tundra, from the Finish word *tunturia,* meaning treeless plain, occurs only in the Northern Hemisphere, north of the temperate forest (or taiga). The growing season is short, only 50–60 days, and

BIOMES: HOW CLIMATE AFFECTS THE DISTRIBUTION OF SPECIES ON EARTH 997

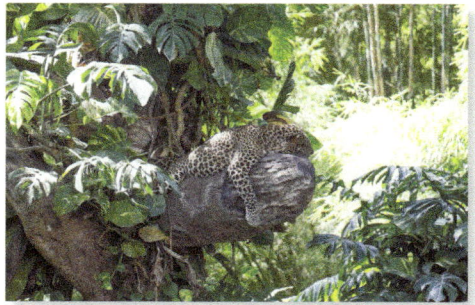
(a) Tropical rain forest, Java

(b) Tropical deciduous forest, India

(c) Temperate rain forest, Washington State

(d) Temperate deciduous forest, Maryland

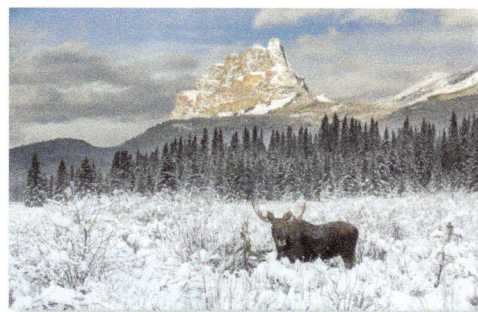
(e) Temperate coniferous forest (or taiga), Canada

(f) Tropical grassland or savanna, Kenya

(g) Temperate grassland (or prairie), Wyoming

(h) Hot desert, Namibia

(i) Cold desert, Mongolia

(j) Tundra, Alaska

Figure 45.13 Examples of terrestrial biomes.
(a) ©Vladislav T. Jirousek/Shutterstock; *(b)* ©Theo Allofs/theoallofs.com; *(c)* ©RGB Ventures/SuperStock/Alamy; *(d)* ©Marie-Ann Daloia/123RF; *(e)* ©John E Marriott/Getty Images; *(f)* ©Jon Arnold Images Ltd/Alamy; *(g)* ©D. Robert & Lorri Franz/Corbis Documentary/Getty Images; *(h)* ©Eye Ubiquitous/Getty Images; *(i)* ©Art Wolfe/Getty Images; *(j)* ©Bruce Coleman/Photoshot/Avalon

temperatures are so cold that water is locked away in **permafrost,** a layer of permanently frozen soil. Vegetation exists as slow-growing mosses, sedges, shrubs, and associated trees that grow close to the ground (**Figure 45.13j**). So little moisture falls that desert conditions exist in some areas. Many animals migrate or hibernate in the winter and those that remain are heavily furred.

Modifications to Biome Classification Although broad terrestrial biomes are a useful way of defining the main large-scale communities on Earth, ecologists acknowledge that not all communities fit neatly into 1 of the 10 major biome types. One biome type often grades into another at biome boundaries, as seen on mountain ranges. Soil conditions can also influence biome vegetation. In

California, temperate grassland with isolated trees is abundant—except on serpentine soils, which are dry and nutrient-poor and support only sparse vegetation adapted to this extreme environment. In the eastern U.S., most of New Jersey's coastal plain, called the Pine Barrens, consists of sandy, nutrient-poor soil that cannot support the surrounding deciduous forest and instead contains grasses and low shrubs growing among open stands of pygmy pine and oak trees.

Aquatic Biomes Are Distinguished by Temperature, Salinity, Oxygen Content, Depth, Current Strength, and Light

Within aquatic environments, several biome types are also recognized, including marine aquatic biomes (intertidal zone, coral reef, and open ocean) and freshwater habitats (lakes, rivers, and wetlands) (Figure 45.14). These biomes are distinguished primarily by differences in temperature, salinity, oxygen content, depth, current strength, and availability of light. Freshwater habitats are traditionally divided into lentic, or standing-water, habitats (from the Latin *lenis,* meaning calm) and lotic, or running-water, habitats (from the Latin *lotus,* meaning washed).

Open Ocean The open ocean is the world's most extensive biome, covering 70% of the Earth's surface with water averaging 4,000 m in depth. Microscopic, photosynthetic organisms called **phytoplankton** are the most common element of plant life, accounting for half the photosynthetic activity on Earth. These are fed upon by **zooplankton,** minute, drifting animal organisms. Larger, free-swimming animals include fishes, squid, sea turtles, and marine mammals.

Intertidal Zone The **intertidal zone,** where the land meets the sea, is alternately submerged and exposed by the daily cycle of the tides (Figure 45.14a). Intertidal areas can consist of rocky shores, sandy shores, or mudflats. In rocky shores, plant life consists of green algae and seaweeds, and animal life can be quite diverse. Sea anemones, snails, hermit crabs, and small fishes live in tide pools; limpets, mussels, sea stars, sea urchins, and chitons occur on the rock face. In tropical areas, mangrove trees colonize the mudflats and may form extensive forests. In temperate areas, salt marsh grasses colonize muddy areas. Sandy or muddy shores also contain a variety of burrowing worms and crabs.

Coral Reef In warm waters of at least 20°C, coral reefs occur in shallow areas where light penetrates and allows photosynthesis to occur (Figure 45.14b). Sunlight is important because many corals harbor photosynthetic algae, or dinoflagellates, that contribute nutrients to the animals and require light to live. An immense variety of fishes, perhaps 40% of the fish species on Earth, are found on coral reefs. There is also high diversity of invertebrates such as snails, sea stars, octopuses, and sea urchins. The largest coral reef in the world is the Great Barrier Reef off the Australian coastline, but other coral reefs are found in the Atlantic Ocean, the Red Sea, and the Pacific and Indian Oceans.

Lentic Habitats **Lentic** habitats consist of still, sometimes deep water in lakes (Figure 45.14c). In addition to phytoplankton, lentic habitats may have rooted vegetation that extends above the water surface, such as cattails. Frogs, turtles, fishes, crayfish, and aquatic insects may be common, and in tropical and subtropical lakes, alligators and crocodiles can be found.

Lotic Habitats In **lotic** habitats, flowing water prevents nutrient accumulations and phytoplankton blooms (Figure 45.14d). The current also mixes the water, providing a well-oxygenated habitat. In fast-moving rivers, leaves from the surrounding forest are the primary food source for animals. In slow-moving streams and rivers, algae and rooted plants may be present. Many invertebrates are flat and attach themselves to rocks to avoid being swept away. Fish such as trout may be present in clear, cool, oxygen-rich rivers, while in murkier waters catfish and carp are present.

Wetlands At the margins of both lentic and lotic habitats, wetlands may develop. These areas are regularly saturated by surface water or groundwater and include marshes (treeless areas) and swamps (wet areas dominated by trees). Many wetlands are seasonally flooded. These are among the most productive areas in the world, with floating plants, such as lilies, and rooted plants, such as sedges, cypress trees, and gum trees. Wetlands contain abundant animals with many insect species, wading and diving birds, amphibians, reptiles, otters, and alligators.

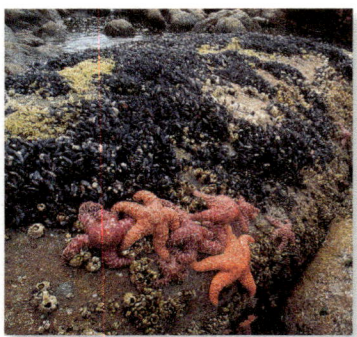
(a) Intertidal, rocky shore, Washington State

(b) Coral reef, Caribbean Sea

(c) Lentic, or lake, habitat

(d) Lotic, or river, habitat

Figure 45.14 Aquatic biomes.
(a) ©Nature Picture Library/Alamy; *(b)* ©adokon/Getty Images; *(c)* ©imageBROKER/Alamy; *(d)* ©Ron Crabtree/Getty Images

EVOLUTIONARY CONNECTIONS

Continental Drift and Biogeography Help to Explain Species Distribution

Knowledge of biomes does not tell us everything about the distribution of plant and animal life on Earth. An understanding of evolution and geological change over large areas and time periods helps to explain some of the patterns seen in modern biomes. For example, South America, Africa, and Australia have similar biomes, ranging from tropical to temperate, yet each continent has distinctive animal and plant life. South America is inhabited by sloths, anteaters, armadillos, and monkeys with prehensile tails. Africa possesses a wide variety of antelopes, zebras, giraffes, lions, baboons, okapis, and aardvarks. Australia, which has no native placental mammals except bats, is home to a variety of marsupials such as kangaroos, koala bears, Tasmanian devils, and wombats, as well as the egg-laying monotremes—the duck-billed platypus and echidna. Most continents also have distinct species of plants; for example, eucalyptus trees are native only in Australia. In South American deserts, succulent plants belong to the family Cactaceae, the cacti; in Africa, they belong to the genus *Euphorbia*, the spurges.

One explanation for these species distributions is that abiotic factors are of paramount importance and that each region supports the flora and fauna best adapted to it. However, the spread of introduced species has proved this explanation incorrect: European rabbits introduced into Australia proliferated rapidly, and eucalyptus from Australia grows well in California. Based on biogeography, a better explanation is that different flora and fauna are the result of the independent evolution of separate, unconnected populations, which have generated different species in different places. A knowledge of continental drift and evolution is therefore vital to understanding modern distributions of species.

The relative location of the landmasses on Earth has changed enormously over time as a result of continental drift, the slow shifting of the Earth's landmasses over time (refer back to Figure 22.5). The current distribution of the essentially flightless bird family, the ratites, in the Southern Hemisphere is the result of continental drift. The common ancestor of these birds occurred on a supercontinent that included what are now South America, Africa, and Australia. As the continent split apart, genera evolved separately so that today there are ostriches in Africa, emus in Australia, and rheas in South America.

Continental drift is not the only mechanism that creates widely separate populations of closely related species. The distributions of many present-day species are relics of once much broader distributions. For example, there are currently four living species of tapir: three in Central and South America and one in Malaysia (**Figure 45.15**). Fossil records reveal a more widespread distribution over much of Europe, Asia, and North America. The oldest fossils come from Europe, making it likely that this was the center of the

Biology Principle

Populations of Organisms Evolve from One Generation to the Next and Are Related by an Evolutionary History

Knowing that all tapir species share a common European ancestor helps to explain their current distribution pattern.

Figure 45.15 Tapir distribution. This map shows the geographic ranges of the four modern tapir species (see the key below the map). Fossil evidence suggests a European origin and a more widespread distribution, with tapirs dying out in other regions.

Tapirus indicus   *Tapirus pinchaque*   *Tapirus terrestris*   *Tapirus bairdi*

Figure 45.16 The biogeographic regions proposed by A. R. Wallace. Note that the borders do not always demarcate continents.

- The location of many fossils and living organisms can be explained by evolution in one supercontinent followed by subsequent continental drift. Other distinct distribution patterns are relics of once much broader distributions (Figure 45.15).
- Six major biogeographic regions—Nearctic, Palearctic, Neotropical, Ethiopian, Oriental, and Australian—each of which contains distinct fauna and flora, are recognized today (Figure 45.16).

45.2 Test Your Knowledge

1. Which terrestrial biome type is present where rainfall is substantial, temperatures are hot year round, and there is a distinct dry season?
   a. tropical rain forest
   b. tropical deciduous forest
   c. temperate forest
   d. savanna
   e. prairie

2. The unique marsupial fauna of Australia is largely a result of
   a. adiabatic cooling.
   b. climate.
   c. continental drift and evolution.
   d. rain shadows.
   e. biogeographic regions.

origin of tapirs. The migrations of later-evolving tapirs resulted in a more widespread distribution. However, global cooling resulted in the extinction of tapirs in all areas except the two tropical locations.

British explorer Alfred Russel Wallace was one of the earliest scientists to realize that certain plant and animal taxa were restricted to certain geographic areas of the Earth (see Chapter 19). For example, the distribution patterns of guinea pigs, anteaters, and other groups are confined to Central and South America, from central Mexico southward. Based on this and other observations, Wallace divided the world's biota into six major **biogeographic regions**: Nearctic, Palearctic, Neotropical, Ethiopian, Oriental, and Australian (Figure 45.16). These regions are still widely accepted today, though debate continues about the exact location of the boundary lines. Biogeographic regions correspond largely to continents but more exactly to areas bounded by major barriers to dispersal, such as oceans, deserts, and mountain ranges.

45.2 Reviewing the Concepts

- Biomes are large-scale communities characterized by distinctive plant and animal life and located in a particular climate (Figures 45.11, 45.12).
- Terrestrial biomes are generally named for their climate and vegetation type and include tropical rain forest, tropical deciduous forest, temperate rain forest, temperate deciduous forest, temperate coniferous forest (taiga), tropical grassland (savanna), temperate grassland (prairie), hot and cold deserts, and tundra. In mountain ranges, biome type may change on an elevation gradient (Figure 45.13).
- Within aquatic environments, biomes include marine aquatic biomes (the intertidal zone, coral reef, and open ocean) and freshwater lakes, rivers, and wetlands. These are distinguished by differences in temperature, salinity, oxygen content, depth, current strength (lentic versus lotic), and availability of light (Figure 45.14).

Assessing Your Knowledge and Skills

Foundational Knowledge

Multiple Choice

1. The equinoxes occur on
   a. December 21 and September 22.
   b. March 21 and December 21.
   c. June 21 and December 21.
   d. June 21 and March 21.
   e. March 21 and September 22.

2. Tidal range is maximal in
   a. a neap tide.
   b. a spring tide.
   c. syzygy.
   d. both a and c.
   e. both b and c.

3. The largest biome type, by area occupied, is the
   a. open ocean.
   b. tropical rain forest.
   c. tundra.
   d. hot desert.
   e. lentic habitat.

4. Which of the following characteristics is (are) commonly used to identify the terrestrial biomes of the Earth?
   a. temperature
   b. precipitation
   c. vegetation
   d. all of the above
   e. a and b only

5. What is the driving force that determines the circulation of the atmospheric air?
   a. temperature differences of the Earth
   b. winds
   c. ocean currents
   d. mountain ridges
   e. All of the above determine the circulation of atmospheric air.

Short Answer

1. If mountains are closer to the Sun than valleys are, why aren't mountains hotter?
2. **PRINCIPLES** A principle of biology is that living organisms interact with their environment. In most locations on Earth at about 20–30° latitude, air cools and descends, and hot deserts occur. Florida is situated between 31° north and 24° north. Why doesn't it support a desert biome?

Critical-Thinking Skills

Multiple Choice

1. You have come across an old dataset in your research lab. The researcher recorded temperature, precipitation, and plant species over a time span of 5 years but apparently forgot to record the location! Upon further examination, you find that the average temperature in the mystery location is 10°C, the average precipitation is 25 cm, and plant matter is sparse, made up mostly of short, scrubby species. From what biome was the dataset most likely recorded?
   a. tundra
   b. temperate grassland
   c. cold desert
   d. hot desert
   e. taiga (temperate coniferous forest)
2. You have survived a plane crash. When you regain consciousness, you cannot remember what your original destination was, but you make several observations about your surroundings that may help you to figure out where you are. First, you notice that tall trees are plentiful and the ground is covered with fallen leaves. The air temperature is very hot, and you notice that the soil is quite dry underneath the leaf litter. Into what biome has your plane most likely crashed?
   a. hot desert
   b. tropical deciduous forest
   c. temperate rain forest
   d. tropical rain forest
   e. tropical grassland
3. How would you best explain the existence of a mountain range that has abundant conifer trees, ferns, and mosses on the east side but only sparse, shrubby plants on the west side?
   a. Adiabatic cooling leads to temperature gradients that produce the differing plant communities.
   b. Precipitation differences due to a rain shadow on the west side produce the differing plant communities.
   c. Differences in the specific heat of the two sides of the mountain produce the differing plant communities.
   d. Differences in solar intensity produce the differing plant communities.
4. What factor would you expect to show the largest difference between the lotic and intertidal biomes?
   a. salinity
   b. plant species diversity
   c. water pH
   d. animal species diversity
   e. temperature variation
5. A genus of lizards contains two extinct species and five living species. The oldest fossils from the genus belong to the two extinct species and are known only from the Ethiopian biogeographic region. Two of the living species are found only in the Neotropical biogeographic region, and the other three are found only in the Australian biogeographic region. Fossil records confirm that all living species are native to the areas they currently inhabit. What is the most reasonable conclusion about the evolution of these lizard species?
   a. The Neotropical and Australian lineages share a common ancestor that originated in the Neotropical region.
   b. The Neotropical and Australian lineages share a common ancestor that originated in the Australian region.
   c. The Neotropical and Australian lineages share a common ancestor that originated in the Ethiopian region.
   d. The Neotropical and Australian lineages do not share a common ancestor but evolved independently through convergent evolution.
   e. The Neotropical and Australian lineages share a common ancestor, which originated in the Palearctic region.

Short Answer

1. If the Earth was not tilted 23.5° on its axis, how would climate and life be affected?
2. Water is most dense at 4°C. As water warms or cools from 4°C, it becomes less dense. At 0°C, water freezes and is in its least dense state. Describe how these density differences affect aquatic organisms. Now imagine that ice were the most dense state for water. How would this affect freshwater life?

46 The Age of Humans

Source: National Park Service, Everglades National Park

Burmese python in the Florida Everglades. This species of snake was introduced into the Florida Everglades by humans and has dramatically decreased the populations of several native species.

Chapter Outline

- 46.1 Human Population Growth
- 46.2 Global Warming and Climate Change
- 46.3 Pollution and Human Influences on Biogeochemical Cycles
- 46.4 Pollution and Biomagnification
- 46.5 Habitat Destruction
- 46.6 Overexploitation
- 46.7 Invasive Species

Assessing Your Knowledge and Skills

In January 2016, British geoscientist Colin Waters and colleagues proposed the formal recognition of a new geological epoch, the Anthropocene, the last and latest epoch of the quaternary period (refer back to Figure 22.4). Some scientists have argued that Anthropocene is a pop-culture term and a political statement, based on the word **anthropogenic,** meaning resulting from human activity, rather than a new epoch. But support is growing to embrace the Anthropocene, a term coined in the 1980s by American ecologist Eugene Stoermer and later popularized by Dutch atmospheric chemist Paul Crutzen. In the 1970s and 1980s, Crutzen showed how human pollutants can damage the ozone layer, a discovery for which he won a Nobel Prize in 1995, along with Mexican chemist Mario Molina and American chemist Frank Rowland.

While agreement is widespread that humans have a profound effect on the global environment, the formal recognition of a new geological time unit is subject to specific criteria. Global-scale changes must be recorded in geological stratigraphic (layered) material, usually rock, and in ice or marine sediments. Waters and colleagues noted the appearance of manufactured items in these sediments, including plastics, concrete, aluminum, and particulates from fossil-fuel combustion. Perhaps the most distinctive new, identifiable items in sediments were radioactive elements from the 500 aboveground nuclear blasts that occurred between 1945 and 1963, which created globe-encircling debris.

Whether or not the Anthropocene gains wide acceptance in the scientific literature, there is no disputing the immense impacts of humans on natural systems. In this chapter, we will examine some of these impacts, which include global warming, effects on biogeochemical cycles, habitat destruction such as deforestation, the overexploitation of various species via overfishing and overhunting, and the introduction of invasive species to all corners of the globe (see the chapter-opening photo). We will begin the chapter by addressing the huge numbers of humans on the planet and our prodigious rate of population growth.

46.1 Human Population Growth

Learning Outcomes

1. Describe how differences in age structure and human fertility across different countries affect human population growth.
2. Explain the concept of an ecological footprint.

In 2015, the world's population was estimated to be increasing at the rate of 150 people every minute: about 2 per minute in developed nations and 148 per minute in less-developed nations. In this section, we will examine human population growth trends in more detail and discuss how knowledge of the human population's age structure and fertility levels can help to predict its future growth. We will then investigate the carrying capacity of the Earth for humans and explore how the concept of an ecological footprint, which measures human resource use, can help us to determine this carrying capacity.

Human Populations Show Extreme Recent Growth

Until the beginning of agriculture and the domestication of animals, about 10,000 B.C.E., the average rate of human population growth was very low. With the establishment of agriculture, and the advent of modern medicine, the world's population had grown to about 800 million by the year 1750. Between 1750 and 2011, a relatively short period of human history, the world's human population surged from 800 million to 7 billion (**Figure 46.1**). In 2017, the number of humans was estimated at 7.5 billion. Scientists are very interested in determining when and at what size the human population will level off.

Knowledge of a Population's Age Structure Can Help to Predict Its Future Growth

The age structure of a population can help to predict future population growth. In all populations, **age structure** refers to the relative numbers of individuals of each defined age group. This information is commonly displayed as a population pyramid. In West Africa, for example, children younger than age 15 make up nearly half of the population, creating a pyramid with a wide base and narrow top (**Figure 46.2a**). Even if fertility rates decline, the population is expected to significantly increase as these young people move into childbearing age. The age structure of western Europe is much more uniform (**Figure 46.2b**). Even if the fertility rate of young women in western Europe increases to a level higher than that of their mothers, the annual numbers of births will still be relatively low because of the low number of women of childbearing age.

Human Population Fertility Rates Vary Widely Around the World

Global population growth can be estimated by determining the **total fertility rate (TFR),** the average number of live births a woman would have during her lifetime if she passed through her childbearing years according to current age-specific fertility rates (**Figure 46.3**). The total fertility rate differs considerably from one geographic area to another because of differences in women's education, wealth (money for contraception), employment levels, religious beliefs, and other factors. In Africa, the total fertility rate of 4.7 in 2010 had declined substantially since the 1970s, when it was around 6.7 children per woman. In Latin America and Southeast Asia, the rates declined even more from the 1970s and are now at around 2.2. Canada and most countries in Europe have a TFR of less than 2.0. The TFR was about 1.76 in the U.S. in 2017, a 30-year low. In Russia, fertility rates have dropped to 1.61. In China, although the TFR is only 1.6, the population there will still continue to increase until at least 2025 because of the large number of women of reproductive age. Although the global TFR declined from 4.47 in the 1970s to 2.42 in 2016, this is still greater than the average of 2.3 needed for zero population growth. The replacement rate is slightly higher than 2.0, to replace mother and father, due to natural mortality prior to reproduction. The replacement rate varies globally, from 2.1 in developing countries to between 2.5 and 3.3 in less-developed countries.

The wide variation in fertility rates makes it difficult to predict future population growth. The 2010 United Nations report presented world population projections to the year 2100 for three different growth scenarios: low, medium, and high (**Figure 46.4**). The three scenarios are based on three different assumptions about fertility rate. Using a low-fertility rate estimate of only 1.5 children per woman, the population would reach a maximum of about 8 billion people by 2050. A more realistic assumption may be to use the fertility rate estimate of 2.0 or 2.5. With these rates, the population would continue to rise to 10 or almost 16 billion, respectively.

The Concept of an Ecological Footprint Helps to Estimate Carrying Capacity

Recall from Chapter 43 that carrying capacity is the maximum population size that can be sustained by an environment. What is the Earth's carrying capacity for the human population, and when will it be reached? Estimates vary widely. Much of the speculation on the upper boundary of the world's population size centers on lifestyle. To use a simplistic example, if everyone on the planet ate meat extensively and drove large cars, then the carrying capacity would be a lot less than if people were vegetarians and used bicycles as their main means of transportation.

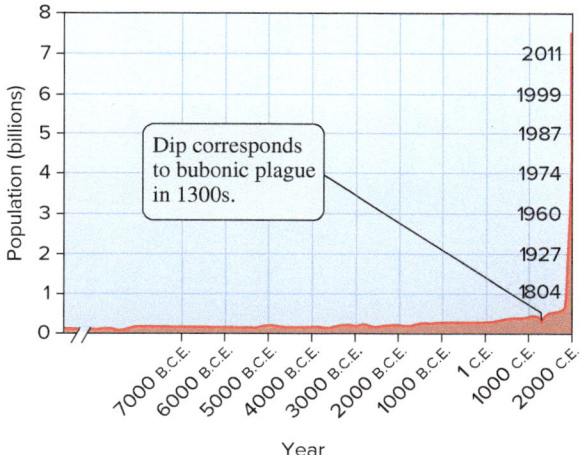

Figure 46.1 The growth pattern of the human population through history.

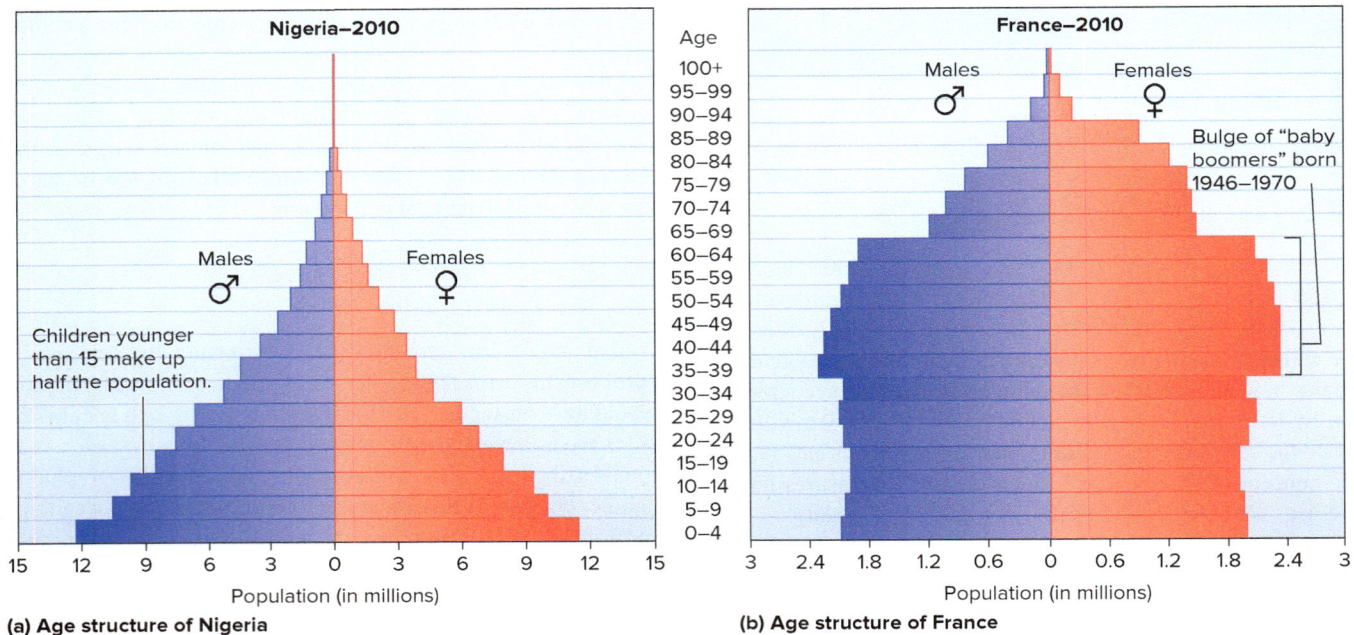

Figure 46.2 The age structure of human populations in Nigeria and France, in 2010. (a) In developing areas of the world such as Nigeria, children are the most abundant age group. Population growth is rapid. (b) In the developed countries of western Europe, the age structure is more evenly distributed. The bulge represents those born in the post–World War II "baby boom," when birth rates climbed due to stabilization of political and economic conditions. Population growth is close to zero.

Concept Check: *If the population pyramid in (part a) were inverted, what would you conclude about the age structure of the population?*

Figure 46.3 Total fertility rates (TFRs) among major regions of the world. Data refer to the average number of children born to a woman during her lifetime.

In the 1990s, Swiss sustainability advocate Mathis Wackernagel and his coworkers calculated how much land is needed to support each person on Earth. Everybody has an effect on the Earth, because we consume the land's resources, including crops, wood, fossil fuels, minerals, and so on. Thus, each person has an **ecological footprint**—the amount of productive land needed to support a person. The average footprint size for each person on the planet is about 3 hectares (1 ha = 10,000 m^2), but wide variation is found around the globe

Biology Principle

Biologists Use Models to Test Predictions and Convey Ideas

How TFR is calculated has a great influence on assumptions about how human global population size will change over the next 80 years.

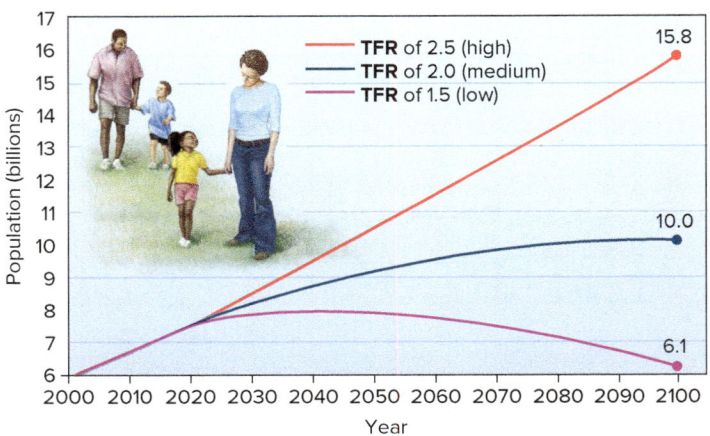

Figure 46.4 Population predictions for 2000–2100, using three different total fertility rates (TFRs).

Biology Principle

Biology Is a Quantitative Science

Detailed measurements of food and water consumption, energy use, and other factors are needed to calculate the ecological footprints of people in different geographic areas.

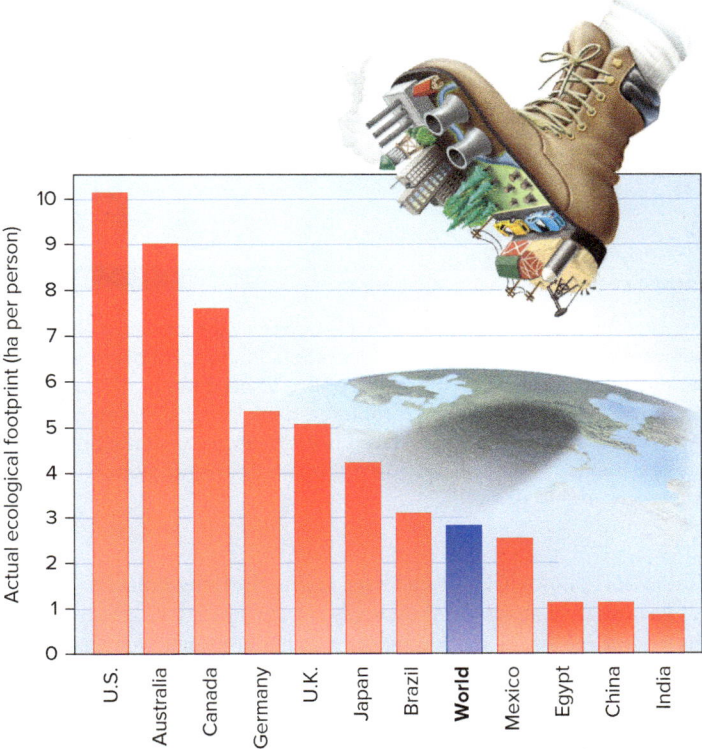

Figure 46.5 Ecological footprints of different countries. An ecological footprint is the amount of productive land needed to support the average individual of that country.

 Concept Check: *What is your ecological footprint?*

(Figure 46.5). The ecological footprint of the average Canadian is 7.5 hectares; it is about 10 hectares for the average American.

In most developed countries, the largest component of land use is for energy, followed by food and then forestry. Much of the land needed for energy absorbs the CO_2 emitted by the use of fossil fuels. If everyone required 10 hectares, as the average American does, we would need three Earths to provide us with the needed resources. Many people in less-developed countries use far fewer resources.

Globally, humans are already beyond the Earth's carrying capacity if we are to live in a sustainable manner. How has this happened? Many people live in an unsustainable manner, using more resources than can be regenerated in any given year. Furthermore, a rapidly growing human population contributes to environmental changes on a broad scale, including habitat loss for wildlife, overhunting and overfishing, and pollution. As described next, pollution is thought to contribute to climate change on a global scale.

46.1 Reviewing the Concepts

- Human population growth has skyrocketed over the last 250 years (Figure 46.1).
- Differences in the age structure of a population, the numbers of individuals in each age group, can influence future patterns of population growth (Figure 46.2).
- Although they have been declining worldwide, total fertility rates (TFRs) differ markedly between developed and less-developed countries. Predicting growth of the human population depends on the total fertility rate that is used (Figures 46.3, 46.4).
- The ecological footprint is the amount of productive land needed to support each person on Earth. Because many people live in a nonsustainable manner, globally we are already in an ecological deficit (Figure 46.5).

46.1 Testing Your Knowledge

1. The average total fertility rate needed for zero population change across the world is
   a. 1.7.   c. 2.0.   e. 2.5.
   b. 1.9.   d. 2.3.

2. The total fertility rate is highest in
   a. Latin America.
   b. Southeast Asia.
   c. Europe.
   d. Africa.
   e. North America.

46.2 Global Warming and Climate Change

Learning Outcomes

1. Describe how the greenhouse effect contributes to global warming.
2. Explain how global warming may affect sea levels and precipitation patterns.
3. Predict how climate change will change species distribution patterns.

Global warming refers to an increase in Earth's average surface temperature. In recent years, ecologists have been investigating how human activities have contributed to global warming. A key impact of global warming is **climate change,** which is a long-term change in Earth's climate or a change in climate in a particular region. In most regions on Earth, global warming is expected to increase surface temperatures. However, because global warming can affect ocean currents and the circulation of the atmosphere, some regions may actually become colder or experience seasonal changes such as colder winters.

To see how global warming causes climate change, we will begin by exploring how Earth's atmosphere acts like the glass panes in a greenhouse, trapping heat inside. By understanding this process, ecologists are better able to predict the future effects of a warming Earth. We will explore how human activities have contributed to global warming and consider some of the potential consequences of climate change, which include rising sea levels, changes in precipitation, and changes in the distributions of species.

The Greenhouse Effect Causes the Earth's Temperature to Rise

The Earth is warmed by the greenhouse effect. In a greenhouse, sunlight penetrates the glass and heats the plants inside. The heat is reradiated but the glass traps the heat inside. Similarly, solar radiation in the form of short-wave energy passes through the atmosphere to heat the surface of the Earth. This energy is then reradiated from the Earth's warmed surface into the atmosphere, but in the form of long-wave infrared radiation. Atmospheric gases absorb much of this infrared energy and radiate it a second time to the Earth's surface, causing its temperature to rise further (**Figure 46.6**). The greenhouse effect is important to life on Earth. Without the greenhouse effect, global temperatures would be much lower than they are, perhaps averaging only −17°C compared with the existing average of +15°C.

The greenhouse effect is caused by a small group of gases, mainly water vapor, that together make up less than 1% of the total volume of the atmosphere. After water vapor, the four most significant greenhouse gases are carbon dioxide, methane, nitrous oxide, and chlorofluorocarbons (**Table 46.1**).

Ecologists are concerned that human activities are increasing the magnitude of the greenhouse effect, resulting in the gradual elevation of the Earth's surface temperature. In particular, the burning of fossil fuels appears to be the main culprit of global warming. Fossil fuels such as oil, coal, and natural gas are high in carbon and release carbon dioxide (CO_2) into the atmosphere when burned. The atmospheric concentrations of most greenhouse gases have increased since the Industrial Revolution over 200 years ago. Of those increasing, the one that is the greatest contributor to global warming is CO_2. As Table 46.1 shows, CO_2 has a lower global warming potential per unit of gas (relative absorption) than any other major greenhouse gas, but its concentration in the atmosphere is much higher. An analysis of air trapped in glaciers shows that concentrations of atmospheric CO_2 increased from about 280 parts per million (ppm) in the preindustrial 19th century to 400 ppm in 2015 (**Figure 46.7a**). Since 1957, air samples have been collected directly at Mauna Loa, Hawaii, a relatively unpolluted site. The data show a 25% increase in atmospheric CO_2 levels in just 58 years, from 313 ppm to 400 ppm during 1957–2015 (**Figure 46.7b**).

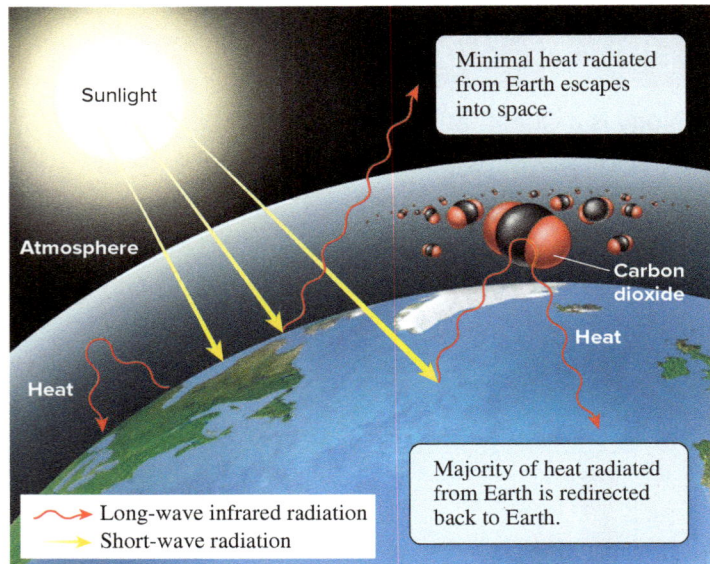

Figure 46.6 The greenhouse effect is caused by the insulating effect of atmospheric carbon dioxide. Solar radiation, in the form of short-wave energy, passes through the atmosphere to heat the Earth's surface. Long-wave infrared energy is radiated back into the atmosphere. Most infrared energy is reradiated back to Earth by atmospheric gases, including carbon dioxide molecules, causing global temperatures to rise.

Table 46.1	The Major Greenhouse Gases and Their Contribution to Global Warming			
	Carbon dioxide (CO_2)	**Methane (CH_4)**	**Nitrous oxide (N_2O)**	**Chlorofluorocarbons (CFCs)**
Relative absorption in ppm of increase*	1	21	310	10,000
Atmospheric concentration (ppm†)	400	1.75	0.315	0.0005
Contribution to global warming	73%	7%	19%	1%
Percentage from natural sources; type of source	20–30%; volcanoes	70–90%; swamps, gas from termites and ruminants	90–100%; soils	0%
Major human-made sources	Fossil-fuel use, deforestation	Rice paddies, landfills, biomass burning, coal and gas exploitation	Cultivated soil, fossil-fuel use, automobiles, industry	Previously manufactured products (for example, aerosol propellants) now banned in the U.S. and the E.U.

*Relative absorption is the warming potential per unit of gas.
†ppm = parts per million.

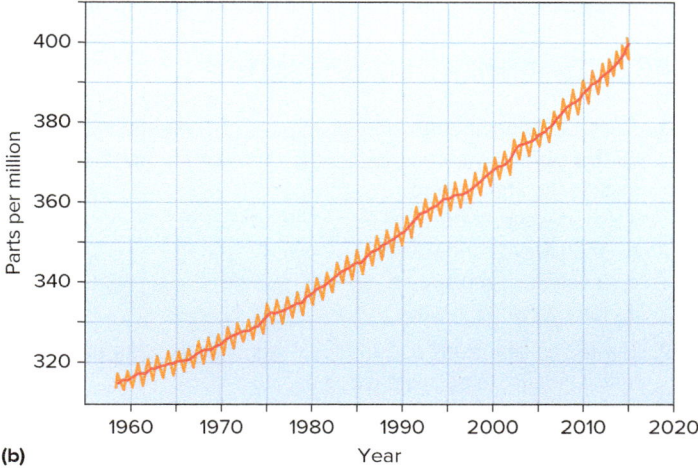

Figure 46.7 The increase of atmospheric carbon dioxide. **(a)** Increases in the burning of fossil fuels have caused an increase in atmospheric CO_2 since the late 19th century. **(b)** Since 1957, CO_2 levels have shown consistent increases in air samples collected at Mauna Loa, Hawaii.

Concept Check: *Why does the amount of CO_2 fluctuate seasonally in the graph in part (b)?*

In addition, seasonal oscillations in CO_2 occur, as seen in the orange line in Figure 46.7b. These types of oscillations differ in the two hemispheres of Earth. The Northern Hemisphere, where Hawaii is located, has greater land area and plant biomass than the Southern Hemisphere. In the northern summer, more CO_2 is absorbed by plants and atmospheric CO_2 level declines slightly. In the northern winter, less CO_2 is absorbed and atmospheric CO_2 levels increase.

To predict the effects of global warming, most scientists focus on a future point, about 2100, when the concentration of atmospheric CO_2 will have doubled—that is, increased to about 700 ppm compared with the late-20th-century level of 350 ppm. Ecologists hypothesize that at that time, average global temperatures will be about 2–4°C warmer than present. This increase in heat might not seem like much, but it is comparable to the warming that ended the last Ice Age. As discussed next, one consequence of global warming is its effect on sea levels.

Global Warming Is Expected to Cause Sea Levels to Rise

The term **sea level** refers to the average level for the surface of one or more of Earth's oceans. The 2014 report of the Intergovernmental Panel on Climate Change (IPCC) suggested that a business-as-usual scenario with the same level of increasing greenhouse gas emissions would lead to a 2–4°C warming by about the end of this century. With a 2–4°C warming, sea levels are expected to rise by about 50–60 cm due to the melting of glaciers and the thermal expansion of water. Sea levels have already risen 10–25 cm over the past century, and seawater entry into coastal forests has killed trees in many areas.

An alarming problem is the long time lag between warming temperatures and the time it takes for ice to melt. Even if future temperature increases were prevented, glaciers would continue to melt for many years. For example, the ice sheets of Greenland will take hundreds of years to melt, even though the increase in temperature needed to melt them could be reached in 50–100 years. You might envision this as being similar to an ice cube on a kitchen counter—it takes a while for the ice cube to melt, even though room temperature is considerably above freezing. Therefore, even if no greenhouse gases were added to the atmosphere after 2020, at least a quarter of the sea-level rise would still be expected by 2100.

One Effect of Global Warming on Climate Change Is to Alter Precipitation Patterns on Land

An increase in global temperatures is also expected to promote climate change and alter global precipitation. Higher temperatures increase evaporation, and warmer air holds more water vapor than cooler air. Global warming will generally lead to more frequent and heavier precipitation. Over the period 1986–2000, the average increase in precipitation over land was 3.5 mm per year. However, over normally dry areas, the level of moisture in the atmosphere may not increase enough to facilitate rain. Therefore, the increased temperatures will likely lead to widespread droughts in desert areas. We can make the analogy that the atmosphere is like a sponge; warm, moist air allows the sponge to absorb more water, but in dry conditions nothing more can be rung out of the sponge. Northern and southern temperate areas have already shown precipitation increases, as have tropical and subtropical areas. However, desert areas such as the Sahel region south of the Sahara have become drier. In essence, the wet get wetter and the dry get drier.

Climate Change Is Expected to Alter the Distribution of Species

Assuming that the previous scenario of climate change is accurate, scientists are beginning to consider the consequences on natural and human-made ecosystems. As most regions on Earth become warmer and a few regions become colder, most plant species cannot easily disperse and move north or south into newly created climatic regions that will be suitable for them. Many tree species take hundreds, even thousands, of years for seed dispersal.

American paleoecologist Margaret Davis was among the first to investigate how species' ranges would be changed in a globally warmed world. She predicted that in the event of a CO_2 doubling, many tree species in North America, such as sugar maple, would suffer range contractions because their southern ranges would retreat northward

Biology Principle

Biologists Use Models to Test Predictions and Convey Ideas

Detailed models taking into account changes in temperature and precipitation are needed to predict the future distributions of different species on a globally warmed Earth.

Figure 46.8 Possible changes in the ranges of sugar maples due to climate change. **(a)** This map takes into account only temperature changes. **(b)** This map takes into account temperature and precipitation changes. In both maps, the yellow shading indicates areas where sugar maples now exist but would not survive due to climate changes associated with a doubling of atmospheric CO_2 levels. The red shading indicates areas where sugar maples do not currently exist but would have a favorable climate to exist if CO_2 levels doubled. The orange shading indicates the overlap where sugar maples currently exist and where they could also exist due to climate change.

46.2 Reviewing the Concepts

- Global warming is an increase in the surface temperature on Earth. It is expected to result in climate change.
- The greenhouse effect is the process by which short-wave solar radiation passes through the atmosphere to warm the Earth and is reradiated back into the atmosphere as long-wave infrared radiation. Much of this radiation is radiated back to Earth's surface a second time, causing its temperature to rise (Figure 46.6).
- An increase in greenhouse gases, particularly of CO_2, is intensifying the greenhouse effect, causing climate change (Table 46.1, Figure 46.7).
- Two consequences of climate change are rising sea levels and changes in precipitation patterns.
- Ecologists predict that climate change will have a large impact on the distribution of the world's organisms (Figure 46.8).

46.2 Testing Your Knowledge

1. Solar radiation in the form of _____ energy heats the Earth's surface, whereas _____ infrared energy is radiated back to the atmosphere.
   a. short-wave, short-wave
   b. short-wave, long-wave
   c. long-wave, short-wave
   d. long-wave, long-wave

2. The rise in sea levels due to global warming over the past century is thought to be in the range of
   a. 1–5 cm.   d. 25–50 cm.
   b. 5–10 cm.   e. 50–100 cm.
   c. 10–25 cm.

46.3 Pollution and Human Influences on Biogeochemical Cycles

Learning Outcomes

1. Outline the steps of the carbon cycle, and describe the environmental effects of elevated atmospheric concentrations of CO_2.
2. Describe the processes of the water cycle, and explain how they are affected by humans.
3. Describe the role of phosphorus in lake eutrophication.
4. List the five main steps of the nitrogen cycle, and identify the human influences on it.

A unit of energy moves through an ecosystem only once, passing through the trophic levels of a food web from producer to consumer and dissipating as heat. In contrast, chemical elements such as carbon or nitrogen follow a cycle, moving from the physical environment to organisms and back to the environment, where the cycle begins again. Because the movements of chemicals through ecosystems involve biological, geological, and chemical transport mechanisms, they are termed **biogeochemical cycles.** Biological mechanisms involve the absorption of chemicals by living organisms and the subsequent

(yellow areas in **Figure 46.8a**). This contraction in the trees' distribution could be offset by the creation of new favorable habitats in Canada (red areas in Figure 46.8a). However, most scientists hypothesize that the climatic zones would shift toward the poles faster than the tree species could either migrate via seed dispersal or evolve to become better adapted to their altered climate. Therefore, the areas shown in red in Figure 46.8a are hypothetical and may not be achieved. In addition, changed precipitation patterns will alter the distribution patterns and densities of organisms still further. When Davis mapped out the future distribution of the sugar maple, *Acer saccharum,* in a globally warmed world and took into account changed precipitation patterns, the predicted future area of distribution was considerably less than the prediction taking into account only temperature changes (**Figure 46.8b**).

release of those chemicals back into the environment. Geological mechanisms include the weathering and erosion of rocks and the transport of elements by surface and subsurface drainage. Chemical transport mechanisms include dissolved matter in rain and snow, atmospheric gases, and dust blown by the wind.

In addition to the basic building blocks of carbon, hydrogen, and oxygen, the elements required in the greatest amounts by living organisms are phosphorus and nitrogen. In this section, we will take a detailed look at the cycles of these nutrients, which can be divided into two broad types: (1) local cycles, such as the phosphorus cycle, which involve elements with no atmospheric mechanism for long-distance transfer, and (2) global cycles, which involve an interchange between the atmosphere and the ecosystem. In our discussion of biogeochemical cycles, we will take a close look at how these cycles are affected by human activities, such as the burning of fossil fuels and the use of fertilizers.

The Carbon Cycle Is Affected by Human Activities Such as the Burning of Fossil Fuels and Deforestation

The movement of carbon from the atmosphere into organisms and back again is known as the carbon cycle. The rates of the processes shown in **Figure 46.9** vary greatly. Some carbon becomes locked up in reservoirs with a low turnover rate. For example, a fraction of the material from primary producers is transformed into deposits of coal, natural gas, and oil, which are collectively known as **fossil fuels.** In addition, much carbon is incorporated into the shells of marine organisms, which eventually form huge limestone deposits on the ocean floor or in terrestrial rocks, where turnover is extremely slow. As a result, rocks and fossil fuels contain the world's largest reserves of carbon.

In contrast, much of the carbon in phototrophs, which include plants, algae, and cyanobacteria, turns over much more rapidly. These organisms acquire CO_2 from the atmosphere and incorporate it into the organic matter of their own biomass via photosynthesis. Each year, plants, algae, and cyanobacteria remove approximately one-seventh of the CO_2 from the atmosphere. At the same time, respiration and the decomposition of phototrophs recycle a similar amount of carbon back into the atmosphere as CO_2. Other natural sources also release significant amounts of CO_2 into the atmosphere. These sources include volcanoes, hot springs, and fires. Animals can return some CO_2 to the atmosphere by respiration and decomposition, but the amount of carbon flowing through this part of the cycle is minimal.

Human activities, primarily the burning of fossil fuels, are increasingly causing large amounts of CO_2 to enter the atmosphere. In addition, deforestation contributes to the elevation of atmospheric CO_2 because

Figure 46.9 The carbon cycle. Each year, plants and algae remove about one-seventh of the CO_2 in the atmosphere. Animal respiration has such a small effect that it is not represented. The width of each arrow indicates the relative contribution of each process to the cycle.

Concept Check: *Where are the greatest reservoirs of global carbon?*

BioConnections: *Refer back to Table 2.1. Carbon is one of just four elements that account for the vast majority of atoms in living organisms. What are the other three, and therefore what biogeochemical cycles might be the most important to us?*

there is less vegetation to absorb CO_2 from the atmosphere. Direct measurements over the past five decades show a steady rise in atmospheric CO_2 (see Figure 46.7), a pattern that shows no sign of slowing.

What are the consequences of elevated levels of atmospheric CO_2? As discussed in Section 46.2, ecologists have identified a link between a rise in atmospheric CO_2 and global warming. Elevated atmospheric CO_2 can also increase the acidity of the oceans, mainly near the surface, because carbonic acid is formed when CO_2 dissolves in water. Higher acidity inhibits shell growth in marine mollusks, crabs, corals, and echinoderms and can cause reproductive disorders in some fish. Also, as described next, higher CO_2 levels may boost plant growth but lower the amount of herbivory.

FEATURE INVESTIGATION

Stiling and Drake's Experiments with Elevated CO_2 Showed an Increase in Plant Growth but a Decrease in Herbivory

How will forests of the future respond to elevated CO_2? To begin to answer such a question, ecologists ideally would enclose large areas of forests with chambers, increase the CO_2 content within the chambers, and measure the responses. This has proved to be difficult for two reasons. First, it is hard to enclose large trees in chambers, and second, it is expensive to increase CO_2 levels over a large area.

In much of Florida's forests, trees are small, only 3–5 m high at maturity, because frequent lightning-initiated fires prevent the growth of larger trees. In a discovery-based investigation, American ecologists Peter Stiling and Bert Drake were able to increase CO_2 levels around patches of forest at the Kennedy Space Center in Cape Canaveral, Florida. In the 1990s, they teamed up with NASA engineers to create 16 circular, open-topped chambers (**Figure 46.10**). In 8 of these, they

Figure 46.10 The effects of elevated atmospheric CO_2 on herbivore abundance and mortality.

(photos) ©Peter Stiling

GOAL To determine the effects of elevated CO_2 on a forest ecosystem; effects on herbivore survival are highlighted here.

STUDY LOCATION Patches of forest at the Kennedy Space Center in Cape Canaveral, Florida.

	Experimental level	Conceptual level
1. Erect 16 open-top chambers around native vegetation. Increase CO_2 levels from 350 ppm to 700 ppm in half of them.		Expected atmospheric CO_2 level is 700 ppm by end of the 21st century. Open-top chambers allow movement of insect herbivores in and out of chambers.
2. Conduct a yearly count of numbers of leaf miners per 200 leaves in each chamber.	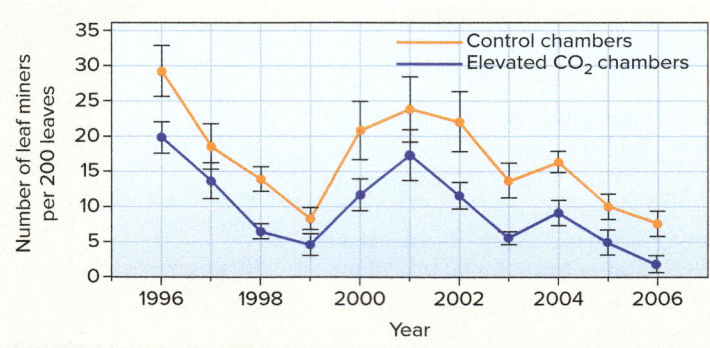	
3. Examine the leaf mines created by leaf miners to determine if the leaf miners survived and emerged, or if they died due to nutritional inadequacy or by the action of predators or parasitoids.	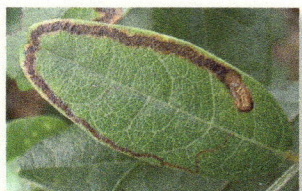	Elevated CO_2 reduces foliar nitrogen, inhibits normal insect development, and prolongs the feeding time of herbivores, allowing natural enemies greater opportunities to attack them.

4. **THE DATA**

Source of mortality*	Elevated CO_2 (% mortality)	Control (% mortality)
Nutritional inadequacy	10.2	5.0
Predators	2.4	2.0
Parasitoids	10.0	3.2

*Data refer only to mortality of larvae within leaves and do not sum to 100%. Mortality of eggs on leaves, pupae in the soil, and flying adults is unknown.

5. **CONCLUSION** Elevated CO_2 increases herbivore mortality and decreases herbivore abundance.

6. **SOURCE** Stiling, P., and Cornelissen, T. 2007. How does elevated carbon dioxide (CO_2) affect plant-herbivore interactions? A field experiment and meta-analysis of CO_2-mediated changes on plant chemistry and herbivore performance. *Global Change Biology* 13: 1823–1842.

increased atmospheric CO_2 to double the ambient level, from around 350 ppm to 700 ppm, the latter of which is the atmospheric concentration predicted by the end of the 21st century. The experiments commenced in 1996 and lasted until 2007.

The experiment shown in Figure 46.10 focuses on the effects of elevated CO_2 on a type of herbivore called a leaf miner, the most common type of herbivore at this site. There is usually only one leaf miner per leaf so that a count of the number of leaf miners per 200 leaves gives an accurate estimate of the level of herbivory. Because the chambers were open-topped, these insect herbivores could come and go. Censuses were conducted on all species of leaf miners (step 2), and mines were examined to assign causes of mortality to the larvae (step 3). Leaf miner larvae are small enough to burrow between the surfaces of plant leaves, creating leaf mines. These mines result in brown areas, as shown in Figure 46.10, step 3. The analysis of a leaf mine reveals whether a larva survived and emerged, died of nutritional inadequacy, was eaten by predators, or was attacked by parasitoids.

As seen in the data in step 4 of Figure 46.10, the mortality of insect leaf miners was higher under conditions of elevated CO_2. This decrease in herbivore survival occurred even though the plants produced more biomass in elevated CO_2, because CO_2 is limiting to plant growth. (Note: The data for plant biomass increases are not shown in the figure.) Lower leaf miner densities could be attributed to higher mortality rates due to nutritional inadequacy, predators, or parasitoids (see step 3).

Stiling and Drake hypothesized that part of the reason for the decline was that even though plants increased in mass, the existing soil nitrogen was diluted over a greater volume of plant material, so the nitrogen level in leaves decreased. This could have resulted in increased insect mortality by two means. First, poorer leaf quality could have directly increased insect death because leaf nitrogen levels may have been too low to support the normal development of the leaf miners. Second, lower leaf quality is expected to increase the amount of time insects need to feed to gain sufficient nitrogen. Increased feeding times, in turn, may have led to increased exposure to natural enemies, such as predatory spiders and ants and parasitoids (refer back to Figure 43.11), so mortality from natural enemies also increased (see the data of Figure 46.10). Thus, in a world of elevated CO_2, plant growth may increase and herbivory may decrease.

Experimental Questions

1. What was the hypothesis of Stiling and Drake's experiment?
2. **SCISKILLS** ▶ Explain the purpose of increasing the CO_2 levels in only half of the chambers in the experiment and not all of the chambers.
3. **SCISKILLS** ▶ Imagine that the percent mortality from parasitoids in the eight elevated CO_2 chambers was 6, 11, 9, 8, 13, 12, 14 and 7 and that in the ambient chambers was 1, 4, 5, 2, 3, 6, 1.6, and 3. Perform a statistical test to see if mortality from parasitoids was different between elevated and ambient CO_2 chambers.

Global Warming and Interruption of Water Flow By Humans Affect the Water Cycle

The water cycle, also called the hydrological cycle, differs from the cycles of other nutrients in that very little of the water that cycles through ecosystems is chemically changed by any of the cycle's components (**Figure 46.11**). This cycle is a physical process, fueled by the Sun's energy, rather than a chemical one, because it consists of essentially two phenomena: evaporation and precipitation. Even so, the water cycle has important biological components. Over land, 90% of the water that reaches the atmosphere is moisture that has passed through plants and exited from the leaves via transpiration. Only about 2% of the total volume of Earth's water is found in the bodies of organisms or is held frozen or in the soil. The rest cycles from bodies of water, to the atmosphere, and then to the land and back to bodies of water again.

Human activities have altered the water cycle in different ways. As discussed in Section 46.2, climate change may have the greatest effect on the water cycle by altering precipitation patterns around the world and causing glaciers to melt. On a smaller scale, humans have built structures that affect the flow of moving bodies of water. To increase the amount of available water and to create hydroelectric power, humans have interrupted the water cycle in many ways, most prominently through the use of dams to create reservoirs. Such dams can greatly interfere with the migration of fishes, such as salmon, and affect their ability to reproduce and survive. Other activities, such as tapping into underground water supplies, or **aquifers,** for drinking

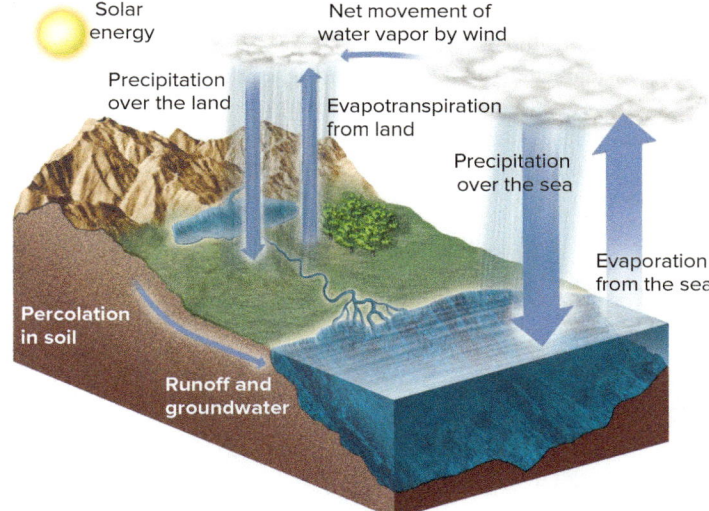

Figure 46.11 The water cycle. This cycle is primarily a physical process, not a chemical one. Solar energy drives the water cycle, causing evaporation of water from the ocean and evapotranspiration from the land. This is followed by condensation of water vapor into clouds and precipitation. The width of each arrow indicates the relative contribution of each process to the cycle.

water removes more water than is put back by rainfall and can cause shallow ponds and lakes to dry up and sinkholes to develop, exacerbating local shortages.

Fertilizers and Other Pollutants Affect the Phosphorus Cycle

Phosphorus has no gaseous phase and thus no atmospheric component; that is, it is not moved by wind. As a result, the phosphorus cycle is viewed as a local cycle, even though phosphorus can be transported over long distances by moving water (Figure 46.12). The Earth's crust is the main storehouse for this element. Weathering and erosion of rocks release phosphorus into the soil, though fertilizers made from mined phosphate rock now release more than this natural process. Plants have the metabolic means to absorb dissolved ionized forms of phosphorus, the most important of which occurs as phosphate (HPO_4^{2-} or $H_2PO_4^-$). Herbivores obtain their phosphorus only from eating plants, and carnivores obtain it by eating herbivores and/or other carnivores. When plants and animals excrete wastes or die, the phosphorus becomes available to decomposers, which release it back to the soil. Leaching and runoff eventually wash much phosphate into aquatic systems, where plants and algae use it. Phosphate that is not taken up into the food chain settles to the ocean floor or lake bottom, forming sedimentary rock.

Phosphorus is commonly limiting in aquatic ecosystems. When more phosphorus is added, the growth of algae and aquatic plants can dramatically increase. In lakes, primary production and total phosphorus concentration appear to follow a linear relationship (Figure 46.13). What is the consequence of high primary production? When the algae and plants die, they sink to the bottom, where bacteria decompose them and consume the dissolved oxygen in the water. Dissolved oxygen concentrations can drop too low for fishes to breathe, killing them. The process by which elevated nutrient levels lead to an overgrowth of algae and the subsequent depletion of water oxygen concentrations is known as **eutrophication.**

Eutrophication is frequently due to the enrichment of water with nutrients derived from human activities, such as fertilizer use and sewage dumping. Lake Washington, near Seattle, received large amounts of lightly treated sewage from 1941 to 1963, at about the same time phosphorus-based detergents came into wide use.

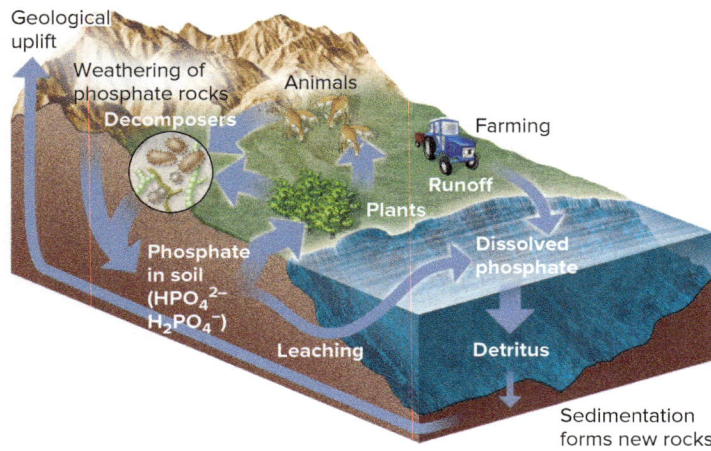

Figure 46.12 **The phosphorus cycle.** Unlike other major biogeochemical cycles, the phosphorus cycle does not have an atmospheric component and thus cycles only locally. The widths of the arrows indicate the relative contribution of each process to the cycle.

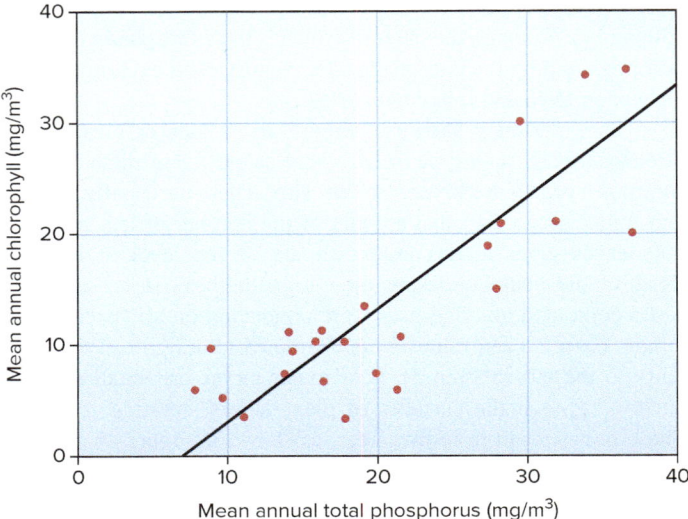

Figure 46.13 **The relationship between primary production and total phosphorus concentration.** As shown in this graph, primary production (measured by chlorophyll concentration) increases linearly with an increase in phosphorus. Each dot represents a different lake.

These two factors were enough to cause severe eutrophication as algae proliferated, darkened the water, and washed ashore to rot and smell. The lake became known locally as Lake Stinko (Figure 46.14a). In the 1960s, sewage flow was diverted from the lake and algae decreased significantly. Water quality improved dramatically and the lake's transparency, as little as 30 inches in 1964, improved to 10 feet in 1968 and as much as 25 feet in 1993 (Figure 46.14b).

Unfortunately, for many other lakes the story is not as rosy. Lake Erie became eutrophic in the 1960s due to the runoff of fertilizer rich in phosphorus from farms and to the industrial and domestic pollutants released from the many cities along its shores. Fish species such as whitefish and lake trout became severely depleted. The U.S. and Canada teamed together to reduce the levels of discharge by 80%, primarily through eliminating phosphorus in laundry detergents and maintaining strict controls on the phosphorus content of wastewater from sewage treatment plants. By the 1980s, Lake Erie experienced fewer algal blooms, clearer water, and a restoration of fish populations. However, since the mid-1990s, high phosphorus concentrations have returned, algae biomass has increased, and the lake has undergone eutrophication once more.

Fertilizers and Industrial Pollutants Affect the Nitrogen Cycle

Nitrogen is an essential component of proteins, nucleic acids, and chlorophyll. Research in 2018, led by scientists at University California, Davis, has shown that in many areas of the world the weathering of bedrock is an important nitrogen source. On average, about 25% of the nitrogen in ecosystems is derived from rocks. However, the vast majority, 75%, comes from the atmosphere.

Because 78% of the Earth's atmosphere consists of nitrogen gas (N_2), it may seem that nitrogen should not be in short supply for organisms. However, nitrogen is often a limiting factor because

(a) Polluted (eutrophication)

(b) Cleared up

Figure 46.14 Phosphorus pollution in Lake Washington. The lake **(a)** in the 1950s, when eutrophic and polluted by sewage and fertilizer runoff, and **(b)** in the 1970s, after eutrophication was reversed.
(a) Source: Seattle Municipal Archives (Item Number: 101715); (b) Source: National Archives and Records Administration (552235)

N_2 molecules must be broken apart before the individual nitrogen atoms can combine with other elements. Because there is a triple bond between its two nitrogen atoms, N_2 is very stable, and only certain bacteria can break it apart into usable forms such as ammonia (NH_3). This process, called nitrogen fixation, is a critical component of the five-part nitrogen cycle (**Figure 46.15**):

1. A few species of bacteria can accomplish **nitrogen fixation,** that is, convert atmospheric N_2 to forms usable by other organisms. The bacteria that fix nitrogen are fulfilling their own metabolic needs, but in the process, they release ammonia (NH_3) or ammonium (NH_4^+), which can be used by some plants. An important group of nitrogen-fixing bacteria, known as rhizobia, live in root nodules of legumes, including peas, beans, lentils, and peanuts, and some woody plants. Cyanobacteria are important nitrogen fixers in terrestrial and aquatic systems (refer back to Figure 30.10).

2. In the process of **nitrification,** soil bacteria convert NH_3 or NH_4^+ to nitrate (NO_3^-), a form of nitrogen commonly used by plants. Bacteria of the genera *Nitrosomonas* and *Nitrococcus* first oxidize the forms of ammonia to nitrite (NO_2^-), after which bacteria of the genus *Nitrobacter* convert NO_2^- to NO_3^-.

3. **Assimilation** is the process by which inorganic substances are incorporated into organic molecules. In the nitrogen cycle, organisms assimilate nitrogen by taking up NH_3, NH_4^+, and NO_3^- formed through nitrogen fixation and nitrification and then incorporating them into organic molecules. Plants take up these forms of nitrogen through their roots, and animals assimilate nitrogen from the plant tissues they ingest.

4. Ammonia can also be formed in the soil through the decomposition of plants and animals and the release of animal waste. **Ammonification** is the conversion of organic nitrogen to NH_3 and NH_4^+. This process is carried out by bacteria and fungi. Most soils are slightly acidic and, because of an excess of H^+, the NH_3 rapidly gains an additional H^+ to form NH_4^+. Because many soils lack nitrifying bacteria, ammonification is the most common pathway for nitrogen to enter the soil.

5. **Denitrification** is the reduction of NO_3^- to N_2. Denitrifying bacteria, which are anaerobic and use NO_3^- in their metabolism instead of O_2, perform the reverse of their nitrogen-fixing counterparts by delivering N_2 to the atmosphere. This process contributes only a relatively small amount of nitrogen to the atmosphere.

Human activities have approximately doubled the rate of nitrogen input to the nitrogen cycle. Industrial fixation of nitrogen for the production of fertilizer makes a major contribution to the pool of nitrogen-containing material in the soils and waters of agricultural regions. As with phosphorus, fertilizer runoff can cause eutrophication of rivers and lakes, and as the resultant algae die, decomposition by bacteria depletes the oxygen level of the water, resulting in fish kills. Excess NO_3^- in surface or groundwater systems used for drinking water is also a health hazard, particularly for infants. In the body, NO_3^- is converted to NO_2^-, which then combines with hemoglobin to form methemoglobin, a type of hemoglobin that does not carry oxygen. In infants, the production of large amounts of NO_2^- can cause methemoglobinemia, a dangerous condition in which the level of O_2 carried through the body decreases. Finally, burning fossil fuels releases not only carbon but also nitrogen in the form of nitrous oxide (N_2O), which contributes to air pollution. N_2O can react with rainwater to form nitric acid (HNO_3), a component of acid rain, which decreases the pH of lakes and streams and increases fish mortality. The release of sulfur from the burning of fossil fuels also contributes to acid rain, because the sulfur reacts with rainwater to form sulfuric acid.

46.3 Reviewing the Concepts

- Elements such as carbon, phosphorus, and nitrogen recycle from the physical environment to organisms and back in what are called biogeochemical cycles.
- In the carbon cycle, phototrophs incorporate CO_2 from the atmosphere into their biomass; decomposition of plants and respiration recycle most of this CO_2 back to the atmosphere.

Biology Principle

Biologists Use Models to Test Predictions and Convey Ideas

Biogeochemical cycles are models that show how chemicals are recycled in ecosystems, yet they are often represented as more complex than the simple cycles they actually are. The Modeling Challenge in the figure legend asks you to redraw Figure 46.15, the nitrogen cycle, as a simple circle or cycle.

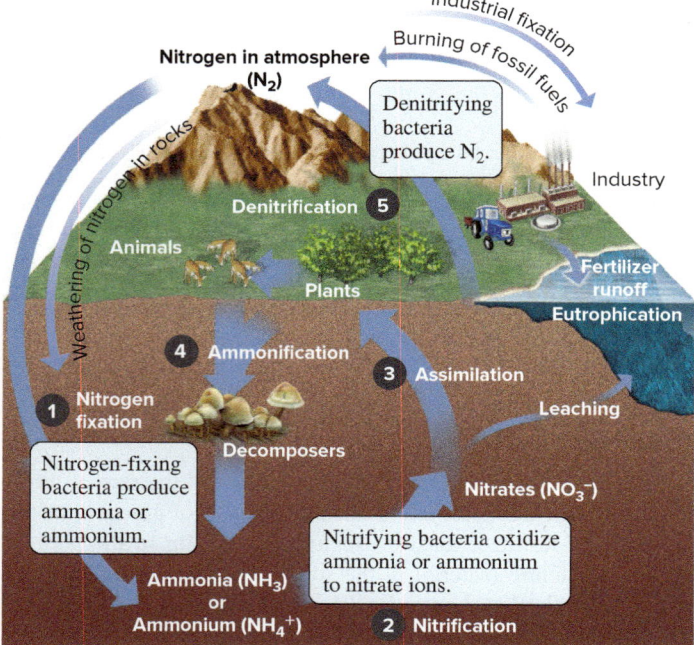

Figure 46.15 **The nitrogen cycle.** The five main parts of the nitrogen cycle are (1) nitrogen fixation, (2) nitrification, (3) assimilation, (4) ammonification, and (5) denitrification. The recycling of nitrogen from dead plants and animals into the soil and then back into plants is of paramount importance because this is an important pathway for nitrogen to enter the soil. The width of each arrow indicates the relative contribution of each process to the cycle.

Modeling Challenge: Biogeochemical cycles can be represented by interactions among two types of components: reservoirs and processes. In this Modeling Challenge, draw a model in which each reservoir is shown within a box and each process is depicted as an arrow that connects two reservoirs. Arrange the five reservoirs listed below into a circle to represent the nitrogen cycle, with atmospheric nitrogen at the 12 o'clock position. Connect the five reservoirs in their correct order with the appropriate six processes so that the nitrogen cycle is drawn correctly in a clockwise sequence. Note: Five arrows can connect the five boxes to form a circle, and a sixth arrow can reach across the circle to connect two boxes.
  Reservoirs: atmospheric nitrogen (N_2), nitrogen in plants, nitrate ions (NO_3^-) in soil, ammonia (NH_3) or ammonium (NH_4^+) in soil, dead organic matter
  Processes: nitrification, assimilation, death, nitrogen fixation, denitrification, ammonification (or decomposition)

Human activities, primarily the burning of fossil fuels, are causing increased amounts of CO_2 to enter the atmosphere (Figures 46.9, 46.10).

- The water cycle is a physical rather than a chemical process because it consists of essentially two phenomena: evaporation and precipitation. Climate change is altering precipitation patterns. In addition, alteration of the water cycle by dams can greatly disrupt the migration of fishes such as salmon and trout (Figure 46.11).
- The phosphorus cycle lacks an atmospheric component and thus is a local cycle. An overabundance of phosphorus can cause the overgrowth of algae and subsequent depletion of oxygen levels, called eutrophication (Figures 46.12, 46.13, 46.14).
- The nitrogen cycle has five parts: nitrogen fixation, nitrification, assimilation, ammonification, and denitrification. In the nitrogen cycle, atmospheric nitrogen is unavailable for use by most organisms and must be converted to usable forms by certain bacteria. The activities of humans, including the production and use of fertilizers, have altered the nitrogen cycle (Figure 46.15).

46.3 Testing Your Knowledge

1. Industrial fixation of nitrogen for the production of fertilizer makes a significant contribution to the pool of nitrogen in
   a. soils.
   b. water.
   c. the atmosphere.
   d. both a and b.
   e. all of the above.

2. The increase of atmospheric CO_2 levels is caused by
   a. the burning of fossil fuels.
   b. deforestation.
   c. ocean acidification.
   d. both a and b.
   e. all of the above.

46.4 Pollution and Biomagnification

Learning Outcome

1. Define biomagnification, and explain its relationship to pollution.

In Section 46.3, we considered how pollution generated by humans can affect biogeochemical cycles. Certain types of pollutants can also accumulate within the bodies of animals and plants. The tendency of certain chemicals to concentrate in organisms at higher trophic levels in food chains, which is called **biomagnification**, can cause health and reproductive problems for certain organisms. The passage of dichlorodiphenyltrichloroethane (DDT), an insecticide used against mosquitoes and agricultural pests, through food chains provides a startling example of biomagnification. DDT was first synthesized by chemists in 1874. In 1939, its insecticidal properties were recognized by Paul Müller, a Swiss scientist who won the 1948 Nobel Prize in Physiology or Medicine for his discovery of and subsequent research on the uses of the chemical. The first important application of DDT was in human health programs during and after World War II, particularly as a means of controlling mosquito-borne malaria; at that time, its use in agriculture also began. The global production of DDT peaked in 1970, when 175 million kilograms of the insecticide was manufactured.

DDT has several chemical and physical properties that profoundly influence its ecological effect. First, DDT is persistent in the environment. It is not rapidly degraded to other, less toxic chemicals by microorganisms or by physical agents such as light and heat. The typical persistence in soil of DDT is about 10 years, which is two to three times longer than the persistence of many other insecticides. Another important characteristic of DDT is its low solubility in water and its high solubility in fats, or lipids. In the environment, most lipids are present in living tissue. Therefore, because of its high lipid solubility, DDT tends to concentrate in biological tissues.

Because biomagnification occurs at each step of the food chain, organisms at higher trophic levels can accumulate especially high concentrations of DDT in their lipids. A typical pattern of biomagnification is illustrated in **Figure 46.16**, which shows the relative amounts of DDT found in a Lake Michigan food chain. The highest concentration of the insecticide was found in gulls, tertiary consumers that feed on fishes, which are the secondary consumers that eat small insects.

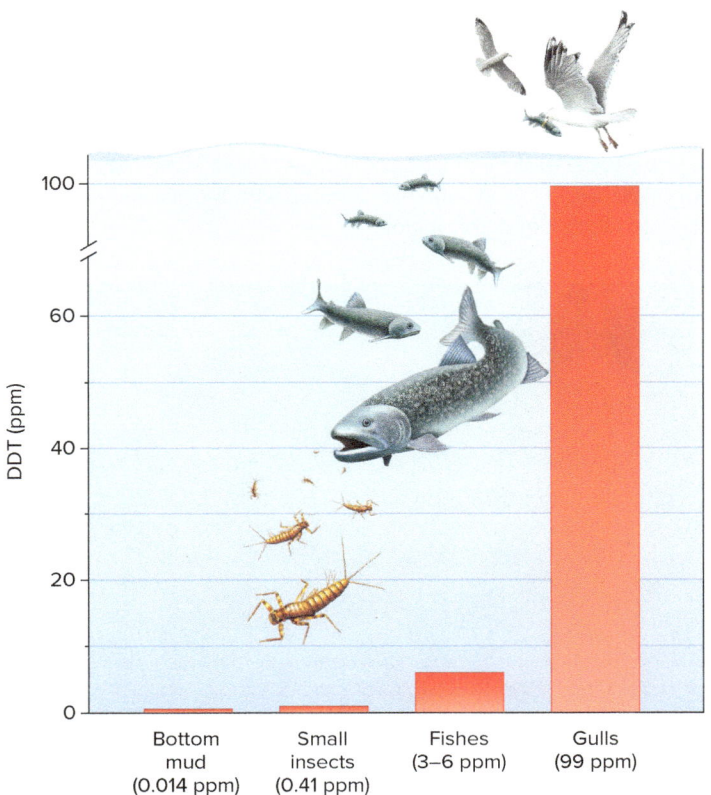

Figure 46.16 Biomagnification in a Lake Michigan food chain. The DDT tissue concentration in gulls, tertiary consumers, was about 240 times that in the small insects sharing the same environment. The biomagnification of DDT in lipids causes its concentration to increase at each successive link in the food chain. A ppm is the equivalent of 1 mg of DDT per kg of biological material.

Biology Principle

Biology Affects Our Society

Discovering how DDT impacts eggshell thickness resulted in the reduction of DDT usage, which thereby increased populations of some bird species.

DDT (dichlorodiphenyltrichloroethane)
- Persists in environment
- High solubility in lipids
- Found in high concentrations at higher trophic levels

Figure 46.17 Thinning of eggshells caused by DDT. Ibis eggs, Texas Gulf Coast, which were crushed by the weight of a parent that was incubating them.
©George Silk/The LIFE Picture Collection/Getty Images

An unanticipated effect of DDT on bird species, particularly birds of prey, was its interference with the metabolic process of eggshell formation. The result was thin-shelled eggs that often broke under the weight of incubating birds (**Figure 46.17**). DDT was responsible for a dramatic decrease in the populations of many birds due to failed reproduction. Relatively high levels of the chemical were also found in some game fishes, which, as a result, became unfit for human consumption.

Because of growing awareness of the adverse effects of DDT, most industrialized countries, including the U.S., had banned the use of the chemical by the early 1970s. The good news is that following the outlawing of DDT, populations of the most severely affected bird species have recovered. However, had scientists initially possessed a more thorough knowledge of how DDT accumulates in food chains, some of the damage to the bird populations might have been prevented.

Ironically, scientists have recently discovered that birds of prey, often the unintended target of pesticides, may actually decrease the need for pesticides. For example, farmers are attracting American kestrels, small falcons, by installing nest boxes. In Michigan cherry orchards, kestrels consume crop pests such as grasshoppers, rodents and starlings, helping to produce fruit crops attractive to consumers, at a cost far lower than by using pesticides.

In addition to DDT, biomagnification occurs for other types of molecules that do not break down quickly in the environment and are not excreted efficiently by living organisms. Such substances include mercury, which is emitted from coal-fired power plants, and persistent organic pollutants, or POPs, which are used in herbicides and pesticides. Underwater mining of the ocean floor to extract minerals releases sulfide and selenium, which can also biomagnify in food chains.

46.4 Reviewing the Concepts

- The increase in the concentration of certain chemicals in living organisms, called biomagnification, can occur at each trophic level of a food web (Figures 46.16, 46.17).

46.4 Testing Your Knowledge

1. The concentration of certain chemicals, such as DDT, in higher trophic levels is known as
    a. eutrophication.
    b. biomagnification.
    c. biogeochemical cycling.
    d. energy transfer.
    e. turnover.

46.5 Habitat Destruction

Learning Outcomes

1. Identify the main causes of habitat destruction by humans.
2. Explain why deforestation is a particularly destructive form of habitat loss.
3. Describe how the impact of agriculture causes ecological changes.

Habitat destruction is usually a human-driven process in which a natural habitat is altered in a way that prevents it from supporting the species that were originally present. Organisms that previously occupied the habitat are displaced or unable to survive, thereby reducing biodiversity. Habitat destruction is the primary cause of species extinction, and a high percentage of existing species are threatened by this process (Figure 46.18). Other human-driven practices that are threatening the survival of many species include pollution, which has already been discussed in this chapter, and overexploitation and invasive species, which are described in the last two sections.

Habitat destruction includes deforestation, conversion of habitat to agricultural land, urbanization, strip mining, quarrying, and many other forms of land modification. Urbanization, the development of cities on previously natural or agricultural areas, is the most human-dominated and fastest-growing type of land use worldwide, and it devastates the land more severely than nearly any other form of habitat destruction. Freshwater habitats have also suffered via dam construction and river channelization. Wetlands have been drained for agricultural purposes and have been filled in for urban or industrial development. In the U.S., as much as 90% of the freshwater marshes have disappeared in states such as Iowa and California, though the

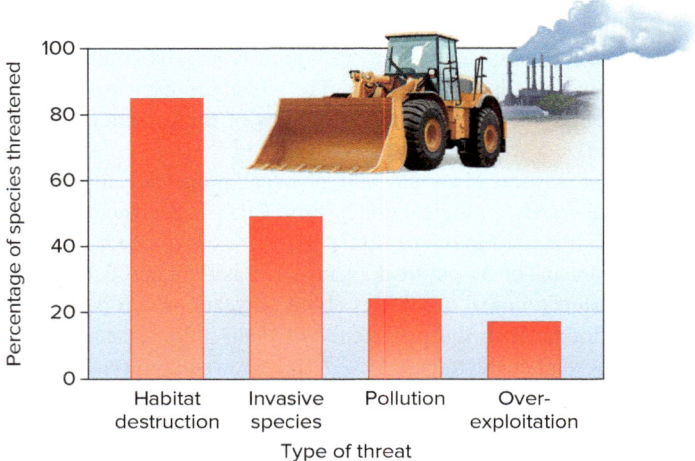

Figure 46.18 Percentages of plant and animal species threatened by various causes in the U.S.

 Concept Check: *Why don't these categories sum to 100%?*

national average is approximately 53%. In this section, we will examine the two most widespread and interrelated types of habitat destruction, at least in terms of their influence on species extinctions: deforestation and the conversion of land for agricultural purposes.

Deforestation by Humans Threatens the Existence of Many Species

Deforestation, the conversion of forested areas to nonforested land, is a prime cause of the extinction of species (Figure 46.19). About one-third of the world's land surface is covered with forests, and much of this area is at risk of deforestation.

Figure 46.19 Deforestation. Cascade Mountains near Seattle, Washington, 1906.
Source: Library of Congress, Prints & Photographs Division [LC-USZ62-67666]

Figure 46.20 Habitat destruction as an important cause of species extinction. The ivory-billed woodpecker, the third-largest woodpecker in the world, was thought to have gone extinct in the southeastern U.S. because of habitat destruction.
©James T. Tanner/Science Source

Many species live in forests. For example, among North American terrestrial wildlife, about one-quarter of the bird species (272 species) and more than 10% of mammal species (49 species) have an obligatory relationship with forest cover, meaning that they depend on trees for food and nesting sites. In terms of wildlife use, oaks are among the most valuable trees in North America. At least 100 species of birds and mammals include acorns in their diets, and for many species of wildlife, the annual acorn crop is a major determinant of their abundance. Most woodpeckers, as well as many other types of birds, nest in holes that they excavate in trees, and their food usually consists of insects collected on or in trees. The ivory-billed woodpecker, *Campephilus principalis,* the largest woodpecker in North America and a former inhabitant of wetlands and forests of the southeastern United States, is presumed to have gone extinct in the 1950s due to the destruction of its habitat by heavy logging (**Figure 46.20**).

Tropical forests, primarily rain forests but also deciduous forests, are among the most species-rich terrestrial habitats, and they exist mainly in three areas of the globe: Africa, Asia, and Latin America. Latin America contains more than the other two areas combined, with Brazil having the greatest amount of tropical forest of any nation. The Amazon rain forest has been termed "the lungs of the planet," because it produces more than 20% of the world's oxygen. Tropical forests, which once covered 14% of the Earth's dry land, now cover only about 7%. The good news is that rates of tropical deforestation decreased from 0.18% per year in the 1990s to 0.08% annually between 2010 and 2015.

Although tropical deforestation occurs for different reasons, clearing land for agriculture has been identified as the prime cause of forest loss. Logging in excess of regrowth is also a significant cause of loss, particularly in Asian forests. Fuelwood collection can also be important but generally is more of a problem in lightly wooded areas. Finally, the construction of mines, dams, and oil installations is a minor cause of direct deforestation, but the discharge of chemicals and silt into rivers can cause much damage indirectly. The roads built into these regions often open them up to further development.

Conservation of tropical forests would not only save many rare species from extinction but also benefit humans in several ways. First, many of the world's crops, including oranges, lemons, bananas, cacao (chocolate), coffee, and vanilla, evolved in tropical rain forests. Rain forests remain a repository of genetic variation that could be used in future breeding of these crops. In addition, rain forests are particularly rich in plants with unusual chemicals that they use in defense (see Chapter 43). Many of these chemicals also have medicinal value. Although many prescription drugs sold worldwide come from plant-derived sources, less than 1% of tropical plants have been tested for their medicinal properties. One estimate has proposed lifetime values for tropical forests at $6,330 per hectare if renewable and sustainable resources such as fruit, rubber, and nuts are harvested. In comparison, forests are worth only $1,000 per hectare for timber and $2,960 per hectare when converted to agricultural grazing land. For these reasons, it seems to make good economic and ecological sense to slow the rate of tropical deforestation.

The Development of Land for Agriculture Causes Adverse Ecological Changes

No other human endeavor on Earth requires as much land as agriculture. More land has been converted to agriculture since 1945 than in the 18th and 19th centuries combined. The average area of land under crop cultivation worldwide averages 11.5%, with an additional 25.8% given over to rangeland for grazing. However, this amount varies substantially among regions (**Table 46.2**).

The planting of crops and the grazing of livestock have produced perhaps the most far-reaching ecological effects on Earth (**Figure 46.21**). The destruction of habitat to plant agricultural crops and to provide pasture for grazing livestock can create soil erosion, increased flooding, declining soil fertility, silting of rivers, and desertification as well as remove valuable wildlife habitat. Also, as we have seen, runoff water from agricultural land often contains high levels of nitrogen and phosphorus from fertilizers as well as

Table 46.2 Percentage of Land Area Used for Agricultural Purposes

Continental area	Cropland (%)	Pastures (%)	Total (%)
World	11.5	25.8	37.3
Asia	20.0	33.0	53.0
Central America and the Caribbean	15.6	36.7	52.3
Europe	13.1	7.9	21.0
Middle East and North Africa	7.6	28.1	35.6
North America	11.4	12.5	23.9
Oceania	6.2	48.2	54.4
South America	6.8	28.8	35.6
Sub-Sahara and Africa	8.1	33.9	42.0

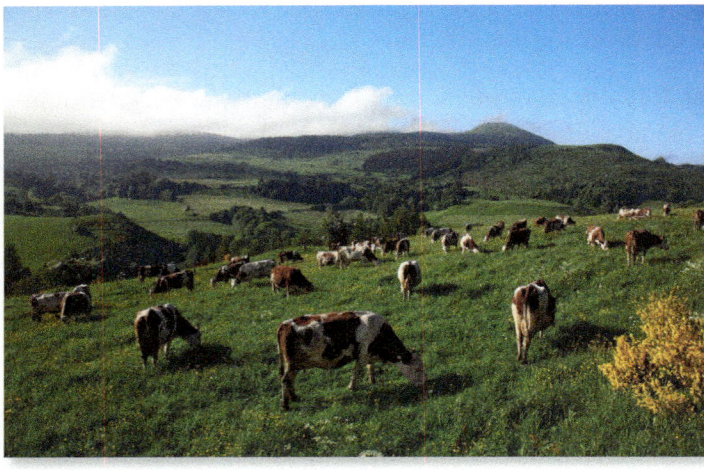

Figure 46.21 Modern agriculture and its impact on the environment. **(a)** This wheat crop could not exist without human help, and at the same time, crops sustain human populations. **(b)** The grazing of livestock has radically altered the ecological landscape.
(a) Source: Scott Bauer/USDA-ARS; *(b)* ©Martial Colomb/Photographer's Choice/Getty Images

residual pesticides and manure, all of which contaminate streams. Some 4 billion tons of topsoil are washed into U.S. waterways each year, severely impacting the organisms that live there. Topsoil can also be whipped away by wind.

46.5 Reviewing the Concepts

- Habitat destruction involves the conversion of habitat to agricultural land, urbanization, the draining of swamps, strip mining, dam destruction, and many other forms of land modification. Habitat destruction is the primary cause of species extinction and is threatening many current species (Figure 46.18).

- Deforestation—in particular, tropical deforestation—threatens many species with extinction because of the high numbers of species that live in forests and the high rate of forest loss (Figures 46.19, 46.20).

- The planting of crops and the grazing of livestock are among the most important drivers of ecological change. Habitat destruction has changed huge areas of the planet into agricultural areas and rangelands for livestock (Figure 46.21, Table 46.2).

46.5 Testing Your Knowledge

1. Habitat destruction includes
    a. deforestation.
    b. the conversion of habitat to agricultural land.
    c. urbanization.
    d. the draining of wetlands.
    e. all the above.

46.6 Overexploitation

Learning Outcomes

1. Outline the main causes of the overexploitation of plant and animal populations.
2. Describe examples of the overexploitation of animals and plants.
3. **SCISKILLS** ▶ Calculate the maximum sustainable yield of a population based on its carrying capacity and per capita rate of increase.

In ecology, **overexploitation** is the practice in which humans harvest a particular species at a rate that is unsustainable, based on its natural rate of mortality and capacity for reproduction. Overexploitation, particularly the hunting of animals, has been the cause of many extinctions in the past. In this section, we will consider several examples of overexploitation and then examine how ecologists make calculations to determine if overexploitation is occurring.

Many Species Have Gone Extinct or Are Currently Threatened Due to Overexploitation

With regard to animals, hunting and fishing have been the main practices that facilitate overexploitation. For other groups, such as plants, humans' ability to identify valuable species and remove them from their native habitats has led to overexploitation. In 2015, a study by Canadian conservation biologist Chris Darimont and colleagues examined the effects of humans as hunters of terrestrial mammals and fishers of marine fishes. They compared 2,125 estimates of exploited animal populations and showed that humans kill adult prey at rates up to 14 times higher than other predators do (**Figure 46.22**). Darimont and colleagues termed humans "super predators" and suggested that in the long run this level of hunting and fishing would not be sustainable. As seen in Figure 46.22, humans are predators of species

THE AGE OF HUMANS

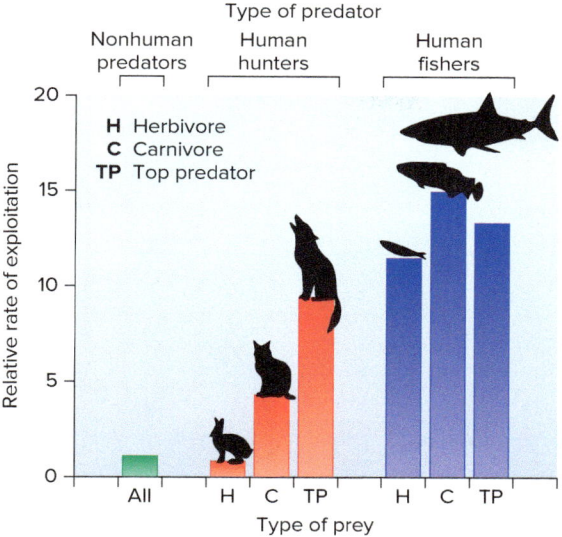

Figure 46.22 Humans as hunters. The rates at which humans exploit adult land mammals and marine fish vastly exceed the impacts of other predators. Marine fish of all trophic levels are similarly affected. In contrast, land predators are exploited at much higher rates than herbivores. A top predator is defined as the most successful predator in a given habitat, which also can be a prey to humans.

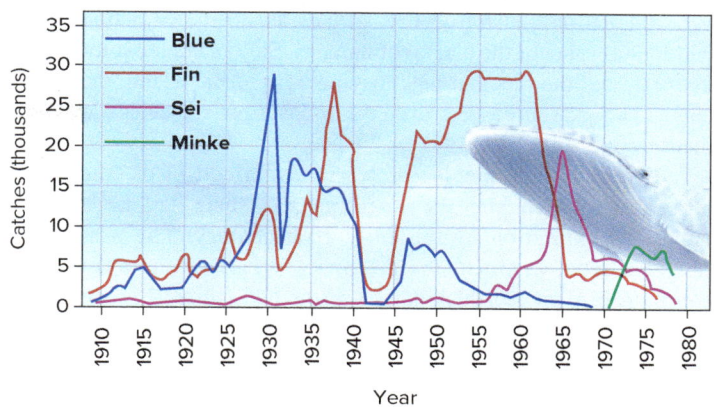

Figure 46.23 Sequential decline of whale catches in the Antarctic, showing the strong effect of human predators.

that are considered the top predators in their native environment. For example, humans have greatly decreased the populations of wolves, which are successful predators in many different habitats.

Let's now consider a few examples in which humans have overexploited animals and plants.

Land Mammals The list of mammals threatened with extinction or actually driven extinct by hunting is long. In the prairies of North America, hunting diminished populations of buffalo from around 70 million in the 18th century to 1,150 by 1899. However, since then, buffalo numbers have rebounded to about 350,000. In Russia, another inhabitant of the prairies—the Eurasian wild horse, also known as the tarpan, *Equus ferus*—had been hunted to extinction by the 1860s.

Even today, poachers threaten many land mammals. Rhinoceroses have a horn and elephants have tusks that some people believe will cure everything from cancer to hangovers. The black market value for rhino horn in Southeast Asia is about $65,000 per kilogram, making it more expensive by weight than gold, diamonds, or cocaine. In 2011 alone, poachers killed roughly 1 in every 12 African elephants. Many experts believe that the last of the great woolly mammoth populations were hunted to extinction at the end of the last glacial period about 12,000 years ago.

Whales In 2014, Roberta Rocha, director of science at the New Bedford Whaling Museum in Massachusetts, estimated the numbers of whales killed by industrial harvesting during the last century, 1900–1999. Rocha's estimate revealed that nearly 3 million whales were harvested in what was the largest cull, in terms of animal biomass, in human history. The history of whaling in general, and Antarctic whaling in particular, has been characterized by a progression from more valuable or more easily caught species to less attractive ones, as populations of the original targets were depleted (**Figure 46.23**). In the Antarctic, blue whales, *Balaenoptera musculus*, dominated the catches through the 1930s, but by the middle 1950s, few were being taken, although the species was not legally protected until 1965. As the populations of blue whales diminished, attention turned to the fin whale, *B. physalus*, which was originally the most abundant of all whales in the Southern Ocean. By the 1960s, the numbers of this species had diminished rapidly. Sei whales, *B. borealis*, were almost ignored by whalers until the bigger species were no longer available. They were hardly taken at all until about 1958, but then catches increased rapidly and reached a peak of about 20,000 in 1964–1965. Catches declined rapidly thereafter, this time due to the introduction of a catch limit of 10. Then, the relatively small minke whales, *B. acutorostrata*, which were ignored in the Southern Ocean until 1971–1972, began to be taken. Since that time, minkes have been the largest component of the southern baleen whale catch. The story was similar in the Northern Hemisphere.

In 1982, the International Whaling Commission (IWC) voted for a moratorium on all commercial whaling. The IWC proposed that commercial whaling be ended in 1985–1986, a proposal that did not actually take effect until 1988. The good news is that following the moratorium, the populations of some whales have increased. Blue whales appeared to quadruple their numbers off the California coast during the 1980s, showing the impact that protection from a predator can have. American molecular geneticist Steve Palumbi and his colleagues estimated past population sizes of North Atlantic whales by examining mitochondrial DNA sequence variation. Population size estimates for fin and humpback whales suggested that pre-whaling populations were 6–20 times higher than present-day population estimates and that full recovery of whale populations will take another 70–100 years.

Birds A poignant example of overexploitation was the dodo, *Raphus cucullatus*, a flightless bird native only to the island of Mauritius that had no known predators. A combination of overexploitation and introduced species led to its extinction within 200 years of the arrival of humans. Sailors hunted it for its meat, and the rats and pigs they brought to the island, the latter as a food source, destroyed the dodos' eggs and chicks in their ground nests.

Two abundant species of North American birds, the passenger pigeon and the Carolina parakeet, had been hunted to extinction by the early 20th century. The passenger pigeon, *Ectopistes migratorius,* was once the most common bird in North America, probably accounting for over 40% of the entire number of birds (Figure 46.24). Flock sizes were estimated to be over 1 billion birds. It may seem improbable that the most common bird on the continent could be hunted to extinction for its meat, but that is just what happened. The flocking behavior of the birds made them relatively easy targets for hunters, who used special firearms to harvest the birds in quantity. In 1876, in Michigan alone, over 1.6 million birds were killed and sent to markets in the eastern U.S. Similarly, the Carolina parakeet, *Conuropsis carolinensis,* the only species of parrot native to the eastern U.S., had been similarly hunted to extinction by the early 1900s.

Fishes Many species of fish are harvested to the degree that the rate of removal exceeds the rate of reproduction. Eventually, these fish populations crash and the fishery collapses. In the case of the Canadian cod fishery, overfishing and collapse came in 1992 after hundreds of years of fishing (Figure 46.25). At that time, the fishery was shut down. There is hope that the Canadian cod fishery will eventually recover, but the recovery will be slow because the fish do not spawn until age 7. A 2015 study showed that populations near Newfoundland and Labrador were still only 25% of their original sizes.

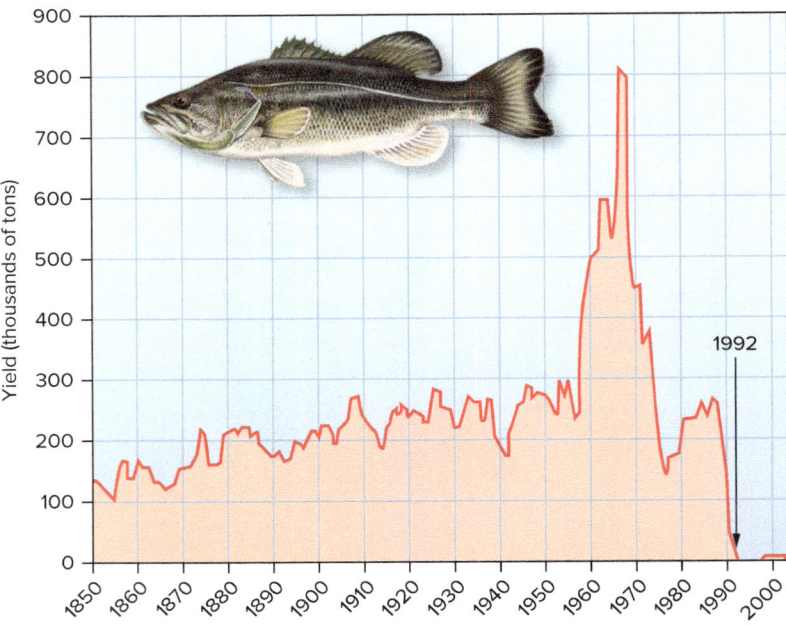

Figure 46.25 Changes in yields in the Canadian cod fishery from 1850 to 2000. The fishery collapsed in 1992 following overfishing.

In the Florida Keys, historical photographs document the reduction of large trophy fish taken over the period 1956–2007 (Figure 46.26). The mean fish size declined from an estimated 19.9 kg in 1956 to 2.3 kg in 2007! There was also a shift in the species caught. In 1956, large groupers, *Epinephellus* spp., were commonly landed together with large sharks greater than 2 m. In 2007, small snappers, *Lutjanus campechanus* and *Ocyurus chrysurus*—average length 34.4 cm—and sharks less than 1 m were landed.

Plants Populations of many species of valuable plants have also been severely reduced for their human uses, including West Indian mahogany, *Swietenia mahogani,* in the Bahamas and Lebanese cedar, *Cedrus libani,* which in Lebanon has been reduced to a few scattered forest remnants. Rare cacti and orchids have also been threatened by collectors, who seek to own a rare organism or to profit from its sale. Rare orchids are very valuable. A single stem of Rothschild's orchid, *Paphiopedilum rothschildianum,* also known as the Gold of Kinabalu, is reported to sell for $5,000 (Figure 46.27). No wonder that after its discovery, this orchid was stripped from the wild in northern Borneo by orchid smugglers, pushing it close to extinction.

Plants may also be collected for their medicinal value, real or supposed. In the U.S., American ginseng, *Panax quinquefolicus,* native to eastern North America, has been sought after to treat the common cold. Plants are so enthusiastically collected by "sang hunters," who also sell it to Chinese traders, that ginseng has become threatened or endangered in some states. The Botanic Gardens Conservation International Organization has stated that over 400 medical plant species are at risk of extinction because of overcollection and deforestation, including many species of yew tree, *Taxus* spp., whose bark is used for cancer drugs.

Figure 46.24 Overexploitation has caused species extinctions. The passenger pigeon, *Ectopistes migratorius,* which was once among the most abundant bird species on Earth, was hunted to extinction for its meat.
©Topham/The Image Works

THE AGE OF HUMANS   1021

Figure 46.26 Change in the sizes of trophy fish caught on Key West charter boats. **(a)** In 1957, **(b)** the early 1980s, and **(c)** 2007.

(a) Source: From the archives of Monroe County Public Library, Key West, Florida. Photo by Wil-Art Studio/Art Stickel; *(b)* Source: From the archives of Monroe County Public Library, Key West, Florida. Photo by Dale McDonald; *(c)* Photo courtesy Loren McClenachan, Ph.D.

Figure 46.27 Rothschild's orchid, *Paphiopedilum rothschildianum*. Overcollection in northern Borneo has pushed this species close to extinction in the wild.
©Matthew Lambley/Alamy

Biology Principle | Biology is a quantitative science

Quantitative Analysis: Ecologists Make Calculations to Determine If Overexploitation Is Occurring

For a population that is being harvested by humans, the **yield** is the number of individuals harvested in a given unit of time. **Maximum sustainable yield (MSY)** is the largest number of individuals that can be harvested without causing long-term decreases in the population. This is like skimming the interest from a bank account but leaving the principal intact to ensure the same interest payment in future years. If the maximum sustainable yield is exceeded on a consistent basis, overexploitation is occurring.

According to the logistic equation described in Chapter 43, population increase, dN/dt, occurs according to the following equation:

$$\frac{dN}{dt} = rN\frac{(K - N)}{K}$$

where

N is the number of individuals
t is a unit of time
r is the per capita rate of population growth
K is the carrying capacity

The greatest population growth, and thus the maximum sustainable yield, occurs at the midpoint of the logistic curve, at $K/2$ (refer back to Figure 43.6). Maximum yield is obtained when populations are maximally increasing.

MSY can thus be estimated as

$$MSY = \frac{rK}{2}$$

As an example, if $K/2 = 10,000$ and $r = 0.14$, then MSY = 1,400.

MSY has been used extensively for fisheries management. However, use of the model has some drawbacks. For example, biologists cannot easily go under water and count the number of available individuals. The model also ignores the age and reproductive status of the individuals being harvested. As an alternative, simple models based on data collection from fisheries may be used, which compare costs and revenues and consider MSY (**Figure 46.28**). The maximum profit (red line) is greater than the profit obtained at MSY (blue line).

Why is MSY important to fisheries (and the harvesting of other species)? Let's assume that total economic costs increase linearly as fishing effort increases. Total revenue also increases with fishing effort, up to a point. After MSY is reached, any increase in effort would result in a decrease in revenue as the fish populations became overfished. The maximum profit is equal to the biggest difference between the cost and revenue curves, and as the red line between the two arrows in Figure 46.28 shows, this occurs below the effort needed to reach the maximum sustainable yield.

Why do fisheries become overfished? Most estimates of MSY are nearly always too high because of overestimates of population size. Furthermore, incremental improvements are made to fishing gear over the years, and the same level of effort results in increased catches. Finally, as fish populations decline, the prices inch upward, making it more likely for commercial fishing to keep harvesting them. In the case of the cod industry, the introduction of radar and sonar allowed crews to pursue fish over huge areas. The crews also caught enormous numbers of noncommercial fish, which were discarded. Among these were important prey species for cod, such as capelin. Finally, although undersized cod, which cannot spawn, were returned to the ocean, such discards do not always survive. As the cod populations went into a tailspin, increased effort resulted in more undersized fish being caught and more discarded.

Crunching the Numbers: In a study of red snapper, *Lutjanus campechanus*, in the Gulf of Mexico, the carrying capacity of a particular population is estimated at 100,000 and the per capita rate of increase is 0.05. What is the maximum sustainable yield?

46.6 Reviewing the Concepts

- Human overexploitation of plants and animals occurs via hunting, fishing, and the removal of species from their native habitat. On average, humans decrease prey populations much more than natural predators do (Figure 46.22).
- Human overexploitation has decreased the populations of many species and driven some of them to extinction. Examples include whales, the passenger pigeon, and many mammals, fishes and plants (Figures 46.23–46.27).
- Maximum sustainable yield is the largest number of individuals that can be harvested without causing long-term decreases in the population. Simple economic models suggest that populations should be harvested at levels of maximum profit, which is well below the maximum sustainable yield (Figure 46.28).

46.6 Testing Your Knowledge

1. Which group of whales was the first to be overharvested in the Southern Hemisphere?
   a. blue
   b. fin
   c. minke
   d. sei
   e. humpback
2. The maximum yield occurs at which point of the logistic curve?
   a. beginning
   b. middle
   c. upper end

46.7 Invasive Species

Learning Outcomes

1. Define introduced species and invasive species.
2. Compare and contrast invasive species as competitors, as predators, and as pathogens.

Introduced species are those species moved by humans from their native habitat to another location. Most often, the species are introduced for agricultural or landscaping purposes or as sources of

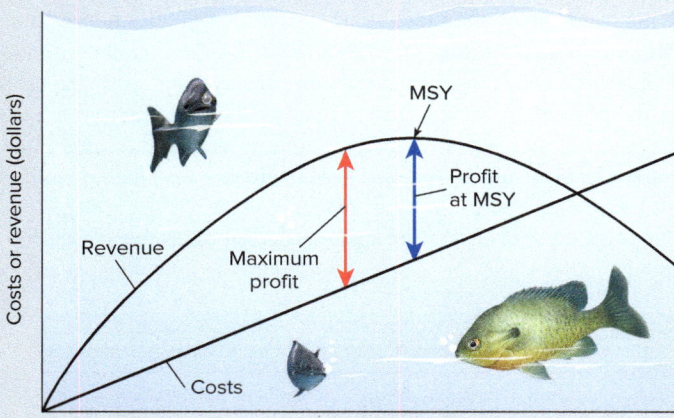

Figure 46.28 Economic fisheries model based on revenue and costs. The model assumes that costs increase linearly with effort but revenue reaches a maximum at maximum sustainable yield (MSY). Maximum profit, however, occurs below MSY.

(a) Feral dog

(b) Purple loosestrife

Figure 46.29 Introduced species as competitors. (a) Feral dogs called dingoes are thought to have outcompeted the thylacine in Australia. (b) Introduced plants can choke out native vegetation, as happens with purple loosestrife in the U.S. Northeast.
(a) ©ksumano/123RF; *(b)* ©Reimar Gaertner/Alamy

timber, meat, or wool, and they may need humans for their continued survival. In some cases, as with plants, insects, or aquatic species, organisms are unintentionally transported via the movement of cargo by ships or planes. Regardless of the way they have been transported, some introduced species become **invasive species,** spreading naturally and impacting native species. Why do invasive species become successful in their new habitat? In most cases, they have one or more of the following characteristics.

- Invasive species may reproduce rapidly.
- They may not have natural enemies in their new habitat.
- Certain native species may lack defenses against the invasive species.
- Invasive species may compete aggressively for resources.
- They may tolerate a wide variety of habitat conditions.

In the U.S. alone, over 4,500 invasive species have been identified, and 15% cause severe ecological or economic harm. Of introduced vertebrates, 142 species have self-sustaining populations in the wild. These include ring-necked pheasants, *Phasianus colchicus,* which were brought over by hunters, and Burmese pythons, *Python molorus,* which were introduced by pet owners. Of the 300 most invasive weeds in the U.S., over half were brought in for gardening, horticulture, or landscape purposes. These include purple loosestrife, *Lythrum salicaria,* and Japanese honeysuckle, *Lonicera japonica,* in the Northeast; kudzu, *Pueraria lobata,* in the Southeast; Chinese tallow, *Sapium sebiferum,* in the South; and leafy spurge, *Euphorbia esula,* in the Great Plains. In this section, we will examine how the interactions between invasive and native species involve competition, predation, and parasitism.

Invasive Species May Compete Against Native Species

Ecologists have discovered many examples in which invasive species outcompete native species. For example, placental mammals have been found to be particularly effective competitors in Australia, where introduced feral dogs—*Canis lupus,* ssp. *dingo*—are thought to have outcompeted the thylacine, *Thylacine cynocephalus,* a native marsupial, wolflike animal, in mainland Australia (**Figure 46.29a**). Sheep introduced for the wool industry appear to compete successfully with a variety of kangaroo species, especially the brush-tailed rock wallaby and larger species, such as the red kangaroo and western gray kangaroo.

Introduced vertebrates are a problem in aquatic environments, too. In California, 48 of 137 species of freshwater fish are nonnative. Of these, 24 are known to have a negative impact on native fish.

Invasive plants are major competitors of native plants throughout the world. In some cases, invasive plants have become so successful that they have produced near monocultures—areas where they exist as the sole plant species. In the U.S. Northeast, purple loosestrife is a strong competitor, choking out other plant species in wetlands (**Figure 46.29b**). In south Florida, three invasives—Brazilian pepper, *Schinus terebinthifolius;* Australian pine, *Casuarina equisetifolia;* and, especially punk-tree, *Melaleuca quinquenervia*—are outcompeting native vegetation in the Everglades.

Invasive Species May Act as Predators

Many striking examples of the powerful effects of invasive species have been provided by predator introductions. Sea lampreys, *Petromyzon marinus,* are primitive, jawless fish that feed by rasping holes in the sides of larger fish species and feeding on their blood (**Figure 46.30a**). They spawn in freshwater streams, and the juveniles return to salt water (or one of the Great Lakes as a substitute) to develop. Sea lampreys found their way into Lake Ontario in the mid-1800s by way of the Erie Canal (**Figure 46.30b**). Improvements to the Welland Canal allowed lampreys to bypass Niagara Falls and colonize the rest of the Great Lakes. In 1921, lampreys were found in Lake Erie, and by 1938 they had been found all the way up into Lake Superior. Lake trout, *Salvelinus namaycush,* were the lampreys' preferred prey in the Great Lakes. The lake trout fishing industry declined somewhat in the 1940s due to overfishing, but the decline was hastened by the arrival of lampreys (**Figure 46.30c**). By the 1960s, lake trout catches had been reduced by over 90% of their historic averages, and the fishery collapsed.

Figure 46.30 Effects of invasive sea lamprey on the lake trout population in the Great Lakes. (a) Sea lamprey attached to a lake trout. (b) Historical passage of lampreys into the Great Lakes from the Atlantic. (c) Effects of lampreys on the trout population in Lake Superior. Note that detailed records of sea lamprey populations were not available before 1956.

(a) ©Dr. Marc E. Gaden, University of Michigan

In the late 1940s, the state of Michigan began looking at control measures for lampreys. Barriers were erected across the mouths of streams to prevent the return of spawning adults, but they were difficult to maintain and were not 100% effective. Attention turned to chemical control of juvenile lampreys, which spend 3–5 years buried in stream sediments, filter feeding. The chemical trifluoromethyl nitrophenol (TFM) proved effective in controlling juvenile lampreys in streams and was first applied in 1958 to streams feeding into Lake Superior. Treatments reduced lamprey densities by 90%, and the population of lake trout increased in the 1980s, aided by an active program of restocking. To maintain lamprey control, streams are treated every 3–5 years. The removal of invasive species is time-consuming and costly but may lead to the recovery of native prey species.

A second example of an invasive species acting as a predator is the Burmese python in the Florida Everglades (see the chapter-opening photo). Pythons were likely introduced prior to 1985, and the population has grown exponentially since then. Pythons in Florida consume a wide range of vertebrate prey, including endangered species such as the wood stork, *Mycteria americana,* and the Key Largo woodrat, *Neotoma floridana.* Survey data gathered by counting animals and roadkills showed that raccoons, opossums, and rabbits were often observed during the period 1993–1999, before pythons became common. In the period 2003–2011, when pythons were common, the decline in observations was 99.3% for raccoons, 98.9% for opossums, 94.1% for white-tailed deer, and 87.5% for bobcats (Figure 46.31). No rabbits or foxes were even observed. In south Florida today, these species exist only in locations peripheral to the Everglades, where pythons are rare.

Some Invasive Species Are Lethal Pathogens

A **pathogen** is a microorganism or virus that causes disease symptoms in its host. Some pathogens become invasive when moved to a new geographic location. For example, chestnut blight, *Cryphonectria parasitica,* is a fungus from Asia that was accidentally introduced to

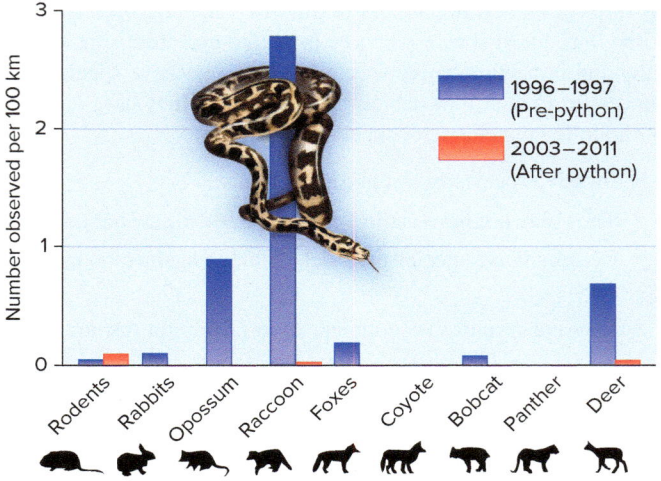

Figure 46.31 Effect of invasive Burmese pythons on wildlife abundance in the Everglades National Park, Florida. According to roadside surveys, a decline in mammal abundance occurred after pythons became common (2003–2011).

BioConnections: *Based on models of population growth outlined in Section 43.3, when do scientists believe the first Burmese pythons were released into the Everglades?*

New York around 1904 from imported Asian chestnut trees, *Castanea crenata.* At that time, the American chestnut tree, *Castanea dentata,* was one of the most common trees in the eastern U.S. It was said that a squirrel could jump from one chestnut to another all the way from Maine to Georgia without touching the ground. The densest populations of chestnuts occurred in the Appalachian Mountains. There, humans and wildlife alike feasted on the bountiful supply of nuts. The tree was ingrained in American culture, as in "chestnuts roasting

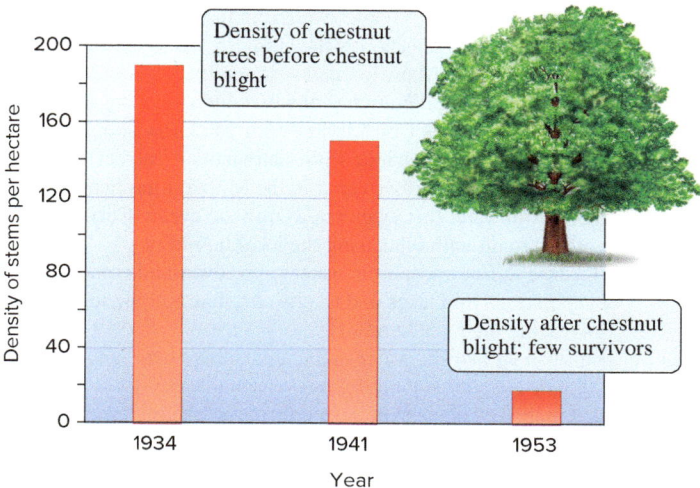

Figure 46.32 Effects of introduced parasites on American chestnut trees. The reduction in density of American chestnut trees in North Carolina following the introduction of chestnut blight disease from Asia shows the severe effect that parasites can have on their hosts. By the 1950s, this once prevalent species had been virtually eliminated.

on an open fire." By the 1950s, however, chestnut blight had significantly reduced the density of American chestnut trees in all areas of the U.S., including North Carolina (**Figure 46.32**). Because at least one tree in four in these forests was a chestnut, the effect was very noticeable, although oaks and hickories replaced chestnuts in the canopy. Eventually, the fungus eliminated nearly all chestnut trees across North America.

46.7 Reviewing the Concepts

- Invasive species are those species introduced into new geographic areas by humans and spread on their own without human support. Invasive species may impact native species via competition, predation, and parasitism (Figures 46.29–46.32).

46.7 Testing Your Knowledge

1. Why are invasive species introduced into areas of the world where they are not normally found?
   a. for agriculture
   b. for timber
   c. for meat
   d. by accident
   e. All of the above are correct.

2. Invasive Burmese pythons
   a. have outcompeted native snakes in many areas of the U.S.
   b. were released in the Great Lakes area to control invasive lampreys.
   c. have severely impacted raccoons, opossums, and rabbits in south Florida.
   d. were important in the spread of chestnut blight in the U.S.
   e. None of the above is correct.

Assessing Your Knowledge and Skills

Foundational Knowledge

Multiple Choice

1. Which anthropogenic gas contributes most to global warming?
   a. methane
   b. chlorofluorocarbons
   c. carbon dioxide
   d. nitrogen oxides
   e. sulfur dioxide

2. The amount of land necessary for survival for each person in a sustainable world is known as
   a. the sustainability level.
   b. the ecological impact.
   c. the ecological footprint.
   d. the survival needs.
   e. the carrying capacity.

3. Human production of fertilizers first impacts which stage of the nitrogen cycle?
   a. nitrogen fixation
   b. nitrification
   c. assimilation
   d. ammonification
   e. denitrification

4. Most recorded extinctions have been caused by
   a. invasive species.
   b. habitat destruction.
   c. overexploitation.
   d. a and b equally.
   e. a, b, and c equally.

5. Invasive species commonly
   a. outcompete native species.
   b. prey on native species.
   c. parasitize native species.
   d. do both a and b.
   e. do a, b, and c.

Short Answer

1. The Earth's atmosphere consists of 78% nitrogen. Why is nitrogen a limiting nutrient, and how can we increase the supply of nitrogen to plants?

2. Why does maximum sustainable yield occur at the midpoint of the logistic curve and not where the population is at its carrying capacity?

Critical-Thinking Skills

Multiple Choice

1. You are inspired by your professor's lecture on human demography to decrease your ecological footprint. Which of the following steps would best help you to accomplish this goal?
   a. Carpool, walk, or bike more often.
   b. Trade in your sedan for an SUV you can use for camping.
   c. Start a high-protein diet that replaces refined carbohydrates with meat.
   d. Heat your home using a fireplace instead of natural gas.
   e. Live out in the countryside, even if your job is in the city.

2. A positive feedback loop is one that tends to amplify a process. Why would the effect of deforestation on the water cycle be considered a positive feedback loop?
   a. Deforestation leads to new types of habitat, which lead to greater species diversity.
   b. Deforestation leads to reduced nutrient uptake during the growing season, which results in more fertile soil.
   c. Deforestation leads to a decrease in precipitation, which results in further loss of vegetation.
   d. Deforestation leads to an increase in the concentration of nutrients in runoff, which then leads to eutrophication.

3. Using a site in northern California, you repeat Stiling and Drake's experiments in elevated and ambient CO_2 chambers and achieve similar results. However, you notice that for one species of herbivore, the mortality rate from predators and parasitoids is equal in the elevated and ambient chambers, while the mortality rate due to nutritional inadequacy is higher in elevated chambers. The herbivore species in question is native to Asia and has only recently arrived in the United States. What is a likely explanation for your findings?
   a. CO_2 becomes toxic to the Asian herbivore at the levels found in the elevated chambers.
   b. The Asian herbivore does not take longer to complete development in elevated CO_2 chambers, so it is not subject to additional enemy attack.
   c. The Asian herbivore takes longer to complete development in the elevated chambers but has very few natural enemies due to its foreign origin.
   d. Individuals of the Asian herbivore species thrive in the elevated CO_2 chamber, growing so large that natural enemies no longer attack them.

4. Lead is known to undergo biomagnification and accumulate in bones. What do you predict will happen to lead taken up by a tiger after it dies in the wild?
   a. The lead will be broken down to a nontoxic form.
   b. The lead will be eliminated from the food web because no species prey upon the tiger.
   c. The lead will re-enter the food web via evaporation and precipitation.
   d. The lead will re-enter the food web via detritivores and decomposers.

5. In the summer of 2016, people sweltering in the midwestern U.S. didn't say, "It's not the heat, it's the humidity." Instead, they said, "It's the cornsweat." Many blamed surrounding acres of cornfields for increasing the local humidity and thereby increasing the perceived temperature. Considering what you know about the water cycle, which of the following statements is correct?
   a. Most atmospheric water vapor in the Midwest has been blown there from water that evaporated from the ocean. Corn has nothing to do with increasing the local humidity.
   b. Because most atmospheric water vapor over land comes from evaporation from lakes and rivers, corn has nothing to do with increasing the local humidity.
   c. Because 90% of the water in the atmosphere over land arrived there by evapotranspiration, it is reasonable to conclude that increased acreage of corn cultivation can increase local humidity.
   d. Because local temperatures are based on climate, and plants capture carbon dioxide, the fields of corn plants should actually decrease local temperatures.

Short Answer

1. **PRINCIPLES** A principle of biology is that biologists use models to test predictions and convey ideas. In one family, the parents, who were born in 1900, have twin daughters at age 20 but then have no more children. Their children, grandchildren, and so on behave in the same way. All descendants live to age 100. In another family, the parents, who were also born in 1900, delay reproduction until age 33 but have triplet daughters. Their children and grandchildren behave in the same way. Which mother has the most female descendants by 2000? What can you conclude?

2. A species of invasive herbivore is inadvertently introduced to a new continent. Which type of organisms may be affected and why?

Biodiversity and Conservation Biology

47

Spix's macaw (*Cyanopsitta spixii*). Fewer than 100 individuals of this species are known to exist in the rain forests of Brazil.

Chapter Outline

47.1 Genetic, Species, and Ecosystem Diversity

47.2 Value of Biodiversity to Human Welfare

47.3 Biodiversity and Ecosystem Function

47.4 Conservation Strategies

Assessing Your Knowledge and Skills

In 2009, Jeff Corwin, an American conservationist and host for programs on Animal Planet and other television networks, published a book entitled *100 Heartbeats: The Race to Save the Earth's Most Endangered Species*. The Hundred Heartbeat Club had been created earlier by biologist E. O. Wilson to highlight the plight of animal species, such as Spix's macaw (*Cyanopsitta spixii*) in Brazil (see the chapter-opening photo), the Chinese river dolphin (*Lipotes vexillifer*), and the Philippine eagle (*Pithecophaga jefferyi*), that have 100 or fewer individuals left alive (and hence that number of heartbeats away from extinction). Saving species from extinction is important in its own right, but as we will see, conservation of biological diversity also has great economic and social value to humankind.

Biological diversity, or **biodiversity**, encompasses the genetic diversity of species, the variety of species, and the different ecosystems they form. The field of **conservation biology** uses principles and knowledge from molecular biology, genetics, and ecology to protect and sustain biodiversity. Because conservation biology draws from nearly all chapters of this textbook, a discussion of it is a fitting way to conclude our study of biology. In this chapter, we will begin by examining the questions of what biodiversity is and why it should be conserved. We will explore how much biodiversity is needed for ecosystems to function properly.

Later, we will consider what is being done to help conserve the world's endangered plant and animal life. This includes identifying global areas rich in species and establishing parks and refuges of the appropriate size, number, and connectivity. We will also discuss the conservation of particularly important types of species and how to restore damaged habitats to a more natural condition. We will then examine how captive-breeding programs help to build populations of rare species prior to their release back into the wild. Some programs have also used modern genetic techniques such as cloning to help breed and perhaps eventually increase populations of endangered species. Finally, we will discuss the concept of sustainable development and whether this organizing principle can help global conservation efforts.

47.1 Genetic, Species, and Ecosystem Diversity

Learning Outcome

1. List and describe the three levels of biodiversity.

Biodiversity can be examined on three levels: genetic diversity, species diversity, and ecosystem diversity. Each level of biodiversity provides valuable benefits to humanity.

Genetic diversity is the amount of genetic variation occurring within and between populations. Without such variation, populations cannot respond so quickly to changes in environmental conditions, resulting in population decline and even extinction. Maintaining genetic variation in the wild relatives of crops may be vital to the continued success of crop-breeding programs. For example, the café marron (*Ramosmania rodriguesii*), a wild relative of the coffee plant that is native to a tiny island off the coast of Mauritius, was assumed to be extinct until 1979, when one surviving tree was identified. Today, cuttings from the tree are being cultured in London's Kew Gardens (**Figure 47.1**). Researchers are hoping to identify café marron genes that could be transferred to coffee plants, thereby allowing them to be grown in a wider range of soils and elevations.

The second level of biodiversity is **species diversity**, the number and relative abundance of species in a community (refer back to Section 44.1). In 1973, the U.S. Endangered Species Act (ESA) was enacted, which was designed to protect both endangered and threatened species. **Endangered species** are those species that are in danger of extinction throughout all or a significant portion of their range. **Threatened species** are those species likely to become endangered in the foreseeable future. Many species are currently threatened. According to the International Union for Conservation of Nature (IUCN), more than 25% of fish species that live on coral reefs, 22% of all mammals, 12% of birds, and 31% of amphibians are threatened with extinction. In 2000, the World Wildlife Foundation placed Atlantic cod, *Gadus morhua*, on the endangered species list as a result of overfishing. Nine of 17 populations of commercially important Chinook salmon, *Oncorhynchus tshawytscha*, in California and Oregon are listed as endangered or threatened.

The third level of biodiversity is ecosystem diversity, the diversity of structure and function within an ecosystem. Conservation has largely focused attention on species-rich ecosystems such as tropical rain forests. Over 120 prescription drugs used to treat malaria, cancer, and other diseases were developed from rain forest plants, yet less than 1% of such plants have been tested for medicinal properties. However, some ecologists have argued that relatively species-poor ecosystems, such as prairies, are also highly threatened and in equal need of conservation. More than 99% of the original tallgrass prairie in the United States has been converted to agricultural land.

47.1 Reviewing the Concepts

- Biodiversity represents diversity at three levels: genetic diversity, species diversity, and ecosystem diversity. Conservation biology uses principles and knowledge from molecular biology, genetics, and ecology to protect biological diversity (Figure 47.1).

47.1 Testing Your Knowledge

1. In 1977, Rafael Guzman, a Mexican biologist, discovered a previously unknown wild relative of corn that is resistant to many of the viral diseases that infect domestic corn. Agriculturalists believe cross-breeding could improve current corn crops. In this case, biodiversity is important at which level(s)?
    a. ecosystem
    b. species
    c. genetic
    d. community
    e. both a and b

47.2 Value of Biodiversity to Human Welfare

Learning Outcome

1. Outline the benefits of biological diversity to human welfare.

Why should biodiversity be a concern? American biologists Paul Ehrlich and E. O. Wilson have suggested that the loss of biodiversity should be an area of great concern for at least three reasons:

1. Humans depend on plants, animals, and microorganisms for a wide range of food, medicine, and industrial products.
2. Ecosystems provide an array of essential services, such as clean air and water.
3. Humans have an ethical responsibility to protect what are our only known living companions in the universe.

In this section, we will examine some of the primary reasons it is so important to preserve biodiversity.

Figure 47.1 Café marron being cultured in London's Kew Gardens. These plants are derived from just one surviving individual found in Mauritius.
©Florapix/Alamy

Society Benefits Economically from Biodiversity

Many different sectors of society benefit from biodiversity. Three key sectors are the pharmaceutical, agricultural, and natural products industries.

Pharmaceutical Industry The pharmaceutical industry is heavily dependent on plant, animal, fungal, and bacterial products for source material. Worldwide prescription drug sales are forecast to be $1 trillion in 2020.

An estimated 50,000–70,000 plant species are used in traditional and modern medicine. About 25% of the prescription drugs in the U.S. alone are derived from plants, and the 2015 market value of such drugs was estimated to be $374 billion, accounting for a little less than half the global pharmaceutical market. Many medicines come from plants found only in tropical rain forests. These include quinine, a drug from the bark of the cinchona tree (*Cinchona officinalis*) (**Figure 47.2a**), which is used for treating malaria, and vincristine, a drug derived from rosy periwinkle (*Catharanthus roseus*), which is a treatment for leukemia and Hodgkin's disease.

Animal products have also been important to the pharmaceutical industry. For example, the venom of the gila monster (*Heloderma suspectum*), one of only two venomous lizards in the world, is being used to treat people who are resistant to conventional treatment for type 2 diabetes, a disease that may affect 30% of Americans at some point in their lives. And a protein from the South American pit viper (*Bothrops jacara*) may help to control human blood pressure. Tarantula venom may be helpful in treating neurological disorders such as Parkinson's disease.

Fungi and soil bacteria have provided an important reservoir of drugs. In 1928, Scottish scientist Alexander Fleming discovered that a fungus of the genus *Penicillium* produces penicillin, one of the first and most widely used antibiotics. In 1964, a group of Canadian scientists traveled to Easter Island and identified a drug produced by a soil bacterium that suppresses immune reactions in humans. The drug was named rapamycin after the indigenous name of Easter Island, which is Rapa Nui. Rapamycin is now used to prevent organ rejection in transplants and as a coating on heart stents.

Agriculture Although humans depend on only about 20 plant species to provide 90% of the world's food, wild relatives of these crops provide a useful reservoir of genetic material for developing pest-resistant varieties or strains that can grow in marginal areas. A gene from longstamen rice (*Oryza longistaminata*), a grass species from Africa, has been integrated into the genome of commercially grown rice (*O. sativa*) to confer resistance to rice blight disease (**Figure 47.2b**). In the 1970s, infusion of genetic material from wild corn in Mexico was used to protect commercially grown U.S. corn from a leaf fungus, which had killed 15% of the crop. Lake Placid mint (*Dicerandra frutescens*), native only to central Florida, produces a powerful insect-repelling chemical that may have benefits for crop protection. Another endangered species, buffalo clover (*Trifolium stoloniferum*), has high protein content and is a perennial, making it of high potential value as a forage crop.

Natural Products Industry Many products used by humans are harvested from plant species growing in their native environments. Maintenance of these native environments is important to preserve such biodiversity. Natural plant products include wood, rubber, and dietary supplements. The forest products industry manufactures over $200 billion in products in the U.S. annually. Maple syrup, nuts, blueberries and algae are all harvested, in addition to the trees themselves. Interestingly, a relatively new commercial crop is made by the flowering shrub *Parthenium argentartum*. It produces high amounts of natural rubber and grows in the deserts of the U.S. Southwest, adding economic value to marginal lands.

Animals, too, have great commercial value in the natural products industry. Salmon fishing in the Pacific Northwest supports over 60,000 jobs and injects over $1 billion into the economy (**Figure 47.2c**). In the U.K., the sea fish catch is worth over $500 million annually.

(a)

(b)

(c)

Figure 47.2 Societal benefits of biodiversity. (a) The bark of the cinchona tree (*Cinchona officinalis*), a plant found only in tropical rain forests, is used to produce quinine, an effective treatment for malaria. (b) A gene from longstamen rice (*Oryza longistaminata*), native to Africa, was used to reduce rice blight disease in rice, *O. sativa,* native to Asia. (c) Commercial fishing for salmon is an important part of the economy in the U.S. Pacific Northwest.

(a) ©Heather Angel/Natural Visions; *(b)* ©Clement Philippe/Arterra Picture Library/Alamy; *(c)* ©Joshua Roper/Alamy

Exploitation of wild animals, often known as bushmeat, in Africa is a major source of animal protein for more than a billion of the world's poorest people. Game birds, especially pheasants and ducks, are often shot in North America and Europe.

Natural Ecosystems Provide Essential Services to Humans

Beyond the direct economic gains from biodiversity, humans benefit enormously from the essential services that natural ecosystems provide (Table 47.1). For example, forests soak up carbon dioxide, maintain soil fertility, and retain water, helping to prevent or minimize flooding; estuaries provide water filtration and protect rivers and coastal shores from excessive erosion. *Prochlorococcus*, an abundant, ocean-dwelling genus of cyanobacteria, was discovered only in 1986, yet it is estimated to produce about 20% of the oxygen we breathe. Another function of ecosystems is the maintenance of populations of natural predators to regulate pest outbreaks and reservoirs of pollinators to pollinate crops and other plants. In addition, approximately 75% of the 100,000 chemicals released into the environment can be degraded by living organisms. The loss of biodiversity can disrupt an ecosystem's ability to carry out such functions.

In the 1990s, farmers in India began using the anti-inflammatory drug diclofenac to reduce pain and fever in their livestock. They could hardly anticipate that vultures scavenging on dead carcasses would accumulate large doses of the drug and die of renal failure. But the consequences did not stop there. Following a 97% reduction in vulture numbers over a 14-year period, the population of feral dogs exploded, buoyed by the availability of uneaten carcasses. The incidence of rabies in humans increased, with estimates of an additional 48,000 people dying over the 14-year time span. The loss of the scavenging services of the vultures was estimated to cost India $24 billion.

A 2014 paper in the journal *Global Environmental Change* by economist Robert Costanza and colleagues made an attempt to calculate the monetary value of ecosystems to various economies (in 2007 dollars). They came to the conclusion that at the time, the world's ecosystems were worth more than $124 trillion a year, nearly twice the gross national product of the world's economies combined ($75.2 trillion in 2007 dollars).

Preservation of Natural Ecosystems May Be Performed for Noneconomic Reasons

Arguments can also be made against the loss of biodiversity on ethical grounds. As only one of millions of species on Earth, many people have argued that humans have no right to destroy other species and the environment around us. In the U.S., California condors (*Gymnogyps californianus*) and grizzly bears (*Ursus arctos*) have little economic value to people, yet we assign them enough value that we preserve the habitats they live in. American philosopher Tom Regan suggests that animals should be treated with respect because they have a life of their own and therefore have value apart from anyone else's interests. American law professor Christopher Stone, in an influential 1972 article titled "Should Trees Have Standing?" has argued that entities such as nonhuman natural objects, like trees or lakes, should be given legal rights just as corporations are treated as individuals for certain purposes. As biologist E. O. Wilson proposed in a 1984 concept known as biophilia, humans have innate attachments with species and natural habitats because of our close association for over millions of years.

John Muir, the founder of the Sierra Club, thought that natural areas had spiritual value and should be preserved rather than used as a source of natural products. This idea, which became known as the preservationist ethic, contrasts with the resource conservation ethic, which focuses on management of natural areas to allow the wisest current and future use of natural resources.

Table 47.1	Examples of the World's Ecosystem Services
Service	**Example**
Atmospheric gas supply	Regulation of carbon dioxide, ozone, and oxygen levels
Climate regulation	Regulation of carbon dioxide, nitrogen dioxide, and methane levels
Disturbance regulation	Storm protection; flood control
Water regulation	Regulation of hydrological flows
Water supply	Irrigation; water for industry
Erosion control	Retention of topsoil; reduction of accumulation of sediments in lakes
Soil formation	Soil formation processes
Nutrient cycling	Nitrogen, phosphorus, and carbon cycles
Waste treatment	Sewage purification
Pollination	Pollination of crops
Biological control	Pest population regulation
Wilderness and refuges	Habitat for wildlife
Food production	Crops; livestock
Raw materials	Fossil fuels; timber
Genetic resources	Medicines; genes for plant resistance
Recreation	Ecotourism; outdoor recreation
Culture	Aesthetic and educational value

47.2 Reviewing the Concepts

- The preservation of biodiversity has been justified because of its economic value, because of the value that ecosystem services provide, and on ethical grounds (Figure 47.2, Table 47.1).

47.2 Testing Your Knowledge

1. Biodiversity benefits
   a. the pharmaceutical industry.
   b. agriculture.
   c. the natural products industry.
   d. the world's ecosystems services.
   e. all of the above.

47.3 Biodiversity and Ecosystem Function

Learning Outcomes

1. **SCISKILLS** ▶ Create graphical representations of possible relationships between biodiversity level and ecosystem function.
2. Describe experimental evidence that shows how species diversity and ecosystem function are linked.

Because biodiversity affects the health of ecosystems, ecologists have explored the question of how much diversity is needed for ecosystems to function properly. In this section, we will look at several models that explore the relationships between biodiversity and ecosystem function and examine an experimental approach used to study this relationship.

Ecologists Have Described Several Models for the Relationship Between Ecosystem Function and Biodiversity

In the 1950s, British ecologist Charles Elton proposed the **diversity-stability hypothesis,** which suggests that species-rich communities are more stable than those with fewer species (refer back to Section 44.2). If we use stability as a measure of ecosystem function, Elton's hypothesis suggests a linear correlation between diversity and ecosystem function; as diversity increases, ecosystem function increases proportionately (**Figure 47.3a**).

Australian ecologist Brian Walker proposed an alternative to this idea, termed the **redundancy hypothesis** (**Figure 47.3b**). According to this hypothesis, ecosystem function increases rapidly at fairly low levels of diversity, but then it levels off because most additional species are functionally redundant.

Two alternative hypotheses relating biodiversity and ecosystem services have also been proposed. The **keystone hypothesis** (**Figure 47.3c**) proposes that ecosystem function dramatically rises as biodiversity approaches its natural levels. Finally, the **idiosyncratic hypothesis** addresses the possibility that although ecosystem function can change as the number of species increases or decreases, the amount and direction of change are unpredictable (**Figure 47.3d**).

Determining the correct model(s) is very important, as our understanding of the effect of species loss on ecosystem function can greatly affect the way we manage our environment. Only relatively recently has the link between ecosystem function and biodiversity, usually measured as species richness, been studied.

Field Experiments Have Explored the Relationship Between Species Richness and Ecosystem Function

In the mid-1990s, American plant ecologist David Tilman and colleagues performed experiments in the field to determine how species richness affected ecosystem functioning. Tilman's previous experiments had suggested that species-rich grasslands were more stable (that is, they were more resistant to the ravages of drought and

Biology Principle

Biologists Use Models to Test Predictions and Convey Ideas

Many different models have been proposed that relate ecosystem function to species richness. Such models can be tested experimentally to determine if they are correct, if they are incorrect, or if they need refining.

Figure 47.3 Graphical representations of possible relationships between ecosystem function and biodiversity. The two solid dots represent the end points of a continuum of species richness. The first solid dot is at the origin, where there are no species and no community services. The second solid dot represents natural levels of species diversity. The relationship is strongest in (a) and weakest in (d).

recovered from drought more quickly) than species-poor grasslands (refer back to Figure 44.4). In subsequent experiments, Tilman's group sowed multiple plots, each 3 m × 3 m and on comparable soils, with seeds of 1, 2, 4, 6, 8, 12, or 24 species of prairie plants. Exactly which species were sown into each plot was determined randomly from a pool of 24 native species. The treatments were replicated 21 times, for a total of 147 plots.

The results showed that the plots with more species (more diverse plots) had increased productivity, expressed as a percentage of plant cover (the amount of ground covered by leaves of plants) than the plots with fewer species (less diverse plots) (**Figure 47.4a**). This occurred because of a greater variety of plant growth forms that could

Biology Principle

Biologists Use Models to Test Predictions and Convey Ideas

Figure 47.4 shows three graphical models that depict the relationships between ecosystem function and species richness in prairie plant communities. The Modeling Challenge in the figure legend asks you to redraw each of these graphs to represent different possible relationships between ecosystem function and species richness, as shown in Figure 47.3.

(a) Plant cover increased with more species.

(b) Available nitrate decreased with more species.

(c) Invasive species decreased with more species.

Figure 47.4 The relationship of species richness to ecosystem function.

Modeling Challenge: The goal of this Modeling Challenge is to relate three of the hypotheses described in Figure 47.3 to the experimental data described in Figure 47.4. Begin with the graphs shown in Figure 47.3a, c, and d, and erase the labels on the *x*-axis and *y*-axis. Then, relabel the axes according to the experimental data shown in Figure 47.4. Relate Figure 47.3a, c, and d to Figure 47.4b, a, and c, respectively. Based on your graphs, what predictions do the three hypotheses (diversity-stability, keystone, and idiosyncratic) make with regard to the experimental data shown in Figure 47.4?

---

use light at different levels of the canopy. The more species-rich plots also used more nutrients, such as nitrate (NO_3^-), than the less diverse plots because a greater variety of plant root lengths could use nutrients at different levels of the soil (**Figure 47.4b**). Furthermore, the frequency of invasive plant species (species not originally planted in the plots) decreased with increased plant species richness (**Figure 47.4c**).

Although Tilman's experiments show a relationship between species richness and ecosystem function, they also suggest that most of the advantages of increasing diversity come with the first 5–10 species, beyond which adding more species appears to have little to no effect. These data support the redundancy hypothesis (compare Figure 47.4a with Figure 47.3b). Confirmation of the redundancy hypothesis is also observed on a larger scale. The productivity of temperate forests on different continents is roughly the same, despite different numbers of tree species—729 in East Asia, 253 in North America, and 124 in Europe. The presence of more tree species may ensure a supply of "backups," should some of the most productive species die off from insect attack or disease. This can happen, as was seen in the demise of the American chestnut tree. Diseases devastated this species, and its presence in forests dramatically had decreased by the mid-20th century (refer back to Figure 46.32). The forests filled in with other species and continued to function as before in terms of nutrient cycling and gas exchange. However, although the forests continued to function without the American chestnuts, some important changes occurred. For example, the loss of chestnuts deprived bears and other animals of an important source of food and may have affected their reproductive capacity and hence the size of their populations.

47.3 Reviewing the Concepts

- Four models describe the relationship between biodiversity and ecosystem function: the diversity-stability, redundancy, keystone, and idiosyncratic hypotheses (Figure 47.3).
- Field experiments have shown that increased biodiversity results in increased ecosystem function, supporting the redundancy hypothesis (Figure 47.4).

47.3 Testing Your Knowledge

1. The idea that there is a linear correlation between diversity and ecosystem function is known as the _____ hypothesis.
   a. keystone
   b. diversity-stability
   c. linear
   d. redundancy
   e. idiosyncratic

2. Experimental evidence suggests that
   a. ecosystem function levels off at fairly low levels of diversity.
   b. ecosystem function plummets as soon as biodiversity decreases from natural levels.
   c. there is a linear correlation between diversity and ecosystem function.
   d. there is an unpredictable relationship between diversity and ecosystem function.
   e. there is no relationship between diversity and ecosystem function.

47.4 Conservation Strategies

Learning Outcomes

1. List and describe the criteria that conservation biologists use to identify areas for protection.
2. Explain how the principles of island biogeography and landscape ecology are used to create nature preserves.
3. Describe different approaches that conservation biologists use to protect individual species.
4. Define restoration ecology, and describe the approaches used to restore degraded ecosystems and populations of species.
5. Describe the advantages and disadvantages of using cloning as a conservation strategy to help save endangered species.
6. Define sustainable development, and describe how this principle can help biologists set aside land to conserve species.

As discussed in Chapter 46, human activities have had many negative effects that threaten biodiversity and ecosystems. The goal of conservation biologists is to manage natural resources with the aim of protecting species, their habitats, and ecosystems. In their efforts to maintain the diversity of life on Earth, these biologists are active on many fronts.

We will begin this section by discussing how conservation biologists identify the global areas most in need of conservation. We will then examine some of the factors that must be considered when designing nature preserves, also called parks. These factors include the size and shape of the preserve and the ability of species to move from one nature preserve to another. In some cases, efforts are aimed at protecting species that have a great ecological impact; in other cases, they are aimed at protecting species, such as the panda bear, that are recognizable and garner support from humans. We will then explore the topic of **habitat restoration**—the full or partial repair or replacement of biological habitats and/or their populations that have been degraded or destroyed. Such restoration may involve captive breeding programs to re-establish populations of threatened species in the wild. We will also discuss how genetic cloning may be used as an additional tool to help conserve endangered species. Finally, we will examine the concept of sustainable development and explore how correct valuation of our living and nonliving natural resources promotes conservation.

Conservation Seeks to Establish Protected Areas

Currently, about 15.4% of global land area and 3.4% of the global ocean area are under some form of environmental protection. More than 217,000 separate areas are protected, with more being added daily. Conservation biologists often must make decisions regarding which habitats should be protected. Many conservation efforts have focused on saving habitats in so-called megadiversity countries, because they often have the greatest number of species. However, more recent strategies have promoted the preservation of certain key areas with the highest levels of unique species or the preservation of representative areas of all types of habitat, even relatively species-poor areas.

Megadiversity Countries One strategy of targeting areas for conservation is to identify **megadiversity countries,** those countries with the greatest numbers of species. Using the number of plants, vertebrates, and selected groups of insects as criteria, American biologist Russell Mittermeier and colleagues determined that just 17 countries are home to nearly 70% of all known species. Brazil, Indonesia, and Colombia top the list, followed by Australia, Peru, Mexico, Madagascar, China, and 9 other countries. The megadiversity country approach suggests that conservation efforts should be focused on the most biologically rich countries. However, although megadiversity areas may contain the most species, they do not necessarily contain the greatest number of unique species. The mammal species list for Peru is 344, and for Ecuador, it is 271; of these, however, 208 species are common to both.

Areas Rich in Endemic Species Another approach to setting conservation priorities—one adopted by the organization Conservation International—takes into account the number of species that are **endemic;** that is, they are found only in a particular place or region. This approach suggests that conservationists should focus their efforts on **biodiversity hot spots,** regions that are biologically diverse and under threat of destruction. To qualify as a

Figure 47.5 Locations of major biodiversity hot spots around the world. Hot spots have high numbers of endemic species.

biodiversity hot spot, a region must meet two criteria: (1) It must contain at least 1,500 species of endemic vascular plant species, and (2) it must have lost at least 70% of its original habitat. Vascular plants were chosen as the primary group of organisms to determine whether or not an area qualifies as a hot spot, mainly because most other terrestrial organisms depend on them to some extent.

Conservationists Norman Myers, Russell Mittermeier, and colleagues identified 34 biodiversity hot spots that together occupy a mere 2.3% of the Earth's surface but contain 150,000 endemic plant species, or 50% of the world's total (**Figure 47.5**). This approach proposes that protecting geographic hot spots will prevent the extinction of a larger number of endemic species than would protecting areas of a similar size elsewhere. The main argument against using hot spots as the criterion for targeting conservation efforts is that the areas richest in endemic species—tropical forests—would receive the majority of attention and funding, perhaps at the expense of protecting other areas.

Representative Habitats and Crisis Ecoregions In a third approach to prioritizing areas for conservation, some scientists have recently argued that we need to conserve representatives of all major habitats. Prairies are a case in point. An example is the Pampas region of South America, which is arguably the most threatened habitat on the continent because of rapid conversion of its natural grasslands to ranch land and agriculture (**Figure 47.6**). The Pampas does not compare well in richness or endemics with the rain forests, but it is a unique area that without preservation could disappear. By selecting habitats that are most distinct from

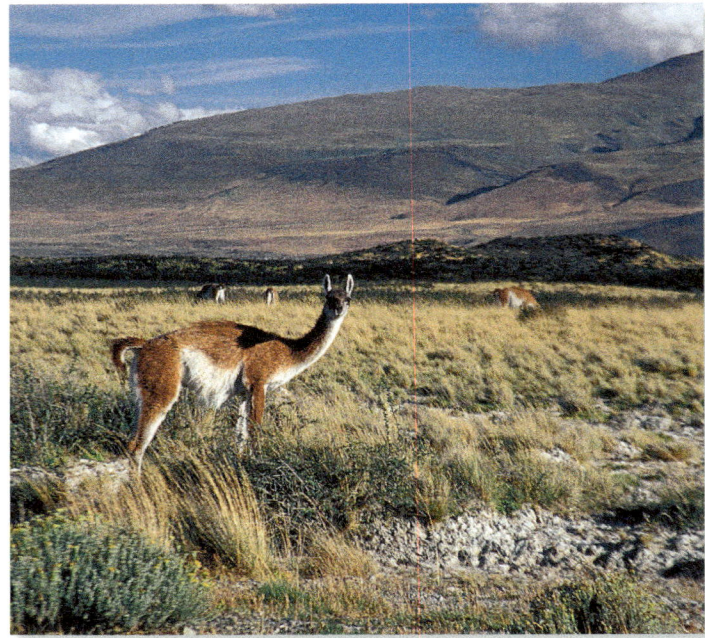

Figure 47.6 The Pampas, Argentina. This habitat is not rich in species but is threatened due to conversion to ranch land and agriculture. The guanaco, shown here, is a characteristic grazer of pampas grass.
©Vick Fisher/Alamy

those already preserved, many areas that are threatened but not biologically rich may be preserved in addition to the less immediately threatened, but richer, tropical forests.

Within these areas, some regions are seen as more critical than others. Habitat loss has been most extensive in temperate grasslands and tropical and temperate deciduous forests, where, in each case, over 50% of the land has been converted to other uses such as agriculture. At the same time, very little, less than 10%, of these agriculturally rich areas has been protected. Such habitats are thus termed **crisis ecoregions.**

Last of the Wild The final approach to conservation involves the preservation of regions of the world relatively untouched by humans. Scientists have mapped out the extent of the human footprint on the globe. The areas of the Earth that fall within the lowest 10% of the human-affected areas have been termed the "last of the wild." Such areas, because they are relatively pristine, offer a great opportunity for conservationists because of their relatively intact communities. These wild areas include tundra and boreal forests of Russia and Canada, as well as some desert biomes and tropical forests.

Preserve Design Incorporates Principles of Island Biogeography and Landscape Ecology

After identifying areas to preserve, conservationists must determine the size, arrangement, and management of the protected land. Among the questions conservationists ask is whether one large preserve is preferable to an equivalent area composed of smaller preserves. Ecologists also need to determine whether nature preserves should be close together or far apart and whether or not they should be connected by strips of suitable habitat to allow the movement of plants and animals between them.

The Role of Island Biogeography According to the equilibrium model of island biogeography (refer back to Section 44.4), the biodiversity on an island tends toward an equilibrium number that is determined by the balance between two factors: immigration rates and extinction rates. This theory has been applied to nature preserves because, in essence, they are islands in a sea of human-altered habitat.

One question for conservationists is how large a protected area should be (**Figure 47.7a**). According to island biogeography, the number of species should increase with increasing area (the species-area effect). Thus, a larger area would mean that a larger number of species would be protected. In addition, larger preserves have other benefits. For example, they are beneficial for organisms that require large spaces, including migrating species and species with extensive territories, such as lions and tigers.

A related question is whether it is preferable to protect a single large preserve or several smaller ones (**Figure 47.7b**). This question is called the **SLOSS debate** (for single large or several small). Proponents of the single large preserve claim that a larger preserve is better able to protect more and larger populations than an equal area divided into small areas. According to island biogeography, a larger block of habitat would support more species than would several smaller blocks.

However, many empirical studies suggest that multiple small sites of equivalent area will contain more species, because a series of small sites is more likely to contain a broader variety of habitats than one large site. Looking at a variety of sites, American conservationists Jim Quinn and Susan Harrison concluded that animal life was richer in collections of small preserves than in a smaller number of larger ones. In their study, having more habitat types outweighed the effect of larger area on biodiversity. In addition, another benefit of a series of smaller preserves is a reduction of extinction risk by a single event such as a wildfire or the spread of disease.

The Role of Landscape Ecology Landscape ecology is an area of ecology that examines the spatial arrangement of communities and ecosystems in a geographic area. Landscape ecologists have suggested that small preserves should be linked together by **movement corridors,** thin strips of land that permit the movement of species between patches (**Figure 47.7c**). Such corridors ideally facilitate the movements of organisms that are vulnerable to predation outside their natural habitat or those that have poor powers of dispersal between habitat patches. In this way, if a population in one small preserve experiences a disaster, immigrants from neighboring populations can more easily recolonize it. This avoids the need for humans to physically move new plants or animals into an area.

Several types of habitat function as movement corridors, including hedgerows (linear patches of shrubs and small trees), which facilitate the movement and dispersal of species between forest fragments (**Figure 47.8**). In China, corridors of habitat have been established to link small, adjacent populations of giant pandas. However,

(a) Large or small (b) Single large or several small (c) Linked or separate (d) Round or oblong

Figure 47.7 The theoretical design of nature preserves. **(a)** A larger preserve holds more species and has low extinction rates. **(b)** A given area should be fragmented into as few pieces as possible. **(c)** Maintaining or creating corridors between fragments may also enhance dispersal. **(d)** Circular-shaped areas minimize the amount of edge effects. The labels better and worse refer to theoretical principles generated by the equilibrium model of island biogeography, but empirical data have not supported all the predictions.

Concept Check: *What are some of the potential risks in connecting areas via movement corridors?*

Figure 47.8 Movement corridors.
©Andrew Parker/Alamy

✓ **Concept Check:** *How do these European hedgerows act as movement corridors?*

disadvantages are associated with movement corridors, which can facilitate the spread of disease, invasive species, and fire between small preserves.

Finally, nature preserves are often designed to minimize **edge effects**, the special physical conditions that exist at the boundaries, or edges, of ecosystems. Habitat edges, particularly those between a natural habitat, such as a forest, and developed land, are often different in physical characteristics from the habitat core. For example, the center of a forest is shaded by trees and has less wind and light than the forest edge, which is unprotected. Many forest-adapted animals therefore shy away from forest edges and prefer forest centers. For this reason, circular preserves are generally preferable to oblong ones, because the amount of edge is minimized (see **Figure 47.7d**).

The Single-Species Approach Focuses Conservation Efforts on Particular Types of Species

As we have seen, many conservation biologists consider broad, global issues when deciding which habitats are more important to preserve, as well as more local issues concerning the sizes, shapes, and interconnectedness of preserves. When establishing and managing nature preserves, they may take a single-species approach in their conservation efforts, which focuses on saving species that are deemed particularly important. As with habitat conservation, various approaches are used to identify the species that are most important.

Indicator Species Some conservation biologists have suggested that certain organisms can be monitored as **indicator species,** those species whose status provides information on the overall health of an ecosystem. For example, corals are good indicators of marine processes such as siltation—the accumulation of sediments transported by water. Because siltation reduces the availability of light, the abundance of many marine organisms decreases in such situations, with corals among the first to display a decline in health. Coral bleaching is also an indicator of climate change.

Polar bears (*Ursus maritimus*) are thought to be a mammalian indicator species of global climate change (**Figure 47.9a**). Most scientists are in agreement that global warming is causing the ice in the Arctic to melt earlier in the spring than it did in the past. Because polar bears rely on the ice to hunt for seals, the earlier breakup of the ice is leaving the bears less time to feed and build the fat that enables them to sustain themselves and their young. A U.S. Geological Survey study concluded that a future reduction of Arctic ice could result in a loss of two-thirds of the world's polar bear population within 50 years. In May 2008, the polar bear was listed as a threatened species under the U.S. Endangered Species Act (ESA).

Umbrella Species **Umbrella species** are those species whose habitat requirements are so large that protecting them would protect many other species existing in the same habitat. The northern spotted owl (*Strix occidentalis*) of the Pacific Northwest is considered to be an important umbrella species (**Figure 47.9b**). A pair of birds needs at least 800 hectares of old-growth forest to survive and reproduce, so maintaining healthy owl populations helps to ensure the survival of many other forest-dwelling species. In the southeastern U.S., the red-cockaded woodpecker (*Picoides borealis*) is often seen as that region's equivalent of the northern spotted owl because it requires large tracts of old-growth long-leaf pine (*Pinus palustris*), including old, diseased trees in which it can excavate its nests.

Flagship Species In the past, conservation resources were often allocated to a **flagship species,** a single large or instantly recognizable species. Such species were typically chosen because they were attractive and thus more readily engendered support from the public for their conservation. The concept of the flagship species, typically a charismatic vertebrate such as the American buffalo (*Bison bison*), has often been used to raise awareness for conservation in general. The giant panda (*Ailuropoda melanoleuca*) is the World Wildlife Fund's emblem for endangered species, and the Florida panther (*Puma concolor*) has become a symbol of the state's conservation campaign (**Figure 47.9c**).

Keystone Species A different conservation strategy focuses on **keystone species,** those species within a community that have a role out of proportion to their abundance or biomass. The beaver, a relatively small animal, can have a drastic impact on a community by building a dam and flooding an entire river valley (**Figure 47.9d**). The resultant lake may become a home to fish species, wildfowl, and aquatic vegetation. A decline in the number of beavers may have serious ramifications for the remaining community members, promoting fish die-offs, waterfowl loss, and the death of vegetation adapted to waterlogged soil.

American ecologist John Terborgh considers tropical palm nuts and figs to be keystone plant species, because they produce fruit

Figure 47.9 Indicator, umbrella, flagship, and keystone species. (a) Polar bears have been called an indicator species of global climate change. (b) The northern spotted owl is considered an umbrella species for the old-growth forest in the Pacific Northwest. (c) The Florida panther has become a flagship species for Florida. (d) The American beaver, a keystone species, creates large dams across streams, and the resultant lakes provide habitats for a great diversity of species.

(a) ©Robert E. Barber/Alamy; *(b)* Source: John & Karen Hollingsworth/U.S. Fish & Wildlife Service; *(c)* Source: Larry Richardson/U.S. Fish & Wildlife Service; *(d)* ©Robert McGouey/Wildlife/Alamy

(a) Indicator species: polar bear

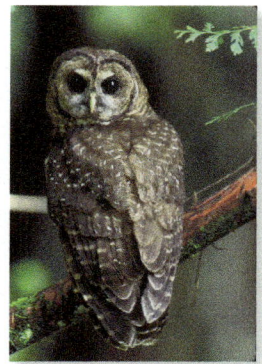
(b) Umbrella species: northern spotted owl

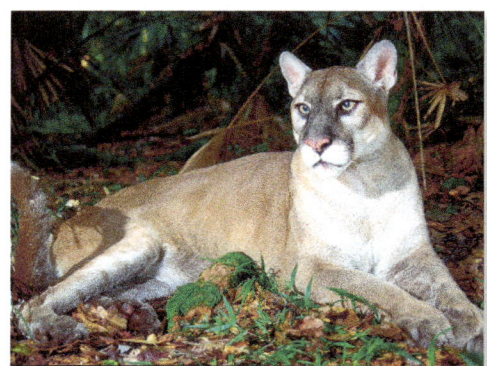
(c) Flagship species: Florida panther

(d) Keystone species: American beaver

during otherwise fruitless times of the year and are thus critical resources for tropical forest fruit-eating animals, including primates, rodents, and many birds. Together, these fruit eaters account for as much as three-quarters of the tropical forest animal biomass. Without the fruit trees, widespread extinction of these animals could occur.

Habitat Restoration Attempts to Improve Degraded Habitats

Another aspect of conservation efforts is improving the quality of a previously degraded habitat. As noted at the beginning of this section, habitat restoration is the full or partial repair or replacement of biological habitats and/or their populations that have been degraded or destroyed. It can focus on complete restoration, rehabilitation, or replacement of a habitat, and it can involve returning species to the wild following captive breeding.

Types of Habitat Restoration In complete restoration, conservation biologists attempt to return a habitat to its composition and condition prior to the disturbance. Under the leadership of American ecologist Aldo Leopold, the University of Wisconsin pioneered the restoration of prairie habitats as early as 1935, converting agricultural land back to species-rich prairies (**Figure 47.10a**).

The second approach, habitat restoration, aims to return the habitat to something similar to, but a little less than, full restoration, a goal called rehabilitation. In Florida, phosphate mining involves removing a layer of topsoil, mining the phosphate-rich layers, returning the topsoil, and then replanting the area. Unfortunately, species such as cogongrass (*Imperata cylindrica*), an invasive Southeast Asian species, may invade these disturbed areas. Even so, the restoration revegetates the area (**Figure 47.10b**).

The third approach to repairing a habitat, termed replacement, makes no attempt to restore what was originally present but instead replaces the original ecosystem with a different one. Ecosystem replacement is particularly useful for places in which the terrain has been substantially altered by past human activities. It would be nearly impossible to re-create the original landscape of an area that was mined for stone or gravel. In these situations, however, wetlands or lakes may be created in the open pits (**Figure 47.10c**).

Reintroductions and Captive Breeding Reintroducing species into areas where they previously existed is a valuable conservation strategy that can re-establish populations in areas where they once occurred (see the chapter-opening photo, Chapter 43). Captive breeding, the propagation of animals and plants outside their natural habitat to produce stock for subsequent release into the wild, has proved valuable in re-establishing breeding populations following extinction or near extinction. Zoos, aquariums, and botanical gardens often play a key role in captive breeding, propagating species that are highly threatened in the wild.

(a) Complete restoration

(b) Rehabilitation

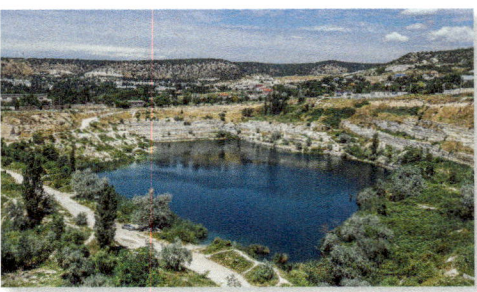
(c) Ecosystem replacement

Figure 47.10 Habitat restoration. **(a)** The University of Wisconsin pioneered the practice of complete restoration of agricultural land to native prairies. **(b)** In Florida, complete restoration after phosphate mining is not usually possible. After topsoil is replaced, invasive species such as cogongrass often grow, resulting in habitat rehabilitation rather than complete restoration. **(c)** This old limestone mine in Crimea has been flooded to provide a valuable freshwater habitat, replacing the ecosystem that was originally present.
(a) ©University of Wisconsin-Madison Arboretum and University Boards of Regents; *(b)* ©D.L. Rockwood, School of Forest Resources and Conservation, University of Florida, Gainesville, FL; *(c)* ©Vladimir Mulder/Shutterstock

Several classic programs illustrate the value of captive breeding and reintroduction. The peregrine falcon (*Falco peregrinus*) had become extinct in nearly all of the eastern U.S. by the mid-1940s, a decline that was linked to the effects of DDT (refer back to Figure 46.16). In 1970, American biologist Tom Cade gathered falcons from other parts of the country to start a captive breeding program at Cornell University. Since then, the program has released thousands of birds into the wild, and in 1999, the peregrine falcon was removed from the list of endangered species. A captive breeding program is also helping to save the California condor (*Gymnogyps californicus*) from extinction. At a cost of $35 million, this species conservation project is the most expensive one ever undertaken in the U.S. In the 1980s, there were only 22 known condors, some in captivity and some in the wild. Scientists made the decision to capture the remaining wild birds in order to protect and breed them (**Figure 47.11a**). By 2016, the captive population numbered 167 individuals, and 268 birds were living in the coastal mountains of California; northern Baja California, Mexico; the Grand Canyon area of Arizona; and Zion National Park, Utah (**Figure 47.11b**). A milestone was reached in 2003, when a pair of captive-reared California condors bred in the wild. Because the number of individuals in any captive breeding program is initially small, care must be exercised to avoid inbreeding. Matings are usually carefully arranged to maximize resultant genetic variation in offspring.

Cloning May Be Used to Save Endangered Species

In 1997, geneticist Ian Wilmut and colleagues at Scotland's Roslin Institute announced to the world that they had cloned a now-famous sheep, Dolly, from mammary cells of an adult ewe. Since then, interest has arisen among conservation biologists about whether the same technology might be used to save species on the verge of extinction. Scientists were encouraged that in January 2001, an Iowa farm cow called Bessie gave birth to a cloned Asian gaur (*Bos gaurus*), an endangered species. The gaur, an oxlike animal native to the jungles of India and Burma, was cloned from a single skin cell taken from a dead animal. To clone the gaur, scientists removed the nucleus from a cow's egg and replaced it with a nucleus from the gaur's cell. The treated egg was then placed into the cow's uterus. Unfortunately, the gaur died from dysentery 2 days after birth, although scientists believe this was unrelated to the cloning procedure.

More recently, other endangered species have been cloned. In 2003, another type of endangered wild cattle, the Javan banteng (*Bos javanicus*), was successfully cloned (**Figure 47.12**). In 2005, clones of the African wildcat (*Felis libyca*) successfully produced

Figure 47.11 Captive breeding programs. The California condor (*Gymnogyps californicus*), the largest bird in the U.S., with a wingspan of nearly 3 m, has been bred in captivity in California. **(a)** A researcher at the San Diego Zoo Safari Park feeds a chick with a puppet so that the bird will not become habituated to the presence of humans. **(b)** This captive-bred condor soars over the Grand Canyon. Note the tag on the underside of its wing.
(a) ©Charlie Neuman/San Diego Union-Tribune/ZUMA Press Inc/Alamy; *(b)* ©Simpson/Photri Images/Alamy

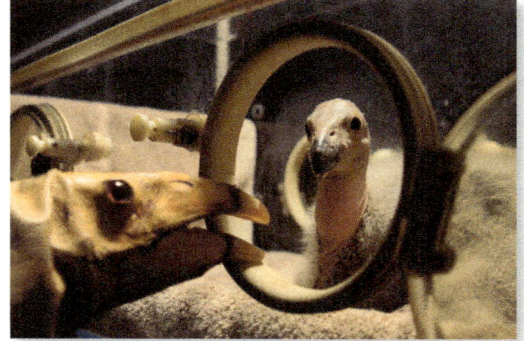
(a) A condor chick being fed using a puppet

(b) A released captive-bred condor

Biology Principle

The Genetic Material Provides a Blueprint That Allows Organisms to Grow, Develop, and Reproduce

Genetic cloning uses a somatic cell, which contains the complete DNA or genetic blueprint of the animal that is to be cloned.

Figure 47.12 Cloning an endangered species. In 2004, this 8-month-old cloned Javan banteng (*Bos javanicus*) made its public debut at the San Diego Zoo.
©Ken Bohn/UPI/Newscom

wildcat kittens. This was the first time that clones of a wild species had bred. In 2009, a cloned Pyrenean ibex (*Capra pyrenaica*) was born but lived only 7 minutes due to physical defects in the lungs. The last wild Pyrenean ibex had died in Spain in 2000. Finally, in 2017, a group of Chinese scientists produced two clones of long-tailed macaque monkeys, the first time a primate had been cloned. Other candidates for cloning include the Sumatran tiger (*Panthera tigris*) and the giant panda. Brazil plans to clone eight of its endangered species. Cloning extinct animals such as the woolly mammoth (*Mammuthus primigenius*) or Tasmanian tiger (*Thylacinus cynocephalus*) would be more difficult due to a lack of fully preserved DNA.

Despite the promise of cloning, a number of issues remain unresolved:

1. Scientists would have to develop an intimate knowledge of different species' reproductive cycles. For sheep and cows, this was routine, based on the vast experience in breeding these species, but eggs of different species, even if they could be harvested, often require different nutritive media in laboratory cultures.
2. Because it is desirable to use available natural mothers for the gestation of cloned animals, scientists will have to identify surrogate females of similar but more common species that can carry the fetus to term.
3. Some argue that cloning does not address the root causes of species loss, such as habitat fragmentation or poaching, and that resources would be better spent elsewhere—for example, in preserving the remaining the habitats of endangered species.
4. Cloning from a limited number of sources might not be able to increase the genetic variation in a species. However, if cells were obtained from many different deceased animals—for example, from their skin—these clones could theoretically reintroduce lost genes back into the population.

Many biologists believe that while cloning may have a role in conservation, it is only part of the solution and we need to address what made the species go extinct before attempting to restore it.

Conservation Efforts May Be Aimed at Sustainability

As we saw in Section 47.2, biodiversity is of great value to human welfare. Therefore, conservation is clearly a matter of great importance, and a failure to value and protect our natural resources adequately could be a grave mistake. Some authors, most recently the American ecologist and geographer Jared Diamond, have investigated why many societies of the past—including Angkor Wat, Easter Island, and the Mayans—collapsed or vanished, leaving behind monumental ruins. Diamond has concluded that the collapse of these societies occurred partly because people inadvertently destroyed the ecological resources on which their societies depended.

Modern nations such as Rwanda face similar issues. The country's population density is the highest in Africa, and it has a limited amount of land that can be used for growing crops. By the late 1980s, the need to feed a growing population had led to the wholesale clearing of Rwanda's forests and wetlands, with the result that little additional land was available to farm. Increased population pressure and food shortages fueled by environmental scarcity were likely contributing factors in igniting the genocide in that country in 1994.

With an increasing human population, a globally changing climate, and losses in forests and other habitats, biologists are faced with severe challenges in helping human societies to set aside land to conserve species. Part of the solution may lie in the concept of **sustainable development,** which is development that meets the needs of the present generation without compromising the ability of future generations to meet their needs. This is an ambitious goal that requires us to combine a knowledge of biology and the environment with social issues and economics.

We must correctly value **natural capital,** the world's stock of natural resources such as soil, water, and living organisms, and include its total value in the manufacture of goods. For example, the market price of timber affects private firms' decisions regarding the planting and harvesting of woodlands. However, the carbon, recreation, and air quality of benefits of woodlands may be much greater than market revenues for timber that accrue to woodland owners. We need to further develop renewable energy sources, such as solar, hydroelectric, and wind energy, and encourage more sustainable

Critical-Thinking Skills—Multiple Choice:
1. a 2. b 3. b 4. e 5. d

Critical-Thinking Skills—Short Answer:
1. This mutation would not affect transcription. RNA polymerase would still transcribe the gene. However, it would affect translation. A polypeptide would be made that was only five amino acids long.
2. The RNA components of spliceosomes and ribosomes perform structural and catalytic roles. In the case of spliceosomes, the RNA is thought to catalyze the splicing reactions. In ribosomes, an rRNA catalyzes the peptidyl transfer reaction. Proteins are needed in spliceosomes and ribosomes to perform a structural role. They hold the RNA in the correct configuration so that its catalytic function is achieved.

Chapter 11

Concept Checks
Figure 11.1 The binding of an ncRNA to DNA or another RNA could be inhibited by the formation of a stem-loop, because it would inhibit base pairing.

Figure 11.5 Telomerase uses the ncRNA called TERC as a template to make the DNA repeat sequence.

Figure 11.6 HOTAIR binds to the target gene because the RNA in HOTAIR is complementary to the target gene.

Figure 11.8 RISC binds to a specific mRNA because the miRNA is complementary to that mRNA. The miRNA and mRNA hydrogen bond to each other.

Figure 11.10 The crRNA directly recognizes the bacteriophage DNA.

BioConnections
Figure 11.4 DNA polymerase cannot copy the 3′ end of a DNA strand because it needs a primer to make DNA and it synthesizes DNA only in the 5′ to 3′ direction.

Figure 11.9 In eukaryotes, proteins that are destined for the ER, Golgi, vacuoles, lysosomes, plasma membrane, and secretion need SRP to reach their correct destination. In bacteria and archaea, proteins that are secreted need SRP.

Feature Investigation Questions
1. The *mex*-3 mRNA corresponds to the sense strand.
2. The researchers began with a copy of the *mex*-3 gene in a plasmid. When RNA polymerase and nucleotides were added, the sense strand of *mex*-3 mRNA was made. They also modified this plasmid by placing the promoter on the opposite side of the *mex*-3 coding sequence. When RNA polymerase and nucleotides were added, the antisense strand of *mex*-3 mRNA was made.
3. A mixture of *mex*-3 sense and antisense RNA, which presumably formed double-stranded RNA, was most effective at causing degradation of the *mex*-3 mRNA.

Testing Your Knowledge
11.1: 1. e 2. b
11.2: 1. e
11.3: 1. e
11.4: 1. e
11.5: 1. e
11.6: 1. d 2. c
11.7: 1. a

Assessing Your Knowledge and Skills

Foundational Knowledge—Multiple Choice:
1. e 2. c 3. d 4. a 5. d

Foundational Knowledge—Short Answer:
1. HOTAIR: scaffold, guide
   RNA of SRP: scaffold, alters protein function
   microRNA: guide
   crRNA: guide

2. The structure of HOTAIR provides a scaffold for the binding of two histone-modifying complexes. Its structure also is complementary to GA-rich sequences next to certain target genes, such as the *HoxD* gene, and thereby guides the histone-modifying complexes to those genes. Such genes are then covalently modified, which causes them to be repressed.

Critical-Thinking Skills—Multiple Choice:
1. b 2. a 3. e 4. d 5. c

Critical-Thinking Skills—Short Answer:
1. The rRNAs within the small and large ribosomal subunits act as scaffolds for the binding of many different ribosomal proteins. The 16S rRNA acts as a guide that binds to mRNA and causes the mRNA to bind to the ribosome. An ncRNA within the peptidyltransferase center acts as a ribozyme.
2. The ncRNAs that are associated with translation may have arisen during the RNA world. They are necessary for the synthesis of proteins, which arose after the RNA world. The ncRNAs that bind directly to DNA or bind directly to a specific type of protein may have arisen after the RNA world because their function would not have been needed prior to the existence of DNA or proteins within cells.

Chapter 12

Concept Checks
Figure 12.2 Gene regulation causes each cell type to express its own unique set of proteins, which, in turn, are largely responsible for the morphologies and functions of the cells.

Figure 12.6 The *lacZ*, *lacY*, and *lacA* genes are under the control of the *lac* promoter.

Figure 12.9 The lac repressor keeps the *lac* operon turned off unless lactose is present in the environment. The activator allows the bacterium to choose between glucose and lactose.

Figure 12.14 Some histone modifications may alter chromatin structure in a way that promotes transcription, whereas others may inhibit transcription.

Figure 12.17 The advantage of alternative splicing is that it allows a single gene to encode two or more polypeptides. This enables organisms to have smaller genomes, which are more efficient and easier to package into a cell.

Figure 12.18 When iron levels rise in the cell, the iron binds to IRP and removes it from the mRNA that encodes ferritin. This results in the rapid translation of ferritin protein, which can store excess iron. Unfortunately, ferritin storage does have limits, so iron poisoning can occur if too much is ingested.

BioConnections
Figure 12.8 In eukaryotic cells, cAMP acts as a second messenger in signal transduction pathways.

Figure 12.15 A nucleosome is composed of DNA wrapped around an octamer of histone proteins.

Testing Your Knowledge
12.1: 1. d 2. a
12.2: 1. b
12.3: 1. c 2. e
12.4: 1. d 2. b
12.5: 1. d 2. a

Assessing Your Knowledge and Skills

Foundational Knowledge—Multiple Choice:
1. b 2. c 3. c 4. d 5. d

Foundational Knowledge—Short Answer:
1. (a) regulatory protein, (b) small effector molecule, (c) segment of DNA, (d) small effector molecule, (e) regulatory protein

2. Gene regulation offers key advantages such as (1) proteins are made only when they are needed, (2) proteins are made in the correct cell type, and (3) proteins are made at the correct stage of development. These advantages are important for reproduction and sustaining life.

Critical-Thinking Skills—Multiple Choice:

1. c 2. b 3. c 4. e 5. a

Critical-Thinking Skills—Short Answer:

1. Garcinol inhibits the acetylation of histones. Histones that are not acetylated bind more tightly to DNA. This tighter binding inhibits transcriptional initiation and elongation, both of which require the removal of histones.
2. Perhaps the DNA methyltransferase is responsible for methylating and inhibiting a gene that causes a cell to become a muscle cell. The methyltransferase is inactivated by the mutation.

Chapter 13

Concept Checks

Figure 13.1 Only germ-line cells give rise to gametes (sperm or egg cells). A somatic cell cannot give rise to a gamete, and therefore a somatic mutation cannot be passed to offspring.

Figure 13.2 This is a mutation that occurred in a somatic cell, so it cannot be transmitted to an offspring.

Figure 13.5 A thymine dimer is harmful because it can cause errors in DNA replication.

Figure 13.6 First, we subtract the number of colonies formed in the absence of mutagen (2) from those formed in the presence of mutagen (44) and obtain a value of 42 for the number of mutations caused by the mutagen. If we divide 42 by 2 million, the rate is 2.1×10^{-5}.

Figure 13.7 UvrC and UvrD are responsible for removing the damaged DNA. UvrC makes cuts on both sides of the damage, and then UvrD removes the damaged region.

Figure 13.9 The growth factor turns on a signaling pathway that ultimately leads to cell division.

Figure 13.12 The type of cancer associated with this fusion is leukemia, which is a cancer of blood cells. The gene fusion produces a gene that is expressed in blood cells because it has the *bcr* promoter. The abnormal fusion protein promotes cancer in these cells.

Figure 13.13 Checkpoints prevent cell division if a genetic abnormality is detected. This helps to properly maintain the genome, thereby minimizing the possibility that a cell harboring a mutation will divide to produce two daughter cells.

BioConnections

Figure 13.9 Drugs that inhibit protein kinases may be used to combat cancer if they target the protein kinases that are overactive in certain forms of cancer.

Feature Investigation Questions

1. Some biologists believed that heritable traits could be altered by physiological events. This suggests that mutations could be stimulated by certain needs of the organism. Others believed that mutations are random. If a mutation had a beneficial effect that improved survival and/or reproductive success, the mutation would be maintained in the population through natural selection.
2. The Lederbergs were testing the hypothesis that mutations are random events. By subjecting the bacteria to a type of environmental stress, the bacteriophage, the researchers would be able to see if the stress induced mutations or if mutations occurred randomly.
3. When looking at the number and location of colonies that were resistant to viral infection, the pattern was consistent among the secondary plates. This indicates that the mutation that allowed the colonies to be resistant to viral infection occurred on the master plate. The secondary plates introduced the selective agent that allowed the resistant bacteria colonies to survive and reproduce while the other colonies were destroyed. Thus, mutations occurred randomly in the absence of any selective agent.

Testing Your Knowledge

13.1: 1. b 2. d
13.2: 1. b 2. d
13.3: 1. e
13.4: 1. d 2. d

Assessing Your Knowledge and Skills

Foundational Knowledge—Multiple Choice:

1. d 2. e 3. d 4. b 5. c

Foundational Knowledge—Short Answer:

1. Spontaneous mutations are due to natural biological processes, whereas induced mutations are due to environmental agents. Both can be harmful. Some induced mutations are avoidable.
2. Mutations may alter the expression of a gene and/or alter the function of a protein encoded by a gene. In many cases, such changes are harmful because a gene may not be expressed at the correct level, or the protein may not function as well as the normal (nonmutant) protein.

Critical-Thinking Skills—Multiple Choice:

1. b 2. e 3. d 4. b 5. a

Critical-Thinking Skills—Short Answer:

1. The mutagen is removing an amino group from cytosine, which converts it to uracil. Nitrous acid is an example of a mutagen that can deaminate a base; it can change a cytosine to uracil.
2. The role of *p53* is to sense DNA damage and prevent damaged cells from proliferating. Perhaps, prior to birth, the fetus is in a protected environment, so that DNA damage may be minimal. In other words, the fetus may not really need *p53*. After birth, agents such as UV light may cause DNA damage. At this point, *p53* is important. A *p53* knockout is more sensitive to UV light because it cannot repair its DNA properly in response to this DNA-damaging agent, and it cannot kill cells that have become irreversibly damaged.

Chapter 14

Concept Checks

Figure 14.1 Chromosomes are readily seen when they are compacted in a dividing cell. By adding such a drug, you increase the percentage of cells that are actively dividing.

Figure 14.2 Interphase consists of the G_1, S, and G_2 phases of the cell cycle.

Figure 14.4 Cancer would not occur if both copies of the *Rb* gene and both copies of the *E2F* gene were rendered inactive due to mutations. An active copy of the *E2F* gene is needed to promote cell division.

Figure 14.5 As shown in the inset, each object is a pair of sister chromatids.

Figure 14.6 The astral microtubules, which extend away from the chromosomes, are important for positioning the spindle apparatus within the cell. The polar microtubules project into the region between the two poles. Polar microtubules that overlap with each other play a role in the separation of the two poles. Kinetochore microtubules are attached to kinetochores at the centromeres and are needed to sort the chromosomes.

Figure 14.8 Cytokinesis in both animal and plant cells separates a mother cell into two daughter cells. In animal cells, cytokinesis involves the formation of a cleavage furrow, which constricts like a drawstring to separate the cells. In plants, the two daughter cells are separated by the formation of a cell plate, which forms a cell wall between the two daughter cells.

Figure 14.11 The original cell is diploid with two sets of chromosomes, whereas the four resulting cells are haploid with one set of chromosomes.

Figure 14.12 The purpose of meiosis in animals is to produce gametes. These gametes combine during fertilization to produce a diploid organism. Following fertilization, the purpose of mitosis is to produce a multicellular organism.

Figure 14.14 Inversions and the translocations shown here do not affect the total amount of genetic material.

BioConnections

Figure 14.3 Checkpoints prevent cancer by checking the integrity of the genome. If abnormalities in DNA structure are detected or if a chromosome is not properly attached to the spindle, the checkpoint will delay cell division until the problem is fixed. If it cannot be fixed, the checkpoint will initiate the process of apoptosis, thereby killing a cell that may harbor mutations. This prevents the proliferation of cells that have the potential to be cancerous.

Testing Your Knowledge

14.1: 1. d 2. b 3. d
14.2: 1. c 2. b 3. d
14.3: 1. a 2. a 3. c
14.4: 1. d 2. e 3. c

Assessing Your Knowledge and Skills

Foundational Knowledge—Multiple Choice:

1. b 2. b 3. c 4. d 5. c

Foundational Knowledge—Short Answer:

1. In diploid species, chromosomes are present in pairs, one from each parent, and contain similar gene arrangements. Such chromosomes are homologous. When DNA is replicated, two identical copies are created and are sister chromatids.

2. Mitosis is a process that produces two daughter cells with the same genetic material as the original cell. In the case of plants and animals, this allows a fertilized egg to develop into a multicellular organism composed of many genetically identical cells.

Critical-Thinking Skills—Multiple Choice:

1. c. Note that the question says that you counted the number of chromatids. If it had said sister chromatids, you would have counted 24 of them, which equals the number of chromosomes in the original cell prior to cell division. 2. c. In a karyotype, you are looking at condensed chromosomes that have already replicated. Each of the two homologs is composed of a pair of sister chromatids that has two copies of the gene. 3. c. If the cell had been in mitosis, 14 sister chromatids would have been aligned in a single row. If it had been in meiosis I, 14 sister chromatids would have been aligned in a double row. 4. e. The chromosome with the deletion would not be able to be attached to the mitotic spindle. 5. b. The normal gamete contained 23 chromosomes.

Critical-Thinking Skills—Short Answer:

1. One reason is that the chromosomes might get tangled up with each other at metaphase and anaphase, which would prevent their proper segregation. A second possible reason is that compact chromosomes are less likely to become damaged when they are moving during mitosis.

2. This abnormal crossover would produce a reciprocal translocation.

Chapter 15

Concept Checks

Figure 15.2 Having blue eyes is a variant (also called a trait). A character is a more general term, which in this case would refer to eye color.

Figure 15.4 In this procedure, stamens are removed from the purple flower to prevent self-fertilization.

Figure 15.5 The reason offspring of the F_1 generation exhibit only one variant of each character is because one trait is dominant over the other.

Figure 15.6 The ratio of the alleles (T to t) is 1:1. The reason the phenotypic ratio is 3:1 is that T is dominant to t.

Figure 15.7 If the linked hypothesis had been correct, the ratio would have been 3 round, yellow to 1 wrinkled, green.

Figure 15.11 No. If two parents are affected with the disease, they must be homozygous for the mutant allele if it's recessive. Two homozygous parents would have to produce all affected offspring. If they don't, then the inheritance pattern is not recessive.

Figure 15.12 In a dominant pattern of inheritance, all affected offspring have at least one affected parent.

Figure 15.15 No. You need a genetically homogenous population to study the norm of reaction. A wild population of squirrels is not genetically homogenous, so it could not be used.

Figure 15.16 The person would be a female. The presence of the Y chromosome determines maleness in humans.

BioConnections

Figure 15.8 Sexual reproduction is the process in which two haploid gametes (for example, sperm and egg) combine with each other to begin the life of a new individual. Each gamete contributes one set of chromosomes. The resulting zygote has chromosomes that occur in pairs (one from each parent). The members of each pair are called homologs of each other; they carry the same types of genes.

Figure 15.9 Alleles segregate from each other during the process of meiosis. Meiosis begins with a diploid cell that has pairs of genes, which may be found in different alleles. During meiosis, these pairs of genes separate and end up in different haploid cells. Therefore, each haploid cell has only one copy of each gene. In other words, each haploid cell has only one allele of a given gene.

Testing Your Knowledge

15.1: 1. c 2. d
15.2: 1. b 2. b
15.3: 1. b
15.4: 1. b 2. d
15.5: 1. b 2. d

Assessing Your Knowledge and Skills

Foundational Knowledge—Multiple Choice:

1. e 2. c 3. a 4. d 5. d

Foundational Knowledge—Short Answer:

1. The genotype is the genes (alleles) that an individual carries, whereas the phenotype is the characteristics that an individual displays. It is possible for two individuals to have the same phenotype and different genotypes. For example, TT and Tt pea plants have different genotypes, but they have the same phenotype (tall).

2. The environment is needed so that genes can be expressed. For example, organic molecules and energy are needed for transcription and translation. In addition, environmental factors influence the outcome of traits. For example, sunlight can cause a tanning response, thereby affecting the darkness of the skin.

Critical-Thinking Skills—Multiple Choice:

1. b 2. d 3. c 4. c 5. d

Critical-Thinking Skills—Short Answer:

1. Recessive alleles are often loss-of-function alleles. It would generally be more likely for a recessive allele to eliminate a trait or function rather than create one. Therefore, the peach carries the dominant allele, whereas the nectarine has a loss-of-function allele that prevents fuzz formation. The recessive allele is in a gene that is necessary for fuzz formation.

2. Though the patterns have some similarities, there are certain features to distinguish them. If a female showing the recessive trait was crossed to a male showing the dominant trait, all of the male offspring would show the recessive trait if it was X-linked. If it was not X-linked, half or all of the male offspring would show the dominant trait if the father was heterozygous or homozygous, respectively. If the father showed the dominant trait and the mother was heterozygous, half of the male offspring would show the recessive trait if it was X-linked, but none of the female offspring would.

Chapter 16

Concept Checks

Figure 16.1 Only the allele inherited from the father would be expressed in the offspring, Because he is heterozygous, half of the offspring would be normal size and half would be dwarf.

Figure 16.3 The Barr body is much more compact than the other X chromosome in the cell. This compaction prevents most of the genes on the Barr body from being expressed.

Figure 16.7 The gene is located in the chloroplast DNA. In this species, chloroplasts are transmitted from parent to offspring via eggs but not via sperm.

Figure 16.10 Crossing over occurred during oogenesis in the heterozygous female of the F_1 generation to produce the recombinant offspring of the F_2 generation.

BioConnections

Figure 16.6 The evolutionary origin of these organelles is an ancient endosymbiotic relationship. Mitochondria are derived from purple bacteria, and chloroplasts are derived from cyanobacteria.

Figure 16.10 Crossing over occurs during prophase of meiosis I.

Feature Investigation Questions

1. Bateson and Punnett were testing the hypothesis that the gene pairs that influence flower color and pollen shape would assort independently of each other. The two traits were expected to show a pattern consistent with Mendel's law of independent assortment.
2. The expected results were a phenotypic ratio of 9:3:3:1. The researchers expected 9/16 of the offspring would have purple flowers and long pollen, 3/16 of the offspring would have purple flowers and round pollen, 3/16 of the offspring would have red flowers and long pollen, and 1/16 of the offspring would have red flowers and round pollen.
3. To conduct the test, you hypothesize that the genes independently assort, because this allows you to calculate the expected values as shown in the data of Figure 16.9. The calculated chi square value is very high, and you would reject the hypothesis that the genes independently assort. As an alternative, you would accept the hypothesis that the genes are linked.

Testing Your Knowledge

16.1: 1. e
16.2: 1. a 2. c
16.3: 1. a 2. c
16.4: 1. a 2. d
16.5: 1. a 2. d
16.6: 1. c 2. b

Assessing Your Knowledge and Skills

Foundational Knowledge—Multiple Choice:

1. e 2. d 3. a 4. d 5. c

Foundational Knowledge—Short Answer:

1. Epigenetics is the study of mechanisms that lead to changes in gene expression that can be passed from cell to cell and are reversible but do not involve a change in the sequence of DNA. Not all epigenetic changes are passed from parent to offspring. For example, a cigarette smoker could acquire an epigenetic change in a lung cell; this change would not be passed to offspring.
2. Epigenetics alters the way that genes are expressed, and thereby can affect an individual's traits. Such traits, in turn, may affect reproduction.

Critical-Thinking Skills—Multiple Choice:

1. c 2. a 3. e 4. e 5. e

Critical-Thinking Skills—Short Answer:

1. The agent in cigarette smoke may cause methylation near the promoter of the *p53* gene, thereby inhibiting transcription. Another possibility is that it may affect covalent histone modifications or chromatin remodeling so that a closed conformation results and the gene is not transcribed.
2. The feeding of 5-azocytidine would increase the frequency of mice with yellow and slightly mottled coats due to the overproduction of yellow pigment. On the other hand, the frequency of pseudo-agouti mice (as well as those with heavily mottled coats) would decrease. This is expected to occur because the methylation of the promoter would be prevented in the offspring, leading to the overexpression of the *Agouti* gene.

Chapter 17

Concept Checks

Figure 17.4 There appear to be two nucleoids in this bacterial cell.

Figure 17.5 The loop domains are held in place by proteins that bind to the DNA at the bases of the loops. The proteins also bind to each other.

Figure 17.6 Bacterial chromosomes and plasmids are similar in that they typically contain circular DNA molecules. However, bacterial chromosomes are usually much longer than plasmids and carry many more genes. Also, bacterial chromosomes tend to be more compacted due to the formation of loop domains and supercoiling.

Figure 17.7 16 hours is the same as 32 doublings. So, $2^{32} = 4,294,967,296$. (The actual number would be much less because the cells would deplete the growth media and grow more slowly than the maximal rate.)

Figure 17.9 During conjugation, only one strand of the DNA from an F factor is transferred from the donor to the recipient cell. The single-stranded DNA in both cells is then used as a template to create double-stranded F factor DNA in both cells.

Figure 17.11 Transduction is not a normal part of the phage life cycle. It is a mistake in which a piece of the bacterial chromosome is packaged into a phage coat and is then transferred to another bacterial cell.

BioConnections

Figure 17.2 Viral release occurs as a budding process in which a membrane vesicle is formed that surrounds the capsid. Similarly, exocytosis involves the formation of a membrane vesicle that encloses some type of cargo.

Figure 17.4 A nucleoid is not a membrane-bound organelle. It is simply a region where a bacterial chromosome is found. A cell nucleus in a eukaryotic cell has an envelope with a double membrane.

Figure 17.10 Griffiths was able to show that genetic material was transferred to type R bacteria, which converted them to type S. This occurred via transformation. Later, Avery, MacLeod, and McCarty determined that DNA was the material that was being transferred.

Testing Your Knowledge

17.1: 1. d
17.2: 1. a
17.3: 1. c 2. d 3. e
17.4: 1. b 2. a

Assessing Your Knowledge and Skills

Foundational Knowledge—Multiple Choice:

1. e 2. b 3. b 4. d 5. c

Foundational Knowledge—Short Answer:

1. Conjugation: The process involves a direct physical contact between two bacterial cells in which a donor cell transfers a strand of DNA to a recipient cell. Transformation: This occurs when a living bacterium takes up genetic information that has been released from a dead bacterium. Transduction: When a virus infects a donor cell, it incorporates a fragment of bacterial

chromosomal DNA into a newly made virus particle. The virus then transfers this fragment of DNA to a recipient cell.

2. Horizontal gene transfer is the transfer of genes from one organism to another organism that is not its offspring. These acquired genes sometimes increase survival and therefore may have an evolutionary advantage. Such genes may even promote the formation of new species. From a medical perspective, an important example of horizontal gene transfer is when one bacterium acquires an antibiotic-resistance gene from another bacterium and then itself becomes resistant to that antibiotic. This phenomenon is making it increasingly difficult to treat a wide variety of bacterial diseases.

Critical-Thinking Skills—Multiple Choice:

1. c 2. c 3. a 4. a. When conjugation occurs, an F⁻ cell becomes F⁺ and the F⁺ cell remains F⁺. Therefore, the F⁺ cells increase over time. 5. e. Horizontal gene transfer can occur between members of the same bacterial species and between different species via conjugation, transduction, and transformation.

Critical-Thinking Skills—Short Answer:

1. Drugs that affect spike glycoproteins can interfere with the ability of a virus to bind to its host cell. This prevents the virus from entering the host cell.
2. It is not a form of sexual reproduction, in which two distinct parents produce gametes that unite to form a new individual. However, conjugation is similar to sexual reproduction in the sense that the genetic material from two cells is somewhat mixed. In conjugation, there is not the mixing of two genomes, one from each gamete. Instead, there is a transfer of genetic material from one cell to another. This transfer can alter the combination of genetic traits in the recipient cell.

Chapter 18

Concept Checks

Figure 18.2 No. A recombinant vector has been made, but it has not been cloned. In other words, copies of the recombinant vector have not been made yet.

Figure 18.3 The insertion of chromosomal DNA into the vector disrupts the *lacZ* gene, thereby preventing the expression of β-galactosidase. The functionality of *lacZ* can be determined by providing the growth medium with a colorless compound, X-Gal, which is cleaved by β-galactosidase into a blue dye. Bacterial colonies containing recircularized vectors form blue colonies, whereas colonies containing recombinant vectors carrying a segment of chromosomal DNA will be white.

Figure 18.5 The 600-bp piece would be closer to the bottom. Smaller pieces travel faster through the gel.

Figure 18.6 The primers are complementary to sequences at each end of the DNA region to be amplified.

Figure 18.8 If a dideoxynucleotide ddNTP is added to a growing DNA strand, the strand can no longer grow because the 3′ —OH group, the site of attachment for the next nucleotide, is missing.

Figure 18.9 A fluorescent spot identifies a cDNA that is complementary to a particular DNA sequence. Because the cDNA was generated from mRNA, this technique identifies a gene that has been transcribed in a particular cell type under a given set of conditions.

Figure 18.10 The sgRNA is composed of two different components of the bacterial defense system, crRNA and tracrRNA, which have been linked together.

Figure 18.11 One reason is that more complex species tend to have more genes. A second reason is that species vary with regard to the amount of repetitive DNA present in their genome.

Figure 18.17 Retrotransposons. A single retrotransposon can be transcribed into multiple copies of RNA, which can be converted to DNA by reverse transcriptase, and inserted into multiple sites in the genome.

BioConnections

Figure 18.2 Plasmids are small, circular molecules of DNA that exist independently of the bacterial chromosome. They have their own origin of replication. Many plasmids carry genes that convey some type of selective advantage to the host cell, such as antibiotic resistance.

Figure 18.6 Primers are needed in a PCR experiment because DNA polymerase cannot begin DNA replication on a bare template strand.

Figure 18.13 The family members that are expressed at early stages of development (embryonic and fetal stages) have a higher affinity for oxygen than the adult form. This allows the embryo and fetus to obtain oxygen from its mother's bloodstream.

Testing Your Knowledge

18.1: 1. d 2. b 3. b
18.2: 1. e 2. c
18.3: 1. e
18.4: 1. d 2. d
18.5: 1. d 2. c

Assessing Your Knowledge and Skills

Foundational Knowledge—Multiple Choice:

1. b 2. b 3. e 4. a 5. c

Foundational Knowledge—Short Answer:

1. a. yes
   b. no, it's only one chromosome in the nuclear genome
   c. yes
   d. yes
2. The genome contains the information for the production of cellular proteins and non-coding RNAs; it is a blueprint. The production of proteins is largely responsible for determining cellular characteristics, which, in turn, determine an organism's traits and allow it to reproduce.

Critical-Thinking Skills—Multiple Choice:

1. c. Remember that GC base pairs form three hydrogen bonds, whereas AT base pairs form only two. The sticky end produced by *Hae*II is composed of two Gs and two Cs, whereas the other restriction enzymes have sticky ends with some As and Ts. 2. b. This procedure provides the most definitive evidence that the vector has a segment of DNA inserted into it. 3. c. The primers used in PCR need to be complementary to ends of the gene of interest. If the gene of interest has never been sequenced, you have no way of knowing what the sequence of the primers should be. 4. b. Nucleotides are added to the 3′ end of the primer, which is 20 nucleotides long. T is complementary to A, so ddTs will be incorporated opposite to As in the sequence that is to the right of the primer-annealing site. For example, if a ddT is incorporated at the first A, the DNA segment will be 21 nucleotides long. 5. d. A DNA microarray is used to analyze gene expression at the whole genome level.

Critical-Thinking Skills—Short Answer:

1. See Figure 18.7. A ddNTP is missing an oxygen at the 3′ position. This prevents the further growth of a DNA strand, thereby causing chain termination.
2. You would conclude that the glutamic acid is probably important. The only amino acid substitution that gave a substantial amount of functional activity was an aspartic acid. Glutamic acid and aspartic acid have very similar amino acid side chains (see Chapter 3); they both contain a carboxyl (COOH) group. Based on these results, you would suspect that a carboxyl group at this location in the protein might be important for its function.

Chapter 19

Concept Checks

Figure 19.2 A single organism does not evolve. Populations may evolve from one generation to the next.

Figure 19.5 Due to a changing global climate, the island fox became isolated from the mainland species. Over time, natural selection resulted in adaptations in the population on the island and eventually resulted in a new species with characteristics that are somewhat different from the mainland species.

Figure 19.6 Many answers are possible. One example is the wing of a bird and the wing of a bat.

Figure 19.9 Rhesus and green monkeys = 0, Congo puffer fish and European flounder = 2, and Rhesus monkey and Congo puffer fish = 10. Pairs that are closely related evolutionarily have fewer differences than do pairs that are more distantly related.

Figure 19.11 If C^R is 0.4, then C^W must be 0.6, because the allele frequencies add up to 1.0. The heterozygote frequency ($2pq$) equals $2(0.4)(0.6)$, which equals 0.48, or 48%.

Figure 19.12 Over the short run, alleles that confer better fitness would be favored and increase in frequency, perhaps enhancing diversity. Over the long run, however, an allele that confers high fitness in the homozygous state may become monomorphic, thereby reducing genetic diversity.

Figure 19.13 Stabilizing selection eliminates alleles that give phenotypes that deviate significantly from the average phenotype. For this reason, it tends to decrease genetic diversity.

Figure 19.15 If malaria was eradicated, there would be no selective advantage for the heterozygote. The H^S allele would eventually be eliminated because the H^SH^S homozygote has a lower fitness. Directional selection would occur.

Figure 19.18 This trait is likely to be involved in intersexual selection. Such traits often play a role in mate choice.

Figure 19.20 The bottleneck effect decreases genetic diversity. This may eliminate adaptations that promote survival and reproductive success. Therefore, the bottleneck effect makes it more difficult for a population to survive.

Figure 19.22 Migration promotes gene flow, which tends to make the allele frequencies in neighboring populations more similar to each other. It also promotes genetic diversity by introducing new alleles into populations.

BioConnections

Figure 19.2 Both natural selection and chemical selection involve processes in which the relative proportions of something in a population increase compared with something else. In natural selection, it is the relative proportions of individuals with certain traits that increase. In chemical selection, molecules with certain characteristics increase their relative numbers compared with other molecules.

Figure 19.21 There are lots of possibilities. The idea is that you are changing one codon to another codon that specifies the same amino acid. For example, changing a codon from GGA to GGG is likely to be neutral because both codons specify glycine.

Figure 19.23 Inbreeding tends to produce homozygotes. Initially, inbreeding would result in more homozygotes in a population. Over the long run, however, if a homozygote has a lower fitness, inbreeding would accelerate the elimination of the allele from the population.

Feature Investigation Questions

1. The island has a moderate level of isolation but is located near enough to the mainland to have some migrants. The island is an undisturbed habitat, so the researchers would not have to consider the effects of human activity on the study. Finally, the island had an existing population of ground finches that would serve as the study organism over many generations.

2. First, the researchers were able to show that beak depth is a genetic trait that has variation in the population. Second, the depth of the beak is an indicator of the types of seeds the birds can eat. The birds with larger beaks can eat larger and drier seeds; therefore, changes in the types of seeds available could act as a selective force on the bird population. During the study period, annual changes in rainfall occurred, which affected the seed sizes produced by the plants on the island. In the drier year, fewer small seeds were produced, so the birds would have to eat larger, drier seeds.

3. The researchers found that following the drought in 1978, the average beak depth in the finch population increased. This indicated that birds with larger beaks were better able to adapt to the environmental changes due to the drought and produce more offspring. This is direct evidence of the phenomenon of natural selection.

Testing Your Knowledge

19.1: 1. e 2. e
19.2: 1. e 2. b
19.3: 1. e 2. e
19.4: 1. c 2. a 3. a
19.5: 1. b 2. b
19.6: 1. e 2. a

Assessing Your Knowledge and Skills

Foundational Knowledge—Multiple Choice:

1. d 2. b 3. d 4. d 5. b

Foundational Knowledge—Short Answer:

1. In convergent evolution, two different species from different lineages show similar characteristics because they occupy similar environments. An example is the long snout and tongue of both the giant anteater, found in South America, and the echidna, found in Australia. These structures enable the animals to feed on ants, but the structures evolved independently. These observations support the idea that evolution results in adaptations to particular environments.

2. Homologous structures are two or more structures that are similar because they are derived from a common ancestor. An example is the same set of bones found in the human arm, turtle arm, bat wing, and whale flipper. The forearms in these species have been modified to perform different functions. This supports the idea that all of these animals evolved from a common ancestor by descent with modification.

Critical-Thinking Skills—Multiple Choice:

1. d. This is an example of convergent evolution, not homology. It is not due to descent from a common ancestor. 2. c 3. d. This is an example of heterozygote advantage. 4. c 5. c

Critical-Thinking Skills—Short Answer:

1. The frequency of the disease is a genotype frequency because it represents individuals with the disease. If you let q^2 represent the genotype frequency, then q equals the square root of 0.04, which is 0.2. If $q = 0.2$, then $p = 1 - q$, which is 0.8. The frequency of heterozygous carriers is $2pq$, which is $2(0.8)(0.2) = 0.32$, or 32%.

2. Evolution is a heritable change in one or more characteristics in a population or species over time. Evolution means change. Evolution can occur on a small scale in the case of a single gene or on a larger scale in the case of the formation of a new species. By comparison, natural selection is one process that causes evolution to happen. Natural selection occurs because some individuals (with a higher fitness) are more likely to survive and reproduce, and others are not. Over many generations, natural selection leads to adaptation.

Chapter 20

Concept Checks

Figure 20.1 There are a lot of possibilities. Certain grass species look quite similar. Elephant species look very similar. And so on.

Figure 20.3 Temporal isolation is an example of a prezygotic isolating mechanism. Because the species breed at different times of the year, hybrid zygotes are not produced between the two species.

Figure 20.5 Hybrid sterility is a type of postzygotic isolating mechanism. A hybrid between the two species is produced, but it is sterile.

Figure 20.10 The insects on a given type of host plant would tend to breed with each other, and natural selection would favor the development of traits that are an advantage for feeding on that host. Over time, the accumulation of genetic changes may lead to reproductive isolation between the populations of insects that have different host preferences.

Figure 20.11 If the *Gremlin* gene was underexpressed, less gremlin protein would be produced. Because gremlin protein inhibits apoptosis, more cell death would occur, and the result would probably be smaller feet, and maybe they would not be webbed.

Figure 20.12 By comparing the number of *Hox* genes in many different animal species, a general trend is observed that animals with more complex body structures have a greater number of *Hox* genes.

BioConnections

Figure 20.2 Female choice is a prezygotic isolating mechanism.

Figure 20.7 The Hawaiian Islands have many different ecological niches that can be occupied by birds. The first founding bird inhabitants evolved to occupy those niches, thereby evolving into many different species.

Feature Investigation Questions

1. Podos hypothesized that the morphological changes in the beak would also affect the birds' songs. A bird's song is an important factor in mate choice. If changes in the beak alter the song of the bird, reproductive ability would be affected. Podos suggested that changes in the beak morphology could thus lead to reproductive isolation among the birds.
2. Podos first caught male birds in the field and collected data on beak size. The birds were banded for identification and released. Later, the banded birds' songs were recorded and analyzed for range of frequencies and trill rates. The results were then compared with similar data from other species of birds to determine if beak size constrained the frequency range and trill rate of the song.
3. The results of the study indicated that natural selection on beak size due to changes in diet can lead to changes in song. Considering the importance of bird song to mate choice, the changes in the song can also lead to reproductive isolation. A by-product of adaptation refers to a change in the phenotype that is not directly acted on by natural selection. In the case of the Galápagos finches, the changes in beak size were directly related to diet; however, as a consequence of that selection, the song pattern was also altered. The change in song pattern was a by-product.

Testing Your Knowledge

20.1: 1. d 2. a
20.2: 1. b 2. b
20.3: 1. a 2. e
20.4: 1. d

Assessing Your Knowledge and Skills

Foundational Knowledge—Multiple Choice:

1. b 2. c 3. a 4. b 5. c

Foundational Knowledge—Short Answer:

1. Prezygotic isolating mechanisms prevent the formation of the zygote. An example is mechanical isolation, the incompatibility of genitalia. Postzygotic isolating mechanisms act after the formation of the zygote. An example is inviability of the hybrid that is formed. (Other examples shown in Figure 20.2 would also be correct.) Postzygotic mechanisms are more costly because some energy is spent in the formation of a zygote and its subsequent growth.
2. One example involves the *Hox* genes, which control morphological features along the anteroposterior axis in animals. An increase in the number of *Hox* genes during evolution is associated with an increase in body complexity and may have spawned many different animal species.

Critical-Thinking Skills—Multiple Choice:

1. c 2. e 3. b. It is temporal because the flowers in the two species do not open at the same time. 4. d 5. b

Critical-Thinking Skills—Short Answer:

1. Allopatric speciation and sympatric speciation are distinguished by the level of geographic isolation. In allopatric species, two groups of the same species become very isolated from each other and then develop into separate species. This typically involves the accumulation of many small genetic changes that leads to adaptations and reproductive isolation. In sympatric speciation, there may not be a physical separation. For example, polyploidy can cause reproductive isolation and result in new species. Also, species may become adapted to local environments and develop into new species.
2. a. Allopatric speciation may occur because the offspring are geographically isolated from the original population.
   b. Sympatric speciation may occur if polyploid plants are formed that cannot successfully interbreed with members of the original population.
   c. Allopatric speciation may occur due to geographic isolation, though some limited inbreeding may occur in hybrid zones where the narrow streams are found.

Chapter 21

Concept Checks

Figure 21.2 A phylum is broader than a family.

Figure 21.3 Yes. They can have many common ancestors, depending on how far back you go in the tree. For example, dogs and cats have a common ancestor that gave rise to mammals, and an older common ancestor that gave rise to vertebrates. The most recent common ancestor is the point at which two species diverged from each other.

Figure 21.4 An order is a smaller taxon that would have a more recent common ancestor.

Figure 21.7 A hinged jaw is the character common to the salmon, lizard, and rabbit, but not to the lamprey.

Figure 21.11 Gorillas and humans would be expected to have fewer genetic differences because their common ancestor (labeled C in the figure) is more recent than that of orangutans and gorillas, which is ancestor B.

Figure 21.12 Monophyletic groups are based on the concept that a particular group of species descended from a common ancestor. When horizontal gene transfer occurs, not all of the genes in a species were inherited from the common ancestor, so this muddles the concept of monophyletic groups.

BioConnections

Figure 21.1 The domains Bacteria and Archaea have organisms with prokaryotic cells.

Figure 21.10 There are lots of possibilities. The idea is that you are changing one codon to another codon that specifies the same amino acid. For example, changing a codon from GGA to GGG is likely to be neutral because both codons specify glycine.

Testing Your Knowledge

21.1: 1. c 2. b
21.2: 1. d 2. a 3. b
21.3: 1. b 2. c
21.4: 1. a 2. d
21.5: 1. c

Assessing Your Knowledge and Skills

Foundational Knowledge—Multiple Choice:

1. d 2. d 3. b 4. b 5. b

Foundational Knowledge—Short Answer:

1. If neutral mutations occur at a relatively constant rate, they act as a molecular clock on which to measure evolutionary time. Genetic diversity between species that is due to neutral mutations gives an estimate of the time elapsed since the last common ancestor. A molecular clock can provide a timescale to a phylogenetic tree.
2. Morphological analysis focuses on morphological features of extinct and modern species. Many traits are analyzed to obtain a comprehensive picture of two species' relatedness. Convergent evolution leads to similar traits that arise independently in different species as they adapt to similar environments. Convergent evolution can, therefore, cause errors in a phylogenetic

tree if a researcher assumes that a particular trait arose only once and that all species having the trait have the same common ancestor.

Critical-Thinking Skills—Multiple Choice:

1. c. The branch point that gave rise to these two species is the most recent.
2. e 3. e 4. b 5. b

Critical-Thinking Skills—Short Answer:

1. An example of a phylogenetic tree is shown below. Note: More than one correct tree is possible. The nodes can be rotated 180° and still give a correct answer. For example, the European flounder and Congo puffer fish could be swapped so that the Congo puffer fish is second from the left and the European flounder is third from the left.

2. Taxonomy is the field of biology that is concerned with the theory, practice, and rules of classifying living and extinct organisms and viruses. In taxonomy, scientists place organisms into distinct groups based on similarities and differences. This allows scientists to more easily appreciate the similarities and differences between groups of species. With regard to applications, many are possible. In the field of conservation, knowing that a group is a unique species may intensify conservation efforts if its population becomes small. In agriculture, knowing that two species are or are not closely related may influence whether or not a breeder will attempt to produce interspecies hybrids.

Figure 22.22   In areas where dairy products were consumed by children and adults, there would be a selective advantage for lactose tolerance. An example would be most of Europe. In areas where dairy products were not consumed, lactose intolerance would provide a selective advantage because it would be more energy efficient to turn off the lactase gene after weaning.

BioConnections

Figure 22.7   First, the process of membrane invagination created the nuclear envelope. Second, endocytosis may have enabled an ancient archaeon to take up a bacterial cell. Over time, bacterial genes were transferred to the nucleus, which gave rise to the eukaryotic nuclear genome. An engulfed bacterial cell eventually became a mitochondrion, and an engulfed cyanobacterial cell became a chloroplast in algae and plants.

Figure 22.12   Two key features are mammary glands and hair. They also have specialized teeth, external ears, and enlarged skulls that harbor highly developed brains. Mammals are typically endothermic.

Testing Your Knowledge

22.1: 1. b 2. e
22.2: 1. c 2. a 3. b
22.3: 1. e 2. a

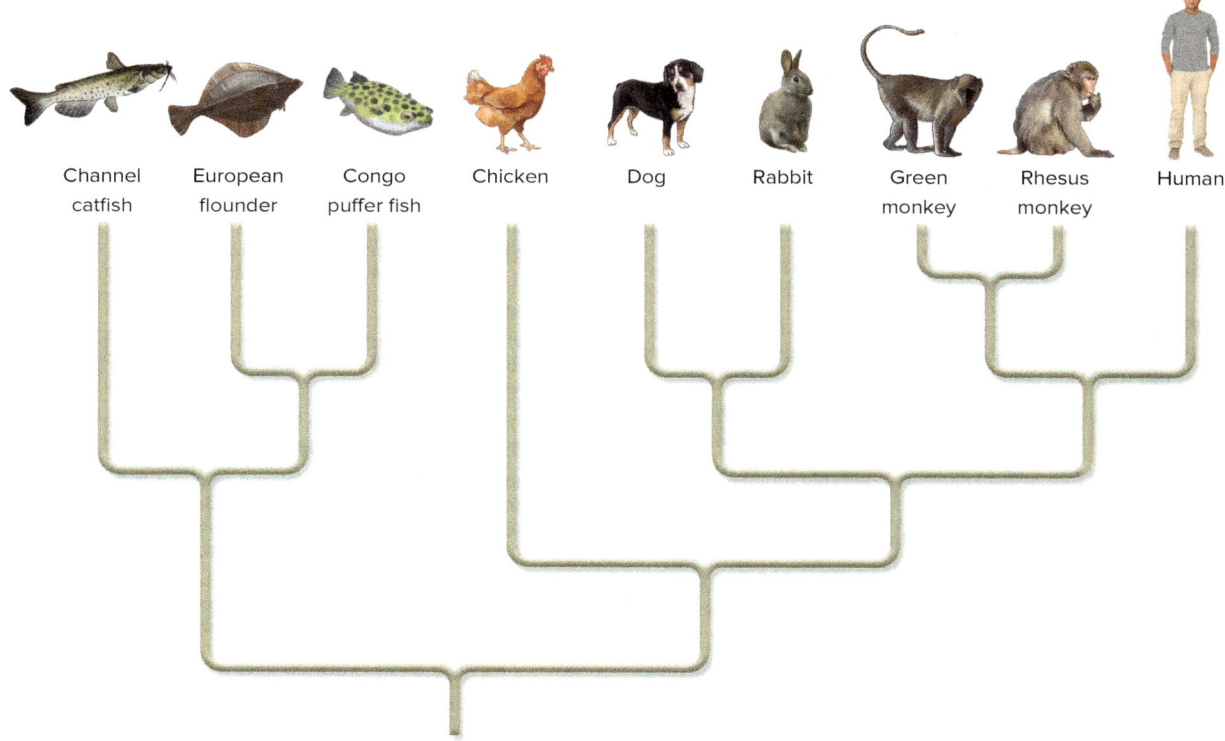

Chapter 22

Concept Checks

Figure 22.2   In a sedimentary rock formation, the layer at the bottom is usually the oldest.

Figure 22.9   Most animal species, including fruit flies, fishes, and humans, exhibit bilateral symmetry.

Figure 22.15   Defining features of primates are grasping hands; eyes situated on the front of the head to facilitate binocular vision; a large brain; digits with flat nails instead of claws; and complex social behavior including well-developed parental care.

Figure 22.21   Humans originated in Africa. To get to South America, they had to migrate to the Middle East, then Asia, then North America, and finally to South America.

Assessing Your Knowledge and Skills

Foundational Knowledge—Multiple Choice:

1. c 2. e 3. b 4. b 5. d

Foundational Knowledge—Short Answer:

1. The relative ages of fossils can be determined by their locations in sedimentary rock formation. Older fossils are found in lower layers. A common way to determine the ages of fossils is via radiometric dating, which is often conducted using a piece of igneous rock from the vicinity of the fossil. A radioisotope is an unstable isotope of an element that decays spontaneously, releasing radiation at a constant rate. The half-life is the length of time required for a radioisotope to decay to exactly one-half of its initial value. To determine the age of a rock (and that of a nearby fossil), scientists can measure the amount of a given radioisotope as well as the amount of the decay product.

Figure 26.8 Echinodermata and chordate are the two phyla that have endoskeletons.

Figure 26.19 No, technically it pumps hemolymph through vessels into sinuses.

Figure 26.29 These organs, called statocysts, are located at the base of the antennules.

Figure 26.33 The shared derived character is the notochord.

Testing Your Knowledge

26.1: 1. b
26.2: 1. a 2. c 3. c
26.3: 1. d
26.4: 1. d 2. b
26.5: 1. c
26.6: 1. c 2. c 3. d
26.7: 1. b 2. d 3. a
26.8: 1. a 2. e 3. c

Assessing Your Knowledge and Skills

Foundational Knowledge—Multiple Choice:

1. e 2. d 3. d 4. d 5. c

Foundational Knowledge—Short Answer:

1. Gametes would dry out on land, and internal fertilization prevents this from happening. Also, water facilitates the movement of gametes, reducing the need for internal fertilization.

2. Positive impacts: food sources, pets, transportation, drug testing, guide dogs, jewelry, clothes, pollination of flowers and agricultural crops, ecotourism, decomposition, nutrient cycling, and biological control of pests, among others.

   Negative impacts: parasites, pests, vectors of disease, competitors for food, some potentially harmful to humans, among others.

Critical-Thinking Skills—Multiple Choice:

1. d 2. b 3. d 4. a 5. e

Critical-Thinking Skills—Short Answer:

1. They are polyphyletic. These three taxa do not share a common ancestor. The annelids are members of the Lophotrochozoa. The arthropods are members of the Ecdyszoa. Both are protostomes. The Chordates are deuterostomes.

2. Insects possess a number of critical innovations. They evolved wings but retained their walking legs. This means they can fly yet thrive in a variety of habitats. They have evolved a variety of mouthparts, which enables them to feed on a variety of foods, from foliage and plant sap to rotting fruits and vertebrate blood. Many insects undergo complete metamorphosis, allowing larvae to occupy completely different habitats from the adults.

Chapter 27

Concept Checks

Figure 27.6 Ray-finned fishes (but not sharks) have a (1) a bony skeleton; (2) a mucus-covered skin; (3) a swim bladder; and (4) an operculum covering the gills.

Figure 27.7 Both lungfishes and coelocanths are Sarcopterygians, having lobe fins.

Figure 27.9 No, some species give birth to live young.

Figure 27.11 Besides the amniotic egg, other critical innovations in amniotes are thoracic breathing; internal fertilization; a thicker, less permeable skin; and more efficient kidneys.

Figure 27.12 Snakes evolved from tetrapod ancestors but subsequently lost their limbs. Some species have tiny vestigial limbs.

Figure 27.15 Adaptations in birds to reduce body weight for flight include a lightweight skull, a reduction of organ size, no urinary bladder, and a reduction of organs outside of breeding season. Also female birds have one ovary and relatively few eggs.

BioConnections

Figure 27.5 Ions and water move across gills by diffusion and osmosis respectively. This can create imbalances in osmolarity and volume of fluids inside the body.

Figure 27.10 Bright or warning coloration advertises the poisonous nature of many species and deters predators.

Figure 27.14 Both classes have four-chambered hearts and care for their young.

Figure 27.19 The Echidna is another species of prototherian.

Feature Investigation Questions

1. The researchers were interested in determining the method in which *Hox* genes controlled limb development.

2. The researchers bred mice that were homozygous for certain mutations in specific *Hox* genes. This allowed the researchers to determine the function of individual genes.

3. The researchers found that homozygous mutants would develop limbs of shorter lengths compared to the wild-type mice. The reduced length was due to the lack of development of particular bones in the limb, specifically the radius, ulna, and some carpels. These results indicated that simple mutations in a few genes could lead to dramatic changes in limb development.

Testing Your Knowledge

27.1: 1. b
27.2: 1. e
27.3: 1. c 2. d 3. e
27.4: 1. e 2. d
27.5: 1. c 2. d 3. e
27.6: 1. b

Assessing Your Knowledge and Skills

Foundational Knowledge—Multiple Choice:

1. a 2. a 3. c 4. d 5. c

Foundational Knowledge—Short Answer:

1. Both taxa have external limbs that move when the attached muscles contract or relax. The difference is that arthropods have external skeletons with the muscles attached internally, whereas vertebrates have internal skeletons with the muscles attached externally.

2. Endothermy (warm-bloodedness) probably evolved independently in both birds and mammals. If the common ancestor of reptiles and birds were endothermic, the chances are that all reptiles would be endothermic.

Critical-Thinking Skills—Multiple Choice:

1. d 2. b 3. d 4. c 5. c

Critical-Thinking Skills—Short Answer:

1. Hinged jaws changed the nature of how vertebrates acquired food. They permitted predators to bite prey and remove chunks of flesh. This in turn stimulated the evolution of prey defenses against predators, including faster swimming or body armor. Predators themselves had to swim faster to catch their prey and develop more powerful jaws and teeth to penetrate the prey's armor. This evolutionary arms race between predator and prey was continued as vertebrates moved onto land and was especially apparent in dinosaurs, some of which developed thick scales and body plates to defend themselves against powerful carnivorous species.

2. (i) Because of their large body size, ancestral amphibians likely faced challenges supporting themselves out of the aquatic environment.

(ii) Although there was far more oxygen available to gills in air than in water, the delicate structure of gills couldn't be supported in the terrestrial environment, so a new method of breathing had to evolve.

(iii) The energetic costs of locomotion on land and greater oxygen demand from enlarged muscles required modification of the heart and circulatory system.

(iv) Even though food and oxygen were abundant on land, early amphibians could not venture too far away from water because egg deposition was restricted to moist environments.

Chapter 28

Concept Checks

Figure 28.6 As in the case of shoots, the capacity to divide the root into two equal pieces by means of a line drawn from opposite sides through the center would indicate that a root has superficial radial symmetry. In order to determine that an organ has radial symmetry at the cellular level, you would have to compare the microscopic views of randomly chosen, wedge-shaped pieces of cross-sectional slices. If the structure of the wedges is similar, the organ has radial symmetry at the microscopic level.

Figure 28.8 Locating stomata on the darker and cooler lower leaf surface helps reduce water loss from the leaf.

Figure 28.12 A twig having five sets of bud scale scars is likely to be approximately 6 years old.

Figure 28.16 Cactus stems are green and photosynthetic, playing the role served by the leaves of most plants.

Figure 28.20 A woody stem builds up a thicker layer of wood than inner bark, in part because older tracheids and vessel element walls are not lost during shedding of bark, which is the case for secondary phloem. In addition, plants typically produce a greater volume of xylem than phloem tissue per year, in part because vessel elements are relatively wide. A large volume of water-conducting tissue helps plants maintain a large amount of internal water.

Figure 28.23 Lateral roots are produced from internal meristematic pericycle tissue because roots do not produce axillary buds like those from which shoot branches develop. Internal production of branch roots helps to prevent them from shearing off as the root tip grows through abrasive soil.

BioConnections

Figure 28.1 The flowering plant life cycle is like that of other plants in that a haploid gametophyte generation alternates with a diploid sporophyte generation, but the size of flowering plant gametophytes is smaller than those of mosses and ferns.

Feature Investigation Questions

1. The advantages of using natural plants include the opportunity to avoid influencing plants with unnatural environmental factors, such as artificial light, and the exposure of all experimental plants to similar growth conditions. In addition, the investigators studied the leaves of some large trees, which would be hard to accommodate in a greenhouse.
2. Pinnately veined leaves were splinted to prevent their breaking, since they were cut at the single main vein, which has both support and conducting functions.
3. Sack and associates measured leaf water conduction at two or more places on each leaf because the effect of cutting a vein might have affected some portions of leaves more than others.

Testing Your Knowledge

28.1: 1. d 2. a
28.2: 1. e 2. b
28.3: 1. d 2. e
28.4: 1. d 2. c

Assessing Your Knowledge and Skills

Foundational Knowledge—Multiple Choice:
1. d 2. b 3. a 4. c 5. a

Foundational Knowledge—Short Answer:
1. If leaves were generally radially symmetric (shaped like spheres or cylinders), leaves would not have maximal ability to absorb sunlight, and they would not be able to optimally disperse excess heat from their surfaces.
2. Although tall herbaceous plants exist (palms and bamboo are examples) the additional support and water-conducting capacity that are provided by secondary xylem allow woody plants to grow tall.

Critical-Thinking Skills—Multiple Choice:
1. d 2. d 3. d 4. d 5. b

Critical-Thinking Skills—Short Answer:
1. Plant leaves arise from shoot apical meristems, which in the case of the water lilies are present on the bottom-growing stem (the rhizome). After petioles are cut, new leaves will continue to arise unless the rhizomes are removed.
2. Pneumatophores are aerial roots that take up oxygen from the air and transmit the oxygen to submerged roots, whose watery sediment habitats are typically deficient in oxygen. Without oxygen, mangrove roots would not be able to conduct cellular respiration to generate ATP needed for root growth and mineral uptake, so the mangrove plants would grow less well and possibly die.

Chapter 29

Concept Checks

Figure 29.4 Auxin efflux carriers could be located on the upper sides of root cells, thereby allowing auxin to move upward in roots.

Figure 29.6 The triple response of dicot seedlings to internally produced ethylene allows them to protect the delicate apical meristem from damage as the seedling emerges through the soil.

Figure 29.7 The active conformation of phytochrome absorbs far-red light. Such absorption causes the active conformation of phytochrome to change to the inactive conformation and to move out of the nucleus and into the cytosol.

Figure 29.8 The inactive conformation of phytochrome would absorb the red portion of sunlight, thereby converting phytochrome into the active conformation.

Figure 29.9 Exposing plants to brief periods of darkness during the daytime will have no effect on flowering because flowering is determined by night length.

Figure 29.11 Yes, just as shoots exhibit negative gravitropism in upward growth, roots are capable of using negative phototropism to grow downward, because light decreases with depth in the soil.

Figure 29.13 Predators are more likely to be able to find their prey if the latter are concentrated and exposed while feeding on plants. Plants benefit when predation removes herbivores, a process that lessens damage to plants.

Figure 29.15 Similar suites of protective plant hormones, such as jasmonic acid, are used in both types of defenses.

BioConnections

Figure 29.3 The activation of the defense hormone receptor is the plant signaling process most similar to the animal cell communication shown in Figure 8.4.

Testing Your Knowledge

29.1: 1. a
29.2: 1. d 2. b 3. a 4. c 5. e 6. d
29.3: 1. a 2. c 3. b 4. c
29.4: 1. c
29.5: 1. e

Assessing Your Knowledge and Skills

Foundational Knowledge—Multiple Choice:

1. c 2. c 3. a 4. e 5. d

Foundational Knowledge—Short Answer:

1. Behavior is defined as the response of a living thing to a stimulus. Therefore, because plants display many kinds of responses to diverse stimuli, they display behavior.
2. Many kinds of disease-causing bacteria and fungi occur in nature, and these organisms evolve very quickly, producing diverse elicitors. Thus, plants must maintain a stock of resistance genes, each having many alleles.

Critical-Thinking Skills—Multiple Choice:

1. c; because phytochrome responds most strongly to red- and far-red-light changes 2. d; because gravity affects both root and shoot systems of Earth's plants 3. d; because at least some plants and fungi are known to exhibit these particular responses 4. d. In the absence of gravity, which normally induces downward growth of most plant roots, you would expect the seedlings to generate embryonic roots in various directions. 5. a. NO affects the cell death process that helps to keep disease from spreading through plant tissues.

Critical-Thinking Skills—Short Answer:

1. As an evolutionary response to diverse types of herbivores, different plants have evolved complex suites of chemicals that decrease edibility.
1. You could collect wild relatives of the crop from diverse locales, grow each type in the presence of the herbivores of interest, then assess the ability of each plant type to grow despite herbivore presence. Plant types that grow best could be subjected to additional rounds of selection to identify those most resistant to the herbivore of concern.

Chapter 30

Concept Checks

Figure 30.8 Mineral nutrient leaching occurs more readily from sandy soils than from clay soils.

Figure 30.10 Soil crusts containing nitrogen-fixing cyanobacteria increase soil fertility, fostering the growth of larger plants that stabilize soils against erosion and provide forage for animals.

Figure 30.16 Your drawing should show that equivalent amounts of water will move into and out of a turgid cell having a water potential of 0, when the cell is placed into pure water. When placed in pure water, a plasmolyzed cell having a water potential of –1.0 MPa will gain water. Your drawing should show that more water moves into the cell than out of it. When placed in pure water, a flaccid cell having a water potential of –0.5 MPa will gain water. This is because water moves from a region of higher water potential to a region of lower water potential, and 0 is greater than –0.5. Your drawing should show that more water moves into the flaccid cell than out of it.

Figure 30.22 The evaporation of water has a powerful cooling effect because it disperses heat so effectively. Water has the highest heat of vaporization of any known liquid.

Figure 30.25 In the ocotillo's desert habitat, times of drought and contrasting availability of water sufficient to support the development and photosynthetic function of leaves do not occur at predictable times, as is the case for temperate forests. For this reason, ocotillo leaf abscission is not amenable to the evolution of genetic mechanisms that allow leaf drop to be precisely timed in anticipation of the onset of drought.

BioConnections

Figure 30.19 Tight junctions of intestinal epithelium and Casparian strips of endodermal cells of plant roots both incorporate materials that form a tight seal, preventing movement of materials from one location to another.

Testing Your Knowledge

30.1: 1. e 2. b
30.2: 1. c 2. c
30.3: 1. c 2. d
30.4: 1. d 2. d
30.5: 1. d 2. c

Assessing Your Knowledge and Skills

Foundational Knowledge—Multiple Choice:

1. e 2. a 3. b 4. d 5. a

Foundational Knowledge—Short Answer:

1. In the case of plant fertilizers, more is not better, because the ion concentration of overfertilized soil may become so high as to draw water from plant cells. In this case, the cells would be bathed in a hypertonic solution and would likely lose water to the solution. If plant cells lose too much water, they will die.
2. When the natural vegetation is removed, transpiration stops, so water is not transported from the ground to the atmosphere, where it may be an important contributor to local rainfall. Extensive removal of plants actually changes local climate in ways that reduce agricultural productivity and human survival.

Critical-Thinking Skills—Multiple Choice:

1. e. Water readily enters intercellular spaces of roots up to the endodermal cylinder, and the entry of ions into root cells increases osmotic flow of water through the symplast. Roots that are active in transport generally use shoot-supplied sugars in cell respiration to power growth and ion uptake. In particular cases such as the storage roots produced by sugar beet, sugars might accumulate. 2. a. Thickened inner cells walls are less extensible than are thinner-walled outer cell walls having radial microfibrils, with the result that when guard cells are turgid, the outer walls expand outward while inner wall shape does not change much, thereby opening the stomatal pore. 3. c. Companion cells are key to production of P protein, which helps to prevent excess loss of fluid from damaged sieve-tube elements. 4. d. Lush foliage would generate high levels of photosynthate, which would be loaded into the phloem for transfer to developing fruit. 5. b. Sand is silicon dioxide, so it contains an abundance of Si.

Critical-Thinking Skills—Short Answer:

1. You might start by evaluating crop production in controlled temperature conditions, to determine optimal temperature and temperature range allowing plant survival. You might then grow plants at optimal temperatures under different watering regimes that mimic rainfall amounts and patterns occurring in nature. Relative water content can be measured as a way of inferring overall plant vulnerability to water stress.
2. Agricultural experts are concerned that adding excess fertilizer to crop fields increases the costs of crop production. Ecologists are concerned that excess fertilizers will wash from crop fields into natural waters and cause harmful overgrowths of cyanobacteria, algae, and aquatic plants. Entry of excess fertilizers into coastal waters is associated with the development of "dead zones" in which animal life has been lost, affecting coastal fisheries.

Chapter 31

Concept Checks

Figure 31.2 Because gametophytes are haploid, they lack the potential for allele variation at each gene locus that is present in diploid sporophytes. Hence, gametophytes are more vulnerable to environmental stresses. By living within the diploid tissues of flowers, flowering plant gametophytes are protected to some extent, and the plant does not lose its gamete-producing life cycle stage.

Figure 31.3 Some flowers lack some of the major flower parts.

Figure 31.4 By clustering its stamens around the pistil, the hibiscus flower increases the chance that a pollinator will both pick up pollen and deliver pollen from another hibiscus flower on the same trip.

Figure 31.6 The absence of showy petals often correlates with wind pollination, because large petals would interfere with the shedding of pollen in the wind.

Figure 31.9 The maximum number of cells in a mature male gametophyte of a flowering plant is three: a tube cell and two sperm cells.

Figure 31.10 Female gametophytes are not photosynthetic and cannot produce their own food. Enclosed within ovules, female gametophytes lack direct access to the outside environment. Carpels contain veins of vascular tissue that bring nutrients from sporophytic tissue to ovules.

Figure 31.13 During their maturation, the cotyledons of eudicot seeds absorb the nutrients originally present in endosperm.

BioConnections

Figure 31.1 The female gametophytes of a flowering plant are produced within ovules that lie within ovaries of flowers. The male gametophytes of flowering plants are produced by pollen grains that originate within the stamen anthers of a flower.

Figure 31.11 The plant pollen tube is analogous in function to a human penis in that both structures accomplish internal fertilization. The plant pollen tube grows long enough to deposit sperm at the micropyle within the body of the female gametophyte, much as an animal penis deposits sperm within the female's body. In both cases, sperm are more likely to survive and accomplish fertilization than if they were deposited outside the female body.

Feature Investigation Questions

1. The large flowers of this lily enabled the investigators to more easily mark petals and record the positions of marks over time.
2. Time-lapse video reduced the amount of time the investigators would have to spend recording changes in the positions of petal marks.
3. Results obtained by using mathematical models can be compared with actual measurements to assess the accuracy of the models. An accurate mathematical model indicates a relatively full understanding of the physical processes involved. If models are sufficiently accurate, they can be utilized in other situations.

Testing Your Knowledge

31.1: 1. d 2. a
31.2: 1. c
31.3: 1. b
31.4: 1. b
31.5: 1. d

Assessing Your Knowledge and Skills

Foundational Knowledge—Multiple Choice:
1. a 2. b 3. d 4. c 5. a

Foundational Knowledge—Short Answer:
1. Pollen grains are vulnerable to mechanical damage and microbial attack during the journey through the air from the anthers of a flower to a stigma. Sporopollenin is an extremely tough polymer that helps to protect pollen cells from these dangers. The function of the beautiful sculptured patterns of sporopollenin on pollen surfaces is unclear.
2. Flower diversity is an evolutionary response to diverse pollination circumstances. For example, plants such as oak and corn that are wind-pollinated produce flowers having a poorly developed perianth. If such wind-pollinated flowers had large, showy perianths, they would get in the way of pollen dispersal or acquisition. On the other hand, flowers that are pollinated by animals often have diverse shapes and attractive petals of differing colors or fragrances that have coevolved with different types of animal pollinators.

Critical-Thinking Skills—Multiple Choice:
1. b. The coordinated fertilization of an egg by one of the two sperm delivered to an ovule by a pollen tube, and fusion of the second sperm with two other ovule nuclei, represents the evolution of an efficient process that provides young embryos with a rich food source in the form of embryos. This efficient process underlies extensive use by humans of cultivated grain endosperm as a major source of food. 2. d. Although many flowering plants employ a sexual process to produce seeds, some plants are able to produce seeds by an asexual process. This difference is important because sexually produced seeds are commonly genetically diverse, whereas asexually produced seeds are likely to be more genetically uniform. 3. b. Environmental signals, such as day length, induce leaf vein companion cells to make the protein FT, which is then transported in the phloem to the shoot apex, stimulating the production of flowers. 4. b. Complete flowers are defined as flowers that include all four basic flower parts: sepals and petals (which in many flowers look similar and are thus known as tepals), pollen-producing anthers, and ovule-producing carpels. Because perfect flowers are those having both anthers and carpels, complete flowers are also perfect. 5. c. Each pollen tube that successfully reaches an ovule may release two sperm, one of which may fertilize the egg (the other may participate in endosperm formation).

Critical-Thinking Skills—Short Answer:
1. Orchid flowers are typically bilaterally symmetric, a trait that is commonly associated with animal pollination. Bees, for example, are strong enough to move through complex orchid flowers to accomplish pollination. The gene *CYCLOIDEA* is associated with the development of bilaterally symmetric flowers, such as snapdragons and orchids.
2. The embryos within seeds are vulnerable to mechanical damage and microbial attack after they are dispersed. Seed coats protect embryos from these dangers and help to prevent seeds from germinating until conditions are favorable for seedling survival and growth.

Chapter 32

Concept Checks

Figure 32.6 No. The brain, for example, does not contain muscle tissue (although the blood vessels supplying the brain do contain smooth muscle).

Figure 32.7 Surface area is important to any living organism that needs to exchange materials with the environment. A good example of a high surface area/volume ratio is that of most tree leaves. This makes leaves ideally suited for such processes as light absorption (required for photosynthesis) and the exchange of gases and water with the environment.

Figure 32.9 Humans are endotherms and homeotherms.

Figure 32.10 The spider is a heterotherm. Its body temperature changes with that of the environment.

Figure 32.12 Heat moves from the warmer artery to the cooler veins. This prevents heat from reaching the extremities and being lost to the environment.

Figure 32.13 Blood, including plasma and blood cells, would leak out of the blood vessel into the interstitial space. The fluid level of the bloodstream would decrease, and that of the interstitial space near the site of the injury would increase. Eventually, the blood that entered the interstitial space would be degraded by enzymes, resulting in the characteristic skin appearance of a bruise. If the injury were very severe, the fluid level in the blood could decrease to a point where the various tissues and organs of the body would not receive sufficient nutrients and oxygen to function normally.

Figure 32.14 No, obligatory exchanges must always occur, but animals can minimize obligatory losses through modifications in behavior. For example, terrestrial animals that seek shade on a hot, sunny day reduce evaporative water loss. As another example, reducing activity minimizes water loss due to respiration.

BioConnections

Figure 32.6 Note that both the stomach and the intestine depicted in Figures 32.6 and 37.6 contain layers of muscle wrapped around the lumen. Although you will learn later that the stomach and intestine have many different functions, this similarity in anatomy suggests that both of these organs may perform the similar activity of mechanically breaking apart chunks of food and propelling the contents from one region to another.

Figure 32.8 Negative feedback is a common way in which the product of a chemical reaction inhibits the rate of that reaction, thus maintaining homeostatic levels of the product in a cell.

Feature Investigation Questions

1. Symptoms of prolonged, heavy exercise include fatigue, muscle cramps, and even occasionally seizures. Fatigue results from the reduction in blood flow to muscles and other organs. Muscle cramps and seizures are the result of imbalances in plasma electrolyte levels. Cade and his colleagues hypothesized that maintaining proper water and electrolyte levels would prevent these problems, and that if water and ion levels were maintained, athletic performance should not decrease as rapidly with prolonged exercise.

2. To test their hypothesis, the researchers created a drink that would restore the correct proportions of lost water and ions within the athletes. If the athletes consumed the drink during exercise, they should not experience as much fatigue or muscle cramping, and thus their performance should be enhanced compared with a control group of athletes that drank only water.

3. Since perspiration (and thus loss of ions and water) would usually be minor under such conditions, a reasonable hypothesis is that a drink containing ions such as those in sports drinks would be of little or no benefit to a person who was not exercising with at least some effort.

Testing Your Knowledge

32.1: 1. b 2. (see following text)

supports the body and produces locomotion—musculoskeletal system

absorbs nutrients into the body—digestive system

regulates growth, metabolism, reproduction, mineral balance—endocrine system

distributes solutes throughout the body—circulatory system

provides a barrier against pathogens entering the body—integumentary system

32.2: 1. d

32.3: 1. a 2. b

32.4: 1. c 2. e

32.5: 1. e 2. e

Assessing Your Knowledge and Skills

Foundational Knowledge—Multiple Choice:

1. b 2. b 3. b 4. b 5. c

Foundational Knowledge—Short Answer:

1. Homeostatic control systems include a sensor that monitors a variable such as body temperature, a set point that is the normal value for the variable, an integrator (usually in the brain) that compares the set point to the value detected by the sensor, and an effector that initiates a change in the variable to return it to a homeostatic value whenever the variable is disrupted.

2. Negative feedback loop—In a negative feedback loop, the system is working to return to homeostasis. Sometimes there is a physiological disturbance that pushes a system away from homeostasis. When this occurs, the organism must bring this system back to homeostasis. An example of this is when the human body temperature exceeds about 37°C. When this occurs, compensatory mechanisms such as perspiring will occur until temperature homeostasis (37°C) is achieved.

   Positive feedback loop—In a positive feedback loop, instead of moving toward homeostasis, the system is moving away from homeostasis. Positive feedback is much less common in nature than negative feedback. The process of blood clotting in mammals is one example of positive feedback.

Critical-Thinking Skills—Multiple Choice:

1. e 2. e 3. c 4. d 5. c

Critical-Thinking Skills—Short Answer:

1. No, not necessarily. Body temperature, for example, is maintained at different set points in birds and mammals. Other vertebrates and most invertebrates do not have temperature set points; their body temperature simply conforms closely to that of the environment. As another example, a giraffe has a set point for blood pressure that is higher than that of a human being, because a giraffe's circulatory system must generate enough pressure to pump blood up its long neck to its brain.

2. Maintaining homeostasis requires continual supplies of energy. Animals consume food, and the energy from that food helps to sustain activities that maintain variables, such as body temperature and pH, and synthesis of complex molecules. Without this energy, it would be difficult or impossible for animals to maintain many important biological processes within a narrow range despite changes in the environment.

Chapter 33

Concept Checks

Figure 33.3 Many reflexes, such as the knee-jerk reflex, cannot be prevented once started. Others, however, can be controlled to an extent. Open your eyes widely and gently touch your eyelashes. A reflex that protects your eye will tend to make you close your eyelid. However, you can overcome this reflex with a bit of difficulty if you need to, for example, when you are putting in contact lenses.

Figure 33.5 Yes, the flow of K^+ down its concentration gradient does create an electrical gradient because K^+ is electrically charged. The net flow of K^+ will stop when the concentration gradient balances the electrical gradient. This occurs at the equilibrium potential.

Figure 33.10 Action potentials can propagate more quickly along an axon. This is especially important for long axons, such as those that carry signals from the spinal cord to distant muscles.

Figure 33.12 In the absence of such enzymes, neurotransmitters remain in the synapse for longer periods, and the postsynaptic cell can become overstimulated. In addition, the ability of the postsynaptic cell to respond to multiple, discrete inputs from the presynaptic cell will be compromised.

Figure 33.20 No, these divisions are present in all vertebrates.

Figure 33.21 The occipital lobe is part of the cerebral cortex, and therefore also part of the cerebrum.

Figure 33.23 Thinking requires energy! Even daydreaming requires energy; imagine how much energy the brain uses when you concentrate for 60 minutes on a difficult exam. In fact, you just expended energy thinking about this question!

BioConnections

Figure 33.5 Electrochemical gradients are used in transporting solutes across a cell membrane. They are also important in the production of ATP in mitochondria, in muscle contraction, and osmotic regulation of cell volume.

Figure 33.6 Water molecules move through membrane channels called aquaporins.

Figure 33.14 Animals are multicellular heterotrophs (cannot make their own food) whose cells lack a cell wall. Most animals have a nervous system, muscles, the ability to move about during at least some phase of their life cycle, and the ability to reproduce sexually.

Feature Investigation Questions

1. Gaser and Schlaug hypothesized that repeated exposure to musical training would increase the size of certain areas of the brain associated with motor, auditory, and visual skills. All three skills are commonly used in reading and performing musical pieces. The researchers used MRI to examine the areas of the brain associated with motor, auditory, and visual skills in three groups of individuals: professional musicians, amateur musicians, and nonmusicians. The researchers found that certain areas of the brain were larger in the professional musicians compared to the other groups and were larger in the amateur musicians compared to the nonmusicians.

2. Schmithorst and Holland found that when exposed to music, certain regions of the brains of musicians were activated differently compared with the brains of nonmusicians. This study supports the hypothesis that there is a

difference in the brains of musicians versus nonmusicians. The experiment conducted by Gaser and Schlaug compared the size of certain regions of the brain among professional musicians, amateur musicians, and nonmusicians. Schmithorst and Holland, however, were also able to detect functional differences between musicians and nonmusicians.

3. A trained athlete might be expected to have greater development of brain areas devoted to motor functions. For example, athletes who engage in repetitive activities such as tennis and badminton have been observed to have greater volumes of motor and balance regions of the brain, including the cerebellum. Someone deaf from birth might be predicted to have decreased development of auditory regions such as those in the temporal lobes. However, it is not necessarily this simple. For example, MRI studies sometimes but not always reveal differences in cell number of the temporal lobes of deaf versus hearing subjects, but there is generally significantly less development of fiber tracts (white matter) in this region in the brains of deaf persons. Therefore, the size of a brain structure (that is, the number of neurons) by itself may not always be predictive of function.

Testing Your Knowledge

33.1: 1. d 2. c

33.2: 1. d 2. e

33.3: 1. d 2. a

33.4: 1. c 2. c

33.5: 1. (see following text)

nerve net—cnidarian

two ventral nerve cords—*Planaria*

nerve ring and radial nerves—sea star (echinoderm)

brain with several well-defined subdivisions—*Drosophila*

33.5: 2. (see following text)

cerebral cortex—thinking and learning

midbrain—sensory processing and alertness

ganglia—local signaling in a body part

hindbrain—control of homeostatic functions such as breathing

33.6: 1. c

33.6: 2. (see following text)

thalamus—filtering incoming sensory information

hypothalamus—controlling body temperature and circadian rhythms

basal nuclei—fine tuning, starting, or inhibiting movement

cerebral cortex—decision making and exercising judgment

pons—controlling breathing/relay between cerebellum and rest of brain

33.7: 1. d 2. b

Assessing Your Knowledge and Skills

Foundational Knowledge—Multiple Choice:

1. c 2. d 3. d 4. e 5. b

Foundational Knowledge—Short Answer:

1. In a graded potential, a weak stimulus causes a small change in the membrane potential, whereas a strong stimulus produces a greater change. Graded potentials occur along the dendrites and cell body. If a graded potential reaches the threshold potential at the axon hillock, an action potential results. This is a change in the membrane potential that is of a constant value and is propagated from the axon hillock to the axon terminals.

2. The activities of animal nervous systems provide many examples of new properties emerging from complex interactions. For example, you learned about reflexes in this chapter, which are behaviors that emerge from interactions between individual neurons that form communication circuits between the peripheral and central nervous systems. You also learned about such higher properties as conscious thought, which also emerges from the interactions between many individual cells, each of which is in communication with up to hundreds of thousands of other cells. Individually, the cells cannot "think," but networked together in elaborate ways, a person like yourself can think, remember, plan ahead, and interpret your environment.

Critical-Thinking Skills—Multiple Choice:

1. a 2. b 3. c 4. d 5. a

Critical-Thinking Skills—Short Answer:

1. Not necessarily, because it is not simply the mass of the brain that determines how intelligent an animal may be or of what it may be capable. The total surface area of the brain, including ridges and folds, is also very important.

2. An increase in extracellular Na^+ concentration would slightly depolarize neurons, thereby changing the resting membrane potential. This effect would be minimal, however, because the resting membrane is not very permeable to Na^+. However, the shape of the action potentials in such neurons would be a little steeper, and the peak a little higher, because the electrochemical gradient favoring Na^+ entry into the cell through voltage-gated channels would be greater.

Chapter 34

Concept Checks

Figure 34.9 The advantage of having heat-sensing organs on both sides of the head is that it provides information about the source of the heat, much like our ability to detect the origin of a sound based on differences in arrival times to our two ears.

Figure 34.14 Because red-green color blindness is a sex-linked recessive gene, males require only a single defective allele on an X chromosome, whereas females require two defective alleles, one on each X chromosome.

Figure 34.17 Photoreceptors (rods and cones) synapse with bipolar cells, which in turn synapse with ganglion cells.

Figure 34.20 The receptor proteins are located in the membranes of microvilli extending from the sensory receptor cells.

BioConnections

Figure 34.1 The term sensory receptor refers to a type of cell that can respond to a particular type of stimulus. The term membrane receptor refers to a protein within a cell membrane that binds a ligand, thereby generating signals that initiate a cellular response.

Figure 34.3 Cilia are cell extensions that contain, in their internal structure, microtubules and motor proteins that cause the cilia to beat, or move, in a coordinated fashion. Stereocilia are membrane projections that are not motile but instead are deformed by the movements of surrounding fluids.

Figure 34.7 Statoliths are also found in the roots and shoots of plants. They serve as a gravity-detection mechanism that results in roots growing downward and shoots growing upward.

Feature Investigation Questions

1. One hypothesis was that many different types of odor molecules might bind to one or just a few types of receptor proteins, with the brain responding differently depending on the number or distribution of the activated receptors. The second hypothesis was that organisms can make a large number of receptor proteins, each type binding a particular odor molecule or group of odor molecules. According to this hypothesis, it is the *type* of receptor protein, and not the number or distribution of receptors, that is important for olfactory sensing. The researchers extracted RNA molecules from the olfactory receptor cells of the nasal epithelium. They then used this RNA to identify genes that encoded G-protein-coupled receptor proteins.

2. The results of the experiment conducted by Buck and Axel support the hypothesis that animals discriminate between different odors based on having a variety of receptor proteins that recognize different odor molecules. Current research suggests that each olfactory receptor cell has a single type of receptor protein that is specific to particular odor molecules. Because most odors are due to multiple chemicals that activate many different types of odor receptor proteins, the brain detects odors based on the combination of the activated receptor proteins. Odor seems to be discriminated by many olfactory receptor proteins, which are in the membrane of separate olfactory receptor cells.

3. Among other things, animals use odors to detect potential sources of food, shelter, mates, and danger. Discerning between the odors of food versus those of potentially toxic material is key to the survival of many animals (think, for example, how we recognize fresh milk and spoiled milk). Many animals also use olfaction in finding shelter; some birds, for example can distinguish the odors produced by particular beneficial plants that they use to build their nests. These plants have natural pathogen-killing properties and help maintain a healthier nest site. Odor preference also plays an important role in helping animals find mates via the release of pheromones, particularly animals that live in dark environments such as nocturnal animals. The ability to discern the odors of different predators even when the predator may not be close enough to hear, or may not be in sight, is a huge survival advantage for many animals, who generally can differentiate such odors from those of nonthreatening species.

Testing Your Knowledge

34.1: 1. b
34.2: 1. c 2. c
34.3: 1. c
34.4: 1. e 2. d
34.5: 1. c 2. c
34.6: 1. b 2. c

Assessing Your Knowledge and Skills

Foundational Knowledge—Multiple Choice:

1. e 2. b 3. a 4. d 5. d

Foundational Knowledge—Short Answer:

1. Sensory transduction—the process by which incoming stimuli are converted into neural signals. An example is the signals generated in the retina when a photon of light strikes a photoreceptor. Perception—an awareness of the sensations that are experienced. An example is an awareness of what a particular visual image is.
2. These orientations permit animals to detect circular or angular movement of the head in three different planes. The fluid in a canal that is oriented in the same plane as the plane of movement will respond maximally to the movement. For example, the canal that is oriented horizontally would respond to horizontal movements, while the other two canals would not. Overall, by comparing the signals from the three canals, an animal's brain can interpret the motion in three dimensions.

Critical-Thinking Skills—Multiple Choice:

1. b 2. b 3. c 4. c 5. c

Critical-Thinking Skills—Short Answer:

1. In both types of eyes, there are photoreceptors, which contain structural specializations such as microvilli or flat discs that provide increased surface area for light reception. Both types of eyes also possess a cornea and lens, as well as a pigmented region to absorb scattered light. With both types, objects can be seen, motion can be detected, and color can be sensed.
2. Of the various senses, the sense of olfaction (smell) is least important for the survival of humans. As diurnal animals, we rely largely on our visual sense. Sound is a critical way to learn about impending danger, such as a car horn, but also is our major means of communication. Other senses, such as the ability to sense pain, have acutely important functions from time to time. Olfaction, though often a pleasurable sense and at times a protective one (think of the smell of spoiled food), nonetheless provides little survival advantage to us. In fact, many people spend much of their lives with greatly diminished olfactory abilities, whether from chronic allergies or other problems, and are not hindered in any significant way. The story is very different for animals such as nocturnal mammals, which rely very heavily on olfaction to find food, locate mates, and avoid predators.

Chapter 35

Concept Checks

Figure 35.1 In addition to not having a requirement to shed their skeletons periodically, animals with endoskeletons can use their skin as an efficient means of heat transfer (refer back to Figure 32.11). In addition, the body surface of such animals is often a highly sensitive sensory organ.

Figure 35.3 Without the connection to bone that a tendon provides, a muscle could not cause movement of that bone.

Figure 35.11 Na^+ enters the muscle cell through ion channels. The acetylcholine receptor on skeletal muscle cells is also a ligand-gated ion channel; when acetylcholine binds the receptor, it induces a shape change that opens the channel. This allows the entry of Na^+ into the cell down its electrochemical gradient.

BioConnections

Figure 35.10 Voltage-gated Ca^{2+} channels exist in the terminals of all axons that communicate by chemical signaling (neurotransmitter release). In those cases, depolarization of the axon terminal opens Ca^{2+} channels, allowing Ca^{2+} to enter the terminal and trigger exocytosis of stored vesicles containing neurotransmitter molecules.

Figure 35.11 Myelin is a lipid-rich membrane sheath that speeds up conduction of action potentials along an axon. Action potentials are regenerated at discrete intervals along the axon wherever the myelin sheath is interrupted by a node of Ranvier. This is known as saltatory conduction.

Testing Your Knowledge

35.1: 1. b 2. b
35.2: 1. e 2. d
35.3: 1. c
35.4: 1. e 2. c

Assessing Your Knowledge and Skills

Foundational Knowledge—Multiple Choice:

1. d 2. e 3. e 4. d 5. a

Foundational Knowledge—Short Answer:

1. (1) The cycle begins with the binding of an energized myosin cross-bridge to an actin molecule on a thin filament.
   (2) The cross-bridge moves, and the thin filaments slide past the thick filaments.
   (3) The ATP binds to myosin, causing the cross-bridge to detach.
   (4) The ATP bound to myosin is hydrolyzed by ATPase, re-forming the energized state of myosin.
2. Exoskeletons are found in many invertebrates such as arthropods. They are found on the outside of an animal's body, and they protect and support the animal. Endoskeletons are internal skeletons made of bone or cartilage and are found in echinoderms and vertebrates. They provide support and protection of internal organs, and they are important for movement and locomotion. They also are the site of blood cell development and contribute to the homeostasis of certain ions such as calcium and phosphate.

Critical-Thinking Skills—Multiple Choice:

1. d 2. c 3. b 4. e 5. c

Critical-Thinking Skills—Short Answer:

1. Electrical events must precede mechanical (contractile) events in skeletal muscle. Acetylcholine released into the synapse of a neuromuscular junction binds to its receptor on the postsynaptic cell, the muscle fiber. It is the electrical activity of the resulting action potential that triggers an increase in cytosolic Ca^{2+} in the fiber. This, in turn, activates the sliding-filament mechanism. Without electrical excitation, therefore, contraction cannot occur.
2. The use of energy released by the hydrolysis of ATP is fundamental to muscle function and locomotion. Recall that ATP must be hydrolyzed during the cross-bridge cycle for skeletal muscle cells to shorten. Energy is also used to maintain calcium ion balance in the sarcoplasmic reticulum and is used in all forms of locomotion. The amount of energy expended by an animal during locomotion reflects how well it is adapted to the environment in which it must move.

Chapter 36

Concept Checks

Figure 36.1 Open circulatory systems evolved prior to closed systems. However, this does not mean that open systems are "primitive" compared to closed circulatory systems. It is better to think of open systems as being ideally suited to the needs of those animals that have them. Arthropods are an incredibly successful order of animals, with the greatest number of species and inhabiting virtually every ecological niche on the planet. Clearly, their type of circulatory system has not prevented arthropods from achieving their great success.

Figure 36.6 The aorta and all arteries branching from it carry oxygenated blood, as do the pulmonary veins returning from the lungs.

Figure 36.8 The left and right ventricles pump blood through the semilunar valves into the aorta and the pulmonary trunk, respectively.

Figure 36.9 Body fluids, both extracellular and intracellular, contain large amounts of charged ions, which are capable of conducting electricity. The slight electric currents generated by the beating heart muscle cells are conducted through the surrounding body fluids by the movements of ions in those fluids. This is recorded by the surface electrodes and amplified by the recording machine.

Figure 36.13 The valves open toward the heart. When the head is upright, the valves are open, and blood drains from the head to the right atrium by gravity. When the giraffe lowers its head to drink, however, gravity would prevent the venous blood from reaching the heart; instead, blood would pool in the head and could raise pressure in the head and brain. The valves in the neck veins prevent this by closing so that blood does not flow backwards.

Figure 36.16 Imagine holding several thin sheets of a wet substance, such as paper towels. If you wave them in the air, what happens? The sheets stick to one another because of surface tension and other properties of moist surfaces. This is what happens to the lamellae in gills when they are in air. When the lamellae stick to each other, the surface area available for gas exchange is reduced and the fish suffocates.

Figure 36.17 Several factors probably limit insect body size, but the respiratory system most likely is one such factor. If an insect grew to the size of a human, for example, the trachea and tracheoles would be so large and extensive that there would be little room for any other internal organs in the body! Also, the mass of the animal's body and the forces generated during locomotion would probably collapse the tracheoles. Finally, diffusion of oxygen from the surface of the body to the deepest regions of a human-sized insect would take far too long to support the metabolic demands of internal structures.

Figure 36.18 As the lungs expand, the pressure within them decreases, as defined by Boyle's law. This permits air to flow into the lungs.

BioConnections

Figure 36.4 Immune defenses are found in most living organisms. Many bacteria produce antibacterial secretions that kill other bacteria. Plants, as shown in Figures 29.14 and 29.15, have a wide array of pathogen-fighting mechanisms.

Figure 36.16 Countercurrent exchange is an efficient means of heat transfer between arteries and veins, such as those near the skin surface of the legs of a wading bird. Heat from the descending arteries is transferred to surrounding veins, which return the warm blood to the heart, preventing heat loss through the skin to the water.

Figure 36.24 The brainstem includes the midbrain, pons, and medulla oblongata. See Figure 33.20 for an illustration of the major parts of the human brain.

Testing Your Knowledge

36.1: 1. c 2. a
36.2: 1. e 2. a
36.3: 1. e 2. b
36.4: 1. c 2. e
36.5: 1. c 2. e
36.6: 1. b
36.7: 1. b 2. b
36.8: 1. c 2. e
36.9: 1. b 2. b
36.10: 1. c
36.11: 1. e 2. c 3. a

Assessing Your Knowledge and Skills

Foundational Knowledge—Multiple Choice:
1. b 2. a 3. b 4. b 5. b

Foundational Knowledge—Short Answer:

1. *Closed circulatory system*—In a closed circulatory system, the blood and interstitial fluid are contained within tubes called blood vessels and are transported by a pump called the heart. All of the nutrients and oxygen that tissues require are delivered directly to them by the blood vessels. Advantages of closed circulatory systems are that different parts of an animal's body can receive blood flow in proportion to that body part's metabolic requirements at any given time. Due to its efficiency, a closed circulatory system allows organisms to become larger. *Open circulatory system*—In an open circulatory system, the organs are bathed in hemolymph that ebbs and flows into and out of the heart(s) and body cavity, rather than blood being directed to all cells. As in a closed circulatory system, there are a pump and blood vessels, but these two structures are less developed and less complex compared to those in a closed circulatory system. Partly as a result, organisms such as mollusks and arthropods are generally limited to being relatively small, although exceptions do exist.

2. Each time the heart beats, it progresses through a relaxation and filling phase (diastole) and a contraction and ejection phase (systole). Systole is preceded by electrical stimulation beginning in the pacemaker cells of the right atrium (or single atrium of animals without double circulations). Heart valves separate atria from ventricles and ventricles from arteries. They open in one direction only, which permits blood to flow only in a forward direction. If valves opened backwards, then blood could flow from ventricles to atria, and arteries to ventricles.

Critical-Thinking Skills—Multiple Choice:
1. c 2. b 3. b 4. d 5. e

Critical-Thinking Skills—Short Answer:

1. Carbon dioxide, hydrogen ions, and heat are produced by metabolism; the more active a cell is, the more these products are generated. The products, in turn, reduce the ability of hemoglobin to bind oxygen. In this way, more active regions of an animal's body obtain more oxygen in proportion to the metabolic demand at that time.

2. Hemoglobin is a protein with quaternary structure (see Figure 36.21) in which the four different subunits can bind up to a total of four oxygen molecules. The association of iron and a polypeptide forms a new molecule with complex ligand-binding properties. It is the structure of the subunits and their relationship to each other that contributes to their ability to bind O_2 and to the nonlinear relationship of the oxygen-hemoglobin dissociation curve. In addition, however, interactions of hemoglobin with other molecules, such as CO_2, change the structure of hemoglobin in such a way that its properties change. Under such conditions, hemoglobin is less able to bind O_2 and consequently it releases the gas. Any molecule that binds to hemoglobin will alter its structure and change its properties; these revert to the original state once the bound molecules are released.

Chapter 37

Concept Checks

Figure 37.2 Extracellular digestion protects the interior of cells from enzyme activity and allows for the intake of larger amounts of food, which can be slowly digested and absorbed gradually.

Figure 37.8 Active transport requires energy provided by ATP. Thus, absorption of nutrients by this mechanism is an energy-requiring event, and some portion of an animal's regular nutrient consumption is used to provide the energy required to absorb the nutrients.

Figure 37.11 By resynthesizing triglycerides from absorbed fatty acids and monoglycerides, a steep diffusion gradient is maintained for further diffusion of the smaller molecules into the cell.

Figure 37.14 The time required for the vesicles to move to the plasma membrane and fuse with it is much shorter than the time required for new GLUTs to be synthesized by activation of GLUT-encoding genes. Thus, the action of insulin on cells is very quick, because the GLUTs are already synthesized.

Figure 37.15 The glycerol and fatty acids used to make glucose are the breakdown products of triglycerides that were stored in adipose tissue during the absorptive state. The amino acids used to make glucose are derived from the breakdown of protein in muscle and other tissue.

Figure 37.17 Secretion of substances into excretory organ tubules is advantageous because it increases the amount of a substance that gets removed from the body by the excretory organs. This is important, because many substances that get secreted are potentially toxic. Filtration, though efficient, is limited by the volume of fluid that can leave the capillaries and enter the excretory tubule.

Figure 37.18 Insects remove soluble wastes by the process of secretion, unlike many other animals such as the annelid shown in part (a), which removes soluble wastes by filtration.

BioConnections

Table 37.2 Plants require many of the same minerals animals do, particularly potassium, phosphorus, iron, magnesium, nitrogen, and others. They also require water and a source of carbon. Microorganisms also require minerals, water, and carbon sources; see Chapter 23 for details.

Figure 37.8 Transmembrane transport processes are not unique to animals, and one or more types are found in virtually all cells.

Figure 37.14 Exocytosis, a feature characteristic of animal cells, involves the fusion of intracellular vesicles with the plasma membrane, resulting in the release of the vesicle contents into the extracellular fluid. See Figure 5.21 for a general description and Figure 33.12 for a specific example unique to animal cells.

Figure 37.23 Insulin-sensitive cells such as those of adipose tissue contain glucose transporters (GLUTs) in the membranes of cytosolic vesicles. When an adipose cell is stimulated by insulin, the GLUT-containing vesicles fuse with the cell's plasma membrane, making it possible for the cell to transport glucose from the extracellular fluid into the cell cytosol.

Figure 37.26 Countercurrent exchange promotes heat retention in the extremities of some animals, and it facilitates gas exchange across the gills of fishes.

Feature Investigation Questions

1. The surprising observation that some people with gastritis or ulcers have living bacteria (*H. pylori*) in their stomachs, and that administering bacteria-killing compounds provided some relief from the symptoms, led to the hypothesis that *H. pylori* infection is a cause of ulcers in humans.
2. The results did support the hypothesis; however, the results also clearly indicate that not all ulcers are due to *H. pylori* infection.
3. A combined treatment with bismuth and an antibiotic is the most effective treatment. It is apparent, however, that even in individuals with continued *H. pylori* infection, some ulcers will heal on their own. In the absence of bismuth/antibiotic therapy, though, the likelihood of a recurrence of an ulcer is much greater.

Testing Your Knowledge

37.1: 1. c 2. b
37.2: 1. c
37.3: 1. a 2. b 3. d
37.4: 1. c 2. a
37.5: 1. a 2. b
37.6: 1. a 2. c
37.7: 1. d 2. c
37.8: 1. d 2. d

Assessing Your Knowledge and Skills

Foundational Knowledge—Multiple Choice:
1. d 2. b 3. a 4. e 5. e

Foundational Knowledge—Short Answer:

1. The three processes are filtration, reabsorption, and secretion. During filtration, an organ acts as a sieve or filter, removing some of the water and its small solutes from the blood, interstitial fluid, or hemolymph, while retaining blood cells and large solutes such as proteins. Reabsorption is the process whereby epithelial cells of an excretory organ recapture useful solutes that were filtered. Secretion is the process whereby epithelial cells of an excretory organ transport unneeded or harmful solutes from the blood to the excretory tubules for elimination. Some substances, such as glucose and amino acids, are reabsorbed but not secreted, whereas some other substances, such as toxic compounds, are not reabsorbed and are secreted. Still other substances, namely proteins, are not filtered at all.
2. The mammalian nephron consists of the renal corpuscle (the glomerulus and Bowman's capsule), where filtration takes place, and the renal tubule. The latter includes the proximal tubule, where reabsorption of most solutes and secretion occur; the loop of Henle, where reabsorption of water occurs; and the distal tubule, where fine-tuning of water and ion reabsorption and secretion occur. From the distal tubule, the filtrate enters the collecting duct, where additional water may be reabsorbed. Although the degree of reabsorption and the specific solutes being affected change along the length of the nephron, reabsorption can occur throughout the renal tubule. Secretion is limited to the proximal and distal tubules, and filtration is limited to the region of Bowman's capsule.

Critical-Thinking Skills—Multiple Choice:
1. b 2. a 3. d 4. e 5. e

Critical-Thinking Skills—Short Answer:

1. The crop is a dilation of the esophagus, which stores and softens food. The gizzard contains swallowed pebbles that help to pulverize food. Both of these functions are adaptations that assist digestion in birds, which do not have teeth and therefore do not chew food. Humans, like many animals, can chew food before swallowing and therefore there was no selection process for the evolution of a crop or gizzard. Sauropod dinosaurs were herbivores that probably contained a gizzard-type stomach in which stones helped to grind coarse vegetation. Such stones would have become smooth after months or even years of rumbling around in the gizzard. Some of these sauropods are known to have lacked the sort of grinding teeth characteristic of modern mammalian herbivores, and thus a gizzard would have aided in their digestion much as it does in modern birds.
2. By functioning in fluid and ion balance, the kidneys help to regulate blood pressure, solute composition (and thereby electrical activity of nervous and muscle tissue), the pH of body fluids (which in turn is vital for many homeostatic processes), and osmolarity (important for maintaining cell shape and function, among other things). Healthy kidneys are important for all the organ systems of vertebrates.

Chapter 38

Concept Checks

Figure 38.10 Not all mammals use the energy of sunlight to synthesize vitamin D. Many animals, such as those that inhabit caves or that are strictly nocturnal, rarely are exposed to sunlight. Some of these animals get their vitamin D from dietary sources. How others maintain calcium balance without dietary or sunlight-derived active vitamin D remains uncertain.

Figure 38.14 Because 20-hydroxyecdysone is a steroid hormone, you would predict that its receptor would be intracellular. All steroid hormones interact with receptors located either in the cytosol or, more commonly, in the nucleus. The hormone-receptor complex then acts to promote or inhibit transcription of one or more genes. The receptor for 20-hydroxyecdysone is indeed found in cell nuclei.

BioConnections

Figure 38.1 Plants also make hormones, which are secreted into the extracellular fluid and travel via the vascular tissue of the plant. As in animals, hormones have numerous functions in plants related to growth, differentiation, responses to stress, reproduction, and others.

Figure 38.3 When dopamine is secreted from an axon terminal into a synapse, where it diffuses to a postsynaptic cell, it is considered a neurotransmitter. When it is secreted from an axon terminal into the extracellular fluid, from where it diffuses into the blood, it is considered a hormone.

Figure 38.7 In addition to the pancreas, certain other organs in an animal's body may contain both exocrine and endocrine tissue or cells. For example, you learned in Chapter 37 that the vertebrate alimentary canal is composed of several types of secretory cells. Some of these cells release hormones into the blood that regulate the activities of the pancreas and other structures, such as the gallbladder. Other cells of the alimentary canal secrete exocrine products such as acids or mucus into the gut lumen that directly aid in digestion or act as a protective coating, respectively.

Feature Investigation Questions

1. Banting and Best based their procedure on a medical condition that results when pancreatic ducts are blocked. The exocrine cells will deteriorate in a pancreas that has obstructed ducts; however, the islet cells are not affected. The researchers proposed to experimentally replicate the condition to isolate the cells suspected of secreting the glucose-lowering factor. From these cells, they assumed they would be able to extract the substance of interest without contamination or degradation due to exocrine products.

2. The extracts obtained by Banting and Best did contain insulin, the glucose-lowering factor, but were of low strength and purity. Collip developed a procedure to obtain a more purified extract with higher concentrations of insulin.

3. Because insulin is a polypeptide, it cannot be absorbed across the intestines; moreover, acid and digestive enzymes would degrade it in the stomach and small intestine. Thus, oral administration of insulin would be ineffective.

Testing Your Knowledge

38.1: 1. c 2. b
38.2: 1. a 2. b
38.3: 1. b 2. d
38.4: 1. b 2. a
38.5: 1. b 2. b
38.6: 1. c
38.7: 1. b

Assessing Your Knowledge and Skills

Foundational Knowledge—Multiple Choice:
1. b 2. e 3. b 4. e 5. d

Foundational Knowledge—Short Answer:

1. Type 1 diabetes mellitus is characterized by insufficient production of insulin due to the immune system destroying the insulin-producing cells of the pancreas. In type 2 diabetes mellitus, insulin is still produced by the pancreas, but adipose and muscle cells do not respond normally to insulin. Type 2 diabetes mellitus is far more common in humans.

2. Two hormones, 20-hydroxyecdysone and juvenile hormone, play major roles in insect development. Ecdysone is secreted by a pair of glands called the prothoracic glands. 20-Hydroxyecdysone stimulates rapid differentiation and causes the larva to molt. Juvenile hormone is secreted by the corpus allatum, and its main function is to control the nature of the molt induced by 20-hydroxyecdysone. Not until the levels of juvenile hormone decline to nearly zero does the molt result in the transition from a larva to a pupa. Many hormones are required for normal growth and development in vertebrates, notably thyroid hormone during developmental stages and growth hormone and IGF-1 during growth phases of life (such as during adolescence and puberty in humans).

Critical-Thinking Skills—Multiple Choice:
1. e 2. e 3. d 4. b 5. e

Critical-Thinking Skills—Short Answer:

1. Leptin is produced by adipose cells in proportion to the amount of stored triglyceride. In this way, an animal's appetite and metabolic rate can be altered to match its energy stores. In addition, leptin stimulates the secretion of reproductive hormones such as LH and FSH; when adipose stores are low, leptin secretion decreases and this removes some of the stimulation to produce LH and FSH. The benefit of this relationship is that it helps ensure that fertility is linked with adequate energy reserves in females.

2. Insulin acts to lower blood glucose concentrations (for example, after a meal), whereas glucagon elevates blood glucose (for example, during fasting). Insulin acts by stimulating the insertion of glucose transporter proteins into the cell membrane of muscle and fat cells. Glucagon acts in part by stimulating glycogenolysis in the liver. If a high dose of glucagon were injected into a nonfasting animal, the blood concentration of glucagon would increase rapidly. This would stimulate increased glycogenolysis, resulting in an above-normal blood glucose concentration.

Chapter 39

Concept Checks

Figure 39.9 FSH and LH concentrations do not surge in males, but instead remain fairly steady, because the testes do not show cyclical activity. Sperm production in males is relatively unchanging throughout much of life after puberty.

Figure 39.12 Pregnancy and subsequent lactation require considerable energy and, therefore, nutrient ingestion. Consuming the placenta provides the female with a rich source of protein and other important nutrients.

BioConnections

Figure 39.1 Like animals, most plants reproduce by sexual reproduction, but some reproduce asexually. Asexual reproduction is also seen in prokaryotes and protists.

Figure 39.10 In addition to its other functions, the placenta must serve the function of the lungs for the fetus, because the fetus's lungs are not breathing air during this time. Arteries always carry blood away from the heart; veins carry blood to the heart. Consequently, blood leaving the heart of the fetus and traveling through arteries to the placenta is deoxygenated. As blood leaves the placenta and returns to the heart, the blood has become oxygenated as oxygen diffuses from the maternal blood into fetal blood. That oxygenated blood then gets pumped from the fetal heart through other arteries to the rest of the fetus's body.

Figure 39.11 Positive feedback also occurs during ovulation in the ovarian cycle (see Figure 39.9). Stimulation of an ovarian follicle by LH causes the growth of the follicle and the release of estradiol, which further stimulates LH, which causes more follicle activity, and so on until ovulation occurs.

Feature Investigation Questions

1. Using *Daphnia,* Paland and Lynch compared the accumulation of mitochondrial mutations between sexually reproducing populations and asexually reproducing populations.

2. Of all the mutations observed in the populations, 17.7% were either moderately or mildly deleterious, and all of these persisted in the asexually reproducing species, whereas only 4.4% persisted in the sexually reproducing species. Thus, the asexually reproducing species were about four times more likely to retain a deleterious mutation.

3. Sexual reproduction allows for mixing of the different alleles of genes with each generation, thereby increasing genetic variation within the population. This can prevent the accumulation of deleterious alleles in the population.

Testing Your Knowledge

39.1: 1. d
39.2: 1. c 2. d
39.3: 1. b 2. b
39.4: 1. c 2. b
39.5: 1. d 2. c
39.6: 1. b

Assessing Your Knowledge and Skills

Foundational Knowledge—Multiple Choice:

1. c 2. e 3. a 4. c 5. b

Foundational Knowledge—Short Answer:

1. Cells of the hypothalamus produce two important hormones that regulate reproduction. GnRH stimulates the anterior pituitary gland to release two gonadotropic hormones, LH and FSH. These two hormones regulate the production of gonadal hormones and development of gametes in both sexes. In addition, increased secretion of GnRH contributes to the initiation of puberty. The hypothalamus also produces oxytocin, a hormone that is stored in the posterior pituitary gland and stimulates milk release during lactation.

2. Cleavage is the stage of animal development in which the zygote divides and forms a mass of smaller cells, creating a hollow ball called a blastula or blastocyst. In gastrulation, the blastula develops into a highly organized structure with three germ layers—endoderm, mesoderm, and ectoderm. During this event, the archenteron, which will become the digestive tract, is also formed. Neurulation is the next stage and includes the formation of the neural tube within the embryo. The final major event in development is organogenesis, during which tissues become organized into functional organs: the heart, lungs, gills, and so on.

Critical-Thinking Skills—Multiple Choice:

1. b 2. a 3. a 4. c 5. e

Critical-Thinking Skills—Short Answer:

1. External fertilization results in the exposure of gametes to predation and other environmental dangers. Many animals have evolved the ability to lay enormous numbers of eggs to compensate for these dangers.

2. Sexual reproduction requires that males and females of a species produce different gametes and that these gametes come into contact with each other. This requires males and females to expend energy to locate mates. It also may require the production of very large numbers of gametes to increase the likelihood that the eggs are fertilized. These costs are outweighed by the genetic diversity afforded by sexual reproduction.

Chapter 40

Concept Checks

Figure 40.2 Although swelling is one of the most obvious manifestations of inflammation, it has no significant adaptive value of its own. It is a consequence of fluid leaking out of blood vessels into the interstitial space. It can, however, contribute to pain sensations, because the buildup of fluid may cause distortion of connective tissue structures such as tendons and ligaments. Pain, while obviously unpleasant, is an important signal that alerts many animals to the injury and serves as a reminder to protect the injured site.

Figure 40.4 Recall from Chapter 36 that as blood circulates, a portion of the plasma—the fluid part of blood—exits venules and capillaries and enters the interstitial fluid. Most of the plasma is reabsorbed back into the capillaries, but a portion gets left behind. That excess fluid is drained away by lymphatic vessels and becomes lymph. Without lymphatic vessels, fluid would accumulate outside the blood, in the interstitial fluid.

Figure 40.6 Helper T cells contribute to both humoral and cell-mediated immunity.

Figure 40.9 Both B- and T-cell receptors have transmembrane domains, a constant region, and a variable region that binds a specific antigen.

Figure 40.13 Because an animal may encounter the same type of pathogen many times during its life, having a secondary immune response means that future infections will be fought off much more efficiently.

BioConnections

Figure 40.1 There are far more erythrocytes in blood than all leukocytes combined (in humans, for example, there are at least 1,000 times more erythrocytes).

Feature Investigation Questions

1. The amino acid sequence of Toll protein shared similarities with a portion of a protein known to be involved in immune responses in vertebrates. In addition, activation of Toll protein and the vertebrate immune protein (a cytokine receptor) resulted in the generation of some of the same intracellular signals. This suggested that in addition to its characterized role in embryonic development, Toll may also be important in immune functions in flies.

2. No, Toll protein is not a receptor that recognizes pathogen-associated molecular patterns (PAMPs) expressed on microbial surfaces, and thus it is distinguishable from Toll-like receptors in vertebrates. Toll is, however, a transmembrane protein that binds to extracellular signals; these signals arise, however, not from the microbes themselves but rather from proteins that are endogenous to fruit flies and that are generated during infections.

3. Yes, the results of the survival study clearly implicated Toll as a protein required for the induction of antimicrobial proteins and the ability to withstand fungal infection. Thus, the investigators' hypothesis was supported.

Testing Your Knowledge

40.1: 1. d
40.2: 1. a 2. c
40.3: 1. e 2. e
40.4: 1. c 2. b

Assessing Your Knowledge and Skills

Foundational Knowledge—Multiple Choice:

1. c 2. c 3. e 4. a 5. d

Foundational Knowledge—Short Answer:

1. Pathogens are disease-causing viruses and microorganisms, including certain bacteria, and eukaryotic parasites, such as certain protists, fungi, and small worms. Bacteria are single-celled prokaryotes that lack a true nucleus but are capable of reproducing on their own, whereas viruses are nucleic acids packaged in a protein coat that require a host cell to reproduce.

2. Lymphocytes include B cells, cytotoxic T cells, helper T cells, NK cells, and regulatory T cells. When stimulated by antigen, B cells differentiate into plasma cells and secrete antibodies. Cytotoxic T cells directly bind to antigen-bearing cells and kill them by secreting chemicals into the extracellular fluid. Helper T cells are involved in numerous aspects of immune defense. These include binding to antigen fragments complexed to class II MHC proteins on the surface of antigen-presenting cells, such as macrophages. This process activates the helper T cell, stimulating it to divide and enabling it to secrete IL-2 and other cytokines. The cytokines, in turn, stimulate B cells and cytotoxic T cells. Helper T cells, once activated, also bind to B cells and stimulate B-cell clonal selection. NK cells are similar to cytotoxic T cells but also function in some innate immune responses. Regulatory T cells are inhibitory; they act as a check on the activity of other T cells.

Critical-Thinking Skills—Multiple Choice:

1. d 2. b 3. b 4. d 5. c

Critical-Thinking Skills—Short Answer:

1. Immunoglobulins are proteins that consist of several structural domains, one of which is highly variable in its amino acid sequence. It is this variability that imparts upon a given immunoglobulin the ability to bind to a specific antigen among the many that an animal might encounter. Some immunoglobulins possess a domain that anchors the protein in a plasma membrane, allowing the protein to act as a receptor on, say, a T cell. Other immunoglobulins are lacking this domain and instead are secreted by a cell where they enter the body fluids and serve as antibodies.

2. Innate immunity is present at birth and is found in all animals. These defenses recognize general, conserved features common to a wide array of pathogens and include external barriers, such as the skin, and internal defenses involving phagocytes and other cells. Adaptive immunity develops *after* an animal has been exposed to a *particular* antigen. The responses include humoral and cell-mediated defenses. Adaptive immunity appears to be largely restricted to vertebrates. Unlike innate immunity, in adaptive immunity, the response to an antigen is greatly increased if an animal is exposed to that antigen again at some future time.

Chapter 41

Concept Checks

Figure 41.5 If the plasma protein concentration decreases, Starling forces favor a net outflow of fluid from a capillary because there is less osmotic force favoring fluid movement into the capillary.

Bioconnections

Figure 41.8 Positive feedback also occurs during the ovarian cycle in mammals at the time of ovulation (see Figure 39.9), and during the process of birth in mammals (see Figure 39.11).

Testing Your Knowledge

41.1: 1. b
41.2: 1. b 2. e
41.3: 1. b 2. b
41.4: 1. d

Assessing Your Knowledge and Skills

Foundational Knowledge—Multiple Choice:

1. b 2. d 3. c 4. e 5. e

Foundational Knowledge—Short Answer:

1. Light-headedness can occur in some people when donating blood, which is essentially a carefully controlled hemorrhage. Initially, as the homeostatic processes described in this chapter are just beginning, there is a period of instability with respect to blood pressure control. While lying or sitting down, this is rarely noticeable, but upon standing gravity counteracts the movement of blood through limb veins back to the heart, causing a sudden decrease in pressure. Fainting does not usually occur, however, because the baroreceptor reflex responds immediately to this sudden change in pressure, and within seconds the heart rate and cardiac output are increased due to the actions of the sympathetic nervous system. This phenomenon can actually happen anytime, even under ordinary circumstances, but is more noticeable when a person's blood volume is reduced such as when donating blood.

2. Hemorrhage is a loss of blood. Blood volume and blood pressure are directly related; that is, if blood volume decreases, so does blood pressure. Should blood pressure fall significantly, there will not be sufficient pressure to deliver a rate of blood flow that maintains the functions of cells in an animal's body. For instance, active organs, like an animal's brain, will not receive sufficient oxygen and nutrients to sustain the activity of brain cells. When this happens, cells may begin to die.

Critical-Thinking Skills—Multiple Choice:

1. a 2. a 3. c 4. c 5. a

Critical-Thinking Skills—Short Answer:

1. Animals in nature are confronted with many types of homeostatic challenges that often require integrated responses in multiple organ systems. For example, a bird flying at high altitudes, such as over a mountain range, faces the challenge of obtaining sufficient oxygen from the environment. You might predict that in such situations a bird may increase its breathing rate, adjust its cardiac output, or both; indeed, both of these changes and several others do occur. Likewise, imagine a fish that migrates between fresh and salt water, such as a salmon. These animals alter the function of their respiratory and urinary systems to help compensate for the changes in ion and water movement across the gills when moving from one environment to another. Yet another common example is starvation, during which the nervous, endocrine, urinary, and digestive systems help to maintain glucose homeostasis by processes such as gluconeogenesis.

2. All of the homeostatic responses to hemorrhage described in this chapter require energy. Increasing the activity of any muscle, including that of the heart and the respiratory muscles, requires a considerable increase in expenditure of energy (ATP). Activity of the nervous system, so vital to the compensatory response to hemorrhage, requires continual ATP production and hydrolysis to maintain ion concentration gradients across the plasma membrane of neurons; without this, there could be no flow of current along a neuron. Many of the transport processes in the kidneys also require ATP to drive the ion pumps that move ions across membranes and that create osmotic gradients for the movement of water.

Chapter 42

Concept Checks

Figure 42.2 In operant conditioning, an animal's behavior is reinforced by a consequence, either a reward or a punishment. In classical conditioning, an involuntary response comes to be associated with a stimulus that did not originally elicit the response, as with Pavlov's dog salivating at the sound of a metronome.

Figure 42.3 Tinbergen manipulated pinecones, but not all digger wasp nests are surrounded by pinecones. You could manipulate branches, twigs, stones, and leaves to determine the necessary size and dimensions of objects that digger wasps use as landmarks.

Figure 42.7 The individuals in the center of the group are less likely to be attacked than those on the edge of the group. This is referred to as the geometry of the selfish herd.

Figure 42.11 Because of the genetic benefit, the answer is nine cousins. Consider Hamilton's rule, expressed in the formula $rB > C$. Using cousins, $B = 9$, $r = 0.125$, and $C = 1$, and $1.125 > 1$. Using sisters, $B = 1$, $r = 0.5$, and $C = 0.5$. Because rB would not be greater than C, there would be no net genetic benefit in self-sacrifice.

Figure 42.12 All the larvae in the group are likely to be the progeny of one egg mass from one adult female moth. The death of the one caterpillar teaches a predator to avoid the pattern and benefits the caterpillar's close kin.

BioConnections

Figure 42.2 Toxic or bad-tasting prey species converge on the same color patterns to reinforce the basic distasteful design and enhance operant conditions.

Feature Investigation Questions

1. Tinbergen observed the activity of digger wasps as they prepared to leave the nest. Each time, the wasp hovered and flew around the nest for a period of time before leaving. Tinbergen suggested that during this time, the wasp was making a mental map of the nest site. He hypothesized that the wasp was using characteristics of the nest site, particularly landmarks, to help relocate it.

2. Tinbergen placed pinecones around the nest of the wasps. When the wasps left the nest, he removed the pinecones from the nest site and set them up in the same pattern a distance away, constructing a sham nest. For each trial, the wasps went directly to the sham nest, which had the pinecones around it. This indicated to Tinbergen that the wasps identified the nest based on the pinecone landmarks.

3. Yes. Tinbergen also conducted an experiment to determine if the wasps were responding to the visual cue of the pinecones or the chemical cue of the pinecone scent. The results of this experiment indicated that the wasps responded to the visual cue of the pinecones and not their scent.

Testing Your Knowledge

42.1: 1. b 2. e 3. c
42.2: 1. e
42.3: 1. d 2. b 3. c
42.4: 1. c 2. e
42.5: 1. c 2. b

Assessing Your Knowledge and Skills

Foundational Knowledge—Multiple Choice:

1. d 2. e 3. b 4. c 5. c

Foundational Knowledge—Short Answer:

1. The donation of the male's body to the female is the ultimate nuptial gift. It is possible that this meal enables the females to produce more eggs. In this way, the male's genes will be passed on to future generations.

ALS (amyotrophic lateral sclerosis), 777
Alternation of generations, 301, 302f, 529, 535f
　in flowering plants, 673, 673f, 674f
Alternative splicing, 215, 263–264, 263f
Altitude, gas pressure and, 796
Altruism, 941–943, 942f, 943f
Alveolata, 494–497, 494f, 495f, 496f, 499t
Alveoli, 800, 801f
Alzheimer, Alois, 740
Alzheimer's disease (AD), 245, 245t, 740
AM (arbuscular mycorrhizae), 499t, 500–501, 501f, 520–521, 520f
Amanita muscaria, 505f
Amanita virosa, 505
Amazon rain forest, 1017
Amblyomma hebraeum, 576f
Amborella trichopoda, 545, 545f
Ambulocetus, 393f, 394
Amenorrhea, 870
Amensalism, 957, 957t
American avocet, 602f
American buffalo, 1019, 1036
American chestnut, 1024–1025, 1025f, 1032–1033
American ginseng, 1020
American Lung Association, 809–810
Ames, Bruce, 274
Ames test, 274–275, 274f
Amine hormones, 849
Amino acids
　essential, 814–815
　found in living organisms, 48f
　gene mutations and, 268–269, 268t
　genetic code and, 216, 216t
　neurotransmitters, 728
　overview of, 47–48, 48f, 49f
　protein digestion into, 825, 825f
　use and storage of, 828–830, 828f, 829f
Amino functional group, 39t
Amino group, 47
Aminoacyl site (A site), 223, 223f
Aminoacyl tRNA, 222
Aminoacyl-tRNA synthetases, 221–222, 222f
Ammonia, 28f
Ammonia (NH_3), 833–834, 833f
Ammonification, 1013, 1014f
Ammonium ions (NH_4^+), 833–834, 833f
Amnion, 598, 599f
Amniotes. *See also* Mammals
　birds, 589t, 599–602, 601f, 602f, 782–783, 783f, 820, 820f, 886–887, 887f, 966, 966f, 968, 969t
　as clade of vertebrates, 588f
　eggs and terrestrial life adaptations of, 598–599, 599f
　reptiles, 589t, 599–601, 600f, 601f, 751, 751f, 886–887, 887f
Amniotic egg, 598–599, 599f
Amoebae, 490, 490f, 497
Amoebozoa, 497, 499t
Amphibia, 589t
Amphibians, 589t, 596–598, 597f, 733f, 797, 886–887, 887f, 889f
Amphipathic molecules, 31, 31f, 99
Amphiprion ocellaris, 10–11, 10f
Amplicon analysis, 512f, 513
Amplicons, 513
amp^R gene, 366, 367f
Ampulla, echinoderms, 580f, 581

Amylase, 820, 823, 824f
β-Amyloid, 245, 740
Amyotrophic lateral sclerosis (ALS), 777
Anabolic reactions, 129
Anabolism, 75
Anaerobes, 488
Anaerobic respiration, 146, 147f
Anagenesis, 437, 437f
Analogous structures, 395, 395f
Anaphase (mitosis), 293, 295f
Anaphase I (meiosis), 299f, 300
Anaphase II (meiosis), 299f
Anatomical homologies, 392t, 396–397, 396f, 397t
Anatomy, 13, 13f
　fossil record and, 453t
Anchoring junctions, 116–118, 117f
Androgen receptors, 53, 53f
Androgens, 851f, 867
Aneuploid conditions in humans, 306, 306t
Aneuploidy, 305, 305f, 306, 306t
Anfinsen, Christian, 51–53, 52f
Angina pectoris, 808
Angiography, 808
Angioplasty, 808, 808f
Angiosperms. *See also* Flowering plant reproduction and development
　ancestry and diversity of, 530f, 531, 534, 536f
　architecture of, 614, 614f
　attack responses, 643–646, 644f, 645f, 646f
　behavioral responses, 631–633, 631f, 632f, 633f
　cell communication, 632–633, 633f
　cellular level transport, 656–661, 657f, 658f, 659f
　definition of, 534, 608, 610
　distinctive features of, 549t
　diversity of modern, 542–549, 542f, 543f, 544f, 545f, 546f, 547f, 548f–549f
　drought stress adaptations, 648, 660, 667, 667f, 668f
　flowers of, 542–544, 542f, 543f, 544f, 673–675, 674f, 677–680, 677f, 678f, 679f, 680f–681f
　fossils, 537f
　fruits, 534, 536f, 546, 546f, 576, 612, 675–676, 675f, 685–686, 686f
　gravity and touch responses, 641–643, 641f, 642f, 643f
　growth and development, 614–618, 614f, 615f, 615t, 616f, 617f, 618f
　hormones, 619, 631, 634–638, 634t, 635f, 637f
　human influences on, 550–551, 550f
　introduction to, 609, 630, 648
　life cycle, 543, 543f, 610–613, 610f, 611f, 612f, 613f, 673–676, 673f, 674f, 675f
　light responses, 638–640, 639f, 640f, 641f
　long-distance transport, 657f, 663–670, 664f, 665f, 666f, 667f, 668f, 669f
　monocots and eudicots, 545, 545f
　nutritional requirements, 649–652, 649f, 649t, 650f, 651f, 652f
　phyla, 530f, 531, 545f
　reproductive development of, 612
　rise of, 537
　root system, 612f, 616–617, 626, 626f, 627f
　secondary metabolites, 546–548, 547f, 548f–549f
　seeds, 545–546, 545f, 546f, 576, 610–611, 610f, 611f, 639, 639f, 640f, 675–676, 675f, 685–687, 685f, 687f, 689
　sexual cycle, 610, 610f, 673, 673f, 674f

　shoot system, 612f, 618–625, 619f, 620f, 621f–622f, 623f, 624f, 625f
　soil nutrition, 652–656, 653f, 654f, 655f, 656f
　structure and life span variations of, 613, 613t
　structure of, 542f
　tissue level transport, 661–663, 661f, 662f
　tissues and specialized cells, 615, 615f, 615t
Angiotensin II (AII), 924, 925f
Animal cell
　compartmentalization of, 71f, 75f
　complexity of, 95t
　extracellular matrix (ECM) in, 89–94, 91f
　gap junctions of, 113, 113f
　nucleus and endomembrane system, 80f
　osmosis in, 105, 106f
　polysaccharide secretion by, 92, 92f
　structure of, 71f
　tight junctions of, 118, 118f
Animal classification, 552
　body cavity, 556–557, 558f
　body plans, 554f, 555–556, 555f, 556f, 557f
　choanoflagellate-like ancestor, 554–555, 554f
　embryonic development, 555–556, 557f
　germ layers, 555
　phyla, 554, 554f
　protostome clades, 557–558, 559f
　segmentation, 557, 558f
Animal pole, 887, 887f
Animals. *See also* Invertebrates; Vertebrates
　body organization, 694–699, 694f, 695f, 696f, 697f, 698f, 698t
　body temperature regulation in, 703–707, 704f, 705f, 706f
　characteristics of, 553, 553f
　chromosome number changes in, 305–306, 306t
　diversity of, 552, 554–555
　embryonic development in, 555–556, 557f, 885–891, 886f, 887f, 888f, 889f, 890f
　as eukaryotes, 9f
　evolution of, 554–555, 554f
　excretory systems, 833–836, 833f, 834f, 835f
　feeding habits, 815
　form and function relationship in, 699–700, 700f
　homeostasis principles in, 700–703, 702f
　internal fluid homeostasis in, 693, 707–712, 708f, 709f, 710f–711f
　introduction to, 693
　microbiome engineering for, 524–526, 525f
　microbiomes, 521–523, 522f–523f
　nutrients required by, 814–815, 815t, 816t
　nutrition, 553, 553f, 814–817, 814f, 815t, 816t
　sampling of, 949–950, 949f, 950f
　sexual reproduction in, 871
　skeletons of, 766–768, 766f, 767f
　in web of life, 447f
Anions, 27
Anna's hummingbird, 602f
Annelids, 556–557, 558f, 570–572, 570f, 571f, 730, 730f, 835, 835f
Annuals, 613
Anodorhynchus hyacinthinus, 602f
Anopheles mosquitoes, 496–497, 496f
Antagonists, 769, 769f
Antarctic krill, 579f
Anteaters, 3f, 4, 395, 395f
Anterior, 555
Anterior pituitary gland, 850f, 852f, 853, 853t

Anther, 674, 674f, 675
Anthozoa, 562–563, 563f, 563t
Anthrax, 484
Anthropocene, 1002
Anthropoidea, 463, 464f, 465f
Antibiotic resistance, 487
Antibiotic resistance, acquired, 360–361
Antibiotics, 222–223
Antibodies, 902, 905–906, 905f, 906f
Anticodon, 217, 218f
Antidiuretic hormone (ADH), 840–841, 841f, 853–854, 863, 863f, 926
Antigen-presenting cells (APCs), 907–908, 908f
Antigens, 902, 904, 904f
  recognition of, 906–908, 908f
Antiparallel, 193, 194f, 196
Antiporter, 110, 110f
Antipredator adaptations, 959–961, 960f
Antirrhinum majus, 678, 679f
Antisense RNA, 237–238, 237f–238f
Ants, 577t, 578
Anura, 597, 597f
Anus, 819, 819f, 827
Aorta, 786, 787f
APCs (antigen-presenting cells), 907–908, 908f
Apical meristems, 529
Apical-basal polarity, 614, 614f
Apicomplexa, 494, 496–497, 496f, 499t
Aplysina archeri, 561f
Apoda, 597, 597f
Apomixis, 689
Apoplastic transport, 661, 661f, 662f
Apoptosis, 280–281, 290
Aposematic coloration, 960f, 961
Appendicular skeleton, 767, 767f
Apple maggot fly, 426
Aquaporins, 108–109, 108f, 840–841
Aquatic biomes, 998, 998f
Aquatic ecosystems, 983–984
  food chain in, 979f
Aqueous solutions, 30
Aquifers, 1011
Aquila verreauxii, 602f
Arabidopsis, 684f, 685
Arabidopsis thaliana, 525, 611f, 626, 643–646
  alternative splicing in, 264, 264t
  genome of, 376
  as model organism, 13
Arachnida, 575, 576f
Arber, Werner, 364
Arbuscles, 500
Arbuscular mycorrhizae (AM), 499t, 500–501, 501f, 520–521, 520f
Archaea
  ancient lineage, 478–479, 478f, 479f
  distinguishing features of, 435t
  diversity and ecological importance of, 481–482, 481f
  genomes of, 375, 376t
  horizontal gene transfer in evolution of, 479
  overview of, 8, 9f, 10, 434, 434f
  phyla, 481
  phylogenetic tree, 478, 478f
  prokaryotic cells of, 8, 69
  structure of, 481, 481f
  in web of life, 447f
Archaean eon, 453, 454f, 456

Archaeopteryx lithographica, 462, 462f, 601, 601f
Archenteron, 888, 889f
Arid climate, 648, 660, 667, 667f, 668f
Armor, 960f, 961
Arnold, William, 158–159
Arnon, Daniel, 157
ART (assisted reproductive technologies), 892
Arteries, 782, 783f, 790, 791f
  disease of, 808, 809f
Arterioles, 790, 791f
  hemorrhage response, 924, 925f
  resistance, 793–794
Arthrobotrys anchonia, 506f
Arthropods
  body plans, 573–574, 574f
  Chelicerata, 574–575, 575t, 576f
  Crustacea, 574, 575t, 578–579, 579f
  diversity, 574, 575t, 577–578, 577t
  excretory systems, 835, 835f
  Hexapoda, 574, 575t, 576–578, 577t, 578f
  Myriapoda, 574–576, 575t, 576f
  nervous system, 730, 730f
  respiratory system, 798–799, 799f
  segmentation, 557, 558f, 574, 574f
  skeletons of, 766, 766f
  subphyla and characteristics of, 574, 575t
Artificial selection, 395–396, 396f
  microbiomes, 524f, 525–526, 525f
  plants, 550, 550f
Artiodactyla, 605t
Asarum canadensis, 520f
Ascaris lumbricoides, 573
Ascomycetes, 499t, 501, 501f
Ascorbic acid, 816t
Asexual reproduction
  advantages and disadvantages, 872
  annelids, 571
  definition of, 291, 871
  flowering plants, 688–689, 688f
  fungi, 503–504, 504f
  mollusks, 568–569
  overview of, 871–872, 871f
  sponges, 562
Asian gaur, 1038
Aspergillus fumigatus, 899–901, 900f–901f
Aspergillus versicolor, 504f
Aspirin, 1, 2f
Assimilation, 1013, 1014f
Assisted reproductive technologies (ART), 892
Association, 335–336
Associative learning, 932
Asteroidea, 581, 582t
Asthma, 808–809, 810f
AT/GC rule, 193, 194f, 196, 198–199, 199f, 200
Atherosclerosis, 808, 808f
Athlete's foot, 504
Athletic performance, 710–712, 710f–711f
Atlantic cod, 1028
Atlas moth, 552
Atmospheric circulation, 989–990, 990f
Atmospheric composition, 455
Atmospheric pressure, 796
Atomic mass, 23–24
Atomic nucleus, 22, 22f
Atomic number, 23
Atoms, 5, 6f, 22–24, 22f
ATP synthase, 140–145, 142f, 143f, 144f

Atrioventricular (AV) node, 787–789, 788f, 789f
Atrioventricular (AV) valves, 786
Atrium, 782, 783f, 787–788, 787f, 788f
Attachment, 349, 350f
Attack, plant responses to, 643–646, 644f, 645f, 646f
Attenborough, David, 11
Atticus atlas, 552
Audition, 747–748, 748f, 749f
Auditory communication, 936
Australian pine, 1023
Australopithecus, 465–467, 467f
Australopithecus afarensis, 466, 466f
Australopithecus africanus, 466–467, 466f
Australopithecus garhi, 466f, 467
Autocrine signaling, 171, 172f
Autoimmune disease, 765, 911
Autonomic nervous system, 732–735, 734f
  hemorrhage response, 919–921, 920f, 924–926, 925f
Autoradiography, 163f–164f
Autosomes, 286, 320
Autotomy, echinoderms, 581
Autotrophs, 151, 456–457, 456f, 488–489, 979
AUX1 (auxin influx carrier), 635, 635f
Auxin efflux carriers, 635, 635f
Auxin influx carrier (AUX1), 635, 635f
Auxin-responsive genes, 635
Auxins, 619, 634t, 635–636, 635f
AV (atrioventricular) node, 787–789, 788f, 789f
AV (atrioventricular) valves, 786
Avery, Oswald, 189–191
Aves, 589t, 599–602, 601f, 602f
Axel, Richard, 759–761, 760f
Axial skeleton, 767, 767f
Axillary buds, 619, 619f
Axon hillock, 715, 716f
Axon terminals, 715, 716f, 724, 725f
  posterior pituitary, 853–854
Axoneme, 78
Axons, 715, 716f
  transmission of electrical signals along, 722–725, 723f, 725f
Azidothymidine (AZT), 353

B

B cells, 902–903, 903f, 904f
  activation of, 908
  immunoglobulins produced by, 905–906, 905f, 906f
  plasma cells produced by, 906, 907f
  self recognition by, 911
Bacillariophyceae, 489, 490f, 497
Bacilli, 486, 486f
Bacillus anthracis, 376t, 484
Bacillus subtilis, 224, 224f, 354f
Backbone, 193, 193f
  chordates with. *See* Vertebrates
Bacskai, Brian, 183
Bacteria
  ancient lineage, 478–479, 478f, 479f
  antibiotics, 222–223
  binary fission of, 293, 295, 296f, 356, 357f, 483–484, 484f
  cell complexity of, 95t
  cell division in, 293, 295, 296f
  cell structure of, 70f

Bacteria, *continued*
  cell walls, 486–487, 486f, 487f
  cells of, 94–95
  cellular adaptations, 487–488, 488f
  cellular surface coating, 487
  chromosomal loops and DNA supercoiling, 354–355, 355f
  circular chromosomes of, 354, 375
  colony, 356, 356f
  counts, 483–484, 484f
  cyanobacteria as photosynthetic, 482–483, 483f
  distinguishing features of, 435t
  diversity and ecological importance of, 482–485, 482t, 483f, 484f
  diversity in cells and metabolism of, 485–489, 485f, 486f, 487f, 488f
  DNA microarrays and, 373t
  domain of, 434, 434f
  elongation stage in translation in, 226, 226f
  flagella, 487–488, 488f
  gene regulation in, 249–250, 249f, 251f
  gene transfer between, 357–361, 358f, 358t, 359f, 360f
  genetic properties of, 354–356, 354f, 355f, 356f
  genome of, 375, 376t
  grouped by nutrition and response to oxygen, 488–489
  horizontal gene transfer, 479
  initiation of translation in, 225f
  introduction to, 345
  microbiomes, 517–518
  molecular gene expression, 210f
  motility, 487–488, 488f
  nitrogen fixation, 655, 656f
  overview of, 8, 9f, 10
  pathogens, 896
  phyla included in, 482t
  phylogenetic tree, 478, 478f
  plasma membrane, 481, 481f, 485f, 486
  plasmids of, 355, 355f
  prokaryotic cells of, 69
  Proteobacteria, 482t, 483
  regulation of transcription in, 250, 251f, 252–256
  reproduction of, 483–484, 484f
  ribosomes of, 222, 223f, 223t
  shapes of, 486, 486f
  as single cells, 479, 479f
  structure of, 481, 481f, 485–489, 485f, 486f, 487f, 488f
  termination of translation in, 227, 227f
  transformation, 359, 359f
  transformation experiments, 188–189, 188f
  translation in, 226–227, 226f, 227f, 227t
  ulcers with, 843–845, 844f
  in web of life, 447f
Bacterial mRNA, 216–217, 216f
Bacteriophages, 347, 348f
  defense against, 242–244, 242f–243f
  latency in, 352
  reproductive cycle of, 349, 350f–351f, 352
Bacteroides, 517
Bacteroidetes, 482t
Baculoviruses, 346t
Baker's yeast. *See Saccharomyces cerevisiae*
Balance, sense of, 748–751, 749f, 750f
Balanced polymorphism, 405
Balancing selection, 405–406, 406f

Balloon angioplasty, 808, 809f
Banting, Frederick, 858–861, 859f–860f
Bark, 624–625, 624f, 625f
Barnacles, 578–579
Baroreceptor reflex, 919–921, 920f, 927–928
Baroreceptors, 919–921, 920f
Barr, Murray, 331
Barr body, 331–334, 332f
Bartel, David, 65
Basal bodies, 78
Basal cells, 282, 282f
Basal lamina, 696
Basal metabolic rate (BMR), 704
Basal nuclei, 735f, 736, 736f
Basal transcription, 257
Base analogues, 273
Base excision, 276t
Base pairs, 193
Base substitution, 268, 268t
Basement membrane, 282, 282f
Bases, 34, 35f, 54, 55f
Basidiomycetes, 499t, 501, 501f
Basilar membrane, 748, 749f
Basophils, 897f, 898
Batesian mimicry, 960f, 961
Bateson, William, 340–341
HMS *Beagle*, 389, 389f
Beaks, 602, 602f
Bears, 11
Beavers, 604f, 951, 952t, 1036, 1037f
Bedrock, 652, 653f
Bees, 525–526, 525f, 577t, 578
Beetles, 577–578, 577t
Behavior, 931
  genetics and learning influences on, 932–935, 932f, 933f–934f, 935f
Behavioral adaptations, heat gain and loss through, 706
Behavioral ecology
  altruism, 941–943, 942f, 943f
  communication, 936–938, 937f
  genetics and learning influences on behavior, 932–935, 932f, 933f–934f, 935f
  introduction to, 931–932
  living in groups, 938–941, 938f, 939f, 940f, 940t
  mating behavior, 936, 944–946, 944f, 945f, 946f
  optimality theory, 938–941, 938f, 939f, 940f, 940t
Behavioral isolation, 420, 420f, 421f
Beijerinck, Martinus, 345
Belding's ground squirrels, 942
Bendall, Fay, 159
Bending of filament, 77f, 78
Benign tumor, 277f, 278
Benson, Andrew Adam, 161
Bernal, John, 62
Bertram, Ewart, 331
Best, Charles, 858–861, 859f–860f
β pleated sheet, 49, 50f
β subunits, 142–143, 143f
Betta splendens, 593
Bidirectional replication, 200, 200f
Bienertia cycloptera, 651, 652f
Biennials, 613
Bifidobacterium longum, 521
Big Bang, 450
Bilateral symmetry, 459, 459f, 555, 555f
Bilateria, 555–557, 555f, 556f, 557f, 558f

Bile, 825–827, 826f
Binary fission, 483–484, 484f
  in bacteria, 293, 295, 296f, 356, 357f
  in mitochondria and chloroplasts, 87–88, 87f
Binocular vision, 463, 464f
Binomial nomenclature, 11, 435–436
Biochemical regulation, 131
Biochemistry, 37, 134
Biodiversity
  economic benefits of, 1029–1030, 1029f
  ecosystem function and, 1031–1033, 1031f, 1032f
  ethical reasons for conservation of, 1030
  genetic, species, and ecosystem diversity, 1028
  introduction to, 1027
  value of, to human welfare, 1028–1030, 1030t
Biodiversity hot spots, 1033–1034, 1034f, 1035f
Biofilms, 487, 518, 518f
Biogenic amines, 728
Biogeochemical cycles, 1008–1014
Biogeographic regions, 1000, 1000f
Biogeography, 392t, 394, 999
Biological control, 961f
Biological evolution, 4
Biological membranes, 98, 98t. *See also* Membranes
Biological species concept, 419
Biological stimuli, plant responses to, 631, 632f
Biology
  classification of living organisms, 8–11, 9f, 10f
  definition of, 1
  discovery-based science and hypothesis testing, 14–16, 15f, 16f
  evolution, study of, 11
  evolutionary history, 7–8
  as experimental science, 3f, 4
  hypothesis versus theory, 14
  levels of organization, 5, 6f
  levels of organization, investigating life at, 13–14, 13f
  model-based learning of, 17
  models and simulations for, 3f, 5
  observation and experimentation, 17
  principles of, 2–5
  as quantitative science, 3f, 4–5
  as scientific discipline, 12–18
  as social discipline, 18
  society, influence on, 3f, 5
  unity and diversity of life, 7–12
Bioluminescence, 560
Biomagnification, 1014–1016, 1015f
Biomass
  ecosystem production of, 983–986, 983f, 984f, 985f, 986f
  pyramid of, 980, 982f
Biomes, 988, 995–998, 995f, 996f, 997f, 998f
Biophilia, 1030, 1040
Biosphere, 5, 6f, 151
Biotic interactions, 930
Biotin, 816t
Biparental inheritance, 339
Bipedalism, 464–466, 466f
Birds, 589t, 599–600, 601f
  adaptations, 601–602, 602f
  DDT and, 1014–1015, 1015f
  digestive system, 820, 820f
  dinosaurs and, 462, 462f
  double circulation, 782–783, 783f
  meroblastic cleavage, 886–887, 887f

overexploitation of, 1019–1020
species richness, 966, 966f, 968, 969t
Birds' nest fungi, 504, 505f
Birds-of-paradise, 11, 12f
Birks, H. John, 966
Birth, 883–885, 883f, 884f, 885f
Biting teeth, 604f
Bivalent, 297, 298f
Bivalvia, 569, 569f, 569t
Black grouse, 946f
Black tip grouper, 593f
Black widow spider, 575, 576f
Blackberries, 686, 686f
Blade, 620, 620f
Blanton, Laura, 522–524, 522f–523f
Blastocoel, 886
Blastocyst, 880, 880f, 887, 887f, 888f
Blastoderm, 887, 887f
Blastomeres, 886
Blastomyces dermatitidis, 506
Blastopore, 555–556, 556f, 557f, 888, 889f
Blastula, 555, 556f, 886, 887f
Blocker function, 232
Blood, 697f, 782
composition of, 784–786, 785f, 786f
gas transport by, 803–806, 803f, 804f, 805f
redistribution of, 921–922, 922f
Blood clotting, 785, 786f
Blood flow, 793–795, 794t, 795f
skin, 705, 706f
Blood flukes, 566
Blood glucose, 830–832, 832f, 856–857, 858f
Blood loss. *See* Hemorrhage
Blood pressure
baroreceptor reflex and, 919–921, 920f, 927–928
blood flow and resistance, relationship to, 793–795, 794t, 795f
blood volume and, 917–918, 918f
definition of, 789
hemorrhage effects on, 917–918, 918f, 927t
redistribution of blood with low, 921–922, 922f
respiratory system aids, 922
table salt and, 21
Blood vessels, 781f, 782–783, 790–793, 791f, 792f, 793f
disease of, 808, 809f
resistance, 793–795, 794t, 795f
Blood volume, 917–918, 918f, 927t
Blooming flower, 679–680, 680f–681f
Blooms
algal, 983–984
cyanobacterial, 483
phytoplankton, 491, 491f
Blue-green blooms, 483
BMP4 (bone morphogenetic protein 4), 427–428, 428f
BMP4 gene, 427–428, 428f
BMR (basal metabolic rate), 704
Bobcat, 604f
Body cavity, 556–557, 558f
Body fluids, homeostasis of, 693, 707–712, 708f, 709f, 710f–711f
Body plans
animal, classification, 554f, 555–556, 555f, 556f, 557f
animal, organization, 694–699, 694f, 695f, 696f, 697f, 698f, 698t
annelids, 570–571, 570f
arthropods, 573–574, 574f
chordates, 582, 582f
cnidarians, 562, 563f
crustaceans, 578, 579f
echinoderms, 580–581, 580f
flatworms, 564, 564f
insects, 574, 574f, 576–577
mollusks, 568, 568f
Body surface
gas exchange across, 797
immune defense, 897
Body temperature, homeostatic regulation of, 703–707, 704f, 705f, 706f
Bodyguard wasp, 643
Bohr effect, 804–805, 805f, 922
Bolotin, Alexander, 242
Bonds, chemical. *See* Chemical bonds
Bone, 697f
bird, 602
components of, 767
public health impacts and, 777, 777f
skeletal muscle movement of, 768–769, 768f, 769f
Bone diseases, 777, 777f
Bone morphogenetic protein 4 (BMP4), 427–428, 428f
Bonobos, 944
Bont ticks, 576f
Bony fishes, 588f, 589t, 590, 591f, 592–593, 593f, 594f
Boore, Jeffrey, 574
Boron, 649t
Borrelia burgdorferi, 375
Bosch, Carl, 656
Botanic Gardens Conservation International Organization, 1020
Botox, 484
Botrychium lunaria, 533f
Bottjer, David, 459
Bottleneck effect, 410–411, 411f
Botulinum toxin, 484
Boveri, Theodor, 316
Bowman's capsule, 836–838, 837f, 838f
Box jellies, 562, 563t
Boyer, Paul, 143
Boyle, Robert, 799
Boyle's law, 799, 799f
Brachiopod, 460, 460f, 567, 568f
Brachychiton, 612f
Bracts, 623, 623f
Brain
cerebellum, 735, 735f
cerebral cortex, 730–731, 731f, 735f, 736, 736f
cerebral ganglia, 564, 564f
cerebrum, 730–731, 731f, 735f, 736, 736f
definition of, 715
divisions of, 730, 731f
evolution of, 729–730, 730f
forebrain, 730, 730f, 735–736, 735f, 736f
fruit fly, 714, 730, 730f
hindbrain, 730, 730f, 735, 735f
human, 735–739, 735f, 736f, 737f, 738f, 771, 771f
imaging studies of, 737–739, 737f, 738f
jaw muscles and, 771, 771f
midbrain, 730, 730f, 735, 735f
of musicians versus nonmusicians, 737–739, 738f
size of, 463
Brainstem, 735
Branched filaments, 483, 483f
Branching tubes in lungs, 800, 801f
Brassinosteroids, 634t, 638
Brazilian pepper, 1023
Breakdown, of organic molecules, 40, 40f
Brittle star, 581, 581f, 582t
Bronchi, 800, 801f
Bronchioles, 800, 801f
hyperreactive, 808–809, 810f
Bronchodilators, 809
Brooker, Robert, 109
Brown adipose tissue, 706–707
Brown recluse spider, 575
Bruneau, Benoit, 783
Brush border, 822f, 823
Brush border enzymes, 823–825, 824f
Bryophytes, 530f, 531, 531f, 532f, 549t
in evolutionary history of land plants, 537
Bryozoa, 567, 568f
Bubonic plague, 895, 1003f
Buccal pumping, 596
Buck, Linda, 759–761, 760f
Bud scales, 623, 623f
Budding, 871, 871f
Buds, 611
Buffalo, 1019, 1036
Buffalo clover, 1029
Buffenstein, Rochelle, 586
Buffers, 35
Buffon, Georges, 388
Bugs, 577t
Bulk feeders, 553f
Bulk flow, 663
Bundle-sheath cells, 165–166, 166f
Burmese pythons, 955, 955f, 1024, 1024f
Bursera simaruba, 949
Burying beetles, 945
Bush, Guy, 426
Butterflies, 552, 577–578, 577t, 578f, 755, 960f, 961
Buttress roots, 626, 626f

C

C (cytosine), 54, 55f, 192, 192f, 193, 193f
C_3 plants, 165, 165f
C_4 plants, 165–166, 166f, 167f
Cactoblastis cactorum, 961, 961f
Cactus, 623, 623f
Cactus moth, 961, 961f
Cade, Robert, 710–712, 710f–711f
Cade, Tom, 1038
Cadherins, 117
Caecilians, 597–598, 597f
Caenorhabditis elegans, 13, 573
alternative splicing in, 264, 264t
Café marron, 1028, 1028f
Caffeine, 547, 547f
Calamospiza melanocorys, 945
Calcitonin, 862–863
Calcium
cross-bridge cycle, 771–772, 772f
electrical excitation, 773, 774f
fertilization and, 876, 876f
life and, 24, 24t
muscle contraction regulation, 772–773, 773f
as plant nutrient, 649t
regulation of, 861–863, 862f
smooth ER storage of, 82
Calcium-dependent protein kinases (CDPKs), 632

Cisternae, 81, 81f, 82, 83f
Cisternal maturation model, 82
CITES (Convention on International Trade in Endangered Species of Wild Fauna and Flora), 539
Citric acid cycle, 132, 133f, 138, 139f, 140f
Clade, 438
Cladistics
 ancestral and derived characters, 440–441, 441f
 shared derived characters and, 441–444, 441f, 442f, 443f
Cladogenesis, 422, 437, 437f
Cladograms, 440–444, 441f, 442f, 443f
Cladonia, 519f
Cladophora, 493f, 511f, 518, 518f
Clams, 567–570, 569f, 569t
Class, 10f, 11, 435, 436f
Classical conditioning, 932–933
Claudin, 118, 118f
Claviceps purpurea, 505, 505f
Cleavage, 555–556, 556f, 557f, 886–887, 886f, 887f, 888f
Cleavage furrow, 293, 296f
Cleavage phase, 134
Clements, Frederic, 971
Climate
 atmospheric circulation, 989–990, 990f
 carbon dioxide and, 651, 1006–1007, 1006f, 1006t, 1007f
 definition of, 989
 elevation and proximity of landmass, 992–993, 992f, 993f
 introduction to, 988
 major biomes, 995–998, 995f, 996f, 997f, 998f
 ocean currents, waves, and tides, 993–994, 993f, 994f
 overview of, 989–994
 plant transpiration effects on, 664–665, 665f
 tilt and rotation of Earth, 990–992, 991f
Climate change, 1005–1008, 1036
Climax community, 971
Cloaca, 819, 820f
Clonal deletion, 911
Clonal inactivation, 911
Clonal selection, 906, 907f
Cloning, 1038–1039, 1039f
Clonorchis sinensis, 566, 566f
Closed circulatory system, 570, 781f, 782–784, 783f, 784f, 791f
Closed conformation, 259
Clostridium, 488
Clostridium botulinum, 484
Clostridium difficile, 484f
Clostridium perfringens, 488
Clostridium symbiosum, 523
Clostridium tetani, 484
Clot formation, 785, 786f
Clownfish, 10–11, 10f
Club mosses, 533f
CML (chronic myelogenous leukemia), 280, 280f
Cnidaria, 562–563, 563f, 563t, 729, 730f, 817, 817f
Cnidocil, 562
Cnidocytes, 562, 563f
CNS. *See* Central nervous system
CO (cardiac output), 794, 794t, 795f, 919, 927t
CO_2. *See* Carbon dioxide
CoA (Coenzyme A), 137, 138f

Coacervates, 63, 63f
Coactivators, 258, 259f
Coal Age, 537, 538f
Coal deposits, 537, 538f
Cocci, 486, 486f
Coccidioides immitis, 506
Cochlea, 748, 748f, 749f, 763
Cod, 1020, 1020f, 1022, 1028
Coding sequence, 216t, 217, 218f
Coding strand, 212
Codons, 216, 216f, 217, 225, 225f, 226–227, 227f
Coefficient of relatedness, 941, 941f
Coelacanths, 589t, 593, 594f
Coelom, 556
Coelomates, 556, 558f
Coenzyme A (CoA), 137, 138f
Coenzymes, 128
Cofactors, 128
Coffea arabica, 547f
Cogongrass, 1037, 1038f
Cohesion, 33, 33f
Cohesion-tension theory, 663, 664f
Cold desert, 995f, 996, 996f, 997f
Coleoptera, 577t
Coleoptile, 687, 687f
Collagen, 91, 91t, 92
 nematode, 573
Collecting duct, 837, 837f, 840–841, 840f, 841f
Collenchyma tissue, 615, 615f
Colligative properties, 32
Collins, Francis, 16
Collip, James, 858–861, 859f–860f
Colon, 827
Colonial bentgrass, 404–405, 405f
Colony, bacterial, 356, 356f
Colony of cells, 483, 483f
Color, of flowers, 678
Color blindness, 755–756, 755f
Color vision, 755–756, 755f
Comb jellies, 560, 560f
Combinatorial control, 257
Comma-shaped bacteria, 486, 486f
Commensalism, 957, 957t, 962
Commercial fisheries, 1020, 1020f, 1022, 1022f, 1029–1030, 1029f
Common ancestor, homologies and, 396–398, 396f, 397f
Communal courting, 945, 946f
Communication, 936–938, 937f. *See also* Cell communication
Community, 5, 6f, 965
 diversity of, 966–968, 967f, 969t
 succession, 971–974, 972f, 973f, 974f
Community ecology, 965
 community stability, 970, 970f
 food webs and energy flow, 979–983, 979f, 981f, 982f
 island biogeography, 974–979, 975f, 976f, 977f–978f
 species richness, 966–970, 966f, 967f, 968f, 969t, 970f, 976f, 977f
 succession, 971–974, 972f, 973f, 974f
Community stability, 970, 970f
Compaction, of chromosomes, 205, 206f, 207
Compartmentalization, 70, 71f, 73f, 75f
Compensatory reflex, 919–921, 920f
Competent, 358, 366

Competition, 956–958, 957f, 957t, 958f, 959f
Competitive exclusion principle, 957–958
Competitive inhibition, 127f, 128
Complement proteins, 899
Complementary DNA (cDNA), 367, 372, 759–761, 760f
Complete flowers, 677
Complete metamorphosis, 578, 578f
Complex interactions, 3f, 4, 51f
Compound, 25
Compound eyes, 744, 752, 753f
Compound microscope, 66
Concentration, 31
Concentration gradient, 104, 104f
 neuron, 718–721, 719f, 720f
Condoms, 892, 893f
Conduction, heat, 705
Cones, 754–756, 754f
Confocal fluorescence microscopy, 68f
Congenital insensitivity to pain, 744
Conidia, 504, 504f
Conifers, 530f, 531, 534, 536f, 539f, 540–542, 540f, 541f, 549t
Conjugation, 357–359, 358f, 358t
Connective tissues, 694f, 696, 697f, 698f
Connell, Joseph, 958, 972–973
Connexon, 113
Conservation biology
 cloning, 1038–1039, 1039f
 habitat restoration, 1037–1038, 1038f
 introduction to, 1027
 preserve design, 1035–1036, 1035f, 1036f
 protected areas, 1033–1035, 1034f
 single-species approaches to, 1036–1037, 1037f
 strategies for, 1033–1040
 sustainability and, 1039–1040
Conservative mechanism, 197, 197f
Constant region, 905f, 906
Constitutive genes, 248
Contact-dependent signaling, 171, 172f
Contigs, 513, 513f
Continental drift, 455, 455f, 999
Contraception, 892–893, 893f
Contractile tubes, 783–784
Contractile vacuoles, 84, 84f, 105, 106f
Contraction of skeletal muscle, 768–775, 768f, 769f, 770f, 771f, 772f, 773f, 774f
Contrast, 66
Control group, 15, 15f
Convection, 705
Convention on International Trade in Endangered Species of Wild Fauna and Flora (CITES), 539
Convergent evolution, 392t, 395, 395f
Cope's giant salamander, 598
Copper, 649t
Coral crab, 579f
Coral reef, 998, 998f
Corals, 562, 563t
Core promoter, 257, 258f
Coriolis, Gaspard Gustave de, 991
Coriolis force, 991–992, 992f
Cork cambium, 624f, 625, 625f
Corn, 550, 550f, 664, 677, 678f, 687f
Cornea, 753, 753f
Corona, 567, 567f
Coronary artery bypass, 808
Coronary artery disease, 808

Corpus luteum, 881, 881f
Correns, Carl, 310, 322, 337–338
Cortex, 615
Cortical collecting duct, 837, 837f, 840–841, 840f, 841f
Cortical reaction, 876, 876f
Cortisol, 856, 864
Corwin, Jeff, 1027
Costanza, Robert, 1030
Cotesia rubecula, 643
Cotranslational sorting, 89, 90f
Cotyledons, 684f, 685, 685f, 687, 687f
Countercurrent exchange, 705, 706f, 797
Countercurrent multiplication system, 840
Courtship, 936
Covalent bonds
   carbon and, 38, 38f
   common element formation of, 26f
   electrons and, 25–26
   formation of, 25f
   nucleotides, 192–193, 193f
   polar or nonpolar, 26
Covalent histone modification, 329t
Covalent modifications, 260, 260f, 272–273, 273f
CpG islands, 261
Crabs, 578–579, 579f
Cracking beak, 602f
Cranial nerves, 732, 733f
Cranium, 587
Crenarchaeota, 481
Cretaceous period, 462
Cretaceous-Paleogene (K/T) event, 537
Crick, Francis, 196, 196f, 199f, 210
Crinoidea, 581, 582t
Crisis ecoregions, 1034–1035, 1034f
CRISPR RNA (crRNA), 233t, 372–373, 374f
CRISPR-Cas system, 242–244, 242f–243f, 372–373, 374f, 552
Cristae, 86, 86f, 133
Critical period, 935
Crocodiles, 589t, 599, 601, 601f
   double circulation, 782–783, 783f
Crocodilia, 589t, 599, 601, 601f
Crop, 820, 820f
Cross-bridge cycle, 771–772, 772f
   regulation of, 772–773, 773f
Cross-bridges, 769f, 770
Crosslinking glycans, 93, 93f
Cross-pollination, 312, 312f, 675
crRNA (CRISPR RNA), 233t, 372–373, 374f
Crustacea, 574, 575t, 578–579, 579f
Crustose lichens, 519, 519f
Crutzen, Paul, 1002
Cryosphere, 515, 515f
Cryptelytrops albolabris, 751f
Cryptococcus neoformans, 506
Cryptomonad, 490f
Cryptosporidium parvum, 489
Ctenophores, 560, 560f
C-terminus, 48, 49f, 217
Cubozoa, 562–563, 563f, 563t
Currents, ocean, 993–994, 993f
Currie, David, 967
Cuticle
   ecdysozoan, 572–574
   plant, 533, 534f, 615
Cuvier, Georges, 388

Cyanobacteria
   body types, 483, 483f
   earliest fossils and, 450, 451f
   early emergence of, 456–457, 456f
   in microbiomes, 515–516, 516f, 517–518
   nitrogen fixation, 483, 488–489, 655, 656f
   overview of, 482t
   as photosynthetic bacteria, 482–483, 483f
   structure of, 485–487, 485f
Cycads, 530f, 531, 534, 536f, 539, 539f, 549t
Cyclic AMP (cAMP)
   lac operon and, 255, 256f
   production of, 180, 181f
   reversal of cellular response and, 182
   as second messenger, 632
   signal amplification and speed with, 183, 183f
   signal transduction pathways, 180–182, 181f
Cyclic electron flow, 156–157, 158f
Cyclic photophosphorylation, 157, 158f
Cyclin-dependent kinases (cdks), 289–290, 290f
Cyclins, 289–290, 290f
CYCLOIDEA genes, 678, 679f
Cyclostomata, 587–590, 589f, 589t, 590f
Cylindrospermum, 479f
Cystic fibrosis (CF), 16, 16f, 319–320, 320f, 321t, 322
Cytogenetics, 286
Cytokines, 898
Cytokinesis, 288f, 289, 293, 296f, 299f
Cytokinins, 634t, 636
Cytoplasm, 69, 70f, 95t
Cytosine (C), 54, 55f, 192, 192f, 193, 193f
Cytoskeleton, 71f, 73f, 75–77, 76t, 102, 102f
Cytoskeleton filaments, 75–77, 76t, 77f
Cytosol
   in animals, 71f
   cilia and flagella, 78, 78f
   cytoskeleton, 75–77, 76t
   definition of, 74
   in eukaryotic cells, 95f
   motor proteins, 77–78, 77f
   in plant defense, 644
   in plants, 73f
   synthesis and breakdown of molecules, 75
Cytosolic leaflet, 99, 99f
Cytotoxic T cells, 902–904, 903f, 904f, 908–910, 909f

D

Daeschler, Ted, 392
Daisy, 672, 675
Dalton (Da), 23
Dams, 1011, 1012f
Damselflies, 577t
Daphnia pulex, 872, 873f
Darimont, Chris, 1018
Dart, Raymond, 466
Darwin, Charles, 170, 387, 388–390, 389f, 408, 416, 542, 571, 631
Darwin, Erasmus, 388
Darwin, Francis, 170, 631
Dasypeltis inornata, 600f
Datana ministra, 942, 942f
Daughter strands, 197, 197f
Davis, A. P., 595–596, 595f–596f
Davis, Margaret, 1007–1008
Day-neutral plants, 639–640
ddGTP (dideoxyguanosine triphosphate), 370f

ddNTPs (dideoxynucleotide triphosphates), 370f, 371
DDT (dichlorodiphenyltrichloroethane), 1014–1015, 1015f
Dead zones, 654
Deafness, 763
Decapoda, 579, 579f
Decomposers, 489
   fungi, 506, 506f
Decoy function, 232, 233f
Deep-sea vent hypothesis, 61–62, 62f
Deer, 604f
Defecation, 827
Defense proteins, 47t
Defenses
   plant, 643–646, 644f, 645f, 646f
   against predation, 959–961, 960f, 961f
   sponges, 561
Deficiency, nutrient, 815t
Deforestation, 1016–1017, 1016f
Degenerate, 216
Dehydration, 708
Dehydration reactions, 40, 40f, 42, 42f, 44, 44f
Deinococcus-Thermus, 482t
Deletion, 304, 304f
DELLA proteins, 636, 637f
Demodex brevis, 575
Demodex canis, 575
Demography, 948, 951–953, 952f, 952t
Dendrites, 715, 716f
Dendritic cells, 897f, 898
Denisovans, 468, 468f, 470
Denitrification, 1013, 1014f
Dense connective tissue, 697f
Density, population, 949–950, 949f, 950f, 962, 962f
Density-dependent factors, 962, 962f
Density-independent factor, 962
Deoxynucleoside triphosphates, 201
Deoxyribonucleic acid (DNA). *See* DNA (deoxyribonucleic acid)
Deoxyribose, 54, 192
Depolarization, 721, 723f
Depression, 714, 740
Desiccation-resistant skin, 598
Desmidium, 493f
Desmosomes, 117, 117f
Determinate cleavage, 556, 557f
Determinate growth, 612
Detritivores, 980
Detritus, 980
Deuterostomes, 556, 557f
Deuterostomia
   Chordata, 557, 558f, 580–584, 582f, 583f, 584f
   Echinodermata, 580–581, 580f, 581f, 582t, 729, 730f, 766f
Development
   animal, 553, 555–556, 557f
   definition of, 427
   hormonal control of, 864–866, 865f, 866f
   insects, 578, 578f
Developmental disorders, 892
Developmental homologies, 392t, 397
Devonian period, 460, 594
Diabetes, 174, 857–861, 859f–860f
Diamond, Jared, 1039
Diaphragm, 801–802, 801f, 802f
Diastole, 788–789
Diatoms, 489, 490f, 497

Dicamptodon copei, 598
Dichlorodiphenyltrichloroethane (DDT), 1014–1015, 1015f
Diclofenac, 1030
Dictyostelia, 499t
Dictyostelium discoideum, 497
Dideoxy chain-termination method, 371–372, 372f
Dideoxy sequencing, 371–372, 371f
Dideoxyguanosine triphosphate (ddGTP), 370f
Dideoxynucleotide triphosphates (ddNTPs), 370f, 371–372, 371f
Diencephalon, 735–736, 735f, 736f
Diet
  epigenetic changes and, 334–335, 335f
  essential nutrients from, 814–815, 815t, 816t
Differential gene regulation, 72
Differential interference contrast (Nomarski) microscopy, 68f
Diffusion, 103–104, 104f
Digestion, 814, 814f
  carbohydrate, 820, 823–825, 824f
  lipid, 825–827, 826f
  in mouth, 820
  principles of, 817–818, 817f
  protein, 820–821, 825, 825f
  in small intestine, 821–823, 822f
  in stomach, 820–821, 821f
Digestive system
  absorptive states, 828–832, 828f, 831f, 832f
  alimentary canal, 814, 814f, 817–820, 819f, 820f, 842–845, 843f, 844f
  in animals, 698t
  cellulose, processing of, 821, 821f
  definition of, 818
  esophagus, 819–820, 819f, 820f
  food processing in, 820–821
  introduction to, 813
  lactose intolerance, 471, 471f, 824–825
  large intestine, 819, 819f, 827
  liver, 819, 819f, 825–827, 826f
  mouth, 819–820, 819f, 820f
  pancreas, 819, 819f, 823, 823f, 825–827, 826f, 830–831, 832f
  postabsorptive states, 828–832, 829f, 831f, 832f
  principles of digestion, 817–818, 817f
  public health impacts and, 842–845, 843f, 844f
  small intestine, 819, 819f, 821–827, 822f, 824f, 825f, 826f
  stomach, 698f, 819–821, 819f, 821f
  vertebrate, 818–827, 819f, 820f, 821f, 822f, 823f, 824f, 825f, 826f
Digger wasps, 933–935, 933f–934f
Dihybrid, 315f, 316
1,25-Dihydroxyvitamin D, 851f, 861–863, 862f
Dingoes, 1023, 1023f
Dinoflagellates, 494, 494f
Dinosaurs, 461–462
Dinozoa, 494, 494f, 499t
Diploblastic animals, 555
Diploid, 286, 301
Diploid sporophyte, 534, 535f
Diploid-dominant species, 301, 302f
Dipnoi, 593, 594f
Diprotodontia, 605t
Diptera, 577t
Direct intercellular signaling, 170–171, 172f
Direct repair, 275, 276t

Directional selection, 403, 404f
Disaccharides, 41–42, 42f, 823, 824f
Discovery-based science, 14–16, 15f, 16f
Diseases and disorders
  autoimmune, 765, 911
  bone and skeletal muscles, 777–778, 777f, 778f
  digestive system, 842–845, 843f, 844f
  endocrine system, 867–868
  epigenetics and, 335–336, 336t
  excretory system, 844–845, 846f
  fungal, 506, 506f
  hemorrhage, 927–928, 928f
  immune system, 912–913, 913f
  mitochondrial diseases, 339t
  nervous system, 739–740, 740t
  non-coding RNAs (ncRNAs) and, 244–245, 245t
  parasitic flatworms, 565–566, 565f, 565t, 566f
  parasitic nematodes, 573, 574f
  reproductive, 891–893, 893f
  respiratory and circulatory systems, 807–810, 808f, 809f, 810f, 811f
  sensory systems, 762–763, 762f
Dispersive mechanism, 197, 197f
Dissociation constant, 174
Distal tubule, 837, 837f, 840–841, 840f, 841f
Distribution of species, climate change and, 1007–1008, 1008f
Disulfide bridge, 51
Diversifying selection, 403–405, 405f
Diversity of species, 966–969, 966f, 967f, 968f, 969t
Diversity-stability hypothesis, 970, 970f, 1031, 1031f
DNA (deoxyribonucleic acid)
  base composition of, 196t
  definition of, 4, 187
  double helix of, 54–55, 55f, 194f, 195–196
  identification of, 189–191, 189f–190f
  information storage by, 211, 217
  mutagens and, 272–273, 273f
  nucleotides and, 54, 55f, 192, 192f
  repair of, 275–277, 276t
  replication of. *See* DNA replication
  sequence changes of, 445, 446f
  sequences of, 511–513, 512f, 513f
  sequencing of, 371–372, 371f, 381f
  strand of, 192–193, 193f
  structure of, 55f, 191–195, 191f
  synthesis of, 201–202, 201f, 202f
  theory of, 14
  X-ray diffraction of, 195–196, 195f
DNA helicase, 200, 200f
DNA library, 367–368, 368f
DNA ligase, 203, 203f, 366f
DNA methylation, 259, 261–262, 328, 329t, 330, 331f, 334–335, 335f
DNA methyltransferase, 261
DNA microarray, 372, 373f, 373t
DNA polymerase, 201–203, 201f, 202f, 203f, 369, 371
DNA primase, 201–202, 201f
DNA replication
  accuracy of, 203
  AT/GC rule and, 198, 199f
  definition of, 188, 197
  leading and lagging strands, 202–203, 202f, 203f
  molecular mechanism of, 200–204
  ncRNA role in, 234–235
  origin of replication and, 200, 200f

  proposed mechanisms of, 197–198, 197f, 198f
  protein requirements of, 200–202, 200f, 201f, 202f
DNA supercoiling, 355, 355f
DNA topoisomerase, 200, 200f
DNA transposons, 382, 382f
DNA-binding domain, 53, 53f
DNase, 189f–190f, 190–191
Dobzhansky, Theodosius, 417
Dodo birds, 1019–1020
Dog breeds, 395–396, 396f
Dogfish shark, 592f
Dolphins, 604f
Domains
  definition of, 8, 10f, 434
  distinguishing features of, 435t
  of life on Earth, 478f
  overview of, 434–435, 434f
  of proteins, 53–54, 53f
  in taxonomic hierarchy, 435, 436f
Domestication, 550, 550f
Dominance, protein function and, 321–322, 321f, 321t
Dominant traits, 312–313, 320f
Dormancy
  bacteria, 484, 484f
  plant, 638, 685, 685f
Doron, Shany, 899
Dorsal hollow nerve cord, 582, 582f
Dorsal side, 555
Dorudon, 393f, 394
Dosage compensation, 332
Double bonds, 25–26, 101
Double circulation, 782–783, 783f
Double fertilization, 543, 543f, 676, 682–683, 683f
Double helix
  ball-and-stick model of, 196, 196f
  within chromosomes, 191f
  definition of, 192
  discovery of, 195–196
  grooves of, 193, 195f
  mutagens and distortion of, 273
  nucleotides of, 54–55
  structure of, 55f, 194f
Dove strategy, 940–941, 940f, 940t
Down, John Langdon, 306
Down syndrome, 306, 306t
Dragonfly, 559f, 577, 577t
Drake, Bert, 1009–1011
Dreissena polymorpha, 567–568, 970
Droplet organelles, 70–71
Drosophila melanogaster
  alternative splicing in, 264, 264t
  brain, 714, 730, 730f
  chromosomes of, 304–305, 305f
  compound eyes, 753f
  genome of, 376
  as model organism, 12
  Toll protein, 899–901, 900f–901f
  transmission patterns of genes in, 341
Drought, 648, 660, 667, 667f, 668f
Drugs
  nervous system effects of, 740–742, 741t
  from plants, 541f, 546–548, 548f–549f
Dry fruits, 546, 546f
Dry-land agriculture, 648
Dubois, Eugene, 467
Duchenne muscular dystrophy, 777
Duck foot, 427–428, 428f

Duck-billed platypus, 604, 606f
Duckweed, 612, 612f, 678
Duodenal ulcer, 842, 843f
Duodenum, 823, 823f
Duplication, 304, 304f
Duve, Christian de, 84
Dynamic instability, 75
Dynein, 77f, 78, 79f
Dysplasia, 282, 282f

E

E site (exit site), 223, 223f
Ears, 748, 748f, 749f
Earth
   Big Bang, 450
   history of life on. *See* Life on earth, history of
   temperature variation of, 989, 989f
   tilt and rotation of, 990–992, 991f
Earthworms, 556–557, 558f, 570–571, 570f, 571f
Ebola virus, 347t
Ecdysis, 558, 559f, 572
Ecdysozoa, 557–558, 559f, 572. *See also* Arthropods
   Nematoda, 556, 558f, 573, 573f, 574f
ECG (electrocardiogram), 789, 789f
Echidna, 395, 395f
Echinodermata
   body plans, 580–581, 580f
   diversity of, 581, 581f, 582t
   endoskeleton, 766f
   major classes and characteristics of, 582t
   nervous system, 729, 730f
Echinoidea, 581, 582t
Echinops sp., 377, 377f
Ecological factors, in speciation, 418
Ecological footprint, 1003–1005, 1005f
Ecological function, 483–484, 484f
Ecological organization
   biomass production in ecosystems, 983–986, 983f, 984f, 985f, 986f
   community stability, 970, 970f
   food webs and energy flow, 979–983, 979f, 981f, 982f
   introduction to, 965
   island biogeography, 974–979, 975f, 976f, 977f–978f
   species richness, 966–970, 966f, 967f, 968f, 969t, 970f, 976f, 977f
   succession, 971–974, 972f, 973f, 974f
Ecological pyramids, 980, 982f
Ecological species concept, 419
Ecology, 13, 13f, 930. *See also* Behavioral ecology; Community ecology; Population ecology
   ecosystems, 965
   landscape, 1035–1036, 1036f
   seedless plant role in, 537, 538f
*Eco*RI, 364–365, 366f, 366t, 367–368
Ecosystems
   biomass production in, 983–986, 983f, 984f, 985f, 986f
   definition of, 5, 6f, 965
   diversity of, 1028, 1029–1030
   function of, 1031–1032, 1031f, 1032f
   replacement of, 1038f
   services of, 1030, 1030t
Ecosystems ecology, 965
Ecotypes, 417

Ectoderm, 555, 556f, 888, 889f
Ectomycorrhizae, 521, 521f
Ectotherms, 704, 704f, 705f
Edge effects, 1036
Edidin, Michael, 102
Edward syndrome, 306t
Effective number of species, 968
Effectors
   homeostatic control system, 701, 702f, 919
   plant cell communication, 632–633, 633f
Efferent arterioles, 925f
Egestion, 814, 814f
EGF (epidermal growth factor), 178, 179f
Eggs
   amniote, 598–599, 599f
   animal, 871, 874–877, 875f, 876f
   flowering plant, 674–675
Egler, Frank, 973
Ehrlich, Paul, 1028
Einstein, Albert, 154
Eisner, Tom, 960
Ejaculation, 878–879, 878f
Elastic fibers, 92, 92f
Elasticity, 92
Elastin, 92, 92f
Electrical excitation
   contraction coupling with, 773, 774f
   heart, 787–788, 788f
Electrical properties of neuron, 718–721, 719f, 720f
Electrical signals of neuron, 721–725, 722f, 723f, 725f
Electrical synapse, 726
Electrical/ion gradient, 122t
Electrocardiogram (ECG), 789, 789f
Electrochemical gradient, 719–721, 720f
   ATP-driven ion pump generation of, 111, 112f
   electron transport chain (ETC) and, 139–140, 141f
   ion, 111, 111t
   overview of, 104, 104f
Electrogenic pump, 111, 112f
Electromagnetic radiation, 153–154, 154f
Electromagnetic spectrum, 153–154, 154f
Electron microscope, 66, 67f
Electron shell, 22–23, 23f
Electron transport chain (ETC), 139–140, 141f, 155–156
Electronegativity, 26
Electrons
   covalent bonds and, 25–26, 25f, 26f
   definition of, 22
   light energy absorption by, 154, 154f
   orbitals and, 22–23, 23f
   overview of, 22–23, 22f
   photosystems and energy levels of, 159–160, 160f
   redox reactions and, 130
Electrophoresis, gel, 368–369, 368f
Elements, 22, 24, A–1
Elephantiasis, 573, 574f
Elephants, 604f, 1019
Elicitors, 644, 645f
Ellis-van Creveld syndrome, 411
Elongation factors, 221t, 226
Elongation stage
   of transcription, 212, 213f, 262f
   of translation, 226–227, 226f
Elton, Charles, 970, 980, 1031
Embryogenesis, 684–685, 684f
   somatic, 688–689, 688f

Embryonic development
   animal classification based on, 555–556, 557f
   in animals, 555–556, 557f, 885–891, 886f, 887f, 888f, 889f, 890f
   cleavage-stage embryo, 886–887, 886f, 887f, 888f
   flowering plant, 684–685, 684f
   gastrulation, 886f, 888–889, 889f
   general events of, 885–886, 886f
   germ layers, 888–889, 889f
   neurulation, 886f, 890–891, 890f
   organogenesis, 891
   specific features of, 555–556, 556f
Embryophytes, 549t
Embryos
   in animals, 871
   cleavage-stage, 886–887, 886f, 887f, 888f
   flowering plant, 610–611, 610f, 611f, 675–676, 675f, 684–685, 684f, 685f, 688–689, 688f
   in humans, 882, 882f
   uterine preparation for, 882, 882f
Emergency contraception pills, 892–893
Emergent properties
   mammalian internal organization, 694f
   of multicellularity, 458f
   principle of, 3f, 4
   protein-protein interaction, 51f
Emerging viruses, 352–353, 353f
Emerson, Robert, 158–159
Emperor scorpion, 576f
Emphysema, 809, 811f
Empirical thought, 388
Encephalartos laurentianus, 539
Encyclopedia of DNA Elements (ENCODE) Project, 381
Endangered species, 1028, 1038–1039
Endangered Species Act, U.S. (ESA), 1028, 1036
Endemic plants and animals, 394, 394f
Endemic species, 1033–1034
Endergonic reactions, 123, 124
Endocrine disruptors, 868
Endocrine glands, 849
Endocrine signaling, 171, 172f
Endocrine system
   growth and development, 864–866, 865f, 866f
   hemorrhage response, 924–926, 925f
   hormone types and mechanisms, 849–850, 850f, 851f
   introduction to, 848–849
   metabolism and energy balance, 848, 854–861, 854f, 855f, 856f, 857f, 858f, 859f–860f
   mineral balance, 861–864, 862f, 863f
   nervous system, links with, 852–854, 852f, 853t
   overview of, 698t, 850f
   public health impacts and, 867–868
   reproduction, 867
Endocytosis, 114–116, 115t, 116f, 849
Endoderm, 555, 556f, 888, 889f
Endodermis, 661–662, 662f
Endomembrane system
   of cells, 75f, 79, 80f
   endoplasmic reticulum, 81–82, 81f
   in eukaryotic cells, 95t
   eukaryotic nucleus, 80–81, 80f, 81f
   Golgi apparatus, 82
   lysosomes, 82–83
   peroxisomes, 84, 85f
   in plant and animal cells, 95t
   vacuoles, 83–84, 84f

Endometrium, 882, 882f
Endomycorrhizae, 520, 520f
Endoplasmic reticulum (ER), 81–82, 81f, 239, 241, 241f
Endoskeleton, 766–767, 766f
   chordate, 581–582
   echinoderm, 580, 580f
   vertebrate, 587
Endosperm, 534, 676
Endospores, 484, 484f
Endosymbiont, 492
Endosymbiosis, 88, 457, 457f, 492, 494, 495f
Endosymbiosis theory, 88, 88f
Endothelium, 790, 791f
Endotherms, 704, 704f, 705–706, 705f, 706f
Energy
   anaerobic respiration, 146, 147f
   carbohydrate, protein, and fat metabolism, 145–146, 145f
   cellular functions and, 69
   chemical reactions and, 122–124
   citric acid cycle, 138, 139f, 140f
   definition of, 22, 122
   enzymes, 125–129
   fermentation, 146–147, 148f
   forms of, 122–123, 122t
   free energy and chemical reactions, 123
   glycolysis, 134–135, 135f
   introduction to, 121
   metabolism and cellular respiration, 129–134
   nutrients for, 814, 828–830, 828f, 829f
   oxidative phosphorylation, 138–145
   pyramid of, 980, 982f
   pyruvate, breakdown of, 137, 138f
   types of, 122, 122f, 122t
   use of, 42f
Energy balance, 848, 854–861, 854f, 855f, 856f, 857f, 858f, 859f–860f
Energy flow, 979–983, 979f, 981f, 982f
Energy intermediates, 130
Energy investment phase, 134
Energy liberation phase, 134
Energy use by living organisms. *See also* Metabolism; Photosynthesis
   in food web, 982f
   glucose energy storage, 42f
   heterotrophic protozoan, 490f
   Indian pipe, 650f
   lactose metabolism, 249f
   nitrogenous wastes, 833f
   principle of, 2, 3f, 4
   resting membrane potential, 719f
   root-hair membrane ion uptake, 658f
English ivy, 395, 395f
Enhancers, 258
Enterobius vermicularis, 573
Enterococcus faecalis, 361
Enterokinase, 825, 825f
Enthalpy (H), 123
Entomology, 576
Entropy, 123, 123f
Entry, 349, 350f
Environment
   adaptation to, 426–427, 426f
   animal homeostasis and, 701
   through earth's history, 453, 455–456
   fossil record and, 453t
   heat exchange with, 704–705, 705f
   natural selection and, 471
Environment, organism interaction with
   blackberry flower and fruit, 686f
   nervous tissue, 695f
   norm of reaction and, 323f
   principle of, 3f, 4
   yeast cells and glucose, 170f
Environmental agents, effects of, 334–336, 335f, 336t
Environmental science, 930
Environmental stimuli, plant responses to, 631–632, 632f
Environmental stress, plant response to, 638
Environmental systems, 325
Enzyme-linked receptors, 175, 175f, 178–180
Enzymes
   activation energy and, 125–126, 125f
   additional factors on function of, 128, 129f
   chemical reactions and, 125–126, 126f
   coacervates and, 63
   definition of, 26, 125
   Golgi apparatus and, 82
   in metabolism, 75
   restriction, 364, 366f, 366t
   substrates and, 126–128, 127f
   temperature and, 128, 129f
   velocity and substrate concentration, 127f
Enzyme-substrate complex, 126–127, 126f
Eosinophils, 897f, 898
Ephedra, 541–542, 541f, 546
Ephedrine, 541f
Epicotyl, 685, 685f, 687, 687f
Epidermal growth factor (EGF), 178, 179f
Epidermis, 615
Epidermophyton floccosum, 504
Epigenetic inheritance, 310t, 328
Epigenetic silencing, 331
Epigenetics
   chromatin modifications and, 328–329, 329t
   definition of, 328
   environmental agents, effects of, 334–336
   genomic imprinting, 329–330, 330f, 331f
   overview of, 328–329
   X-chromosome inactivation (XCI), 331–334, 332f, 332t, 333f
Epimutation, 328
Epinephelus fasciatus, 593f
Epinephrine, 180–182, 181f, 182f
Episome, 352
Epithalamus, 735f, 736, 736f
Epithelial cells, 818–819, 821–827, 822f, 824f, 825f, 826f
Epithelial tissues, 694f, 695–696, 696f, 698f
Epitope, 907
EPO (erythropoietin), 868, 923, 924f
Eppley, Sarah, 531
EPSP (excitatory postsynaptic potential), 726
Epstein-Barr virus, 346t, 347t
Equilibrium, population, 953
Equilibrium model of island biogeography, 975–979, 975f, 976f, 977f–978f
   preserve design, 1035–1036, 1035f, 1036f
Equilibrium potential, 719–721, 720f
Equisetum telmateia, 533f
ER (endoplasmic reticulum), 81–82, 81f, 239, 241, 241f
ER lumen, 81f, 82
ER signal sequence, 241, 241f
Ergot, 505, 505f
Errantia, 571, 571f
Erythrocytes, 785, 785f, 792f, 803f, 917, 923–924, 924f
Erythropoietin (EPO), 868, 923, 924f
ESA (Endangered Species Act, U.S.), 1028, 1036
Escherichia coli
   alternative splicing and, 264, 264t
   anaerobic respiration by, 146, 147f
   bacterial colony of, 356, 356f
   base composition of DNA in, 196t
   binary fission of, 356, 357f
   chromosomal loops of, 354, 355f
   chromosomes of, 354, 355f, 375
   conjugation in, 357–359, 358f
   DNA repair in, 276
   DNA replication in, 197–198, 198f, 202, 203f
   DNA supercoiling of, 354, 355f
   as donor strains, 357–358, 358f
   evolutionary relationships of, 224, 224f
   gene mutations in, 269, 271, 271f–272f
   gene regulation, 249–250, 249f
   genome of, 376t
   Gram-negative bacteria, 486f
   horizontal gene transfer in, 360, 479
   lac operon and, 254–256, 254f–255f
   lactose permease, 109
   as model organism, 13
   nucleotide excision repair (NER), 276, 276f
   reproduction of, 356, 356f
   ribosomes of, 222
   structure of, 70f
   transduction in, 360, 360f
   translation in, 220
   viral reproductive cycle in, 349, 350f–351f, 352
   as virus host, 346t
Esophagus, 819–820, 819f, 820f
   lining of, 696f
Essay on the Principle of Population (Malthus), 389
Essential amino acids, 814–815
Essential fatty acids, 815
Essential nutrients, 814–815, 815t, 816t
Estradiol, 867, 880
Estrogen receptors, 53, 53f, 177, 177f
Estrogens, 851f, 867
ETC (electron transport chain), 139–140, 141f, 155–156
Ethology, 931–932
Ethylene, 634t, 636–638, 637f
Euchromatin, 205
Eudicots, 545, 545f, 613, 613t, 677–678, 685, 685f
Euglenoids, 492–493, 493f
Eukarya (domain), 9f, 10, 434, 434f, 435t, 447f, 478f
   phylogenetic tree, 492f
Eukaryotes
   alternative splicing in, 264, 264t
   definition of, 70
   evolution, 491–492, 492f
   fungi as plastidless, 479, 480f
   gene regulation in, 250–251, 251f
   genomes of, 376–380, 377f, 377t
   molecular gene expression, 210f
   multicellular, 458–459, 458f
   parasites, 896
   protists as early diverging, 479, 480f
   regulation of transcription in, 256–262

RNA modifications in, 213–215, 214f
RNA splicing and translation, regulation in, 262–265
transcription in, 212
Eukaryotic cells
  chromosome structure and number of, 303–307
  chromosomes of, 80–81, 81f
  cilium or flagellum structure, 79f
  cycle of, 286–290, 288f, 290f
  definition of, 10
  examples of, 9f
  gene regulation in, 250, 250f
  interacting parts of, 94, 95f, 95t
  meiosis and sexual reproduction, 297–301
  mitotic cell division, 291–296
  morphology of, 72, 72f
  non-coding RNAs' role in DNA replication, 234–235
  organelles of, 70, 71f, 73f
  origin of, 457–458, 457f
  protein sorting, 90f
  ribosomes of, 222, 223t
  translation in, 226–227, 227t
  types of cytoskeletal filaments found in, 75–77, 76t
Eukaryotic chromosomes
  cell cycle and, 281f, 286–290, 288f, 290f
  introduction to, 285
  meiosis and sexual reproduction, 297–301
  mitotic cell division and, 291–296
  molecular structure of, 204–207
  replication and compaction of, 291, 292f
  variation in structure and number of, 303–307
Eukaryotic supergroups, 491–492, 492f, 499t
Eulipotyphla, 605t
Euphausiacea, 579, 579f
Euploid organisms, 304
Eurasian wild horse, 1019
Europe, fertility rate, 1003
European common frog, 597f
Euryarchaeota, 481
Eusociality, 943, 943f
Eutherians, 604–605, 605t, 606f
Eutrophication, 1012–1013, 1013f
Evaporation, 33, 33f, 705
  ion and water exchange due to, 709, 709f
Evaporative heat loss, 705–706
Evapotranspiration rate, 967
Evo-devo (evolutionary developmental biology), 427–431, 428f
Evolution. *See also* Human evolution; Plant evolution
  animal, 554–555, 554f
  of baroreceptors, 921
  of color vision, 755–756, 755f
  convergent, 392t, 395, 395f
  definition of, 4, 387
  eukaryote, 491–492, 492f
  evidence of evolutionary change, 391–398, 392t
  evolutionary developmental biology, 427–431, 428f
  of feathers, 601
  of flowers, 544f
  fossil record and, 451–453
  of fungi, 499, 500f
  genes in populations, 398–402
  genetic drift, 409–412
  of heart, 783–784, 784f
  of hormones, 864

  identification of species, 417–419, 418f
  of immune system, 899
  introduction to, 387–388, 416–417
  of jaws, 590, 591f
  lactose intolerance and, 824–825
  microbial, 479
  migration and nonrandom mating, 413–414
  of myosins, 771, 771f
  natural selection, 390, 391f, 402–409
  of nervous system, 729–731, 730f, 731f
  overview of, 388–391
  reproductive isolation and, 419–421, 420f
  speciation, mechanisms of, 422–427
  study of, 11
  vertical, 7, 8f
Evolutionarily conserved, 224, 224f
Evolutionary change
  biogeography, 394
  convergent evolution, 395, 395f
  evidence of, 391–398, 392t
  fossils, 392, 393f, 394
  genes and timing of, 444–445, 444f
  homology, 396–398, 396f, 397f
  introduction to, 391, 392t
  selective breeding, 395–396, 396f
Evolutionary connections
  cellular adaptations to drought stress, 660
  choanoflagellates, 178, 180f
  continental drift and biogeography, 999–1000, 999f, 1000f
  eukaryotic cell origins, 457–458, 457f
  FtsZ protein, 293, 295, 296f
  Hox genes, 428–430, 429f, 430f
  mitochondria and chloroplasts, 88, 88f
  polymorphism, 399, 399f
  small subunit rRNAs, 224, 224f
  taxonomic hierarchy, 435
Evolutionary developmental biology (evo-devo), 427–431, 428f
Evolutionary history
  of bears, 11
  connection of life forms, 7
  examples of, 7f
  horizontal gene transfer, 7–8
  principle of, 3f, 4
  small subunit rRNAs, comparisons of, 224, 224f
  vertical descent with mutation, 7, 8f
Evolutionary lineage concept, 419
Evolutionary relationships, 418
  cladogram and, 442, 442f
  molecular systematics and, 440
  phylogenetic tree and, 437–438, 437f
  taxonomic groups and, 438, 438f
Excavata, 492–493, 492f, 493f, 499t
Excitation-contraction coupling, 773, 774f
Excitatory postsynaptic potential (EPSP), 726
Excretion, 834, 834f
Excretory system
  definition of, 833
  in different animal groups, 833–836, 833f, 834f, 835f
  introduction to, 813
  kidneys, 599, 836–842, 837f, 838f, 839f, 840f, 841f
  large intestine and anus, 827
  nitrogenous wastes removal by, 833–836, 833f, 834f, 835f
  overview of, 698t

  public health impacts and, 844–845, 846f
Excurrent siphon, 583
Exercise, 776, 831–832
Exergonic reaction, 123, 124f, 125f
Exit site (E site), 223, 223f
Exocytosis, 114–116, 115f, 115t
Exons, 213, 214–215, 215f
Exoskeletons, 558, 559f, 572–574, 766, 766f
Expansins, 617
Expansion
  lungs, 801–802, 802f
  plant cell, 617–618, 617f, 618f, 636–637, 637f
  stretch receptors' detection of, 747
Experimental group, 15, 15f
Experimentation, 17
  inhibition, 974f
  logistic growth of yeast cells, 956f
  nitrogen and phosphorus limitation of primary production, 983f
  principle of, 3f, 4
"Experiments on Plant Hybrids" (Mendel), 310
Exponential growth, 954–955, 954f, 955f, 956f
Expression, 243f, 244
Extant species, 433, 434f
Extensors, 768–769, 769f
External fertilization, 877
External gills, 797, 797f
Extinct species, 433, 434f
Extinction
  habitat destruction and, 1017, 1017f
  in island biogeography model, 975–979, 975f, 976f, 977f–978f
  mass extinctions, 455–456
  overexploitation, 1018–1022, 1020f
Extracellular digestion, 817, 817f
Extracellular fluid, 707–708, 708f, 863–864, 863f
Extracellular leaflet, 99, 99f
Extracellular matrix (ECM)
  adhesive and structural proteins in, 91–92, 92f
  anchoring junctions and, 116–118, 117f
  cell signaling and, 91, 91f
  diagram of, 91f
  epithelial tissues, 696
  polysaccharides in, 92, 92f
  transmembrane proteins and, 102, 102f
Extracellular polymeric substance, 487
Extranuclear inheritance, 310t, 337–339, 337f, 338f, 339f, 339t
Extraterrestrial hypothesis, 61
Extremophiles, 481
Eye color, 472, 472f
Eyes, 744, 752–758, 753f, 754f, 755f, 756f, 757f

F

F factor, 357–359, 358f
F_1 generation, 312, 313f, 315–316, 315f
F_2 generation, 312, 313f, 315–316, 315f
Facilitated diffusion, 103, 103f, 107–109, 108f, 657–658, 657f
  in digestive system, 818, 823, 824f, 825, 825f
Facilitation, 971–972, 972f, 973f, 974f
Facultative anaerobes, 488
Falsifiable, 14
Family, 10f, 11, 435, 436f
Fast fibers, 776
Fast-glycolytic fibers, 776

HGT. *See* Horizontal gene transfer
Hibiscus flower, 545*f*
Hierarchical models, 17
Highly active antiretroviral therapy (HAART), 913
Highly repetitive sequences, 380
Hill, Robin, 159
Hillig, Karl, 547–548, 548*f*–549*f*
Hindbrain, 730, 730*f*, 735, 735*f*
Hinge of motor proteins, 77, 77*f*
Hippocampus, 736
Hirudo medicinalis, 571*f*
Histone acetyltransferase, 260
Histone code hypothesis, 260
Histone variants, 259, 260*f*
　localization of, 329*t*
Histones, 204, 204*f*
　gene transcription and, 260–261, 260*f*, 262*f*
Histoplasma capsulatum, 506
HIV. *See* Human immunodeficiency virus
Hoffman, Jules, 899–901, 900*f*–901*f*
Holland, Scott, 737–739, 738*f*
Holobiont, 517
Holoblastic cleavage, 887, 887*f*
Hologenome, 517
Holothuroidea, 581, 582*t*
Homeostasis. *See also* Hemorrhage
　animal principles, 700–703, 702*f*
　autonomic nervous system regulation of, 732–735, 734*f*
　blood clotting, 785, 786*f*
　blood glucose, 830–832, 832*f*
　body temperature, 703–707, 704*f*, 705*f*, 706*f*
　calcium, 861–863, 862*f*
　challenge to, 916–917
　contractile vacuoles, 105, 106*f*
　control system, 701–702, 702*f*
　definition of, 4, 693
　external versus internal environments, 701
　feedback inhibition, 132*f*
　feedforward regulation, 702
　gas and solute exchange between capillaries and cells, 792*f*
　glucose, 830–832, 831*f*, 832*f*, 856–857, 858*f*
　glycolysis and, 136*f*–137*f*
　hindbrain regulation of, 735, 735*f*
　hormones and, 848
　internal fluids, 693, 707–712, 708*f*, 709*f*, 710*f*–711*f*
　internal volume, 105, 106*f*
　kidney disease, 845
　leaf abscission, 668*f*
　negative feedback, 702, 702*f*
　in plants, 534*f*
　potassium, 863–864, 863*f*
　principle of, 3*f*, 4, 33*f*
　respiratory system, 792*f*
　sodium, 840, 840*f*, 863–864, 863*f*
　vertebrates, 701
　water and, 33*f*
Homeostatic control system, 701–702, 702*f*, 919
Homeostatic response
　rapid phase of, 919–922, 920*f*, 922*f*, 927*t*
　secondary phase of, 923–926, 924*f*, 925*f*, 927*t*
Homeotherms, 704
Hominid, 468*f*
Hominidae, 465*f*
Hominins, 462, 464–466, 466*f*

Hominoidea, 463, 464*f*, 465*f*
Hominoids, 462
Homo erectus, 466*f*, 467–468, 468*f*, 469
Homo ergaster, 466*f*, 467, 468*f*, 469
Homo floresiensis, 469
Homo genus, 467–468
Homo habilis, 466*f*, 467
Homo heidelbergensis, 466*f*, 467, 468, 468*f*, 469
Homo neanderthalensis, 466*f*, 467, 468, 468*f*, 469, 469*f*, 470
Homo sapiens
　alternative splicing in, 264, 264*t*
　family Hominidae, 465*f*
　genetic diversity in, 473, 473*f*
　human evolution, 466*f*
　interbreeding of, 470
　origin of, 467, 468–469, 468*f*, 470*f*
　in Quaternary period, 462
Homologous chromosomes, 286–287, 301, 317, 317*f*
Homologous genes, 378, 378*f*
Homologous structures, 396, 396*f*
Homologs, 286
Homology, 286, 396–398, 396*f*, 397*f*, 438–440
Homozygote, 313–314
Honeybees, 327, 936–938, 937*f*
Honeycreepers, 423, 423*f*
Honeysuckle vine, 631*f*
Hooke, Robert, 66
Hookworms, 573
Horizontal gene transfer (HGT), 7–8, 8*f*, 360–361, 401*t*, 446–448, 447*f*, 479
Hormones
　birth, 884–885, 884*f*, 885*f*
　blood flow regulation, 794
　concentrations of, 849
　definition of, 171, 848
　evolution of, 864
　growth and development control by, 864–866, 865*f*, 866*f*
　leaf development, 619
　male reproductive function, 879, 879*f*
　metabolism control by, 848, 854–861, 854*f*, 855*f*, 856*f*, 857*f*, 858*f*, 859*f*–860*f*
　mineral balance control by, 861–864, 862*f*, 863*f*
　misuse of, 868
　ovarian cycle, 881–882, 882*f*
　plant, 619, 631, 634–638, 634*t*, 635*f*, 637*f*
　receptors for, 849, 864
　reproduction control by, 867
　structures of, 849, 851*f*
　synthesis and removal of, 849
　therapeutic uses, 868
　types and mechanisms of, 849–850, 850*f*, 851*f*
　uterus preparation for embryo, 882, 882*f*
Hornworts, 529, 530*f*, 531, 531*f*, 549*f*
Horse lineage, 7, 8*f*
Horses, evolution of, 439–440, 439*f*
Horsetails, 532, 533*f*
Horvath, Philippe, 242
Host cell, 346, 346*t*
Host range, 346, 347*f*
Host-associated microbiomes, 516, 517*f*
　acquisition of, 517, 518*f*
　animal, 521–523, 522*f*–523*f*
　bacterial, 517–518
　functions of, 517–519, 518*f*, 519*f*

　fungal, 518–521, 519*f*, 520*f*, 521*f*
　host capabilities extended by, 517, 518*f*
　plant, 519–521, 520*f*, 521*f*
　protist, 518, 518*f*
Hosts, 516
Hot desert, 995*f*, 996, 996*f*, 997*f*
HOTAIR (*Hox* transcript antisense intergenic RNA), 233*t*, 235–236, 236*f*, 245, 245*t*
Hox genes
　in animal genomes, 553
　body pattern evolution and, 428–429, 429*f*
　ctenophores, 560
　evolution of, 555
　expression of, 428–429, 430*f*
　gnathostomes, 590
　organ development and function, 699
　organogenesis, 891
　origins of, 429–430
　tetrapods, 594–596, 595*f*–596*f*
　vertebrates, 587
Hox transcript antisense intergenic RNA (HOTAIR), 233*t*, 235–236, 236*f*, 245, 245*t*
HR (hypersensitive response), 645, 645*f*
H_S (Shannon diversity index), 968, 969*t*
Human evolution
　Australopithecus and *Paranthropus*, 465–467
　bipedalism, 464–466, 466*f*
　continued evolution, 471–472, 471*f*, 472*f*
　Homo genus, 467–468
　Homo sapiens, 462, 465*f*, 467–469, 468*f*, 470, 470*f*
　human compared with chimpanzee genomes, 464
　interbreeding and, 470
　phenotypic variation and, 472–473, 473*f*
　possible scenario for, 466*f*
　primate evolution and, 463–464, 464*f*
Human Genome Project, 379–380
Human immunodeficiency virus (HIV)
　characteristics of, 346*t*, 347*t*
　drugs for, 353
　as emerging virus, 353, 353*f*
　micrograph of, 353*f*
　public health impact of, 913, 913*f*
　reproductive cycle of, 349, 350*f*–351*f*, 352
Human impacts
　global warming and climate change, 1005–1008, 1036
　habitat destruction, 1016–1018, 1016*f*, 1017*f*
　human population growth, 1003–1005, 1003*f*
　introduction to, 1002
　invasive species, 1016*f*, 1022–1025
　overexploitation, 1016*f*, 1018–1022, 1020*f*
　pollution and biomagnification, 1014–1016, 1015*f*
　pollution and human influences on biogeochemical cycles, 1008–1014
Human Microbiome Project, 521
Humans
　angiosperm diversification influenced by, 550–551, 550*f*
　biodiversity, value to, 1028–1030, 1030*t*
　brains of, 735–739, 735*f*, 736*f*, 737*f*, 738*f*, 771, 771*f*
　gametogenesis, 874, 875*f*, 878–881, 878*f*, 879*f*, 881*f*
　as hunters, 1018–1019, 1019*f*
　jaw muscles of, 771, 771*f*

microbiome of, 510, 516–517, 517f, 518f, 521–523, 522f–523f
nervous system structure and function in, 732–739, 733f, 734f, 735f, 736f, 737f, 738f
neurons, 72, 72f
pedigree analysis of, 319–320, 320f
population growth, 1003–1005, 1003f
pregnancy and birth in, 883–885, 883f, 884f, 885f
reproductive structure and function, 877–882, 878f, 879f, 880f, 881f, 882f
skin cells, 72, 72f
viruses and, 346, 347t, 352
Humboldt Current, 993, 993f
Humoral immunity, 902–906, 904f, 905f, 907f
Humus, 652–653
Hundred Heartbeat Club, 1027
Huntington disease, 320, 320f
Hutton, James, 388
Hyacinth macaw, 602f
Hybrid breakdown, 420f, 421
Hybrid inviability, 420, 420f
Hybrid sterility, 420–421, 420f, 421f
Hybridization, 312, 312f
Hydra, 562, 563t, 871f
Hydrocarbons, 38
Hydrochloric acid, 820–821
Hydrogen, 22f, 24, 24t, 26f
as plant nutrient, 649t
Hydrogen bonds, 26–27, 27f, 50
Hydrogen fluoride, 25, 25f
Hydrogen ions, 34
Hydrological cycle, 1011, 1011f
Hydrolysis, 124f
Hydrolysis reactions, 32, 33f, 40, 40f
Hydrophilic molecules, 30–31
Hydrophilic pocket, 109, 109f
Hydrophilic regions, 31, 31f
Hydrophilic solutes, 103–104, 104f
Hydrophobic effect, 50–51
Hydrophobic molecules, 31
Hydrophobic regions, 31, 31f
Hydrostatic pressure in kidneys, 838, 838f
Hydrostatic skeleton, 556
Hydroxide ions, 34
20-Hydroxyecdysone, 851f, 865, 866f
Hydroxyl functional group, 39t
Hydrozoa, 562–563, 563f, 563t
Hymenoptera, 577t
Hyperplasia, 282, 282f
Hyperpolarization, 721
in photoreceptor cells, 756, 756f
Hypersensitive response (HR), 645, 645f
Hypertension, 808
Hyperthyroidism, 855
Hypertonic, 105, 105f
Hyphae, 501–502, 502f
tip growth of, 503, 503f
Hypocotyl, 685, 685f, 687, 687f
Hypothalamus, 735f, 736, 736f, 850f
pituitary gland and, 852–854, 852f, 853t
Hypothesis, 14
Hypothesis testing, 14–16, 15f, 16f
Hypothyroidism, 855
Hypotonic, 105, 105f
Hyracotherium, 7, 8f
Hystrix africaeaustralis, 604f

I

I band, 769f, 770
Ice, 32, 32f
Ice, microbiomes in, 515–516, 515f
Idiosyncratic hypothesis, 1031, 1031f
Ig (immunoglobulin), 905–906, 905f, 906f
IGF-1 (insulin-like growth factor-1), 865
Illicit drugs, 741–742, 741t
Immigration, in island biogeography model, 975–979, 975f, 976f, 977f–978f
Immune system
acquired immunity, 896, 902–912, 903f, 904f, 905f, 907f, 908f, 909f, 910f, 911f, 913
definition of, 895
hemorrhage response, 926
innate immunity, 895, 897–901, 897f, 898f, 900f–901f
introduction to, 895–896
overview of, 698t
pathogens and, 896
public health impacts and, 912–913, 913f
Immune tolerance, 911
Immunity, 895
Immunoglobulin (Ig), 905–906, 905f, 906f
Immunological memory, 911–912, 911f
Imperfect flowers, 677, 678f
Implantation, 883–884, 887, 888f
Imprinted genes, 329, 330f
Imprinting, 935, 935f
In vitro translation system, 217
Inactivation gate, 724
Inbreeding, 413–414, 414f
Inclusive fitness, 942
Incomplete dominance, 310t, 322–323, 322f
Incomplete flowers, 677
Incomplete metamorphosis, 578, 578f
Incurrent siphon, 583
Independent assortment, law of, 315–316, 315f, 317, 318f, 319, 328, 340–343, 340f
Indeterminate cleavage, 556, 557f
Indeterminate growth, 611–612, 612f
Indian pipe, 650, 650f
Indicator species, 1036, 1037f
Indoleacetic acid, 634t, 635, 635f
Induced fit, 126–127
Induced mutations, 272, 272t
Inducer, 255
Inducible operon, 255
Industrial nitrogen fixation, 656
Infection, activated cytotoxic T cells, 908–910, 909f
Infertility, 891–892
Inflammation, 898, 898f, 928, 928f
Inflorescences, 546
Influenza virus, 346t, 347t, 348f
Information, 69, 188
Information storage, 65, 211, 217
Infundibular stalk, 853
Ingestion, 814, 814f
Ingroup, 441f, 442
Inheritance
of acquired characteristics, 388
chromosome theory of, 316–319
definition of, 186, 309
environmental agents, effects of, 334–336
epigenetics, 328–329
extranuclear, 337–339, 337f, 338f, 339f, 339t
genomic imprinting, 329–330, 330f, 331f
introduction to, 309–310, 327–328
linkage of genes on the same chromosome, 340–343
Mendel's laws of, 310–316
pedigree analysis of human traits, 319–320, 320f
sex chromosomes and X-linked patterns of, 324–325, 324f
types of and their molecular basis, 310t
variations in, and their molecular basis, 320–323, 321f, 322f
X-chromosome inactivation (XCI), 331–334, 332f, 332t, 333f
Inhibition, 972, 974f
Inhibitors, 127–128, 127f
Inhibitory postsynaptic potential (IPSP), 726
Initiation factors, 221t, 225
Initiation stage
in transcription, 212, 213f
in translation, 225–226, 225f
Initiator tRNA, 225–226, 225f
Innate behavior, 932, 932f
Innate immunity
body surface as initial line of defense, 897
definition of, 895
inflammation, 898, 898f
interferons and complement proteins, 899
phagocytic cells, 897–898, 897f
Toll protein and, 899–901, 900f–901f
toll-like receptors, 899
Inner bark, 624–625, 624f, 625f
Inorganic fertilizers, 654
Insects
body plans, 574, 574f, 576–577
diversity of, 552, 574, 577–578, 577t
excretory systems, 835, 835f
hormonal control of development, 865–866, 866f
orders and characteristics of, 577t
reproduction and development, 578, 578f
respiratory system, 798–799, 799f
Instar, 578
Insulin
diabetes and, 174, 857–861, 859f–860f
isolation of, 858–861, 859f–860f
in metabolism, 830–832, 831f, 832f, 856–857, 858f
pancreas secretion of, 856–857, 858f
Insulin-like growth factor-1 (IGF-1), 865
Integral membrane proteins, 100, 100f
Integrase, 349
Integration, 349, 350f
Integrators, 701, 702f
Integrins, 117
Integumentary system, 698t, 926
Interbreeding, 470
Intercellular channels, 113–114, 113f, 114f
Interference, 243f, 244
Interferons, 899
Intergovernmental Panel on Climate Change (IPCC), 1007
Interleukin 1, 908, 908f, 911
Intermediate filaments, 75–76, 76t
Internal environment, homeostatic control system regulating, 701–702, 702f
Internal fertilization, 599, 877
Internal fluids, homeostasis of, 693, 707–712, 708f, 709f, 710f–711f
Internal gills, 797–798, 798f

Mother of Thousands, 688–689, 688f
Moths, 552, 577, 577t, 961, 961f
Motility
    bacteria, 487–488, 488f
    protists, 490, 490f
Motor neurons, 716, 717f, 718f
Motor proteins, 47t, 77–78, 77f
Mount St. Helens, 965, 971–972
Mountains, 992–993, 992f
Mouse. See Mus musculus
Mouth, 819–820, 819f, 820f
Movement
    animal, 553
    bacteria, 487–488, 488f
    of bones, 768–769, 768f, 769f
    of cargo, 77, 77f
    of filament, 77f, 78
    flagella and cilia, 78, 78f
    hair cells and, 747, 747f
    motor proteins and cytoskeleton filaments, 77f
    plant, 631, 631f
    protists, 490, 490f
    skeletal muscle fibers, 775–776
Movement corridors, 1035, 1036f
MRI (magnetic resonance imaging), 737–739, 737f, 738f
mRNA. See Messenger RNA
mRNA degradation, 236–241
MRSA (methicillin-resistant *Staphylococcus aureus*), 361, 487
MSY (maximum sustainable yield), 1021–1022, 1022f
Mucilage, 483, 483f, 487
Mucoromycota, 499t, 500, 501f
Mucosa, digestive system, 819, 823
Mud salamander, 597f
Muir, John, 1030
Müller, Paul, 1014
Müllerian mimicry, 960f, 961
Mullis, Kary, 369
Multicellular choanoflagellate, 180f
Multicellularity, 458–459, 458f
Multiple fruits, 546, 546f
Multiregional hypothesis, 468
Mus musculus
    genome of, 376
    as model organism, 13
Muscle activity, heat production through, 706–707
Muscle cells, fermentation in, 147, 148f
Muscle contractions
    electrical excitation coupling with, 773, 774f
    force generation for, 768–775, 768f, 769f, 770f, 771f, 772f, 773f, 774f
    regulation of, 772–773, 773f
Muscle fibers, 768–770, 768f, 769f
Muscle tissues, 694–695, 694f, 695f, 698f, 768
Muscular dystrophy, 777–778
Musculoskeletal system, 698t
Musicians, brains of, 737–739, 738f
Mussels, 567–569, 569t, 970
Mutagens, 272–275, 273f, 273t, 274f
Mutations
    causes of, 270–275, 272t
    chromosome structure and, 304, 304f
    consequences of, 268–270
    definition of, 7, 267
    DNA sequence and, 268
    frameshift, 268t, 269

gene expression and, 269, 269t
in germ-line or somatic cells, 269, 270f
induced, 272, 272t
introduction to, 267–268
missense, 268–269, 268t, 279, 279f
nonsense, 268t, 269
oncogenes and, 278–279, 278f, 279f
point, 268, 268t
as random events, 271, 271f–272f
sexual reproduction in elimination of, 872, 873f
silent, 268, 268t
spontaneous, 272, 272t
tetrapod limbs, 594–596, 595f–596f
tumor-suppressor genes and, 281
vertical descent with, 7, 8f
Mutualism, 957, 957f, 957t, 962
Myasthenia gravis, 765
Mycelium, 501–503, 502f, 504f
Mycobacterium tuberculosis, 376t
Mycorrhizae, 500, 520–521, 520f, 521f
Myelin sheath, 715, 716f, 724, 725f
Myers, Norman, 1034
MYH16 gene, 771, 771f
Myocardial infarction (MI), 780, 808, 809f
Myoclonic epilepsy, 339t
Myofibrils, 768f, 769–770, 769f, 774f
Myogenic heart, 787
Myosin, 77f, 78
Myosins, 769–771, 769f, 770f, 771f
    cross-bridge cycle, 771–772, 772f
Myostatin, 778, 778f
Myriapoda, 574–576, 575t, 576f
Mysticeti, 393f, 394
Myxine glutinosa, 589f

N

NAD^+ (nicotinamide adenine dinucleotide), 130, 131f, 146–147, 148f
NADH, 130, 131f
    fermentation and, 146–147, 148f
    oxidation of, 142
NADPH (nicotinamide adenine dinucleotide phosphate), 152, 153f, 156, 157f, 159–160, 160f
Nägeli, Karl, 188
Na^+/K^+-ATPase, 111, 112f
    in neurons, 718–719, 719f
Naked mole rats, 586, 943, 943f
Nanopore sequencing, 372
Nardoo, 528
Nasal passage, 696f
Nash, John, 939
Nathans, Daniel, 364
National Institutes of Health (NIH), 379
Natural capital, 1039–1040
Natural enemies, 959–961, 960f, 961f
Natural killer (NK) cells, 897f, 898, 902–903, 903f, 910
Natural products industry, 1029–1030
Natural selection
    balancing selection, 405–406, 406f
    continued human evolution, 471–472, 471f, 472f
    definition of, 7, 390
    directional selection, 403, 404f
    diversifying selection, 403–405, 405f
    evolutionary adaptation, 391f
    fitness and, 403

Grants' observations of, 406–408, 407f–408f
introduction to, 402
microevolution and, 401t
overview of, 390, 391f
patterns of, 403–406
reproductive success and, 402–403
sexual selection, 408, 409f
stabilizing selection, 403, 404f
studies of, 392t
Nature preserve design, 1035–1036, 1035f, 1036f
Nautiluses, 569t
ncRNAs. See Non-coding RNAs
Neanderthals, 466f, 467, 468, 468f, 469, 469f, 470
Necator americanus, 573
Nectar, 545, 545f
Nectar-feeding beak, 602f
Nectarinia reichenowi, 939, 940f
Negative control, 252, 254–255, 254f–255f
Negative feedback loop, 702, 702f, 879, 879f
Negative frequency-dependent selection, 406
Negative pressure filling, 799–800, 799f, 801–802, 802f
Negative regulation of cell division, 281
Nelumbo nucifera, 686
Nematocysts, 562, 563f
Nematodes, 556, 558f, 573, 573f, 574f. See also *Caenorhabditis elegans*
Neoceratodus forsteri, 594f
Neotestudina rosatii, 504f
Nephila clavipes, 945f
Nephrons, 836–838, 837f, 838f, 839f
NER (nucleotide excision repair), 275–277, 276f, 276t
Nernst, Walther, 720
Nernst equation, 720–721
Nerve cords, 582, 582f, 730, 730f
Nerve nets, 729, 730f
Nerves, 715, 732, 733f
Nervous system
    blood flow regulation by, 793–794
    cellular components of, 715–717, 716f, 717f, 718f
    communication at synapses in, 726–729, 726f, 727f, 728f
    definition of, 714
    electrical properties of neurons in, 718–721, 719f, 720f
    endocrine system, links with, 852–854, 852f, 853t
    evolution and development of, 729–731, 730f, 731f
    generation and transmission of electrical signals in, 721–725, 722f, 723f, 725f
    hemorrhage response, 919–921, 920f, 924–926, 925f
    overview of, 698t
    public health impacts and, 739–742, 740t, 741t
    redistribution of blood by, 921–922, 922f
    resting membrane potential in, 718–721, 719f, 720f
    structure and function of, 732–739, 733f, 734f, 735f, 736f, 737f, 738f
    ventilation control by, 806–807, 806f, 807t
Nervous tissues, 694f, 695, 695f, 698f
Net primary production (NPP), 983
Net reproductive rate, 953
Neural crest, 890–891, 890f
Neural tube, 890, 890f
Neurogenic heart, 787
Neurogenic muscle weakness, 339t
Neurohormone, 853

Neurological disorders, 245, 245t
Neuromuscular junction, 773–775, 774f
Neurons, 695, 695f
　definition of, 714–715
　electrical properties of, 718–721, 719f, 720f
　function of, 715–717, 716f, 717f, 718f
　generation and transmission of electrical signals along, 721–725, 722f, 723f, 725f
　morphology of, 72, 72f
　specialization of, 715–717, 717f, 718f
　structure of, 715, 716f
Neuropeptides, 728
Neuroscience, 714–715. *See also* Nervous system; Sensory systems
　public health and, 739–742, 740t, 741t, 762–763, 762f
Neurotransmission
　disorders of, 740
　drugs disrupting, 741–742, 741t
Neurotransmitters, 714, 726–728, 727f, 728f
Neurulation, 886f, 890–891, 890f
Neutral theory of evolution, 412, 412f, 444
Neutral variation, 412, 412f
Neutrons, 22, 22f, 24
Neutrophils, 897f, 898
New Guinea, 976, 977f
NFR (nucleosome-free region), 261, 261f, 262f
NH_3 (ammonia), 833–834, 833f
NH_4^+ (ammonium ions), 833–834, 833f
Niacin, 816t
Niche, 957–958
Nickel, 649t
Nicolson, Garth, 99
Nicotinamide adenine dinucleotide (NAD^+), 130, 131f, 146–147, 148f
Nicotinamide adenine dinucleotide phosphate (NADPH), 152, 153f, 156, 157f, 159–160, 160f
Nicrophorus defodiens, 945
Nierhaus, Knud, 223
Nigeria, age structure of, 1004f
NIH (National Institutes of Health), 379
Nirenberg, Marshall, 217, 218–220
Nitric oxide, 631, 638
Nitrification, 1013, 1014f
Nitrogen
　covalent bonds and, 26f
　diagram of, 23f
　in living organisms, 24, 24t
　as plant nutrient, 649t
　plant requirement for, 654–656, 656f
　in terrestrial ecosystems, 983, 983f
Nitrogen cycle, 1012–1013, 1014f
Nitrogen fixation, 483, 488–489, 1013, 1014f
　bacterial, 655, 656f
　industrial, 656
　microbiome, 513–514
　for plants, 654–656, 656f
Nitrogenous wastes, 709, 709f, 833–836, 833f, 834f, 835f
Nitrous oxide, 272–273, 273f, 1006, 1006t
NK (natural killer) cells, 897f, 898, 902–903, 903f, 910
Nociception, 752
Nociceptors, 752
Nodes, 437, 437f, 618–619, 619f
Nodes of Ranvier, 715, 716f, 724, 725f
Nodules, 655, 656f

Non-coding RNAs (ncRNAs)
　chromatin structure and transcription, effects of, 235–236
　definition of, 209, 230
　eukaryotic DNA replication, role in, 234–235
　examples of, 233t
　functions of, 230–232, 231f, 232f, 233f
　genome defense, 242–244, 242f–243f
　human disease and plant health, role in, 244–246, 245t, 246t
　importance of, 211
　introduction to, 229–230
　molecular binding of, 230, 231f
　overview of, 230–234, 231f
　protein sorting and, 239, 241, 241f
　translation, mRNA degradation, and RNA modification, effects on, 236–241
Noncompetitive inhibition, 127f, 128
Nondisjunction, 305, 306f
Nonionizing radiation, 273, 273f
Nonpolar bonds, 38f
Nonpolar covalent bonds, 26
Nonpolar molecules, 26
Nonrandom mating, 401t, 413–414, 414f
Nonrecombinant types, 341–343, 342f
Nonrecombinants, 343
Nonsense mutation, 268t, 269
Nonshivering thermogenesis, 707
Norm of reaction, 323, 323f
Northern spotted owl, 1036, 1037f
Norwalk virus, 347t
Noses, 758, 758f
Notochord, 582, 582f, 889, 889f
Nowell, Peter, 280
NPP (net primary production), 983
N-terminus, 48, 49f, 217
Nuclear envelope, 71f, 73f, 80, 80f
Nuclear genome, 87, 337f, 376–377, 377t, 381f, 457–458, 457f
Nuclear lamina, 205, 205f
Nuclear matrix, 81, 205, 205f
Nuclear pore, 71f, 73f, 80, 80f
Nuclear receptors, 53–54, 53f
Nuclearia, 498–499, 498f, 500f
Nucleic acids
　comparisons with, 41t
　definition of, 54, 191
　double helix, 54–55, 55f
　nucleotides, 54, 55f
　RNA single strand, 55, 55f
　sources, functions, and deficiency of, 815t
　structure of, 191–195
Nucleoid, 69, 70f, 354, 354f, 485f, 486
Nucleolus, 71f, 73f, 81, 81f
Nucleosome, 204, 204f
Nucleosome-free region (NFR), 261, 261f, 262f
Nucleotide excision repair (NER), 275–277, 276f, 276t
Nucleotides, 54, 55f, 191, 191f, 192–193, 192f, 193f
Nucleus
　in animal cells, 71f
　atomic, 22, 22f
　of cells, 79–85, 80f
　chromosome territory, 81, 81f
　definition of, 10
　electrons and, 22–23, 23f
　in eukaryotic cells, 95f

　in plant cells, 73f
Nudibranchs, 569f, 569t, 570, 797f
Numbers, pyramid of, 980, 982f
Nüsslein-Volhard, Christiane, 899
Nutation, 631, 631f
Nutrients
　absorption of, 814, 814f, 817–818, 817f
　capillary exchange, 790–791, 791f, 792f
　definition of, 649, 813
　for energy and molecular synthesis, 814, 828–830, 828f, 829f
　essential, 814–815, 815t, 816t
　plant, 649, 649f, 649t, 650f
　sources, functions, and deficiency of, 815t
　in terrestrial ecosystems, 983, 983f
　use and storage of, 828–830, 828f, 829f
Nutrition
　animal, 553, 553f, 814–817, 814f, 815t, 816t
　bacteria, 488–489
　definition of, 813
　for energy and molecular synthesis, 814
　essential nutrients, 814–815, 815t, 816t
　feeding habits, 815
　fungi, 501–503, 502f, 503f
　infertility and, 892
　nutrient absorption, 814, 814f, 817–818, 817f
　nutrient use and storage, 828–830, 828f, 829f
　plant requirements, 649–652, 649f, 649t, 650f, 651f, 652f
　regulation of absorptive and postabsorptive states, 830–832, 831f, 832f
　reproduction and hormones, 867
Nuts, 546

O

O_2. *See* Oxygen
Obesity, 813
Obligate aerobes, 488
Obligate anaerobes, 488
Observation, 17
Oca2 gene, 472, 472f
Occipital lobe, 736, 736f
Occludin, 118, 118f
Ocean currents, waves, and tides, 993–994, 993f, 994f
Oceans, microbiomes in, 515–516, 515f
Ocellaris clownfish, 10–11, 10f
Ochoa, Severo, 217
Ocotillo, 667, 667f
Octet rule, 25
Octopuses, 567–570, 569f, 569t
Odobenus rosmarus, 604f
Odocoileus virginianus, 945f
Odonata, 577t
Odontoceti, 393f, 394
Odor molecules, 758–761, 758f, 760f
Odum, Howard, 981
Oils, 44, 45f
Okazaki fragments, 202–203, 202f, 203f
Oleic acid, 44, 45f
Olfaction, 758–761, 758f, 760f
Olfactory receptors, 758–761, 758f, 760f
Omnivores, 815
On the Origin of Species (Darwin, C.), 389–390
Oncogene, 278–280, 278f, 279f
100 Heartbeats: The Race to Save the Earth's Most Endangered Species (Corwin), 1027

1000 Genomes Project, 472–473
Onions, 687*f*
Oocyte, 874, 875*f*
Oogenesis, 874, 875*f*, 880–881, 881*f*
Oogonia, 874, 875*f*
Oparin, Alexander, 60, 63
Open circulatory system, 568, 781–782, 781*f*
Open complex, 212
Open conformation, 259
Open ocean, 998
Operant conditioning, 933, 933*f*
Operator, 254
Operculum, 592–593, 797, 798*f*
Operon, 253
Ophiuroidea, 581, 582*t*
Opisthokonta, 491, 497–498, 498*f*, 499*t*
Opposable thumbs, 464*f*
Opsin, 754–755
Opsonization, 906
Optimality theory, 938–941, 938*f*, 939*f*, 940*f*, 940*t*
Opuntia stricta, 961, 961*f*
Oral contraceptives, 892–893, 893*f*
Orbitals, 22–23, 23*f*
Orchids, 1020, 1021*f*
Order, 10*f*, 11, 435, 436*f*
Ordovician period, 460, 460*f*
Organ fusion, flower, 678
Organ of Corti, 748, 749*f*
Organ systems
   animal, 694*f*, 696–699, 698*f*, 698*t*
   plant, 611, 612*f*
Organelles. *See also specific organelles*
   compartmentalization of, 70, 71*f*, 73*f*, 75*f*
   definition of, 70
   droplet, 70–71
   genomes of, 337–339
   protein sorting to, 89, 90*f*
   semiautonomous, 85–88, 85*f*, 95*f*
Organic fertilizers, 654
Organic matter, 652–653
Organic molecules
   breakdown of for energy, 130
   carbohydrates, 41–43, 42*f*, 43*f*
   carbon atom and carbon-containing molecules, 38, 38*f*, 39*t*
   definition of, 37
   fermentation of, 146–147, 148*f*
   formation of, prior to existence of cells, 60–62, 62*f*
   four major classes of, 40–41, 41*t*
   lipids, 44–46, 44*f*, 45*f*, 46*f*
   nonpolar and polar bonds in, 38*f*
   nucleic acids, 54–55, 55*f*
   proteins, 47–54
   synthesis and breakdown of, 40, 40*f*
Organic polymers, 62–63
Organisms
   animals, 694*f*
   classification of living, 8–11, 9*f*, 10*f*
   definition of, 2, 5, 6*f*
   energy usage of. *See* Energy use by living organisms
   environmental interaction of, 3*f*, 4
   evolution from one generation to the next, 3*f*, 4
   gene expression and characteristics of, 211
   homeostasis, 3*f*, 4
   model, 12–13
Organization, 5, 6*f*, 13–14, 13*f*, 69

Organogenesis, 891
Organs
   animal, 694*f*, 696–699, 698*f*, 698*t*
   bird, 602
   blood redistribution to, 921–922, 922*f*
   definition of, 5, 6*f*, 696
   hemorrhage effects on, 917–918, 918*f*
   lymphoid, 902, 903*f*
   plant, 611, 612*f*, 615–617, 615*f*, 615*t*, 616*f*
Orgel, Leslie, 62
Origin of Eukaryotic Cells (Margulis), 88
Origin of replication, 200, 200*f*
Ornithoptera alexandrae, 552
Ornithorhynchus anatinus, 604, 606*f*
Orthoptera, 577*t*
Oryza nivara, 550
Oryza rufipogon, 550
Oryza sativa, 550
Oscillatoria, 483*f*
Osmolarity, 708, 710
   of kidney filtrate, 839–840, 839*f*, 841*f*
Osmosis
   in animal cells, 105, 106*f*
   definition of, 105
   facilitated diffusion, 107–109, 108*f*
   homeostasis and, 693
   in kidney, 839–840, 840*f*
   across membrane, 105, 105*f*, 106*f*
   overview of, 657–658, 657*f*, 658*f*
   in plant cells, 105, 106*f*
   pressure-flow hypothesis, 667–670, 669*f*
Osmotic lysis, 105, 106*f*
Osteichthyans, 588*f*, 589*t*, 590, 591*f*, 592–593, 593*f*, 594*f*
Osteomalacia, 777
Osteoporosis, 777, 777*f*
Otoliths, 750, 750*f*
Outer bark, 624–625, 624*f*, 625*f*
Outgroup, 441*f*, 442
Out-of-Africa hypothesis, 468–469, 470*f*
Ova, 871
Ovarian cycle, 880–882, 881*f*, 882*f*
Ovaries
   animal reproduction, 874, 880–881, 880*f*, 881*f*
   in endocrine system, 850*f*, 851*f*
   flowering plant, 674, 674*f*, 685–686, 686*f*
Overexploitation, 1016*f*, 1018–1022, 1020*f*
Oviduct, 880, 880*f*
Oviparous species, 592
Ovoviparous species, 592
Ovulation, 874
Ovules, 539, 674, 674*f*, 682, 683*f*
Oxidation, 130
Oxidative fibers, 776
Oxidative phosphorylation
   ATP synthase, 140–145, 142*f*, 143*f*, 144*f*
   chemiosmosis and, 132–133, 133*f*, 140–142
   electrochemical gradient, 139–140, 141*f*
   electron transport chain, 139–140, 141*f*
   introduction to, 139
   NADH oxidation, 142
   overview of, 141*f*
   regulation of, 142
Oxygen (O_2)
   atmospheric composition and, 455
   bacterial response to, 488–489
   blood transport of, 803–806, 803*f*, 804*f*, 805*f*

   covalent bonds and, 26*f*
   cyanobacteria and production of, 482–483
   exchange of, 781, 792*f*, 797–800, 797*f*, 798*f*, 799*f*
   in gills, 797–798, 797*f*, 798*f*
   in living organisms, 24, 24*t*
   in lungs, 799–800, 799*f*
   photosynthesis and, 153*f*, 155–156, 157*f*
   as plant nutrient, 649*t*
   respiratory center response to, 806*f*, 807
   respiratory pigment binding of, 803–805, 803*f*, 804*f*, 805*f*
   in tracheal systems, 798–799, 799*f*
Oxygen-hemoglobin dissociation curve, 804–805, 804*f*, 805*f*
Oxytocin, 853–854, 884–885, 884*f*
Oysters, 567–570, 569*t*

P

P generation, 312, 313*f*, 315–316, 315*f*
P site (peptidyl site), 223, 223*f*
P wave, 789, 789*f*
p53 protein, 289–290
P680 pigment, 159, 159*f*
Pääbo, Svante, 464
Pacemaker, 787–788, 788*f*
Pain
   insensitivity to, 744
   sense of, 752
Pakicetus, 392, 393*f*
Paland, Susanne, 872, 873*f*
Paleontologists, 451, 453*t*
Paleozoic era, 459–461
Palmate venation, 620, 621*f*–622*f*
Palumbi, Steve, 1019
PAMPs (pathogen-associated molecular patterns), 899
Panamic porkfish, 422*f*, 423
Pancreas
   in digestive system, 819, 819*f*, 823, 823*f*, 825–827, 826*f*, 830–831, 832*f*
   in endocrine system, 850*f*, 856–861, 857*f*, 858*f*, 859*f*–860*f*
Pancreatic duct, 823*f*, 826*f*
Pancreatic lipase, 825, 826*f*
Pancrustacea, 574
Pandinus imperator, 576*f*
Panther chameleon, 600*f*
Pantothenic acid, 816*t*
Paper chromatography, 163*f*
Papilio xuthus, 755
Papillomavirus, 347*t*
Paracrine signaling, 171, 172*f*
Paralogs, 378
Paramecium caudatum, 106*f*
Paramecium species, 490*f*, 957–958, 958*f*
Paranthropus, 466–467
Paranthropus boisei, 466–467, 466*f*
Parapodia, 570
Parasites
   eukaryotic, 896
   flatworms, 565–566, 565*f*, 565*t*, 566*f*
   invasive species as, 1024–1025, 1025*f*
   nematodes, 573, 574*f*
Parasitism, 957, 957*f*, 957*t*, 959, 959*f*
Parasitoidism, 957, 957*t*, 959, 959*f*
Parasympathetic division, 733, 734*f*, 919, 920*f*
Parathyroid glands, 850*f*

Parathyroid hormone (PTH), 861–863, 862*f*
Parenchyma cells/tissue, 615, 615*f*
Parental care, 463
Parental strands, 197, 197*f*
Parietal lobe, 736, 736*f*
Parsimony, principle of, 443, 443*f*
Parthenium argentatum, 1029
Parthenogenesis, 871
Partial pressure, 796
Partial pressure of O_2 (P_{O_2}), 804–805, 804*f*, 805*f*, 807
Particulate inheritance, 309
Parturition, 884–885, 884*f*, 885*f*
Passenger pigeon, 1020, 1020*f*
Passive absorption, 818
Passive transport, 103, 103*f*
Patau, Klaus, 332
Patau syndrome, 306*t*
Paternal inheritance, 339
Pathogen-associated molecular patterns (PAMPs), 899
Pathogenic fungi, 506, 506*f*
Pathogens
  plant defenses against, 644–646, 645*f*, 646*f*
  types of, 896
Pattern formation, 427
Pauling, Linus, 196, 440
Pavlov, Ivan, 932
PCR. *See* Polymerase chain reaction
Pea aphids, 426–427, 426*f*
Pea plants, 311–312, 311*f*, 312*f*, 636, 687*f*
Peat moss, 532*f*, 537
Pectins, 93, 93*f*
Pedigree analysis of human traits, 319–320, 320*f*
Peer-review process, 18
Pelecanus onocrotalus, 602*f*
Pelomyxa, 490*f*
Penicillin, 1029
Penicillium, 1029
Penis, 877, 878, 878*f*
Pentoses, 41–42
Pepsin, 820–821
Peptide bonds, 47, 49*f*, 217
Peptidoglycans, 42, 43, 486–487, 487*f*
Peptidyl site (P site), 223, 223*f*
Peptidyl transfer reaction, 226, 226*f*
Per capita growth rate, 954
Perception, 745, 746*f*
Peregrine falcon, 1038
Perennials, 613
Perfect flowers, 677
Perianths, 545–546, 545*f*, 677
Pericarp, 676
Periderm, 625
Periodic table, 23, A–1
Peripheral membrane proteins, 100, 100*f*
Peripheral nervous system (PNS), 715, 732–735, 733*f*, 734*f*
Periphyton, 491, 491*f*
Perissodactyla, 605*t*
Peristalsis, 820, 820*f*
Peritubular capillaries, 837, 837*f*
Permeability, membrane, 719
Permian period, 461*f*
Peroxisomes, 71*f*, 73*f*, 84, 85*f*
Perspiration, 709–712, 710*f*–711*f*
PET (positron-emission tomography), 135, 135*f*
Petals, 542–544, 542*f*, 543*f*, 544*f*, 673, 674*f*

Petiole, 620, 620*f*
Petrogale assimilis, 604, 606*f*
Petromyzon marinus, 590*f*
pH, 34–35, 35*f*, 128
  hemoglobin and, 804–805, 805*f*
pH scale, 35*f*
Phage λ, 346*t*
  latency in, 352
  reproductive cycle of, 349, 350*f*–351*f*
Phage P1, 360, 360*f*
Phage T4, 346*t*, 347, 348*f*, 349
Phages. *See* Bacteriophages
Phagocytes, 897–898, 897*f*
Phagocytosis, 116, 492, 897
Phallus impudicus, 505*f*
Phanerozoic eon, 453, 454*f*, 459–462
Pharmaceutical industry, 1029
Pharyngeal slits, 582, 582*f*
Pharynx, 800, 801*f*, 819–820, 819*f*
Phase contrast microscopy, 68*f*
Phenolics, 546–547, 547*f*, 643
Phenotypes
  as characteristics, 314–315
  environment and, 323, 323*f*
  environmental agents, effects of, 334–335, 335*f*
  intermediate, 322–323, 322*f*
  recessive alleles, 321–322, 321*t*
  spatial expression of genes and, 427–428, 428*f*
  variation in modern humans, 472–473, 473*f*
Phenylketonuria (PKU), 321*t*
Pheromones, 936
Philanthus triangulum, 933–935, 933*f*–934*f*
Philippine eagle, 1027
Philodina roseola, 567*f*
Phloem, 532–533, 615–616, 615*f*, 623–625, 624*f*, 625*f*
  transport in, 667–670, 669*f*
Phosphate, 124, 124*f*
Phosphate functional group, 39*t*
Phosphate mining, 1037, 1038*f*
Phosphates
  cross-bridge cycle, 771, 772*f*
  of DNA, 193, 193*f*
Phosphatidylcholine, 46*f*
Phosphodiester linkage, 193, 193*f*
Phosphodiesterase, 182
Phosphoenolpyruvate, 130
Phospholipid bilayer
  in biological membranes, 99, 99*f*
  simple diffusion and, 103–104, 104*f*
Phospholipids, 45, 46*f*
Phosphorus, 24, 24*t*, 983, 983*f*
  as plant nutrient, 649*t*
Phosphorus cycle, 1012, 1012*f*
Phosphorylation, 124
Photoautotrophs, 151, 488
Photoheterotrophs, 488
Photons, 154
  photoreceptor activity and, 756, 756*f*
Photoperiodism, 639–640, 639*f*, 640*f*, 641*f*
Photophosphorylation, 157
Photoreception, 752–758, 753*f*, 754*f*, 755*f*, 756*f*, 757*f*
Photoreceptors, 752, 754–756, 754*f*, 755*f*, 756*f*, 757*f*
Photorespiration, 165–166, 166*f*, 167*f*
Photosynthesis
  biosphere and, 151, 151*f*
  Calvin cycle, 160–164

  cellular respiration and, 151, 151*f*
  chloroplasts and, 86–87, 87*f*, 152, 152*f*
  cyanobacteria and, 482–483, 483*f*
  definition of, 3*f*, 4, 150
  introduction to, 150
  molecular features of photosystems, 158–160
  overview of, 151–152
  photorespiration and, 165–166, 166*f*, 167*f*
  reactions that harness light energy, 153–157
  stages of, 152, 153*f*
  variations in, 165–167
Photosynthetic pigments, 155, 155*f*, 156*f*
Photosystem I (PSI)
  cyclic electron flow, 156–157, 158*f*
  electron energy levels and, 159–160, 160*f*
  linear electron flow, 155–156, 157*f*
Photosystem II (PSII)
  electron energy levels and, 159–160, 160*f*
  linear electron flow, 155–156, 157*f*
  molecular features of, 158–159, 159*f*
Photosystems, molecular features of, 158–160
Phototropism, 170, 171*f*, 632, 636
Phthiraptera, 577*t*
Phylogenetic trees
  animal, 554*f*
  Archaea and Bacteria, 478, 478*f*
  eukaryotic supergroups, 492*f*
  evolutionary relationships and, 437–438, 437*f*
  horizontal gene transfer, 447*f*
  introduction to, 437
  parsimony, principle of, 443, 443*f*
  plants, 530*t*, 539*f*, 545*f*
  rooted, 443
  shared characteristics and, 440–442, 441*f*
  taxonomic groups and, 438, 438*f*
  vertebrates, 588*f*
Phylogeny, 437, 445, 446*f*
Phylum, 10*f*, 11, 435, 436*f*
Physcomitrella patens, 536
Physical Basis of Life, The (Bernal), 62
Physical mutagens, 272
Physical stimuli, plant responses to, 631, 632*f*
Physical systems
  definition of, 515
  microbiomes of, 514–516, 515*f*, 516*f*
Physiological adaptation hypothesis, 271
Physiology, 13, 13*f*
Phytochrome, 639–640, 639*f*, 640*f*
Phytophthora infestans, 480*f*, 489, 497
Phytoplankton, 491, 491*f*
Pigments, 154–155, 155*f*, 156*f*
Pili, 70, 70*f*, 487
Pill bug, 579*f*, 597
Pine Barrens, 998
Pineal gland, 850*f*
Pineapples, 546, 546*f*
Pinnate venation, 620, 621*f*–622*f*
Pinocytosis, 115–116
Pinus, 536*f*, 540, 540*f*
Pinus resinosa, 521*f*
Pinworms, 573
Pipefish, 946
Pistils, 542*f*, 543, 544*f*, 674, 674*f*, 682
Pisum sativum, 311–312, 311*f*, 312*f*
Pitfall trap, 949–950, 949*f*
Pituitary gland, 850*f*
  hypothalamus and, 852–854, 852*f*, 853*t*

PKA (protein kinase A), 181–183, 181f, 182f, 183f
PKU (phenylketonuria), 321t
Placenta, 605, 883–884, 883f
Placoderms, 591f
Planarian, 564f, 565t
Planctomycetes, 482t, 485f, 486
Plant cells
  cell wall of, 89–94, 93f
  communication, 632–633, 633f
  compartmentalization of, 73f, 75f
  complexity of, 95t
  expansion of, 617–618, 617f, 618f, 636–637, 637f
  nucleus and endomembrane system, 80f
  osmosis in, 105, 106f
  plasmodesmata of, 113–114, 114f
  structure of, 73f
Plant evolution
  ancestry of land plants, 529–536, 529f, 530f, 531f, 532f, 533f, 534f, 535f, 536f
  angiosperm diversification, human influences on, 550–551, 550f
  diversity of modern angiosperms, 542–549, 542f, 543f, 544f, 545f, 546f, 547f, 548f–549f
  diversity of modern gymnosperms, 539–542, 539f, 540f, 541f
  evolutionary relationships, 529, 530f
  history of land plants, 536–538, 537f, 538f
  human influences on, 550–551, 550f
  introduction to, 528
Plants. *See also* Flowering plants
  algae and, 493–494, 493f, 499t, 529, 530f, 549t
  ancestry and diversity of, 529–536, 529f, 530f, 531f, 532f, 533f, 534f, 535f, 536f
  behavioral responses, 631–633, 631f, 632f, 633f
  cell communication, 632–633, 633f
  chromosome number changes in, 306–307, 307f
  distinctive features of, 529, 549t
  drought stress adaptations, 648, 660, 667, 667f, 668f
  drugs from, 541f, 546–548, 548f–549f
  as eukaryotes, 9f
  evolutionary history of, 536–538, 537f, 538f
  gymnosperms, 530f, 531, 534, 536f, 537, 539–542, 539f, 540f, 541f, 549t
  hormones, 619, 631, 634–638, 634t, 635f, 637f
  invasive species, 1023, 1023f
  life cycle, 529, 532f, 535f, 540f, 543, 543f
  liverworts and hornworts, 529, 530f, 531, 531f, 532f, 537, 549t
  lycophytes and pteridophytes, 529, 530f, 531–534, 533f, 534f, 535f, 537, 538f, 539f, 549t
  microbiome engineering for, 524–525, 524f
  microbiomes, 519–521, 520f, 521f
  mosses, 529, 530f, 531, 531f, 532f, 537, 549t, 673, 674f
  movement, 631, 631f
  non-coding RNAs (ncRNAs), 245, 246t
  organs, 611, 612f, 615–617, 615f, 615t, 616f
  overexploitation of, 1020, 1021f
  photosynthesis in. *See* Photosynthesis
  photosynthetic pigments of, 155, 155f, 156f
  phyla, 529–531, 530f, 539f, 545f
  sampling of, 949, 949f
  transpiration, 663–667, 664f, 665f, 666f, 667f, 668f
  vascular tissues, 532–533, 542, 615–616, 615f, 623–625, 623f, 624f, 625f
  in web of life, 447f

Plaques, atherosclerotic, 808, 808f
Plasma, 707, 785, 785f, 786f, 923, 924f
Plasma cells, 902, 903f, 904f, 906, 907f
Plasma membrane
  in animal cells, 71f, 95t
  in bacteria, 95t, 481, 481f, 485f, 486
  hormone receptors, 849
  definition of, 98
  functions of, 98t
  in plant cells, 73f, 95t
  in plant defense, 644
  structure of, 69, 70f
Plasmids, 355, 355f, 364
Plasmodesmata, 113–114, 114f
Plasmodium, 489, 496–497, 496f
Plasmodium falciparum, 496–497, 496f
Plasmolysis, 105, 106f, 658, 659f
Plasmolyzed cell, 658, 659f
Plastids, 87, 339f
  primary and secondary, 494, 495f
Platalea ajaja, 602f
Platelets, 785, 786f
Platyhelminthes, 564–566, 564f, 565f, 565t, 566f, 729, 730f
Pleiotropy, 322
Pleodorina californica, 458f, 459
Pleural sacs, 800–801
Pneumatophores, 626, 626f
Pneumocystis jiroveci, 353, 506
PNS (peripheral nervous system), 715, 732–735, 733f, 734f
P_{O_2} (partial pressure of O_2), 804–805, 804f, 805f, 807
Podos, Jeffrey, 424–425
Poinsettia, 623, 623f
Point mutation, 268, 268t
Poiseuille, Jean Marie Louis, 793
Poiseuille's law, 793
Polar bears, 11, 1036, 1037f
Polar bonds, 38f
Polar covalent bonds, 26, 27f
Polar molecules, 26
Polar transport, 635–636, 635f
Pole, of spindle apparatus, 292, 293f
Pollen, 534
  germination of, 675–676, 675f, 682
Pollen grains, 674, 681–682, 682f
Pollen tube, 676, 682–683, 683f
Pollination, 543, 545–546, 545f, 675, 675f
Pollinators, 545–546, 545f
Pollution
  biomagnification and, 1014–1016, 1015f
  human influences on biogeochemical cycles, 1008–1014
  species threatened by, 1016f
Poly A tail, 214, 214f
Polyandrous mating systems, 944f, 945f, 946
Polycistronic mRNA, 253
Polygonum, 543f
Polygynous mating systems, 944f, 945, 945f, 946f
Polymerase chain reaction (PCR), 369–370, 369f, 759–761, 760f
  microbiome analysis, 512f, 513
  polymerase used in, 507
Polymerization, 234, 235f
Polymers, 40, 40f

Polymorphic gene, 399
Polymorphism, 399, 399f
Polyp, 562, 563f
Polypeptide hormones, 849
Polypeptides
  definition of, 4, 47, 210
  formation of, 49f
  gene mutations and, 268–269, 268t
  sorting of, 239, 241, 241f
  synthesis of, 226–227, 226f
Polyplacophora, 569, 569f, 569t
Polyploidy, 305, 305f, 306–307, 307f, 426
Polysaccharides, 42–43, 43f, 92, 92f, 823, 824f
Polyunsaturated fatty acids, 44
Pons, 735, 735f
Population
  definition of, 948
  density and size of, 949–950, 949f, 950f, 962, 962f
  growth of, 952–958, 954f, 955f, 956f, 958f, 959f
Population density, 949–950, 949f, 950f, 962, 962f
Population ecology
  definition of, 948
  demography, 948, 951–953, 952f, 952t
  measurement of population size and density, 949–950, 949f, 950f
  population growth, 952–958, 954f, 955f, 956f, 958f, 959f
  species interactions, 956–963, 957f, 957t, 958f, 959f, 960f, 961f, 962f
Population genetics, 387–388, 399–401, 401f
Populations
  definition of, 5, 6f, 388
  as dynamic units, 398
  genes in, 398–402
  genetic drift and size of, 410–411, 410f, 411f
  human population growth, 1003–1005, 1003f
Populations, evolution and
  allopatric speciation, 422f
  animal phylogeny, 554f
  choanoflagellates and, 180f
  eukaryotic supergroups and, 492f
  evolutionary relationship of plant phyla, 530f
  fungi relationship, 500f
  genetic drift, 411f
  gibberellin-DELLA system evolution, 637f
  plants in hot and dry environments, 167f
  principle of, 3f, 4
  sympatric speciation, 426f
  taxonomy and, 436f
  vertebrate nervous system, 733f
  whales, 393f
Porcupine, 604f
Porifera, 561–562, 561f
Porkfish, 422f, 423
Portal veins, 853
Portuguese man-of-war, 562–563, 563f, 563t
Positive control, 252, 255–256, 256f
Positive feedback loop, 702, 928, 928f
  birth, 884–885, 884f, 885f
Positron-emission tomography (PET), 135, 135f
Postabsorptive states, 828–830, 829f
  regulation of, 830–832, 831f, 832f
Postanal tail, 582, 582f
Posterior, 555
Posterior pituitary gland, 850f, 852f, 853–854
Postsynaptic cells, 726–728, 726f, 727f, 728f

Postsynaptic membrane receptors, 728, 728f
Post-translation, 251, 251f
Post-translational sorting, 89, 90f
Postzygotic isolating mechanisms, 419, 420f, 421
Potassium
   kidney reabsorption of, 840, 840f
   as plant nutrient, 649t
   regulation of, 863–864, 863f
Potential energy, 122, 122f
Power of Movement in Plants, The (Darwin, C., and Darwin, F.), 631
Power stroke, 771, 772f
Prairie, 995f, 996, 996f, 997f
Prairie dogs, 942–943, 943f
Prebiotic soup, 60
Precambrian era, 45, 454f
Precipitation, 648
Precipitation patterns, 1007
Predation
   groups for success of, 939, 939f
   many-eyes hypothesis and, 938, 938f
   species interactions, 957, 957f, 957t, 959–961, 959f, 960f, 961f
Predators, invasive species as, 1023–1024
Predatory fungi, 506, 506f
Predictions, 14
Pregnancy, 883–885, 883f, 884f, 885f
   prevention of, 892–893, 893f
Preinitiation complex, 258, 258f, 262f
Pre-mRNA, 210, 212f, 213, 214f, 215f
   alternative splicing of, 263–264, 263f
Presbytis entellus, 605, 606f
Prescription drugs, 1029
Preservationist ethic, 1030
Preserve design, 1035–1036, 1035f, 1036f
Pressure
   gas, 796
   skin receptors detecting, 746–747
Pressure potential, 659, 659f
Pressure-flow hypothesis, 667–670, 669f
Presynaptic cells, 726, 726f, 727f
Prey
   defenses of, 959–961, 960f, 961f
   predation effects on, 961, 961f
Prezygotic isolating mechanisms, 419–421, 420f
Prickly pear cactus, 961, 961f
Primary active transport, 110, 110f
Primary cell wall, 93, 93f
Primary consumers, 979–980, 979f, 981f
Primary electron acceptor, 159, 159f
Primary endosymbiosis, 494, 495f
Primary growth, 615, 615t
Primary immune responses, 911, 911f
Primary lymphoid organs, 902, 903f
Primary meristem, 615–617, 615f, 616f
Primary oocytes, 874, 875f
Primary plastids, 494, 495f
Primary producers, 979–980, 979f, 981f
Primary production
   in aquatic ecosystems, 983–984
   consumption of, 984–985, 985f
   across earth, 984, 984f
   in terrestrial ecosystems, 983, 983f
Primary spermatocytes, 874, 875f
Primary structure of proteins, 49, 50f
Primary succession, 971, 972f, 973f
Primary tissues, 615, 615t

Primary vascular tissues, 615–616, 615f, 623–624, 623f, 624f
Primates, 605t, 755
   classification of, 465f
   evolution of, 445, 446f, 463–464, 464f
   evolutionary tree of, 464f
Primer, 201, 202, 202f
PRL (prolactin), 853t
Probing beak, 602f
Probiotic treatment, 525
Proboscidea, 605t
Proceedings of the Linnean Society of London, 389
Prochlorococcus, 1030
Producers, 489
Products, 29
Progesterone, 851f, 867
Proglottids, 565–566, 565f
Prokaryotes, 478, 896
Prokaryotic cells
   definition of, 8, 10
   examples of, 9f
   origin of, 456
   shapes of, 486, 486f
   structure of, 69–70, 70f
Prolactin (PRL), 853t
Prometaphase, 293, 294f
Prometaphase I, 297, 298f, 300
Prometaphase II, 298f
Promiscuous mating systems, 944, 944f
Promoter, 211, 212f, 253
   core, 257, 258f
Proofreading, 203
Prophage, 349
Prophase, 293, 294f
Prophase I, 297, 298f
Prophase II, 298f
Propionibacterium acnes, 488
Proplastids, 87
Prostaglandins, 884–885, 884f
Prosthetic groups, 128
Proteases, 82, 189f–190f, 190, 825, 825f
Protected areas, 1033–1035, 1034f
Protein kinase A (PKA), 181–183, 181f, 182f, 183f
Protein kinase cascade, 178, 179f
Protein kinases, 175, 175f
Protein phosphatases, 182
Protein subunits, 50, 50f
Protein-encoding genes, 209
   microbiome analysis, 513–514
Protein-protein interactions, 51, 51f, 69
Proteins
   amino acids, 47–48, 48f, 49f
   categories and functions of, 47t
   cell characteristics determined by, 71–72
   comparisons with, 41t
   definition of, 4, 47, 210
   digestion of, 820–821, 825, 825f
   folding of, 51–53, 52f
   functional domains of, 53–54, 53f
   hierarchy of structure of, 49–50, 50f
   membranes and, 99–100, 99f, 100f
   metabolism of, 145–146, 145f
   microbiome analysis, 513–514
   ncRNA and function or stability of, 232
   organism characteristics and, 211
   polypeptides and, 210
   processing of, 82

   sorting of, 82, 89, 90f, 239, 241, 241f
   sources, functions, and deficiency of, 815t
   structure, determining factors of, 50–51
   temperature effects on, 703
   transmembrane movement of, 102, 102f
   transport. See Transport proteins
   use and storage of, 828–830, 828f, 829f
Proteobacteria, 88, 88f, 482t, 483
Proteoglycans, 92, 92f
Proteolysis, 82
Proteomes, 72, 250
Proterozoic eon, 453, 454f, 457, 458–459
Prothoracicotropic hormone (PTTH), 865, 866f
Protists
   algae, 479, 489, 490f, 493–494, 493f, 499t, 511f, 515, 515f, 518, 518f, 529, 530f, 549t, 654
   Alveolata, 494–497, 494f, 495f, 496f, 499t
   Amoebozoa, 497, 499t
   animal evolution from, 554–555, 554f
   competition, 957–958, 958f
   cytoplasmic extensions, 497, 497f
   definition of, 479
   diversity and ecological importance of, 489–498, 490f, 491f, 492f, 493f, 494f, 495f, 496f, 497f, 498f, 499t
   ecological roles, 489, 490f
   as eukaryotes, 9f
   eukaryotic supergroups, 491–492, 492f, 499t
   examples of, 480f
   Excavata, 492–493, 492f, 493f, 499t
   feeding groove, 492–493, 492f, 493f
   flagella, 490, 490f, 497–498, 497f, 498f
   fungus-like, 489
   habitats, 490–491, 491f
   land plants and, 493–494, 493f, 499t
   membrane sacs, 494, 494f
   microbiomes, 518, 518f
   motility, 490, 490f
   Opisthokonta, 491, 497–498, 498f, 499t
   protozoa, 479, 489, 490f
   pseudopodia, 490, 490f, 497
   Rhizaria, 494, 497, 497f, 499t
   Stramenopila, 494, 497, 497f, 499t
Protobiont, 63, 63f
Proton-motive force, 140
Protons, 22, 22f, 23
Proto-oncogenes, 279–280, 279f
Protostomes, 556–558, 557f, 559f
Prototherians, 603–604, 605t, 606f
Protozoa, 479, 489, 490f
Proventriculus, 820f, 821
Provirus, 349
Proximal tubule, 837–838, 838f, 839f
Proximate causes, 932
Psaronius, 460, 461f
Pseudoceros bifurcus, 565f
Pseudocoelomates, 556, 558f
Pseudoephedrine, 541f
Pseudomonas simiae, 525
Pseudopodia, 490, 490f, 497
Pseudosalix handleyi, 537f
Pseudotriton montanus, 597f
PSI. See Photosystem I
PSII. See Photosystem II
Psilocybe, 505
Psilotum nudum, 533f, 534f

Solution, 30
Solvent, 30, 103
Somatic cells, 269, 270f, 331–332, 332f
Somatic embryogenesis, 688–689, 688f
Somatic nervous system, 732
Somniosus microcephalus, 592
Sorting signals, 89
Sound
    detection of, 747–748, 748f, 749f, 763
    hair cells and, 747, 747f, 749f
Source pool, 975–976, 977f
Sousa, Wayne, 972
South American pit viper, 1029
Southeast Asia, fertility rate, 1003
Soybean plant, 656f
Spatial expression of genes, 427–428, 428f
Spatial learning, 933
Speciation, 417
    allopatric speciation, 422–423, 422f
    mechanisms of, 422
    reproductive isolation in finches, 424–425, 424f, 425f
    sympatric speciation, 426–427, 426f
Species
    definition of, 5, 388, 416
    distribution of, 999–1000, 999f, 1000f, 1007–1008, 1008f
    domains of, 434–435, 434f, 435t
    identification of, 417–419, 418f
    in taxonomic hierarchy, 435, 436f
Species concept, 418–419
Species diversity, 1028
    definition of, 968
    patterns of, 966–969, 966f, 967f, 968f, 969t
Species interactions
    amensalism, 957, 957t
    commensalism, 957, 957t, 962
    competition, 956–958, 957f, 957t, 958f, 959f
    herbivory, 957, 957f, 957t, 959, 959f
    mutualism, 957, 957f, 957t, 962
    natural enemies, 959–961, 960f, 961f
    parasitism, 957, 957f, 957t, 959, 959f
    parasitoidism, 957, 957t, 959, 959f
    predation, 957, 957f, 957t, 959–961, 959f, 960f, 961f
Species richness
    birds, 966, 966f, 968, 969t
    community stability and, 970, 970f
    definition of, 966
    ecosystem function, 1031–1033, 1032f
    island size and, 976f
    patterns of, 966–969, 966f, 967f, 968f, 969t
    source pool, distance from, 977f
    species-area hypothesis of, 966–967, 967f
    species-productivity hypothesis of, 967–968, 967f
    species-time hypothesis of, 966, 967f
Species turnover, 975–976
Species-area effect, 967
Species-area hypothesis, 966–967, 967f
Species-area relationships, 975–976, 975f, 976f
Species-distance relationships, 975–976, 975f, 977f
Species-productivity hypothesis, 967–968, 967f
Species-time hypothesis, 966, 967f
Specific heat, 32
Sperm
    animal, 871, 874–879, 875f, 876f, 878f
    flowering plants, 674, 682–683, 683f

Spermatids, 874, 875f
Spermatocyte, 874, 875f
Spermatogenesis, 874, 875f, 878–879, 878f, 879f
Spermatogonia, 874, 875f
Spermatophytes, 534, 549t
Spermophilus beldingi, 942
Sphagnum, 532f, 537
Sphere-shaped bacteria, 486, 486f
Spicules, 561
Spiders, 575, 576f, 945f
Spike mosses, 533f
Spinal cord, 715
Spinal nerves, 732, 733f
Spines, 623, 623f
Spiracles, 574, 574f, 798, 799f
Spiral cleavage, 556, 557f
Spiral-shaped bacteria, 486, 486f
Spirilli, 486
Spirochaetes, 482t, 486, 486f
Spix's macaw, 1027
Spleen, 926
Spliceosome, 214–215, 215f
Splicing, 214–215, 215f
Sponges, 180f, 561–562, 561f
Spongin, 561
Spontaneous movements, 101f
Spontaneous mutations, 272, 272t
Sporangia, 529, 674
Spores, 502–504, 502f, 504f, 506, 506f
    plant, 529
Sporophytes, 301, 529, 534, 535f, 610, 610f
    development of, 611–612, 612f, 675–676, 675f
    in sexual cycle of flowering plants, 673, 673f, 674f
Sporopollenin, 529
Spotted sandpiper, 946
Squamata, 589t, 599–601, 600f
Squid, 567–570, 569t
Squirrels, 942
SRP (signal recognition particle), 239, 241, 241f
SSRIs (selective serotonin reuptake inhibitors), 740
SSU rRNA (small subunit rRNA), 224, 224f, 445
Stability, community, 970, 970f
Stabilizing selection, 403, 404f
Stahl, Franklin, 197–198, 198f
Stamens, 542–544, 542f, 543f, 544f, 673–674, 674f
Staminate flowers, 677, 678f
Standard fluorescence microscopy, 68f
Standard light microscopy, 68f
Staphylococcus aureus, 361
Starch, 42, 43, 43f
Starling, Ernest, 922–923
Starling forces, 923, 924f
Start codon, 216t, 217, 225, 225f
Statocysts, 748–750, 749f
Statoliths, 641–642, 641f, 642f, 749f, 750
STDs (sexually transmitted diseases), 891–892
Stearic acid, 44, 45f
Stedman, Hansell, 771
Stem cells, 617
Stems, 532–533, 534f, 611, 611f
    modified, 625
    primary structure and development, 615–616, 615f
Stents, 808, 809f
Stereocilia, 747, 747f, 749f
Steroid hormones, 177, 849, 851f
Steroids, 45, 47f

Steward, F. C., 688
Stigma, 542f, 543, 674f
Stigonema, 483f
Stiling, Peter, 1009–1011
Stimulus, sensory, 745, 746f
Stingrays, 592f
Stinkhorn fungi, 504, 505f
Stoermer, Eugene, 1002
Stomach, 698f, 819–821, 819f, 821f
    in endocrine system, 850f
Stomach acid, 813, 820–821, 842–845, 843f, 844f
Stomata, 152, 152f, 533–534, 534f, 616, 616f, 651, 665–666, 665f, 666f
    closure of, 638
Stone, Christopher, 1030
Stop codon, 216t, 217, 226–227, 227f
Stork, Nigel, 576
Stovepipe sponge, 561f
Strahl, Brian, 260
Strain, 312, 357
Straining reactants, 126
Stramenopila, 494, 497, 497f, 499t
Strand, of DNA, 191, 191f, 192–193, 193f
Strasburger, Eduard, 316
Strawberries, 546, 546f
Strepsirrhini, 463, 464f, 465f
Streptococcus pneumoniae, 486f, 487
    base composition of DNA in, 196t
    Griffith's bacterial transformation experiments on, 188–189, 188f
    transformation in, 359
Streptophytes, 529, 530f, 549t
Stretch receptors, 747
Stroke volume (SV), 794, 794t, 927t
Stroma, 87, 87f, 152, 152f
Stromatolites, 456, 456f
Strong, Donald, 966
Strong acid, 34
Struck, Torsten, 571
Structural formula, 25
Structural models, 17
Structural proteins, 47t, 91t, 92, 92f
Structure determines function
    AT/GC rule and, 199f
    bird beak variety, 602f
    cerebral cortex folding, 731f
    conformational change, 109f
    elastin protein, 92f
    external gills, 797f
    gametogenesis, 875f
    hemoglobin, 803f
    horse evolution, 439f
    hydrophilic pocket, 109f
    polyp and medusa forms of cnidarians, 563f
    principle of, 3f, 4
    renal corpuscle, 838f
    shade- and sun-leaves, 651f
    small intestine tissue arrangement, 822f
    steroid hormones, 851f
    transfer RNA, 221f
    vertebrate jaw evolution, 591f
    visual communication in fireflies, 937f
    visual pigment, 755f
    xylem, 623f
Stylets, nematode, 573
Subatomic particles, 22
Submetacentric, 303, 303f

Subsidence zones, 989, 990f
Subsoil, 652, 653f
Subspecies, 417
Substrate-level phosphorylation, 132
Substrates, 126–128, 127f
Subterranean root microbiomes, 519–520
Succession
   definition of, 971
   facilitation, 971–972, 972f, 973f, 974f
   inhibition, 972, 974f
   island biogeography model, 975–979, 975f, 976f, 977f–978f
   models of, 973, 974f
   tolerance, 972–973, 974f
Sugar loading, 667–670, 669f
Sugar maple, 667, 1007–1008, 1008f
Sugar-phosphate backbone, 193, 193f, 195f
Sulfate functional group, 39t
Sulfhydryl functional group, 39t
Sulfolobus solfataricus, 376t
Sulfur, 649t
Sumatran tiger, 1039
Sun tracking, 642
Sun-leaves, 651, 651f
Supergroups, 10, 10f
   definition of, 434, 491
   eukaryotic, 434–435, 434f, 436f, 491–492, 492f, 499t
Surface area/volume (SA/V) ratio, 72–73, 74f, 700, 700f
Surface tension, 33, 33f
   lung, 802
Surfactant, 802
Survivorship curves, 951, 952f
Survivorship rate, age-specific, 952–953
Suspension feeders, 553f
Sustainability, 1039–1040
Sustainable development, 1039
Sutter, Nathan, 396
Sutton, Walter, 316
SV (stroke volume), 794, 794t, 927t
Svedberg, Theodor, 222
Sweat, 709–712, 710f–711f
Swim bladder, 593
Swine flu (H1N1), 352
Sycamore, 623, 623f
Symbiotic relationship, 88, 88f
Symmetry
   animal, 555, 555f
   flower, 678, 679f
   plant, 614, 614f
Sympathetic division, 733, 734f, 919–921, 920f, 924–926, 925f
Sympatric speciation, 426–427, 426f
Symplast, 661
Symplastic transport, 661, 661f, 662f
Symporter, 110–111, 110f
Synapses, 726–729, 726f, 727f, 728f
Synapsis, 297, 298f
Synaptic cleft, 726–727, 727f
Synechococcus, 515
Syngnathus typhle, 946
Synthesis of organic molecules, 40, 40f
Synthetic microbiomes, 525
Synthetic RNA, 217–218, 218f–219f, 220
Systematics, 433, 438–440, 438f, 439f
Systemic acquired resistance (SAR), 645–646, 646f

Systemic circulation, 782, 783f
Systemin, 638, 644, 644f
Systems biology, 13–14, 13f, 94, 95f, 95t
Systole, 788–789
Szostak, Jack, 63, 65

T

T (thymine), 54, 55f, 192, 192f, 193, 193f
T_3 (triiodothyronine), 854f, 855, 855f
T cells, 902–904, 903f, 904f
   activation of, 908
   antigens recognition, 906–908, 908f
   cells killed by, 908–910, 909f
   HIV infection, 913
   self recognition by, 911
T wave, 789, 789f
T1DM (type 1 diabetes mellitus), 174, 857–861, 859f–860f
T2DM (type 2 diabetes mellitus), 857
Table salt, 21, 27, 28f, 30f
Tactile communication, 936–938, 937f
Tadpoles, 597, 597f
Taenia pisiformis, 565f
Tagmata, 574, 574f
Taiga, 995f, 996, 996f, 997f
Tail of motor proteins, 77, 77f
Tangl, Eduard, 113
Tansley, A. G., 965
Tapeworms, 565, 565f, 565t
Tapir distribution, 999, 999f
Taproot system, 616, 626
Taq polymerase, 369–370, 369f
Tarantula venom, 1029
Tarpan, 1019
Tasmanian tiger, 1039
Taste, sense of, 761, 761f
Taste buds, 761, 761f
TATA box, 257, 258f
Taxol, 546
Taxon, 434
Taxonomy
   binomial nomenclature, 11, 435–436
   cladistics, 440–444, 441f, 442f, 443f
   definition of, 8, 433
   evolutionary relationships and, 438, 438f
   examples of, 10f, 434f
   horizontal gene transfer, 446–448, 447f
   introduction to, 433–434
   molecular clocks, 444–445, 444f, 446f
   overview of, 434–436
   phylogenetic trees and, 437–438, 437f, 438f
   taxonomic hierarchy, 435
Tay-Sachs disease, 321t
Tbx5 gene, 783–784, 784f
Teal, John, 984
Tearing beak, 602f
Teeth, 603, 604f
Telocentric, 303, 303f
Telomerase, 234, 235f
Telomerase reverse transcriptase (TERT), 234
Telomerase RNA component (TERC), 233t, 234, 235f
Telomeres, 234, 234f
Telophase, 293, 295f
Telophase I, 299f, 300
Telophase II, 299f

TEM. *See* Transmission electron microscopy
Temperate coniferous forest, 995f, 996, 996f, 997f
Temperate deciduous forest, 995–996, 995f, 996f, 997f
Temperate grassland, 995f, 996, 996f, 997f
Temperate rain forest, 529f, 995, 995f, 996f, 997f
Temperature
   atmospheric circulation and, 989–990
   detection of, 751, 751f
   Earth and historical changes in, 455
   Earth's, variation in, 989, 989f
   effects of, 703
   enzymes and, 128, 129f
   gas solubility and, 796
   heat exchange and, 704–705, 705f
   heat gain and loss in endotherms, 705–706, 706f
   hemoglobin and, 804–805, 805f
   homeostasis and, 3f, 4
   homeostatic regulation of body, 703–707, 704f, 705f, 706f
   membranes and, 703
   metabolic rate and, 703–704, 704f
   muscle activity and brown adipose tissue in heat production, 706–707
Template strands, 198, 199f, 212
Temporal isolation, 419, 420f
Temporal lobe, 736, 736f
Temporal models, 17
Tendrils, 623, 623f
Tensile strength, 92
Teosinte, 550, 550f
Terborgh, John, 1036–1037
TERC (telomerase RNA component), 233t, 234, 235f
Terebratulina septentrionalis, 568f
Termination stage
   of transcription, 213f
   of translation, 226–227, 227f
Terminator, 211, 212, 212f
Termites, 577t, 578, 943
Terpenes, 546, 547f, 643
Terpenoids, 546, 547f, 643
Terrestrial biomes, 995–998, 995f, 996f, 997f
Terrestrial ecosystems
   food chain in, 979f
   primary production in, 983, 983f
Terrestrial mammals, 392, 393f, 394
Terrestrial tetrapods, 392, 392f
Territory, 936
   defense of, 939–941, 940f, 940t
TERT (telomerase reverse transcriptase), 234
Tertiary consumers, 980
Tertiary endosymbiosis, 494, 495f
Tertiary period, 462
Tertiary structure of proteins, 50, 50f
TEs (transposable elements), 381–383, 382f, 383f
Testable, 14
Testcross, 343
Testes
   in endocrine system, 850f, 851f
   reproduction, 874, 878–879, 878f, 879f
Testosterone, 867, 879, 879f
Testudines, 589t, 599–600, 600f
Tetanus, 484
Tetrahydrocannabinol (THC), 547–548, 548f–549f
Tetrao tetrix, 946f
Tetraploid organisms, 305, 305f

T

Tetrapods. *See also* Mammals
  amniotes, 598–603, 599f, 600f, 601f, 602f
  amphibians, 589t, 596–598, 597f, 733f, 797, 886–887, 887f, 889f
  clades of vertebrates, 588f
  limbs of, 594–596, 595f–596f
  origin of, 594
TFIID (transcription factor IID), 258, 259f
TFR (total fertility rate), 1003, 1004f
Thalamus, 735f, 736, 736f
Thaumarchaeota, 481
THC (tetrahydrocannabinol), 547–548, 548f–549f
Theory, 14
Theria, 604–605, 605t, 606f
Therman, Eeva, 332
Thermodynamics
  first law of, 122, 153
  second law of, 123
Thermoreception, 751, 751f
Thermoreceptors, 751, 751f
Thermus aquaticus, 507
Thiamine, 816t
Thick filaments, 769–771, 769f, 770f
Thin filaments, 769–771, 769f, 770f
30-nm fiber, 205, 205f, 206f
Thoracic breathing, 598
Thoraco-abdominal pump, 922
Thornton, Joseph, 864
Threatened species, 1028
Three-cell model, 989–990, 990f
Threshold potential, 722, 723f
Thylacine, 1023
Thylakoid lumen, 152, 152f
Thylakoid membrane, 87, 87f
Thylakoids, 152, 152f, 485–486, 485f
Thymine (T), 54, 55f, 192, 192f, 193, 193f
Thymine dimer, 273, 273f
Thymus, 902, 903f
Thyroid gland, 850f, 854–855, 854f, 855f
Thyroid hormone, 848, 854–855, 854f, 855f
  in growth and development, 865, 865f
Thyroid-stimulating hormone (TSH), 853t, 855, 855f
Thyrotropin-releasing hormone (TRH), 854–855, 855f
Ticks, 575, 576f
Tidal ventilation, 802
Tides, 993–994, 994f
Tight junctions, 118, 118f
Tiktaalik roseae, 392, 392f
Tilman, David, 970, 970f, 1031–1032
Tinbergen, Niko, 933–935, 933f–934f
Tissues, 5, 6f
  animal, 694–696, 694f, 695f, 696f, 697f, 698f, 768
  plant, 529, 532–533, 542, 615–616, 615f, 615t, 623–625, 623f, 624f, 625f
TLRs (toll-like receptors), 899
TMV (tobacco mosaic virus), 345, 346t, 348f, 352
TNF (tumor necrosis factor), 908, 908f, 911
Toads, 589t, 597–598, 597f
Tobacco mosaic virus (TMV), 345, 346t, 348f, 352
Tobacco smoke, 809–810
Tolerance, 972–973, 974f
Toll protein, 899–901, 900f–901f
Toll-like receptors (TLRs), 899
Tonicella lineata, 569f
Topsoil, 652, 653f
Total fertility rate (TFR), 1003, 1004f
Total peripheral resistance (TPR), 794

Touch, 746–747
  plant responses to, 641–643, 641f, 642f, 643f
Toxins
  arthropod, 575–576
  bacteria, 483–484
  fungal, 505, 505f
  microbiomes producing, 516, 516f, 518
TPR (total peripheral resistance), 794
Trace elements, 24
Trachea, 798, 799f, 800, 801f
Tracheal system, 574, 798–799, 799f
Tracheids, 532, 534f
Tracheophytes, 532–534, 534f, 549t
tracrRNA, 372–373, 374f
Tracts, 732
Traits, 311, 313f
Transcription
  definition of, 210
  diagram of, 212f
  DNA methylation and, 261–262
  elongation stage of, 212, 213f, 262f
  in eukaryotes, 212
  gene regulation and, 250–251, 251f
  overview of, 211–212, 212f
  regulation of in bacteria, 250, 251f, 252–256
  regulation of in eukaryotes, 251, 251f, 256–262
  RNA polymerase and DNA template during, 212, 213f
  stages of, 213f
Transcription factor IID (TFIID), 258, 259f
Transcription factors, 173, 178, 179f, 212
  general, 258, 258f, 259f
  regulatory, 252, 252f
  in transcription regulation in eukaryotes, 256–258, 258f, 259f
Transcriptional activation, 261, 262f
Transcriptional start site, 257, 258f
Transcriptome, 514
Transduction, 357, 357t, 360, 360f
Transects, 949, 949f
Transepithelial transport, 118
Transfer RNA (tRNA)
  aminoacyl-tRNA synthetases and, 221–222, 222f
  anticodons in, 217, 218f
  binding to ribosomes, 218–220
  common structural features, 220–221
  definition of, 211
  elongation stage, 226–227, 226f
  information and, 217, 218f
  introns in, 215
  as ncRNA, 233t
  overview of, 55
  in ribosome structure, 223f
  RNA triplets, 217–218, 218f–219f, 220
  structure of, 221f
  translation and, 221t, 225–226, 225f
Transformation, 189, 357, 357t, 359, 359f
Transformation experiments, on bacteria, 188–189, 188f
Transformation principle, 189, 189f–190f
Transition state, 126
Transitional form, 392, 392f
Translation
  aminoacyl-tRNA synthetases, 221–222, 222f
  bacterial and eukaryotic comparison, 227t
  definition of, 210
  elongation stage of, 226–227, 226f

  in eukaryotes, regulation of, 262–264, 265f
  functional sites for, 223–224, 223f
  gene regulation and, 251, 251f
  genetic code, 216–220
  initiation of, 225–226, 225f
  machinery of, 220–224, 221t
  non-coding RNAs, effects of, 236–241
  overview of, 210–211
  ribosomes, 222–223
  stages of, 225–227
  termination of, 226–227, 227f
  transfer RNA structural features, 220–221
Translation factors, 221t
Translocation, 234–235, 235f, 304, 304f
Transmembrane gradient, 104, 104f
Transmembrane proteins, 100, 100f, 102, 102f
Transmembrane transport, 661, 661f
Transmission, 188
Transmission electron microscopy (TEM), 66–67, 69f, 481f
Transpiration, 663, 664f
  global climate effects of, 664–665, 665f
  plant adaptations reducing, 665–667, 665f, 666f, 667f, 668f
Transplantation
  of gut microbiomes, 524–525
  of kidneys, 845–846
Transport, 103–106
  at cellular level in plants, 656–661, 657f, 658f, 659f
  gases in blood, 803–806, 803f, 804f, 805f
  long-distance in plants, 657f, 663–670, 664f, 665f, 666f, 667f, 668f, 669f
  at tissue level in plants, 661–663, 661f, 662f
Transport proteins
  active transport, 110–111, 110f
  CFTR gene and, 16
  channels for solute movement, 107, 107f
  ion electrochemical gradients, 111, 111t, 112f
  in plants, 657–658, 657f
  transporters, 47t, 109–110, 109f, 110f
Transposable elements (TEs), 381–383, 382f, 383f
Transposase, 382, 382f
Transposition, 381–383, 382f, 383f
Transverse tubules (T-tubules), 773, 774f
Traps, 949–950, 949f
"Tree of life," 8, 448
Tree python, 600f
Tree trunk, 624–625, 624f, 625f
Trematodes, 565–566, 565t, 566f
TRH (thyrotropin-releasing hormone), 854–855, 855f
Trial-and-error learning, 933, 933f
Triassic period, 461, 461f
Trichoderma, 480f
Trichomes, 622, 622f
Trichomonas vaginalis, 477, 493
Triglycerides
  digestion and absorption of, 825–827, 826f
  fatty acids, 44, 45f
  formation of, 44, 44f
  use and storage of, 828–830, 828f, 829f
Triiodothyronine (T_3), 854f, 855, 855f
Trilobita, 574
Trilobite, 460, 460f
Triple response, 637, 637f
Triple X syndrome, 306, 306t, 332
Triplet, 217–218, 218f–219f, 220

Triploblastic animals, 555
Triploid organisms, 305, 305f
Trisomy, 305, 305f, 306t
Triticum aestivum, 306, 307f, 550
Triticum boeoticum, 550
Triticum dicoccoides, 550
tRNA. *See* Transfer RNA
Trochophore larva, 558, 559f
Trombicula alfreddugesi, 575, 576f
Trophectoderm, 887, 888f
Trophic level, 979–980, 979f, 981f, 982f
Trophy fish, 1020, 1021f
Tropical deciduous forest, 995, 995f, 996f, 997f
Tropical forests, 1017
Tropical grassland, 995f, 996, 996f, 997f
Tropical rain forest, 995, 995f, 996f, 997f
Tropomyosin, 772–773, 773f
α-Tropomyosin, 263, 263f
Troponin, 772–773, 773f
True-breeding line, 312
Truffles, 504, 505f
Trunk, tree, 624–625, 624f, 625f
Trypsin, 825, 825f
Tschermak, Erich von, 310
TSH (thyroid-stimulating hormone), 853t, 855, 855f
Tsui, Lap-Chee, 16
T-tubules (transverse tubules), 773, 774f
Tubal ligation, 892, 893f
Tube cell, 682
Tube feet, 580–581, 580f
Tube worms, 571, 571f
Tuber melanosporum, 505f
Tubulin, 293, 295
Tumor necrosis factor (TNF), 908, 908f, 911
Tumors, 277f, 278
    DNA microarrays and, 373t
Tumor-suppressor genes, 278, 280–282
Tundra, 995f, 996–997, 996f, 997f
Tunic, 583
Tunicates, 583, 584f
Turbellarians, 565, 565f, 565t
Turgid cell, 658, 659f
Turgidity, 33, 33f
Turgor pressure, 658, 659f
Turkana boy, 467
Turner syndrome, 306, 306t, 332
Turtles, 589t, 599–600, 600f
Tusks, 604f
Twelve-wired bird-of-paradise, 12f
Two-factor cross, 315–316, 315f
Type 1 diabetes mellitus (T1DM), 174, 857–861, 859f–860f
Type 2 diabetes mellitus (T2DM), 857
Tyrannosaurus rex, 462
Tyrosine kinases, 175, 175f

U

U (uracil), 55, 55f, 192, 192f
Ulcer, 842–845, 843f, 844f
Ultimate causes, 932
Ultraviolet (UV) light, 154, 170, 272–273, 273f
Umbilicaria, 519f
Umbrella species, 1036, 1037f
Unbranched filaments, 483, 483f
Unicells, 483, 483f
Unicellular choanoflagellate, 180f

Uniformitarianism, 388
Uniporter, 110, 110f
Unsaturated, 101
Unsaturated fatty acids, 44, 45f
Upwellings, 983
Uracil (U), 55, 55f, 192, 192f
Urea, 833f, 834
Uremia, 845
Urey, Harold, 60–61
Uric acid, 833f, 834
Urinary system, 694f, 696f, 836, 837f
    hemorrhage response, 924–926, 925f
Urine, 834, 840–841, 840f, 841f, 927t
Urochordata, 583, 584f
Urodela, 597–598, 597f
Ursus arctos, 521
Uterine cycle, 882, 882f
Uterus, 880, 880f
    pregnancy, 883–884, 883f
    preparation for embryo, 882, 882f
UV (ultraviolet) light, 154, 170, 272–273, 273f

V

Vaccinations, 911
Vacuoles, 83–84, 84f, 617, 617f
Vagina, 877
Vaginal diaphragms, 892, 893f
Valence electrons, 23
Valves
    heart, 786, 789f
    vein, 791–792, 793f
Van der Waals dispersion forces, 26–27, 44, 51
Van Helmont, Jan Baptista, 151
Van Valen, Leigh, 419
Variable region, 905f, 906
Variants, 311
Variation, genetic material and, 188
Varicella-zoster virus, 352
Variola virus, 347t
Vascular bundles, 615–616, 615f
Vascular cambium, 624, 624f, 625f
Vascular seedless plants, 529, 530f, 531–534, 533f, 534f, 535f, 539f, 549t
    in evolutionary history of land plants, 537, 538f
Vascular tissues, 532–533, 542, 615–616, 615f, 623–625, 623f, 624f, 625f
Vasectomy, 892, 893f
Vasoconstriction, 793, 924, 925f
Vasodilation, 793
Vectors, 364, 367–368
Vegetal pole, 886–887, 887f
Vegetative growth, 611–612, 612f
Veins, 782, 783f, 791–792, 791f, 793f
    leaf, 616, 616f, 620, 621f–622f
Velocity, of chemical reaction, 127, 127f
Venom, 575–576
Ventilation, 797, 800, 801f
    control of, 806–807, 806f, 807t
    tidal, 802
Ventral side, 555
Ventricles, 782, 783f, 787–788, 787f, 788f
Venules, 791–792, 791f
Vergnaud, Giles, 242
Vernalization, 328
Vernanimalcula guizhouena, 459, 459f
Verreaux's eagle, 602f

Vertebrae, 587
Vertebrates, 581–582, 582f. *See also* Mammals
    amniote eggs and terrestrial life adaptations, 598–599, 599f
    amphibians, 589t, 596–598, 597f, 733f, 797, 886–887, 887f, 889f
    birds, 589t, 599–602, 601f, 602f, 782–783, 783f, 820, 820f, 886–887, 966, 966f, 968, 969t
    brains of, 730, 731f
    characteristics of, 553, 553f, 587, 588f, 589t
    chondrichthyans, 589t, 591–592, 591f, 592f
    circulatory system in, 782–784, 783f, 784f
    clades, 587, 588f, 589t
    cyclostomes, 587–590, 589f, 589t, 590f
    definition of, 586
    digestive system, 818–827, 819f, 820f, 821f, 822f, 823f, 824f, 825f, 826f
    eyes of, 752–754, 753f
    growth regulation in, 865, 865f
    heart of, 786–790, 787f, 788f, 789f
    homeostasis, 701
    introduction to, 586–587
    jaws of, 587, 590, 591f
    nervous system structure and function in, 732–739, 733f, 734f, 735f, 736f, 737f, 738f
    osteichthyans, 588f, 589t, 590, 591f, 592–593, 593f, 594f
    reptiles, 589t, 599–601, 600f, 601f, 751, 751f, 886–887
    respiratory system, 799–800, 799f
    Sarcopterygii, 589t, 590, 592–593, 594f
    sense of balance in, 748–751, 749f, 750f
    skeletons of, 767, 767f
    tetrapod limbs, 594–596, 595f–596f
Vertical evolution, 7, 8f
Vesicles, 79
Vesicular transport model, 82
Vessels, 542
Vestibular system, 748–751, 748f, 749f, 750f
Vestigial structures, 397, 397t
Via, Sara, 426
Vibrio cholerae, 375, 486f
Vibrio parahaemolyticus, 479f
Vibrios, 486, 486f
Vigilance, 938, 938f
Villi, 821, 822f
Vincristine, 1029
Viral assembly, 351f, 352
Viral component synthesis, 351f, 352
Viral envelope, 347, 348f
Viral genome, 347
Viral reproductive cycles
    attachment, 349, 350f
    definition of, 349
    entry, 349, 350f
    integration, 349, 350f
    latency in bacteriophages, 352
    latency in human viruses, 352
    release, 351f, 352
    synthesis of viral components, 351f, 352
    viral assembly, 351f, 352
Viral vectors, 364
Virchow, Rudolf, 59, 91, 286
Viruses
    characteristics of, 346, 346t
    definition of, 346
    drugs for, 353

INDEX

Viruses, *continued*
  emerging viruses, 352–353, 353*f*
  human hosts, 346, 347*t*, 352
  introduction to, 345
  latency in, 352
  pathogens, 896
  structural differences in, 346–347, 348*f*
  variation in, 346–347, 348*f*
  viral reproductive cycles, 349, 350*f*–351*f*, 352
Visceral mass, mollusk, 568, 568*f*
Visual communication, 936, 937*f*
Visual disorders, 762–763, 762*f*
Visual pigments, 754–756, 755*f*, 756*f*
Vitamin A, 816*t*
Vitamin B_1, 816*t*
Vitamin B_2, 816*t*
Vitamin B_6, 816*t*
Vitamin B_{12}, 816*t*
Vitamin C, 816*t*
Vitamin D, 816*t*, 861–863, 862*f*
Vitamin D deficiency, 472
Vitamin E, 816*t*
Vitamin K, 816*t*
Vitamins, 815, 816*t*
Viviparous species, 592
Volcanic eruptions, 455
Voltage-gated ion channels, 721, 722*f*, 724
Volvocine algae, 458–459, 458*f*
Volvox aureus, 458*f*, 459
Vries, Hugo de, 310
Vulpes vulpes, 931

W

Wächtershäuser, Günter, 61–62
Wackernagel, Mathis, 1004
Waddington, Conrad, 328
Waggle dance, 936–938, 937*f*
Walker, Brian, 1031
Walker, John, 143
Wallace, Alfred, 389, 1000, 1000*f*
Wallin, Ivan, 88
Walrus, 604*f*
Warburg, Otto, 135
Warren, J. Robin, 843–845, 844*f*
Wasps, 577*t*, 578, 933–935, 933*f*–934*f*
Waste. *See* Nitrogenous wastes
Water
  athletic performance and, 710–712, 710*f*–711*f*
  concentration, 31
  evaporation of, 709, 709*f*
  hemorrhage response, 923, 924*f*
  homeostasis and, 33*f*
  hydrophilic and hydrophobic regions, 31, 31*f*
  ion balance with, 708–712, 709*f*, 710*f*–711*f*
  ions and polar molecules dissolving in, 30–31, 30*f*
  kidney reabsorption of, 839–841, 839*f*, 841*f*
  plant cell expansion with, 617–618, 617*f*, 618*f*, 636–637, 637*f*
  plant loss of, 663–667, 664*f*, 665*f*, 666*f*, 667*f*, 668*f*
  as plant nutrient, 650
  polar covalent bonds in, 27*f*
  properties of, 30–33
  seed germination and, 686
  shape of, 28*f*
  structural formula of, 25
  table salt dissolving in, 30*f*
  tasks of in living organisms, 32–33, 33*f*
  three states of, 32, 32*f*
Water channels, 107–109, 108*f*
Water cycle, 1011, 1011*f*
Water ferns, 528
Water flea, 872, 873*f*
Water flow, interruption of, 1011
Water potential, 658–660, 659*f*
Water stress, 660
Water vapor, 32
  plant-transpired, 664–665, 665*f*
Water vascular systems, 580–581, 580*f*
Waterland, Robert, 334
Waters, Colin, 1002
Water-soluble vitamins, 816*t*
Watson, James, 196, 196*f*, 199*f*, 379
Wavelength, 153, 154*f*
Waves, ocean, 993–994, 993*f*
Waxes, 45
Weak acid, 34
Weaponry, 960*f*, 961
"Web of life," 8, 447*f*, 448
Weevils, 577*t*
Weight loss, 813
Weinberg, Wilhelm, 400–401
Weismann, August, 188
Welland Canal, 1023, 1024*f*
Welwitschia mirabilis, 541*f*, 542
West Indian mahogany, 1020
Wetlands, 998
Whales, 392, 393*f*, 394, 1019, 1019*f*
Wheat, 306, 307*f*, 550
Wheat rust, 506, 506*f*
Whisk fern, 532, 533*f*
White matter, 732, 733*f*
White pelican, 602*f*
White willow, 1, 2*f*
White-bearded manakin, 945
White-lipped pit viper, 751*f*
White-tailed deer, 945*f*, 1024, 1024*f*
Whittaker, Robert, 995
Whittaker plot, 995, 995*f*
Whole metagenomic sequencing (WMS), 512*f*, 513
Whorls, 677, 678*f*
Wieschaus, Eric, 899
Wilkins, Maurice, 196
Wilmut, Ian, 1038
Wilson, Allan, 469–470
Wilson, E. O., 417, 975–978, 975*f*, 977*f*–978*f*, 1027, 1028, 1030, 1040
Wilson's bird-of-paradise, 12*f*
Wings, insects, 552, 576–578
Wintercreeper, 395, 395*f*
WMS (whole metagenomic sequencing), 512*f*, 513
WntA gene, 552
Wöhler, Friedrich, 37
Wolf, Larry, 939
Wolves, 604*f*, 948, 954, 955*f*
Wood, 624–625, 624*f*, 625*f*
Wood stork, 1024
Woodpigeon, 938, 938*f*
Woody plants, 623
Woolly mammoth, 1019, 1039
World Wildlife Foundation, 1028
Wright, Sewall, 409
Wuchereria bancrofti, 573, 574*f*

X

X inactivation center (Xic), 332, 334
X-0 system, 324–325, 324*f*
X-chromosome inactivation (XCI), 331–334, 332*f*, 332*t*, 333*f*
Xic (X inactivation center), 332, 334
X-linked genes, 325
X-linked inheritance patterns, 310*t*, 324–325, 324*f*
X-linked recessive traits, 325, 363
X-ray diffraction, 195–196, 195*f*
X-Y system, 324, 324*f*
Xylem, 532–533, 615–616, 615*f*, 623–625, 623*f*, 624*f*, 625*f*
  transport in, 663, 664*f*

Y

Yasumura, Yuki, 636
Yeast cells, 170*f*
  fermentation in, 147, 148*f*
  logistic growth of, 956*f*
Yeasts, 504, 956*f*. *See also Saccharomyces cerevisiae*
Yersinia pestis, 895
Yew tree, 1020
Yolk sac, 598, 599*f*
Yoshida, Masasuke, 143–145
Yucca plants, 418

Z

Z line, 769, 769*f*
Z scheme, 159–160, 160*f*
Zea mays, 550, 550*f*, 664, 677, 678*f*, 687*f*
Zeatin, 634*t*, 636
Zebra, 918*f*
Zebra mussel, 567–568, 970
Zika virus, 352–353
Zinc, 649*t*
Zinnia flower, 545*f*, 546
Zona pellucida, 874
Zone of elongation, 626, 627*f*
Zone of maturation, 626, 627*f*
Zuckerkandl, Emile, 440
Z-W system, 324*f*, 325
Zygote, 297, 676, 871
  embryonic development from, 684–685, 684*f*